Study Guide for

Organic Chemistry
Structure and Reactivity

Third Edition

Seyhan N. Eğe
The University of Michigan

Roberta W. Kleinman
Lock Haven University

Marjorie L. C. Carter
Eastern Michigan University

D. C. Heath and Company   LEXINGTON, MASSACHUSETTS / TORONTO

About the cover: Chemists deal every day with questions of structure and reactivity in their work. The picture you see on the cover was generated using molecular modeling programs and computer graphics in the laboratories of the Parke-Davis Pharmaceutical Research Division of Warner-Lambert as part of their ongoing search for more effective medications. The blue and white structure represents a molecule of tacrine (Cognex™), which was recently approved for the treatment of Altzheimer's disease. It is shown docked into the active site of acetylcholinesterase, an enzyme known to be involved in the processes of learning and memory. Scientists, in designing such medications, work with knowledge of the three-dimensional structures (shown in magenta) of the protein molecules that are the enzymes and of the drugs that complex with and inhibit the reactivity of these enzymes. In designing new and more potent drugs, chemists take into account all of the spatial and electronic properties of both the enzymes and of the drug under consideration. Computer modeling is a powerful tool that enables chemists to examine possible structures of drugs and choose the ones most likely to have the right structure and reactivity interactions and, therefore, to be active before starting the extensive experimentation work that eventually leads to clinical testing.

The chemistry discussed in this book is aimed at developing the understanding that chemists have of the structure of molecules. The book focuses on how the reactivity of such compounds is determined by their structures and how chemists use such understanding in creating new substances.

Cover image courtesy of Daniel Ortwine, Parke-Davis Pharmaceutical Research, a Division of Warner-Lambert Company.

Published simultaneously in Canada.

Printed in the United States of America.

International Standard Book Number: 0-669-34162-2

10 9 8 7 6 5 4 3 2

Contents

To the Student   **viii**

Chapter 1   **An Introduction to Structure and**
Bonding in Organic Compounds   **1**

Map 1.1   Ionic compounds and ionic bonding 1
Map 1.2   Covalent compounds 2
              Workbook Exercises 5
Map 1.3   Covalent bonding 14
Map 1.4   Isomers 17
Map 1.5   Polarity of covalent molecules 22
Map 1.6   Nonbonding interactions between molecules 26

Chapter 2   **Covalent Bonding and Chemical Reactivity**   **37**

              Workbook Exercises 37
Map 2.1   Molecular orbitals and covalent bonds 39
Map 2.2   Hybrid orbitals 44
Map 2.3   Bond lengths and bond strengths 47
Map 2.4   Functional groups 53

Chapter 3   **Reactions of Organic Compounds as**
Acids and Bases   **60**

              Workbook Exercises 60
Map 3.1   The Brønsted-Lowry theory of acids and bases 62
Map 3.2   The Lewis theory of acids and bases..64
Map 3.3   Relationship of acidity to position of the central element in the
              periodic table 72
Map 3.4   The relationship of pK_a to energy and entropy factors 79
Map 3.5   The effects of structural changes on acidity and basicity 85

Chapter 4   **Reaction Pathways**   **98**

              Workbook Exercises 98
Map 4.1   The factors that determine whether a chemical reaction between a given
              set of reagents is likely 99
Map 4.2   A nucleophilic substitution reaction 101
Map 4.3   The factors that determine the equilibrium constant for a reaction 102
Map 4.4   The factors that influence the rate of a reaction as they appear in the rate
              equation 102
Map 4.5   Some of the factors other than concentration that influence the rate
              of a reaction 103
Map 4.6   Factors that influence the rate of a reaction, and the effect of
              temperature 103
Map 4.7   An electrophilic addition reaction to an alkene 104
Map 4.8   Markovnikov's rule 105
Map 4.9   Carbocations 107
Map 4.10   Energy diagrams 108

Chapter 5   **Alkanes and Cycloalkanes**   124

                                 Workbook Exercises 124
Map   5.1   Conformation 129
Map   5.2   Representations of organic structures 131
Map   5.3   Conformation in cyclic compounds 138

Chapter 6   **Stereochemistry**   153

                                 Workbook Exercises 153
Map   6.1   Stereochemical relationships 156
Map   6.2   Chirality 156
Map   6.3   Definition of enantiomers 156
Map   6.4   Optical activity 160
Map   6.5   Formation of racemic mixtures in chemical reactions 161
Map   6.6   Configurational isomers 166
Map   6.7   Diastereomers 171
Map   6.8   The process of resolution 172

Chapter 7   **Nucleophilic Substitution and**
                                 Elimination Reactions   183

                                 Workbook Exercises 183
Map   7.1   A typical S_N2 reaction 191
Map   7.2   The S_N1 reaction 192
Map   7.3   The factors that are important in determining nucleophilicity 194
Map   7.4   The factors that determine whether a substituent is a good leaving
                                 group 199
Map   7.5   A comparison of E1 and E2 reactions 202
Map   7.6   Chapter summary 206-207

Chapter 8   **Alkenes**   225

                                 Workbook Exercises 225
Map   8.1   Reactions that form carbocations 230
Map   8.2   Reactions of carbocations 231
Map   8.3   The hydroboration reaction 237
Map   8.4   Catalytic hydrogenation reactions 238
Map   8.5   The addition of bromine to alkenes 245
Map   8.6   The oxidation reactions of alkenes 248

Chapter 9   **Alkynes**   273

                                 Workbook Exercises 273
Map   9.1   Outline of the synthesis of a disubstituted alkyne from a terminal
                                 alkyne 274
Map   9.2   Electrophilic addition of acids to alkynes 276
Map   9.3   Reduction reactions of alkynes 278

Chapter 10   **Infrared Spectroscopy**   293

Chapter 11   **Nuclear Magnetic Resonance Spectroscopy**   298

Workbook Exercises 298

Chapter 12   **Alcohols, Diols, and Ethers**   **315**

Map 12.1   Conversion of alkenes to alcohols 319
Map 12.2   Conversion of alcohols to alkyl halides 323
Map 12.3   Preparation of ethers by nucleophilic substitution reactions 327
Map 12.4   Ring-opening reactions of oxiranes 331
Map 12.5   Oxidation and reduction at carbon atoms 334
Map 12.6   Reactions of alcohols with oxidizing agents 339
Map 12.7   Summary of the preparation of and reactions of alcohols and ethers 344

Chapter 13   **Aldehydes and Ketones. Reactions at**
                            Electrophilic Carbon Atoms   **371**

                            Workbook Exercises 371
Map 13.1   Some ways to prepare aldehydes and ketones 380
Map 13.2   The relationship between carbonyl compounds, alcohols and alkyl
                       halides 381
Map 13.3   Organometallic reagents and their reactions with compounds
                       containing electrophilic carbon atoms 385
Map 13.4   Some ways to prepare alcohols 386
Map 13.5   Hydrates, acetals, ketals 391
Map 13.6   Reactions of carbonyl compounds with compounds related to
                       ammonia 392
Map 13.7   Reduction of carbonyl groups to methylene groups 399
Map 13.8   Protecting groups 403

Chapter 14   **Carboxylic Acids and Their Derivatives I.**
                            Nucleophilic Substitution Reactions
                            at the Carbonyl Group   **445**

                            Workbook Exercises 445
Map 14.1   Relative reactivities in nucleophilic substitutions 451
Map 14.2   Preparation of carboxylic acids 456
Map 14.3   Hydrolysis reactions of acid derivatives 460
Map 14.4   Mechanism of nucleophilic substitution reactions of acid
                       derivatives. Acyl transfer reactions 462
Map 14.5   Reactions of acids and acid derivatives with alcohols 470
Map 14.6   Reactions of acids and acid derivatives with ammonia or amines 472

Chapter 15   **Carboxylic Acids and Their Derivatives II.**
                            Synthetic Transformations and
                            Compounds of Biological Interest   **491**

Map 15.1   Reactions of organometallic reagents with acids and acid derivatives 495
Map 15.2   Reduction of acids and acid derivatives 499

Chapter 16   **Enols and Enolate Anions as Nucleophiles I.**
                            Halogenation, Alkylation, and
                            Condensation Reactions   **532**

                            Workbook Exercises 532

Map 16.1   Enolization 536
Map 16.2   Reactions of enols and enolates with electrophiles 540
Map 16.3   Thermodynamic and kinetic enolates 541
Map 16.4   Alkylation reactions 544
Map 16.5   The aldol condensation 547
Map 16.6   Acylation reactions of enolates 552

Chapter 17   **Polyenes**   **585**

Map 17.1   Different relationships between multiple bonds 585
Map 17.2   A conjugated diene 586
Map 17.3   Addition to dienes 586
Map 17.4   The Diels-Alder reaction 589
Map 17.5   Visible and ultraviolet spectroscopy 601

Chapter 18   **Enols and Enolate Anions as Nucleophiles II.**
                    Conjugate Addition Reactions; Ylides   **633**

Map 18.1   Electrophilic alkenes 633
Map 18.2   Reactions of electrophilic alkenes 643
Map 18.3   Ylides 644
Map 18.4   The Wittig reaction 646
Map 18.5   Dithiane anions 648

Chapter 19   **The Chemistry of Aromatic Compounds I.**
                    Electrophilic Aromatic Substitution   **686**

            Workbook Exercises 686
Map 19.1   Aromaticity 689
Map 19.2   Electrophilic aromatic substitution 691
Map 19.3   Essential steps of an electrophilic aromatic substitution 693
Map 19.4   Reactivity and orientation in electrophilic aromatic substitution 693
Map 19.5   Electrophiles in aromatic substitution reactions 702

Chapter 20   **Free Radicals**   **727**

Map 20.1   Chain reactions 727
Map 20.2   Halogenation of alkanes 729
Map 20.3   Selective free radical halogenations 733
Map 20.4   Free radical addition reactions of alkenes 737
Map 20.5   Oxidation reactions as free radical reactions 744

Chapter 21   **Mass Spectrometry**   **766**

Chapter 22   **The Chemistry of Amines**   **779**

Map 22.1   Preparation of amines 788
Map 22.2   Nitrosation reactions 793
Map 22.3   Rearrangements to nitrogen atoms 795
Map 22.4   Reactions of quaternary nitrogen compounds 800

Chapter 23   **The Chemistry of Aromatic Compounds II.**
                    Synthetic Transformations   **828**

Map 23.1   Diazonium ions in synthesis 829
Map 23.2   Reactions of diazonium ions 832
Map 23.3   Nucleophilic aromatic substitution 836

Chapter 24   The Chemistry of Heterocyclic Compounds   867

Map 24.1   Classification of cyclic compounds 868
Map 24.2   Synthesis of heterocycles from carbonyl compounds 874
Map 24.3   Substitution reactions of heterocycles 880

Chapter 25   Carbohydrates   919

Map 25.1   Classification of carbohydrates 919
Map 25.2   Structure of glucose 924
Map 25.3   Reactions of monosaccharides 936
Map 25.4   The Kiliani-Fischer synthesis 939

Chapter 26   Amino Acids, Peptides, and Proteins   966

Map 26.1   Amino acids, polypeptides, proteins 966
Map 26.2   Acid-base properties of amino acids 967
Map 26.3   Proof of structure of peptides and proteins 977
Map 26.4   Protection of functional groups in peptide synthesis 978
Map 26.5   Activation of the carboxyl group in peptide synthesis 979
Map 26.6   Conformation and structure in proteins 984

Chapter 27   Macromolecular Chemistry   1025

Map 27.1   Polymers 1025
Map 27.2   Properties of polymers 1026
Map 27.3   Stereochemistry of polymers 1033
Map 27.4   Types of polymerization reactions 1039

Chapter 28   Concerted Reactions   1053

Map 28.1   Concerted reactions 1053
Map 28.2   Cycloaddition reactions 1060
Map 28.3   Electrocyclic reactions 1064
Map 28.4   Woodward-Hoffmann rules 1066
Map 28.5   Sigmatropic rearrangements 1072
Map 28.6   Carbenes 1074

            Chapter 2 triangles to construct tetrahedra ....endpage

To the Student

The *Study Guide* that accompanies your text has been prepared to help you study organic chemistry. The textbook contains many problems designed to assist you in reviewing the chemistry that you need to know. The *Study Guide* contains the answers to these problems worked out in great detail to help you to develop the patterns of thought and work that will enable you to complete a course in organic chemistry successfully. In addition, notes that clarify points that may give you difficulty are provided in many answers.

The *Study Guide* also contains *Workbook Exercises,* created by Brian Coppola of the University of Michigan. These exercises are designed to help you review previous material and to introduce you to the problem-solving skills you will need for new material. They are found only in the *Study Guide,* and no answers are given for them. Many of the exercises can be explored with other students in your class.

Suggestions for the best way to study organic chemistry are given on pages 2 - 4 of the textbook. Before you work with the *Study Guide,* please review those pages in the text. There you will find emphasis on four things necessary for success in organic chemistry:

1. steady consistent studying,
2. working with problems as the most effective way to learn organic chemistry,
3. training your eye to look for the structural features in molecules that determine their chemical and physical properties, and
4. training your hand to draw correct structural formulas for reacting species and products.

This *Study Guide* contains two features to help you to study more effectively. The sections on Problem-Solving Skills in the textbook show you how to analyze problems in a systematic way by asking yourself questions about the structural changes in and the reactivity of the reagents shown in the problem. These same questions are used in arriving at the answers shown in the *Study Guide* for some of the problems. If you follow the reasoning shown in these answers, you will review the thinking patterns that are useful in solving problems.

The *Study Guide* also contains concept maps, which are summaries of important ideas or patterns of reactivity presented in a two-dimensional outline form. The textbook has notes in the margins telling you when a concept map appears in the *Study Guide*. The concept maps are located among the answers to the problems. The Table of Contents of the *Study Guide* on pp. iii - vi will tell you where each concept map is. The concept maps will be the most useful to you if you use them as a guide to making your own. For example, when you review your lecture notes, you will learn the essential points much more easily if you attempt to summarize the contents of the lecture in the form of a concept map. At a later time you may want to combine the contents of several lectures into a different concept map. Your maps need not look like the ones in the *Study Guide*. What is important is that you use the format to try to see relationships among ideas, reactions, and functional groups in a variety of ways.

The *Study Guide* will be most helpful to you if you make every attempt to solve each problem completely before you look at the answer. Recognition of a correct answer is much easier than being able to produce one yourself, so if you simply look up answers in the book to see whether you "know how to do the problem" or "understand" a principle, you will probably decide that you do. In truth, however, you will not have gained the practice in writing structural formulas and making the step-by-step decisions about reactivity that you will need when faced by similar questions on examinations. Work out all answers in detail, writing correct, complete formulas for all reagents and products. Build molecular models to help you draw correct three-dimensional representations of molecules. Consult the models whenever you are puzzled by questions of stereochemistry.

If you do not understand the answer to a problem, study the relevant sections of the text again, and then try to do the problem once more. The problems will tell you what you need to spend most of your time studying. As you solve the many review problems that bring together material from different chapters, your knowledge of the important reactions of organic chemistry will solidify.

We hope that the *Study Guide* will serve you as a model for the kind of disciplined care that you must take with your answers if you wish to train yourself to arrive at correct solutions to problems with consistency. We hope also that it will help you to develop confidence in your ability to master organic chemistry so that you enjoy your study of a subject that we find challenging and exciting.

Seyhan N. Eğe
Roberta W. Kleinman
Marjorie L. C. Carter

1

An Introduction to Structure and Bonding in Organic Compounds

Concept Map 1.1   Ionic compounds and ionic bonding.

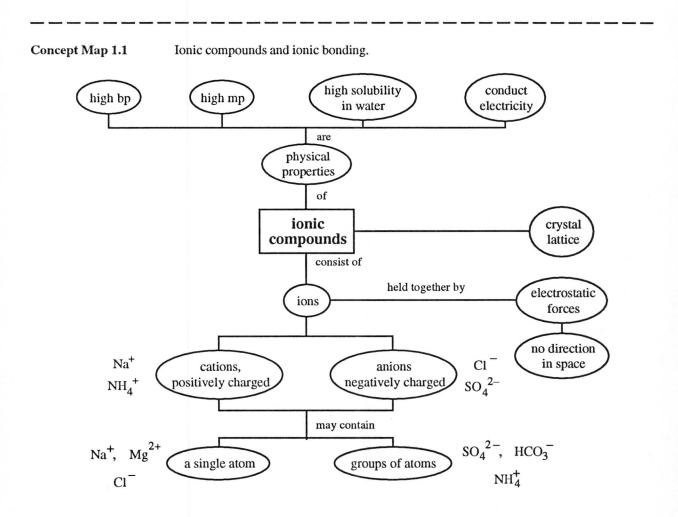

Concept Map 1.2   Covalent compounds.

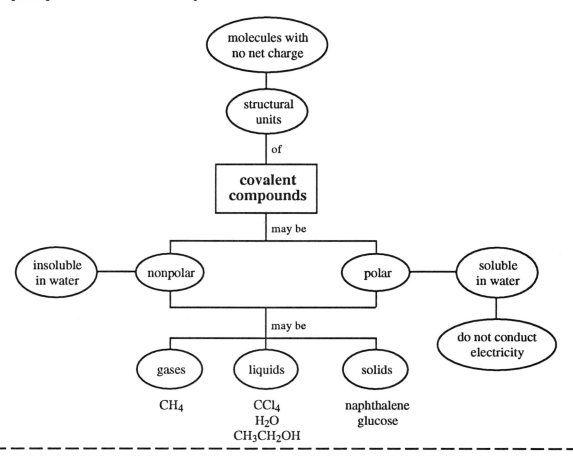

gases   liquids   solids

CH₄   CCl₄   naphthalene
                H₂O   glucose
            CH₃CH₂OH

- -

1.1   Note: The names of the compounds shown below are given for information. You are not yet expected to know how to name the compounds, but an examination of the names to see if you recognize an emerging pattern is fun.

CH₃CH₂CH₂CH₂OH

Lewis structure for 1-butanol   condensed formula for 1-butanol

connectivity: four carbon atoms in a row with the oxygen atom at the end of the row

CH₃CH₂CHCH₃ or CH₃CH₂CHOHCH₃
            |
            OH

Lewis structure for 2-butanol   condensed formula for 2-butanol

connectivity: four carbon atoms in a row with the oxygen atom on the second carbon atom of the row

1.1 (cont)

Lewis structure for 2-methyl-2-propanol

condensed formula for 2-methyl-2-propanol

CH_3CCH_3 or $(CH_3)_3COH$

connectivity: three carbon atoms in a row with one carbon atom and one oxygen atom on the second carbon
atom of the row

Lewis structure for methyl isopropyl ether

condensed formula for methyl isopropyl ether

CH_3OCHCH_3 or $CH_3OCH(CH_3)_2$

connectivity: one carbon atom bonded to an oxygen that is bonded to two more carbon atoms in a row, with
a third carbon atom attached to the first of these carbons

Lewis structure for diethyl ether

condensed formula for diethyl ether

$CH_3CH_2OCH_2CH_3$

connectivity: two carbon atoms in a row bonded to an oxygen atom that is bonded to two other carbon atoms
in a row

1.2   The names (and the connectivities) of the compounds in this problem are related to some of those in Problem
1.1. See if you can find the pattern.

Lewis structure for 1-bromobutane

condensed formula for 1-bromobutane

$CH_3CH_2CH_2CH_2Br$

connectivity: four carbon atoms in a row with the bromine atom at the end of the row

1.2 (cont)

```
     H   H   H   H
     |   |   |   |
 H — C — C — C — C — H
     |   |   |   |
     H   H  :Br: H
              ••
```

Lewis structure for 2-bromobutane

$CH_3CH_2CHCH_3$ or $CH_3CH_2CHBrCH_3$ with Br below

condensed formula for 2-bromobutane

connectivity: four carbon atoms in a row with the bromine atom on the second carbon atom of the row

```
         H
         |
     H — C — H
     H       H
     |       |
 H — C — C — C — H
     |       |
     H  :Br: H
         ••
```

Lewis structure for 2-bromo-2-methylpropane

CH_3 over CH_3CCH_3 with Br below or $(CH_3)_3CBr$

condensed formula for 2-bromo-2-methylpropane

connectivity: three carbon atoms in a row with the bromine atom and a carbon atom on the second carbon atom of the row

```
         H
         |
     H — C — H
     H       H    ••
     |       |    ••
 H — C — C — C — Br:
     |   |   |    ••
     H   H   H
```

Lewis structure for 1-bromo-2-methylpropane

CH_3 over CH_3CHCH_2Br or $(CH_3)_2CHCH_2Br$

condensed formula for 1-bromo-2-methylpropane

connectivity: three carbon atoms in a row with the bromine atom at the end of the row and a carbon atom on the second carbon atom of the row

1.3

```
     H   H   H   ••
     |   |   |
 H — C — C — C — N — H
     |   |   |   |
     H   H   H   H
```

Lewis structure for propylamine

$CH_3CH_2CH_2NH_2$

condensed formula for propylamine

connectivity: three carbon atoms in a row with the nitrogen atom at the end of the row

1.3 (cont)

Lewis structure for methylethylamine   condensed formula for methylethylamine

$CH_3CH_2NHCH_3$

connectivity: two carbon atoms in a row attached to a nitrogen atom that is bonded to one other carbon atom

Lewis structure for trimethylamine   condensed formula for trimethylamine

CH_3NCH_3 or $(CH_3)_3N$

with CH_3 above

connectivity: a nitrogen atom with three single carbon atoms bonded to it

---

Workbook Exercises

In most chemical reactions, molecules undergo changes in the connectivity, that is, the bonding, of their atoms. Learning how to identify these changes rapidly is a skill you should master.

EXERCISE I.   Identify these features in the following chemical reactions:

   (1)   bonds broken

   (2)   bonds formed

   (3)   redistribution of bonding and nonbonding electrons

EXAMPLE

Na^+ $^-$:C≡N: + $CH_3CH_2CH_2$—Br: ⟶ $CH_3CH_2CH_2$—C≡N: + Na^+ :Br: $^-$

SOLUTION

While you are becoming comfortable with molecular structures, it is a good idea to use complete Lewis structures to follow the changes in connectivity. You should also examine the formal charges and electron configurations of the atoms involved in those changes.

Workbook Exercises (cont)

One bond is broken: the 2-electron C—Br bond breaks, and this electron pair becomes the fourth nonbonding electron pair on the bromide ion that forms. One bond is formed: the nonbonding electron pair on the carbon atom of cyanide ion (CN⁻) is used to form the new C—C bond.

EXERCISE I.   Identify the bonds broken, the bonds formed, and the way electrons have been redistributed in the processes of chemical change for the following reactions.

1. $(CH_3)_2CHCH_2\overset{..}{\underset{..}{O}}H$ + $Na^+\ H\!:\!^-$ $\longrightarrow$ $(CH_3)_2CHCH_2\overset{..}{\underset{..}{O}}\!:^-\ Na^+$ +   H—H

2.

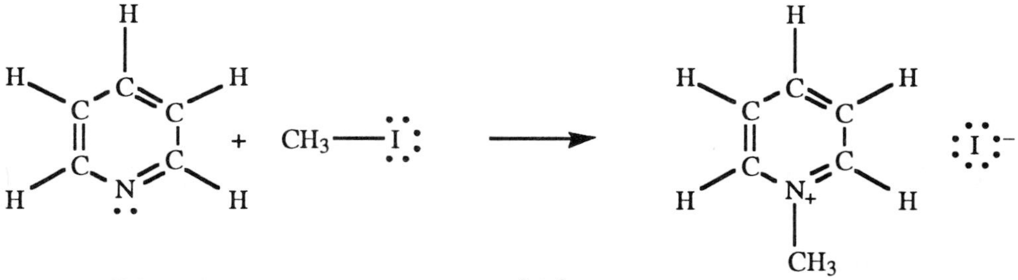

3.

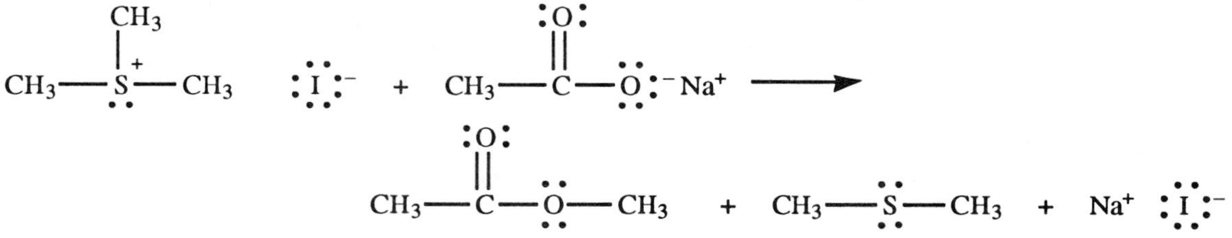

The next two examples involve double bonds, introduced in Section 1.5 of the text. Can you also describe these changes?

4.

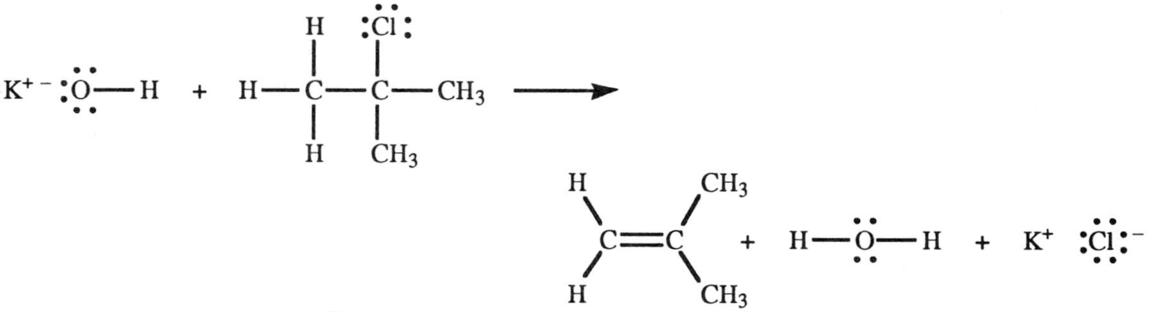

5.

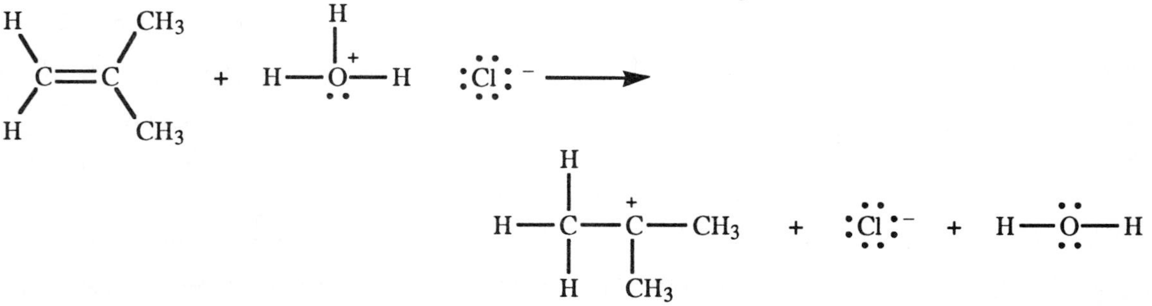

Workbook Exercises (cont)

EXERCISE II.   Complete each of the equations by providing the structure of the molecule that will balance the chemical equation. In a balanced equation, there must be an equal number of each kind of atom on both sides of the equation. The overall charge must also be the same on both sides of a chemical reaction equation. All of the atoms in the compounds in these exercises have closed shell configurations.

EXAMPLE

$$Li^+ \; ^-\!:\!\ddot{S}H \quad + \quad (CH_3)_2CH\ddot{C}\!\!:\!\!\longrightarrow \quad A \quad + \quad Li^+ \; :\!\ddot{C}\!l\!:^-$$

SOLUTION

As in Exercise I, we can identify some of the bonding changes. From this information, and the fact that the equation must balance, we determine the structure of the unknown substance **A**.

What we see: the C—Cl bond breaks, and the electron pair from the single bond becomes the fourth nonbonding electron pair of chloride ion on the right side of the equation. The unknown compound **A** must (a) incorporate the atoms from the left hand side of the equation that are not shown on the right, (b) have an overall neutral charge in order to keep the charges balanced, and (c) be comprised of atoms with closed shell configurations.

So, compound **A** needs to include (a) the SH atoms derived from the ionic compound LiSH, and (b) the atoms from $(CH_3)_2CHCl$, except for chloride ion which appears as a product. The product ion, Cl^-, must come from the uncharged molecule $(CH_3)_2CHCl$ because there is no other source of chlorine atoms on the left hand side of the equation. Although we cannot identify the structure of **A** with certainty at this point, we can account for the atoms from $(CH_3)_2CHCl$ that remain when the chloride ion, Cl^-, is removed. Perhaps only temporarily, we can imagine the presence of a positively

$$charged\ fragment,\ CH_3\!-\!\underset{+}{\overset{\overset{\displaystyle H}{|}}{C}}\!-\!CH_3\ ,\ that\ comes\ from\ removing\ Cl^-\ from\ the\ uncharged\ starting\ compound.\ In\ our$$

$$imagination,\ the\ fragments\ we\ can\ use\ to\ make\ \mathbf{A}\ are\ the\ cation,\ CH_3\!-\!\underset{+}{\overset{\overset{\displaystyle H}{|}}{C}}\!-\!CH_3\ ,\ and\ the\ anion,\ ^-\!:\!\ddot{S}H\ .\ There\ are$$

two possible ways to create a compound from these fragments:

the first way gives us an ionic compound: the second way gives us a covalent compound:

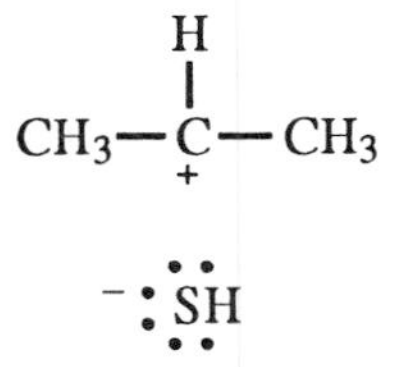

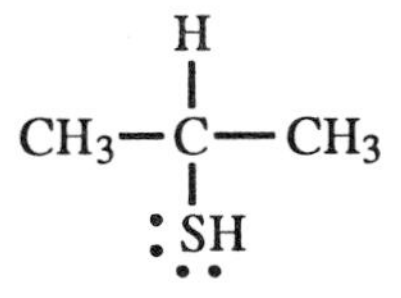

$$CH_3\!-\!\underset{+}{\overset{\overset{\displaystyle H}{|}}{C}}\!-\!CH_3 \qquad\qquad\qquad CH_3\!-\!\underset{\underset{\displaystyle :\ddot{S}H}{|}}{\overset{\overset{\displaystyle H}{|}}{C}}\!-\!CH_3$$

$$^-\!:\!\ddot{S}H$$

Is there a way to decide between these two possible structures for **A**? Or are they both acceptable answers? When solving an open-ended problem such as this one, either in the homework or on an examination, it is important to check the assumptions and information given in the problem. In this case, rereading the question tells us that "All of the atoms in these compounds have closed shell configurations." Therefore, the covalent structure is the only one that satisfies this criterion. In the ionic structure, the positively charged carbon has an open shell.

EXERCISE II.   Complete each of the following equations, as demonstrated in the example above.

1. $$CH_3CH_2\ddot{B}r\!: \quad + \quad B \quad \longrightarrow \quad CH_3CH_2\overset{+}{N}H_3 \quad + \quad :\!\ddot{B}r\!:^-$$

Workbook Exercises (cont)

2. CH_3—C—Cl: + Li$^+$ $^-$:O—CH_3 $\longrightarrow$ **C** + Li$^+$:Cl:$^-$

3.   **D** + CH_3—O—CH_3 $\longrightarrow$ CH_3—O$^+$—CH_3 + :F:$^-$

The next two examples in this exercise involve double bonds. What are the structures of **E** and **F**?

4.   **E** + [H, CH_3 / C=C / H, H] $\longrightarrow$ H—C—C—CH_3 (with Br, Br above; H, H below)

5.   Na$^+$ $^-$:O—H + H—C—C—Br (with Cl, Br above; Cl, Br below) $\longrightarrow$ H—O—H + **F** + Na$^+$:Br:$^-$

- -

1.4   The molecular formulas for the following compounds match the C_nH_{2n+2} rule, either right away or after hydrogen atoms are mentally substituted for halogen atoms.

(c) $C_2HBrClF_3$ has 5 halogen atoms, therefore it is the equivalent of C_2H_6.

(e) $C_{20}H_{42}$

(g) $CHCl_3$ is the equivalent of CH_4

The structural formulas of these three compounds will contain only singly bonded atoms. None of them can have a ring structure.

1.5   Formal charge = number of valence electrons – number of nonbonding electrons – $\frac{1}{2}$ number of bonding electrons.

(a) CCl_4

:Cl:
:Cl: C :Cl:
:Cl:

:Cl:
:Cl—C—Cl:
:Cl:

Cl $7 - 6 - \frac{1}{2}(2) = 0$

C $4 - 0 - \frac{1}{2}(8) = 0$

Each chlorine atom has 6 nonbonding electrons and 1 electron from a pair of bonding electrons. Therefore each chlorine atom has 7 valence electrons, the number it needs, and no formal charge. Carbon shares 8 bonding electrons. It effectively has 4 electrons around it, the number it needs to have no formal charge.

(b) CH_3Br

H
H :C: Br:
H

H
H—C—Br:
H

H $1 - 0 - \frac{1}{2}(2) = 0$

C $4 - 0 - \frac{1}{2}(8) = 0$

Br $7 - 6 - \frac{1}{2}(2) = 0$

1.5 (cont)

(c) $CH_3\overset{+}{O}H_2$

$$\begin{array}{c} H\ \ H \\ \overset{\cdot\cdot}{}\ \ \overset{\cdot\cdot}{}\ + \\ H\!:\!C\!:\!\underset{\cdot\cdot}{O}\!:\!H \\ H \end{array}$$

Structure with $H-C-\overset{+}{\underset{\cdot\cdot}{O}}-H$

(deficiency of one electron)

H	$1 - 0 - \frac{1}{2}(2) = 0$
C	$4 - 0 - \frac{1}{2}(8) = 0$
O	$6 - 2 - \frac{1}{2}(6) = +1$

(d) $^-NH_2$

$$H\!:\!\underset{\cdot\cdot}{N}\!:\!H\ \ ^-$$

$$H-\underset{\cdot\cdot}{N}-H\ \ ^-$$

H	$1 - 0 - \frac{1}{2}(2) = 0$
N	$5 - 4 - \frac{1}{2}(4) = -1$

(excess of one electron)

Each hydrogen atom shares 2 bonding electrons and therefore effectively has 1 electron and no formal charge. The nitrogen atom has 4 nonbonding electrons, and shares 4 bonding electrons. It effectively has 6 electrons around it, one more than the 5 electrons it needs to be uncharged. Nitrogen therefore has a formal charge of –1.

(e) $CH_3\overset{+}{N}H_3$

$$\begin{array}{c} H\ \ H \\ \overset{\cdot\cdot}{}\ \ \overset{\cdot\cdot}{}+ \\ H\!:\!C\!:\!N\!:\!H \\ H\ \ H \end{array}$$

$$H-C-\overset{+}{N}-H$$

(deficiency of one electron)

H	$1 - 0 - \frac{1}{2}(2) = 0$
C	$4 - 0 - \frac{1}{2}(8) = 0$
N	$5 - 0 - \frac{1}{2}(8) = +1$

(f) H_2NNH_2

$$\begin{array}{c} H\ \ H \\ \overset{\cdot\cdot}{}\ \ \overset{\cdot\cdot}{} \\ H\!:\!\underset{\cdot\cdot}{N}\!:\!\underset{\cdot\cdot}{N}\!:\!H \end{array}$$

$$H-\underset{\cdot\cdot}{N}-\underset{\cdot\cdot}{N}-H$$

H	$1 - 0 - \frac{1}{2}(2) = 0$
N	$5 - 2 - \frac{1}{2}(6) = 0$

(g) PH_3

$$H\!:\!\underset{\cdot\cdot}{P}\!:\!H$$

$$H-\underset{\cdot\cdot}{P}-H$$

P	$5 - 2 - \frac{1}{2}(6) = 0$
H	$1 - 0 - \frac{1}{2}(2) = 0$

(h) H_2S

$$H\!:\!\overset{\cdot\cdot}{\underset{\cdot\cdot}{S}}\!:\!H$$

$$H-\overset{\cdot\cdot}{\underset{\cdot\cdot}{S}}-H$$

H	$1 - 0 - \frac{1}{2}(2) = 0$
S	$6 - 4 - \frac{1}{2}(4) = 0$

(i) CH_3CH_2OH

$$\begin{array}{c} H\ \ H \\ \overset{\cdot\cdot}{}\ \ \overset{\cdot\cdot}{}\ \ \overset{\cdot\cdot}{} \\ H\!:\!C\!:\!C\!:\!\underset{\cdot\cdot}{O}\!:\!H \\ H\ \ H \end{array}$$

$$H-C-C-\overset{\cdot\cdot}{\underset{\cdot\cdot}{O}}-H$$

H	$1 - 0 - \frac{1}{2}(2) = 0$
C	$4 - 0 - \frac{1}{2}(8) = 0$
O	$6 - 4 - \frac{1}{2}(4) = 0$

1.5 (cont)

(j) $HOCH_2CH_2OH$

$$H : \overset{..}{\underset{..}{O}} : \overset{\overset{H}{|}}{\underset{\overset{|}{H}}{C}} : \overset{\overset{H}{|}}{\underset{\overset{|}{H}}{C}} : \overset{..}{\underset{..}{O}} : H$$

$$H-\overset{..}{\underset{..}{O}}-\overset{\overset{H}{|}}{\underset{\overset{|}{H}}{C}}-\overset{\overset{H}{|}}{\underset{\overset{|}{H}}{C}}-\overset{..}{\underset{..}{O}}-H$$

H $1 - 0 - \frac{1}{2}(2) = 0$

C $4 - 0 - \frac{1}{2}(8) = 0$

O $6 - 4 - \frac{1}{2}(4) = 0$

(k) $CH_3\overset{+}{\underset{\underset{CH_3}{|}}{O}}CH_3$

$$H : \overset{\overset{H}{..}}{\underset{\overset{..}{H}}{C}} : \overset{\overset{+}{..}}{\underset{..}{O}} : \overset{\overset{H}{..}}{\underset{\overset{..}{H}}{C}} : H$$
$$H : \overset{}{\underset{\overset{..}{H}}{C}} : H$$

$$H-\overset{\overset{H}{|}}{\underset{\overset{|}{H}}{C}}-\overset{\overset{..}{+}}{\underset{|}{O}}-\overset{\overset{H}{|}}{\underset{\overset{|}{H}}{C}}-H$$
$$H-\overset{}{\underset{\overset{|}{H}}{C}}-H$$

H $1 - 0 - \frac{1}{2}(2) = 0$

C $4 - 0 - \frac{1}{2}(8) = 0$

O $6 - 2 - \frac{1}{2}(6) = +1$

(deficiency of one electron)

Each hydrogen atom shares 2 bonding electrons and therefore effectively has 1 electron and no formal charge. Each carbon atom shares 8 bonding electrons and thus effectively has 4 electrons and no formal charge. The oxygen atom has 2 nonbonding electrons and shares 6 bonding electrons. It effectively has 5 electrons, 1 fewer than the 6 electrons it needs to be uncharged, and therefore has a formal charge of +1.

1.6   (a)

$$CH_3 - \overset{\overset{CH_3}{|}}{\underset{\underset{CH_3}{|}}{\overset{+}{N}}} - CH_3$$

N $5 - 0 - \frac{1}{2}(8) = +1$

Each hydrogen atom shares 2 bonding electrons, and therefore effectively has 1 electron and no formal charge. Each carbon atom shares 8 bonding electrons, and thus effectively has 4 electrons and no formal charge. The nitrogen atom shares 8 bonding electrons. It effectively has 4 electrons, 1 fewer than the 5 electrons it needs to be uncharged, and therefore has a formal charge of +1.

(b)

$$: \overset{..}{\underset{..}{Br}} - \overset{}{\underset{}{C}} - \overset{..}{\underset{..}{Br}} :$$

C $4 - 2 - \frac{1}{2}(4) = 0$

Each bromine atom has 6 nonbonding electrons and shares 2 bonding electrons, and therefore effectively has 7 electrons and no formal charge. The carbon atom has 2 nonbonding electrons and shares 4 bonding electrons, and thus effectively has 4 electrons and no formal charge.

(c)

$$CH_3 - \overset{\overset{H}{|}}{\underset{\underset{..}{}}{\overset{+}{O}}} - CH_3$$

O $6 - 2 - \frac{1}{2}(6) = +1$

(d)

$$CH_3 - \overset{..}{\underset{..}{N}} - H$$

N $5 - 4 - \frac{1}{2}(4) = -1$

1.6 (cont)

(e)

$$\text{Cl}_2\text{C}\text{Cl}_2^{-}$$

C $4 - 2 - \frac{1}{2}(6) = -1$

(f)

$$CH_3 - C^{+} - CH_3 \text{ with } CH_3$$

C $4 - 0 - \frac{1}{2}(6) = +1$

Each chlorine atom has 6 nonbonding electrons and shares 2 bonding electrons, and therefore effectively has 7 electrons and no formal charge. The carbon atom has 2 nonbonding electrons and shares 6 bonding electrons. It effectively has 5 electrons around it, one more than the 4 electrons it needs to be uncharged, and therefore has a formal charge of –1.

1.7

$$F - B^{-} - N^{+} - H$$

F $7 - 6 - \frac{1}{2}(2) = 0$

H $1 - 0 - \frac{1}{2}(2) = 0$

B $3 - 0 - \frac{1}{2}(8) = -1$ (excess of one electron)

N $5 - 0 - \frac{1}{2}(8) = +1$ (deficiency of one electron)

Each fluorine atom has 6 nonbonding electrons and shares 2 bonding electrons. Each fluorine atom therefore effectively has 7 electrons and no formal charge. Each hydrogen atom shares 2 bonding electrons and therefore effectively has 1 electron and no formal charge. The nitrogen atom shares 8 bonding electrons. It effectively has 4 electrons, 1 fewer than the 5 electrons it needs to be uncharged, and therefore has a formal charge of +1. The boron atom shares 8 bonding electrons. It effectively has 4 electrons around it, one more than the 3 electrons it needs to be uncharged and therefore has a formal charge of –1.

1.8 BF_3 has room in its orbitals to accept a pair of electrons. We expect it to react with any uncharged or negatively charged species that has nonbonding electrons.

$$F - B + H - O - H \longrightarrow F - B^{-} - O^{+} - H$$

$$F - B + H - N - H \longrightarrow F - B^{-} - N^{+} - H$$

1.8 (cont)

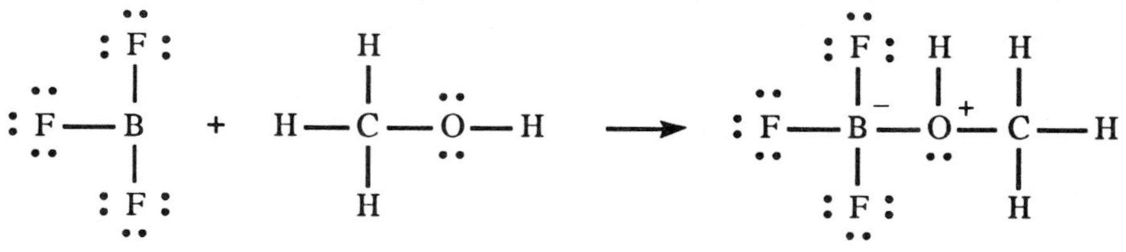

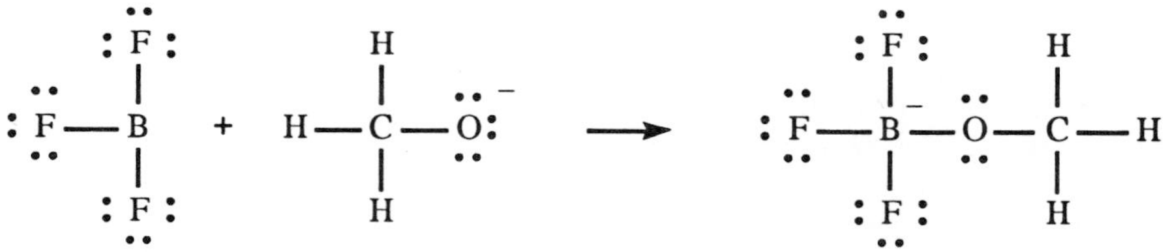

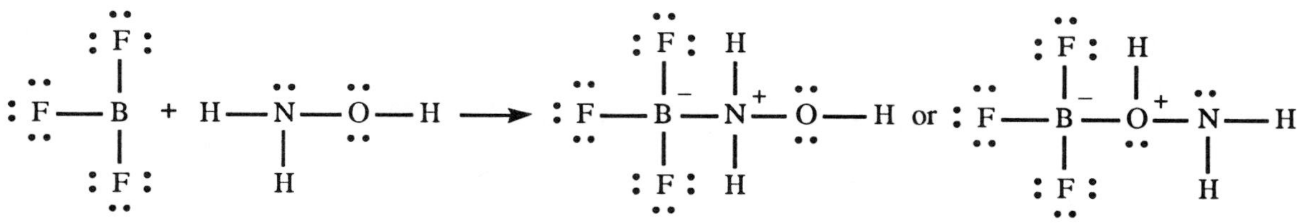

$$:F-B + H-C-O-H \longrightarrow :F-B-O^+-C-H$$

$$:F-B + H-C-O: \longrightarrow :F-B-O-C-H$$

$$:F-B + H-N-O-H \longrightarrow :F-B-N^+-O-H \text{ or } :F-B-O^+-N-H$$

Even though the hydronium ion has a pair of nonbonding electrons, the positive charge on the ion makes them unlikely to participate in further bonding. The electrons on any positively charged species are tightly held, and not easily donated to another atom. Exploring the result of the reaction of BF_3 with H_3O^+ is useful. The resulting species is quite unstable because of the double positive charge on the oxygen atom next to the negative charge on the boron.

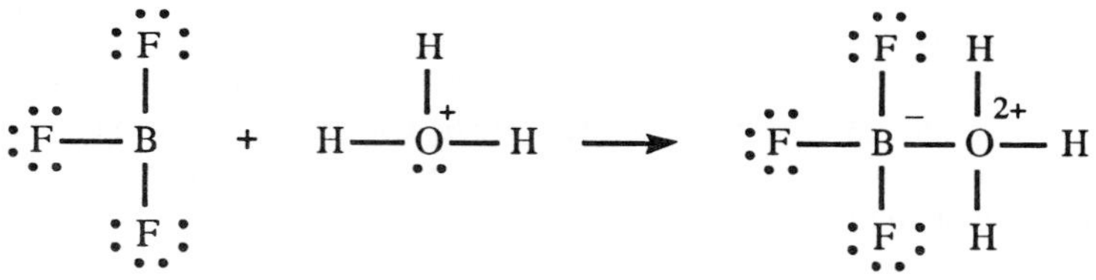

$$:F-B + H-O^+-H \longrightarrow :F-B-O^{2+}-H$$

unstable

1.9

Condensed Formula		Lewis Structures

(a) CH_3CCH_3 (with O double bonded)

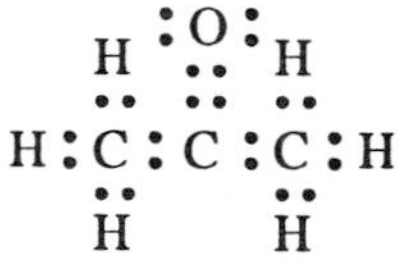

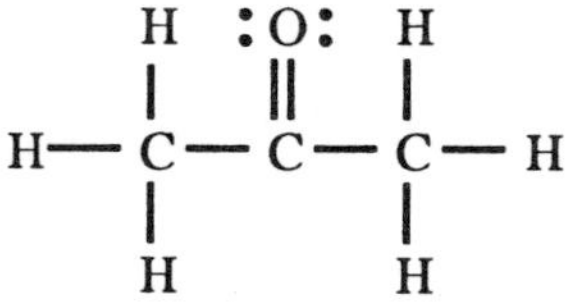

(b) $CH_3C\equiv CH$

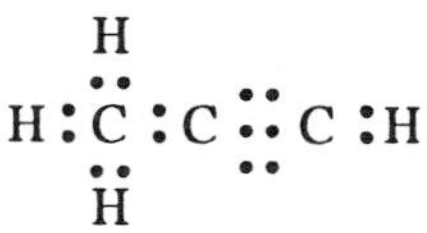

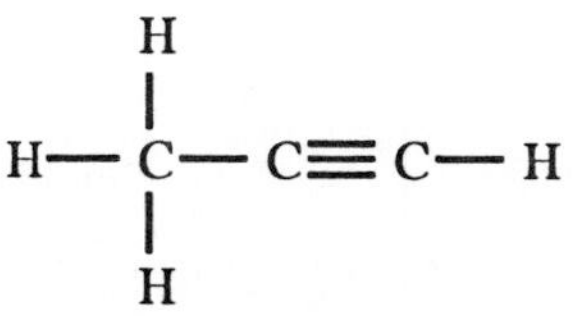

1.9 (cont)   <u>Condensed Formula</u>   <u>Lewis Structures</u>

(c) $\underset{\displaystyle \text{HCNH}_2}{\overset{\displaystyle \overset{O}{\|}}{}}$

(d) $\underset{\displaystyle \text{HCOH}}{\overset{\displaystyle \overset{O}{\|}}{}}$

(e) $\underset{\displaystyle \text{CH}_3\text{COCH}_3}{\overset{\displaystyle \overset{O}{\|}}{}}$

(f) $\text{HON}\!=\!\text{O}$

or

Note that the nonbonding electrons may be placed around a doubly bonded oxygen in two ways, as illustrated in the structural formulas above.

(g) $\text{CH}_3\text{N}\!=\!\text{NCH}_3$

(h) $\text{CH}_2\!=\!\text{CHCl}$

1.10   (a)   The saturated compound with 2 carbons would be C_2H_6. C_2H_4 is missing 2 hydrogens; therefore, it has 1 unit of unsaturation.

(b) $C_{14}H_9Cl_5$ is the equivalent of $C_{14}H_{14}$. The saturated compound with 14 carbons would be $C_{14}H_{30}$. $C_{14}H_{14}$ is missing $(30 - 14) = 16$ hydrogens; therefore, it has $16/2 = 8$ units of unsaturation.

(c) $C_2HBrClF_3$ has 0 units of unsaturation (see Problem 1.4).

(d) $C_{10}H_{16}$ has 3 units of unsaturation ($C_{10}H_{22}$ is saturated).

(e) $C_{20}H_{42}$ is saturated (see Problem 1.4).

(f) C_2H_3Cl is the equivalent of C_2H_4 (see part (a).

(g) $CHCl_3$ is saturated (see Problem 1.4)

(h) C_4H_7Cl is the equivalent of C_4H_8. The saturated compound with 4 carbons is C_4H_{10}. It therefore has 1 unit of unsaturation.

Concept Map 1.3   Covalent bonding.

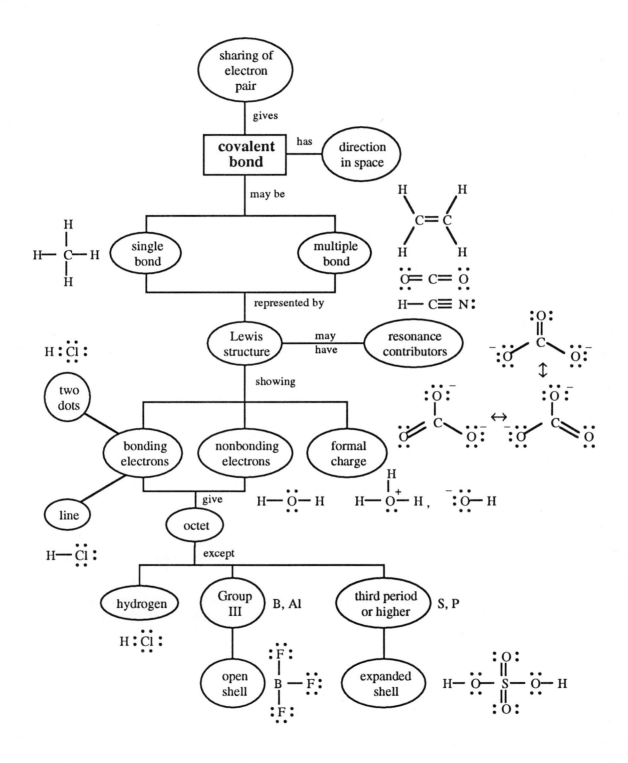

1.11

(a)

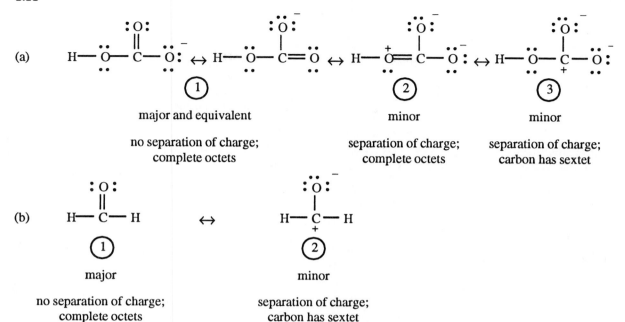

1 major and equivalent

no separation of charge;
complete octets

2 minor

separation of charge;
complete octets

3 minor

separation of charge;
carbon has sextet

(b)

1 major

no separation of charge;
complete octets

2 minor

separation of charge;
carbon has sextet

Note that such a minor resonance contributor may nevertheless be significant in the chemical reactivity of a species (see Problems 1.12 and 1.13 for examples).

(c)

1 major and equivalent

no separation of charge;
complete octets

2 minor

separation of charge;
nitrogen has sextet

(d)

1 major

no separation of charge;
complete octets;
negative charge on more
electronegative atom

2 minor

no separation of charge;
complete octets;
negative charge on less
electronegative atom

3 minor

separation of charge;
carbon has sextet

(e)

equivalent resonance contributors

complete octets;
no contributor with complete octets that does
not have separation of charge can be written

1.11 (cont)

(f)

$$:\ddot{O}-\overset{+}{\ddot{O}}=\ddot{O} \leftrightarrow \ddot{O}=\overset{+}{\ddot{O}}-\ddot{O}: \leftrightarrow :\ddot{O}-\ddot{O}-\overset{+}{\ddot{O}} \leftrightarrow \overset{+}{\ddot{O}}-\ddot{O}-\ddot{O}:$$

①   ②

major and equivalent   minor and equivalent
complete octets   one oxygen has sextet

all contributors have separation of charge

(g)

$$\ddot{O}=\overset{+}{N}=\ddot{O} \leftrightarrow \ddot{O}=N-\overset{+}{\ddot{O}} \leftrightarrow \overset{+}{\ddot{O}}-N=\ddot{O}$$

①   ②

major   minor and equivalent

complete octets;   oxygen has sextet;
positive charge on less   positive charge on more
electronegative atom   electronegative atom

(h)

$$^-:\ddot{O}-\overset{\overset{\displaystyle :\ddot{O}:^-}{|}}{\underset{2+}{Cl}}-\ddot{O}:^- \leftrightarrow {}^-:\ddot{O}-\overset{\overset{\displaystyle :\ddot{O}:}{\|}}{\underset{+}{Cl}}-\ddot{O}:^- \leftrightarrow {}^-:\ddot{O}-\overset{\overset{\displaystyle :\ddot{O}:^-}{|}}{\underset{+}{Cl}}=\ddot{O} \leftrightarrow \ddot{O}=\overset{\overset{\displaystyle :\ddot{O}:^-}{|}}{\underset{+}{Cl}}-\ddot{O}:^-$$

②   ①

minor   major and equivalent

large separation of charge   less separation of charge;
                                        more covalent bonds

Note that chlorine is in the third period of the periodic table and may, therefore, have more than eight electrons around it.

1.12   The boron atom in boron trifluoride has only six electrons and can accept an electron pair from the nitrogen atom in ammonia to complete the octet.

$$\begin{array}{cc} :\ddot{F}: & H \\ | & | \\ :\ddot{F}-B & :N-H \\ | & | \\ :\ddot{F}: & H \end{array}$$

formation of covalent bond

Likewise the reaction between carbon dioxide and hydroxide ion is rationalized by looking at a resonance contributor in which the carbon in carbon dioxide has only six electrons and can accept an electron pair from the oxygen atom of the hydroxide ion to complete the octet.

$$\ddot{O}=\overset{+}{C}-\ddot{O}:^- \qquad :\ddot{O}-H$$

formation of covalent bond

1.13   Ammonia will react with formaldehyde. In the minor resonance contributor of formaldehyde (Problem 1.6b), the carbon atom has a sextet and can accept an electron pair from the nitrogen atom in ammonia to complete the octet.

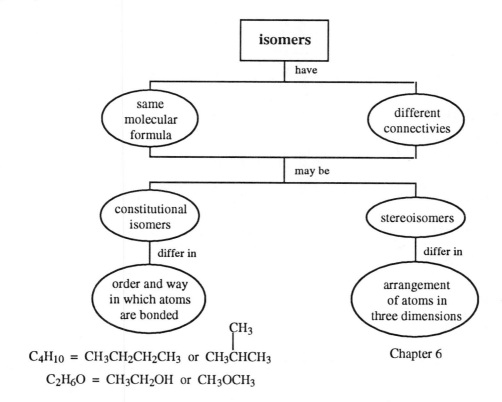

formation of covalent bond

- -

Concept Map 1.4   Isomers.

isomers

have

same molecular formula

different connectivies

may be

constitutional isomers

stereoisomers

differ in

differ in

order and way in which atoms are bonded

arrangement of atoms in three dimensions

Chapter 6

C_4H_{10} = $CH_3CH_2CH_2CH_3$ or $CH_3\overset{\underset{\displaystyle |}{CH_3}}{C}HCH_3$

C_2H_6O = CH_3CH_2OH or CH_3OCH_3

- -

1.14 C_3H_7Cl

$CH_3CH_2CH_2Cl$

$CH_3CHClCH_3$

1.14 (cont)

C_3H_8O

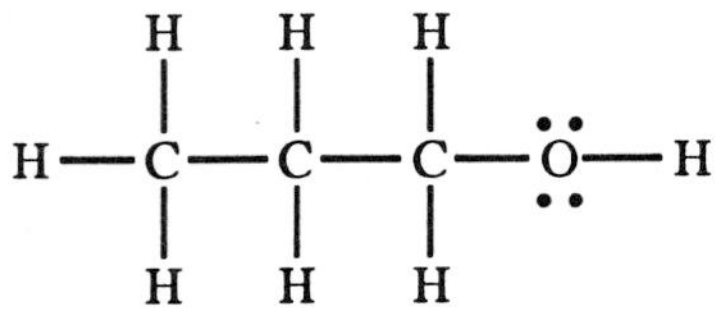

$CH_3CH_2CH_2OH$ 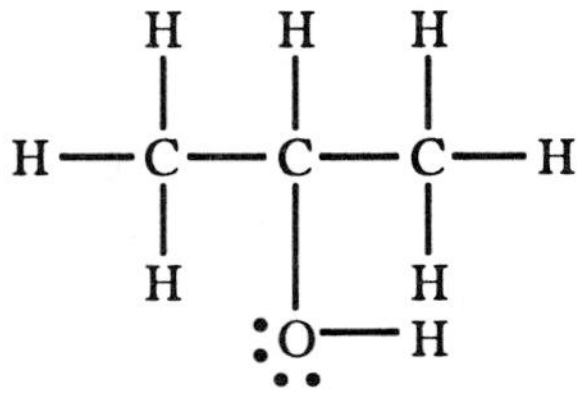 $CH_3CHOHCH_3$ 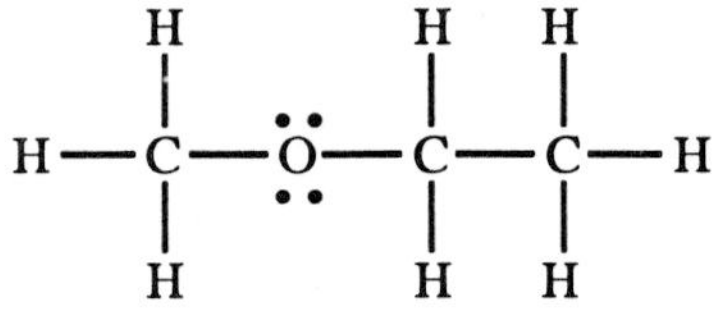 $CH_3OCH_2CH_3$

$C_4H_8Cl_2$

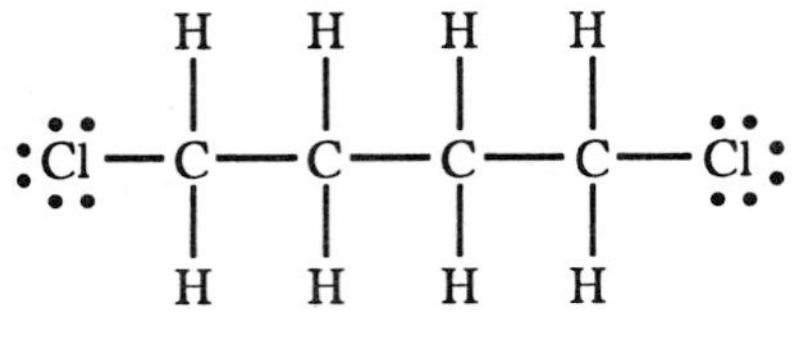

$ClCH_2CH_2CH_2CH_2Cl$ 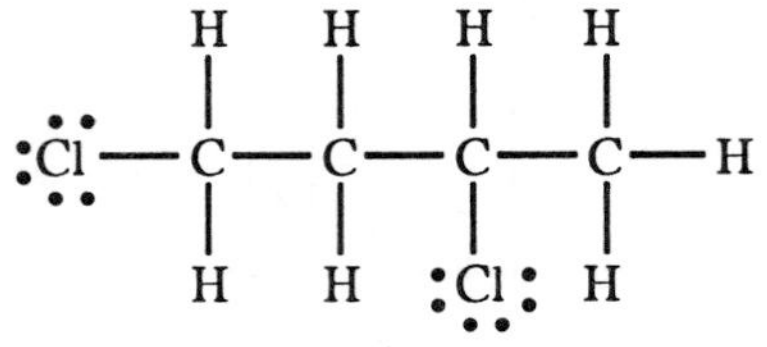 $ClCH_2CH_2CHClCH_3$ 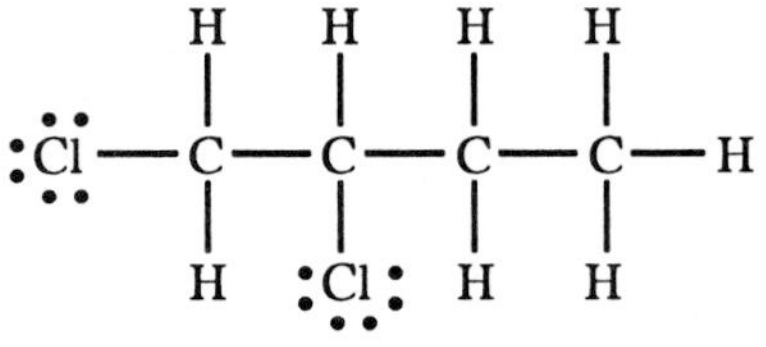 $ClCH_2CHClCH_2CH_3$

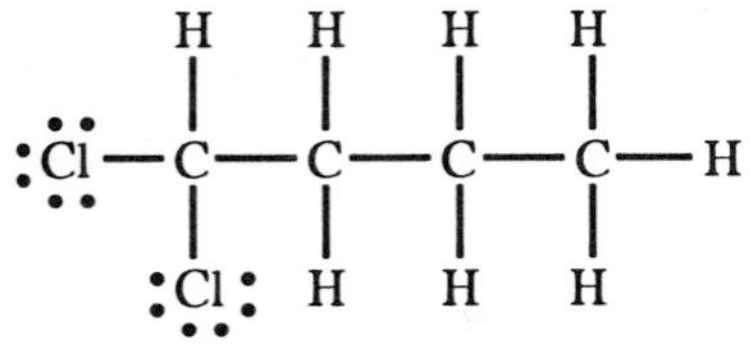

$CHCl_2CH_2CH_2CH_3$ 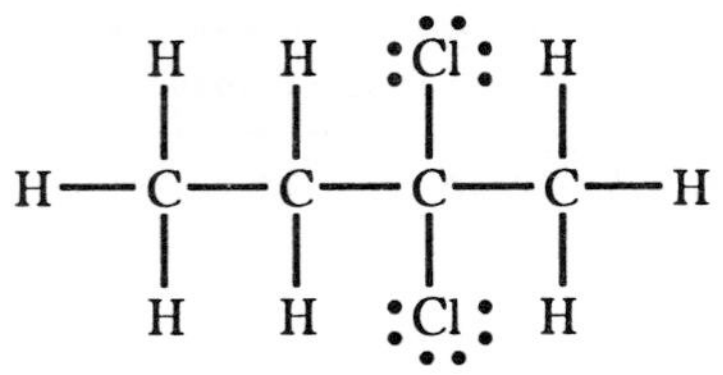 $CH_3CH_2CCl_2CH_3$ 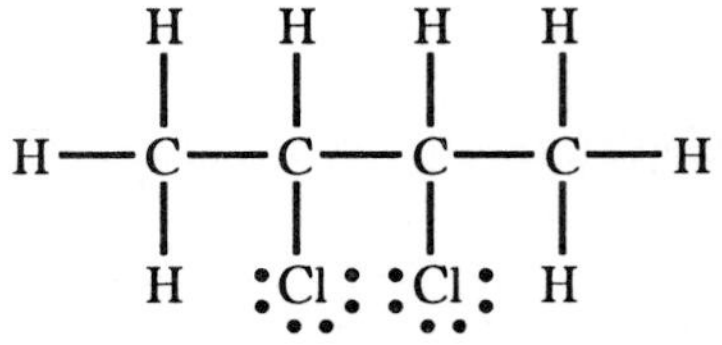 $CH_3CHClCHClCH_3$

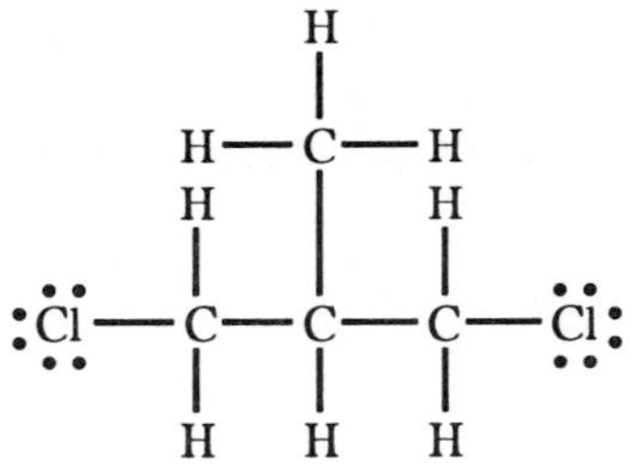

$ClCH_2CH(CH_3)CH_2Cl$ 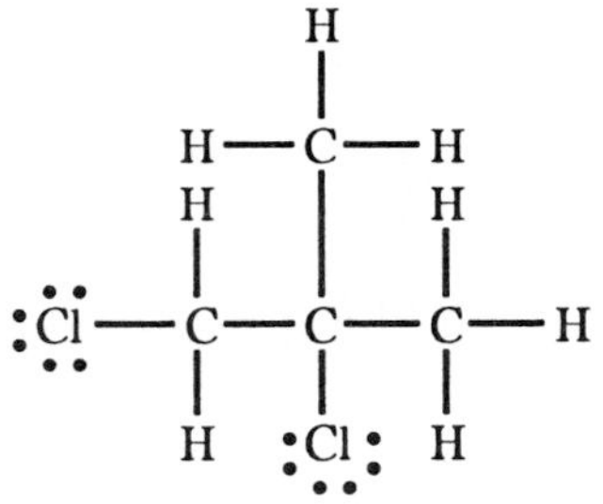 $ClCH_2CCl(CH_3)CH_3$ 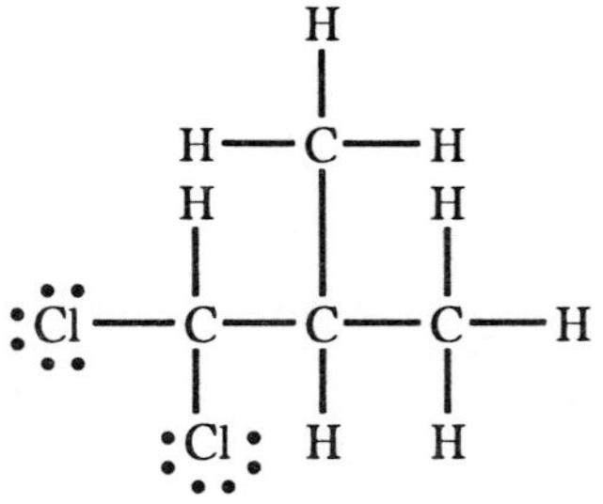 $CHCl_2CH(CH_3)CH_3$

C_2H_4O

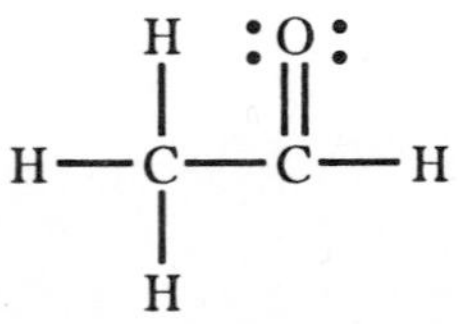

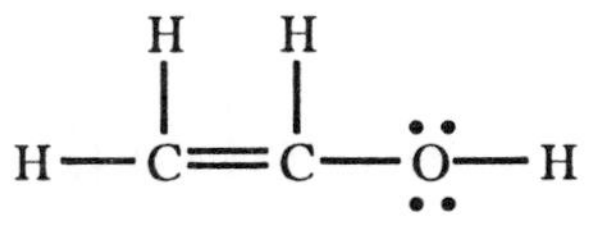

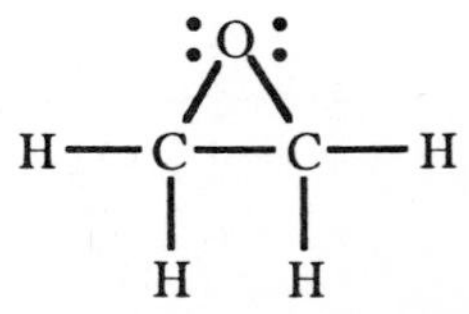

1.14 (cont)

$$\underset{CH_3\overset{\displaystyle O}{\overset{\displaystyle \|}{C}}H}{}$$

$$CH_2 = CHOH$$

enol

$$\underset{CH_2 - CH_2}{\overset{O}{\triangle}}$$

1.15　Different three-dimensional representations of the compounds and the relationships between them are shown below and on the next page. Note that there are many more possibilities for each compound besides those that are shown. It will be helpful to work with molecular models to see the different orientations in space that molecules and parts of molecules may have.

(a)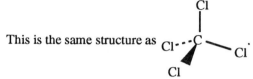

This is the same structure as

The formula representing the molecule has been rotated around an axis through the top chlorine atom and the carbon atom.

(b)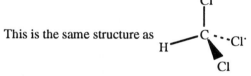

This is the same structure as

The formula representing the molecule has been rotated in the plane of the paper.

(c)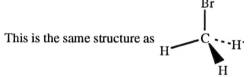

This is the same structure as

The formula representing the molecule was first rotated around an axis going through the top hydrogen atom and the carbon atom and then rotated in the plane of the paper.

(d)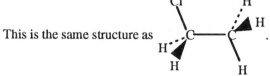

This is the same structure as

The formula representing the molecule was rotated around an axis perpendicular to the carbon-carbon bond.

(e)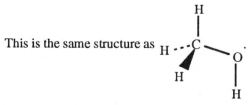

This is the same structure as

1.15 (cont)

The formula representing the molecule was changed by rotating the O—H bond around an axis going through the carbon atom and the oxygen atom without rotating the carbon atom. This type of transformation will be discussed in greater detail in Chapter 5.

(f)

This is the same structure as

The formula representing the molecule was rotated around an axis going through the carbon atom and the nitrogen atom.

(g)

This is the same structure as

The formula representing the molecule was changed by rotating the CH_3 (methyl) group on the right about an axis through the carbon-oxygen bond on the left.

(h)

This is the same structure as

The formula representing the molecule was changed by rotating the CH_3 (methyl) group on the right about an axis through the carbon-sulfur bond on the left.

1.16   (a)

molecule perpendicular to plane of paper

molecule in plane of paper

(b)

Bond angles and bond lengths in this CH_3 are the same as in the other one.

1.16 (cont)

(c)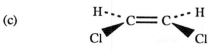

Chlorine atoms are both on
the same side of the paper.

1.09 Å
~120°
1.34 Å

Chlorine atoms are on the
opposite side of the paper.

H   ~120°   Cl
1.09 Å
1.34 Å

(d)

1.09 Å ———▶

H   180°   180°
C—C≡C—H

This bond is expected
to be less than 1.54 Å

1.20 Å 1.06 Å

(e)

H···
C══O
H

molecule perpendicular
to plane of paper

1.09 Å ———▶

H   ~120°
C══O
H

molecule in plane of paper

1.17   (a)

H ··· N—H δ+
         |
         H
        δ−

(b)

O δ−
H   H δ+

(c)

Br δ−
|
C δ+
H ···   H
    H

(d)

δ+   δ−
O══C══O

(e)

F
|
F··· C δ+
| F δ−
F

(f)

H   Cl δ−
 C══C δ+
Cl   H

(g)

H
H   C δ+—O δ−
    | |
    H   H δ+

Concept Map 1.5   Polarity of covalent molecules.

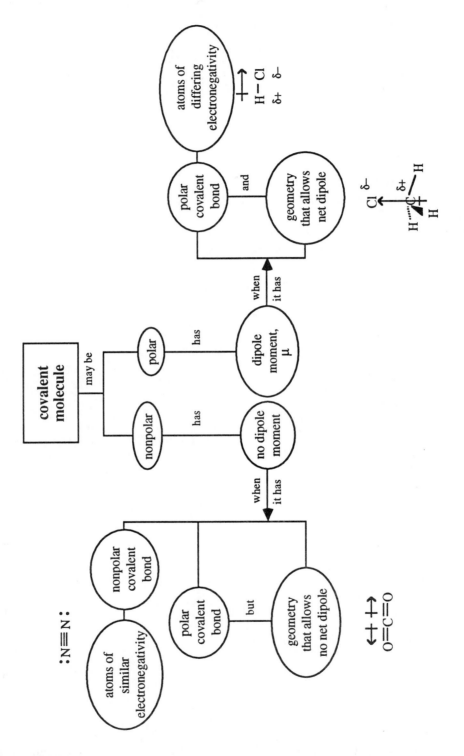

1.18   The molecular dipole is the **vector sum** of the individual bond dipoles in a molecule. A bond dipole is a **vector** quantity which by definition has both **magnitude** and **direction.** An arrow is generally used to represent a vector. The length of the arrow represents the magnitude of the charge separation and the direction of the arrow points toward the negative end of the dipole. When two vectors are added together, the direction of the dipole must be considered as well as the magnitude. Thus two vectors of equal magnitude and direction will yield a **resultant** vector pointing in the same direction with twice the magnitude as the original vectors, but two vectors of equal magnitude and opposite direction will exactly cancel out. For example, carbon dioxide is a linear molecule that has no dipole moment because it has equal bond dipoles that point in opposite directions.

vectors of individual bond
dipoles exactly cancel because
the angle between them is 180°

Water is a bent molecule that has two equal bond dipoles pointing toward the oxygen atom away from the hydrogen atoms. The angle between the two bond dipoles is 104∞. The resultant vector, the dipole moment, points toward the oxygen atom and bisects the angle between the two hydrogen atoms. The magnitude of the dipole moment is more than that of each individual bond dipole but less than the sum.

vectors of individual
bond dipoles at an angle
result in a vector sum

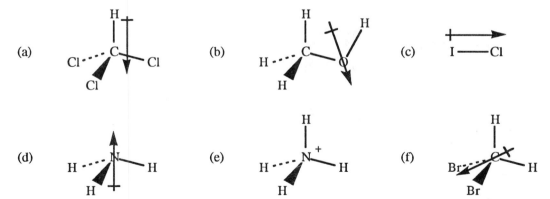

Ions have full charges, either positive
or negative, and thus do not have dipole moments, which are partial positive and negative charges separated
within the same molecule.

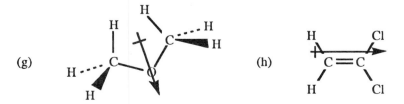

1.19   (a) $CH_3CH_2CH_2CH_2OH$ has the highest boiling point because of hydrogen bonding.

(b) CH_3OH has the highest boiling point because of hydrogen bonding.

(c) CH_3OH has the highest boiling point [see(b)]. The hydrogen bonding in CH_3SH is not significant (see p. 32 in the text).

1.20   (a) CH_3CH_2OH

1.20 (cont)

(b) $CH_3CH_2CH_2OH$

see (a) above and substitute $CH_3CH_2CH_2$— for CH_3CH_2—

(c) $HOCH_2CH_2CH_2CH_2CH_2OH$

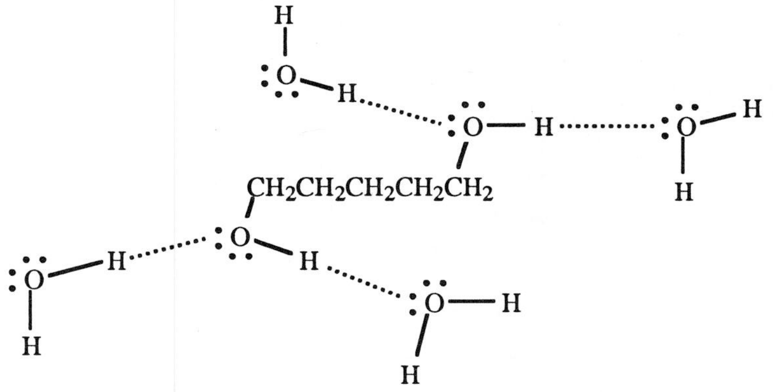

$CH_3CH_2CH_2CH_2CH_2OH$ has one half as many hydrogen bonds as $HOCH_2CH_2CH_2CH_2CH_2OH$ since it has only one —OH group.

(d) $CH_3CH_2\overset{\overset{\displaystyle O}{\|}}{C}OH$

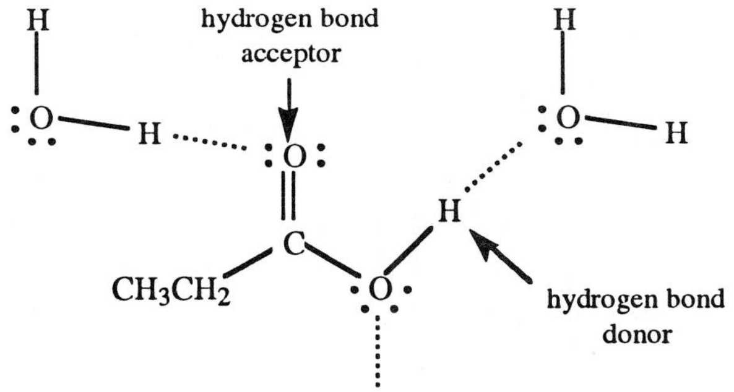

hydrogen bond acceptor

hydrogen bond donor

$CH_3\overset{\overset{\displaystyle O}{\|}}{C}OCH_3$ is only a hydrogen bond acceptor.

(e) $CH_3CH_2CH_2NH_2$

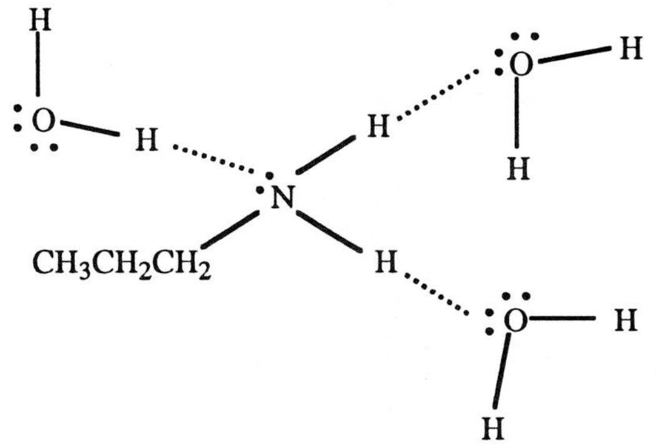

$CH_3CH_2CH_2Cl$ does not participate in hydrogen bonding to any significant extent and has little solubility in water.

Concept Map 1.6   Nonbonding interactions between molecules.

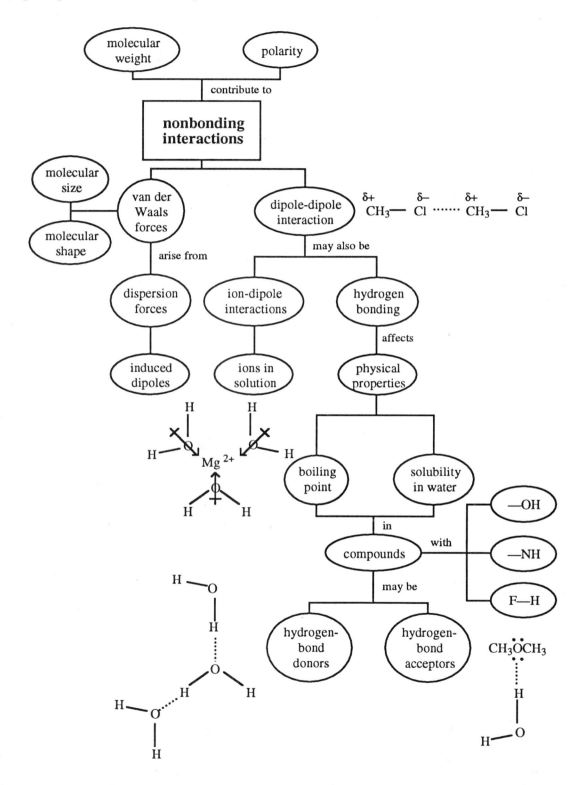

1.21 (a) MgF_2, ionic

(b) SiF_4, covalent
Si—F, polar bond

(c) NaH, ionic
(the anion is $H\!:^-$)

(d) ClF, covalent
polar bond

(e) SCl_2, covalent
S—Cl, polar bond

(f) OF_2, covalent
O—F, polar bond

(g) SiH_4, covalent
Si—H, polar bond

(h) PH_3, covalent

(i) $NaOCH_3$, ionic
(the anion, CH_3—$\ddot{\underset{\cdot\cdot}{O}}\!:^-$,
contains covalent bonds)

(j) CH_3Na, ionic
(the anion, $^-\!:CH_3$,
contains covalent bonds)

(k) Na_2CO_3, ionic
(the anion, CO_3^{-2},
contains covalent bonds)

(l) BrCN, covalent
Br—C and C≡N,
polar bonds)

1.22

(a) A Lewis structure showing $:\!\ddot{F}$—N—$\ddot{F}\!:$ with an $:\!\ddot{F}\!:$ below N.

(b) A Lewis structure showing $:\!\ddot{Cl}$—Al—$\ddot{Cl}\!:$ with a $:\!\ddot{Cl}\!:$ below Al.

(c) A Lewis structure showing H—C—S—C—H with H atoms; S has two lone pairs.

(d) A Lewis structure showing H—C—N—H with H atoms and a lone pair on N.

(e) A Lewis structure showing H—C—C—C—H with H atoms and $:\!\ddot{Cl}\!:$ below the central C.

(f) A Lewis structure showing $^-:\!\ddot{O}$—H.

(g) A Lewis structure showing H—C—C—C—$\ddot{O}$—H with H atoms.

(h) A Lewis structure showing H—$\ddot{O}$—$\ddot{O}$—H.

(i) A Lewis structure showing H—C—$\ddot{N}$—$\ddot{O}$—H with H atoms.

(j) A Lewis structure showing H—C—$\ddot{S}$—H with H atoms.

(k) A Lewis structure showing H—Si—H with H atoms.

(l) A Lewis structure showing H—$\ddot{O}$—S—$\ddot{O}$—H with $:\!O\!:$ double-bonded above and $:\!O\!:$ double-bonded below S.

Here sulfur has 12 electrons around it but a zero formal charge.

S $6 - 0 - \frac{1}{2}(12) = 0$

(m) A Lewis structure showing H—$\ddot{O}$—$\overset{+}{N}$—$\ddot{O}\!:^-$ with $:\!O\!:$ double-bonded above N.

1.23

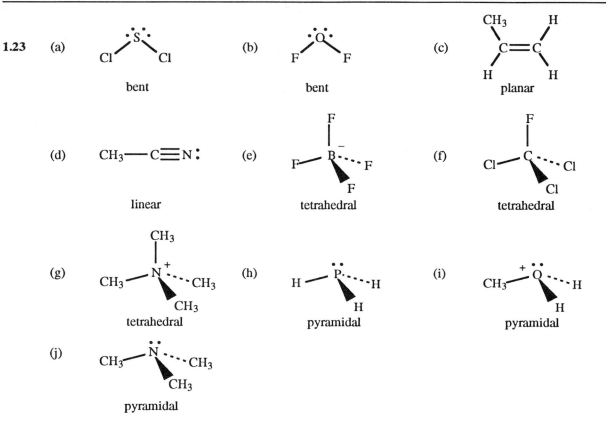

(a) bent

(b) bent

(c) planar

(d) linear

(e) tetrahedral

(f) tetrahedral

(g) tetrahedral

(h) pyramidal

(i) pyramidal

(j) pyramidal

Note that the pyramidal shapes are derived from the tetrahedron. The distribution in space of the bonding and nonbonding electron pairs around these atoms is tetrahedral. The geometry around these atoms appears pyramidal because our experimental methods detect only the positions of the atoms and not those of the electrons.

1.24 (a)

(b)

(c) does not have a significant dipole moment

(d) C—F bond dipole is much greater than C—Br bond dipoles

1.24 (cont)

(e)

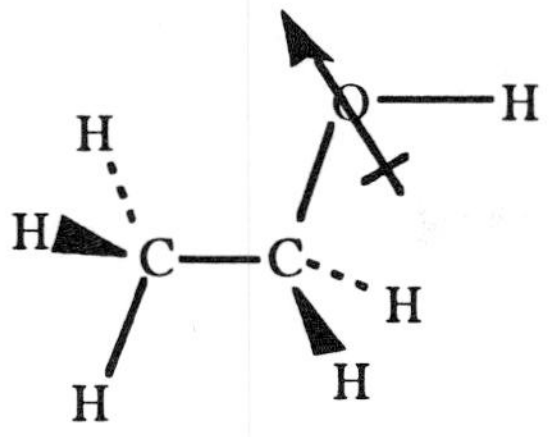

(f)

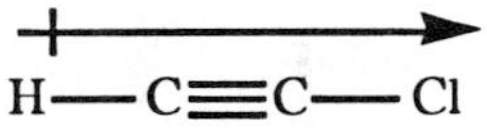

(g)

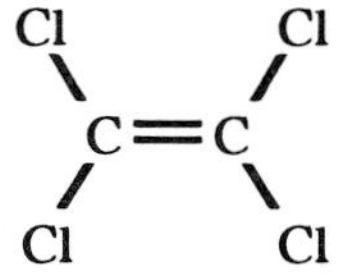

individual bond dipoles cancel

1.25 (a) Hydrogen atoms are attached to carbon, not oxygen, nitrogen or fluorine. Hydrogen bonding is not important.

(b) There is a hydrogen atom attached to oxygen, so hydrogen bonding is important.

(c) No hydrogen atoms are on oxygen, nitrogen or fluorine, so no hydrogen bonding.

(d) There is a hydrogen atom attached to oxygen, so hydrogen bonding is important.

(e) Hydrogen atoms are attached to nitrogen, so hydrogen bonding does occur.

(f) Here, too, hydrogen bonding occurs because of presence of N—H bonds.

(g) The hydrogen attached to sulfur does not participate significantly in hydrogen bonding.

1.26

(a) Yes

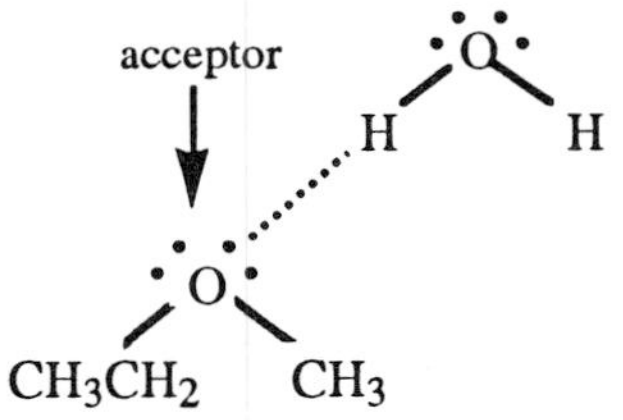

(b) Yes

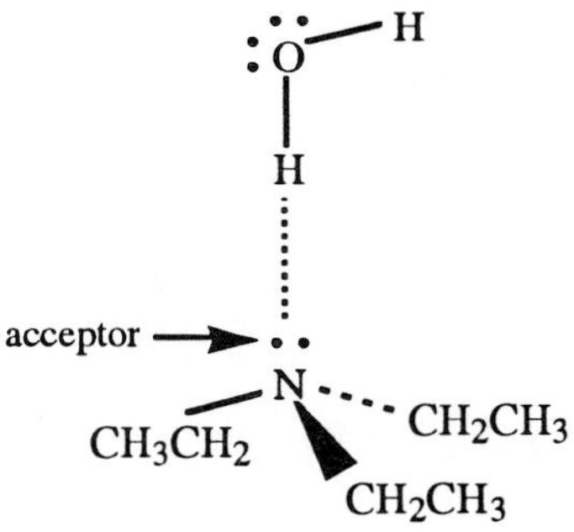

(c) Yes

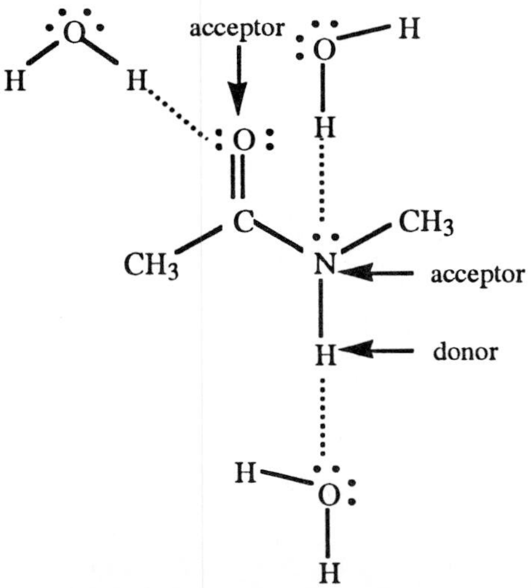

(d) Yes

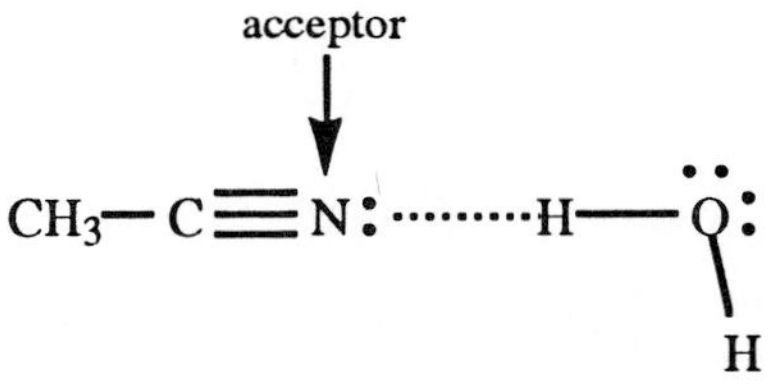

1.26 (cont)

(e)   Yes

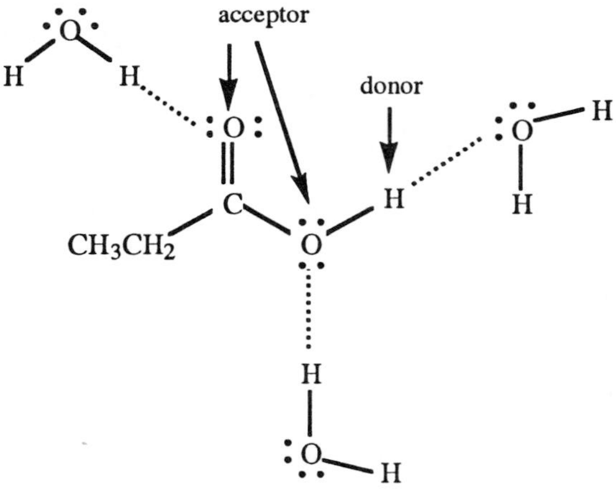

(f)   No

(g)   Yes

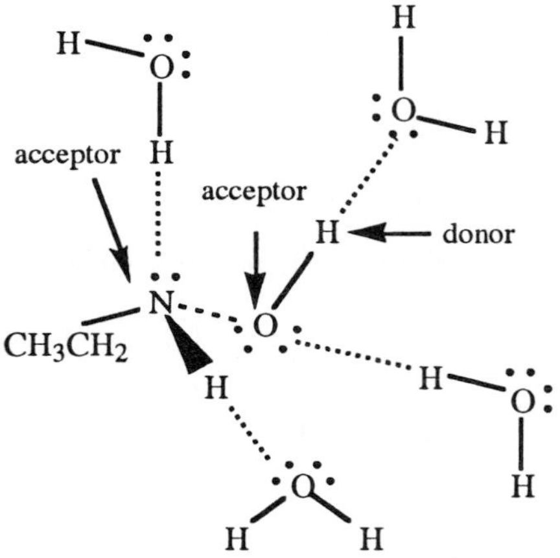

(h)   No

(i)   Yes

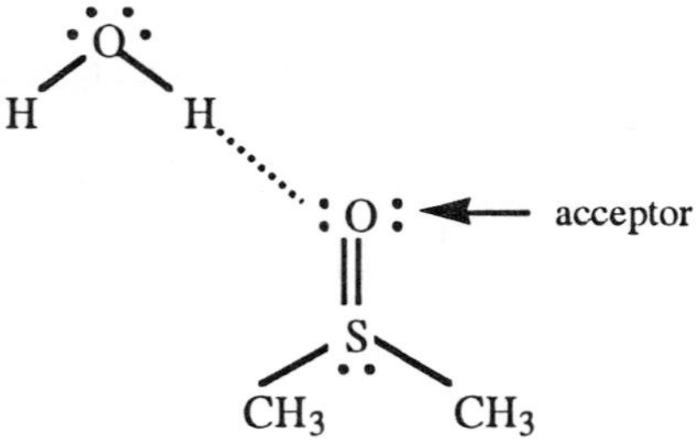

Sulfur is not a significant hydrogen
bond acceptor

1.27   (a) NaCl is more soluble (ionic compound, strong ion-dipole interactions)

(b)  $\overset{\overset{\displaystyle O}{\|}}{\text{H}}\text{COH}$ (does not have nonpolar carbon chain)

(c)  $\overset{\overset{\displaystyle O}{\|}}{\text{H}}\text{COH}$ (hydrogen bond acceptor and donor, does not have nonpolar carbon chain)

1.27 (cont)

    (d) $CH_3CH_2OCH_2CH_3$ (larger dipole, oxygen more electronegative than sulfur; also hydrogen bond acceptor)

    (e) $HOCCH_2CH_2COH$ (two hydrogen bond donor and four acceptor sites)

1.28

In water, all atoms are involved in hydrogen bonding. A strong, three-dimensional network results.

In hydrogen fluoride, only two atoms are involved in hydrogen bonding, giving rise to a weaker linear network.

In ethanol, only two atoms are involved in hydrogen bonding. The rest of the molecule is nonpolar and involved in van der Waals interactions, which are much weaker.

1.29   :F — B — F: ↔   F = B — F: ↔   :F — B = F   ↔   :F — B — F:
              | | | ‖
             :F: :F: :F: :F:

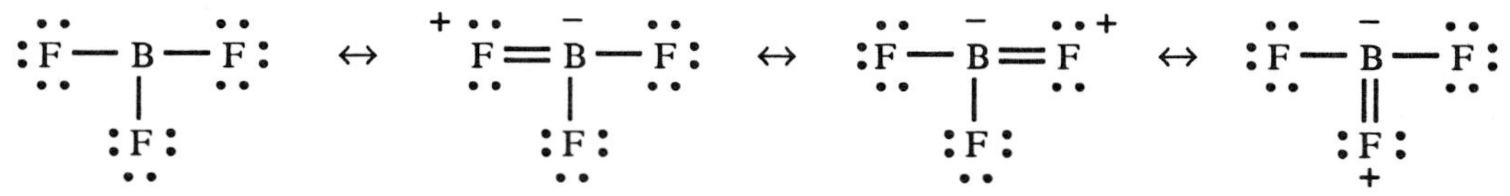

The shorter-than-expected B—F bond lengths suggest that the compound has resonance contributors in which there is a double bond between the boron and each of the fluorine atoms. Such resonance contributors have the advantage of having an octet around each atom and additional bonding, even though they also put a positive charge on the electronegative fluorine atom.

1.30   (a)

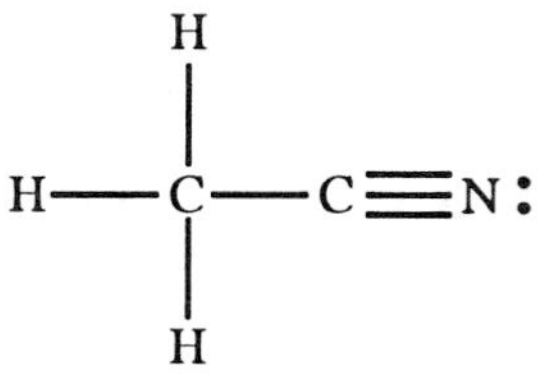

          H
          |
    H —— C —— C ≡≡ N :   Lewis structure
          |
          H

       (b) 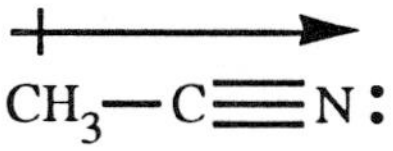   CH₃ — C ≡≡ N :   dipole moment

       (c) 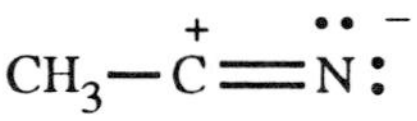   CH₃ — C ≡≡ N :   Resonance contributor with separation of charge and negative formal charge on the more electronegative element. A separation of charge increases the dipole moment.

1.31

(a) 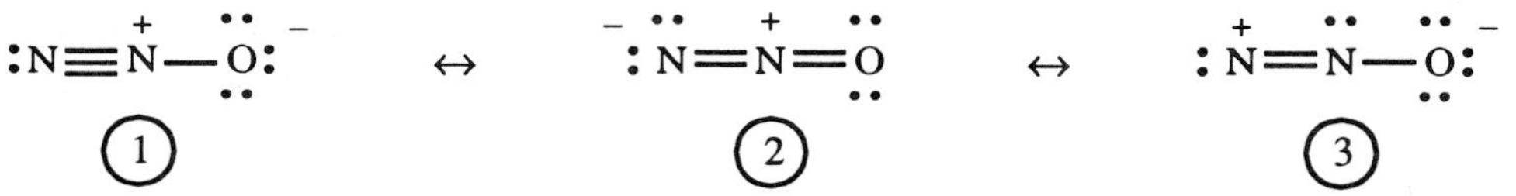   :N ≡ N — O: ↔   : N = N = O   ↔   : N = N — O:

          ①   ②   ③

        major   minor
    complete octets;   complete octets;   nitrogen has sextet;
  negative charge on more   negative charge on less large separation of
    electronegative atom   electronegative atom   charge

(b)

         : O: : O: : O:
          ‖   ‖   ‖
    : O — S — O: ↔   O = S — O: ↔   : O — S = O
          ‖ | |
         : O: : O: : O:
                                                         ↕

         : O: : O: : O:
          | | |
    O = S = O   ↔   O = S — O: ↔   : O — S = O
          | ‖   ‖
         : O: : O: : O:

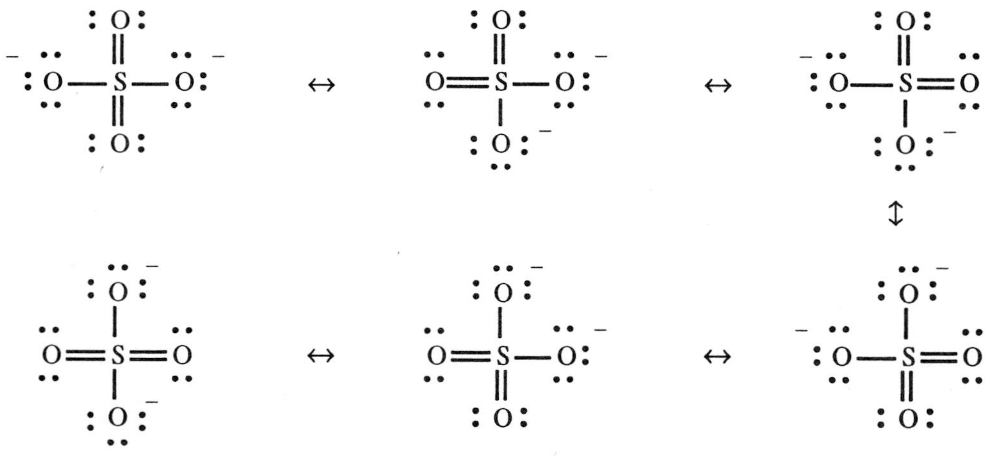

                              major
   all equivalent with no separation of charge and at least an octet around each atom

1.31 (b) (cont)

$$\begin{array}{ccc} \text{(structures)} & \leftrightarrow & \text{(structures)} \end{array} \qquad \leftrightarrow \qquad \text{(three others)};$$

minor
all have separation of charge

(c) $CH_3 — C — O — H$ (1) $\leftrightarrow$ $CH_3 — C = O — H$ (2) $\leftrightarrow$ $CH_3 — C — O — H$ (3)

major	minor	
complete octets;	complete octets;	carbon has sextet;
no separation of charge	large separation of charge	separation of charge

1.32   (a) $:C = O$ $^-:C \equiv O:^+$ $^+:C — O:^-$

| carbon does not have octet | both atoms have octets but negative charge is on less electronegative atom | carbon does not have an octet but negative charge is on more electronegative atom |

(b) Fe^{2+} is an electron pair **acceptor.** The carbon atom of carbon monoxide bonds to iron; it must be an electron pair **donor.** Therefore, the resonance contributor in which carbon has a negative charge and each atom has a complete octet is more important than the other contributors. The resonance contributor with separation of charge and a positive charge on the carbon atom would lead to electrostatic repulsion between Fe^{2+} and the carbon atom and must, therefore, not make a significant contribution to the structure.

1.33 $\overset{-}{N} = \overset{+}{N} = \overset{-}{N}$ (1) $\leftrightarrow$ $^{2-}:N — \overset{+}{N} \equiv N:$ $\leftrightarrow$ $:N \equiv \overset{+}{N} — N:^{2-}$ (2)

major	minor
all atoms have a completeoctet;	all atoms have a complete octet;
separation of charge with one	separation of charge with both
negative charge on each end nitrogen	negative charges on one atom

1.34

$$
\begin{array}{ccc}
\underset{\substack{(1)\\ \text{major}\\ \text{complete octets;}\\ \text{no separation of charge}}}{
\chemfig{O=C(-[6]N(-)(-))-}}
&\longleftrightarrow&
\underset{\substack{(2)\\ \text{minor}\\ \text{complete octets; separation}\\ \text{of charge with negative charge}\\ \text{on more electronegative atom}}}{
\chemfig{^-O-C(=[6]N^+)}}
&\longleftrightarrow&
\underset{\substack{(3)\\ \text{carbon has sextet;}\\ \text{separation of charge}}}{
\chemfig{^-O-C^+(-[6]N)}}
\end{array}
$$

Structure (1): $:\overset{\cdot\cdot}{\underset{}{O}}:$ double-bonded to C, C bonded to CH$_3$ and N(CH$_3$) with lone pair on N.
— **major; complete octets; no separation of charge**

Structure (2): $\overset{\cdot\cdot}{\underset{\cdot\cdot}{O}}{}^{-}$ single bond to C, C double bond to N^{+}.
— **minor; complete octets; separation of charge with negative charge on more electronegative atom**

Structure (3): $\overset{\cdot\cdot}{\underset{\cdot\cdot}{O}}{}^{-}$ single bond to C^{+}, N with lone pair.
— **carbon has sextet; separation of charge**

1.35

(a)

$$\text{H}-\overset{\overset{\textstyle H}{|}}{\underset{\underset{\textstyle H}{|}}{\text{C}}}-\overset{\cdot\cdot}{\text{N}}=\text{C}=\overset{\cdot\cdot}{\underset{\cdot\cdot}{\text{S}}}$$

(b)

$$\text{H}-\overset{\overset{\textstyle H}{|}}{\underset{\underset{\textstyle H}{|}}{\text{C}}}-\overset{\cdot\cdot}{\text{N}}\overset{+}{=}\overset{\cdot\cdot}{\text{C}}-\overset{\cdot\cdot}{\underset{\cdot\cdot}{\text{S}}}:^{-}
\quad\longleftrightarrow\quad
\text{H}-\overset{\overset{\textstyle H}{|}}{\underset{\underset{\textstyle H}{|}}{\text{C}}}-\overset{-}{\underset{\cdot\cdot}{\text{N}}}-\overset{+}{\text{C}}=\overset{\cdot\cdot}{\underset{\cdot\cdot}{\text{S}}}$$

(c)

$$\text{H}-\overset{\overset{\textstyle H}{|}}{\underset{\underset{\textstyle H}{|}}{\text{C}}}-\overset{\cdot\cdot}{\underset{\cdot\cdot}{\text{S}}}-\text{C}\equiv\text{N}:$$

1.36

(a)

$$
{}^{-}:\overset{\overset{\textstyle H}{|}}{\underset{\underset{\textstyle H}{|}}{\text{C}}}-\overset{+}{\underset{}{\text{N}}}\overset{\overset{\textstyle :\overset{\cdot\cdot}{O}:^{-}}{|}}{=}\overset{\cdot\cdot}{\underset{\cdot\cdot}{\text{O}}}
\;\longleftrightarrow\;
{}^{-}:\overset{\overset{\textstyle H}{|}}{\underset{\underset{\textstyle H}{|}}{\text{C}}}-\overset{+}{\underset{}{\text{N}}}\overset{\overset{\textstyle :O:}{\|}}{-}\overset{\cdot\cdot}{\underset{\cdot\cdot}{\text{O}}}:^{-}
\;\longleftrightarrow\;
\overset{\overset{\textstyle H}{|}}{\underset{\underset{\textstyle H}{|}}{\text{C}}}=\overset{+}{\underset{}{\text{N}}}\overset{\overset{\textstyle :\overset{\cdot\cdot}{O}:^{-}}{|}}{-}\overset{\cdot\cdot}{\underset{\cdot\cdot}{\text{O}}}:^{-}
$$

    minor   minor   major

All have the same separation of charge but the one above has the negative charge localized on the most electronegative atoms, the oxygens.

(b) The two minor resonance contributors above account for the reactivity of the nitromethane ion with a positively charged species, which forms a bond to the carbon atom.

1.37 SO_2 is an 18-electron molecule.

Note that this structure is possible because S is in the third period of the periodic table.

These resonance contributors would lead us to expect the bond lengths is SO_2 to be somewhere between 1.49 Å and 1.70 Å. The experimental bond length (1.43 Å) is shorter than expected. Chemists point to the polarity of the bonds (the strong attraction between the positive charge on sulfur and the negative charge on oxygen, for example) as a further factor in rationalizing such observations.

1.38

(a)

The nonbonded electron pair on the nitrogen has become a bond between nitrogen and carbon and is not as available for bonding.

(b)

One nonbonded electron pair on nitrogen has become a bond between the two nitrogens. One of the bonding electron pairs between carbon and nitrogen is now a nonbonding pair on carbon and is available for bonding to reagents that are deficient in electrons.

(c)

One of the nonbonded pairs on the oxygen atom is now a bonding pair between the oxygen atom and carbon 1. The bonding pair between carbons 1 and 2 has become a nonbonding pair first on carbon 2, then on carbon 4 and finally on carbon 6 with the negative charge that was on oxygen now on these carbons.

1.39   (a)

(b)   Each has a resonance contributor in which the carbon atom marked *C has a positive charge. Thus each compound has a partial positive character at that carbon atom, making it reactive toward a negatively charged ion such as RS:⁻ .

2

Covalent Bonding and Chemical Reactivity

Workbook Exercises

Simple geometrical relationships are an important part of chemistry. Some of these can be demonstrated with the two tetrahedral blocks that you can construct using the two large cardboard triangles at the back of the *Study Guide*. Fold each along the three dotted lines to bring the three darkened corners together, tuck the flaps inside, and then tape the sides together to make a tetrahedral block.

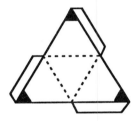

Hold the two tetrahedral blocks as indicated in the pictures below. Make special note of the spatial relationships between the vertices that are *not* touching.

(1) Sharing a single vertex
(touch A to A)

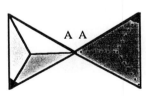

the remaining 6 vertices
can have many orientations,
one of which is shown

(2) Sharing an edge
(A to A and B to B)

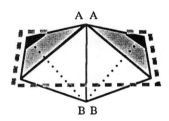

the remaining 4 vertices
are coplanar
(What happens if you connect
A to B and B to A instead?)

(3) Sharing a face
(A to A, B to B, C to C)

the remaining 2 vertices
are colinear

There are many good sets of molecular models available commercially. You should develop the habit of using molecular models to help visualize the three-dimensional relationships of atoms in molecules. Using your set of molecular models, assemble 2 tetrahedral atoms and then join them together in a singly bonded connection. You will find that a vertex or bond of one tetrahedral atom will need to be connected directly to the other tetrahedral atom, rather than to a vertex of the other atom. Learn how to construct double and triple bonds with your model set and compare the geometry of the atoms in those constructions with those of the tetrahedral blocks. Keep the cardboard tetrahedra that you have constructed. You will need them for the exercises in later chapters.

	1s	2s	2p			3s	3p		
2.1 sodium	⇅	⇅	⇅	⇅	⇅	↑			
magnesium	⇅	⇅	⇅	⇅	⇅	⇅			
aluminum	⇅	⇅	⇅	⇅	⇅	⇅	↑		
silicon	⇅	⇅	⇅	⇅	⇅	⇅	↑	↑	
phosphorus	⇅	⇅	⇅	⇅	⇅	⇅	↑	↑	↑
sulfur	⇅	⇅	⇅	⇅	⇅	⇅	⇅	↑	↑
chlorine	⇅	⇅	⇅	⇅	⇅	⇅	⇅	⇅	↑
argon	⇅	⇅	⇅	⇅	⇅	⇅	⇅	⇅	⇅

2.2

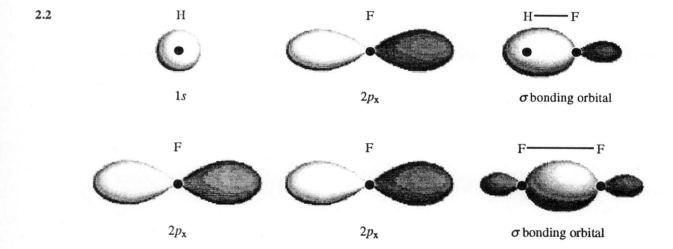

2.3

(a) No overlap possible; wrong orientation of *p* orbital in space. The *s* orbital is approaching the nodal plane of the *p* orbital where there is no electron density.

(b) No overlap possible; wrong orientation of one *p* orbital in space.

(c) No bonding will occur. The orbitals are out of phase, and only an antibonding interaction will occur.

(d) Bonding will occur. The orbitals are in phase and are approaching each other so that overlap is possible.

Concept Map 2.1 Molecular orbitals and covalent bonds.

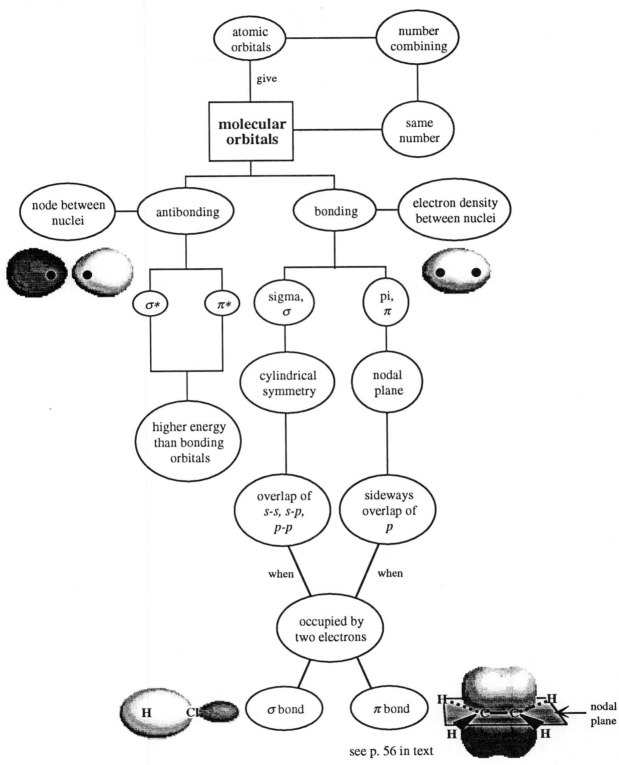

see p. 56 in text

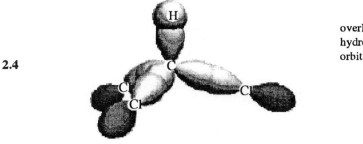

2.4

overlap of 1s orbital on
hydrogen and sp³ hybrid
orbital on carbon

overlap of 3p orbital on
chlorine and sp³ hybrid
orbital on carbon

overlap of 1s orbital on
hydrogen and sp³ hybrid
orbital on carbon

overlap of two sp³ hybrid
orbitals, one on each carbon

2.5

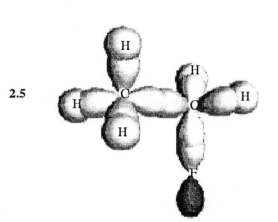

overlap of 2p orbital on
fluorine and sp³ hybrid
orbital on carbon

2.6   The three 2p orbitals on nitrogen are 90° apart. Bonds formed by the overlap of these orbitals would have bond
angles close to 90°. The actual bond angles in ammonia have been experimentally determined to be 107°.

2.7   nonbonding electrons

2.8   nonbonding electrons

methanol

2.8 (cont)

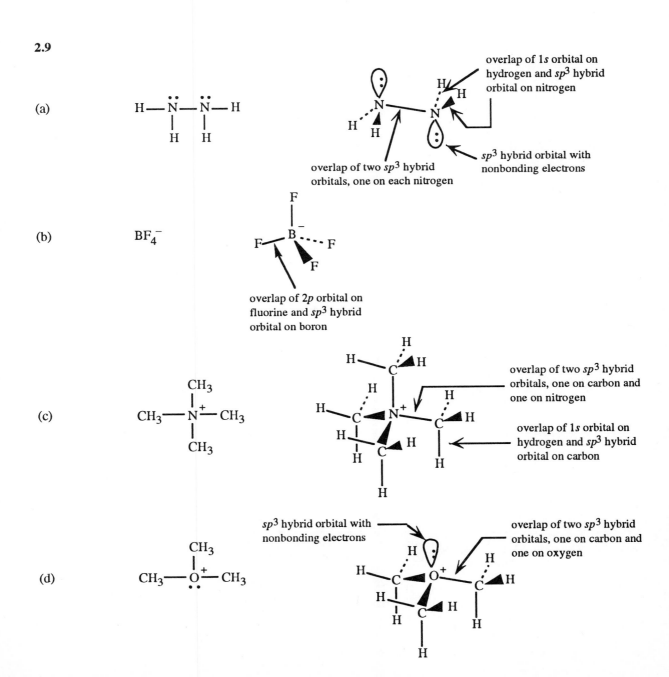

2.9

(a) H—N̈—N̈—H

overlap of 1s orbital on hydrogen and sp³ hybrid orbital on nitrogen

overlap of two sp³ hybrid orbitals, one on each nitrogen

sp³ hybrid orbital with nonbonding electrons

(b) BF₄⁻

overlap of 2p orbital on fluorine and sp³ hybrid orbital on boron

(c)

CH₃—N⁺—CH₃ with CH₃ above and CH₃ below

overlap of two sp³ hybrid orbitals, one on carbon and one on nitrogen

overlap of 1s orbital on hydrogen and sp³ hybrid orbital on carbon

(d)

CH₃—O⁺—CH₃ with CH₃ above

sp³ hybrid orbital with nonbonding electrons

overlap of two sp³ hybrid orbitals, one on carbon and one on oxygen

2.10 (a) alcohol   (b) ether   (c) alkane

(d) amine   (e) alcohol   (f) amine

2.11   Reaction with H_2SO_4?

(a)

$CH_3CH_2CH_2CH_2CH_2$ H   Yes

(b)

$CH_3CH_2CH_2$ CH_3 Yes

(c)   No

(d) CH_3CH_2 —N····H   Yes

(e)

CH_3CH_2CH H   Yes

CH_3

(f) CH_3CH_2—N····H   Yes

CH_2CH_3

2.12 (a)

Oxygen is not hybridized; spectro-
scopic evidence shows the two pairs
of nonbonding electrons on oxygen to
be different in energy. Therefore we
use a model that has the electrons in
different orbitals, the $2s$ and $2p_y$.

overlap of an sp^2 hybrid
orbital on carbon and a
$2p_x$ orbital on oxygen

sp^2

overlap of an sp^2 hybrid
orbital on carbon with s
orbital on hydrogen

overlap of an sp^3 hybrid orbital
on carbon with sp^2 hybrid
orbital on the second carbon

sp^3

overlap of an sp^3 hybrid
orbital on carbon with s
orbital on hydrogen

σ bond backbone

2.12 (a) (cont)

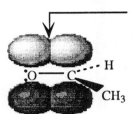

overlap of two $2p_z$ orbitals, one on carbon and one on oxygen, in phase with each other

π bond

(b)   nonbonding electrons in sp^2 hybrid orbitals $\Big\{$

$\leftarrow p_z$ orbital

sp^2

In this model, the energy levels of the nonbonding electrons should be the same.

2.13

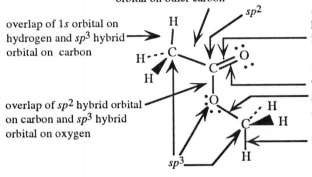

overlap of sp^3 hybrid orbital on one carbon and sp^2 hybrid orbital on other carbon

overlap of 1s orbital on hydrogen and sp^3 hybrid orbital on carbon

overlap of sp^2 hybrid orbital on carbon and sp^3 hybrid orbital on oxygen

σ bond, overlap of sp^2 hybrid orbital on carbon and $2p_x$ orbital on oxygen

π bond, overlap of $2p_z$ orbitals, one on carbon and one on oxygen

overlap of two sp^3 hybrid orbitals, on one carbon and one on oxygen

overlap of 1s orbital on hydrogen and sp^3 hybrid orbital on carbon

2.14   (a) alcohol   (b) alkene   (c) aldehyde

(d) ketone   (e) amine   (f) alkane

(g) carboxylic acid   (h) ether   (i) ester

2.15   (a) alcohol and carboxylic acid   (b) amine and carboxylic acid   (c) ketone and carboxylic acid

(d) alcohol and aldehyde   (e) alkene and carboxylic acid

2.16

(a)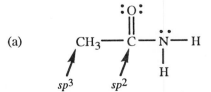

(b)
$$CH_3CH_2-\overset{\displaystyle :\overset{..}{O}:}{\overset{\|}{C}}-\overset{..}{\underset{..}{O}}-\overset{H}{\underset{}{C}}=\overset{H}{\underset{}{C}}-H$$

sp^3 sp^2

2.16 (cont)

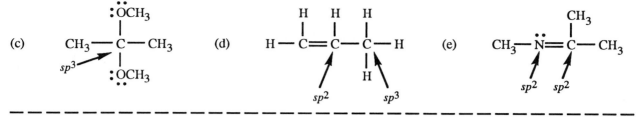

(c)   CH₃—C—CH₃   (d)   H—C=C—C—H   (e)   CH₃—N=C—CH₃

- -

Concept Map 2.2   Hybrid orbitals.

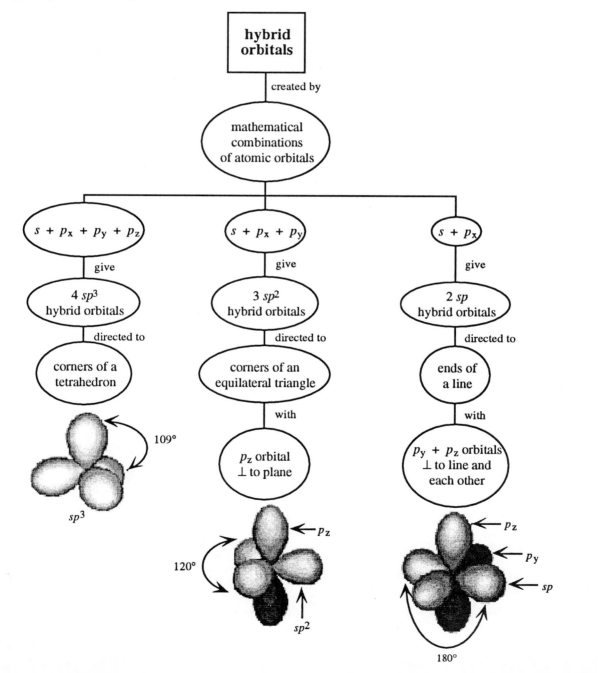

2.17   (a) alkyne   (b) alkene   (c) ketone   (d) ester

         (e) nitrile   (f) aldehyde   (g) carboxylic acid (h) amine

2.18   (a)

$$H—C=O$$
$$|$$
$$H$$

$3\sigma, 1\pi$

(b)

H   H
| |
H—C—C—O—H
| |
H   H

8σ

(c)

$$H—C\equiv C—C\equiv C—H$$

$5\sigma, 4\pi$

(d)

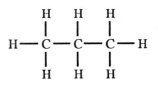

$6\sigma, 3\pi$

(e)

H   O   H
| || |
H—C—C—O—C—H
| |
H   H

$10\sigma, 1\pi$

(f)

H   H   H
| | |
H—C—C—C—H
| | |
H   H   H

10σ

2.19

overlap of 1s orbital
on hydrogen and *sp*
hybrid orbital on carbon

overlap of two *sp* hybrid
orbitals, one on carbon and
one on nitrogen

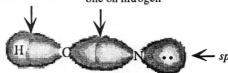

← *sp*

σ bond backbone

overlap of two sets of *p*
orbitals, two each on
nitrogen and carbon,
which are at angles of
90° to each other

π bonds

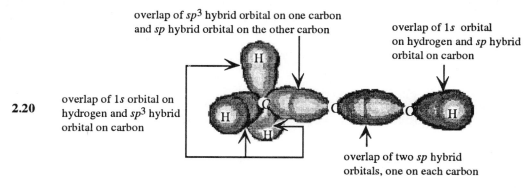

overlap of sp^3 hybrid orbital on one carbon
and sp hybrid orbital on the other carbon

overlap of $1s$ orbital
on hydrogen and sp hybrid
orbital on carbon

2.20   overlap of $1s$ orbital on
hydrogen and sp^3 hybrid
orbital on carbon

overlap of two sp hybrid
orbitals, one on each carbon

σ bond backbone

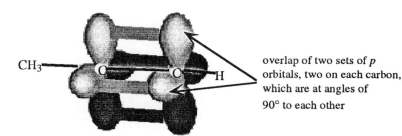

CH₃

overlap of two sets of p
orbitals, two on each carbon,
which are at angles of
90° to each other

π bonds

2.21

(a)   CH₃⊰H   +   Cl⊰Cl   ⟶   CH₃⊰Cl   +   H⊰Cl

Bond energies (kcal/mol)   99   58   81   103

$$\Delta H_r = (99 + 58) - (81 + 103) = -27 \text{ kcal/mol}$$

(b)   H₂C⊰O   +   H⊰C≡N   ⟶   H⊰O⊰CH₂⊰C≡N

Bond energies (kcal/mol)   176   99   111   86   83

$$\Delta H_r = (176 + 99) - (111 + 86 + 83) = -5 \text{ kcal/mol}$$

(Note that to calculate ΔH_r in a molecule in which a π bond is broken, both the π
bond and the σ bond are broken and the σ bond is reformed in the product.)

(c)   H₂C⊰CH₂   +   H⊰Br   ⟶   H⊰CH₂⊰CH₂⊰Br

146   87   99   83   68

Bond energies (kcal/mol) $\Delta H_r = (146 + 87) - (99 + 83 + 68) = -17 \text{ kcal/mol}$

                                           83   146   111

(d)   CH₃—CH⊰CH₂   ⟶   CH₃—CH⊰CH₂   +   H⊰O—H

         86 ⌇⌇ ⌇⌇ 99
            OH   H

Bond energies (kcal/mol) $\Delta H_r = (86 + 83 + 99) - (146 + 111) = +11 \text{ kcal/mol}$

Concept Map 2.3   Bond lengths and bond strengths.

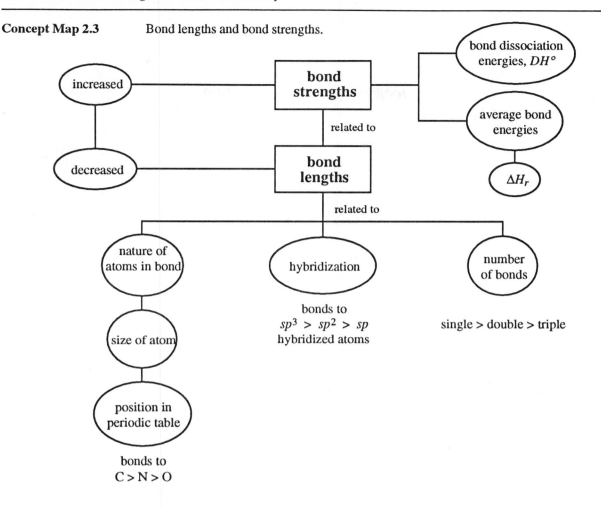

2.22   (a)

chloroethane   chloroethene

(b)

A resonance contributor of chloroethene has a double bond between carbon and chlorine, making the bond shorter (1.72 Å) than expected for a carbon-chlorine single bond (1.78 Å). A bond with partial double bond character is also stronger than a single bond, and therefore has a higher bond dissociation energy.

2.22 (cont)

(c)

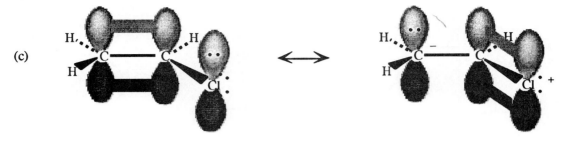

(d)   The carbon atom to which the chlorine is bonded in chloroethene is an sp^2-hybridized carbon atom. An sp^2-hybridized carbon atom forms a shorter and stronger bond to chlorine than the sp^3-hybridized carbon atom of chloroethane (p. 74 in the text).

2.23

(a) $CH_3CH_2CH = CHCH_3$ + Br_2 $\xrightarrow[\substack{\text{carbon} \\ \text{tetrachloride} \\ \text{dark, 25 °C}}]{}$ $CH_3CH_2\underset{\underset{Br}{|}}{C}H\underset{\underset{Br}{|}}{C}HCH_3$

(b) $CH_3C \equiv CCH_3$ + $2\,Br_2$ $\xrightarrow[\substack{\text{carbon} \\ \text{tetrachloride} \\ \text{dark, 25 °C}}]{}$ $CH_3\underset{\underset{Br}{|}}{\overset{\overset{Br}{|}}{C}}-\underset{\underset{Br}{|}}{\overset{\overset{Br}{|}}{C}}CH_3$

Br_2 arrow down to:

$\underset{Br}{\overset{CH_3}{C}} = \underset{CH_3}{\overset{Br}{C}}$ $\xrightarrow{Br_2}$

(c) $HC \equiv CCH_2C \equiv CH$ + $4\,Br_2$ $\xrightarrow[\substack{\text{carbon} \\ \text{tetrachloride} \\ \text{dark, 25 °C}}]{}$ $HC\underset{\underset{Br}{|}}{\overset{\overset{Br}{|}}{}}-\underset{\underset{Br}{|}}{\overset{\overset{Br}{|}}{C}}CH_2\underset{\underset{Br}{|}}{\overset{\overset{Br}{|}}{C}}-\underset{\underset{Br}{|}}{\overset{\overset{Br}{|}}{C}}H$

Br_2 arrow down to:

$\underset{Br}{\overset{H}{C}} = \underset{CH_2C \equiv CH}{\overset{Br}{C}}$ $\xrightarrow{Br_2}$ $\underset{Br}{\overset{H}{C}} = \overset{Br}{\underset{\underset{\underset{Br}{\overset{C}{|}}}{CH_2}}{C}} = C$ ... $\xrightarrow{Br_2}$ $HCBr_2CBr_2CH_2 \quad C = C \quad Br / H$ $\uparrow Br_2$

2.23 (cont)

(d) 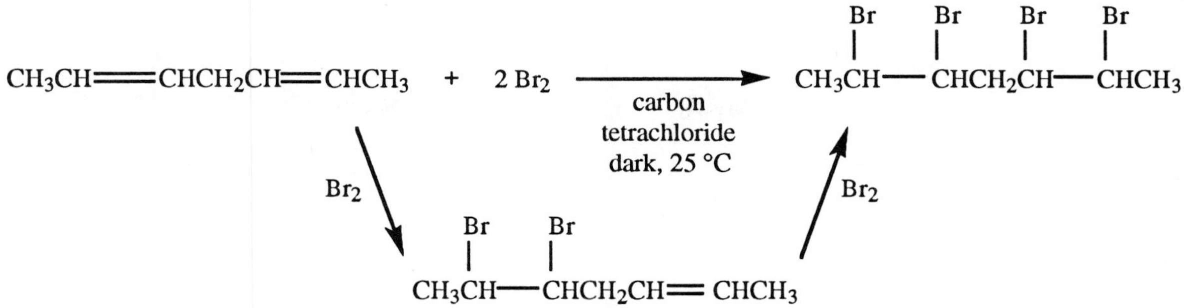 CH_3CH═══$CHCH_2CH$═══$CHCH_3$ +   2 Br_2 $\xrightarrow[\substack{\text{carbon} \\ \text{tetrachloride} \\ \text{dark, 25 °C}}]{}$ $\underset{\underset{|}{Br}}{CH_3CH}$—$\underset{\underset{|}{Br}}{CHCH_2CH}$—$\underset{\underset{|}{Br}}{CHCH_3}$

$\Big\downarrow Br_2$ $\Big\uparrow Br_2$

$\underset{\underset{|}{Br}}{CH_3CH}$—$\underset{\underset{|}{Br}}{CHCH_2CH}$═══$CHCH_3$

2.24   Alkanes: Tetrahedral carbon atoms bonded only to other tetrahedral carbon or hydrogen atoms

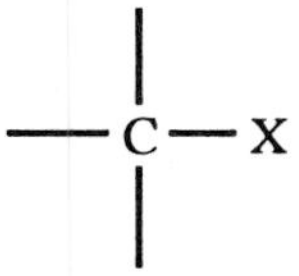

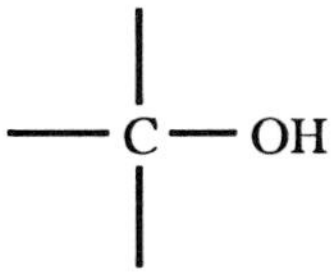

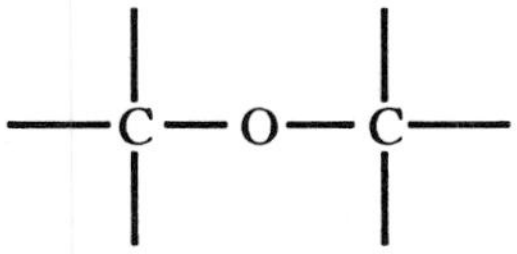

Alkenes: —C═C—   Alkynes: —C≡C—

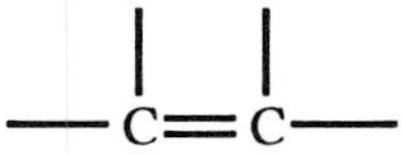

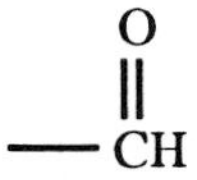

Alkyl halides: —C—X   Alcohols: —C—OH

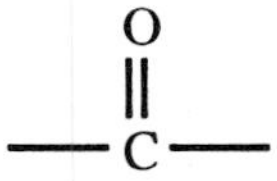

Ethers: —C—O—C—   Aldehydes: —CH (═O)

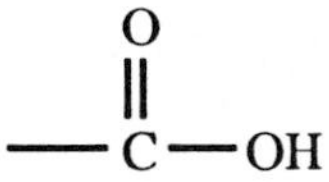

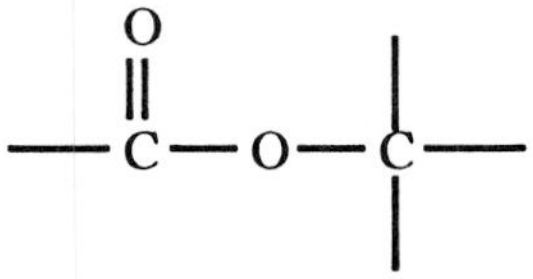

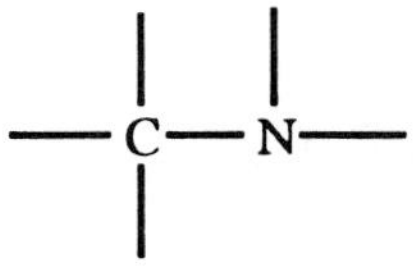

Ketones: —C— (═O)   Carboxylic acids: —C—OH (═O)

Esters: —C—O—C— (═O)   Amines: —C—N—

Nitriles: —C≡N   Aromatic hydrocarbons:

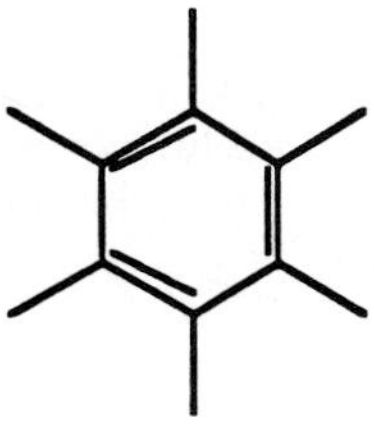

2.25

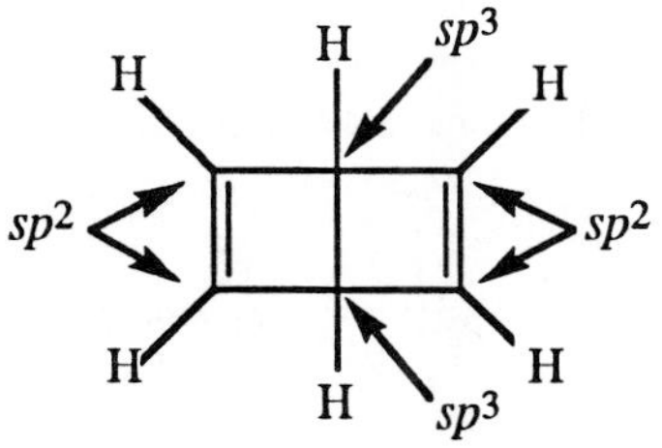

No delocalization of electrons in this structure.
We would expect the compound to behave like an alkene.

2.25 (cont)

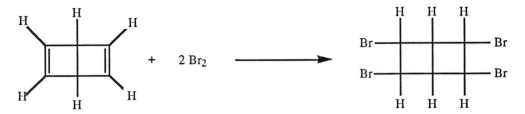

- -

Concept Map 2.4   (see p. 53)

- -

2.26   (a)   H : Be : H

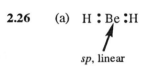

sp, linear

(b)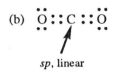

sp, linear

(c)

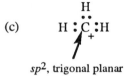

sp², trigonal planar

(d)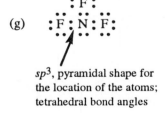

sp³, tetrahedral

(e)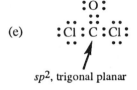

sp², trigonal planar

(f)

sp³, pyramidal shape for
the location of the atoms;
tetrahedral bond angles

(g)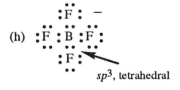

sp³, pyramidal shape for
the location of the atoms;
tetrahedral bond angles

(h)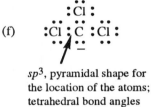

sp³, tetrahedral

2.27   (a) B_2H_6 cannot be adequately represented by a conventional Lewis structure. All twelve electrons
contributed by the six hydrogen atoms and two boron atoms are needed to bond the hydrogen atoms to
the boron atoms. There are no electrons left to hold the two boron atoms together.

$$\begin{array}{ccc} & H & H \\ & | & | \\ H - & B & B - H \\ & | & | \\ & H & H \end{array}$$

(b) $B_2H_6{}^{2-}$ is analogous to ethane, C_2H_6.

2.27 (cont)

(c)
$$\left[\begin{array}{c} \overset{H}{\underset{H}{|}} \quad \overset{H}{\underset{H}{|}} \\ H-B-B-H \end{array} \right]^{2-}$$
or
$$\begin{array}{c} \overset{H}{\underset{H}{|}} \quad \overset{H}{\underset{H}{|}} \\ H-\overset{-}{B}-\overset{-}{B}-H \end{array}$$

This structure is more correct
because it localizes formal charges.

(d)

(B and B structure with wedge and dash bonds to H atoms)

2.28   (a) $CH_3C\equiv CH$ $H_2C=CHCH_3$ (benzene ring) $CH_3CH_2-CH_3$

shortest   longest

(b) $H_2C=O$ $\overset{O}{\overset{||}{HC}}-OCH_3$ CH_3-OH

shortest   longest

(c) $CH_3C\equiv N$ $H_2C=NCH_3$ CH_3-NH_2

shortest   longest

2.29

(a) $CH_3-C\equiv C-\overset{CH_3}{\underset{OH}{\overset{|}{\underset{|}{C}}}}-CH_3$

alkyne

alcohol

(b) $CH_3-\overset{CH_3}{\overset{|}{C}}=CH-\overset{O}{\overset{||}{C}}-CH_3$

alkene   ketone

(c)   (benzene ring)$-CH_2-\overset{CH_3}{\overset{|}{N}}-CH_3$

aromatic hydrocarbon

amine

(d) $CH_3CH_2-\overset{|}{\underset{Br}{CH}}-CH=CH_2$

alkyl halide   alkene

(e) $\overset{O}{\overset{||}{HOC}}-CH_2CH_2-\overset{O}{\overset{||}{COH}}$

carboxylic   carboxylic
acid   acid

(f) $CH_3-\overset{O}{\overset{||}{CO}}-CH_2-CH_2-O-CH_3$

ester   ether

2.29 (cont)

(g)

aromatic hydrocarbon | aldehyde

(h) CH_3—CH—CH_2—CH
O
aldehyde
OH
alcohol

(i)
aromatic hydrocarbon | CH_2OH alcohol

(j) $CH_3CH_2CH_2$—C≡N
nitrile

2.30 (a)
CH₃
CH
CH₂   CH—CH   aldehyde
O
CH₂—CH   C   alkene
CH   C≡C   CH₃
CH₃   ketone   H   C
O
ketone

(b)
CH₂   CH₂   CH₂
CH₂   CH   CH   CH₂   CH₃
CH₂   CH₂   CH₃
N
amine
CH₂CH₃

(c)

alkyl halide — Cl / CH₃ — C — ... alcohol HO / CH₃ ... vinyl halide

alkyl halide — Br / H   alkyl halide — Br   alkene

(d)
H   H
C=C
alkene
O
CH₃CO(CH₂)₁₀   CH₂CH₃
ester

(e)
H   H H   H
C=C   C=C
alkene   alkene
CH₃(CH₂)₄   CH₂   (CH₂)₂C≡C(CH₂)₂COH
O
alkyne   carboxylic
acid

Concept Map 2.4   Functional Groups.

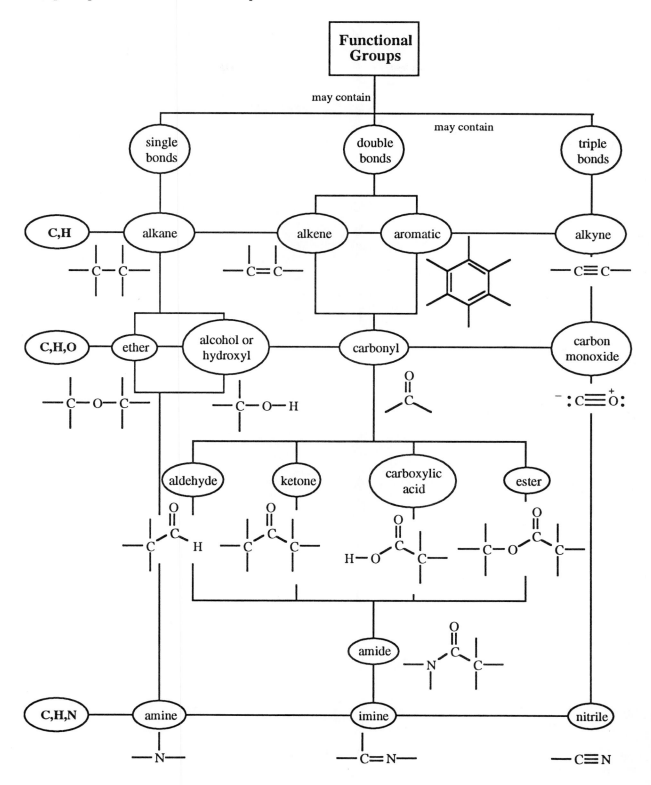

2.31

(a) $H_2C\!\!=\!\!O$

↑

sp^2

(b) CH_3NHOH

sp^3

(c) $CH_3CH\!\!=\!\!CHCH_3$

sp^3 sp^2 sp^3

one of two isomers

(d) $CH_3C\!\!\equiv\!\!CCH_3$

sp^3 sp sp^3

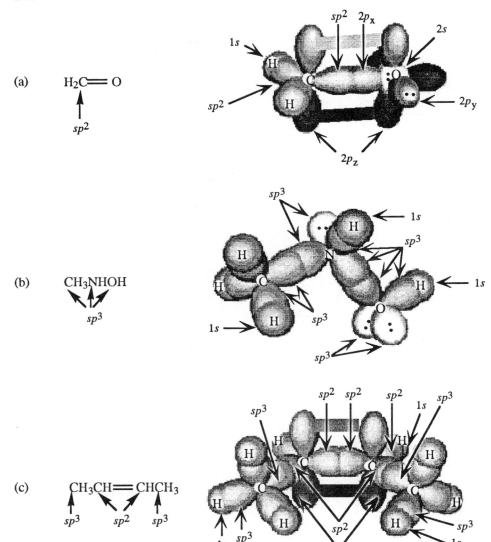

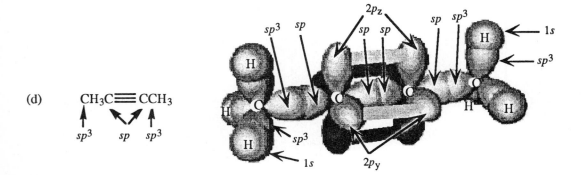

2.32   (a)   H —N̈ = N̈— H

(b) sp^2

(c)   *sp*² hybrid orbital ⟶   ⟵ $2p_z$ orbital

*sp*² hybrid orbital ⟶
in *xy* plane

*sp*² hybrid orbitals ⟵
in *xy* plane

(d)   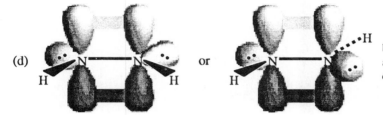   or   better because lone pairs
are farther from each other;
electron pairs repel each other

2.33

(a) $CH_3C\equiv N\ddot{:}$ + H— OSOH ⟶ $CH_3C\equiv \overset{+}{N}— H$ HSO_4^-

with O double bonded above and below the central S.

(b) $CH_3\overset{\ddot{O}\ddot{}}{\underset{\|}{C}}CH_3$ + H— OSOH ⟶ $CH_3\overset{\overset{+}{\ddot{O}}— H}{\underset{\|}{C}}CH_3$ HSO_4^-

(c) $CH_3\overset{CH_3}{\underset{\|}{C}}= \underset{\cdot\cdot}{N}CH_3$ + H— OSOH ⟶ $CH_3\overset{CH_3}{\underset{\underset{H}{\|}}{C}}= \overset{+}{N}CH_3$ HSO_4^-

(d) $CH_3\overset{:\ddot{O}:}{\underset{\|\cdot\cdot}{C}}OCH_3$ + H— OSOH ⟶ $CH_3\overset{:\overset{+}{O}— H}{\underset{\|\cdot\cdot}{C}}OCH_3$ HSO_4^-

or

$CH_3\overset{:\ddot{O}:}{\underset{\underset{\overset{+}{H}}{\|\cdot\cdot}}{C}}O— CH_3$ HSO_4^-

2.34

(a) $CH_3CH=CHCH_3$ + Br_2 $\xrightarrow[\text{dark, 25 °C}]{\text{carbon tetrachloride}}$ $CH_3CH\!-\!CHCH_3$
                                                                                                              $\quad\quad\;\; |\quad\;\; |$
                                                                                                              $\quad\quad\;\; Br\quad Br$

(b) $CH_3CH_2CH=CH_2$ + H_2 $\xrightarrow[\text{catalyst}]{}$ $CH_3CH_2CH\!-\!CH_2$
                                                                                                    $\quad\quad\;\;\; |\quad\;\; |$
                                                                                                    $\quad\quad\;\;\; H\quad H$

(c) $CH_3CH_2CH_2CH_3$ + Br_2 $\xrightarrow[\text{dark, 25 °C}]{\text{carbon tetrachloride}}$ no reaction

(d)   ⬡$-CH=CH_2$ + Br_2 $\xrightarrow[\text{dark, 25 °C}]{\text{carbon tetrachloride}}$ ⬡$-CH\!-\!CH_2$
                                                                                                              $\quad\quad\;\; |\quad\;\; |$
                                                                                                              $\quad\quad\;\; Br\quad Br$

(e) $CH_3C\equiv CH$ + Br_2 (excess) $\xrightarrow[\text{dark, 25 °C}]{\text{carbon tetrachloride}}$
$$\begin{array}{cc} Br & Br \\ | & | \\ CH_3C\!-\!CH \\ | & | \\ Br & Br \end{array}$$

(f) $CH_3C\equiv CH$ + H_2 (excess) $\xrightarrow[\text{catalyst}]{}$
$$\begin{array}{cc} H & H \\ | & | \\ CH_3C\!-\!CH \\ | & | \\ H & H \end{array}$$

2.35

(a) $H_2C\!\!\Vert\!\!CH_2$ + $Br\!\!\Vert\!\!Br$ $\longrightarrow$ $Br\!\!\Vert\!\!CH_2\!\!\Vert\!\!CH_2\!\!\Vert\!\!Br$

Bond energies (kcal/mol)   146   46   68   83   68

$$\Delta H_r = (146 + 46) - (68 + 83 + 68) = -27 \text{ kcal/mol}$$

(b)
$$\begin{array}{c} CH_3 \\ | \\ CH_3\!-\!C\!\!\Vert\!\!OH \\ | \\ CH_3 \end{array} + H\!\!\Vert\!\!Cl \longrightarrow \begin{array}{c} CH_3 \\ | \\ CH_3\!-\!C\!\!\Vert\!\!Cl \\ | \\ CH_3 \end{array} + H\!\!\Vert\!\!OH$$

          86   103   81   111

$$\Delta H_r = (86 + 103) - (81 + 111) = -3 \text{ kcal/mol}$$

(c)
$$\begin{array}{c} O \\ \| \\ CH_3\!-\!C\!-\!CH_3 \end{array} + H\!\!\Vert\!\!O\!-\!H \longrightarrow \begin{array}{c} O\!\!\Vert\!\!H \\ | \\ CH_3\!-\!C\!-\!CH_3 \\ | \\ O\!-\!H \end{array}$$

179   111   111
                                                                86
                                                                86

$$\Delta H_r = (179 + 111) - (111 + 86 + 86) = +7 \text{ kcal/mol}$$

2.36

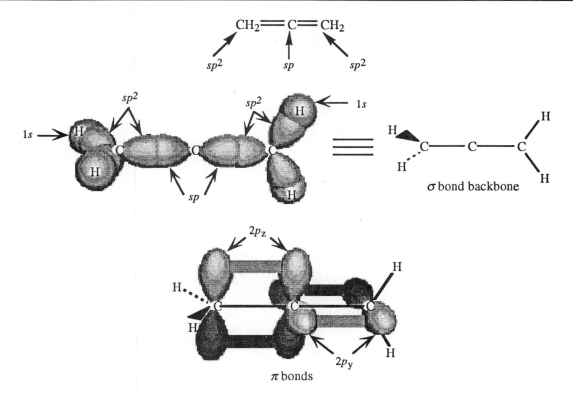

$$CH_2 = C = CH_2$$

$sp^2 \qquad sp \qquad sp^2$

σ bond backbone

$2p_z$

$2p_y$

π bonds

2.37   **(a)**

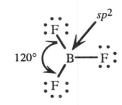

120°

sp^2

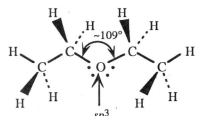

~109°

sp^3

(b)

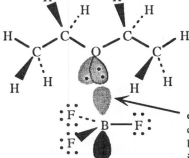

empty *p* orbital on boron overlaps easily with non-bonding electrons on oxygen in diethyl ether

(c)   Boron trifluoride etherate has a higher boiling point than either of the starting materials because the molecular weight of the etherate is higher and because the etherate is a more polar compound than either of the starting materials.

2.37 (c) (cont)

$$F - B^{-} - \overset{+}{O} \begin{array}{l} CH_2CH_3 \\ CH_2CH_3 \end{array}$$

with F on top and bottom of B

separation of charge in boron trifluoride etherate makes it polar;
dipole-dipole interactions between molecules leads to higher boiling point

2.38

(a) $H - \overset{H}{\underset{H}{\overset{|}{C}}} - \ddot{N} = \overset{+}{N} = \ddot{N}^{-} \longleftrightarrow H - \overset{H}{\underset{H}{\overset{|}{C}}} - \overset{..}{\underset{..}{N}}{}^{-} - \overset{+}{N} \equiv N: \longleftrightarrow H - \overset{H}{\underset{H}{\overset{|}{C}}} - \overset{+}{N} \equiv \overset{+}{N} - \overset{..}{\underset{..}{N}}:{}^{2-}$

This resonance contributor has the greatest separation of charge and makes the smallest contribution.

(b)   The N—N—N bond angle is 180° with *sp* hybridization for the central nitrogen atom.

2.39   (a)   (b)   *sp*^{2} (c)   *sp*

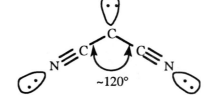

~120°

(d)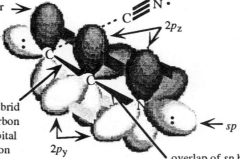

empty 2*p*$_z$ orbital overlapping with *p*$_z$ orbital on other carbon

sp^2

2*p*$_z$

overlap of *sp*^{2} hybrid orbital on one carbon and *sp* hybrid orbital on the other carbon

2*p*$_y$

sp

overlap of *sp* hybrid orbital on carbon and *sp* hybrid orbital on nitrogen

CH 2 *Covalent Bonding and Chemical Reactivity* 59

2.40 (a)

$$O = C = C = C = O$$

$$\uparrow \quad \uparrow \quad \uparrow$$
$$sp \quad sp \quad sp$$

(b)

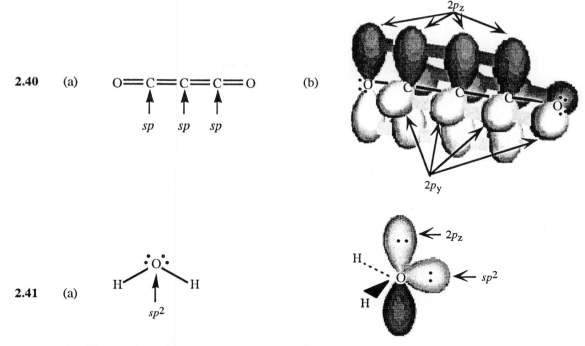

2.41 (a)

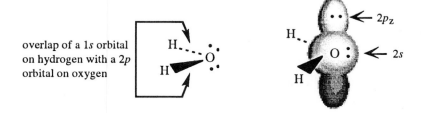

(b) The one in which the oxygen atom is sp^2-hybridized. The electron pair in the p orbital should have a different energy than the electron pair in the sp^2 hybrid orbital. The model of water in which the oxygen atom is sp^3-hybridized has both pairs of nonbonding electrons in sp^3 hybrid orbitals. These electrons should, therefore, be of the same energy. Experimentally we should not see them as occupying two different energy levels.

(c) Yes, lack of hybridization at the oxygen atom also results in two different energy levels for the two pairs of nonbonding electrons.

overlap of a $1s$ orbital on hydrogen with a $2p$ orbital on oxygen

An sp^2-hybridized oxygen atom would be expected to have a H—O—H bond angle of 120°. If oxygen were not hybridized, the H—O—H bond angle would be expected to be 90°. The experimentally determined value for the bond angle is 105°, between 120° and 90°.

(d) The oxygen atom would have sp hybridization.

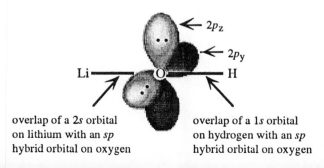

overlap of a $2s$ orbital on lithium with an sp hybrid orbital on oxygen

overlap of a $1s$ orbital on hydrogen with an sp hybrid orbital on oxygen

3
Reactions of Organic Compounds as Acids and Bases

Workbook Exercises

Much of Chapter 3 is devoted to learning how to rank acidities. It is important for you to be able to draw and recognize, both quickly and accurately, the form of a molecule resulting from the loss of a proton. Such a reaction is called a deprotonation reaction.

The connectivity change that occurs in deprotonation may be imagined in the following way:

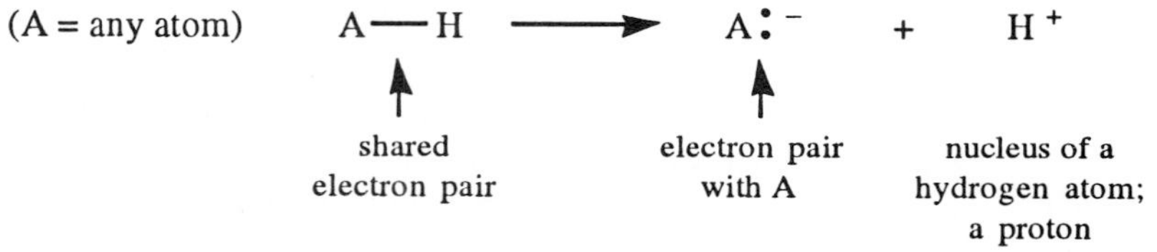

$$(A = any \ atom) \qquad A{-}H \longrightarrow A\!:^{-} \quad + \quad H^{+}$$

| shared electron pair | electron pair with A | nucleus of a hydrogen atom; a proton |

EXERCISE I. Draw all of the products which can result from a single deprotonation of the molecule shown below.

EXAMPLE

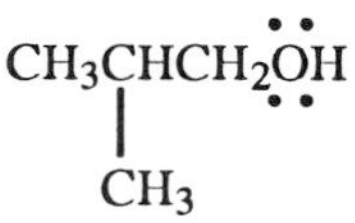

$$CH_3CHCH_2\ddot{O}H$$
$$\qquad\quad |$$
$$\qquad CH_3$$

SOLUTION

(1) Redraw the compound as a complete Lewis structure and then decide, based on connectivity, how many different types of hydrogen atoms there are.

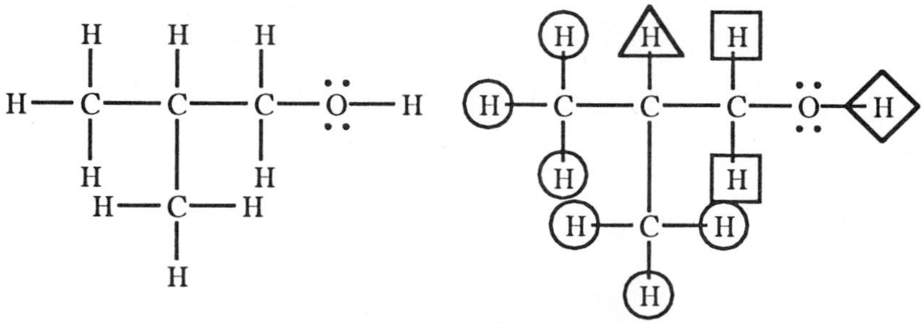

There are 4 different types of hydrogen atoms

(Note that regardless of how the molecule is drawn, the set of 6 hydrogens in circles are of the same type. That is, they are equally situated with respect to the other atoms in the molecule.)

Workbook Exercises (cont)

(2)   There are 4 different deprotonations possible.

loss of ⬦ H⁺

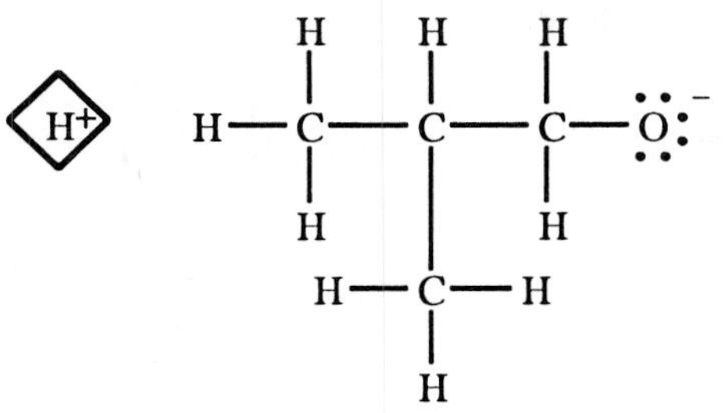

loss of △ H⁺

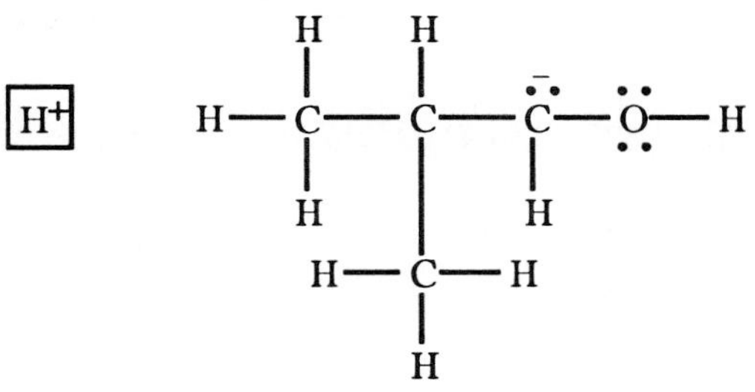

loss of ▢ H⁺

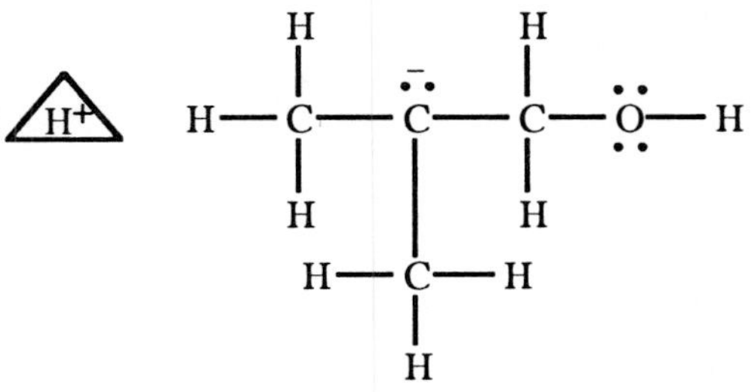

loss of ◯ H⁺

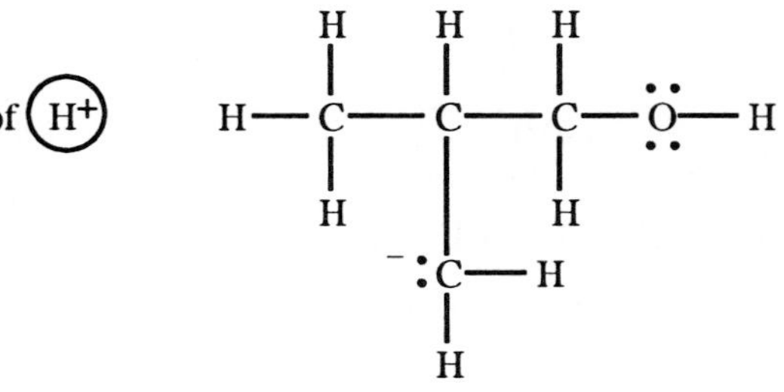

EXERCISE I.   Draw all of the products which can result from a single deprotonation of the molecules shown below.

(a)

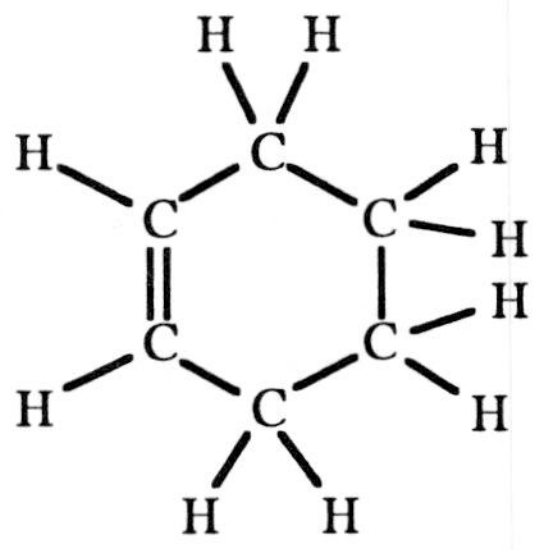

(b) $HOCCH_2CH_3$

(c) H_2

(d)

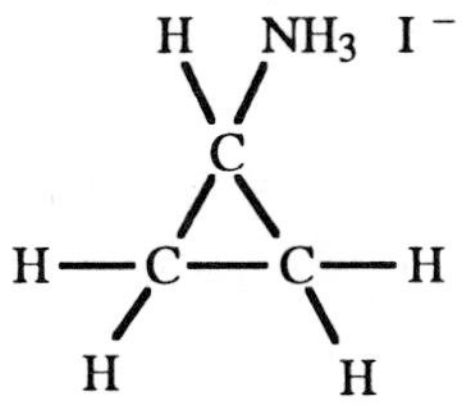

(e) $ClCH_2CN$

(f) $HC{\equiv}CCH_2CH$

(g) $H_2OCH_2CH_3$

(h) $(CH_3)_3S^{+\,-}BF_4$

EXERCISE II.   The reverse of a deprotonation reaction is a protonation reaction. Protonation is the attaching of a proton, H^+, to a molecule or ion. In order to form a bond with a proton, another atom must contribute both of the electrons to the bond (H^+ has no electrons to contribute!). Nonbonding electron pairs from uncharged or negatively charged atoms are often the source of electrons for forming a new bond to a proton.

The connectivity change that occurs in protonation may be drawn in the following way:

(A = any atom)   uncharged A: + a proton (H^+) source ⟶ $\overset{+}{A}$—H

                         anionic A:⁻ + a proton (H^+) source ⟶   A—H

                                                                          in both cases, the non-
                                                                          bonding electron pair
                                                                          provides the A—H bond

Workbook Exercises (cont)

Draw all of the products which can result from a single deprotonation of the molecules shown below. Locating the nonbonding electrons in each molecule is a good way to start.

(a) $CH_3CH_2CH_2O^-\ Na^+$

(b) $(CH_3)_2CHNCH_2CH_2NH_2$

(c) $\overset{\displaystyle O}{\overset{\|}{CH_3OCNH_2}}$

(d)

(e) $Li^+\ ^-C\!\equiv\!CCH_2CH_2O^-\ Li^+$

(f)

(g) $\overset{\displaystyle OCH_3}{\underset{\displaystyle OH}{CH_3CCH_2CH_2CH_3}}$

(h) $\overset{\displaystyle O}{\overset{\|}{^-OC}}\!\!-\!\!\underset{\displaystyle NH_2}{CHCH_2OH}$

Concept Map 3.1   The Brønsted-Lowry theory of acids and bases.

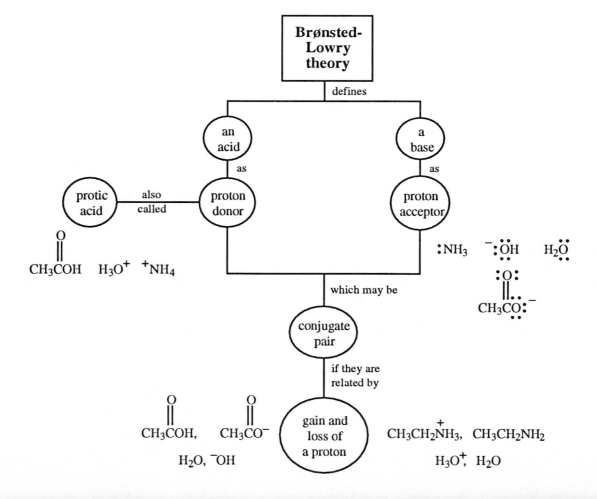

3.1   (a) $(CH_3)_2O$ $(CH_3)_2OH^+$

base   conjugate acid

(b) $H_2SO_4 = (HO)_2SO_2$ $HSO_4^- = HOSO_3^-$

acid   conjugate base
(one resonance contributor)

(c) $^-NH_2$ NH_3

base   conjugate acid

(d) $CH_3OH_2^+$ CH_3OH

acid   conjugate base

(e) $H_2C=O$ $H_2C=OH^+$

base   conjugate acid

(f) CH_3OH CH_3O^-

acid   conjugate base

Concept Map 3.2   The Lewis theory of acids and bases.

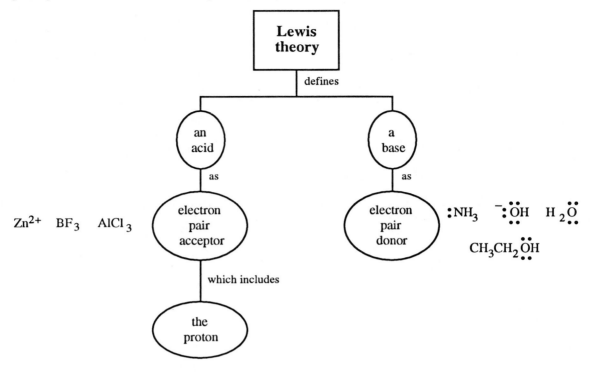

Lewis theory

defines

an acid

a base

as   as

Zn^{2+} BF_3 $AlCl_3$

electron pair acceptor

electron pair donor

$:NH_3$ $^-:\overset{..}{\underset{..}{O}}H$ $H_2\overset{..}{\underset{..}{O}}$

$CH_3CH_2\overset{..}{\underset{..}{O}}H$

which includes

the proton

- -

3.2

(a) Cu^{2+} +   4 H—N—H ⇌
          Lewis acid   Lewis base

(b)   Lewis acid   Lewis base

3.2 (cont)

(c)

$$:\!F\!-\!B\!-\!F\!: \;+\; CH_3CH_2CH=O \;\rightleftharpoons\; CH_3CH_2CH=\overset{+}{O}\!-\!\overset{-}{B}F_3$$

Lewis acid   Lewis base

(d)

$$Hg^{2+} \;+\; CH_3CH_2\ddot{S}\!-\!H \;\rightleftharpoons\; \left[\,CH_3CH_2\overset{Hg}{\underset{}{S}}\!-\!H\,\right]^{2+}$$

Lewis acid   Lewis base

3.3

(a)

$$CH_3\ddot{N}HCH_3 \;+\; AlCl_3 \;\rightleftharpoons\; CH_3\overset{+}{N}HCH_3 \text{ with } \overset{-}{AlCl_3} \text{ and } H$$

$$CH_3\ddot{N}HCH_3 \;+\; H_2SO_4 \;\rightleftharpoons\; CH_3\overset{+}{N}CH_3 \text{ (with 2 H)} \;+\; HSO_4^-$$

(b)

$$HCHO \;+\; AlCl_3 \;\rightleftharpoons\; :\overset{+}{O}\!-\!AlCl_3 \text{ (HCH)}$$

$$HCHO \;+\; H_2SO_4 \;\rightleftharpoons\; :\overset{+}{O}\!-\!H \text{ (HCH)} \;+\; HSO_4^-$$

(c)

$$HC\equiv CH \;+\; AlCl_3 \;\rightleftharpoons\; H\!-\!\overset{+}{C}\!=\!C\!-\!H \text{ with } \overset{-}{AlCl_3}$$

$$HC\equiv CH \;+\; H_2SO_4 \;\rightleftharpoons\; H\!-\!\overset{+}{C}\!=\!C\!-\!H \text{ (with H)} \;+\; HSO_4^-$$

3.3 (cont)

(d)

$$CH_3CH_2-\overset{\overset{\displaystyle CH_2CH_3}{|}}{\underset{\cdot\cdot}{P}}-CH_2CH_3 \;+\; AlCl_3 \;\rightleftharpoons\; CH_3CH_2-\overset{\overset{\displaystyle CH_2CH_3}{|}}{\underset{\underset{\displaystyle \overset{-}{AlCl_3}}{|}}{\overset{+}{P}}}-CH_2CH_3$$

$$CH_3CH_2-\overset{\overset{\displaystyle CH_2CH_3}{|}}{\underset{\cdot\cdot}{P}}-CH_2CH_3 \;+\; H_2SO_4 \;\rightleftharpoons\; CH_3CH_2-\overset{\overset{\displaystyle CH_2CH_3}{|}}{\underset{\underset{\displaystyle H}{|}}{\overset{+}{P}}}-CH_2CH_3 \;+\; HSO_4^{-}$$

(e)

$$CH_3CH_2-\underset{\cdot\cdot}{\overset{\cdot\cdot}{O}}H \;+\; AlCl_3 \;\rightleftharpoons\; CH_3CH_2-\underset{\cdot\cdot}{\overset{\overset{\displaystyle {}^{-}AlCl_3}{\overset{|}{+}}}{O}}H$$

$$CH_3CH_2-\underset{\cdot\cdot}{\overset{\cdot\cdot}{O}}H \;+\; H_2SO_4 \;\rightleftharpoons\; CH_3CH_2-\underset{\cdot\cdot}{\overset{\overset{\displaystyle H}{\overset{|}{+}}}{O}}H \;+\; HSO_4^{-}$$

(f)

$$CH_3CH_2\underset{\cdot\cdot}{\overset{\cdot\cdot}{S}}-CH_2CH_3 \;+\; AlCl_3 \;\rightleftharpoons\; CH_3CH_2\underset{\cdot\cdot}{\overset{\overset{\displaystyle {}^{-}AlCl_3}{\overset{|}{+}}}{S}}-CH_2CH_3$$

$$CH_3CH_2\underset{\cdot\cdot}{\overset{\cdot\cdot}{S}}-CH_2CH_3 \;+\; H_2SO_4 \;\rightleftharpoons\; CH_3CH_2\underset{\cdot\cdot}{\overset{\overset{\displaystyle H}{\overset{|}{+}}}{S}}-CH_2CH_3 \;+\; HSO_4^{-}$$

3.4   (a)   ethyl acetate

$$H-\overset{\overset{\displaystyle H}{|}}{\underset{\underset{\displaystyle H}{|}}{C}}-\overset{\overset{\displaystyle :\!O\!:}{\|}}{C}-\underset{\cdot\cdot}{\overset{\cdot\cdot}{O}}-\overset{\overset{\displaystyle H}{|}}{\underset{\underset{\displaystyle H}{|}}{C}}-\overset{\overset{\displaystyle H}{|}}{\underset{\underset{\displaystyle H}{|}}{C}}-H$$

$$CH_3\overset{\overset{\displaystyle :O:}{\|}}{C}-\underset{\cdot\cdot}{\overset{\cdot\cdot}{O}}-CH_2CH_3 \;+\; H_2SO_4 \;\rightleftharpoons\; CH_3\overset{\overset{\displaystyle :\overset{+}{O}-H}{\|}}{C}-\underset{\cdot\cdot}{\overset{\cdot\cdot}{O}}-CH_2CH_3 \;+\; HSO_4^{-}$$

$$CH_3\overset{\overset{\displaystyle :O:}{\|}}{C}-\underset{\cdot\cdot}{\overset{\cdot\cdot}{O}}-CH_2CH_3 \;+\; H_2SO_4 \;\rightleftharpoons\; CH_3\overset{\overset{\displaystyle :O:}{\|}}{C}-\underset{\underset{\displaystyle H}{\overset{|}{+}}}{\overset{\cdot\cdot}{O}}-CH_2CH_3 \;+\; HSO_4^{-}$$

(b)   2-(*N,N*-dimethylamino)ethanol

$$H-\underset{\cdot\cdot}{\overset{\cdot\cdot}{O}}-\overset{\overset{\displaystyle H}{|}}{\underset{\underset{\displaystyle H}{|}}{C}}-\overset{\overset{\displaystyle H}{|}}{\underset{\underset{\displaystyle H}{|}}{C}}-\underset{\cdot\cdot}{N}\Big(\overset{\overset{\displaystyle H-\overset{\overset{\displaystyle H}{|}}{\underset{\underset{\displaystyle H}{|}}{C}}-H}{}}{}\Big)-\overset{\overset{\displaystyle H}{|}}{\underset{\underset{\displaystyle H}{|}}{C}}-H$$

3.4 (b) (cont)

$$\underset{\overset{\displaystyle H-\overset{\cdot\cdot}{\underset{\cdot\cdot}{O}}-CH_2CH_2-\overset{\overset{\displaystyle CH_3}{|}}{\underset{\cdot\cdot}{N}}-CH_3}{}}{} + H_2SO_4 \;\rightleftharpoons\; H-\overset{\cdot\cdot}{\underset{\cdot\cdot}{O}}-CH_2CH_2-\overset{\overset{\displaystyle CH_3}{|}}{\underset{\overset{|}{H}}{\overset{+}{N}}}-CH_3 \;+\; HSO_4^-$$

$$H-\overset{\cdot\cdot}{\underset{\overset{+}{\underset{|}{H}}}{O}}-CH_2CH_2-\overset{\overset{\displaystyle CH_3}{|}}{\underset{\cdot\cdot}{N}}-CH_3 \;+\; H_2SO_4 \;\rightleftharpoons\; H-\overset{\cdot\cdot}{\underset{\overset{+}{\underset{|}{H}}}{O}}-CH_2CH_2-\overset{\overset{\displaystyle CH_3}{|}}{\underset{\cdot\cdot}{N}}-CH_3 \;+\; HSO_4^-$$

(c)   2-propen-1-ol

$$H-C\!\!=\!\!C-\overset{\overset{\displaystyle H}{|}}{\underset{\overset{|}{H}}{C}}-\overset{\cdot\cdot}{\underset{\cdot\cdot}{O}}-H$$

(with H's on the first two carbons)

$$CH_2\!\!=\!\!CHCH_2-\overset{\cdot\cdot}{\underset{\cdot\cdot}{O}}-H \;+\; H_2SO_4 \;\rightleftharpoons\; CH_2\!\!=\!\!CHCH_2-\overset{\cdot\cdot}{\underset{\overset{+}{\underset{|}{H}}}{O}}-H \;+\; HSO_4^-$$

$$CH_2\!\!=\!\!CHCH_2-\overset{\cdot\cdot}{\underset{\cdot\cdot}{O}}-H \;+\; H_2SO_4 \;\rightleftharpoons\; \overset{\overset{\displaystyle H}{|}}{CH_2}-\overset{+}{CHCH_2}-\overset{\cdot\cdot}{\underset{\cdot\cdot}{O}}-H \;+\; HSO_4^-$$

(d)   methyl isocyanate

$$H-\overset{\overset{\displaystyle H}{|}}{\underset{\overset{|}{H}}{C}}-\overset{\cdot\cdot}{N}\!\!=\!\!C\!\!=\!\!\overset{\cdot\cdot}{\underset{\cdot\cdot}{O}}$$

$$CH_3-\overset{\cdot\cdot}{N}\!\!=\!\!C\!\!=\!\!\overset{\cdot\cdot}{\underset{\cdot\cdot}{O}} \;+\; H_2SO_4 \;\rightleftharpoons\; CH_3-\overset{\overset{+}{\underset{|}{H}}}{N}\!\!=\!\!C\!\!=\!\!\overset{\cdot\cdot}{\underset{\cdot\cdot}{O}} \;+\; HSO_4^-$$

$$CH_3-\overset{\cdot\cdot}{N}\!\!=\!\!C\!\!=\!\!\overset{\cdot\cdot}{\underset{\cdot\cdot}{O}} \;+\; H_2SO_4 \;\rightleftharpoons\; CH_3-\overset{\cdot\cdot}{N}\!\!=\!\!C\!\!=\!\!\overset{\cdot\cdot}{\underset{\overset{+}{\underset{|}{H}}}{O}} \;+\; HSO_4^-$$

3.5   <u>base</u>   <u>conjugate acid</u>

(a)

$$\underset{\underset{\displaystyle H}{}}{\overset{\overset{\displaystyle CH_3CH_2}{}}{C}}\!\!=\!\!\underset{\underset{\displaystyle CH_2CH_3}{}}{\overset{\overset{\displaystyle H}{}}{C}}$$

$$\underset{\underset{\displaystyle H}{}}{\overset{\overset{\displaystyle CH_3CH_2}{}}{\overset{+}{C}}}-\underset{\underset{\displaystyle CH_2CH_3}{}}{\overset{\overset{\displaystyle H}{}}{C}}-H$$

(b) $CH_3CH_2-\overset{\cdot\cdot}{\underset{\cdot\cdot}{O}}-CH_2CH_3$ $CH_3CH_2-\overset{\cdot\cdot}{\underset{\overset{+}{\underset{|}{H}}}{O}}-CH_2CH_3$

3.5

	base	conjugate acid

(c) $CH_3CH_2\overset{\cdot\cdot}{N}CH_2CH_3$ with H below $CH_3CH_2\overset{+}{N}CH_2CH_3$ with H above and below

(d) $CH_3CH_2CH_2\!-\!\overset{\cdot\cdot}{\underset{\cdot\cdot}{S}}\!-\!CH_3$ $CH_3CH_2CH_2\!-\!\overset{\cdot\cdot}{\underset{+}{S}}\!-\!CH_3$ with H below

(e) $CH_3CH_2\overset{\displaystyle :\overset{\cdot\cdot}{O}:}{\overset{\|}{CH}}$ $CH_3CH_2\overset{\displaystyle :\overset{+}{O}\!-\!H}{\overset{\|}{CH}}$

(f) $CH_3CH_2CH_2\!-\!\overset{\cdot\cdot}{\underset{\cdot\cdot}{O}}H$ $CH_3CH_2CH_2\!-\!\overset{\cdot\cdot}{O}H$ with $\overset{+}{H}$ below

(g) $CH_3CH_2\overset{\displaystyle :\overset{\cdot\cdot}{O}:}{\overset{\|}{C}}CH_2CH_3$ $CH_3CH_2\overset{\displaystyle :\overset{+}{O}\!-\!H}{\overset{\|}{C}}CH_2CH_3$

3.6   Equations from Problem 3.3 rewritten:

(a) $CH_3\overset{CH_3}{\underset{H}{\overset{|}{N}}}\!:$ $:\!\overset{\cdot\cdot}{\underset{:Cl:}{\underset{|}{Al}}}\!-\!\overset{\cdot\cdot}{\underset{\cdot\cdot}{Cl}}\!:$ $\longrightarrow$ $CH_3\overset{CH_3}{\underset{H}{\overset{|}{\overset{+}{N}}}}\!-\!\overset{-}{AlCl_3}$

$CH_3\overset{CH_3}{\underset{H}{\overset{|}{N}}}\!:$ $H\!-\!\overset{\cdot\cdot}{\underset{\cdot\cdot}{O}}\!-\!SO_3H$ $\longrightarrow$ $CH_3\overset{CH_3}{\underset{H}{\overset{|}{\overset{+}{N}}}}\!-\!H$ $^-\!:\!\overset{\cdot\cdot}{\underset{\cdot\cdot}{O}}\!-\!SO_3H$

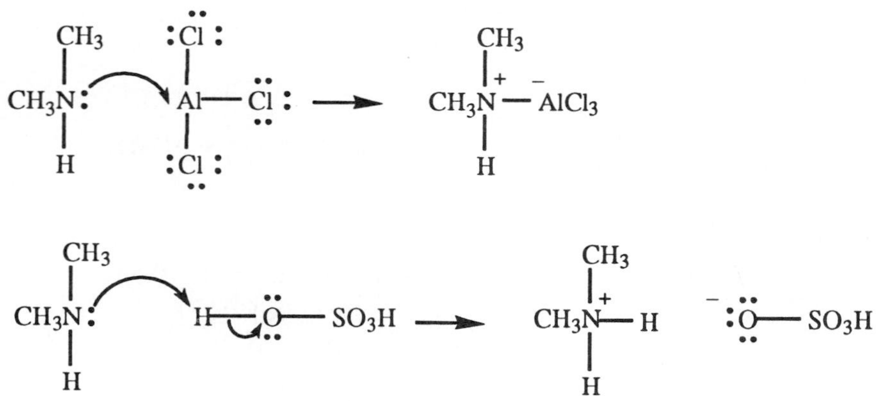

(b) $\overset{:O:}{\underset{HCH}{\overset{\|}{}}}$ $:\!\overset{\cdot\cdot}{\underset{:Cl:}{\underset{|}{Al}}}\!-\!\overset{\cdot\cdot}{\underset{\cdot\cdot}{Cl}}\!:$ $\longrightarrow$ $\overset{:\overset{+}{O}\!-\!AlCl_3}{\underset{HCH}{\overset{\|}{}}}$

$\overset{:O:}{\underset{HCH}{\overset{\|}{}}}$ $H\!-\!\overset{\cdot\cdot}{\underset{\cdot\cdot}{O}}\!-\!SO_3H$ $\longrightarrow$ $\overset{:\overset{+}{O}\!-\!H}{\underset{HCH}{\overset{\|}{}}}$ $^-\!:\!\overset{\cdot\cdot}{\underset{\cdot\cdot}{O}}\!-\!SO_3H$

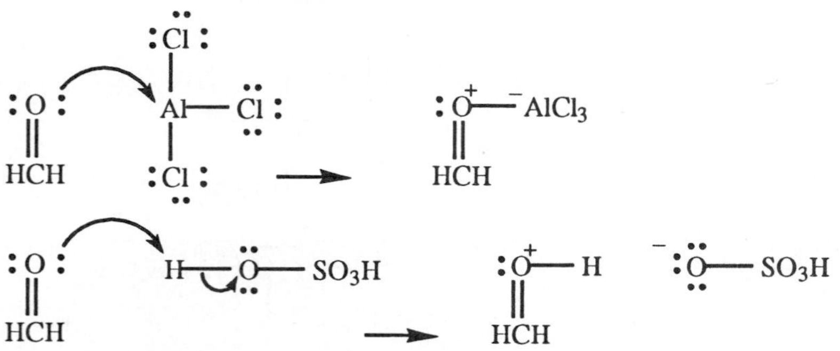

3.6 (cont)

(c) $HC \equiv CH$ + AlCl$_3$ (with lone pairs) $\longrightarrow$ $H - \overset{+}{C} = C - H$, $^-AlCl_3$

$HC \equiv CH$ $H - \overset{..}{O} - SO_3H$ $\longrightarrow$ $H - \overset{+}{C} = \overset{H}{\underset{}{C}} - H$ $\overset{..}{:}\overset{..}{O} - SO_3H$

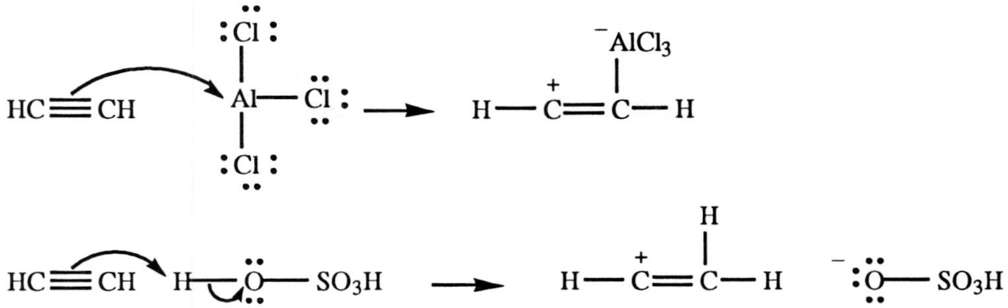

(d) $CH_3CH_2 - \overset{CH_2CH_3}{\underset{..}{P}} - CH_2CH_3$ + AlCl$_3$ $\longrightarrow$ $CH_3CH_2 - \overset{CH_2CH_3}{\underset{AlCl_3}{\overset{+}{P}}} - CH_2CH_3$

$CH_3CH_2 - \overset{CH_2CH_3}{\underset{..}{P}} - CH_2CH_3$ $H - \overset{..}{O} - SO_3H$ $\longrightarrow$ $CH_3CH_2 - \overset{CH_2CH_3}{\underset{H}{\overset{+}{P}}} - CH_2CH_3$ $^-\overset{..}{:}\overset{..}{O} - SO_3H$

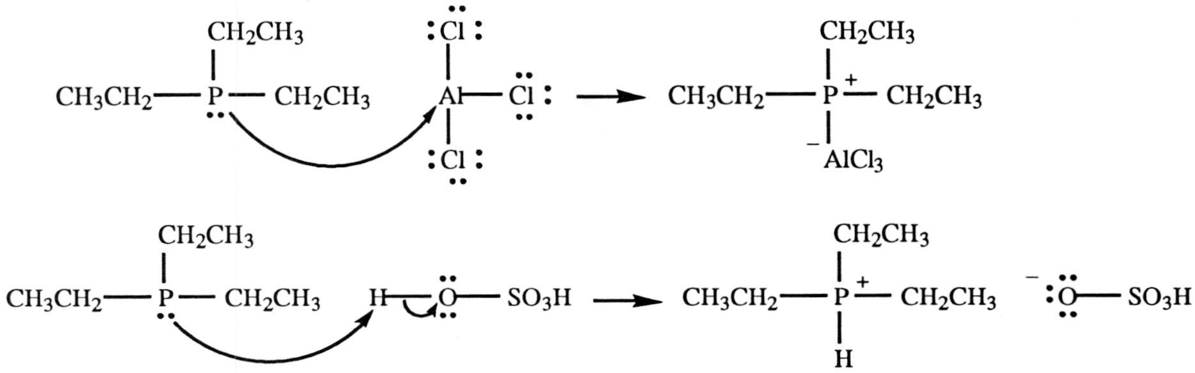

(e) $CH_3CH_2 - \overset{..}{O}H$ + AlCl$_3$ $\longrightarrow$ $CH_3CH_2 - \overset{^-AlCl_3}{\overset{+}{O}H}$

$CH_3CH_2 - \overset{..}{O}H$ $H - \overset{..}{O} - SO_3H$ $\longrightarrow$ $CH_3CH_2 - \overset{H}{\overset{+}{O}H}$ $^-\overset{..}{:}\overset{..}{O} - SO_3H$

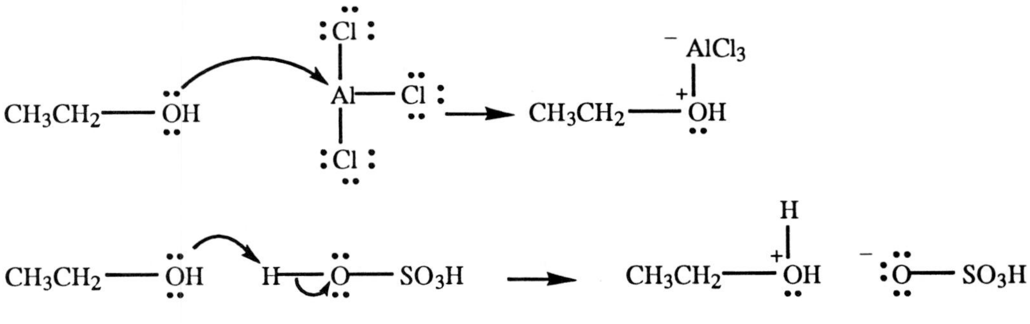

(f) $CH_3CH_2\overset{..}{S} - CH_2CH_3$ + AlCl$_3$ $\longrightarrow$ $CH_3CH_2\overset{^-AlCl_3}{\overset{+}{S}} - CH_2CH_3$

$CH_3CH_2\overset{..}{S} - CH_2CH_3$ $H - \overset{..}{O} - SO_3H$ $\longrightarrow$ $CH_3CH_2\overset{H}{\overset{+}{S}} - CH_2CH_3$ $^-\overset{..}{:}\overset{..}{O} - SO_3H$

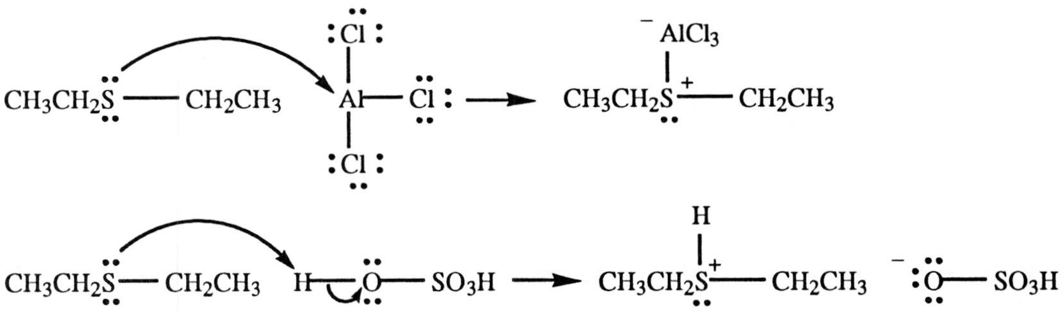

Equations from Problem 3.4 rewritten:

(a) $\overset{:O:}{\underset{CH_3C - \overset{..}{O} - CH_2CH_3}{\parallel}}$ $H - \overset{..}{O} - SO_3H$ $\longrightarrow$ $\overset{:\overset{+}{O} - H}{\underset{CH_3C - \overset{..}{O} - CH_2CH_3}{\parallel}}$ $^-\overset{..}{:}\overset{..}{O} - SO_3H$

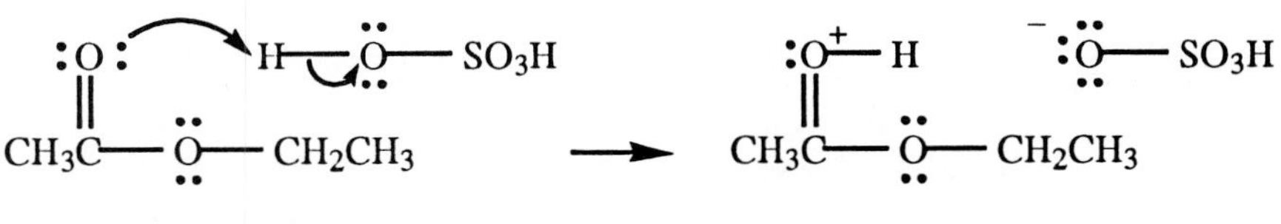

3.6 (a) (cont)

$$CH_3C(=\ddot{O})-\ddot{O}-CH_2CH_3 + H-\ddot{O}-SO_3H \longrightarrow CH_3C(=\ddot{O})-\overset{+}{\underset{H}{\ddot{O}}}-CH_2CH_3 \quad \ ^-\!:\!\ddot{O}-SO_3H$$

(b)

$$H-\ddot{O}-CH_2CH_2-\underset{\ddot{\,}}{N}(-CH_3)(-CH_3) + H-\ddot{O}-SO_3H \longrightarrow H-\ddot{O}-CH_2CH_2-\overset{+}{\underset{H}{N}}(-CH_3)(-CH_3) \quad \ ^-\!:\!\ddot{O}-SO_3H$$

$$H-\ddot{O}-CH_2CH_2-\underset{\ddot{\,}}{N}(-CH_3)(-CH_3) + H-\ddot{O}-SO_3H \longrightarrow H-\overset{+}{\underset{H}{\ddot{O}}}-CH_2CH_2-\underset{\ddot{\,}}{N}(-CH_3)(-CH_3) \quad \ ^-\!:\!\ddot{O}-SO_3H$$

(c)

$$CH_2=CHCH_2-\ddot{O}-H + H-\ddot{O}-SO_3H \longrightarrow CH_2=CHCH_2-\overset{+}{\underset{H}{\ddot{O}}}-H \quad \ ^-\!:\!\ddot{O}-SO_3H$$

$$CH_2=CHCH_2-\ddot{O}-H + H-\ddot{O}-SO_3H \longrightarrow CH_2-\overset{+}{C}HCH_2-\ddot{O}-H \quad \ ^-\!:\!\ddot{O}-SO_3H$$
$$\underset{H}{|}$$

(d)

$$CH_3-\underset{\ddot{\,}}{N}=C=\ddot{O} + H-\ddot{O}-SO_3H \longrightarrow CH_3-\overset{+}{\underset{H}{N}}=C=\ddot{O} \quad \ ^-\!:\!\ddot{O}-SO_3H$$

$$CH_3-\underset{\ddot{\,}}{N}=C=\ddot{O} + H-\ddot{O}-SO_3H \longrightarrow CH_3-\underset{\ddot{\,}}{N}=C=\overset{+}{\underset{H}{\ddot{O}}} \quad \ ^-\!:\!\ddot{O}-SO_3H$$

3.7   (a)

$$CH_3CH_2 - \overset{\underset{|}{H}}{\overset{..}{O}}: \quad H - \overset{..}{\underset{..}{I}}: \longrightarrow CH_3CH_2 - \overset{\underset{|}{H}}{\overset{+}{\underset{..}{O}}} - H \quad : \overset{..}{\underset{..}{I}}: ^-$$

(b)

$$CH_3 - \overset{\underset{|}{CH_3}}{\underset{CH_3}{N}}: \quad H - \overset{..}{\underset{..}{Cl}}: \longrightarrow CH_3 - \overset{\underset{|}{CH_3}}{\overset{+}{\underset{CH_3}{N}}} - H \quad : \overset{..}{\underset{..}{Cl}}: ^-$$

(c)

$$CH_3CH_2 - \overset{\underset{|}{CH_3CH_2}}{\underset{CH_3CH_2}{O}}: \quad \overset{:\overset{..}{F}:}{\underset{:\overset{..}{F}:}{B}} - \overset{..}{\underset{..}{F}}: \longrightarrow CH_3CH_2 - \overset{\underset{|}{CH_3CH_2}}{\overset{+}{O}} - \overset{:\overset{..}{F}:}{\underset{:\overset{..}{F}:}{B^-}} - \overset{..}{\underset{..}{F}}:$$

(d)

$$CH_3C \equiv CCH_3 \longrightarrow CH_3\overset{+}{C} = CCH_3$$
$$\underset{H - \overset{..}{\underset{..}{Br}}:}{} \qquad \underset{\overset{|}{H}}{} \qquad : \overset{..}{\underset{..}{Br}}: ^-$$

(e)

$$CH_3 - \overset{\underset{|}{CH_3}}{\overset{+}{C}} - CH_3 \longrightarrow CH_3 - \overset{\underset{|}{CH_3}}{\underset{CH_2 - CH_2}{C}} - CH_3$$
$$\underset{CH_2 = CH_2}{} \qquad \underset{+}{}$$

3.8   <u>most basic</u>   <u>least basic</u>

(a) CH_3NHCH_3 > CH_3OCH_3

amine   ether
(nonbonding electrons   (nonbonding electrons
on nitrogen)   on oxygen)

(b) CH_3^- > NH_2^- > OH^- > F^-

methyl anion   amide ion   hydroxide ion   fluoride ion
a carbanion   (nonbonding electrons on
(nonbonding electrons   fluorine; EN 4.0)
on carbon; EN 2.5)

EN = electronegativity

(c) $CH_3\overset{-}{N}CH_3$ > CH_3O^-

amide anion   alkoxide anion

Concept Map 3.3   Relationship of acidity to position of the central element in the
                           periodic table.

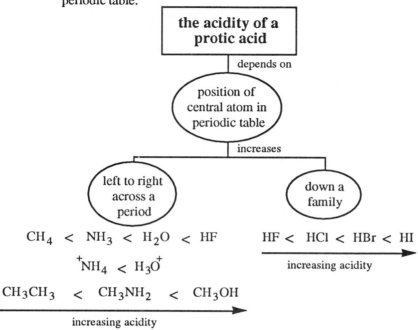

$$CH_4 \;<\; NH_3 \;<\; H_2O \;<\; HF$$

$$^+NH_4 \;<\; H_3O^+$$

$$CH_3CH_3 \;<\; CH_3NH_2 \;<\; CH_3OH$$

increasing acidity

$$HF \;<\; HCl \;<\; HBr \;<\; HI$$

increasing acidity

3.9   H₂S is the stronger acid.

$$H_2S + HO^- \rightleftharpoons HS^- + H_2O$$

Hydrogen sulfide ionizes more readily than does water because the hydrogen sulfide anion is more stable
relative to its conjugate acid than is hydroxide ion. The hydrogen sulfide anion is a weaker base than hydroxide
anion for the same reason that methanethiolate anion is a weaker base than the hydroxide anion. The sulfur atom
is larger than an oxygen atom. Therefore the negative charge on the hydrogen sulfide anion is more spread out.
The anion is more stable relative to its conjugate acid than is hydroxide ion, and less likely to be protonated.

3.10 $$CH_3CH_2SH + CH_3CH_2O^- \rightleftharpoons CH_3CH_2S^- + CH_3CH_2OH$$

Ethanethiolate ion is a weaker base than methoxide ion because the negative charge is more spread out on the
sulfur atom than on the oxygen atom. (See the answer to Problem 3.9.) In the reaction of ethanethiol and
methoxide ion, the equilibrium lies far to the right.

3.11   (a) One of the nonbonding electron pairs on the oxygen atom can react with the empty *p*-orbital of the
                 carbocation.

3.11 (cont)

(b) The product formed is a protonated alcohol, an oxonium ion, which has a pK_a ~ −2. It will transfer a proton to the oxygen atom of water in an equilibrium reaction. With a large excess of water, equilibrium lies toward the right.

(excess)

3.12   least acidic   most acidic

(a) $CH_3CH_2CH_3$ < $CH_3CH_2NH_2$ < CH_3CH_2OH < CH_3CH_2SH

           alkene   amine   alcohol   thiol

(b) CH_3OCH_3 < CH_3NHCH_3 < $CH_3\overset{+}{O}{-}CH_3$ (with H above)

        (proton on a carbon atom)   (proton on a nitrogen atom)   (proton on the oxygen atom
                                                                              of an oxonium ion)

(c) $CH_3CH_2CH_3$ < $CH_3\overset{+}{N}{-}CH_3$ (with H above and below)   < $CH_3\overset{+}{O}{-}CH_3$ (with H above)

           alkane   ammonium ion   oxonium ion

3.13   acid   conjugate base

(a)

(b) $CH_3CH_2CH_2\ddot{S}H$ $CH_3CH_2CH_2\ddot{S}{:}^-$

(c)

3.13 (cont)   <u>acid</u>   <u>conjugate base</u>

(d)

(e)

(f)

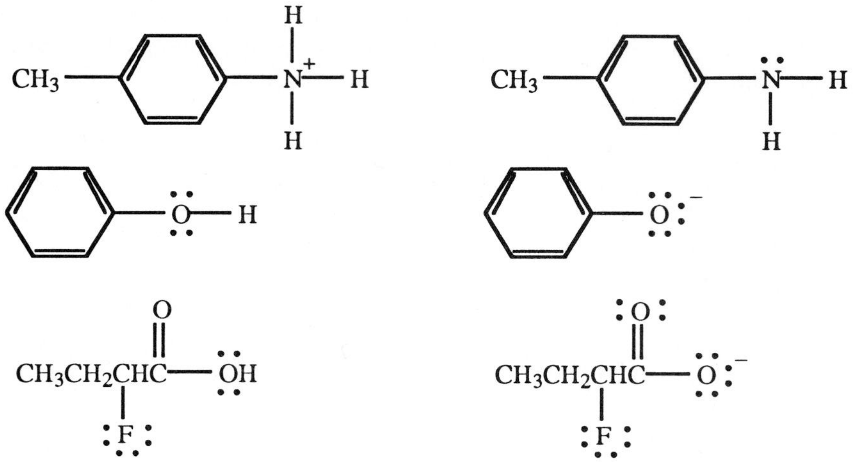

3.14 HO_3SO—H   + H_2O ⇌ HO_3SO^- + H_3O^+

sulfuric acid   water   hydrogen   hydronium
                                                sulfate anion   ion

$K_a = 10^9$ $K_a = 10^{1.7}$

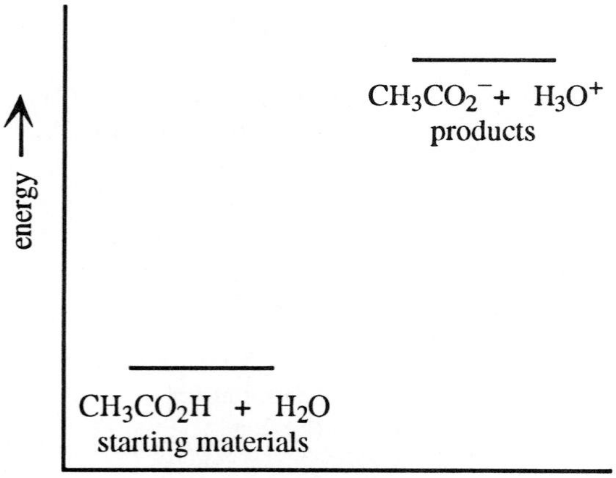

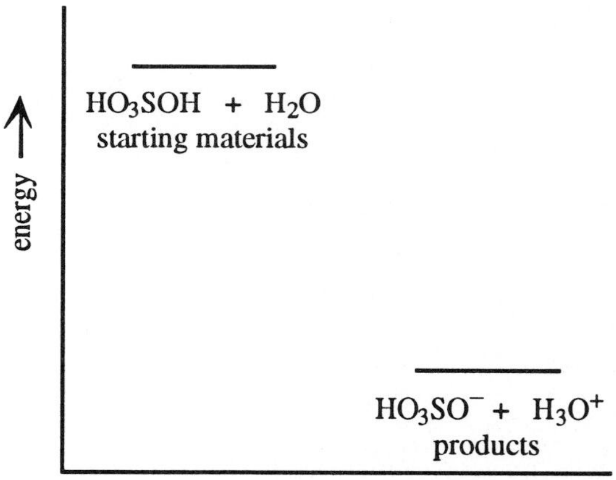

3.15   The acidity constant is

$$K_a = \frac{[H^+]\,[\colon\!B]}{[HB^+]}$$

where $\colon\!B$ and HB^+ are the conjugate base and conjugate acid forms respectively.

pK_a is defined as $-\log K_a$

For formic acid $pK_a = -\log(1.99 \times 10^{-4})$

$= -(\log 1.99 + \log 10^{-4})$

$= -(0.299 - 4)$

3.15 (cont)

$$pK_a = -(-3.701)$$

$$pK_a = +3.701$$

For acetic acid $pK_a = +4.76$

As the pK_a gets larger (more positive), K_a gets smaller and the amount of dissociation of the acid decreases. Formic acid has the smaller pK_a, and thus the larger K_a, and is a stronger acid than acetic acid.

3.16

 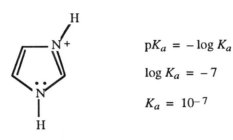

base   conjugate acid
                     (oneresonance
                     contributor)

$$pK_a = -\log K_a$$

$$\log K_a = -7$$

$$K_a = 10^{-7}$$

3.17

$$ClCH_2\overset{O}{\overset{\|}{C}}OH \ + \ H_2O \ \rightleftharpoons \ ClCH_2\overset{O}{\overset{\|}{C}}O^- \ + \ H_3O^+$$

$$K_{eq} = \frac{[ClCH_2CO_2^-]\,[H_3O^+]}{[ClCH_2CO_2H]\,[H_2O]}$$

$$K_{eq}[H_2O] = K_a = \frac{[ClCH_2CO_2^-]\,[H_3O^+]}{[ClCH_2CO_2H]}$$

$$ClCH_2\overset{O}{\overset{\|}{C}}OH \ + \ CH_3\overset{O}{\overset{\|}{C}}O^- \ \rightleftharpoons \ ClCH_2\overset{O}{\overset{\|}{C}}O^- \ + \ CH_3\overset{O}{\overset{\|}{C}}OH$$

$$K_{eq} = \frac{[ClCH_2CO_2^-]\,[CH_3CO_2H]}{[ClCH_2CO_2H]\,[CH_3CO_2^-]}$$

$$K_{eq} = \frac{K_a \text{ of } ClCH_2CO_2H}{K_a \text{ of } CH_3CO_2H} = \frac{\dfrac{[ClCH_2CO_2^-]\,[H_3O^+]}{[ClCH_2CO_2H]}}{\dfrac{[CH_3CO_2^-]\,[H_3O^+]}{[CH_3CO_2H]}}$$

$$K_{eq} = \frac{[ClCH_2CO_2^-]\,\cancel{[H_3O^+]}\,[CH_3CO_2H]}{[ClCH_2CO_2H]\,[CH_3CO_2^-]\,\cancel{[H_3O^+]}}$$

$$K_{eq} = \frac{[ClCH_2CO_2^-]\,[CH_3CO_2H]}{[ClCH_2CO_2H]\,[CH_3CO_2^-]} = K_{eq} \text{ for the reaction of chloroacetic acid with acetate ion}$$

3.18 $\Delta G° = -2.303\, RT \log K_{eq}$

$$-2.59 \text{ kcal/mol} = (-2.303)(1.987 \times 10^{-3}\,\text{kcal/mol·K})(298\,\text{K})(\log K_{eq})$$

$$\log K_{eq} = 1.899$$

$$K_{eq} = 10^{1.899} = 79.3$$

$$K_{eq} = \frac{K_a \text{ of ClCH}_2\text{CO}_2\text{H}}{K_a \text{ of CH}_3\text{CO}_2\text{H}}$$

$$79.3 = \frac{K_a \text{ of ClCH}_2\text{CO}_2\text{H}}{(1.75 \times 10^{-5})}$$

$$K_a (\text{ClCH}_2\text{CO}_2\text{H}) = (79.3)(1.75 \times 10^{-5}) = 1.39 \times 10^{-3}$$

$$pK_a (\text{ClCH}_2\text{CO}_2\text{H}) = -\log K_a = -\log (1.39 \times 10^{-3})$$

$$pK_a = -(-3 + 0.143) = 2.86$$

This is very close to the value, 2.8, given in the front cover of the textbook for chloroacetic acid.

3.19 $\underline{pK_a}$

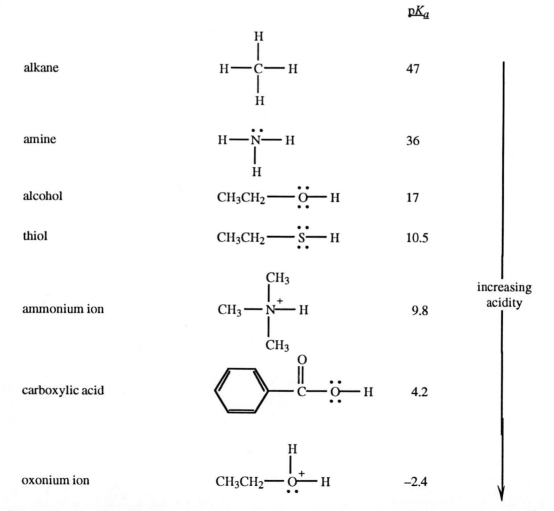

		$\underline{pK_a}$
alkane		47
amine		36
alcohol		17
thiol		10.5
ammonium ion		9.8
carboxylic acid		4.2
oxonium ion		-2.4

increasing acidity

(Note that in this table acidity increases down the table, unlike the pK_a table inside the front cover of the textbook.)

3.20

(a) CH_3CH_2OH + K^+OH^- $\rightleftharpoons$ $CH_3CH_2O^-\,K^+$ + H_2O

        pK_a 17 pK_a 15.7

(b) $CH_3CH_2\overset{+}{N}H_2\;Cl^-$ + Na^+OH^- $\rightleftharpoons$ CH_3CH_2NH + H_2O + Na^+Cl^-
        $\underset{CH_3CH_2}{\big|}$ $\underset{CH_3CH_2}{\big|}$
        pK_a 10 pK_a 15.7

(c) $CH_2F\overset{O}{\overset{\|}{C}}OH$ + $Na^+\,HCO_3^-$ $\rightleftharpoons$ $CH_2F\overset{O}{\overset{\|}{C}}O^-\,Na^+$ + H_2CO_3 $\rightleftharpoons$ H_2O + $CO_2\uparrow$

        pK_a 2 pK_a 6.5

(d) $CH_3-\!\!\bigcirc\!\!-OH$ + $Na^+\,HCO_3^-$ $\rightleftharpoons$ $CH_3-\!\!\bigcirc\!\!-O^-\,Na^+$ + H_2CO_3

        pK_a 10 pK_a 6.5

(e) $\bigcirc\!\!-NH_2$ + HCl $\rightleftharpoons$ $\bigcirc\!\!-\overset{+}{N}H_3\;Cl^-$

        pK_a −7 pK_a 4.6

(f) $CH_3CH_2CH_2SH$ + $CH_3CH_2O^-\,Na^+$ $\rightleftharpoons$ $CH_3CH_2CH_2S^-\,Na^+$ + CH_3CH_2OH

        pK_a 10.5 pK_a 17

(g) $CH_3-\!\!\bigcirc\!\!-\overset{O}{\overset{\|}{C}}OH$ + $Na^+\,CN^-$ $\rightleftharpoons$ $CH_3-\!\!\bigcirc\!\!-\overset{O}{\overset{\|}{C}}O^-\,Na^+$ + HCN

        pK_a 4 pK_a 9.1

(h) $CH_3\overset{O}{\overset{\|}{C}}CH_2\overset{O}{\overset{\|}{C}}CH_3$ + $CH_3O^-\,Na^+$ $\rightleftharpoons$ $CH_3\overset{O}{\overset{\|}{C}}\overset{..}{C}HCCH_3$ + CH_3OH
                                           Na^+

        pK_a 9.0 pK_a 15.5

3.21 $\Delta G^\circ = -RT \ln K_{eq}$

$\qquad = -2.303 RT \log K_{eq}$

$pK_{eq} = -\log K_{eq}$

therefore $\Delta G^\circ = 2.303 RT\, pK_{eq}$

At 298 K with $R = 1.987 \times 10^{-3}$ kcal/mol·K

$\Delta G^\circ = (2.303)\,(1.987 \times 10^{-3}\ \text{kcal/mol·K})\,(298\ \text{K})\, pK_{eq}$

$\Delta G^\circ = 1.36$ kcal/mol $\times pK_{eq}$

3.22 $pK_{eq} = (pK_a \text{ of starting acid}) - (pK_a \text{ of product acid})$

starting materials	products

$CH_3CH_2^- \quad + \quad CH_3NH_2$ $\qquad\qquad\qquad$ $CH_3CH_3 \quad + \quad CH_3NH^-$

$\qquad\qquad pK_a \sim 36$ $\qquad\qquad\qquad\qquad\qquad\qquad\qquad$ $pK_a \sim 49$

$pK_{eq} = 36 - 49 = -13;\ \Delta G^\circ = 1.4$ kcal/mol $\times (-13) = -18.2$ kcal/mol

There is a large decrease in free energy so the reaction goes essentially to completion as written.

$CH_3NH^- \quad + \quad CH_3OH$ $\qquad\qquad\qquad$ $CH_3NH_2 \quad + \quad CH_3O^-$

$\qquad\qquad pK_a\ 15.5$ $\qquad\qquad\qquad\qquad\qquad\qquad\qquad$ $pK_a \sim 36$

$pK_{eq} = 15.5 - 36 = -20.5;\ \Delta G^\circ = 1.4$ kcal/mol $\times (-20.5) = -28.7$ kcal/mol

There is a large decrease in free energy so the reaction essentially to completion as written.

$CH_3OH \quad + \quad OH^-$ $\qquad\qquad\qquad$ $CH_3O^- \quad + \quad H_2O$

$pK_a\ 15.5$ $\qquad\qquad\qquad\qquad\qquad\qquad\qquad\qquad\qquad$ $pK_a\ 15.7$

$pK_{eq} = 15.5 - 15.7 = -0.2;\ \Delta G^\circ = 1.4$ kcal/mol $\times (-0.2) = -0.28$ kcal/mol

There is only a small decrease in free energy. Approximately equal amounts of both starting material and products are present at equilibrium.

$CH_3SH + OH^-$ $\qquad\qquad\qquad\qquad\qquad$ $CH_3S^- + H_2O$

$pK_a \sim 10.5$ $\qquad\qquad\qquad\qquad\qquad\qquad\qquad\qquad$ $pK_a\ 15.7$

$pK_{eq} = 10.5 - 15.7 = -5.2;\ \Delta G^\circ = 1.4$ kcal/mol $\times (-5.2) = -7.3$ kcal/mol

There is a large decrease in free energy so the reaction goes essentially to completion as written.

$CH_3NH_3^+ + OH^-$ $\qquad\qquad\qquad\qquad$ $CH_3NH_2 + H_2O$

$pK_a\ 10.6$ $\qquad\qquad\qquad\qquad\qquad\qquad\qquad\qquad\quad$ $pK_a\ 15.7$

$pK_{eq} = 10.6 - 15.7 = -5.1;\ \Delta G^\circ = 1.4$ kcal/mol $\times (-5.1) = -7.1$ kcal/mol

There is a large decrease in free energy so the reaction goes essentially to completion as written.

$CH_3OH_2^+ + H_2O$ $\qquad\qquad\qquad\qquad\qquad$ $CH_3OH + H_3O^+$

$pK_a \sim -1.7$ $\qquad\qquad\qquad\qquad\qquad\qquad\qquad\qquad$ $pK_a -1.7$

3.22 (cont)

$pK_{eq} = -1.7 - (-1.7) = 0; \Delta G° = 1.4 \text{ kcal/mol} \times (0) = 0$

There is essentially no decrease in free energy. Approximately equal amounts of both starting material and products are present at equilibrium.

Concept Map 3.4   The relationship of pK_a to energy and entropy factors.

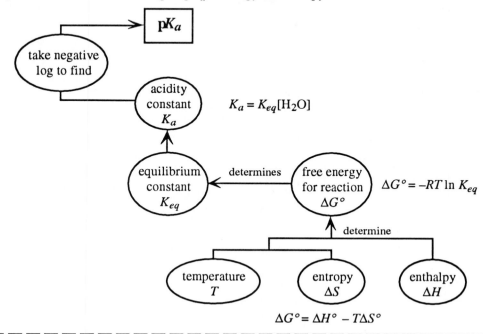

3.23 CF_3CH_2OH ionizes more readily than CH_3CH_2OH because the $CF_3CH_2O^-$ ion is more stable relative to its conjugate acid than is the ethoxide ion. The electron-withdrawing effect (negative inductive effect) of the three fluorine atoms reduces the negative charge on the oxygen atom in 2,2,2-trifluoroethoxide ion relative to that in ethoxide ion. Withdrawal of negative charge from the oxygen atom stabilizes the anion.

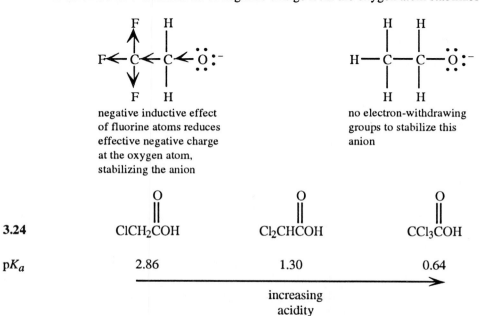

negative inductive effect
of fluorine atoms reduces
effective negative charge
at the oxygen atom,
stabilizing the anion

no electron-withdrawing
groups to stabilize this
anion

3.24

$ClCH_2\overset{O}{\overset{\|}{C}}OH$	$Cl_2CH\overset{O}{\overset{\|}{C}}OH$	$CCl_3\overset{O}{\overset{\|}{C}}OH$

pK_a 2.86   1.30   0.64

increasing
acidity

3.24 (cont)

Inductive effects are additive. The more electron-withdrawing chlorine atoms there are in the molecule, the more stable its conjugate base will be relative to the acid, and, therefore, the more acidic the conjugate acid will be.

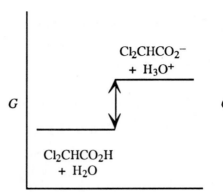

electron-withdrawing effect of one chlorine atom stabilizes the anion relative to acetate ion, making the conjugate acid more acidic than acetic acid (pK_a 4.8)

electron-withdrawing effect of two chlorine atoms leads to still further stabilization of the anion, and further strengthening of the conjugate acid

electron-withdrawing effect of three chlorine atoms makes this anion the most stable of the three compared here and, thus, its conjugate acid is the strongest

large difference in energy between between acid and its conjugate base

much smaller difference in energy between acid and its conjugate base

3.25

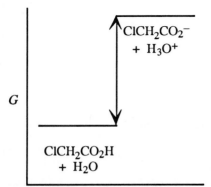

cyanoacetic acid   cyanoacetate anion

3.25 (cont)

Cyanoacetic acid is stronger than acetic acid because its conjugate base is a weaker base than acetate ion. The atom attached to the α-carbon atom has a positive charge in one important resonance contributor. The cyano group is therefore electron withdrawing, reducing electron density at the carboxylate ion by its inductive effect. The withdrawal of charge stabilizes the cyanoacetate ion relative to its acid, shifting the equilibrium toward greater ionization for cyanoacetic acid than for acetic acid.

nitroacetic acid   nitroacetate anion

Resonance involving the nitro group

Resonance involving the carboxylate group
(There are three more resonance contributors in which the nitro group
looks as it does in the resonance contributor at the upper left.)

Nitroacetic acid is a stronger acid than acetic acid because its anion is a weaker base than acetate ion. The nitrogen attached to the α–carbon atom is more electronegative than hydrogen and always has a formal positive charge. The nitro group, with one nitrogen atom and two oxygen atoms, is a strongly electron-withdrawing group. Nitroacetic acid is a stronger acid than cyanoacetic acid because the atom attached to the α-carbon atom in nitroacetic acid carries a full positive charge (all resonance contributors have a positively charged nitrogen atom). Only one resonance contributor in cyanoacetic acid has a positive carbon atom attached to the α-carbon atom.

3.26   In ethoxide anion, the basic oxygen atom is attached to an sp^3-hybridized carbon atom with no great polarity. In the acetate anion, the basic oxygen atom is attached to an sp^2-hybridized carbon atom which is more electronegative than an sp^3-hybridized carbon atom (see p. 64 in the text). In addition, this sp^2-hybridized carbon atom is attached to the still more electronegative oxygen atom, so it has considerable positive character. The positive nature of the carbon atom of the carbonyl group attracts electron density from the negatively charged oxygen atom of the acetate anion, resulting in stabilization of the anion.

sp^3-hybridized carbon
atom; no stabilization of
the negative charge

sp^2-hybridized carbon atom
with partial positive charge;
attracts electron density
from the negatively charged
oxygen atom

3.27   Triiodomethane (CHI_3) would be expected to be a weaker acid than trichloromethane ($CHCl_3$), because iodine is less electronegative than chlorine and cannot stabilize the conjugate base of triiodomethane as effectively as chlorine stabilizes the trichloromethyl anion.

the strongly electronegative (3.0) chlorine atoms withdraw electron density from the carbon atom, stabilizing the anion

The iodine atoms have about the same electronegativity as carbon (2.4 vs. 2.5), and, therefore, do not really have a negative inductive effect. Any stabilization comes from the presence of atoms larger than hydrogen, which help to spread the charge away from the carbon and over the whole ion in a way that the small hydrogen cannot.

3.28   The carbanion formed upon deprotonation of acetone is stabilized by resonance. The conjugate base of propane has no such stabilization.

stabilization of carbanion by
resonance delocalization of charge

no resonance delocalization
of charge possible

3.29

(a)

The anion produced by removal of the proton of the carboxyl group is stabilized by resonance. The anion that would be produced by removal of the proton on the hydroxyl group would not be stabilized by resonance.

(b)

The anion produced by the removal of the α-hydrogen atom is stabilized by resonance.

3.29 (cont)

(c) $HOCH_2CH_2CH_2$—S̈—H   :Ö—H   ⟶ $HOCH_2CH_2CH_2$—S̈:⁻   + H_2O

The removal of the proton from the thiol group is easier than the removal of a proton from the hydroxyl group. The larger size of the sulfur atom means that the negative charge is more spread out and, therefore, better stabilized than if it were on an oxygen atom.

(d)

:B̈r: :O:
 | ‖
:B̈r—C—CCH₃   ⟶   [:B̈r—C̈—CCH₃   ⟷   :B̈r—C=CCH₃]
 | + H_2O
 H
  ⟍ :Ö—H

The hydrogen atom on the α-carbon atom with the bromine atoms is more acidic than the hydrogen on the other α-carbon atom because the electron-withdrawing effect of the bromine atoms, as well as resonance, stabilizes the anion that results from its loss.

- -

Concept Map 3.5   (see p. 85)

- -

3.30

(a)

[piperidinium ion] + Na⁺ OH⁻ ⟶ [piperidine] + H_2O + Na⁺ + Cl⁻

pK_a ~10.6 pK_a 15.7

$pK_{eq} = 10.6 - 15.7 = -5.1; \Delta G° = 1.4$ kcal/mol $\times (-5.1) = -7.1$ kcal/mol
There is a large decrease in free energy so the reaction goes essentially to completion as written.

(b) H_2O + $CH_3CH_2\overset{+}{O}CH_2CH_3$ ⇌ $CH_3CH_2OCH_2CH_3$ + H_3O^+
                         |
                         H

           pK_a –3.6 pK_a –1.7

$pK_{eq} = -3.6 - (-1.7) = -1.9; \Delta G° = 1.4$ kcal/mol $\times (-1.9) = -2.7$ kcal/mol
There is an increase in free energy. More starting material than product will be present at equilibrium.

(c)
       O   O
       ‖   ‖
      CH_3COH + $CH_3CH_2S^-$ Na⁺ ⟶ CH_3CO^- Na⁺ + CH_3CH_2SH

        pK_a 4.8 pK_a 10.5

$pK_{eq} = 4.8 - 10.5 = -5.7; \Delta G° = 1.4$ kcal/mol $\times (-5.7) = -8.0$ kcal/mol
There is a large decrease in free energy so the reaction goes essentially to completion as written.

3.30 (cont)

(d) $CCl_3\overset{\displaystyle O}{\overset{\|}{C}}OH$ + $Na^+ HCO_3^-$ ⇌ $CCl_3\overset{\displaystyle O}{\overset{\|}{C}}O^- Na^+$ + H_2CO_3 (⇌ H_2O + $CO_2\uparrow$)

   pK_a 0.6 pK_a 6.5

pK_{eq} = 0.6 – 6.5 = –5.9; $\Delta G°$ = 1.4 kcal/mol × (–5.9) = –8.3 kcal/mol
There is a large decrease in free energy so the reaction goes essentially to completion as written.

(e) $CH_3CH_2\overset{\displaystyle O}{\overset{\|}{C}}OH$ + NH_3 ⇌ $CH_3CH_2\overset{\displaystyle O}{\overset{\|}{C}}O^-$ + NH_4^+

   pK_a ~4.9 pK_a 9.4

pK_{eq} = 4.9 – 9.4 = –4.5; $\Delta G°$ = 1.4 kcal/mol × (–4.5) = –6.3 kcal/mol
There is a large decrease in free energy so the reaction goes essentially to completion as written.

(f) CH_3NO_2 + $CH_3CH_2O^- Na^+$ ⇌ $Na^+ {}^-CH_2NO_2$ + CH_3CH_2OH

   pK_a 10.2 pK_a 17

pK_{eq} = 10.2 – 17 = –6.8; $\Delta G°$ = 1.4 kcal/mol × (–6.8) = –9.5
There is a large decrease in free energy so the reaction goes essentially to completion as written.

(g) CH_3CH_2SH + $NaNH_2$ ⇌ $CH_3CH_2S^- Na^+$ + NH_3

   pK_a 10.5 pK_a 36

pK_{eq} = 10.5 – 36 = –25.5; $\Delta G°$ = 1.4 kcal/mol × (–25.5) = –35.7
There is a large decrease in free energy so the reaction goes essentially to completion as written.

(h) $HC\equiv CH$ + $Na^+ HCO_3^-$ ⇌ $HC\equiv C^- Na^+$ + H_2CO_3

   pK_a 26 pK_a 6.5

pK_{eq} = 26 – 6.5 = +19.5; $\Delta G°$ = 1.4 kcal/mol × (19.5) = +27.3
There is a large increase in free energy so no reaction takes place. Only starting material will be present.

Concept Map 3.5   The effects of structural changes on acidity and basicity.

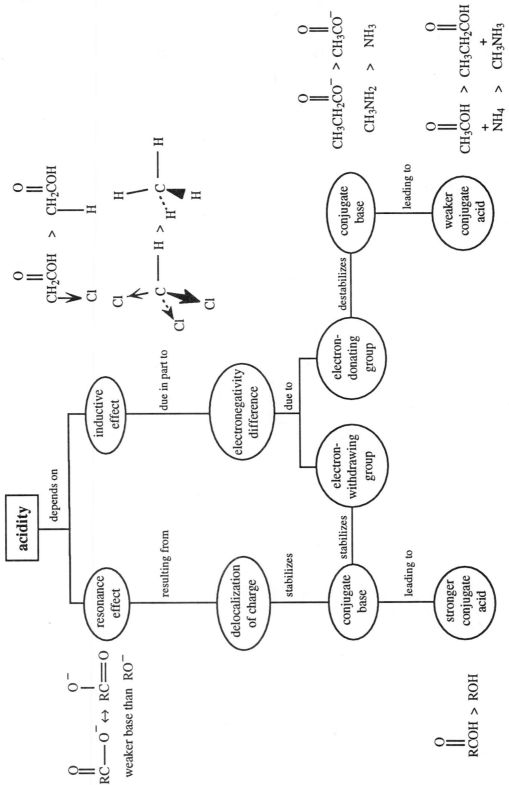

3.31   <u>least acidic</u>   <u>most acidic</u>

$$NaH \quad < \quad AlH_3 \quad < \quad H_2S \quad < \quad HCl$$

The order of relative acidity of the hydrogen atoms increases as the electronegativity of the atom to which the hydrogen atom is attached increases. This is due to the increasing stability of the conjugate base as the electronegativity of the central atom increases.

3.32   <u>least acidic</u>   <u>most acidic</u>

(a)
$$CH_3CH_2\overset{O}{\overset{\|}{C}}OH \quad < \quad ClCH_2\overset{O}{\overset{\|}{C}}OH \quad < \quad Cl^- \; \overset{+}{H_3}NCH_2\overset{O}{\overset{\|}{C}}OH$$

(b)
$$CH_3\overset{O}{\overset{\|}{C}}OH \quad < \quad CCl_3\overset{O}{\overset{\|}{C}}OH \quad < \quad CF_3\overset{O}{\overset{\|}{C}}OH$$

(c) $\quad CH_3CH_2OH \quad < \quad ClCH_2CH_2OH \quad < \quad FCH_2CH_2OH$

(d) $\quad CH_3CH_2NH_2 \quad < \quad CH_3CH_2OH \quad < \quad CH_3CH_2SH$

(e)
$$CH_3\overset{O}{\overset{\|}{C}}OCH_3 \quad < \quad CH_3\overset{O}{\overset{\|}{C}}OH \quad < \quad CH_3\overset{\overset{+}{O}-H}{\overset{\|}{C}}OH$$

(f)
$$CH_3\overset{O}{\overset{\|}{C}}OCH_3 \quad < \quad CH_3\overset{O}{\overset{\|}{C}}NH_2 \quad < \quad CH_3\overset{O}{\overset{\|}{C}}OH$$

3.33   <u>least basic</u>   <u>most basic</u>

(a)
$$CH_3\overset{..}{\underset{..}{O}}CH_2 \quad < \quad CH_3\overset{CH_3}{\underset{..}{\overset{|}{N}}}CH_3 \quad < \quad Cl_3C\overset{..}{:}{}^-$$

(b)
$$\overset{..}{:}NF_3 \quad < \quad :NH_2\overset{..}{\underset{..}{O}}H \quad < \quad :NH_3$$

(c)
$$\overset{+}{NH_4} \quad < \quad :NH_3 \quad < \quad \overset{-\;..}{:NH_2}$$

(this cannot act as a base
because there are no
nonbonding electrons)

$$CH_3\overset{CH_3}{\underset{CH_3}{\overset{|}{\underset{|}{Si}}}}CH_3 \quad < \quad CH_3\overset{..}{\underset{..}{S}}CH_3 \quad < \quad CH_3\overset{CH_3}{\underset{..}{\overset{|}{P}}}CH_3$$

nbonding electrons)

3.34

(a) F^- + BF_3 $\rightleftharpoons$ BF_4

   Lewis base   Lewis acid

(b) Ag^+ + $2\ NH_3$ $\rightleftharpoons$ $Ag(NH_3)_2^+$

   Lewis acid   Lewis base

(c) $Al(H_2O)_6^{3+}$ + OH^- $\rightleftharpoons$ $Al(OH)(H_2O)_5^{2+}$ + H_2O

   Brønsted-Lowry   Brønsted-Lowry   conjugate base   conjugate acid
        acid   base

(d) $CH_3CH_2\overset{\displaystyle O}{\overset{\displaystyle \|}{C}}OH$ + OH^- $\rightleftharpoons$ $CH_3CH_2\overset{\displaystyle O}{\overset{\displaystyle \|}{C}}O^-$ + H_2O

   Brønsted-Lowry   Brønsted-Lowry   conjugate base   conjugate acid
        acid   base

(e) $CH_3CH_2\overset{\displaystyle CH_2CH_3}{\overset{\displaystyle |}{N}}CH_2CH_3$ + $CF_3\overset{\displaystyle O}{\overset{\displaystyle \|}{C}}OH$ $\rightleftharpoons$ $CH_3CH_2\overset{\displaystyle CH_2CH_3}{\underset{\displaystyle H}{\overset{\displaystyle |}{\overset{+}{N}}}}CH_2CH_3$ + $CF_3\overset{\displaystyle O}{\overset{\displaystyle \|}{C}}O^-$

     Brønsted-Lowry   Brønsted-Lowry   conjugate acid   conjugate base
          base   acid

(f) $CH_3CH_2SCH_2CH_3$ + BF_3 $\rightleftharpoons$ $CH_3CH_2\overset{\overset{\displaystyle ^-BF_3}{\displaystyle |}}{\underset{+}{S}}CH_2CH_3$

      Lewis base   Lewis acid

(g) $ClCH_2\overset{\displaystyle O}{\overset{\displaystyle \|}{C}}OH$ + HCO_3^- $\rightleftharpoons$ $ClCH_2\overset{\displaystyle O}{\overset{\displaystyle \|}{C}}O^-$ + H_2CO_3

   Brønsted-Lowry   Brønsted-Lowry   conjugate base   conjugate acid
        acid   base

(h) $H_2PO_4^-$ + OH^- $\rightleftharpoons$ HPO_4^{2-} + H_2O

   Brønsted-Lowry   Brønsted-Lowry   conjugate base   conjugate acid
        acid   base

(i) $CH_3CH_2CH_2SH$ + $CH_3CH_2O^-$ $\rightleftharpoons$ $CH_3CH_2CH_2S^-$ + CH_3CH_2OH

      Brønsted-Lowry   Brønsted-Lowry   conjugate base   conjugate acid
           acid   base

3.35   Basicity is determined by the availability of the electron pair that will remove the proton from the acid. $CH_3OCH_2CH_2CH_2NH_2$ is a weaker base than $CH_3CH_2CH_2CH_2NH_2$ because the electron-withdrawing oxygen atom of the methoxy group is pulling electron density away from the nitrogen atom. In $(CH_3O)_2CHCH_2NH_2$ there are two electron-withdrawing methoxy groups closer to the nitrogen atom than in the first case, which further reduce the availability of the electron pair. In $N{\equiv}CCH_2CH_2NH_2$ the cyano group is even more electron-withdrawing than the methoxy group because of the positive charge on the carbon atom of the cyano group in one resonance contributor (see Problem 3.25).

3.36

(a)

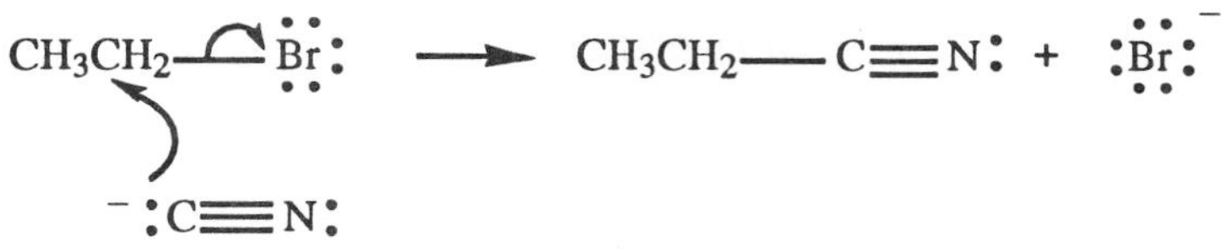

(b)

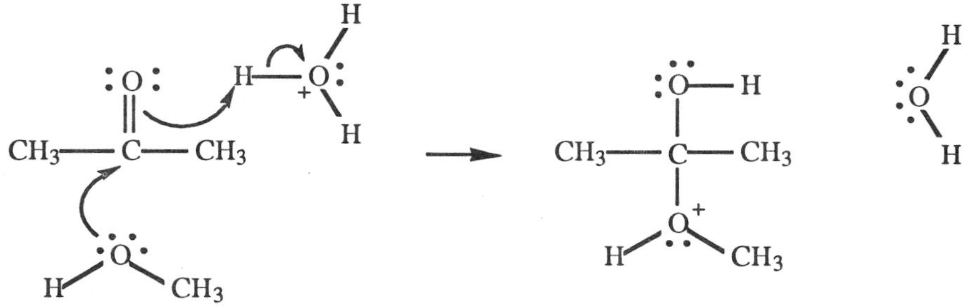

(c)

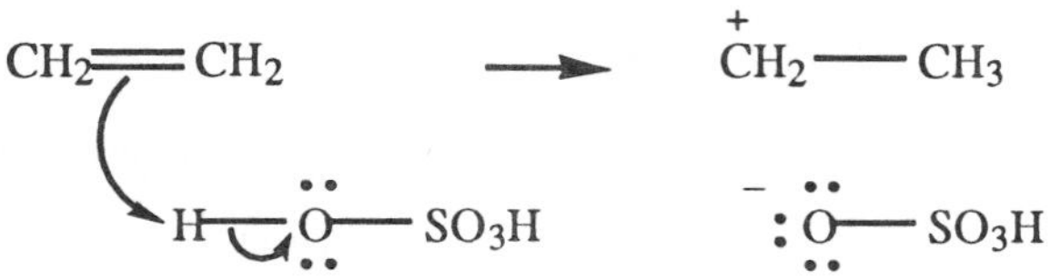

(d)

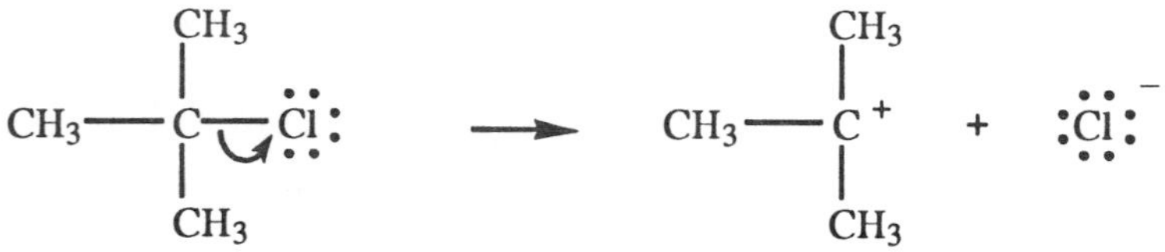

(e)

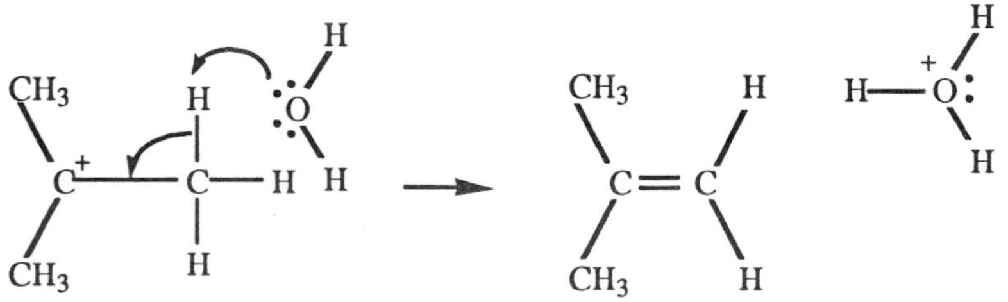

(f)

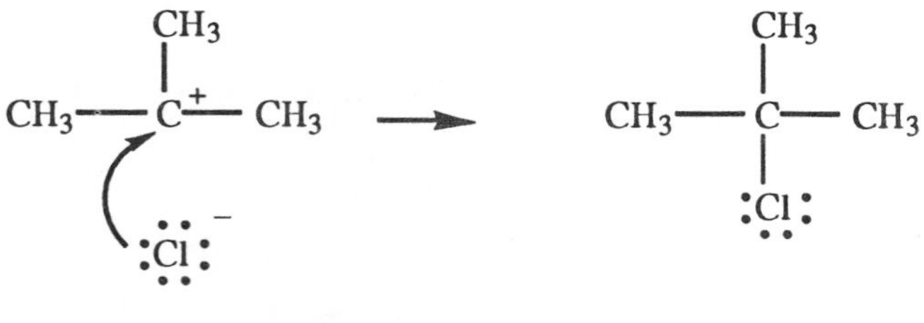

3.36 (cont)

(g) 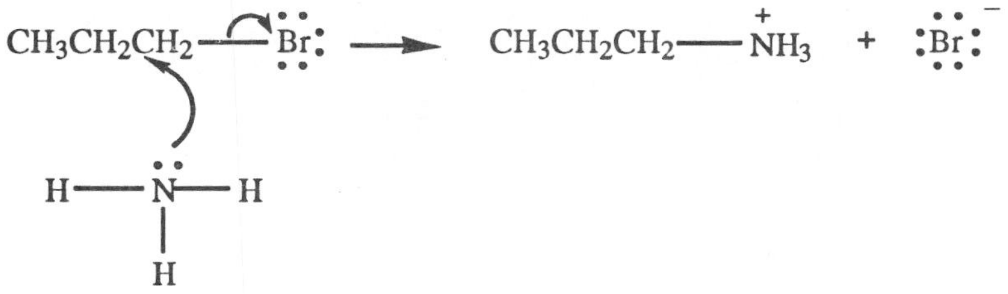   CH₃CH₂CH₂—$\overset{\cdot\cdot}{\underset{\cdot\cdot}{Br}}$: ⟶   CH₃CH₂CH₂—$\overset{+}{N}$H₃   +   :$\overset{\cdot\cdot}{\underset{\cdot\cdot}{Br}}$:⁻

H—$\overset{\cdot\cdot}{N}$—H
|
H

(h)   H—$\overset{\underset{|}{H}}{\underset{|}{C}}$—$\overset{:\overset{\cdot\cdot}{O}\cdot}{C}$—CH₃   ⟶   H—C＝C—CH₃ with $\overset{H}{|}$ and :$\overset{\cdot\cdot}{\underset{\cdot\cdot}{O}}$:⁻

H—$\overset{\cdot\cdot}{\underset{\cdot\cdot}{O}}$—H

:$\overset{\cdot\cdot}{\underset{\cdot\cdot}{O}}$—H⁻

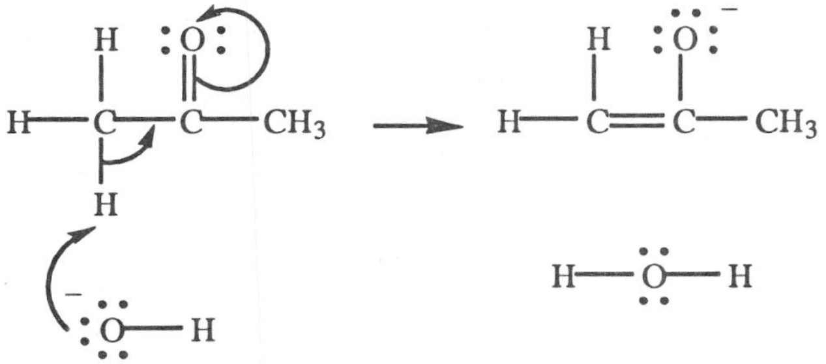

(i) 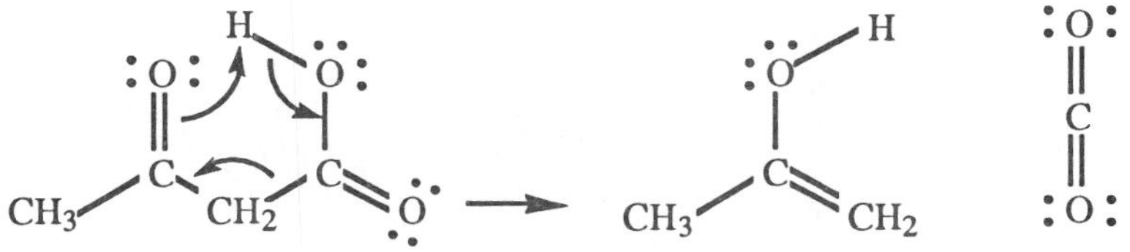   CH₃—C(=:$\overset{\cdot\cdot}{\underset{\cdot\cdot}{O}}$:)—CH₂—C(=$\overset{\cdot\cdot}{\underset{\cdot\cdot}{O}}$:)—$\overset{H}{\overset{|}{:\overset{\cdot\cdot}{O}}}$:   ⟶   CH₃—C(=$\overset{\cdot\cdot}{\underset{\cdot\cdot}{O}}$—H)＝CH₂   +   :$\overset{\cdot\cdot}{O}$:＝C＝$\overset{\cdot\cdot}{\underset{\cdot\cdot}{O}}$:

(j)    CH₃—C(=:$\overset{\cdot\cdot}{O}$:)—CH₃   +   H—$\overset{+}{\underset{H}{O}}$—H   ⟶   CH₃—C(=$\overset{+}{O}$—H)—CH₃   + $\overset{H}{\underset{H}{\overset{\cdot\cdot}{O}}}$

(k) 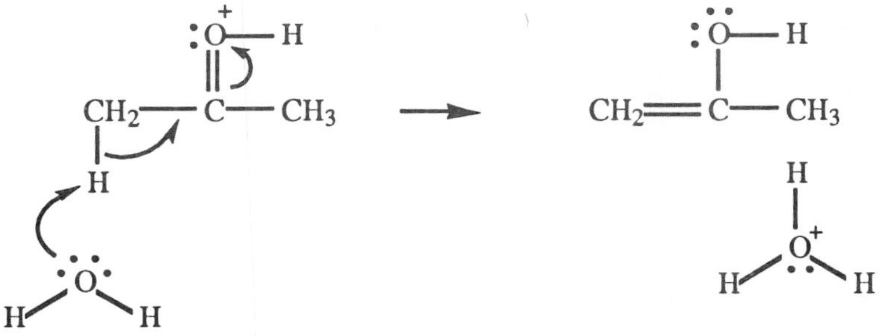   CH₂—C(=$\overset{+}{O}$—H)—CH₃ with $\overset{|}{H}$ and :$\overset{\cdot\cdot}{\underset{H\,\,\,\,H}{O}}$ ⟶   CH₂＝C(—$\overset{\cdot\cdot}{O}$—H)—CH₃   +   H—$\overset{+}{\underset{H}{O}}$—H

3.37

(a)    CHCl₂$\overset{O}{\overset{||}{C}}$OH

$pK_a = 1.3$
$K_a = 10^{-pK_a} = 10^{-1.3}$
$= 10^{+0.7} \times 10^{-2}$
$K_a = 5 \times 10^{-2}$

3.37 (cont)

(b) $CH_3\overset{+}{N}H_3$

$$pK_a = 10.4$$
$$K_a = 10^{-pK_a} = 10^{-10.4}$$
$$= 10^{+0.6} \times 10^{-11}$$
$$K_a = 4 \times 10^{-11}$$

(c) CCl_3CH_2OH

$$pK_a = 12.2$$
$$K_a = 10^{-pK_a} = 10^{-12.2}$$
$$= 10^{+0.8} \times 10^{-13}$$
$$K_a = 6 \times 10^{-13}$$

3.38

(a) CH_3CH_2OH

$$K_a = 10^{-17} \quad = 10^{-pK_a}$$
$$pK_a = -\log K_a \quad = +17$$

(b) CH_3CH_2SH

$$K_a = 3.16 \times 10^{-11}$$
$$pK_a = -\log (3.16 \times 10^{-11})$$
$$= (\log 3.16 + \log 10^{-11})$$
$$= (0.5 - 11) = -(-10.50)$$
$$pK_a = +10.50$$

(c)

$$\overset{\overset{\text{H}}{|}}{CH_3CH_2\underset{+}{O}CH_2CH_3}$$

$$K_a = 3.98 \times 10^{+3}$$
$$pK_a = -\log (3.98 \times 10^{+3})$$
$$= (\log 3.98 + \log 10^{+3})$$
$$= (0.69 + 3) = -(3.60)$$
$$pK_a = -3.60$$

3.39   (a) $H\!:^-$

(b) $Na^+ H^- \quad + \quad H_2O \quad \longrightarrow \quad Na^+ OH^- \quad + \quad H_2\uparrow$

(c)   Hydride ion, H^-, is a base. Hydrogen, H_2, is its conjugate acid.

(d) $CH_3CH_2OH \quad + \quad Na^+ H^- \quad \longrightarrow \quad CH_3CH_2O^- Na^+ \quad + \quad H_2\uparrow$

3.40
(a)

$$CH_3CH_2\overset{..}{\underset{\overset{|}{H}}{\overset{+}{O}}}CH_2CH_3 \; + \; CH_3\overset{..}{\underset{\overset{|}{H}}{N}}CH_3 \longrightarrow CH_3CH_2\overset{..}{\underset{..}{O}}CH_2CH_3 \; + \; CH_3\overset{\overset{H}{|}}{\underset{\overset{|}{H}}{\overset{+}{N}}}CH_3$$

A   B

$pK_a -3.6$ $pK_a \sim 10$

3.40 (a) (cont)

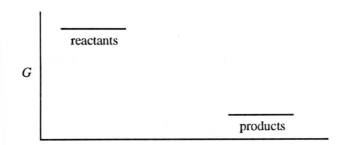

(b) $CH_3\overset{O}{\overset{\|}{C}}\overset{\cdot\cdot}{C}H\overset{O}{\overset{\|}{C}}OCH_2CH_3$ + $CH_3CH_2\overset{\cdot\cdot}{O}H$ ⇌ $CH_3\overset{O}{\overset{\|}{C}}CH_2\overset{O}{\overset{\|}{C}}OCH_2CH_3$ + $CH_3CH_2\overset{\cdot\cdot}{\underset{\cdot\cdot}{O}}\!:^-\,Na^+$

$\overset{-}{Na^+}$ C   D

pK_a 17   pK_a 11.0

(energy diagram: reactants low, products high, labeled G vertical axis)

(c) $[(CH_3)_3Si]_2\overset{\cdot\cdot}{N}H$ + $CH_3CH_2CH_2CH_2:^-\,Li^+$ ⇌ $[(CH_3)_3Si]_2\overset{\cdot\cdot}{N}{}^-\,Li^+$ + $CH_3CH_2CH_2CH_3$

pK_a ~10   E   F

pK_a ~49

(energy diagram: reactants high, products low, labeled G vertical axis)

(d)

$CH_3CH_2-\overset{\overset{:\overset{\cdot\cdot}{O}:}{\|}}{\underset{+}{N}}-\overset{\cdot\cdot}{\underset{\cdot\cdot}{O}}:^-$ + $(CH_3)_3C-\overset{\cdot\cdot}{\underset{\cdot\cdot}{O}}:^-$ ⇌ $CH_3\overset{\cdot\cdot}{C}H-\overset{\overset{:\overset{\cdot\cdot}{O}:}{\|}}{\underset{+}{N}}-\overset{\cdot\cdot}{\underset{\cdot\cdot}{O}}:^-$ + $(CH_3)_3C-\overset{\cdot\cdot}{O}H$

G   H

pK_a ~10.2   pK_a 19

3.40 (d) (cont)

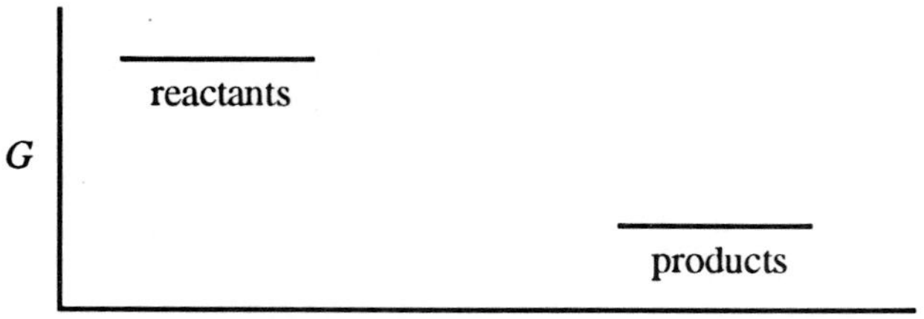

(e)

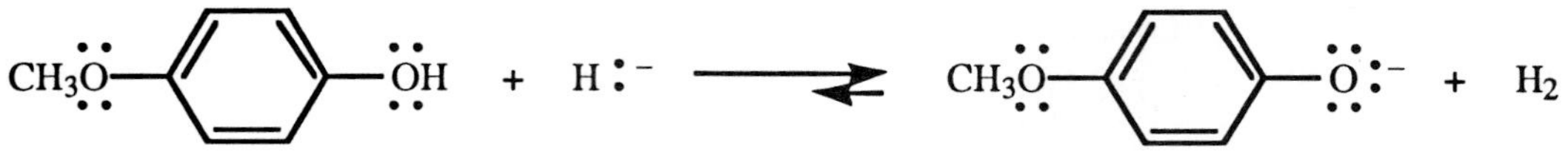

$$CH_3\ddot{O}-\langle\rangle-\ddot{O}H \;+\; H\!:^{-} \;\rightleftharpoons\; CH_3\ddot{O}-\langle\rangle-\ddot{O}\!:^{-} \;+\; H_2$$

I   J

$pK_a \sim 10$ $pK_a\ 35$

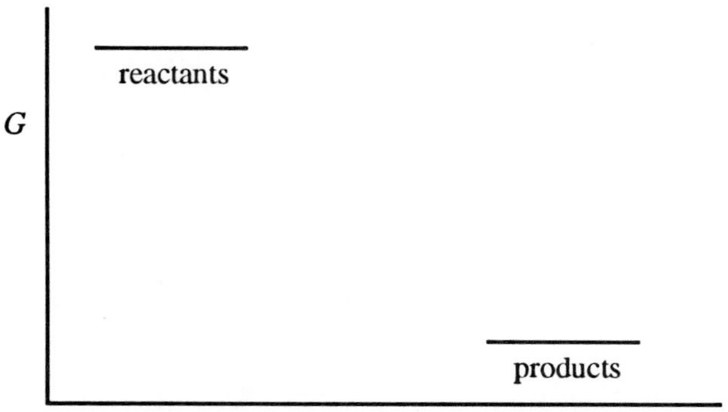

3.41

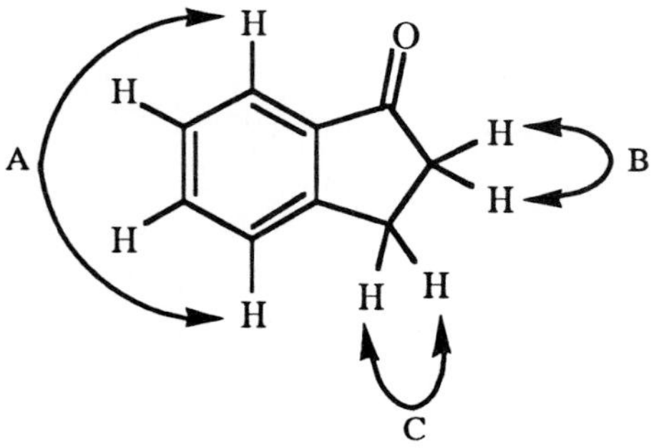

1-indanone

(a)   A $pK_a \sim 43$, analogous to benzene

       B $pK_a \sim 19$, analogous to acetone

       C $pK_a \sim 41$ analogous to toluene

(b)

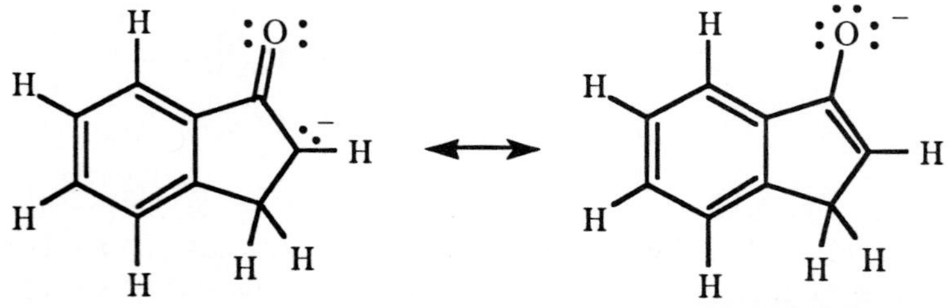

major resonance contributor

3.41 (cont)

(c)

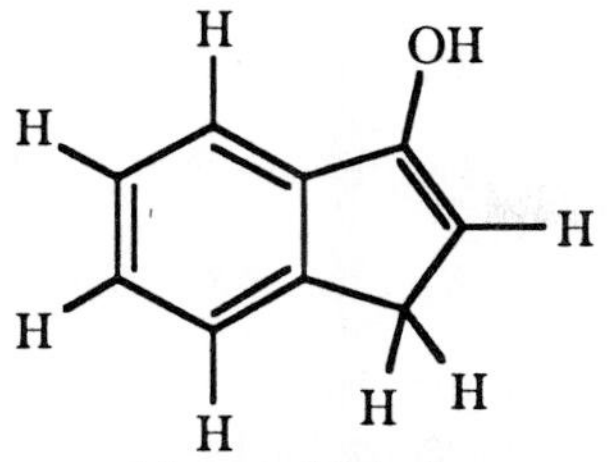

enol form of 1-indanone

3.42   (a)

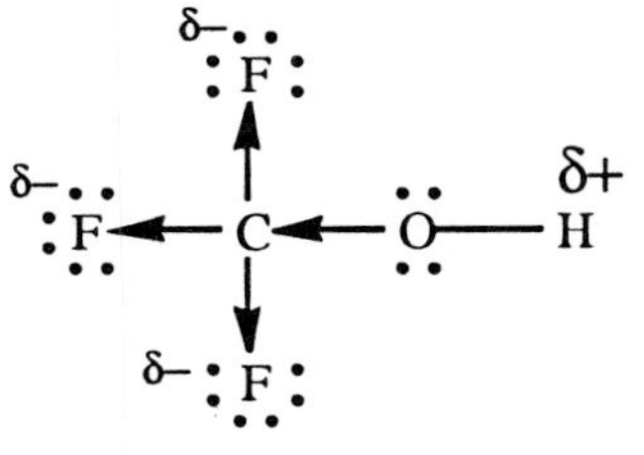

acid 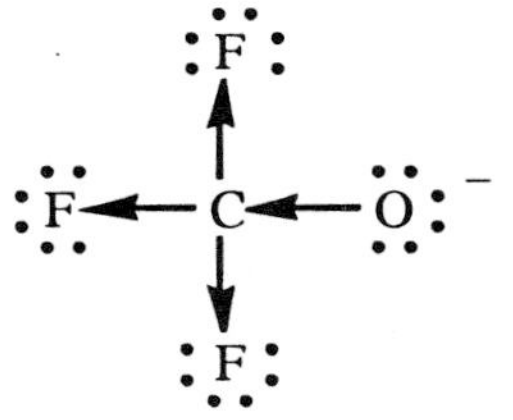   conjugate base

The inductive effect of the fluorines makes the proton on the hydroxyl group more acidic and the anion less basic. The inductive effect of the fluorines also reduces the availability of the electron pairs on the oxygen atom so they are not as readily involved in hydrogen bonding, decreasing the boiling point.

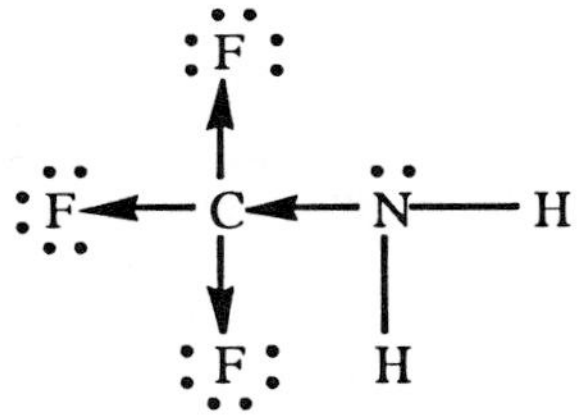

The basicity of the nitrogen atom is also decreased by the electron-withdrawing effect of the fluorines, which makes the electron pair on the nitrogen less available for protonation.

(b) $CF_3\overset{+}{N}H_3$ Cl^- + $CH_3\underset{CH_3}{\overset{CH_3}{N:}}$ $\rightleftharpoons$ $CF_3\overset{..}{N}H_2$ + $CH_3\underset{CH_3}{\overset{CH_3}{\overset{+}{N}}}-H$ Cl^-

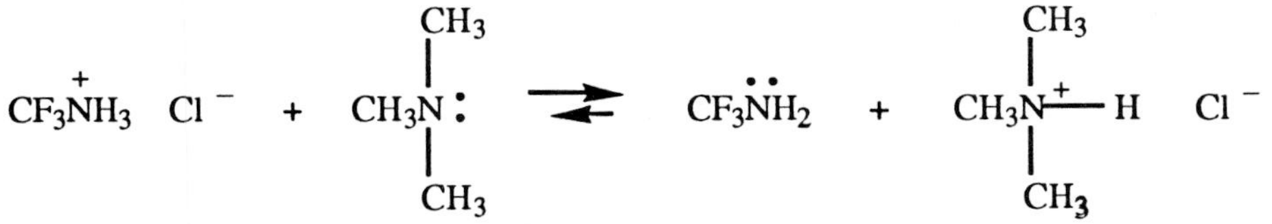

We would expect trimethylamine to be a stronger base than trifluoromethylamine [see part (a)], and to remove a proton from the trifluoromethylammonium ion.

3.43   When a strong acid such as H—Cl is added to water, the following reaction takes place.

$$H-Cl + H_2O \longrightarrow Cl^- + H_3O^+$$

$pK_a -7$ $pK_a -1.7$

The equilibrium constant for this reaction is greater than 10^{+5}, and the reaction proceeds essentially to completion. Similar reactions take place in water with other strong acids such as sulfuric, nitric, and phosphoric acids. A solution of strong acid in water thus consists of the hydronium ion, (the strongest acid which can exist in water) and the conjugate base of the strong acid.

3.43 (cont)

When a strong base such as amide ion is added to water, a similar reaction takes place.

$$NH_2^- + H_2O \longrightarrow NH_3 + OH^-$$

$$pK_a\ 15.7 \quad pK_a\ 36$$

Again the equilibrium constant is very large ($>10^{+20}$), and the reaction proceeds essentially to completion. A solution of a strong base in water thus consists of the hydroxide ion (the strongest base that can exist in water) and the conjugate acid of the strong base.

3.44

$$pK_a\ \overset{+}{R}NH_3 \sim 10.6 \qquad\qquad pK_a\ RCOH \sim 4.0$$

(a)　The carboxylate anion, —CO⁻, is a weaker base than the amino group, —NH₂. Therefore, a proton will be transferred from the carboxyl group to the amino group. Glycine is best represented as H₃NCH₂CO⁻.

(b)　The high melting point for a compound with such a low molecular weight is indicative of an ionic compound. The high water solubility is also indicative of strong electrostatic forces. Both observations are consistent with the charged structure, H₃NCH₂CO⁻, for glycine.

3.45

(a)

3.45 (a) (cont)

$$H_2N-C(=NH_2^+)-NH-CH_2CH_2CH_2CH(NH_3^+)-C(=O)-O^-$$

(b)   Arginine is more basic than lysine because the conjugate acid of arginine is stabilized by resonance. The positive charge is delocalized to three nitrogen atoms in arginine. In lysine it is localized on a single nitrogen atom.

3.46   (a) $$K_a = \frac{[A^-]\,[H_3O^+]}{[HA]}$$

taking the log of both sides

$$\log K_a = \log\left(\frac{[A^-]\,[H_3O^+]}{[HA]}\right)$$

$$\log K_a = \log [H_3O^+] + \log \frac{[A^-]}{[HA]}$$

multiplying both sides by −1

$$-\log K_a = -\log [H_3O^+] - \log \frac{[A^-]}{[HA]}$$

rearranging the equation by adding $\log \frac{[A^-]}{[HA]}$ to both sides

$$-\log [H_3O^+] = -\log K_a + \log \frac{[A^-]}{[HA]}$$

substituting the following relationships

$$pK_a = -\log K_a$$

$$pH = -\log [H^+] = -\log [H_3O^+]$$

we get the Henderson-Hasselbach equation

$$pH = pK_a + \log \frac{[A^-]}{[HA]}$$

(b)   when $pH = pK_a$, $\log \frac{[A^-]}{[HA]} = 0$ and $\frac{[A^-]}{[HA]} = 1$

or the concentrations of the acid and its conjugate base are equal.

$$[HA] = [A^-]$$

3.46 (cont)

(c)   pH $\cong 6.5$

pK$_a$ = 7

$$pH = pK_a + \log \frac{[A^-]}{[HA]}$$

$$6.5 = 7 + \log \frac{[A^-]}{[HA]}$$

$$\log \frac{[A^-]}{[HA]} = -0.5$$

$$\frac{[A^-]}{[HA]} = 0.316$$

$$[A^-] = 0.316\,[HA]$$

and $[HA] = \sim3\,[A^-]$

There is more protonated than unprotonated imidazole present at pH 6.5.

3.47 (a)

H—C—C—H with H, Cl on top and H, Br on bottom

pK$_a$ 23.8

H—C—C—H with H, H on top and H, H on bottom

pK$_a$ ~49

The conjugate base of haloethane is stabilized by the electron-withdrawing effect of the halogen atoms. No such stabilization is available to the conjugate base of ethane.

stabilization of anion by electron-withdrawing effect of halogen atoms

no comparable stabilization of anion of ethane

3.47 (cont)

(b)

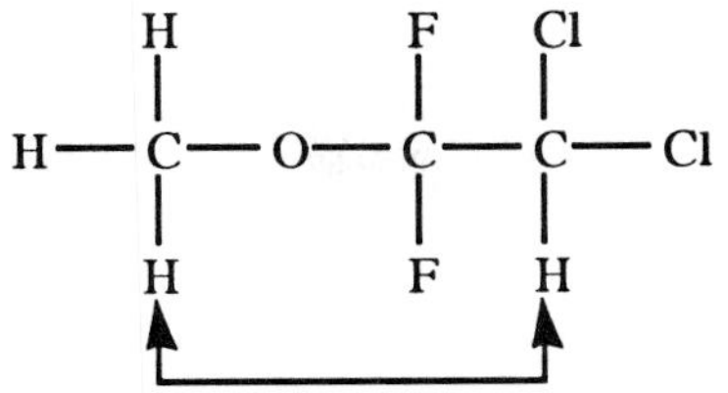

potential sites of
deprotonation

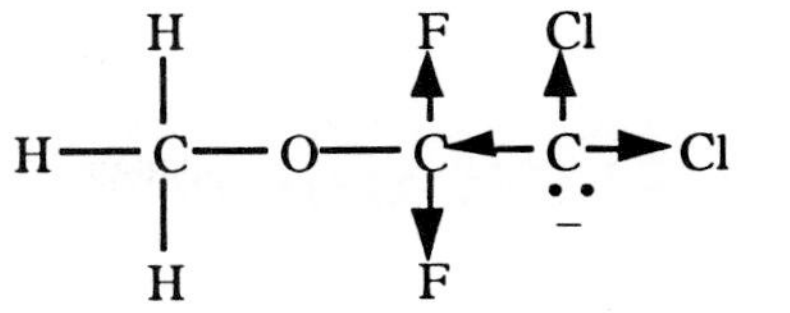

more stable of the two possible conjugate
bases; anion stabilized by the electron-
withdrawing effect of halogen atoms

(c)

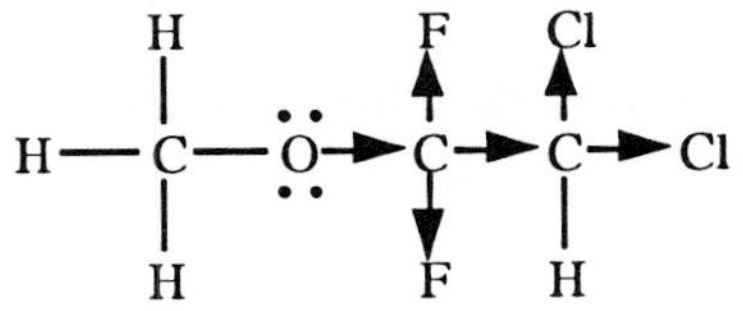

methoxyflurane

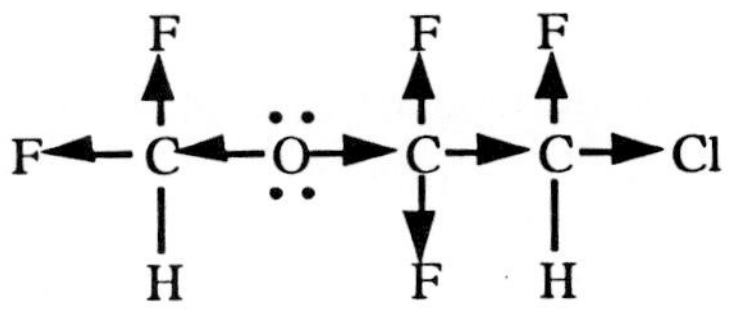

enflurane

The oxygen atom in methoxyflurane is more basic than the oxygen atom in enflurane. Enflurane has more fluorine atoms withdrawing electron density from the oxygen atom than does methoxyflurane.

4

Reaction Pathways

Workbook Exercises

EXERCISE I. A number of examples of connectivity changes were given as Workbook Exercises I and II in Chapter 1. Use the curved arrow notation for mechanisms that was introduced in Chapter 3 to restate, with this symbolism, the bonding changes you have already described with words.

EXERCISE II. Draw the structures of the products implied by each of the following mechanisms.

(a)

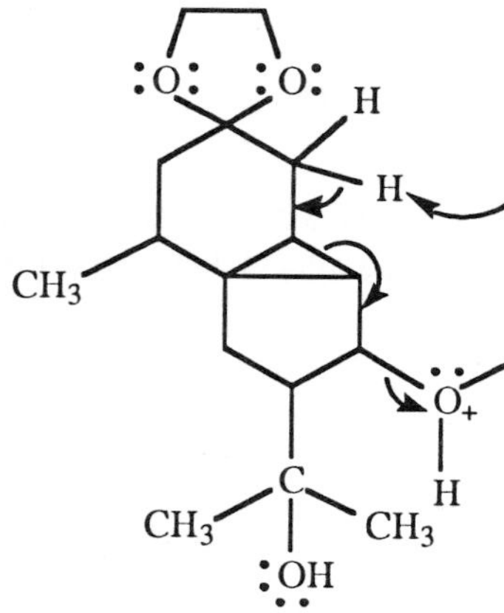

(b)

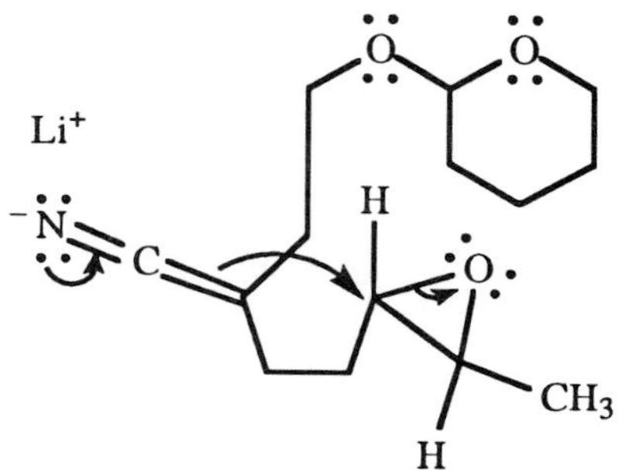

(c)

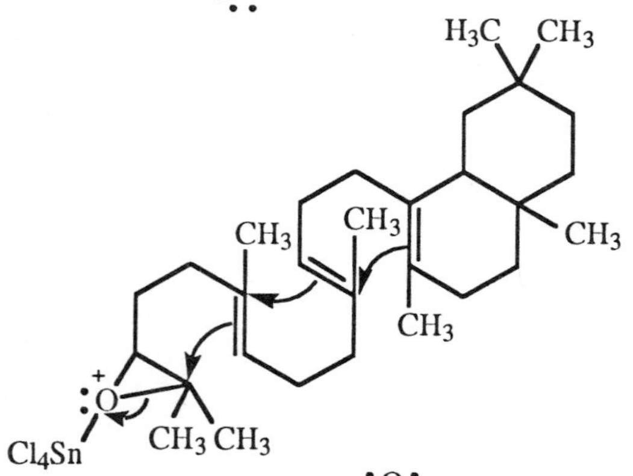

(d)

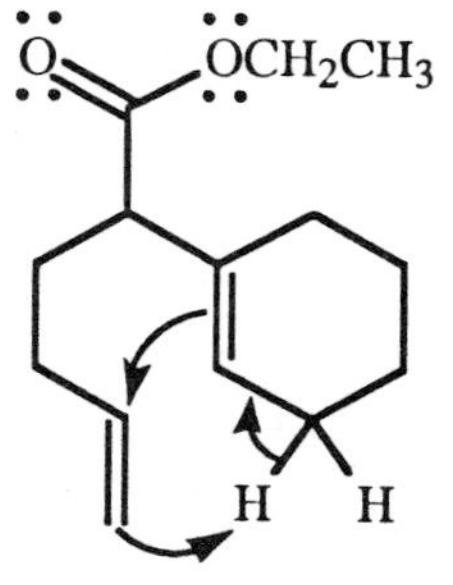

(e)

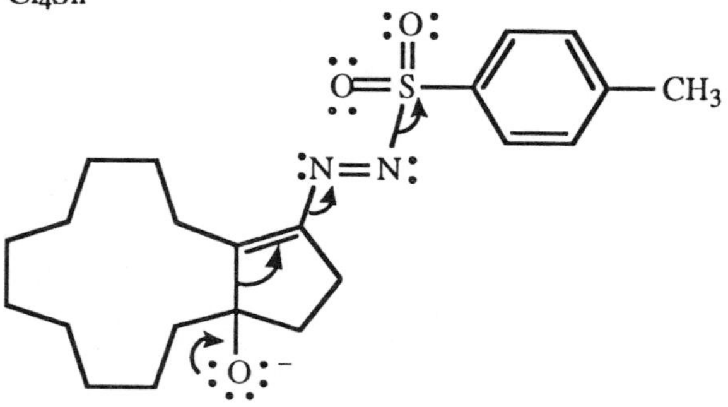

(f)

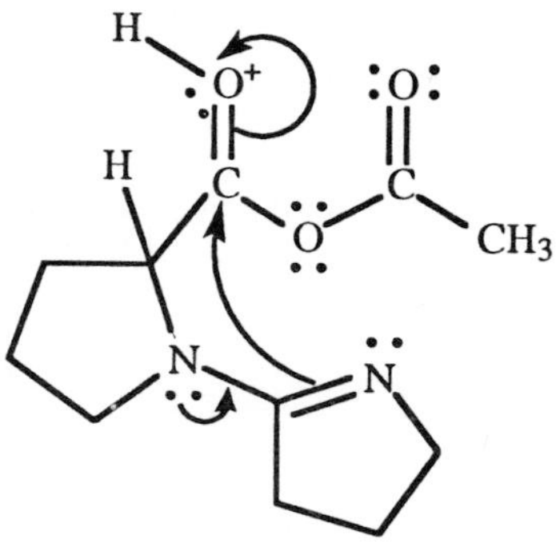

Workbook Exercises (cont)

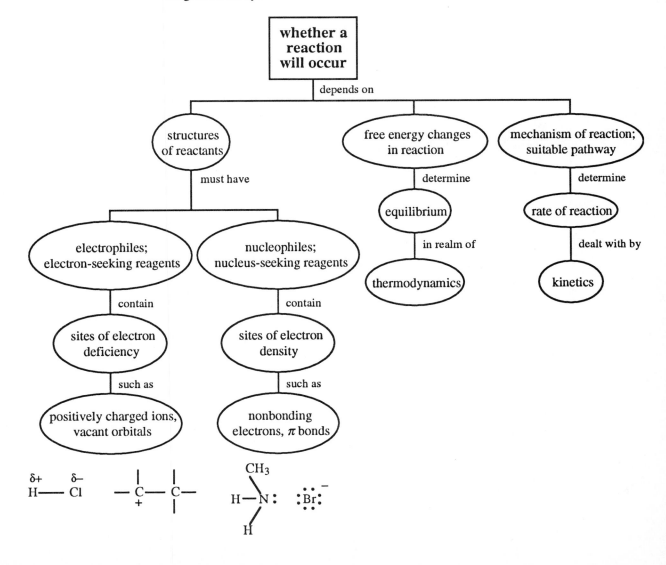

Concept Map 4.1   The factors that determine whether a chemical reaction between a given set of reagents is likely.

4.1   For those species in which more than one electrophilic or nucleophilic sites are present, the stronger site is indicated.

(a) CH_3—$\overset{..}{\underset{..}{O}}:^-$ (b) $:PH_3$ (c) Cu^{2+} (d) $\overset{\delta+}{H}$—$\overset{\delta-}{\underset{..}{Br}}:$ (e) $\overset{\delta+}{CH_3}$—$\overset{\delta-}{\underset{..}{Cl}}:$

nucleophile   nucleophile   electrophile   electrophile   electrophile

(f) $CH_3\overset{..}{N}H_2$ (g) $H:^-$ (h) $(CH_3)_3B$ (i) $H\overset{..}{O}$—$\overset{..}{N}H_2$ (j) $HC\equiv CH$

nucleophile   nucleophile   electrophile   nucleophile   nucleophile

(k) $AlCl_3$ (l) $\overset{H}{\underset{H}{\diagdown}}\overset{\delta+}{C}=\overset{\delta-}{\underset{..}{O}}$ (m) $:\overset{..}{\underset{..}{I}}:^-$ (n) CH_3CH_2—$\overset{..}{\underset{..}{S}}:^-$ (o) Hg^{2+}

electrophile   electrophile   nucleophile   nucleophile   electrophile

4.2

(a)   electrophile ⟶ $CH_3\frown\overset{..}{\underset{..}{Br}}:$ ⟶ CH_3—$\overset{..}{\underset{..}{O}}H$ + $:\overset{..}{\underset{..}{Br}}:^-$

nucleophile ⟶ $^-:\overset{..}{\underset{..}{O}}$—H

(b)   electrophile ⟶ $CH_3\frown\overset{..}{\underset{..}{I}}:$ ⟶ CH_3—$\overset{..}{\underset{..}{O}}H$ + $:\overset{..}{\underset{..}{I}}:^-$

nucleophile ⟶ $^-:\overset{..}{\underset{..}{O}}$—H

(c)   electrophile ⟶ $CH_3CH_2\frown\overset{..}{\underset{..}{Cl}}:$ ⟶ CH_3CH_2—$\overset{..}{\underset{..}{O}}H$ + $:\overset{..}{\underset{..}{Cl}}:^-$

nucleophile ⟶ $^-:\overset{..}{\underset{..}{O}}$—H

(d)   electrophile ⟶ $CH_3\frown\overset{..}{\underset{..}{I}}:$ ⟶ CH_3—$\overset{+}{N}H_3$ $:\overset{..}{\underset{..}{I}}:^-$

nucleophile
$:NH_3$

4.3   Sodium ion, Na⁺, has a complete valence shell (see Problem 2.1 for the electronic configuration of elemental sodium), and the energies of the empty orbitals in the $n=3$ shell are too high to be available for accepting an electron pair. Both the mercury(II), Hg^{2+}, and copper(II), Cu^{2+}, ions are transition metals that have lower energy empty orbitals available to accept an electron pair.

Concept Map 4.2   A nucleophilic substitution reaction.

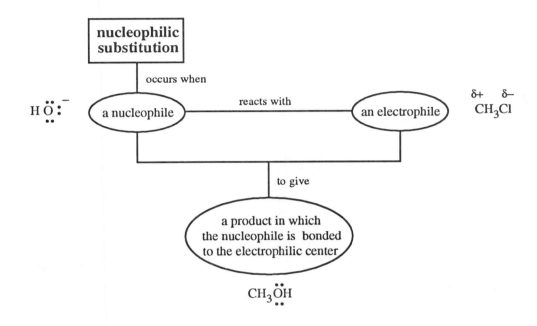

4.4   (a)   Overall equation must be:

$$Cl_2 \ + \ NH_3 \ \longrightarrow \ H_3\overset{+}{N}\!-\!Cl \ + \ Cl^-$$

$$H_3\overset{+}{N}\!-\!Cl \ + \ Cl^- \ + \ Na^+OH^- \ \longrightarrow \ H_2N\!-\!Cl \ + \ Na^+Cl^- \ + \ H_2O$$

(b)   Mechanism

Concept Map 4.3   The factors that determine the equilibrium constant for a reaction.

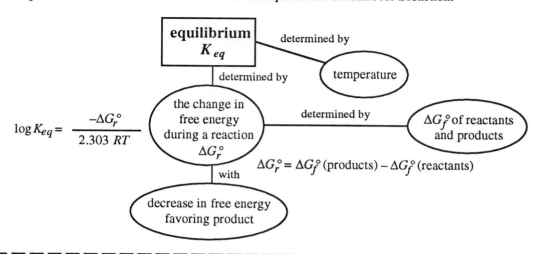

$$\log K_{eq} = \frac{-\Delta G_r^\circ}{2.303 \, RT}$$

$$\Delta G_r^\circ = \Delta G_f^\circ \, (\text{products}) - \Delta G_f^\circ \, (\text{reactants})$$

Concept Map 4.4   The factors that influence the rate of a reaction as they appear in the rate equation.

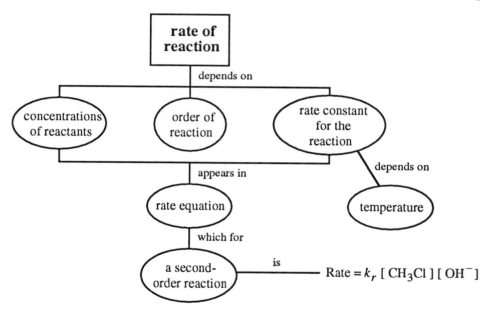

$$\text{Rate} = k_r \, [\, CH_3Cl \,] \, [\, OH^- \,]$$

4.5   At 311 K, $k_r = 3.55 \times 10^{-5}$ L/mol·sec.

Initial rate $= k_r[CH_3Cl][OH^-]$; when $[OH^-]$ 0.05 M.

Initial rate $= (3.55 \times 10^{-5}$ L/mol·sec$)(0.003$ mol/L$)(0.05$ mol/L$)$.

$= 5.33 \times 10^{-9}$ mol/L·sec, which is five times the original rate (see p. 130 in the text).

When $[CH_3Cl] = 0.001$ M, the rate will be one-third the rate when $[CH_3Cl] = 0.003$ M, or 3.55×10^{-10} mol/L·sec.

Concept Map 4.5   Some of the factors other than concentration that influence the rate of a reaction.

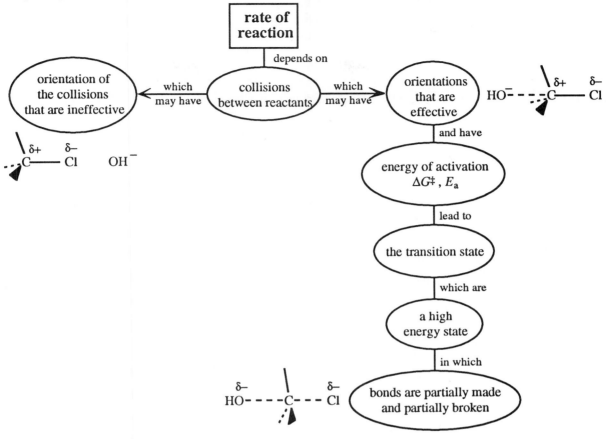

Concept Map 4.6   Factors that influence the rate of a reaction, and the effect of temperature.

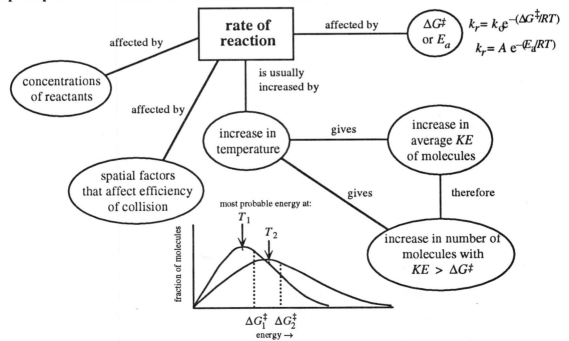

4.6   At 308 K, $k_r = 5.3 \times 10^{-4}$ L/mol·sec.

Initial rate = $(5.3 \times 10^{-4}$ L/mol·sec$)(0.001$ mol/L$)(0.01$ mol/L$)$

= 5.3×10^{-9} mol/L·sec

Concept Map 4.7   An electrophilic addition reaction to an alkene.

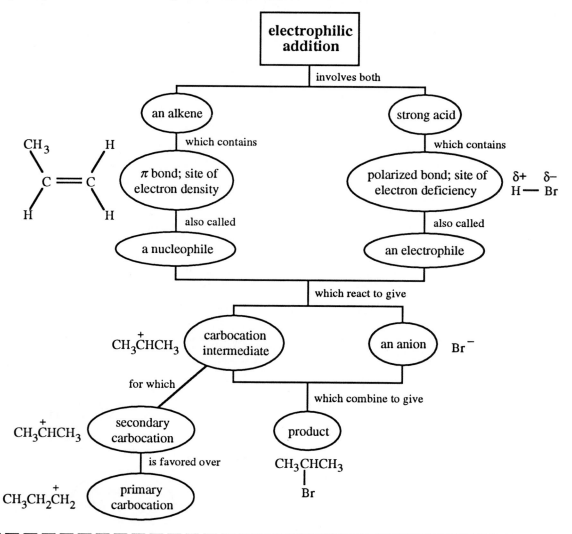

4.7   The double bond in propene is another possible nucleophile in the reaction mixture.

$$CH_3CH\!=\!CH_2 \longrightarrow CH_3CH\!-\!CH_2\!-\!\overset{CH_3}{\underset{+}{CHCH_3}}$$

$$\overset{\curvearrowright}{CH_3\underset{+}{CHCH_3}}$$

a new secondary
carbocation

Concept Map 4.8   Markovnikov's Rule.

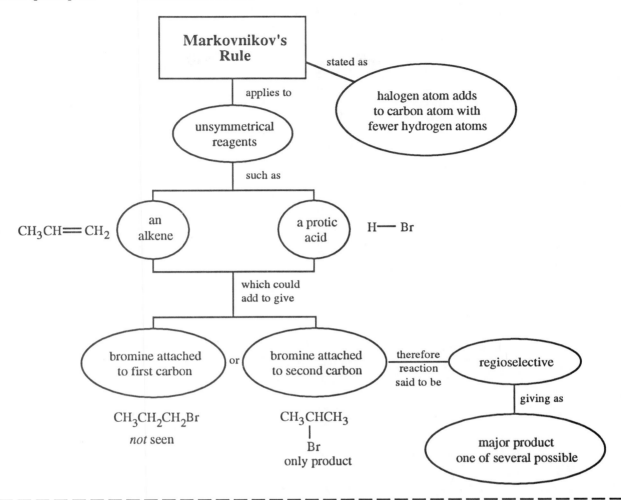

- -

4.8

(a) $CH_3\overset{\overset{\displaystyle CH_3}{|}}{C}=CH_2$ + HBr $\longrightarrow$ $CH_3\overset{\overset{\displaystyle CH_3}{|}}{\underset{\underset{\displaystyle Br}{|}}{C}}CH_3$

(b) $CH_3CH_2CH=CH_2$ + HBr $\longrightarrow$ $CH_3CH_2\underset{\underset{\displaystyle Br}{|}}{C}HCH_3$

(c) $CH_3CH_2CH=CHCH_3$ + HBr $\longrightarrow$ $CH_3CH_2\underset{\underset{\displaystyle Br}{|}}{C}HCH_2CH_3$ + $CH_3CH_2CH_2\underset{\underset{\displaystyle Br}{|}}{C}HCH_3$

4.9   The methyl cation (CH_3^+) is less stable than the *n*-propyl cation ($CH_3CH_2CH_2^+$) because the methyl cation has no alkyl groups to stabilize the positive charge.

4.10   The *tert*-butyl cation is more stable than the isopropyl cation because the *tert*-butyl cation has three alkyl groups
to stabilize the positive charge and the isopropyl group has only two.

tert-butyl cation	$CH_3{-}\overset{\displaystyle CH_3}{\underset{\displaystyle CH_3}{\overset{\displaystyle |}{\underset{\displaystyle |}{C}}}}{}^+$	most stable
isopropyl cation	$CH_3{-}\overset{\displaystyle CH_3}{\underset{\displaystyle H}{\overset{\displaystyle |}{\underset{\displaystyle |}{C}}}}{}^+$	
n-propyl cation	$CH_3CH_2{-}\overset{\displaystyle H}{\underset{\displaystyle H}{\overset{\displaystyle |}{\underset{\displaystyle |}{C}}}}{}^+$	
methyl cation	$H{-}\overset{\displaystyle H}{\underset{\displaystyle H}{\overset{\displaystyle |}{\underset{\displaystyle |}{C}}}}{}^+$	least stable

4.11 $\log K_{eq} = -(\Delta G°/2.303\,RT)$

At 25 °C (298 K),

for 1-bromopropane

$\log K_{eq} = -(-7.63\ \text{kcal/mol})/[(2.303)(1.99 \times 10^{-3}\ \text{kcal/mol·K})(298\ \text{K})]$

$\log K_{eq} = 5.59$

$K_{eq} = 10^{5.59} = 10^{0.59} \times 10^5 = 3.89 \times 10^5$

for 2-bromopropane

$\log K_{eq} = -(-8.77\ \text{kcal/mol})/[(2.303)(1.99 \times 10^{-3}\ \text{kcal/mol·K})(298\ \text{K})]$

$\log K_{eq} = 6.42$

$K_{eq} = 10^{6.42} = 10^{0.42} \times 10^6 = 2.63 \times 10^6$

Note that the answers that you get may not correspond exactly to the numbers obtained here, depending on
whether you round off to the correct number of significant figures at each stage of the calculation or carry all
figures through and round off at the end of the calculation.

Concept Map 4.9   Carbocations.

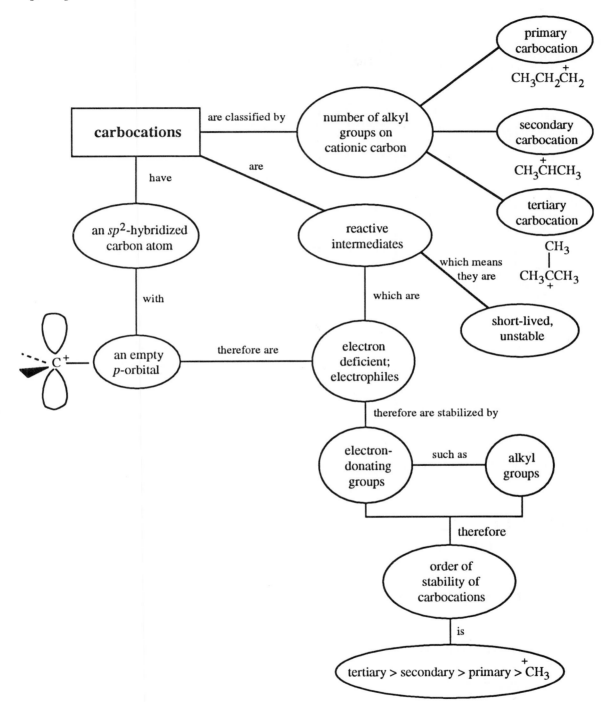

Concept Map 4.10   Energy diagrams.

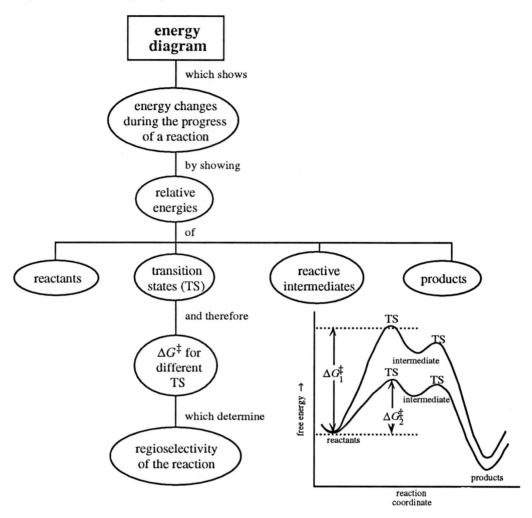

- -

4.12   (a) $CH_3CH = CH_2$ (g)   +   HI (g) $\longrightarrow$ CH_3CHCH_3 (g)

$$\qquad\qquad\qquad\qquad\qquad\qquad\qquad\qquad\qquad\qquad\qquad\qquad\qquad\qquad\qquad | \atop I$$

(b)   Rate $= k$ [$CH_3CH=CH_2$][HI]

4.13   Concentration is defined as the number of moles of a substance in a given volume. For an ideal gas, the concentration at constant temperature and volume is proportional to the pressure.

$$P = \frac{n}{V} RT$$

$P_{propene} = P_{HI} = (45 \text{ mm Hg})/(760 \text{ mm Hg/atm})$

Rate $= k (P_{propene})(P_{HI})$

Rate $= (1.66 \times 10^{-6}/\text{atm·sec})(45/760)^2 \text{atm}^2$

Rate $= 5.8 \times 10^{-9}$ atm/sec

4.14   For $CH_3CH \!=\! CH_2$ +   HI $\longrightarrow$ CH_3CHCH_3
$$\underset{\text{I}}{|}$$

$k_r = 2.61 \times 10^{-3}$ L/mol·sec

with $E_a = 22.4$ kcal/mol and $A = 2.5 \times 10^7$ L/mol·sec

For $CH_3CH \!=\! CH_2$ +   HI $\longrightarrow$ $CH_3CH_2CH_2I$

$k_r = Ae^{-E_a/RT}$

with $E_a = 38$ kcal/mol and $A = 2.5 \times 10^7$ L/mol·sec

$k_r = (2.5 \times 10^7$ L/mol·sec$)\exp(-38$ kcal/mol$)/[(1.99 \times 10^{-3}$ kcal/mol·K$)(490$ K$)]$

$k_r = (2.5 \times 10^7$ L/mol·sec$)\exp(-39.0)$

$k_r = (2.5 \times 10^7$ L/mol·sec$)(1.2 \times 10^{-17})$

$k_r = 3.0 \times 10^{-10}$ L/mol·sec

Rate for the formation of the isopropyl cation = $k_{\text{isopropyl}}[CH_3CH\!=\!CH_2][HI]$

Rate for the formation of the n-propyl cation = $k_{n\text{-propyl}}[CH_3CH\!=\!CH_2][HI]$

$$\frac{k_{\text{isopropyl}}}{k_{n\text{-propyl}}} = \frac{2.61 \times 10^{-3} \text{ L/mol·sec}}{3.0 \times 10^{-10} \text{ L/mol·sec}}$$

$$\frac{k_{\text{isopropyl}}}{k_{n\text{-propyl}}} = 8.7 \times 10^6$$

In the reaction of propene with hydrogen iodide, 8.7 million isopropyl cations are formed for every n-propyl cation.

4.15

(a)

(b)

4.15 (b) (cont)

Note that even though both carbon and nitrogen have nonbonding electron pairs, the electrons on the carbon are more available for bonding in nucleophilic substitution reactions because carbon bears the negative charge and is less electronegative than nitrogen and, therefore, holds its electrons less tightly.

(c)

(d)

(e)

4.16   <u>reagent</u>   <u>type</u>   <u>reason for choice</u>

(a) $CH_3O^- Na^+$ nucleophile   The carbon atom bonded to the iodine atom in ethyl iodide is an electrophilic carbon and will react with a reagent that is a nucleophile. The reaction is a nucleophilic substitution. CH_3O^- is the nucleophile. The reagent is shown as also containing a sodium ion, because it is not possible to have a reagent with only anions in it. The sodium ions are called "spectator" ions. They do not participate directly in the reaction, but serve to neutralize the negative charges on the anions before and after the reaction takes place.

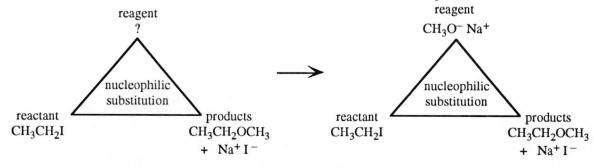

4.16 (cont)

(b)   H₂SO₄   electrophile   The double bond in 1-butene is a nucleophile and reacts with the electrophilic proton in sulfuric acid. The carbocation produced in the first step of the reaction is an electrophile and will react with the conjugate base of sulfuric acid, a nucleophile, which is also produced in the first step of the reaction. The overall reaction is an electrophilic addition.

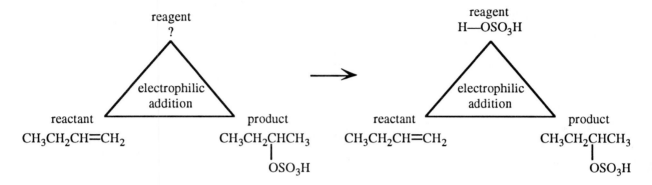

(c)   Br₂   electrophile   The nucleophilic double bond in 1-butene will react with the electrophile bromine (see Chapter 8 for the mechanism of the addition of bromine to alkenes). The overall reaction is an electrophilic addition.

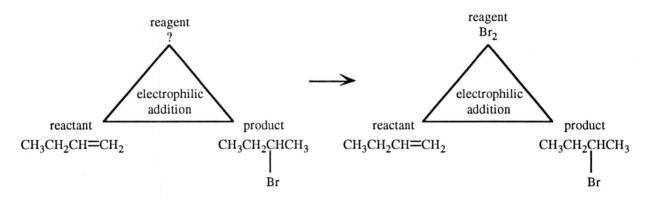

(d)   NH₃   nucleophile   The carbon atom bonded to the bromine atom in *n*-propyl bromide is an electrophilic carbon and will react with a reagent that is a nucleophile. The reaction is a nucleophilic substitution.

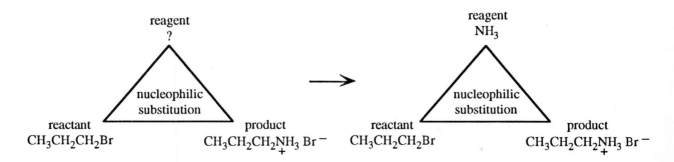

4.17

(a) $CH_3CH_2CH{=}CH_2 \xrightarrow{?} CH_3CH_2CHCH_3$
                                                                         |
                                                                        SH

The elements of hydrogen sulfide, H_2S, have added to the double bond. The double bond is a nucleophile and is protonated by strong Brønsted acids. H_2S, with a $pK_a \sim 10.5$, is not sufficiently acidic (hydrogen sulfide is similar to ethanethiol, CH_3CH_2SH, and can be expected to have a similar pK_a); a hydrogen halide such as HBr ($pK_a - 8$) is. In the product that is formed, the bromine atom will be attached to an electrophilic carbon atom and will leave when the hydrogen sulfide anion is used as a nucleophile.

$CH_3CH_2CH{=}CH_2 \xrightarrow{\text{HBr}} CH_3CH_2CHCH_3$
                                                                              |
                                                                             Br

$CH_3CH_2CHCH_3 \xrightarrow{\text{Na}^+ \text{SH}^-} CH_3CH_2CHCH_3 + \text{Na}^+ \text{Br}^-$
| |
Br   SH

(b) $CH_3CH{=}CHCH_3 \xrightarrow{?} CH_3CHCH_2CH_3$
                                                                         |
                                                                        Cl

The elements of hydrogen chloride, HCl, have added to the double bond. Unlike H_2S, hydrogen chloride is a strong enough acid ($pK_a - 7$) to protonate the double bond. The cation produced by protonation will then react with the conjugate base of hydrogen chloride to give the desired product.

$CH_3CH{=}CHCH_3 \xrightarrow{\text{HCl}} CH_3\overset{+}{C}H{-}CHCH_3 + \text{Cl}^- \longrightarrow CH_3CHCH_2CH_3$
                                                                              | |
                                                                             H   Cl

(c) $CH_2{=}CH_2 \xrightarrow{?} CH_3CH_2\overset{+}{N}H_3 \ \text{I}^-$

The elements of the ammonium ion, NH_4^+, have added to the double bond. NH_4^+ is not sufficiently acidic ($pK_a\,9.4$) but HI (pK_a-9) is. Ammonia can then be used as a nucleophile to react with the electrophilic carbon atom in ethyl iodide, the product of the first reaction.

$CH_2{=}CH_2 \xrightarrow{\text{HI}} CH_3CH_2I \xrightarrow{\text{NH}_3} CH_3CH_2\overset{+}{N}H_3 \ \text{I}^-$

(d) $CH_3CHCH_2CH_2CH_3 \xrightarrow{?} CH_3CHCH_2CH_2CH_3$
                      | |
                     Br   CN

Cyanide ion has substituted for bromide ion. This is a nucleophilic substitution reaction on 2-bromopentane, which requires only one species, NaCN.

$CH_3CHCH_2CH_2CH_3 \xrightarrow{\text{Na}^+ \text{CN}^-} CH_3CHCH_2CH_2CH_3 + \text{Na}^+ \text{Br}^-$
| |
Br   CN

4.18

(a)

nucleophile

$CH_3CH = CH_2$ → electrophile $CH_3\overset{+}{C}HCH_3$ → CH_3CHCH_3

$H - \overset{..}{\underset{..}{O}} - SO_3H$

electrophile

nucleophile → $\overset{-}{\underset{..}{:O}} - SO_3H$

$\overset{..}{\underset{..}{:O}} - SO_3H$

(b)

electrophile

$CH_3CH_2CH_2 - \overset{..}{\underset{..}{Br}}:$ → $CH_3CH_2CH_2 - \overset{+}{P}H_3$

$:\overset{..}{\underset{..}{Br}}:^-$

nucleophile → $:PH_3$

(c)

electrophile

$CH_3CH_2 - \overset{..}{\underset{..}{I}}:$ → $CH_3CH_2 - \overset{..}{\underset{..}{S}} - CH_2CH_3$ + $:\overset{..}{\underset{..}{I}}:^-$

$CH_3CH_2 - \overset{..}{\underset{..}{S}}:^-$ ← nucleophile

(d)

nucleophile

$\underset{CH_3}{\overset{CH_3}{C}} = CH_2$ → electrophile $CH_3 - \overset{CH_3}{\underset{}{\overset{+}{C}}} - CH_3$ → $CH_3 - \overset{CH_3}{\underset{:\overset{..}{\underset{..}{Cl}}:}{C}} - CH_3$

electrophile → $H - \overset{..}{\underset{..}{Cl}}:$

nucleophile → $:\overset{..}{\underset{..}{Cl}}:^-$

(e) $CH_3CH_2CH = CHCH_3$ + Br_2 → $CH_3CH_2CH - CHCH_3$

nucleophile   electrophile Br Br

4.19   <u>reagent</u>   <u>type</u>   <u>reason for choice</u>

(a) $Na^+ SH^-$ nucleophile   The carbon atom bonded to the bromine atom in methyl bromide is an electrophilic carbon and will react with a reagent that is a nucleophile. The reaction is a nucleophilic substitution.

4.19 (cont)

(b)   Na$^+$ CN$^-$ nucleophile   The carbon atom bonded to the bromine atom in ethyl bromide is an electrophilic carbon and will react with a reagent that is a nucleophile. The reaction is a nucleophilic substitution.

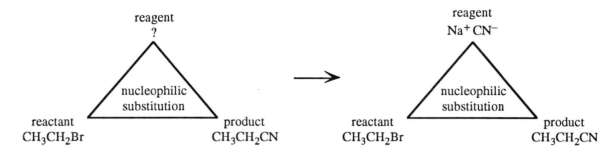

(c)   Cl$_2$ electrophile   The nucleophilic double bond in ethylene will react with the electrophile chlorine (see Chapter 8 for the mechanism of the addition of chlorine to alkenes). The reaction is an electrophilic addition.

(d)   CH$_3$CH$_2$O$^-$ Na$^+$ nucleophile   The carbon atom bonded to the bromine atom in *n*-propyl bromide is an electrophilic carbon and will react with a reagent that is a nucleophile. The reaction is a nucleophilic substitution.

(e)   HI   electrophile   The nucleophilic double bond in 1-butene will react with the electrophilic proton of hydrogen iodide and the electrophilic carbocation that is produced will then react with the iodide ion. The reaction is an electrophilic addition.

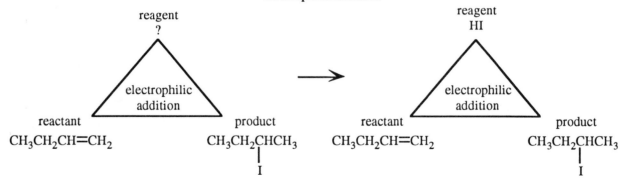

(f)   Na$^+$ I$^-$ nucleophile   The carbon atom bonded to the bromine atom in isopropyl bromide is an electrophilic carbon and will react with a reagent that is a nucleophile. The reaction is a nucleophilic substitution.

(g)   Na$^+$ CN$^-$ nucleophile   The carbon atom bonded to the chlorine atom in 7-cyano-3-heptyn-1-ol is an electrophilic carbon and will reaction with a nucleophilic species. The reaction is a nucleophilic substitution.

(h)   HBr   electrophile   The nucleophilic double bond in methylcyclopentene will react with the electrophilic proton of hydrogen bromide and the electrophilic carbocation that is produced will then react with the bromide ion. The reaction is an electrophilic addition.

4.19 (cont)

(i)   First step:
        $Na^+ NH_2^-$ base   The alcohol has been converted into its conjugate base in the first step. We need a base that is much stronger than the conjugate base of the alcohol to carry out the deprotonation. Amide ion, NH_2^-, is such a base (see the pK_a table).

        Second step:
        CH_3I electrophile   The oxygen atom of the conjugate base of the alcohol is connected to a methyl group in the product. The oxygen anion is a nucleophile; therefore, we need an electrophilic carbon atom bearing a leaving group to react with it.

4.20

(a) $CH_2\!\!=\!\!CH_2 \xrightarrow{\ ?\ } CH_3CH_2OCH_3$

The elements of methanol, CH_3OH, have added to the double bond. The double bond is a nucleophile and is protonated by strong Brønsted acids. Methanol is not sufficiently acidic (pK_a 15.5), but a hydrogen halide such as HBr (pK_a –8) is. In the product that is formed, the bromine atom will be attached to an electrophilic carbon atom and will leave when sodium methoxide is used as a nucleophile.

$$CH_2\!\!=\!\!CH_2 \xrightarrow{HBr} CH_3CH_2Br \xrightarrow{CH_3O^- Na^+} CH_3CH_2OCH_3$$

(b) $CH_3CH\!\!=\!\!CH_2 \xrightarrow{\ ?\ } CH_3\underset{\underset{I}{|}}{C}HCH_3$

The elements of hydrogen iodide, HI, have added to the double bond. Hydrogen iodide is a strong enough acid (pK_a –9) to protonate the double bond. The cation produced by protonation will then react with iodide ion, the conjugate base of hydrogen iodide, to give the desired product.

$$CH_3CH\!\!=\!\!CH_2 \xrightarrow{HI} CH_3\underset{\underset{I}{|}}{C}HCH_3$$

(c) $CH_3CH_2CH\!\!=\!\!CH_2 \xrightarrow{\ ?\ } CH_3CH_2\underset{\underset{OH}{|}}{C}HCH_3$

The elements of water, H_2O, have added to the double bond. Water is not sufficiently acidic (pK_a 15.7) but the hydronium ion (pK_a –1.7), formed when a strong acid such as sulfuric acid is added to water, is (See Problem 3.43). The solvent, water, will then act as a nucleophile to react with the carbocation to give the final product.

$$CH_3CH_2CH\!\!=\!\!CH_2 \xrightarrow[H_2SO_4]{H_2O} CH_3CH_2\underset{\underset{OH}{|}}{C}HCH_3$$

4.20 (cont)

(d) $CH_2{=}CH_2$ $\xrightarrow{\;?\;}$ $CH_3CH_2\overset{+}{N}H_2CH_3$ Br^-

The elements of the methylammonium ion, $CH_3NH_3^+$, have added to the double bond. $CH_3NH_3^+$ is not sufficiently acidic (pK_a 10.6) but HBr (pK_a –8) is. Methylamine can then be used as a nucleophile to react with the electrophilic carbon atom in ethyl bromide, the product of the first reaction.

$CH_2{=}CH_2$ $\xrightarrow{\;HBr\;}$ CH_3CH_2Br $\xrightarrow{\;CH_3NH_2\;}$ $CH_3CH_2\overset{+}{N}H_2CH_3$ Br^-

4.21

(a)   electrophile $\longrightarrow$ $CH_3{-}\overset{..}{\underset{..}{Cl}}:$ $\longrightarrow$ $:\overset{..}{\underset{..}{I}}{-}CH_3$ $:\overset{..}{\underset{..}{Cl}}:^-$

nucleophile $\longrightarrow$ $:\overset{..}{\underset{..}{I}}:^-$

(b)   acid$\longrightarrow$ $H{-}\overset{..}{\underset{..}{Br}}:$ $:\overset{..}{\underset{..}{Br}}:^-$

$CH_3{-}\underset{CH_3}{\overset{CH_3}{\underset{|}{\overset{|}{C}}}}{-}\overset{..}{\underset{..}{O}}H$ $\longrightarrow$ $CH_3{-}\underset{CH_3}{\overset{CH_3}{\underset{|}{\overset{|}{C}}}}{-}\overset{+}{\underset{..}{O}}H_2$ $\longrightarrow$ $CH_3{-}\underset{CH_3}{\overset{CH_3}{\underset{|}{\overset{|}{C^+}}}}$ $:\overset{..}{\underset{..}{O}}{-}H$

base

(c) $:N{\equiv}C:^-$ electrophile $\overset{H}{\underset{H}{C}}{=}\overset{..}{\underset{..}{O}}$ $\longrightarrow$ $:N{\equiv}C{-}\overset{H}{\underset{H}{C}}{-}\overset{..}{\underset{..}{O}}:^-$

nucleophile

(d)   nucleophile $CH_3{-}\overset{..}{\underset{..}{I}}:$ $\longrightarrow$ $:\overset{..}{\underset{..}{I}}:^-$ CH3

electrophile

4.21 (cont)

(e)

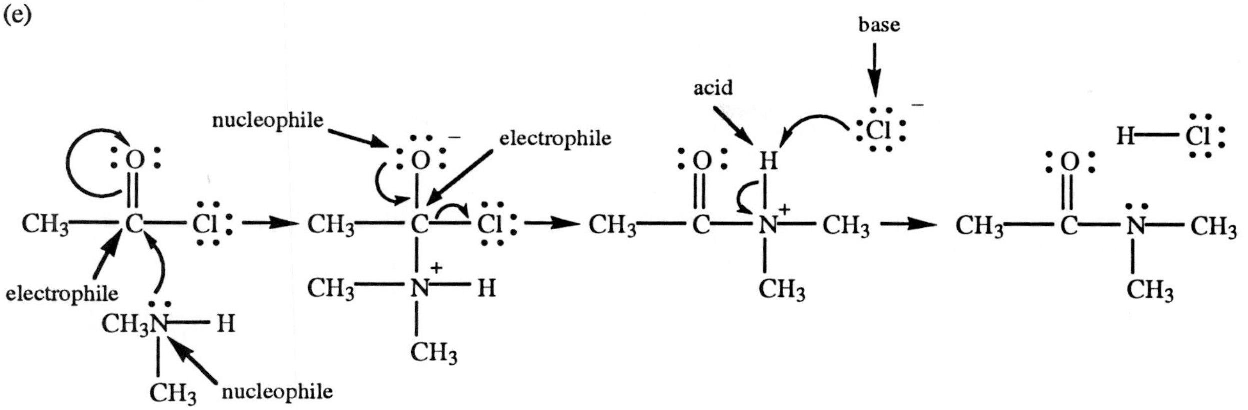

(If excess $(CH_3)_2NH$ is used, it will act as a base in the second step of the reaction.)

(f)

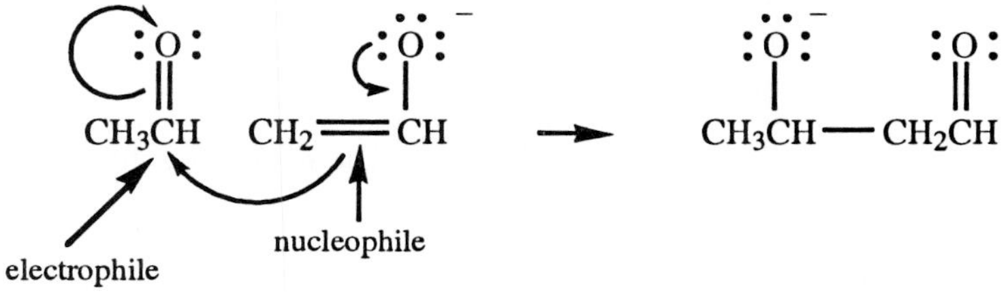

(g)

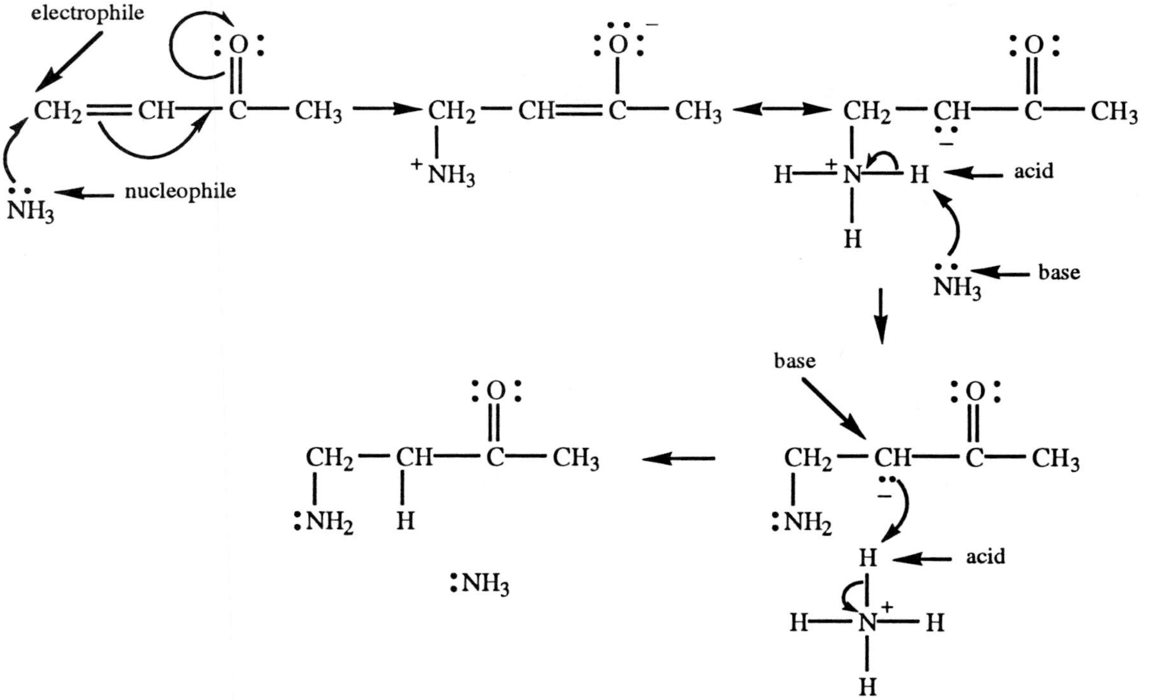

4.22   No, the rate of reaction will decrease as the reaction progresses. The rate is a function of the concentration of starting material and it decreases as starting material is converted into product.

4.23   At 308 K, $k = 5.3 \times 10^{-4}$ L/mol·sec. When the reaction is half over, the concentrations of hydroxide ion and of methyl bromide will each be half the initial concentrations.

$$\text{Rate} = k\,[CH_3Br][OH^-]$$

$$= k\,(0.5[CH_3Br]_{initial})(0.5[OH^-]_{initial})$$

$$= 0.25\,(\text{initial rate}) = 0.25(5.3 \times 10^{-9}\,\text{L/mol·sec})$$

$$= 1.3 \times 10^{-9}\,\text{L/mol·sec}$$

4.24   When a reaction has gone to 99% completion, the final concentration of product will be 99% that of the initial starting material concentration.

$$[\text{product}]_{99\%\,\text{completion}} = 0.99[\text{starting material}]_{initial}$$

and the concentration of starting material will be reduced to 1% of the initial concentration.

$$[\text{starting material}]_{99\%\,\text{completion}} = 0.01[\text{starting material}]_{initial}$$

The reaction is over when it reaches equilibrium and

$$K_{eq} = [\text{product}]_{eq}/[\text{starting material}]_{eq} = 0.99[\text{starting material}]_{initial}/0.01[\text{starting material}]_{initial}$$

$$= 0.99/0.01 = 99$$

$$\Delta G_r = -RT \ln K_{eq}$$

At 25 °C (298 K)

$$\Delta G_r = -(1.99 \times 10^{-3}\,\text{kcal/mol·K})(298\,\text{K}) \ln 99$$

$$= -2.7\,\text{kcal/mol}$$

4.25
(a) $CH_3Cl + H_2O \rightarrow CH_3OH + H_3O^+ + Cl^-$

The nucleophile in the reaction is water. The other products of the reaction are the hydronium ion and chloride ion.

(b)

(c) $\text{Rate} = k\,[CH_3Cl][H_2O]$

4.25 (cont)

(d)   The density of water at room temperature is approximately 1 g/mL. One liter of water contains 1000 g or 55.6 mol (1000 g)/(18 g/mol). The concentration of water in pure water is thus 55.6 M. The concentration of chloromethane that dissolves in one liter of water is only 0.003 moles, which is not enough to significantly change the concentration of water in an aqueous solution of chloromethane. According to the stoichiometry of the reaction, one equivalent of chloromethane reacts with one equivalent of water. At the end of the reaction, assuming all the starting material is converted to product, the final concentration of water will be (55.6 – 0.003) M or, to the correct number of significant figures, 55.6 M. It would not be possible to see a change in the concentration of water as the reaction proceeds. In other words, the concentration of water remains essentially constant as the reaction proceeds, and can be incorporated into the rate constant.

Rate = $k'[CH_3Cl]$ where $k' = k[H_2O]$

This type of reaction, in which the solvent is also the nucleophile, is an example of **a solvolysis reaction.**

4.26 (a)
$$CH_3\overset{\displaystyle CH_3}{\underset{}{C}}=CH_2 \text{ (g)} \quad + \quad H_2O \text{ (l)} \quad \longrightarrow \quad CH_3\overset{\displaystyle CH_3}{\underset{\displaystyle OH}{C}}CH_3 \text{ (l)}$$

$\Delta G_r° = \Delta H_r° - T\Delta S_r°$

$\Delta H_r° = -85.87 \text{ kcal/mol} - (-4.04 \text{ kcal/mol} - 68.32 \text{ kcal/mol}) = -13.51 \text{ kcal/mol}$

$\Delta S_r° = 46.30 \text{ cal/mol·K} - (16.72 \text{ cal/mol·K} + 70.17 \text{ cal/mol·K}) = -40.59 \text{ cal/mol·K}$

$\qquad = -4.059 \times 10^{-2} \text{ kcal/mol·K}$

$\Delta G_r° = (-13.51 \text{ kcal/mol}) - (298 \text{ K})(-4.06 \times 10^{-2} \text{ kcal/mol·K})$

$\qquad = -1.41 \text{ kcal/mol}$

$\Delta G_r° = -RT \ln K_{eq}$

$\ln K_{eq} = -\Delta G_r°/RT$

$K_{eq} = \exp(-\Delta G_r°/RT) = \exp[-(-1.41 \text{ kcal/mol})/(1.99 \times 10^{-3} \text{ kcal/mol·K})(298 \text{ K})]$

$\qquad = 10.8$

(b) When sulfuric acid is added to water, the proton is transferred completely to form the hydronium ion and the hydrogen sulfate anion.

$$H_2SO_4 + H_2O \rightarrow H_3O^+ + HSO_4^-$$

$pK_a -9 \qquad\qquad\qquad pK_a -1.7$

In a dilute solution of H_2SO_4 in water, the electrophile is the hydronium ion, not sulfuric acid. The nucleophile is the double bond of the alkene, which requires a strong acid for protonation. H_2O (pK_a 15.7) is not strong enough, but H_3O^+ is. The hydronium ion is regenerated in the last step of the mechanism, so only a catalytic amount of sulfuric acid is necessary (see Section 8.4 for the mechanism).

4.27 (a) $\Delta S_r° = (74.27 \text{ cal/mol·K} + 45.11 \text{ cal/mol·K}) - (78.32 \text{ cal/mol·K}) = +41.06 \text{ cal/mol·K}$

$\qquad = +4.106 \times 10^{-2} \text{ kcal/mol·K}$

The entropy increases during the reaction because two independent molecules, water and cyclohexene, are formed from each molecule of cyclohexanol.

$\Delta H_r° = (-57.80 \text{ kcal/mol} - 1.28 \text{ kcal/mol}) - (-70.40 \text{ kcal/mol}) = +11.32 \text{ kcal/mol}$

4.27 (cont)

(b) $\Delta G_r^\circ = \Delta H_r^\circ - T\Delta S_r^\circ$

$\Delta G_r^\circ = (11.32 \text{ kcal/mol}) - (298 \text{ K})(4.106 \times 10^{-2} \text{ kcal/mol·K})$

$= -0.916 \text{ kcal/mol}$

$\Delta G_r^\circ = -RT \ln K_{eq}$

$\ln K_{eq} = -\Delta G_r^\circ / RT$

$K_{eq} = \exp(-\Delta G_r^\circ / RT) = \exp[-(-0.916 \text{ kcal/mol})/(1.99 \times 10^{-2} \text{ kcal/mol·K})(298 \text{ K})]$

$= 4.69$, therefore the forward reaction is favored at this temperature.

(c) Because the number of molecules is the same on both sides of the first equation, the change in entropy will be small, and the change in free energy will approximately equal the change in enthalpy. The evolution of heat means that $\Delta H_r^\circ < 0$ (products have a lower enthalpy than starting materials) and therefore that the free energy of the products (protonated alcohol and dihydrogen phosphate ion) is lower than the free energy of the starting materials (alcohol and phosphoric acid).

(d) The formation of the very high energy carbocation is the rate-determining step in the addition of an electrophile to an alkene. It is likely that step (2), the formation of the carbocation, is the rate-determining step in the dehydration. Therefore, even though overall energy changes favor the formation of the product at room temperature [part (b)], no product is formed until enough heat is supplied to get over the energy barrier represented by the free energy of activation of the rate-determining step.

4.28

(a) $CH_3CH_2CH_2CH_2CH_2 \!-\! C\!\equiv\!C\!-\!H$ + HBr $\xrightarrow[\substack{\text{silica gel} \\ 24 \text{ h}}]{H_2O}$ $CH_3CH_2CH_2CH_2CH_2 \!-\! C\!=\!CH_2$

1 equivalent of HBr $\overset{|}{Br}$

(b) $CH_3CH_2CH_2CH_2CH_2 \!-\! \overset{+}{C}\!=\!CH_2$ $CH_3CH_2CH_2CH_2CH_2 \!-\! \overset{+}{C}H\!-\!CH_2$

reactive intermediate   reactive intermediate
from protonation of alkyne   from protonation of alkene

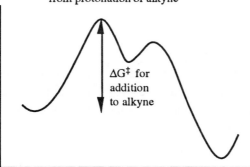

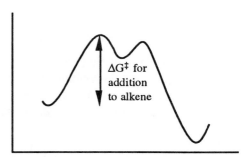

positive charge on *sp* carbon, therefore less stable carbocation because an *sp* carbon is more electronegative than an *sp*²-carbon (p. 64 in text); has higher $\Delta G^\ddagger$, therefore forms with greater difficulty in the rate-determining step of the reaction

positive charge on *sp*² carbon, therefore more stable cation; has lower $\Delta G^\ddagger$, therefore forms more easily in the rate-determining step of the reaction

4.29

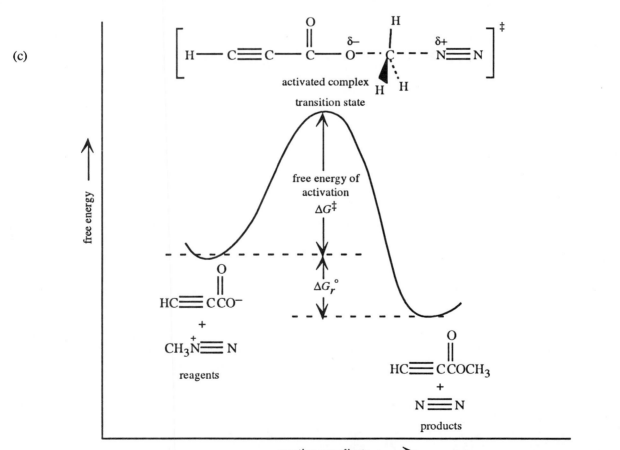

(a)

conjugate base   conjugate acid

(b)

(c)

activated complex
transition state

free energy of
activation
$\Delta G^{\ddagger}$

ΔG_r°

free energy

$HC\!\!\equiv\!\!CCO^-$
+
$CH_3\overset{+}{N}\!\!\equiv\!\!N$
reagents

$HC\!\!\equiv\!\!CCOCH_3$
+
$N\!\!\equiv\!\!N$
products

reaction coordinate ⟶

4.30

(a)   The strongest acid that is present when a small amount of sulfuric acid is added to methanol is the methyl–oxonium ion, $CH_3OH_2^+$.

$$CH_3OH \quad + \quad H_2SO_4 \quad \xrightleftharpoons{\qquad\qquad} \quad CH_3\overset{+}{O}-H \quad + \quad HSO_4^-$$
$$\qquad\qquad\qquad\qquad\qquad\qquad\qquad\qquad\qquad | $$
$$\qquad\qquad\qquad\qquad\qquad\qquad\qquad\qquad\quad H$$

excess   pK_a –9   pK_a ~–2.4
                                                                      strongest acid present

(b)

reused to
protonate alkene

(c)   Step 1, in which the tertiary carbocation is formed, is the most likely to be the rate-determining step.

(d)

$CH_3-\overset{\overset{\displaystyle CH_3}{|}}{\underset{\underset{\displaystyle H}{|}}{C}}-\overset{+}{C}H_2$, the isobutyl cation that would be needed to form methyl isobutyl ether, is a primary cation

and much less stable than the *tert*-butyl cation shown forming in part (b). The energy of activation for the formation of the *tert*-butyl cation is much lower than that for the formation of the isobutyl cation. Therefore the *tert*-butyl cation forms faster and most of the product comes from this cation.

4.30 (cont)

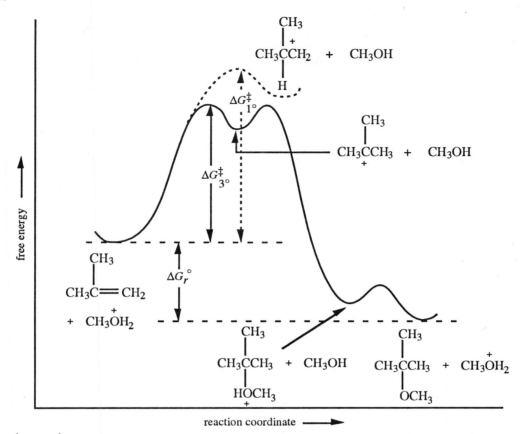

$\Delta G_{1°}^{\ddagger} > \Delta G_{3°}^{\ddagger}$, therefore formation of the 3° cation is much faster than formation of the 1° cation, and the product is the one that comes from the 3° cation.

5

Alkanes and Cycloalkanes

Workbook Exercises

EXERCISE I. Insight into molecular structure comes from a detailed examination of the numbers and kinds of atoms in a molecular formula. For each of the following molecular formulas, determine the units of unsaturation (see Section 1.5).

    (a) $C_{10}H_{16}$, limonene from lemon oil

    (b) $C_{17}H_{21}NO_4$, cocaine, a local anesthetic

    (c) $C_{22}H_{37}NO_2$, anandiamide, a natural ligand for the receptor in the brain for tetrahydrocannabinol, the active ingredient of marijuana

    (d) $C_{23}H_{46}$, muscalure, an insect attractant

    (e) $C_5H_9NO_2$, proline, an amino acid

    (f) $C_4H_{12}N_2$, putrescine, found in decayed tissues

    (g) $C_6H_8O_6$, ascorbic acid

    (h) C_3H_6O, propylene oxide, a sterilant

    (i) $C_{43}H_{78}N_6O_{13}$, muroctasin, an immunostimulant

EXERCISE II. What are some structural features that could account for the units of unsaturation in each of the compounds in Exercise I? Try to give up to three examples of such structural features for each. For example, proline [part (e)], has two units of unsaturation. These units could be represented by two rings, a ring and a carbon-oxygen double bond, or a carbon-nitrogen triple bond. Other structural features are also possible based on the information given.

EXERCISE III. The structures for most of the compounds listed in Exercise I are found in chemistry textbooks. Find as many of these structures as you can, identify the units of unsaturation actually present, and compare this count to your answers to Exercise II.

———

5.1

(a)

5.1 (cont)

(b) 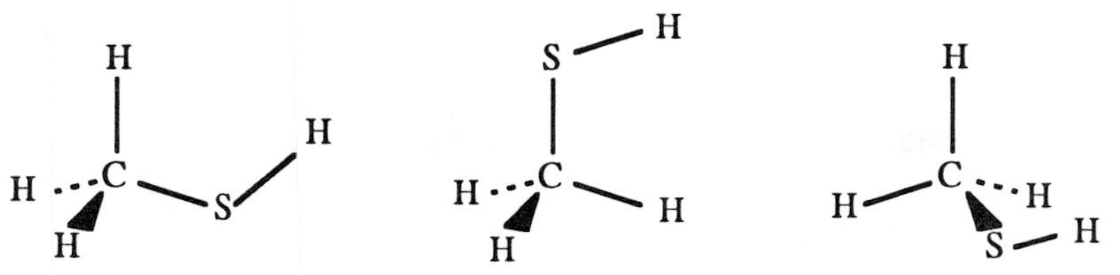 etc.

(c) 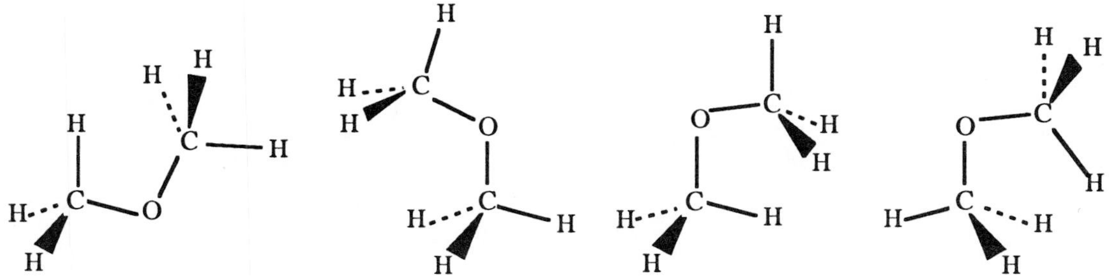 etc.

(d) 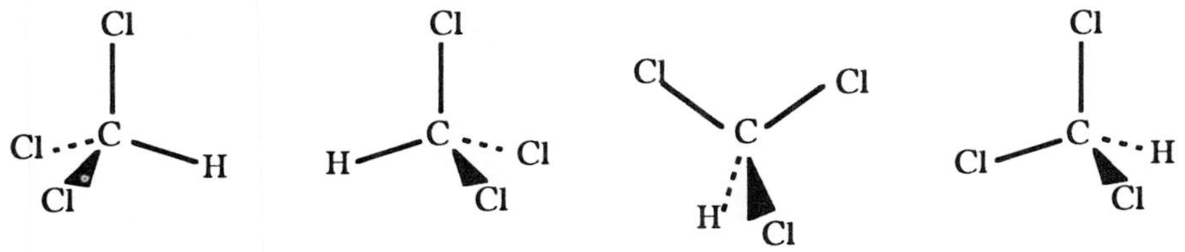 etc.

(e) 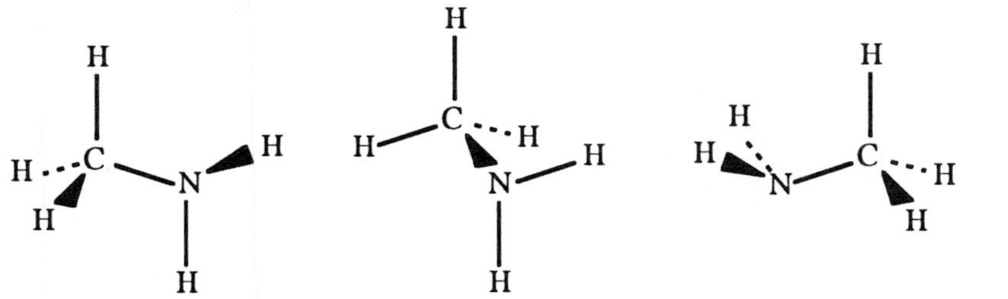 etc.

5.2 $C_2H_3Cl_3$: CH_3CCl_3 $ClCH_2CHCl_2$

Lewis structures

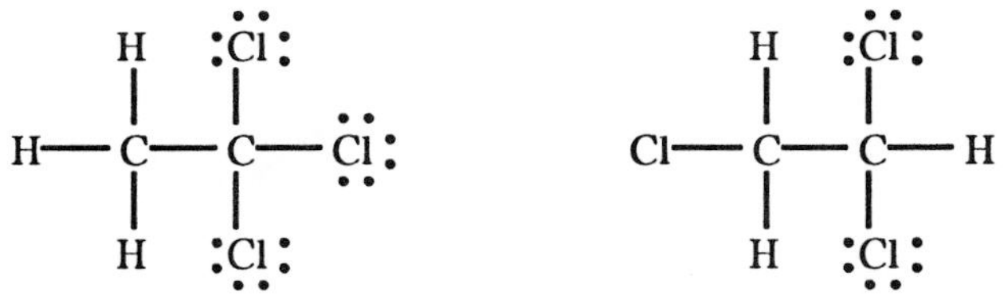

three-dimensional representations

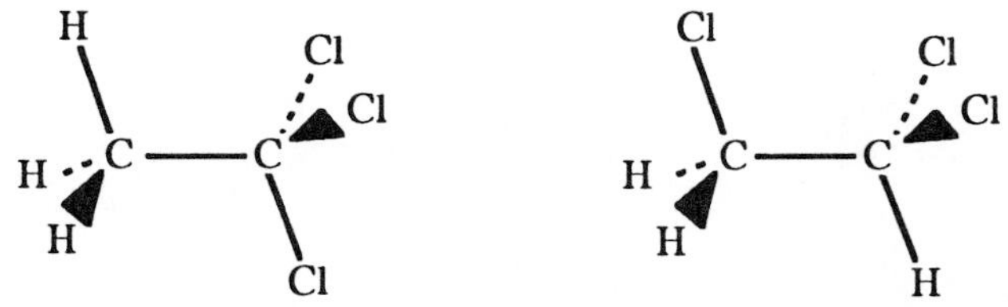

5.2 (cont)

$C_2H_2Cl_4$: $ClCH_2CCl_3$ $Cl_2CHCHCl_2$

Lewis structures

three-dimensional
representations

5.3 $C_3H_6Cl_2$:

condensed   (1) $CH_3CH_2CHCl_2$ (2) $CH_3CCl_2CH_3$
formulas

            (3) $CH_3CHClCH_2Cl$ (4) $ClCH_2CH_2CH_2Cl$

Lewis structures (1)   (2)

                (3)   (4)

three-dimensional (1)   (2)
representations

                (3)   (4)

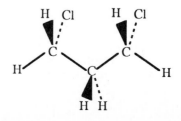

5.4   (a) ethyl   (b) *sec*-butyl

(c) isobutyl   (d) methyl

(e) isopropyl   (f) propyl

(g) *tert*-butyl   (h) ethyl; butyl

5.5   (a) Constitutional isomers

(b) Not constitutional isomers; both formulas have the same connectivity.

(c) Constitutional isomers

(d) Constitutional isomers

(e) Constitutional isomers

(f) Constitutional isomers

(g) Not constitutional isomers; both formulas have the same connectivity.

5.6

(a)

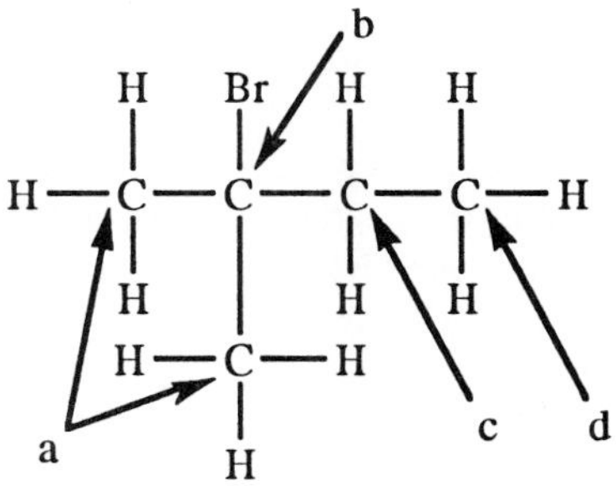

(b)

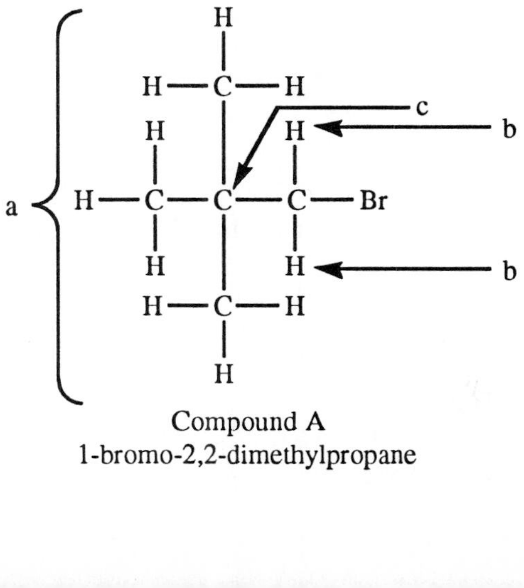

Compound A
1-bromo-2,2-dimethylpropane

5.6 (b) (cont)

The hydrogen atoms belong in two groups, 9 in group *a* and two in group *b*. There are three groups of carbon atoms, those bonded to the *a*-hydrogen atoms, the one bonded to the *b*-hydrogen atom, and the one labeled *c*, with no hydrogen atoms on it. There are no hydrogen atoms that are 3-bond neighbors of each other in this molecule.

Compound B
1-bromopentane

There are five groups of hydrogen atoms, labeled *a*-, *b*-, *c*-, *d*-, and *e*-hydrogen atoms. Each group of hydrogen atoms is bonded to the correspondingly labeled carbon atoms. The *a*-hydrogen atoms are 3-bond neighbors to the *b*-hydrogen atoms. The *b*-hydrogen atoms are 3-bond neighbors to the *a*- and *c*-hydrogen atoms. The *c*-bond hydrogen atoms are 3-bond neighbors to the *b*- and *d*-hydrogen atoms. The *d*-hydrogen atoms are 3-bond neighbors to the *c*- and *e*-hydrogen atoms. The *e*-hydrogen atoms are 3-bond neighbors to the *d*-hydrogen atoms.

Compound C
3-bromopentane

There are three groups of hydrogen atoms and three groups of carbon atoms in the molecule, the *a*-hydrogen atoms (and carbon atoms), the *b*-hydrogen atoms (and carbon atoms), and the *c*-hydrogen atom (and the carbon atom to which it is bonded). The *a*-hydrogen atoms are 3-bond neighbors to the *b*-hydrogen atoms. The *b*-hydrogen atoms are 3-bond neighbors to the *a*- and *c*-hydrogen atoms. The *c*-bond hydrogen atom is a 3-bond neighbor to the *b*-hydrogen atom.

5.7

two molecular models of 2-bromo-2-methylbutane

front H of group b
replaced by Cl

back H of group b
replaced by Cl

5.7 (cont)

These two new compounds are not identical. They are stereoisomers of each other (p. 22 in the text and Chapter 6). They have the same connectivity, but differ in spatial relationships. We cannot pick one of these molecules off the page, move it over, and make it fit exactly on the other molecule. The hydrogens in group b, therefore, are equivalent only if we do nothing to distinguish them spatially. Living organisms, for example, have the capacity to make such distinctions in chemical reactions.

5.8

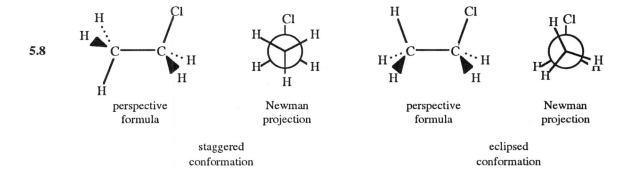

| perspective formula | Newman projection | perspective formula | Newman projection |

staggered
conformation

eclipsed
conformation

Concept Map 5.1   Conformation.

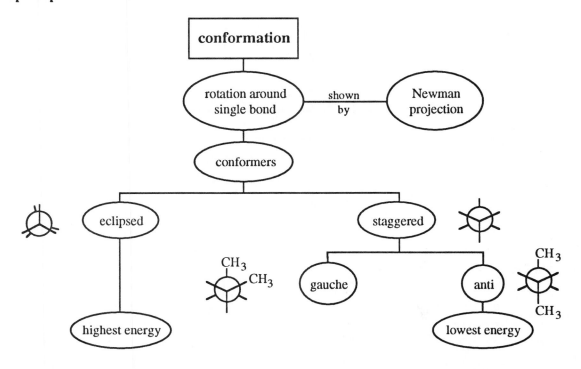

5.9

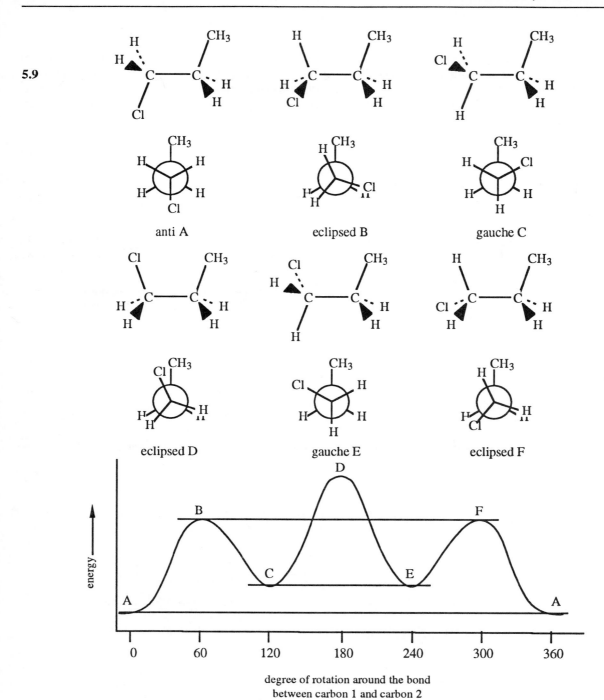

anti A   eclipsed B   gauche C

eclipsed D   gauche E   eclipsed F

degree of rotation around the bond
between carbon 1 and carbon 2

5.10

no dipole moment
C—Cl bond dipoles cancel

5.10 (cont)

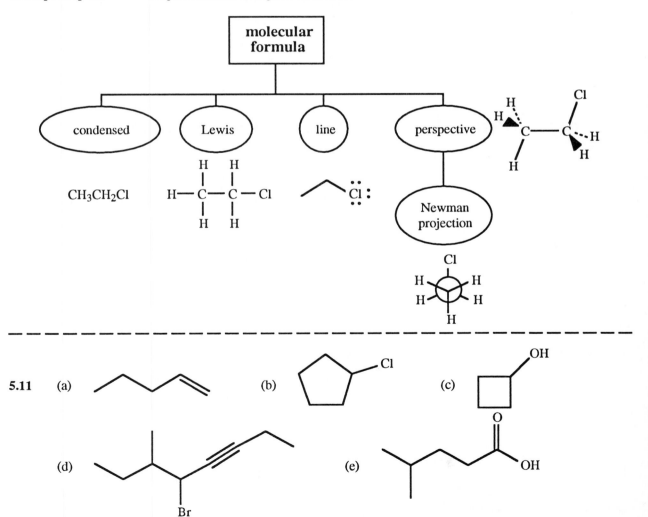

dipole moment
C—Cl bond dipoles do not cancel

At room temperature in the liquid state, a significant portion of 1,2-dichloroethane molecules exist in the gauche conformation.

Even though the negative ends of the C—Cl bond dipoles are close enough in the gauche conformation to repel each other, considerable van der Waals attraction between the chlorine atoms also exists. There is no van der Waals attraction between the chlorine atoms in the anti conformation.

— —

Concept Map 5.2   Representations of organic structures.

5.12

(a)

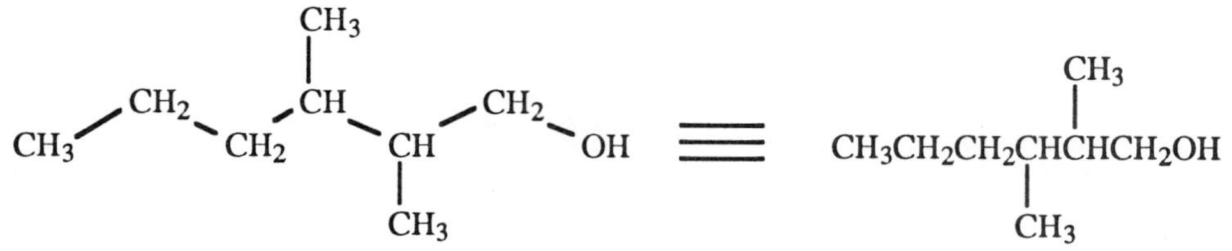

$$CH_3CH_2CH_2CHCHCH_2OH$$

$$CH_3CH_2CH_2CH(CH_3)CH(CH_3)CH_2OH$$

(b)

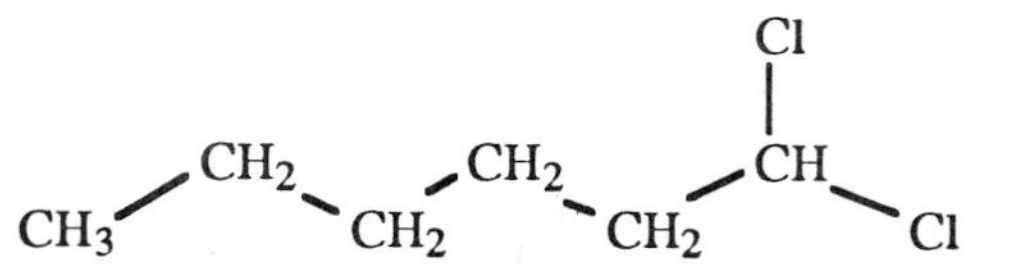

$$CH_3CH_2CH_2CH_2CH_2CHCl_2$$

(c)

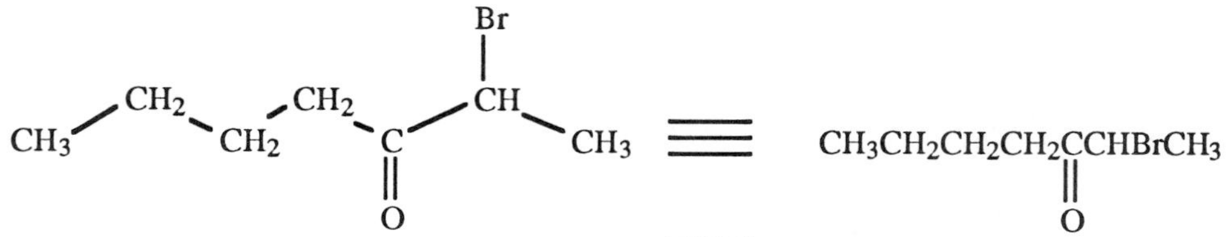

$$CH_3CH_2CH_2CH_2CCHBrCH_3$$

$$CH_3CH_2CH_2CH_2COCHBrCH_3$$

(d)

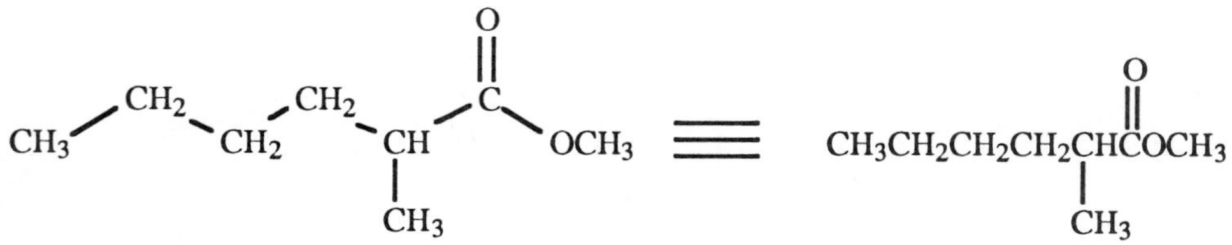

$$CH_3CH_2CH_2CH_2CHCOCH_3$$

$$CH_3CH_2CH_2CH_2CH(CH_3)COOCH_3$$

(e)

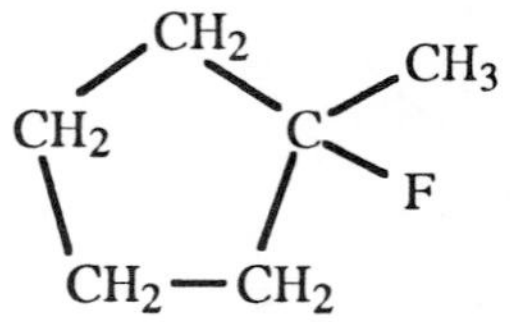

5.13

(a) 2,2,6,7-tetramethyloctane

(b) 5-ethyl-2,6-dimethyloctane

(c) 2,3-dimethylpentane

(d) 4-*tert*-butylheptane

(e) 2,2,5-trimethyl-6-propyldecane

(f) 6-isopropyl-3-methylnonane

5.14   (a) $CH_3CCH_2CH_2CH_2CH_2CH_3$

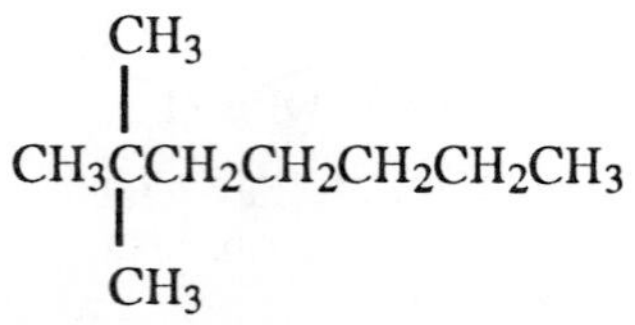

(b) $CH_3CHCH_2CH_2CH_2CHCH_2CH_2CH_2CH_3$

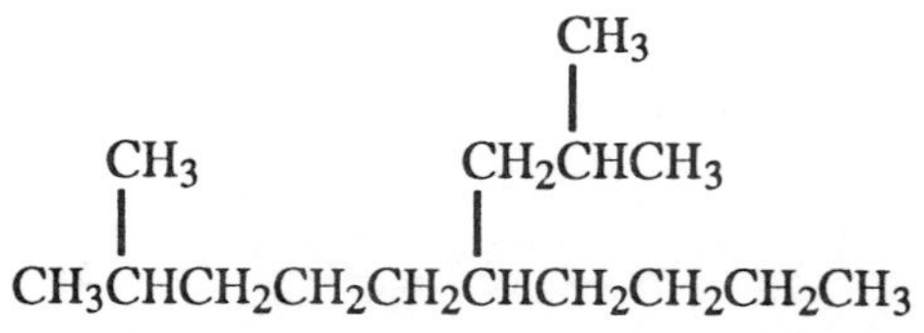

5.14 (cont)

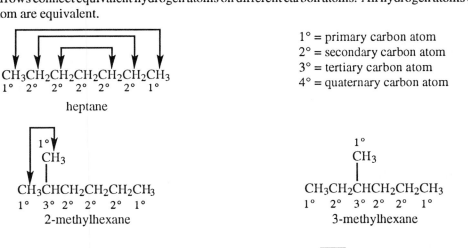

(c) CH₃CHCH₂CH₂CCH₂CH₂CHCH₃   (d) CH₃CH₂CHCHCH₂CH₂CH₂CH₃

with substituents CH₃, CH₃CHCH₃, CH₃ above and CH₃CHCH₃ below for (c); CH₂CH₃ above and CH₃CCH₃ with CH₃ below for (d)

(e) CH₃CCH₂CHCH₂CHCH₂CH₂CH₂CH₂CH₂CH₃

with substituents CH₃, CH₃, CH₂CH₃ above and CH₃ below

5.15   Arrows connect equivalent hydrogen atoms on different carbon atoms. All hydrogen atoms on the same carbon atom are equivalent.

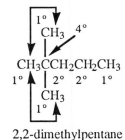

$1°$ = primary carbon atom
$2°$ = secondary carbon atom
$3°$ = tertiary carbon atom
$4°$ = quaternary carbon atom

CH₃CH₂CH₂CH₂CH₂CH₂CH₃
$1°$ $2°$ $2°$ $2°$ $2°$ $2°$ $1°$
heptane

$1°$
CH₃
CH₃CHCH₂CH₂CH₂CH₃
$1°$ $3°$ $2°$ $2°$ $2°$ $1°$
2-methylhexane

$1°$
CH₃
CH₃CH₂CHCH₂CH₂CH₃
$1°$ $2°$ $3°$ $2°$ $2°$ $1°$
3-methylhexane

$1°$
CH₃ $4°$
CH₃CCH₂CH₂CH₃
$1°$ $2°$ $2°$ $1°$
CH₃
$1°$
2,2-dimethylpentane

$1°$ $1°$
CH₃   CH₃
CH₃CH — CHCH₂CH₃
$1°$ $3°$ $3°$ $2°$ $1°$
2,3-dimethylpentane

$1°$
CH₃
$4°$
CH₃CH₂ — C — CH₂CH₃
$1°$ $2°$ $2°$ $1°$
CH₃
$1°$
3,3-dimethylpentane

$1°$ $1°$
CH₃   CH₃
CH₃CHCH₂CHCH₃
$1°$ $3°$ $2°$ $3°$ $1°$
2,4-dimethylpentane
all $1°$ H are equivalent
all $2°$ H are equivalent
all $3°$ H are equivalent

5.15 (cont)

$$\underset{1°}{\overset{1°}{CH_3}} \quad \underset{1°}{\overset{1°}{CH_3}}$$

CH₃C——CHCH₃
1° |4°   3° 1°
CH₃
1°

2,2,3-trimethylbutane

2° 1°
CH₂CH₃
|
CH₃CH₂CHCH₂CH₃
1° 2° 3° 2° 1°

3-ethylpentane

all 1° H are equivalent

all 2° H are equivalent

5.16   (a) 2-bromo-2,3-dimethylpentane   (b) 3-pentanol

(c) 4-chloro-3-hexanol   (d) 4,4-dichloro-2,2-dimethylhexane

(e) 2,3-dibromo-2,3-dimethylbutane   (f) 3-iodononane

(g) 1,1,1-trichlorobutane   (h) 4-*tert*-butyl-3-nonanol

5.17   (a)   CH₃CHCH₂CH₂CH₂CH₂CH₂CH₃
|
I

(b)   CH₃CH₂CHCH₂CH₂CH₃
|
OH

(c)   CH₃CH₂OH

(d)
CH₃
|
ClCH₂CHCH₂CCH₃
| |
OH   CH₃

(e)   HOCH₂CHCH₂OH
|
OH

(f)   CH₃CH₂CH₂CH₂CH₂OH

(g)   HOCH₂CH₂OH

(h)
CH₂CH₃
|
CH₃CHCCH₂CH₂CH₂CH₃
| |
Cl Cl

(i)
CH₃
|
CH₃CH₂C——CHCH₂CH₂CH₃
| |
CH₃   Br

(j)
CH₂CH₃
|
CH₃CHCHCH₂CH₂CH₂CH₃
|
Br

5.18   C₆H₁₃Cl

CH₃CH₂CH₂CH₂CH₂CH₂Cl   1° 1-chlorohexane

CH₃CH₂CH₂CH₂CHCH₃
|
Cl   2° 2-chlorohexane

5.18 (cont)

$$CH_3CH_2CH_2CHCH_2CH_3$$
$$\overset{|}{Cl}$$

2°   3-chlorohexane

$$\overset{CH_3}{\underset{|}{}}$$
$$CH_3CHCH_2CH_2CH_2Cl$$

1°   1-chloro-4-methylpentane

$$\overset{CH_3}{\underset{|}{}}$$
$$CH_3CHCH_2CHCH_3$$
$$\overset{|}{Cl}$$

2°   2-chloro-4-methylpentane or
     4-chloro-2-methylpentane

$$\overset{CH_3}{\underset{|}{}}$$
$$CH_3CHCHCH_2CH_3$$
$$\overset{|}{Cl}$$

2°   3-chloro-2-methylpentane

$$\overset{CH_3}{\underset{|}{}}$$
$$CH_3CCH_2CH_2CH_3$$
$$\overset{|}{Cl}$$

3°   2-chloro-2-methylpentane

$$\overset{CH_3}{\underset{|}{}}$$
$$ClCH_2CHCH_2CH_2CH_3$$

1°   1-chloro-2-methylpentane

$$\overset{CH_3}{\underset{|}{}}$$
$$CH_3CH_2CHCH_2CH_2Cl$$

1°   1-chloro-3-methylpentane

$$\overset{CH_3}{\underset{|}{}}$$
$$CH_3CH_2CHCHCH_3$$
$$\overset{|}{Cl}$$

2°   2-chloro-3-methylpentane

$$\overset{CH_3}{\underset{|}{}}$$
$$CH_3CH_2CCH_2CH_3$$
$$\overset{|}{Cl}$$

3°   3-chloro-3-methylpentane

$$\overset{CH_2Cl}{\underset{|}{}}$$
$$CH_3CH_2CHCH_2CH_3$$

1°   3-(chloromethyl)pentane or
     1-chloro-2-ethylbutane

5.18 (cont)

$$CH_3CCH_2CH_2Cl$$ with CH_3 above and CH_3 below the second carbon   1°   1-chloro-3,3-dimethylbutane

$$CH_3C\!-\!CHCH_3$$ with CH_3 above CH_3C, and CH_3 and Cl below   2°   3-chloro-2,2-dimethylbutane

$$CH_3CCH_2CH_3$$ with CH_2Cl above and CH_3 below   1°   1-chloro-2,2-dimethylbutane

$$CH_3CHCHCH_2Cl$$ with CH_3 above and CH_3 below   1°   1-chloro-2,3-dimethylbutane

$$CH_3CH\!-\!CCH_3$$ with CH_3 and CH_3 above, and Cl below   3°   2-chloro-2,3-dimethylbutane

5.19   (1) 1,1-dichlorobutane   (2) 2,2-dichlorobutane

        (3) 1,2-dichlorobutane   (4) 1,3-dichlorobutane

        (5) 2,3-dichlorobutane   (6) 1,4-dichlorobutane

        (7) 1,1-dichloro-2-methylpropane   (8) 1,2-dichloro-2-methylpropane

        (9) 1,3-dichloro-2-methylpropane

5.20   (a) ethylbenzene   (b) *sec*-butylbenzene

        (c) isobutylbenzene   (d) 1,4-dimethylbenzene (*p*-xylene)

        (e) isopropylbenzene   (f) propylbenzene

        (g) *tert*-butylbenzene   (h) 1-butyl-4-ethylbenzene

Note that if the substituent groups are simple alkyl groups, the compounds are named as alkylbenzenes. When the alkyl groups become large or are complicated by having other substituents on them (Problem 5.21), the benzene ring is treated as one of the substituents.

5.21   (a) 1-chloro-3-phenylpropane   (b) 3-iodo-3-phenylhexane

       (c) 1,2-dichloro-1-phenylpropane   (d) 2-methyl-2-phenylhexane

       (e) 3-phenyl-1-propanol   (f) 3-methyl-1-phenylbutane

5.22   (a) 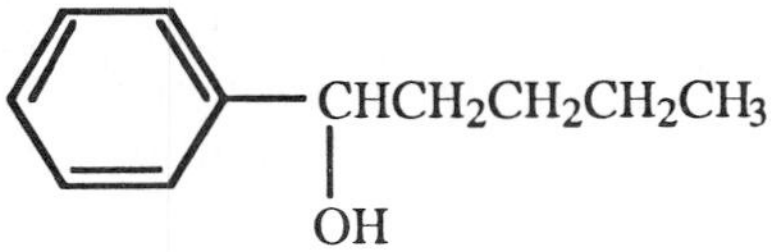 $-CHCH_2CH_2CH_2CH_3$ with OH below

       (b) 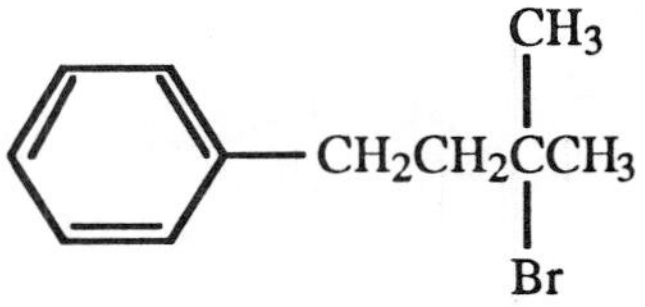 $-CH_2CH_2CCH_3$ with CH_3 above and Br below

       (c) 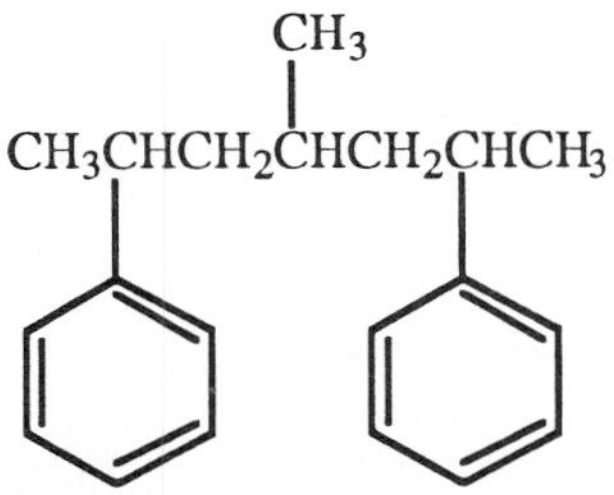 $CH_3CHCH_2CHCH_2CHCH_3$ with two phenyl groups and CH₃

       (d) 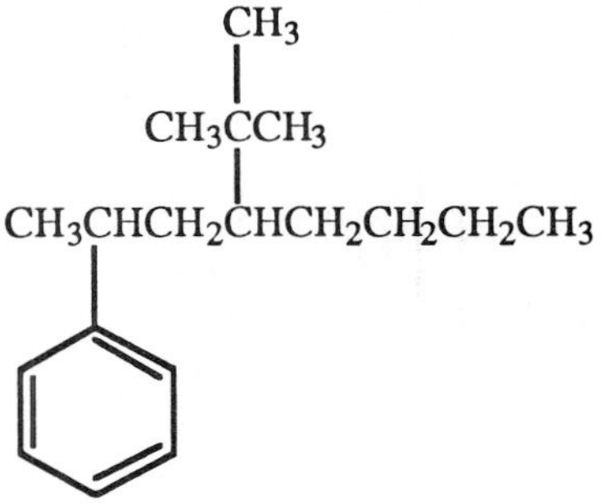 $CH_3CHCH_2CHCH_2CH_2CH_2CH_3$ with CH_3CCH_3 / CH_3 and phenyl

       (e) 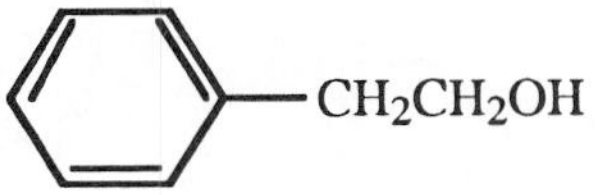 $-CH_2CH_2OH$

5.23   (a) 1-bromo-1-methylcyclobutane   (b) ethylcyclopropane

       (c) 1-ethyl-1-methylcyclopentane   (d) iodocyclopentane

       (e) cyclobutanol   (f) bromocyclopropane

5.24   (a) 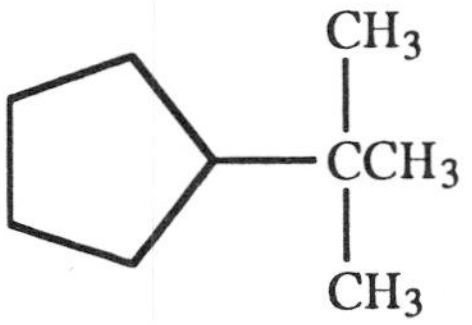   (b)

       (c) 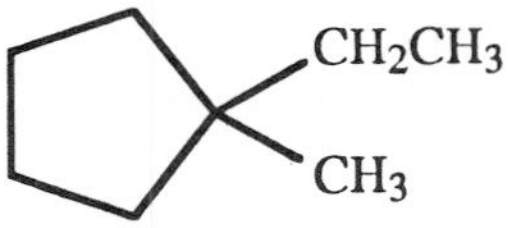   (d)

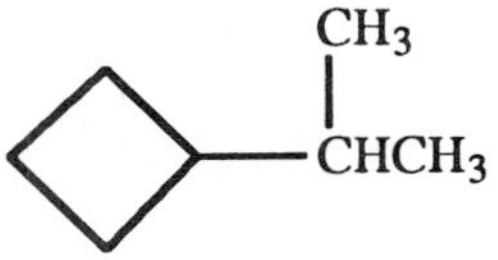

       (e) 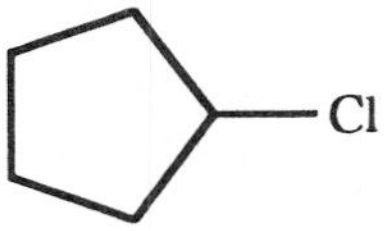 —Cl   (f) 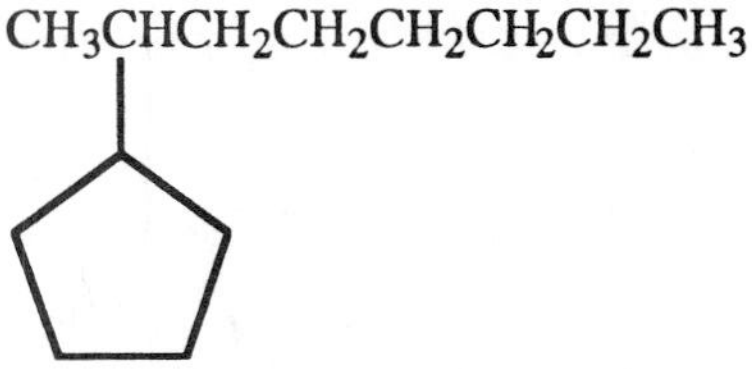 $CH_3CHCH_2CH_2CH_2CH_2CH_2CH_3$

       (g)

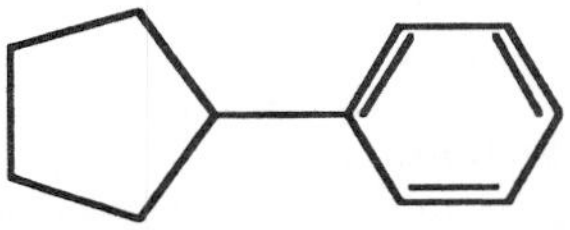

Concept Map 5.3   Conformation in cyclic compounds.

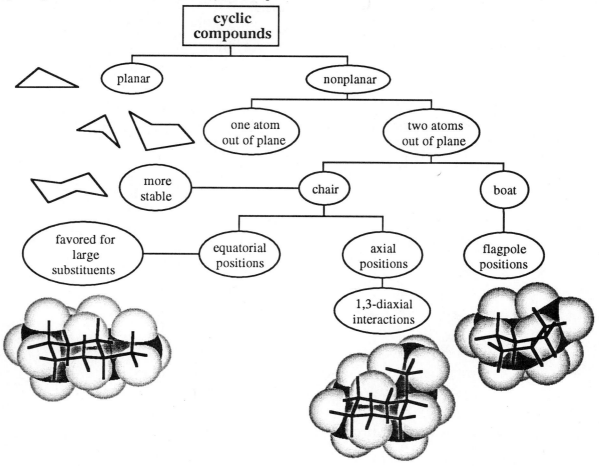

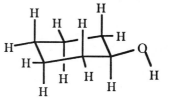

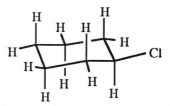

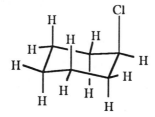

5.25

planar form

chair conformations
axial chlorine

equatorial chlorine

planar form

chair conformations
axial hydroxyl group

equatorial hydroxyl group

5.26

$$CH_3\underset{\underset{CH_3}{|}}{\overset{\overset{CH_3}{|}}{C}} \cdot \quad > \quad CH_3\underset{\underset{H}{|}}{\overset{\overset{CH_3}{|}}{C}} \cdot \quad > \quad CH_3\underset{\underset{H}{|}}{\overset{\overset{H}{|}}{C}} \cdot \quad > \quad H\underset{\underset{H}{|}}{\overset{\overset{H}{|}}{-C}} \cdot$$

5.27
(a) 1-bromo-1-methylcyclopropane
(b) 2,2,5,5-tetramethylhexane
(c) 1,1-dimethylcyclobutane
(d) 2-chloro-2-methylpentane
(e) isopropylcyclohexane or 2-cyclohexylpropane
(f) 1-ethyl-1-cyclopentanol
(g) 1-chloro-1-ethylcyclohexane
(h) 1-phenyl-1-cyclohexanol
(i) 3-bromo-3-ethyl-1-heptanol
(j) 6-*tert*-butyl-2,3-dichlorooctane or 6,7-dichloro-3-ethyl-2,2-dimethyloctane

5.28

$$\underset{\underset{Cl}{|}}{Br\,CHCF_3}$$

2-bromo-2-chloro-1,1-trifluoroethane

5.29 (a)
$$CH_3\underset{\underset{OH}{|}}{\overset{\overset{CH_3}{|}}{C}}CH_2CH_2CH_2CH_3$$

(b)

(c)

(d)
$$CH_3CH_2CH_2\underset{\underset{\underset{CH_3}{|}}{\overset{CH_3CCH_3}{}}}{\overset{CH_3}{C}}HCH_2CH_2CH_3$$

(e)
$$CH_3CH_2CH_2CH_2CHCH_2CH_2CH_2CH_3$$

(f)
$$-CH_2\underset{\underset{}{}}{\overset{\overset{CH_3}{|}}{C}}HCH_3$$

(g)
$$CH_3CH_2CHCH_2CH_2CH_2CH_3$$

(h)

(i)
$$CH_3\underset{\underset{Cl}{|}}{\overset{\overset{CH_3}{|}}{C}}CH_2CH_2CH_2CH_2\underset{\underset{OH}{|}}{\overset{\overset{CH_3}{|}}{C}}CH_3$$

5.30

(a) $C_5H_{12}O$

$CH_3CH_2CH_2CH_2CH_2OH$

1-pentanol

$CH_3CHCH_2CH_2CH_3$
|
OH

2-pentanol

$CH_3CH_2CHCH_2CH_3$
|
OH

3-pentanol

CH_3
|
$CH_3CHCH_2CH_2OH$

3-methyl-1-butanol

CH_3
|
$CH_3CHCHCH_3$
|
OH

3-methyl-2-butanol

CH_3
|
$CH_3CCH_2CH_3$
|
OH

2-methyl-2-butanol

CH_3
|
$CH_3CH_2CHCH_2OH$

2-methyl-1-butanol

CH_3
|
CH_3CCH_2OH
|
CH_3

2,2-dimethyl-1-propanol

$CH_3CH_2CH_2CH_2OCH_3$

methyl butyl ether

CH_3
|
$CH_3CHCH_2OCH_3$

methyl isobutyl ether

CH_3
|
$CH_3CH_2CHOCH_3$

methyl *sec*-butyl ether

CH_3
|
CH_3COCH_3
|
CH_3

methyl *tert*-butyl ether

$CH_3CH_2CH_2OCH_2CH_3$
ethyl propyl ether

CH_3
|
$CH_3CHOCH_2CH_3$
ethyl isopropyl ether

(b) $C_3H_5Cl_3$

$CH_3CH_2CCl_3$	$CH_3CHClCHCl_2$	$ClCH_2CH_2CHCl_2$	$CH_3CCl_2CH_2Cl$	$ClCH_2CHClCH_2Cl$
1,1,1-trichloro-propane	1,1,2-trichloro-propane	1,1,3-trichloro-propane	1,2,2-trichloro-propane	1,2,3-trichloro-propane

5.31   (a)

CH_3
|
$CH_3CH_2CCH_2CH_3$
|
CH_3

most spherical of the three constitutional isomers, least van der Waals attraction between the molecules

(b)

CH_3
|
$CH_3CCH_2CH_3$
|
CH_3

same as (a)

5.31 (cont)

$$CH_3$$
(c) 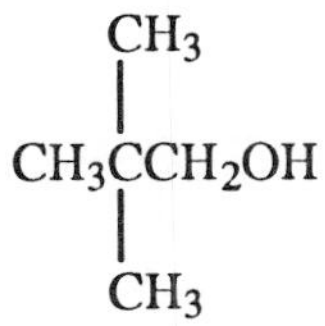 CH_3CCH_2OH same as (a)
$$CH_3$$

$$CH_3$$
(d) 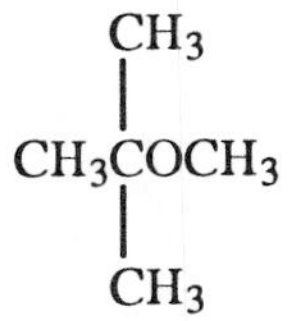 CH_3COCH_3 least polar (the other two are alcohols and form hydrogen bonds)
$$CH_3$$

(e)    [cyclopentane ring with O]   least polar

5.32   (a) C_5H_9Br

Each halogen atom counts as a hydrogen atom in determining the units of unsaturation.
Units of unsaturation $= [(2 \times 5) + 2 - 10]/2 = (12\text{-}10)/2 = 1$
Some structural formulas are shown on the next page:

alkenes: $CH_3CH_2CH\!=\!CHCH_2Br$ $CH_3CH_2CH\!=\!CCH_3$ $CH_3\overset{\displaystyle CH_3}{C}\!=\!CCH_3$
                                                                          Br Br

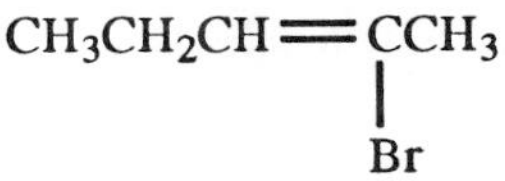

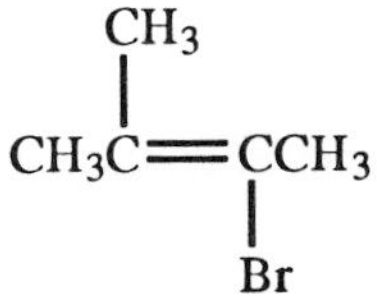

cyclic
compounds: [cyclopentane]—Br   [cyclobutane with CH_3 and Br]   [cyclopropane]—CHCH_3 with Br   [cyclopropane with CH_3]—CH_2Br

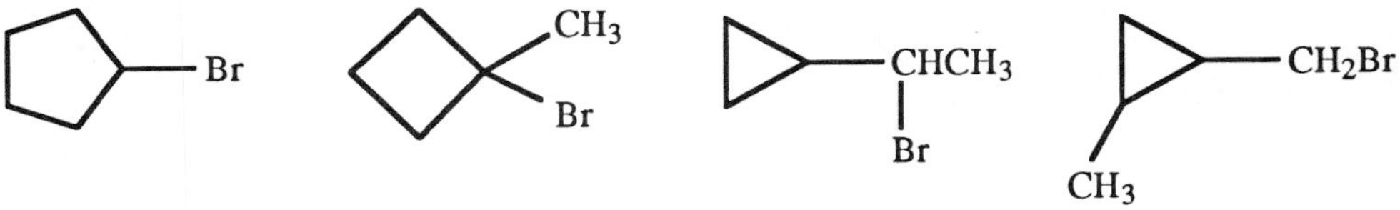

(b) $C_5H_{10}O$

Oxygen atoms do not affect the calculation
Units of unsaturation $= [(2 \times 5) + 2 - 10]/2 = (12\text{-}10)/2 = 1$

Some structural formulas are shown below:

aldehydes $\overset{\displaystyle O}{\overset{\displaystyle \|}{CH_3CH_2CHCH}}$ $\overset{\displaystyle O}{\overset{\displaystyle \|}{CH_3CH_2CCH_2CH_3}}$
and ketones: CH_3

unsaturated $CH_2\!=\!CHCH_2CH_2CH_2OH$ $CH_3CHCH\!=\!CHCH_3$
alcohols: OH

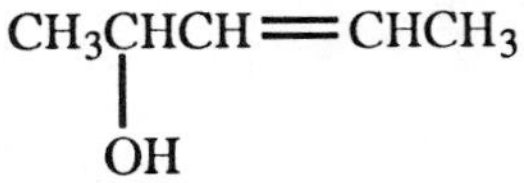

5.32 (b) (cont)

unsaturated ethers:

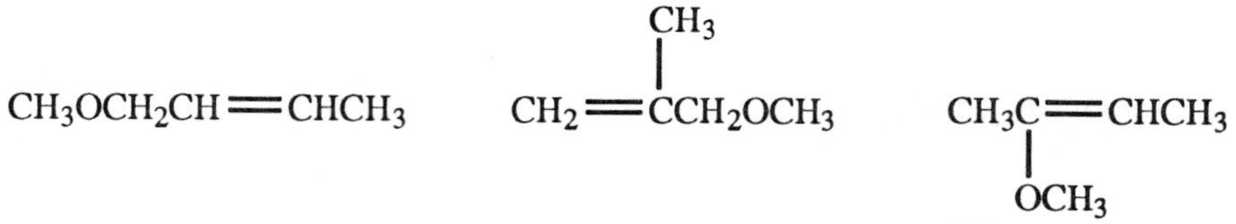

$CH_3OCH_2CH=CHCH_3$

$CH_2=\overset{\overset{\displaystyle CH_3}{|}}{C}CH_2OCH_3$

$CH_3C=CHCH_3$
$\overset{|}{O}CH_3$

cyclic alcohols:

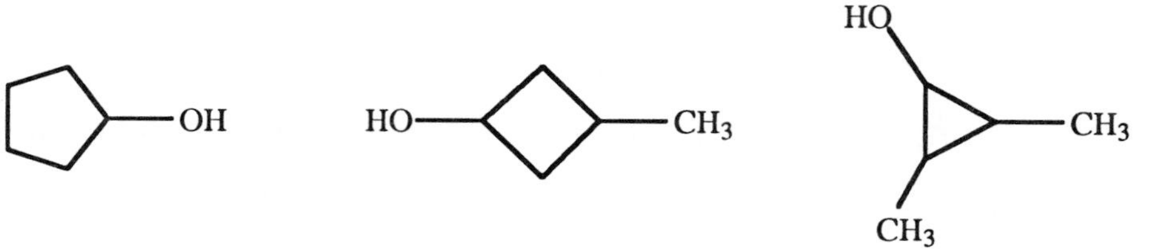

—OH

HO— —CH₃

—CH₃

cyclic ethers:

—CH₃

—CH₂CH₃

—CHCH₃
$\overset{|}{C}H_3$

—OCH₃

—OCH₃

—CH₂OCH₃

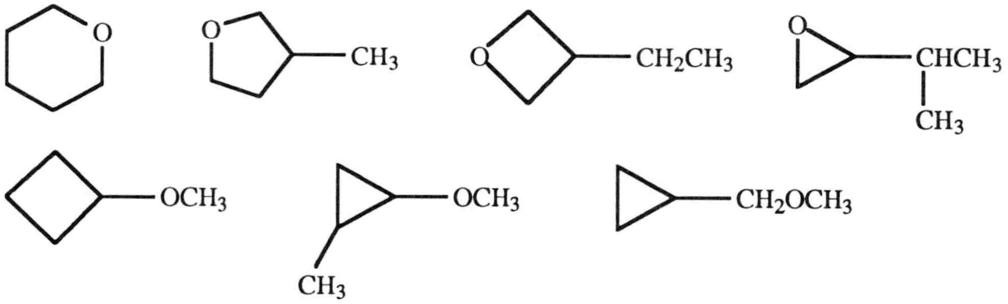

(c) C_5H_8O

Units of unsaturation = $[(2 \times 5) + 2 - 8]/2 = (12-8)/2 = 2$

Some structural formulas are shown below:

unsaturated aldehydes and ketones:

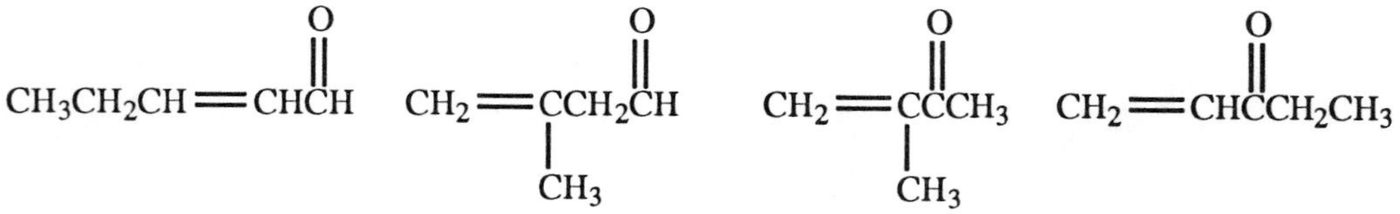

$CH_3CH_2CH=CH\overset{\overset{\displaystyle O}{||}}{C}H$

$CH_2=C\overset{\overset{\displaystyle O}{||}}{C}H_2\overset{\overset{\displaystyle O}{||}}{C}H$
$\overset{|}{C}H_3$

$CH_2=\overset{\overset{\displaystyle O}{||}}{C}\overset{\overset{\displaystyle }{}}{C}CH_3$
$\overset{|}{C}H_3$

$CH_2=CH\overset{\overset{\displaystyle O}{||}}{C}CH_2CH_3$

unsaturated alcohols and ethers:

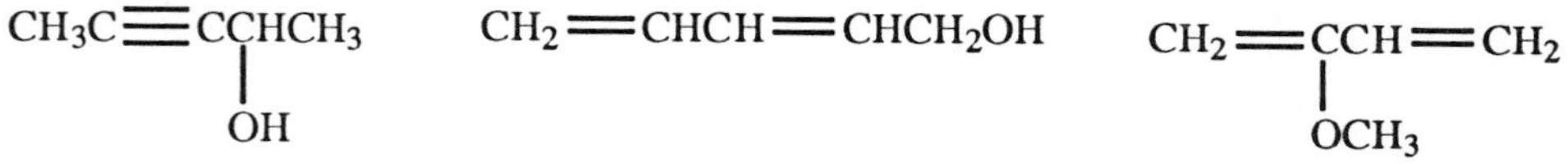

$CH_3C\equiv CCHCH_3$
$\overset{|}{O}H$

$CH_2=CHCH=CHCH_2OH$

$CH_2=CCH=CH_2$
$\overset{|}{O}CH_3$

cyclic compounds:

=O

—$\overset{\overset{\displaystyle O}{||}}{C}CH_3$

—OCH₃

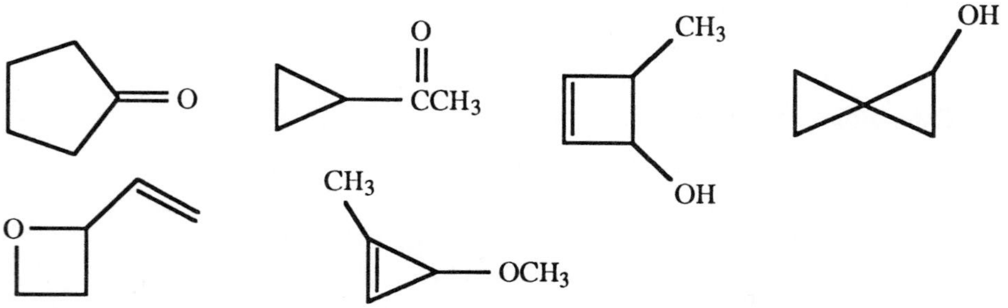

5.32 (cont)

(d)   Cyclic compounds that have rings with four or more carbon atoms will not react easily with bromine in carbon tetrachloride. The ones with the molecular formula C_5H_9Br are:

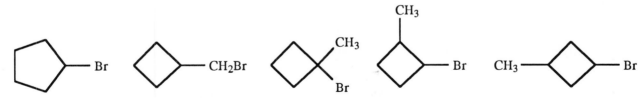

(e)   The hydrogenation data tells us that the original compound was an alkene with the carbon skeleton of 3-methyl-1-butanol.

$$\overset{\overset{\displaystyle CH_3}{\displaystyle |}}{CH_3CHCH_2CH_2OH}$$

The following structures are possible.

$$\overset{\overset{\displaystyle CH_3}{\displaystyle |}}{CH_3C}\!=\!CHCH_2OH \qquad\qquad \overset{\overset{\displaystyle CH_3}{\displaystyle |}}{CH_2}\!=\!CCH_2CH_2OH$$

5.33   (a)

triple bond: $CH_3CH_2CH_2CH_2CH_2C\!\equiv\!CH$ $\overset{\overset{\displaystyle CH_3}{\displaystyle |}}{CH_3CHCH_2C}\!\equiv\!CCH_3$

two double bonds: $CH_3CH\!=\!CHCH\!=\!CHCH_2CH_3$ $\overset{\overset{\displaystyle CH_3}{\displaystyle |}}{\underset{\underset{\displaystyle CH_3}{\displaystyle |}}{CH_2\!=\!CCHCH\!=\!CH_2}}$

a ring and
double bond:

two rings:

(b)   The original compound must have the carbon skeleton present in 2-methylhexane.

$$\overset{\overset{\displaystyle CH_3}{\displaystyle |}}{CH_3CHCH_2CH_2CH_2CH_3}$$

The possible structures are shown below and on the next page.

$$\overset{\overset{\displaystyle CH_3}{\displaystyle |}}{CH_3CHCH_2CH_2C}\!\equiv\!CH \qquad \overset{\overset{\displaystyle CH_3}{\displaystyle |}}{CH_3CHCH_2C}\!\equiv\!CCH_3 \qquad \overset{\overset{\displaystyle CH_3}{\displaystyle |}}{CH_3CHC}\!\equiv\!CCH_2CH_3$$

5.33 (b) (cont)

$$CH_3CHCH_2CH = C = CH_2 \quad\text{(with } CH_3 \text{ above CH)}$$

$$CH_3CHCH = CHCH = CH_2 \quad\text{(with } CH_3 \text{ above CH)}$$

$$CH_3C = CHCH_2CH = CH_2 \quad\text{(with } CH_3 \text{ above C)}$$

$$CH_2 = CCH_2CH_2CH = CH_2 \quad\text{(with } CH_3 \text{ above C)}$$

$$CH_3CHCH = C = CHCH_3 \quad\text{(with } CH_3 \text{ above CH)}$$

$$CH_3C = CHCH = CHCH_3 \quad\text{(with } CH_3 \text{ above C)}$$

$$CH_2 = CCH_2CH = CHCH_3 \quad\text{(with } CH_3 \text{ above C)}$$

$$CH_2 = CCH = CHCH_2CH_3 \quad\text{(with } CH_3 \text{ above C)}$$

5.34   Isomers of $C_3H_8O_2$

$CH_3OCH_2OCH_3$	$CH_3OCH_2CH_2OH$	$HOCH_2CH_2CH_2OH$
two ether groups	ether alcohol	2 hydroxyl groups
no hydrogen bonding	hydrogen bonding	more hydrogen bonding
bp 45 °C	bp 124 °C	bp 215 °C

5.35

(a)

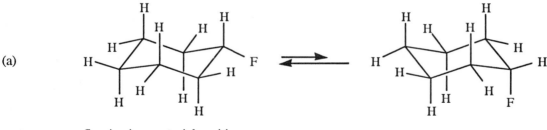

fluorine in equatorial position
less crowded
lower in energy

(b) The equilibrium constant, $K_{eq} = \dfrac{[\text{axial}]}{[\text{equatorial}]}$, is less than 1. The conformation with the fluorine in the equatorial position predominates.

(c) The energy difference between the two conformations of bromocyclohexane would be expected to be larger than that for the fluoro compound. Bromine is considerably larger than fluorine and would experience even more crowding in the axial position than fluorine would.

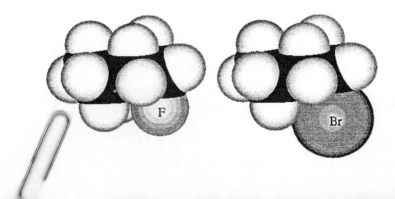

5.36

(a)

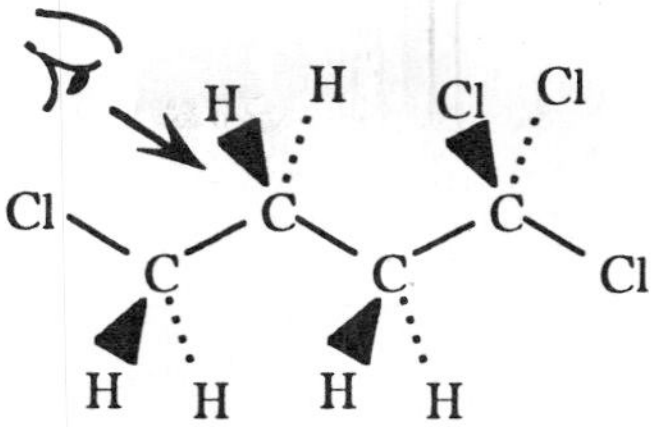

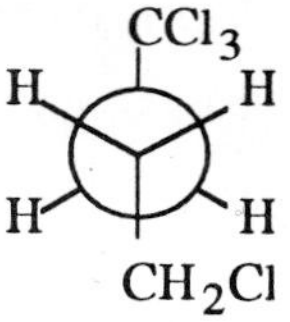

1,1,1,4-tetrachlorobutane
most stable conformation A

(b)

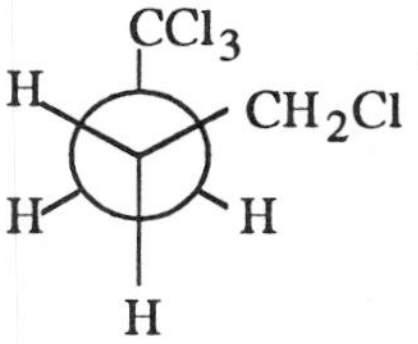

C
one of two
gauche forms

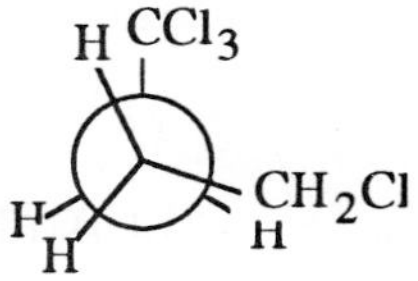

B
one of the two more
stable eclipsed forms

(c)

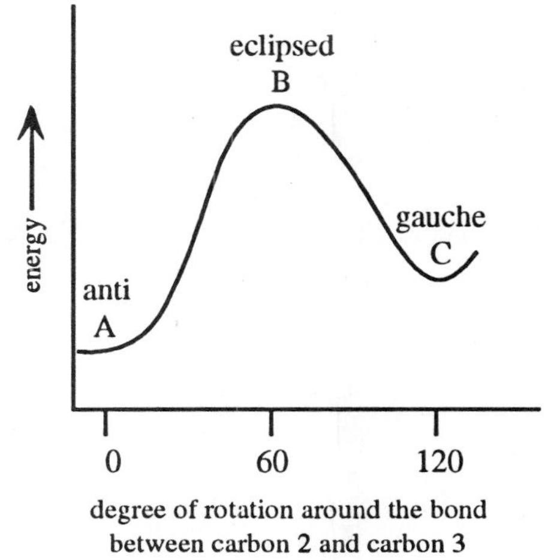

degree of rotation around the bond
between carbon 2 and carbon 3

5.37

(a) $CH_3CH_2NH_2$ + HCl $\longrightarrow$ $CH_3CH_2NH_3^+$ Cl$^-$
               base   acid

(b) $CH_3CH_2C\equiv CH$ + $NaNH_2$ $\longrightarrow$ $CH_3CH_2C\equiv C^- Na^+$ + NH_3
                    acid   base

(c) $(CH_3CH_2CH_2)_2NH$ + CH_3CH_2Br $\longrightarrow$ $(CH_3CH_2CH_2)_2\overset{+}{N}HCH_2CH_3$ Br$^-$
                nucleophile   electrophile

(d) $CH_3CH_2CH\!=\!CHCH_2CH_3$ + HBr $\longrightarrow$ $CH_3CH_2CH_2CHBrCH_2CH_3$
                nucleophile   electrophile

5.37 (cont)

(e) $CH_3CH_2CH_2SH$ + NaOH $\longrightarrow$ $CH_3CH_2CH_2S^- Na^+$ + H_2O
      acid   base

(f)   ⬡—$\overset{+}{N}H_2CH_3$ Cl^- + NaOH $\longrightarrow$ ⬡—$NHCH_3$ + NaCl + H_2O
              acid   base

(g) $CH_3CH_2CH = CHCH_3$ + Br_2 $\longrightarrow$ $CH_3CH_2CHBrCHBrCH_3$
       nucleophile   electrophile

(h)   ⬡—SNa + $CH_3CH_2CH_2Br$ $\longrightarrow$ ⬡—$SCH_2CH_2CH_3$ + NaBr
         nucleophile   electrophile

(i)   Br—⬡—$\overset{O}{\overset{\|}{C}}OH$ + $NaHCO_3$ $\longrightarrow$ Br—⬡—$\overset{O}{\overset{\|}{C}}O^- Na^+$ + H_2O + $CO_2\uparrow$
           acid   base

(j) $CH_3CH_2CH_2C \equiv CCH_3$ + 2 H_2 $\xrightarrow{PtO_2}$ $CH_3CH_2CH_2CH_2CH_2CH_3$

(k) CH_3CH_3 + Cl_2 $\xrightarrow{hv}$ CH_3CH_2Cl + CH_3CHCl_2 + $ClCH_2CH_2Cl$ + HCl

Patterns are not obvious for reactions (j) and (k).

5.38

(a)   ⬡—SH $\xrightarrow[H_2O]{NaOH}$ ⬡—$S^- Na^+$ $\xrightarrow{CH_3CHCH_2CH_2Br}$

electrophile with carbon
skeleton found in product

$\overset{CH_3}{|}$
$CH_3CHCH_2CH_2Br$

lost in product
therefore need OH^-

nucleophile

⬡—$CH_2CH_2CHCH_3$ with $\overset{CH_3}{|}$ on the CH

5.38 (cont)

(b)   nucleophile
$$CH_3NH_2$$
$$CH_3CH_2Br \xrightarrow{\hspace{2cm}} CH_3CH_2\overset{+}{N}H_2CH_3 \; Br^-$$

this portion of the product
comes from the nucleophile;
the rest must be from the
electrophile it reacts with

(c)
$$CH_2\!=\!CH_2 \xrightarrow{\text{HBr}} CH_3CH_2Br \xrightarrow{CH_3O^- Na^+} CH_3CH_2OCH_3$$
nucleophile

looks as though CH_3OH has added; CH_3OH
is too weak an acid to add to double bond;
must use two consecutive reactions

(d)
$$CH_3CH_2CH_2CH_2NH_2 \xrightarrow{\text{HCl}} CH_3CH_2CH_2CH_2\overset{+}{N}H_3 \; Cl^-$$
base   conjugate acid of base

(e)   electrophile
2 Br$_2$
$$CH_3CH_2C\!\equiv\!CH \xrightarrow{\hspace{2cm}} CH_3CH_2CBr_2CHBr_2$$
nucleophile with carbon   Br$_2$ added twice
skeleton of product; triple bond
will react two times with Br$_2$

(f)   electrophile with carbon
skeleton found in product
$$CH_3CH_2CH_2CH_2OH \xrightarrow{NH_2^-} CH_3CH_2CH_2CH_2O^- \xrightarrow{CH_3Br} CH_3CH_2CH_2CH_2OCH_3$$

proton lost in product   nucleophile
therefore need base

(g)
$$CH_3CH\!=\!CHCH_3 \xrightarrow{\text{HBr}} CH_3CHCH_2CH_3 \xrightarrow{Na^+ CN^-} CH_3CHCH_2CH_3$$
nucleophile | |
Br   CN

looks as though CN has added; HCN is
too weak an acid to add to double bond;
must use two consecutive reactions

(h)
$$CH_3CHCH_2CH\!=\!CH_2 \xrightarrow[\text{PtO}_2]{\text{H}_2} CH_3CHCH_2CH_2CH_3$$
| |
OH   OH

has two more hydrogens than starting
material, therefore reagent is H$_2$

5.38 (cont)

(i)

hydrogen substituted by halogen,
therefore need halogen with light

5.39

(a)   1,2-ethanediol

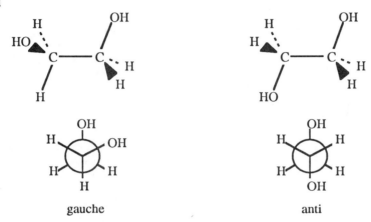

1,2-dimethoxyethane

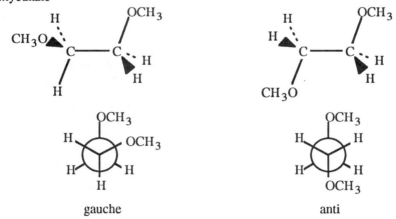

(b)   Hydrogen bonding can take place in the gauche conformation of 1,2-ethanediol but not in the anti conformation.

(c)   The order of stability is A > C > B.

5.39 (cont)

(d)

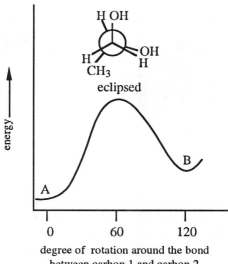

eclipsed

energy →

A

B

0   60   120

degree of rotation around the bond
between carbon 1 and carbon 2

5.40 (a) The free energy difference between the two conformations can be determined from the equilibrium constant, which in turn can be determined from the ratios of the axial and equatorial conformations.

$$\Delta G° = -2.303 \, RT \log K_{eq} \text{ and } K_{eq} = \frac{[\text{axial}]}{[\text{equatorial}]}$$

$$\Delta G° = -2.303 \, RT \log \frac{[\text{axial}]}{[\text{equatorial}]} = +2.303 \, RT \log \frac{[\text{equatorial}]}{[\text{axial}]}$$

The equilibrium with the greatest amount of the equatorial conformation will have the largest $\Delta G°$. Equilibrium **C** has a ratio of 35:1 in favor of the equatorial conformation and therefore has the greatest difference in free energy.

(b) The bulkier the group attached to the ring, the greater will be the interaction of that group with other substituents on the ring. In particular, the larger the groups are that are attached to the atom directly bonded to the ring, the greater the interaction with other ring substituents when the bulky group is in the axial position.

In **A**, the C(ring)—N—C bond angle is 180° and there are no bulky methyl groups on the substituent. There is little interaction between the group and the ring hydrogens. The conformation in which the substituent is equatorial is only slightly favored.

H

N≡C

180°

H

180°

N

C

5.40 (b) (cont)

In **B**, the C(ring)—N—C bond angle is ~120° and there are two bulky methyl groups on the carbon atom of the substituent. There is some interaction between the methyl groups and the ring hydrogen atoms in both conformations, but the interaction is greater when the substituent is in the axial position. In the axial conformation shown below, there are 1,3-diaxial interactions between the axial ring hydrogen atoms and one of the methyl groups. There are other conformations in which the methyl group is not interacting. The conformation in which the substituent is equatorial is more favored for **B** than the corresponding conformer is for **A**.

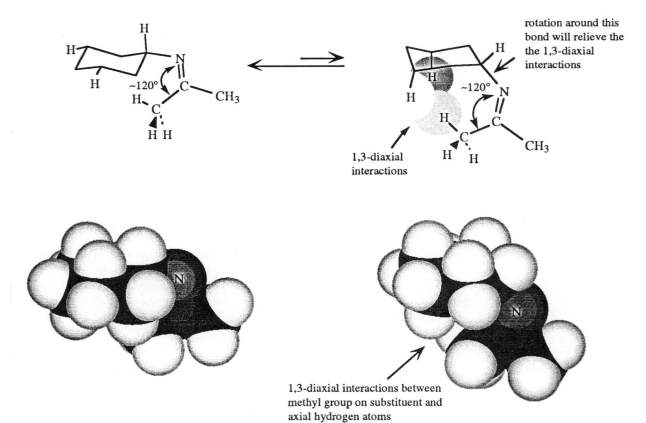

5.40 (b) (cont)

In **C**, the C(ring)—N—C bond angle is ~ 109°. The methyl groups are attached to the nitrogen atom and are thus closer to the ring than in **B**. There are much greater interactions between the methyl groups and the axial hydrogen atoms on carbons 3 and 5 in the axial conformation of **C** than in the axial conformation of **B**. The equatorial conformation in **C** is thus much more favored than in the other two equilibria.

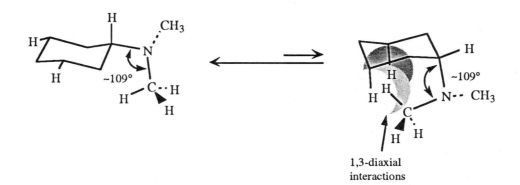

1,3-diaxial
interactions

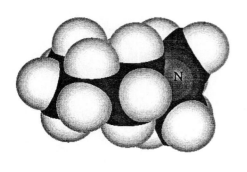

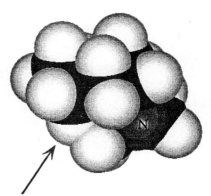

1,3-diaxial interactions. The methyl group in C is closer to the axial hydrogen atoms than in B.

(c) $K_{eq} = \dfrac{[\text{axial}]}{[\text{equatorial}]}$

For **C**, the equilibrium constant is $1/35$.

(d) cyclohexane—N(CH$_3$)$_2$ $\xrightarrow{\text{CH}_3\text{I}}$ cyclohexane—$\overset{+}{\text{N}}$(CH$_3$)$_3$ I$^-$

5.40 (d) (cont)

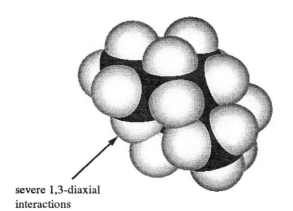

conformation with equatorial substituent;
no 1,3-diaxial interactions

conformation with axial substituent;
severe 1,3-diaxial interactions;
less favorable than conformation with
substituent in equatorial position

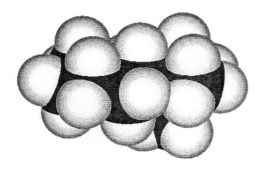

severe 1,3-diaxial
interactions

The equilibrium constant for the conformational change for this new compound will be even smaller than that for **C**. The substituent on the new compound has an even greater bulk and thus greater crowding in the axial position, especially from the axial hydrogens on carbons 3 and 5 (1,3 diaxial interactions). Unlike in **B** and **C**, rotation around the C(ring)—N bond will not relieve the interactions.

6
Stereochemistry

Workbook Exercises

Before you begin to learn about some of the three-dimensional relationships that can exist in molecules, you should realize that these relationships are not unique to molecules, but also exist in the macroscopic world of common objects. To get the most benefit from the following exercises, you will need the two cardboard tetrahedra you constructed for the workbook exercises in Chapter 2. Use your model set to construct two tetrahedra with four atoms of different colors attached to each center.

In earlier chapters you learned the symbolism used to represent a tetrahedral arrangement of atoms around a central atom, for example, methane, CH_4.

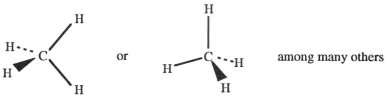

These are two-dimensional representations, called perspective drawings, of a three-dimensional object. Your study of stereochemistry in Chapter 6 will be easier if you become comfortable relating these two-dimensional pictures with the three-dimensional structures they represent.

EXERCISE I.   (a) Place one of the tetrahedra on the table so that the face with only letters on it is toward you, and one of the edges is parallel to an edge of the table.

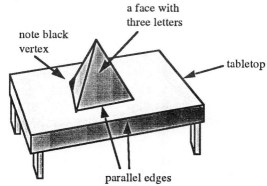

(b) The overall arrangement of the three letters on the face of the cardboard tetrahedron pointed towards you should be

, if you use lines to point to each vertex from the center.

Add the letters A, B, and C
as they appear to you on the
face of the tetrahedron:

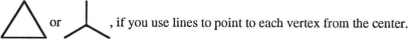

Workbook Exercises (cont)

(c) You can also use lines, dashes and wedges to draw the arrangement of the four vertices of your cardboard model. If you imagine a point at the center of the cardboard model and then lines originating from there to each of the four vertices, the picture would look as follows:

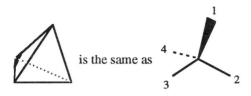

is the same as

The face defined by the numbers 1-2-3,  , is still directed toward you.

Draw a representation of the cardboard model you have on the table using lines, dashes, and wedges, and add the A, B, C, and ● labels as they appear to you on the model.

(d) If you let your eyes travel from number 1 to number 2 to number 3 on the front face of the tetrahedron drawn in part (c), your eyes would be moving the way the hands of a clock do, or "clockwise."

counterclockwise   clockwise

Look at your own cardboard model and decide what direction your eyes must travel in going from A to B to C.

(e) Place your other cardboard model on the table with one of its edges parallel to the tabletop and with its A-B-C face toward you. How do your eyes travel in going from A to B to C? Draw the figure using the symbolism of lines, dashes, and wedges to indicate three dimensionality.

(f) Place a molecular model of a tetrahedron with four different attached atoms on the table so that two of the attached atoms line up with the edge of the tabletop in a way that is similar to the way the front edge of your cardboard model is situated on the table. The four atoms on the molecular model should correspond in space to the vertices of your cardboard model. Draw the molecular model using three-dimensional symbolism.

(g) Select one of your cardboard models. There are five other edges on the cardboard model besides the one parallel to the edge of the table. Reposition the model so that each edge, in turn, is parallel to the edge of the table. Draw a three-dimensional representation for each one. Note that all six pictures you will have drawn for the model represent the same object. For those with an A-B-C face toward you, is the A→B→C direction clockwise or counterclockwise?

(h) You now have six three-dimensional representations for one of your cardboard models and one for the second one. Do any of the six representations of one model exactly match one for the second model?

EXERCISE II.   We will now arbitrarily assign a set of priorities to the four vertices on your cardboard models:

(1) = highest priority   = B
(2) = second highest   = A
(3) = third highest   = ●
(4) = lowest priority   = C

Workbook Exercises (cont)

(a) With one of your hands, grasp the cardboard model by the lowest priority vertex (C). (Lightly pinch the vertex between your thumb, index and middle fingers.) Hold the model so that the B-A-● face that is opposite this vertex is facing you. On this face, do your eyes turn clockwise or counterclockwise in looking from (1) to (2) to (3) (or B→A→●)? Using the same set of priorities, answer the same question for your other cardboard model. Now assign a set of priorities to the four colored atoms on one of your molecular models. Assign a clockwise or counterclockwise direction in going from (1) to (2) to (3) by grasping the group with priority (4) and holding the model so that the 1-2-3 face is pointing toward you.

(b) Using the same molecular model from part (a), disconnect and reconnect any two of the atoms, exchanging their positions. Keeping the same priorities that you assigned in part (a), decide whether your eyes move clockwise or counterclockwise when you look from (1) to (2) to (3) when atom (4) is held away from you. Exchange any two other atoms and examine the (1) to (2) to (3) direction again. Repeat this exercise once more. What generalization can you make about the number of different arrangements possible for four different atoms in a tetrahedral arrangement?

EXERCISE III. When you hold the molecular model so that atom (4) is towards you, what direction do your eyes move in looking from (1) to (2) to (3)?

EXERCISE IV. Arbitrarily assign priorities to the following four symbols: P, β, Σ, and q.

(1) = highest priority = _____   (2) = second highest = _____

(3) = third highest   = _____   (4) = lowest priority = _____

Using your molecular models as a guide, decide whether your eyes turn clockwise or counterclockwise in going from (1) to (2) to (3) when the priority (4) symbol in the following representations is held away from you.

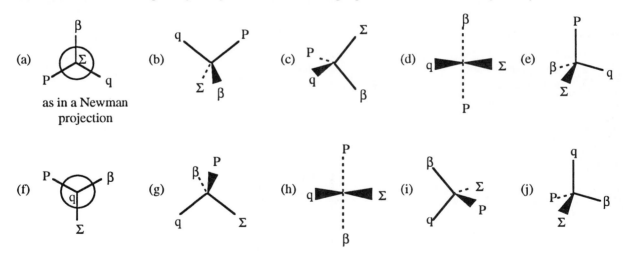

Which of the representations above correspond to the same arrangement in space?

Concept Map 6.1   Stereochemical relationships.

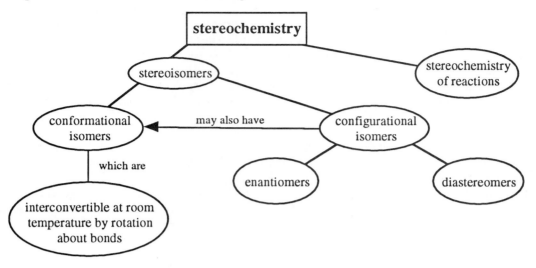

Concept Map 6.2   Chirality.

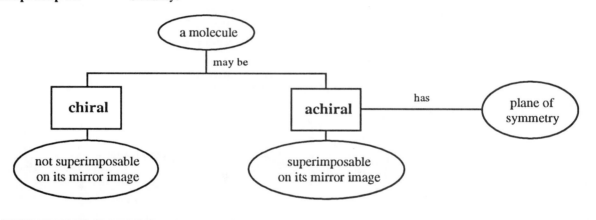

Concept Map 6.3   Definition of enantiomers.

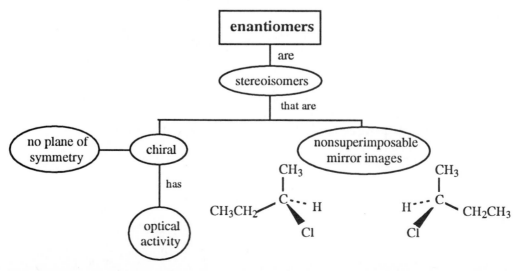

6.1   (a)

If the mug had the word "MOTHER" written on one side, there would no longer be any planes of symmetry. The mug would now be chiral.

    plane of symmetry

If the mug had the word "MOM" instead of "MOTHER" written on one side, it would still be chiral. If the word "MOM" were written opposite the handle, it would be achiral.

(b)

Thread is helically wound around a spool. A helix has no planes of symmetry. A spool with helically wound thread has no plane of symmetry and is chiral.

(c)

   plane of symmetry

An empty spool has many planes of symmetry. One is shown.

6.2   (a)

helical thread; chiral

(b)   plane of symmetry

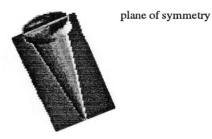

achiral

(c)

prongs equal length; achiral
prongs not equal length; chiral

(d)   plane of synnetry

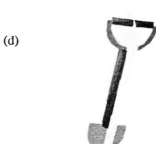

achiral

mirror

(e)   plane of symmetry

socks are interchangeable; achiral

(f)

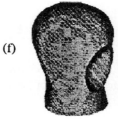

some mittens have a plane of symmetry;
the mittens shown do not. Each mitten
is chiral and is the mirror image isomer of
the other mitten.

(g)

                                plane of symmetry

achiral

6.2 (cont)

(h)

Notice buttons and buttonholes; chiral

(i)

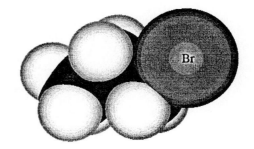

chiral

6.3   (a)   No

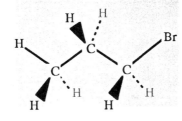

The molecule has a plane of symmetry.
The atoms that are behind the plane
are shown as lighter letters

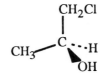

(b)   Yes

CH$_2$Cl

H····C—CH$_3$

HO

CH$_2$Cl

CH$_3$—C····H

OH

(c)   Yes

CH$_2$Cl

H····C—CH$_2$CH$_3$

H$_3$C

CH$_2$Cl

CH$_3$CH$_2$—C····H

CH$_3$

(d)   No

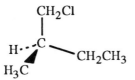

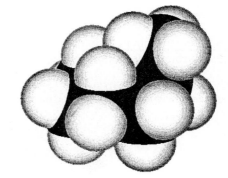

The molecule has a plane of symmetry.
[See part (a)]

6.3 (cont)

(e)   Yes

$$CH_3CH_2CH_2 \overset{\overset{\displaystyle CH_3}{|}}{\underset{Br}{C}} \cdots H \qquad\qquad H \cdots \overset{\overset{\displaystyle CH_3}{|}}{\underset{Br}{C}} - CH_2CH_2CH_3$$

(f)   Yes

$$CH_3CHCH_2 \overset{\overset{\displaystyle CH_3}{|}}{\underset{OH}{C}} \cdots H \qquad\qquad H \cdots \overset{\overset{\displaystyle CH_3}{|}}{\underset{HO}{C}} - CH_2CHCH_3$$
$$\qquad\;\;|\qquad\qquad\qquad\qquad\qquad\qquad\qquad\qquad\quad |$$
$$\qquad\;\;CH_3\qquad\qquad\qquad\qquad\qquad\qquad\qquad\;CH_3$$

Concept Map 6.4   Optical activity.

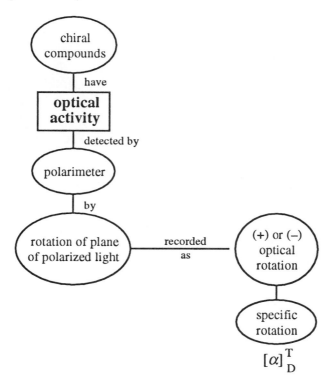

6.4 $\dfrac{\text{optical rotation of unknown solution}}{\text{optical rotation of known concentration}} = \dfrac{+13.3°}{+53.2°} = \dfrac{1}{4}$

therefore the concentration of the unknown solution $= \dfrac{1}{4} \times$ the concentration of the original solution

$$= \dfrac{1}{4} \times (20 \text{ g}/100 \text{ mL}) = 5.0 \text{ g}/100 \text{ mL} = 0.05 \text{ g/mL}$$

6.5 $[\alpha]_D^{20°} = \dfrac{+2.46°}{(1.0 \text{ dm}) (5 \text{ g}/100 \text{ mL})} = +49.2° \ (c \ 0.05, \text{ ethanol})$

6.6 The equation for specific rotation can be rearranged.

$$\alpha = [\alpha]_D^{20°} \, lc$$

The actual rotation, α, is directly proportional to the concentration. If the solution is diluted, for example, to twice the original volume, the concentration, and therefore the optical rotation, will be cut in half. Thus if the original specific rotation is +90°, the actual rotation for the diluted solution will be 45° in a clockwise direction. If the specific rotation is –270°, the actual rotation will be –135°, or 225° in a clockwise direction. In general, decreasing the concentration will cause a positive rotation to decrease in the clockwise direction and a negative rotation to increase in the clockwise direction.

Concept Map 6.5   Formation of racemic mixtures in chemical reactions.

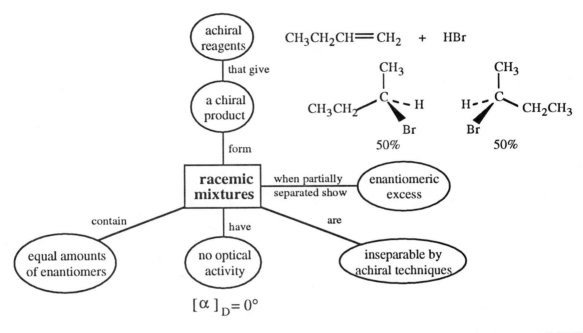

6.7   (a) $\text{enantiomeric excess} = \dfrac{\text{actual rotation}}{\text{rotation of pure enantiomer}} \times 100$

$$\text{enantiomeric excess} = \frac{-14.70°}{-39.5°} \times 100 = 37.2\%$$

(b)

(*S*)-(–)-2-bromobutanoic acid   (*R*)-(+)-2-bromobutanoic acid
68.6%   31.4%

6.7 (cont)

The mixture containing a 37.2% enantiomeric excess of the levorotatory compound is 100 − 37.2 or 62.8% racemic. The racemic portion of the mixture contains equal amounts (62.8/2 = 31.4%) of each enantiomer. Therefore the solution contains 31.4% of the dextrorotatory and (31.4 + 37.2)% or 68.6% of the levorotatory isomer.

6.8   (a) (*S*)-3-hexanol   (b) (*S*)-2-bromopentane

(c) (*R*)-3-iodo-3-methylhexane   (d) (*R*)-1-bromo-1-chloropropane

(e) (*R*)-1,2-dichlorobutane   (f) (*S*)-4-chloro-2-methyl-1-butanol

(g) (*S*)-1-chloro-3,4-dimethylpentane   (h) (*R*)-4-bromo-2-butanol

6.9   (a)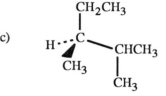

(b)

(c)

(d)

(e)

(f)

6.10   (a)

(*R*)-3-methylhexane   (*S*)-3-methylhexane

These are enantiomers

(b)

(2*R*, 4*R*)-2-bromo-4-methylhexane   (2*S*, 4*S*)-2-bromo-4-methylhexane

These enantiomers are the diastereomers of the two enantiomers on the next page.

6.10 (b) (cont)

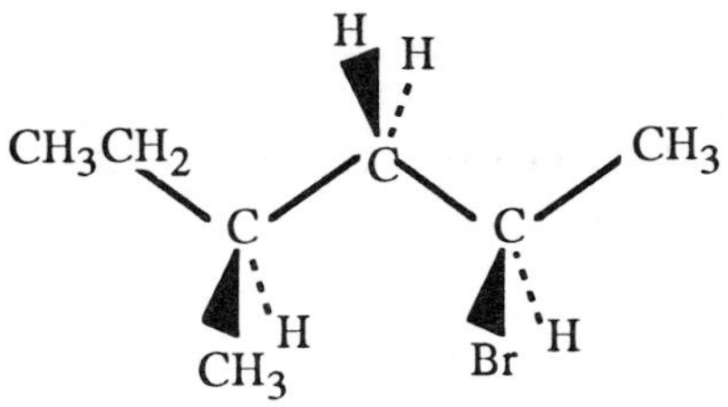

(2*R*, 4*S*)-2-bromo-4-methylhexane

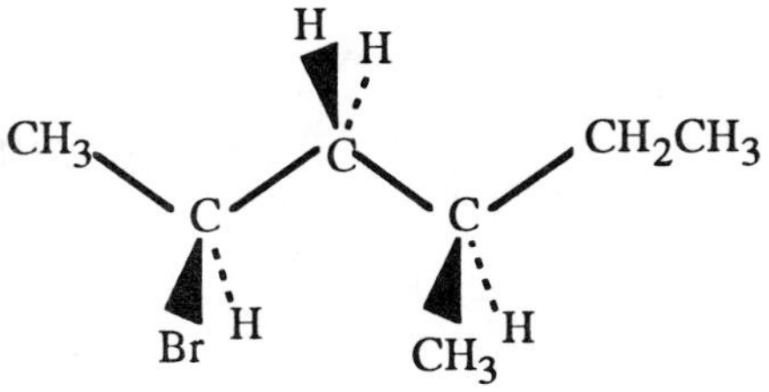

(2*S*, 4*R*)-2-bromo-4-methylhexane

These enantiomers are the diastereomers of the two enantiomers on the previous page.

(c)

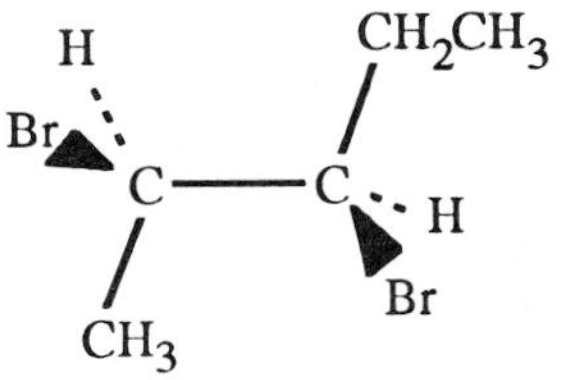

(2*R*, 3*R*)-2,3-dibromopentane

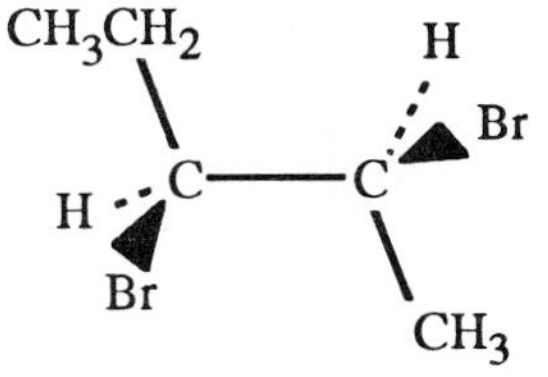

(2*S*, 3*S*)-2,3-dibromopentane

These enantiomers are the diastereomers of the two enantiomers below.

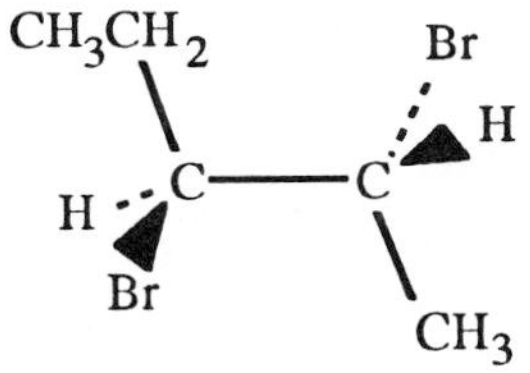

(2*R*, 3*S*)-2,3-dibromopentane

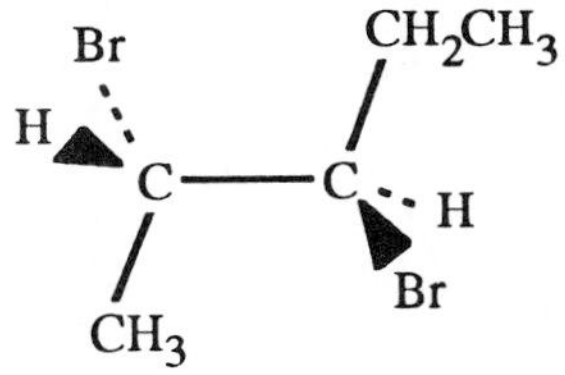

(2*S*, 3*R*)-2,3-dibromopentane

These enantiomers are the diastereomers of the two enantiomers above.

6.11   3-bromo-2-butanol

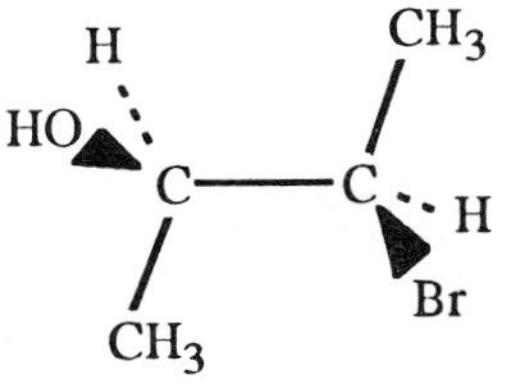

(2*R*, 3*R*)-3-bromo-2-butanol

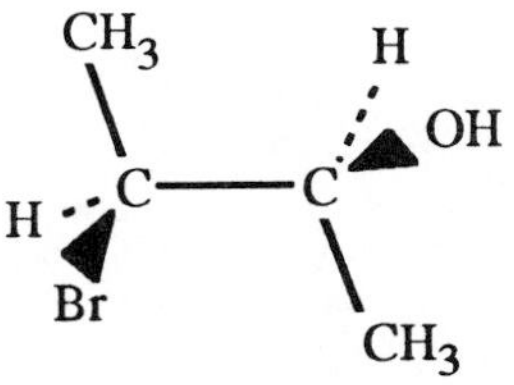

(2*S*, 3*S*)-3-bromo-2-butanol

These enantiomers are the diastereomers of the two enantiomers on the next page.

6.11 (cont)

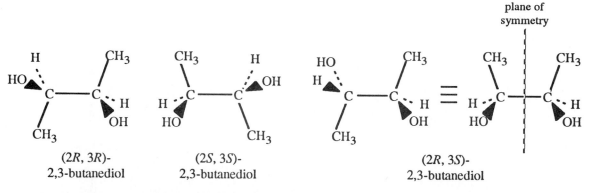

(2R, 3S)-3-bromo-2-butanol   (2S, 3R)-3-bromo-2-butanol

These enantiomers are the diastereomers of the two enantiomers on the previous page.

2-chloro-3-methylheptane

(2R, 3R)-2-chloro-3-methylheptane   (2S, 3S)-2-chloro-3-methylheptane

These enantiomers are the diastereomers of the two enantiomers below.

(2R, 3S)-2-chloro-3-methylheptane   (2S, 3R)-2-chloro-3-methylheptane

These enantiomers are the diastereomers of the two enantiomers above.

- -

Concept Map 6.6   (see p. 166)

- -

6.12   2,3-butanediol

plane of
symmetry

(2R, 3R)-   (2S, 3S)-   (2R, 3S)-
2,3-butanediol   2,3-butanediol   2,3-butanediol

These enantiomers are the diastereomers of the meso compound.   meso compound

6.12 (cont)

2,3-pentanediol

(2*R*, 3*R*)-2,3-pentanediol   (2*S*, 3*S*)-2,3-pentanediol

These enantiomers are the diastereomers of the two enantiomers below.

(2*R*, 3*S*)-2,3-pentanediol   (2*S*, 3*R*)-2,3-pentanediol

These enantiomers are the diastereomers of the two enantiomers above.

2,4-pentanediol

(2*R*, 4*R*)-2,4-pentanediol   (2*S*, 4*S*)-2,4-pentanediol   (2*R*, 4*S*)-2,4-pentanediol

These enantiomers are the diastereomers of the meso compound.   meso compound

6.13   2,3-dihydroxybutanedioic acid

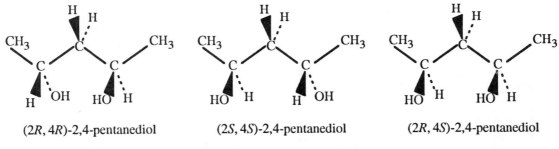

(2*R*, 3*R*)-2,3-dihydroxybutanedioic acid   (2*S*, 3*S*)-2,3-dihydroxybutanedioic acid
(+)-tartaric acid   (–)-tartaric acid

6.13 (cont)

These enantiomers on the previous page are the diastereomers of the meso compound.

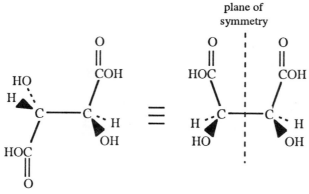

(2*R*, 3*S*)-2,3-dihydroxybutanedioic acid

meso-tartaric acid

Concept Map 6.6   Configurational isomers.

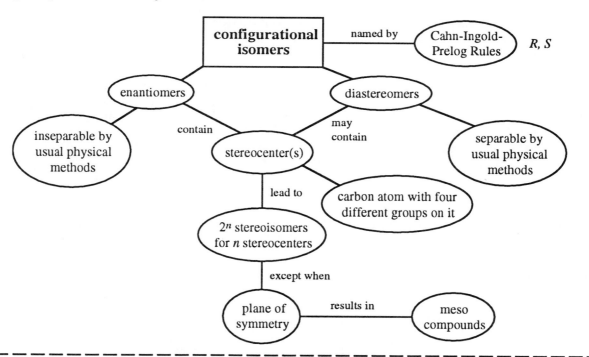

6.14   For a compound with *n* stereocenters, there can be a maximum of 2^n stereoisomers. 2,3,4-Tribromohexane has 3 stereocenters and thus a maximum of 8 stereoisomers. The number of different isomers will be reduced if there are meso compounds.

6.14 (cont)

(2*R*,3*R*,4*R*)-2,3,4-tribromohexane

(2*S*,3*S*,4*S*)-2,3,4-tribromohexane

These two compounds are enantiomers and are diastereomers of the other six stereoisomers.

(2*R*,3*S*,4*R*)-2,3,4-tribromohexane

(2*S*,3*R*,4*S*)-2,3,4-tribromohexane

These two compounds are enantiomers and are diastereomers of the other six stereoisomers.

(2*R*,3*R*,4*S*)-2,3,4-tribromohexane

(2*S*,3*S*,4*R*)-2,3,4-tribromohexane

These two compounds are enantiomers and are diastereomers of the other six stereoisomers.

(2*S*,3*R*,4*R*)-2,3,4-tribromohexane

(2*R*,3*S*,4*S*)-2,3,4-tribromohexane

These two compounds are enantiomers and are diastereomers of the other six stereoisomers.

6.15   (a)

(2*S*, 3*S*)-3-methoxy-2-methyl-4-phenyl-1-butanol

(b)

(2*R*, 3*S*)-3-methoxy-2-methyl-4-phenyl-1-butanol

This is a diastereomer of the original compound. The configuration at carbon 2 was changed by reversing the methyl group and the hydrogen atom.

We could also have 2*R*, 3*R*, the enantiomer of the original compound, and 2*S*, 3*R*, the enantiomer of the structure shown above.

6.16

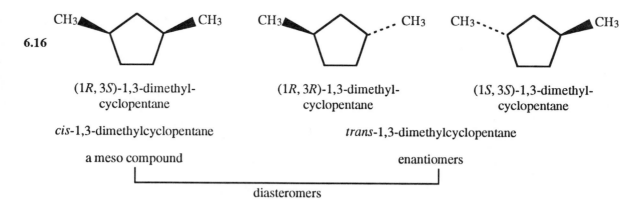

(1*R*, 3*S*)-1,3-dimethyl-cyclopentane	(1*R*, 3*R*)-1,3-dimethyl-cyclopentane	(1*S*, 3*S*)-1,3-dimethyl-cyclopentane
cis-1,3-dimethylcyclopentane	*trans*-1,3-dimethylcyclopentane	
a meso compound	enantiomers	

diasteromers

Note that in the meso compound the stereocenters have opposite configurations. Also, once an assignment of configuration is carefully made for one of the stereoisomers, the configurations of stereocenters can be assigned by looking to see if they are the same, or opposite, in configuration. For example, once the stereocenters in *cis*-1,3-dimethylcyclopentane have been assigned (see explanation that follows), the other two stereoisomers are easy. The first trans isomer has the same configuration (methyl up) at carbon 1 and the opposite configuration (methyl down) at carbon 3 as the cis isomer. Therefore, it is the (1*R*, 3*R*) isomer.

6.17

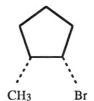
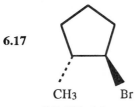

(1R, 2R)-1-bromo-
2-methylcyclopentane

(1S, 2S)-1-bromo-
2-methylcyclopentane

(1R, 2S)-1-bromo-
2-methylcyclopentane

(1S, 2R)-1-bromo-
2-methylcyclopentane

trans-1-bromo-2-methylcyclopentane

cis-1-bromo-2-methylcyclopentane

enantiomers   enantiomers

diastereomers

6.18

trans-1-bromo-2-methylcyclohexane   *cis*-1-bromo-2-methylcyclohexane

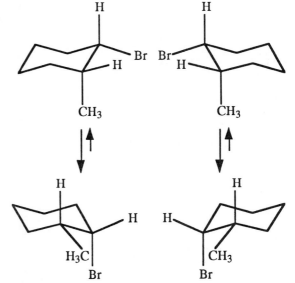

enantiomers   enantiomers

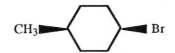

trans-1-bromo-4-methylcyclohexane   *cis*-1-bromo-4-methylcyclohexane

6.18 (cont)

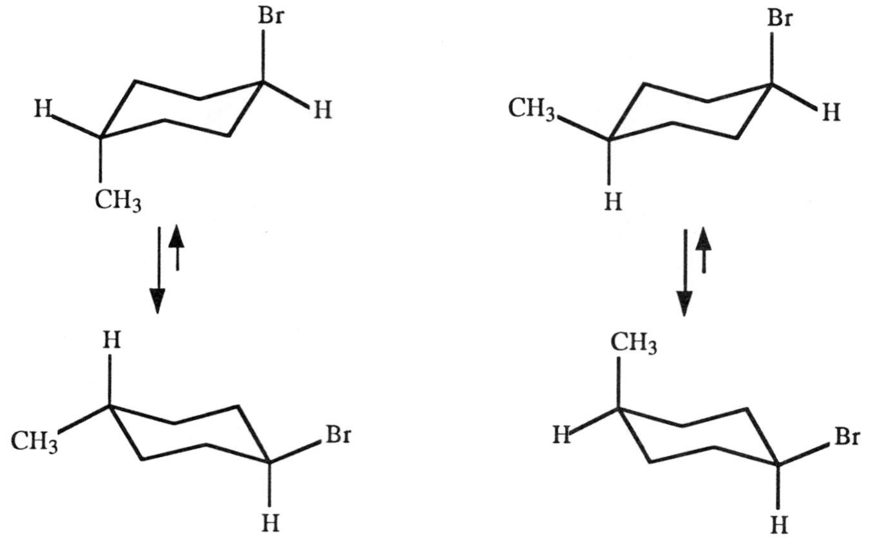

6.19

(a)

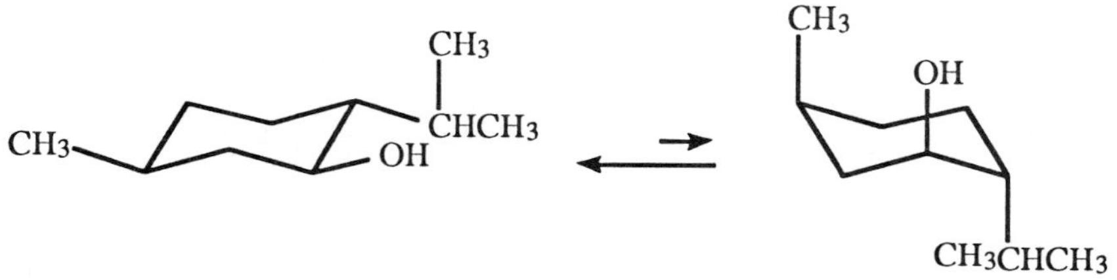

all substituents equatorial;
the most stable isomer of
2-isopropyl-5-methylcyclohexanol

all substituents axial

(b)

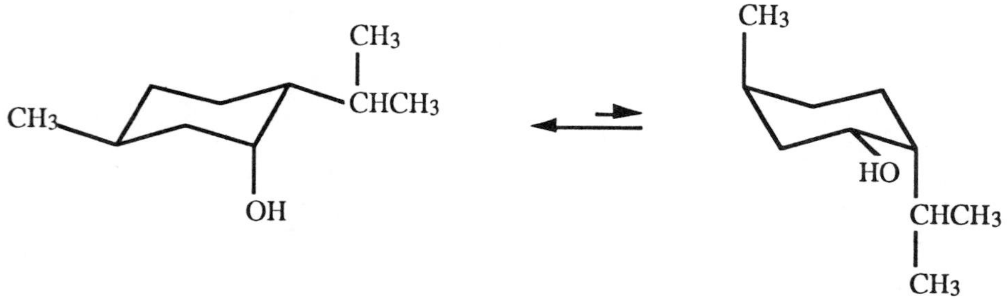

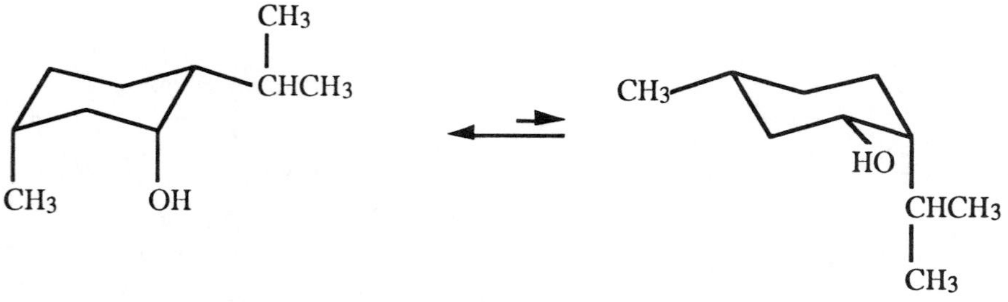

some other possible stereoisomers

Concept Map 6.7   Diastereomers.

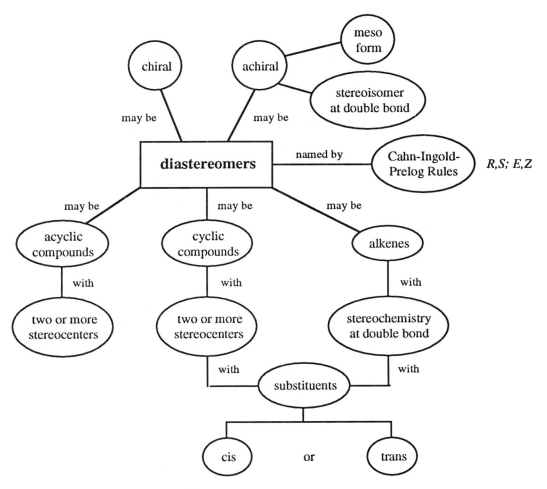

6.20

(E)-3-methyl-3-hexene

(Z)-1,3-dibromo-1-chloro-2-
methylpropene

(Z)-5-chloro-3-ethyl-
2-pentenoic acid

(Z)-3-ethyl-3-hexen-1-yne

Concept Map 6.8   The process of resolution.

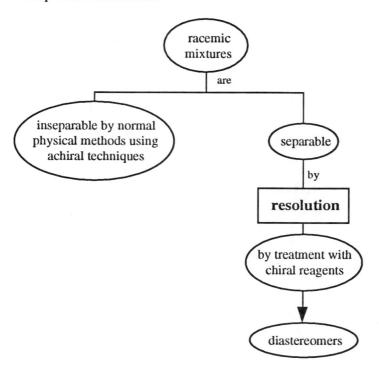

6.21

water-soluble salt

water-insoluble amine   water-soluble salt

6.22

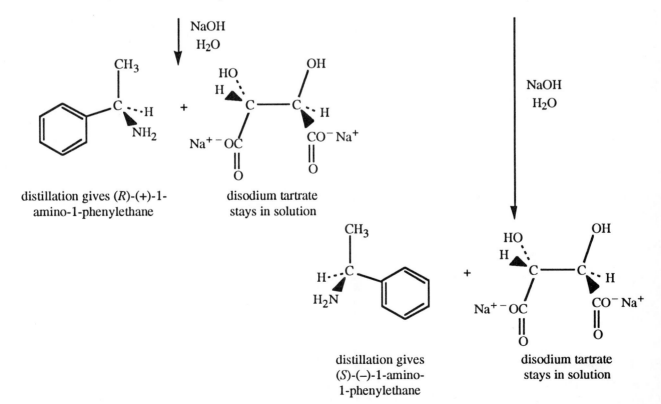

(R)-(+)-1-amino-1-phenylethane   (S)-(–)-1-amino-1-phenylethane   (2R,3R)-(+)-tartaric acid

methanol

salt of (R)-(+)-1-amino-1-phenylethane
with (2R,3R)-(+)-tartaric acid; more
soluble in methanol; stays in solution

salt of (S)-(–)-1-amino-1-phenylethane
with (2R,3R)-(+)-tartaric acid; less
soluble in methanol; crystallizes out

NaOH
H₂O

NaOH
H₂O

distillation gives (R)-(+)-1-
amino-1-phenylethane

disodium tartrate
stays in solution

distillation gives
(S)-(–)-1-amino-
1-phenylethane

disodium tartrate
stays in solution

6.23   (a) (S)-3-bromo-2-methylpentane   (b) (S)-1-phenyl-1-butanol

(c) butane   (d) (S)-1,2-dibromopentane

(e) (R)-3-methyl-3-hexanol   (f) (S)-3-methyl-3-hexanol

(g) (2R, 3S)-2,3-dibromobutane   (h) (1S, 2S, 4S)-1,2-dibromo-4-methylcyclohexane
     meso-2,3-dibromobutane

(i) (1R, 2R)-2-bromocyclopentanol   (j) (2R, 3S)-2,3-dichloropentane

The assignment of configuration for (h) is easiest if the chair form is converted to the planar form using wedges and dashed lines to indicate the stereochemistry of the substituents. Assignment of configuration at carbon-2, for example, is shown below.

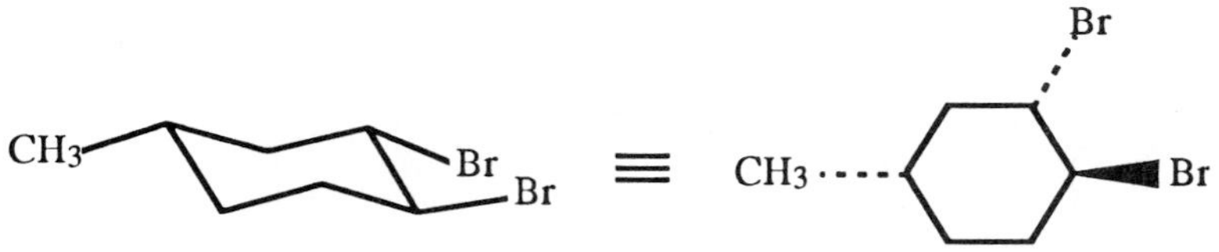

(1S, 2S, 4S)-1,2-dibromo-4-methylcyclohexane

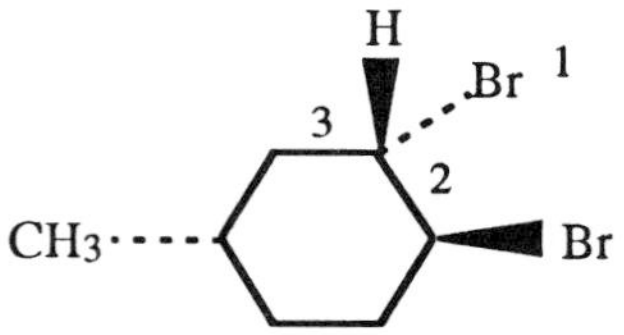

The bromine is pointing down, therefore the group of lowest priority, the hydrogen, is pointing up. Our eye travels clockwise from priority 1 to priority 2 to priority 3, therefore the stereocenter *looks R*, but is really *S* because we are looking at it from the wrong face of the molecule.

6.24   b, c, e, h, i, and k have stereocenters.

(b)

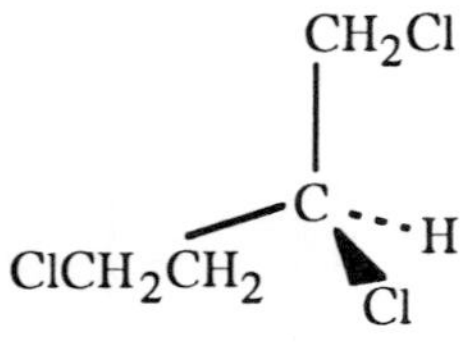

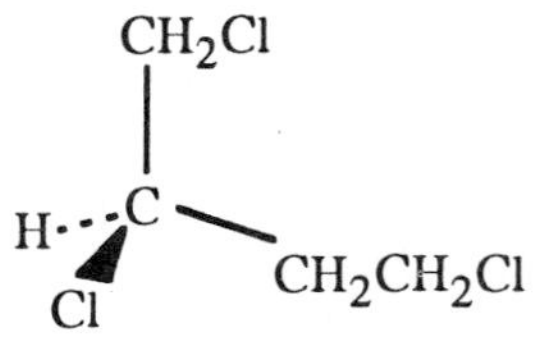

(c)

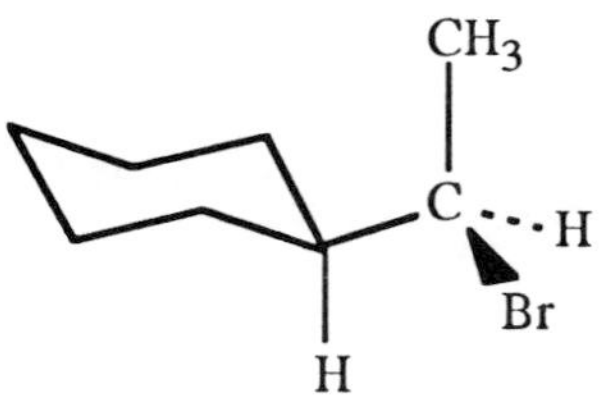

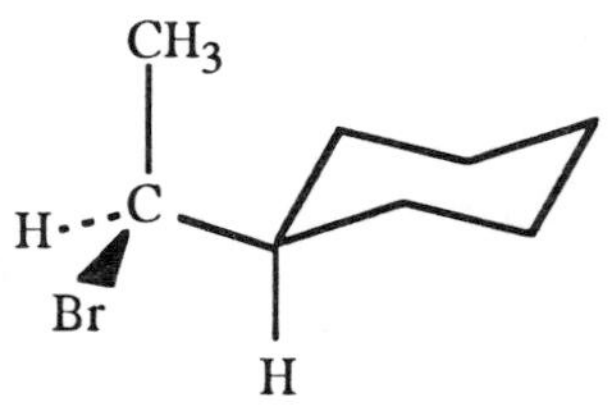

(e)

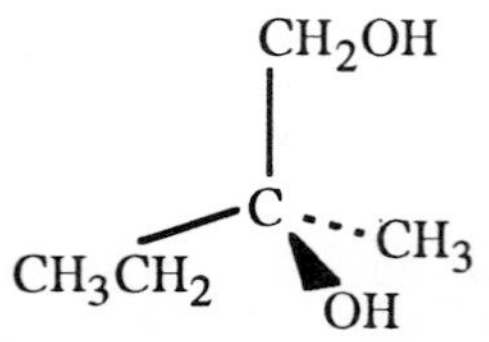

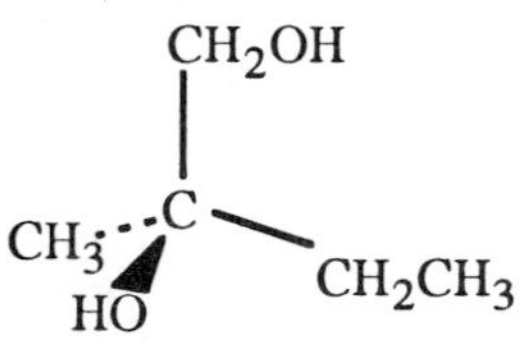

6.24 (cont)

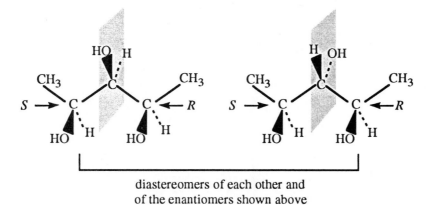

(h)

(i)

(k)

enantiomers

The central atom in each of these enantiomers is not a stereocenter because two of the substituents on the carbon are identical (including their configurations).

diastereomers of each other and
of the enantiomers shown above

The central carbon atom in each of these meso forms is a stereocenter but because of the symmetry of the molecule, the resulting compounds are not optically active. The two groups containing stereocenters that are attached to the center carbon atom appear identical, but they have opposite configurations and are therefore not identical.

6.25   (a) enantiomers   (The molecule on the left has the *S* configuration, the one on the right, the *R* configuration.)

(b) identical

(c) identical

(d) identical

(e) identical

(f) enantiomers   (The molecule on the left has the *R* configuration; the other, the *S* configuration.)

(g) enantiomers   (The molecule on the left has the *R* configuration; the other, the *S* configuration. Note also that the molecule on the right is not the mirror image of the molecule on the left but is a conformer of the mirror image.)

(h) constitutional isomers   [(*R*)-3-bromo-1-chlorobutane and (2*R*,3*S*)-2-bromo-3-chlorobutane]

(i) identical   (Both have the *R* configuration.)

(j) conformers

(k) constitutional isomers   [(*R*)-2-bromo-1-chlorobutane and (*S*)-3-bromo-1-chlorobutane]

(l) conformers   (The chlorine atom is equatorial in the structure on the left and axial in the one on the right.)

6.26

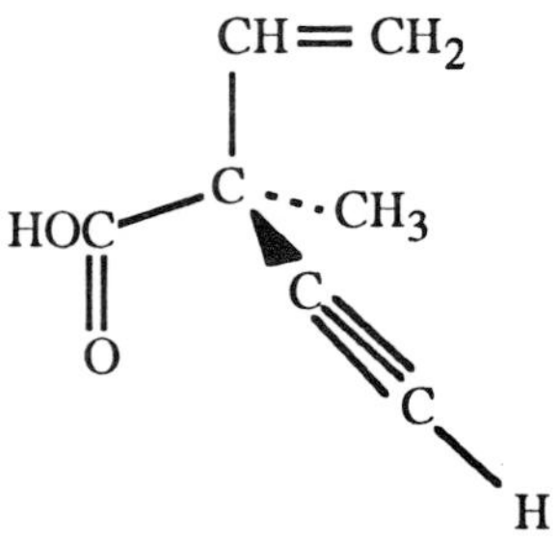

6.27

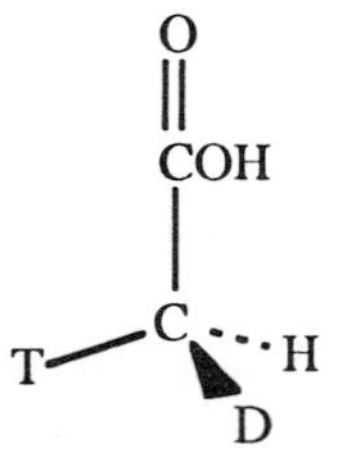

(*S*)-isomer 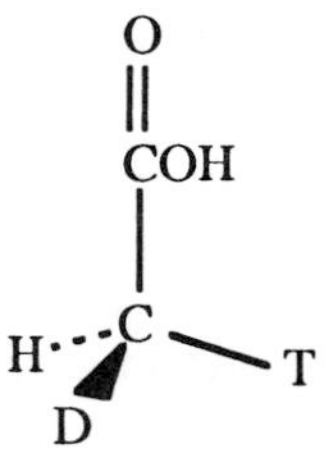   (*R*)-isomer

6.28 $\text{enantiomeric excess} = \dfrac{\text{actual rotation}}{\text{rotation of pure enantiomer}} \times 100$

$88 = \dfrac{-4.1°}{\text{rotation of pure enantiomer}} \times 100$

rotation of pure enantiomer $= -4.7°$

(*S*)-(–)-citronellol

6.29   (a)

(*R*)-(–)-1-amino-2-propanol	(*S*)-(+)-1-amino-2-propanol
hydrochloride	hydrochloride

(b)   The *S*-enantiomer was recovered with higher optical purity (higher absolute specific rotation).

(c) $\dfrac{\text{enantiomeric excess}}{\text{for the } R\text{-isomer}} = \dfrac{-31.5°}{-35°} \times 100 = 90\%$

6.30 $\alpha = [\alpha]_D^{20°} \; lc$

$\alpha = (-92°)(0.5 \text{ dm})(1 \text{ g}/100 \text{ mL})$

$\alpha = -0.46°$

6.31 $\text{enantiomeric excess} = \dfrac{\text{actual rotation}}{\text{rotation of pure enantiomer}} \times 100$

$[\alpha]_D^{20°}$ for pure (+) enantiomer $= +13.90°$, therefore $[\alpha]_D^{20°}$ for pure (–) enantiomer $= -13.90°$.

$\text{enantiomeric excess} = \dfrac{-3.5°}{-13.90°} \times 100 = 25\%$

The mixture contains a 25% enantiomeric excess of the levorotatory compound and is 100 – 25 or 75% racemic. The racemic portion of the mixture contains equal amounts (37.5%) of each enantiomer. Therefore the solution contains 37.5% of (+)-2-butanol and (37.5 + 25)% or 62.5% of (–)-2-butanol.

6.32

CH₃
CH₃NCH₂
OH
CH₃—C—C—CH₂—⟨phenyl⟩
H
⟨phenyl⟩

(2S, 3R)-4-dimethylamino-1,2-
diphenyl-3-methyl-2-butanol

medically useful form

CH₃
HO
CH₃
⟨phenyl⟩—CH₂—C—C—NCH₃
CH₂
CH₃
H
⟨phenyl⟩

(2R, 3S)-4-dimethylamino-1,2-
diphenyl-3-methyl-2-butanol

CH₃
CH₃NCH₂
OH
H—C—C—CH₂—⟨phenyl⟩
CH₃
⟨phenyl⟩

(2S, 3S)-4-dimethylamino-1,2-
diphenyl-3-methyl-2-butanol

CH₃
HO
CH₃
⟨phenyl⟩—CH₂—C—C—NCH₃
CH₂
H
CH₃
⟨phenyl⟩

(2R, 3R)-4-dimethylamino-1,2-
diphenyl-3-methyl-2-butanol

6.33

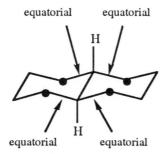

equatorial   equatorial
H

equatorial   H   equatorial

In *trans*-decalin, both of the methylene
groups on the ring junction are
equatorial in each case; more stable

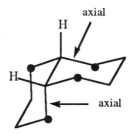

axial
H

H

axial

In *cis*-decalin, one of the methylene
groups on the ring junction is
axial in each case; less stable

6.34

(a)

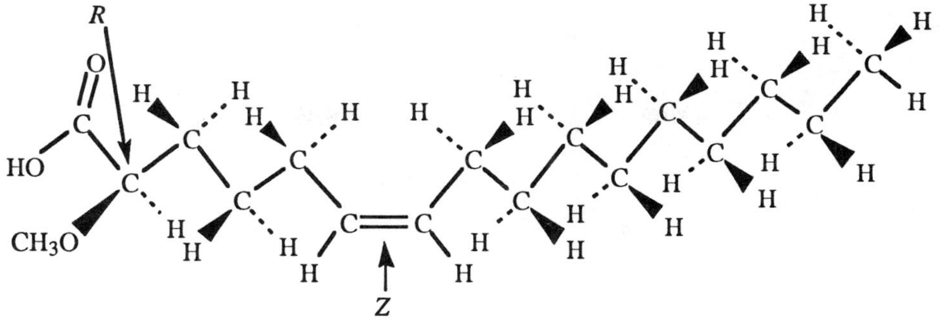

(b)   (*R*)-(*Z*)-2-methoxy-6-hexadecenoic acid

(c)

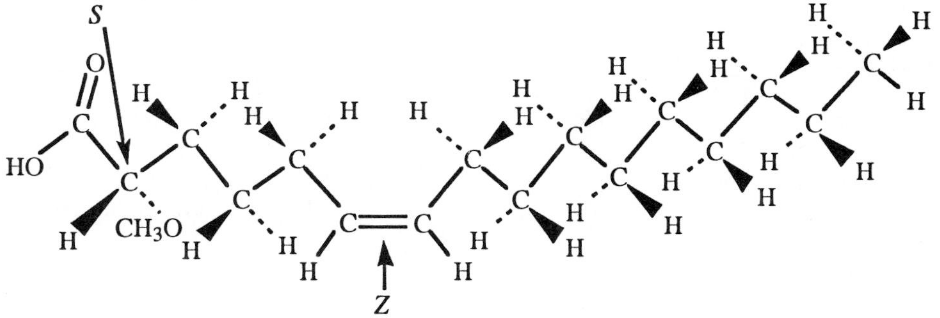

(*S*)-(*Z*)-2-methoxy-6-hexadecenoic acid

This is the enantiomer of the compound shown in (a). The *R*, *E* compound and the *S*, *E* compound would be diastereomers of the compound shown in (a).

6.35

(a)

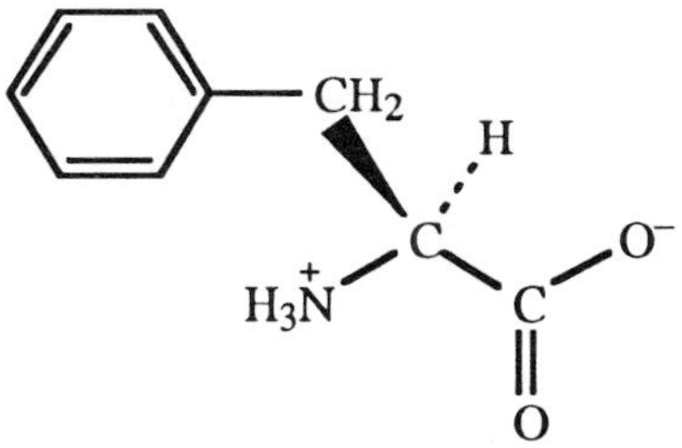

(*S*)-phenylalanine

(b)   A racemic mixture of phenylalanine would not be optically active. Therefore the optical rotation would be zero.

(c)   The sign of rotation cannot be predicted for (*S*)-phenylalanine from its structural formula, nor from its *S* configuration. It will, however, be optically active.

(d)

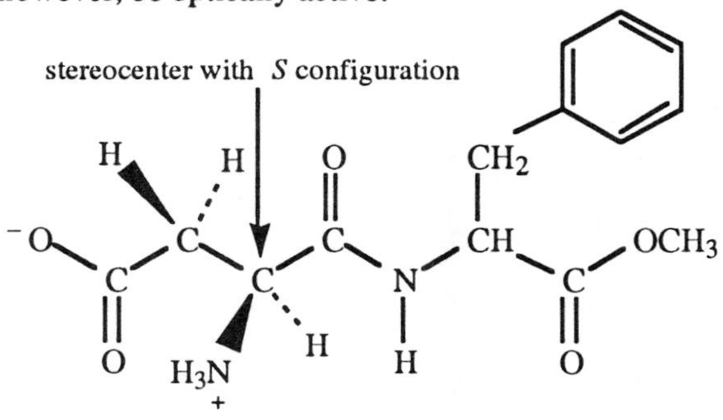

6.36   (a)

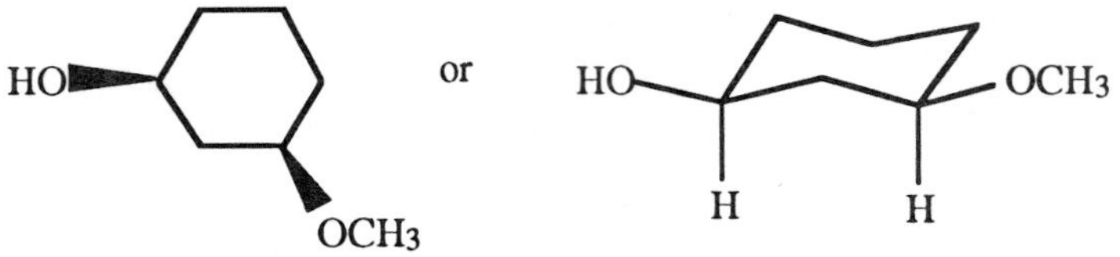

(b)

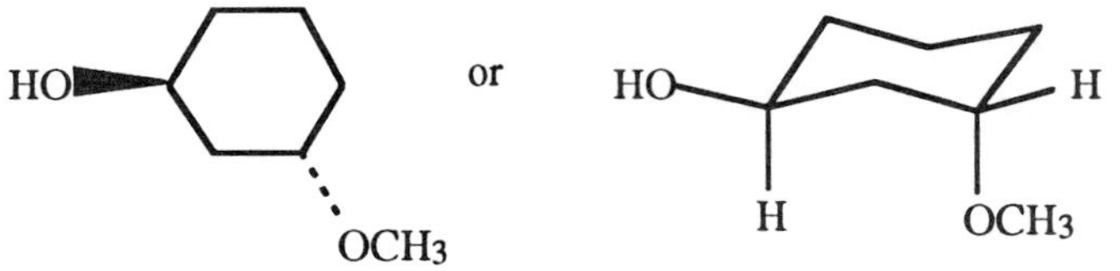

(c)

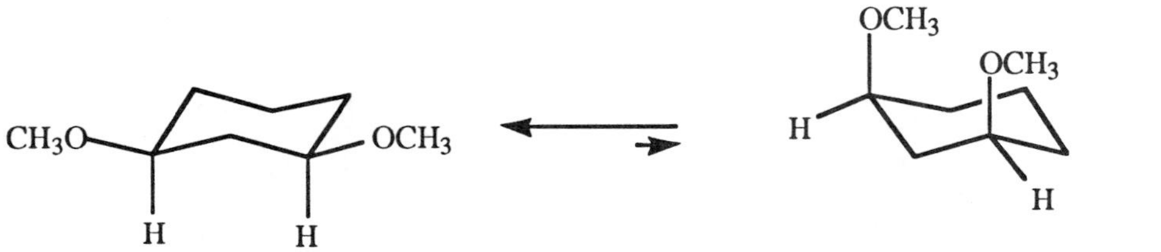

diequatorial conformer of *cis*-
1,3-dimethoxycyclohexane;
most stable conformer of the
four shown

diaxial conformer of *cis*-
1,3-dimethoxycyclohexane

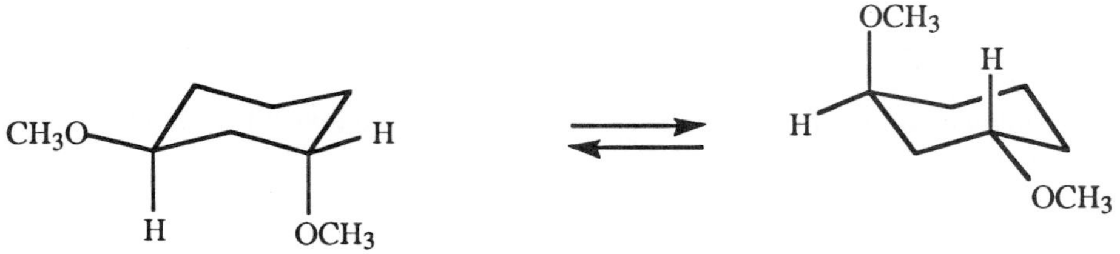

both conformers of *trans*-1,3-dimethoxycyclohexane
have one axial and one equatorial methoxy group

6.37   (a)

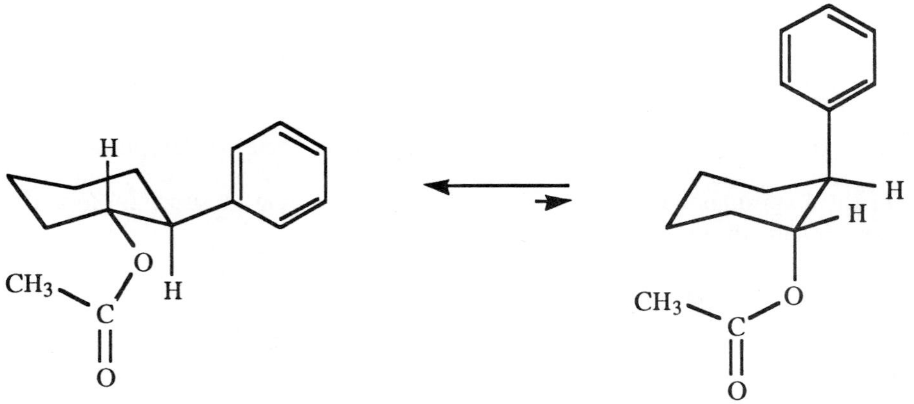

more stable conformer

chair conformations of Compound **A**

(b)   Compound **B** and Compound **C** are diastereomers.

6.37 (cont)

(c)

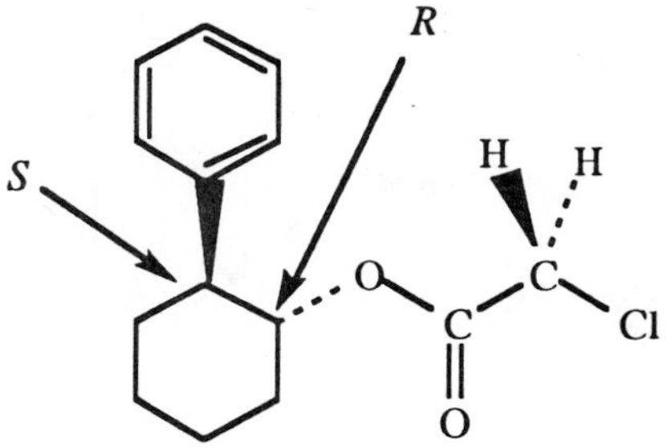

(d)

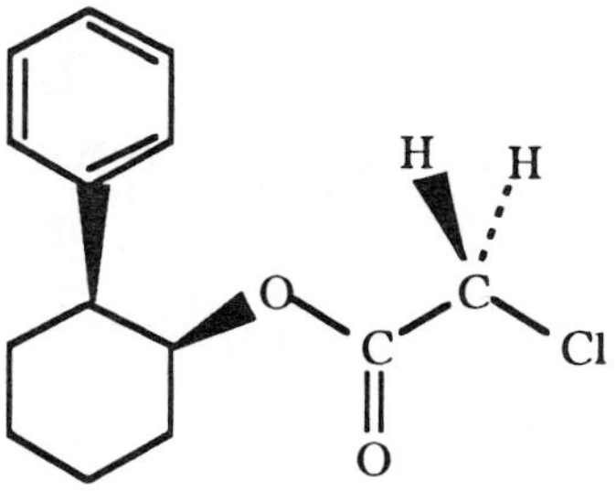

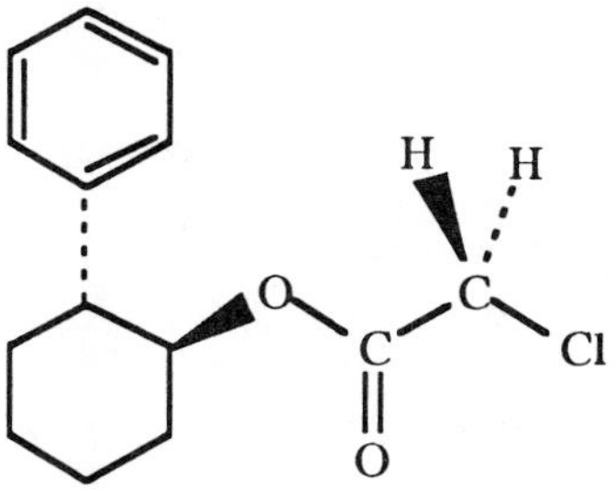

the *S, S* stereoisomer, which
is a diastereomer of Compound **A**.
The *R, R* diastereomer, the enantiomer
of the compound above, is also a
stereoisomer of Compound **A**.

the enantiomer of Compound **A**

6.38

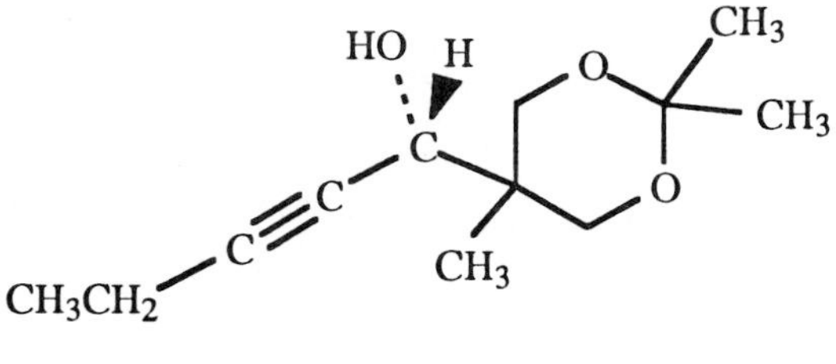

or

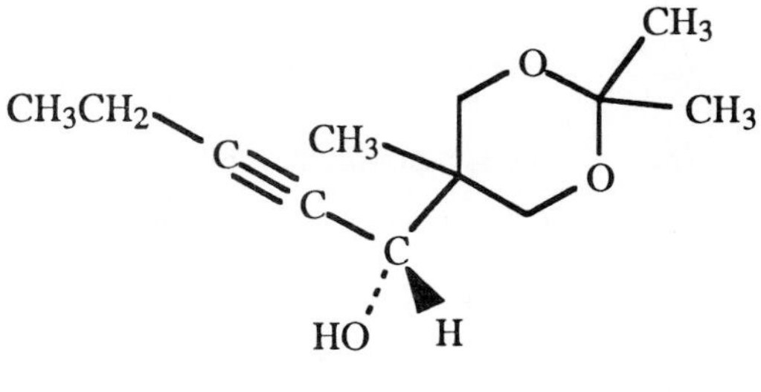

6.39

(a)

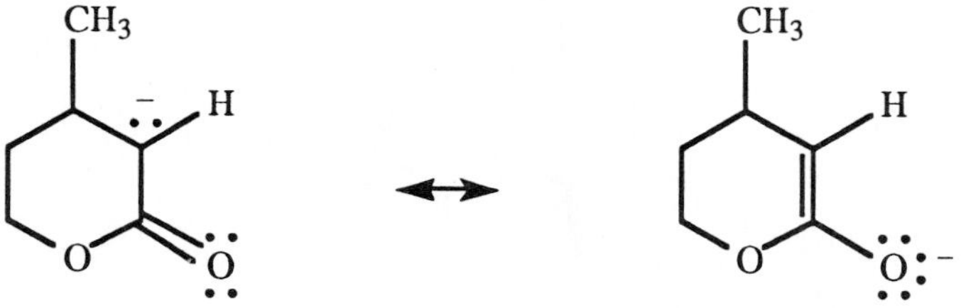

conjugate base of 3-methylpentanolide
only one stabilized by resonance

6.39 (cont)

(b)

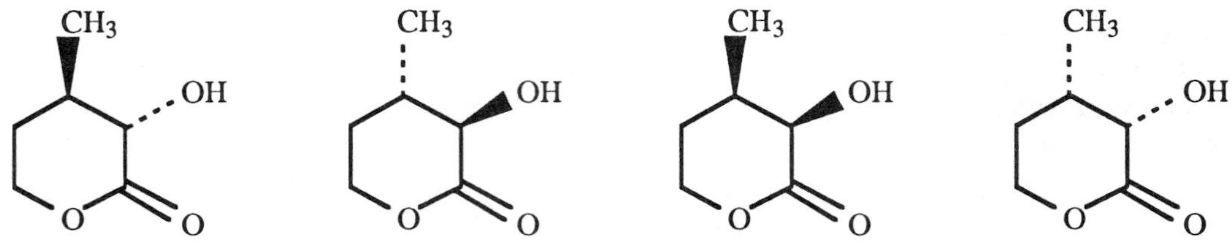

(2S, 3R)-(–)-2-
hydroxy-3-methyl-
pentanolide

6.40

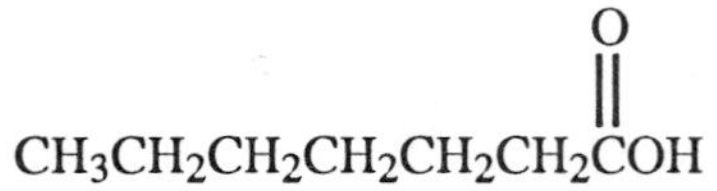

$$CH_3CH_2CH_2CH_2CH_2CH_2COH$$

heptanoic acid

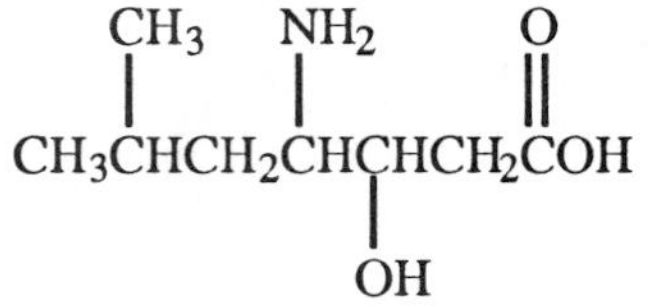

$$CH_3CHCH_2CHCHCH_2COH$$

4-amino-3-hydroxy-6-methylheptanoic acid

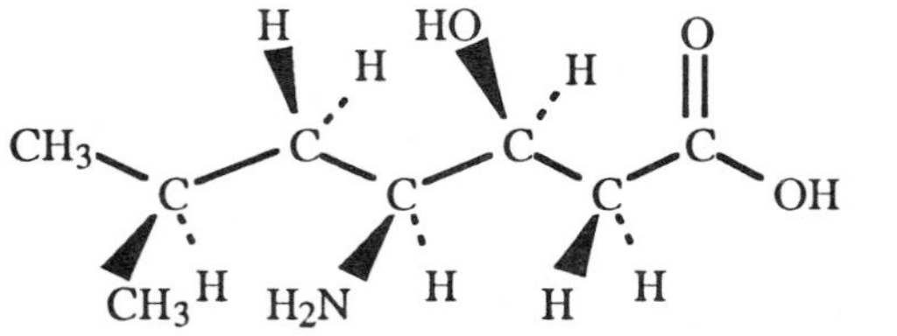

(3S, 4S)-4-amino-3-hydroxy-6-methylheptanoic acid
statine

6.41   (a)   It has 5 stereocenters.

(b)   It has a double bond.

(c)

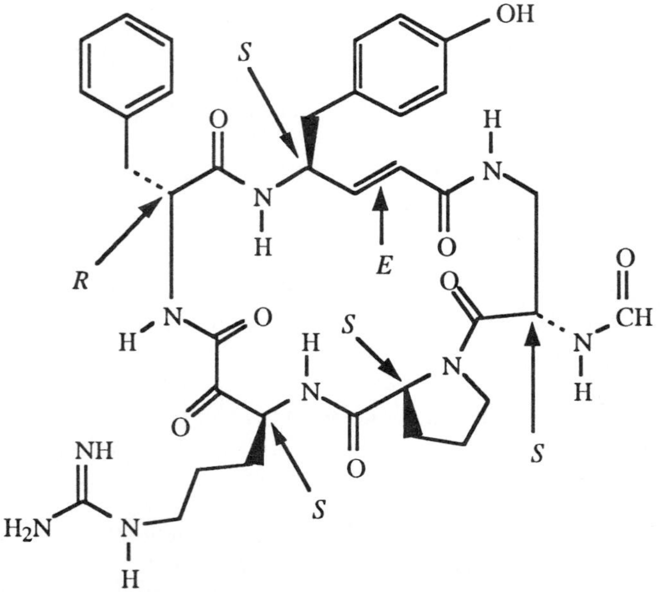

7

Nucleophilic Substitution and Elimination Reactions

Workbook Exercises

The structures in this text are comprised of single (σ), double ($\sigma + \pi$) and/or triple ($\sigma + 2\pi$) bonds. It is not surprising, then, that chemical reactions involve changes in σ and/or π bonding. In this chapter, you will be introduced to two important ways in which certain σ bonds can undergo changes. Many σ bonds to carbon are polar, or they are weak, meaning that they can be induced to break relatively easily. The ones shown below usually break heterolytically, in the same way that hydrogen chloride reacts with a base:

$$H_3N: \quad H \quad Cl: \longrightarrow H_3\overset{+}{N}—H \quad :Cl:^-$$

base

Table 1   Examples of the direction of ionization of polar or weak σ bonds between carbon and elements of groups 15, 16, and 17 in the periodic table.

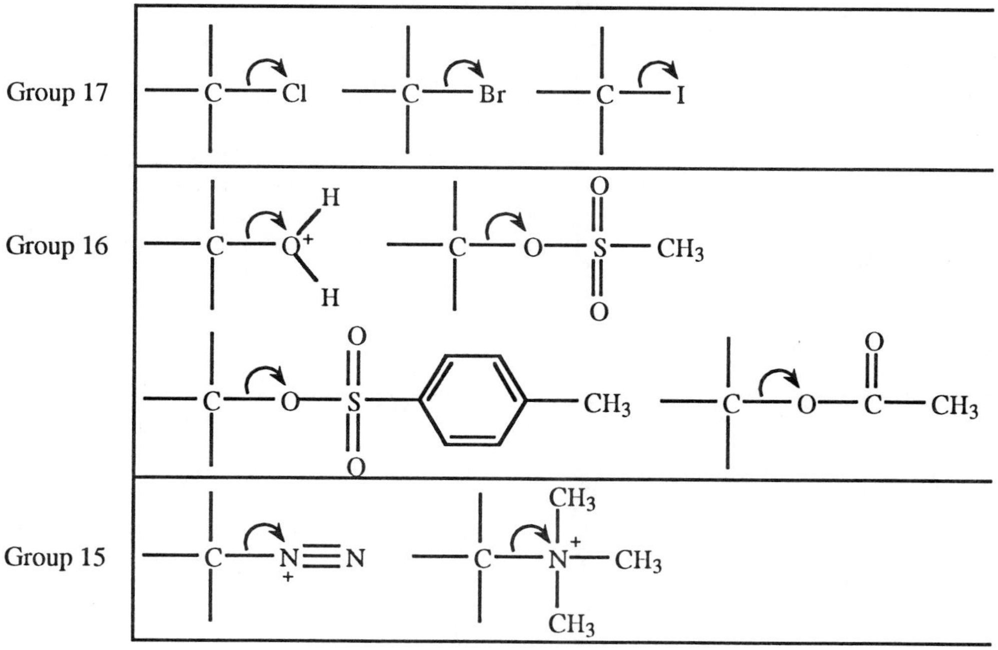

The carbon atoms shown in the table above are all losing a pair of electrons to the group, called the leaving group, that leaves when the bond breaks. In order to maintain or regain the closed shell configuration at these atoms, a new bond derived from an incoming electron pair is required. Two sources of such electron pairs are possible: (1) A nonbonding electron pair of some atom forms a new σ bond in a substitution reaction, and (2) the electrons from an adjacent C—H bond, which is deprotonated, gives a new π bond in an elimination reaction.

Workbook Exercises (cont)

For example, when trimethylamine is combined with 2-chloropentane, a mixture of products is observed:

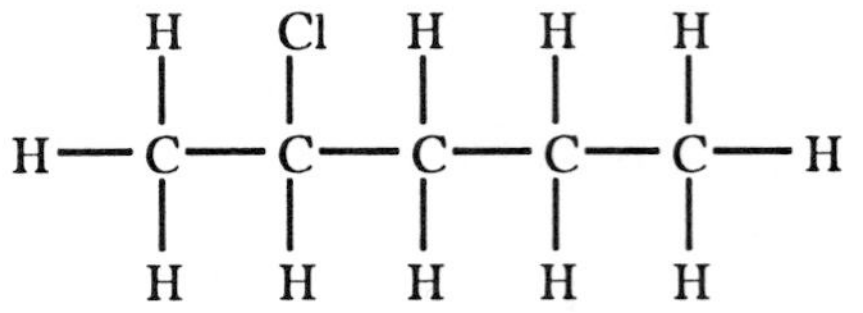

2-chloropentane
(the C—Cl bond can break
to form a chloride ion)

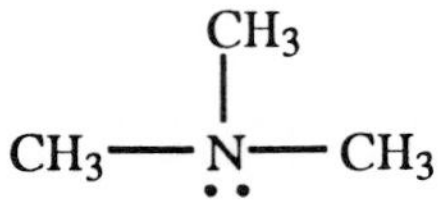

trimethylamine
(the N atom electron pair can
be used to (1) form a new C—N
bond or (2) deprotonate a C—H
adjacent to the C—Cl)

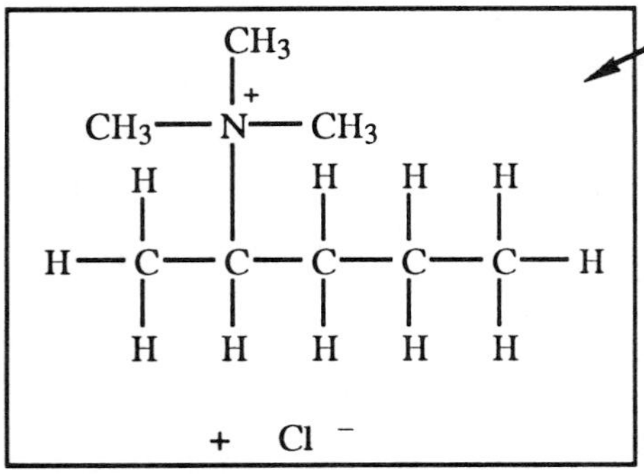

the connectivity resulting from **substitution**
(because the carbon atom attached to the
nitrogen atom is a stereocenter,
stereoisomerism is possible)

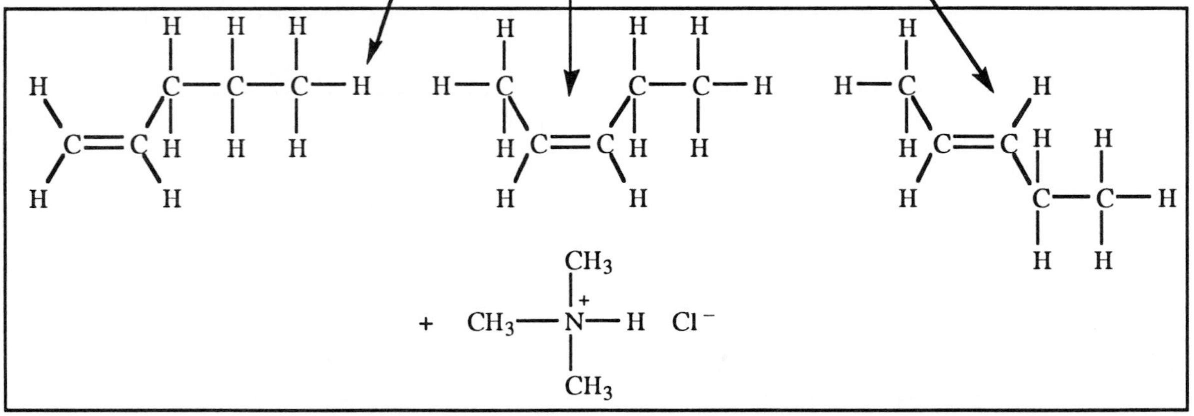

the connectivity resulting from **elimination**
(loss of the chloride ion as well as the proton from an adjacent
C—H bond to give a π bond. Sometimes stereoisomers are possible.)

Workbook Exercises (cont)

EXERCISE I.   Predict the connectivities of the possible substitution and elimination product(s) derived from the following chemical reactions. Think about the possibility of the formation of stereoisomers and show them when appropriate. Use Table 1 to help you decide which atoms or groups can be lost easily.

(a)

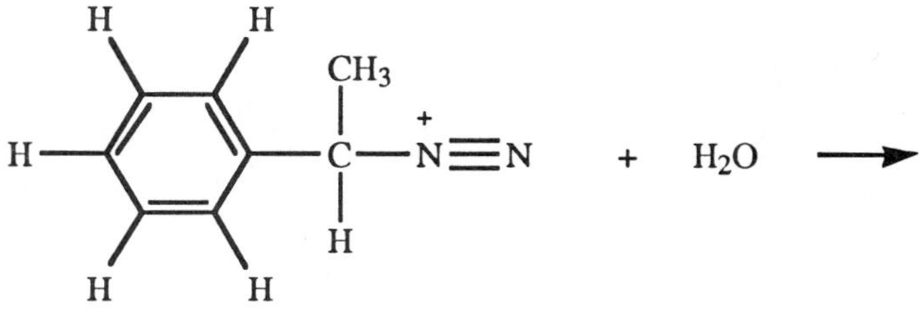

(b)

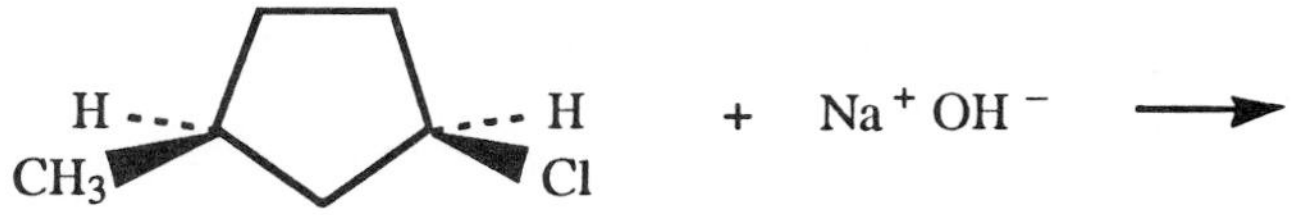

(c)

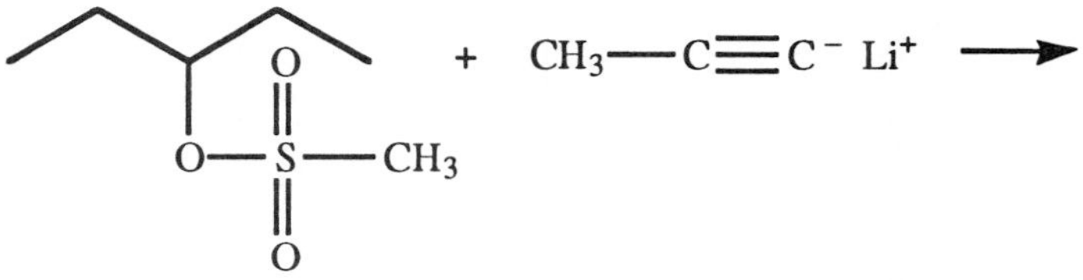

(d)

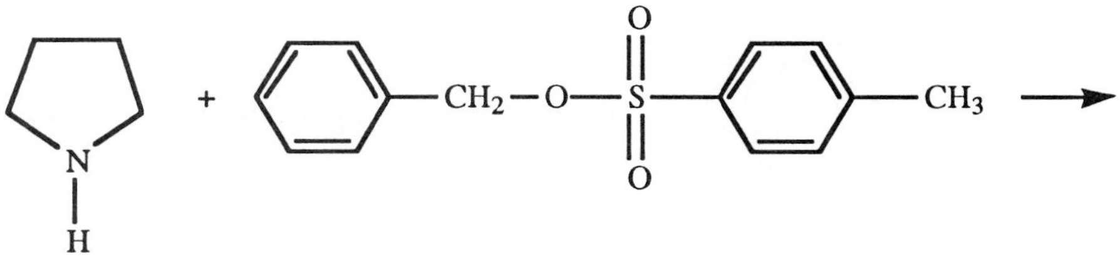

(e)

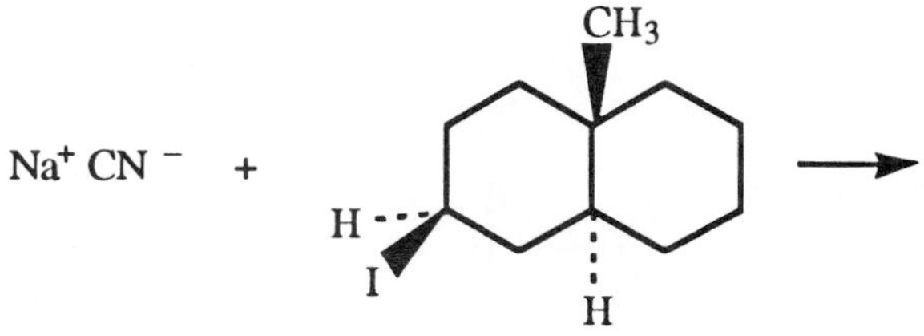

(f)

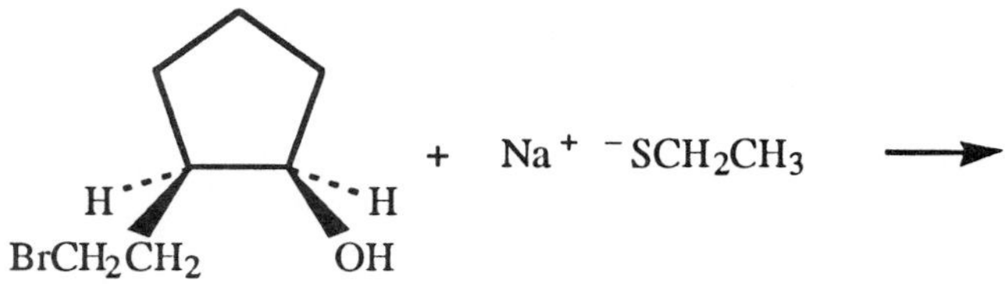

Workbook Exercises (cont)

EXERCISE II.   Complete the following chemical equations based on the information provided. There may be more than one solution to each problem.

(a)   A   + Na^+ $^-OCH_2CH_3$ $\longrightarrow$ [structure: $(CH_3)(CH_3)C=C(H)(H)$]   + Na^+ Cl^- + CH_3CH_2OH

(b)   [structure: allyl–$SiCl_3$]   +   3 CH_3^- Li^+ $\longrightarrow$ B   +   3 Li^+ Cl^-

(c)   C   + $CH_3CH_2CH_2I$ $\longrightarrow$ $CH_3-\overset{O}{\overset{\|}{C}}-\underset{\underset{CH_2CH_2CH_3}{|}}{CH}-\overset{O}{\overset{\|}{C}}-OCH_3$ + Na^+ I^-

(d) Li^+ $^-SCH_3$ +   D $\longrightarrow$ [cyclohexene with SCH₃ substituent]   + Li^+ $^-O-\overset{O}{\underset{O}{\overset{\|}{\underset{\|}{S}}}}-$[benzene ring]$-CH_3$

         $\left(\ +\ \text{[1,3-cyclohexadiene]}\ +\ HSCH_3\ +\ Li^+\,^-O-\overset{O}{\underset{O}{\overset{\|}{\underset{\|}{S}}}}-\text{[benzene ring]}-CH_3\ \right)$

(e)   [structure: $CH_3O-C(=O)-C(H)(CH_2Ph)-N\overset{+}{\equiv}N$]   +   E $\longrightarrow$ [structure: $CH_3O-C(=O)-C(H)(CH_2Ph)-\overset{+}{O}H_2$]   + $N\equiv N$

         +   F (a stereoisomer of the substitution product)   +   G and H (2 elimination products)

(f)   [bicyclic decalin structure with Br—O, CH₃, H substituents]   + Na^+ OH^- $\longrightarrow$ I   + H_2O + Na^+ Br^-

         (Hint: This is an elimination reaction in which bromide ion is the leaving group.)

7.1   (a)

(b)

first product;
acid with $pK_a \sim 10$

is a base that has a conjugate acid with $pK_a \sim 10$. In the presence of excess base, the following

equilibrium will be established.

| base | acid | | conjugate acid | conjugate base |
| in large excess | $pK_a \sim 10$ | | $pK_a \sim 10$ | |

Because the two acids have roughly the same acidity, the relative amounts of the different species at equilibrium will be determined by the relative concentrations of the different reagents. Because there is a large excess of the amine used as a starting material in the reaction, the proton will spend most of its time with that amine and the product amine will be mostly in the unprotonated form.

(c)

7.2   The answers to part (b) analyze the problem as shown in the Problem Solving Skills section of the chapter. The rest of the problems show only the equations, written so as to emphasize the necessity of reasoning backwards from the structure of the desired product to the structure of the starting material.

(a)

7.2 (cont)

(b) $CH_3CH_2CH_2CH_2CH_2Br \longrightarrow CH_3CH_2CH_2CH_2CH_2CH(\overset{\overset{\displaystyle O}{\displaystyle \|}}{C}OCH_2CH_3)_2$

1. What are the connectivities of the two compounds? How many carbon atoms does each contain? Are there any rings? What are the positions of branches and functional groups on the chains?

    The starting material has five carbon atoms in a row bonded to a bromide atom. In the product the bromine has been replaced by a carbon atom bearing two ester groups.

2. How do the functional groups change in going from starting material to product? Does the starting material have a good leaving group?

    The bromine atom in the starting material is a good leaving group. The alkyl halide function in the starting material has been converted to an extra carbon with two ester functions on it.

3. Is it possible to dissect the structures of the starting material and product to see which bonds must have been broken and which formed?

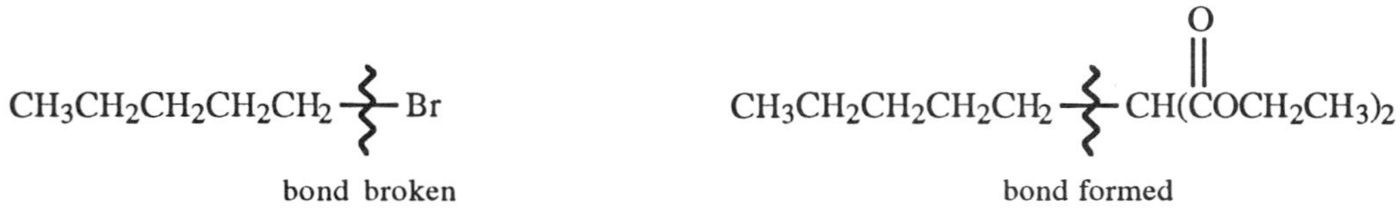

    $CH_3CH_2CH_2CH_2CH_2 \overset{\displaystyle \}}{\underset{\displaystyle \}}{|}} Br$ $CH_3CH_2CH_2CH_2CH_2 \overset{\displaystyle \}}{\underset{\displaystyle \}}{|}} CH(\overset{\overset{\displaystyle O}{\displaystyle \|}}{C}OCH_2CH_3)_2$

    bond broken   bond formed

4. New bonds are created when an electrophile reacts with a nucleophile. Do we recognize any part of the product molecule as coming from a good nucleophile or an electrophilic addition?

    Because the bromine atom is a good leaving group, and the carbon to which it is bonded is an electrophile, we suppose that the new fragment that bonds to it is a nucleophile. The pieces we need are:

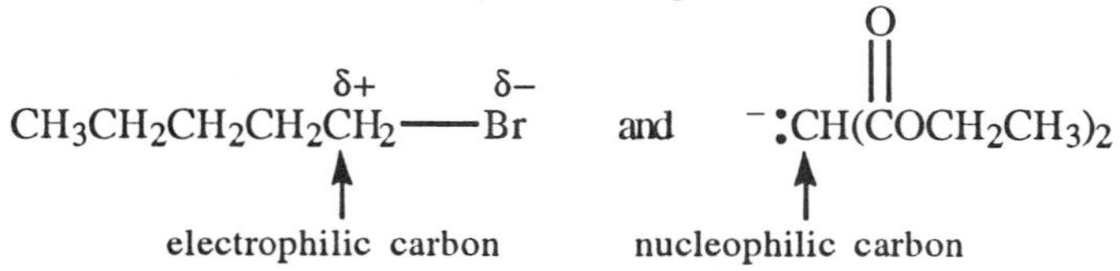

    $CH_3CH_2CH_2CH_2\overset{\delta+}{C}H_2 \overset{\delta-}{\text{——}} Br$ and $^-{:}CH(\overset{\overset{\displaystyle O}{\displaystyle \|}}{C}OCH_2CH_3)_2$

         electrophilic carbon   nucleophilic carbon

    We find the nucleophile in Table 7.1.

5. What type of compound would be a good precursor to the product?

    Displacement of the bromide ion by an incoming nucleophilic carbon atom will give us the product we want. We need a nucleophile with the structure shown above.

6. After this step, do we see how to get from starting material to product? If not, do we need to analyze the structure obtained in step 5 by applying questions 4 and 5 to it?

    We can now complete the synthesis.

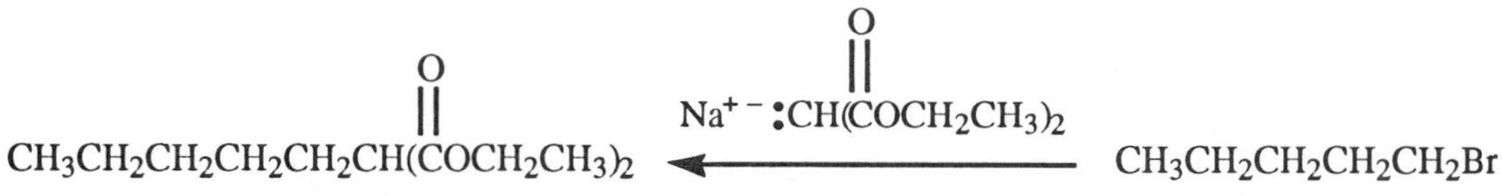

    $CH_3CH_2CH_2CH_2CH_2CH(\overset{\overset{\displaystyle O}{\displaystyle \|}}{C}OCH_2CH_3)_2 \xleftarrow{\quad Na^{+\,-}{:}CH(\overset{\overset{\displaystyle O}{\displaystyle \|}}{C}OCH_2CH_3)_2 \quad} CH_3CH_2CH_2CH_2CH_2Br$

7.2 (cont)

(c)

7.3 The initial rate of the reaction will be

rate $= k\,[CH_3CH_2Br]\,[OH^-]$

rate $= (1.7 \times 10^{-3}\,L/mol\cdot s)(0.05\,mol/L)(0.07\,mol/L)$

rate $= 6 \times 10^{-6}\,mol/L\cdot s$

If the concentration of bromoethane is doubled (increased from 0.05 mol/L to 0.1 mol/L), the rate will double. The new rate would be $\sim 1 \times 10^{-5}\,mol/L\cdot s$.

7.4 For the reaction

$$CH_3CH_2CH_2CH_2Br \;+\; K^+\,I^- \xrightarrow[\text{acetone}]{} CH_3CH_2CH_2CH_2I \;+\; K^+\,Br^- \downarrow$$

The rate at which *n*-butyl bromide is used up is:

rate $= k\,[CH_3CH_2CH_2CH_2Br]\,[I^-]$

rate $= (1.09 \times 10^{-1}\,L/mol\cdot min)(0.06\,mol/L)(0.02\,mol/L)$

rate $= 1.3 \times 10^{-4}\,mol/L\cdot min$

7.5 For the reaction

rate $= k\,[tert\text{-butyl chloride}]$

rate $= (0.145/h)(0.0824\,mol/L)$

rate $= 1.19 \times 10^{-2}\,mol/L\cdot h$

The units in which this rate is expressed are mol/L·h. The rate constant will remain the same, but the rate of the reaction will decrease as the *tert*-butyl chloride is used up in the reaction. When half of the *tert*-butyl chloride is gone, the rate of the reaction will be half of what it was initially or $5.95 \times 10^{-3}\,mol/L\cdot h$.

7.6   The following species can act as bases (electron donors).

$$CH_3CH_2\!\!-\!\!\ddot{\underset{\cdot\cdot}{O}}\!\!-\!\!H \qquad CH_3\underset{\underset{CH_3}{|}}{\overset{\overset{CH_3}{|}}{C}}\!\!-\!\!\ddot{\underset{\cdot\cdot}{O}}\!\!-\!\!CH_2CH_3 \qquad CH_3\underset{\underset{CH_3}{|}}{\overset{\overset{CH_3}{|}}{C}}\!\!-\!\!\ddot{\underset{\cdot\cdot}{O}}\!\!-\!\!H$$

$$CH_3\overset{\overset{CH_3}{|}}{C}\!\!=\!\!CH_2 \qquad :\overset{\cdot\cdot}{\underset{\cdot\cdot}{Cl}}:^- \qquad H_2\overset{\cdot\cdot}{\underset{\cdot\cdot}{O}} \qquad ^-:\overset{\cdot\cdot}{\underset{\cdot\cdot}{O}}H$$

7.7   (S)-(+)-2-Bromobutane will racemize by undergoing reversible bimolecular nucleophilic substitution reactions with inversion of configuration. The optical rotation will gradually decrease to 0° as the system reaches equilibrium.

7.8   The fact that the rate of racemization is twice the rate of substitution of the radioactive iodine, I^*, shows that every substitution occurs with inversion of configuration.

                   (S)-(+)-2-iodooctane   (R)-(−)-2-iodooctane

                                      inversion of configuration

Each S_N2 reaction incorporates *one* radioactive iodine atom into 2-iodooctane, but results in the loss of optical activity equal to that of two molecules of (S)-(+)-2-iodooctane: the one that had been converted to the (R)-isomer, and the other, the optical activity of which is cancelled by the presence of the (R)-isomer. Therefore the rate of racemization, the rate at which optical activity is lost, is *twice* the rate of the substitution reaction.

7.9

Concept Map 7.1   A typical S_N2 reaction.

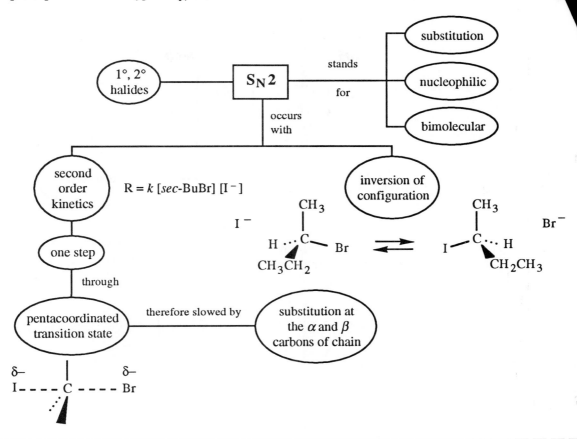

- -

7.10 S_N1 mechanism

The S_N1 reaction.

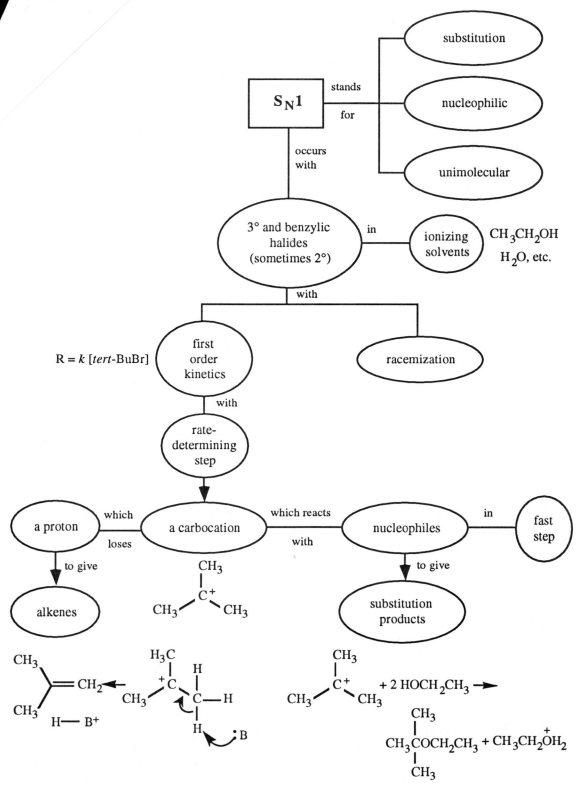

The factors that are important in determining nucleophilicity.

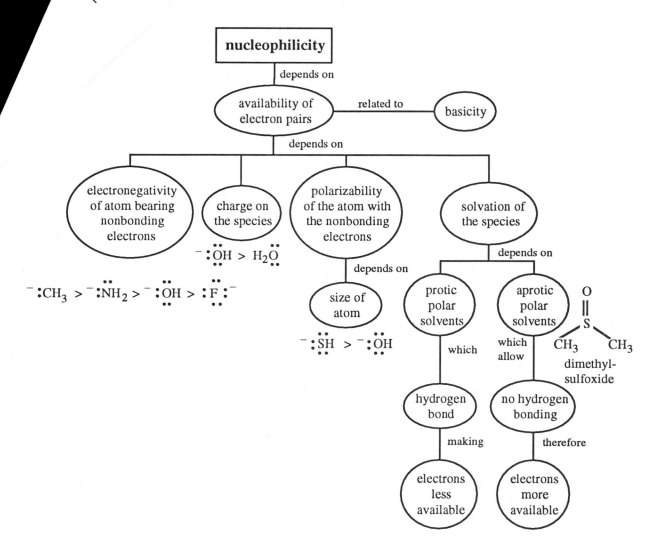

- -

7.14   (a)

$H_2\ddot{N}\!\!-\!\!\ddot{O}H \xrightarrow{\;H\ddot{C}l\!:\;} H_3\overset{+}{N}\!\!-\!\!\ddot{O}H \qquad :\ddot{C}l:^-$

$H_2\ddot{N}\!\!-\!\!\ddot{O}H \xrightarrow{\;CH_3\ddot{I}:\;} CH_3\overset{H}{\underset{H}{\overset{|}{\underset{|}{N^+}}}}\!\!-\!\!\ddot{O}H \qquad :\ddot{I}:^-$

7.11   (a)   S⁻   >   ⟨benzene⟩—O⁻   (b)   OH⁻   >   CH₃CO⁻ (with C=O)

(Sulfur is more polarizable than oxygen.)   (The hydroxide ion is more basic than the
                                                      acetate ion. See Problem 7.12.)

(c)   OH⁻   >   NO₃⁻

(The hydroxide ion is more basic than the
nitrate ion. See Problem 7.12.)

7.12

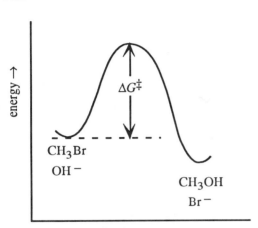

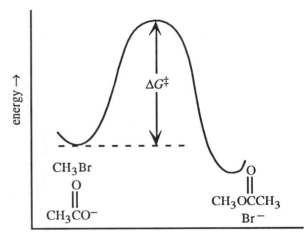

Hydroxide ion is more basic than acetate ion. The pK_a of water, the conjugate acid of hydroxide ion, is
15.7. The pK_a of acetic acid, the conjugate acid of acetate ion, is 4.8. The greater basicity of the oxygen
atom in hydroxide ion means that its electrons are more loosely held than those in acetate ion. Bonding
between hydroxide ion and the carbon atom in methyl bromide starts more easily, and it requires less
energy to get to the transition state.

Concept Map 7.3   (see p. 194)

7.13   The increase in the relative rates of the substitution reaction in this series of solvents is related to the decreasing
ability of the solvent to hydrogen bond to the nucleophile, chloride ion, and to the steric hindrance of the positive
end of the dipole of the carbonyl group.

CH₃OH	HCNH₂ (with C=O)	HCN(CH₃)₂ (with C=O)	CH₃CN(CH₃)₂ (with C=O)
methanol	formamide	*N,N*-dimethyl- formamide	*N,N*-dimethyl- acetamide

Methanol hydrogen bonds strongly and reduces the nucleophilicity of the chloride anion. Formamide has
N—H bonds, so it also hydrogen bonds with chloride anion, but not as strongly as methanol does. Neither *N,N*-
dimethylformamide nor *N,N*-dimethylacetamide can hydrogen bond at all, therefore the nucleophile behaves
as a strong nucleophile in both solvents. The methyl substituent on the carbonyl group of *N,N*-dimethylacet-
amide prevents the chloride anion from getting as close to the positive end of the dipole as it can in *N,N*-
dimethylformamide, making the chloride anion an even stronger nucleophile.

7.14 (cont)

(b)   The pK_as of the oxonium ions are all less than zero while the pK_as of the ammonium ions are all between 4 and 11. The pK_a of the conjugate acid of hydroxylamine falls in the range of values for the ammonium ion.

$$p K_a \quad \overset{+}{N} H_4 \qquad 9.4$$

$$p K_a \quad H_3\overset{+}{N}OH \qquad 5.97$$

The presence of an electron-withdrawing oxygen atom on the nitrogen atom makes the nonbonding electrons on nitrogen less available. Hydroxylamine is therefore a weaker base than ammonia, and the conjugate acid of hydroxylamine is a stronger acid than ammonium ion.

7.15   The hydronium ion comes from protonation of the solvent, water, by the acid, HBr, that was added to the reaction mixture.

$$H_2O \ + \ HBr \ \rightarrow \ H_3O^+ \ + \ Br^-$$

7.16   The other nucleophiles that are present are the starting alcohol and water.

Two additional substitution reactions are possible. The carbocation can react with water to regenerate the starting alcohol or with another alcohol molecule to produce an ether.

Nucleophiles are also bases and can remove a proton from the carbocation to give an alkene, the elimination product.

7.16 (cont)

7.17 CH_3NH_2 + $CH_3CH_2CH_2CH_2\overset{+}{O}\!\!-\!\!H$ with an H above the O ⇌ $CH_3\overset{+}{N}H_3$ + $CH_3CH_2CH_2CH_2O\!\!-\!\!H$

     base   acid   acid   conjugate base

                    $pK_a \sim -2.4$ pK_a 10.6

7.18   All of the nucleophiles in Table 7.1 are too strongly basic to be used with oxonium ions formed by protonating an alcohol.

nucleophile	conjugate acid of nucleophile	pK_a of conjugate acid
HO^-	H_2O	15.7
RO^-	ROH	~16
ArO^-	$ArOH$	~10
HS^-	H_2S	~10
RS^-	RSH	~10.5
ArS^-	$ArSH$	~7.8
NH_3	$NH_4{}^+$	9.4
RNH_2	$RNH_3{}^+$	~10.6
$N_3{}^-$	HN_3	4.7
CN^-	HCN	9.1
$RC\!\equiv\!C^-$	$RC\!\equiv\!CH$	~26
$(CH_3CH_2O\overset{O}{\overset{\|}{C}})_2CH^-$	$(CH_3CH_2O\overset{O}{\overset{\|}{C}})_2CH_2$	~11

7.19

protonation of ethanol   ethyloxonium ion with a good leaving group

17.9 (cont)

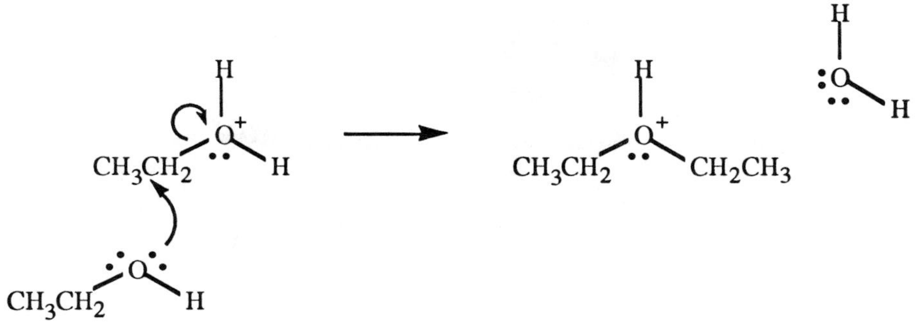

*nucleophile displacing
leaving group*

diethyloxonium ion

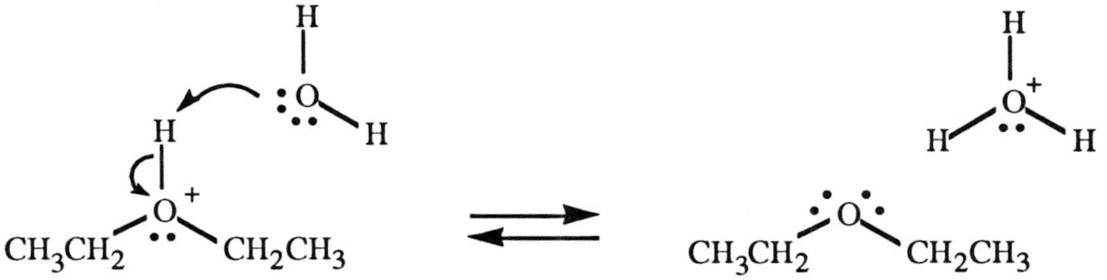

*deprotonation of the
diethyloxonium ion*

7.20

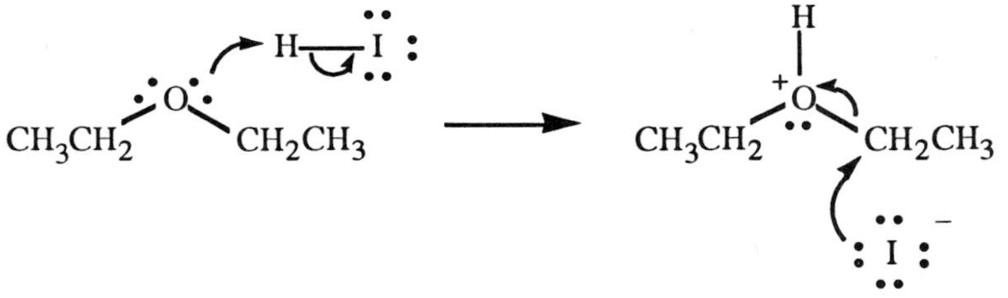

*protonation of
diethyl ether*

*nucleophile displacing
good leaving group*

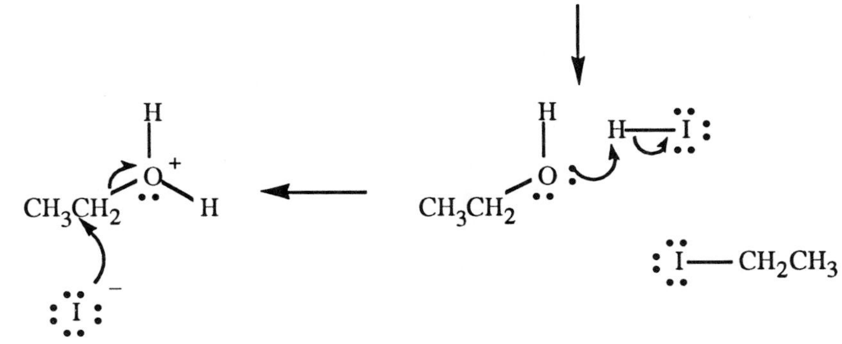

*nucleophile displacing
good leaving group*

*protonation of
ethanol*

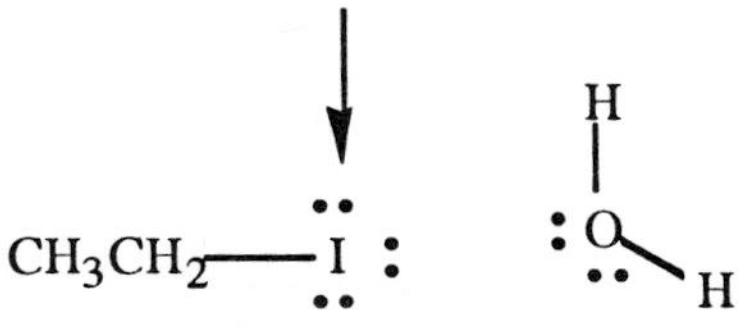

7.21   (1)

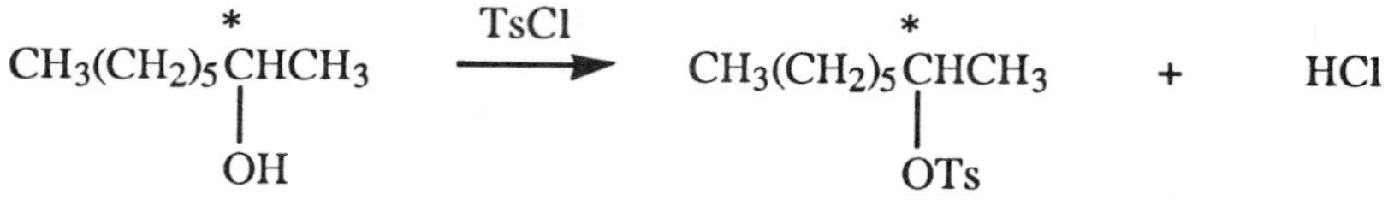

$$CH_3(CH_2)_5\overset{*}{C}HCH_3 \xrightarrow{\text{TsCl}} CH_3(CH_2)_5\overset{*}{C}HCH_3 \quad + \quad HCl$$

with OH below on left, OTs below on right

(+)-2-octanol   (+)-2-octyl tosylate

No bonds are broken at the chiral carbon atom; the reaction goes with retention of configuration.

(2)

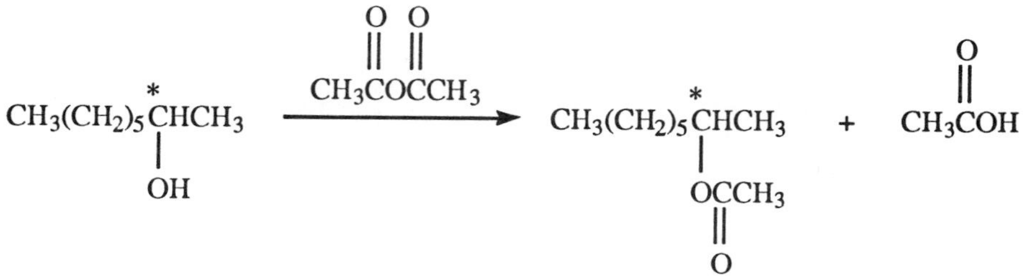

$$CH_3(CH_2)_5\overset{*}{C}HCH_3 \xrightarrow{\text{CH}_3\text{COCCH}_3} CH_3(CH_2)_5\overset{*}{C}HCH_3 \quad + \quad CH_3COH$$

with OH below on left; OCCH$_3$ (O) below center; the reagent is $\underset{\quad}{CH_3COCCH_3}$ with two C=O groups

(–)-2-octanol   (–)-2-octyl acetate

The starting alcohol here is of opposite configuration to the alcohol we had in equation 1; we know this because it has the opposite sign of rotation. The reaction is not one that we recognize, and the sign of rotation does not tell us whether the configuration of the acetate is inverted or not. Let us reserve judgement until we look at the next reaction.

(3)

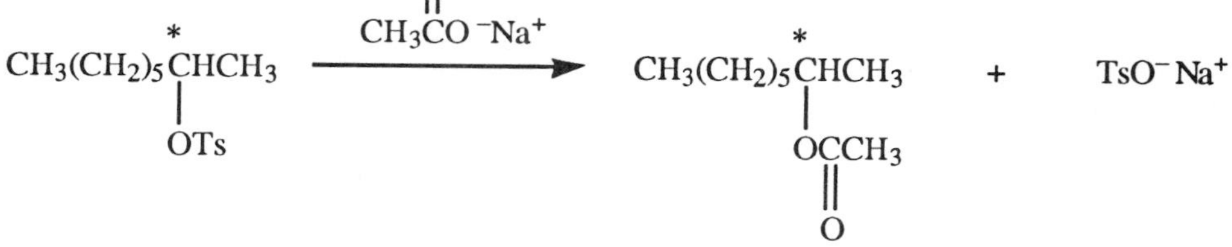

$$CH_3(CH_2)_5\overset{*}{C}HCH_3 \xrightarrow{\text{CH}_3\text{CO}^-\text{Na}^+} CH_3(CH_2)_5\overset{*}{C}HCH_3 \quad + \quad TsO^-Na^+$$

with OTs below on left; OCCH$_3$ (O) below center

(+)-2-octyl tosylate   (–)-2-octyl acetate

We recognize this as an S_N2 reaction, therefore it goes by inversion of configuration. If the (–)-acetate has the opposite configuration from that of the (+)-tosylate, it must also have a configuration opposite to that of the (+)-alcohol because the formation of the tosylate from the (+)-alcohol did not involve inversion of configuration [see part (1)]. Therefore, the configuration of the (–)-acetate must be the same as that of the (–)-alcohol. Reaction 2 must proceed with retention of configuration.

7.22

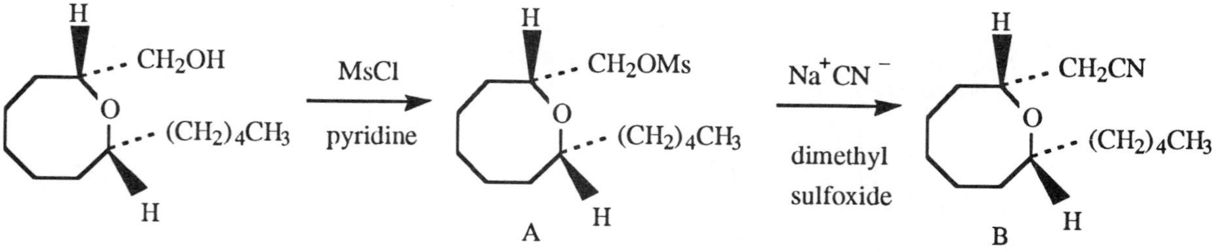

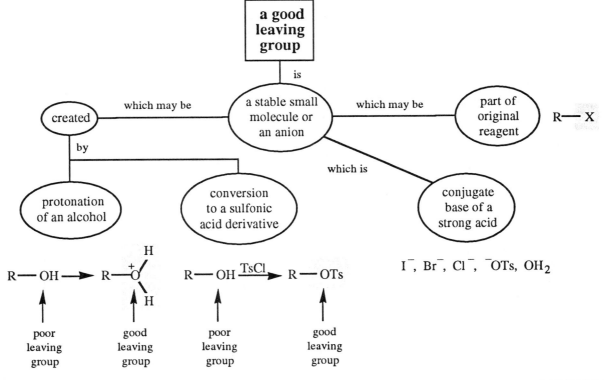

7.23 The reaction of CH₃C≡C—C(H)(OH)—CH₂OH with TsCl (1–2 equiv), pyridine, dichloromethane.

primary carbon; less hindered OH group; reaction with TsCl favored → CH₂OTs

secondary carbon; OH group more hindered therefore slower to react with TsCl

---

Concept Map 7.4   The factors that determine whether a substituent is a good leaving group.

a good leaving group is a stable small molecule or an anion

which may be created by
— protonation of an alcohol
— conversion to a sulfonic acid derivative

which may be part of original reagent   R—X

which is conjugate base of a strong acid

I⁻, Br⁻, Cl⁻, ⁻OTs, OH₂

R—OH → R—O⁺(H)(H)

poor leaving group   good leaving group

R—OH →(TsCl) R—OTs

poor leaving group   good leaving group

7.24   (1) S_N2

(2) S_N1 slow

7.24 (cont)

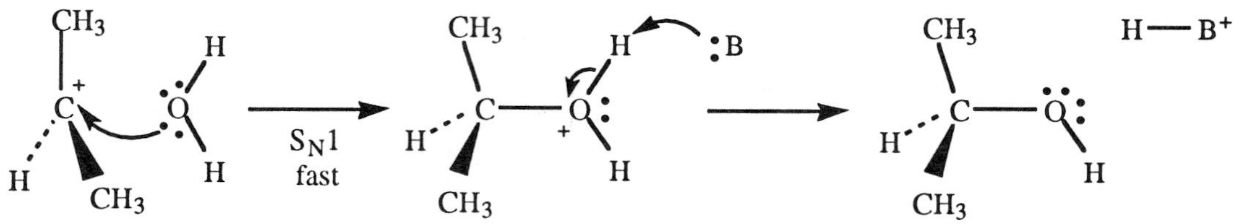

Note that attack of a nucleophile on the electrophilic carbocation can occur at either the front face (where the leaving group was attached) or at the back face. The next equation shows attack of a nucleophile at the back face.

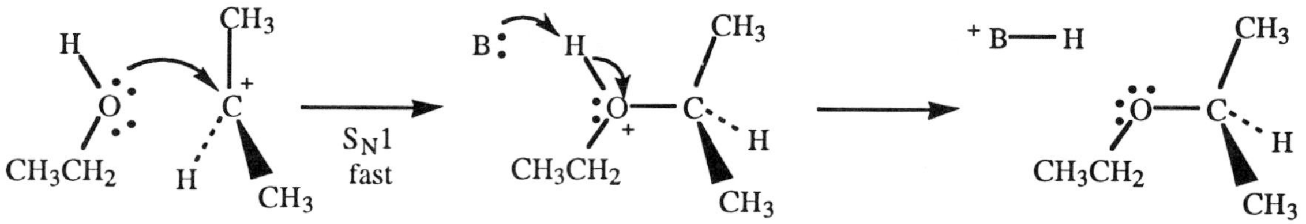

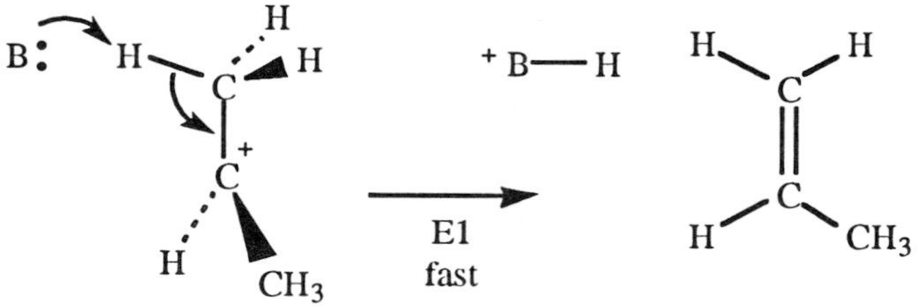

(3)

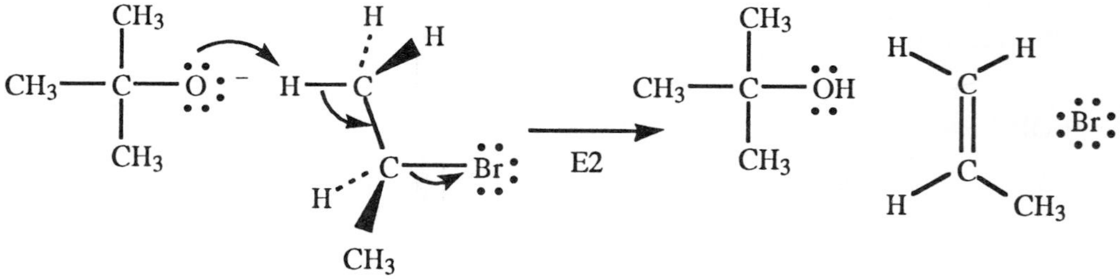

7.25 Yes

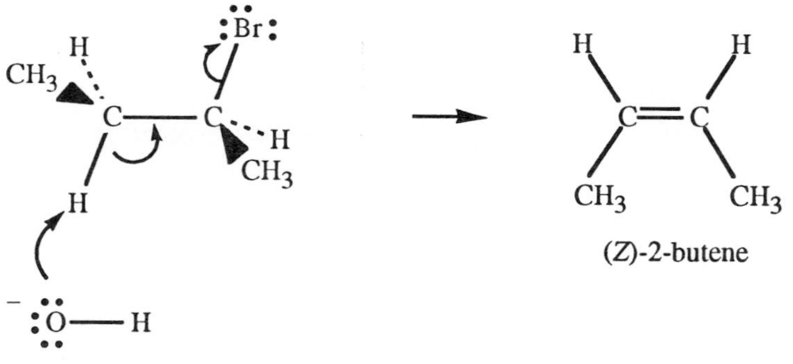

(Z)-2-butene

The transition state for the formation of (E)-2-butene is less crowded, more favorable, and lower in energy than the transition state for the formation of (Z)-2-butene where the two methyl groups are close together.

7.26

(a) $CH_3CH_2CH_2CH_2CH_2Br$ $\xrightarrow[\Delta]{\underset{\text{ethanol}}{KOH}}$ $CH_3CH_2CH_2CH{=}CH_2$ + H_2O + K^+Br^-

(b)

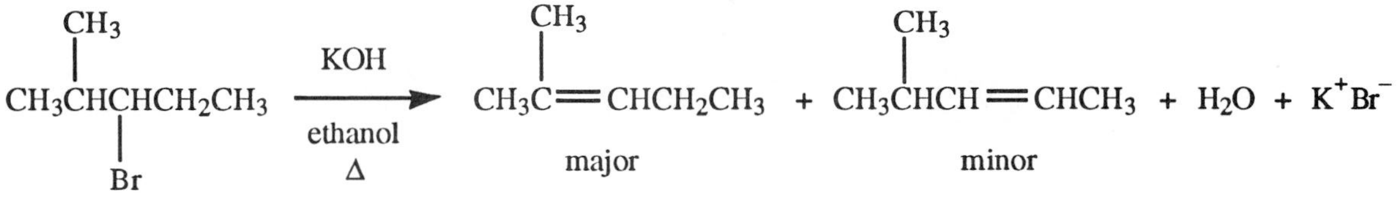

$CH_3\overset{\displaystyle CH_3}{\underset{\displaystyle |}{C}}HCHCH_2CH_3$ $\xrightarrow[\Delta]{\underset{\text{ethanol}}{KOH}}$ $CH_3C{=}CHCH_2CH_3$ + $CH_3\overset{\displaystyle CH_3}{\underset{\displaystyle |}{C}}HCH{=}CHCH_3$ + H_2O + K^+Br^-

major   minor

(c)

major   minor   very minor

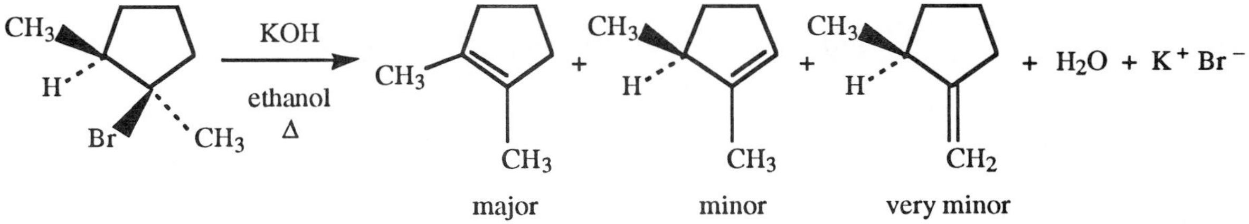

(d)

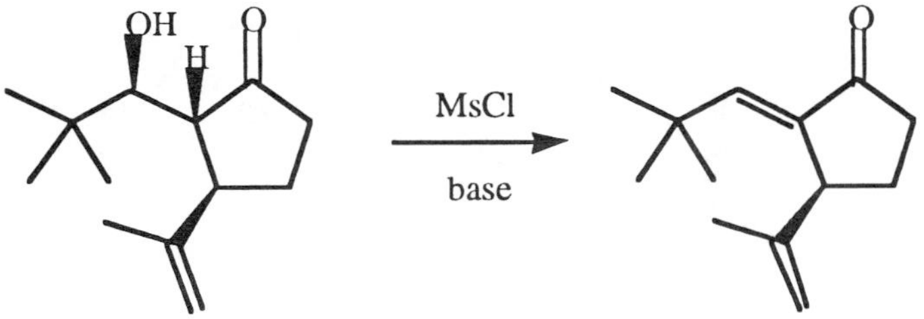

7.27

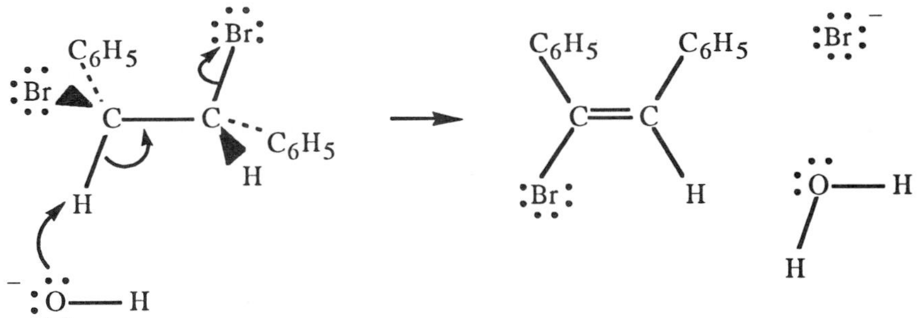

(1R,2S)-1,2-dibromodiphenylethane   (E)-1-bromo-1,2diphenylethene

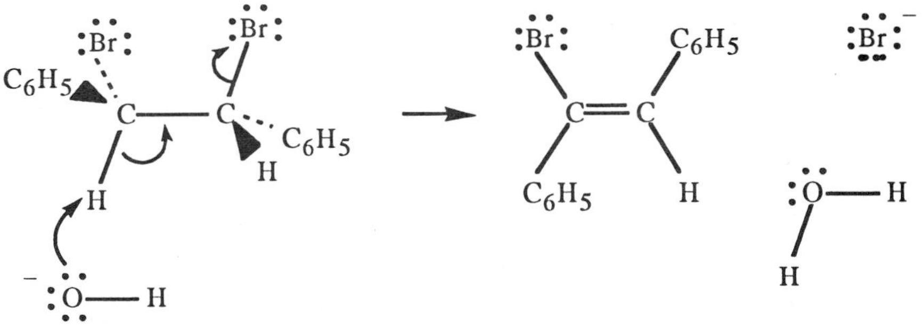

(1R,2R)-1,2-dibromodiphenylethane   (Z)-1-bromo-1,2diphenylethene

Concept Map 7.5   A comparison of E1 and E2 reactions.

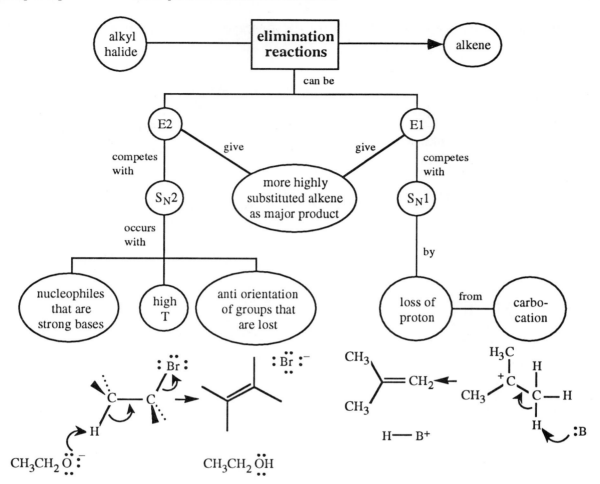

- -

Concept Map 7.6   (See pp. 206 - 207)

- -

7.28

(a) $CH_3CH_2CH_2CH_2Br$ $\xrightarrow[\substack{\text{ethanol} \\ \Delta}]{\text{KOH}}$ $CH_3CH_2CH{=}CH_2$ $+$ $CH_3CH_2CH_2CH_2OH$ $+$ H_2O $+$ $K^+ Br^-$

major product with strong
base at high temperature

(b) $CH_3CH_2CH_2CH_2Br$ $\xrightarrow[\substack{50\% \text{ aqueous} \\ \text{ethanol}}]{0.1 \text{ M NaOH}}$ $CH_3CH_2CH_2CH_2OH$ $+$ $Na^+ Br^-$

(c) $CH_3CH_2CH_2CH_2Br$ $\xrightarrow{NH_3}$ $CH_3CH_2CH_2CH_2\overset{+}{N}H_3$ Br^- $\xrightarrow{NH_3 \text{ (excess)}}$

$CH_3CH_2CH_2CH_2NH_2$ $+$ $NH_4^+ Br^-$

7.28 (cont)

(d) $CH_3CH_2CH_2CH_2Br$ $\xrightarrow{Na^+ N_3^-}$ $CH_3CH_2CH_2CH_2N_3$ $+$ $Na^+ Br^-$

(e) $CH_3CH_2CH_2CH_2Br$ $\xrightarrow{Na^+ CN^-}$ $CH_3CH_2CH_2CH_2CN$ $+$ $Na^+ Br^-$

(f) $CH_3CH_2CH_2CH_2Br$ $\xrightarrow{CH_3CH_2S^- Na^+}$ $CH_3CH_2CH_2CH_2SCH_2CH_3$ $+$ $Na^+ Br^-$

(g) $CH_3CH_2CH_2CH_2Br$ $\xrightarrow{CH_3\overset{\displaystyle O}{\overset{\|}{C}}O^- Na^+}$ $CH_3CH_2CH_2CH_2O\overset{\displaystyle O}{\overset{\|}{C}}CH_3$ $+$ $Na^+ Br^-$

(h) $CH_3CH_2CH_2CH_2Br$ $\xrightarrow{CH_3CH_2C\equiv C^- Na^+}$ $CH_3CH_2CH_2CH_2C\equiv CCH_2CH_3$ $+$ $Na^+ Br^-$

(i) $CH_3CH_2CH_2CH_2Br$ $\xrightarrow{Na^+ \ ^-CH(\overset{\displaystyle O}{\overset{\|}{C}}OCH_2CH_3)_2}$ $CH_3CH_2CH_2CH_2CH(\overset{\displaystyle O}{\overset{\|}{C}}OCH_2CH_3)_2$ $+$ $Na^+ Br^-$

7.29

(a)

leaving group

$-CH_2CH_2CH_2OTs$ $+$ N_3^- $\longrightarrow$ $-CH_2CH_2CH_2N_3$ $+$ TsO^-

nucleophile

(b)

leaving group   nucleophile

$-CH_2CH_2Cl$ $+$ Na^+CN^- $\longrightarrow$ $-CH_2CH_2CN$ $+$ $NaCl$

$^-:C\equiv N:$ $\longleftrightarrow$ $:C\overset{\displaystyle ..}{=}N:^-$

more nucleophilic

(c) $CH_3CH_2CH_2CH_2Br$ $+$ $CH_3CH_2S^- Na^+$ $\longrightarrow$ $CH_3CH_2CH_2CH_2SCH_2CH_3$ $+$ $NaBr$

leaving group   nucleophile

7.29 (cont)

(d) $CH_3CH_2CH_2Cl$ + $CH_3C \equiv C^- Na^+$ $\longrightarrow$ $CH_3CH_2CH_2C \equiv CCH_3$ +   NaCl

         $\underset{\text{leaving group}}{\uparrow}$ $\underset{\text{nucleophile}}{\uparrow}$

(e) $\underset{\underset{\text{leaving group}}{\uparrow}}{CH_3\overset{\overset{CH_3}{|}}{C}HCH_2CH_2Br}$ + $\underset{\underset{\text{nucleophile}}{\uparrow}}{NH_3}$ (excess) $\longrightarrow$ $CH_3\overset{\overset{CH_3}{|}}{C}HCH_2CH_2NH_2$ + NH_4Br

(f) CH_3I + $Na^+ \ ^-CH(\overset{\overset{O}{\|}}{C}OCH_2CH_3)_2$ $\longrightarrow$ $CH_3CH(\overset{\overset{O}{\|}}{C}OCH_2CH_3)_2$ +   NaI

         $\underset{\text{leaving group}}{\uparrow}$ $\underset{\text{nucleophile}}{\uparrow}$

(g) CH_3CH_2I + $P(CH_2CH_2CH_2CH_3)_3$ $\longrightarrow$ $CH_3CH_2\overset{+}{P}(CH_2CH_2CH_2CH_3)_3 \ I^-$

         $\underset{\text{leaving group}}{\uparrow}$ $\underset{\text{nucleophile}}{\nwarrow}$

(h) $HOCH_2CH_2Cl$ + $CH_3S^- Na^+$ $\longrightarrow$ $HOCH_2CH_2SCH_3$ +   NaCl

         $\underset{\text{leaving group}}{\uparrow}$ $\underset{\text{nucleophile}}{\uparrow}$

7.30   (a) $SH^- > Cl^-$

     (b) $(CH_3)_3P > (CH_3)_3B$ (The boron atom has an empty orbital and is an electrophile; the phosphorus atom has nonbonding electrons.)

     (c) $CH_3NH^- > CH_3NH_2$

     (d) $CH_3SCH_3 > CH_3OCH_3$

7.31

(a)

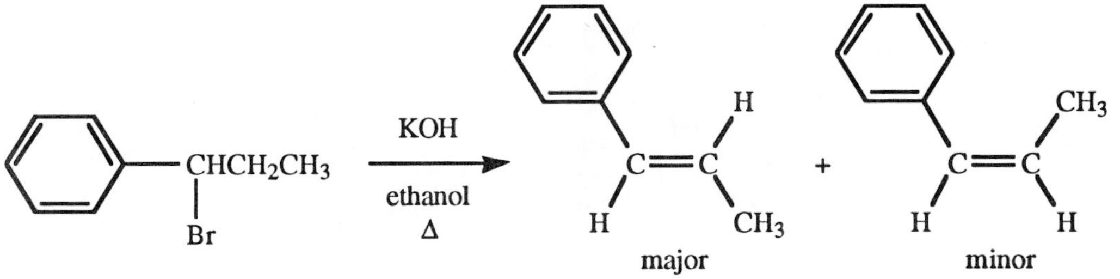

(b)

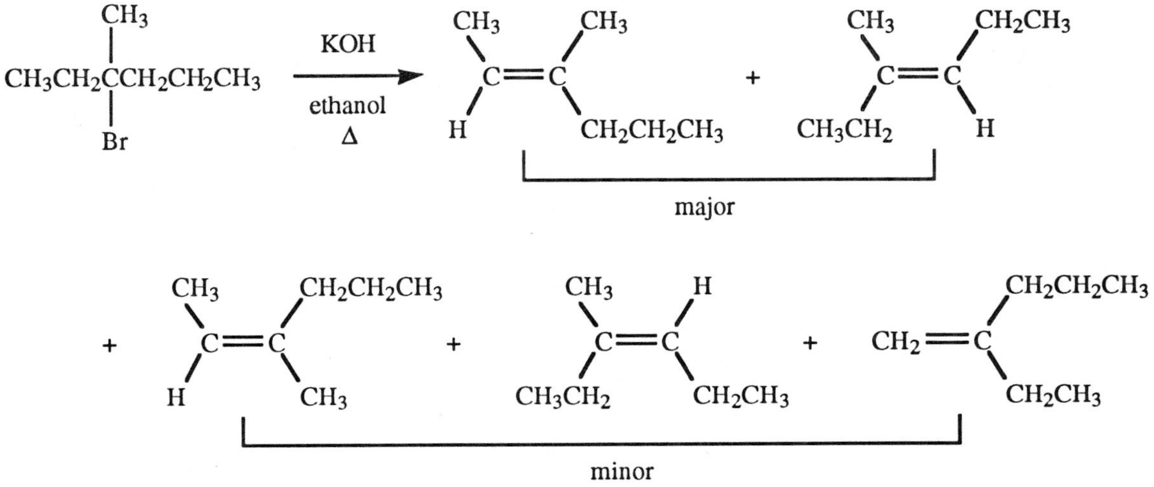

(c)

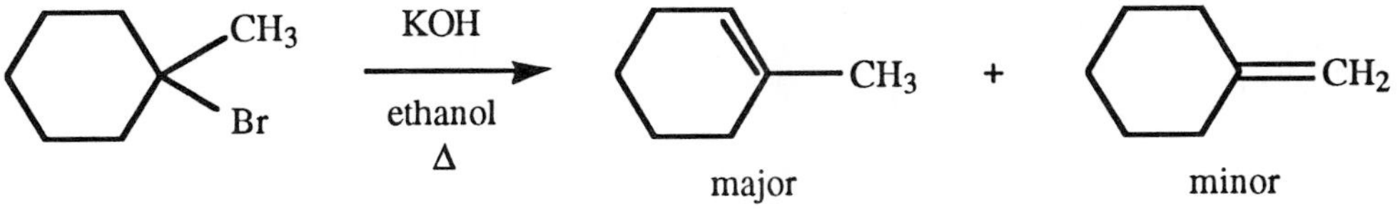

(d)

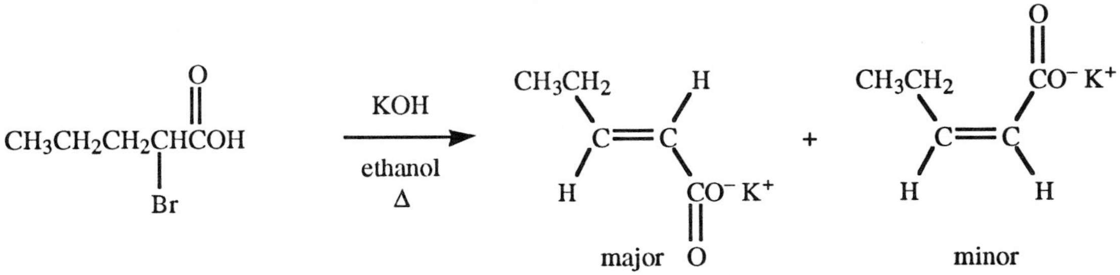

(e)

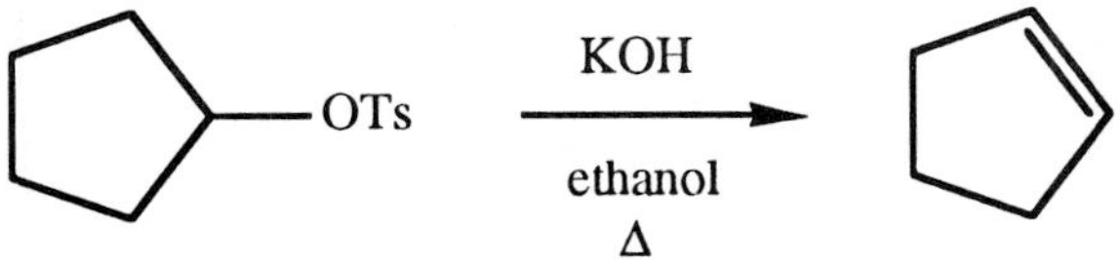

Concept Map 7.6   Chapter summary.

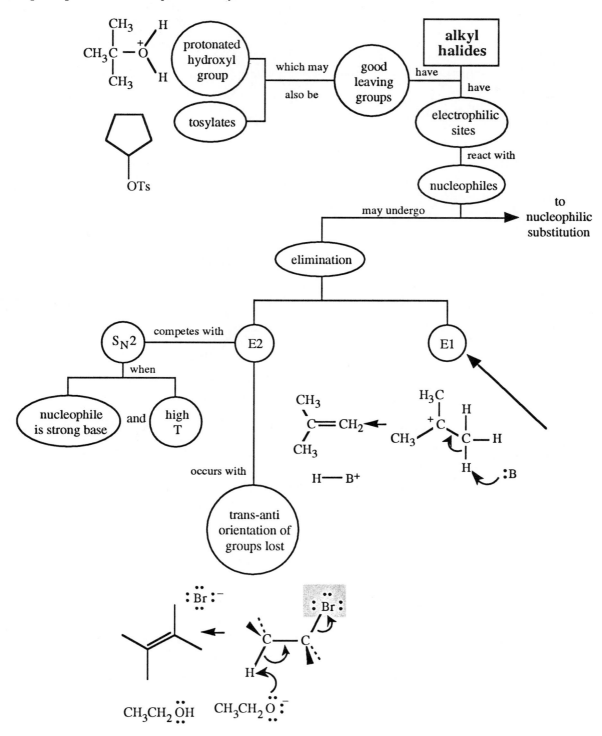

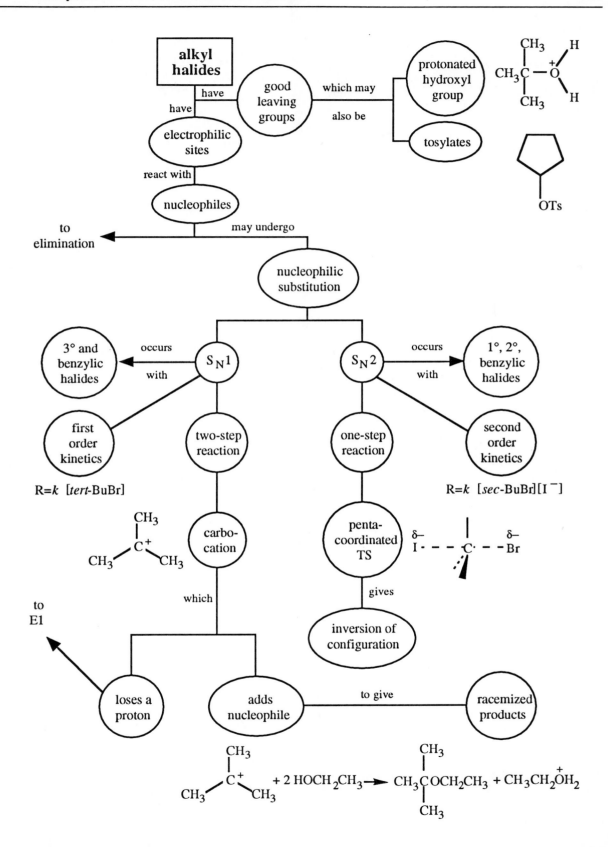

7.32

(a)

$$CH_3CHCH_3 \xrightarrow[\text{pyridine}]{\text{TsCl}} CH_3CHCH_3 \xrightarrow{\text{NaN}_3} CH_3CHCH_3$$

with OH below the first, OTs below the second (A), and N₃ below the third (B).

(a)

CH₃CHCH₃ with OH
→ TsCl / pyridine → CH₃CHCH₃ with OTs (A)
→ NaN₃ → CH₃CHCH₃ with N₃ (B)

(b)

$$\underset{\overset{|}{CH_3}}{CH_3CHCH_2CH_2CH_2CH_2Br} \xrightarrow[\text{ethanol}]{\text{NaCN}} \underset{\overset{|}{CH_3}}{CH_3CHCH_2CH_2CH_2CH_2CN}$$

C

(c)

cyclopentyl bromide $\xrightarrow[\substack{\text{ethanol}\\ \Delta}]{\text{KOH}}$ cyclopentene D

(d)

$$\underset{\overset{|}{CH_3}}{CH_3CH_2CH_2C}=CHCH_2CH_2OH \xrightarrow[\text{pyridine}]{\text{TsCl}}$$

$$\underset{\overset{|}{CH_3}}{CH_3CH_2CH_2C}=CHCH_2CH_2OTs \xrightarrow[\text{acetone}]{\text{NaI}} \underset{\overset{|}{CH_3}}{CH_3CH_2CH_2C}=CHCH_2CH_2I$$

E   F

(e)

$\xrightarrow[\substack{\text{ethanol}\\ \Delta}]{\text{KOH}}$

major product G (1-ethylcyclohexene) + minor product (3-ethylcyclohexene)

G $\xrightarrow{\text{HI}}$ H (1-ethyl-1-iodocyclohexane, I)

(f)

$\xrightarrow[\text{pyridine}]{\text{TsCl}}$ I $\xrightarrow[\substack{N,N\text{-dimethyl-}\\ \text{formamide}}]{\text{NaCN}}$ J

7.33

(a)

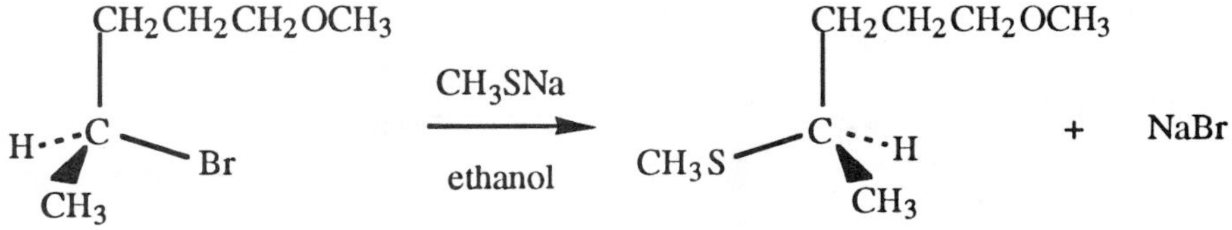

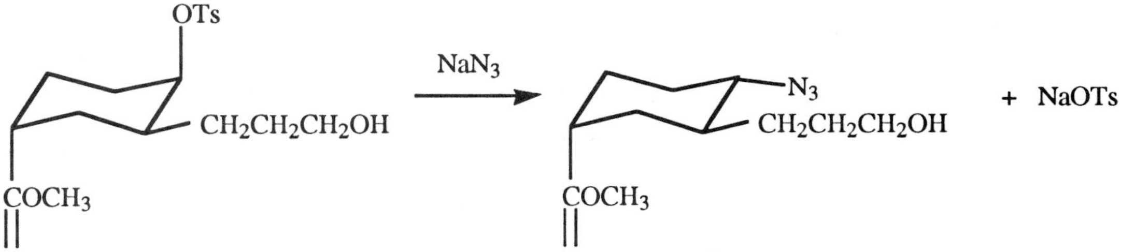

(b)

(c)

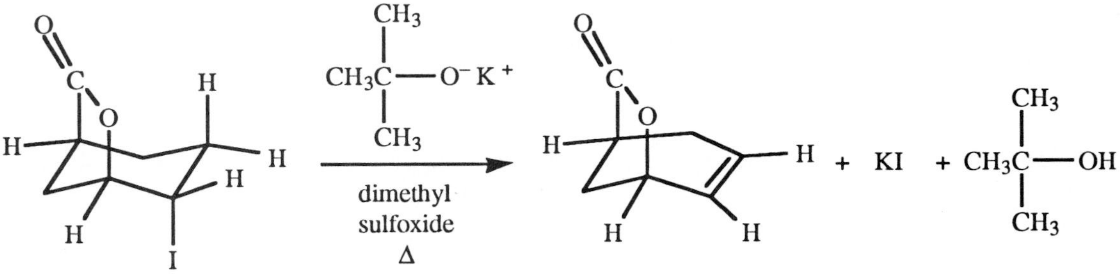

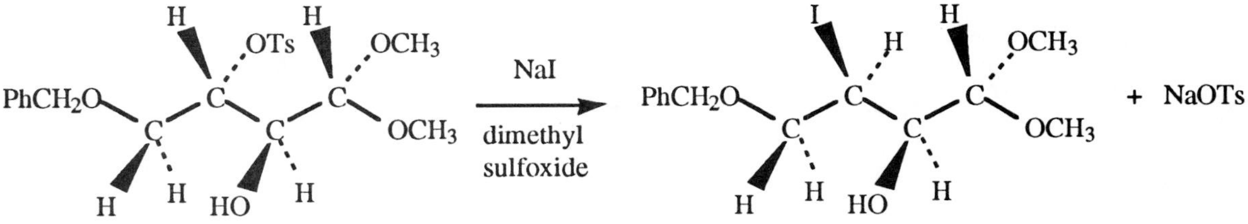

(d)

(e)

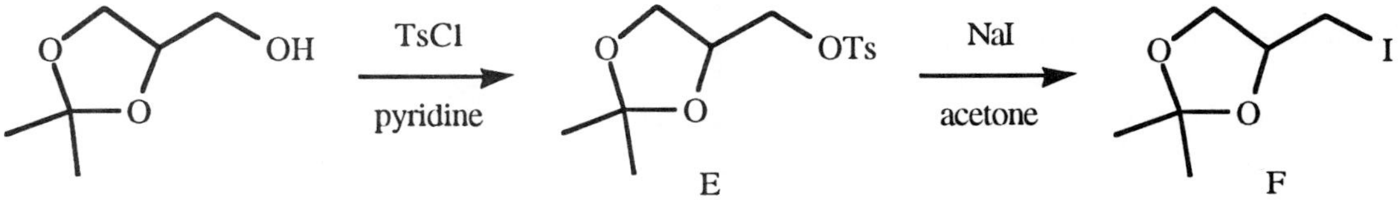

(f)

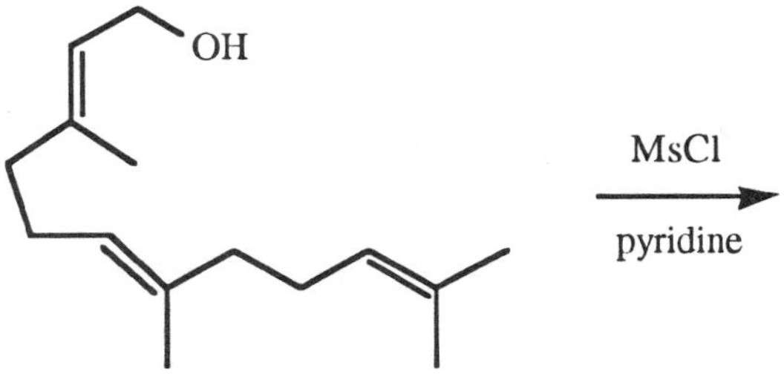

7.33 (f) (cont)

G   H

7.34

(a) The reaction of the 1° alkyl halide, $CH_3CH_2CH_2Br$, will be faster than that of the 2° alkyl halide, $CH_3CHBrCH_3$. There is greater steric crowding in the transition state for the S_N2 reaction of the 2° alkyl halide with the good nucleophile, CN^-.

(b) The S_N2 reaction of the more nucleophilic CH_3S^- ion will be faster than that of the CH_3O^- ion.

(c) The *tert*-butyl bromide reacts faster in the S_N1 solvolysis reaction than *n*-butyl bromide because a 3° carbocation intermediate is much more stable than a 1° carbocation intermediate. We know that this is an S_N1 reaction because there is a good ionizing solvent (H_2O), and no strong nucleophile present.

(d) These reactions are electrophilic addition reactions, and go through carbocation intermediates. The second reaction, which gives a 3° carbocation intermediate, goes faster than the first reaction, which gives a 2° carbocation intermediate.

7.35

1. What are the connectivities of the two compounds? How many carbon atoms does each contain? Are there any rings? What are the positions of branches and functional groups on the carbon skeletons?

   There is one more carbon atom in the product. It has been added to carbon 1 of the starting material and is part of the cyano group. The starting material has two hydroxyl groups. One hydroxyl group is on a primary carbon atom and the other is on a tertiary carbon atom. The product contains a cyano group on the primary carbon atom and a hydroxyl group on the tertiary carbon atom.

2. How do the functional groups change in going from starting material to product? Does the starting material have a good leaving group?

   The primary hydroxyl group has been replaced by a cyano group. There is no good leaving group in the starting material but a hydroxyl group can be converted into one.

3. Is it possible to dissect the structures of the starting material and product to see which bonds must be broken and which formed?

7.35 (cont)

4. New bonds are created when an electrophile reacts with a nucleophile. Do we recognize any part of the product molecule as coming from a good nucleophile or an electrophilic addition?

    The cyano group, —CN, comes from CN¯, a good nucleophile.

5. What type of compound would be a good precursor to the product?

    Something with a good leaving group where the hydroxyl group is, such as water, a halide, or a tosylate group, would be a good precursor. We cannot use strong acid to convert the hydroxyl group into a better leaving group because the cyanide anion is a better base than the alcohol and would deprotonate the oxonium ion.

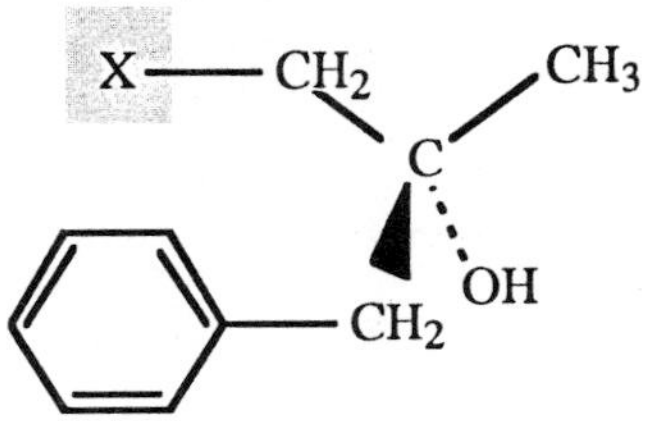

6. After this last step, do we see how to get from starting material to product? If not, we need to analyze the structure obtained in step 5 by applying questions 4 and 5 to it.

    The precursor containing the tosylate group can easily be made from the starting material and tosyl chloride. A primary alcohol will react faster with tosyl chloride than a tertiary alcohol (see Problem 7.23). We can now complete the problem.

Complete correct answer:

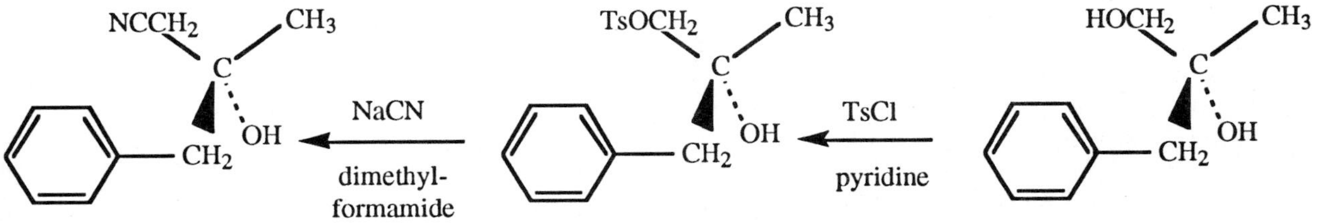

7.36   The answers to parts (a), (e) and (f) analyze the problems as shown in the Problem Solving Skills section of the chapter. The rest of the problems show only the equations, written so as to emphasize the necessity of reasoning backward from the structure of the desired product to the structure of the starting material.

7.36 (cont)

(a)

1. What are the connectivities of the two compounds? How many carbon atoms does each contain? Are there any rings? What are the positions of branches and functional groups on the carbon skeletons?

    The product has two more carbon atoms than the starting material. These carbon atoms are on the sulfur atom. The thioether group in the product is on the same carbon atom that the hydroxyl group was attached to in the starting material.

2. How do the functional groups change in going from starting material to product? Does the starting material have a good leaving group?

    The hydroxyl group has been replaced by a thioether. The hydroxyl group is not a good leaving group but can be converted into one.

3. Is it possible to dissect the structures of the starting material and product to see which bonds must be broken and which formed?

          bond broken   bond formed

4. New bonds are created when an electrophile reacts with a nucleophile. Do we recognize any part of the product molecule as coming from a good nucleophile or an electrophilic addition?

    The ethanethiolate anion, $CH_3CH_2S^-$, is a good nucleophile.

5. What type of compound would be a good precursor to the product?

    The displacement of a leaving group (halide or tosylate ion) by the nucleophile will give us the product we want.

6. After this last step, do we see how to get from starting material to product? If not, we need to analyze the structure obtained in step 5 by applying questions 4 and 5 to it.

    We now can complete the synthesis.

7.36 (a) (cont)

The compound could also have been synthesized using an S_N1 reaction with ethanethiol in an ionizing solvent such as water, but the product of the reaction with the solvent would also form. When possible, it is better to use the S_N2 reaction for synthesis, instead of an S_N1 reaction, because there will be fewer competing side reactions to lower the yield of the desired product.

(b) $CH_3CH_2CH_2CH_2CH_2I$ $\xleftarrow{\quad HI \quad}$ $CH_3CH_2CH_2CH_2CH_2OH$

(c) $CH_3CH_2CH_2CH_2C{\equiv}CH$ $\xleftarrow{\quad HC{\equiv}C^-Na^+ \quad}$ $CH_3CH_2CH_2CH_2Br$

(d)

1. What are the connectivities of the two compounds? How many carbon atoms does each contain? Are there any rings? What are the positions of branches and functional groups on the carbon skeletons?

    The carbon skeletons of the two compounds are the same. The compounds have opposite stereochemistry.

2. How do the functional groups change in going from starting material to product? Does the starting material have a good leaving group?

    A hydroxyl group has been replaced by an amino group. The hydroxyl group is not a good leaving group but can be converted into one.

3. Is it possible to dissect the structures of the starting material and product to see which bonds must be broken and which formed?

bond broken   bond formed

7.36 (e) (cont)

4. New bonds are created when an electrophile reacts with a nucleophile. Do we recognize any part of the product molecule as coming from a good nucleophile or an electrophilic addition?

    The amino group comes from ammonia, NH_3, a good nucleophile.

5. What type of compound would be a good precursor to the product?

    Displacement of a leaving group (halide or tosylate ion) by the nucleophile, NH_3, will give the product we want.

    This compound must have the same stereochemistry as the alcohol, so that an S_N2 reaction with ammonia gives the amine with the inverted configuration.

6. After this last step, do we see how to get from starting material to product? If not, we need to analyze the structure obtained in step 5 by applying questions 4 and 5 to it.

    We know how to convert a hydroxyl group into a leaving group with retention of configuration. We can now finish the synthesis.

(f)

1. What are the connectivities of the two compounds? How many carbon atoms does each contain? Are there any rings? What are the positions of branches and functional groups on the carbon skeleton?

    The carbon skeletons are identical in starting material and products. The azide functional group in the product is on the second carbon atom of the chain, carbon two of the double bond in the starting material.

2. How do the functional groups change in going from the starting material to the product? Does the starting material have a good leaving group?

    The double bond has disappeared. The product is an azide. There is no good leaving group in the starting material.

3. Is it possible to dissect the structure to see which bonds must be broken and which formed?

7.36 (f) (cont)

bond broken   bonds formed

4. New bonds are created when an electrophile reacts with a nucleophile. Do we recognize any part of the product molecule as coming from a good nucleophile or an electrophilic addition?

    The azide group, $-N_3$, comes from N_3^-, a good nucleophile.

5. What type of compound would be a good precursor to the product?

    We need a compound that has a good leaving group where the azide group is in the product. Displacement of a leaving group (halide or tosylate ion) by the azide nucleophile will give us the product we want.

6. After this last step, do we see how to get from starting material to product? If not, we need to analyze the structure obtained in step 5 by applying questions 4 and 5 to it.

    Since the starting material does not have a leaving group, we need to repeat steps 4 and 5.

4. (again) New bonds are created when an electrophile reacts with a nucleophile. Do we recognize any part of the product molecule as coming from a good nucleophile or an electrophilic addition?

    A good leaving group such as a halide can come from an electrophilic addition of a hydrogen halide to the double bond.

5. (again) What type of compound would be a good precursor to the product?

    The electrophilic addition of a hydrogen halide such as hydrogen bromide to the double bond will give us the intermediate we need. We can now complete the problem.

(g)

7.37   (a)   Rate $= k$ [$CH_3CH_2CH_2CH_2Br$] [NaOH]

       (b)   The problem does not specify whether the energy diagram should show differences in free energy or enthalpy. The diagram below is drawn showing the experimental energy of activation, E_a, and the enthalpy of reaction, ΔH_r.

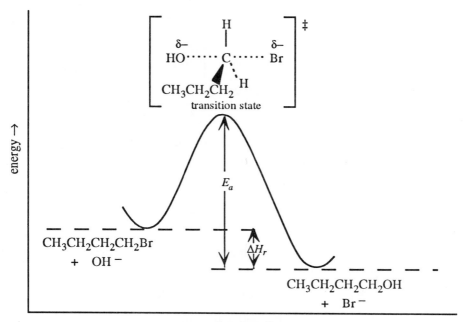

       (c)   The rate will be doubled.

       (d)   The rate will be halved.

       (e)   If some of the ethanol is replaced with water, there will be only a slight effect on the rate since there is only a slight build-up of positive charge in the transition state on the carbon atom undergoing the substitution.

$$\begin{array}{c} C_6H_5 \\ | \end{array}$$

7.38   (a)   Rate $= k$ [$C_6H_5CCH_3$]

$$\begin{array}{c} | \\ Br \end{array}$$

       (b)   See the note to the energy diagram in Problem 7.37. This diagram is drawn showing free energy of activation and the free energy of reaction.

7.38 (cont)

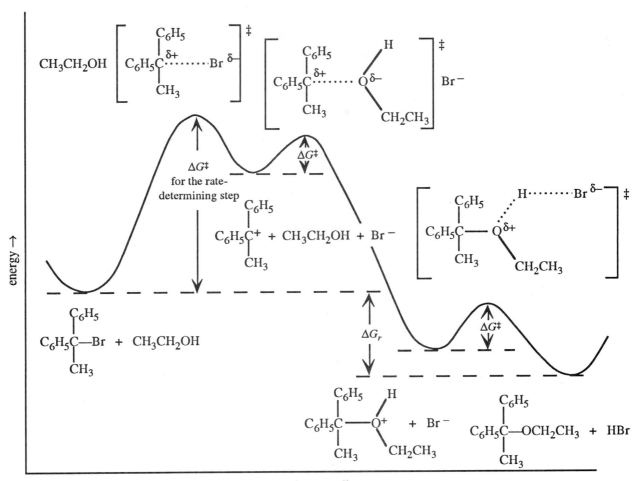

(c)   The rate will double.

(d)   Water is a better ionizing solvent than ethanol. The reaction involves the formation of a carbocation, resulting in the separation of charge in the rate-determining step, so an increase in the water content of the solvent will cause an increase in the rate of the reaction by stabilizing the activated complex at the transition state.

7.39

$$CH_3CH{=}CHCH_2 \overset{\curvearrowleft}{\,}\underset{\cdot\cdot}{\overset{\cdot\cdot}{Cl}}: \longrightarrow [CH_3CH{=}CH{-\!-}\overset{+}{C}H_2 \longleftrightarrow CH_3\overset{+}{C}H{-\!-}CH{=}CH_2] \; :\underset{\cdot\cdot}{\overset{\cdot\cdot}{Cl}}:^-$$

resonance stabilized carbocation, an allylic cation, has
partial positive character at carbon 1 and carbon 3

7.39 (cont)

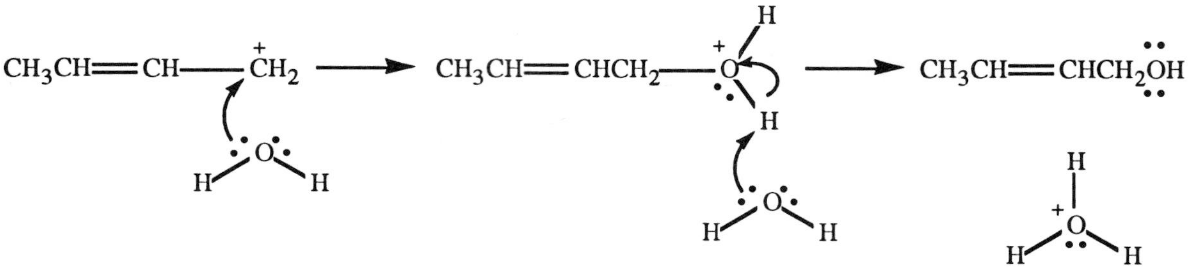

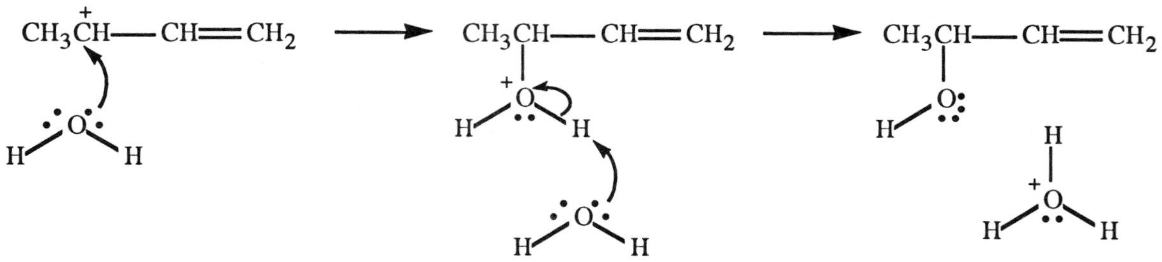

$$CH_3CH\!=\!CH\!-\!\overset{+}{C}H_2 \longrightarrow CH_3CH\!=\!CHCH_2\!-\!\overset{\overset{H}{|}}{\underset{|}{\overset{+}{O}}}\!\!-\!H \longrightarrow CH_3CH\!=\!CHCH_2\ddot{O}H$$

7.40

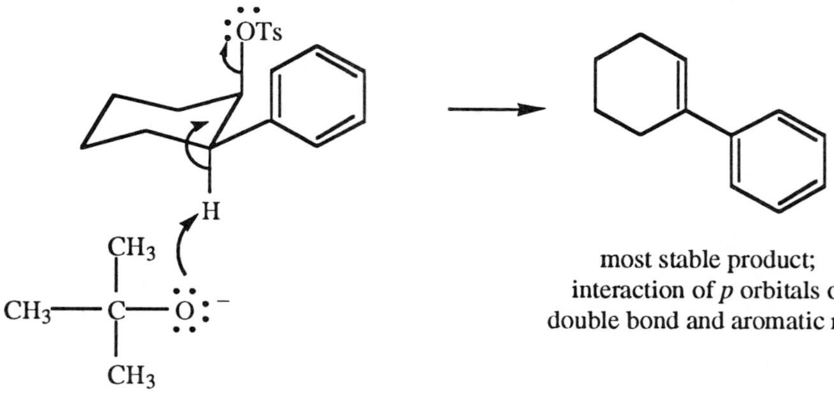

most stable product;
interaction of *p* orbitals of
double bond and aromatic ring

For *cis*-2-phenylcyclohexyl tosylate, the most stable product arises from a conformation of the cyclohexane ring that has the phenyl group in the equatorial position, and the leaving group, the tosyl group, in the axial position required for the E2 reaction to occur (see p. ~ in the text).

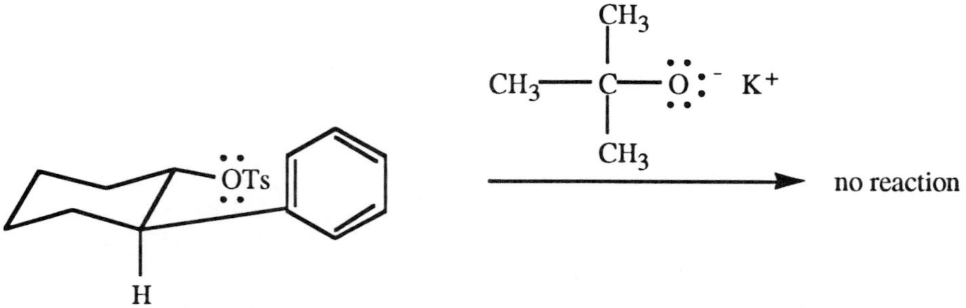

no reaction

In *trans*-2-phenylcyclohexyl tosylate the bulky phenyl ring prevents the molecule from achieving the diaxial conformation required for elimination, and no reaction takes place.

7.41

(a)

menthyl chloride;
all of the substituents are equatorial
in this chair conformation of the
cyclohexane ring

only one hydrogen is axial
to the axial chlorine

a

neomenthyl chloride;
two hydrogens axial to
the axial chlorine

b

(b)   When the reaction with menthyl chloride is carried out in aqueous ethanol, a solvent which supports ionization better than pure ethanol, a planar carbocation is formed, and a hydrogen on either side of the cationic center is eliminated to give the two products observed.

aqueous
ethanol

two products

planar carbocation

7.42

(a)

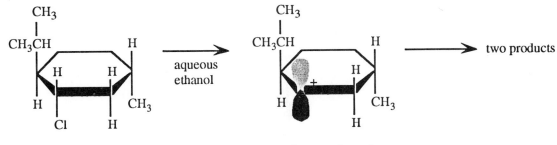

TsCl

pyridine

$(CH_3)_2NH$ (excess)

7.42 (cont)

(b)   The hydroxyl group on the 1° carbon atom is less hindered than the hydroxyl group on the 3° carbon atom and will react faster with tosyl chloride.

7.43   (a) The similarity of the products from the two reactions suggests that both reactions proceed through the same intermediate, a carbocation.

(b) The fact that they proceed at different rates suggests that the formation of the carbocation is the rate-determining step. The higher rate for the first reaction suggests that chloride ion is a better leaving group than dimethyl sulfide.

7.44

thiocyanate ion

cyanate ion

In the thiocyanate ion, the electrons on the sulfur are more polarizable than those on the nitrogen, and that end of the ion is the better nucleophile. In the cyanate ion, the electrons on the nitrogen are more polarizable than those on the oxygen, and the nitrogen is a better nucleophile.

7.45   (a)

(S)-2-bromopentane

(R)-2-bromopentane
inversion of configuration

These compounds are enantiomers.

(b)   The results of Experiment B, in which the reaction rate increases with an increase in the concentration of bromide ion, indicates that the reaction proceeds by an S_N2 reaction. The rate of an S_N2 reaction depends on the concentration of the nucleophile as well as on the concentration of the alkyl halide.

7.45 (cont)

(c) The starting material is reversibly converted to product (Statement 3).

The starting material and the product are at the same level with respect to energy, therefore neither form is favored. The reaction proceeds randomly, converting each enantiomer to the other, until the mixture becomes optically inactive. The interconversion of the enantiomers continues, but can no longer be observed by a change in the optical activity of the solution.

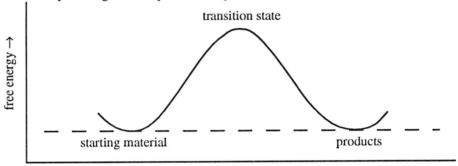

7.46

(a) The base is
$$CH_3CH \overset{\overset{\displaystyle CH_3}{\underset{\displaystyle CHCH_3}{|}}}{\underset{\displaystyle }{|}}\text{—}\overset{..}{\underset{..}{N}}\text{:}^{-}$$
The pK$_a$ of its conjugate acid is ~36, analogous to $NH_3 \rightarrow NH_2^-$.

most acidic proton; pK$_a$ ~ 10

(b) analogous to phenol,

anion from the first deprotonation of 2-hydroxyacetophenone.

(c)

7.46 (c) (cont)

This is a nucleophilic substitution reaction. Silicon is a much larger atom than carbon, so an S_N2 reaction at a 3° Si atom is not as unlikely as it is on C.

(d)

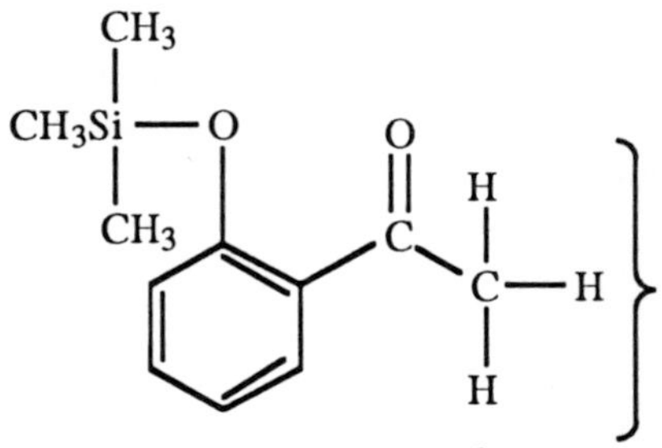

second most acidic protons in 2-hydroxyacetophenone; $pK_a \sim 20$,

analogous to acetone, CH_3CCH_3.

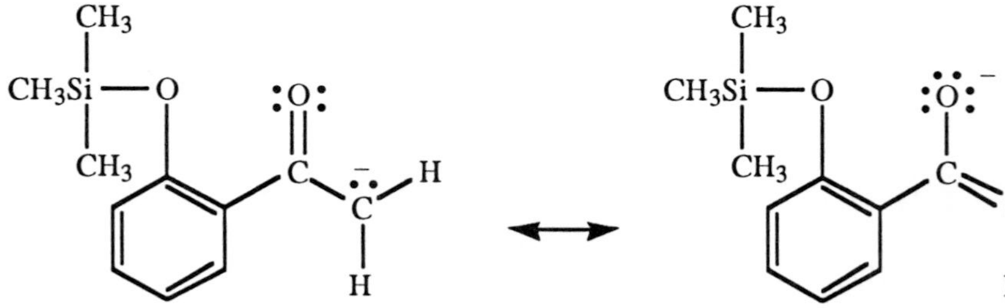

7.47

(a)

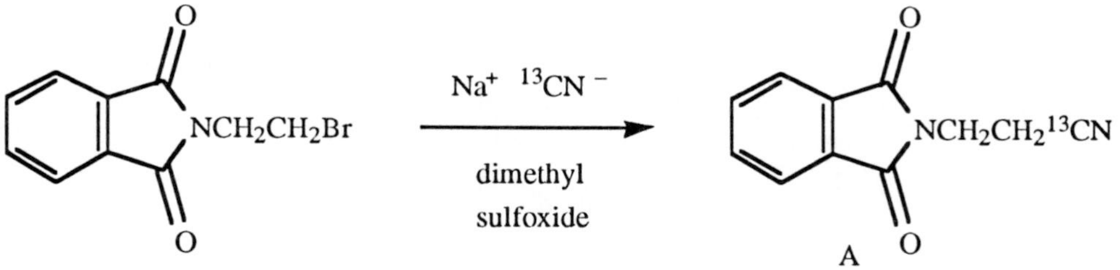

(b)

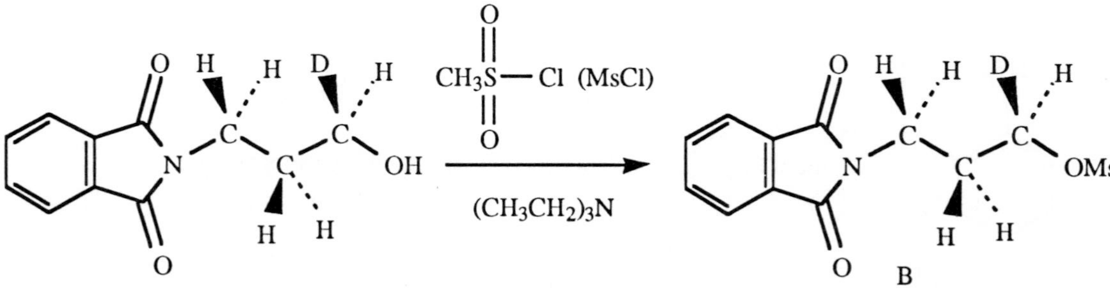

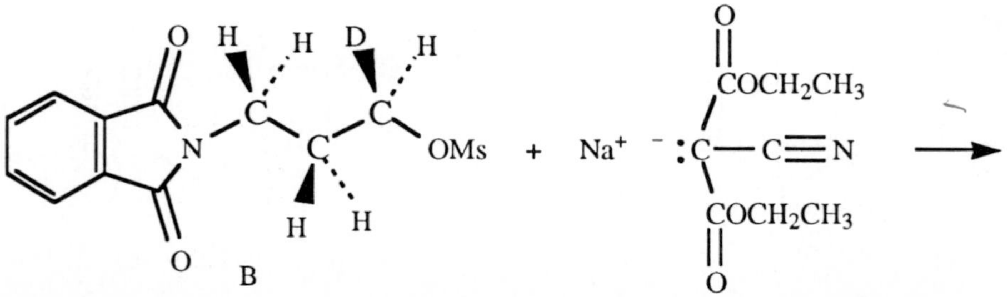

7.47 (b) (cont)

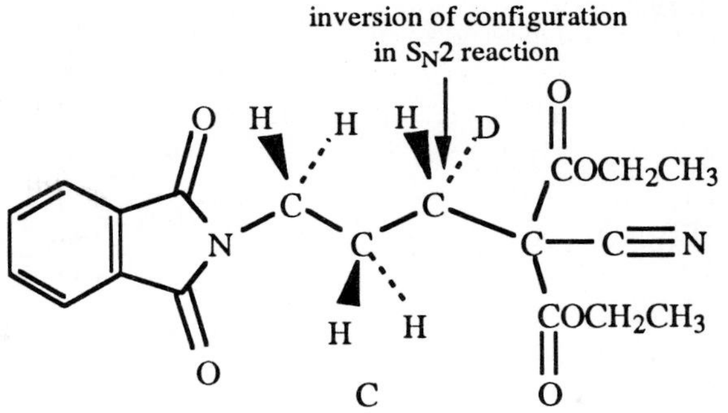

inversion of configuration
in S_N2 reaction

7.48

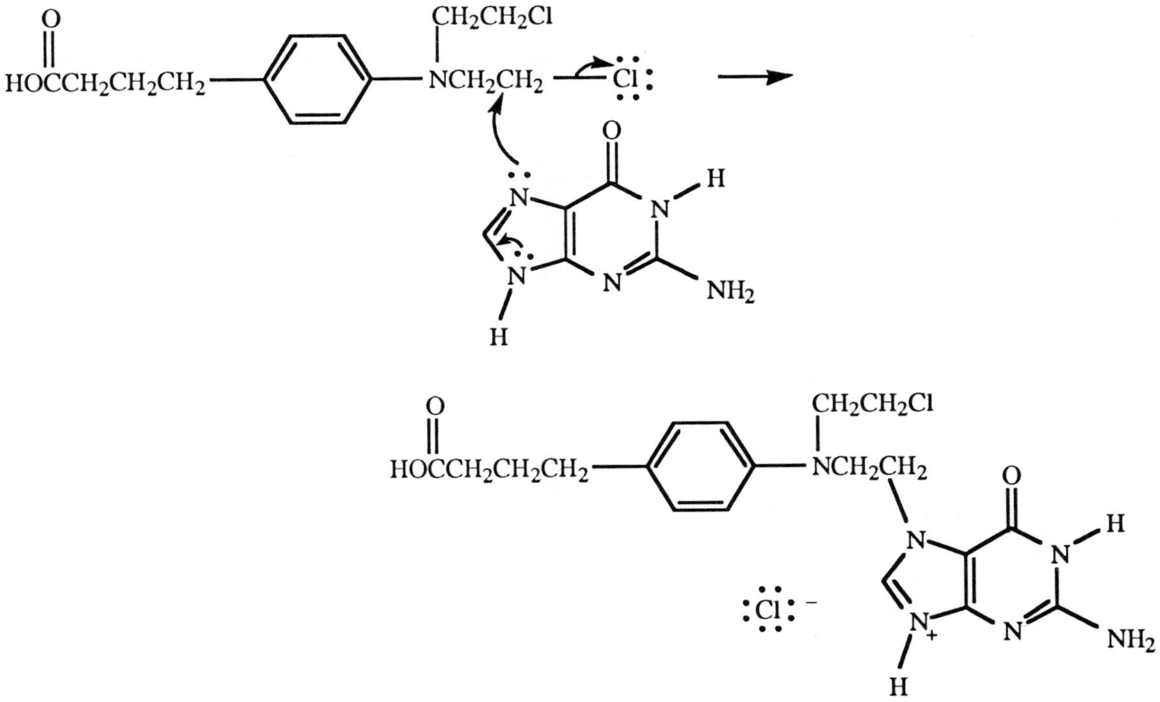

7.49

(a)

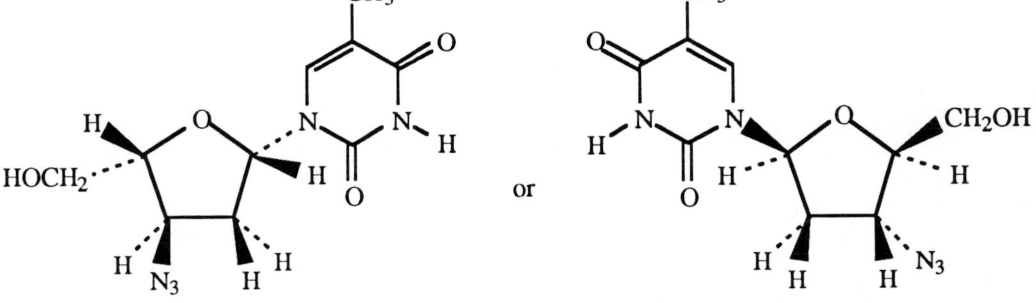

(–)-AZT   or   (–)-AZT

7.49 (cont)

(b)

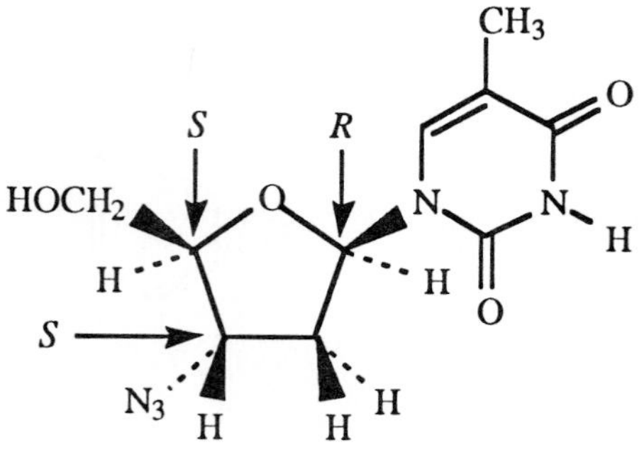

(+)-AZT

(c)   −56°

(d) The number of stereoisomers possible for a compound with n stereocenters is 2^n. There are 3 stereocenters so the total number of stereoisomers possible is 2^3 or 8.

(e) The reaction gives a mixture of stereoisomers at the stereocenter that undergoes substitution. This implies that the reaction went by way of a carbocation intermediate, an S_N1 reaction.

8

Alkenes

Workbook Exercise

Addition to π bonds was first encountered in Chapter 4. Use the questions below to probe your understanding of the structural changes that occur in an addition reaction.

Q1.   Under what experimental conditions does the addition occur?

Q2.   What molecule is adding; what are the groups that add?

Q3.   What is the regioselectivity (for example, the Markovnikov orientation)?

Q4.   What is the stereochemistry? The two groups can add on the same side of the π bond (called **syn** addition), or on opposite sides of the π bond (called **anti** addition).

EXAMPLE

Using a familiar example:

(D = deuterium, ^{2}H)

a mixture of stereoisomers

SOLUTION

Answers to Q1, Q2, and Q3: HCl adds H and Cl with the Markovnikov orientation.

Answer to Q4: Visualizing the changes is often easier if the alkene is represented as perpendicular to the plane of the paper.

Workbook Exercises (cont)

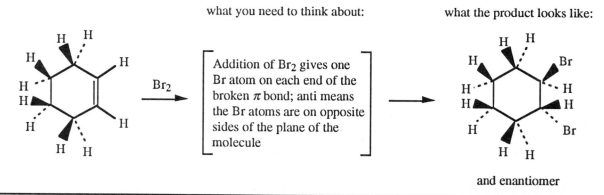

a pair of enantiomers representing
syn (same side) addition

a pair of enantiomers representing
anti (opposite side) addition

An important skill for you to develop is to connect a written (and/or verbal) description of a structural change with a molecular representation of that transformation. The following example will help you develop this skill.

EXAMPLE

Draw the structure corresponding to the anti addition of bromine to cyclohexene (the six-membered hydrocarbon ring with one carbon-carbon double bond).

SOLUTION

Restatement of the verbal description in structural terms:

what you need to think about:

what the product looks like:

Br$_2$

> Addition of Br$_2$ gives one Br atom on each end of the broken π bond; anti means the Br atoms are on opposite sides of the plane of the molecule

and enantiomer

EXERCISE I.   In each case, provide structural formulas for the starting materials, reagents, or products needed to complete the equation. Examine all of the information carefully.

(a)

syn addition of H$_2$

A + B

a pair of diastereomers
derived from the addition to
each face of the π bond

Workbook Exercises (cont)

(b)   C or D or E $\xrightarrow{\text{Markovnikov addition of HBr}}$

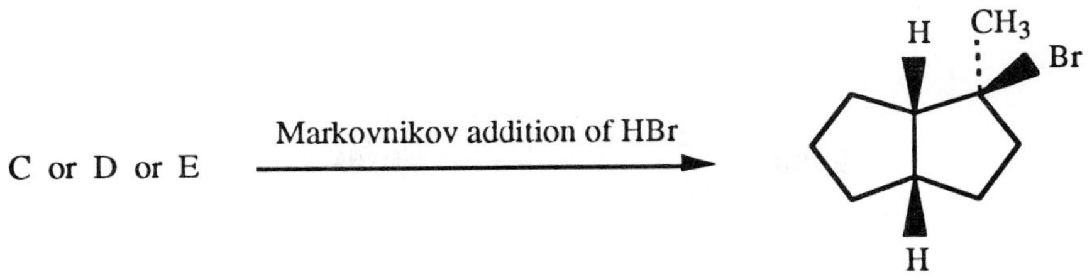

and other isomers

(c)   F $\xrightarrow{\text{syn addition of an oxygen atom}}$

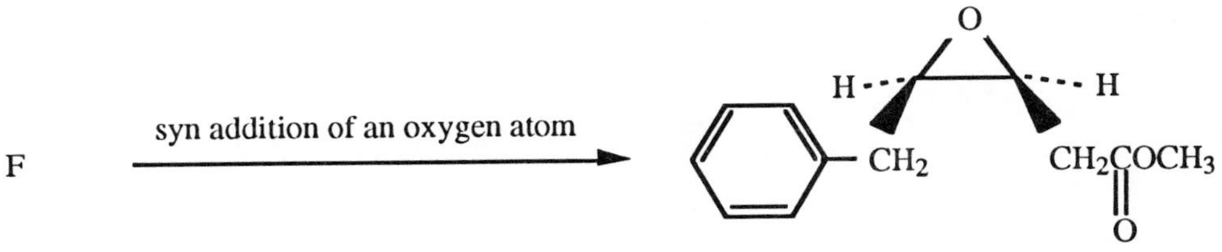

and enantiomer

EXERCISE II. Give a complete description of the changes in connectivity that accompany each of the following reactions.

(a)

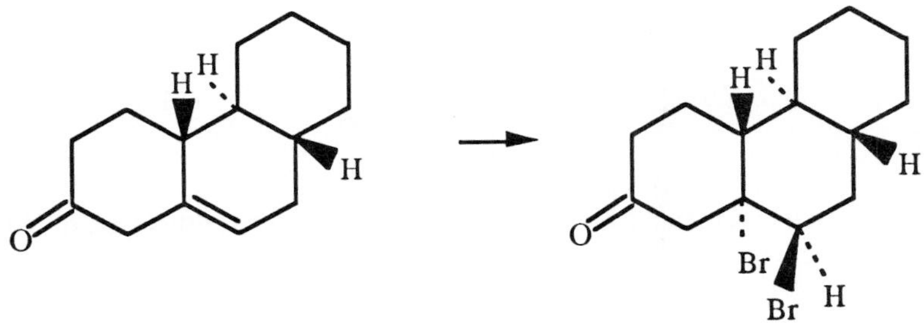

(b)

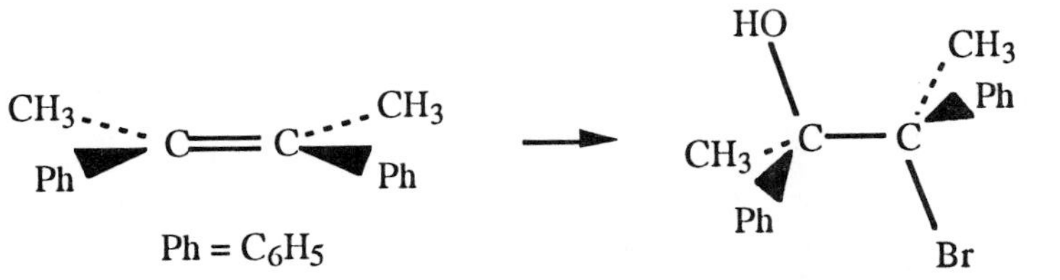

Ph = C$_6$H$_5$

and enantiomer

(c)

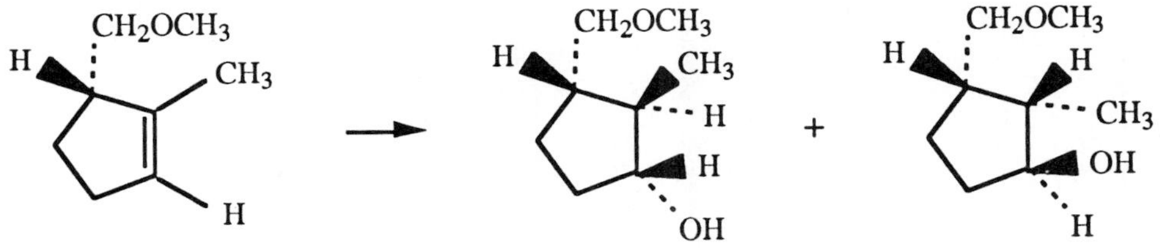

8.1 C_5H_{10}

(1)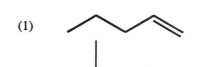

(2)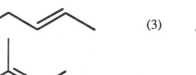

(3)

(4)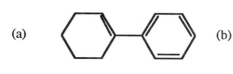

(5)

(6)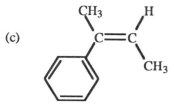

8.2   (1) 1-pentene   (2) (*E*)-2-pentene   (3) (*Z*)-2-pentene

(4) 3-methyl-1-butene   (5) 2-methyl-2-butene   (6) 2-methyl-1-butene

8.3

(a)   (b)   (c)

(d)   (e)   (f)

8.4   No. Each carbon atom of the double bond must have two different substituents for this type of isomerism to be possible.

8.5

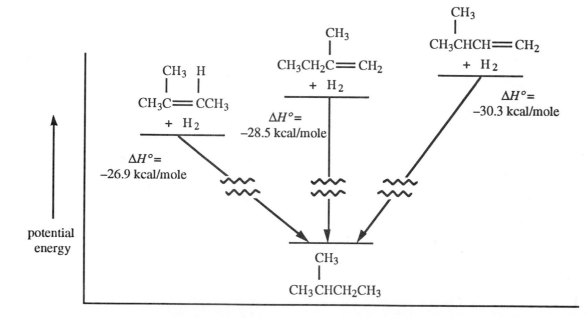

8.6   A concept triangle will be used to answer part (a). Only the equations will be shown for the rest of the problem.

(a)

$$\text{(cyclopentene)} \xrightarrow[\text{H}_2\text{SO}_4]{\text{H}_2\text{O}} \text{?}$$

The starting material is an alkene. The reagent is a mixture of water with a strong acid; therefore, the most abundant species in solution will be the hydronium ion, H_3O^+, an electrophile. The type of reaction is an electrophilic addition to the double bond. Because the electrophile in H_2O is H_3O^+, the overall addition is that of H—OH to the double bond.

The alkene is symmetric so we do not have to worry about regioselectivity in this case.

(b)

(c) $CH_3CH_2CH_2CH_2CH=CH_2 \xrightarrow[\text{H}_2\text{SO}_4]{\text{H}_2\text{O}} CH_3CH_2CH_2CH_2CHCH_3$

with OH below the CH

racemic mixture

(d)

$$\underset{\underset{OH}{|}}{\overset{\overset{CH_3}{|}}{CH_3CCH_3}} \xrightarrow[\Delta]{\text{H}_3\text{PO}_4} \overset{\overset{CH_3}{|}}{CH_3C}=CH_2$$

(e) $\underset{\underset{OH}{|}}{CH_3CHCH_3} \xrightarrow[\Delta]{\text{H}_3\text{PO}_4} CH_3CH=CH_2$

8.6 (cont)

(f) $CH_2 \!=\! CH_2$ $\xrightarrow[\text{H}_2\text{SO}_4]{\text{H}_2\text{O}}$ CH_3CH_2OH

(g)
$$\underset{\underset{CH_3}{|}}{CH_3C}\!=\!CHCH_2CH_3 \xrightarrow[\text{H}_2\text{SO}_4]{\text{H}_2\text{O}} \underset{\underset{OH}{|}}{\overset{\overset{CH_3}{|}}{CH_3C}}CH_2CH_2CH_3$$

(h)

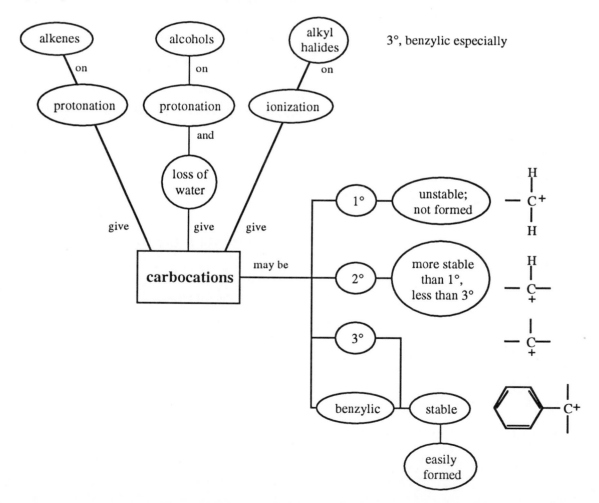

Concept Map 8.1   Reactions that form carbocations.

3°, benzylic especially

Concept Map 8.2   Reactions of carbocations.

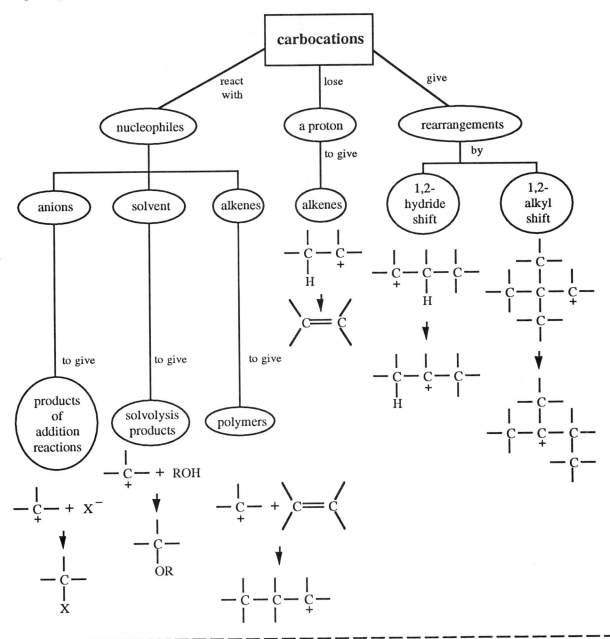

8.7

8.7 (cont)

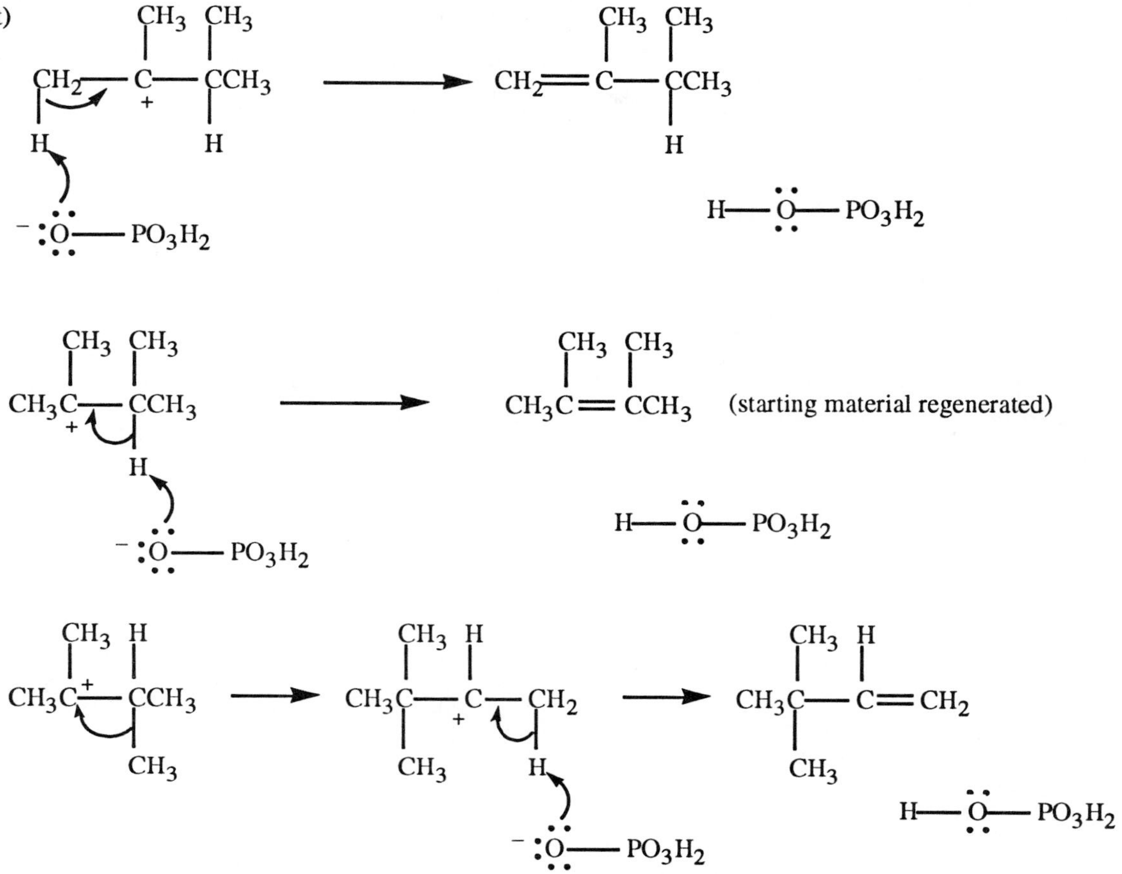

$$CH_3C = CCH_3 \quad \text{(starting material regenerated)}$$

Note that very little of this product is seen. It arises from a secondary cation, which is less stable than a tertiary cation.

8.8

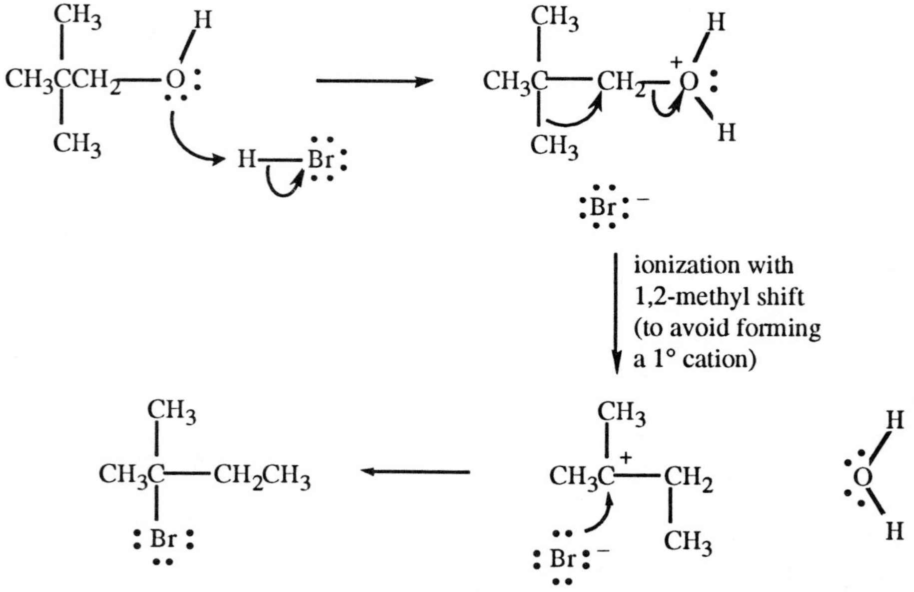

ionization with
1,2-methyl shift
(to avoid forming
a 1° cation)

8.9   Note: H—B$^+$ stands for a Brønsted-Lowry acid and :B stands for the conjugate base of the acid. They are used in mechanisms when more than one acid can transfer a proton and more than one base can accept a proton.

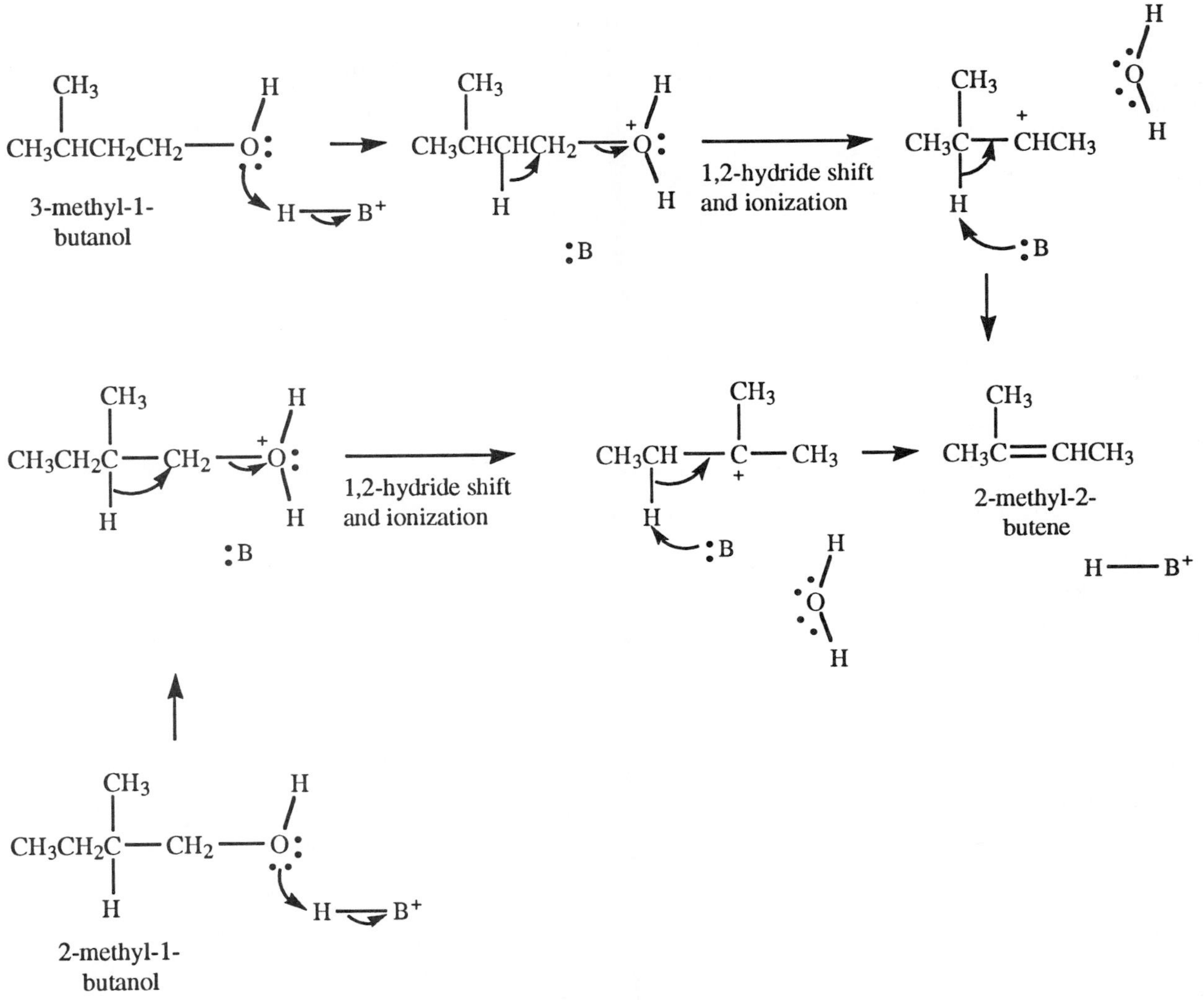

3-methyl-1-butanol

1,2-hydride shift and ionization

2-methyl-1-butanol

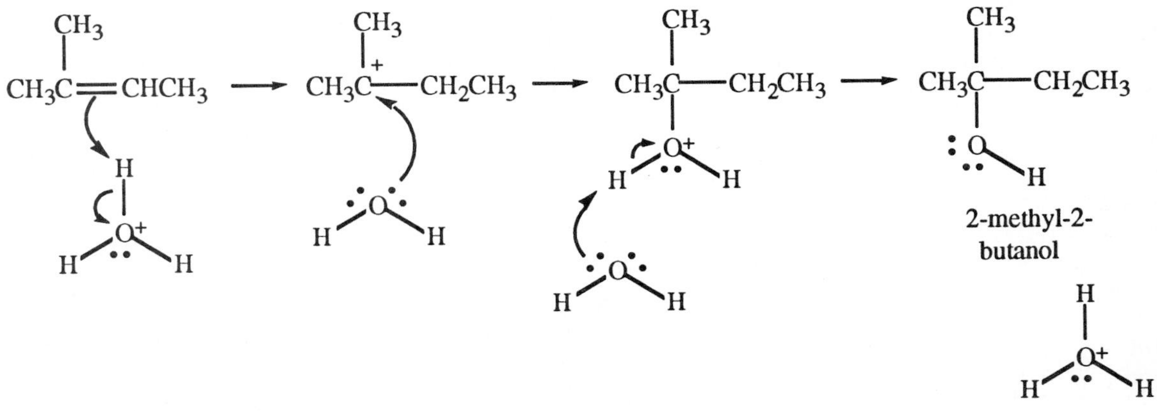

1,2-hydride shift and ionization

2-methyl-2-butene

2-methyl-2-butanol

8.10

$$\text{H} \ddot{\underset{\cdot\cdot}{\text{B}}} \text{H}$$

with H above B

shape is trigonal planar;
hybridization of the
boron atom is *sp²*

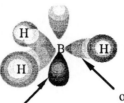

empty *p*
orbital

overlap of *sp²* hybrid
orbital on boron and
1 *s* orbital on hydrogen

8.11

(cyclopentene) $\xrightarrow{\text{BH}_3}$ (tricyclopentylborane)

$$\text{CH}_3(\text{CH}_2)_7\text{CH}=\text{CH}_2 \xrightarrow{\text{BH}_3} \text{CH}_3(\text{CH}_2)_7\text{CH}_2\text{CH}_2\text{B}\underset{\overset{|}{\text{CH}_2\text{CH}_2(\text{CH}_2)_7\text{CH}_3}}{\text{CH}_2\text{CH}_2(\text{CH}_2)_7\text{CH}_3}$$

with CH₂CH₂(CH₂)₇CH₃ above B

$$\underset{\overset{|}{\text{CH}_3}}{\overset{\overset{\text{CH}_3}{|}}{\text{CH}_3\text{C}}}-\text{CH}=\text{CH}_2 \xrightarrow{\text{BH}_3} \underset{\overset{|}{\text{CH}_3}}{\overset{\overset{\text{CH}_3}{|}}{\text{CH}_3\text{C}}}\text{CH}_2\text{CH}_2-\text{B}-\text{CH}_2\text{CH}_2\underset{\overset{|}{\text{CH}_3}}{\overset{\overset{\text{CH}_3}{|}}{\text{C}}}\text{CH}_3$$

with CH₂CH₂C(CH₃)₂CH₃ group above B

8.12

(a) $\text{CH}_3\text{CH}_2\text{CH}_2\underset{\overset{|}{\text{CH}_3}}{\text{C}}=\text{CH}_2 \longrightarrow \text{CH}_3\text{CH}_2\text{CH}_2\underset{\overset{|}{\text{CH}_3}}{\text{CH}}\text{CH}_2-\text{B}$

8.12 (cont)

(b)

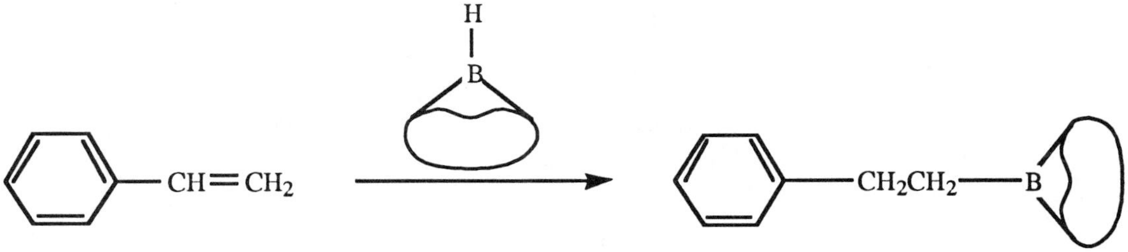

(c)

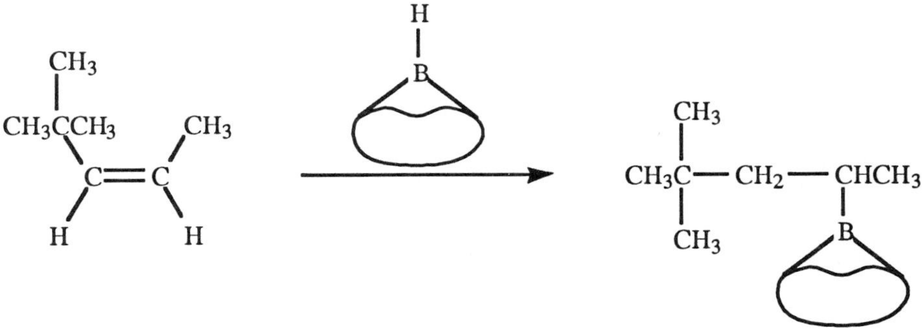

(d)

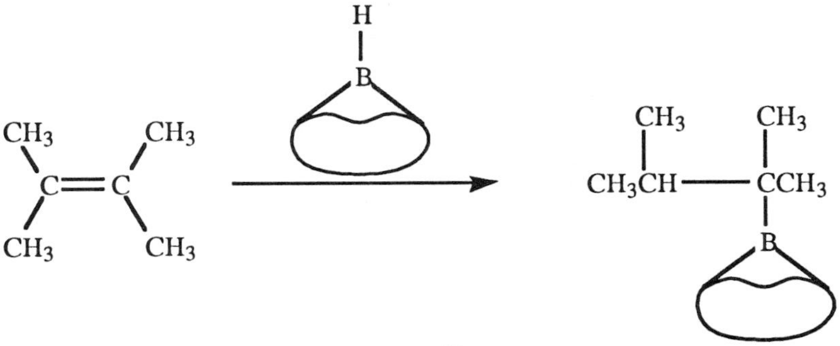

(e)

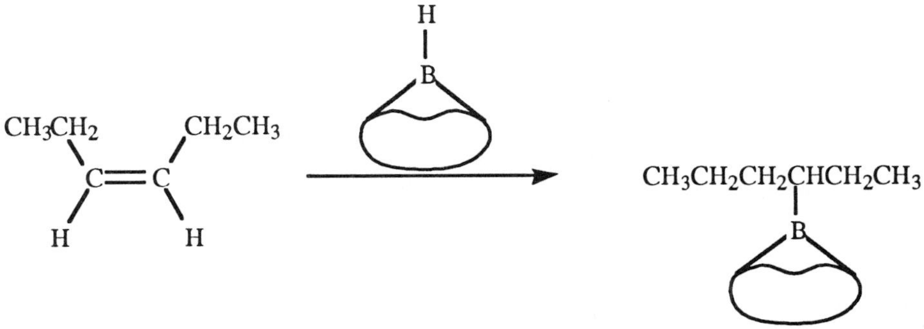

Concept Map 8.3   (see p. 237)

8.13

(a)

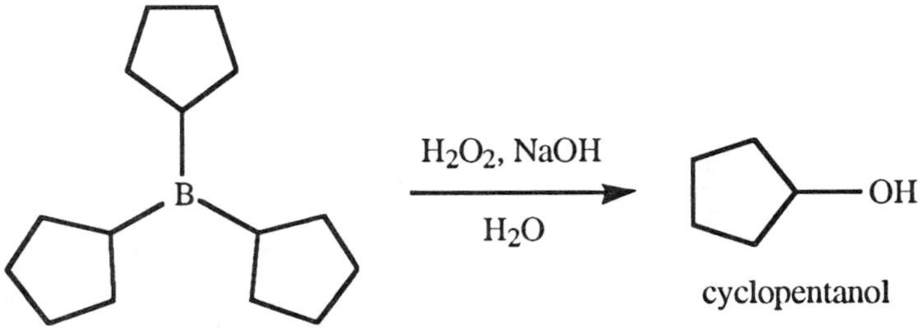

cyclopentanol

8.13 (cont)

(b)

$$CH_2CH_2(CH_2)_7CH_3$$
$$|$$
$$CH_3(CH_2)_7CH_2CH_2BCH_2CH_2(CH_2)_7CH_3 \quad \xrightarrow[\text{H}_2\text{O}]{\text{H}_2\text{O}_2,\ \text{NaOH}} \quad CH_3(CH_2)_8CH_2OH$$

1-decanol

(c)

$$\xrightarrow[\text{H}_2\text{O}]{\text{H}_2\text{O}_2,\ \text{NaOH}}$$

3,3-dimethyl-1-butanol

(a')

$$\xrightarrow[\text{H}_2\text{O}]{\text{H}_2\text{O}_2,\ \text{NaOH}}$$

$$CH_3CH_2CH_2CHCH_2OH$$

2-methyl-1-pentanol

(b')

$$\xrightarrow[\text{H}_2\text{O}]{\text{H}_2\text{O}_2,\ \text{NaOH}}$$

2-phenylethanol

(c')

$$\xrightarrow[\text{H}_2\text{O}]{\text{H}_2\text{O}_2,\ \text{NaOH}}$$

$$CH_3C - CH_2 - CHCH_3$$

4,4-dimethyl-2-pentanol

(d')

$$\xrightarrow[\text{H}_2\text{O}]{\text{H}_2\text{O}_2,\ \text{NaOH}}$$

$$CH_3CH - CCH_3$$

2,3-dimethyl-2-butanol

(e')

$$CH_3CH_2CH_2CHCH_2CH_3 \quad \xrightarrow[\text{H}_2\text{O}]{\text{H}_2\text{O}_2,\ \text{NaOH}} \quad CH_3CH_2CH_2CHCH_2CH_3$$
$$| $$
$$OH$$

3-hexanol

Concept Map 8.3 The hydroboration reaction.

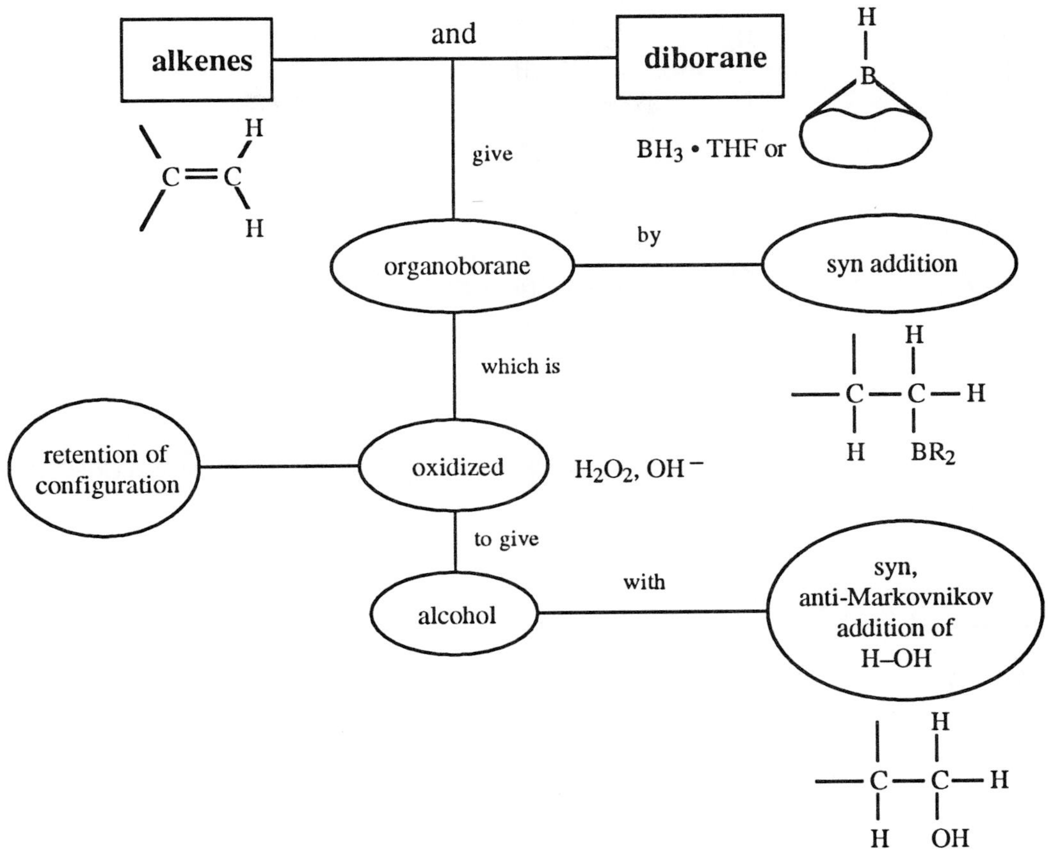

8.14

(a)

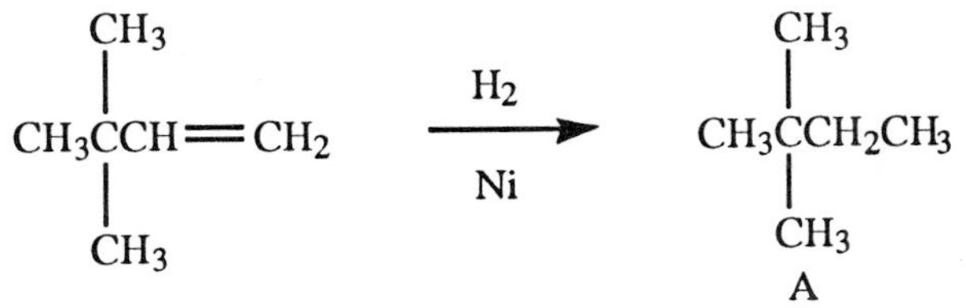

(b)

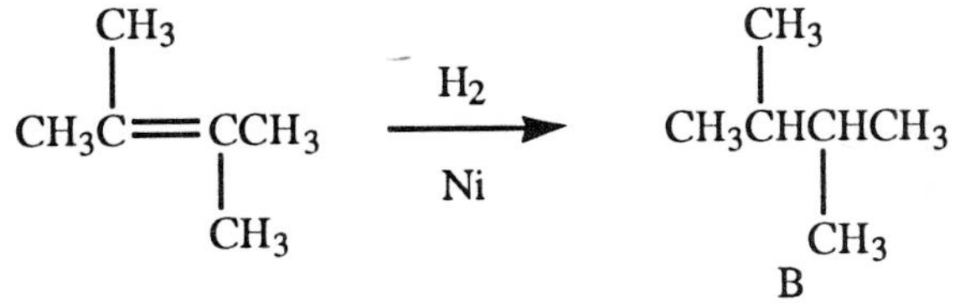

(c)

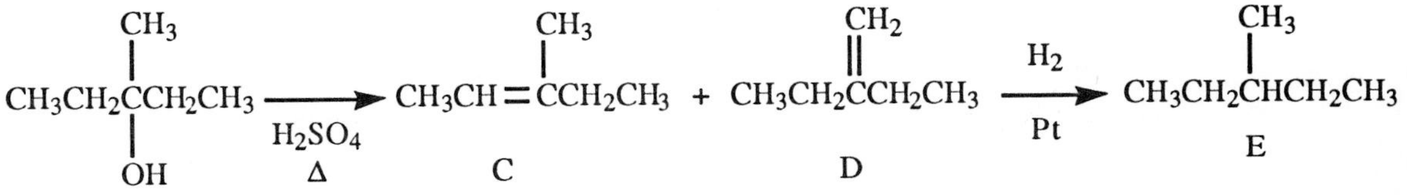

8.14 (cont)

(d)

$$\begin{array}{c} CH_3 \\ | \\ CH_3CHC{=}CH_2 \\ | \\ CH_3 \end{array} \quad \xrightarrow[\text{Ni}]{H_2} \quad \begin{array}{c} CH_3 \\ | \\ CH_3CHCHCH_3 \\ | \\ CH_3 \end{array}$$

F

(e)

$$\begin{array}{c} CH_3 \\ | \\ CH_3CCH_2CH_2CH_3 \\ | \\ OH \end{array} \quad \xrightarrow[\Delta]{H_3PO_4} \quad \begin{array}{c} CH_3 \\ | \\ CH_3C{=}CHCH_2CH_3 \end{array} \quad + \quad \begin{array}{c} CH_3 \\ | \\ CH_2{=}CCH_2CH_2CH_3 \end{array} \quad \xrightarrow[\text{Pt}]{H_2}$$

G   H

$$\begin{array}{c} CH_3 \\ | \\ CH_3CHCH_2CH_2CH_3 \end{array}$$

I

- -

Concept Map 8.4   Catalytic hydrogenation reactions.

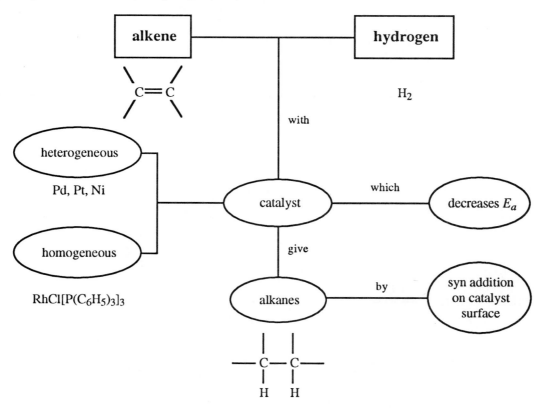

8.15

(a) $CH_3CH_2CH=CHCH_2CH_3$ $\xrightarrow[\text{Pt}]{\text{H}_2}$ $CH_3CH_2CH_2CH_2CH_2CH_3$

(b) $CH_3CH_2CH_2\overset{\displaystyle CH_3}{\underset{\displaystyle}{C}}=CH_2$ $\xrightarrow[\substack{\text{RhCl}[\text{P}(\text{C}_6\text{H}_5)_3]_3 \\ \text{benzene}}]{\text{H}_2}$ $CH_3CH_2CH_2\overset{\displaystyle CH_3}{\underset{\displaystyle}{CH}}CH_3$

(c)   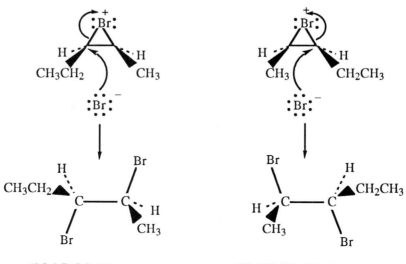 — structure with CH$_3$ $\xrightarrow[\text{Pd/C}]{\text{H}_2}$ — structure with CH$_3$

(d) $\xrightarrow[\text{Pd/C}]{\text{H}_2 \text{ (excess)}}$

(e) $CH_3\overset{\displaystyle CH_3}{\underset{\displaystyle CH_3}{C}}CH=CH_2$ $\xrightarrow[\substack{\text{RhCl}[\text{P}(\text{C}_6\text{H}_5)_3]_3 \\ \text{benzene}}]{\text{H}_2}$ $CH_3\overset{\displaystyle CH_3}{\underset{\displaystyle CH_3}{C}}CH_2CH_3$

(f) $\xrightarrow[\substack{\text{RhCl}[\text{P}(\text{C}_6\text{H}_5)_3]_3 \\ \text{benzene}}]{\text{H}_2}$

(g) $CH_3CH_2CH_2CH_2CH_2CH=CH_2$ $\xrightarrow[\substack{\text{RhCl}[\text{P}(\text{C}_6\text{H}_5)_3]_3 \\ \text{benzene}}]{\text{H}_2}$ $CH_3CH_2CH_2CH_2CH_2CH_2CH_3$

8.16   1.   Attack at carbon 3 of the enantiomeric bromonium ions derived from (Z)-2-pentene.

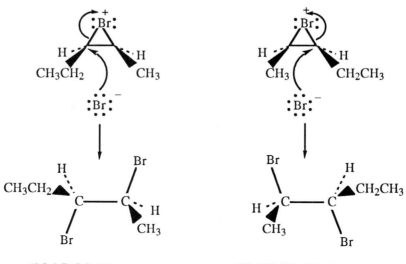

(2*S*, 3*S*)-2,3-dibromopentane   (2*R*, 3*R*)-2,3-dibromopentane

8.16 (1) (cont)

This is the same mixture as that obtained by attack at carbon 2 shown on p. ~ in the text.

2. Attack at carbon 2 of the enantiomeric bromonium ions derived from (*E*)-2-pentene.

(2*S*, 3*R*)-2,3-dibromopentane   (2*R*, 3*S*)-2,3-dibromopentane

Attack at carbon 3 of the enantiomeric bromonium ions derived from (*E*)-2-pentene.

(2*R*, 3*S*)-2,3-dibromopentane   (2*S*,3*R*)-2,3-dibromopentane

This is the same mixture as that obtained by attack at carbon 2.

8.17   Part (a) will be analyzed in detail. Only the equations will be shown for the rest of the problem.

(a)

$$CH_3CH_2 \quad\quad H$$
$$C = C$$
$$H \quad\quad CH_2CH_3$$

$\xrightarrow[\text{tetrachloride}]{\text{carbon}}^{Br_2}$?

1.   To what functional group class do the reactants belong?

One reactant is an alkene. The double bond of the alkene is a nucleophile. The other reagent is a halogen, an electrophile.

2.   Does either reactant have a good leaving group?

No.

3.   Are any of the reactants good acids, bases, nucleophiles, or electrophiles? Is there an ionizing solvent present?

Bromine, above the arrow, is an electrophile. The solvent, carbon tetrachloride, is not an ionizing solvent.

4.   What is the most likely first step for the reaction? Most common reactions can be classified either as protonation-deprotonation, or as reactions of a nucleophile with an electrophile.

Attack by the nucleophilic double bond on the electrophilic bromine leading to the formation of a bromonium ion is the first step.

5.   What are the properties of the species present in the reaction mixture after the first step? Is any further reaction likely?

The species present in the reaction mixture are the bromonium ion, an electrophile, and bromide ion, a nucleophile. The bromonium ion will react with bromide ion to give the dibromide product.

6.   What is the stereochemistry of the reaction?

The opening of the bromonium ion by bromide leads to an overall anti addition and the creation of two stereocenters. The product, however, has a plane of symmetry and is therefore a meso compound.

8.17 (a) (cont)

Complete correct answer:

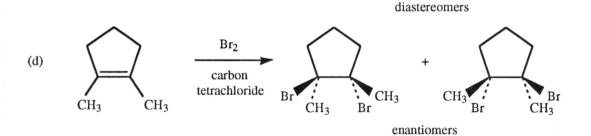

(3*R*,4*S*)-3,4-dibromohexane
meso-3,4-dibromohexane

(b)

racemic mixture

(c)

diastereomers

(d)

enantiomers

8.18   Formation of product from an unsymmetrical bromonium ion intermediate.

(1*S*, 2*R*)-2-bromo-1-phenyl-1-propanol

8.18 (cont)

(1*R*, 2*S*)-2-bromo-1-phenyl-1-propanol

Formation of products from a carbocation.

(1*S*, 2*S*)-2-bromo-1-
phenyl-1-propanol

(1*R*, 2*S*)-2-bromo-1-
phenyl-1-propanol

8.18 (cont)

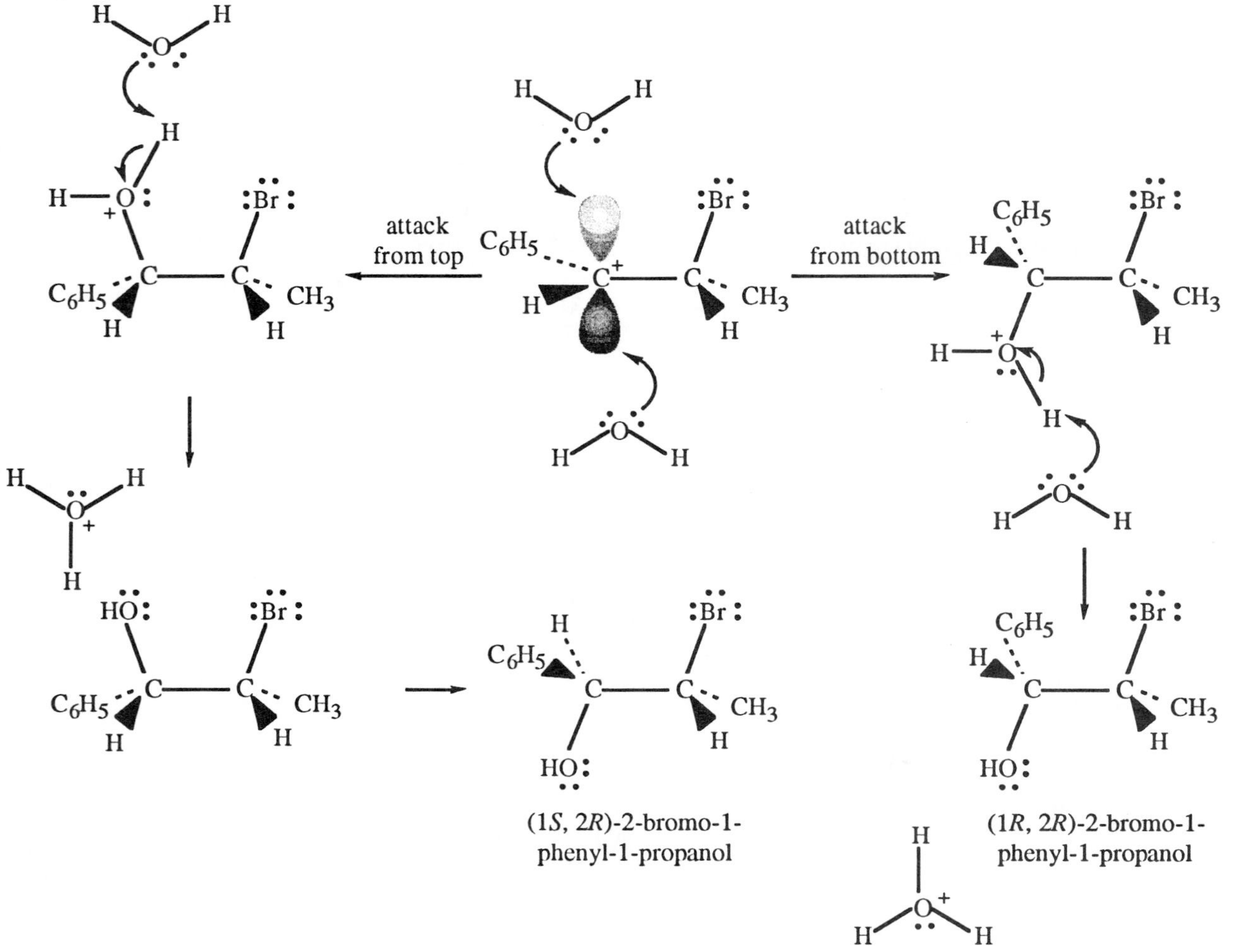

(1*S*, 2*R*)-2-bromo-1-
phenyl-1-propanol

(1*R*, 2*R*)-2-bromo-1-
phenyl-1-propanol

8.19

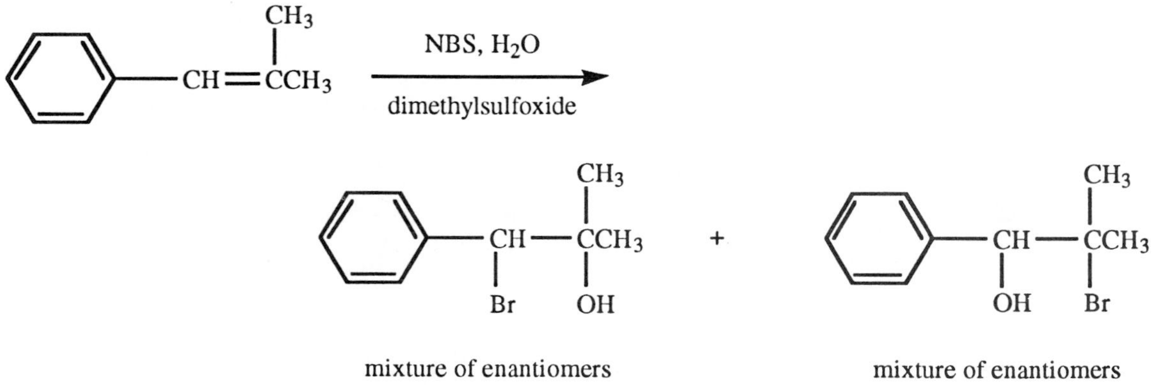

$$\text{Ph—CH}=\text{C(CH}_3\text{)CH}_3 \xrightarrow[\text{dimethylsulfoxide}]{\text{NBS, H}_2\text{O}}$$

mixture of enantiomers   +   mixture of enantiomers

Benzylic and tertiary carbocations are both relatively stable intermediates. Therefore, both the benzylic carbon atom and the tertiary carbon atom have about the same amount of partial charge in the bromonium ion. Attack by the nucleophile can occur at either carbon.

Concept Map 8.5   The addition of bromine to alkenes.

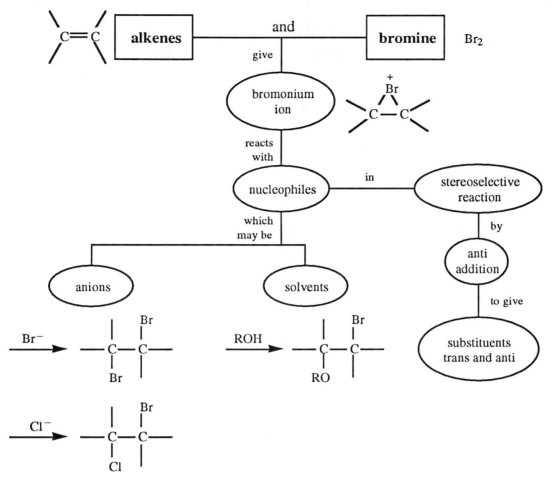

8.20

8.21

(a)

$$CH_3C(CH_3)=CHCH_3 \xrightarrow{O_3} \xrightarrow[H_2O]{Zn} CH_3\overset{\displaystyle O}{\overset{\|}{C}}CH_3 \;+\; CH_3\overset{\displaystyle O}{\overset{\|}{C}}H$$

(b)

$$CH_3\overset{\displaystyle CH_3}{\underset{|}{CH}}CH=\overset{\displaystyle CH_2CH_3}{\underset{|}{C}}CH_2CH_3 \xrightarrow{O_3,\ CH_3OH} \xrightarrow{(CH_3)_2S} CH_3\overset{\displaystyle CH_3}{\underset{|}{CH}}-\overset{\displaystyle O}{\overset{\|}{C}}H \;+\; CH_3CH_2\overset{\displaystyle O}{\overset{\|}{C}}CH_2CH_3$$

(c)

$$CH_3\overset{\displaystyle CH_3}{\underset{|}{CH}}CH=\overset{\displaystyle CH_2CH_3}{\underset{|}{C}}CH_2CH_3 \xrightarrow[H_2O]{O_3} \xrightarrow{H_2O_2,\ NaOH} \xrightarrow{H_3O^+} CH_3\overset{\displaystyle CH_3}{\underset{|}{CH}}-\overset{\displaystyle O}{\overset{\|}{C}}OH \;+\; CH_3CH_2\overset{\displaystyle O}{\overset{\|}{C}}CH_2CH_3$$

(d)

1-methylcyclohexene $\xrightarrow{O_3,\ CH_3OH} \xrightarrow{(CH_3)_2S} CH_3\overset{\displaystyle O}{\overset{\|}{C}}CH_2CH_2CH_2CH_2\overset{\displaystyle O}{\overset{\|}{C}}H$

(e)

cyclododecadiene $\xrightarrow[\substack{\text{dichloromethane} \\ 0\ °C}]{O_3} \xrightarrow{H_2O_2,\ NaOH} 2\ Na^+\ {}^-O\overset{\displaystyle O}{\overset{\|}{C}}CH_2CH_2CH_2CH_2\overset{\displaystyle O}{\overset{\|}{C}}O^-\ Na^+$

(f)

$$CH_3CH_2\overset{\displaystyle CH_3}{\underset{|}{C}}=CH_2 \xrightarrow{O_3} \xrightarrow[H_2O]{Zn} CH_3CH_2\overset{\displaystyle O}{\overset{\|}{C}}CH_3 \;+\; H\overset{\displaystyle O}{\overset{\|}{C}}H$$

(g)

$$CH_3(CH_2)_{13}CH=CH_2 \xrightarrow[H_2O]{O_3} \xrightarrow{H_2O_2,\ NaOH} \xrightarrow{H_3O^+} CH_3(CH_2)_{13}\overset{\displaystyle O}{\overset{\|}{C}}OH \;+\; H\overset{\displaystyle O}{\overset{\|}{C}}OH$$

8.22

$$CH_3CH=\overset{\displaystyle CH_3}{\underset{\underset{\displaystyle CH_3}{|}}{\underset{|}{C}}}CH_2\overset{\displaystyle CH_3}{\underset{|}{C}}CH_3 \;+\; CH_3CH_2\overset{\displaystyle CH_3}{\underset{\underset{\displaystyle CH_3}{|}}{\underset{|}{C}}}=CH\overset{}{C}CH_3 \xrightarrow{O_3} \xrightarrow[H_2O]{Zn}$$

90%   10%

$$CH_3\overset{\displaystyle O}{\overset{\|}{C}}H \;+\; CH_3CH_2\overset{\displaystyle O}{\overset{\|}{C}}CH_2\overset{\displaystyle CH_3}{\underset{\underset{\displaystyle CH_3}{|}}{\underset{|}{C}}}CH_3 \;+\; CH_3CH_2\overset{\displaystyle O}{\overset{\|}{C}}CH_2CH_3 \;+\; CH_3\overset{\displaystyle CH_3}{\underset{\underset{\displaystyle CH_3}{|}}{\underset{|}{C}}}-\overset{\displaystyle O}{\overset{\|}{C}}H$$

~45%   ~45%   ~5%   ~5%

8.23 $CH_3CH_2\overset{\displaystyle O}{\overset{\|}{C}}CH_2CH_3$ + $CH_3\overset{\displaystyle O}{\overset{\|}{C}}H$ $\xleftarrow{\text{ozonolysis}}$ $CH_3CH\!=\!\overset{\displaystyle CH_2CH_3}{C}CH_2CH_3$ $\xleftarrow{\text{dehydration}}$ $CH_3CH_2\overset{\displaystyle CH_2CH_3}{\underset{OH}{C}}CH_2CH_3$

original alcohol

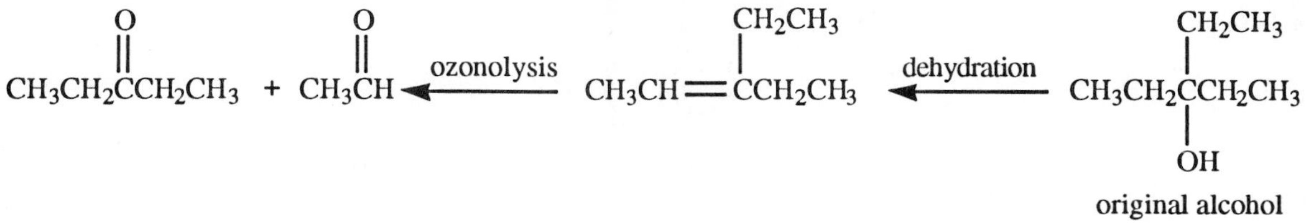

8.24

(a)

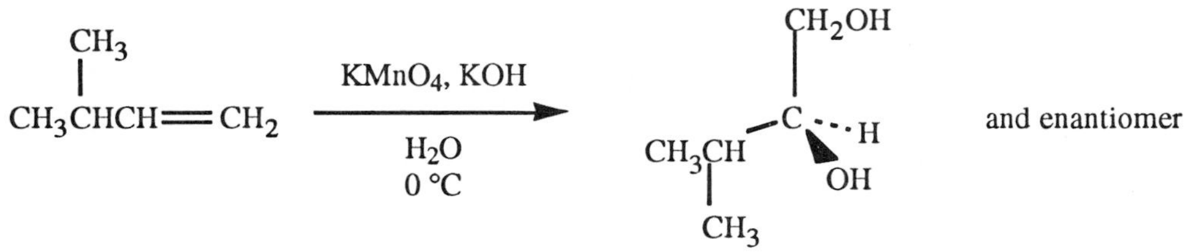

(b)

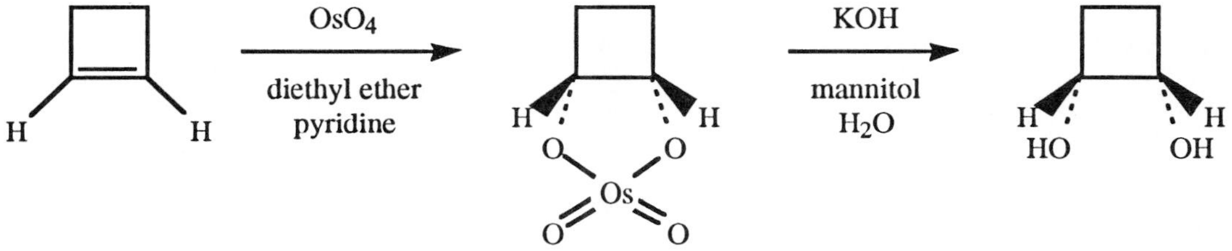

(c)

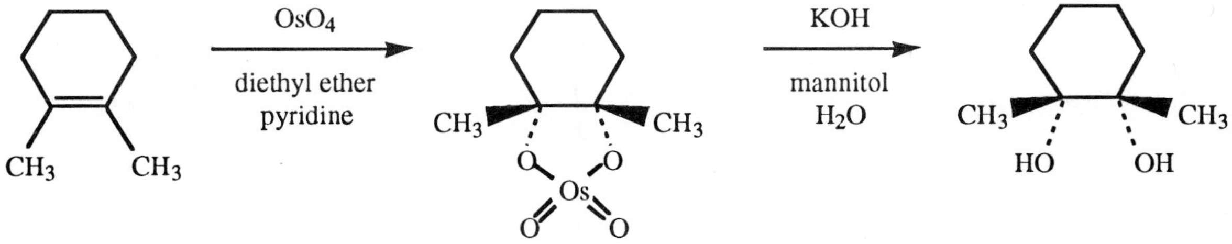

(d)

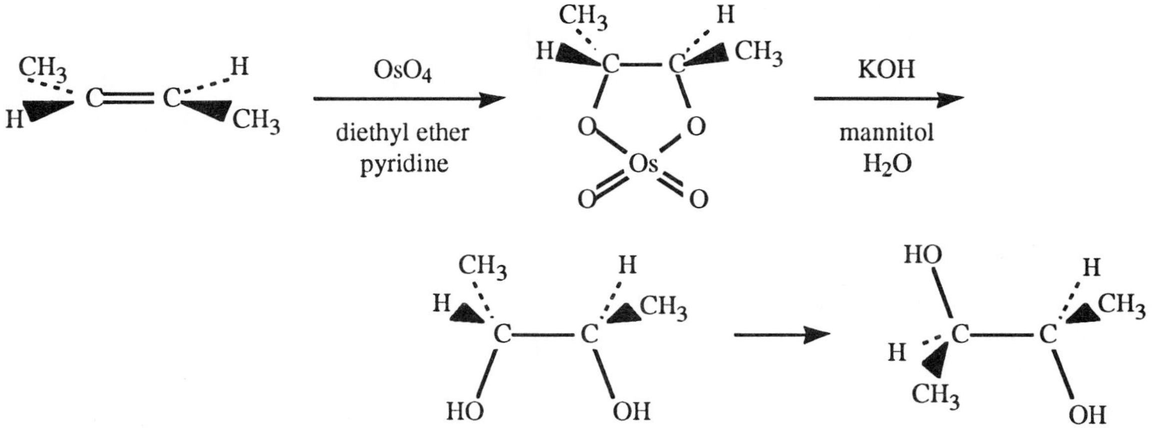

and enantiomer
(higher energy conformation)

and enantiomer
(lower energy conformation)

8.24 (cont)

(e)

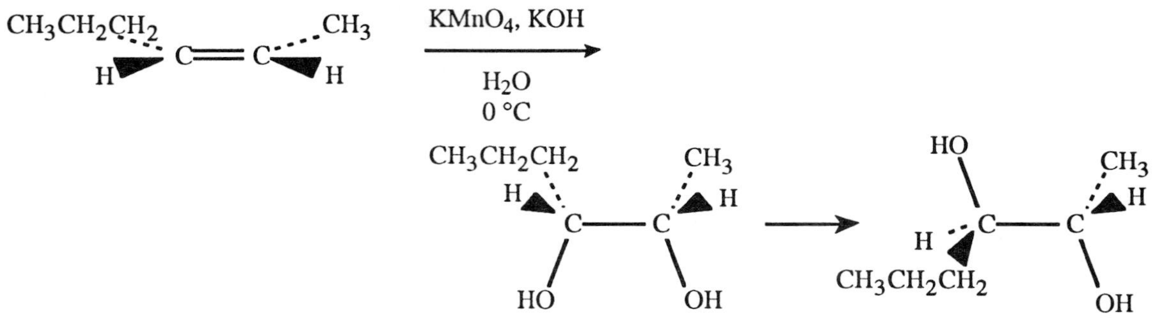

CH₃CH₂CH₂ and CH₃ groups shown on alkene reacting with KMnO₄, KOH, H₂O, 0 °C

and enantiomer

(higher energy
conformation)

and enantiomer

(lower energy
conformation)

Concept Map 8.6   The oxidation reactions of alkenes.

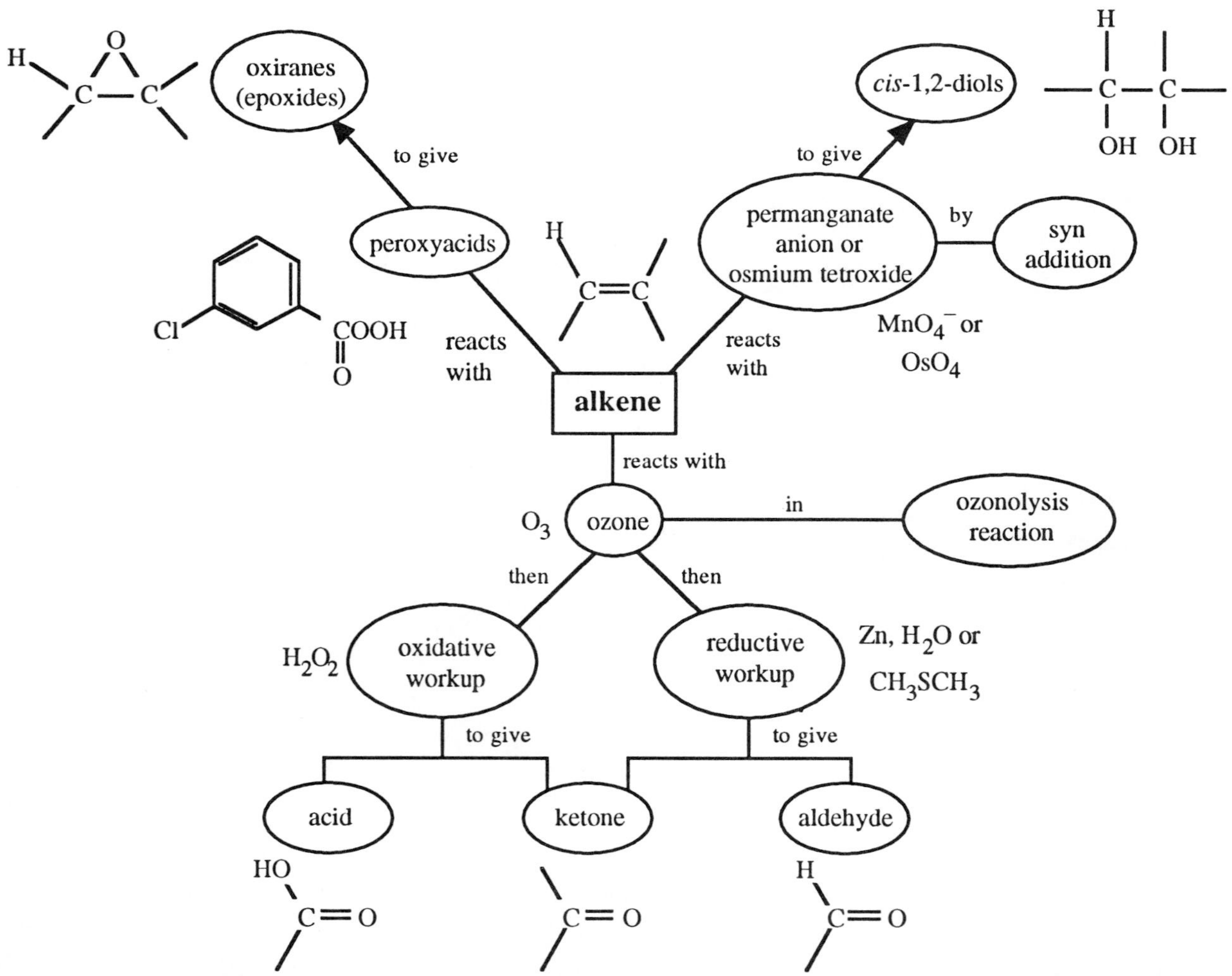

8.25

(a)

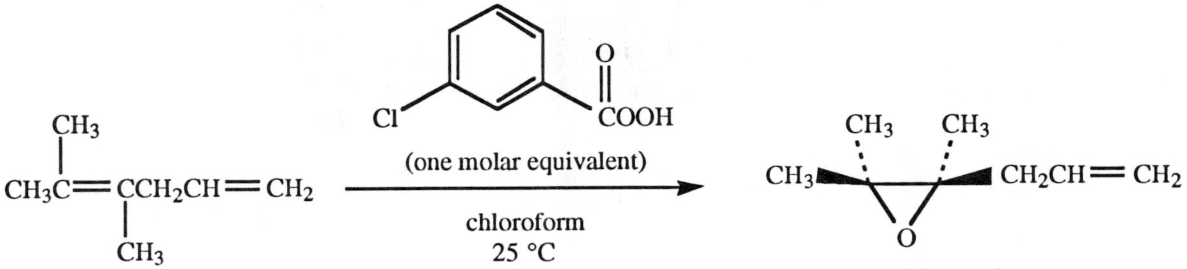

and enantiomer

(b)

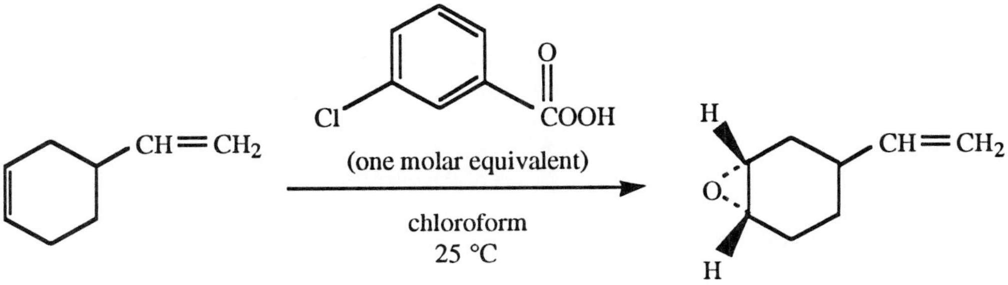

mixture of diastereomers

Note that in parts (a) and (b), the more highly substituted double bond is more nucleophilic and therefore reacts faster with the peroxyacid than the less substituted double bond.

(c) $CH_3CH_2CH_2CH=CH_2$

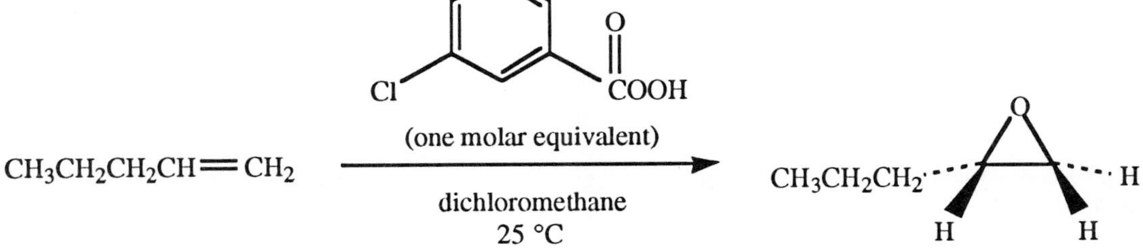

and enantiomer

(d)

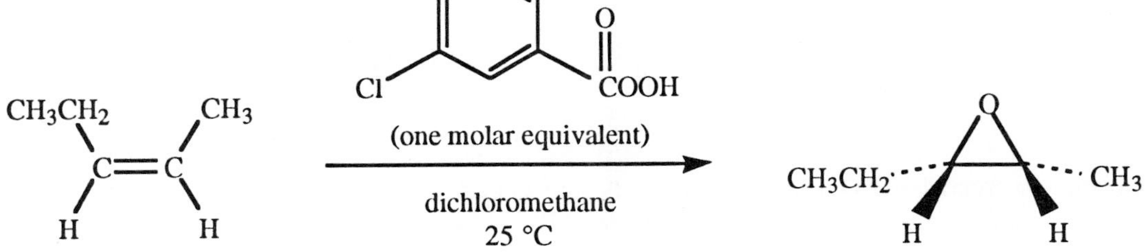

and enantiomer

(e)

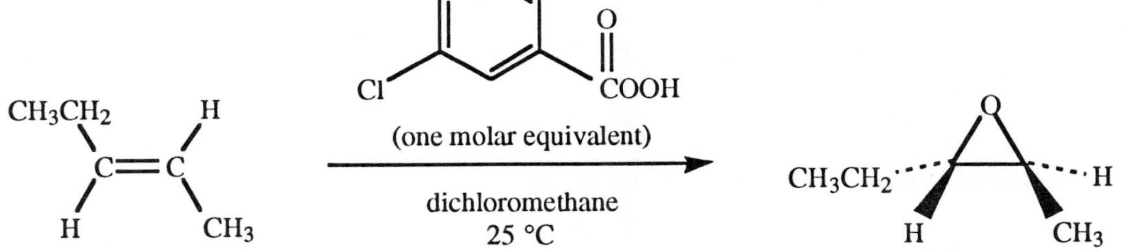

and enantiomer

8.26

1. What are the connectivities of the two compounds? How many carbon atoms does each contain? Are there any rings? What are the positions of branches and functional groups on the carbon skeletons?

(E)-2-butene

(2S, 3S)-2,3-butanediol
drawn in the eclipsed
conformation. No plane
of symmetry is present.

more stable anti
conformation of
(2S, 3S)-2,3-butanediol

one of the enantiomeric products

The carbon skeletons of the two compounds are identical. Each contains four carbon atoms. The two carbon atoms that were doubly bonded in the starting material have hydroxyl groups on them in the product. The methyl groups in the starting alkene and in the eclipsed conformation of the product diol are on the opposite side of the two reacting carbon atoms.

2. How do the functional groups change in going from starting material to product? Does the starting material have a good leaving group?

The starting material is an alkene. It contains a double bond. The product is a diol with the two hydroxyl groups on adjacent carbon atoms. There is no leaving group present in the starting material.

3. Is it possible to dissect the structures of the starting material and product to see which bonds must be broken and which formed?

4. New bonds are created when an electrophile reacts with a nucleophile. Do we recognize any part of the product molecule as coming from a nucleophile or an electrophilic addition reaction?

A double bond is converted to a diol by reaction with an oxidizing agent (an electrophile) such as permanganate ion or osmium tetroxide. The reaction of permanganate ion or osmium tetroxide is a syn addition so would give the stereochemistry we need to get from an alkene with methyl groups on opposite sides of the double bond to a diol, which also has methyl groups on the opposite sides of the molecule in its eclipsed conformation.

5. Do we see how to get from starting material to the product?

The starting material can be converted directly to product.

8.26 (cont)

Complete correct answer:

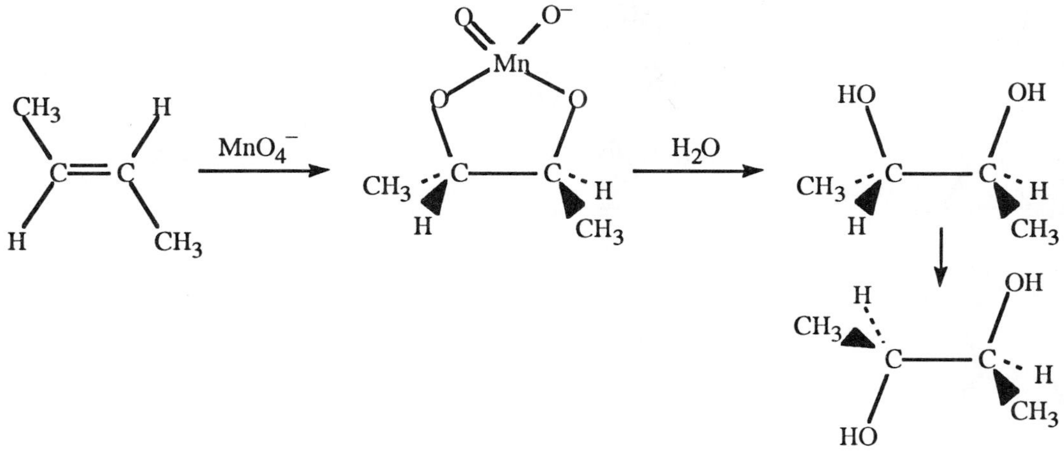

and enantiomer

and enantiomer

8.27 (a) (*E*)-3-methyl-3-heptene (b) 3-ethyl-2,5-dimethyl-2-hexene

(c) (*E*)-3,4-dichloro-3-heptene (d) (*Z*)-3,4-dimethyl-3-heptene

(e) 2-methyl-1-hexene (f) (*Z*)-2-chloro-3-ethyl-3-hexene

(g) 3-bromocyclohexene (h) (*R*)-1,3-dichlorobutane

(i) (*Z*)-4-bromo-3-ethyl-2-pentene (j) 3-chloro-2-methyl-2-pentene

(k) *cis*-1,2-dibromocyclobutane
meso-1,2-dibromocyclobutane
(1*R*, 2*S*)-1,2-dibromocyclobutane
(1*S*, 2*R*)-1,2-dibromocyclobutane

(l) (2*R*, 3*S*)-2-bromo-3-chlorobutane

8.28 (a)

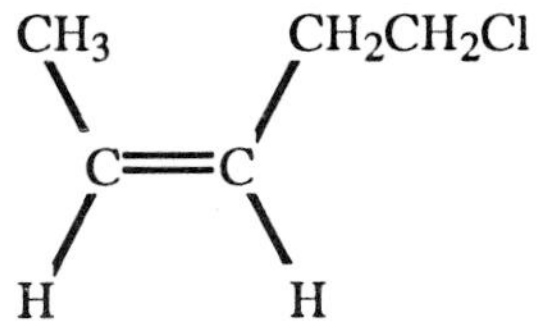

(b)

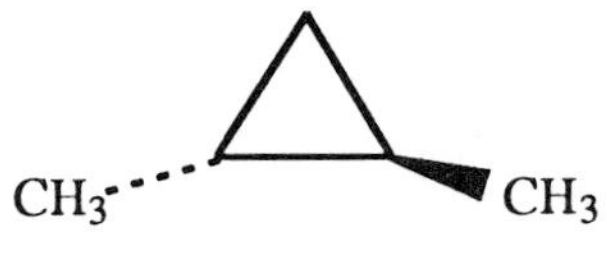

and enantiomer

(c)

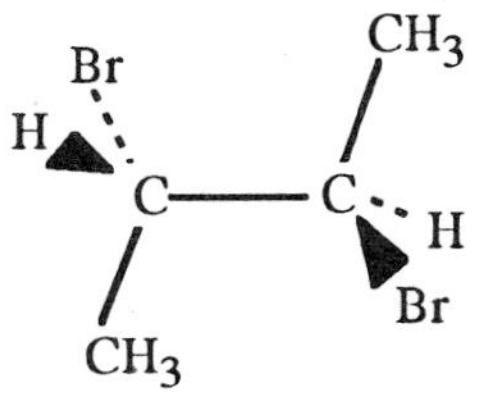

(d)

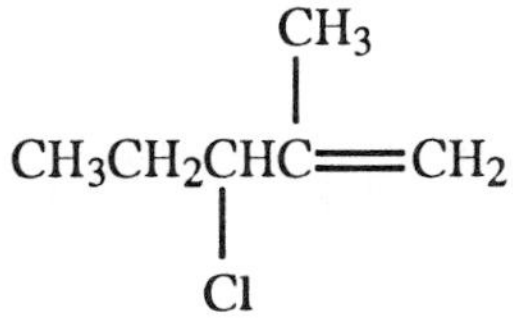

(e)

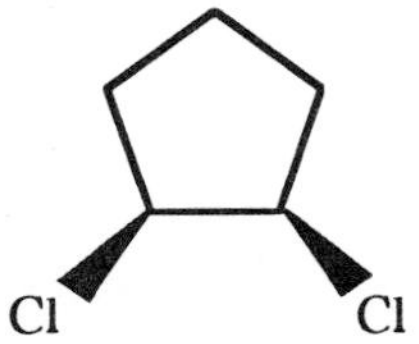

(f)

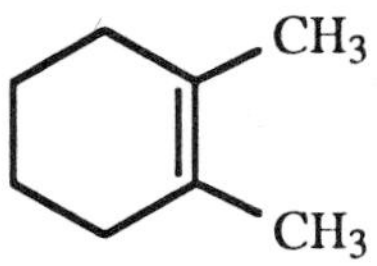

8.29

(a)

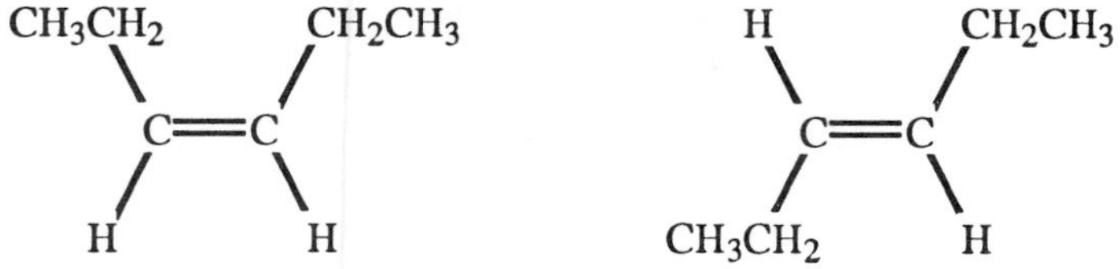

CH₃CH₂, CH₂CH₃, C=C, H, H   H, CH₂CH₃, C=C, CH₃CH₂, H

(b)

CH₃ ··· CH₃   CH₃ ··· CH₃   CH₃ ··· CH₃

(c)

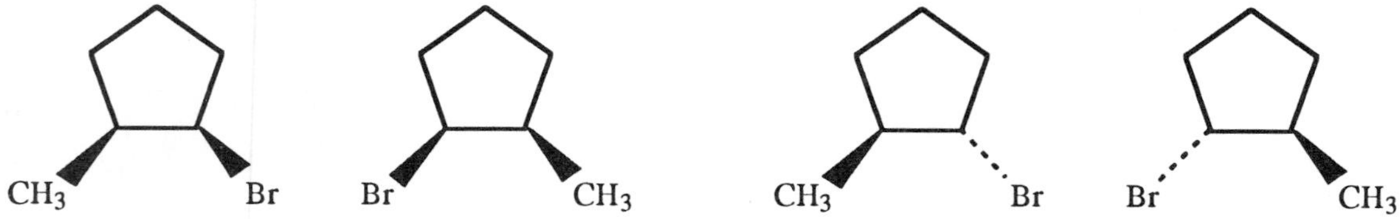

CH₃ ··· Br   Br ··· CH₃   CH₃ ··· Br   Br ··· CH₃

(d)

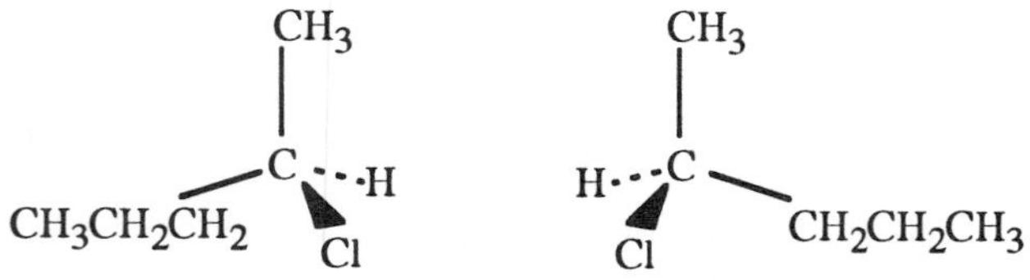

CH₃, C, H, CH₃CH₂CH₂, Cl   CH₃, C, H, Cl, CH₂CH₂CH₃

(e)

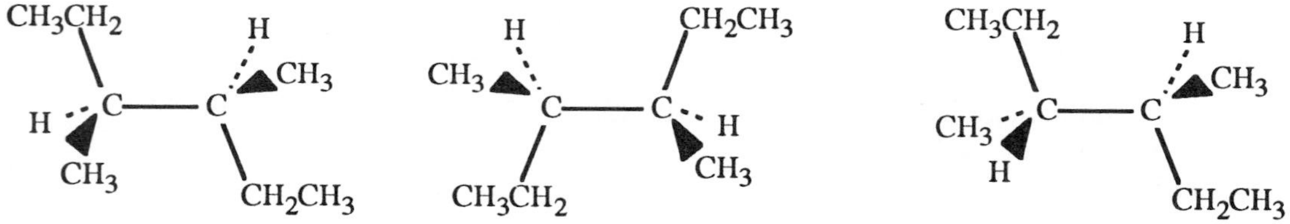

CH₃CH₂, H, CH₃, C, C, H, CH₃, CH₂CH₃   H, CH₂CH₃, CH₃, C, C, H, CH₃CH₂, CH₃   CH₃CH₂, H, CH₃, C, C, CH₃, H, CH₂CH₃

(f)

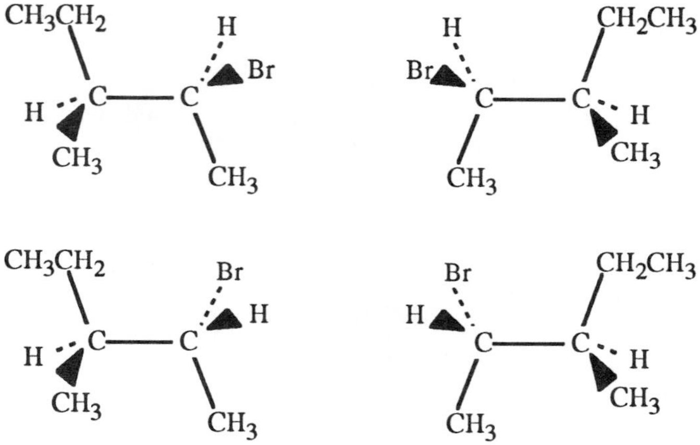

CH₃CH₂, H, H, C, C, Br, CH₃, CH₃   H, CH₂CH₃, Br, C, C, H, CH₃, CH₃

CH₃CH₂, Br, H, C, C, H, CH₃, CH₃   Br, CH₂CH₃, H, C, C, H, CH₃, CH₃

8.29 (cont)

(g)

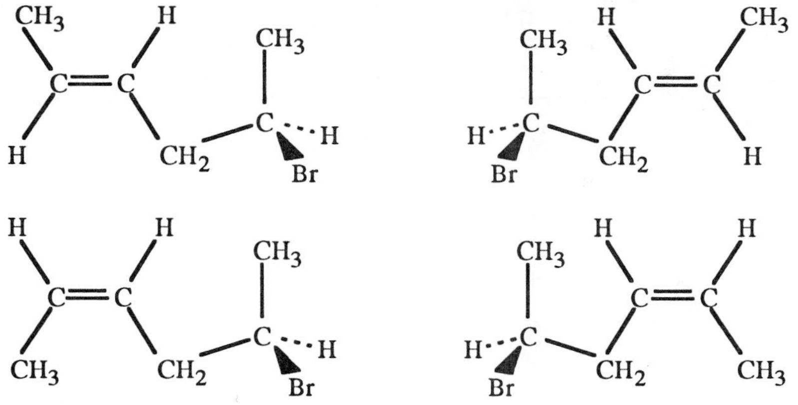

8.30

(a)

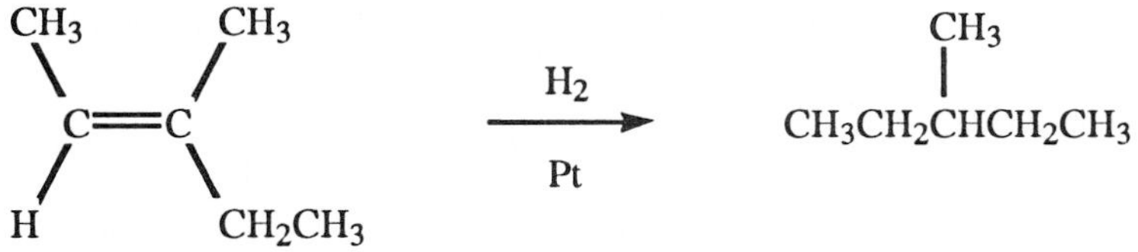

$$\xrightarrow[\text{Pt}]{\text{H}_2}$$

(b)

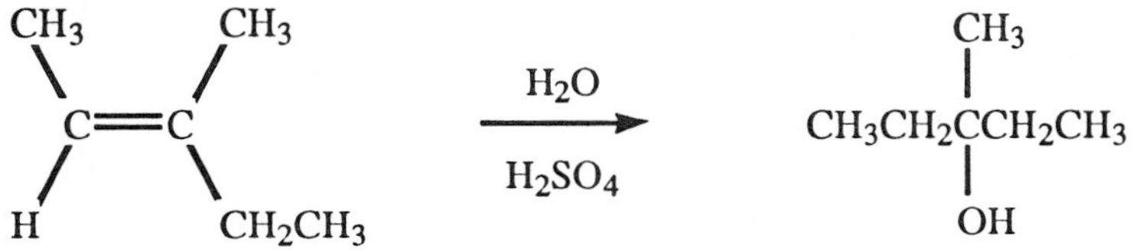

$$\xrightarrow[\text{H}_2\text{SO}_4]{\text{H}_2\text{O}}$$

(c)

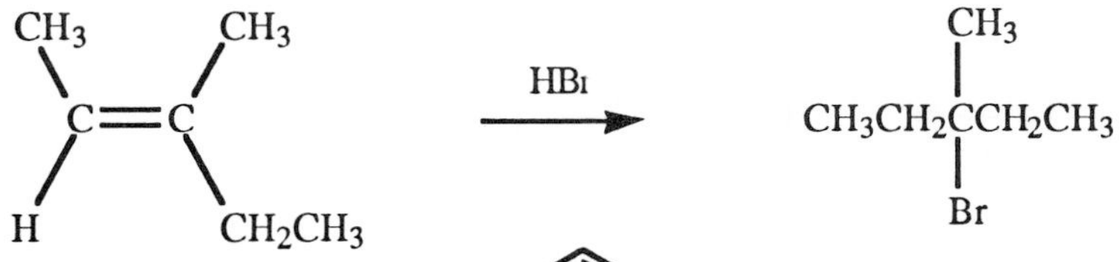

$$\xrightarrow{\text{HBl}}$$

(d)

$$\xrightarrow{}$$

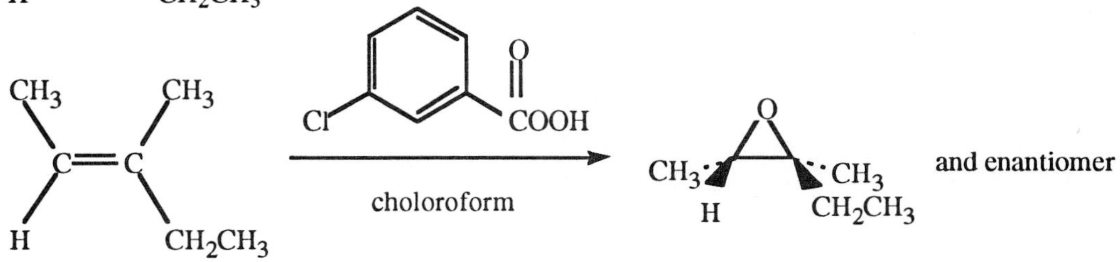

choloroform

and enantiomer

(e)

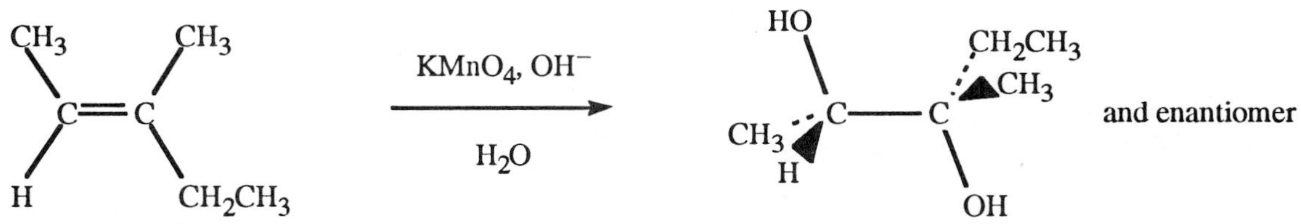

$$\xrightarrow[\text{H}_2\text{O}]{\text{KMnO}_4,\ \text{OH}^-}$$

and enantiomer

(f)

$$\xrightarrow{\text{O}_3,\ \text{CH}_3\text{OH}} \xrightarrow{(\text{CH}_3)_2\text{S}}$$

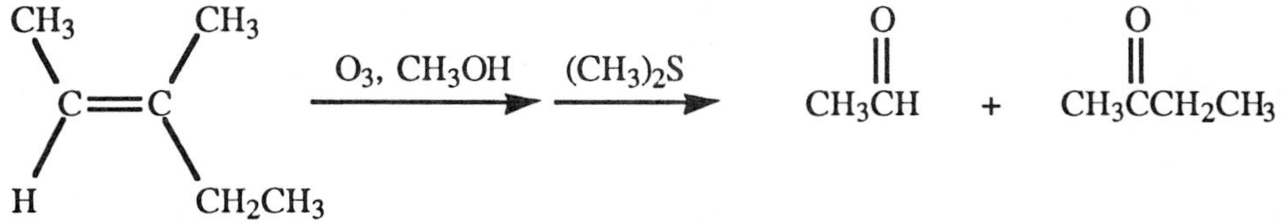

8.30 (cont)

(g) 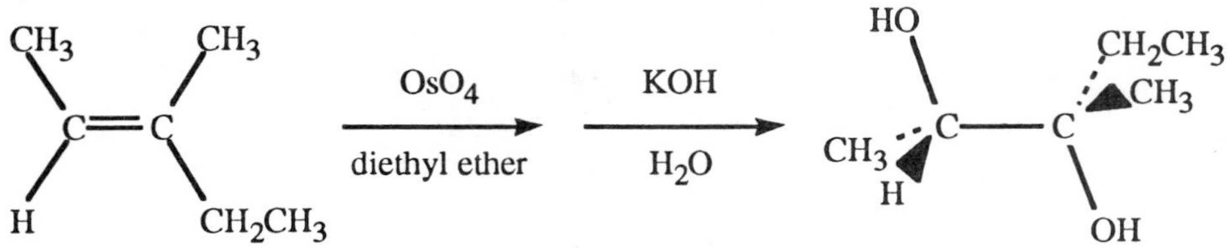 and enantiomer

(h)

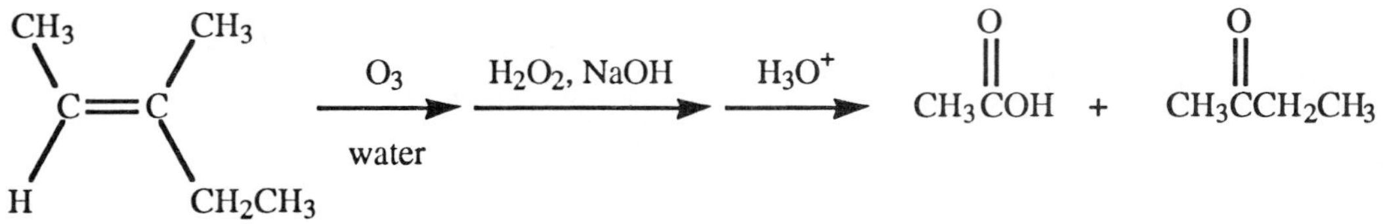

(i) 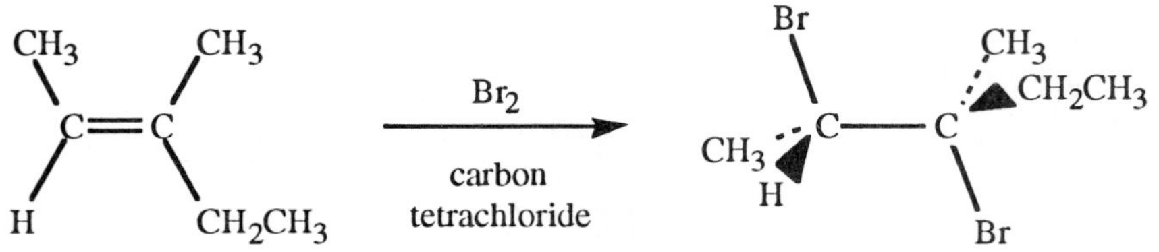 and enantiomer

(j) 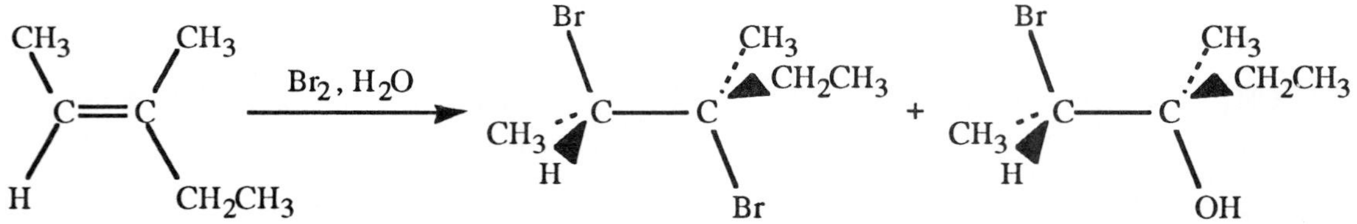 and enantiomer   and enantiomer

(k)

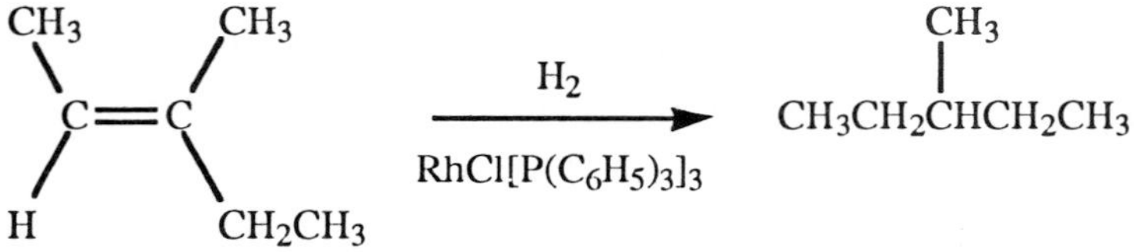

8.31

(a) 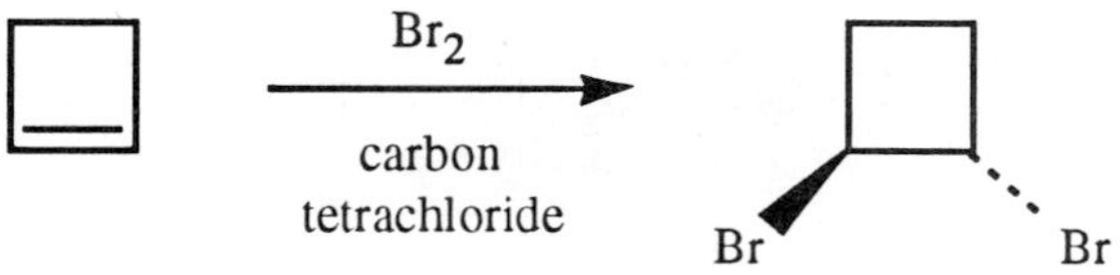 and enantiomer

(b)

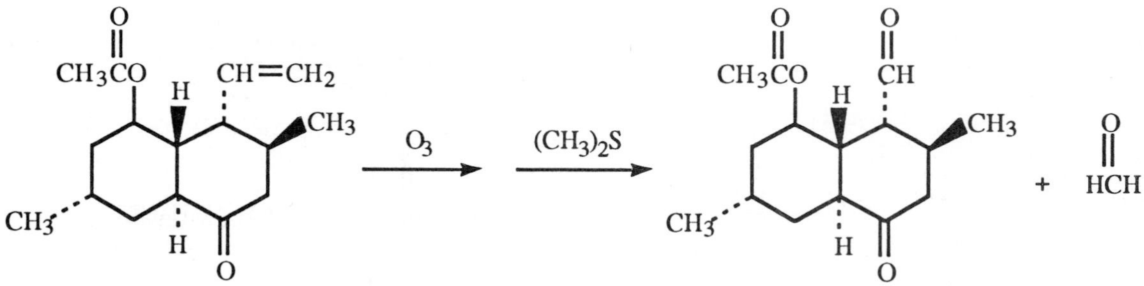

8.31 (cont)

(c) 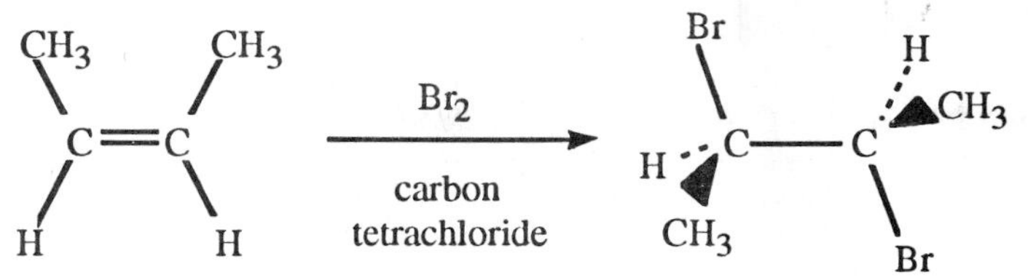 and enantiomer

(d) 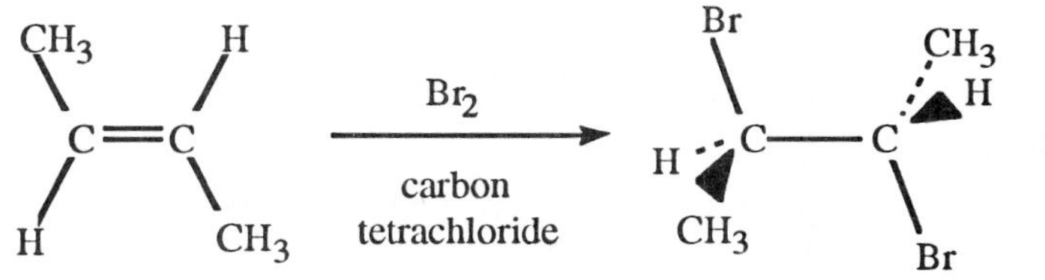 meso compound

(e) 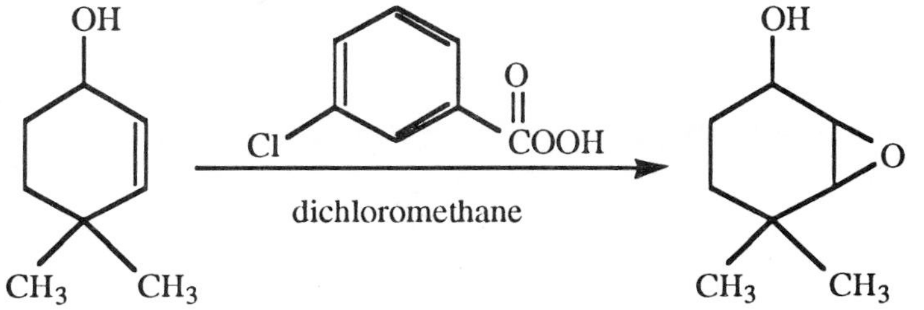 mixture of diastereomers

(f) 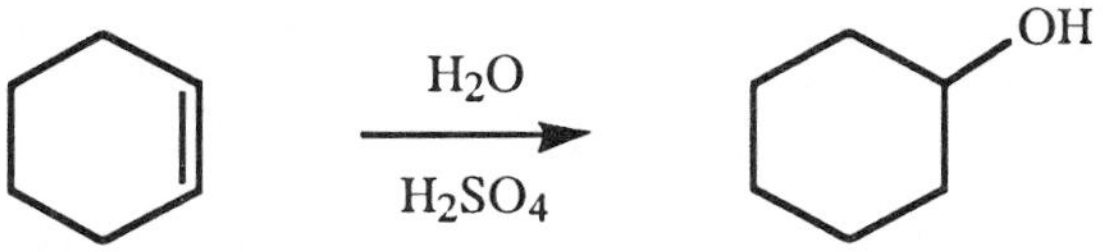

8.32

(a) 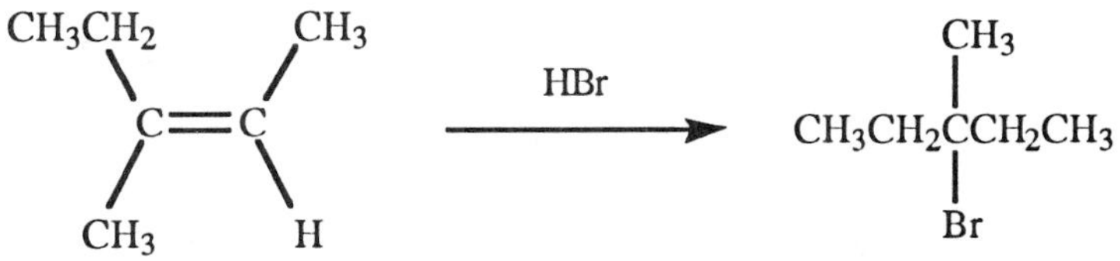

Note that (*E*)-3-methyl-2-pentene also gives 3-bromo-3-methylpentane [Prob. 8.30 (c)].

(b) 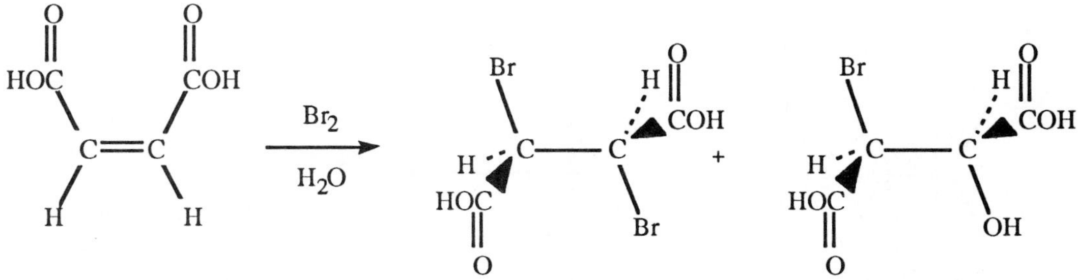

and enantiomer and enantiomer

8.32 (cont)

(c)

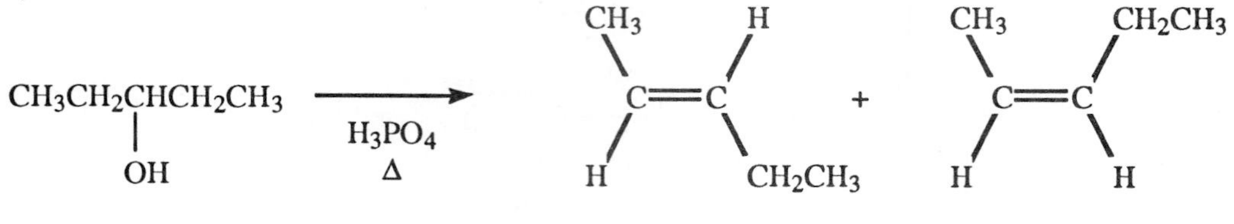
CH3CH2CHCH2CH3
OH
H3PO4
Δ
CH3
H
C=C
H
CH2CH3
+
CH3
CH2CH3
C=C
H
H

major product

(d)

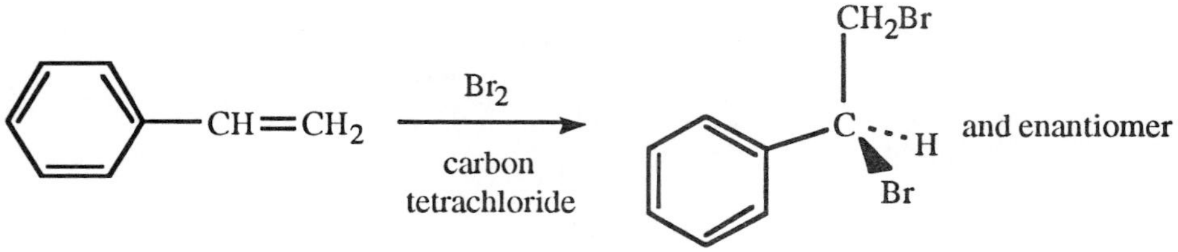
CH=CH2
Br2
carbon
tetrachloride
CH2Br
C
H
Br
and enantiomer

(e)

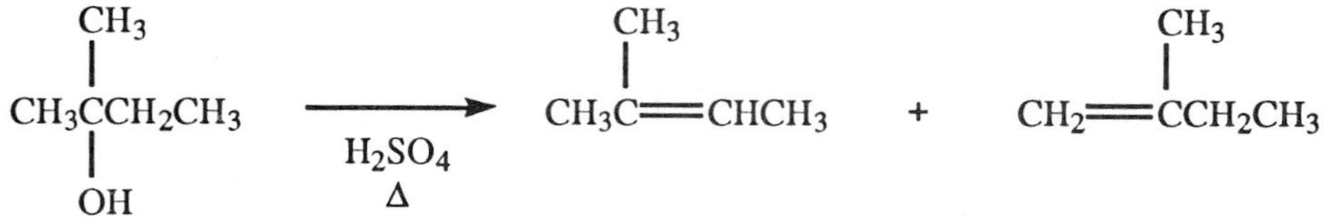
CH3
CH3CCH2CH3
OH
H2SO4
Δ
CH3
CH3C=CHCH3
+
CH3
CH2=CCH2CH3

(f)

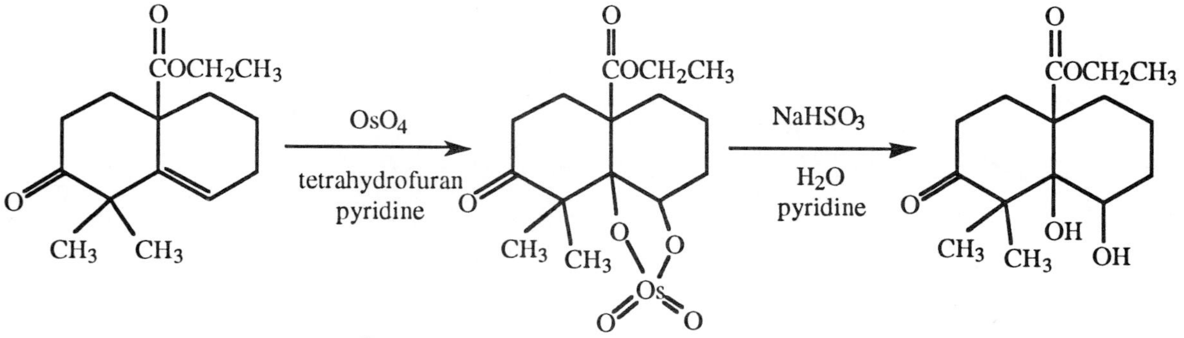
O
COCH2CH3
O
CH3
CH3
OsO4
tetrahydrofuran
pyridine
Os
NaHSO3
H2O
pyridine
OH
OH

(g)

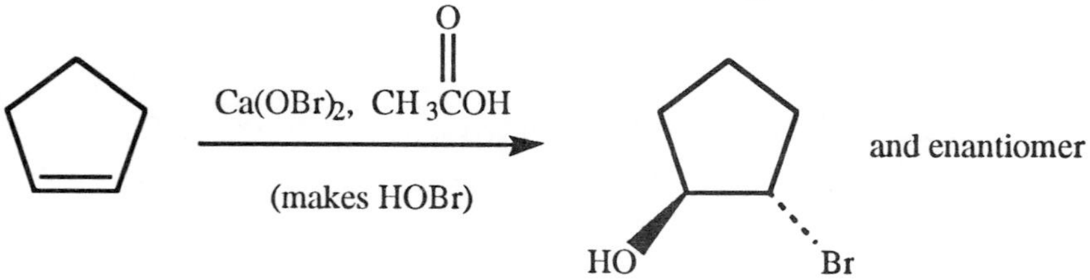
O
Ca(OBr)2, CH3COH
(makes HOBr)
HO
Br
and enantiomer

8.33

(a)

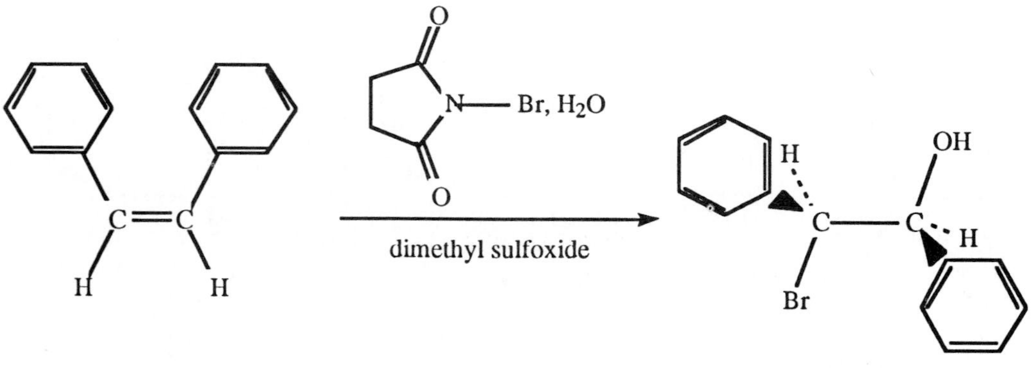
C=C
H
H
O
N— Br, H2O
O
dimethyl sulfoxide
H
OH
C
C
H
Br

and enantiomer
A

8.33 (cont)

(b)

$$\xrightarrow[\text{Pt}]{\text{H}_2}$$

B

(c)

$$\xrightarrow{\text{CH}_3(\text{CH}_2)_{10}\overset{\displaystyle O}{\overset{\|}{C}}\text{OOH}}$$

C   +   D

(d)

$$\xrightarrow{\text{O}_3, \text{CH}_3\text{OH}} \xrightarrow{(\text{CH}_3)_2\text{S}}$$

E   +   F

(e)

$$\xrightarrow{\text{BD}_3}$$

$$\left(\text{CH}_3\text{COCH}_2\text{CH}_2\text{CH}_2\text{CHDCH}\underset{\underset{\text{CH}_3}{|}}{}\right)_3\text{B} \quad + \quad \left(\text{CH}_3\text{COCH}_2\text{CH}_2\text{CH}_2\text{CH}\underset{\underset{\text{CHDCH}_3}{|}}{}\right)_3\text{B} \xrightarrow{\text{H}_2\text{O}_2, \text{NaOH}}$$

G   H

$$\text{CH}_3\text{COCH}_2\text{CH}_2\text{CH}_2\text{CHDCH}\underset{\underset{\text{CH}_3}{|}}{}\text{—OH} \quad + \quad \text{CH}_3\text{COCH}_2\text{CH}_2\text{CH}_2\text{CH}\underset{\underset{\text{CHDCH}_3}{|}}{}\text{—OH}$$

I   J

(f)

$$\xrightarrow[\substack{\text{RhCl[P(C}_6\text{H}_5)_3]_3 \\ \text{benzene}}]{\text{H}_2}$$

K

8.33 (cont)

(g)

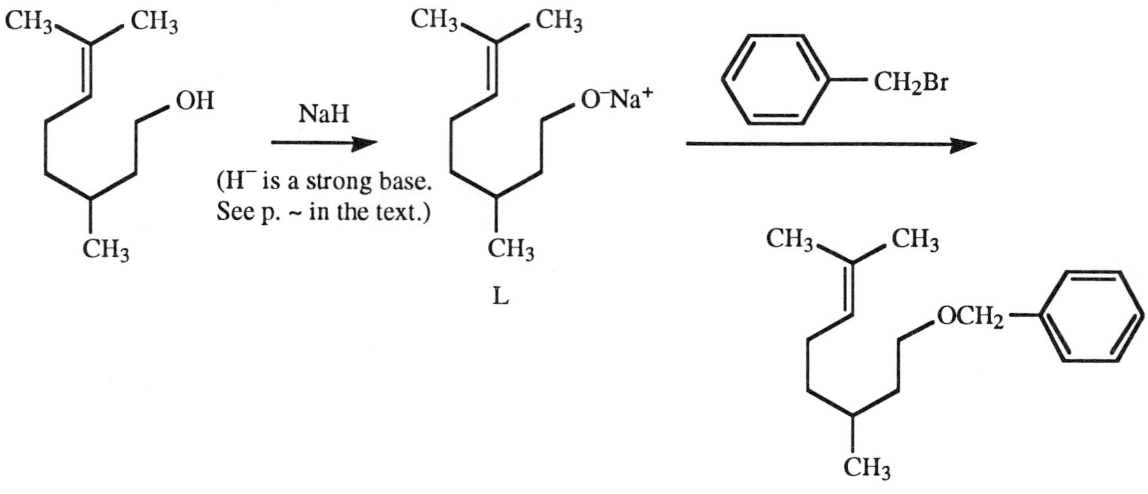

L

M

8.34

(a)

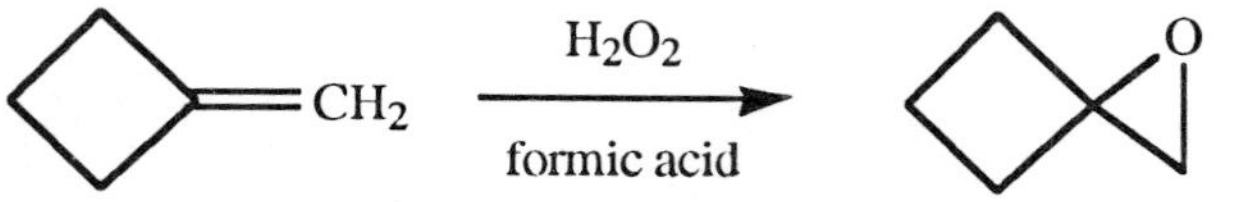

A

(b)

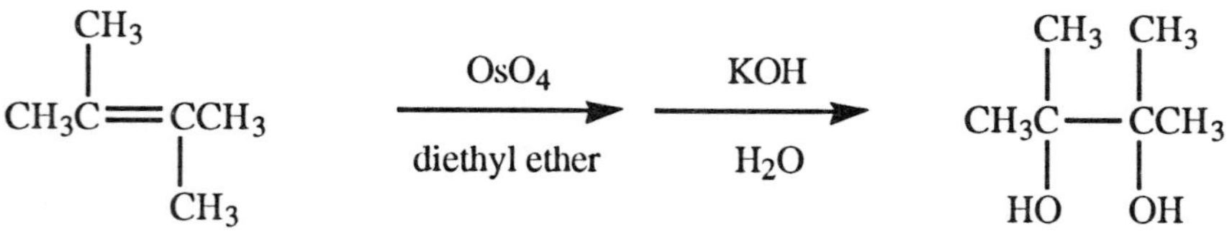

B

(c)

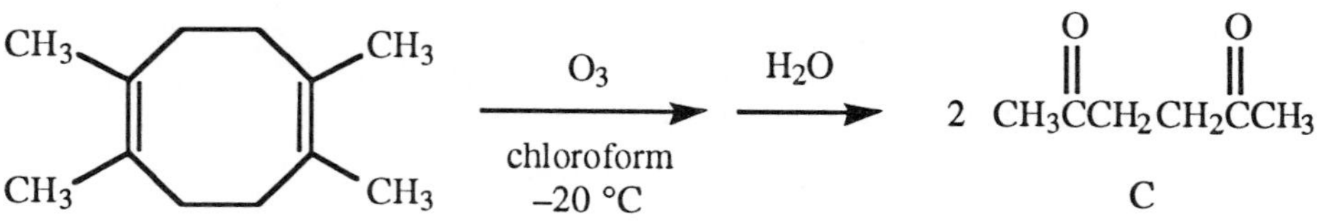

C

(d)

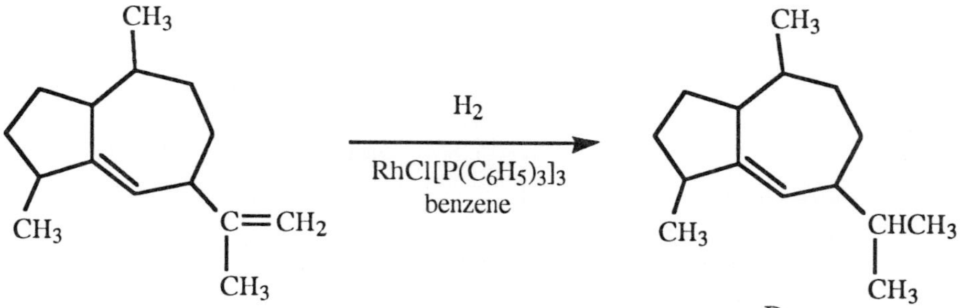

D

8.34 (cont)

(e)

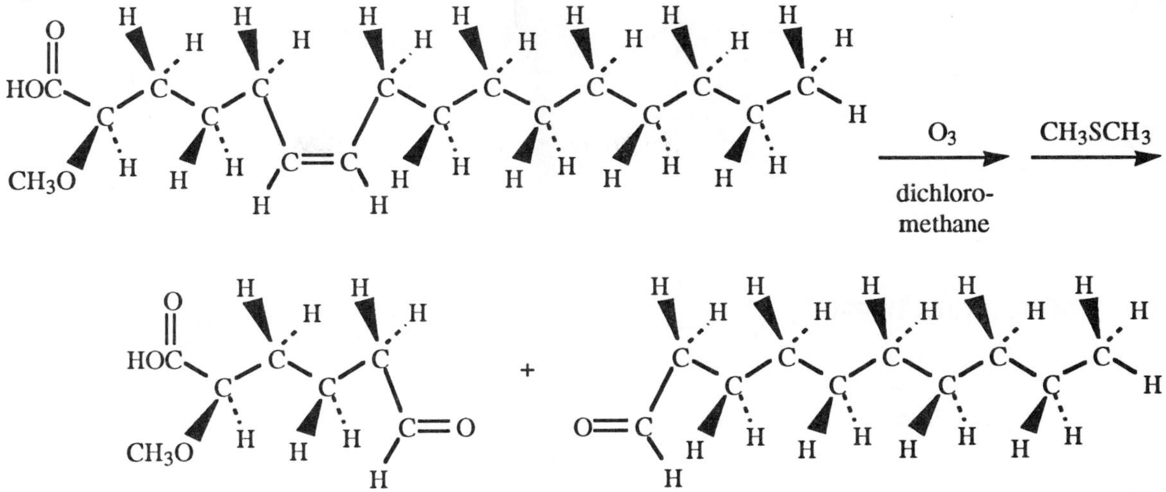

O_3 → CH_3SCH_3 →
dichloro-
methane

E
optically active

+

F
optically inactive

(f)

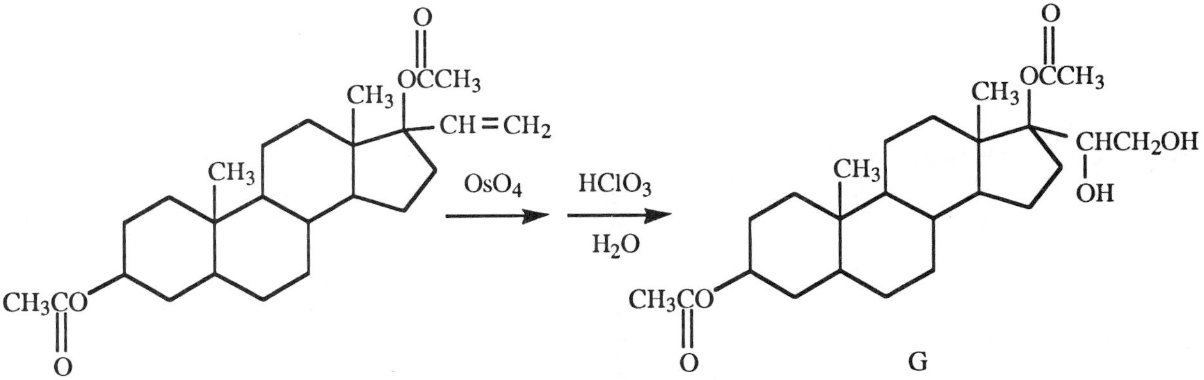

OsO_4 $HClO_3$ →
H_2O

G

(g) CH_3C=$CHCH_2CH_2COCH_3$ (with CH_3 and O substituents) → $\dfrac{O_3}{\text{dichloromethane} \ -78\,°C}$ → $(CH_3)_2S$ → CH_3CCH_3 + $HCCH_2CH_2COCH_3$

H   I

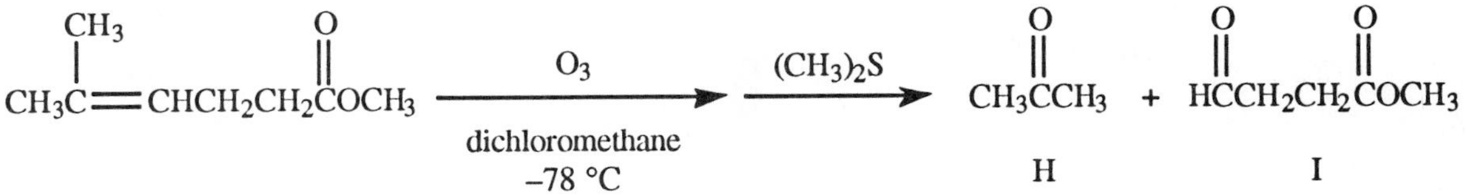

8.35

(a)

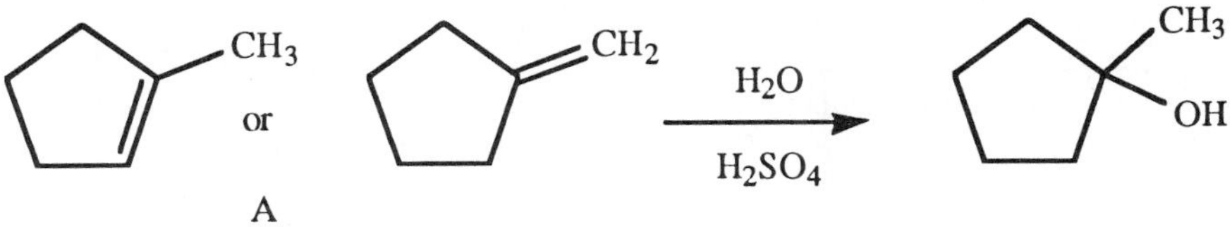

A

$\dfrac{H_2O}{H_2SO_4}$ →

8.35 (cont)

(b)

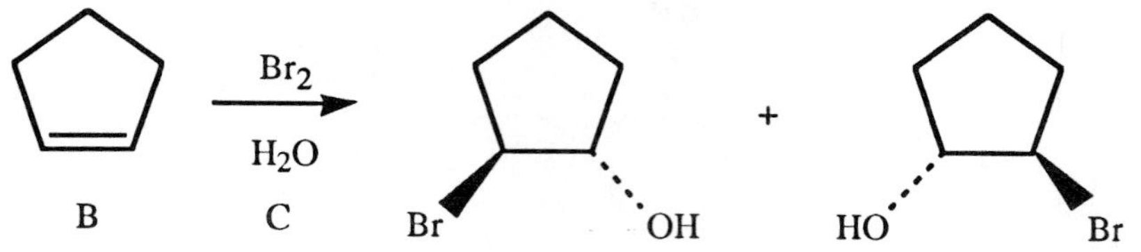

(c)

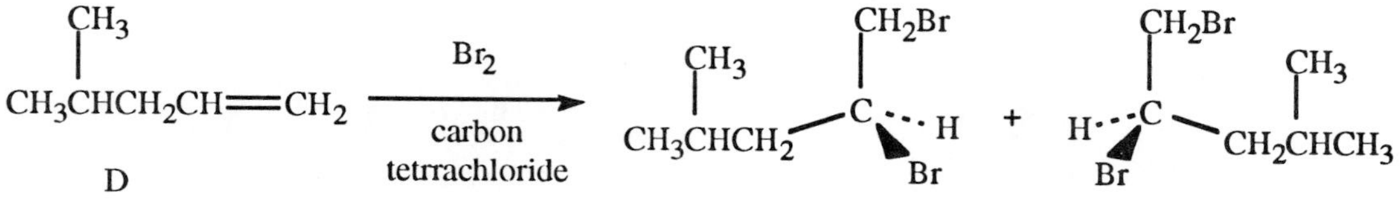

(d)

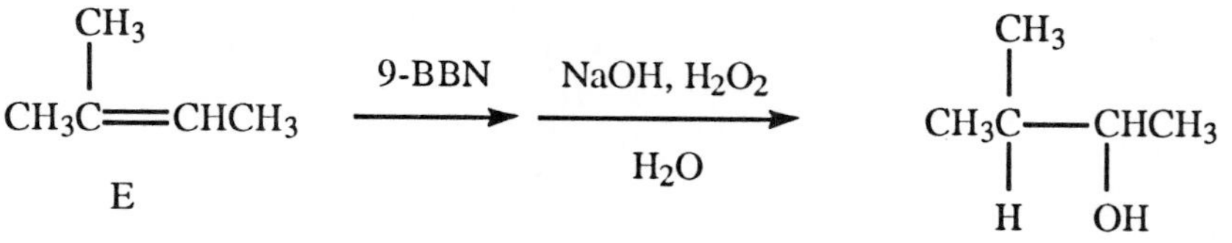

(e)

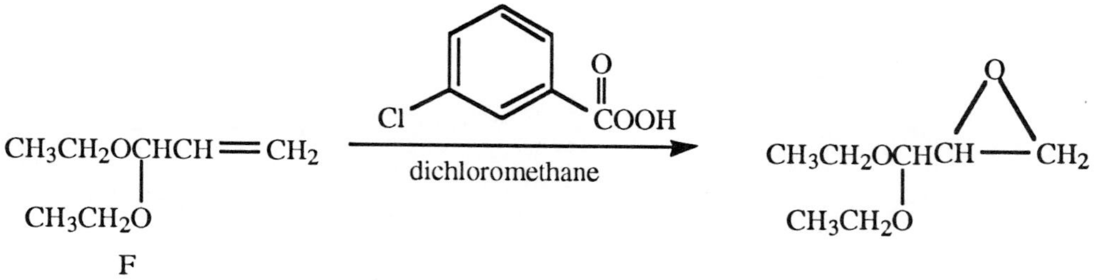

(f)

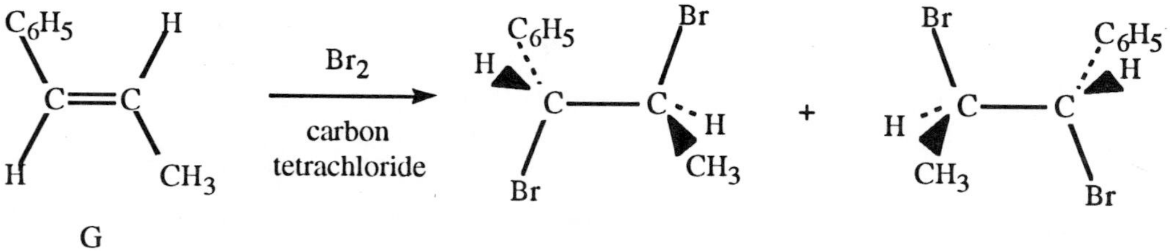

(g)

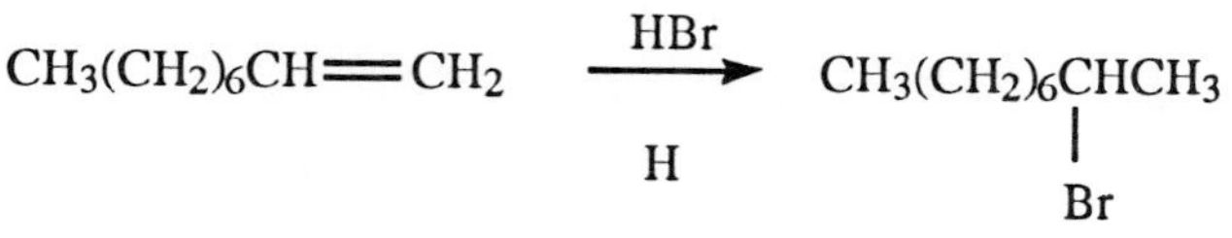

$$CH_3(CH_2)_6CH=CH_2 \xrightarrow[\;\;\;\;\;\;\;]{HBr} CH_3(CH_2)_6CHCH_3$$

H

Br

(h)

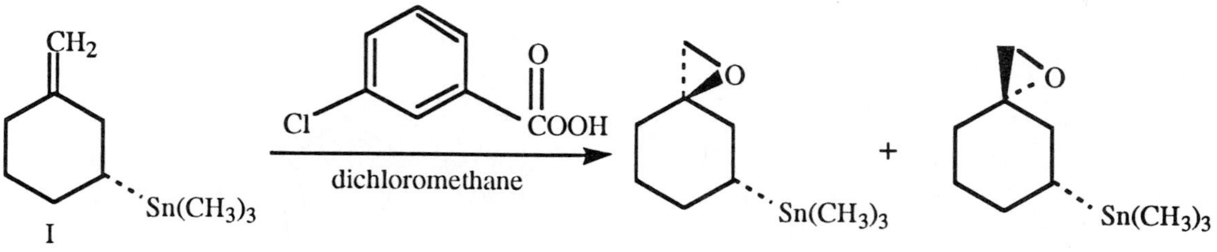

8.35 (cont)

(i)

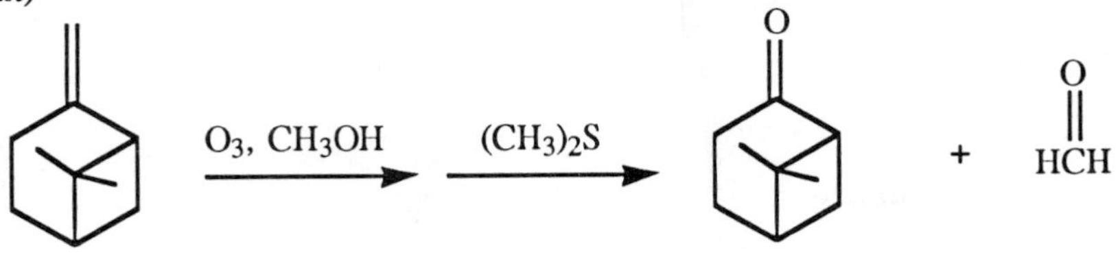

O3, CH3OH
(CH3)2S
O
+
O
HCH

J

(j)

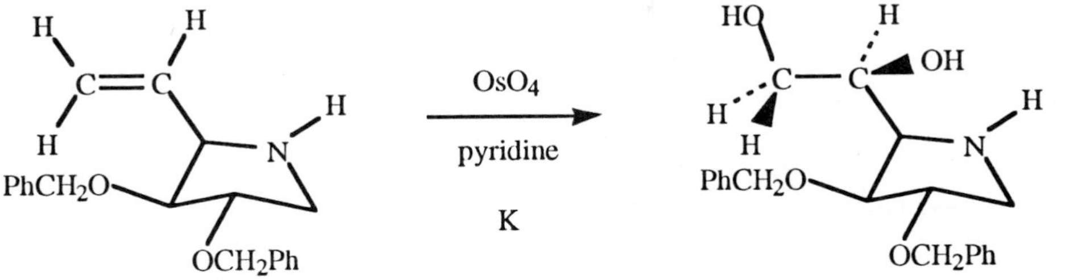

OsO4
pyridine
K
PhCH2O
OCH2Ph
HO
OH

8.36

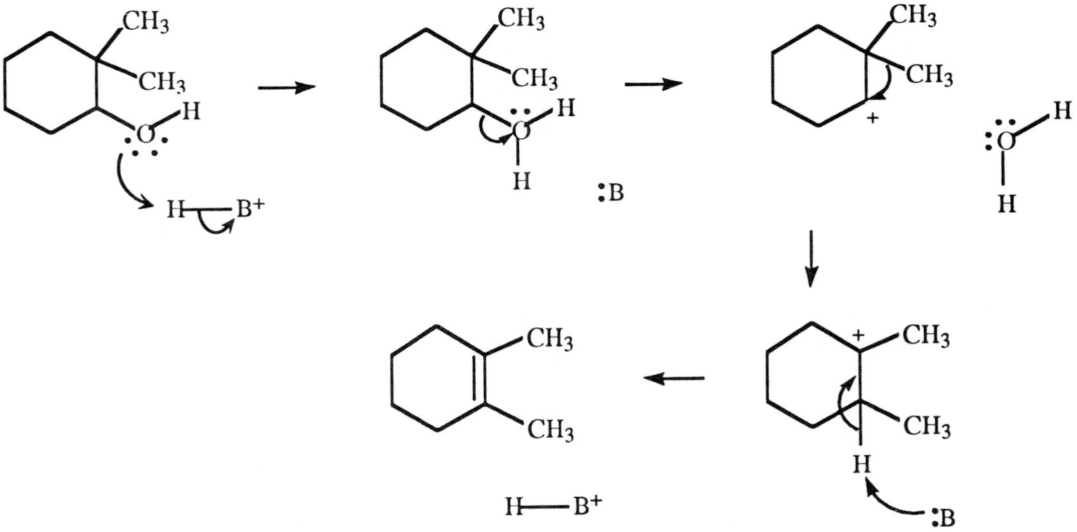

CH3
:B
H—B+

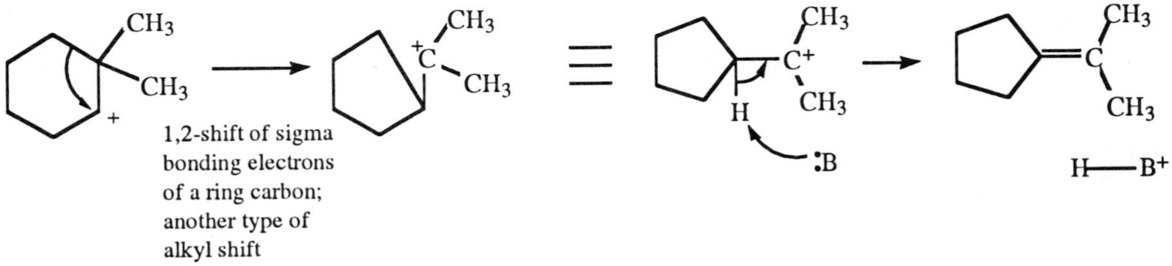

CH3
1,2-shift of sigma bonding electrons of a ring carbon; another type of alkyl shift
:B
H—B+

8.37

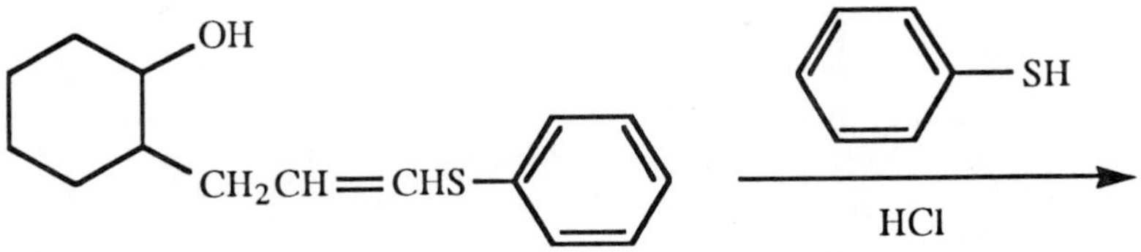

OH
CH2CH=CHS
SH
HCl

8.37 (cont)

The most likely first step for the reaction is protonation of the double bond by hydrogen chloride. The most stable carbocation is the one adjacent to the sulfur, which can stabilize the positive charge by delocalization.

The species now present in the reaction mixture is an electrophile. It will react with thiophenol, which is a better nucleophile than chloride ion.

The protonated thioether will now transfer a proton to a base.

8.38

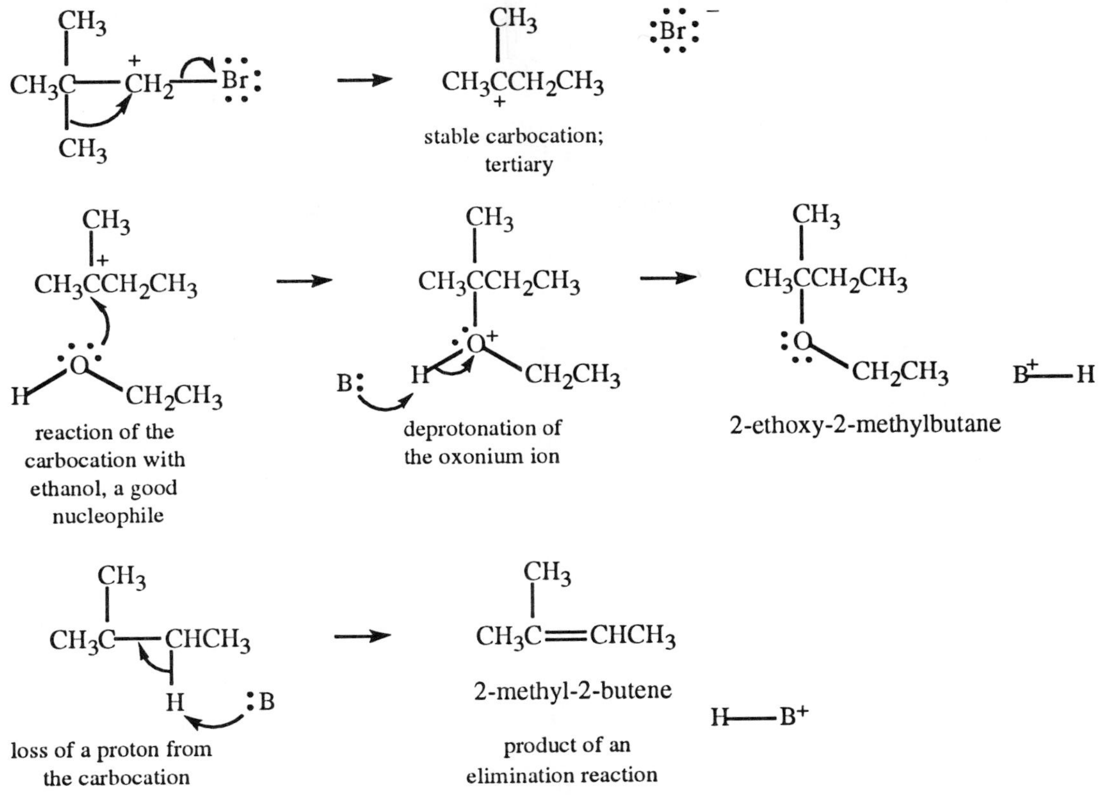

stable carbocation;
tertiary

reaction of the
carbocation with
ethanol, a good
nucleophile

deprotonation of
the oxonium ion

2-ethoxy-2-methylbutane

loss of a proton from
the carbocation

2-methyl-2-butene

product of an
elimination reaction

8.39   Formation of 2,3-dimethyl-1,3-butadiene from pinacol

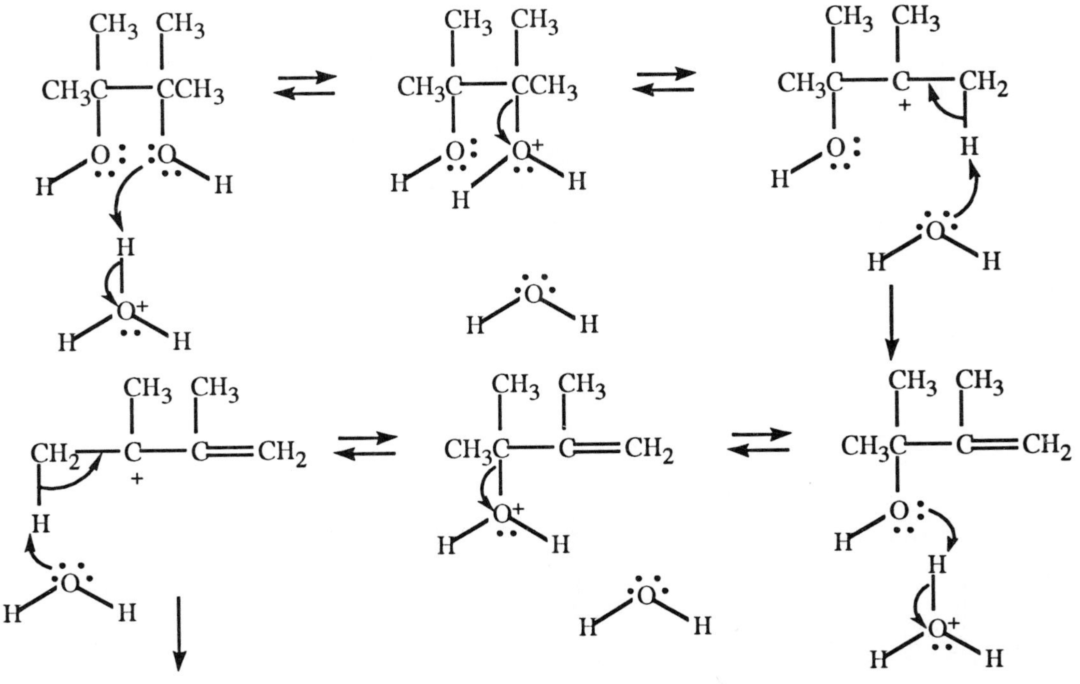

8.39 (cont)

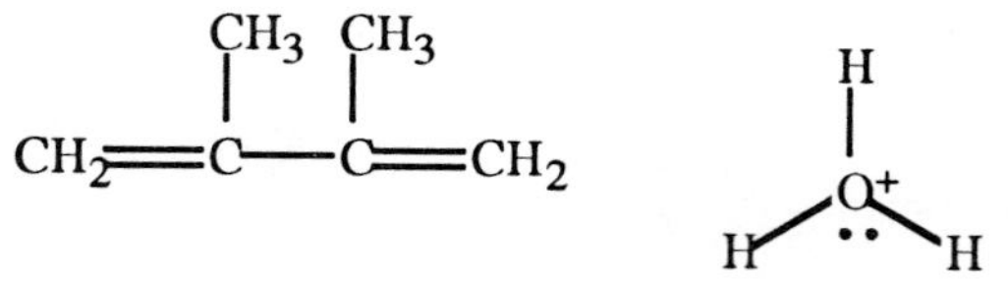

$$CH_2=\overset{\overset{\displaystyle CH_3}{|}}{C}-\overset{\overset{\displaystyle CH_3}{|}}{C}=CH_2$$

$$\overset{\displaystyle H}{\underset{H}{\overset{|}{\underset{\ddot{\cdot}}{O}}^{+}}}{}_{H}$$

2,3-dimethyl-1,3-butadiene

Formation of pinacolone

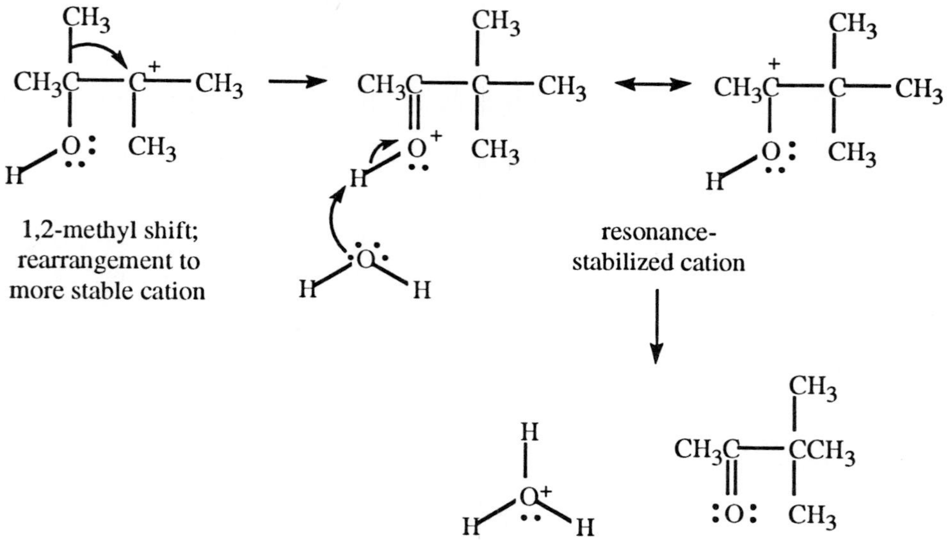

1,2-methyl shift;
rearrangement to
more stable cation

resonance-
stabilized cation

pinacolone

8.40

(a)

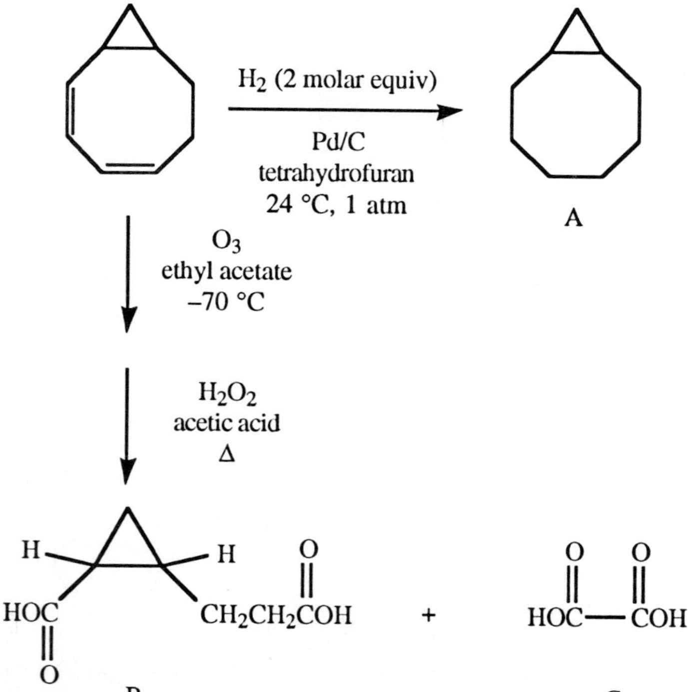

H₂ (2 molar equiv)

Pd/C
tetrahydrofuran
24 °C, 1 atm

A

O₃
ethyl acetate
−70 °C

H₂O₂
acetic acid
Δ

$$\underset{B}{\overset{\displaystyle H}{\underset{\displaystyle HOC}{\overset{|}{\triangle}}}\ \ \overset{\displaystyle H}{\underset{\displaystyle CH_2CH_2COH}{|}}}$$

+

$$\underset{C}{\overset{O\ \ \ \ O}{\underset{HOC-COH}{\|\ \ \ \ \|}}}$$

8.40 (cont)

(b)

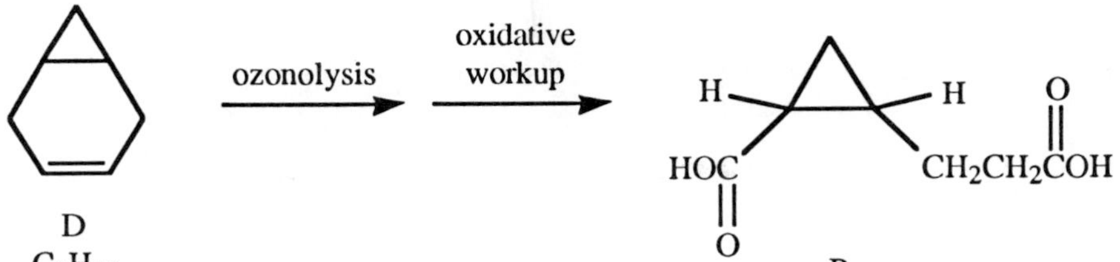

D
C_7H_{10}
3 units of unsaturation:
2 rings, 1 double bond

8.41

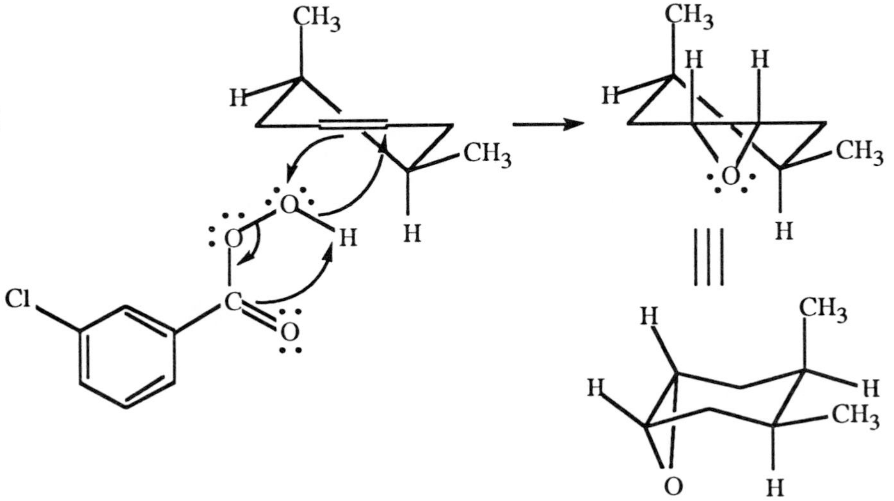

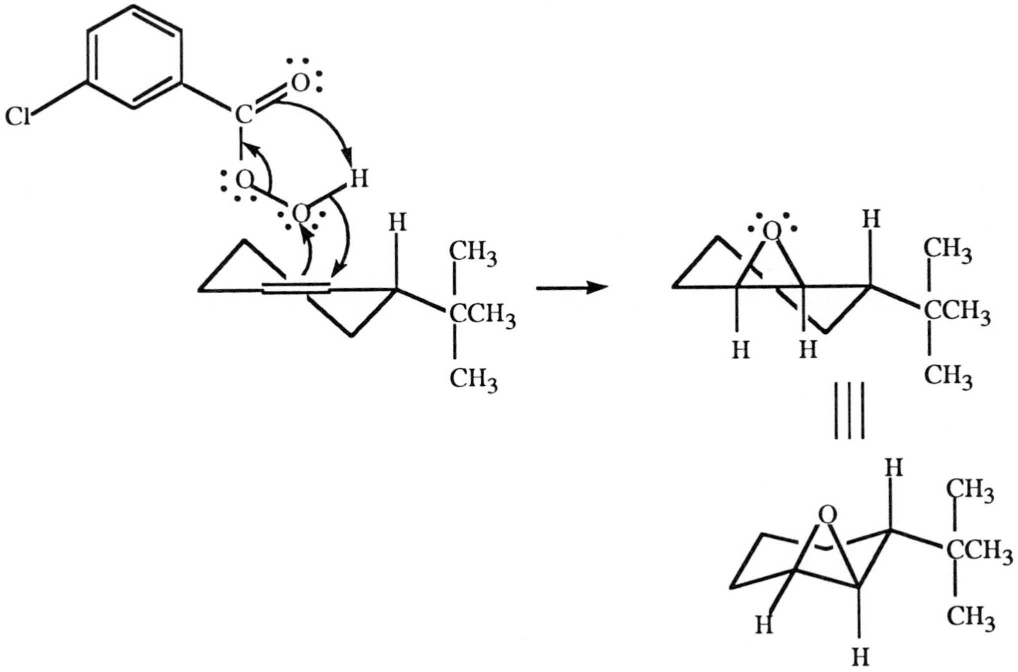

In each case the peroxyacid approaches the double bond from the less hindered side. If the starting material is racemic, the product will also be racemic. None of the diastereomeric product will form in either reaction.

8.42

(a)

$$
\begin{array}{c}
\underset{|}{\overset{CH_3}{\underset{OH}{C}}}\text{---}\underset{|}{\overset{CH_3}{\underset{OH}{C}}}CH_3 \\
CH_3
\end{array}
$$

(b)

$$CH_3(CH_2)_7CH\!\!=\!\!CH_2 \quad \xrightarrow[(CH_3CH_2)_4N^+\,OH^{\,-}]{(CH_3)_3COOH,\ OsO_4} \quad \xrightarrow[H_2O]{NaHSO_3}$$

and enantiomer

(c)

$$\xrightarrow[(CH_3CH_2)_4N^+\,OH^{\,-}]{(CH_3)_3COOH,\ OsO_4} \quad \xrightarrow[H_2O]{NaHSO_3}$$

and enantiomer

(d)

$$\xrightarrow[(CH_3CH_2)_4N^+\,OH^{\,-}]{(CH_3)_3COOH,\ OsO_4} \quad \xrightarrow[H_2O]{NaHSO_3}$$

meso compound

8.43

$$\underset{\substack{A \\ C_{11}H_{14}O_4}}{CH_3\overset{O}{\overset{\|}{C}}CH\!\!=\!\!CHCH_2CH_2\,\overset{O}{\overset{\|}{C}}CH\!\!=\!\!CHCOCH_3} \quad \xrightarrow[\substack{Pd/C \\ ethanol}]{H_2} \quad \underset{\substack{B \\ C_{11}H_{18}O_4 \\ 3\ units\ of\ unsaturation}}{CH_3\overset{O}{\overset{\|}{C}}CH_2CH_2CH_2CH_2\,\overset{O}{\overset{\|}{C}}CH_2CH_2\,\overset{O}{\overset{\|}{C}}OCH_3}$$

C
$C_{11}H_{18}O_2$
3 units of unsaturation

8.44

(a)

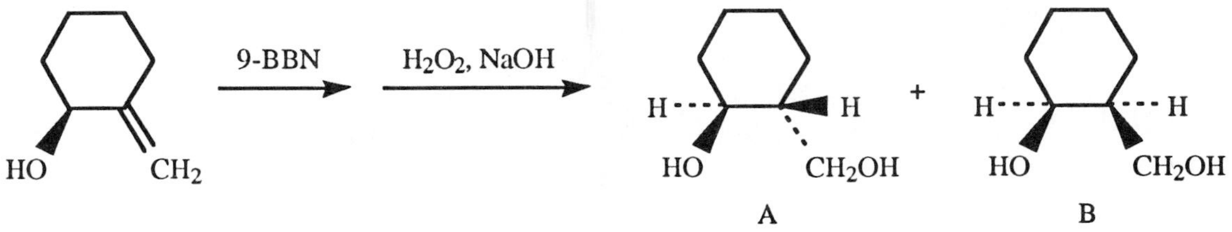

Compound A, in which the large groups are trans to each other on the cyclohexane ring, is expected to be thermodynamically more stable. Compound B, however, is formed by way of a sterically less hindered transition state.

(b)

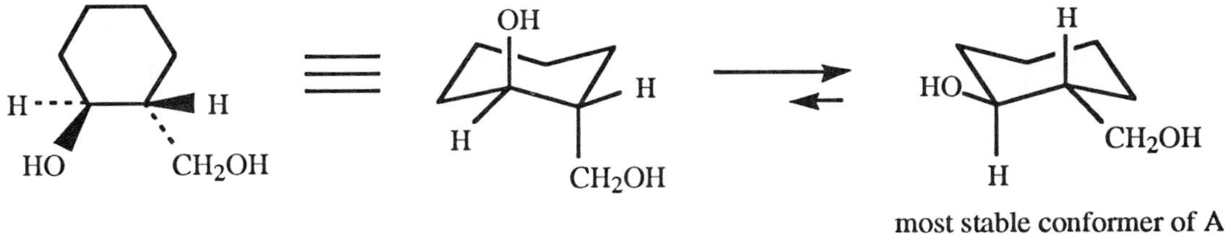

most stable conformer of A

8.45   The reaction causes a change in the connectivities of the carbon atoms. In the starting material the phenyl ring is attached to an unlabeled carbon atom. In the product, it is bonded to the C-14 labeled carbon atom. A rearrangement has taken place. Such rearrangements require carbocation intermediates.

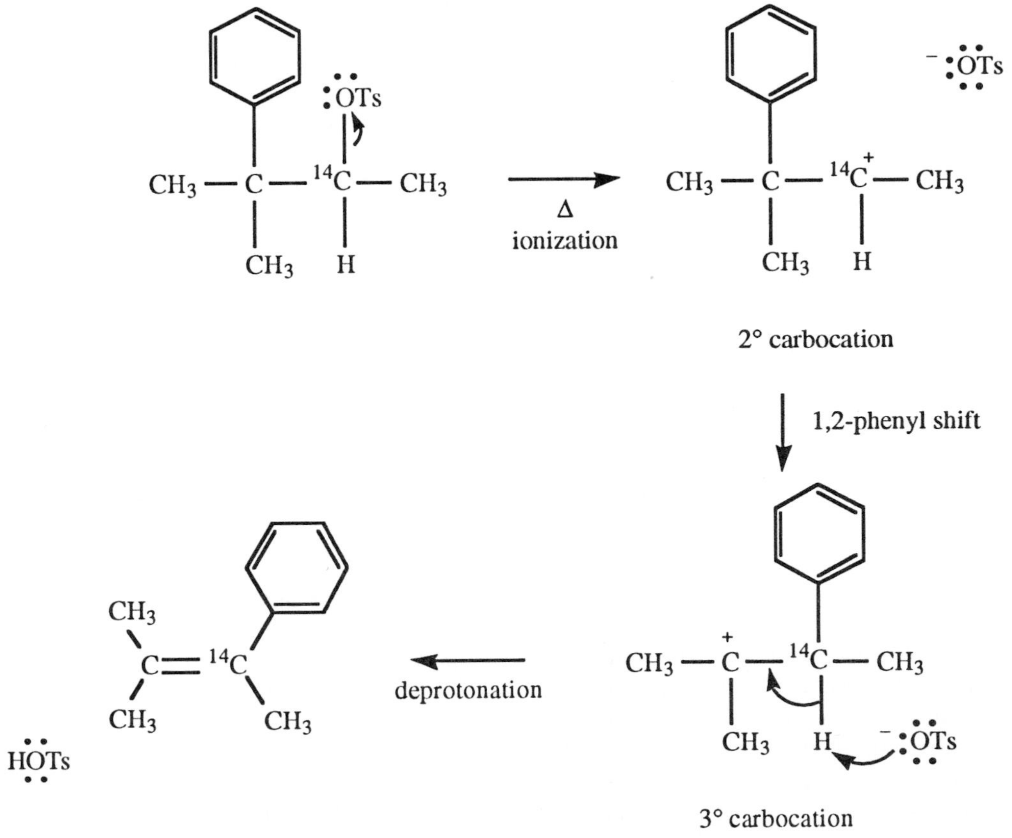

2° carbocation

1,2-phenyl shift

3° carbocation

deprotonation

8.46

(a)   (Z)-4-methyl-3-hexen-1-ol

(b)

$$CH_3CH_2, \ CH_2CH_2OH \xrightarrow[\substack{\text{pyridine} \\ A}]{TsCl} CH_3CH_2, \ CH_2CH_2OTs \xrightarrow[B]{Na^+I^-} CH_3CH_2, \ CH_2CH_2I$$

with $C=C$ bearing CH_3 and H substituents throughout.

(c)

[Mechanism scheme: protonation of the hydroxyl group by H—I, forming an oxonium ion that is displaced by iodide to give the **desired product** (the alkene with CH₂CH₂—I and regenerated H₂O and HI).]

desired product

[Lower scheme: protonation of the alkene C=C by H—I gives a carbocation]

$$^+C(CH_3)(CH_3CH_2)-C(H)(CH_2CH_2-O-H)-H$$

[which is attacked by iodide to give]

$$CH_3CH_2-C(I)(CH_3)-C(H)(CH_2CH_2-O-H)-H$$

unavoidable side reaction

It is important to remember that competing side reactions are unavoidable even if we use one equivalent of each reagent. The smallest quantity of material that we can experiment with contains billions of molecules of each reagent, randomly colliding with each other. The protonation of the hydroxyl group, especially, creates another acid, the oxonium ion, which is also strong enough to protonate the alkene, creating a carbocation that will react with the iodide ion. When a molecule contains more than one functional group, both may become involved in reactions unless we choose our reagents very carefully.

8.47

(a)

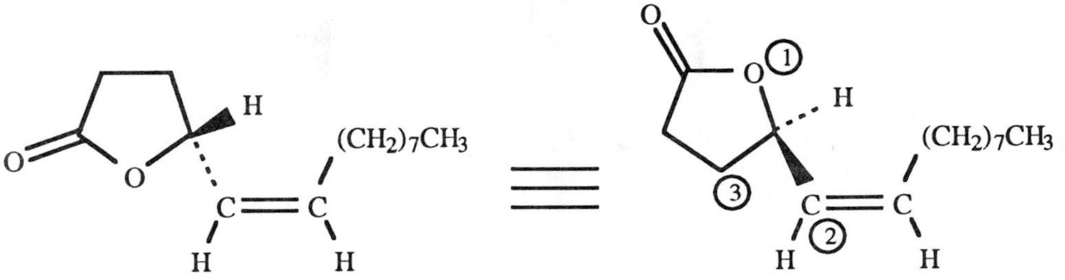

(R)-$(-)$-(Z)-isomer

(b)   The name of the alcohol without indication of stereochemistry is 6-methyl-5-hepten-1,2-diol.

two methyl groups
on double bond,
therefore no stereo-
chemistry at double bond

only one stereocenter;
R-configuration

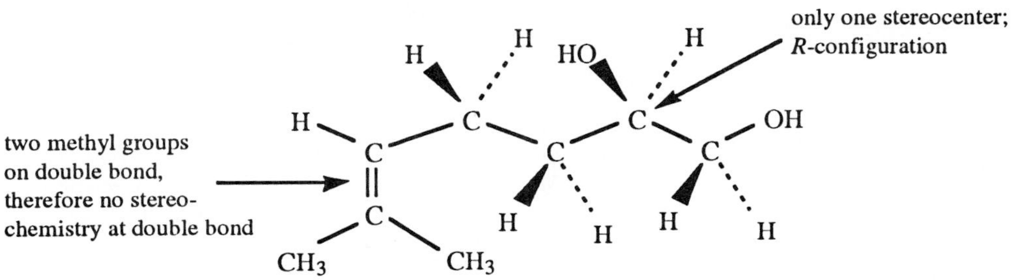

therefore the full correct name is (R)-6-methyl-5-hepten-1,2-diol.

(c)   The carbon-carbon double bond has been cleaved in this reaction, and a carbon-oxygen double bond in an aldehyde function appears on the product we see, which contains most of the carbon atoms of the starting material. This appears to be an ozonolysis reaction with a reductive work-up.

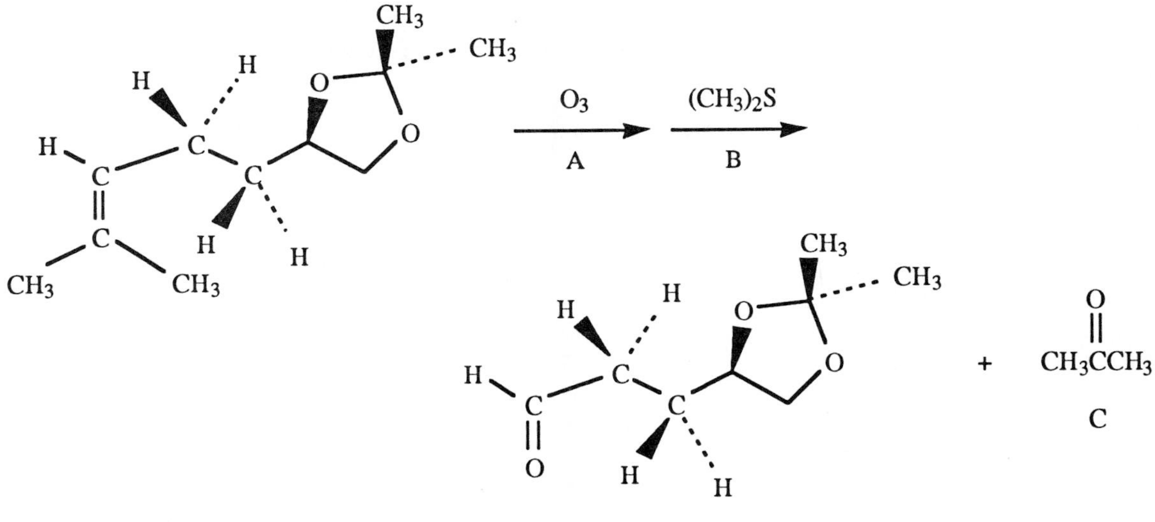

8.48

(a)   An anti-Markovnikov addition of H_2O appears to have taken place. It is a hydroboration-oxidation reaction.

8.48 (cont)

(b)

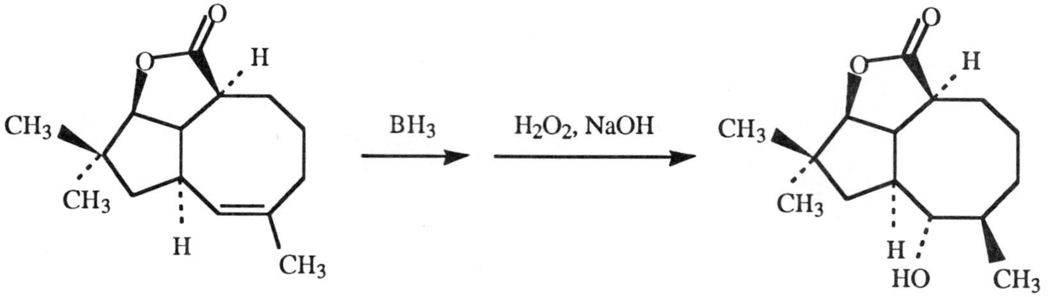

the approach of the BH₃
from the side away from the
substituents at the top
of the ring is favored

8.49

(a)

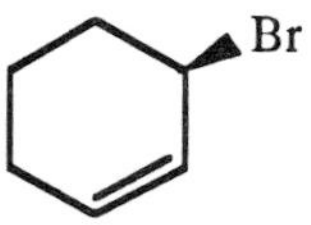

(*R*)-3-bromocyclohexene

(b)

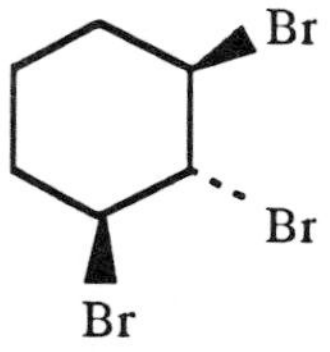

Compound A
a meso form,
therefore optically inactive

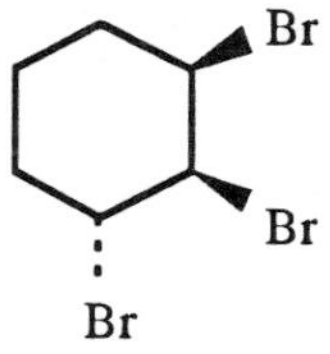

Compound B
chiral,
therefore optically active

Both products are formed by anti addition of Br₂ to the double bond.

8.50

(a)

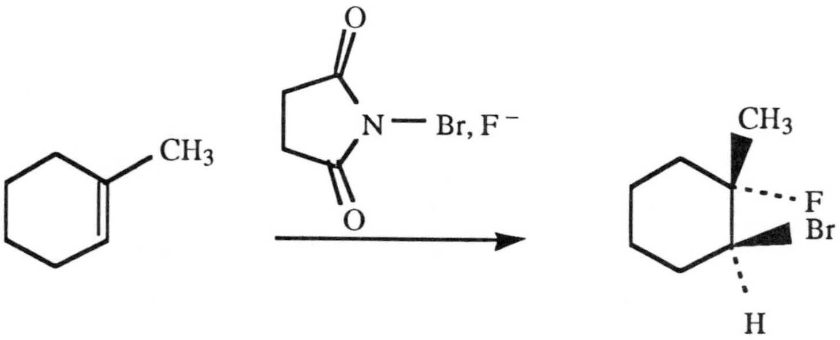

and enantiomer

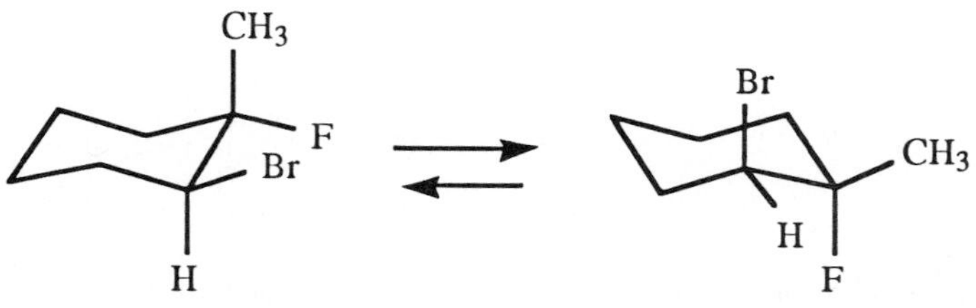

8.50 (cont)

(b)   *S* at each stereocenter
       (1*S*, 2*S*)-2-bromo-1-fluoro-1-methylcyclohexane

(c)

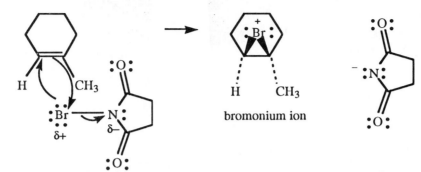

bromonium ion

If a bromonium ion were the intermediate in the reaction, we would expect attack of fluoride ion at the less hindered carbon atom, which is not the regioselectivity observed. Therefore, the intermediate must be the partially open bromonium ion with cationic character at the 3° carbon atom.

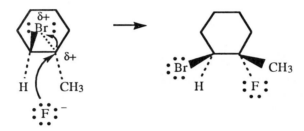

8.51

(a)

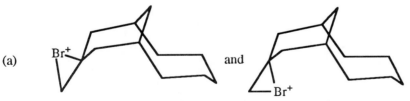

and

(b)   The one with the bromine pointing away from the other ring system is the one proposed to form in this reaction. The other one is sterically hindered. The bromine molecule would not be able to get to the underface of the double bond.

(c)

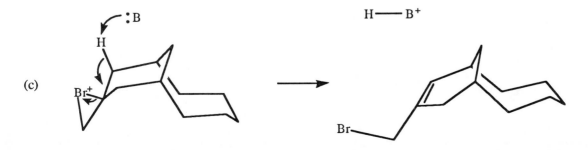

8.52

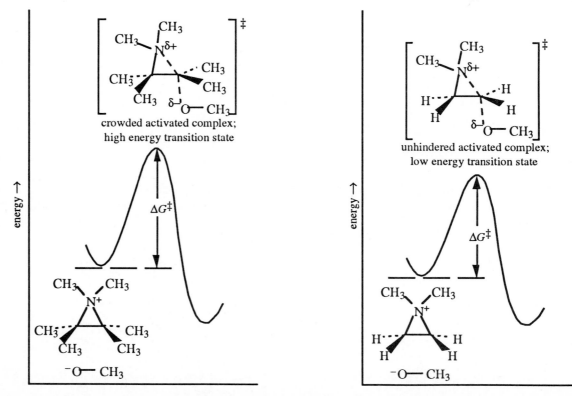

(a)

A

B

(b)   Methanol is a strong enough nucleophile to attack and open the strained three-membered ring with a positively charged nitrogen atom as the leaving group. Remember the reactions of the bromonium ion with water or alcohols. Methanol is not a strong enough base to deprotonate a carbon atom to form the alkene.

(c)   The rate of the reaction increases as crowding at the carbon atom undergoing the S_N2 reaction decreases.

crowded activated complex;
high energy transition state

unhindered activated complex;
low energy transition state

9

Alkynes

Workbook Exercises

Your study of the structure and reactivity of organic compounds has included a number of different ways to represent a molecule based upon the type of information that is known or needed. One way of organizing this information is under the general heading of "isomerism," which can be viewed according to the hierarchy presented below.

Isomerism (or, ways to represent molecules)

A **molecular formula** provides the possibility of constitutional isomers that differ in their connectivity (see Chapter 1).

A particular **connectivity** provides the possibility that there are stereoisomers that differ in the three-dimensional arrangement of bonded atoms (see Chapter 6).

A particular **stereoisomer** (or a particular connectivity, if no stereoisomers are possible) may have conformational isomers that differ in nonbonded three-dimensional arrangements (see Chapter 5).

EXERCISE I.   For each of the three molecular formulas (a, b, and c) shown below,

(a) $C_6H_{12}O_2$ (b) C_7H_{10} (c) $C_5H_{11}NO$

1.   Create 10 different constitutional isomers (for a total of 30 molecules). The molecules that you create should be comprised of uncharged, closed-shell atoms.

2.   Decide how many stereoisomers are possible for each constitutional isomer that you create.

3.   Choose five examples of single bonds between any 2 tetrahedral atoms in the isomers you create. Draw Newman projections for each example showing different possible conformations. For any cyclohexane or closely related six-membered ring, create the two appropriate chair forms.

(Note: This is an ideal problem to work with other students. You should also save your responses to part 1 of this exercise for the workbook exercise in Chapter 11.)

— —

9.1   (a) 6,6-dichloro-2-methyl-3-heptyne   (b) 3,3-dimethyl-1-pentyne

(c) 4-bromo-2-hexyne   (d) 2,6-dimethyl-3-octyne

9.2 (a) $CH_3CH_2CH_2C{\equiv}CCCH_3$ with CH_3 above and Cl below   (b) $CH_3C{\equiv}CCH_2CHCH_2CH_2CH_3$ (with phenyl group)   (c) $CH_3C{-}C{\equiv}CCH_2CH_3$ with CH_3 above and CH_3 below

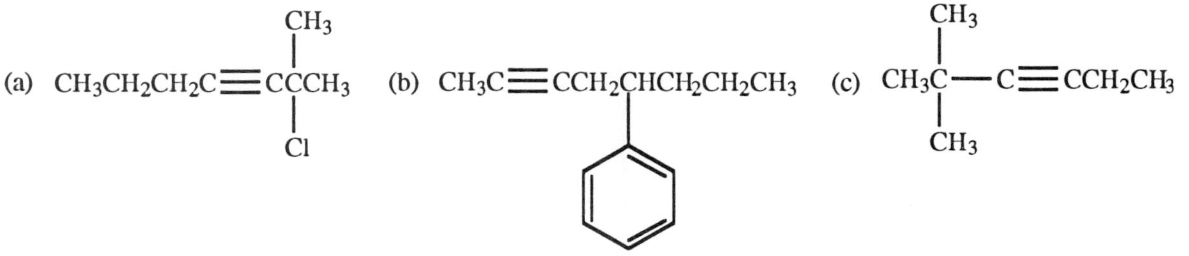

Concept Map 9.1   Outline of the synthesis of a disubstituted alkyne from a terminal alkyne.

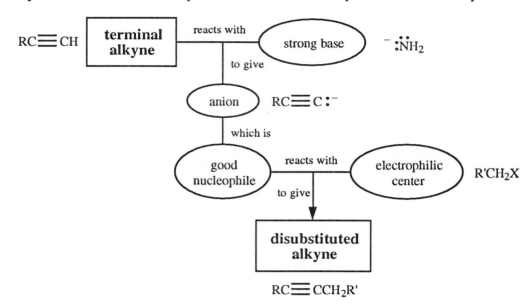

9.3   (a) $CH_3CH_2CH_2C \equiv CH$ $\xrightarrow[\text{NH}_3 \text{ (liq)}]{\text{NaNH}_2}$ $CH_3CH_2CH_2C \equiv C:^- Na^+$ $\xrightarrow{CH_3CH_2CH_2Br}$

                                                                                                    A

$CH_3CH_2CH_2C \equiv CCH_2CH_2CH_3$

                                                                                                    B

(b) $HC \equiv C:^- Na^+$ $\xrightarrow[\text{NH}_3 \text{ (liq)}]{\overset{\displaystyle CH_3}{\overset{|}{CH_3CHCH_2CH_2Br}}}$ $CH_3\overset{\displaystyle CH_3}{\overset{|}{CH}}CH_2CH_2C \equiv CH$

        C

                          D

(c) $CH_3CH_2CH_2C \equiv CH$ $\xrightarrow[\text{NH}_3 \text{ (liq)}]{\text{NaNH}_2}$ $CH_3CH_2CH_2C \equiv C:^- Na^+$ $\xrightarrow{(CH_3)_2SO_4}$

                          E   F

$CH_3CH_2CH_2C \equiv CCH_3$

                                                                                                    G

The leaving group in dimethyl sulfate, $(CH_3)_2SO_4$, is the methyl sulfate anion, $^-OSO_3CH_3$.

9.4   (a) $\xrightarrow{\text{HBr}}$

9.4 (cont)

(b) $CH_3C{\equiv}CCH_3$ $\xrightarrow{\text{HBr (excess)}}$ $CH_3CH_2\overset{\displaystyle Br}{\underset{\displaystyle Br}{C}}CH_3$

(c) $CH_3CH_2CH_2CH{=}CH_2$ $\xrightarrow{\text{HCl}}$ $CH_3CH_2CH_2\underset{\displaystyle Cl}{C}HCH_3$

(d) $\xrightarrow{\text{HI}}$

(e) $HC{\equiv}CH$ $\xrightarrow{\text{HBr (excess)}}$ CH_3CHBr_2)

(f)   —$CH_2CH{=}CH_2$ $\xrightarrow{\text{HBr}}$ —$CH_2\underset{\displaystyle Br}{C}HCH_3$

9.5 $CH_3C{\equiv}C(CH_2)_7\overset{\displaystyle O}{C}OH$ $\xrightarrow[\substack{80\% \\ H_2SO_4}]{}$ $CH_3\overset{\displaystyle O}{C}CH_2(CH_2)_7\overset{\displaystyle O}{C}OH$ + $CH_3CH_2\overset{\displaystyle O}{C}(CH_2)_7\overset{\displaystyle O}{C}OH$

- -

Concept Map 9.2   (see p. 276)

- -

9.6

Concept Map 9.2   Electrophilic addition of acids to alkynes.

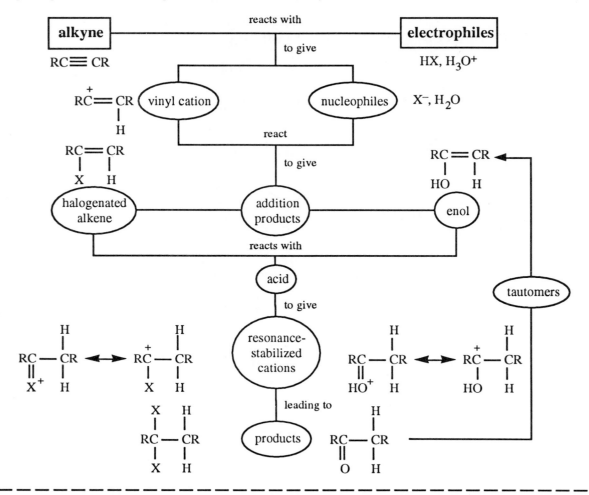

9.8

1. What are the connectivities of the starting material and the products? How many carbon atoms does each contain? Are there any rings? What are the positions of branches and functional groups on the carbon skeletons?

$$CH_3(CH_2)_{11}C \equiv CH \quad \xrightarrow{?} \quad CH_3(CH_2)_{11}C \equiv CCH_2(CH_2)_7CH_3$$

1-tetradecyne   10-tricosyne

The starting material has a linear chain of 14 carbon atoms; the product has a linear chain of 23 carbon atoms. The starting material is a 1-alkyne and the product is a 10-alkyne.

2. How do the functional groups change in going from starting material to product? Does the starting material have a good leaving group?

A nonyl group in the product has replaced the terminal hydrogen in the starting material.

3. Is it possible to dissect the structures of the starting material and product to see which bonds must be broken and which formed?

bond broken   bond formed

$$CH_3(CH_2)_{11}C \equiv C \text{\textbraceleft} H \qquad\qquad CH_3(CH_2)_{11}C \equiv C \text{\textbraceleft} CH_2(CH_2)_7CH_3$$

Pieces that we need: $CH_3(CH_2)_{11}C \equiv C \text{---}$ and $\text{---} CH_2(CH_2)_7CH_3$

4. New bonds are created when an electrophile reacts with a nucleophile. Do we recognize any part of the product molecule as coming from an electrophile or a nucleophile?

The new carbon-carbon bond must have come from a nucleophilic substitution reaction. The terminal alkyne can be converted into a nucleophile. The alkyl group, attached to a leaving group, must supply the electrophile.

$$CH_3(CH_2)_{11}C \equiv C\text{:}^- \quad \xrightarrow{CH_3(CH_2)_7CH_2Br} \quad CH_3(CH_2)_{11}C \equiv CCH_2(CH_2)_7CH_3$$

5. After this last step, do we see how to get from starting material to product? If not, we need to analyze the structure obtained in step 4 by applying questions 3 and 4 to it.

Complete correct answer:

$$CH_3(CH_2)_{11}C \equiv CH \quad \xrightarrow[\text{NH}_3 \text{ (liq)}]{\text{NaNH}_2} \quad CH_3(CH_2)_{11}C \equiv C\text{:}^- Na^+ \quad \xrightarrow{CH_3(CH_2)_7CH_2Br}$$

$$CH_3(CH_2)_{11}C \equiv CCH_2(CH_2)_7CH_3$$

9.9 $CH_3(CH_2)_{11}C \equiv CCH_2(CH_2)_7CH_3 \quad \xrightarrow[\substack{PtO_2 \\ \text{hexane}}]{H_2 \text{ (1 equivalent)}}$

9.9 (cont)

$$CH_3(CH_2)_{11} \quad CH_2(CH_2)_7CH_3$$
$$C=C \qquad + \quad CH_3(CH_2)_{21}(CH_3) \quad + \quad CH_3(CH_2)_{11}C\equiv CCH_2(CH_2)_7CH_3$$
$$H \qquad H$$

$$\qquad\qquad \text{alkene} \qquad\qquad\qquad\qquad \text{alkane} \qquad\qquad\qquad\qquad \text{alkyne}$$

In the presence of one equivalent of hydrogen, the first product to form will be the alkene. As the concentration of alkene builds up in the reaction mixture, it will start to react with hydrogen to give the alkane. To the extent that this second reaction occurs, some of the alkyne will remain unchanged, because the system will run out of hydrogen before it runs out of alkyne.

— —

Concept Map 9.3   Reduction reactions of alkynes.

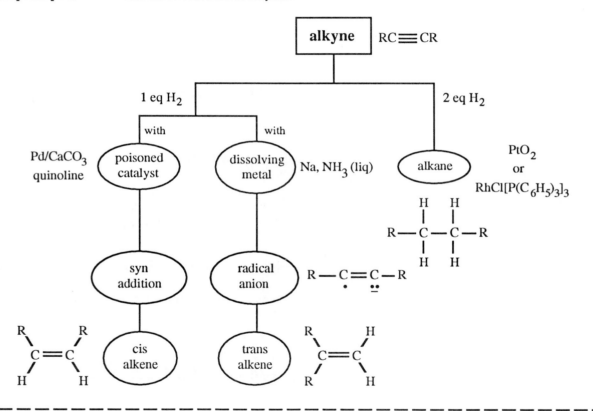

— —

9.10

(a) $CH_3CH_2CH_2CH_2C\equiv CH \xrightarrow[\substack{Pd/BaSO_4 \\ \text{quinoline}}]{H_2} CH_3CH_2CH_2CH_2CH=CH_2$

(b) $CH_3CH_2CH_2CH_2C\equiv CCH_3 \xrightarrow[\substack{Pd/BaSO_4 \\ \text{quinoline}}]{H_2}$

$$CH_3CH_2CH_2CH_2 \qquad CH_3$$
$$C=C$$
$$H \qquad\qquad H$$

9.10 (cont)

(c) 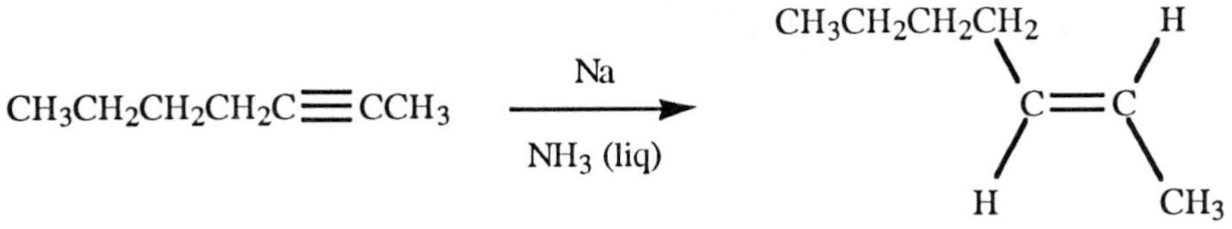 $CH_3CH_2CH_2CH_2C\equiv CCH_3$ $\xrightarrow[\text{NH}_3 \text{ (liq)}]{\text{Na}}$

(d) $CH_3CH_2CH_2C\equiv CCH_3$ $\xrightarrow[\text{RhCl[P(C}_6\text{H}_5)_3]_3]{\text{H}_2}$ $CH_3CH_2CH_2CH_2CH_2CH_3$

(e) $CH_3CH_2CH_2CH_2C\equiv CCH_3$ $\xrightarrow[\substack{\text{Pt} \\ \text{acetic acid}}]{\text{H}_2 \text{ (excess)}}$ $CH_3CH_2CH_2CH_2CH_2CH_2CH_3$

9.11 $CH_3C\equiv CCH_3$ $\xrightarrow[\text{Pd/Al}_2\text{O}_3]{\text{D}_2}$

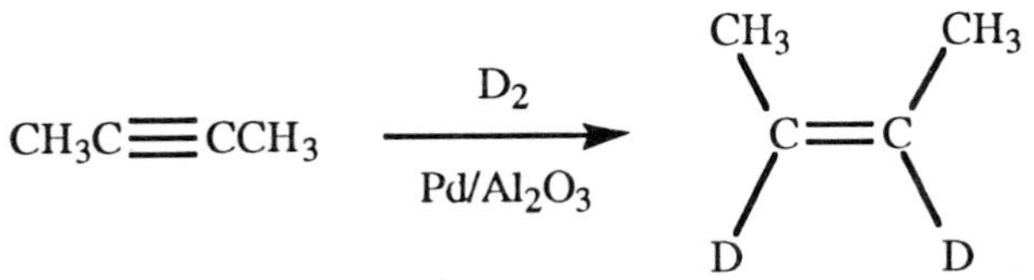

9.12

(a)

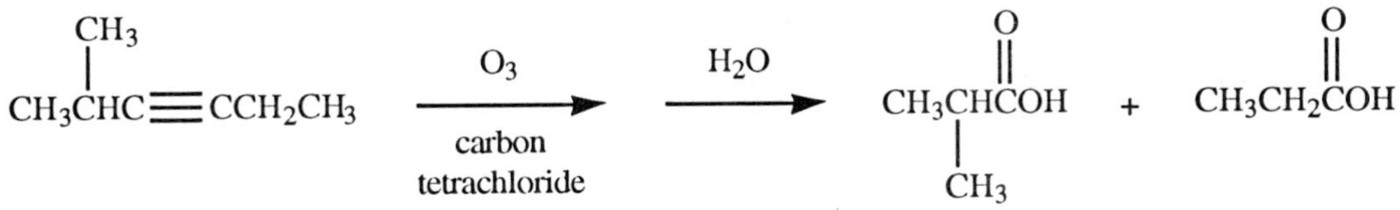

(b)

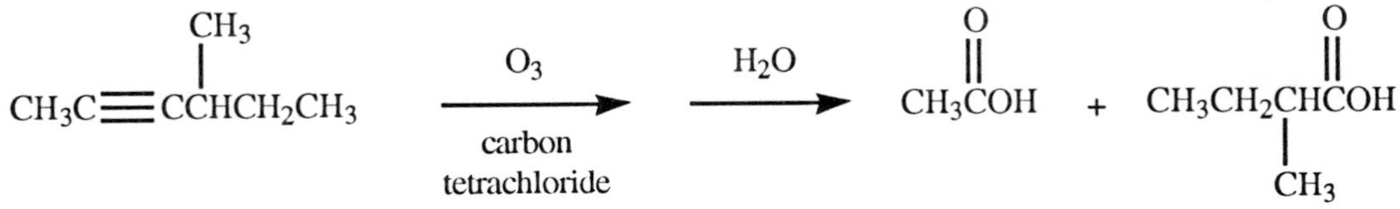

9.13

(a)

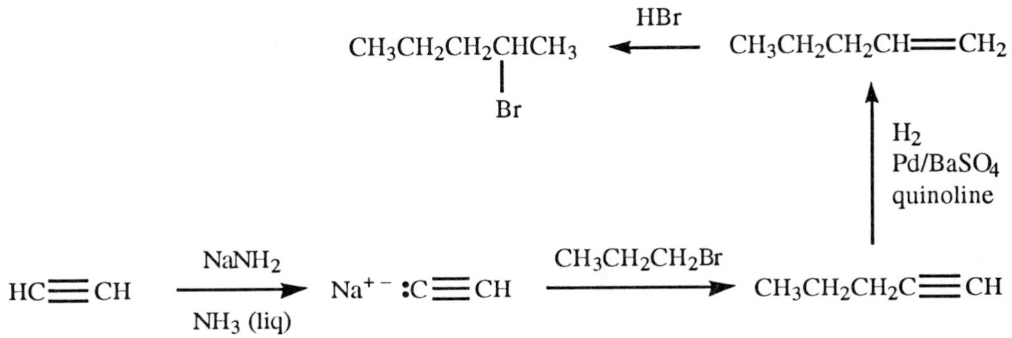

9.13 (cont)

(b)

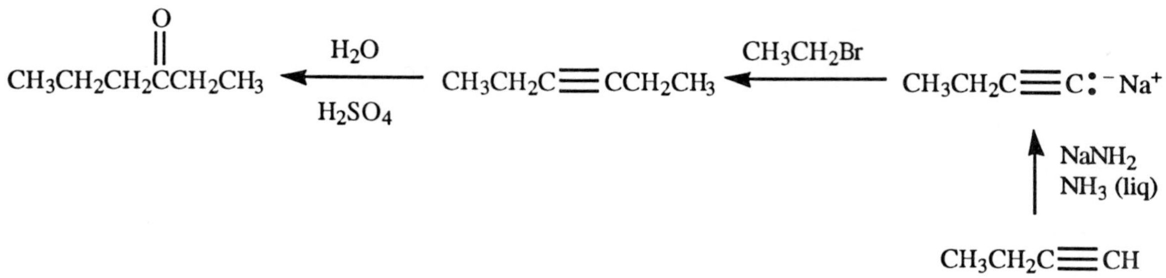

$$CH_3CH_2CH_2\overset{\overset{\displaystyle O}{\|}}{C}CH_2CH_3 \xleftarrow[\text{H}_2\text{SO}_4]{\text{H}_2\text{O}} CH_3CH_2C\equiv CCH_2CH_3 \xleftarrow{CH_3CH_2Br} CH_3CH_2C\equiv C\!:^- Na^+$$

$$\Big\uparrow \begin{array}{l} NaNH_2 \\ NH_3 \text{ (liq)} \end{array}$$

$$CH_3CH_2C\equiv CH$$

(c)

$$\xleftarrow[\text{H}_2\text{O}]{KMnO_4,\ KOH}$$

(Z-alkene structure)

$$\Big\uparrow \begin{array}{l} H_2 \\ Pd/BaSO_4 \\ quinoline \end{array}$$

$$CH_3CH_2CH_2C\equiv C\!:^- Na^+ \xrightarrow{CH_3CH_2Br} CH_3CH_2CH_2C\equiv CCH_2CH_3$$

$$\Big\uparrow \begin{array}{l} NaNH_2 \\ NH_3 \text{ (liq)} \end{array}$$

$$CH_3CH_2CH_2C\equiv CH$$

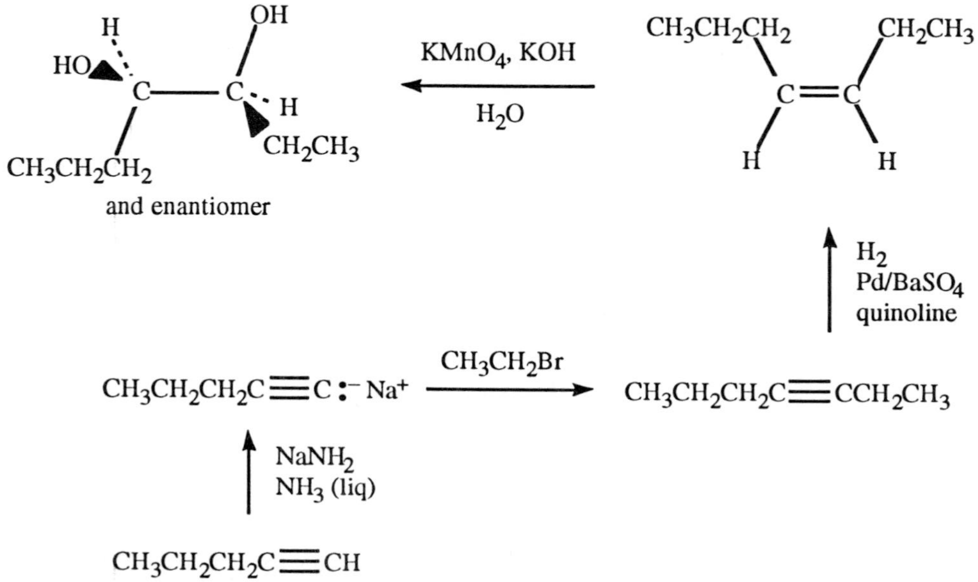

9.14   (a) (*R*)-2-chloro-4-octyne   (b) (*Z*)-3-methyl-3-heptene

(c) 2-methyl-4-octyn-2-ol   (d) *trans*-1-methyl-2-vinylcyclohexane

9.15   (a)

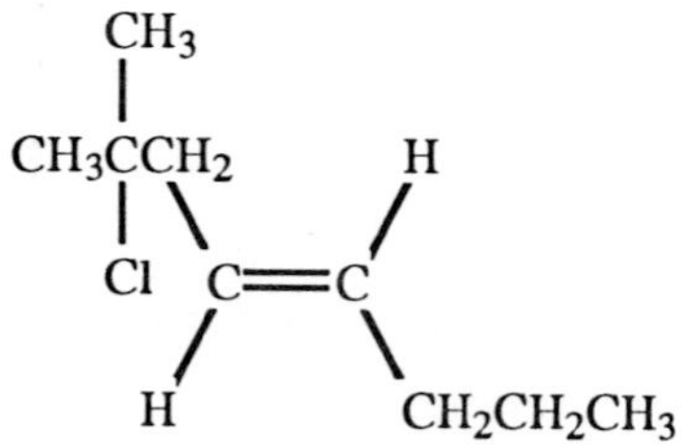

(b)

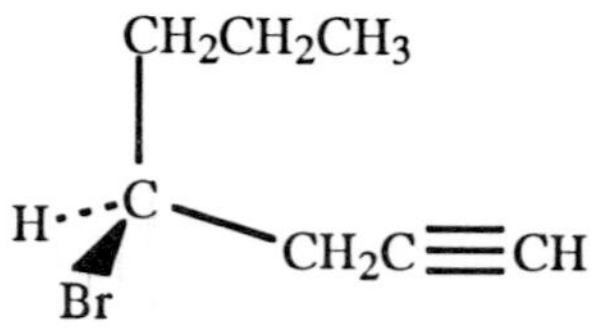

(c)

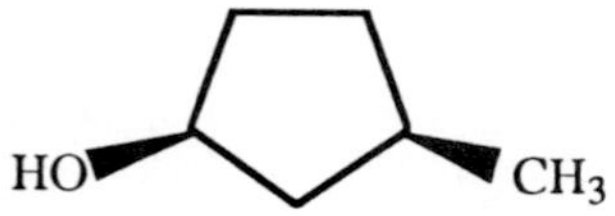

and enantiomer

(d) $CH_3CH_2C\equiv CCH_2CH_2OH$

9.16

(a) 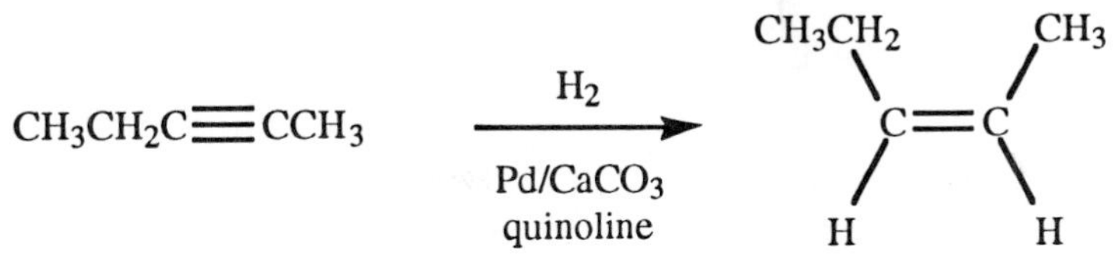 $CH_3CH_2C\equiv CCH_3$ $\xrightarrow[\text{quinoline}]{\text{H}_2 \atop \text{Pd/CaCO}_3}$

(b) $CH_3CH_2C\equiv CCH_3$ $\xrightarrow[\text{Pt}]{\text{H}_2 \text{ (excess)}}$ $CH_3CH_2CH_2CH_2CH_3$

(c) 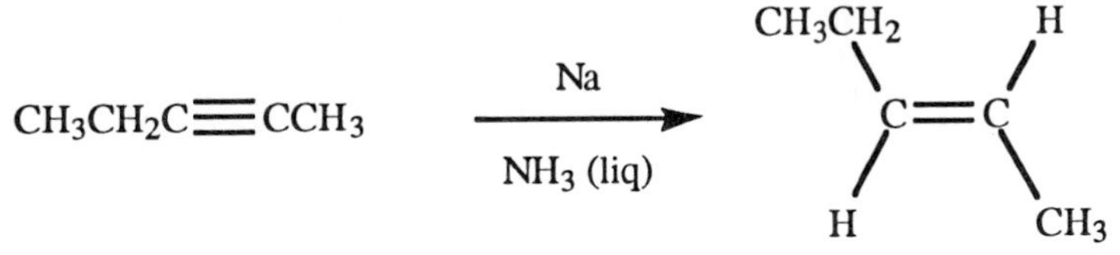 $CH_3CH_2C\equiv CCH_3$ $\xrightarrow[\text{NH}_3 \text{ (liq)}]{\text{Na}}$

(d) 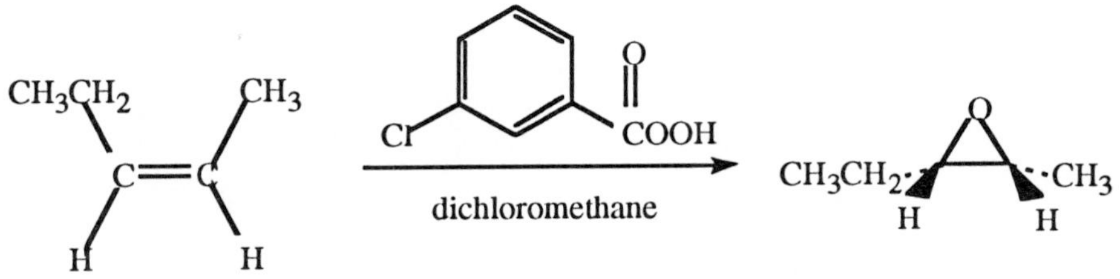 $\xrightarrow{\text{dichloromethane}}$ $CH_3CH_2\cdots\overset{O}{\triangle}\cdots CH_3$ and enantiomer

(e) 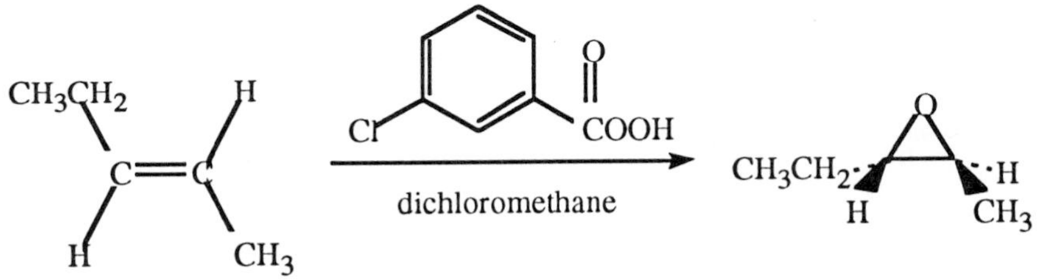 $\xrightarrow{\text{dichloromethane}}$ $CH_3CH_2\cdots\overset{O}{\triangle}\cdots H$ and enantiomer

(f) 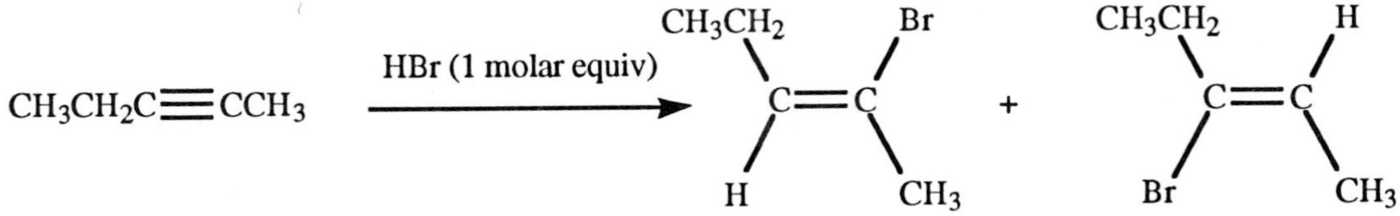 $CH_3CH_2C\equiv CCH_3$ $\xrightarrow{\text{HBr (1 molar equiv)}}$ $+$

major products

(g) $CH_3CH_2C\equiv CCH_3$ $\xrightarrow{\text{HBr (2 molar equiv)}}$ $CH_3CH_2CBr_2CH_2CH_3$ $+$ $CH_3CH_2CH_2CBr_2CH_3$

(h) 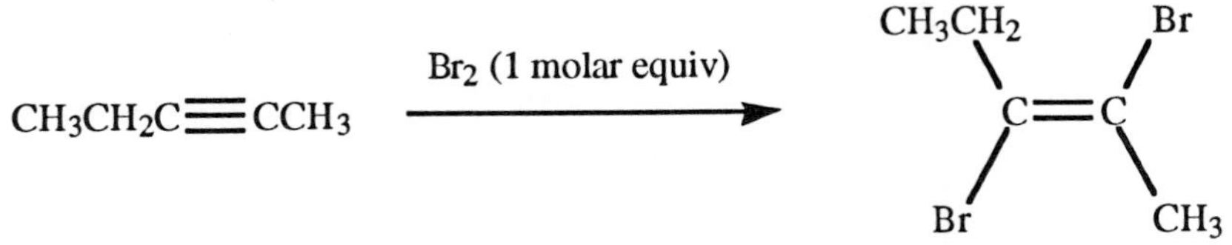 $CH_3CH_2C\equiv CCH_3$ $\xrightarrow{\text{Br}_2 \text{ (1 molar equiv)}}$

9.16 (cont)

(i) $CH_3CH_2C\equiv CCH_3$ $\xrightarrow{\text{Br}_2 \text{ (2 molar equiv)}}$ $CH_3CH_2CBr_2CBr_2CH_3$

(j) $CH_3CH_2C\equiv CCH_3$ $\xrightarrow[\text{carbon}\atop\text{tetrachloride}]{\text{O}_3}$ $\xrightarrow{\text{H}_2\text{O}}$ $CH_3CH_2\overset{\displaystyle O}{\overset{\|}{C}}OH$ + $CH_3\overset{\displaystyle O}{\overset{\|}{C}}OH$

(k) $CH_3CH_2C\equiv CCH_3$ $\xrightarrow[\text{H}_2\text{SO}_4 \atop \text{HgSO}_4]{\text{H}_2\text{O}}$ $CH_3CH_2\overset{\displaystyle O}{\overset{\|}{C}}CH_2CH_3$ + $CH_3CH_2CH_2\overset{\displaystyle O}{\overset{\|}{C}}CH_3$

(l) $CH_3CH_2C\equiv CCH_3$ $\xrightarrow[\text{RhCl}[\text{P}(\text{C}_6\text{H}_5)_3]_3]{\text{H}_2}$ $CH_3CH_2CH_2CH_2CH_3$

9.17

(a)

A

(b) $CH_2{=}CHBr$ $\xrightarrow{\text{HBr}}$ CH_3CHBr_2

                                             B

(c) $CH_2{=}CHCH_2Br$ $\xrightarrow[\text{carbon}\atop\text{tetrachloride}]{\text{Br}_2}$ $BrCH_2CHBrCH_2Br$

                                                  C

(d)

D

(e)

E

9.17 (cont)

(f)
$$CH_3(CH_2)_7C\equiv C(CH_2)_7\overset{O}{\overset{\|}{C}}OH \xrightarrow[\text{carbon tetrachloride}]{O_3} \xrightarrow{H_2O} CH_3(CH_2)_7\overset{O}{\overset{\|}{C}}OH + HO\overset{O}{\overset{\|}{C}}(CH_2)_7\overset{O}{\overset{\|}{C}}OH$$

                                                                        F   G

(g)

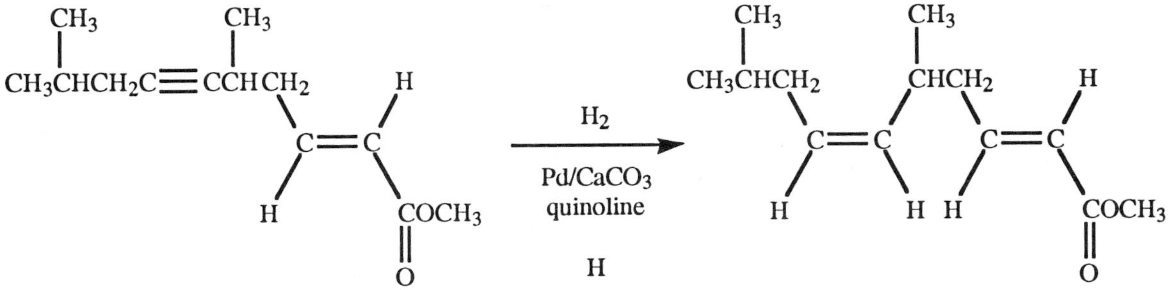

(h)

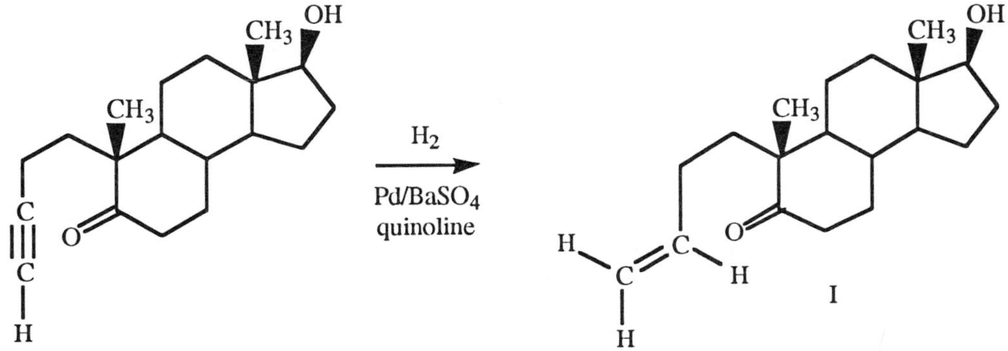

9.18

(a)
$$CH_3C\equiv CCH_3 \xrightarrow[H_2SO_4]{H_2O} CH_3\overset{O}{\overset{\|}{C}}CH_2CH_3$$

                                    A

(b)

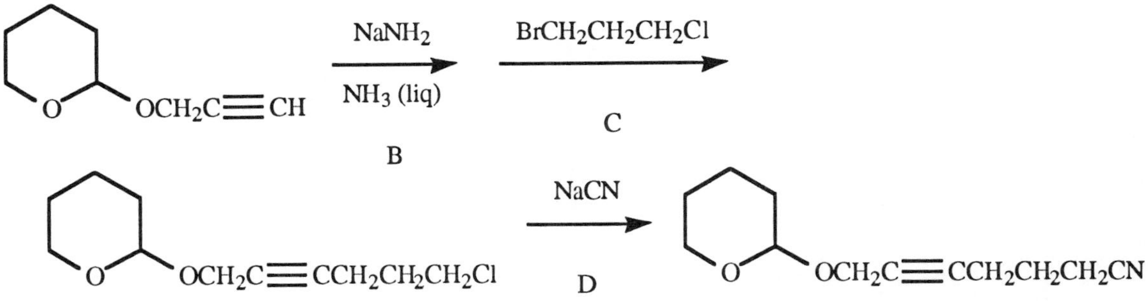

(c)
$$CH_2=CHCH_2CH_2Br \xrightarrow[NH_3 \text{ (liq)}]{HC\equiv C:^- Na^+} CH_2=CHCH_2CH_2C\equiv CH$$

                                                                    E

9.18 (cont)

(d) 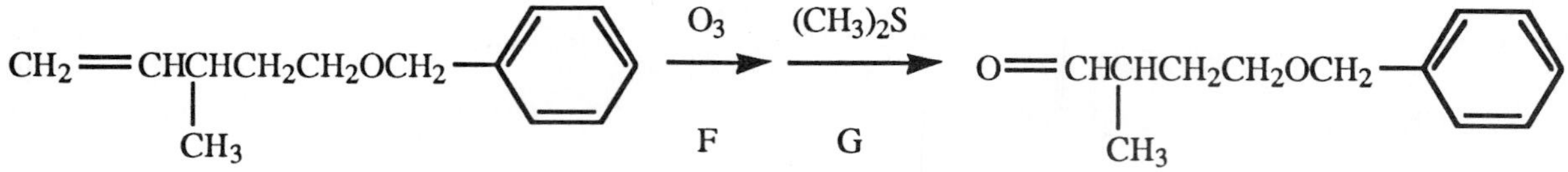 CH_2=CHCHCH$_2$CH$_2$OCH$_2$—⟨phenyl⟩ $\xrightarrow{O_3}$ $\xrightarrow{(CH_3)_2S}$ O=CHCHCH$_2$CH$_2$OCH$_2$—⟨phenyl⟩
        | |
        CH$_3$ CH$_3$
                                          F   G

(e) HC≡CH $\xrightarrow{Na^+ NH_2^-}$ HC≡C:$^-$Na$^+$ $\xrightarrow{\text{⟨phenyl⟩—CH}_2\text{OCH}_2\text{CH}_2\text{Br}}$
            H   I   J

HC≡CCH$_2$CH$_2$OCH$_2$—⟨phenyl⟩

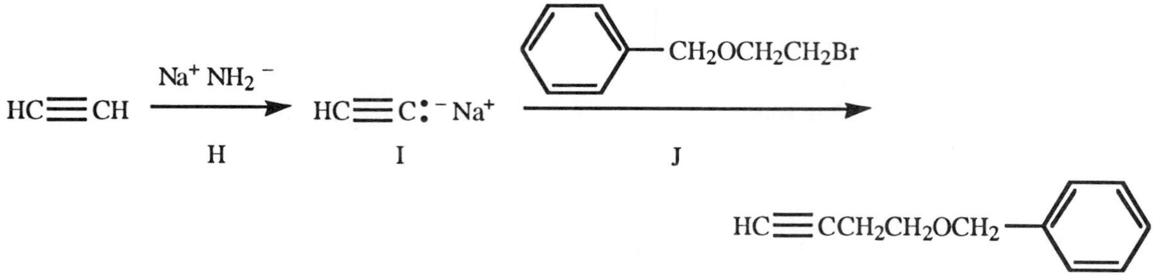

(f) ⟨phenyl⟩—C≡C—COCH$_2$CH$_3$ $\xrightarrow[\substack{Pd/BaSO_4 \\ quinoline}]{H_2}$

K

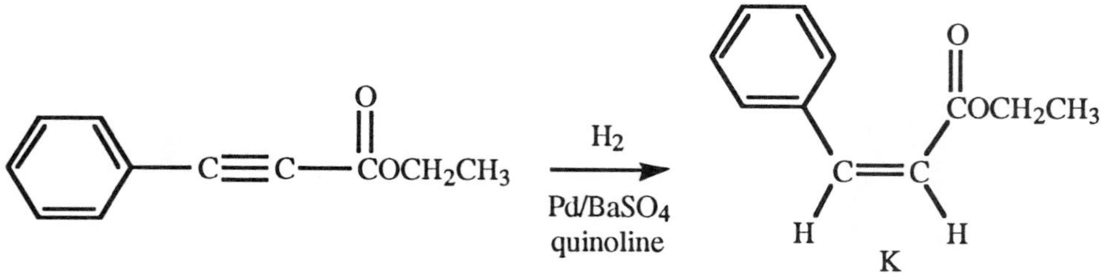

9.19

(a)

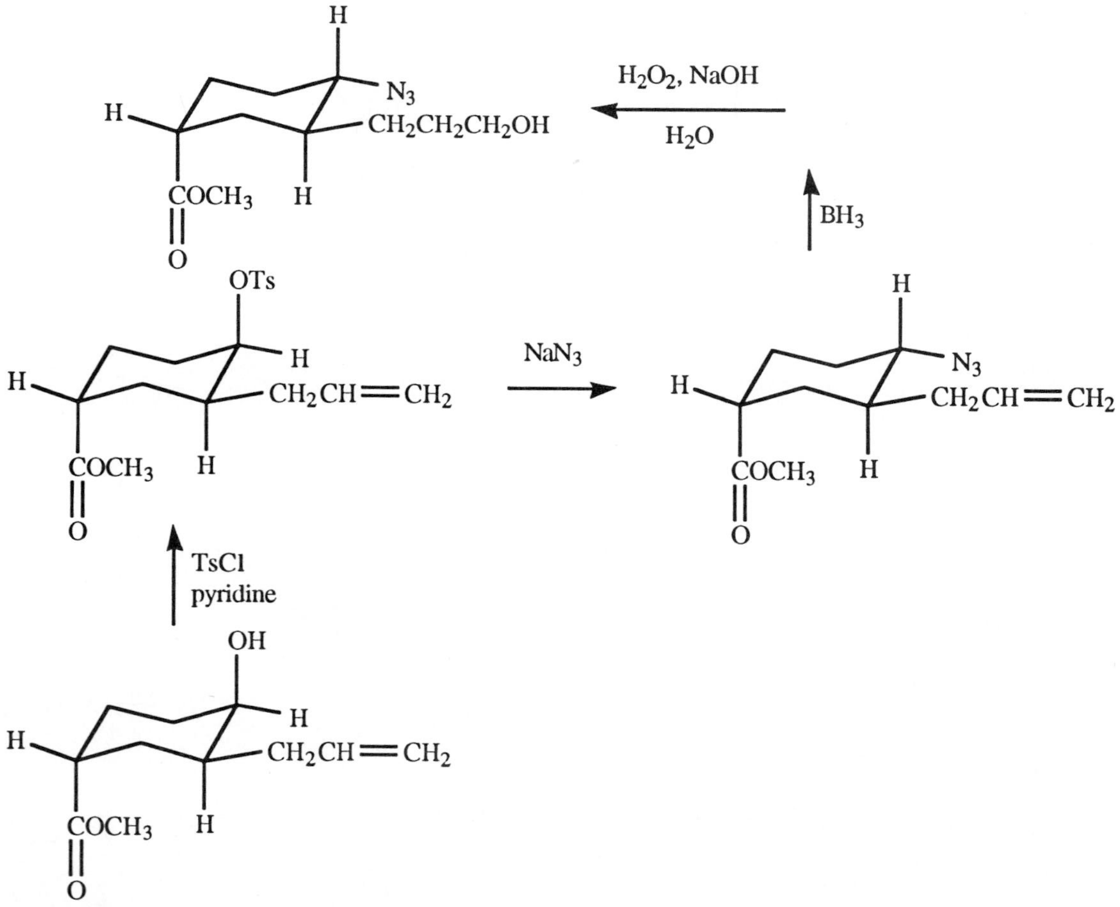

9.19 (cont)

(b)

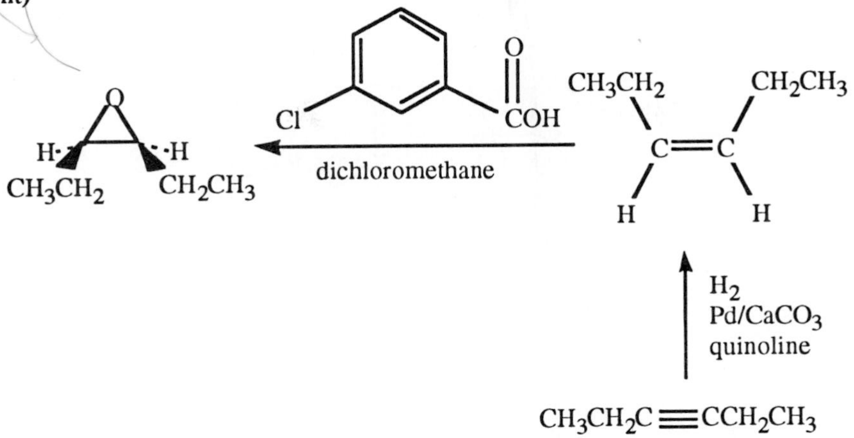

(c)

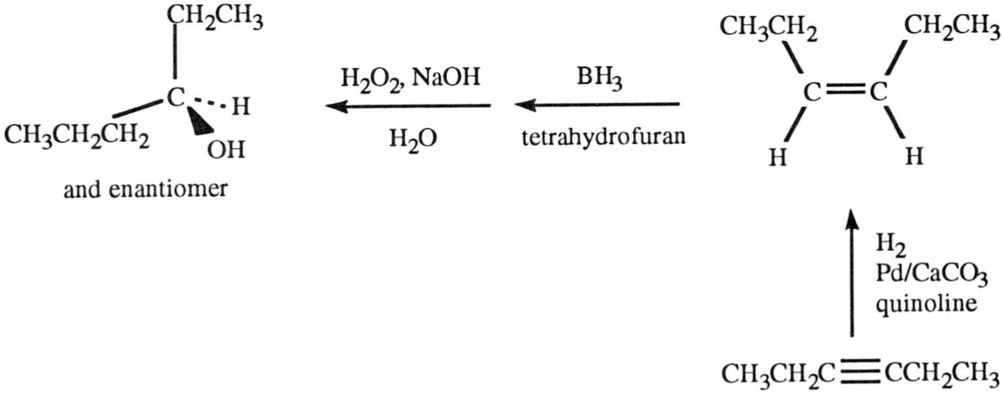

and enantiomer

An alternate synthesis is given below.

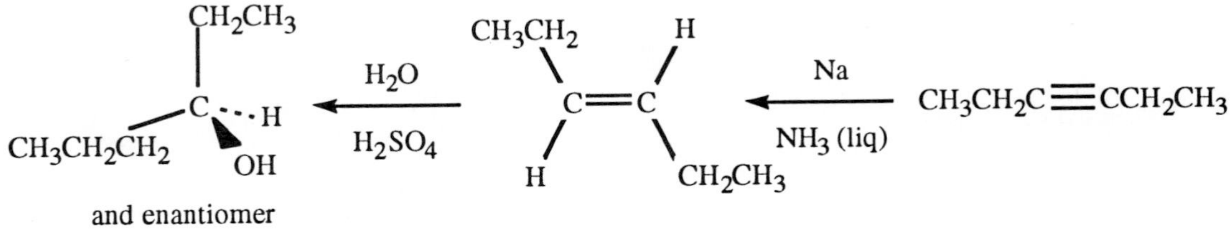

and enantiomer

(d)

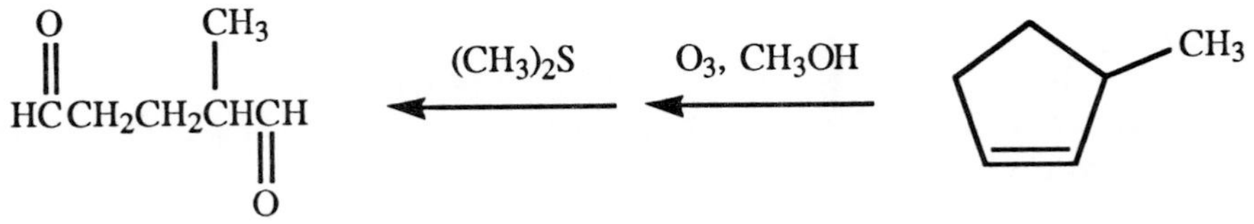

(e)

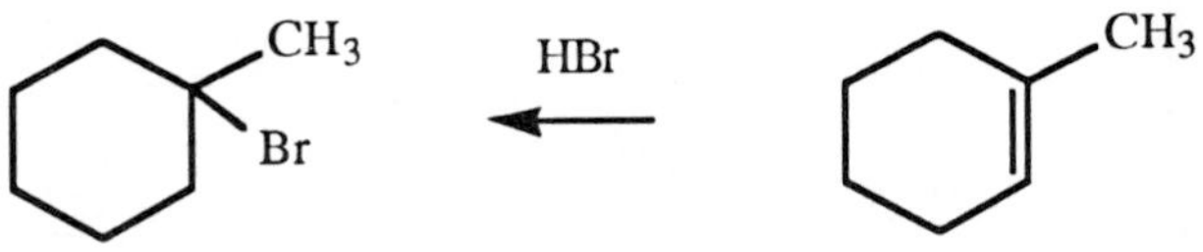

9.19 (cont)

(f)

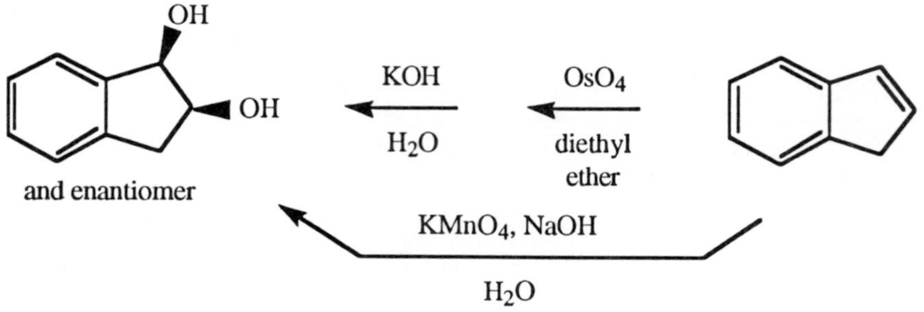

and enantiomer

KOH / H$_2$O ← OsO$_4$ / diethyl ether

KMnO$_4$, NaOH / H$_2$O

9.20

(a)

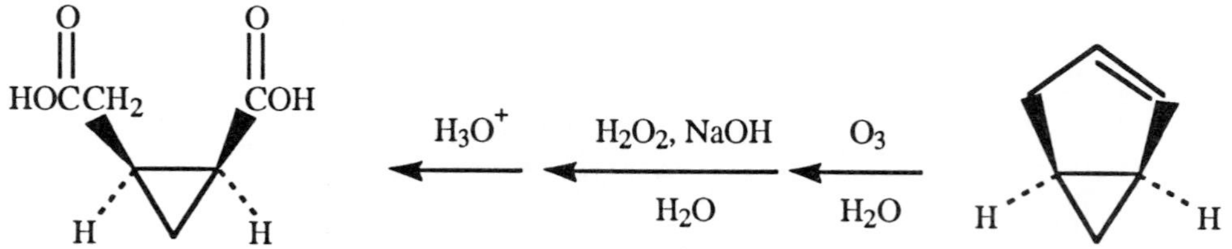

H$_3$O$^+$ ← H$_2$O$_2$, NaOH / H$_2$O ← O$_3$ / H$_2$O

(b)

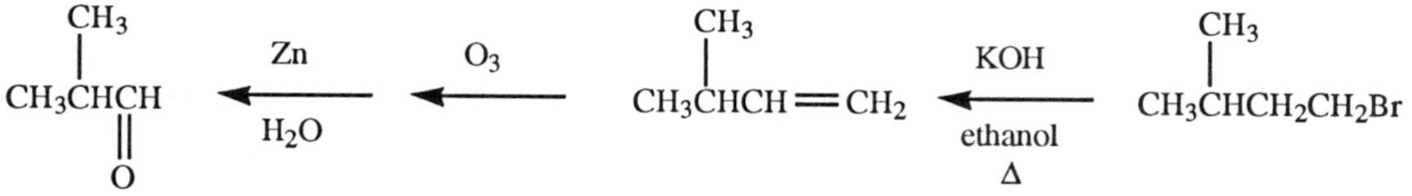

$$CH_3CHCH \overset{O}{\underset{\|}{}}$$ Zn / H$_2$O ← O$_3$ ← CH$_3$CHCH=CH$_2$ KOH / ethanol / Δ ← CH$_3$CHCH$_2$CH$_2$Br

(c)

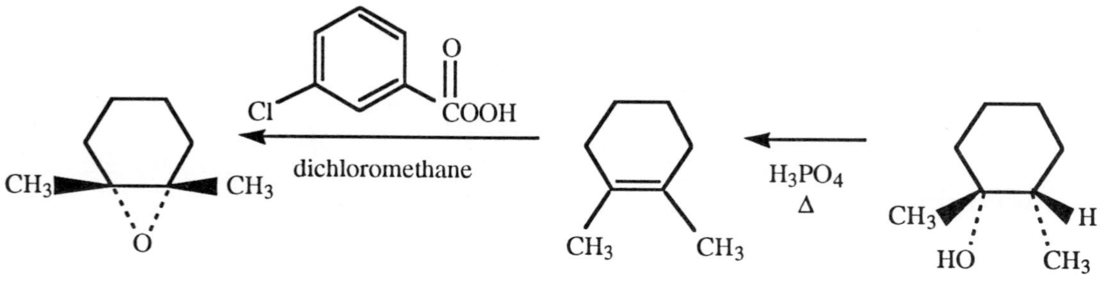

dichloromethane ← H$_3$PO$_4$ / Δ

(d)

H$_2$ / Pd/BaSO$_4$ / quinoline ← CH$_3$CH$_2$C≡CCH$_2$CH$_3$ ← CH$_3$CH$_2$Br CH$_3$CH$_2$C≡C:$^-$ Na$^+$

NaNH$_2$ / NH$_3$ (liq)

CH$_3$CH$_2$Br → HC≡C:$^-$Na$^+$ / NH$_3$ (liq) → CH$_3$CH$_2$C≡CH

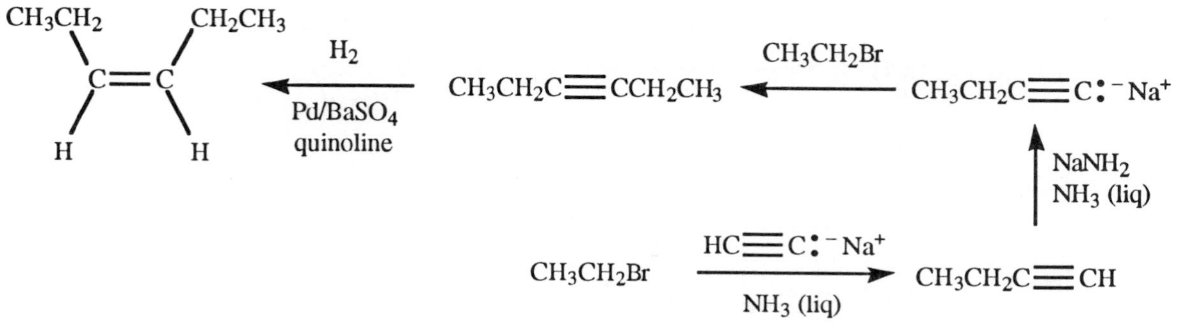

9.20 (cont)

(e)

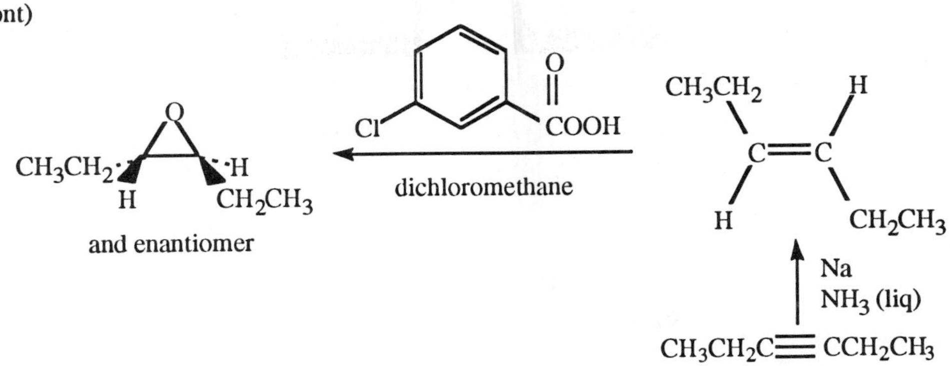

and enantiomer

from (d)

9.21   (a) $C_{13}H_{10}O \xrightarrow[\text{cat}]{H_2} C_{13}H_{14}O$ The corresponding saturated hydrocarbon would have the formula $C_{13}H_{28}O$.

$C_{13}H_{10}O$ has 9 units of unsaturation $\left(\dfrac{28-10}{2}=9\right)$.

After hydrogenation, 7 units of unsaturation remain $\left(\dfrac{28-14}{2}=7\right)$.

Two units of unsaturation $(9-7=2)$ correspond to multiple bonds that are easily hydrogenated.

Possible structures for Carlina-oxide are

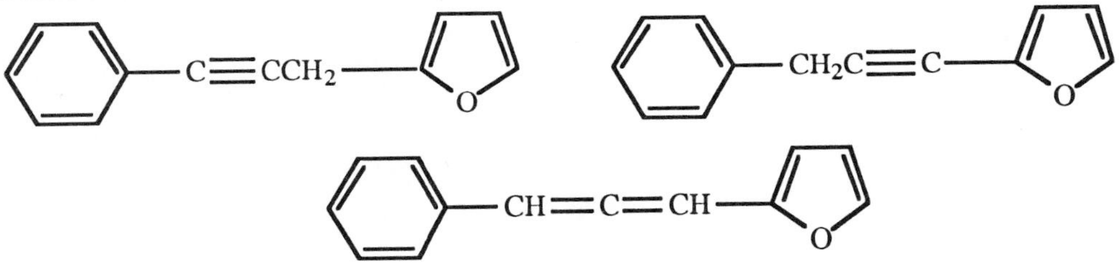

(b)

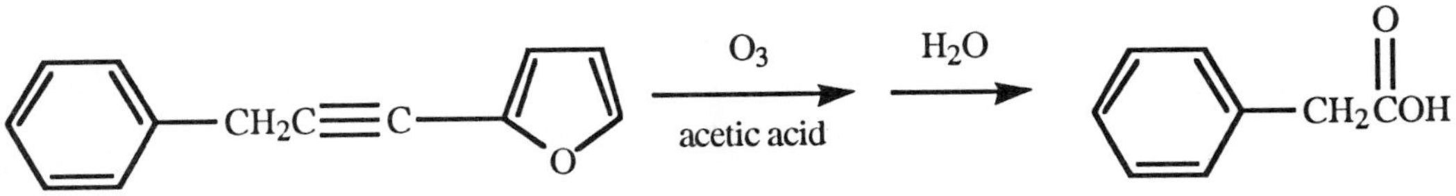

Carlina-oxide

+ unisolated compounds

9.22

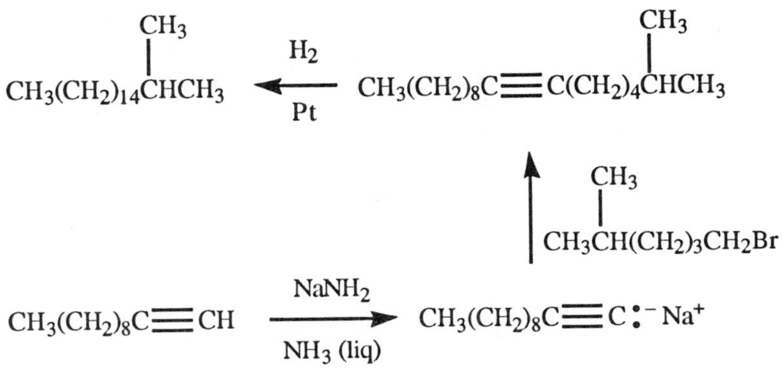

9.23

(a) $HOCH_2(CH_2)_7C\equiv CH$ $\xrightarrow[NH_3 \text{ (liq)}]{Na^+ NH_2^- \text{ (excess)}}$ $Na^{+\,-}OCH_2(CH_2)_7C\equiv C:^- Na^+$ $\xrightarrow{CH_3CH_2CH_2CH_2Br}$

                                                                   A

$\xrightarrow{H_3O^+}$ $HOCH_2(CH_2)_7C\equiv CCH_2CH_2CH_2CH_3$ $\xrightarrow[\text{poisoned catalyst}]{H_2}$

B

a carbon anion is more
nucleophilic than an oxygen anion

spodoptol
C

(b)

(c)

$CH_3CH_2CH_2CH_2\overset{\overset{\textstyle O}{\|}}{C}H$

Aldehyde D

$HOCH_2(CH_2)_7\overset{\overset{\textstyle O}{\|}}{C}H$

Aldehyde E

9.24   (a)

disparlure

dichloromethane

$CH_3(CH_2)_9C\equiv C(CH_2)_4\overset{\overset{\textstyle CH_3}{|}}{C}HCH_3$ $\xrightarrow[\substack{Pd/CaCO_3 \\ \text{quinoline}}]{H_2}$

$\overset{\overset{\textstyle CH_3}{|}}{CH_3CH}(CH_2)_3CH_2Br$

$CH_3(CH_2)_9C\equiv C:^- Na^+$ $\xleftarrow[NH_3 \text{ (liq)}]{NaNH_2}$ $CH_3(CH_2)_9C\equiv CH$

9.24 (cont)

(b)

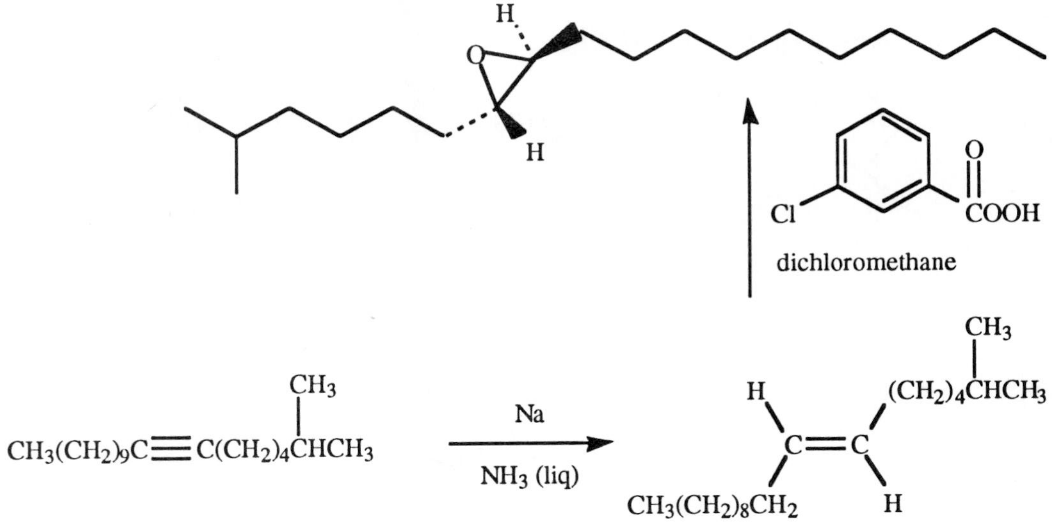

9.25 $CH_3CH_2C \equiv CCH_2CH_3$ $\xrightarrow{BH_3}$ $\xrightarrow[H_2O]{H_2O_2, NaOH}$

$\xrightarrow{\text{tautomerization}}$ $CH_3CH_2CH_2CCH_2CH_3$

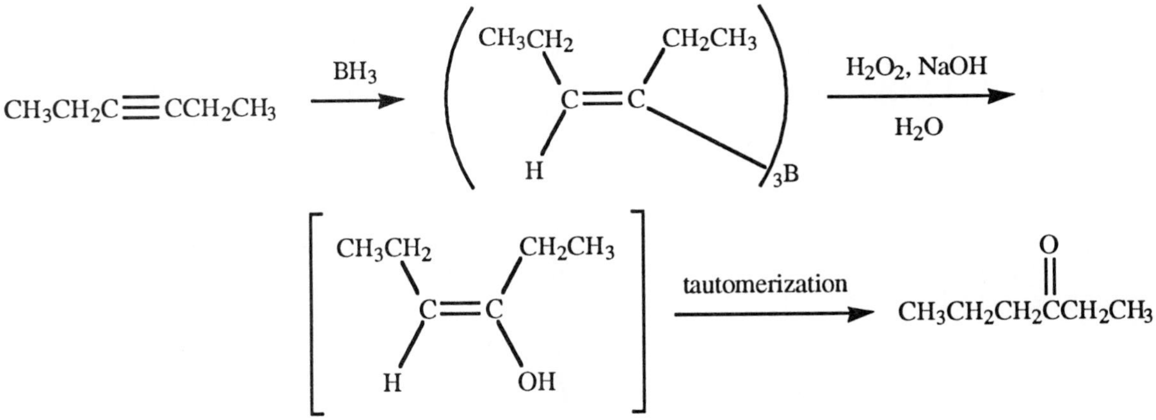

9.26

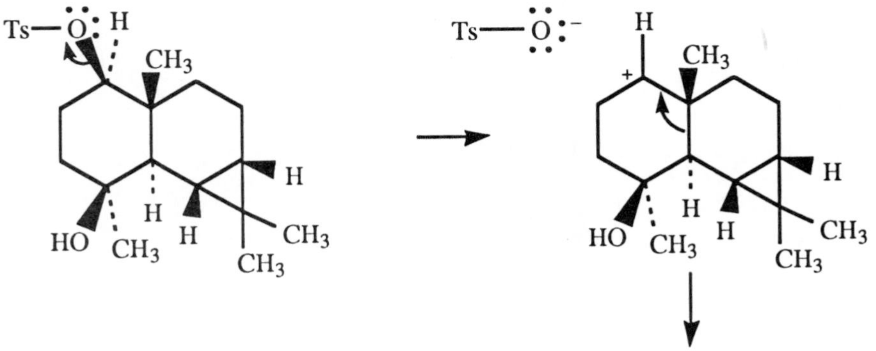

9.26 (cont)

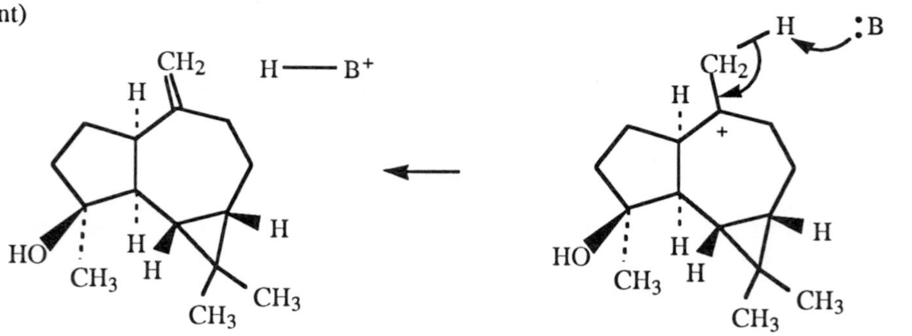

9.27

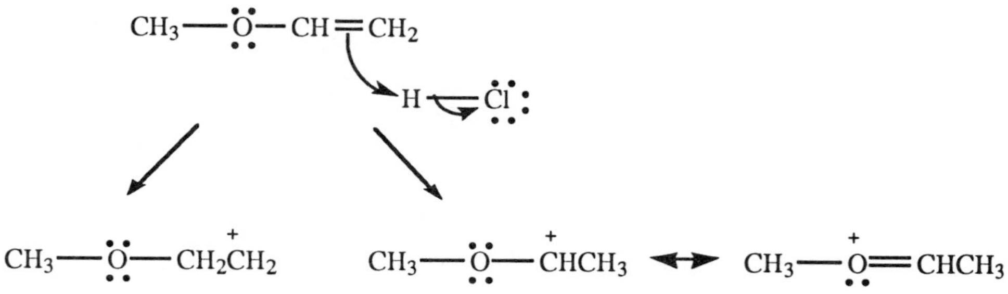

primary carbocation;
electron-withdrawing
group on α-carbon;
less stable; does not form

resonance stabilized secondary
carbocation; more stable

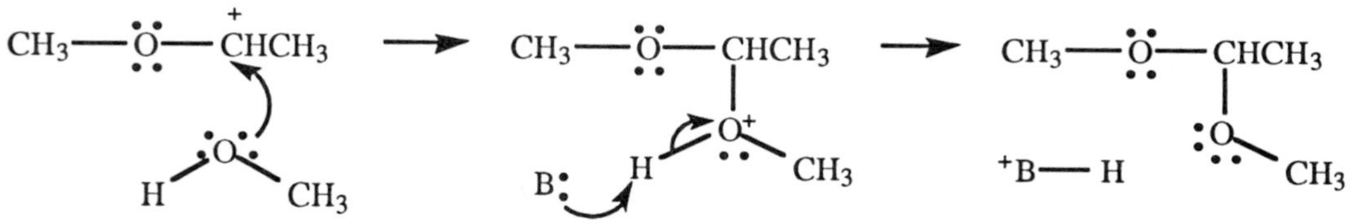

9.28

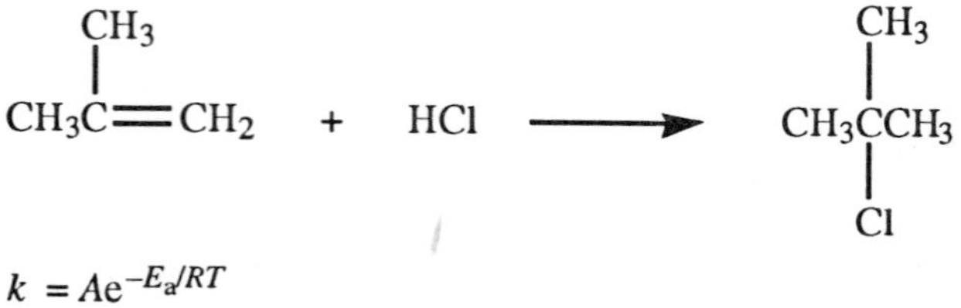

$$CH_3C = CH_2 \ + \ HCl \ \longrightarrow \ CH_3\overset{CH_3}{\underset{Cl}{C}}CH_3$$

$k = Ae^{-E_a/RT}$

$$k = (1 \times 10^{11} \ \text{mL/mol·s})(\exp\frac{-28.8 \ \text{kcl/mol}}{(1.99 \times 10^{-3} \ \text{kcal/mol·K})(298 \ \text{K})})$$

$k = (1 \times 10^{11} \ \text{mL/mol·s})(\exp -48.6)$

$k = (1 \times 10^{11} \ \text{mL/mol·s})(8.1 \times 10^{-22})$

$k = 8 \times 10^{-11} \ \text{mL/mol·s}$

9.28 (cont)

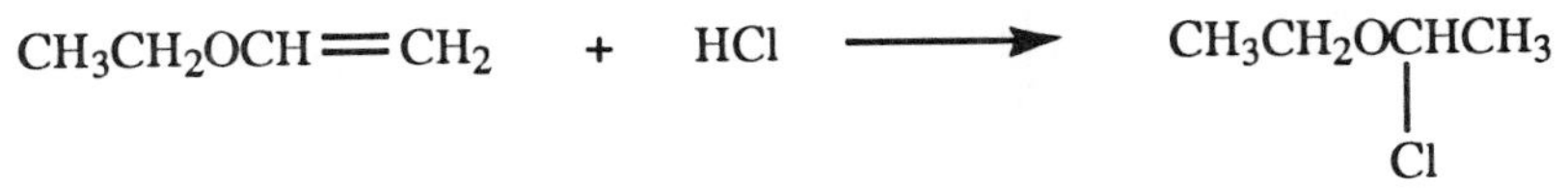

$$CH_3CH_2OCH=CH_2 \quad + \quad HCl \quad \longrightarrow \quad CH_3CH_2OCHCH_3$$
$$\underset{Cl}{|}$$

$$k = (5 \times 10^8 \text{ mL/mol·s})(\exp\frac{-14.7 \text{ kcal/mol}}{(1.99 \times 10^{-3} \text{ kcal/mol·K})(298 \text{ K})})$$

$$k = (5 \times 10^8 \text{ mL/mol·s})(\exp -24.8)$$

$$k = (5 \times 10^8 \text{ mL/mol·s})(1.7 \times 10^{-11})$$

$$k = 9 \times 10^{-3} \text{ mL/mol·s}$$

In the reaction with 2-methylpropene, the intermediate is a tertiary carbocation that is stabilized by the electron-donating inductive effect of the three methyl groups. In the reaction with ethyl vinyl ether, the intermediate is a carbocation in which the positive charge can be delocalized over the carbon atom and the adjacent oxygen atom.

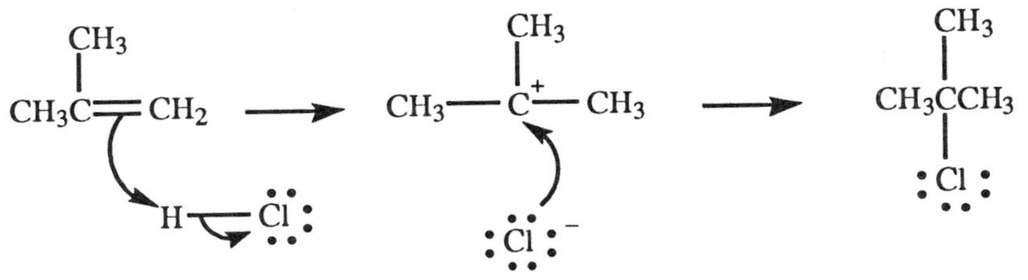

tertiary carbocation;
less stable than the resonance-
stabilized carbocation from
ethyl vinyl ether

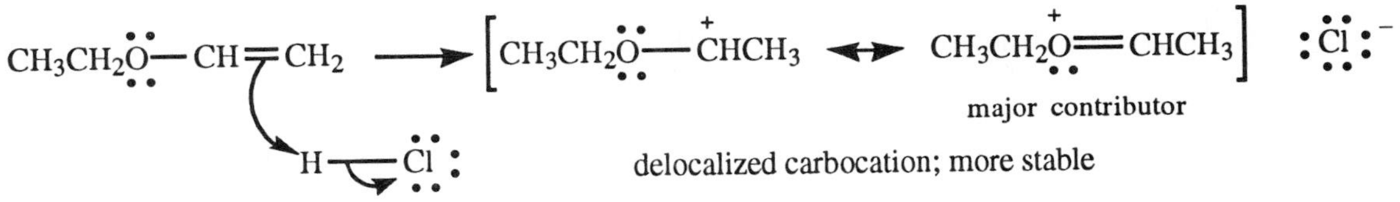

major contributor

delocalized carbocation; more stable

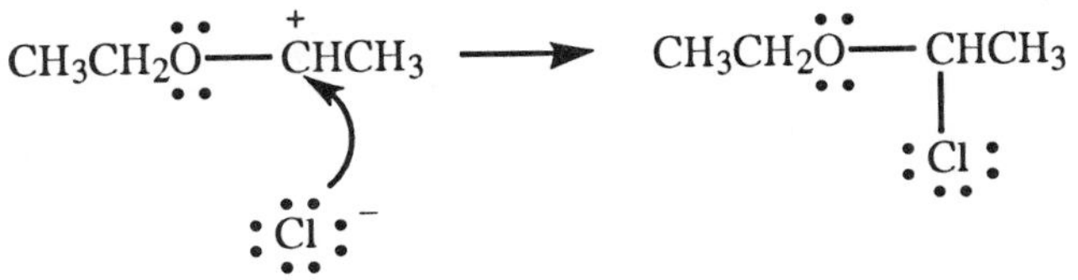

9.28 (cont)

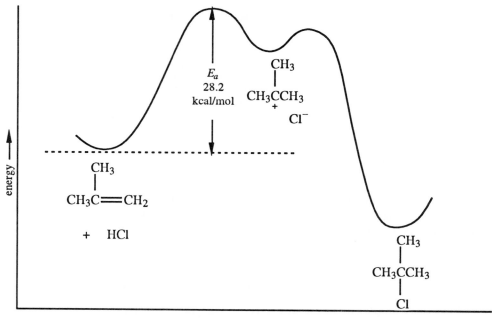

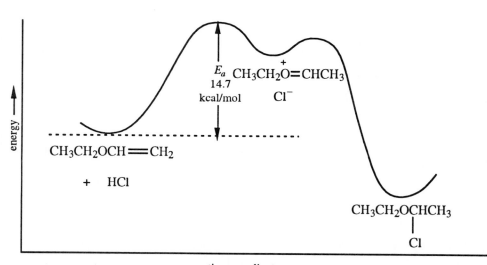

The reactive intermediate formed in this reaction is more highly stabilized, and therefore, closer in energy to the energy level of the starting materials than is the case for 2-methylpropene. The E_a for this reaction is therefore lower, and the reaction is faster.

10

Infrared Spectroscopy

10.1   Compound A has bands at
3090 cm^{-1} (=C—H)
1650 cm^{-1} (C=C)
therefore compound A is an alkene.

Compound B has bands at
3300 cm^{-1} (≡C—H)
2100 cm^{-1} (C≡C)
therefore compound B is an alkyne.

Compound C has bands at
3330 cm^{-1} (—O—H)
1060 cm^{-1} (C—O)
therefore compound C is an alcohol.

Compound D has bands at
3080 cm^{-1} (=C—H)
1640 cm^{-1} (C=C)
therefore compound D is an alkene.

10.2   Compound E, $C_4H_{10}O$, has no units of unsaturation. It has bands at
3325 (broad) cm^{-1} (—O—H)
1040 cm^{-1} (C—O)
therefore it is an alcohol.
Possible structures are:

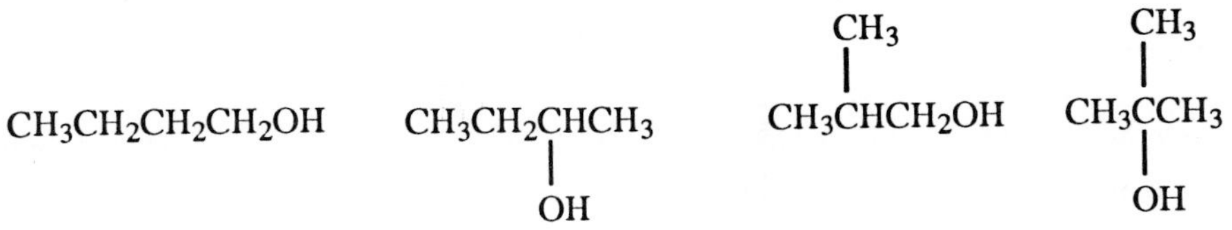

$CH_3CH_2CH_2CH_2OH$ $CH_3CH_2\underset{\underset{OH}{|}}{C}HCH_3$ $CH_3\underset{\overset{|}{CH_3}}{C}HCH_2OH$ $CH_3\underset{\underset{OH}{|}}{\overset{\overset{CH_3}{|}}{C}}CH_3$

Compound F, $C_6H_{12}O$, has one unit of unsaturation. It has bands at
3325 (broad) cm^{-1} (—O—H)
1060 cm^{-1} (C—O)
therefore it must be an alcohol. Since there is one unit of unsaturation and no alkene stretching frequency, it must be a cyclic alcohol.

10.2 (cont)

Possible structures are:

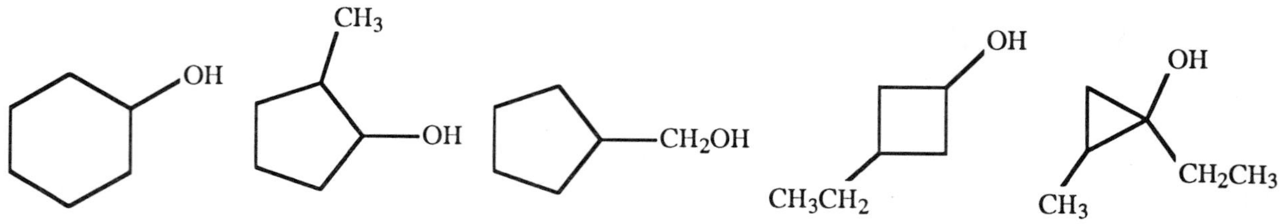

Compound G, $C_6H_{12}O$, has one unit of unsaturation. It has bands at
1700 cm^{-1} (C=O).
The unit of unsaturation corresponds to the carbonyl group; therefore, it cannot have a ring.
Possible structures are:

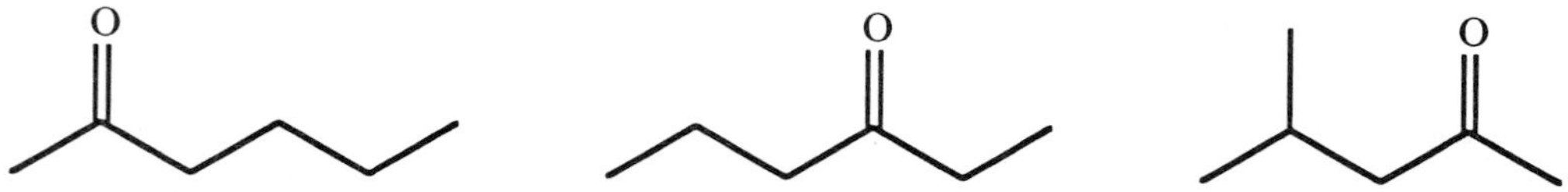

Compound H, C_4H_8O, has one unit of unsaturation. It has bands at
2700 cm^{-1} (O=C—H). (The band at 2900 cm^{-1} is hidden in the alkane band.)
1740 cm^{-1} (C=O)

Possible structures are:

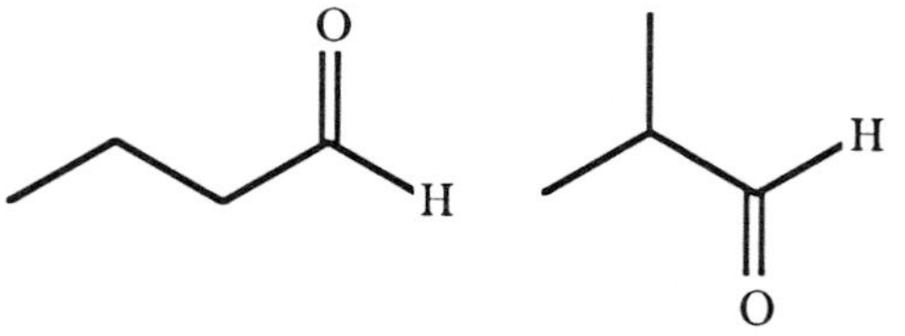

10.3   The best approach to identifying which spectrum belongs to which compound is to first determine what functional groups are present in each compound and then to look for the specific bands in the spectra that correspond to the functional groups.

There is only one alcohol, which should have a broad band at ~3350 cm^{-1} and no band in the carbonyl region. The only spectrum that fits is spectrum 4. Spectrum 4, therefore, must belong to 1-octanol.
        3325 (broad) cm^{-1} (—O—H)
        1060 cm^{-1} (C—O)

Two of the compounds are acids. Acids should have a broad band at ~3300–2400 cm^{-1} and a band in the carbonyl region. Spectra 5 and 6 have the bands that correspond to a carboxylic acid. 3-Butenoic acid is also an alkene, and its spectrum should have a band at ~1600 cm^{-1}. Spectrum 5 has a band at 1645 cm^{-1}, but spectrum 6 does not have a band in this region; therefore, spectrum 5 belongs to 3-butenoic acid and spectrum 6 belongs to octanoic acid.

10.3 (cont)

Spectrum 5 (3-butenoic acid).

3100–2500 (broad) cm^{-1} (—O—H)

  1715 cm^{-1} (C=O)

  1645 cm^{-1} (C=C)

Spectrum 6 (octanoic acid).

  3100–2500 (broad) cm^{-1} (—O—H)

  1710 cm^{-1} (C=O)

One of the compounds is an aldehyde, which should have a band in the carbonyl region and a band at 2700 cm^{-1}. Spectrum 2 has these two bands and must, therefore, belong to 10-undecenal, an unsaturated aldehyde. Spectrum 2 also has a band at ~1600 cm^{-1}, which confirms the presence of an alkene.

  2715 cm^{-1} (O=C—H)

  1725 cm^{-1} (C=O)

  1640 cm^{-1} (C=C)

Of the last two spectra to assign (1 and 3), one must belong to an ester (methyl pentanoate) and the other to a ketone (3-methylcyclohexanone). Both have bands in the carbonyl region, but spectrum 1 has broad bands between 1000-1300 cm^{-1}, suggesting that this spectrum belongs to an ester. Spectrum 3 must, therefore, belong to the ketone.

Spectrum 1 (methyl pentanoate)

  1740 cm^{-1} (C=O)

  1260, 1175 cm^{-1} (C—O)

Spectrum 3 (3-methylcyclohexanone)

  1710 cm^{-1} (C=O)

10.4   Compound I has bands at

1740 cm^{-1} (C=O)

1240, 1050 cm^{-1} (C—O)

It is therefore an ester.

Compound J has bands at

3325 (broad) cm^{-1} (—O—H)

1060 cm^{-1} (C—O)

There is no carbonyl absorption; it is therefore an alcohol.

Compound K has bands at

3100 cm^{-1} (=C—H)

1640 cm^{-1} C=C

It is therefore an alkene.

Compound L has bands at

2720 cm^{-1} (O=C—H)

1730 cm^{-1} (C=O)

It is therefore an aldehyde.

10.5   Compound M, C_3H_3Cl, has 2 units of unsaturation. It has bands at

   3300 cm^{-1} ($\equiv$C—H)

   2130 cm^{-1} (C$\equiv$C)

   The two units of unsaturation correspond to the triple bond. It is a terminal alkyne.
   There is only one possible structure.

   HC$\equiv$CCH$_2$Cl

   Compound N, $C_5H_{10}O$, has one unit of unsaturation. It has bands at

   1715 cm^{-1} (C=O)

   The absorption at 1715 cm^{-1} indicates that it has a carbonyl group. With only one unit of unsaturation, it cannot
   contain a ring. No absorption at ~2700 cm^{-1} indicates that it is a ketone and not an aldehyde.
   Possible structures are:

   Compound O, $C_5H_{10}O_2$, has one unit of unsaturation. It has bands at

   3200–2500 (broad) cm^{-1} (—O—H)

   1710 cm^{-1} (C=O)

   These two bands indicate a carboxylic acid. With one unit of unsaturation in the carbonyl group, the compound
   cannot contain a ring.
   Possible structures are:

   Compound P, $C_8H_8O_2$, has five units of unsaturation. It has bands at

   1725 cm^{-1} (C=O)

   1600 cm^{-1} (C=C (aromatic))

   1280, 1110 cm^{-1} (C—O)

   When a compound has at least four units of unsaturation, an aromatic ring is usually present. This is confirmed
   by the band at 1600 cm^{-1}. The fifth unit of unsaturation is the carbonyl group. The bands at 1280 and 1110
   cm^{-1}, along with the presence of a carbonyl, indicate an aromatic ester.
   There is only one possible structure.

10.6   Compound Q has bands at

3335 (broad) cm^{-1} (—O—H)
3080 cm^{-1} (C=C—H)
1650 cm^{-1} (C=C)
1048 cm^{-1} (C—O)

It is an alcohol and an alkene.

Compound R has bands at

2200 cm^{-1} (C≡C)
1675 cm^{-1} (C=O)

No band is seen at 3300 cm^{-1}; therefore, the compound is not a terminal alkyne. The low value for the carbonyl absorption suggests that the carbonyl group is conjugated with the multiple bond. The absence of a band at 2700 cm^{-1} indicates that it is a conjugated ketone and not a conjugated aldehyde.

Compound S has bands at

3460 (broad) cm^{-1} (—O—H)
1710 cm^{-1} (C=O)

There is no band at 2700 cm^{-1}; therefore, the compound is a ketone and not an aldehyde. The other functional group in the compound is an alcohol.

Compound T has bands at

3100 cm^{-1} (C=C—H)
1740 cm^{-1} (C=O)
1640 cm^{-1} (C=C)
1240, 1040 cm^{-1} (C—O)

The bands at 1240 and 1040 cm^{-1}, along with the presence of a carbonyl, indicate an ester. The other functional group is an alkene.

10.7   (+)-17-Methyltestosterone

11

Nuclear Magnetic Resonance Spectroscopy

Workbook Exercises

In Section 5.6 of your text, you solved a set of problems designed to make you think about molecular symmetry. The ideas about structure represented in these problems apply directly to one of the most useful techniques for determining molecular structure, called nuclear magnetic resonance (NMR) spectroscopy, the topic of Chapter 11.

A brief review of the ideas and terminology introduced in Section 5.6. (You should re-examine that section of the text for a more complete look at this topic.)

There are three structural descriptions for this molecule that will relate to NMR spectroscopy:

I. **The number of sets of atoms and their relationships within a molecule.**
There are three distinct, nonequivalent sets, or groups, of hydrogen atoms (a, b and c) based on overall molecular symmetry (for example, the distance from the oxygen atom).

II. **The relative sizes of groups of atoms.**
The total of 10 hydrogen atoms that comprise sets a, b and c are in the ratio of 3:1:6, respectively.

III. **The environment of neighboring atoms as related by the number of intervening bonds.**
Any hydrogen atom in set c is related to hydrogen atom b by 3 intervening bonds (H_b—C—C—H_c). H_b is a 3-bond neighbor for any of the H_c atoms.

As you begin to study Chapter 11, remember that NMR spectroscopy simply provides graphical information that can be interpreted in terms of:
    1.   the number of groups of equivalent atoms and their connectivity relationships within a molecule.
    2.   the relative numbers of atoms within each of the groups.
    3.   the environment of groups of atoms based on the number of bonds intervening between them and other atoms.

The following two workbook exercises are extensions of the problems from Section 5.6. They will focus your attention on structural ideas important to your study of NMR spectroscopy.

Workbook Exercises (cont)

EXERCISE I.   You created a number of molecules with different connectivities for part (1) of the workbook exercise in Chapter 9. Identify the number and relative size of the sets of equivalent hydrogen atoms for each of your molecules. Do the same for sets of carbon atoms. What differences in the infrared (IR) spectra (Chapter 10) of these isomers could be used to distinguish them from each other?

EXERCISE II.   In this problem, groups of equivalent atoms are ranked according to their proximity to an electronegative group in the molecule (for example, of groups a, b and c, group a is closest to the most electronegative group).

Molecules A, B, and C have the same molecular formula, $C_6H_{12}O_2$. Molecule A has a strong IR absorption between $3200-3400$ cm^{-1}, among others, although the area between $1600-2200$ cm^{-1} is clear of significant absorptions. Molecule A has three groups of equivalent hydrogen atoms (a, b and c) in the ratio of 2:1:3, respectively. Set a has two nonequivalent 3-bond hydrogen neighbors (hereafter referred to as "neighbors" unless a different description is required). Set b has four neighbors, and set c has no neighbors. Molecule A has four groups of carbon atoms.

Molecule B has a broad IR absorption between $2500-3200$ cm^{-1} as well as a strong absorption at 1720 cm^{-1}, among others. Molecule B has four groups of equivalent hydrogen atoms (a, b, c, and d) in the ratio of 1:2:6:3, respectively. Set a has no neighbors. Set b has three neighbors. Set c has no neighbors, and set d has two neighbors. There are five groups of carbon atoms in molecule B.

Molecule C has a broad IR absorption between $3450-3550$ cm^{-1}, and an intense absorption at 1715 cm^{-1}, among others. There are five groups of hydrogen atoms in molecule C (a, b, c, d, and e) in the ratio of 1:3:2:3:3, respectively. Set a has no neighbors. Set b has no neighbors. Set c has three neighbors. Set d has no neighbors, and set e has two neighbors. There are six groups of carbon atoms in molecule C.

Draw structural formulas for molecules A, B, and C that are compatible with the spectral data (and the molecular formula) that you have for them.

— —

11.1   Arrows point to chemical-shift equivalent hydrogen atoms. Hydrogen atoms on the same methyl group are always chemical-shift equivalent. Hydrogen atoms on the same methylene group are usually, but not always, chemical-shift equivalent.

(a) $CH_3CH_2OCH_2CH_3$ (b) $CH_3CH_2CH_2Br$ (c)

(d) CH_3CHCH_2Cl with CH_3 (e) $CH_3CH_2CCH_2CH_3$ (with O double bonded)   (f) $CH_3CH_2CH_2CH_2CH_3$

11.2   Compound A is 1,1,2-trichloroethane, $ClCH_2CHCl_2$. This is the only compound of the three for which we expect to see more than one band. The hydrogen atoms attached to the two carbon atoms of 1,1,2-trichloroethane are not chemical-shift equivalent. One is bonded to a carbon atom with two chlorine atoms and absorbs farther from TMS than the other two, which are bonded to a carbon atom with only one chlorine atom.

Acetone and 1,2-dichloroethane will each have only a single band. We can distinguish them from each other by the position of the band in the spectrum.

$$\underset{\displaystyle \parallel}{\overset{\displaystyle O}{}}$$

Compound B is acetone, CH_3CCH_3. All six hydrogen atoms in acetone are chemical-shift equivalent. A single absorption band is seen in the spectrum at approximately the same place as one of the bands in the spectrum of methyl acetate (Figure 11.1).

Compound C is 1,2-dichloroethane, $ClCH_2CH_2Cl$. All four hydrogen atoms in 1,2-dichloroethane are chemical-shift equivalent and give rise to a singlet in the spectrum. The position of the band in the spectrum of 1,2-dichloroethane is similar to that of one of the bands in the spectrum of 1,1,2-trichloroethane.

11.3   (a) $^{28}_{14}Si$ even atomic and mass number; no nuclear magnetic moment

(b) $^{15}_{7}N$ odd mass and atomic number; has nuclear magnetic moment

(c) $^{19}_{9}F$ odd mass and atomic number; has nuclear magnetic moment

(d) $^{31}_{15}P$ odd mass and atomic number; has nuclear magnetic moment

(e) $^{11}_{5}B$ odd mass and atomic number; has nuclear magnetic moment

(f) $^{32}_{16}S$ even atomic and mass number; no nuclear magnetic moment

11.4   The band which moves progressively upfield as the solution of the alcohol is diluted is the band for the hydrogen atom of the hydroxyl group. As the alcohol is diluted, the amount of hydrogen bonding decreases.

Hydrogen bonding further deshields a hydrogen atom because the hydrogen atom involved in hydrogen bonding is partially bonded to another electronegative atom (O, N, or F) and has even less electron density around it than a hydrogen atom which is not involved in hydrogen bonding.

R——O——H   R——O——H - - - - O——R
                                                               /
                                                              H

dilute solution   concentrated solution

no hydrogen bonding;   hydrogen bonding; the
the hydrogen atom is less   hydrogen atom is more
deshielded and absorbs   deshielded and absorbs
farther upfield   farther downfield

11.5

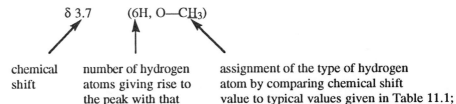

The absorption bands between δ 1.0 and 3.0 in the spectrum of cyclohexene fall in the region for hydrogen atoms bonded to *sp³*-hybridized carbon atoms. The four hydrogen atoms absorbing farther downfield (δ ~2) are the allylic hydrogen atoms on carbon atoms 3 and 6. These hydrogens are more deshielded than the hydrogens on carbon atoms 4 and 5 because they are closer to the double bond. The four hydrogen atoms farther upfield (δ~1.5) are those on carbon atoms 4 and 5.

11.6   In this problem, in which we assign the structure of a compound from its molecular formula and spectral data, the proton magnetic resonance data are analyzed in two ways. The chemical shift values for groups of different hydrogen atoms are read off the spectrum, these values are compared with the chemical shift values given in Table 11.1 for different types of hydrogen atoms, and a tentative assignment of partial structure is made. The integration of the spectrum is examined by measuring the relative heights of the steps in the integration curves. The two types of information are summarized in the answers to these problems in the following way:

δ 3.7   (6H, O—C$\underline{H}_3$)

chemical shift	number of hydrogen atoms giving rise to the peak with that chemical shift	assignment of the type of hydrogen atom by comparing chemical shift value to typical values given in Table 11.1; the results of the integration also taken into account here

Compound D, $C_8H_{10}O_2$, has four units of unsaturation. Integration of the spectrum gives a ratio of 1.6:1 or 3:2 for the relative areas under the peaks. The molecular formula tells us that there are ten hydrogen atoms that absorb so the actual ratio is 6:4. Therefore, we have six hydrogen atoms that absorb at δ 3.7 and four that absorb at δ 6.8. The results of this reasoning and of our inspection of Table 11.1 are summarized below.

δ 3.7   (6H, —OC$\underline{H}_3$)
δ 6.8   (4H, Ar$\underline{H}$)

The compound is 1,4-dimethyoxybenzene.

Note that the symmetry of the structure of the compound is reflected in the simplicity of the spectrum.

CH_3O———⟨ ⟩———OCH_3

Compound E, C_8H_{10}, has four units of unsaturation. Integration of the spectrum gives a ratio of 1.6:1 or 3:2 (or 6:4 for a total of ten hydrogen atoms in E) for the relative areas under the peaks.

δ 2.3   (6H, ArC$\underline{H}_3$)
δ 6.9   (4H, Ar$\underline{H}$)

11.6 (cont)

As with the spectrum of Compound D, the appearance of the spectrum of Compound E suggests symmetry in the structure. The compound is 1,4-dimethylbenzene (*p*-xylene).

$$CH_3 \text{---} \langle \text{benzene ring} \rangle \text{---} CH_3$$

Compound F, C_7H_7Cl, has four units of unsaturation. (Each halogen atom counts as one hydrogen atom in determining units of unsaturation.) Integration of the spectrum gives a ratio of 2.6:1 or 5:2 (for a total of seven hydrogen atoms in F) for the relative areas under the peaks.

δ 4.4	(2H, Y—$\underline{CH_2}$—X, where both X and Y deshield the methylene group)
δ 7.3	(5H, Ar$\underline{H}$)

The compound is benzyl chloride.

$$\langle \text{benzene ring} \rangle \text{---} CH_2Cl$$

Note that in using Table 11.1, we must realize that the values given will not match exactly the chemical shift values that we get from the spectra. We must put all the information together to make assignments of structure. In this problem, once we know that an aryl ring is present, we understand why the hydrogen atoms of the methylene group absorb at lower field than a typical RCH_2X group (δ 3.5 in Table 11.1).

Compound G, C_8H_8O, has five units of unsaturation. Integration of the spectrum gives a ratio of 1.7:1 or 5:3 (for a total of eight hydrogen atoms in G) for the relative areas under the peaks.

δ 2.5	$(3H, -\overset{\displaystyle O}{\overset{\displaystyle \|}{C}}C\underline{H_3})$
δ 7–8	(5H, Ar$\underline{H}$)

The compound is acetophenone (1-phenylethanone).

$$\langle \text{benzene ring} \rangle \text{---} \overset{\displaystyle O}{\overset{\displaystyle \|}{C}}CH_3$$

The presence of the carbonyl group in the molecule was deduced from the chemical shift of the methyl group, the total number of carbon atoms in the molecule, and the one unit of unsaturation that had to be accounted for once the aromatic ring, with four units of unsaturation, was identified.

11.7   The spectrum of bromoethane has two groups of peaks with an integration giving a ratio of 1.5:1 or 3:2 for the relative areas under the groups of small peaks.

11.7 (cont)

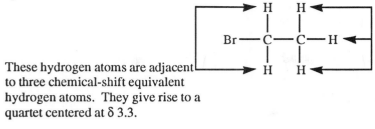

These hydrogen atoms are adjacent to three chemical-shift equivalent hydrogen atoms. They give rise to a quartet centered at δ 3.3.

These hydrogen atoms are adjacent to two chemical-shift equivalent hydrogen atoms. They give rise to a triplet centered at δ 1.8.

There are four possible spin orientations of the two hydrogen atoms of the methylene group. Two of these are indistinguishable. The magnetic field sensed by the adjacent hydrogen atoms varies slightly with a probability of 1:2:1, giving rise to similar relative intensities of the small peaks in the triplet.

(The value of the coupling constant is read off the spectrum.)

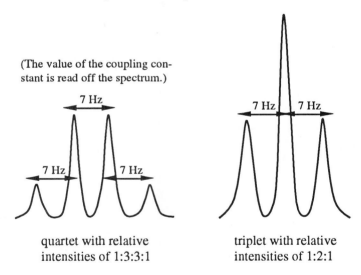

quartet with relative intensities of 1:3:3:1

triplet with relative intensities of 1:2:1

The data from the spectrum may be summarized as follows:

	δ 1.8	(3H, triplet, J 7 Hz, C$\underline{H}_3$CH$_2$—)
	δ 3.3	(2H, quartet, J 7 Hz, CH$_3$C$\underline{H}_2$—)
or as	δ 1.8	(3H, t, J 7 Hz, C$\underline{H}_3$CH$_2$—)
	δ 3.3	(2H, q, J 7 Hz, CH$_3$C$\underline{H}_2$—)

11.8   In the problems that follow, a third type of information is added to the two that were used to make structural assignments in Problem 11.6. The splitting patterns are analyzed, coupling constants are measured if necessary, and a determination is made of the number of hydrogen atoms coupling with the ones that give rise to the peak we are analyzing. The information is presented in the way shown at the end of Problem 11.7.

11.8 (cont)

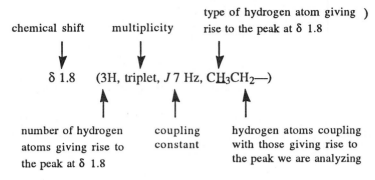

chemical shift   multiplicity   type of hydrogen atom giving)
                                  rise to the peak at δ 1.8

δ 1.8   (3H, triplet, *J* 7 Hz, C$\underline{H}_3$C$\underline{H}_2$—)

number of hydrogen   coupling   hydrogen atoms coupling
atoms giving rise to   constant   with those giving rise to
the peak at δ 1.8   the peak we are analyzing

Compound H, C_9H_{12}, has four units of unsaturation. The spectrum has bands at

δ 1.3   (3H, doublet, *J* 8 Hz, C$\underline{H}_3$—CH—C$\underline{H}_3$)
δ 2.8   (1H, septet, *J* 8 Hz, CH$_3$—C$\underline{H}$—CH$_3$)
δ 6.8   (5H, singlet, Ar$\underline{H}$)

The compound is isopropylbenzene (cumene).

11.9

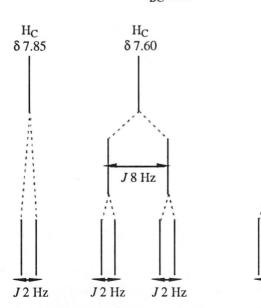

J_{AB} 8 Hz
J_{BC} 2 Hz

| H$_C$ | H$_C$ | H$_C$ |
| δ 7.85 | δ 7.60 | δ 7.00 |

J 8 Hz

J 2 Hz   *J* 2 Hz   *J* 2 Hz   *J* 8 Hz

11.10 $CH_3CH_2CH_2Br$
                                            a   b   c
                                            $J_{ab} = J_{bc} = 7$ Hz

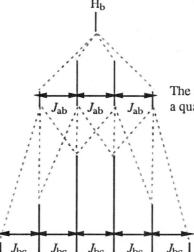

The methylene hydrogen atoms are split into a quartet by the adjacent methyl hydrogen atoms.

Each peak of the quartet is further split into a triplet by the other methylene hydrogen atoms. If the coupling constants for the two types of interactions are the same, the triplets overlap to give a sextet.

11.11   The top spectrum is that of 4-methylbenzonitrile. The symmetrical splitting pattern for the hydrogen atoms on the aromatic ring, similar in appearance to a doublet of doublets, is typical of an aromatic ring that is substituted in the para position with two substituents of differing electronegativity.

The bottom spectrum is that of 3-methylbenzonitrile. An aromatic ring with two substituents of differing electronegativity meta to each other will not give the characteristic doublet of doublets.

11.12   Compound I, $C_5H_{10}O$, has one unit of unsaturation. The spectrum has bands at

δ 1.2	(6H, singlet, $(C\underline{H}_3)_2C$)
δ 3.2	(1H, singlet, —O$\underline{H}$)
δ 4.6–5.0	(2H, multiplet, $C\underline{H}_2$=CH—)
δ 5.2–5.9	(1H, multiplet, CH_2=C$\underline{H}$—)

The compound is 2-methyl-3-buten-2-ol. The assignment of the band at δ 3.2 could not be made until it became clear that the unit of unsaturation is a double bond (bands at δ 4.6–5.9).

$$CH_2{=\!\!=}CH—\overset{\displaystyle CH_3}{\underset{\displaystyle OH}{\overset{|}{\underset{|}{C}}}}CH_3$$

Compound J, C_4H_8O, has one unit of unsaturation. The spectrum has bands at

δ 1.1	(6H, doublet, *J* 8 Hz, $(C\underline{H}_3)_2CH$—)
δ 2.3	(1H, multiplet, *J* 8 Hz, $(CH_3)_2C\underline{H}$—)
δ 9.0	(1H, multiplet, —$\overset{\displaystyle O}{\overset{\|}{C}}\underline{H}$)

11.12 (cont)

The compound is 2-methylpropanal (isobutyraldehyde). The hydrogen atom on the carbonyl group is coupled with a very small coupling constant to the α–hydrogen, which appears as a septet with further fine splitting.

$$\begin{array}{c} CH_3 \\ | \\ CH_3CHCH \\ \| \\ O \end{array}$$

Compound K, $C_8H_{10}O$, has four units of unsaturation. The spectrum has bands at

δ 2.2	(3H, singlet, ArC$\underline{H}_3$)
δ 3.6	(3H, singlet, C$\underline{H}_3$O—)
δ 6.2–6.8	(4H, multiplet, Ar$\underline{H}$)

The compound is 1-methoxy-4-methylbenzene (*p*-methylanisole).

11.13   Compound L, $C_5H_{12}O$, has no units of unsaturation. The infrared spectrum has bands at

3350 cm^{-1}	(—O—H)
3000, 1460 cm^{-1}	(C—H)
1120 cm^{-1}	(C—O)

The proton magnetic resonance spectrum has bands at

δ 0.9	(6H, triplet, J 6 Hz, C$\underline{H}_3$CH$_2$)
δ 1.3	(4H, quintet, J 6 Hz, CH$_3$C$\underline{H}_2$CH)
δ 3.3	(1H, quintet, J 6 Hz, CH$_2$C$\underline{H}$CH$_2$)
δ 3.6	(1H, singlet, —O$\underline{H}$)

The compound is 3-pentanol. Note how distorted the patterns at δ 0.9 and δ 1.3 are. Distorted patterns can occur when coupled hydrogen atoms have similar values for their chemical shifts.

$$\begin{array}{c} CH_3CH_2CHCH_2CH_3 \\ | \\ OH \end{array}$$

11.14   Compound M, $C_5H_8O_2$, has two units of unsaturation. The infrared spectrum has bands at

3100 cm^{-1}	(=C—H)
1760 cm^{-1}	(C=O)
1675 cm^{-1}	(C=C)
1200 cm^{-1}	(C—O)

11.14 (cont)

The proton magnetic resonance spectrum has bands at

δ 1.8	(3H, multiplet, C$\underline{H}_3$C=C)
δ 1.9	(3H, singlet, C$\underline{H}_3$C=O)
δ 4.3	(2H, multiplet, C$\underline{H}_2$=C)

The compound is 2-propenyl ethanoate (isopropenyl acetate).

$$\begin{array}{c} \quad\quad\; O \\ \quad\quad\; \| \\ CH_3CO\,C\!=\!\!=\!CH_2 \\ \quad\quad | \\ \quad\quad CH_3 \end{array}$$

Note that at first glance the two peaks around δ 2.0 appear to be a doublet, but since there are no other bands in the spectrum showing the same separation, the two peaks cannot have arisen because of coupling to another hydrogen atom. We assign the peaks to two nonequivalent methyl groups, one on a double bond and one α to a carbonyl group. One of the peaks for a methyl group is not as high as the other one. The band at δ 4.3 is also thickened. This indicates that the methyl group on the double bond and the vinylic hydrogen atoms are coupled to each other with a small coupling constant.

11.15   Isomer 1 has only two sets of carbon atoms and is 2,3-dimethylbutane.

$$\begin{array}{c} CH_3 \quad CH_3 \\ | \quad\quad | \\ CH_3CH - CHCH_3 \end{array}$$

Isomer 2 has four sets of carbon atoms and is 3-methylpentane.

$$\begin{array}{c} CH_3 \\ | \\ CH_3CH_2CHCH_2CH_3 \end{array}$$

Isomer 3 has five sets of carbon atoms and is 2-methylpentane.

$$\begin{array}{c} CH_3 \\ | \\ CH_3CHCH_2CH_2CH_3 \end{array}$$

11.16   (a)

Spectrum 3 has bands for four different types of carbon atoms, one (~50 ppm) for a carbon atom bonded to a nitrogen atom.

11.16 (cont)

(b)

$$CH_3CH_2CHCH_3$$
$$|$$
$$OH$$

Spectrum 2 has bands for four different types of carbon atoms, one (~70 ppm) for a carbon atom bonded to an oxygen atom.

(c)

$$\overset{O}{\underset{\|}{}} \overset{O}{\underset{\|}{}}$$
$$CH_3CH_2COCCH_2CH_3$$

Spectrum 5 has bands for three different types of carbon atoms, one (10 ppm) for an unsubstituted carbon atom, one (~30 ppm) for a carbon a to a carbonyl group, and one (170 ppm) for the carbon atom of the carbonyl group in a carboxylic acid derivative.

(d)

Spectrum 4 has bands for two different types of carbon atoms, one (~70 ppm) for a carbon atom bonded to an oxygen atom.

(e)

Spectrum 1 has bands for eight different types of carbon atoms, four of which (110-150 ppm) are in the alkene region.

11.17 Compound N, C_5H_8O has two units of unsaturation. Three absorption bands in the spectrum for five carbon atoms suggests that the compound has symmetry. One unit of unsaturation is a carbonyl group (219.6 ppm), and the other one must be a ring because no bands appear in the region of the spectrum where the carbon atoms of an alkene absorb. The compound is cyclopentanone.

11.18 Compound O, $C_8H_8O_2$, has five units of unsaturation. The spectrum (note the expanded scale) has bands at

δ 3.7	(3H, multiplet, —OC$\underline{H}_3$)
δ 6.8–8	(4H, multiplet, Ar$\underline{H}$)
δ 9.6	(1H, singlet, O=C$\underline{H}$)

The compound is 4-methoxybenzaldehyde,

11.18 (cont)

Compound P, $C_4H_8O_2$, has one unit of unsaturation. The spectrum has bands at

δ 1.2	(3H, triplet, *J* 7 Hz, C$\underline{H}_3$CH$_2$)
δ 2.0	(3H, singlet, C$\underline{H}_3$C=O)
δ 4.1	(2H, quartet, *J* 7 Hz, CH$_3$C$\underline{H}_2$—O)

$$\overset{\displaystyle O}{\overset{\displaystyle \|}{}}$$

The compound is ethyl acetate, CH$_3$COCH$_2$CH$_3$.

Compound Q, $C_5H_8O_2$, has two units of unsaturation. The spectrum has bands at

δ 2.2	(3H, singlet, C$\underline{H}_3$—C=C)
δ 2.5	(3H, singlet, C$\underline{H}_3$—C=C)
δ 5.8	(1H, singlet, $\underline{H}$C=C)
δ 12.0	(1H, singlet, —CO$\underline{H}$)

The compound is 3-methyl-2-butenoic acid,

$$\begin{array}{ccc} CH_3 & & H \\ & \diagdown \quad \diagup & \\ & C{=}C & \\ & \diagup \quad \diagdown & \\ CH_3 & & \underset{\displaystyle \|}{\underset{\displaystyle O}{COH}} \end{array}$$

Note that at first glance the two peaks around δ 2.0 appear to be a doublet, but since there are no other bands in the spectrum showing the same separation, the two peaks cannot have arisen because of coupling to another hydrogen atom. We assign the peaks to two nonequivalent methyl groups, one cis and one trans to the carboxylic acid group. The band at δ 5.8 shows some evidence of coupling (with a very small coupling constant) with the methyl groups.

Compound R, C_7H_9N, has four units of unsaturation. (When determining the units of unsaturation in a compound containing nitrogen, eliminate the nitrogens from the molecular formula and then subtract one hydrogen for each nitrogen atom to determine the number of hydrogen atoms that corresponds to the saturated formula. The adjusted molecular formula for compound R is C_7H_8) The spectrum has bands at

δ 2.8	(3H, singlet, C$\underline{H}_3$N)
δ 3.4	(1H, singlet, N$\underline{H}$)
δ 6.4–7.4	(5H, multiplet, Ar$\underline{H}$)

The compound is *N*-methylaniline, ⬡—NCH$_3$ with $\overset{|}{H}$. The broad band at δ 3.4 results from coupling

between the nucleus of $^{14}_{7}N$ and the hydrogen atom on it.

11.18 (cont)

Compound S, $C_7H_{12}O_3$, has two units of unsaturation. The spectrum has bands at

δ 1.2	(3H, triplet, *J* 7 Hz, C<u>H</u>$_3$CH$_2$)
δ 2.1	(3H, singlet, C<u>H</u>$_3$C=O)
δ 2.6	(4H, multiplet, —C<u>H</u>$_2$C<u>H</u>$_2$—) The chemical shift value suggests that each of the methylene groups is α to a carbonyl group.
δ 4.1	(2H, quartet, *J* 7 Hz, CH$_3$C<u>H</u>$_2$—O)

The fragments listed above account for six of the seven carbon atoms, two of the three oxygen atoms, and one of the two units of unsaturation. The remaining carbon and oxygen atoms and one unit of unsaturation make

$$\overset{O}{\overset{\|}{}} \qquad \overset{O}{\overset{\|}{}}$$

up a carbonyl group. The compound is ethyl 4-oxopentanoate, $CH_3CCH_2CH_2COCH_2CH_3$.

Compound T, $C_6H_{15}N$, has no units of unsaturation. The spectrum has bands at

δ 0.9	(3H, triplet, *J* 7 Hz, C<u>H</u>$_3$CH$_2$)
δ 2.2	(2H, quartet, *J* 7 Hz, CH$_3$C<u>H</u>$_2$N)

The fragments listed above account for five hydrogen atoms and two carbon atoms in an ethyl group. The compound has fifteen hydrogen atoms and six carbon atoms, and must be a symmetric compound containing three ethyl groups. The compound is triethylamine, $(CH_3CH_2)_3N$.

11.19   Compound U, $C_{10}H_{12}O$, has five units of unsaturation. The infrared spectrum has bands at 1717 cm^{-1} (C=O). There are no bands in the infrared at about 2700 and 2900 cm^{-1}; therefore, the compound must be a ketone.

The proton magnetic resonance spectrum has bands at

δ 1.8	(3H, singlet, C<u>H</u>$_3$C=O)
δ 2.4–2.8	(4H, multiplet, —C<u>H</u>$_2$C<u>H</u>$_2$—)
δ 6.8	(5H, singlet, Ar<u>H</u>)

The compound is 4-phenyl-2-butanone, —CH$_2$CH$_2$$\overset{O}{\overset{\|}{C}}CH_3$.

The peaks around δ 2.5 may be seen as a pair of overlapping triplets.

11.20   Compound V, $C_5H_{12}O$, has no units of unsaturation. The infrared spectrum has bands at

3320 cm^{-1}	(—O—H)
1060 cm^{-1}	(C—O)

The infrared spectrum indicates that V is an alcohol.

11.20 (cont)

The proton magnetic resonance spectrum has bands at

δ 0.8	(6H, doublet, $(CH_3)_2CH$)
δ 1.0–2.0	(3H, multiplet, —$CHCH_2$—)
δ 2.9	(1H, singlet, —OH)
δ 3.4	(2H, triplet, —CH_2CH_2—O)

The compound is 3-methyl-1-butanol, $CH_3CHCH_2CH_2OH$.

$$\overset{\displaystyle CH_3}{\underset{\displaystyle |}{}}$$

11.21

δ 0.8 (3H, triplet, J_{ac} 7 Hz)

These are the methyl hydrogen atoms labeled a. They are split into a triplet by the two hydrogen atoms on the adjacent methylene group.

δ 1.1 (3H, doublet, J_{bd} 7 Hz)

These are the methyl hydrogen atoms labeled b. They are split into a doublet by the hydrogen atom on the adjacent benzylic carbon atom.

δ 1.4 (2H, multiplet, $J_{ac} = J_{cd} = 7$ Hz)

These are the methylene hydrogen atoms labeled c. They are split by the hydrogen atom on the adjacent benzylic carbon atom and the hydrogen atoms on the adjacent methyl group.

δ 2.4 (1H, sextet, $J_{bd} = J_{cd} = 7$ Hz)

This is the benzylic hydrogen atom labeled d. It is split into a sextet by the hydrogen atoms on the adjacent methylene group and the hydrogen atoms on the adjacent methyl group.

δ 6.8 (5H, singlet)

These are the aromatic hydrogen atoms labeled e. Even though they are not chemical-shift equivalent, the alkyl substituent does not change the magnetic environment enough to cause them to have different chemical shifts. Hydrogen atoms that have the same chemical shift will not couple.

11.22 Compound W has bands in its infrared spectrum at 1685 cm^{-1} (conjugated C=O). The infrared spectrum indicates that W is a ketone.

The proton magnetic resonance spectrum has bands at

δ 1.1	(6H, doublet, J 7 Hz, $(CH_3)_2CH$—)
δ 3.3	(1H, septet, $(CH_3)_2CH$—)
δ 7.0–7.7	(5H, multiplet, ArH)

11.22 (cont)

The chemical shift of the methine hydrogen atom and the fact that the aryl hydrogen atoms are split, indicate that the carbonyl group is between the aryl and the isopropyl groups. This is confirmed by the presence of a band for a conjugated carbonyl group in the infrared. The compound is 2-methyl-1-phenyl-1-propanone

(isobutyrophenone),

$$\begin{array}{c} O \\ \parallel \\ -CCHCH_3 \\ | \\ CH_3 \end{array}$$

11.23

H_c   Br

Br —   — NH$_2$

H_b   H_a

The bands are analyzed below:

δ 3.8	(2H, broad multiplet, —N$\underline{H}_2$)
δ 6.3	(1H, doublet, J_{ab} 8 Hz, Ar$\underline{H}_a$)
δ 6.8	(1H, doublet of doublets, J_{ab} 8 Hz, J_{bc} 2 Hz, Ar$\underline{H}_b$)
δ 7.1	(1H, doublet, J_{bc} 2 Hz, Ar$\underline{H}_c$)

H_a is the farthest upfield of the aryl hydrogen atoms because it is next to the amino group, which is strongly electron-donating (see Section 22.3 in the text), and distant from the electron-withdrawing bromine atoms. It is split into a doublet by H_b with J 8 Hz.

H_b has the next highest chemical shift. It is next to only one bromine atom. A diagram showing the splitting pattern for H_b is shown below.

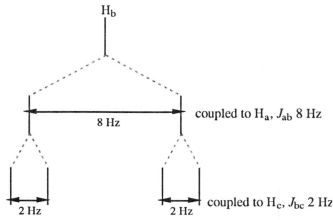

H_c is the most deshielded of the aryl hydrogen atoms because it is between two bromine atoms. It is also split into a doublet by H_b with J 2 Hz.

11.24   The two peaks observed for the methyl groups cannot be due to spin-spin coupling because there are no other bands in the spectrum that also show splitting.

The two separate absorption bands for the two methyl groups tell us that there is partial double bond character between the carbon atom of the carbonyl group and the nitrogen atom, which prevents free rotation. If free rotation were possible, the two methyl groups would become chemically equivalent, and only one chemical shift would be observed for both. Because there is not free rotation, the two methyl groups are in different environments and thus have two different chemical shifts.

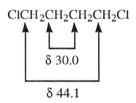

Resonance contributors for the amide group showing the existence of partial double bond character for the bond between the carbon atom of the carbonyl group and the nitrogen atom.

11.25   Compound X, $C_4H_8Cl_2$, has no units of unsaturation. We know that there are only two different types of carbon atoms in X because the carbon-13 magnetic resonance spectrum has only two bands. X must be 1,4-dichlorobutane.

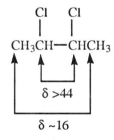

2,3-Dichlorobutane also has the symmetry required for this spectrum, but we would expect the methyl groups to absorb at higher field than the chemical shift values observed (Table 11.4).

$$\text{CH}_3\text{CH—CHCH}_3$$

2,3-Dichlorobutane

Also note that the carbon atoms bearing the chlorine atoms in this compound are even more highly substituted than in 1,4-dichlorobutane. We would expect them to absorb at lower field.

11.26   The infrared spectrum of compound Y shows the presence of an ester (1738 cm^{-1} for the C=O stretching frequency and 1256 and 1173 cm^{-1} for the C—O stretching frequencies).

The proton magnetic resonance spectrum shows the presence of hydrogen atoms bonded to a carbon atom a to a carbonyl group (δ 1.96). The carbon-13 magnetic resonance spectrum shows the presence of at least four different carbon atoms.

11.26 (cont)

δ 22.3   *sp*³-carbon atom

δ 28.1   *sp*³-carbon atom a to a carbonyl group

δ 79.9   *sp*³-carbon atom bonded to an oxygen atom

δ 170.2   ester carbonyl carbon atom

The proton magnetic resonance integration (3:1) has too few hydrogen atoms for the number of carbon atoms indicated by the carbon-13 spectrum. The actual ratio must be a multiple, such as 6:2 or 9:3. We cannot distribute eight hydrogen atoms on a minimum of four carbon atoms in such a way as to give only two chemical shifts. Compound Y must therefore have a total of twelve hydrogen atoms, nine on three methyl groups which are part of a *tert*-butyl group and three on a methyl group which is a to the ester group. Compound Y is *tert*-butyl acetate.

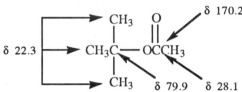

11.27   The infrared spectrum of Compound Z shows the presence of an alcohol (3400 cm⁻¹ for the O—H stretching frequency and 1030 cm⁻¹ for the C—O stretching frequency) and a terminal alkyne (3300 cm⁻¹ for the ≡C—H stretch and 2100 cm⁻¹ for the C≡C stretch). Therefore, Compound Z must have at least three carbon atoms, one bonded to the alcohol group and the two carbon atoms of the triple bond. The carbon-13 nuclear magnetic resonance spectrum shows only three bands, confirming that Compound Z has three different types of carbon atoms. There is no carbonyl absorption in the infrared spectrum and no band at ~ δ 200 in the carbon-13 nuclear magnetic resonance spectrum, so the molecule does not contain a carbonyl group. The carbon-13 chemical shifts and the splitting patterns are consistent with the following structure:

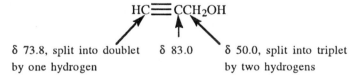

The structure is supported by the infrared spectrum (discussed above) and the proton magnetic resonance spectrum, analyzed below:

δ 2.5   (1H, triplet, *J* 3 Hz, — CH₂C≡C$\underline{H}$)

δ 3.1   (1H, broad singlet, —O$\underline{H}$)

δ 4.3   (2H, doublet, *J* 3 Hz, O—C$\underline{H}_2$C≡CH)

12

Alcohols, Diols, and Ethers

12.1   1,2-Ethanediol (ethylene glycol) is a very polar compound and hydrogen bonds strongly to water. It is infinitely soluble in water. A solution of ethylene glycol in water depresses the freezing point of water so that the mixture does not freeze at temperatures usually seen in the winter time. This keeps pipes and radiators from bursting from the expansion of the water in them as it forms ice. The high boiling point of ethylene glycol ensures that it will not boil out of the radiator of a car when the engine is hot.

12.2

polar   nonpolar

Cholesterol is not very soluble in water even though it has a hydroxyl group because the rest of the molecule is nonpolar and is too large to be brought into solution by the small polar hydroxyl group (Section 1.8B).

12.3   Cholesterol is an alkene as well as an alcohol.

12.3 (cont)

12.4   (a)   The location of the hydroxyl group on the chain affects the solubility in water of the isomeric alcohols. *n*-Butyl alcohol has the polar hydroxyl group at one end of the molecule and a long hydrophobic chain stretching away from it. In *sec*-butyl alcohol all parts of the hydrocarbon portion of the molecule are closer to, and more under the influence of, the polar hydroxyl group. In *tert*-butyl alcohol there is no large hydrophobic chain that is far away from the polar, water-soluble portion of the molecule. The hydroxyl group is able to carry the rest of the molecule into solution with it.

(b)   Hydrogen bonding of the alcohol is the most important factor in determining water solubility. The bromo compound is less soluble than the slightly more polar chloro compound. The dipole is weaker for the bromo compound so even those interactions are not helpful for the bromo compound.

12.4   (c) Hydrogen bonding among molecules of the diol give it a very high boiling point. The alcohol also has hydrogen bonding, but with only one hydroxyl group to participate in the hydrogen bonding, it has a lower boiling point than the diol. The ether cannot have hydrogen bonding among its molecules, therefore it has the lowest boiling point of the three.

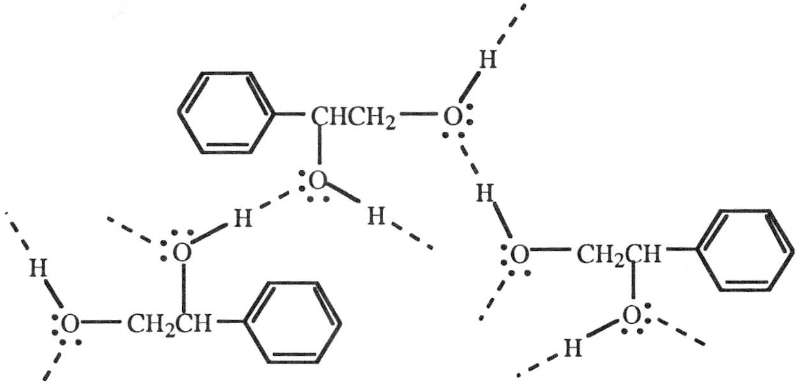

extensive hydrogen
bonding in diol

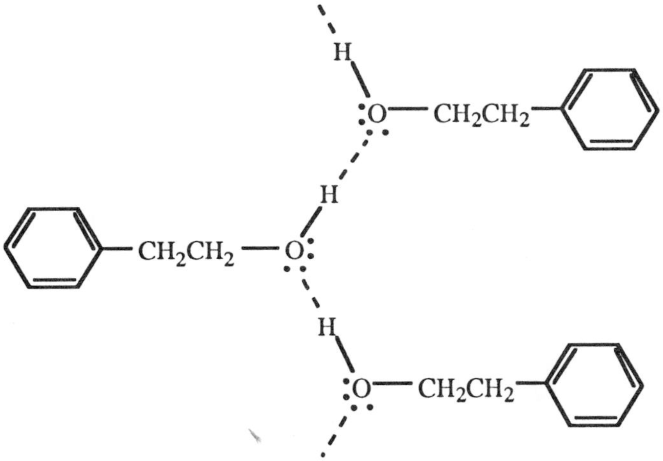

less extensive hydrogen
bonding in alcohol

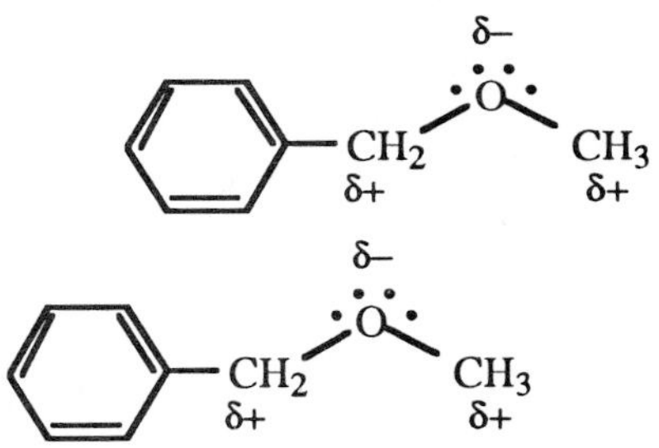

only dipole-dipole
interactions in ether

12.5

(a)   3-methyl-1-butanol

(b)   (*E*)-4-hexen-3-ol

(c)   2-pentyn-1-ol

(d)   4,5-dichloro-1-hexanol

(e)   1-chloro-3-ethoxypropane

(f)   diphenyl ether
       (phenoxybenzene)

(g)   cyclohexanol

(h)   *trans*-2-methylcyclopentanol

(i)   *cis*-4-*tert*-butylcyclohexanol

(j)   (*R*)-5-hexen-3-ol

(k)   1,3-dimethoxypropane

(l)   1,2-dimethoxybenzene

12.6   (a)

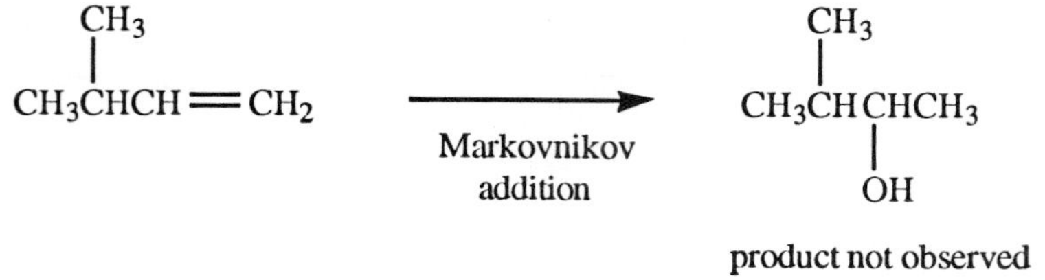

$$CH_3\overset{\overset{\textstyle CH_3}{|}}{CH}CH=CH_2 \xrightarrow[\text{addition}]{\text{Markovnikov}} CH_3\overset{\overset{\textstyle CH_3}{|}}{CH}\underset{\underset{\textstyle OH}{|}}{CH}CH_3$$

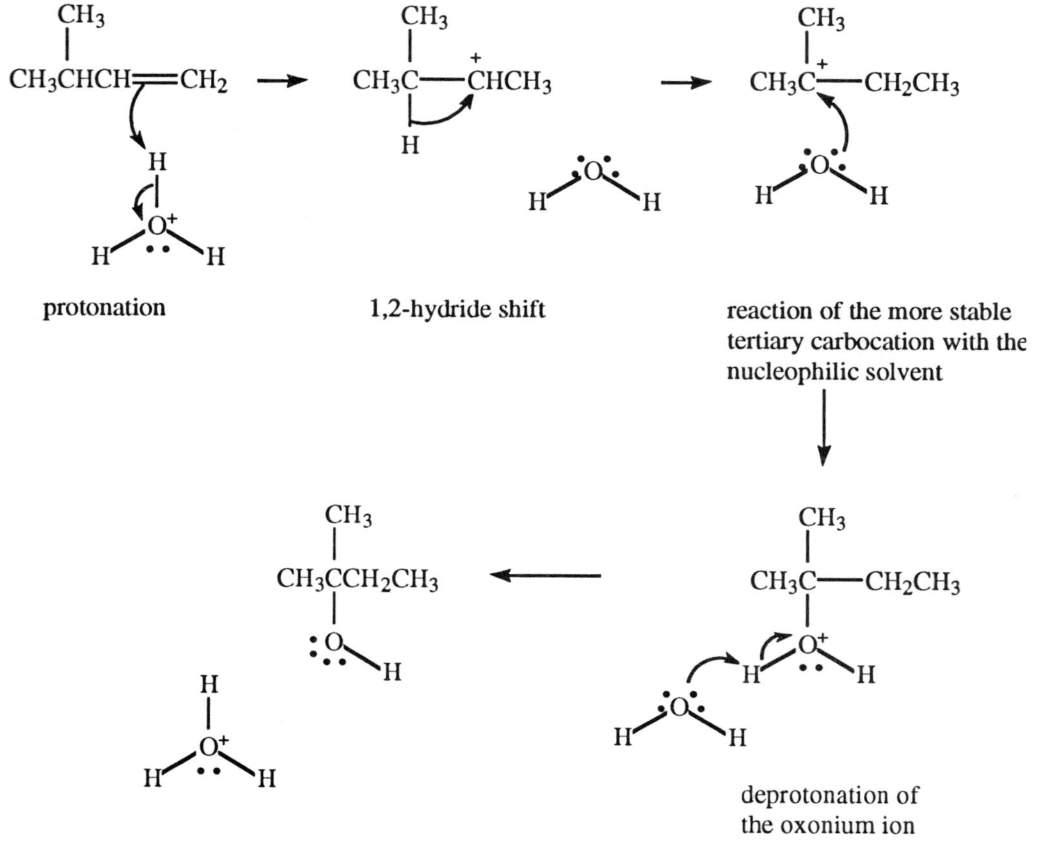

product not observed

(b)

protonation   1,2-hydride shift   reaction of the more stable
                                                tertiary carbocation with the
                                                nucleophilic solvent

deprotonation of
the oxonium ion

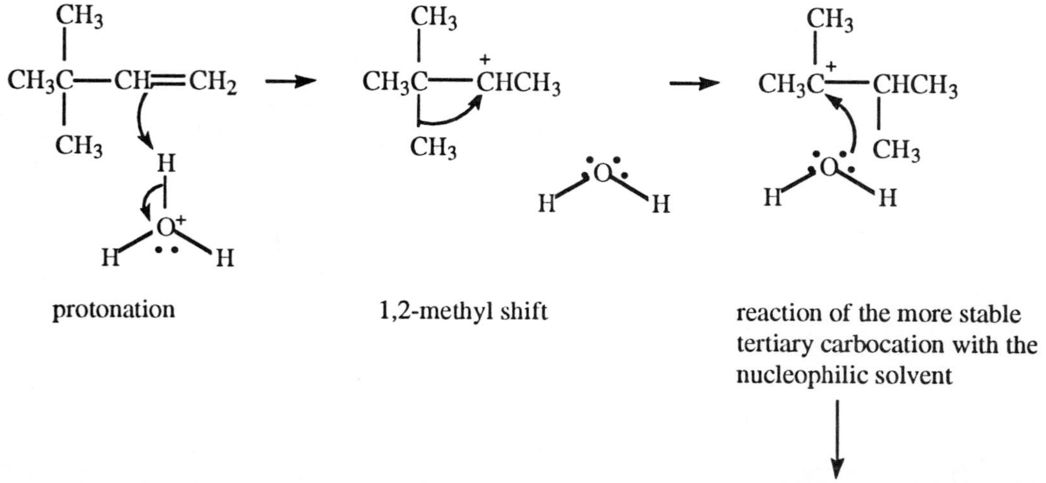

12.7

protonation   1,2-methyl shift   reaction of the more stable
                                                tertiary carbocation with the
                                                nucleophilic solvent

12.7 (cont)

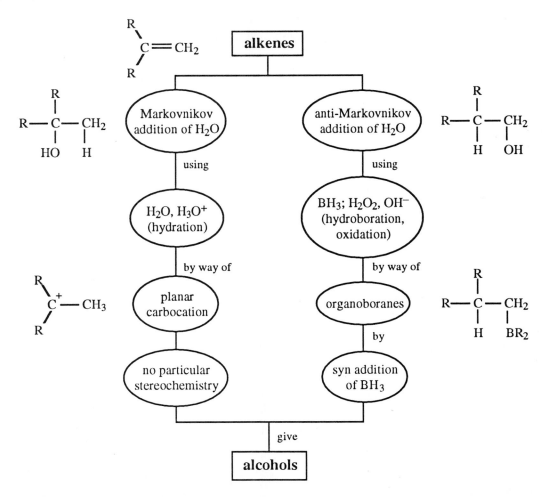

12.8 $CH_2{=}CHCH_2CH_2CH_2OH \xrightarrow{\quad H_3O^+ \quad} CH_3CHCH_2CH_2CH_2OH$
                                                                    |
                                                                   OH

Dehydration would also occur if the temperature were raised.

- -

Concept Map 12.1   Conversion of alkenes to alcohols.

12.9

$CH_3CH_2CHCH_2CH_3 \longrightarrow CH_3CH_2CHCH_2CH_3 \longrightarrow CH_3CH_2\overset{+}{C}HCH_2CH_3$

$CH_3CH_2CHCH_2CH_3$

3-bromopentane
from an S$_N$1 reaction

$CH_3CH_2CHCH_2CH_3 \longrightarrow CH_3CH_2CHCH_2CH_3$

3-bromopentane
from an S$_N$2 reaction

$CH_3CH_2\overset{+}{C}H\!\!-\!\!CHCH_3 \xrightarrow[\text{shift}]{\text{1,2-}\atop\text{hydride}} CH_3CH_2CH_2\overset{+}{C}HCH_3 \longrightarrow CH_3CH_2CH_2CHCH_3$

2-bromopentane
from the secondary cation
formed by rearrangement
of the 3-pentyl cation

12.10

$CH_3CH_2CHCH_2CH_3$

ionization assisted by ZnCl$_2$

$CH_3CH_2CH_2\overset{+}{C}HCH_3 \rightleftharpoons CH_3CH_2\overset{+}{C}HCH_2CH_3 \rightleftharpoons CH_3\overset{+}{C}HCH_2CH_2CH_3$

2-pentyl cation	**3-pentyl cation**	**2-pentyl cation**

with "1,2-hydride shift" labels between the cations.

12.10 (cont)

$$CH_3CH_2CH_2CHCH_3$$

$$:\overset{..}{\underset{..}{Cl}}:$$

2-chloropentane

$$CH_3CH_2CHCH_2CH_3$$

$$:\overset{..}{\underset{..}{Cl}}:$$

3-chloropentane

$$CH_3CHCH_2CH_2CH_3$$

$$:\overset{..}{\underset{..}{Cl}}:$$

2-chloropentane

There are two equivalent positions on the five-carbon chain that give rise to 2-pentyl cations, and thus to 2-chloropentane, and only one position that leads to the formation of 3-chloropentane. If the two carbocations are roughly equal in stability, we would expect that statistically twice as much 2-chloropentane as 3-chloropentane will form. This is observed experimentally.

Zinc chloride is a Lewis acid. It complexes with the chloride ion assisting with the ionization of the secondary alkyl halides, allowing equilibrium to be established between the cations.

12.11

12.12 (1) With zinc chloride in concentrated hydrochloric acid, we have reaction conditions that allow for ionization of the alkyl halides that are formed as products and for equilibrium among different ionic species. As we saw in Problem 12.9, zinc chloride assists in the ionization of alkyl halides.

12.12 (1) (cont)

$$CH_2CH_3$$
$$CH_3CH_2CH—CH_2—\overset{..}{\underset{..}{Cl}}:$$

1-chloro-2-ethylbutane

O with two H

$$CH_2CH_3$$
$$CH_3CH_2CH—CH_2 \overset{\delta^+}{\underset{..}{Cl}}: ---- \overset{\delta^-}{ZnCl_2} \longrightarrow CH_3CH_2\overset{+}{C}HCH_2CH_2CH_3 \longrightarrow CH_3CH_2CHCH_2CH_2CH_3$$

rearrangement of developing
carbocation (postulated to
avoid writing a primary carbocation as an intermediate)

Lewis acid
assisting in
the departure
of chloride ion

$$:\overset{..}{\underset{..}{Cl}}:^-$$

$$:\overset{..}{\underset{..}{Cl}}:$$
3-chlorohexane

$$CH_2CH_3$$
$$CH_3CH_2\overset{}{\underset{H}{C}}—CH_2 \overset{\delta^+}{\underset{..}{Cl}}: ---- \overset{\delta^-}{ZnCl_2} \longrightarrow CH_3CH_2\overset{+}{\underset{}{C}}CH_3 \longrightarrow CH_3CH_2\overset{}{\underset{}{C}}CH_3$$

rearrangement of
developing carbocation

Lewis acid
assisting in
the departure
of chloride ion

$$:\overset{..}{\underset{..}{Cl}}:^-$$

$$:\overset{..}{\underset{..}{Cl}}:$$
3-chloro-3-methylpentane

$$CH_3CH—\overset{+}{C}HCH_2CH_2CH_3 \longrightarrow CH_3\overset{+}{C}HCH_2CH_2CH_2CH_3 \longrightarrow CH_3CHCH_2CH_2CH_2CH_3$$
$$\underset{H}{}$$

rearrangement of one
secondary carbocation
into another one

$$:\overset{..}{\underset{..}{Cl}}:^-$$

$$:\overset{..}{\underset{..}{Cl}}:$$
2-chlorohexane

(2) With thionyl chloride in pyridine, an S_N2 reaction takes place. The alcohol is converted into the corresponding halide without rearrangement.

$$CH_2CH_3 \quad H$$
$$CH_3CH_2CH—CH_2—\overset{..}{\underset{..}{O}}:$$

$$:\overset{..}{\underset{..}{Cl}} \quad \overset{..}{\underset{..}{Cl}}:$$
$$S$$
$$\overset{||}{\underset{..}{O}}:$$

$$CH_2CH_3 \quad H$$
$$CH_3CH_2CH—CH_2—\overset{+}{\underset{}{O}}: \quad N (pyridine)$$
$$\underset{S}{} \quad \overset{..}{\underset{..}{Cl}}:$$
$$:\overset{..}{\underset{..}{O}}:^- \quad \overset{..}{\underset{..}{Cl}}:$$

12.12 (2) (cont)

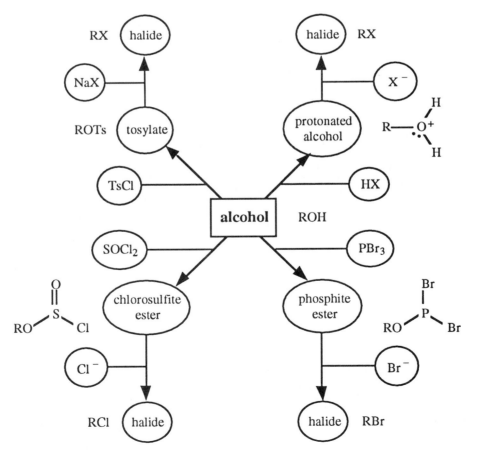

CH₃CH₂CH(CH_2CH_3)—CH₂—Cl:
1-chloro-2-ethylbutane

- -

Concept Map 12.2   Conversion of alcohols to alkyl halides.

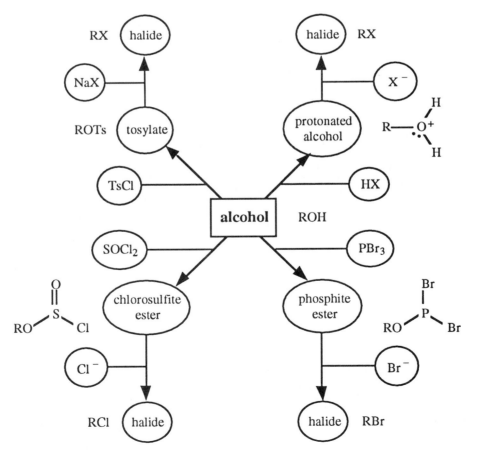

12.13 (a)

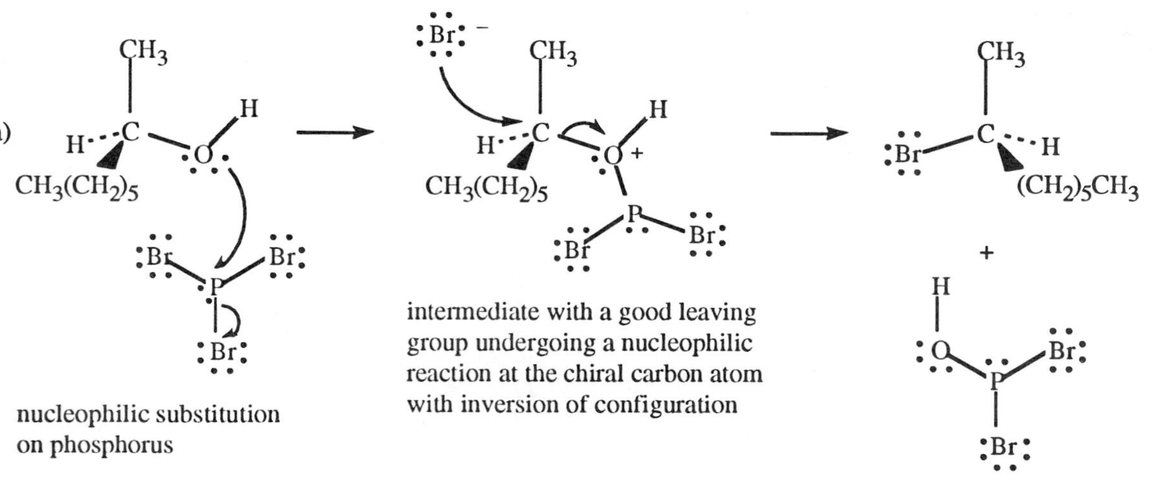

nucleophilic substitution
on phosphorus

intermediate with a good leaving
group undergoing a nucleophilic
reaction at the chiral carbon atom
with inversion of configuration

(b)

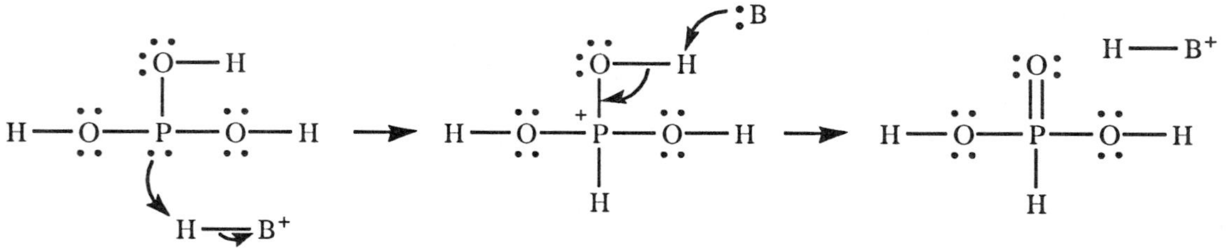

12.14 (a)

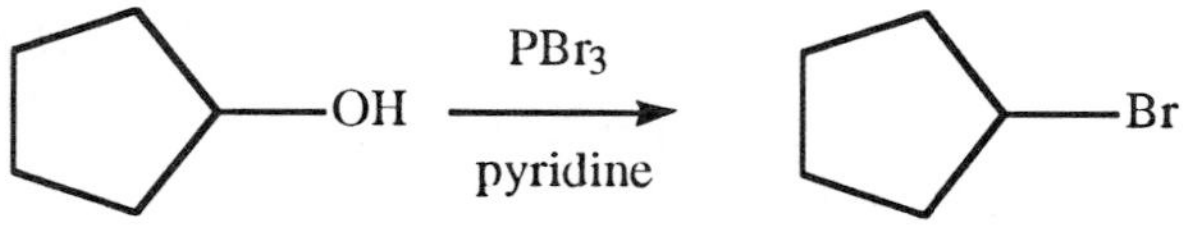

(b)

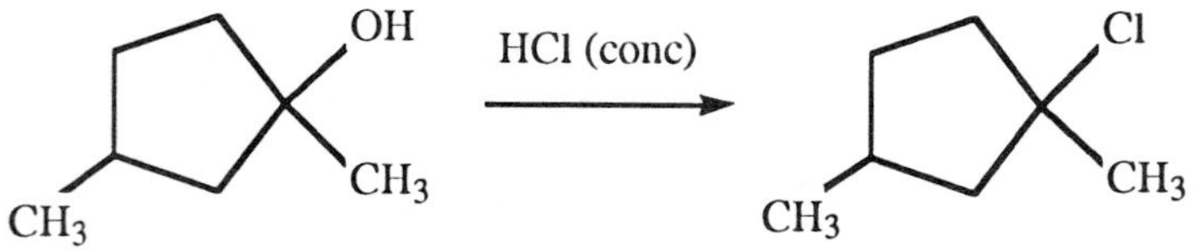

(c) $CH_3(CH_2)_9CH_2OH$ $\xrightarrow[\Delta]{SOCl_2}$ $CH_3(CH_2)_9CH_2Cl$

(d) $CH_3(CH_2)_8CH_2OH$ $\xrightarrow[\Delta]{HBr}$ $CH_3(CH_2)_8CH_2Br$

(e) $CH_3SCH_2CH_2OH$ $\xrightarrow[\text{chloroform}]{SOCl_2}$ $CH_3SCH_2CH_2Cl$

(f) $CH_3CH_2OCH_2CH_2OH$ $\xrightarrow[\text{pyridine}]{PBr_3}$ $CH_3CH_2OCH_2CH_2Br$

12.14 (cont)

(g)

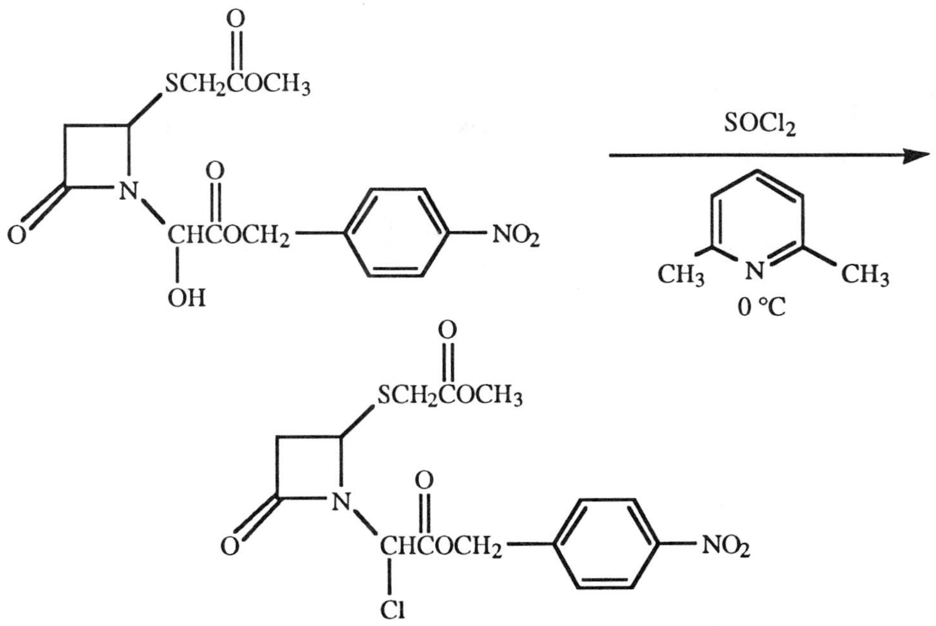

12.15

(a)

$$CH_3CHOH \xrightarrow{Na} CH_3CHO^-Na^+ \xrightarrow{CH_3CH_2CH_2Br} CH_3CHOCH_2CH_2CH_3$$

(b)

$$CH_3COH \xrightarrow{K} CH_3CO^-K^+ \longrightarrow CH_3COCH_2-\text{(phenyl)}$$

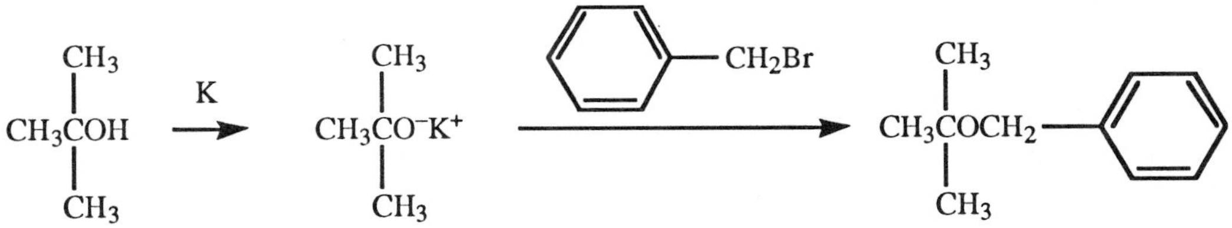

(c) $CH_3CH_2OH \xrightarrow{Na} CH_3CH_2O^-Na^+ \xrightarrow{CH_2=CHCH_2Br} CH_3CH_2OCH_2CH=CH_2$

(d) $CH_3CH_2CH_2CH_2OH \xrightarrow{Na} CH_3CH_2CH_2CH_2O^- Na^+ \xrightarrow{CH_3I} CH_3CH_2CH_2CH_2OCH_3$

or $CH_3OH \xrightarrow{Na} CH_3O^- Na^+ \xrightarrow{CH_3CH_2CH_2CH_2Br} CH_3CH_2CH_2CH_2OCH_3$

(e) $CH_3CH_2CH_2CH_2CH_2CH_2OH \xrightarrow{Na} CH_3CH_2CH_2CH_2CH_2CH_2O^- Na^+ \xrightarrow{CH_3CH_2Br}$

$$CH_3CH_2CH_2CH_2CH_2CH_2OCH_2CH_3$$

or $CH_3CH_2OH \xrightarrow{Na} CH_3CH_2O^- Na^+ \xrightarrow{CH_3CH_2CH_2CH_2CH_2CH_2Br}$

$$CH_3CH_2CH_2CH_2CH_2CH_2OCH_2CH_3$$

12.15 (cont)

(f)

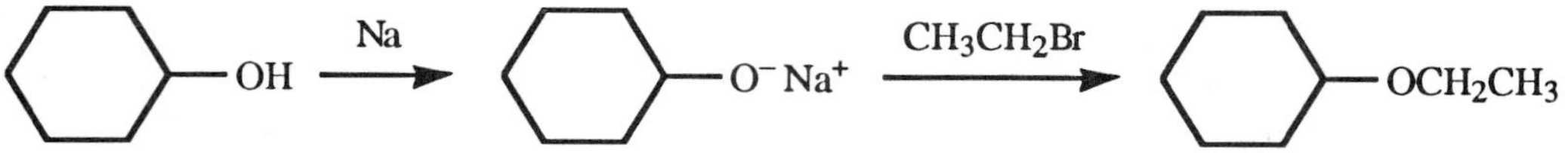

12.16

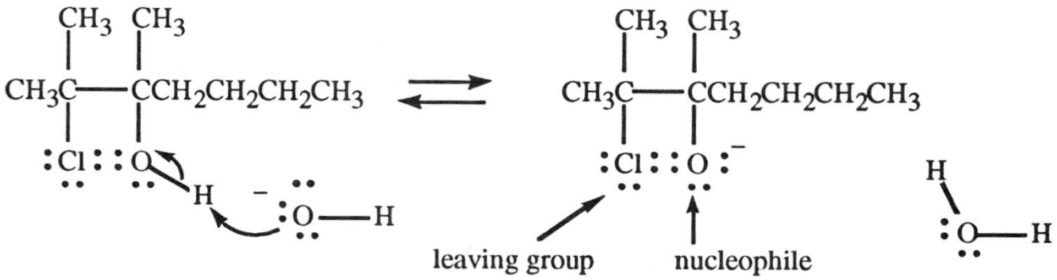

leaving group   nucleophile

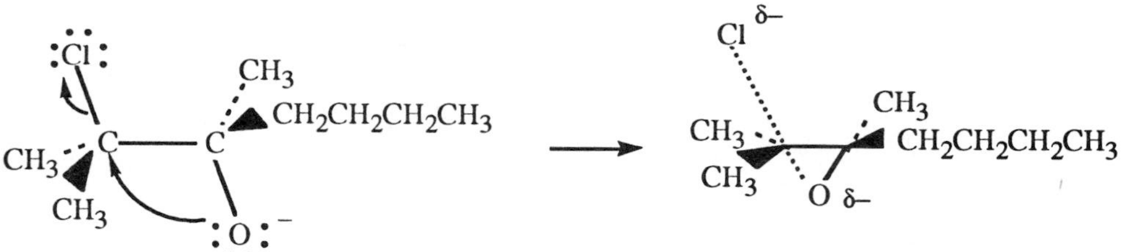

trans, anti orientation of the
nucleophile and the leaving group

transition state for an intramolecular S_N2 reaction

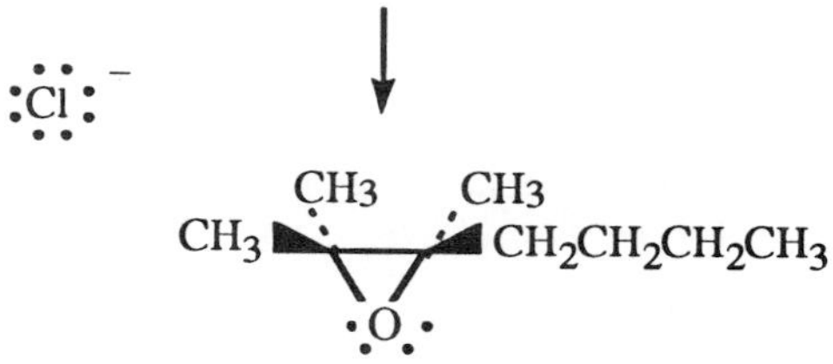

one of two possible enantiomers

12.17

(a)

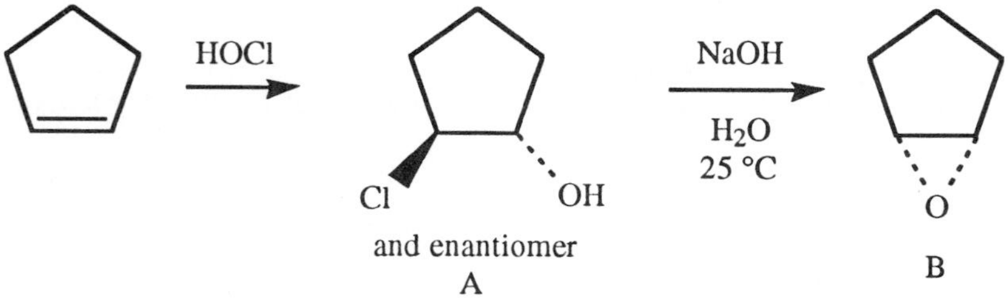

HOCl

NaOH

H_2O
25 °C

Cl   OH
and enantiomer
A

B

(b)

Cl
|
$CH_3CH_2CH_2CHCHCH_2CH_2CH_3$
|
OH

NaOH

H_2O
25 °C

$CH_3CH_2CH_2CH$———$CHCH_2CH_2CH_3$
\   /
O

C

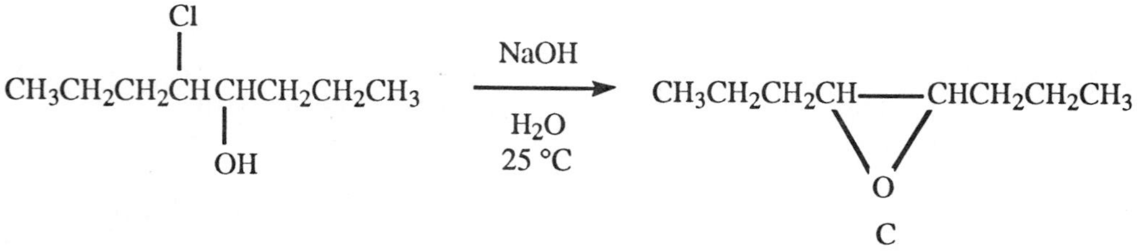

12.17 (cont)

(c)

$$CH_3$$
$$ClCH_2\overset{|}{\underset{|}{C}}CH_2CH_2CH_2CH_3 \quad \xrightarrow[\substack{H_2O \\ 25\,°C}]{NaOH} \quad CH_2\overset{CH_3}{\underset{\underset{D}{\diagdown O \diagup}}{-}}\overset{|}{C}CH_2CH_2CH_2CH_3$$
$$OH$$

(d)

$$BrCH_2CH_2CH_2CH_2CH_2OH \quad \xrightarrow[\substack{H_2O \\ 25\,°C}]{NaOH}$$

E

(e)

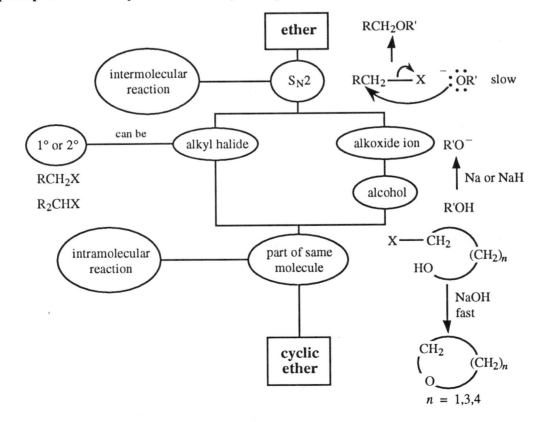

Concept Map 12.3   Preparation of ethers by nucleophilic substitution reactions.

12.18

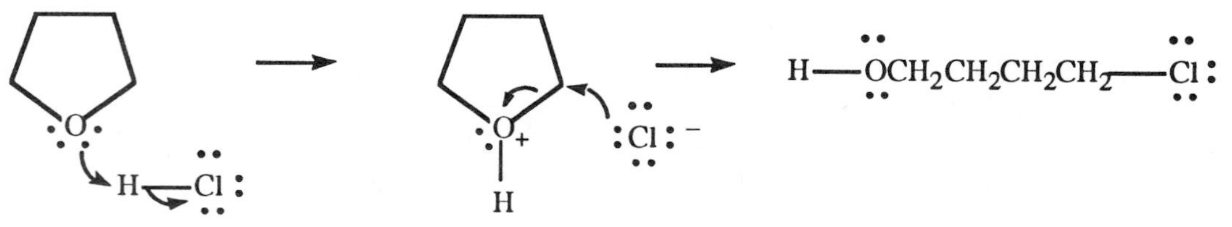

protonation of
the ether

nucleophilic attack on
the oxonium ion

12.19

(1)

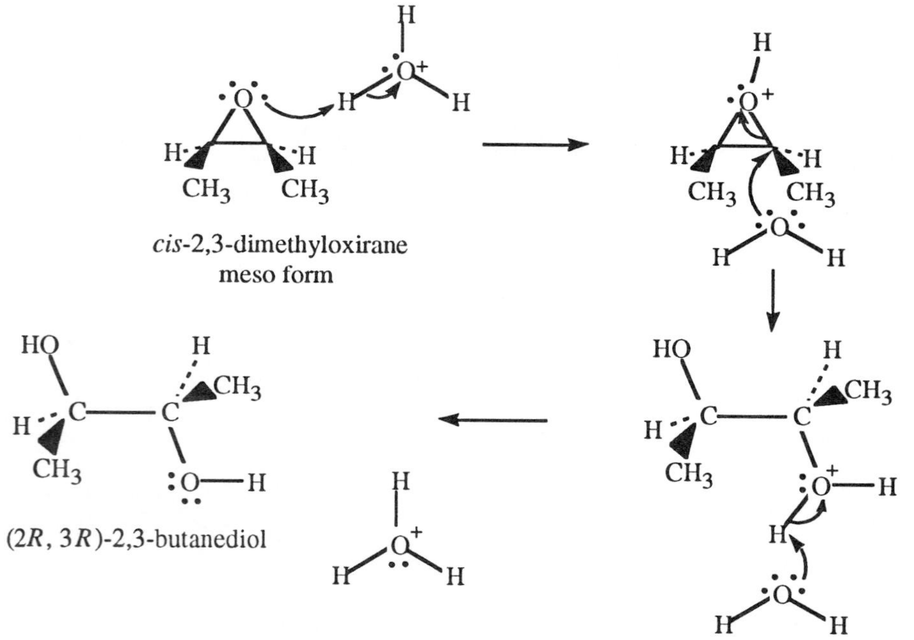

cis-2,3-dimethyloxirane
meso form

(2*R*, 3*R*)-2,3-butanediol

(2)

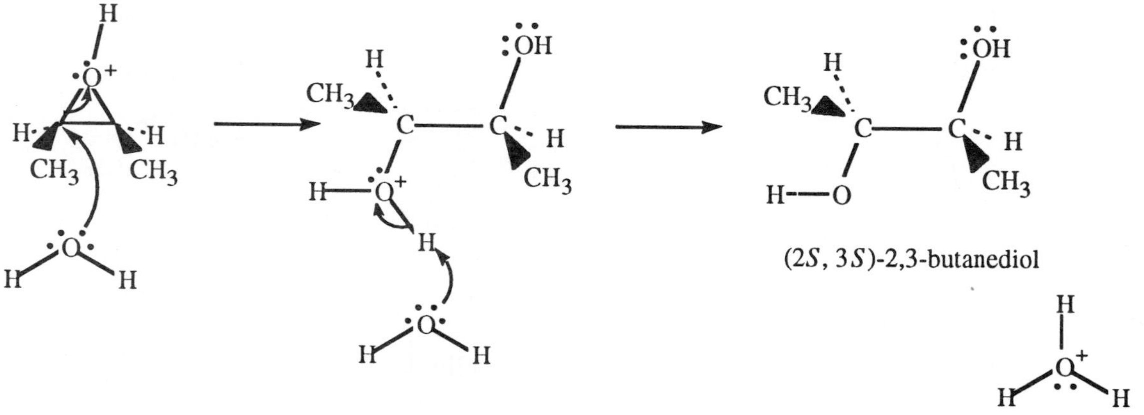

(2*S*, 3*S*)-2,3-butanediol

12.19 (cont)

(3)

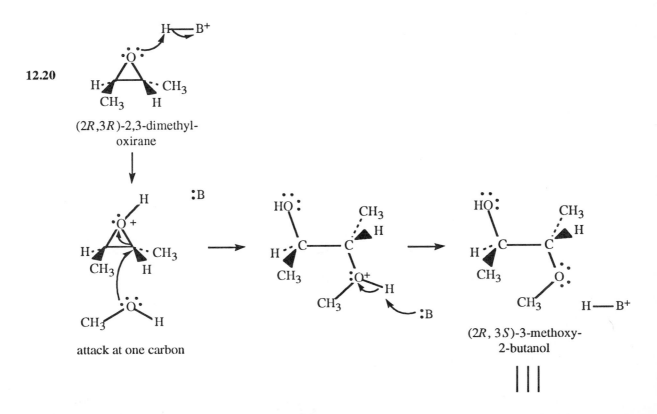

trans-2,3-dimethyloxirane
(2*R*, 3*R*)-enantiomer

(2*R*, 3*S*)-2,3-butanediol
meso-2,3-butanediol

Note that attack at the other carbon atom also leads to the meso compound. (2*S*, 3*S*)-2,3-Dimethyloxirane, the enantiomer of the trans isomer shown above, also gives the same compound.

12.20

(2*R*,3*R*)-2,3-dimethyl-
oxirane

attack at one carbon

(2*R*, 3*S*)-3-methoxy-
2-butanol

12.20 (cont)

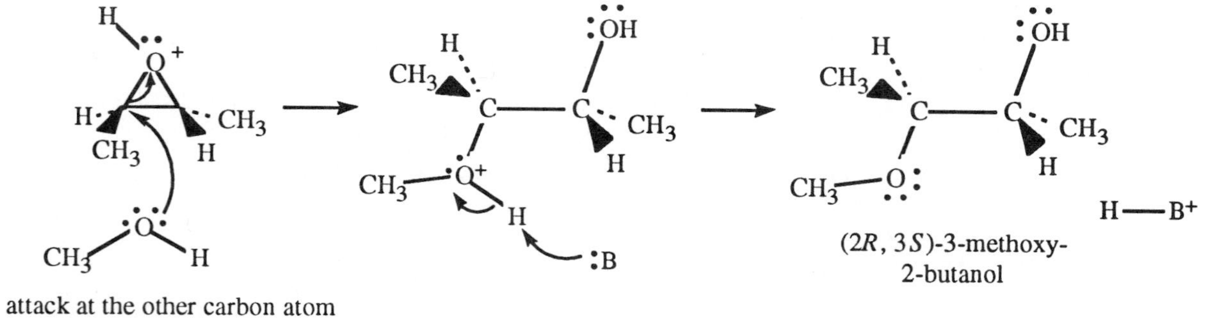

attack at the other carbon atom

(2R, 3S)-3-methoxy-
2-butanol

H—B$^+$

12.21

(a)

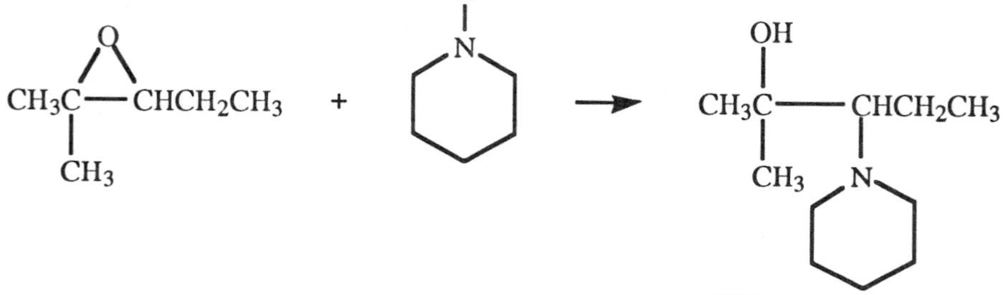

(b)

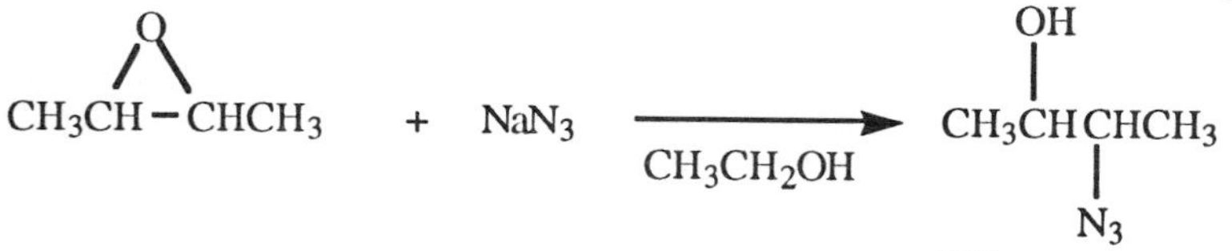

(c)

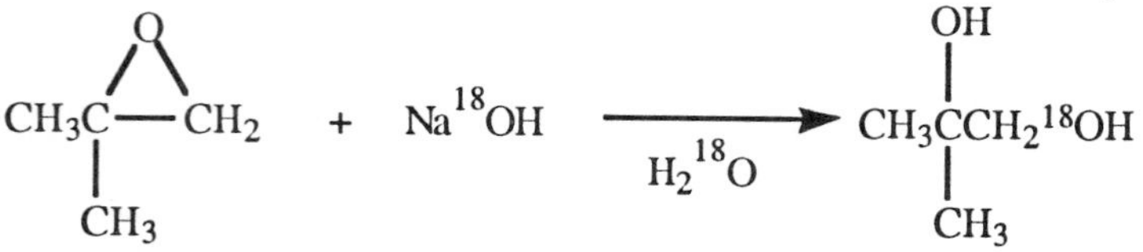

(d)

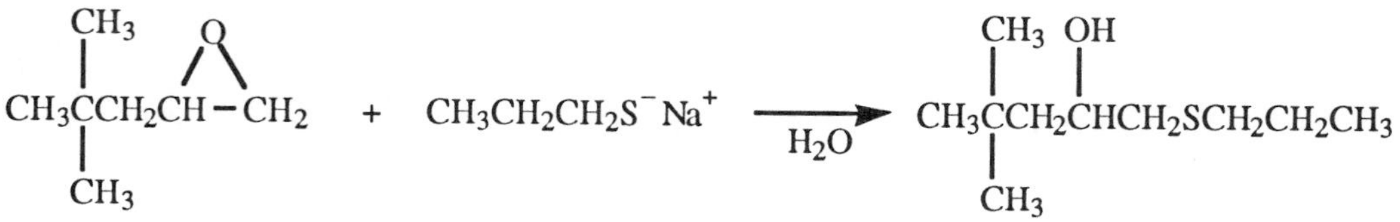

(e)

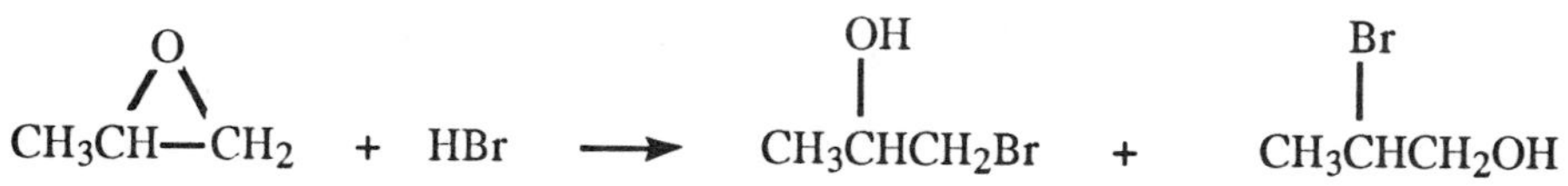

(f)

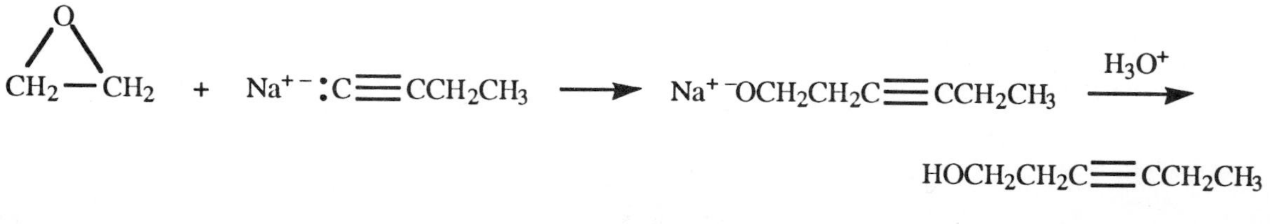

Concept Map 12.4   Ring-opening reactions of oxiranes.

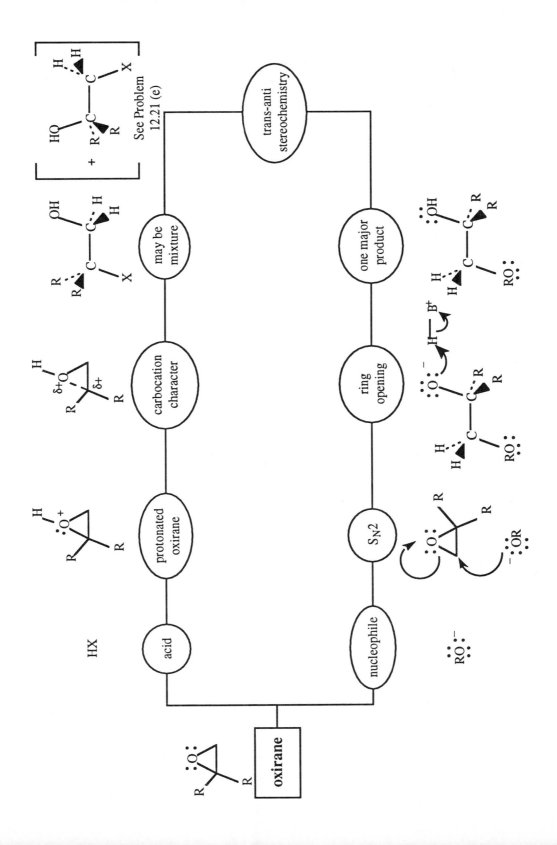

12.22 $CH_3SH + CH_3CH_2O^-Na^+ \longrightarrow CH_3S^-Na^+ + CH_3CH_2OH$

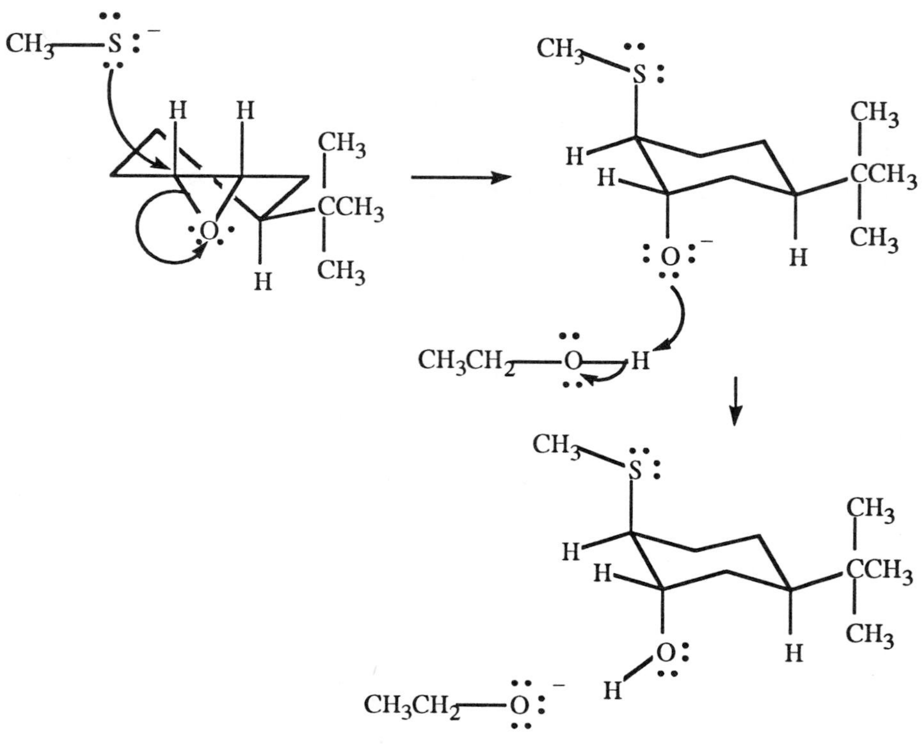

12.23

(a)

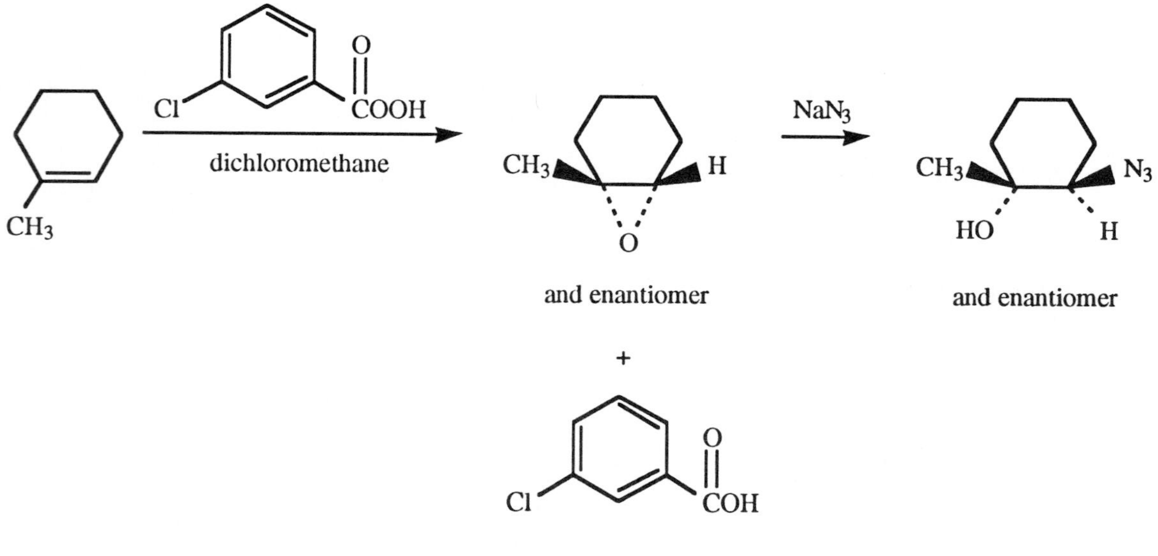
Cl
COOH
dichloromethane
CH3
NaN3
H
N3
HO
and enantiomer
and enantiomer
+
COH

(b)

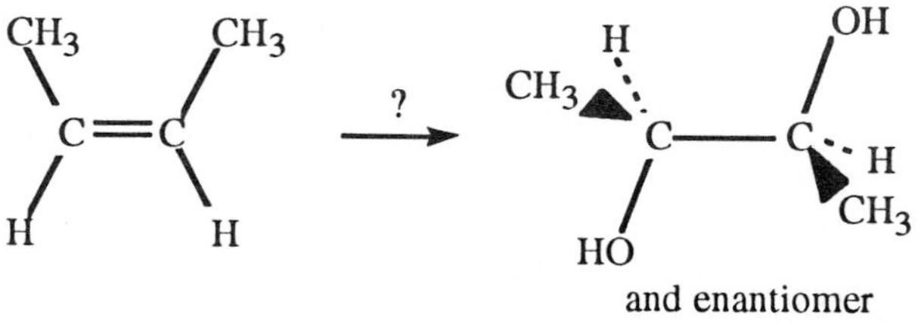
?
and enantiomer

12.23 (b) (cont)

1. What are the connectivities of the two compounds? How many carbon atoms does each contain? Are there any rings? What are the positions of branches and functional groups on the carbon skeletons?

    The carbon skeletons are the same. The alkene has two methyl groups cis to each other on the double bond. The methyl groups are also on the same side of the plane of the page which contains the hydroxyl groups, carbon-2, and carbon-3.

    The two hydroxyl groups are anti to each other.

2. How do the functional groups change in going from starting material to product? Does the starting material have a good leaving group?

    The double bond of the alkene has disappeared. Two hydroxyl groups have been attached to the carbon atoms that were part of the double bond.

3. Is it possible to dissect the structures of the starting material and product to see which bonds must be broken and which formed?

bond broken	bonds formed

    A π bond was broken; two carbon-oxygen bonds were formed anti to each other. The anti orientation of the two hydroxyl groups suggests the opening of an oxirane ring by hydroxide ion or water.

4. New bonds are created when an electrophile reacts with a nucleophile. Do we recognize any part of the product molecule as coming from a good nucleophile or an electrophilic addition?

    An alkene can be converted into a compound containing one oxygen by oxidation with the electrophilic reagent, *m*-chloroperoxybenzoic acid. A hydroxyl group can come from water, a nucleophile.

5. What type of compound would be a good precursor to the product?

    An oxirane can be converted to the 1,2-diol.

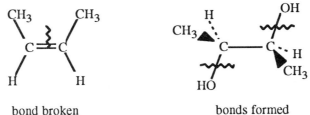

12.23 (b) (cont)

6.   After this last step, do we see how to get from starting material to product? If not, we need to analyze the structure
     obtained in step 5 by applying questions 4 and 5 to it.

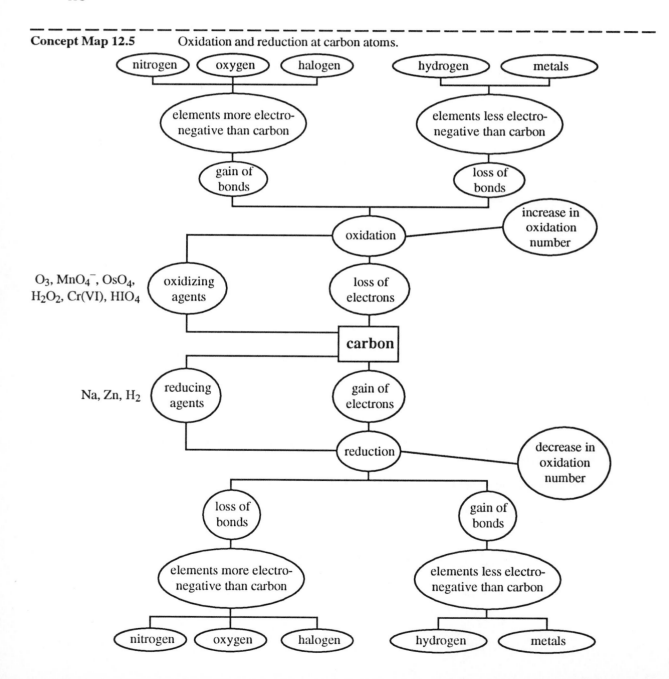

Concept Map 12.5   Oxidation and reduction at carbon atoms.

12.24 (a)

loss of two hydrogens and increase in oxidation number; oxidation; oxidizing agent

(b)

an increase in oxidation number at each carbon; oxidation; oxidizing agent

(c) $CH_3CH_2Br \longrightarrow CH_3CH_2NH_2$

$-1 \nearrow \qquad\qquad -1 \nearrow$

no change in oxidation state; not an oxidation-reduction reaction

(d)

gain of two oxygens and increase in oxidation number at each carbon; oxidation; oxidizing agent

(e)

no change in oxidation state; not an oxidation-reduction reaction

(f)

$$CH_3CH_2\overset{\overset{\textstyle O}{\|}}{C}H \longrightarrow CH_3CH_2CH_2OH$$

$+1 \qquad\qquad -1$

gain of two hydrogens and decrease in oxidation number; reduction; reducing agent

(g) $CH_3CH\!\!=\!\!CHCH_3 \longrightarrow$

$$CH_3CH\!\!-\!\!CHCH_3$$
$$\;\;|\qquad\;\;|$$
$$OH\quad OH$$

$-1 \quad -1 \qquad\qquad 0 \quad 0$

increase in oxidation number at each carbon; oxidation; oxidizing agent

(h) $CH_3C\!\!\equiv\!\!CCH_2CH_3 \longrightarrow CH_3CH\!\!=\!\!CHCH_2CH_3$

$0 \quad 0 \qquad\qquad\qquad -1 \quad -1$

gain of two hydrogens and decrease in oxidation number at each carbon; reduction; reducing agent

12.24 (cont)

(i)

$$CH_3C \equiv CCH_3 \longrightarrow CH_3\overset{O}{\overset{\|}{C}} - \overset{O}{\overset{\|}{C}}CH_3$$

0   0   +2 +2

gain of two oxygens and increase in oxidation number; oxidation; oxidizing agent

(j)

$$CH_3CH_2CH_2OH \longrightarrow CH_3CH_2\overset{O}{\overset{\|}{C}}OH$$

−1   +3

loss of two hydrogens and gain of oxygen; oxidation; oxidizing agent

(k)

$$CH_3\overset{O}{\overset{\|}{C}}H \longrightarrow CH_3 - \overset{OCH_3}{\underset{OCH_3}{\overset{|}{\underset{|}{C}}}} - H$$

+1   +1

no change in oxidation state; not an oxidation-reduction reaction

12.25

methane	chloromethane methyl chloride	dichloromethane methylene chloride
tetrachloromethane carbon tetrachloride	trichloromethane chloroform	

12.26

(a)

$$CH_3\overset{CH_3}{\underset{OH}{\overset{|}{\underset{|}{C}}H}}CH_2CHCH_3 + CrO_3 \xrightarrow[H_2SO_4]{H_2O} CH_3\overset{CH_3}{\overset{|}{C}H}CH_2\overset{O}{\overset{\|}{C}}CH_3 + Cr^{3+}$$

                                     orange   blue-green
                                                                      (positive test)

12.26 (cont)

(b)

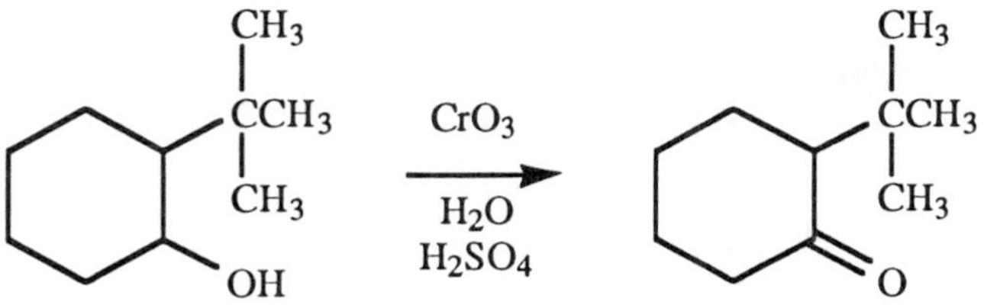

$$+ \quad Cr^{3+}$$
(positive test)

(c)

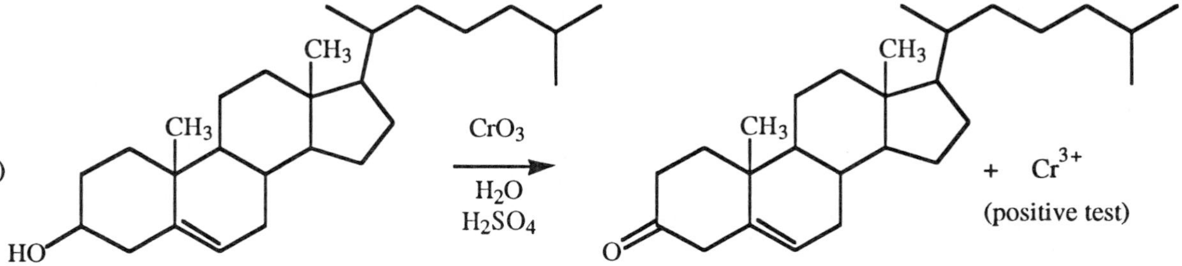

$$+ \quad Cr^{3+}$$
(positive test)

(d)

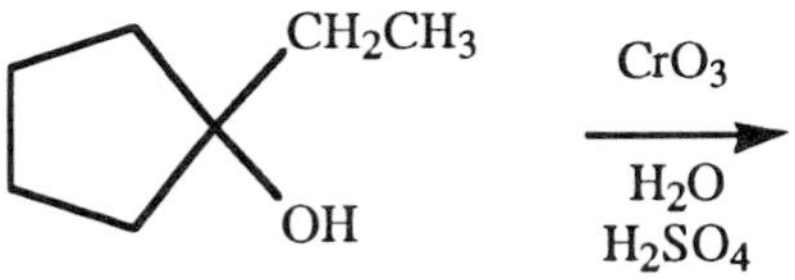

$\xrightarrow[\text{H}_2\text{SO}_4]{\text{CrO}_3 \quad \text{H}_2\text{O}}$ 3° alcohol, solution remains orange (negative test)

(e)

$$CH_3CH_2CH_2CH_2CH_2OH \xrightarrow[\substack{H_2O \\ H_2SO_4}]{CrO_3} CH_3CH_2CH_2CH_2\overset{\displaystyle O}{\overset{\|}{C}}OH \quad + \quad Cr^{+3} \text{ (positive test)}$$

(f)

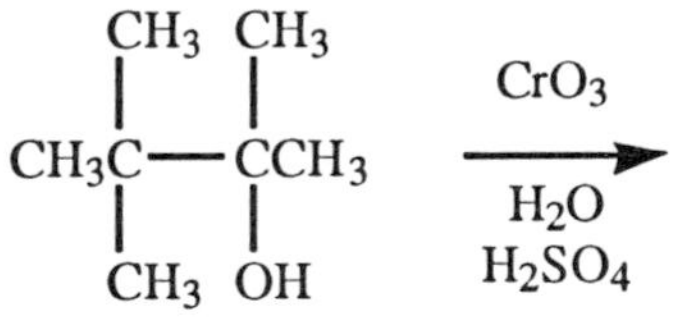

$\xrightarrow[\substack{\text{H}_2\text{O} \\ \text{H}_2\text{SO}_4}]{\text{CrO}_3}$ 3° alcohol, solution remains orange (negative test)

12.27

(1)

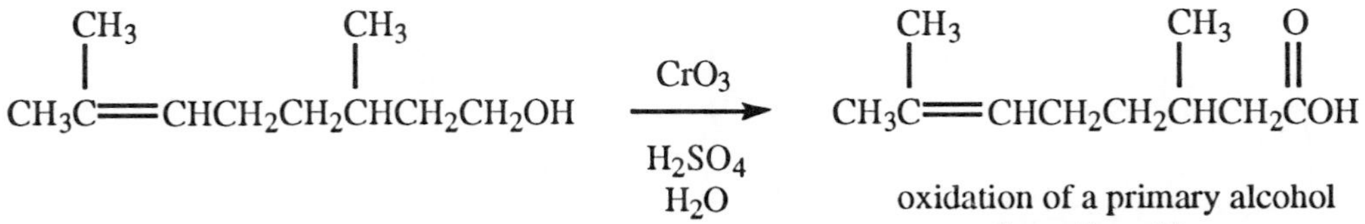

oxidation of a primary alcohol
to a carboxylic acid

(2)

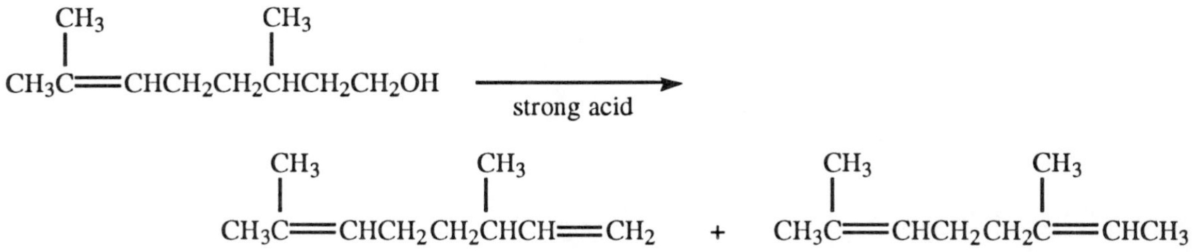

dehydration of alcohol

12.27 (cont)

(3)

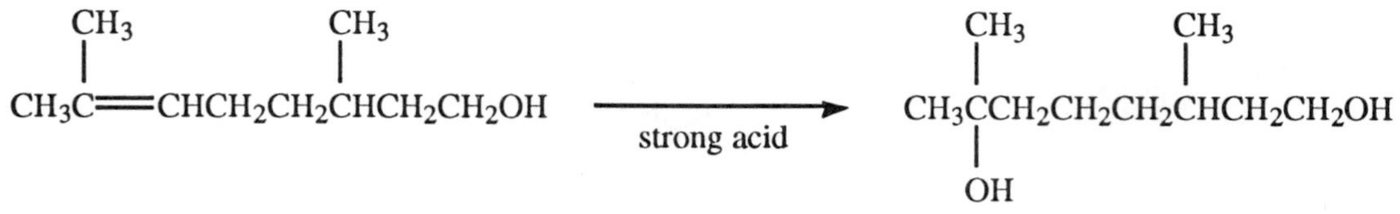

hydration of alkene

In addition, polymerization reactions (Section 8.4B) of alkenes occur in the presence of strong acids.

12.28

(a)

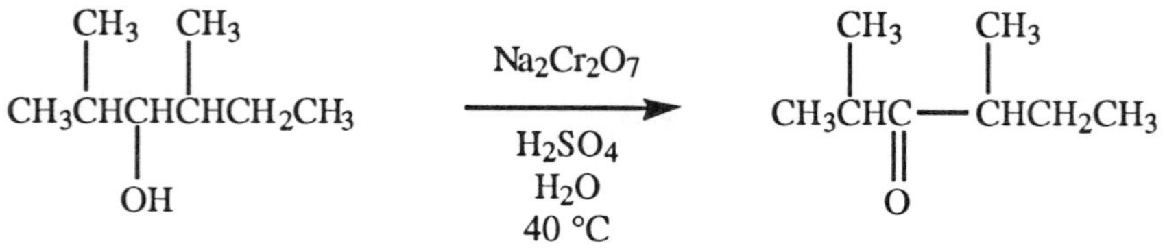

(b)

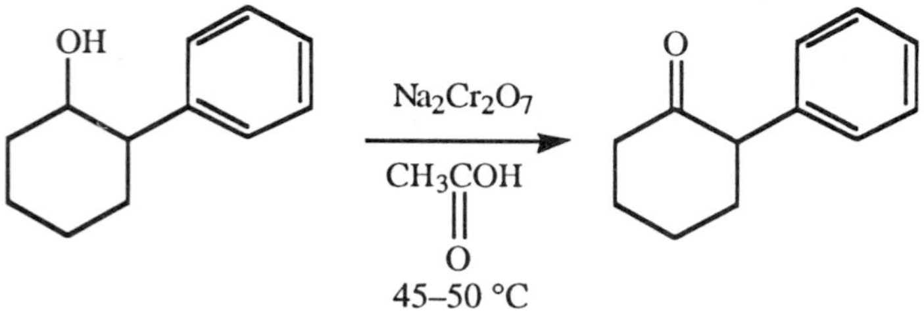

(c)

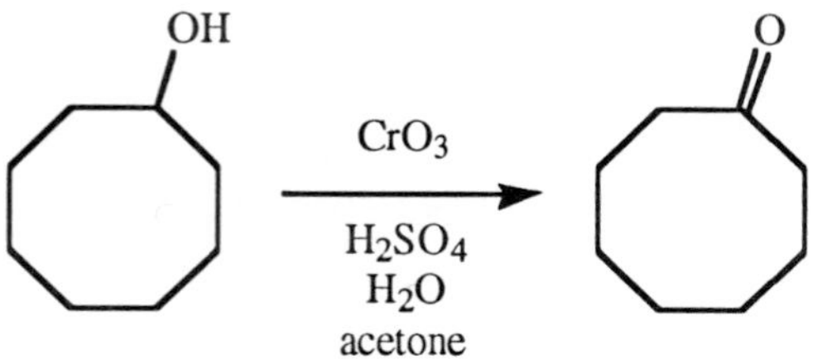

(d)

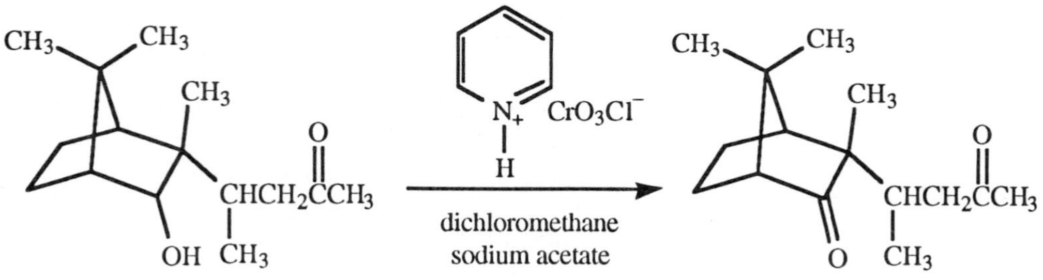

(e)

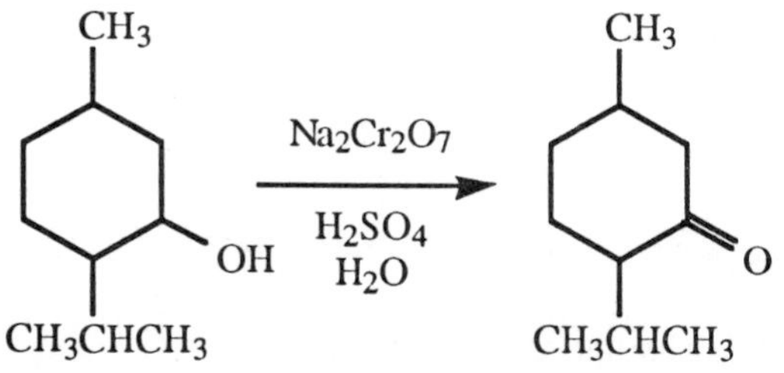

12.28 (cont)

(f)

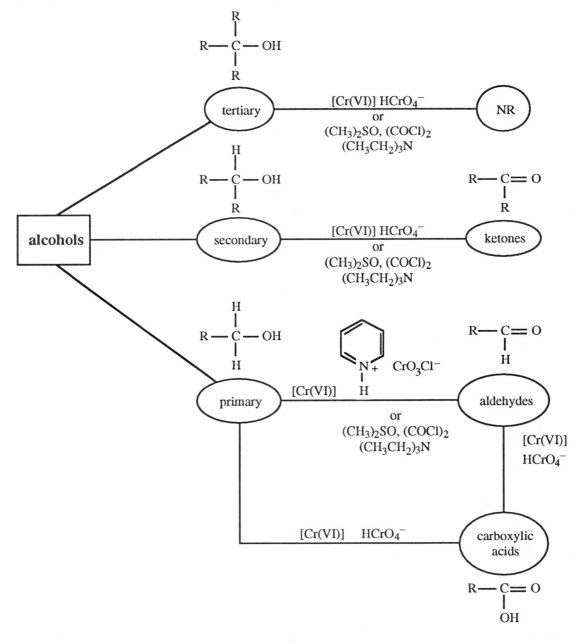

- -

Concept Map 12.6   Reactions of alcohols with oxidizing agents.

12.29   (a)

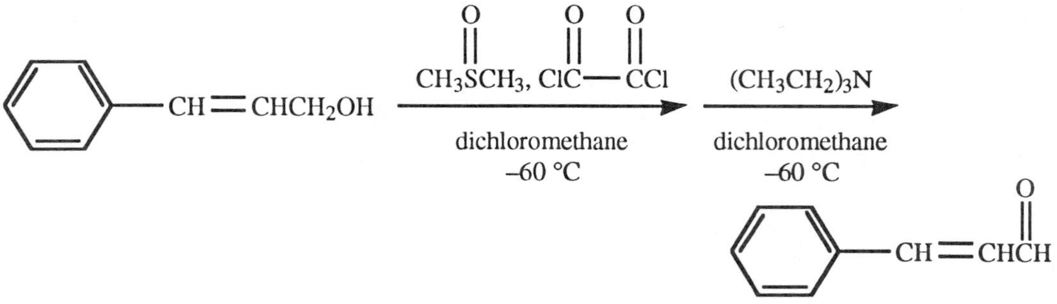

(b)

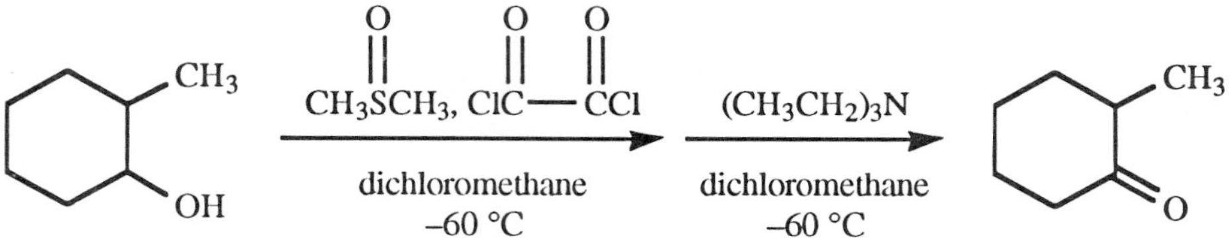

(c)

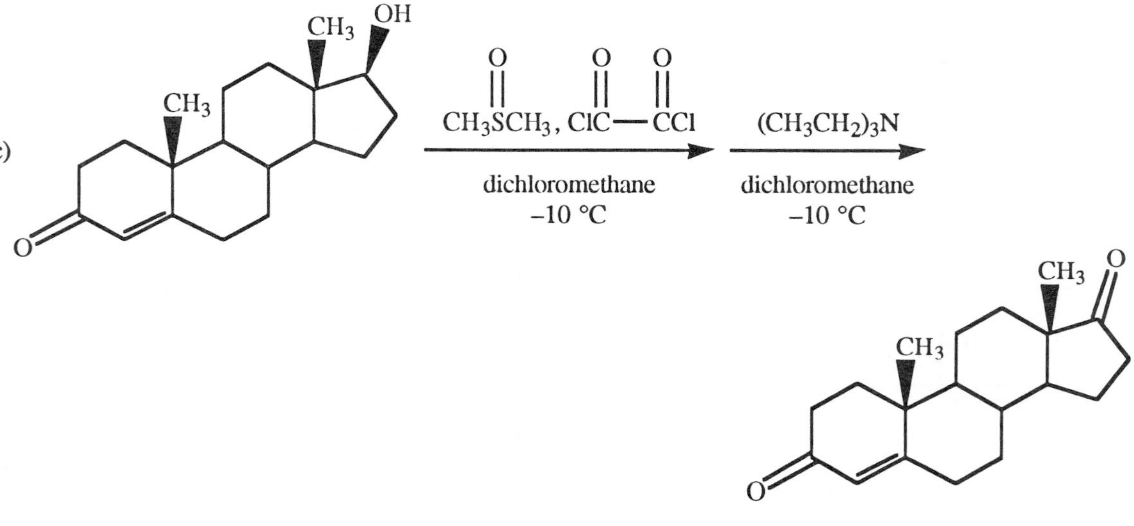

12.30

Experiment 1. The mechanism for the second stage of the Swern reaction has the base deprotonating the carbon atom next to the sulfur atom and not the one next to the oxygen atom. Dimethyl sulfoxide with deuterium substituting for hydrogen makes it possible to see the location of the deprotonation.

(a)

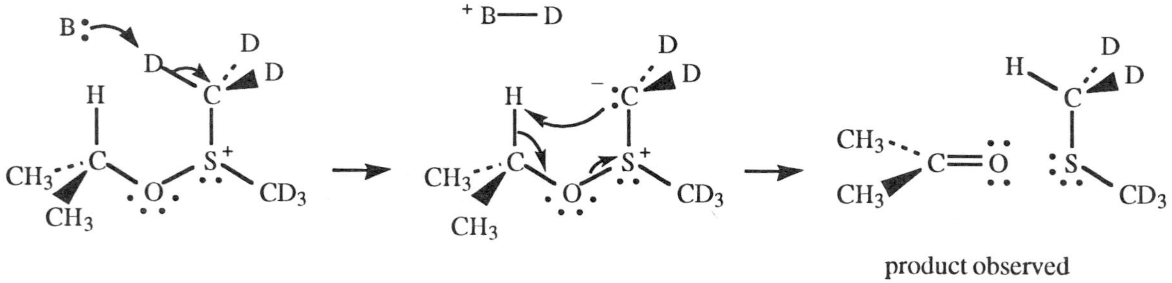

product observed

An alternative mechanism would have been the direct deprotonation of the carbon atom adjacent to the oxygen atom.

12.30 (cont)

(b)

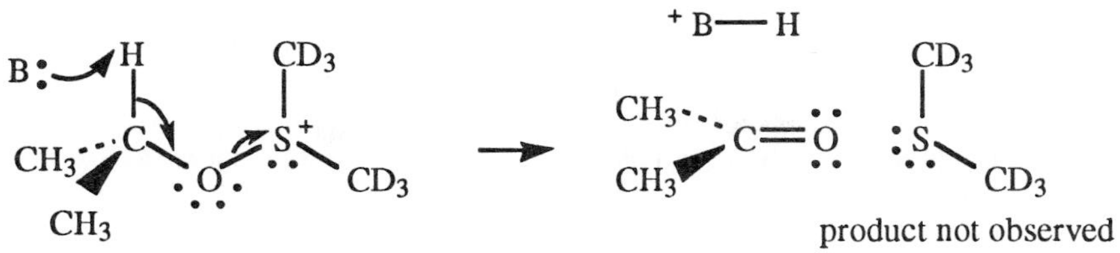

product not observed

Such a mechanism results in the formation of dimethyl sulfide with six deuterium atoms, instead of the five actually observed.

Experiment 2. This experiment also supports the mechanism shown above. If there are no hydrogen atoms on the carbon atom attached to the sulfur atom, the carbonyl compound does not form, again demonstrating that pathway (b) shown above does not occur. The products observed come from an S_N2 reaction.

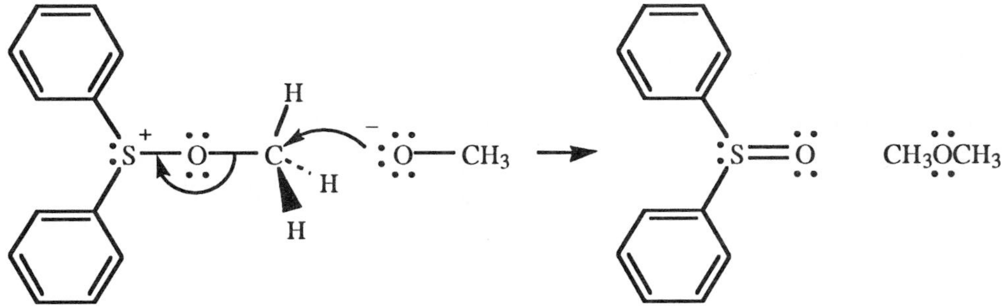

12.31

(a)

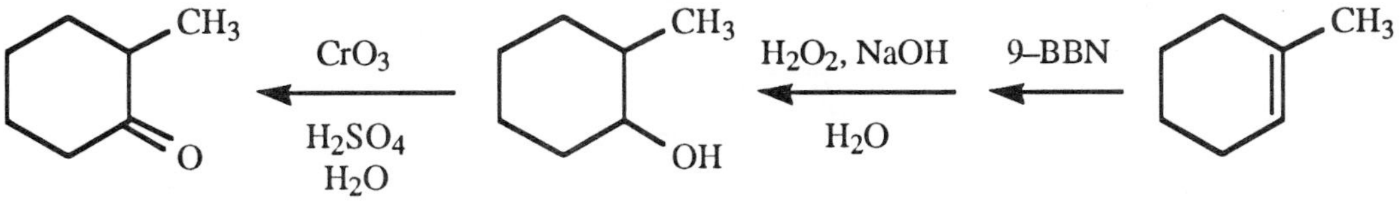

Note that this problem closely resembles the problem in Problem-Solving Skills Section 12.6. Practice applying the same questions to it.

(b)

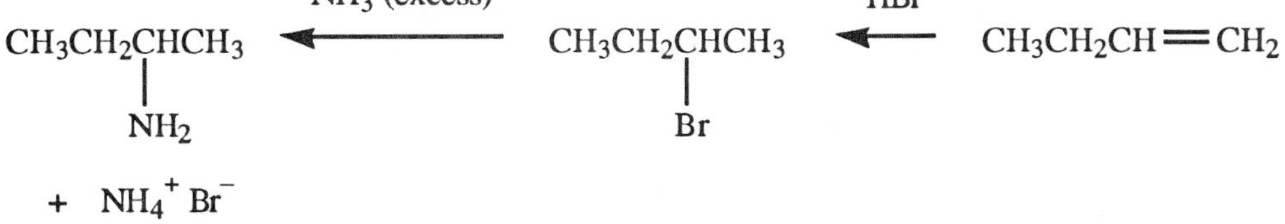

$$CH_3CH_2CHCH_3 \quad \xleftarrow{\text{NH}_3 \text{ (excess)}} \quad CH_3CH_2CHCH_3 \quad \xleftarrow{\text{HBr}} \quad CH_3CH_2CH=CH_2$$

with NH_2 and Br substituents

$+ \ NH_4^+ \ Br^-$

(c)

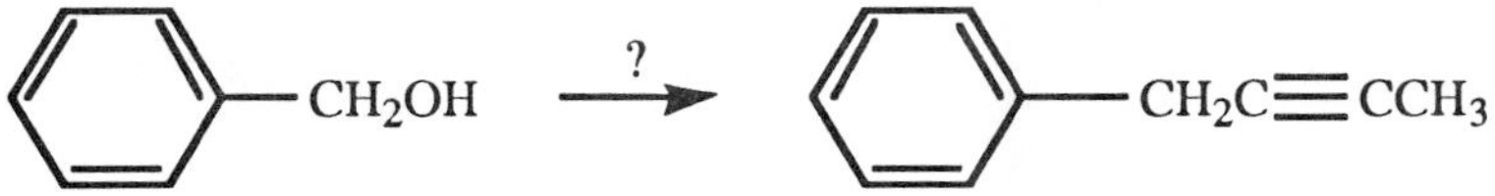

1. What are the connectivities of the two compounds? How many carbon atoms does each contain? Are there any rings? What are the positions of branches and functional groups on the carbon skeletons?

Both compounds contain an aromatic ring. The alcohol group on the benzylic carbon atom in the starting material has been replaced by a three-carbon chain containing a triple bond (a 1-propynyl group) in the product.

12.31 (c) (cont)

2. How do the functional groups change in going from starting material to product? Does the starting material have a good leaving group?

    The benzylic carbon atom is bonded to the first carbon atom of the triple bond. There is no good leaving group present in the starting material.

3. Is it possible to dissect the structures of the starting material and product to see which bonds must be broken and which formed?

    A carbon-oxygen bond was broken. A carbon-carbon bond was formed.

4. New bonds are created when an electrophile reacts with a nucleophile. Do we recognize any part of the product molecule as coming from a good nucleophile or an electrophilic addition?

    The 1-propynyl group can come from an S_N2 reaction using an anion derived from an alkyne as the nucleophile.

5. What type of compound would be a good precursor to the product?

    A compound with a good leaving group, such as an alkyl halide or a tosylate, would be a good precursor to the product.

    An alkyl bromide can easily be made from the starting material by reaction with concentrated hydrobromic acid.

6. After this last step, do we see how to get from starting material to product? If not, we need to analyze the structure obtained in step 5 by applying questions 4 and 5 to it.

12.32

(a)

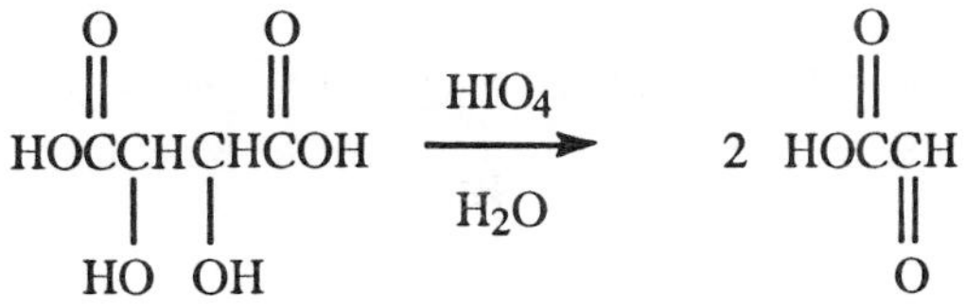

(b)

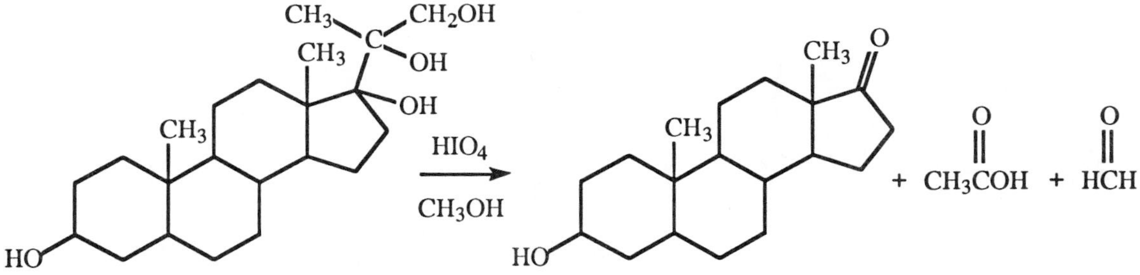

(c)

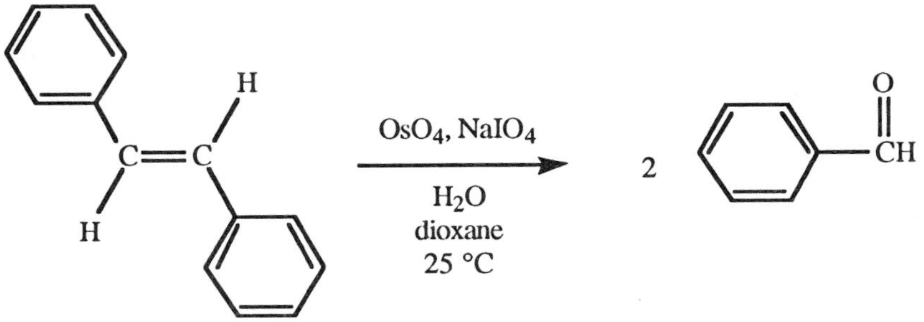

- -

Concept Map 12.7   (see p. 344)

(see p. 344)

- -

12.33

(a) 1,3-propanediol

(b) 1,2-dimethoxyethane

(c) 3-ethoxy-2,2-dimethyl-1-propanol

(d) 1-bromo-2-methoxyethane
(ethyl phenyl ether)

(e) ethoxybenzene

(f) 2-ethoxyethanol

(g) cyclobutanol

(h) (2S)(E)-5-methyl-4-hepten-2-ol

(i) (1R, 2R)-2-bromocyclopentanol

(j) benzyloxyphenylmethane
(dibenzyl ether)

(k) (1S, 2S)-2-methoxycyclohexanol

(l) 1,1,1-triethoxyethane

Concept Map 12.7   Summary of the preparation of and reactions of alcohols and ethers.

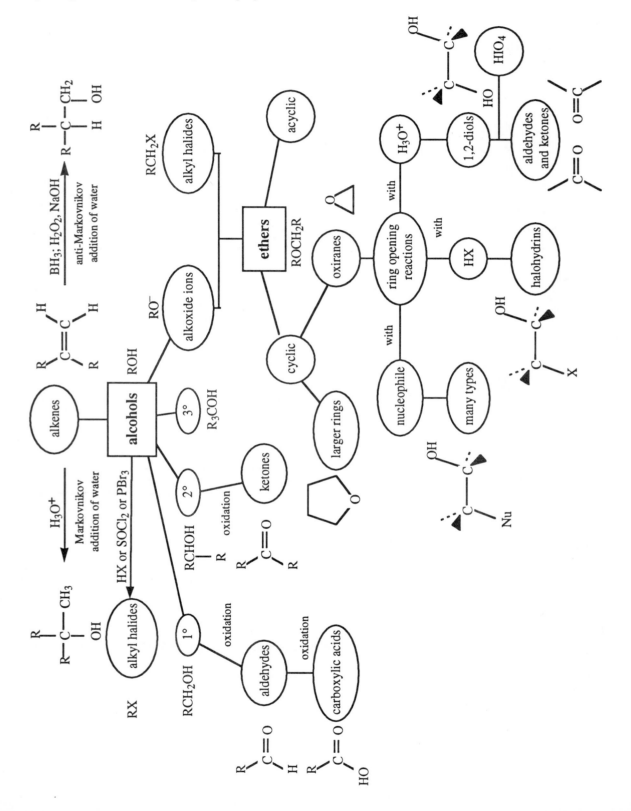

12.34

(a) $CH_3CH_2OC=CH_2$
$\quad\quad\quad\quad\;\;|$
$\quad\quad\quad\quad\;\;Cl$

(b)

$$\underset{\displaystyle H\quad\quad\quad OCH_3}{\overset{\displaystyle CH_3CH_2CH_2O\quad\quad H}{C=C}}$$

(c) $ClCH_2CHCH_2Cl$
$\quad\quad\quad\quad\;|$
$\quad\quad\quad\quad OH$

(d) ⬛—CH_2OH

(e)
$$\underset{\displaystyle CH_3CH_2}{\overset{\displaystyle OH}{CH_3\cdots \overset{|}{C}-CH_2CH=CH_2}}$$

(f)
$$\underset{\displaystyle CH_3}{\overset{\displaystyle CH_2OH}{H\cdots \overset{|}{C}-Cl}}$$

(g)
$$\underset{\displaystyle CH_3CH_2}{\overset{\displaystyle C\equiv CH}{CH_3\cdots \overset{|}{C}-OH}}$$

(h) $O_2NCH_2CH_2OH$

(i)
$$\underset{\displaystyle CH_3CH_2}{\overset{\displaystyle OH}{H\cdots \overset{|}{C}-CH_2\underset{\displaystyle CH_3}{\overset{\displaystyle CH_3}{\overset{|}{C}CH_2CH_3}}}}$$

(j)
$$\underset{\displaystyle HO\quad\quad\quad CH_2CH_3}{\overset{\displaystyle CH_3CH_2\quad\quad CH_3}{H\cdots \overset{|}{C}-\overset{|}{C}\cdots H}}$$

12.35

(a) $CH_3CH_2CH_2OH \xrightarrow[\Delta]{HBr} CH_3CH_2CH_2Br \;+\; H_2O$

(b) $CH_3CH_2CH_2OH \xrightarrow[\Delta]{PBr_3} CH_3CH_2CH_2Br \;+\; H_3PO_3$

(c) $CH_3CH_2CH_2OH \xrightarrow[\text{pyridine}]{SOCl_2} CH_3CH_2CH_2Cl \;+\; SO_2\uparrow \;+$

$\quad\quad N^+ H \;\; Cl^-$ (pyridinium chloride)

(d) $CH_3CH_2CH_2OH \xrightarrow{NaH} CH_3CH_2CH_2O^-Na^+ \;+\; H_2\uparrow$

(e) $CH_3CH_2CH_2OH \xrightarrow[\text{cold}]{H_3O^+} CH_3CH_2CH_2\overset{+}{O}H_2 \;+\; H_2O$

(f) $CH_3CH_2CH_2OH \xrightarrow[\substack{H_2SO_4 \\ \Delta}]{} CH_3CH=CH_2 \;+\; H_2O$

(g) $CH_3CH_2CH_2OH \xrightarrow{Na} CH_3CH_2CH_2O^-Na^+ \xrightarrow{CH_3CH_2CH_2CH_2Br}$

$\quad\quad\quad CH_3CH_2CH_2OCH_2CH_2CH_2CH_3 \;+\; NaBr$

(h) $CH_3CH_2CH_2OH \xrightarrow[\substack{H_2SO_4 \\ \Delta}]{Na_2Cr_2O_7} CH_3CH_2\overset{\displaystyle O}{\overset{\displaystyle \|}{C}}OH$

12.35 (cont)

(i) $CH_3CH_2CH_2OH$ $\xrightarrow[\text{dichloromethane}]{}$ $CH_3CH_2CH{\overset{\text{O}}{\parallel}}$

reagent above arrow: pyridinium $+ CrO_3Cl^-$ (with N–H)

(j) $CH_3CH_2CH_2OH$ $\xrightarrow{CH_3SCH_3,\ ClC(=O)-CC(=O)l}$ $\xrightarrow{(CH_3CH_2)_3N}$ $CH_3CH_2CH{\overset{\text{O}}{\parallel}}$

12.36

(a) $CH_3CH_2CH_2CH_2CHCH_3$ (with OH) $\xrightarrow[\Delta]{HBr}$ $CH_3CH_2CH_2CH_2CHCH_3$ (with Br) $+$ $CH_3CH_2CH_2CHCH_2CH_3$ (with Br)

under the two products: **major**

$+$ H_2O $+$ some elimination products

(b) $CH_3CH_2CH_2CH_2CHCH_3$ (with OH) $\xrightarrow[\Delta]{PBr_3}$ $CH_3CH_2CH_2CH_2CHCH_3$ (with Br) $+$ H_3PO_3

(c) $CH_3CH_2CH_2CH_2CHCH_3$ (with OH) $\xrightarrow[\text{pyridine}]{SOCl_2}$ $CH_3CH_2CH_2CH_2CHCH_3$ (with Cl) $+$ $SO_2\uparrow$ $+$ pyridinium chloride

(d) $CH_3CH_2CH_2CH_2CHCH_3$ (with OH) $\xrightarrow{NaH}$ $CH_3CH_2CH_2CH_2CHCH_3$ (with O^-Na^+) $+$ $H_2\uparrow$

(e) $CH_3CH_2CH_2CH_2CHCH_3$ (with OH) $\xrightarrow[ZnCl_2]{HCl}$ $CH_3CH_2CH_2CH_2CHCH_3$ (with Cl) $+$ $CH_3CH_2CH_2CHCH_2CH_3$ (with Cl)

(f) $CH_3CH_2CH_2CH_2CHCH_3$ (with OH) $\xrightarrow[\Delta]{H_2SO_4}$ $CH_3CH_2CH_2CH=CHCH_3$ $+$

$CH_3CH_2CH_2CH_2CH=CH_2$ $+$ $CH_3CH_2CH=CHCH_2CH_3$ $+$ H_2O

12.36 (cont)

(g) $CH_3CH_2CH_2CH_2CHCH_3$ $\xrightarrow[\substack{H_2SO_4 \\ \Delta}]{Na_2Cr_2O_7}$ $CH_3CH_2CH_2CH_2\overset{\displaystyle O}{\overset{\|}{C}}CH_3$
                         |
                        OH

(h) $CH_3CH_2CH_2CH_2CHCH_3$ $\xrightarrow[\text{dichloromethane}]{}$ $CH_3CH_2CH_2CH_2\overset{\displaystyle O}{\overset{\|}{C}}CH_3$
                         |
                        OH

(reagent above arrow: pyridinium, N–H$^+$ CrO$_3$Cl$^-$)

(i) $CH_3CH_2CH_2CH_2CHCH_3$ $\xrightarrow[]{CH_3SCH_3,\ ClC\overset{O}{\overset{\|}{}}\!-\!\overset{O}{\overset{\|}{}}CCl}$ $\xrightarrow[]{(CH_3CH_2)_3N}$ $CH_3CH_2CH_2CH_2\overset{\displaystyle O}{\overset{\|}{C}}CH_3$
                         |
                        OH

12.37 (a) cyclohexanol (ring with —OH)

          2. H_2SO_4, cold, conc
          4. CrO_3, H_2O, H_2SO_4

(b) alkene with CH_3 and H on one carbon, H and CH_2CH_3 on the other

          2. H_2SO_4, cold, conc
          3. Br_2, carbon tetrachloride

(c) alkene with CH_3 and H on one carbon, H and CH_2Cl on the other

          2. H_2SO_4, cold, conc
          3. Br_2, carbon tetrachloride
          5. $AgNO_3$, CH_3CH_2OH
          6. NaI, acetone

(d) cyclopentane ring with CH_3 and OH on same carbon

          2. H_2SO_4, cold, conc

(e) benzene ring with —CH_2Cl

          5. $AgNO_3$, CH_3CH_2OH
          6. NaI, acetone

(f) $CH_3CH_2CH_2CH_2CH_2OH$

          2. H_2SO_4, cold, conc
          4. CrO_3, H_2O, H_2SO_4

12.37 (cont)

(g) $CH_3C \equiv CCH_2CH_3$

2. H_2SO_4, cold, conc
3. Br_2, carbon tetrachloride

(h)

(i) $HOCH_2CH_2OH$

1. H_2O, cold
2. H_2SO_4, cold, conc
4. CrO_3, H_2O, H_2SO_4

(j)
$$CH_3CH_2\overset{\displaystyle CH_3}{\underset{\displaystyle Br}{C}}CH_3$$

5. $AgNO_3$, CH_3CH_2OH

(k) $-OCH_2CH_3$

2. H_2SO_4, cold, conc

(l) $CH_3CH_2CH_2CH_2CH_2CH_2Cl$

6. NaI, acetone

12.38   (a)   The amino alcohol with two hydrogen bonding groups is most soluble in water. The alcohol, which can act as a donor as well as an acceptor of hydrogen bonds, is more soluble in water than the ester, which is only an acceptor of hydrogen bonds.

(b)   The more stable the conjugate base (an anion in this example), the stronger is the acid (the alcohol). The *tert*-butoxide anion is destabilized by the bulk of the methyl groups that interferes with solvation (and therefore stabilization) of the anion. The trifluoroethoxide anion, by contrast, is stabilized by the electron-withdrawing effect of the fluorine atoms.

12.38 (b) (cont)

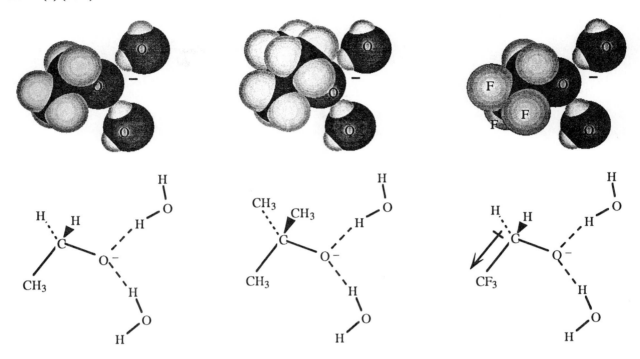

(c) The compound that has the highest boiling point and the highest solubility in water is the one with two hydroxyl groups, which leads to increased hydrogen bonding, both between molecules of the same kind (giving rise to a higher boiling point) and between the compound and water (giving rise to higher solubility in water). Of the compounds with only one hydroxyl group, the one with the higher molecular weight and the longer carbon chain has the higher boiling point (from increased van der Waals attraction) and the lower solubility in water (because the hydrophobic part of the molecule is too large to be brought into solution by the polar hydroxyl group). See Sections 1.10B and 1.10C for review.

(d) The hydroxyl group hydrogen bonds more strongly than the thiol group, and the alcohol is therefore more soluble than the corresponding sulfur compound.

12.38 (d) (cont)

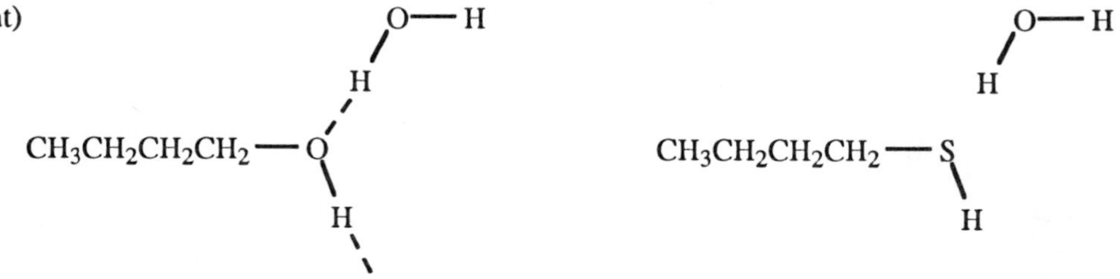

12.39

(a)

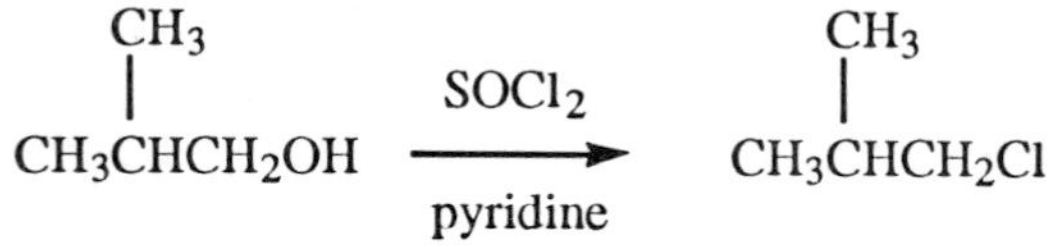

$$CH_3CHCH_2OH \xrightarrow[\text{pyridine}]{SOCl_2} CH_3CHCH_2Cl$$

(with CH$_3$ substituents on both structures)

(b)

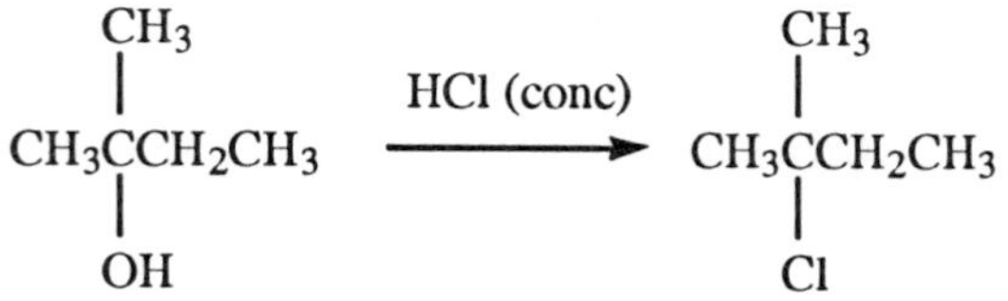

$$CH_3CCH_2CH_3 \text{ (CH}_3\text{, OH)} \xrightarrow{HCl \text{ (conc)}} CH_3CCH_2CH_3 \text{ (CH}_3\text{, Cl)}$$

(c)

$$CH_3CH_2CHCH_3 \text{ (OH)} \xrightarrow{PBr_3} CH_3CH_2CHCH_3 \text{ (Br)}$$

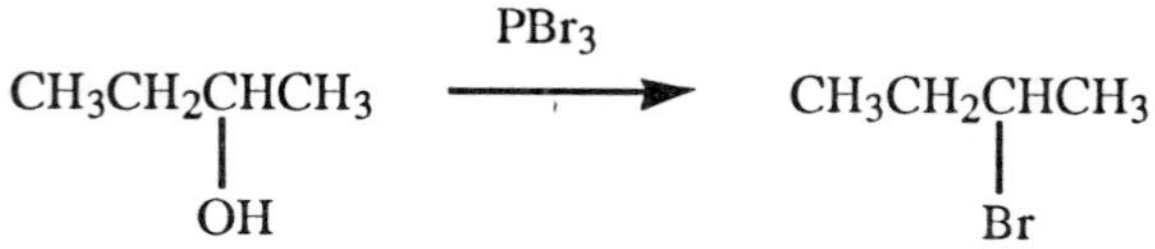

(d)

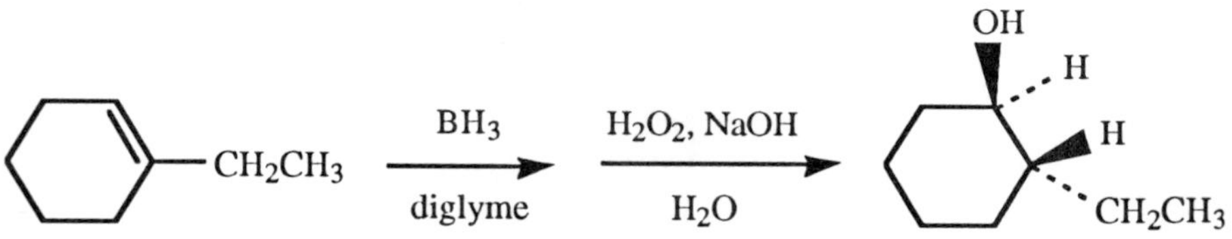

and enantiomer

(e)

$$HOCH_2(CH_2)_4CH_2OH \xrightarrow{HBr \text{ (g) (excess)}} BrCH_2(CH_2)_4CH_2Br$$

(f)

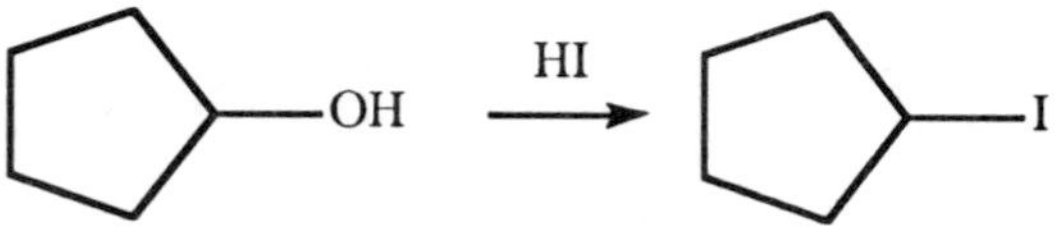

(g)

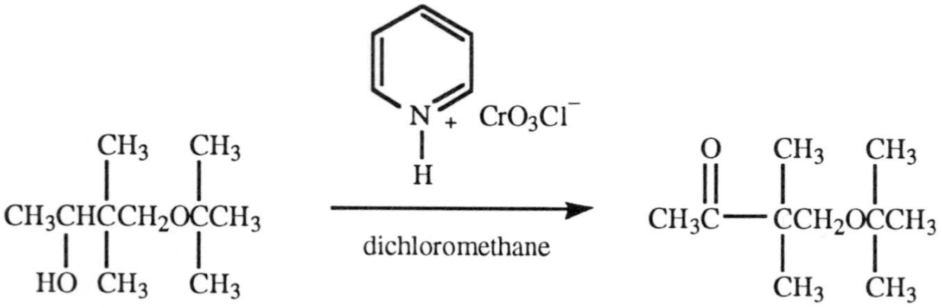

12.39 (cont)

(h) 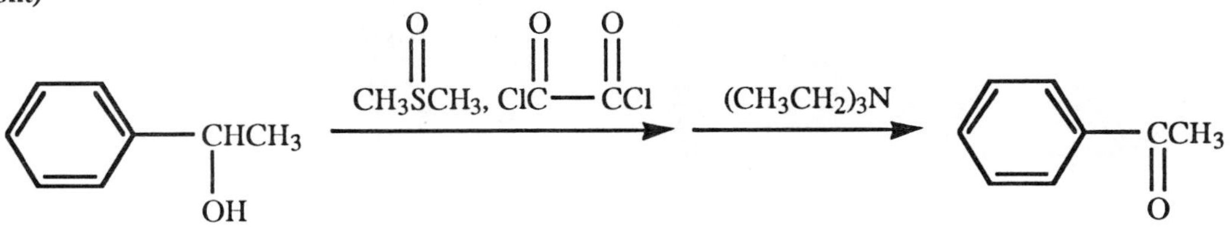
CHCH3
OH
CH3SCH3, ClC—CCl
(CH3CH2)3N
CCH3
O

12.40

(a) 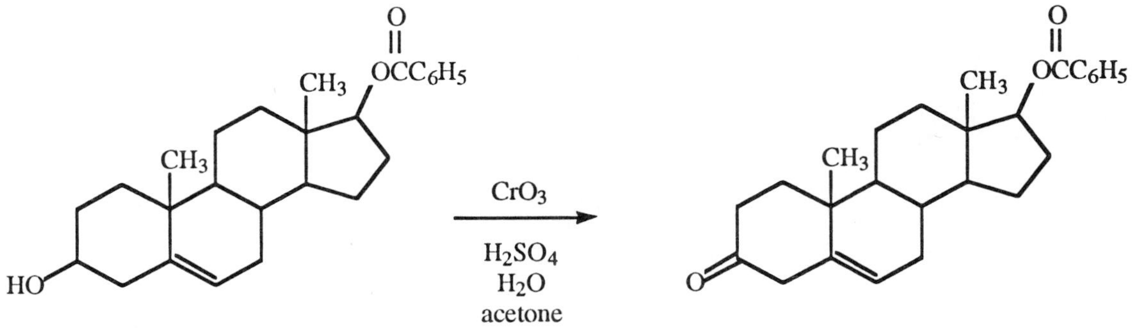
OCC6H5
CH3
CrO3
H2SO4
H2O
acetone
HO
O

(b) 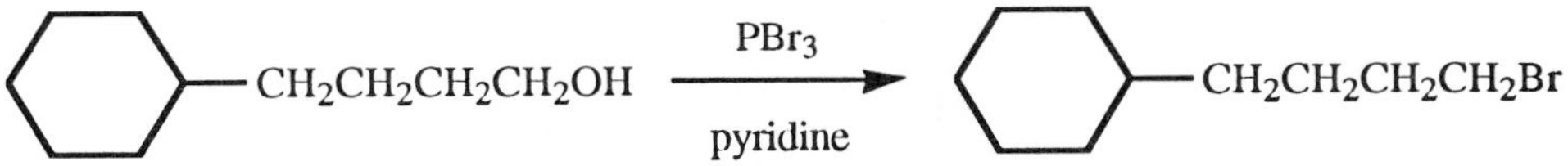
CH2CH2CH2CH2OH
PBr3
pyridine
CH2CH2CH2CH2Br

(c) 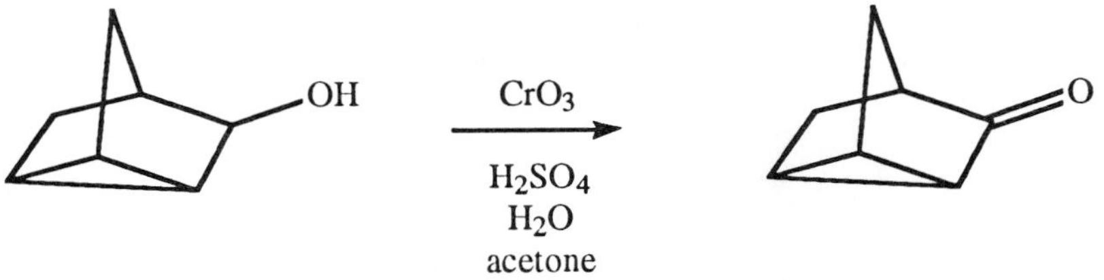
OH
CrO3
H2SO4
H2O
acetone
O

(d) 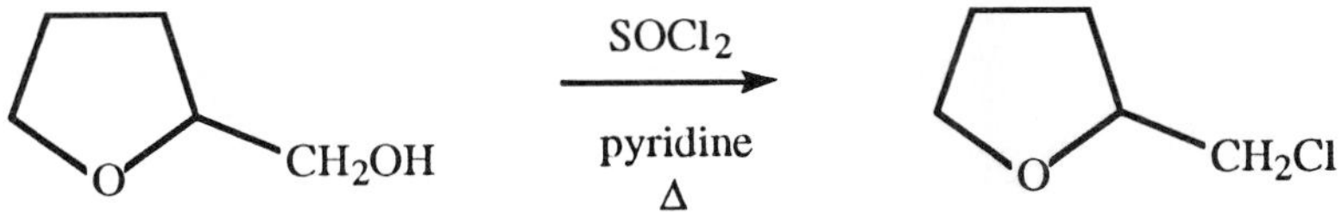
SOCl2
pyridine
Δ
CH2OH
CH2Cl
O

(e) 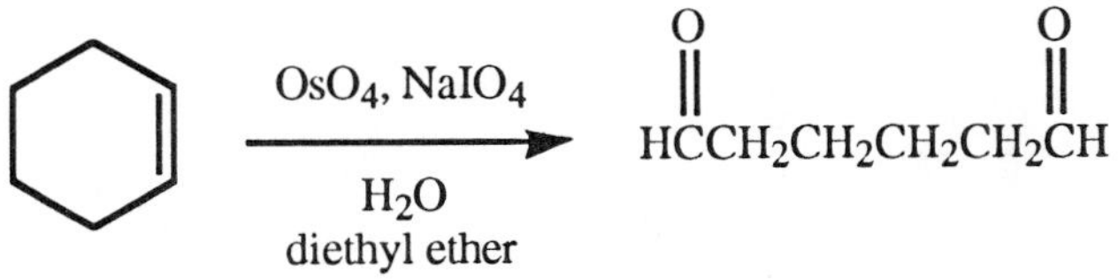
OsO4, NaIO4
H2O
diethyl ether
HCCH2CH2CH2CH2CH

(f) 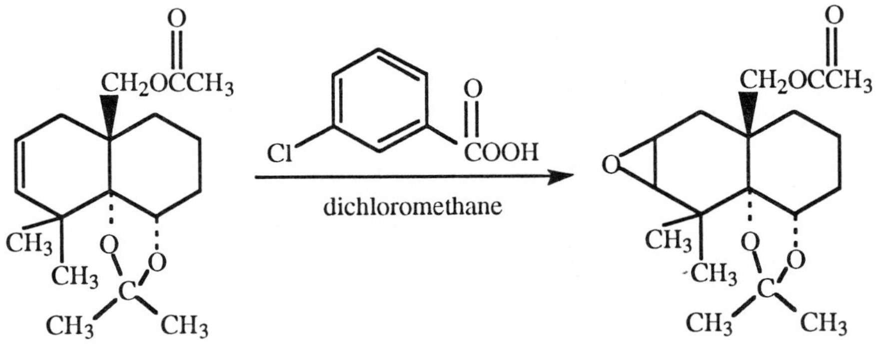
CH2OCCH3
Cl
COOH
dichloromethane
CH3
C
O

two diastereomers

12.40 (cont)

(g)

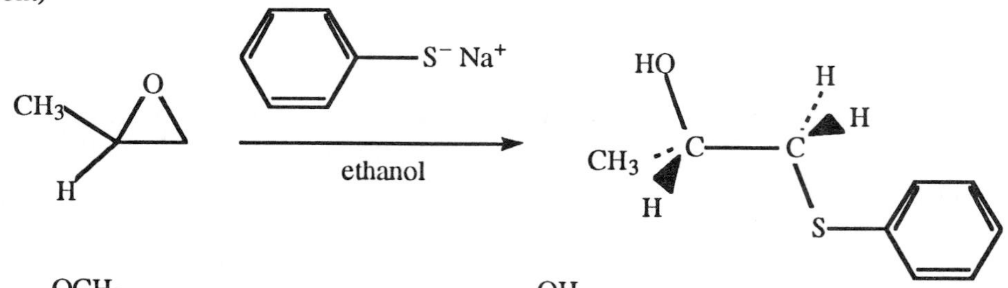

(h)

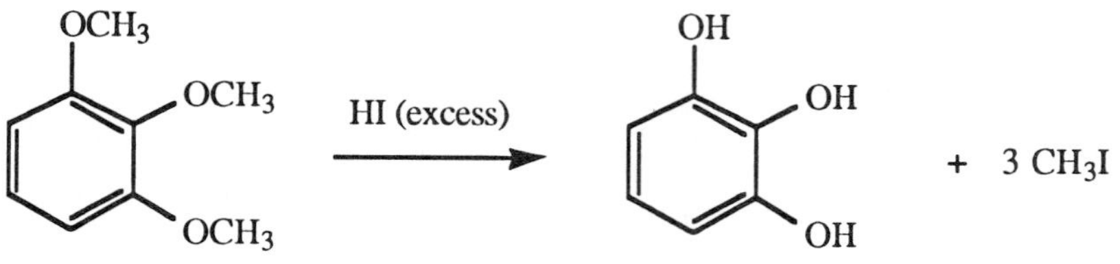

12.41

(a)

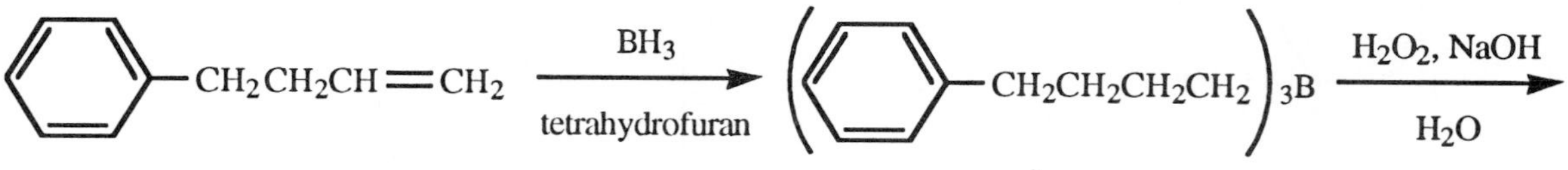

A

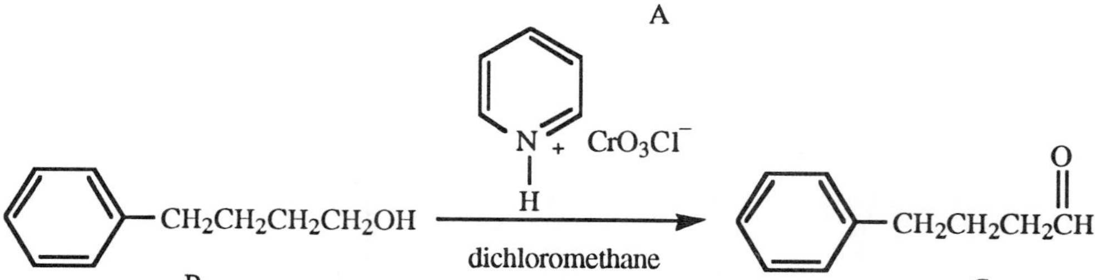

B

C

(b)

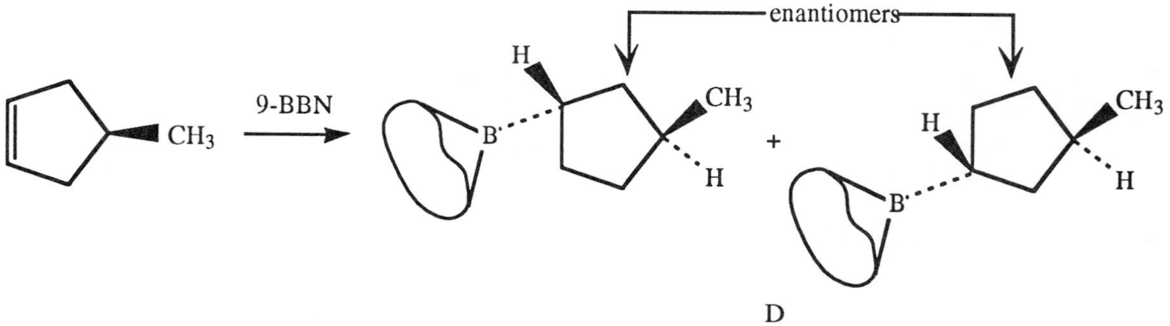

D

major isomers from less hindered transition state

+

12.41 (b) (cont)

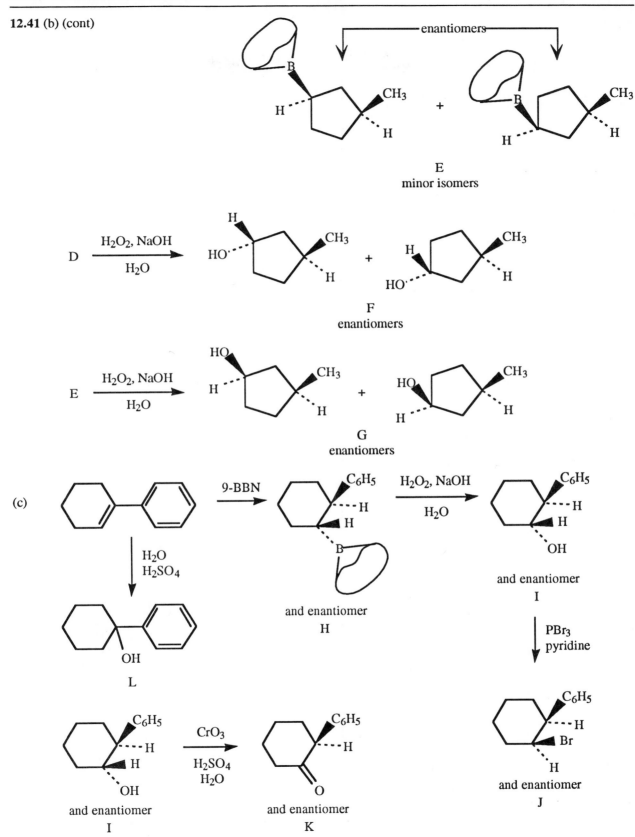

E
minor isomers

F
enantiomers

G
enantiomers

(c)

L

and enantiomer
H

and enantiomer
I

and enantiomer
I

and enantiomer
K

and enantiomer
J

12.41 (cont)

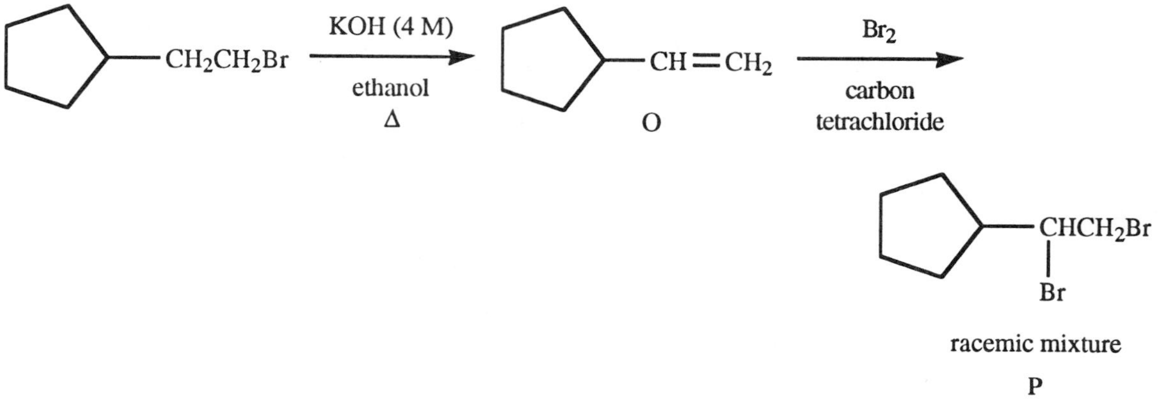

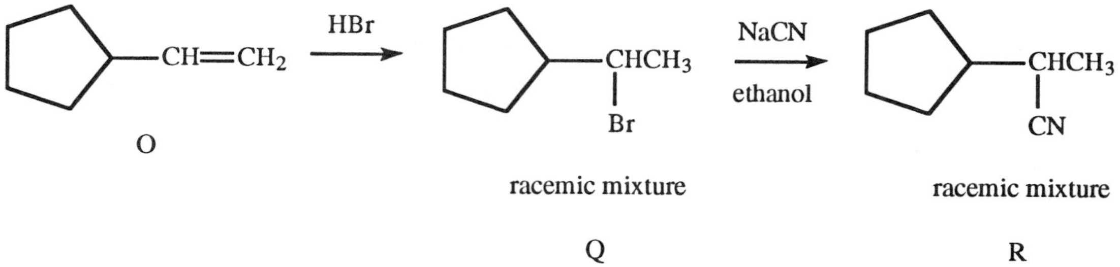

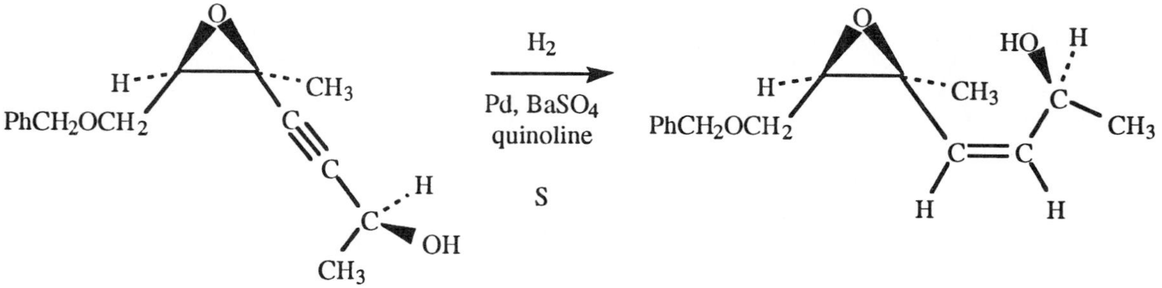

(d)

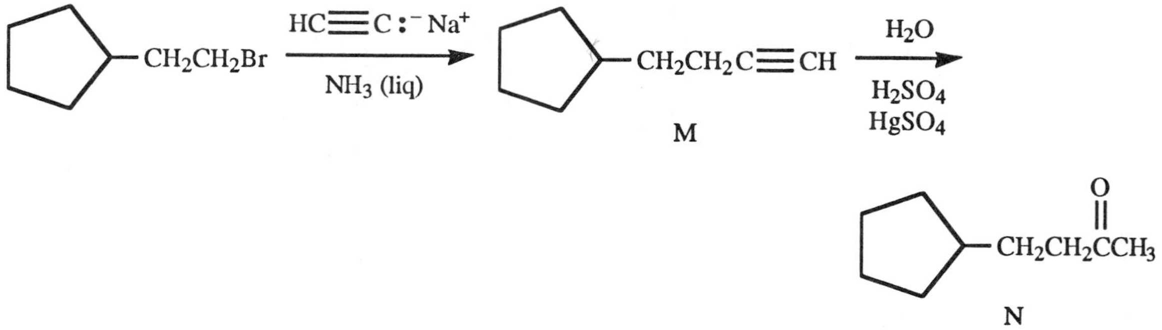

(e)

12.42

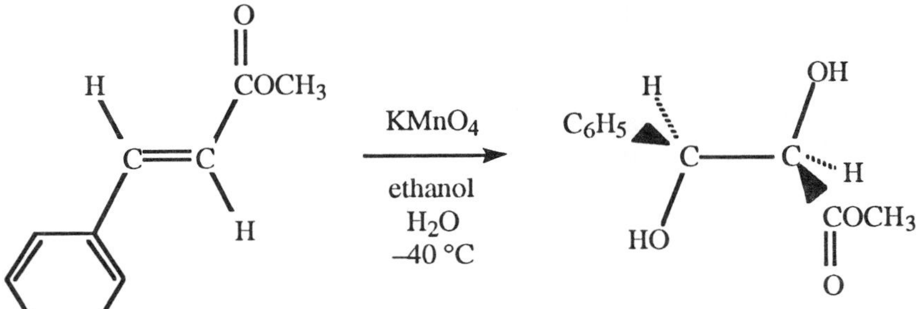

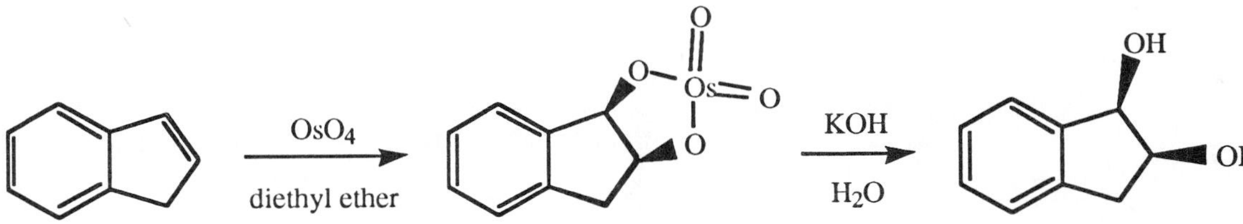

(a)

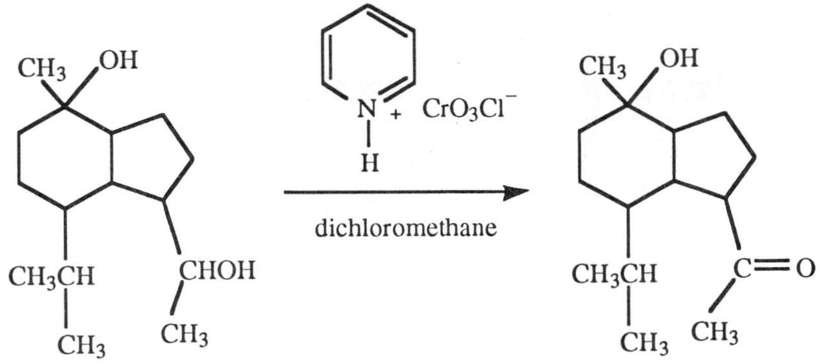

(b)

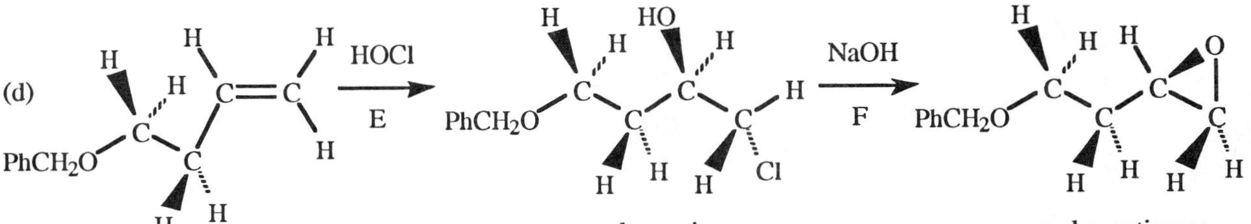

(c)

(d)

(e)

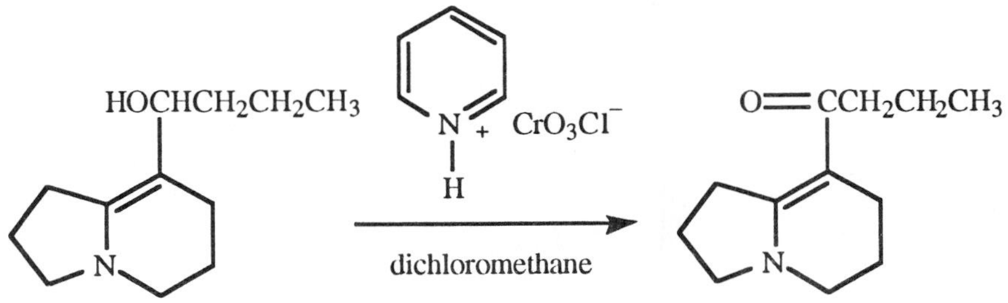

12.42 (cont)

(f)

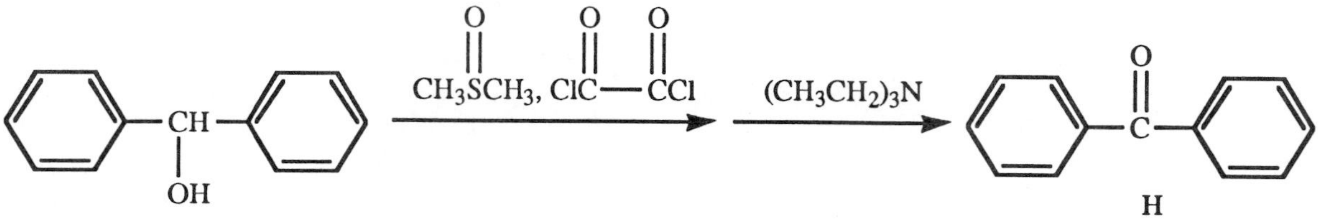

12.43

(a)

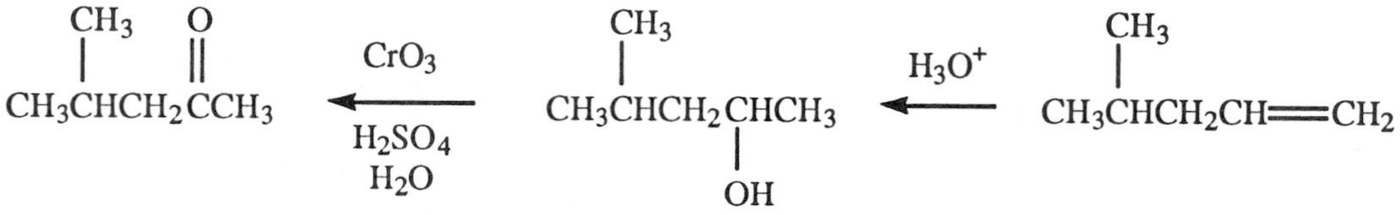

(b)

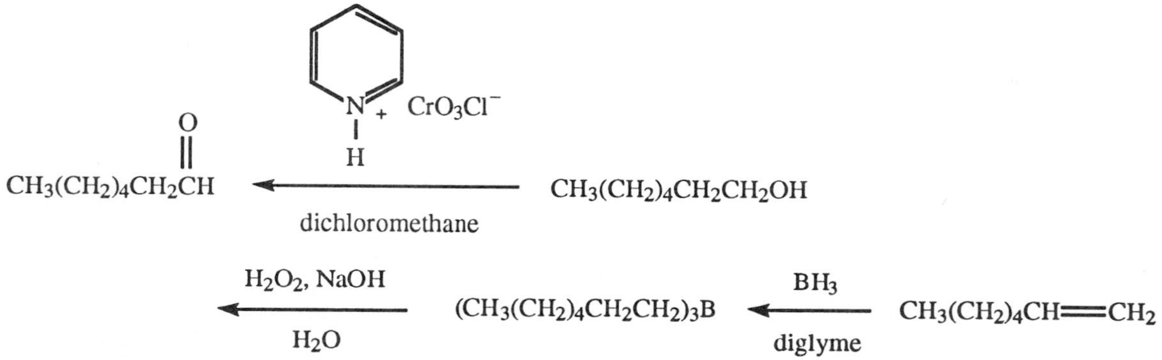

(c)

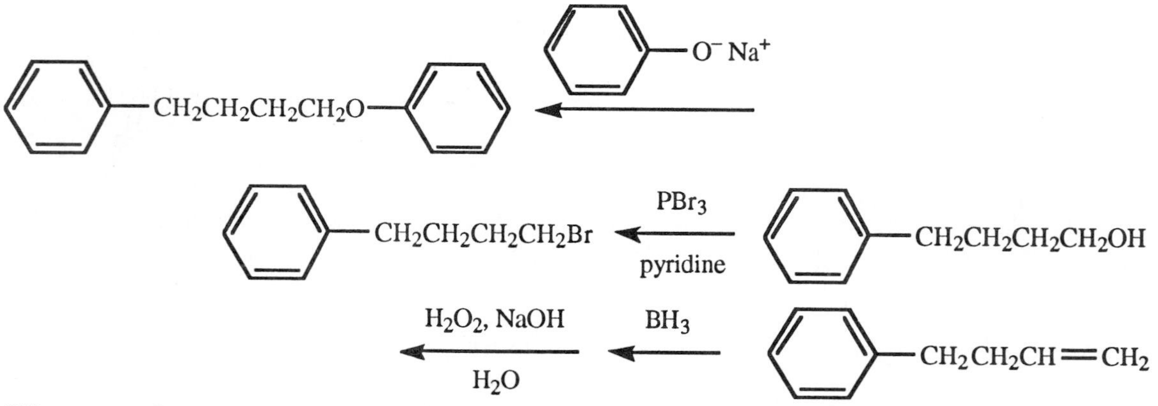

(d)

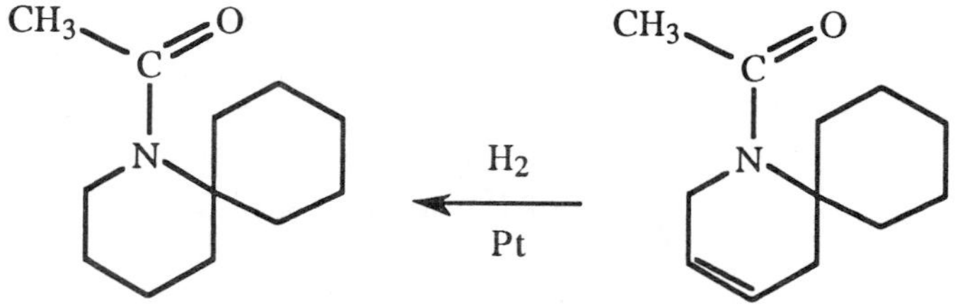

12.43 (cont)

(e)

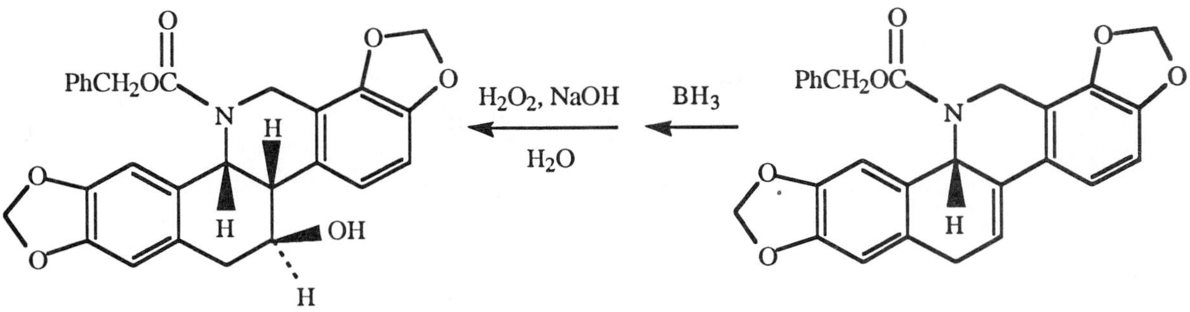

(f)

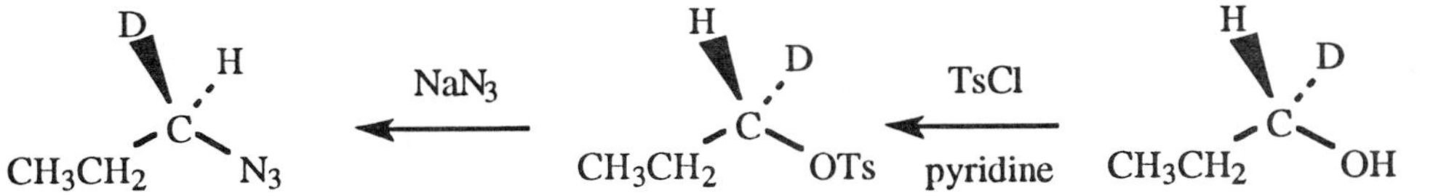

(g)

$$CH_3(CH_2)_4CH_2OCH_3 \xleftarrow{CH_3I} CH_3(CH_2)_4CH_2O^-Na^+ \xleftarrow{Na} CH_3(CH_2)_4CH_2OH$$

$$\xleftarrow[H_2O]{H_2O_2,\ NaOH} \xleftarrow{BH_3} CH_3(CH_2)_3CH{=}CH_2$$

(h)

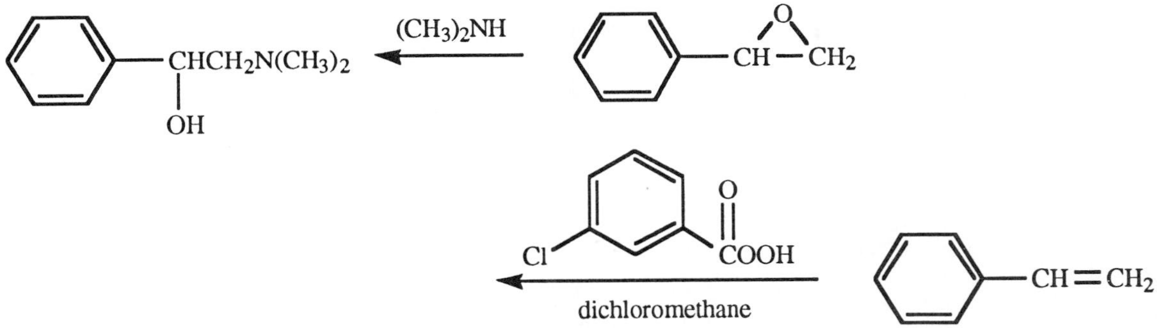

12.44

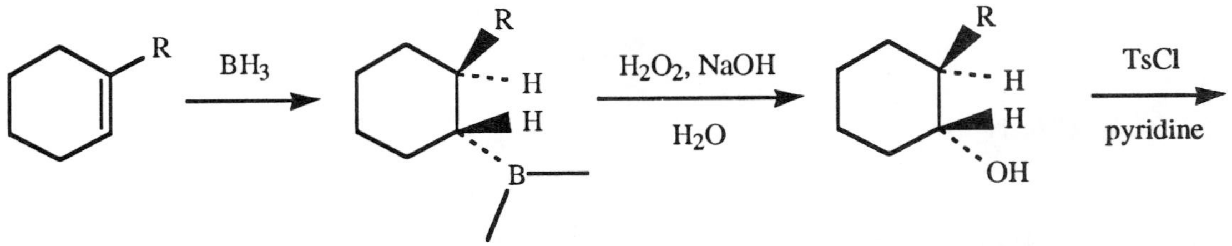

syn addition of BH_3

migration of carbon atom with retention of configuration

12.44 (cont)

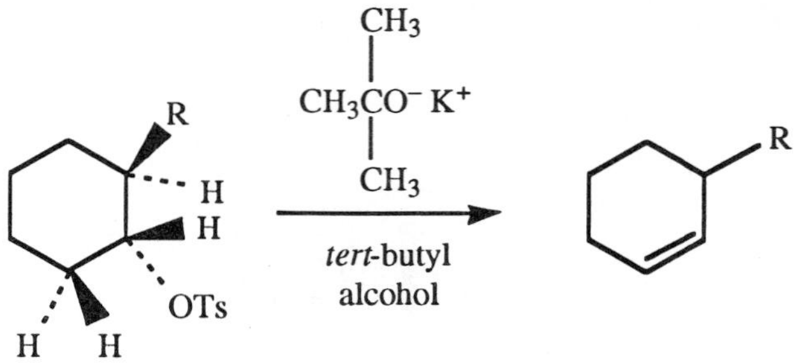

preparation of
tosylate with
retention of
configuration

elimination of the
tosylate with hydrogen
atom that is trans and
anti to it

12.45   (a)

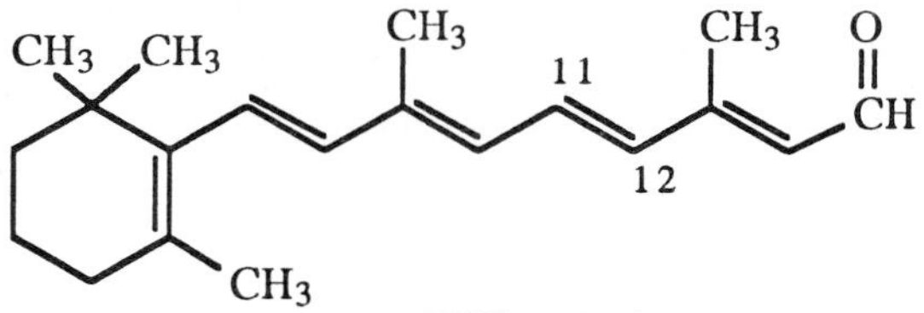

(11*E*)-retinal

(b)

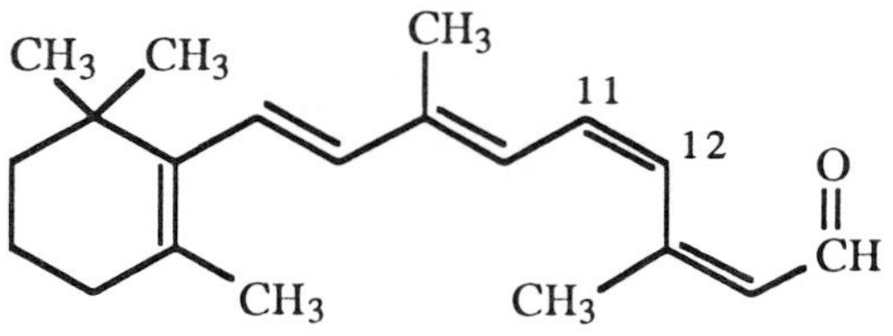

(11*Z*)-retinal

(c)

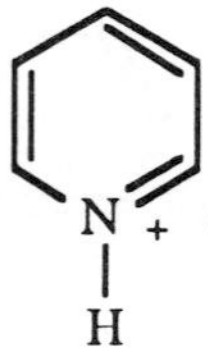

CrO_3Cl^- in dichloromethane

12.46

(a) (1)

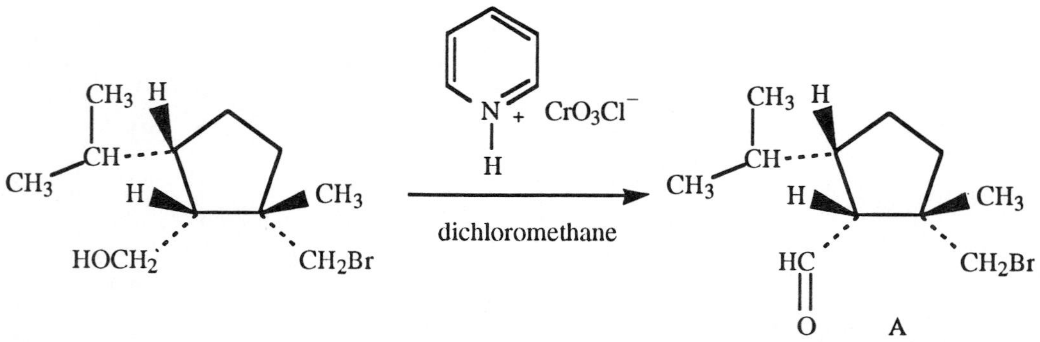

12.46 (a) (cont)

(2)

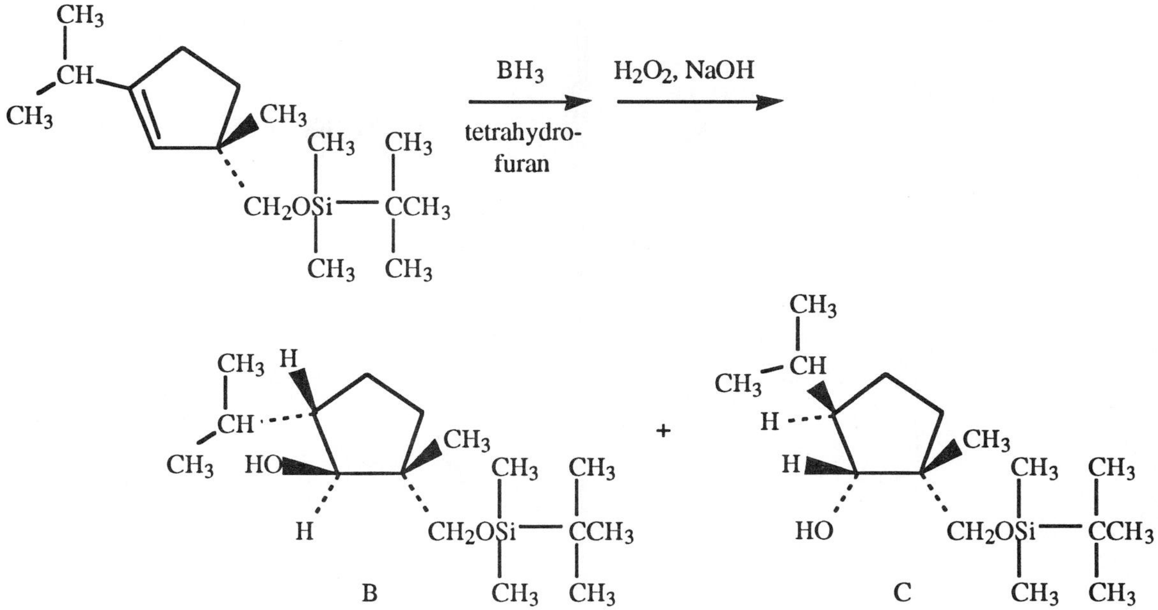

BH$_3$ H$_2$O$_2$, NaOH

tetrahydro-furan

B   +   C

(3)

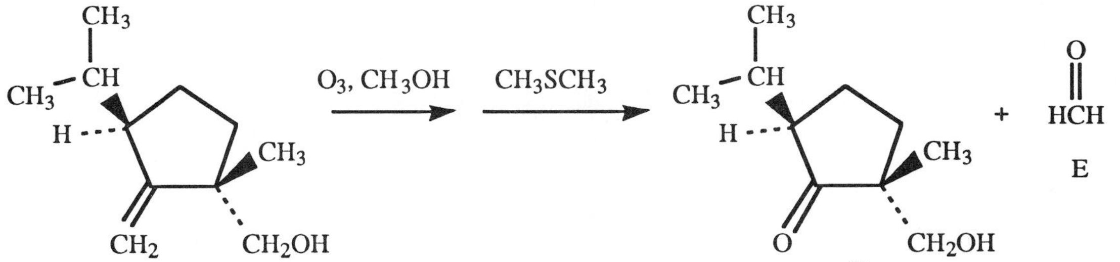

O$_3$, CH$_3$OH   CH$_3$SCH$_3$

D   +   HCH

E

(4)

$$CH_3CO^- K^+$$

methanol

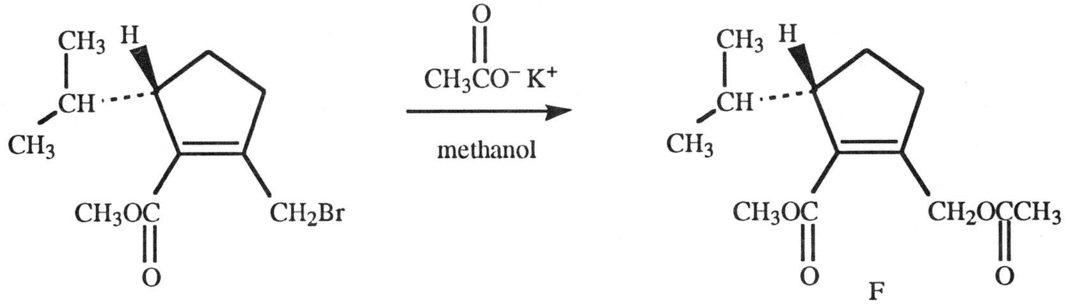

F

(b)

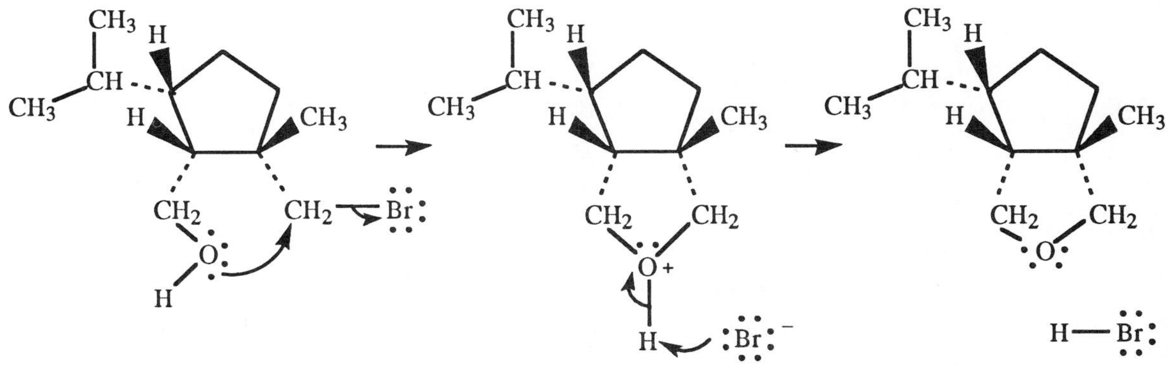

12.47

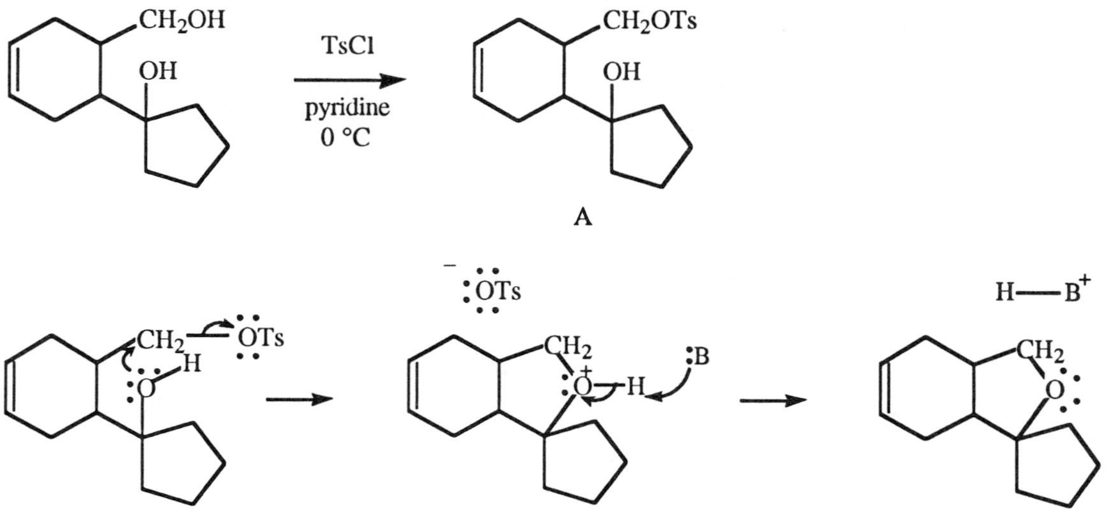

A

12.48 $CH_3C \equiv CCH_2C \equiv CCH_2CH_2CH_2OH$

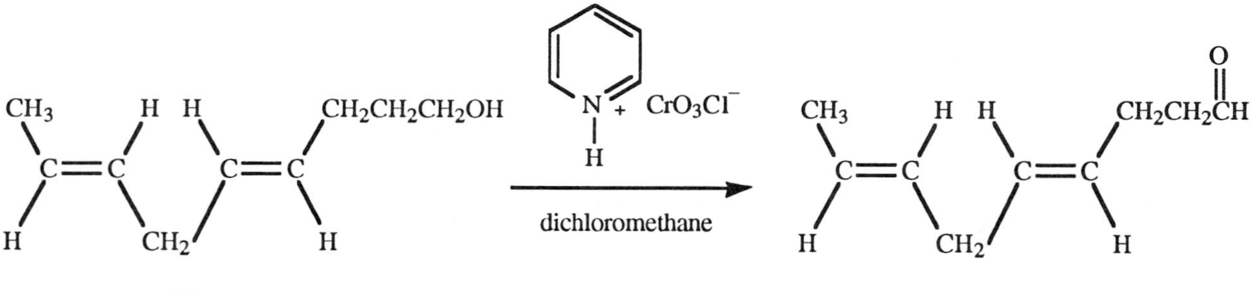

X

Y

12.49 $C_6H_5CH \overset{O}{-} CH_2$ $\xrightarrow[\text{H}_2\text{SO}_4]{\text{CH}_3\text{OH}}$ $C_6H_5\underset{\underset{OCH_3}{|}}{C}HCH_2OH$

$C_6H_5CH \overset{O}{-} CH_2$ $\xrightarrow[\text{CH}_3\text{OH}]{\text{CH}_3\text{O}^-\text{Na}^+}$ $C_6H_5\underset{\underset{OH}{|}}{C}HCH_2OCH_3$

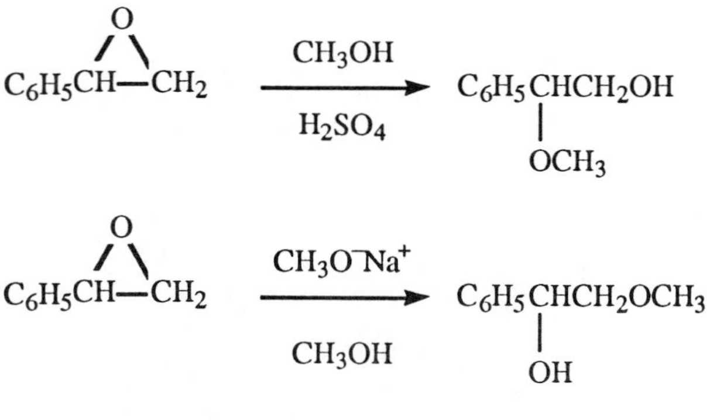

12.49 (cont)

(1)

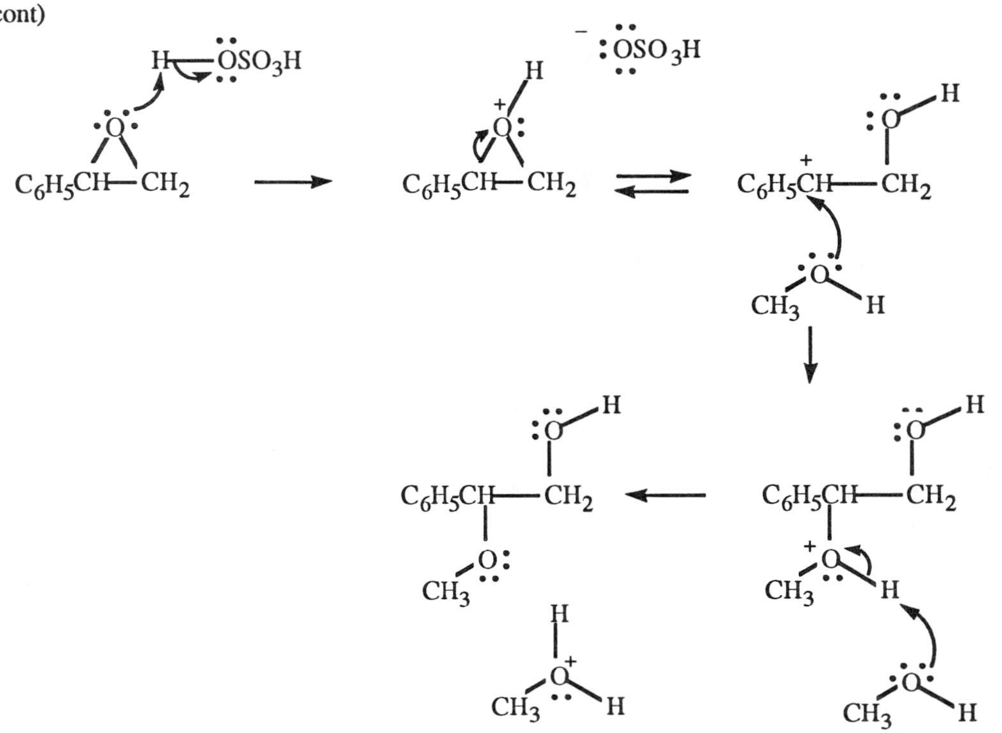

(2)

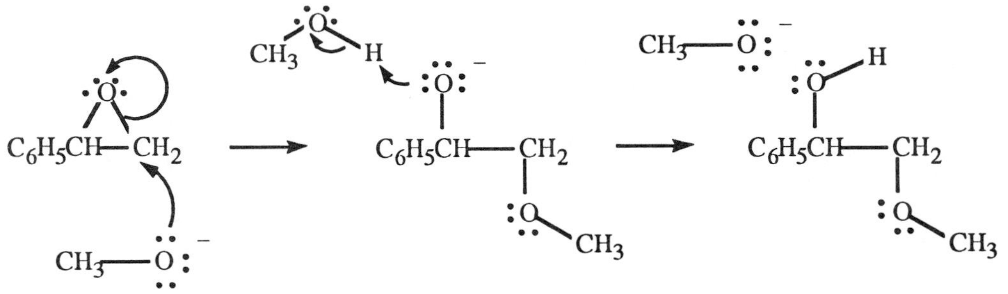

12.50

(a)

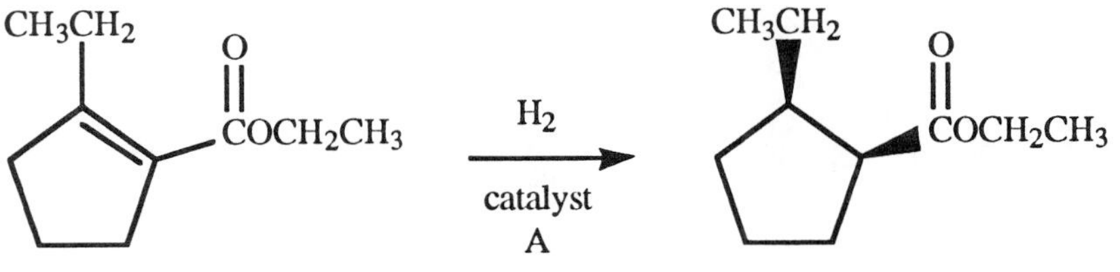

and enantiomer

(b)

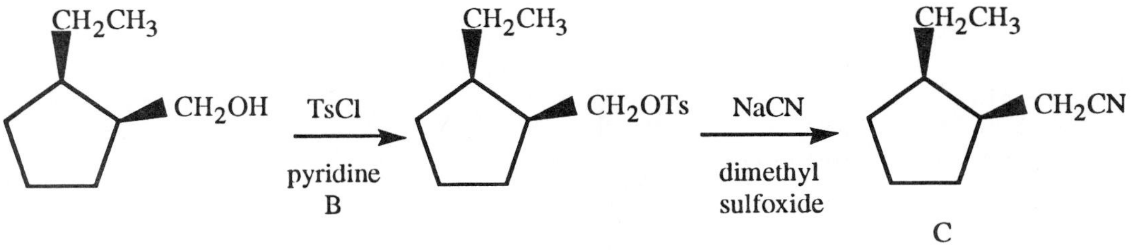

12.50 (cont)

(c)

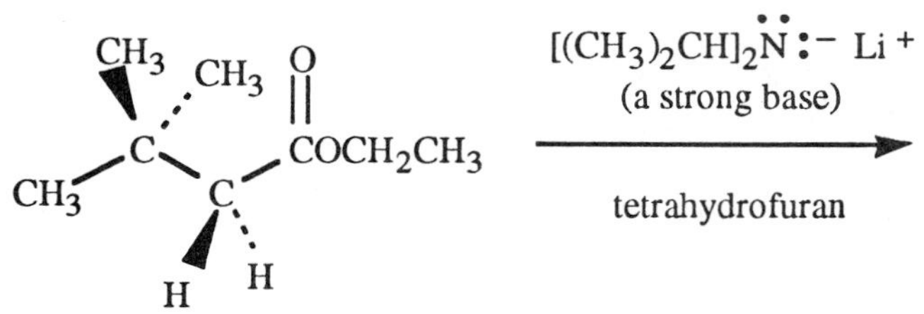

CH3
CH3
O
CH3
C
C
COCH2CH3
H
H
[(CH3)2CH]2N:− Li+
(a strong base)
tetrahydrofuran

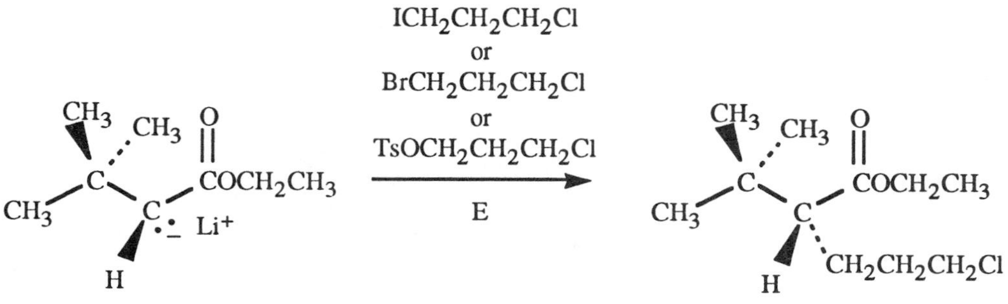

ICH2CH2CH2Cl
or
BrCH2CH2CH2Cl
or
TsOCH2CH2CH2Cl
E
Li+
COCH2CH3
CH2CH2CH2Cl

D
a good nucleophile

and enantiomer

(d)

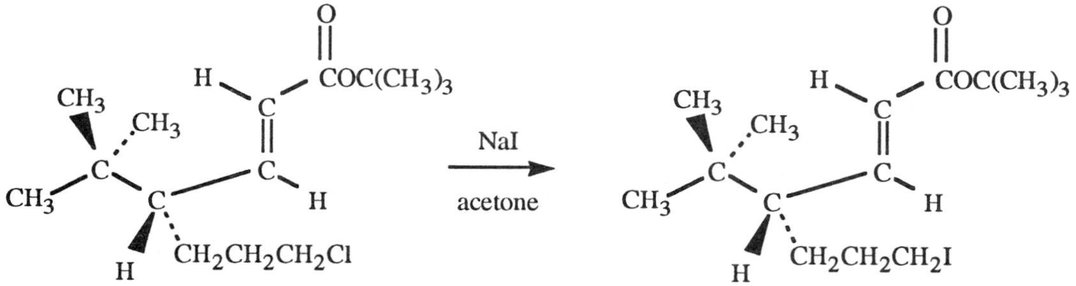

COC(CH3)3
CH2CH2CH2Cl
NaI
acetone
CH2CH2CH2I

F

12.51

(a)

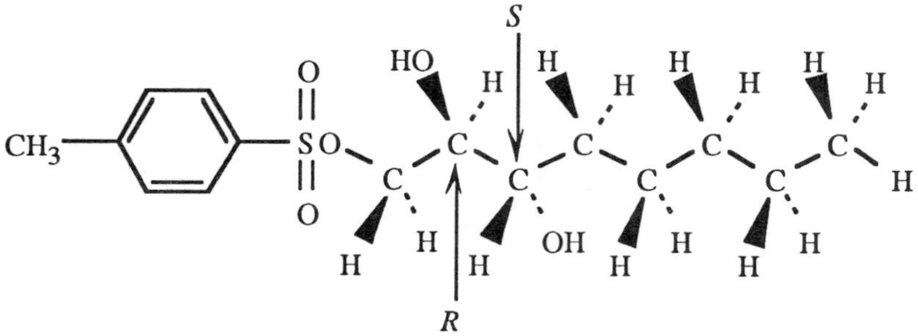

S
R
CH3
SO
HO
OH

(b)

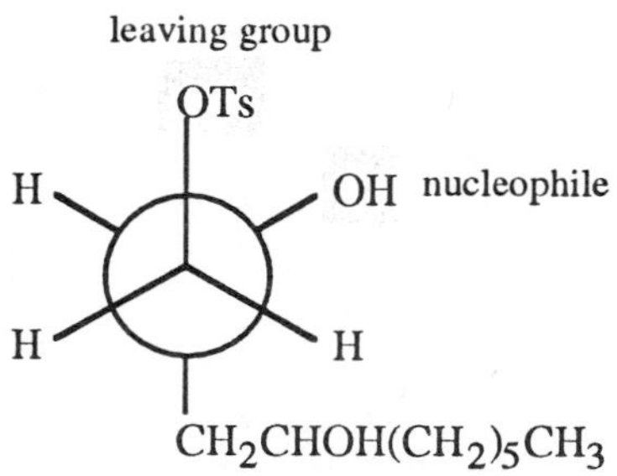

leaving group
OTs
OH nucleophile
H
H
H
CH2CHOH(CH2)5CH3

12.51 (b) (cont)

This conformation will not give oxirane A because the hydroxyl group that will serve as the nucleophile in the displacement reaction is not anti to the leaving group, the tosylate.

(c)   Potassium carbonate is a source of the base, carbonate ion, CO_3^{-2}. The pKa of the conjugate acid of CO_3^{-2} is bicarbonate ion, HCO_3^-, pK_a 10.2. The carbonate ion does not completely deprotonate the hydroxyl group, pK_a ~ 17. Because the displacement reaction is intramolecular, and, therefore, very fast, it is sufficient to create only a low concentration of alkoxide ions, which are then rapidly converted into the oxirane.

(d)

12.52 $ClCH_2CH_2CH_2CCH_3$ (with O double bond) $\xrightarrow{\text{reduction}}$

(*S*)-5-chloro-2-pentanol   (*S*)-2-methyltetrahydrofuran

12.53

(a) $HC\equiv CCH_2OH$ $\xrightarrow[\text{NH}_3 \text{ (liq)}]{2 \text{ LiNH}_2}$ $Li^+ {}^-:C\equiv CCH_2O^- Li^+$

A

$Li^+ {}^-:C\equiv CCH_2O^- Li^+$ $\xrightarrow[\text{H}_2\text{O}]{CH_3(CH_2)_3CH_2Br \quad H_3O^+}$ $CH_3(CH_2)_3CH_2C\equiv CCH_2OH$

A   2-octyn-1-ol

(b) $Li^+ {}^-:C\equiv CCH_2OCH_2(CH_2)_3CH_3$

12.53 (cont)

(c)   An acetylide anion (charge on carbon) is more nucleophilic than an alkoxide ion (charge on oxygen) and will
react faster with the electrophilic alkyl halide.

(d) $HC \equiv CCH_2O^- Li^+$

12.54

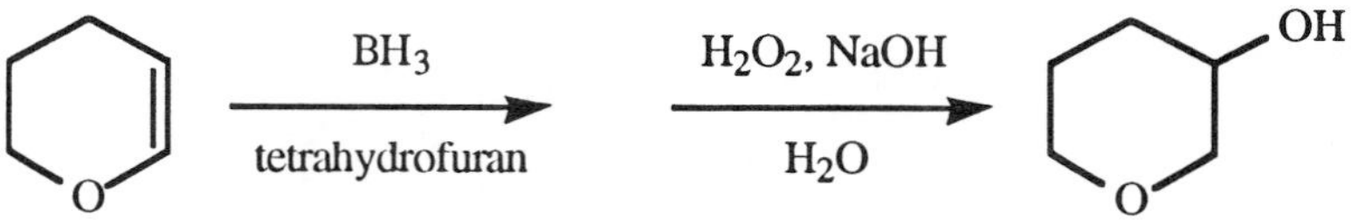

The unsymmetrical alkene can react with diborane by way of two different transition states.

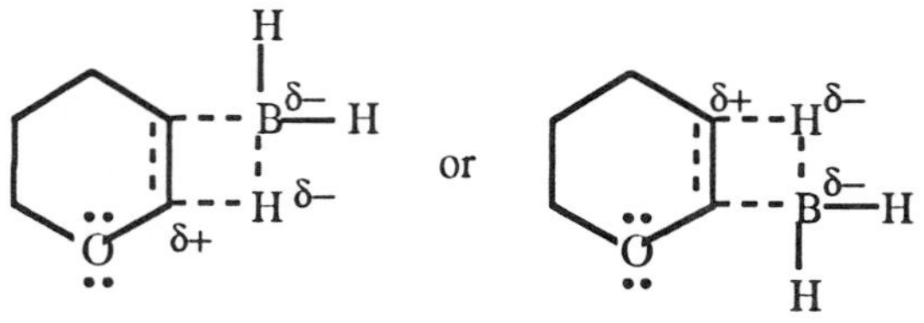

The one with the partial positive charge next to the oxygen is more stable, because the charge can be delocalized
to oxygen, shown by resonance contributors written as though an intermediate with a full positive charge forms.

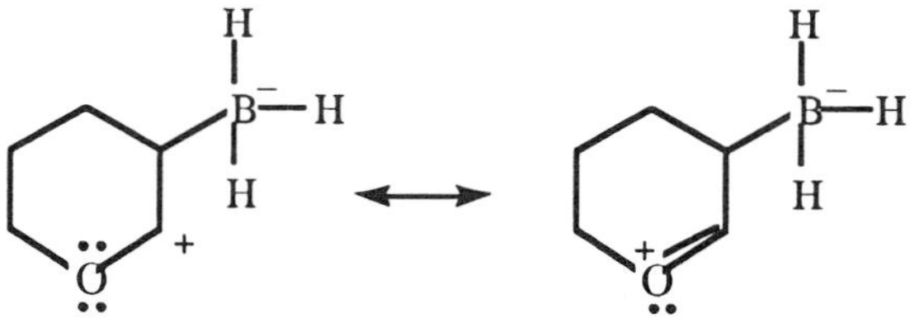

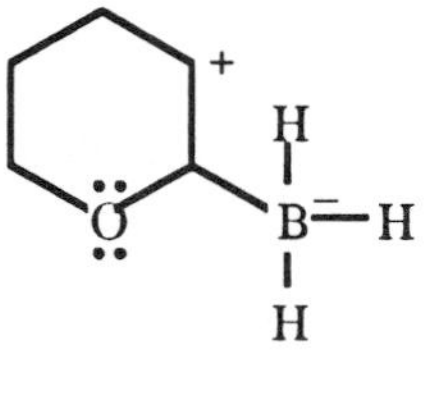

has no additional stabilization

The reaction goes by way of the transition state of lower energy corresponding to the more stable of the two
reaction intermediates.

12.55

(a)

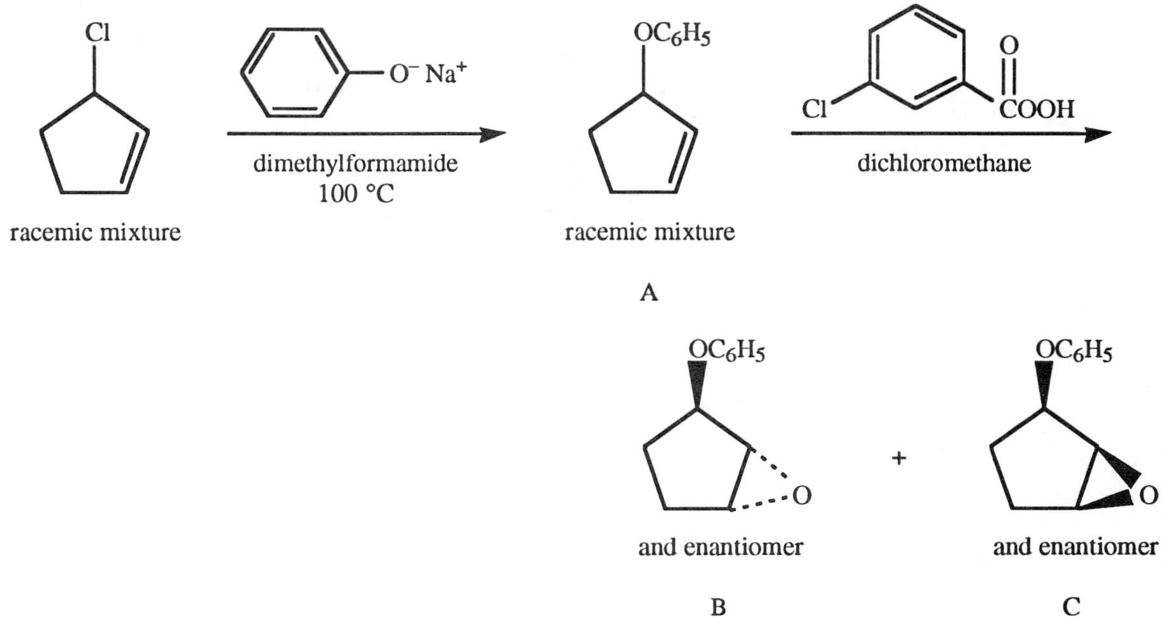

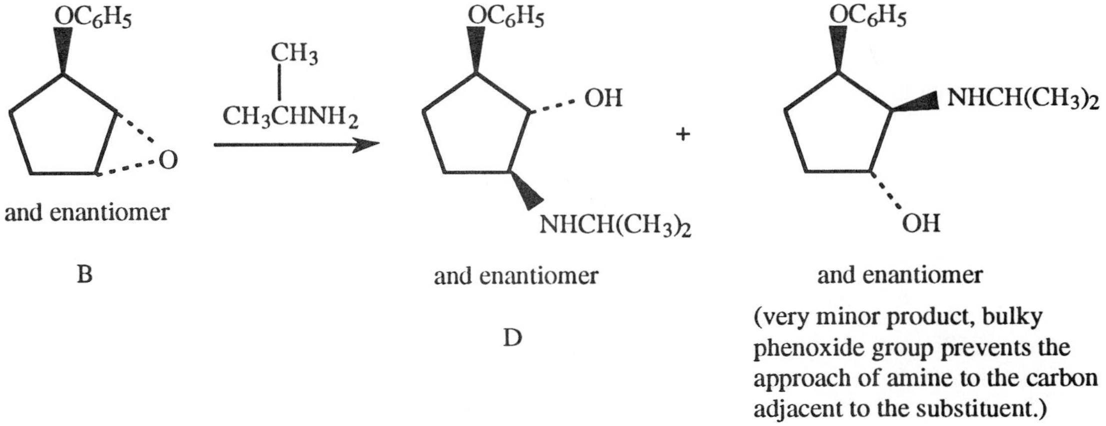

racemic mixture → (with $C_6H_5O^-Na^+$, dimethylformamide, 100 °C) → **A** racemic mixture (with 3-chlorobenzoic acid, $COOH$, dichloromethane) →

B (and enantiomer) + **C** (and enantiomer)

B (and enantiomer) → (with CH_3CHNH_2, CH_3) → **D** (and enantiomer) + (very minor product, bulky phenoxide group prevents the approach of amine to the carbon adjacent to the substituent.)

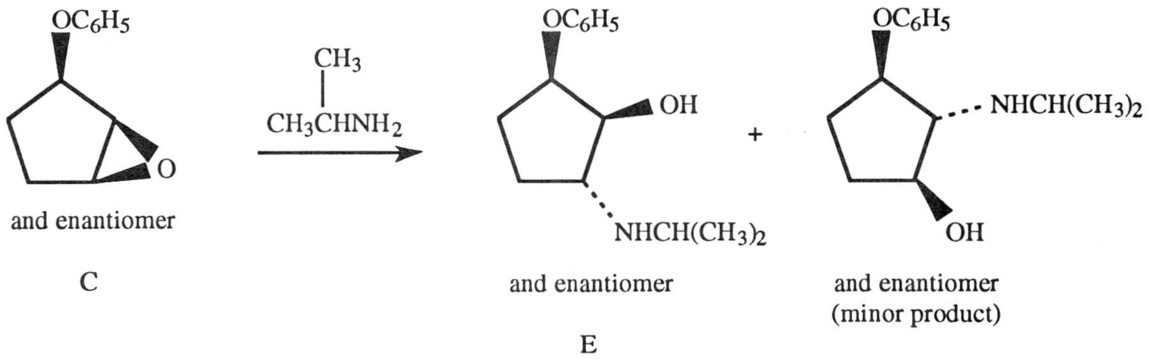

C (and enantiomer) → (with CH_3CHNH_2, CH_3) → **E** (and enantiomer) + (and enantiomer) (minor product)

(b)   D and E are diastereomers. D is a racemic mixture.

12.56

(a)

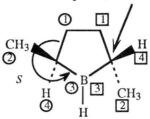

looks *R*, because we are looking at the stereocenter from the side on which the group of lowest priority comes up at us, therefore it is really *S*

(2*S*, 5*S*)-2,5-dimethyl-1-boracyclopentane

The numbers in the circles indicate priorities of groups on the left-hand stereocenter for the purpose of assigning configuration; the numbers in the squares indicate priority of groups on the right-hand stereocenter. Boron has a lower atomic number than carbon and so comes third in the order of priorities.

(b)

(*S*)-(+)-2-methyl cyclopentanone

or

(*R*)-(–)-2-methyl cyclopentanone

(c)

H_2O_2, NaOH

syn addition of H—OH to give the *R* configuration at the carbon atom bearing the methyl group

12.57

(a)   **(1)**

BH_3 H_2O_2, NaOH

A

12.57 (cont)

(2)

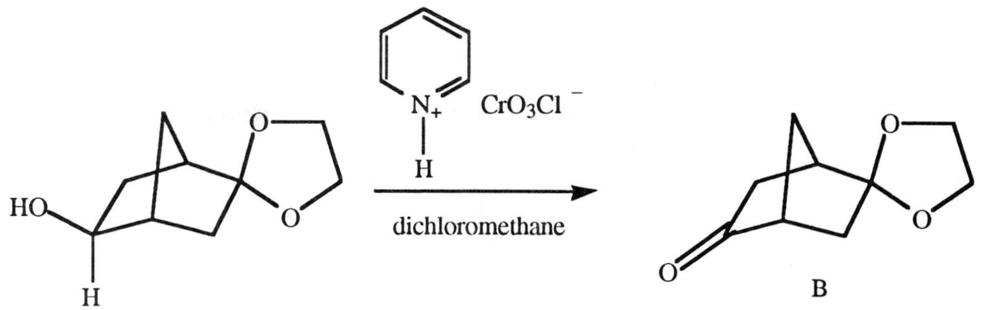

(3)

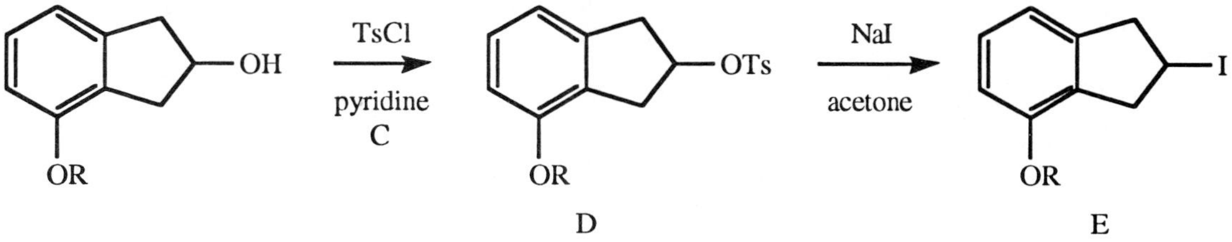

(b)   (1)

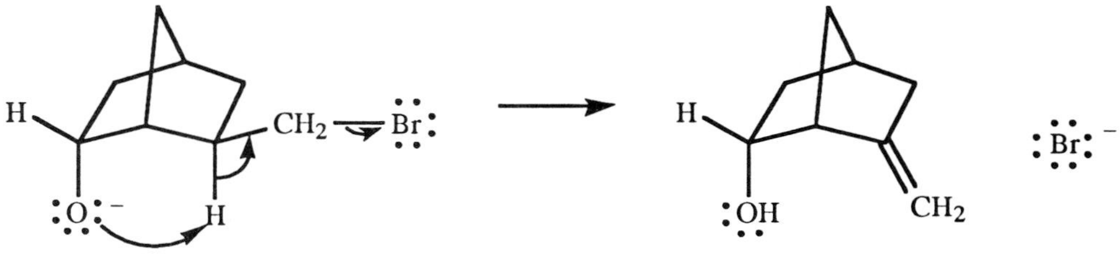

(2)

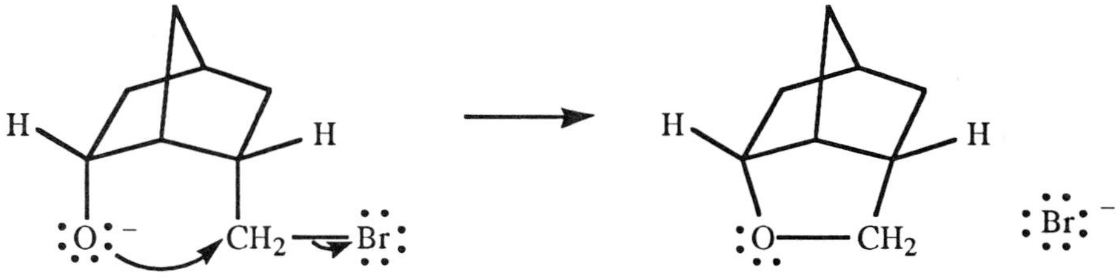

cyclic ether

12.58   The following equations can be used to calculate the free energy of activation from the heat and entropy of activation.

$$\Delta G^{\ddagger} = \Delta H^{\ddagger} - T\Delta S^{\ddagger}$$

For the formation of oxirane

$$\Delta G^{\ddagger} = (23.2 \text{ kcal/mol} - (303 \text{ K})(9.9 \times 10^{-3} \text{ kcal/mol·K})$$

$$\Delta G^{\ddagger} = 20.2 \text{ kcal/mol}$$

12.58 (cont)

For the formation of tetrahydrofuran

$\Delta G^{\ddagger} = (19.8 \text{ kcal/mol} - (303 \text{ K})(-5 \times 10^{-3} \text{ kcal/mol·K})$

$\Delta G^{\ddagger} = 21.3 \text{ kcal/mol}$

The formation of oxirane has the smaller free energy of activation and therefore the faster rate. Even though the heat of activation is smaller for the formation of tetrahydrofuran than for oxirane, the reaction giving rise to tetrahydrofuran has a negative (unfavorable) entropy of activation whereas the entropy of activation for the formation of oxirane is positive, contributing to a smaller overall free energy of activation for the reaction.

12.59   Compound A, $C_5H_{10}O$, has one unit of unsaturation and Compound B, C_5H_8O, has two units of unsaturation. From the infrared spectrum, A is an alcohol (O—H stretching frequency at 3330 cm^{-1}) and B is a ketone (C=O stretching frequency at 1750 cm^{-1} and no band at 2700 cm^{-1}). The permanganate and bromine tests are negative, therefore neither A nor B has a double bond. They must therefore each contain a ring. The carbonyl stretching frequency is outside of the range given in the table for ketones. This is because the carbonyl stretching frequency for cyclic ketones increases as the size of the ring decreases; 1750 cm^{-1} is typical for five-membered ring ketones while four-membered ring ketones absorb at 1789 cm^{-1}. Compound A must be cyclopentanol and Compound B must be cyclopentanone.

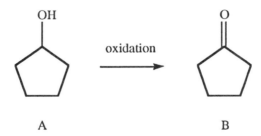

A   B

12.60   Compound C, $C_7H_{16}O$, has no units of unsaturation. The infrared spectrum shows the presence of an alcohol (3300 cm^{-1} for the O—H stretch and 1110 cm^{-1} for the C—O stretch).

Compound D, $C_7H_{14}O$, has one unit of unsaturation. The infrared spectrum shows the presence of an aldehyde (2700 cm^{-1} for the C—H stretch and 1725 cm^{-1} for the C=O stretch).

Analysis of the proton magnetic resonance spectrum of Compound C is shown below:

δ 0.9	(3H, triplet, C$\underline{\text{H}}_3$CH$_2$—)	
δ 1.3	(10H, multiplet, —(C$\underline{\text{H}}_2$)$_5$—)	(A linear chain of methylene groups that have similar chemical shifts and are at least one carbon away from electron-withdrawing atoms or π bonds will usually appear as a complex multiplet in this region.)
δ 2.2	(1H, singlet, —O$\underline{\text{H}}$)	
δ 3.6	(2H, triplet, —CH$_2$C$\underline{\text{H}}_2$—O—)	

12.60 (cont)

The carbon-13 magnetic resonance spectrum shows six alkyl carbon atoms (14.2, 23.1, 26.4, 29.7, 32.4, and 33.2 ppm) suggesting a linear chain. The seventh carbon atom (62.2) falls in the range of a carbon atom bonded to an oxygen atom. Compound C is 1-heptanol, $CH_3(CH_2)_5CH_2OH$. It must be oxidized by pyridinium chlorochromate to heptanal.

Analysis of the proton magnetic resonance spectrum of Compound D confirms that it is heptanal.

δ 0.9	(3H, triplet, $C\underline{H}_3CH_2$—)
δ 1.3	(8H, multiplet, —$(C\underline{H}_2)_4$—)
δ 2.4	(2H, multiplet, —$C\underline{H}_2C=O$)
δ 9.8	(1H, triplet, —$C\underline{H}_2\overset{\displaystyle O}{\overset{\displaystyle \|}{C}}H$)

The carbon-13 magnetic resonance spectrum confirms the presence of an aldehyde functional group (202.2 ppm).

$$CH_3(CH_2)_5CH_2OH \qquad CH_3(CH_2)_5\overset{\displaystyle O}{\overset{\displaystyle \|}{C}}H$$

$$\text{C} \qquad\qquad\qquad \text{D}$$

12.61 Compound E, $C_5H_{12}O$, has no units of unsaturation. The infrared spectrum shows the presence of an alcohol (3400 cm^{-1} for the O—H stretch). Compound E does not react with chromium trioxide so it must be a tertiary alcohol. Only one structural formula is possible for a five-carbon tertiary alcohol. Compound E must be 2-methyl-2-butanol (*tert*-amyl alcohol).

$$\overset{\displaystyle CH_3}{\underset{\displaystyle OH}{\overset{\displaystyle |}{\underset{\displaystyle |}{CH_3CCH_2CH_3}}}} \quad \xrightarrow[\substack{H_2SO_4 \\ H_2O}]{CrO_3} \quad \text{solution remains orange}$$

$$\text{E}$$

Analysis of the magnetic resonance spectra of Compound E confirms the structure. The proton magnetic resonances are given below:

δ 0.9	(multiplet, ($C\underline{H}_3CH_2$—)
δ 1.2	(singlet, $(C\underline{H}_3)_2C$—)
δ 1.5	(multiplet, —$C\underline{H}_2CH_3$)

There are no bands for a hydrogen atom on a carbon bonded to an oxygen atom (~4 ppm), which also confirms that Compound E is a tertiary alcohol. The proton on the oxygen atom is hidden in one of the multiplets.

The carbon-13 magnetic resonance spectrum shows only four types of carbon atoms. Three of the types are alkyl carbon atoms (8.8, 28.9, and 36.8 ppm) and one is a tertiary carbon atom bonded to an oxygen atom (70.6 ppm).

Compound F, $C_5H_{12}O$, also has no units of unsaturation. The infrared spectrum again shows the presence of an alcohol (3400 cm^{-1} for the O—H stretch). A positive reaction with chromium trioxide points to a secondary or primary alcohol. The possible structures are shown on the next page.

12.61 (cont)

Primary alcohols:

$$CH_3CH_2CH_2CH_2CH_2OH \qquad \overset{\overset{\displaystyle CH_3}{|}}{CH_3CHCH_2CH_2OH} \qquad \overset{\overset{\displaystyle CH_3}{|}}{CH_3CH_2CHCH_2OH} \qquad \overset{\overset{\displaystyle CH_3}{|}}{\underset{\underset{\displaystyle CH_3}{|}}{CH_3CCH_2OH}}$$

Secondary alcohols:

$$\underset{\underset{\displaystyle OH}{|}}{CH_3CH_2CH_2CHCH_3} \qquad \underset{\underset{\displaystyle OH}{|}}{CH_3CH_2CHCH_2CH_3} \qquad \overset{\overset{\displaystyle CH_3}{|}}{\underset{\underset{\displaystyle OH}{|}}{CH_3CHCHCH_3}}$$

The carbon-13 magnetic resonance spectrum shows five bands. The three alcohols that have five different types of carbon atoms are 1-pentanol, 2-pentanol, and 2-methyl-1-butanol. Of these, only 2-pentanol will have the observed doublet for the hydrogen atoms on carbon-1 and a multiplet for the hydrogen atom on the carbon atom bearing the hydroxyl group.

Analysis of the rest of the magnetic resonance spectra of Compound F confirms the structure. The proton magnetic resonance is given below:

δ 0.9-1.4   (complex multiplet, C$\underline{H}_3$C$\underline{H}_2$C$\underline{H}_2$—)

δ 1.2   (doublet, C$\underline{H}_3$CH—)

δ 3.7   (multiplet, —CH$_2$C$\underline{H}$CH$_3$)

$$\overset{\displaystyle |}{O—}$$

$$\underset{\underset{\displaystyle OH}{|}}{CH_3CH_2CH_2CHCH_3} \quad \xrightarrow[\substack{H_2SO_4 \\ H_2O}]{CrO_3} \quad CH_3CH_2CH_2\overset{\overset{\displaystyle O}{\|}}{C}CH_3$$

$$F$$

Note that the hydrogen atom on the hydroxyl group is not seen as a separate band in the spectrum of either Compound E or Compound F.

13

Aldehydes and Ketones. Addition Reactions at Electrophilic Carbon Atoms

Workbook Exercises

The reactions in the next set of chapters combine the mechanistic steps of substitution, addition, and elimination. If you feel unsure about your understanding of the general use of these terms, please take some time to review.

Like carbon-carbon π bonds, carbon-oxygen and carbon-nitrogen π bonds undergo addition reactions. Notice the similar changes in connectivity associated with the addition of methanol, under acidic conditions, to a $\diagdown C = C \diagup$, a $\diagdown C = N \diagup$, and a $\diagdown C = O$ functional group.

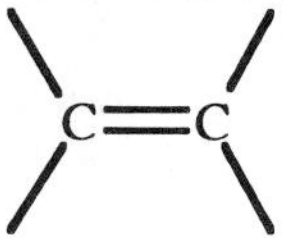

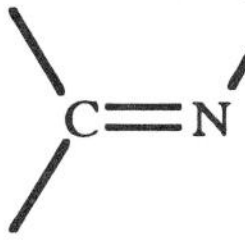

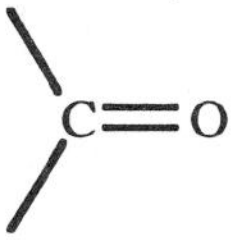

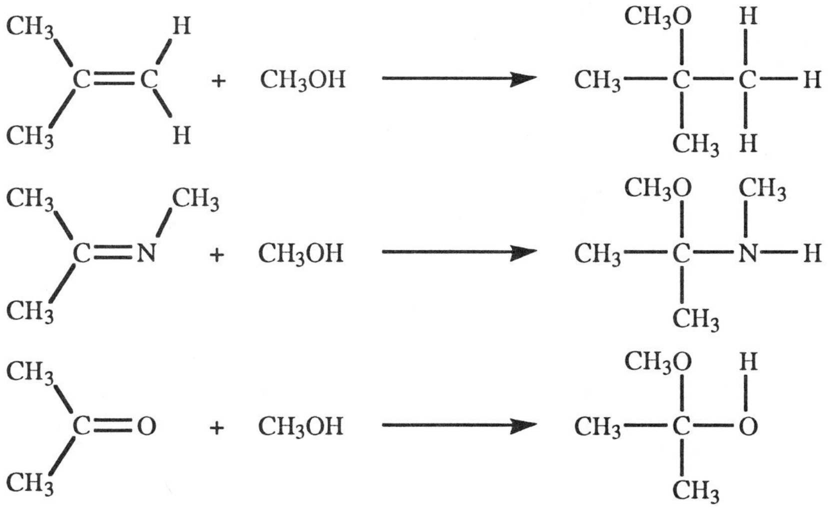

The orientation (regioselectivity) of the addition reaction to π bonds where one of the atoms is nitrogen or oxygen is highly predictable in the direction shown above. The reaction is also reversible.

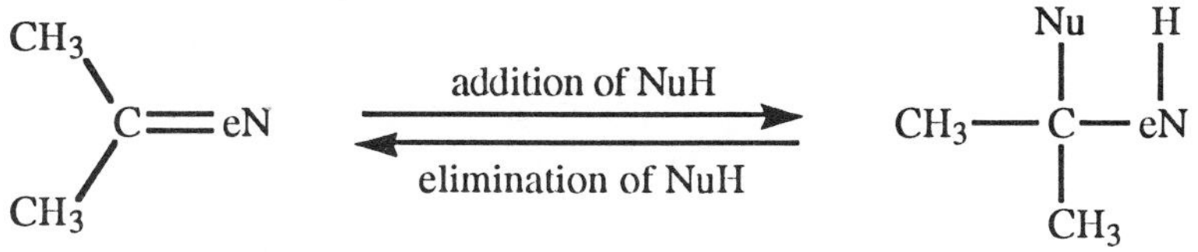

Workbook Exercises (cont)

The simple addition reaction products shown on the previous page are generally not observed when the nucleophilic group is a nitrogen (RNH_2), an alcohol (ROH), or a thiol (RSH). Instead, the observed product, in many cases, can be understood in terms of an overall loss of water (H_2O) between the reactants:

Connectivity changes that show the overall stoichiometry:

For nucleophiles involving RNH_2:

For nucleophiles involving ROH or RSH:

The ways of visualizing the reactions shown above are useful for quickly predicting structural relationships.

EXERCISE I.   Predict the product of the reaction, where water is also formed, when each of the compounds shown below is combined in separate reactions with (1) ethanol (CH_3CH_2OH), (2) methylamine (CH_3NH_2), (3) hydrazine (NH_2NH_2), and (4) ethanethiol (CH_3CH_2SH).

(a) 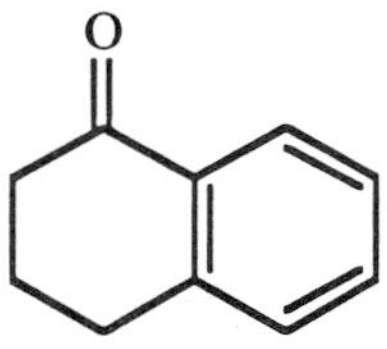   (b) 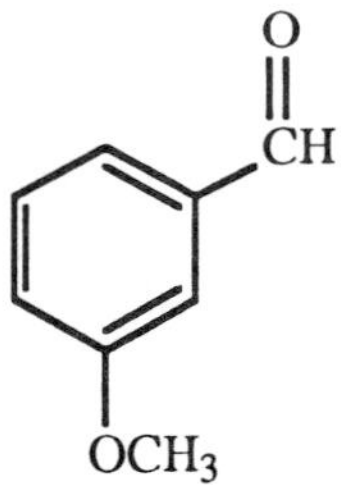   (c)  $CH_3CH_2CCH_2CH_3$

EXERCISE II.   The reactions shown above are reversible when an excess of water is present. What are the structures of the carbonyl compounds (the compounds with a carbon-oxygen π bond) and the nucleophile(s) that result from re-introducing water into the compounds shown below?

(a) 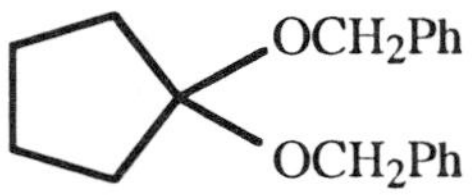   (b) 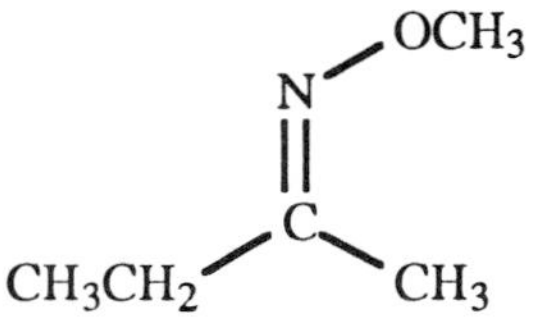   (c)

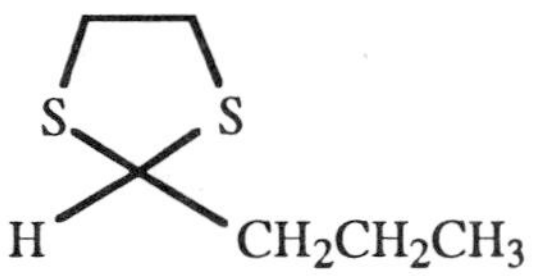

(d) 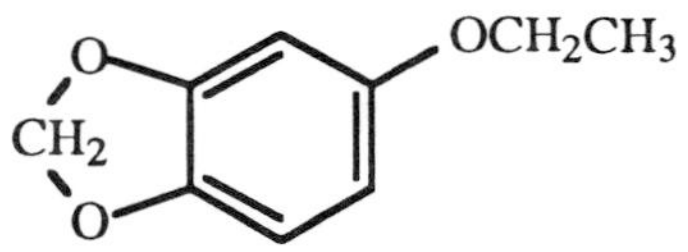   (e)

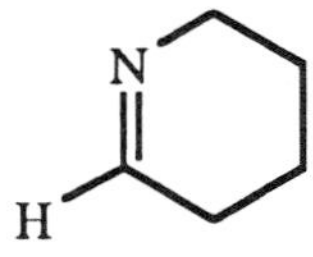

Workbook Exercises (cont)

The transformations shown above may be seen in detail as the combination of two fundamental mechanistic steps. In the case of the RNH_2 nucleophiles, the formation of the carbon-nitrogen double bond is understood to be an addition to the carbon-oxygen π bond followed by the elimination of water. In the case of the ROH and RSH nucleophiles, the first step is also an addition to the carbon-oxygen π bond. This is followed by a substitution of the hydroxyl group, which, under acidic conditions, leaves as a molecule of water. The reverse reactions also have addition-elimination or substitution-elimination steps.

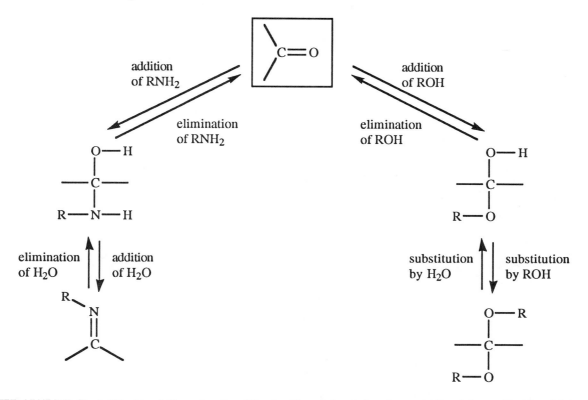

EXERCISE III. Redo Workbook Exercises I and II using the mechanistic scheme outlined above. Problem I (c), using ethanethiol, is solved for you here as an example.

SOLUTION

13.1

2-Butanol is an alcohol and can serve as both a hydrogen bond donor and as an acceptor. Strong hydrogen bonding among molecules of 2-butanol accounts for its high boiling point.

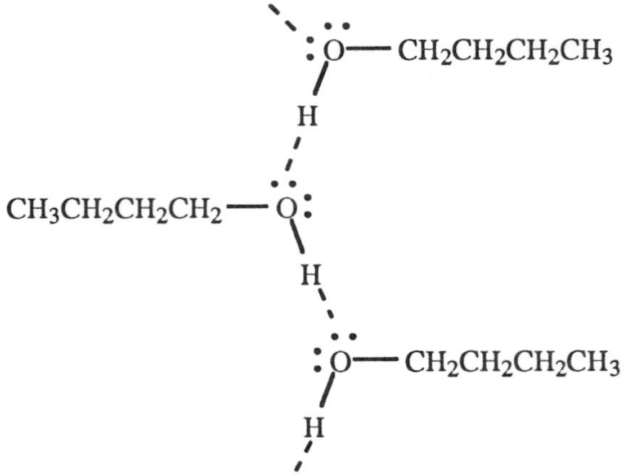

There is no hydrogen bonding among molecules of 2-butanone and of diethyl ether. They can serve only as hydrogen bond acceptors, not donors. Therefore, the intermolecular forces acting on the molecules of 2-butanone and of diethyl ether are dipole-dipole interactions. The strong polarity of the carbonyl group results in stronger dipole-dipole interactions between molecules of 2-butanone than for molecules of diethyl ether.

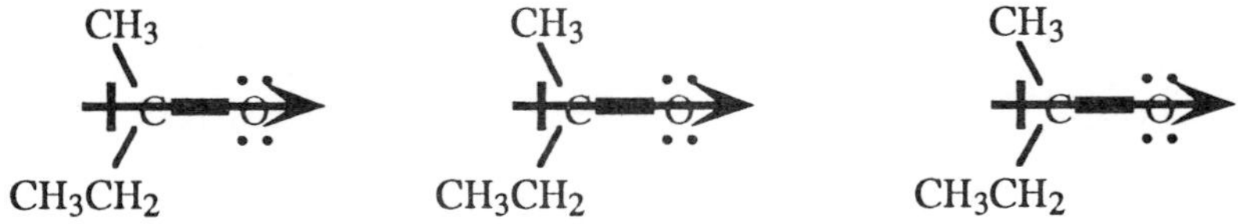

strong dipole-dipole interactions,
high boiling point

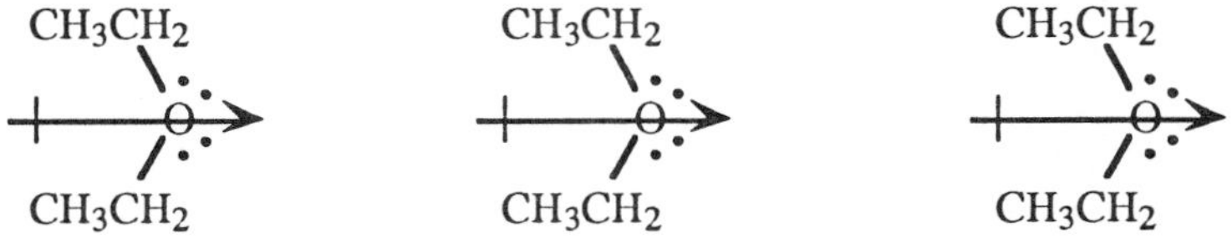

weak dipole-dipole interactions,
low boiling point

13.2   (a)

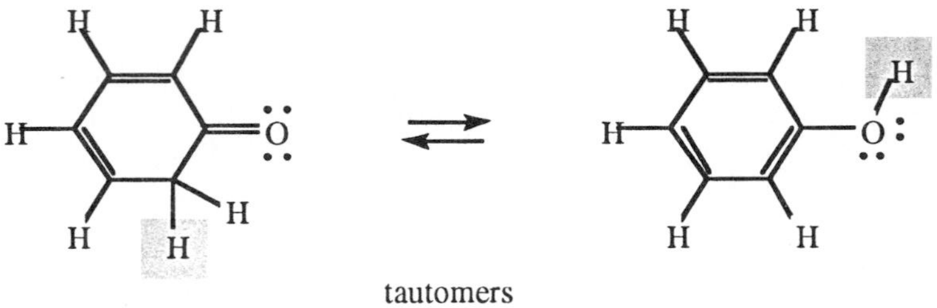

tautomers

(b)

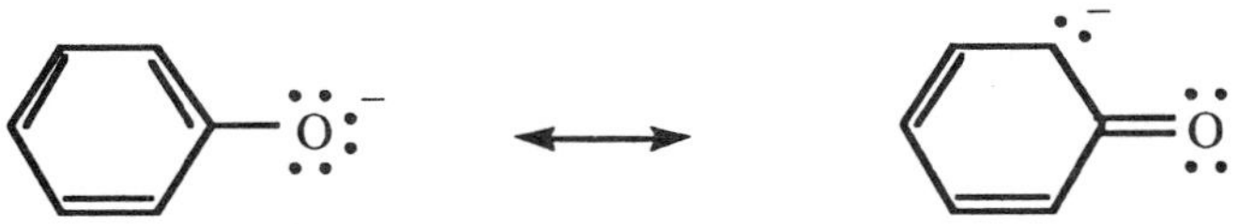

resonance contributors

13.2 (cont)

(c)

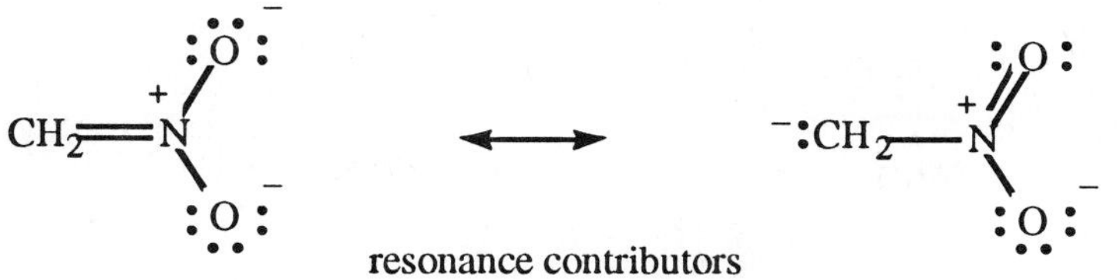

resonance contributors

(d)

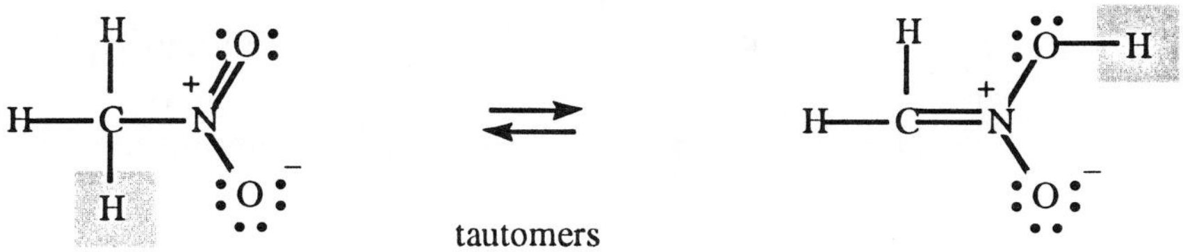

tautomers

13.3

(a)

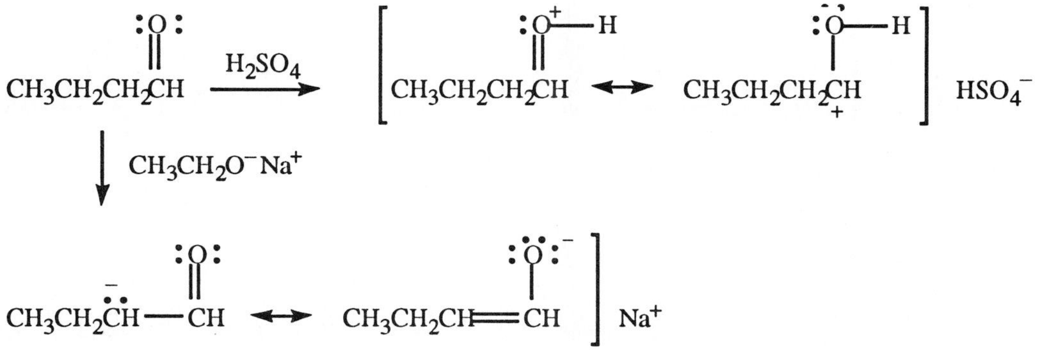

(b)

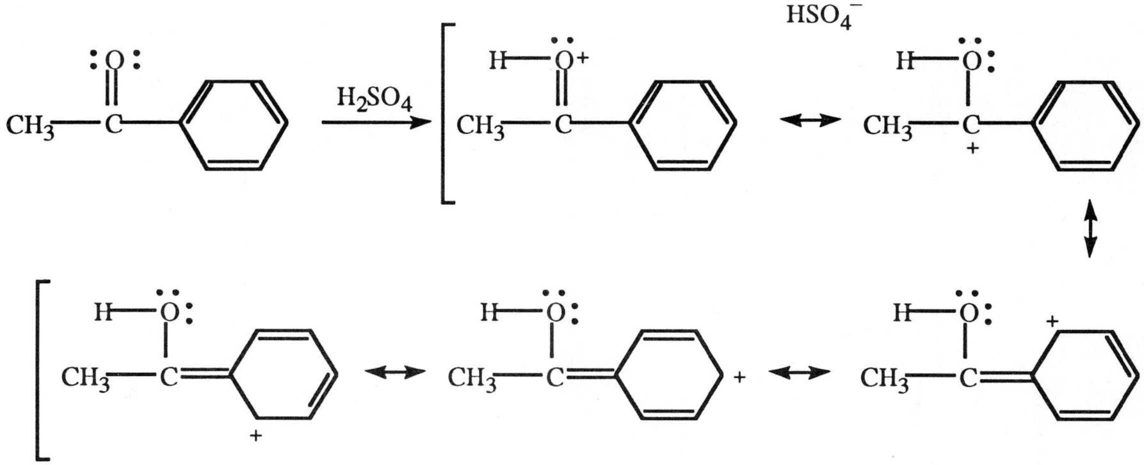

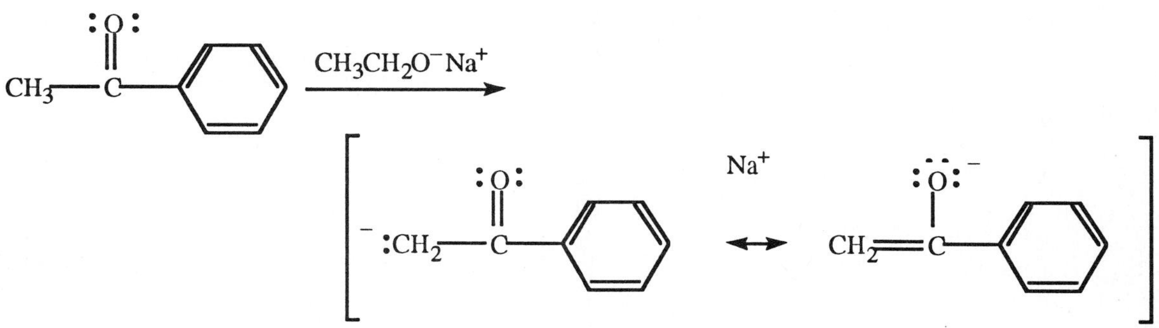

13.3 (cont)

(c)

$$
\begin{array}{c}
\underset{\underset{|}{C}H_3}{} \quad \overset{:\ddot{O}:}{\underset{\|}{}} \\
CH_3C\!\!=\!\!CHCH_2CH_2CCH_3 \xrightarrow{H_2SO_4}
\end{array}
$$

$$
\left[
\begin{array}{cc}
\underset{\overset{|}{C}H_3}{} & \overset{+}{\underset{\|}{:\ddot{O}\!-\!H}} \\
CH_3C\!\!=\!\!CHCH_2CH_2CCH_3 & \longleftrightarrow & CH_3C\!\!=\!\!CHCH_2CH_2\overset{+}{C}CH_3
\end{array}
\right] HSO_4^-
$$

nonbonding electrons become protonated more readily than π electrons do

$$
CH_3C\!\!=\!\!CHCH_2CH_2CCH_3 \xrightarrow{CH_3CH_2O^- Na^+}
$$

$$
\left[
\begin{array}{cc}
CH_3C\!\!=\!\!CHCH_2CH_2\overset{:O:}{\underset{}{C}}\!-\!\ddot{C}H_2 & \longleftrightarrow & CH_3C\!\!=\!\!CHCH_2CH_2C\!\!=\!\!CH_2
\end{array}
\right] Na^+
$$

or

$$
\left[
\begin{array}{cc}
CH_3C\!\!=\!\!CHCH_2\overset{..}{C}H\!-\!CCH_3 & \longleftrightarrow & CH_3C\!\!=\!\!CHCH_2CH\!\!=\!\!CCH_3
\end{array}
\right] Na^+
$$

(d)

$$
\xrightarrow{H_2SO_4}
\left[
\quad \longleftrightarrow \quad
\right] HSO_4^-
$$

$$
\downarrow CH_3CH_2O^- Na^+
$$

$$
\left[
\quad \longleftrightarrow \quad
\right] Na^+
$$

13.4

(a)   3-hydroxy-2-
       methylpentanal

(b)   2-pentanone

(c)   4-chlorobenzaldehyde

(d)   4-heptyn-3-one
       propen-1-one

(e)   (*E*)-1,3-diphenyl-2-
       nonen-2-one

(f)   (*Z*)-6-methyl-5-

(g)   2,2-dibromocyclohexanone

(h)   trichloroethanal
       (trichloroacetaldehyde)

(i)   3,3-dimethylcyclohexane-
       carbaldehyde

(j)   2,4-pentanedione

(k)   2-ethyl-2-methylcyclo-
       butanone

(l)   *cis*-2-propylcyclopentane-
       carbaldehyde

13.5

(a)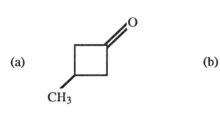

(b)   (structure)

(c)   (structure)

(d)   (structure)

(e) O_2N—(ring)—$\overset{O}{\overset{\|}{C}}H$

(f) $CH_3CH_2\overset{O}{\overset{\|}{C}}CH_2CH_2OH$

(g) $HC\overset{O}{\overset{\|}{C}}CH_2CH_2CH_2CH_2\overset{O}{\overset{\|}{C}}H$

(h)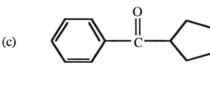

(i)   O=(ring)=O

- -

Concept Map 13.1   (see p. 380)

- -

13.6

(a) $HC\equiv CCH_2CH_2CH=CHCH_2OH$ $\xrightarrow{\overset{O}{\overset{\|}{CH_3SCH_3}} \ \overset{O}{\overset{\|}{ClC}}—\overset{O}{\overset{\|}{CCl}}} \xrightarrow{(CH_3CH_2)_3N}$

$HC\equiv CCH_2CH_2CH=CHC\overset{O}{\overset{\|}{\ }}H$

A

13.6 (cont)

(b)

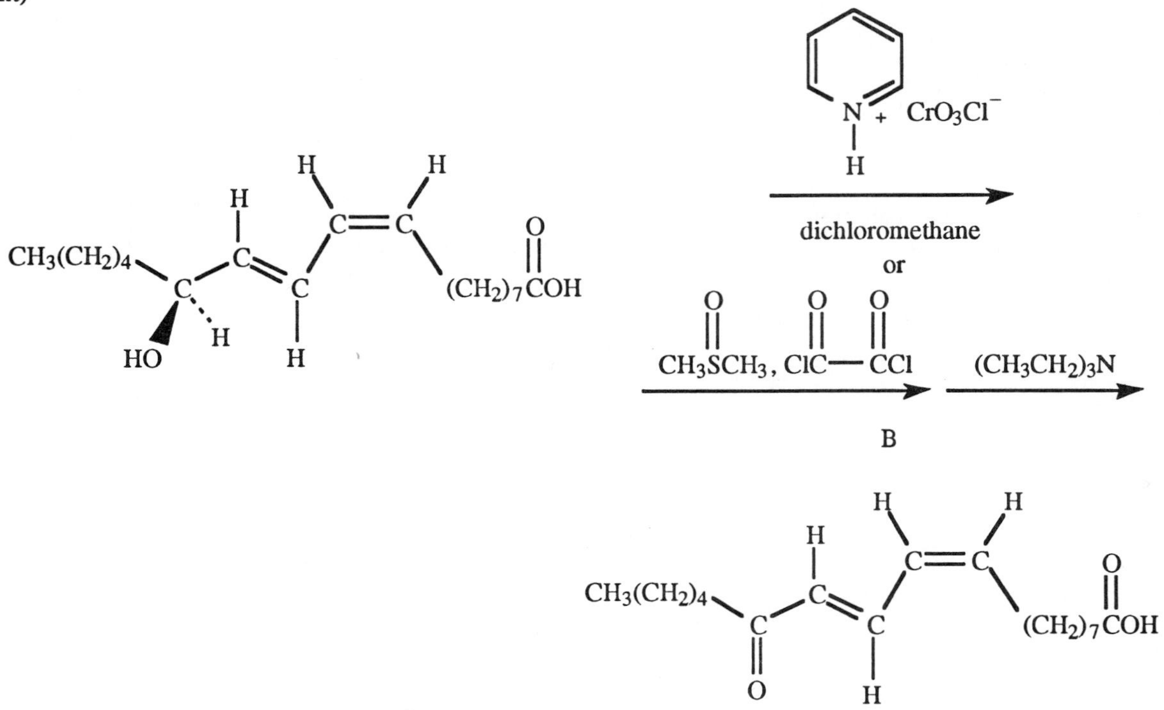

B

(c)

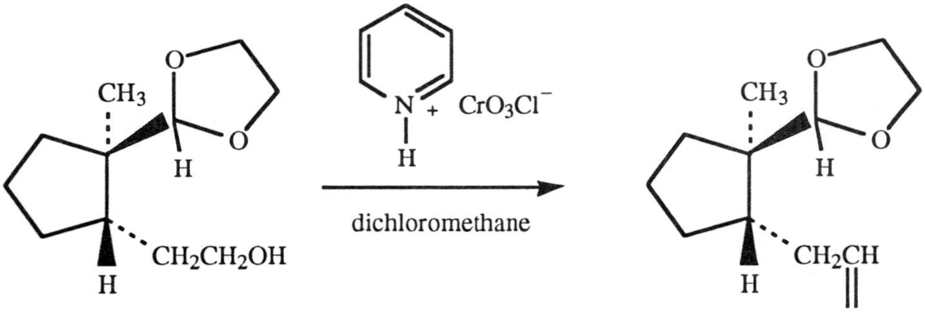

C

- -

Concept Map 13.2   (see p. 381)

- -

13.7

(a)

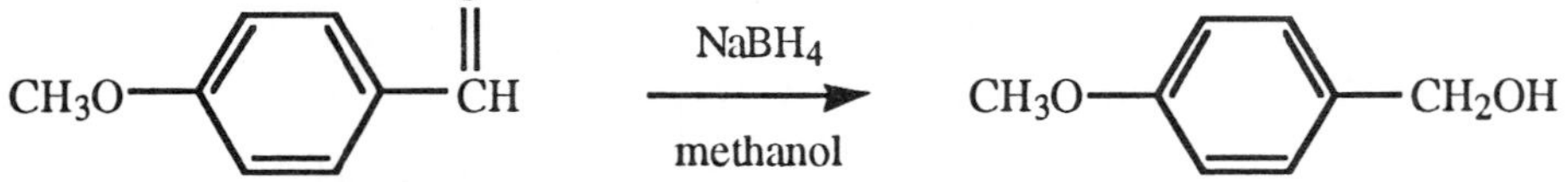

(b)

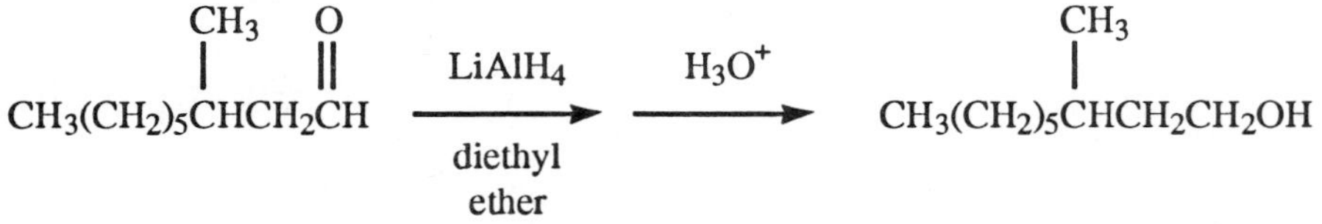

13.7 (cont)

(c)

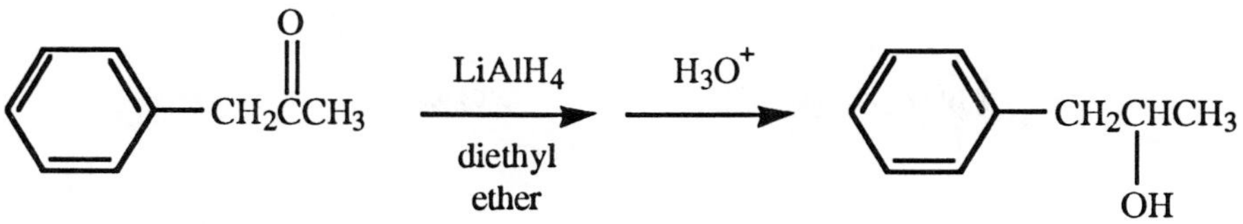

(d)

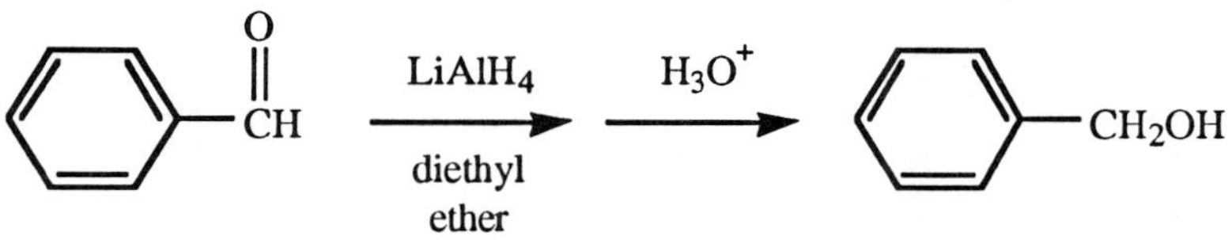

(e)

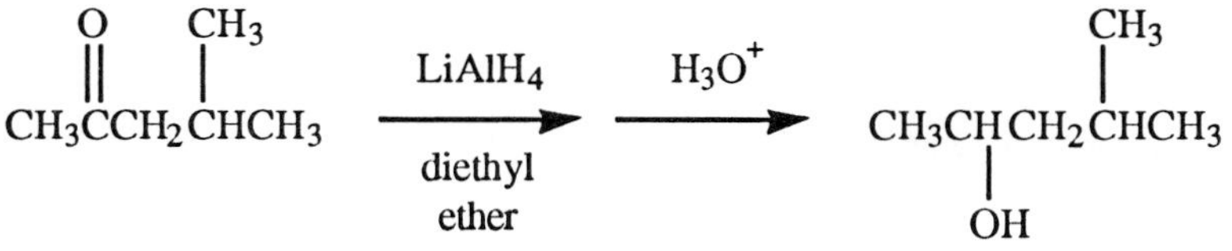

(f)

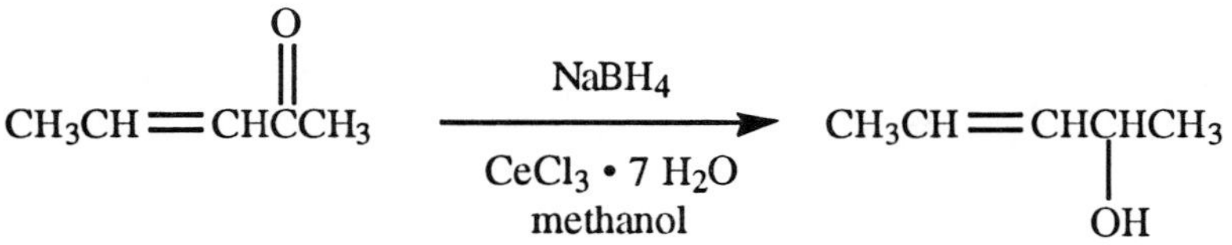

(g)

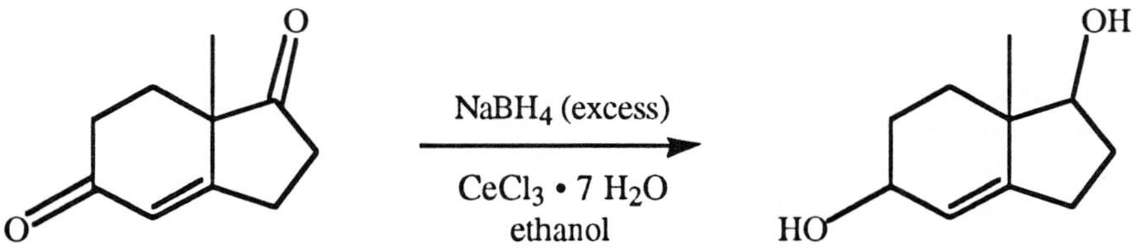

(h)

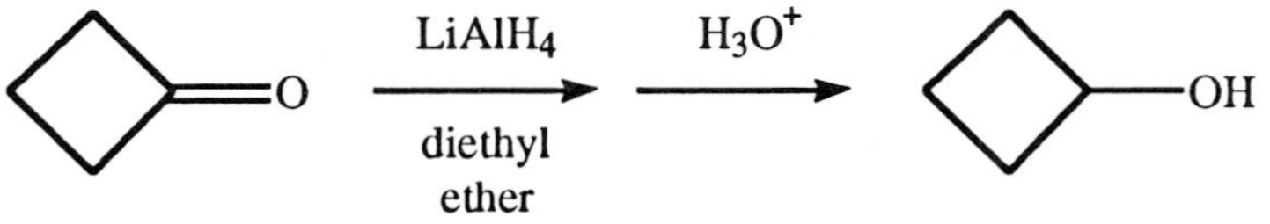

13.8

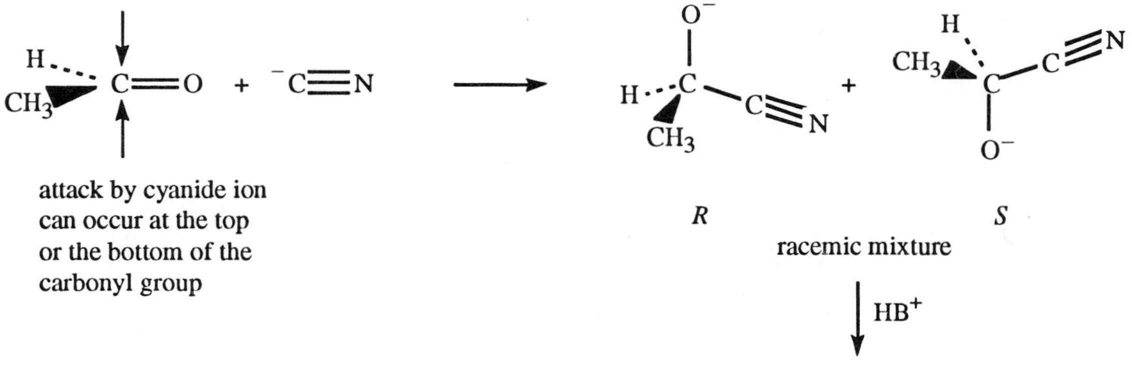

attack by cyanide ion
can occur at the top
or the bottom of the
carbonyl group

13.8 (cont)

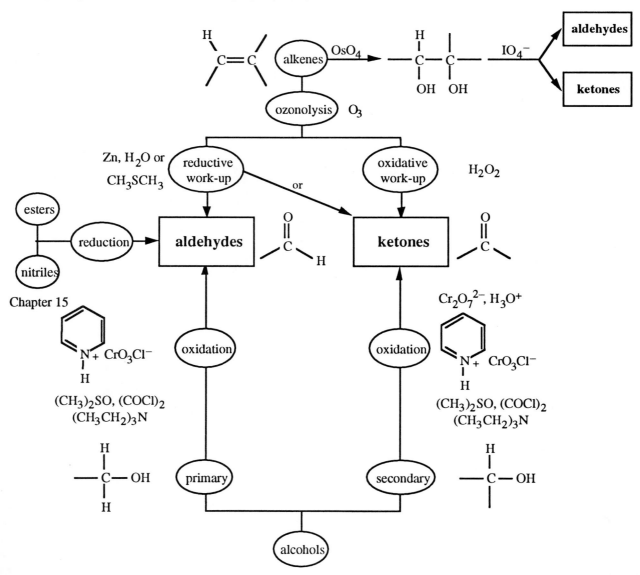

(*R*)-lactic acid   (*S*)-lactic acid   *R*   *S*

also racemic mixture   racemic mixture

Note that the hydrolysis of each nitrile proceeds with retention
of configuration, but hydrolysis of the racemic mixture of
nitriles gives rise to a racemic mixture of hydroxyacids.

– –

Concept Map 13.1   Some ways to prepare aldehydes and ketones.

Concept Map 13.2   The relationship between carbonyl compounds, alcohols, and alkyl halides.

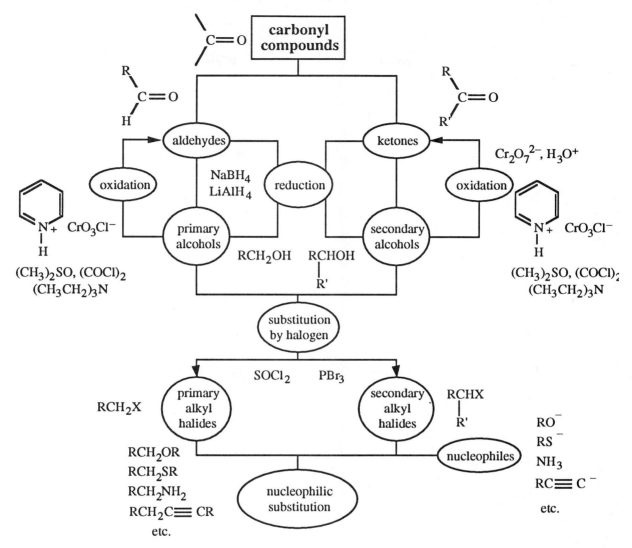

13.9

(a) $CH_3CH_2CH_2CH_2Br$ $\xrightarrow[\text{tetrahydro-furan}]{\text{Li}}$ $CH_3CH_2CH_2CH_2Li$ + Li^+Br^-

(b)   [benzene ring]—CH_2CH_2Br $\xrightarrow[\text{diethyl ether}]{\text{Mg}}$ [benzene ring]—CH_2CH_2MgBr

(c) $CH_2{=\!=}CHCl$ $\xrightarrow[\substack{\text{tetrahydro-furan} \\ \Delta}]{\text{Mg}}$ $CH_2{=\!=}CHMgCl$

13.9 (cont)

(d)

$$\text{C}\equiv\text{CH} \xrightarrow[\text{diethyl ether}]{\text{CH}_3\text{CH}_2\text{MgBr}} \text{C}\equiv\text{CMgBr} + \text{CH}_3\text{CH}_3\uparrow$$

(e)

$$\text{—MgBr} + \underset{\text{CH}_3\overset{\text{O}}{\overset{\|}{\text{C}}}\text{OH}}{} \longrightarrow \text{—H} + \text{CH}_3\overset{\text{O}}{\overset{\|}{\text{C}}}\text{O}^-\text{Mg}^{2+}\text{Br}^-$$

(f) $\text{CH}_3\text{CH}_2\text{C}\equiv\text{C}^-\text{Na}^+ + \text{H}_2\text{O} \longrightarrow \text{CH}_3\text{CH}_2\text{C}\equiv\text{CH}\uparrow + \text{Na}^+\text{OH}^-$

(g)

$$\text{—CH}_2\text{Br} \xrightarrow[\substack{\text{tetrahydro-}\\\text{furan}}]{\text{Li}} \text{—CH}_2\text{Li} + \text{Li}^+\text{Br}^-$$

(h) $\text{CH}_3\text{CH}_2\text{CH}_2\text{Li} + \text{NH}_3 \longrightarrow \text{CH}_3\text{CH}_2\text{CH}_3\uparrow + \text{Li}^+\text{NH}_2^-$

(i) $\text{CH}_3\text{CH}_2\text{MgI} + \text{CH}_3\text{OH} \longrightarrow \text{CH}_3\text{CH}_3\uparrow + \text{CH}_3\text{O}^-\text{Mg}^{2+}\text{I}^-$

(j)

$$\text{C}\equiv\text{CH} \xrightarrow[\text{NH}_3\text{ (liq)}]{\text{Na}^+\text{NH}_2^-} \text{C}\equiv\text{C}^-\text{Na}^+ + \text{NH}_3$$

13.10

(a)

$$\text{=O} + \text{CH}_3\text{CH}_2\text{CH}_2\text{CH}_2\text{Li} \longrightarrow$$

A

$$\xrightarrow{\text{H}_3\text{O}^+}$$

B

(b)

$$\text{—Br} \xrightarrow[\substack{\text{diethyl}\\\text{ether}}]{\text{Mg}} \text{—MgBr} \xrightarrow[\text{CH}_2-\text{CH}_2]{\overset{\text{O}}{\triangle}}$$

C

$$\text{—CH}_2\text{CH}_2\text{O}^-\text{Mg}^{2+}\text{Br}^- \xrightarrow{\text{H}_3\text{O}^+} \text{—CH}_2\text{CH}_2\text{OH}$$

D   E

13.10 (cont)

(c)

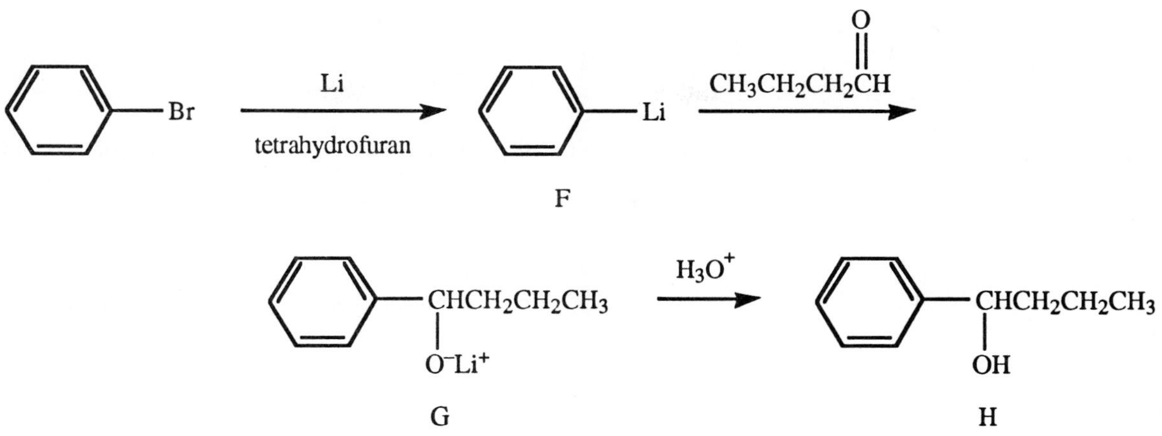
Br
Li
tetrahydrofuran
Li
F
O
CH3CH2CH2CH
CHCH2CH2CH3
O⁻Li⁺
G
H3O⁺
CHCH2CH2CH3
OH
H

(d)

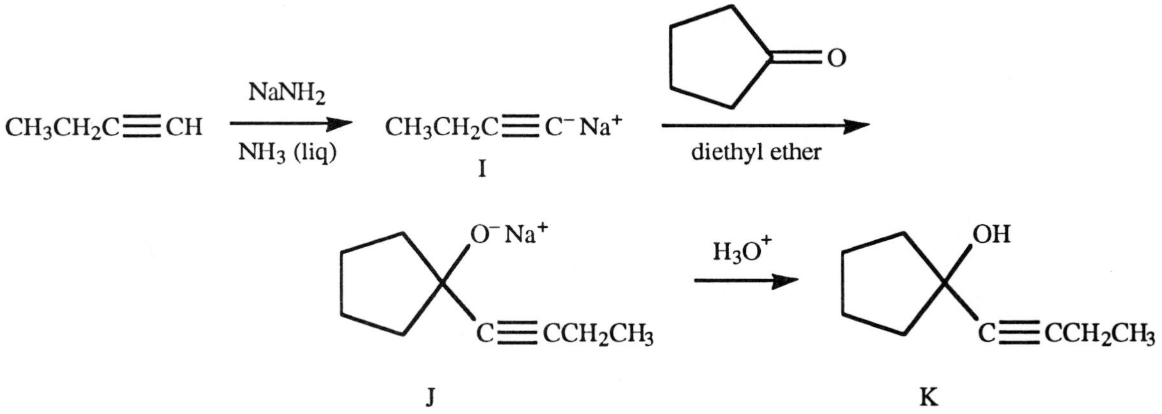
CH3CH2C≡CH
NaNH2
NH3 (liq)
CH3CH2C≡C⁻ Na⁺
I
O
diethyl ether
O⁻ Na⁺
C≡CCH2CH3
J
H3O⁺
OH
C≡CCH2CH3
K

(e)

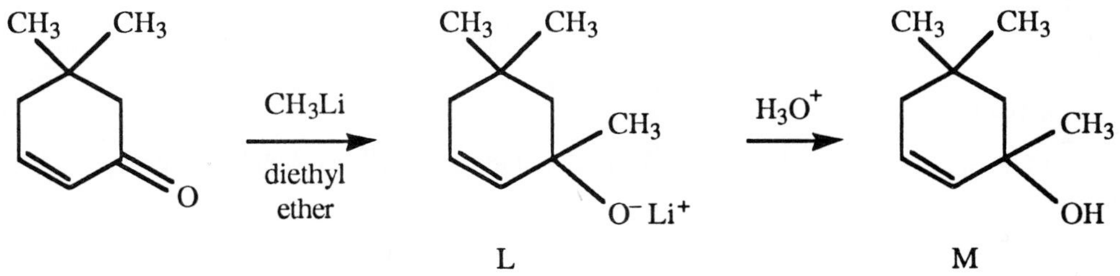
CH3 CH3
O
CH3Li
diethyl ether
CH3 CH3
CH3
O⁻ Li⁺
L
H3O⁺
CH3 CH3
CH3
OH
M

13.11

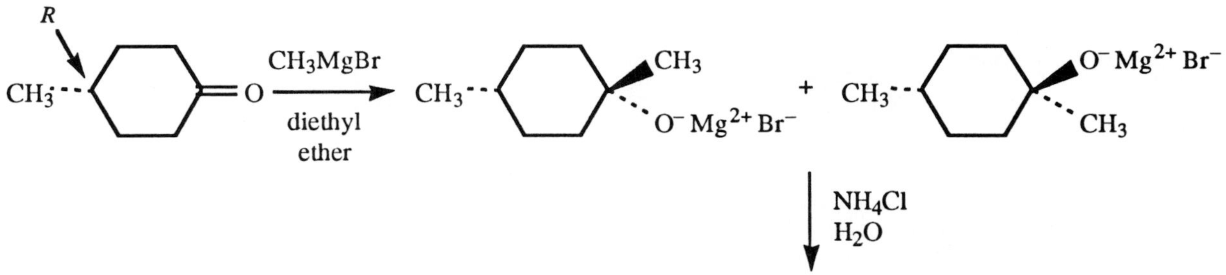
R
CH3
O
CH3MgBr
diethyl ether
CH3
CH3
O⁻ Mg²⁺ Br⁻
+
CH3
O⁻ Mg²⁺ Br⁻
CH3
NH4Cl
H2O

13.11 (cont)

(1*S*, 3*R*)-1,3-dimethyl-
cyclohexanol,
enantiomer of (1*R*, 3*S*)-
1,3-dimethylcyclohexanol
prepared from
(*S*)-3-methylcyclohexanone

+

(1*R*, 3*R*)-1,3-dimethyl-
cyclohexanol,
enantiomer of (1*S*, 3*S*)-
1,3-dimethylcyclohexanol
prepared from
(*S*)-3-methylcyclohexanone

13.12   The racemic mixture (*R*)- and (*S*)-3-methylcyclohexanone would react with methylmagnesium bromide to give
two sets of enantiomers (2 stereocenters give rise to 4 stereoisomers unless the symmetry of the system reduces
that number). Each enantiomer within a pair would form in the same amount because each arises from a
transition state that is the mirror image of the transition state leading to the other enantiomer. The mixture,
therefore, would not be optically active.

- -

Concept Map 13.4   (see p. 386)

- -

13.13

The Grignard reagent approaches the molecule best from the side away from the ether group. This can be seen
most easily in a conformational drawing.

best approach of the
Grignard reagent
to the carbonyl group

Concept Map 13.3 Organometallic reagents and their reactions with compounds containing electrophilic carbon atoms.

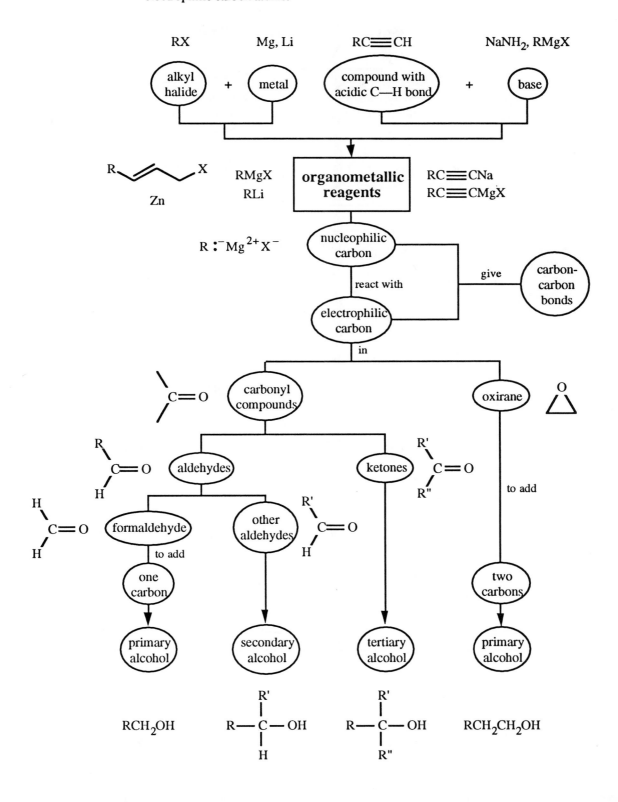

Concept Map 13.4   Some ways to prepare alcohols.

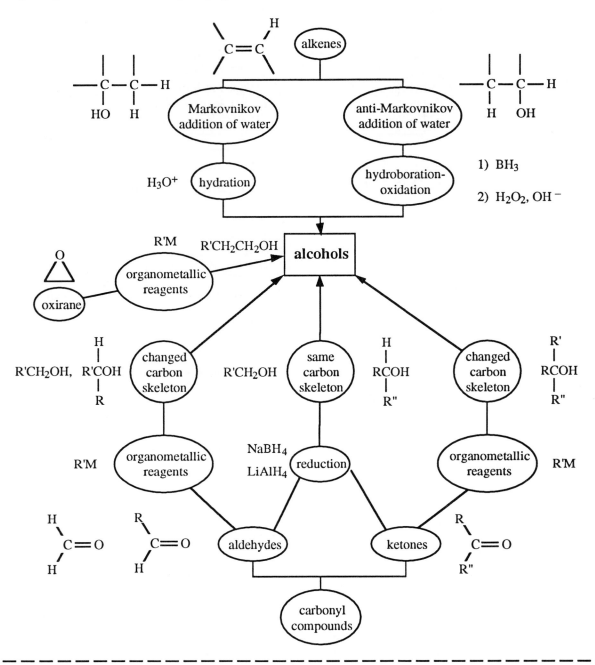

13.14

13.14 (cont)

$$BrMgC\equiv CCH_2CH_2 \qquad H$$
$$C=C \qquad + \qquad$$
$$H \qquad MgBr$$

$$HC\equiv CCH_2CH_2 \qquad H$$
$$C=C \qquad \rightleftharpoons \longrightarrow$$
$$H \qquad H$$
$$\uparrow$$
$$pK_a \sim 36$$

$$BrMgC\equiv CCH_2CH_2 \qquad H \qquad\qquad BrMgC\equiv CCH_2CH_2 \qquad H$$
$$C=C \qquad + \qquad\qquad\qquad C=C$$
$$H \qquad H \qquad\qquad\qquad\qquad\qquad H \qquad H$$

The Grignard reagent that would predominate in the solution
after multiple equilibria.

13.15   (a)   This is a tertiary alcohol. It can be prepared in three ways corresponding to the three dissections shown
below. Each will lead to an organometallic reagent and a ketone. The ketone will be the fragment
containing the oxygen.

$$\begin{array}{cc}
CH_3 & \overset{③}{\quad} \quad CH_2CH_2CH_2CH_3 \\
CH_3C=CH\!-\!\{\!-\!C\!-\!\}\!-\!CH_3 \\
& \underset{①}{\quad}\ OH\ \underset{②}{\quad}
\end{array}$$

Reagents for dissection 1:

$$CH_3 \qquad\qquad CH_3 \qquad\qquad\qquad CH_3 \qquad\qquad O$$
$$CH_3C=CHMgBr\ \text{or}\ CH_3C=CHLi\ (\text{prepared from}\ CH_3C=CHBr)\ \text{and}\ CH_3CCH_2CH_2CH_2CH_3$$

Reagents for dissection 2:

$$CH_3 \qquad O$$
$$CH_3MgBr\ \text{or}\ CH_3Li\ (\text{from}\ CH_3Br)\ \text{and}\ CH_3C=CHCCH_2CH_2CH_2CH_3$$

Reagents for dissection 3:

$$CH_3 \qquad O$$
$$CH_3CH_2CH_2CH_2MgBr\ \text{or}\ CH_3CH_2CH_2CH_2Li\ (\text{from}\ CH_3CH_2CH_2CH_2Br)\ \text{and}\ CH_3C=CHCCH_3$$

(b)   This is a secondary alcohol. Two dissections are possible leading to an aldehyde and an organometallic
reagent. A third dissection indicates that a reduction reaction, hydride addition, will also give this alcohol.

$$\begin{array}{c}
CH_3 \qquad\qquad CH_3 \qquad\qquad ③\ H \\
CH_3C=CH\ CH_2CH_2C=CHCH_2CH_2\!-\!\{\!-\!C\!-\!\}\!-\!C\equiv CH \\
\underset{①}{\quad}\ O\!-\!\{\!-\!H\ \underset{②}{\quad} \\
\underset{③}{\quad}
\end{array}$$

13.15 (b) (cont)

Reagents for dissection 1:

$$\underset{\text{CH}_3}{\text{CH}_3\text{C}}=\text{CHCH}_2\text{CH}_2\overset{\text{CH}_3}{\text{C}}=\text{CHCH}_2\text{CH}_2\text{MgBr} \quad \text{or} \quad \underset{\text{CH}_3}{\text{CH}_3\text{C}}=\text{CHCH}_2\text{CH}_2\overset{\text{CH}_3}{\text{C}}=\text{CHCH}_2\text{CH}_2\text{Li} \quad \text{(prepared}$$

from $\underset{\text{CH}_3}{\text{CH}_3\text{C}}=\text{CHCH}_2\text{CH}_2\overset{\text{CH}_3}{\text{C}}=\text{CHCH}_2\text{CH}_2\text{Br})$ and $\text{H}\overset{\text{O}}{\text{C}}\text{C}\equiv\text{CH}$. This aldehyde may give problems depending on whether nucleophilic addition to the carbonyl group will compete with deprotonation of the alkyne.

Reagents for dissection 2:

$\text{HC}\equiv\text{CMgBr}$ or $\text{HC}\equiv\text{CNa}$ (from $\text{HC}\equiv\text{CH}$) and $\underset{\text{CH}_3}{\text{CH}_3\text{C}}=\text{CHCH}_2\text{CH}_2\overset{\text{CH}_3}{\text{C}}=\text{CHCH}_2\text{CH}_2\overset{\text{O}}{\text{C}}\text{H}$

This is the better choice for an aldehyde and an organometallic reagent.

Reagents for dissection 3:

NaBH_4 and $\underset{\text{CH}_3}{\text{CH}_3\text{C}}=\text{CHCH}_2\text{CH}_2\overset{\text{CH}_3}{\text{C}}=\text{CHCH}_2\text{CH}_2\overset{\text{O}}{\text{C}}-\text{C}\equiv\text{CH}$

(c)   This is a primary alcohol. A primary alcohol is prepared 1) by reduction of an aldehyde, 2) by addition of an organometallic reagent to formaldehyde or, sometimes, by addition of an organometallic reagent to oxirane. The dissections corresponding to these three routes are shown below.

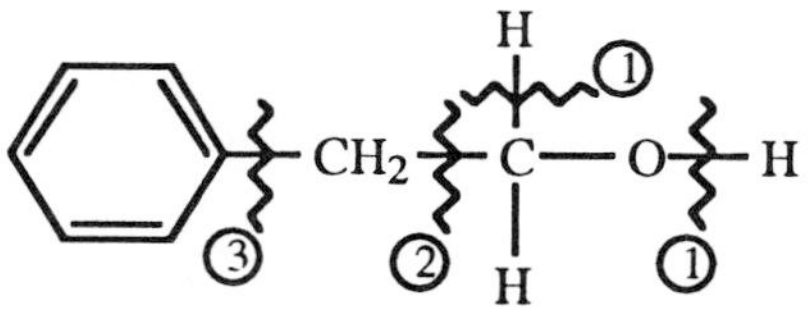

Reagents for dissection 1:

NaBH_4 and

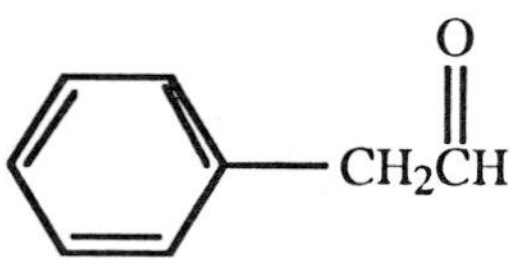

Reagents for dissection 2:

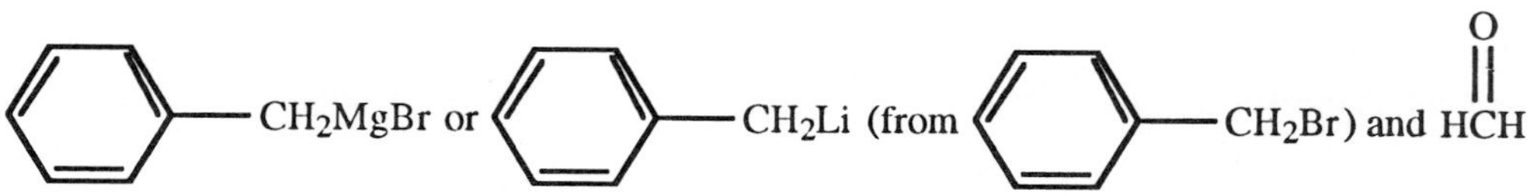

Reagents for dissection 3:

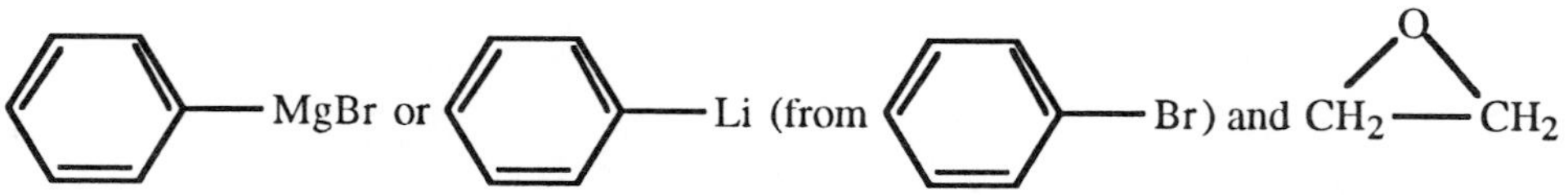

13.16

Two stereoisomeric hemiacetals are possible, resulting from the nucleophilic attack of the hydroxyl group at the bottom or the top of the planar carbonyl group. The two hemiacetals are enantiomers of each other.

13.17   Essentially all the ^{18}O will wind up in the oxygen of the carbonyl group.

13.18

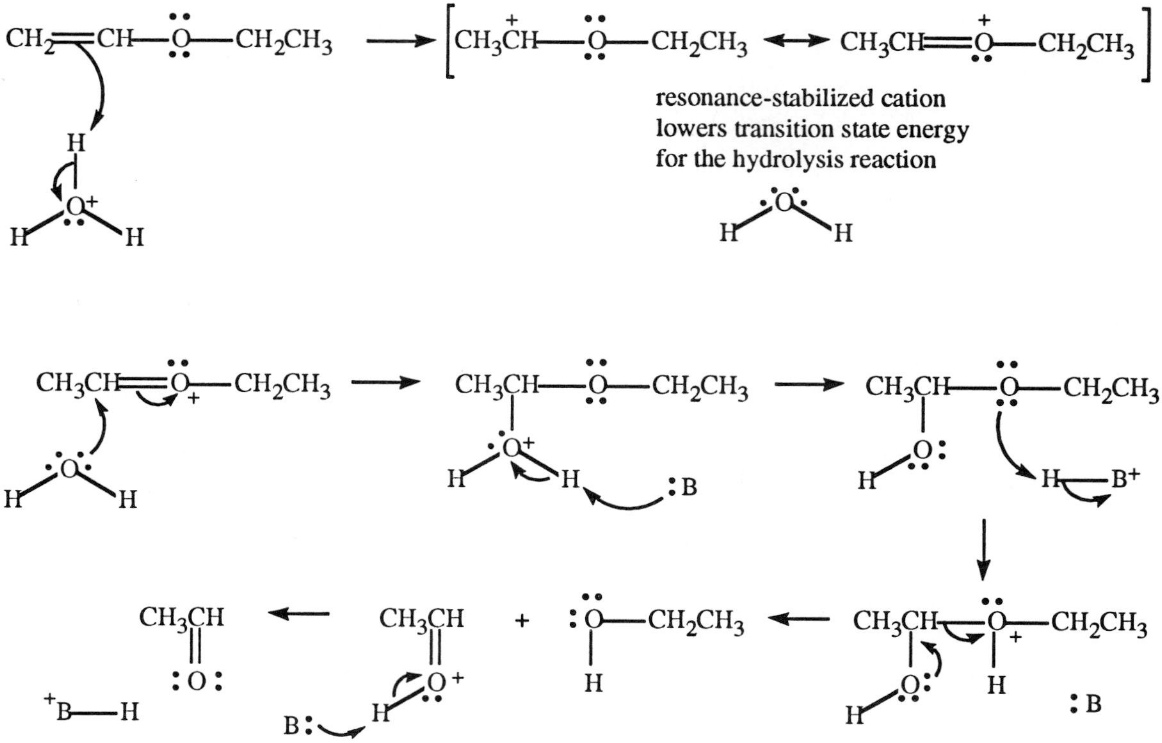
resonance-stabilized cation
lowers transition state energy
for the hydrolysis reaction

13.19

(a)

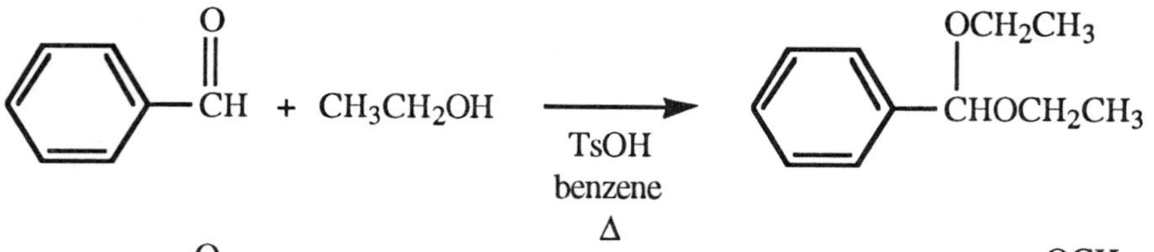
TsOH
benzene
Δ

(b)

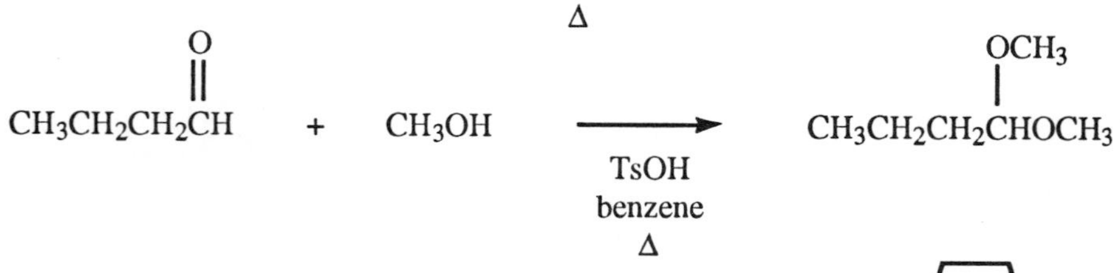
TsOH
benzene
Δ

(c)

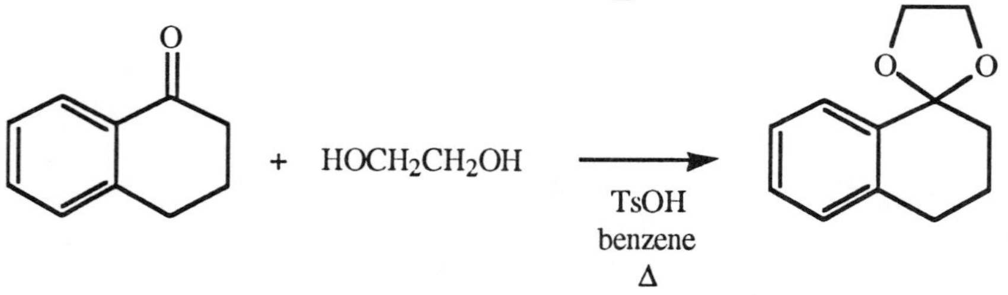
TsOH
benzene
Δ

Concept Map 13.5   Hydrates, acetals, ketals.

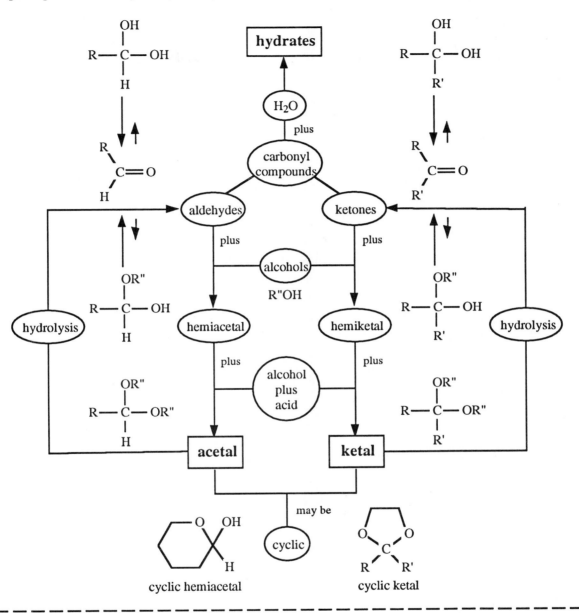

cyclic hemiacetal   cyclic ketal

- -

13.20 $CH_2\text{--}CH\text{--}CH_2$ + $CH_3\overset{\overset{\displaystyle O}{\|}}{C}CH_3$ $\longrightarrow$ $C_6H_{12}O_3$
       | | |
      OH OH OH

$C_3H_8O_3 + C_3H_6O = C_6H_{14}O_4$

$C_6H_{14}O_4 - C_6H_{12}O_3 = H_2O$

One molecule of water has been lost in the reaction.

13.20 (cont)

$HOCH_2CHCH_2OH$ + CH_3CCH_3 $\xrightarrow{\text{TsOH}}$

a cyclic ketal

Concept Map 13.6   Reactions of carbonyl compounds with compounds related to ammonia.

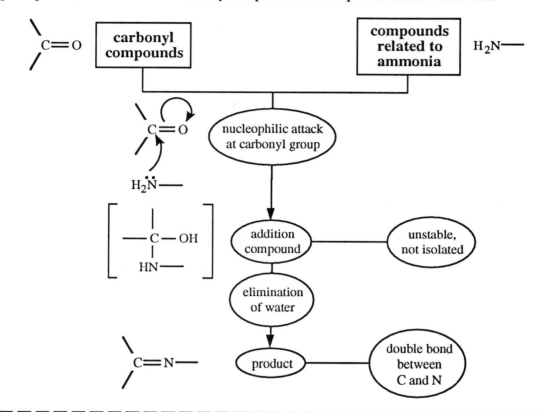

13.21

pyridoxamine-5'-phosphate

2-oxopropanoic acid

13.21 (cont)

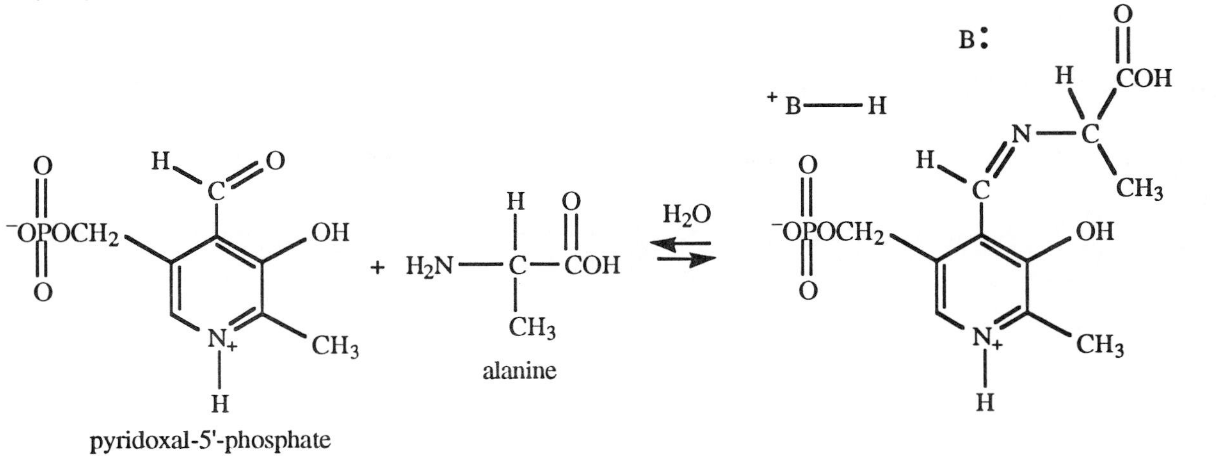

pyridoxal-5'-phosphate

In the presence of an enzyme, which is a chiral reagent, these reactions occur stereoselectively to give (*S*)-alanine.

13.22

(a)

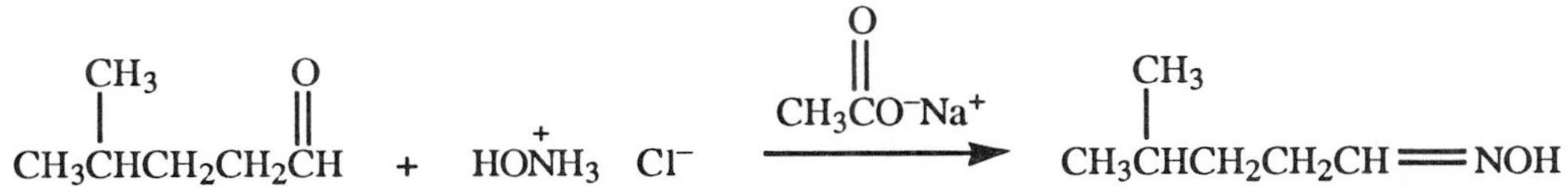

(b)

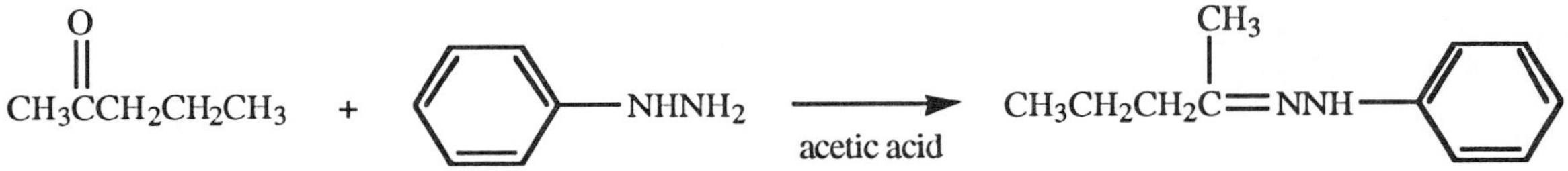

(c)

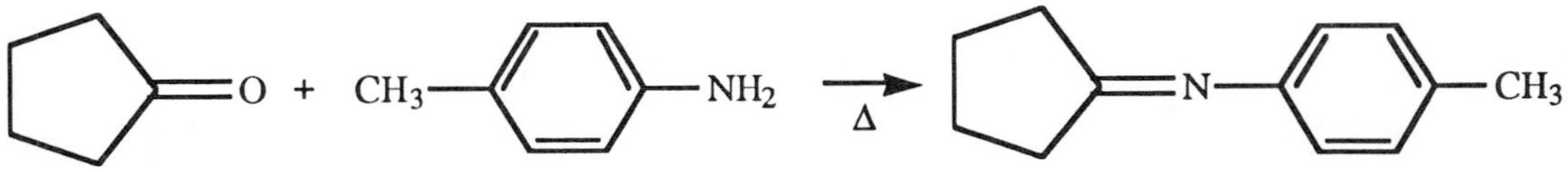

(d)

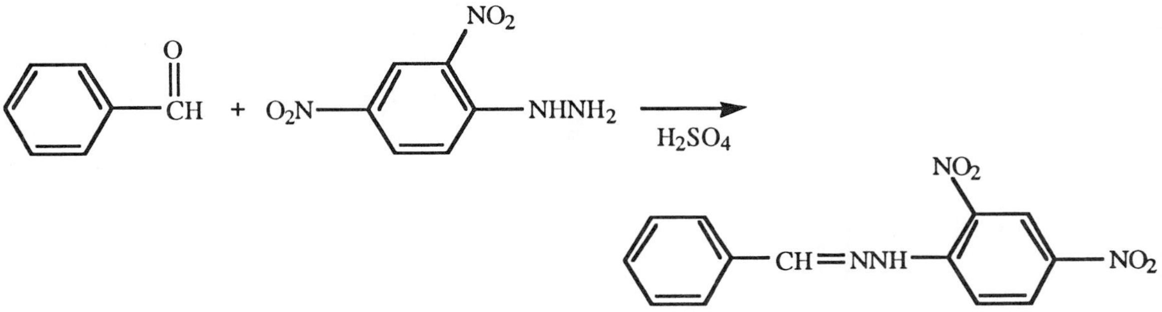

(e)

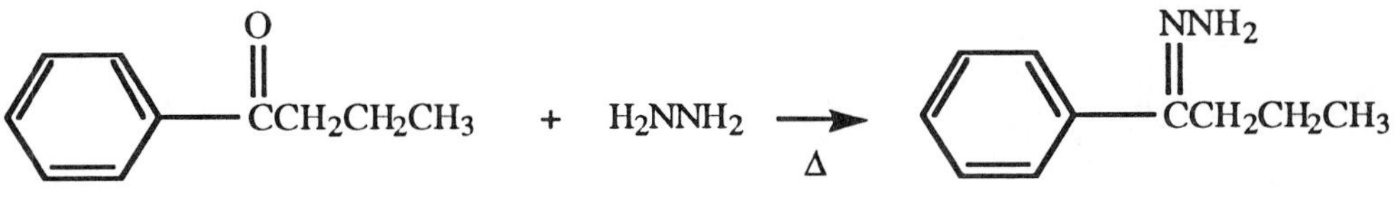

13.22 (cont)

(f) 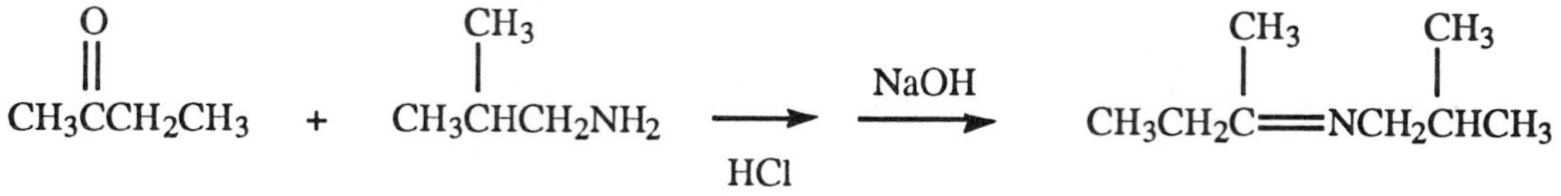 $CH_3CCH_2CH_3$ + $CH_3CHCH_2NH_2$ $\xrightarrow{\text{HCl}}$ $\xrightarrow{\text{NaOH}}$ $CH_3CH_2C{=}NCH_2CHCH_3$

(g)

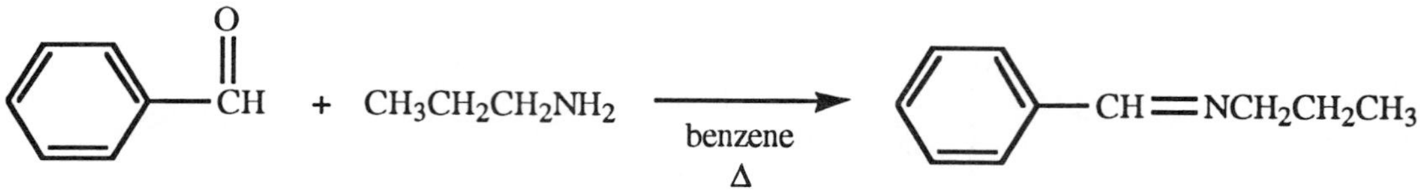

13.23

addition of nucleophilic nitrogen to electrophilic carbonyl group

deprotonation

protonation

loss of water

protonation

deprotonation

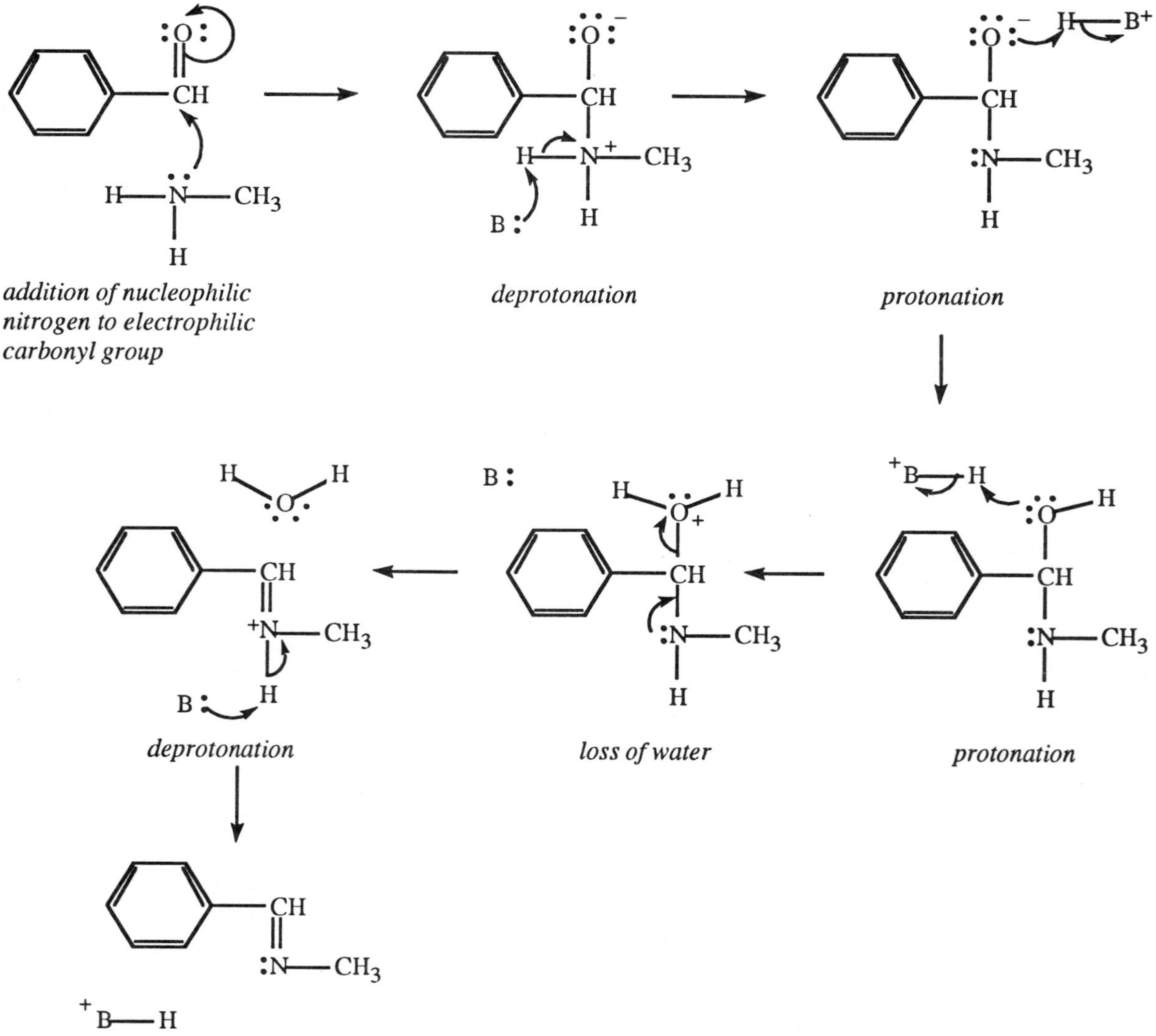

13.24

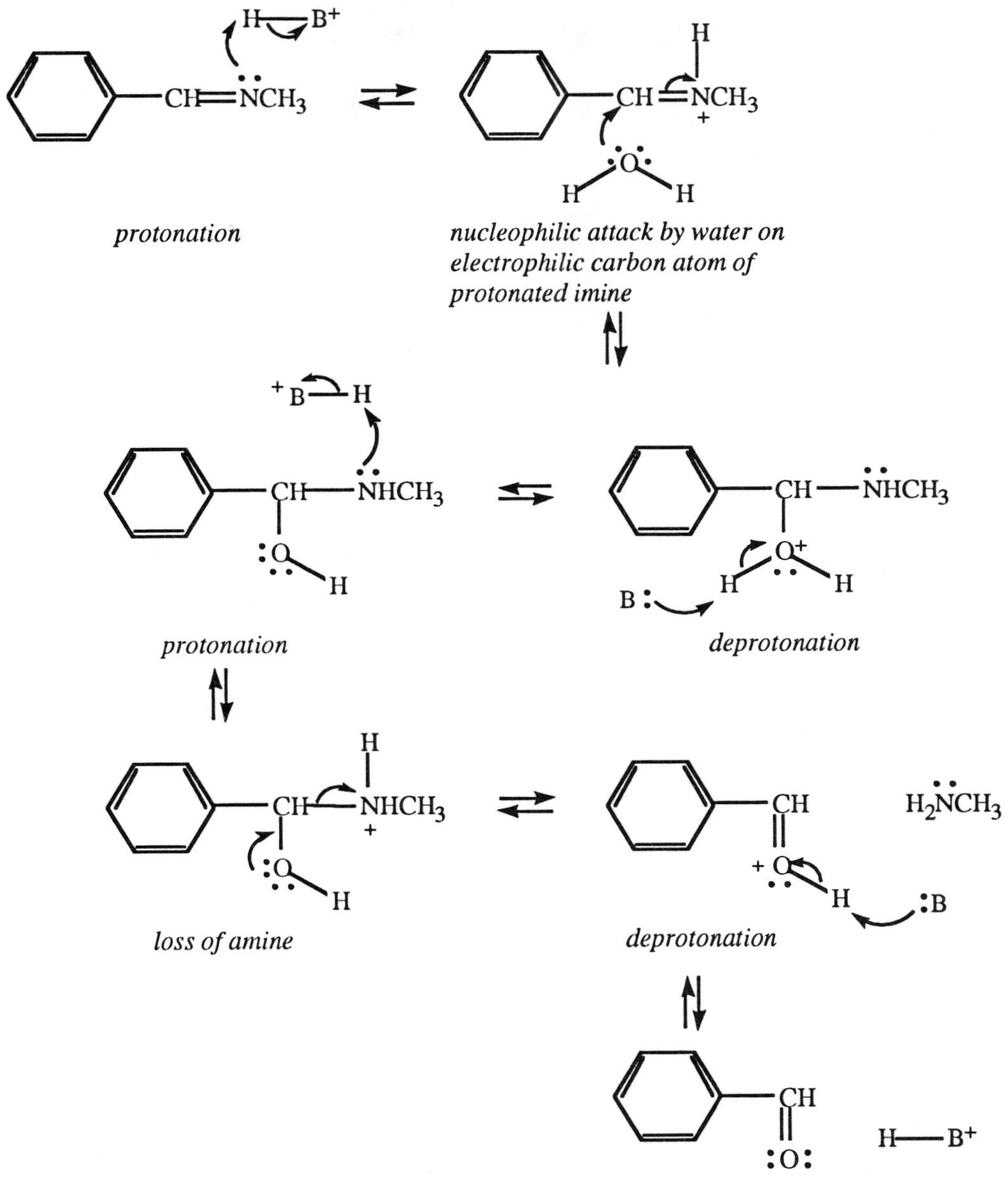

protonation

nucleophilic attack by water on
electrophilic carbon atom of
protonated imine

protonation

deprotonation

loss of amine

deprotonation

13.25

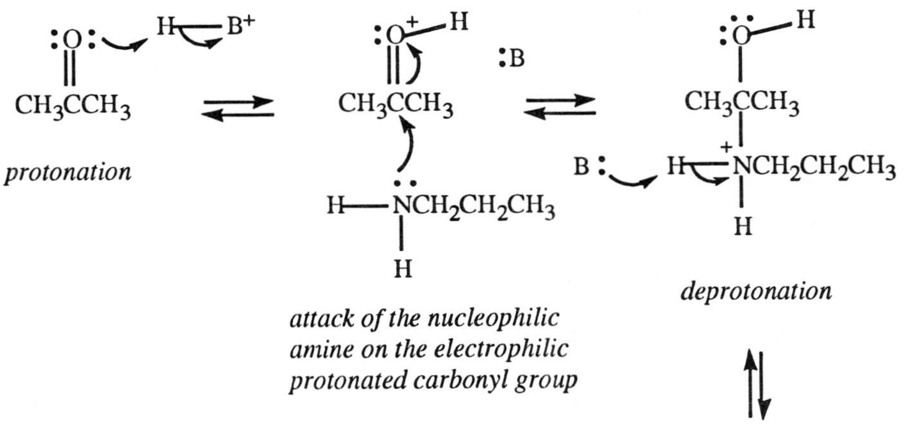

CH_3CCH_3

protonation

attack of the nucleophilic
amine on the electrophilic
protonated carbonyl group

deprotonation

13.25 (cont)

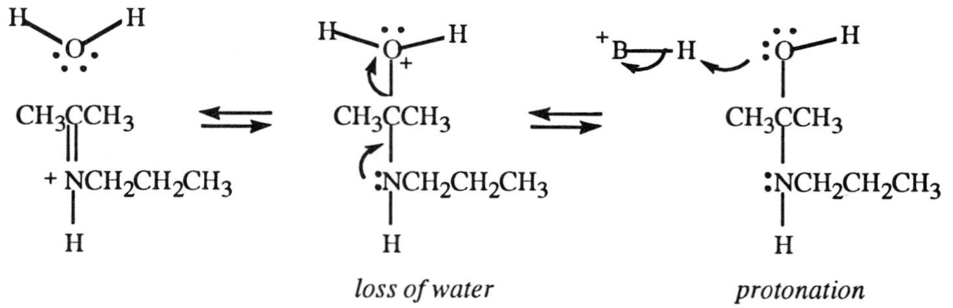

loss of water   *protonation*

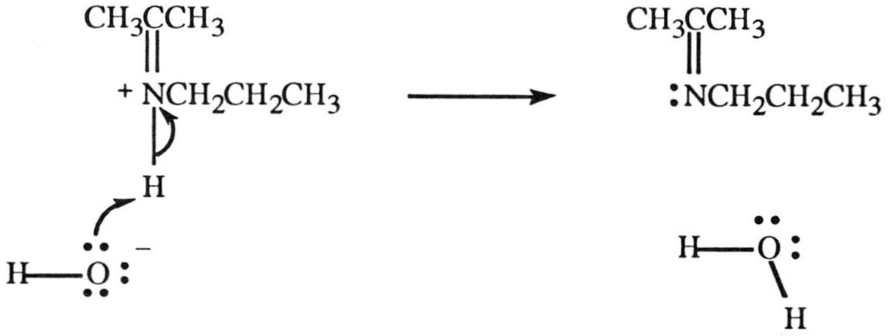

The catalytic amount of hydrochloric acid begins the reaction by protonating the oxygen of the carbonyl group. Sodium hydroxide is added at the end to neutralize the acid in order to prevent catalysis of the reverse reaction, the hydrolysis of the imine (Problem 13.24).

13.26 The major side reaction in the presence of hydrochloric acid would be hydrolysis of the ketal group.

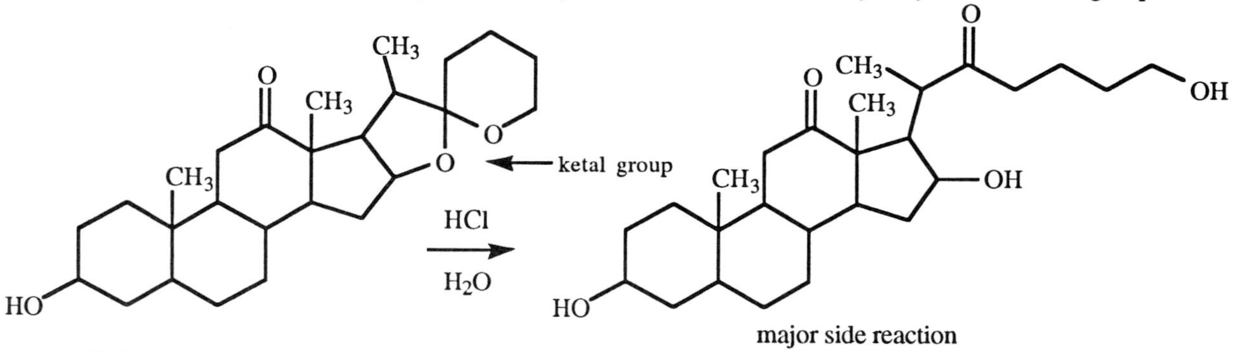

major side reaction

Dehydration of the alcohol function and substitution of a chlorine for the hydroxyl group are also possible side reactions.

Concept Map 13.7   (see p. 399)

13.27

(a)

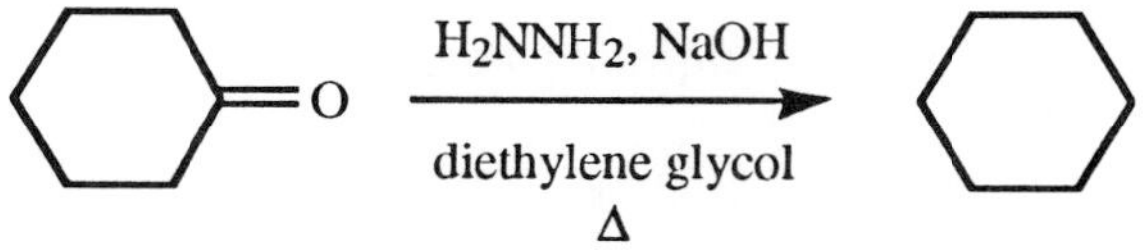

13.27 (cont)

(b)

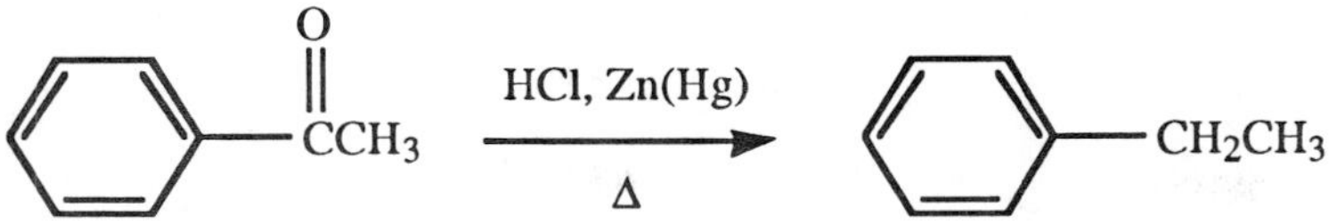

$$\text{(phenyl)}-\overset{\overset{\displaystyle O}{\|}}{C}CH_3 \xrightarrow[\Delta]{HCl,\ Zn(Hg)} \text{(phenyl)}-CH_2CH_3$$

(c)

$$HO-\text{(benzene)}-\overset{\overset{\displaystyle O}{\|}}{C}H \xrightarrow[\Delta]{HCl,\ Zn(Hg)} HO-\text{(benzene)}-CH_3$$

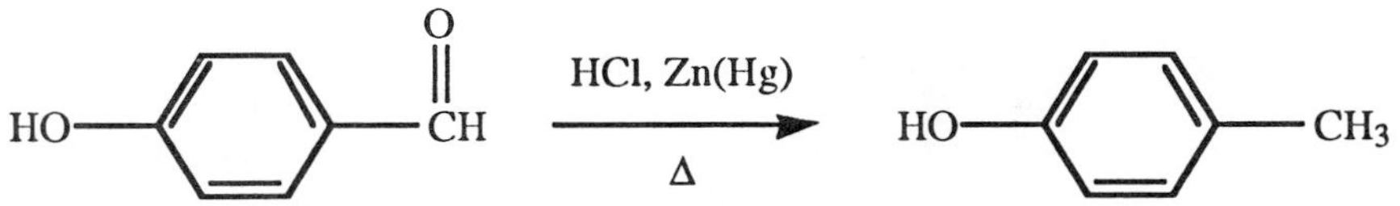

(d)

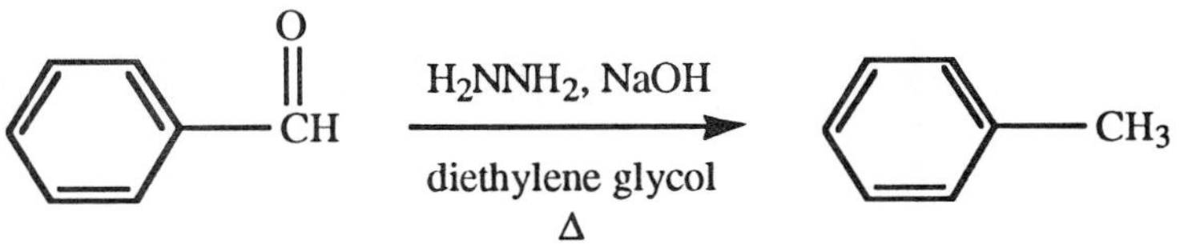

$$\text{(phenyl)}-\overset{\overset{\displaystyle O}{\|}}{C}H \xrightarrow[\substack{\text{diethylene glycol} \\ \Delta}]{H_2NNH_2,\ NaOH} \text{(phenyl)}-CH_3$$

(e)

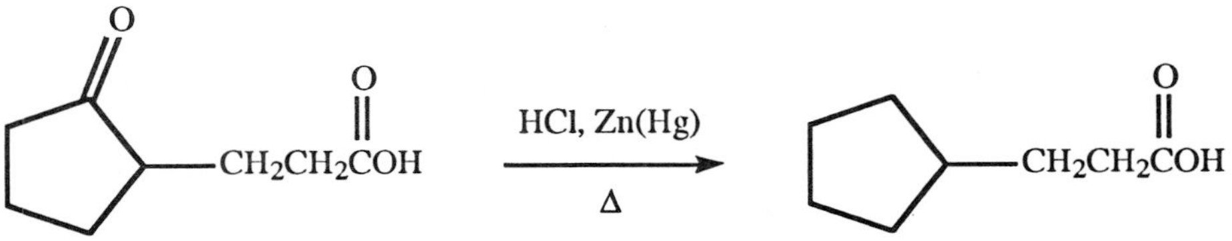

$$\text{(cyclopentanone)}-CH_2CH_2\overset{\overset{\displaystyle O}{\|}}{C}OH \xrightarrow[\Delta]{HCl,\ Zn(Hg)} \text{(cyclopentane)}-CH_2CH_2\overset{\overset{\displaystyle O}{\|}}{C}OH$$

(f)

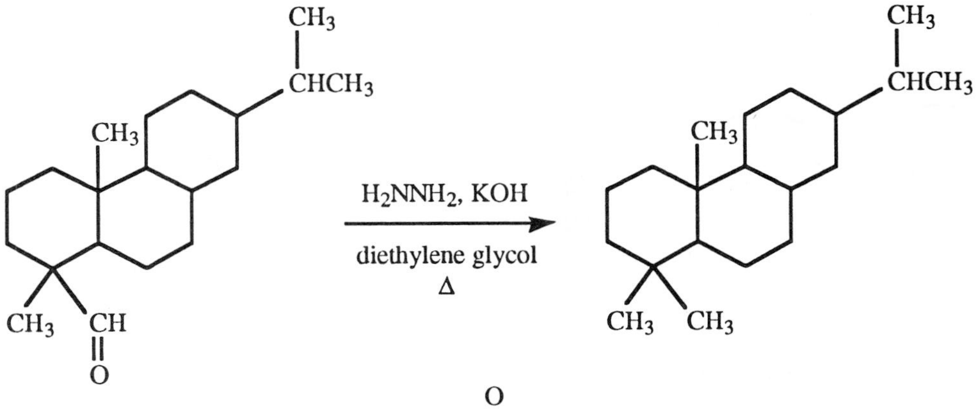

$$\xrightarrow[\substack{\text{diethylene glycol} \\ \Delta}]{H_2NNH_2,\ KOH}$$

(g)

$$\xrightarrow[\substack{ZnCl_2 \\ Na_2SO_4}]{2\ CH_3CH_2SH}$$

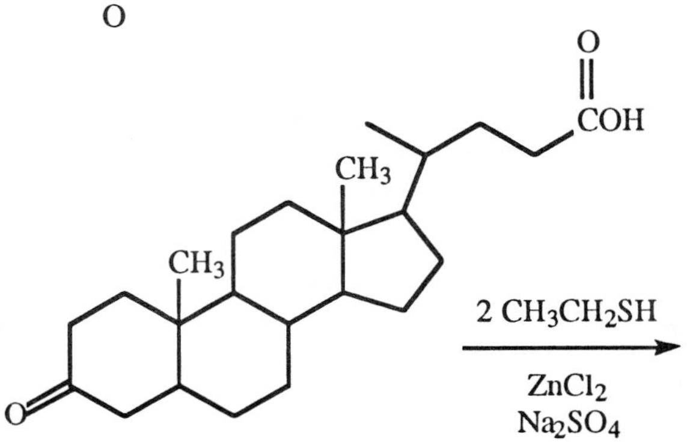

13.27 (g) (cont)

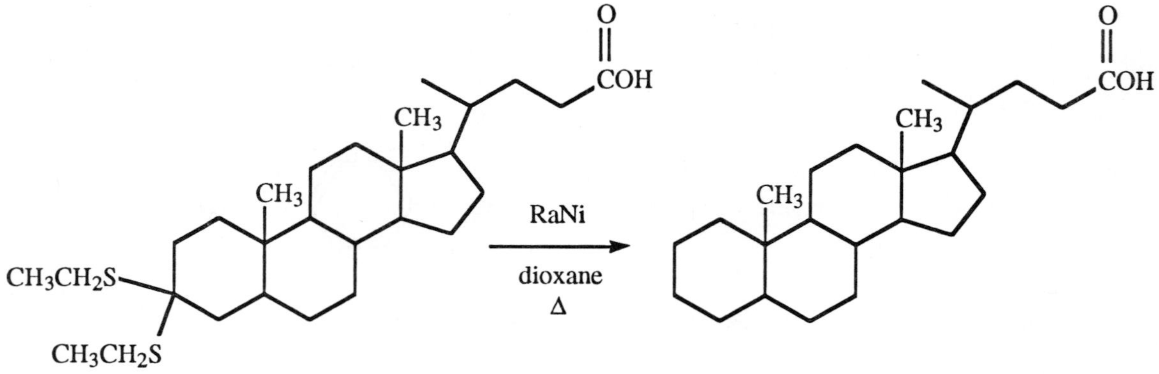

(h)

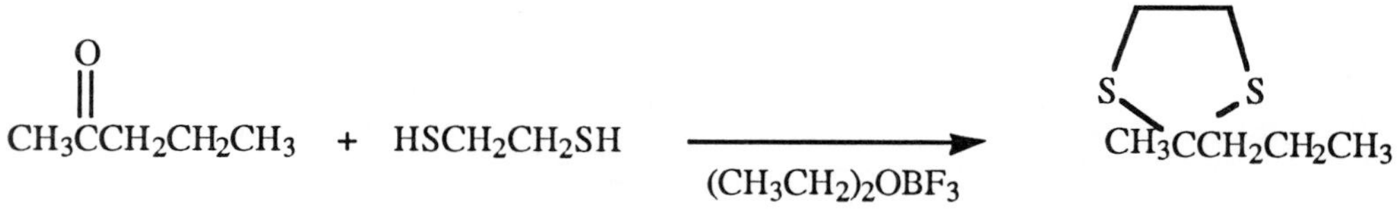

$$CH_3CCH_2CH_2CH_3 \; + \; HSCH_2CH_2SH \xrightarrow[(CH_3CH_2)_2OBF_3]{} CH_3CCH_2CH_2CH_3$$

(i)

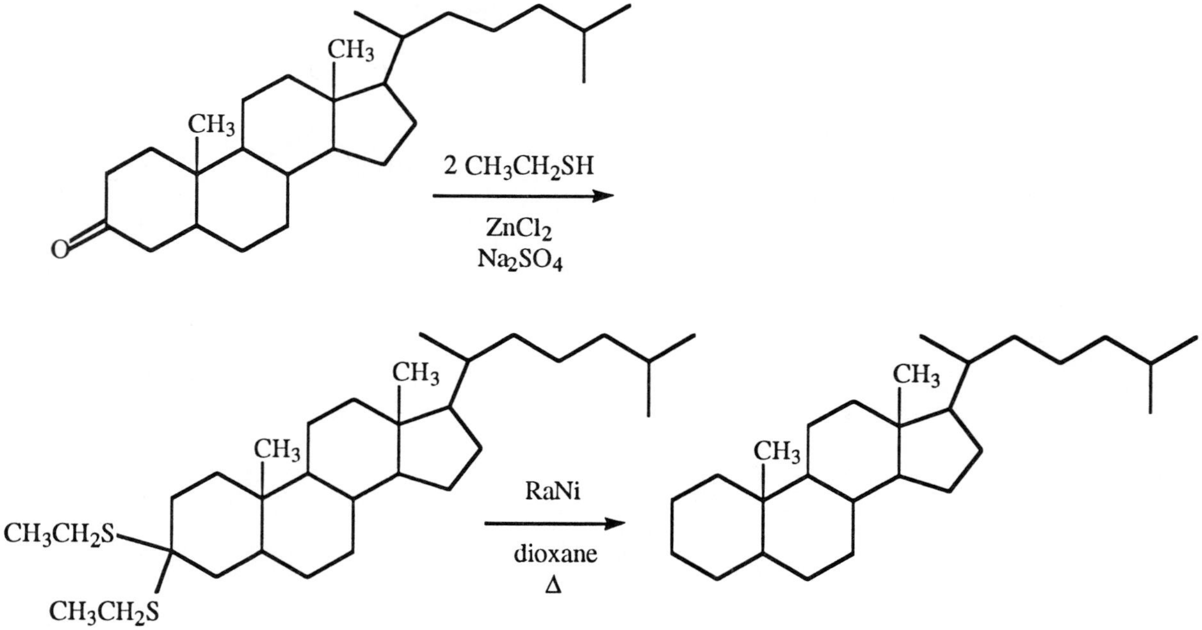

Concept Map 13.7   Reduction of carbonyl groups to methylene groups.

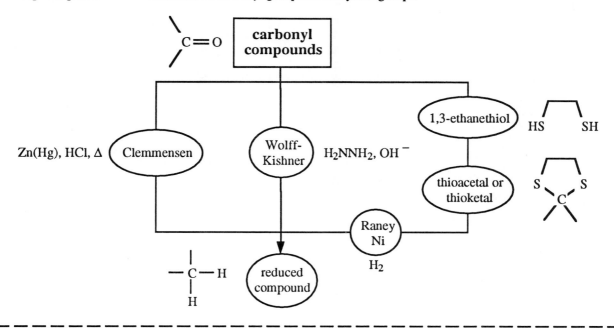

1. What are the connectivities of the two compounds? How many carbon atoms does each contain? Are there any rings? What are the positions of branches and functional groups on the carbon skeletons?

    The atoms of the starting material and product molecules that are the same are shown below. The starting material has a triple bond, an alcohol, and two ketal functional groups; the product has a cis double bond, a ketone and three alcohol functional groups. The rings that are present are cyclic ketals.

2. How do the functional groups change in going from starting material to product? Does the starting material have a good leaving group?

13.28 (cont)

The triple bond in the starting material has been converted to a cis double bond. One of the ketal groups in the starting material has been converted to a ketone; the other ketal group has been converted into two alcohol functional groups. There is no good leaving group in the starting material.

3. Is it possible to dissect the structures of the starting material and product to see which bonds must be broken and which formed?

$CH_3CCH_2CHC \equiv\kern-2pt C-$ ... CH_3 / CH_3 (with OH)

bonds broken

$C=C$ ... CH_3CCH_2CH (OH), $CHCH_2O-H$, CH_2O-H

bonds formed

4. New bonds are created when an electrophile reacts with a nucleophile. Do we recognize any part of the product molecule as coming from a good nucleophile or an electrophilic addition?

No.

5. What type of compound would be a good precursor to the product?

An alkyne which can be converted to a cis alkene by reduction with hydrogen and the appropriate catalyst.

$$CH_3CCH_2CH(OH)\;C=C\;CHCH_2OH\,(CH_2OH) \xleftarrow[\text{quinoline}]{\substack{H_2 \\ Pd/CaCO_3}} CH_3CCH_2CHC \equiv CCHCH_2OH\;(OH)(CH_2OH)$$

The ketal groups in the starting material can be removed by acid hydrolysis.

$$CH_3CCH_2CHC \equiv CCHCH_2OH\;(OH)(CH_2OH) \xleftarrow{H_3O^+} CH_3CCH_2CHC \equiv C-\;(OH)\cdots\;CH_3/CH_3$$

6. After this last step, do we see how to get from starting material to product? If not, we need to analyze the structure obtained in step 5 by applying questions 4 and 5 to it.

$$CH_3CCH_2CH(OH)\;C=C\;CHCH_2OH\,(CH_2OH) \xleftarrow[\text{quinoline}]{\substack{H_2 \\ Pd/CaCO_3}} CH_3CCH_2CHC \equiv CCHCH_2OH\;(OH)(CH_2OH)$$

$\uparrow H_3O^+$

Concept Map 13.8   Protecting groups.

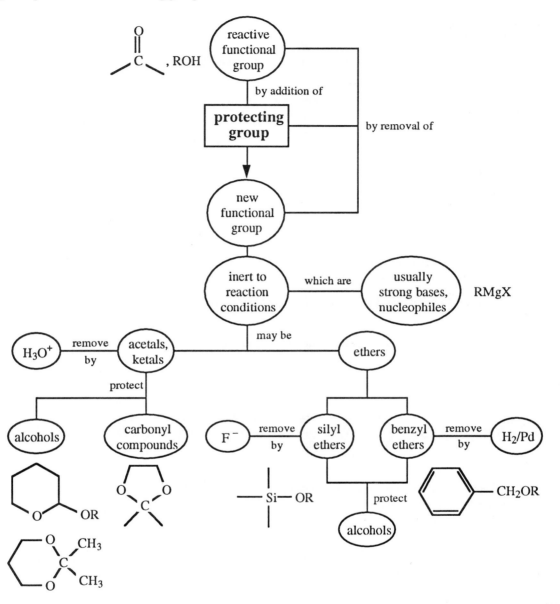

- -

13.31

$$CH_2{=}CCH_2CH_2OCH_2{-}\bigcirc \xrightarrow{?} CH_3CH_2C{-}CHCHCH_2CH_2OCH_2{-}\bigcirc$$

with CH_3 branch and CH_2 / CH_3 branches and OH group.

1. What are the connectivities of the two compounds? How many carbon atoms does each contain? Are there any rings? What are the positions of branches and functional groups on the carbon skeletons?

    The starting material has a double bond and an ether functional group. The product has a double bond, and ether and secondary alcohol functional groups. Carbon 2 of a four carbon segment has added to carbon 1 of the starting material (see structure in part 3).

13.31 (cont)

2. How do the functional groups change in going from starting material to product? Does the starting material have a good leaving group?

    A four carbon segment has added to carbon 1 of the double bond. There is no good leaving group.

3. Is it possible to dissect the structures of the starting material and product to see which bonds must be broken and which formed?

         bonds broken   bonds formed

4. New bonds are created when an electrophile reacts with a nucleophile. Do we recognize any part of the product molecule as coming from a good nucleophile or an electrophilic addition?

    A Grignard reagent is a good nucleophile, which can add to the electrophilic carbon atom of an aldehyde to give a secondary alcohol.

5. What type of compound would be a good precursor to the product?

    An aldehyde would be a good precursor to the product we want.

6. After this last step, do we see how to get from starting material to product? If not, we need to analyze the structure obtained in step 5 by applying questions 4 and 5 to it.

    We need to prepare the aldehyde so we must apply questions 4 and 5 again.

4 (repeated). Do we recognize any part of the precursor molecule as coming from a good nucleophile or an electrophilic addition?

    No.

5 (repeated). What type of compound would be a good precursor to the product?

    An aldehyde can be prepared by the oxidation of a primary alcohol.

13.31 (cont)

A primary alcohol can be made by the electrophilic addition of diborane to the terminal double bond in the starting material, followed by oxidation of the alkylborane.

6 (repeated). After this last step, do we see how to get from starting material to product? If not, we need to analyze the structure obtained in step 5 by applying questions 4 and 5 to it.

The complete synthesis is shown below:

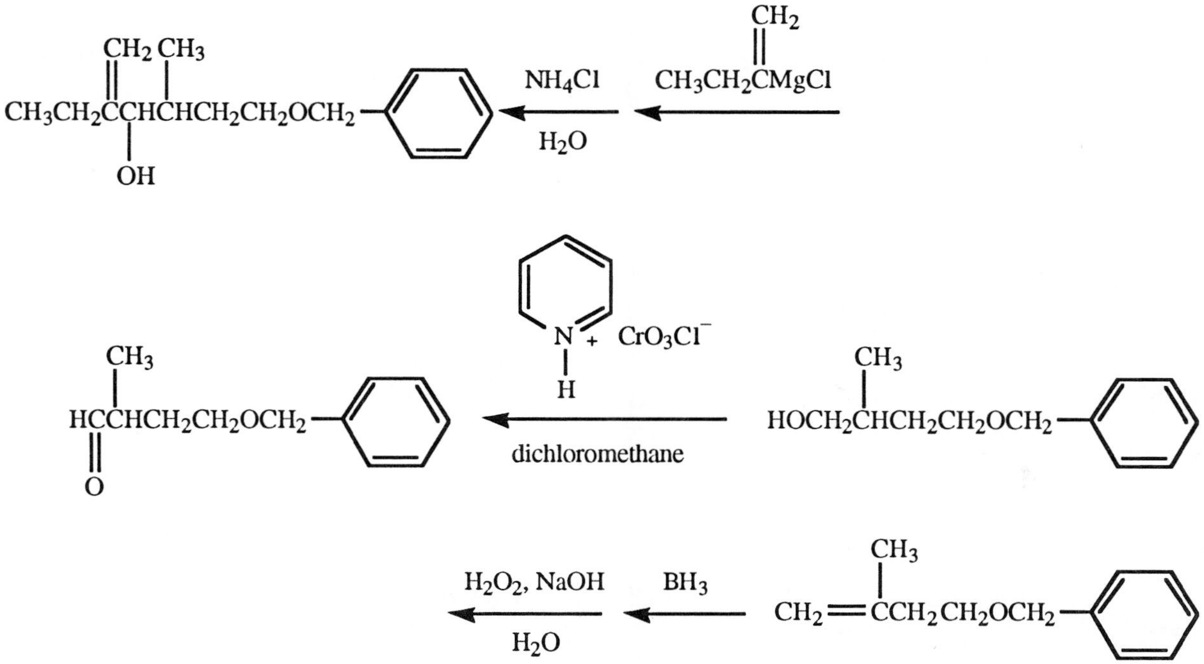

13.32

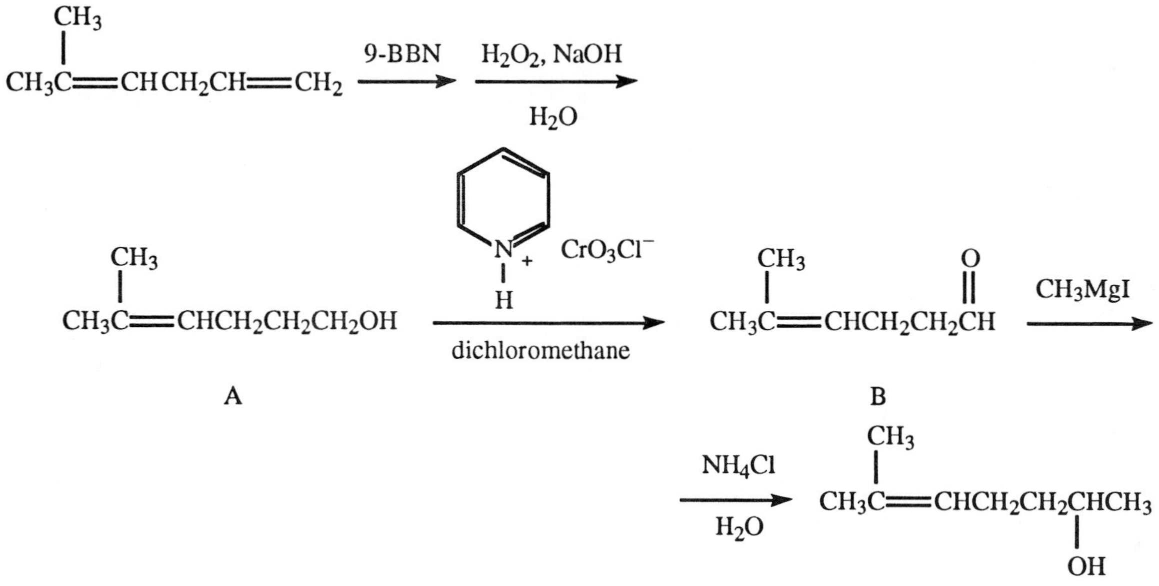

13.33

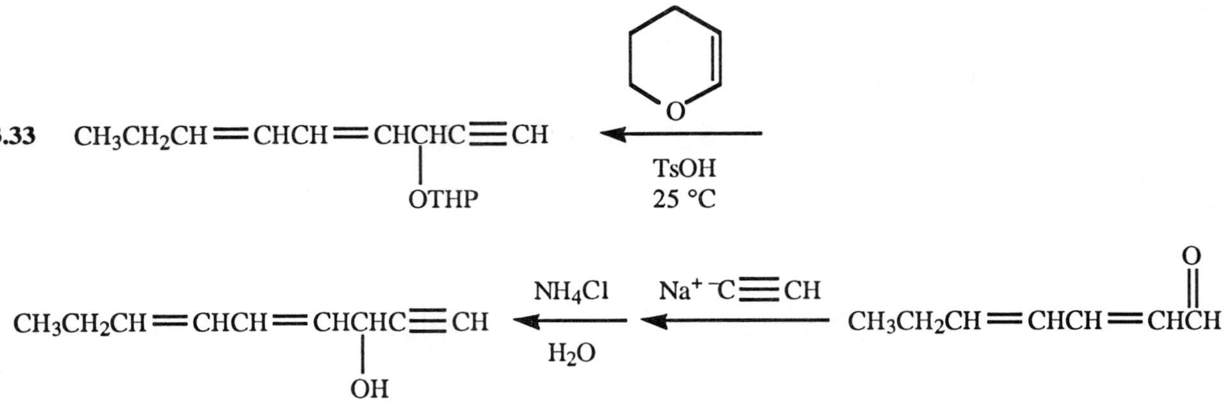
CH3CH2CH=CHCH=CHCHC≡CH
OTHP
O
TsOH
25 °C
CH3CH2CH=CHCH=CHCHC≡CH
OH
NH4Cl
H2O
Na+ −C≡CH
O
CH3CH2CH=CHCH=CHCH

13.34

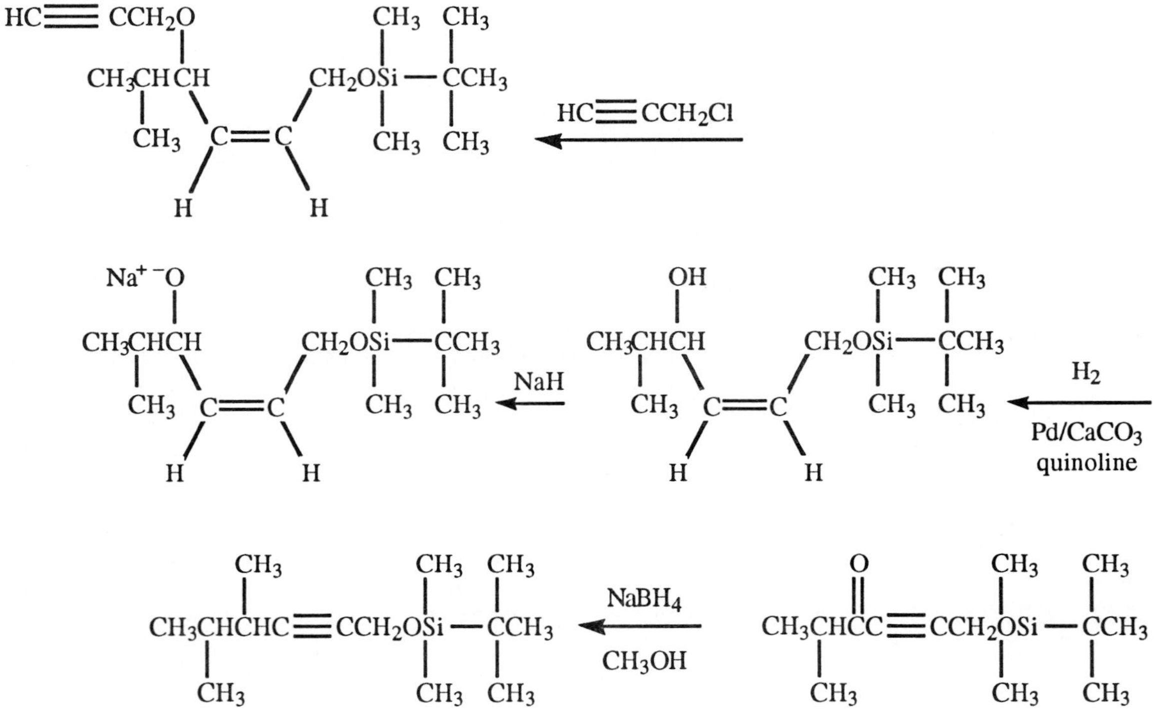
HC≡CCH2O
CH3CHCH
CH3
C=C
H
H
CH2OSi—CCH3
CH3 CH3
CH3 CH3
HC≡CCH2Cl
Na+ −O
NaH
OH
H2
Pd/CaCO3
quinoline
NaBH4
CH3OH
CH3CHCHC≡CCH2OSi—CCH3
CH3CHCC≡CCH2OSi—CCH3

13.35

(a) cyclopentanecarbaldehyde (b) cyclobutanone (c) (*E*)-2-hexenal (d) (*S*)-2-hydroxypropanal

(e) 2-methylcyclohexanone (f) 5-methyl-2-hexanone (g) 4-methoxybenzaldehyde (anisaldehyde)

(h) 3-methyl-2-cyclohexenone (i) (4*S*, 5*S*)-4-chloro-5-methyl-2-heptanone (j) 4-phenylbutanal

13.36

(a)

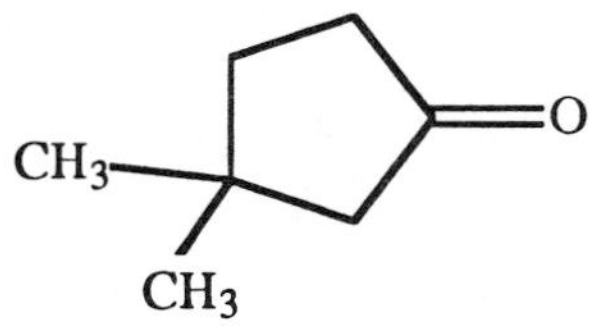

(b)

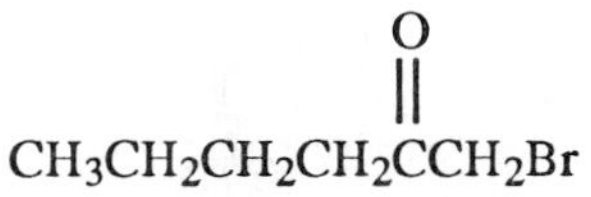

$$\underset{\substack{\| \\ O}}{CH_3CH_2CH_2CH_2CCH_2Br}$$

(c)

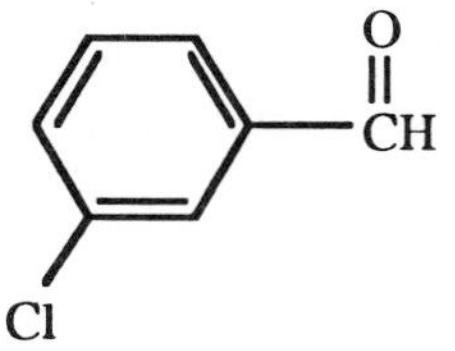

(d)

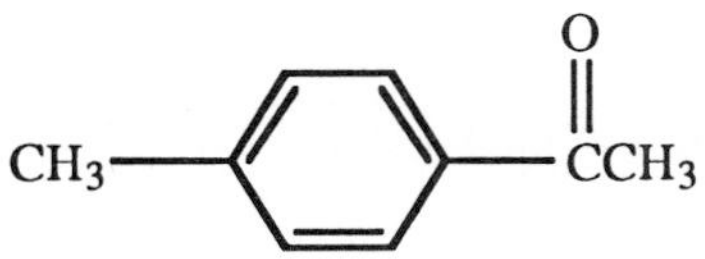

(e)

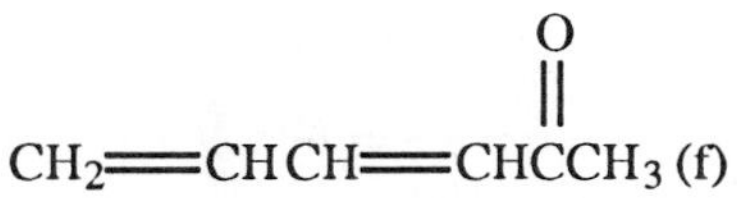

$$CH_2\!\!=\!\!CHCH\!\!=\!\!CHCCH_3$$

(f)

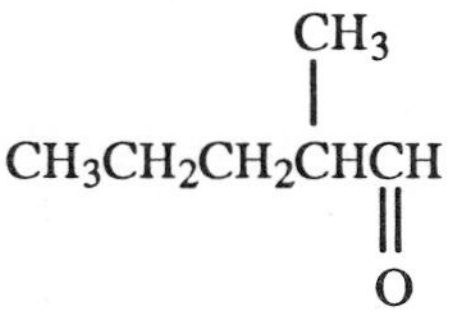

$$CH_3CH_2CH_2CHCH$$

(g)

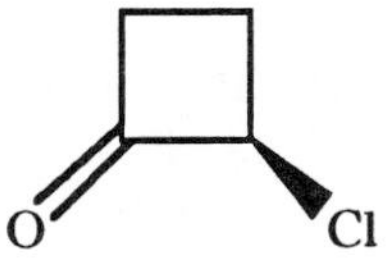

(h)

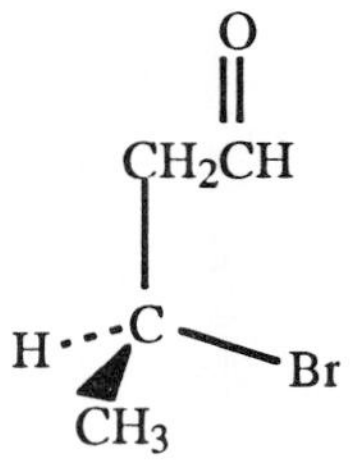

13.37

(a)

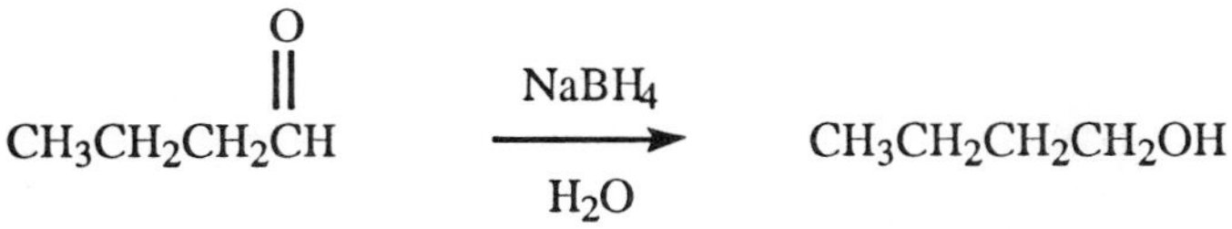

$$CH_3CH_2CH_2CH \xrightarrow[H_2O]{NaBH_4} CH_3CH_2CH_2CH_2OH$$

(b)

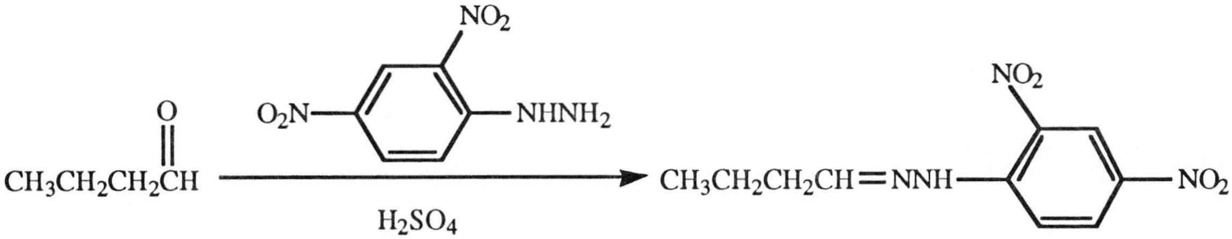

$$CH_3CH_2CH_2CH \xrightarrow[H_2SO_4]{} CH_3CH_2CH_2CH\!\!=\!\!NNH$$

(c)

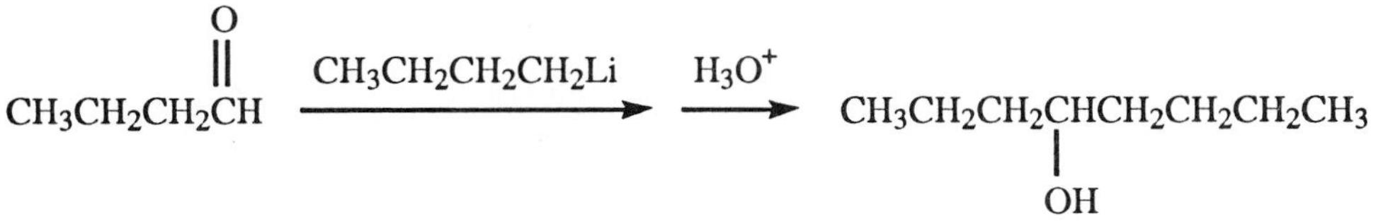

$$CH_3CH_2CH_2CH \xrightarrow[]{CH_3CH_2CH_2CH_2Li \quad H_3O^+} \underset{\substack{| \\ OH}}{CH_3CH_2CH_2CHCH_2CH_2CH_2CH_3}$$

(d)

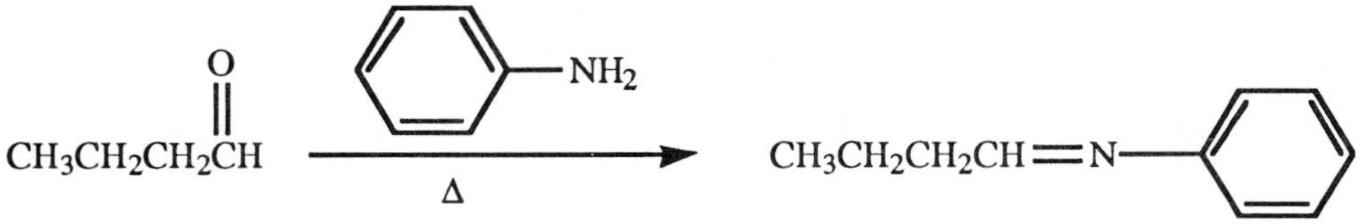

$$CH_3CH_2CH_2CH \xrightarrow[\Delta]{} CH_3CH_2CH_2CH\!\!=\!\!N$$

(e)

$$CH_3CH_2CH_2CH \xrightarrow[\substack{H_2O \\ H_2SO_4}]{NaCN} \underset{\substack{| \\ OH}}{CH_3CH_2CH_2CHCN}$$

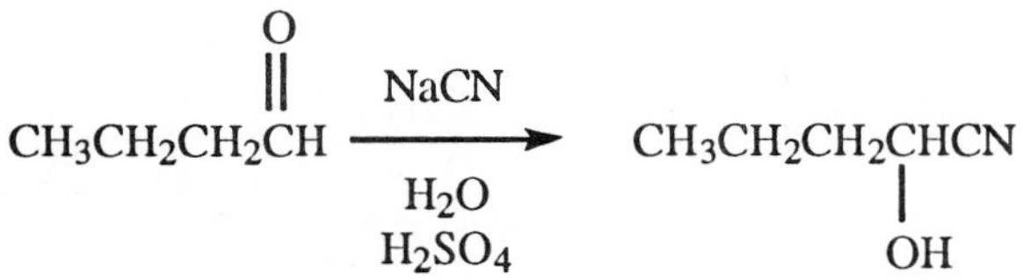

13.37 (cont)

(f)
$$CH_3CH_2CH_2\overset{O}{\underset{\|}{C}}H \xrightarrow[\substack{H_3O^+ \\ H_2O}]{Na_2Cr_2O_7} CH_3CH_2CH_2\overset{O}{\underset{\|}{C}}OH \quad + \quad Cr^{+3}$$

(g)
$$CH_3CH_2CH_2\overset{O}{\underset{\|}{C}}H \xrightarrow[\substack{CH_3CO^-Na^+ \\ \| \\ O}]{HONH_3^+Cl^-} CH_3CH_2CH_2CH{=}NOH$$

(h)
$$CH_3CH_2CH_2\overset{O}{\underset{\|}{C}}H \xrightarrow[\text{diethyl ether}]{CH_3MgI} CH_3CH_2CH_2\underset{\underset{O^-Mg^{2+}I^-}{|}}{C}HCH_3 \xrightarrow{H_3O^+} CH_3CH_2CH_2\underset{\underset{OH}{|}}{C}HCH_3$$

(i)
$$CH_3CH_2CH_2\overset{O}{\underset{\|}{C}}H \xrightarrow[\substack{\text{diethyl} \\ \text{ether}}]{LiAlH_4 \quad H_3O^+} CH_3CH_2CH_2CH_2OH$$

(j)
$$CH_3CH_2CH_2\overset{O}{\underset{\|}{C}}H \xrightarrow[\substack{\text{diethylene glycol} \\ \Delta}]{H_2NNH_2,\ KOH} CH_3CH_2CH_2CH_3$$

(k)
$$CH_3CH_2CH_2\overset{O}{\underset{\|}{C}}H \xrightarrow[\Delta]{HCl,\ Zn(Hg)} CH_3CH_2CH_2CH_3$$

13.38

(a)
$$CH_3CH_2CH_2\overset{O}{\underset{\|}{C}}CH_3 \xrightarrow[H_2O]{NaBH_4} CH_3CH_2CH_2\underset{\underset{OH}{|}}{C}HCH_3$$

(b)
$$CH_3CH_2CH_2\overset{O}{\underset{\|}{C}}CH_3 \xrightarrow[H_2SO_4]{} CH_3CH_2CH_2\underset{\underset{CH_3}{|}}{C}{=}NNH{-}$$

with reagent 2,4-dinitrophenylhydrazine and product bearing the 2,4-dinitrophenyl group (NO_2 substituents).

13.38 (cont)

(c)

$$CH_3CH_2CH_2\overset{\displaystyle O}{\overset{\|}{C}}CH_3 \xrightarrow[\Delta]{\text{⬡—NH}_2} CH_3CH_2CH_2\overset{CH_3}{C}=N\text{—⬡}$$

(d)

$$CH_3CH_2CH_2\overset{\displaystyle O}{\overset{\|}{C}}CH_3 \xrightarrow[\text{tetrhydrofuran}]{\text{⬡—Li}} CH_3CH_2CH_2\overset{CH_3}{\underset{O^-Li^+}{C}}\text{—⬡} \xrightarrow[H_2O]{NH_4Cl}$$

$$CH_3CH_2CH_2\overset{CH_3}{\underset{OH}{C}}\text{—⬡}$$

(e)

$$CH_3CH_2CH_2\overset{\displaystyle O}{\overset{\|}{C}}CH_3 \xrightarrow[\substack{H_2O \\ H_2SO_4}]{NaCN} CH_3CH_2CH_2\overset{CH_3}{\underset{OH}{C}}CN$$

(f)

$$CH_3CH_2CH_2\overset{\displaystyle O}{\overset{\|}{C}}CH_3 \xrightarrow[\substack{CH_3\overset{\|}{\underset{O}{C}}O^-Na^+}]{HONH_3^+Cl^-} CH_3CH_2CH_2\overset{CH_3}{C}=NOH$$

(g)

$$CH_3CH_2CH_2\overset{\displaystyle O}{\overset{\|}{C}}CH_3 \xrightarrow[\text{diethyl ether}]{CH_3CH_2MgI} CH_3CH_2CH_2\overset{CH_3}{\underset{O^-Mg^{2+}Br^-}{C}}CH_2CH_3 \xrightarrow{H_3O^+} CH_3CH_2CH_2\overset{CH_3}{\underset{OH}{C}}CH_2CH_3$$

(h)

$$CH_3CH_2CH_2\overset{\displaystyle O}{\overset{\|}{C}}CH_3 \xrightarrow{CH_3C\equiv C^-Na^+} CH_3CH_2CH_2\overset{CH_3}{\underset{O^-Na^+}{C}}C\equiv CCH_3 \xrightarrow{NH_4Cl} CH_3CH_2CH_2\overset{CH_3}{\underset{OH}{C}}C\equiv CCH_3$$

(i)

$$CH_3CH_2CH_2\overset{\displaystyle O}{\overset{\|}{C}}CH_3 \xrightarrow[\substack{\text{diethylene glycol} \\ \Delta}]{H_2NNH_2, KOH} CH_3CH_2CH_2CH_2CH_3$$

(j)

$$CH_3CH_2CH_2\overset{\displaystyle O}{\overset{\|}{C}}CH_3 \xrightarrow[\Delta]{HCl, Zn(Hg)} CH_3CH_2CH_2CH_2CH_3$$

13.39

(a)

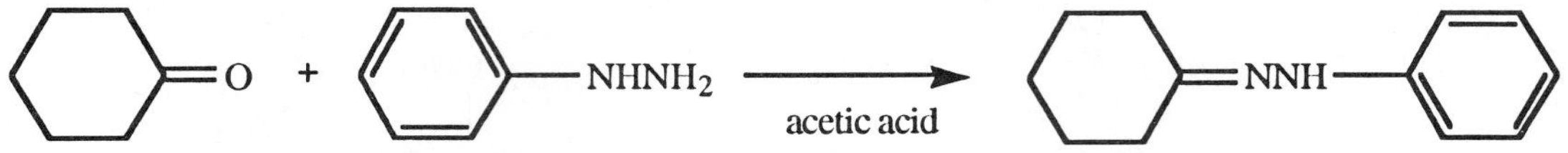

(b)

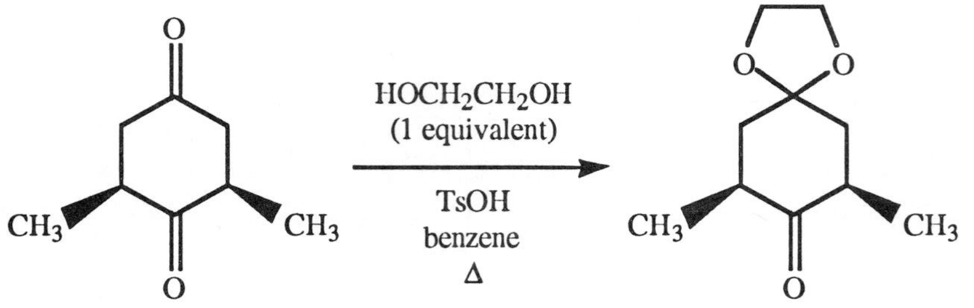

(c)

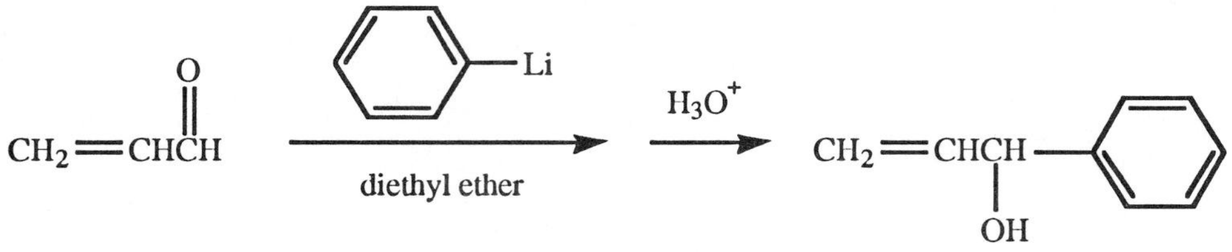

(d)

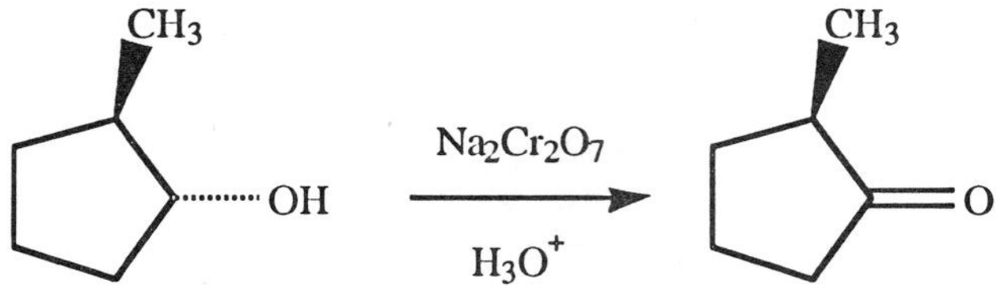

(e)

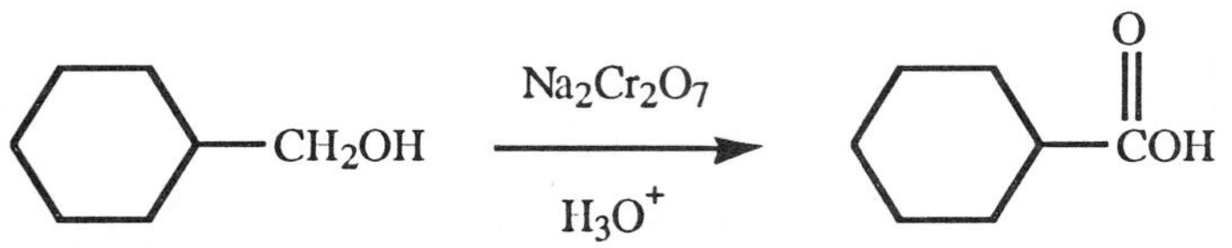

(f)

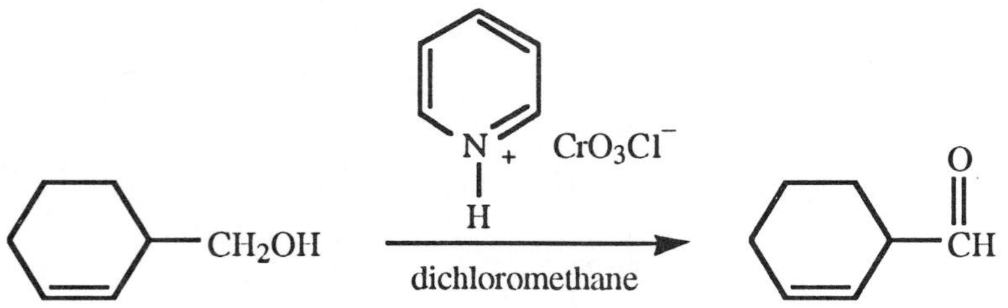

(g)

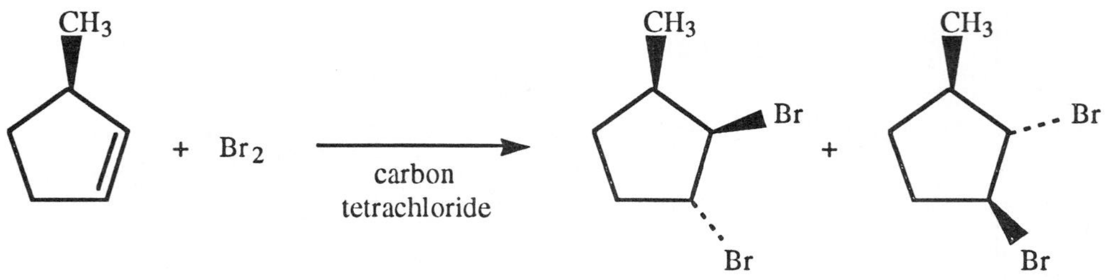

13.39 (cont)

(h)

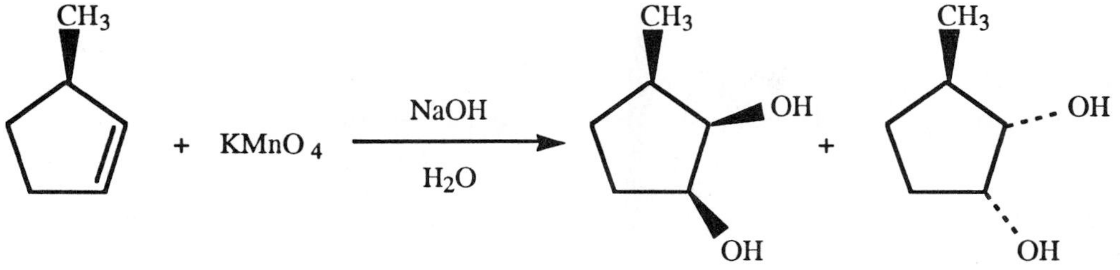

(i)

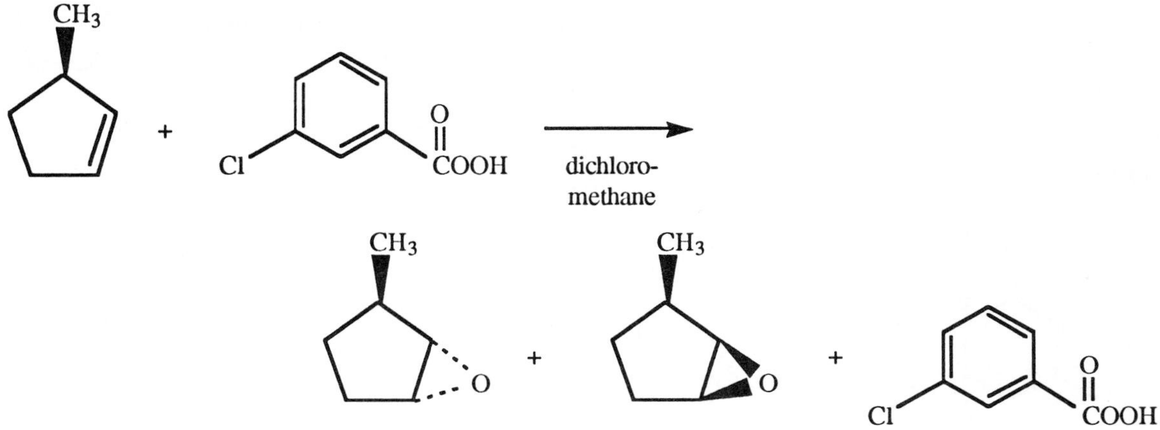

(j)

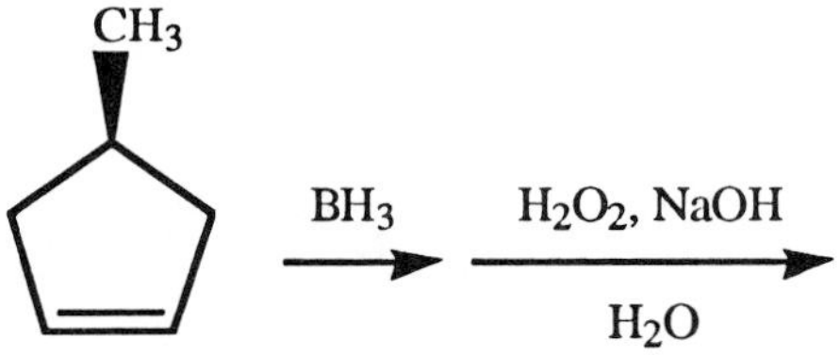

[See Problem 12.41 (b)]

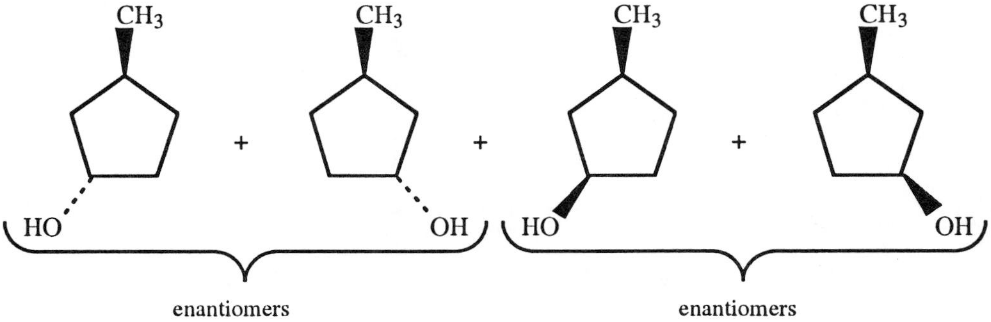

enantiomers   enantiomers

(k)

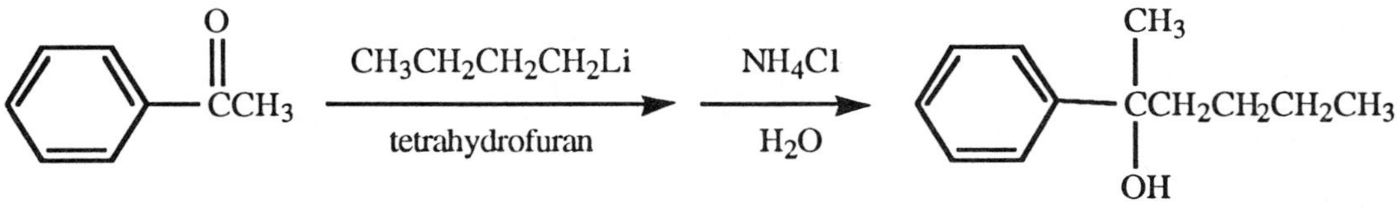

13.39 (cont)

(l)

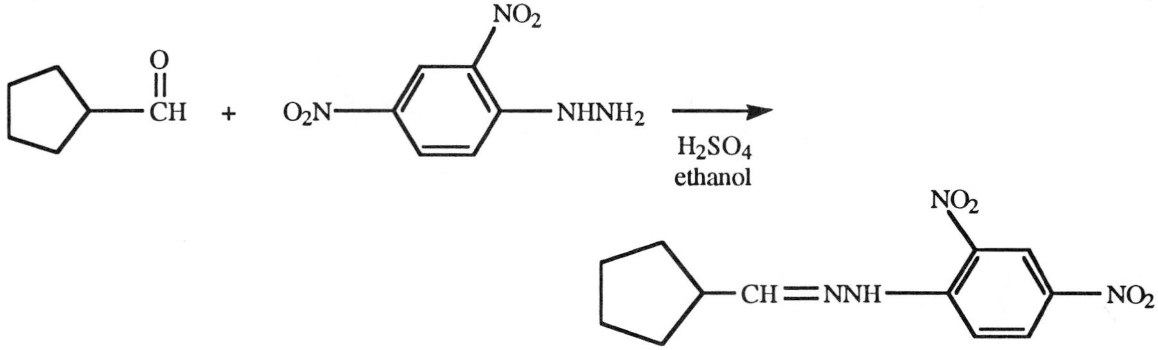

13.40

(a)

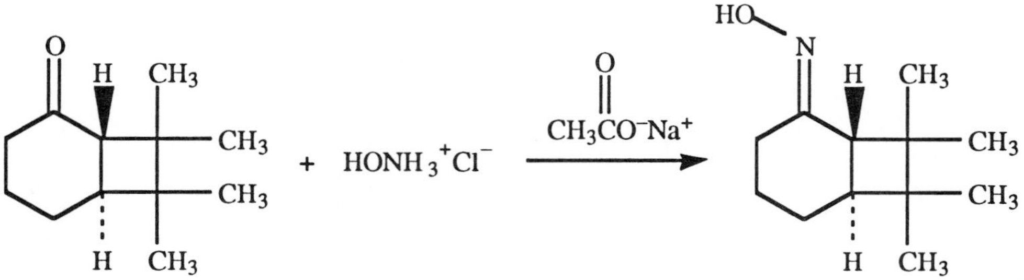

(b)

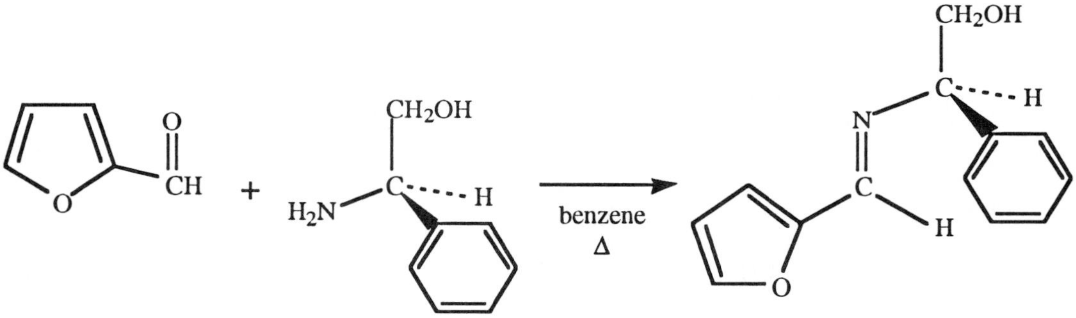

(c) $CH_3CH_2CH_2C\equiv CH$ + $NaNH_2$ $\longrightarrow$ $CH_3CH_2CH_2C\equiv C^-\,Na^+$

NH$_3$ (liq)

(d) $CH_3CH_2CH_2C\equiv CH$ + CH_3CH_2MgBr $\longrightarrow$ $CH_3CH_2CH_2C\equiv CMgBr$ + $CH_3CH_3\uparrow$

diethyl ether

(e)

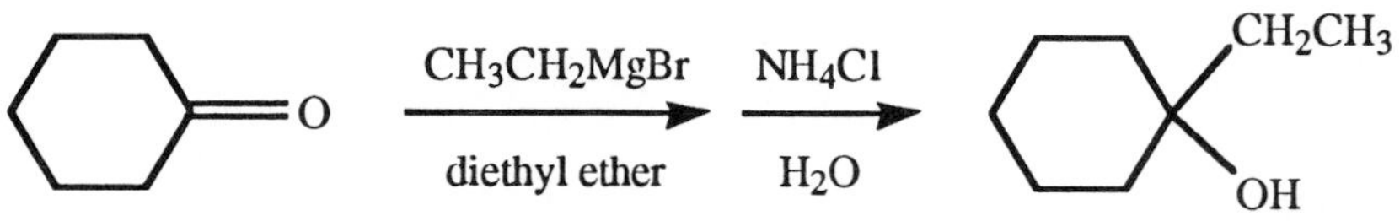

(f)

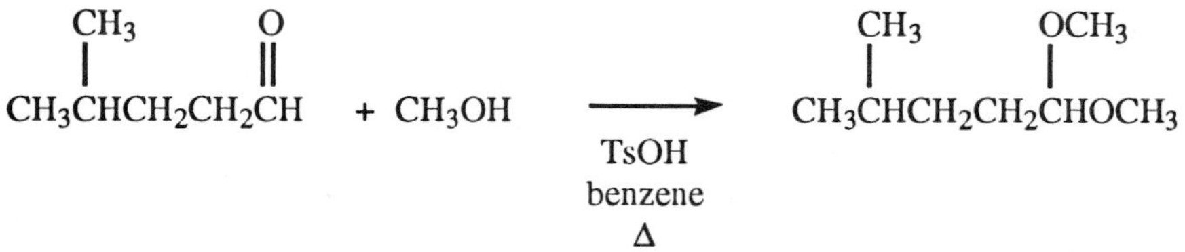

13.40 (cont)

(g)

CH$_3$—⬡—CCH$_2$CH$_3$ (with =O)   +   HOCH$_2$CH$_2$OH $\xrightarrow[\substack{\text{benzene} \\ \Delta}]{\text{TsOH}}$ CH$_3$—⬡—C(O-CH$_2$CH$_2$-O)CH$_2$CH$_3$

(h)

⬡=O   +   HSCH$_2$CH$_2$OH $\xrightarrow{(CH_3CH_2)_2OBF_3}$ [cyclohexane spiro ring with O and S]

(i)

$$\underset{\text{O}}{\overset{\parallel}{CH_3C}}CH_2\underset{\overset{|}{CH_3}}{CH}CH_3 \;+\; CH_3\underset{\overset{|}{CH_3}}{CH}CH_2NH_2 \;\xrightarrow[\text{HCl}]{\text{NaOH}}\; CH_3\underset{\overset{\parallel}{NCH_2\underset{\overset{|}{CH_3}}{CH}CH_3}}{\overset{}{C}}CH_2\underset{\overset{|}{CH_3}}{CH}CH_3$$

(j)

$$CH_3CH_2C\equiv C\underset{\overset{|}{\text{OTHP}}}{CH}CH_2CH_2CH=CH\underset{\overset{\parallel}{O}}{C}CH_3 \;\xrightarrow[\text{base}]{CH_3ONH_3^+ \, Cl^-}$$

$$CH_3CH_2C\equiv C\underset{\overset{|}{\text{OTHP}}}{CH}CH_2CH_2CH=CH\underset{\overset{\parallel}{NOCH_3}}{C}CH_3$$

(k)

$$HOCH_2CH_2\underset{\overset{|}{CH_3}}{\overset{\overset{\displaystyle CH_3}{|}}{C}}CH_2CH=CH\underset{\overset{\parallel}{O}}{C}OCH_3 \;\xrightarrow[\text{dichloromethane}]{\substack{[\text{pyridinium}]\,CrO_3Cl^-}}\; HC\underset{\overset{|}{CH_3}}{\overset{\overset{\displaystyle O \quad CH_3}{\parallel \quad |}}{C}}CH_2CH=CH\underset{\overset{\parallel}{O}}{C}OCH_3$$

(l)

$$\underset{\overset{\parallel}{O}}{CH_3C}(CH_2)_8CH_3 \;\xrightarrow[\Delta]{\text{HCl, Zn(Hg)}}\; CH_3CH_2(CH_2)_8CH_3$$

(m)

$$\underset{\overset{|}{CH_3}}{CH_3CH}CH_2\underset{\overset{\parallel}{O}}{C}CH_2\underset{\overset{|}{CH_3}}{CH}CH_3 \;\xrightarrow[\substack{\text{diethylene glycol} \\ \Delta}]{H_2NNH_2, \text{ NaOH}}\; \underset{\overset{|}{CH_3}}{CH_3CH}CH_2CH_2CH_2\underset{\overset{|}{CH_3}}{CH}CH_3$$

13.41

(a)

⬡—CCH$_2$CH$_2$CH$_2$Cl (with =O) $\xrightarrow[\substack{\text{TsOH} \\ \text{benzene, }\Delta}]{HOCH_2CH_2OH}$

13.41 (a) (cont)

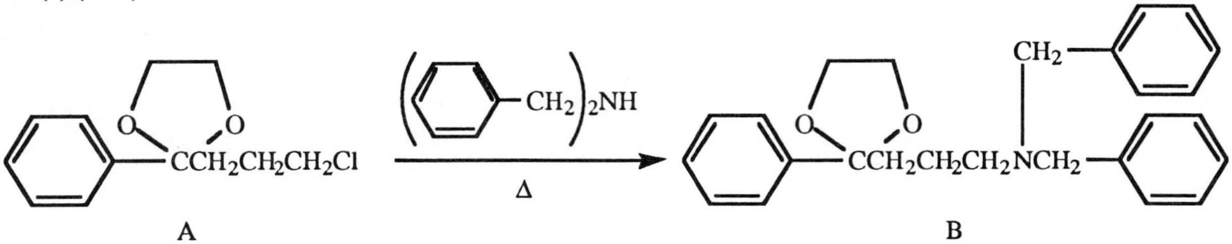

A

B

(b)

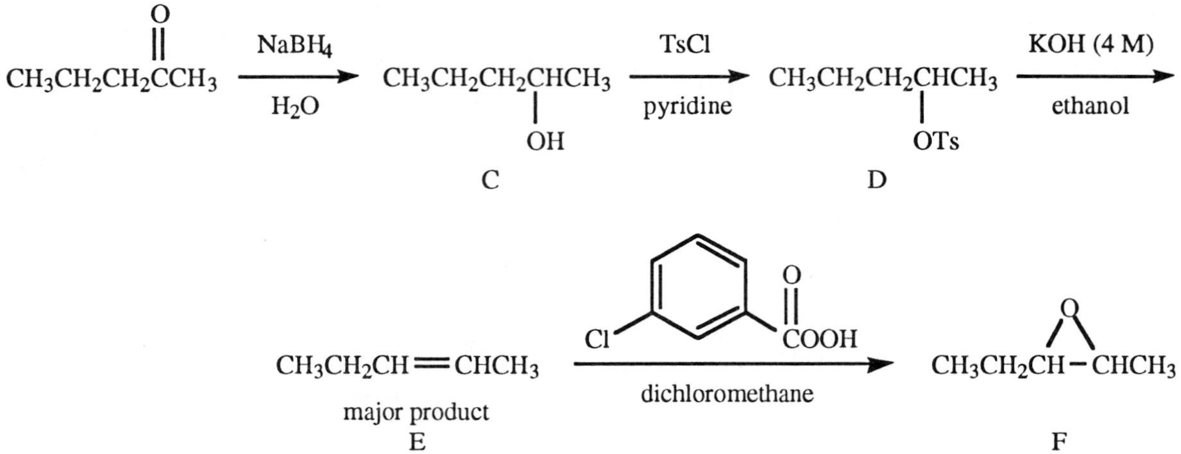

(c)

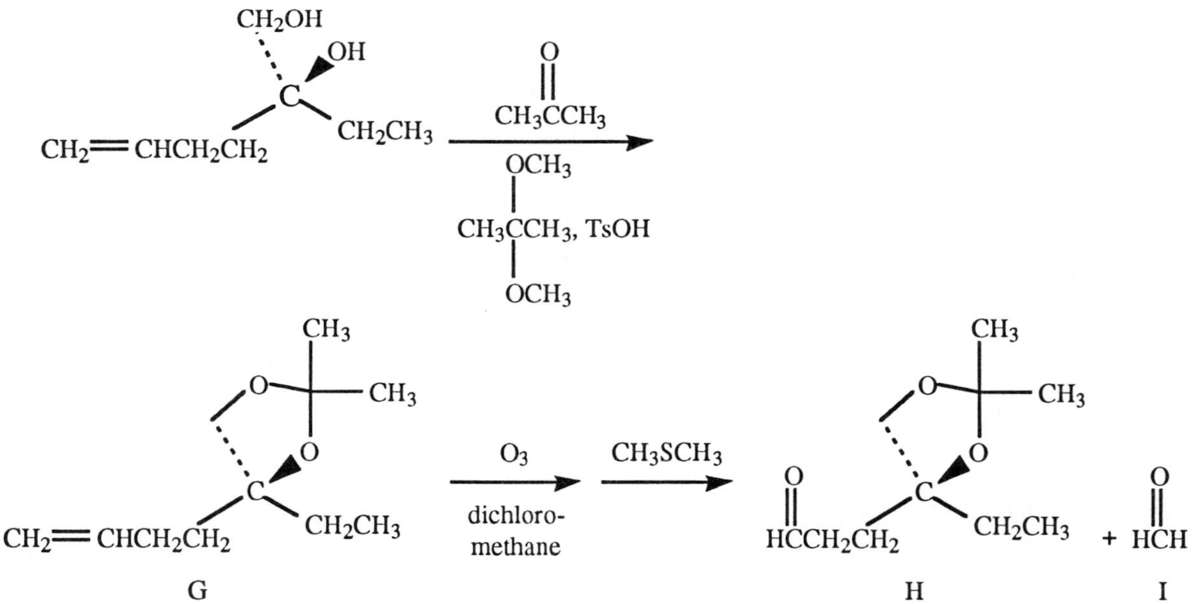

(d)

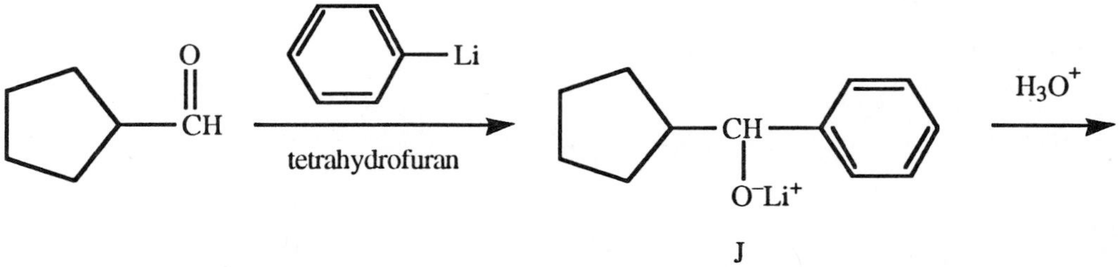

13.41 (d) (cont)

K

$\xrightarrow[\text{H}_3\text{O}^+]{\text{CrO}_3}$

L

$\xrightarrow[\text{acetic acid}]{\text{—NHNH}_2}$

M

(e)

$\text{CH}_3\text{C}=\text{CHCH}_3$ with CH_3 substituent $\xrightarrow[\text{tetrahydro-furan}]{\text{BH}_3}$ N $\xrightarrow[\text{H}_2\text{O}]{\text{H}_2\text{O}_2, \text{NaOH}}$ O $\xrightarrow[\substack{\text{H}_2\text{SO}_4 \\ \text{H}_2\text{O}}]{\text{NaCr}_2\text{O}_7}$

N: $\text{CH}_3\text{C}-\text{CHCH}_3$ with CH_3, H, BR_2

O: $\text{CH}_3\text{CHCHCH}_3$ with CH_3, OH

$\text{CH}_3\text{CHCCH}_3$ with CH_3 and O (P) $\xrightarrow{\text{NaCN}}$ $\text{CH}_3\text{CH}-\text{CCN}$ with CH_3, CH_3, O^-Na^+ (Q) $\xrightarrow{\text{H}_3\text{O}^+}$ $\text{CH}_3\text{CH}-\text{CCN}$ with CH_3, CH_3, OH (R)

(f)

$\text{HC}\equiv\text{CH}$ $\xrightarrow[\text{NH}_3 \text{ (liq)}]{\text{NaNH}_2}$ $\text{HC}\equiv\text{C}^-\text{Na}^+$ (S) $\xrightarrow[\text{diethyl ether}]{\text{—CH}_2\text{Br}}$

T: —$\text{CH}_2\text{C}\equiv\text{CH}$ $\xrightarrow[\text{diethyl ether}]{\text{CH}_3\text{CH}_2\text{MgBr}}$ U: —$\text{CH}_2\text{C}\equiv\text{CMgBr}$ $\longrightarrow$

V: —$\text{CH}_2\text{C}\equiv\text{C}$ with cyclopentane $\text{Br}^-\text{Mg}^{2+}-\text{O}$ $\xrightarrow[\text{H}_2\text{O}]{\text{NH}_4\text{Cl}}$ W: —$\text{CH}_2\text{C}\equiv\text{C}$ with cyclopentane HO

13.41 (cont)

(g)

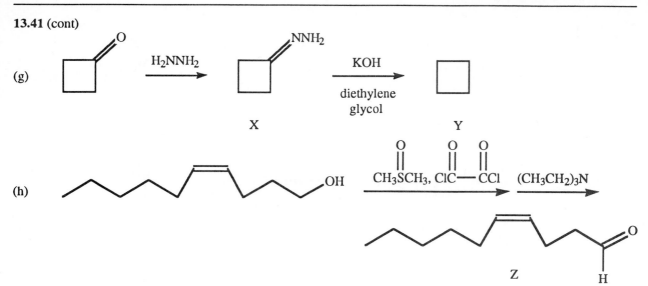

(h)

13.42

(a)   HC≡CCH₂CH₂OH   →(TsOH, Δ)→   HC≡CCH₂CH₂OTHP   →(CH₃CH₂MgBr)→
                                                    A

BrMgC≡CCH₂CH₂OTHP   →(CH₃(CH₂)₅CHO)→   CH₃(CH₂)₅CHC≡CCH₂CH₂OTHP   →(H₃O⁺)→
                                                       |
                                                   O⁻ Mg²⁺ Br⁻
         B   C

CH₃(CH₂)₅CHC≡CCH₂CH₂OH   →(H₂, Pd/CaCO₃, quinoline, ethyl acetate)→   E
        |
        OH
         D

(b)   C₆H₅—Br   →(Mg, diethyl ether)→   C₆H₅—MgBr   →(epoxide)→
                                              F

13.42 (b) (cont)

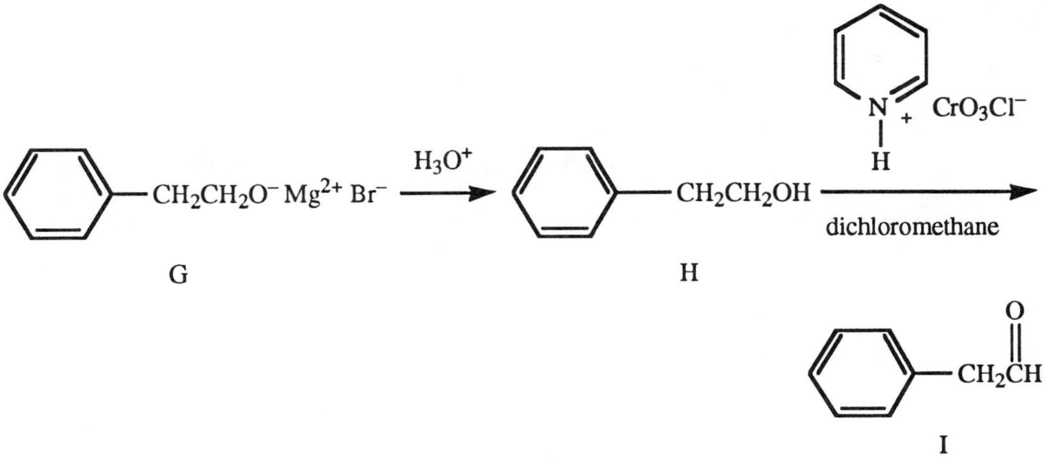

G → H

(c)

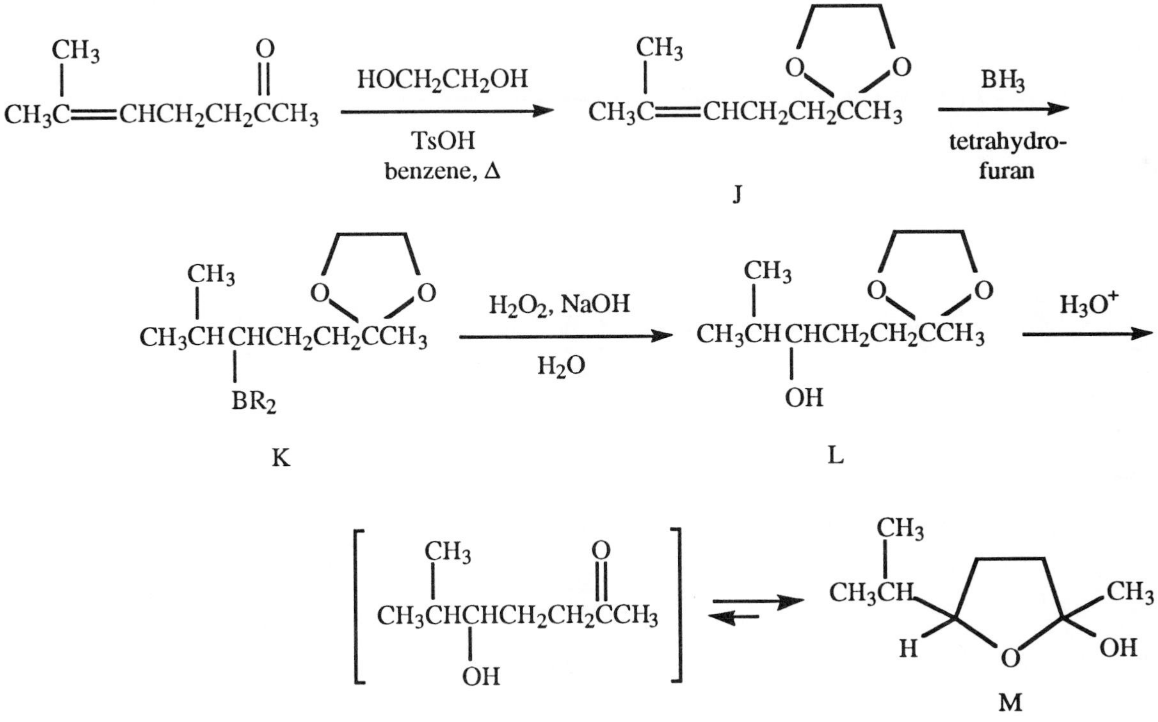

J → K → L

M

(d)

N

O → P → Q

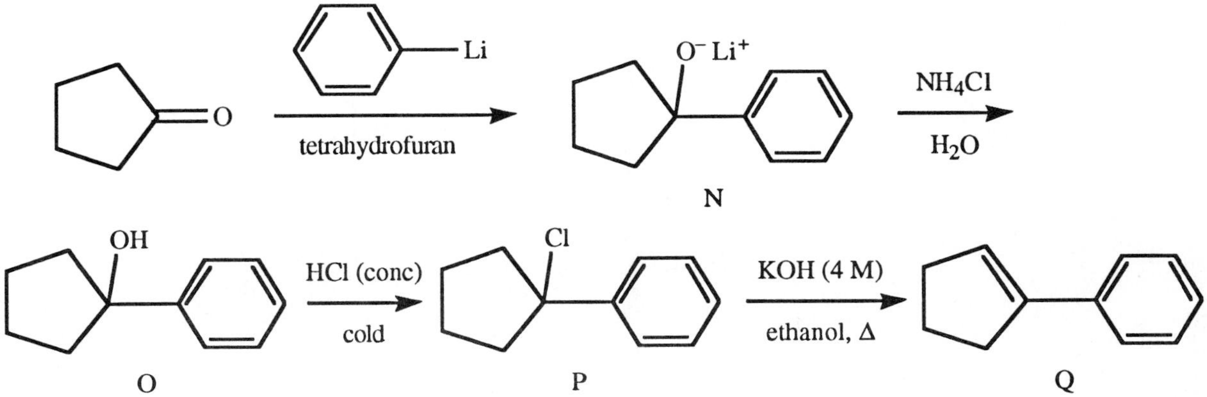

13.42 (cont)

(e)

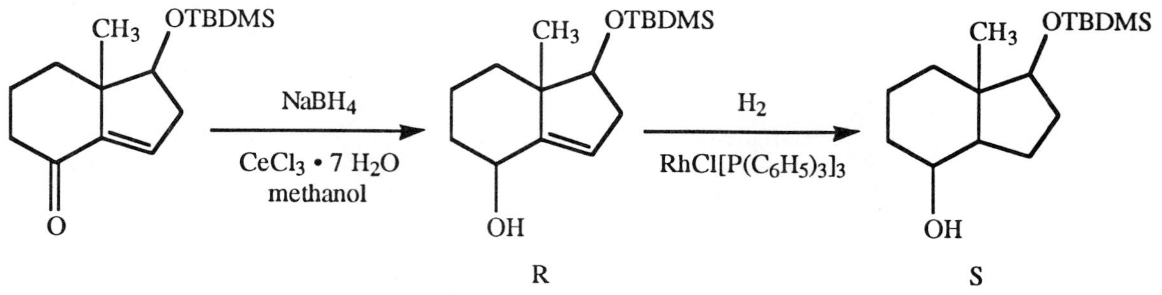

(f)

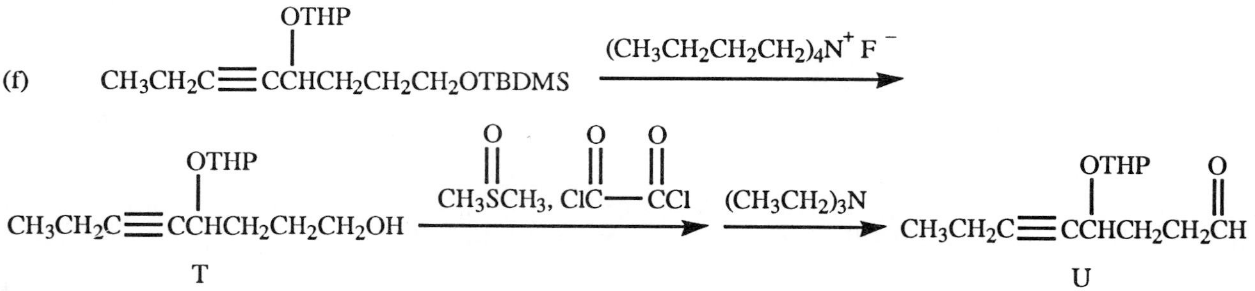

$$CH_3CH_2C\equiv CCHCH_2CH_2CH_2OTBDMS \xrightarrow{(CH_3CH_2CH_2CH_2)_4N^+\,F^-}$$

with OTHP on the carbon

$$CH_3CH_2C\equiv CCHCH_2CH_2CH_2OH \xrightarrow[\quad]{\begin{array}{c}O \quad O \quad O\\ \| \quad \| \quad \|\\ CH_3SCH_3,\ ClC\!-\!CCl\end{array}} \xrightarrow{(CH_3CH_2)_3N} CH_3CH_2C\equiv CCHCH_2CH_2CH$$

with OTHP, T under the left structure, OTHP and O on the right structure (U)

(g)

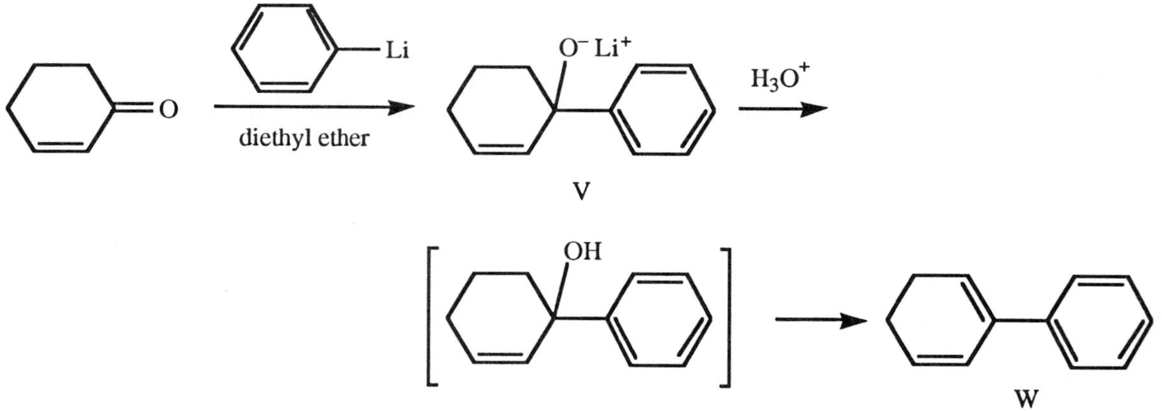

Note that a tertiary alcohol is easily dehydrated, especially when the double bond that is formed is conjugated with an aromatic ring. If we wish to have the alcohol as the product, we use a weak acid, such as NH_4Cl, to protonate the alkoxide.

(h)

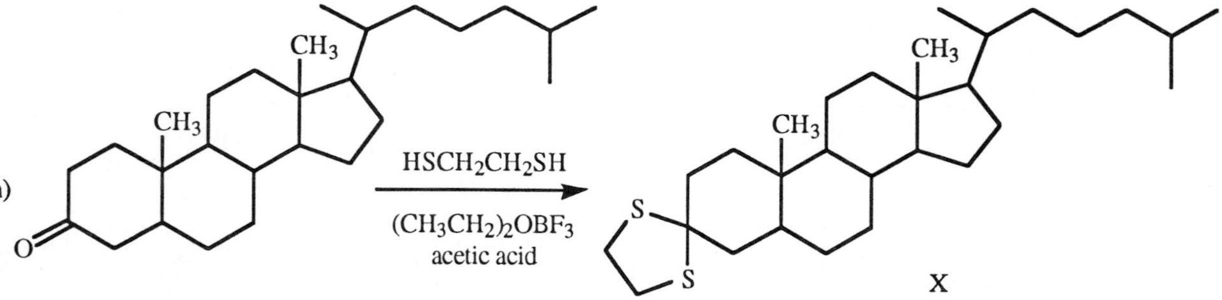

13.42 (cont)

(i)

$$CH_2=CHCH_2CH_2-\!\!\!\bigcirc\!\!=O \xrightarrow[\text{chloroform}]{Br_2}$$

$$BrCH_2CHCH_2CH_2-\!\!\!\bigcirc\!\!=O \underset{Br}{} \xrightarrow[\substack{\text{TsOH}\\\text{benzene, }\Delta}]{HOCH_2CH_2OH} BrCH_2CHCH_2CH_2-\!\!\!\bigcirc\!\!\bigcirc\!\!\bigcirc$$

Y   Br   Z

(j)

$$\underset{CH_3CHCH_2CH_2CCH_2CH_2COH}{\overset{CH_3\quad\quad O\quad\quad O}{|\quad\quad\parallel\quad\quad\parallel}} \xrightarrow[\substack{\text{diethylene glycol}\\\Delta}]{H_2NNH_2,\ NaOH}$$

$$\underset{CH_3CHCH_2CH_2CH_2CH_2CH_2CO^-Na^+}{\overset{CH_3}{|}} \xrightarrow{H_3O^+} \underset{CH_3CHCH_2CH_2CH_2CH_2CH_2COH}{\overset{CH_3\quad\quad\quad\quad\quad\quad\quad O}{|\quad\quad\quad\quad\quad\quad\quad\parallel}}$$

AA   BB

(k)

$$\underset{CH_3CH_2CH_2CH_2C-\!\!\!\bigcirc}{\overset{O}{\parallel}} \xrightarrow[\Delta]{HCl,\ Zn(Hg)} CH_3CH_2CH_2CH_2CH_2-\!\!\!\bigcirc$$

CC

(l)

13.43

(a)

13.43 (a) (cont)

$$CH_2\overset{\displaystyle O}{-\!\!-}CHCH_3$$

Ph–CH₂CHCH₃ with O⁻ Mg²⁺ Br⁻ ← Ph–MgBr ←(Mg, diethyl ether)— Ph–Br

(b)

$$\underset{H}{\overset{CH_3CH_2}{>}}C=C\underset{H}{\overset{CH_2CH_2OH}{<}}$$ ←(H₂, Pd/CaCO₃, quinoline)— $CH_3CH_2C\equiv CCH_2CH_2OH$ ←(H₃O⁺)—

$$CH_3CH_2C\equiv CCH_2CH_2O^-Mg^{2+}Br^-$$ ←(epoxide)— $CH_3CH_2C\equiv CMgBr$ ←(CH₃CH₂MgBr)—

$CH_3CH_2C\equiv CH$ ←(CH₃CH₂Br)— $Na^+\,{}^-C\equiv CH$ ←(NaNH₂, NH₃ (liq))— $HC\equiv CH$

(c)

$$\underset{}{CH_3CH_2CH_2\overset{\displaystyle O}{\overset{\|}{C}}CH_2CH_3}$$ ←(Na₂Cr₂O₇, H₃O⁺)— $$CH_3CH_2CH_2\underset{OH}{CH}CH_2CH_3$$ ←(H₃O⁺)—

$$CH_3CH_2CH_2\underset{O^-Mg^{2+}Br^-}{CH}CH_2CH_3$$ ←($\overset{\displaystyle O}{\overset{\|}{CH_3CH_2CH}}$)— $CH_3CH_2CH_2MgBr$ ←(Mg, diethyl ether)— $CH_3CH_2CH_2Br$

(d)

$$\underset{}{\overset{CH_3}{|}}CH_3C=CHCH_2CH_2CH_3$$
major product
←(H₃PO₄, Δ)—
$$\underset{OH}{\overset{CH_3}{|}}CH_3CCH_2CH_2CH_2CH_3$$ ←(H₃O⁺)—

$$\underset{O^-Mg^{2+}Br^-}{\overset{CH_3}{|}}CH_3CCH_2CH_2CH_2CH_3$$ ←($\overset{\displaystyle O}{\overset{\|}{CH_3CCH_3}}$)— $CH_3CH_2CH_2CH_2MgBr$ ←(Mg, diethyl ether)—

13.43 (d) (cont)

$$CH_3CH_2CH_2CH_2Br \xleftarrow[\text{pyridine}]{PBr_3} CH_3CH_2CH_2CH_2OH \xleftarrow{H_3O^+}$$

$$CH_3CH_2CH_2CH_2O^-Mg^{2+}Br^- \xleftarrow{\triangle O} CH_3CH_2MgBr \xleftarrow[\text{diethyl ether}]{Mg} CH_3CH_2Br$$

(e)

$$\text{C}_6\text{H}_5-CH_2CH_2Br \xleftarrow[\text{pyridine}]{PBr_3} \text{C}_6\text{H}_5-CH_2CH_2OH \xleftarrow{H_3O^+}$$

$$\text{C}_6\text{H}_5-CH_2CH_2O^-Mg^{2+}Br^- \xleftarrow{\triangle O} \text{C}_6\text{H}_5-MgBr \xleftarrow[\text{diethyl ether}]{Mg} \text{C}_6\text{H}_5-Br$$

(f)

$$\xleftarrow[\text{carbon tetrachloride}]{Br_2} \quad \xleftarrow[\text{NH}_2 \text{ (liq)}]{Na}$$

and enantiomer

$$CH_3CH_2CH_2C\equiv CCH_3 \xleftarrow{CH_3I} CH_3CH_2CH_2C\equiv C^-Na^+ \xleftarrow[\text{NH}_3 \text{ (liq)}]{NaNH_2}$$

$$CH_3CH_2CH_2C\equiv CH \xleftarrow{CH_3CH_2CH_2Br} Na^{+-}C\equiv CH \xleftarrow[\text{NH}_3 \text{ (liq)}]{NaNH_2} HC\equiv CH$$

(g)

$$CH_3(CH_2)_8\overset{O}{\overset{\|}{C}}CH=CH_2 \xleftarrow[\text{dichloromethane}]{} CH_3(CH_2)_8\underset{OH}{CH}CH=CH_2 \xleftarrow[\text{H}_2O]{NH_4Cl}$$

$$CH_2=CHCH\overset{O}{\overset{\|}{}} \xleftarrow{} CH_3(CH_2)_7CH_2Li \xleftarrow{Li} CH_3(CH_2)_7CH_2Br \xleftarrow[\text{pyridine}]{PBr_3}$$

13.43 (g) (cont)

$$CH_3(CH_2)_7CH_2OH \xleftarrow{H_3O^+} \xleftarrow{\underset{HCH}{\overset{O}{\parallel}}} CH_3(CH_2)_6CH_2MgBr \xleftarrow[\text{diethyl ether}]{Mg}$$

$$CH_3(CH_2)_6CH_2Br \xleftarrow[\text{pyridine}]{PBr_3} CH_3(CH_2)_6CH_2OH \xleftarrow{H_3O^+} \xleftarrow{\triangle^O} CH_3(CH_2)_4CH_2Br \xleftarrow[\text{diethyl ether}]{Mg}$$

$$CH_3(CH_2)_4CH_2Br \xleftarrow[\text{pyridine}]{PBr_3} CH_3(CH_2)_4CH_2OH \xleftarrow{H_3O^+} \xleftarrow{\triangle^O}$$

$$CH_3CH_2CH_2CH_2MgBr \xleftarrow[\text{diethyl ether}]{Mg} CH_3CH_2CH_2CH_2Br \xleftarrow[\text{pyridine}]{PBr_3} CH_3CH_2CH_2CH_2OH \xleftarrow{H_3O^+}$$

$$\xleftarrow{\triangle^O} CH_3CH_2MgBr \xleftarrow[\text{diethyl ether}]{Mg} CH_3CH_2Br$$

(h)

$$\underset{\text{(benzene ring)}}{} \overset{O}{\underset{\parallel}{C}}CH_2CH_3 \xleftarrow{(CH_3CH_2)_3N} \xleftarrow[\overset{O}{\underset{\parallel}{CH_3SCH_3}}, \overset{O}{\underset{\parallel}{ClC}}-\overset{O}{\underset{\parallel}{CCl}}]{} \underset{\text{(benzene ring)}}{} \underset{\underset{OH}{|}}{CHCH_2CH_3} \xleftarrow[\text{H}_2\text{O}]{NH_4Cl}$$

$$CH_3CH_2\overset{O}{\underset{\parallel}{CH}} \xleftarrow{} \underset{\text{(benzene ring)}}{}-MgBr \xleftarrow[\text{diethyl ether}]{Mg} \underset{\text{(benzene ring)}}{}-Br$$

(i)

and enantiomer   and enantiomer

$H_3O^+ \uparrow$ $\uparrow$

13.43 (i) (cont)

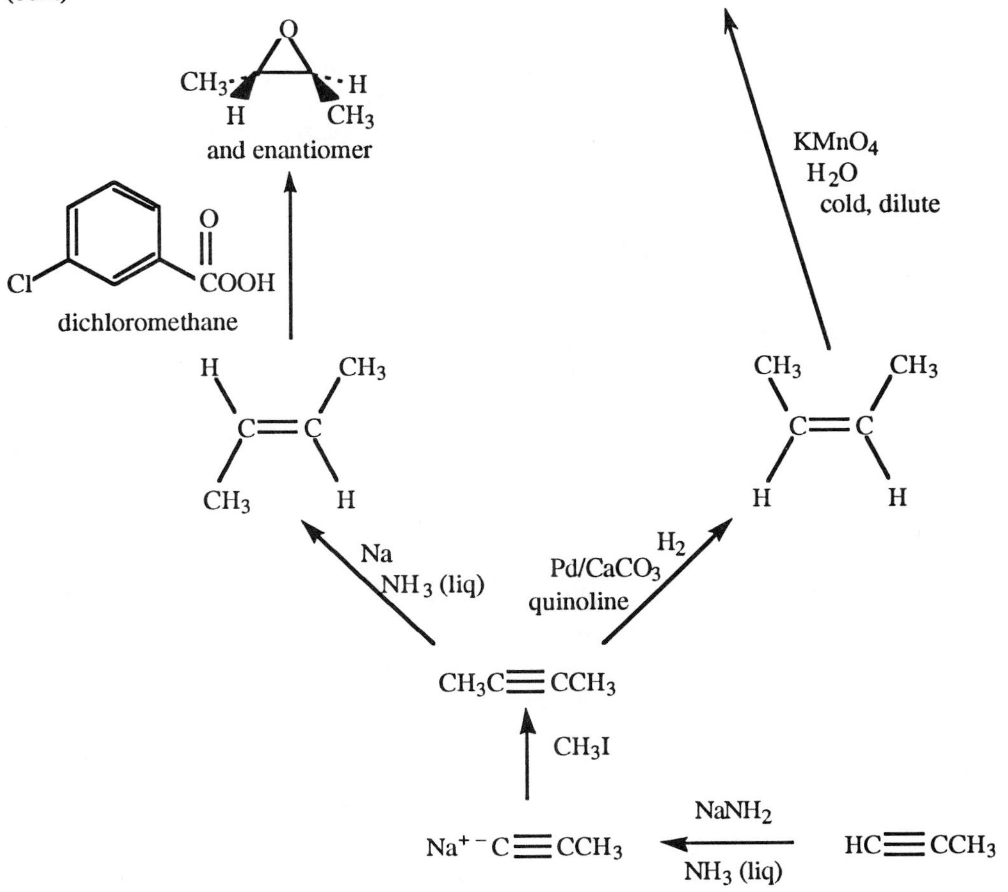

(j)

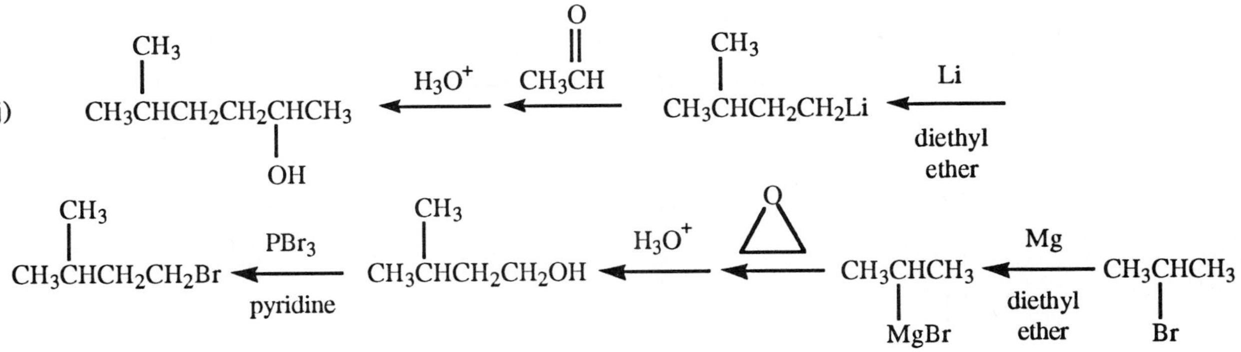

(k)

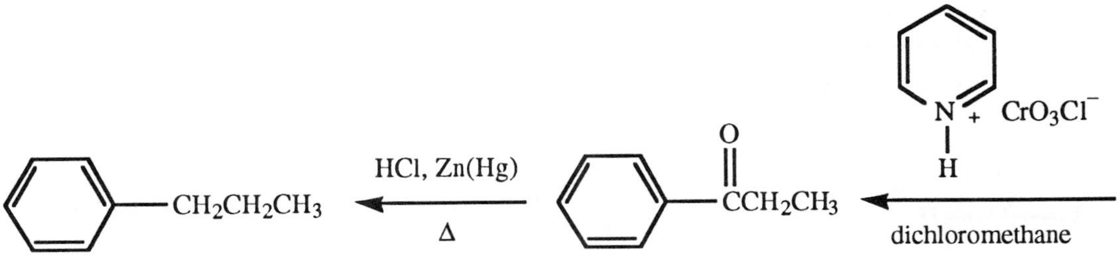

13.43 (k) (cont)

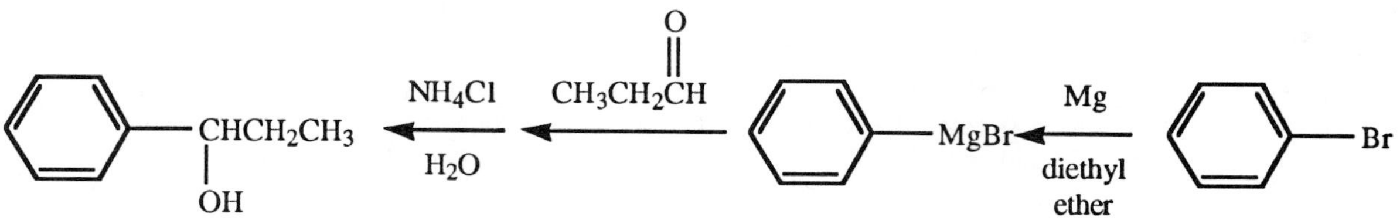

(l)

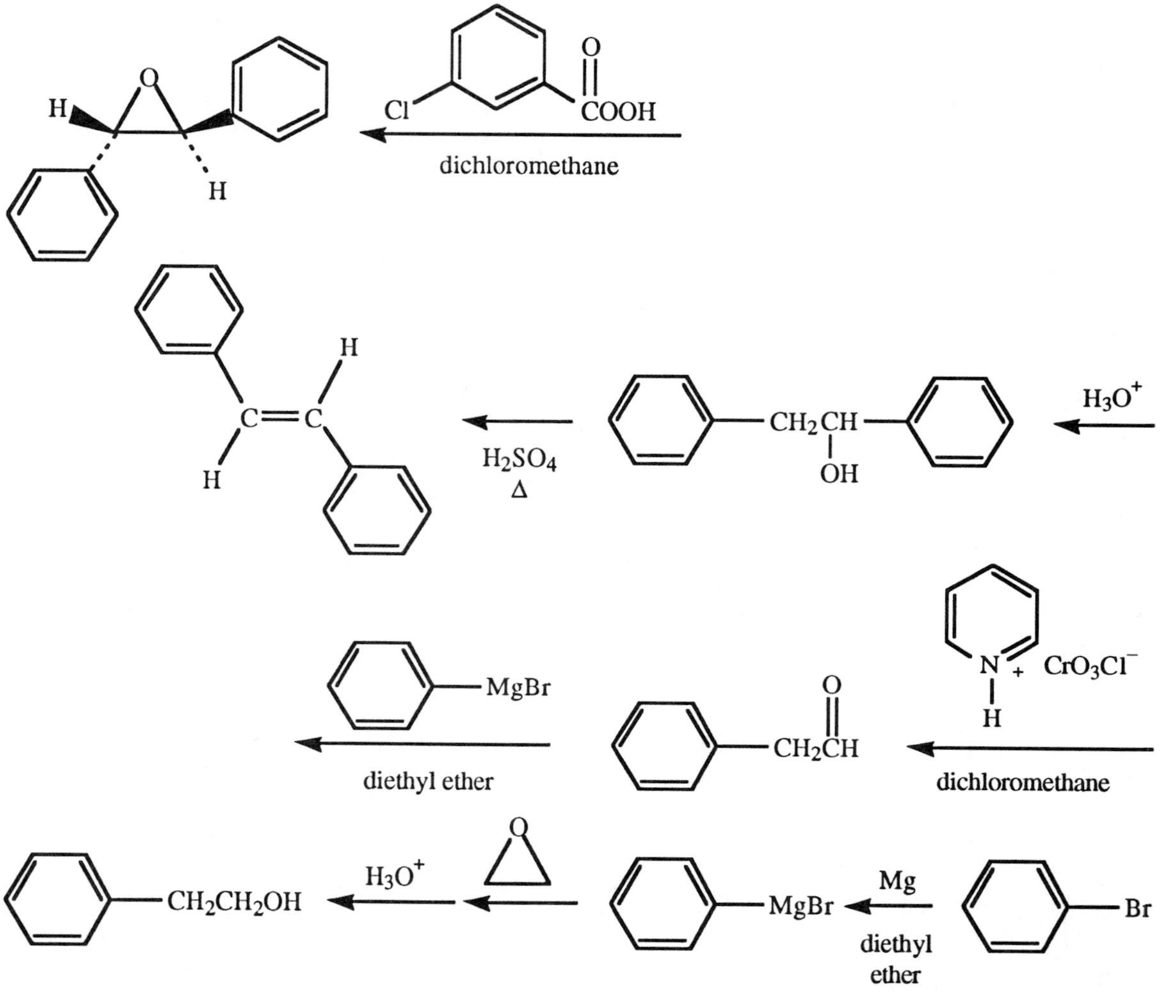

13.44

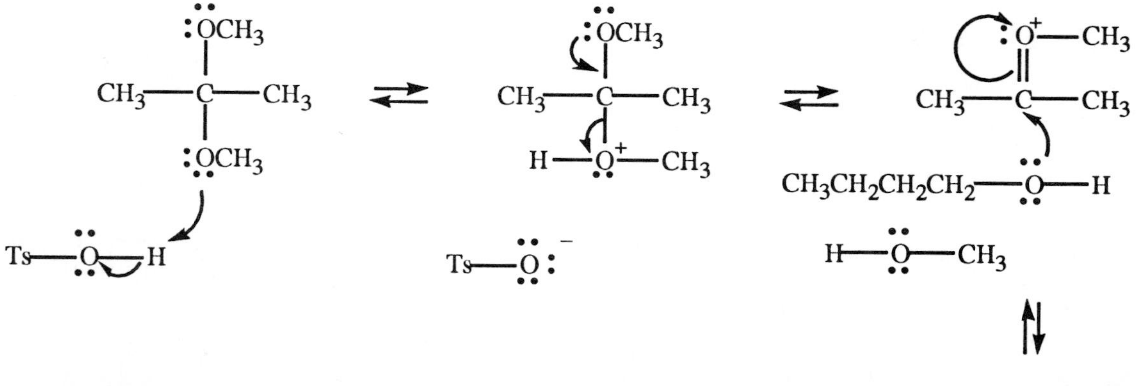

13.44 (cont)

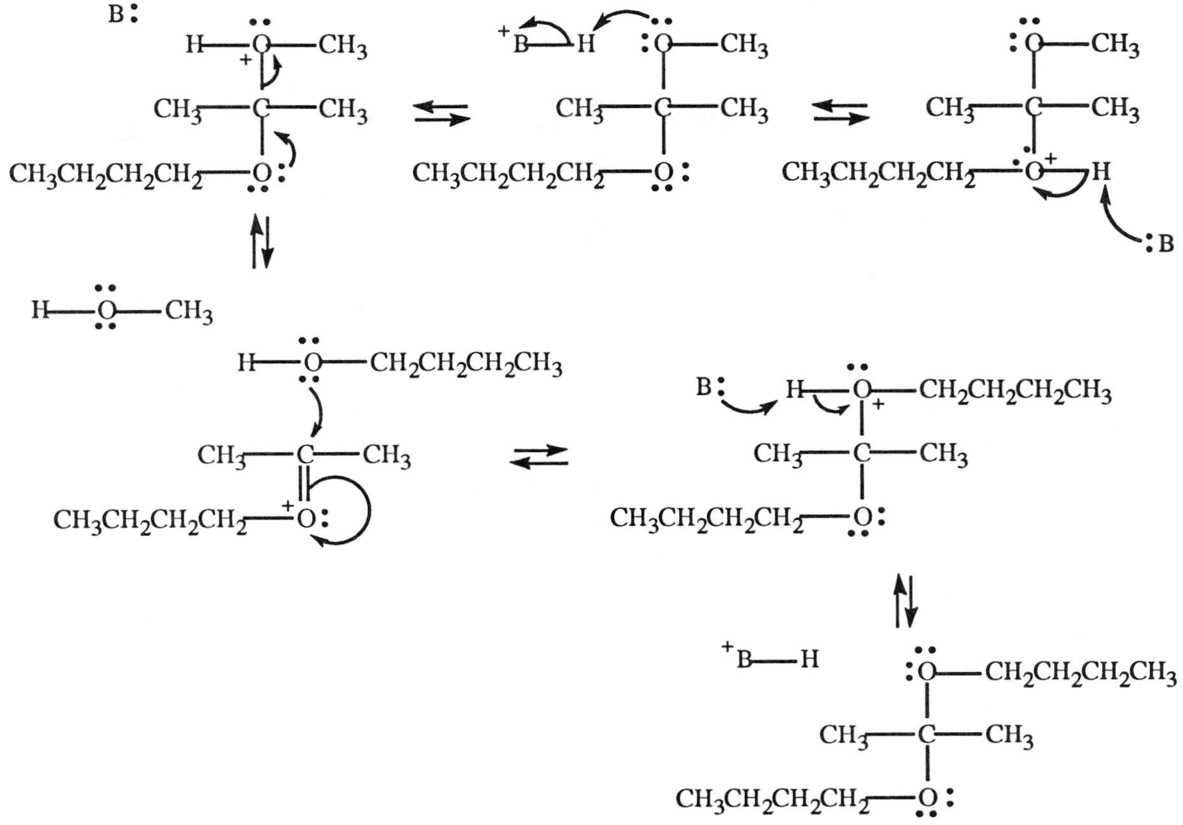

The reaction proceeds by a series of steps, each of which is an equilibrium. Removal by distillation of methanol, the lowest boiling component of the mixture, as it forms, will push the equilibrium towards the dibutoxypropane.

13.45

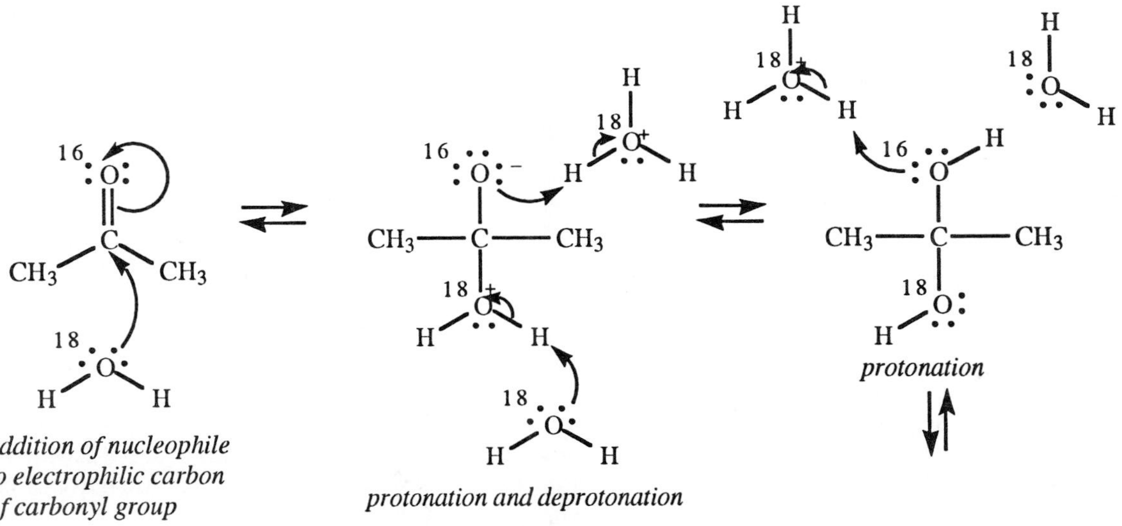

*addition of nucleophile
to electrophilic carbon
of carbonyl group*

protonation and deprotonation

protonation

13.45 (cont)

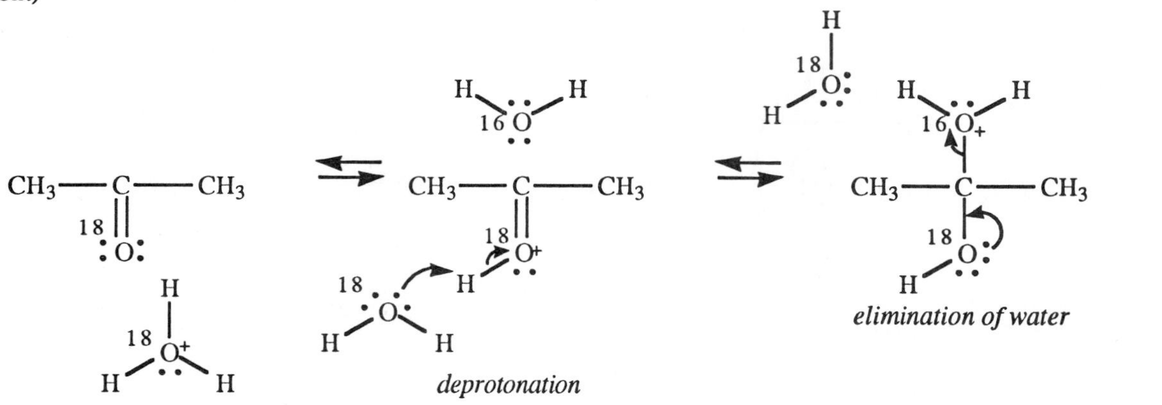

deprotonation   *elimination of water*

13.46

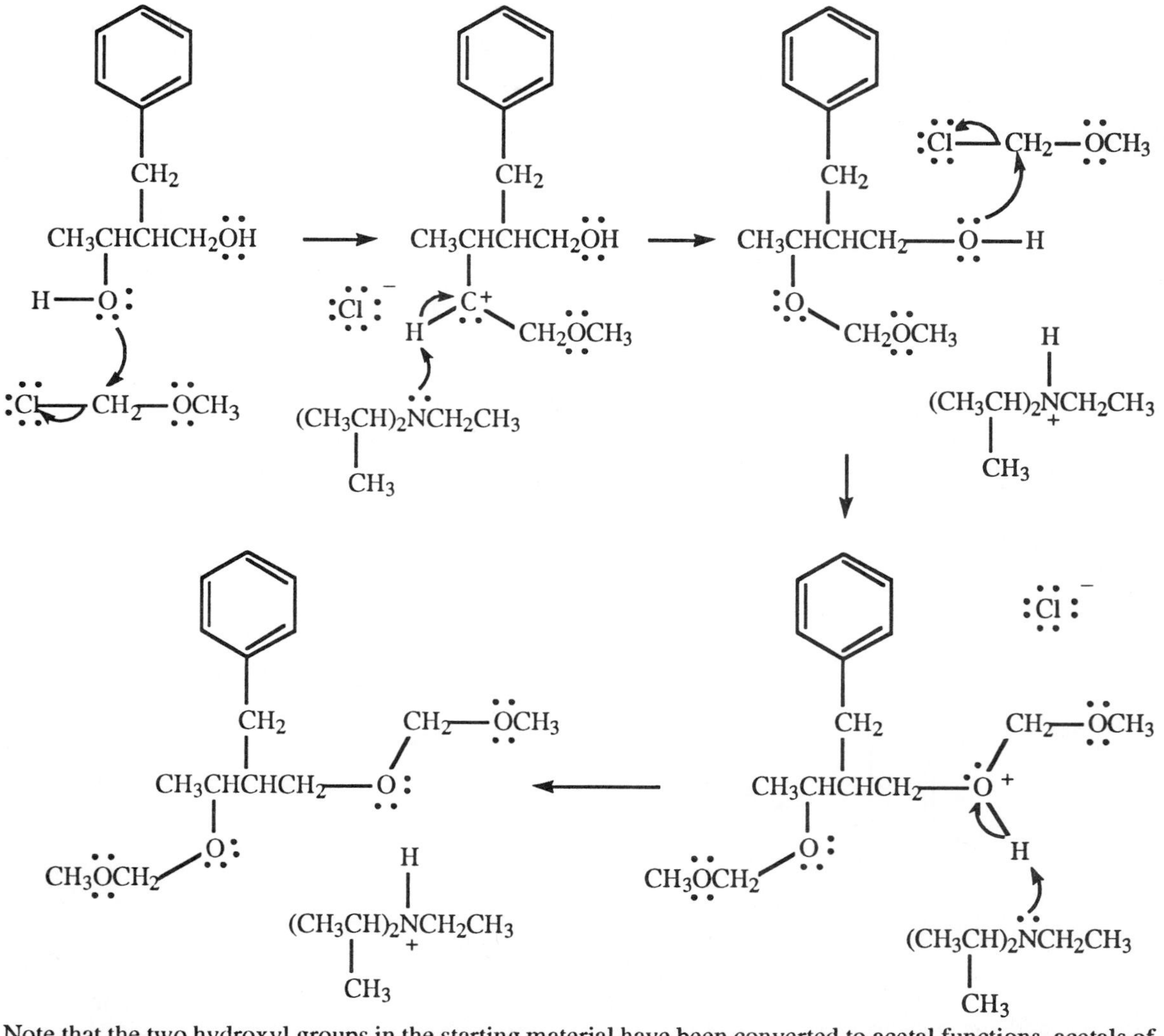

Note that the two hydroxyl groups in the starting material have been converted to acetal functions, acetals of formaldehyde (methanal). The next step is a transacetalization reaction, a conversion of the open-chain acetals to a more stable cyclic acetal of formaldehyde.

13.46 (cont)

13.47 $CH_3(CH_2)_6CH_2$ — Br: rate-determining step ⟶ $CH_3(CH_2)_6CH_2$:Br:⁻

H — $\overline{A}lH_3$

$CH_3(CH_2)_6CH_2$
|
H

:Br:⁻

AlH_3

↓

:Br — $\overline{A}lH_3$

The rate data suggest an S_N2 reaction with hydride ion acting as the nucleophile. The decrease in rate with increased substitution on the β-carbon atom suggests that the transition state for the rate-determining step is more crowded than the starting reagent (Section 7.3B).

carbon with four groups around
it, less crowded starting material

carbon with five groups around
it, more crowded transition state

The dependence of the rate on the leaving group suggests that the bond to the leaving group is also broken in the transition state for the rate-determining step.

13.48

$HC{\equiv}C^-Na^+$ (2 molar equiv) $\xrightarrow{\quad\quad}$ NH_4Cl $\xrightarrow{\quad}$ H_2O

$\xrightarrow[\text{quinoline}]{\begin{array}{c}H_2\\ \hline Pd/CaCO_3\end{array}}$

13.49

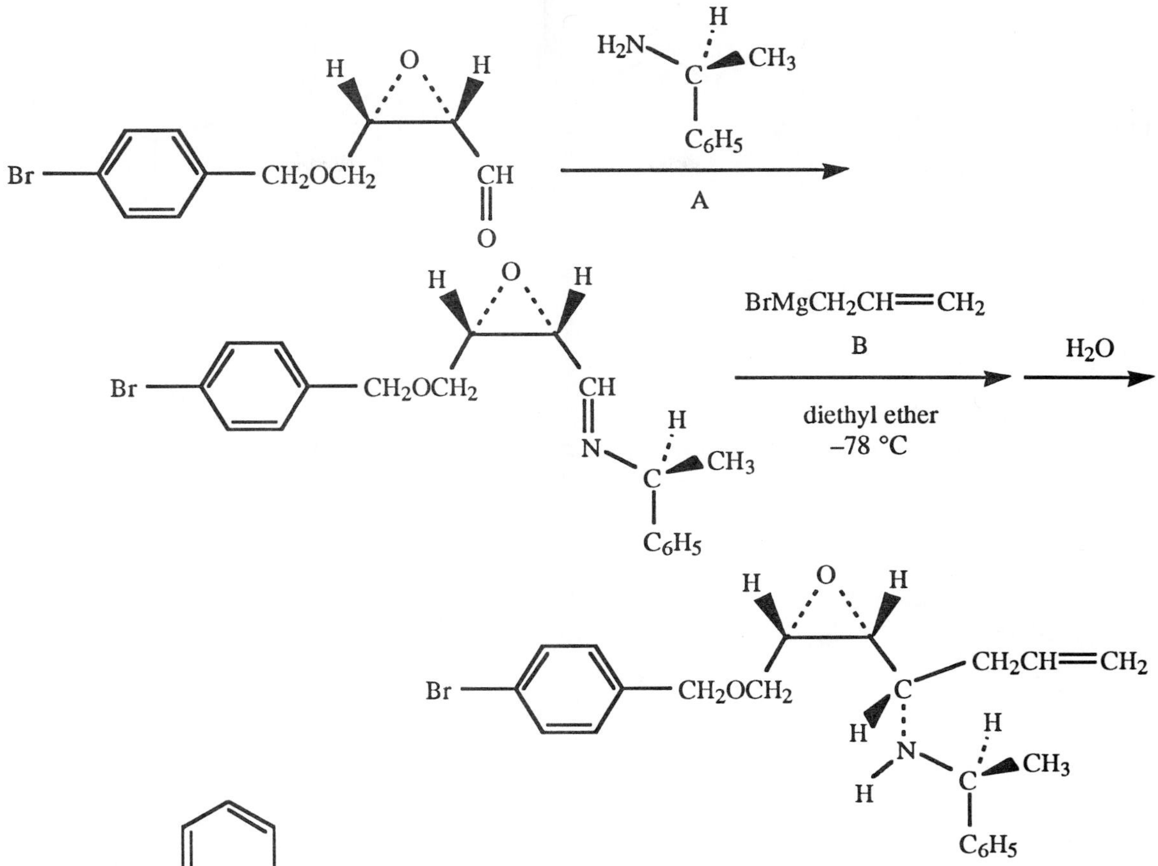

13.50

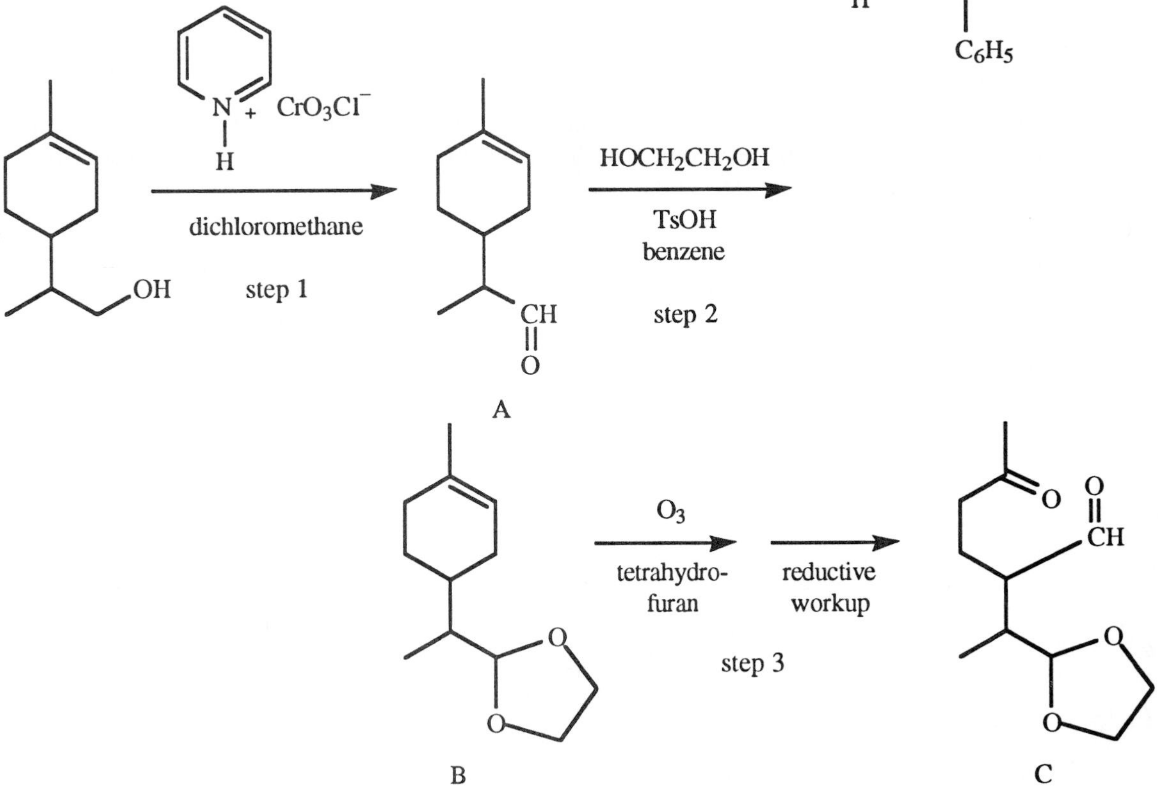

Step 2 was necessary to protect the aldehyde group from oxidation during ozonolysis.

13.51

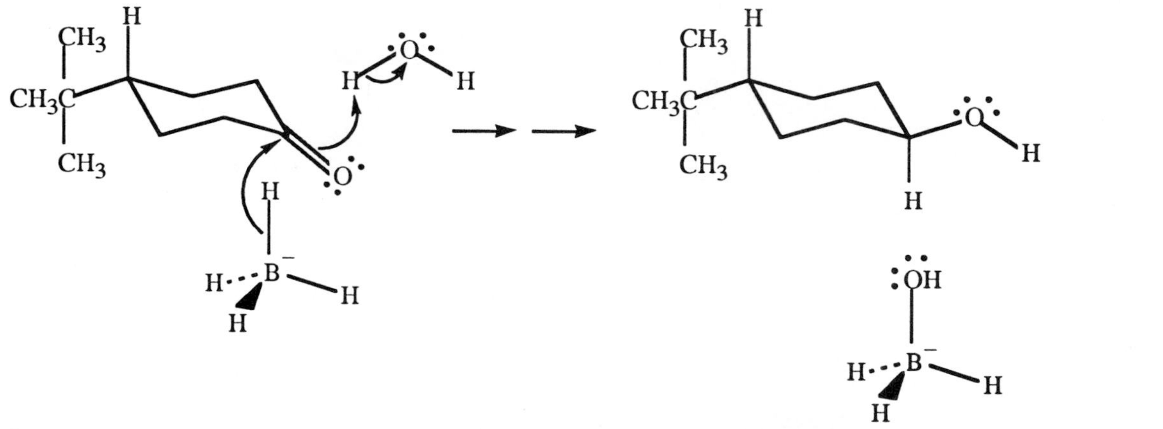

The large *tert*-butyl group occupies an equatorial position on the ring (Section 5.10C). Reduction takes place so that the major product has the hydroxyl group also in the equatorial position. This means that the hydride is delivered to the carbonyl group from the axial direction.

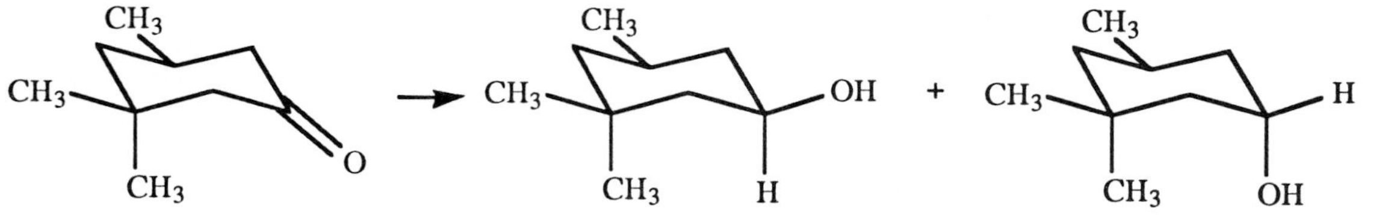

48% product from axial approach of the borohydride anion

52% product from equatorial approach of the borohydride anion

In 3,3,5-trimethylcyclohexanone, one of the methyl groups on carbon 3 is axial and hinders the approach of the borohydride anion from the axial direction. The major product is now the result of reduction of the carbonyl group from the less hindered side of the molecule.

13.52

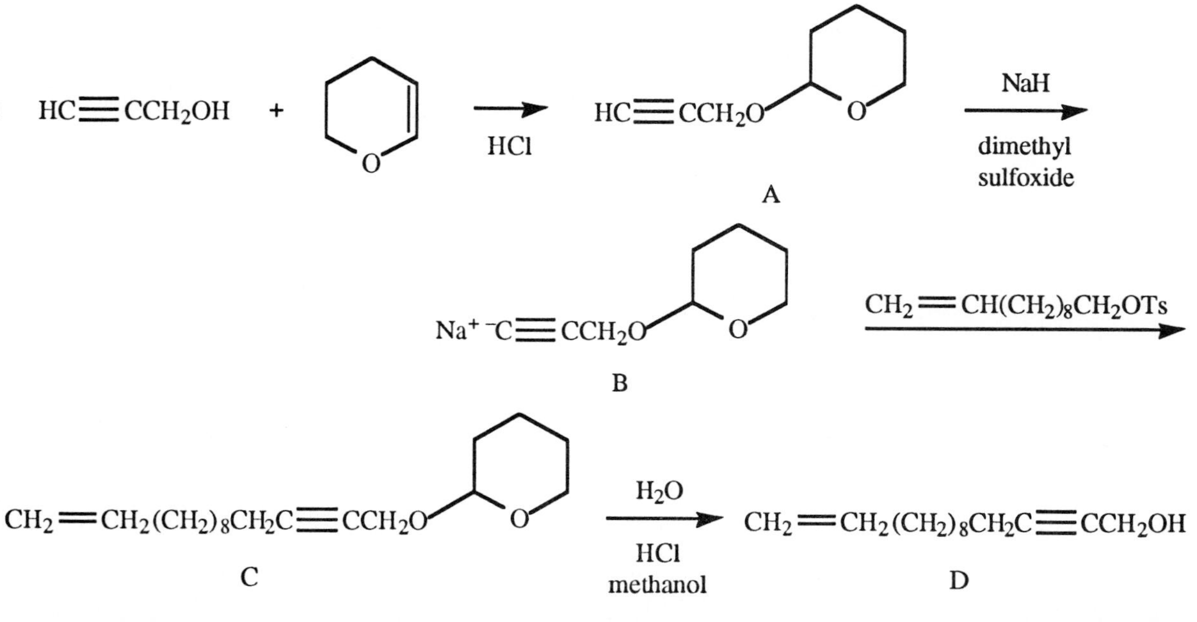
HCl
A
NaH
dimethyl sulfoxide
B
CH2=CH(CH2)8CH2OTs
C
H2O
HCl
methanol
D

13.53 $BrCH_2CH_2CH_2CH_2OH$ $\xrightarrow[\text{dichloromethane}]{}$ $BrCH_2CH_2CH_2\overset{\displaystyle O}{\overset{\|}{C}}H$ $\xrightarrow[\substack{\text{TsOH} \\ \text{benzene, }\Delta}]{HOCH_2CH_2OH}$

with pyridinium CrO_3Cl^- reagent over the first arrow.

A

$BrCH_2CH_2CH_2CH$ (with 1,3-dioxolane ring) $\xrightarrow[\substack{\text{tetrahydro-} \\ \text{furan}}]{Mg}$ $BrMgCH_2CH_2CH_2CH$ (with 1,3-dioxolane ring) $\longrightarrow$

B   **C**

with the enal fragment:
$$\underset{H}{\overset{CH_3}{C}}=\underset{\overset{\displaystyle CH}{\underset{\|}{O}}}{\overset{H}{C}}$$

D

(structure with CH_3/H alkene, $CHCH_2CH_2CH_2CH$ dioxolane, $O^-Mg^{2+}Br^-$) $\xrightarrow{H_3O^+}$ bracketed structure with CH_3/H alkene, CH–OH, $C=O$ (aldehyde H), CH_2, CH_2, CH_2 $\longrightarrow$

E

(structure with CH_3/H alkene attached to tetrahydropyran ring bearing OH and H) $\xrightarrow[\text{dichloromethane}]{\text{pyridinium } CrO_3Cl^-}$ **F** (structure with CH_3/H alkene attached to six-membered lactone ring, $=O$)

$C_8H_{14}O_2$ $C_8H_{12}O_2$

a stable, cyclic hemiacetal   three units of unsaturation
                                               (2 double bonds and a ring)

13.54

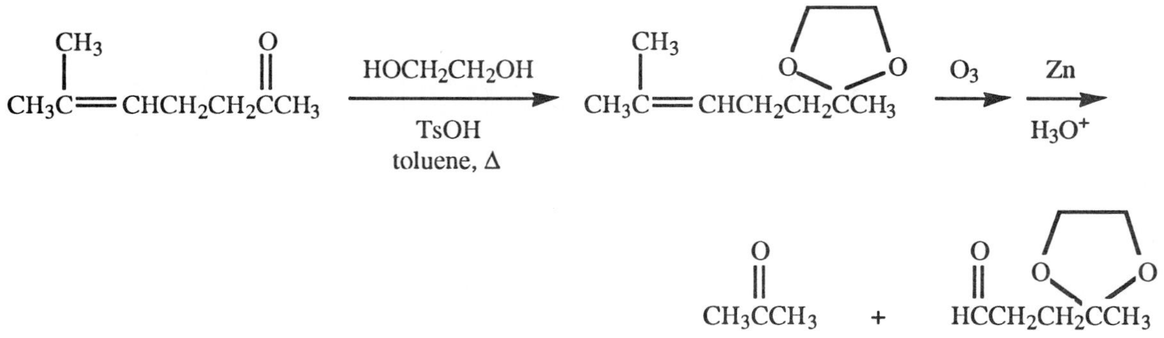

(a)

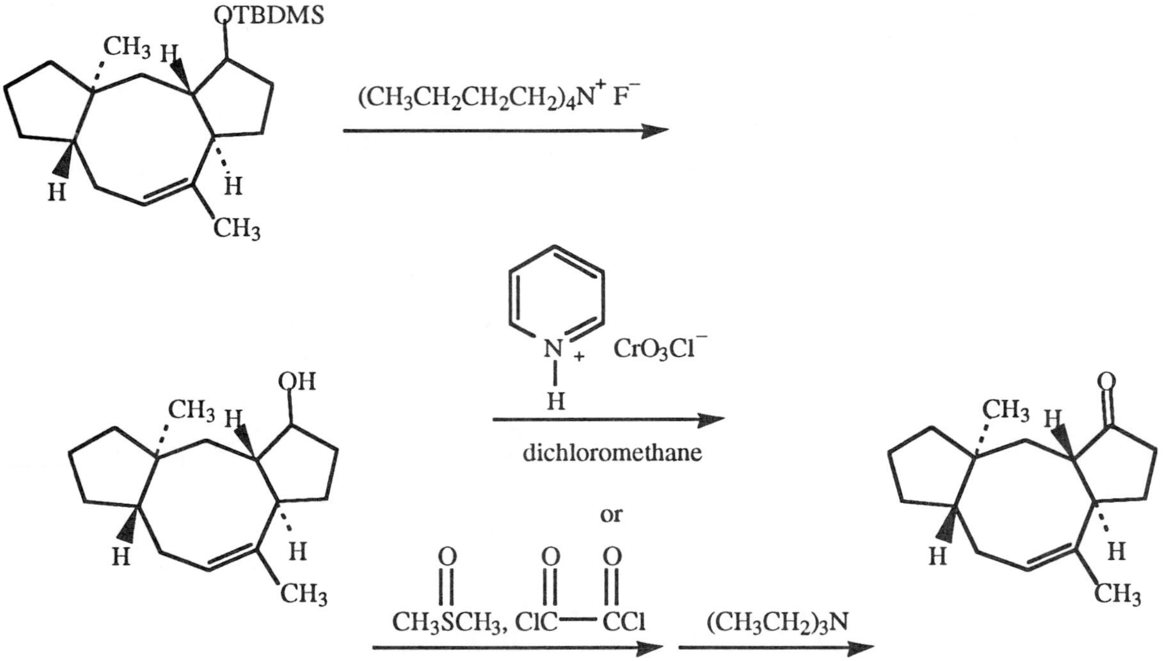

(b)

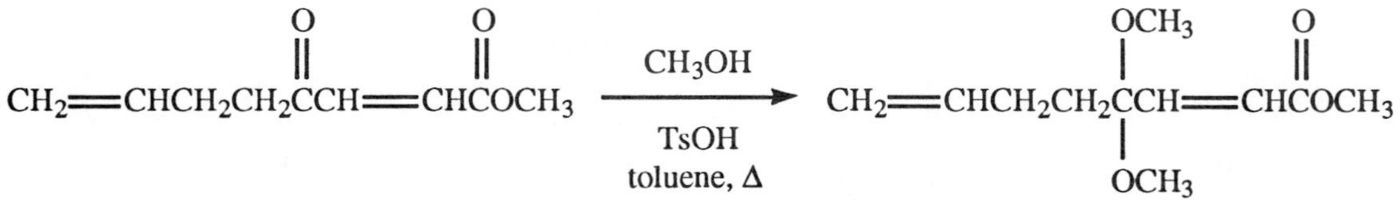

(c)

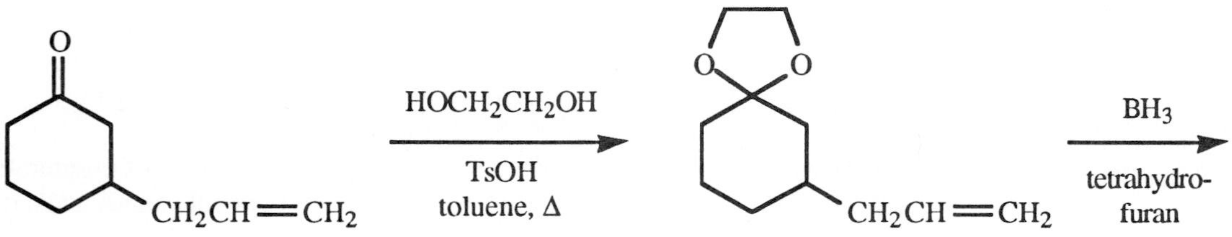

(d)

13.54 (d) (cont)

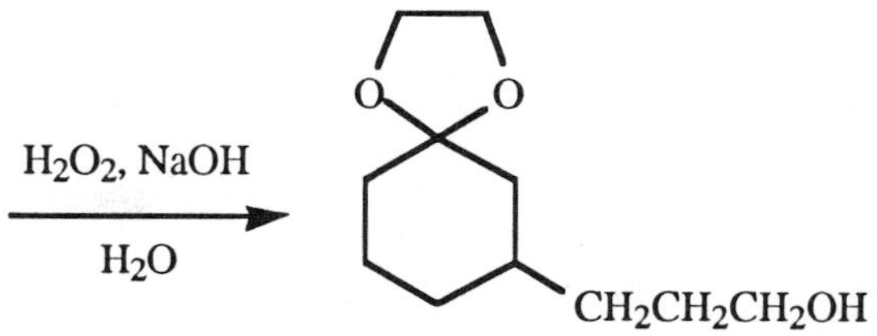

(e)

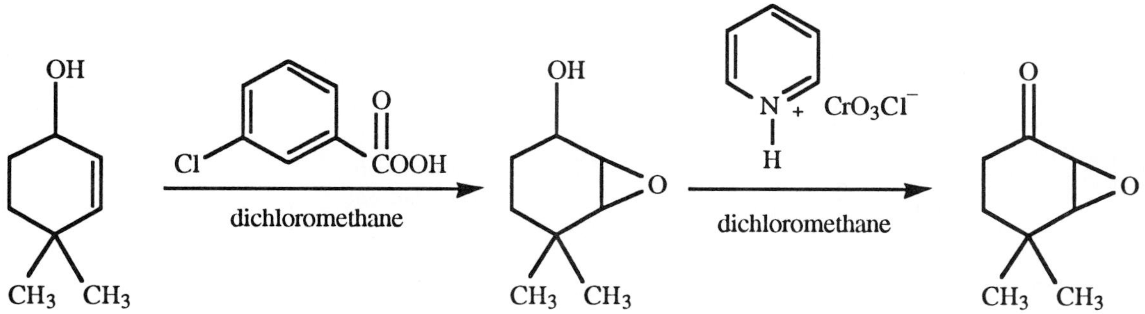

Note that the double bond is oxidized with the peroxyacid before the alcohol is converted to a carbonyl group. A double bond conjugated with a carbonyl group is electron-deficient and not easily attacked by an electron-seeking reagent such as a peroxyacid (Sections 8.9D).

(f)

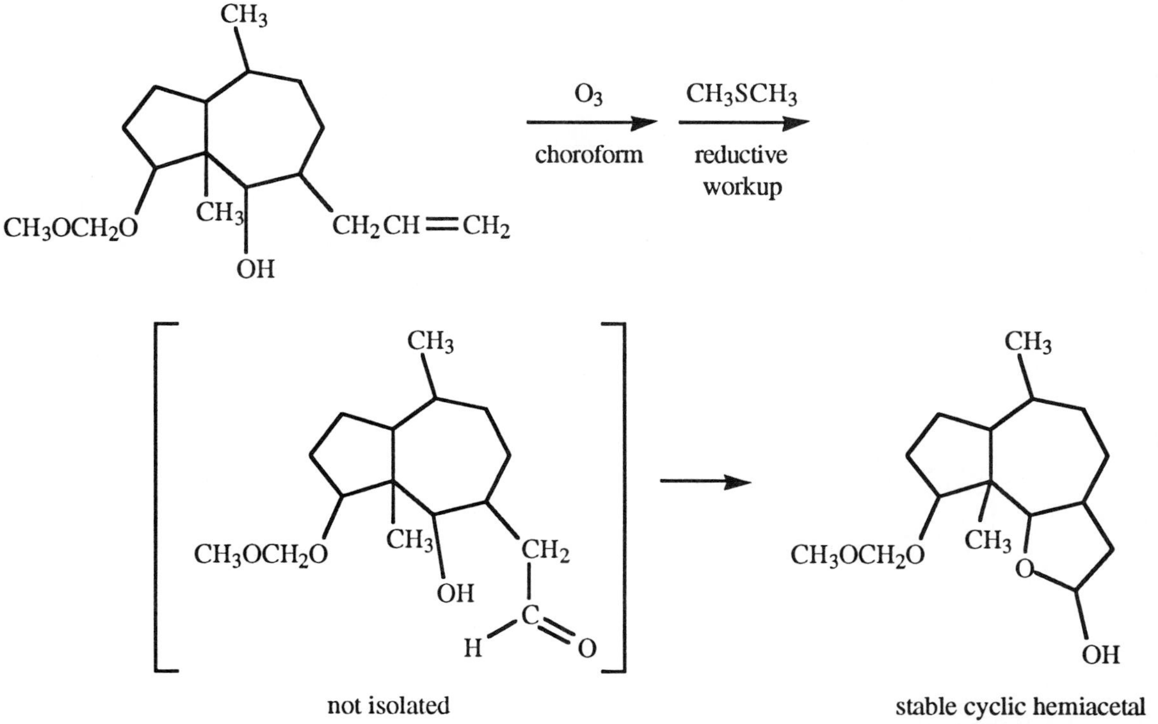

not isolated   stable cyclic hemiacetal

13.54 (cont)

(g)

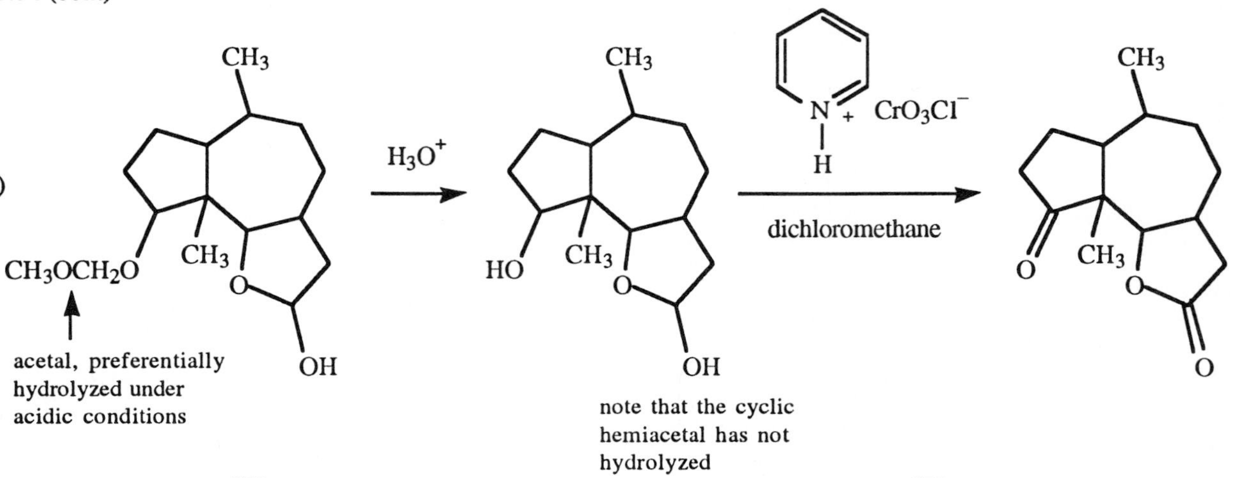
CH3
CH3OCH2O
CH3
O
OH
acetal, preferentially hydrolyzed under acidic conditions
H3O+
HO
note that the cyclic hemiacetal has not hydrolyzed
N+ CrO3Cl−
H
dichloromethane

(h)

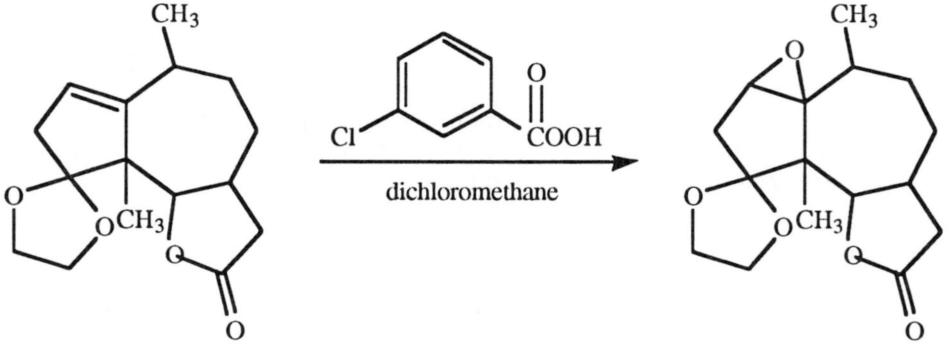
CH3
O
OCH3
O
Cl
COOH
dichloromethane

(i)

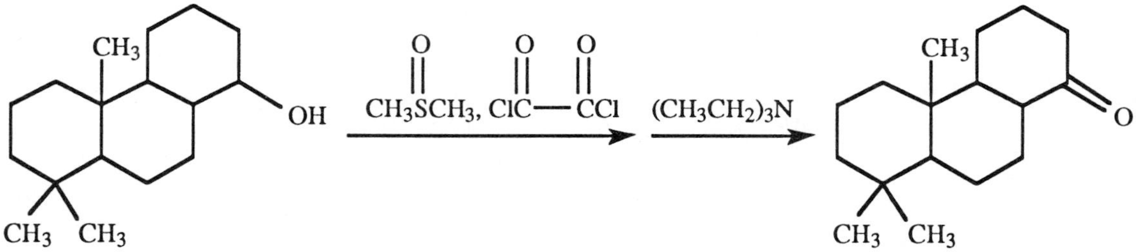
CH3
OH
CH3 CH3
CH3SCH3, ClC—CCl
(CH3CH2)3N
O

(j)

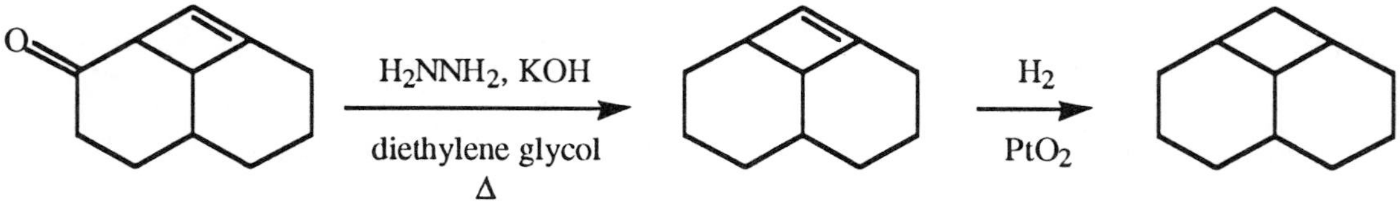
O
H2NNH2, KOH
diethylene glycol
Δ
H2
PtO2

13.55

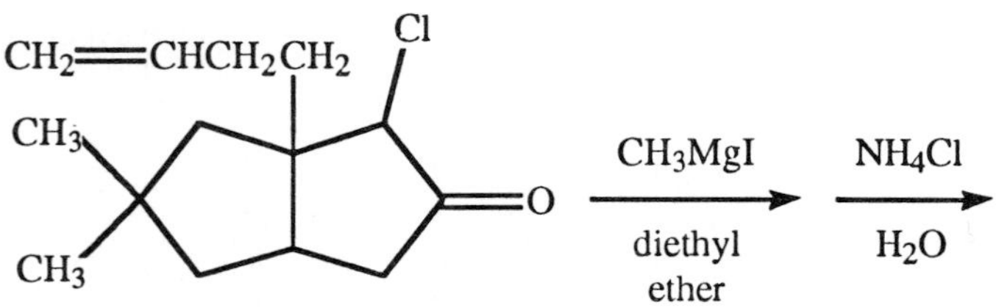
CH2═CHCH2CH2
Cl
CH3
CH3
O
CH3MgI
diethyl ether
NH4Cl
H2O

13.55 (cont)

13.56   The benzylic carbon atom in the starting material that has a hydroxyl group on it is the carbon atom of a carbonyl group in the product. It is also bonded to another carbon atom. The formation of a carbon-carbon σ bond suggests the reaction of an organometallic reagent containing a nucleophilic center with the electrophilic carbon atom of a carbonyl compound. A ketone is formed by oxidation of a secondary alcohol. A secondary alcohol is formed by reaction of an organometallic reagent with an aldehyde. Therefore, we need an aldehyde function where the benzylic alcohol group is in the starting material.

this carbon is the same as the benzylic
carbon with the hydroxyl group in the
starting material shown on the next page

13.56 (cont)

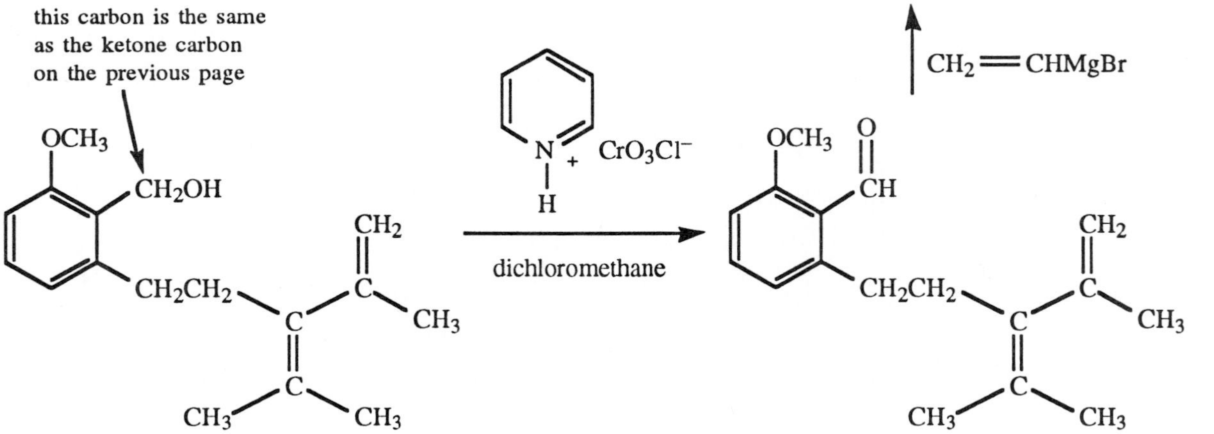
this carbon is the same as the ketone carbon on the previous page
dichloromethane

13.57

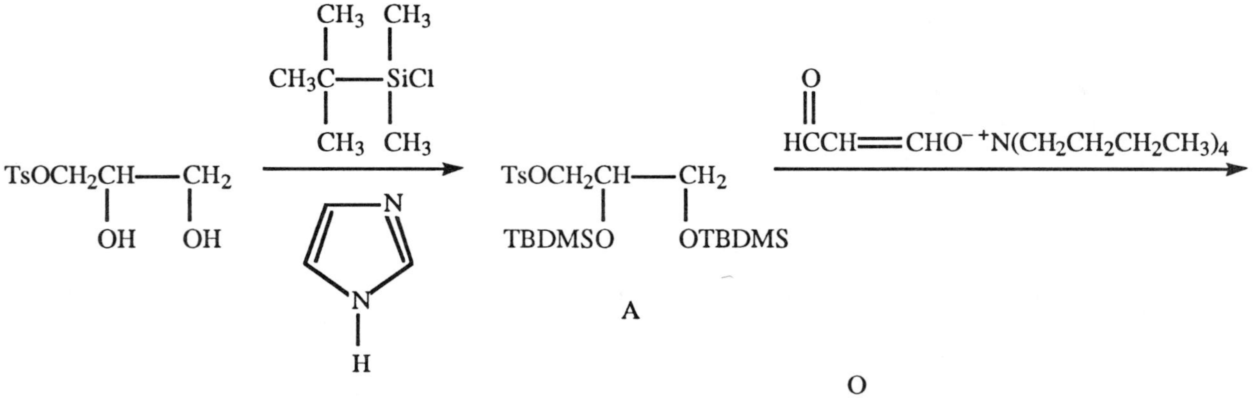

A

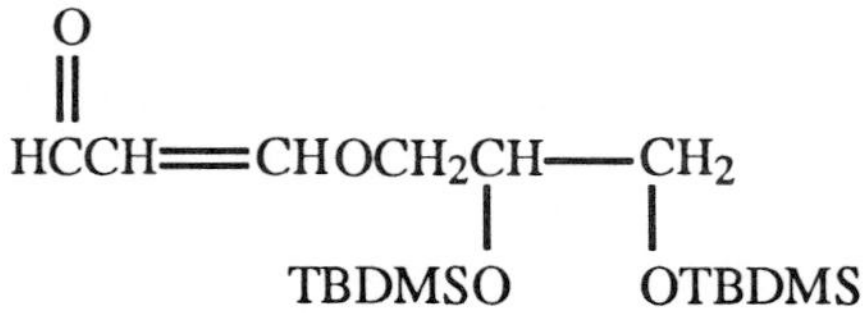

B

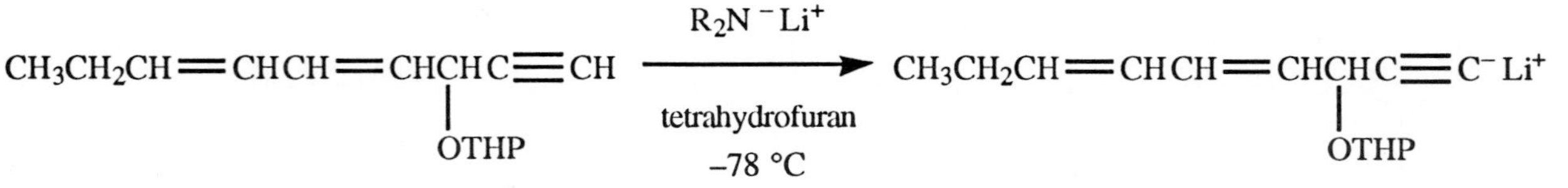
tetrahydrofuran
–78 °C

C

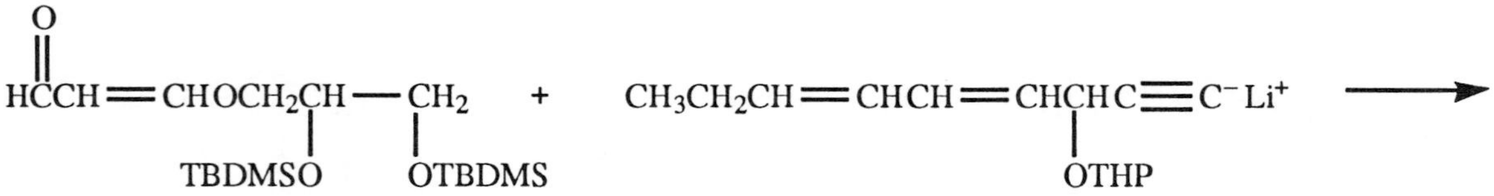

B C

13.57 (cont)

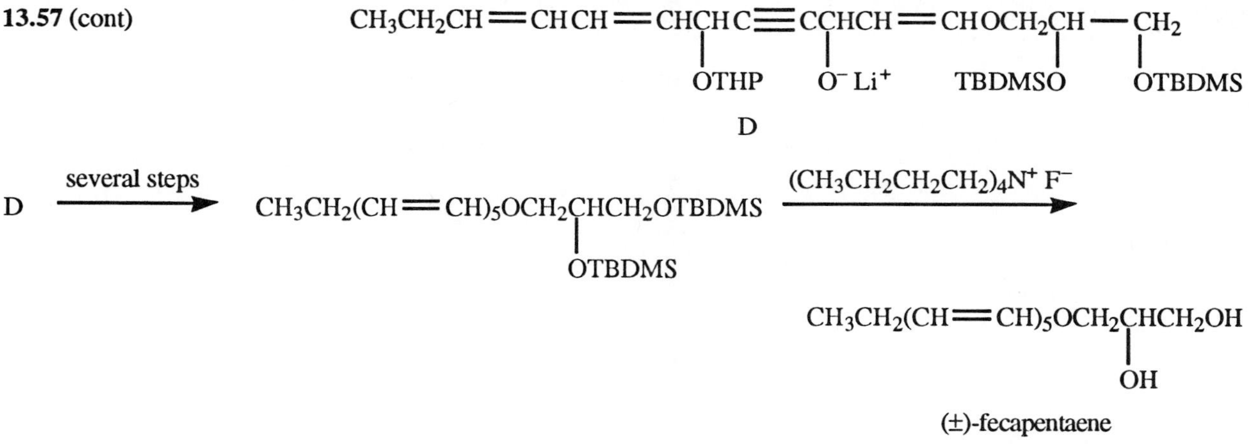
CH3CH2CH=CHCH=CHCHC≡CCHCH=CHOCH2CH—CH2
OTHP
O⁻ Li⁺
TBDMSO
OTBDMS
D
several steps
CH3CH2(CH=CH)5OCH2CHCH2OTBDMS
OTBDMS
(CH3CH2CH2CH2)4N⁺ F⁻
CH3CH2(CH=CH)5OCH2CHCH2OH
OH
(±)-fecapentaene

13.58 (a)

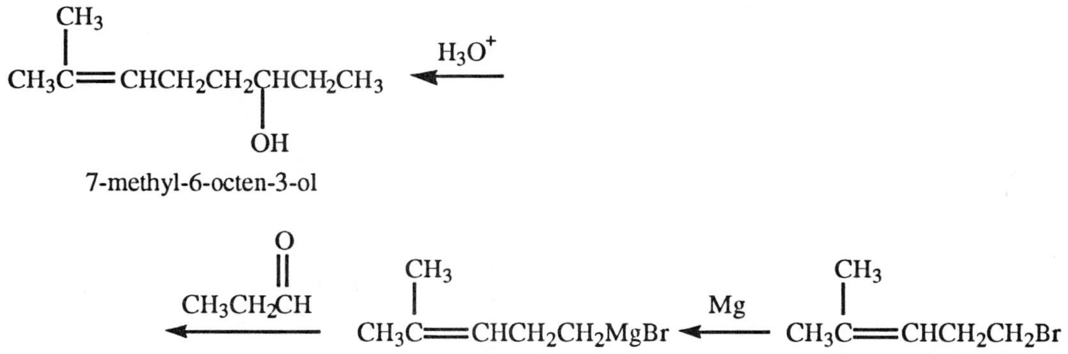
CH3
CH3C=CHCH2CH2CHCH2CH3
OH
H3O⁺
7-methyl-6-octen-3-ol
O
CH3CH2CH
CH3C=CHCH2CH2MgBr
Mg
CH3C=CHCH2CH2Br

(b)

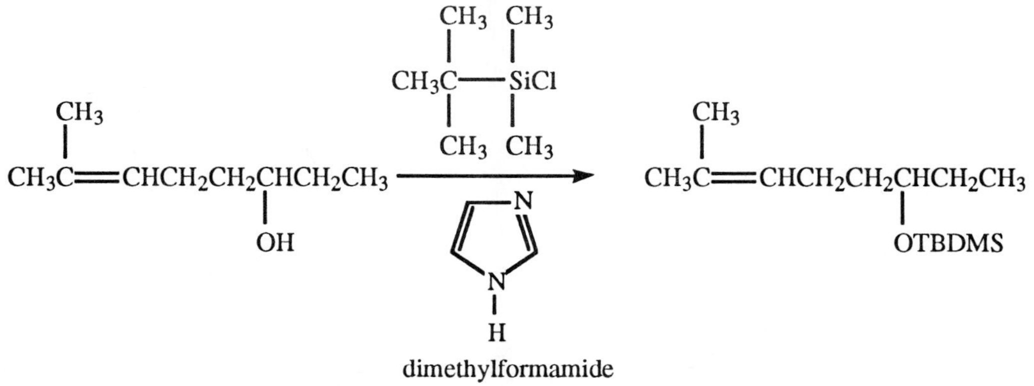
CH3C=CHCH2CH2CHCH2CH3
OH
CH3 CH3
CH3C—SiCl
CH3 CH3
N
H
dimethylformamide
CH3C=CHCH2CH2CHCH2CH3
OTBDMS

(c)

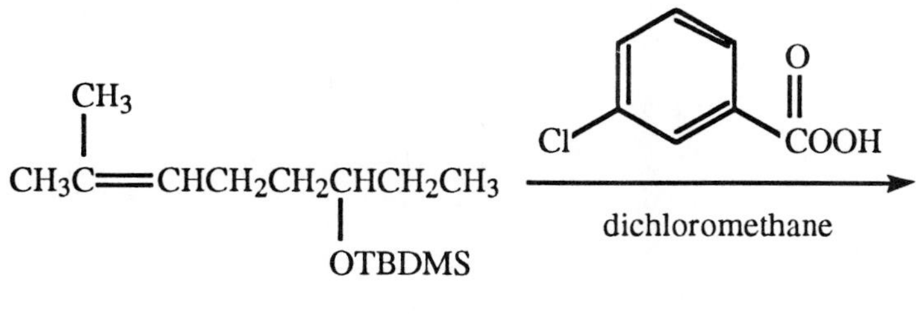
CH3C=CHCH2CH2CHCH2CH3
OTBDMS
Cl
O
COOH
dichloromethane

13.58 (c) (cont)

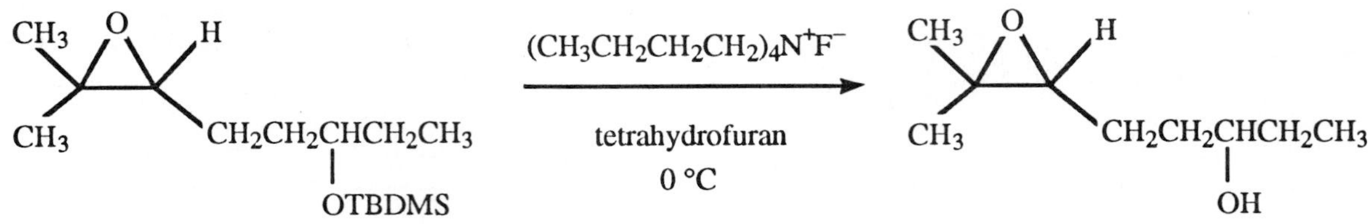

The alcohol functional group is a nucleophile that can open the oxirane ring under acid catalysis (see Section 12.5B).

13.59

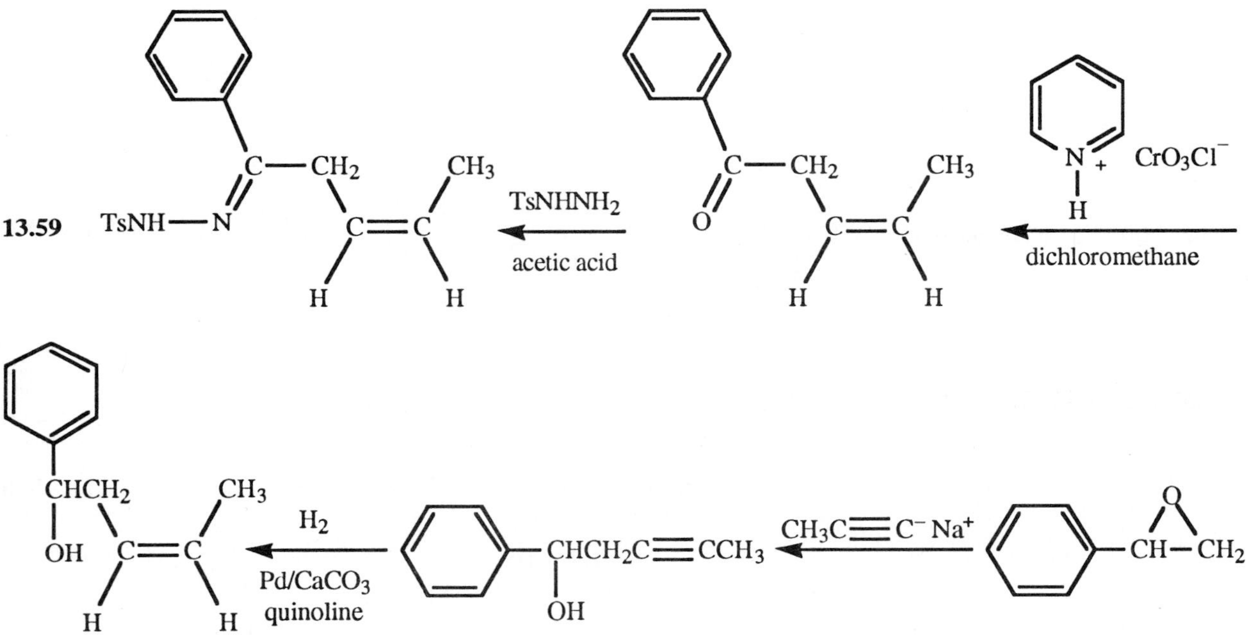

13. 60

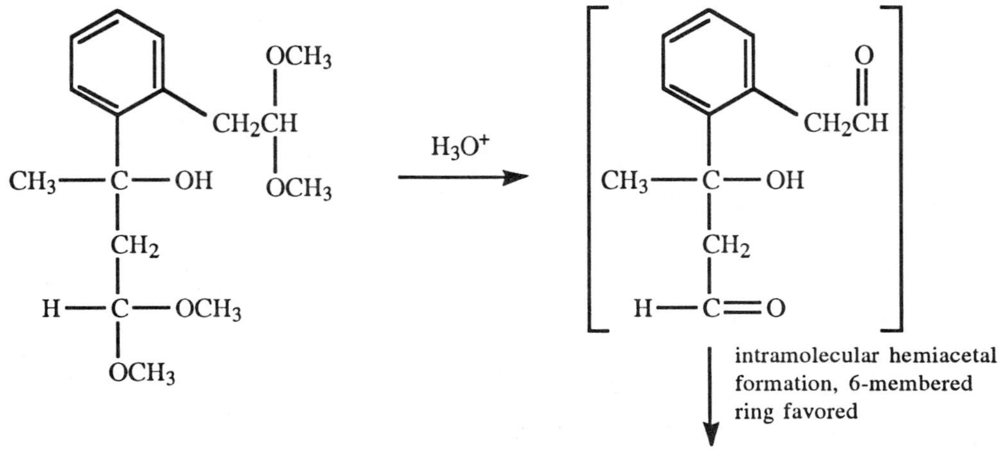

intramolecular hemiacetal formation, 6-membered ring favored

13. 60 (cont)

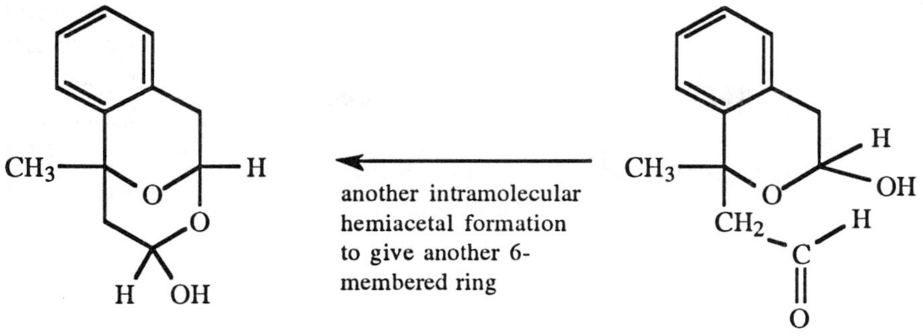

CH3
H
H
OH
another intramolecular hemiacetal formation to give another 6-membered ring
CH3
H
OH
CH2
H
C
O

13.61

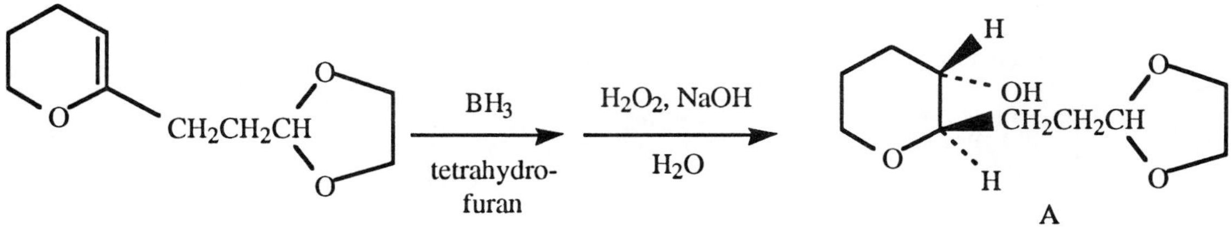

CH2CH2CH
BH3
tetrahydro-furan
H2O2, NaOH
H2O
H
OH
CH2CH2CH
H
A

(See Problem 12.54 for an explanation of the regioselectivity.)

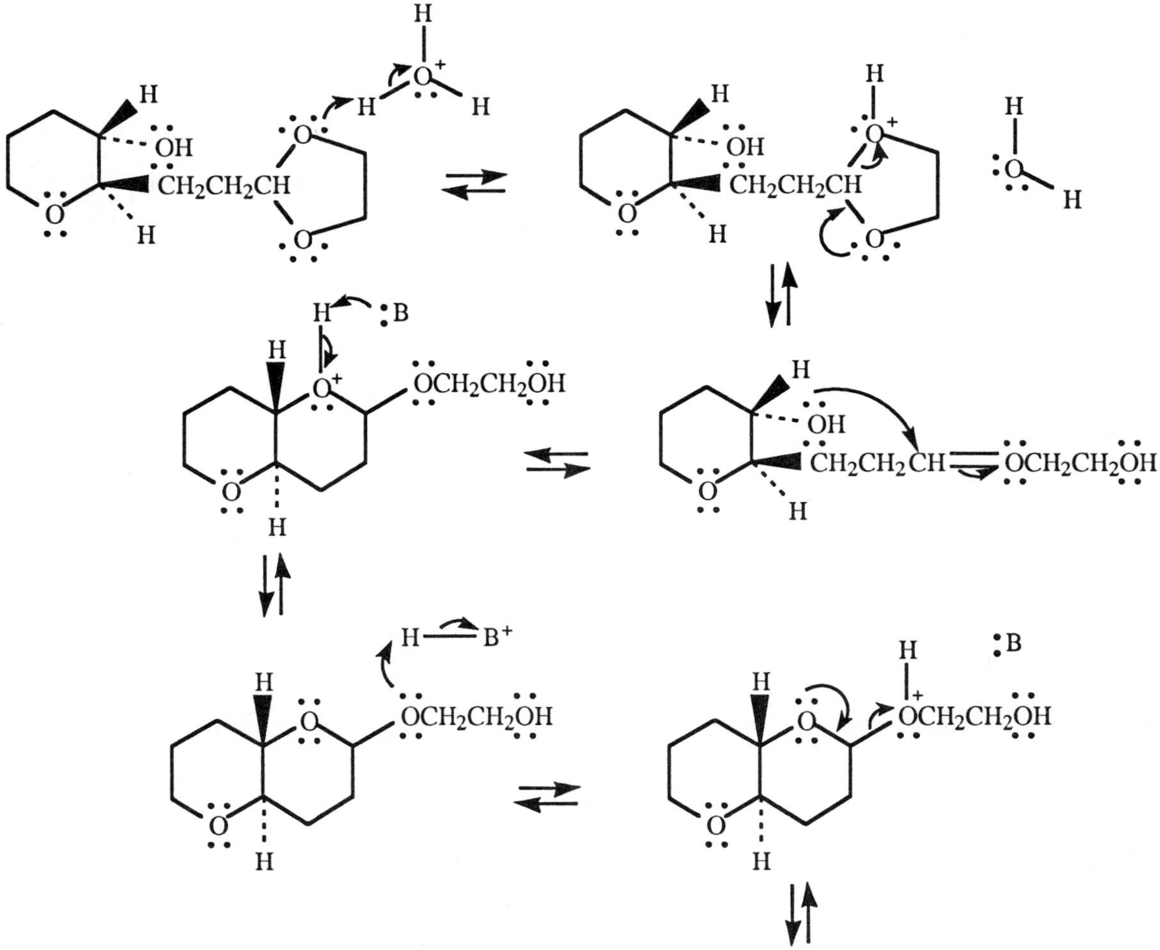

H
OH
CH2CH2CH
H
H
H
H
:B
OCH2CH2OH
H—B+

13.61 (cont)

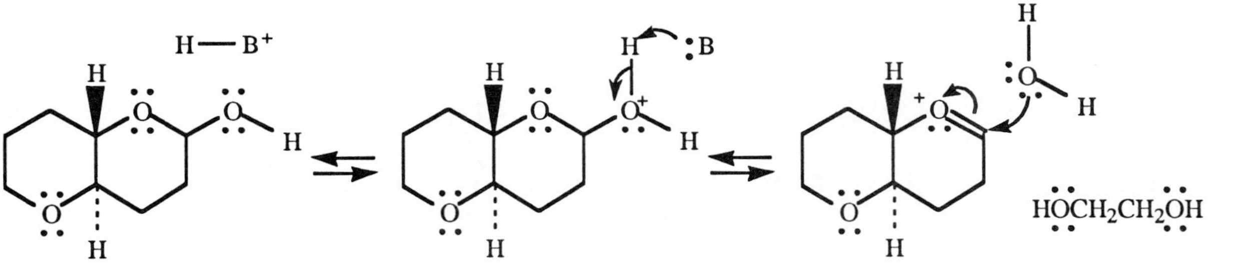

13.62

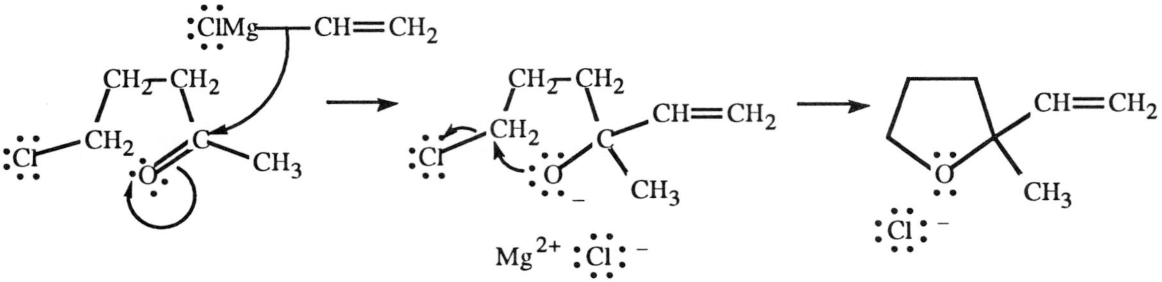

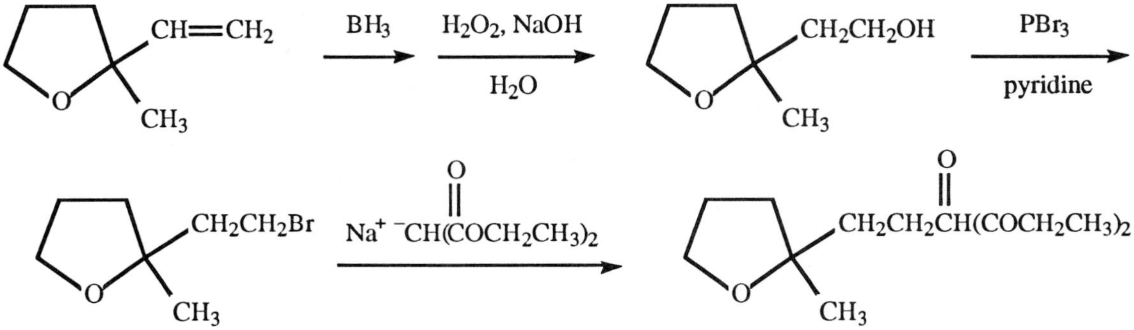

13.63   (a)

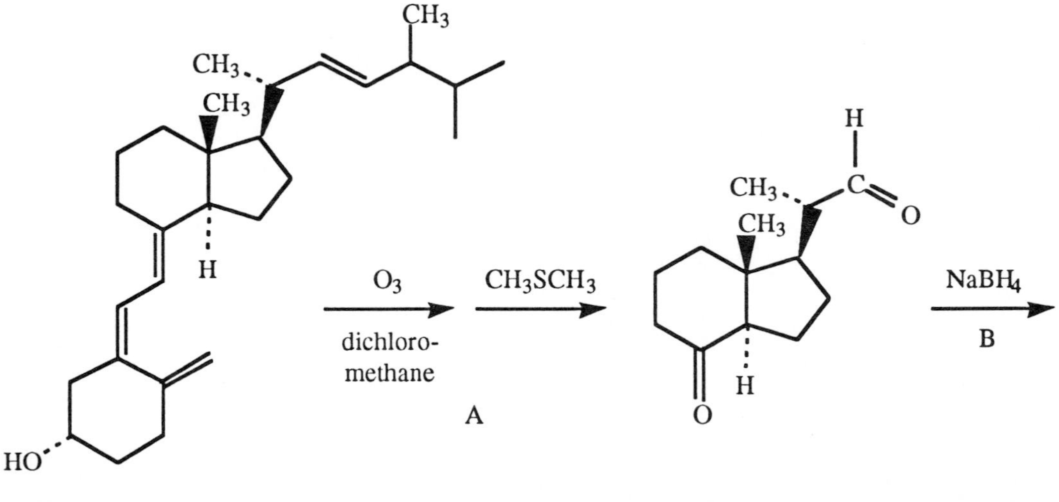

13.63 (a) (cont)

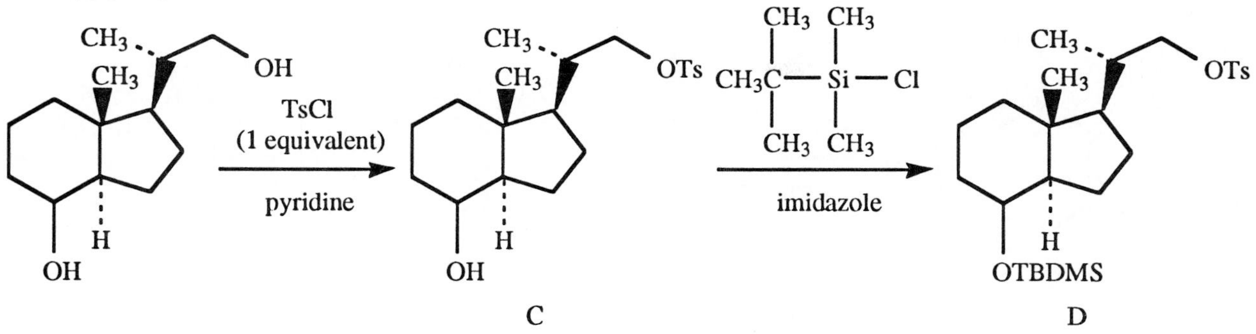

(b)   The only good leaving group in D is the tosylate group. It is the tosyl ester of a primary alcohol, so that must be the carbon atom being oxidized to the aldehyde group. Let RCH_2OTs stand for Compound D.

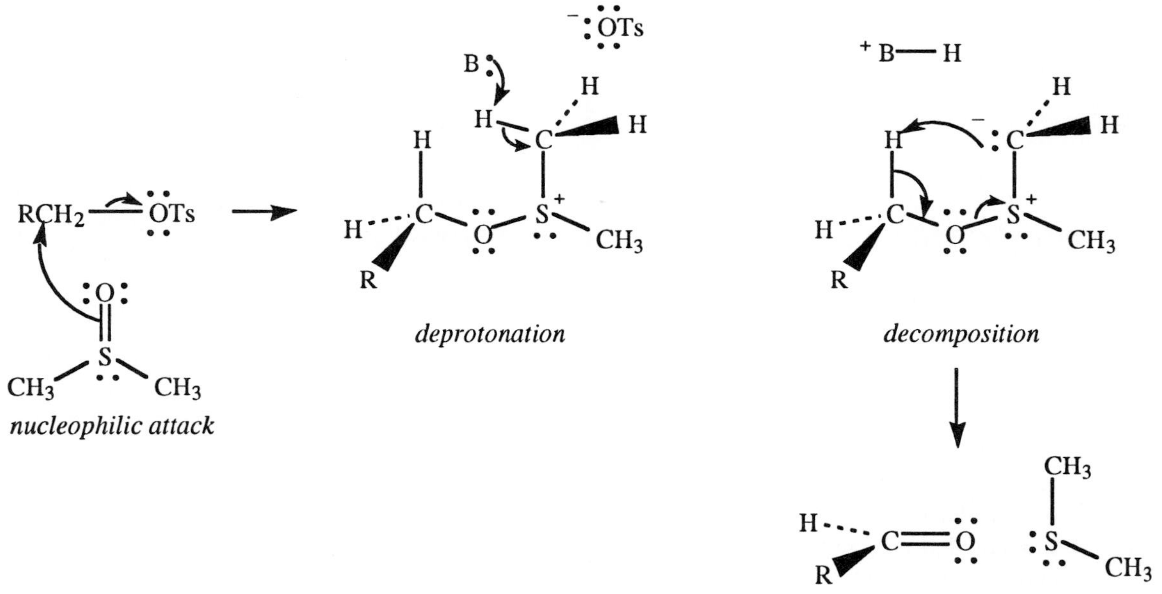

nucleophilic attack   *deprotonation*   *decomposition*

13.64   (a)

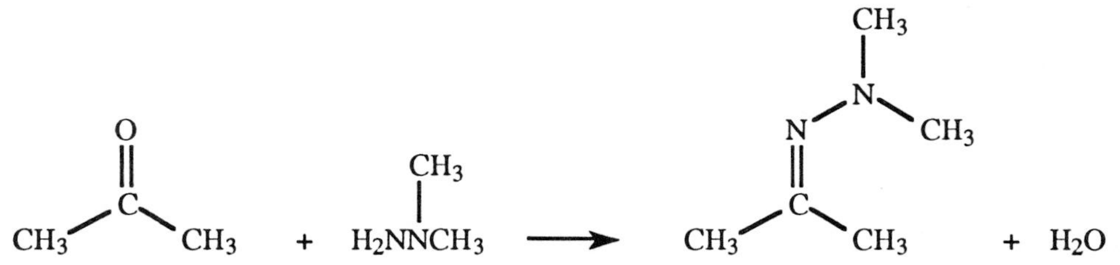

(b)

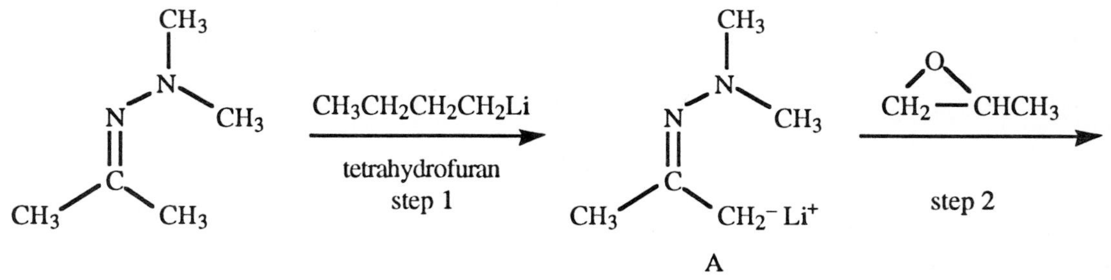

13.64 (b) (cont)

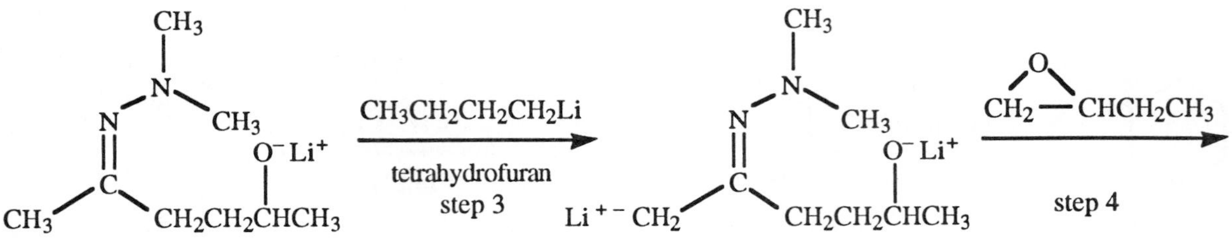

B C

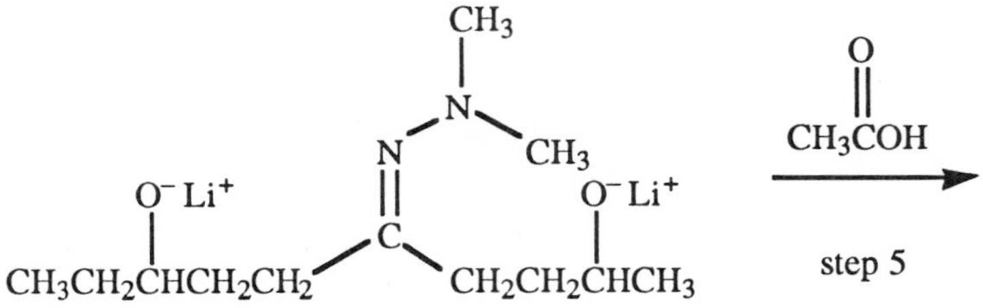

D

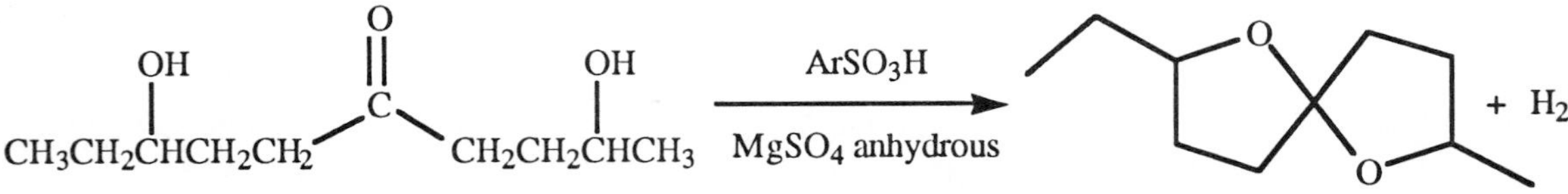

$C_{10}H_{20}O_3$
E

$C_{10}H_{18}O_2$
F

no absorptions above 1500 cm^{-1}
except for strong bands between
2950 and 2990 cm^{-1}

Hydrolysis of the hydrazone functional group in step 5 produces a ketone with a hydroxyl group on each of the two carbon atoms gamma to the carbonyl group. The infrared data tell us that product F no longer has a ketone functional group. In the presence of acid, these two hydroxyl groups form a bicyclic ketal by nucleophilic addition to the carbonyl group followed by elimination of water.

13.65

(a)

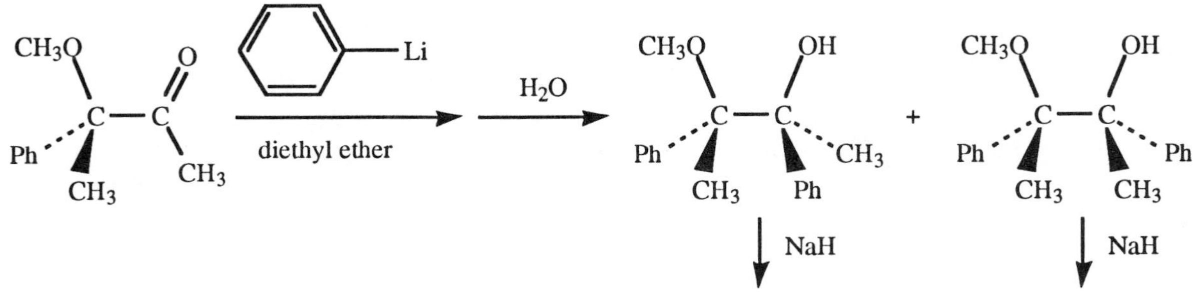

13.65 (a) (cont)

↓ CH₃I   ↓ CH₃I

CH₃O   OCH₃   CH₃O   OCH₃
    S   S   S   R
     C — C   C – – C
Ph   CH₃   Ph   Ph
    CH₃   Ph   CH₃   CH₃

optically active,   meso form,
therefore is ether of   optically inactive,
the major alcohol   therefore is ether of
                             the minor alcohol

(b)

If the lithium ion coordinates with the two oxygen atoms and holds the carbonyl group in the same plane as the ether oxygen, one side of the molecule becomes much more sterically hindered toward the approach of the incoming nucleophile than the other. The major product comes from nucleophilic attack from the side of the carbonyl group away from the bulky phenyl group and close to the smaller methyl group.

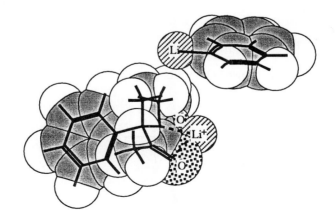

13.66   Compound B, $C_6H_{10}O$, has two units of unsaturation and Compound A, $C_8H_{12}O$, has three units of unsaturation. We can tell from the infrared spectrum that B is a ketone (C=O stretching frequency at 1710 cm^{-1}). A carbonyl group only uses up one unit of unsaturation. The other unit of unsaturation must be a ring or a double bond. No absorption bands around 1600 cm^{-1} or above 3000 cm^{-1} appear in the infrared spectrum, so the compound does not contain a double bond. Cyclic ketones smaller than six carbons absorb at frequencies greater than 1725 cm^{-1} (see Problem 12. ~), therefore Compound B must be cyclohexanone. Compound A must also contain a cyclohexane ring since it is prepared from Compound B. The infrared spectrum of Compound A tells us that it is an alcohol and a terminal alkyne (broad O—H stretching frequency at about 3400 cm^{-1}, ≡C—H at 3308 cm^{-1}, and C≡C at about 2100 cm^{-1}). We can determine what has been added to cyclohexanone by subtracting the two molecular formulas.

$$C_8H_{12}O - C_6H_{10}O = C_2H_2 = \text{the elements of acetylene}$$

The elements of acetylene can be added by reaction of sodium acetylide with a ketone followed by acidification. Compound B is 1-ethynylcyclohexanol. The reaction of B to prepare A is shown below:

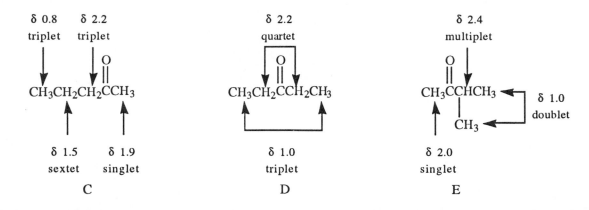

13.67   Compounds C, D, and E, molecular formula $C_5H_{10}O$, all have one unit of unsaturation. The infrared spectra of the three compounds show the presence of a ketone (C=O at 1720 cm^{-1} and no bands at 2900 and 2700 cm^{-1}, which tells us that the compounds are not aldehydes). The presence of a ketone is confirmed by the carbon–13 magnetic resonance spectra in which the ketone carbon atom can be seen for C, D, and E (C at 206.6 ppm, D at 209.3 ppm, and E at 211.8 ppm). There are only three ketones that have the molecular formula $C_5H_{10}O$.

The carbon-13 spectra tell us how many types of carbon atoms each compound has. Compound C has five different carbon atoms. In compound D, there are three sets of carbon atoms, which points to symmetry in the molecule. Compound E has four sets of carbon atoms, and therefore must have two which are the same.

The structural formulas are shown below, along with the proton magnetic resonance data for each one.

δ 0.8	δ 2.2		δ 2.2		δ 2.4
triplet	triplet		quartet		multiplet

$$CH_3CH_2CH_2CCH_3 \qquad CH_3CH_2CCH_2CH_3 \qquad CH_3CCHCH_3$$

δ 1.5	δ 1.9	δ 1.0	δ 2.0
sextet	singlet	triplet	singlet
C		**D**	**E**

δ 1.0 doublet — CH_3

14

Carboxylic Acids and Their Derivatives I. Nucleophilic Substitution Reactions at the Carbonyl Group

Workbook Exercises

There are two common classifications of functional groups that contain the carbonyl group ($-\overset{\overset{\textstyle O}{\|}}{C}-$). These two classifications are based on an observed difference in reactivity. In the case of aldehydes and ketones (see Chapter 13), the fundamental reaction is addition. As demonstrated by the formation of imines and ketals, among others, the addition products of aldehydes and ketones can undergo further reactions. In Chapters 14 and 15, the second type of carbonyl chemistry is presented. In this case, groups other than simple alkyl or aryl (as in ketones) or hydrogen (as in aldehydes) are attached to the carbon atom of a carbonyl group. Instead, halogen, oxygen, nitrogen, or sulfur atoms (symbolized below as the generic electronegative group eN) are bonded to the carbonyl group. The overall reaction of these carbonyl compounds is not addition, as in the case of aldehydes and ketones, but rather substitution of the eN group.

$$R-\overset{\overset{\textstyle O}{\|}}{C}-eN \quad \xrightarrow{\text{nucleophiles} \atop \text{(Nu)}} \quad R-\overset{\overset{\textstyle O}{\|}}{C}-Nu \ + \ eN$$

overall reaction = substitution at the sp^2-hybridized carbon atom
(eN = many O, N, S, and halogen groups, among others)

The functional unit symbolized by $R-\overset{\overset{\textstyle O}{\|}}{C}-$ is called an **acyl group.** The S_N1 and S_N2 substitution reactions of alkyl halides by nucleophiles (R—LG → R—Nu) are called alkylations of the nucleophile, or alkyl transfer reactions.

The substitution of the eN atom group of $R-\overset{\overset{\textstyle O}{\|}}{C}-eN$ by a nucleophile to give $R-\overset{\overset{\textstyle O}{\|}}{C}-Nu$ is called an **acylation** of the nucleophile, or an **acyl transfer reaction.**

A new term, such as acylation, is created in order to express a new idea. The process of acylation can be described as addition followed by elimination. The reference to ideas previously encountered is designed to make an unfamiliar process, such as acylation, more familiar to you. If the reference to addition and elimination reactions is not meaningful to you, your best strategy is to refamiliarize yourself with these previous types of reactions before attempting to master new ones. These types of analogies can be used in many ways to communicate meaning. When referring to the overall change in connectivity in an acylation, the term substitution is used to describe the transformation at the sp^2-hybridized carbon atom.

Workbook Exercises (cont)

EXAMPLE

Compare the four examples of transformations using water as a nucleophile shown below. Each one is accompanied by an equation describing the overall change in connectivity and the mechanistic rationalization of the change observed.

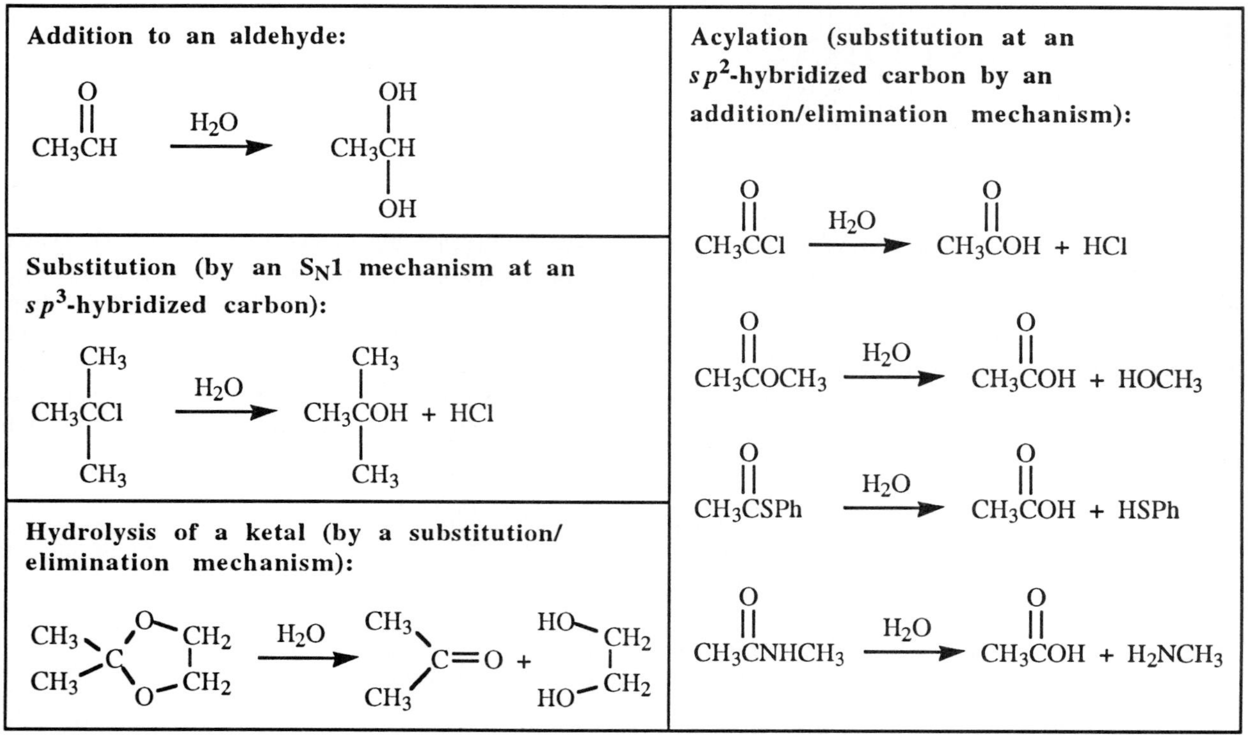
Addition to an aldehyde:
Substitution (by an SN1 mechanism at an sp3-hybridized carbon):
Hydrolysis of a ketal (by a substitution/elimination mechanism):
Acylation (substitution at an sp2-hybridized carbon by an addition/elimination mechanism):
H2O
CH3CH
CH3CCl
CH3COH + HCl
CH3COCH3
CH3COH + HOCH3
CH3CSPh
CH3COH + HSPh
CH3CNHCH3
CH3COH + H2NCH3

EXERCISE I. Complete the following acylation reactions.

(a) 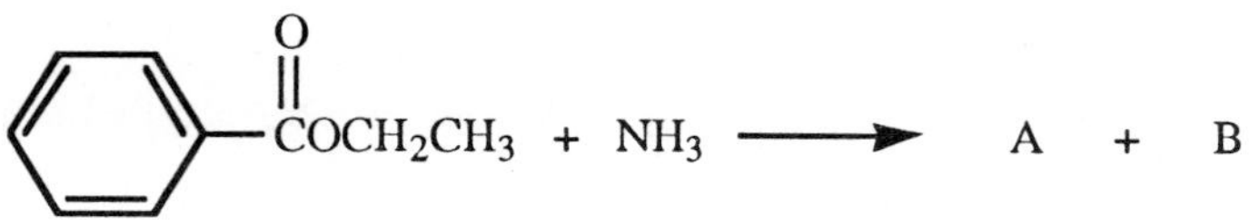

(b) 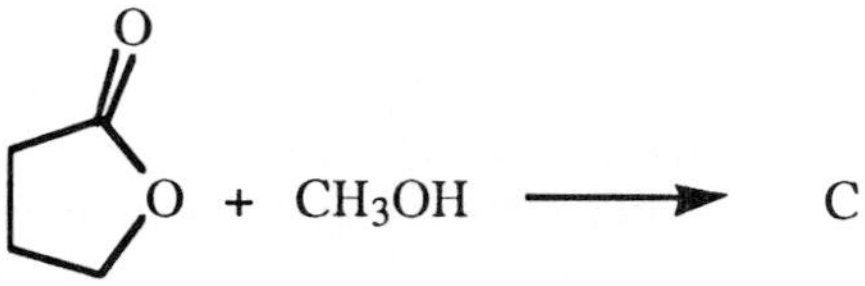

(c) 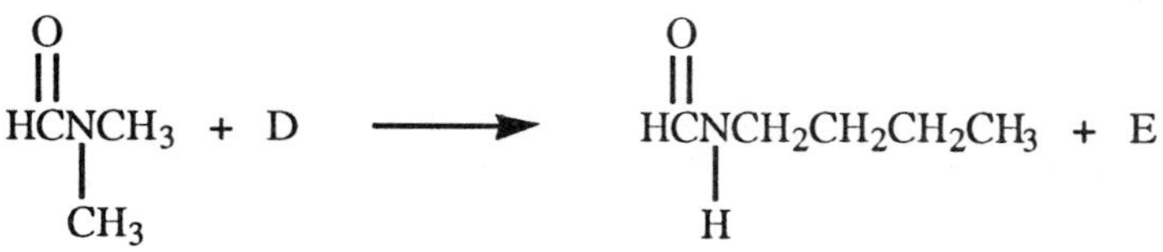

Workbook Exercises (cont)

(d)

$\begin{array}{c} \text{CH}_3 \\ \diagdown \\ \diagup \\ \text{CCCl}_3 \\ \| \\ \text{O} \end{array}$ + NaOH $\longrightarrow$ F* + G*

(*The trichloromethyl group, —CCl$_3$, behaves as a leaving group in this reaction. Under the aqueous base conditions of this reaction, the leaving group is protonated to give an uncharged product, while the acylated nucleophile is deprotonated to give its conjugate base.)

(e)   H   +   I $\longrightarrow$ $\overset{\text{O O}}{\overset{\| \|}{\text{CH}_3\text{COCCH}_3}}$ + NaCl

A MORE DETAILED LOOK AT THE ACYLATION MECHANISM

The overall transformation in an acylation reaction is viewed as the combination of two fundamental mechanistic steps: addition to the carbonyl group followed by an elimination. This is the same way that the reactions of the carbonyl group in Chapter 13 were characterized.

$$\underset{\text{R}-\overset{\overset{\text{O}}{\|}}{\text{C}}-\text{eN}}{} \underset{\substack{\text{elimination} \\ \text{of NuH}}}{\overset{\substack{\text{addition} \\ \text{of NuH}}}{\rightleftharpoons}} \underset{\substack{\text{R}-\overset{\overset{\text{OH}}{|}}{\underset{\underset{\text{eN}}{|}}{\text{C}}}-\text{Nu}}}{} \underset{\substack{\text{addition} \\ \text{of HeN}}}{\overset{\substack{\text{elimination} \\ \text{of HeN}}}{\rightleftharpoons}} \underset{\text{R}-\overset{\overset{\text{O}}{\|}}{\text{C}}-\text{Nu}}{}$$

alcohol intermediate

If the nucleophile is an anion, Nu –, the intermediate will be an alkoxide ion, $\text{R}-\overset{\overset{\text{O}^{-}}{|}}{\underset{\underset{\text{eN}}{|}}{\text{C}}}-\text{Nu}$, rather than an alcohol.

EXAMPLE

This scheme, applied to one of the molecules shown on the previous page, gives the following more detailed description.

previously shown: $\overset{\text{O}}{\overset{\|}{\text{CH}_3\text{CSPh}}}$ $\xrightarrow{\text{H}_2\text{O}}$ $\overset{\text{O}}{\overset{\|}{\text{CH}_3\text{COH}}}$ + HSPh

more detailed view: $\text{CH}_3-\overset{\overset{\text{O}}{\|}}{\text{C}}-\text{SPh}$ $\underset{\substack{\text{elimination} \\ \text{of H}_2\text{O}}}{\overset{\substack{\text{addition} \\ \text{of H}_2\text{O}}}{\rightleftharpoons}}$ $\text{CH}_3-\overset{\overset{\text{OH}}{|}}{\underset{\underset{\text{SPh}}{|}}{\text{C}}}-\text{OH}$ $\underset{\substack{\text{addition} \\ \text{of PhSH}}}{\overset{\substack{\text{elimination} \\ \text{of PhSH}}}{\rightleftharpoons}}$ $\text{CH}_3-\overset{\overset{\text{O}}{\|}}{\text{C}}-\text{OH}$

EXERCISE II. Redo Exercise I using the mechanistic scheme outlined above.

14.1

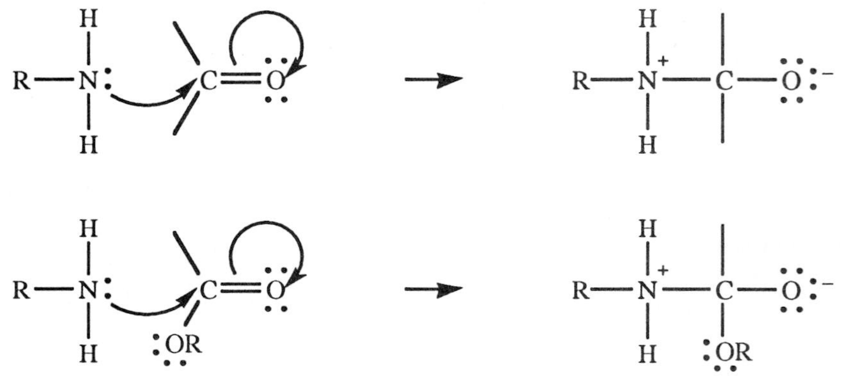

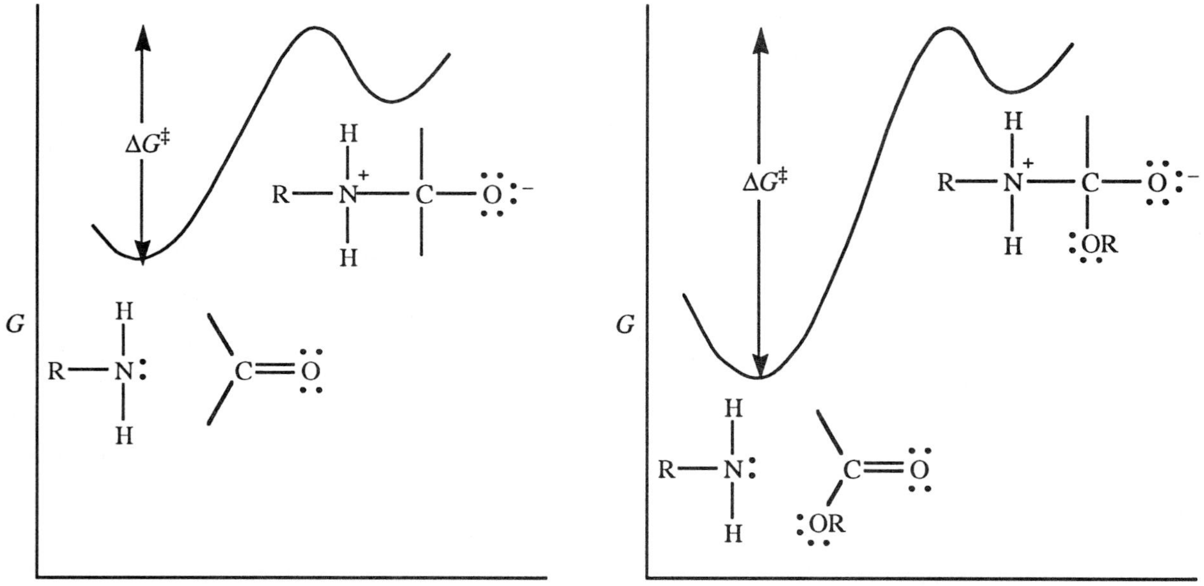

progress of reaction for reaction
of aldehyde or ketone with amine

progress of reaction for reaction
of acid derivative with amine

The carbonyl group of an aldehyde or a ketone is not as highly stabilized as the carbonyl group of an ester is. The free energy of the aldehyde or ketone is thus higher than that of the ester. If we assume that the tetrahedral intermediates, in which we no longer have resonance, are approximately the same energy, then $\Delta G^{\ddagger}$ for an aldehyde or ketone is lower than for an ester. Therefore, an aldehyde or ketone reacts faster with the nucleophile than the ester does.

14.2

(a)

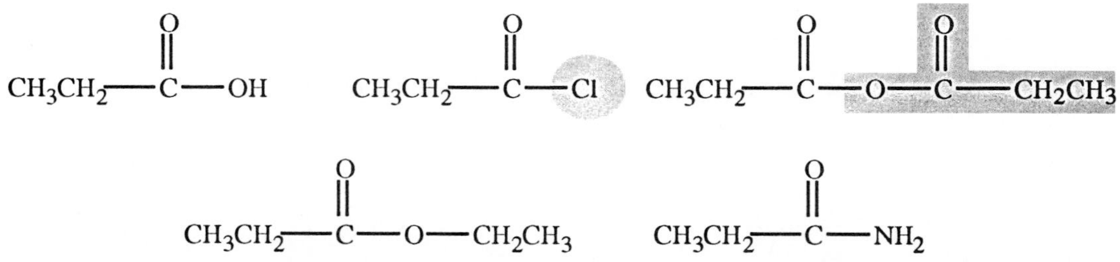

14.2 (cont)

(b) $CH_3CH_2-\overset{\overset{O}{\|}}{C}-\overset{+}{O}H_2$ HSO_4^- $\underset{H_2SO_4}{\overset{\longrightarrow}{\longleftarrow}}$ $CH_3CH_2-\overset{\overset{O}{\|}}{C}-OH$ $\underset{}{\overset{H_2SO_4}{\overset{\longleftarrow}{\longrightarrow}}}$

$CH_3CH_2-\overset{\overset{+O-H}{\|}}{C}-OH$ HSO_4^-

major product

$CH_3CH_2-\overset{\overset{O}{\|}}{C}-\overset{\overset{+}{O}}{\underset{H}{|}}-CH_2CH_3$ $\underset{H_2SO_4}{\overset{\longrightarrow}{\longleftarrow}}$ $CH_3CH_2-\overset{\overset{O}{\|}}{C}-O-CH_2CH_3$ $\underset{}{\overset{H_2SO_4}{\overset{\longleftarrow}{\longrightarrow}}}$

$CH_3CH_2-\overset{\overset{+O-H}{\|}}{C}-O-CH_2CH_3$ HSO_4^-

major product

$CH_3CH_2-\overset{\overset{O}{\|}}{C}-\overset{+}{N}H_3$ $\underset{H_2SO_4}{\overset{\longrightarrow}{\longleftarrow}}$ $CH_3CH_2-\overset{\overset{O}{\|}}{C}-NH_2$ $\underset{}{\overset{H_2SO_4}{\overset{\longleftarrow}{\longrightarrow}}}$ HSO_4^- $CH_3CH_2-\overset{\overset{+O-H}{\|}}{C}-NH_2$

major product

(c)   The good leaving groups are shaded in the formulas shown in parts (a) and (b).

- -

Concept Map 14.1   (see p. 451)

- -

14.3 $CH_3CH_2C-\overset{\overset{:O:}{\|}}{\underset{..}{O}}-H$ $\longleftrightarrow$ $CH_3CH_2\overset{+}{C}-\overset{\overset{:\overset{..}{O}:^-}{|}}{\underset{..}{O}}-H$ $\longleftrightarrow$ $CH_3CH_2C=\overset{\overset{:\overset{..}{O}:^-}{|}}{\underset{..}{\overset{+}{O}}}-H$

$CH_3CH_2C-\overset{\overset{:O:}{\|}}{\underset{..}{C}l}:$ $\longleftrightarrow$ $CH_3CH_2\overset{+}{C}-\overset{\overset{:\overset{..}{O}:^-}{|}}{\underset{..}{C}l}:$ $\longleftrightarrow$ $CH_3CH_2C=\overset{\overset{:\overset{..}{O}:^-}{|}}{\underset{..}{\overset{+}{C}l}}:$

$CH_3CH_2C-\overset{\overset{:O:}{\|}}{\underset{..}{O}}-\overset{\overset{:O:}{\|}}{C}CH_2CH_3$ $\longleftrightarrow$ $CH_3CH_2\overset{+}{C}-\overset{\overset{:\overset{..}{O}:^-}{|}}{\underset{..}{O}}-\overset{\overset{:O:}{\|}}{C}CH_2CH_3$

$\updownarrow$

14.3 (cont)

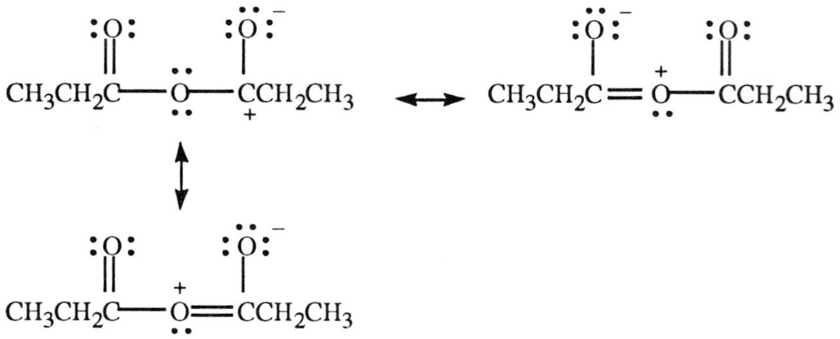

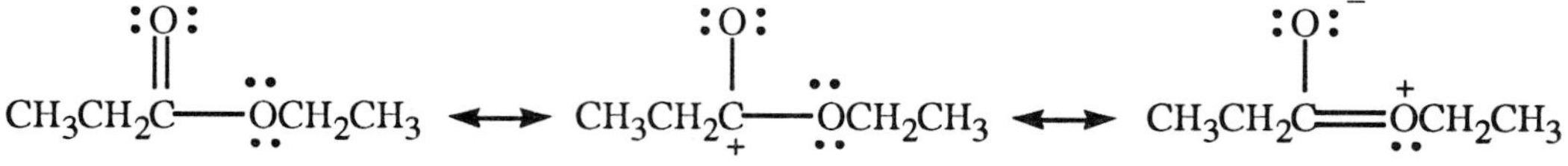

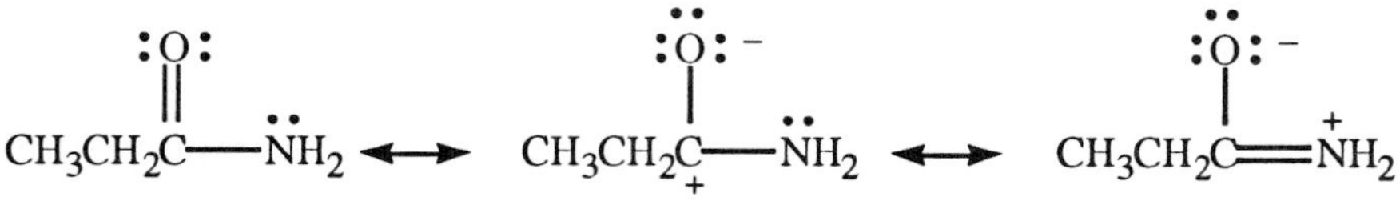

$$CH_3CH_2C\overset{:\overset{\cdot\cdot}{O}:}{\underset{}{\|}}\overset{\cdot\cdot}{\underset{\cdot\cdot}{O}}\overset{+}{\underset{}{}}—\overset{:\overset{\cdot\cdot}{O}:^{-}}{\underset{}{|}}CCH_2CH_3 \longleftrightarrow CH_3CH_2C=\overset{\cdot\cdot}{\underset{\cdot\cdot}{O}}\overset{+}{—}\overset{:\overset{\cdot\cdot}{O}:}{\underset{}{\|}}CCH_2CH_3$$

$$\updownarrow$$

$$CH_3CH_2C\overset{:\overset{\cdot\cdot}{O}:}{\underset{}{\|}}—\overset{+}{\underset{\cdot\cdot}{O}}=\overset{:\overset{\cdot\cdot}{O}:^{-}}{\underset{}{|}}CCH_2CH_3$$

$$CH_3CH_2C\overset{:\overset{\cdot\cdot}{O}:}{\underset{}{\|}}—\overset{\cdot\cdot}{\underset{\cdot\cdot}{O}}CH_2CH_3 \longleftrightarrow CH_3CH_2\overset{+}{\underset{}{C}}\overset{:\overset{\cdot\cdot}{O}:^{-}}{\underset{}{|}}—\overset{\cdot\cdot}{\underset{\cdot\cdot}{O}}CH_2CH_3 \longleftrightarrow CH_3CH_2C=\overset{:\overset{\cdot\cdot}{O}:^{-}}{\underset{}{|}}\overset{+}{\underset{\cdot\cdot}{O}}CH_2CH_3$$

$$CH_3CH_2C\overset{:\overset{\cdot\cdot}{O}:}{\underset{}{\|}}—\overset{\cdot\cdot}{\underset{\cdot\cdot}{N}}H_2 \longleftrightarrow CH_3CH_2\overset{+}{\underset{}{C}}\overset{:\overset{\cdot\cdot}{O}:^{-}}{\underset{}{|}}—\overset{\cdot\cdot}{\underset{}{N}}H_2 \longleftrightarrow CH_3CH_2C=\overset{:\overset{\cdot\cdot}{O}:^{-}}{\underset{}{|}}\overset{+}{\underset{}{N}}H_2$$

The resonance contributors shown above reduce the positive charge on the carbon atom of the carbonyl group. The carbon atom of the carbonyl group in acid derivatives is thus less electrophilic than the carbon atom of the carbonyl group in aldehydes and ketones. The compounds in which the resonance contributors involve oxygen or nitrogen are more stable than the one in which chlorine is involved. The orbitals of oxygen and nitrogen (second row elements of smaller size) overlap with those of carbon better than those of chlorine, therefore their resonance contributors are more important in the stabilization of the compounds. The amide is the most stable of all, because nitrogen is less electronegative than oxygen, and therefore bears a positive charge more easily than oxygen does.

14.4   Propanamide is much less basic than propylamine because the nitrogen atom in propanamide has a partial positive charge (see the resonance contributors for propanamide in Problem 14.3). This means that there is less electron density on the nitrogen atom in propanamide than in propylamine, and, therefore, lower basicity for the amide.

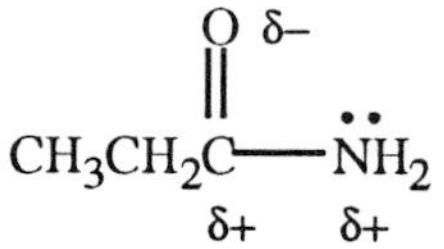

$$CH_3CH_2\underset{\delta+}{C}\overset{\overset{O\ \delta-}{\|}}{—}\underset{\delta+}{\overset{\cdot\cdot}{N}H_2}$$

$$CH_3CH_2CH_2—\overset{\cdot\cdot}{N}H_2$$

The electron pair on the nitrogen atom is less available; the amide nitrogen is less basic.

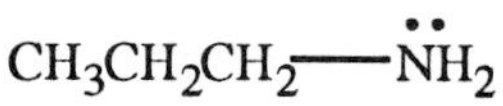

The electron pair on the nitrogen atom is more available; the amine nitrogen is more basic.

$$CH_3CH_2C\overset{:\overset{\cdot\cdot}{O}:}{\underset{}{\|}}—\overset{\cdot\cdot}{\underset{\cdot\cdot}{N}}H^{-} \longleftrightarrow CH_3CH_2C=\overset{:\overset{\cdot\cdot}{O}:^{-}}{\underset{}{|}}\overset{\cdot\cdot}{\underset{}{N}}H$$

$$CH_3CH_2CH_2—\overset{\cdot\cdot}{\underset{\cdot\cdot}{N}}H^{-}$$

delocalization of charge in conjugate base of propanamide; hence stabilization of base relative to its conjugate acid

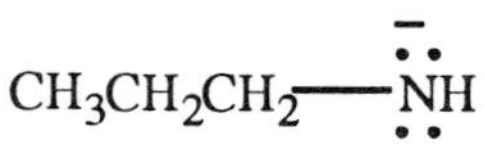

no stabilization of negative charge; therefore strong base

Concept Map 14.1   Relative reactivities in nucleophilic substitutions.

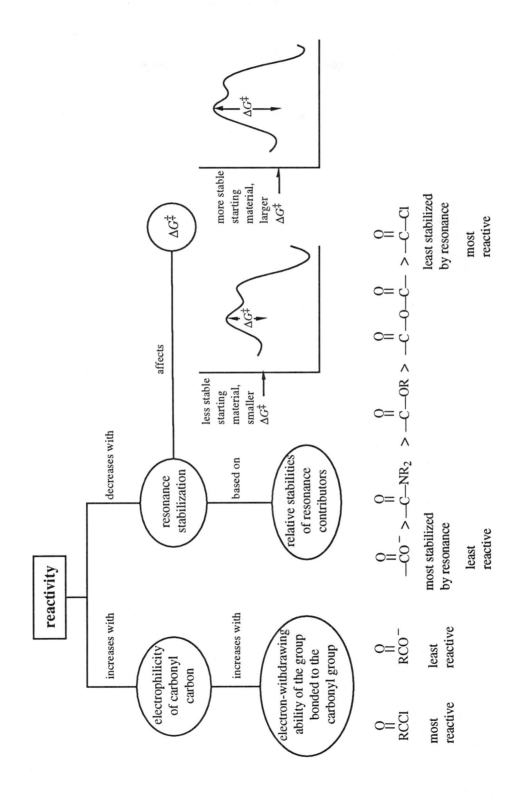

14.5

decreasing solubility →

(a) $CH_3CH_2CH_2CH_2\overset{O}{\overset{\|}{C}}O^-Na^-$ > $CH_3CH_2CH_2CH_2\overset{O}{\overset{\|}{C}}OH$ > $CH_3CH_2\overset{O}{\overset{\|}{C}}OCH_2CH_3$

- ionic
- covalent, but hydrogen bond donor as well as acceptor
- only hydrogen bond acceptor

(b) $CH_3CH_2CH_2\overset{O}{\overset{\|}{C}}OH$ > $CH_3CH_2CH_2CH_2OH$ > $CH_3CH_2\overset{O}{\overset{\|}{C}}OCH_2CH_3$

- can hydrogen bond at the carbonyl group as well as at the hydroxyl group
- only one site for hydrogen bonding
- only hydrogen bond acceptor

(c) $CH_3CH_2CH_2\overset{O}{\overset{\|}{C}}NH_2$ > $CH_3CH_2CH_2\overset{O}{\overset{\|}{C}}OCH_2CH_3$ > $CH_3CH_2CH_2\overset{O}{\overset{\|}{C}}O(CH_2)_4CH_3$

- hydrogen bond donor as well as acceptor
- hydrogen bond acceptor
- hydrogen bond acceptor but nonpolar portion of the molecule now significantly larger than the polar part

14.6

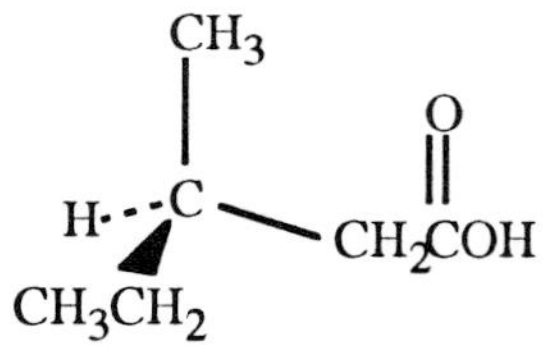
(*R*)-3-methylpentanoic acid

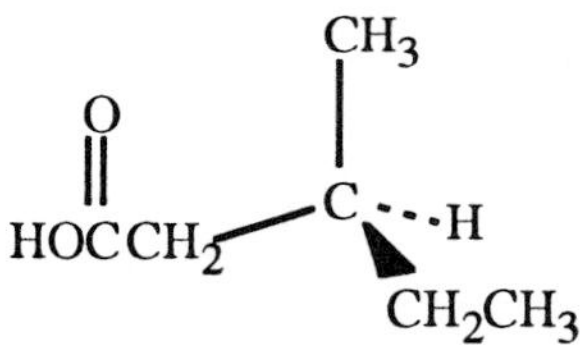
(*S*)-3-methylpentanoic acid

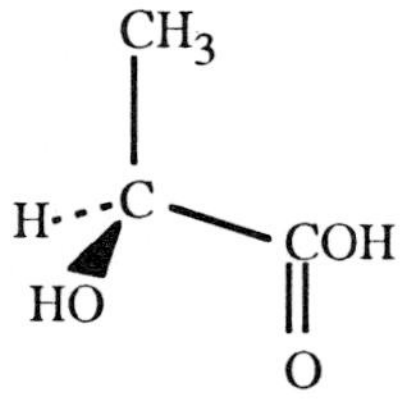
(*S*)-2-hydroxypropanoic acid

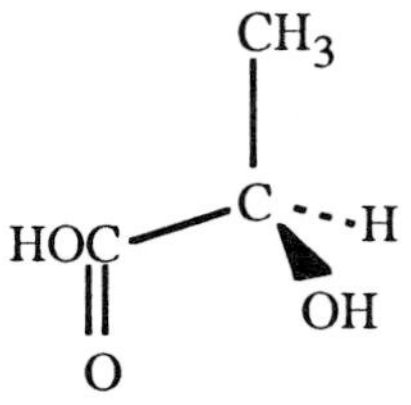
(*R*)-2-hydroxypropanoic acid

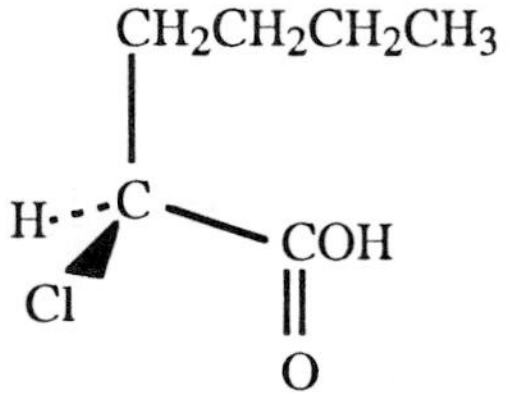
(*S*)-2-chlorohexanoic acid

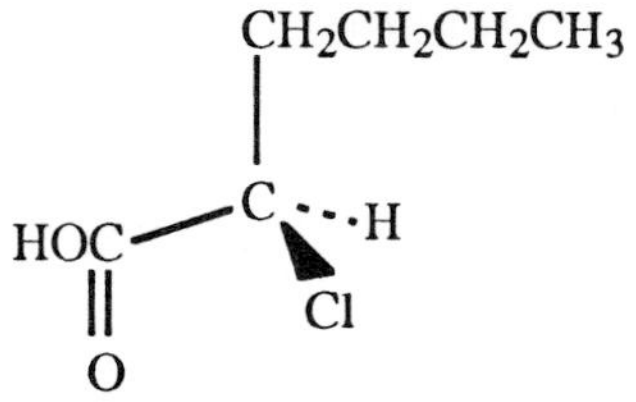
(*R*)-2-chlorohexanoic acid

14.6 (cont)

(R)-4-hydroxy-6-methylheptanoic acid   (S)-4-hydroxy-6-methylheptanoic acid

(S)-alanine   (R)-alanine

(2R, 3S)-tartaric acid   (2S, 3S)-tartaric acid   (2R, 3R)-tartaric acid
meso-tartaric acid

(1R, 2R)-1,2-cyclopentane-   (1S, 2S)-1,2-cyclopentane-   (1R, 2S)-1,2-cyclopentane-
dicarboxylic acid   dicarboxylic acid   dicarboxylic acid
                                                                cis isomer
                                                                meso compound

trans isomers

(1R, 3S)-3-methyl-   (1S, 3R)-3-methyl-   (1R, 3R)-3-methyl-   (1S, 3S)-3-methyl-
cyclohexane-   cyclohexane-   cyclohexane-   cyclohexane-
carboxylic acid   carboxylic acid   carboxylic acid   carboxylic acid

cis isomers   trans isomers

14.7

(a)   undecanoic acid

(b)   (*E*)-2-pentenoic acid

(c)   (*S*)-2-hydroxybutanoic acid

(d)   *m*-bromobenzoic acid

(e)   2-chloro-4-methylpentanoic acid

(f)   5-oxohexanoic acid

(g)   3-hydroxyhexanedioic acid

(h)   (1*R*, 2*R*)-2-hydroxycyclopentanecarboxylic acid
       (*trans*-2-hydroxycyclopentanecarboxylic acid)

14.8

(a)   hexanoyl chloride

(b)   butanoic anhydride

(c)   methyl (*Z*)-4-hexenoate

(d)   isobutyl *p*-chlorobenzoate

(e)   cyclobutanecarbonyl chloride

(f)   *p*-chlorobenzoic anhydride

14.9

(a)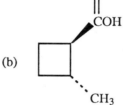

(b)

(c)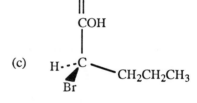

(d) $HOC(CH_2)_6COH$ (with O double bonds)

(e) $CH_3OCCH_2COCH_3$ (with O double bonds)

(f)

(g)

(h) $ClC(CH_2)_3CCl$ (with O double bonds)

(i) $Na^+ {}^-OC—CO^- Na^+$ (with O double bonds)

(j)

14.10

(a)   propyl benzoate

(b)   ethyl 6-oxooctanoate

(c)   decanoic acid

(d)   *m*-bromobenzoyl chloride
       (sodium stearate)

(e)   sodium octadecanoate
       (phenylacetic acid)

(f)   phenylethanoic acid

(g)   *N*-methylacetamide

(h)   *p*-methylbenzamide

(i)   (1*R*, 2*R*)-2-bromocyclo-
       pentanecarboxylic acid

14.10 (cont)

(j)   (*E*)-2-butenoic acid   (k)   ethyl 3-oxopentanoate   (l)   (*R*)-3-hydroxybutanoic acid

(m)   heptanenitrile   (n)   3-methylbutanoic anhydride   (o)   *N*-phenylbutanamide
       (butananilide)

14.11

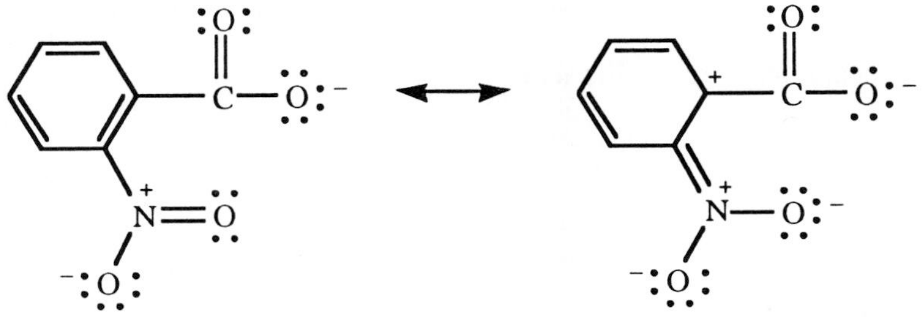

anion of *ortho*-
nitrobenzoic acid

positive charge on the carbon
atom that is adjacent to the
carboxylate group;
stabilization of the anion

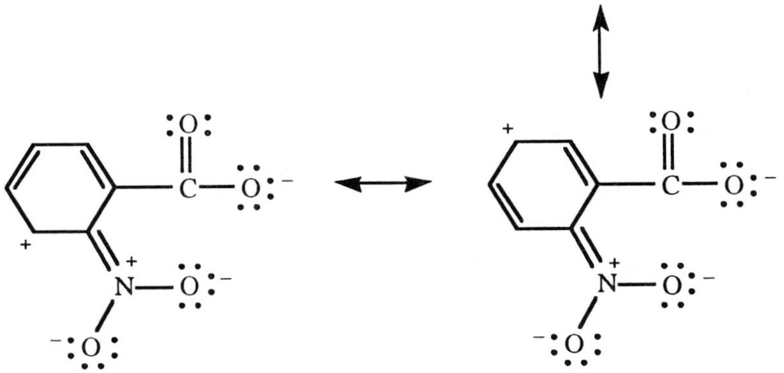

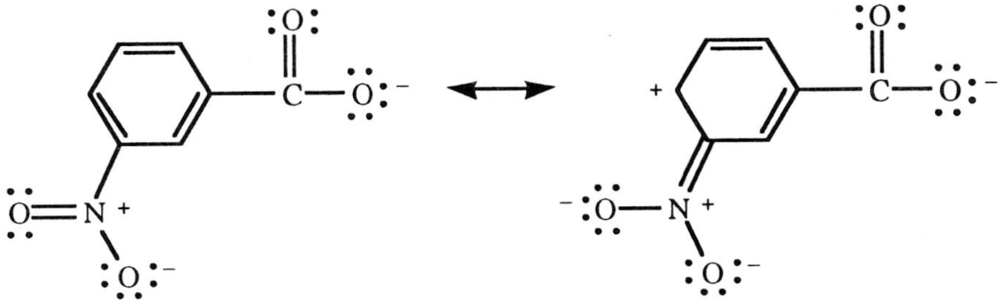

anion of *meta*-
nitrobenzoic acid

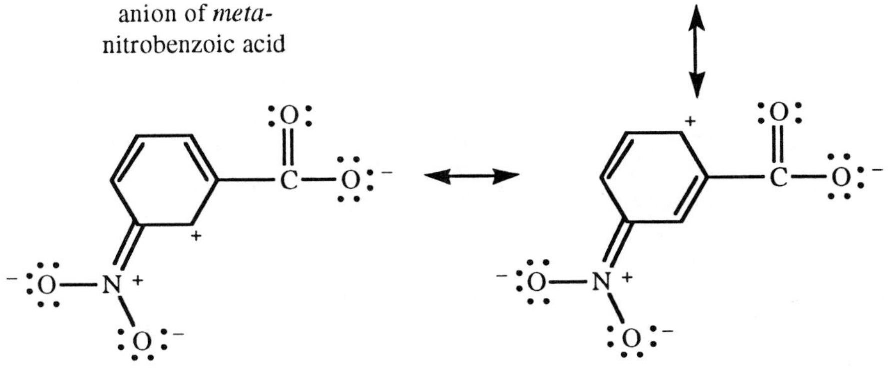

14.11 (cont)

There is no resonance contributor of *meta*-nitrobenzoic acid in which there is a positive charge adjacent to the carboxylate group.

- -

Concept Map 14.2   Preparation of carboxylic acids.

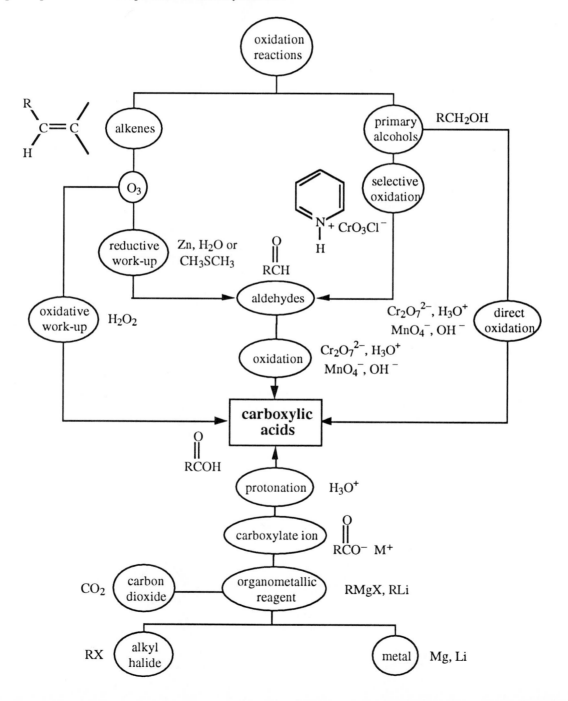

14.12

(a)

CH$_3$—C$_6$H$_4$—Br $\xrightarrow[\text{diethyl ether}]{\text{Mg}}$ CH$_3$—C$_6$H$_4$—MgBr $\xrightarrow{\text{CO}_2}$

A

CH$_3$—C$_6$H$_4$—CO—O$^-$ Mg^{2+} Br$^-$ $\xrightarrow{\text{H}_3\text{O}^+}$ CH$_3$—C$_6$H$_4$—COOH

B   **C**

(b)

$$\underset{}{\text{CH}_3\text{CH}_2\text{CH}_2\overset{\overset{\text{CH}_3}{|}}{\text{CH}}\text{CH}_2\text{CH}_2\text{OH}} \xrightarrow[\substack{\text{H}_2\text{O} \\ \Delta}]{\text{KMnO}_4,\ \text{NaOH}} \text{CH}_3\text{CH}_2\text{CH}_2\overset{\overset{\text{CH}_3}{|}}{\text{CH}}\text{CH}_2\overset{\overset{\text{O}}{\|}}{\text{C}}\text{O}^-\ \text{Na}^+ \xrightarrow{\text{H}_3\text{O}^+}$$

D

$$\text{CH}_3\text{CH}_2\text{CH}_2\overset{\overset{\text{CH}_3}{|}}{\text{CH}}\text{CH}_2\overset{\overset{\text{O}}{\|}}{\text{C}}\text{OH}$$

E

(c)

$\xrightarrow[\text{chloroform}]{\text{O}_3}$ **F** $\xrightarrow[\substack{\text{H}_2\text{O} \\ \Delta}]{\text{Ag}_2\text{O},\ \text{NaOH}}$

$$\text{Na}^+\ {}^-\overset{\overset{\text{O}}{\|}}{\text{O}}\text{C}(\text{CH}_2)_4\overset{\overset{\text{O}}{\|}}{\text{C}}\text{O}^-\ \text{Na}^+ \xrightarrow{\text{H}_3\text{O}^+} \text{HO}\overset{\overset{\text{O}}{\|}}{\text{C}}(\text{CH}_2)_4\overset{\overset{\text{O}}{\|}}{\text{C}}\text{OH}$$

G   **H**

(d)

$\xrightarrow[\substack{\text{H}_2\text{O} \\ \Delta}]{\text{Ag}_2\text{O},\ \text{NaOH}}$ C$_6$H$_5$—CH$_2$CH$_2$CO$^-$ Na$^+$ $\xrightarrow{\text{H}_3\text{O}^+}$

I

C$_6$H$_5$—CH$_2$CH$_2$COOH

J

(e)

$$\text{CH}_3(\text{CH}_2)_5\overset{\overset{\text{O}}{\|}}{\text{C}}\text{H} \xrightarrow[\substack{\text{H}_2\text{SO}_4 \\ \text{H}_2\text{O} \\ \Delta}]{\text{CrO}_3} \text{CH}_3(\text{CH}_2)_5\overset{\overset{\text{O}}{\|}}{\text{C}}\text{OH}$$

K

(f)

CH$_3$(CH$_2$)$_9$CH=CH$_2$ $\xrightarrow[\text{chloroform}]{\text{O}_3}$ $\xrightarrow{\text{H}_2\text{O}_2,\ \text{NaOH}}$ $\xrightarrow{\text{H}_3\text{O}^+}$ $\text{CH}_3(\text{CH}_2)_9\overset{\overset{\text{O}}{\|}}{\text{C}}\text{OH}$ + $\text{H}\overset{\overset{\text{O}}{\|}}{\text{C}}\text{OH}$

L   **M**

14.13

(a)

$$\text{C}_6\text{H}_5\text{—CH}_2\text{CH}_2\overset{\displaystyle O}{\overset{\|}{\text{C}}}\text{OH} \xrightarrow[\Delta]{\text{SOCl}_2} \text{C}_6\text{H}_5\text{—CH}_2\text{CH}_2\overset{\displaystyle O}{\overset{\|}{\text{C}}}\text{Cl}$$

(b)

$$\text{CH}_3\text{CH}_2\text{CH}_2\overset{\displaystyle O}{\overset{\|}{\text{C}}}\text{OH} \xrightarrow{\text{PCl}_3} \text{CH}_3\text{CH}_2\text{CH}_2\overset{\displaystyle O}{\overset{\|}{\text{C}}}\text{Cl}$$

(c)

$$\text{CH}_3\text{CH}_2\text{CH}_2\overset{\displaystyle O}{\overset{\|}{\text{C}}}\text{Cl} \ + \ \text{Na}^+ \ ^-\text{O}\overset{\displaystyle O}{\overset{\|}{\text{C}}}\text{CH}_2\text{CH}_2\text{CH}_3 \longrightarrow \text{CH}_3\text{CH}_2\text{CH}_2\overset{\displaystyle O}{\overset{\|}{\text{C}}}\text{O}\overset{\displaystyle O}{\overset{\|}{\text{C}}}\text{CH}_2\text{CH}_2\text{CH}_3$$

(d)

$$\text{CH}_3(\text{CH}_2)_{10}\overset{\displaystyle O}{\overset{\|}{\text{C}}}\text{OH} \xrightarrow[\Delta]{\text{SOCl}_2} \text{CH}_3(\text{CH}_2)_{10}\overset{\displaystyle O}{\overset{\|}{\text{C}}}\text{Cl}$$

(e)

$$\text{HO}\overset{\displaystyle O}{\overset{\|}{\text{C}}}\text{CH}_2\text{CH}_2\overset{\displaystyle O}{\overset{\|}{\text{C}}}\text{OH} \xrightarrow{\text{CH}_3\overset{\displaystyle O}{\overset{\|}{\text{C}}}\text{O}\overset{\displaystyle O}{\overset{\|}{\text{C}}}\text{CH}_3}$$

(f)

$$\text{CH}_3\text{CH}_2\overset{\displaystyle O}{\overset{\|}{\text{C}}}\text{OH} \xrightarrow[\substack{\text{clay} \\ \Delta \\ 650\ ^\circ\text{C}}]{} \text{CH}_3\text{CH}_2\overset{\displaystyle O}{\overset{\|}{\text{C}}}\text{O}\overset{\displaystyle O}{\overset{\|}{\text{C}}}\text{CH}_2\text{CH}_3$$

(g)

$$\text{HO}\overset{\displaystyle O}{\overset{\|}{\text{C}}}(\text{CH}_2)_5\overset{\displaystyle O}{\overset{\|}{\text{C}}}\text{OH} \xrightarrow[\Delta]{\text{excess SOCl}_2} \text{Cl}\overset{\displaystyle O}{\overset{\|}{\text{C}}}(\text{CH}_2)_5\overset{\displaystyle O}{\overset{\|}{\text{C}}}\text{Cl}$$

14.14   (a)

(b)

14.15

(a)

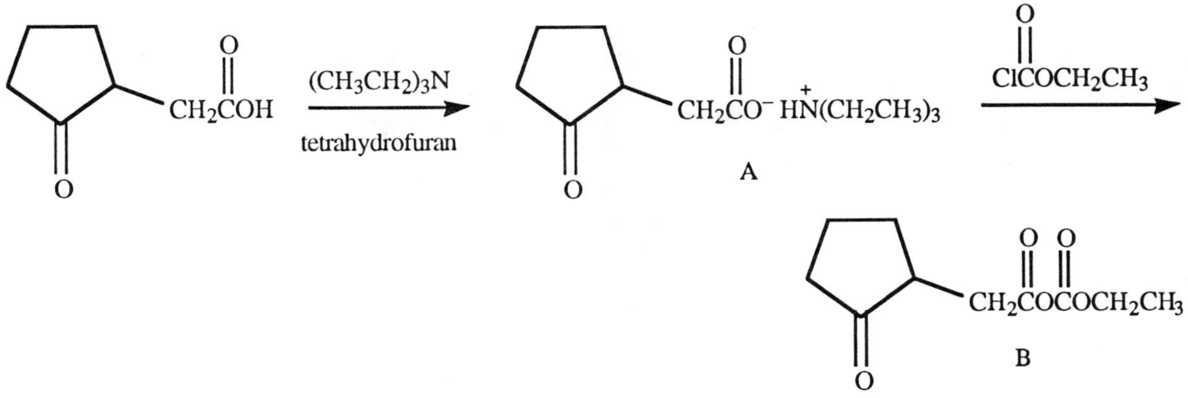

A

B

(b)   Two of the carbonyl groups are acid anhydride-like; the one in the ring is a ketone and does not have a good leaving group on it. Of the two that are acid anhydrides, one is less resonance-stabilized than the other, therefore higher in energy, and therefore more reactive than the other.

two resonance contributors to the
anhydride function that reacts

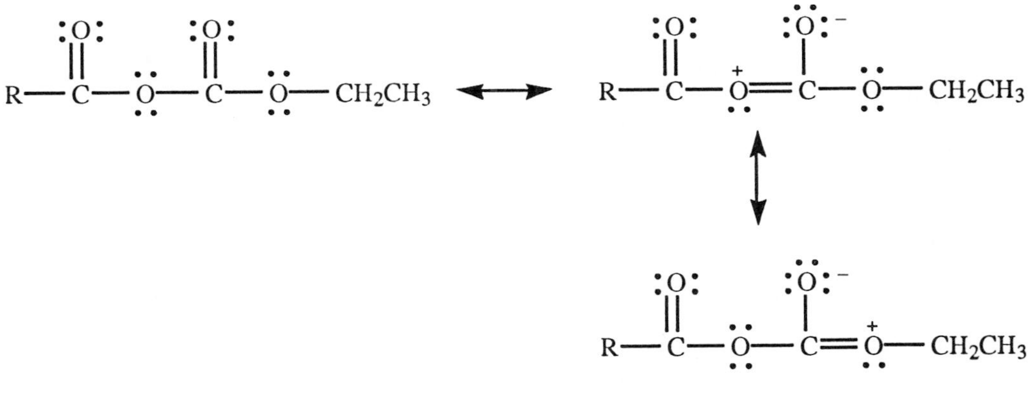

three resonance contributors to the anhydride-
ester function; less reactive than the other anhydride

14.16

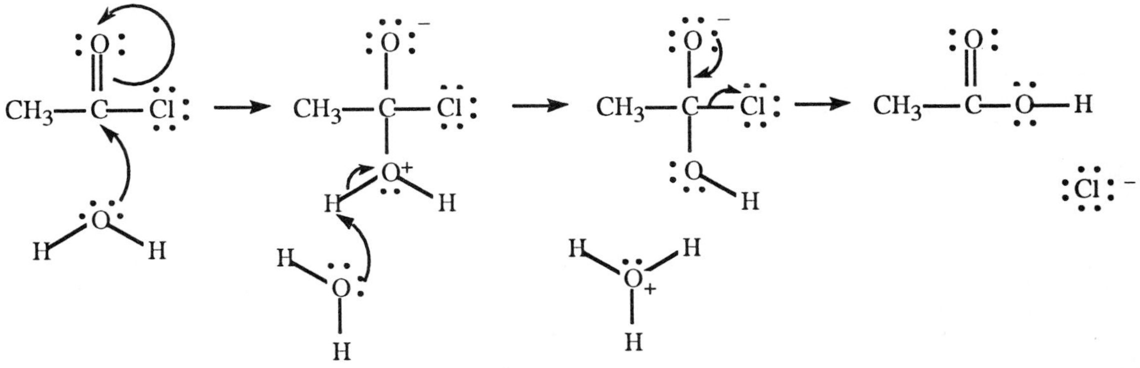

Concept Map 14.3   Hydrolysis reactions of acid derivatives.

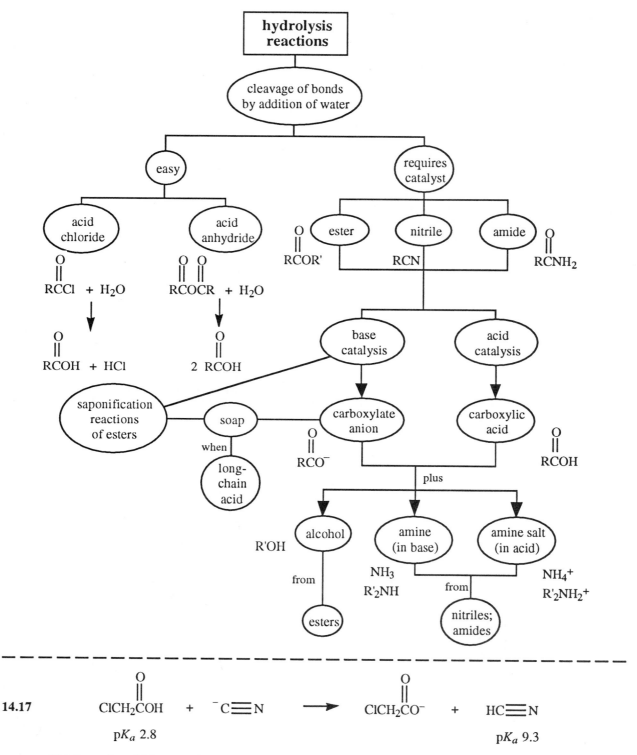

14.17   CICH$_2$COH   + $^-$C≡N   ⟶   CICH$_2$CO$^-$ +   HC≡N

              pK_a 2.8   pK_a 9.3

Chloroacetic acid is a much stronger acid than hydrogen cyanide. The cyanide anion will, therefore, be converted to hydrogen cyanide and will not be available to act as a nucleophile if chloroacetic acid is not first neutralized with base. In addition, even small amounts of hydrogen cyanide are lethal to humans.

14.18

(a)

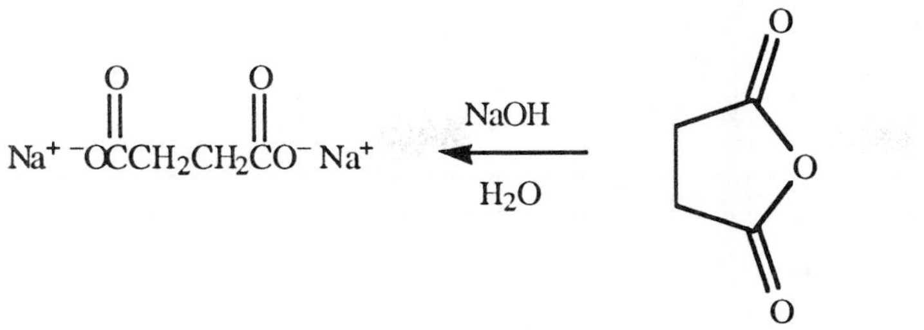

$$\underset{Na^+\ ^-\underset{\|}{\overset{O}{O}}CCH_2CH_2\underset{\|}{\overset{O}{C}}O^-\ Na^+}{} \xleftarrow[\text{H}_2\text{O}]{\text{NaOH}}$$

(b)

$$\underset{\underset{CH_3}{|}}{CH_3CH_2CH_2\underset{\overset{\|}{O}}{\overset{O}{C}}HCOH} \xleftarrow{H_3O^+} \xleftarrow{CO_2} \underset{\underset{CH_3}{|}}{CH_3CH_2CH_2CHMgBr} \xleftarrow{Mg}$$

$$\underset{\underset{CH_3}{|}}{CH_3CH_2CH_2CHBr} \xleftarrow[\text{pyridine}]{PBr_3} \underset{\underset{CH_3}{|}}{CH_3CH_2CH_2CHOH} \xleftarrow{H_3O^+} \xleftarrow{CH_3MgI} \underset{}{CH_3CH_2CH_2\overset{O}{\overset{\|}{C}}H}$$

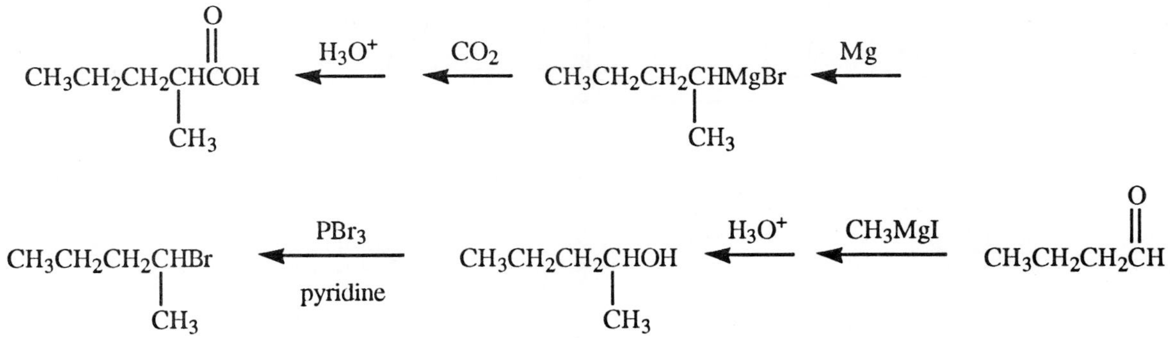

14.19

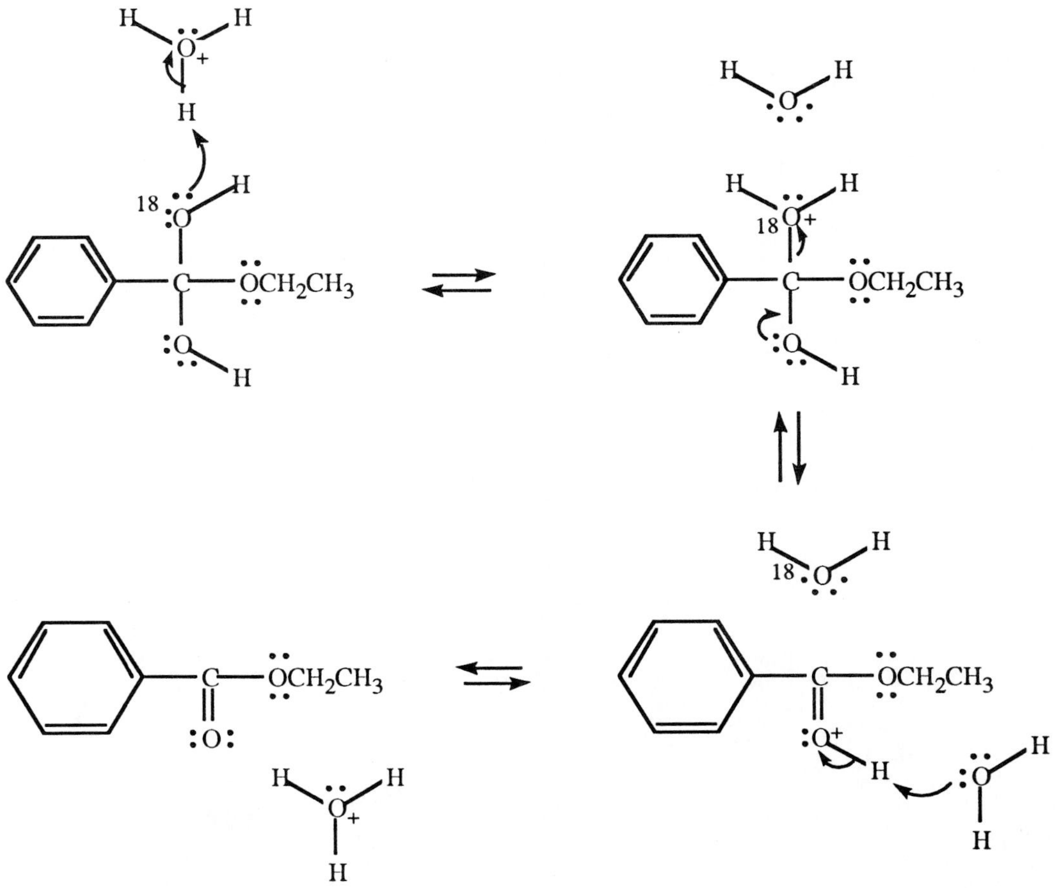

14.20   The two oxygen atoms are equivalent and cannot be distinguished from each other (Section 1.6A).

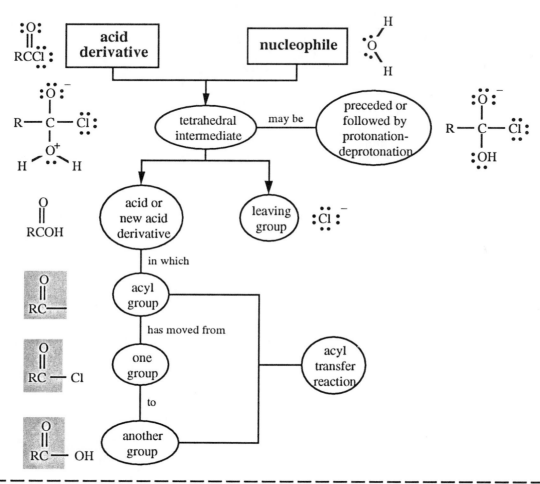

Concept Map 14.4   Mechanism of nucleophilic substitution reactions of acid derivatives. Acyl transfer reactions

14.21   Attack by the hydroxide ion at the carbon atom of the carbonyl group.

S configuration   retains the *S* configuration

14.21 (cont)

retains the *S* configuration   retains the *S* configuration

Attack by the hydroxide ion at the stereocenter.

S configuration   *R* configuration

14.22

1.   How have connectivities been changed in going from reactant to product? How many carbon atoms does each
     contain? What bonds must be broken and formed to transform reactant into product?

bonds broken   *bonds formed*

The reactant and the product have the same number of carbon atoms. Two of the bonds between carbon and nitrogen
have been broken. A carbon-oxygen double bond and two nitrogen-hydrogen single bonds have been formed.

2.   What reagents are present? Are they good acids, bases, nucleophiles, or electrophiles?

H_3O^+ is an acid, H_2O is a weak nucleophile. The carbon atom of the nitrile is electrophilic, and the nitrogen atom
has nonbonding electrons on it.

14.22 (cont)

3.   What is the most likely first step for the reaction: protonation or deprotonation, ionization, attack by a nucleophile, or attack by an electrophile?

An acid is present; therefore, protonation is the most likely first step. Water is a weak nucleophile and is unlikely to attack the unprotonated nitrile.

4.   What are the properties of the species present in the reaction mixture after the first step? What is likely to happen next?

The electrophilicity of the carbon bonded to nitrogen is greatly increased by the protonation of the nitrile; this carbon will now react with water.

4 (repeated).   What are the properties of the species present in the reaction mixture after this step? What is likely to happen next?

The oxonium ion resulting from the attack by water will be deprotonated by the solvent.

Comparing this new species to the desired product shows that it has a carbon-oxygen bond and a nitrogen-hydrogen bond in the right locations. To get to the product requires protonation at nitrogen, and a shifting of the double bond.

bonds broken   *bonds formed*

14.22 (cont)

This is a tautomerization.

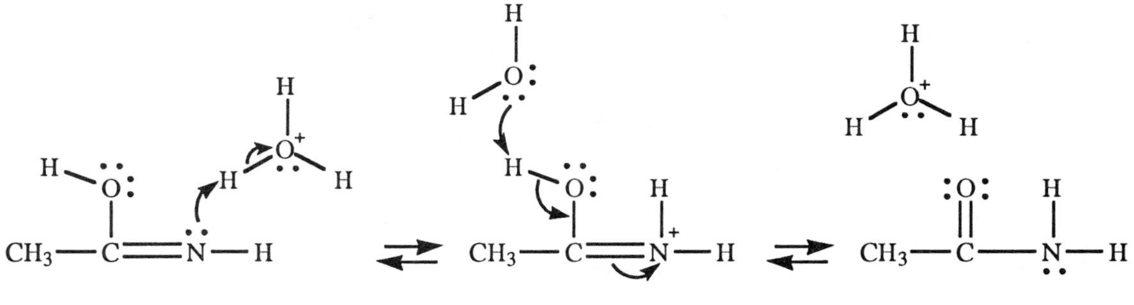

The complete mechanism is shown below:

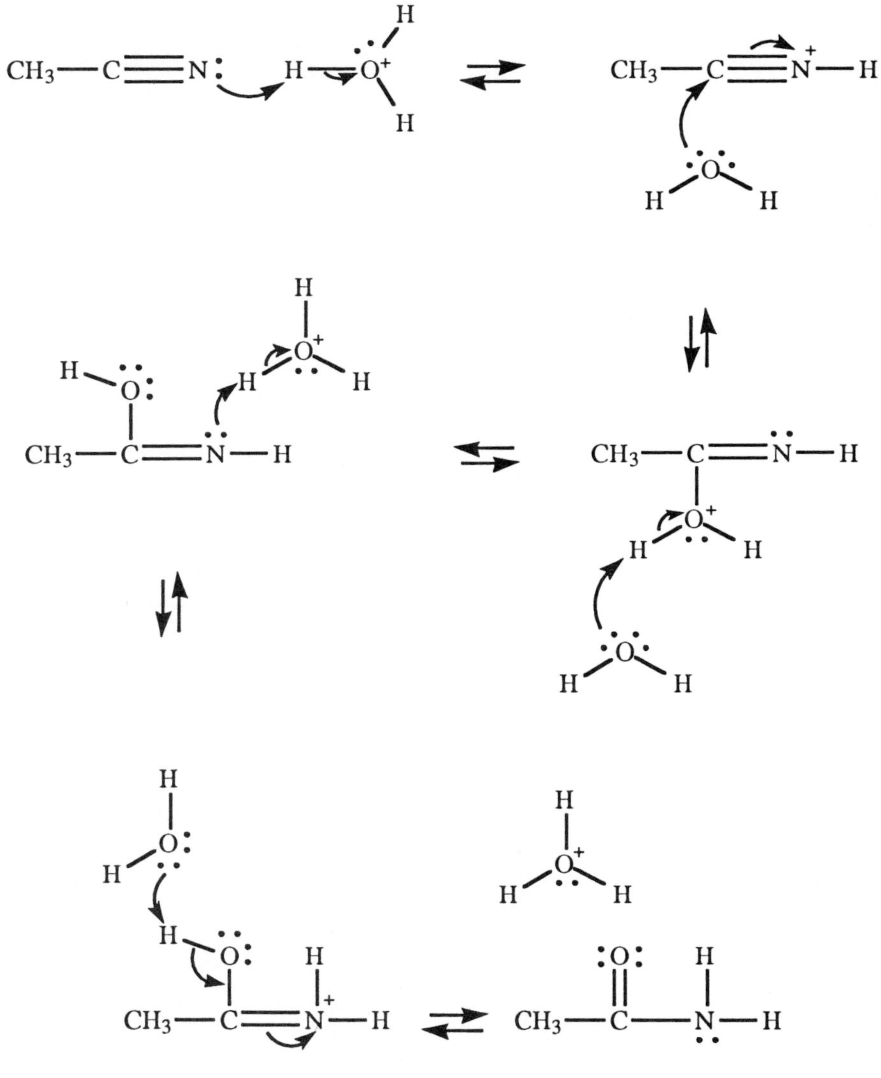

14.23

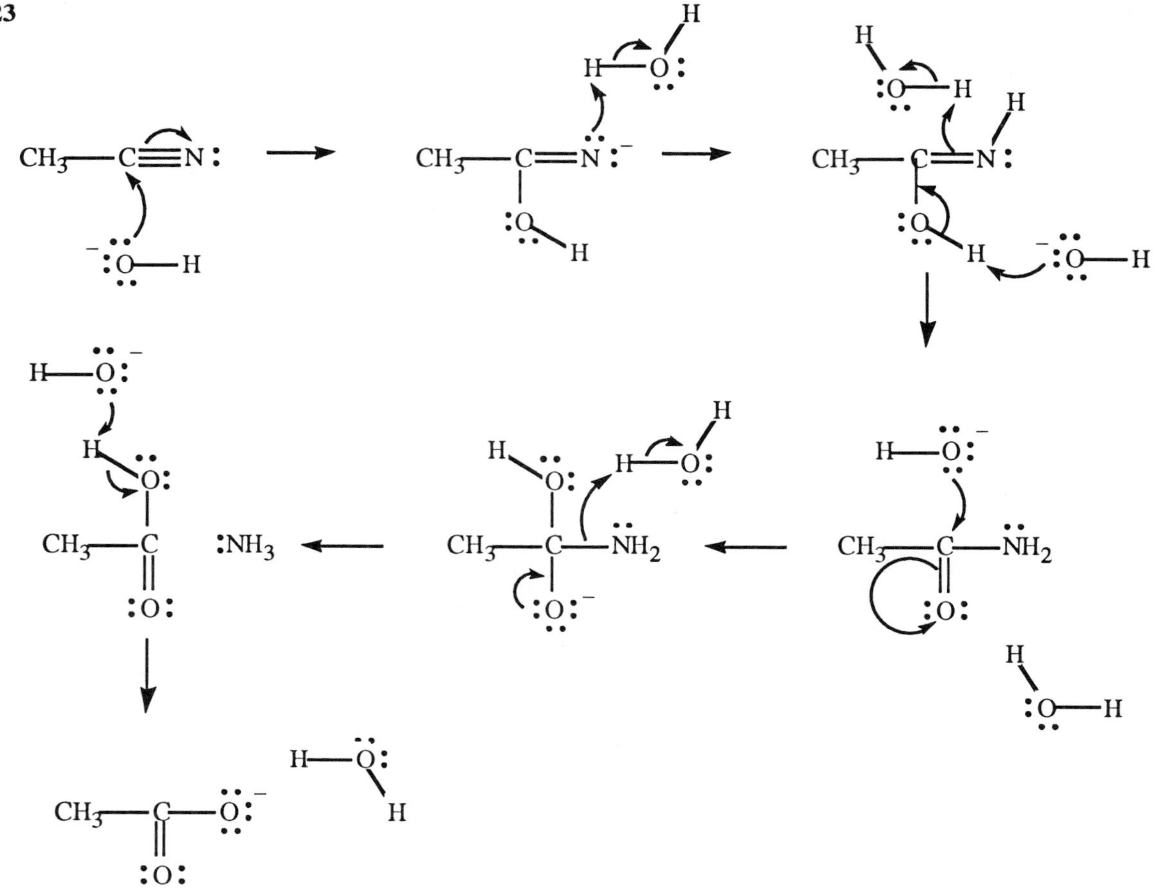

14.24

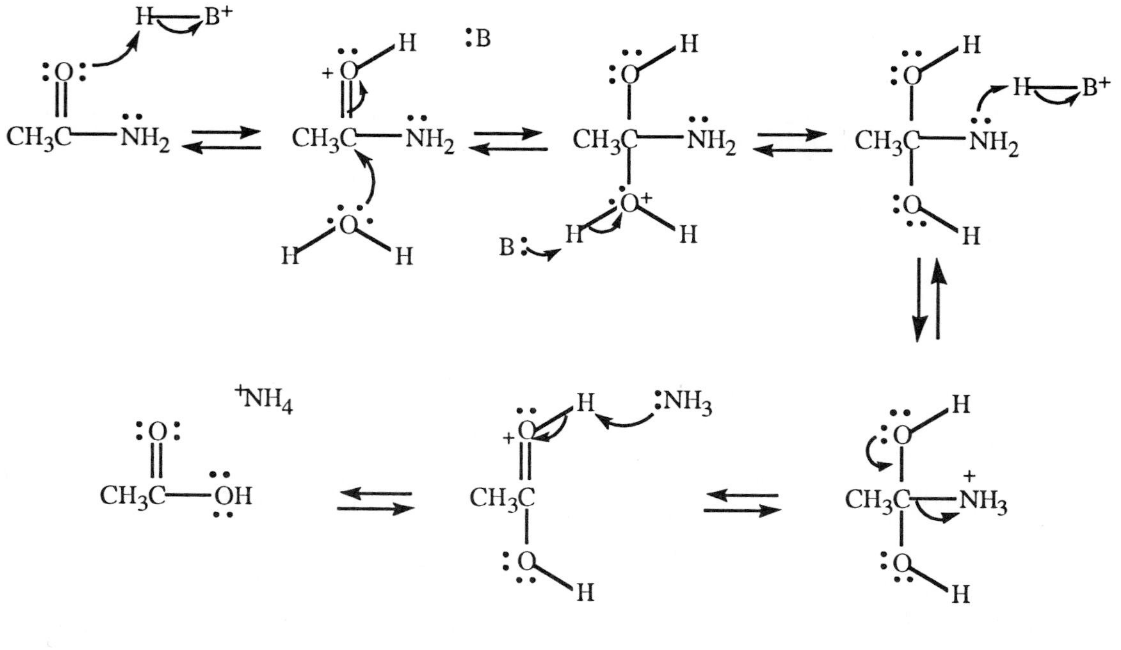

14.25

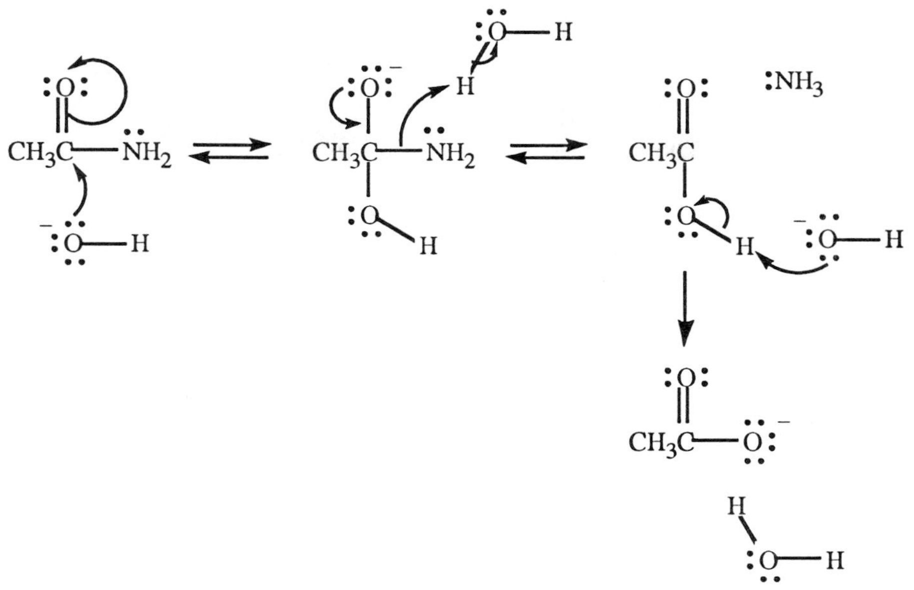

For the carbon-nitrogen bond to break in the basic hydrolysis of acetamide, the leaving group must be the conjugate base of ammonia, amide anion. In basic solution, it is unlikely that protonation of the nitrogen in the tetrahedral intermediate will take place before the carbon-nitrogen bond begins to break. However, amide anion is such a strong base that it is improbable that it has any real existence in water. The mechanism accounts for this by showing the amide ion taking a proton from water as the tetrahedral intermediate breaks up.

14.26 $CH_3CH_2CH_2CH_2OH$ +

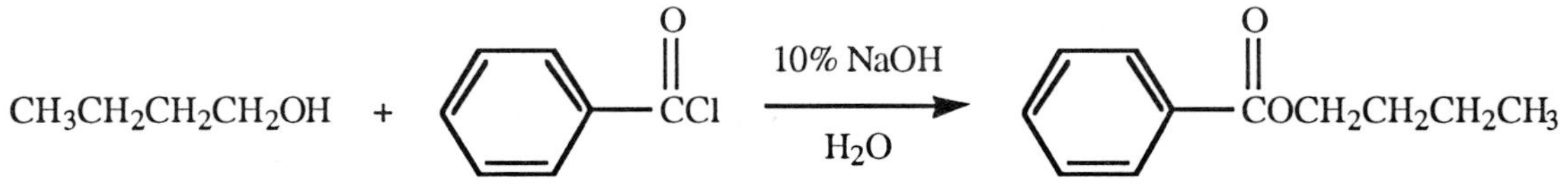

The other organic product formed would be sodium benzoate.

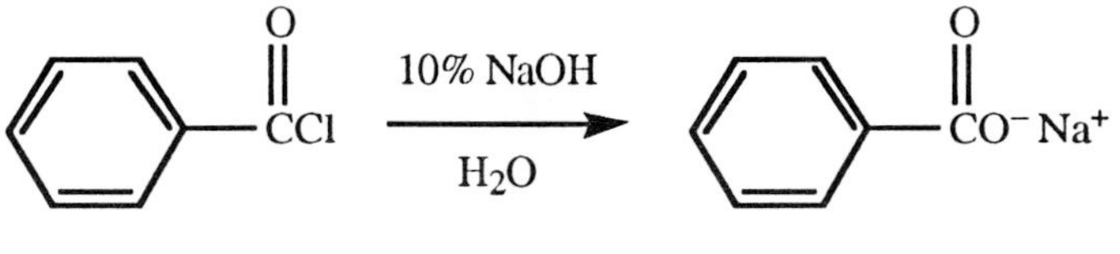

14.27

(a)

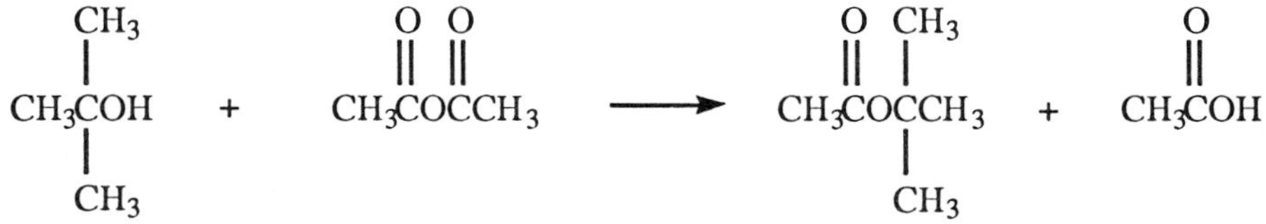

(b) 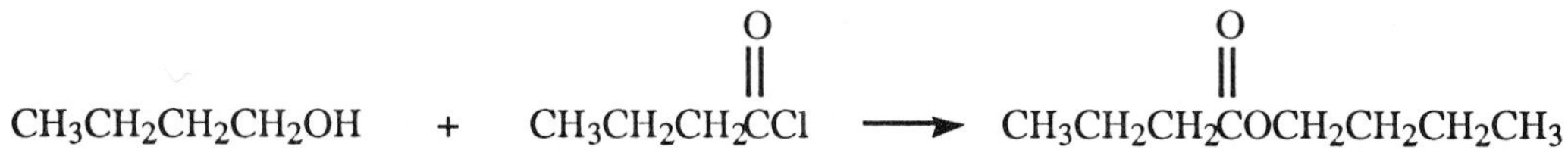 $CH_3CH_2CH_2CH_2OH$ + $CH_3CH_2CH_2CCl$ ⟶ $CH_3CH_2CH_2COCH_2CH_2CH_2CH_3$

14.27 (cont)

(c)

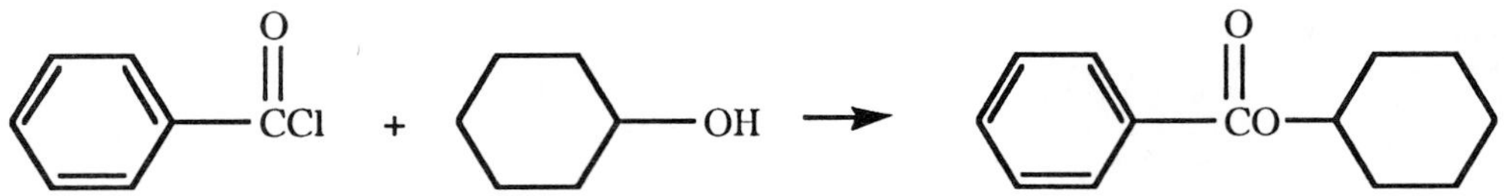

(d)

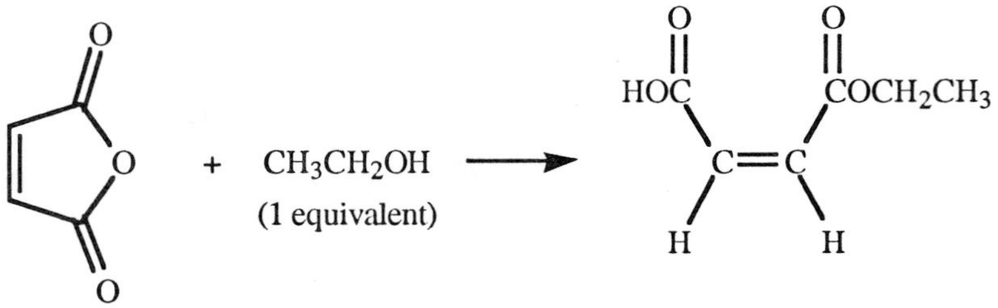

(1 equivalent)

14.28

(a)

$$HOC(CH_2)_9COH \xrightarrow[\Delta]{\begin{array}{c} CH_3OH \text{ (excess)} \\ H_2SO_4 \end{array}} CH_3OC(CH_2)_9COCH_3$$

A

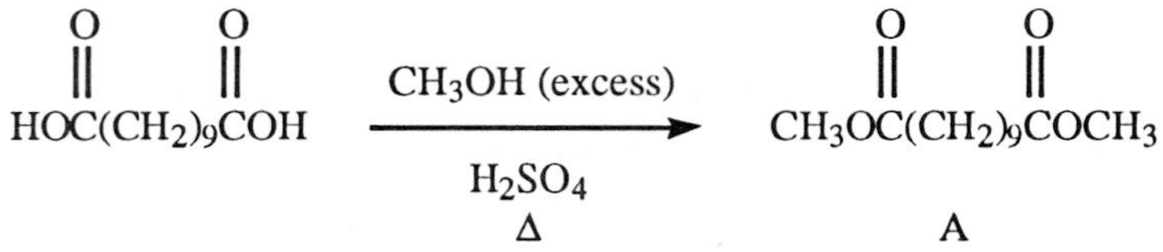

(b)

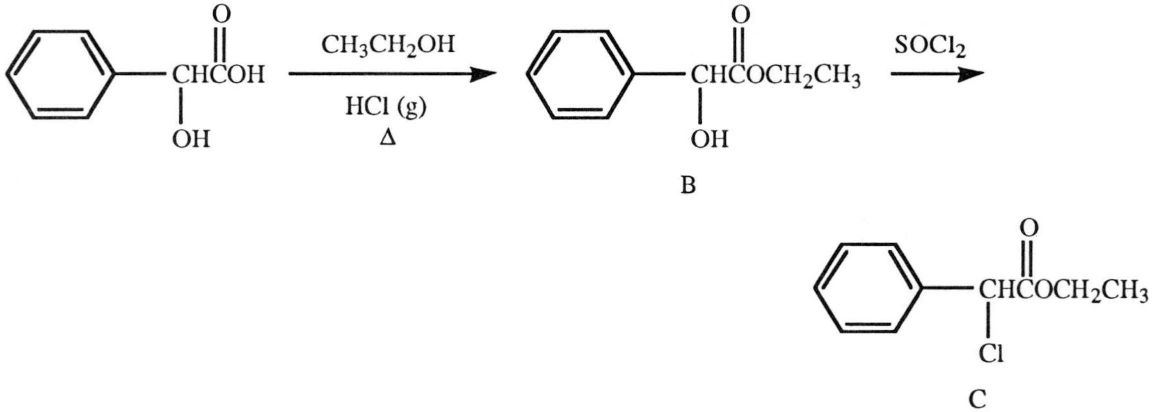

(c)

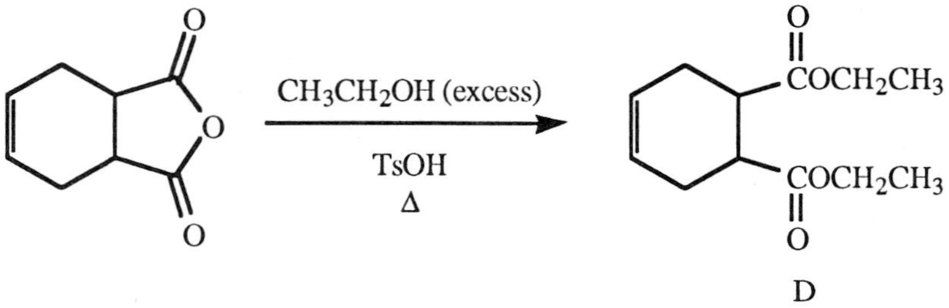

(d)

$$HOCC\equiv CCOH \xrightarrow[\Delta]{\begin{array}{c} CH_3OH \text{ (excess)} \\ H_2SO_4 \end{array}} CH_3OCC\equiv CCOCH_3$$

E

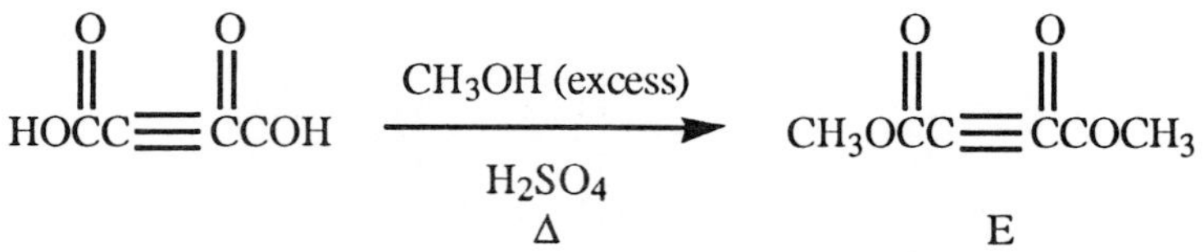

14.28 (cont)

(e)

$$BrCH_2\overset{\overset{\displaystyle O}{\displaystyle \|}}{C}OH \quad \xrightarrow[\substack{H_2SO_4 \\ \Delta}]{CH_3CH_2OH} \quad BrCH_2\overset{\overset{\displaystyle O}{\displaystyle \|}}{C}OCH_2CH_3$$

                                                                    F

(f)

$$CH_3CH{=}CH\overset{\overset{\displaystyle O}{\displaystyle \|}}{C}OH \quad \xrightarrow[\substack{H_2SO_4 \\ benzene \\ \Delta}]{\overset{\displaystyle CH_3CH_2CHCH_3}{\underset{\displaystyle OH}{|}}} \quad CH_3CH{=}CH\overset{\overset{\displaystyle O}{\displaystyle \|}}{C}O\overset{\overset{\displaystyle CH_3}{|}}{C}HCH_2CH_3$$

                                                                              G

14.29

(a)

$$\text{(benzoic acid)} \quad COH \quad + \quad CH_3OH \quad \underset{H_2SO_4}{\rightleftharpoons} \quad COCH_3 \quad + \quad H_2O$$

mp 121 °C	bp 66 °C	bp 198 °C
insoluble in water;	miscible with water;	insoluble in water;
soluble in	soluble in	soluble in
diethyl ether	diethyl ether	diethyl ether

(b)   The reaction mixture consists of excess methanol, unreacted benzoic acid, sulfuric acid, and the products, water and methyl benzoate.

Step 1:   Add more water to the reaction mixture; most of the methanol and the sulfuric acid will go into the water layer. Methyl benzoate is insoluble in water and will appear as an oily layer. Unreacted benzoic acid will remain dissolved in the ester. Add diethyl ether, bp 37 °C, d 0.7, to dissolve the organic compounds. Separate the organic layer from the water layer.

Step 2:   Shake the organic layer (diethyl ether, methyl benzoate, benzoic acid) with saturated aqueous sodium bicarbonate.

$$H_2SO_4 \quad + \quad NaHCO_3 \quad \longrightarrow \quad CO_2\uparrow \quad + \quad H_2O \quad + \quad SO_4{}^{2-} \quad + \quad 2\,Na^+$$

(traces that	(excess)	soluble in water;
remain in the		insoluble in
methyl benzoate)		diethyl ether

$$COH \quad + \quad NaHCO_3 \longrightarrow \quad CO^-Na^+ \quad + \quad CO_2\uparrow \quad + \quad H_2O$$

insoluble in water;   soluble in water; insoluble
soluble in diethyl ether   in diethyl ether

The mixture will separate into two layers. The upper layer will be the ether solution containing methyl benzoate, some of the methanol, and small amounts of water. The lower layer will contain sodium sulfate (the salt formed by the reaction of the remaining sulfuric acid with the sodium bicarbonate), sodium benzoate (the salt from the reaction of benzoic acid with sodium bicarbonate), and some methanol.

14.29 (b) (cont)

Step 3:   Separate the ether layer from the water layer. The ether and remaining methanol can be removed by distillation, leaving the higher boiling methyl benzoate behind. This is usually done after the ether layer is dried by adding a solid, anhydrous inorganic salt to it.

- -

Concept Map 14.5   Reactions of acids and acid derivatives with alcohols.

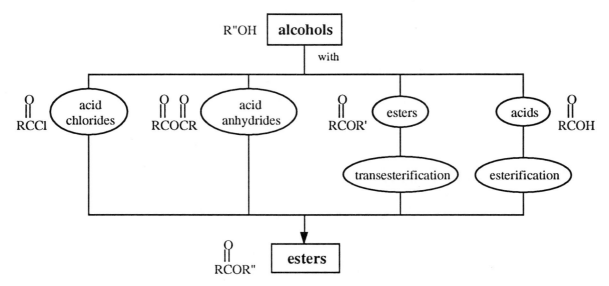

This is a transesterification reaction.

14.31 (cont)

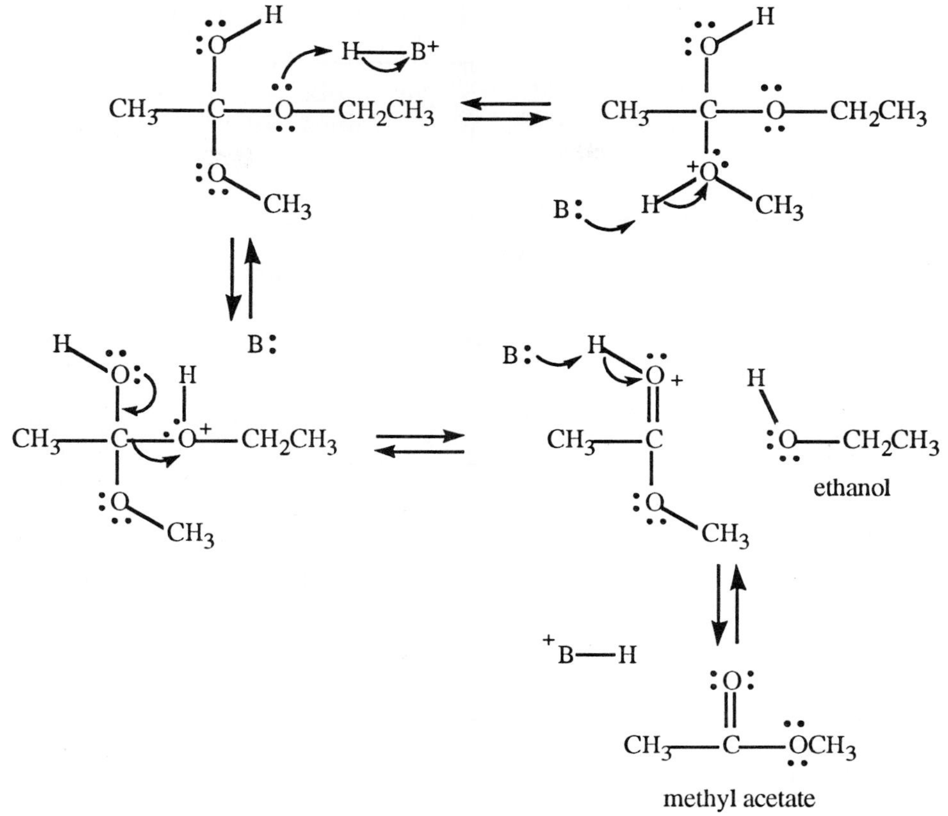

ethanol

methyl acetate

14.32

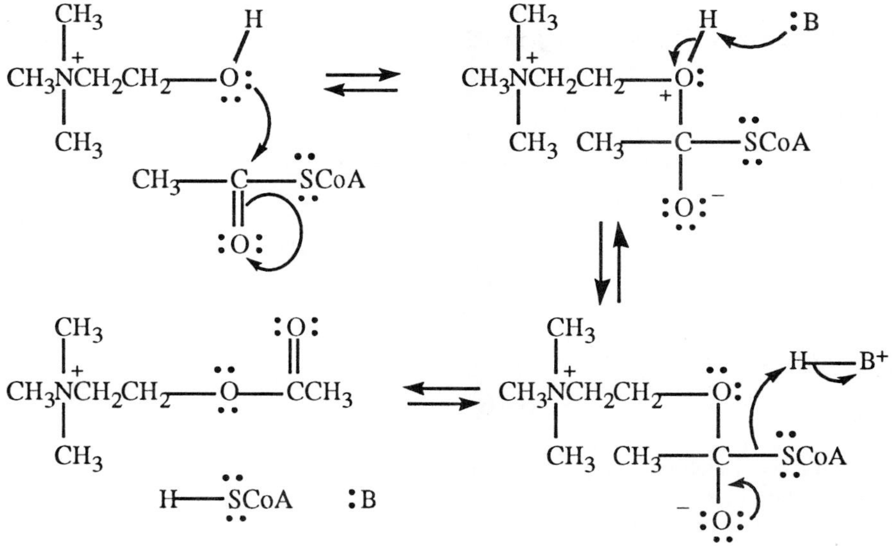

14.33

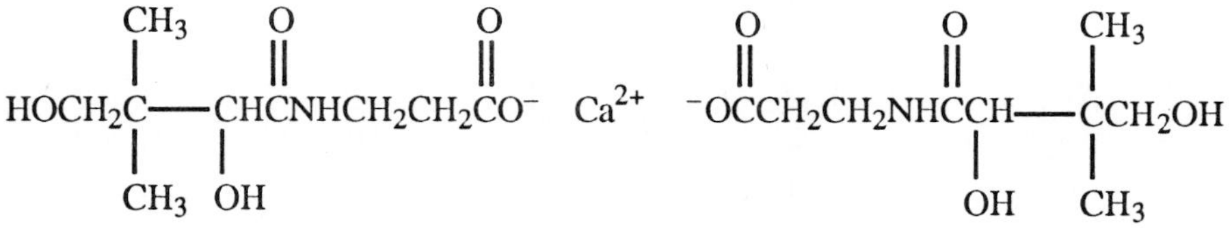

Concept Map 14.6   Reactions of acids and acid derivatives with ammonia or amines.

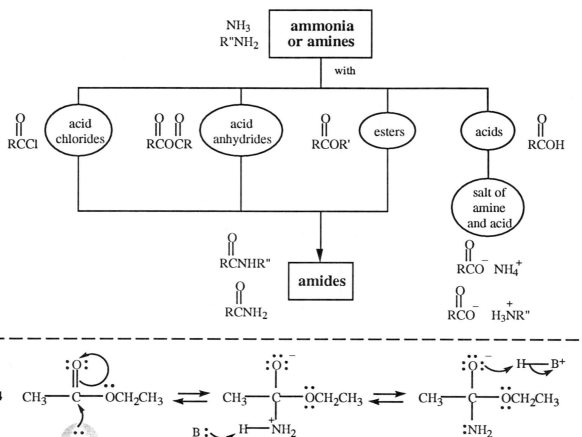

An alkoxide ion is not a good leaving group in an S_N2 reaction, but does serve as a leaving group in acyl transfer reactions. The bond energy of the carbonyl group is so high (p. ~ in the text) that the reaction forming the carbonyl group occurs even when a good leaving group is not present.

14.35

(a) CH₃CH₂OCCH=CHCOCH₂CH₃ $\xrightarrow[\substack{NH_4Cl \\ H_2O}]{NH_3 \text{ (excess)}}$ H₂NCCH=CHCNH₂ + 2 CH₃CH₂OH

$$\text{CH}_3\text{CH}_2\text{OCCH}=\text{CHCOCH}_2\text{CH}_3 \xrightarrow[\substack{\text{NH}_4\text{Cl}\\\text{H}_2\text{O}}]{\text{NH}_3\text{ (excess)}} \underset{\text{A}}{\text{H}_2\text{NCCH}=\text{CHCNH}_2} + 2\,\text{CH}_3\text{CH}_2\text{OH}$$

14.35 (cont)

(b)

B

C

(c) $CH_3CH_2OCCH_2CH_2CH_2CH_2COCH_2CH_3$ $\xrightarrow{H_2NNH_2 \text{ (excess)}}$

$H_2NNHCCH_2CH_2CH_2CH_2CNHNH_2$ + $2\,CH_3CH_2OH$

D

(d) $CH_3(CH_2)_4CHCOH$ $\xrightarrow[\Delta]{SOCl_2}$ $CH_3(CH_2)_4CHCCl$ $\xrightarrow[H_2O]{NH_3}$ $CH_3(CH_2)_4CHCNH_2$

CH_2CH_3 ... CH_2CH_3 ... CH_2CH_3

E ... F

(e)

G

(f) $N\equiv CCH_2COCH_2CH_3$ $\xrightarrow[0\,°C]{NH_3 \atop H_2O}$ $N\equiv CCH_2CNH_2$ + CH_3CH_2OH

H

14.36

(a)   2,4-dichlorobenzoic acid

(b)   4-bromohexanoic acid

(c)   2-methylbutanedioic anhydride (2-methylsuccinic anhydride)

(d)   2-methylpentanenitrile

(e)   ethyl (1*R*, 2*R*)-2-ethylcyclohexane-carboxylate

(f)   *N,N*-dimethylpentanamide

(g)   (*R*)-2-hydroxypentanoic acid

(h)   2-chloropentanoyl chloride

(i)   4-ethylbenzanilide

(j)   methyl 2-bromo-4-methylpentanoate

(k)   ethyl 2-aminobenzoate (2-methyladipic acid)

(l)   2-methylhexanedioic acid

(m) hexanoic anhydride

(n)   (*E*)-4-octenoic acid

(o)   (1*R*, 2*R*)-cyclobutanedicarboxylic acid

14.37

(a)

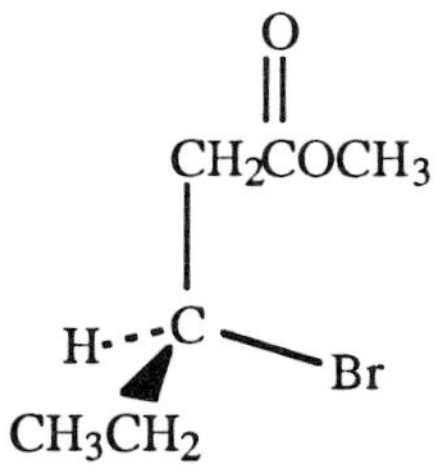

$$CH_2COCH_3$$

H···C—Br

CH_3CH_2

(b)

HO——COH

(c)

Br, Br, CNH_2

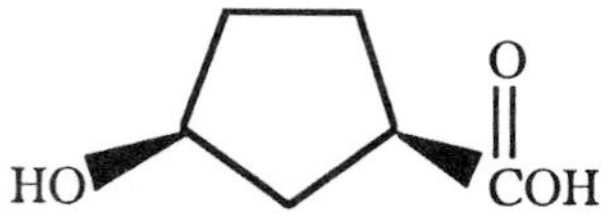

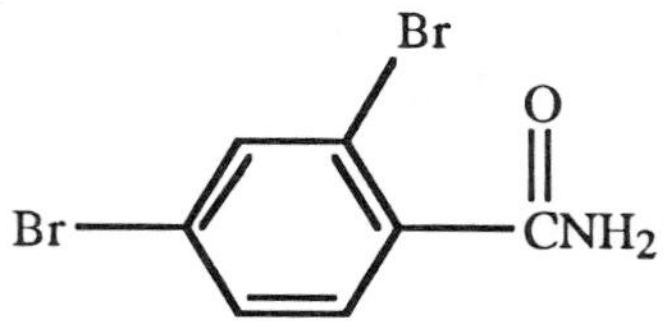

(d)

CH_3CH_2 $CH_2COCH_2CH_3$

C=C

H   H

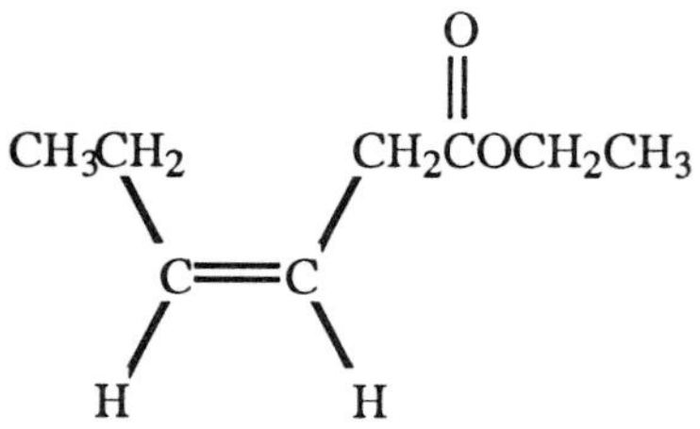

(e) $CH_3CH_2CH_2CH_2CH_2CH_2C\equiv N$

(f) 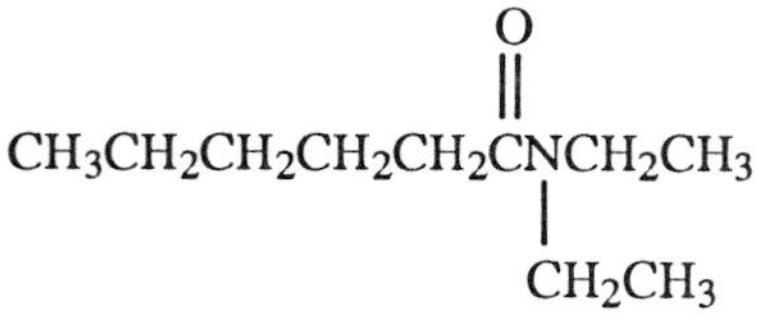 $CH_3CH_2CH_2CH_2CH_2CNCH_2CH_3$

                       CH_2CH_3

(g) 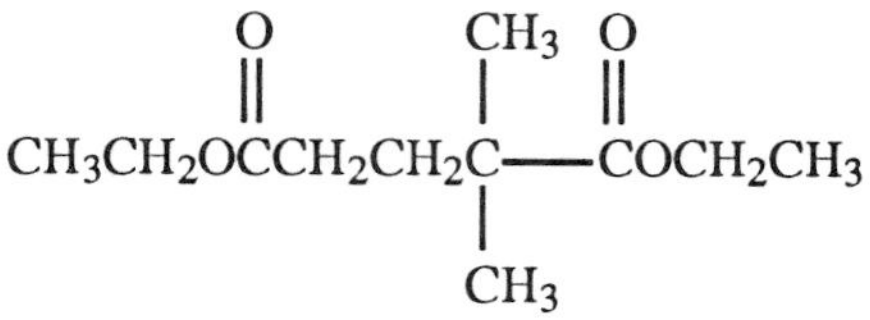 $CH_3CH_2OCCH_2CH_2C$—$COCH_2CH_3$

                            CH_3

(h) 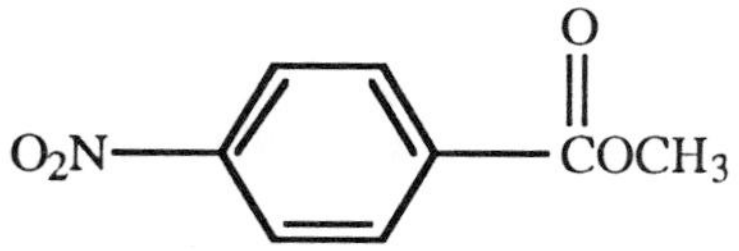 O_2N——$COCH_3$

(i) 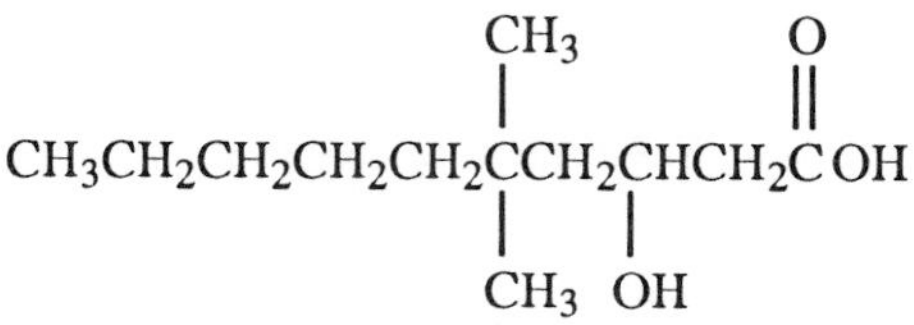 $CH_3CH_2CH_2CH_2CH_2CCH_2CHCH_2COH$

                            CH_3 OH

(j) 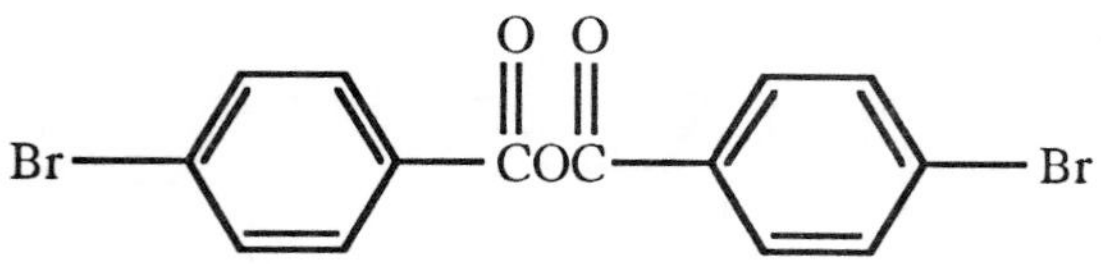   Br——COC——Br

(k) 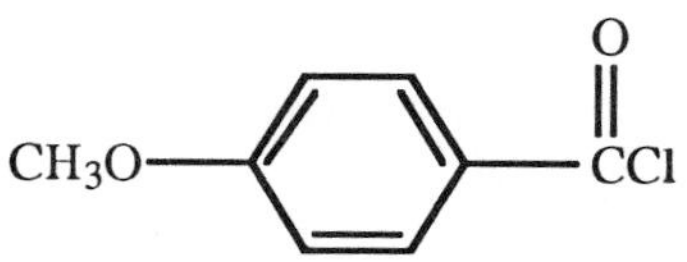 CH_3O——CCl

14.38

(a) 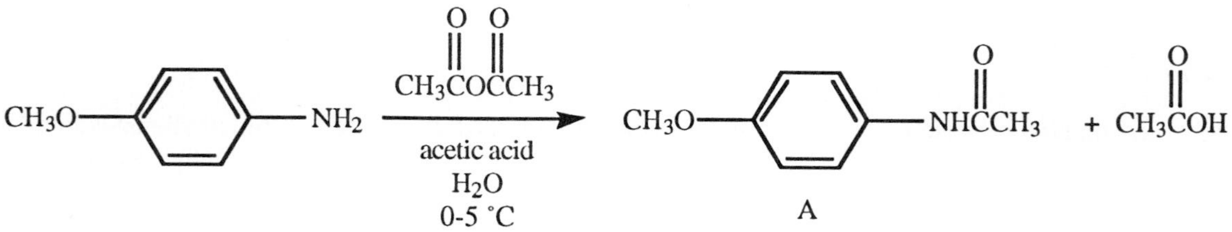 CH_3O——NH_2 $\xrightarrow[\substack{\text{acetic acid} \\ H_2O \\ \text{0-5 °C}}]{CH_3COCCH_3}$ CH_3O——$NHCCH_3$ + CH_3COH

                                                               A

(b) 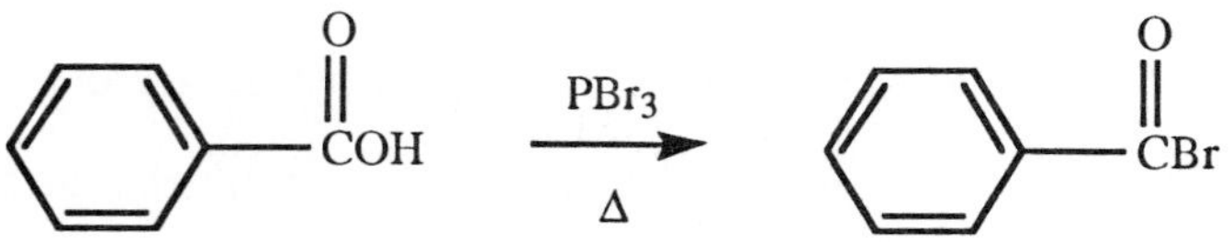   ——COH $\xrightarrow[\Delta]{PBr_3}$ ——CBr

                                             B

14.38 (cont)

(c) 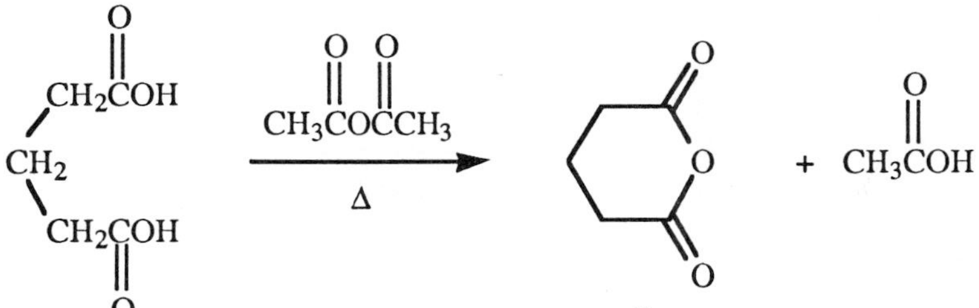

C

(d) 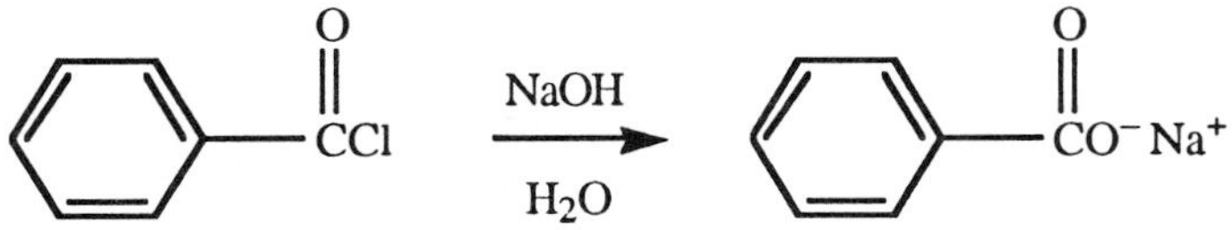

D

(e) 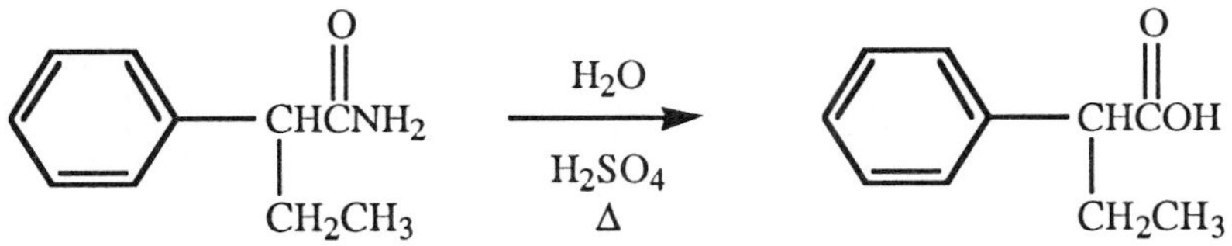

E

(f) 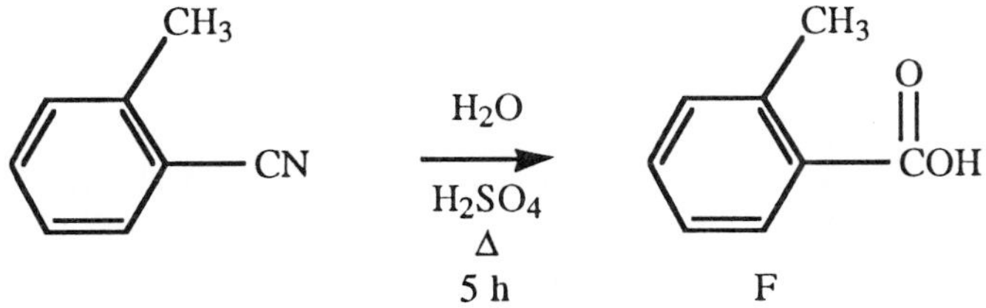

F

(g) $CH_3CH_2\overset{O}{\overset{\|}{C}}Cl \xrightarrow{NH_3\ (excess)} CH_3CH_2\overset{O}{\overset{\|}{C}}NH_2$

G

(h) 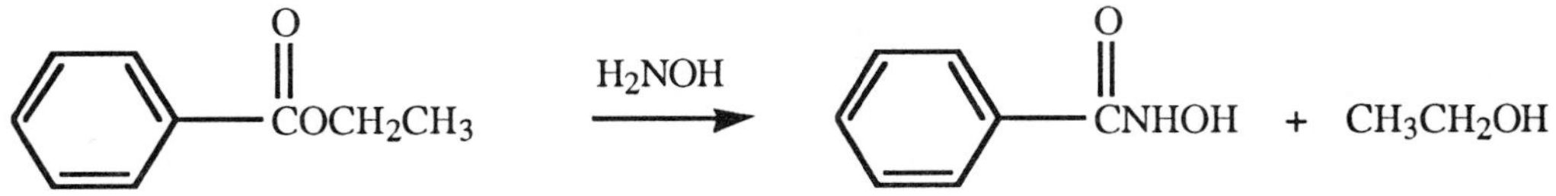

H

(i) $CH_3(CH_2)_5\overset{O}{\overset{\|}{C}}Cl \xrightarrow{CH_3(CH_2)_5\overset{O}{\overset{\|}{C}}O^-Na^+} CH_3(CH_2)_5\overset{O}{\overset{\|}{C}}O\overset{O}{\overset{\|}{C}}(CH_2)_5CH_3$

I

14.39

(a)

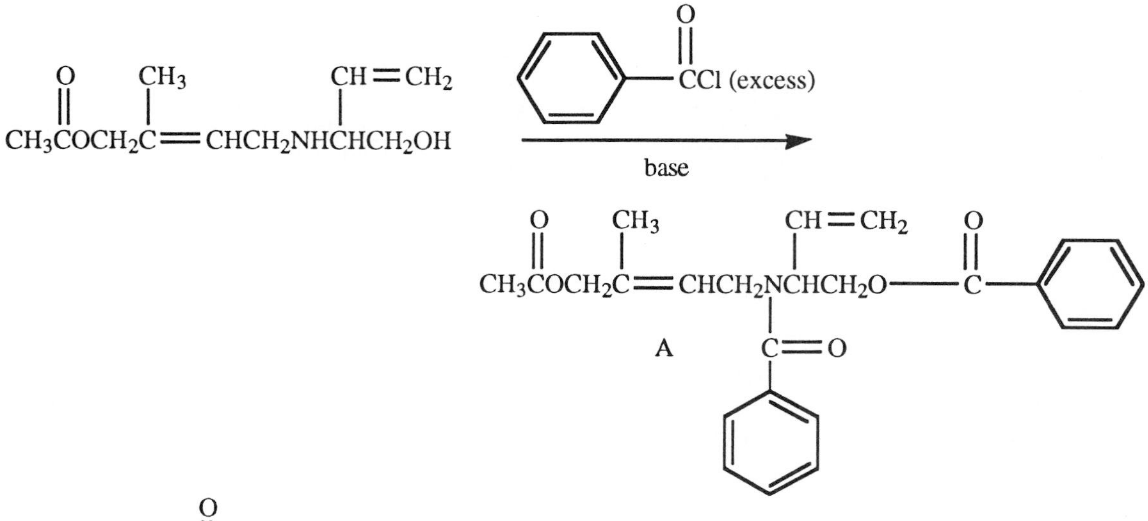

$$CH_3COCH_2C=CHCH_2NHCHCH_2OH \xrightarrow[\text{base}]{\text{C}_6H_5CCl \text{ (excess)}}$$

with substituents CH_3 and $CH=CH_2$

A

$$CH_3COCH_2C=CHCH_2NCHCH_2O-C-C_6H_5$$

(b)

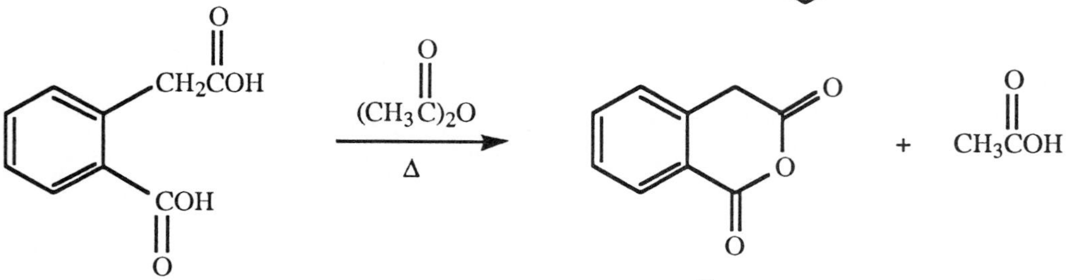

$$\xrightarrow[\Delta]{(CH_3C)_2O}$$

B + CH_3COH

(c) $CH_3(CH_2)_{10}COCH_2(CH_2)_{10}CH_3$ + $CH_3OH \xrightarrow[\Delta]{H_2SO_4} CH_3(CH_2)_{10}COCH_3$ + $CH_3(CH_2)_{10}CH_2OH$

                                                                      C   D

(d)

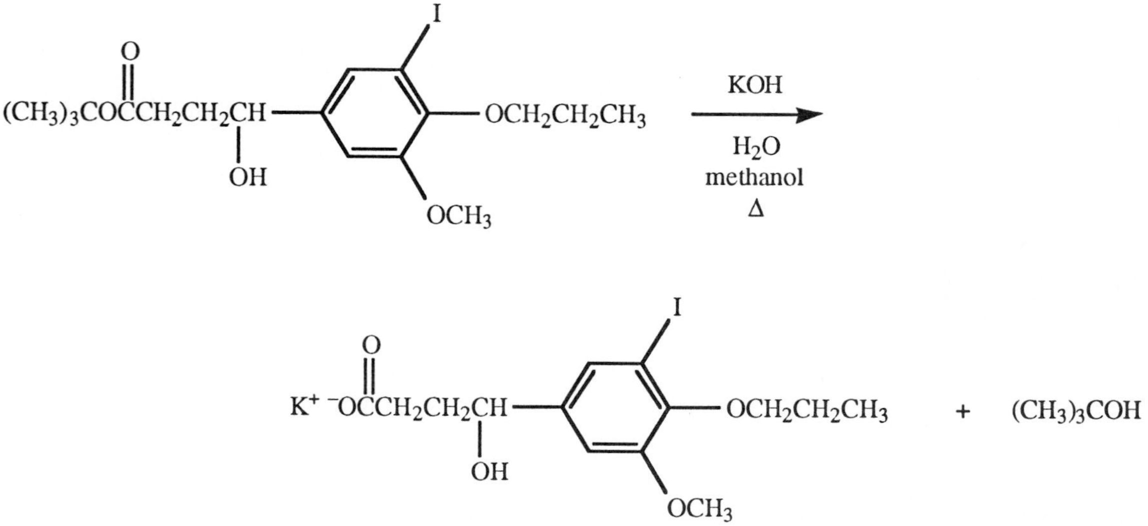

$$(CH_3)_3COCCH_2CH_2CH-\text{(aryl)}-OCH_2CH_2CH_3 \xrightarrow[\substack{H_2O \\ methanol \\ \Delta}]{KOH}$$

with substituents I, OCH$_3$, and OH

$$K^+ {}^-OCCH_2CH_2CH-\text{(aryl)}-OCH_2CH_2CH_3 + (CH_3)_3COH$$

with substituents I, OCH$_3$, and OH

E   F

14.39 (cont)

(e)

$$\underset{HOC(CH_2)_4COH}{\overset{O\quad\quad O}{\|\quad\quad\|}} \quad\xrightarrow{SOCl_2\ (excess)}\quad \underset{ClC(CH_2)_4CCl}{\overset{O\quad\quad O}{\|\quad\quad\|}}$$

G

(f)

$$\underset{ClC(CH_2)_4CCl}{\overset{O\quad\quad O}{\|\quad\quad\|}}$$

G

CH₃ | CH₃COH (excess) | CH₃

$$\begin{array}{c} CH_3 \\ | \\ CH_3COH\ (excess) \\ | \\ CH_3 \end{array}$$

NCH₃ on N-methylaniline reagent

→

$$\underset{CH_3}{\overset{CH_3}{|}}\ \underset{}{\overset{O}{\|}}\ \underset{}{\overset{O}{\|}}\ \underset{CH_3}{\overset{CH_3}{|}}$$
$$CH_3CO-C(CH_2)_4C-OCCH_3$$
$$\underset{CH_3}{|}\qquad\qquad\qquad \underset{CH_3}{|}$$

H

(g)

I   J

(h)

K
major product

+

L
minor product

Nitrogen is more nucleophilic than oxygen, so the major product comes from acylation at the nitrogen atom.

(i)

$$CCl_3CH_2CH_2CH_2CH_2OH \xrightarrow[\substack{H_2O \\ \Delta}]{KMnO_4} \underset{M}{CCl_3CH_2CH_2CH_2\overset{O}{\overset{\|}{C}}O^-K^+} \xrightarrow{H_3O^+} \underset{N}{CCl_3CH_2CH_2CH_2\overset{O}{\overset{\|}{C}}OH}$$

14.40

(a)

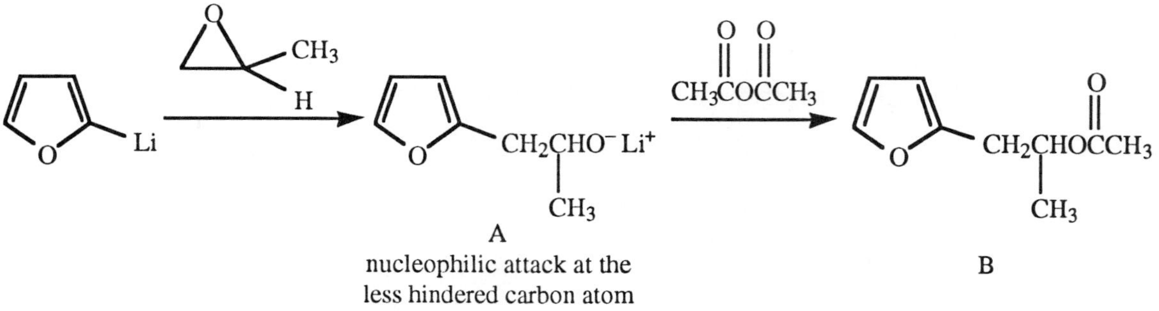
CH3
H
Li
CH2CHO⁻ Li⁺
CH3
A
nucleophilic attack at the less hindered carbon atom
CH3COCCH3
CH2CHOCCH3
CH3
B

(b)

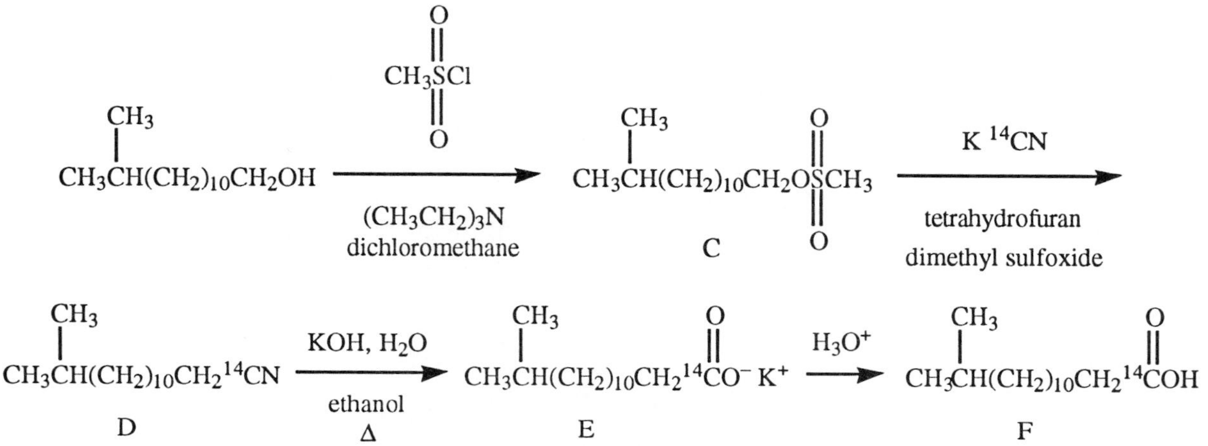
CH3SCl
(CH3CH2)3N
dichloromethane
C
K 14CN
tetrahydrofuran
dimethyl sulfoxide
D
KOH, H2O
ethanol
Δ
E
H3O⁺
F

(c)

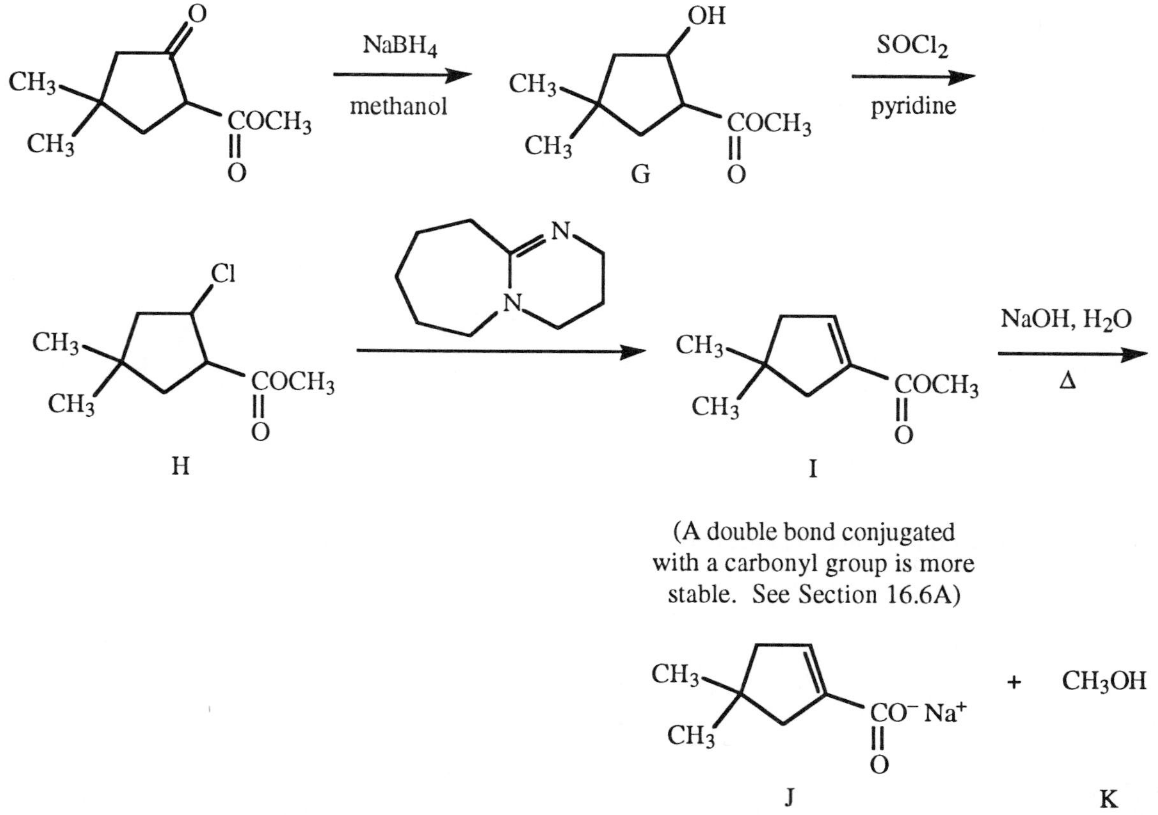
NaBH4
methanol
G
SOCl2
pyridine
H
I
NaOH, H2O
Δ
(A double bond conjugated with a carbonyl group is more stable. See Section 16.6A)
J
+ CH3OH
K

14.40 (cont)

(d)

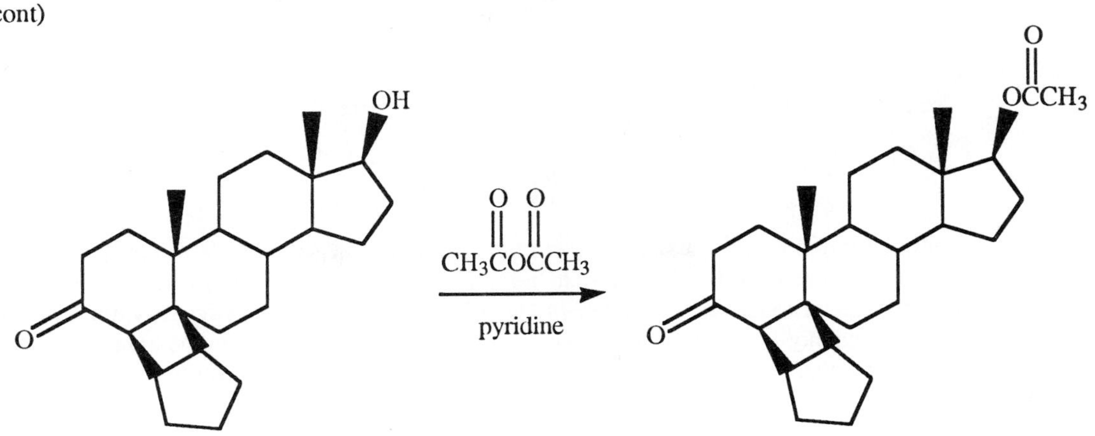

L

(e)

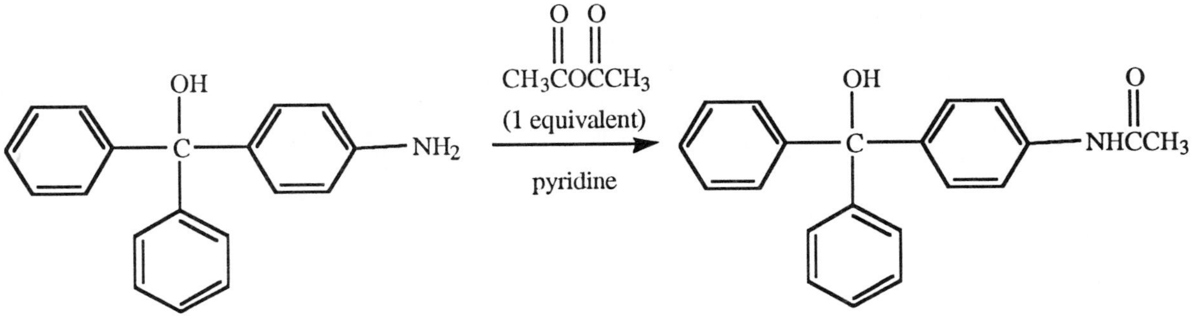

M

(f)

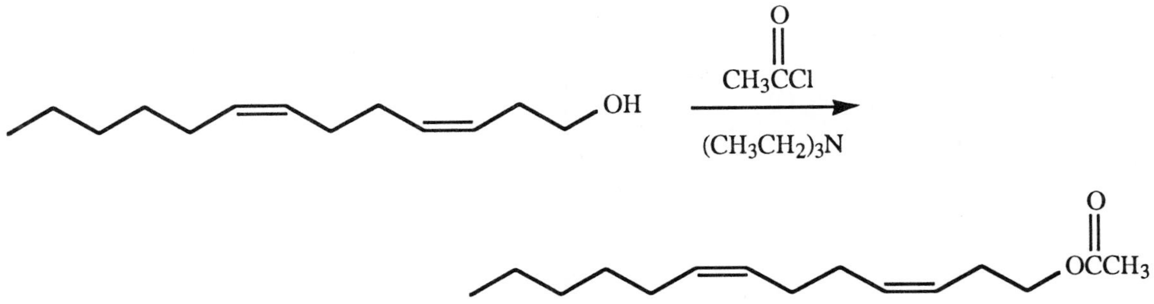

N

(g)

P   Q

14.41 (a)

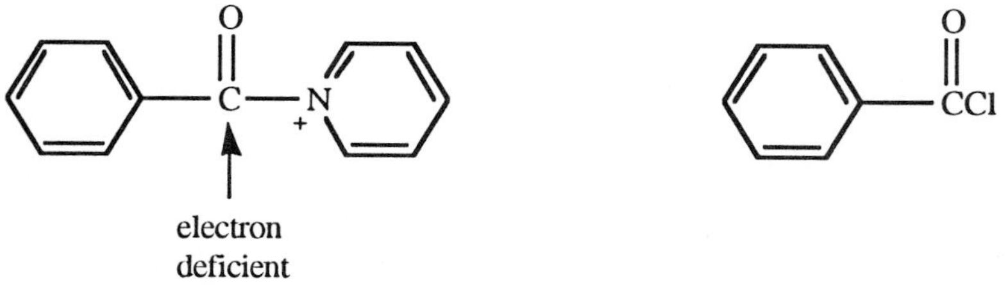

electron
deficient

The intermediate formed when pyridine reacts with benzoyl chloride is an unstable species with a positive charge adjacent to an electron-deficient carbon atom. It has an especially good leaving group because the positive charge on the nitrogen atom decreases in the transition state, lowering the activation energy. When chloride ion is the leaving group, as in benzoyl chloride, charge is building up in the transition state, and the activation energy for that reaction is higher than when pyridine is the leaving group.

(b)

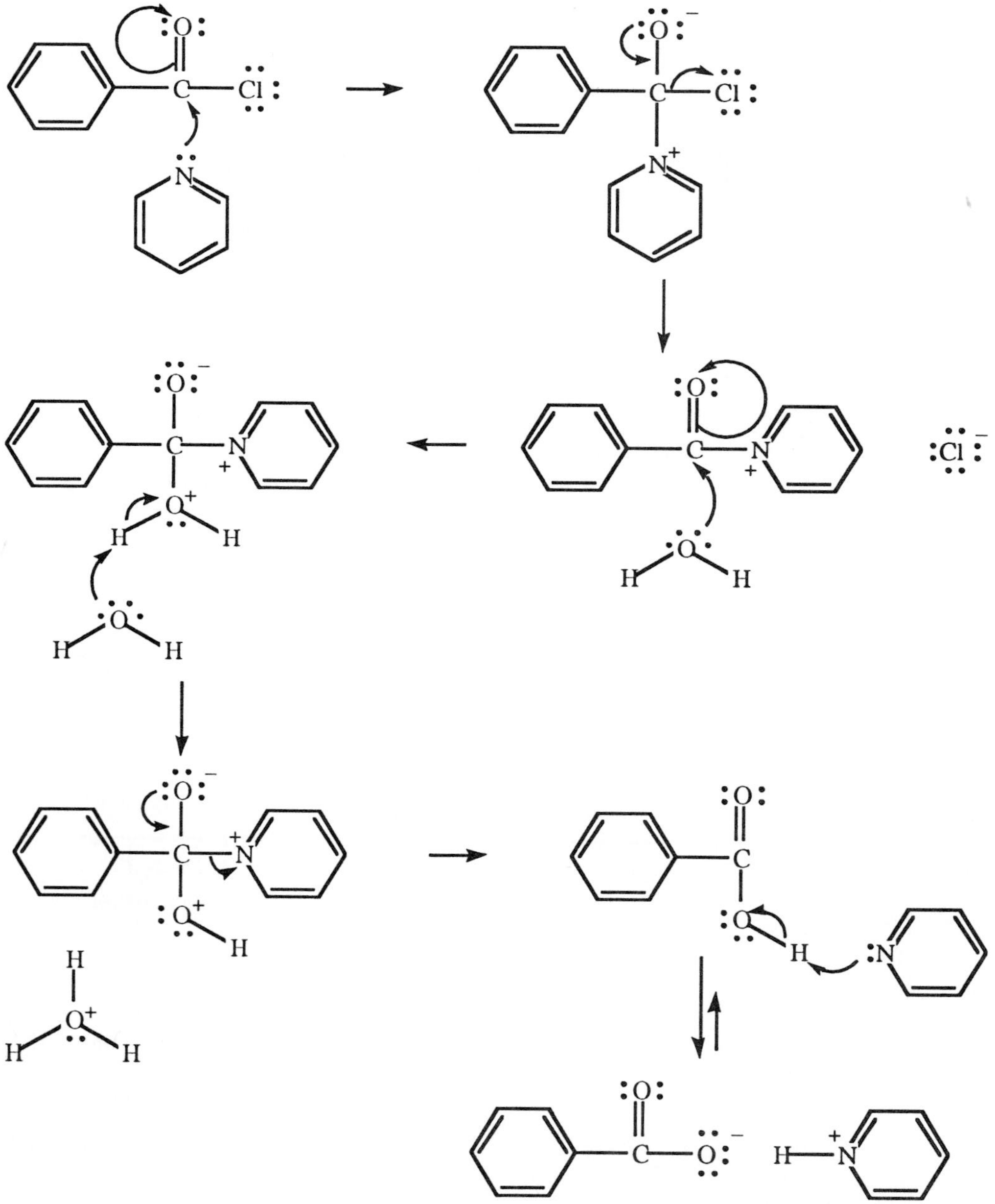

14.41 (cont)

(c)

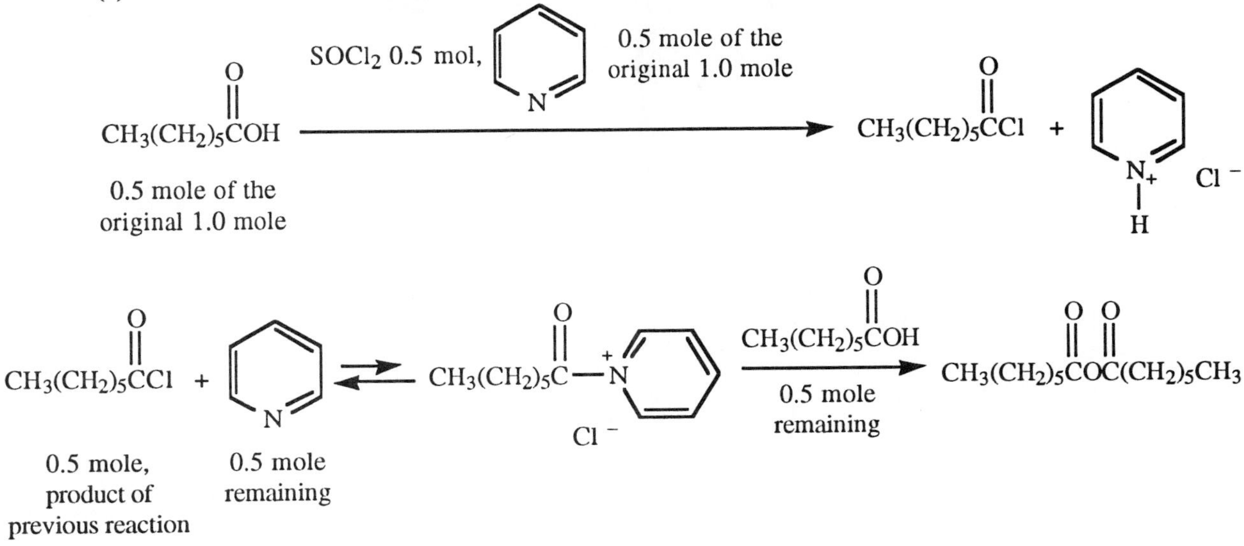

14.42

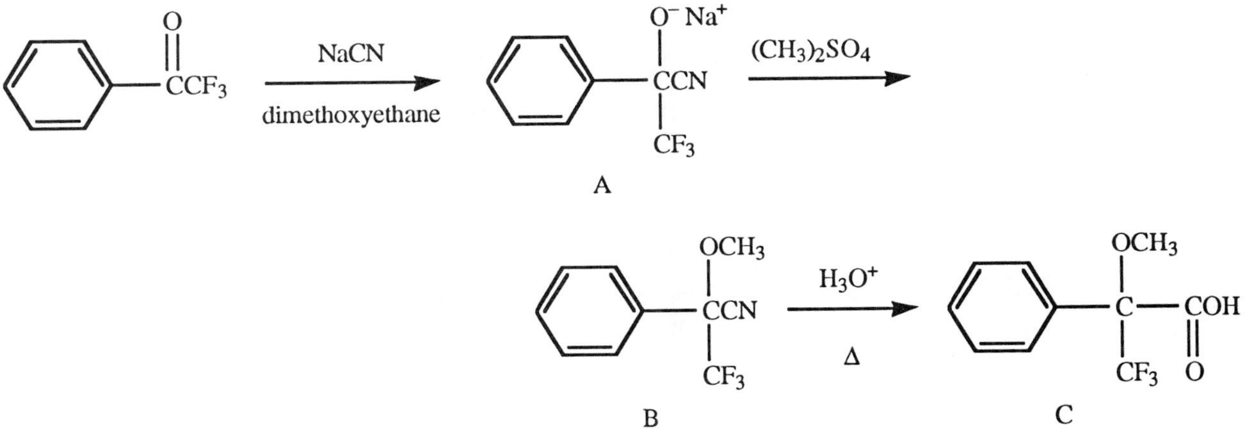

14.43

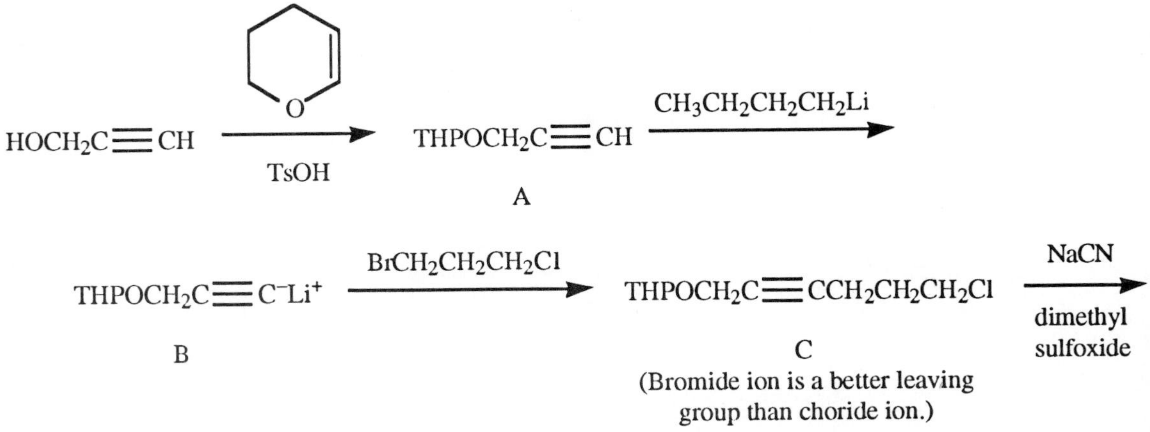

(Bromide ion is a better leaving group than choride ion.)

14.43 (cont)

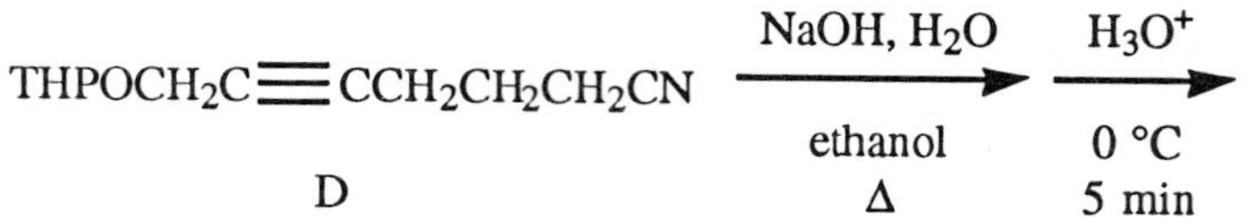

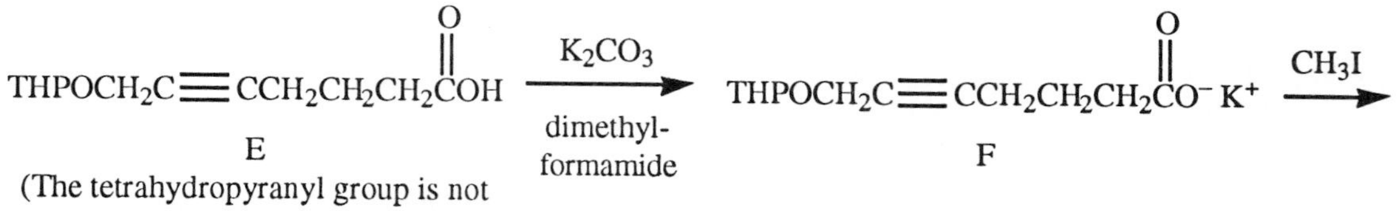

$$\text{THPOCH}_2\text{C}\equiv\text{CCH}_2\text{CH}_2\text{CH}_2\text{CN} \xrightarrow[\substack{\text{ethanol} \\ \Delta}]{\text{NaOH, H}_2\text{O}} \xrightarrow[\substack{0\ °\text{C} \\ 5\ \text{min}}]{\text{H}_3\text{O}^+}$$

D

$$\text{THPOCH}_2\text{C}\equiv\text{CCH}_2\text{CH}_2\text{CH}_2\overset{\overset{\displaystyle O}{\|}}{\text{C}}\text{OH} \xrightarrow[\substack{\text{dimethyl-} \\ \text{formamide}}]{\text{K}_2\text{CO}_3} \text{THPOCH}_2\text{C}\equiv\text{CCH}_2\text{CH}_2\text{CH}_2\overset{\overset{\displaystyle O}{\|}}{\text{C}}\text{O}^-\ \text{K}^+ \xrightarrow{\text{CH}_3\text{I}}$$

E   F

(The tetrahydropyranyl group is not
hydrolyzed under these conditions.)

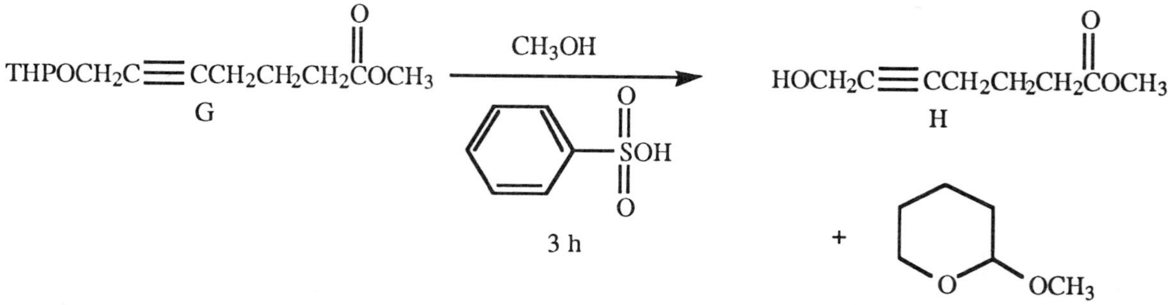

$$\text{THPOCH}_2\text{C}\equiv\text{CCH}_2\text{CH}_2\text{CH}_2\overset{\overset{\displaystyle O}{\|}}{\text{C}}\text{OCH}_3 \xrightarrow[\substack{\text{3 h}}]{\text{CH}_3\text{OH}} \text{HOCH}_2\text{C}\equiv\text{CCH}_2\text{CH}_2\text{CH}_2\overset{\overset{\displaystyle O}{\|}}{\text{C}}\text{OCH}_3$$

G   H

(benzenesulfonic acid reagent drawn as structure with —SOH group; + tetrahydropyranyl methyl ether)

The infrared spectrum of Compound H shows the presence of an alcohol (3450 cm^{-1} for the O—H stretch) and
an ester (1720 cm^{-1} for the C=O stretch). Analysis of the proton magnetic resonance spectrum of Compound
H is shown below:

δ 1.6-2.0   (2H, multiplet, —CH$_2$CH$_2$—)
δ 2.2-2.6   (5H, multiplet, —(≡CCH$_2$CH$_2$CH$_2$C=O and —OH)
δ 3.68   (3H, singlet, —OCH$_3$)
δ 4.2-4.3   (2H, multiplet, —OCH$_2$C≡)

Compound H is methyl 7-hydroxy-5-heptynoate

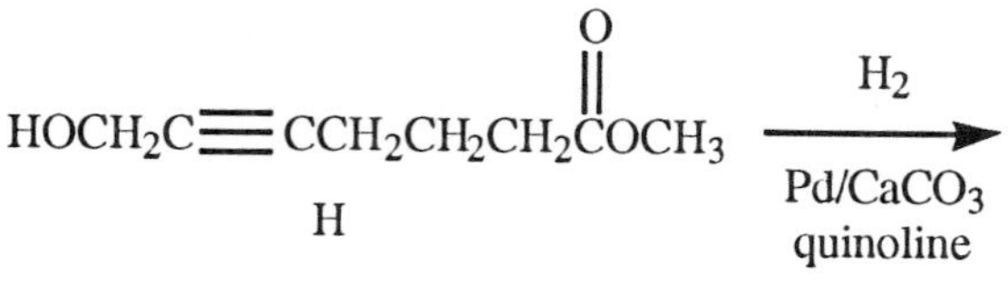

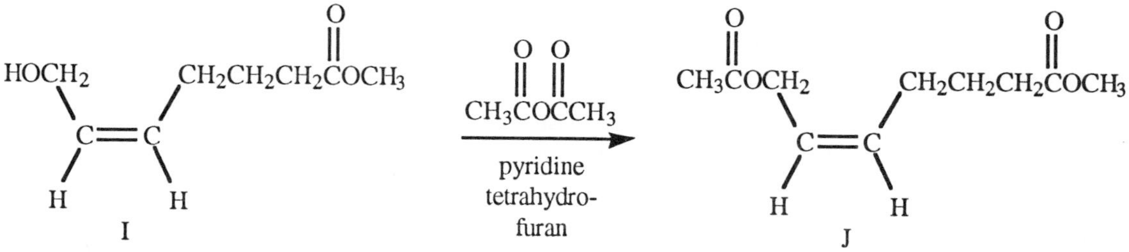

$$\text{HOCH}_2\text{C}\equiv\text{CCH}_2\text{CH}_2\text{CH}_2\overset{\overset{\displaystyle O}{\|}}{\text{C}}\text{OCH}_3 \xrightarrow[\substack{\text{Pd/CaCO}_3 \\ \text{quinoline}}]{\text{H}_2}$$

H

(Compound I: cis alkene structure HOCH$_2$ and CH$_2$CH$_2$CH$_2$COCH$_3$ on C=C with H, H)

$$\xrightarrow[\substack{\text{pyridine} \\ \text{tetrahydro-} \\ \text{furan}}]{\overset{\overset{\displaystyle O\ \ \ O}{\|\ \ \ \|}}{\text{CH}_3\text{COCCH}_3}}$$

(Compound J: cis alkene structure CH$_3$COCH$_2$ and CH$_2$CH$_2$CH$_2$COCH$_3$ on C=C with H, H)

14.44

$$CO_3^{2-} \quad + \quad CH_3OH \quad \rightleftharpoons \quad HCO_3^- \quad + \quad CH_3O^-$$

$$\text{p}K_a \ 15 \qquad\qquad\qquad\qquad \text{p}K_a \ 10.2$$

14.45

$$\xrightarrow[\text{transesterification}]{\substack{CH_3OH \\ HCl}}$$

$$\xrightarrow[\substack{\text{acetone} \\ \text{intramolecular } S_N2 \\ \text{reaction by alkoxide ion}}]{K_2CO_3}$$

14.46

(a)

$$\xrightarrow[\text{pyridine}]{\substack{O \quad O \\ \| \quad \| \\ CH_3COCCH_3 \\ \text{(excess)}}}$$

A

(b)

$$\xrightarrow[\text{TsOH}]{\substack{O \\ \| \\ CH_3CCH_3}}$$

or

$$\xrightarrow[\text{TsOH}]{(CH_3)_2C(OCH_3)_2}$$

B

$$\xrightarrow[\text{dichloromethane}]{Cl-C_6H_4-COOH}$$

C

$$\xrightarrow[\substack{\text{diethyl ether} \\ \text{(a source of hydride} \\ \text{ion, } H:^-, \text{ that reacts} \\ \text{with oxiranes as} \\ \text{a nucleophile)}}]{LiAlH_4}$$

racemic mixture
D

$$\xrightarrow{\substack{O \quad O \quad O \\ \| \quad \| \quad \| \\ CH_3SCH_3, \ ClC-CCl}} \xrightarrow{(CH_3CH_2)_3N}$$

or

$$\xrightarrow[\text{dichloromethane}]{\substack{\text{pyridinium} \\ CrO_3Cl^-}}$$

E

14.47

(a)

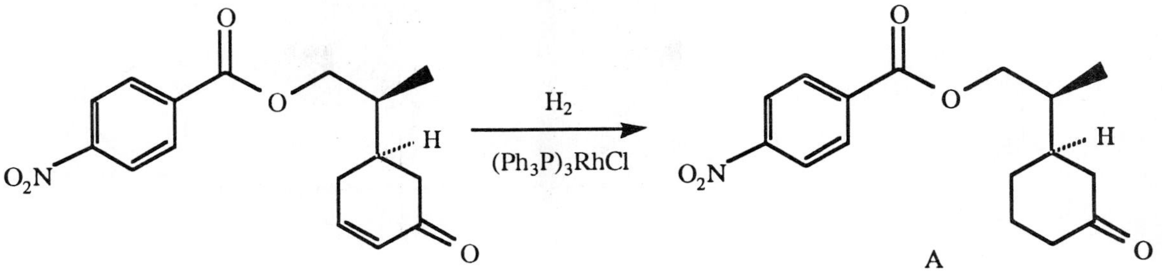
O
O
H
O_2N
O
H_2
$(Ph_3P)_3RhCl$
A

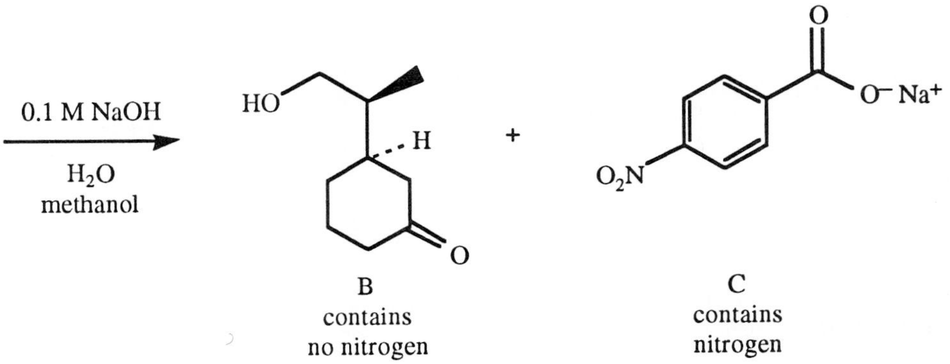
0.1 M NaOH
H_2O
methanol
HO
H
O
+
$O^- Na^+$
O_2N
B
contains
no nitrogen
C
contains
nitrogen

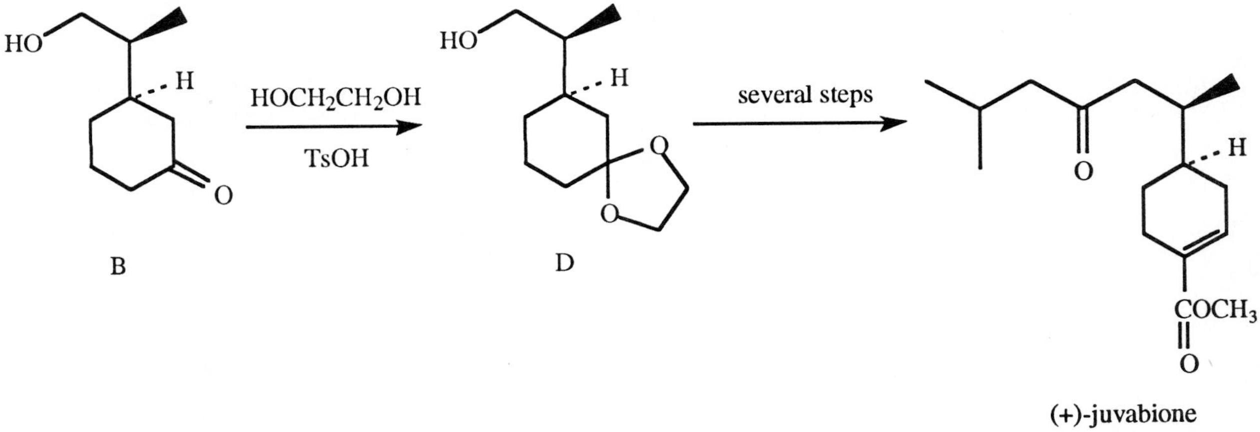
HO
H
O
B
$HOCH_2CH_2OH$
TsOH
D
several steps
$COCH_3$
(+)-juvabione

(b)

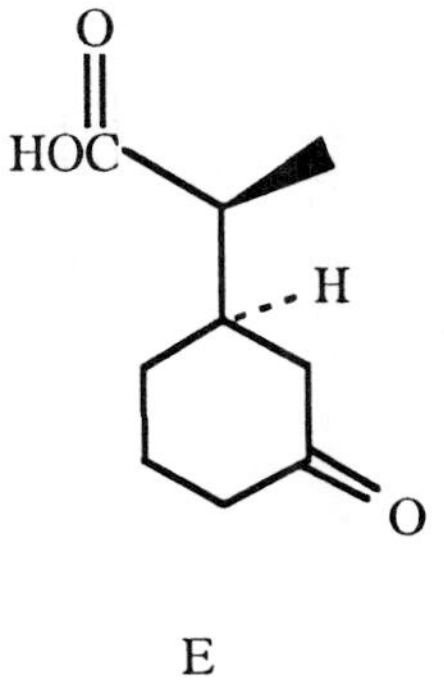
HOC
O
H
E

14.48

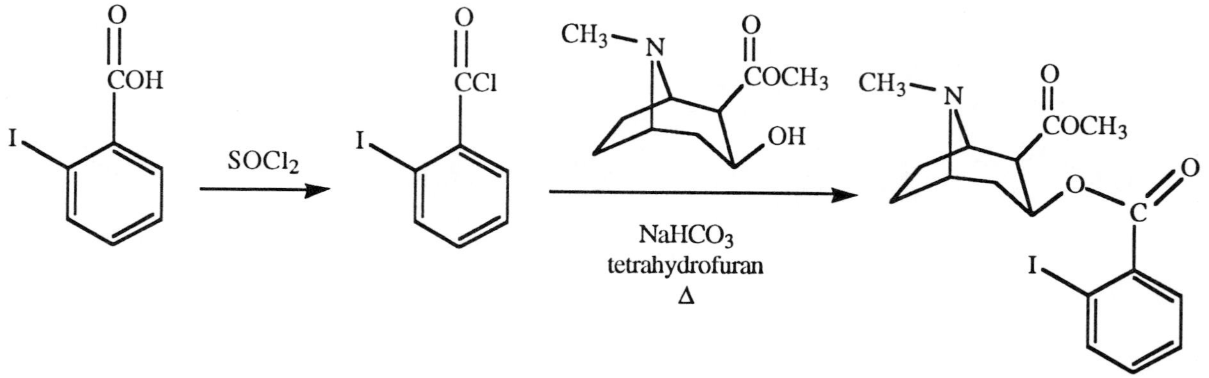

A

B

14.49

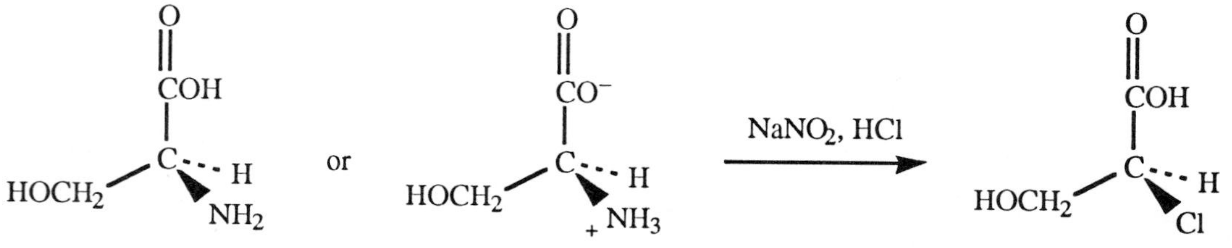

(S)-serine

(S)-2-chloro-3-hydroxy-
propanoic acid

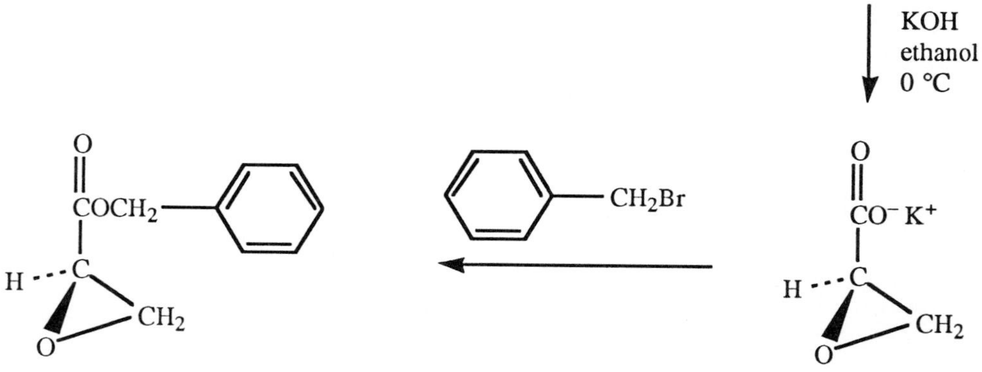

KOH
ethanol
0 °C

benzyl (R)-oxiranecarboxylate

potassium (R)-oxiranecarboxylate

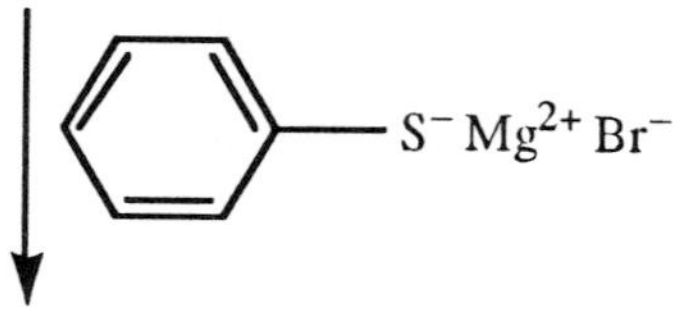

14.49 (cont)

benzyl (*S*)-2-hydroxy-3-(phenylthio)propanoate   (*S*)-2-hydroxy-3-(phenylthio)propanoic acid

14.50 $BrCH_2$ ⌒ $C \equiv CCH_2CH_2CH_2CH_2OTHP$ ⟵ H ⌒ $C \equiv CCH_2CH_2CH_2CH_2OTHP$

bond to be formed   bond to be broken

We need to form a C—C bond; therefore we need an organometallic reagent and a carbonyl compound.

$BrCH_2C \equiv CCH_2CH_2CH_2CH_2OTHP$ ⟵ $\overset{\text{LiBr or NaBr or KBr}}{}$ $TsOCH_2C \equiv CCH_2CH_2CH_2CH_2OTHP$

⟵ $\overset{\text{TsCl}}{\underset{\text{pyridine}}{}}$ $HOCH_2C \equiv CCH_2CH_2CH_2CH_2OTHP$

⟵ $\overset{H_3O^+}{}$ $Br^- Mg^{2+}$ or Na^+ or Li^+ $^-OCH_2C \equiv CCH_2CH_2CH_2CH_2OTHP$

⟵ $\overset{O}{\overset{\|}{HCH}}$ $Br^- Mg^{2+}$ or Na^+ or Li^+ $^-C \equiv CCH_2CH_2CH_2CH_2OTHP$

⟵ $\overset{\text{CH}_3\text{CH}_2\text{CH}_2\text{CH}_2\text{Li or}}{\underset{\text{NaNH}_2 \text{ or CH}_3\text{CH}_2\text{MgBr}}{}}$ $HC \equiv CCH_2CH_2CH_2CH_2OTHP$

14.51

14.51 (cont)

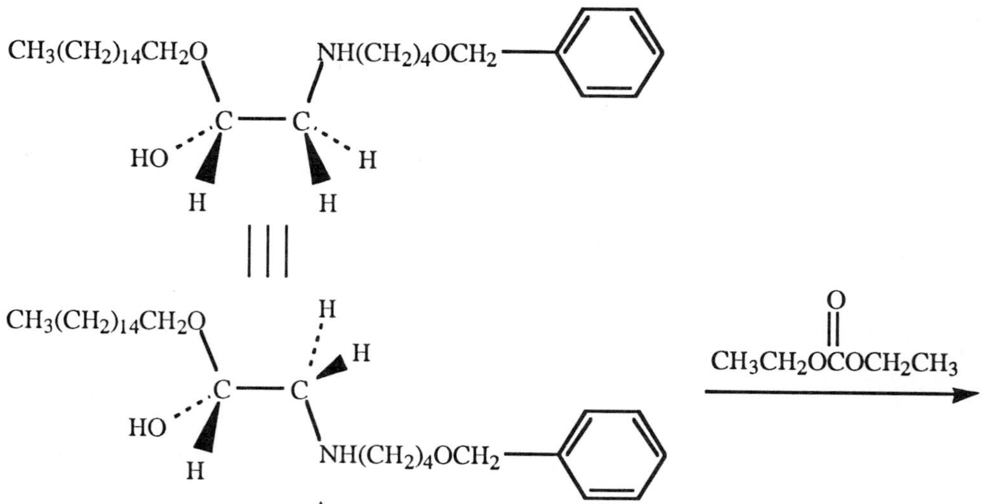

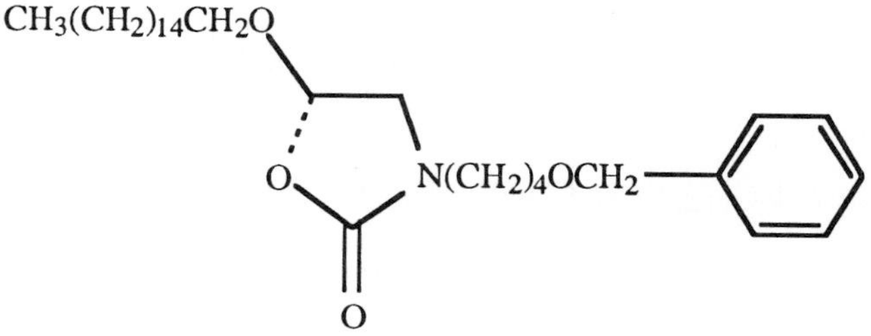

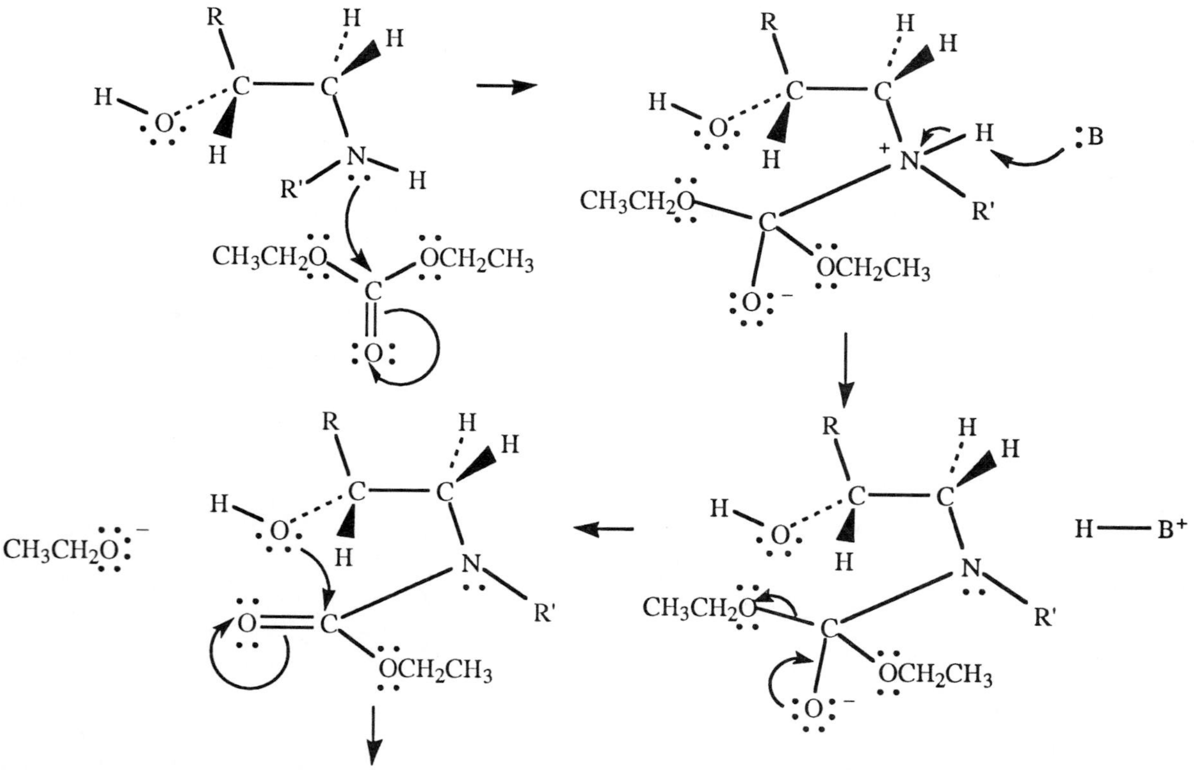

A

B

14.51 (cont)

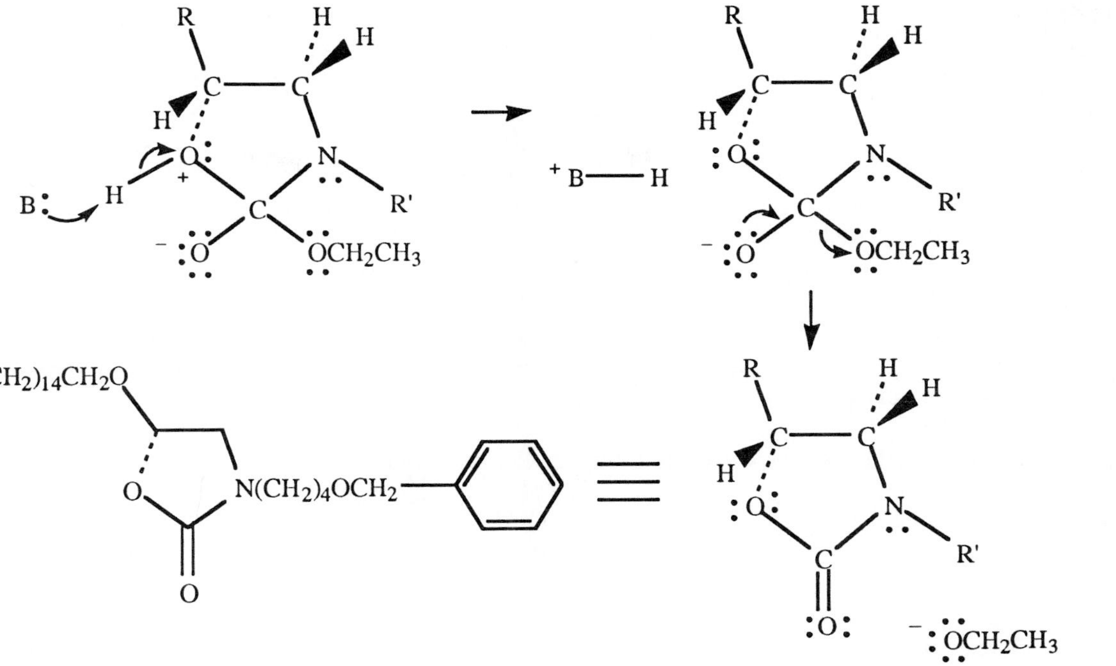

14.52

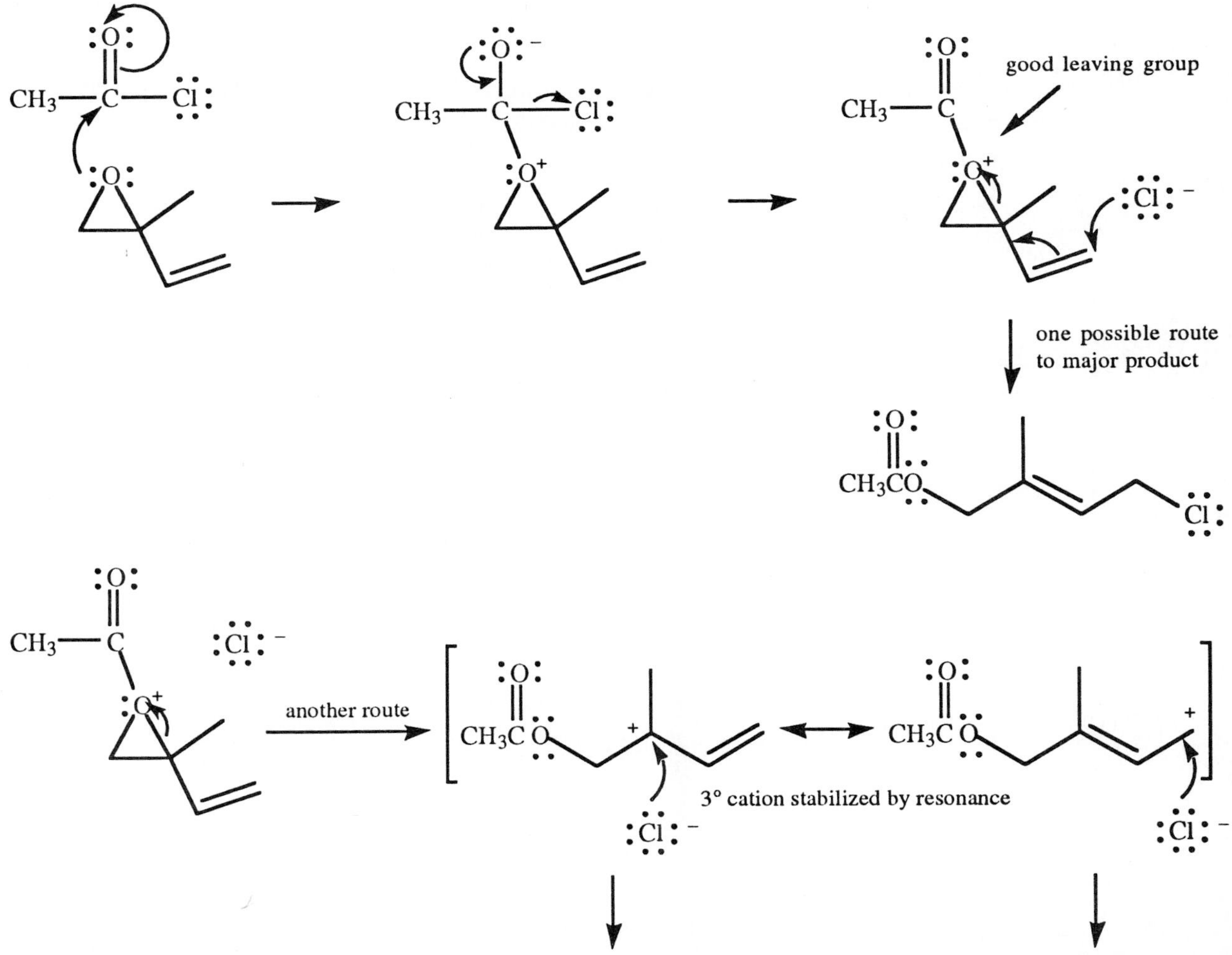

good leaving group

one possible route
to major product

another route

3° cation stabilized by resonance

14.52 (cont)

14.53   Compound A, $C_9H_{10}O_2$, has five units of unsaturation. When a compound has four or more units of unsaturation, we should look for an aromatic ring, which can be seen in both the proton and carbon-13 magnetic resonance spectra. The infrared spectrum shows the presence of an ester (1743 cm^{-1} for the C=O stretch and 1229 and 1118 cm^{-1} for the C—O stretch).

Analysis of the proton spectrum of Compound A is shown below:

δ 2.2	(3H, singlet, C$\underline{H}_3$C=O
δ 5.2	(2H, singlet, ArC$\underline{H}_2$—O—)
δ 7.4	(5H, singlet, Ar$\underline{H}$)

The absence of splitting for the aromatic hydrogen atoms tells us that the group bonded to the aromatic ring does not differ significantly from the aryl carbon atoms in electronegativity and in magnetic anisotropy. A tetrahedral carbon atom meets this requirement.

The carbon-13 magnetic resonance spectrum shows two alkyl carbon atoms (20.7 and 66.1 ppm). There are three types of aromatic carbon atoms (128.1, 128.4, and 136.2 ppm), and one carbonyl carbon atom (170.5 ppm). Compound A is benzyl acetate.

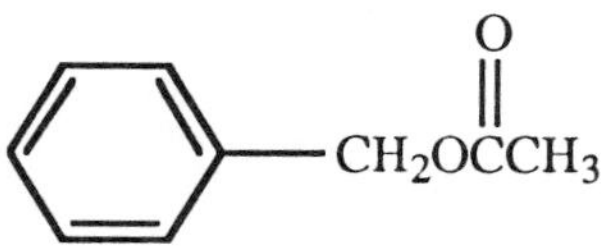

Note that the compound cannot be methyl phenylacetate because the band for the methyl singlet in the proton magnetic resonance spectrum of a methyl ester would be at ~δ 3.7.

15

Carboxylic Acids and Their Derivatives II. Synthetic Transformations and Compounds of Biological Interest

15.1

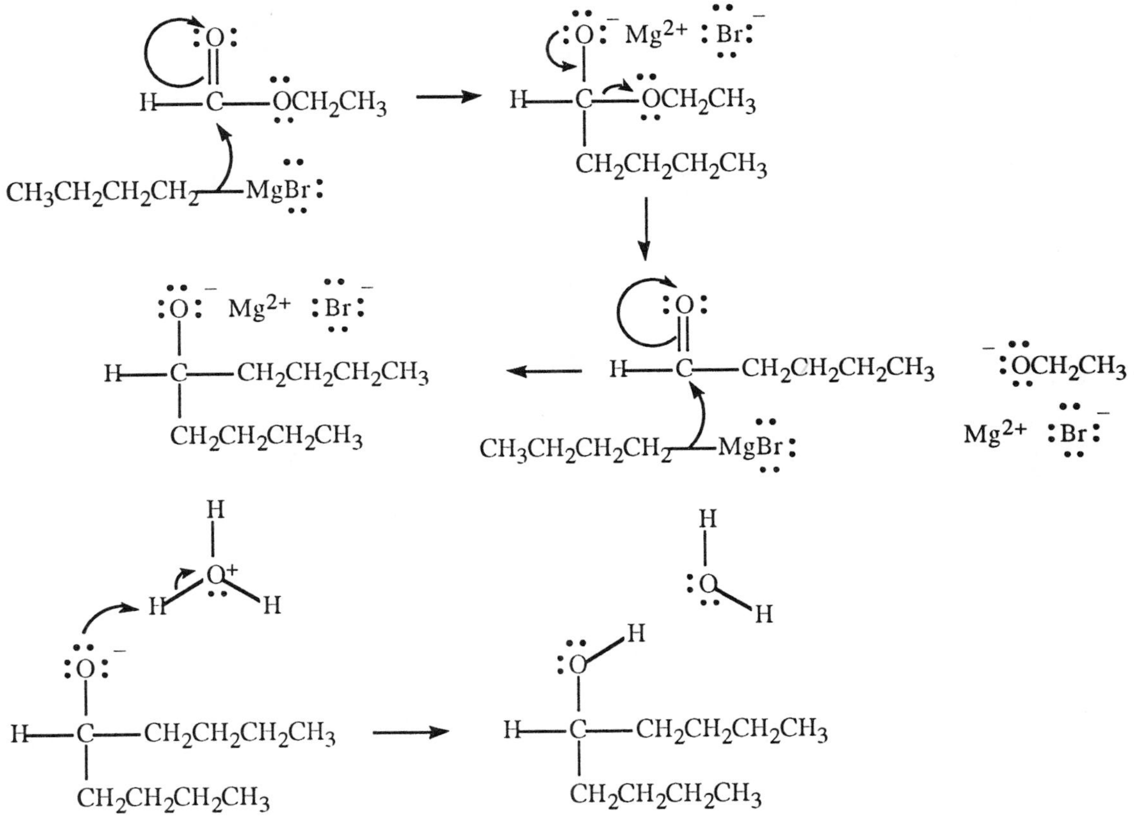

15.2 $CH_3CH_2{}^{14}\underset{\underset{\displaystyle CH_2CH_3}{|}}{\overset{\overset{\displaystyle OH}{|}}{C}}CH_2CH_3$ $\xleftarrow[\text{H}_2\text{O}]{\text{NH}_4\text{Cl}}$ $CH_3CH_2{}^{14}\underset{\underset{\displaystyle CH_2CH_3}{|}}{\overset{\overset{\displaystyle O^-Mg^{2+}Br^-}{|}}{C}}CH_2CH_3$ $\xleftarrow[\text{diethyl ether}]{2\ CH_3CH_2MgBr}$

3-ethyl-3-pentanol-3-^{14}C

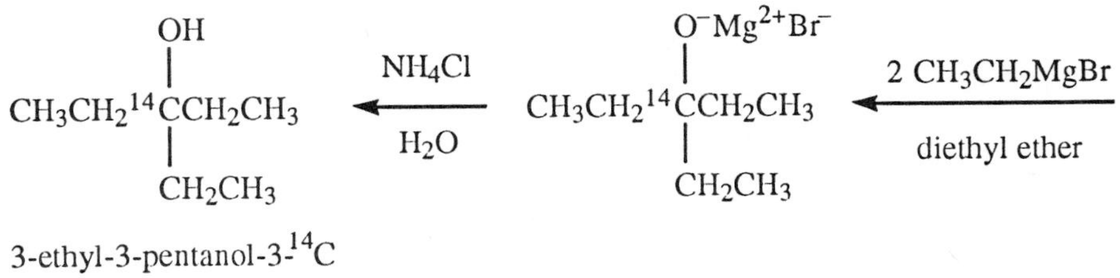

15.2 (cont)

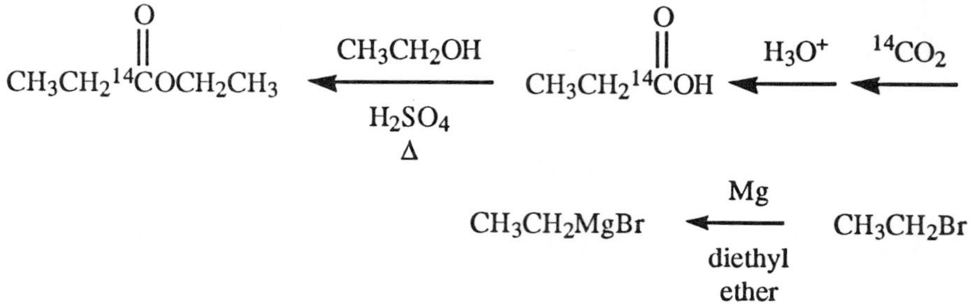

$$CH_3CH_2{}^{14}COCH_2CH_3 \xleftarrow[\substack{H_2SO_4 \\ \Delta}]{CH_3CH_2OH} CH_3CH_2{}^{14}COH \xleftarrow{H_3O^+} \xleftarrow{{}^{14}CO_2}$$

$$CH_3CH_2MgBr \xleftarrow[\substack{\text{diethyl} \\ \text{ether}}]{Mg} CH_3CH_2Br$$

15.3

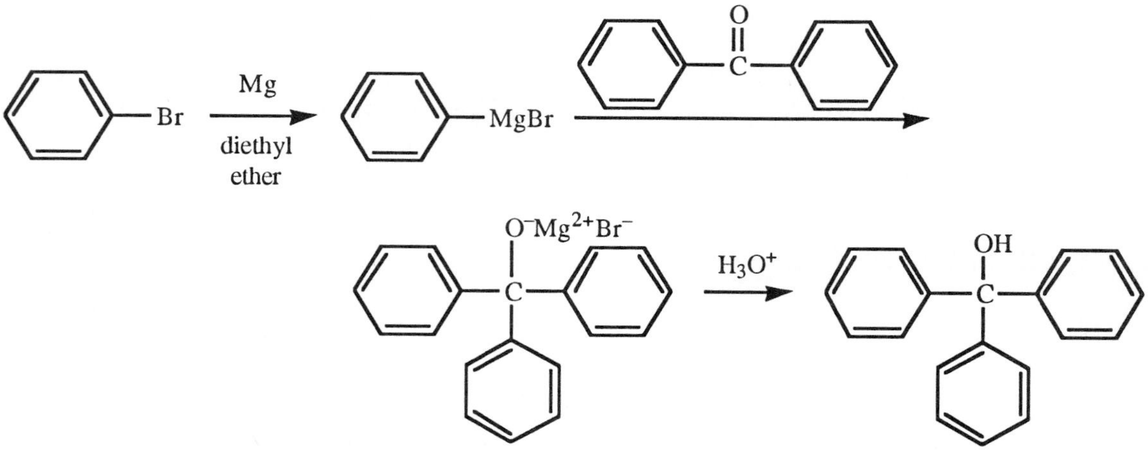

15.4

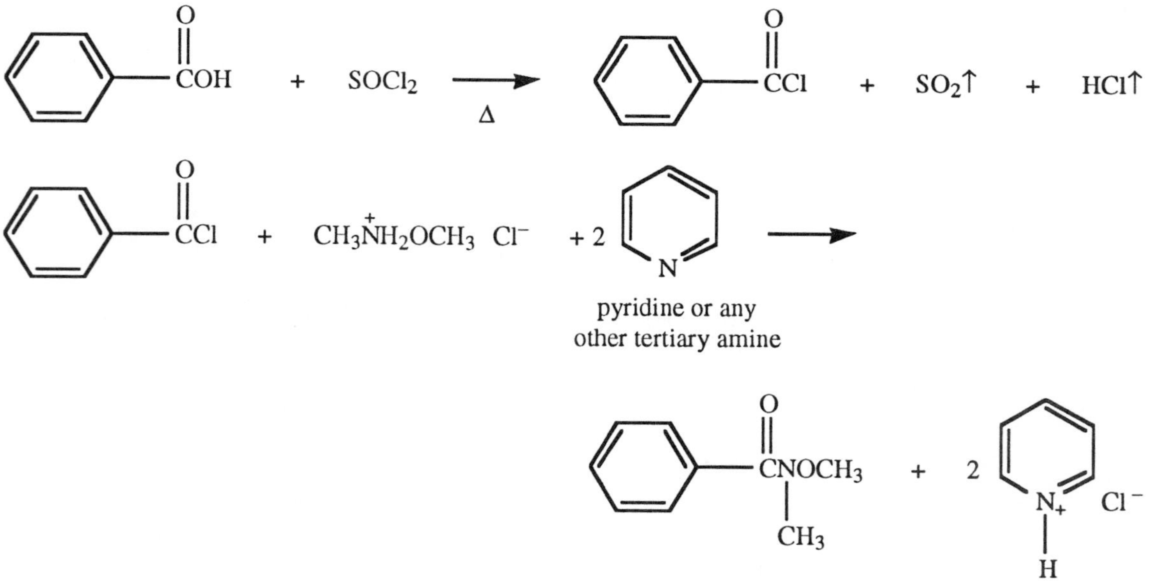

pyridine or any
other tertiary amine

15.5   (a)

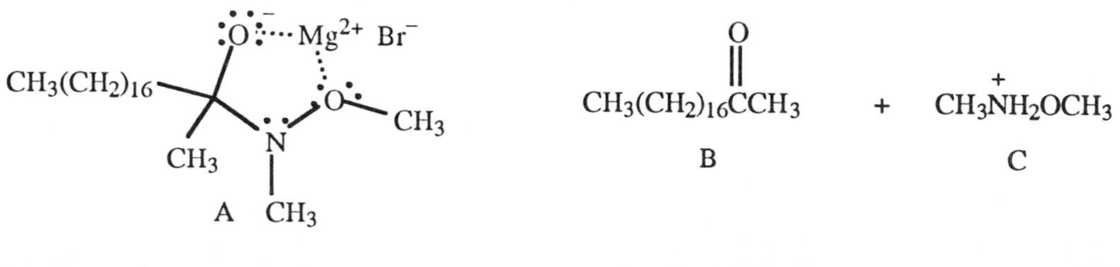

$$CH_3(CH_2)_{16}\overset{O}{\overset{\|}{C}}CH_3 \quad + \quad CH_3\overset{+}{N}H_2OCH_3$$

          B   C

15.5 (cont)

(b)

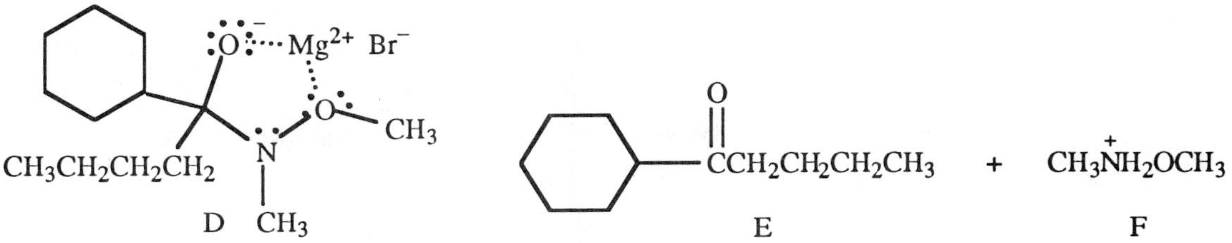

D

E   +   F

(c)

G

H   +   I

- -

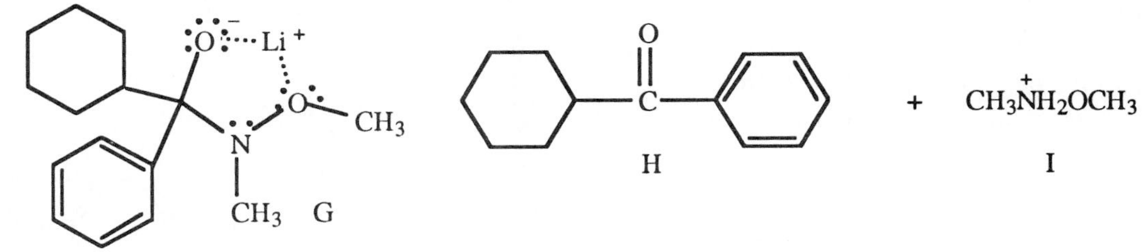

Concept Map 15.1   (see p. 495)

- -

15.6

(a)

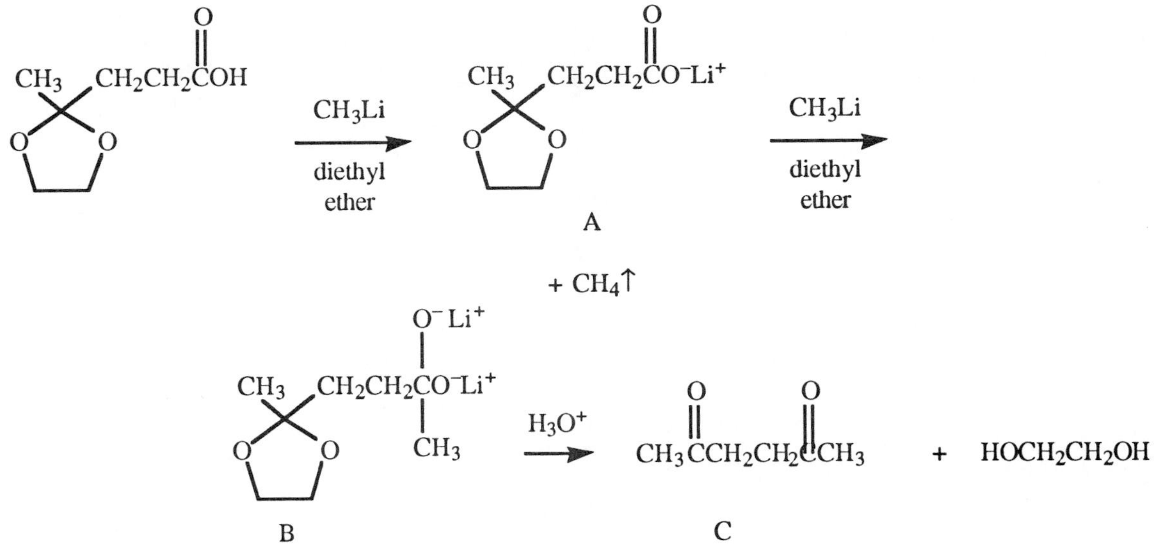

A

$+ CH_4\uparrow$

B 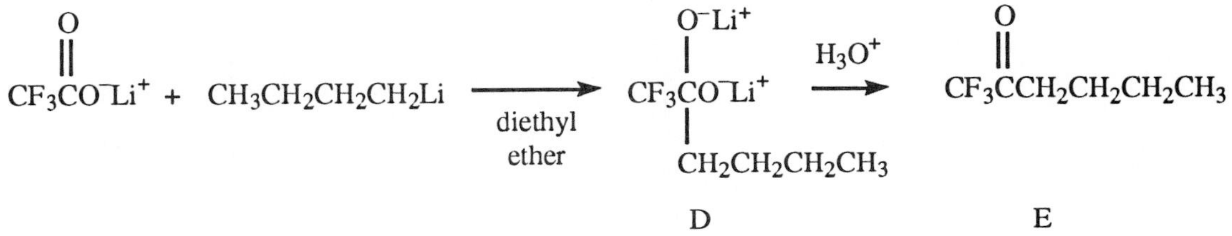   C

(b) $CF_3CO^-Li^+$ + $CH_3CH_2CH_2CH_2Li$ $\xrightarrow{\text{diethyl ether}}$ $CF_3CO^-Li^+$ $\xrightarrow{H_3O^+}$ $CF_3CCH_2CH_2CH_2CH_3$

D   E

(c)

F

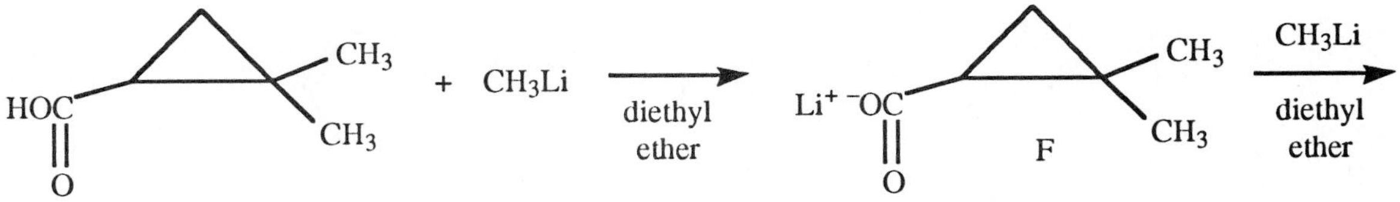

15.6 (c) (cont)

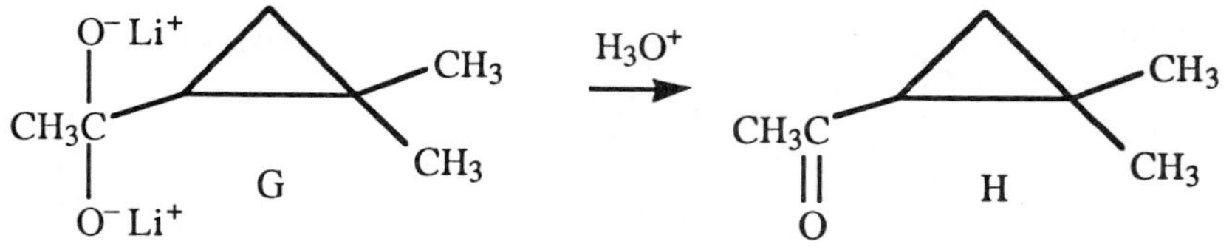

(d)

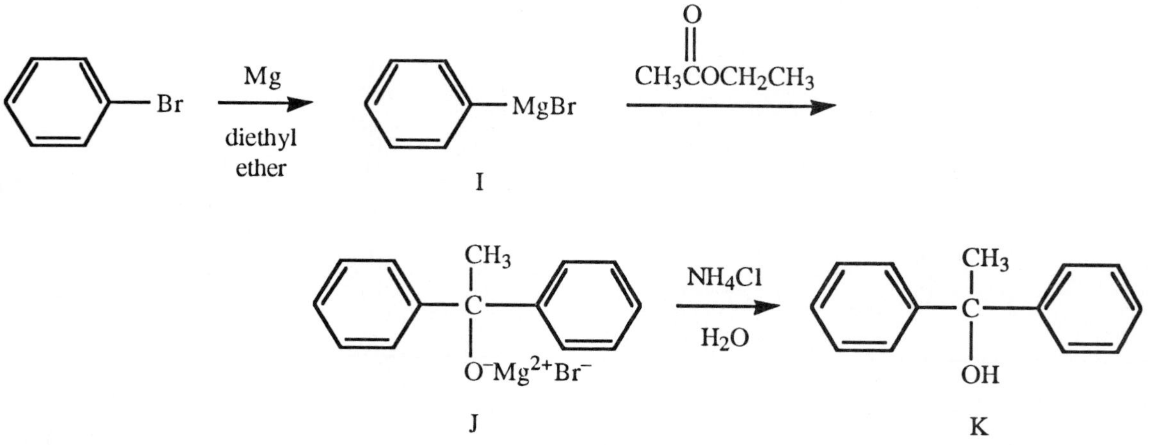

(e)

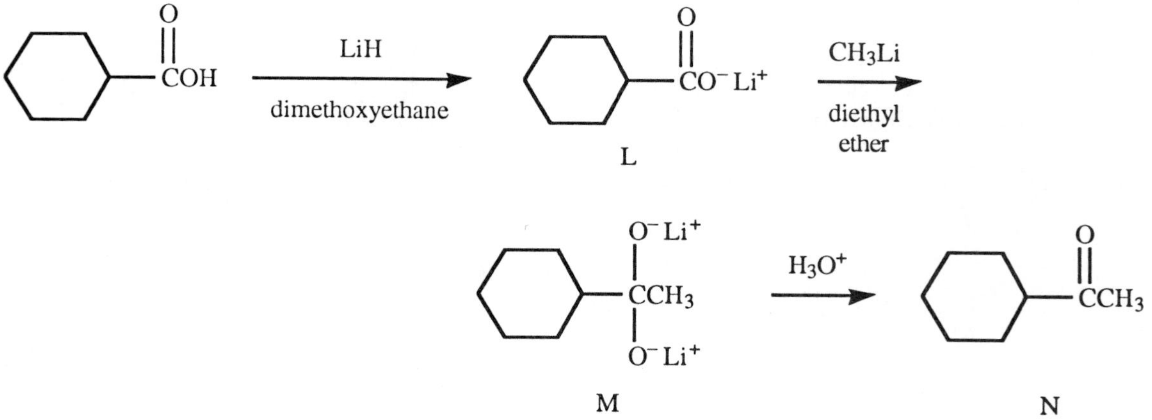

(f)

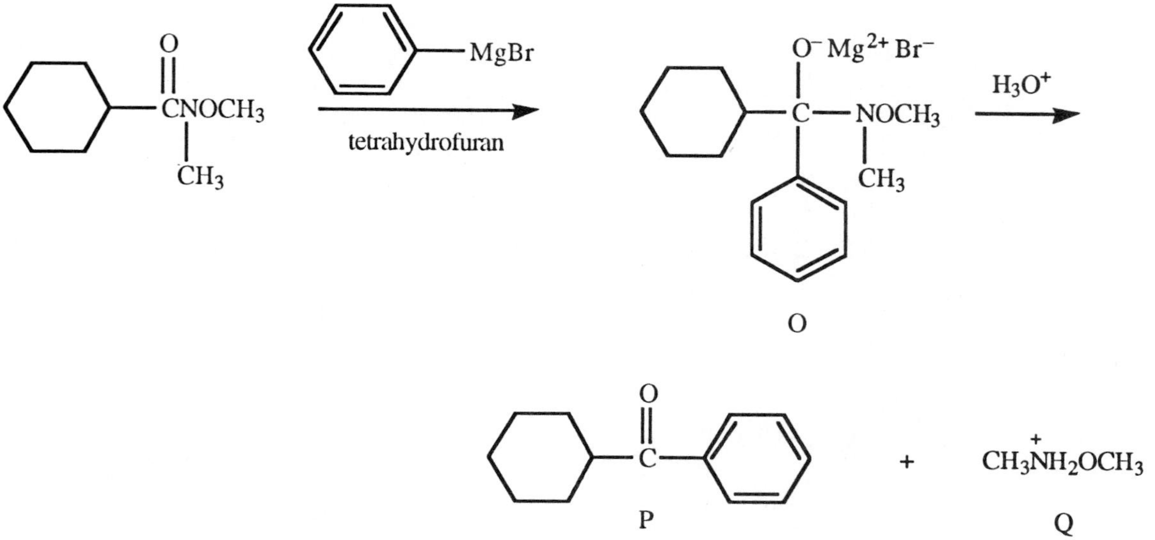

Concept Map 15.1   Reactions of organometallic reagents with acids and acid derivatives.

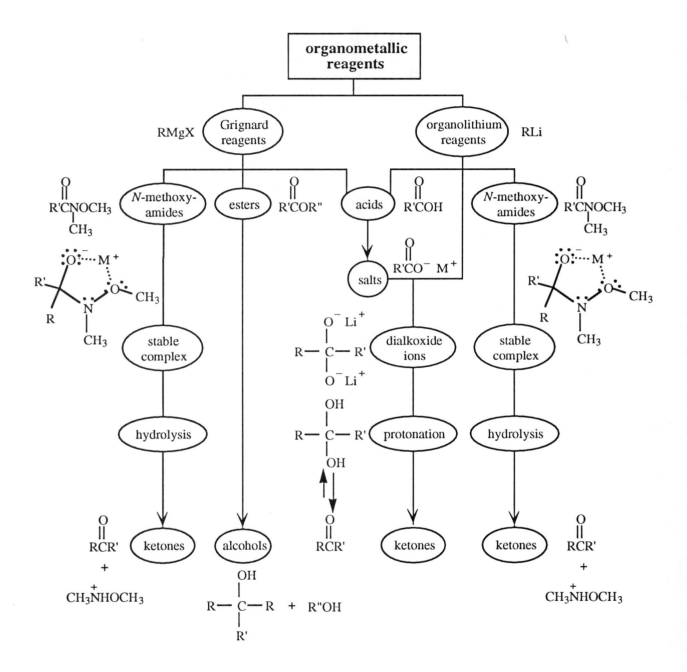

15.7

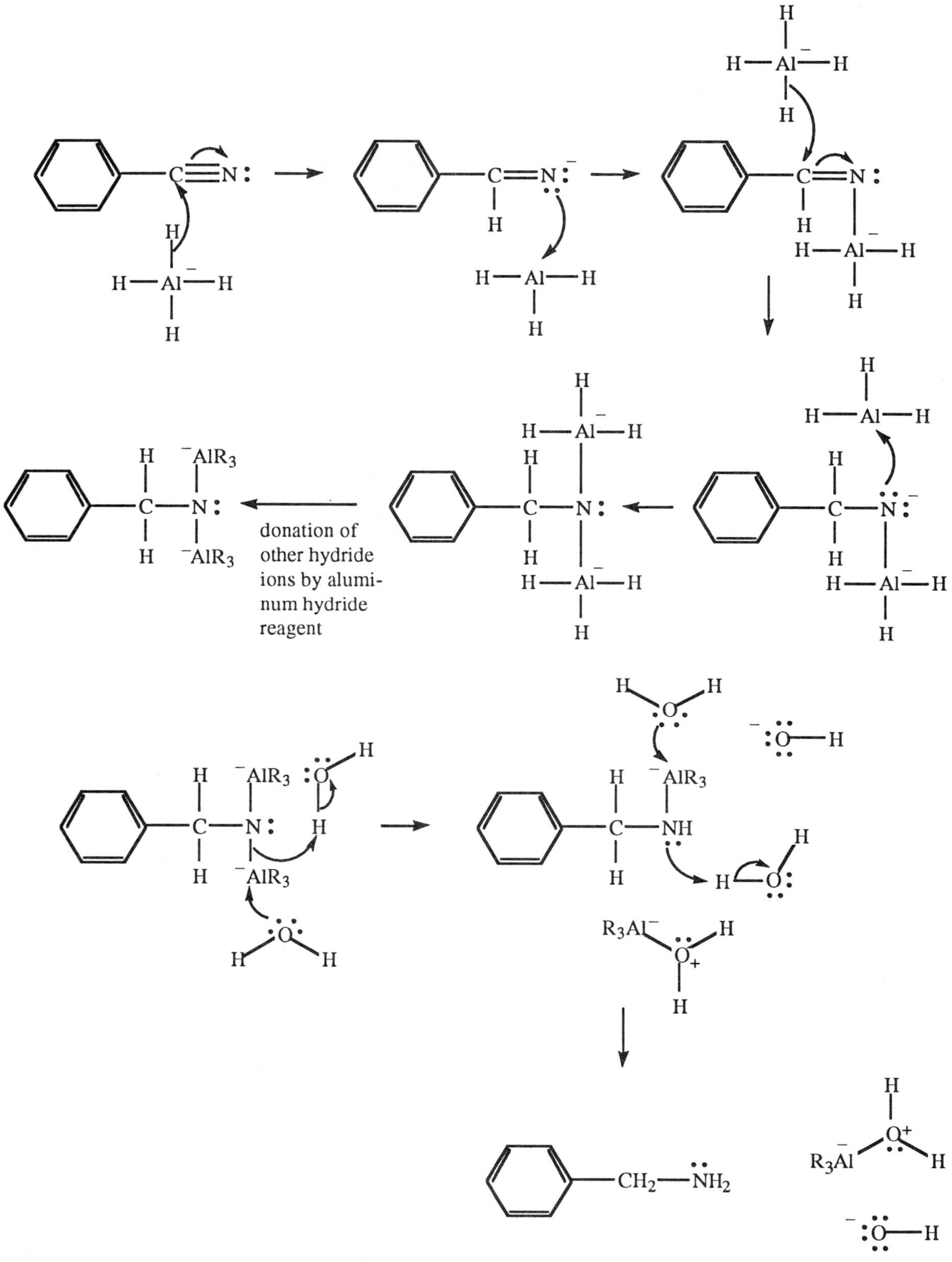

donation of
other hydride
ions by alumi-
num hydride
reagent

15.8

(a)

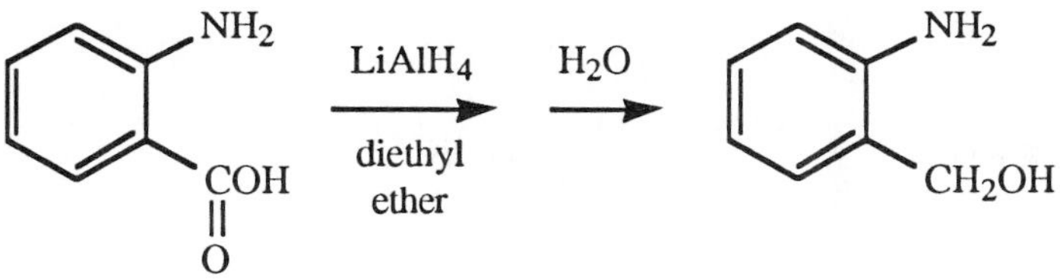

(b)

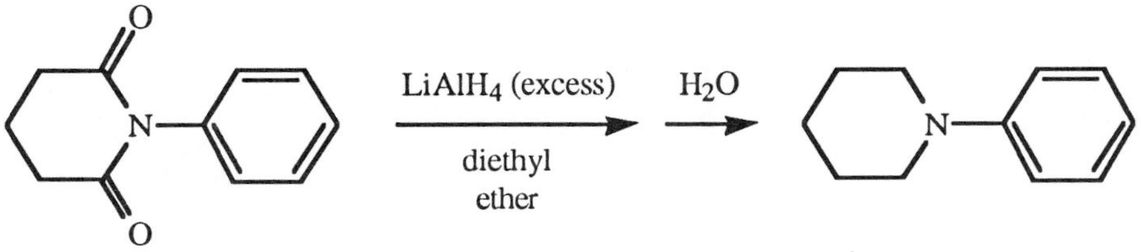

(c)

$$CH_3CH\!=\!CHCH_2CH_2\overset{\displaystyle O}{\overset{\|}{C}}OCH_3 \xrightarrow[\substack{\text{diethyl} \\ \text{ether}}]{\text{LiAlH}_4} \xrightarrow{\text{H}_3\text{O}^+} CH_3CH\!=\!CHCH_2CH_2CH_2OH$$

$$+ \quad CH_3OH$$

(d)

$$CH_3CH_2O\overset{\displaystyle O}{\overset{\|}{C}}(CH_2)_4\overset{\displaystyle O}{\overset{\|}{C}}OCH_2CH_3 \xrightarrow[\substack{\text{diethyl} \\ \text{ether}}]{\text{LiAlH}_4} \xrightarrow{\text{H}_3\text{O}^+} HOCH_2(CH_2)_4CH_2OH \quad + \quad CH_3CH_2OH$$

(e)

$$CH_3CH_2\overset{\displaystyle O}{\overset{\|}{C}}NHCH_2CH_3 \xrightarrow[\substack{\text{diethyl} \\ \text{ether}}]{\text{LiAlH}_4} \xrightarrow{\text{H}_2\text{O}} CH_3CH_2CH_2NHCH_2CH_3$$

(f)

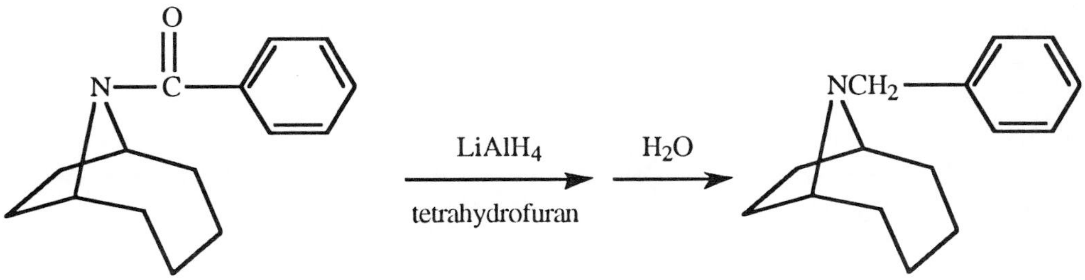

15.9

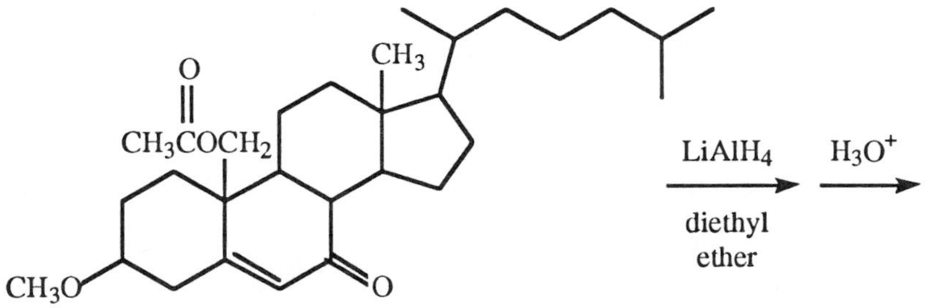

15.9 (cont)

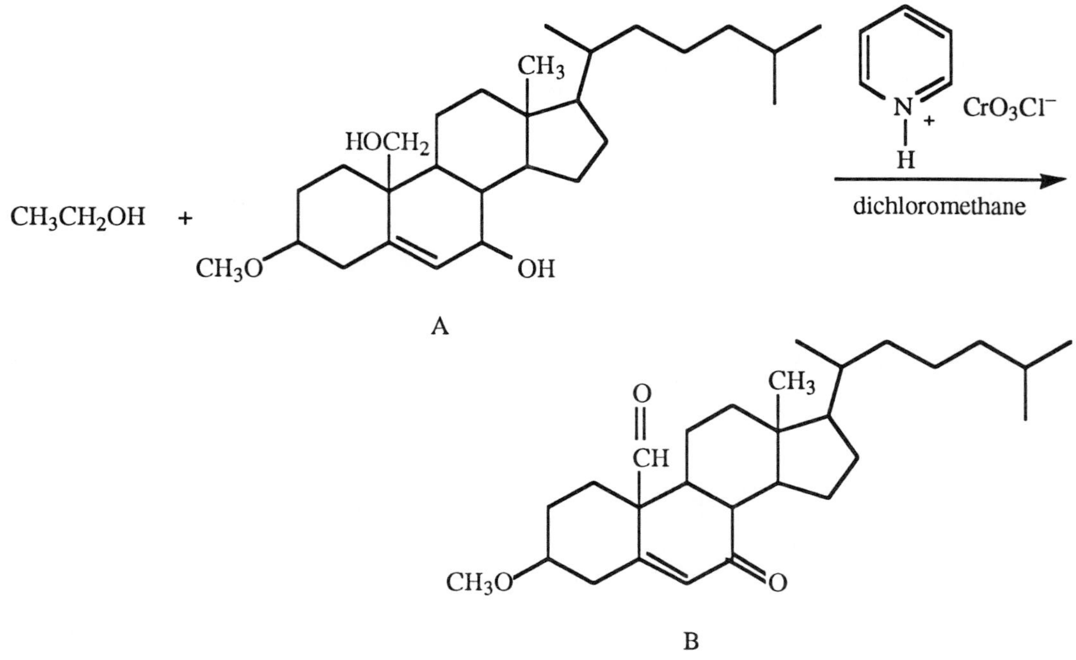

A

B

15.10

(1)

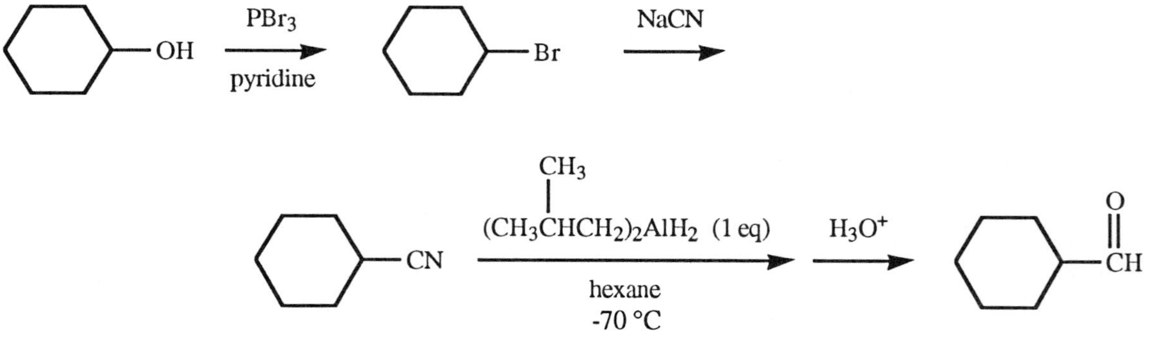

(2)

+ CH_3OH

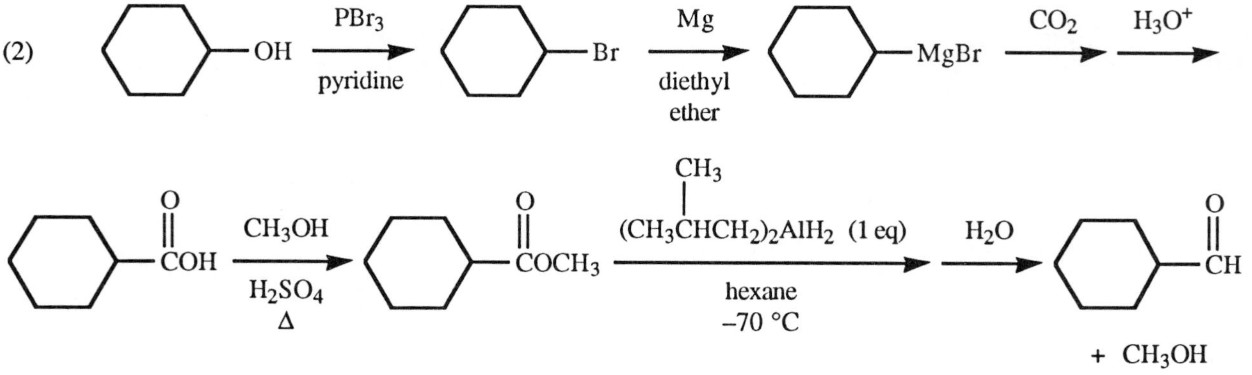

Concept Map 15.2   Reduction of acids and acid derivatives.

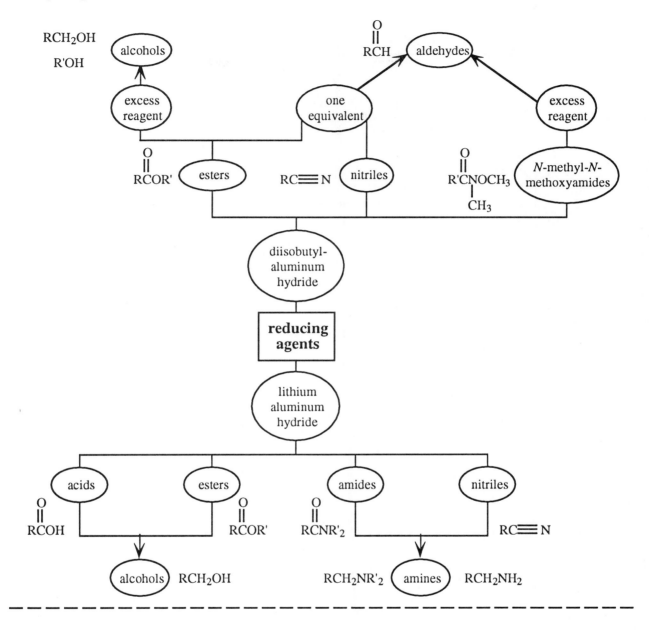

15.11

15.11 (cont)

(b)

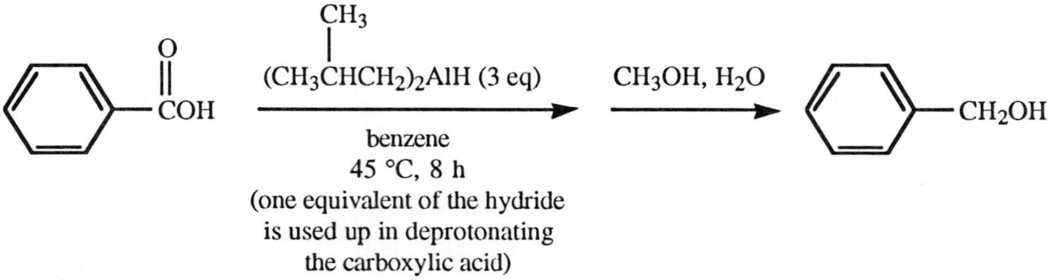

(c)

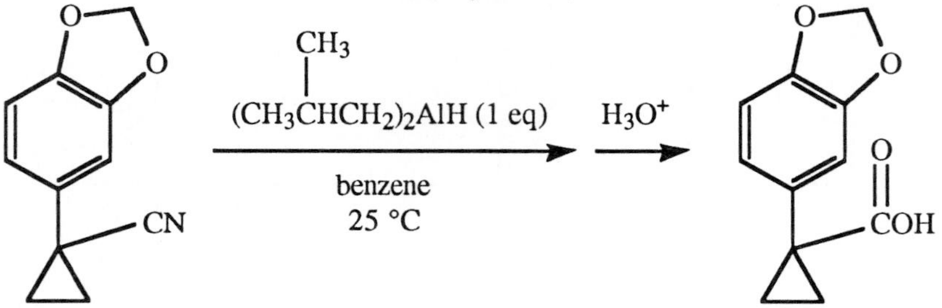

(d)

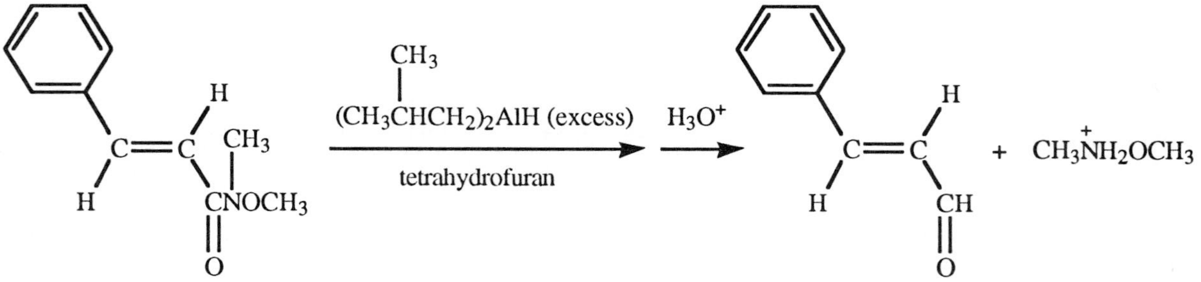

15.12

(a)

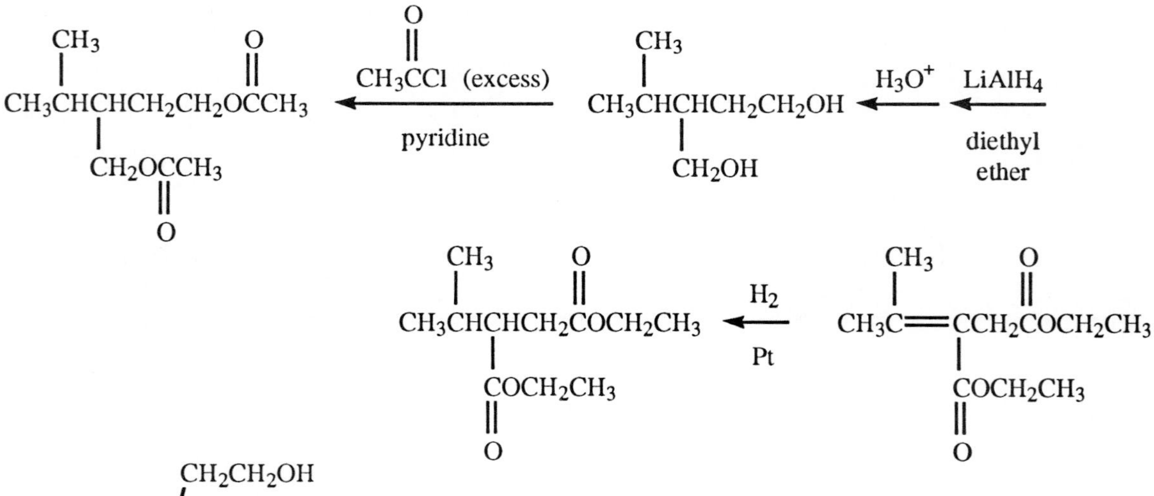

(b)

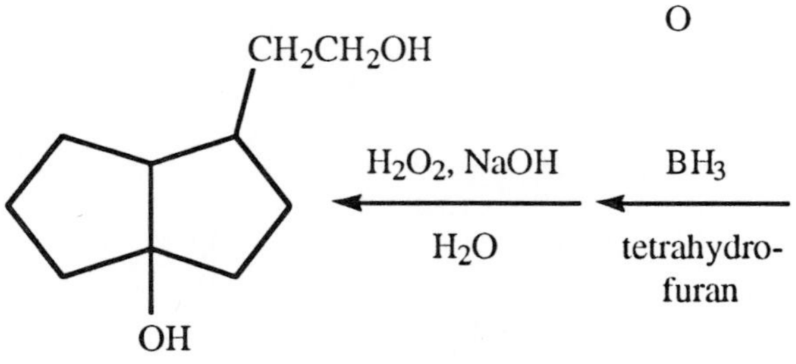

15.12 (b) (cont)

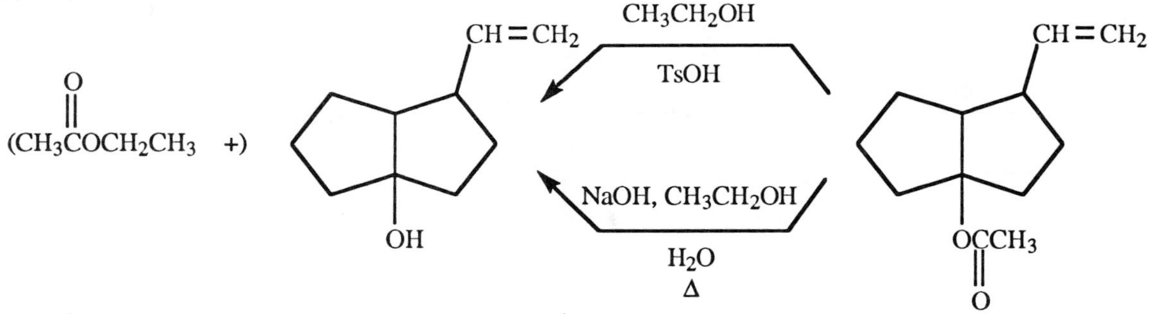

$(CH_3COCH_2CH_3 \ +)$

(c)

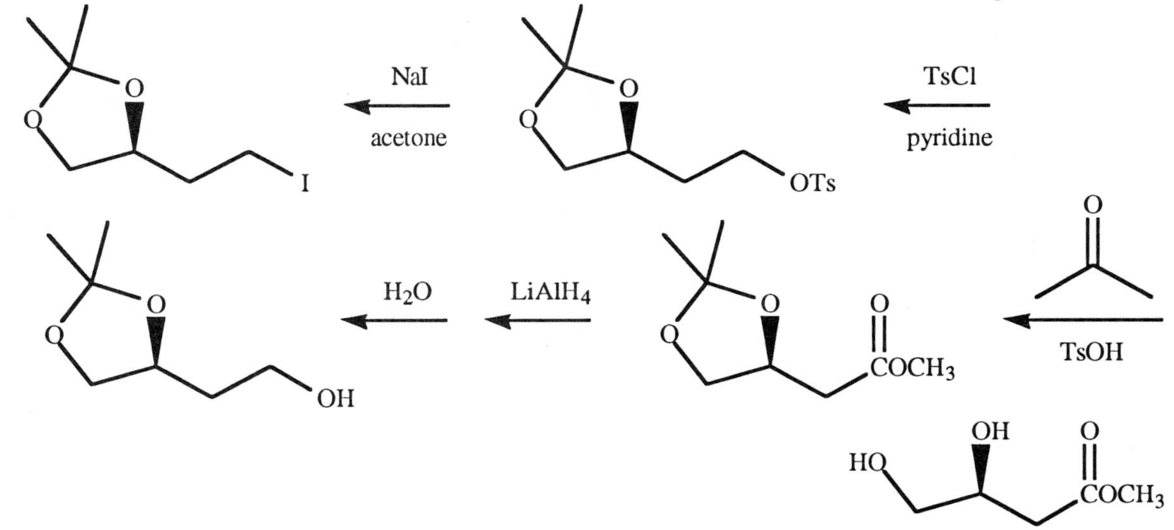

15.13

(a)

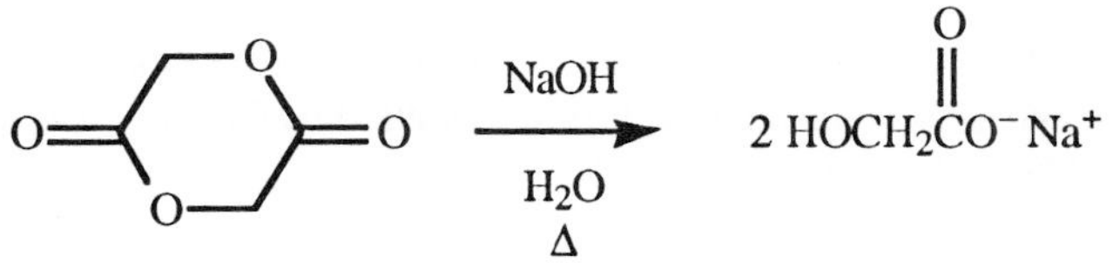

(b)

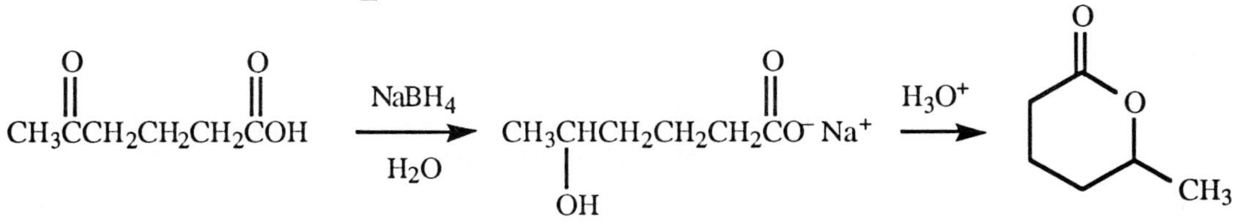

(c)

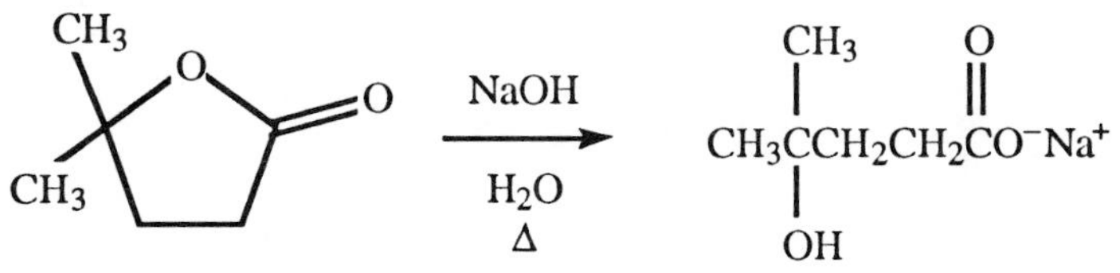

(d)

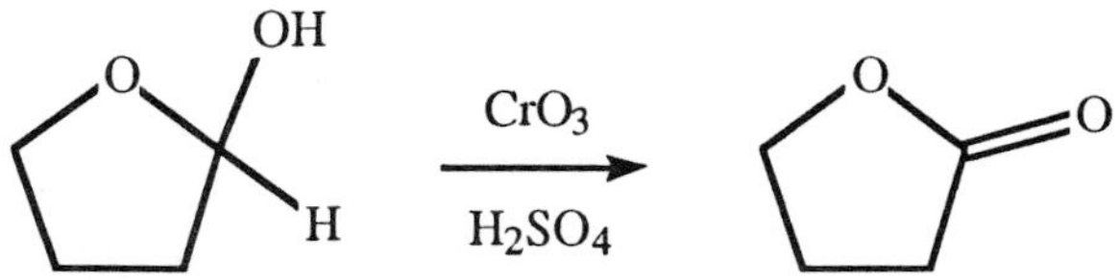

15.13 (cont)

(e)

$$\text{[cyclobutanone]} \xrightarrow[\text{chloroform}]{\text{C}_6\text{H}_5\text{COOH}} \text{[γ-butyrolactone]}$$

15.14

$$\text{[oxepan-2-one]} \xrightarrow{\text{CH}_3\text{OH}} \underset{\text{A}}{\text{HOCH}_2\text{CH}_2\text{CH}_2\text{CH}_2\text{CH}_2\overset{\text{O}}{\overset{\|}{\text{C}}}\text{OCH}_3} \xrightarrow[\text{dichloromethane}]{\underset{\text{H}}{\overset{+}{\text{N}}}\text{CrO}_3\text{Cl}^-}$$

$$\underset{\text{B}}{\text{H}\overset{\text{O}}{\overset{\|}{\text{C}}}\text{CH}_2\text{CH}_2\text{CH}_2\text{CH}_2\overset{\text{O}}{\overset{\|}{\text{C}}}\text{OCH}_3}$$

15.15

$$\xrightarrow{?}$$

1. What are the connectivities of the two compounds? How many carbon atoms does each contain? Are there any rings? What are the positions of branches and functional groups on the carbon skeletons?

    The starting material has eight carbon atoms and is a bicyclic ketone. The product has nine carbon atoms but only has one ring, a five-membered one, which can also be seen in the bicyclic compound. The five-membered ring has a ketone function in it, and is connected by a two-carbon chain to an ester.

2. How do the functional groups change in going from starting material to product? Does the starting material have a good leaving group?

    The ketone of the starting material appears to have been transformed into an ester in the product. A ketone function appears on the five-membered ring in the product. There is no good leaving group in the starting material.

3. Is it possible to dissect the structures of the starting material and product to see which bonds must be broken and which formed?

bond to be broken   bonds to be formed

15.15 (cont)

4.   New bonds are created when an electrophile reacts with a nucleophile. Do we recognize any part of the product molecule as coming from a good nucleophile or an electrophilic addition?

   An ester can be made from an alcohol (a nucleophile) and an acid derivative (an electrophile).

5.   What type of compound would be a good precursor to the product?

   An alcohol can be oxidized to a ketone.

   An alcohol and an ester can be made from a lactone by solvolysis (Problem 15.14).

6.   After this last step, do we see how to get from starting material to product? If not, we need to analyze the structure obtained in step 5 by applying questions 4 and 5 to it.

   We need to prepare the lactone.

4. (repeated) Do we recognize any part of the product molecule as coming from a good nucleophile or an electrophilic addition?

   The alkyl oxygen of a lactone can come from the nucleophilic oxygen of a peroxyacid.

5. (repeated) What type of compound would be a good precursor to the product?

   The Baeyer-Villiger reaction can be used to prepare the lactone from the starting material.

15.15 (cont)

6. **(repeated)** After this last step, do we see how to get from starting material to product? If not, we need to analyze the structure obtained in step 5 by applying questions 4 and 5 to it.

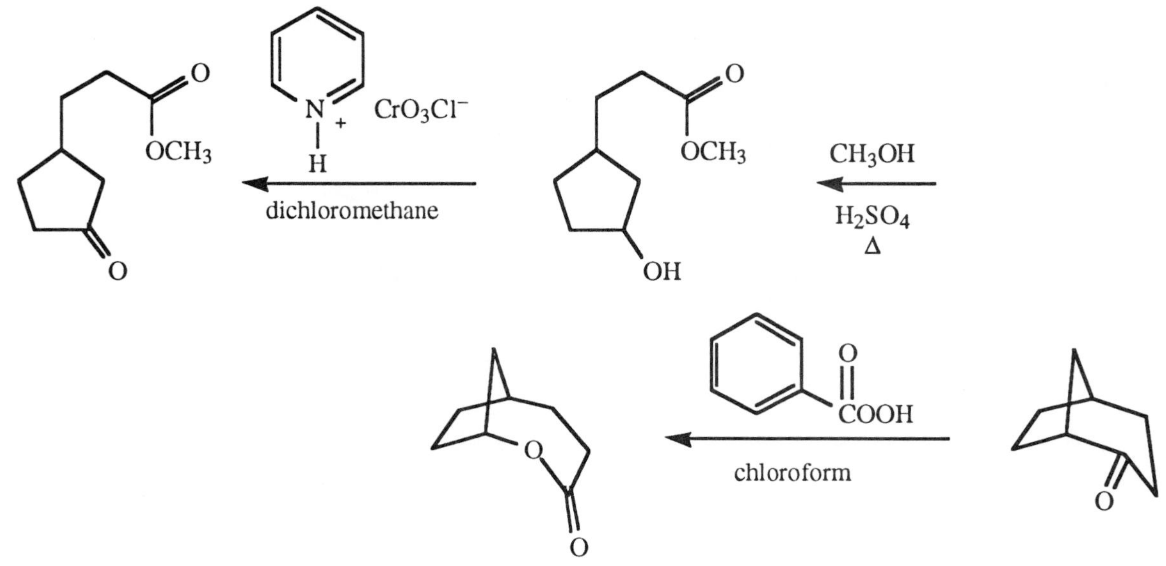

15.16

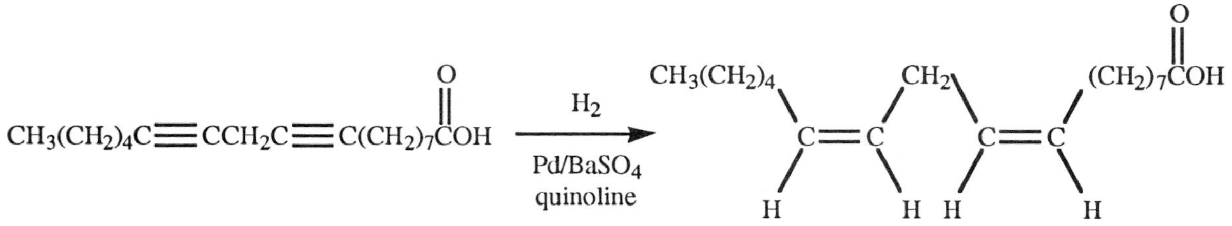

(9Z, 12Z)-9,12-octadecadienoic acid
(linoleic acid)

15.17   Compounds b, c, e, and f are surfactants.

15.18

(a)

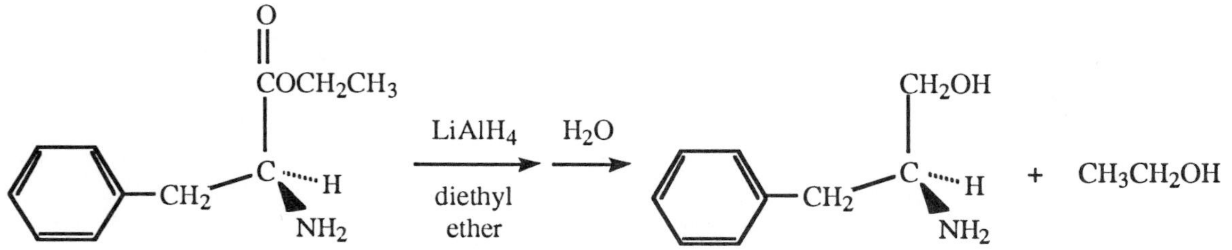

15.18 (cont)

(b)

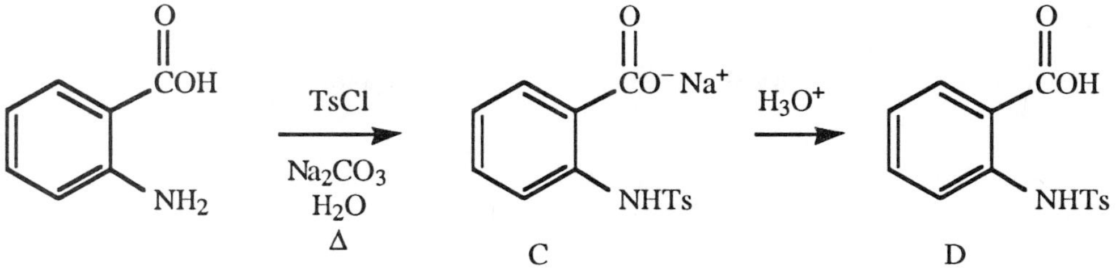

C

D

(c)

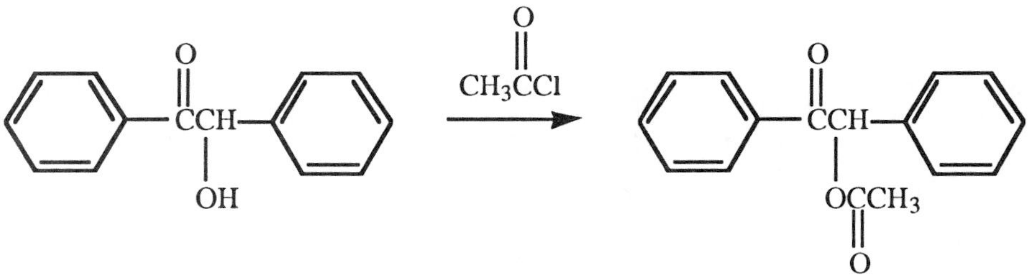

E

(d)

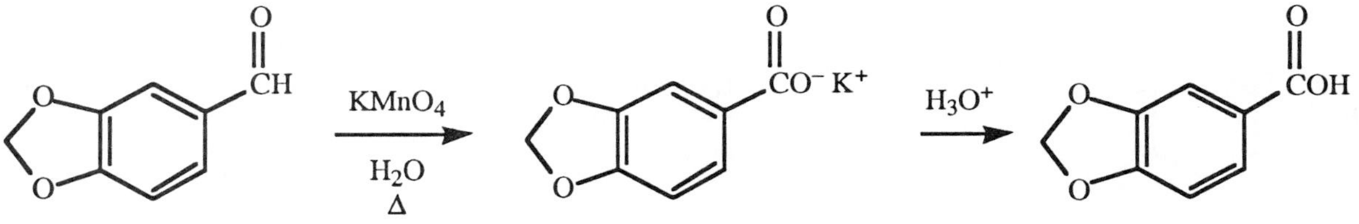

F

G

Note: The cyclic acetal on
the aromatic ring is stable
and does not open during the
protonation of the carboxylate anion.

(e) $CH_3(CH_2)_{14}CH_2OH$ $\xrightarrow[\text{pyridine}]{\text{TsCl}}$ $CH_3(CH_2)_{14}CH_2OTs$

H

(f)

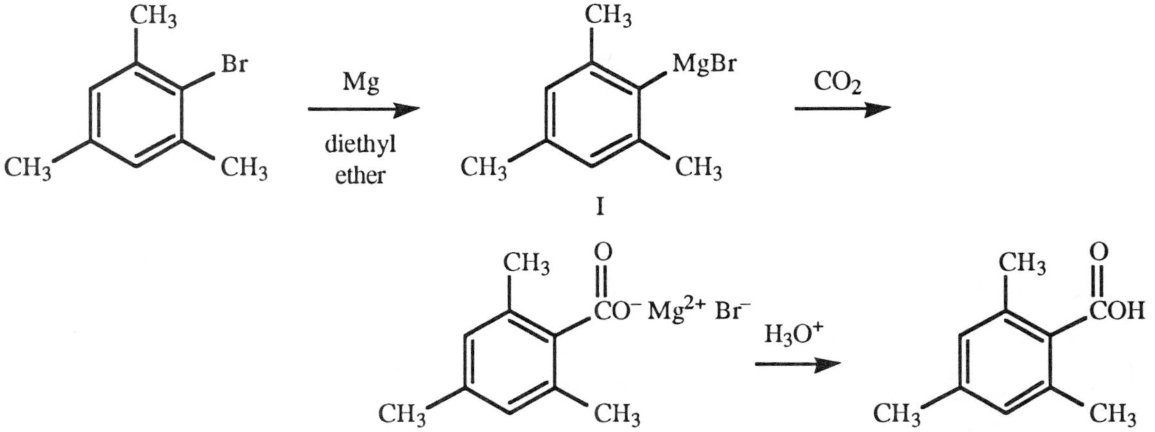

I

J

K

15.18 (cont)

(g)

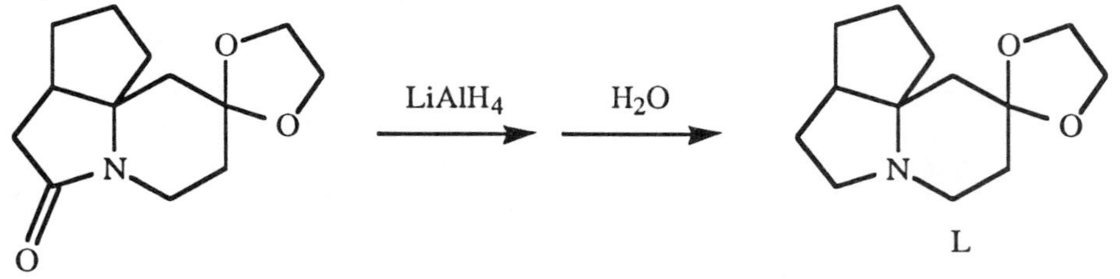
LiAlH$_4$
H$_2$O
L

15.19

(a)

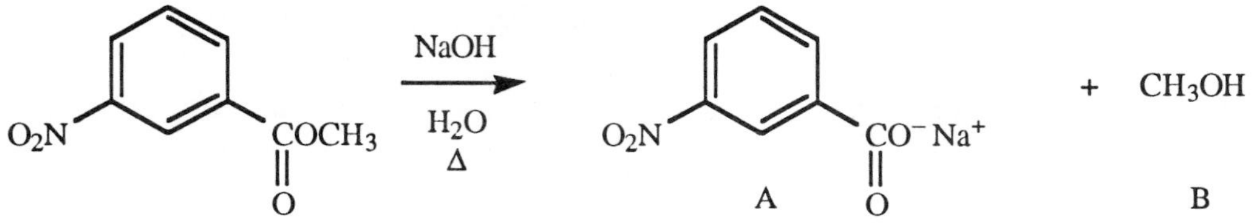
O$_2$N
COCH$_3$
NaOH
H$_2$O
Δ
CO$^-$ Na$^+$
A
+ CH$_3$OH
B

(b)

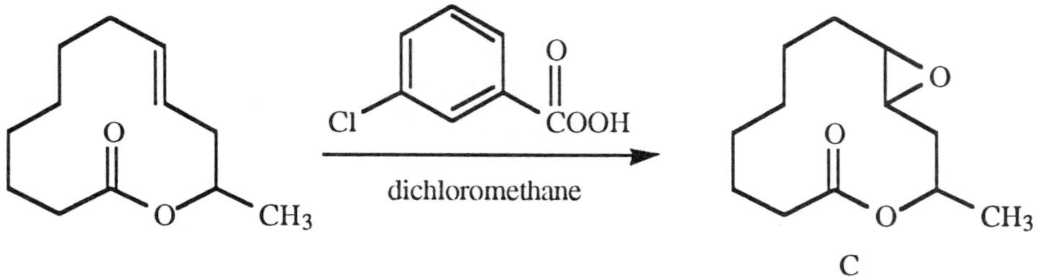
CH$_3$
Cl
COOH
dichloromethane
C

(c)

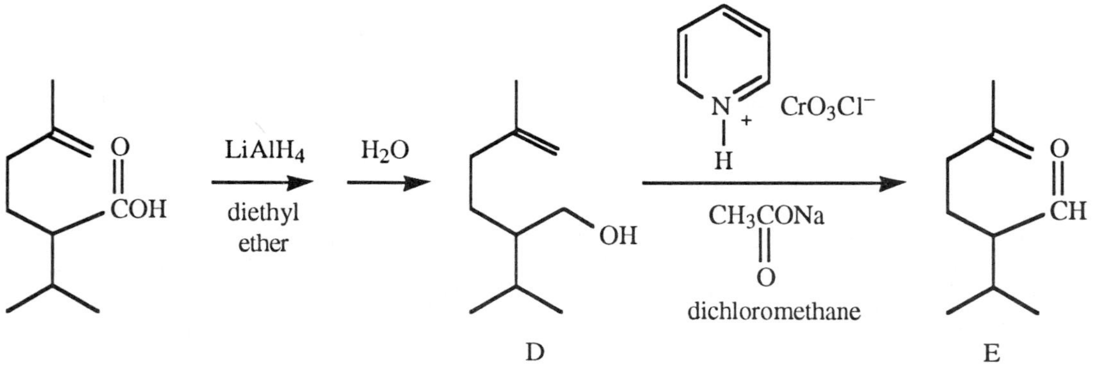
COH
LiAlH$_4$
diethyl ether
H$_2$O
OH
D
CrO$_3$Cl$^-$
CH$_3$CONa
dichloromethane
CH
E

(d)

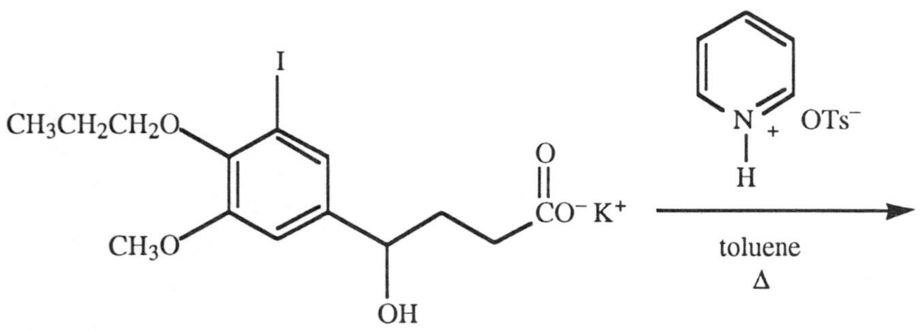
I
CH$_3$CH$_2$CH$_2$O
CH$_3$O
OH
CO$^-$ K$^+$
OTs$^-$
toluene
Δ

15.19 (d) (cont)

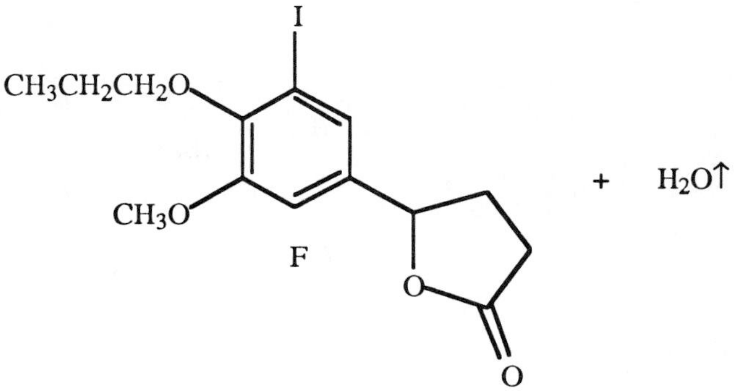

$+$ $H_2O\uparrow$

F

is an acid with pK_a 5.2, close enough to the pK_a of the carboxylic acid ($pK_a \sim 4.8$) to maintain an equilibrium concentration of the acid. The carboxylic acid then forms the lactone under conditions in which water is removed from the reaction mixture.

(e)

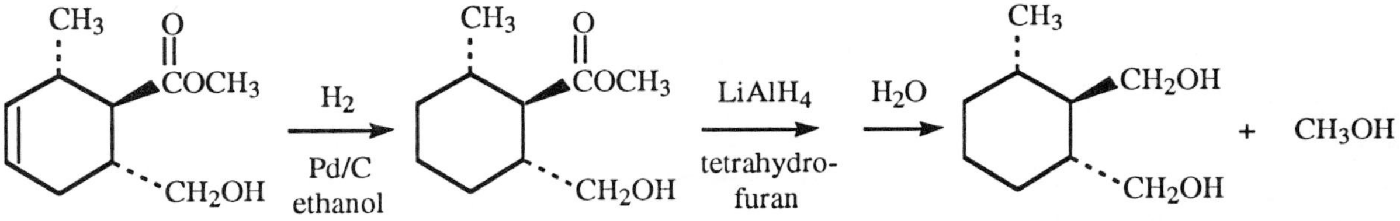

(f)

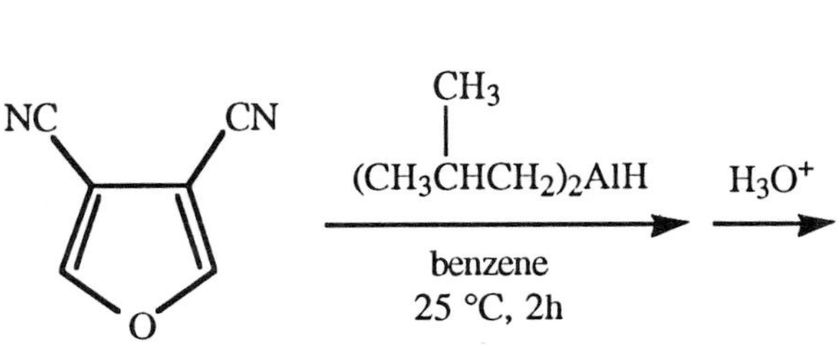

(g)

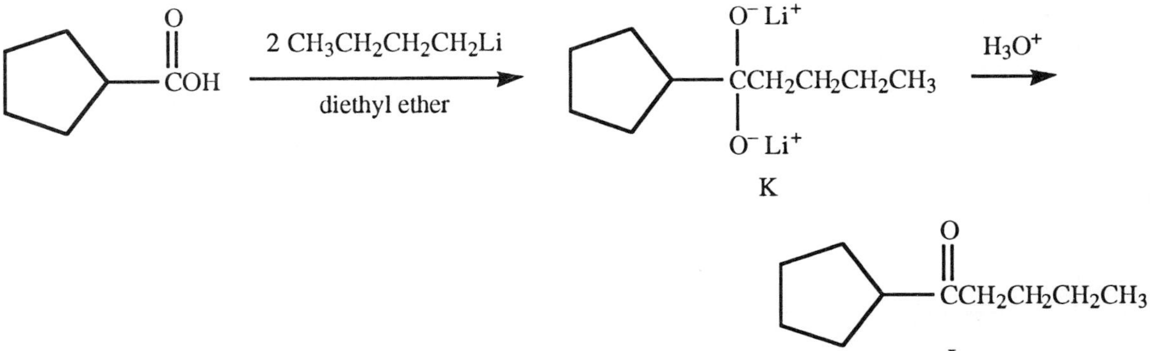

15.19 (cont)

(h)

15.20

(a)

$$CH_2=CHCH=CHCOH \xrightarrow[\substack{\text{diethyl} \\ \text{ether}}]{LiAlH_4} \xrightarrow{H_2O}$$

$$CH_2=CHCH=CHCH_2OH \xrightarrow[\text{dichloromethane}]{} CH_2=CHCH=CHCH$$
$$\qquad\qquad\quad A \qquad\qquad\qquad\qquad\qquad\qquad\qquad\qquad\qquad\qquad B$$

(b)

$$CH_3CH_2CH_2CH_2COCH_2CH_3 \xrightarrow[\text{diethyl ether}]{2\ CH_3CH_2MgBr} CH_3CH_2CH_2CH_2\underset{\underset{O^-Mg^{2+}Br^-}{\overset{CH_2CH_3}{|}}}{\overset{|}{C}}CH_2CH_3 \xrightarrow[H_2O]{NH_4Cl}$$
$$\qquad\qquad\qquad\qquad\qquad\qquad\qquad\qquad\qquad\qquad\qquad\qquad C$$

$$CH_3CH_2CH_2CH_2\underset{\underset{OH}{|}}{\overset{\overset{CH_2CH_3}{|}}{C}}CH_2CH_3$$
$$\qquad\qquad\qquad D$$

(c)

$$CH_3CH_2\overset{\overset{O}{\|}}{C}NHCH_2CH_3 \xrightarrow[\substack{\text{diethyl} \\ \text{ether}}]{LiAlH_4} \xrightarrow{H_2O} CH_3CH_2CH_2NHCH_2CH_3$$
$$\qquad\qquad\qquad\qquad\qquad\qquad\qquad\qquad\qquad\qquad\qquad E$$

(d)

$$CH_3CH=CHCH_2C\equiv CCH_2CH_2COH \xrightarrow[\substack{NH_3\ (liq) \\ \text{ethanol}}]{Li} \xrightarrow{H_3O^+}$$

(e)

$$\xrightarrow[\substack{\text{diethyl} \\ \text{ether}}]{LiAlH_4} \xrightarrow{H_3O^+}$$

15.20 (e) (cont)

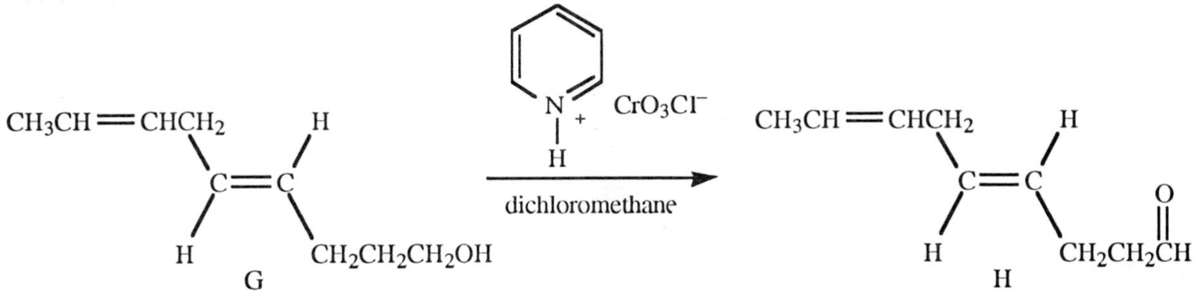

(f)

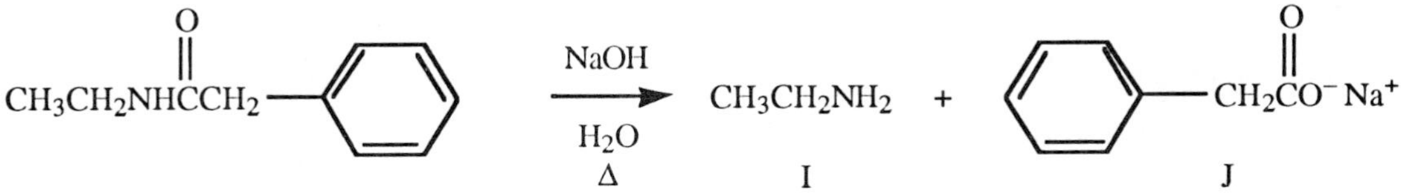

(g)

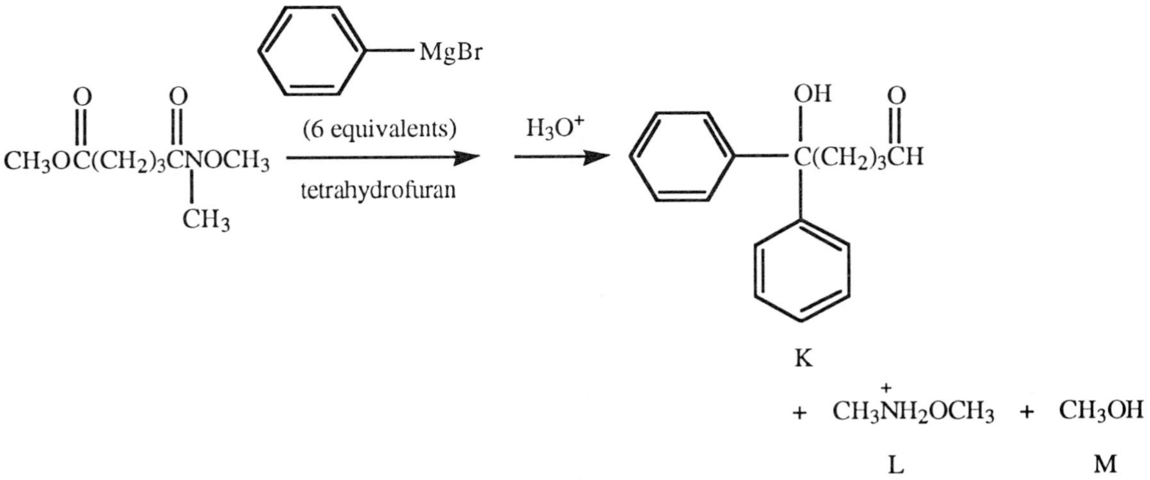

$+$ $CH_3\overset{+}{N}H_2OCH_3$ $+$ CH_3OH

L   M

15.21

(a)

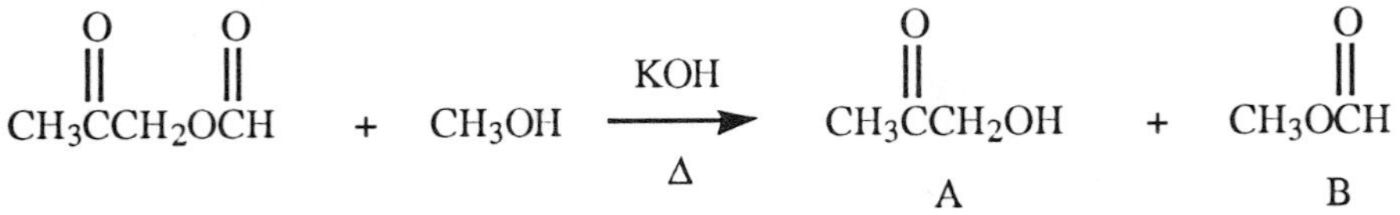

(b)

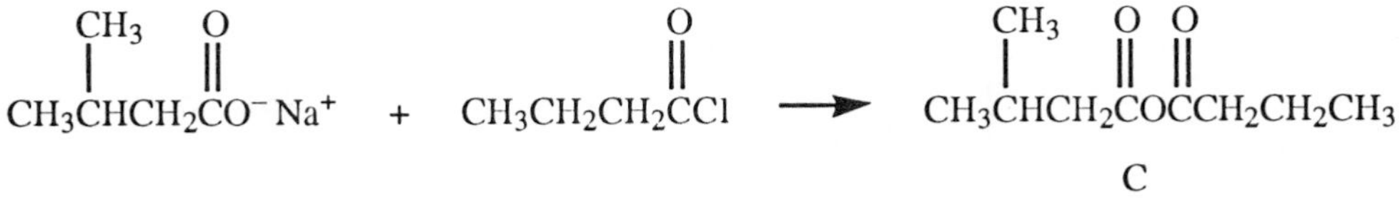

(c)

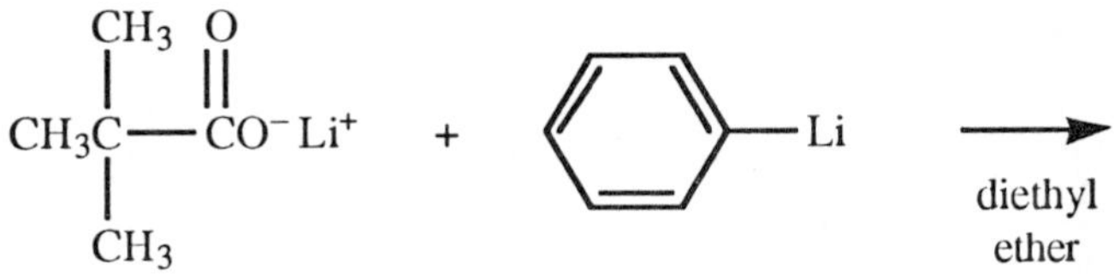

15.21 (c) (cont)

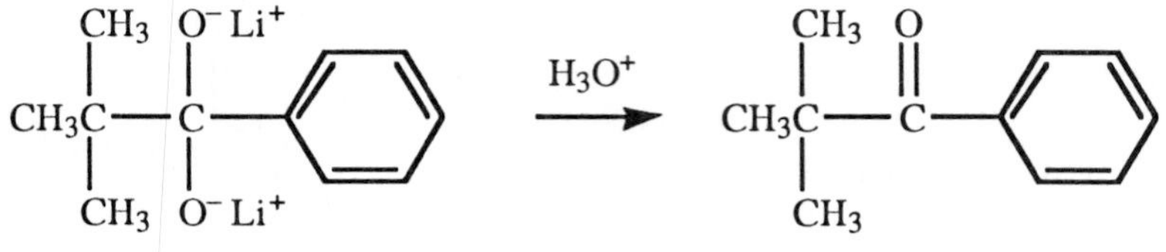

D   →   E

(d)

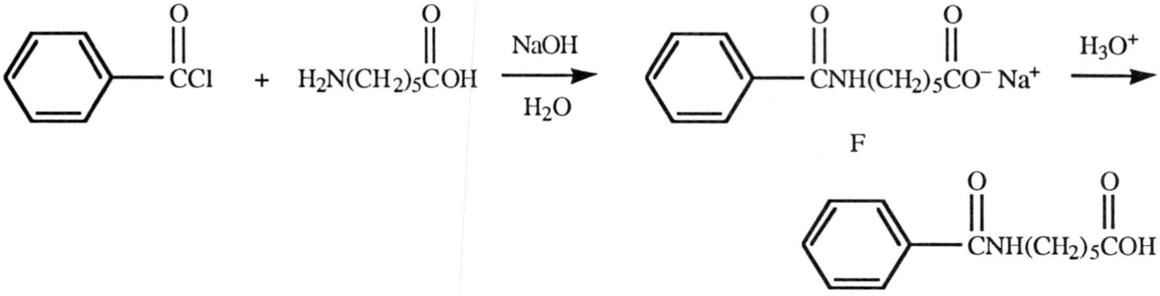

F

G

(e)

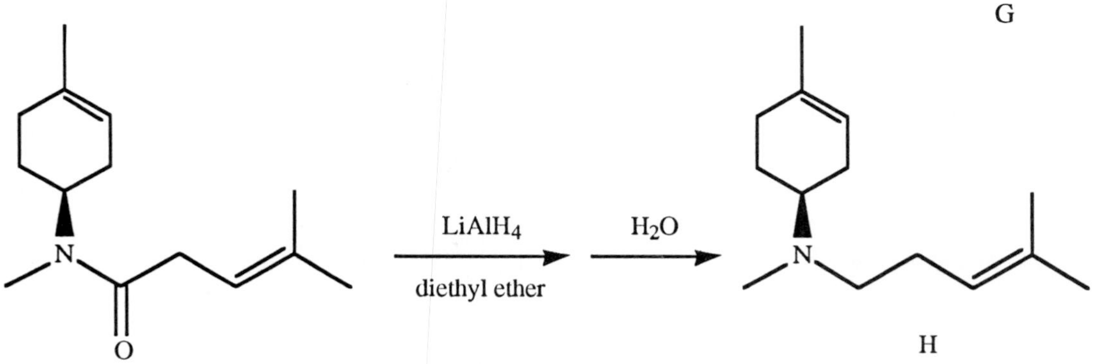

H

(f)

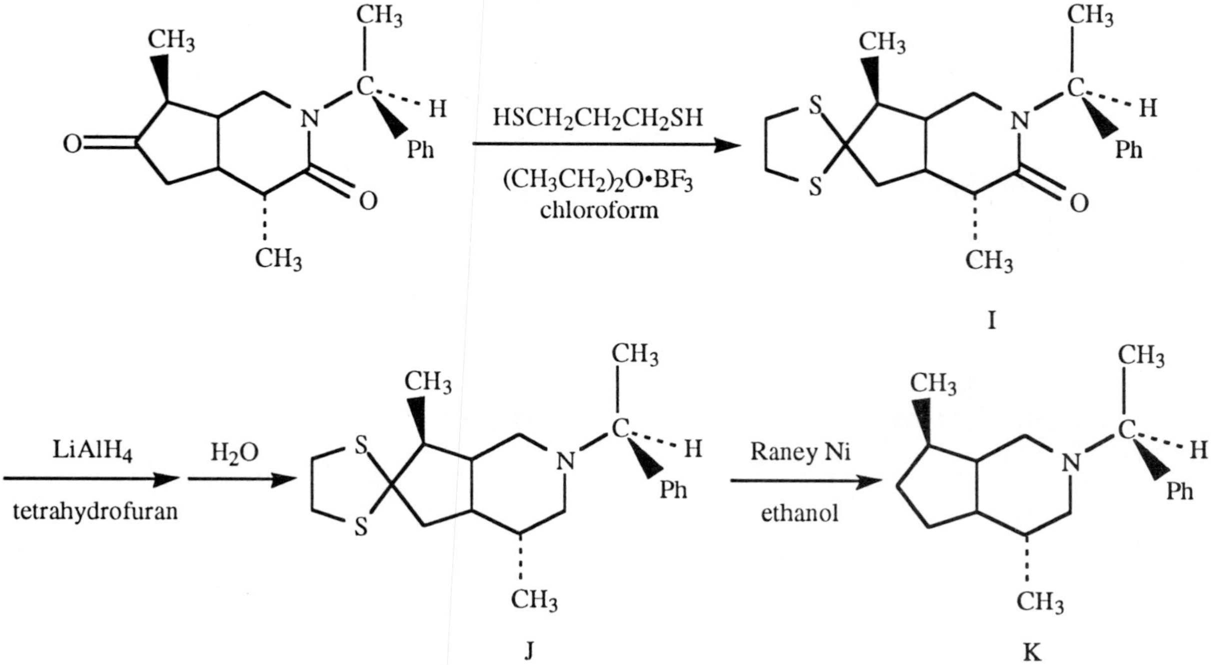

I

J   K

15.21 (cont)

(g)

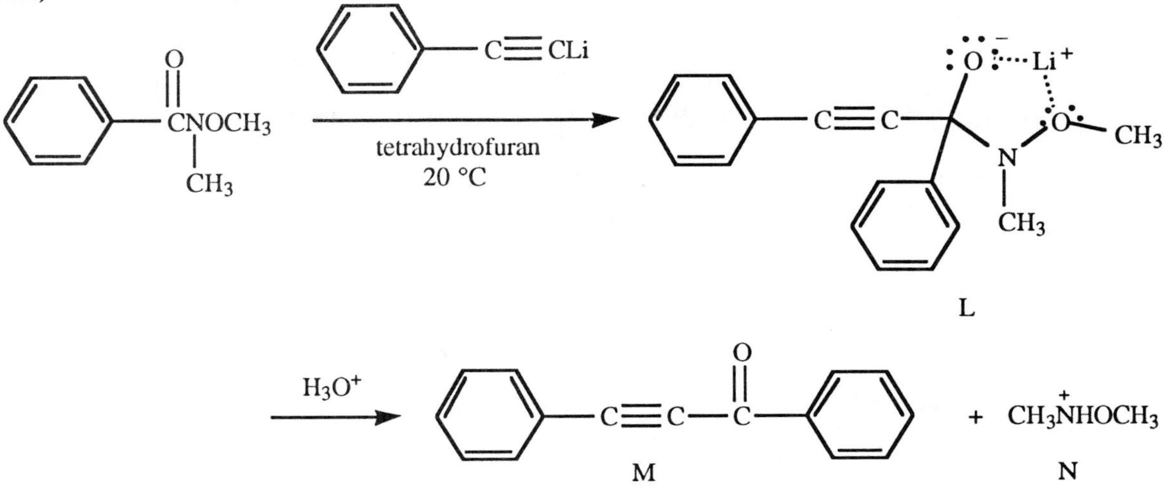

L

M + N

CH₃NHOCH₃

15.22

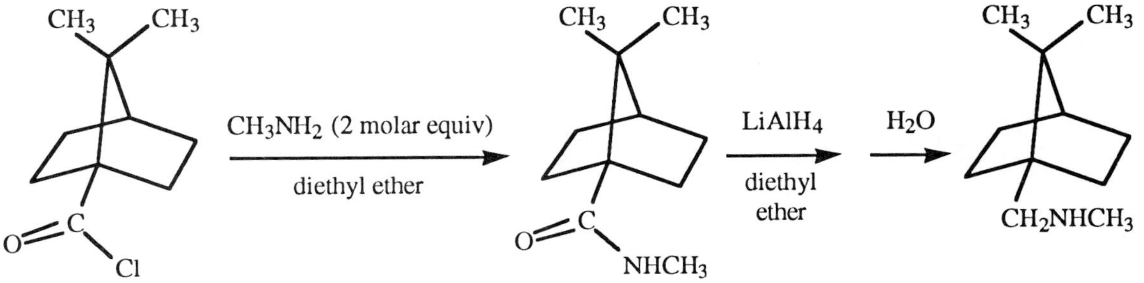

15.23

(a)

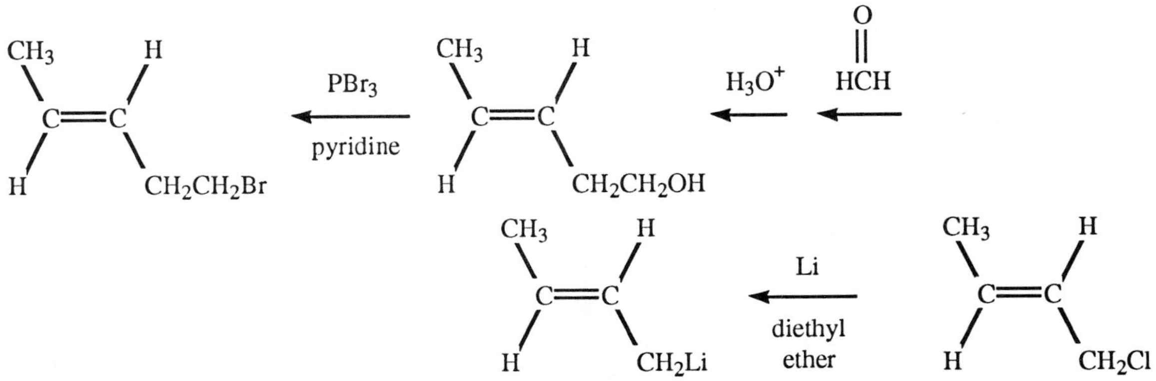

(b)

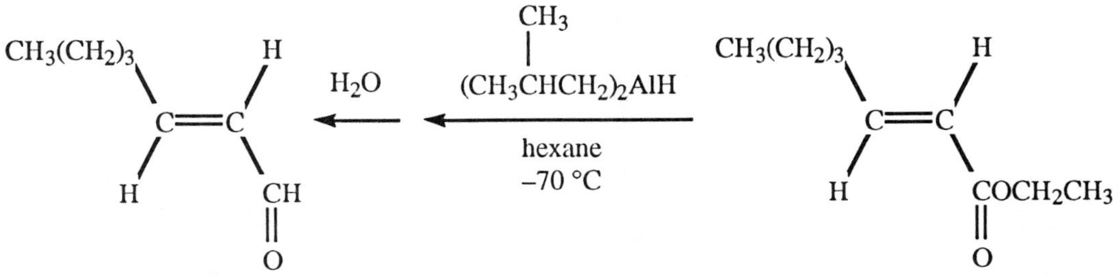

15.23 (cont)

(c)

$$\underset{\overset{|}{\underset{\overset{||}{O}}{OCCH_3}}}{CH_3CHCH_2CH_2CH_2CH} \overset{O}{\Vert} \xleftarrow[\substack{\text{cold} \\ \text{short time}}]{H_3O^+} \underset{\overset{|}{\underset{\overset{||}{O}}{OCCH_3}}}{CH_3CHCH_2CH_2CH_2CHOCH_3} \underset{OCH_3}{|} \xleftarrow[\text{pyridine}]{(CH_3C)_2O \ \overset{O}{\Vert}}$$

$$\underset{OH}{CH_3CHCH_2CH_2CH_2CHOCH_3} \underset{OCH_3}{|}$$

(d)

$$\underset{OH}{\overset{CH_2OH}{\text{cyclohexene ring}}} \xleftarrow[\substack{\text{diethyl} \\ \text{ether}}]{H_2O \quad LiAlH_4} \text{spiro lactone}$$

(e)

$$\underset{\overset{||}{O}}{\overset{O}{\Vert}\text{cyclopentanone}=CH_2} \xleftarrow[\Delta]{\text{base}} \underset{\overset{||}{O}}{\overset{O}{\Vert}\text{cyclopentanone}-CH_2Br} \xleftarrow[\text{pyridine}]{PBr_3}$$

$$\underset{\overset{||}{O}}{\overset{O}{\Vert}\text{cyclopentanone}-CH_2OH \ CNH_2} \xleftarrow[\substack{HCl \\ 40\ ^\circ C \\ \text{short time}}]{H_2O} \text{ketal-cyclopentane } CH_2O\text{-cyclohexyl}, \ C\equiv N$$

(f)

$$\underset{OH}{CH_3(CH_2)_7CHC}\equiv C\overset{O}{\overset{\Vert}{C}}OCH_2CH_3 \xleftarrow[H_2O]{NH_4Cl} \xleftarrow{} CH_3(CH_2)_7\overset{O}{\overset{\Vert}{C}}H$$

$$Na^+ \ ^-C\equiv C\overset{O}{\overset{\Vert}{C}}OCH_2CH_3 \xleftarrow[NH_3\ (liq)]{NaNH_2} HC\equiv C\overset{O}{\overset{\Vert}{C}}OCH_2CH_3$$

15.23 (cont)

(g)

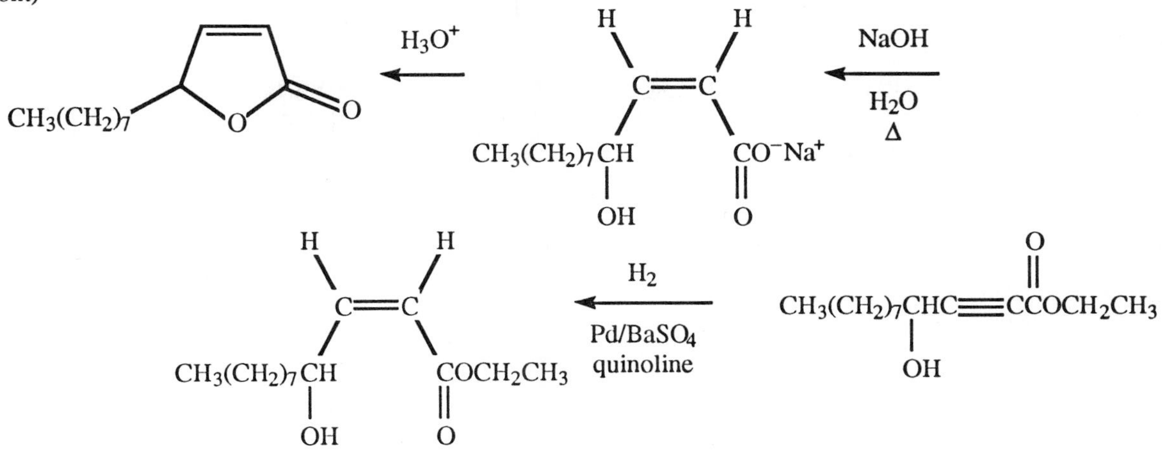

15.24

(a)

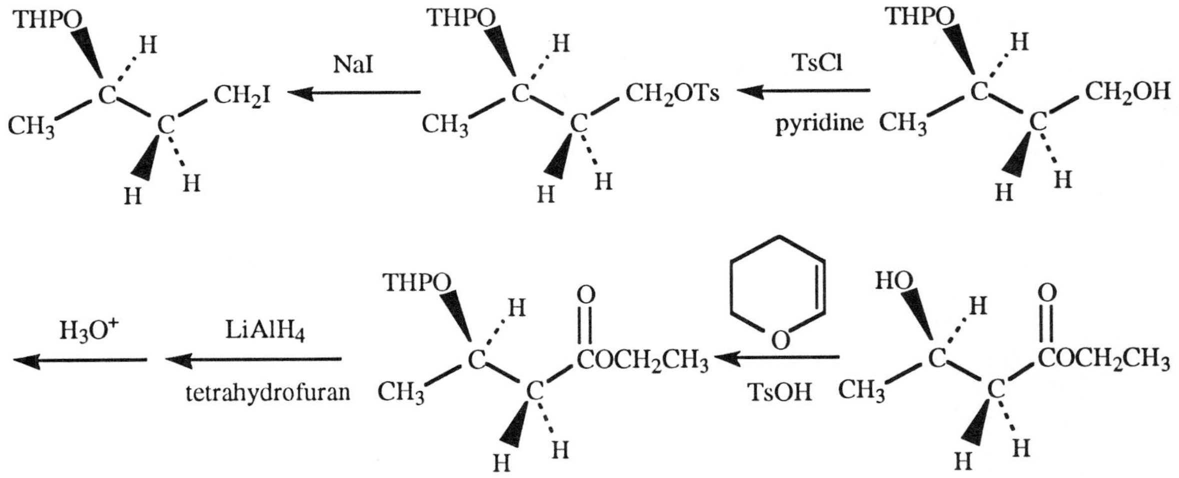

(b)

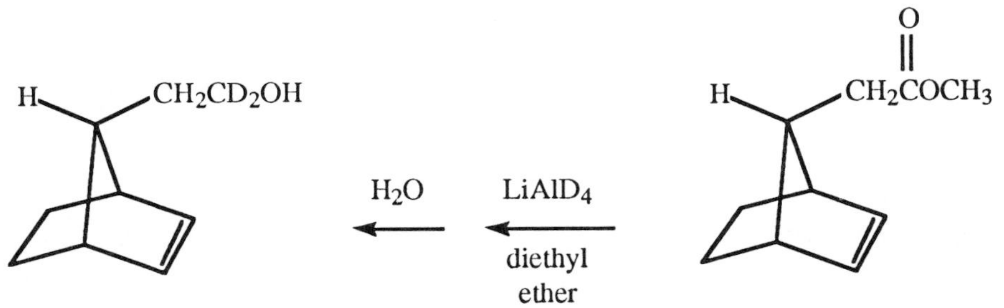

(c)

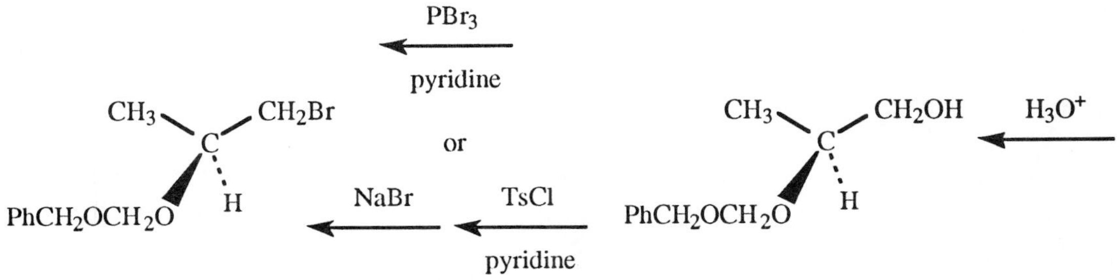

15.24 (c) (cont)

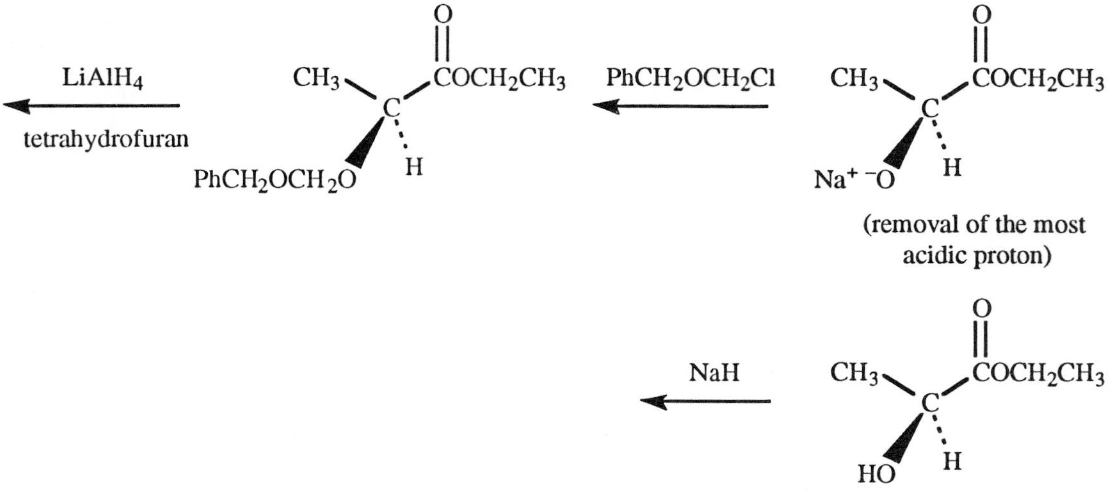

(removal of the most acidic proton)

(d)

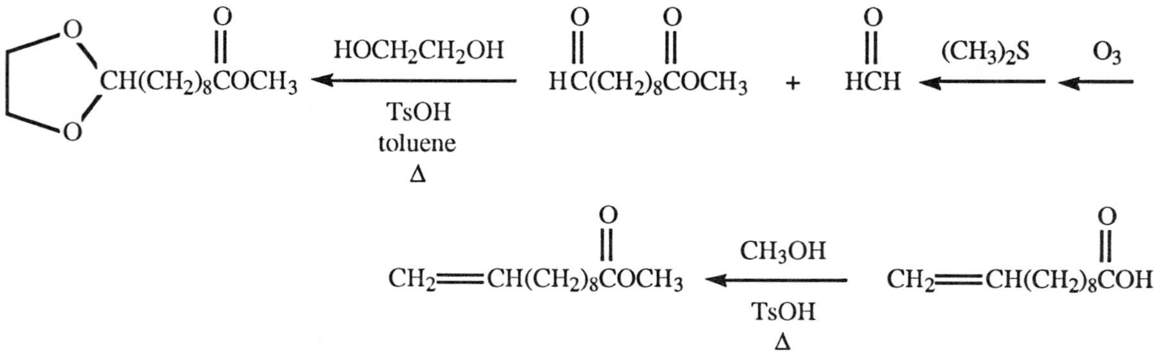

(e)

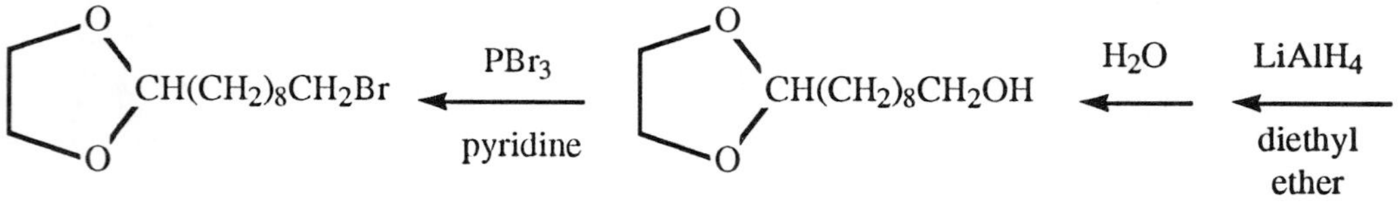

(f)

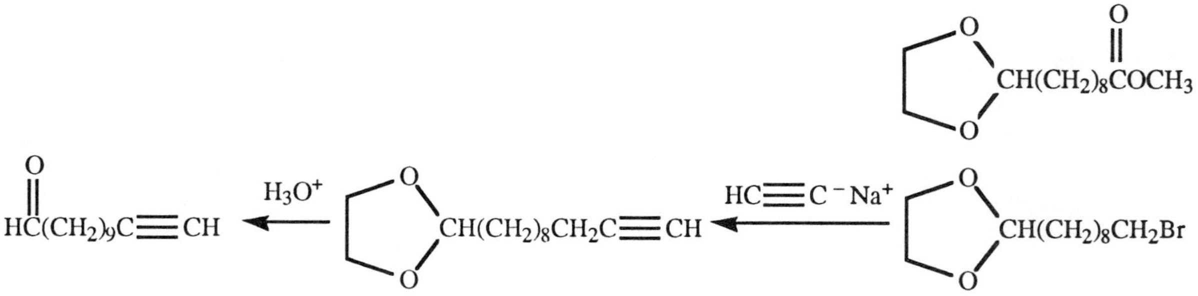

(g)

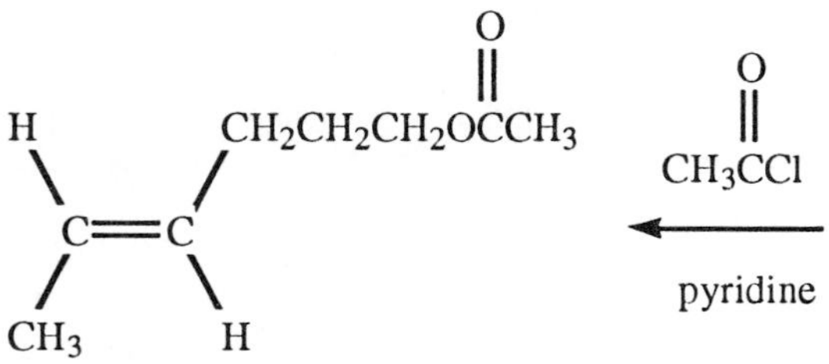

15.24 (g) (cont)

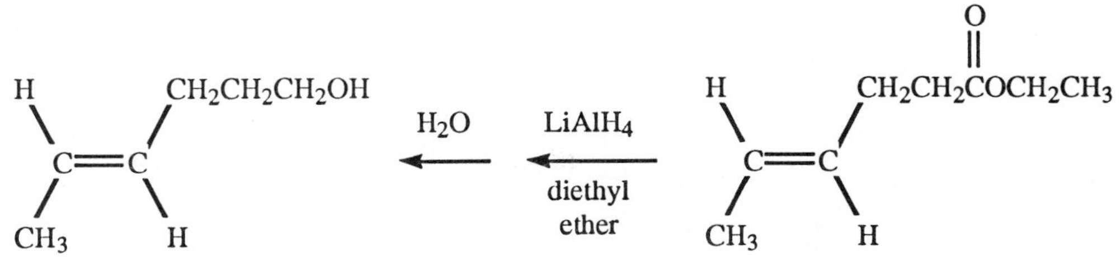

15.25

First reaction:

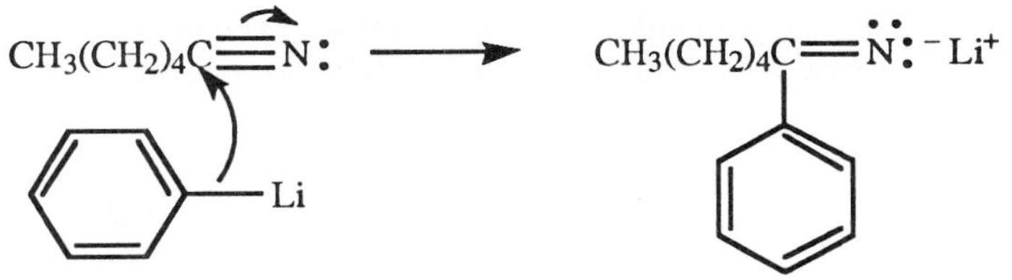

Second reaction:

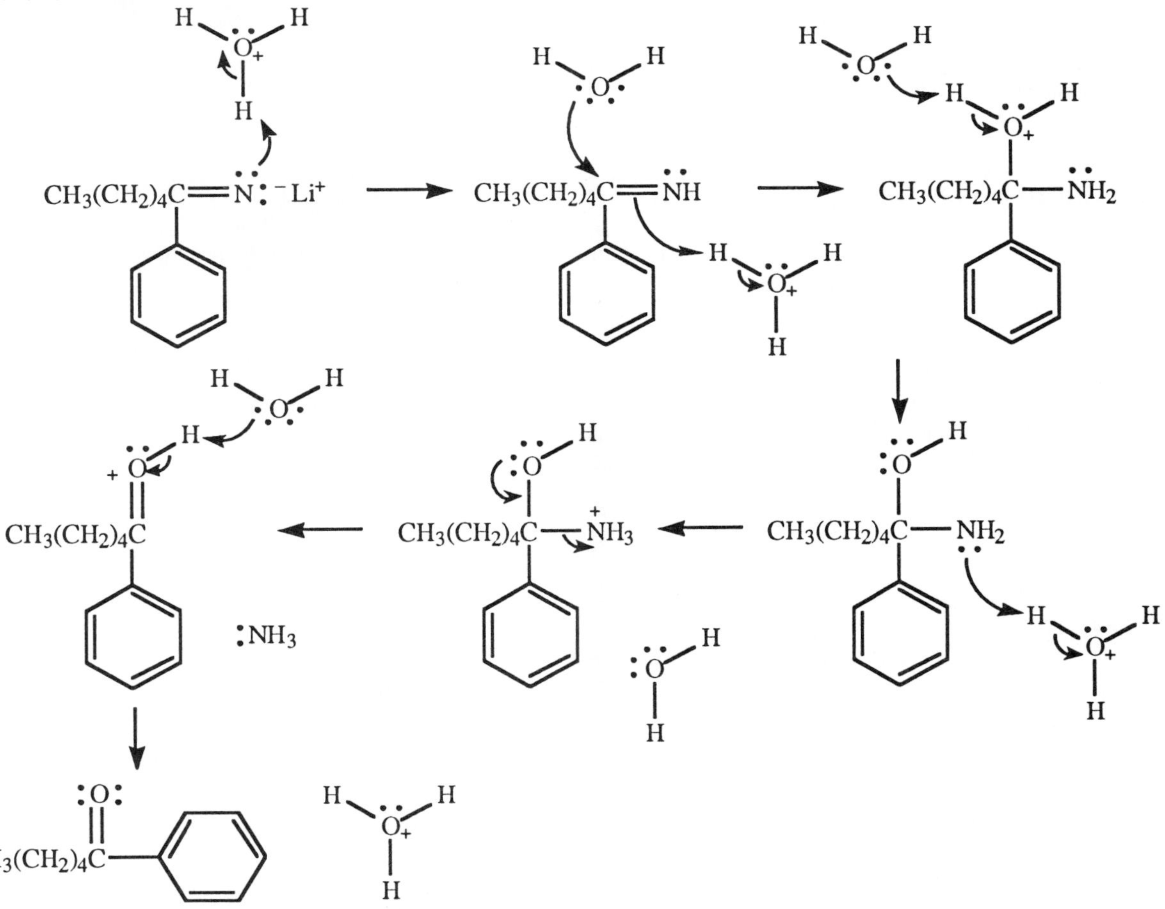

15.26

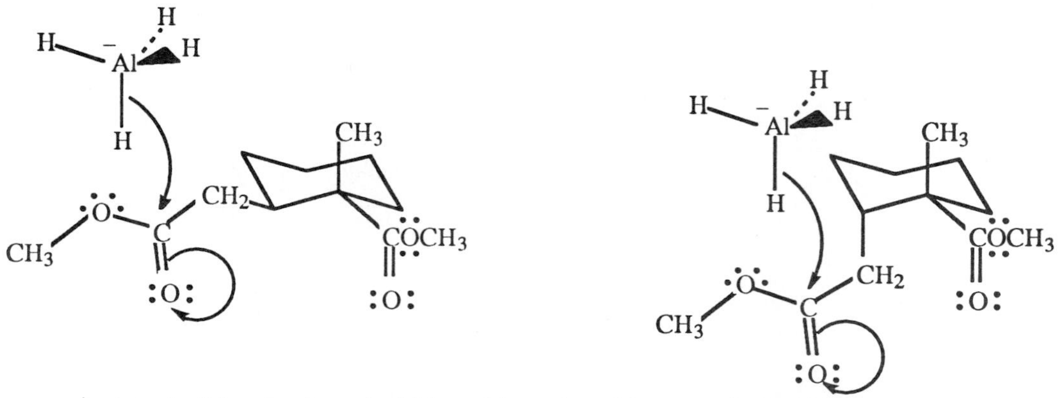

Approach of the aluminum hydride to this carbonyl is not sterically hindered for either the cis or trans isomer. Most of the reaction goes by this pathway.

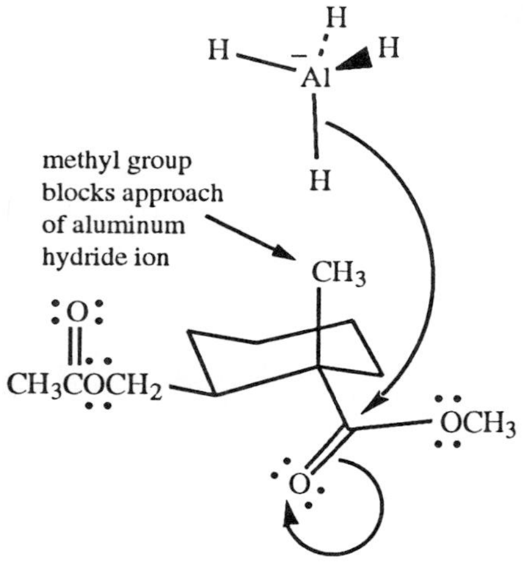

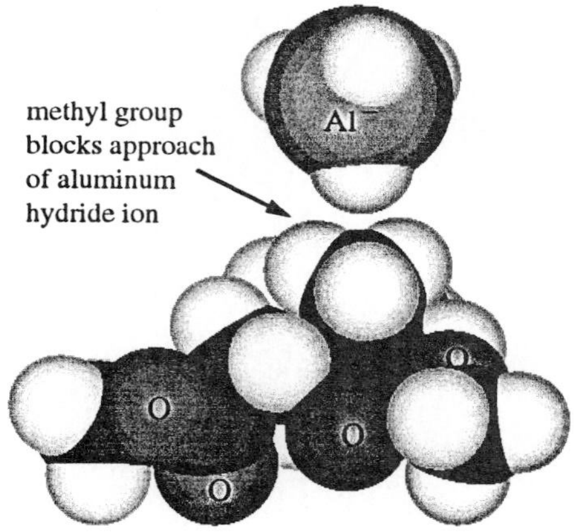

methyl group blocks approach of aluminum hydride ion

methyl group blocks approach of aluminum hydride ion

Approach of the aluminum hydride is sterically hindered for both the cis and trans isomers. This reaction has a higher activation energy than the reaction at the other carbonyl group and, therefore, a slower rate.

15.27

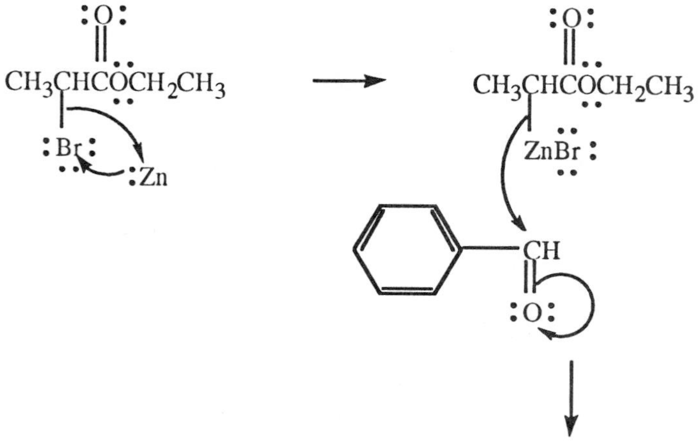

15.27 (cont)

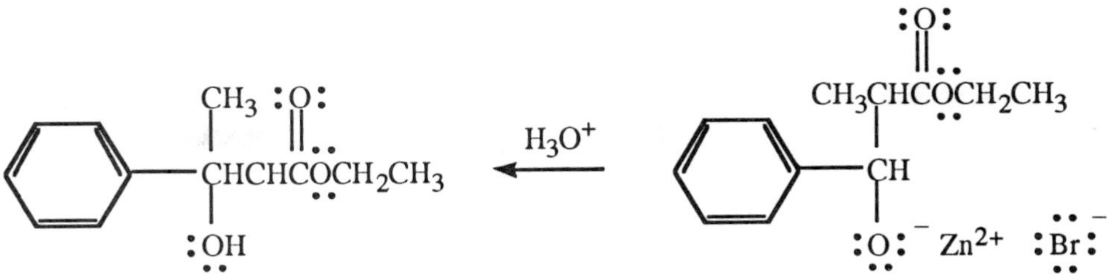

Note the similarity of this reaction to reactions with Grignard and organolithium reagents. The organozinc reagent reacts selectively with the carbonyl group of the aldehyde function in the presence of the ester group.

15.28

(a)

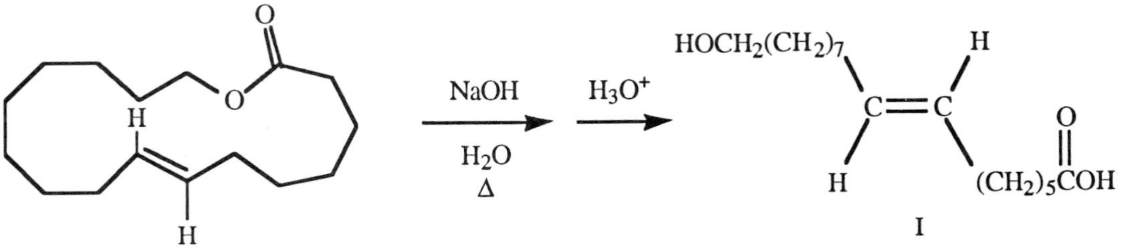

musk ambrette

(b)

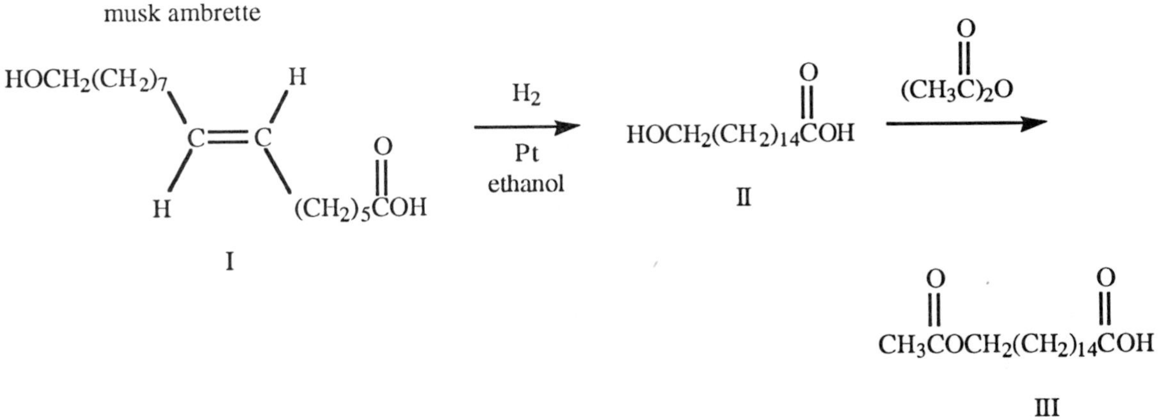

(c)

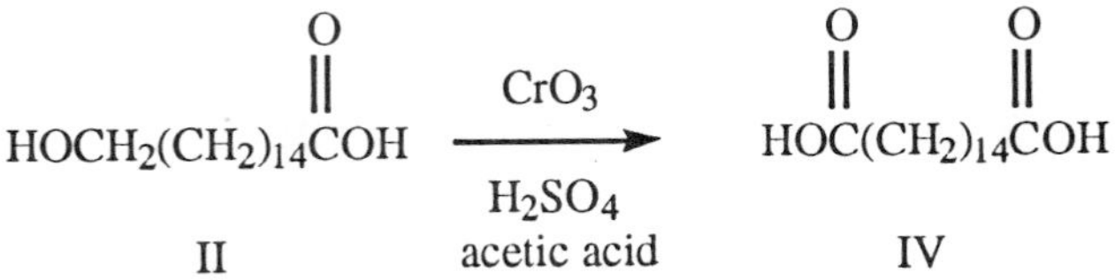

(d)

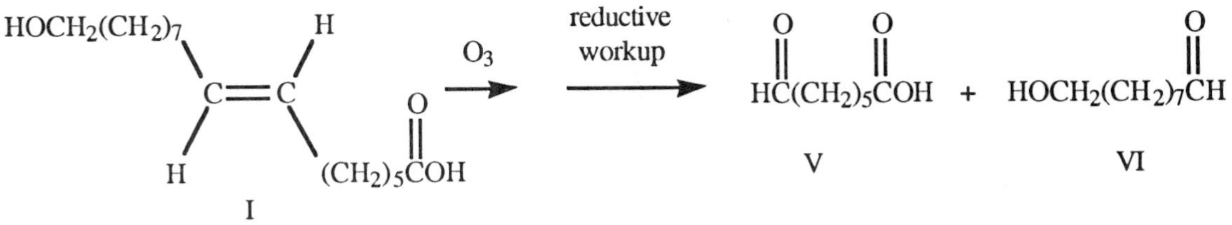

(e)

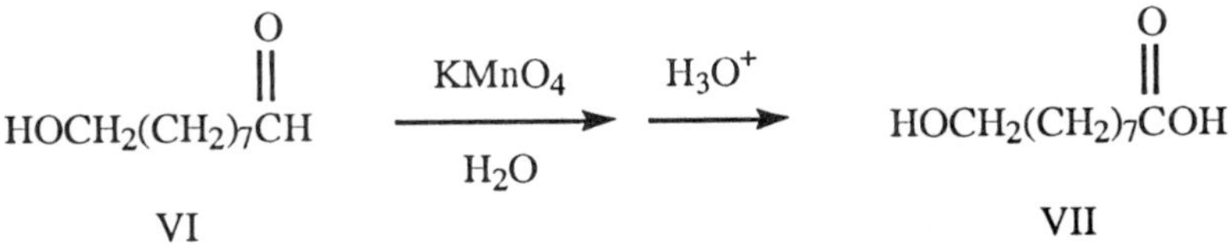

15.28 (cont)

(f) $\underset{\text{VII}}{HOCH_2(CH_2)_7COH}$ $\xrightarrow[\text{H}_2\text{SO}_4]{\text{CrO}_3}$ $\underset{\text{VIII}}{HOC(CH_2)_7COH}$

(g) $\underset{\text{V}}{HC(CH_2)_5COH}$ $\xrightarrow[\text{Na}_2\text{CO}_3]{\overset{+}{\text{HONH}_3}\ \text{Cl}^-}$ $\underset{\text{IX}}{HON{=\!=\!=}CH(CH_2)_5COH}$

(h) $\underset{\text{IX}}{HON{=\!=\!=}CH(CH_2)_5COH}$ $\xrightarrow[\substack{\Delta\\ \text{dehydrating}\\ \text{agent}}]{(CH_3C)_2O}$ $\underset{\text{X}}{N{\equiv}C(CH_2)_5COH}$

(i) $\underset{\text{X}}{N{\equiv}C(CH_2)_5COH}$ $\xrightarrow[\substack{\text{H}_2\text{O}\\ \Delta}]{\text{NaOH}}$ $\xrightarrow{\text{H}_3\text{O}^+}$ $\underset{\text{heptanedioic acid}}{HOC(CH_2)_5COH}$

15.29 $HOCH_2(CH_2)_4COCH_2CH_3$ $\xrightarrow[\text{dichloromethane}]{\text{pyridinium chlorochromate}}$ $\underset{\text{A}}{HC(CH_2)_4COCH_2CH_3}$ $\xrightarrow{CH_2{=\!=}CHMgBr}$

$\xrightarrow[\text{H}_2\text{O}]{\text{NH}_4\text{Cl}}$ $\underset{\text{B}}{\underset{\overset{|}{OH}}{CH_2{=\!=}CHCH}(CH_2)_4COCH_2CH_3}$ $\xrightarrow[\substack{\text{H}_2\text{O}\\ \text{ethanol}}]{\text{KOH}}$ $\underset{\text{C}}{\underset{\overset{|}{OH}}{CH_2{=\!=}CHCH}(CH_2)_4CO^-K^+}$ $\xrightarrow{\text{H}_3\text{O}^+}$

$\underset{\text{D}}{\underset{\overset{|}{OH}}{CH_2{=\!=}CHCH}(CH_2)_4COH}$ $\xrightarrow[\text{pyridine}]{(CH_3C)_2O}$ $\underset{\text{E}}{\underset{\overset{|}{\underset{\overset{||}{O}}{OCCH_3}}}{CH_2{=\!=}CHCH}(CH_2)_4COH}$

The carbonyl group of the aldehyde is less stabilized by resonance than the carbonyl group of the ester and, therefore, is more reactive. (See Section 14.1A.)

15.30

(a)

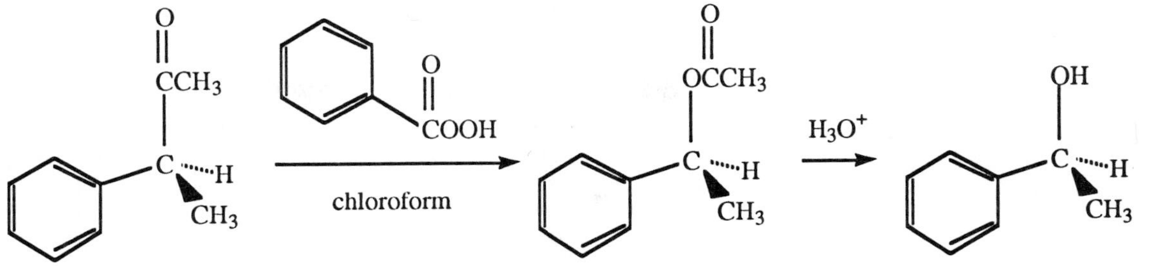

(S)-(+)-3-phenyl-2-butanone   (S)-(–)-1-phenylethanol

The product has the same relative configuration as the starting material. Hydrolysis of an ester proceeds with retention of configuration since the bond to the stereocenter is not broken. Migration of the alkyl group to the oxygen atom in the Baeyer-Villiger reaction must, therefore, also proceed with retention of configuration.

(b)

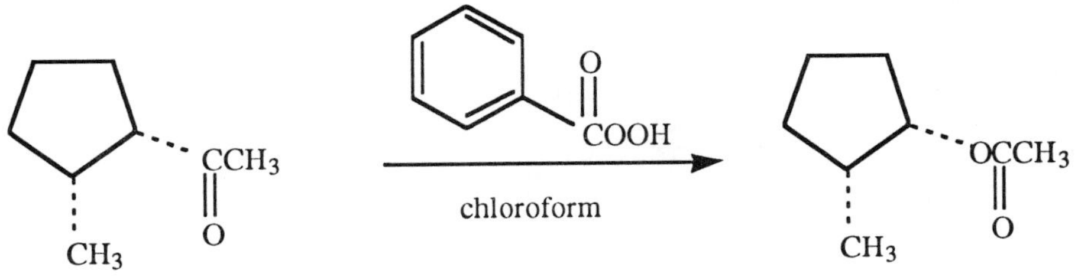

and enantiomer   and enantiomer

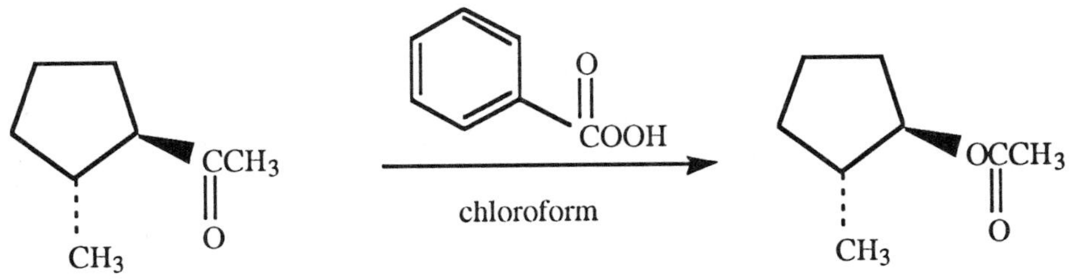

and enantiomer   and enantiomer

15.31

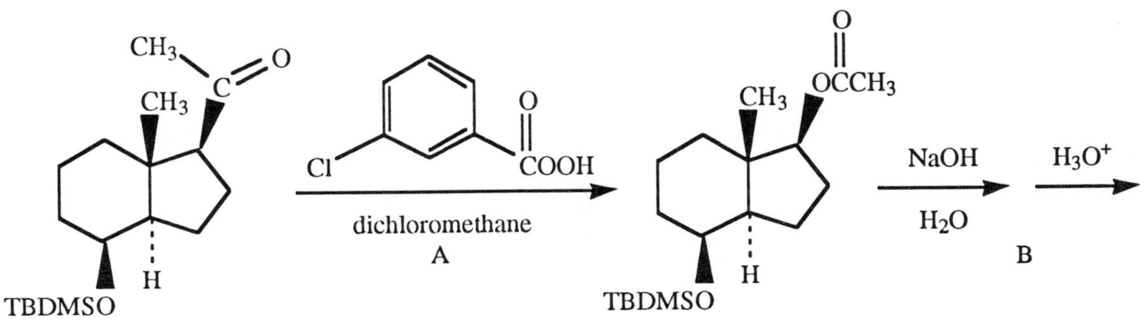

15.31 (cont)

15.32

(a) $CH_2\!=\!CH(CH_2)_9CH_2OH$ $\xrightarrow[\text{acetone}]{}$ $CH_2\!-\!CH(CH_2)_9CH_2OH$

with *m*-chlorobenzoic acid (3-Cl-C$_6$H$_4$-COOH), forming epoxide **A**

$\xrightarrow[\text{tetrahydrofuran}]{CH_3COH,\ LiCl}$ $ClCH_2CH(CH_2)_9CH_2OH$ $\xrightarrow[\substack{H_2SO_4,\ H_2O\\ \text{acetone}}]{CrO_3\ \text{(excess)}}$ $ClCH_2C(CH_2)_9COH$

                                                   B   **C**
                                                                                                        α-chloroketoacid

(b) The product oxirane and *m*-chlorobenzoic acid are both carboxylic acids; they cannot be separated from each other by washing one away from the other with a solution of sodium bicarbonate. Both have similar pK_as; both dissolve in base, and both reprecipitate on acidification. When the products are an alcohol and an acid, separation by washing the acid out with dilute base is possible.

15.33

15.33 (cont)

$$\xrightarrow{\text{H}_2\text{O}}$$

B

$$\xrightarrow[\substack{\text{pyridine} \\ \text{dichloromethane}}]{\text{TsCl}}$$

C

$$\xrightarrow[\substack{\text{tetrahydrofuran} \\ \text{cold} \\ \text{(hydrolysis of the} \\ \text{acetal; removal} \\ \text{of the protecting} \\ \text{group)}}]{\text{H}_3\text{O}^+}$$

D

$$\xrightarrow[\substack{\text{(formation of an} \\ \text{alkoxide and an} \\ \text{intramolecular S}_N2 \\ \text{reaction)}}]{\text{KOH}}$$

E

15.34 $\text{HO(CH}_2)_8\text{OH}$ $\xrightarrow[\text{H}_2\text{O}]{\text{HBr (48\%) (1 eq)}}$ $\text{Br(CH}_2)_8\text{OH}$ $\xrightarrow[\substack{\text{H}_2\text{O} \\ \text{H}_2\text{SO}_4 \\ \text{acetone}}]{\text{CrO}_3}$ $\text{Br(CH}_2)_7\overset{\displaystyle O}{\overset{\|}{\text{C}}}\text{OH}$

A   B

$$\xrightarrow[\substack{\text{NH}_3 \text{ (liq)} \\ \text{tetrahydrofuran}}]{\substack{\text{Li}^+ \text{ NH}_2^- \\ \text{(2 equiv)}}}$$

C

C   + $\text{Br(CH}_2)_7\overset{\displaystyle O}{\overset{\|}{\text{C}}}\text{OH}$ $\longrightarrow$

B

D $\xrightarrow[\text{cold}]{\text{H}_3\text{O}^+}$

15.34 (cont)

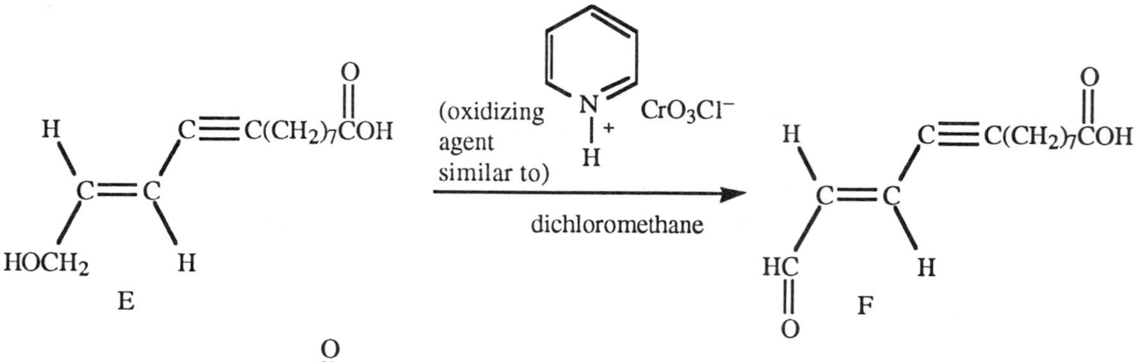

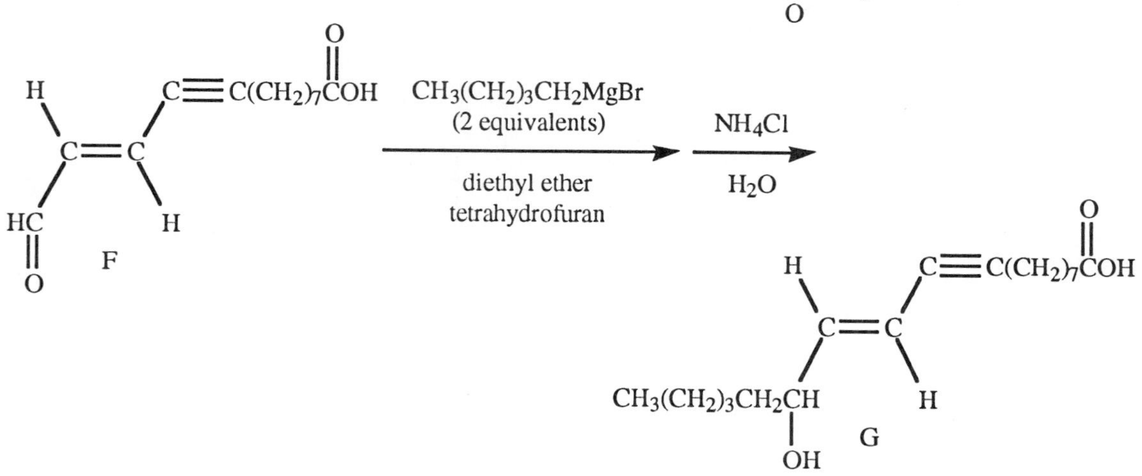

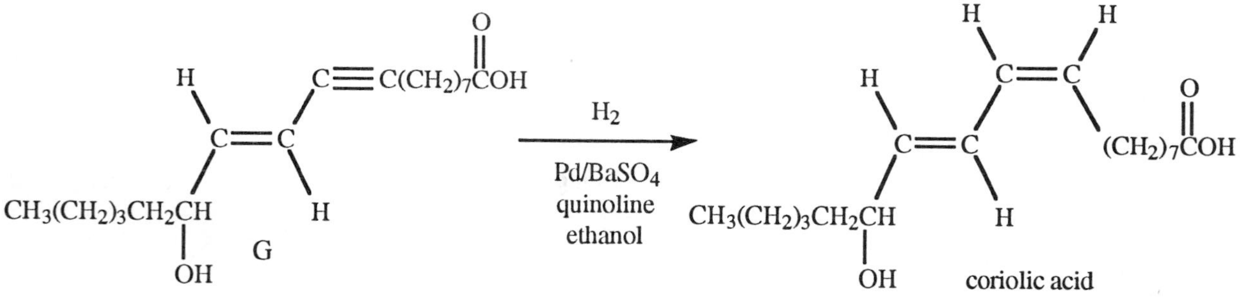

15.35

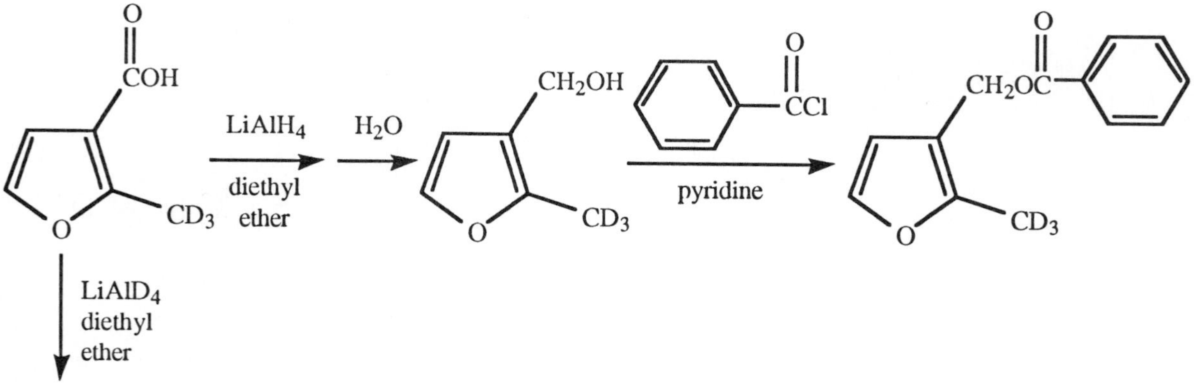

15.35 (cont)

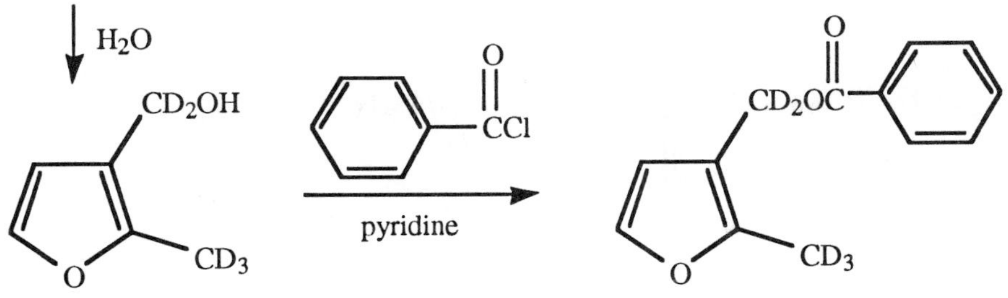

15.36

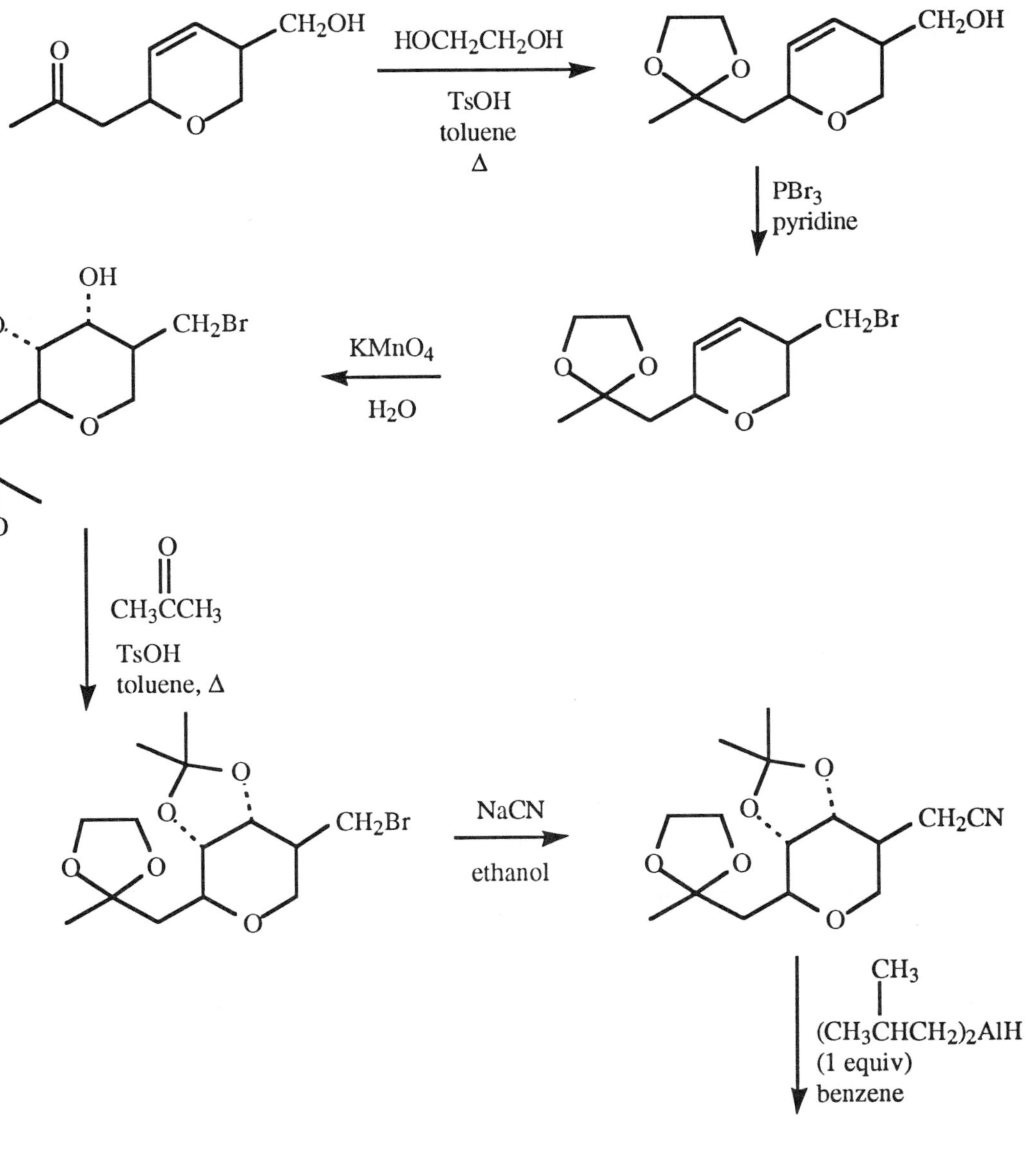

15.36 (cont)

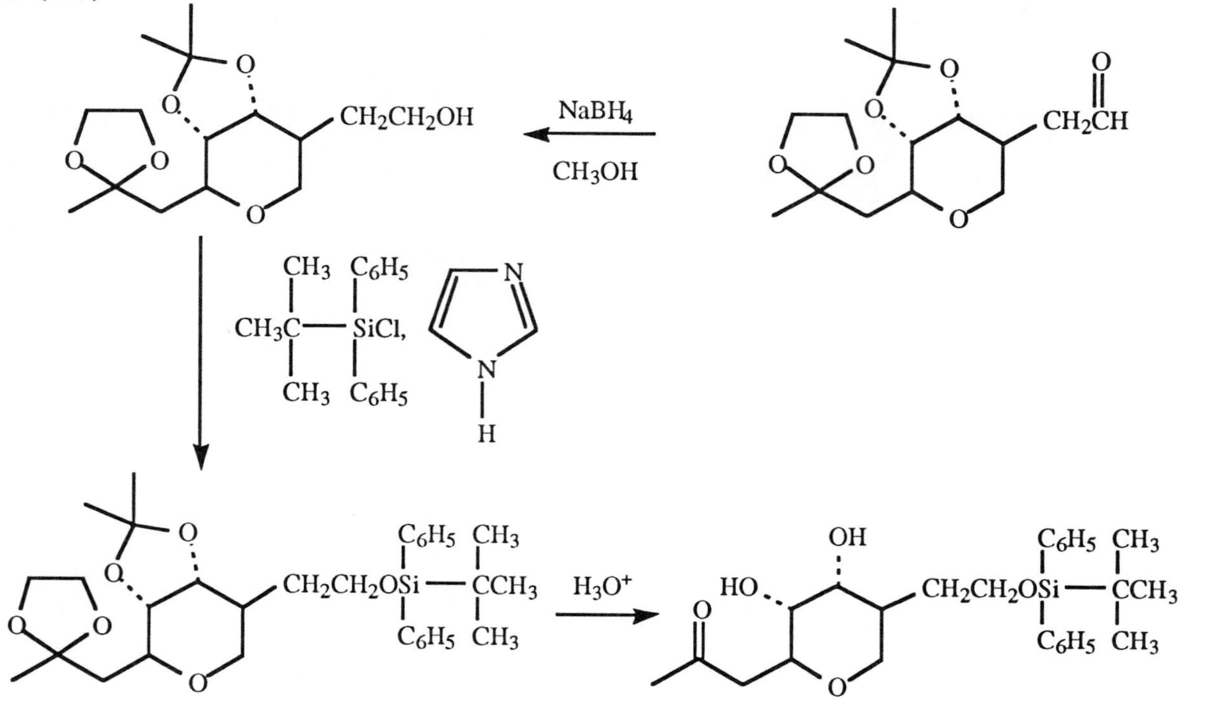

15.37

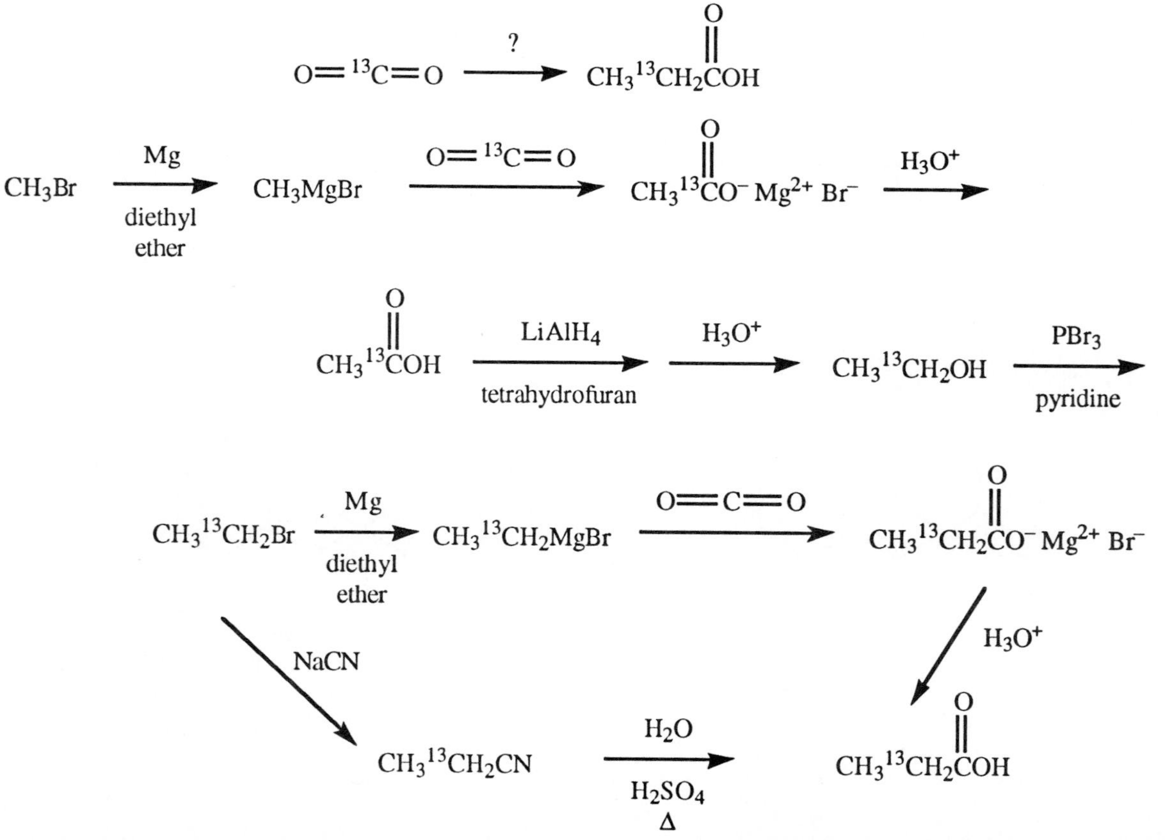

$$O{=}^{13}C{=}O \xrightarrow{\ ?\ } CH_3{}^{13}CH_2\overset{\displaystyle O}{\overset{\|}{C}}OH$$

$$CH_3Br \xrightarrow[\text{diethyl ether}]{Mg} CH_3MgBr \xrightarrow{O{=}^{13}C{=}O} CH_3{}^{13}\overset{\displaystyle O}{\overset{\|}{C}}O^-\,Mg^{2+}\,Br^- \xrightarrow{H_3O^+}$$

$$CH_3{}^{13}\overset{\displaystyle O}{\overset{\|}{C}}OH \xrightarrow[\text{tetrahydrofuran}]{LiAlH_4} \xrightarrow{H_3O^+} CH_3{}^{13}CH_2OH \xrightarrow[\text{pyridine}]{PBr_3}$$

$$CH_3{}^{13}CH_2Br \xrightarrow[\substack{\text{diethyl}\\\text{ether}}]{Mg} CH_3{}^{13}CH_2MgBr \xrightarrow{O{=}C{=}O} CH_3{}^{13}CH_2\overset{\displaystyle O}{\overset{\|}{C}}O^-\,Mg^{2+}\,Br^-$$

NaCN

$$CH_3{}^{13}CH_2CN \xrightarrow[\substack{H_2SO_4\\\Delta}]{H_2O} CH_3{}^{13}CH_2\overset{\displaystyle O}{\overset{\|}{C}}OH$$

$$\xrightarrow{H_3O^+}$$

15.38

15.39   Let RN——H stand for the amine.

15.40

(a)

this carbonyl group more electrophilic now than the other one

Chelation with magnesium ion, an electrophile, activates the carbonyl group of one ester function toward attack by the nucleophile.

(b)   The carbonyl group of an aldehyde is less stabilized by resonance than the carbonyl group of an ester, and therefore is more reactiive (Section 14.1A).

attack at this carbonyl
has to occur from the
bottom face of the ring

Chelation creates a ring in which one side of the carbonyl group is shielded by a large group. It is easiest to see this if you use an *sp*^{3} carbon atom for the atom representing the aldehyde carbonyl group in building your chelate structure with molecular models.

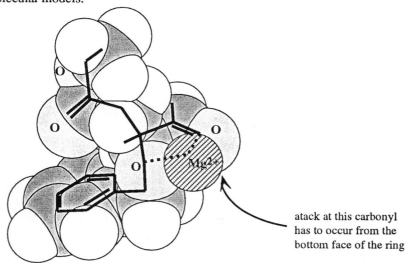

atack at this carbonyl
has to occur from the
bottom face of the ring

15.41   Compounds A and B, $C_6H_{12}O_2$, have one unit of unsaturation. The carbonyl group of an ester function uses up one unit of unsaturation. Therefore, Compounds A and B cannot have any rings or other π bonds.

Analysis of the proton magnetic spectrum of Compound A is shown below:

δ 0.9, 1.5, and	(8H, consisting of two overlapping triplets, (2) C$\underline{H}_3$CH$_2$—,
δ 1.2-1.8	and a multiplet, —C$\underline{H}_2$—)
δ 2.1	(2H, a distorted triplet, —CH$_2$C$\underline{H}_2$C=O)
δ 4.1	(2H, quartet, CH$_3$C$\underline{H}_2$—O—)

Compound A is ethyl butanoate (ethyl butyrate).

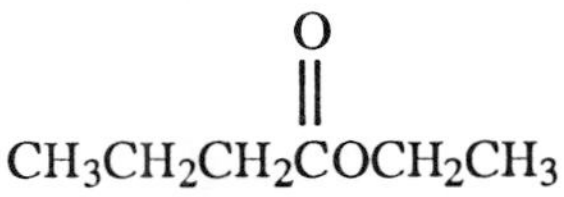

$$\begin{array}{c} O \\ \| \\ CH_3CH_2CH_2COCH_2CH_3 \end{array}$$

There are only two singlets in the spectrum of Compound B. The band at δ 1.1 is three times as intense as the one at δ 3.5. Since there are twelve hydrogen atoms in the molecular formula, the band at δ 1.1 must represent nine hydrogen atoms and that at δ 3.5 must represent three hydrogen atoms. Analysis of the proton magnetic spectrum of Compound B is shown below:

δ 1.1	(9H, singlet, (C$\underline{H}_3$)$_3$C—)
δ 3.5	(3H, singlet, C$\underline{H}_3$—O—)

Compound B is methyl 2,2-dimethylpropanoate.

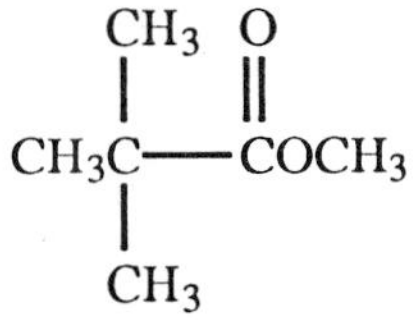

$$\begin{array}{c} CH_3 \quad O \\ | \qquad \| \\ CH_3C\!-\!\!-\!COCH_3 \\ | \\ CH_3 \end{array}$$

15.42   Compound C, $C_9H_{10}O_3$, has five units of unsaturation. The infrared spectrum shows the presence of a carboxylic acid (3300-2600 cm^{-1} for the O—H and 1719 cm^{-1} for the C=O stretching frequencies).

Analysis of the proton magnetic spectrum of Compound C is shown below:

δ 3.5	(2H, singlet, —C$\underline{H}_2$—)	
δ 3.7	(3H, singlet, C$\underline{H}_3$—O—)	
δ 6.7-7.3	(4H, multiplet, Ar$\underline{H}$)	(This is the splitting pattern for a para substituted benzene ring)
δ 11.6	(1H, singlet, O=CO$\underline{H}$)	

Compound C is (4-methoxyphenyl)acetic acid

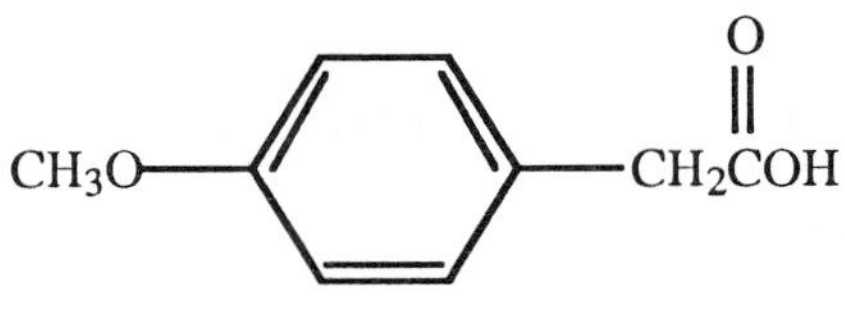

15.43   Compound D has four bands with integrations of 3:2:3:2 going from lower to higher δ values.

Analysis of the proton magnetic spectrum of Compound D is shown below:

δ 1.4	(3H, triplet, *J* 7 Hz, C$\underline{H}_3$CH$_2$—)
δ 4.5	(2H, quartet, J 7 Hz, CH$_3$C$\underline{H}_2$—O—)
δ 7.6-7.9	(3H, multiplet, Ar$\underline{H}$)
δ 8.1-8.4	(2H, multiplet, Ar$\underline{H}$)

Compound D is ethyl benzoate.

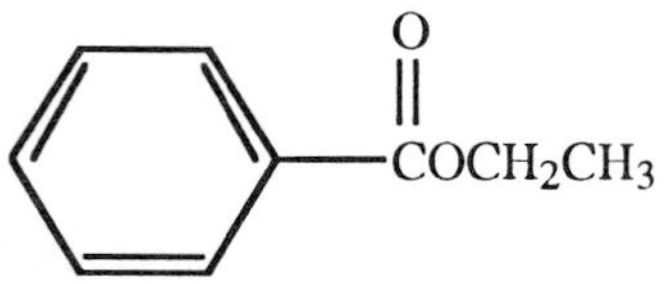

The two hydrogen atoms on carbons 2 and 6 of the aromatic ring are especially deshielded by the carbonyl group of the ester.

Compound E has three bands with integrations of 3:2:3.

Analysis of the proton magnetic spectrum of Compound E is shown below:

δ 1.1	(3H, triplet, *J* 8 Hz, C$\underline{H}_3$CH$_2$—)
δ 2.3	(2H, quartet, J 8 Hz, CH$_3$C$\underline{H}_2$C=O)
δ 3.7	(3H, singlet, C$\underline{H}_3$O—)

Compound E is methyl propanoate (methyl propionate).

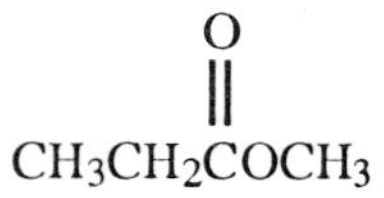

$$\overset{\displaystyle O}{\overset{\displaystyle \|}{CH_3CH_2COCH_3}}$$

15.44   Compound F, C$_5$H$_8$O$_2$, has 2 units of unsaturation. The infrared spectrum shows the presence of an alkene (1635 cm^{-1} for the C=C stretching frequency and 989 cm^{-1} for a C=C—H bending frequency), and an ester (1731 cm^{-1} for the C=O and 1279, 1207, and 1069 cm^{-1} for the C—O stretching frequencies).

Analysis of the proton magnetic spectrum of Compound F is shown below:

δ 1.8	(3H, multiplet, C$\underline{H}_3$C=CH)
δ 3.5	(3H, singlet, C$\underline{H}_3$O—)
δ 5.2	(1H, multiplet, C=C$\underline{H}$)
δ 5.7	(1H, multiplet, C=C$\underline{H}$)

Compound F is methyl 2-methylpropenoate (methyl methacrylate).

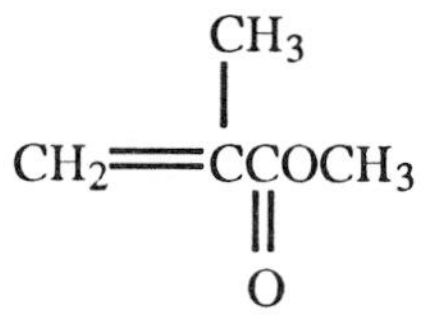

$$\begin{array}{c} CH_3 \\ | \\ CH_2{=\!=}CCOCH_3 \\ \| \\ O \end{array}$$

The allylic and vinyl hydrogen atoms are coupled to each other through the double bond.

15.45 Compound G, $C_8H_{14}O_2$, has 2 units of unsaturation. The infrared spectrum shows the presence of an ester (1736 cm^{-1} for the C=O, and 1160 and 1032 cm^{-1} for the C—O stretching frequencies).

Integration of the proton magnetic resonance spectrum of Compound G gives a 3:2:2 ratio, which adds up to half the number of hydrogen atoms in the molecular formula. This points to symmetry in the molecule.

Analysis of the proton magnetic spectrum of Compound G is shown below:

δ 1.3	(6H, triplet, C$\underline{H}_3$CH$_2$—)
δ 2.6	(4H, singlet, —C$\underline{H}_2$C=O)
δ 4.3	(4H, quartet, CH$_3$C$\underline{H}_2$O—)

Compound G is diethyl butanedioate (diethyl succinate).

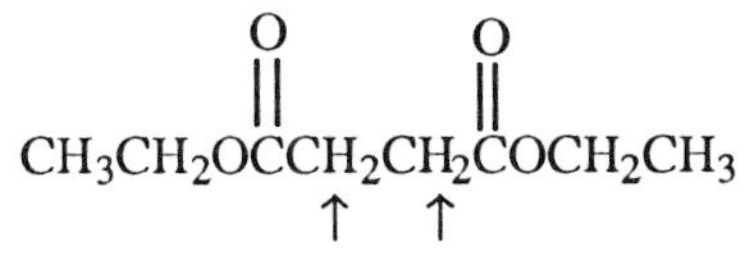

$$\underset{\uparrow\;\;\uparrow}{CH_3CH_2O\overset{\overset{O}{\|}}{C}CH_2CH_2\overset{\overset{O}{\|}}{C}OCH_2CH_3}$$

These hydrogen atoms are chemical-shift equivalent and therefore do not couple.

15.46 The structure shown has the molecular formula $C_{12}H_{20}O_2$ as established by mass spectrometry.

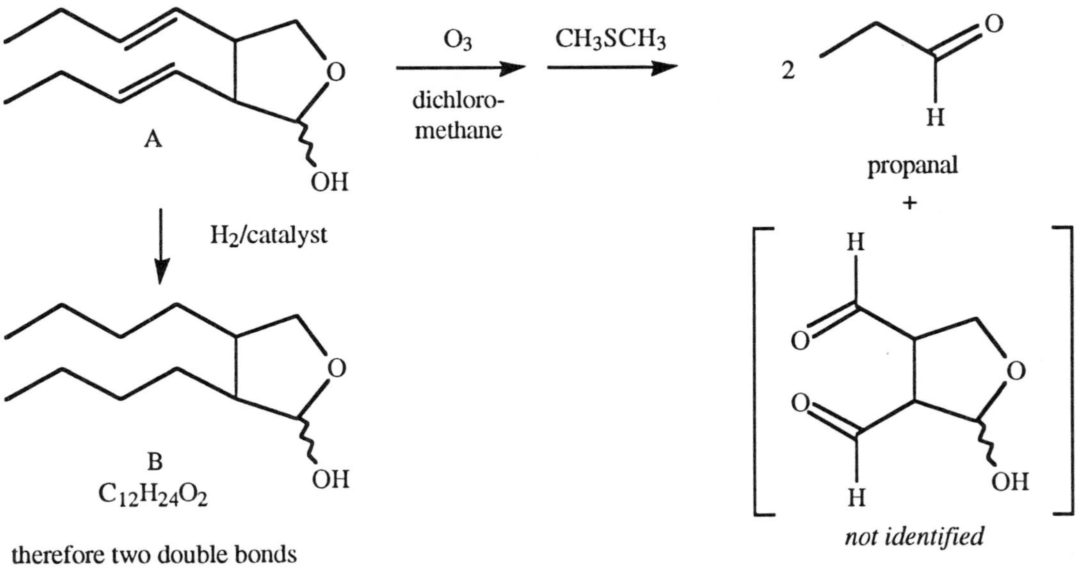

therefore two double bonds or one triple bond in A

3644 cm^{-1} suggests —OH group

1737 cm^{-1} suggests C=O in aldehyde or ketone

968 cm^{-1} suggests —C=C—H

15.46 (cont)

C=O and —OH are compatible with cyclic hemiacetal in equilibrium with open form.

15.47
(a)   Let RMgBr stand for the Grignard reagent.

(b)   Carbon-13 nuclear magnetic resonance spectrum.

The other bands must be the CH_2 groups. It is not possible to assign them exactly with the information we have.

Proton magnetic resonance spectrum.

15.47 (b) (cont)

All the other protons absorb between δ 1.0 – 2.4

(c)

Side product at higher temperatures, formed when the Grignard reagent reacts with the carbonyl group that forms in the first part of the reaction [part (a)]

15.48

(a)

δ 0.76 – 0.97 (d)

3500 cm^{-1}

CH$_3$

1720 cm^{-1}

H—O—CH$_2$—CH$_2$CHCH$_2$—CH$_2$CCH$_3$

δ 2.83 (s)

δ 1.0 – 1.8 (m)

δ 3.40 – 3.68 (t)   δ 2.26 – 2.52 (t)

δ 2.05 (s)

(b)   Compound B no longer has a carbonyl group and has a molecular weight 16 amu larger than that of Compound A; 16 amu ≡ CH$_4$, suggesting that a methyl group has added to the carbonyl group in Compound A. The structure of Compound B is probably:

δ 0.73 – 0.98 (d)   3520 cm^{-1}

3520 cm^{-1}

CH$_3$ O—H ← δ 3.08 (s)

H—O—CH$_2$—CH$_2$CHCH$_2$CH$_2$—C—CH$_3$

δ 3.08 (s)

δ 0.9 – 1.8 (m)   CH$_3$ δ 1.12 (s)

δ 3.50 – 3.75 (t)

(c)   Trimethylchlorosilane destroys any organolithium reagent remaining in the reaction mixture before water is added, liberating the carbonyl group. The fact that so much Compound B is formed in the absence of trimethylchlorosilane indicates that CH$_3$Li survives the addition of H$_2$O long enough to add to the carbonyl group newly liberated in Compound A.

electrophile

(CH$_3$)$_3$Si —— C̈l̈:   →   (CH$_3$)$_3$SiCH$_3$ +   Li$^+$:C̈l̈:$^-$

nucleophile → CH$_3$—Li

16

Enols and Enolate Anions as Nucleophiles I. Alkylation and Condensation Reactions

Workbook Exercises

The most efficient way for you to understand the subject matter in Chapter 16 is to seek out the appropriate analogies in your earlier work, as previously suggested. There are two fundamental ideas in this chapter:

1. Carbon atoms connected to carbonyl groups, as well as to other electron-withdrawing groups, can behave as nucleophiles.

carbonyl group (or some other electron-withdrawing group, such as NO_2, CN, Ph_3P)

carbon atom α to the carbonyl group; the α-carbon atom

hydrogen atom attached to the α-carbon atom; the α-hydrogen atom

The carbon atom α to a carbonyl group can behave as a nucleophile in two ways:

as the uncharged tautomer; the enol form

as the anionic conjugate base; the enolate form

2. These carbon nucleophiles undergo the characteristic reactions that you already associate with nucleophiles: substitution, addition to carbonyl groups, and acylation. As before, reviewing these concepts from earlier chapters will benefit your learning.

EXAMPLE

deprotonation

the enolate anion behaves primarily as a carbon nucleophile

532

Workbook Exercises (cont)

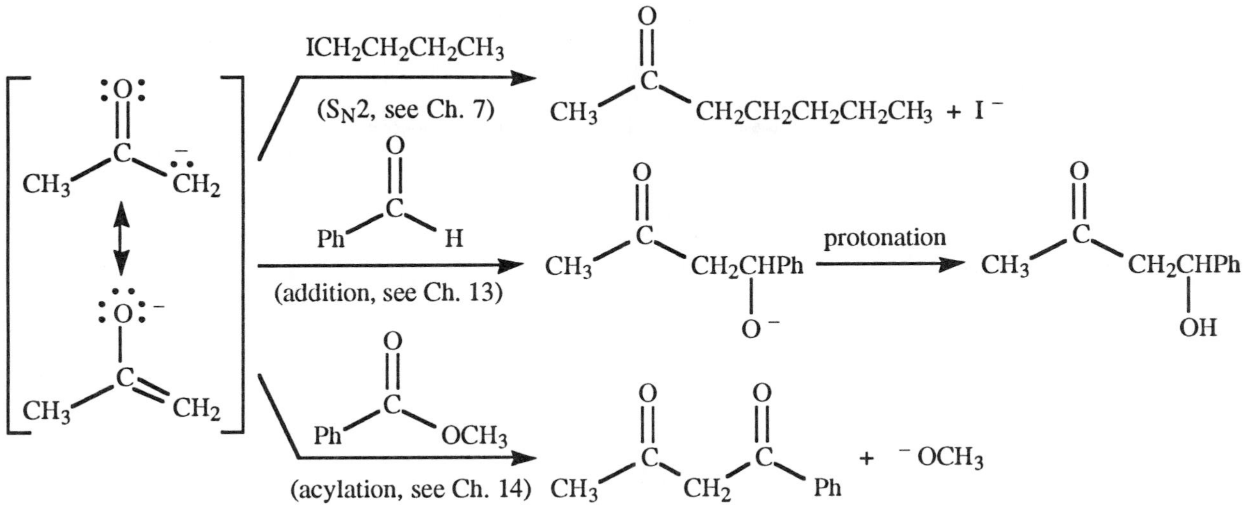

EXERCISE I.   For each of the following compounds, draw the structure of the enol and two resonance forms for the enolate anion . If there is more than one enol/enolate anion possible, show all of the possibilities.

(a)

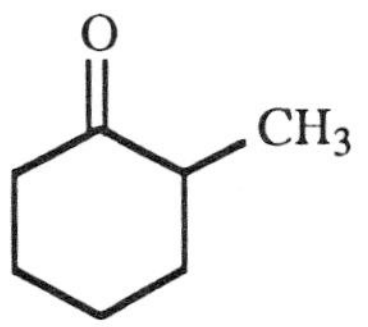

(b)

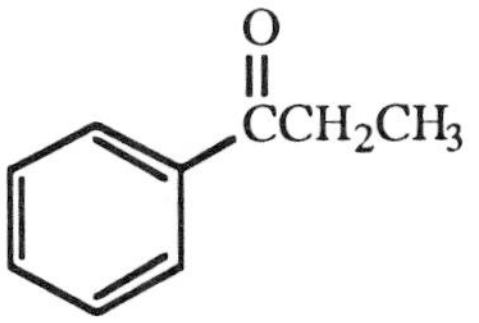

(c) $CH_3CH_2COCH_3$

(d)

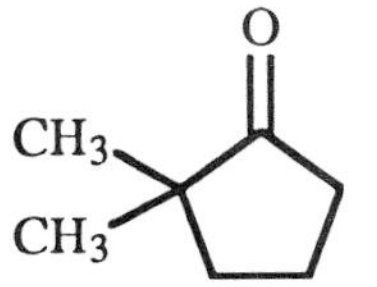

(e) $PhCH_2CH$ (with O above C)

(f) $CH_3CCH_2CCH_3$ (with two O above C)

EXERCISE II.   Use each of the enolate nucleophiles you created in Exercise I in substitution, addition, and acylation reactions with the three electrophiles shown in the example above ($CH_3CH_2CH_2CH_2I$, PhCHO, and $PhCO_2CH_3$). You have seen the mechanisms for each of these types of reactions in earlier chapters, so you should be able to draw mechanisms for the reactions you write.

16.1

1.   ethyl acetate

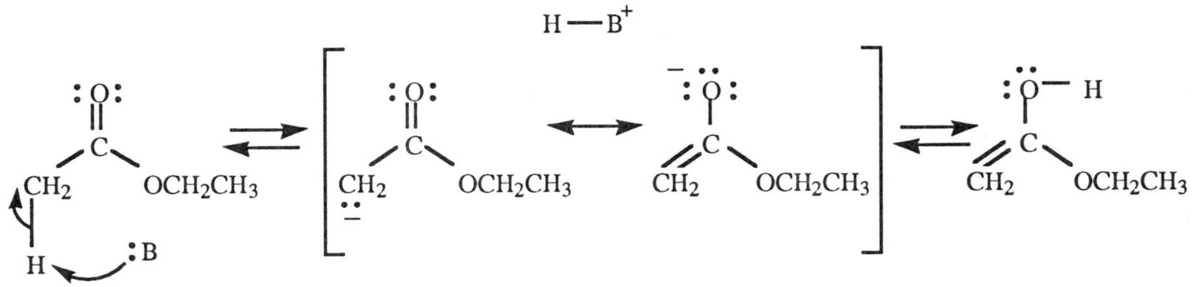

ethyl acetate
pK_a 23

16.1 (cont)

2.   ethyl acetoacetate

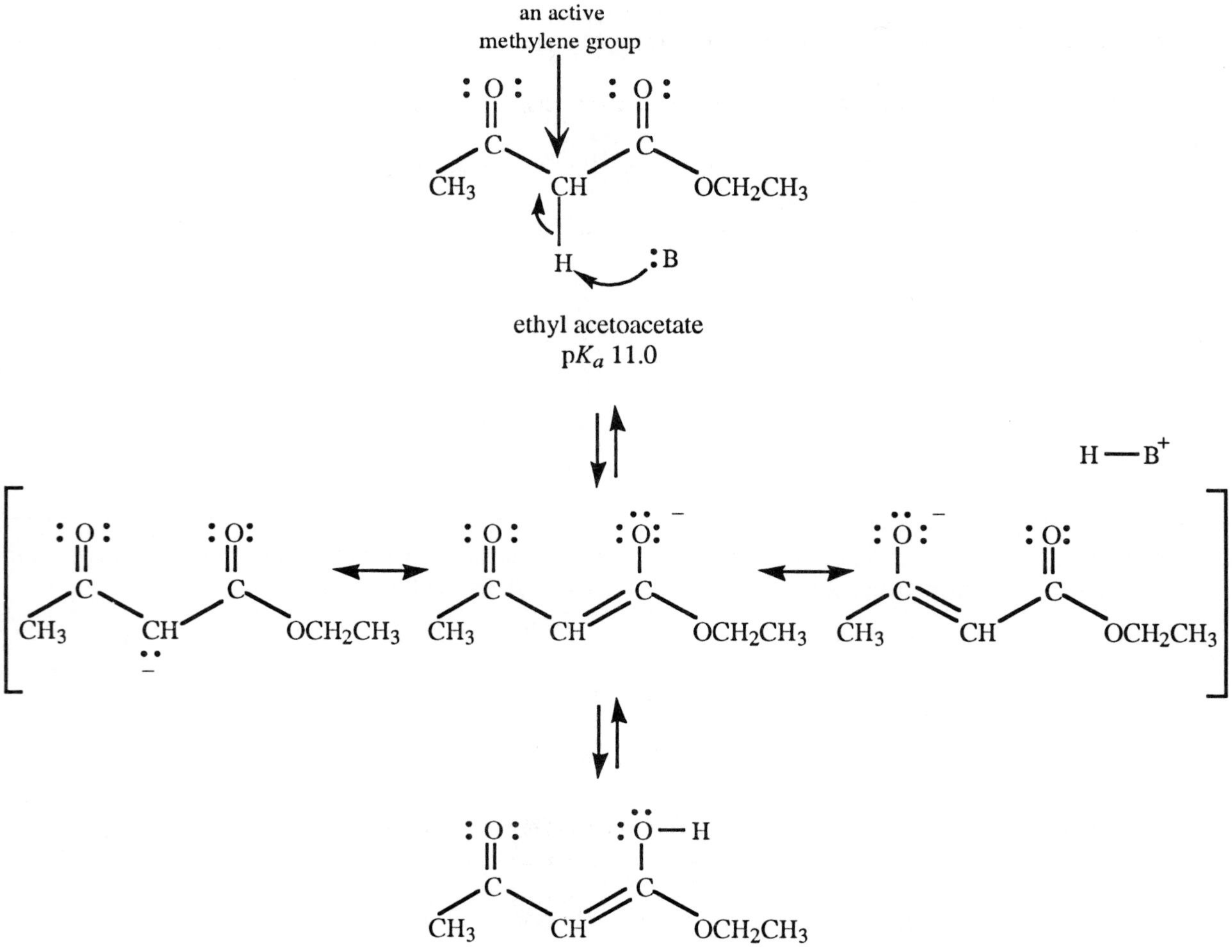

an active
methylene group

ethyl acetoacetate
pK_a 11.0

$H - B^+$

3.   ethyl nitroacetate

an active
methylene group

ethyl nitroacetate
pK_a 5.8

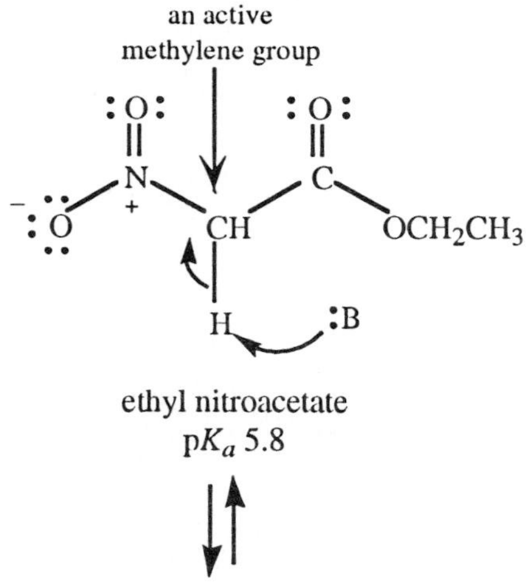

16.1 (3) (cont)

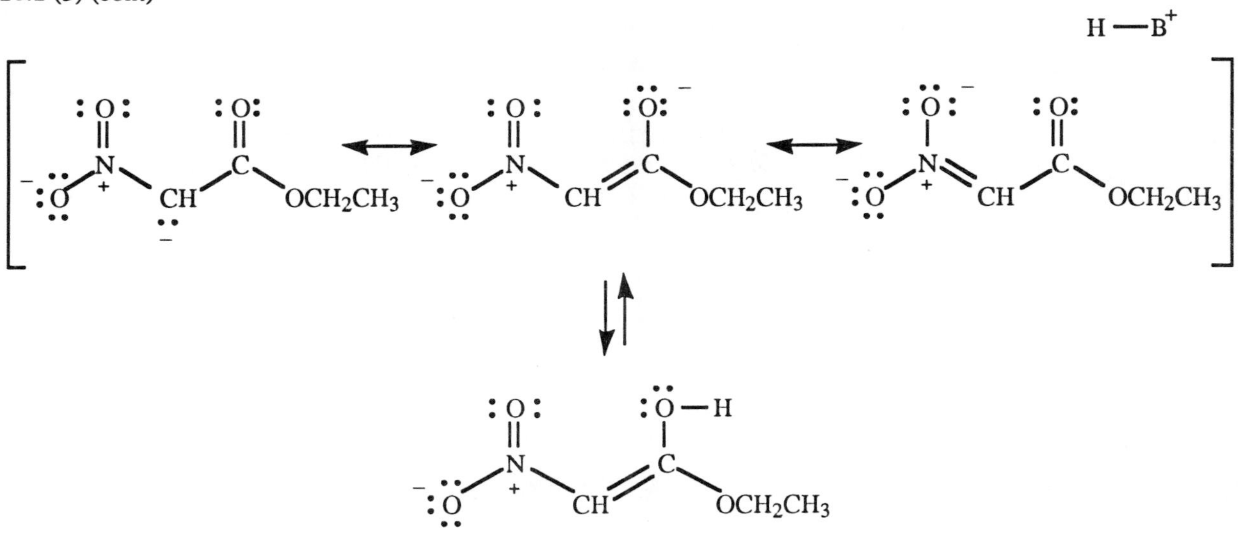

H—B$^+$

$$
\left[
\begin{array}{ccc}
\text{(resonance structure 1)} & \longleftrightarrow & \text{(resonance structure 2)} & \longleftrightarrow & \text{(resonance structure 3)}
\end{array}
\right]
$$

$\updownarrow$

(enol structure)

4.   ethyl cyanoacetate

an active
methylene group

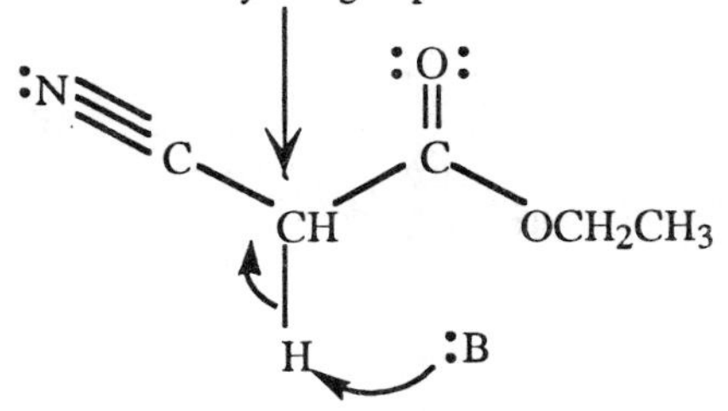

ethyl cyanoacetate
$pK_a > 9$
(cannot be determined accurately
because the compound decomposes)

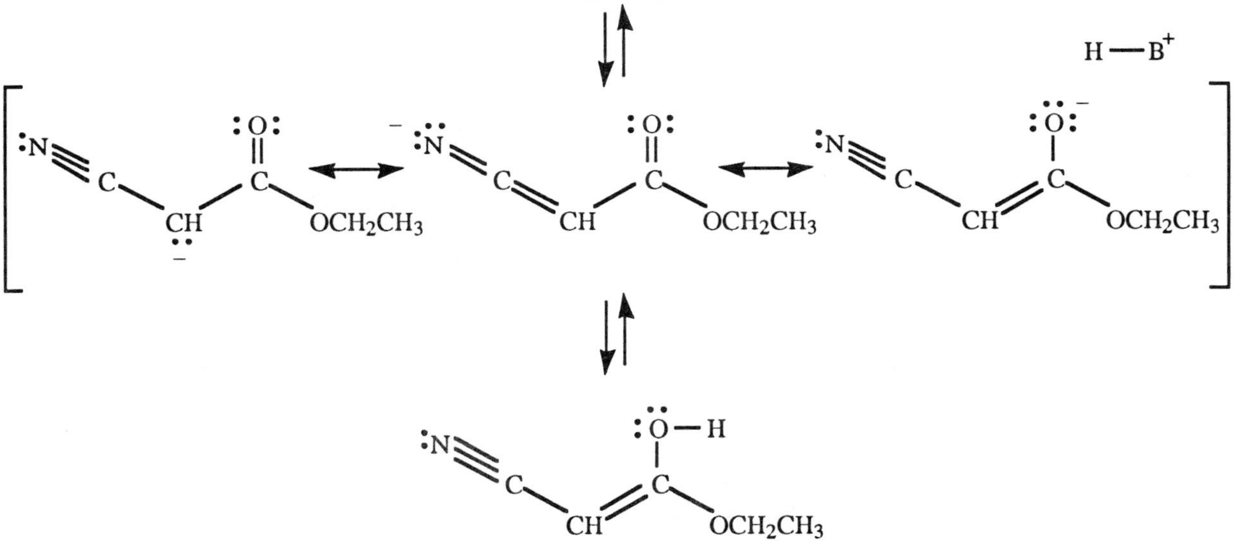

$\updownarrow$

H—B$^+$

$$
\left[
\begin{array}{ccc}
\text{(resonance structure 1)} & \longleftrightarrow & \text{(resonance structure 2)} & \longleftrightarrow & \text{(resonance structure 3)}
\end{array}
\right]
$$

$\updownarrow$

(enol structure)

Concept Map 16.1   Enolization.

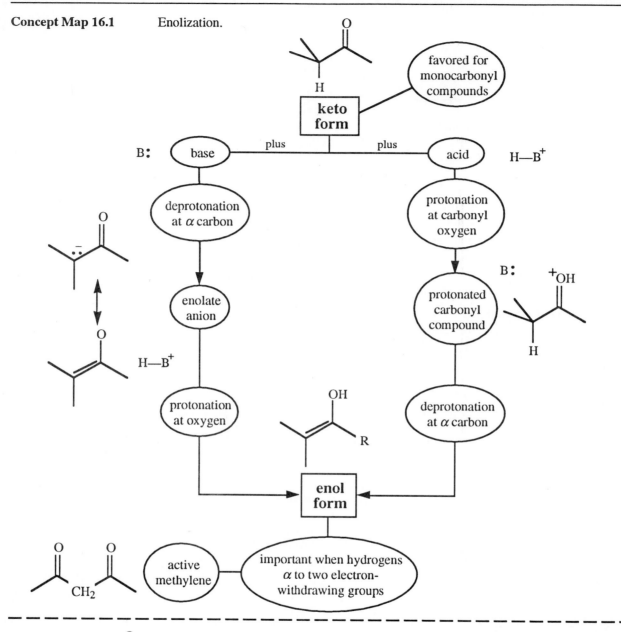

16.2   (a)

(active methylene)

(b)

(active methylene)

(c) $CH_3COCH_2CH_3$

(more electrophilic
carbonyl carbon atom)

(d) $CH_3OCCH_2COCH_3$

(active methylene)

16.3

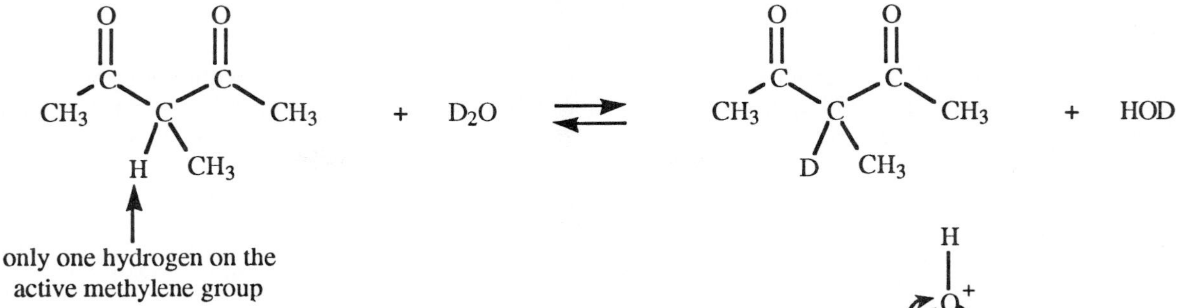
only one hydrogen on the active methylene group

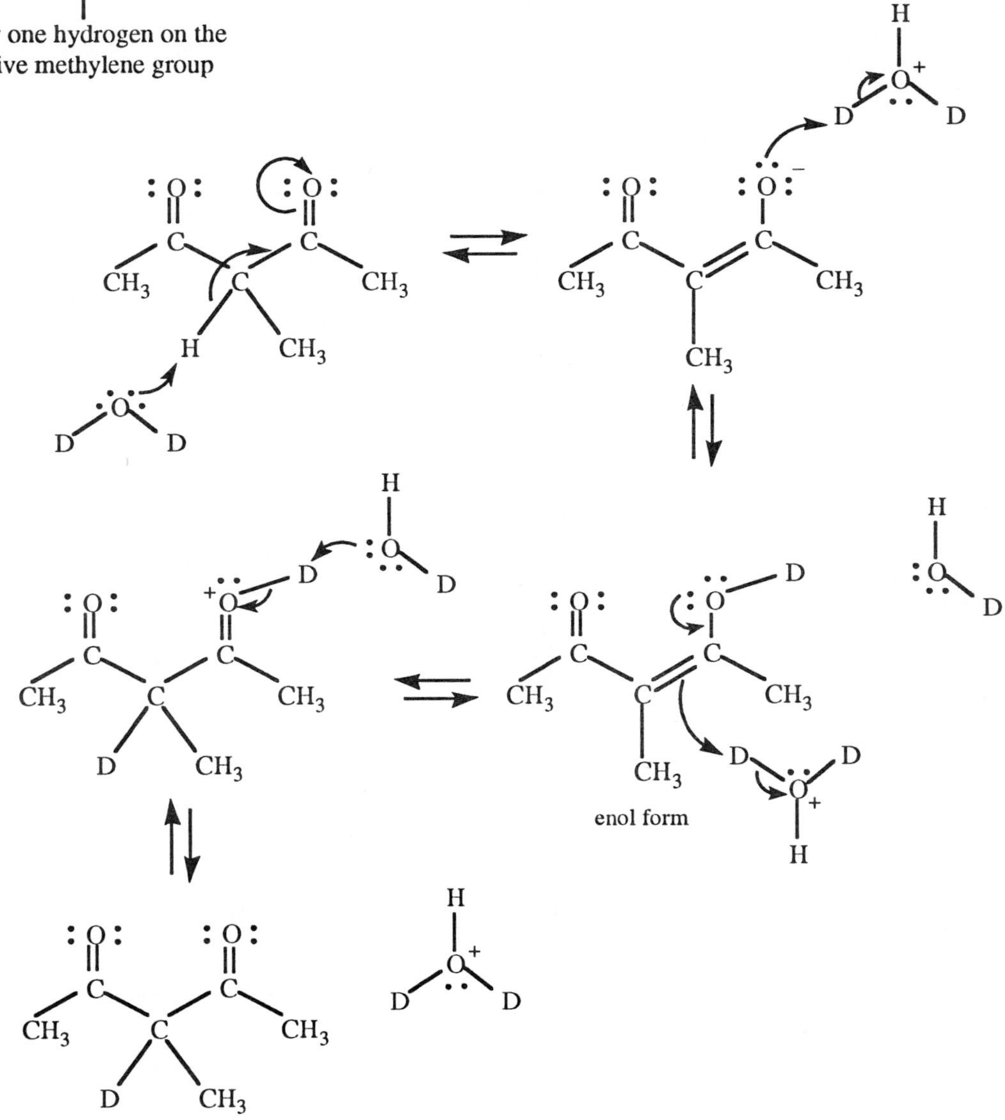
enol form

16.4 (a)

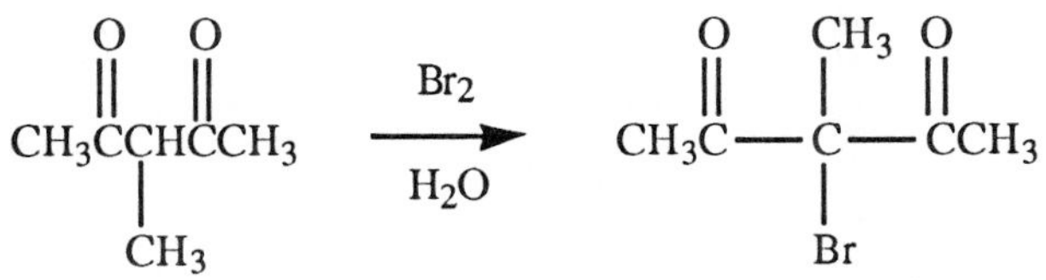
Br2
H2O

16.4 (cont)

(b)   3-Methyl-2,4-pentanedione will enolize in water. This reaction is an equilibrium reaction, and after
         a day, the equilibrium concentration of enol is present. The reaction with bromine is fast because the
         electrophile reacts at the double bond of the enol until the equilibrium concentration of the enol is used
         up.

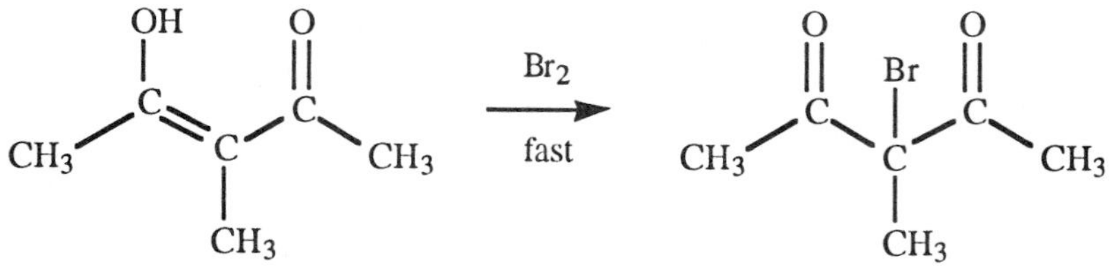

Once the enol has been used up, the rate of the overall reaction is determined by the rate at which the
enol is formed. That is the slow, rate-determining step of this two-step process.

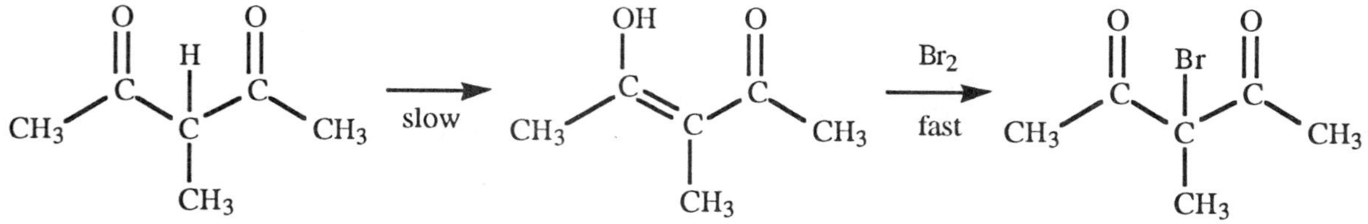

16.5

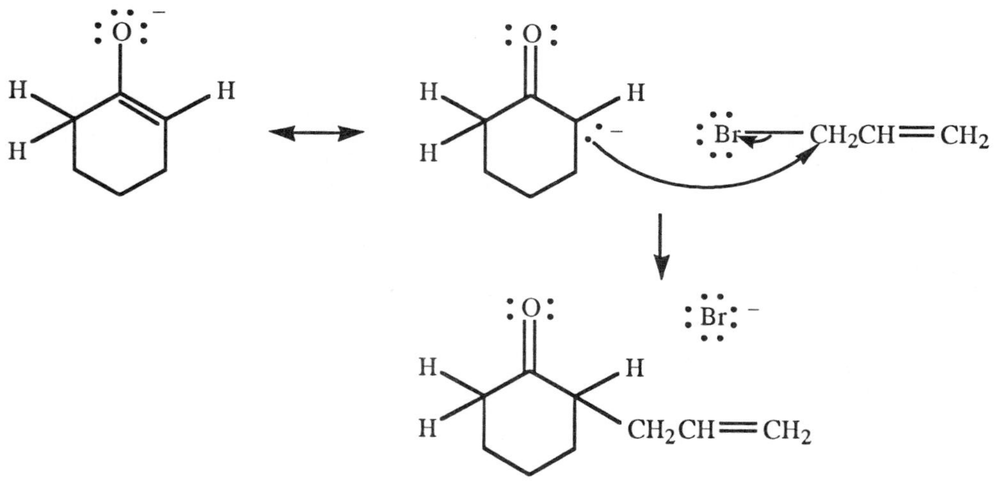

- -

Concept Map 16.2   (see p. 540)

- -

16.6

(a)

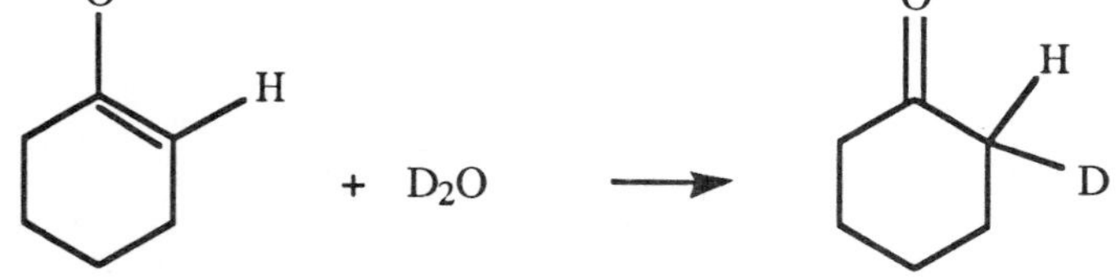

enolate ion   electrophile
(formed in base)

16.6 (a) (cont)

In excess D_2O (solvent) this process will repeat itself until all of the α-hydrogen atoms have been replaced by deuterium.

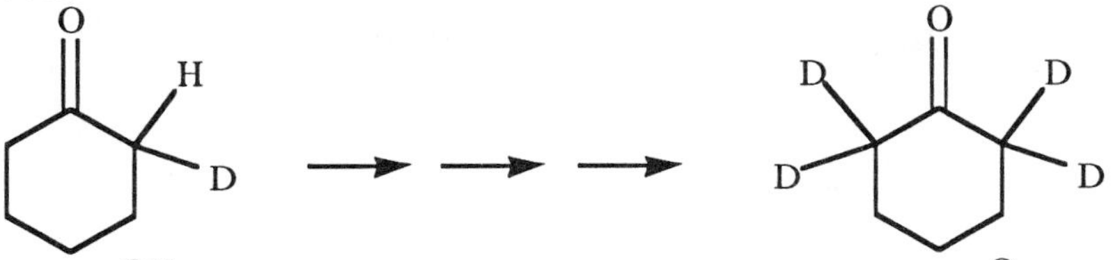

(b)

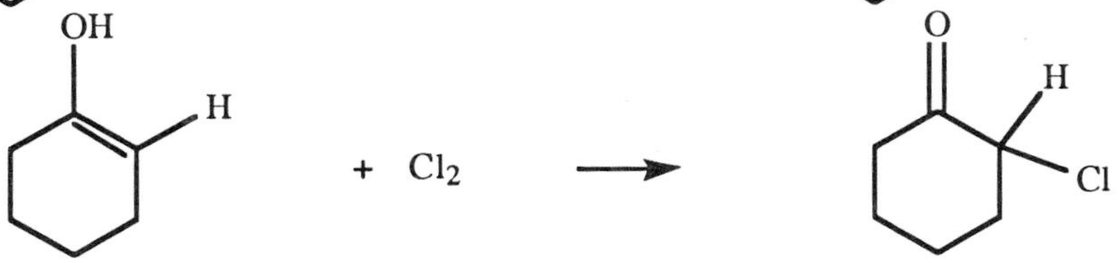

enol
(formed in neutral
or acidic solution)

electrophile

(c) CH_3CH_2C⏐⏐$CHCH_3$ + Br_2 ⟶ $CH_3CH_2CCHCH_3$

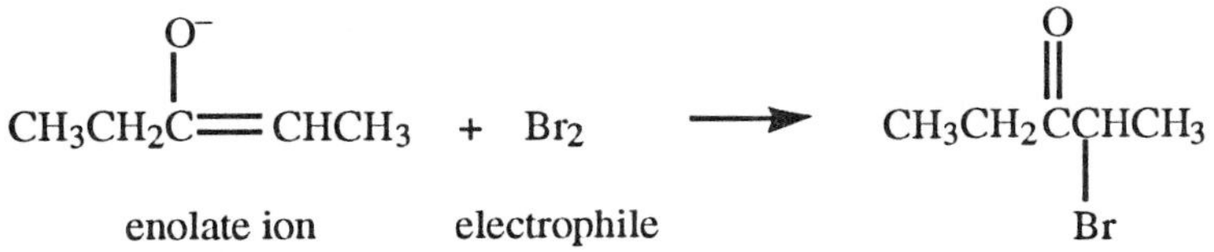

enolate ion   electrophile
(formed in base)   Br

With a second equivalent of bromine, the process will repeat itself. The most acidic proton is the one on the carbon atom bearing the bromine atom.

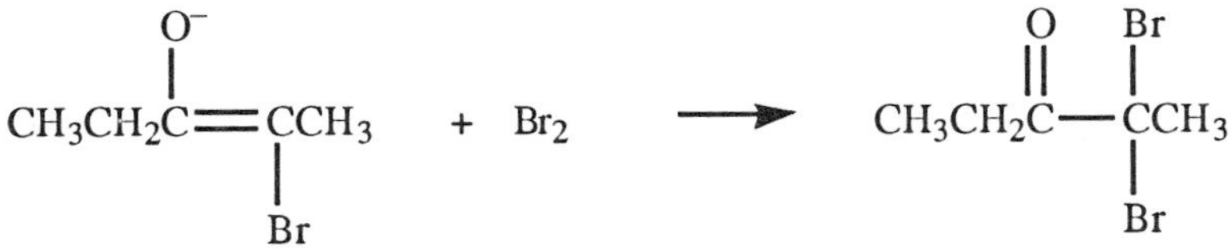

CH_3CH_2C⏐⏐CCH_3 + Br_2 ⟶ CH_3CH_2C—CCH_3

Br
new enolate ion   electrophile

(d)

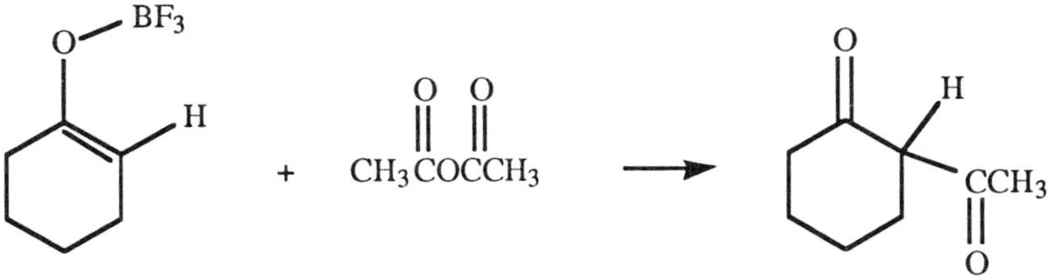

enol form
(BF_3, a Lewis acid
replacing the proton)

+ CH_3COCCH_3

electrophile

Concept Map 16.3   (see p. 541)

16.6 (cont)

(e)

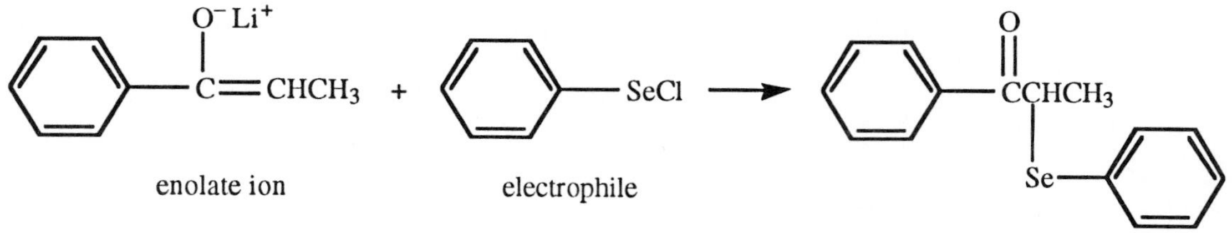

enolate ion   electrophile

(f)

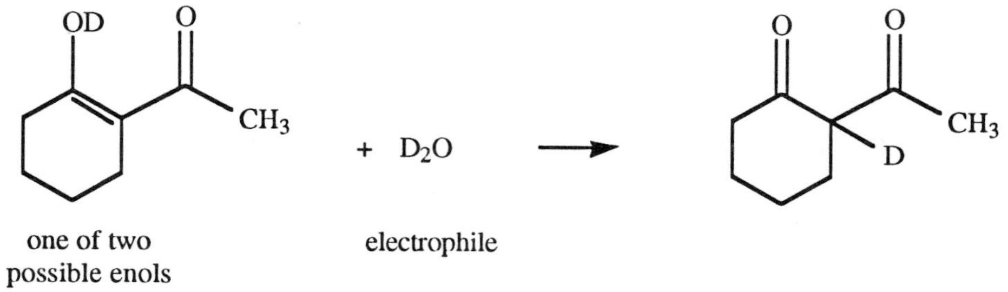

one of two
possible enols   electrophile

Concept Map 16.2   Reactions of enols and enolates with electrophiles.

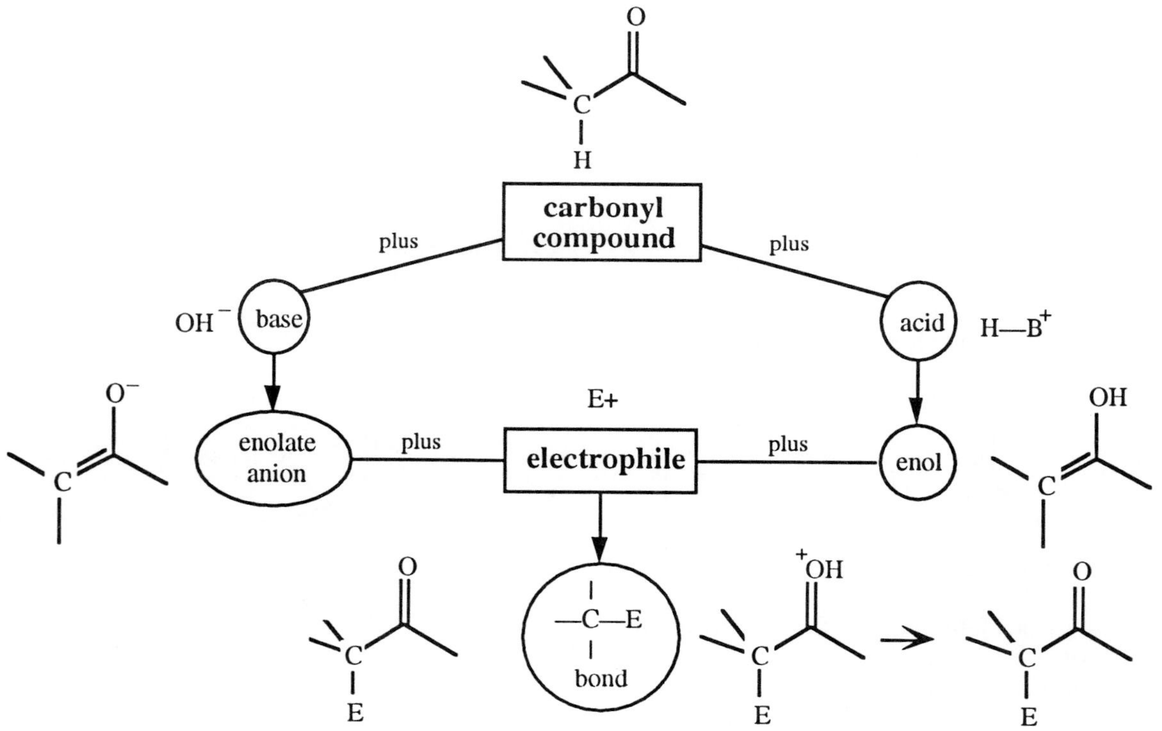

E^+ can be D_2O, X_2, RX, RCX, RCOR', RCOCR

Concept Map 16.3   Thermodynamic and kinetic enolates.

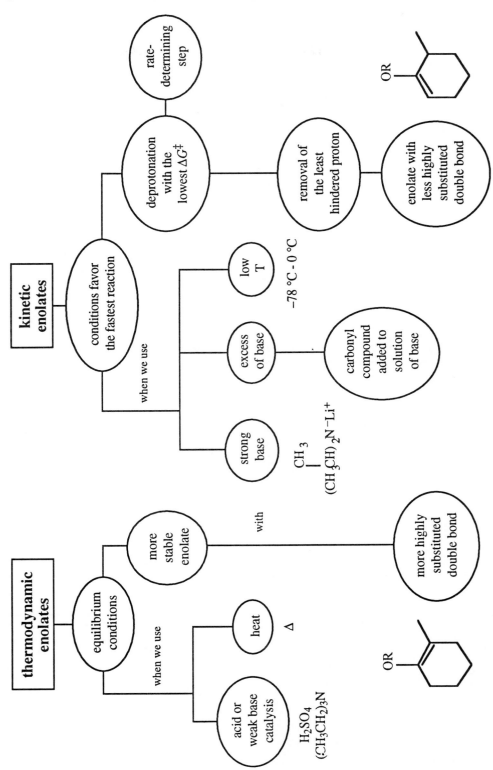

16.7

(a)
$$CH_3CH_2CCH_3 \xrightarrow[\substack{\text{dimethylformamide} \\ \Delta}]{(CH_3)_3SiCl, \ (CH_3CH_2)_3N} CH_3CH=CCH_3$$
with O (double bond) above the leftmost carbonyl, and OSi(CH₃)₃ above the product.

(b)
$$CH_3CH_2CCH_3 \xrightarrow[\substack{\text{1,2-dimethoxyethane}}]{(CH_3CH)_2N^-Li^+} \xrightarrow{(CH_3)_3SiCl, \ (CH_3CH_2)_3N} CH_3CH_2C=CH_2$$
with O (double bond) on the reactant, CH₃ on the reagent, and OSi(CH₃)₃ on the product.

16.8

(a)
A → B

(b)
C

(c)
$$CH_3CH_2CH_2COCH_3 \xrightarrow[\substack{\text{tetrahydrofuran} \\ -78\,°C}]{(CH_3CH)_2N^-Li^+} CH_3CH_2CH=COCH_3 \xrightarrow[\substack{\text{hexamethyl-} \\ \text{phosphoric} \\ \text{triamide} \\ -78\,°C}]{ClCH_2OCH_3} CH_3CH_2CHCOCH_3$$
D   E (O^-Li^+)   F (with CH₂OCH₃ substituent)

Concept Map 16.4   (see p. 544)

16.9

(a)
$$CH_2(COCH_2CH_3)_2 \xrightarrow[\text{ethanol}]{CH_3CH_2ONa} Na^{+ \ -}CH(COCH_2CH_3)_2 \xrightarrow{CH_3CHCH_3} CH_3CHCH(COCH_2CH_3)_2$$
A   B
with Br above the CH₃CHCH₃ reagent and CH₃ above product B.

(b)
$$CH_3CH_2CH(COCH_2CH_3)_2 \xrightarrow[\substack{\textit{tert}\text{-butyl} \\ \text{alcohol}}]{(CH_3)_3CONa} CH_3CH_2C(COCH_2CH_3)_2 \xrightarrow{BrCH_2CH=CH_2}$$
                                              Na⁺
                                               C

16.9 (b) (cont)

$$CH_2\!\!=\!\!CHCH_2C(COCH_2CH_3)_2$$

with structure showing:
$$\underset{CH_2CH_3}{CH_2\!\!=\!\!CHCH_2\overset{\overset{\displaystyle O}{\|}}{C}(COCH_2CH_3)_2}$$

D

(c) $CH_2(COCH_2CH_3)_2$ $\xrightarrow[\text{ethanol}]{CH_3CH_2ONa}$ $Na^+\ {}^-CH(COCH_2CH_3)_2$ $\xrightarrow{}$

E

$\bigcirc$—CH_2Cl

$\bigcirc$—$CH_2CH(COCH_2CH_3)_2$

F

(d) $CH_3\overset{\overset{\displaystyle O}{\|}}{C}CH_2\overset{\overset{\displaystyle O}{\|}}{C}OCH_2CH_3$ $\xrightarrow[\text{ethanol}]{CH_3CH_2ONa}$ $\underset{Na^+}{CH_3\overset{\overset{\displaystyle O}{\|}}{C}\overset{-}{C}H\overset{\overset{\displaystyle O}{\|}}{C}OCH_2CH_3}$ $\xrightarrow{BrCH_2CH_2COCH_2CH_3}$

G

$CH_3\overset{\overset{\displaystyle O}{\|}}{C}\overset{}{C}H\overset{\overset{\displaystyle O}{\|}}{C}OCH_2CH_3$
$\quad\ \overset{|}{CH_2CH_2COCH_2CH_3}$
$\qquad\qquad\qquad\ \overset{\overset{\displaystyle O}{\|}}{}$

H

(e) $CH_3\overset{\overset{\displaystyle O}{\|}}{C}CH_2\overset{\overset{\displaystyle O}{\|}}{C}OCH_2CH_3$ $\xrightarrow[\text{ethanol}]{CH_3CH_2ONa}$ $\underset{Na^+}{CH_3\overset{\overset{\displaystyle O}{\|}}{C}\overset{-}{C}H\overset{\overset{\displaystyle O}{\|}}{C}OCH_2CH_3}$ $\xrightarrow{ClCH_2CH_2CH_2Br}$ $CH_3\overset{\overset{\displaystyle O}{\|}}{C}\overset{}{C}H\overset{\overset{\displaystyle O}{\|}}{C}OCH_2CH_3$

I

$\qquad\qquad\qquad\qquad\qquad\qquad\qquad\qquad\qquad\qquad\overset{|}{CH_2CH_2CH_2Cl}$

J

(Br$^-$ is a better leaving group than Cl$^-$)

(f) $CH_3\overset{\overset{\displaystyle O}{\|}}{C}CH_2\overset{\overset{\displaystyle O}{\|}}{C}OCH_2CH_3$ $\xrightarrow[\text{ethanol}]{CH_3CH_2ONa}$ $\underset{Na^+}{CH_3\overset{\overset{\displaystyle O}{\|}}{C}\overset{-}{C}H\overset{\overset{\displaystyle O}{\|}}{C}OCH_2CH_3}$ $\xrightarrow{CH_2\!\!=\!\!CHCH_2CH_2Br}$

K

$CH_3\overset{\overset{\displaystyle O}{\|}}{C}\overset{}{C}H\overset{\overset{\displaystyle O}{\|}}{C}OCH_2CH_3$
$\quad\ \overset{|}{CH_2CH_2CH\!\!=\!\!CH_2}$

L

$\xrightarrow[\text{toluene}]{Na}$

$CH_3\overset{\overset{\displaystyle O}{\|}}{C}\!\!-\!\!\overset{\overset{\displaystyle Na^+}{}}{\underset{\overset{\displaystyle |}{CH_2CH_2CH\!\!=\!\!CH_2}}{\overset{-}{C}}}\!\!-\!\!\overset{\overset{\displaystyle O}{\|}}{C}OCH_2CH_3$

M

$\xrightarrow{\overset{\overset{\displaystyle I}{|}}{CH_3CHCH_3}}$

structure with central C:
$\underset{\overset{\displaystyle \|}{O}}{CH_3C}\!\!-\!\!\overset{\overset{\displaystyle CH_3CH}{\overset{|}{CH_3}}}{\underset{\underset{\displaystyle CH_2CH_2CH\!\!=\!\!CH_2}{}}{C}}\!\!-\!\!\underset{\overset{\displaystyle O}{}}{\overset{\overset{\displaystyle O}{\|}}{C}OCH_2CH_3}$

16.9 (cont)

(g)

$$\text{(epoxide)} + \text{Na}^+\ ^-\text{CH(COCH}_2\text{CH}_3)_2 \longrightarrow \text{Na}^+\ ^-\text{OCH}_2\text{CH}_2\text{CH(COCH}_2\text{CH}_3)_2$$

P

(h)

$$\text{CH}_3\text{CCH}_2\text{COCH}_2\text{CH}_3 \xrightarrow[\text{tetrahydrofuran}]{\text{NaH}} \text{CH}_3\text{CCH}=\text{COCH}_2\text{CH}_3 \xrightarrow[\text{hexane}]{\text{CH}_3\text{CH}_2\text{CH}_2\text{CH}_2\text{Li}}$$

Q

$$\text{CH}_2=\text{CCH}=\text{COCH}_2\text{CH}_3 \xrightarrow[\text{tetrahydrofuran}]{\text{PhOCH}_2\text{CH}_2\text{Br}} \xrightarrow{\text{H}_3\text{O}^+} \text{PhOCH}_2\text{CH}_2\text{CH}_2\text{CCH}_2\text{COCH}_2\text{CH}_3$$

R

S

$$\xrightarrow{\text{LiAlH}_4} \xrightarrow{\text{H}_3\text{O}^+} \text{PhOCH}_2\text{CH}_2\text{CH}_2\text{CHCH}_2\text{CH}_2\text{OH} + \text{CH}_3\text{CH}_2\text{OH}$$

T   U

Concept Map 16.4   Alkylation reactions.

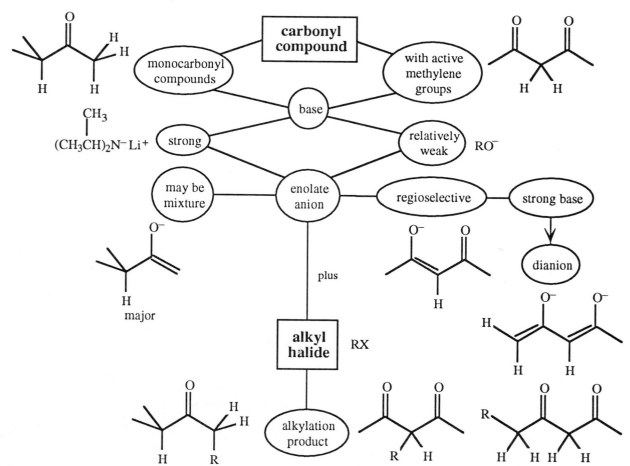

16.10 $CH_3OCCH_2COCH_3$ $\xrightarrow[\text{tetrahydrofuran}]{NaH}$ $CH_3OC=CHCOCH_3$ (A) $\xrightarrow[\text{tetrahydrofuran}]{CH_3CHI}$

$CH_3OCCHCOCH_3$ (B) $\xrightarrow[\text{tetrahydrofuran}]{LiAlH_4}$ $\xrightarrow{H_2O}$ $HOCH_2CHCH_2OH$ (C)

CH_3CHCH_3

(0.3 equivalents)

(a mild acid catalyst)

dichloromethane

$THPOCH_2CHCH_2OH$ (D) $\xrightarrow{CH_3SCH_3,\ ClC—CCl}$ $\xrightarrow{(CH_3CH_2)_3N}$ $THPOCH_2CHCH$ (E)

CH_3CHCH_3

CH_3CHCH_3

16.11

1. $CH_3CHCH_2CH_2Br$ $\xleftarrow{\text{conc HBr}}$ $CH_3CHCH_2CH_2OH$ $\xleftarrow{H_3O^+}$ $\triangle$ $CH_3CHMgBr$ $\xleftarrow[\text{diethyl ether}]{Mg}$ CH_3CHBr

2. $\xleftarrow{H_3O^+}$ $\xleftarrow[\text{methanol}]{H_2NCNH_2,\ CH_3ONa}$

$CH_3CHCH_2CH_2C(COCH_2CH_3)_2$ $\xleftarrow{CH_3CHCH_2CH_2Br}$ $CH_3CH_2C(COCH_2CH_3)_2$ $\xleftarrow[\text{tert-butyl alcohol}]{(CH_3)_3COK}$

CH_2CH_3 K^+

$CH_3CH_2CH(COCH_2CH_3)_2$ $\xleftarrow{CH_3CH_2Br}$ $K^+ {}^-CH(COCH_2CH_3)_2$ $\xleftarrow[\text{tert-butyl alcohol}]{(CH_3)_3COK}$ $CH_2(COCH_2CH_3)_2$

16.12

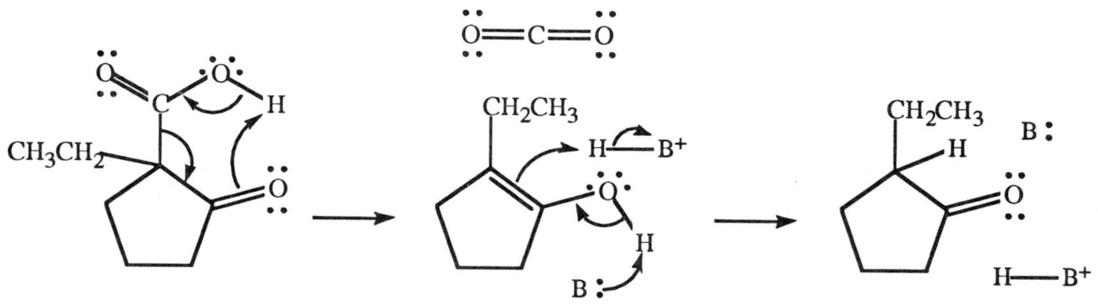

$$O=C=O$$

16.13

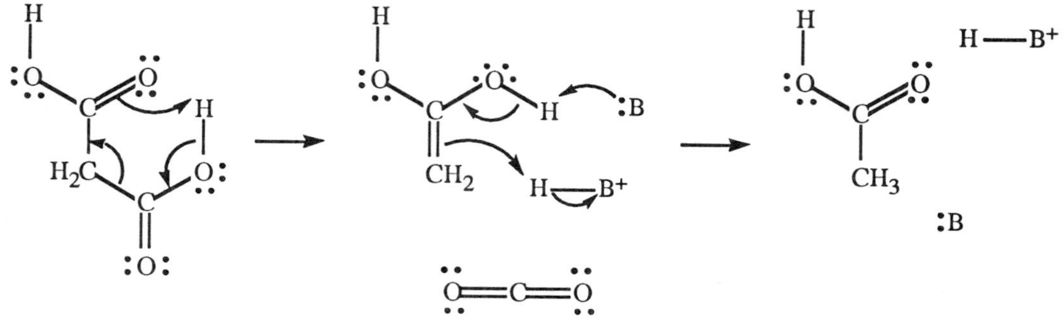

$$O=C=O$$

16.14

(a)
$$CH_3CCH_2COCH_2CH_3 \xrightarrow[\text{ethanol}]{CH_3CH_2ONa} CH_3C\overset{-}{C}HCOCH_2CH_3 \;\; Na^+ \xrightarrow{ClCH_2COCH_2CH_3}$$

A

$$CH_3CCHCOCH_2CH_3 \;\;|\;\; CH_2COCH_2CH_3 \xrightarrow[\Delta]{H_3O^+} CH_3CCH_2CH_2COH \;+\; CO_2\uparrow \;+\; CH_3CH_2OH$$

B   C

(b)
$$C_6H_5CCH_2COCH_2CH_3 \xrightarrow[\text{ethanol}]{CH_3CH_2ONa}$$

$$C_6H_5C\overset{-}{C}HCOCH_2CH_3 \;\; Na^+ \xrightarrow[\Delta]{BrCH_2C\equiv CH} C_6H_5CCHCOCH_2CH_3 \;|\; CH_2C\equiv CH \xrightarrow[\Delta]{H_3O^+}$$

D   E

$$C_6H_5CCH_2CH_2C\equiv CH \;+\; CO_2\uparrow \;+\; CH_3CH_2OH$$

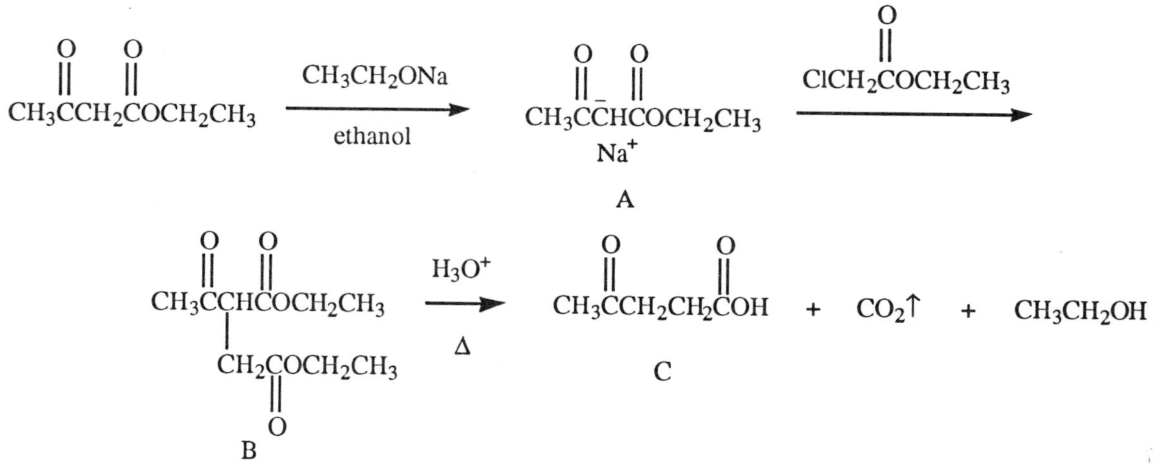

F

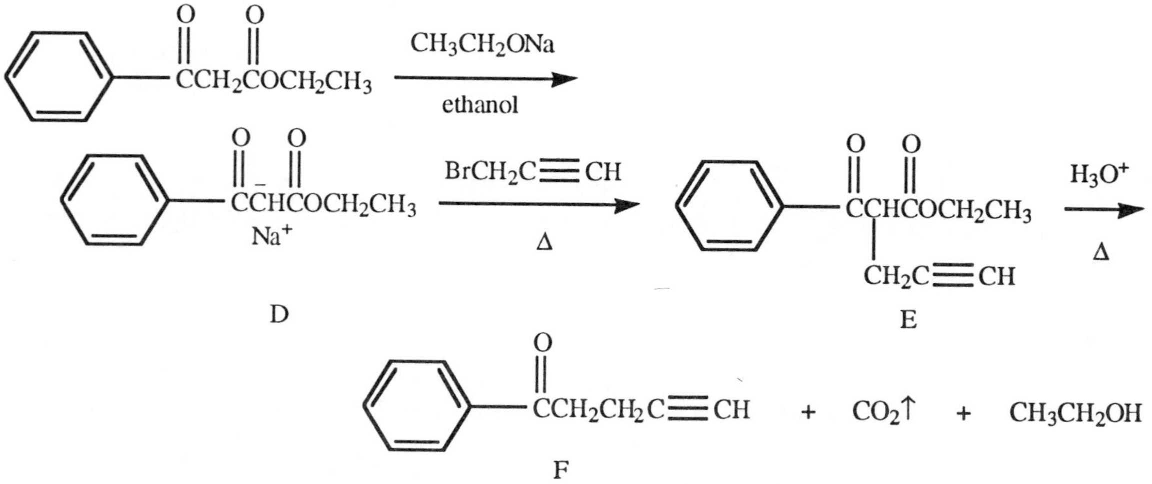

16.14 (cont)

(c) $CH_2(COCH_2CH_3)_2$ $\xrightarrow[\text{ethanol}]{CH_3CH_2ONa}$ $Na^+ \; {}^-CH(COCH_2CH_3)_2$ $\xrightarrow{BrCH_2CH_2CH_2Cl}$

G

$ClCH_2CH_2CH_2CH(COCH_2CH_3)_2$ $\xrightarrow[\text{ethanol}]{CH_3CH_2ONa}$ $\left[ClCH_2CH_2CH_2\bar{C}(COCH_2CH_3)_2 \right] \longrightarrow$

H

(Br⁻ is a better leaving group than Cl⁻)

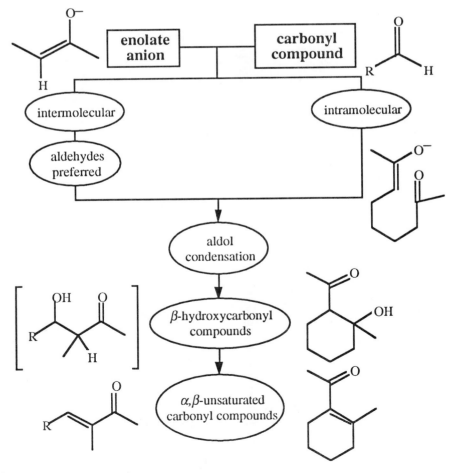

+ CO₂↑ + 2 CH₃CH₂OH

I   J

Concept Map 16.5   The aldol condensation.

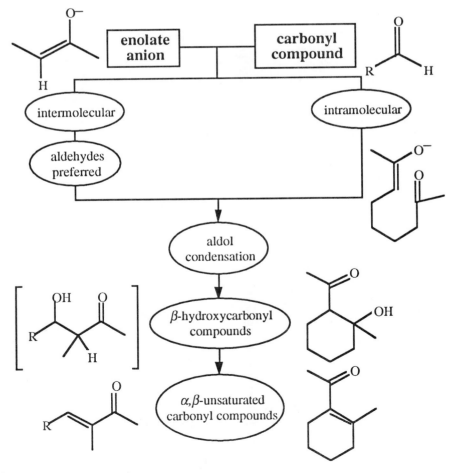

16.15

(a)

$$\text{furan-CH=O} + CH_3CCH_3 \xrightarrow[\text{H}_2\text{O}]{\text{NaOH}} \xrightarrow{\text{H}_3\text{O}^+} \text{furan-CH=CHCCH}_3$$

(b)

$$\text{C}_6\text{H}_5\text{-CH=O} \ (\text{2 equiv}) + CH_3CCH_3 \ (\text{1 equiv}) \xrightarrow[\text{H}_2\text{O}]{\text{NaOH}} \xrightarrow{\text{H}_3\text{O}^+}$$

$$\text{C}_6\text{H}_5\text{-CH=CHCCH=CH-C}_6\text{H}_5$$

(c)

$$\text{C}_6\text{H}_5\text{-CH=O} + CH_3CCH_2CH_3 \xrightarrow{\text{HCl}} \text{C}_6\text{H}_5\text{-CH=CCCH}_3$$
$$\hspace{10cm} | \atop CH_3$$

(The more stable enol from
2-butanone has the more
highly substituted double bond)

(d)

$$\text{C}_6\text{H}_5\text{-CH}_2\text{CH=O} \ (\text{2 equiv}) \xrightarrow[\text{H}_2\text{O}]{\text{NaOH}} \text{C}_6\text{H}_5\text{-C=CHCH}_2\text{-C}_6\text{H}_5$$

with CH group

16.16

(a)

$$CH_3O-\text{C}_6\text{H}_4-\text{CH=O} + CH_3CH_2CH$$

(b)

$$CH_3CH_2CH_2CH_2CH-CHCH \xleftarrow{} \quad 2\ CH_3CH_2CH_2CH_2CH$$
with HO and CH₂CH₂CH₃ substituents

16.16 (cont)

(c)

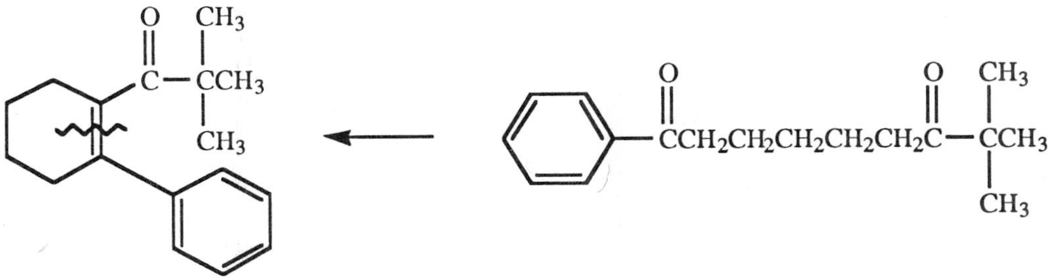

(d)

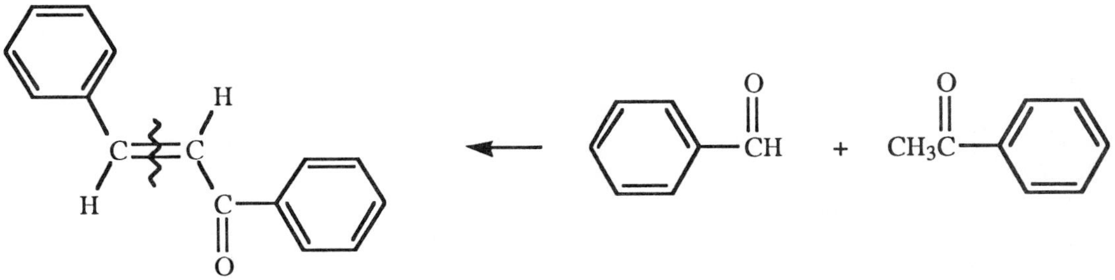

16.17   The aldol condensation product must be going back to being an aldehyde and an enolate ion, which then reacts with the other aldehyde in the reaction mixture.

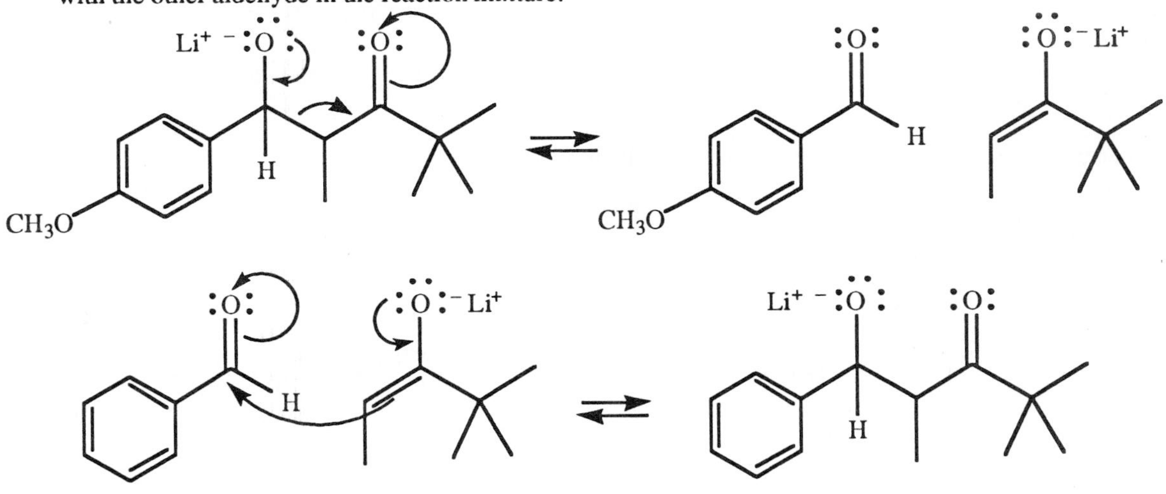

16.18

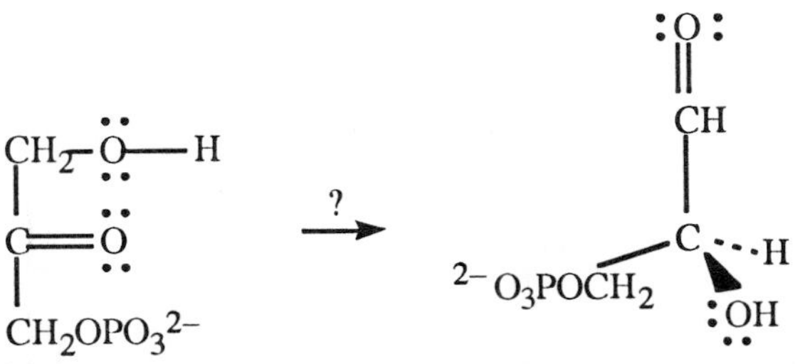

1.   How have connectivities changed in going from reactant to product? How many carbon atoms does each contain? What bonds must be broken and formed to transform starting material into product?

Both reactant and product have three carbon atoms. There is an alcohol function on carbon 1 and a ketone function on carbon 2 in the reactant. There is an aldehyde function on carbon 1 and an alcohol function on carbon 2 in the product. Both have a phosphate ester on carbon 3. The bonds to be broken and formed are shown on the next page.

16.18 (1) (cont)

2. What reagents are present? Are they good acids, bases, nucleophiles, or electrophiles? Is there an ionizing solvent present?

    The triosephosphate enzyme can act as an acid or a base.

3. What is the most likely first step for the reaction: protonation or deprotonation, ionization, attack by a nucleophile, or attack by an electrophile?

    Protonation of the carbonyl group of the ester with later deprotonation of the α hydrogen on carbon 1 is the most likely first step.

4. What are the properties of the species present in the reaction mixture after the first step? What is likely to happen next?

    An enediol is formed, which will tautomerize to the product.

16.19   (a) $\underset{\text{2-oxopentanedioic acid}}{HO\overset{O}{\overset{\|}{C}}CH_2CH_2\overset{O}{\overset{\|}{C}}\overset{O}{\overset{\|}{C}}OH}$

    (b)   We need to hydrolyze the triester and decarboxylate the resulting triacid.

intermediate triacid;
circled carboxylic acid group is β to circled
carbonyl group, so it is lost during the heating

16.20

(a)

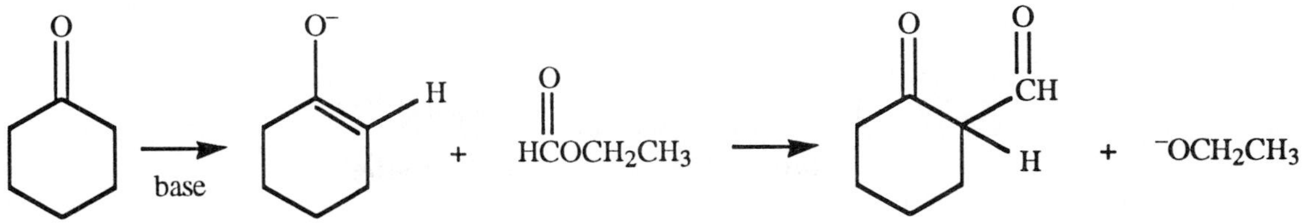

(b)

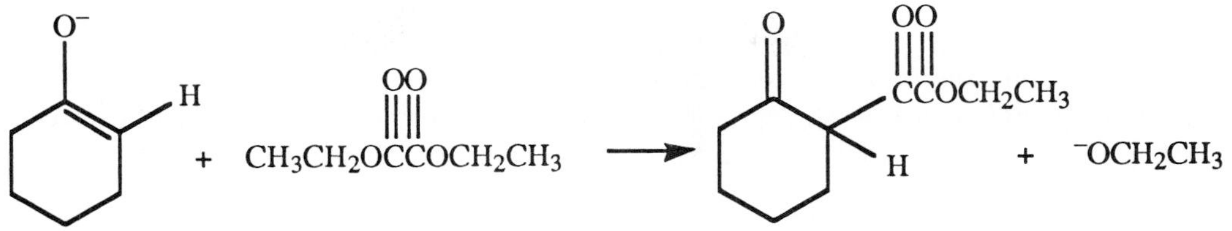

(c)

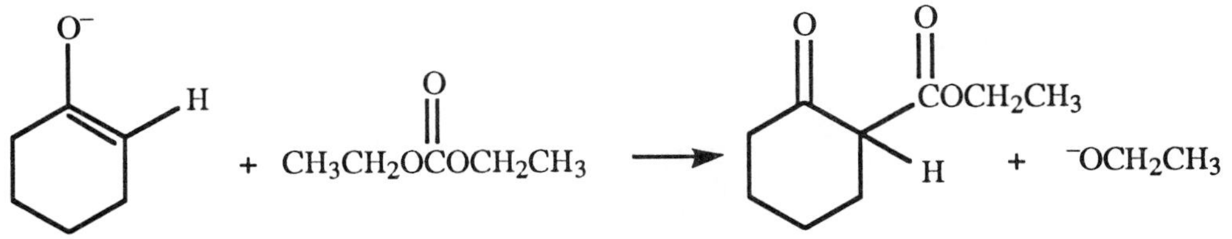

16.21

(a)

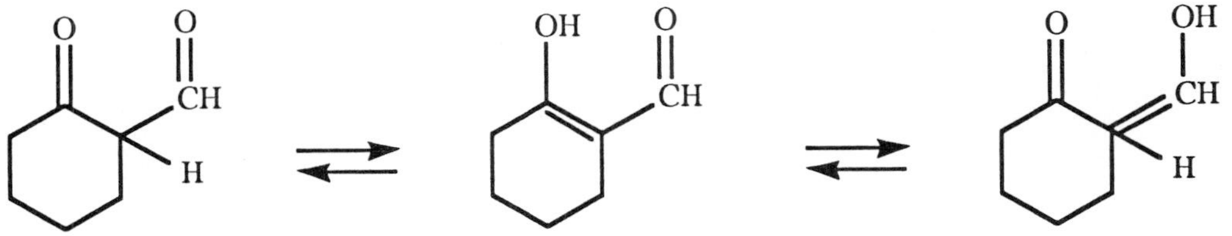

(b)

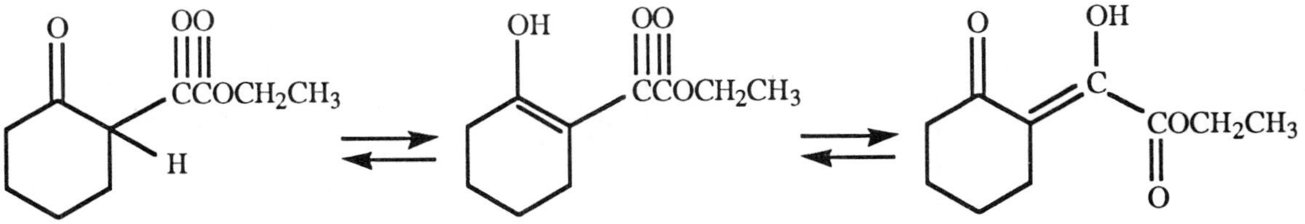

(c)

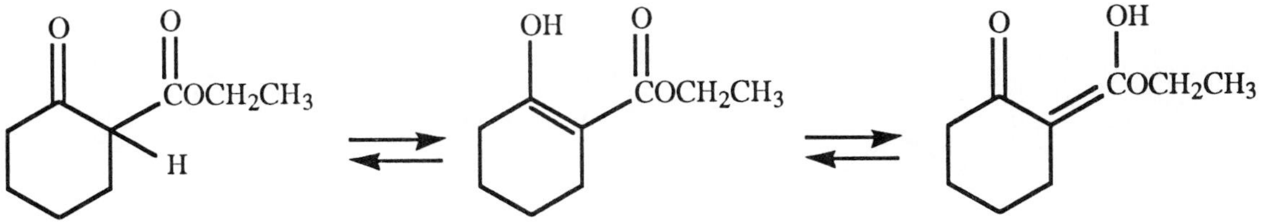

16.22

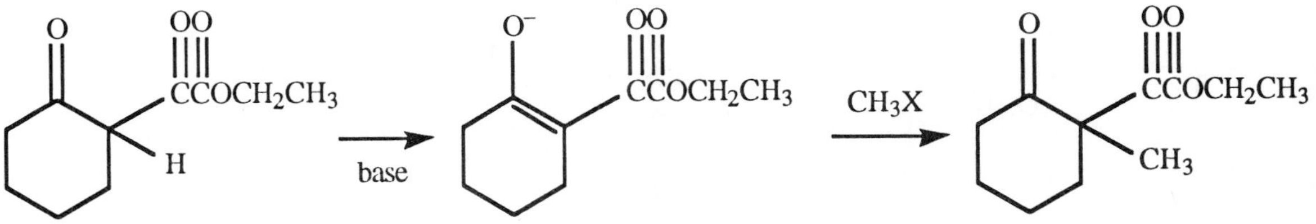

16.23

(structure with curved arrows showing deprotonation)

CH_3C—C—C—$COCH_2CH_3$ with CH_3, $:O:$, CH_3, O, H, $:H$ K^+, CH_3 groups

$\longrightarrow$

K^+

CH_3C=C—C—$COCH_2CH_3$ with CH_3, $:O:^-$, CH_3, O, CH_3 groups, H_2

This is the only enolate ion possible in this system. It does not form under the first set of reaction conditions because the base, ethoxide ion, is not strong enough to deprotonate that carbon atom ($pK_a \sim 23$). When the much stronger base, hydride ion, is used, deprotonation takes place to stabilize the product of the condensation reaction against the reverse reaction.

Concept Map 16.6   Acylation reactions of enolates.

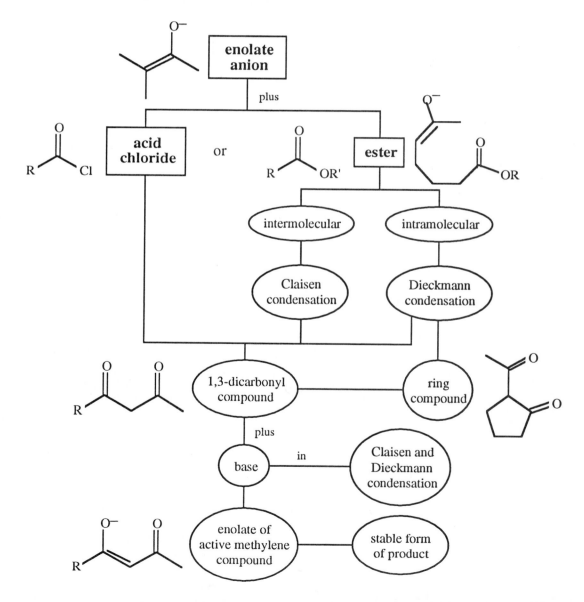

16.24

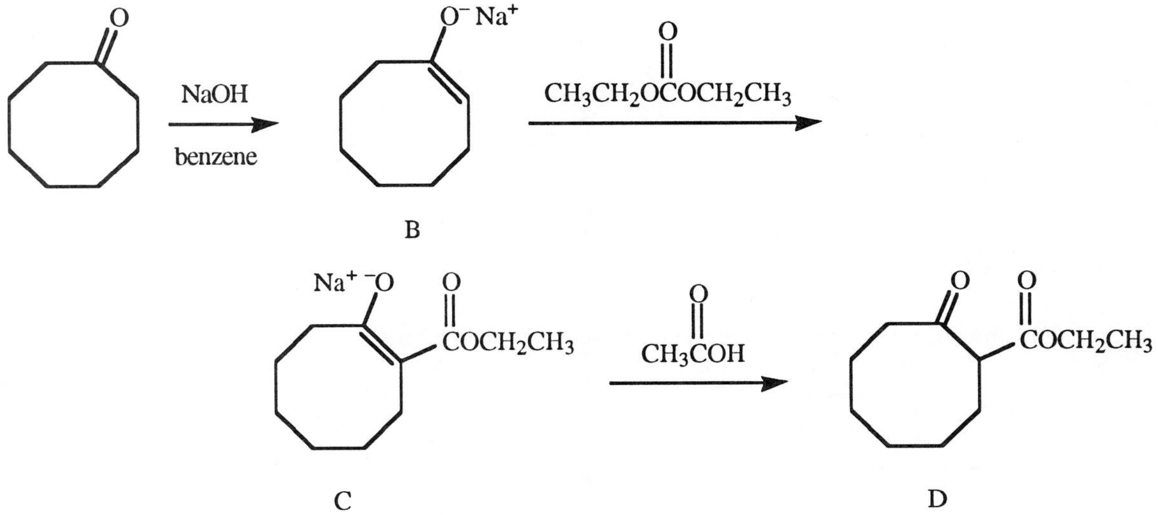

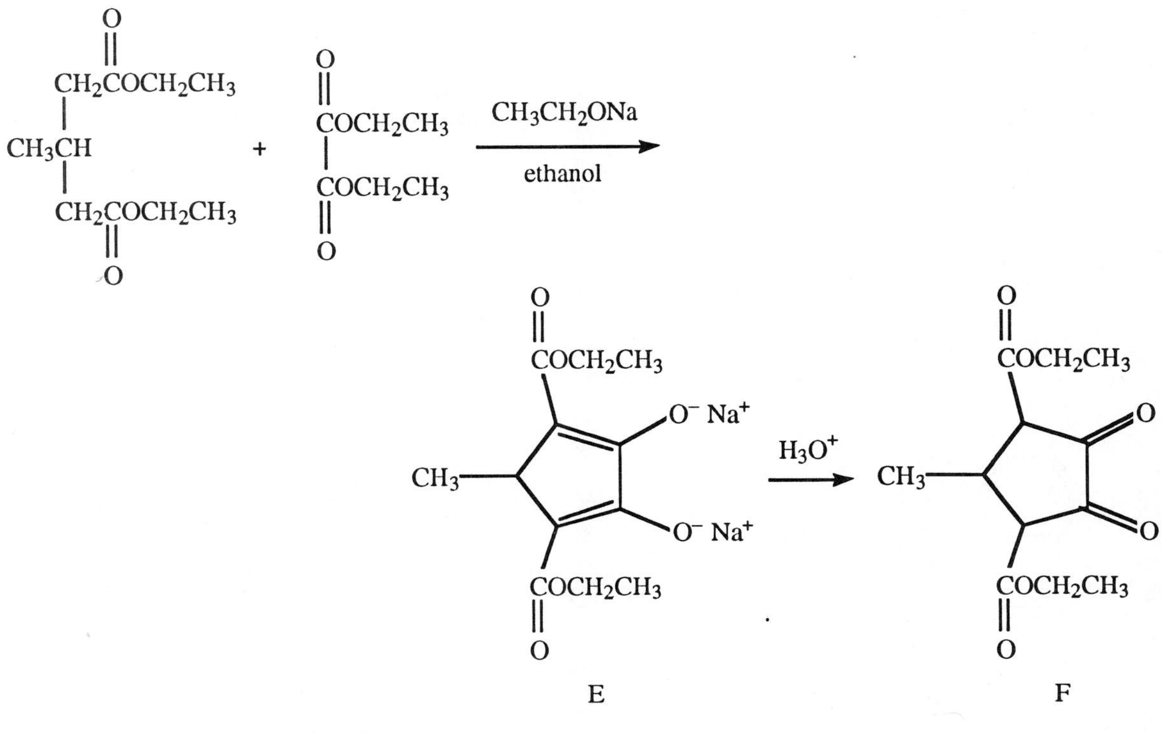

(a)

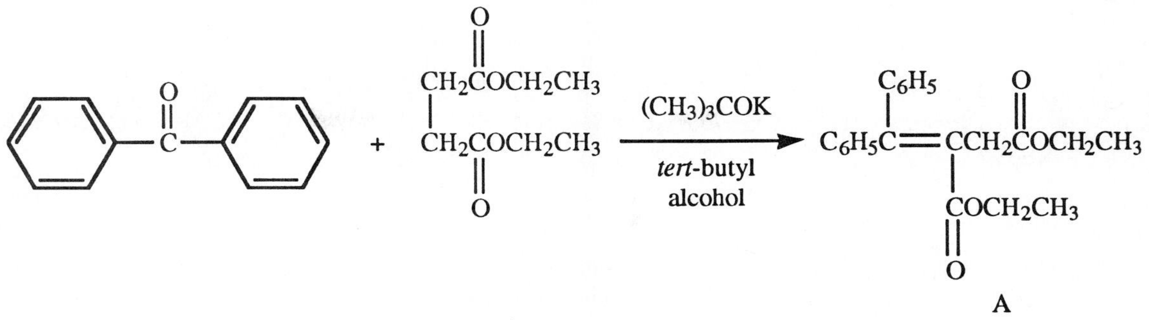

(b)

(c)

16.25

(a)

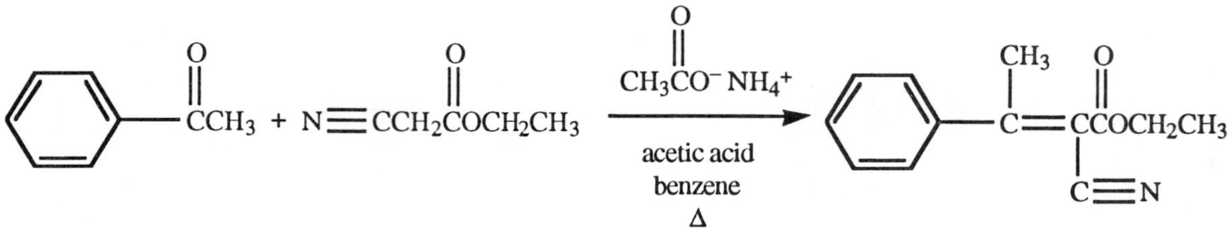
acetic acid
benzene

In the Knoevenagel reaction a key intermediate is the imine (p. 534 in the text) of the ketone, formed from ammonia in equilibrium with ammonium ion in acetic acid.

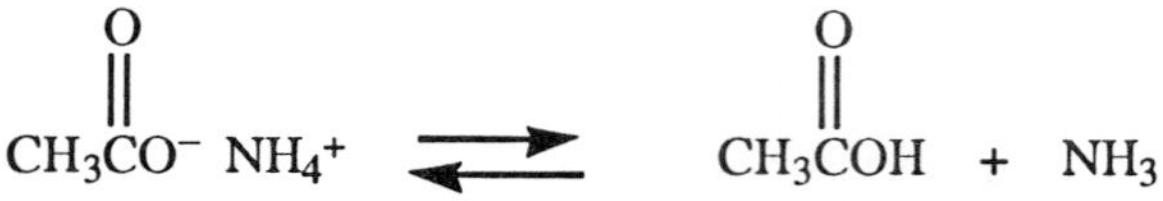

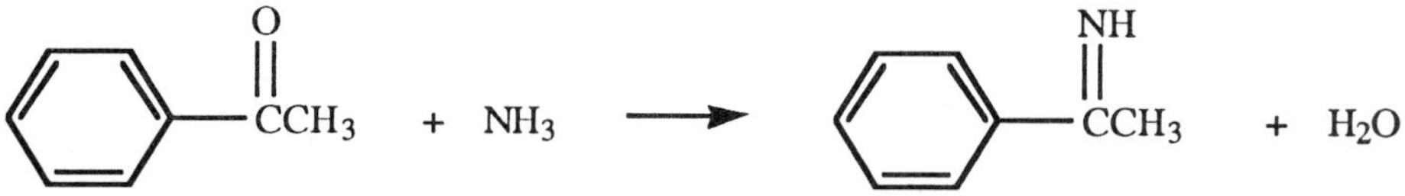

The imine is present in the acidic solution in its protonated form.

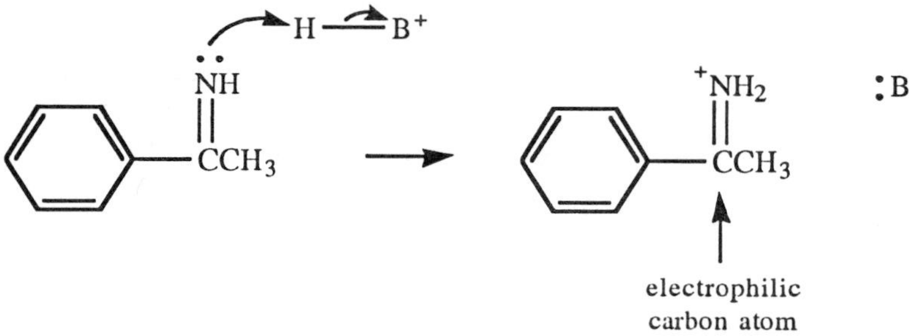
electrophilic
carbon atom

The cyanoester is enolized in the reaction mixture.

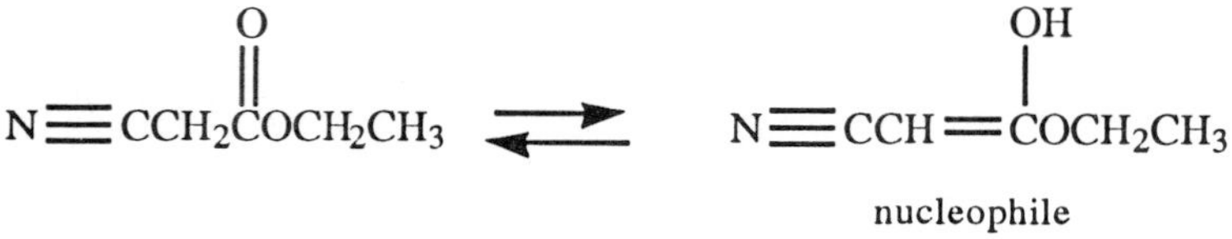
nucleophile

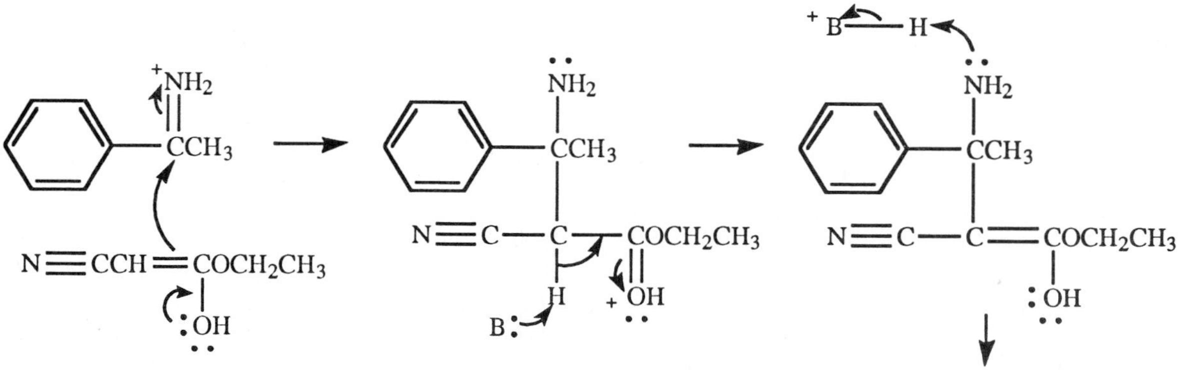

16.25 (a) (cont)

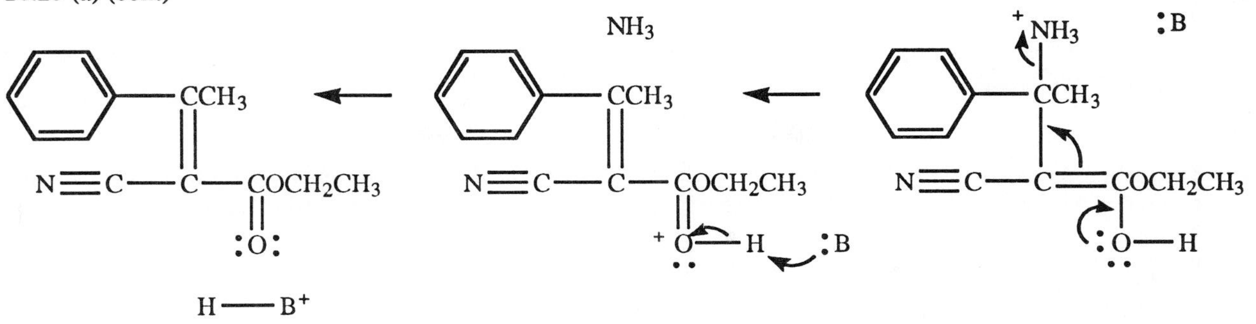

The overall result is a condensation reaction between the carbonyl group of the ketone and the active methylene group of the cyanoester.

(b)

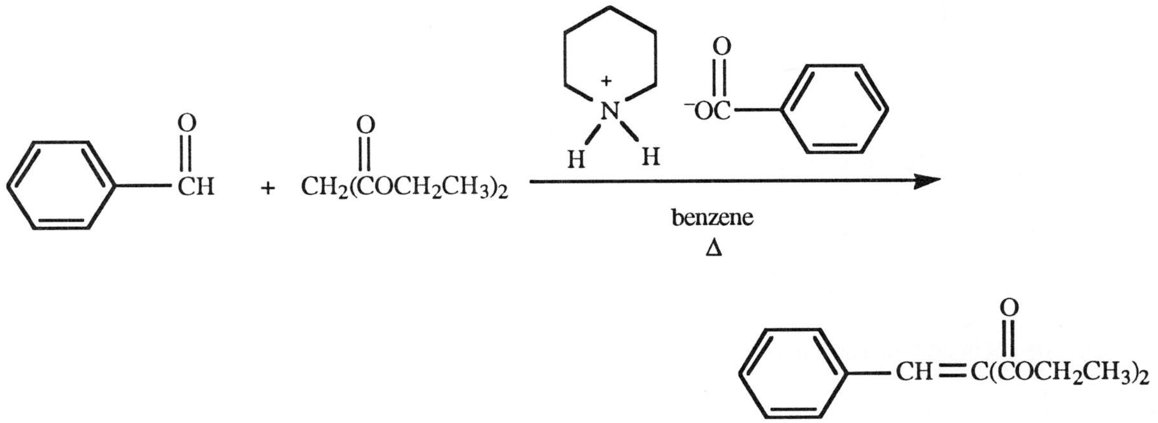

16.26

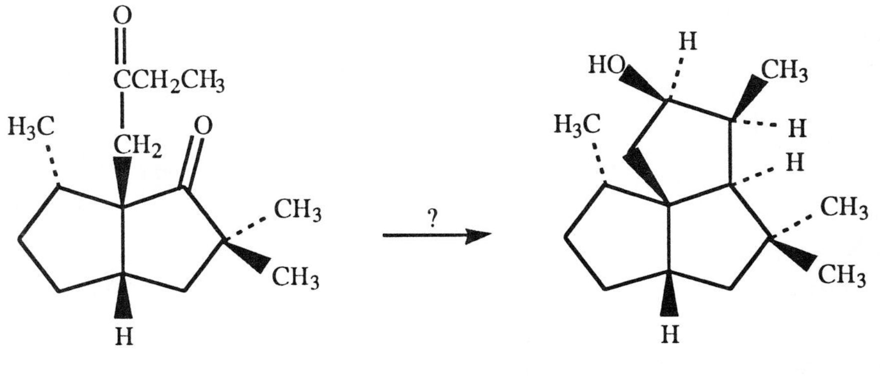

1.   What are the connectivities of the two compounds? How many carbon atoms does each contain? Are there any rings? What are the positions of branches and functional groups on the carbon skeletons?

The product has the same number of carbon atoms as the starting material. There are two five-membered rings in the reactant and three in the product. The carbonyl group of the side chain in the starting material has become an alcohol group in the product and the side chain has been incorporated into the third ring. The new ring in the product is attached at the carbon atom of the carbonyl group in a ring of the starting material.

16.26 (cont)

2. How do the functional groups change in going from starting material to product? Does the starting material have a good leaving group?

    The carbonyl group of the side chain in the starting material has become an alcohol in the product. The carbonyl group that was on a ring has disappeared. There is no good leaving group in the starting material.

3. Is it possible to dissect the structures of the starting material and product to see which bonds must be broken and which formed?

bonds broken   *bonds formed*

    The major transformation is the formation of a bond between one of the carbon atoms α to the carbonyl group of the side chain and the carbon atom of the ring carbonyl group.

4. New bonds are created when an electrophile reacts with a nucleophile. Do we recognize any part of the product molecule as coming from a good nucleophile or an electrophilic addition?

    An enolate anion from a ketone is a good nucleophile and will react with the electrophilic carbon atom of a carbonyl group.

    The enolate from the ketone of the side chain will react with the carbonyl group of the ring ketone in an intramolecular aldol condensation. The product will lose water to form an α,β-unsaturated ketone.

16.26 (cont)

5.   What type of compound would be a good precursor to the product?

A ketone that can be reduced to the alcohol.

The ketone that we need for step 5 is formed by the reduction of the α,β-unsaturated ketone we got in step 4.

6.   After this last step, do we see how to get from starting material to product? If not, we need to analyze the structure obtained in step 5 by applying questions 4 and 5 to it.

16.27

(a)

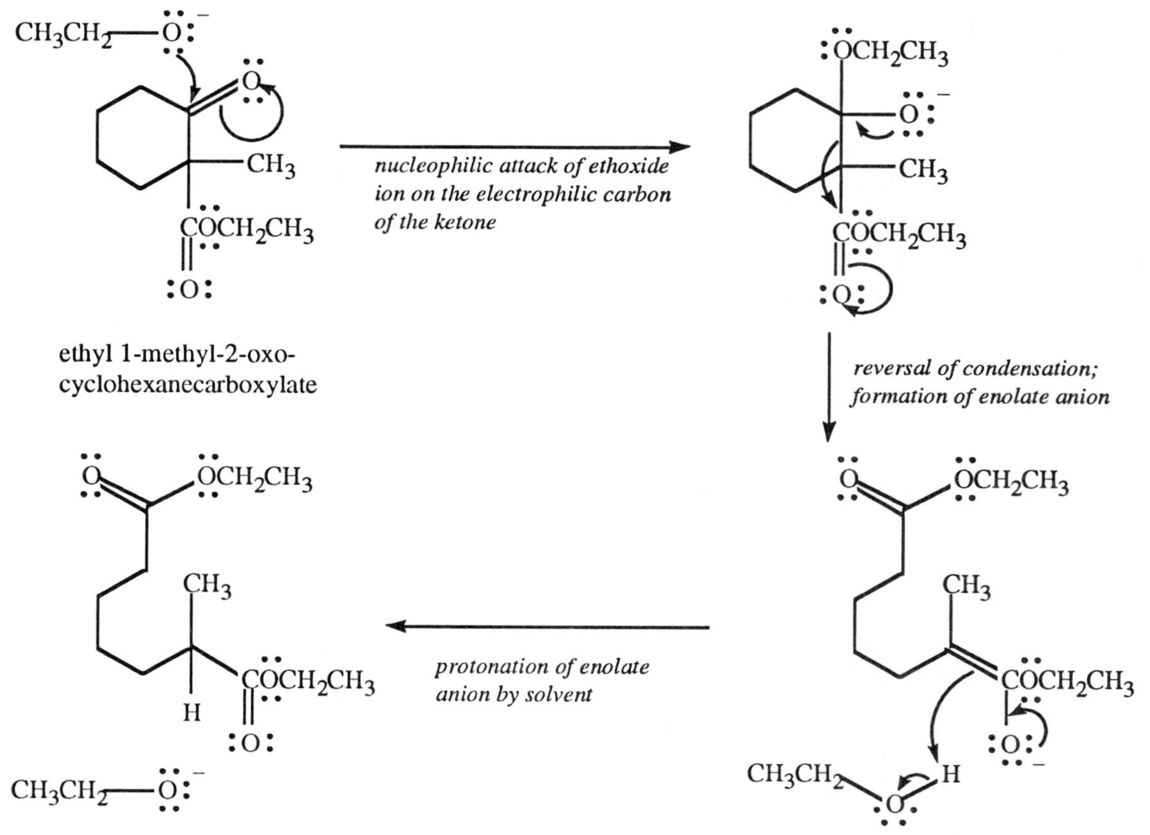

CH3CH2—O:⁻
CH3
COCH2CH3
:O:
ethyl 1-methyl-2-oxo-cyclohexanecarboxylate
nucleophilic attack of ethoxide ion on the electrophilic carbon of the ketone
:OCH2CH3
reversal of condensation; formation of enolate anion
protonation of enolate anion by solvent
H

(b)

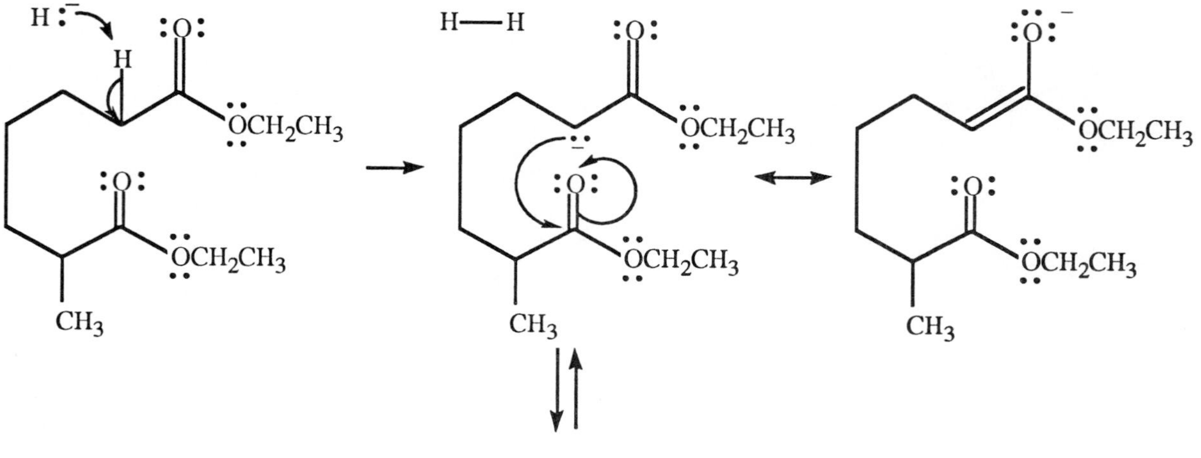

H:⁻
H
:O:
OCH2CH3
CH3
H—H

16.27 (b) (cont)

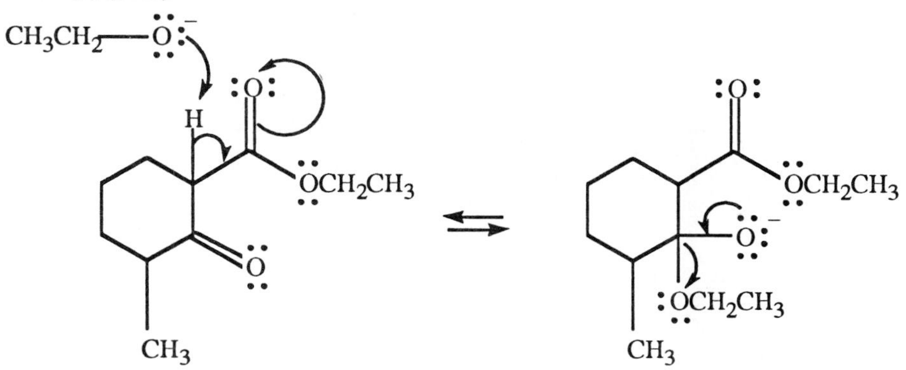

ethyl 3-methyl-2-oxo-
cyclohexanecarboxylate

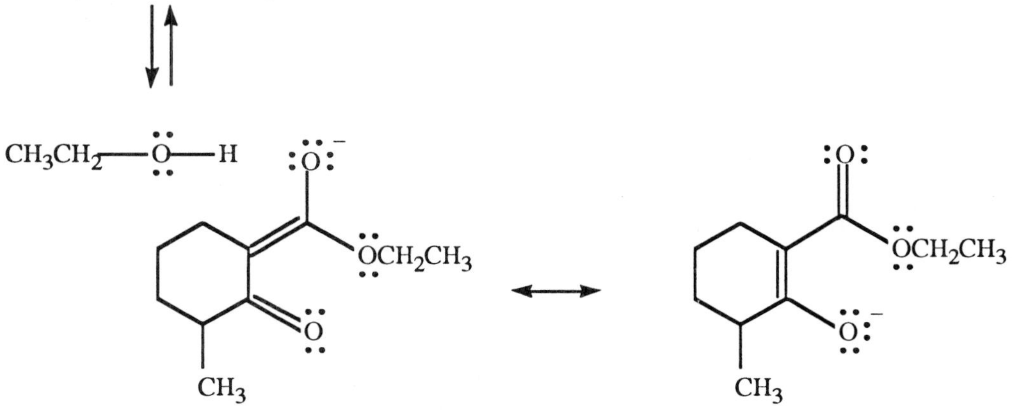

The product β-ketoesters in Claisen condensations are stabilized by deprotonation to the corresponding enolates. Ethyl 1-methyl-2-oxocyclohexanecarboxylate does not have a hydrogen atom on the carbon atom between the two carbonyl groups, and, therefore, cannot be stabilized this way. It undergoes reversal of the cyclization reaction. The product resulting from the other cyclization reaction, ethyl 3-methyl-2-oxocyclohexanecarboxylate, does have an active hydrogen. Under the conditions for the reaction, it is the thermodynamically more stable product.

16.28

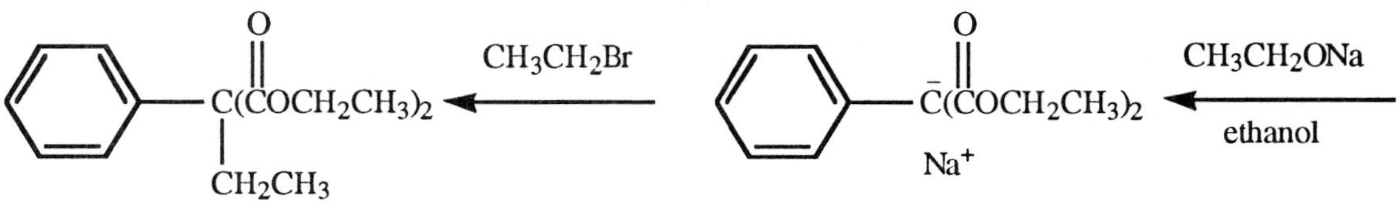

diethyl ethylphenylmalonate

16.28 (cont)

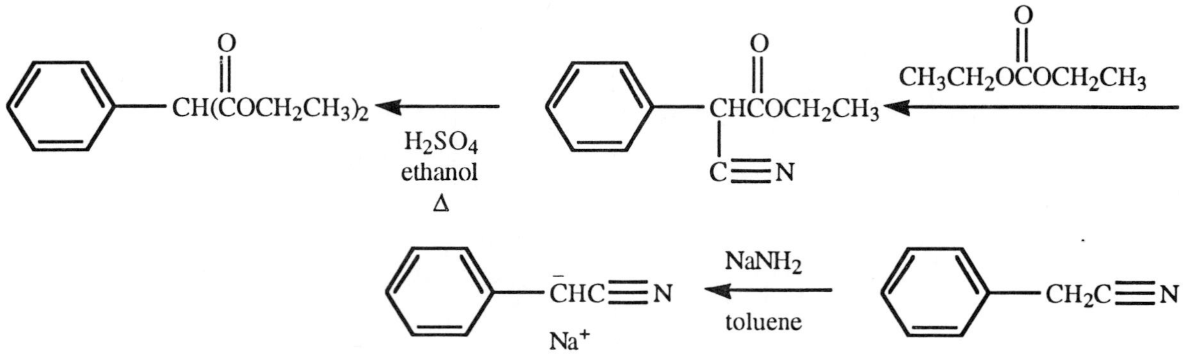

16.29 (a) $CH_3COCH_2COCF_3$ > $CH_3COCH_2COCH_3$

(b) 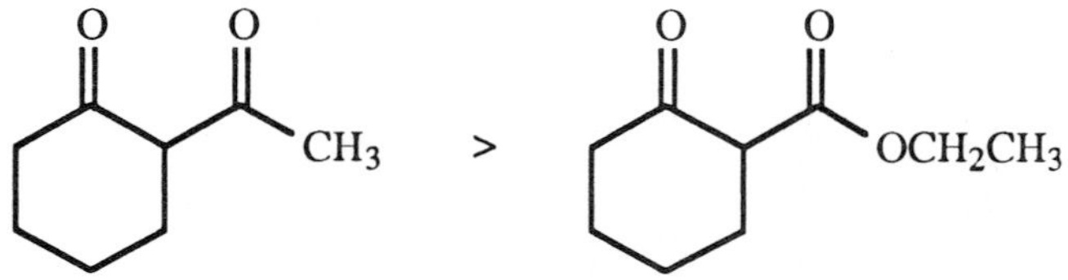

(c) $CH_3COCH_2COOCH_2CH_3$ > $CH_3COCH(CH_2CH_3)COOCH_2CH_3$

(d) $O_2NCH_2NO_2$ > $O_2NCH_2COCH_3$ > $O_2NCH_2COOCH_2CH_3$

(e) $HCOCH_2CHO$ > $CH_3CH_2OCOCH_2COOCH_2CH_3$ > $CH_3CH_2OCOCH(CH_2CH_3)COOCH_2CH_3$

16.30

(a) 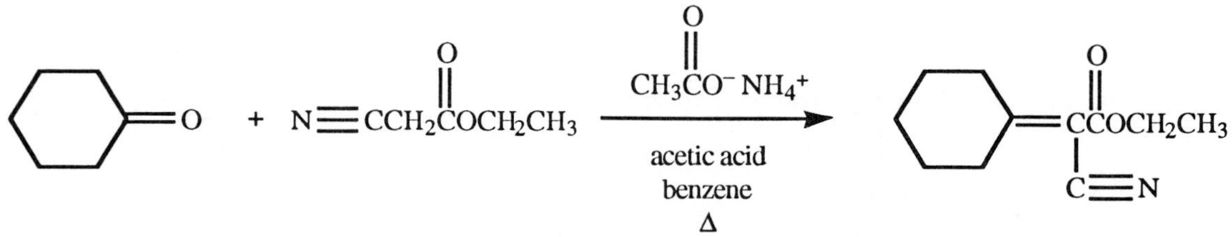

A

16.30 (cont)

(b)

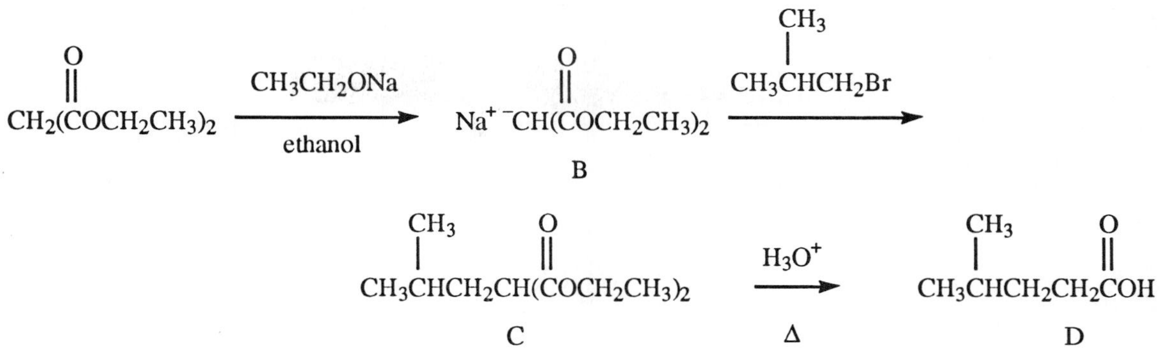

CH$_2$(COCH$_2$CH$_3$)$_2$ $\xrightarrow[\text{ethanol}]{\text{CH}_3\text{CH}_2\text{ONa}}$ Na$^+$ $^-$CH(COCH$_2$CH$_3$)$_2$ $\xrightarrow{\text{CH}_3\text{CHCH}_2\text{Br}}$

B

CH$_3$CHCH$_2$CH(COCH$_2$CH$_3$)$_2$ $\xrightarrow[\Delta]{\text{H}_3\text{O}^+}$ CH$_3$CHCH$_2$CH$_2$COH

C

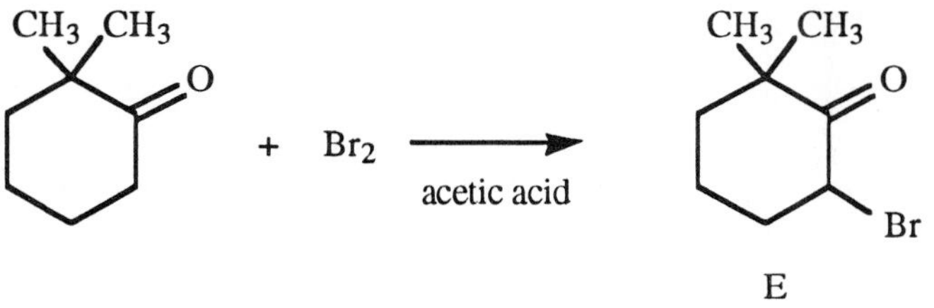

D

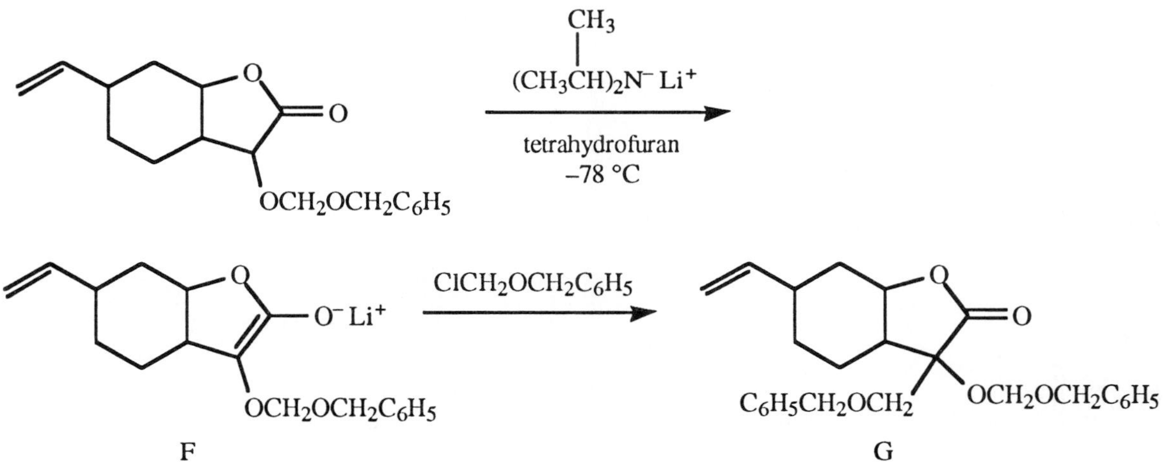

(c)

+ Br$_2$ $\xrightarrow{\text{acetic acid}}$

E

(d)

$\xrightarrow[\substack{\text{tetrahydrofuran} \\ -78\ °C}]{(\text{CH}_3\text{CH})_2\text{N}^- \text{Li}^+}$

F

$\xrightarrow{\text{ClCH}_2\text{OCH}_2\text{C}_6\text{H}_5}$

G

(e) O$_2$N—

—CCH$_3$ + CH$_3$COCH$_3$ $\xrightarrow{\text{CH}_3\text{ONa}}$ $\xrightarrow{\text{H}_3\text{O}^+}$

O$_2$N—

—CCH$_2$CCH$_3$ + CH$_3$OH

H

I

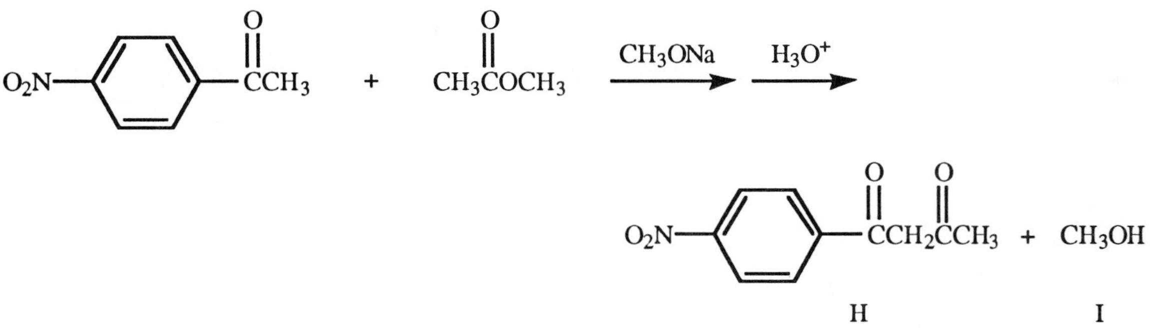

16.30 (cont)

(f)

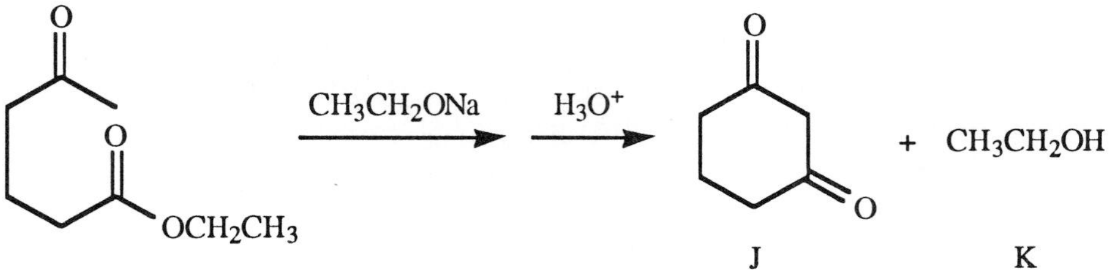
OCH2CH3
CH3CH2ONa
H3O+
+ CH3CH2OH
J
K

(g)

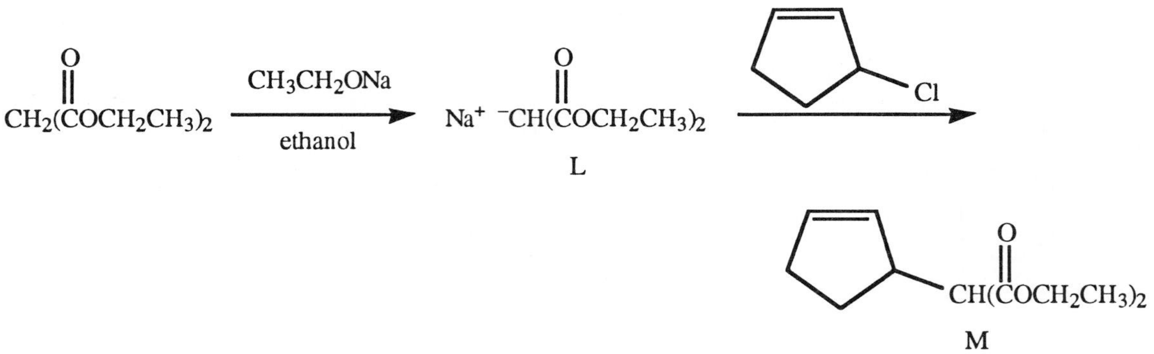
CH2(COCH2CH3)2
CH3CH2ONa
ethanol
Na+ −CH(COCH2CH3)2
L
Cl
CH(COCH2CH3)2
M

16.31

(a)

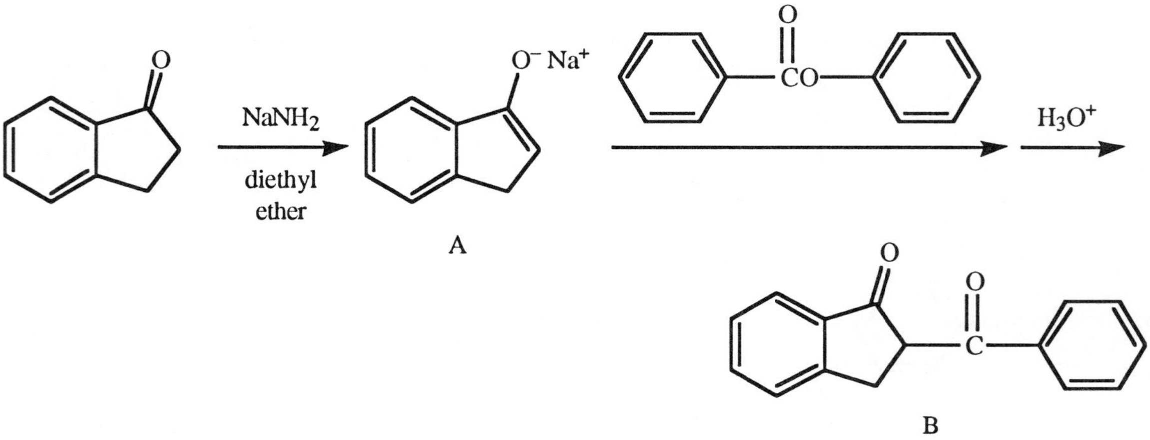
NaNH2
diethyl ether
O− Na+
A
CO
H3O+
C
B

(b)

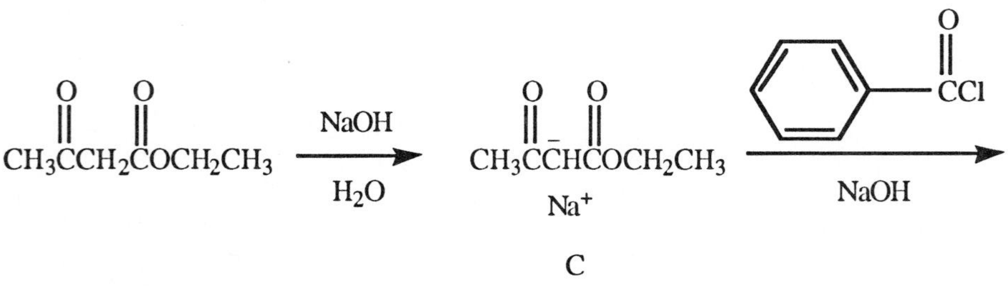
CH3CCH2COCH2CH3
NaOH
H2O
CH3CCHCOCH2CH3
Na+
C
CCl
NaOH

16.31 (b) (cont)

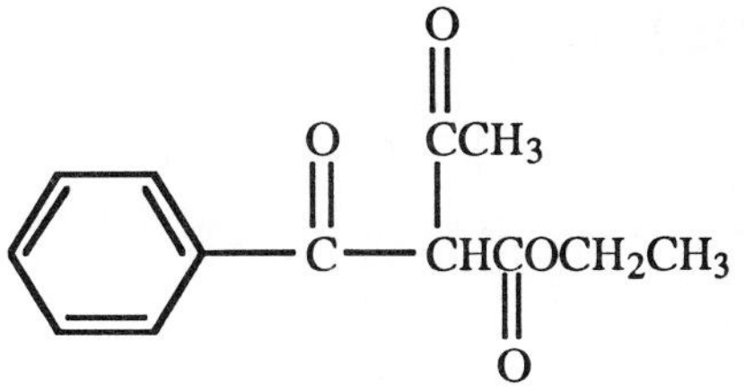

D

(c)

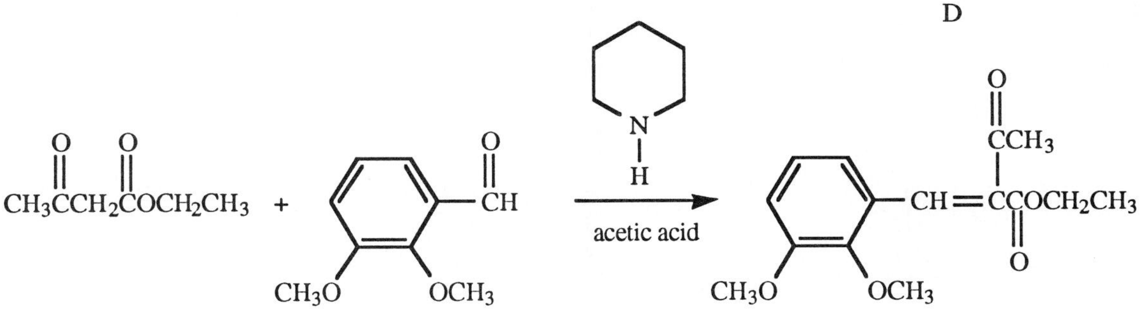

$CH_3CCH_2COCH_2CH_3$ +

E

(d)

$CH_3CCH_2CCH_2CH_2CH_2CH=CH_2$

F

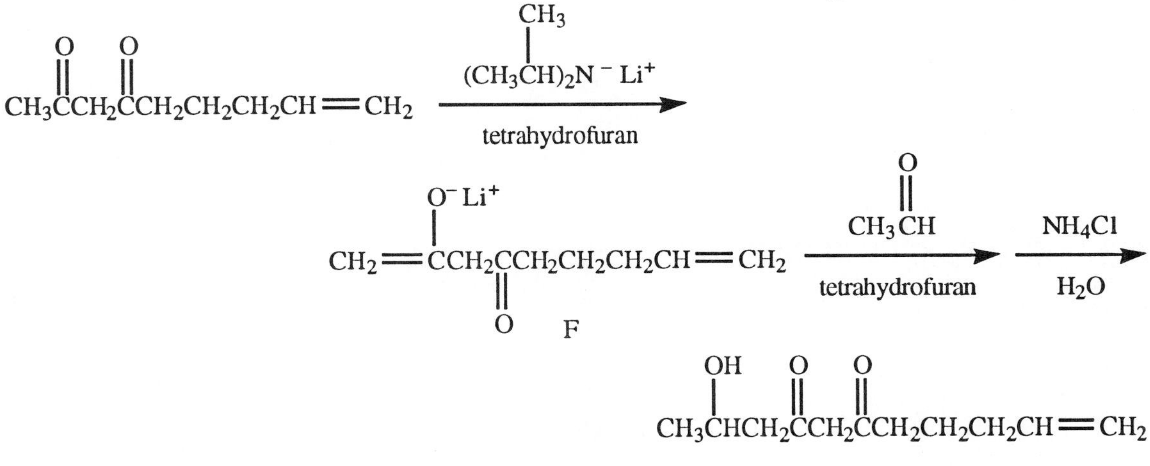

G

(e)

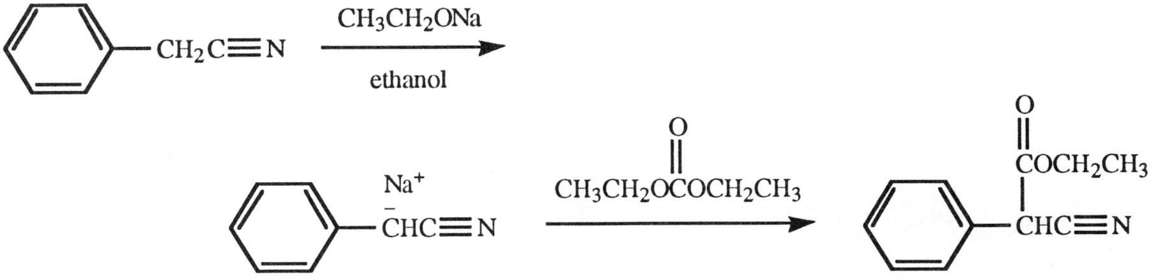

H

I

(f)

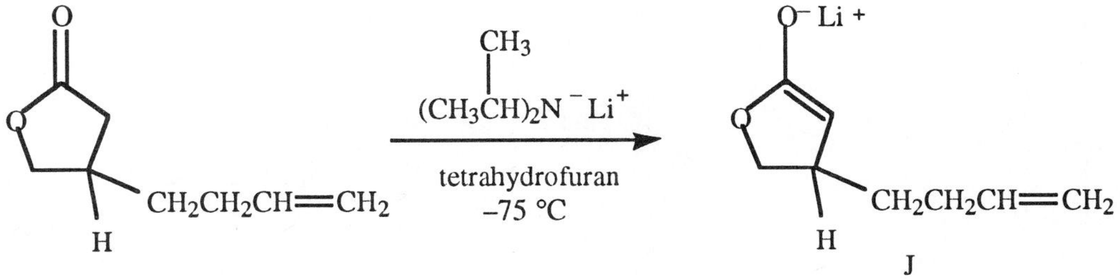

J

16.31 (f) (cont)

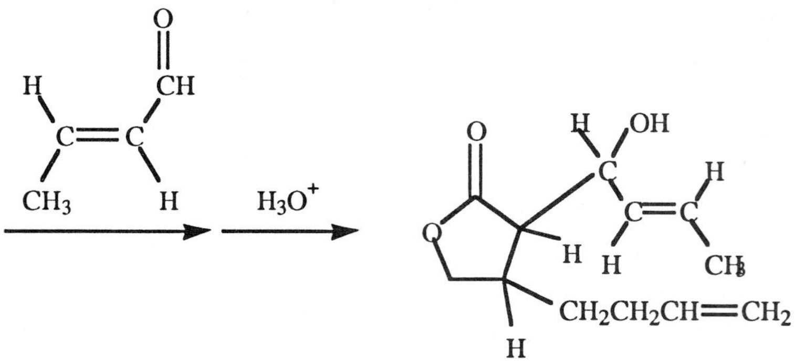

K

(g)

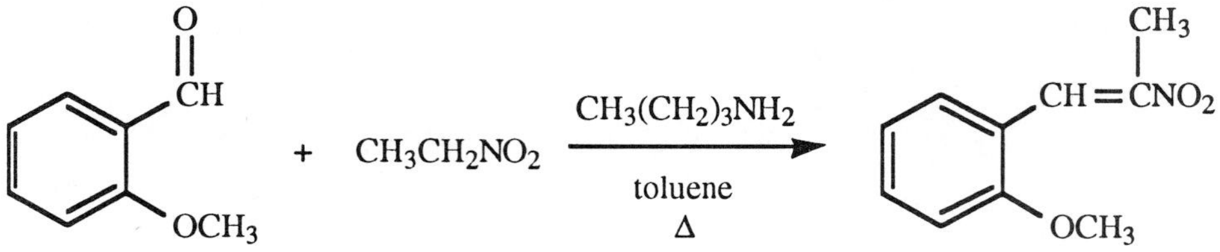

L

(h)

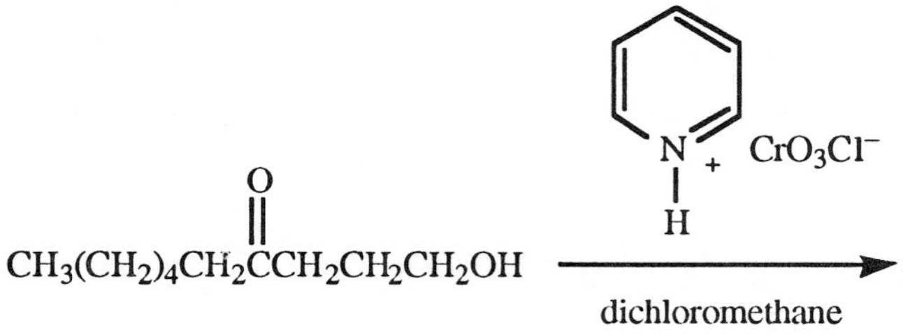

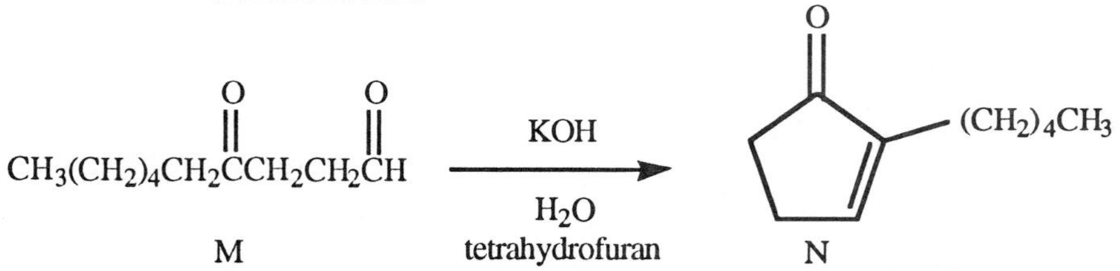

M   N

16.32

(a)

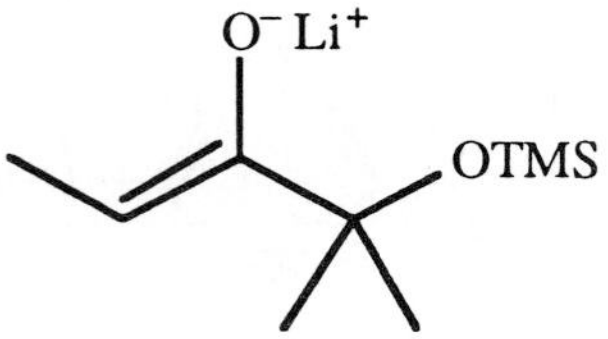

(b)

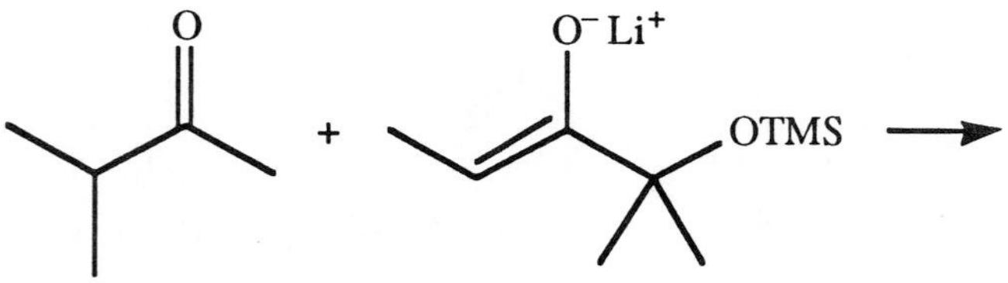

16.32 (b) (cont)

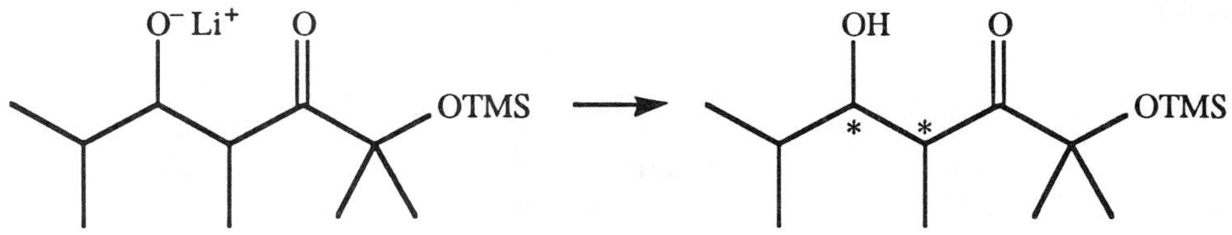

aldol product with
two stereocenters

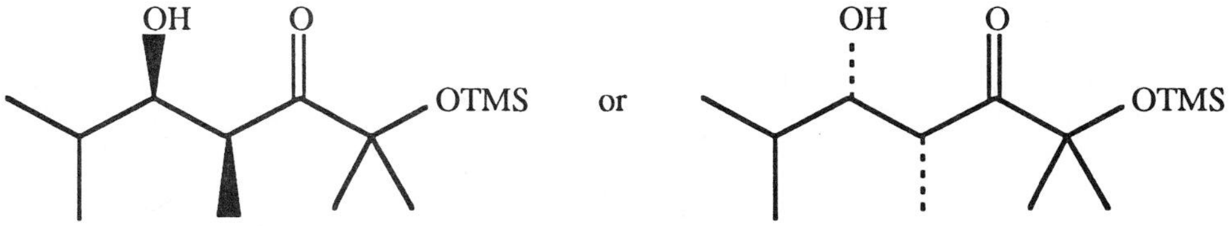

or

16.33

$$CH_3CH_2OCCH_2CCH_2COCH_2CH_3 \xrightarrow{CH_3CH_2O^- Na^+} CH_3CH_2OCCH_2C\!\!=\!\!CHCOCH_2CH_3$$

$$\xrightarrow[\text{ICH}_2\text{CHCH}_3]{\text{OTHP}} CH_3CH_2OCCH_2CCHCH_2CHCH_3 \xrightarrow{CH_3CH_2O^- Na^+}$$

$$CH_3CH_2OCCH\!\!=\!\!CCHCH_2CHCH_3 \xrightarrow{CH_2=CHCH_2CH_2Br} CH_2=CHCH_2CH_2CHCCHCH_2CHCH_3$$

$$\xrightarrow[\substack{H_2O \\ \Delta}]{NaOH} CH_2=CHCH_2CH_2CHCCHCH_2CHCH_3 \xrightarrow[\Delta]{H_3O^+}$$

$$CH_2=CHCH_2CH_2CH_2CCH_2CH_2CHCH_3$$

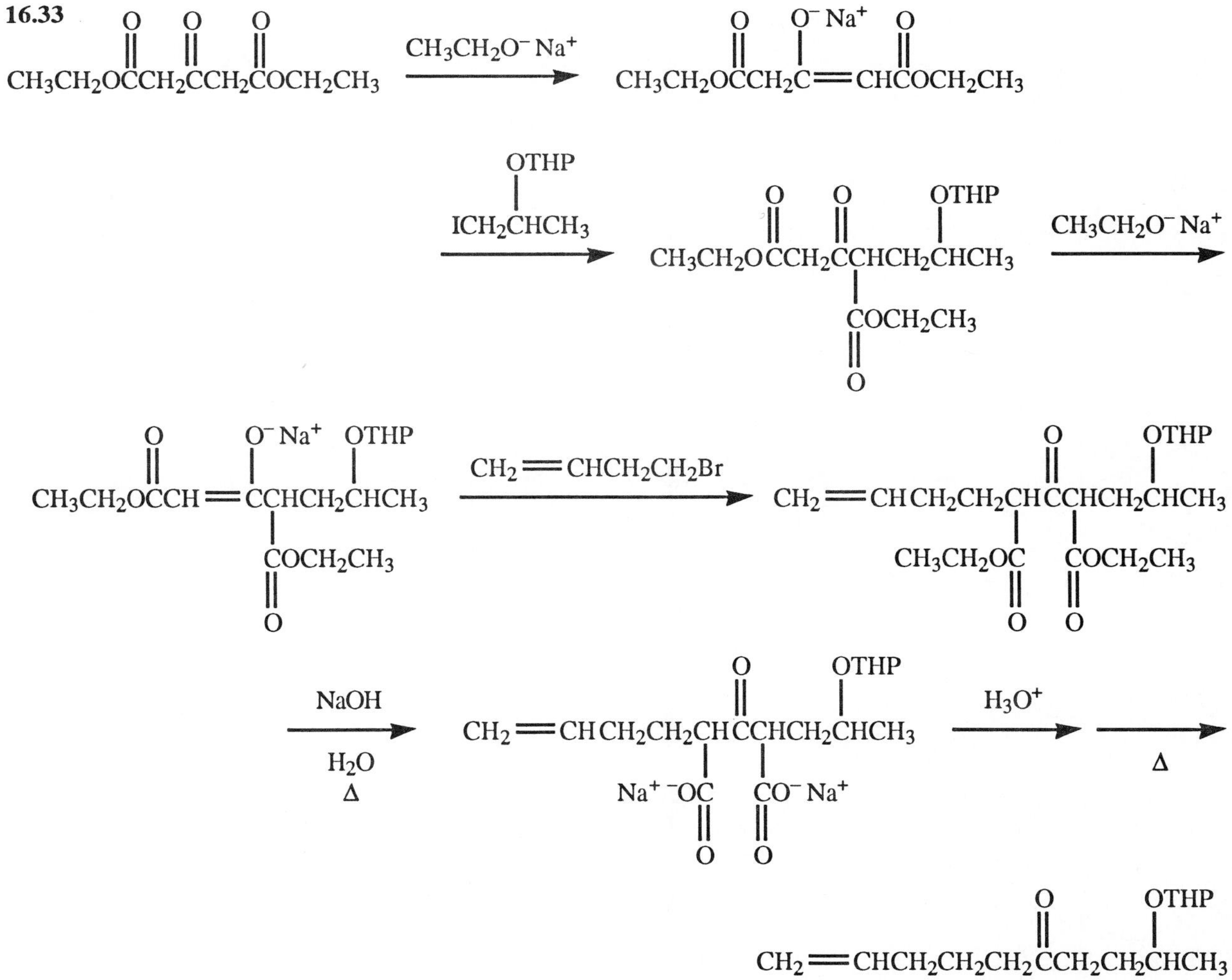

The ester is hydrolyzed with base because long heating with acid would also remove the protecting group.

16.34

Step 1

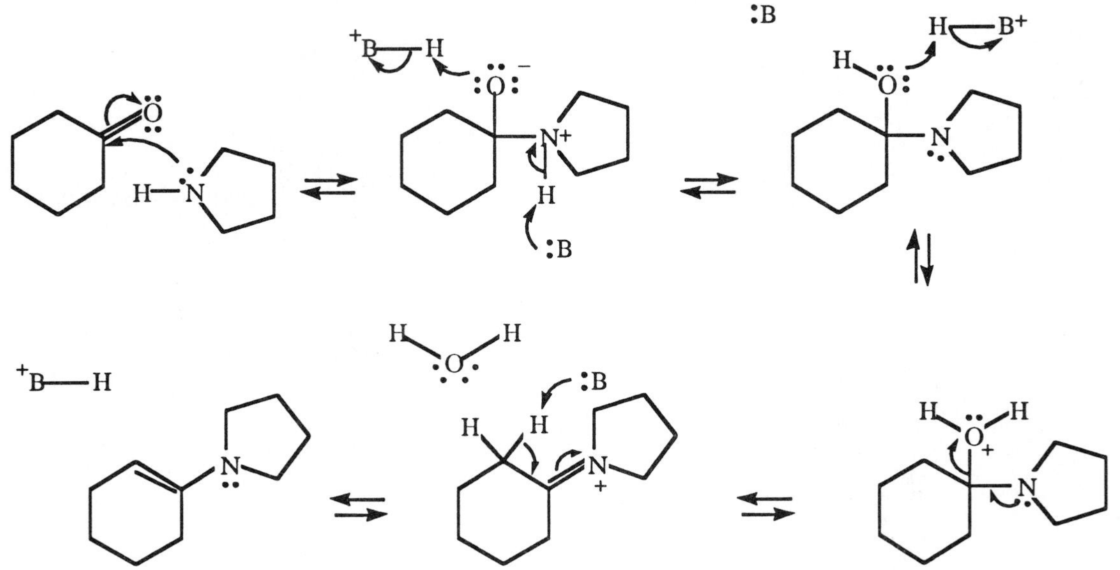

Step 2

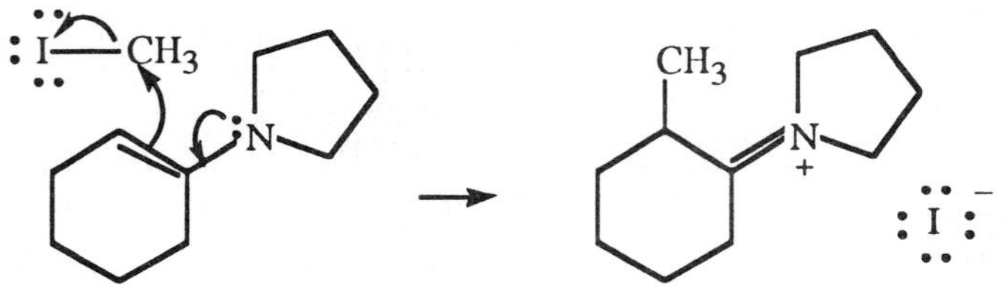

Step 3

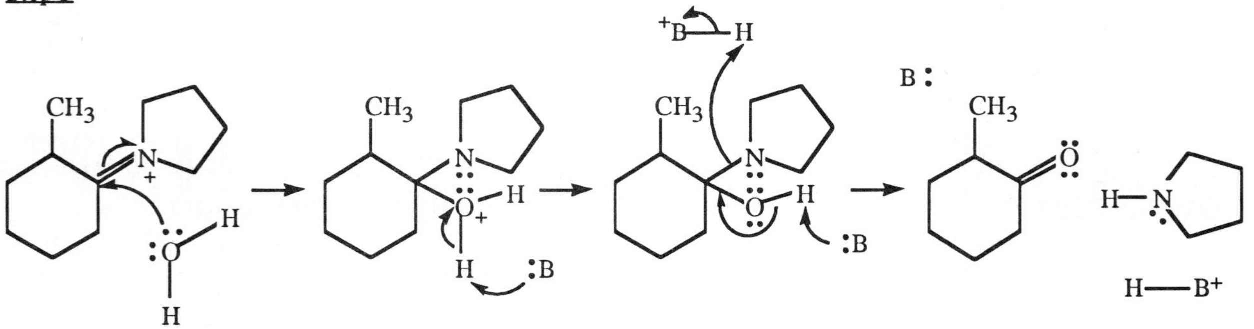

16.35

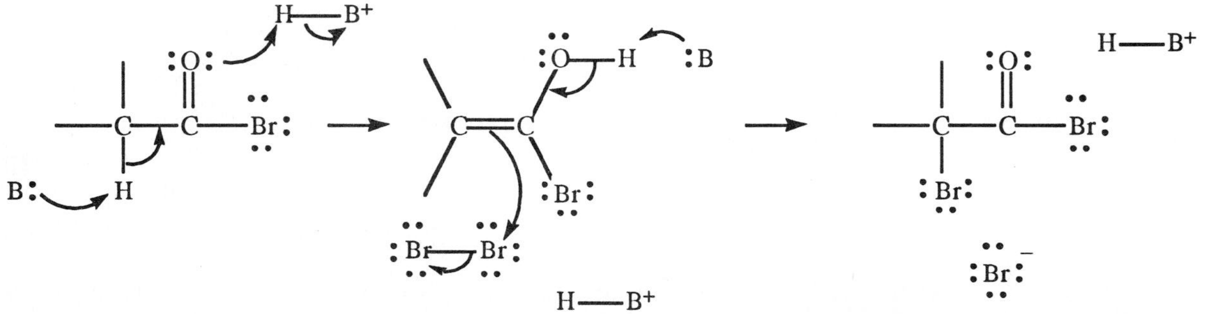

16.35 (cont)

For the acid to react with bromine, enolization must occur at the α-carbon atom of the acid. The carboxylic acid is in equilibrium with the ionized form, the carboxylate anion, which bears a negative charge and, therefore, does not enolize readily. Conversion of the acid to the acid bromide makes the carbon atom of the carbonyl group more electrophilic than it is in the carboxylic acid, thus increasing the acidity of the α-hydrogen atom and promoting enolization. Only a catalytic amount of phosphoric tribromide is used, therefore most of the starting material and the product are present as the carboxylic acid. While the mechanism of the interconversion of the carboxylic acid and acid bromide is not known, it may be seen as taking place by nucleophilic attack of the hydroxyl group of one carboxylic acid on the carbonyl group of the acid bromide, giving an intermediate resembling an acid anhydride in structure. Bromide ion could then displace a carboxylic acid, giving rise to a new acid bromide.

16.36

16.36 (cont)

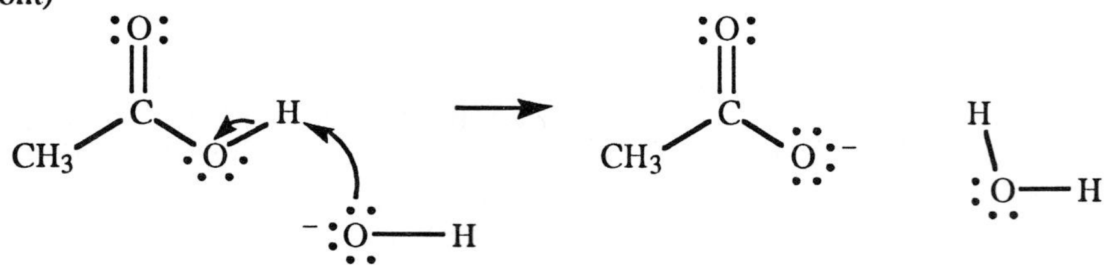

Once one of the α-hydrogen atoms is replaced by a halogen atom, the other hydrogen atoms on that carbon atom become more acidic. Enolization takes place a second and then a third time on the same side. Attack at the carbonyl group by base leads to the loss of the trihalocarbanion, a relatively stable species.

16.37 CO_3^{2-} + H_2O $\rightleftharpoons$ HCO_3^- + OH^-

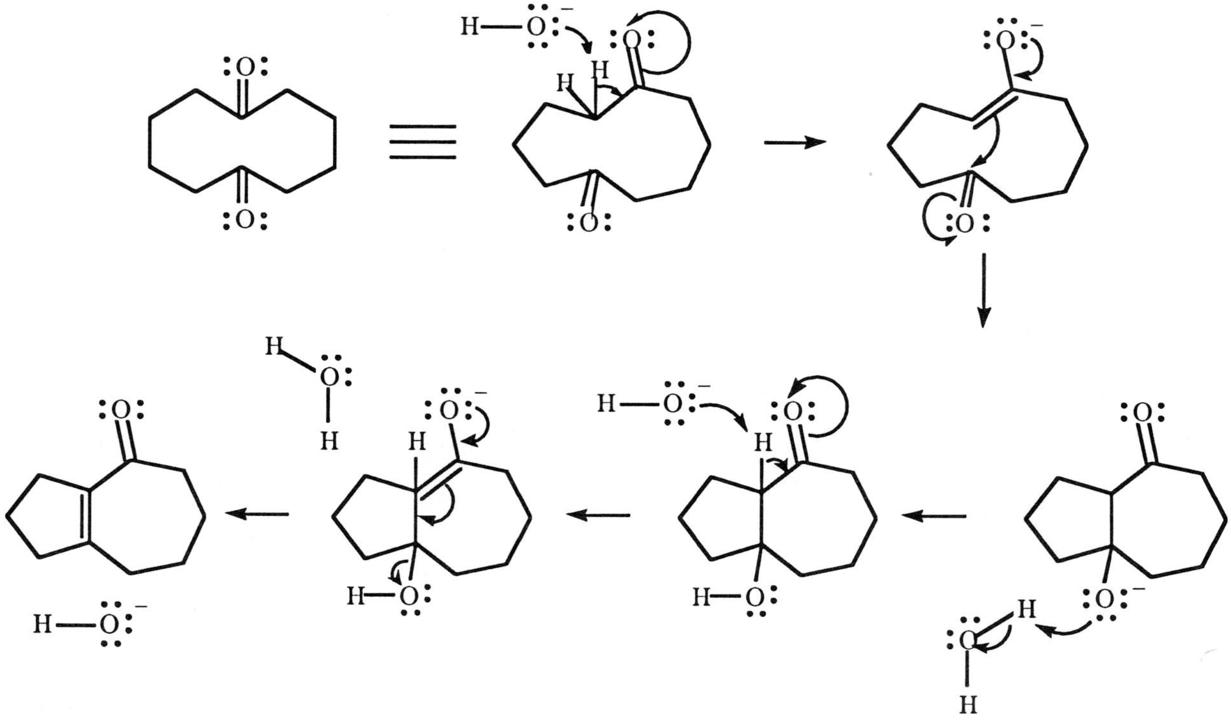

16.38

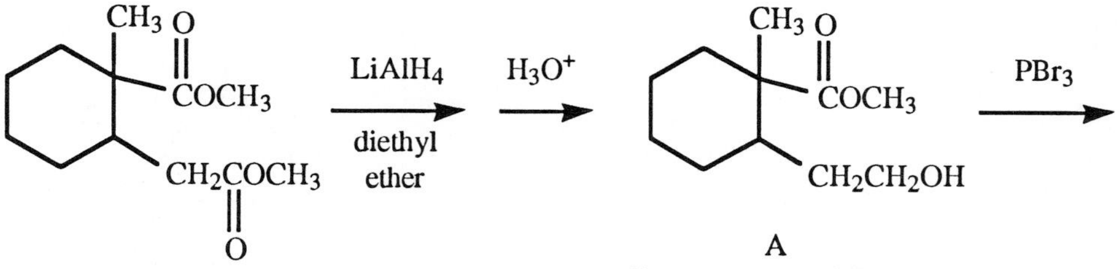

A
(Less sterically hindered
ester is reduced.)

16.38 (cont)

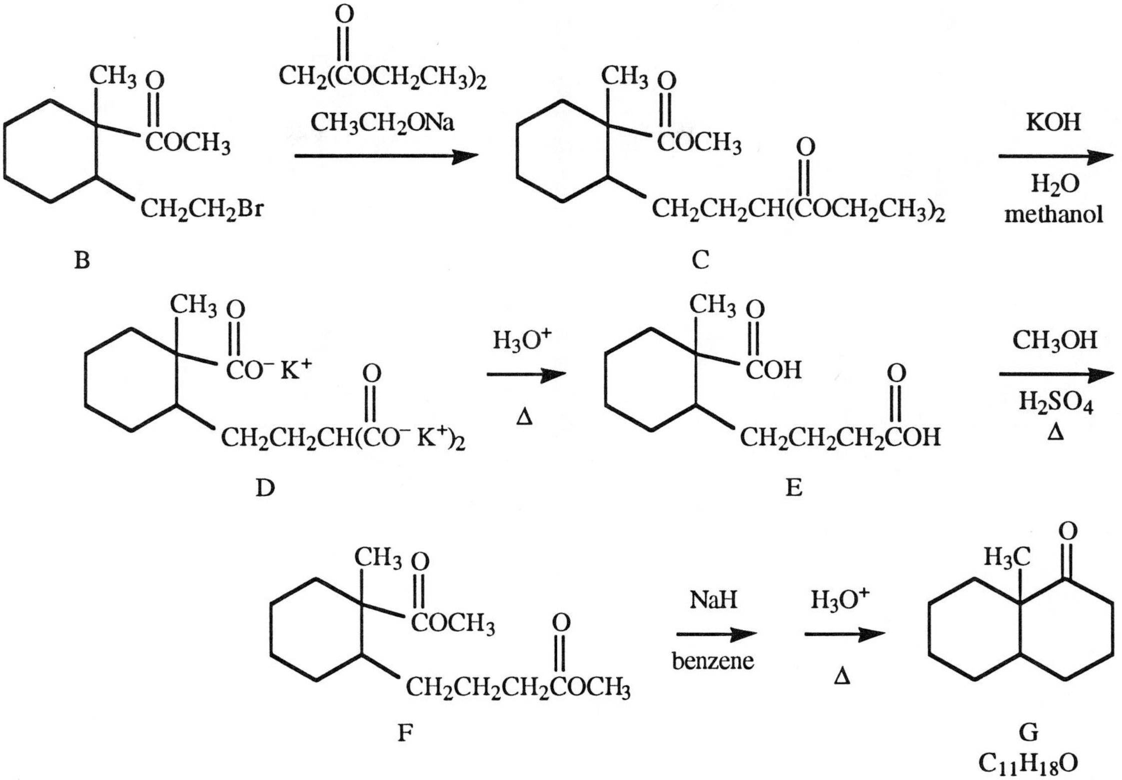

B

C

D

E

F

G
$C_{11}H_{18}O$

16.39

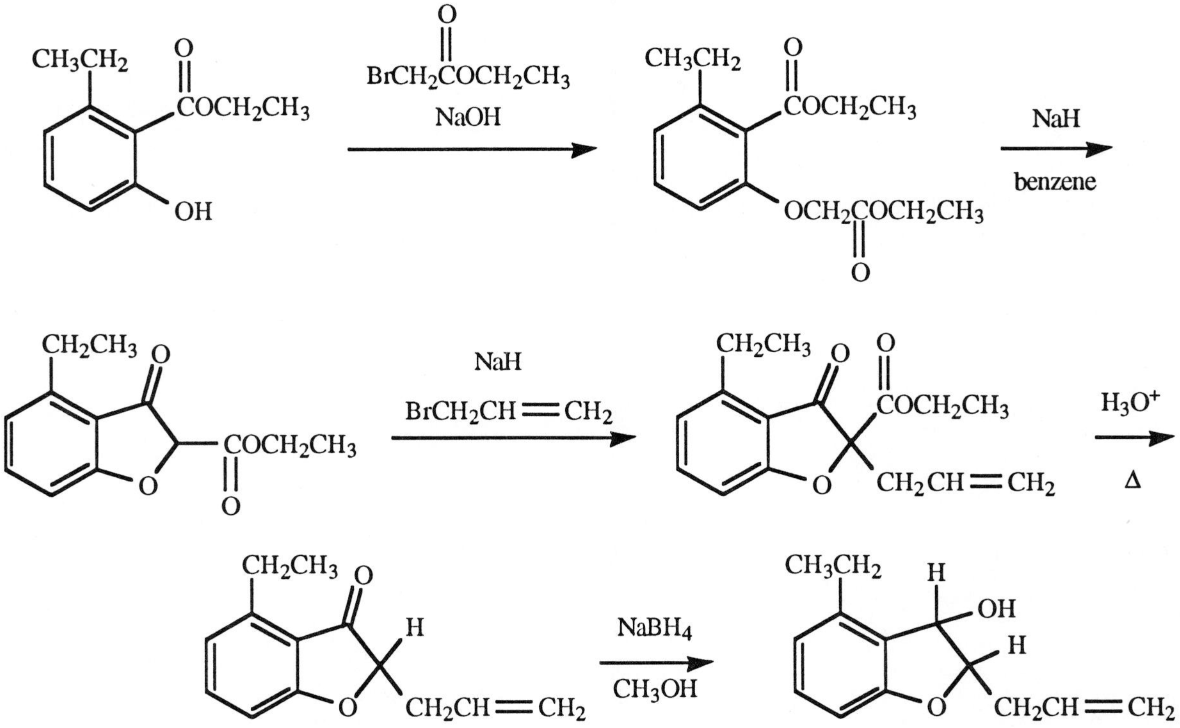

16.40

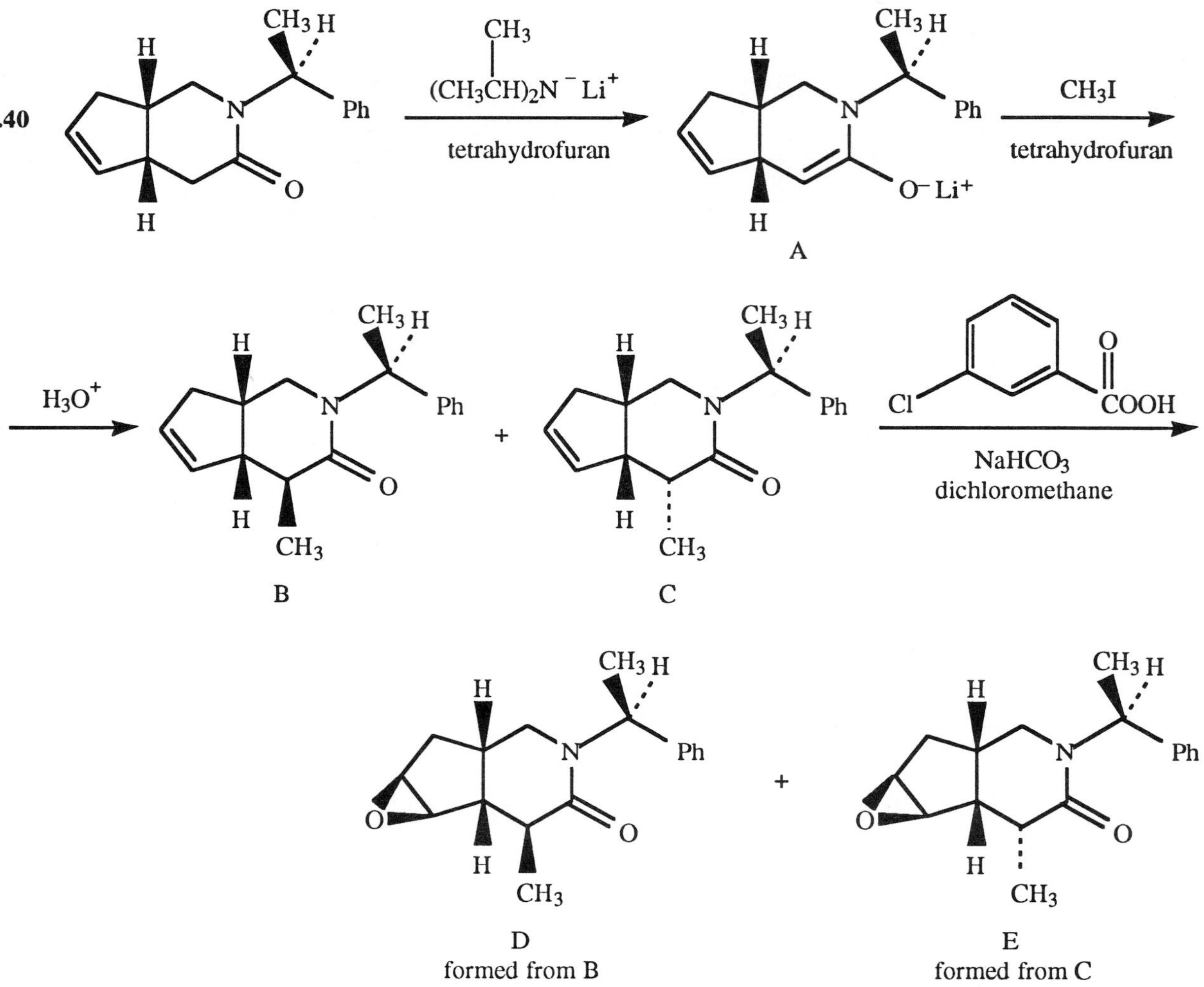

The top face of the ring system, which is on the same side as the hydrogen atoms and away from the bonds to the other ring, is more easily reached by the reagent.

16.41

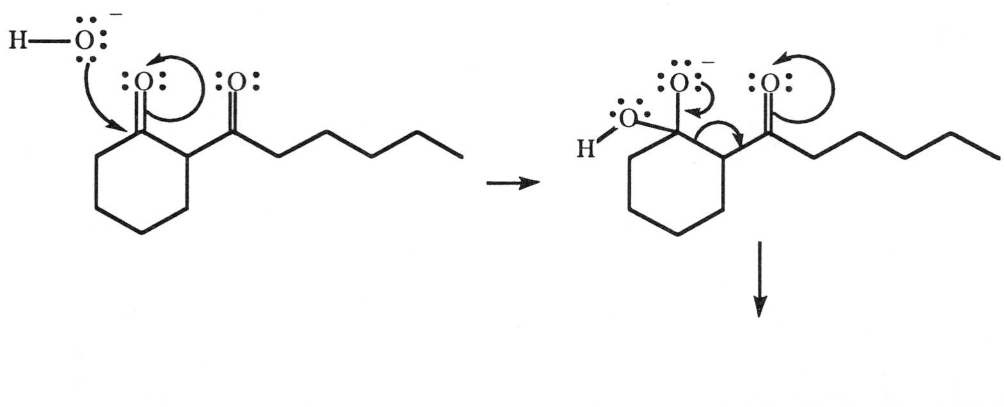

16.41 (cont)

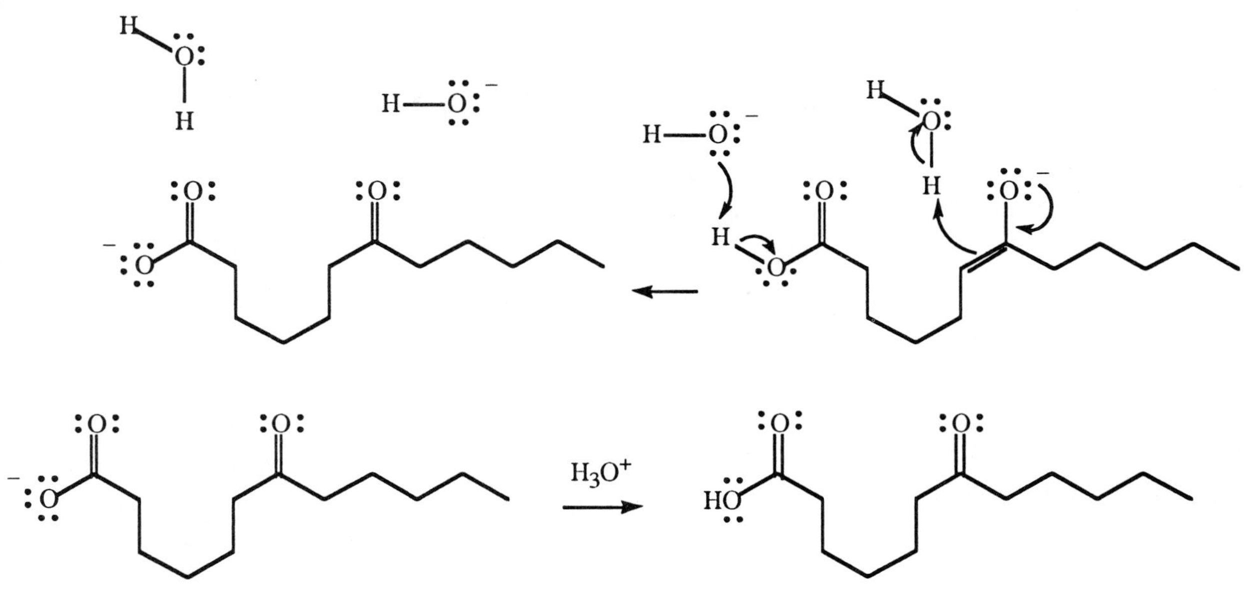

16.42

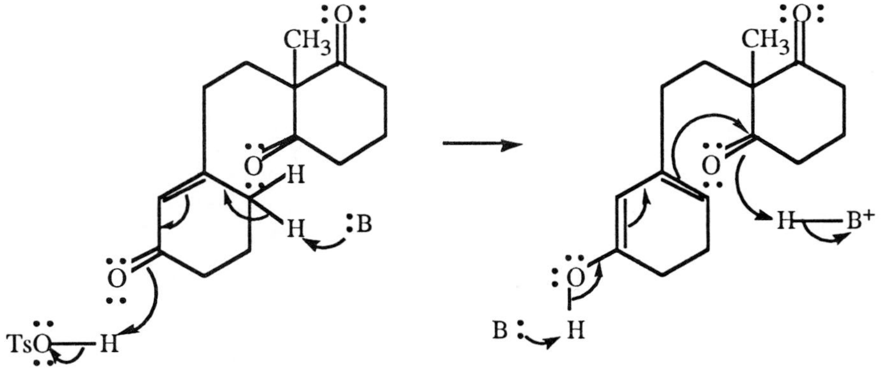

The effect that a carbonyl group
has in increasing the acidity of
an α-hydrogen atom is transmitted
to the γ-position by the conjugated
double bond.

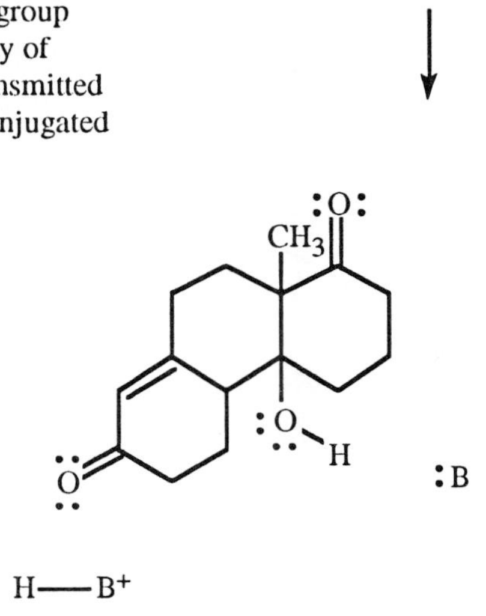

16.43

1.

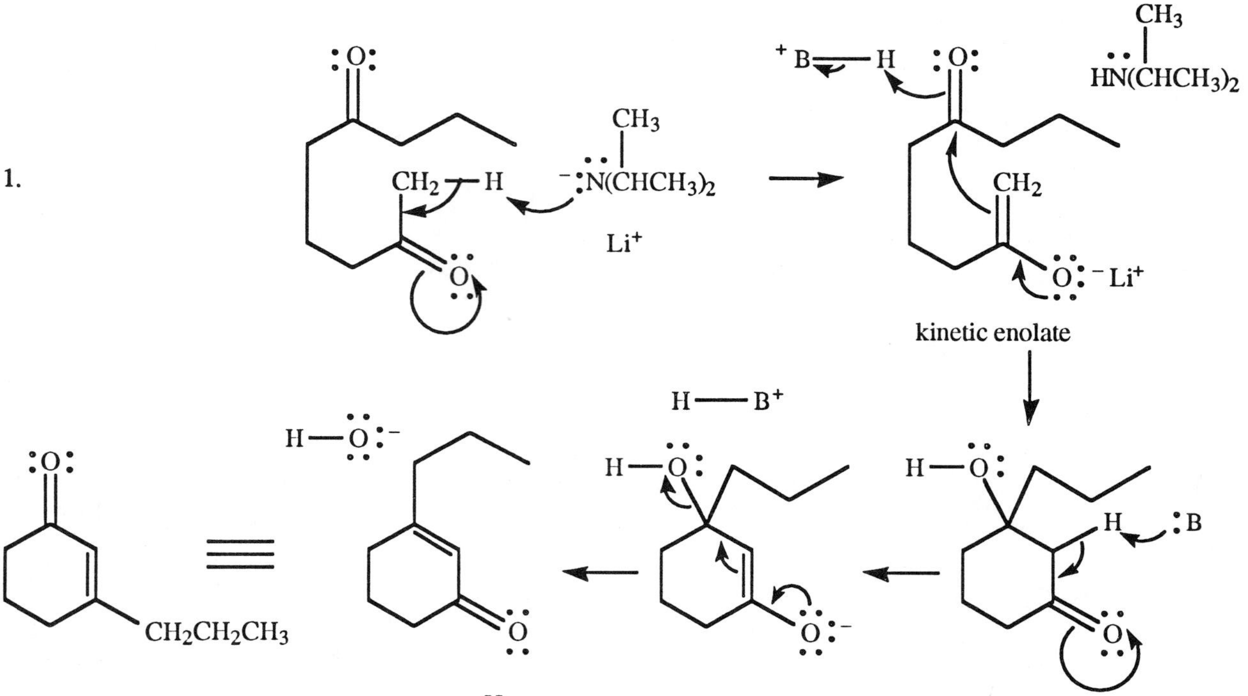
CH₃
HN(CHCH₃)₂
Li⁺
kinetic enolate
H—B⁺
CH₂CH₂CH₃

2.

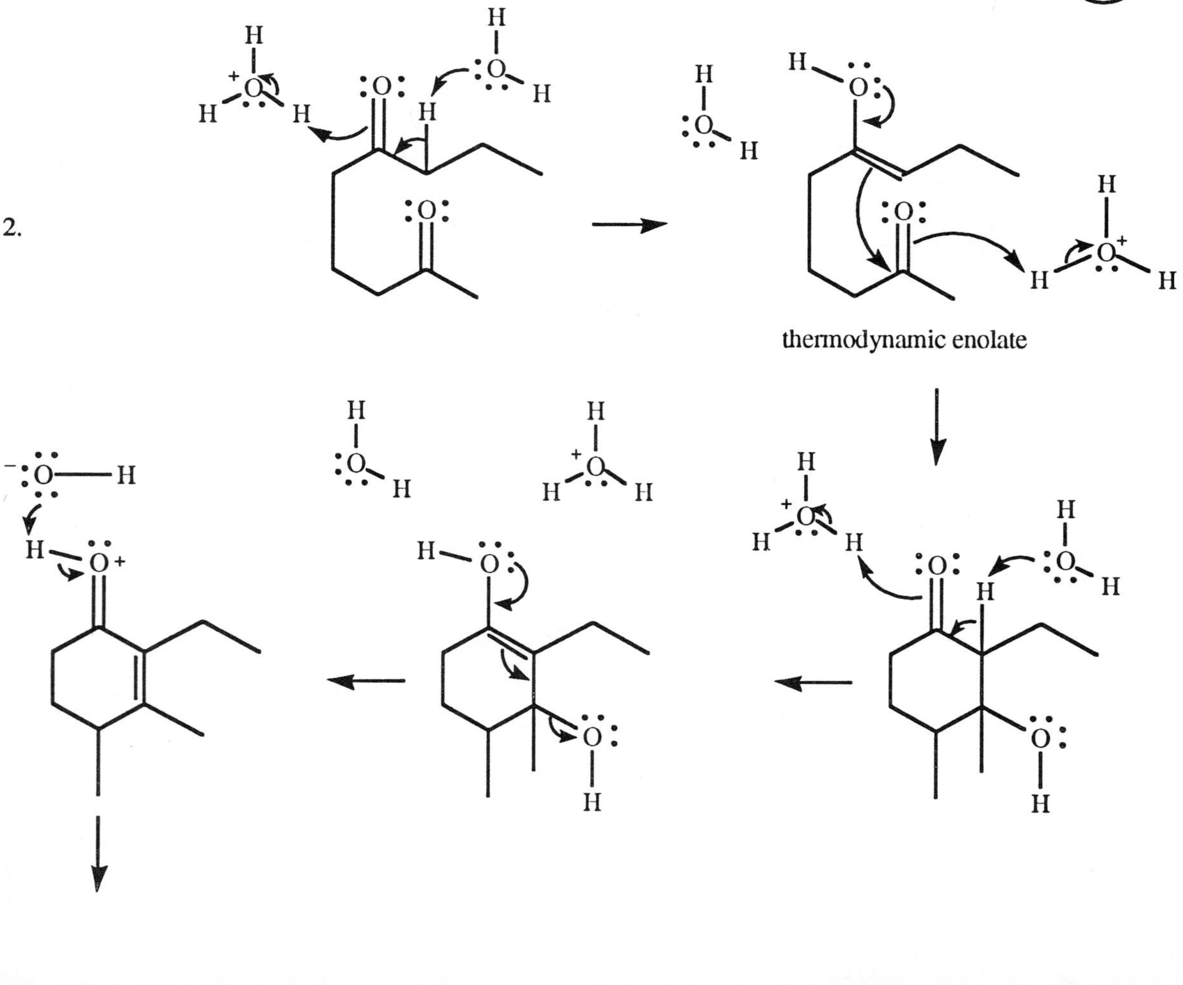
thermodynamic enolate

16.43 (cont)

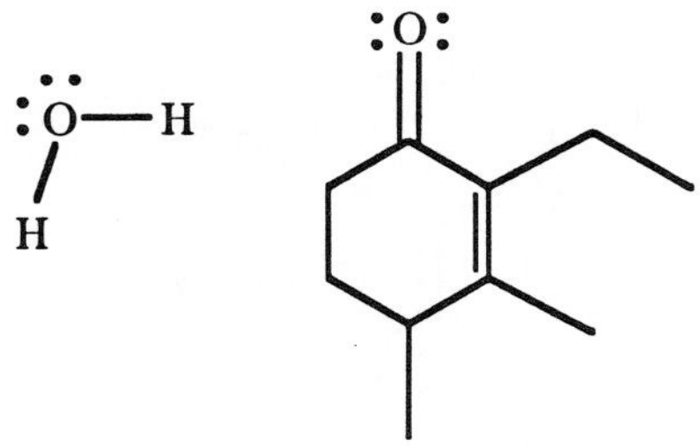

16.44

(a)   lithium diisopropylamide

(b)

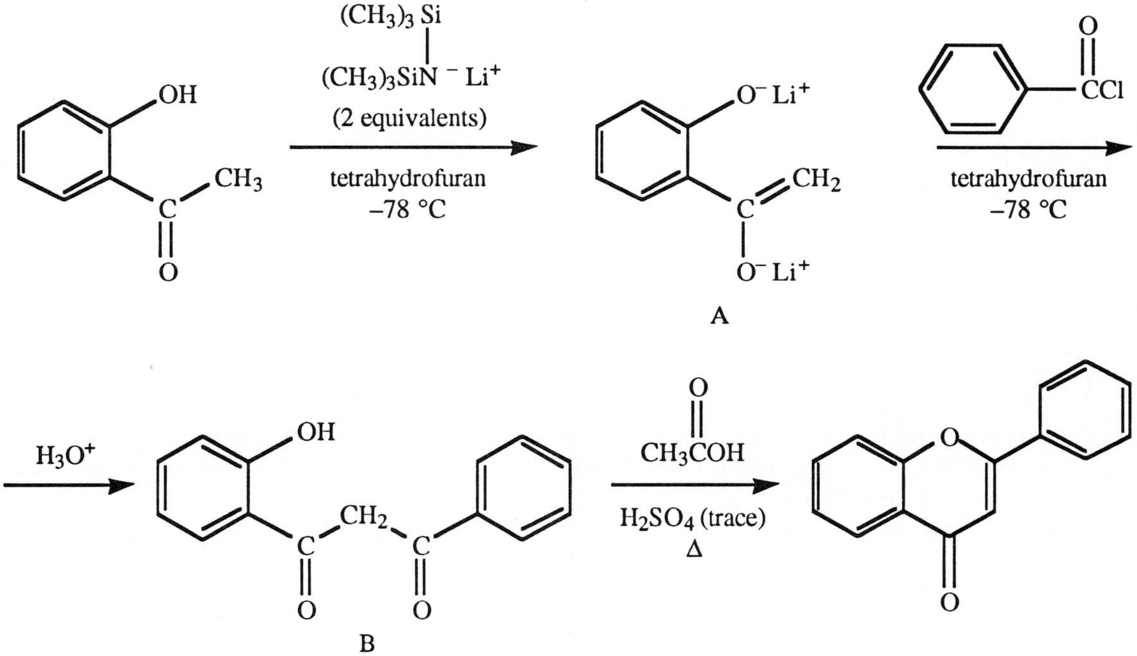

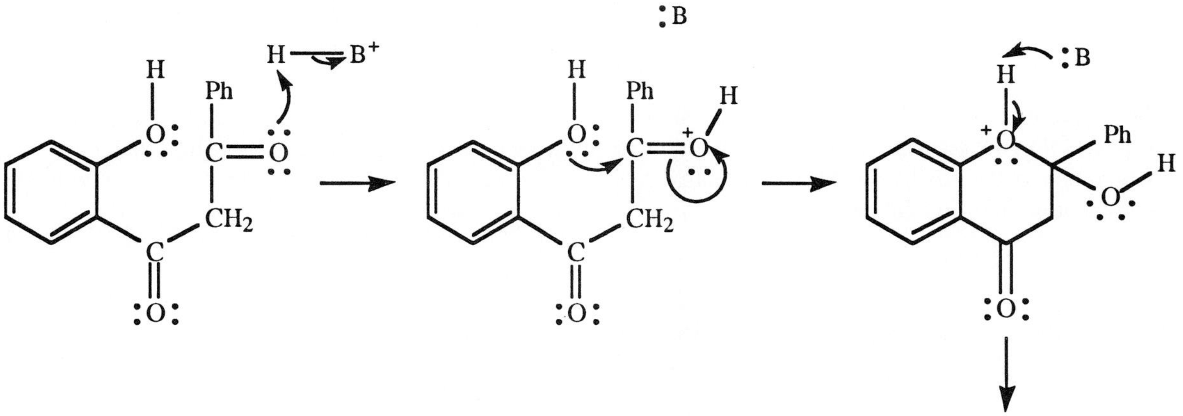

16.44 (cont)

16.45   An examination of the major resonance contributors of the enolate anion from 1-phenyl-2-methyl-1-propanone reveals that there is double bond character to the carbon-carbon bond between carbons 1 and 2 in the structure in which the negative charge is localized on the oxygen atom. To the extent that there is such double bond character, free rotation around that bond is inhibited, and the two methyl groups are no longer chemical-shift equivalent (for example, one is trans and the other cis, to the oxygen atom), and would be expected to have different chemical shifts.

The experimental observation that there are two chemical shifts for the methyl groups in the region δ 1.0–2.0 suggests that the resonance contributor with the negative charge on oxygen is the more important one. We would expect the positively charged lithium ion to be associated with the oxygen atom rather than with the carbon atom in the enolate anion.

16.46 $2 \ CH_3CH_2OH \ + \ Na \ \longrightarrow \ 2 \ CH_3CH_2O^- \ Na^+ \ + \ H_2$

16.46 (cont)

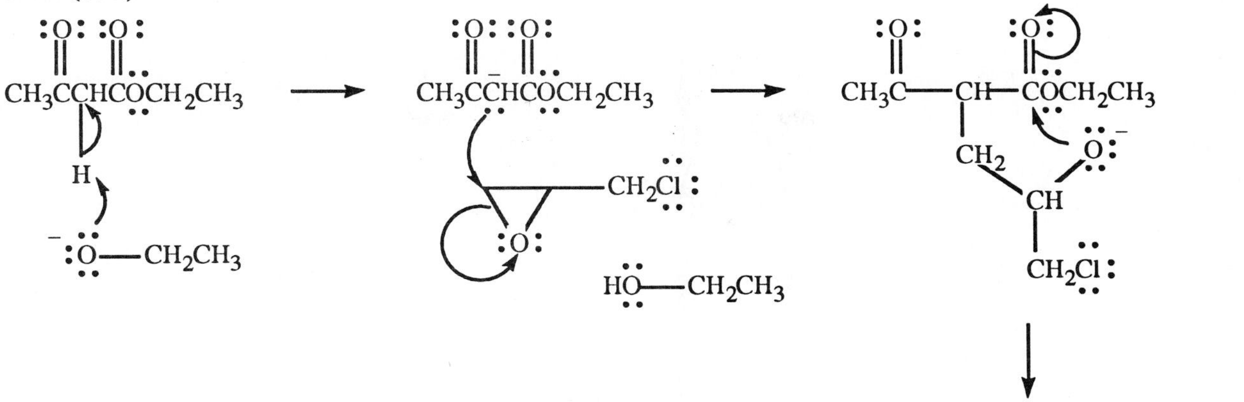

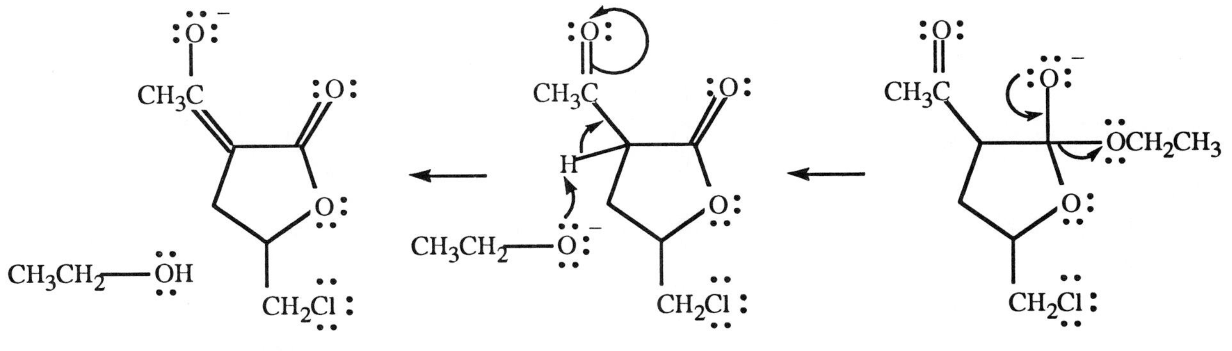

enolate anion,
stable product of
ketolactone in base

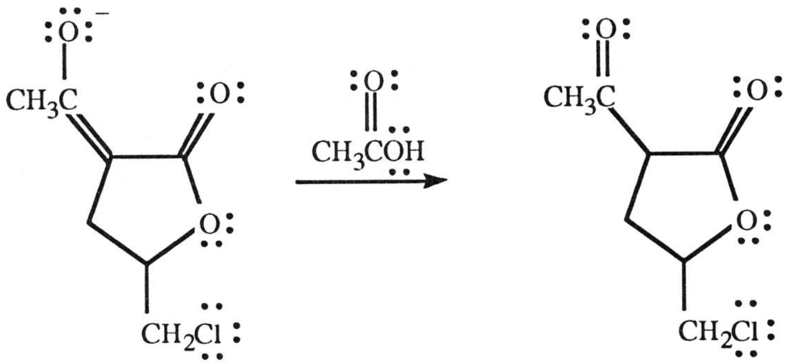

16.47

1.

2.

Note that this reaction is reminiscent of the formation of oxiranes by the intramolecular S_N2 reaction of alkoxide ions derived from halohydrins (Section 12.4C). The thermodynamic enolate of the ketone has the right regiochemistry and stereochemistry to give the three-membered ring.

16.48

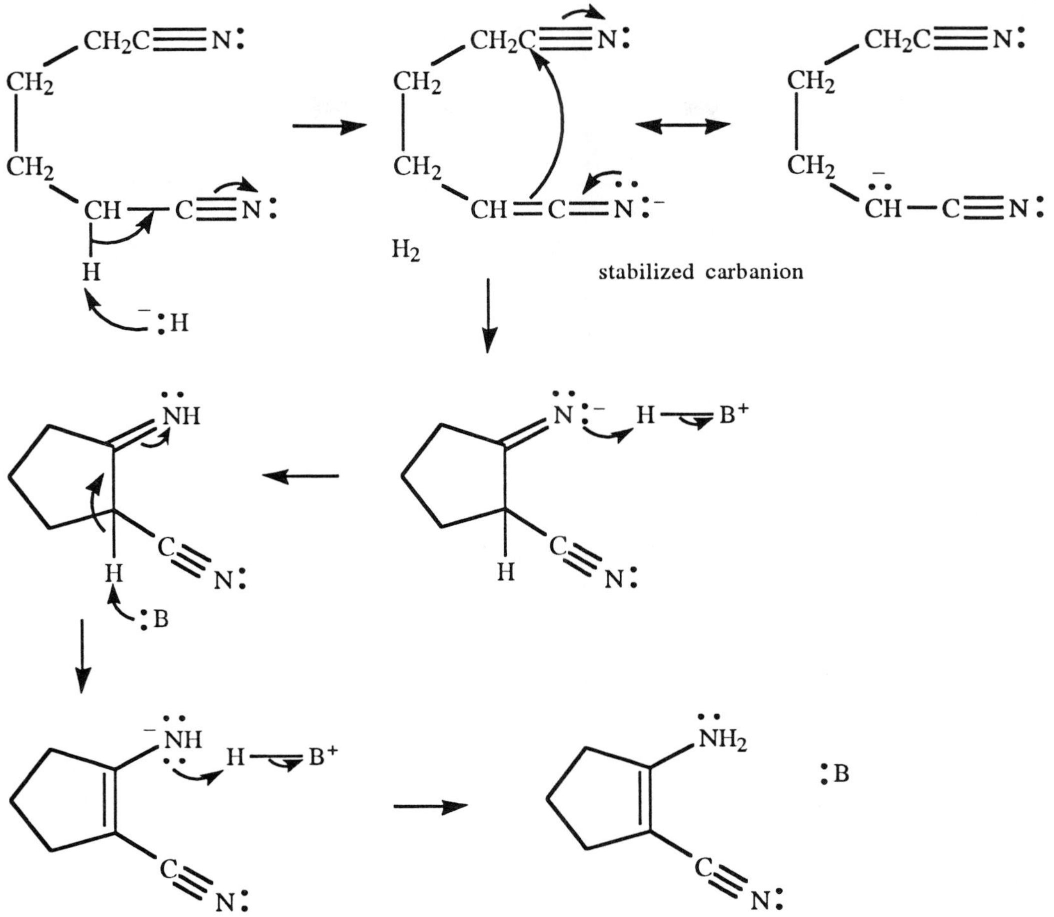
stabilized carbanion

16.49

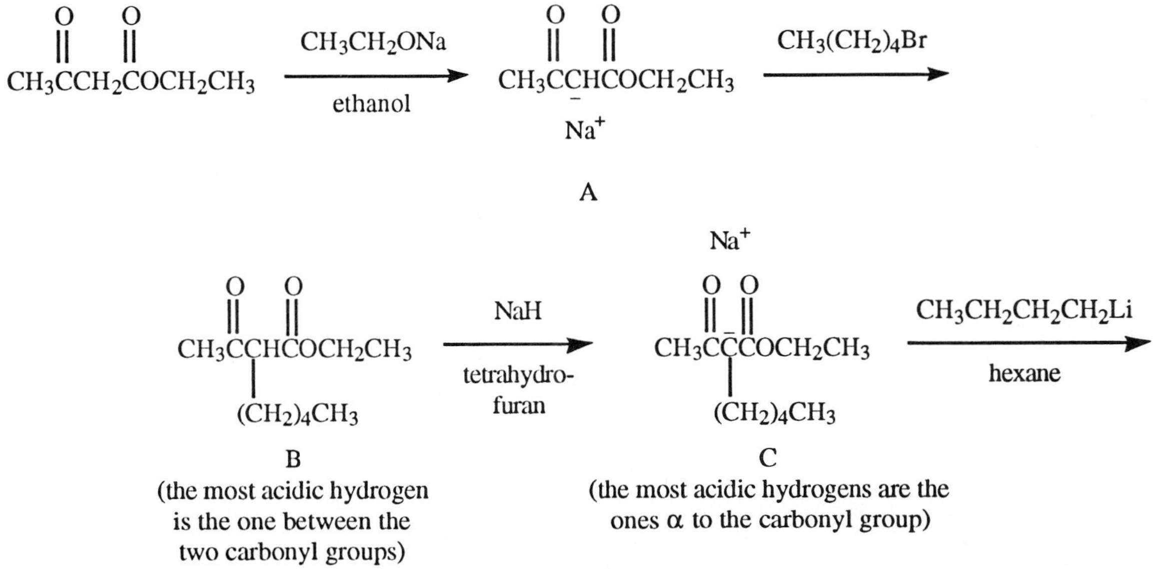
CH3CH2ONa
ethanol
CH3(CH2)4Br
A
NaH
tetrahydro-
furan
CH3CH2CH2CH2Li
hexane
B
(the most acidic hydrogen is the one between the two carbonyl groups)
C
(the most acidic hydrogens are the ones α to the carbonyl group)

16.49 (cont)

Na$^+$

Li—CH$_2$CCCOCH$_2$CH$_3$ $\xrightarrow{\triangle-CH_3}$ CH$_3$CHCH$_2$CH$_2$CCCOCH$_2$CH$_3$ $\xrightarrow[\text{H}_2\text{O}]{\text{NaOH}}$
          (CH$_2$)$_4$CH$_3$ O$^-$Li$^+$ (CH$_2$)$_4$CH$_3$ Δ

D

(the most nucleophilic
carbon is the one α to
only one carbonyl group)

E

Na$^+$

CH$_3$CHCH$_2$CH$_2$CCCO$^-$Na$^+$ $\xrightarrow{\text{H}_2\text{SO}_4}$ CH$_3$CHCH$_2$CH$_2$CCH$_2$(CH$_2$)$_4$CH$_3$ $\xrightarrow{\text{CrO}_3}$
  O$^-$Li$^+$ (CH$_2$)$_4$CH$_3$ OH

F   G

+ CO$_2$

CH$_3$CCH$_2$CH$_2$CCH$_2$(CH$_2$)$_4$CH$_3$ $\xrightarrow[\text{ethanol}]{\text{NaOH} \atop \text{H}_2\text{O}}$ dihydrojasmone

H

dihydrojasmone

CH$_3$CCH$_2$CH$_2$CCH$_2$(CH$_2$)$_4$CH$_3$ $\xrightarrow[\text{ethanol}]{\text{NaOH} \atop \text{H}_2\text{O}}$ CH$_3$CCH$_2$CH$_2$C=CH(CH$_2$)$_4$CH$_3$

more stable enolate anion,
containing the more highly
substituted double bond

16.49 (cont)

$$CH_2\!\!=\!\!\overset{\overset{\displaystyle O^-Na^+}{|}}{C}CH_2CH_2\overset{\overset{\displaystyle O}{\|}}{C}CH_2(CH_2)_4CH_3$$

less stable enolate anion,
with the less highly
substituted double bond

dihydrojasmone

$(CH_2)_5CH_3$

this product does not form

16.50 Compound A, $C_{12}H_{20}O_2$, has three units of unsaturation. The infrared spectrum of Compound A shows the presence of an ester functional group (C=O stretching frequency at 1735 cm^{-1} and C—O stretching frequency at 1225 cm^{-1}), and a double bond (=C—H stretching frequency at 2978 cm^{-1}, and =C—H bending frequency at 975 cm^{-1}). The alkene functional group is not likely to be a terminal double bond because the C=C stretching frequency is not seen in the infrared spectrum. The presence of a double bond is confirmed by the proton magnetic resonance spectrum (δ 5.3, 2H). The proton magnetic resonance spectrum also shows the presence of a methyl group bonded to a carbon atom bearing a single hydrogen atom (δ 1.3, 3H, doublet).

$$\text{Compound A} \xrightarrow[\text{Pt}]{\text{H}_2} \text{Compound B, } C_{12}H_{22}O_2$$

Compound B, $C_{12}H_{22}O_2$, still contains two of the three units of unsaturation, one of which is the ester functional group. The other unit of unsaturation must be a ring.

$$\text{Compound B} \xrightarrow[\substack{\text{H}_2\text{O} \\ \Delta}]{\text{NaOH}} \xrightarrow{\text{H}_3\text{O}^+} \text{Compound C, } C_{12}H_{24}O_3$$

Compound C, $C_{12}H_{24}O_3$, has only one unit of unsaturation, which is most likely the carbonyl group of a carboxylic acid, since treatment with aqueous base followed by acid will hydrolyze an ester to an alcohol and a carboxylic acid. This reaction suggests that the ester functional group was a lactone.

16.50 (cont)

$$\text{Compound C} \xrightarrow[\substack{\text{acetic} \\ \text{acid}}]{\text{CrO}_3} \text{Compound D, } C_{12}H_{24}O_3$$

$$\text{Compound D} \xrightarrow{I_2, \text{ NaOH}} \text{CHI}_3 \ + \ \overset{\displaystyle O}{\overset{\displaystyle \|}{\text{HOC}}}(CH_2)_9\overset{\displaystyle O}{\overset{\displaystyle \|}{\text{C}}}\text{OH}$$

therefore Compound D is $\text{CH}_3\overset{\displaystyle O}{\overset{\displaystyle \|}{\text{C}}}(CH_2)_9\overset{\displaystyle O}{\overset{\displaystyle \|}{\text{C}}}\text{OH}$

and Compound C is $\underset{\displaystyle \overset{|}{\text{OH}}}{\text{CH}_3\text{CH}}(CH_2)_9\overset{\displaystyle O}{\overset{\displaystyle \|}{\text{C}}}\text{OH}$

and Compound B is

The products from the ozonolysis of Compound A allow us to locate the double bond, but do not allow us to determine the stereochemistry of the double bond. If we had more extensive tables correlating infrared absorption frequencies with structure, we would find out that the =C—H bending frequency at 975 cm^{-1} tells us that the double bond is trans.

$$\text{Compound A} \xrightarrow[\substack{\text{ethanol} \\ \text{ethyl acetate}}]{O_3} \xrightarrow[\substack{\text{formic} \\ \text{acid}}]{H_2O_2} \xrightarrow[\text{ethanol}]{\text{KOH}} \xrightarrow{H_3O^+} \underset{\displaystyle \overset{|}{\text{OH}}}{\text{CH}_3\text{CHCH}_2}\overset{\displaystyle O}{\overset{\displaystyle \|}{\text{C}}}\text{OH} \ + \ \overset{\displaystyle O}{\overset{\displaystyle \|}{\text{HOC}}}(CH_2)_6\overset{\displaystyle O}{\overset{\displaystyle \|}{\text{C}}}\text{OH}$$

Compound A must be

16.51

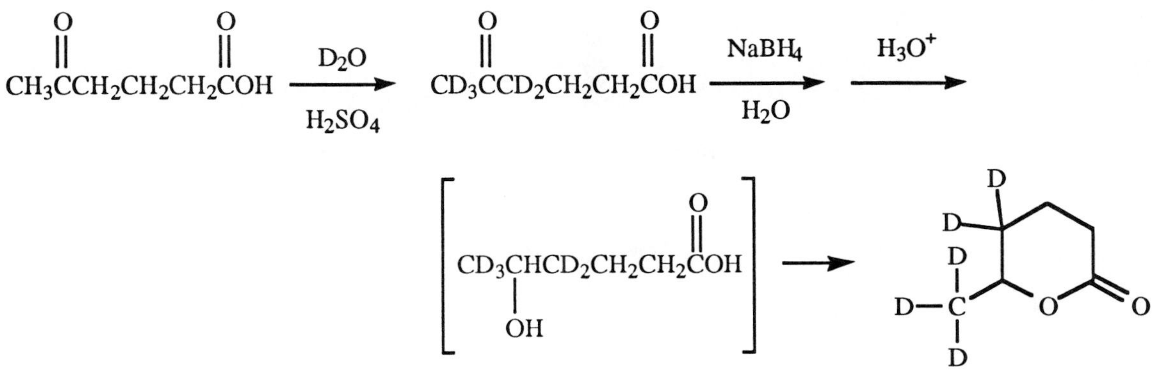

16.52

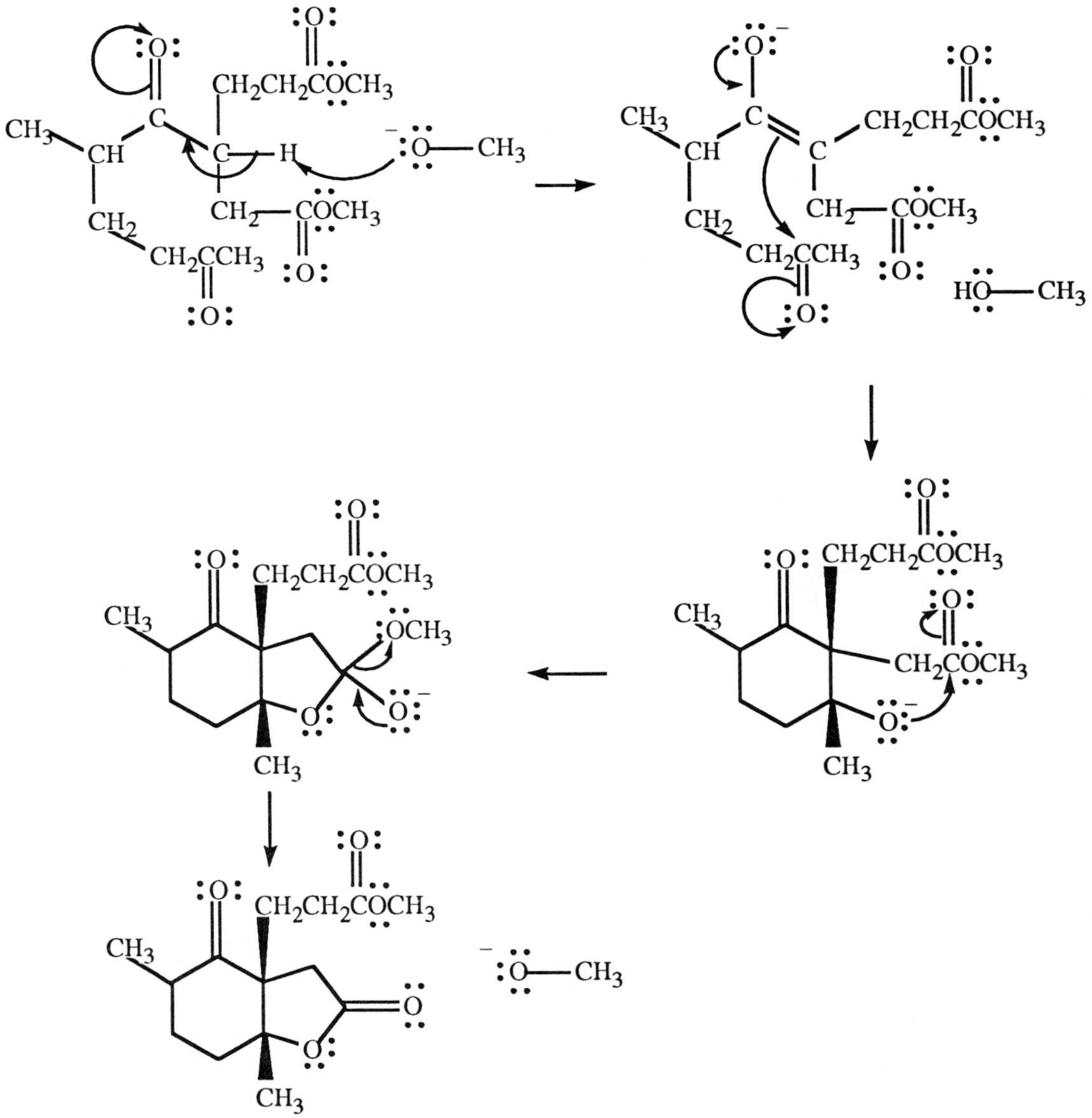

16.53

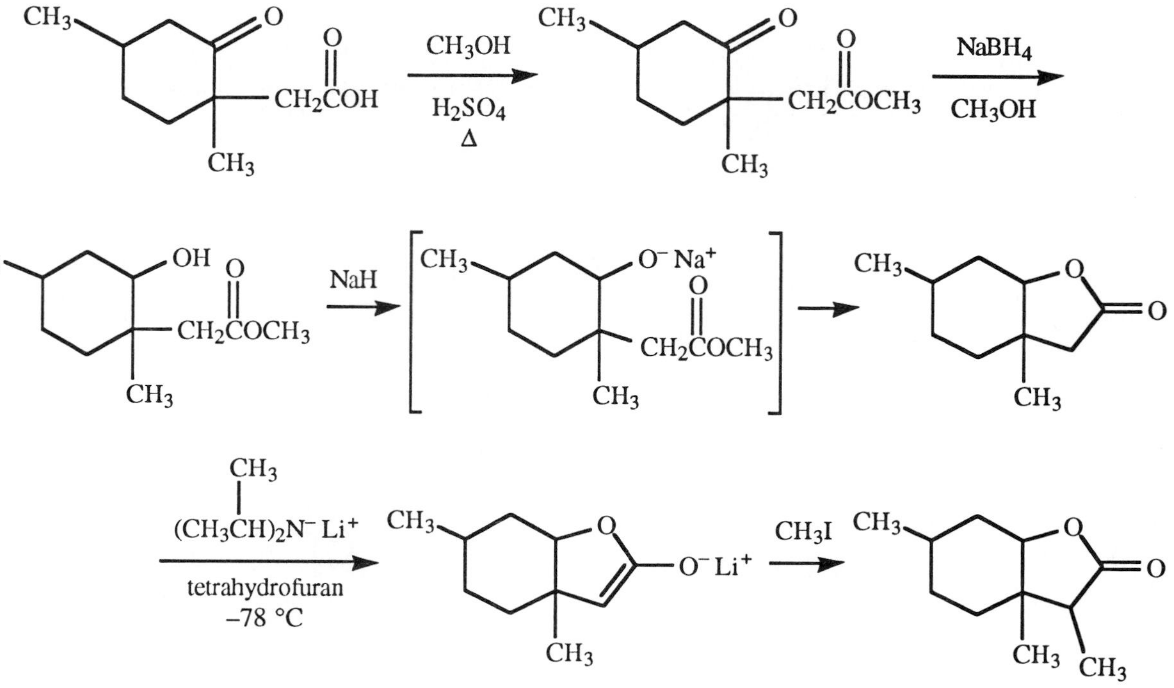

16.54

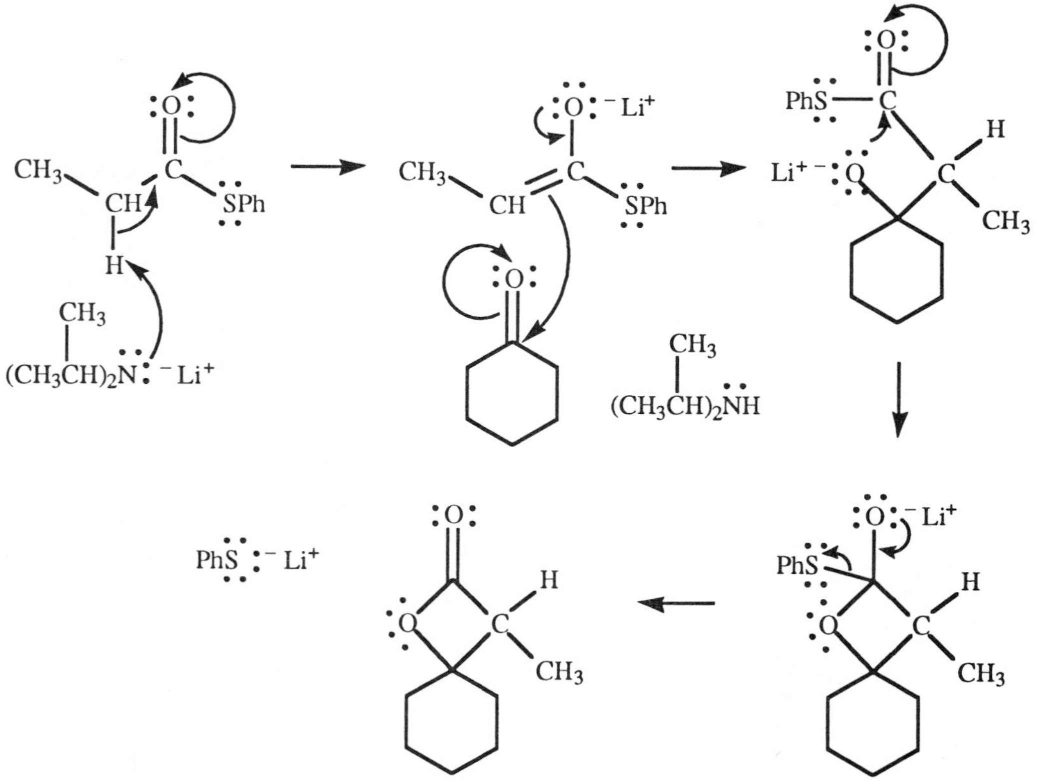

16.55

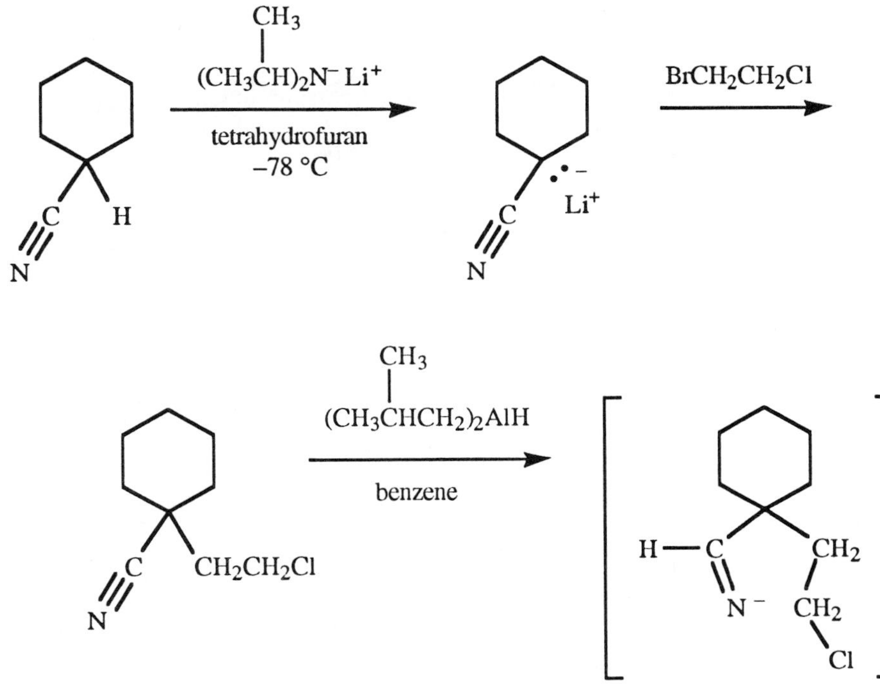

16.56

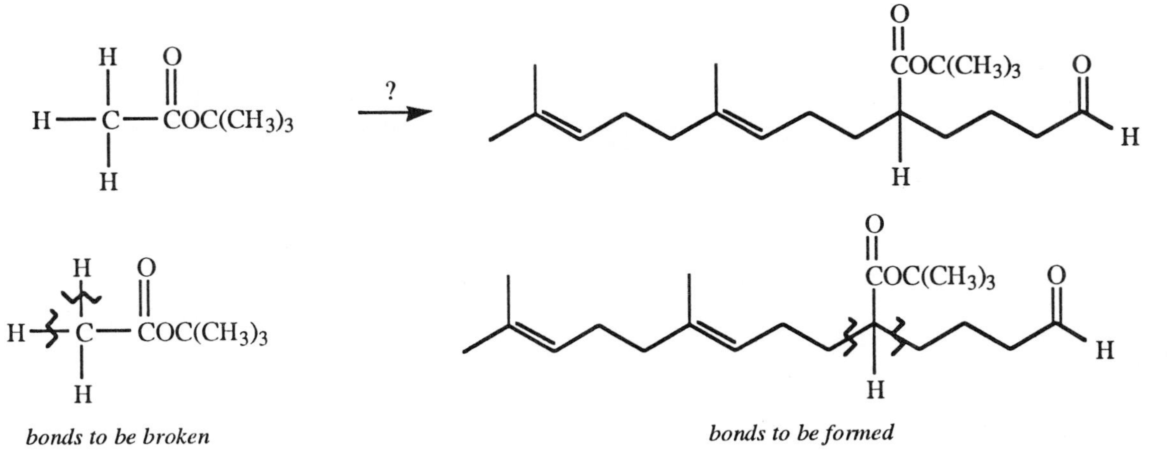

bonds to be broken   *bonds to be formed*

We need two deprotonation and alkylation steps.

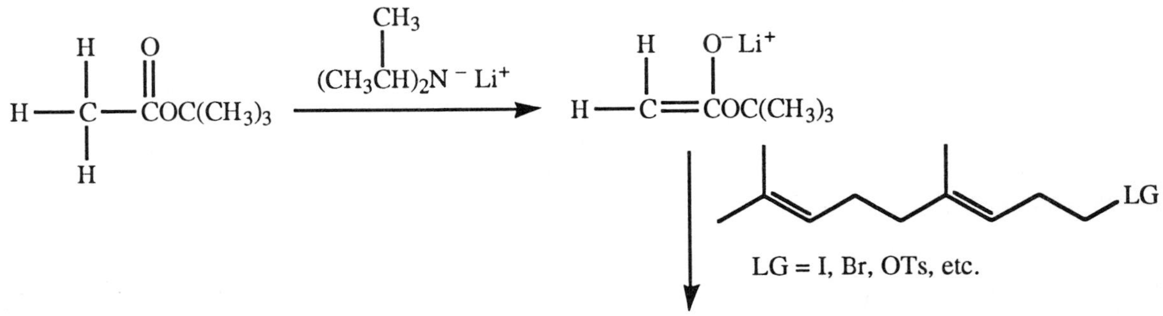

LG = I, Br, OTs, etc.

16.56 (cont)

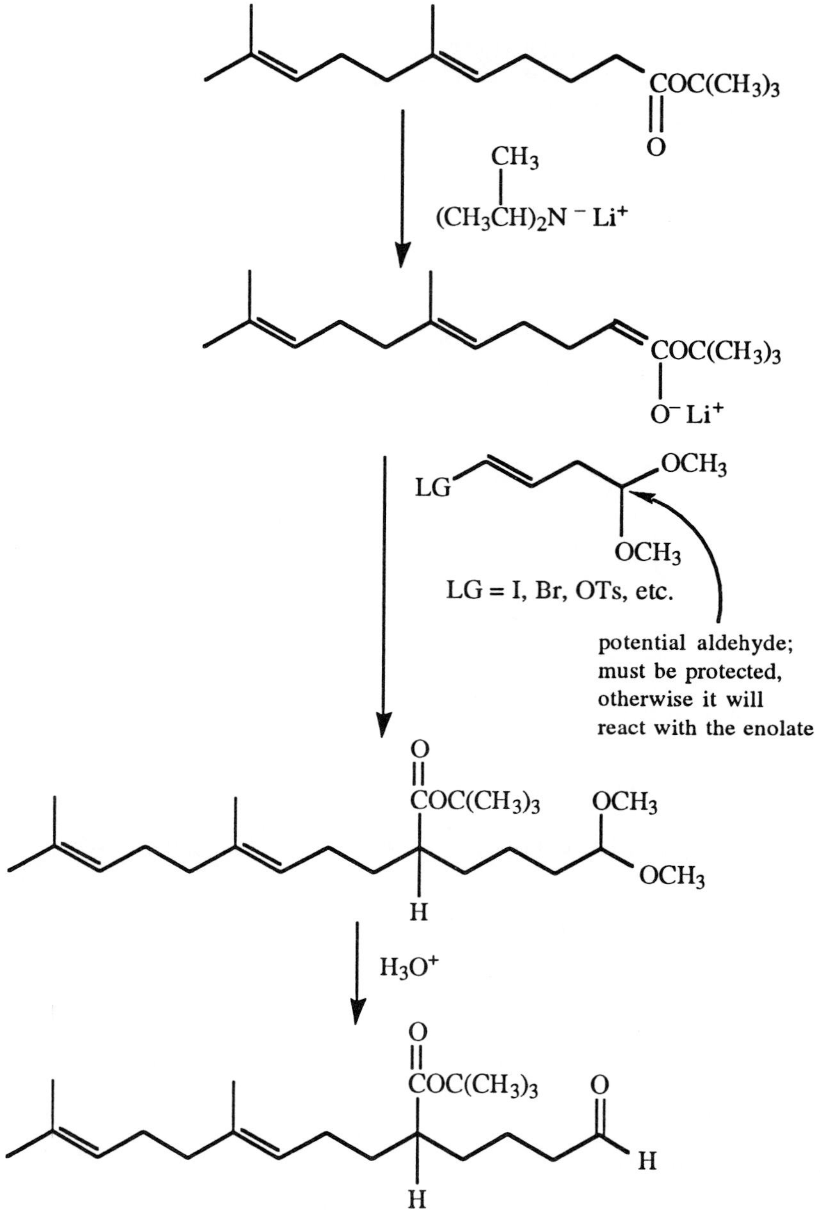
COC(CH3)3
O
CH3
(CH3CH)2N⁻ Li+
COC(CH3)3
O⁻ Li+
LG
OCH3
OCH3
LG = I, Br, OTs, etc.
potential aldehyde; must be protected, otherwise it will react with the enolate
O
COC(CH3)3
OCH3
OCH3
H
H3O+
O
COC(CH3)3
O
H
H

17

Polyenes

Concept Map 17.1   Different relationships between multiple bonds.

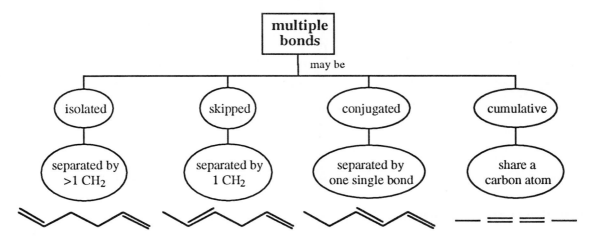

17.1   (a) 1,4-hexadiyne

(b) 2-hexen-4-yne

(c) 1,4-cyclohexadiene

(d) 1,5-cyclooctadiene

(e) (*E*)-1-phenyl-1,4-pentadiene

(f) 1-chloro-2,4-heptadiyne

17.2

(a) $CH_2{=}CHCH_2CH_2\overset{\overset{\textstyle O}{\|}}{C}CH_3$

(b) $HC{\equiv}CCH_2CH_2C{\equiv}CH$

(c) $CH_2{=}CHCH_2CH{=}CH_2$

(d) $CH_3C{\equiv}CCH_2C{\equiv}CCH_3$

(e) $\underset{\displaystyle H}{\overset{\displaystyle CH_3}{{>}}}C{=}C\underset{\displaystyle CH_2\underset{\displaystyle CH_3}{\overset{|}{C}}CH_2CH{=}CH_2}{\overset{\displaystyle H}{<}}$

(f) $HC{\equiv}CCH{=}CHCH{=}CH_2$

Concept Map 17.2   A conjugated diene.

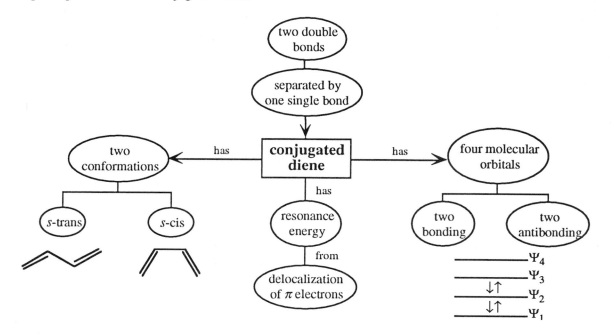

- -

Concept Map 17.3   Addition to dienes.

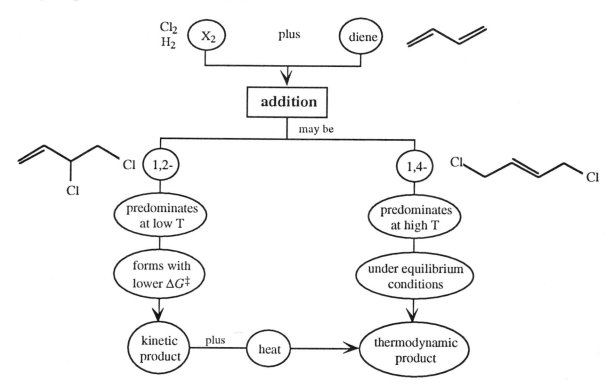

17.3

(a)

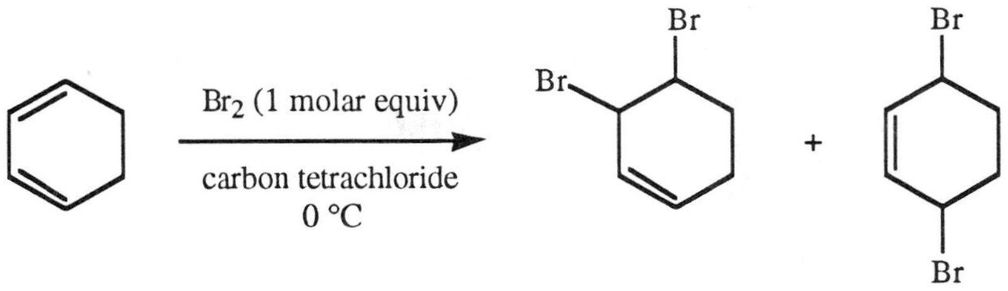
Br$_2$ (1 molar equiv)
carbon tetrachloride
0 °C
Br
Br
+
Br
Br

(b)

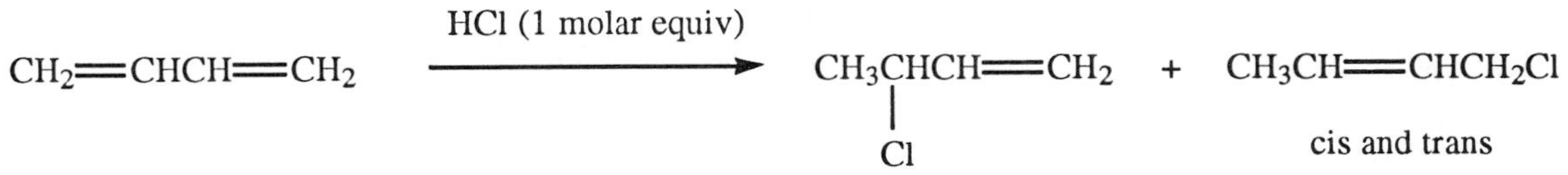
CH$_2$=CHCH=CH$_2$
HCl (1 molar equiv)
CH$_3$CHCH=CH$_2$
Cl
+
CH$_3$CH=CHCH$_2$Cl
cis and trans

(c)

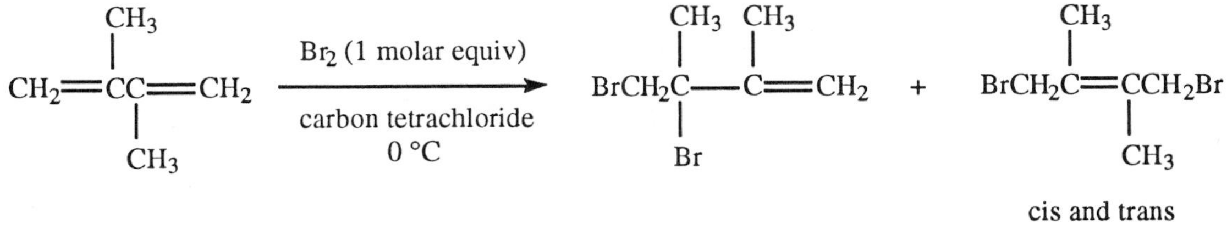
CH$_3$
CH$_2$=CC=CH$_2$
CH$_3$
Br$_2$ (1 molar equiv)
carbon tetrachloride
0 °C
CH$_3$ CH$_3$
BrCH$_2$C—C=CH$_2$
Br
+
CH$_3$
BrCH$_2$C=CCH$_2$Br
CH$_3$
cis and trans

(d)

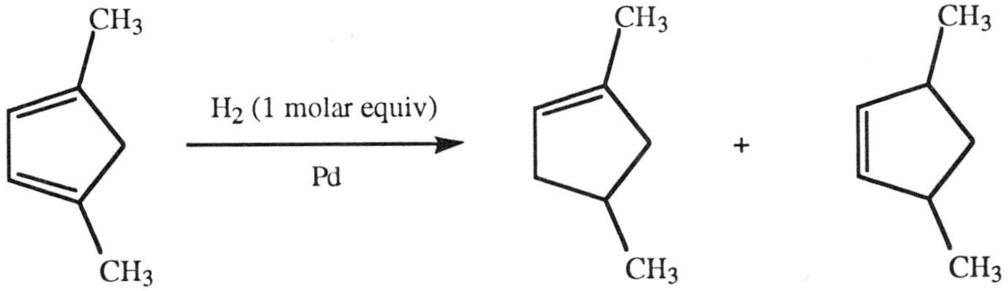
CH$_3$
CH$_3$
H$_2$ (1 molar equiv)
Pd
CH$_3$
CH$_3$
+
CH$_3$
CH$_3$

17.4

(a)

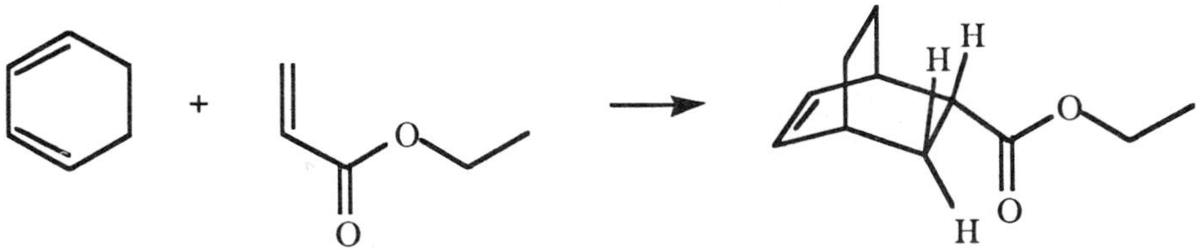
+
O
O
H
H
O
O
H

and enantiomer
A

(b)

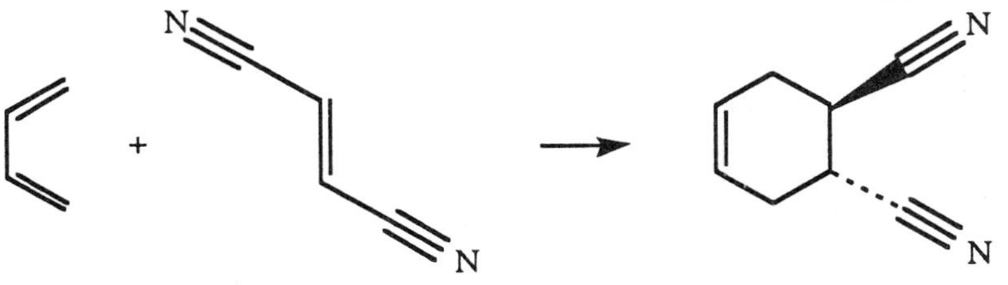
+
N
N
N
N

and enantiomer
B

17.4 (cont)

(c)

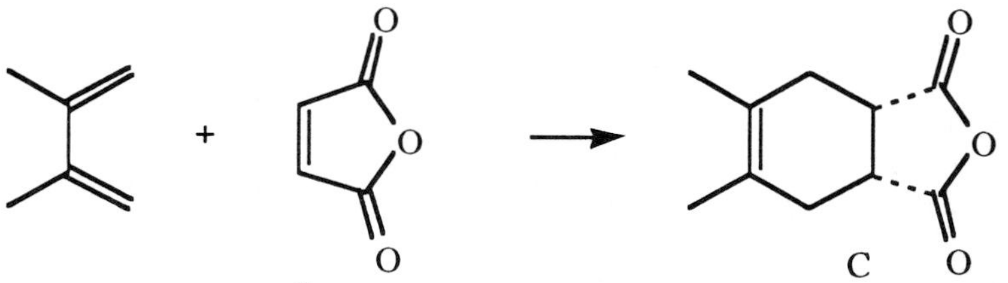

C

(d)

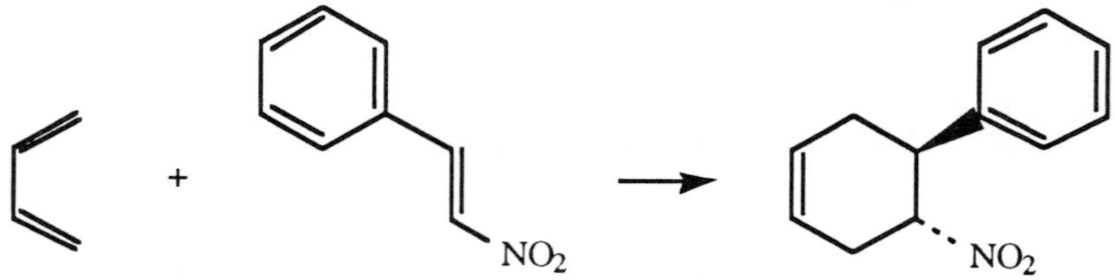

and enantiomer
D

(e)

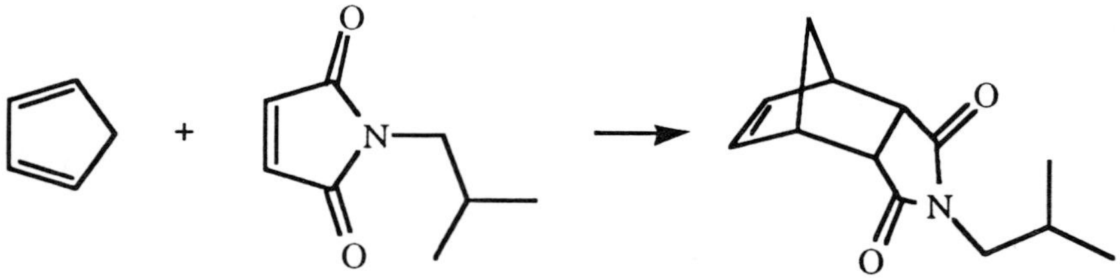

E

(f)

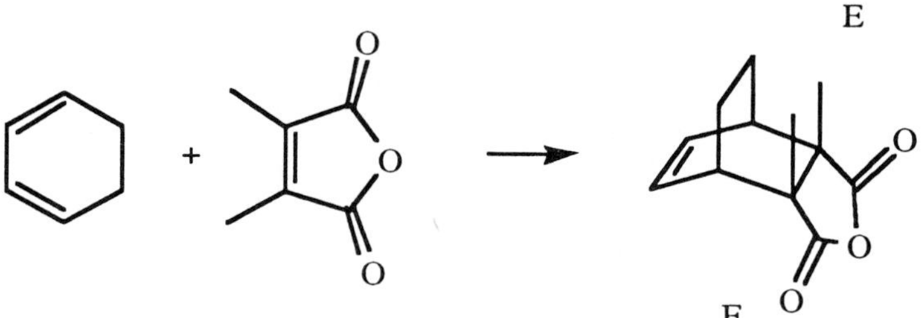

F

(g)

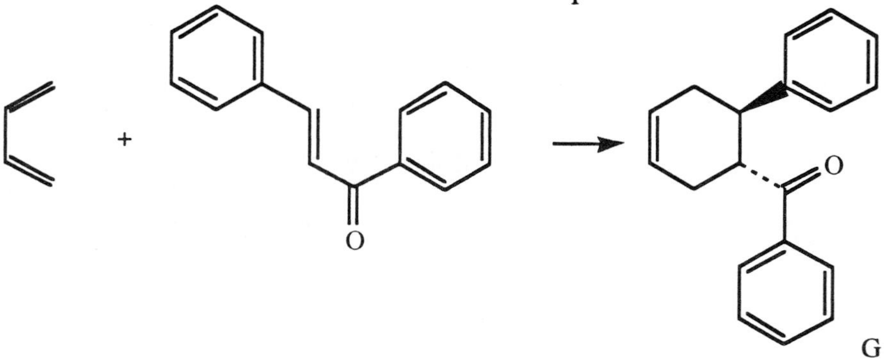

and enantiomer

G

(h)

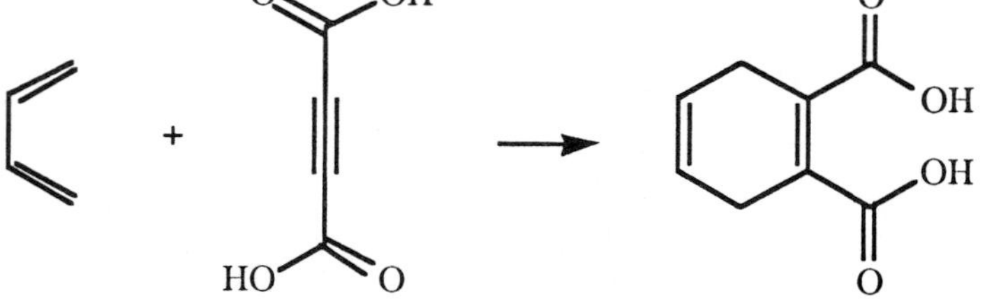

H

17.5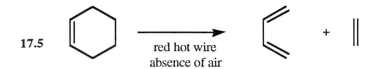

Concept Map 17.4   The Diels-Alder reaction.

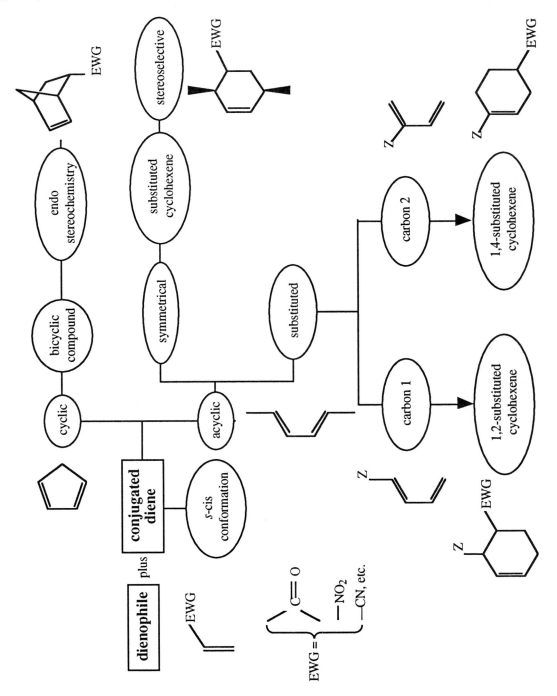

17.6

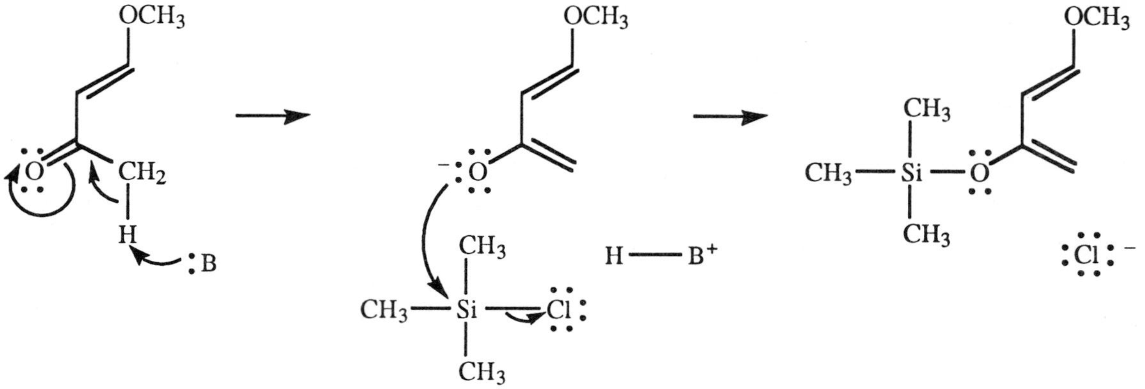

17.7

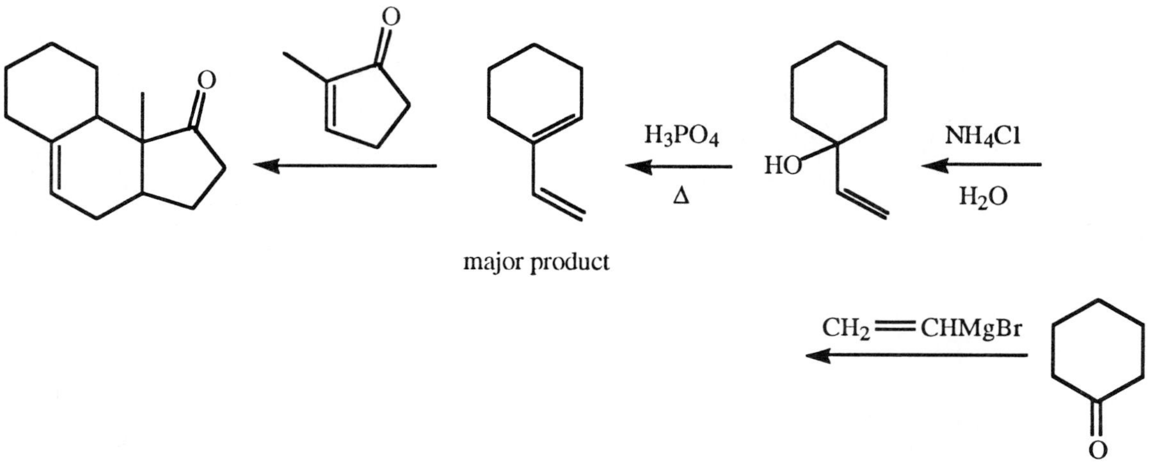

major product

17.8

(a)

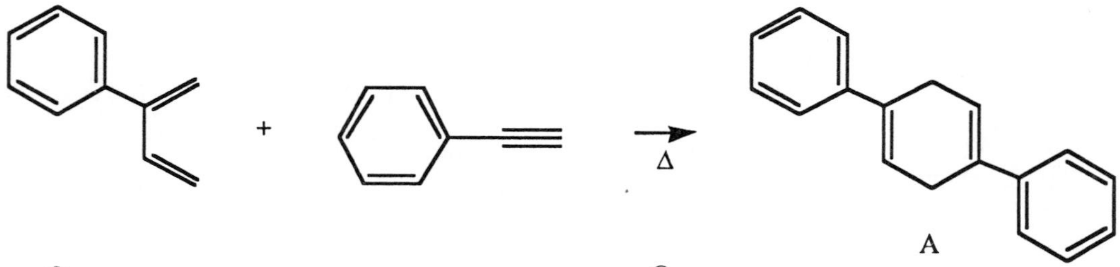

A

(b)

B

(c)

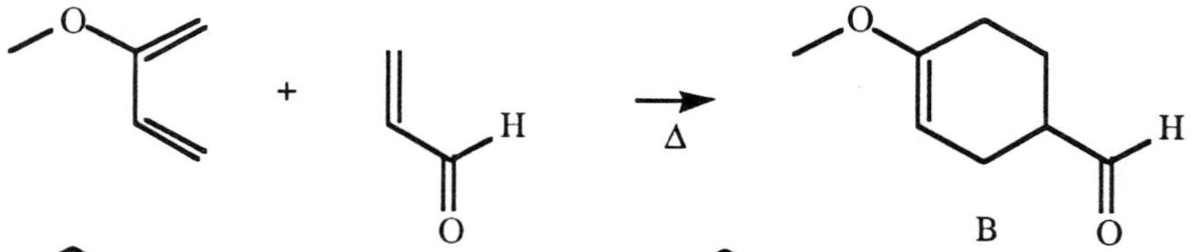

C

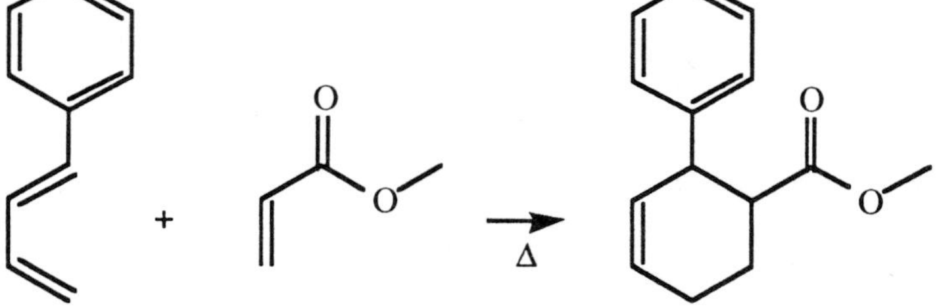

17.8 (cont)

(d) 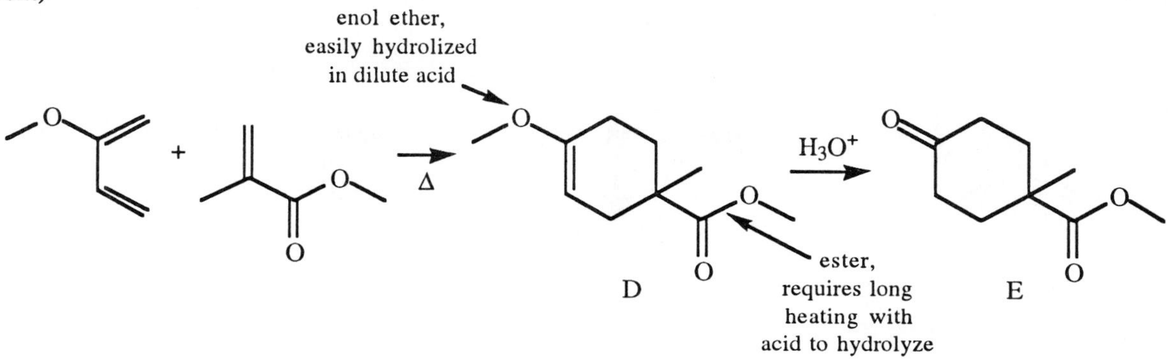
enol ether,
easily hydrolized
in dilute acid
O
O
O
Δ
H3O+
D
E
ester,
requires long
heating with
acid to hydrolyze

(e) 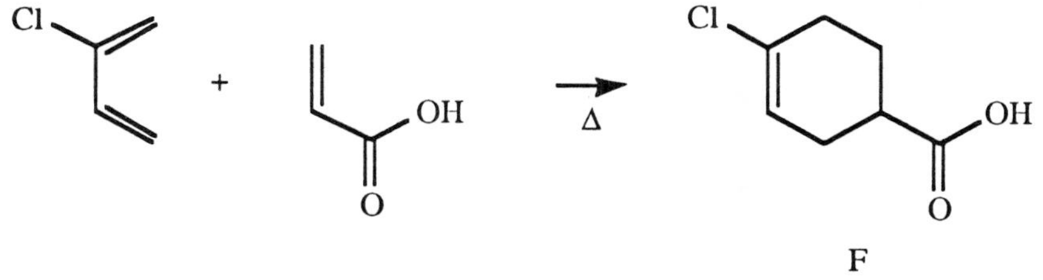
Cl
OH
O
Δ
F

(f) 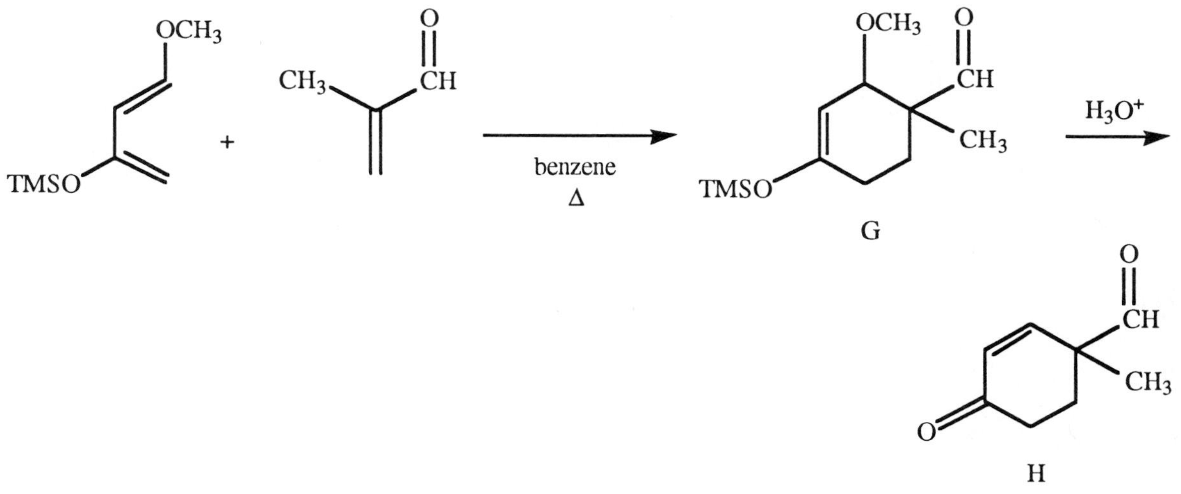
OCH3
TMSO
CH3
CH
O
benzene
Δ
G
H3O+
CH3
H

17.9

(a) 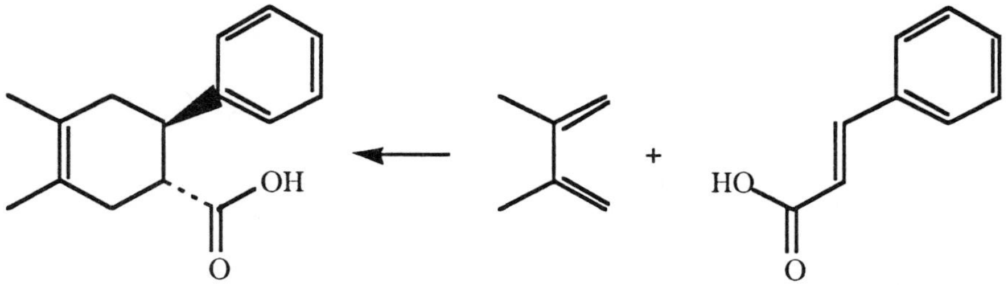
OH
O
HO

17.9 (cont)

(b)

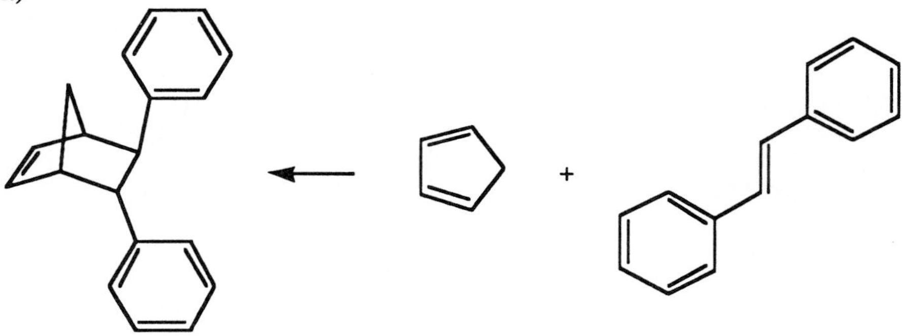

(c)

(d)

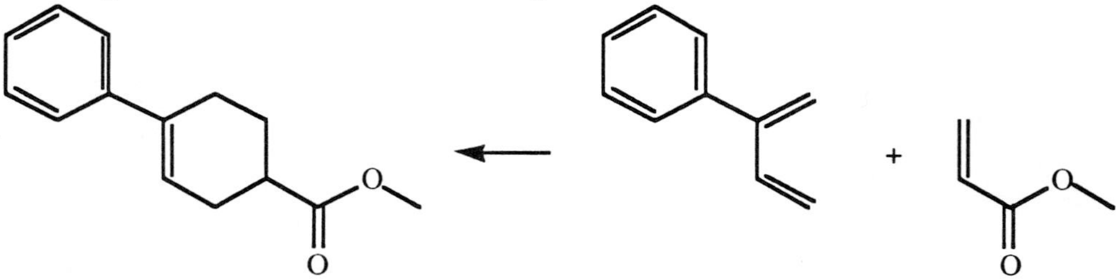

(e)

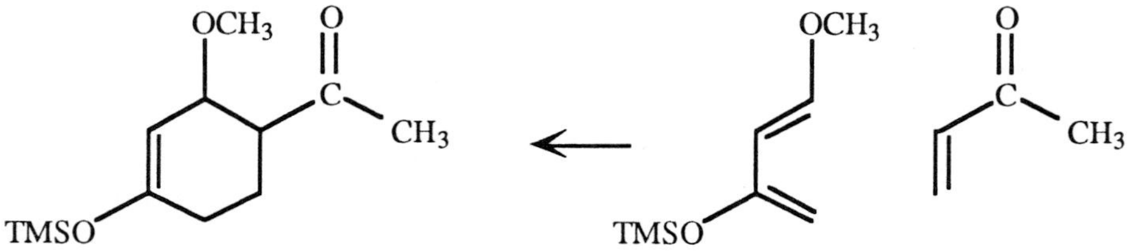

(f)

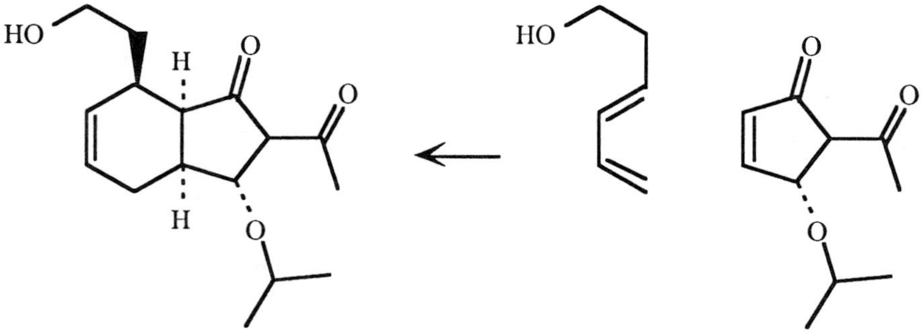

17.10

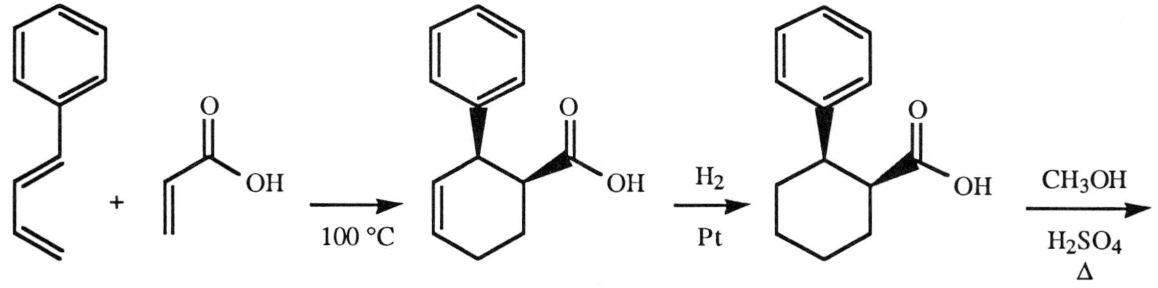
OH
100 °C
OH
H_2
Pt
OH
CH_3OH
H_2SO_4
Δ

A B

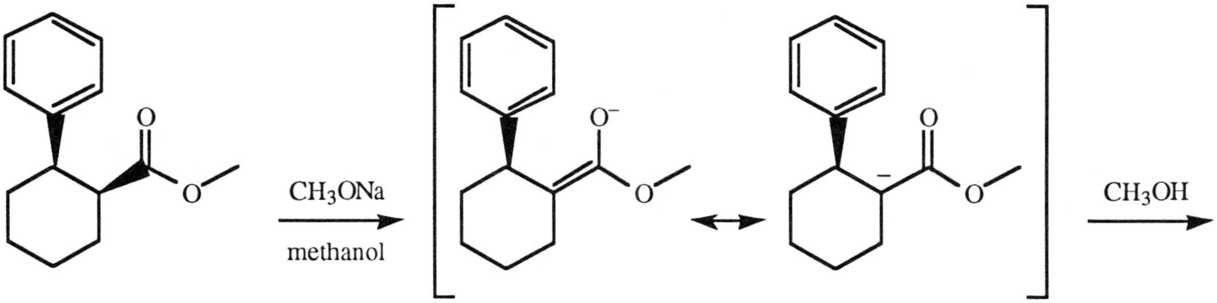
CH_3ONa
methanol
O^-
CH_3OH

C

enolate anion of ester (Section 16.1B); loss of stereochemistry at the α-carbon atom of the ester

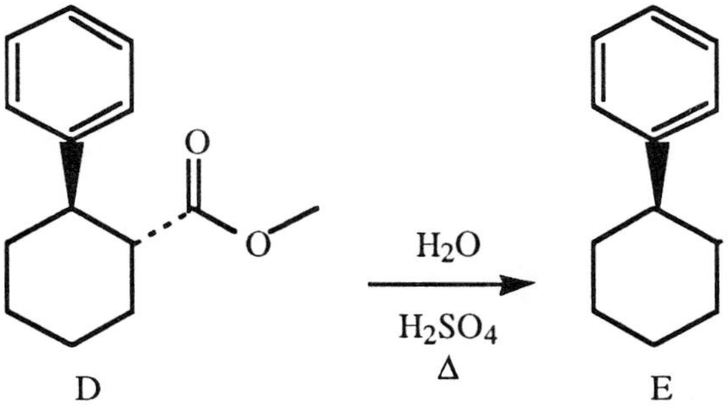
H_2O
H_2SO_4
Δ
OH

D E

Generation of the carbanion from the methyl ester followed by reprotonation is an equilibrium reaction and gives the thermodynamically more stable trans stereoisomer. Compound D is different from Compound C, and Compound E is different from A. Therefore, the initial Diels-Alder product, Compound A, must have been the cis stereoisomer.

17.11

(a)

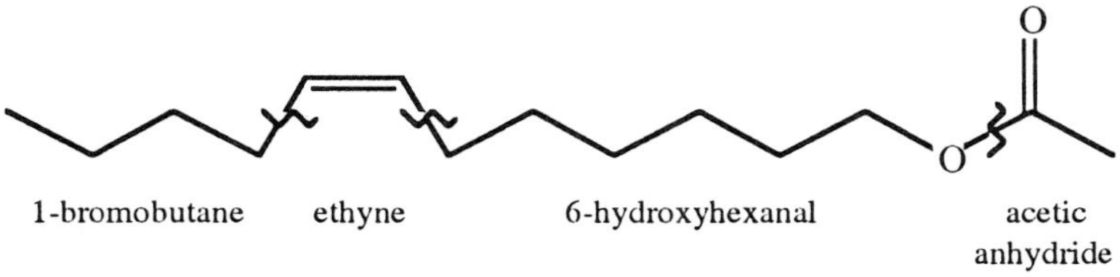
1-bromobutane
ethyne
6-hydroxyhexanal
acetic anhydride

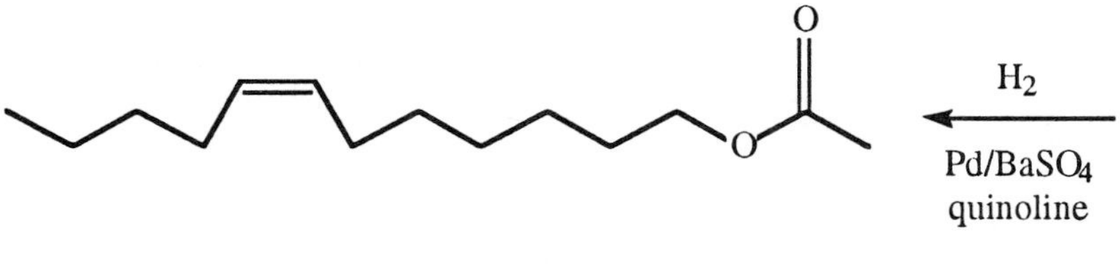
H_2
$Pd/BaSO_4$
quinoline

17.11 (a) (cont)

$$CH_3(CH_2)_3C \equiv C(CH_2)_5CH_2O\overset{\overset{\displaystyle O}{\|}}{C}CH_3 \xleftarrow[\text{pyridine}]{(CH_3\overset{\overset{\displaystyle O}{\|}}{C})_2O} CH_3(CH_2)_3C \equiv C(CH_2)_5CH_2OH \xleftarrow[\text{CH}_3\text{OH}]{\text{NaBH}_4}$$

$$CH_3(CH_2)_3C \equiv C(CH_2)_5\overset{\overset{\displaystyle O}{\|}}{C}H \xleftarrow[\Delta]{H_3O^+} CH_3(CH_2)_3C \equiv C(CH_2)_5\overset{\displaystyle \diagup}{C}H \xleftarrow[\text{1-bromobutane}]{CH_3CH_2CH_2CH_2Br}$$

$$Na^+ \;^-C \equiv CCH_2(CH_2)_4\overset{\displaystyle}{C}H \xleftarrow[\text{NH}_3\text{ (liq)}]{\text{NaNH}_2} HC \equiv CCH_2(CH_2)_4\overset{\displaystyle}{C}H \xleftarrow{HC \equiv C^- Na^+}$$

$$BrCH_2(CH_2)_4\overset{\displaystyle}{C}H \xleftarrow[\text{pyridine}]{PBr_3} HOCH_2(CH_2)_4\overset{\displaystyle}{C}H \xleftarrow[\substack{\text{TsOH}\\\text{benzene}\\\Delta}]{HOCH_2CH_2OH} HOCH_2(CH_2)_4\overset{\overset{\displaystyle O}{\|}}{C}H$$

6-hydroxyhexanal

$$HC \equiv C^- Na^+ \xleftarrow[\text{NH}_3\text{ (liq)}]{\text{NaNH}_2} HC \equiv CH$$

ethyne

Note that the cis double bond, which is the less stable isomer, is introduced as late as possible in the synthesis to prevent isomerization by acidic reagents.

(b)   To get the trans isomer, we would use sodium in liquid ammonia, instead of hydrogen and poisoned catalyst, in the reduction of the triple bond. The reduction in this case would be carried out at the alcohol stage to avoid reaction of the ester with ammonia. Because the trans double bond is the more stable of the two, it is not subject to isomerization as easily as the cis double bond is.

17.12

menthofuran
both are head to tail

or

bisabolol
both are head to tail

17.12 (cont)

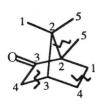

camphor

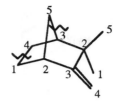

linaloöl

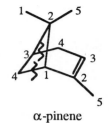

camphene

α-pinene

17.13

(a)

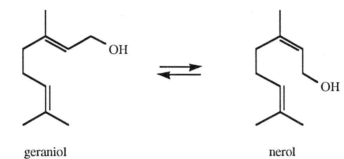

geraniol

⇌

nerol

(*E*)-double bond between
carbon 2 and carbon 3

(*Z*)-double bond between
carbon 2 and carbon 3

protonation of
the double bond

rotation around the
single bond between
carbon 2 and carbon 3

deprotonation

17.13 (cont)

(b)

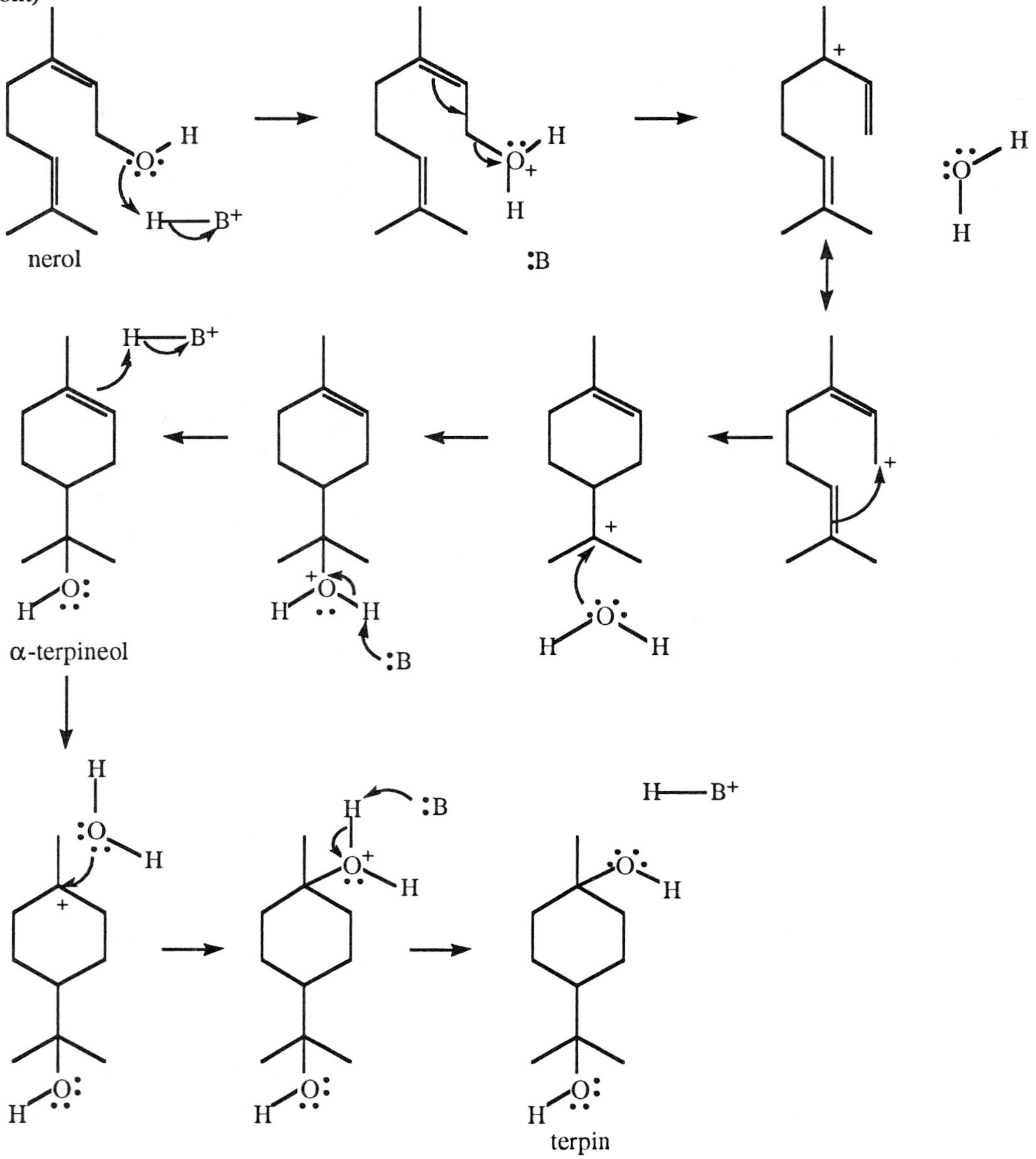

nerol

:B

α-terpineol

terpin

(c)

α-terpineol

:B

limonene

H—B$^+$

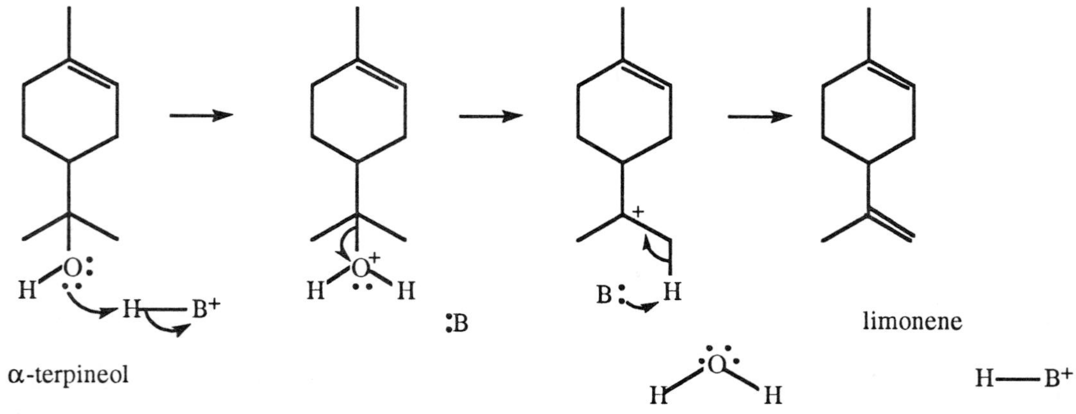

17.13 (c) (cont)

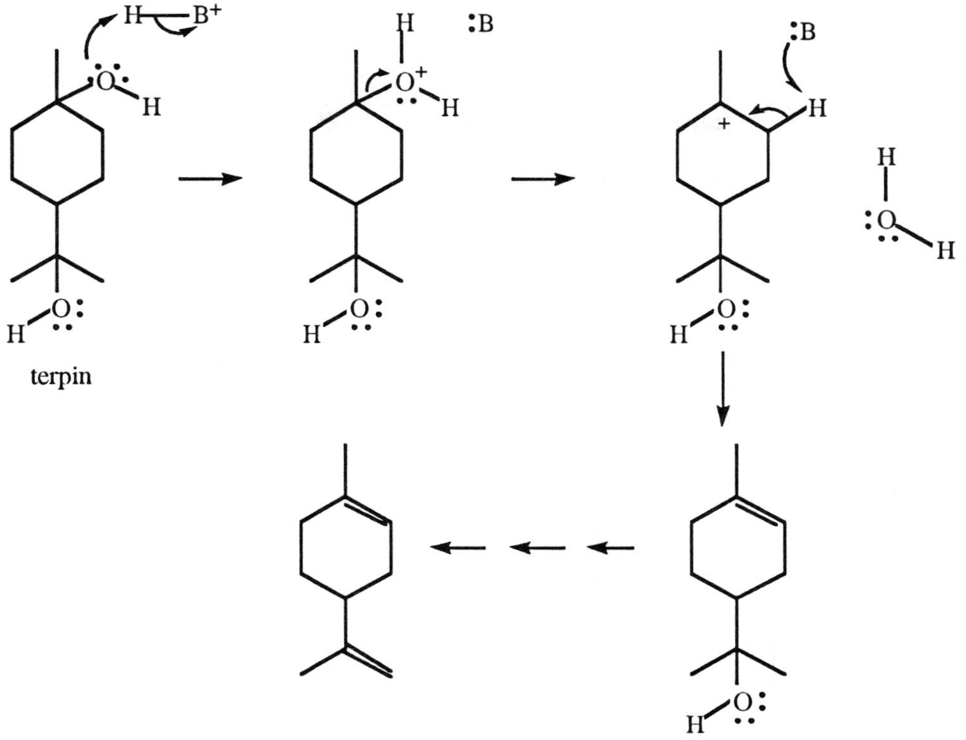

terpin

limonene

α-terpineol

(d)

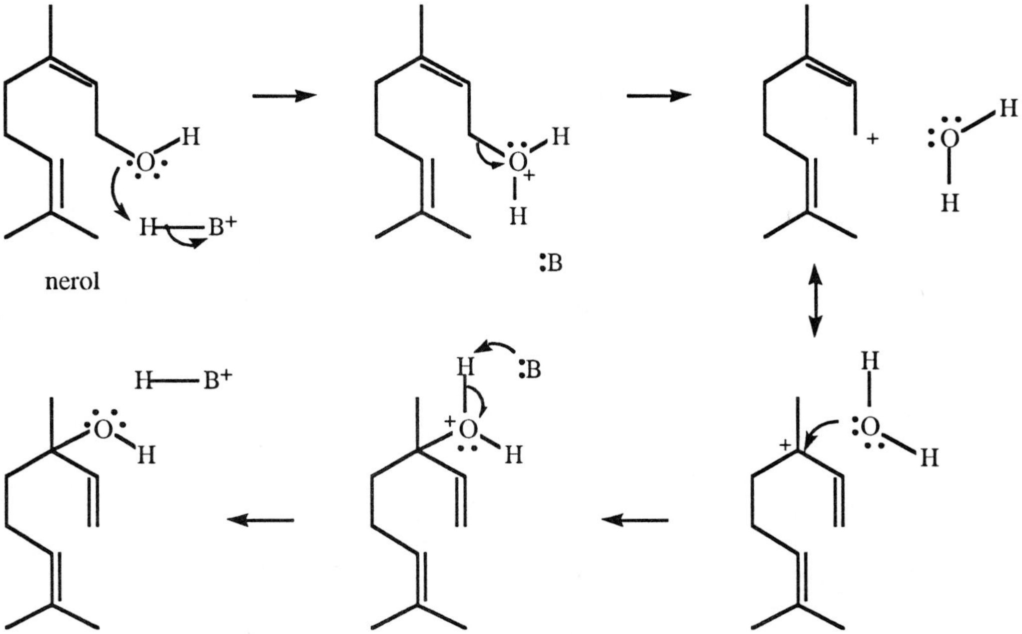

nerol

linaloöl

17.14

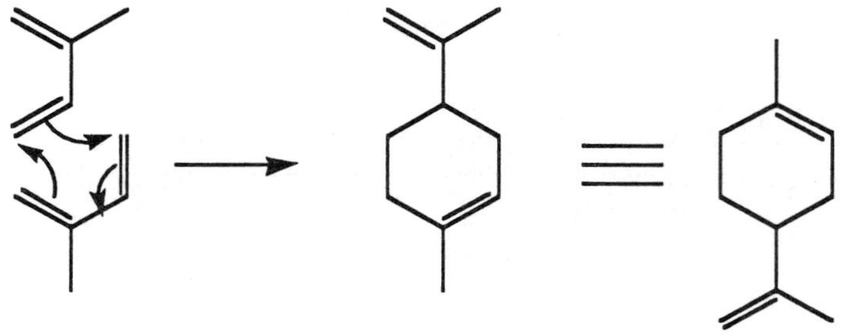

limonene

A Diels-Alder reaction between one isoprene unit and the less hindered double bond of the second one in a head-to-tail fashion gives limonene.

17.15

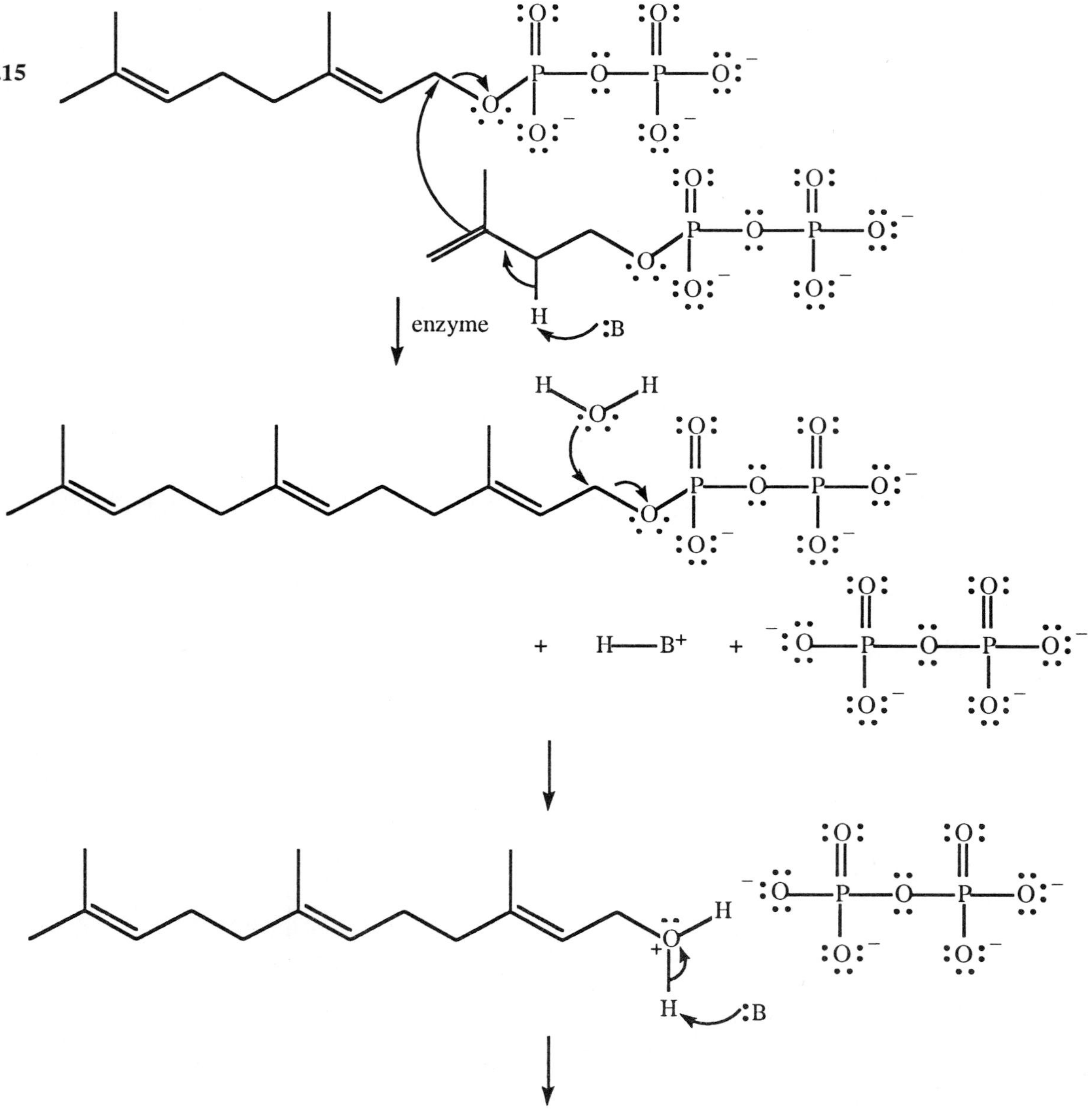

17.15 (cont)

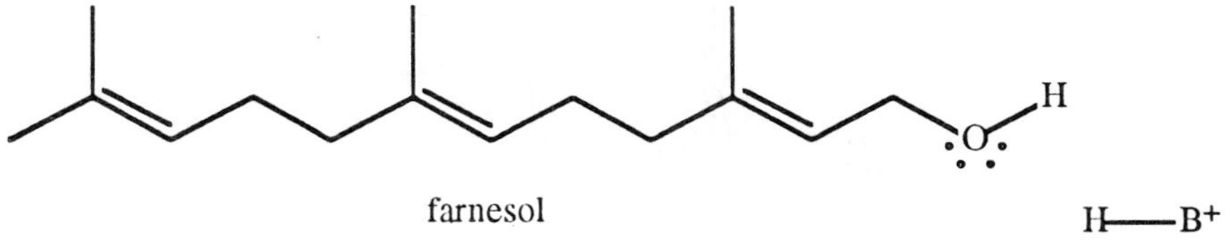

farnesol

H——B⁺

17.16

farnesyl
pyrophosphate

enzyme

geranylgeranyl pyrophosphate

+   H——B⁺   +

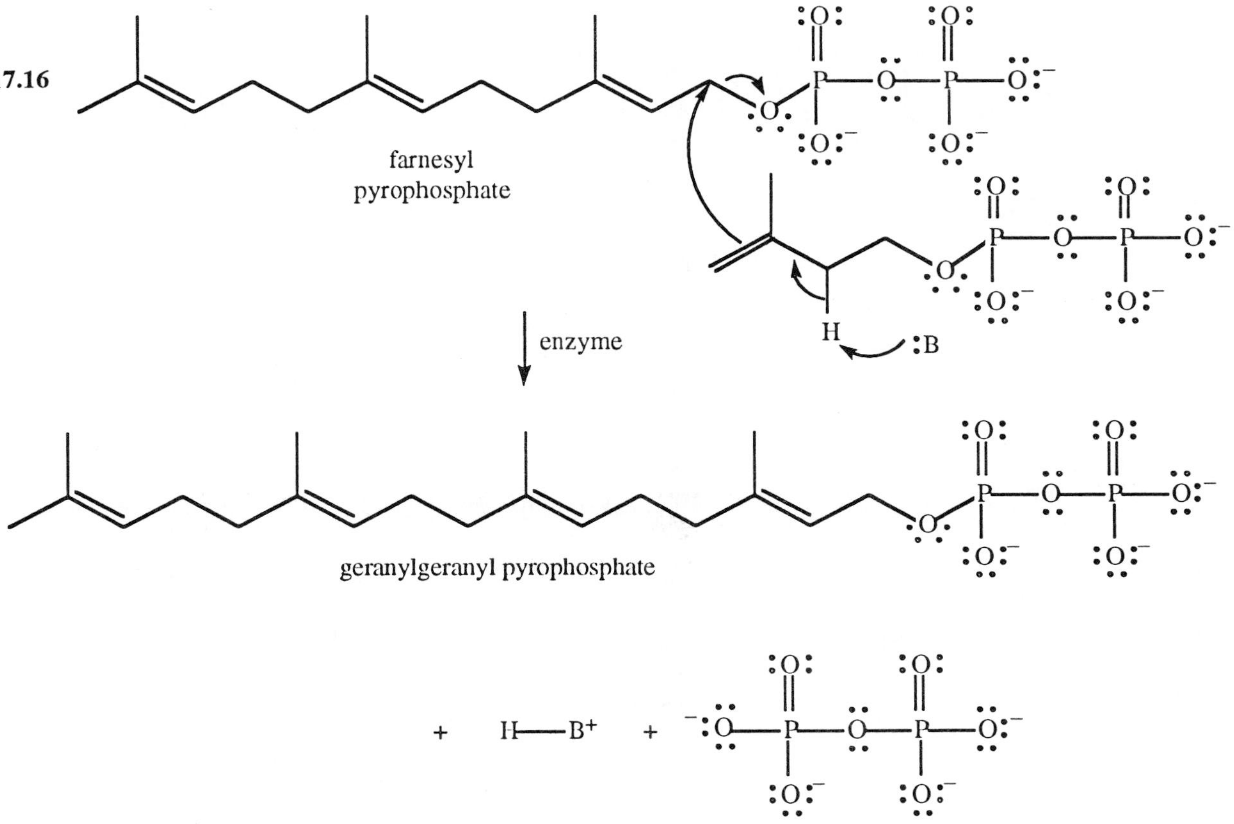

17.17   (a)

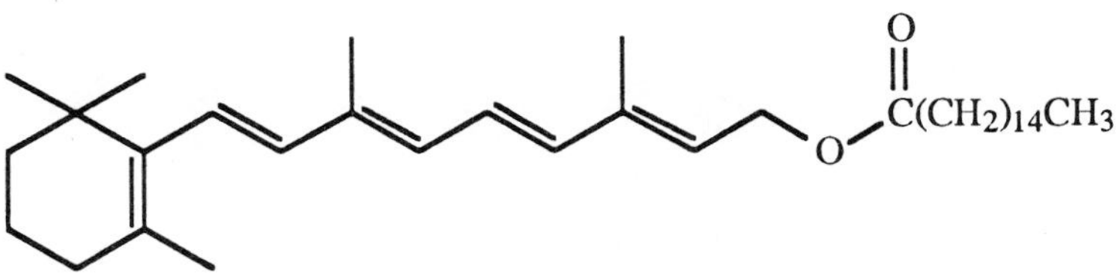

hexadecanoate ester of Vitamin A

(b)

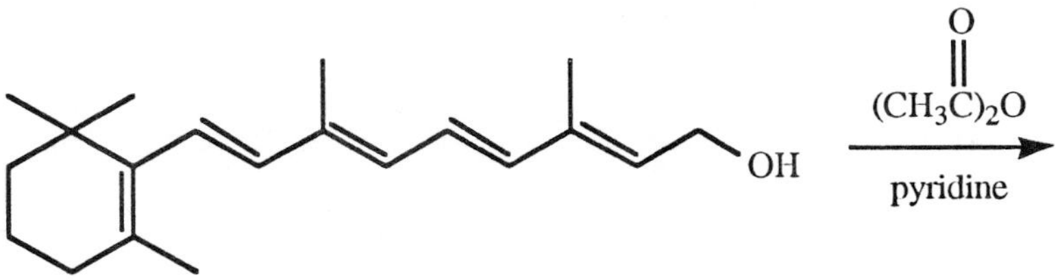

17.17 (b) (cont)

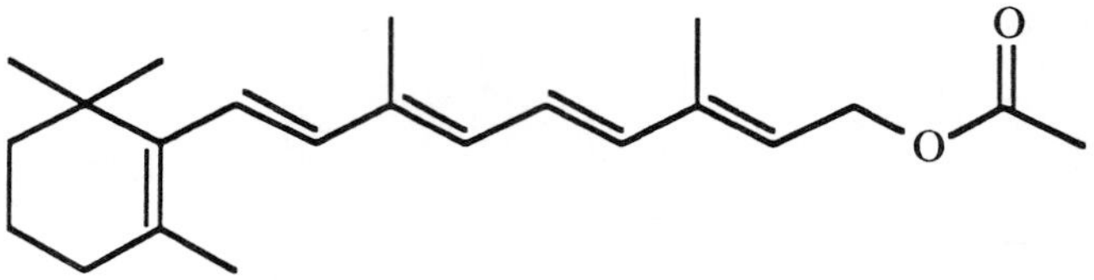

17.18

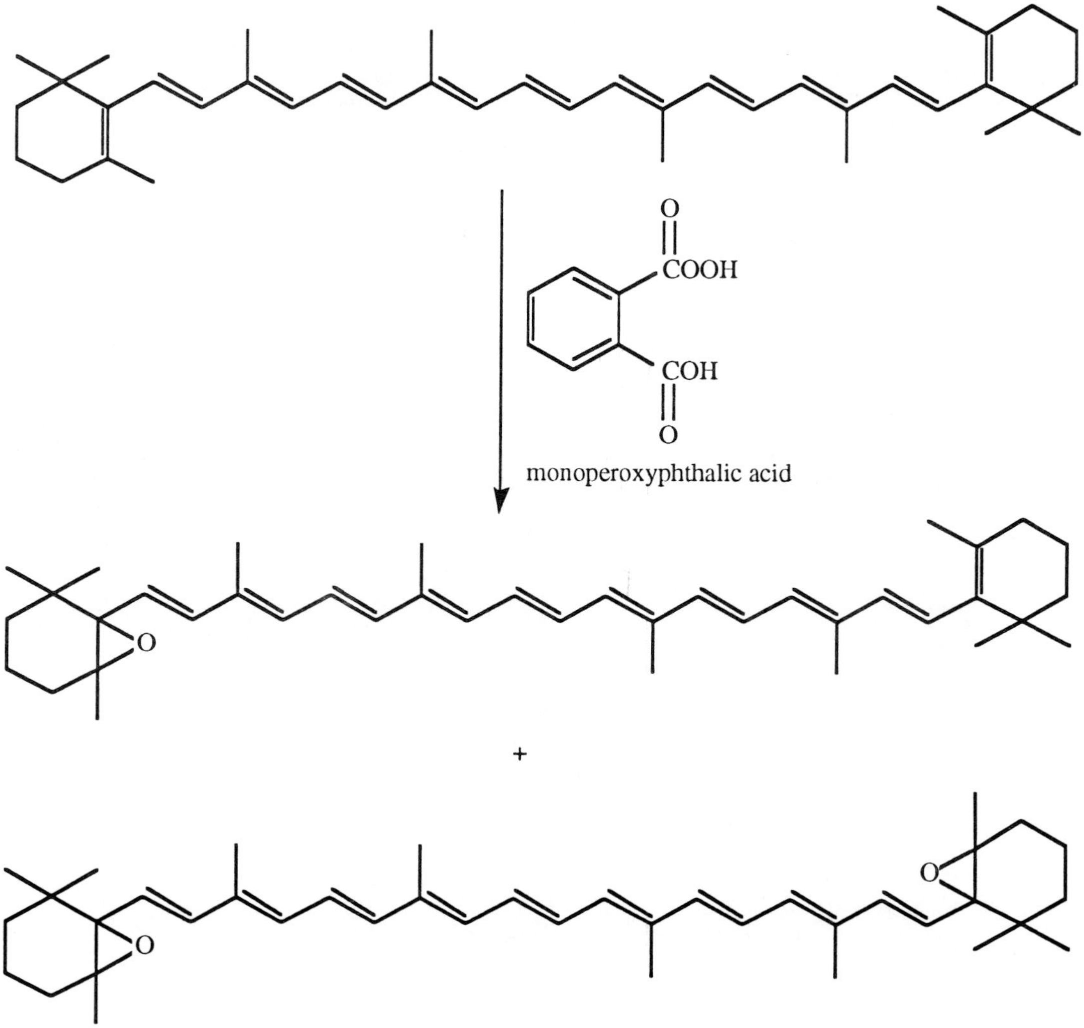

monoperoxyphthalic acid

+

The most highly substituted double bonds are oxidized preferentially by the peroxyacid.

Concept Map 17.5   Visible and ultraviolet spectroscopy.

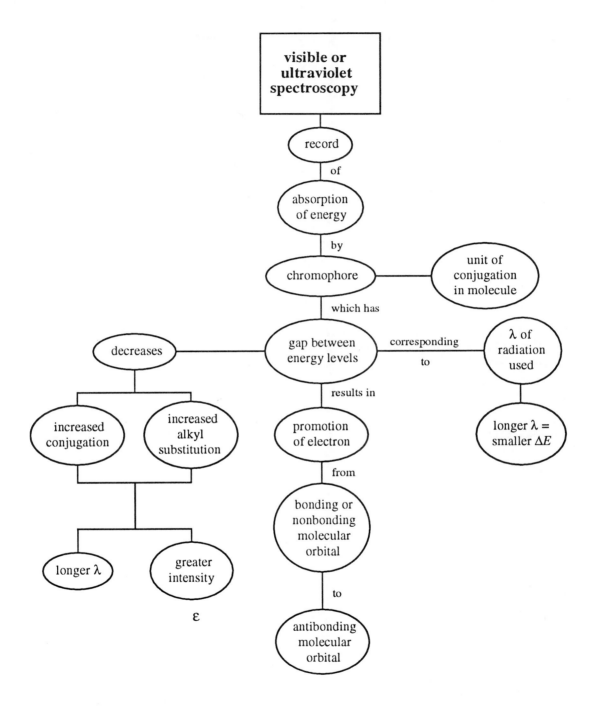

17.19   The compound with the most extended conjugation would be expected to absorb at the longest wavelength.

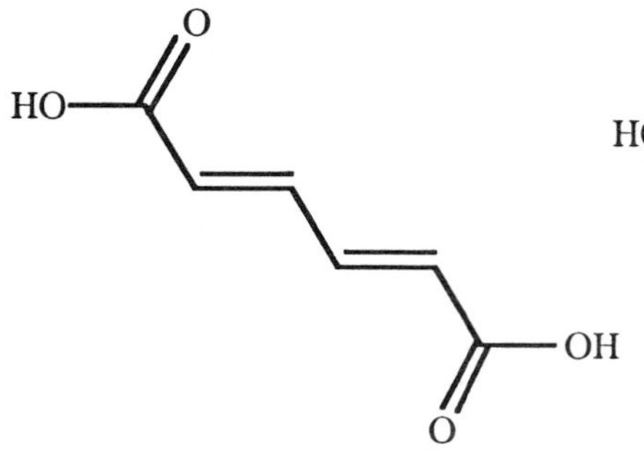

The compound above has the most
extensive conjugation, and would
absorb at the longest wavelength.

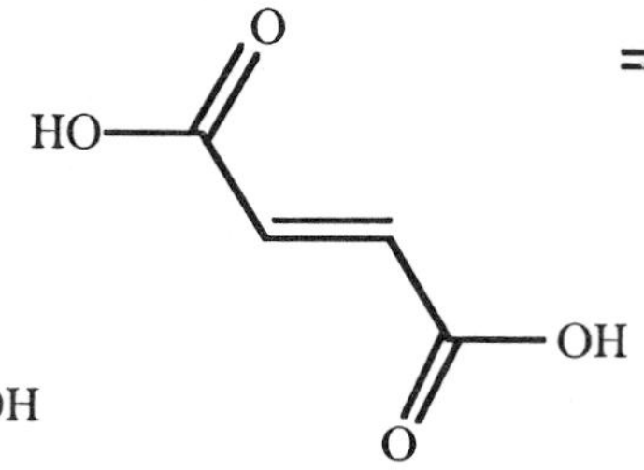

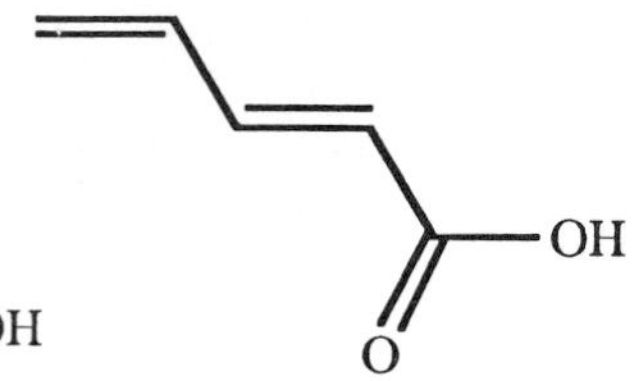

17.20

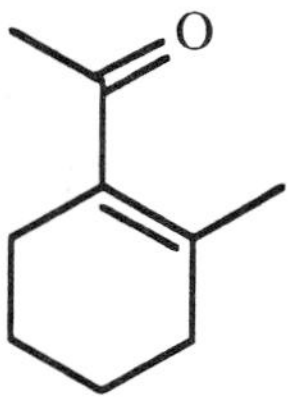

1-Acetyl-2-methyl-1-cyclohexene has two absorption bands, $\lambda_{max}^{ethanol}$ 247 nm (ε 6000) and 305 nm (ε 100). The band at 247 nm corresponds to the $\pi \rightarrow \pi^*$ transition of the conjugated π system, and the band at 305 nm, to the $n \rightarrow \pi^*$ transition of the carbonyl group.

17.21

(a)

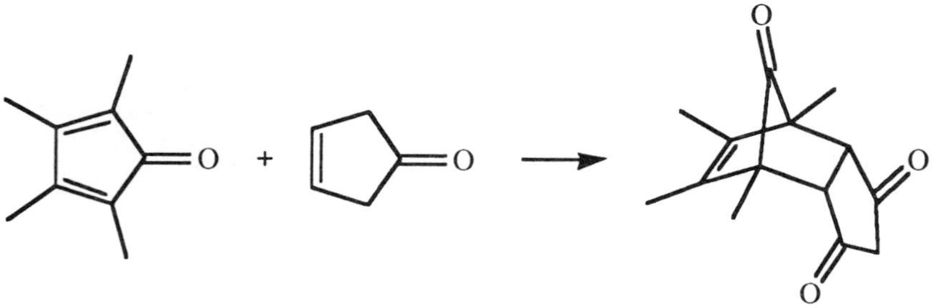

(b)

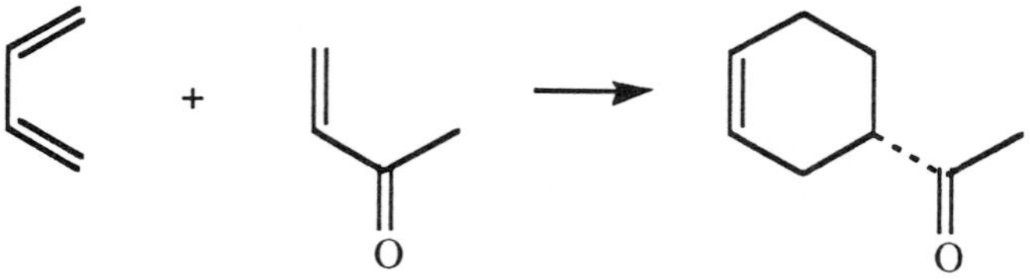

and enantiomer
B

(c)

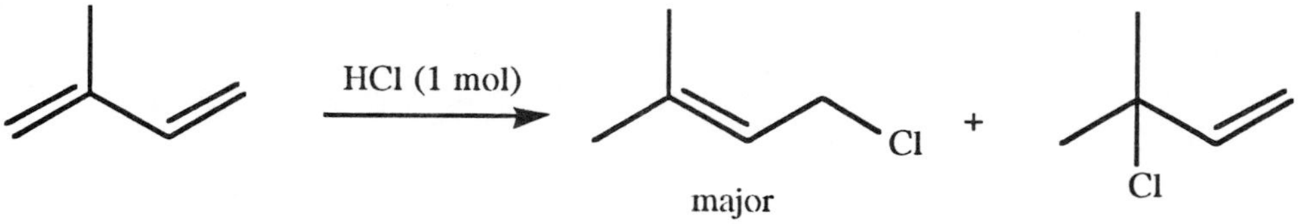

17.21(cont)

(d)

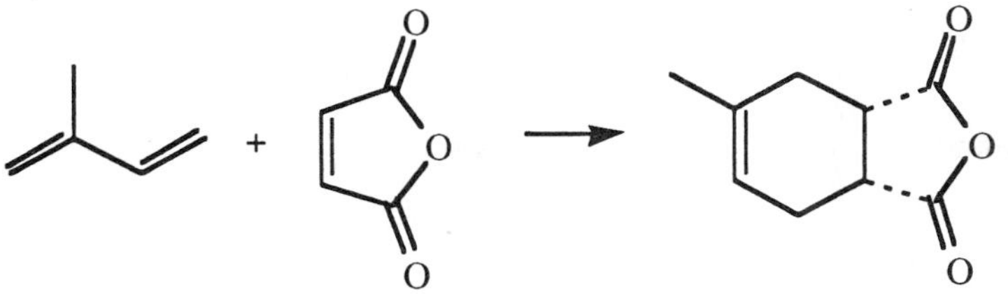

and enantiomer
E

(e)

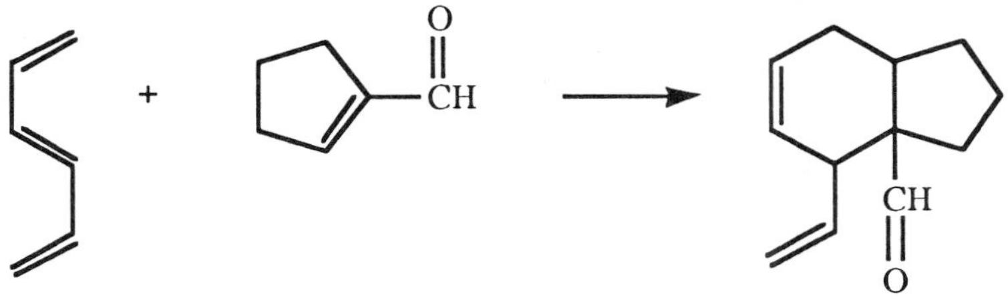

and enantiomer
F

(f) $CH_3C\equiv CCH_2C\equiv CCH_2CH_3$ $\xrightarrow[\substack{Pd/BaSO_4 \\ quinoline}]{H_2}$

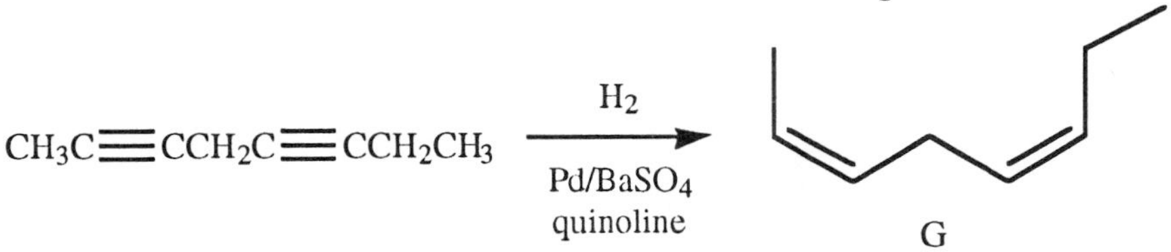

G

(g)

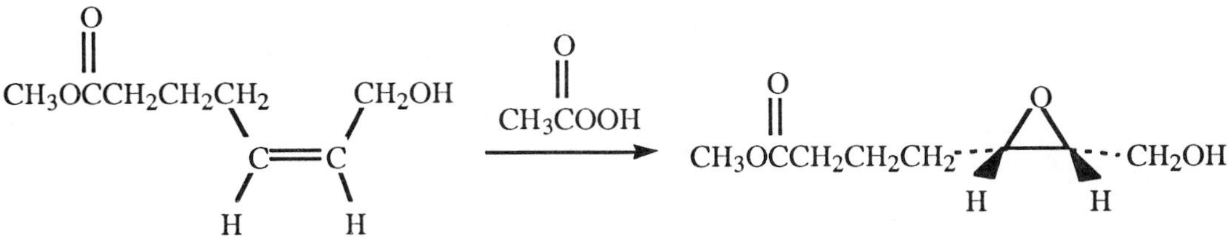

and enantiomer
H

17.22

(a)

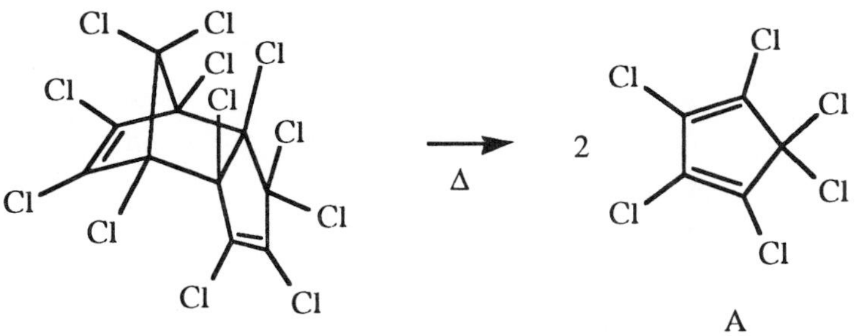

2

A

17.22 (cont)

(b)

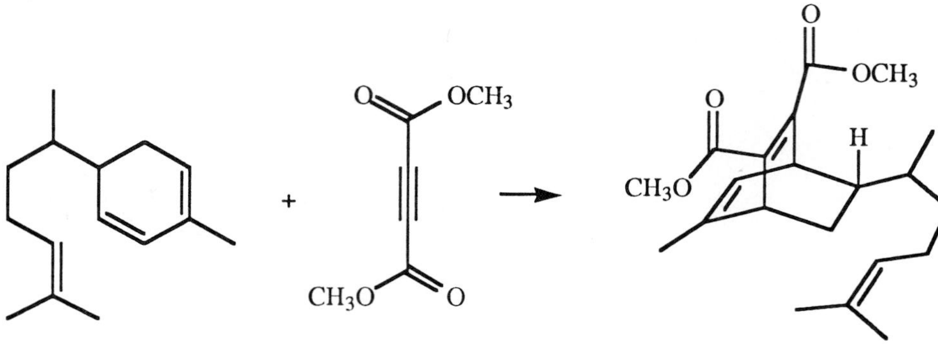

and enantiomer
B

(c)

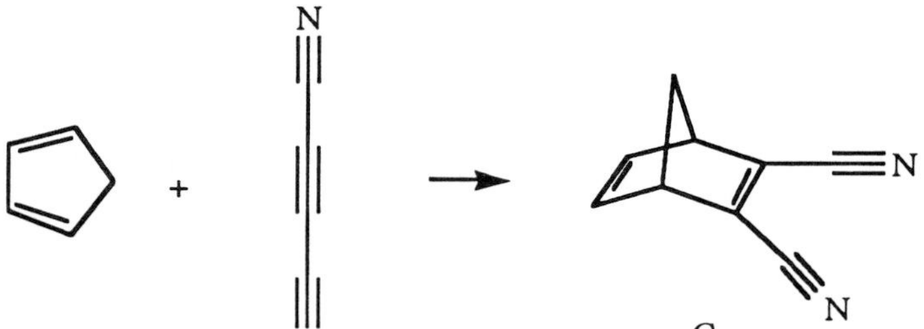

C

(d)

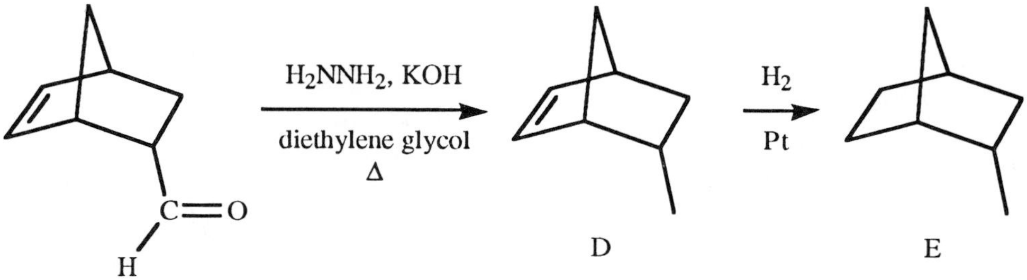

H₂NNH₂, KOH

diethylene glycol
Δ

D

H₂
Pt

E

(e)

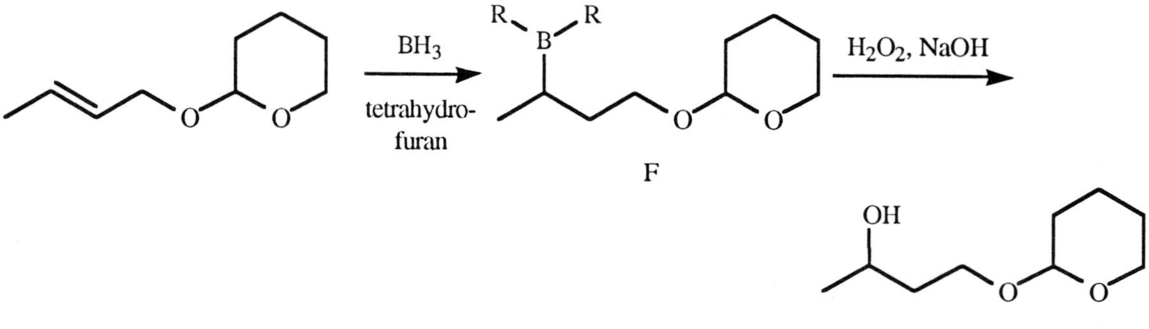

BH₃

tetrahydro-
furan

F

H₂O₂, NaOH

OH

G

17.22 (cont)

(f)

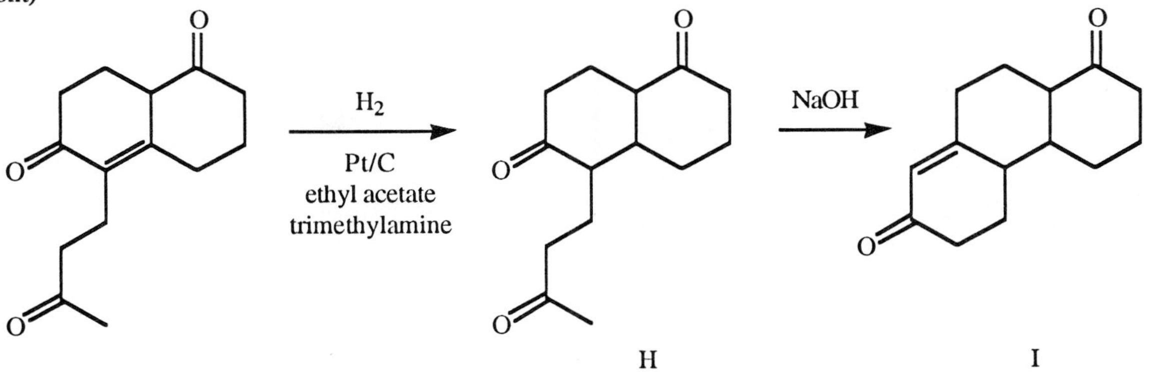

(g)

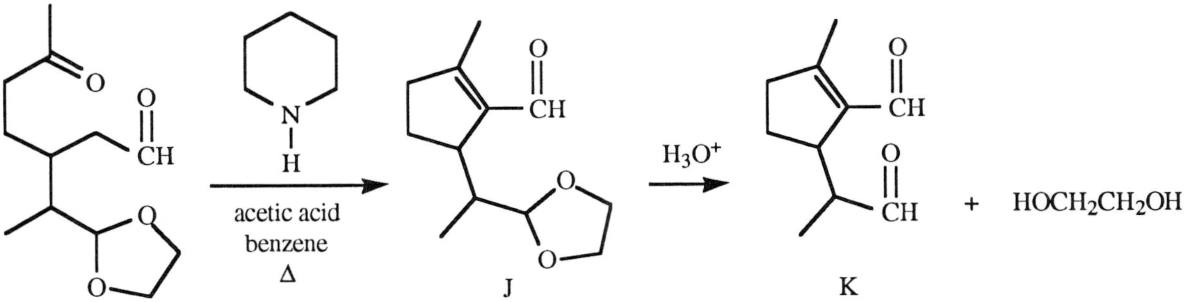

product with the
more highly substituted
double bond favored

(h)

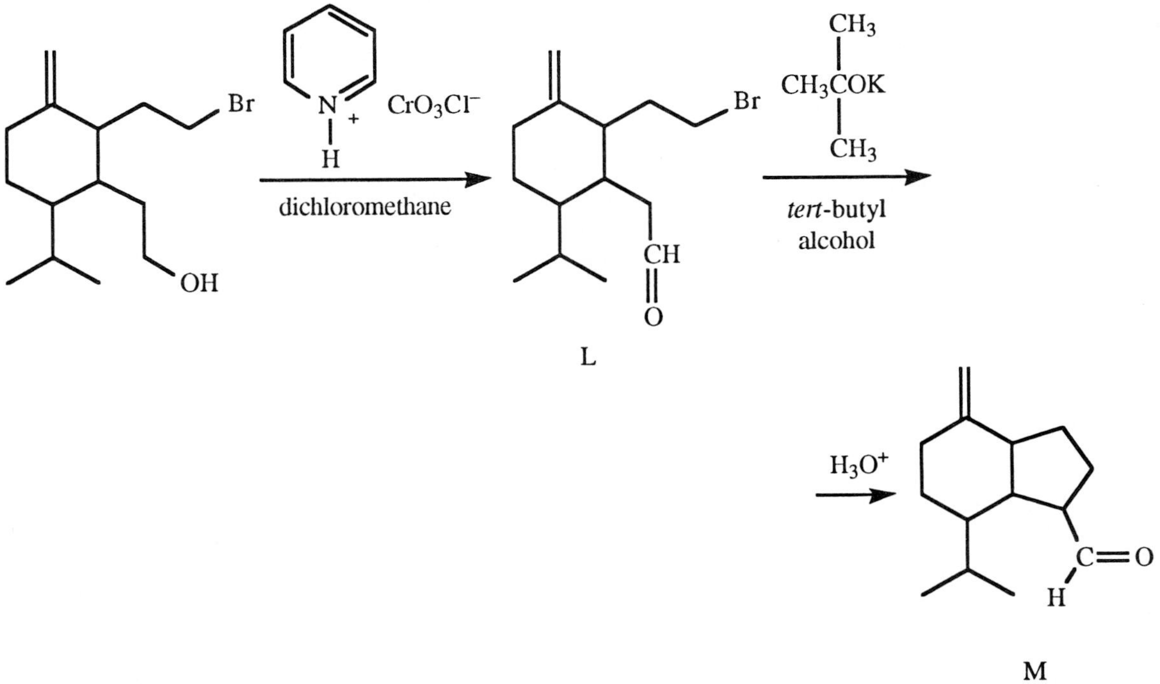

17.22 (cont)

(i)

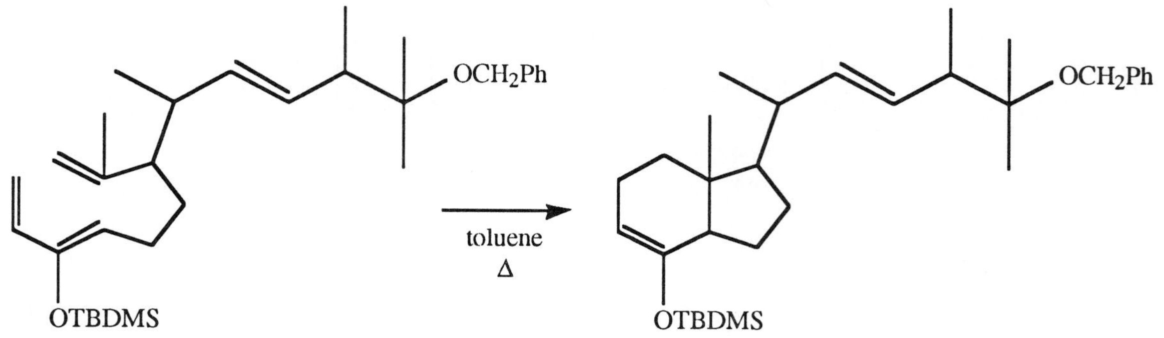
OCH2Ph
OTBDMS
toluene
Δ
OCH2Ph
OTBDMS

N

(j)

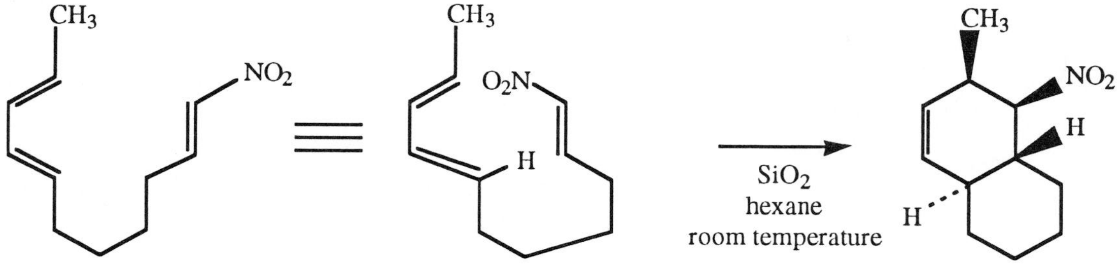
CH3
NO2
CH3
O2N
H
SiO2
hexane
room temperature
CH3
NO2
H
H

O

17.23

(a)

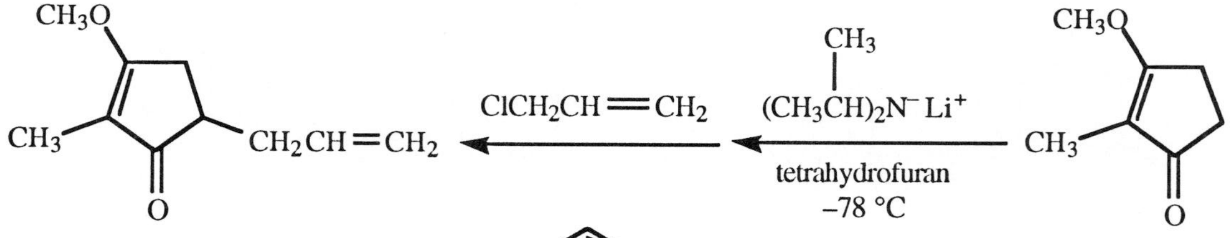
CH3O
CH3
O
CH2CH=CH2
ClCH2CH=CH2
CH3
(CH3CH)2N⁻ Li⁺
tetrahydrofuran
−78 °C
CH3O
CH3
O

(b)

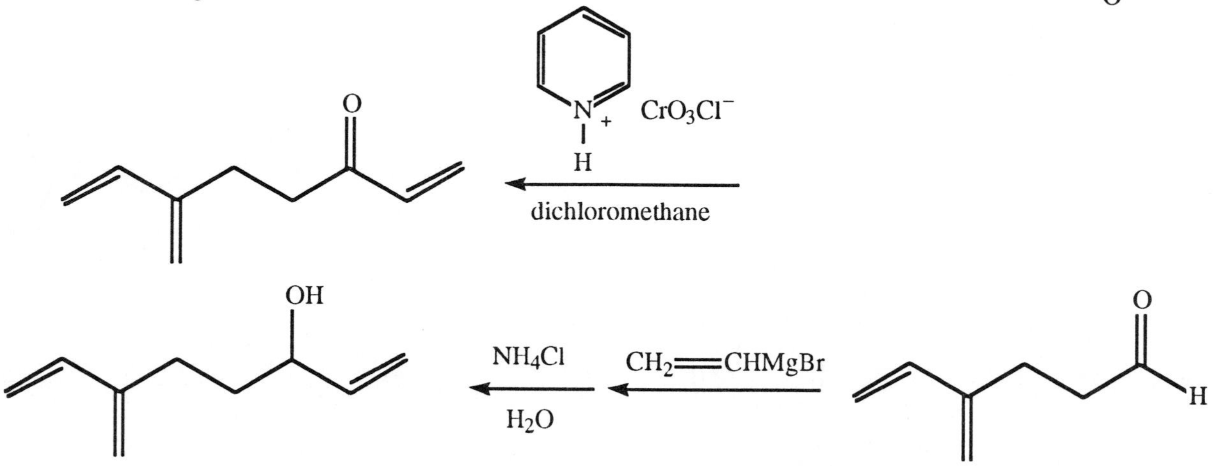
O
N
+
H
CrO3Cl⁻
dichloromethane
OH
NH4Cl
H2O
CH2=CHMgBr
O
H

17.23 (cont)

(c)

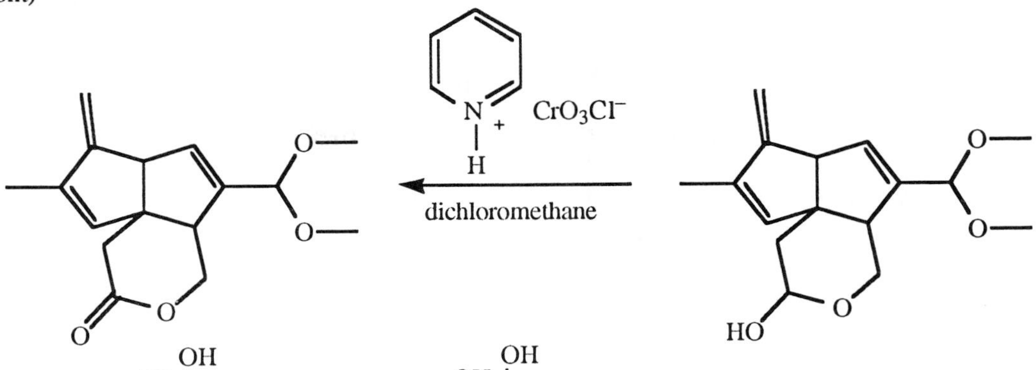

(d)

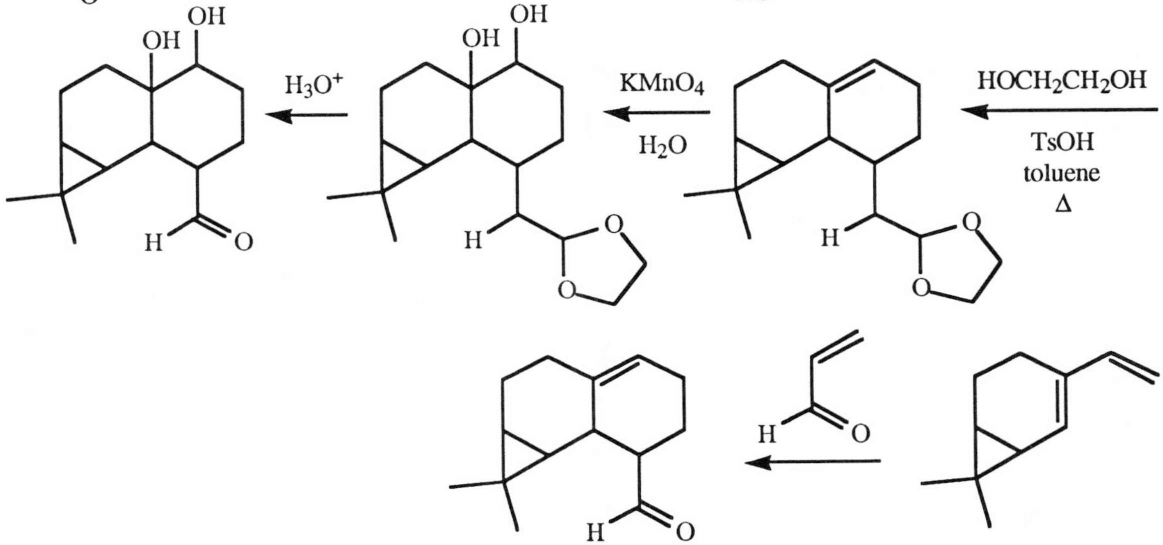

(e)

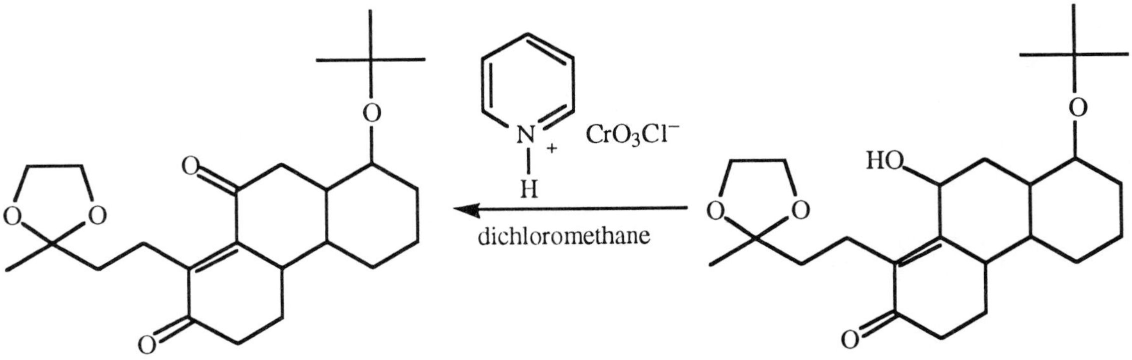

17.24   <u>σ bonds formed</u>

(a)

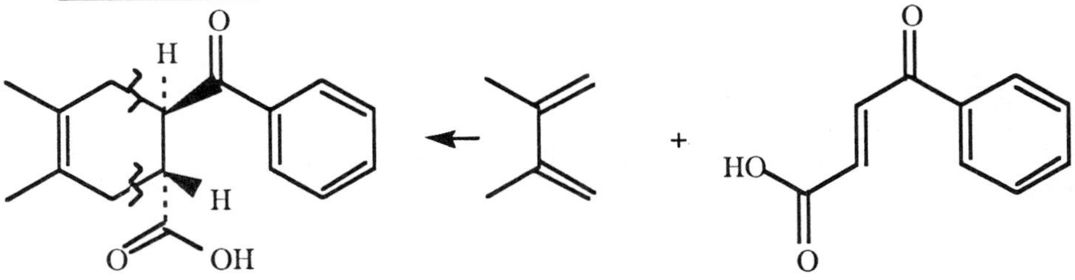

17.24 (cont) <u>σ bonds formed</u>

(b)

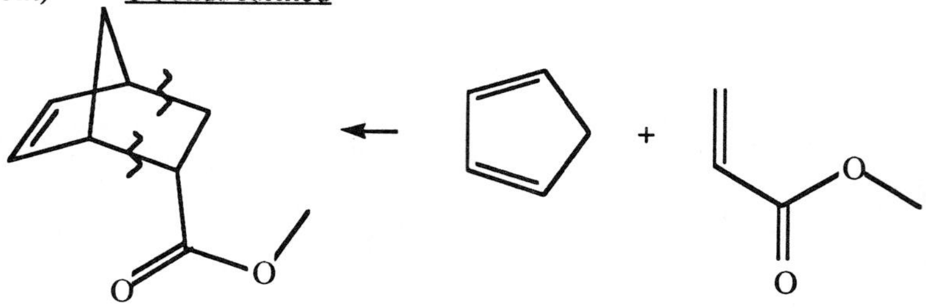

(c)

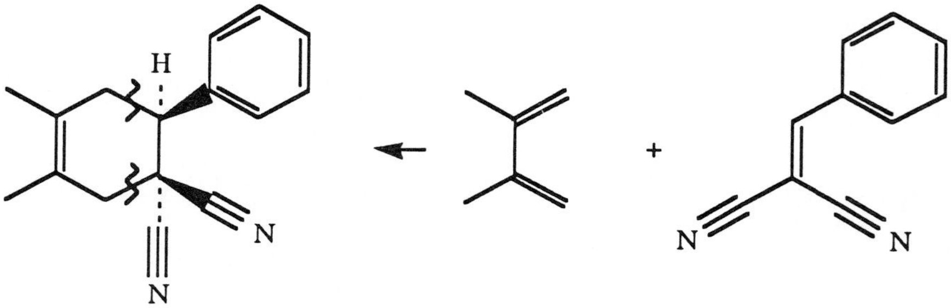

(d)

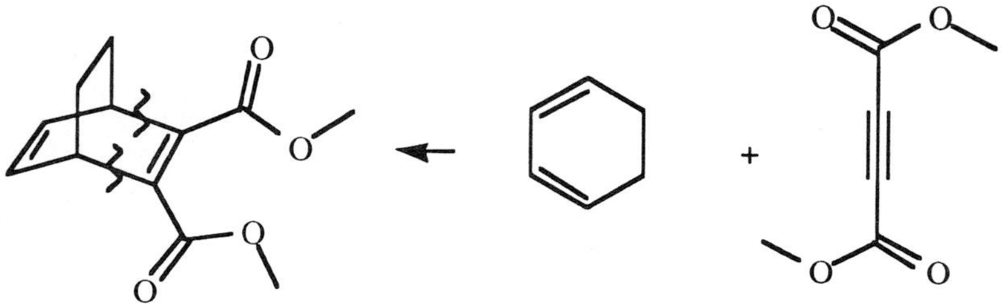

(e)

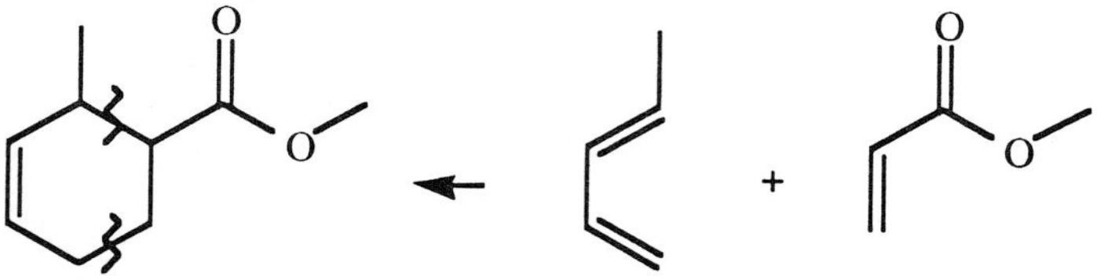

(f)

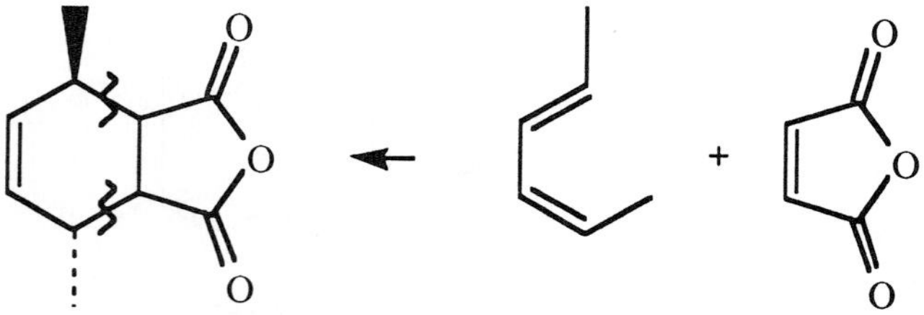

(g)

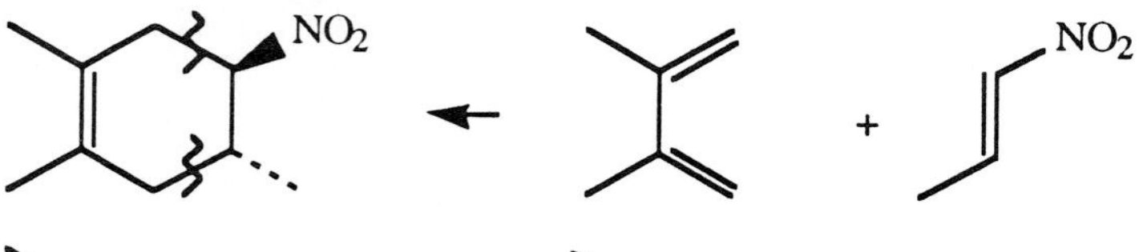

(h)

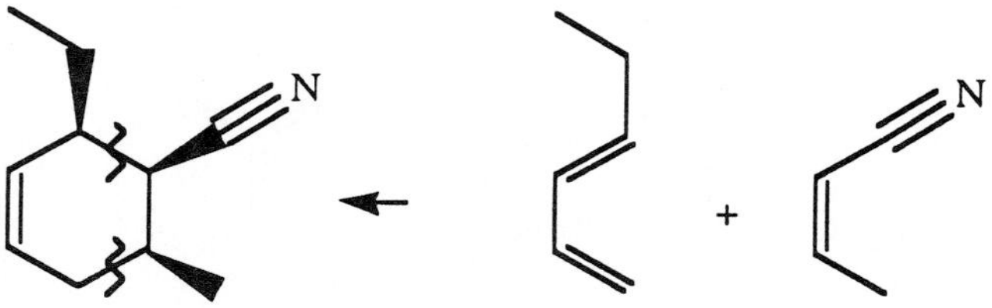

17.25

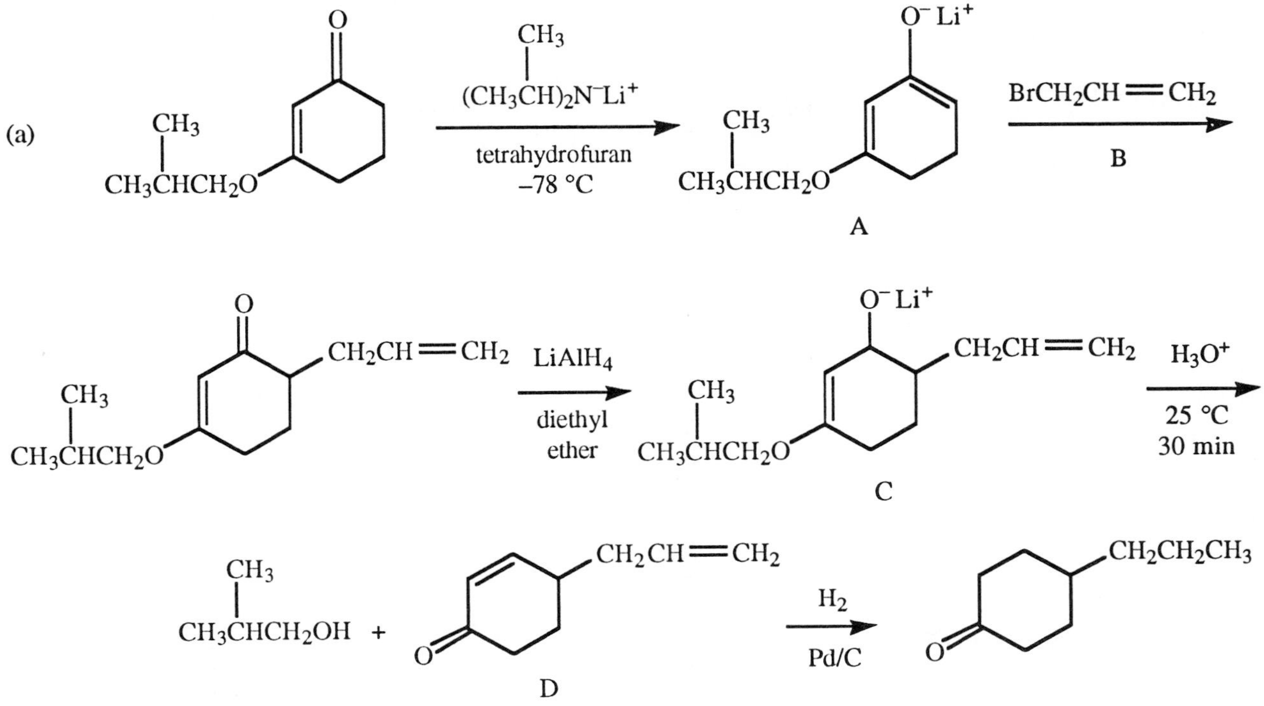
(a)
$(CH_3CH)_2N^-Li^+$
tetrahydrofuran
–78 °C
A
$BrCH_2CH{=}CH_2$
B
$CH_2CH{=}CH_2$
LiAlH$_4$
diethyl
ether
C
H_3O^+
25 °C
30 min
CH_3CHCH_2OH +
D
H_2
Pd/C
$CH_2CH_2CH_3$

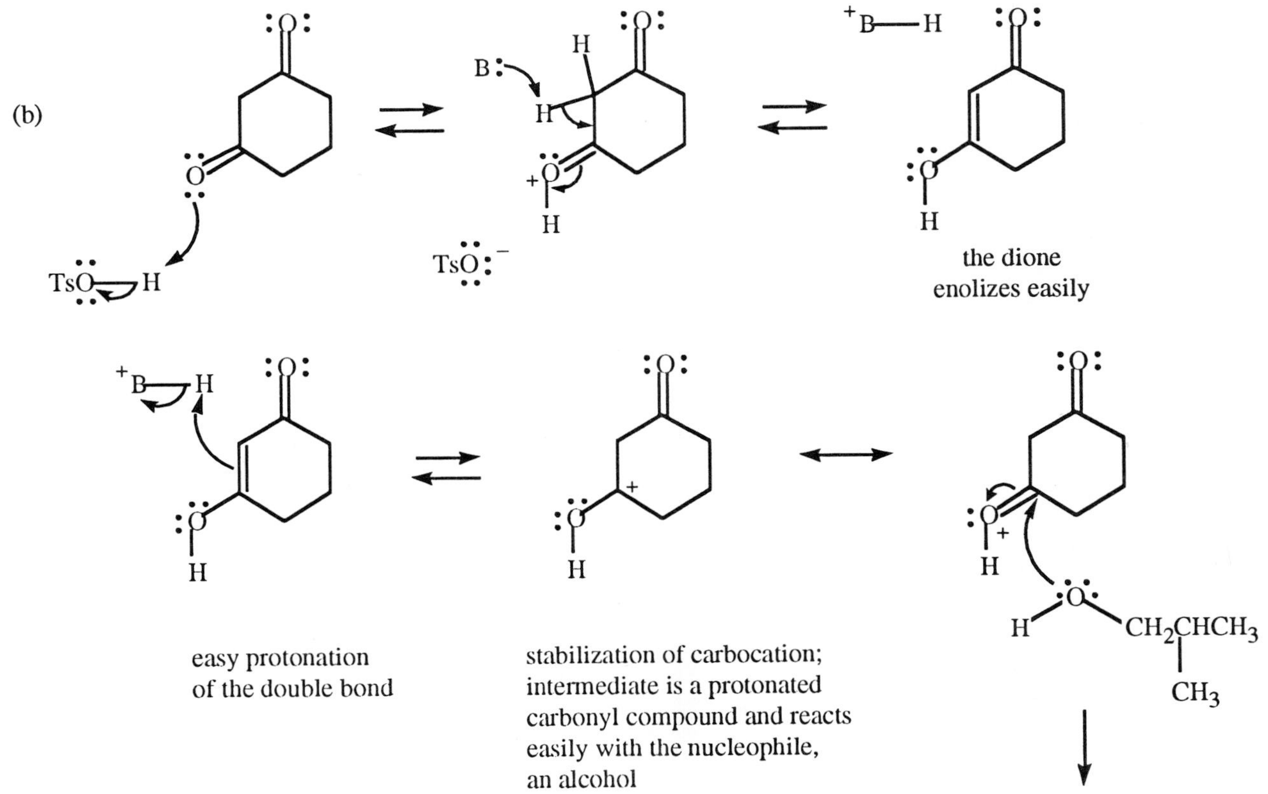
(b)
the dione
enolizes easily
easy protonation
of the double bond
stabilization of carbocation;
intermediate is a protonated
carbonyl compound and reacts
easily with the nucleophile,
an alcohol

17.25 (b) (cont)

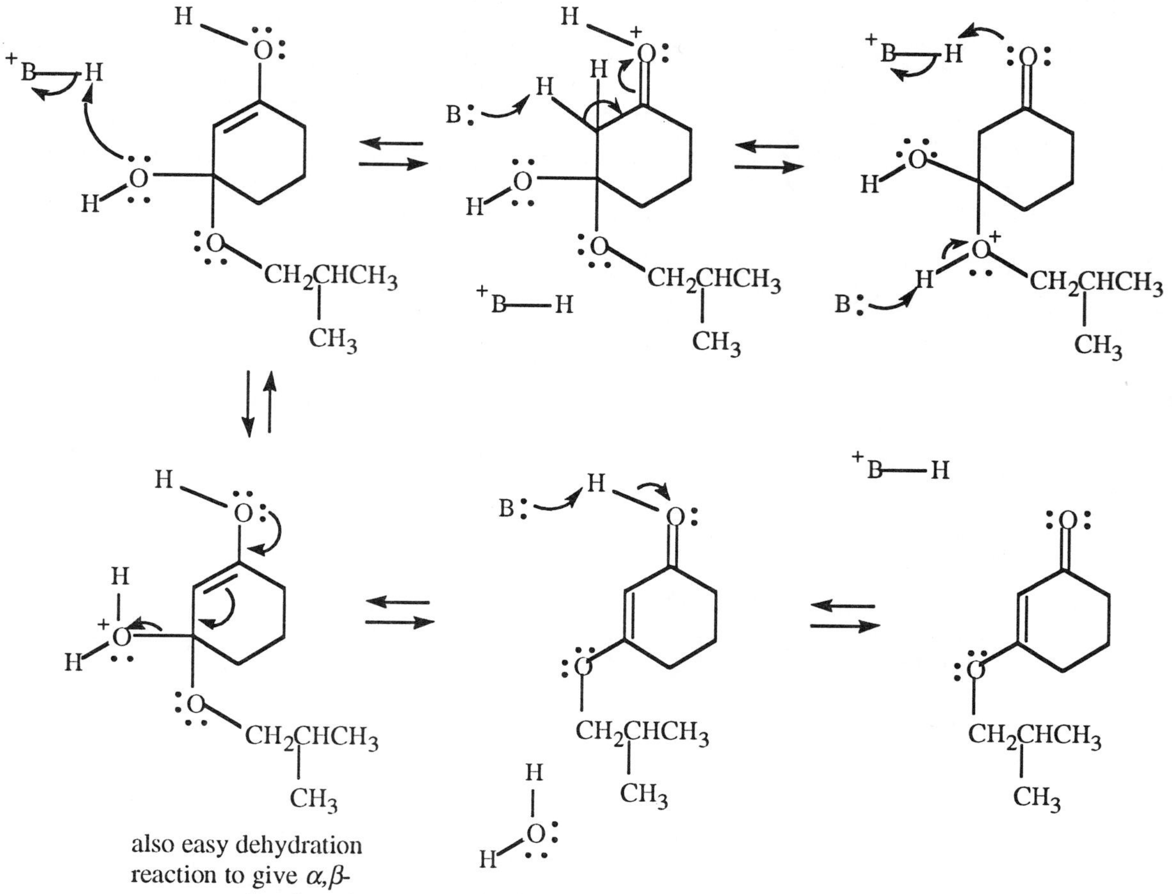

also easy dehydration
reaction to give α,β-
unsaturated carbonyl
compound

17.26

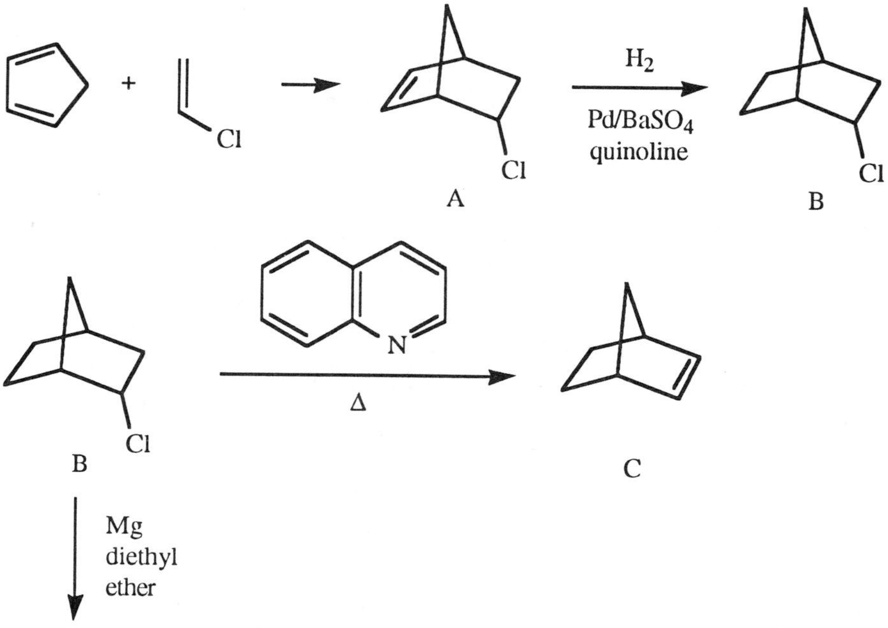

17.26 (cont)

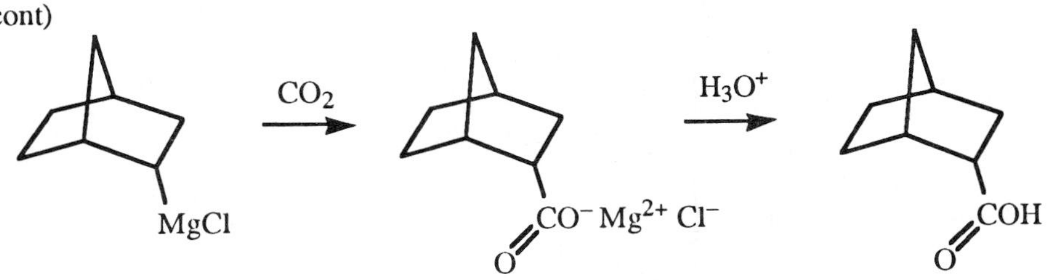

D E F

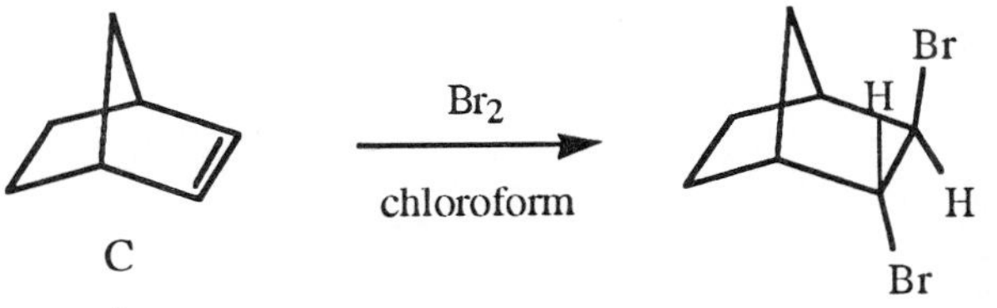

C

and enantiomer
G

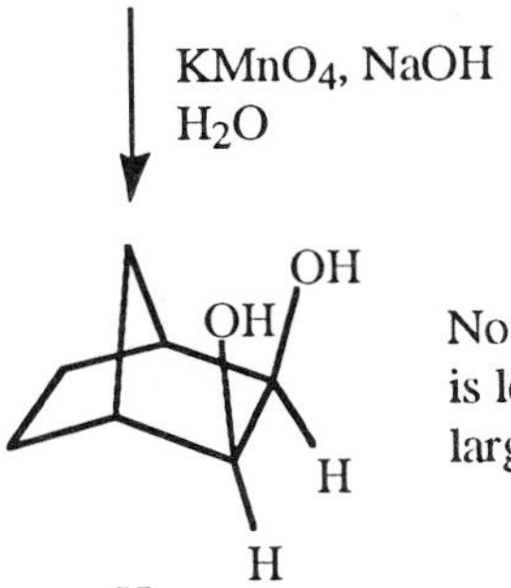

H

Note: the exo side of the bicyclic ring is less hindered to the approach of a large reagent than the endo side.

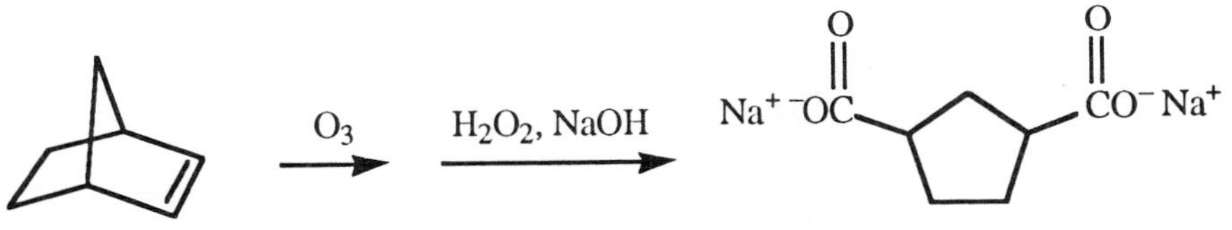

C I

17.27

(a)

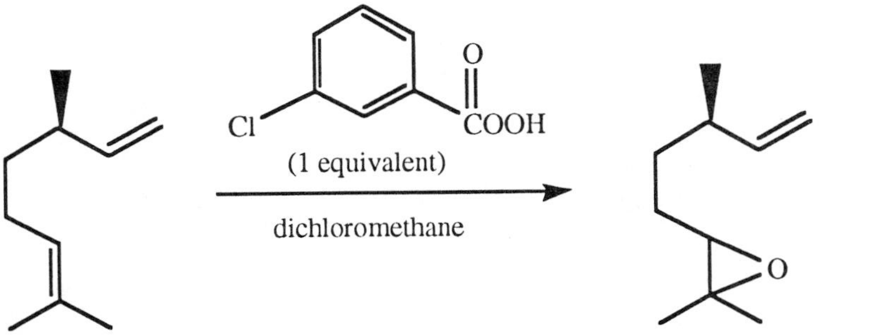

A
(more highly substituted double bond is more nucleophilic)

17.27

(b)

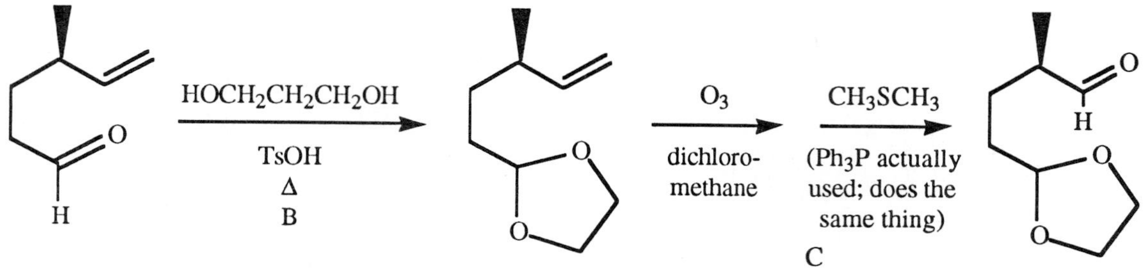
O
H
HOCH2CH2CH2OH
TsOH
Δ
B
O
O
O3
dichloro-
methane
CH3SCH3
(Ph3P actually
used; does the
same thing)
C
O
H
O
O

(c)

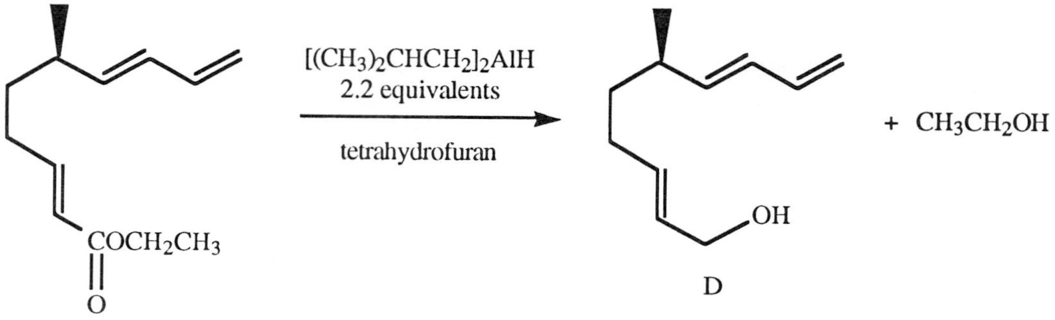
COCH2CH3
O
[(CH3)2CHCH2]2AlH
2.2 equivalents
tetrahydrofuran
OH
D
+ CH3CH2OH

(d)

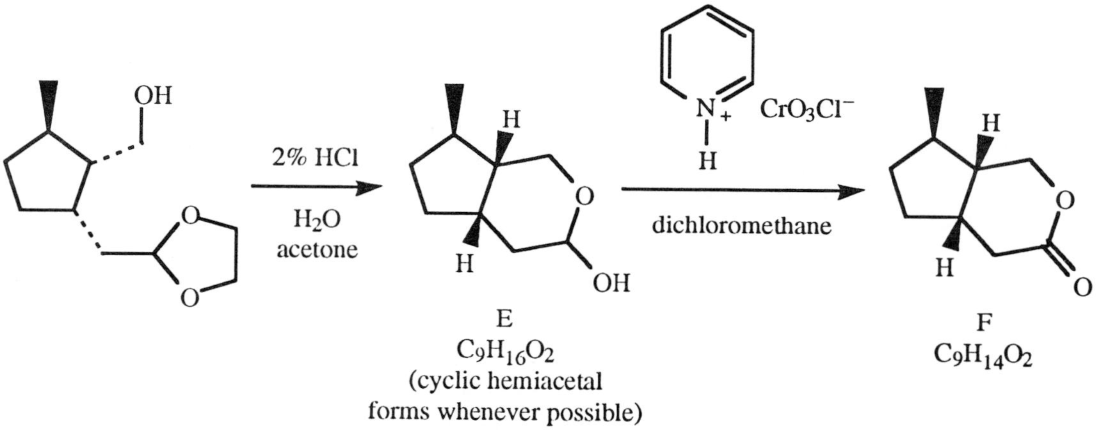
OH
O
O
2% HCl
H2O
acetone
H
O
H
OH
E
C9H16O2
(cyclic hemiacetal
forms whenever possible)
N+
H
CrO3Cl−
dichloromethane
H
O
H
O
F
C9H14O2

(e)

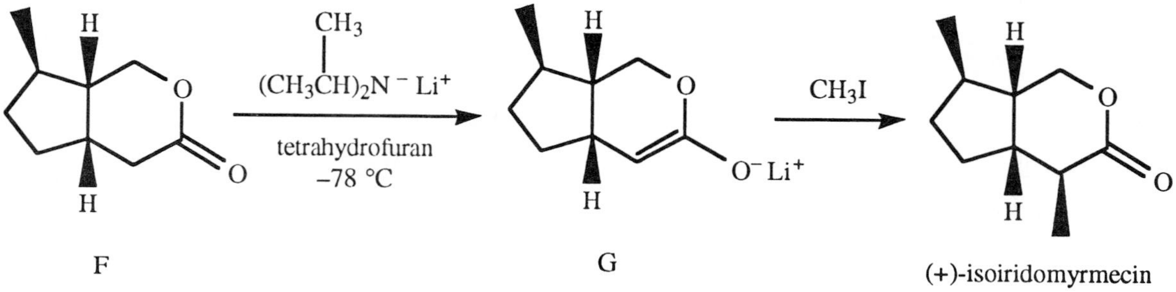
H
O
H
O
F
CH3
(CH3CH)2N− Li+
tetrahydrofuran
−78 °C
H
O
H
O− Li+
G
CH3I
H
O
H
O
(+)-isoiridomyrmecin

17.28

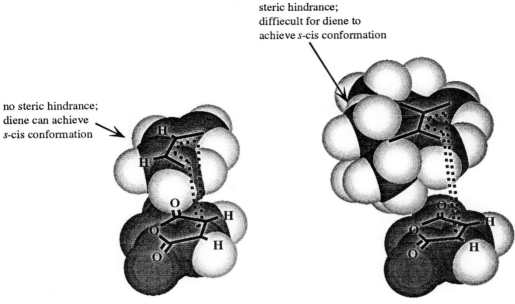

no steric hindrance;
diene can achieve
s-cis conformation

steric hindrance;
diffiecult for diene to
achieve *s*-cis conformation

transition state for the addition
of 1,3-butadiene to maleic anhydride

transition state for the addition
of 2,3-di-*tert*-butyl-1,3-butadiene to
maleic anhydride. The bulkiness of the
tert-butyl groups prevents the diene from
achieving the *s*-cis conformation needed
for the Diels-Alder reaction. Good
overlap of the π system is also difficult,
raising the energy of the transition state to
the point where the reaction does not proceed.

17.29

17.29 (cont)

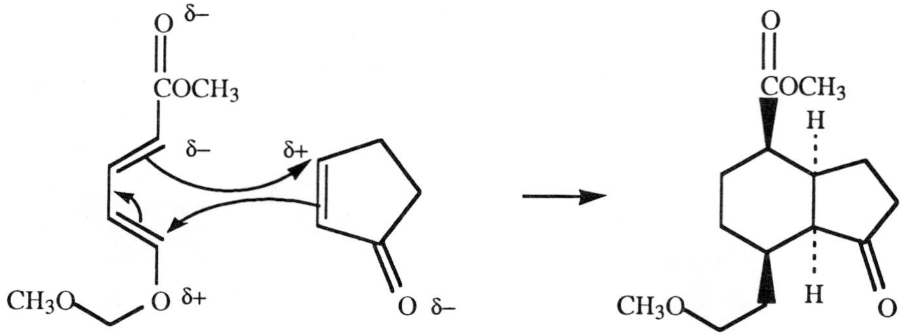

stereochemistry comes from
endo addition of dienophile to diene

17.30

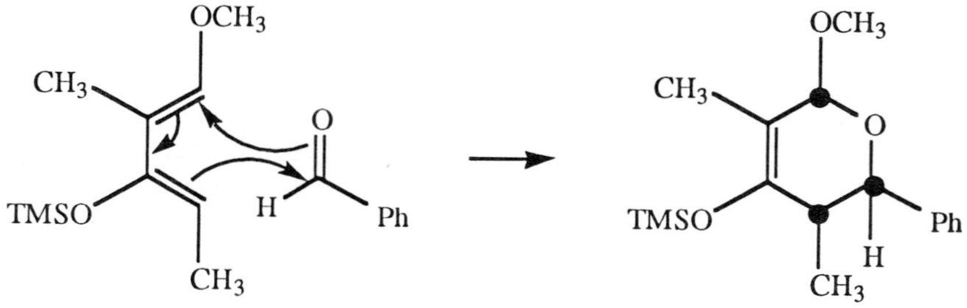

modified   benzaldehyde
Danishefsky's
diene

The Diels-Alder adduct has 3
stereocenters and, therefore, 8
stereoisomers. One stereocenter
is lost on hydrolysis, leaving
4 possible stereoisomers.

In the presence of the bulky magnesium ion–tetrahydrofuran complex (p. 524 of the text) of the carbonyl group, one diastereomer is favored over the other.

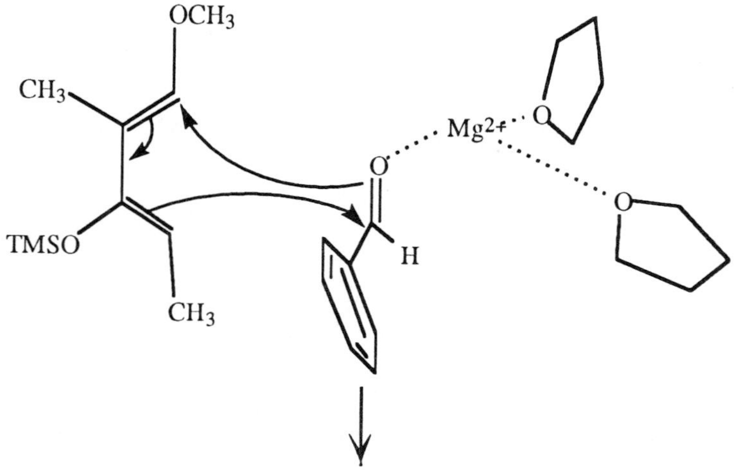

17.30 (cont)

activated complex at transition state

↓

This corresponds to having the bulky magnesium ion-tetrahydrofuran complex exo to the diene and the phenyl group endo. This gives the diastereomer in which the methyl and phenyl groups are cis to each other.

17.31

$$CH_2 = CHCH = CHCH = CH_2$$

1,2-addition   1,4-addition   1,6-addition

$$BrCH_2CHCH = CHCH = CH_2$$
$$|$$
$$Br$$

$$BrCH_2CH = CHCHCH = CH_2$$
$$|$$
$$Br$$
not observed

$$BrCH_2CH = CHCH = CHCH_2Br$$

All three products can be derived from the same intermediate.

$$\overset{+}{BrCH_2CHCH} = CHCH = CH_2 \quad \longleftrightarrow \quad BrCH_2CH = CHCH = \overset{+}{CHCH_2}$$

$$Br^-$$ $$Br^-$$

$$BrCH_2CH = \overset{+}{CHCHCH} = CH_2$$

$$BrCH_2CHCH = CHCH = CH_2$$
$$|$$
$$Br$$
conjugated diene

$$Br^-$$

$$BrCH_2CH = CHCHCH = CH_2$$
$$|$$
$$Br$$
skipped diene

$$BrCH_2CH = CHCH = CHCH_2Br$$
conjugated diene

The products observed have conjugated double bonds and are thermodynamically more stable than the one with the skipped double bonds.

17.32

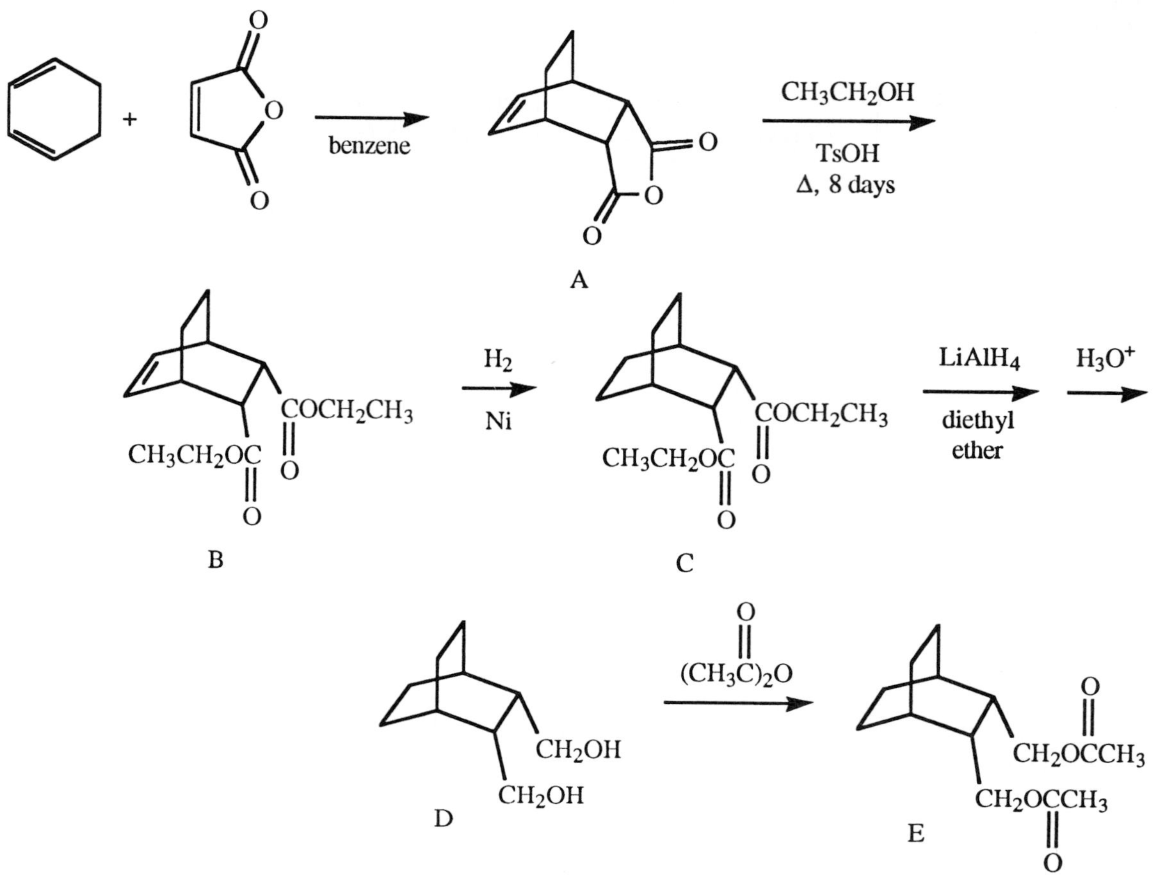

A

B

C

D

E

17.33

(a)

B

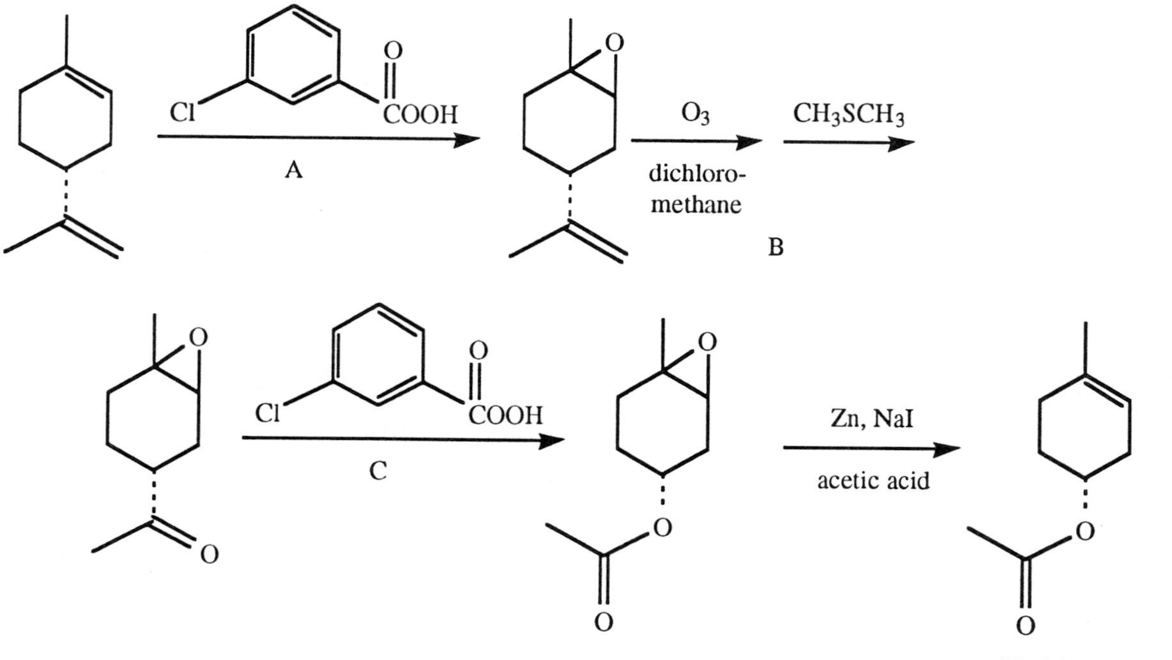

(R)-(+)-acetate
intermediate

17.33 (cont)

(b)

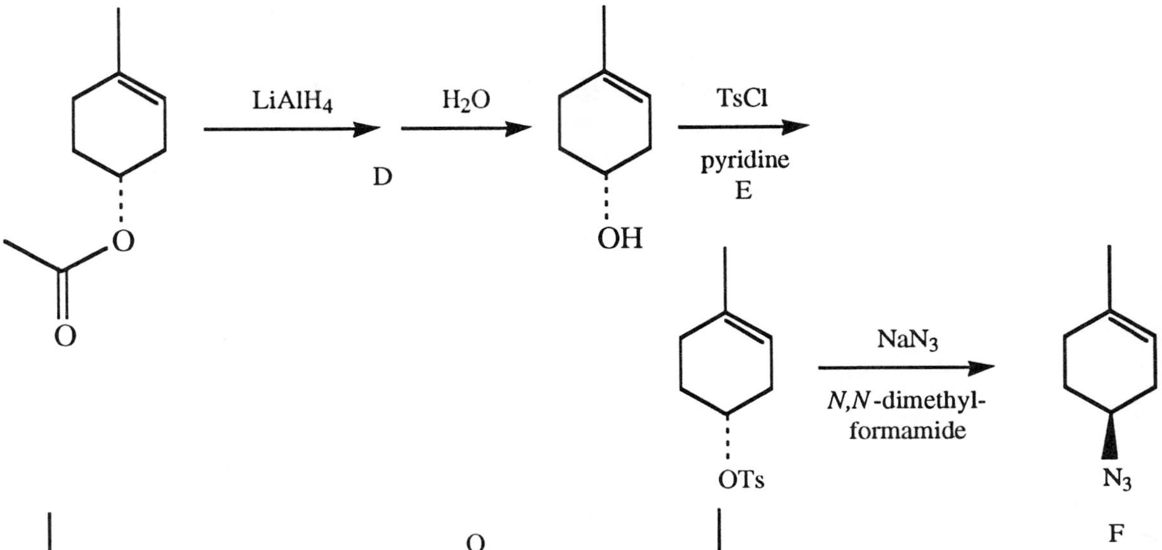

(c)

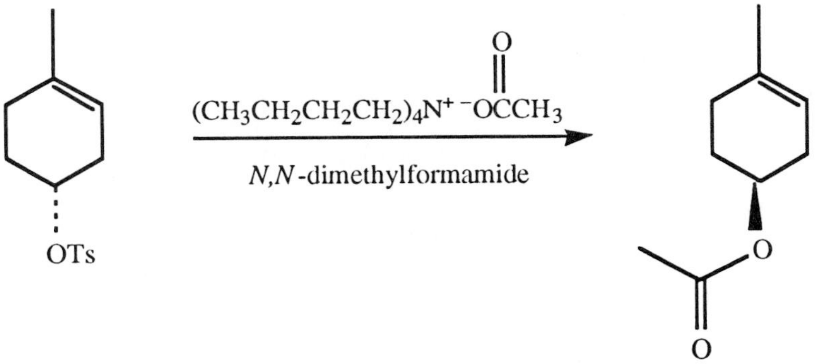

G
stereoisomer of the
(*R*)-(+)-acetate

17.34   (a)

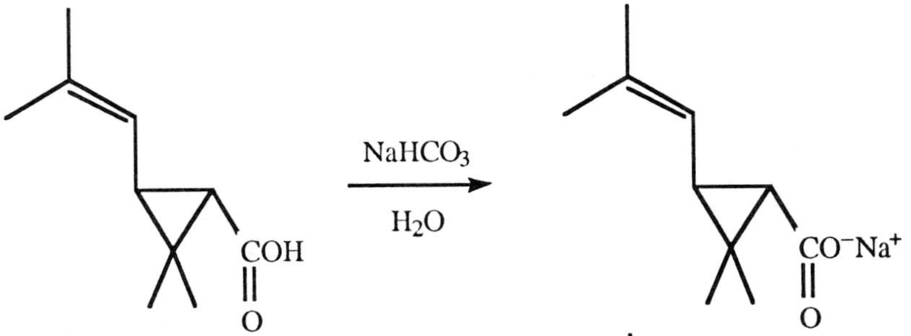

(b)

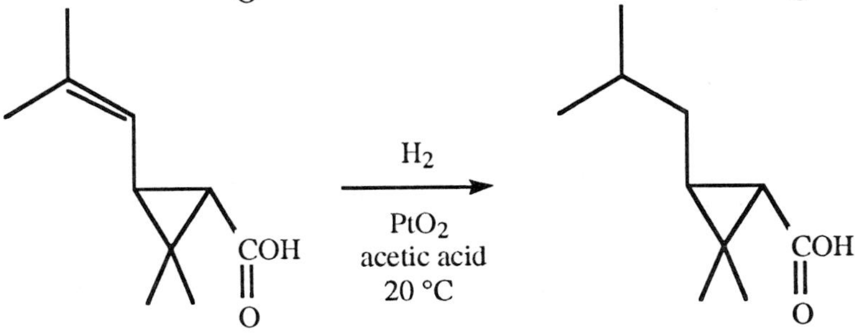

17.34 (cont)

(c)

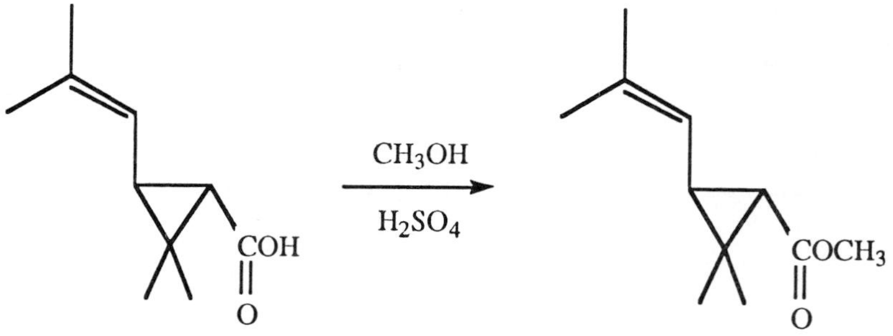

$$\text{(structure)} \xrightarrow[\text{H}_2\text{SO}_4]{\text{CH}_3\text{OH}} \text{(structure)}$$

Starting material: cyclopropane with isopropenyl-type chain and COOH (written as C—OH with O), product with COCH₃.

(d)

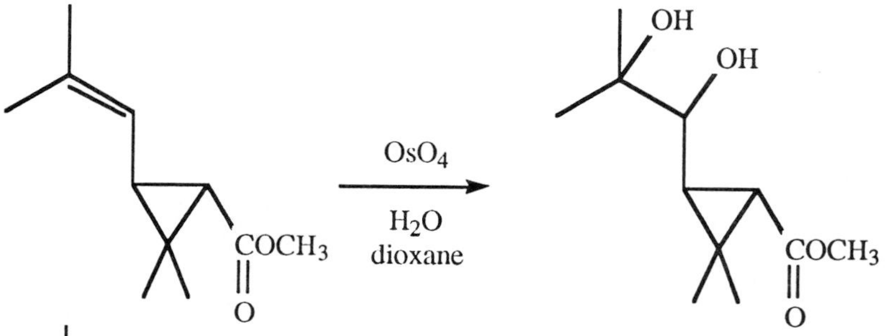

$$\xrightarrow[\substack{\text{H}_2\text{O} \\ \text{dioxane}}]{\text{OsO}_4}$$

Product shows OH, OH diol with COCH₃.

(e)

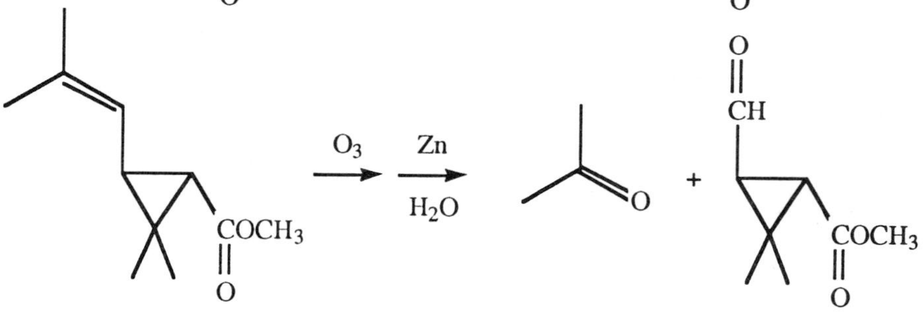

$$\xrightarrow[\text{H}_2\text{O}]{\text{O}_3 \quad \text{Zn}}$$

acetone $+$ aldehyde-cyclopropane with COCH₃.

(f)

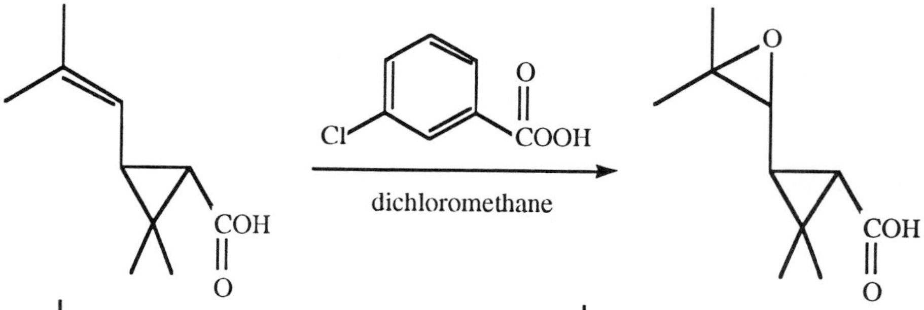

$$\xrightarrow[\text{dichloromethane}]{}$$

reagent: 3-chlorobenzoic acid (Cl-substituted benzene with COOH and O). Product: epoxide with COH.

(g)

$$\xrightarrow[\substack{\text{H}_2\text{O} \\ \Delta}]{\text{CH}_3\text{NHCH}_3}$$

epoxide with COH → product with OH, CH₃N(CH₃) group, CO⁻ H₂N⁺(CH₃)CH₃.

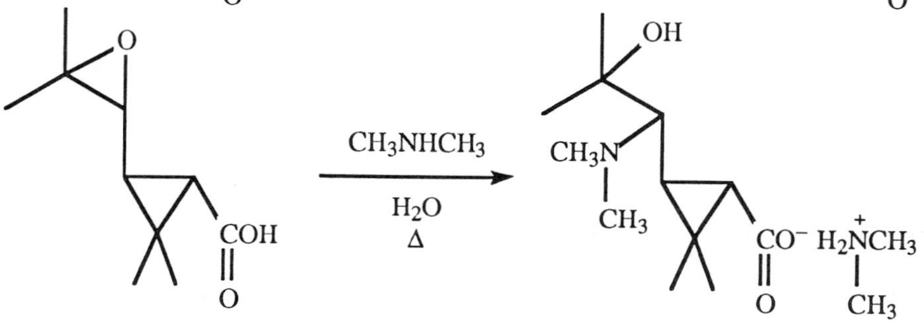

17.35

(a)   Bisabolene has the molecular formula $C_{15}H_{24}$. The corresponding saturated alkane would have the formula $C_{15}H_{32}$. Bisabolene has four units of unsaturation. Hydrogenation of bisabolene gives Compound X, $C_{15}H_{30}$. Therefore, bisabolene contains one ring and three π bonds.

(b)

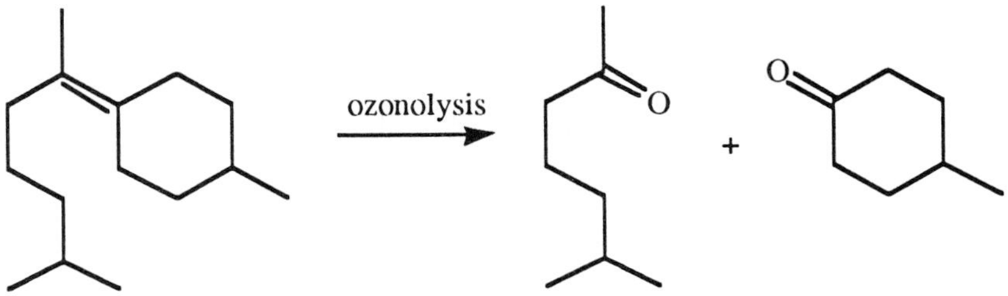

Compound Y
$C_{15}H_{28}$

6-methyl-2-
heptanone

4-methyl-
cyclohexanone

(c)

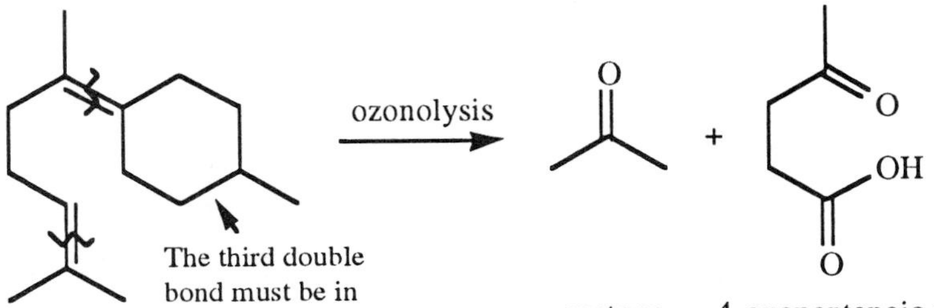

The third double
bond must be in
the ring. The exact
position is uncertain.

acetone   4-oxopentanoic acid

+ other cleavage products

(d)

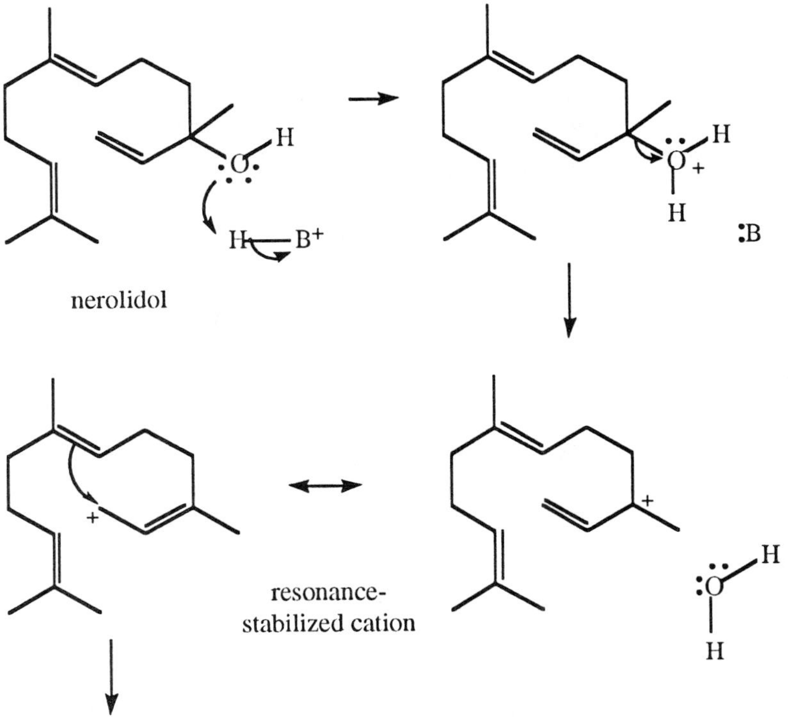

nerolidol

resonance-
stabilized cation

17.35 (d) (cont)

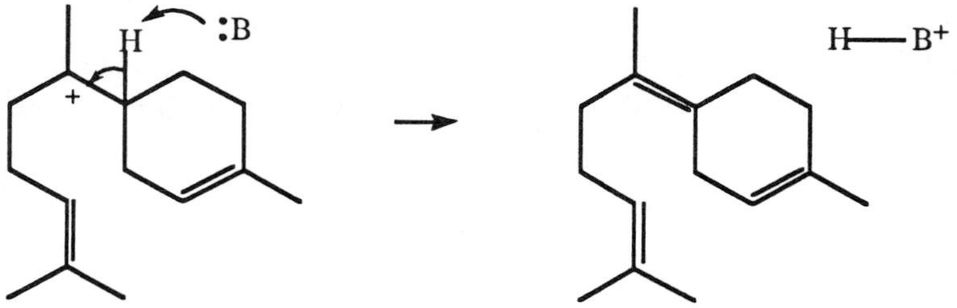

bisabolene
location of third
double bond now known

(e)

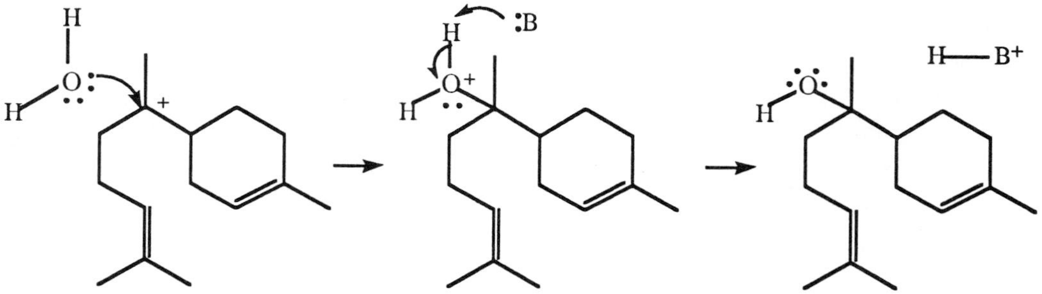

bisabolol

(f)

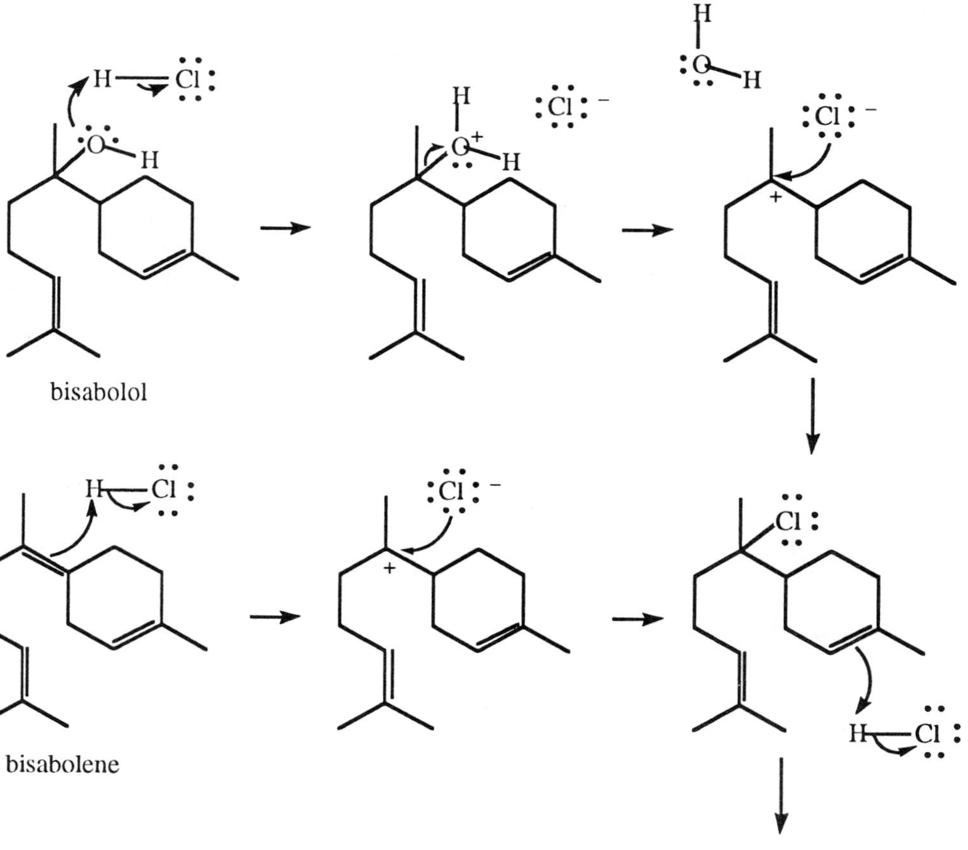

bisabolol

bisabolene

17.35 (f) (cont)

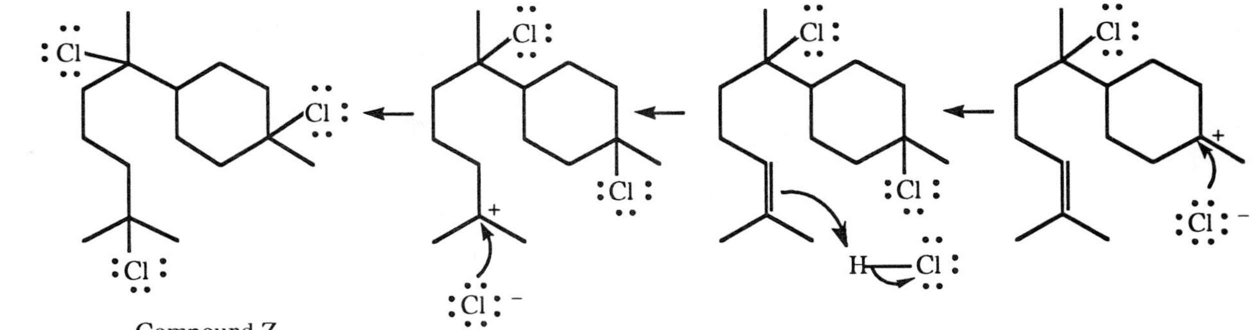

Compound Z

17.36 $HC\equiv CH$ $\xrightarrow[\text{tetrahydrofuran}]{\text{Na}}$ $HC\equiv C^{-}\,Na^{+}$ (A) $\xrightarrow{CH_3(CH_2)_9Br}$

$HC\equiv C(CH_2)_9CH_3$ (B) $\xrightarrow{CH_3CH_2CH_2CH_2Li}$ $Li^{+}\,{}^{-}C\equiv C(CH_2)_9CH_3$ (C) $\xrightarrow[\text{D}]{CH_3CH(CH_3)(CH_2)_4Br}$

$CH_3(CH_2)_9C\equiv C(CH_2)_4CH(CH_3)CH_3$ $\xrightarrow[\text{Pd/BaSO}_4\text{, quinoline; E}]{H_2}$

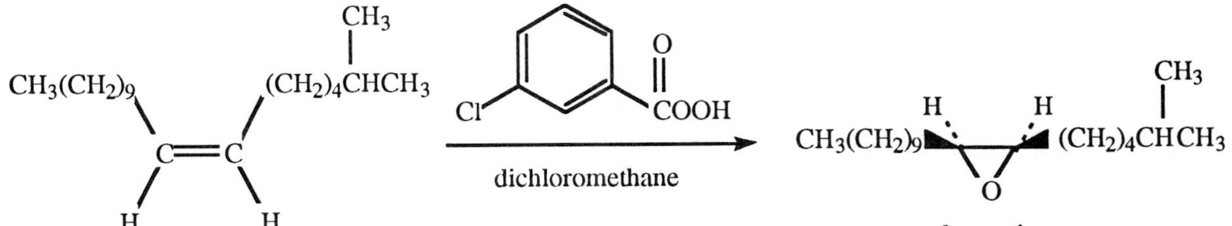

F

G

and enantiomer

disparlure

$CH_3CH(CH_3)(CH_2)_4OH$ $\xrightarrow[\text{pyridine; H}]{PBr_3}$ $CH_3CH(CH_3)(CH_2)_4Br$ (D)

17.37

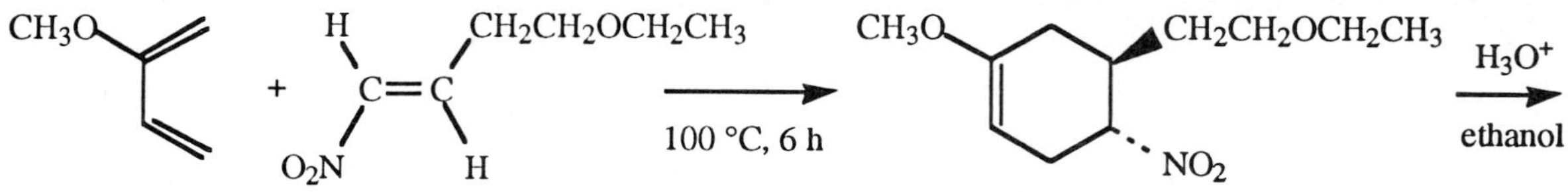

and enantiomer
A

ν_{max} (cm^{-1})
2844 C—H (aliphatic)
1658 C=C
1377 —NO_2
1178, 1166, 1117 C—O—C

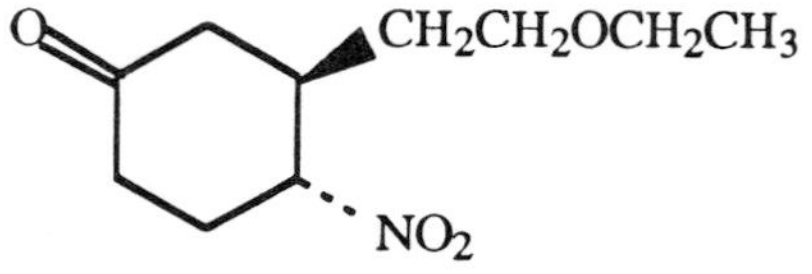

and enantiomer
B

ν_{max} (cm^{-1})
2844 C—H (aliphatic)
1722 C=O
1350 —NO_2
1117 C—O—C

17.38

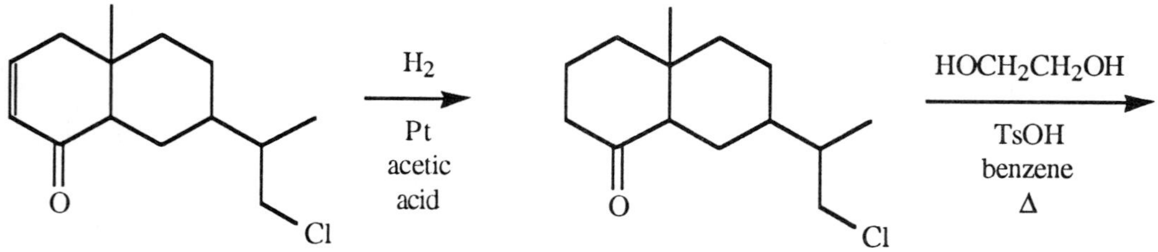

A

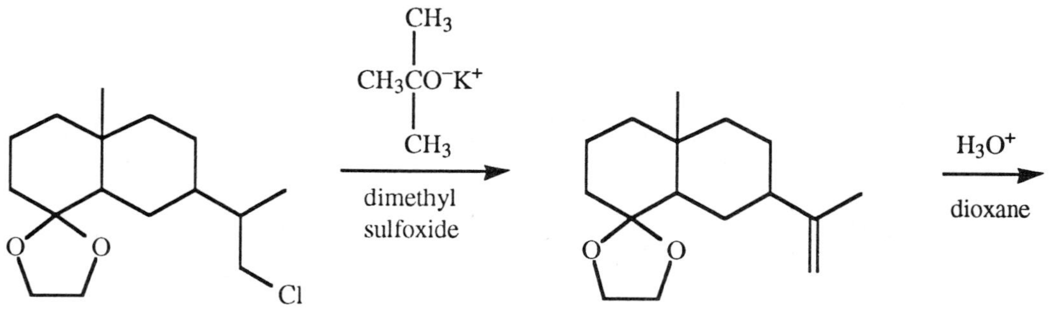

B C

17.38 (cont)

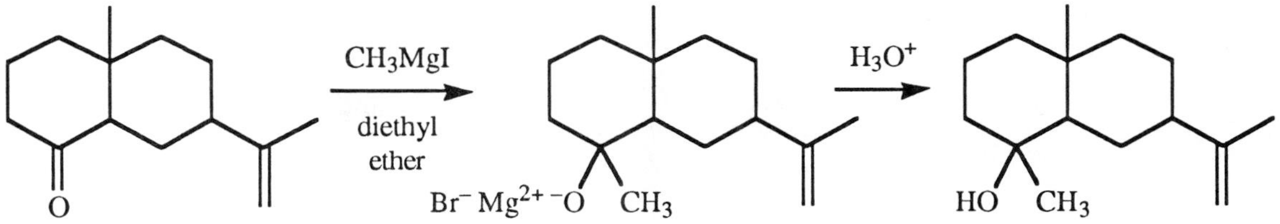

D → (CH₃MgI, diethyl ether) → E → (H₃O⁺) → F
neointermedeol

17.39

R =

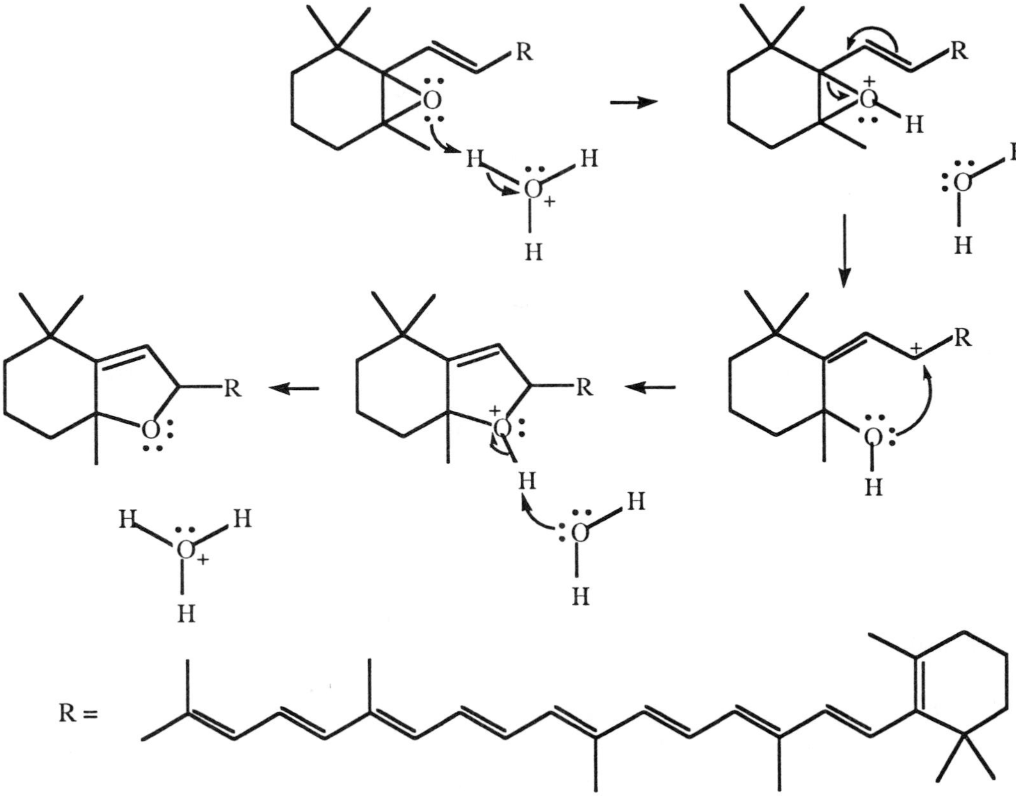

17.40

(a)

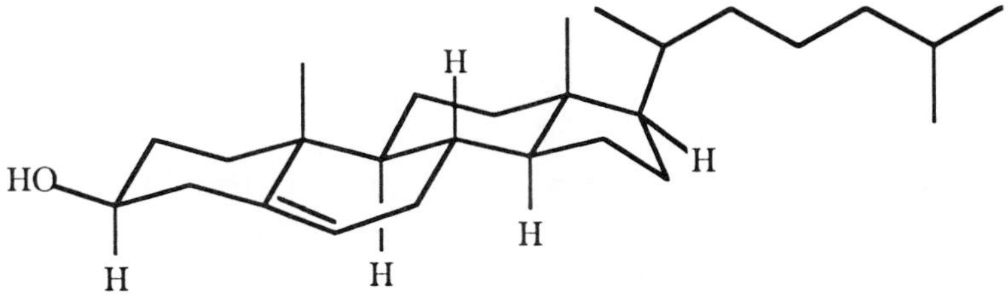

Bromine approaches the molecule from the less
hindered side, away from the methyl group.

↓ Br₂

17.40 (a) (cont)

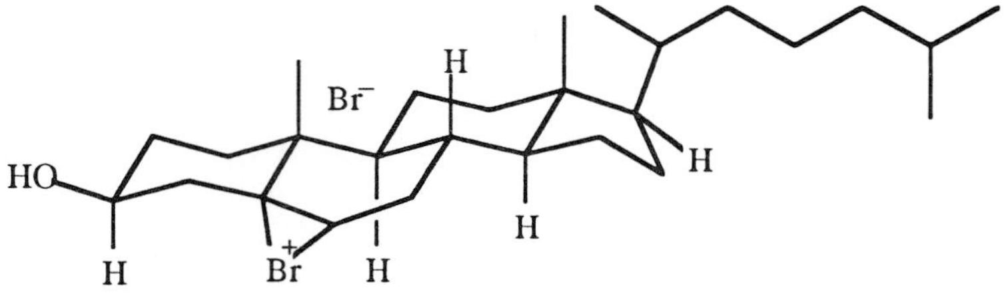

trans-diaxial orientation of the bromine atoms
when the bromonium ion opens

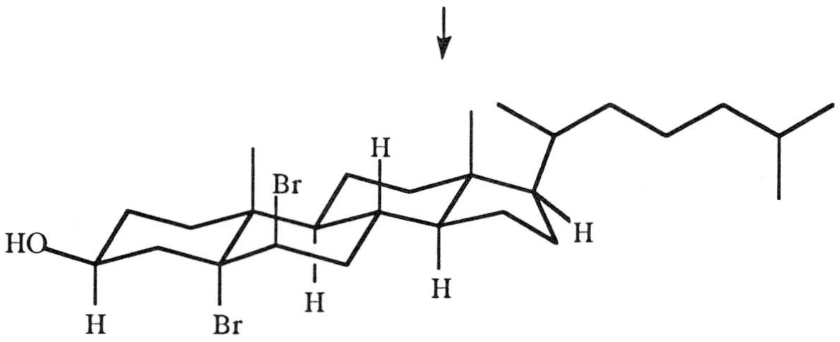

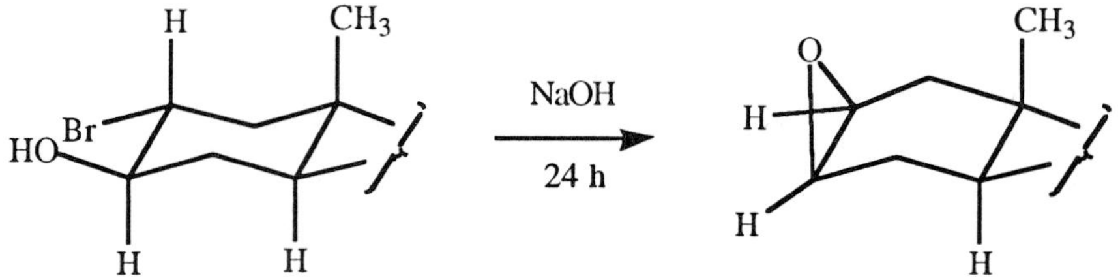

(b)

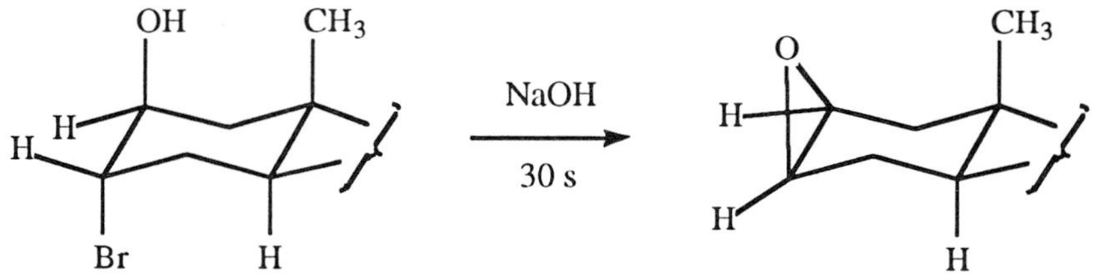

The nucleophile and the leaving group are in a trans-diaxial orientation that is favored for an intramolecular S_N2 reaction.

The nucleophile and the leaving group are in a trans-diequatorial orientation, which is less favorable for an intramolecular S_N2 reaction. Because of the rigidity of the fused ring system, the two groups can have the trans-diaxial orientation to each other only when the cyclohexane ring is in a boat or twist conformation (Section 5.10C). Such a conformation is higher in energy than the chair conformation allowed for the reaction of Compound A, so the rate of reaction is much slower for B than for A.

17.41

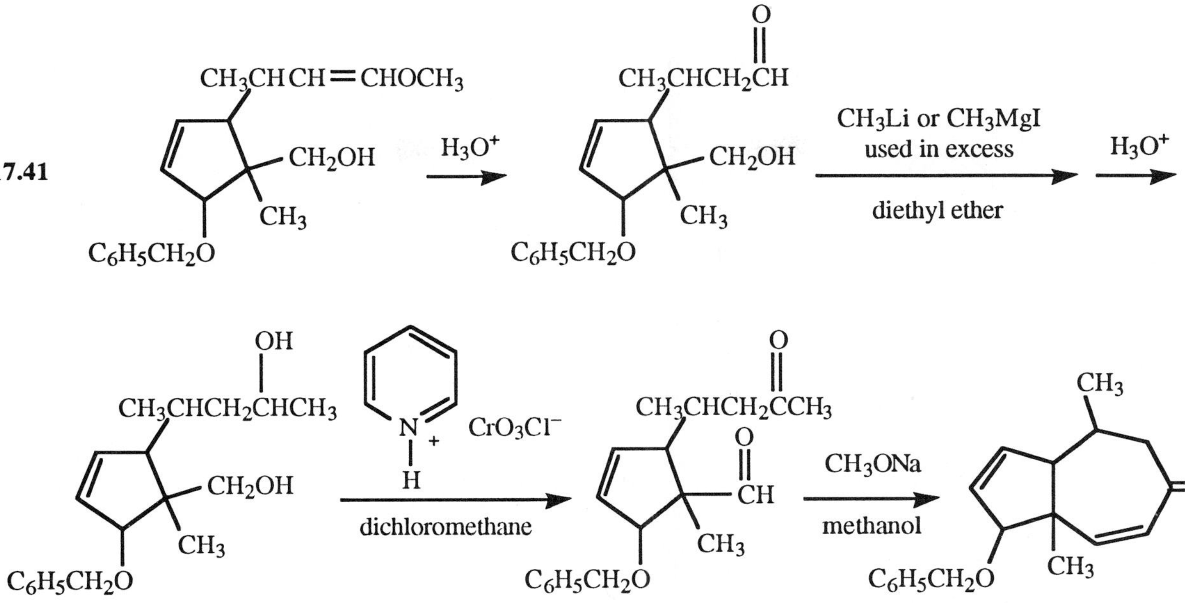

Another product with two fused five-membered rings is theoretically possible. The structure shown for helenalin in the text indicates that the bonds joining the five-membered ring to the seven-membered ring are trans to each other. Such trans stereochemistry at the junction of two fused five-membered rings would create a lot of strain, but is all right at the junction of a five- and seven-membered ring. You may wish to prove this to yourself by building molecular models of both systems.

17.42

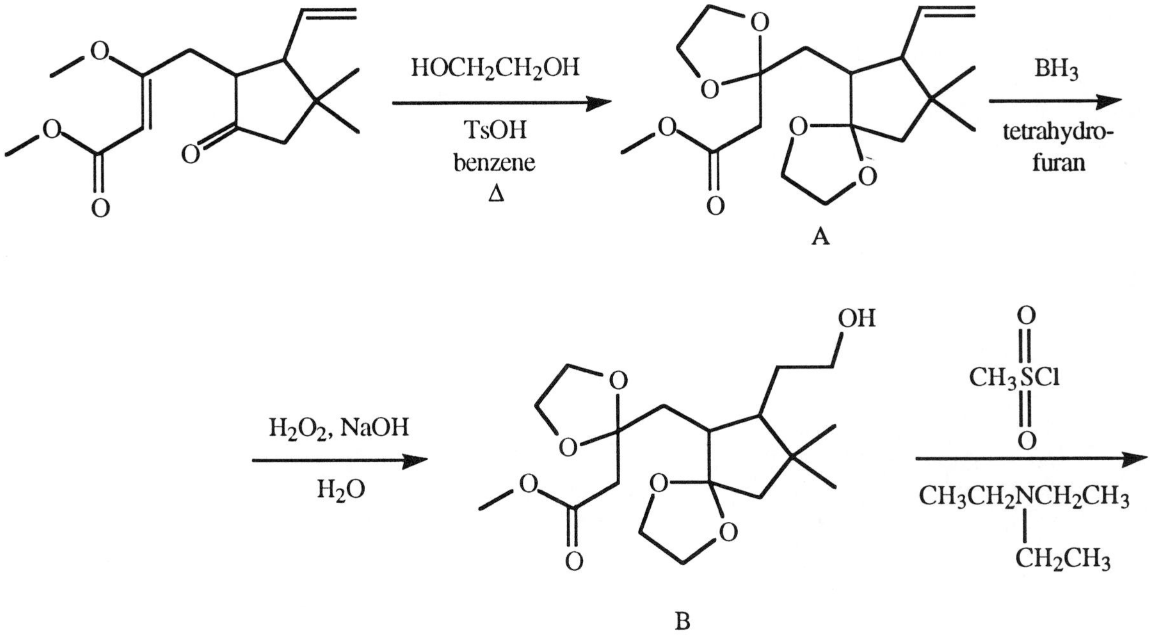

17.42 (cont)

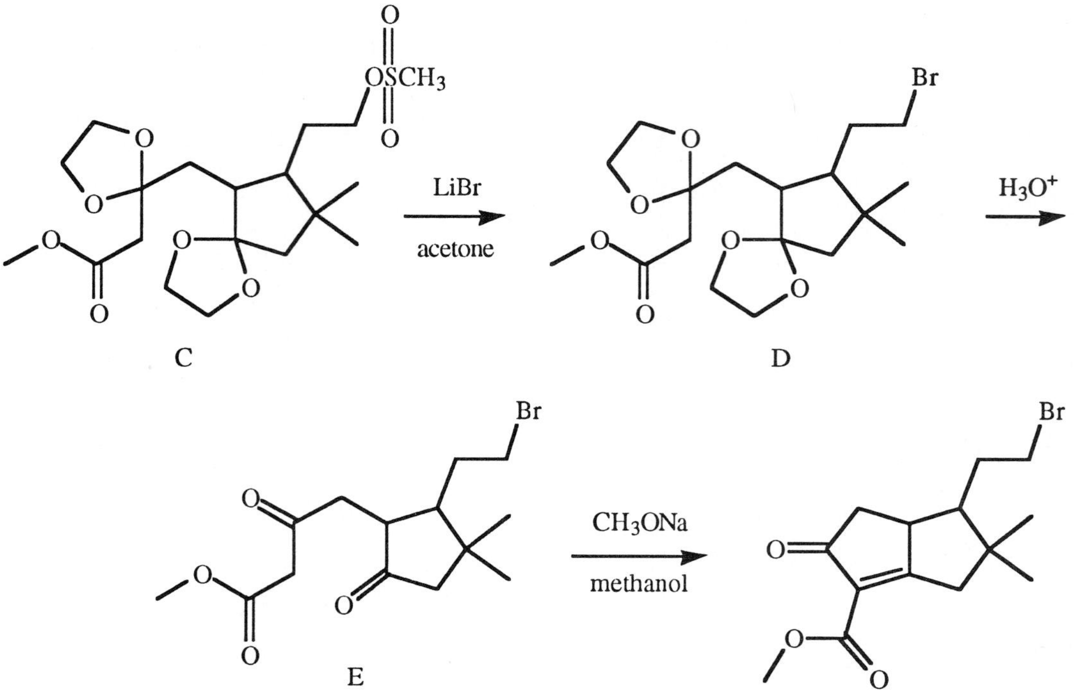

17.43

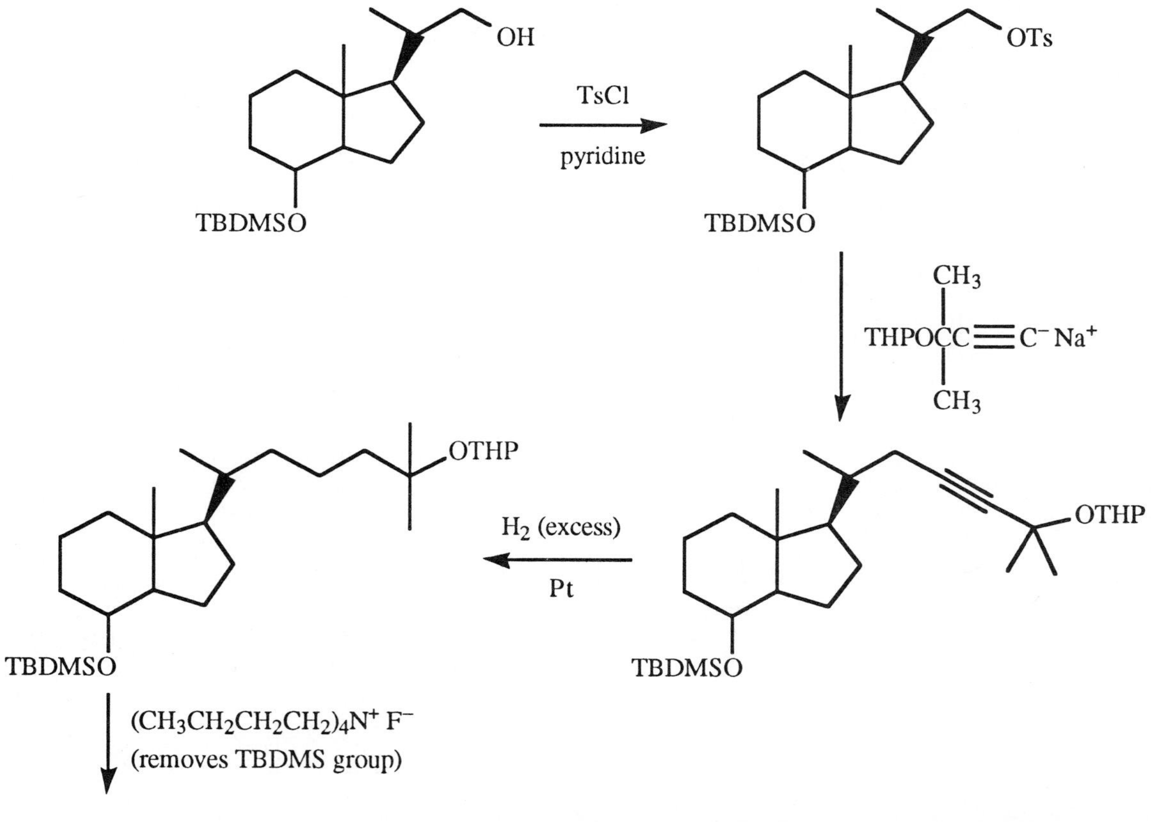

17.43 (cont)

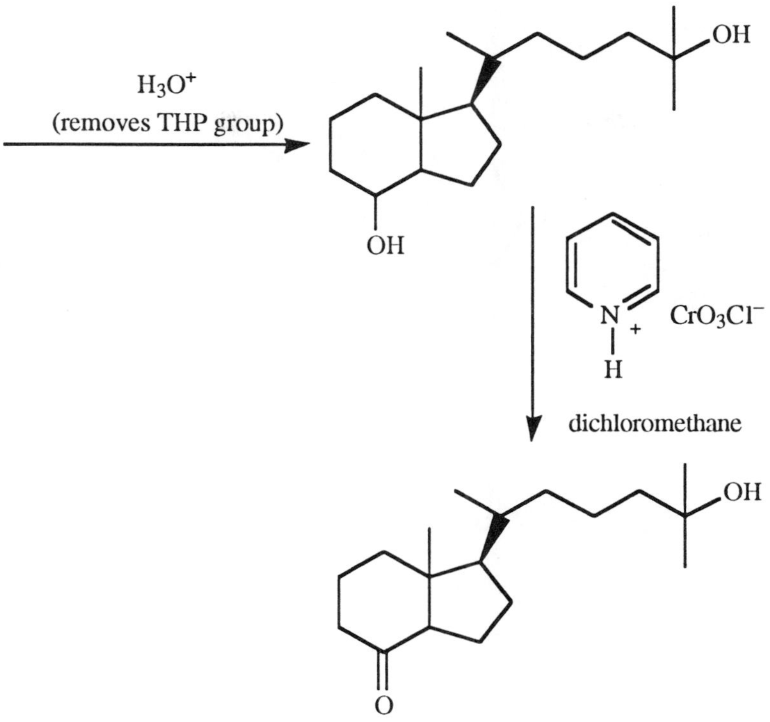

17.44

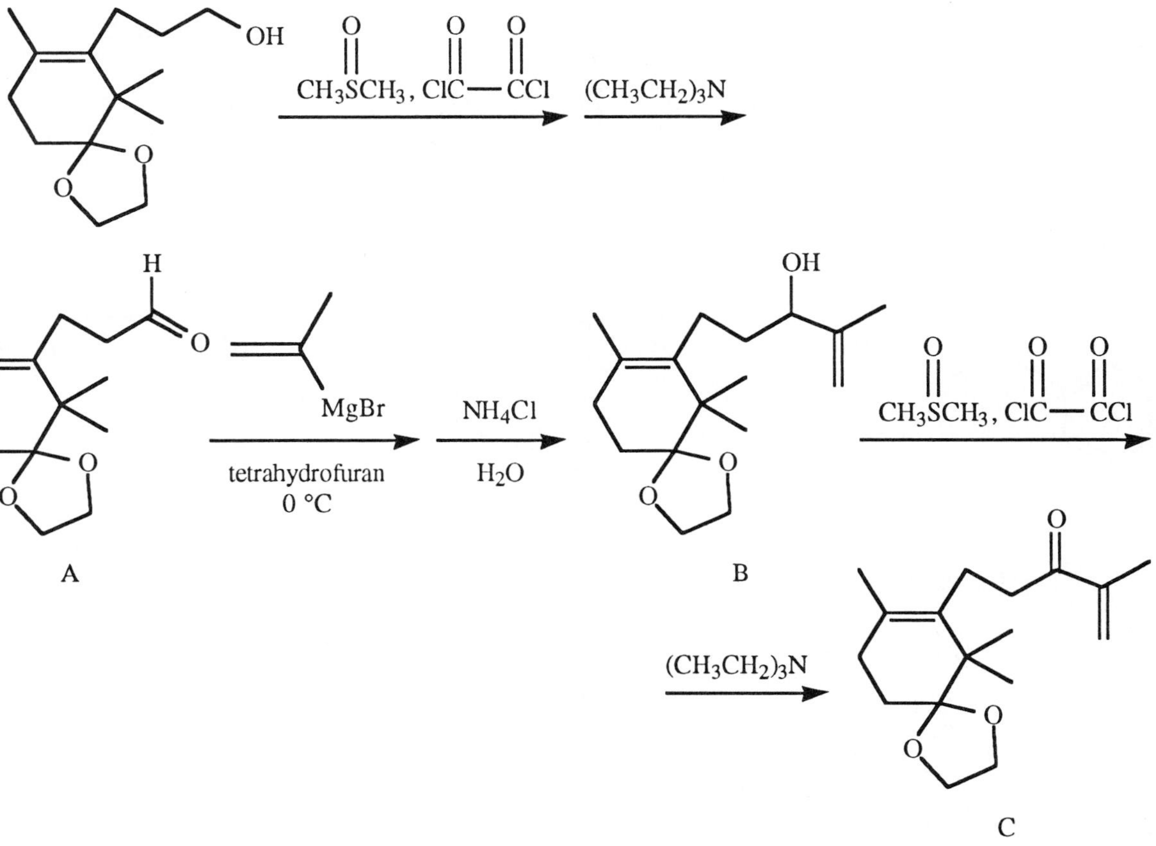

17.44 (cont)

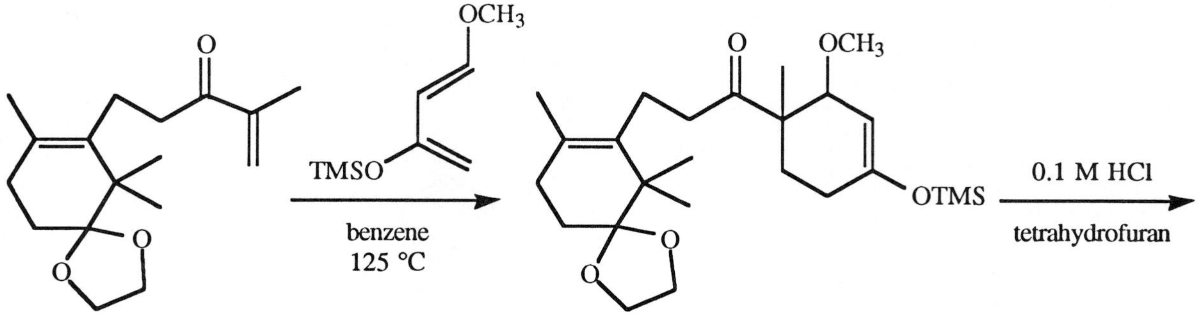

C

D

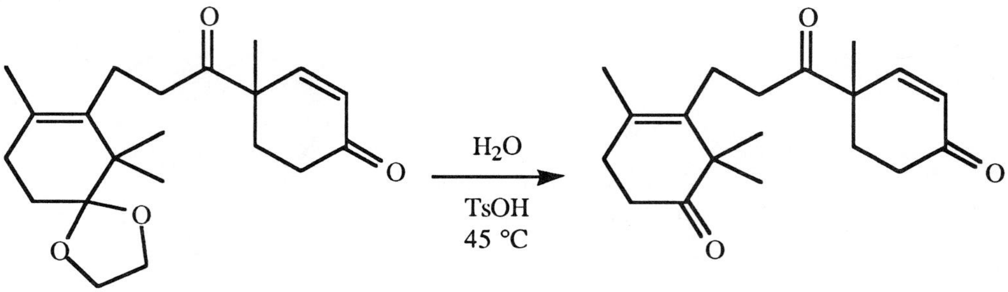

E
$C_{21}H_{30}O_4$

F
$C_{19}H_{26}O_3$

17.45

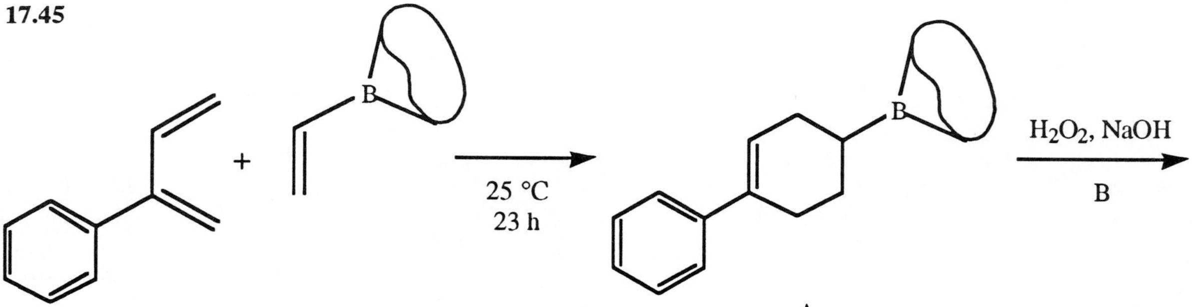

A

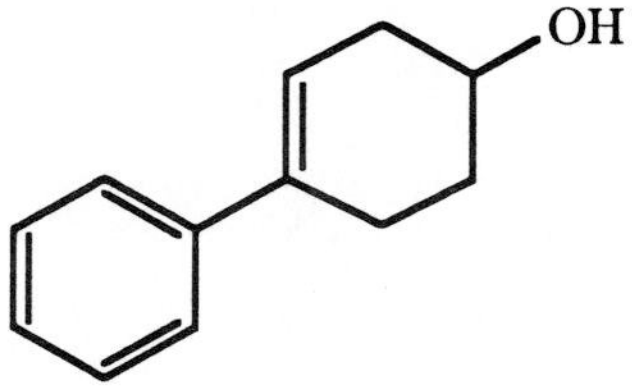

C
a 3-cyclohexen-1-ol

17.46

(a)

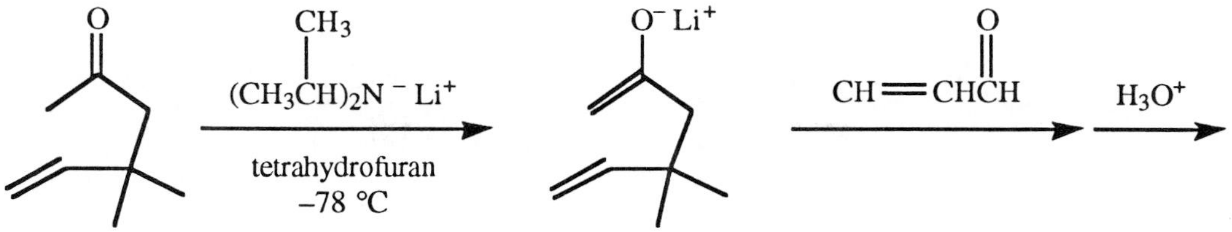

A

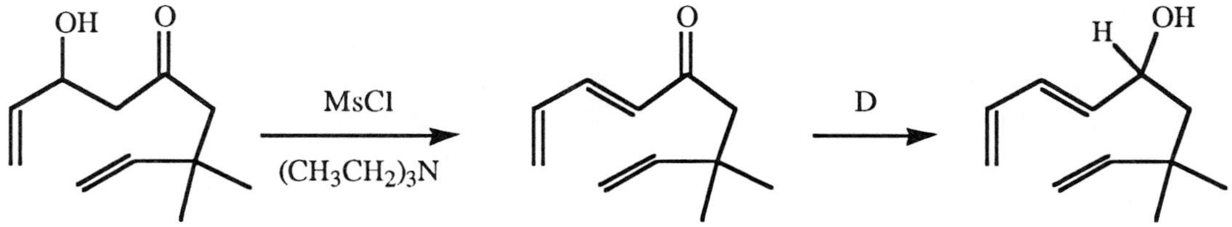

B

MsCl

(CH₃CH₂)₃N

C

the product of an
elimination reaction

D

$D = LiAlH_4$ or $NaBH_4$, $CeCl_3 \cdot 7H_2O$

(b)

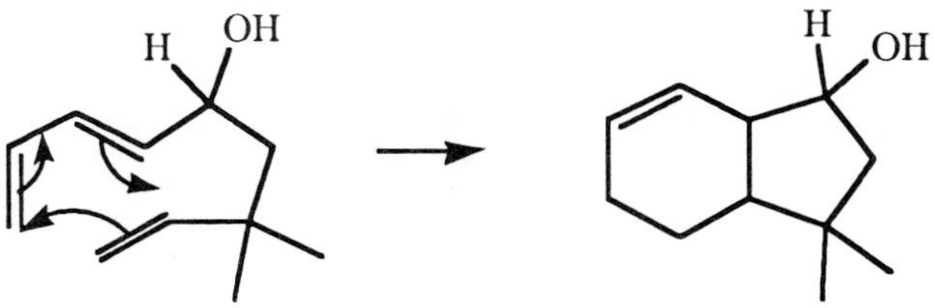

product of Diels-Alder reaction
with no stereochemistry shown

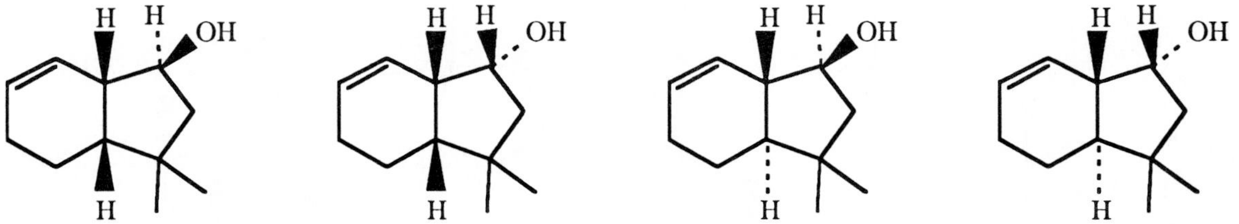

and enantiomer   and enantiomer   and enantiomer   and enantiomer

17.47

(a)

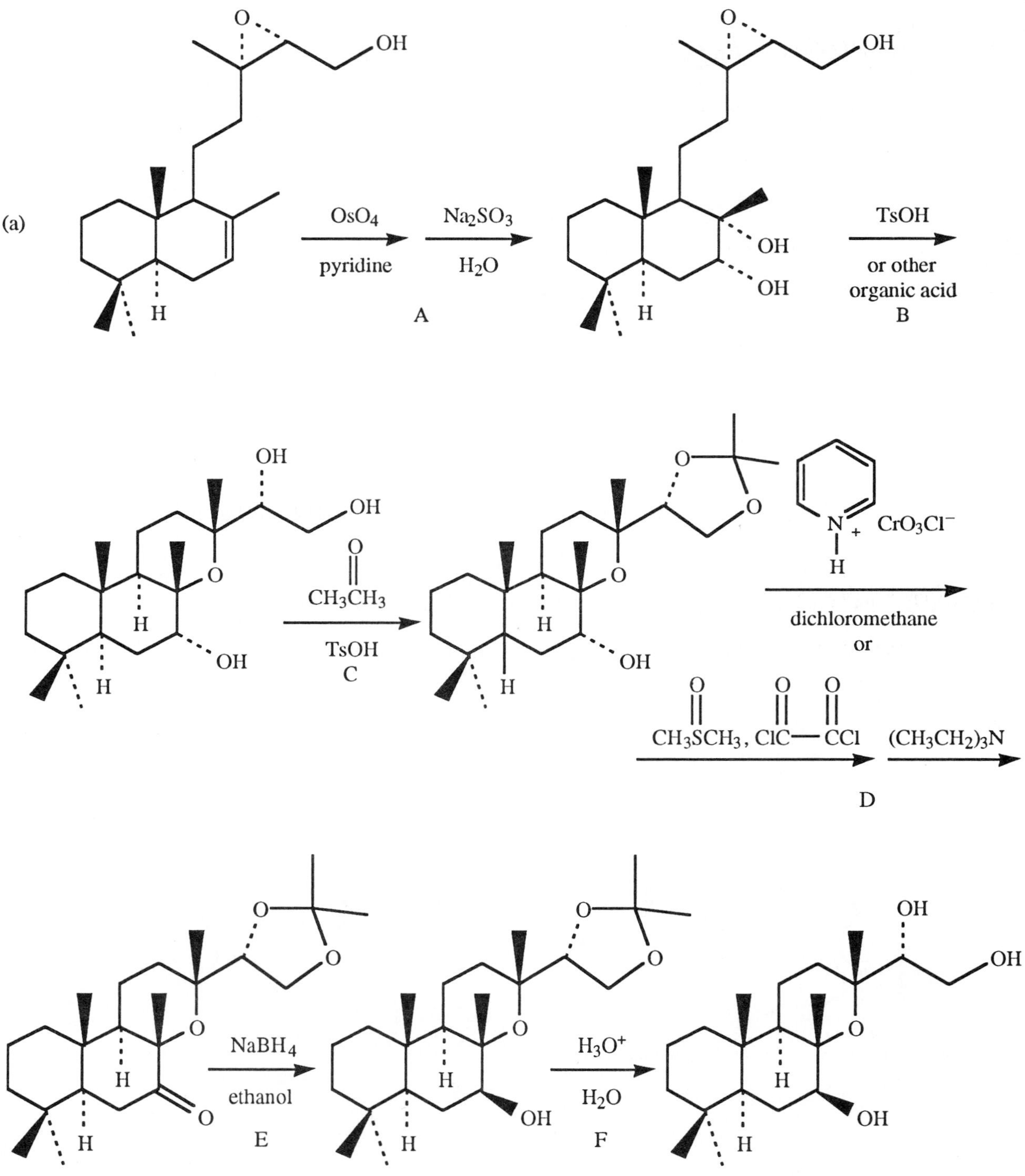

OsO₄ / pyridine → A → Na₂SO₃ / H₂O → B → TsOH or other organic acid

CH₃CH₃ / TsOH → C

pyridinium CrO₃Cl⁻ / dichloromethane or CH₃SCH₃, ClC—CCl / (CH₃CH₂)₃N → D

NaBH₄ / ethanol → E → H₃O⁺ / H₂O → F

(−)-borjatriol

17.47 (cont)

(b)

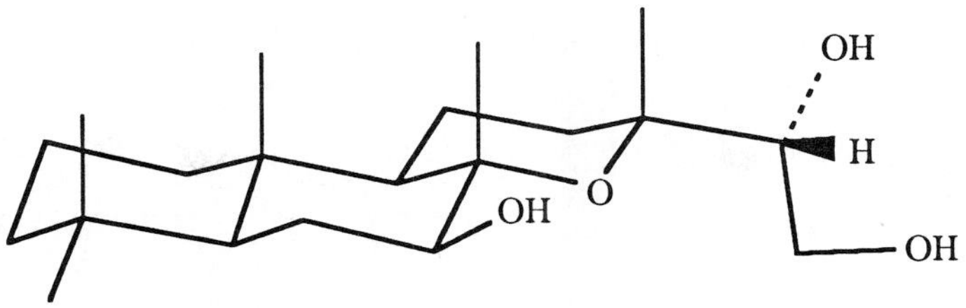

(−)-borjatriol

17.48

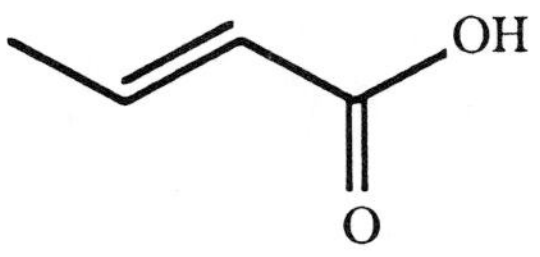

Compound B

λ_{max} 208 nm (ε 12,000)

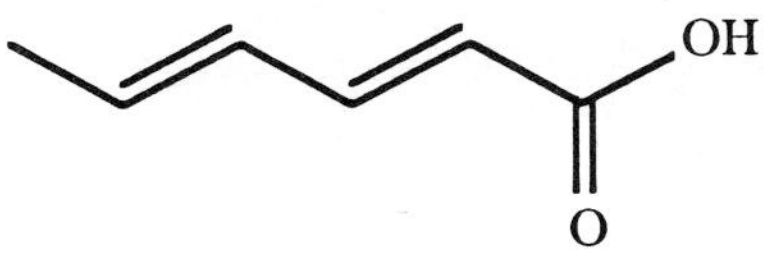

Compound C

λ_{max} 261 nm (ε 25,000)

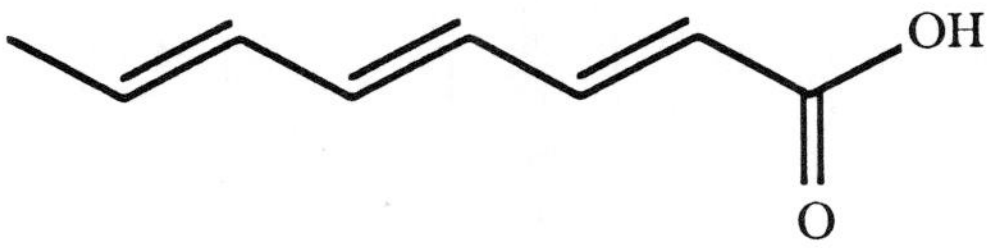

Compound A

λ_{max} 302 nm (ε 36,000)

The wavelength of maximum absorption and the intensity of absorption increases with increasing conjugation.

17.49 (a)   A compound with a molecular formula of C_4H_6O has two units of unsaturation.

(b)   The observed λ_{max} 219 nm (ε 16,6000) is due to the $\pi \rightarrow \pi^*$ transition of a conjugated system, while λ_{max} 318 nm (ε 30) is due to the $n \rightarrow \pi^*$ transition of a carbonyl group. Only three structures fit both the molecular formula and the ultraviolet spectrum.

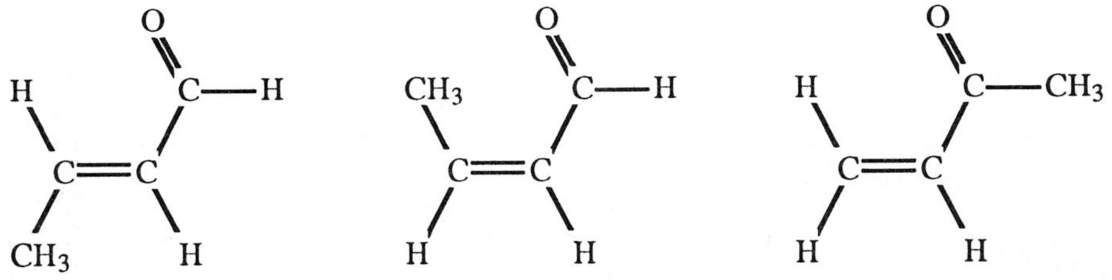

17.49 (cont)

(c)   Some other structures that are not compatible with the data are:

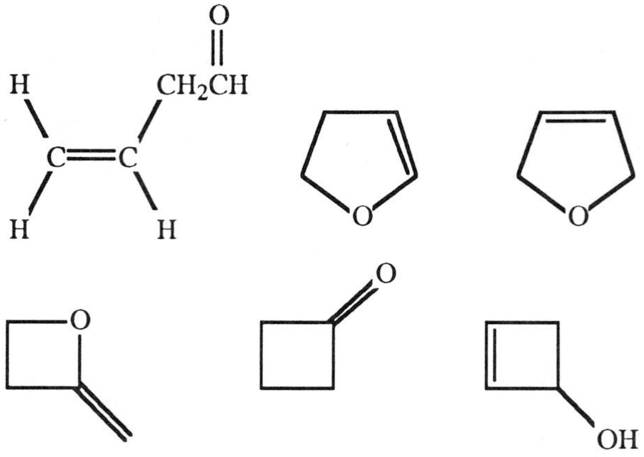

17.50   Of the three structures shown, the one below has the most highly substituted conjugated system and will, therefore, absorb at the highest wavelength.

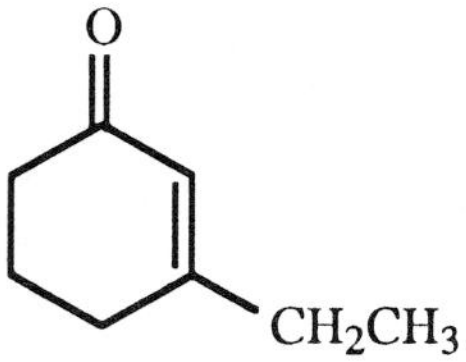

17.51   Of the three irones, β-irone absorbs at the highest wavelength. This suggests that β-irone has the most extended system of conjugation. Of the three structures shown, the one given below has the most extended conjugation and must, therefore, be β-irone.

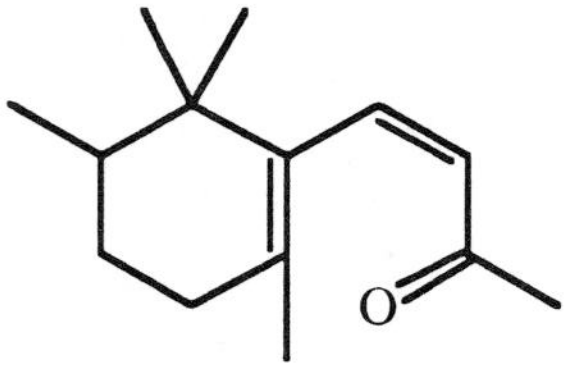

β-irone

17.52

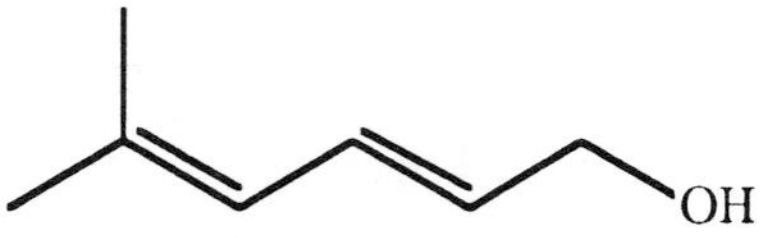

5-Methyl-2,4-hexadien-1-ol has the more highly substituted chromophore and will have the longer wavelength of absorption, 236 nm.

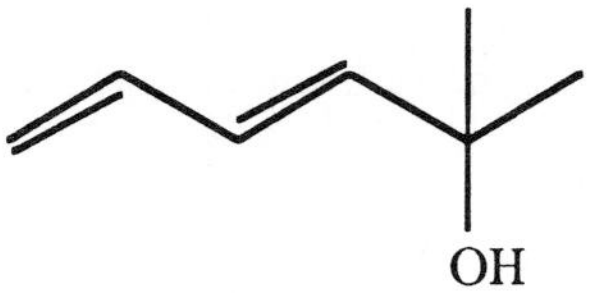

2-Methyl-3,5-hexadien-2-ol will absorb at 223 nm.

18

Enols and Enolate Anions as Nucleophiles II. Conjugate Addition Reactions; Ylides

---

Concept Map 18.1   Electrophilic alkenes.

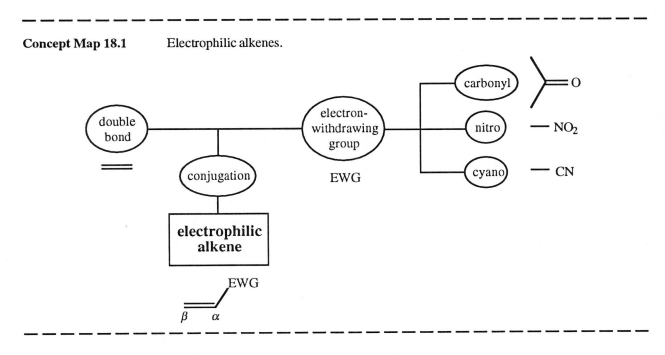

18.1

18.1 (cont)

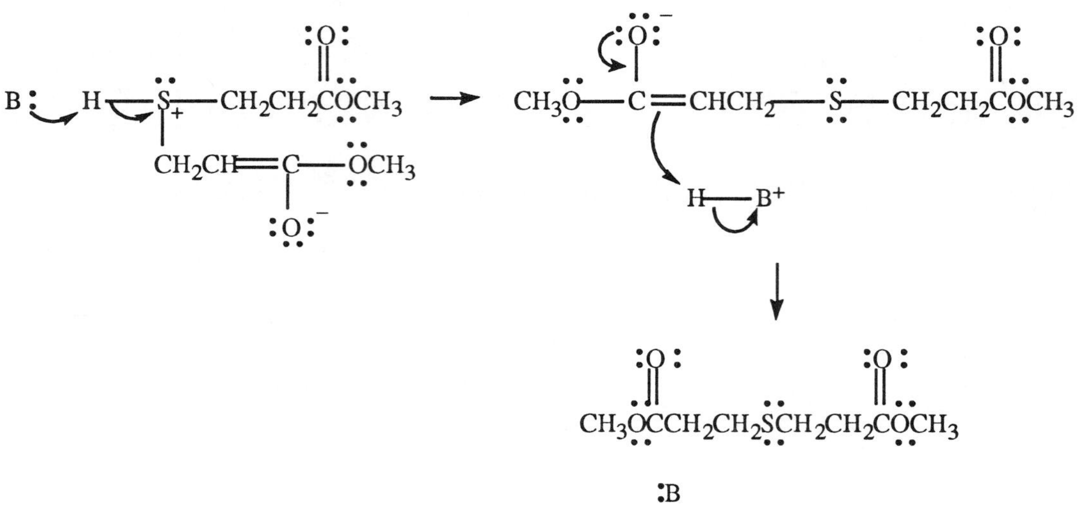

$$CH_3OCCH_2CH_2SCH_2CH_2COCH_3$$

:B

18.2

(*E*)-2-butenedioate
fumarate

attack at C-1 from top   attack at C-1 from bottom   attack at C-2 from top   attack at C-2 from bottom

(*S*)-(–)-malate   (*R*)-(+)-malate   (*S*)-(–)-malate   (*R*)-(+)-malate

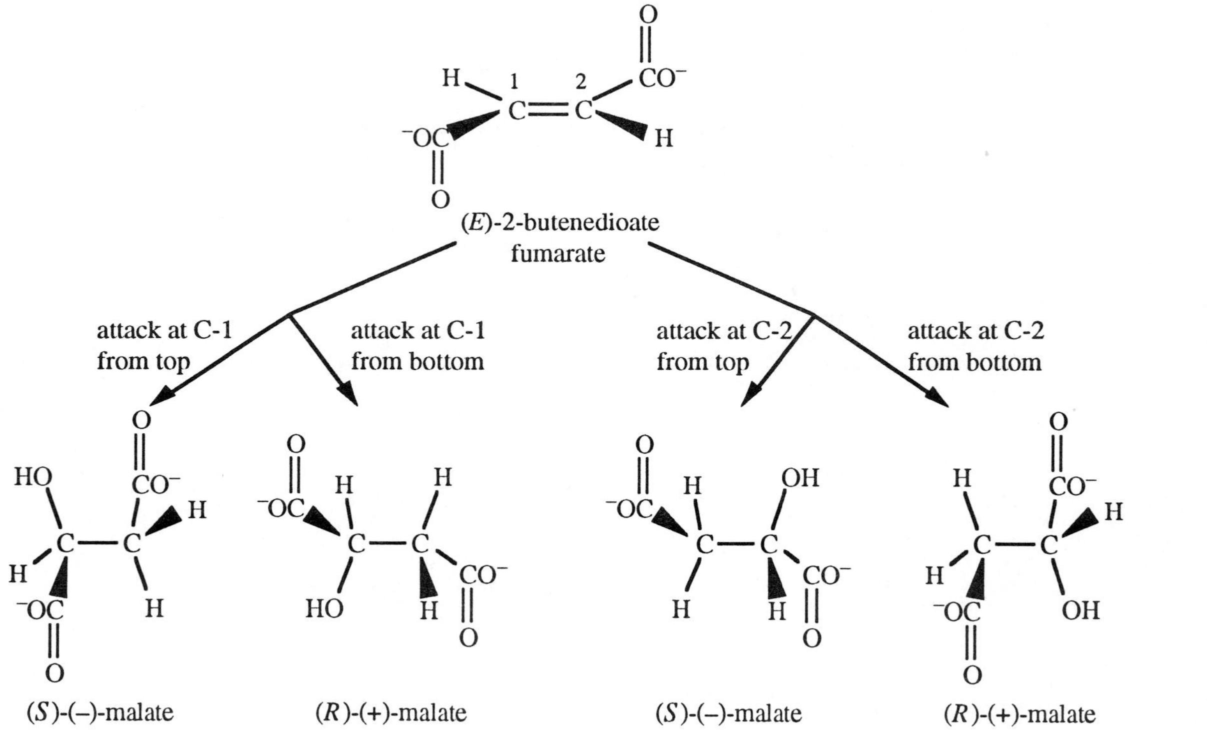

If the reaction were taking place in an achiral environment, we would expect a racemic mixture as a result of all four of these processes occurring with equal probability.

18.3

(a)

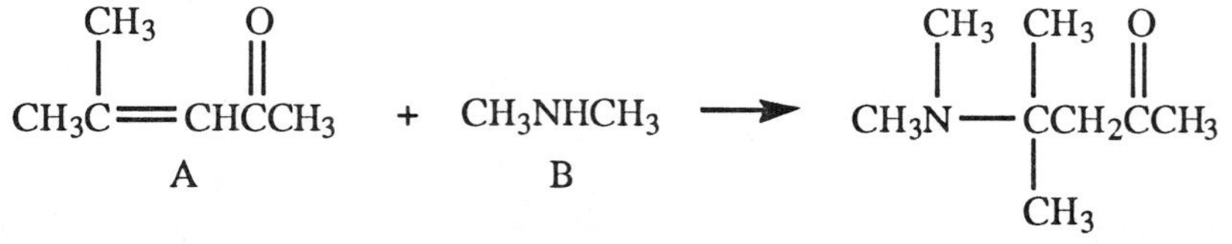

18.3 (cont)

(b) $CH_2{=}CHC{\equiv}N$ + NH_3 $\longrightarrow$ $N{\equiv}CCH_2CH_2NHCH_2CH_2C{\equiv}N$

     (2 equivalents) H_2O C

(c)   —SH + $CH_2{=}CHCOCH_3$ $\xrightarrow{CH_3ONa}$ —SCH_2CH_2COCH_3

       D   E

(d)   + $CH_3CH_2NHCH_2CH_3$ $\longrightarrow$

                  (1 equiv)   diethyl ether

                                          F

18.4

(a) $CH_3CH_2CH_2CH_2Li$ + CuI $\longrightarrow$ $(CH_3CH_2CH_2CH_2)_2CuLi$

                               diethyl ether

                                 0 °C

(b) $CH_3CH_2CH_2CH_2CH_2I$ $\xrightarrow[\substack{\text{diethyl ether} \\ 25\,°C}]{(CH_3)_2CuLi}$ $CH_3CH_2CH_2CH_2CH_2CH_3$

(c) $CH_3(CH_2)_6CH_2Cl$

                      hexamethylphosphoric

                          triamide

                          25 °C

(d) $\xrightarrow[\substack{\text{diethyl ether} \\ 0\,°C}]{(CH_3)_2CuLi}$

18.5 $\xrightarrow[\text{diethyl ether}]{(CH_3)_2CuLi}$ $\xrightarrow[H_2O]{NH_4Cl}$

                               and enantiomer

18.6

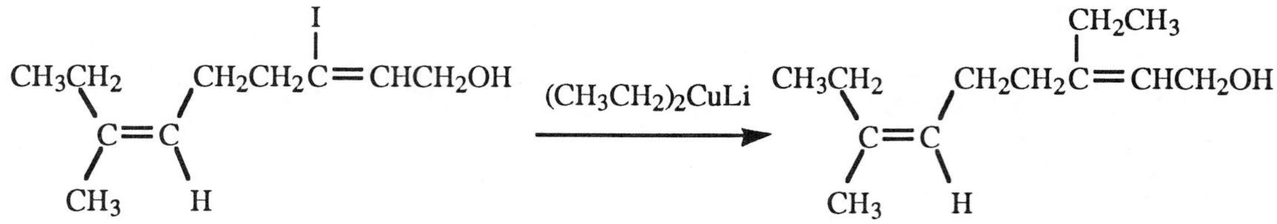

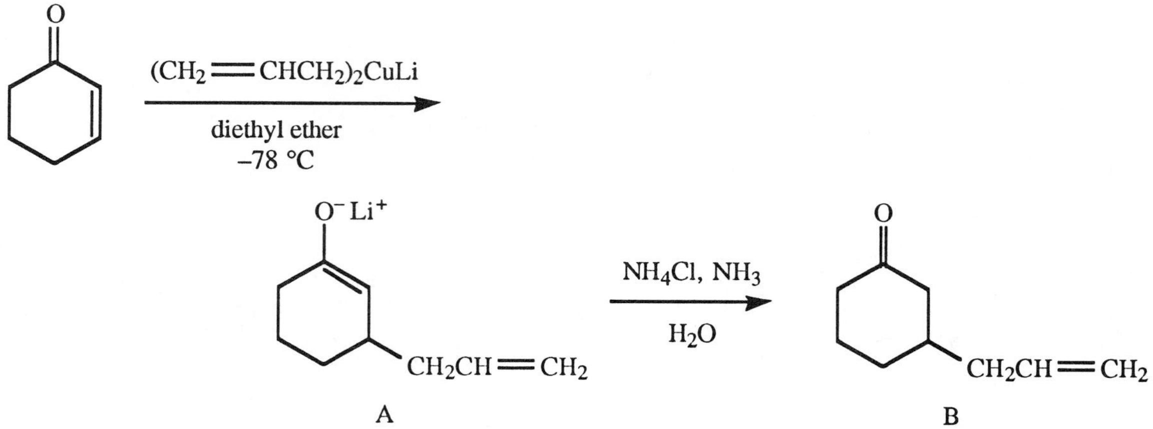

18.7

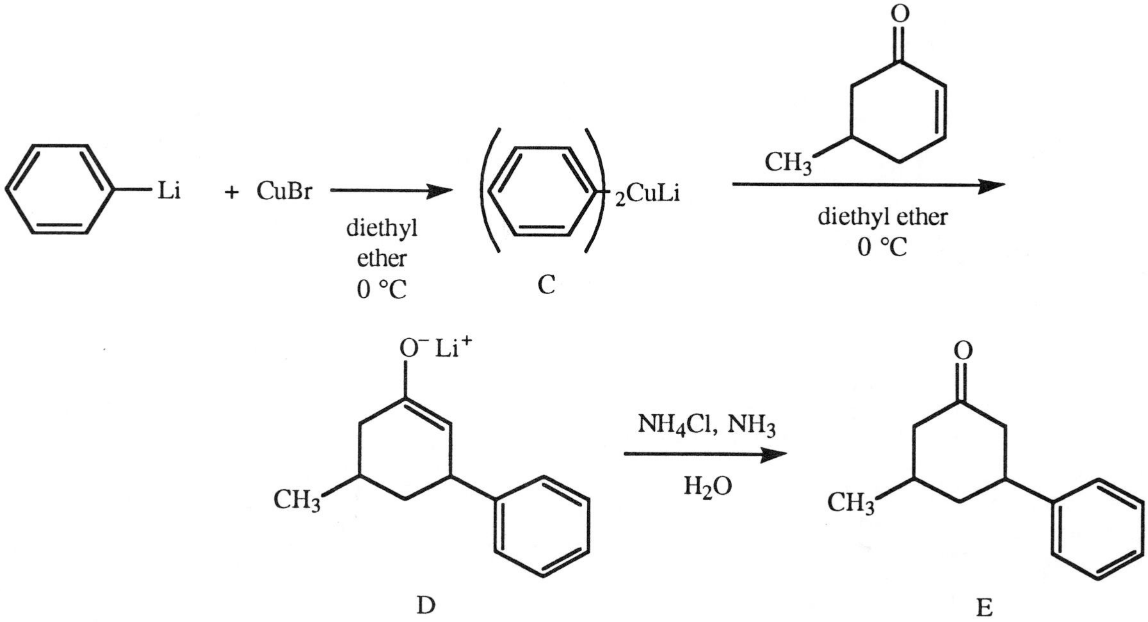

(a)

(b)

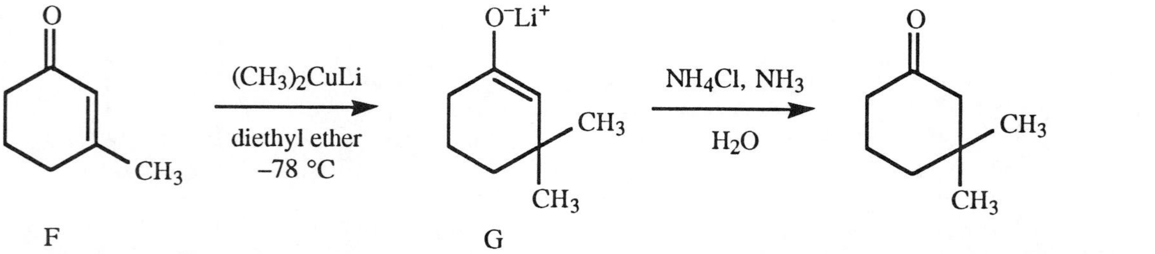

(c)

18.7 (cont)

(d)

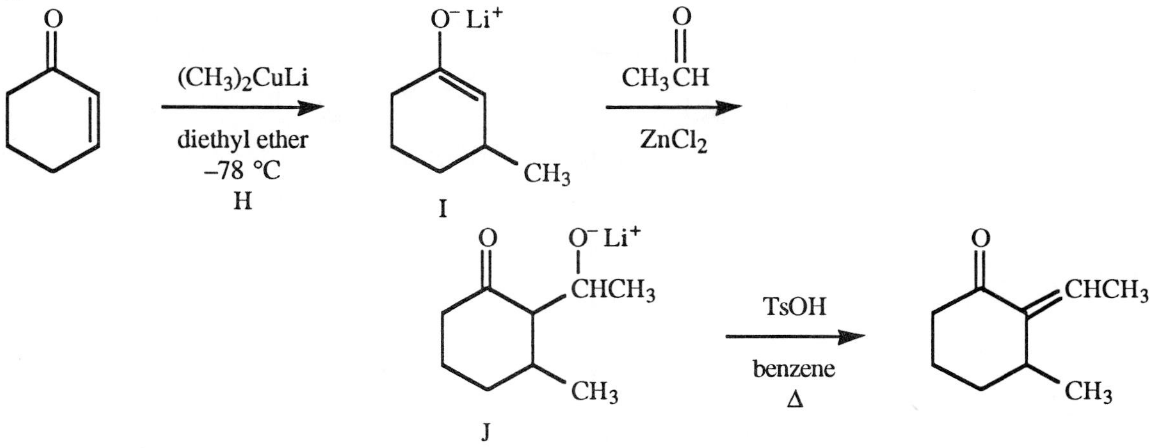

I

J

(e)

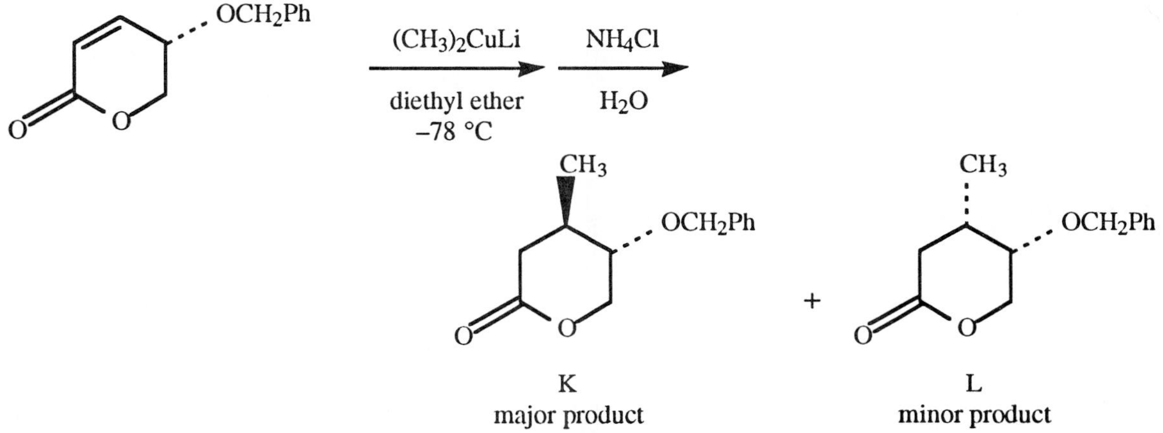

K
major product

L
minor product

18.8

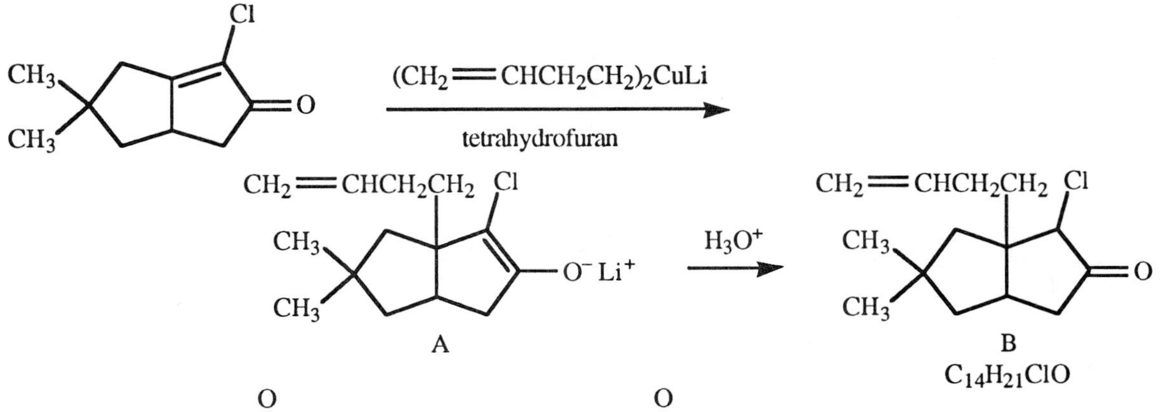

A

B
$C_{14}H_{21}ClO$

18.9 RSH + $CH_3(CH_2)_8\overset{\displaystyle O}{\overset{\|}{C}}CH=CH_2$ $\longrightarrow$ $CH_3(CH_2)_8\overset{\displaystyle O}{\overset{\|}{C}}CH_2CH_2SR$

18.10

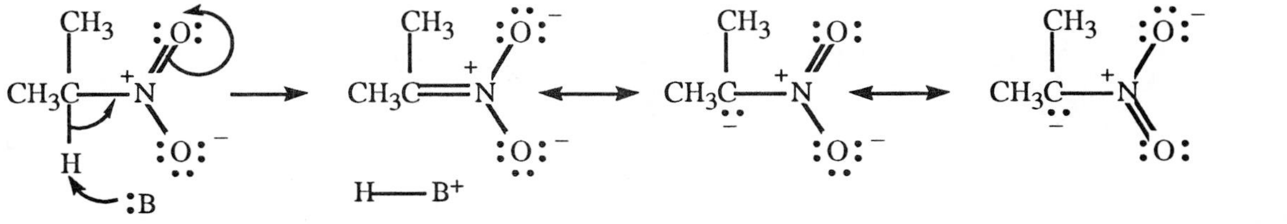

18.11

partial positive charge on β- carbon atom

18.12

(a) $CH_2(C\equiv N)_2$ + $2\ CH_2\!\!=\!\!CHCCH_3$ $\xrightarrow[\text{benzene}]{\text{Na}}$

$$CH_3CCH_2CH_2\underset{\underset{N}{\overset{\overset{N}{|||}}{|}}}{\overset{\overset{|||}{C}}{C}}CH_2CH_2CCH_3$$

(b)

$$\underset{\underset{CH_3}{|}}{CH_3CHCCH_2COCH_2CH_3} + CH_2\!\!=\!\!\underset{\underset{CH_3}{|}}{CCCH_3} \xrightarrow[\text{ethanol}]{\text{KOH}} CH_3\underset{\underset{CH_3}{|}}{CH}\!-\!\underset{\underset{\underset{CH_3\ O}{|}}{CH_2CH-CCH_3}}{CCHCOCH_2CH_3}$$

(c) $N\equiv CCH_2COCH_3$ + $CH_2\!\!=\!\!CHC$— $\xrightarrow[\text{methanol}]{\text{CH}_3\text{ONa}}$

(d)

$\xrightarrow{\text{NaNH}_2}$ $\xrightarrow{CH_2=CHC\equiv N}$

(The more highly substituted enolate anion and the one in which the double bond is conjugated with the ring is the more stable one.)

(e)

+ $CH_2\!\!=\!\!CHCCH_3$ $\xrightarrow[\text{benzene}]{\text{Na}}$

18.12 (cont)

(f) CH_3NO_2 + $\xrightarrow{CH_3CH_2NHCH_2CH_3}$

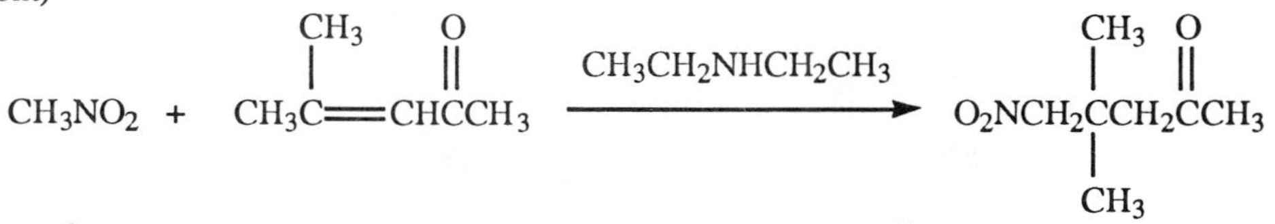

(g)   + $CH_2(COCH_2CH_3)_2$ $\xrightarrow[\text{ethanol}]{CH_3CH_2ONa}$

18.13 $\xrightarrow[\substack{\text{tetrahydrofuran} \\ -67\ °C}]{(CH_3CH)_2N^-Li^+}$ $\xrightarrow[-78\ °C]{CH_2{=}CHCOCH_3}$

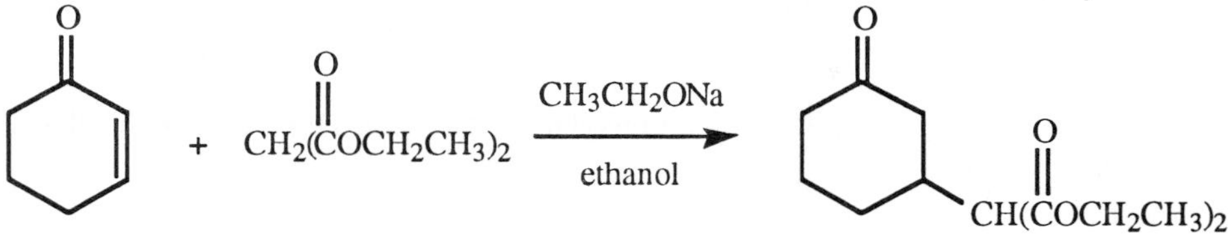

A

$\xrightarrow[-78\ °C]{H_3O^+}$

B

$\xrightarrow[\substack{H_2O \\ \Delta \\ C}]{NaOH}$ $\xrightarrow[\substack{\text{cold} \\ \text{dilute} \\ D}]{H_3O^+}$

B

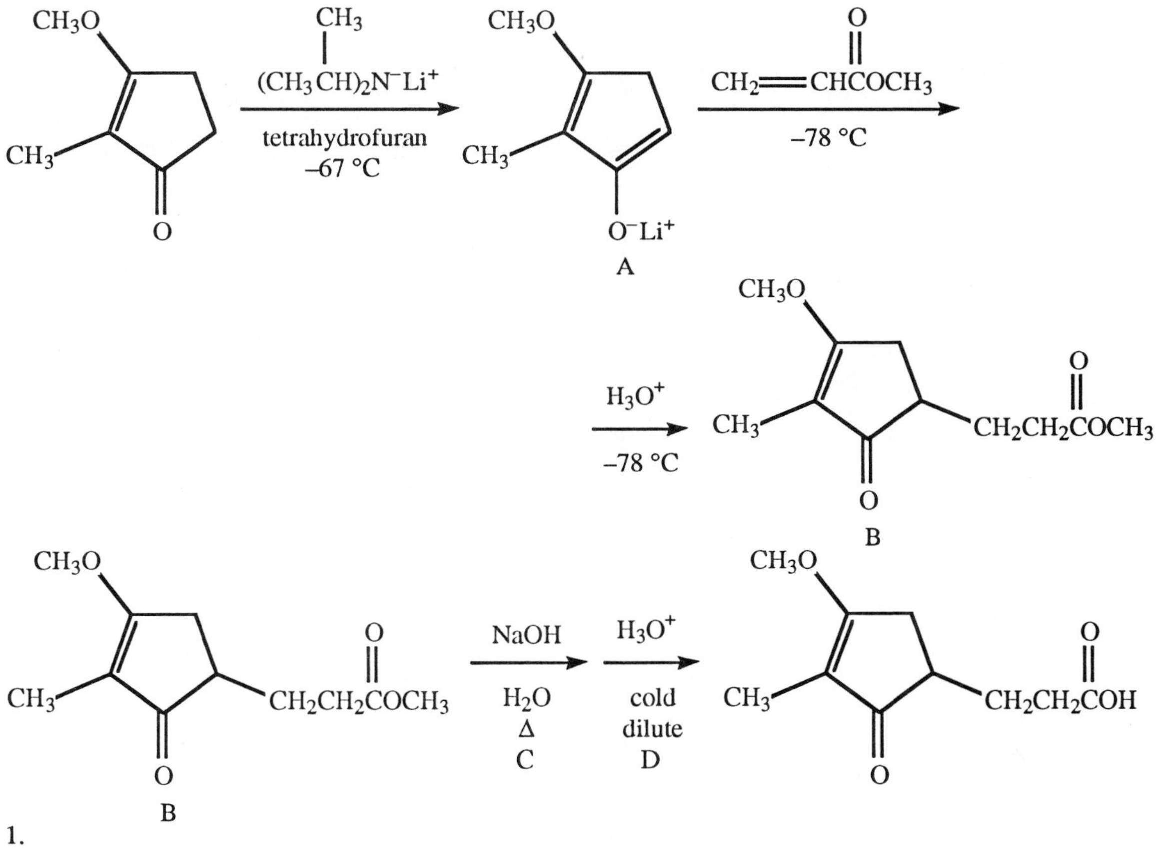

18.14   1.

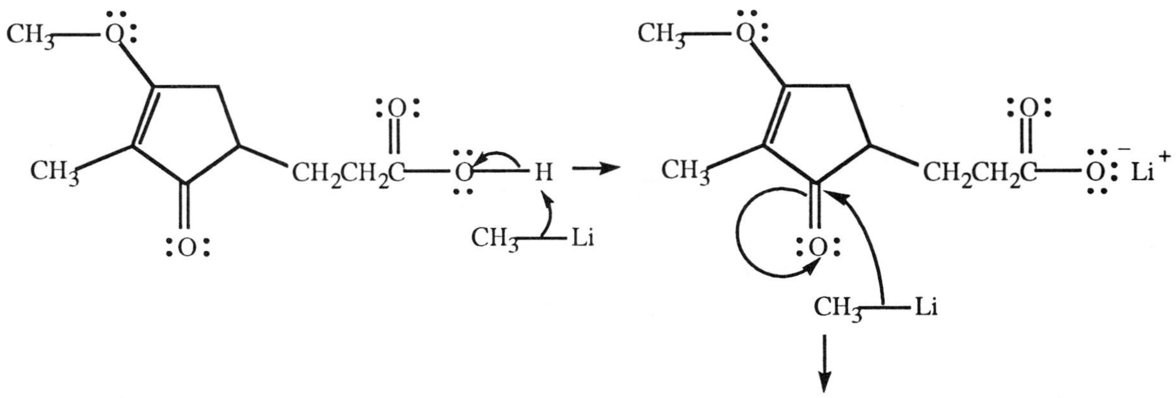

18.14 (1) (cont)

The carbon atom of the carbonyl group in the carboxylate anion is much less electrophilic than the carbon atom of the ketone carbonyl.

2.

The same result can be obtained by hydrolyzing the enol ether (Problem 13.18), then dehydrating the alcohol.

18.15

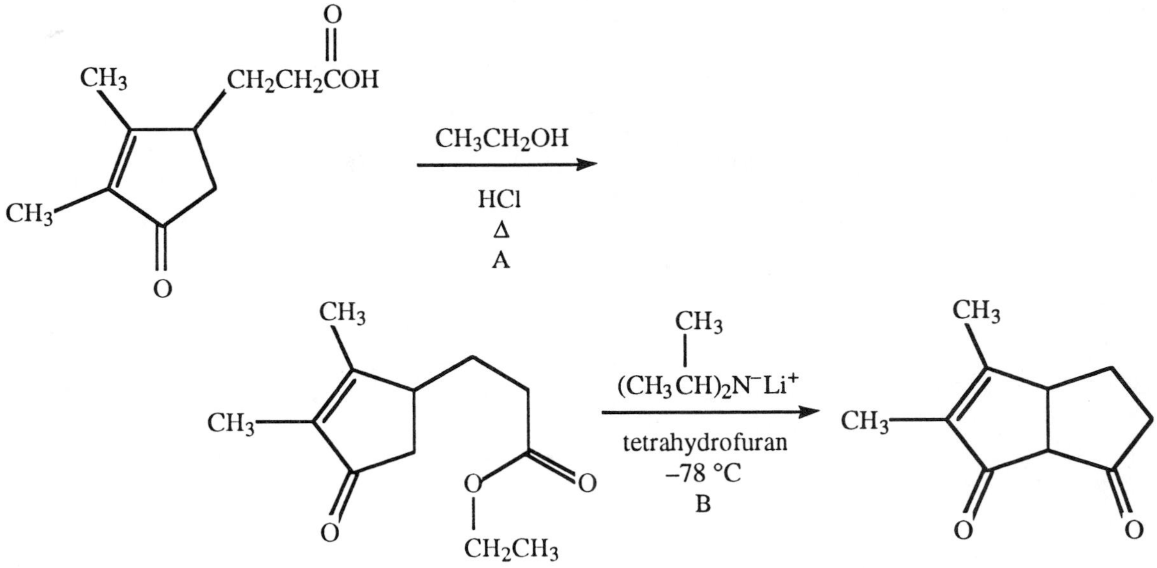

Concept Map 18.2   (see p. 643)

18.16

(a)

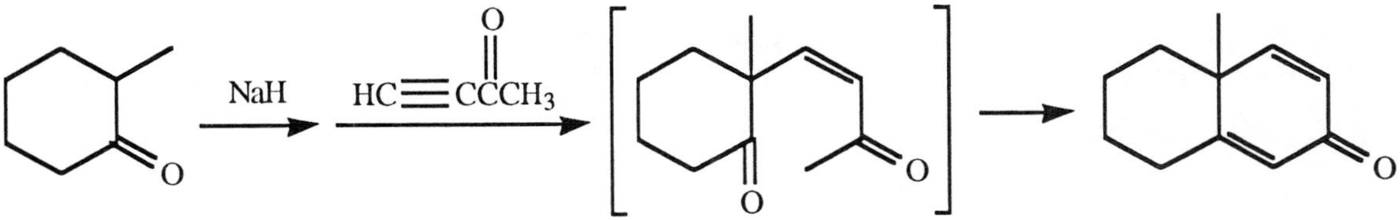

(b)

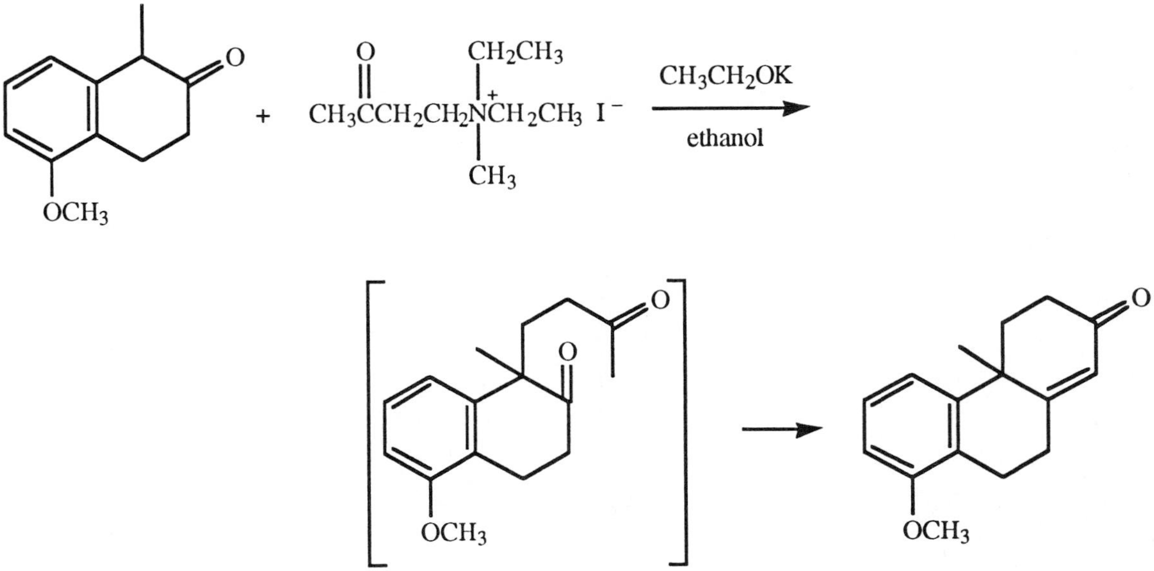

18.16 (cont)

(c)

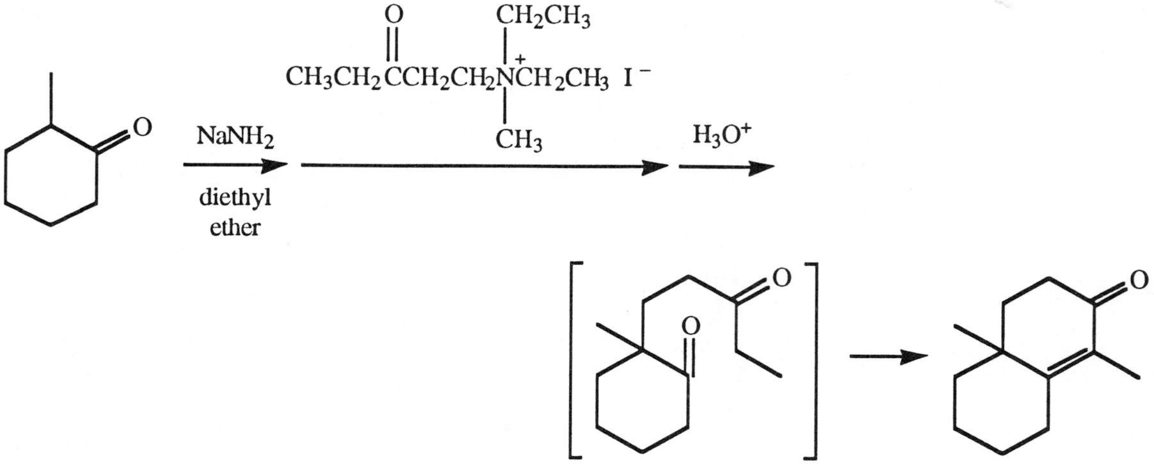

(d)

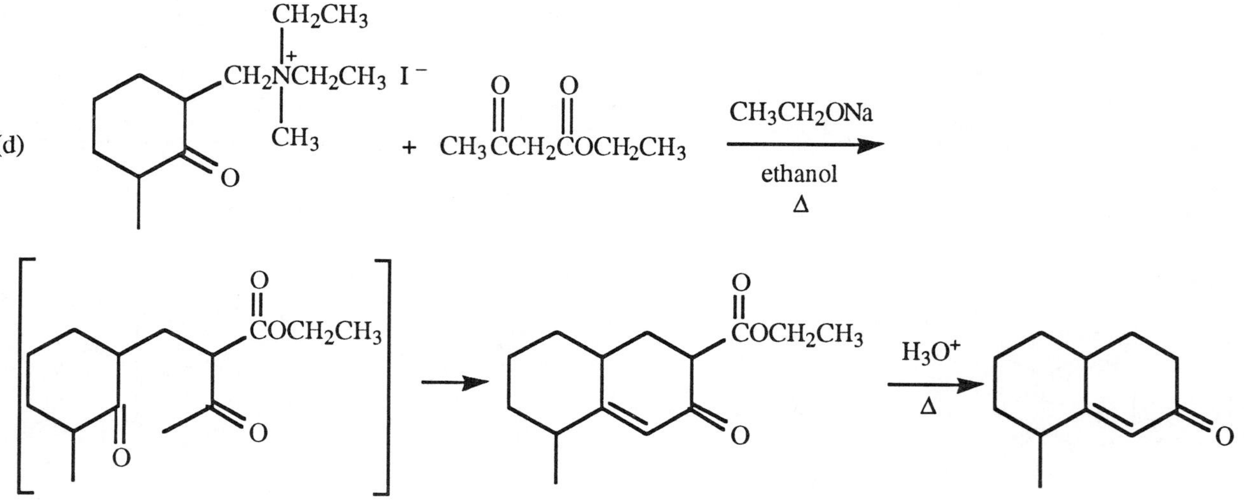

18.17

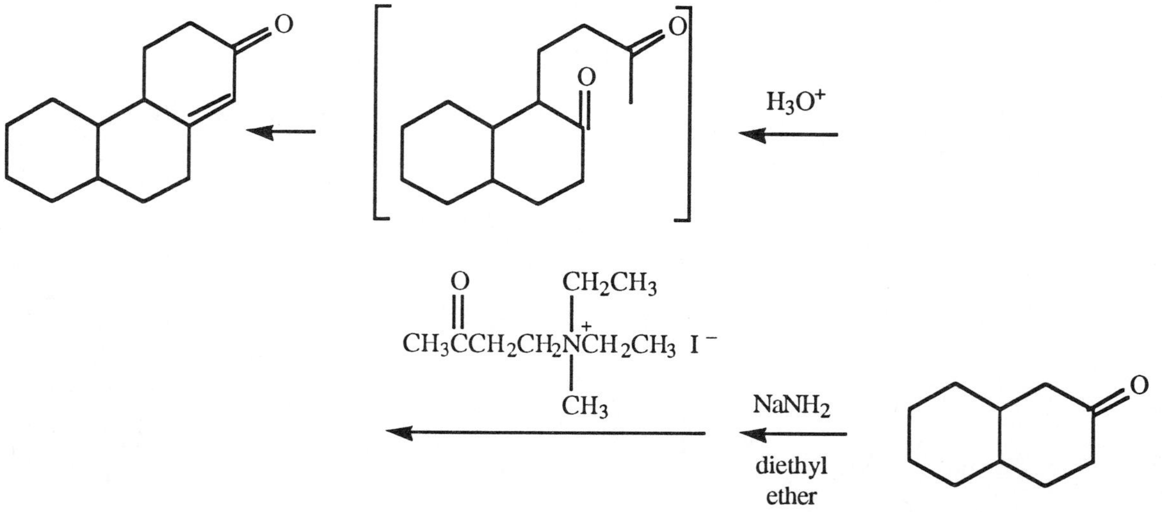

Concept Map 18.2   Reactions of electrophilic alkenes.

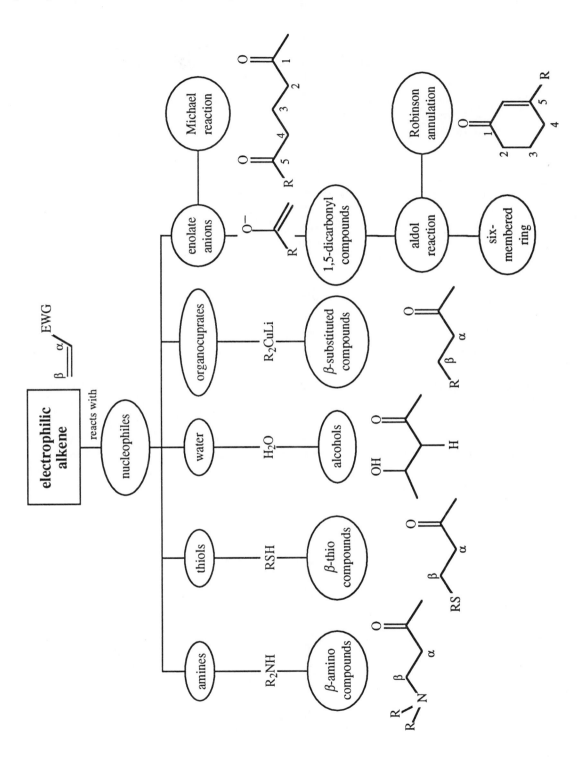

Concept Map 18.3   Ylides.

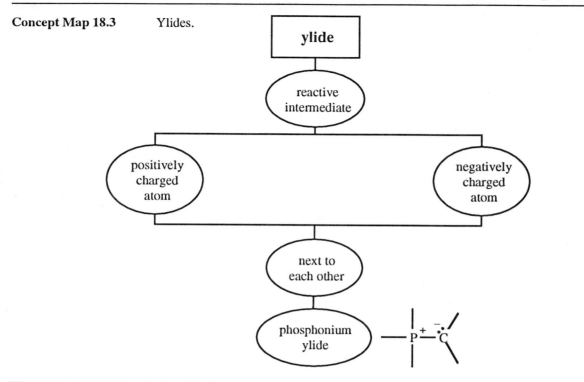

18.18   A carbon atom can be introduced by the reaction of an organometallic reagent such as a Grignard reagent with a carbonyl group (Section 13.6B). A double bond can be introduced via an E2 elimination (Section 7.6)

18.19

(a)

18.19 (a) (cont)

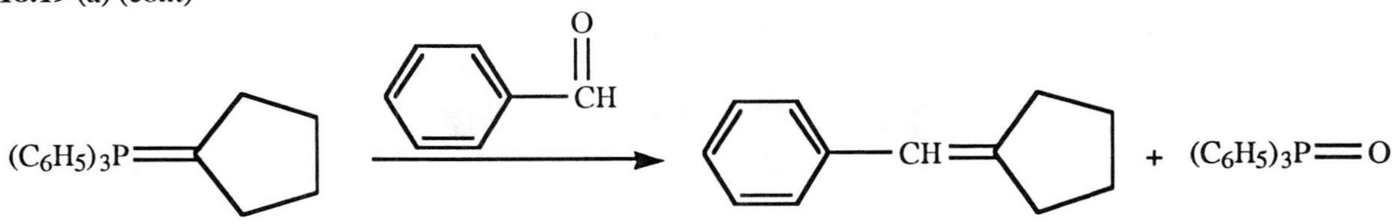

A

B

C
(product in all Wittig
reactions with triphenyl-
phosphonium ylides)

(b) $(C_6H_5)_3\overset{+}{P}CH_2\overset{O}{\overset{\|}{C}}OCH_2CH_3 \ \overset{Br^-}{}$ $\xrightarrow[\text{ethanol}]{CH_3CH_2ONa}$

$(C_6H_5)_3P=CH\overset{O}{\overset{\|}{C}}OCH_2CH_3$ $\xrightarrow{CH_3CH=CH\overset{O}{\overset{\|}{C}}H}$ $CH_3CH=CHCH=CH\overset{O}{\overset{\|}{C}}OCH_2CH_3$

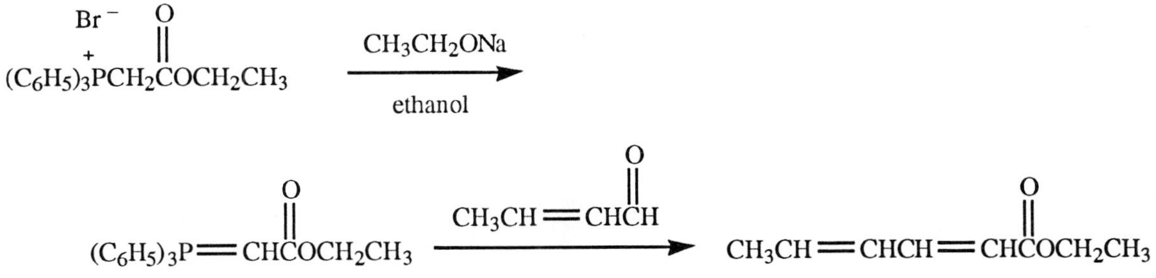

D

E

(c) $(C_6H_5)_3\overset{+}{P}CH_2\overset{Br^-}{}\cdots\overset{O}{\overset{\|}{C}}OCH_2CH_3$ $\xrightarrow[\text{ethanol}]{CH_3CH_2ONa}$

$(C_6H_5)_3P=CH\cdots\overset{O}{\overset{\|}{C}}OCH_2CH_3$ $\xrightarrow{\hspace{3cm}}$

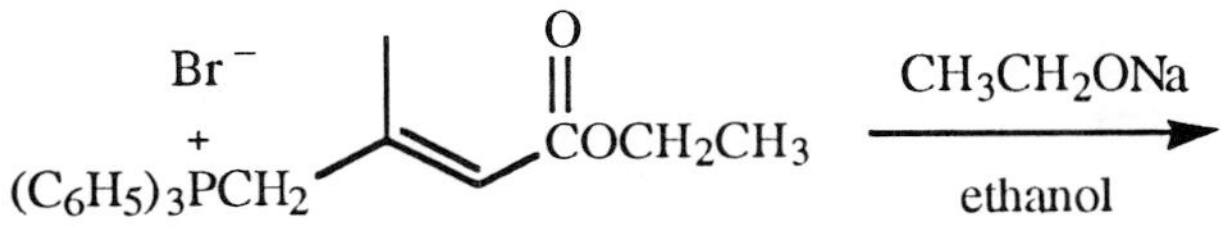

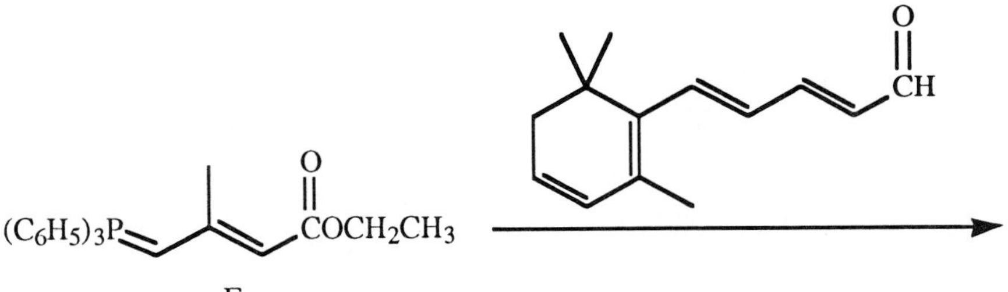

F

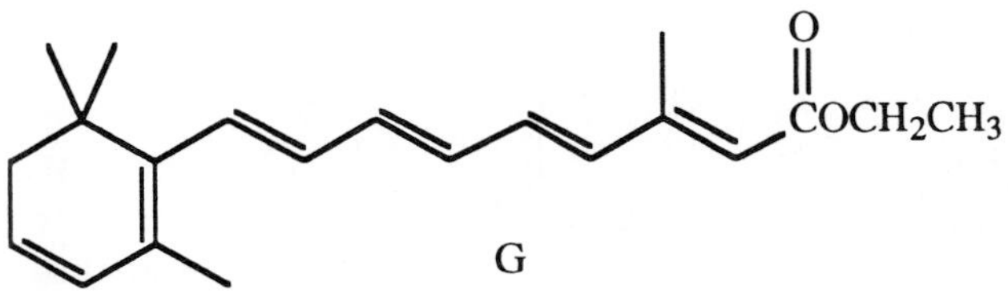

G

(d)

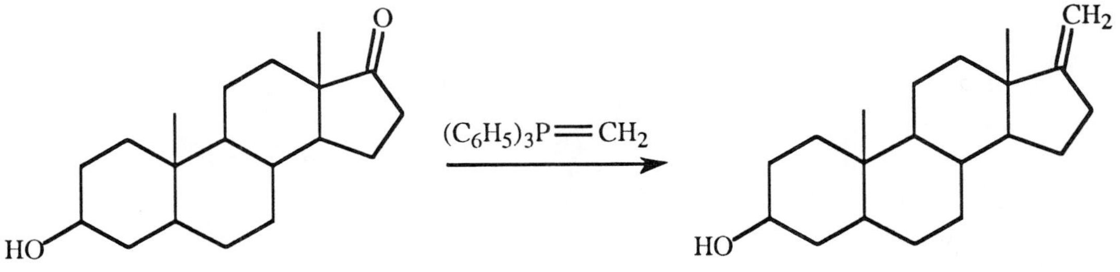

H

Concept Map 18.4   The Wittig reaction.

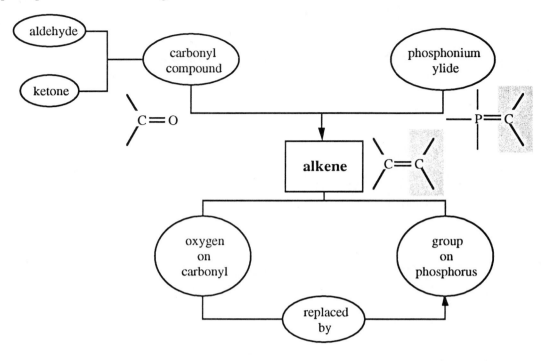

18.20

18.21

(a)

$$CH_2=CHCH_2PPh_2 \xrightarrow[\substack{\text{tetrahydrofuran} \\ \text{hexamethylphosphoric triamide} \\ -78\ °C}]{CH_3CH_2CH_2CH_2Li} CH_2=CHCH=PPh_2$$

A

(b)

$$CH_2=CHCH=PPh_2 \quad + \quad$$

A

B

18.21 (cont)

(c)

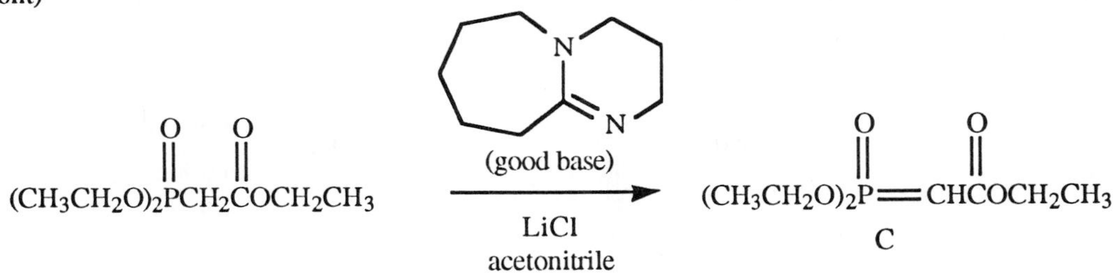

(d)

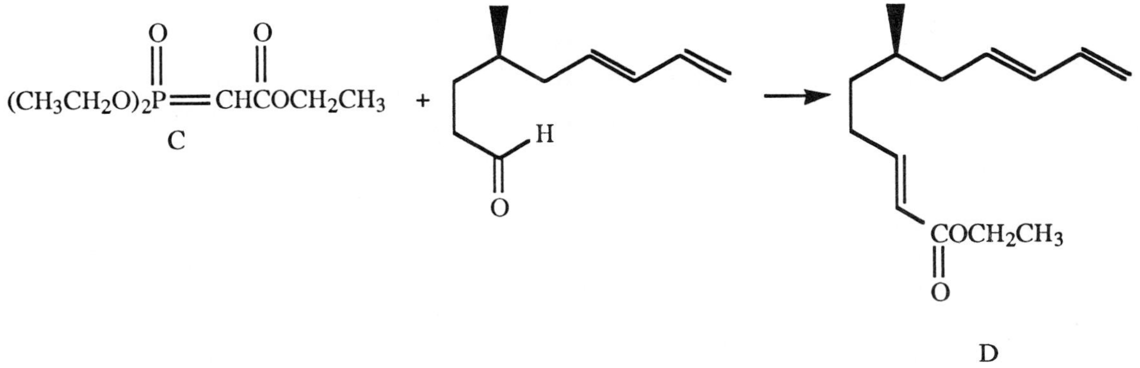

(e)

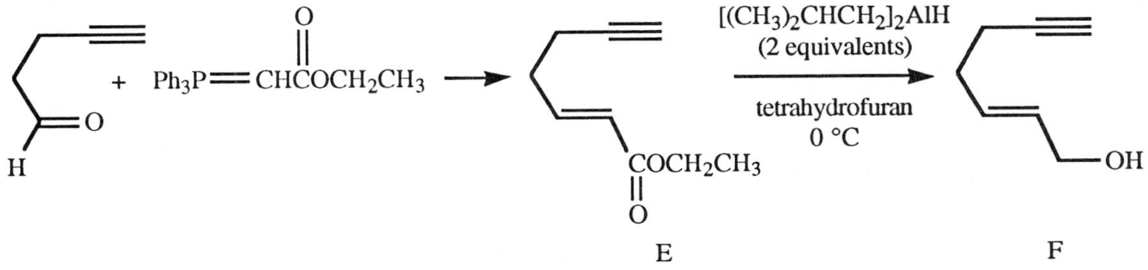

18.22

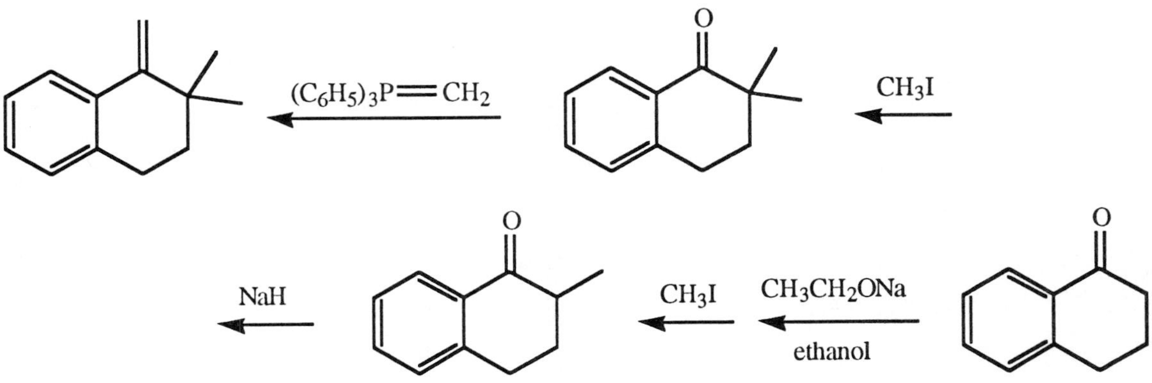

Concept Map 18.5   Dithiane anions.

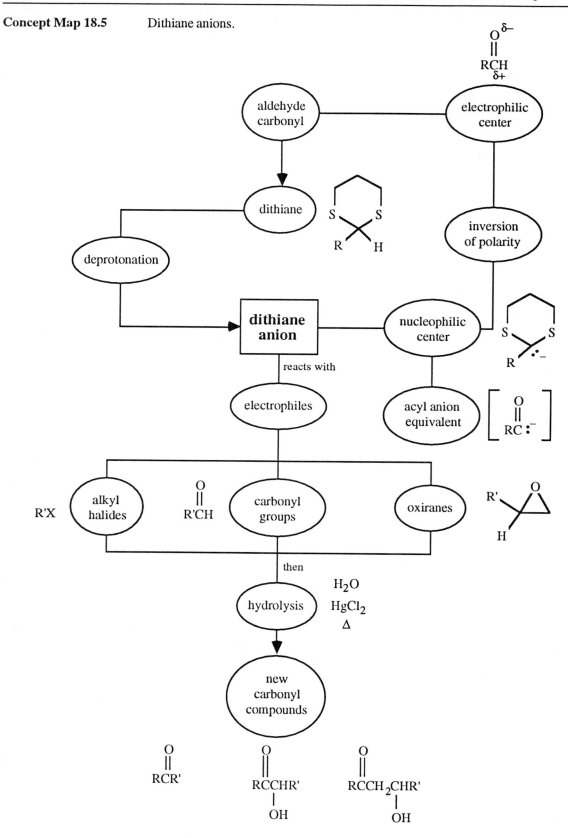

18.23

(a)

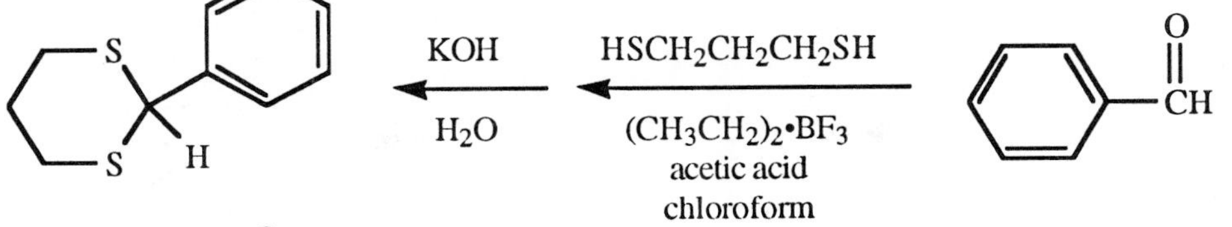

(b)

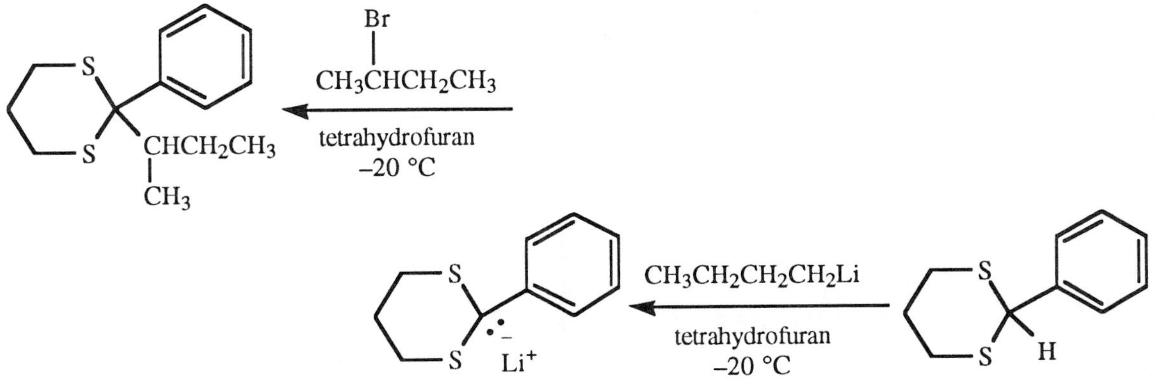

(c)

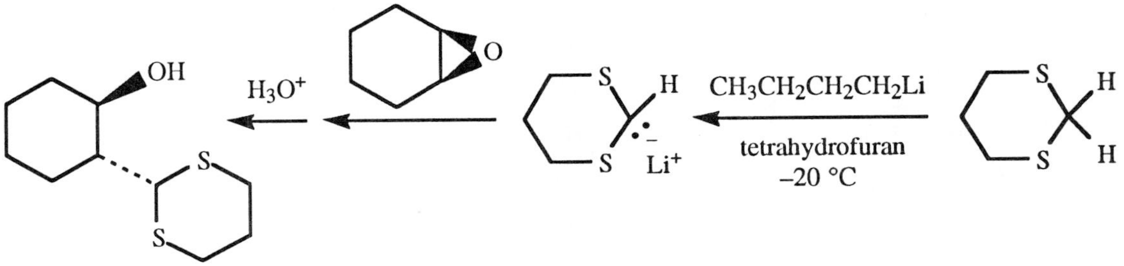

and enantiomer

(d)

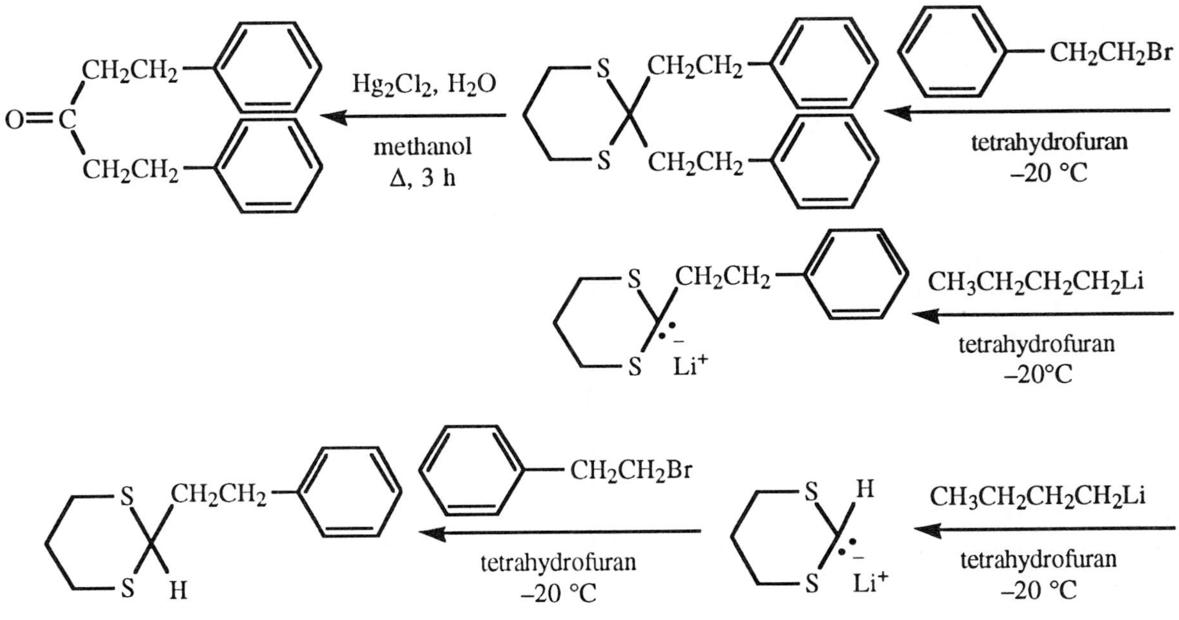

18.23 (d) (cont)

$$\underset{H_2O}{\overset{KOH}{\longleftarrow}} \quad \underset{\substack{(CH_3CH_2)_2 \cdot BF_3 \\ \text{acetic acid} \\ \text{chloroform}}}{\overset{HSCH_2CH_2CH_2SH}{\longleftarrow}} \quad \overset{O}{\overset{\|}{HCH}}$$

18.24   1.

$$CH_3\overset{O}{\overset{\|}{S}}CH_3 \quad + \quad NaH \quad \longrightarrow \quad CH_3\overset{O}{\overset{\|}{S}}\!\!-\!\!\overset{..}{\overset{-}{C}}H_2 \; Na^+ \quad + \quad H_2\!\uparrow$$

2.   The anion formed from dimethyl sulfoxide is stabilized by delocalization of the negative charge to oxygen.

$$CH_3\overset{:O:}{\underset{..}{\overset{\|}{S}}}\!\!-\!\!\overset{-}{\overset{..}{C}}H_2 \quad \longleftrightarrow \quad CH_3\overset{:\overset{..}{O}:^-}{\underset{..}{\overset{|}{S}}}\!\!=\!\!CH_2$$

3.   The anion from dimethyl sulfoxide resembles an enolate anion.

$$CH_3\overset{:\overset{..}{O}:^-}{\overset{|}{C}}\!\!=\!\!CH_2 \qquad\qquad CH_3\overset{:\overset{..}{O}:^-}{\underset{..}{\overset{|}{S}}}\!\!=\!\!CH_2$$

enolate anion   anion from
from acetone   dimethyl sulfoxide

4. $CH_3\overset{O}{\overset{\|}{S}}\overset{..}{\overset{-}{C}}H_2 \; Na^+ \quad +$ [benzophenone structure] $\longrightarrow$ [product structure with O^-Na^+ and CH_2SCH_3]

$CH_3\overset{O}{\overset{\|}{S}}\overset{..}{\overset{-}{C}}H_2 \quad Na^+ \quad +$ [cyclohexyl ketone structure] $-\overset{O}{\overset{\|}{C}}OCH_2CH_3 \longrightarrow$ [product structure] $-\overset{O}{\overset{\|}{C}}CH_2\overset{O}{\overset{\|}{S}}CH_3$

18.25

(a) $CH_3\overset{O}{\overset{\|}{C}}CH\!\!=\!\!\overset{CH_3}{\overset{|}{C}}CH_3 \quad + \quad CH_2(COCH_3)_2 \quad \underset{\text{methanol}}{\overset{CH_3ONa}{\longrightarrow}}$

18.25 (a) (cont)

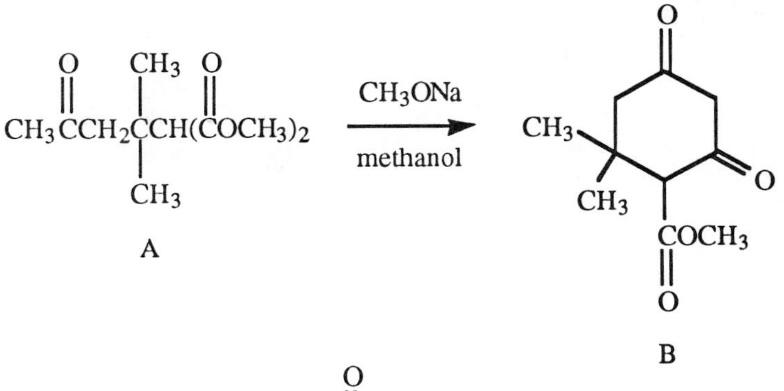

A → B

(b)

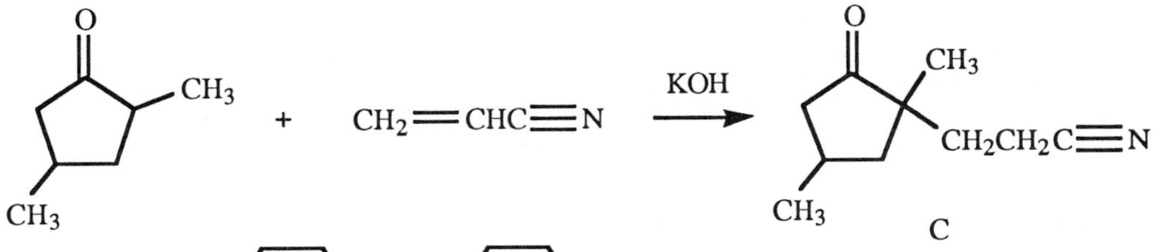

(c)

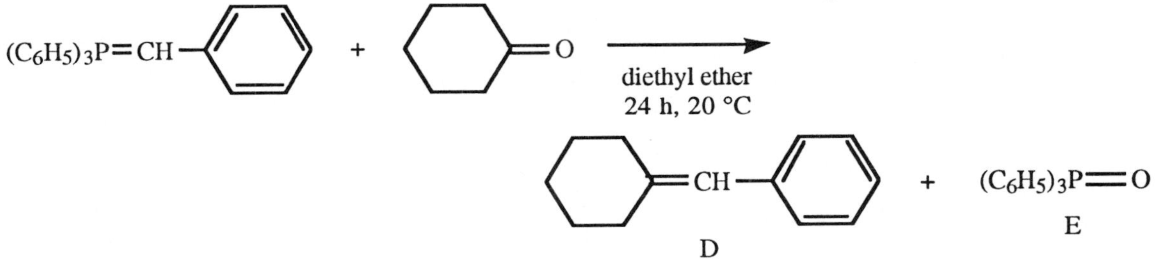

(d)

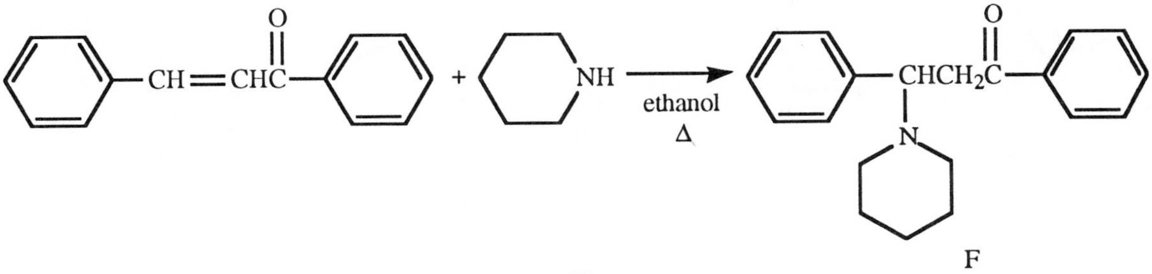

(e)

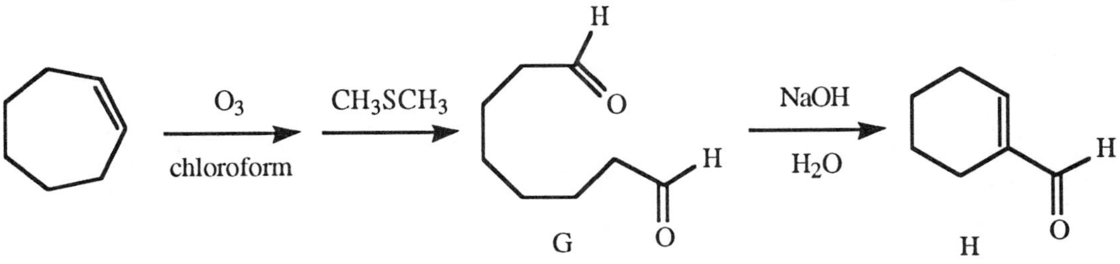

(f)

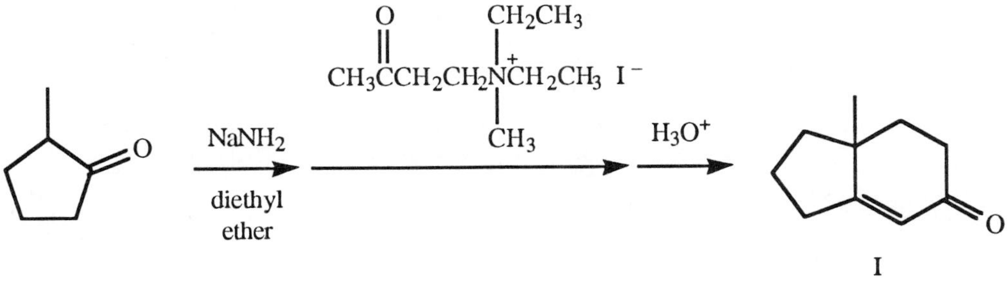

18.25 (cont)

(g)

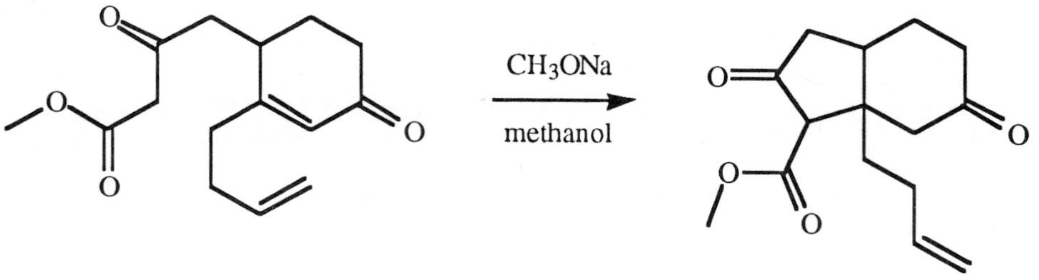

(h)

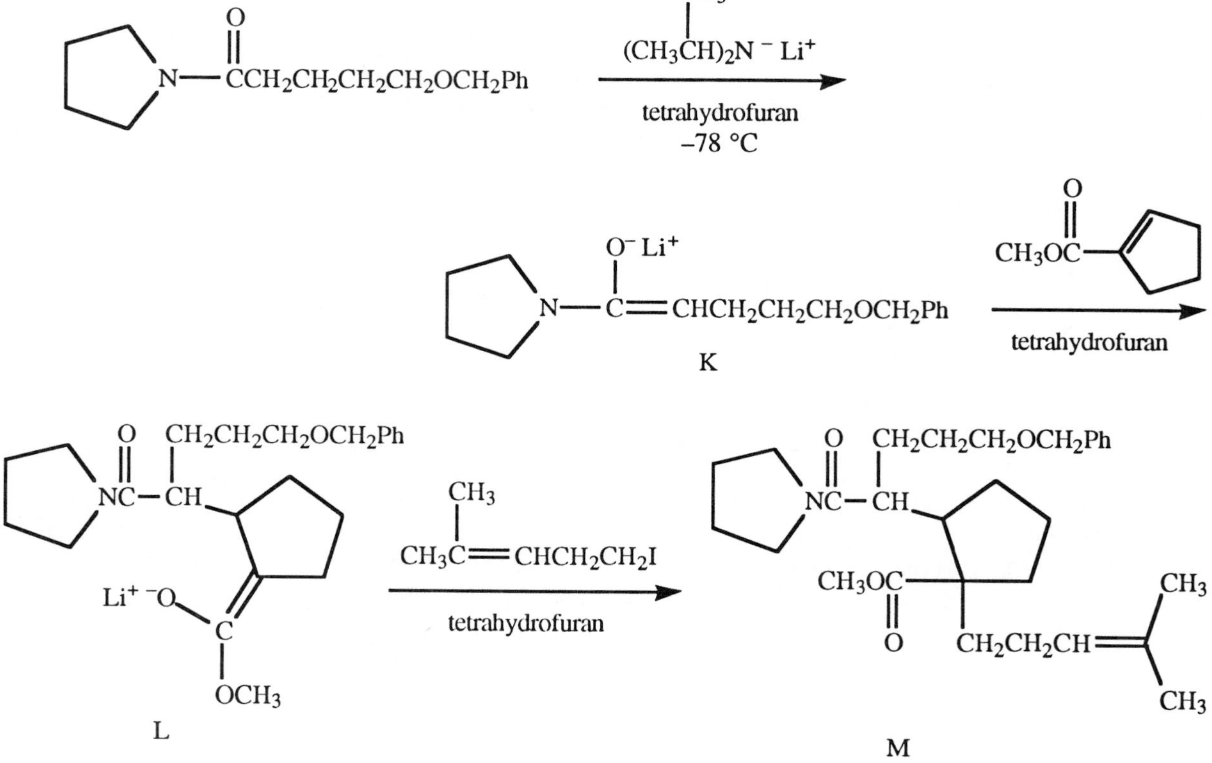

18.26

(a)

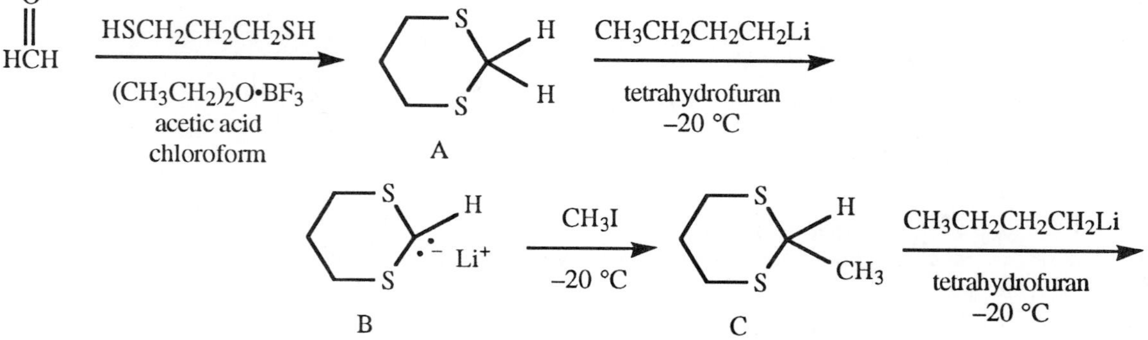

18.26 (a) (cont)

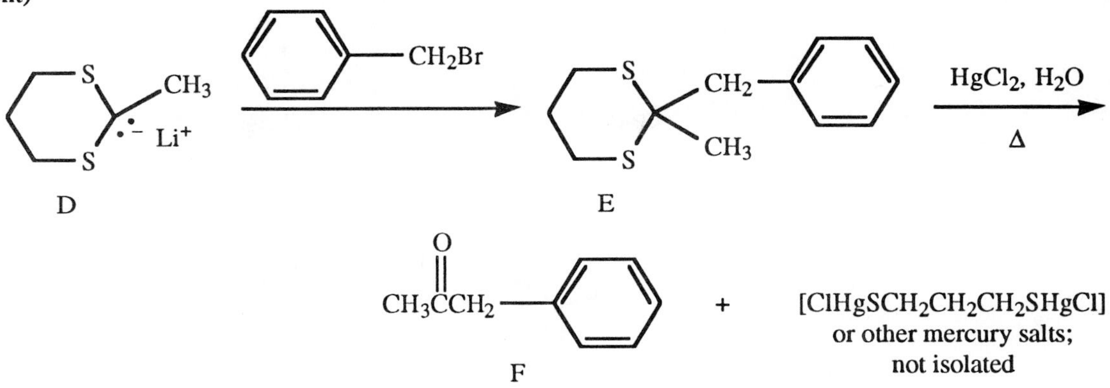

(b) 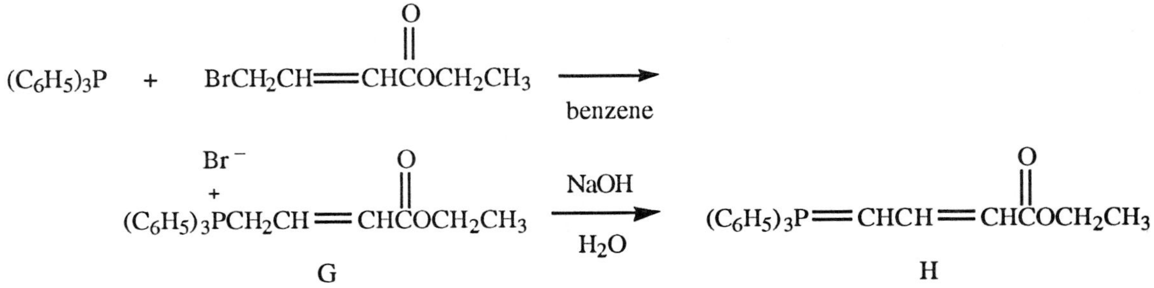 $(C_6H_5)_3P$ + $BrCH_2CH=CHCOCH_2CH_3$ $\xrightarrow{\text{benzene}}$

$$\underset{G}{\overset{Br^-}{\underset{}{(C_6H_5)_3\overset{+}{P}CH_2CH=CHCOCH_2CH_3}}} \xrightarrow[\text{H}_2\text{O}]{\text{NaOH}} \underset{H}{(C_6H_5)_3P=CHCH=CHCOCH_2CH_3}$$

(c)

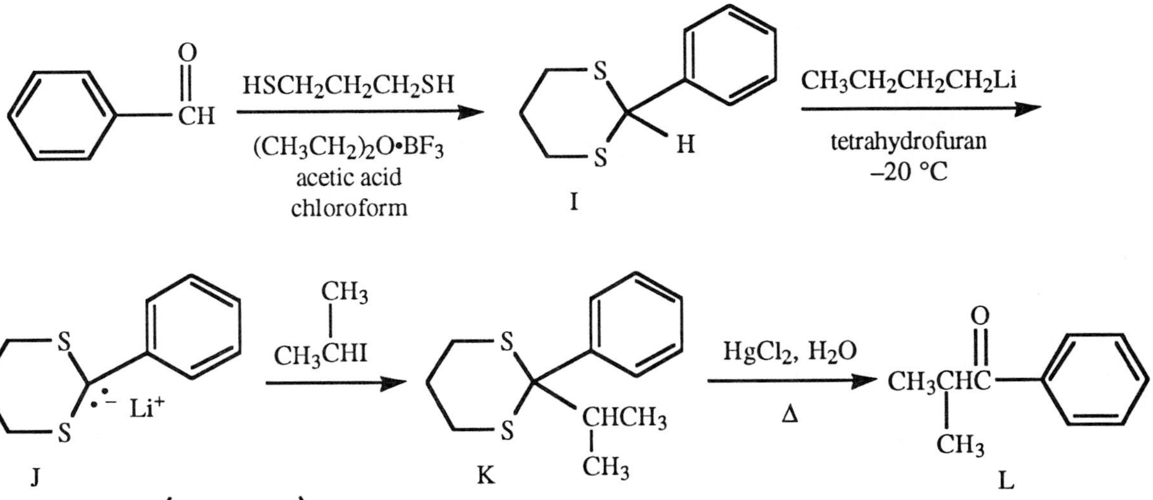

(d)

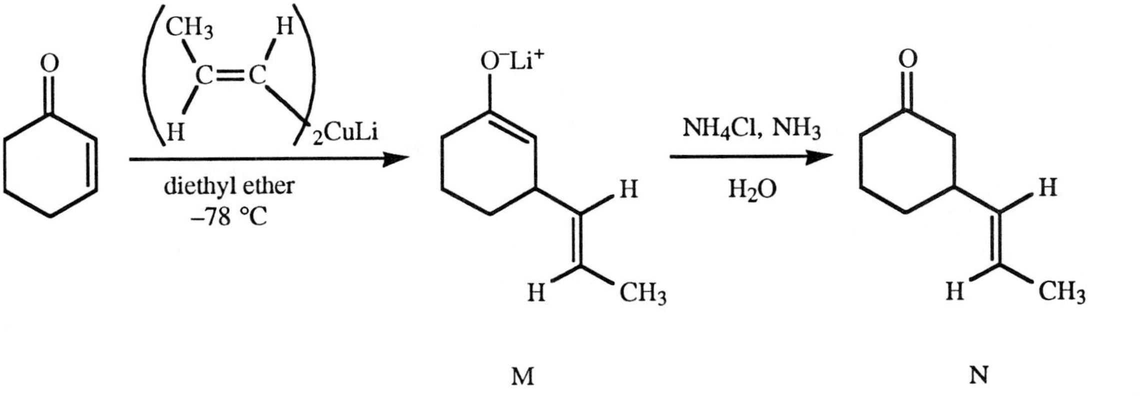

18.26 (cont)

(e)

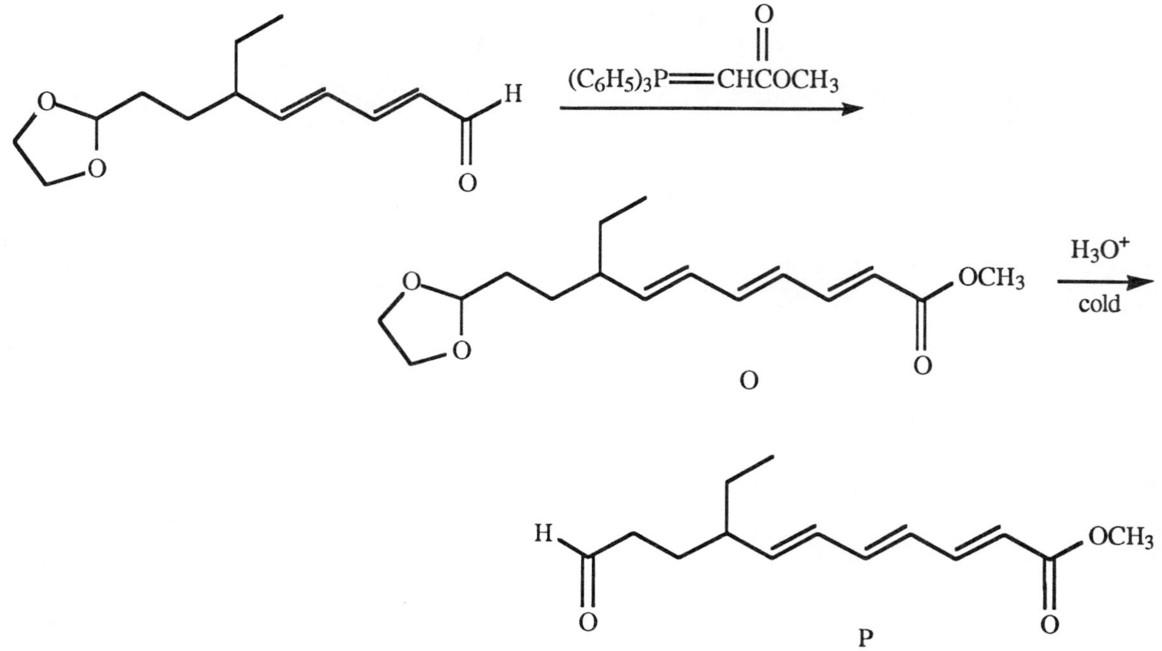

(f)

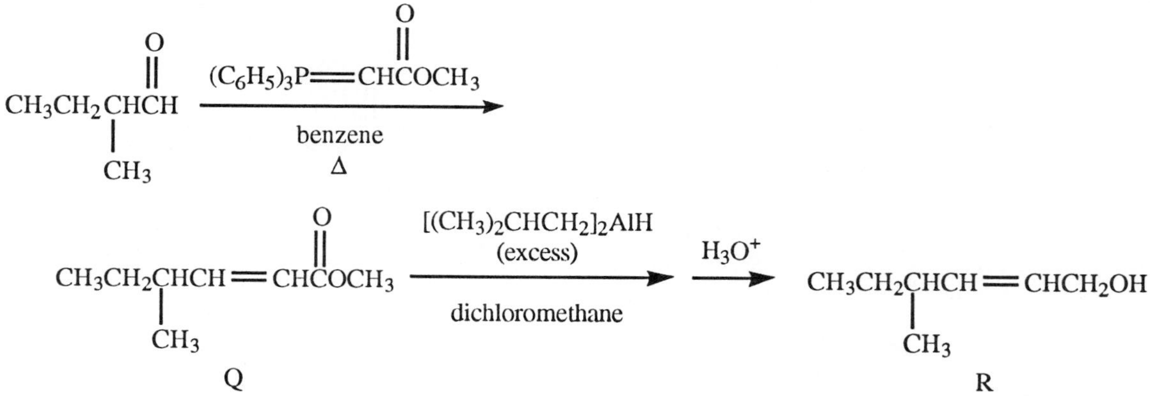

(g)

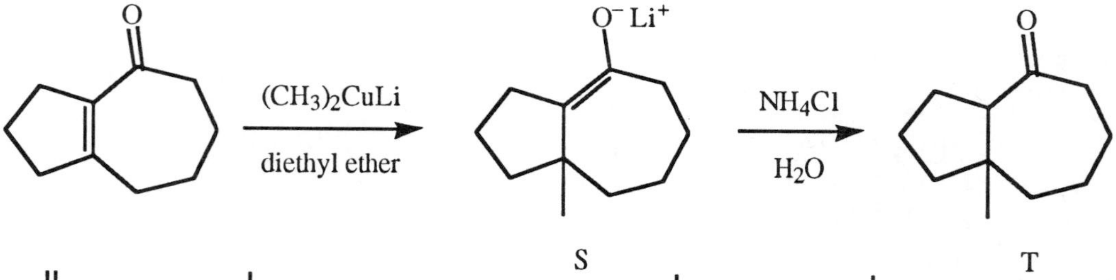

(h)

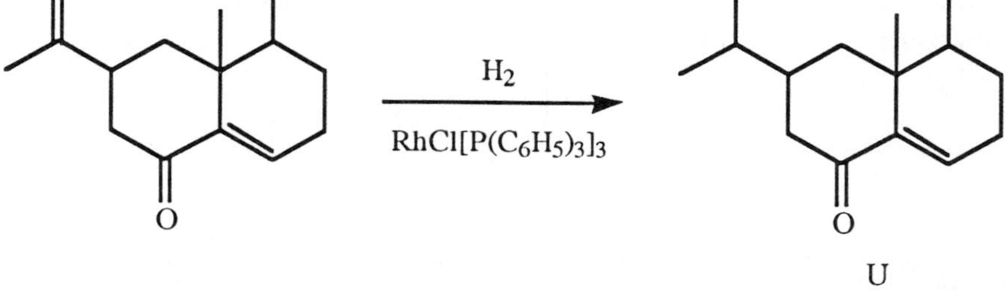

18.26 (cont)

(i)

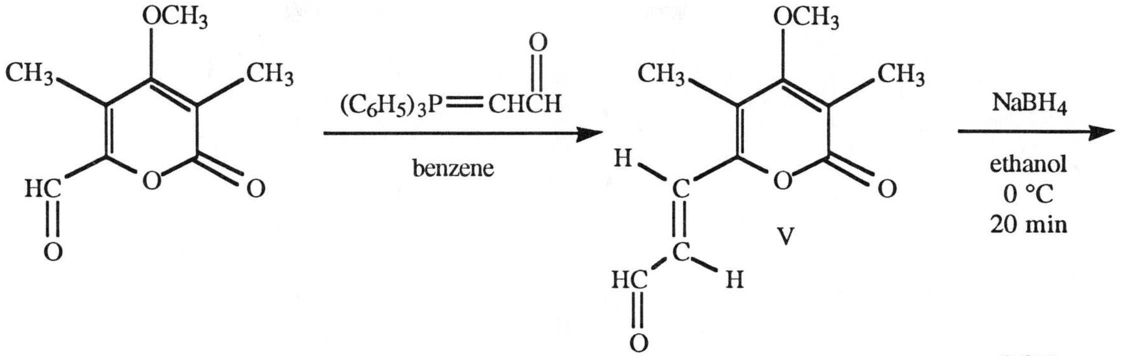

(j)

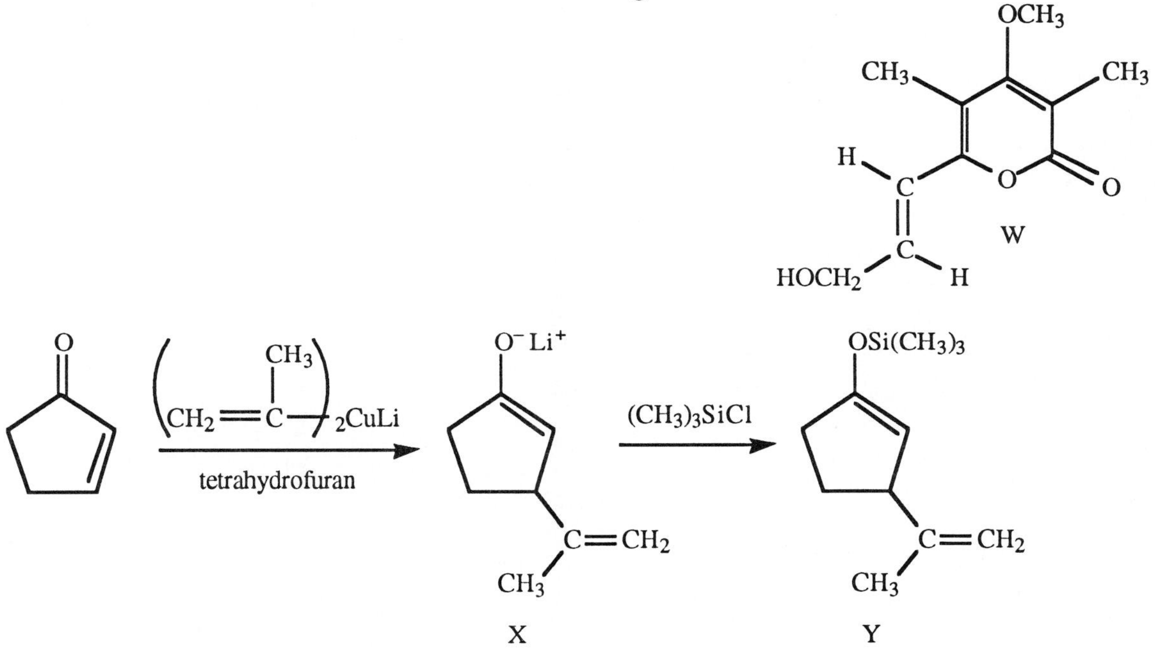

18.27

(a)

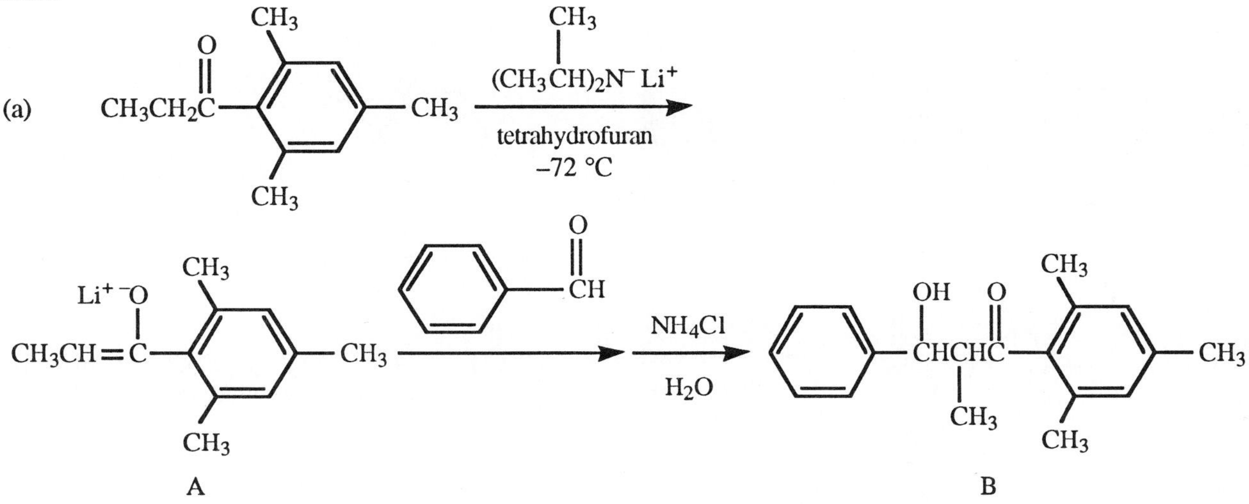

18.27 (cont)

(b)

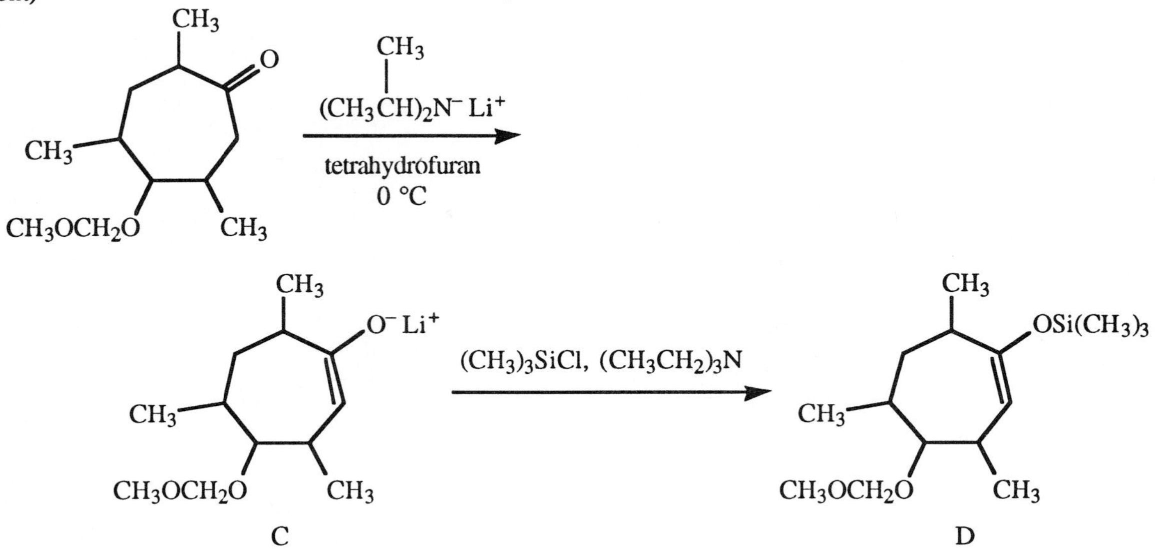

(c)

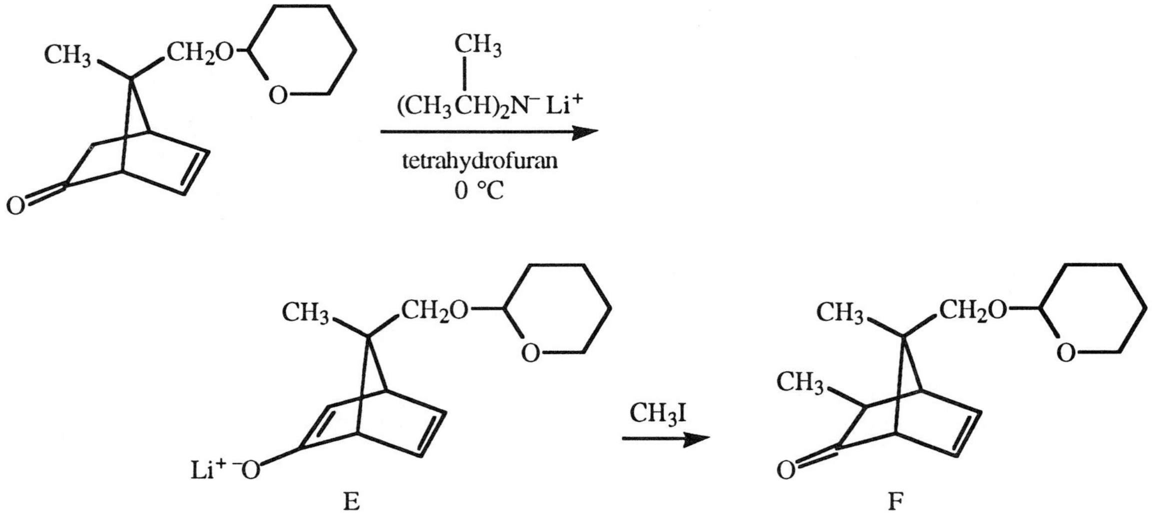

(d)

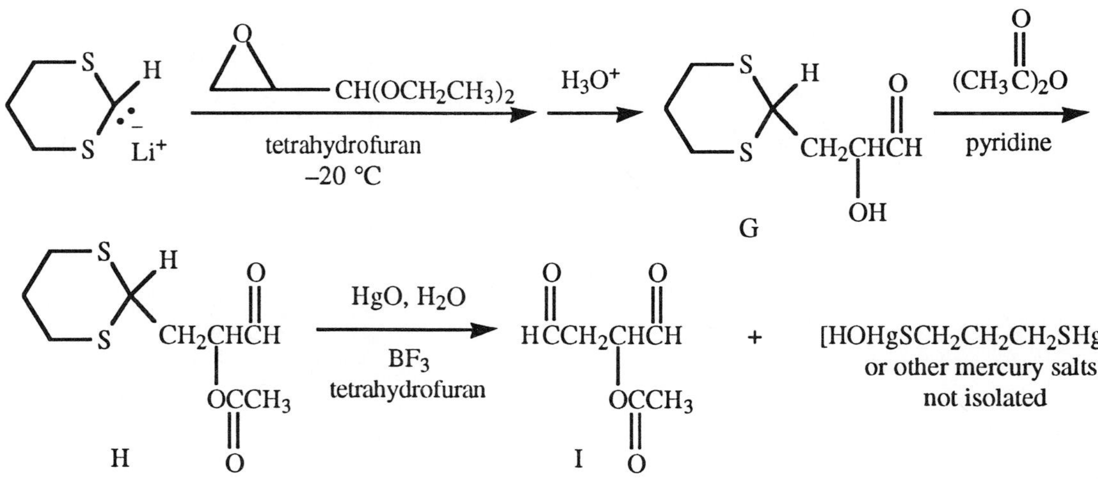

18.27 (cont)

(e)

$$\xrightarrow[\substack{\text{acetonitrile} \\ \text{CaCO}_3, \Delta}]{\text{Hg}_2\text{Cl}_2, \text{H}_2\text{O}}$$

$$\underset{\text{J}}{CH_3\overset{O}{\underset{\|}{C}}CH_2}\text{—}$$

(f)

$$\xrightarrow[\substack{\text{tetrahydrofuran} \\ -78\ °\text{C}}]{\underset{\text{CH}_3}{(CH_3\overset{|}{CH})_2N^-\,Li^+}}$$

K

$$\xrightarrow[\substack{\text{hexamethyl-} \\ \text{phosphoric} \\ \text{triamide} \\ 22\ °\text{C}}]{}$$

L

$$\xrightarrow[\substack{\text{tetrahydrofuran} \\ -78\ °\text{C}}]{\underset{\text{CH}_3}{(CH_3\overset{|}{CH})_2N^-\,Li^+}}$$

M

$$\xrightarrow{\text{CH}_3\text{I}}$$

N

(g)

$$CH_3\overset{O}{\underset{\|}{C}}H \xrightarrow[\substack{\text{HCl (g)} \\ \text{chloroform}}]{\text{HSCH}_2\text{CH}_2\text{CH}_2\text{SH}}$$

O

$$\xrightarrow[\substack{\text{tetrahydrofuran} \\ -20\ °\text{C}}]{\text{CH}_3\text{CH}_2\text{CH}_2\text{CH}_2\text{Li}}$$

P

$$\xrightarrow[]{\underset{\text{CH}_3}{CH_3\overset{|}{C}HBr}}$$

Q

$$\xrightarrow[\text{methanol}]{\text{Hg}_2\text{Cl}_2, \text{H}_2\text{O}} \underset{\text{R}}{CH_3\overset{O}{\underset{\|}{C}}\underset{\underset{CH_3}{|}}{C}HCH_3}$$

(h)

$$\underset{\underset{CH_3}{|}}{\overset{\overset{CH_3}{|}}{CH_3C}}\text{—}\overset{O}{\underset{\|}{C}}CH_2CH_3 \xrightarrow[\substack{\text{tetrahydrofuran} \\ -72\ °\text{C}}]{\underset{\text{CH}_3}{(CH_3\overset{|}{CH})_2N^-\,Li^+}} \underset{\underset{CH_3}{|}}{\overset{\overset{CH_3}{|}}{CH_3C}}\text{—}\overset{O^-\,Li^+}{\underset{\|}{C}}=CHCH_3$$

S

$$\xrightarrow[]{\overset{\overset{\displaystyle O}{\|}}{\underset{}{C}H}\text{-C}_6\text{H}_5}$$

$$\xrightarrow[\text{H}_2\text{O}]{\text{NH}_4\text{Cl}} \underset{\underset{CH_3}{|}}{\overset{\overset{CH_3}{|}}{CH_3C}}\text{—}\overset{O}{\underset{\|}{C}}\underset{\underset{CH_3}{|}}{C}H\overset{OH}{\underset{}{C}}H\text{—C}_6\text{H}_5$$

T

18.27 (cont)

(i)

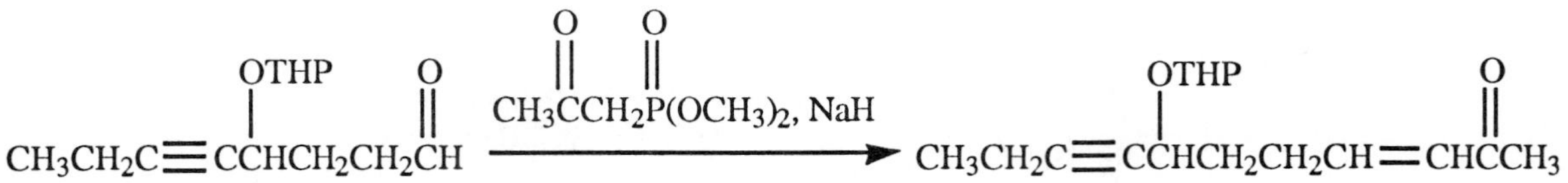

$$CH_3CH_2C\equiv CCHCH_2CH_2CH \xrightarrow{\;\; CH_3CCH_2P(OCH_3)_2,\ NaH \;\;} CH_3CH_2C\equiv CCHCH_2CH_2CH=CHCCH_3$$

U

18.28

(a)

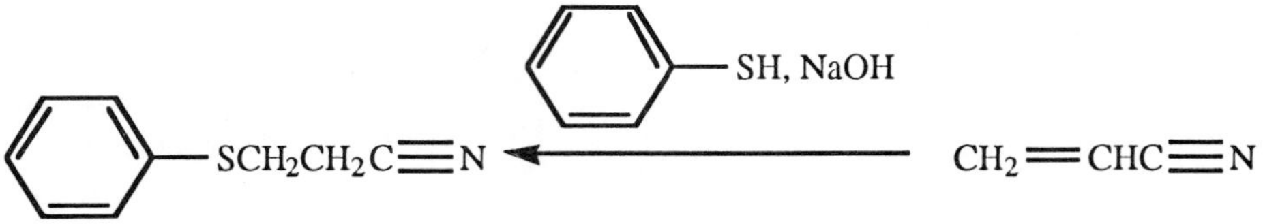

— SCH₂CH₂C≡N   ⟵ — SH, NaOH ⟵   CH₂=CHC≡N

(b)

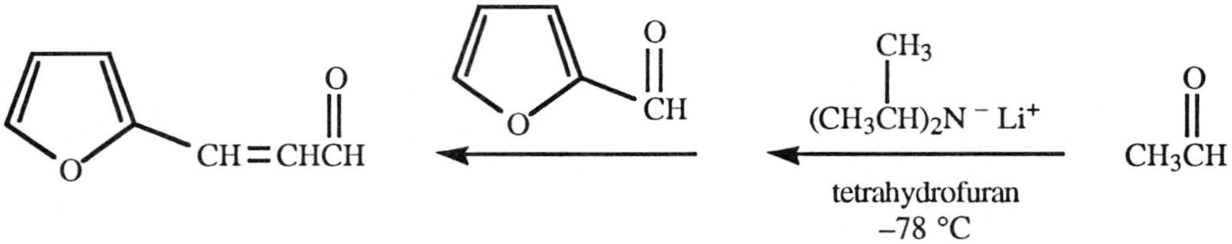

tetrahydrofuran
−78 °C

(c)

$$CH_3CH_2CH_2CH_2CH_2CH_2CH=CH_2 \xleftarrow{\;(CH_2=CHCH_2)_2CuLi\;} CH_3CH_2CH_2CH_2CH_2I$$

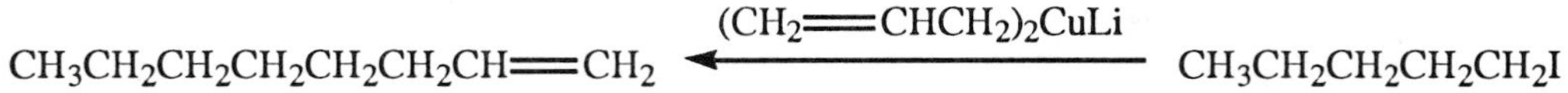

(d)

(CH₃)₂CuLi

diethyl ether

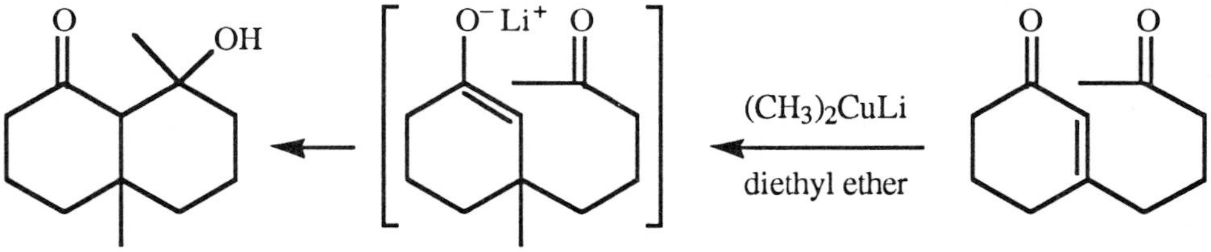

(e)

intermediate from
aldol condensation

Michael addition product

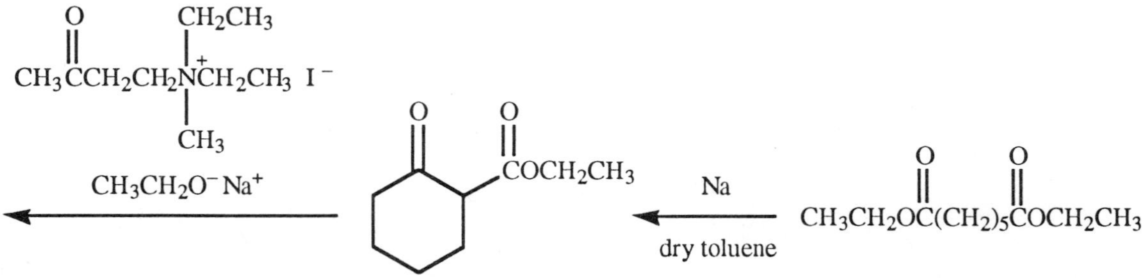

CH₃CH₂O⁻ Na⁺

Na

dry toluene

$$CH_3CH_2OC(CH_2)_5COCH_2CH_3$$

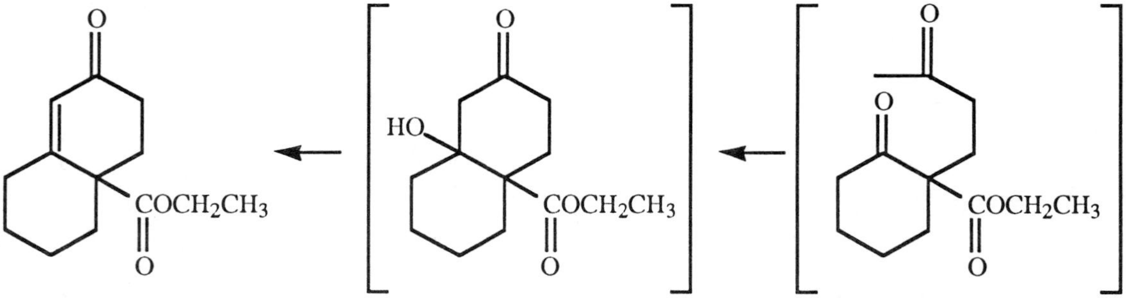

18.28 (cont)

(f)

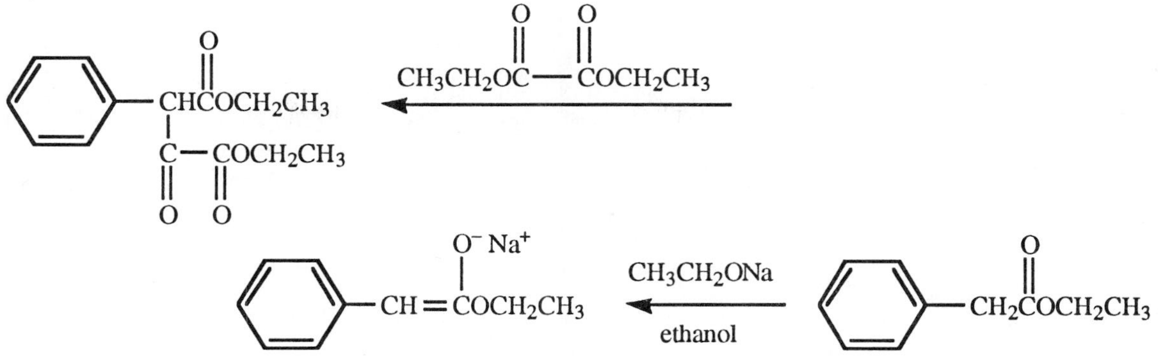
CHCOCH2CH3
C—COCH2CH3
CH3CH2OC—COCH2CH3
O⁻ Na⁺
CH=COCH2CH3
CH3CH2ONa
ethanol
CH2COCH2CH3

18.29

(a)

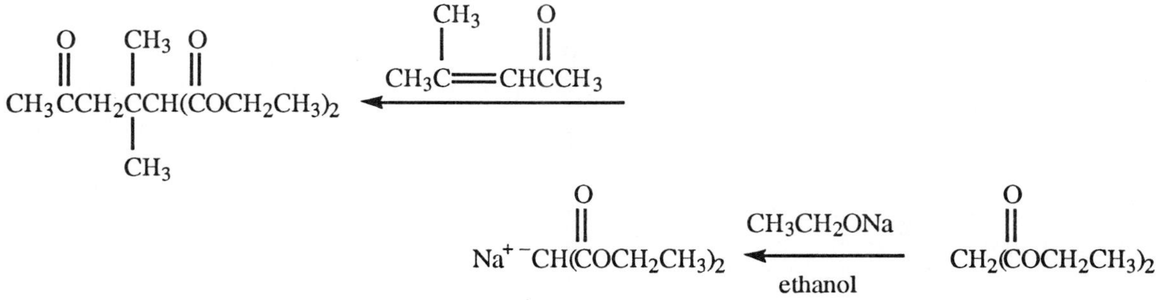
CH3CCH2CCH(COCH2CH3)2
CH3
CH3C=CHCCH3
Na⁺ ⁻CH(COCH2CH3)2
CH3CH2ONa
ethanol
CH2(COCH2CH3)2

(b)

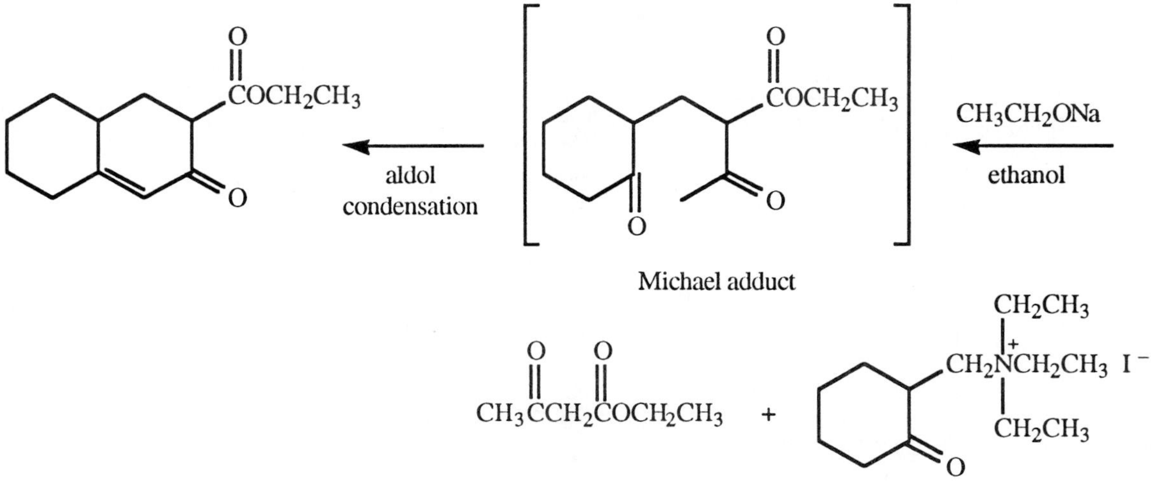
COCH2CH3
aldol
condensation
CH3CH2ONa
ethanol
Michael adduct
CH3CCH2COCH2CH3
CH2CH3
CH2NCH2CH3 I⁻

(c)

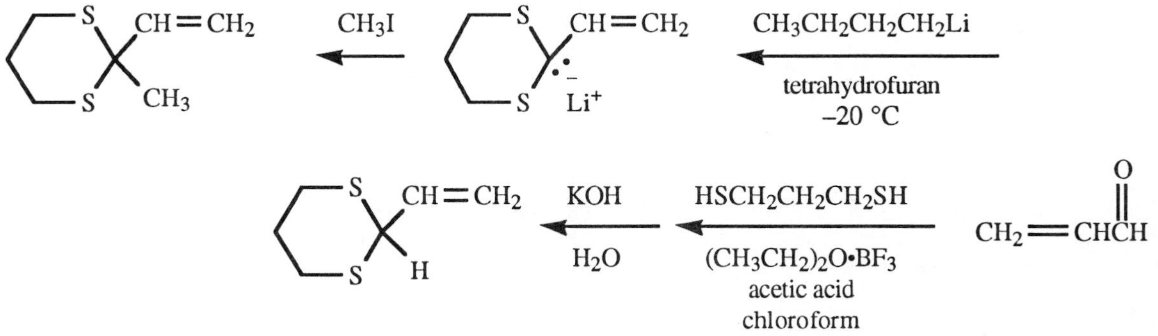
CH=CH2
CH3
CH3I
Li⁺
CH3CH2CH2CH2Li
tetrahydrofuran
−20 °C
KOH
H2O
HSCH2CH2CH2SH
(CH3CH2)2O•BF3
acetic acid
chloroform
CH2=CHCH

18.29 (cont)

(d)

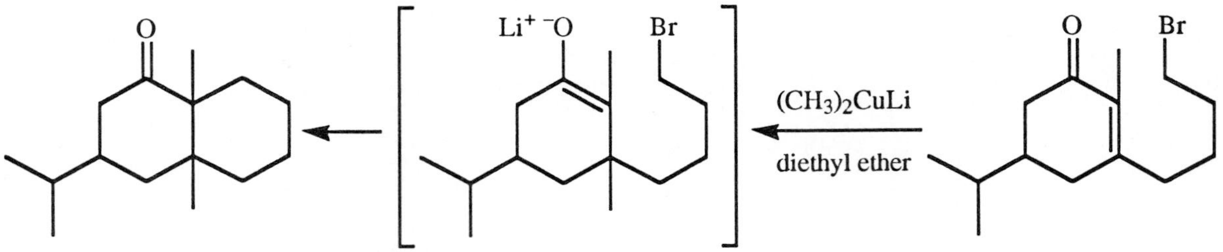

(e)

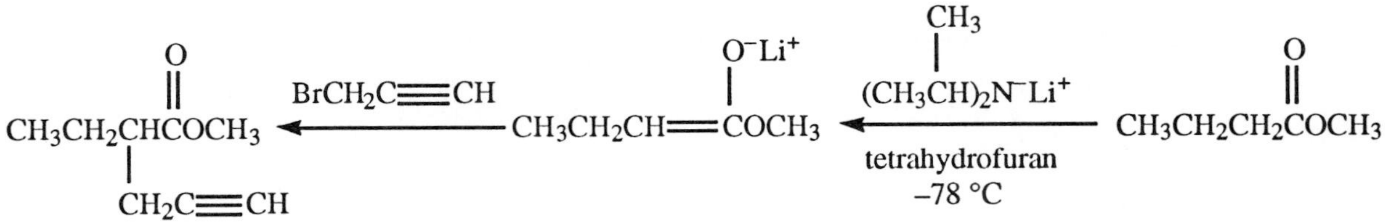

(f)

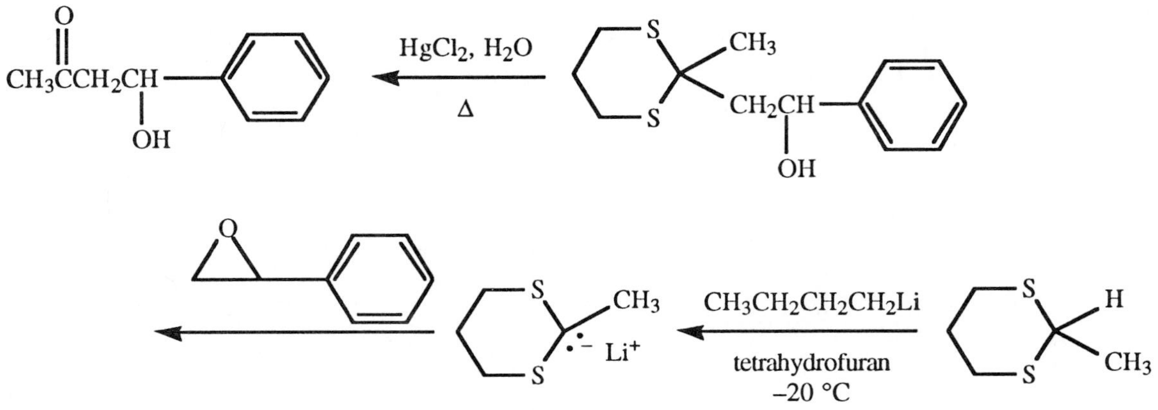

(g)

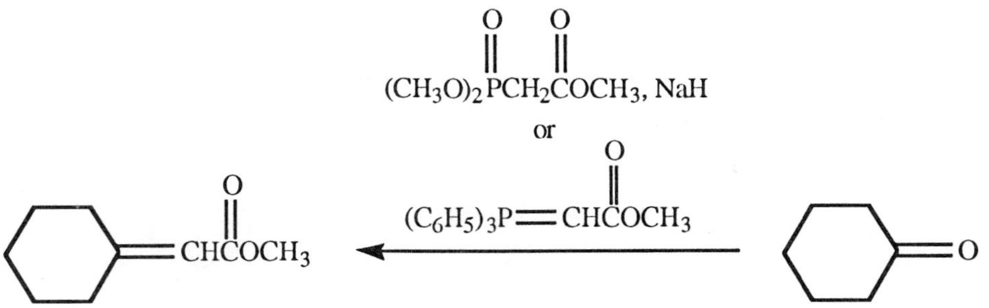

18.30

$$CH_3CH_2\overset{\displaystyle O}{\overset{\displaystyle \|}{C}}N(CH_3)_2 \longrightarrow CH_3CH=\overset{\displaystyle O^-\,Li^+}{\overset{\displaystyle |}{C}}N(CH_3)_2$$

N,N-dimethylpropanamide   enolate ion from
                                     N,N-dimethylpropanamide

$$CH_3\overset{\displaystyle CH_3}{\underset{\displaystyle CH_3}{\overset{\displaystyle |}{\underset{\displaystyle |}{C}}}}-\overset{\displaystyle O}{\overset{\displaystyle \|}{C}}-CH=CHCH_3$$

2,2-dimethyl-4-hexen-3-one

$$CH_3\overset{\displaystyle CH_3}{\underset{\displaystyle CH_3}{\overset{\displaystyle |}{\underset{\displaystyle |}{C}}}}-\overset{\displaystyle O}{\overset{\displaystyle \|}{C}}CH_2-\underset{*}{\overset{\displaystyle CH_3}{\overset{\displaystyle |}{CH}}}-\underset{*}{\overset{\displaystyle CH_3}{\overset{\displaystyle |}{CH}}}-\overset{\displaystyle O}{\overset{\displaystyle \|}{C}}N(CH_3)_2$$

product from the Michael reaction;
2 stereocenters, therefore 4 stereoisomers

one pair of enantiomers

second pair of enantiomers

18.31

resonance contributors for the carbanion from the sulfone

18.31 (cont)

resonance contributors for the carbanion from the sulfoxide

The negative charge on the sulfone carbanion is delocalized to two oxygen atoms while that on the sulfoxide carbanion is delocalized to only one oxygen atom. The conjugate base of the sulfone is, for this reason, more highly stabilized relative to the acid, and is, therefore, a weaker base than is the conjugate base of the sulfoxide. Another way of rationalizing the difference in basicity is to say that the conjugate base of the sulfone has lower electron density at any one site than the conjugate base of the sulfoxide, and is, therefore, less readily protonated.

18.32

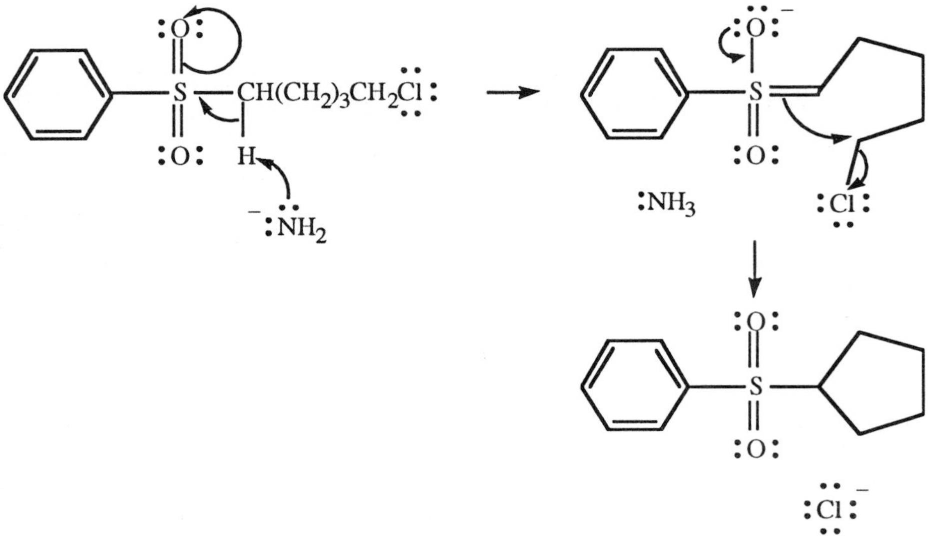

18.33

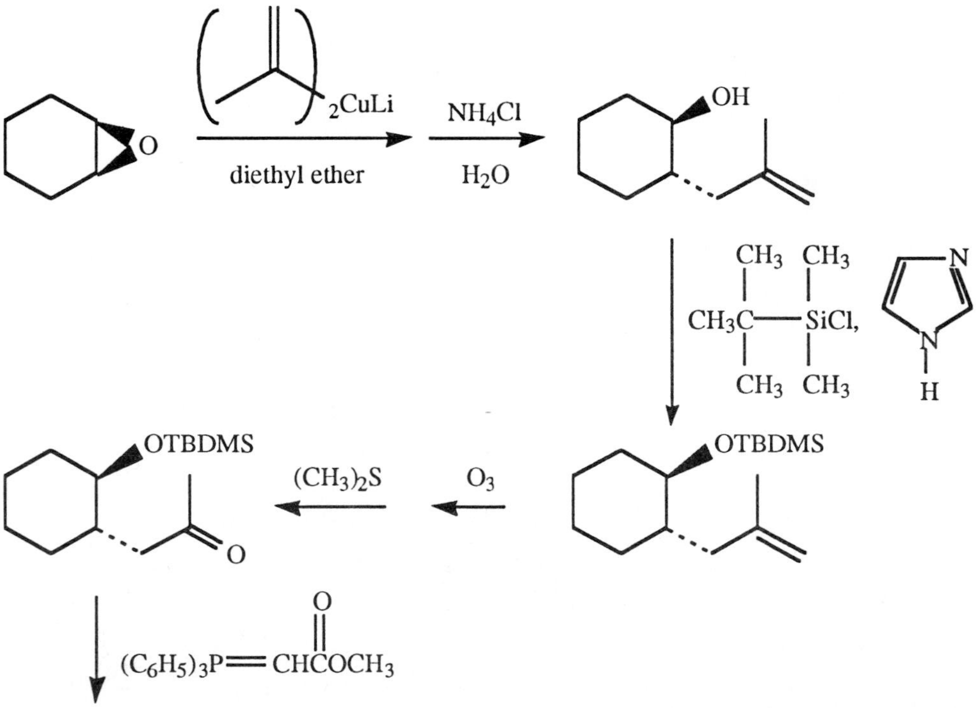

18.33 (cont)

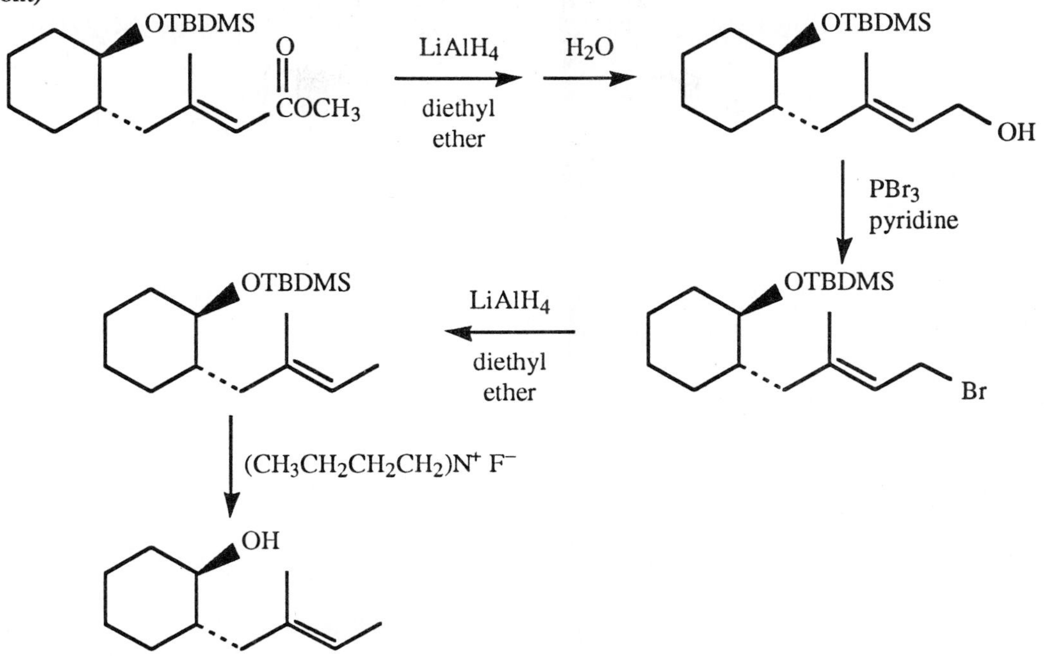

18.34

2-methyl-
butanal

9-bromo-
1-nonanol

potassium
cyanide

$CH_3CH_2CH(CH_2)_9CH_2{}^{14}COH$ $\xleftarrow[\Delta]{H_3O^+}$ $CH_3CH_2CH(CH_2)_9CH_2{}^{14}CN$
|
CH_3

|
CH_3

12-methyltetradecanoic acid

$\uparrow$ $K^{14}CN$
ethanol

$CH_3CH_2CH(CH_2)_9CH_2OH$ $\xrightarrow[\text{pyridine}]{PBr_3}$ $CH_3CH_2CH(CH_2)_9CH_2Br$
|
CH_3

|
CH_3

$\uparrow$ $H_2O, HClO_4$
methanol

18.34 (cont)

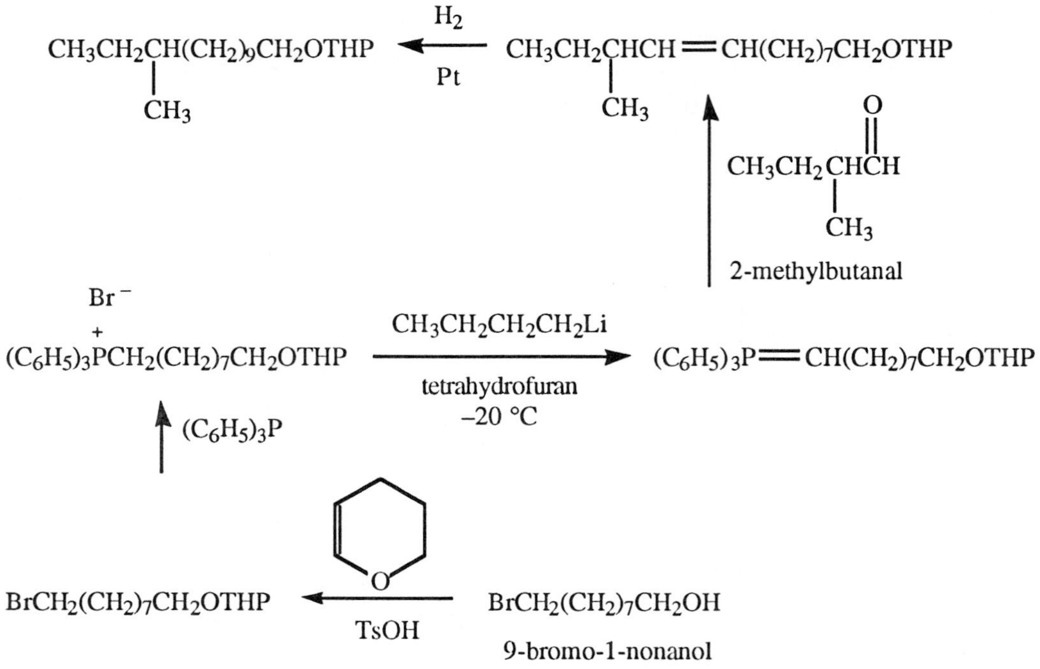

$$CH_3CH_2CH(CH_2)_9CH_2OTHP \xleftarrow[\text{Pt}]{H_2} CH_3CH_2CHCH=CH(CH_2)_7CH_2OTHP$$

with CH_3 branch, and CH_3 branch

2-methylbutanal

$$CH_3CH_2CHCH$$ with CH_3 branch and carbonyl O

$$Br^- \\ + \\ (C_6H_5)_3PCH_2(CH_2)_7CH_2OTHP \xrightarrow[\substack{\text{tetrahydrofuran} \\ -20\ °C}]{CH_3CH_2CH_2CH_2Li} (C_6H_5)_3P=CH(CH_2)_7CH_2OTHP$$

$\uparrow (C_6H_5)_3P$

$$BrCH_2(CH_2)_7CH_2OTHP \xleftarrow[\text{TsOH}]{} BrCH_2(CH_2)_7CH_2OH$$

9-bromo-1-nonanol

18.35

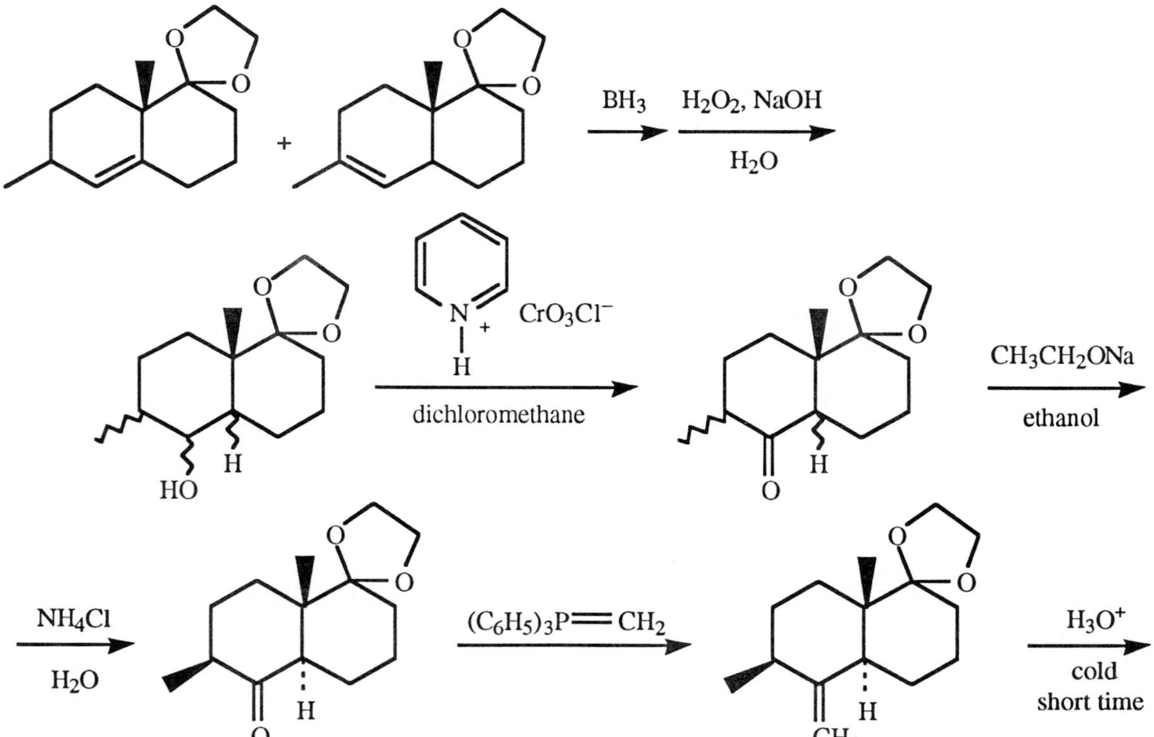

(This is an example of equilibration.
Enolization occurs at both sides of the
carbonyl group. Protonation of the
enolate anion gives the form that is
thermodynamically more stable.)

18.35 (cont)

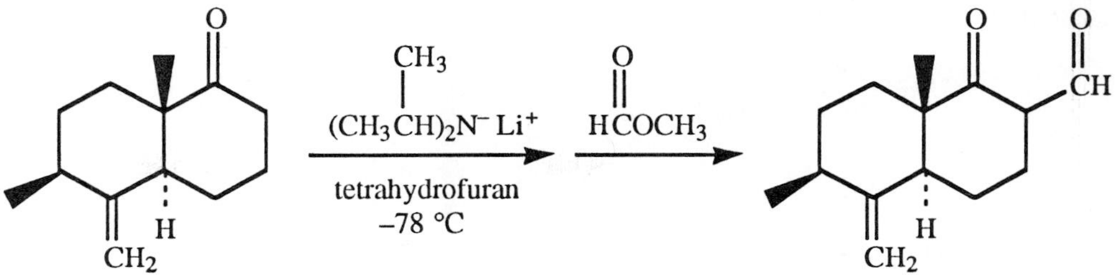

18.36

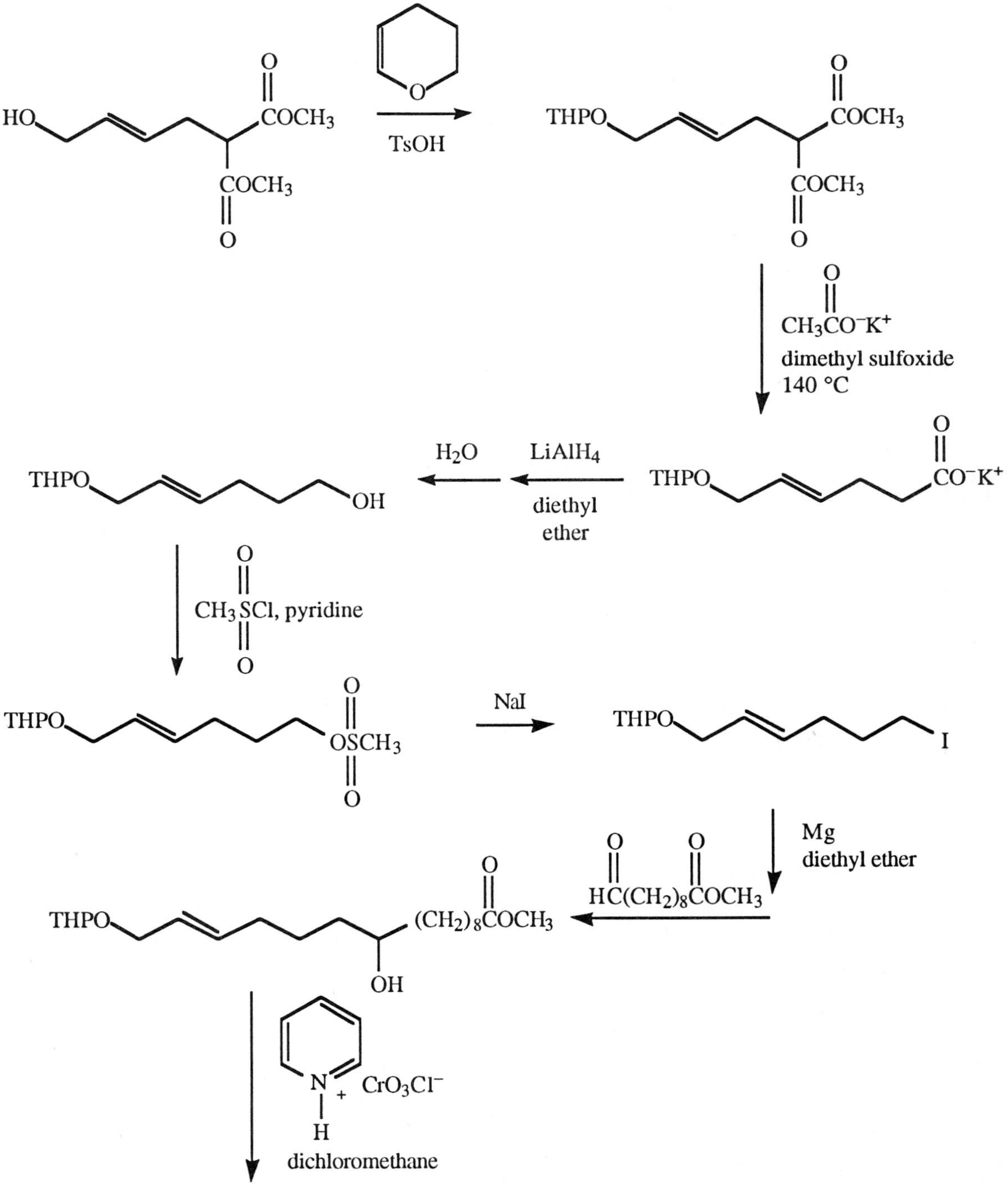

18.36 (cont)

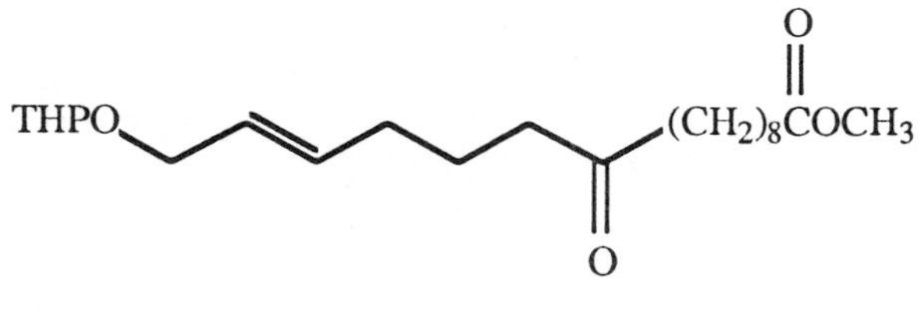

Mechanism of Step 2:

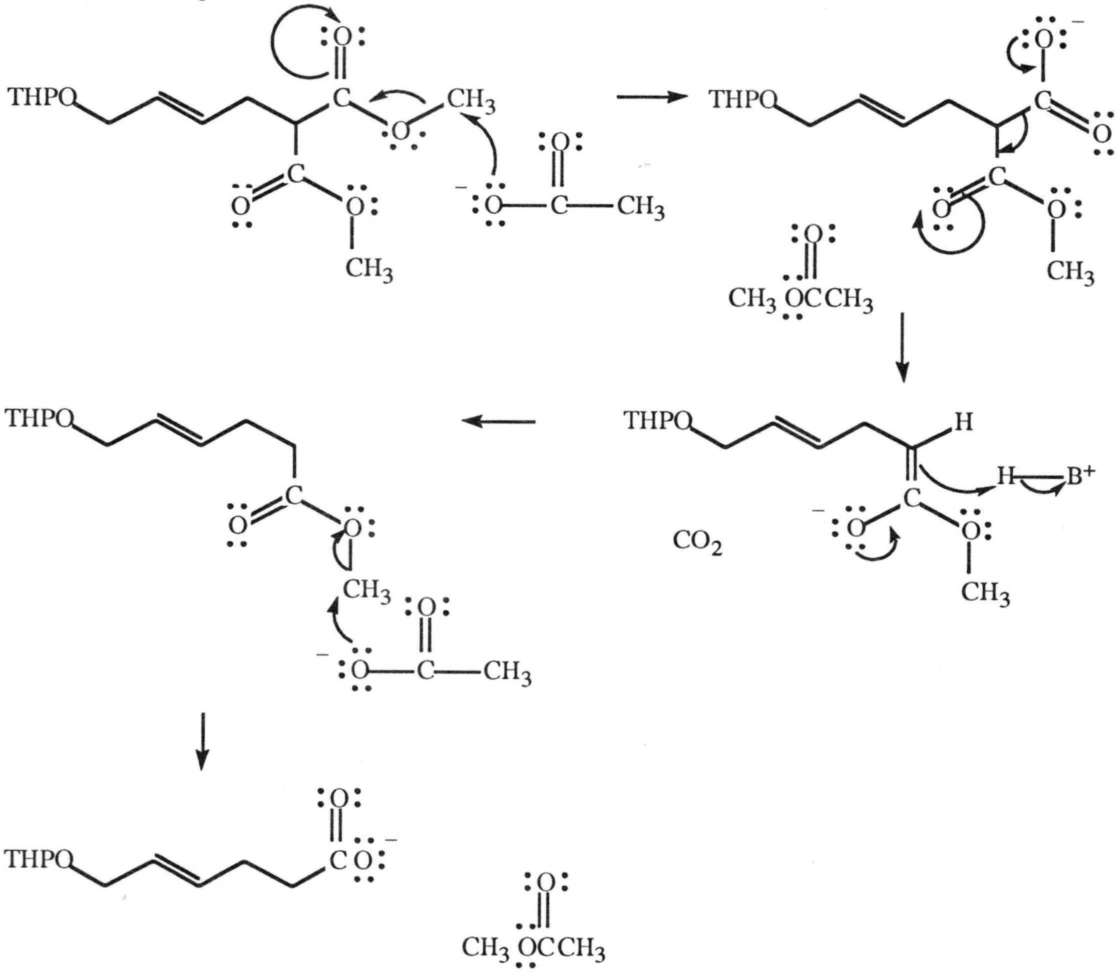

CO$_2$

18.37

(a)

LiI, LiAlH$_4$

diethyl ether
−100 °C

H$_2$O

A
(*S*)-configuration at
the new stereocenter

KH
tetrahydrofuran

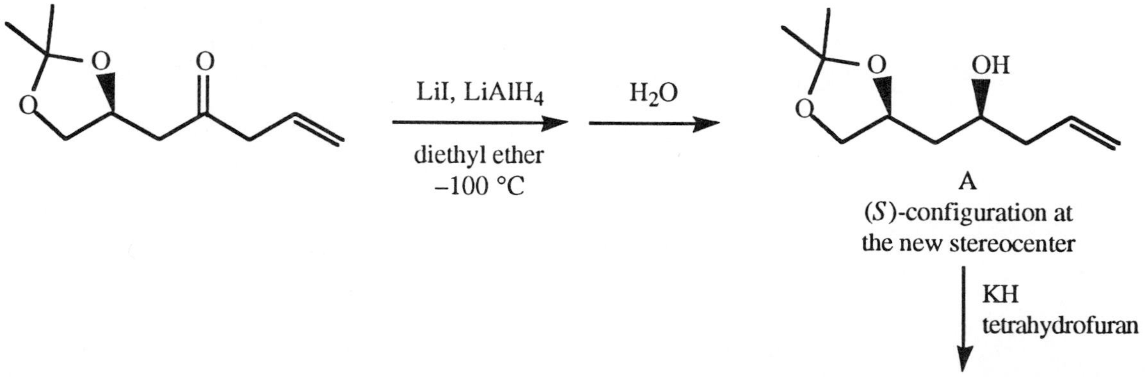

18.37 (a) (cont)

(b)   Lithium ion complexes with the oxygen atom of the carbonyl group and with one of the oxygen atoms of the protected diol to hold the carbonyl group in one conformation. This makes one side of the carbonyl group, the side away from the bulky protected diol function, more accessible to the aluminum hydride anion, the nucleophile in the lithium aluminum hydride reduction.

18.37 (b) (cont)

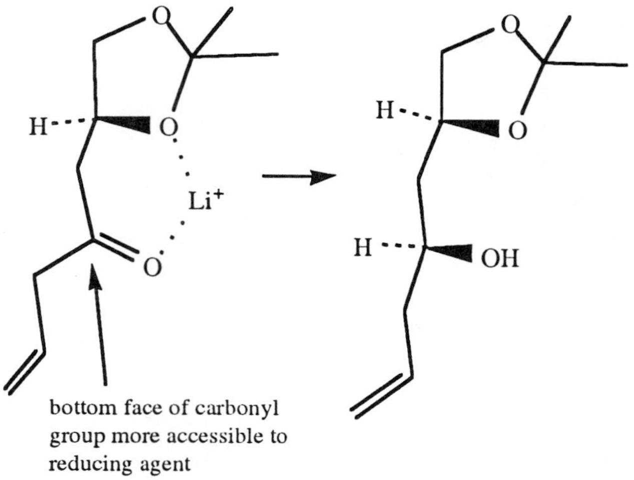

bottom face of carbonyl
group more accessible to
reducing agent

(c)   Let R = 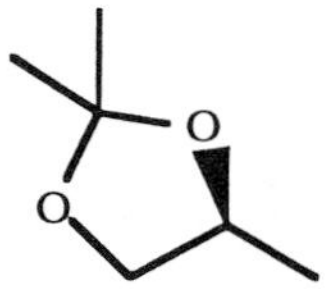   amd R' =

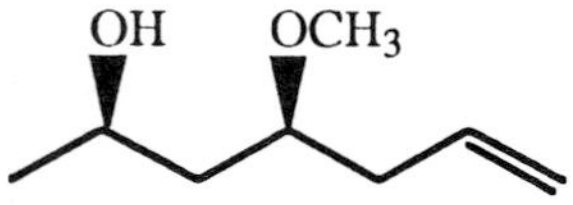

good leaving group

electrophile

nucleophile

sulfur-stabilized
carbocation

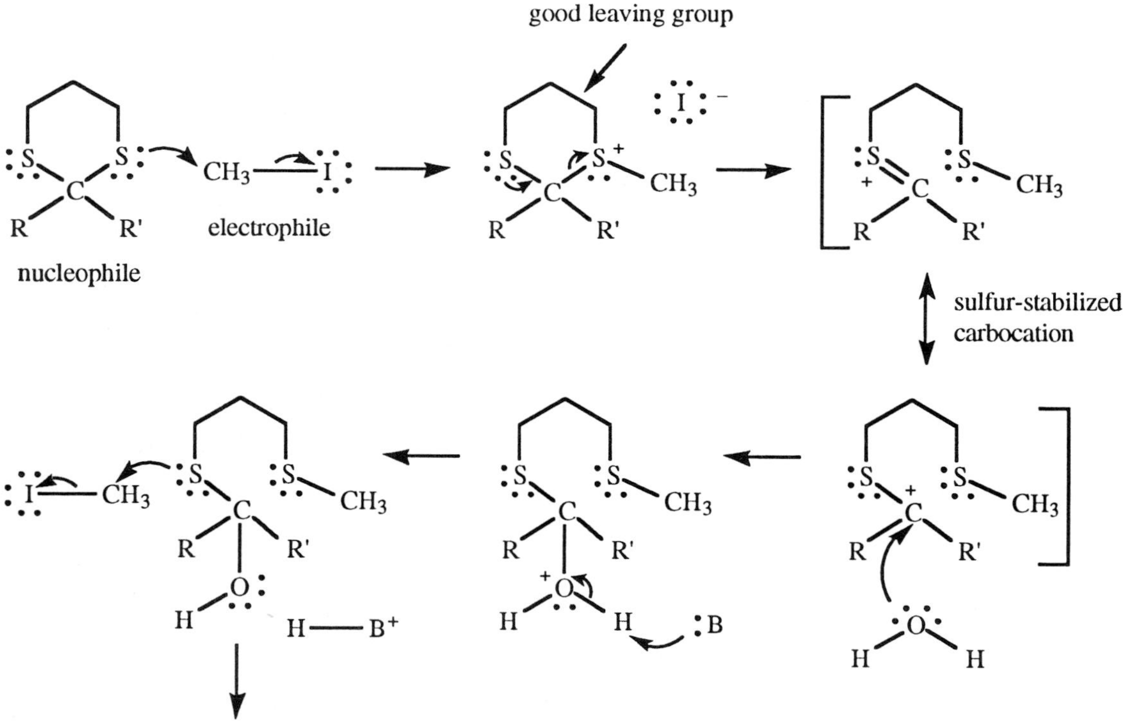

18.37 (c) (cont)

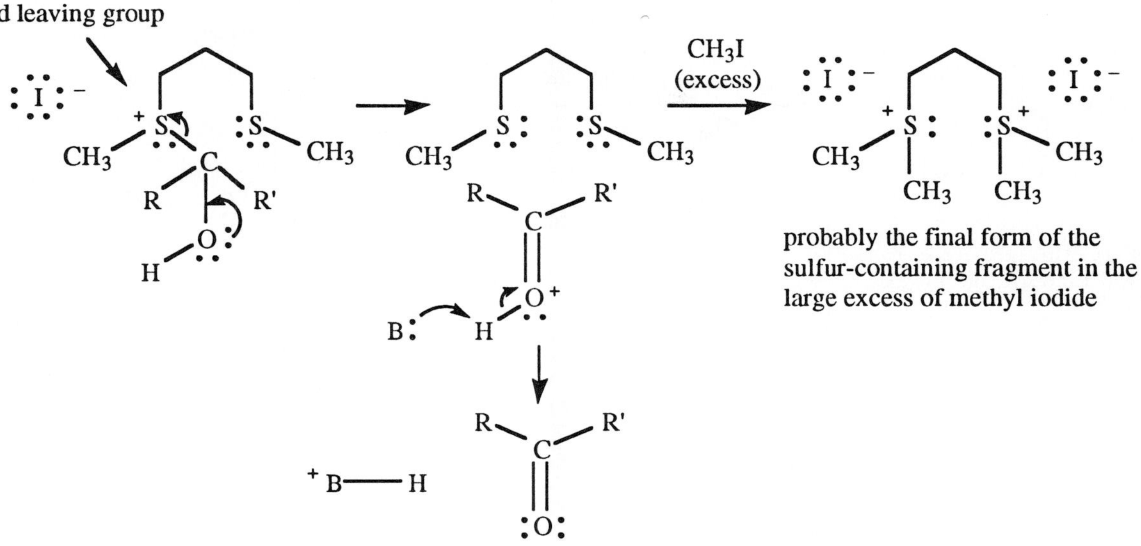

good leaving group

CH₃I
(excess)

probably the final form of the
sulfur-containing fragment in the
large excess of methyl iodide

18.38

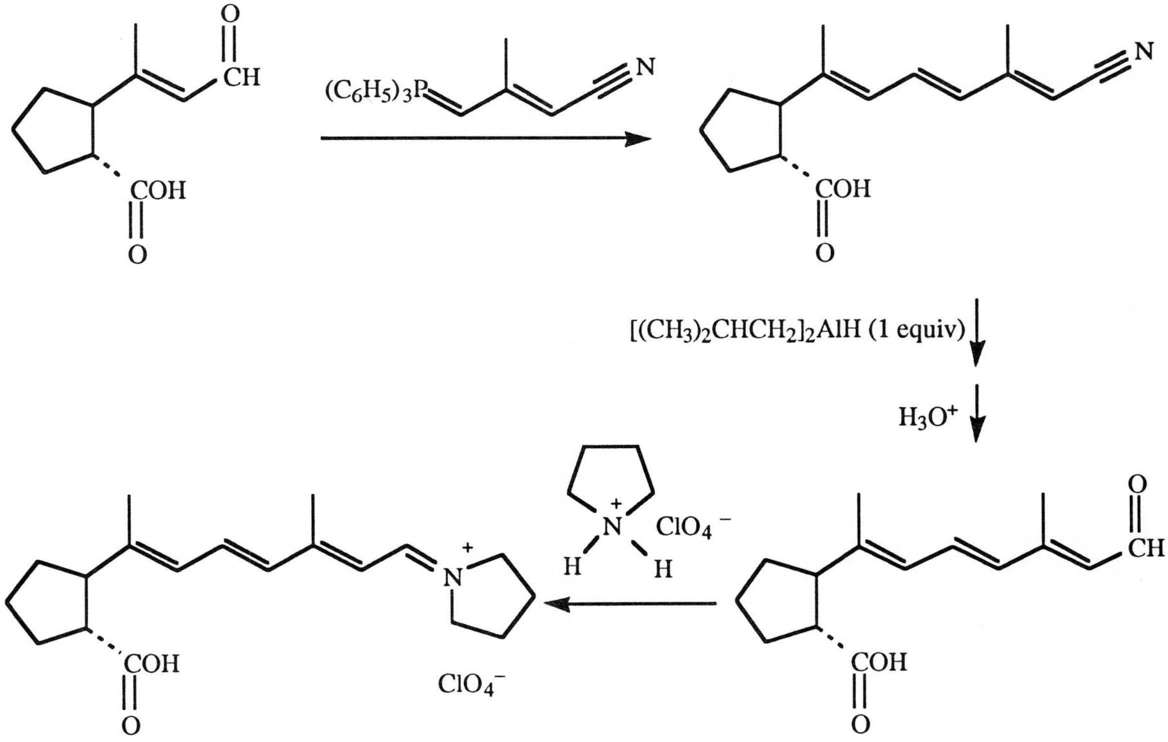

$(C_6H_5)_3P$

$[(CH_3)_2CHCH_2]_2AlH$ (1 equiv)

H_3O^+

ClO_4^-

ClO_4^-

18.39

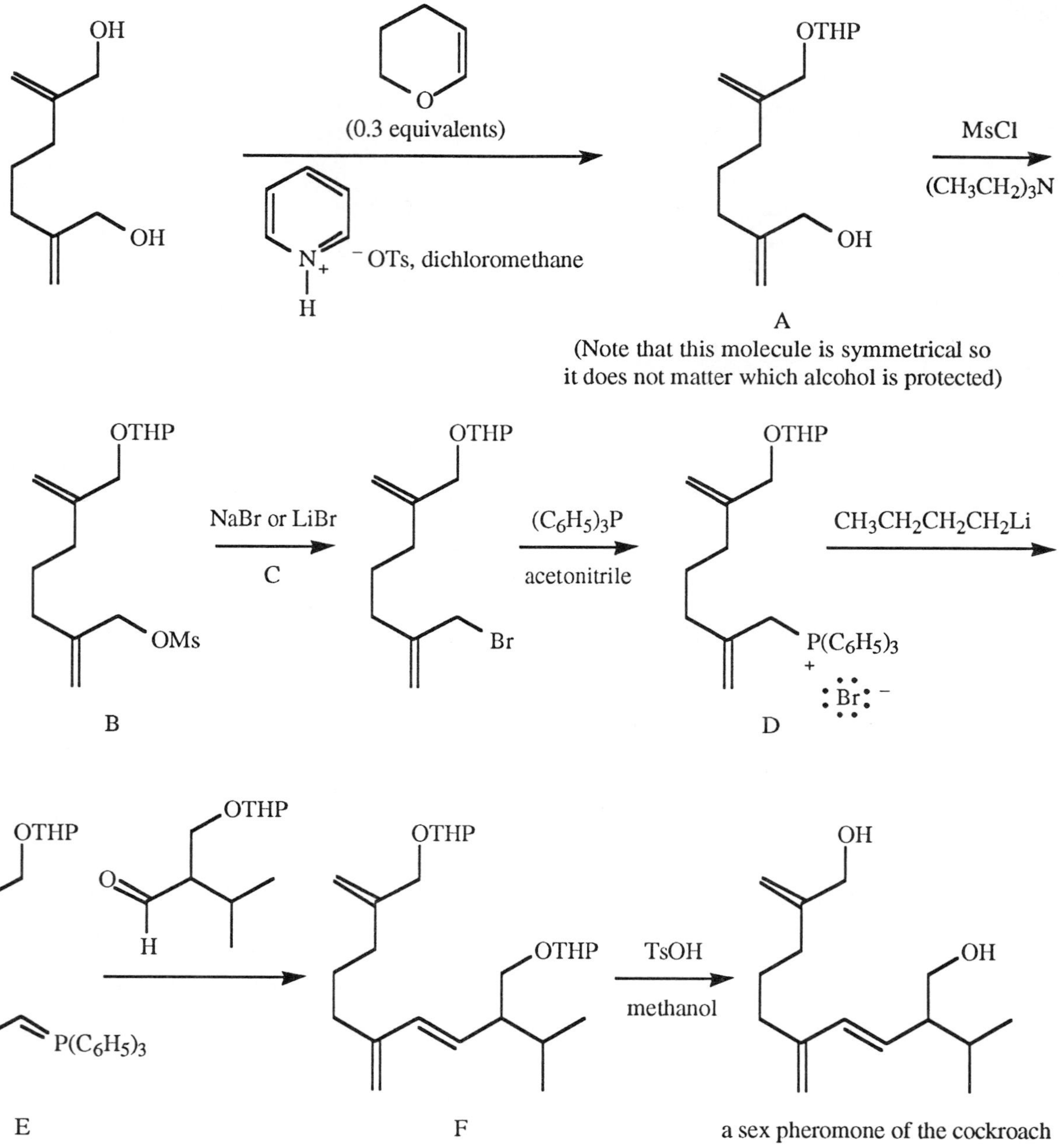

(0.3 equivalents)

${}^{-}$OTs, dichloromethane

$\xrightarrow{\text{MsCl}}$
$(CH_3CH_2)_3N$

A
(Note that this molecule is symmetrical so
it does not matter which alcohol is protected)

B
$\xrightarrow[\text{C}]{\text{NaBr or LiBr}}$

$\xrightarrow[\text{acetonitrile}]{(C_6H_5)_3P}$

D

$\xrightarrow{CH_3CH_2CH_2CH_2Li}$

E

F

a sex pheromone of the cockroach

$\xrightarrow[\text{methanol}]{\text{TsOH}}$

18.40 $HC{\equiv}CH$ + 2 CH_3CH_2MgBr $\xrightarrow[\substack{\text{diethyl ether}\\\text{toluene}}]{}$ $BrMgC{\equiv}CMgBr$ $\xrightarrow{}$

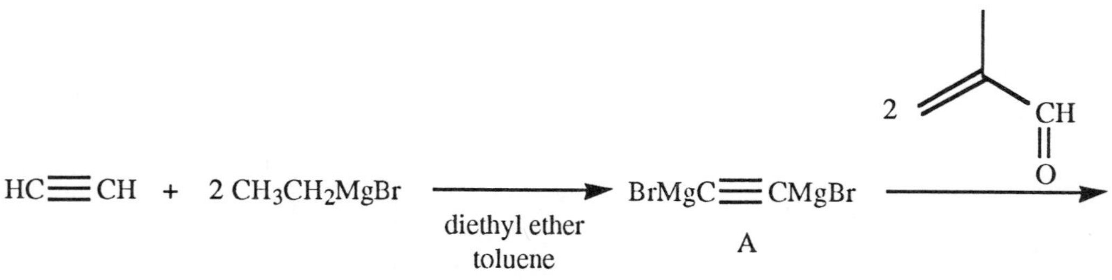

A

18.40 (cont)

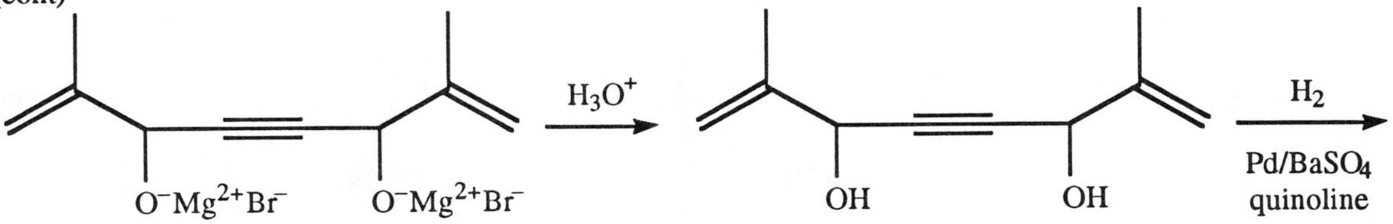

B   C

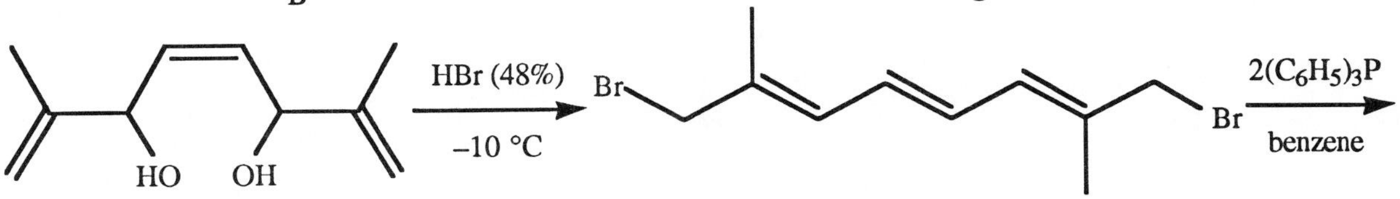

D   E

Allylic carbocations are the intermediates in this reaction. The carbocations, stabilized by delocalization of charge by resonance, react with bromide ion to give the most highly conjugated system. The cis stereochemistry of the central double bond is lost in the conjugated triene, which has the most stable trans configuration at each double bond.

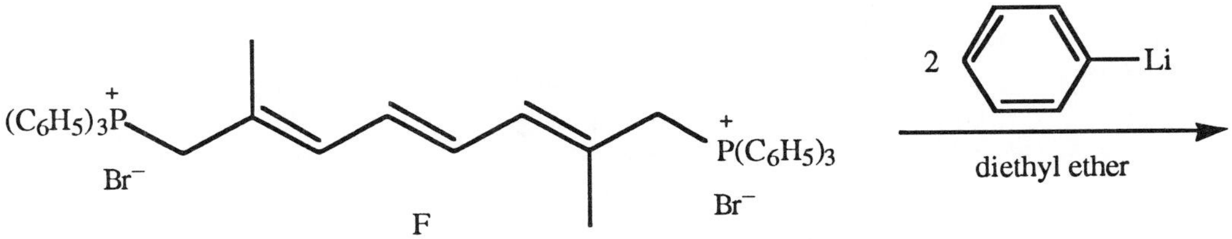

F

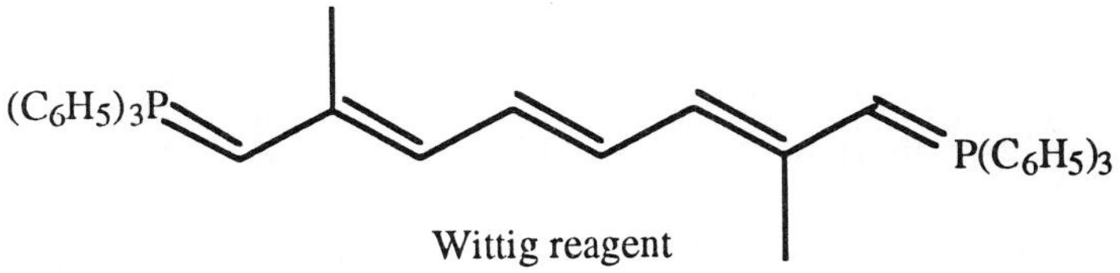

Wittig reagent

18.41

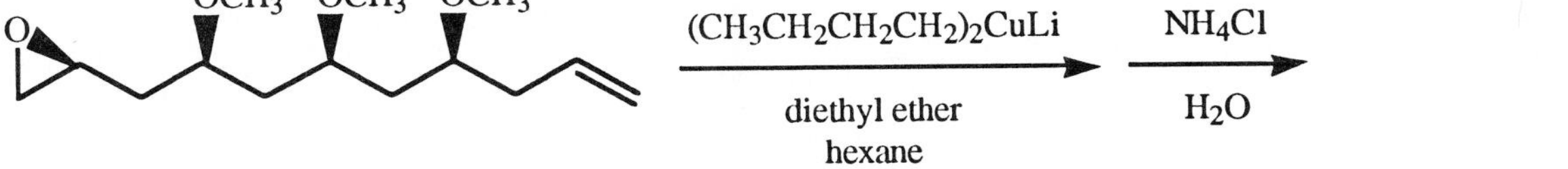

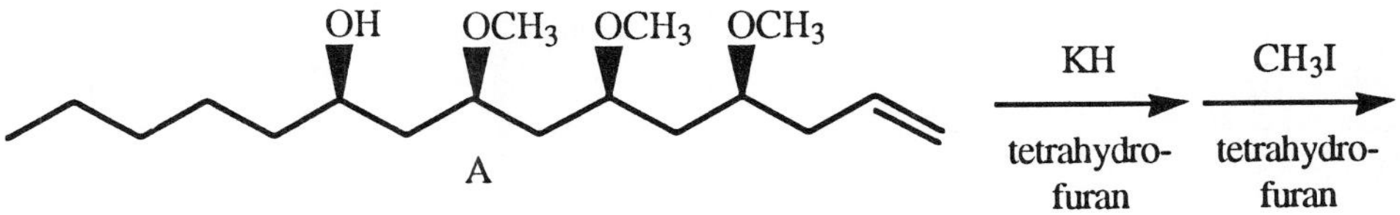

A

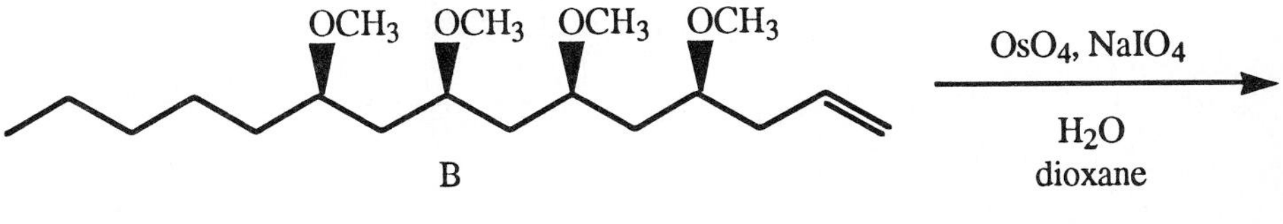

B

18.41 (cont)

$$\text{C} \xrightarrow[\text{(CH}_3\text{CH}_2)_2\text{O·BF}_3]{\text{HSCH}_2\text{CH}_2\text{CH}_2\text{SH}}$$

D

18.42

$$\text{CH}_3\text{CHCH}_2\text{CH}_2\text{MgBr} \xrightarrow[\text{benzene}]{} \text{CH}_3\text{CHCH}_2\text{CH}_2\text{CH}_2\text{CH}_2\text{CH}_2\text{O}^-\text{Mg}^{2+}\text{Br}^- \xrightarrow{\text{H}_3\text{O}^+}$$

A

$$\text{CH}_3\text{CHCH}_2\text{CH}_2\text{CH}_2\text{CH}_2\text{CH}_2\text{OH} \xrightarrow{\text{HBr}} \text{CH}_3\text{CHCH}_2\text{CH}_2\text{CH}_2\text{CH}_2\text{CH}_2\text{Br} \xrightarrow[\substack{N,N\text{-dimethyl-}\\\text{formamide}}]{(\text{C}_6\text{H}_5)_3\text{P}}$$

B   C

$$\text{CH}_3\text{CHCH}_2\text{CH}_2\text{CH}_2\text{CH}_2\text{CH}_2\overset{+}{\text{P}}(\text{C}_6\text{H}_5)_3 \;\text{Br}^- \xrightarrow[\substack{\text{hexamethyl-}\\\text{phosphoric}\\\text{triamide}}]{\text{base}} \text{CH}_3\text{CHCH}_2\text{CH}_2\text{CH}_2\text{CH}_2\text{CH}=\text{P}(\text{C}_6\text{H}_5)_3 \xrightarrow{\text{CH}_3(\text{CH}_2)_9\text{CH}}$$

D   E

$$\text{CH}_3\text{CH(CH}_2)_4 \qquad (\text{CH}_2)_9\text{CH}_3 \xrightarrow[\text{dichloromethane}]{} \text{CH}_3\text{CH(CH}_2)_4 \cdots \text{(CH}_2)_9\text{CH}_3$$

F   and enantiomer

18.43   1.

$$\text{CH}_3\text{CCH}_2\text{COCH}_2\text{CH}_3 \xrightarrow[\text{ethanol}]{\text{CH}_3\text{CH}_2\text{ONa}} \text{CH}_3\text{CCHCOCH}_2\text{CH}_3 \;\;\text{Na}^+ \xrightarrow[\text{(1 molar equiv)}]{\text{Br(CH}_2)_3\text{Br}}$$

A

$$\text{CH}_3\text{CCHCOCH}_2\text{CH}_3 \xrightarrow[\Delta]{\text{HBr}\;(48\%)} \text{CH}_3\text{C(CH}_2)_3\text{CH}_2\text{Br} \xrightarrow[\substack{\text{TsOH}\\\Delta\\\text{benzene}}]{\text{HOCH}_2\text{CH}_2\text{OH}}$$

(CH₂)₂CH₂Br

B   C

18.43 (1) (cont)

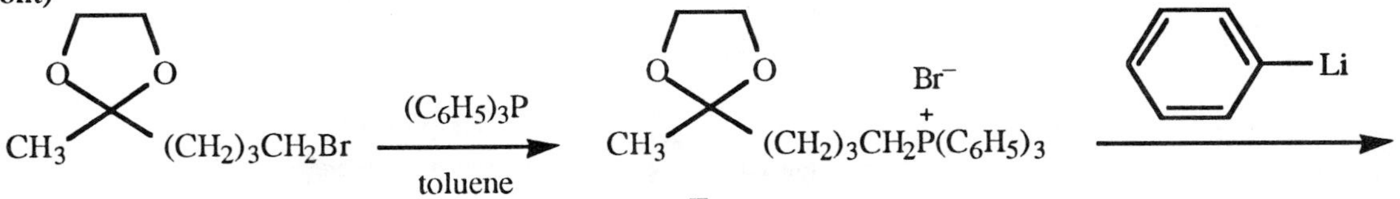

D E

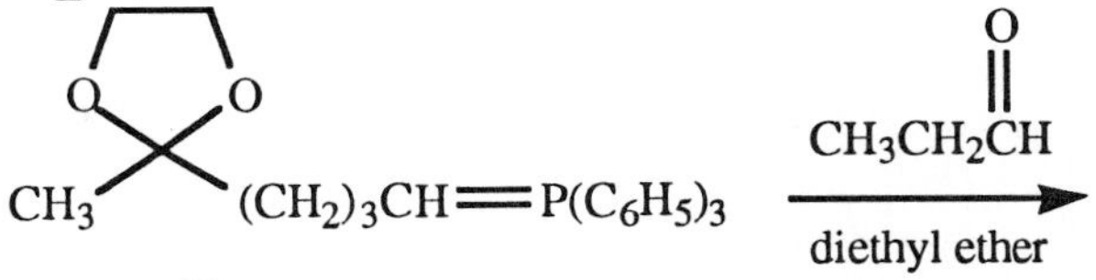

F

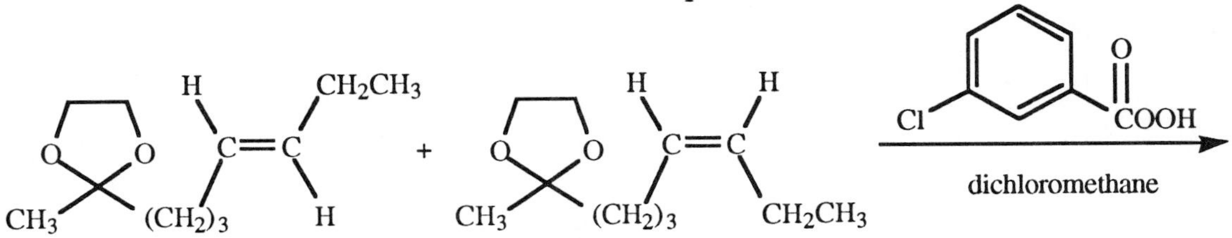

G

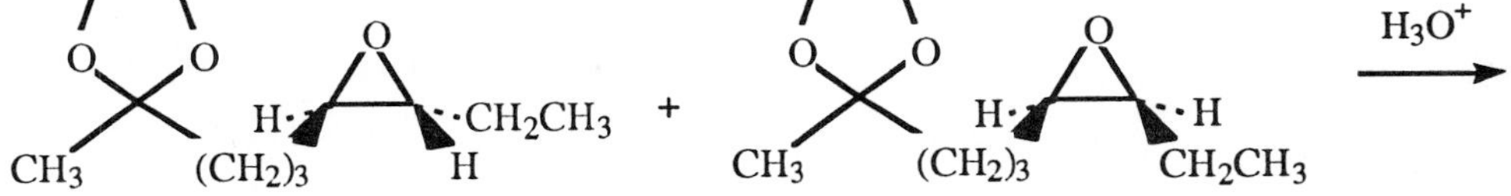

and enantiomer and enantiomer

H (four stereoisomers)

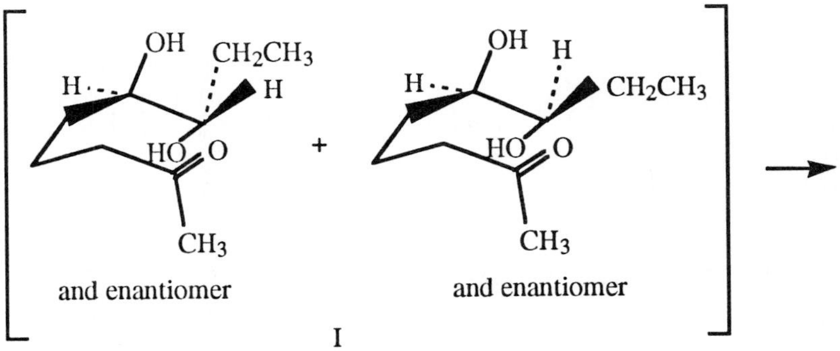

I

Removal of the protecting group from the carbonyl group and opening of the oxirane ring in aqueous acid gives I, which, under acidic conditions, forms the stable cyclic ketal of the resulting ketone-diol.

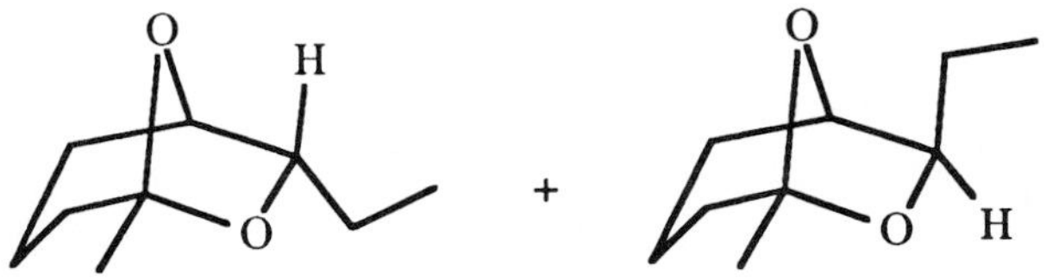

and enantiomer and enantiomer

endo-brevicomin *exo*-brevicomin

18.43 (cont)

2.

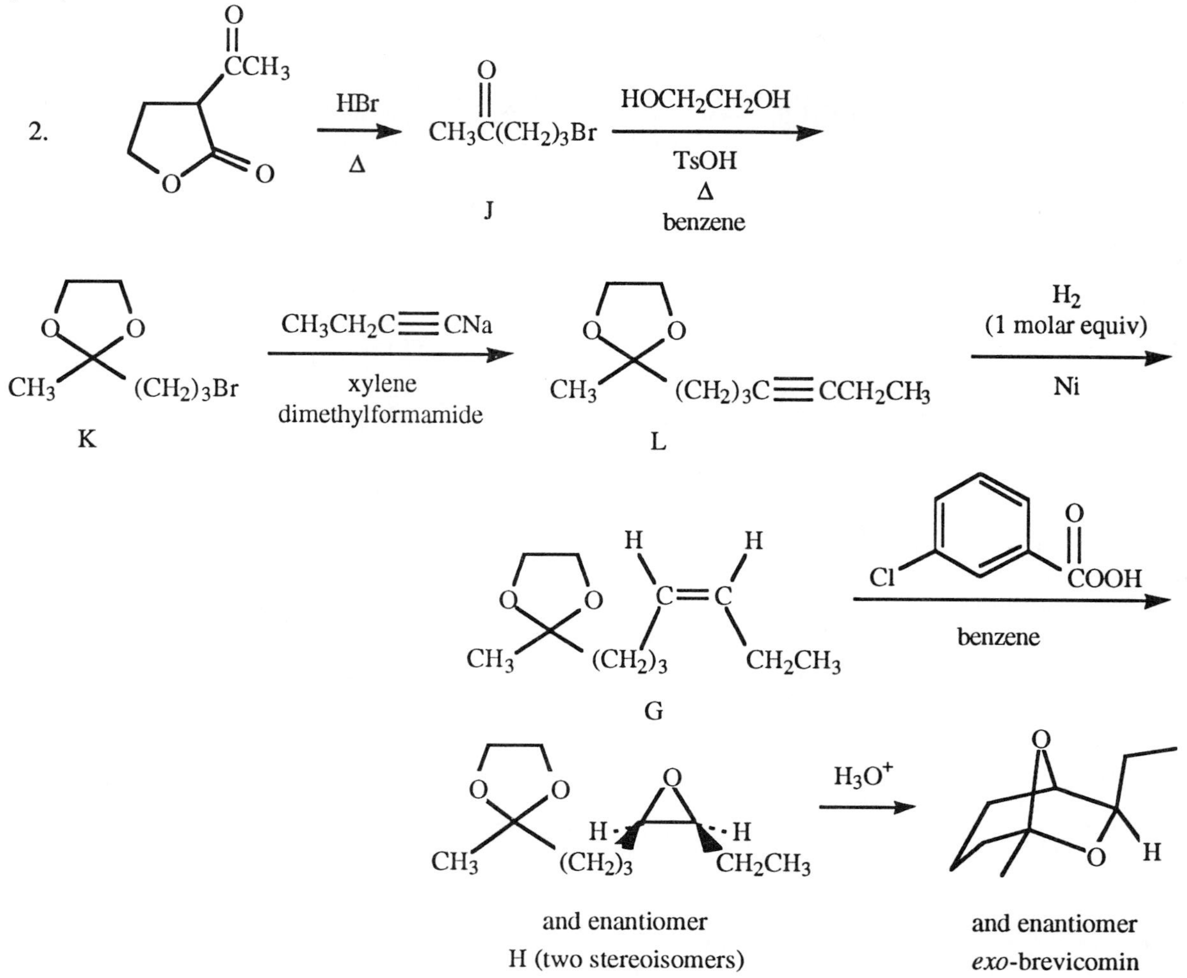

J

HBr
Δ

HOCH₂CH₂OH
TsOH
Δ
benzene

K

CH₃CH₂C≡CNa

xylene
dimethylformamide

L

H₂
(1 molar equiv)

Ni

G

Cl— —COOH

benzene

and enantiomer
H (two stereoisomers)

H₃O⁺

and enantiomer
exo-brevicomin

18.44 (a)

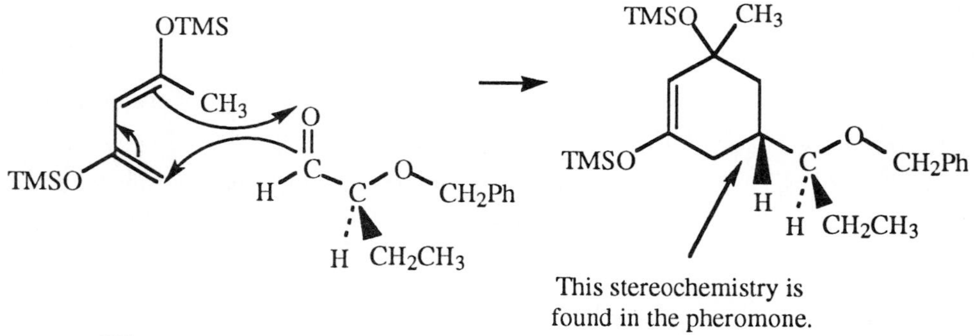

This stereochemistry is
found in the pheromone.

(b)

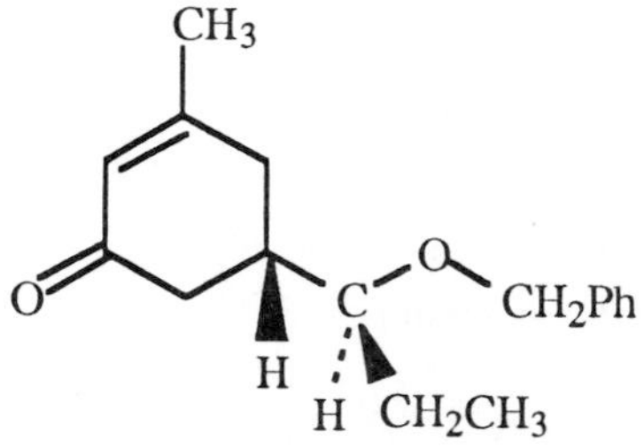

18.45

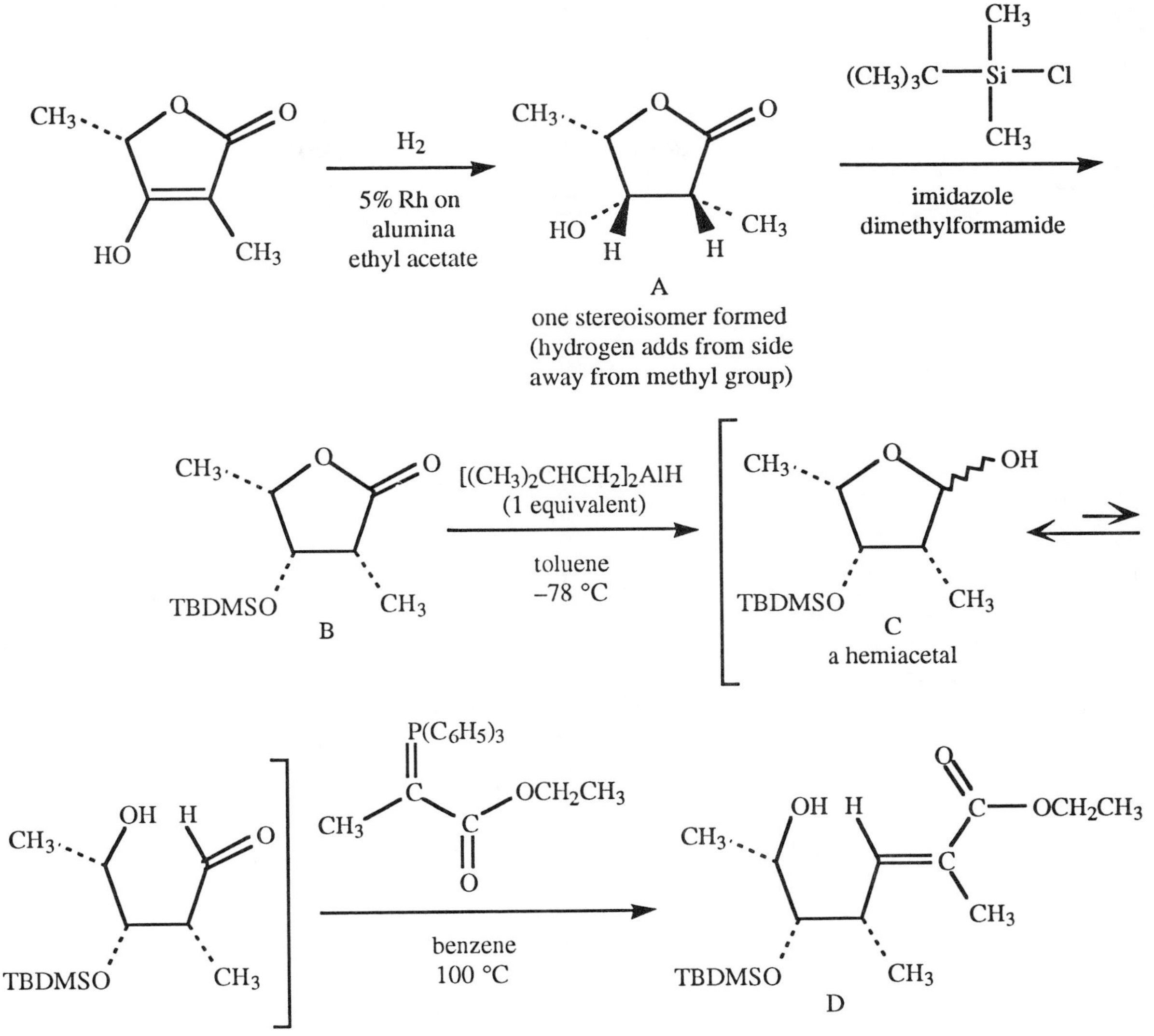

A
one stereoisomer formed
(hydrogen adds from side
away from methyl group)

B

C
a hemiacetal

D

18.46   The kinetic product of the reaction is the 1,2-addition product. It forms almost exclusively when the reaction is run for a short time at low temperatures. The 1,4-addition product is the thermodynamically more stable product. It is the major product when the reaction is run at higher temperatures and for a long period of time allowing for equilibrium to be established.

The reaction products must be formed reversibly, otherwise they could not equilibrate.

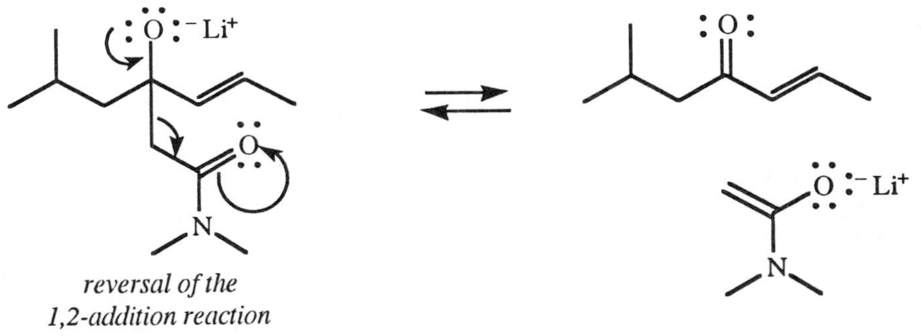

*reversal of the
1,2-addition reaction*

18.46 (cont)

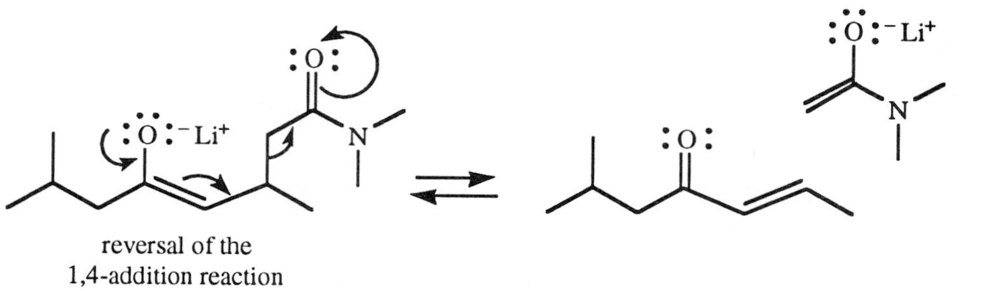
Li+
Li+
N
N
reversal of the
1,4-addition reaction

18.47

(a)

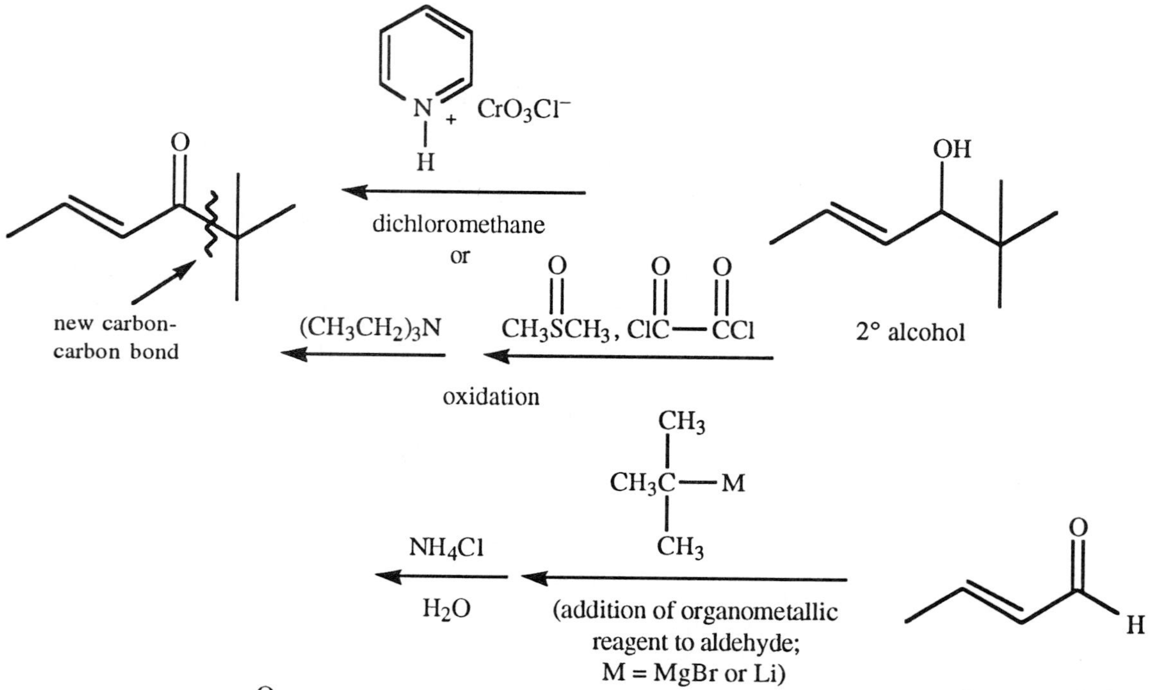
O
new carbon-
carbon bond
N
H
CrO3Cl−
dichloromethane
or
(CH3CH2)3N
CH3SCH3, ClC—CCl
oxidation
OH
2° alcohol
CH3
CH3C—M
CH3
NH4Cl
H2O
(addition of organometallic
reagent to aldehyde;
M = MgBr or Li)
H

(b)

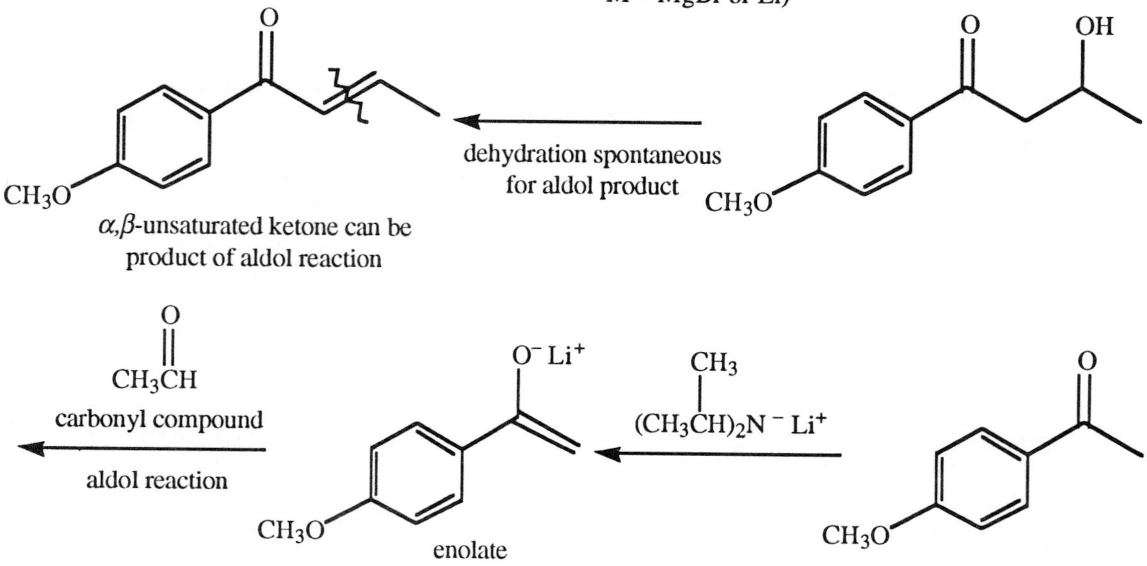
O
CH3O
α,β-unsaturated ketone can be
product of aldol reaction
dehydration spontaneous
for aldol product
OH
CH3CH
carbonyl compound
aldol reaction
O− Li+
enolate
CH3
(CH3CH)2N − Li+

18.47 (cont)

(c)

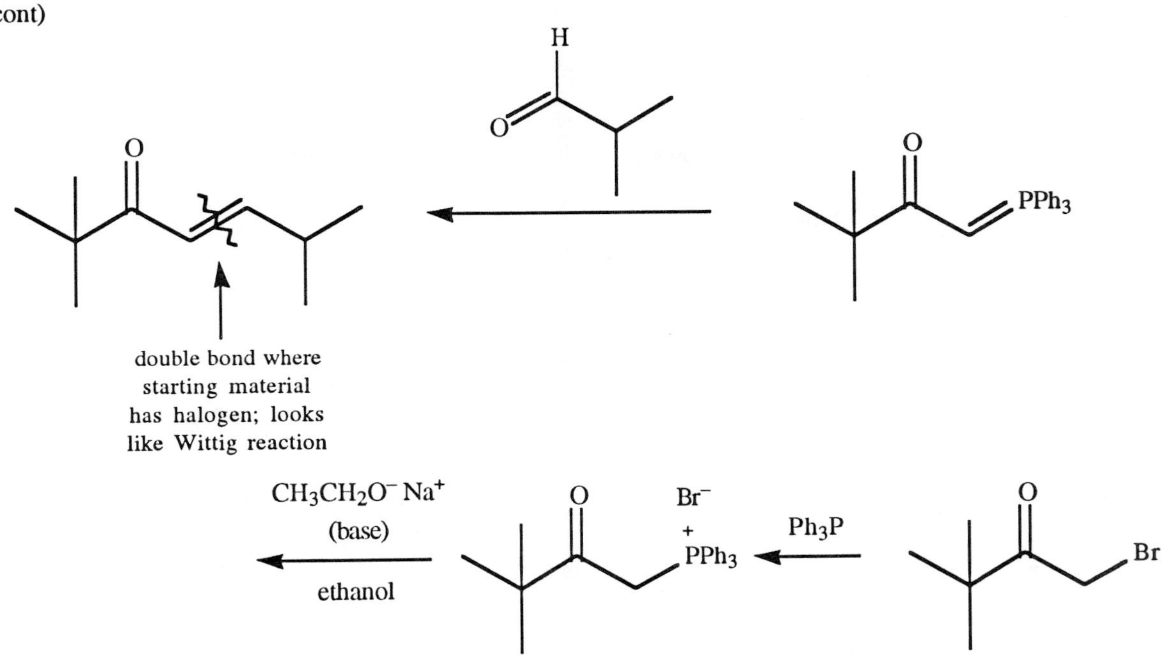
H
O
O
PPh3
O
double bond where
starting material
has halogen; looks
like Wittig reaction
CH3CH2O− Na+
(base)
ethanol
O
Br−
+
PPh3
Ph3P
O
Br

(d)

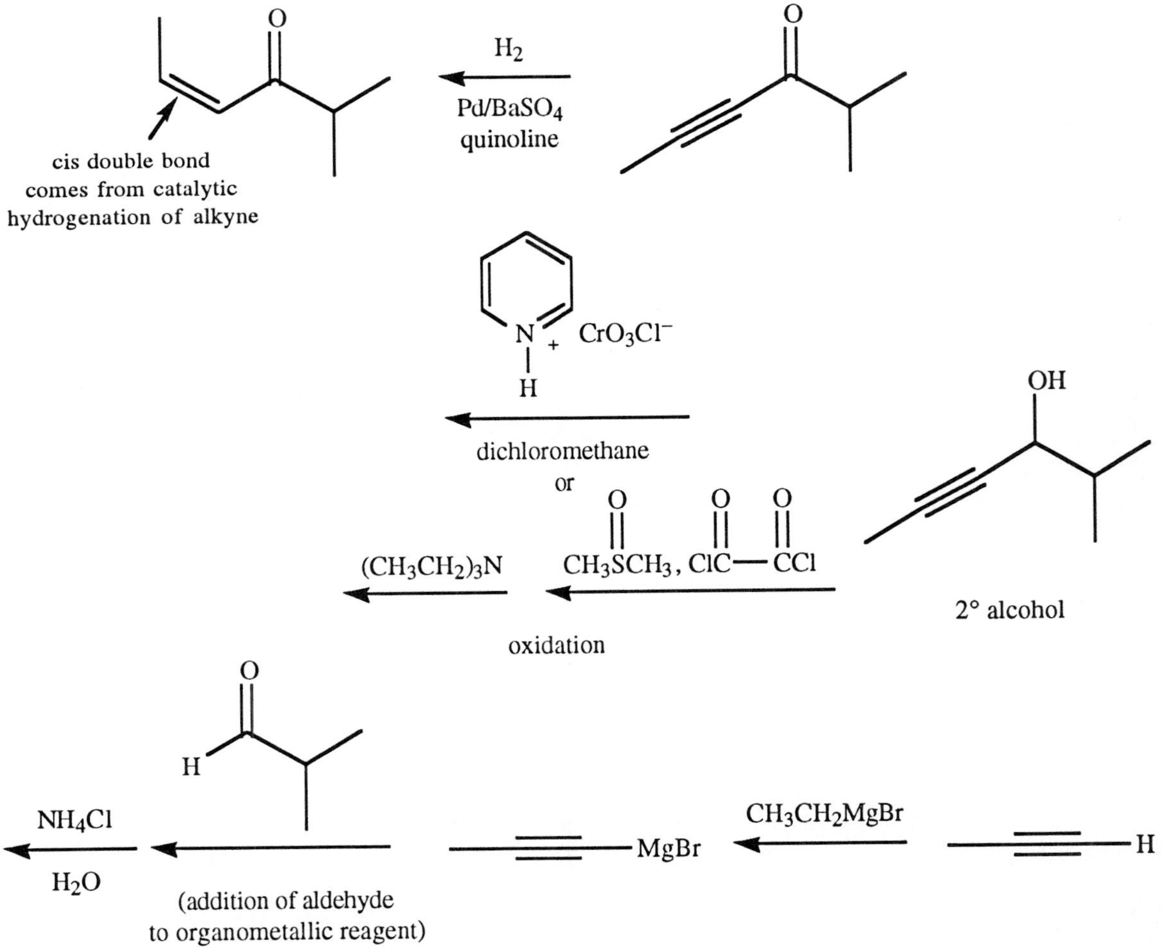
O
H2
Pd/BaSO4
quinoline
O
cis double bond
comes from catalytic
hydrogenation of alkyne
N
+
H
CrO3Cl−
OH
dichloromethane
or
(CH3CH2)3N
CH3SCH3, ClC—CCl
oxidation
2° alcohol
O
H
NH4Cl
H2O
(addition of aldehyde
to organometallic reagent)
MgBr
CH3CH2MgBr
H

18.48

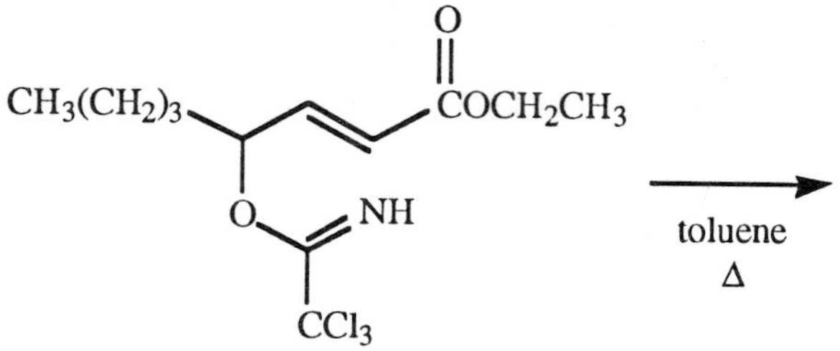

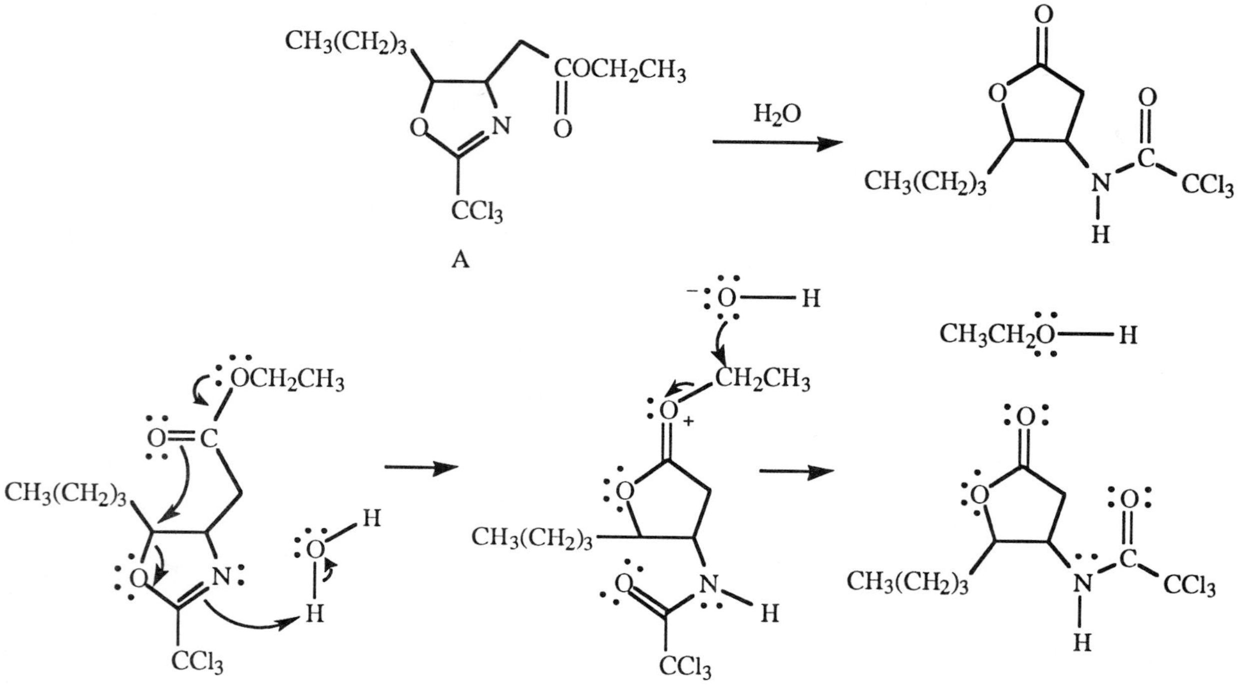

18.49   (a)

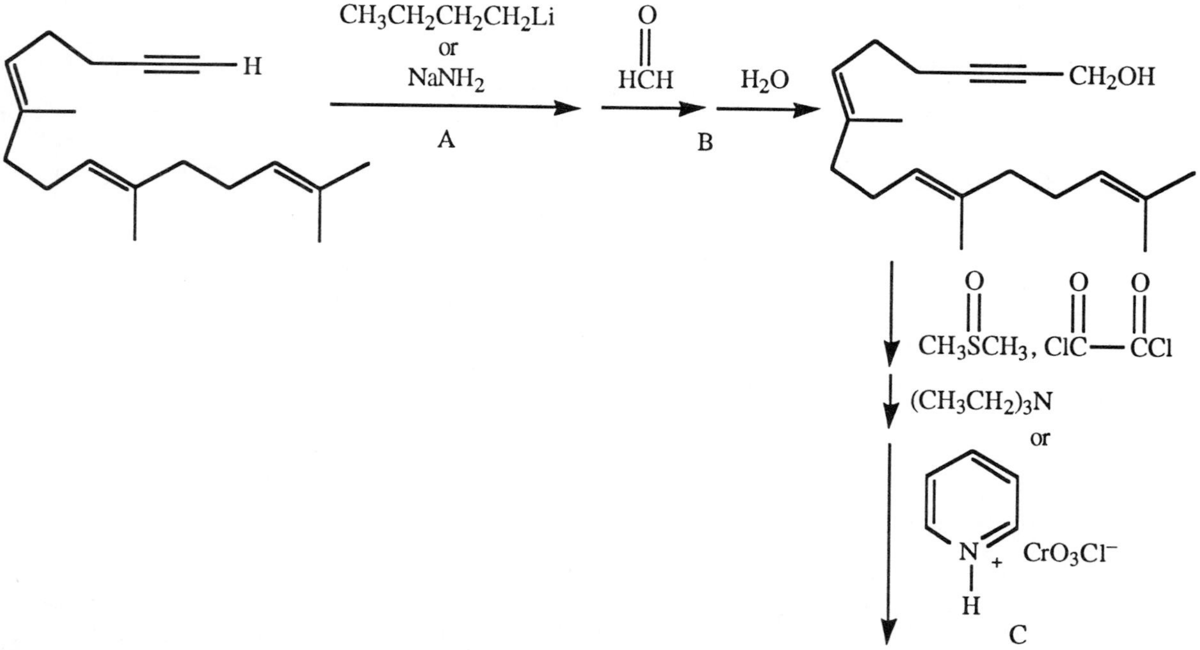

18.49 (a) (cont)

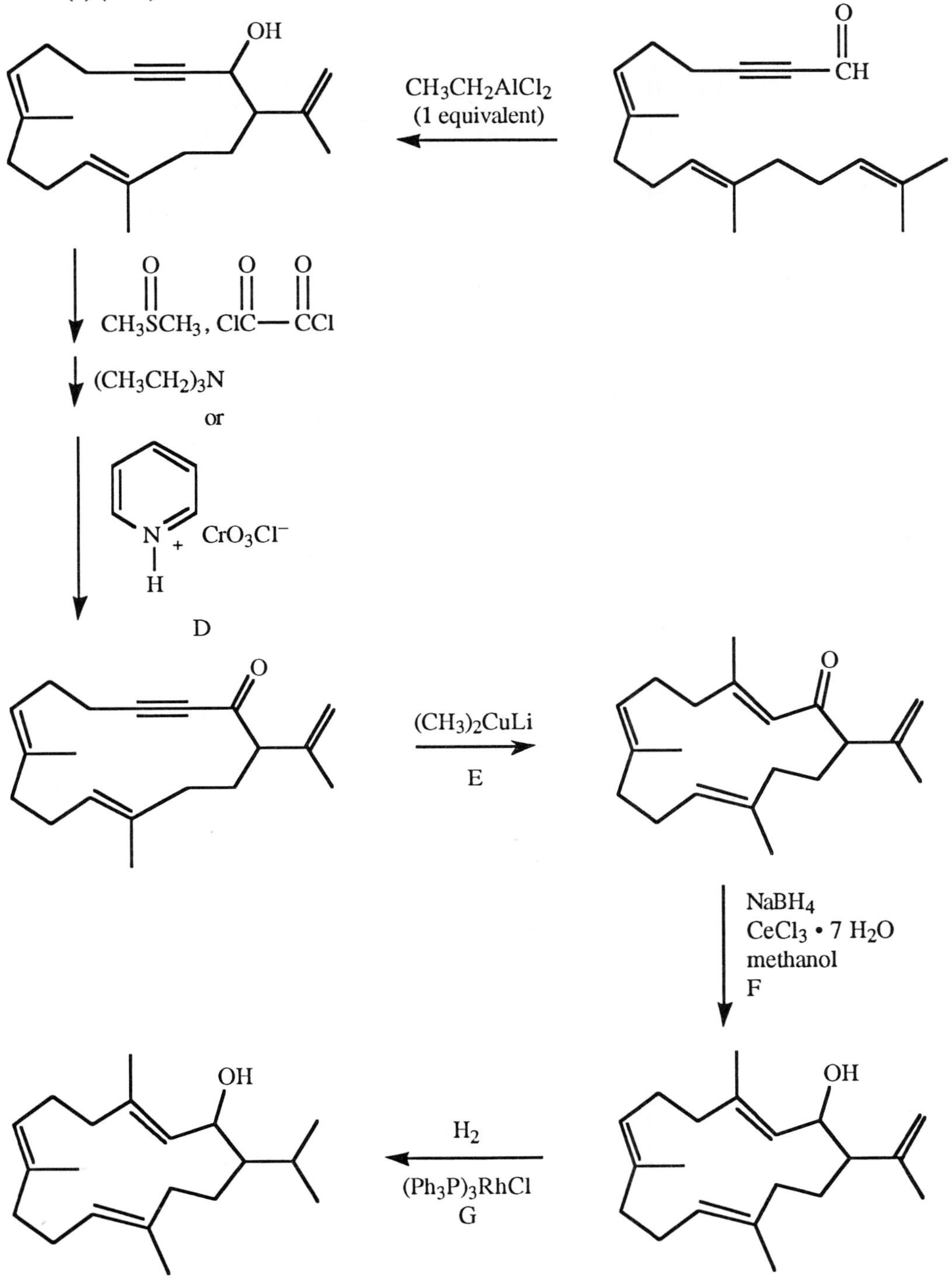
OH
O
CH
CH3CH2AlCl2
(1 equivalent)
O O O
CH3SCH3, ClC—CCl
(CH3CH2)3N
or
N +
H
CrO3Cl−
D
O
(CH3)2CuLi
E
O
NaBH4
CeCl3 • 7 H2O
methanol
F
OH
H2
(Ph3P)3RhCl
G
OH

Hydrogen with palladium or platinum as a catalyst is not used as the reagent in the last step of the synthesis because it is not selective.

18.49 (b)

open shell on aluminum,
therefore Lewis acid

connection made between C-1 and C-14

18.50

(a) 1.   Preparation of dianion from propanoic acid.

2.   Preparation of the Wittig reagent.

18.50 (a) (2) (cont)

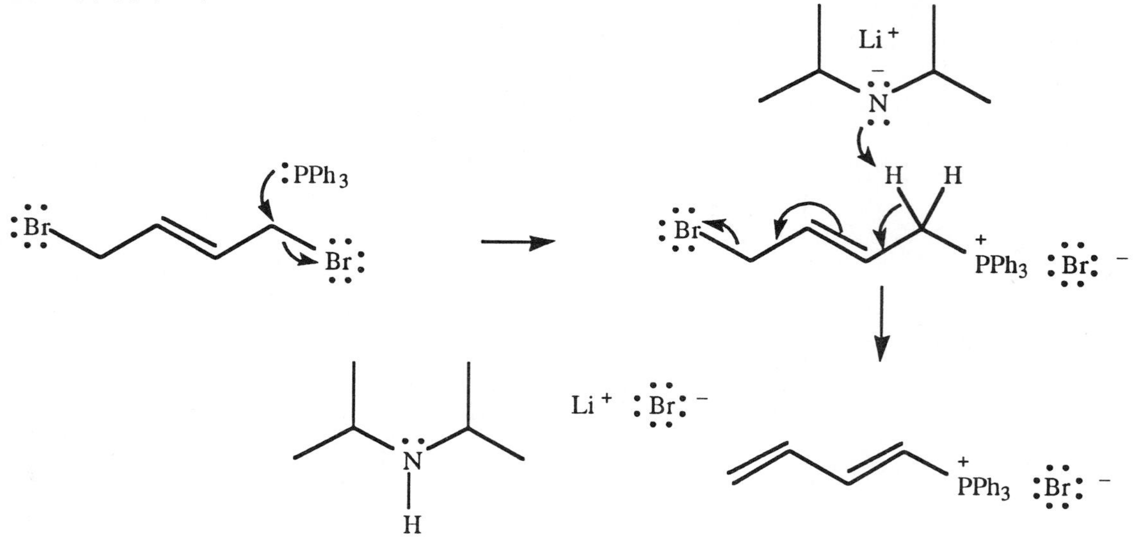

(b)

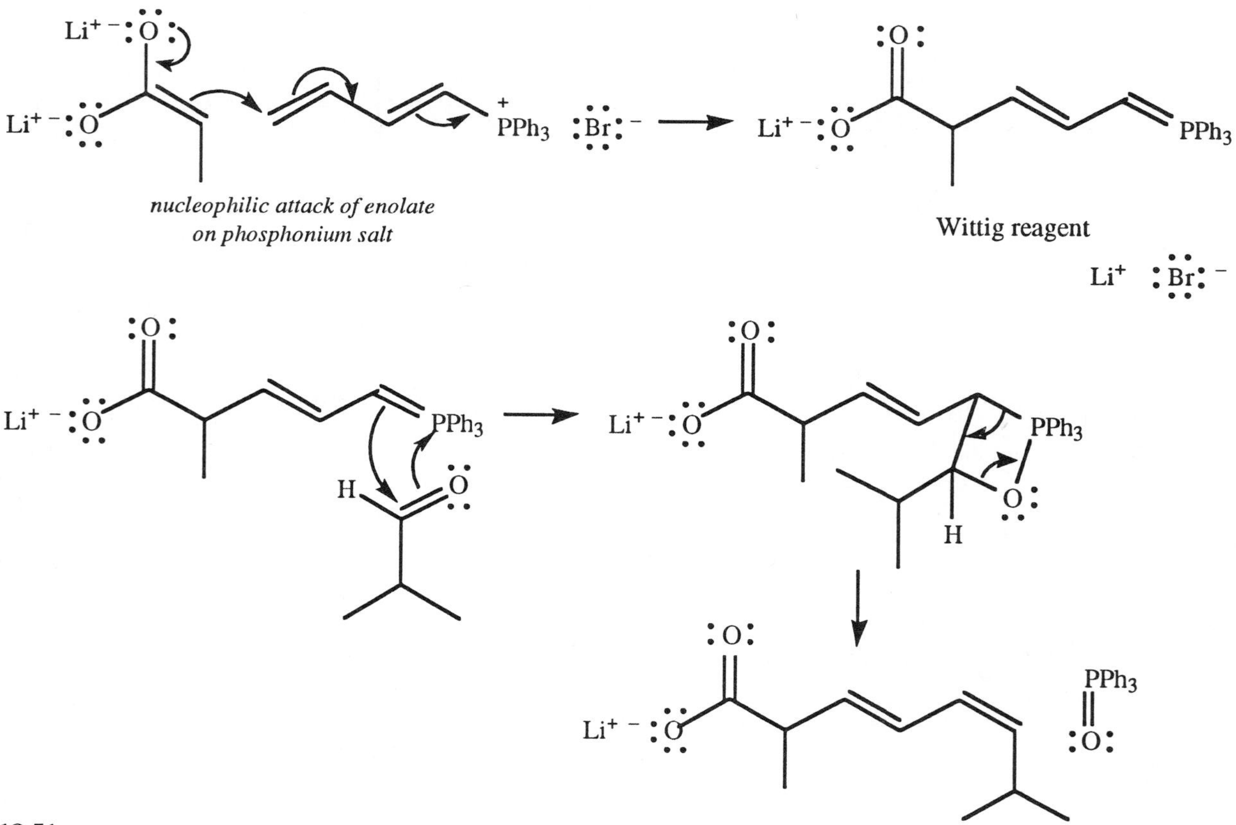

nucleophilic attack of enolate on phosphonium salt

Wittig reagent

18.51

(a)

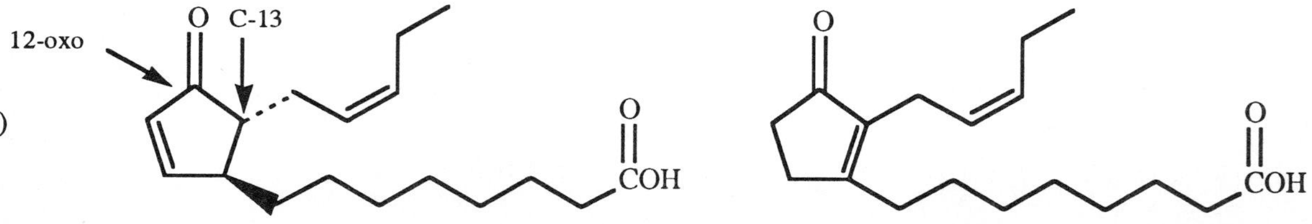

18.51 (cont)

(b)

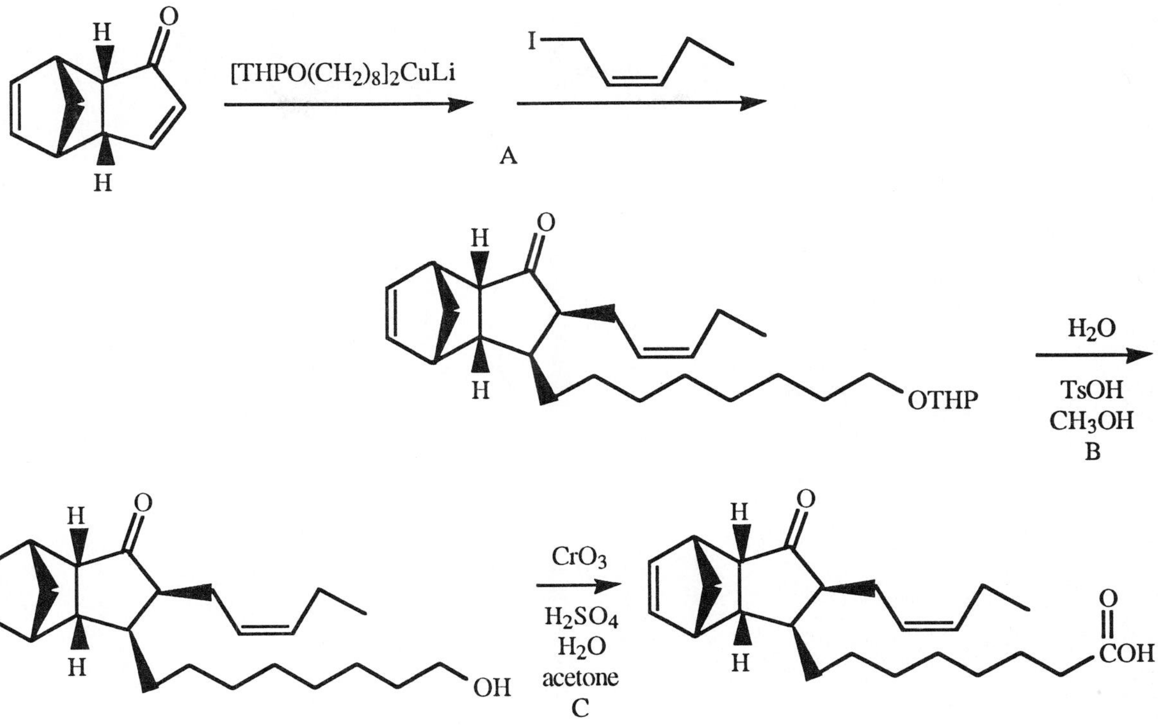

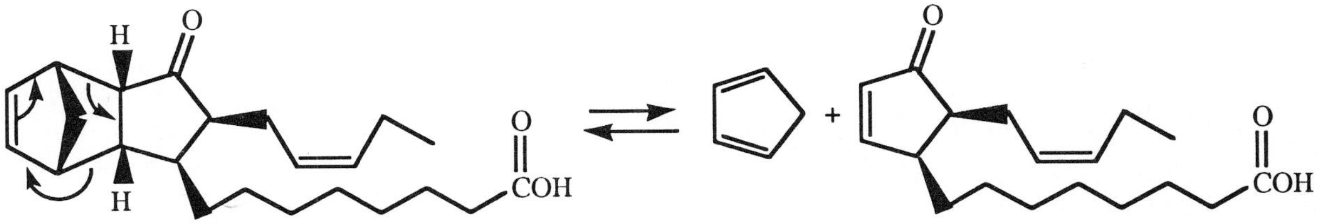

This is a reversible reaction, therefore it is necessary to remove the cyclopentadiene that forms in order to shift the equilibrium toward the desired product, which is the free 12-oxophytodienoic acid.

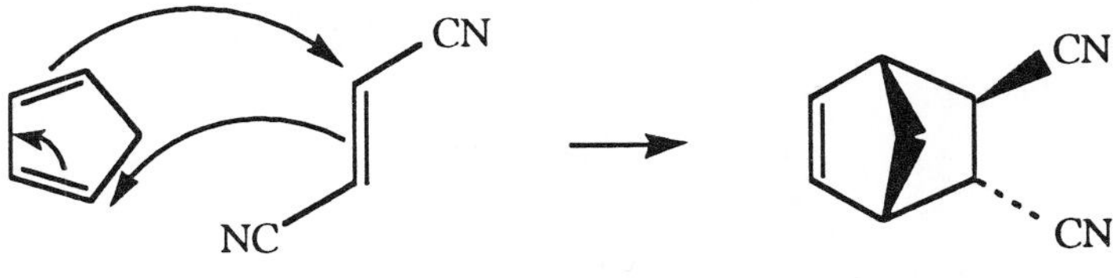

other product
of the reaction

18.52

(a) S_N2 and conjugate addition of an amine to the α,β-unsaturated ester.

18.52 (cont)

(b)

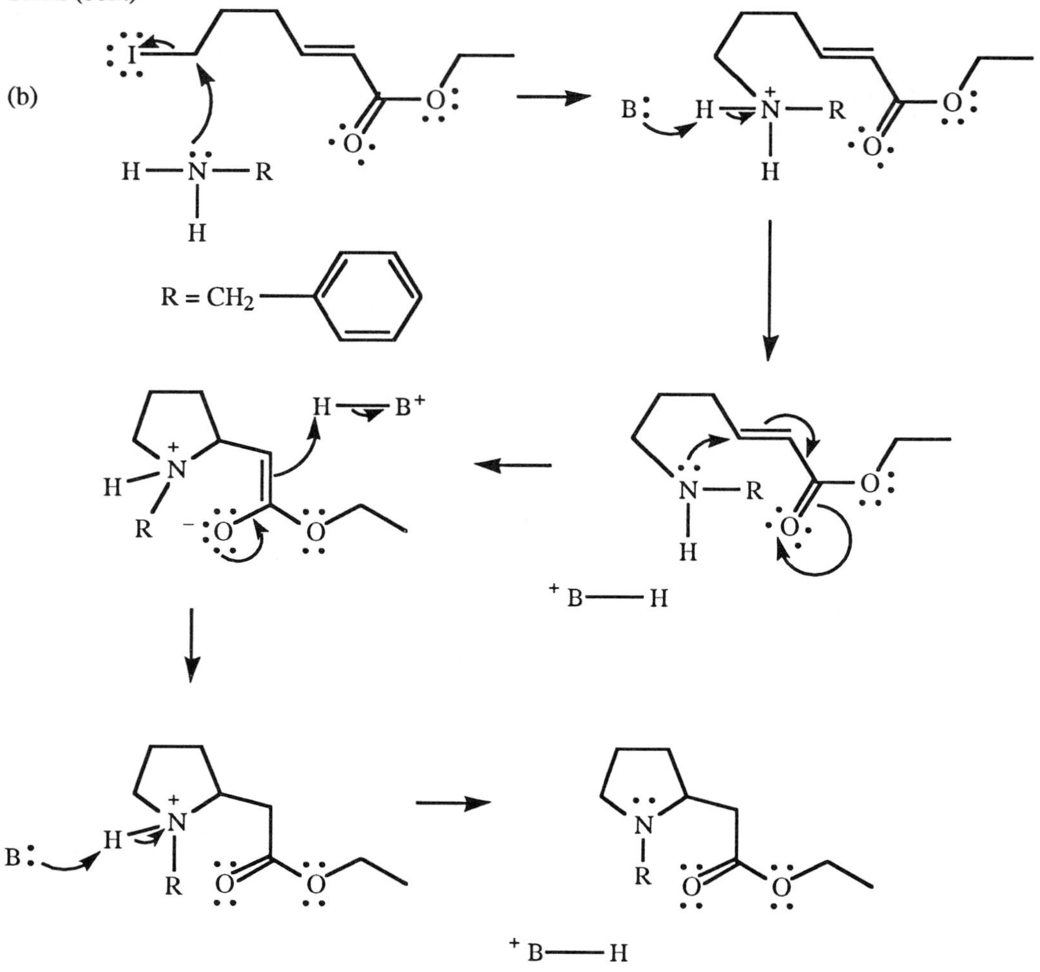

$R = CH_2C_6H_5$

18.53

δ 4.0 (2H, quartet, *J* 8 Hz)

δ 7.2 (1H, doublet, *J* 16 Hz)

δ 1.2 (3H, triplet, *J* 8 Hz)

δ 6.9 - 7.2 (5H, multiplet)

δ 6.0 (1H, doublet, *J* 16 Hz)

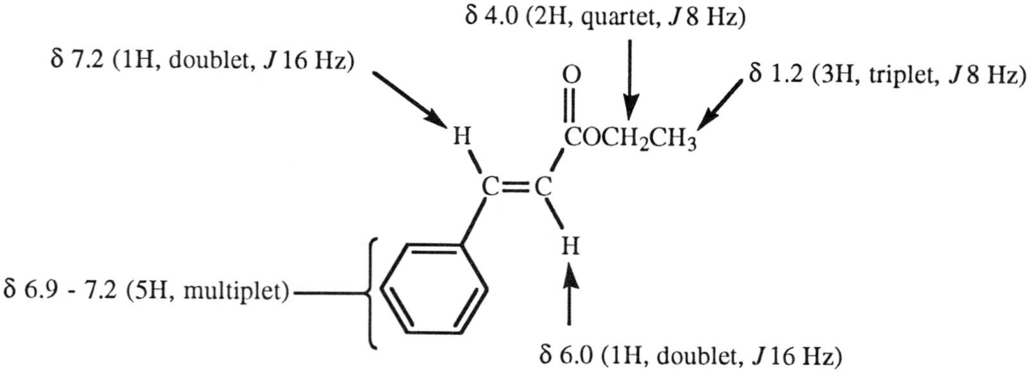

Each vinylic hydrogen atom is coupled only to the other vinylic hydrogen atom and each appears as a doublet with *J* ~ 16 Hz. (The inner peak of the doublet downfield of the aromatic peak is part of the multiplet for the aromatic hydrogens. It is 16 Hz upfield from the peak at d 7.4, which is the outer peak of the doublet.)

18.53 (cont)

position of unsplit vinylic hydrogen

←— 16 Hz —→

The normal range for the chemical shift of vinylic hydrogen atoms is about d 5–6. Both vinylic hydrogen atoms in ethyl (*E*)-3-phenyl propenoate are shifted farther downfield, indicating that they are being deshielded by something other than the carbon-carbon double bond. The higher field (d 6.0) vinylic hydrogen atom is deshielded by the carbonyl group of the ester function. The lower field vinylic hydrogen atom (d 7.2) is even more deshielded because it is b to the carbonyl group and adjacent to the phenyl ring.

The chemical shift of the hydrogen atom on the β carbon atom reflects the lower electron density at this carbon, symbolized by the positive charge shown in the resonance contributor above.

18.54

δ 18.0 —→ CH$_3$ H ←— δ 122.6

δ 147.5 —→ C=C

H   COH ←— δ 172.3   ‖   O

Chemical shifts in C-13 magnetic resonance spectra have essentially the same kind of dependence on electron density as proton chemical shifts. A carbon atom attached to an electron-withdrawing group or atom will be shifted downfield (deshielded) relative to a carbon atom attached to an electron-donating group or atom. The carbon atom on the carboxylic acid group, attached to two oxygens, absorbs farthest downfield and the methyl group farthest upfield. Of the two carbon atoms of the double bond, carbon 3 absorbs farther downfield than carbon 2 because it is *b* to the carbonyl group.

19

The Chemistry of Aromatic Compounds I. Electrophilic Aromatic Substitution

Workbook Exercises

Learning to represent resonance interactions between various functional groups attached to benzene and other benzene-like rings will help you to understand the topics in Chapters 19 and 23. There are two types of resonance interactions with a benzene ring: electron-donating or electron-withdrawing. By examining the nature of the atom(s) attached directly to the benzene ring, you can conclude whether the group is electron-donating or electron-withdrawing when resonance contributors are to be drawn.

(a)   The resonance interaction is electron-donating when the atom attached to the ring has a pair of nonbonding electrons. For example:

(b)   The resonance interaction is electron-withdrawing when the atom attached to the ring is involved with a multiple bond to an electronegative atom such as oxygen or nitrogen. For example:

EXERCISE I.   Draw resonance contributors that show the interaction between the benzene (or benzene-like) ring and the attached group in each of the following examples.

(a) C_6H_5—NHCCH$_3$ (with C=O)

(b)   naphthalene with OH

(c)   naphthalene with OH

(d) C_6H_5—C≡N

(e) C_6H_5—N(O)—O$^-$

(f) C_6H_5—N=O

Workbook Exercises (cont)

EXERCISE II.   Which of the following ions will be resonance-stabilized by one of the groups attached to the benzene ring?

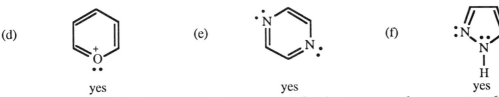

(a)   (b)   (c)   (d)

19.1   The lone pair electrons that are counted in determining aromaticity are those that are needed to generate the conjugated cyclic π system and to bring the total number of π electrons to $4n + 2$.

(a)

yes
planar π system, 6 π electrons,
including one lone pair on oxygen

(b)

no
π system not conjugated

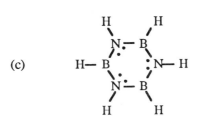

(c)

yes
planar π system, 6 π electrons;
all three lone pairs on nitrogen are needed

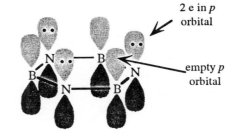

2 e in p orbital

empty p orbital

6 p orbitals; 3 contain 2 electrons,
3 have room to accept electrons

(d)   (e)   (f)

yes
planar π system, 6 π electrons;
the lone pair on oxygen
is not needed

yes
planar π system, 6 π electrons;
the lone pairs on the nitrogens
are not needed

yes
planar π system, 6 π electrons,
including the lone pair on the
nitrogen bonded to the hydrogen

19.2   The hydrogen atoms on the sp^2-hybridized carbon atoms in thiophene absorb at a much lower field than those bound to the sp^2-hybridized carbon atoms in methyl vinyl sulfide. The hydrogen atoms in thiophene are thus much more deshielded than the vinyl hydrogens in methyl vinyl sulfide. The chemical shifts of the thiophene hydrogen atoms fall in the region where hydrogen atoms on aromatic rings absorb, pointing to the presence of a ring current, which is one of the criteria for aromaticity.

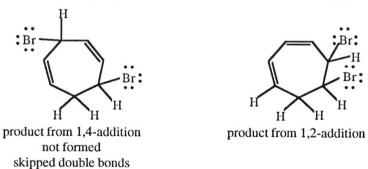

thiophene

CH_3—$\ddot{S}$—$CH{=}CH_2$

methyl vinyl sulfide

$(4n + 2)\,\pi$ electrons
with $n = 1$, therefore
an aromatic system with
delocalization of electrons
over the ring

19.3

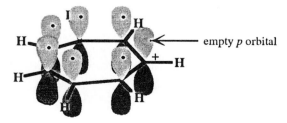

19.4   The dibromide in Problem 19.3 is the 1,6-addition product. Two other possibilities are 1,2- and 1,4-addition products. Either one can lose hydrogen bromide and then ionize to form tropylium bromide.

product from 1,4-addition
not formed
skipped double bonds

product from 1,2-addition

19.5

empty p orbital

σ bonds between the carbon atoms and
between the carbon and hydrogen atoms

Concept Map 19.1 Aromaticity.

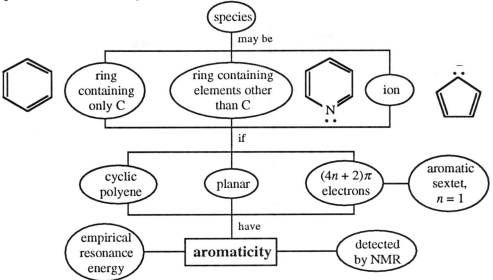

19.6

(a) 1,3-dinitrobenzene
m-dinitrobenzene

(b) 2-methylnaphthalene
β-methylnaphthalene

(c) 3-ethylnitrobenzene
m-ethylnitrobenzene

(d) 4-chlorophenol
p-chlorophenol

(e) 2,4-dichlorotoluene

(f) 4-*tert*-butylnitrobenzene
p-tert-butylnitrobenzene

19.7

(a)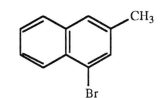

(b)

(c)

(d)

(e)

(f) Br— —Br

(g)

(h)

(i)

19.8

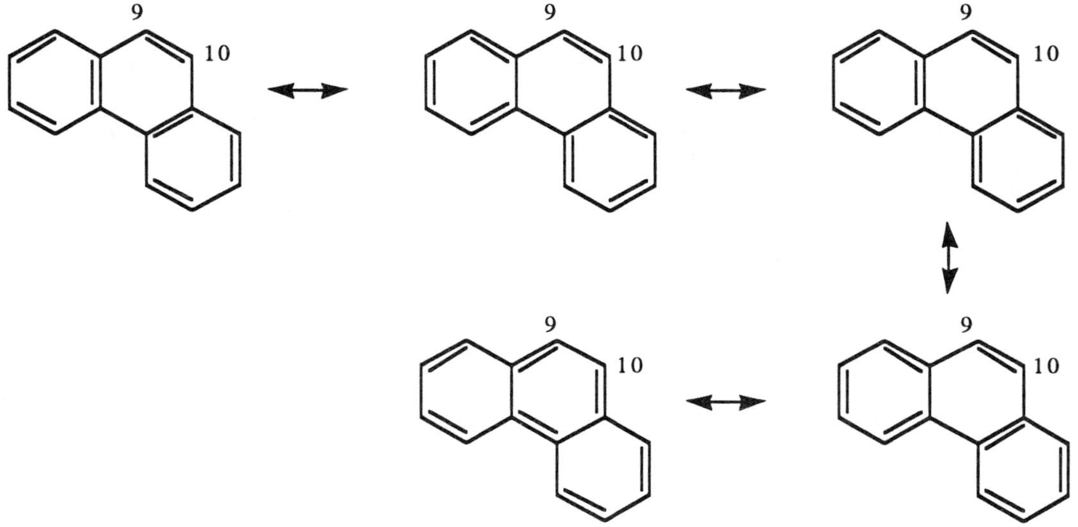

Four of the five resonance contributors of phenanthrene have a double bond at the 9,10 position. That bond, therefore, has more localized double bond character that the other bonds in the molecule.

19.9

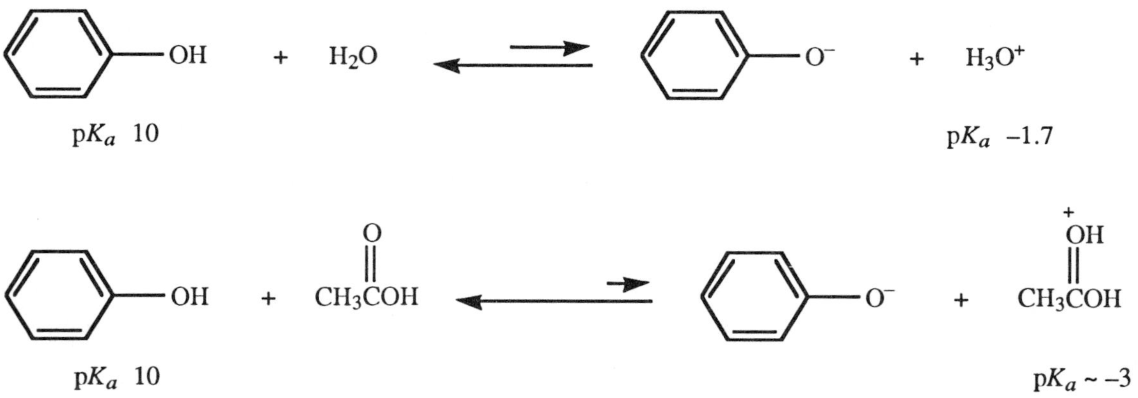

Phenol is much less likely to transfer a proton to acetic acid than it is to water. The water solution will contain a higher concentration of phenolate ion than will the acetic acid solution.

19.10

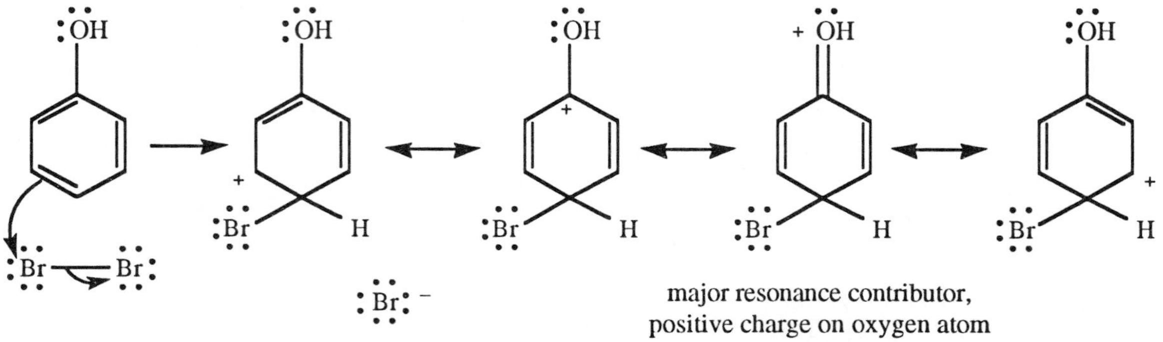

major resonance contributor,
positive charge on oxygen atom

19.10 (cont)

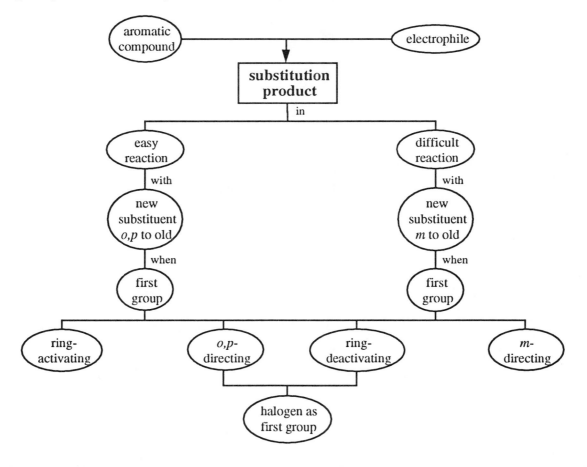

major resonance contributor;
no charge;
more stable than similar
contributor from phenol

The reactive intermediate from the phenolate ion is better stabilized than the one from phenol. The phenolate ion gives rise to a reactive intermediate which has no charge in its major resonance contributor, and hence is more stable than the reactive intermediate from phenol. A more stable reactive intermediate corresponds to a lower transition state and, therefore, to a faster reaction.

- -

Concept Map 19.2   Electrophilic aromatic substitution.

19.11   (a)

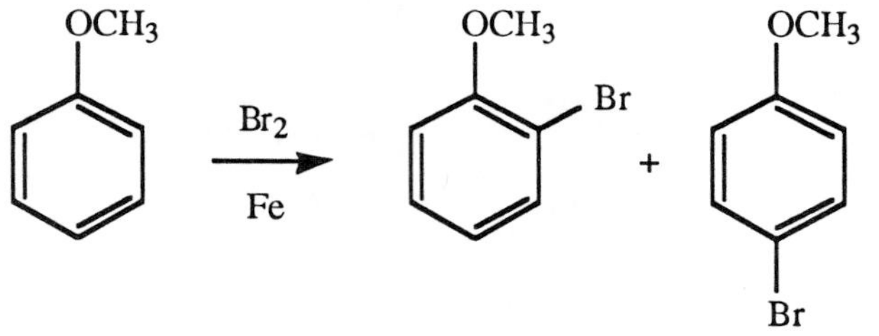

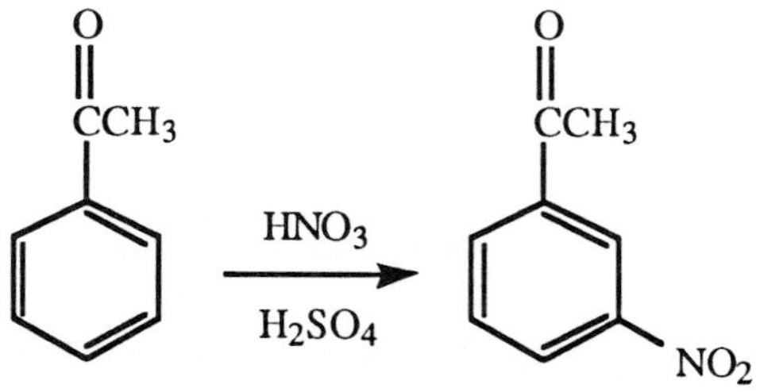

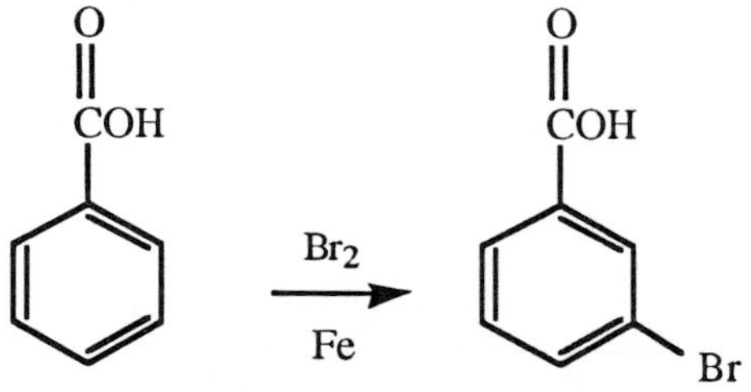

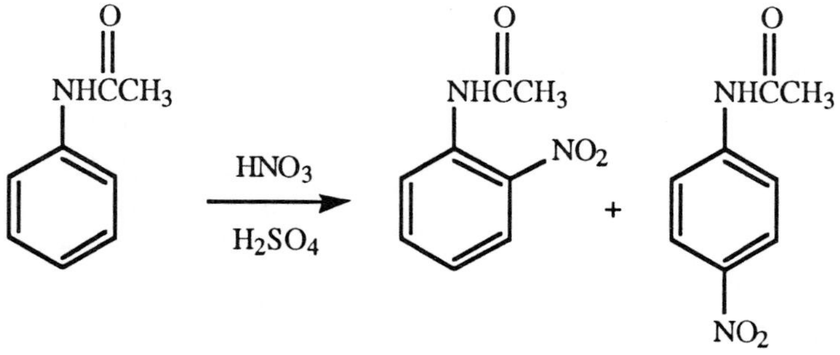

(b)

(c)

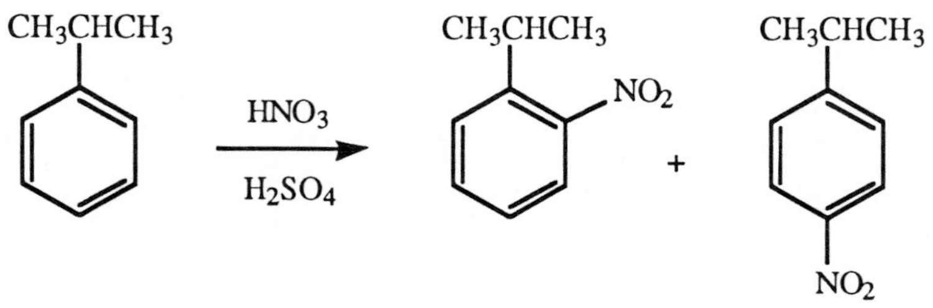

(d)

(e)

Concept Map 19.3   Essential steps of an electrophilic aromatic substitution.

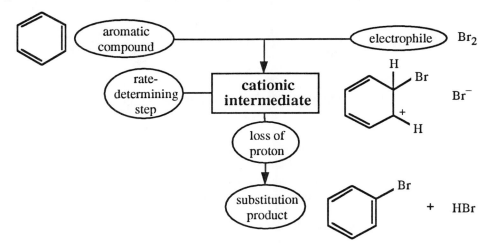

Concept Map 19.4   Reactivity and orientation in electrophilic aromatic substitution.

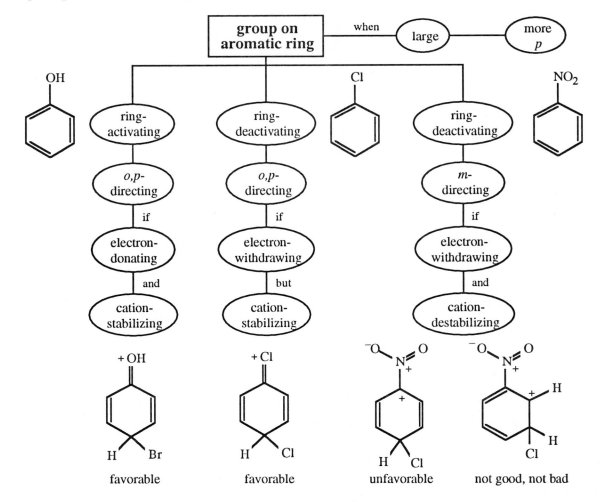

19.12   If electrophilic substitution occurs at the para position of the ring, the intermediate will have the following resonance contributors:

In this resonance contributor
the positive charge on the ring
is next to the positively charged
nitrogen atom; unfavorable

A similar set of resonance contributors can be drawn for ortho substitution. If electrophilic substitution occurs at the meta position, the positive charge on the ring is separated from the positive charge on the nitrogen atom by at least one carbon atom in all resonance contributors.

The intermediate formed on electrophilic attack at the meta position is, therefore, of lower energy than the intermediate formed when reaction occurs at the ortho or para positions.

19.13   (a)   electron-withdrawing substituent; meta directing

(b) $(CH_3)_3P$——   positive charge on atom directly bonded to ring; meta directing

(c)   electron-withdrawing substituent (positively charged nitrogen atom closest to ring); meta directing

(d) $(CH_3)_2\overset{+}{S}$——   positive charge on atom directly bonded to ring; meta directing

(e)   will stabilize a positive charge by resonance; ortho,para directing

(f) $CH_3CH{=}CH{-}$ will stabilize a positive charge by resonance; ortho,para directing

19.14

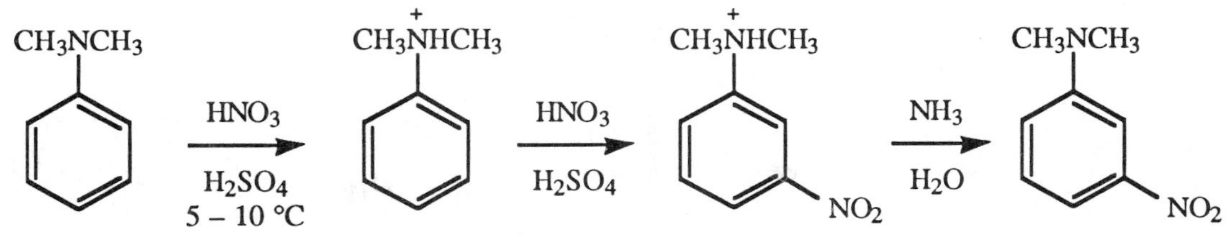

Amines are good Brønsted bases. Strong acid converts the amine into a substituted ammonium ion, which has a positively charged nitrogen atom directly attached to the ring, and is therefore meta directing. The last step of the reaction is the deprotonation of the aromatic ammonium ion by ammonia to generate the free amine.

19.15

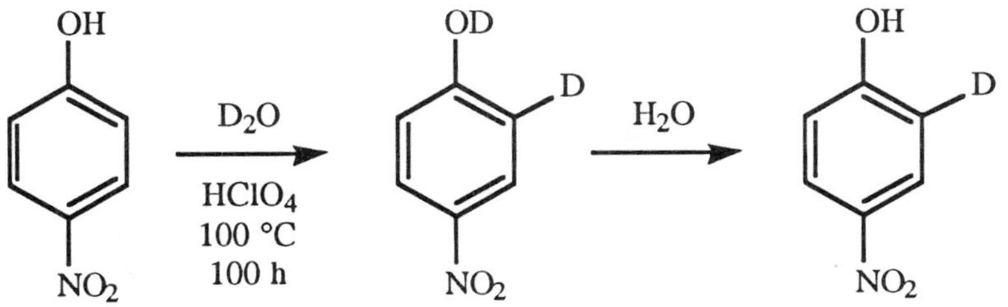

The hydroxyl group hydrogen bonds to water and, therefore, undergoes rapid hydrogen-deuterium exchange. Exchange of the hydrogen atoms on the aromatic ring has a high energy of activation and requires strong acid catalysis because, for exchange to occur, the aromaticity of the ring must be disrupted by the formation of a cationic intermediate. In the absence of strong acid, such exchange is extremely slow.

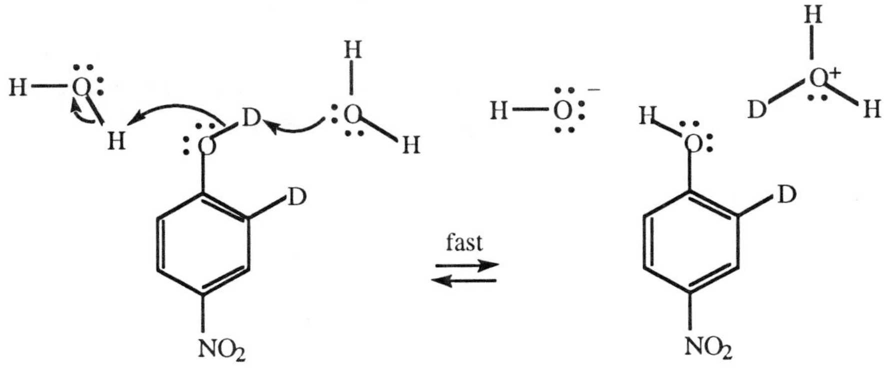

19.16

(a)

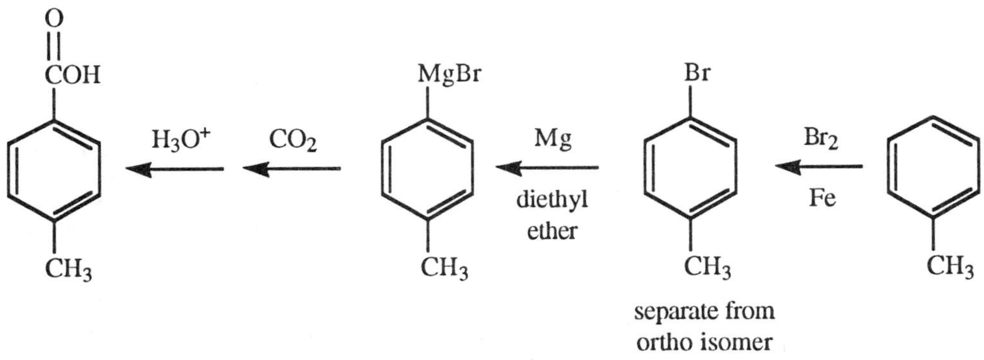

separate from
ortho isomer

19.16 (cont)

(b)

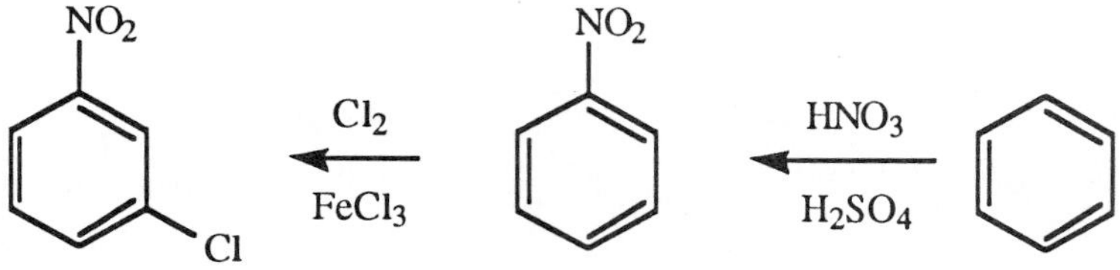

(Note: reversing the order in which these reagents are used would give ortho- and para-substituted products.)

(c)

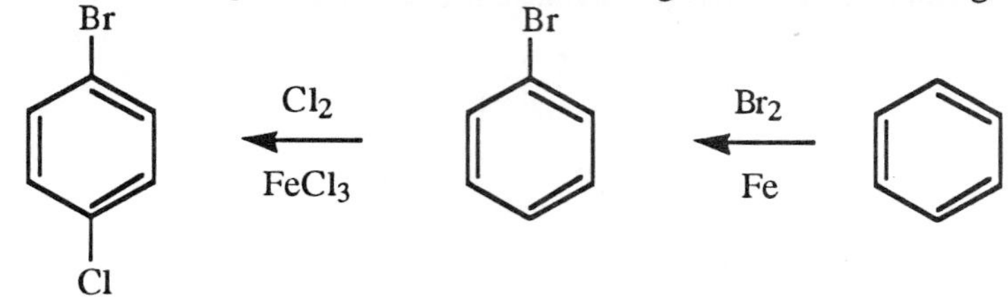

separate from
ortho isomer

(d)

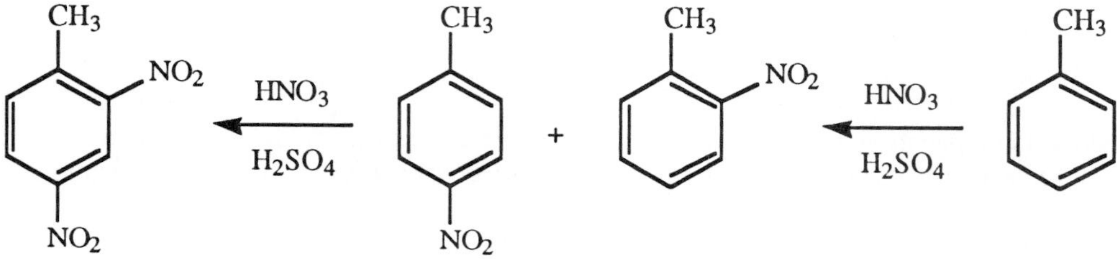

(e)

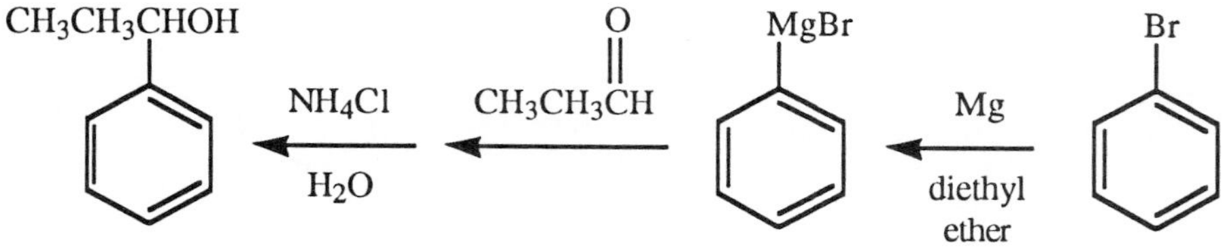

[from (c)]

(f)

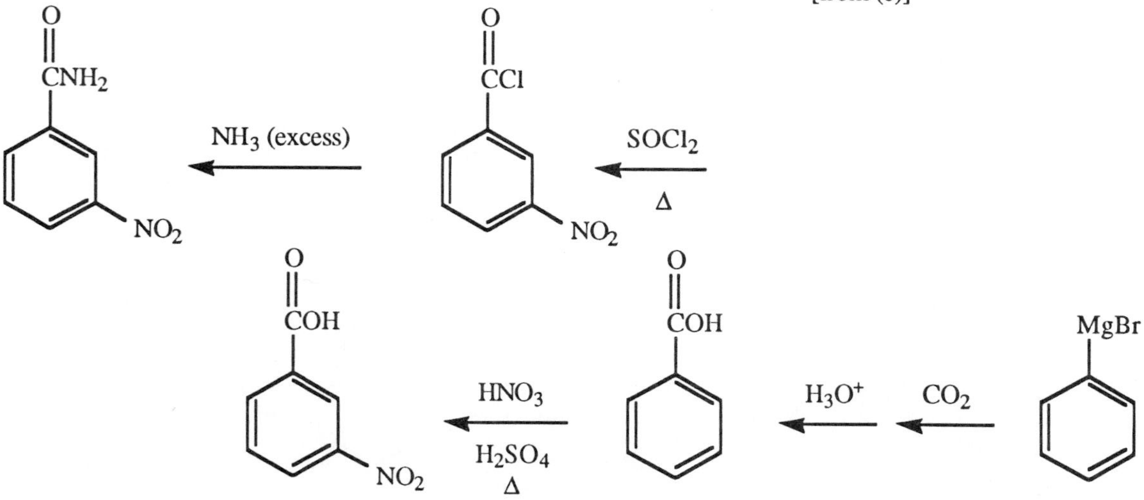

[from (e)]

19.17

(a)

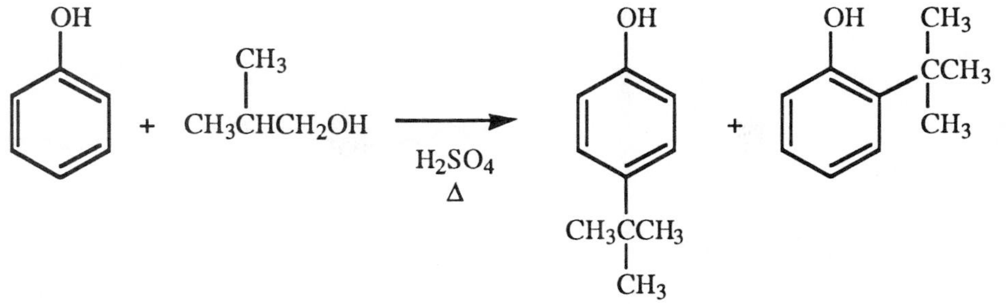

(b)

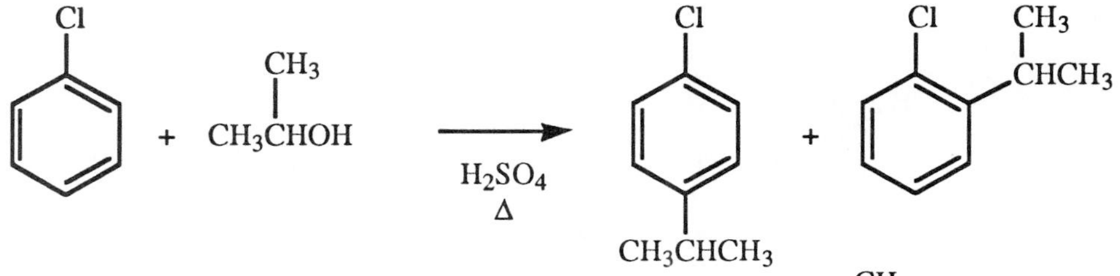

(c)

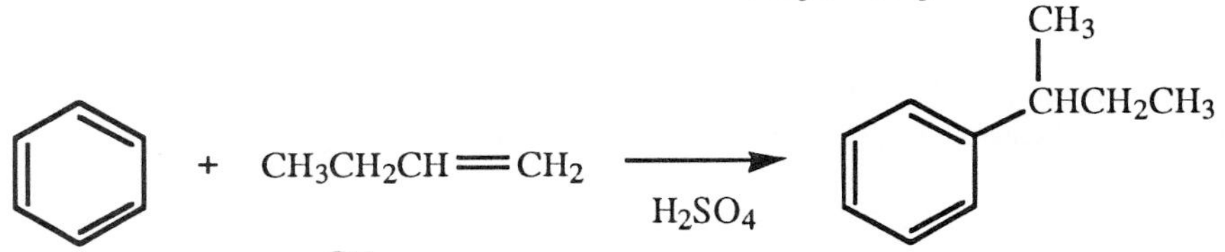

(d)

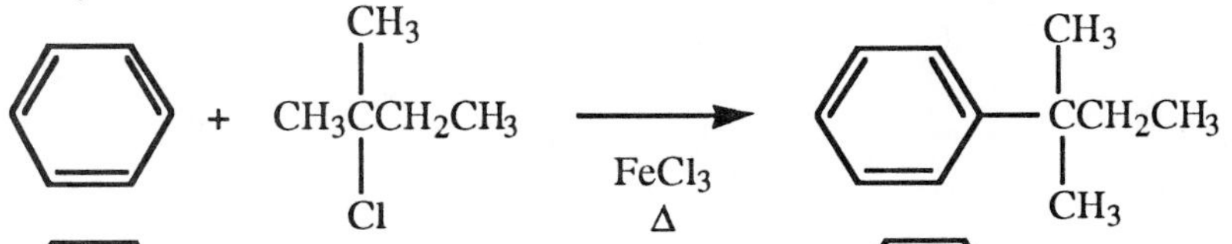

(e)

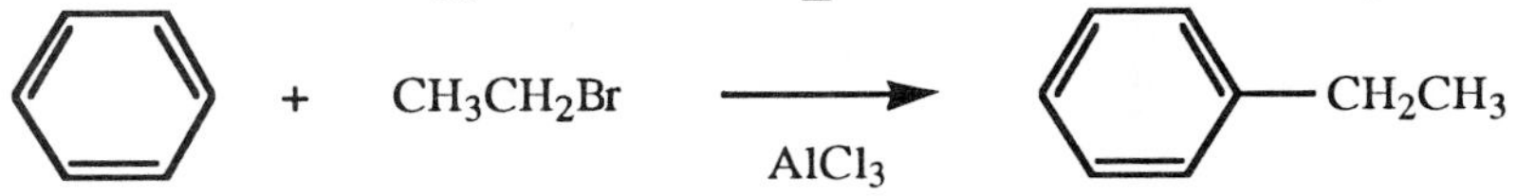

19.18

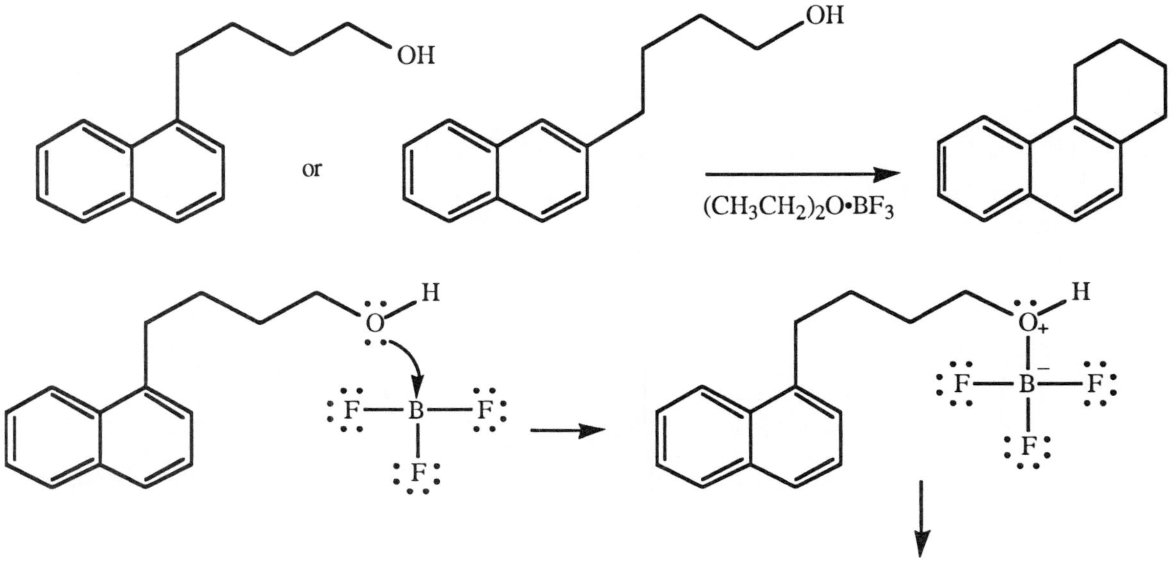

19.18 (cont)

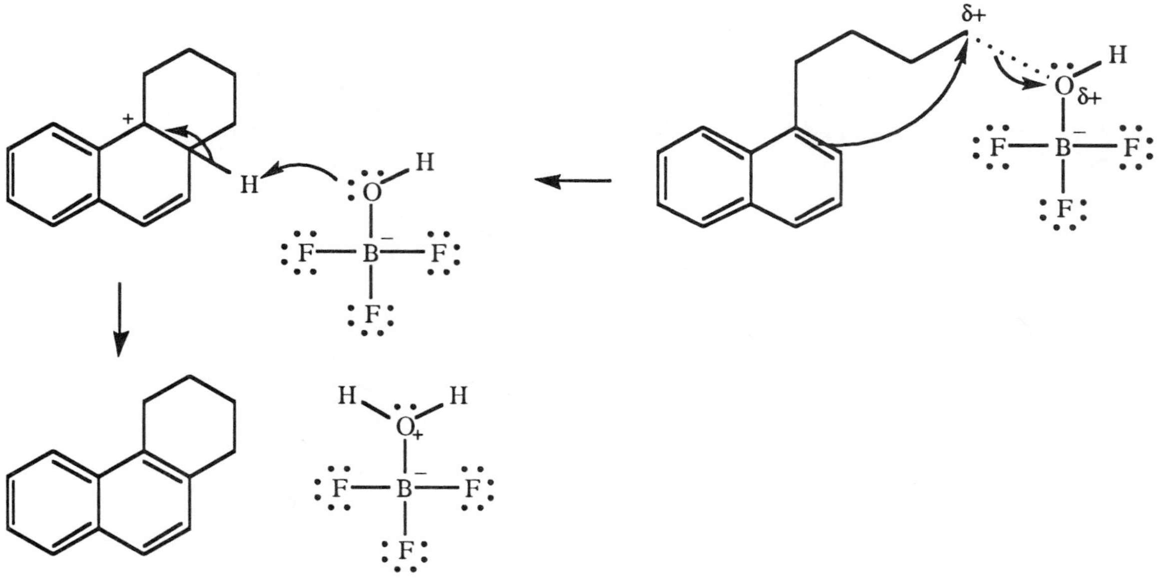

The same process would take place with the other alcohol except that the point of attack of the chain would be different. Attack would be at carbon 1 of the ring instead of carbon 2.

19.19 $C_{60}Cl_n$ + n $\xrightarrow{AlCl_3}$ C_{60}─()$_n$ + n HCl

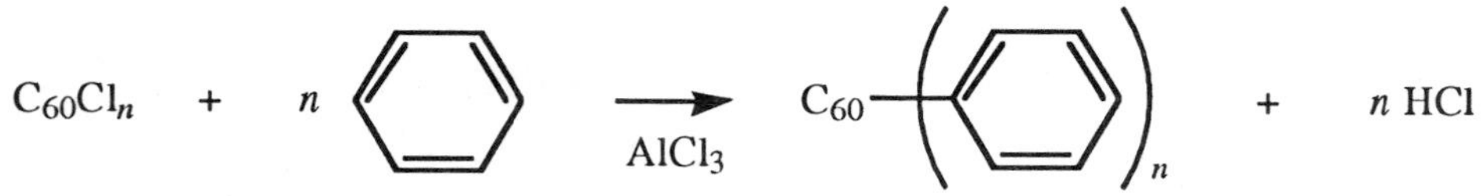

A Friedel-Crafts
reaction takes place

has aromatic hydrogens,
therefore band at δ 7.2
in the NMR

19.20

(a)

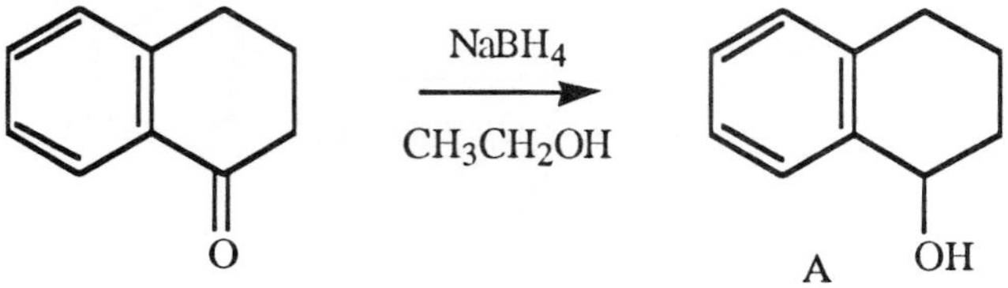

(b)

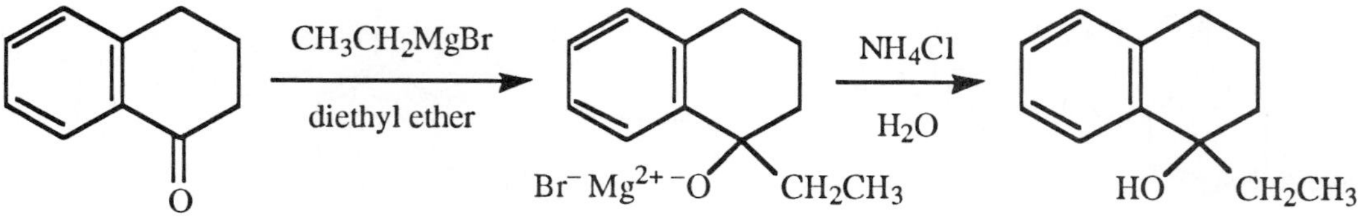

19.20 (b) (cont)

(c)

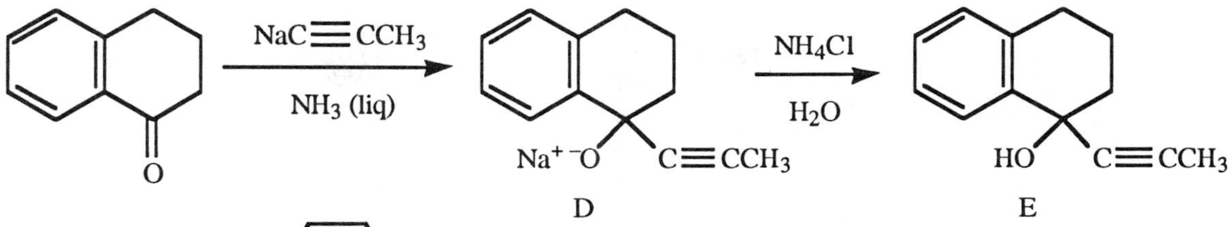

(d)

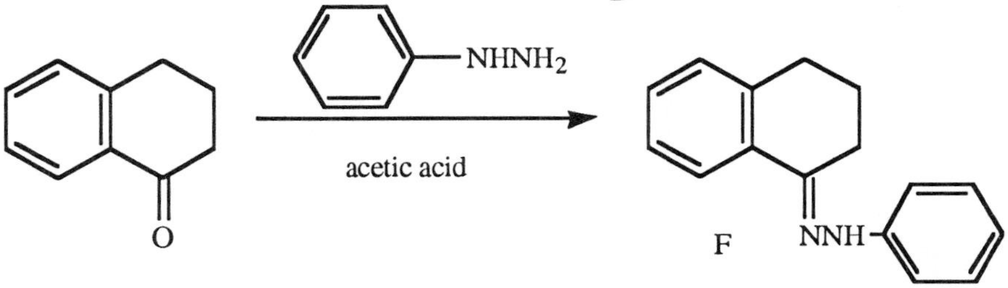

(e)

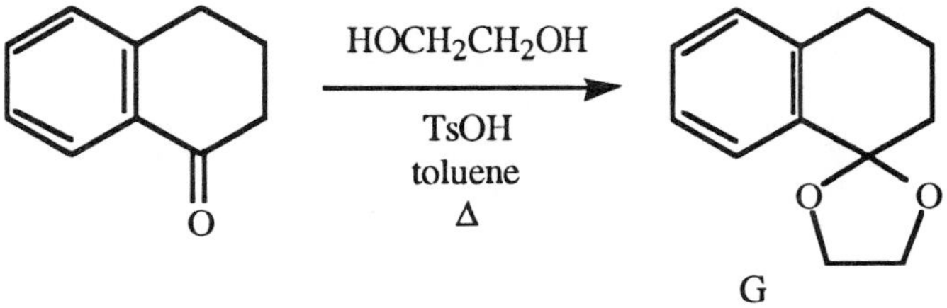

(f)

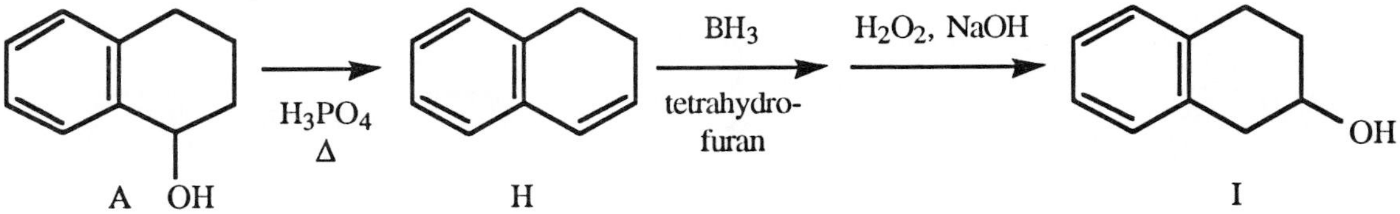

(g)

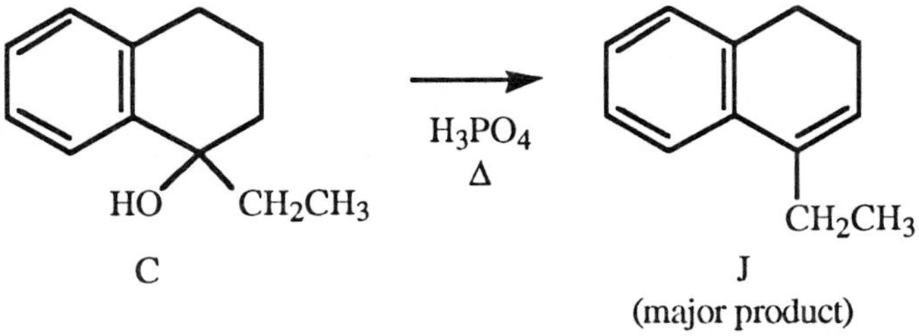

(h)

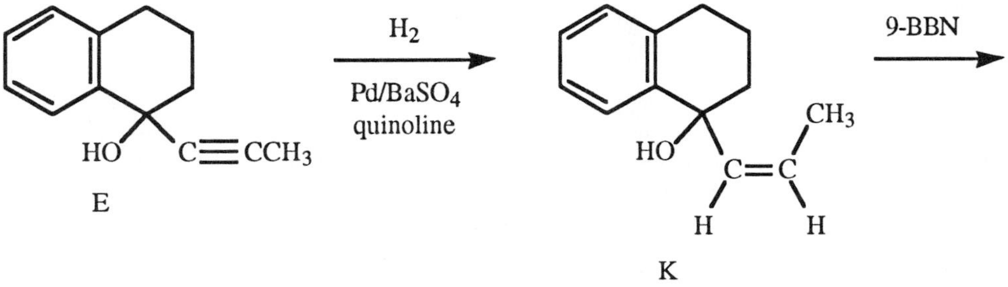

19.20 (h) (cont)

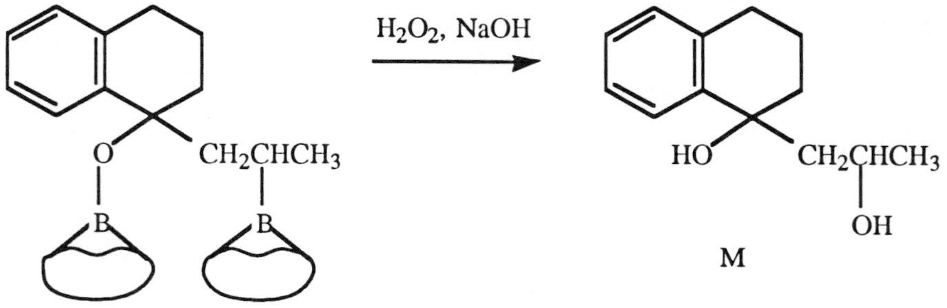

19.21

(a)

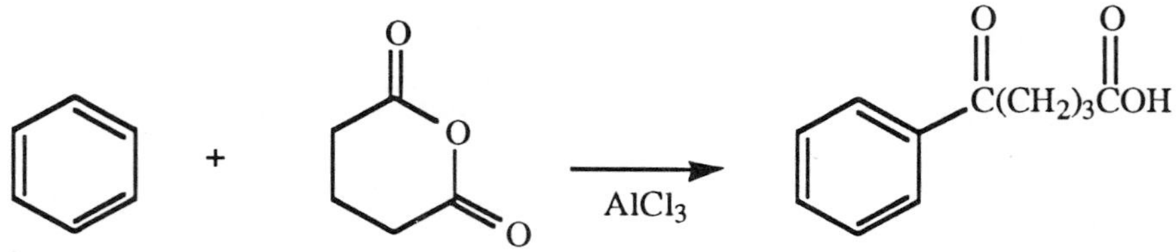

(b)

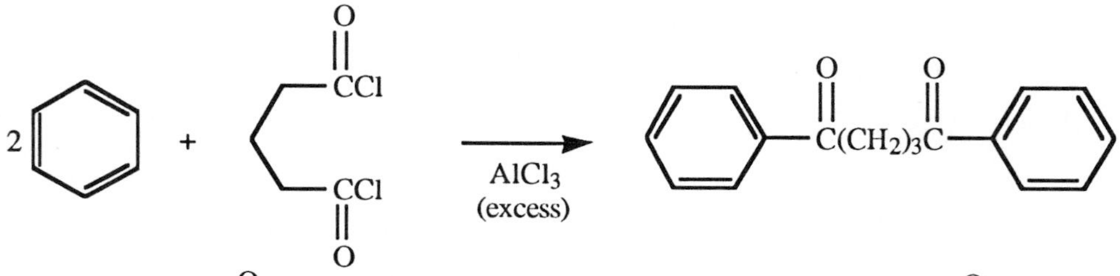

(c)

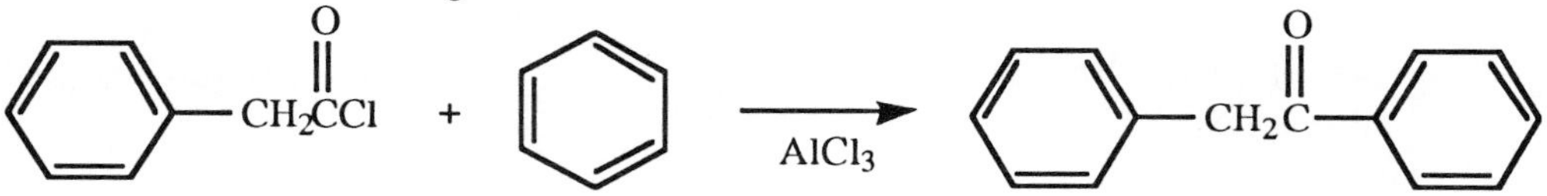

(d)

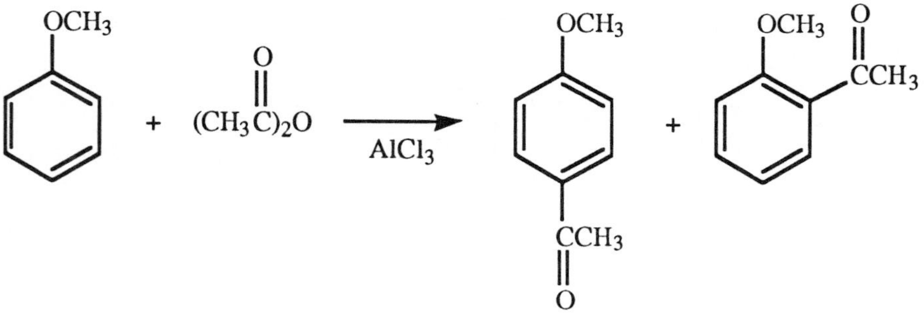

19.22

(a)

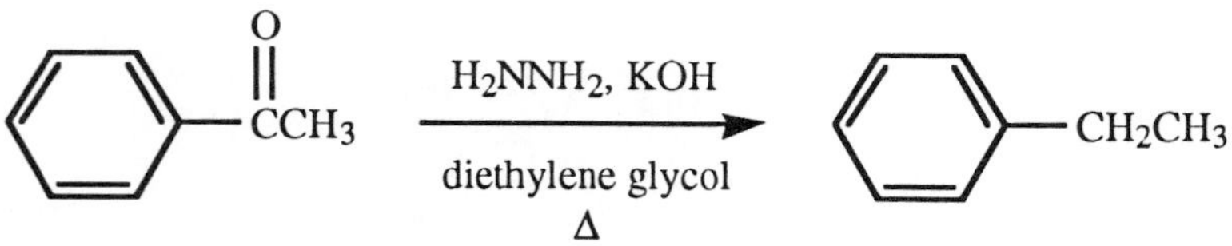

(b)

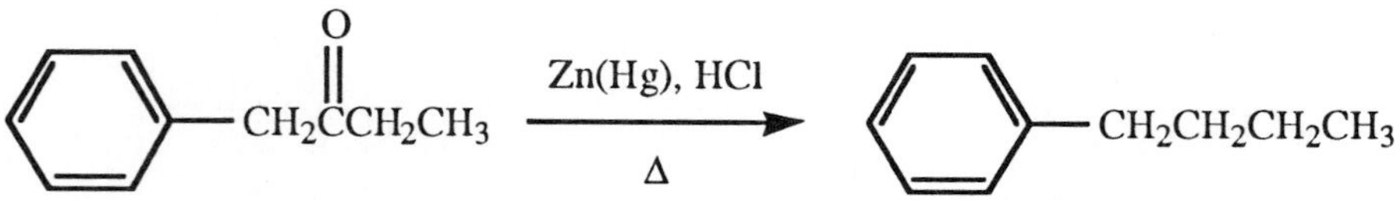

19.22 (cont)

(c)

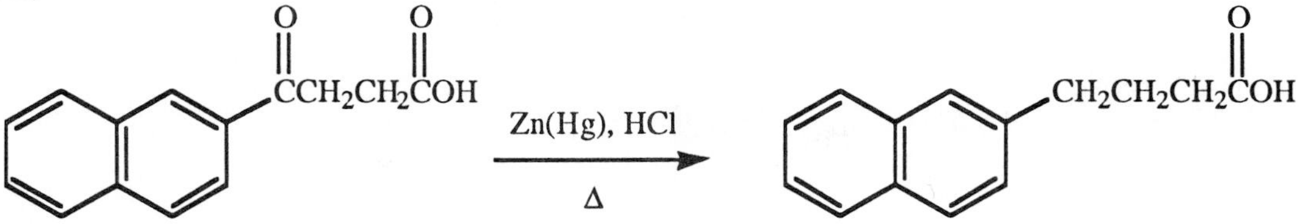

(d)

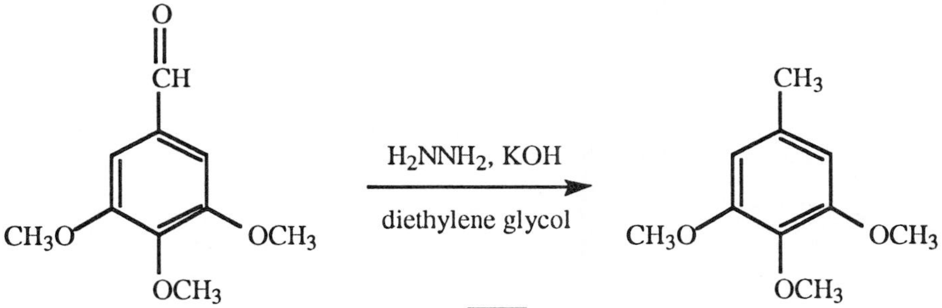

(e)

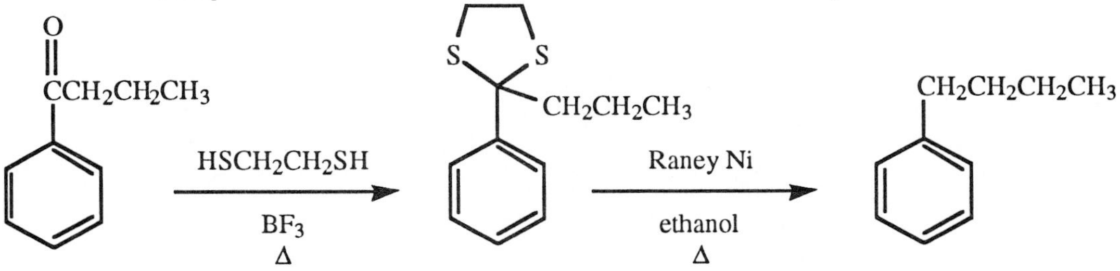

Concept Map 19.5 (see p. 702)

19.23

(a)

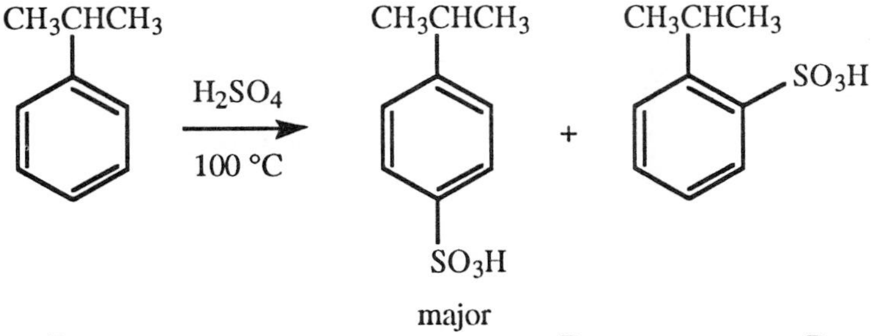

(b)

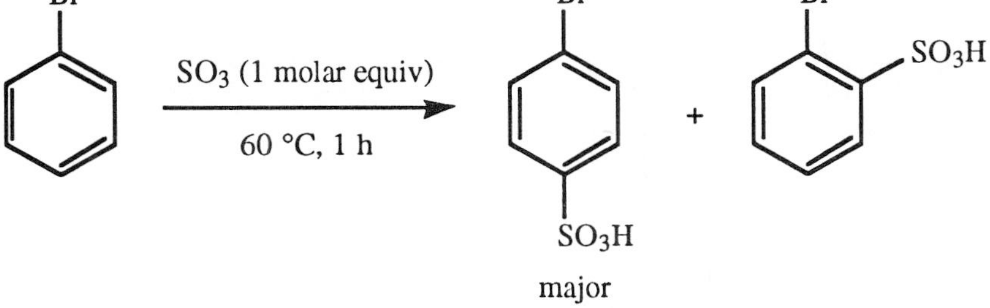

19.23 (cont)

(c)

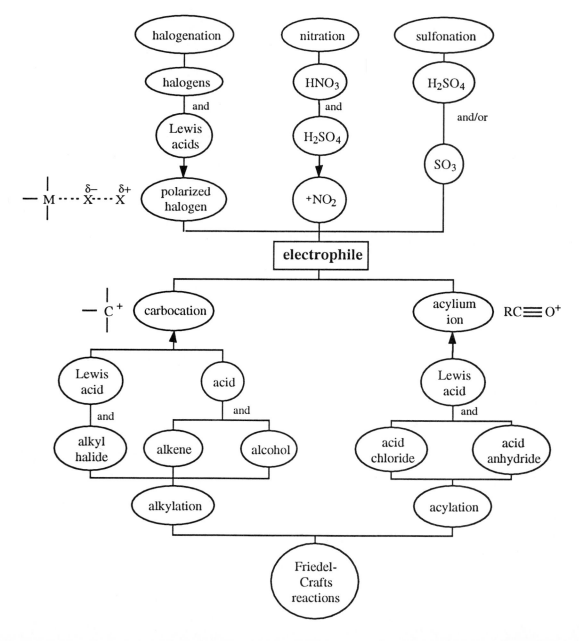

Concept Map 19.5   Electrophiles in aromatic substitution reactions.

19.24   At low temperatures the product mixture reflects the relative stabilities of the different carbocationic intermediates resulting from electrophilic attack at the ortho, meta, and para positions on the ring. The transition states for the different reaction pathways are assumed to be similar to the different intermediates. Therefore, the relative stabilities of the intermediates determine the relative energies of activation for the three reaction pathways, and the rates at which the three products are formed. The reaction at low temperatures is kinetically controlled (see Section 17.3).

The sulfonation reaction is reversible. At high temperatures both the reverse reaction and the forward reaction occur. Equilibrium is established among the different isomers. The ratio of products at the higher temperature represents the relative stabilities of the products. *p*-Toluenesulfonic acid is thermodynamically the most stable of the products and is the major product at high temperatures. Under these conditions the reaction is thermodynamically controlled.

19.25

attack by nucleophile

four groups around the sulfur atom in the starting material (note that this contrasts with three groups around the carbon atom of the carbonyl group in acid derivatives)

intermediate losing the leaving group

five groups around the sulfur atom in the intermediate

deprotonation

19.26

(a)

(b)

19.26 (cont)

(c) [benzenesulfonyl chloride structure] + 2 CH$_3$CH$_2$NHCH$_2$CH$_3$ → [N,N-diethyl benzenesulfonamide structure with CH$_2$CH$_3$ and —NCH$_2$CH$_3$] + CH$_3$CH$_2$NCH$_2$CH$_3$ (with H and H$^+$)

(d) [benzenesulfonyl chloride structure] + CH$_3$CH$_2$CH$_2$CH$_2$NH$_2$ →(NaOH / H$_2$O)→ [sulfonamide sodium salt structure, SNCH$_2$CH$_2$CH$_2$CH$_3$ with O Na$^+$]

(e) [sulfonamide sodium salt structure, SNCH$_2$CH$_2$CH$_2$CH$_3$ with O Na$^+$] →(H$_3$O$^+$)→ [sulfonamide structure, SNHCH$_2$CH$_2$CH$_2$CH$_3$]

(f) [benzenesulfonyl chloride structure] + CH$_3$CH$_2$NCH$_2$CH$_3$ (with CH$_2$CH$_3$) →(NaOH / H$_2$O)→

[benzenesulfonate structure, SO$^-$ Na$^+$] + CH$_3$CH$_2$NCH$_2$CH$_3$ (with CH$_2$CH$_3$) + Na$^+$ Cl$^-$

19.27

(a) [mechanism showing H—O—N=O with curved arrow to H—B$^+$, giving protonated species H—O—N=O with H and :B, then :N=O: ↔ :N≡O: electrophile, and H—O—H]

[N,N-dimethylaniline structure CH$_3$NCH$_3$] + :N=O: → [para-nitroso N,N-dimethylaniline structure CH$_3$NCH$_3$ with :N=O: at para position]

The pK_a of nitrous acid is 3.37 and the pK_a of the *N,N*-dimethylanilinium ion is 5.06. Nitrous acid is not a strong enough acid to fully protonate *N,N*-dimethylaniline and the product will be predominantly para. However, a strong acid such as sulfuric acid would protonate the nitrogen, giving the *N,N*-dimethylanilinium ion. The positively charged ion would be *meta*-directing.

19.27 (cont)

(b)

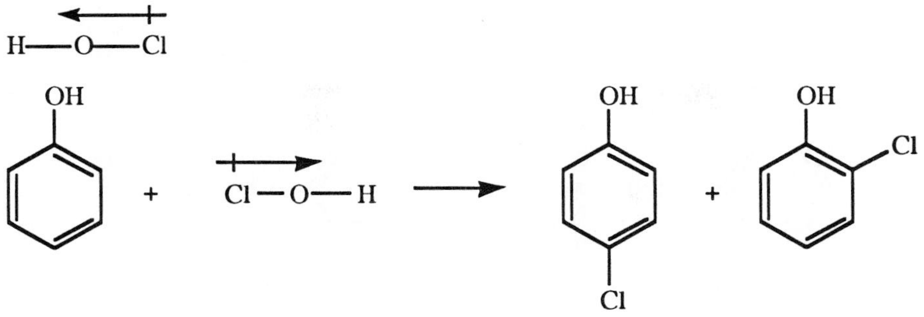

19.28

(a)

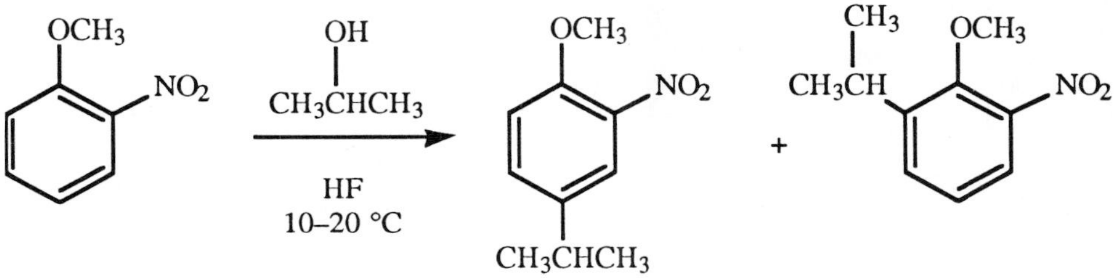

(b)

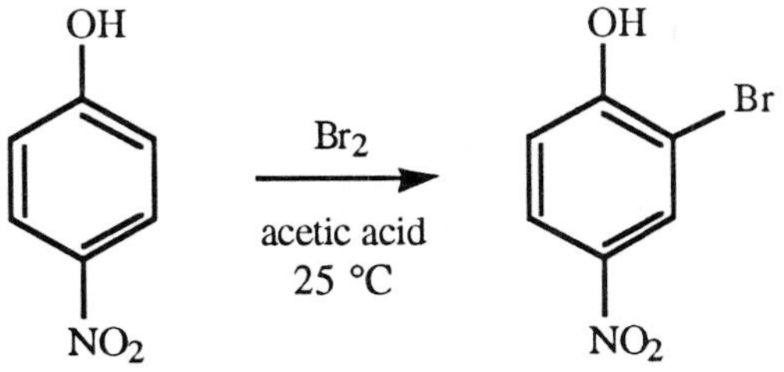

(c)

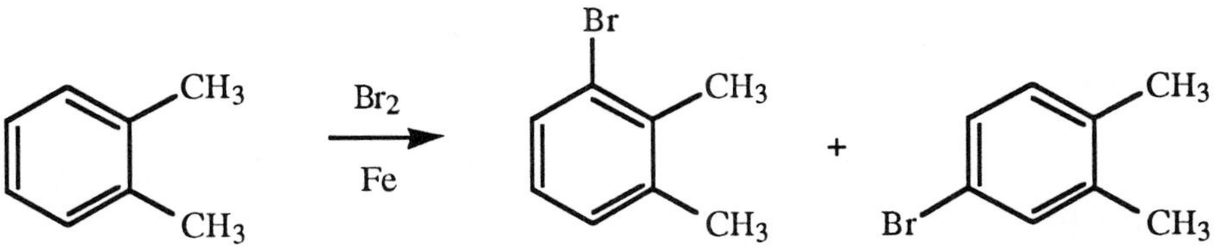

19.29

(a)

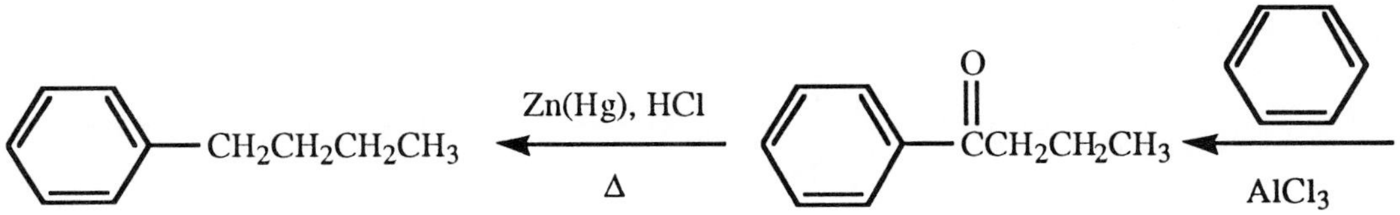

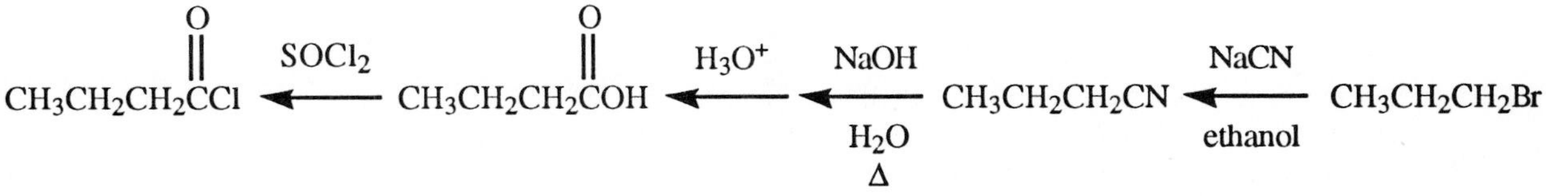

19.29 (cont)

(b)   (1)

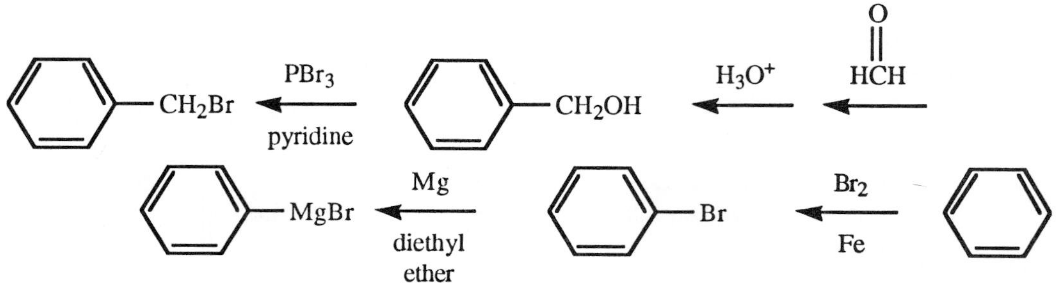

(2)

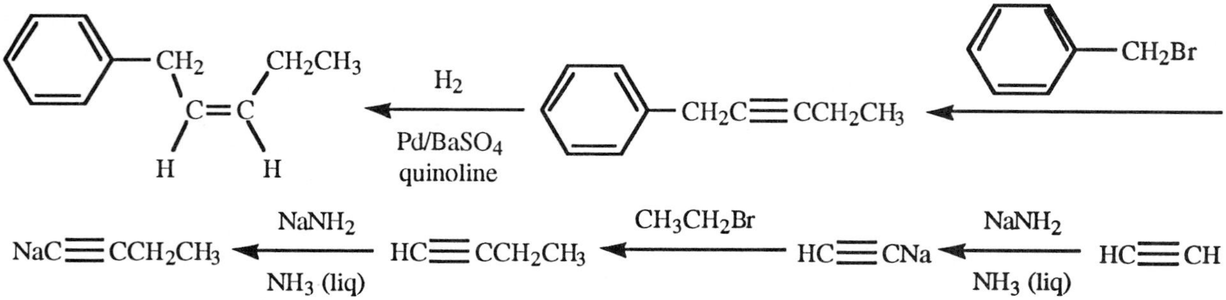

(c)   (1)

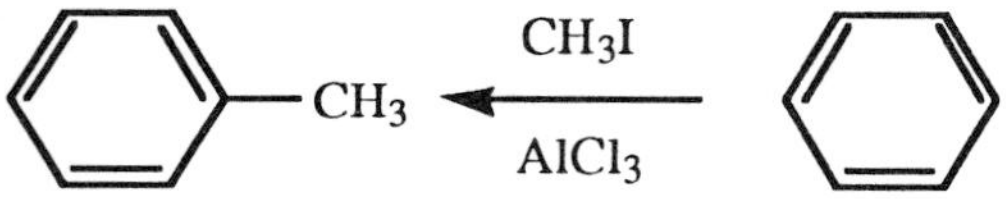

(2)

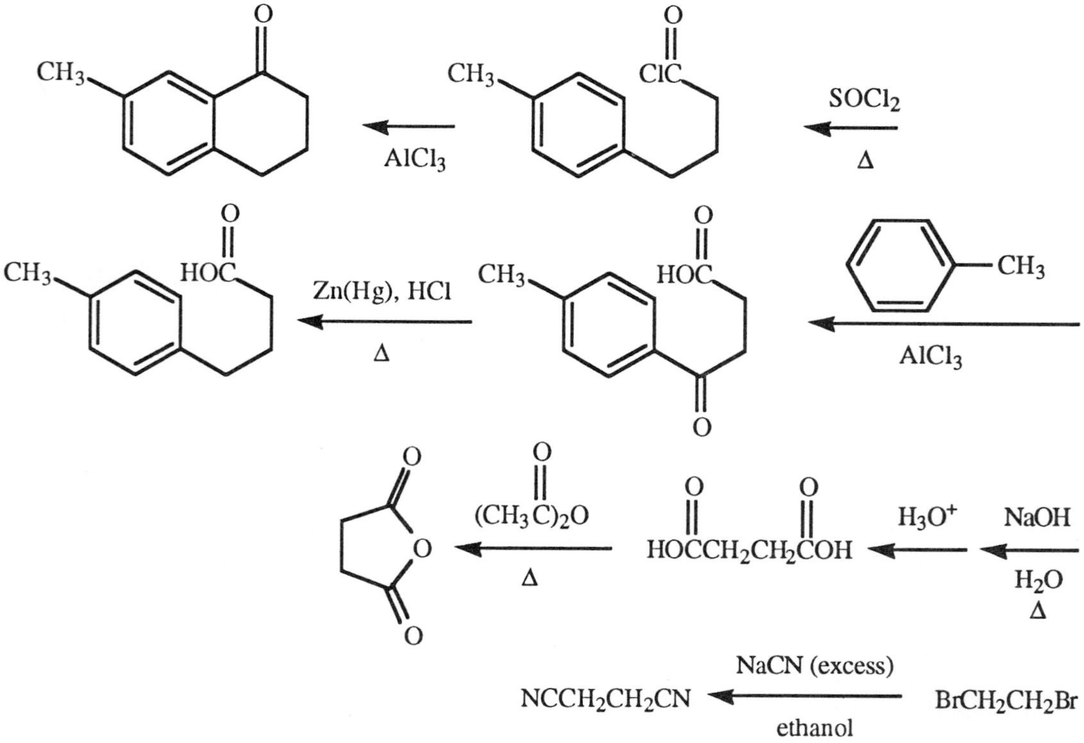

19.29 (cont)

(d)

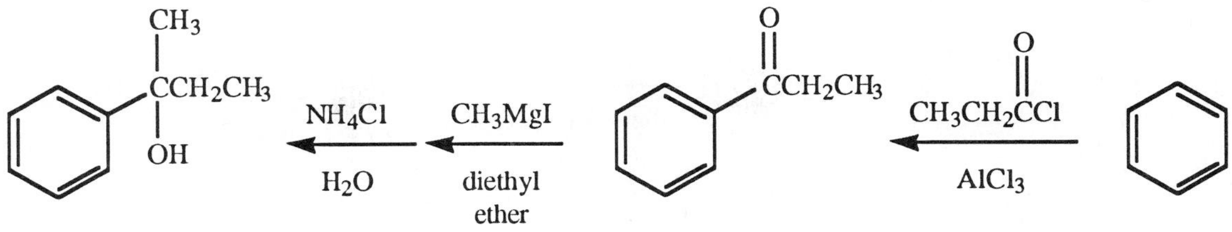

19.30

(a)   3,4-dimethylnitrobenzene   (b)   4-methyl-1-phenylpentane   (c)   2-chloro-6-ethylnaphthalene

(d)   2,4-dinitrotoluene   (e)   *m*-bromobenzoic acid   (f)   methyl 3,5-dimethylbenzoate

(g)   1-(4-methylphenyl)-1-butanone   (h)   *p*-bromobenzaldehyde
                                              4-bromobenzaldehyde

19.31   (a) 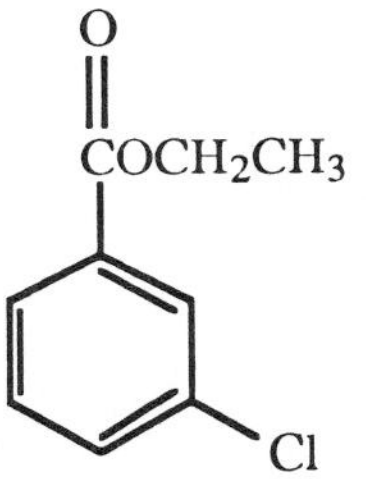   (b) 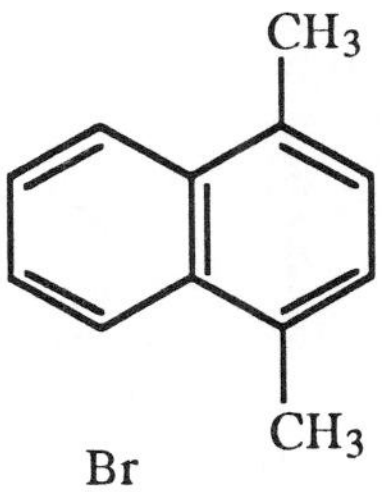   (c)

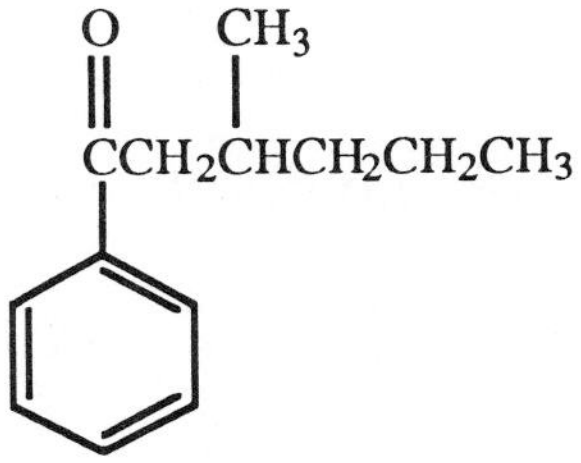

(d) 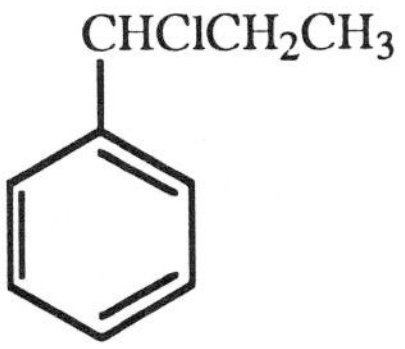   (e) 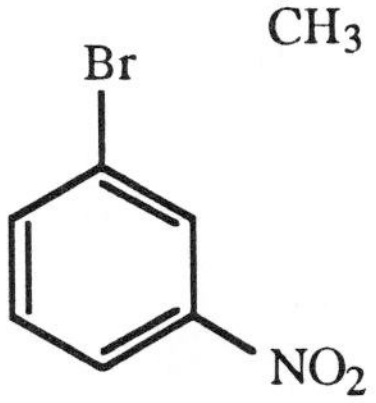   (f)

19.32

(a)

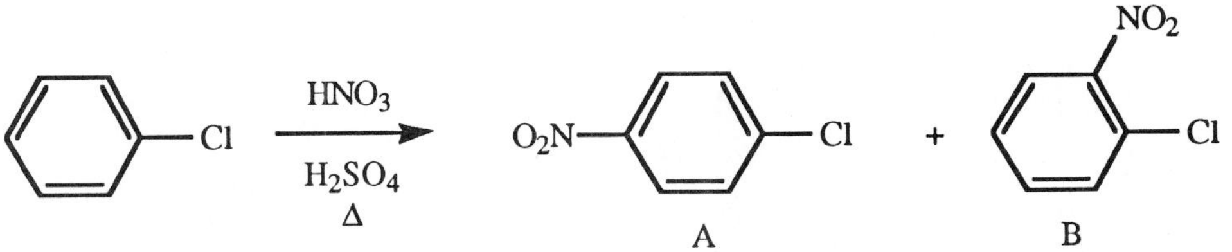

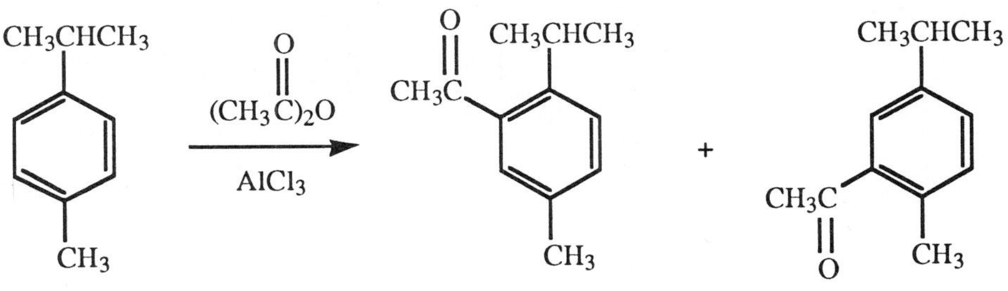

(b)

19.32 (cont)

(c)

$$\text{CH}_3-\bigcirc-\overset{\overset{\displaystyle O}{\|}}{C}\text{CH}_2\text{CH}_2\text{CH}_3 \xrightarrow[\substack{\text{diethylene glycol} \\ \Delta}]{\text{H}_2\text{NNH}_2,\ \text{KOH}} \text{CH}_3-\bigcirc-\text{CH}_2\text{CH}_2\text{CH}_2\text{CH}_3$$

E

(d)

$$\text{CH}_3-\bigcirc-\overset{\overset{\displaystyle O}{\|}}{C}\text{CH}_2\text{CH}_2\text{CH}_3 \xrightarrow[\substack{\text{FeBr}_3 \\ \Delta}]{\text{Br}_2}$$

F

(e)

$$\bigcirc-\overset{\overset{\displaystyle O}{\|}}{C}\underset{\underset{\displaystyle \text{CH}_3}{|}}{C}\text{HCH}_3 \xrightarrow[\Delta]{\text{Zn(Hg),\ HCl}} \bigcirc-\text{CH}_2\underset{\underset{\displaystyle \text{CH}_3}{|}}{C}\text{HCH}_3$$

G

(f)

$$\underset{\bigcirc}{\text{CH}_3\text{CHCH}_3} + \underset{\text{AlCl}_3}{\overset{\text{CH}_3}{\underset{|}{\text{CH}_3\text{CHCl}}}} \longrightarrow$$

H   +   I

(g)

$$\bigcirc + \bigcirc \xrightarrow{\text{HF}}$$

J   +   K

(h)

$$\bigcirc + \bigcirc \xrightarrow{\text{AlCl}_3}$$

L   +   M

19.32 (cont)

(i)

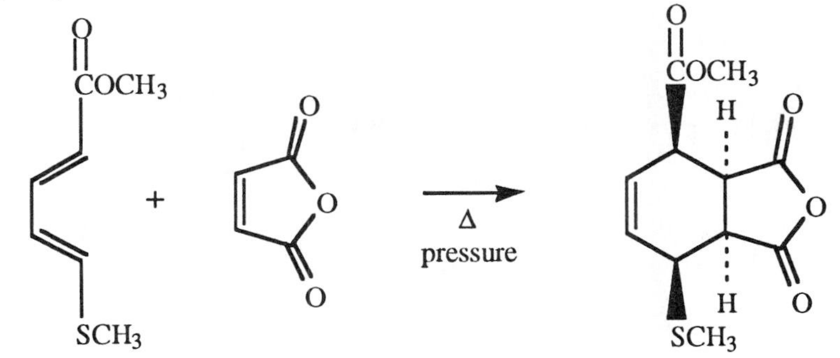

(j)

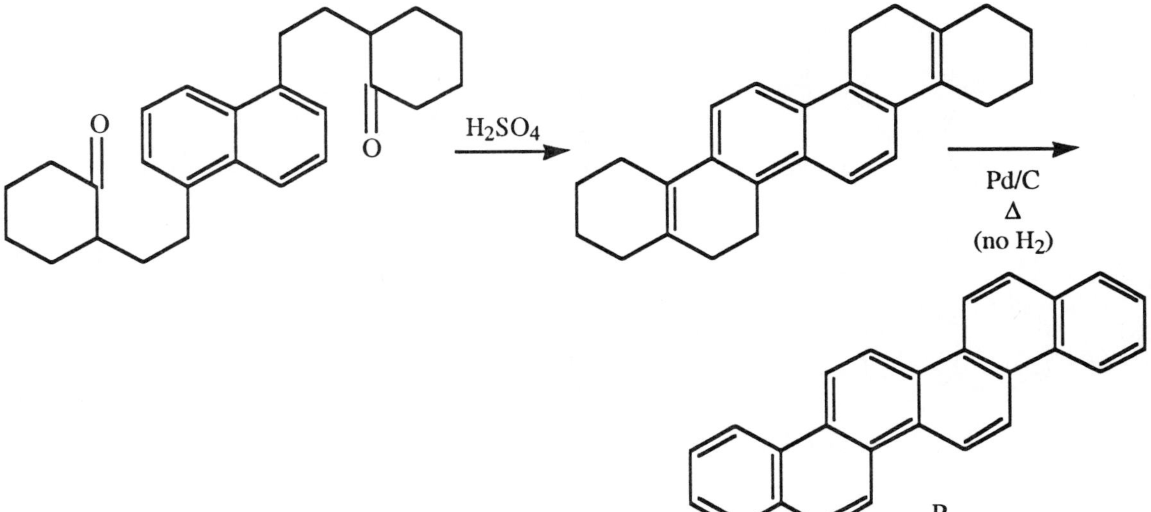

19.33

(a)

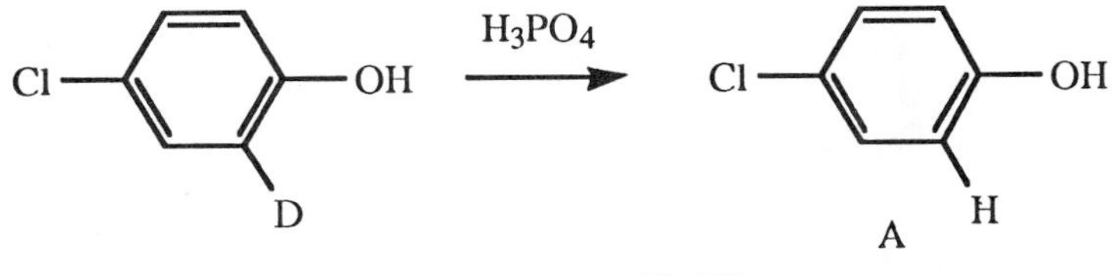

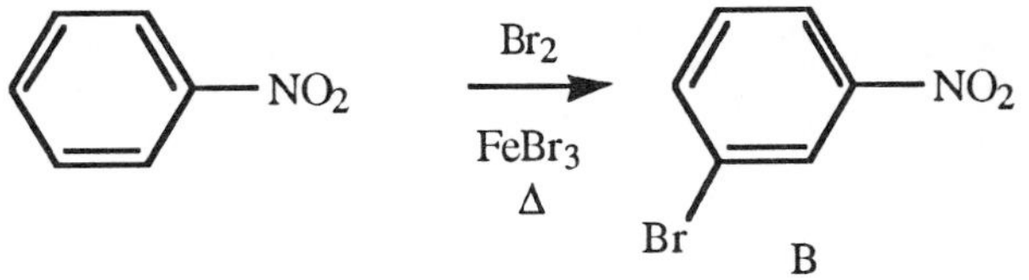

(b)

(c)

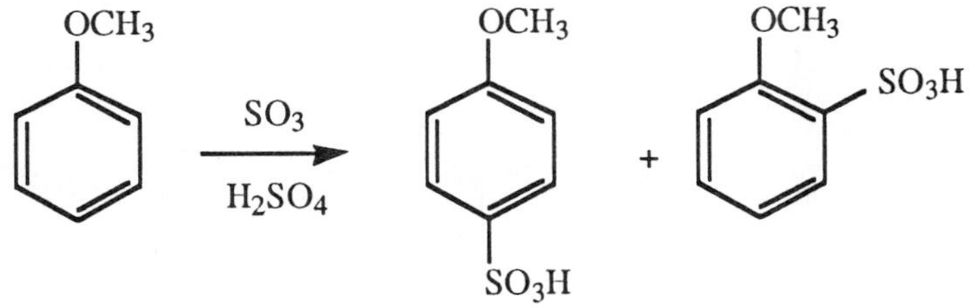

19.33 (cont)

(d)

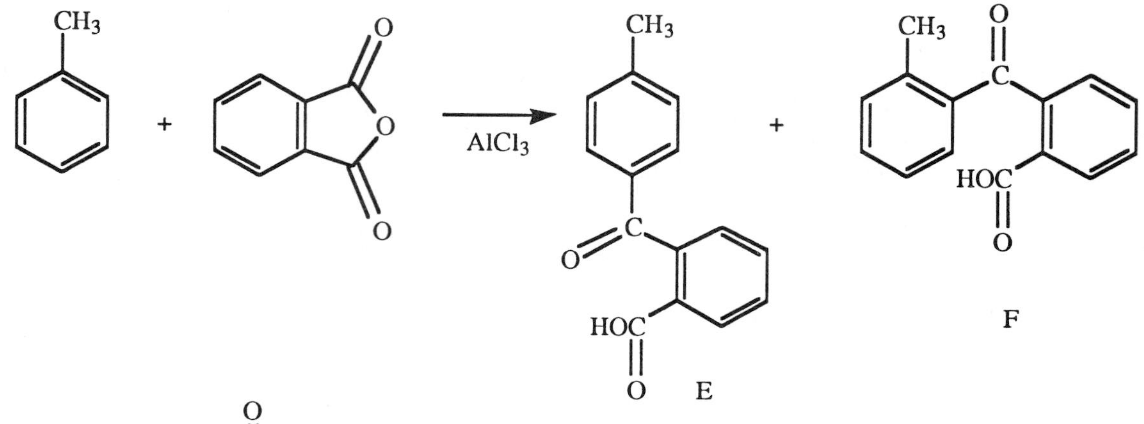

(e)

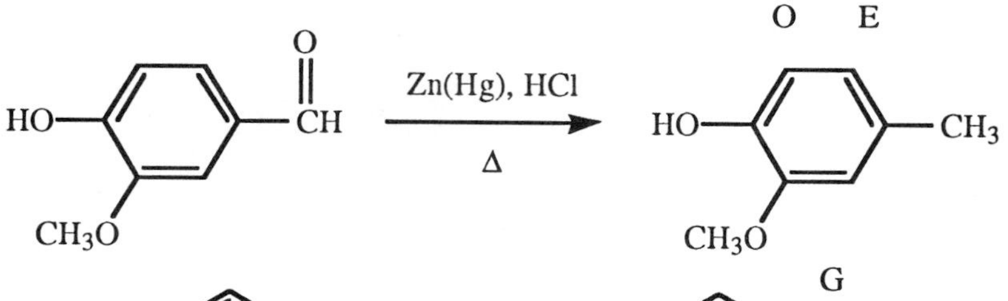

(f)

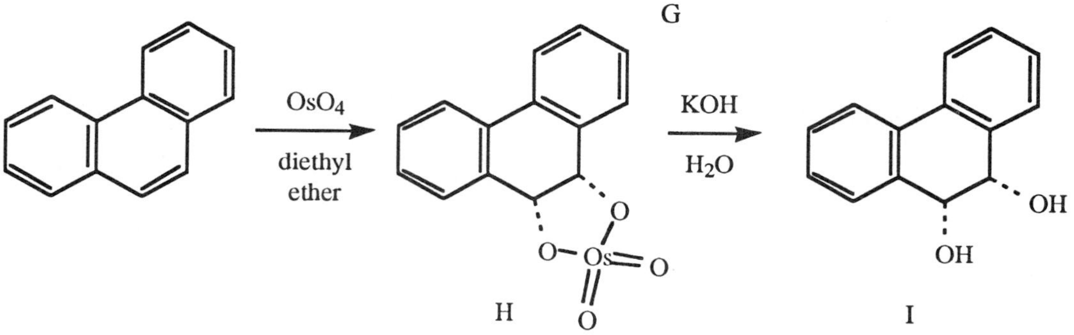

(g)

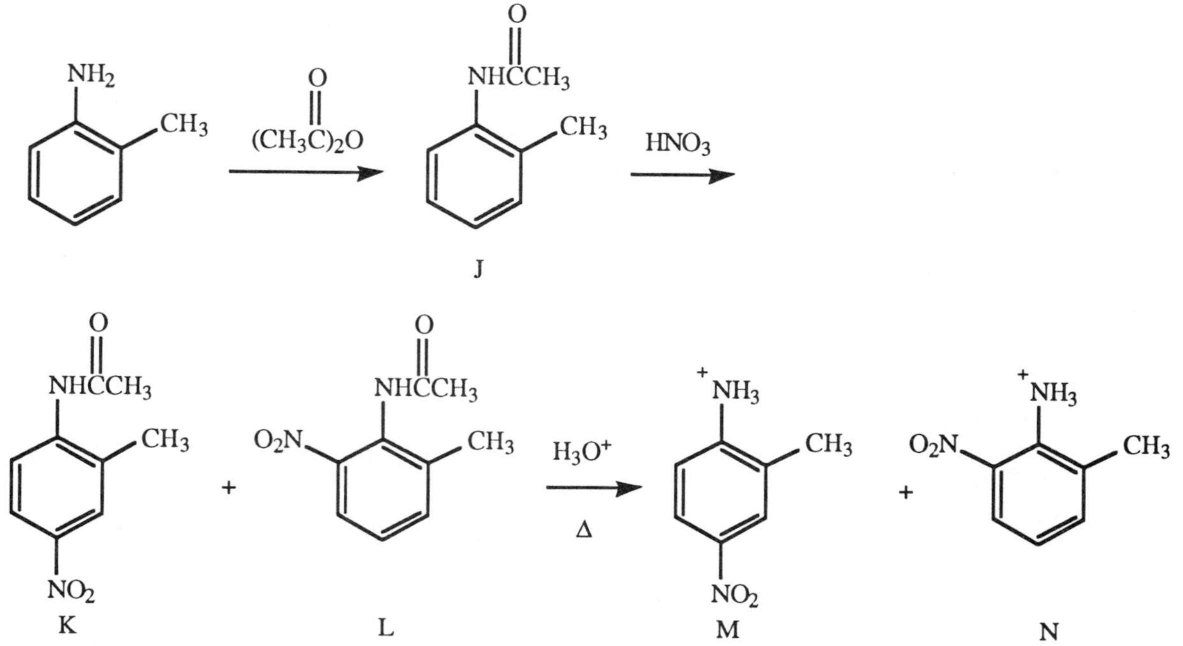

19.33 (cont)

(h)

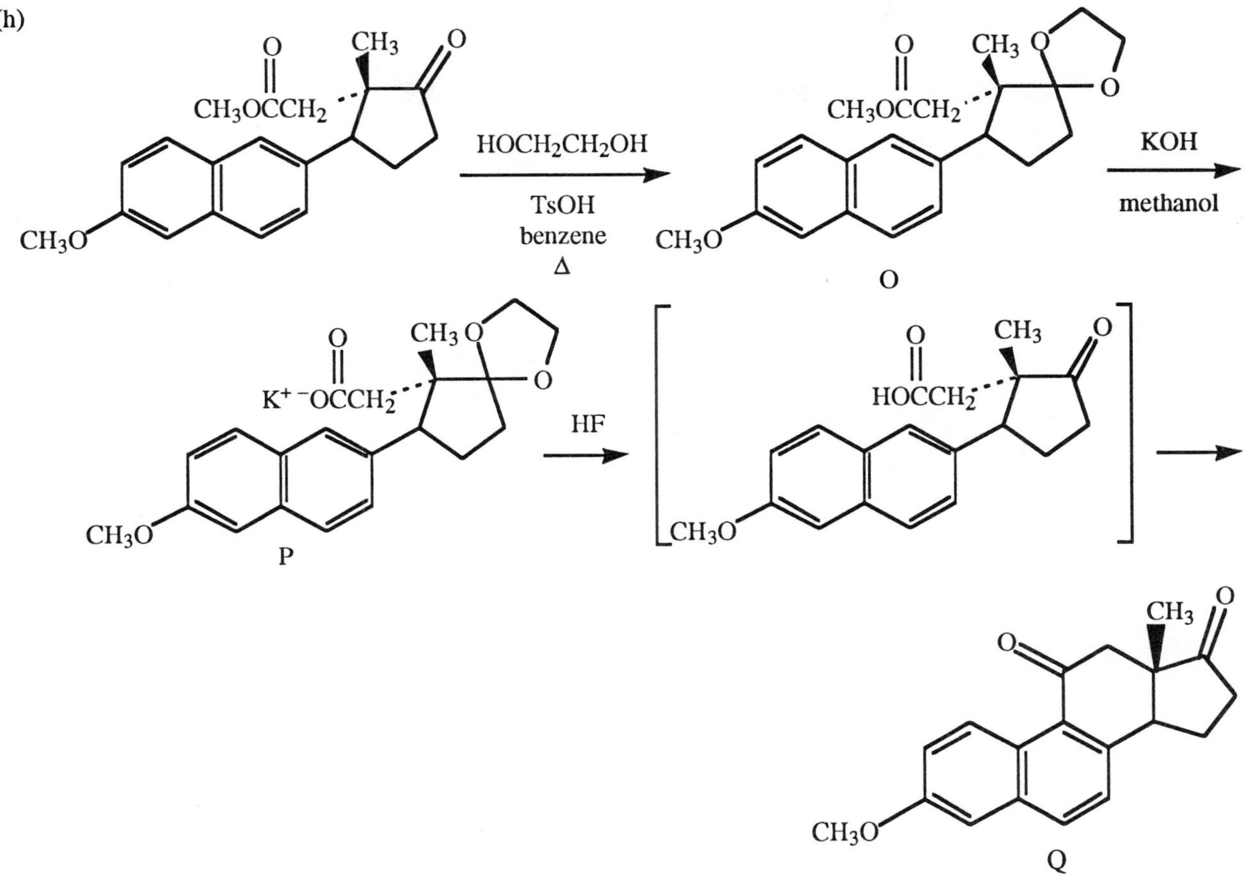

(i)

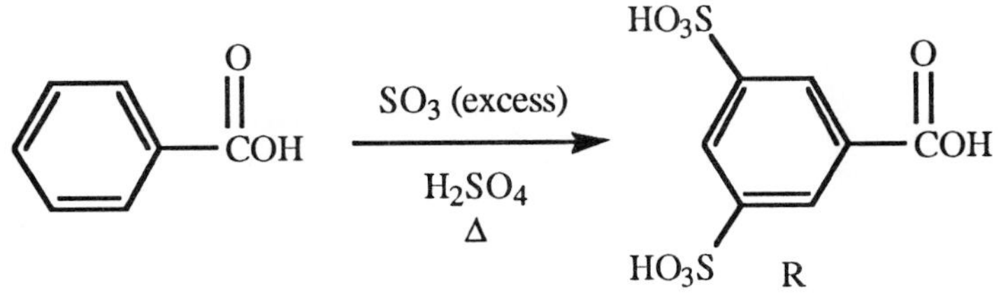

(j)

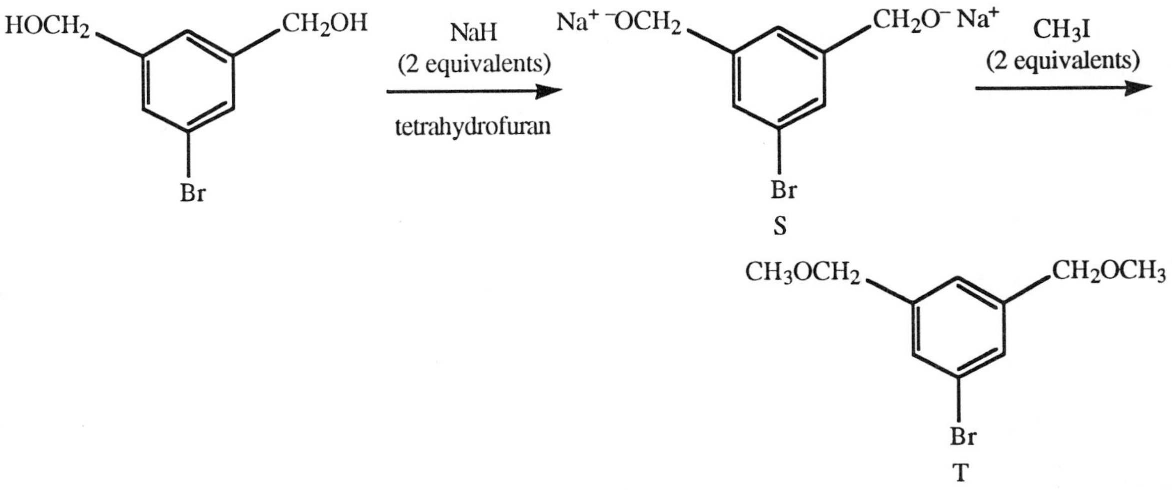

19.34

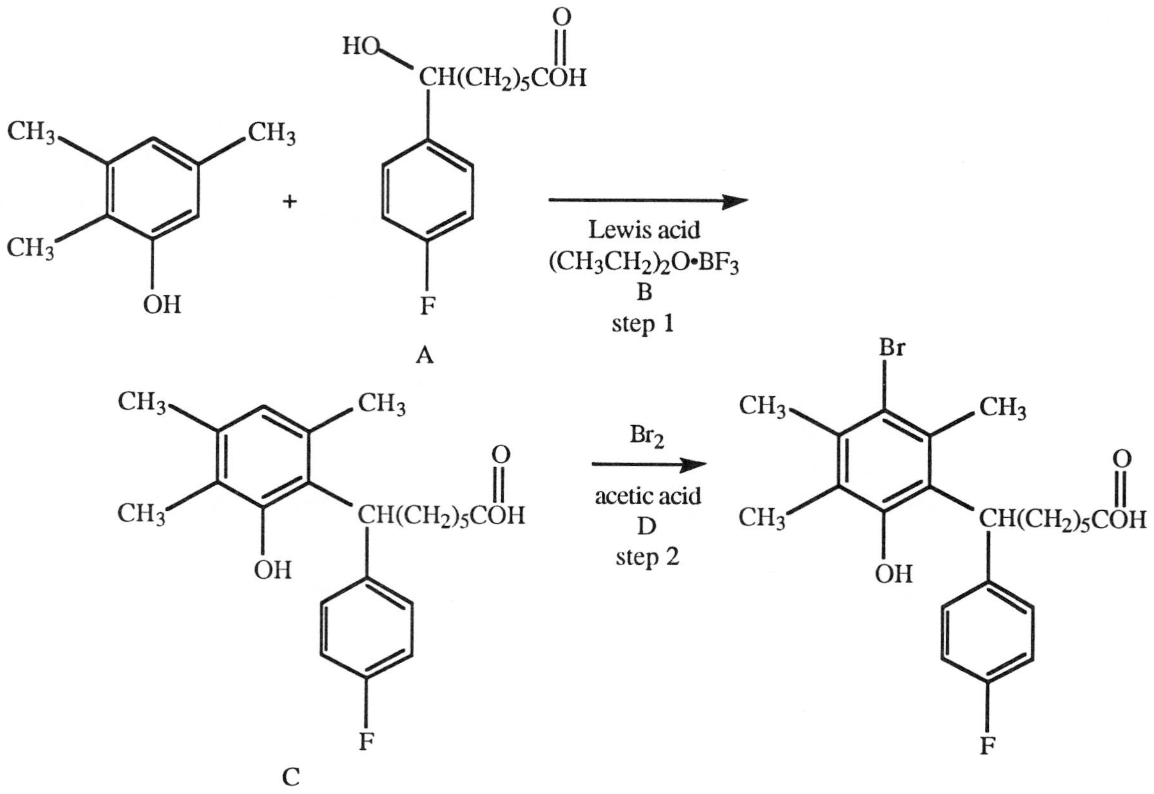

19.35

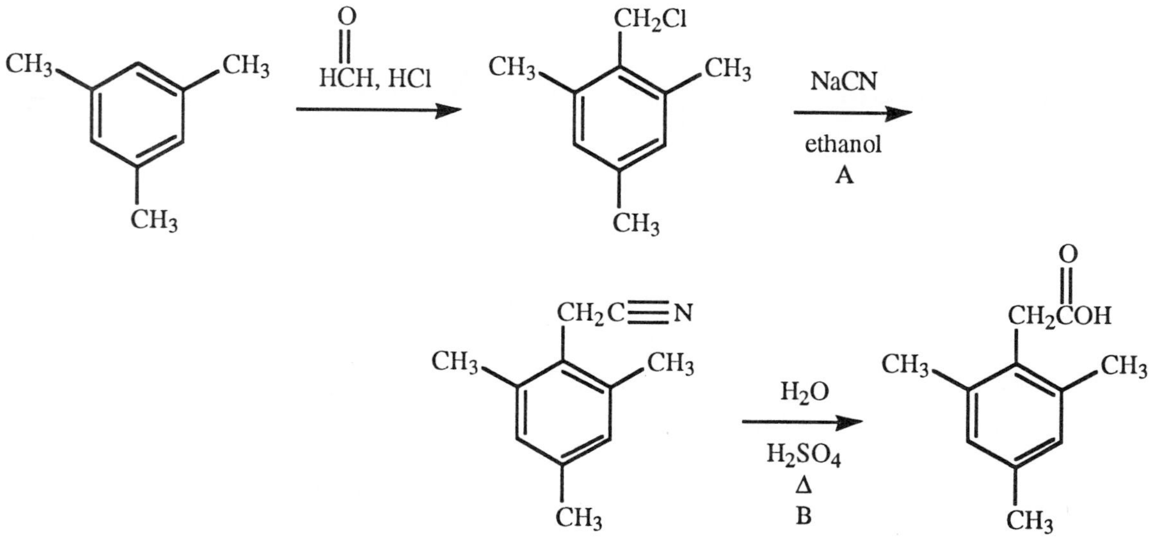

19.35 (cont)

Mechanism:

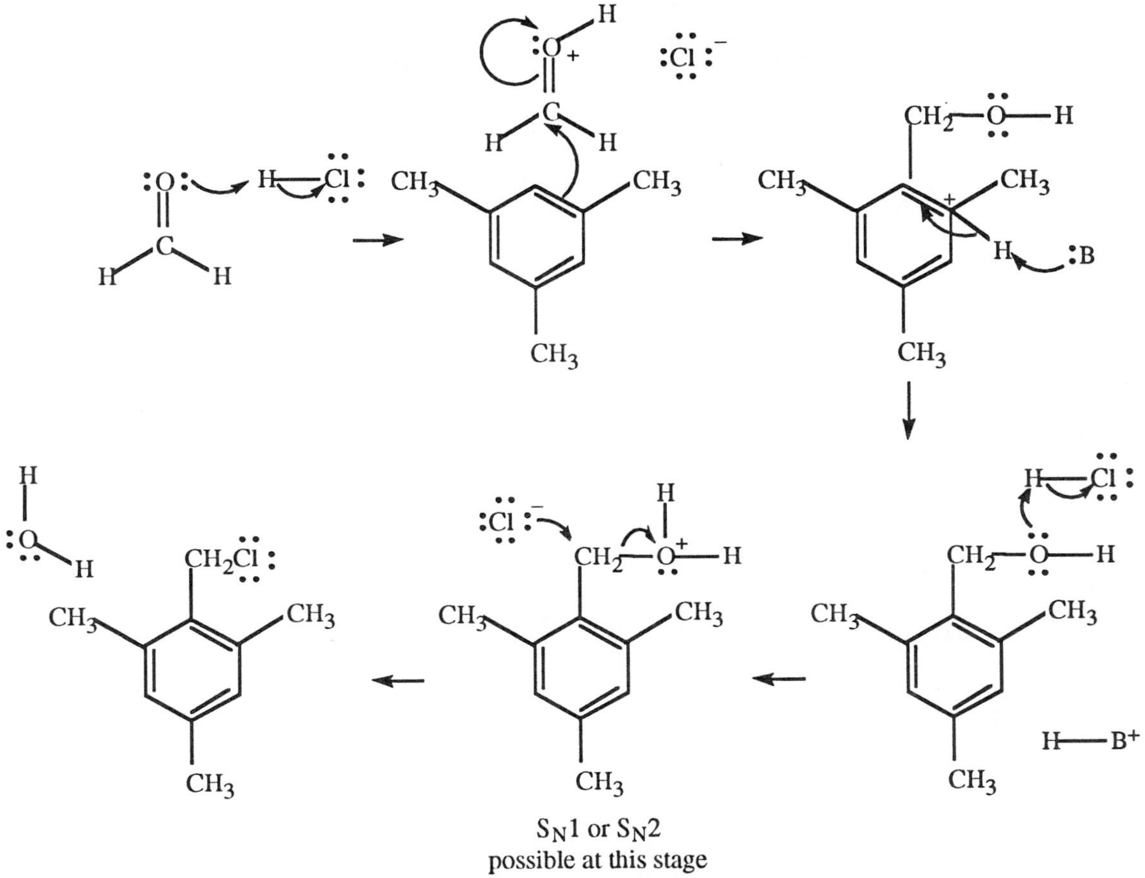

S_N1 or S_N2
possible at this stage

19.36   (a)

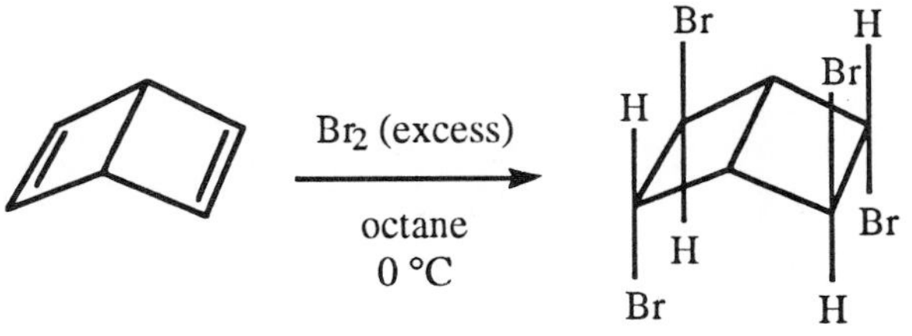

A

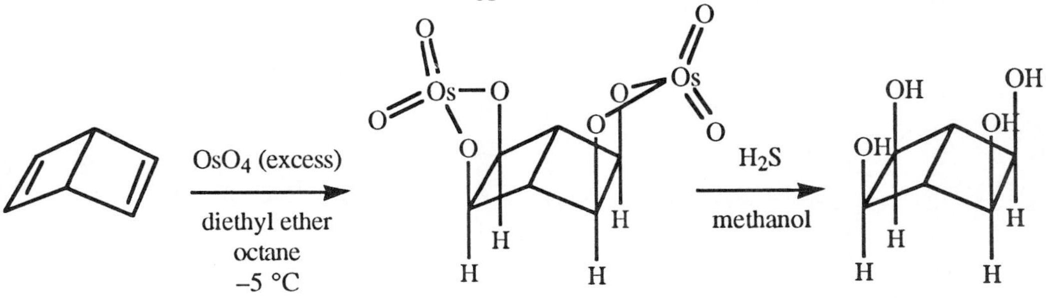

B   C

Only one possible isomer is shown for the reactions above.

19.36 (a) (cont)

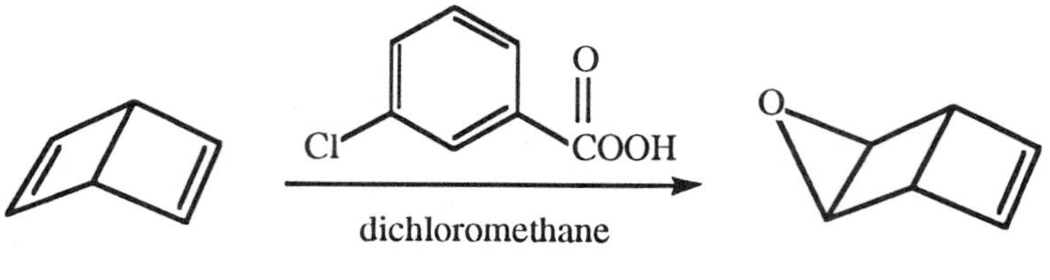

D

(b)

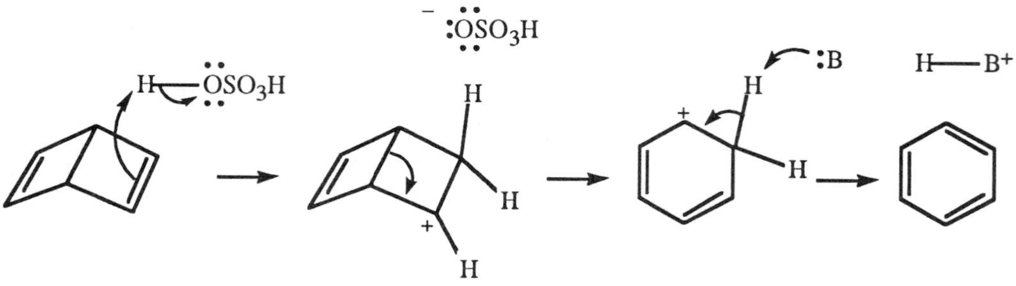

19.37

(a)

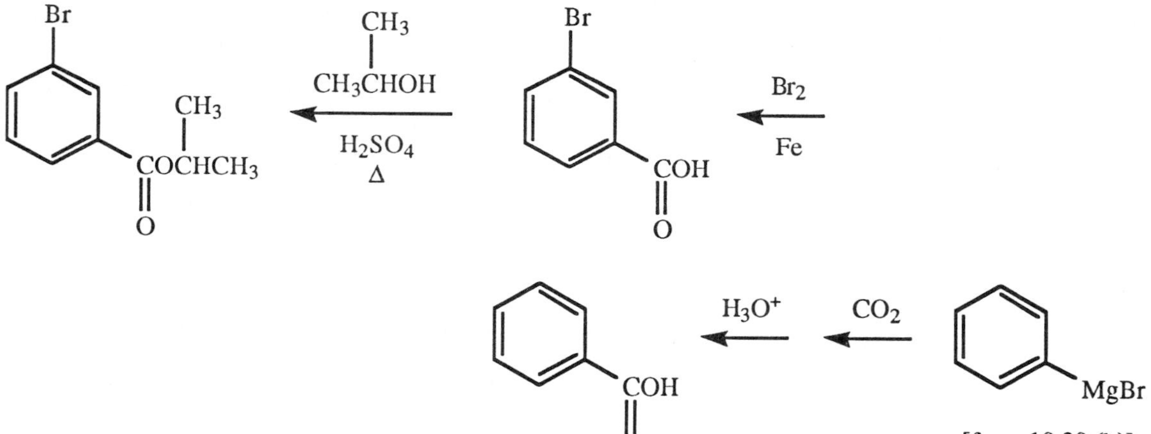

[from 19.29 (b)]

(b)

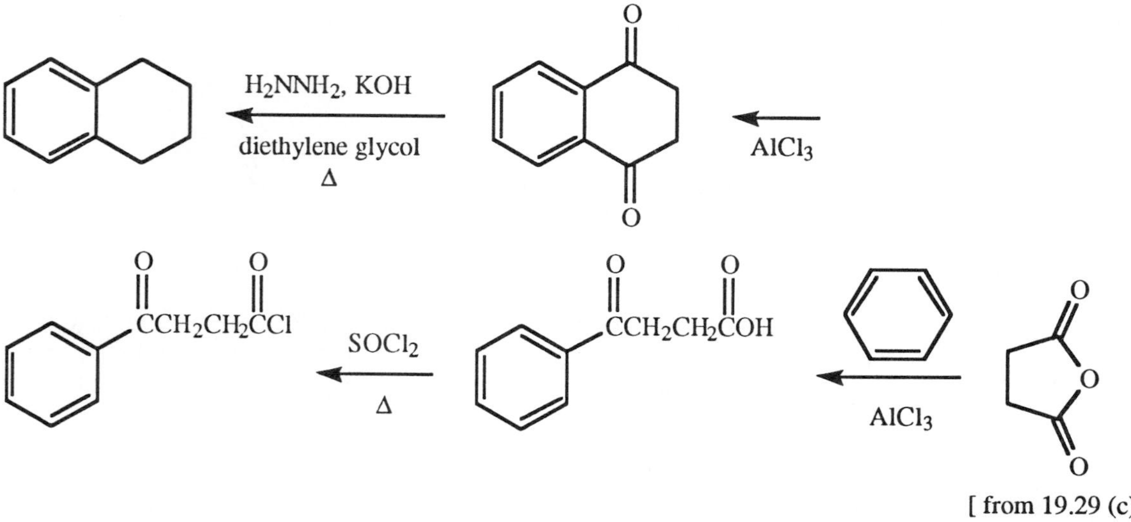

[from 19.29 (c)]

19.37 (cont)

(c)

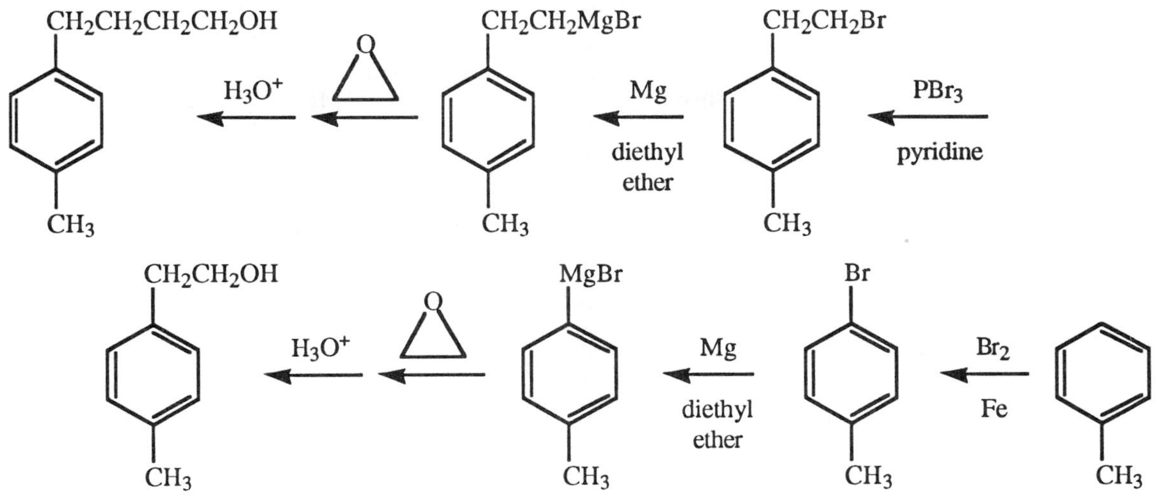

(d)

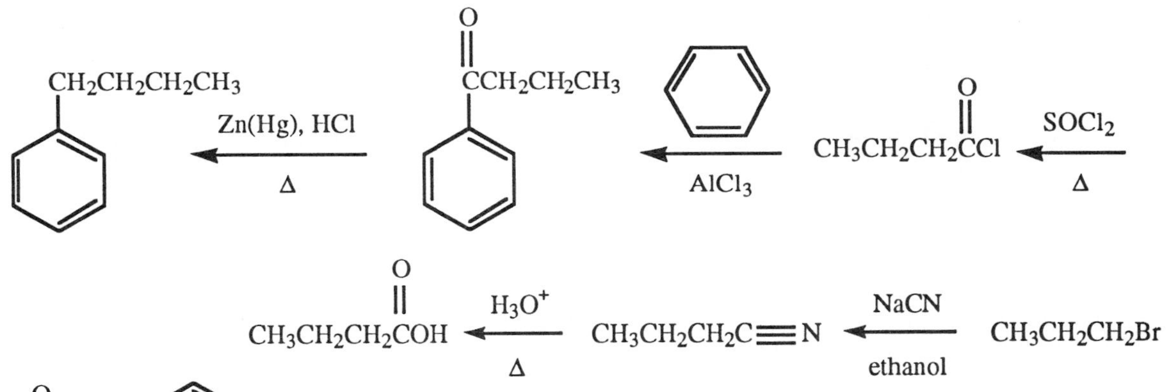

(e) (1)

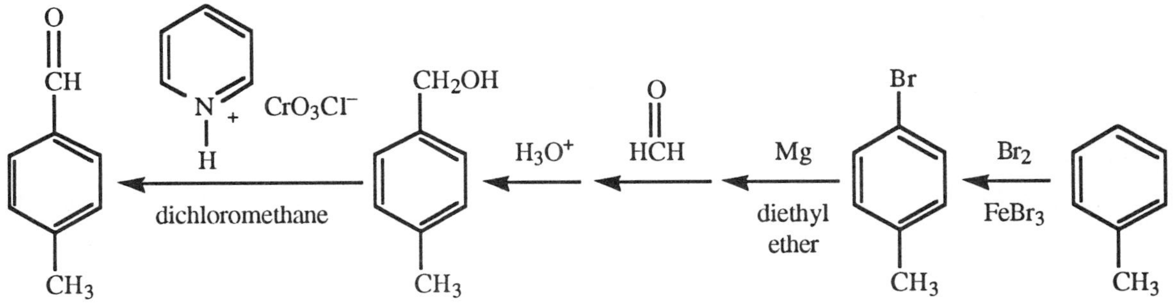

(2)

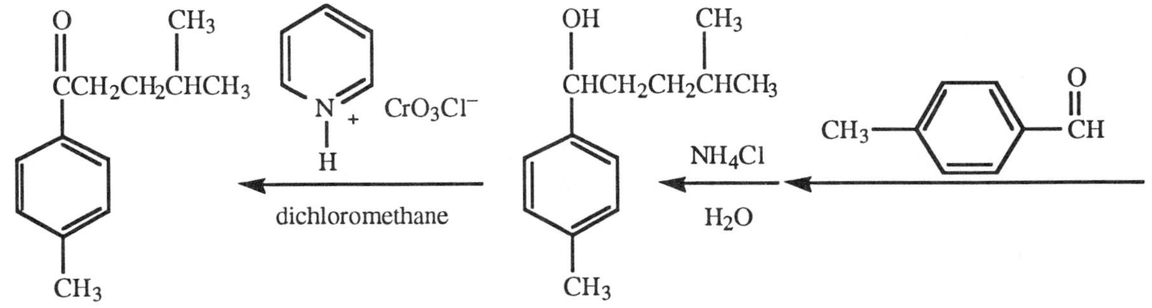

19.37 (e) (2) (cont)

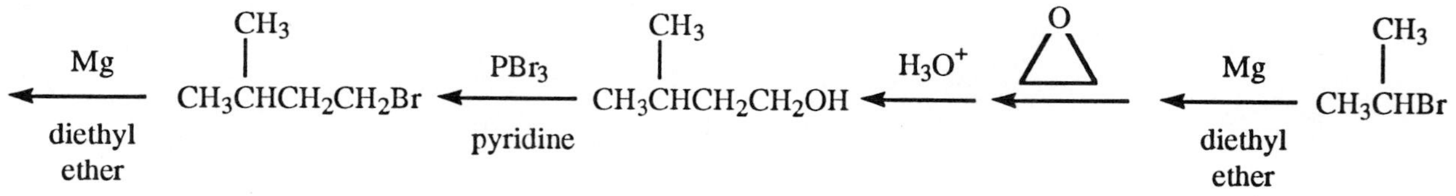

(f)

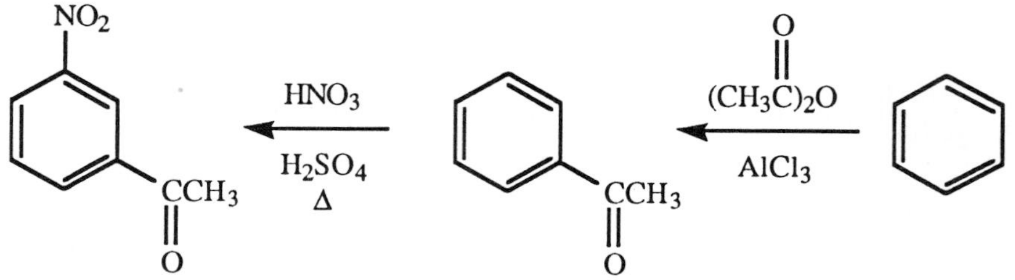

(g)

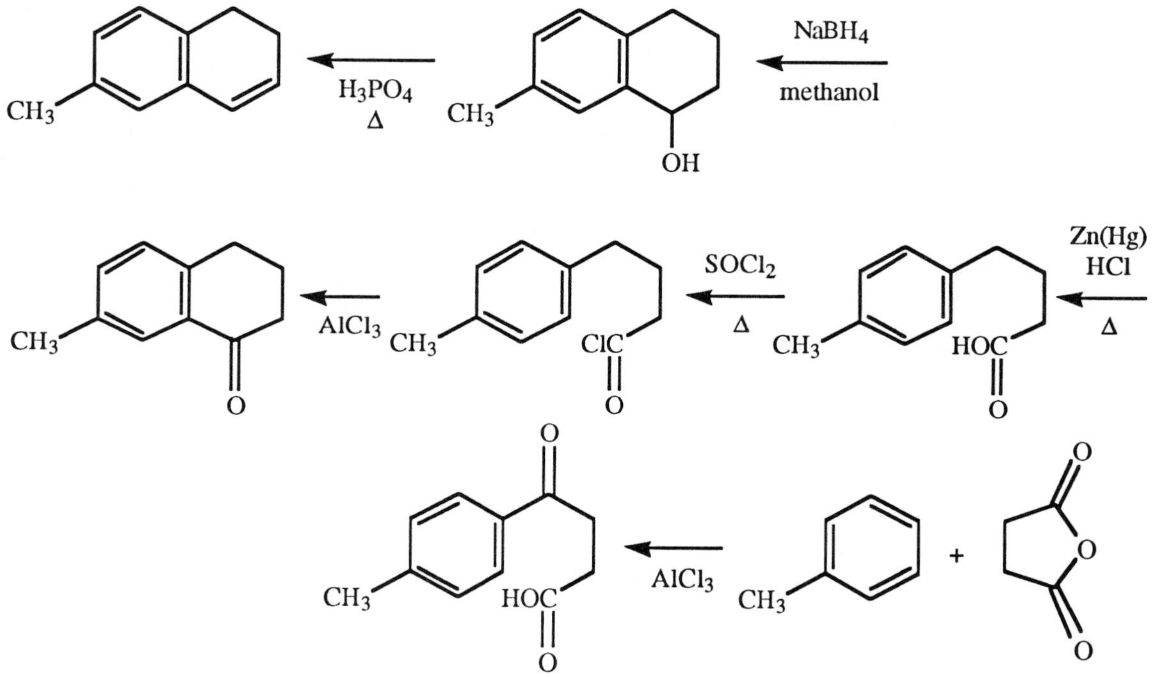

[from 19.29 (c)]

(h)

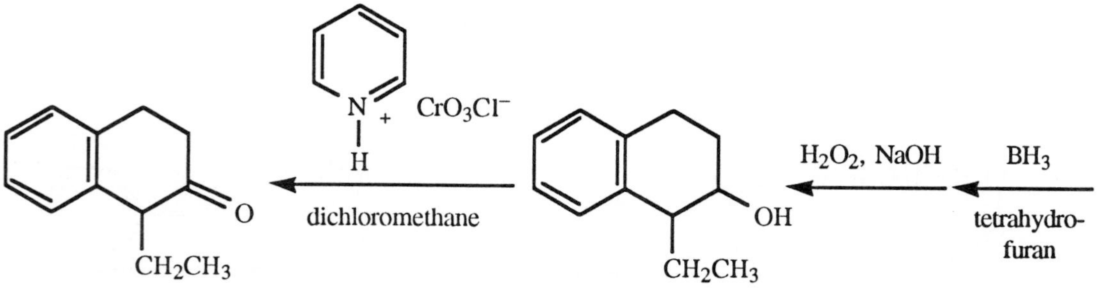

19.37 (h) (cont)

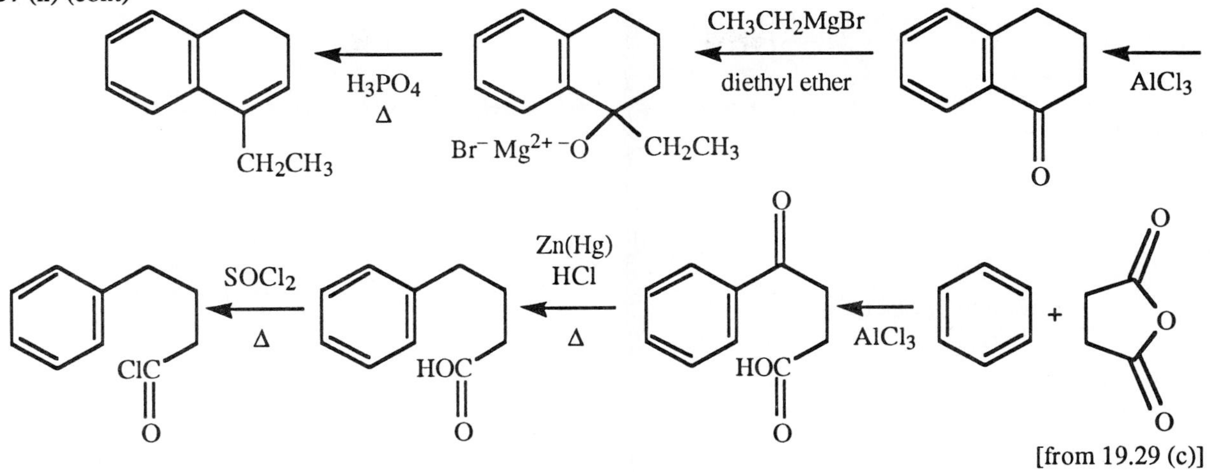

[from 19.29 (c)]

(i)

[from 19.29 (b)]

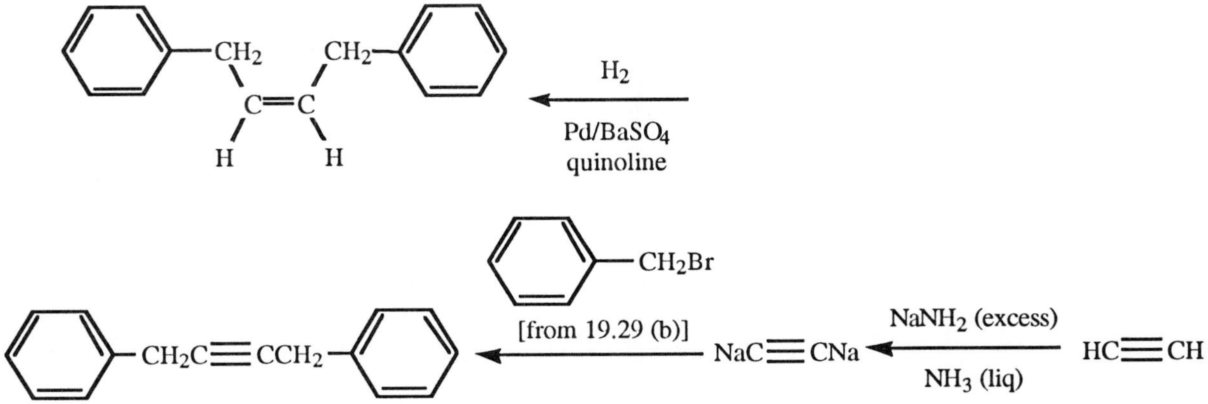

19.38

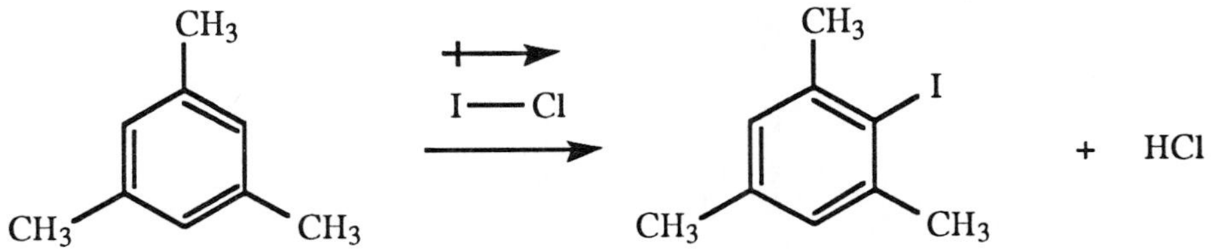

Chlorine is more electronegative than iodine; iodine is, therefore, the electrophile.

19.39

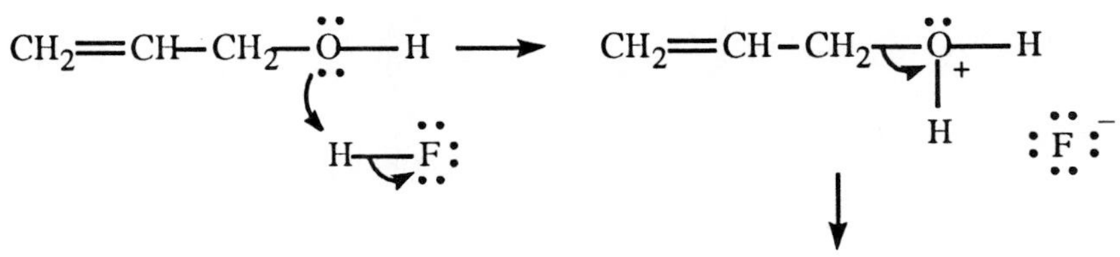

19.39 (cont)

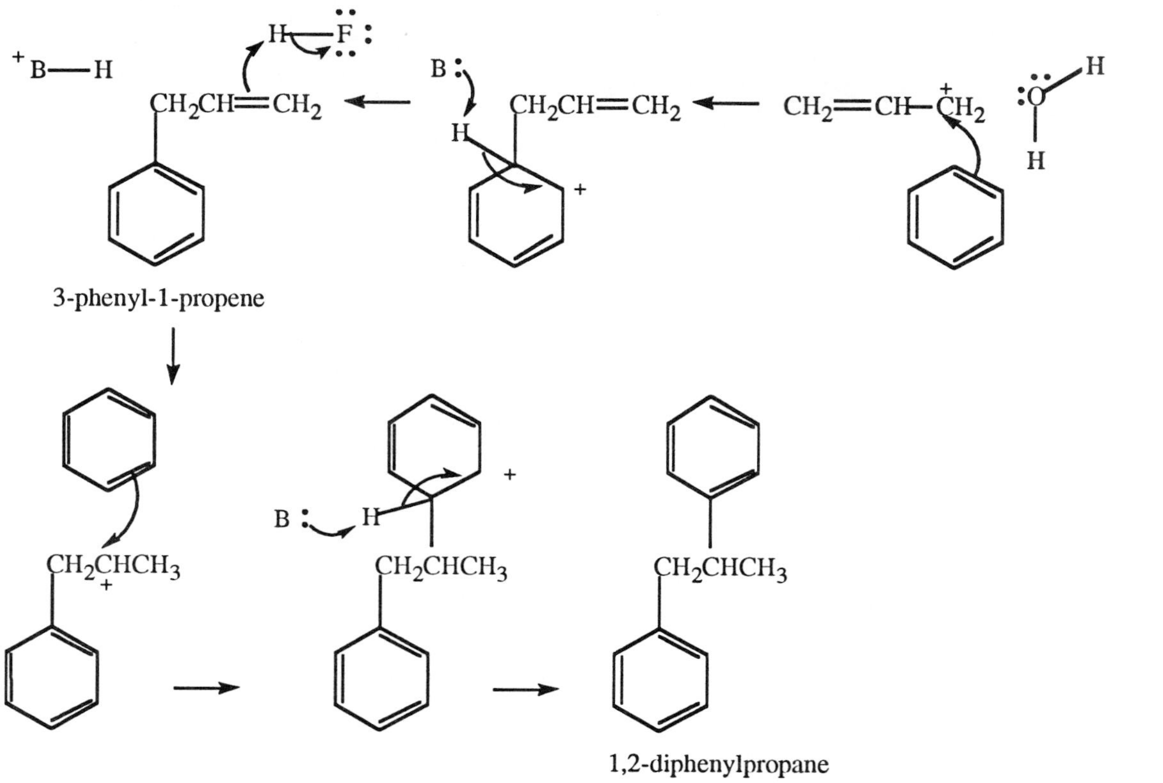

3-phenyl-1-propene

1,2-diphenylpropane

19.40

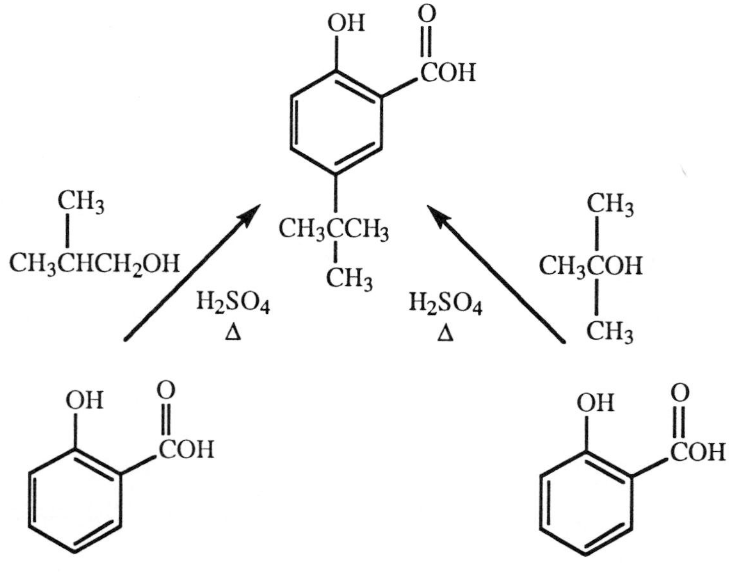

Both alcohols form the same carbocationic intermediate.

19.40 (cont)

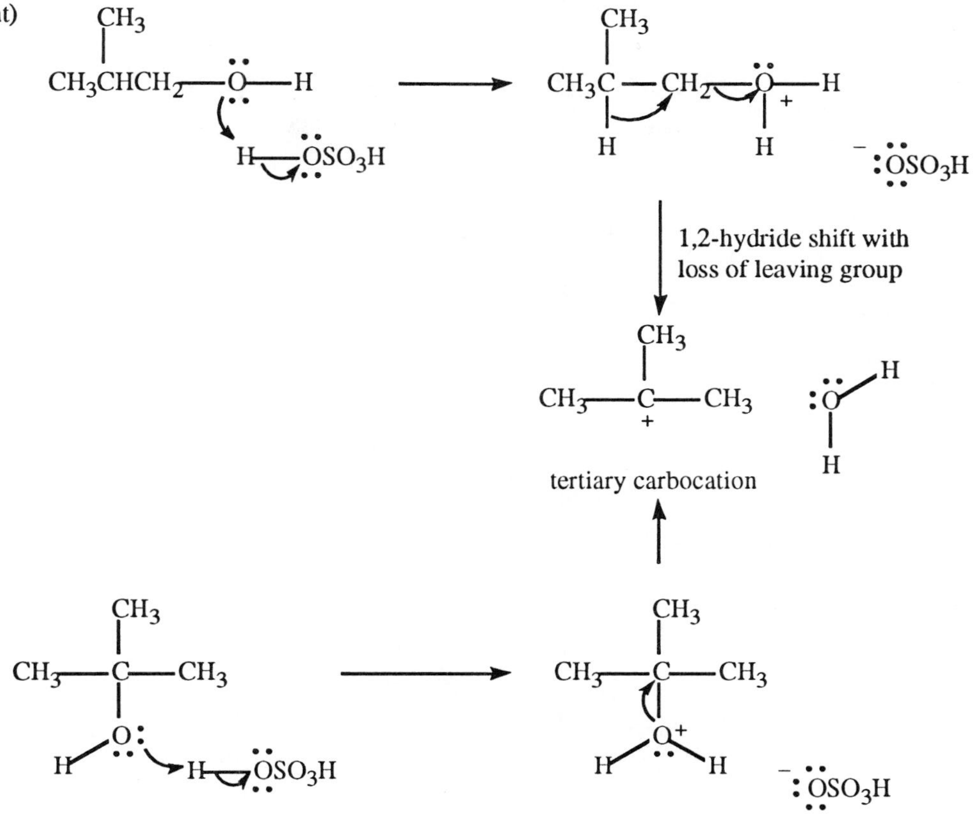

1,2-hydride shift with
loss of leaving group

tertiary carbocation

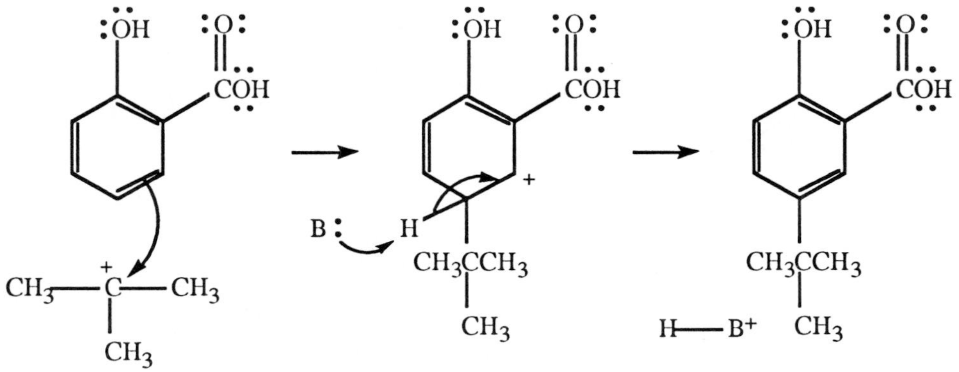

19.41

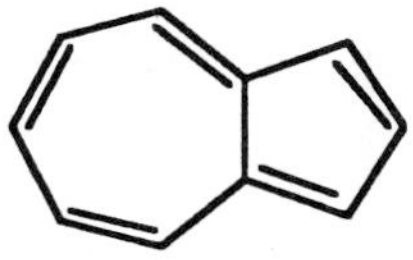

Azulene has a planar π system with 10 π electrons, so we would expect the compound to have aromaticity. It has $4n + 2$ electrons with $n = 2$. One particularly interesting resonance contributor for azulene may be regarded as the juxtaposition of an aromatic tropylium ion with an aromatic cyclopentadienyl anion.

tropylium ion $\longrightarrow$ $\longleftarrow$ cyclopentadienyl anion

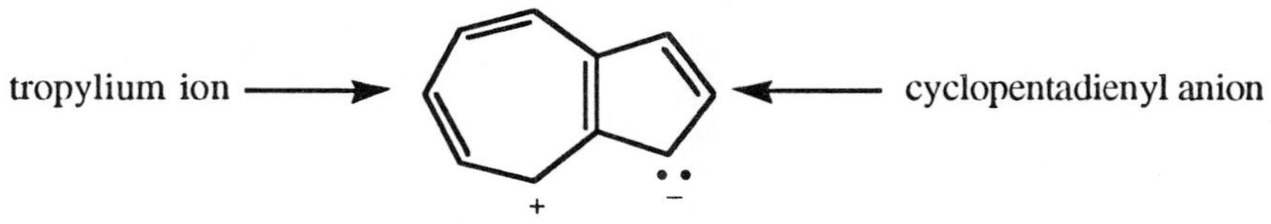

19.42   (a)

19.42 (b) (cont)

The driving force for the reaction is the stability of the aromatic ring, which is why a proton and the benzyloxy group are easily lost.

19.43

The starting material has 24 protons. Two are replaced by bromine atoms in the product, leaving 22 protons. The number of protons in the nuclear magnetic resonance spectrum of the product is 11. The product must, therefore, be symmetrical. One equivalent position on each ring is unhindered by the *tert*-butyl groups.

19.44   The basicity of an amine is determined by the availability of the pair of nonbonding electrons. The nonbonding electrons in ammonia are localized on the nitrogen atom and are readily available for protonation. The nonbonding electrons in pyrrole and indole are part of an aromatic sextet. Protonation of the nitrogen atom in either one of these compounds leads to loss of aromaticity. The conjugate acids of pyrrole and indole are, therefore, of higher energy relative to the bases than the ammonium ion is relative to ammonia.

The acidity of a compound is determined by the relative stabilities of the acid and its conjugate base. When ammonia loses a proton, the amide anion that is formed has the negative charge concentrated on the sp^3 nitrogen atom. Deprotonation of pyrrole or indole gives anions in which the negative charge is concentrated on an sp^2 nitrogen atom, which is more electronegative than an sp^3 nitrogen atom. The conjugate bases of pyrrole and indole are, therefore, weaker bases than the amide anion, and pyrrole and indole are thus stronger acids than ammonia.

$$:NH_3 \longrightarrow \quad {}^{-}:\ddot{N}H_2 \quad \text{anion with localized charge in } sp^3 \text{ hybridized orbital}$$

19.44 (cont)

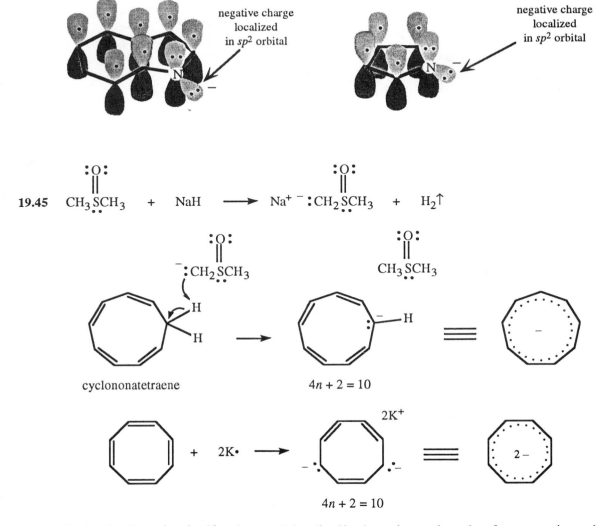

19.45

Both anions have $4n + 2 = 10$ π electrons delocalized in planar rings and are, therefore, aromatic species. They have aromatic stability just as the tropylium ion and the cyclopentadienyl anion do (Section 19.1D).

19.46

Resonance contributors for heptafulvene have the positive charge delocalized over the seven-membered ring in what is a tropylium ion, but the negative charge is localized on a carbon atom. In the dicyano compound, the negative charge can be delocalized onto the two nitrogen atoms. The two electron-withdrawing cyano groups thus stabilize the negative character at the carbon atom outside the ring.

19.46 (cont)

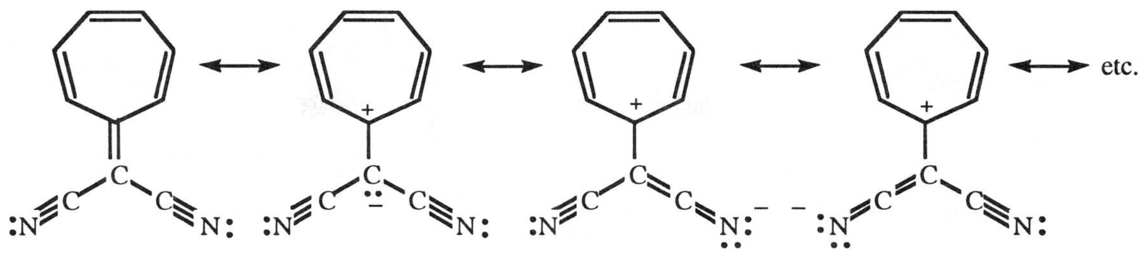

19.47

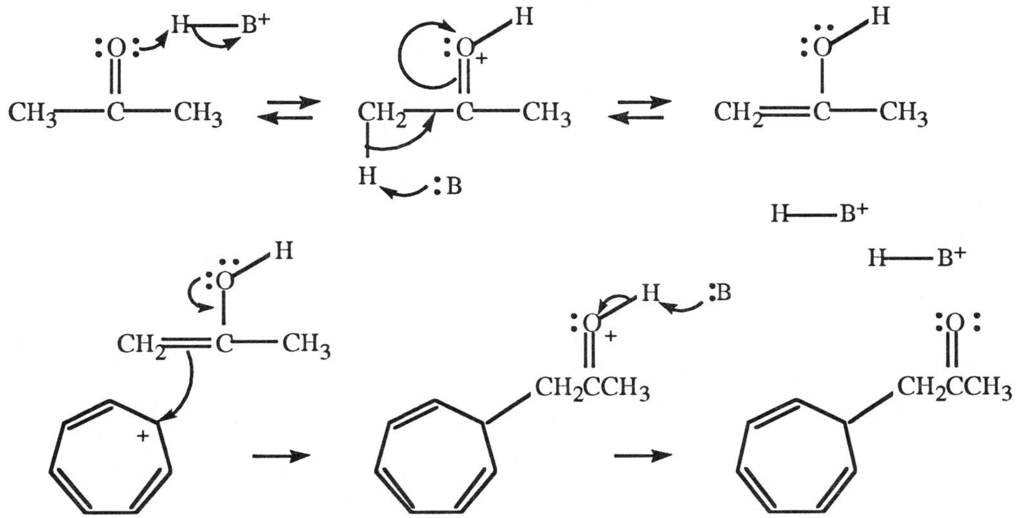

19.48

(a)

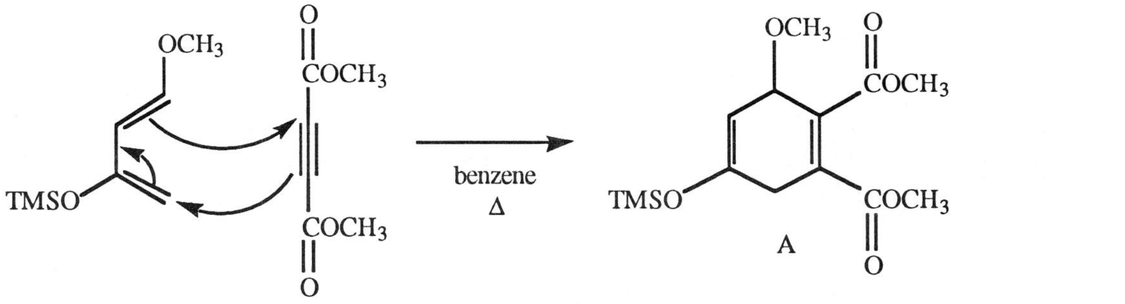

(b)

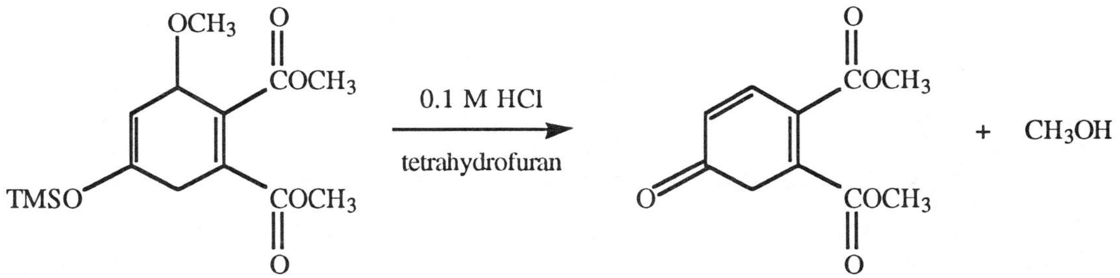

19.48 (b) (cont)

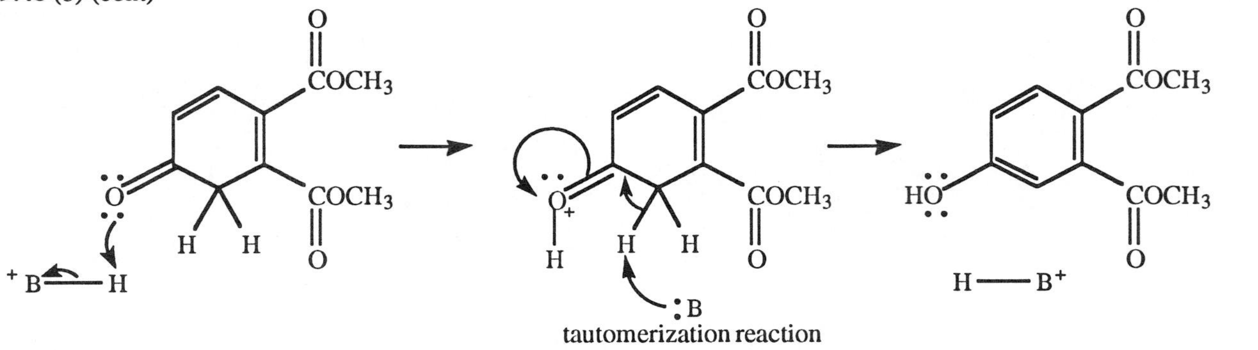

tautomerization reaction

19.49

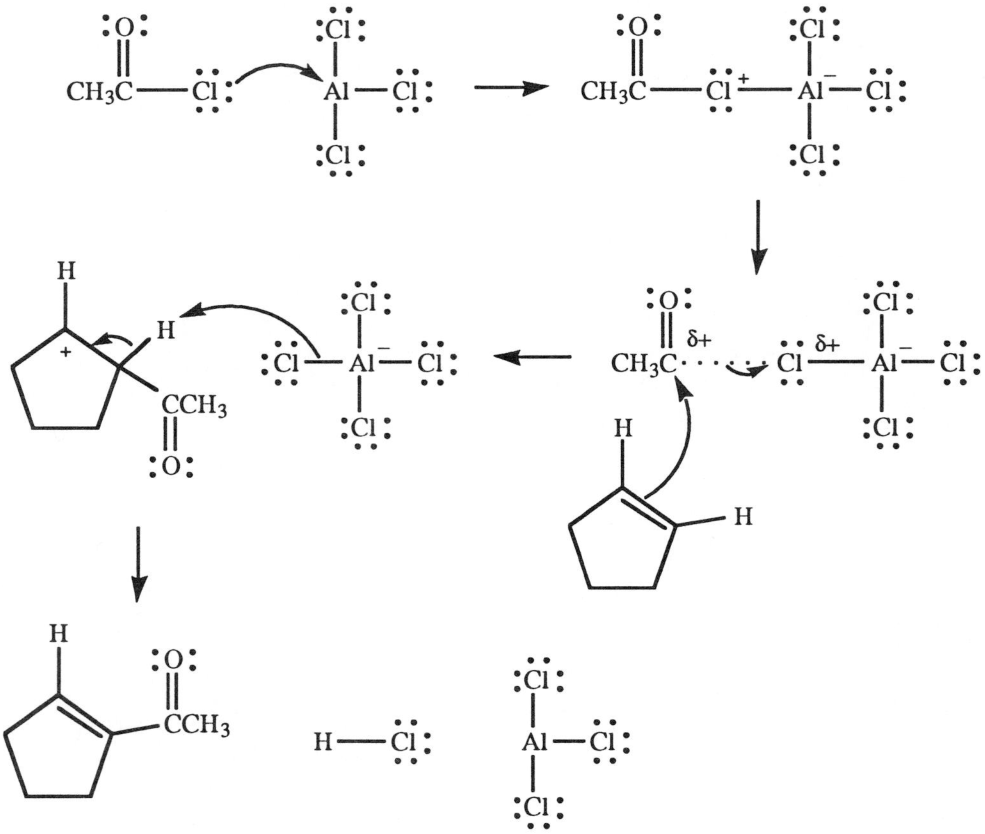

19.50

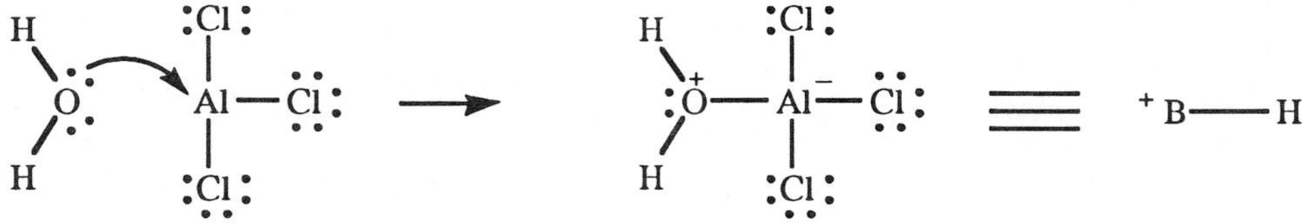

Brønsted acid

19.50 (cont)

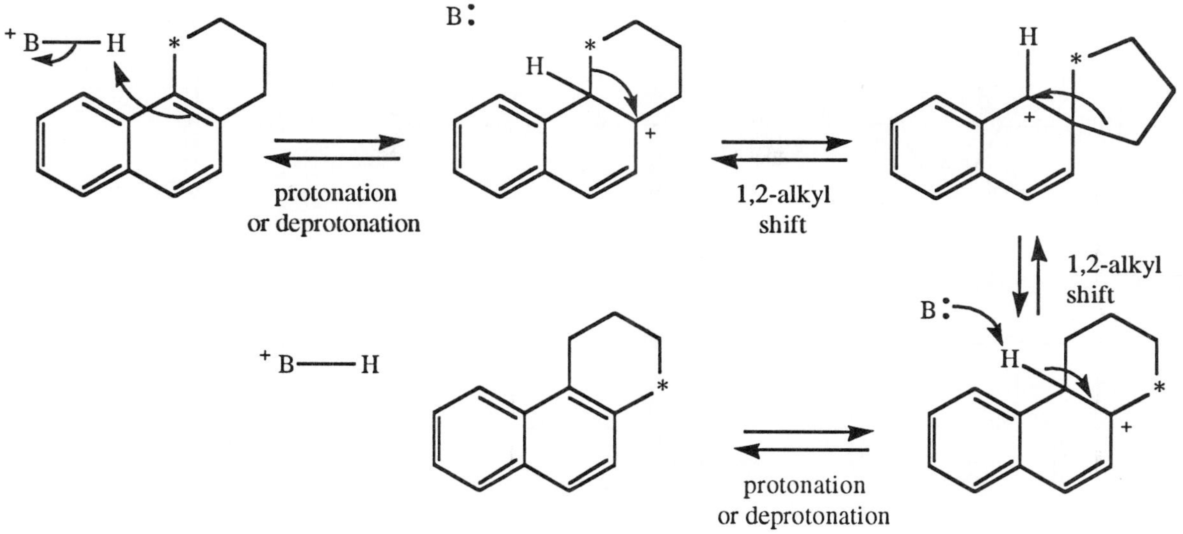

19.51

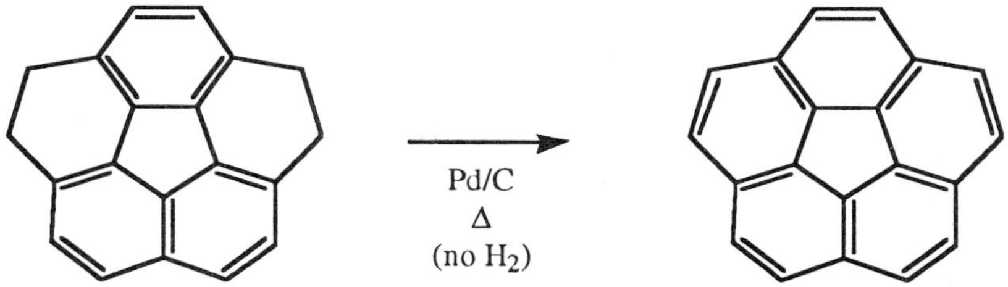

corannulene
$C_{20}H_{10}$

Corannulene has 3 double bonds in each six-membered ring.

19.52

(a)

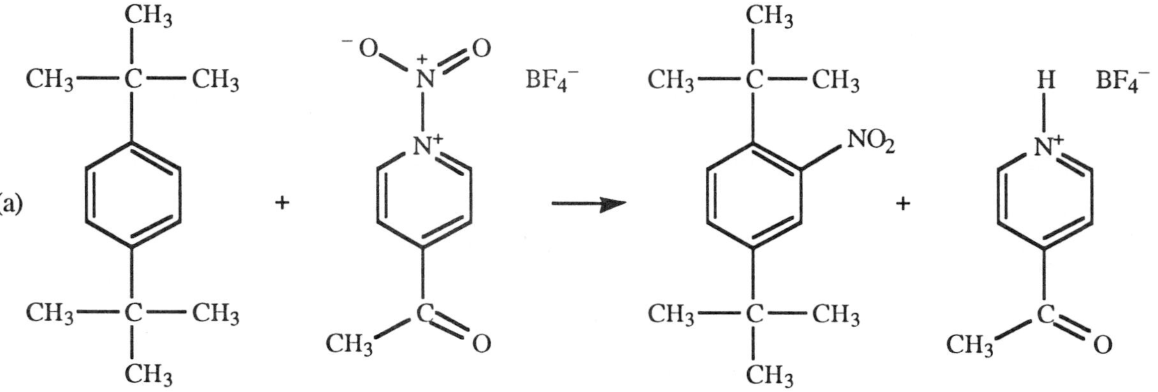

(b)   The minor product is

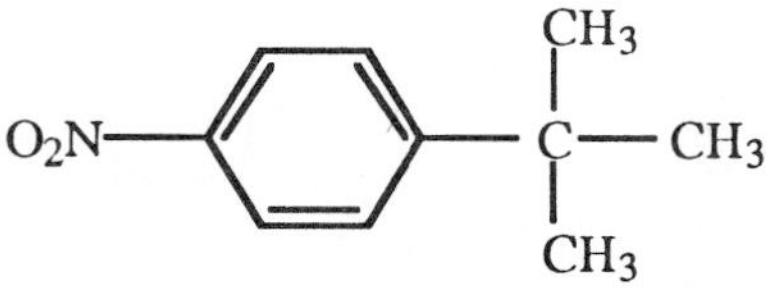

19.52 (b) (cont).

This reaction is possible because of the stability of the tertiary carbocation that is expelled. We would not expect to see the same reaction if a methyl cation were the species being expelled.

19.53   (a) $C_{60} = 60 \times 12 = 720$ amu
1656 − 720 = 936 amu
MW of benzene = 78 amu
936 amu/78 amu = 12 benzene units,
therefore MW 1656 corresponds to $C_{60}(C_6H_6)_{12}$.

(b)   The proton nuclear magnetic resonance spectrum shows that the hydrogen atoms on benzene have been converted into two kinds of hydrogens. Five of them remain as aromatic hydrogens; one has been converted into another type of hydrogen, perhaps more typical of alkene hydrogens.

(c)

20

Free Radicals

Concept Map 20.1   Chain reactions.

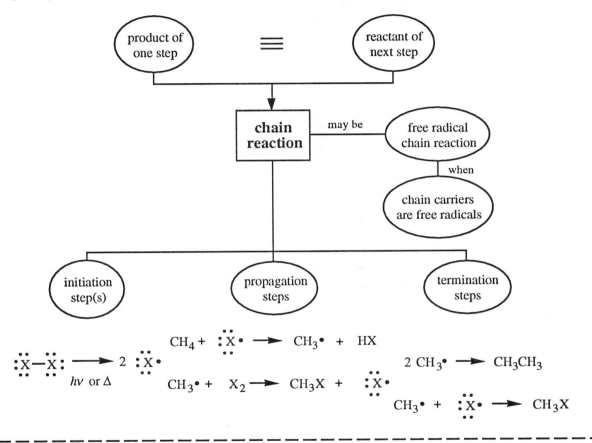

20.1   Propane has six primary hydrogen atoms and two secondary hydrogen atoms for a total of eight hydrogen atoms. If there were no difference in reactivity between primary and secondary hydrogen atoms, for every six primary hydrogen atoms that react, two secondary hydrogen atoms would react, and the product mixture would consist of 75% 1-chloropropane and 25% 2-chloropropane. When the difference in reactivity is considered, for every six primary hydrogen atoms that react, 2×2.5 or five secondary hydrogen atoms would react. The product mixture should thus consist of 55% 1-chloropropane and 45% 2-chloropropane.

$$\frac{6}{6 + 5} \times 100 = 55\%$$

$$\frac{5}{6 + 5} \times 100 = 45\%$$

20.1 (cont)

2-Methylpropane has nine primary hydrogen atoms and one tertiary hydrogen atom. If there were no difference in reactivity between primary and tertiary hydrogen atoms, the product mixture would consist of 90% 1-chloro-2-methylpropane (isobutyl chloride) and 10% 2-chloro-2-methylpropane (*tert*-butyl chloride). When the difference in reactivity is considered, for every nine primary hydrogen atoms that react, 1×4 or four tertiary hydrogen atoms would react. The product mixture should thus consist of 69% 1-chloro-2-methylpropane and 31% 2-chloro-2-methylpropane.

$$\frac{9}{9+4} \times 100 = 69\%$$

$$\frac{4}{9+4} \times 100 = 31\%$$

20.2   The selectivity observed at 300 °C depends on small differences between energies of activation for the abstraction of primary, secondary, and tertiary hydrogen atoms. As the temperature increases, the average kinetic energy of chlorine atoms and of alkane molecules increases, and the number of collisions that have energies in excess of the energy of activation for the abstraction of the different kinds of hydrogen atoms also increases. Differences in energies of activation for different reactions become less and less relevant. We see a leveling off of the selectivity of the reactions. If the temperature is high enough, almost all molecules will have energy equal to or greater than the activation energies for any hydrogen to be abstracted, and the rates of reaction would be the same.

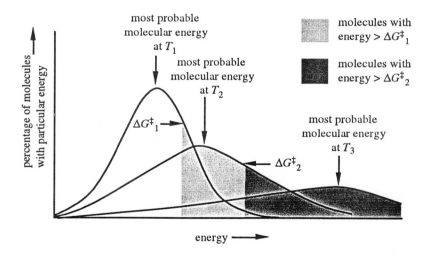

20.3   Neopentane (2,2-dimethylpropane) has only primary hydrogen atoms. The enthalpy for the abstraction of a primary hydrogen atom by a chlorine atom is –5 kcal/mol. The activation energy for this exothermic step is small (~1 kcal/mol, p. 887 in text), and the rate correspondingly fast. The enthalpy for the abstraction of a primary hydrogen atom by a bromine atom, on the other hand, is +12 kcal/mol. The activation energy must be at least as large as the enthalpy, and the rate correspondingly slow.

Concept Map 20.2   Halogenation of alkanes.

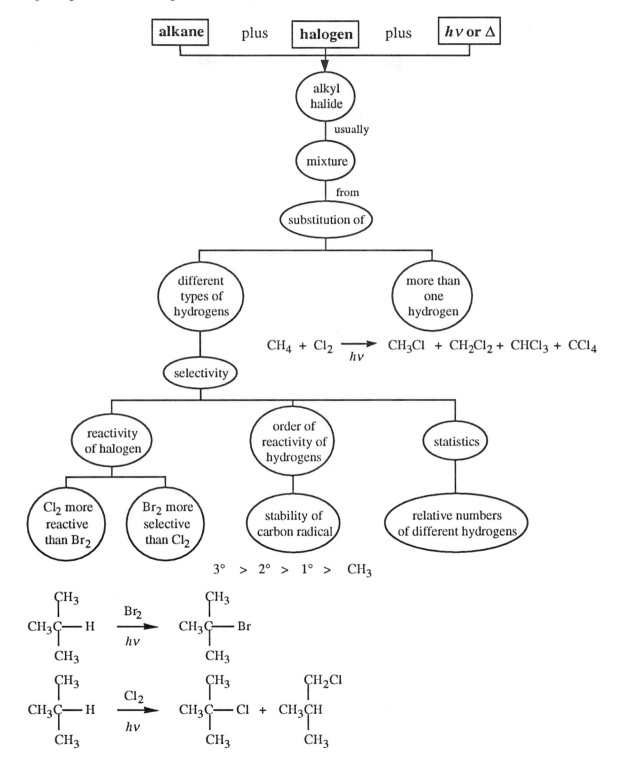

20.4

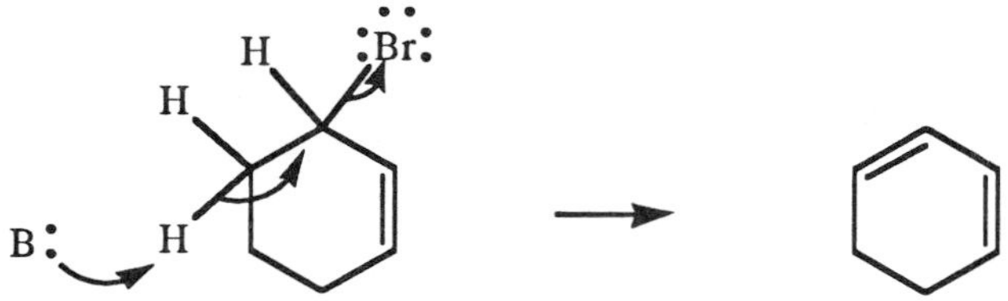

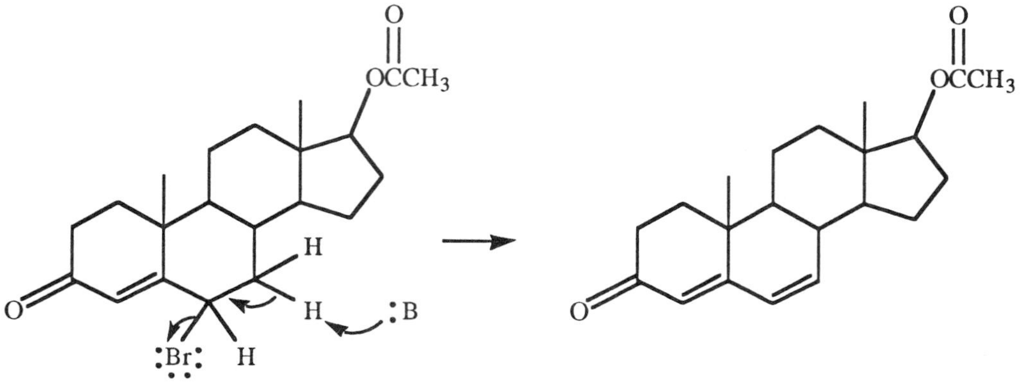

20.5   **(a)**

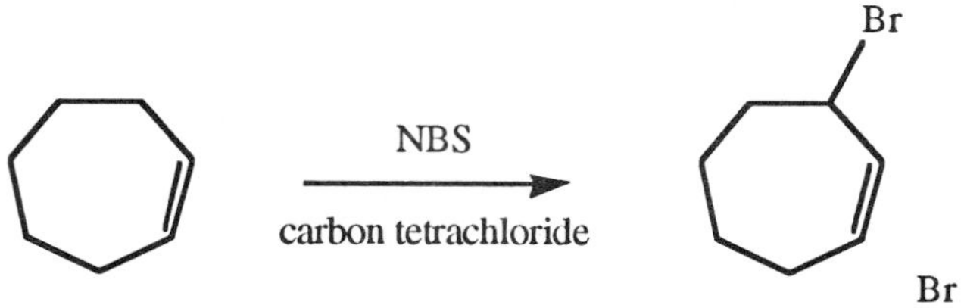

NBS

carbon tetrachloride

(b)

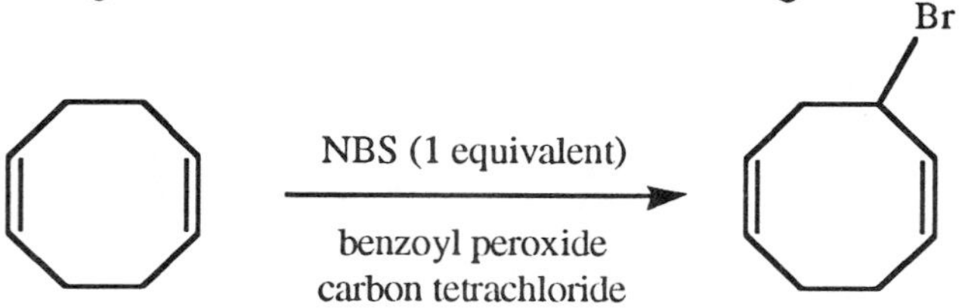

NBS (1 equivalent)

benzoyl peroxide
carbon tetrachloride

(c)

NBS

carbon tetrachloride

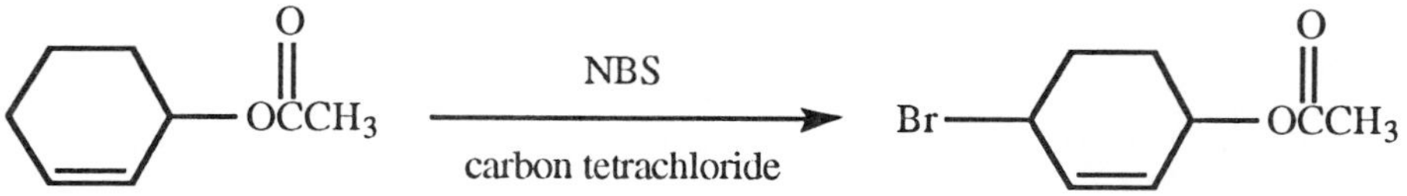

This product is reported to form in 58% yield. Another product is also possible, the one in which the allylic hydrogen atom adjacent to the oxygen atom is replaced.

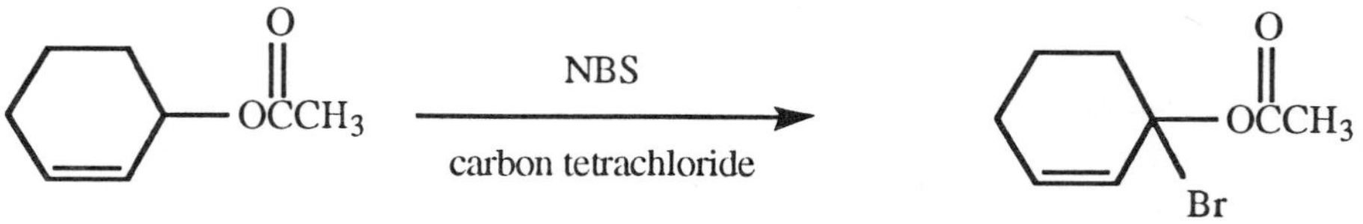

NBS

carbon tetrachloride

Usually an ether oxygen stabilizes a radical at an adjacent carbon atom. The following resonance contributors are written to explain the effect.

In this case, the electron-withdrawing carbonyl group attached to the oxygen atom reduces the electron density at the oxygen atom, making the corresponding resonance contributor unfavorable. The product, with two good leaving groups on the same allylic carbon atom, might also be too unstable to isolate.

20.6

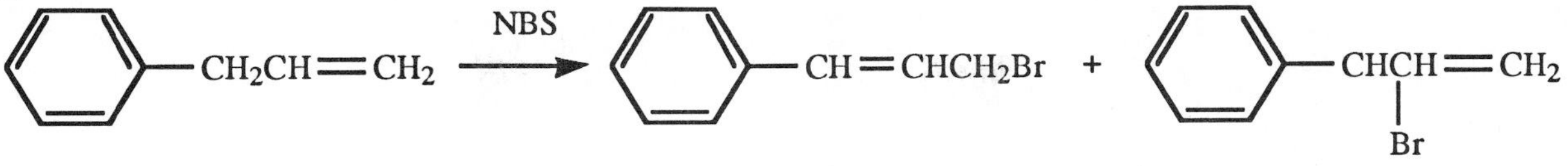
NBS

50% 10%

Initiation step:

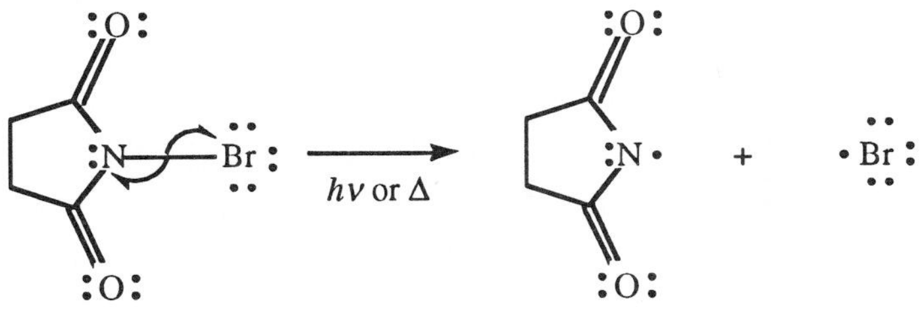
hv or Δ

The role of benzoyl peroxide is not known.

Propagation steps:

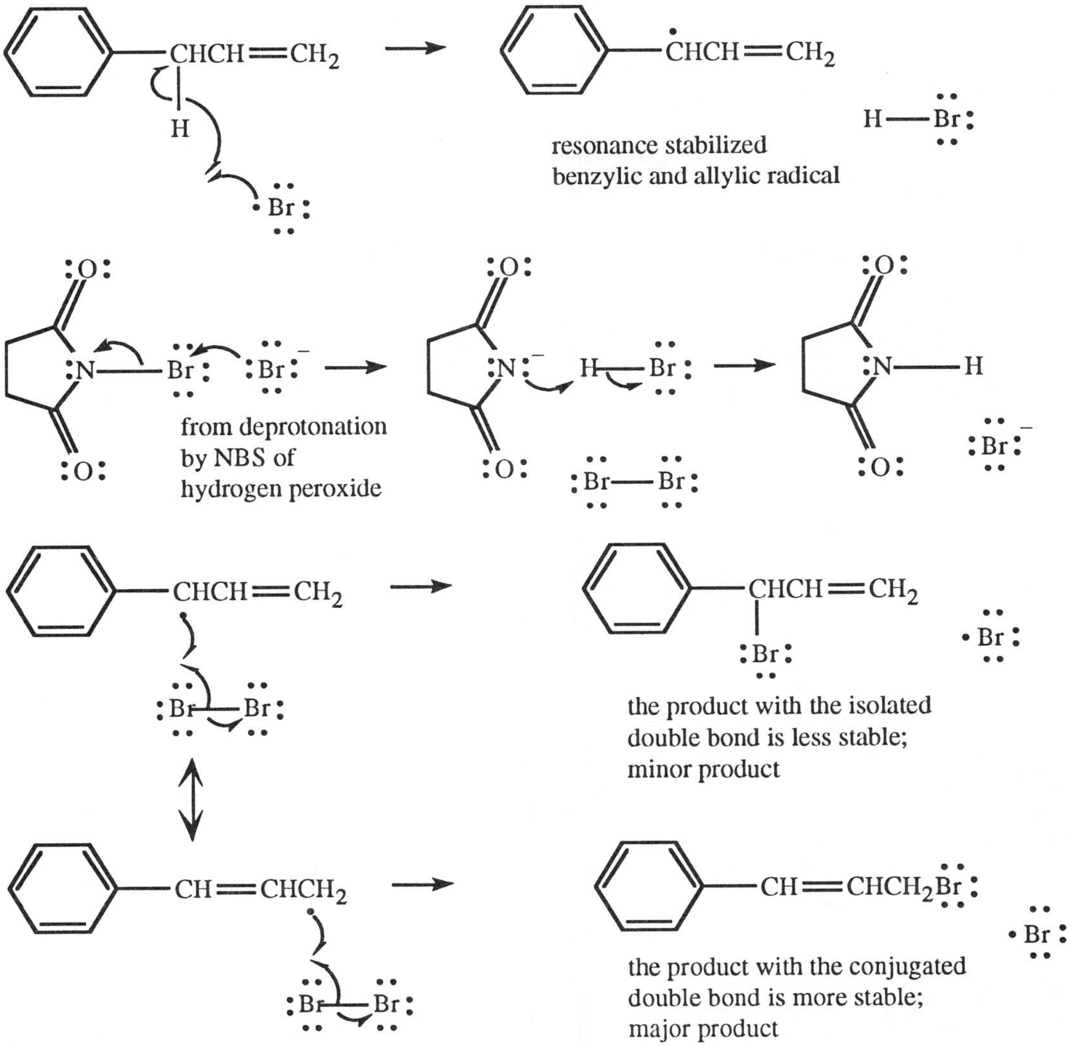
resonance stabilized
benzylic and allylic radical
from deprotonation
by NBS of
hydrogen peroxide
the product with the isolated
double bond is less stable;
minor product
the product with the conjugated
double bond is more stable;
major product

20.7 $CH_3CH_2CH_2CH_2CH = CHCH_3$ $\xrightarrow{\text{NBS}}$ $CH_3CH_2CH_2\overset{\displaystyle|}{\underset{\displaystyle Br}{C}}HCH = CHCH_3$

                                                                                            60%

$CH_3CH_2CH_2CHCH = CHCH_3$

H $\cdot\ddot{B}\ddot{r}\!:$

$CH_3CH_2CH_2\overset{\displaystyle\cdot}{C}HCH = CHCH_3$ $\longleftrightarrow$ $CH_3CH_2CH_2CH = CH\overset{\displaystyle\cdot}{C}HCH_3$

         $\downarrow Br_2$ $\downarrow Br_2$

$CH_3CH_2CH_2\overset{\displaystyle|}{\underset{\displaystyle Br}{C}}HCH = CHCH_3$ $CH_3CH_2CH_2CH = CH\overset{\displaystyle|}{\underset{\displaystyle Br}{C}}HCH_3$

4-bromo-2-heptene   2-bromo-3-heptene

$CH_3CH_2CH_2CH_2CH = CHCH_2$

                H $\cdot\ddot{B}\ddot{r}\!:$

$CH_3CH_2CH_2CH_2CH = CH\overset{\displaystyle\cdot}{C}H_2$ $\longleftrightarrow$ $CH_3CH_2CH_2CH_2\overset{\displaystyle\cdot}{C}HCH = CH_2$

         $\downarrow Br_2$ $\downarrow Br_2$

$CH_3CH_2CH_2CH_2CH = CHCH_2Br$ $CH_3CH_2CH_2CH_2\overset{\displaystyle|}{\underset{\displaystyle Br}{C}}HCH = CH_2$

1-bromo-2-heptene   3-bromo-1-heptene

20.8

(a)

major product

(b)

20.8 (cont)

(c)

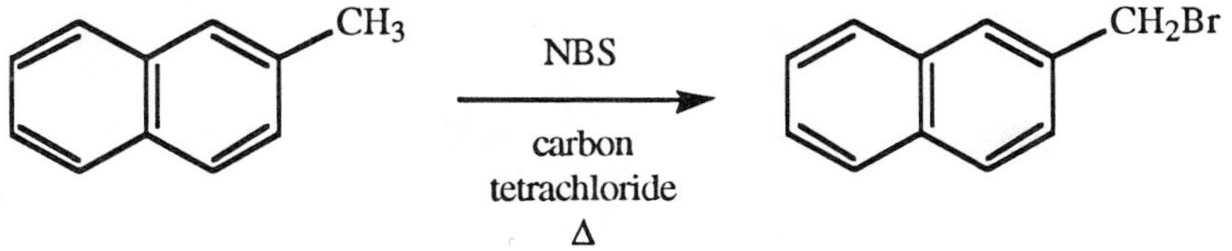

(d)

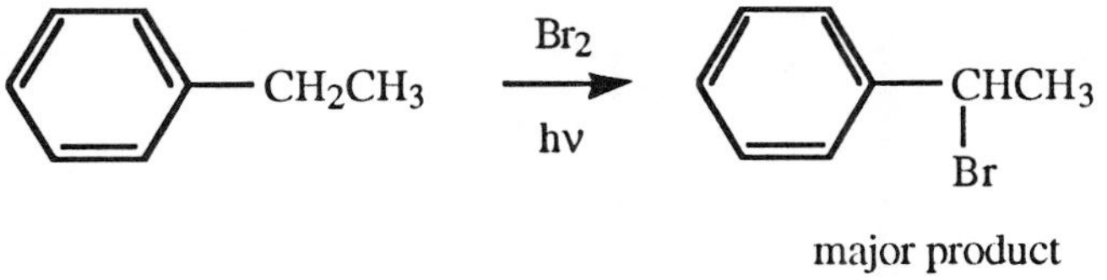

major product

- -

Concept Map 20.3   Selective free radical halogenations.

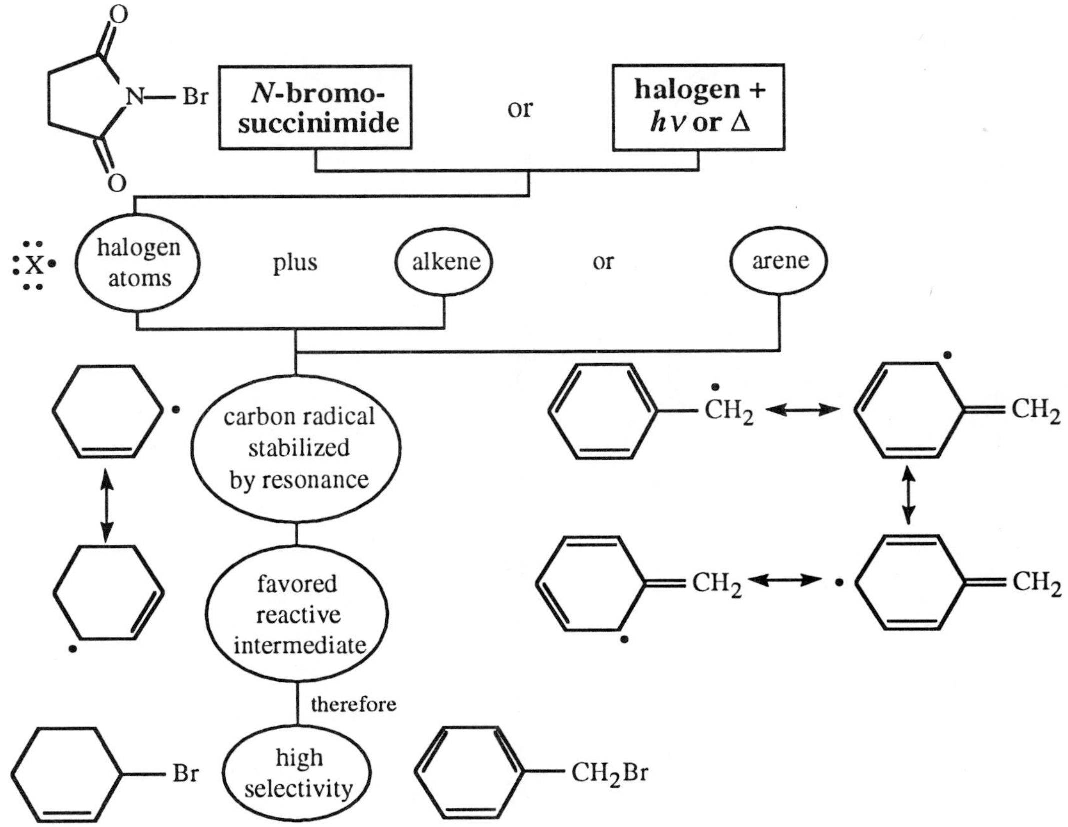

20.9   The unpaired electron can be delocalized over three rings.

20.10

$$CH_3CHCH_2-H \longrightarrow CH_3CHCH_2 + H\cdot \qquad DH° \quad 98 \text{ kcal/mol}$$

(with CH_3 substituent on the CH)

$$CH_3C-H \longrightarrow CH_3C\cdot + H\cdot \qquad DH° \quad 91 \text{ kcal/mol}$$

(with two CH_3 substituents on the C)

$$\Delta(DH°) = 98 - 91 = 7 \text{ kcal/mol}$$

20.10 (cont)

$$CH_3CH_2CH_2\!-\!\!H \longrightarrow CH_3CH_2\underset{\cdot}{CH_2} + H\cdot \qquad DH° \quad 98 \text{ kcal/mol}$$

$$\underset{\displaystyle CH_3CH\!-\!\!H}{\overset{\displaystyle \overset{CH_3}{|}}{}} \longrightarrow \underset{\displaystyle CH_3\underset{\cdot}{CH}}{\overset{\displaystyle \overset{CH_3}{|}}{}} + H\cdot \qquad DH° \quad 95 \text{ kcal/mol}$$

$$\Delta(DH°) = 98 - 95 = 3 \text{ kcal/mol}$$

20.11 The change in mechanism does not change the relative energy levels for propene, hydrogen bromide, *n*-propyl bromide, and isopropyl bromide because the energies of starting material and products are dependent only on the states (i.e., temperature, pressure, physical state, etc.) and not on how the starting material is converted to product. As long as the initial and final states remain the same in the two reactions, the relative energy levels remain the same. What does change is the nature of the intermediates and their relative energies, as well as the activation energies for each step.

20.12

20.12 (cont)

3.

$$\overset{:\overset{\cdots}{O}:}{\underset{}{CH_3C\overset{||}{O}}}\!\!-\!\!\overset{:\overset{\cdots}{O}:}{\underset{\cdots}{OCCH_3}} \quad \xrightarrow{h\nu} \quad 2\ \overset{:\overset{\cdots}{O}:}{\underset{\cdots}{CH_3C\overset{||}{O}}}\cdot$$

$$\overset{:\overset{\cdots}{O}:}{\underset{\cdots}{CH_3C\overset{||}{O}}}\!\!\cdot \quad H\!\!-\!\!SiCl_3 \quad \longrightarrow \quad \cdot SiCl_3 \quad + \quad \overset{:\overset{\cdots}{O}:}{\underset{}{CH_3C\overset{||}{O}H}}$$

$$CH_2\!\!=\!\!CH(CH_2)_5CH_3$$

$$Cl_3SiCH_2CH_2(CH_2)_5CH_3 \quad \longleftarrow \quad Cl_3SiCH_2\overset{\cdot}{C}H(CH_2)_5CH_3$$

$$\cdot SiCl_3 \qquad\qquad\qquad H\!\!-\!\!SiCl_3$$

20.13

(a)

+ CH$_3$SH $\xrightarrow[\text{h}\nu]{\text{acetone}}$

(b) CH$_3$(CH$_2$)$_5$CH=CH$_2$ $\xrightarrow[\text{peroxides}]{\text{HBr}}$ CH$_3$(CH$_2$)$_7$Br

(c) Cl$_3$CBr + $\underset{}{CH_2\!\!=\!\!CCH_2CH_3}\overset{\overset{\textstyle CH_2CH_3}{\textstyle |}}{}$ $\xrightarrow{\Delta}$ $Cl_3CCH_2\overset{\overset{\textstyle CH_2CH_3}{\textstyle |}}{\underset{\underset{\textstyle Br}{\textstyle |}}{C}}CH_2CH_3$

(d) CH$_3$(CH$_2$)$_4$SiH$_3$ + CH$_3$(CH$_2$)$_5$CH=CH$_2$ $\xrightarrow{\text{peroxides}}$ CH$_3$(CH$_2$)$_4$SiH$_2$(CH$_2$)$_7$CH$_3$

Concept Map 20.4   Free radical addition reactions of alkenes.

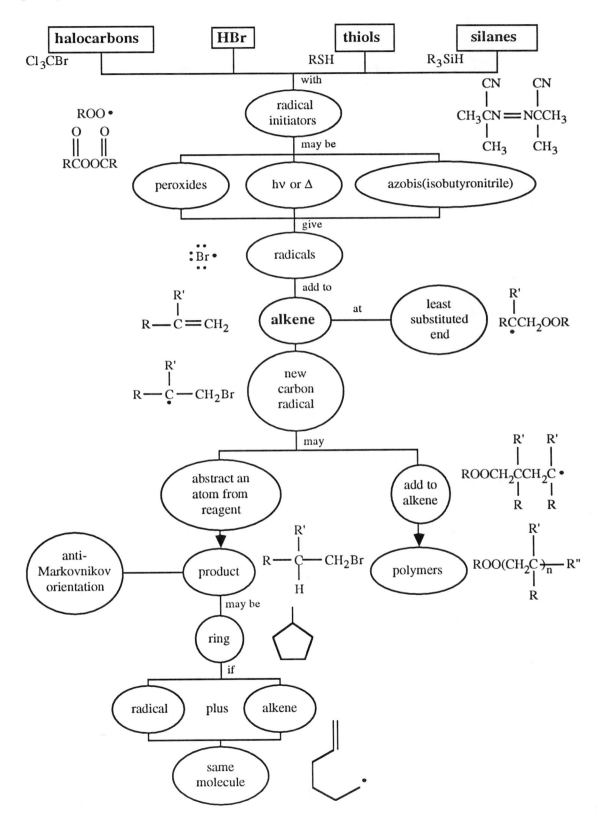

20.14

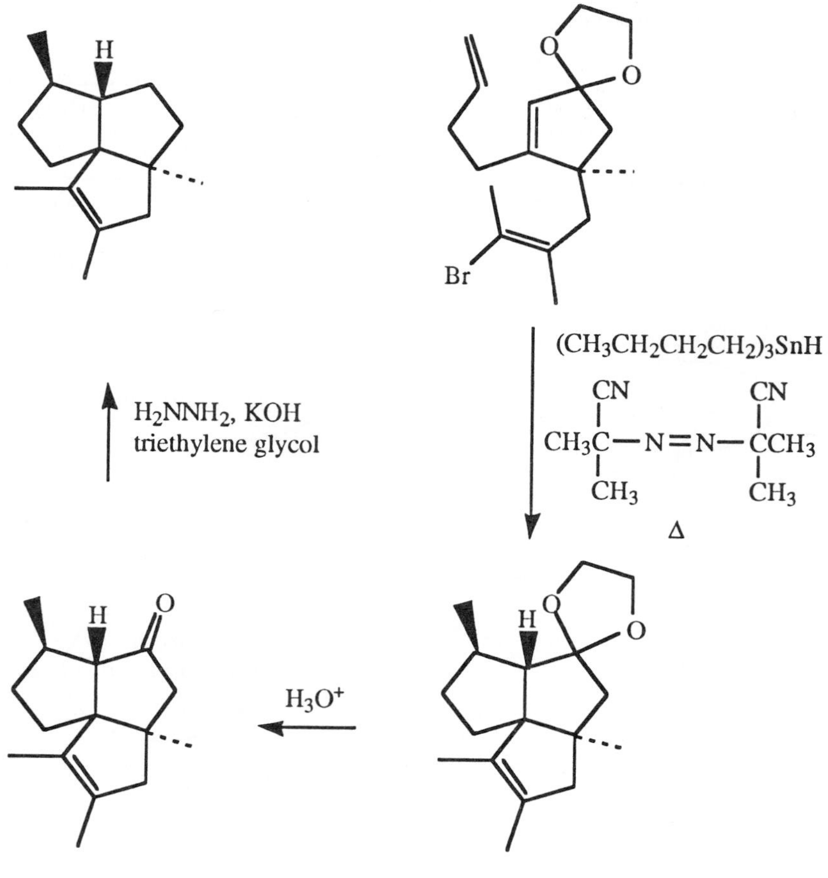
H
O
O
Br
(CH3CH2CH2CH2)3SnH
CN
CN
CH3C—N=N—CCH3
CH3
CH3
Δ
H2NNH2, KOH
triethylene glycol
H3O+
H
O
H
O
O

20.15

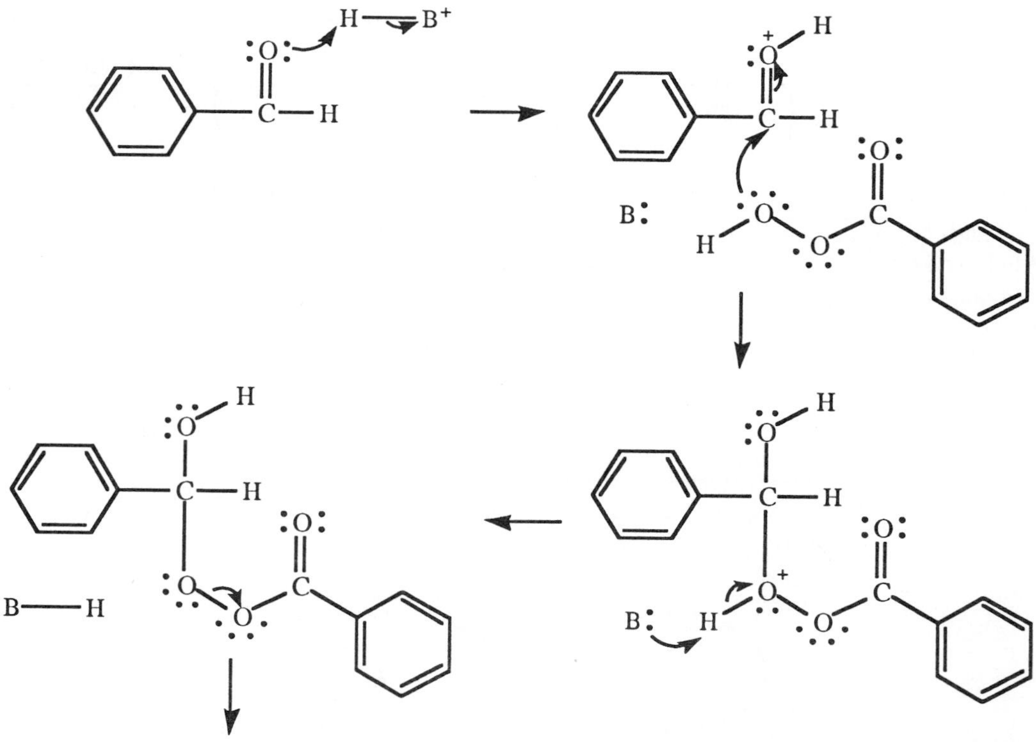
H—B+
O
C—H
O
H
C—H
O
C
B:
O
H
O
H
O
C—H
O
C
+B—H
O
O
H
O
C—H
O
C
B:
H
O
O

20.15 (cont)

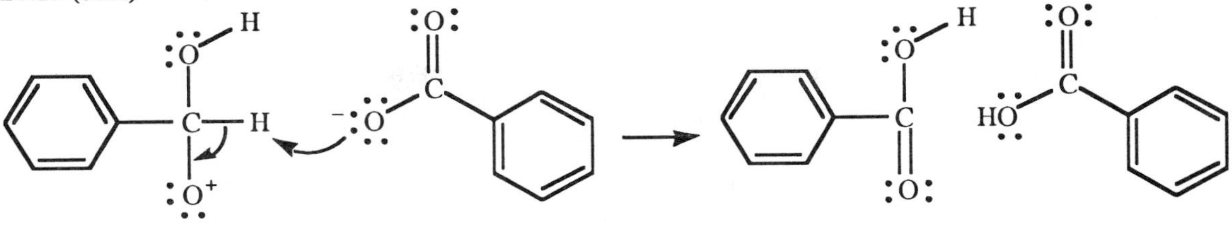

20.16

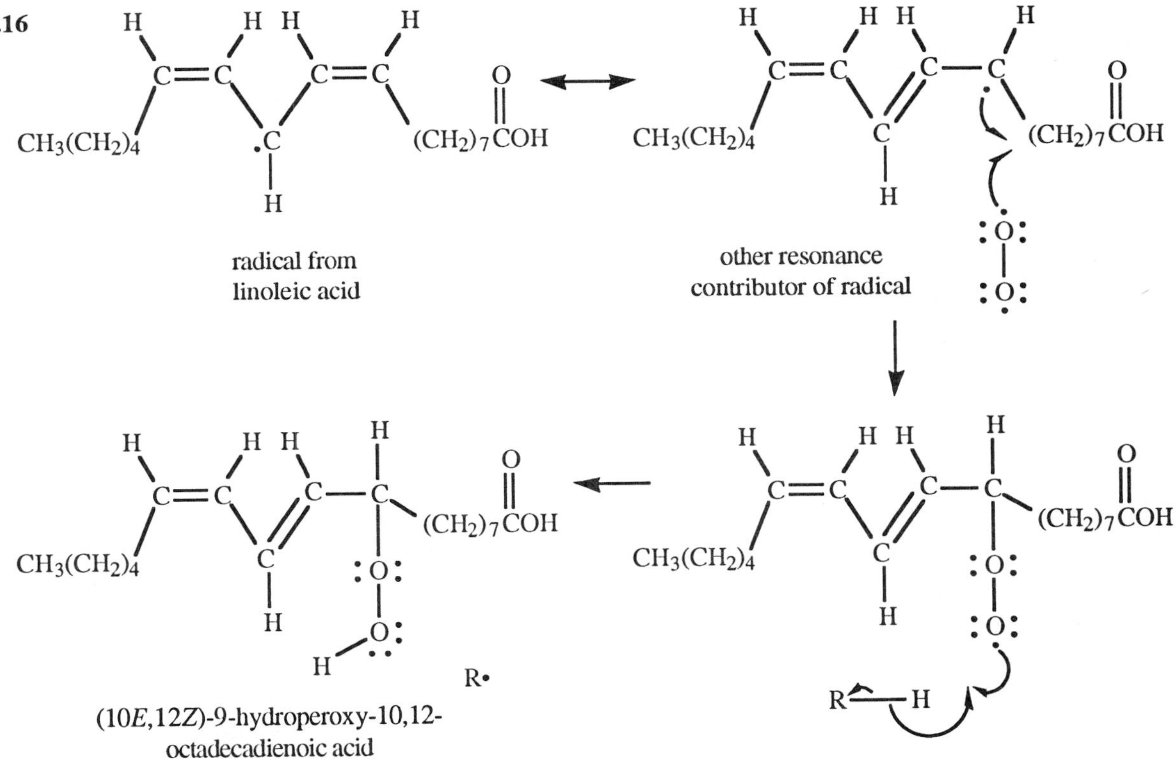

radical from
linoleic acid

other resonance
contributor of radical

(10*E*,12*Z*)-9-hydroperoxy-10,12-
octadecadienoic acid

R•

R—H

20.17

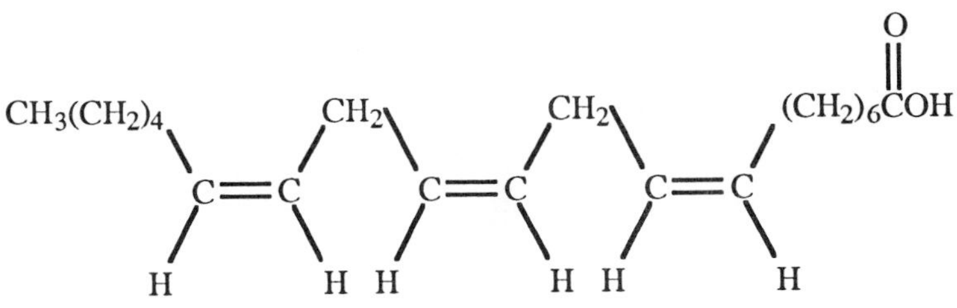

(8*Z*,11*Z*,14*Z*)-8,11,14-icosatrienoic acid

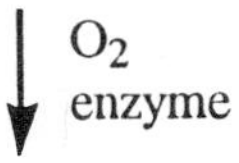

↓ O₂
   enzyme

20.17 (cont)

(8Z,11Z,13E)-15-hydroperoxy-8,11,13-icosatrienoic acid

20.17 (cont)

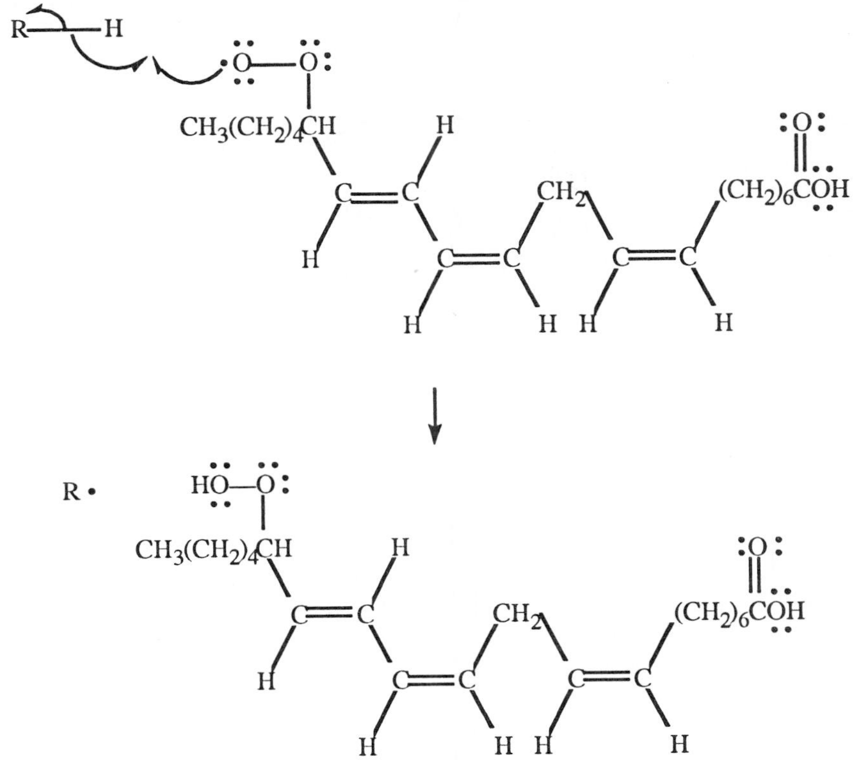

20.18

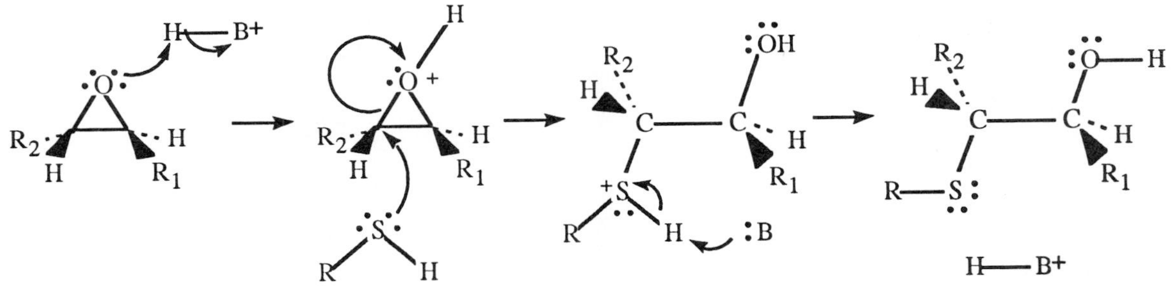

Note that while the opening of the ring in an unsymmetrical oxirane usually gives rise to a mixture of isomers, a reaction in biological systems gives rise to a single product. The mechanism shown above assumes that the reaction is taking place with a preferred orientation at the active site of an enzyme (see Section 26.5 for an example) and that, therefore, only one product is formed.

20.19

(a)

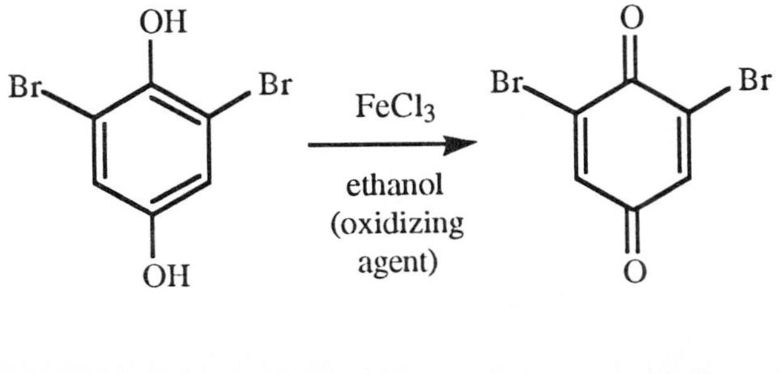

20.19 (cont)

(b)

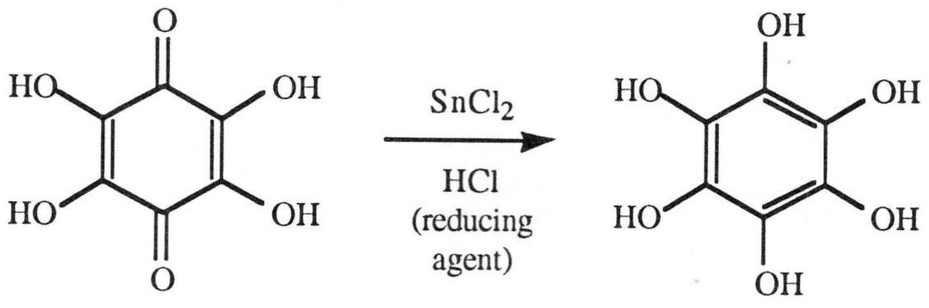

(c)

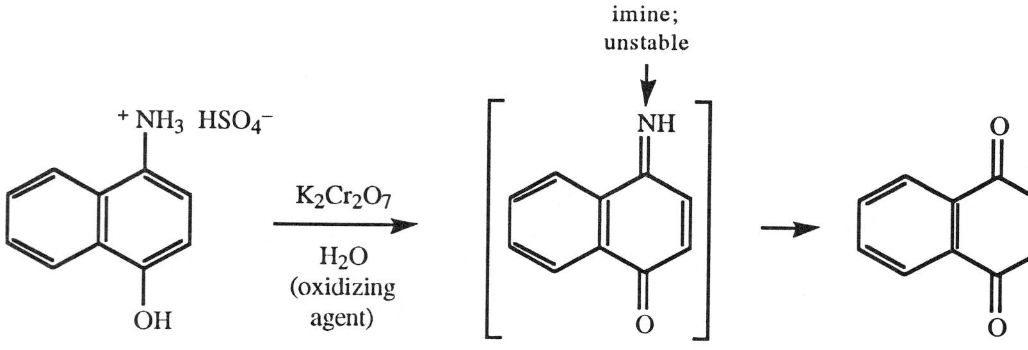

imine;
unstable

unstable, converted in
water to *p*-napthoquinone

(d)

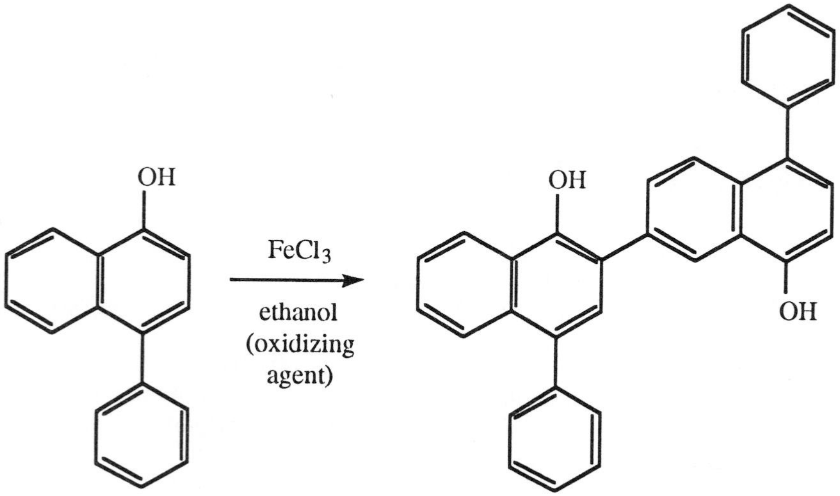

(one of several possible products)

(e)

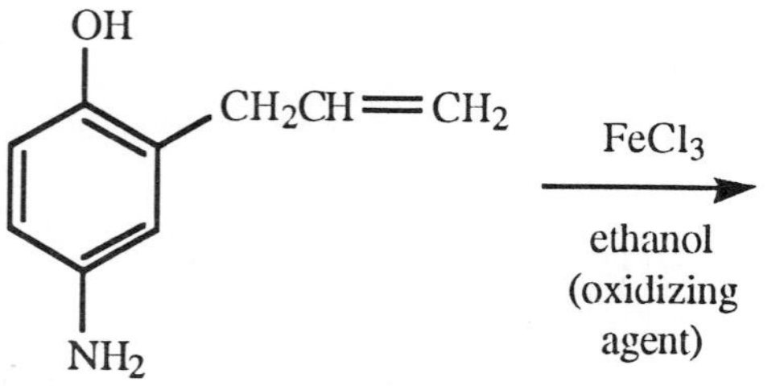

20.19 (e) (cont)

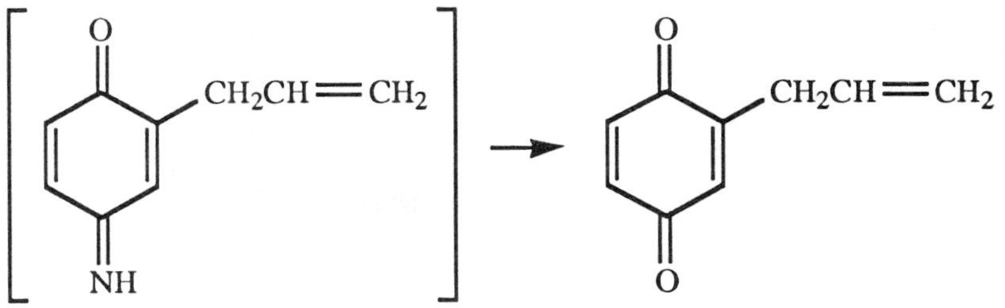

20.20

(a)

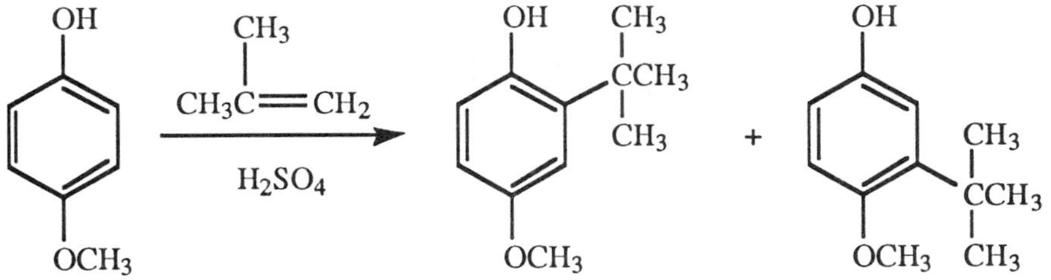

(b)

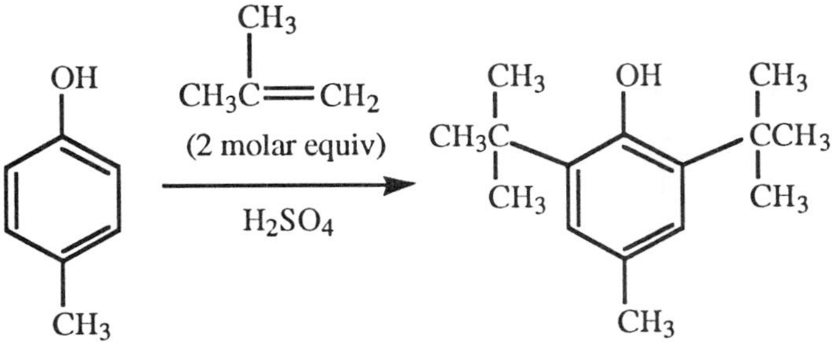

Both syntheses are Friedel-Crafts alkylation reactions. In (a) both the hydroxyl group and the methoxyl group are strong ortho,para directors, leading to attack at positions ortho to each. In (b) the hydroxyl group is a much stronger ortho,para director than the methyl group, and attack is at the two positions ortho to the hydroxyl group.

Concept Map 20.5   (see p. 744)

20.21

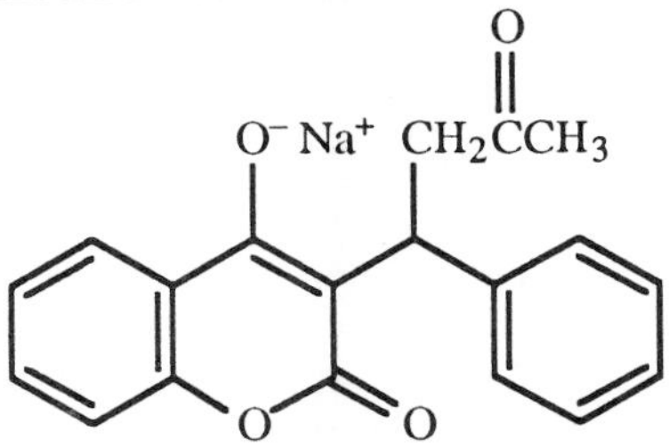

20.22

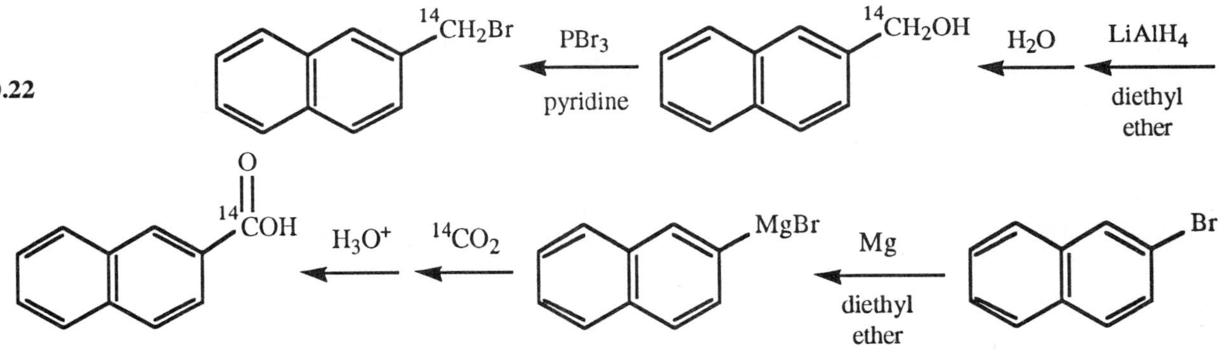

Concept Map 20.5   Oxidation reactions as free radical reactions.

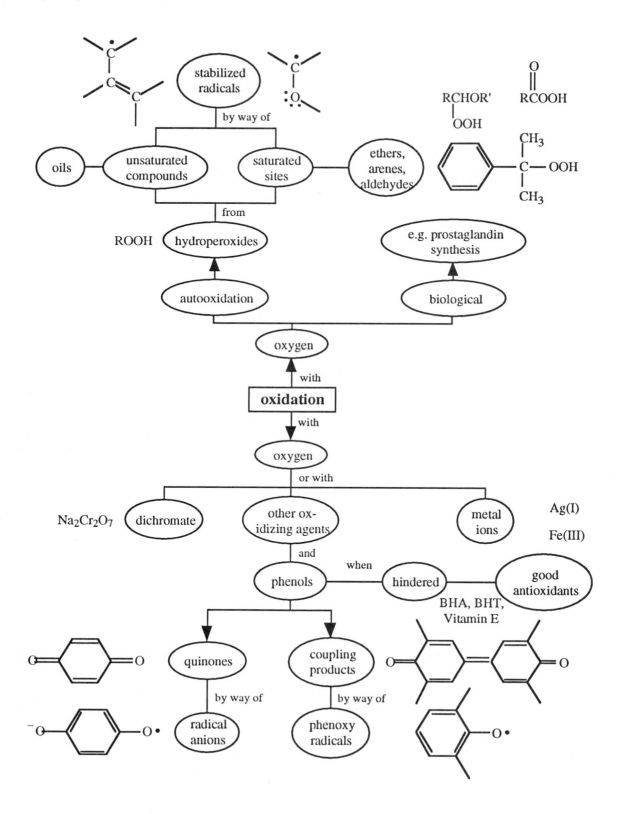

20.23

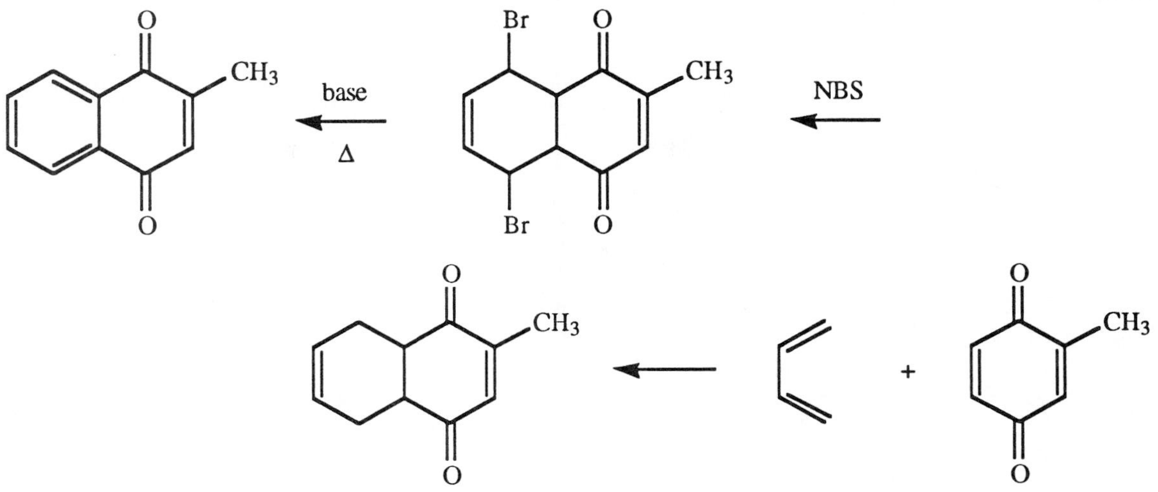

An alternate synthesis is shown below:

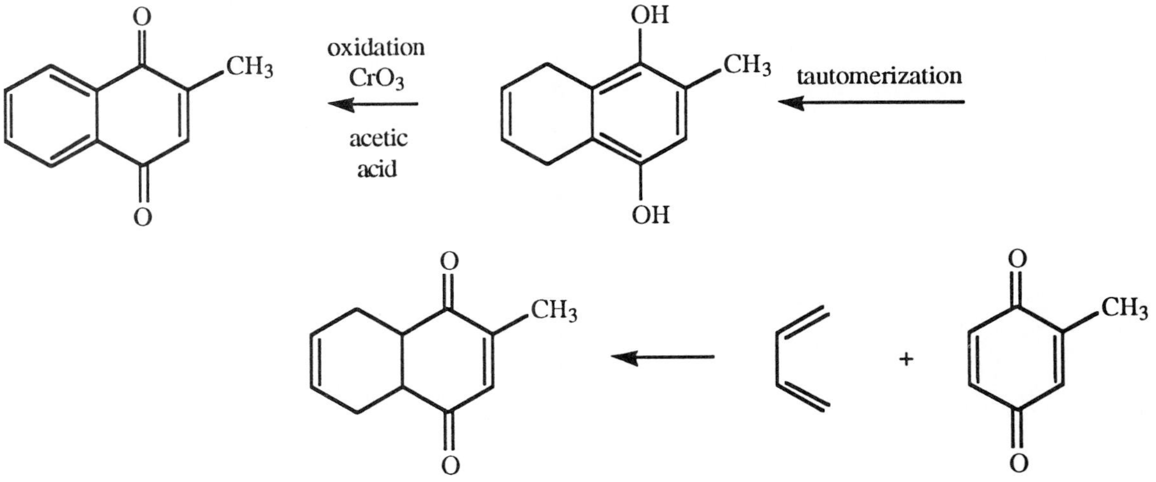

20.24

(a) $CH_3CH= CHCH= CHCH_3$ $\xrightarrow[\text{carbon tetrachloride}]{\text{Br}_2 \text{ (1 molar equiv)}}$

$CH_3CHBrCHBrCH= CHCH_3$ + $CH_3CHBrCH= CHCHBrCH_3$

(b)

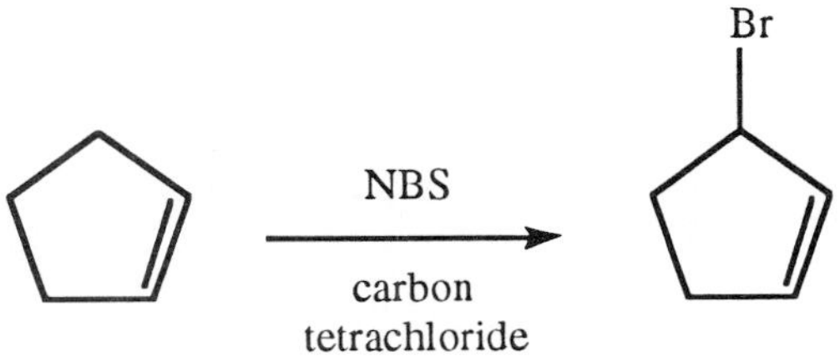

20.24 (cont)

(c)

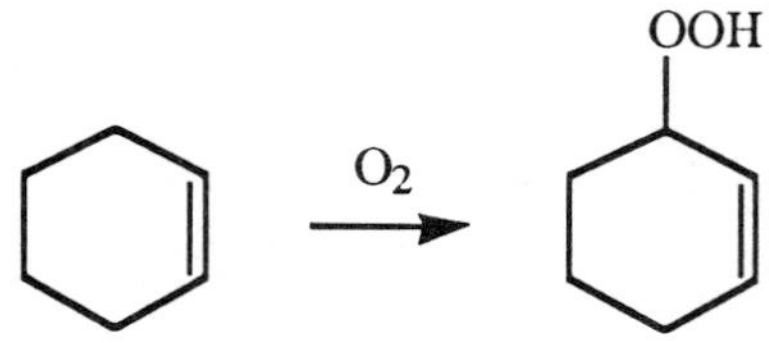

(d) $CH_3CH_2CH_3$ $\xrightarrow[hv]{Cl_2}$ $CH_3CH_2CH_2Cl$ + $CH_3CHClCH_3$ + $CH_3CH_2CHCl_2$ + $CH_3CHClCH_2Cl$

major products   two of the possible minor products

(e)

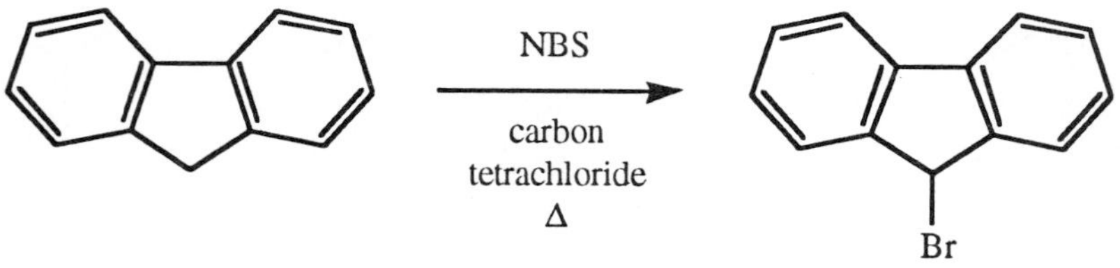

(f)

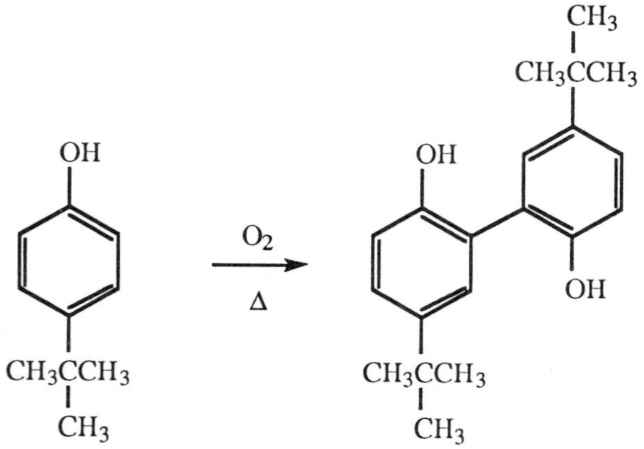

(g)

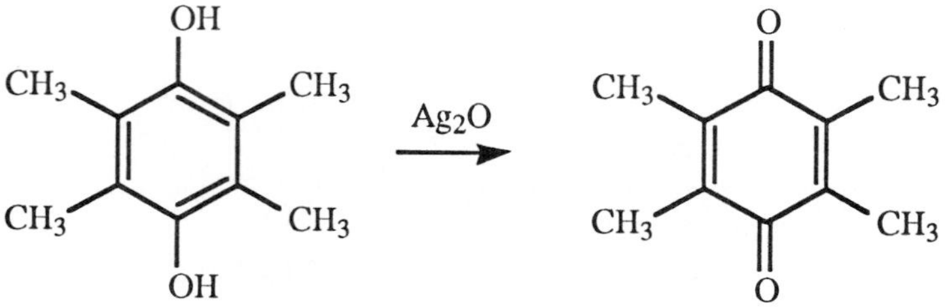

(h)

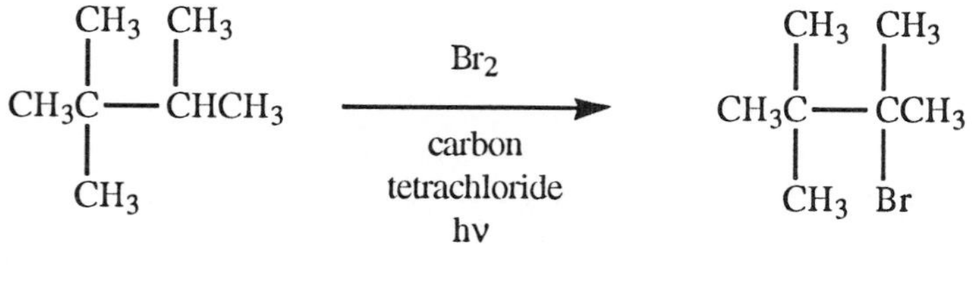

20.24 (cont)

(i)

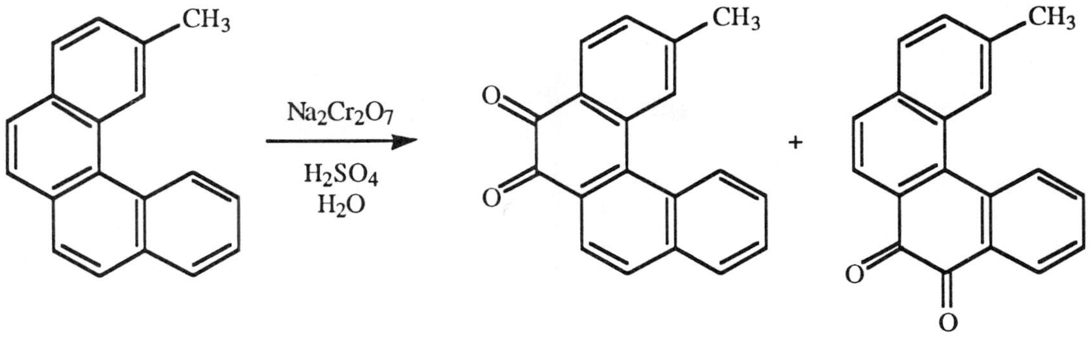

(j)

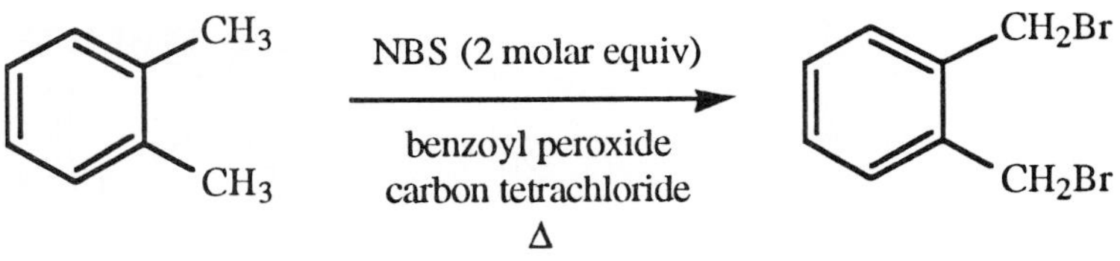

NBS (2 molar equiv)

benzoyl peroxide
carbon tetrachloride
Δ

(k)

$+$ $(CH_3CH_2CH_2CH_2)_3SnH$

90 °C

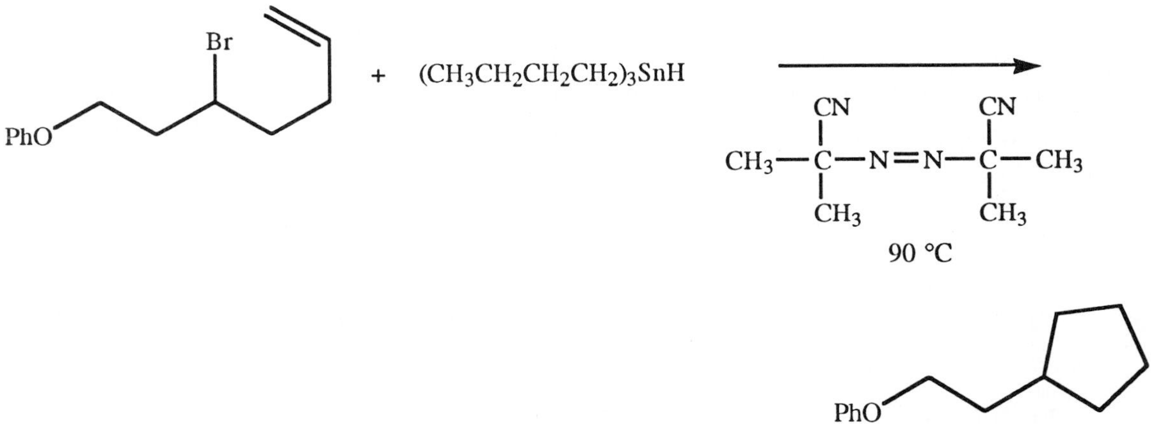

20.25

(a) $CH_3CH_2CH_2CH=CH_2$ $+$ Cl_3SiH $\xrightarrow{280\ °C}$ $CH_3CH_2CH_2CH_2CH_2SiCl_3$

(b) $CH_3(CH_2)_5CH=CH_2$ $+$ CCl_4 $\longrightarrow$ $CH_3(CH_2)_5CHClCH_2CCl_3$

$CH_3COOCCH_3$
$h\nu$

(c)

$\xrightarrow[\text{peroxide}]{\text{HBr}}$

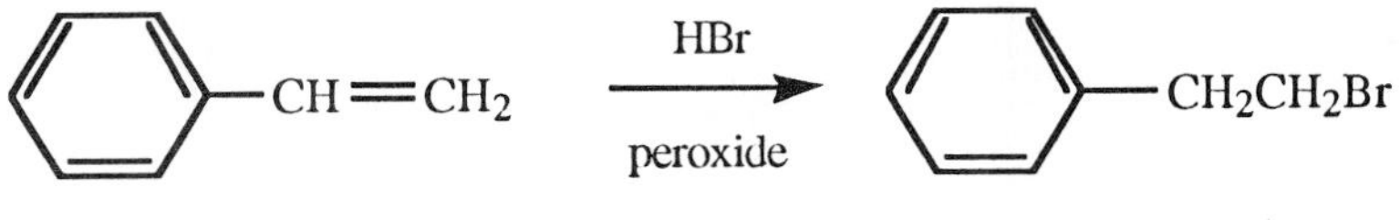

20.25 (cont)

(d)

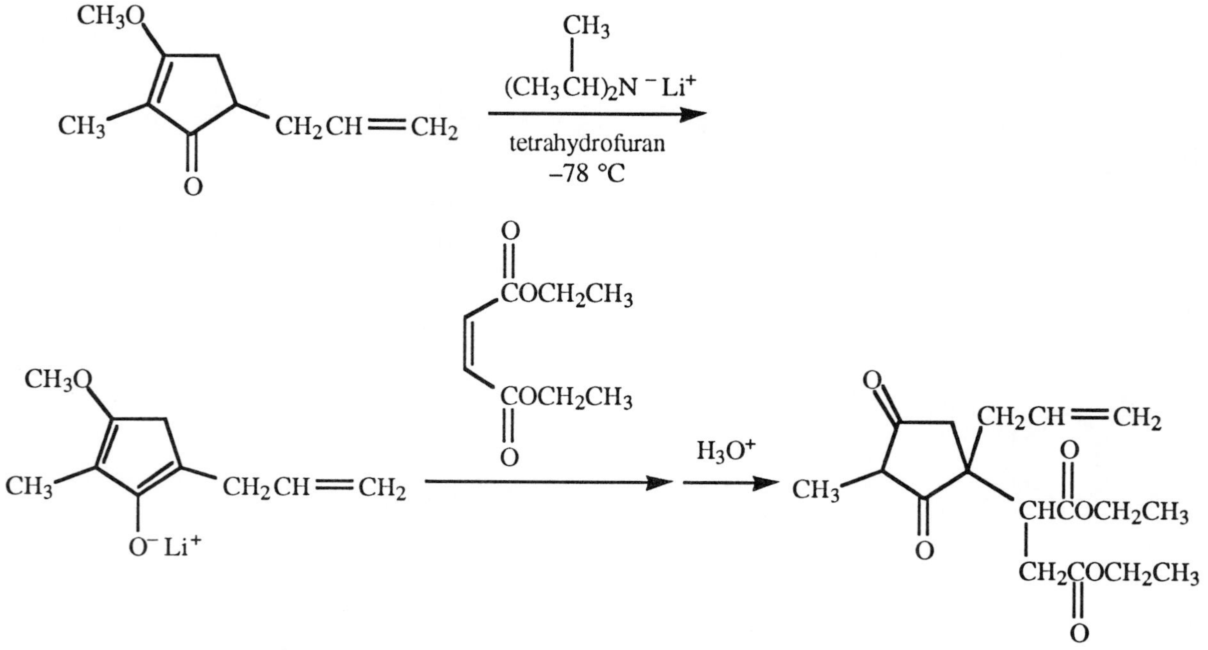

(e) $\underset{\text{(excess)}}{CH_3CH_2CH_2CH_3} \xrightarrow[h\nu]{Cl_2} CH_3CH_2CHClCH_3 + CH_3CH_2CH_2CH_2Cl$ + some polychlorinated products

(f) $CH_3CH(CH_3)CH{=}CH_2 \xrightarrow[\text{peroxide}]{HBr} CH_3CH(CH_3)CH_2CH_2Br$

(g) $CH_3CH_2SH + CH_2{=}CHOCH_2CH_3 \xrightarrow{\Delta} CH_3CH_2SCH_2CH_2OCH_2CH_3$

(h) $BrCH_2CH{=}CH_2 \xrightarrow[O_2]{HBr} BrCH_2CH_2CH_2Br$

(i)

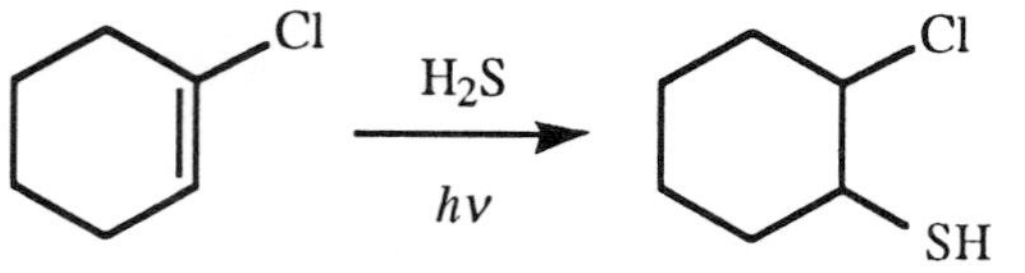

(j)

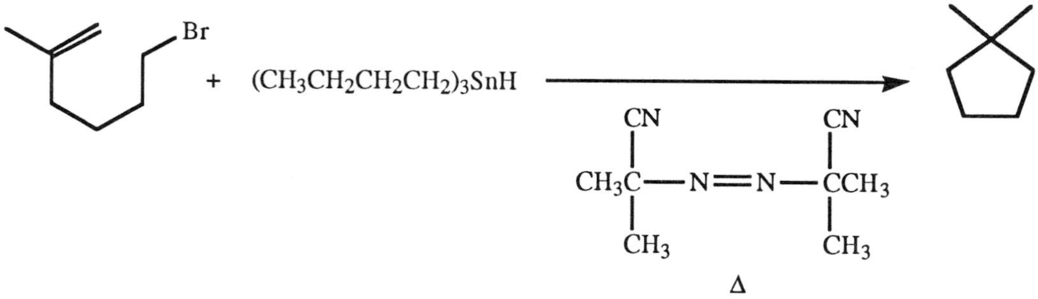

20.25 (cont)

(k)

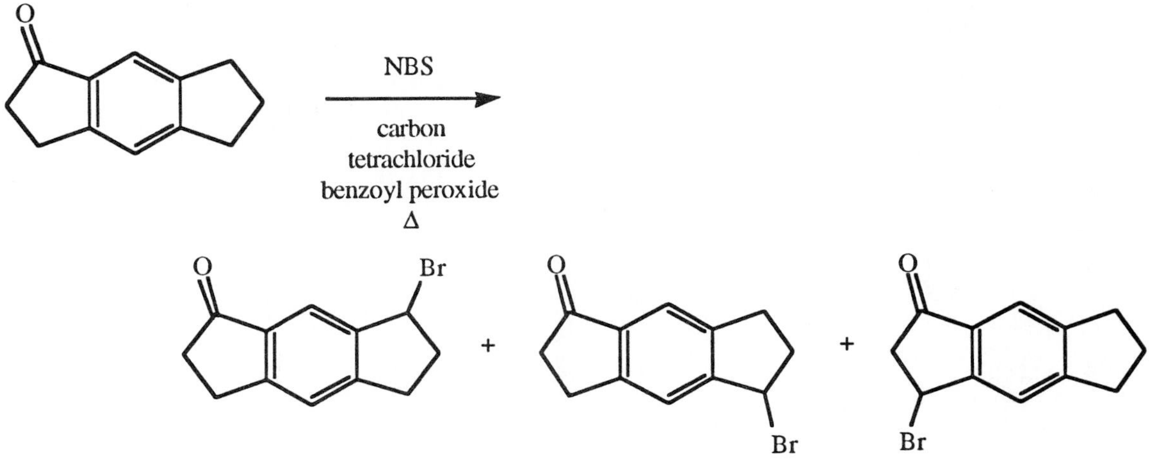

20.26

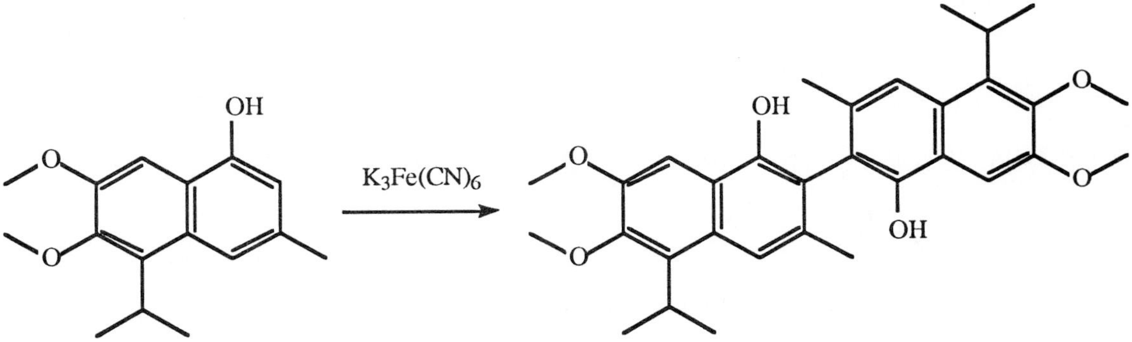

The isopropyl group hinders approach to the position para to the hydroxyl group so we would expect coupling to occur mainly at the ortho position.

20.27

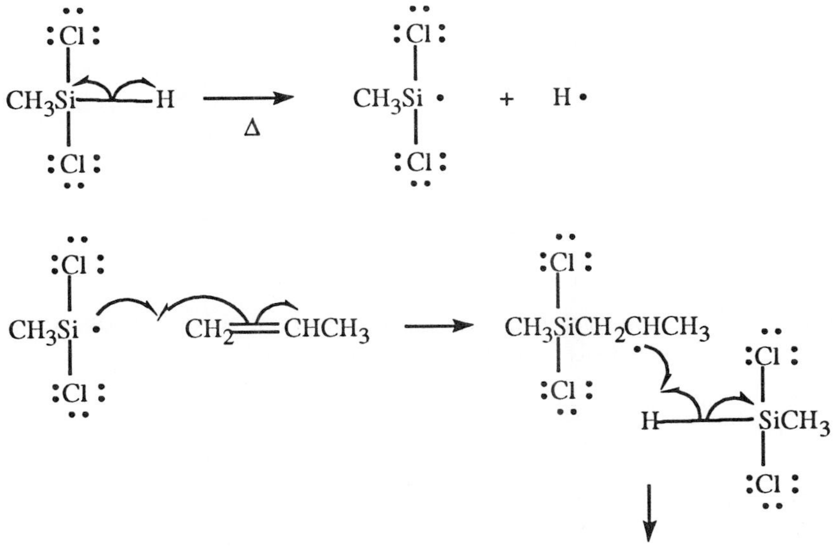

20.27 (cont)

20.28 1. Conditions for the formation of free radicals, such as heat, light, or peroxides, would be recommended for these reactions.

2. *tert*-Butyl alcohol does not have a hydrogen atom attached to the carbon atom bonded to the hydroxyl group and would, therefore, not react with 1-octene under free radical conditions.

20.29

20.30

(a)

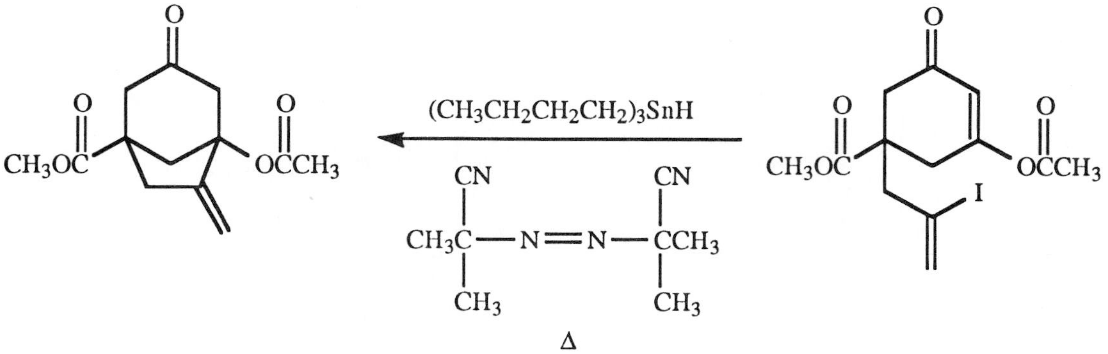

(b)

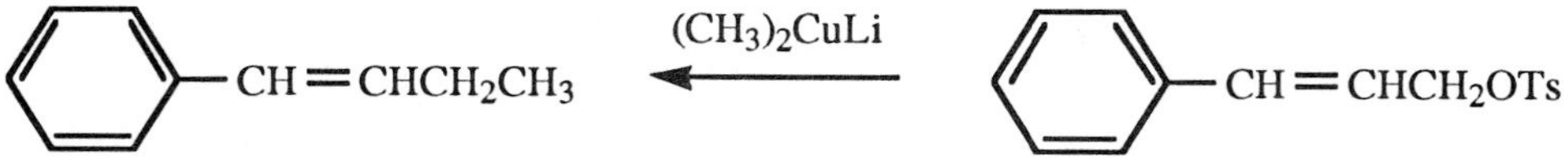

(c)

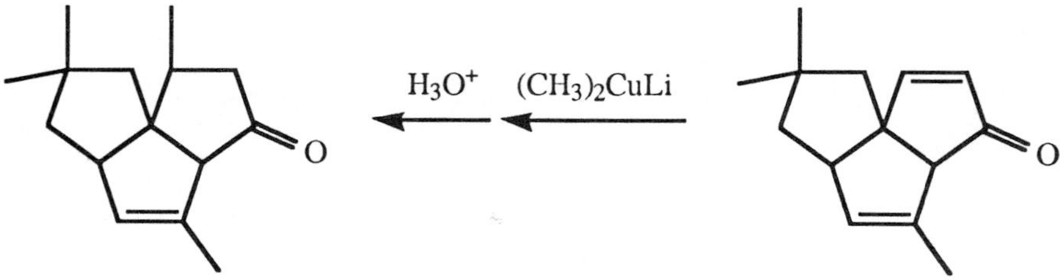

(d)

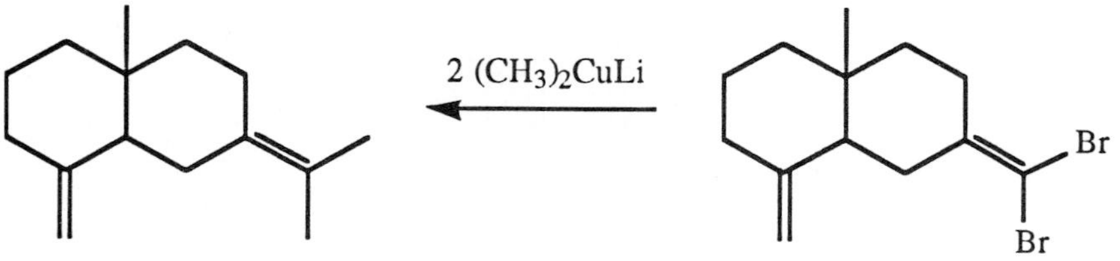

(e)

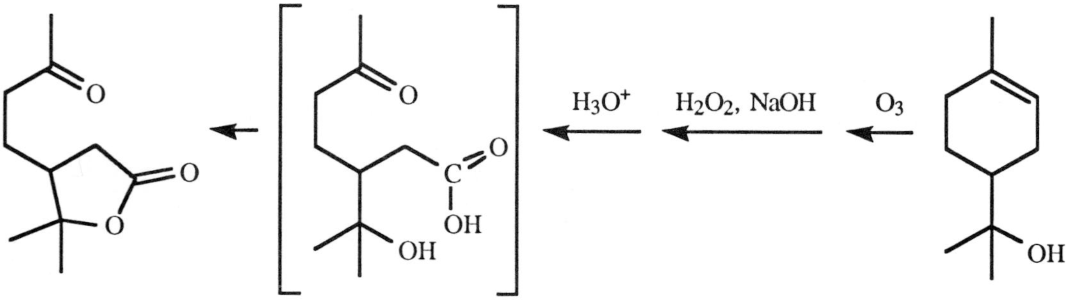

20.31　The final product in this series of reactions is a cyclic hemiketal that contains eleven carbon atoms. That is equal to the number of carbon atoms in the original starting material (except for the three carbon atoms of the protecting group on the diol) plus those of the nitrile and the Grignard reagent. Opening the hemiketal will give the structure of the last intermediate compound in the synthesis.

20.31 (cont)

The first step of the reaction is a free-radical addition to the alkene.

In the next step the Grignard reagent must add to the nitrile.

20.31 (cont)

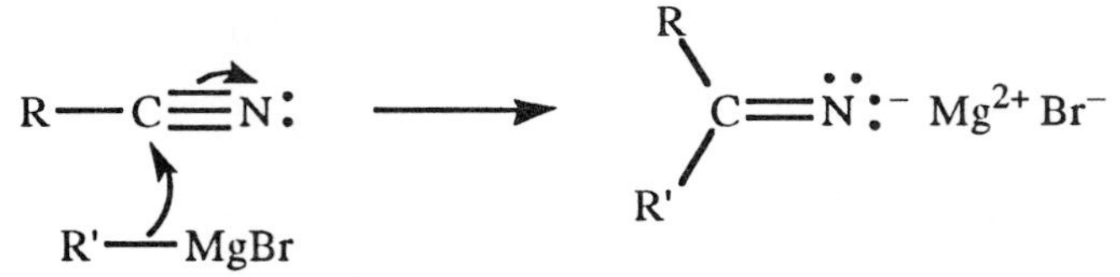

The product of this step is the conjugate base of an imine, which will hydrolyze in dilute acid to a ketone.

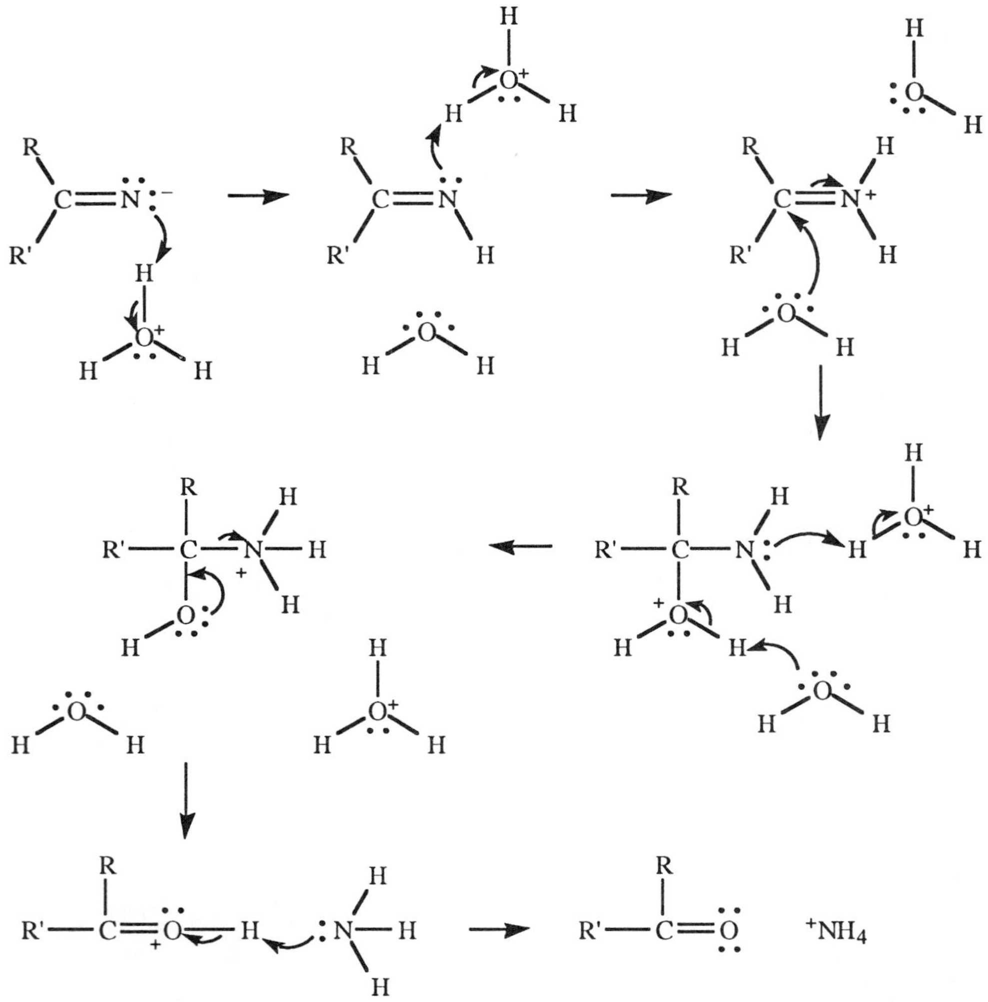

The cyclic ketal protecting group on the diol part of the molecule will be hydrolyzed by the same dilute acid solution as hydrolyzes the imine, giving the dihydroxyketone shown at the beginning of this answer. The hydroxyketone cyclizes spontaneously to the six-membered ring ketal.

20.32

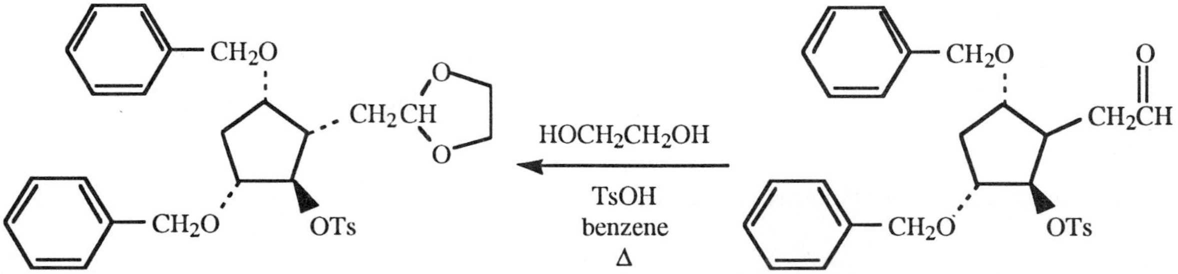

20.32 (cont)

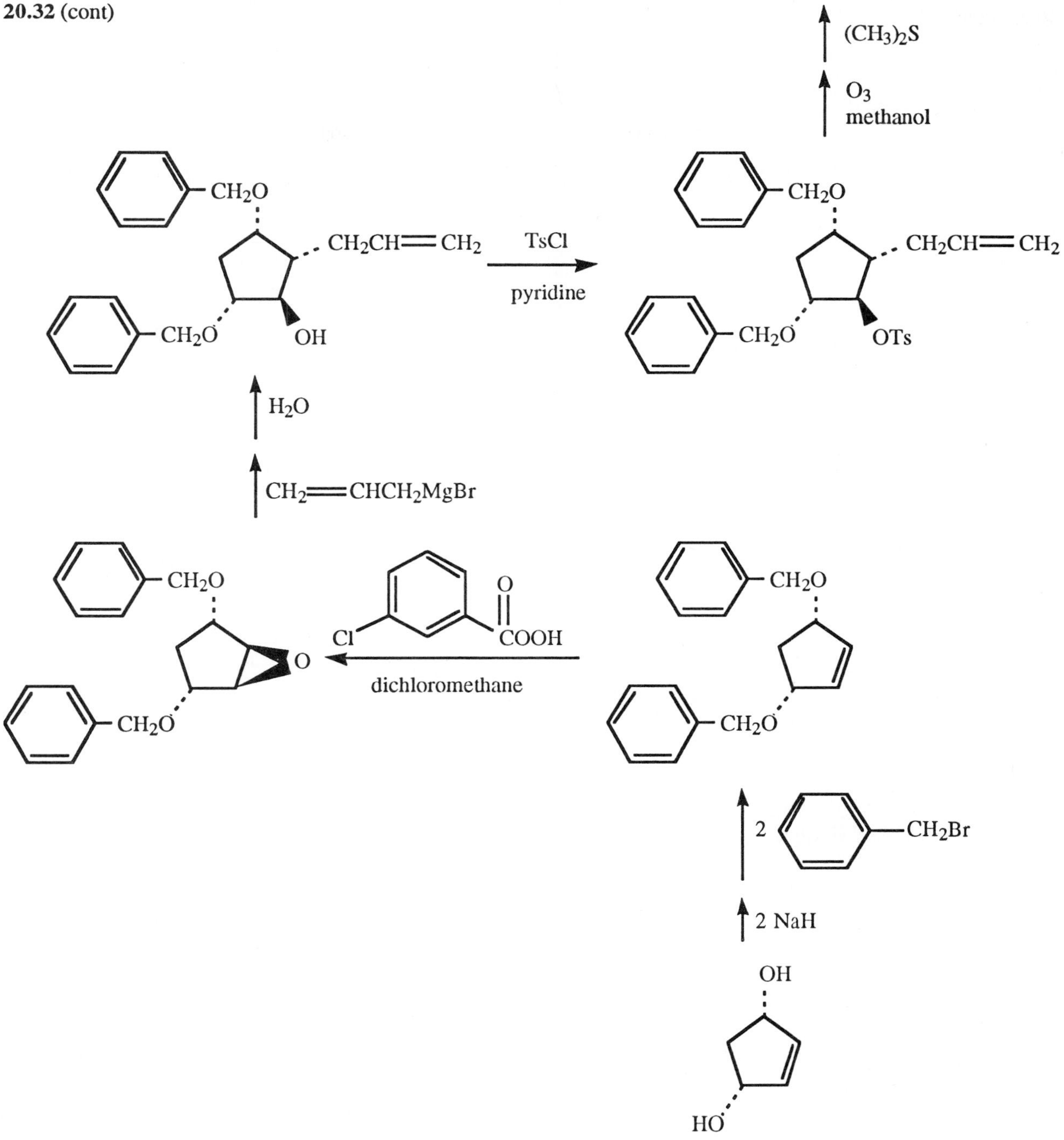

20.33

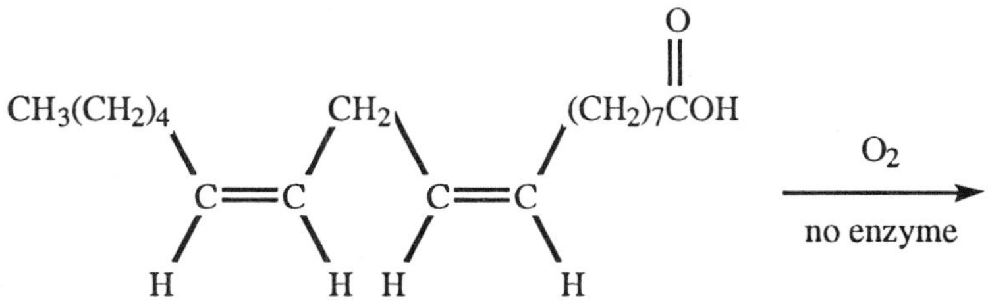

20.33 (cont)

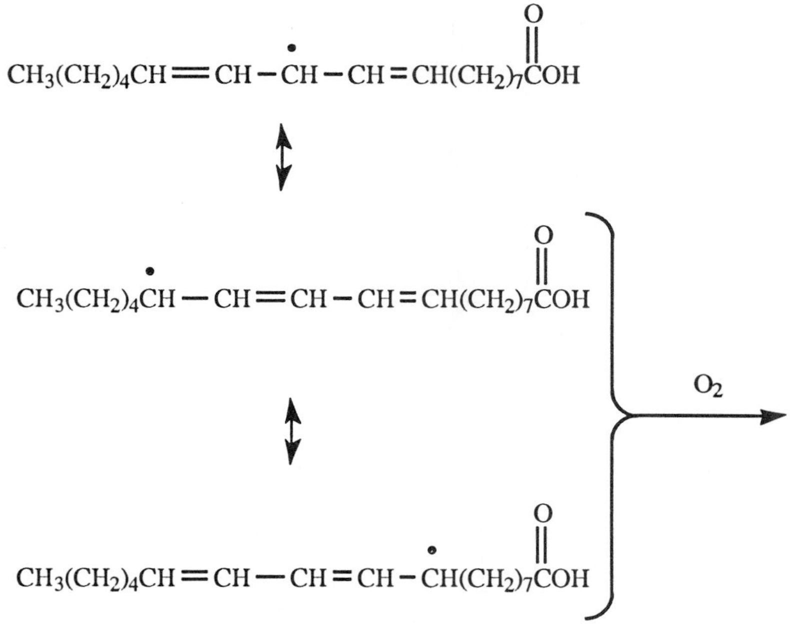

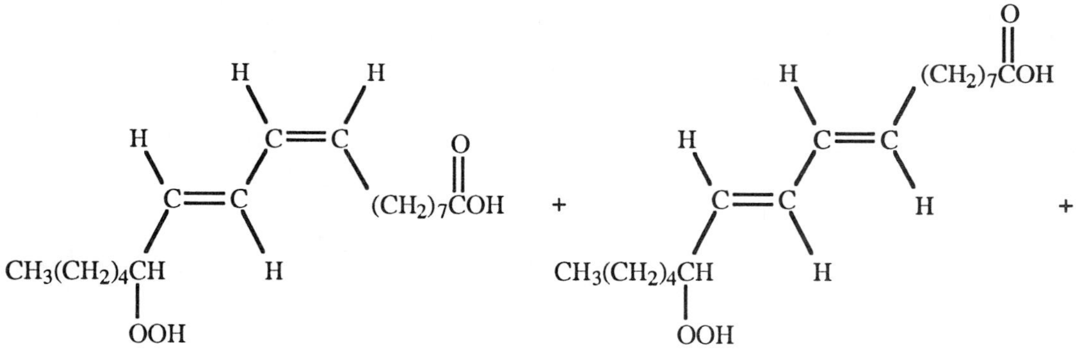

$$CH_3(CH_2)_4CH=CH-\overset{\bullet}{C}H-CH=CH(CH_2)_7\overset{O}{\overset{\|}{C}}OH$$

$\updownarrow$

$$CH_3(CH_2)_4\overset{\bullet}{C}H-CH=CH-CH=CH(CH_2)_7\overset{O}{\overset{\|}{C}}OH$$

$\updownarrow$

$$CH_3(CH_2)_4CH=CH-CH=CH-\overset{\bullet}{C}H(CH_2)_7\overset{O}{\overset{\|}{C}}OH$$

$\xrightarrow{O_2}$

biological product

+

+

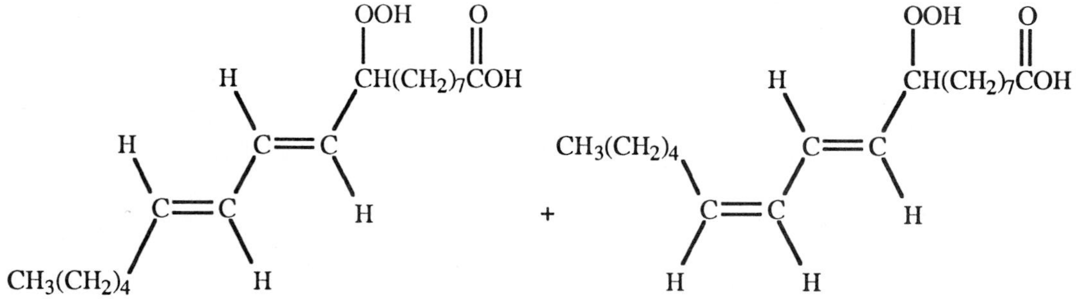

from Problem 20.16

20.34 (1) $CH_3CH \!=\! CH_2$ + $X \cdot$ $\longrightarrow$ $CH_3\overset{\cdot}{C}HCH_2X$ $\underline{\Delta H_r \, (kcal/mol)}$

$C\!=\!C \, (\pi)$ ~59 kcal/mol

C—Cl	~81 kcal/mol	−22
C—Br	~68 kcal/mol	−9
C—I	~52 kcal/mol	+7

(2) $CH_3\overset{\cdot}{C}HCH_2X$ +   HX $\longrightarrow$ $CH_3CH_2CH_2X$ $\underline{\Delta H_r \, (kcal/mol)}$

H—Cl	103 kcal/mol	C—H (2°)	95 kcal/mol	+8
H—Br	87 kcal/mol			−8
H—I	71 kcal/mol			−24

The first chain-propagating step of the free radical chain reaction of hydrogen halides with propene is the addition of the hydrogen atom to the double bond. This step is exothermic for chlorine and bromine, but endothermic for iodine. The second step, the abstraction of a hydrogen atom from the hydrogen halide by the carbon radical, is endothermic for hydrogen chloride, but exothermic for hydrogen bromide and hydrogen iodide. Therefore, only for hydrogen bromide are both steps of the reaction exothermic and expected to have small energies of activation. For the other two hydrogen halides, one or the other step has a high energy of activation. For such systems, the free radical chain reaction is slow, and other reaction pathways, such as reactions by way of carbocation intermediates, are favored.

20.35

20.35 (cont)

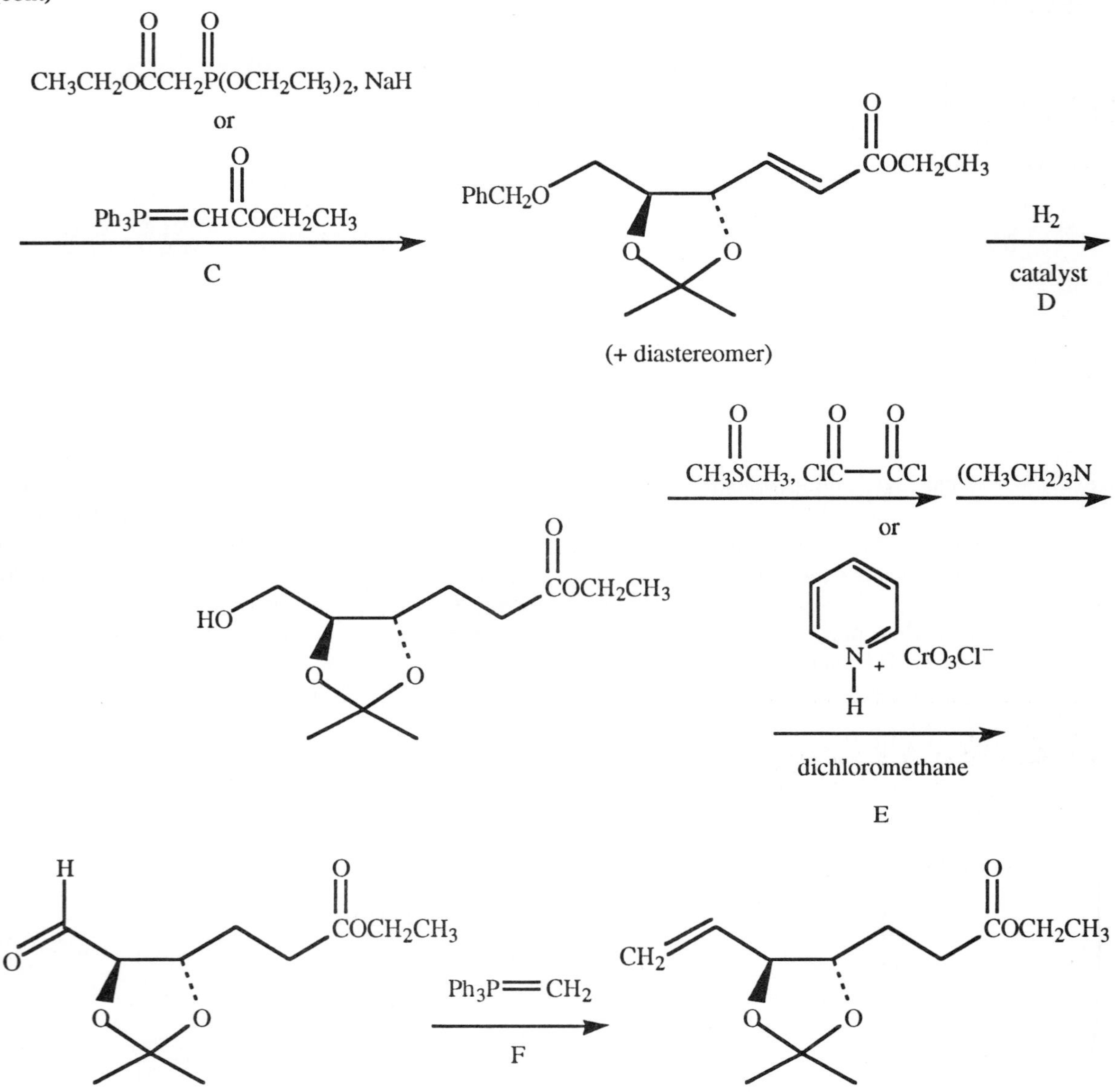

(+ diastereomer)

20.36

1.

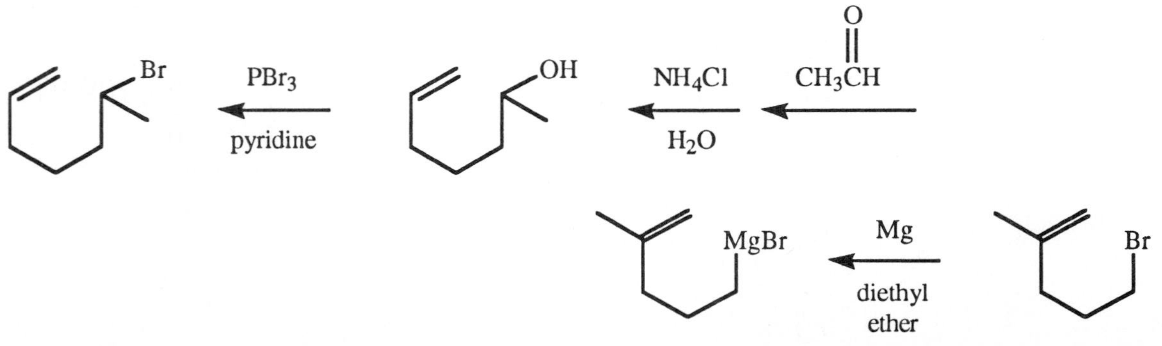

20.36 (cont)

2.

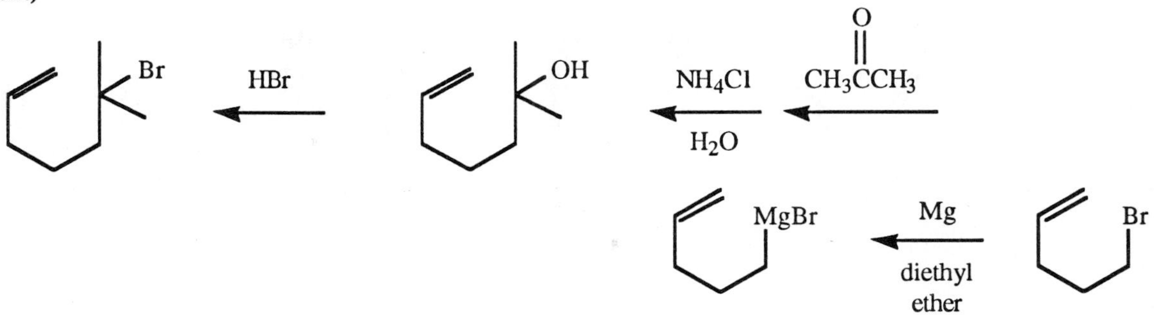

3.

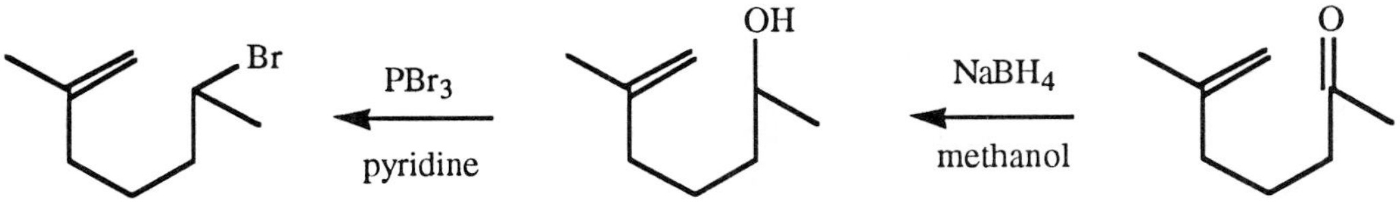

4.

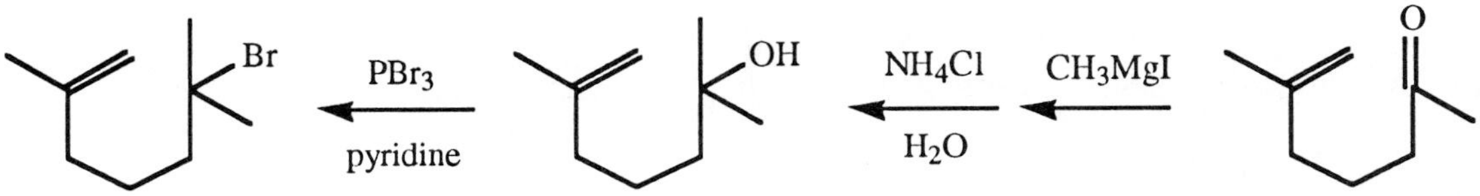

20.37

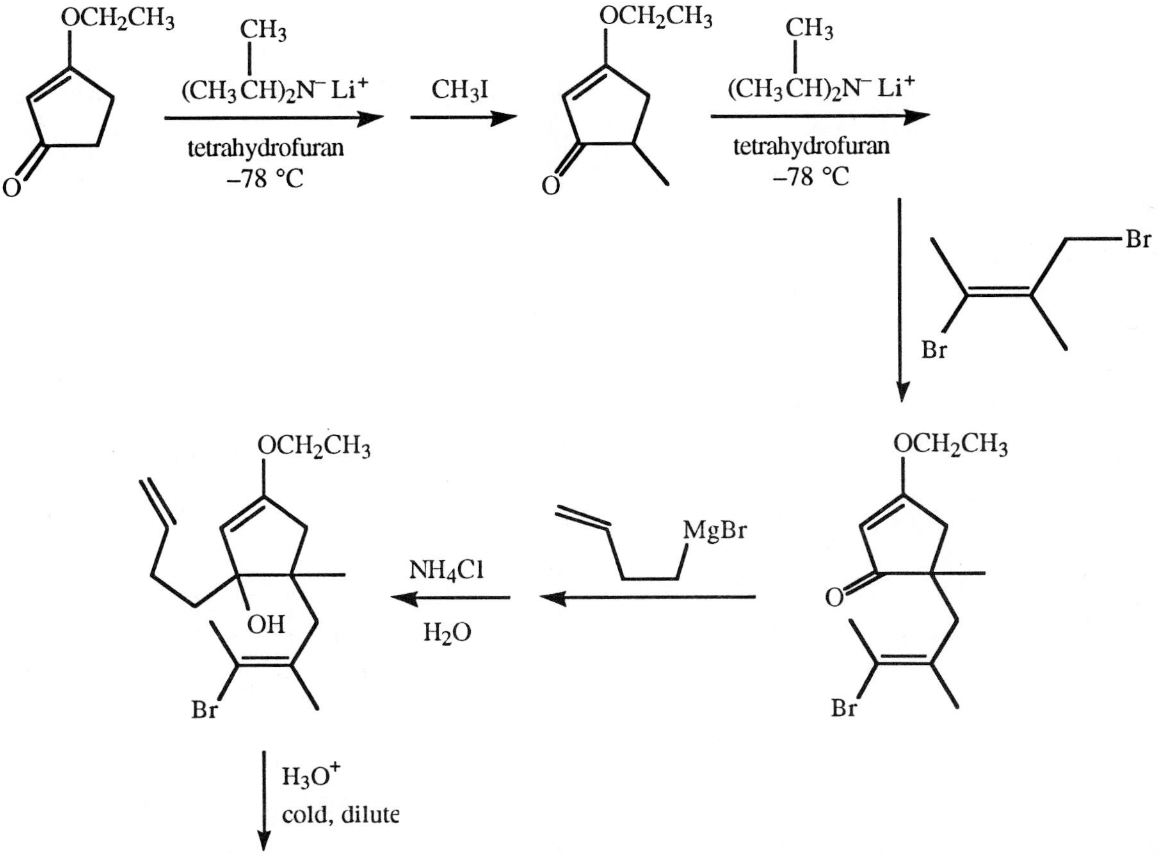

20.37 (cont)

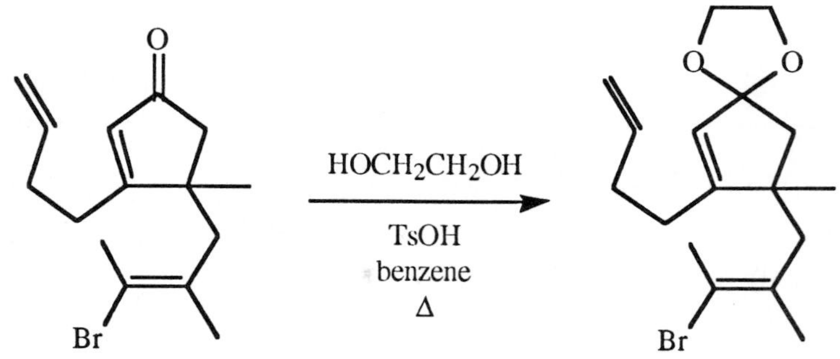

20.38

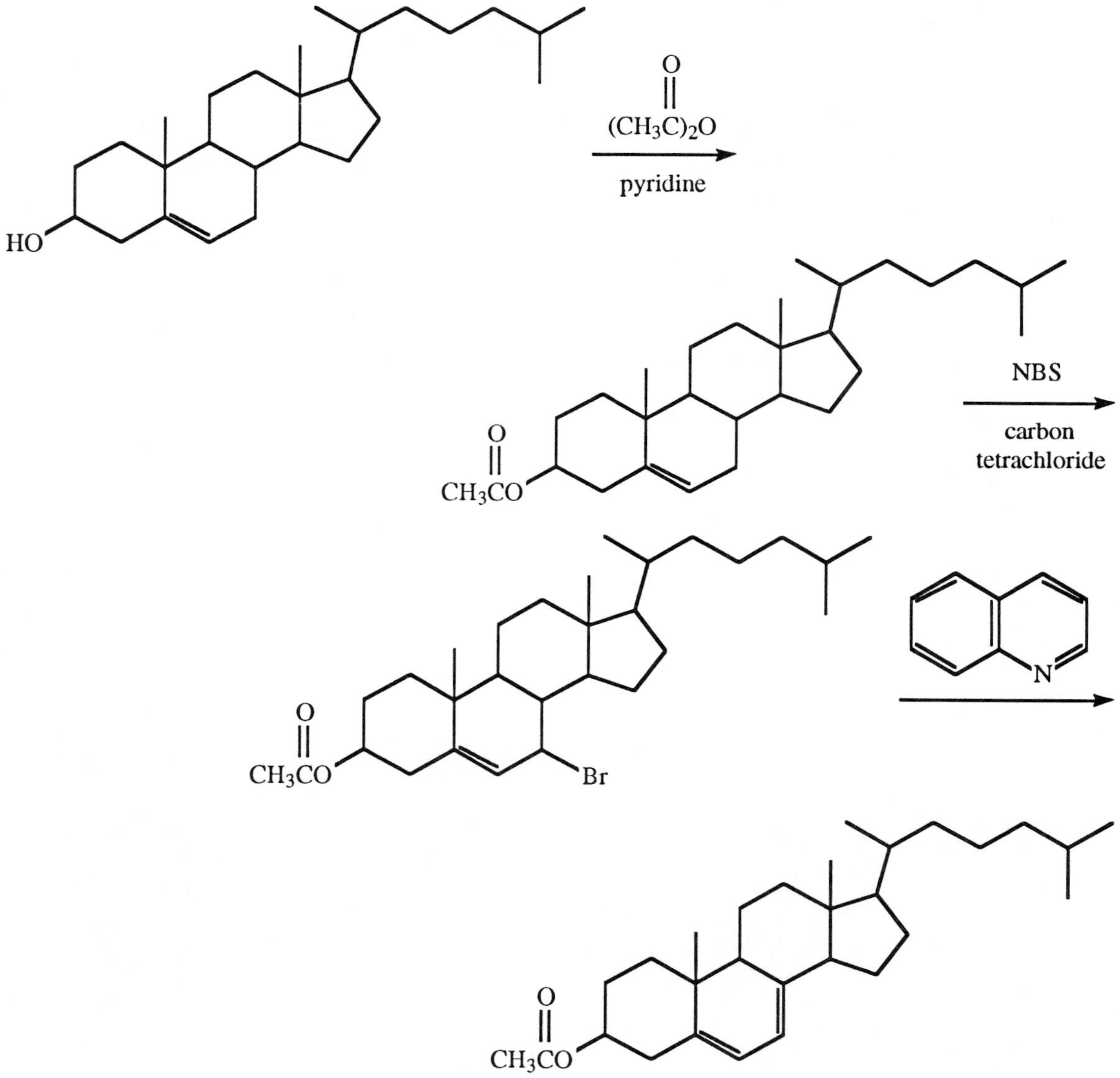

7-dehydrocholesteryl acetate

20.39

20.39 (cont)

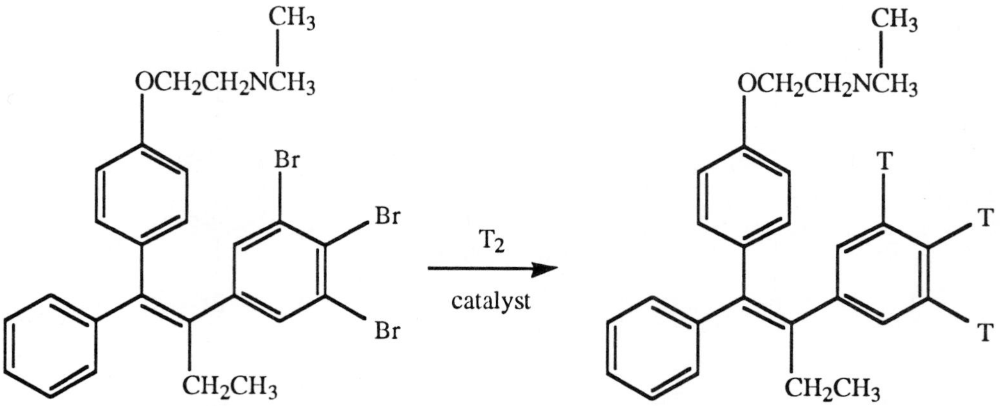

and diastereomer   radioactively labeled tamoxifen

20.40

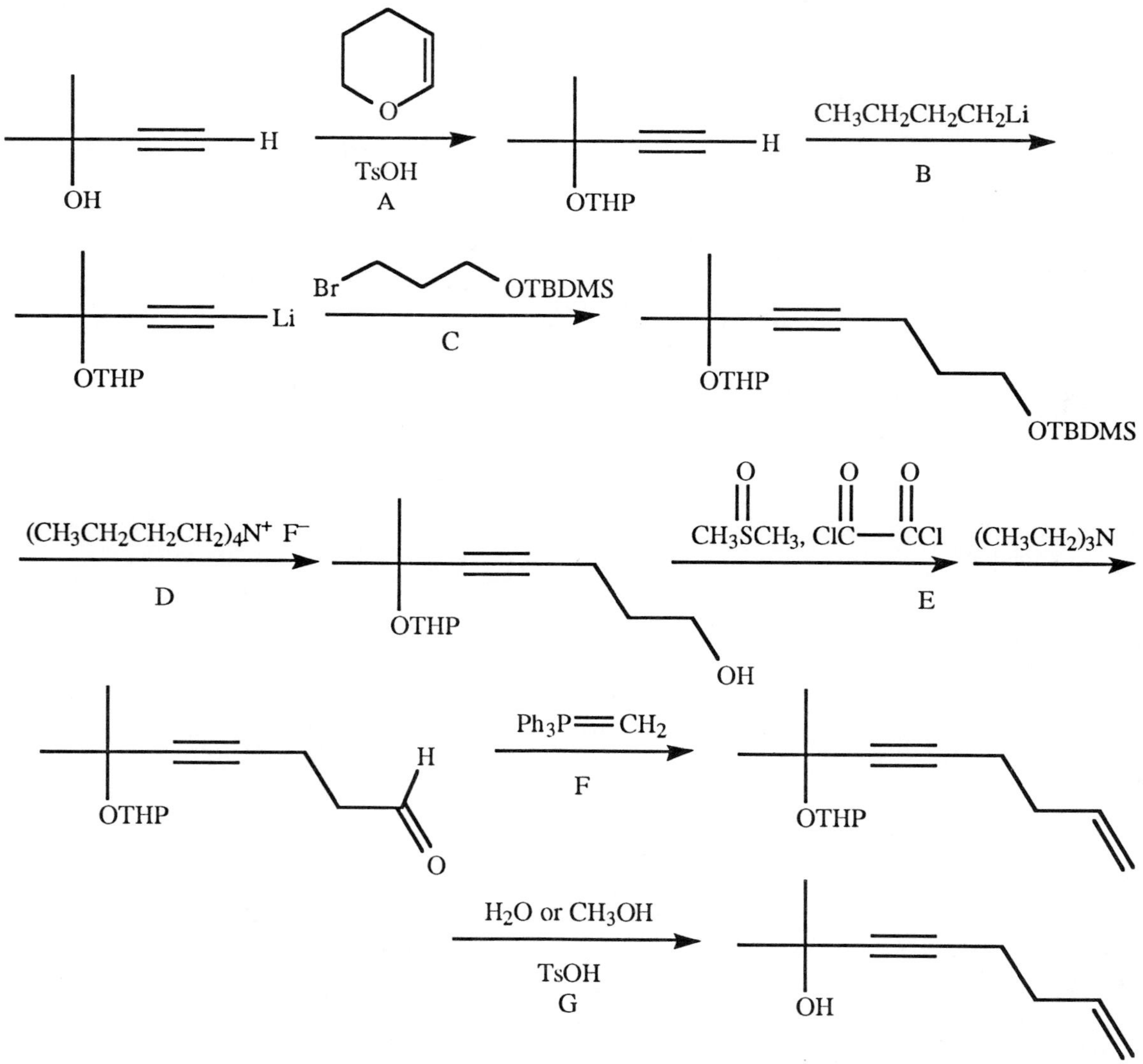

20.41

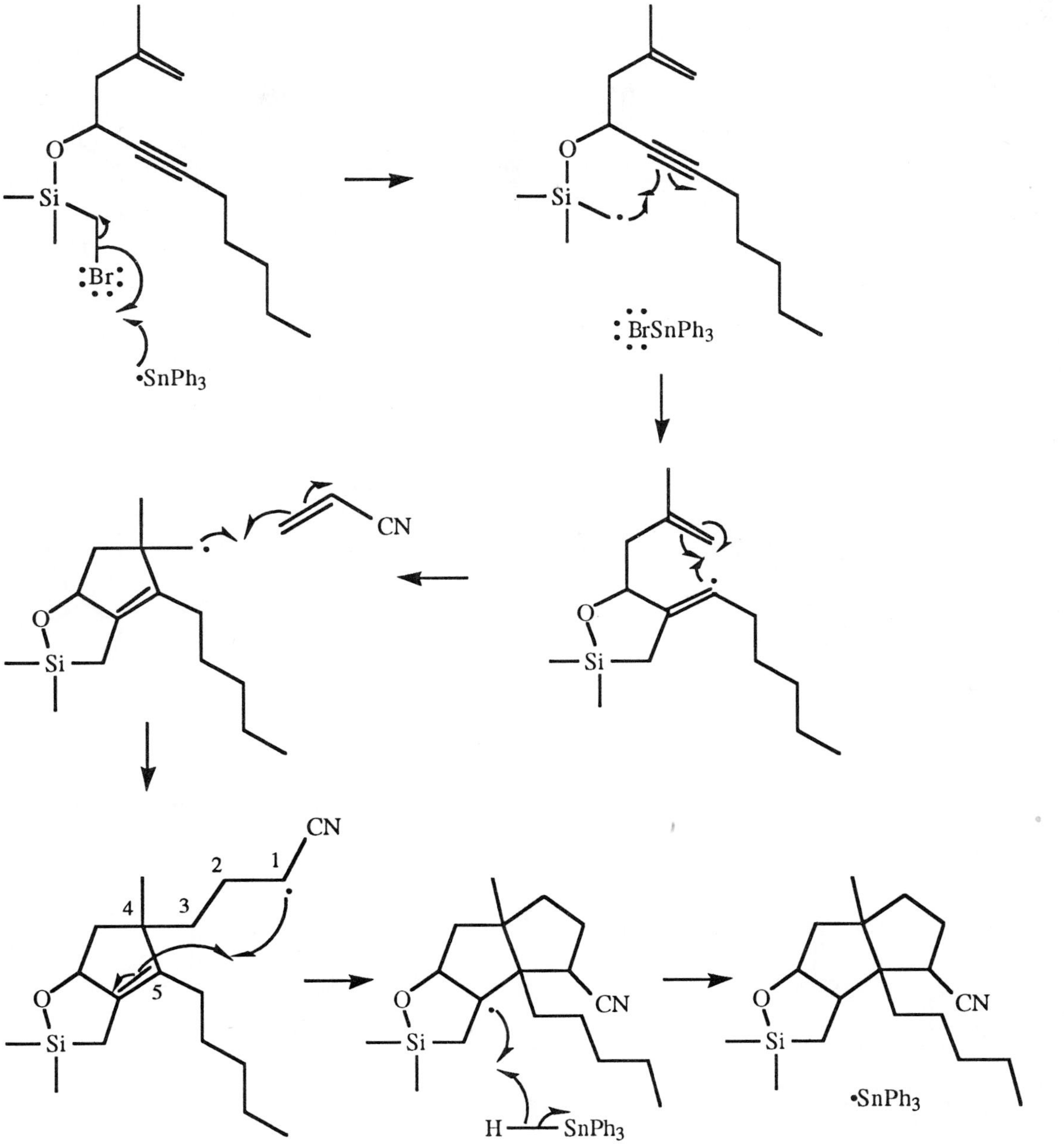

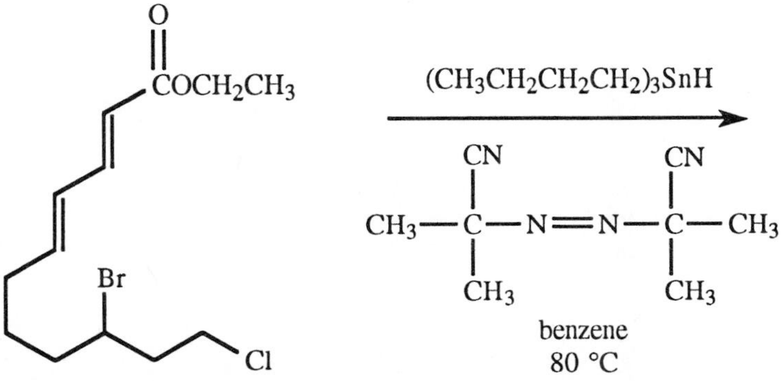

20.42

(a)

20.42 (cont)

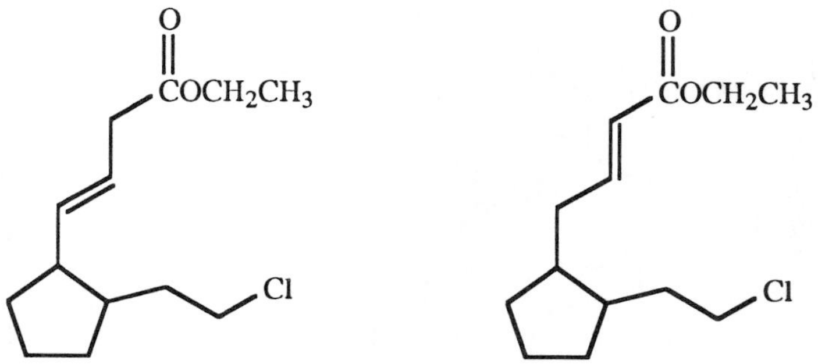

A
two possible products;
the one actually isolated
is the one on the left

(b)

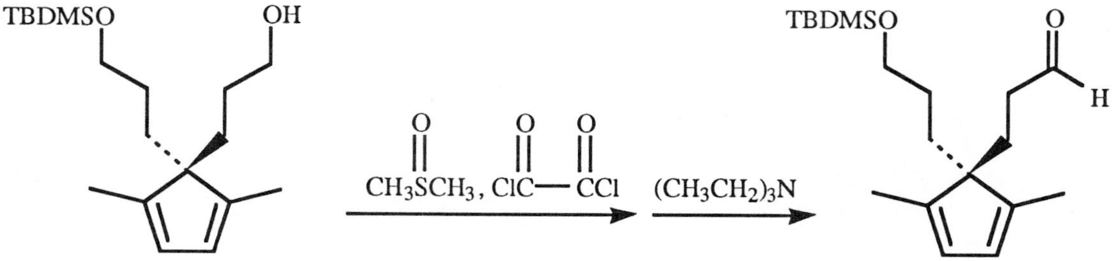

$$CH_3SCH_3, \; ClC-CCl \qquad (CH_3CH_2)_3N$$

B

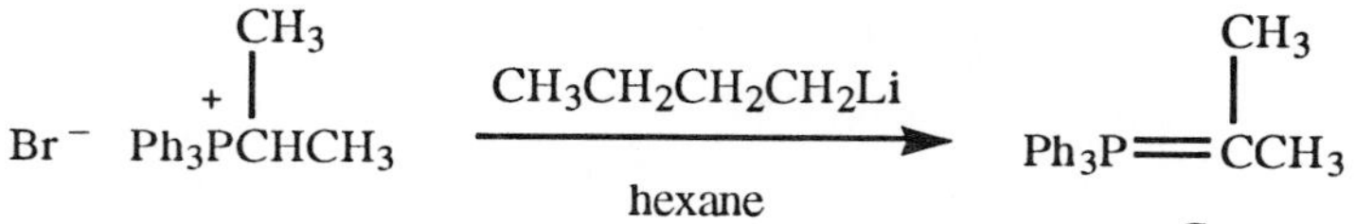

$$Br^- \quad Ph_3\overset{+}{P}CHCH_3 \xrightarrow[\text{hexane}]{CH_3CH_2CH_2CH_2Li} \quad Ph_3P=CCH_3$$

C

B   + $Ph_3P=CCH_3$

C

D

$$\xrightarrow[\text{tetrahydrofuran}]{(CH_3CH_2CH_2CH_2)_4N^+ \; F^-} \xrightarrow[H_2O]{NH_4Cl}$$

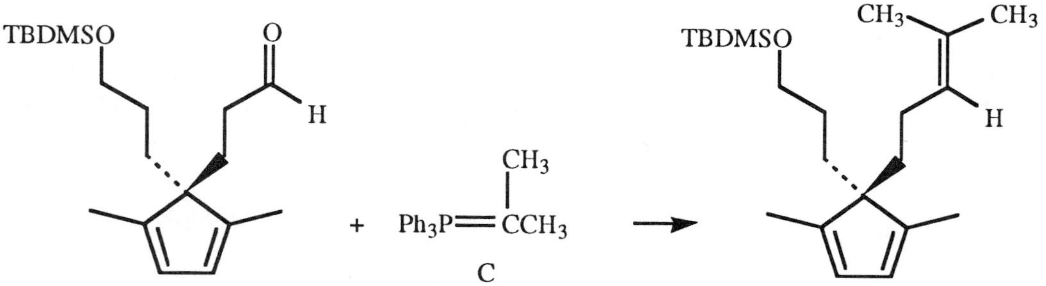

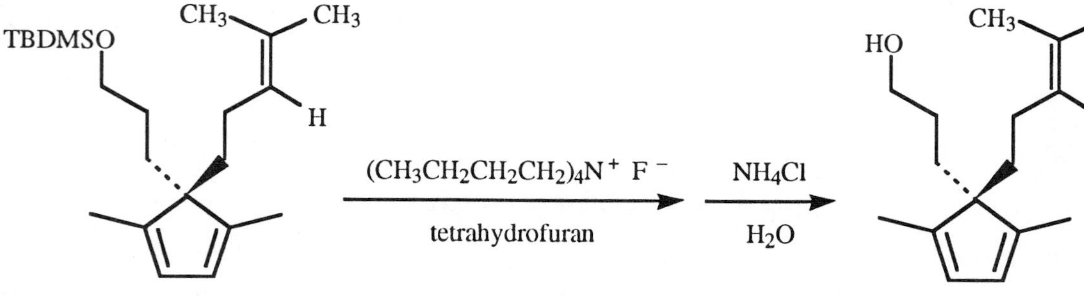

D

E

20.42 (b) (cont)

isopropyl group at this
stereocenter can be in the plane
or down; one is Compound H,
the other Compound I

21

Mass Spectrometry

21.1   If we subtract the m/z value for the base peak from that for the peak for the molecular ion, we can determine the mass of the neutral fragment which is lost.

$$58 \;-\; 29 \;=\; 29$$

molecular ion	base peak	neutral fragment

The fragment is most likely an ethyl radical, $CH_3\overset{\bullet}{C}H_2$, formed by homolytic cleavage of the bond between the carbon atom of the carbonyl group and the α-carbon atom. The fragment that is charged, with an m/e of 29, is a resonance-stabilized acylium ion, $HC{\equiv}O^+$. Putting these two fragments together gives us the identity of the unknown, propanal.

propanal

molecular ion
m/z 58

M$\overset{+}{\cdot}$
m/z 58

ethyl radical
not seen in mass spectrum

an acylium ion
base peak m/z 29

21.2   A hydrogen atom is lost in going from the molecular ion at m/z 46 to the peak at m/z 45. The loss of a hydroxyl radical (m/z 17) from the molecular ion leads to the base peak at m/z 29. Both of these fragments come from homolytic cleavage of the two bonds to the carbonyl group. Putting these fragments together gives us the structure of the unknown acid, which is formic acid.

formic acid

molecular ion
m/z 46

21.2 (cont)

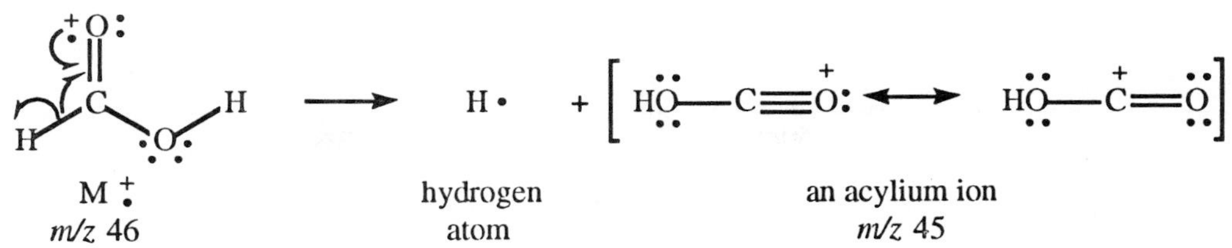

M $\overset{+}{\cdot}$
m/z 46

hydrogen
atom

an acylium ion
m/z 45

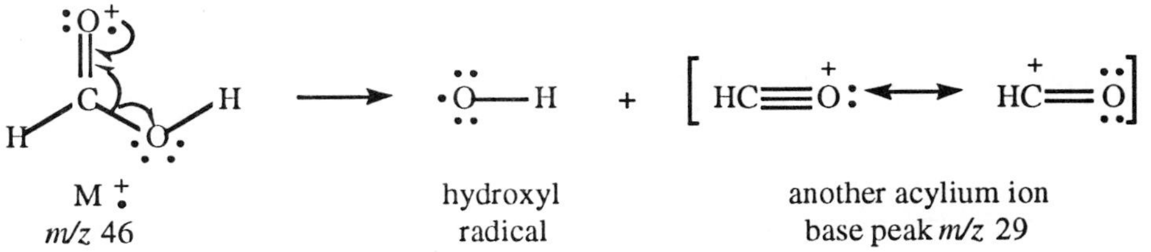

M $\overset{+}{\cdot}$
m/z 46

hydroxyl
radical

another acylium ion
base peak *m/z* 29

21.3   Two possible cleavages of the molecular ion of 3-pentanone at the bonds to the α-carbon atom are shown below:

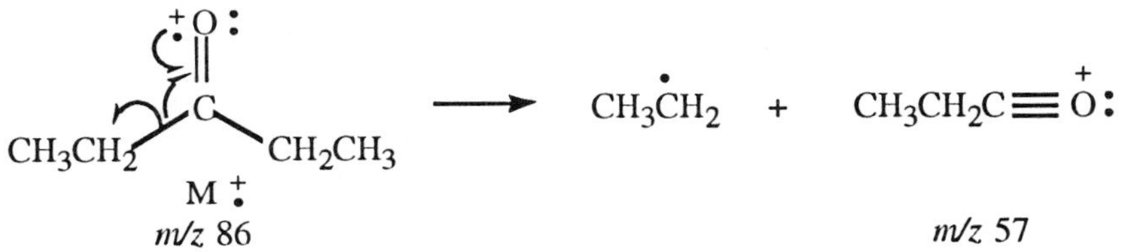

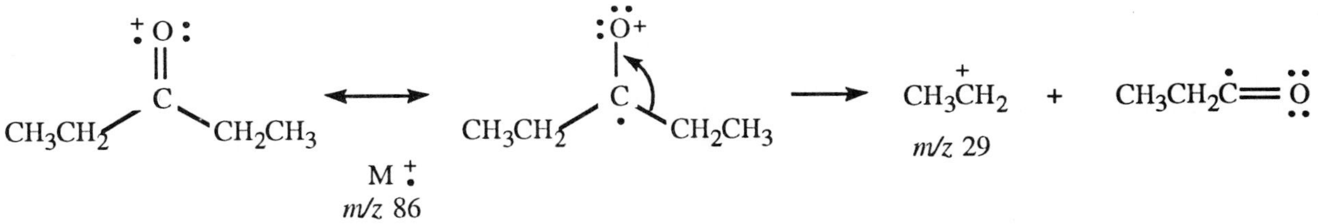

$CH_3\overset{\cdot}{C}H_2$ + $CH_3CH_2C\equiv\overset{+}{O}:$

m/z 57

$CH_3\overset{+}{C}H_2$ + $CH_3CH_2\overset{\cdot}{C}=\overset{\cdot\cdot}{\underset{\cdot\cdot}{O}}$

m/z 29

The expected cleavage of the molecular ion of 2-pentanone at the bonds to the two different α-carbon atoms are shown below:

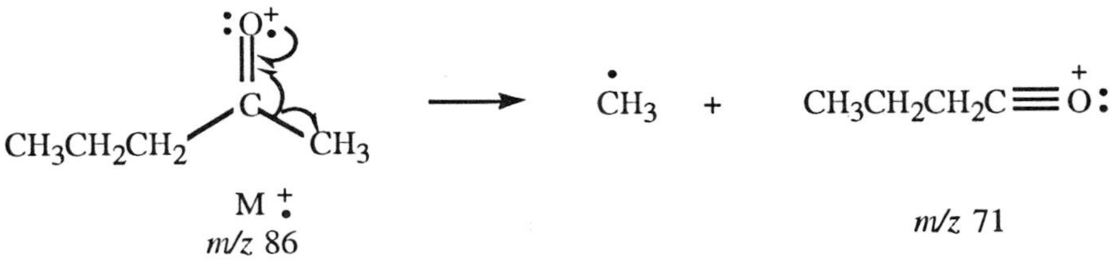

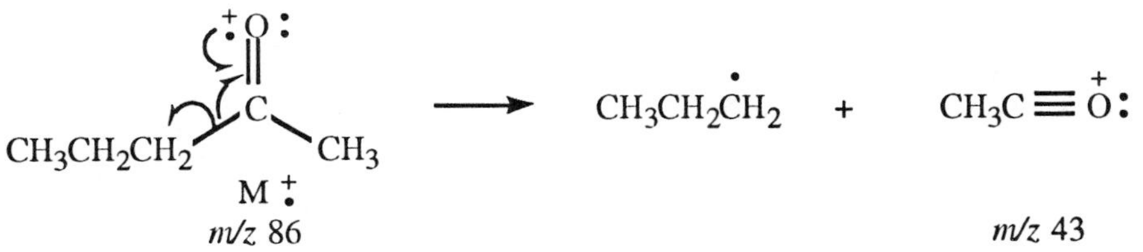

$\overset{\cdot}{C}H_3$ + $CH_3CH_2CH_2C\equiv\overset{+}{O}:$

m/z 71

$CH_3CH_2\overset{\cdot}{C}H_2$ + $CH_3C\equiv\overset{+}{O}:$

m/z 43

The mass spectrum with base peaks at *m/z* 29 and *m/z* 57 is the spectrum of 3-pentanone. The mass spectrum with the base peak at *m/z* 43 and a small peak at *m/z* 71 is the spectrum of 2-pentanone.

21.4   The formation of the base peak in the spectrum of 2,2-dimethylpropanal from the molecular ion is shown below:

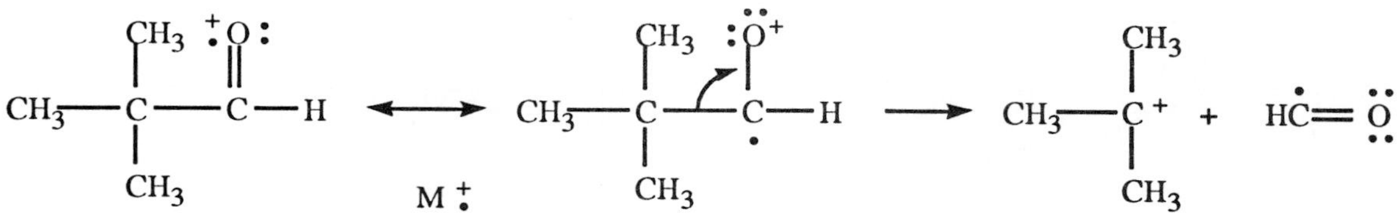

M $\overset{+}{\cdot}$
m/z 86

tert-butyl cation
m/z 57

The equivalent fragmentation from the molecular ion of acetaldehyde is shown below:

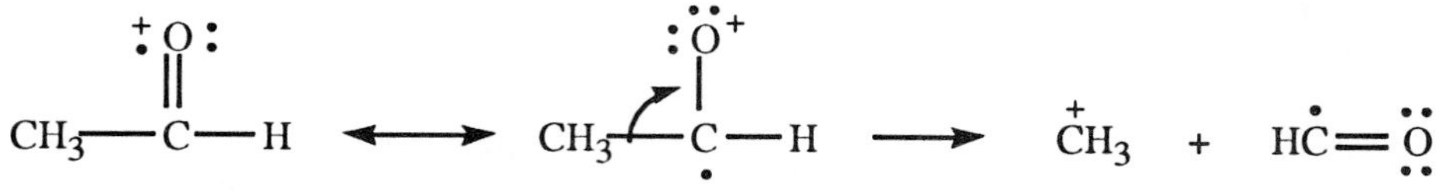

M $\overset{+}{\cdot}$
m/z 44

methyl cation
m/z 15

The base peak in the fragmentation of the molecular ion of 2,2-dimethylpropanal is the stable tertiary cation, the *tert*-butyl cation. The comparable species in the fragmentation of the molecular ion of acetaldehyde would be the unstable methyl cation, *m/z* 15, which does not appear in the spectrum of acetaldehyde.

21.5

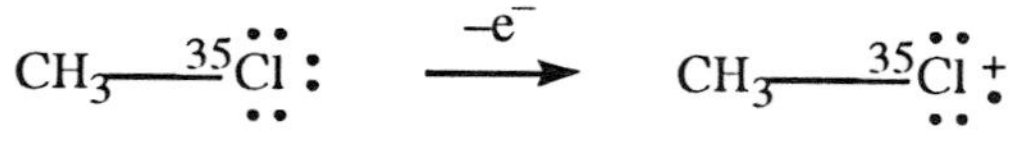

M $\overset{+}{\cdot}$
m/z 50

m/z 15

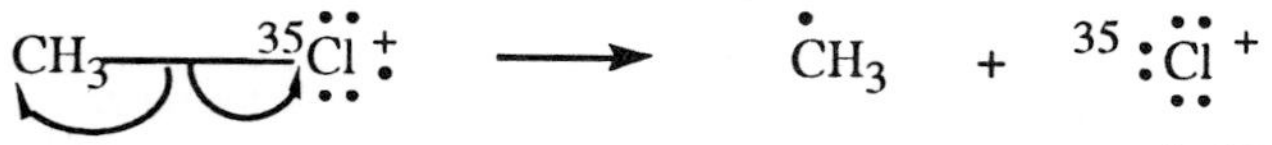

m/z 35
not evident in
the spectrum

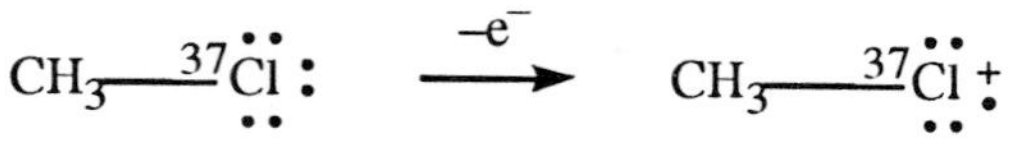

M $\overset{+}{\cdot}$
m/z 52

m/z 15

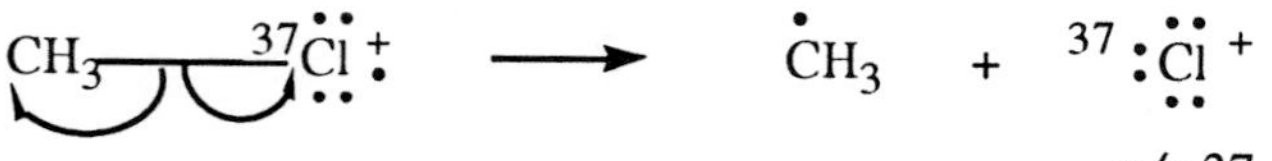

m/z 37
not evident in
the spectrum

21.6   The two ions with the highest *m/e* values appear in this spectrum at *m/z* 64 and *m/z* 66. The one at the higher mass is one-third the intensity of the other, suggesting that these are molecular ions for a compound containing chlorine.

$$M\overset{+}{\cdot} \qquad m/z\ 64 \quad (^{35}Cl)$$

$$M\overset{+}{\cdot} \qquad m/z\ 66 \quad (^{37}Cl)$$

An important fragmentation of the molecular ion gives rise to the peak at *m/z* 29. This is the ethyl cation and results from loss of chlorine from the molecular ion. Putting these fragments together, we identify the haloalkane as ethyl chloride (chloroethane).

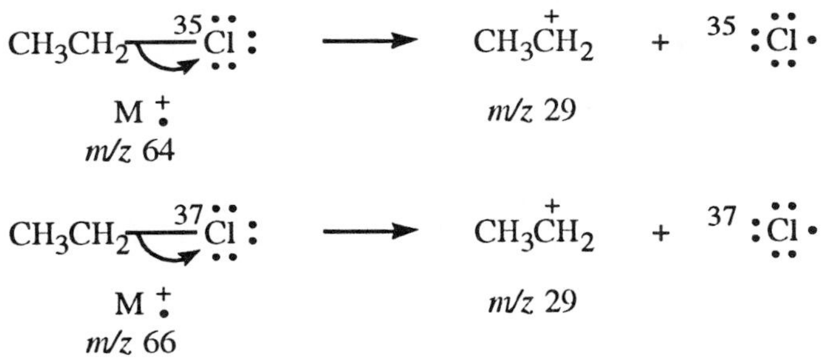

$$CH_3CH_2 \overset{35}{\longrightarrow} \ddot{C}l\!: \longrightarrow CH_3\overset{+}{C}H_2 \ + \ ^{35}\!:\!\ddot{C}l\cdot$$

M $\overset{+}{\cdot}$ m/z 29
m/z 64

$$CH_3CH_2 \overset{37}{\longrightarrow} \ddot{C}l\!: \longrightarrow CH_3\overset{+}{C}H_2 \ + \ ^{37}\!:\!\ddot{C}l\cdot$$

M $\overset{+}{\cdot}$ m/z 29
m/z 66

21.7   The two ions at the highest *m/z* values in this mass spectrum are at *m/z* 148 and at *m/z* 150 and are of equal intensity. This suggests that one of the two halogens in the unknown is bromine. The presence of bromine is confirmed by the two peaks of equal intensity at *m/z* 79 and 81.

The first fragmentation, *m/z* 148 to *m/z* 129, (or *m/z* 150 to *m/z* 131) is the loss of a species with a mass of 19, which corresponds to the atomic weight of fluorine. The base peak, which is at *m/z* 69, results from the loss of a bromine atom. The compound can have only one bromine atom, because the molecular weight is too low for more than one. Therefore, there must be three fluorine atoms. The compound is bromotrifluoromethane, CF_3Br.

The fragmentations are shown below. The molecular ion is shown as being formed by loss of a nonbonding electron from bromine. Because the electronegativity of bromine is lower than that of fluorine, and the bromine atom is much larger than a fluorine atom, it would be easier to lose the electron from bromine than from fluorine.

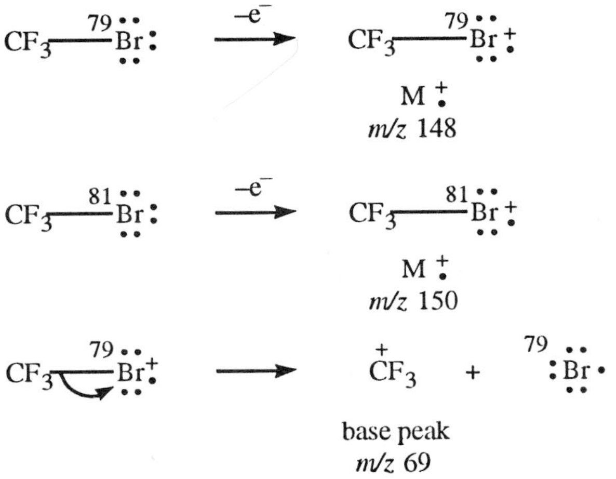

$$CF_3 \overset{79}{\longrightarrow} \ddot{Br}\!: \quad \overset{-e^-}{\longrightarrow} \quad CF_3 \overset{79}{\longrightarrow} \ddot{Br}\overset{+}{\!:}$$

M $\overset{+}{\cdot}$
m/z 148

$$CF_3 \overset{81}{\longrightarrow} \ddot{Br}\!: \quad \overset{-e^-}{\longrightarrow} \quad CF_3 \overset{81}{\longrightarrow} \ddot{Br}\overset{+}{\!:}$$

M $\overset{+}{\cdot}$
m/z 150

$$CF_3 \overset{79}{\longrightarrow} \ddot{Br}\overset{+}{\cdot} \longrightarrow \overset{+}{C}F_3 \ + \ ^{79}\!:\!\ddot{Br}\cdot$$

base peak
m/z 69

21.7 (cont)

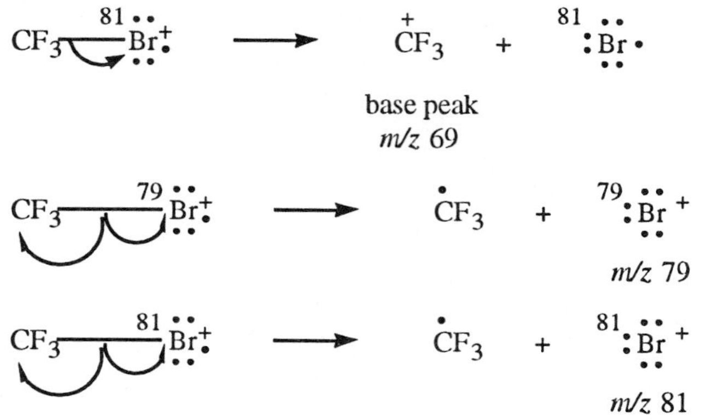

$$CF_3\overset{81}{\underset{}{\overset{\cdot\cdot}{Br}}}\overset{+}{\cdot} \longrightarrow \overset{+}{CF_3} + \overset{81}{\underset{\cdot\cdot}{:Br}}\cdot$$

base peak
m/z 69

$$CF_3\overset{79}{\underset{}{\overset{\cdot\cdot}{Br}}}\overset{+}{\cdot} \longrightarrow \overset{\cdot}{CF_3} + \overset{79}{\underset{\cdot\cdot}{:Br}}{}^{+}$$

m/z 79

$$CF_3\overset{81}{\underset{}{\overset{\cdot\cdot}{Br}}}\overset{+}{\cdot} \longrightarrow \overset{\cdot}{CF_3} + \overset{81}{\underset{\cdot\cdot}{:Br}}{}^{+}$$

m/z 81

Even though the majority of the molecular ions results from loss of an electron from the bromine atom, a certain number of the molecular ions will result from loss of an electron from a fluorine atom. A better way of thinking about this, of course, is to remember that all of the electrons belong to the whole molecule. When an electron is lost, the fragments that result from this molecular ion may be rationalized by picturing the deficiency of the electron as being localized at one site or another. The relative intensities of the bands at *m/z* 69 and at *m/z* 129 and 131 in the spectrum are a measure of the ease with which bromine and fluorine accommodate the deficiency of an electron. Localization of the deficiency of an electron on fluorine leads to the following fragmentations:

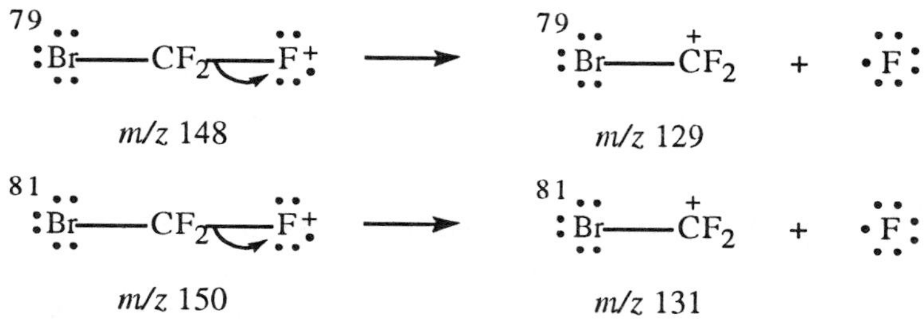

$$\overset{79}{\underset{\cdot\cdot}{:Br}}-CF_2\overset{\cdot\cdot}{\underset{\cdot\cdot}{F}}{}^{+} \longrightarrow \overset{79}{\underset{\cdot\cdot}{:Br}}-\overset{+}{CF_2} + \cdot\overset{\cdot\cdot}{\underset{\cdot\cdot}{F}}:$$

m/z 148   *m/z* 129

$$\overset{81}{\underset{\cdot\cdot}{:Br}}-CF_2\overset{\cdot\cdot}{\underset{\cdot\cdot}{F}}{}^{+} \longrightarrow \overset{81}{\underset{\cdot\cdot}{:Br}}-\overset{+}{CF_2} + \cdot\overset{\cdot\cdot}{\underset{\cdot\cdot}{F}}:$$

m/z 150   *m/z* 131

21.8 The formation of the molecular ion for isobutylamine and the pathway for its fragmentation to the base peak are shown below:

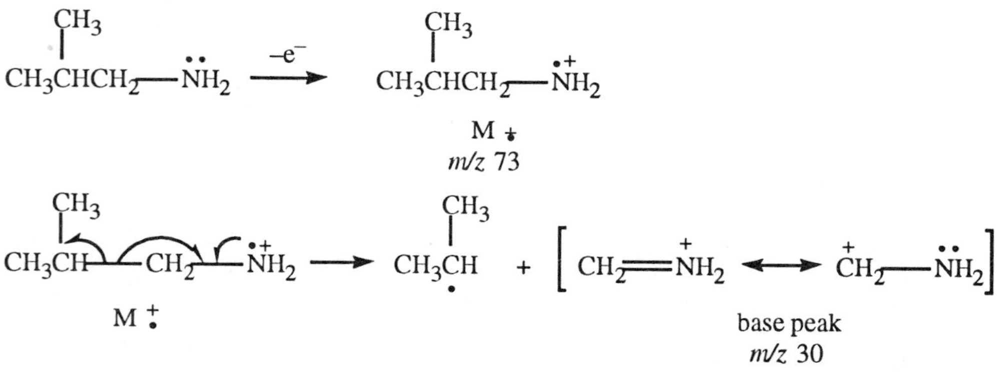

$$\underset{CH_3CHCH_2-\overset{\cdot\cdot}{NH_2}}{\overset{\overset{CH_3}{|}}{}} \overset{-e^-}{\longrightarrow} \underset{CH_3CHCH_2-\overset{\cdot+}{NH_2}}{\overset{\overset{CH_3}{|}}{}}$$

M $\overset{+}{\cdot}$
m/z 73

$$\underset{\underset{M\,\overset{+}{\cdot}}{CH_3CH-CH_2-\overset{\cdot+}{NH_2}}}{\overset{\overset{CH_3}{|}}{}} \longrightarrow \underset{CH_3\overset{\cdot}{CH}}{\overset{\overset{CH_3}{|}}{}} + \left[CH_2\!\!=\!\!\overset{+}{NH_2} \longleftrightarrow \overset{+}{CH_2}-\overset{\cdot\cdot}{NH_2} \right]$$

base peak
m/z 30

21.8 (cont)

The corresponding reaction for *sec*-butylamine is:

$$CH_3CH_2\overset{\overset{\displaystyle CH_3}{|}}{CH}\!-\!\ddot{N}H_2 \xrightarrow{-e^-} CH_3CH_2\overset{\overset{\displaystyle CH_3}{|}}{CH}\!-\!\overset{\bullet+}{N}H_2$$

$$M\overset{+}{\bullet}$$
$$m/z\ 73$$

$$CH_3CH_2\!-\!\overset{\overset{\displaystyle CH_3}{|}}{CH}\!-\!\overset{\bullet+}{N}H_2 \longrightarrow CH_3\overset{\bullet}{C}H_2 \;+\; \left[CH_3CH\!=\!\overset{+}{N}H_2 \longleftrightarrow CH_3\overset{+}{C}H\!-\!\ddot{N}H_2 \right]$$

$$M\overset{+}{\bullet}$$

base peak
$m/z\ 44$

Similarly, for *tert*-butylamine:

$$CH_3\overset{\overset{\displaystyle CH_3}{|}}{\underset{\underset{\displaystyle CH_3}{|}}{C}}\!-\!\ddot{N}H_2 \xrightarrow{-e^-} CH_3\overset{\overset{\displaystyle CH_3}{|}}{\underset{\underset{\displaystyle CH_3}{|}}{C}}\!-\!\overset{\bullet+}{N}H_2$$

$$M\overset{+}{\bullet}$$
$$m/z\ 73$$

$$CH_3\!-\!\overset{\overset{\displaystyle CH_3}{|}}{\underset{\underset{\displaystyle CH_3}{|}}{C}}\!-\!\overset{\bullet+}{N}H_2 \longrightarrow \overset{\bullet}{C}H_3 \;+\; \left[CH_3\overset{\overset{\displaystyle CH_3}{|}}{C}\!=\!\overset{+}{N}H_2 \longleftrightarrow CH_3\overset{\overset{\displaystyle CH_3}{|}}{\underset{+}{C}}\!-\!\ddot{N}H_2 \right]$$

$$M\overset{+}{\bullet}$$

base peak
$m/z\ 58$

The spectrum with the small molecular ion at *m/z* 73 and the base peak at *m/z* 30 is that of isobutylamine. The spectrum with the barely visible molecular ion at *m/z* 73 and the base peak at *m/z* 44 is that of *sec*-butylamine. The spectrum of *tert*-butylamine has a base peak of *m/z* 58. No molecular ion is seen in the spectrum of *tert*-butylamine.

21.9 $CH_3CH_2CH_2CH_2CH\!=\!CH_2 \longrightarrow CH_3CH_2CH_2CH_2CH\overset{\bullet\;+}{-\!\!-}CH_2$

1-hexene

$$M\overset{+}{\bullet}$$
$$m/z\ 84$$

$$CH_3CH_2CH_2\!-\!CH_2\!-\!CH\overset{\bullet\;+}{-\!\!-}CH_2 \longrightarrow CH_3CH_2\overset{\bullet}{C}H_2 \;+\; CH_2\!=\!CH\overset{+}{C}H_2$$

$$M\overset{+}{\bullet}$$
$$m/z\ 84$$

allyl cation
base peak
$m/z\ 41$

21.10   The expected fragmentations of the molecular ion from 2,2-dimethyl-1-phenylpropane are shown below:

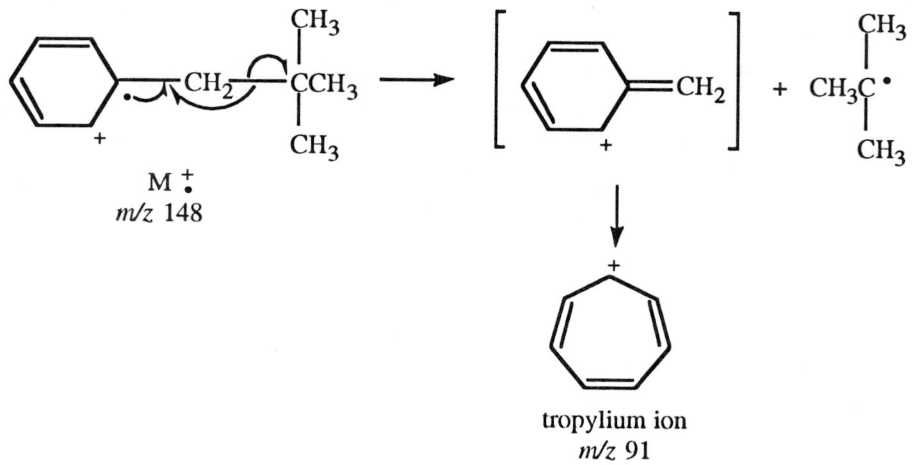

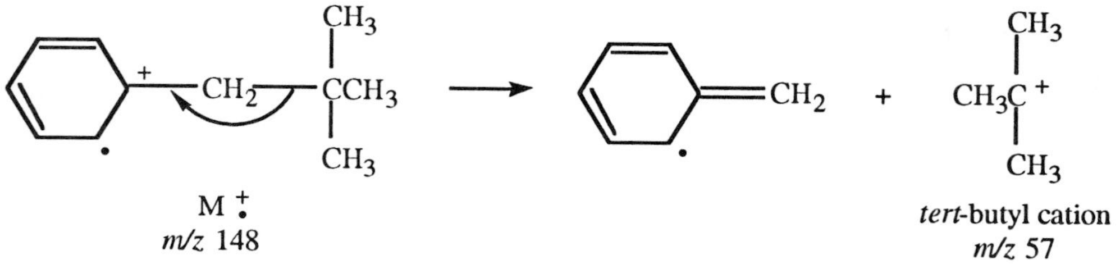

M $\overset{+}{\cdot}$
m/z 148

tropylium ion
m/z 91

M $\overset{+}{\cdot}$
m/z 148

tert-butyl cation
m/z 57

The expected fragmentation of the molecular ion from 2-methyl-3-phenylbutane is shown below:

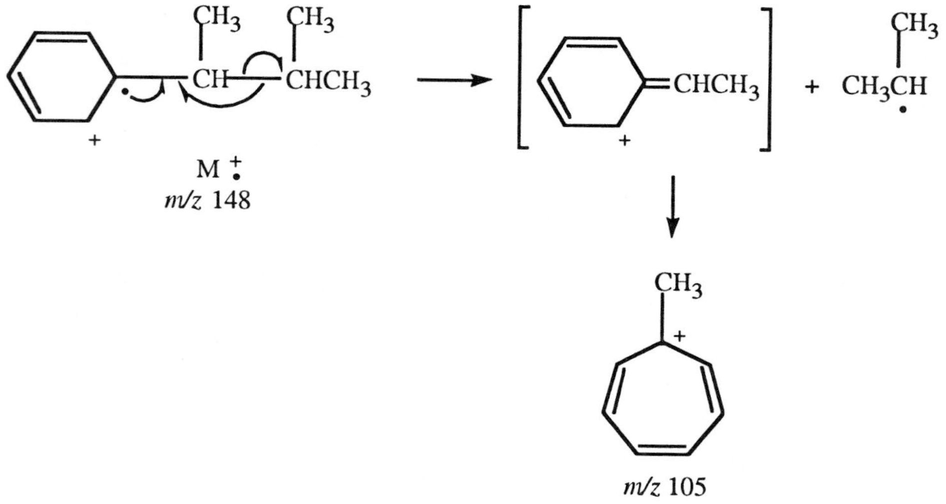

M $\overset{+}{\cdot}$
m/z 148

m/z 105

The spectrum with a peak at *m/z* 91 and the base peak at *m/z* 57 is that of 2,2-dimethyl-1-phenylpropane. The spectrum with the base peak at *m/z* 105 is that of 2-methyl-3-phenylbutane.

21.11

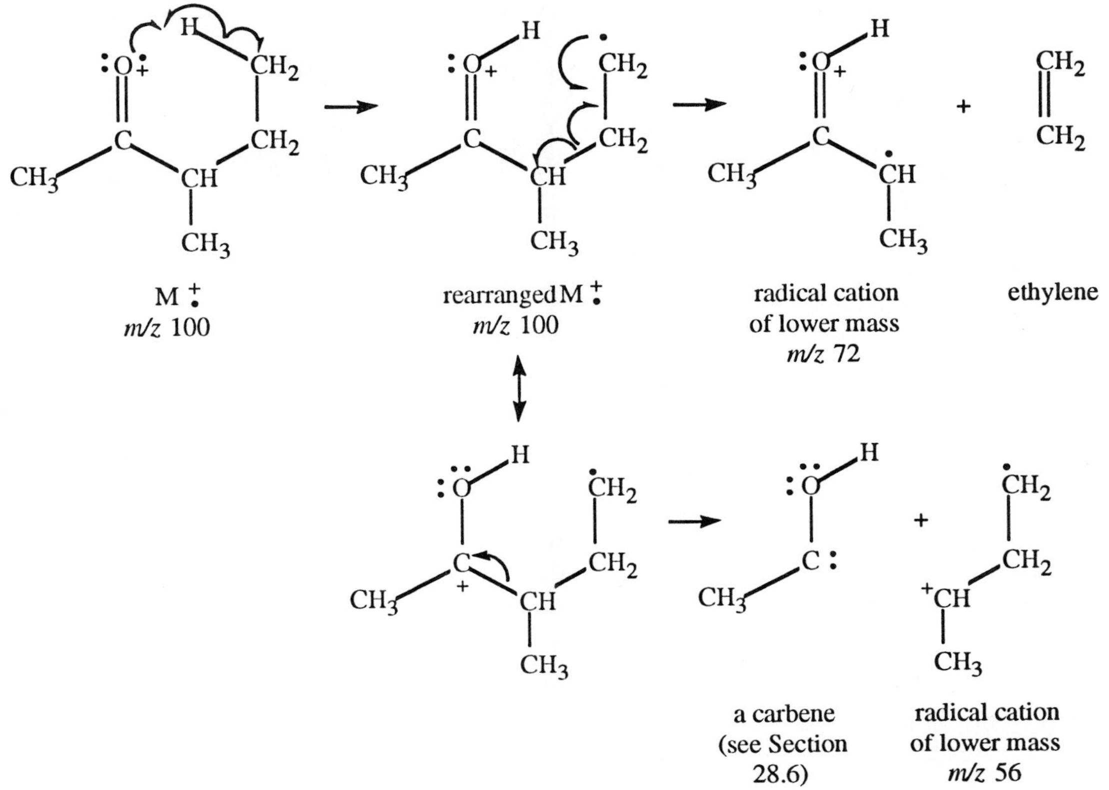

M $\overset{+}{\cdot}$
m/z 100

rearranged M $\overset{+}{\cdot}$
m/z 100

radical cation
of lower mass
m/z 72

ethylene

a carbene
(see Section
28.6)

radical cation
of lower mass
m/z 56

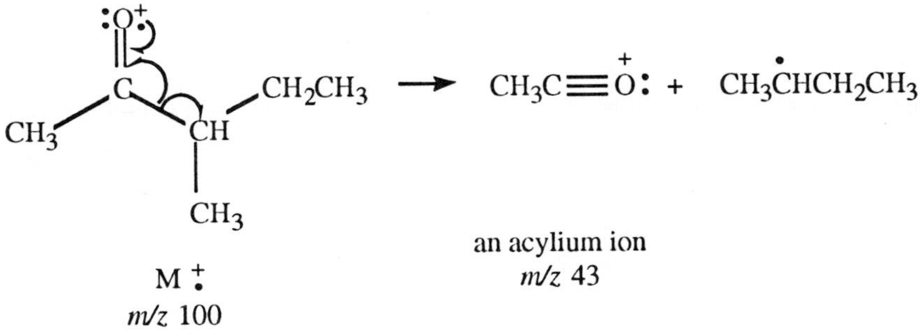

M $\overset{+}{\cdot}$
m/z 100

$$CH_3C\equiv\overset{+}{O}: \;+\;\; CH_3\overset{\cdot}{C}HCH_2CH_3$$

an acylium ion
m/z 43

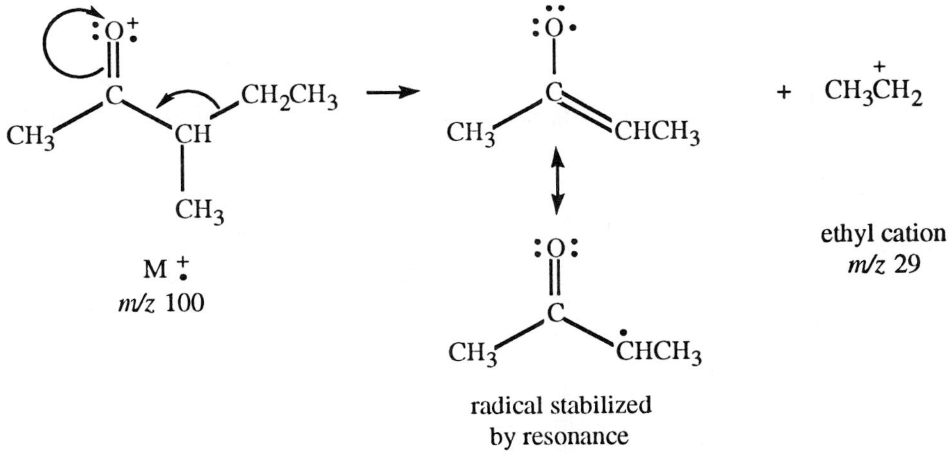

M $\overset{+}{\cdot}$
m/z 100

$$+\;\; CH_3\overset{+}{C}H_2$$

ethyl cation
m/z 29

radical stabilized
by resonance

21.12   One possible rearrangement and fragmentation of the molecular ion from 1-hexene is shown below:

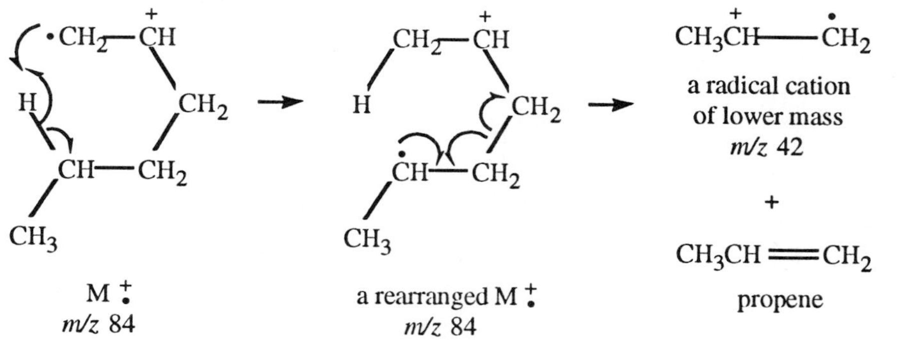

Another possible rearrangement and fragmentation of the molecular ion from 1-hexene is shown below:

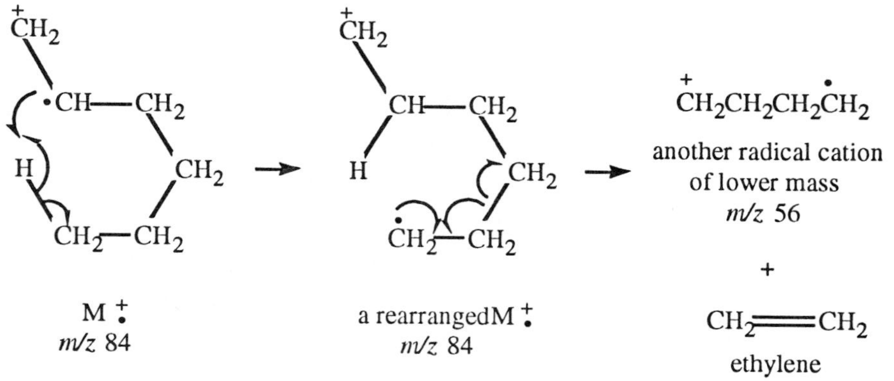

21.13   The infrared spectrum indicates the presence of an alcohol (3338 cm^{-1} is the O—H stretching frequency and 1031 cm^{-1} is the C—O stretching frequency). The proton magnetic resonance spectrum is analyzed below:

δ 1.7	(2H, multiplet, $J \sim 6$ Hz, —CH$_2$CH$_2$CH$_2$—)
δ 2.5	(2H, triplet, J 6 Hz, ArCH$_2$CH$_2$—)
δ 3.3	(2H, triplet, J 6 Hz, —CH$_2$CH$_2$O—)
δ 4.0	(1H, singlet, —OH)
δ 6.7	(5H, singlet, J 6 Hz, ArH)

Analysis of subunits in the proton magnetic resonance spectrum gives us a molecular formula of C$_9$H$_{12}$O, which corresponds to a molecular weight of 136. The molecular ion in the mass spectrum is at m/z 136. We therefore have all the atoms. The base peak (m/z 91) is the tropylium ion, which results from loss of a C$_2$H$_5$O unit and is the result of the following fragmentation.

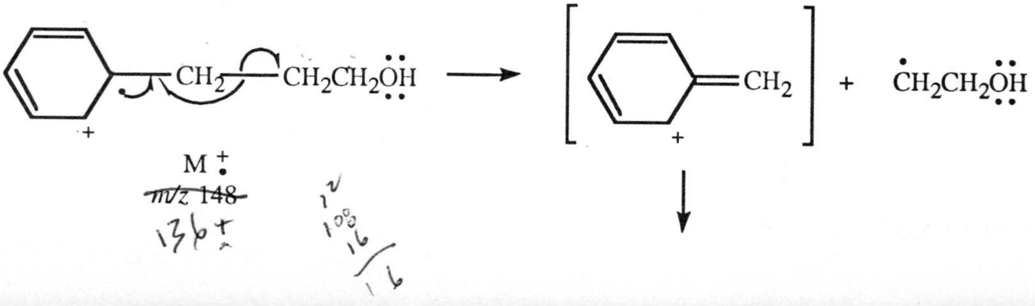

21.13 (cont)

tropylium ion
m/z 91

The compound is 3-phenyl-1-propanol. The C-13 nuclear magnetic resonance data are analyzed below:

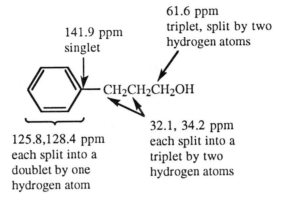

61.6 ppm
triplet, split by two
hydrogen atoms

141.9 ppm
singlet

CH₂CH₂CH₂OH

125.8,128.4 ppm
each split into a
doublet by one
hydrogen atom

32.1, 34.2 ppm
each split into a
triplet by two
hydrogen atoms

21.14   The mass spectrum of Compound B has two molecular ions at *m/z* 60 and 62 having a ratio of intensities of 3:1. This ratio tells us that the halogen in Compound B is chlorine. If we subtract 35 from 60, we get a mass of 25, which corresponds to the molecular formula C_2H, and the structure $HC\equiv C$—. Compound B is chloroethyne, $HC\equiv CCl$.

21.15   Compound C is an ester and must, therefore, contain the structural unit $-CO-$ with double-bond O, which accounts for 44 mass units. The molecular ion appears at *m/z* 60. Subtraction of 44 from 60 leaves 16 mass units, which corresponds to CH_3 and H. The structure of Compound C must be:

$$\underset{\text{methyl formate}}{HCOCH_3}$$

(with C=O double bond)

The presence of a methoxyl group is confirmed by the presence of a band in the proton magnetic resonance spectrum at δ 3.69. The fragmentations of the molecular ion leading to the ions with *m/z* 31 and 29 are shown on the next page.

$$HC-OCH_3 \longrightarrow HC\equiv \overset{+}{O}: + \cdot \overset{..}{O}CH_3$$

M⁺·
m/z 60

an acylium ion
m/z 29

$$HC-OCH_3 \longrightarrow HC + {}^+\overset{..}{O}CH_3$$

m/z 31

21.16   The two peaks of equal size in the mass spectrum of Compound D at m/z 79 and 81 tell us that bromine is present. The fact that the molecular ion fragments before it can be recorded suggests that the compound gives rise to a very stable cation, m/z 43. An acylium ion , $CH_3C\equiv O^+$, fits this description. Compound D is acetyl bromide.

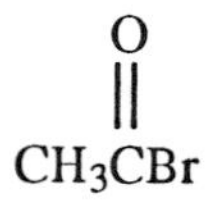

$$\underset{CH_3CBr}{\overset{\displaystyle \overset{O}{\parallel}}{}}$$

(Other possibilities that may come to mind are isopropyl or *n*-propyl bromide. Neither alkyl cation is as stable as the resonance-stabilized acylium ion, and we would therefore expect to see the molecular ion for either of these compounds.)

21.17   The molecular ion of Compound E has m/z 60. There is a difference of 29 mass units between the molecular ion and the base peak at m/z 31, which corresponds to the loss of an ethyl radical and the formation of the oxonium ion, $CH_2\overset{+}{=}OH$, which is usually the base peak in a primary alcohol. Compound E is *n*-propyl alcohol.

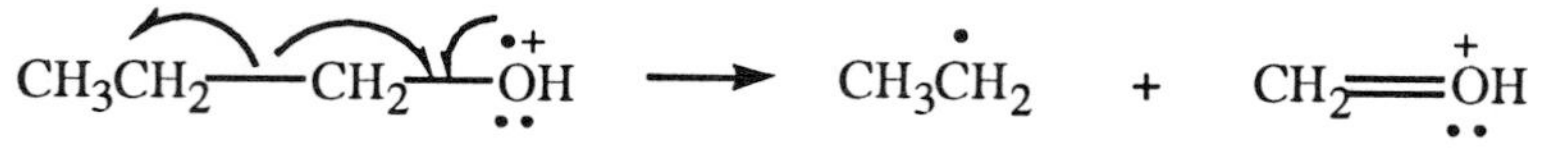

Compound F is isomeric with *n*-propyl alcohol; therefore, it must be isopropyl alcohol. The fragmentation leading to the base peak is shown below:

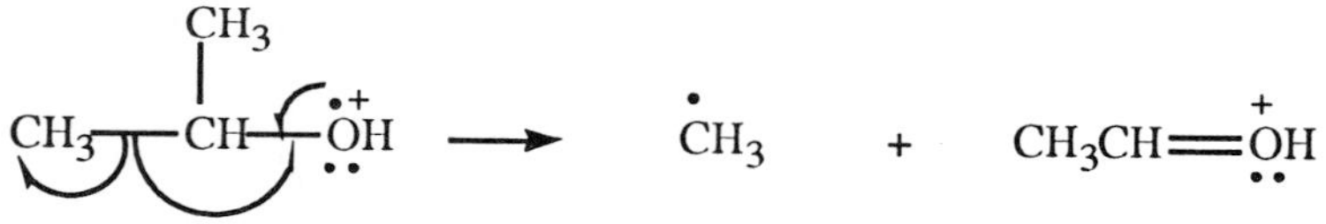

21.18

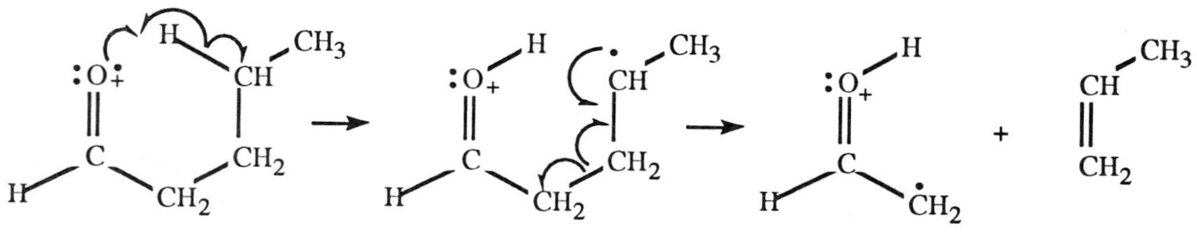

21.19   The mass spectrum of Compound G has two molecular ions of equal intensity at m/z 120 and 122. The presence of these two peaks tells us that Compound G contains bromine. The presence of bromine is confirmed by the two very weak peaks of equal intensity at m/z 79 and 81. The base peak at m/z 41 is the allyl cation, which results from loss of a bromine atom from the molecular ion. Compound G is 3-bromo-1-propene (allyl bromide). The formation of the base peak in the mass spectrum results from loss of a bromine atom.

21.19 (cont)

$$CH_2\!\!=\!\!CHCH_2\overset{79}{\cdots}\overset{..}{\underset{..}{Br}}: \longrightarrow CH_2\!\!=\!\!CHCH_2^+ + \overset{79}{} \cdot\overset{..}{\underset{..}{Br}}:$$

M $\overset{+}{\cdot}$	base peak
m/z 120	*m/z* 41

The proton and C-13 magnetic resonance spectra are analyzed below:

δ 3.7	(2H, doublet, BrCH₂CH—)
δ 4.9 and 5.1	(2H, two doublets of doublets, —CH=CH₂)
δ 5.4 – 6.1	(1H, multiplet, —CH₂—CH=CH₂)

120 ppm   33 ppm

$$CH_2\!\!=\!\!CHCH_2Br$$

135 ppm

21.20   The infrared spectrum tells us that Compound H is a conjugated ketone (1676 cm⁻¹ is the C=O stretching frequency for a conjugated carbonyl group). The proton magnetic resonance spectrum is analyzed below:

δ 2.3	(3H, singlet, CH₃C=O)
δ 3.6	(3H, singlet, —OCH₃)
δ 6.4 – 7.5	(4H, para substitution pattern, ArH)

The fragments add up to C₉H₁₀O₂, which corresponds to a molecular weight of 150. The molecular ion of Compound H is at *m/z* 150. Therefore, all atoms are accounted for. Compound H is *p*-methoxyacetophenone.

The base peak, *m/z* 135, results from loss of a methyl radical from the molecular ion, leaving a resonance-stabilized acylium ion with *m/z* 135.

M $\overset{+}{\cdot}$	
m/z 150	*m/z* 135

21.21   The infrared spectrum tells us that Compound I is an aldehyde (1703 cm^{-1} is the C=O stretching frequency and 2730 cm^{-1} distinguishes an aldehyde group from a ketone group).

δ 2.4	(3H, singlet, ArC$\underline{H}_3$)
δ 7.1 – 7.9	(4H, para substitution pattern, Ar$\underline{H}$)
δ 9.9	(1H, singlet, —C$\underline{H}$=O)

These fragments add up to C_8H_8O corresponding to a molecular weight of 120, which is the *m/z* of the molecular ion of Compound I. Putting the subunits from the proton magnetic resonance together gives us *p*-tolualdehyde. The first base peak, *m/z* 119, results from the loss of a hydrogen atom from the molecular ion, giving an acylium ion. The second base peak, *m/z* 91, is the tropylium ion, which most likely comes from a rearrangement of the methylphenyl cation formed by loss of carbon monoxide from the acylium ion.

M $\overset{+}{\cdot}$
m/z 120

an acylium ion
m/z 119

m/z 119

tropylium ion
m/z 91

The C-13 nuclear magnetic resonance is analyzed below:

134.4 ppm
singlet

145.3 ppm
singlet

21.6 ppm
split into
quartet by three
hydrogen atoms

191.4 ppm
split into
doublet by one
hydrogen atom

129.6 ppm
each split into
doublet by one
hydrogen atom

22

The Chemistry of Amines

22.1

(a) 2-methyl-1-propanamine
isobutylamine

(b) *cis*-2-ethyl-1-cyclopentanamine
cis-1-amino-2-ethylcyclopentane

(c) *N,N*-diethylbutanamine

(d) 3-cyclopropyl-1-propanamine
3-cyclopropyl-1-propylamine

(e) 3-nitroaniline
m-nitroaniline

(f) 2,4-dibromoaniline

22.2

(a)
$$K_{diss} = \frac{[R_3N\!:\,][BR'_3]}{[R_3N^+\!:^-BR'_3]}$$

(b) The dissociation constants for the amine-borane complexes become smaller as the number of substituents on the nitrogen atom increases from none for ammonia to two for dimethylamine. When there are three substituents on the nitrogen atom, the dissociation constant is larger than it is for the other two complexes. The dissociation constant measures the strength of the bond between the atoms of the Lewis base, the amine, and the Lewis acid, the borane. The bond strength is influenced by two factors: (1) electronic factors, mainly the availability of the nonbonding electron pair on the nitrogen atom; and (2) steric factors. The values observed for the dissociation constants may be interpreted to mean that alkyl substitution on the nitrogen atom increases the availability of the nonbonding electrons on the nitrogen atom, but that the presence of three substituents on nitrogen leads to steric hindrance that interferes with bonding in the complex between trimethylamine and trimethylborane. The steric factor is more important in this case than in the reaction of the amine with a Brønsted-Lowry acid because trimethylborane is larger than a proton.

22.3 The acidity of a conjugate acid is related to the availability of the nonbonding electron pair on the base. The observed trend for the conjugate acids of the given amines is due to the decreasing availability of the electron pair on the nitrogen. This decreasing availability is caused by the electron-withdrawing ether and cyano substituents which pull electron density away by induction. The inductive effect is greatest when the electron-withdrawing group is closest to the amino group. The cyano group is more electron-withdrawing than the oxygen because of resonance.

22.4

22.4 (cont)

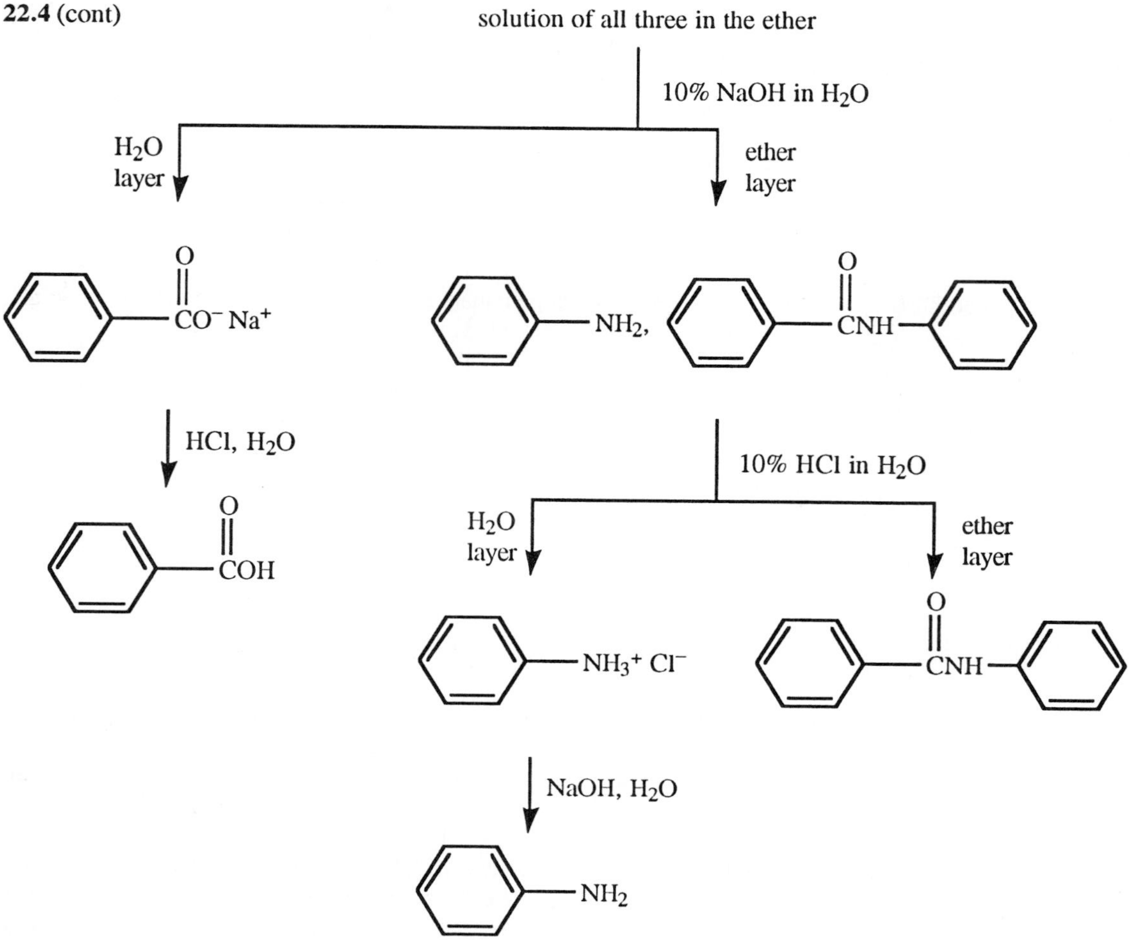
solution of all three in the ether
10% NaOH in H_2O
H_2O layer
ether layer
CO⁻ Na⁺
NH₂,
CNH
HCl, H_2O
COH
10% HCl in H_2O
H_2O layer
ether layer
NH₃⁺ Cl⁻
CNH
NaOH, H_2O
NH₂

22.5

(a) All of the cyanoanilines are weaker bases than aniline itself because the cyano group is an electron-withdrawing group. The effect of the cyano group is greatest when it is ortho, and smallest when it is meta, to the amino group. Resonance effects are important for the ortho- and para-substituted compounds, but only the inductive effect is important in *m*-cyanoaniline.

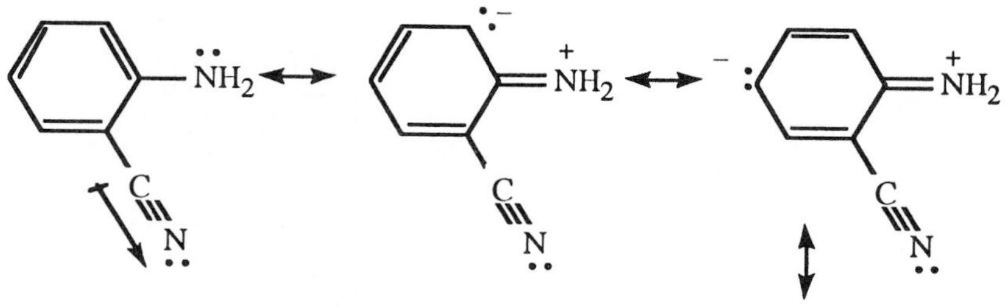

22.5 (a) (cont)

nonbonding electrons of the amino group are delocalized to
the ring and to the cyano group

no resonance delocalization of the nonbonding electrons of the
amino group to the cyano group; only the inductive effect

delocalization of the nonbonding electrons of the amino group to the
ring and to the cyano group; inductive effect of the cyano group in the
para position weaker than when it is in the ortho position

(b)   *p*-Toluidine is a slightly stronger base than aniline. This is attributed to the methyl group. 4-Aminobenzo-
phenone is a weaker base than aniline because the carbonyl group is an electron-withdrawing group.

22.5 (b) (cont)

The carbonyl group decreases the availability of the nonbonding electrons on the amino group by the inductive effect and by resonance. In both series of compounds the trends can be rationalized by looking at the availability of the nonbonding electrons on the amino group in the conjugate base. The more available the electrons, the stronger the base, the weaker is the conjugate acid, and the larger the pK_a.

(c)   The aromatic amine is less basic than the alkyl amines. The nonbonding electrons on the nitrogen atom are delocalized to the aromatic ring in *p*-toluidine; they are localized on the nitrogen atom in diethylamine and ethylamine.

The lower availability of electrons on the nitrogen atom of the aromatic amine compared to the alkyl amine is used to rationalize its lower basicity and the increased acidity of its conjugate acid. Diethylamine, with two alkyl groups on the nitrogen, is slightly more basic than ethylamine, which has only one alkyl group on the nitrogen.

(d)   The less positive (or more negative) a pK_a is, the stronger the conjugate acid and the weaker is its conjugate base. In aniline, the nonbonding electrons of the nitrogen atom are delocalized to one aromatic ring. The conjugate acid of diphenylamine, with a pK_a of 0.8, is a strong acid, and diphenylamine is thus a weak base. The ability of a base to remove a proton from an acid depends on how available the electron pair of the base is, i.e., how much electron density is available. In diphenylamine, the electron pair on the nitrogen atom is delocalized into both phenyl rings, reducing the amount of electron density on the nitrogen atom even more than in aniline.

22.5 (d) (cont)

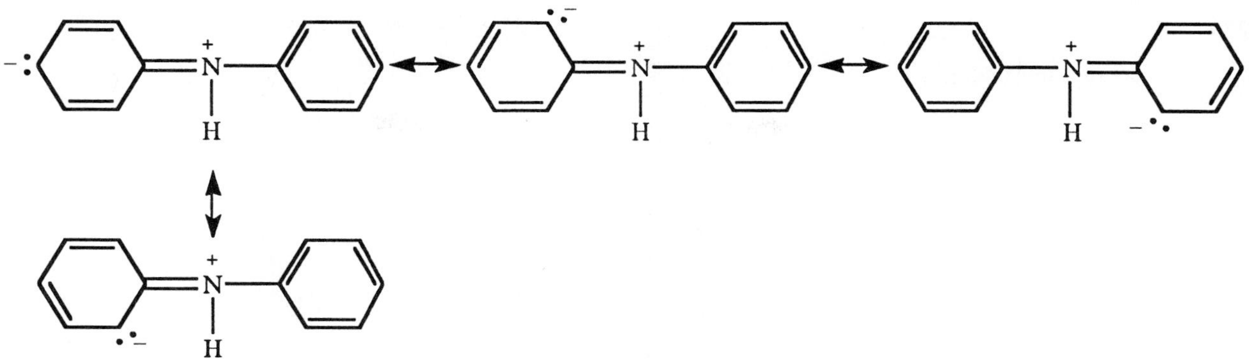

22.6

(a)

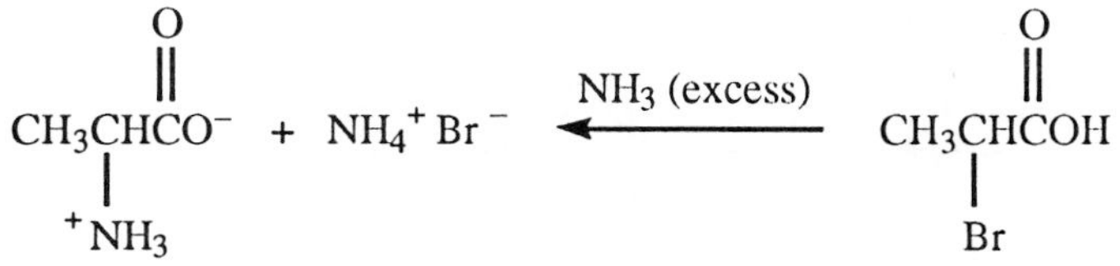

$$CH_3\overset{\displaystyle |}{\underset{\displaystyle +NH_3}{CHCO^-}} + NH_4^+\,Br^- \xleftarrow{\;NH_3\text{ (excess)}\;} CH_3\overset{\displaystyle |}{\underset{\displaystyle Br}{CHCOH}}$$

(b)

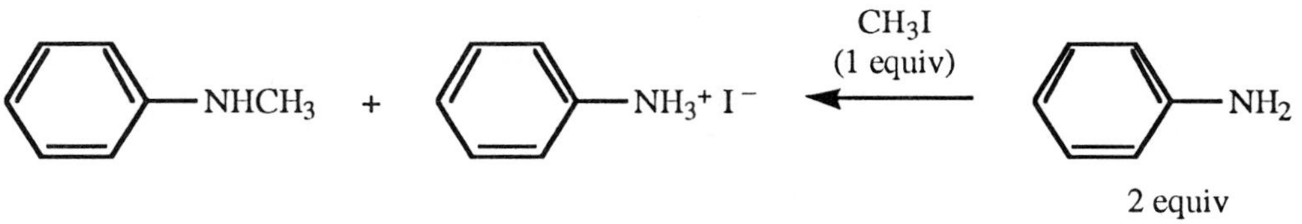

(c)

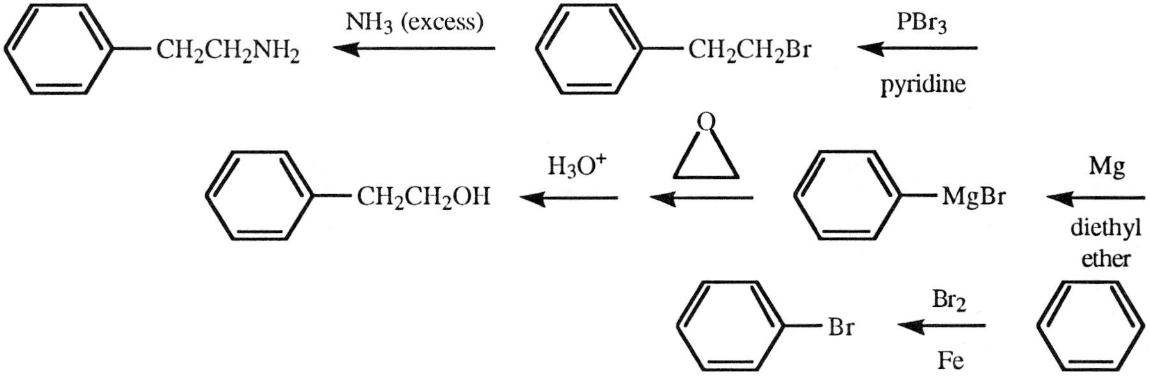

22.7   4-Chlorobutanoic acid is a strong enough acid to protonate the phthalimidate anion, converting it to phthalimide. The nitrogen atom in phthalimide is not nucleophilic because the nonbonding electrons on nitrogen are delocalized to two carbonyl groups.

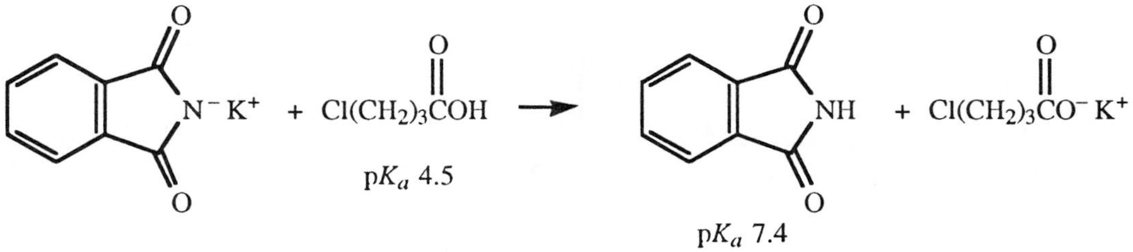

22.8 $ClCH_2CH_2CH_2C\equiv N$ $\xleftarrow[\text{pyridine}]{SOCl_2}$ $HOCH_2CH_2CH_2C\equiv N$ $\xleftarrow[\text{ethanol}]{NaCN}$

4-chlorobutanenitrile

$HOCH_2CH_2CH_2Cl$ $\xleftarrow[\text{pyridine}]{SOCl_2 \text{ (1 molar equiv)}}$ $HOCH_2CH_2CH_2OH$

22.9 $\overset{+}{H_3}NCH_2CH_2CH_2\overset{\displaystyle O}{\overset{\|}{C}}OH$ $+$ CO_3^{2-} $\underset{\longrightarrow}{\longleftarrow}$ $\overset{+}{H_3}NCH_2CH_2CH_2\overset{\displaystyle O}{\overset{\|}{C}}O^-$ $+$ HCO_3^-

$pK_a \sim 10.6$ $pK_a \sim 4.8$ pK_a 10.2

The carboxylic acid group is more acidic than the conjugate acid of the amino group on the amino acid. As long as a large excess of carbonate ion is not used, the amino acid will be in the form shown.

22.10   (a)

(b) $CH_3(CH_2)_4C\equiv C(CH_2)_8N_3$ $\xrightarrow[\substack{Pd/CaCO_3 \\ quinoline}]{H_2}$

22.11

22.12

22.12 (cont)

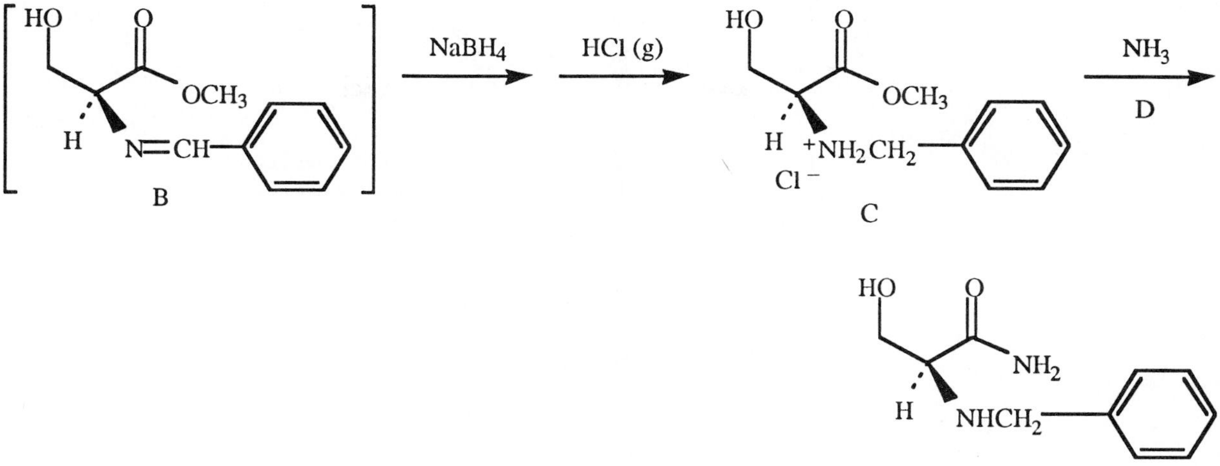

22.13

(a)

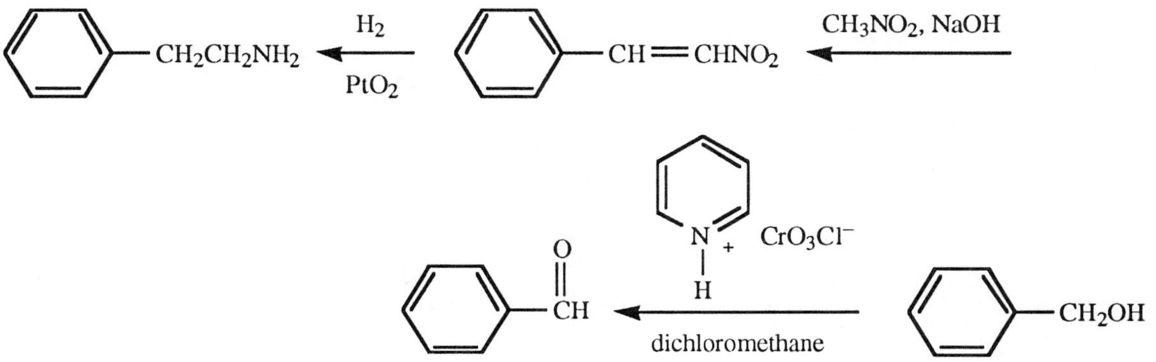

(b)

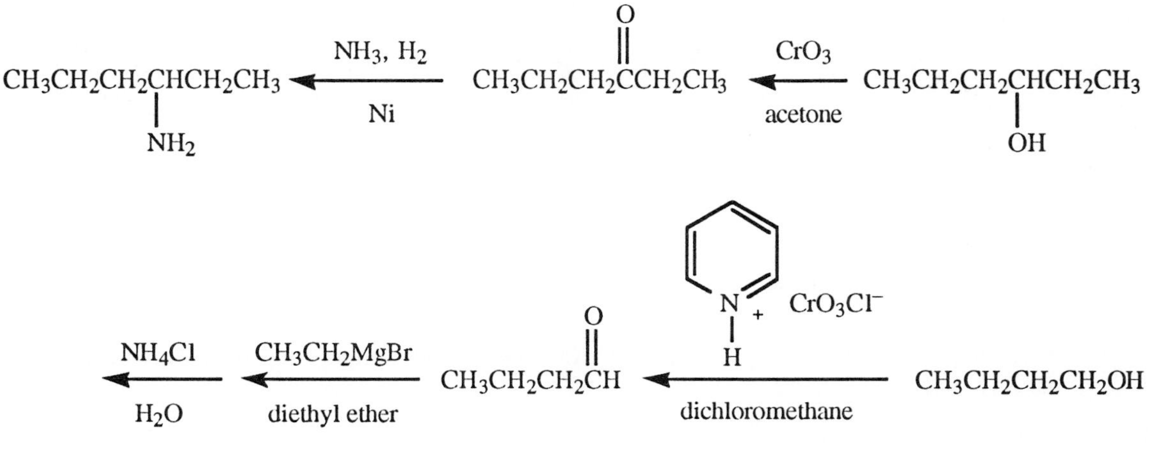

(c)

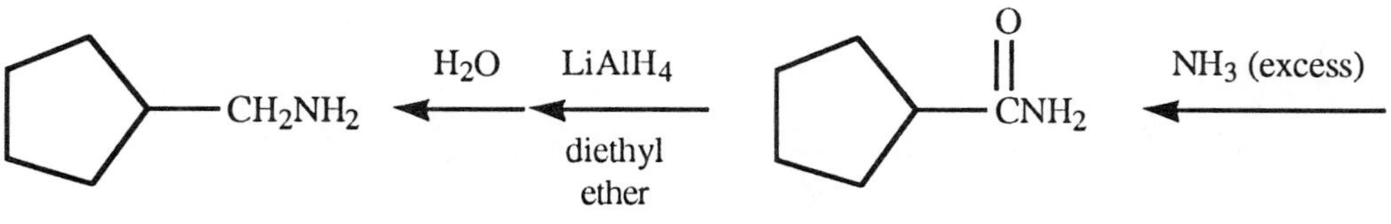

22.13 (c) (cont)

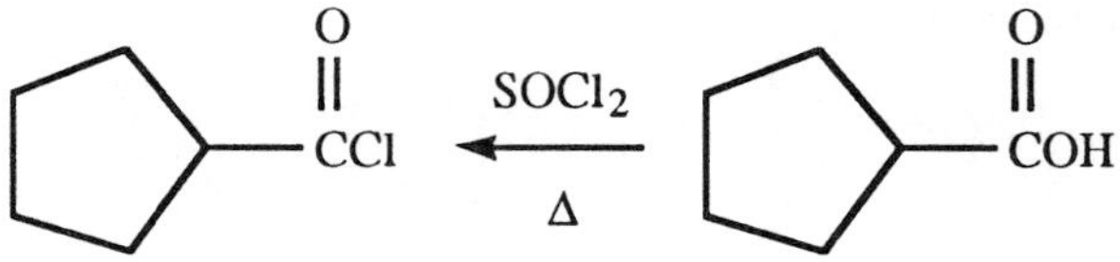

(d)

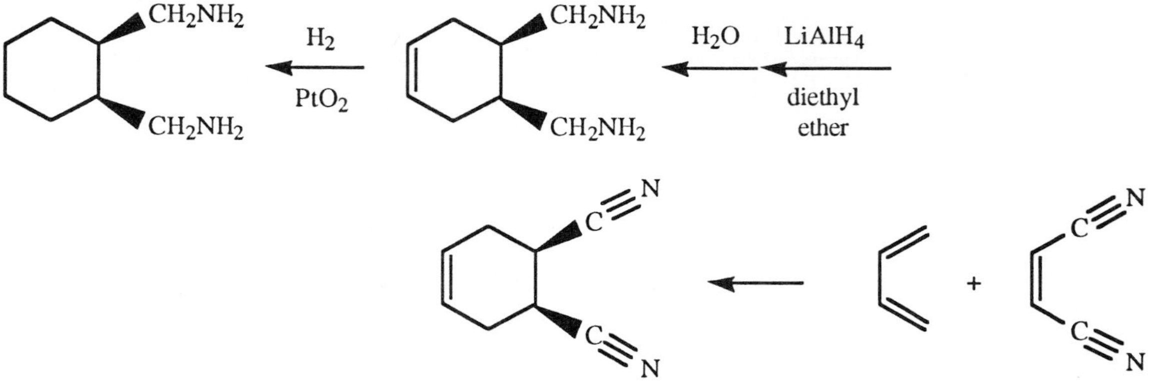

22.14

(a)

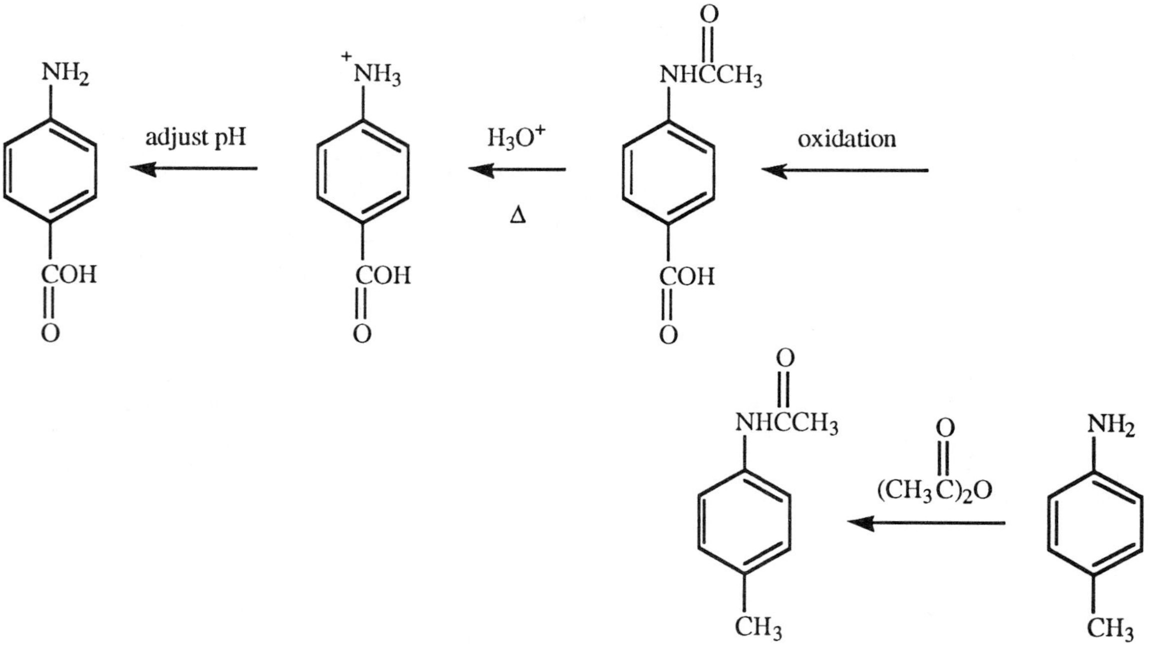

Protection of the amino group is necessary because an aromatic amine is sensitive to oxidizing agents (Section 20.7A).

(b)   If the pH of the solution is too low, deprotonation of the amine function would not occur, and the protonated *p*-aminobenzoic acid would remain in solution.

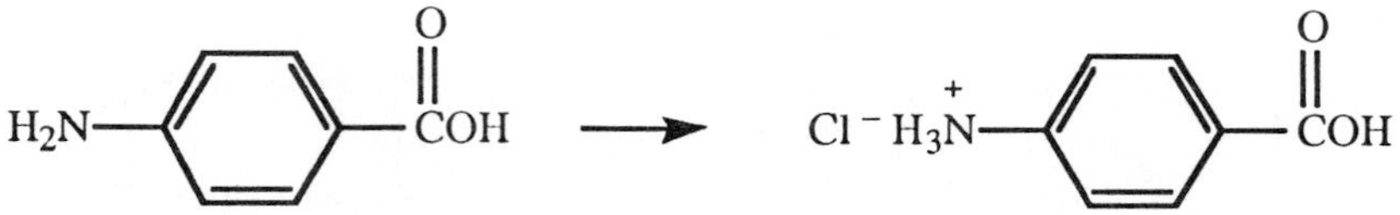

22.14 (b) (cont)

If the pH is too high, deprotonation of the carboxylic acid group would occur, and the *p*-aminobenzoate salt would remain in solution.

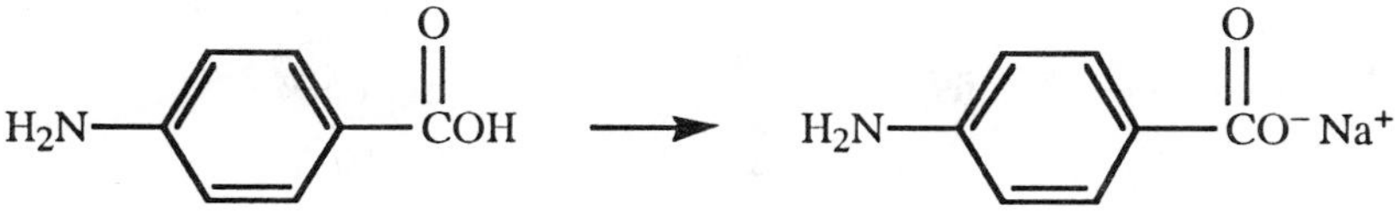

22.15

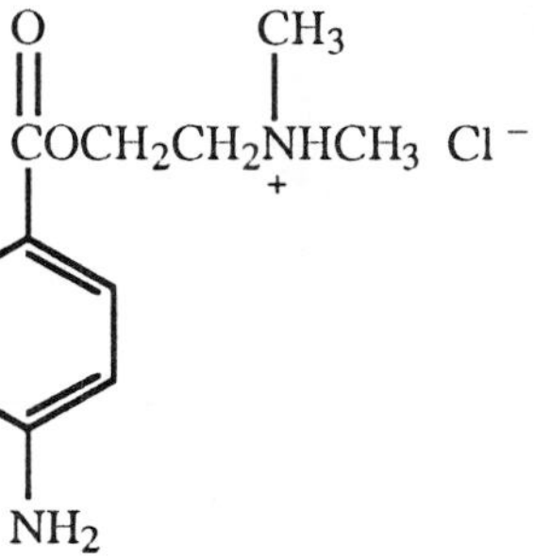

Procaine has two amine functions, an amino group on the aromatic ring, and a tertiary alkyl amine function. The alkyl amine is much more basic (pK_a of a substituted ammonium ion, ~10) than an aryl amine (pK_a of the anilinium ion, 4.6). Therefore procaine hydrochloride has the structure shown above.

——

Concept Map 22.1   (see p. 788)

——

22.16

(a)

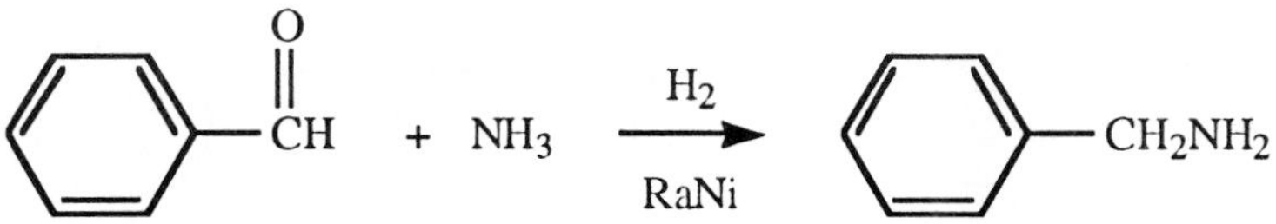

(b)

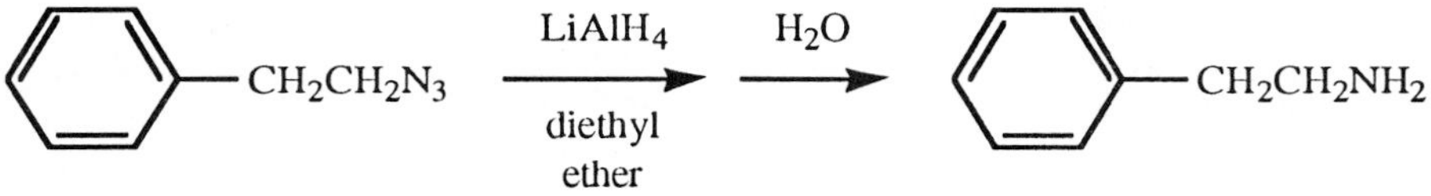

(c)

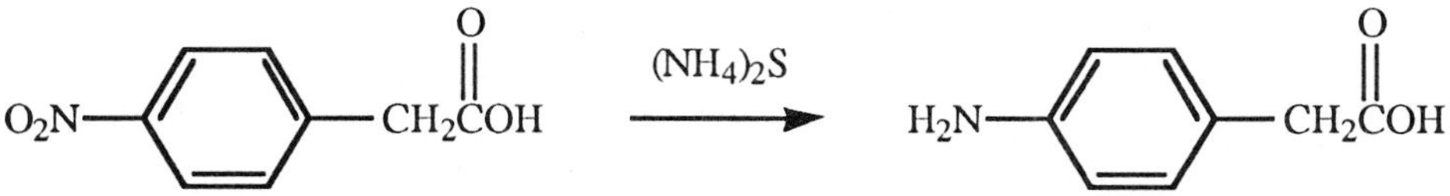

(d)

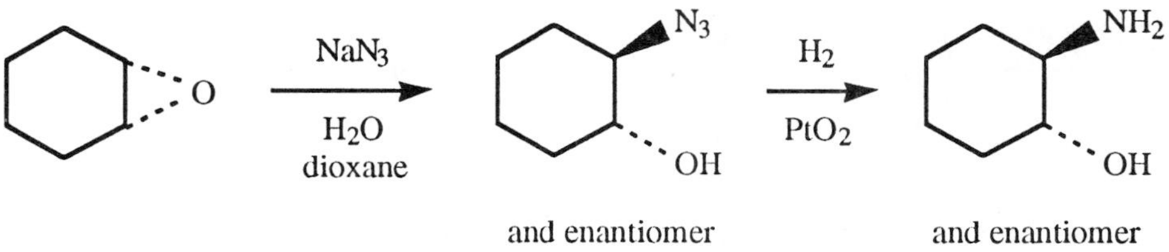

and enantiomer   and enantiomer

(e)

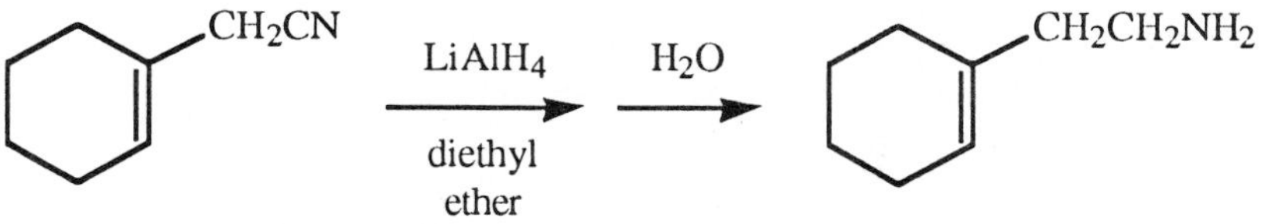

Concept Map 22.1   Preparation of amines.

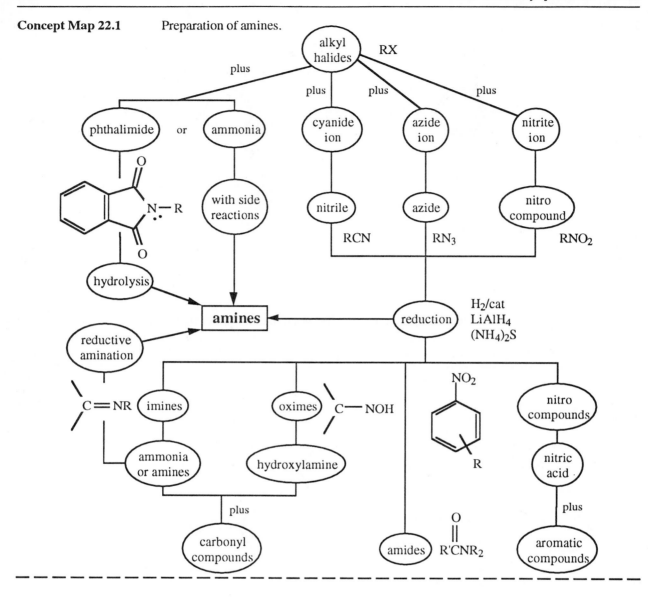

22.17

22.17 (cont)

(c)

$$\text{C}_6\text{H}_5\text{N}(\text{CH}_2\text{CH}_2\text{CH}_2\text{CH}_3)(\text{CH}_2\text{CH}_2\text{CH}_2\text{CH}_3) \xleftarrow[\text{H}_2\text{O}]{\text{NaOH}} \xleftarrow{\text{CH}_3\text{CH}_2\text{CH}_2\text{CH}_2\text{Br}}$$

$$\text{C}_6\text{H}_5\text{—NHCH}_2\text{CH}_2\text{CH}_2\text{CH}_3 \xleftarrow[\text{H}_2\text{O}]{\text{NaBH}_4}$$

$$\text{C}_6\text{H}_5\text{—N}\!=\!\text{CHCH}_2\text{CH}_2\text{CH}_3 \xleftarrow{} \text{C}_6\text{H}_5\text{—NH}_2 \qquad \text{CH}_3\text{CH}_2\text{CH}_2\overset{\text{O}}{\overset{\|}{\text{CH}}}$$

(d)   (1) $\text{CH}_3(\text{CH}_2)_6\text{CH}_2\text{NH}_2 \xleftarrow{\text{NH}_3 \text{ (excess)}} \text{CH}_3(\text{CH}_2)_6\text{CH}_2\text{Br} \xleftarrow[\text{pyridine}]{\text{PBr}_3}$

$$\text{CH}_3(\text{CH}_2)_6\text{CH}_2\text{OH} \xleftarrow{\text{H}_3\text{O}^+} \overset{\triangle\!\!\!^{\text{O}}}{\xleftarrow{}} \text{CH}_3(\text{CH}_2)_4\text{CH}_2\text{MgBr} \xleftarrow[\substack{\text{diethyl} \\ \text{ether}}]{\text{Mg}} \text{CH}_3(\text{CH}_2)_4\text{CH}_2\text{Br} \xleftarrow[\text{pyridine}]{\text{PBr}_3}$$

$$\text{CH}_3(\text{CH}_2)_4\text{CH}_2\text{OH} \xleftarrow{\text{H}_3\text{O}^+} \overset{\triangle\!\!\!^{\text{O}}}{\xleftarrow{}} \text{CH}_3\text{CH}_2\text{CH}_2\text{CH}_2\text{MgBr} \xleftarrow[\substack{\text{diethyl} \\ \text{ether}}]{\text{Mg}} \text{CH}_3\text{CH}_2\text{CH}_2\text{CH}_2\text{Br}$$

(2) $\text{CH}_3(\text{CH}_2)_6\text{CH}_2\text{NH}_2 \xleftarrow[\substack{\text{Pd} \\ \text{acetic acid}}]{\text{H}_2} \text{CH}_3(\text{CH}_2)_5\text{CH}\!=\!\text{CHNO}_2 \xleftarrow{\text{CH}_3\text{NO}_2, \text{NaOH}}$

$$\text{CH}_3(\text{CH}_2)_5\overset{\text{O}}{\overset{\|}{\text{CH}}} \xleftarrow[\substack{\text{dichloromethane} \\ \text{or}}]{\underset{\text{H}}{\overset{+}{\text{pyridinium}}} \text{CrO}_3\text{Cl}^-} \text{CH}_3(\text{CH}_2)_5\text{CH}_2\text{OH} \xleftarrow{\text{H}_3\text{O}^+} \overset{\square\!\!\!^{\text{O}}}{\xleftarrow{}} \text{CH}_3\text{CH}_2\text{CH}_2\text{CH}_2\text{MgBr}$$

[from (d) (1)]

$$(\text{CH}_3\text{CH}_2)_3\text{N} \quad \text{CH}_3\text{SCH}_3, \text{ClC}\overset{\text{O}}{\overset{\|}{}}\!\!-\!\!\overset{\text{O}}{\overset{\|}{}}\text{CCl} \xleftarrow{} \xleftarrow{}$$

22.17 (d) (cont)

(3) $CH_3(CH_2)_6CH_2NH_2$ $\xleftarrow{NH_3, H_2 \atop Ni}$ $CH_3(CH_2)_6CH\!\!\!\overset{O}{\underset{}{\|}}$ $\xleftarrow{\text{dichloromethane}}$ $CH_3(CH_2)_6CH_2OH$

[from (d) (1)]

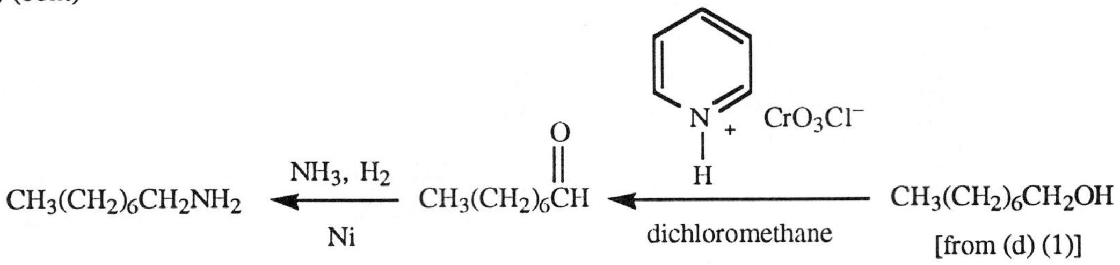

22.18

(a) CH_3NCH_3 $\xleftarrow{NaOH \atop H_2O}$ $\overset{+}{CH_3}NHCH_3$ I^- $\xleftarrow{2\ CH_3I}$ NH_2 $\xleftarrow{NaOH}$ $\xleftarrow{Sn,\ HCl}$

$\xleftarrow{HNO_3 \atop H_2SO_4}$ (benzene with NO_2)

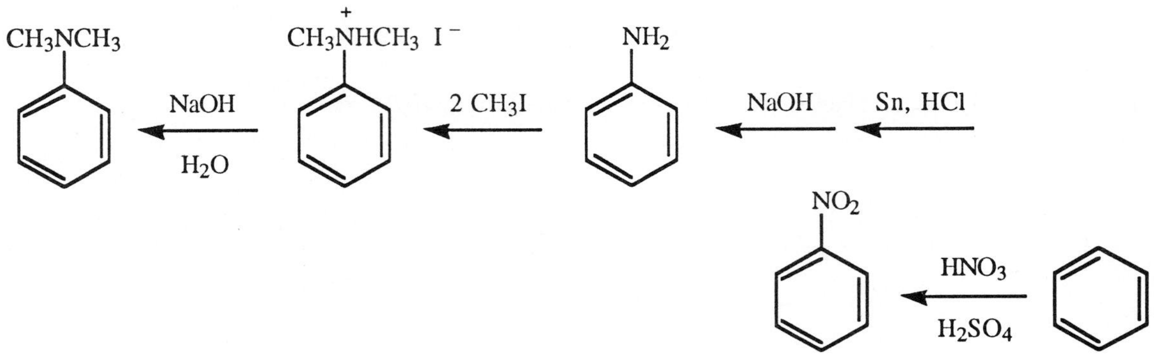

(b) NH_2 / NO_2 $\xleftarrow{NaOH \atop H_2O \atop \Delta}$ $NHCCH_3$ with NO_2 $\xleftarrow{HNO_3 \atop H_2SO_4}$ $NHCCH_3$ $\xleftarrow{(CH_3C)_2O \atop pyridine}$

NH_2 $\xleftarrow{NaOH \atop H_2O}$ $\xleftarrow{Fe,\ HCl}$ NO_2 $\xleftarrow{HNO_3 \atop H_2SO_4 \atop \Delta}$ (benzene)

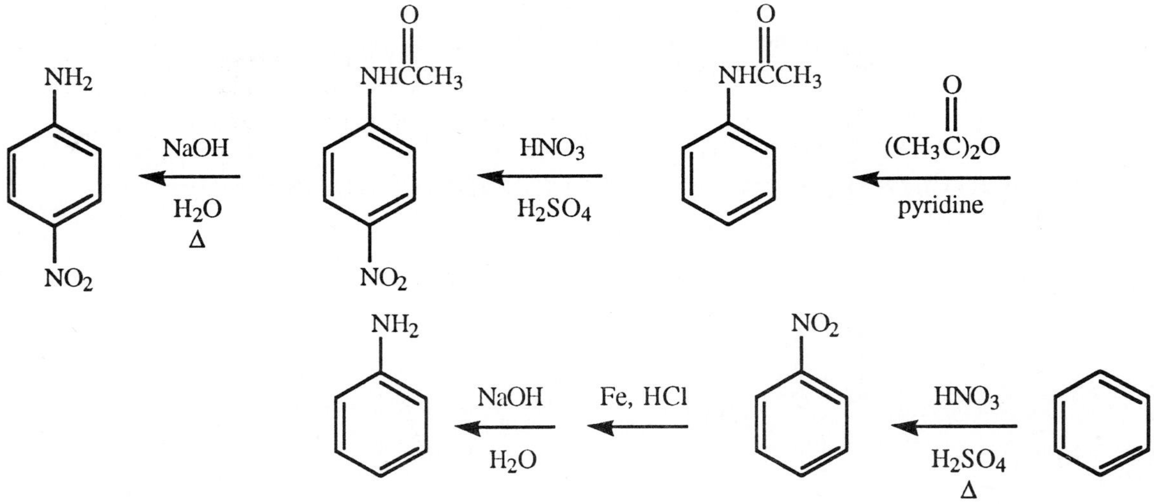

(c) $CH_3CHCH_2CH_2NH_2$ with CH_3 $\xleftarrow{H_2 \atop Pd \atop \text{acetic acid}}$ $CH_3CHCH=CHNO_2$ with CH_3 $\xleftarrow{CH_3NO_2,\ NaOH}$

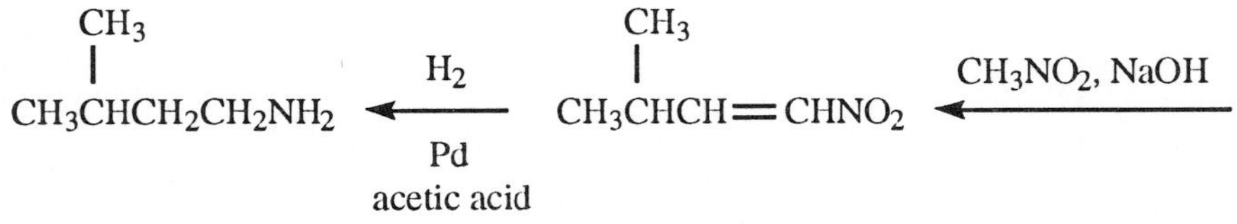

22.18 (c) (cont)

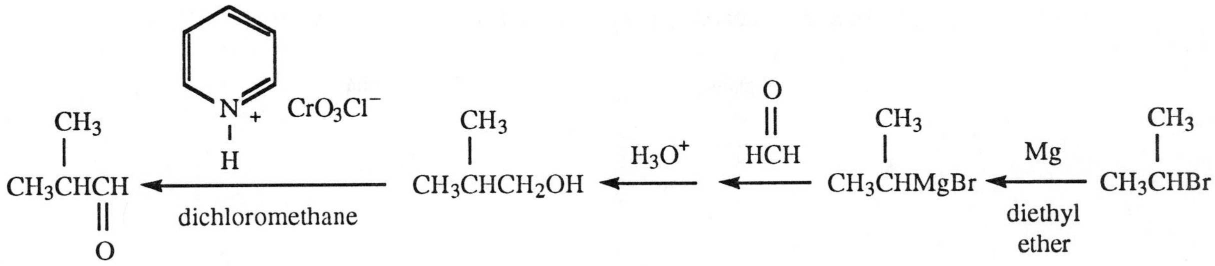

$$CH_3CHCH \xleftarrow[\text{dichloromethane}]{\text{pyridinium } CrO_3Cl^-} CH_3CHCH_2OH \xleftarrow{H_3O^+} \xleftarrow[\text{HCH}]{O} CH_3CHMgBr \xleftarrow[\substack{\text{diethyl} \\ \text{ether}}]{Mg} CH_3CHBr$$

22.19

(a)

$$CH_3CH_2CH_2CHCH_3 \xrightarrow[\substack{H_2O \\ 0\,°C}]{NaNO_2,\ HCl} CH_3CH_2CH_2CHCH_3 + CH_3CH_2CH_2CHCH_3$$

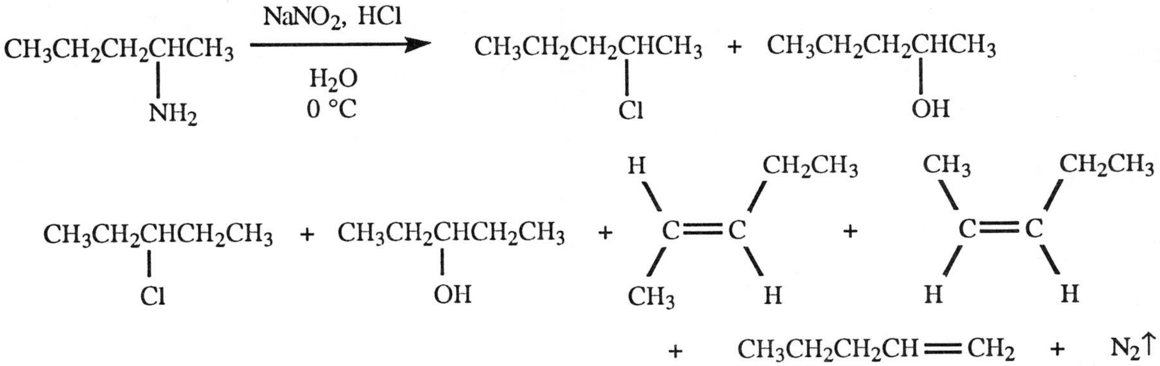

(b)

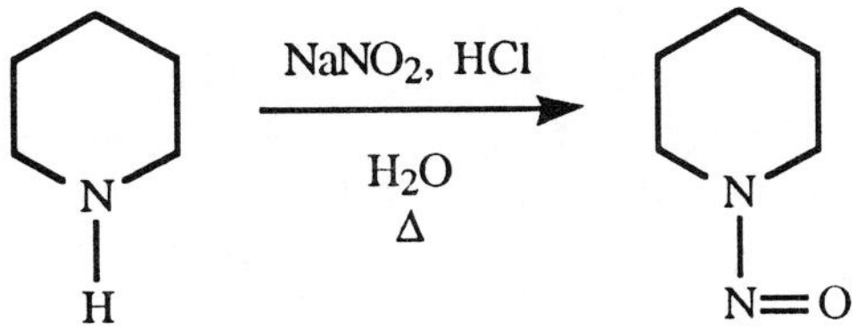

(c)

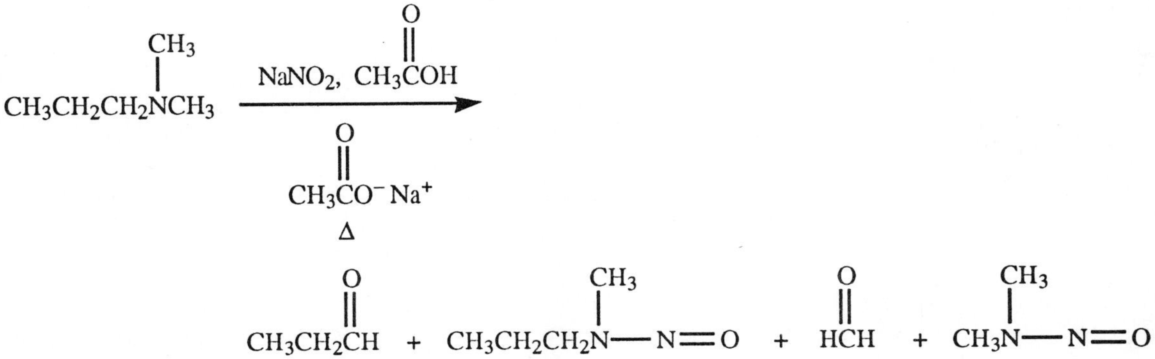

(d)

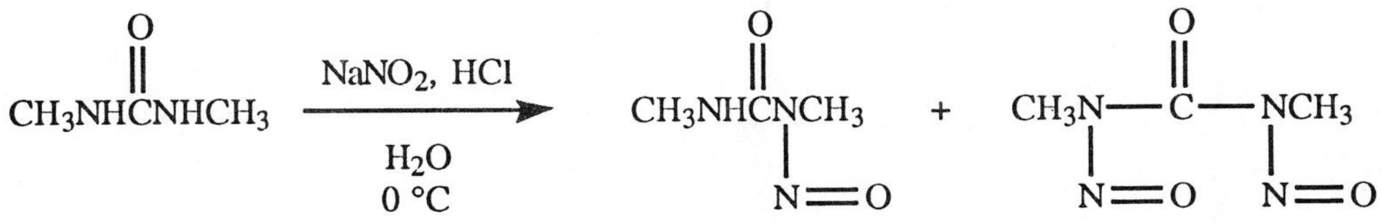

22.19 (cont)

(e)

$$CH_3\underset{\underset{CH_3}{|}}{\overset{\overset{CH_3}{|}}{C}}NH_2 \xrightarrow[\substack{H_2O \\ 0\ °C}]{NaNO_2,\ HCl} CH_3\underset{\underset{Cl}{|}}{\overset{\overset{CH_3}{|}}{C}}CH_3 + CH_3\underset{\underset{OH}{|}}{\overset{\overset{CH_3}{|}}{C}}CH_3 + CH_3\overset{\overset{CH_3}{|}}{C}{=}CH_2 + N_2\uparrow$$

22.20 R_2NH + HCl $\longrightarrow$ $R_2\overset{+}{N}H_2\ Cl^-$

water soluble

R_2NH + HNO_2 $\xrightarrow{HCl}$ $R_2\ddot{N}-N{=}\ddot{O} \longleftrightarrow R_2\overset{+}{N}{=}N-\ddot{\underset{..}{O}}:^-$

The nonbonding electrons on the amine nitrogen atom are delocalized to the nitroso group. The nitrosoamine is therefore much less basic than the original amine and will not form a salt. A nitrosamine resembles an amide in basicity.

22.21   (a)

(S)-[1-²H]-1-butanol
>99% S

(S)-tosylate

(R)-azide

(R)-[1-²H]-1-butanamine
98% R

(S)-[1-²H]-1-butanol
97% S

(b)   The experiment shows that the replacement of —NH₂ by —OH is going cleanly by inversion of stereochemistry, which means that the loss of nitrogen from the intermediate diazonium ion happens by the S_N2 mechanism. Ionization to give a primary cation, which would lose its stereochemical identity, cannot be taking place.

Concept Map 22.2   Nitrosation reactions.

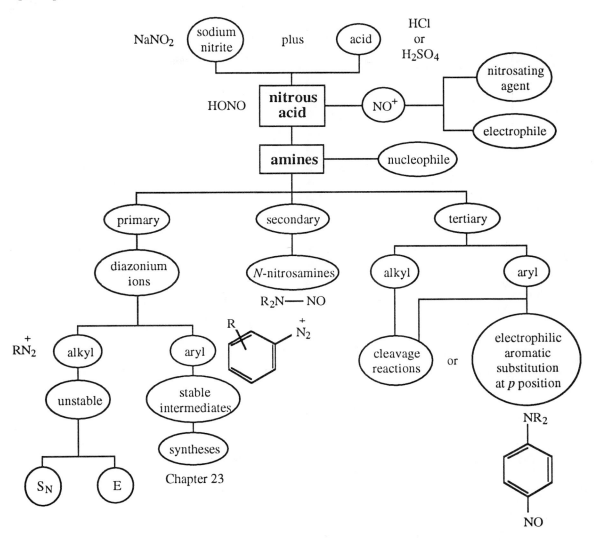

- -

22.22

(a)

22.22 (cont)

(b)

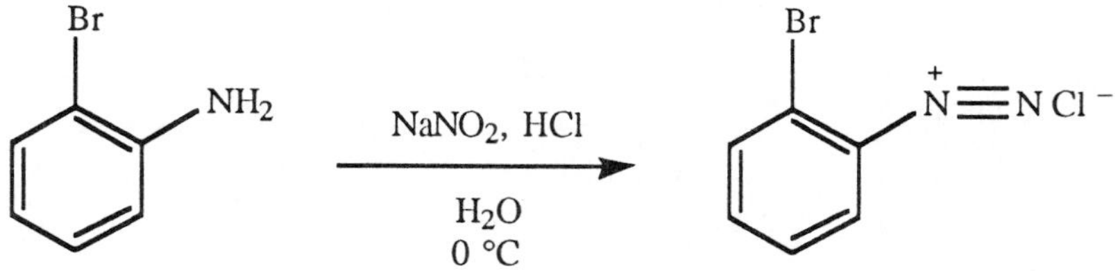

(c)

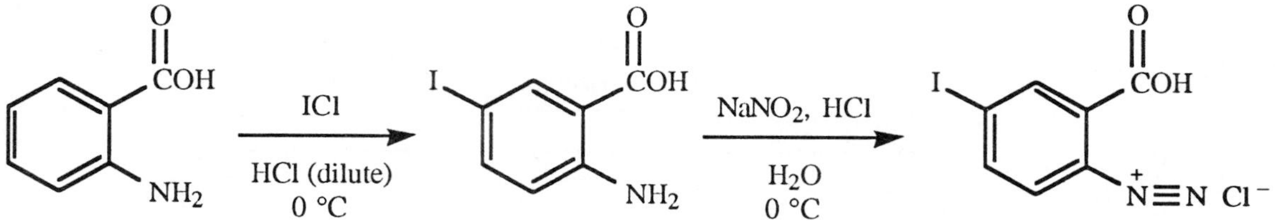

(d)

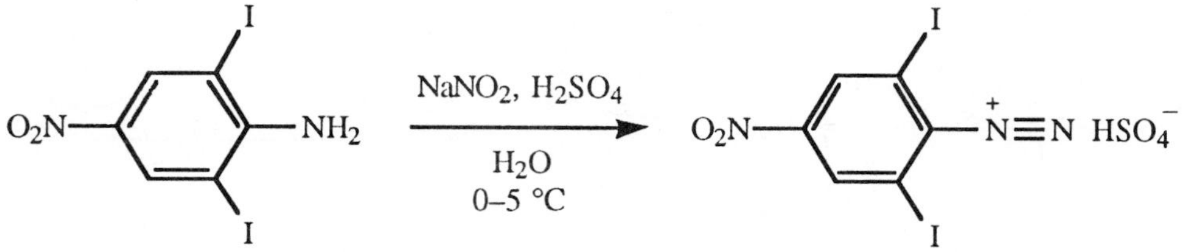

22.23

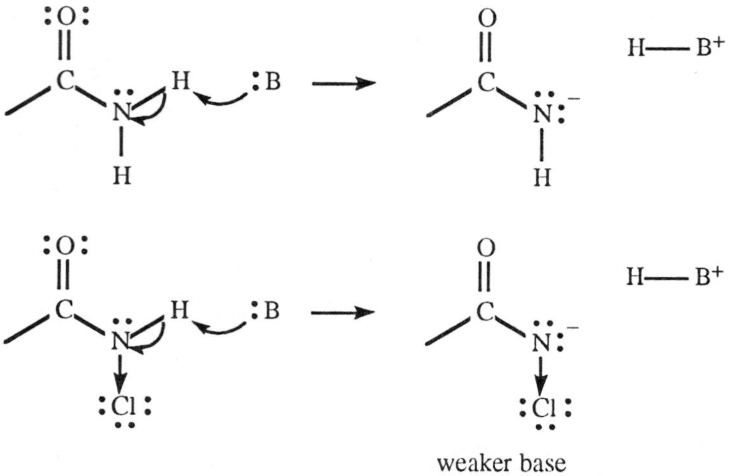

weaker base
than amide anion

Because chlorine is more electronegative than hydrogen, it can stabilize a negative charge more effectively. The anion from the *N*-chloroamide is thus a weaker base than the anion from the amide, and its conjugate acid is a stronger acid than the original unsubstituted amide.

Concept Map 22.3   Rearrangements to nitrogen atoms.

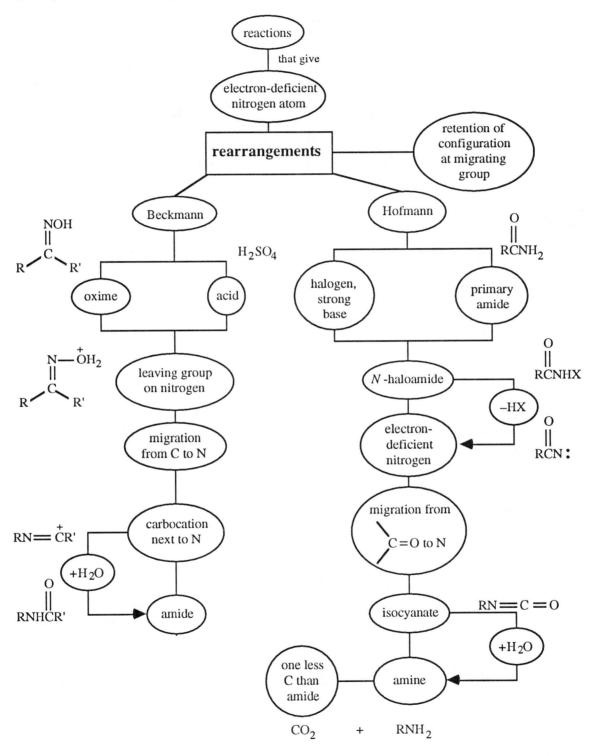

22.24

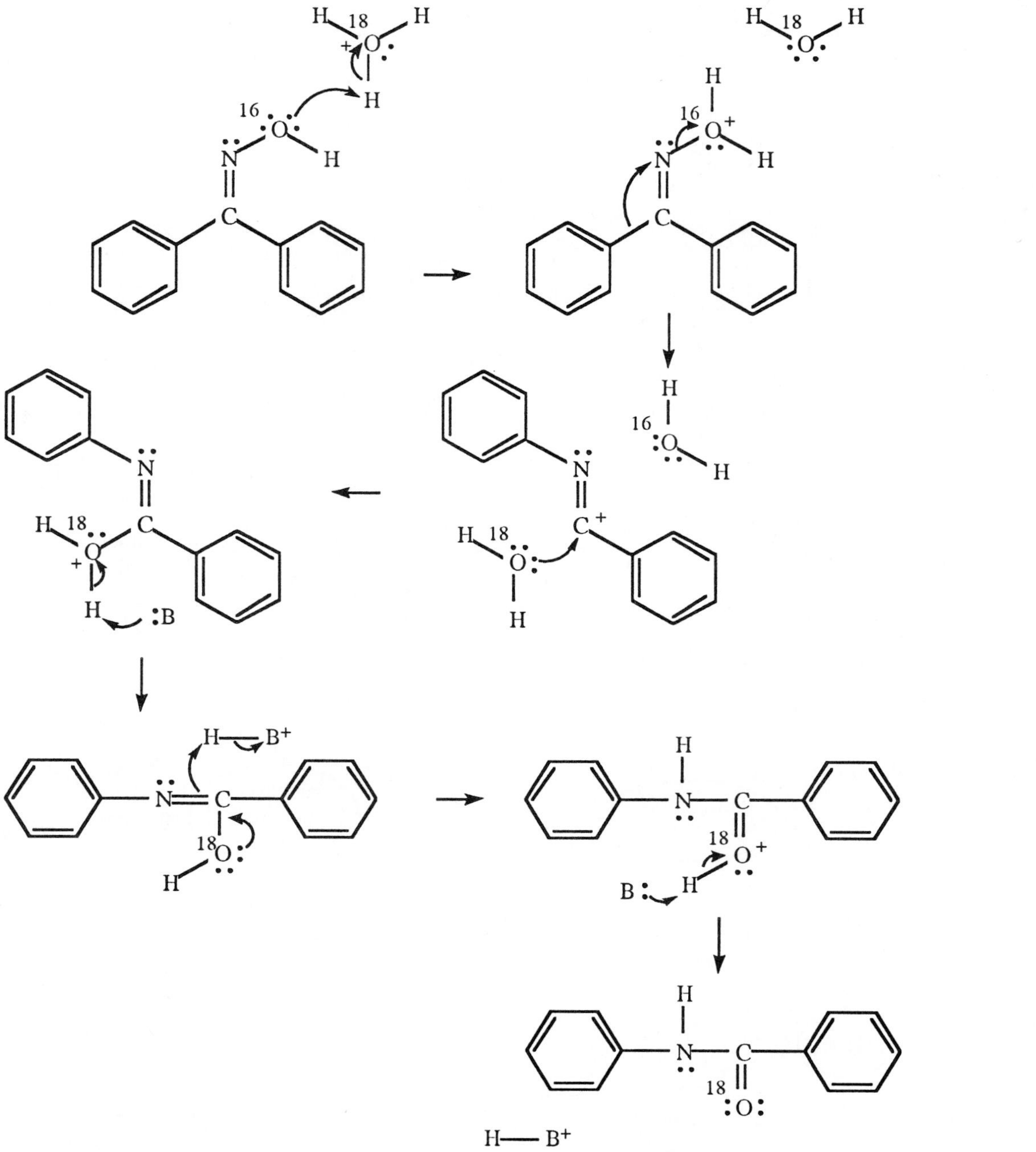

If an intramolecular reaction with water had occurred, the amide would contain ^{16}O instead of ^{18}O.

22.25

(a)

$$\underset{ClCCl}{\overset{O}{\|}} \quad + \quad NH_3 \text{ (excess)} \quad \longrightarrow \quad \underset{H_2NCNH_2}{\overset{O}{\|}}$$

22.25 (cont)

(b) 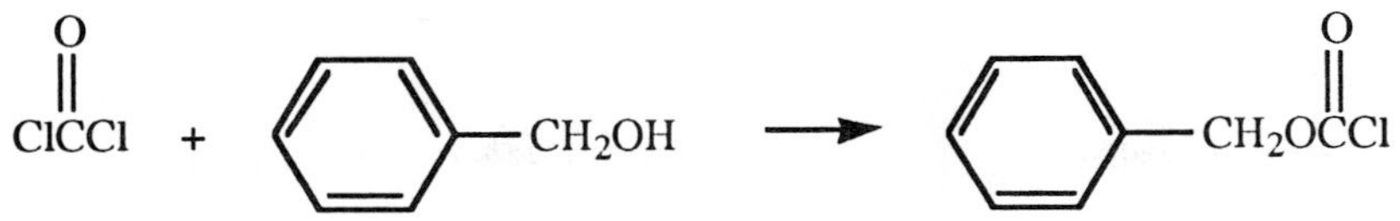   ClCCl +   [phenyl]—CH₂OH   ⟶   [phenyl]—CH₂OCCl

                (1 molar equiv)

(c) 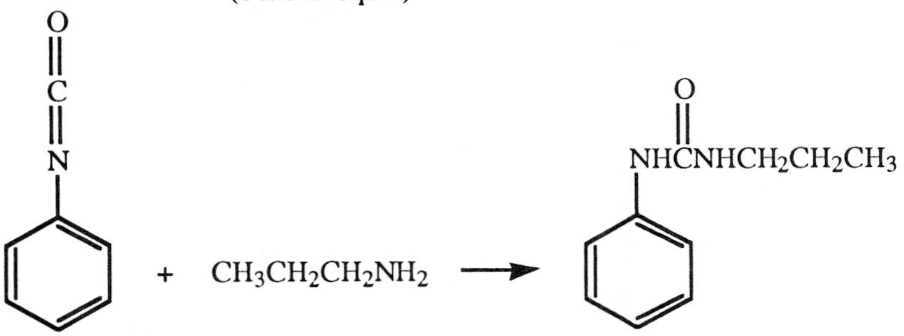   [phenyl isocyanate]   +   CH₃CH₂CH₂NH₂   ⟶   [phenyl-NHCNHCH₂CH₂CH₃]

(d) 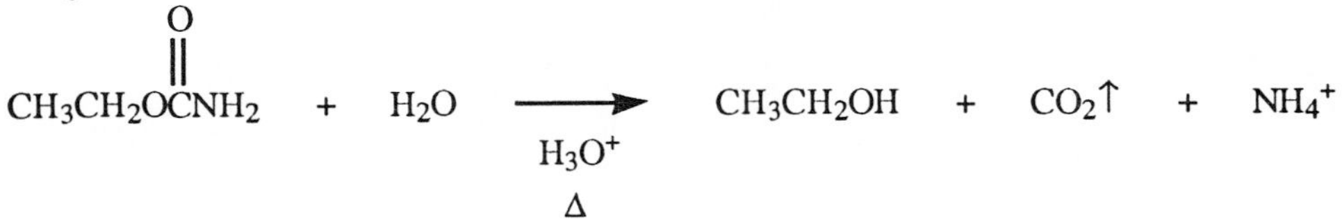   CH₃CH₂OCNH₂   +   H₂O $\xrightarrow[\Delta]{H_3O^+}$ CH₃CH₂OH   +   CO₂↑   +   NH₄⁺

(e) 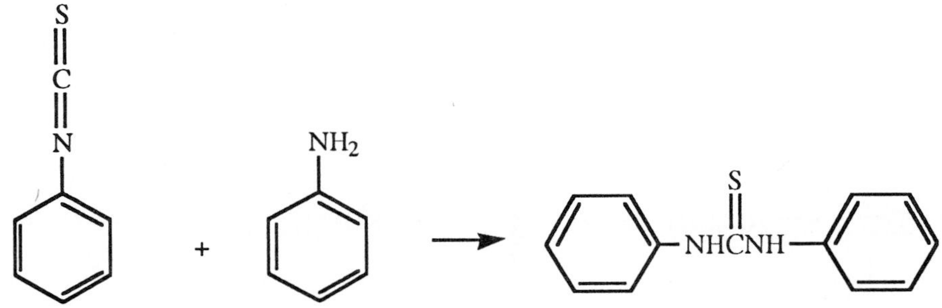   [phenyl isothiocyanate]   +   [aniline, NH₂]   ⟶   [phenyl—NHCNH—phenyl]

    phenyl isothiocyanate   a substituted thiourea

22.26   1.   Degradation in acid

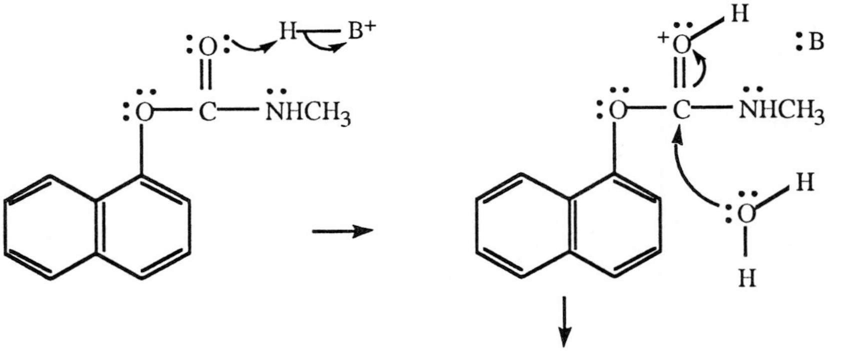

22.26 (1) (cont)

22.26 (cont)

2. Degradation in base

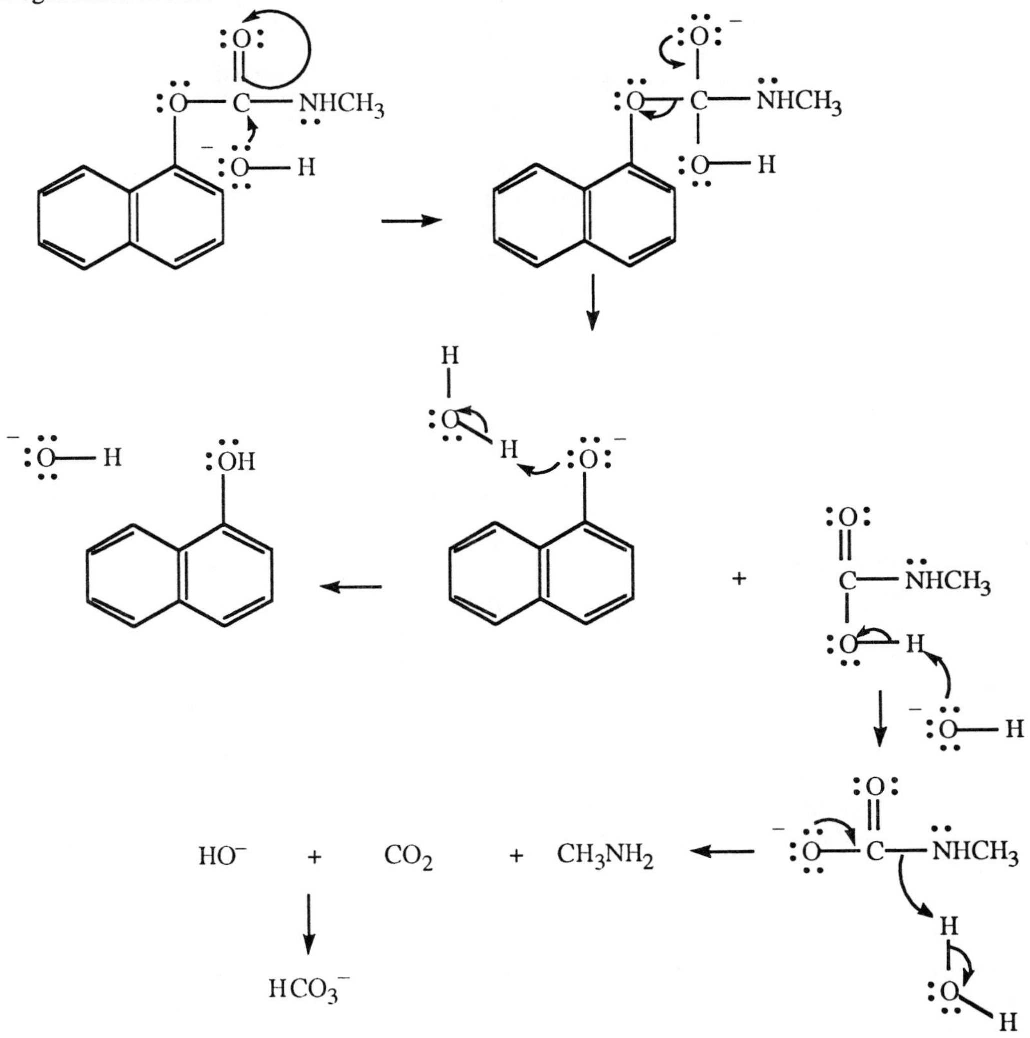

Concept Map 22.4 (see p. 800)

22.27

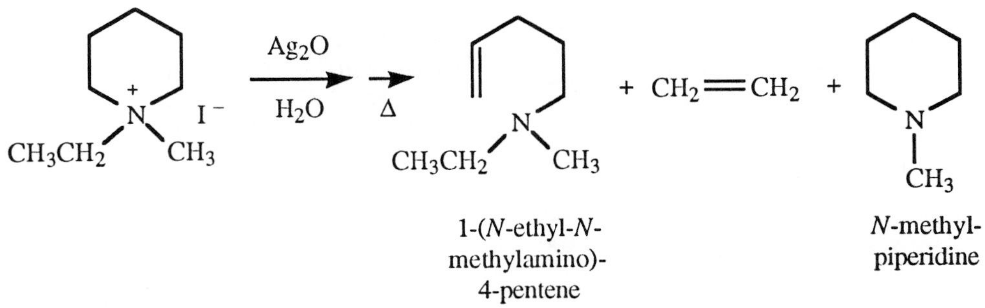
Ag2O
H2O
Δ
+ CH2=CH2 +
1-(N-ethyl-N-methylamino)-4-pentene
N-methylpiperidine

Concept Map 22.4   Reactions of quaternary nitrogen compounds.

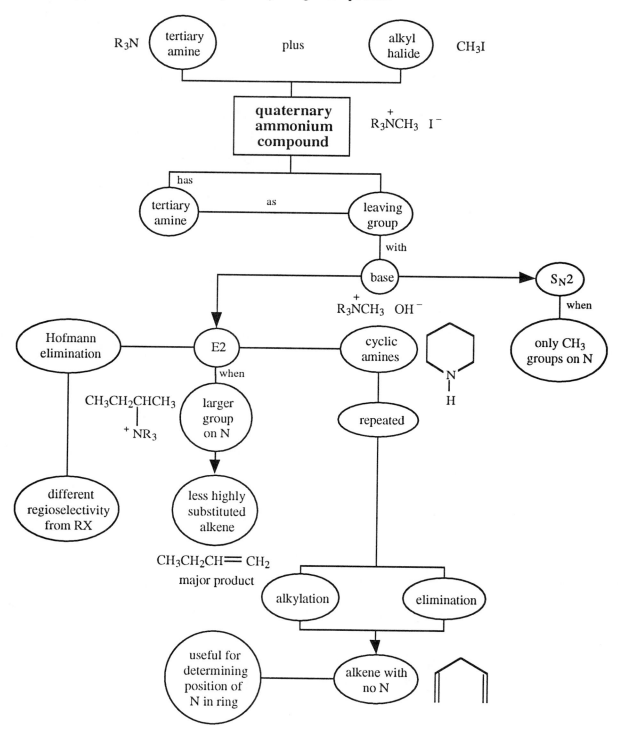

22.28   The Hofmann elimination reaction would have to be carried out three times to free the nitrogen atom from the compound.

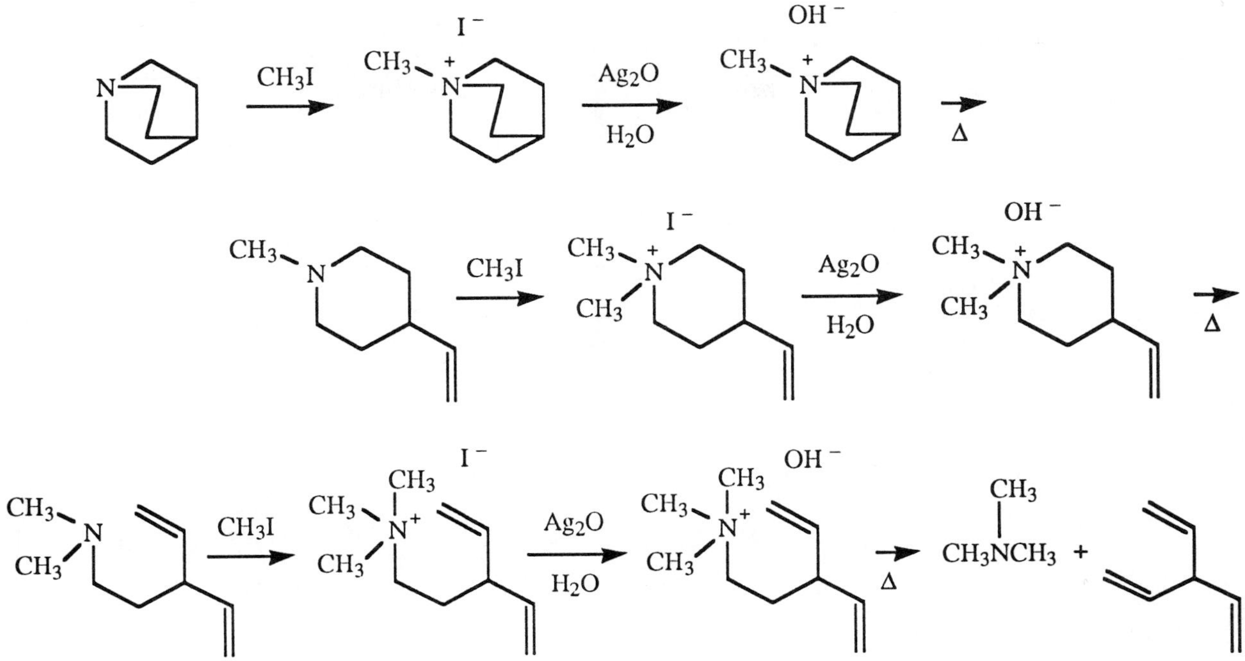

22.29

(a)

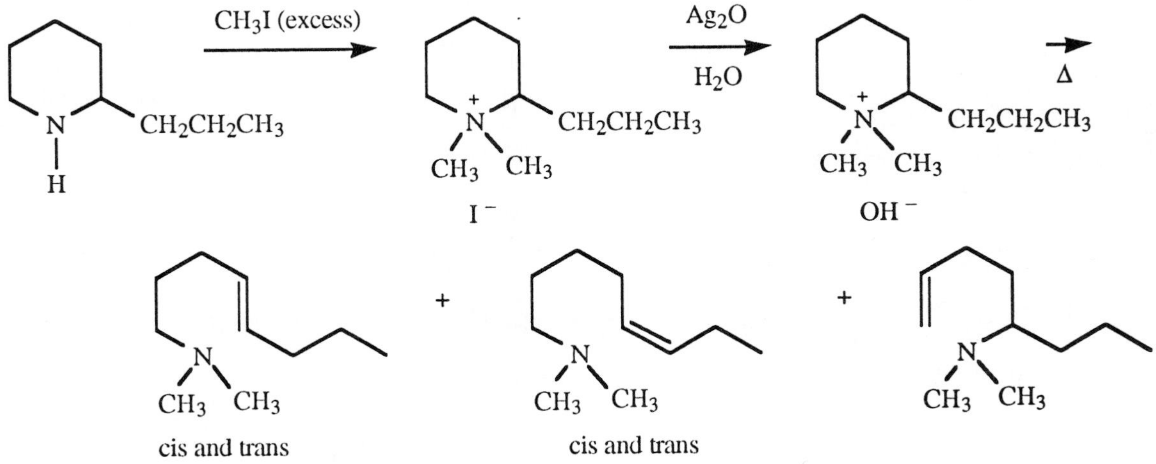

cis and trans   cis and trans

(b)

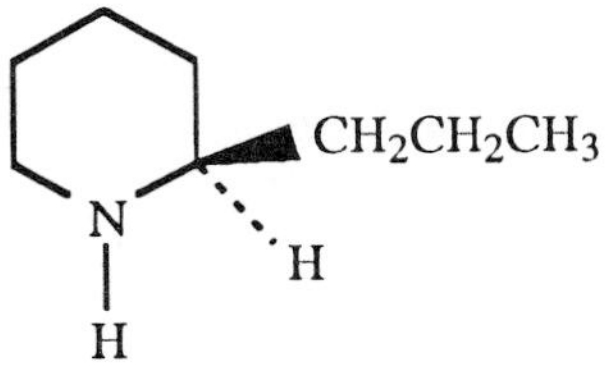

22.30

(a)

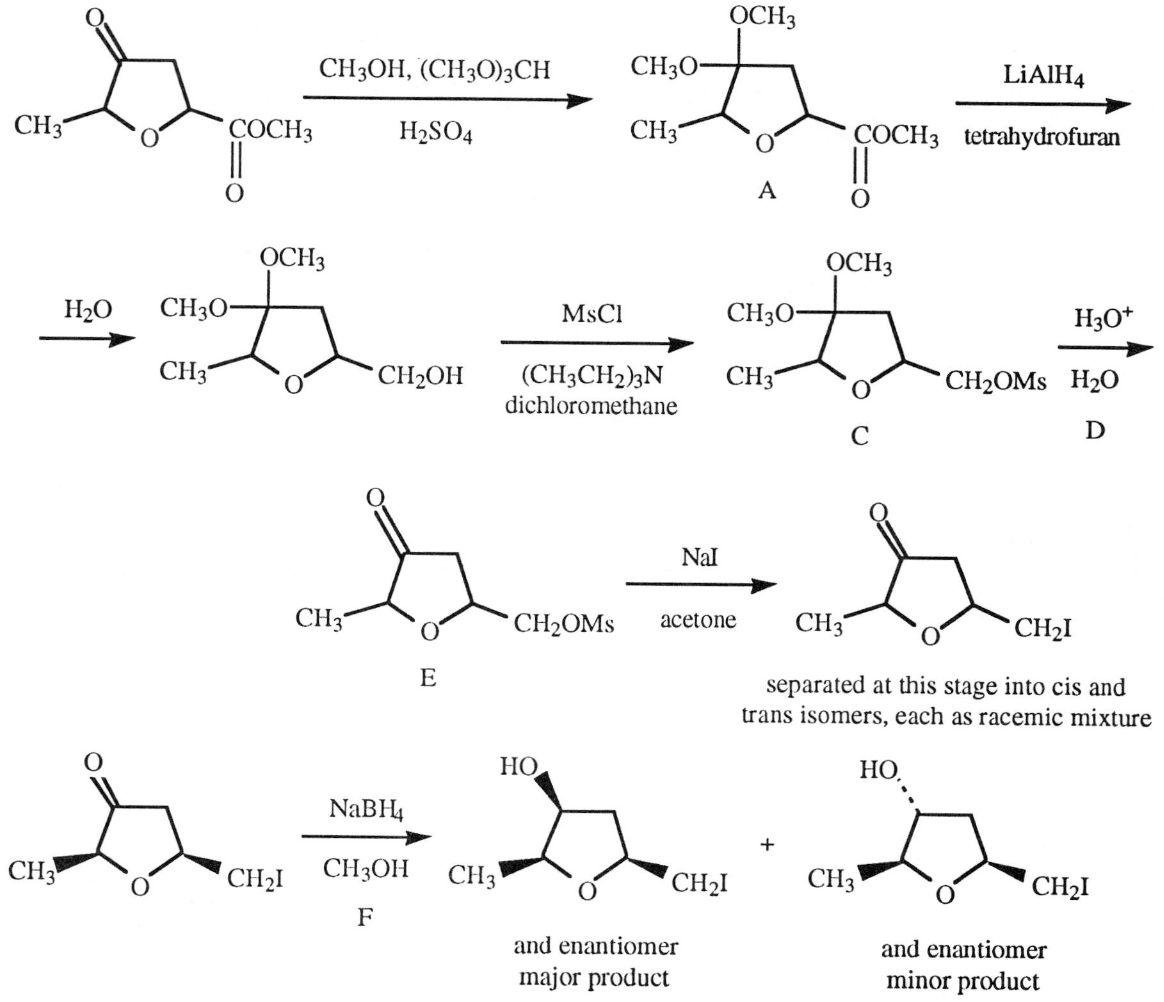

(b)   The minor product can be converted into (+)-muscarine.

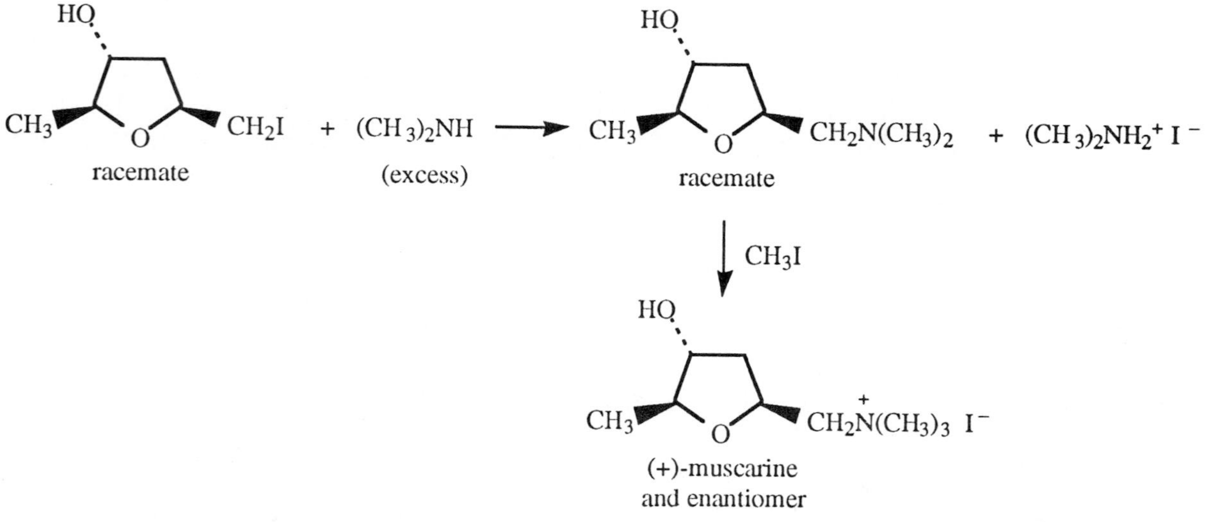

22.31

(a)   *trans*-2-ethyl-1-(*N*-methylamino)cyclobutane   (b)   2-methylpyridine

(c)   triethylamine   (d)   2-amino-4-methylbenzoic acid   (e)   4-aminobutanoic acid

(f)   2,4,6-trichloroaniline   (g)   5-methyl-2-hexanamine   (h)   *cis*-2-aminocyclohexanol
                                            2-amino-5-methylhexane

(i)   *N,N*-diethyl-4-nitroaniline
       N,N-diethyl-*p*-nitroaniline

22.32

(a) 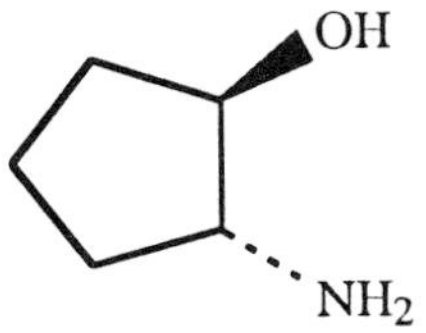   and enantiomer

(b)

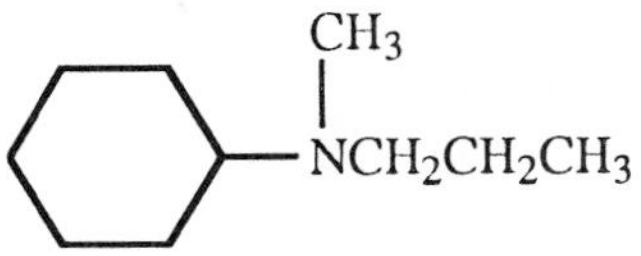

(c)

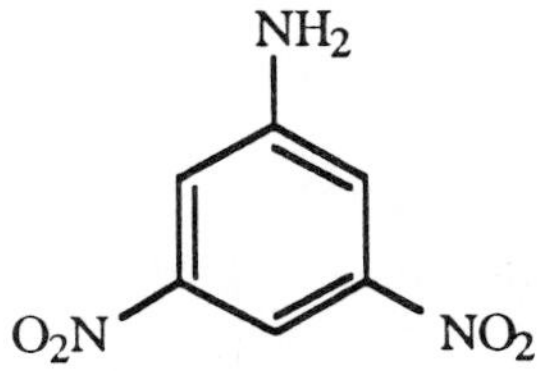

(d)

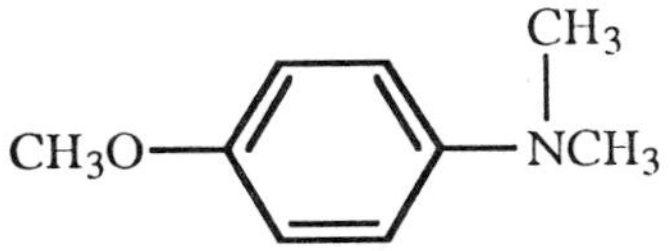

(e)

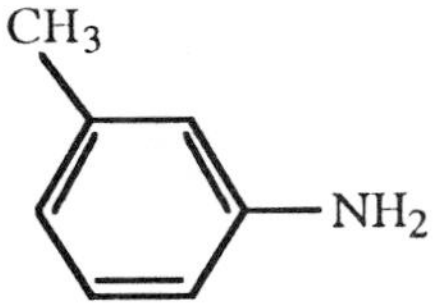

(f) 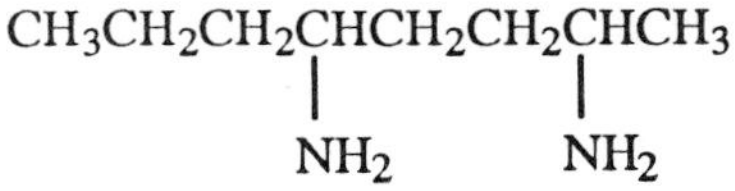 $CH_3CH_2CH_2CHCH_2CH_2CHCH_3$
       with NH_2 and NH_2

(g) 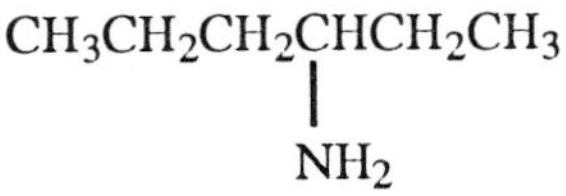 $CH_3CH_2CH_2CHCH_2CH_3$
       with NH_2

(h)

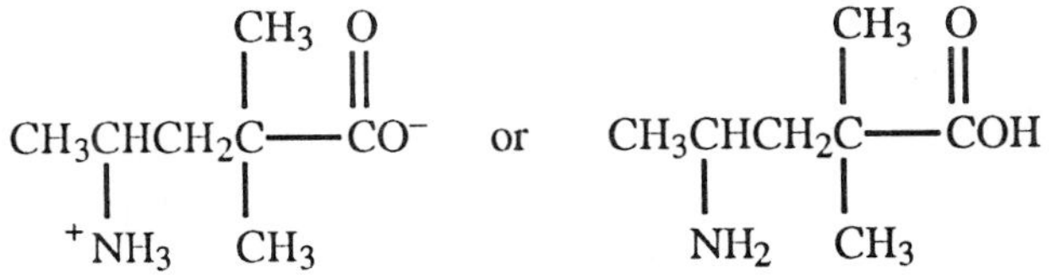

22.33

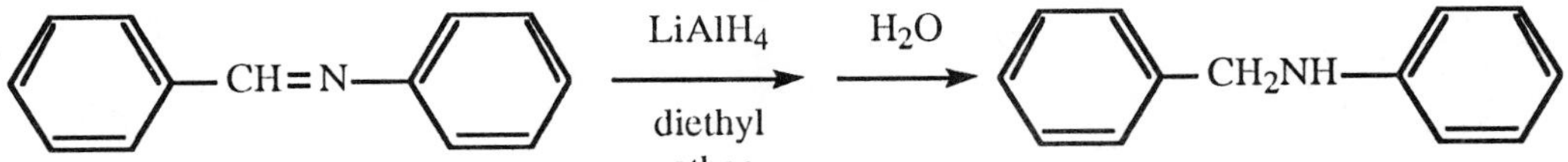

(a)

(b)

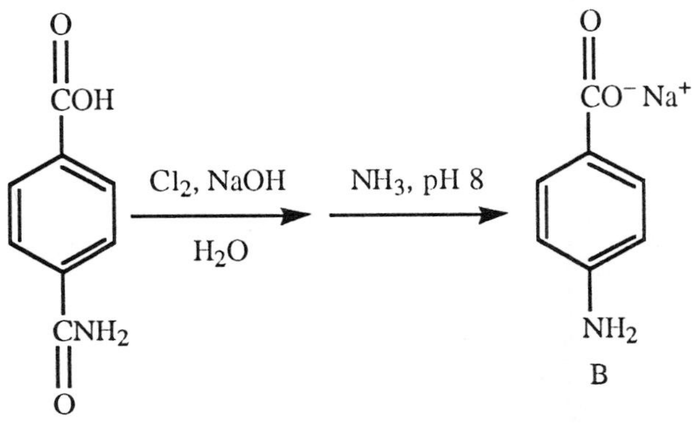

22.33 (cont)

(c)

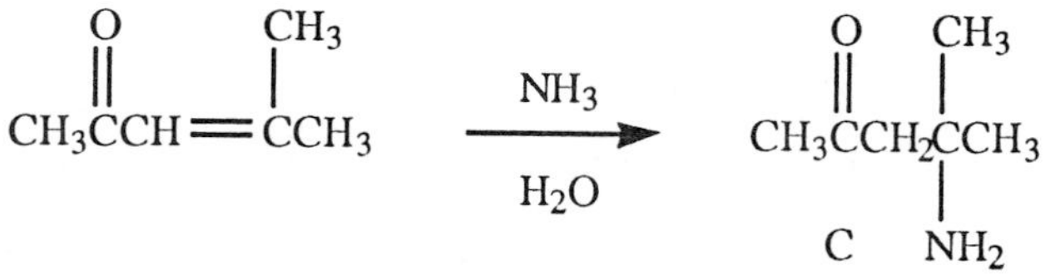

$$CH_3CCH=CCH_3 \xrightarrow[\text{H}_2\text{O}]{\text{NH}_3} CH_3CCH_2CCH_3$$

C

(d)

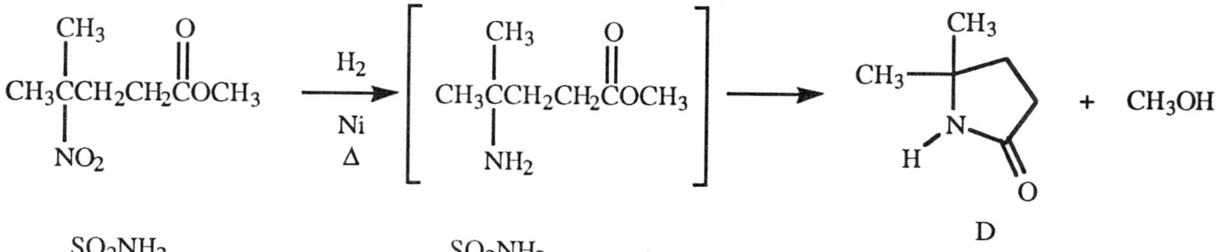

$$CH_3CCH_2CH_2COCH_3 \xrightarrow[\Delta]{\substack{H_2 \\ Ni}} \left[CH_3CCH_2CH_2COCH_3 \right] \longrightarrow D \quad + \quad CH_3OH$$

(e)

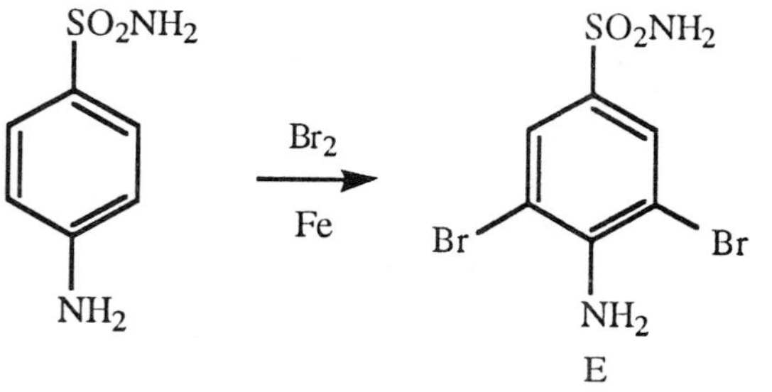

$$\xrightarrow[\text{Fe}]{\text{Br}_2} E$$

(f)

$$CH_3CH_2CCH_2CH_2CH_3 \xrightarrow[\text{Na}_2\text{CO}_3]{\overset{+}{\text{HONH}_3} \ \text{HSO}_4^-}$$

F 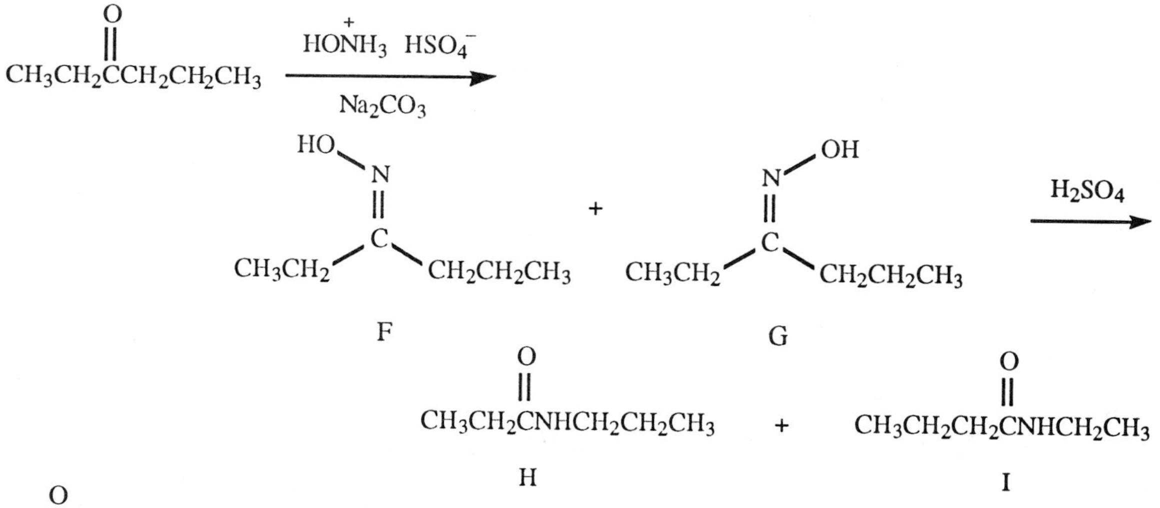   +   G 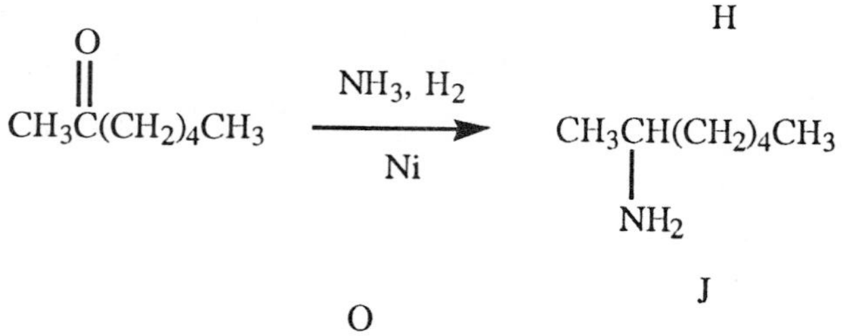 $\xrightarrow{\text{H}_2\text{SO}_4}$

$$CH_3CH_2CNHCH_2CH_2CH_3 \quad + \quad CH_3CH_2CH_2CNHCH_2CH_3$$

H   I

(g)

$$CH_3C(CH_2)_4CH_3 \xrightarrow[\text{Ni}]{\text{NH}_3, \text{H}_2} CH_3CH(CH_2)_4CH_3$$

J

(h)

$$CH_3CH_2CH_2NO_2 \xrightarrow[]{\overset{\text{O}}{\overset{\|}{\text{HCH}}}, \text{NaOH}} CH_3CH_2CHCH_2OH \xrightarrow{\text{H}_2\text{SO}_4, \text{Fe}}$$

K

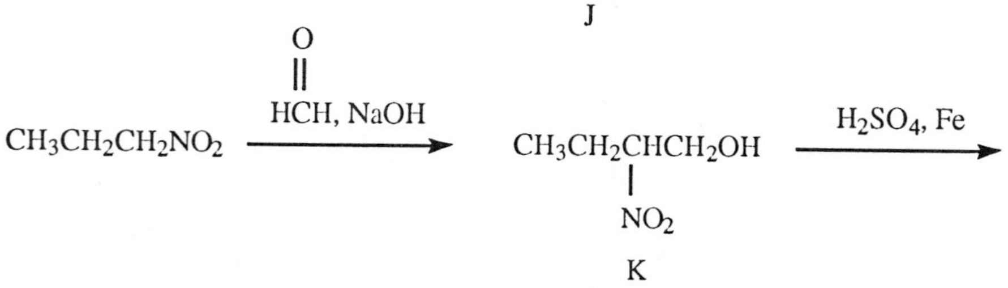

22.33 (h) (cont)

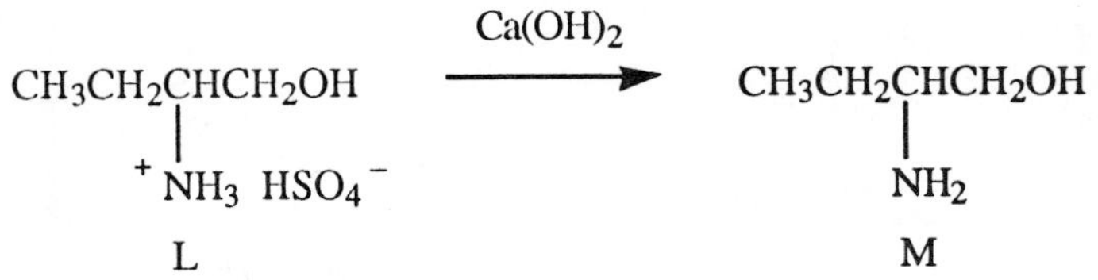

$$CH_3CH_2CHCH_2OH \xrightarrow{Ca(OH)_2} CH_3CH_2CHCH_2OH$$

with $\overset{+}{N}H_3 \ HSO_4^-$ (L) giving NH_2 (M)

22.34

(a)

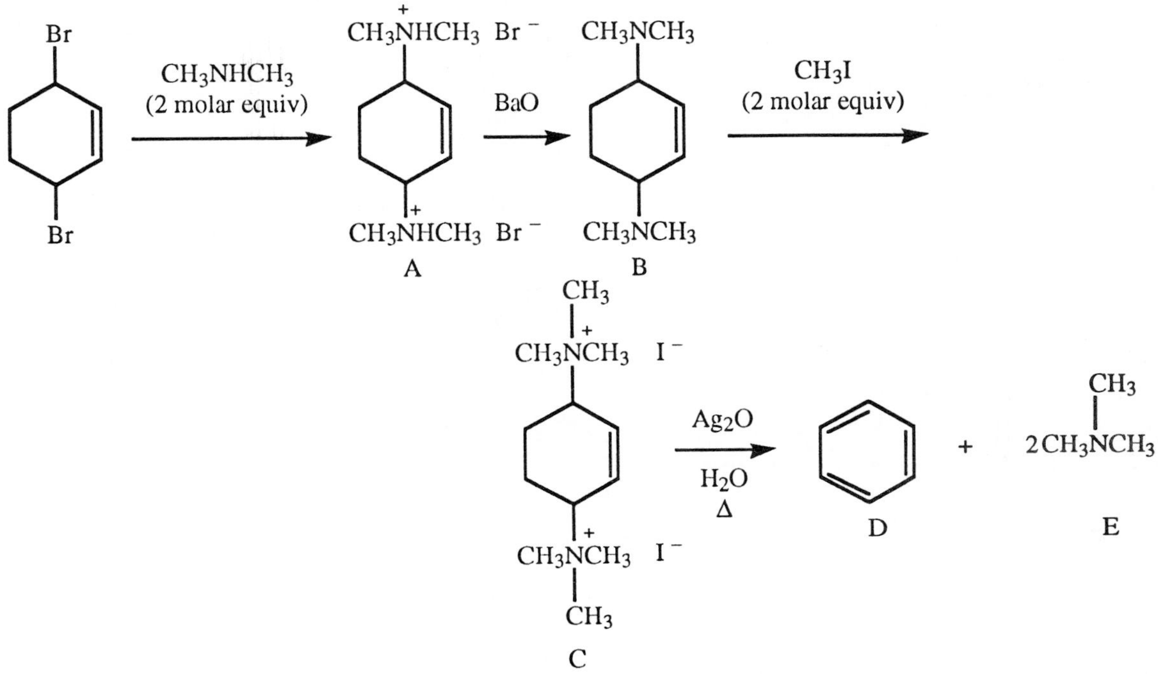

(b)

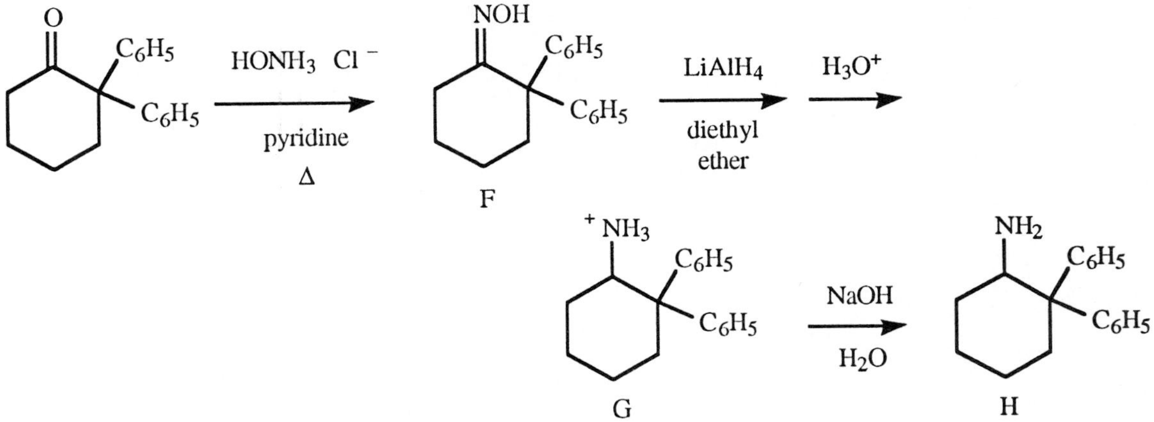

(c)

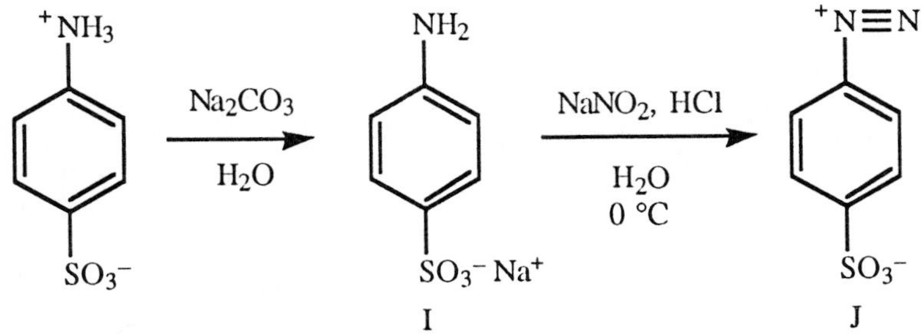

22.34 (cont)

(d)

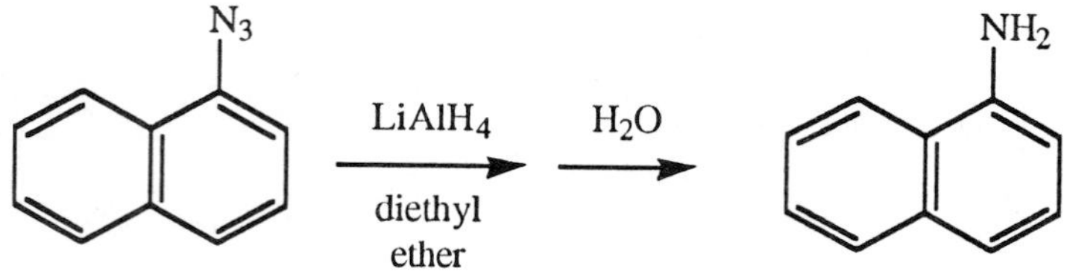

(e)

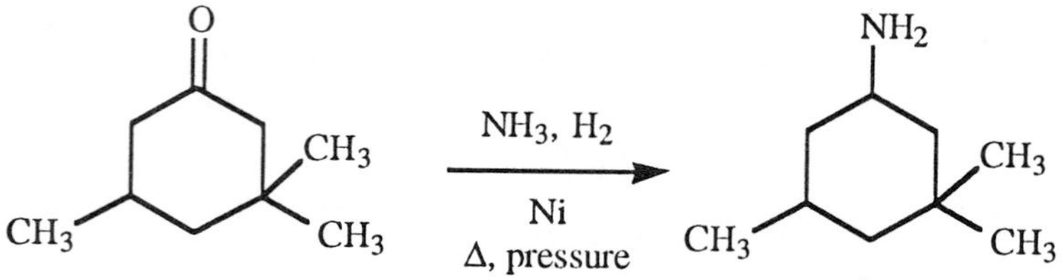

(f)

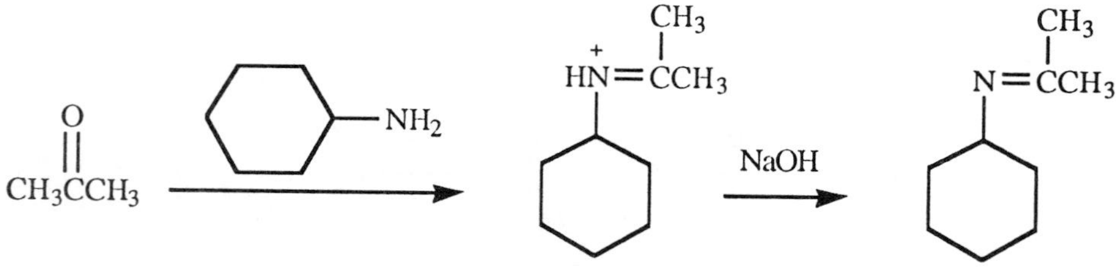

(g)

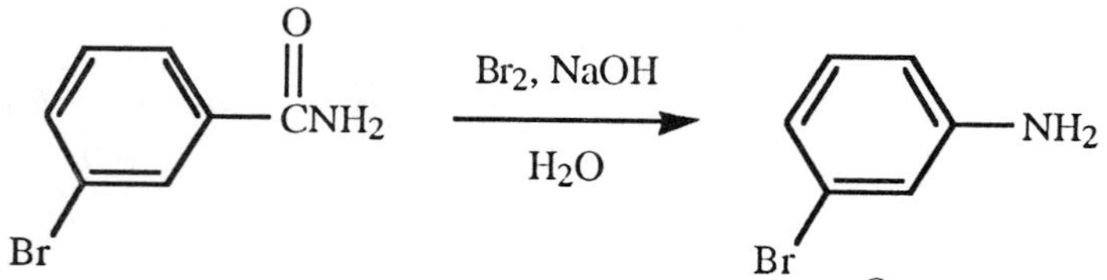

(h)

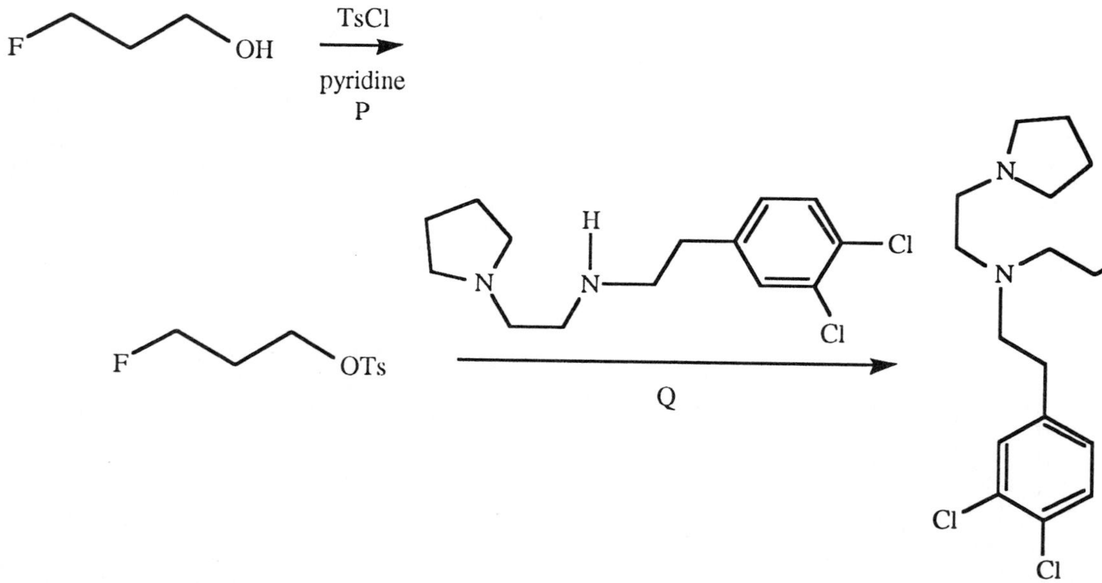

22.35

(a)

CH₃CHCOCH₂CH₃ → $\xrightarrow[\text{dimethyl sulfoxide}]{\text{NaNO}_2}$ → CH₃CHCOCH₂CH₃

with Cl and NO₂ substituents, product A

(b)

$\xrightarrow[\text{diethyl ether}]{\text{LiAlH}_4}$ $\xrightarrow{\text{H}_2\text{O}}$

product B

(c)

$$\text{CH}_3\text{CNH}_2 \xrightarrow[\text{H}_2\text{O}]{\text{Br}_2, \text{NaOH}} \text{CH}_3\text{NH}_2 \;+\; \text{HCO}_3^-$$

C

(d)

$\xrightarrow[\text{Ni}]{\text{H}_2}$

D   E

(e)

CH₃CH₂CHCH₂CH₂CH₃ $\xrightarrow[\text{pyridine}]{\text{TsCl}}$ CH₃CH₂CHCH₂CH₂CH₃ $\xrightarrow{\text{CH}_3\text{NHCH}_3}$ $\xrightarrow{\text{NaOH}}$

with OH and OTs

F

CH₃CH₂CHCH₂CH₂CH₃ $\xrightarrow[\text{acetonitrile}]{\text{CH}_3\text{I}}$ CH₃CH₂CHCH₂CH₂CH₃

CH₃NCH₃

G   H (CH₃NCH₃ I⁻ with CH₃)

(f)

Br(CH₂)₁₀Br + CH₃NCH₃ $\xrightarrow[\substack{\text{methanol} \\ 25\,°\text{C} \\ \text{2 weeks}}]{}$ Br⁻ CH₃N(CH₂)₁₀NCH₃ Br⁻ $\xrightarrow[\text{H}_2\text{O}]{\text{Ag}_2\text{O}}$

(1 equiv)   (2 equiv)

I

22.35 (f) (cont)

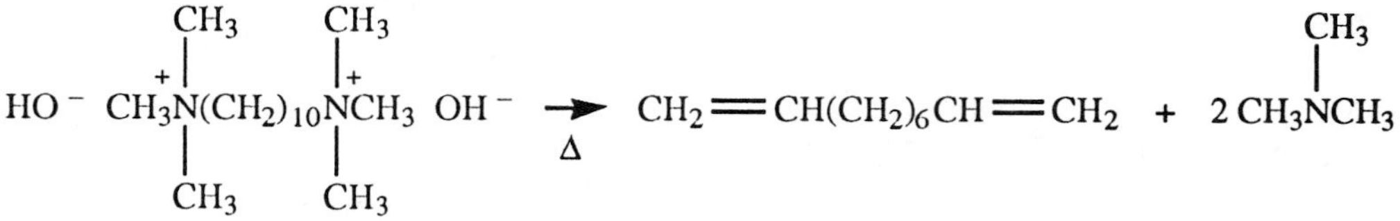

$$HO^- \quad CH_3\overset{\overset{\displaystyle CH_3}{|}}{\underset{\underset{\displaystyle CH_3}{|}}{N^+}}(CH_2)_{10}\overset{\overset{\displaystyle CH_3}{|}}{\underset{\underset{\displaystyle CH_3}{|}}{N^+}}CH_3 \quad OH^- \quad \xrightarrow{\Delta} \quad CH_2{=}CH(CH_2)_6CH{=}CH_2 \quad + \quad 2\ CH_3\overset{\overset{\displaystyle CH_3}{|}}{N}CH_3$$

J   K

(g)

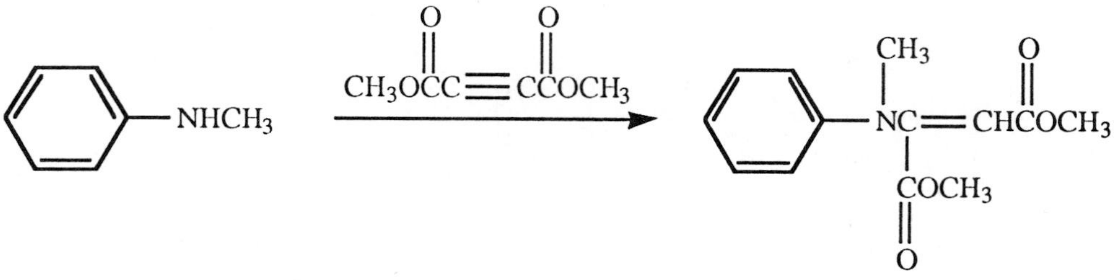

(h)

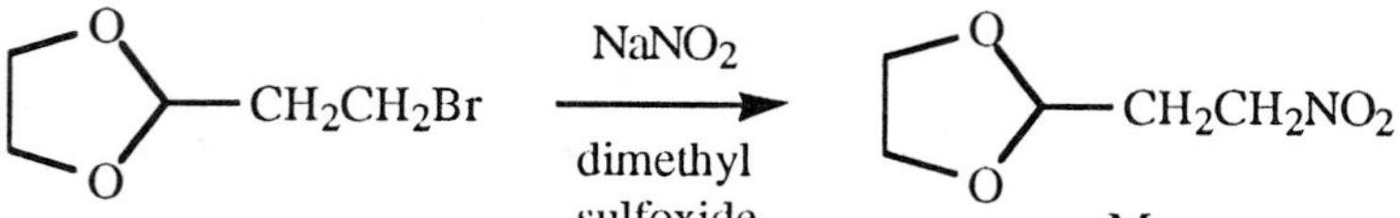

(i)

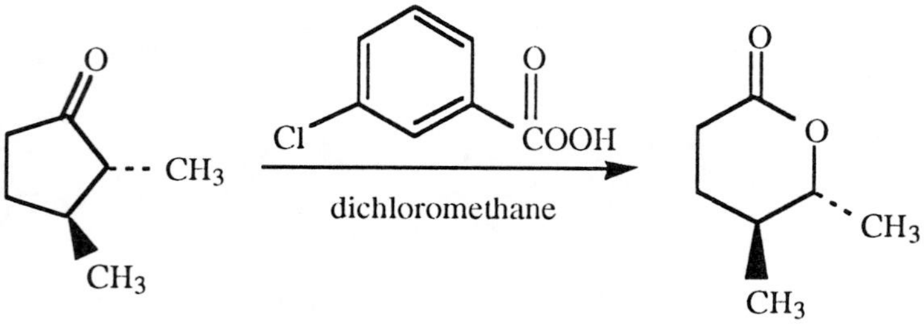

(j)

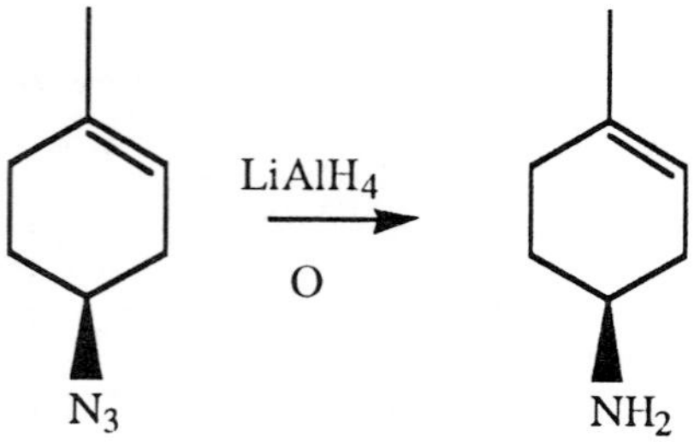

22.36

(a)

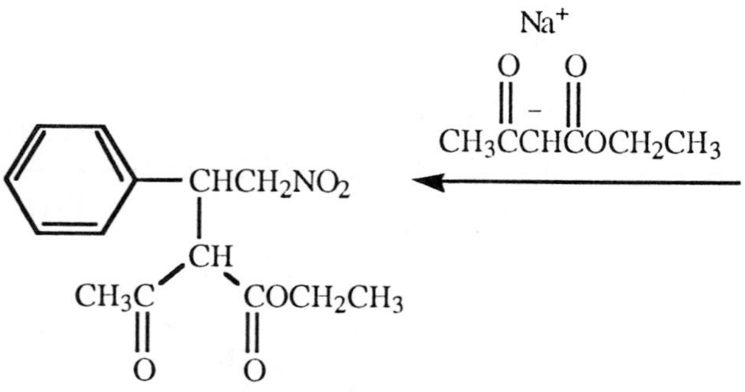

22.36 (a) (cont)

Ph—CH=CHNO₂ ←(HCl)← ←(CH₃NO₂, NaOH / H₂O)← Ph—CHO

(b) H₂NCH₂CHCH₂CH₂NH₂ (with CH₃ branch) ←(Br₂, NaOH)← H₂NC(=O)CH₂CHCH₂CH₂C(=O)NH₂ (with CH₃ branch) ←(NH₃ (excess))←

ClC(=O)CH₂CH(CH₃)CH₂CH₂C(=O)Cl ←(SOCl₂ (2 molar equiv))← HOC(=O)CH₂CH(CH₃)CH₂CH₂C(=O)OH

(c) CH₃CH₂C(=O)NHCH₂CH₃ ←(H₂SO₄, Δ)← CH₃CH₂C(=NOH)CH₂CH₃ ←(HONH₃⁺ Cl⁻ / Na₂CO₃)← CH₃CH₂C(=O)CH₂CH₃

(d) 1. Ph—CH₂CH₂CHCH₂CHCH₃ (NH₂, NH₂) ←(NH₃, H₂ (2 molar equiv) / Ni)← Ph—CH₂CH₂C(=O)CH₂C(=O)CH₃

2. Ph—CH₂CH₂CHCH₂CHCH₃ (NH₂, NH₂) ←(H₂O)← ←(LiAlH₄ / diethyl ether)←

Ph—CH₂CH₂CHCH₂CHCH₃ (N₃, N₃) ←(NaN₃)← Ph—CH₂CH₂CHCH₂CHCH₃ (OTs, OTs) ←(TsCl / pyridine)←

Ph—CH₂CH₂CHCH₂CHCH₃ (OH, OH) ←(NaBH₄ / methanol)← Ph—CH₂CH₂C(=O)CH₂C(=O)CH₃

(pyridinium CrO₃Cl⁻ / dichloromethane)

(e) 1. Ph—CH₂OCH₂CH₂CH₂NH₂ ←(NH₃, H₂ / Ni)← Ph—CH₂OCH₂CH₂CH(=O) ←

22.36 (e 1) (cont)

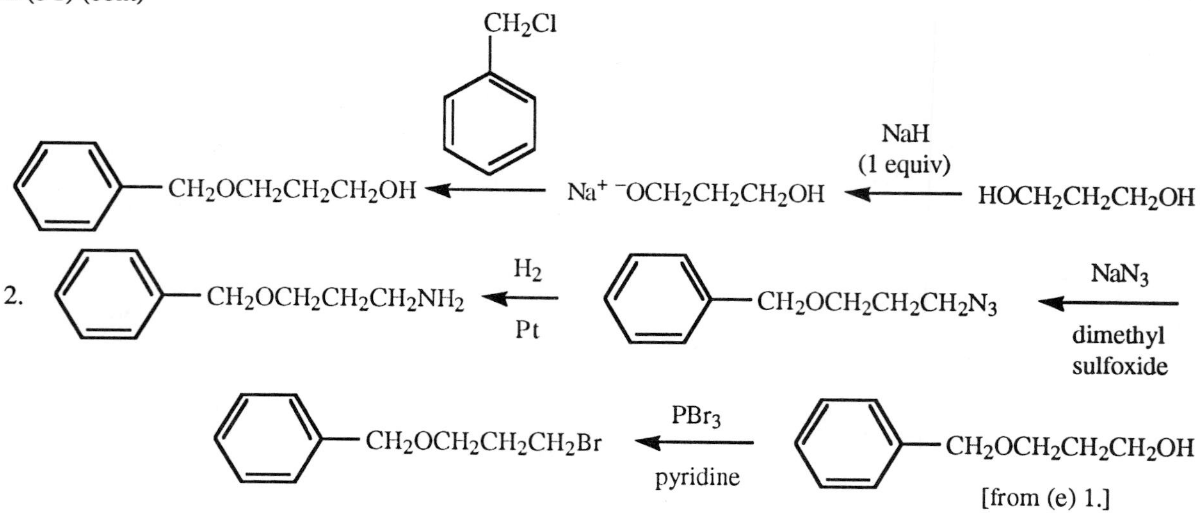

22.37

1.

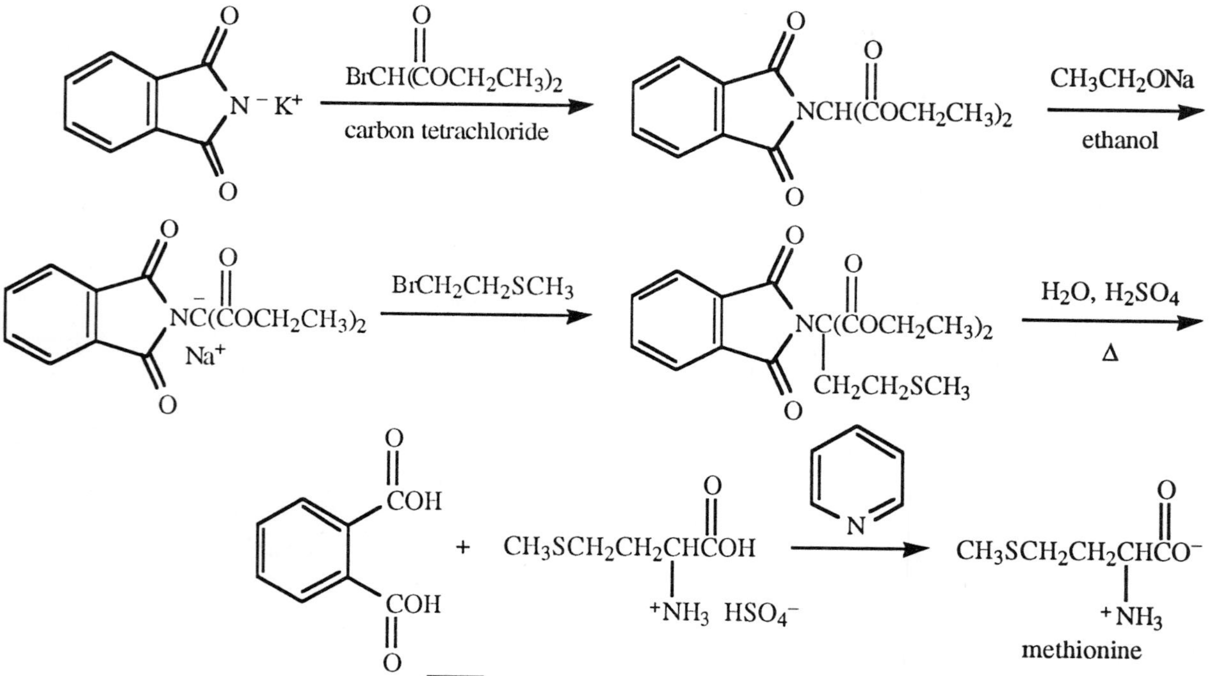

2.

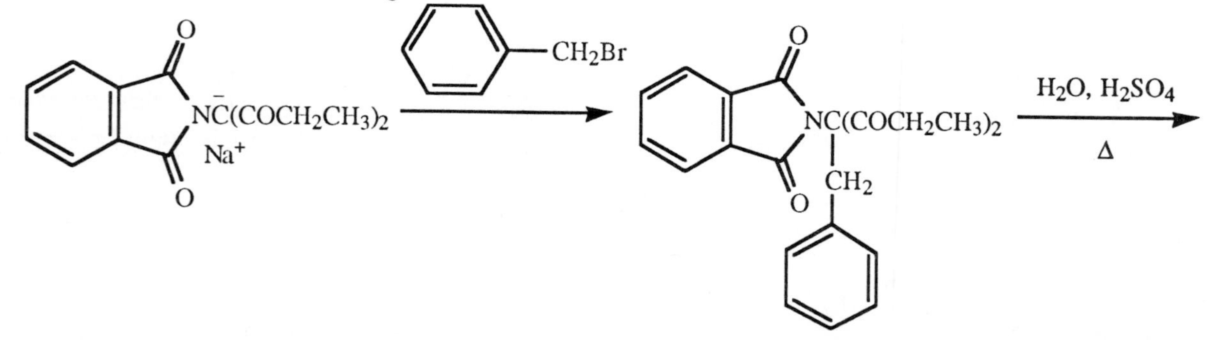

22.37 (2) (cont)

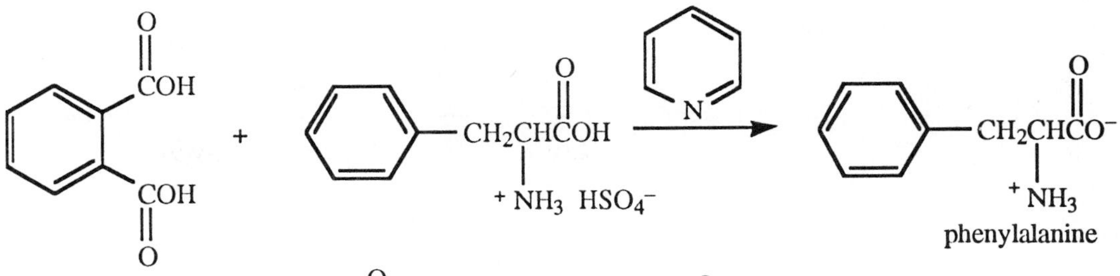

3.

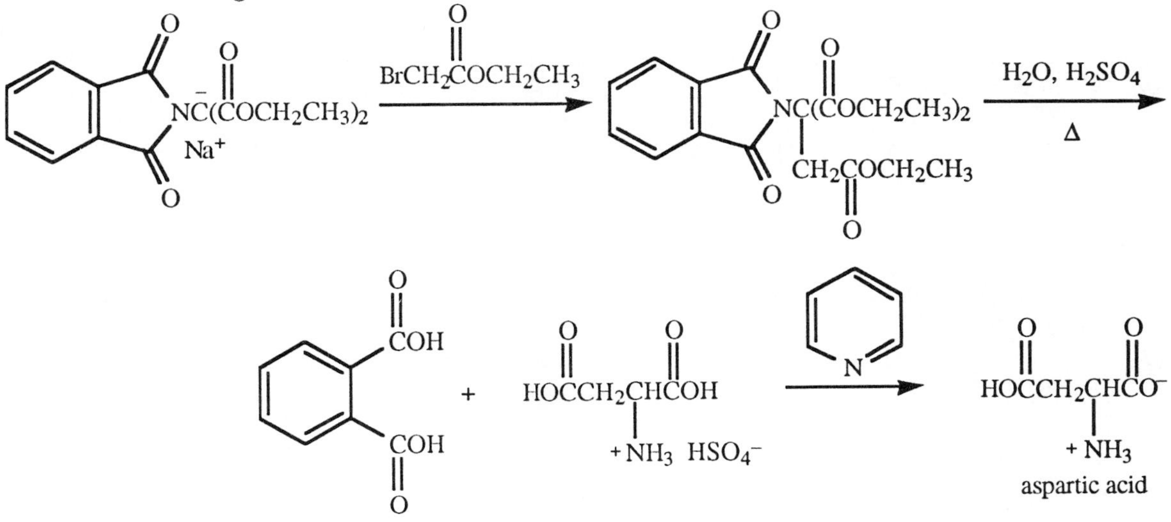

22.38

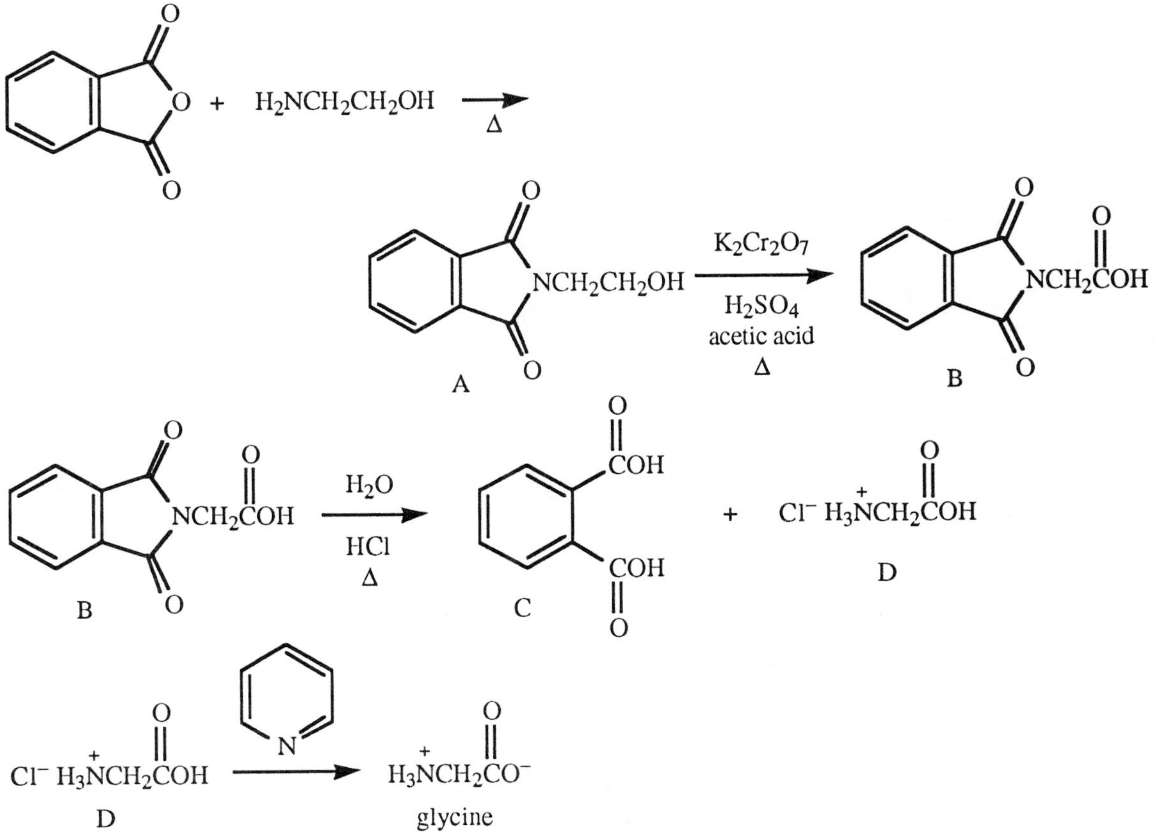

22.39

1.

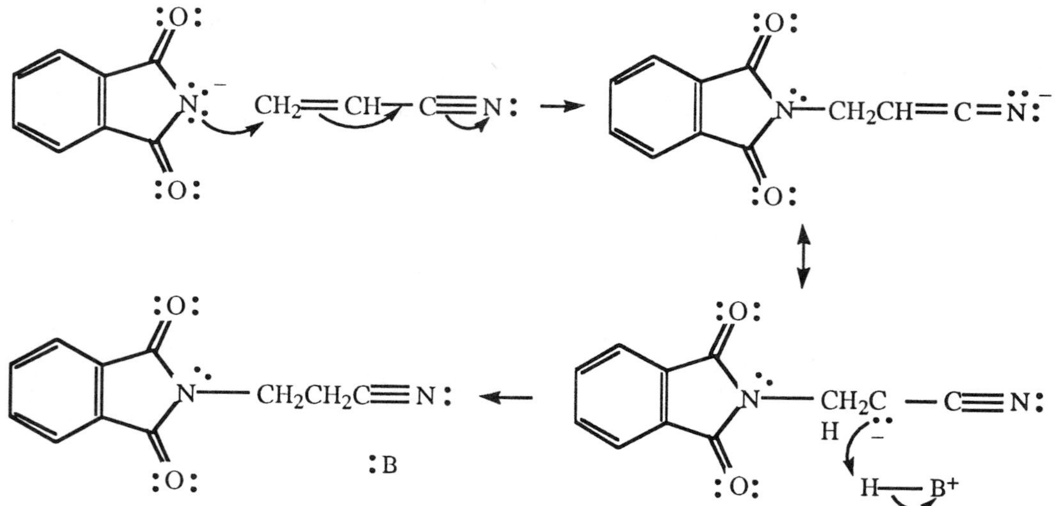

2.

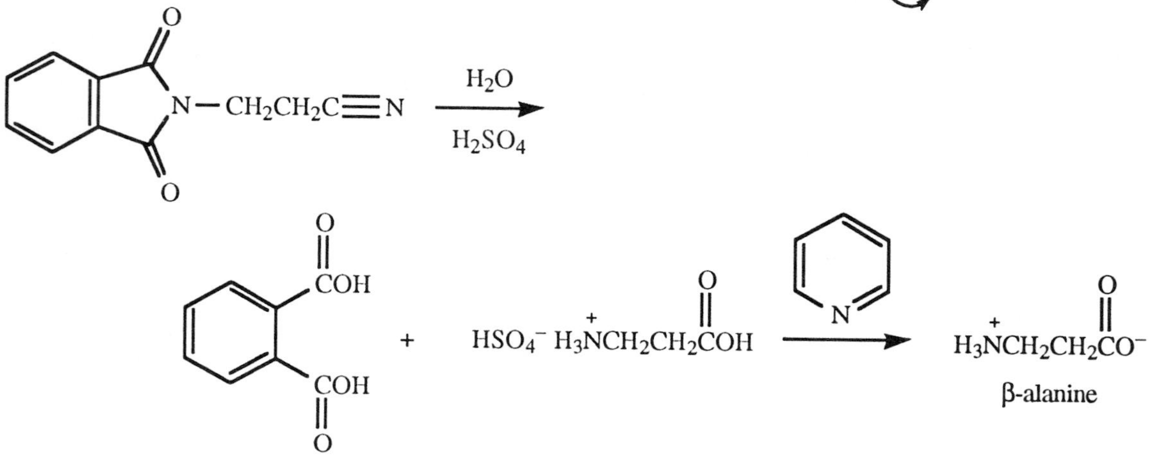

22.40

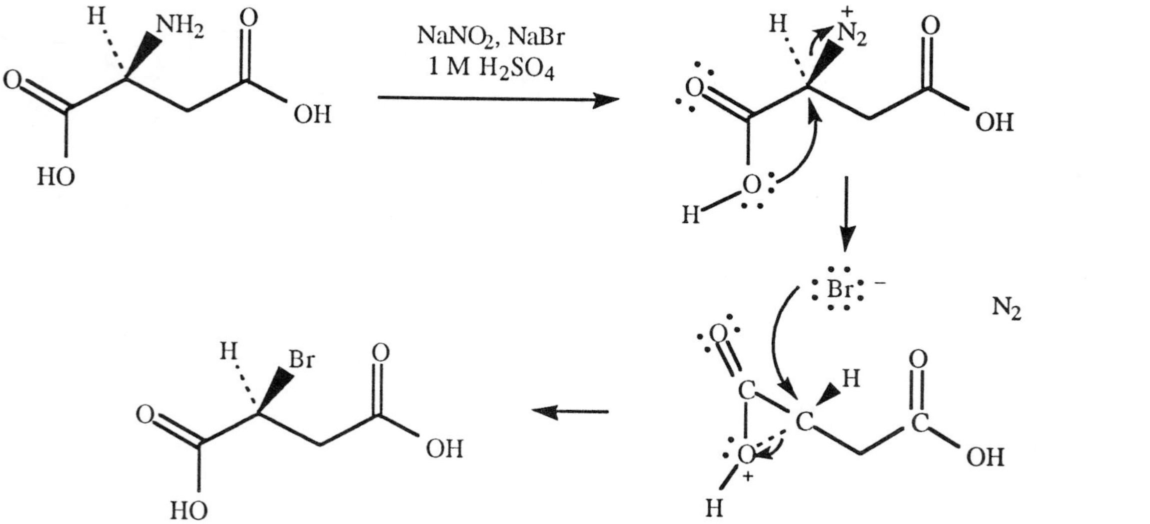

The observed stereochemistry is explained by intramolecular displacement of nitrogen by the carboxylic acid group, followed by attack by bromide ion.

22.41

ionization with
1,2-hydride shift

22.41 (cont)

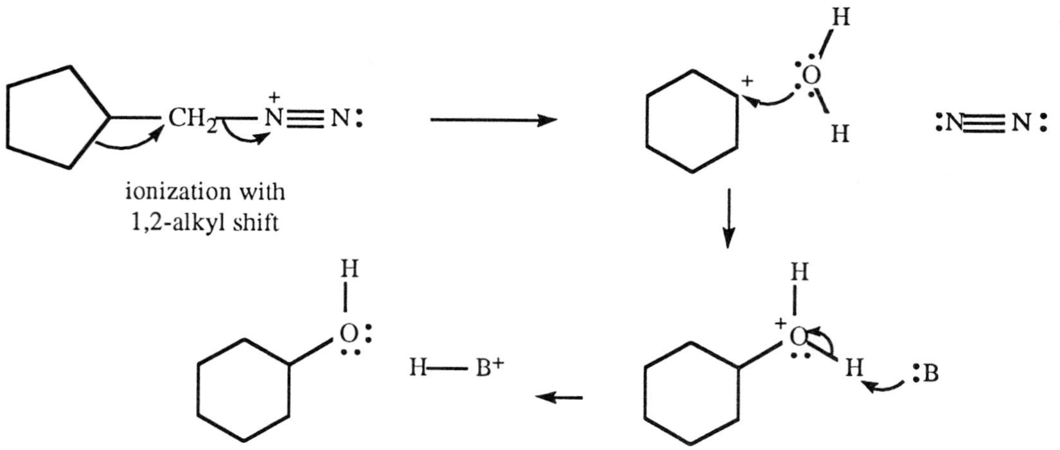

ionization with
1,2-alkyl shift

22.42

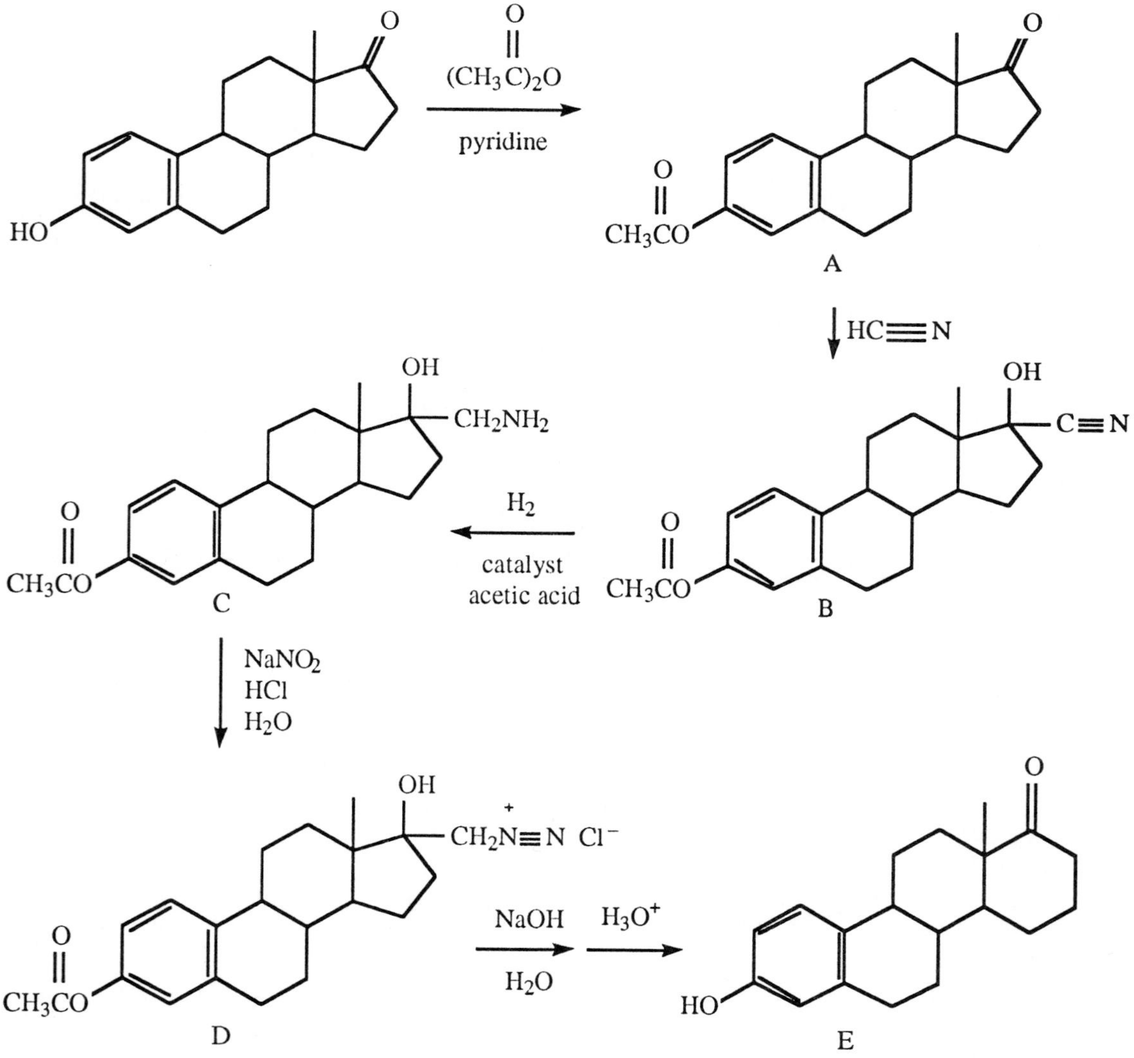

$(CH_3C)_2O$

pyridine

A

$HC\equiv N$

B

H_2

catalyst
acetic acid

C

NaNO$_2$
HCl
H$_2$O

D

NaOH   H$_3$O$^+$

H$_2$O

E
(Note that the acetyl group is
also removed in the last step.)

22.42 (cont)

Mechanism of ring expansion

(See the previous problem for the mechanism of the formation of the diazonium ion.)

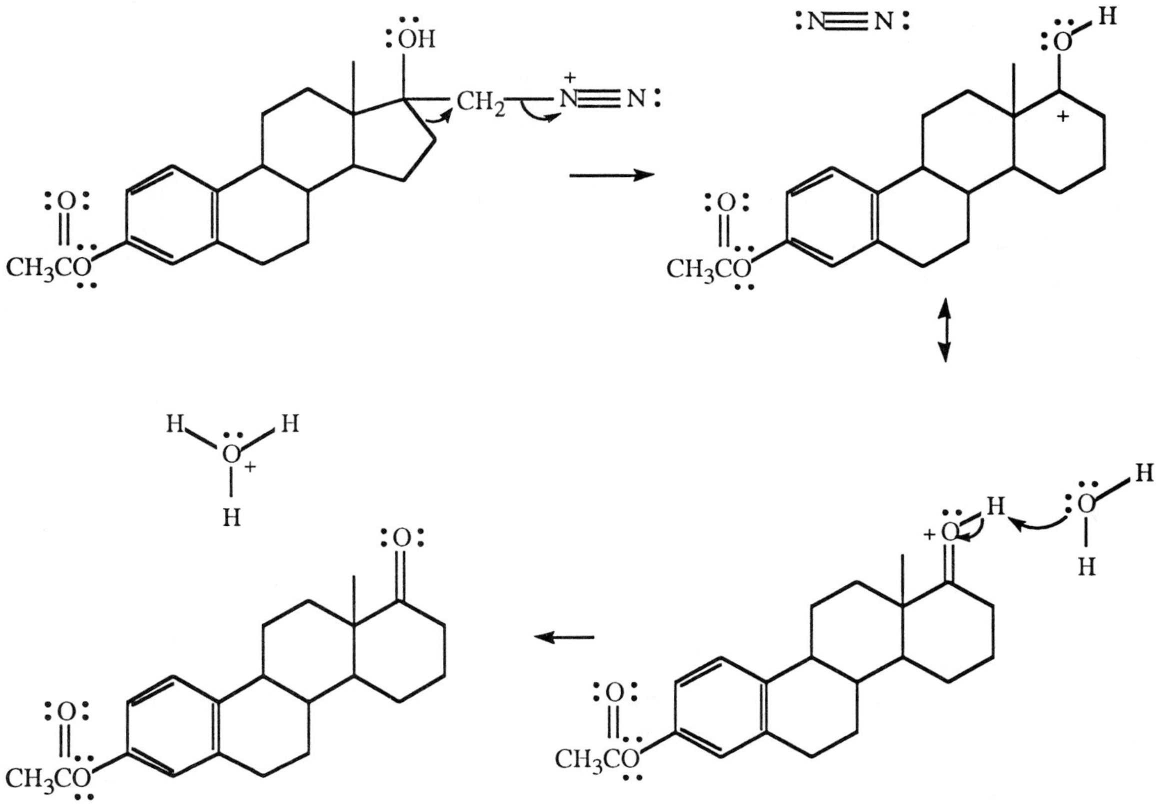

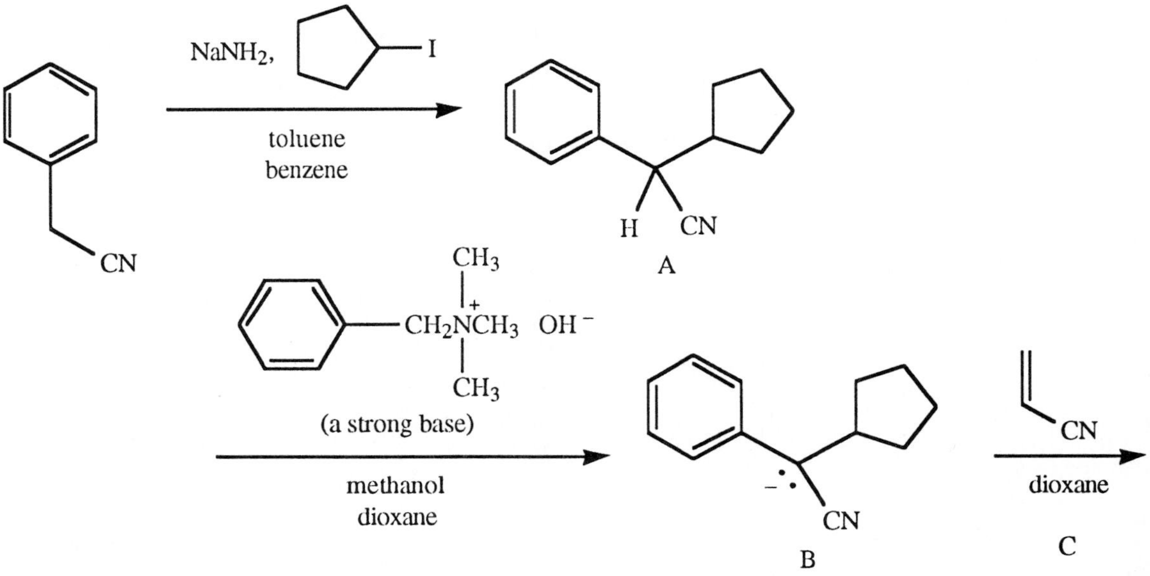

22.43 (cont)

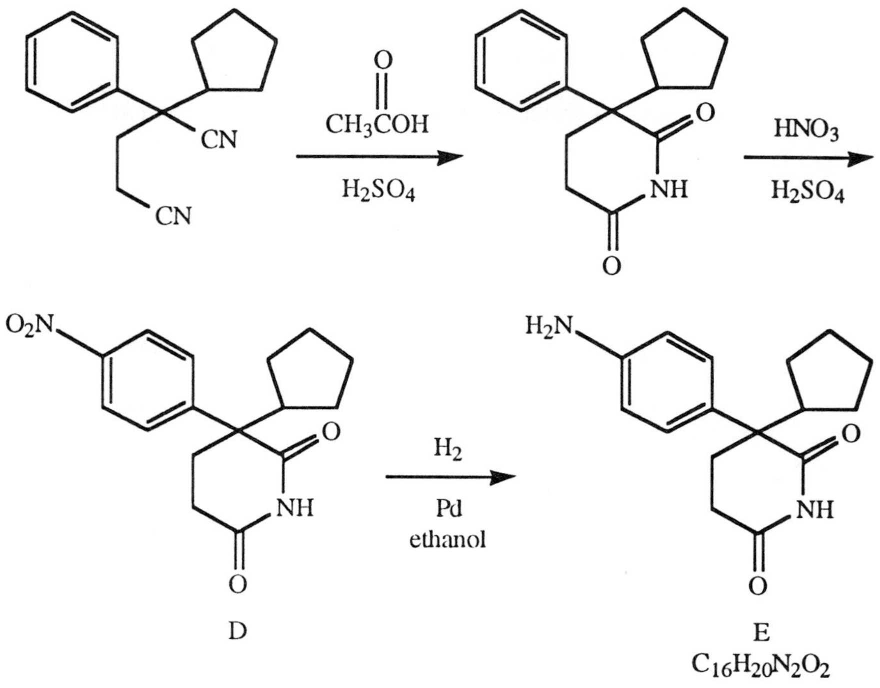

22.44

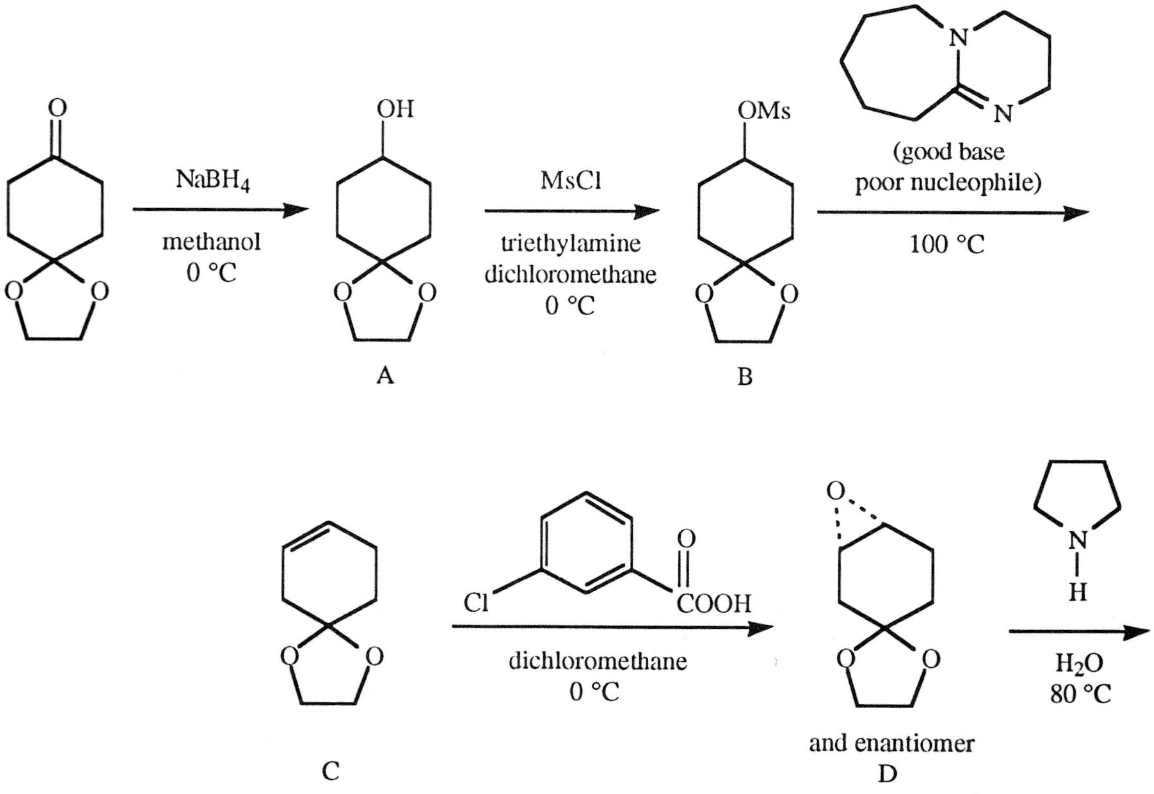

22.44 (cont)

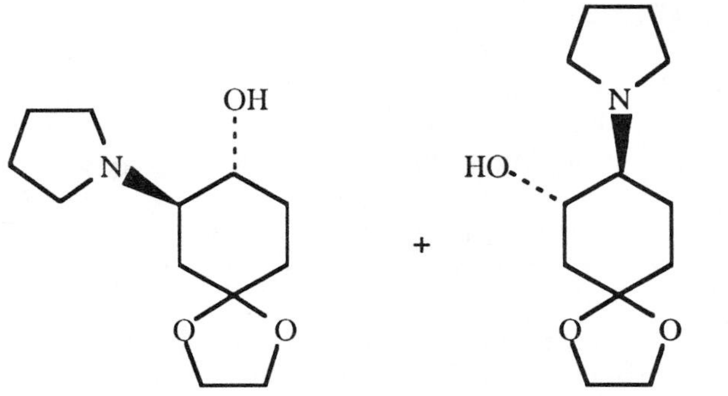
OH
N
HO
N
O
O
+

and enantiomer E $C_{12}H_{21}NO_3$ and enantiomer F $C_{12}H_{21}NO_3$

22.45

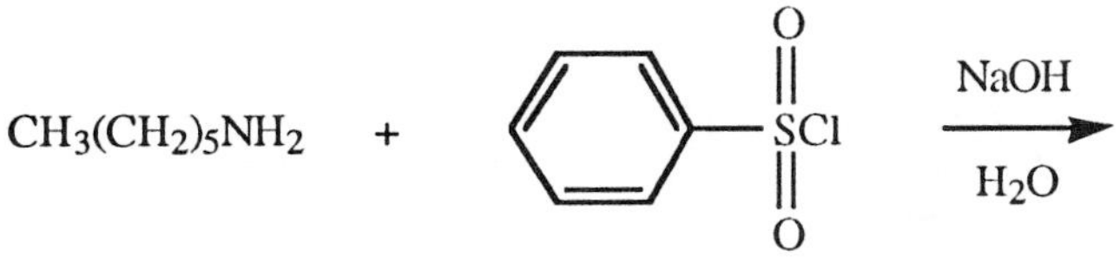
CH3(CH2)5NH2
+
SCl
NaOH
H2O

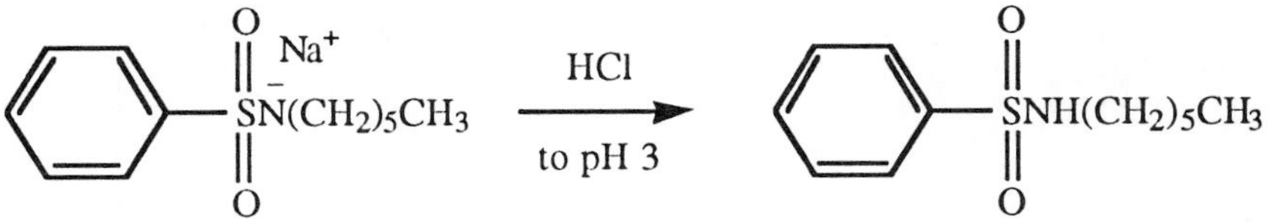
Na+
SN(CH2)5CH3
HCl
to pH 3
SNH(CH2)5CH3

water soluble water-insoluble precipitate

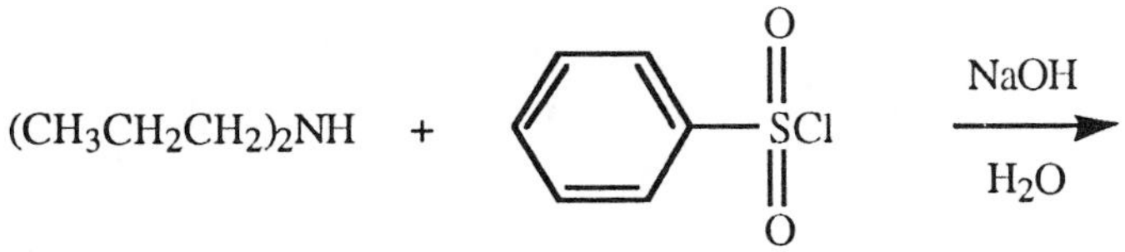
(CH3CH2CH2)2NH
+
SCl
NaOH
H2O

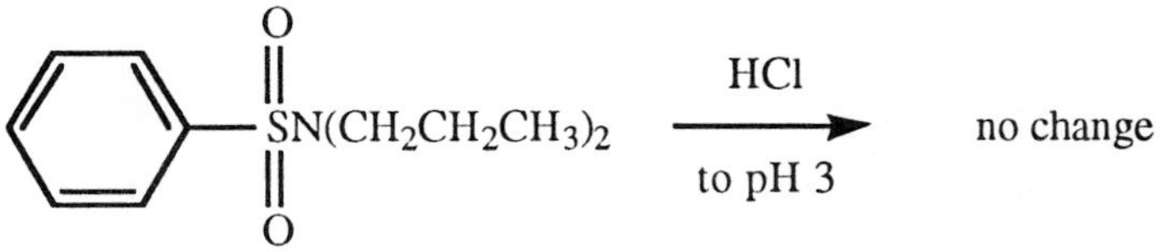
SN(CH2CH2CH3)2
HCl
to pH 3
no change

water-insoluble precipitate

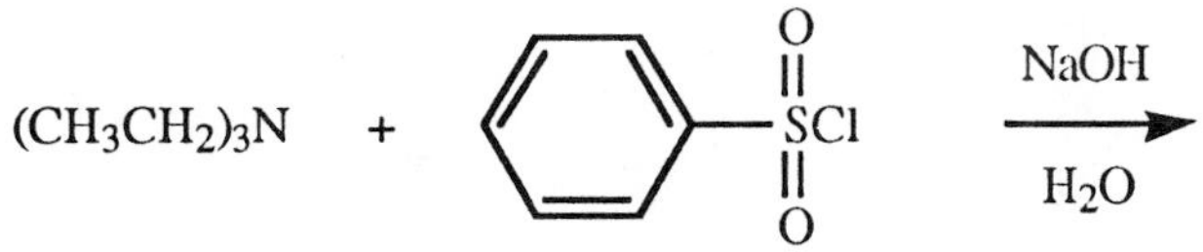
(CH3CH2)3N
+
SCl
NaOH
H2O

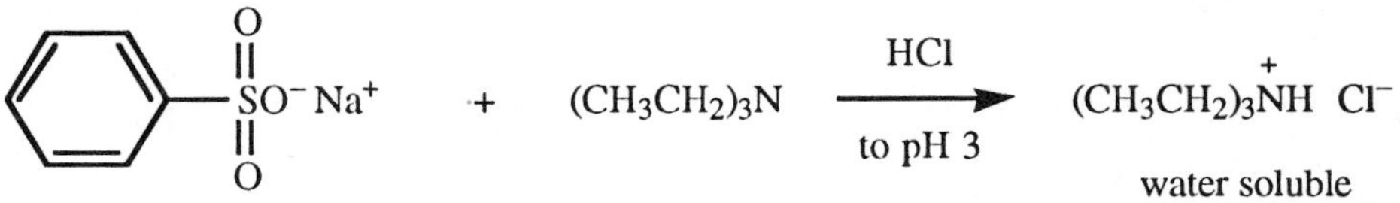
SO− Na+
+
(CH3CH2)3N
HCl
to pH 3
(CH3CH2)3NH Cl−

pK_a of conjugate acid is –0.6 water soluble

22.45 (cont)

A primary amine is converted into a sulfonamide that has an acidic hydrogen atom on the nitrogen and thus is deprotonated by the base. The salt of the sulfonamide is soluble in water, but the solid sulfonamide precipitates when the solution is acidified.

The sulfonamide of a secondary amine has no acidic hydrogen atom on the nitrogen and is not affected by the basic solution. We observe a precipitate in the test tube, and no change is seen when the solution is acidified.

A tertiary amine does not react with benzenesulfonyl chloride, which is converted slowly by base into sodium benzenesulfonate, a water soluble salt. What we observe depends upon the physical properties of the amine. If the amine has a low molecular weight and is soluble in water, we will have a clear basic solution that remains clear upon acidification. If the amine is not soluble in water, we will see the amine as a second phase in base but it will dissolve when we acidify the solution. The pK_a of the conjugate acid of sodium benzenesulfonate is –0.6. The salt remains in solution at a pH that will protonate an amine. Benzenesulfonic acid is also quite soluble in water, so no precipitate is seen under acidic conditions.

22.46

22.47

22.48

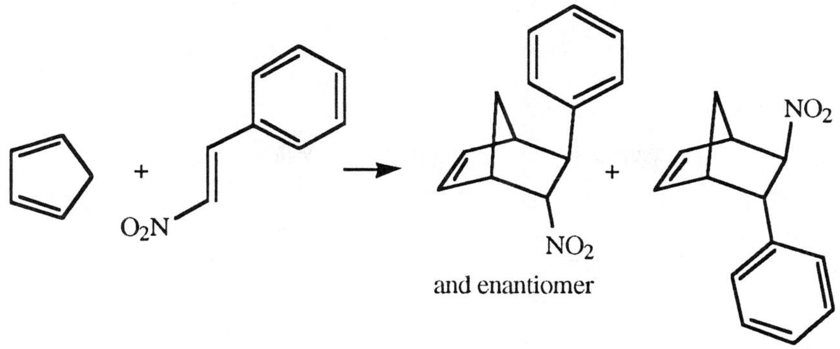

A
(mixture of stereoisomers)

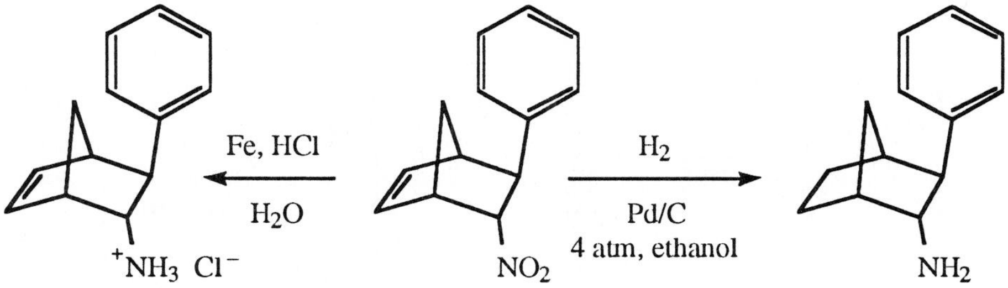

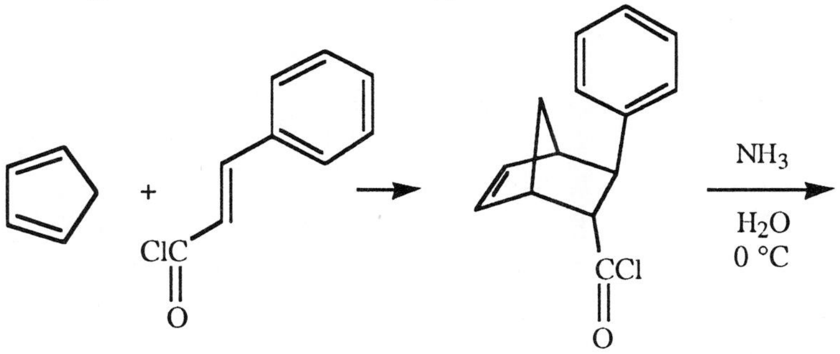

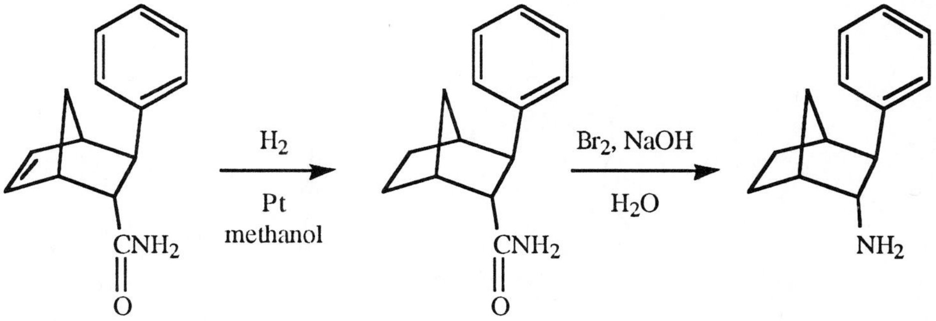

22.48 (cont)

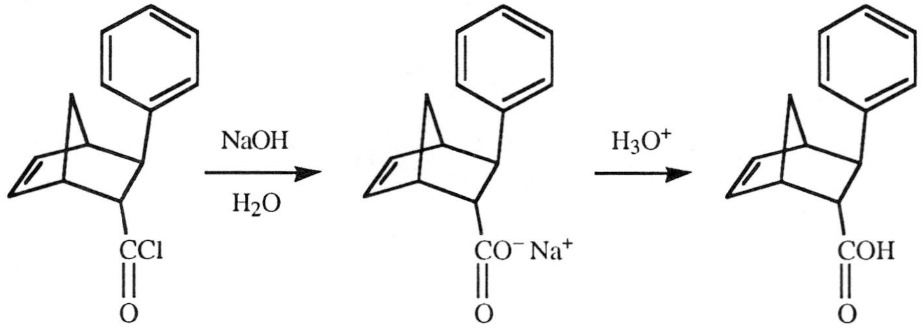

and enantiomer
D

and enantiomer
G

and enantiomer
H

22.49

(a)

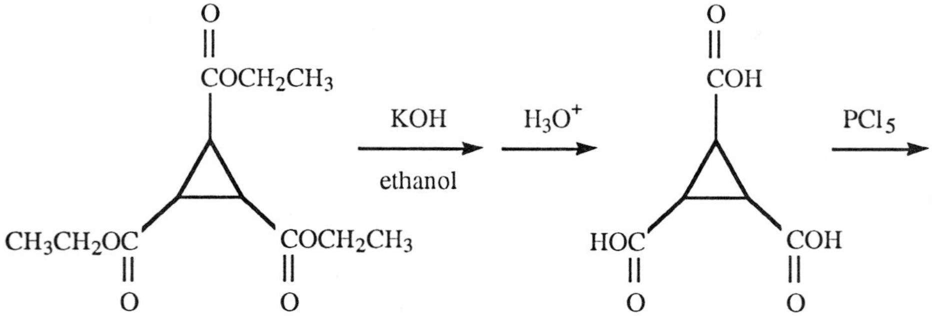

KOH

ethanol

H_3O^+

PCl$_5$

A

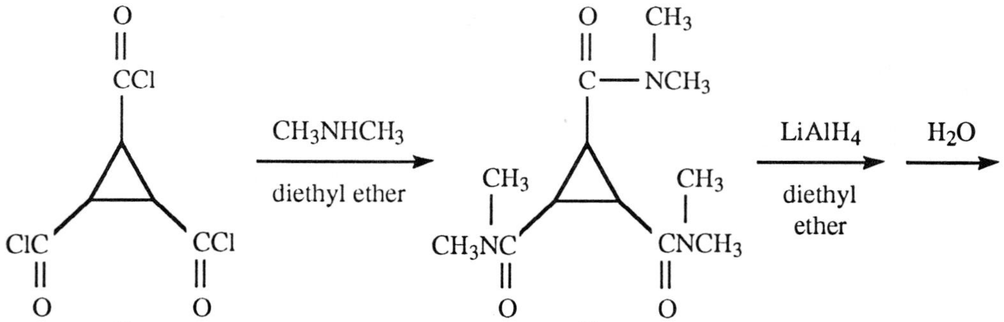

CH_3NHCH_3

diethyl ether

B

LiAlH$_4$

diethyl
ether

H_2O

C

CH_3I
(excess)

ethanol

Ag$_2$O

H_2O

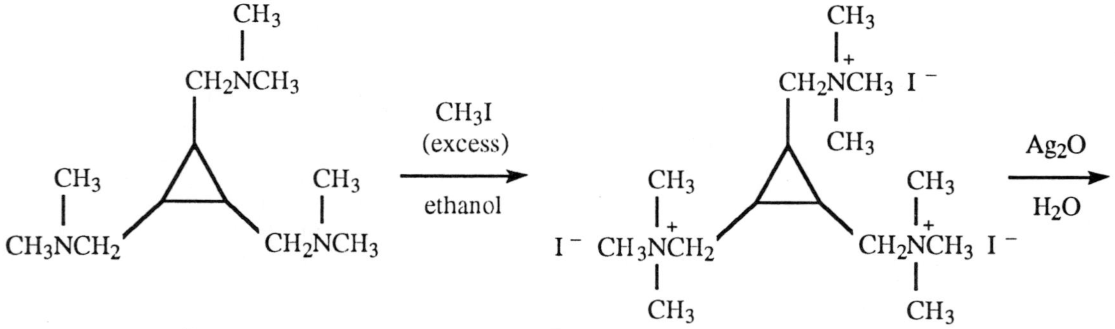

D

E

22.49 (a) (cont)

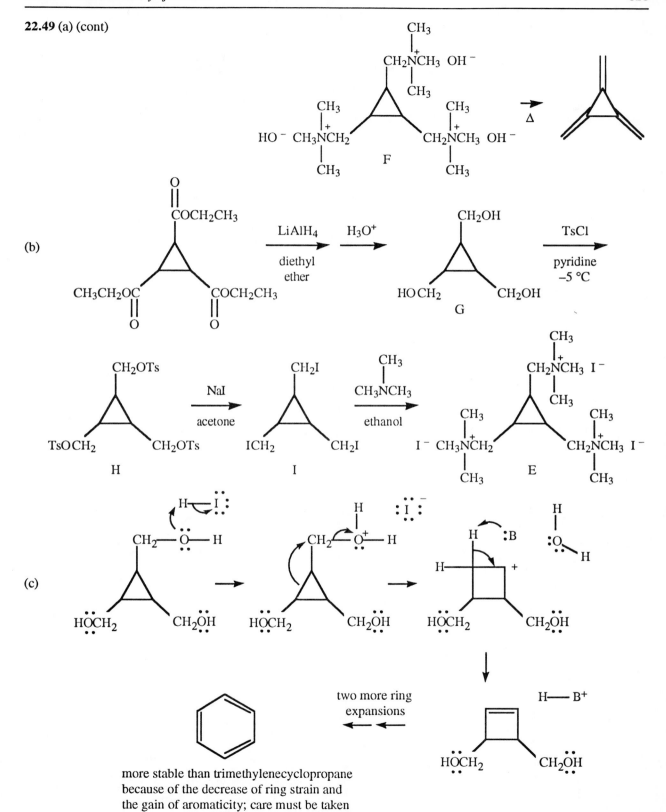

more stable than trimethylenecyclopropane
because of the decrease of ring strain and
the gain of aromaticity; care must be taken
to prevent the formation of carbocations.

22.49 (cont)

(d)   Cis-trans stereoisomerism is possible but no isomers with optical activity are seen because of the symmetry of
the molecule.

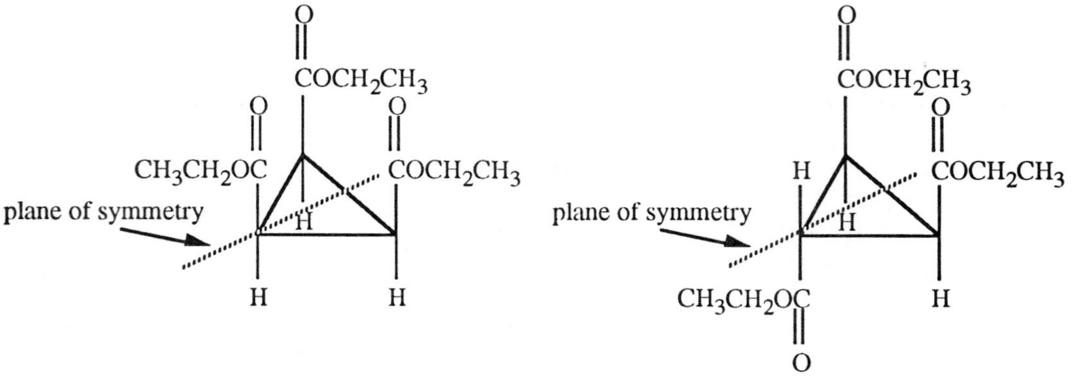

(e)   An E2 elimination could be used.

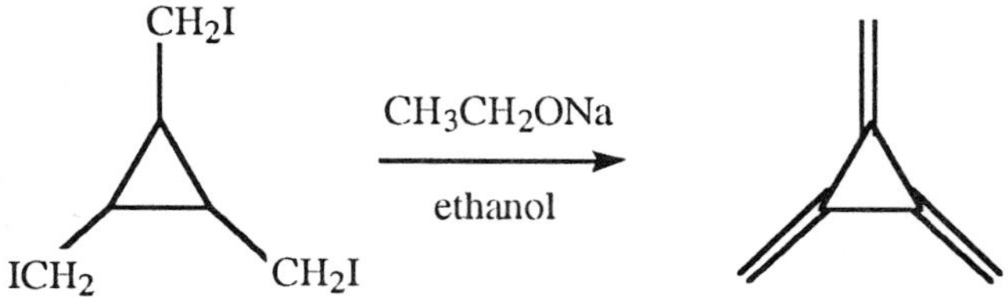

22.50

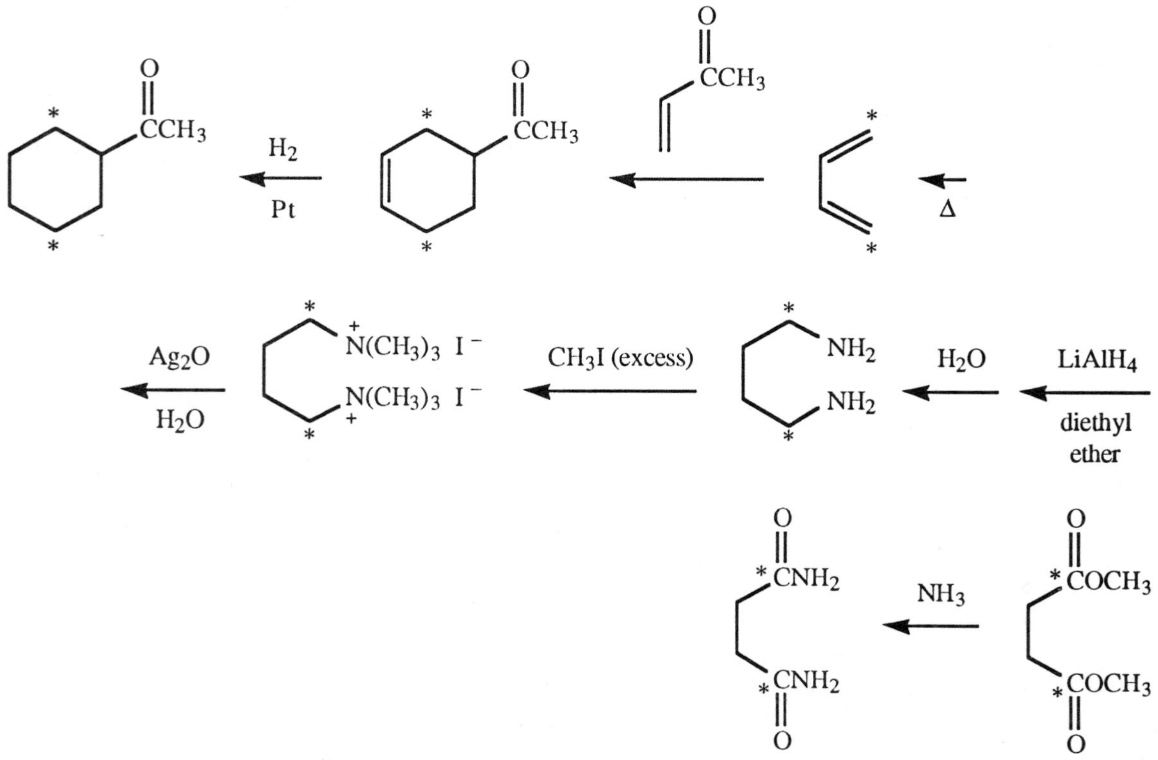

22.51

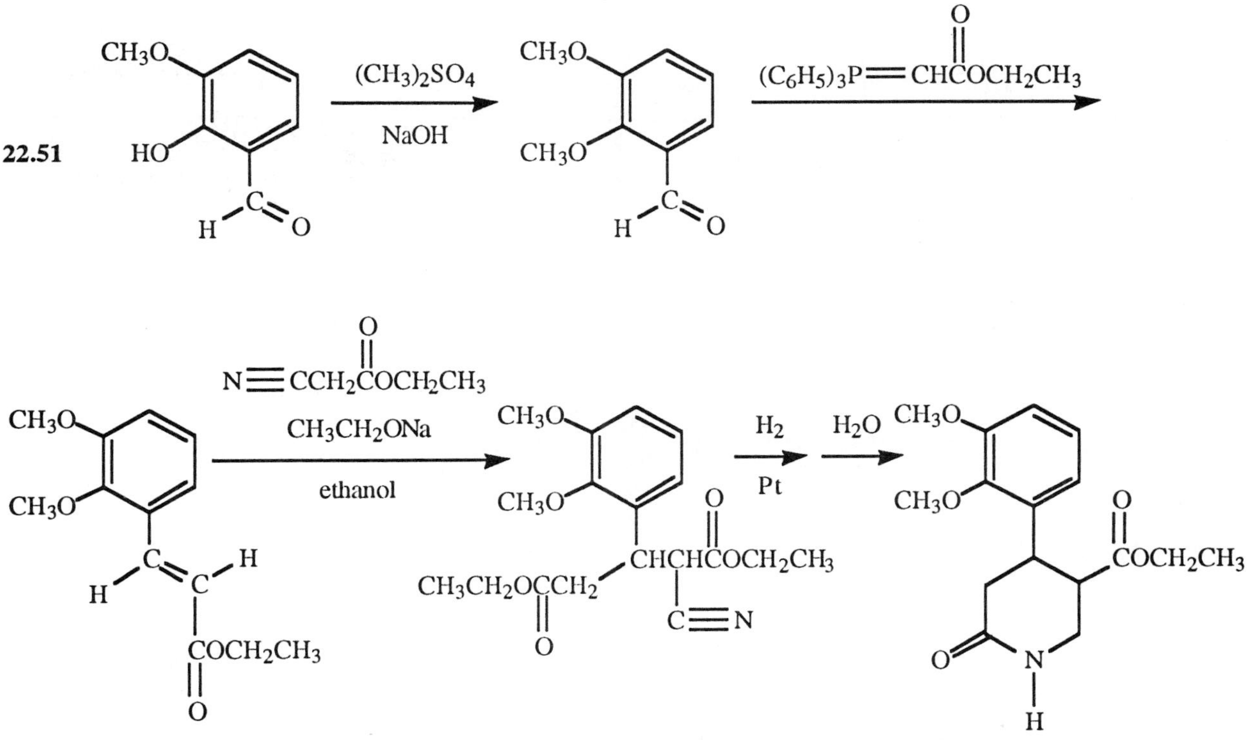

22.52

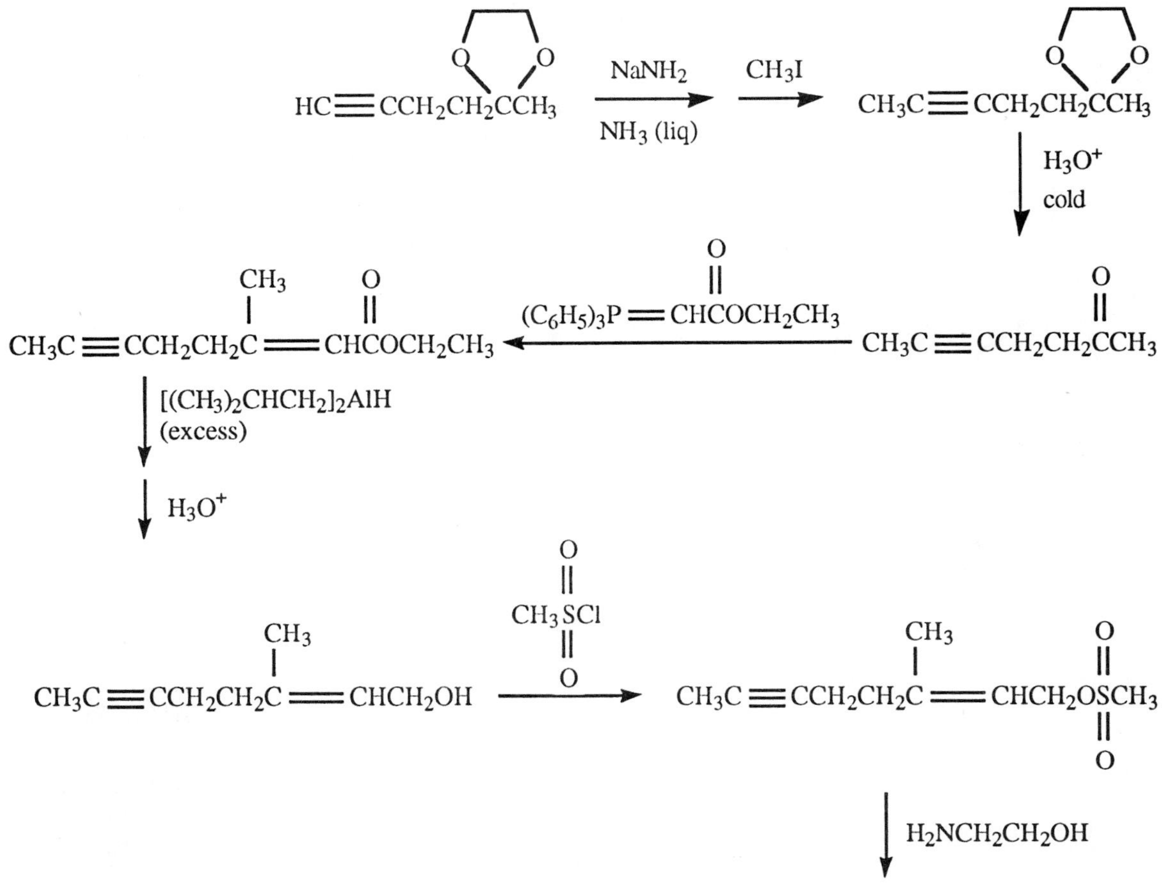

22.52 (cont)

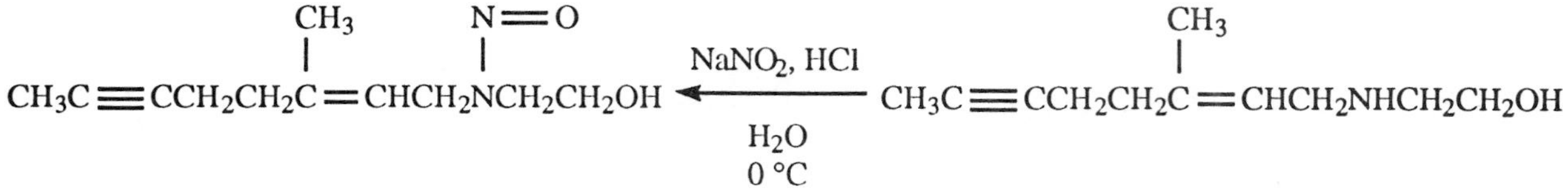

$$CH_3C \equiv CCH_2CH_2C=CHCH_2NCH_2CH_2OH \xleftarrow[\substack{H_2O \\ 0\,°C}]{NaNO_2,\ HCl} CH_3C \equiv CCH_2CH_2C=CHCH_2NHCH_2CH_2OH$$

with CH₃ substituent and N=O on the nitrogen (left), CH₃ substituent (right)

22.53

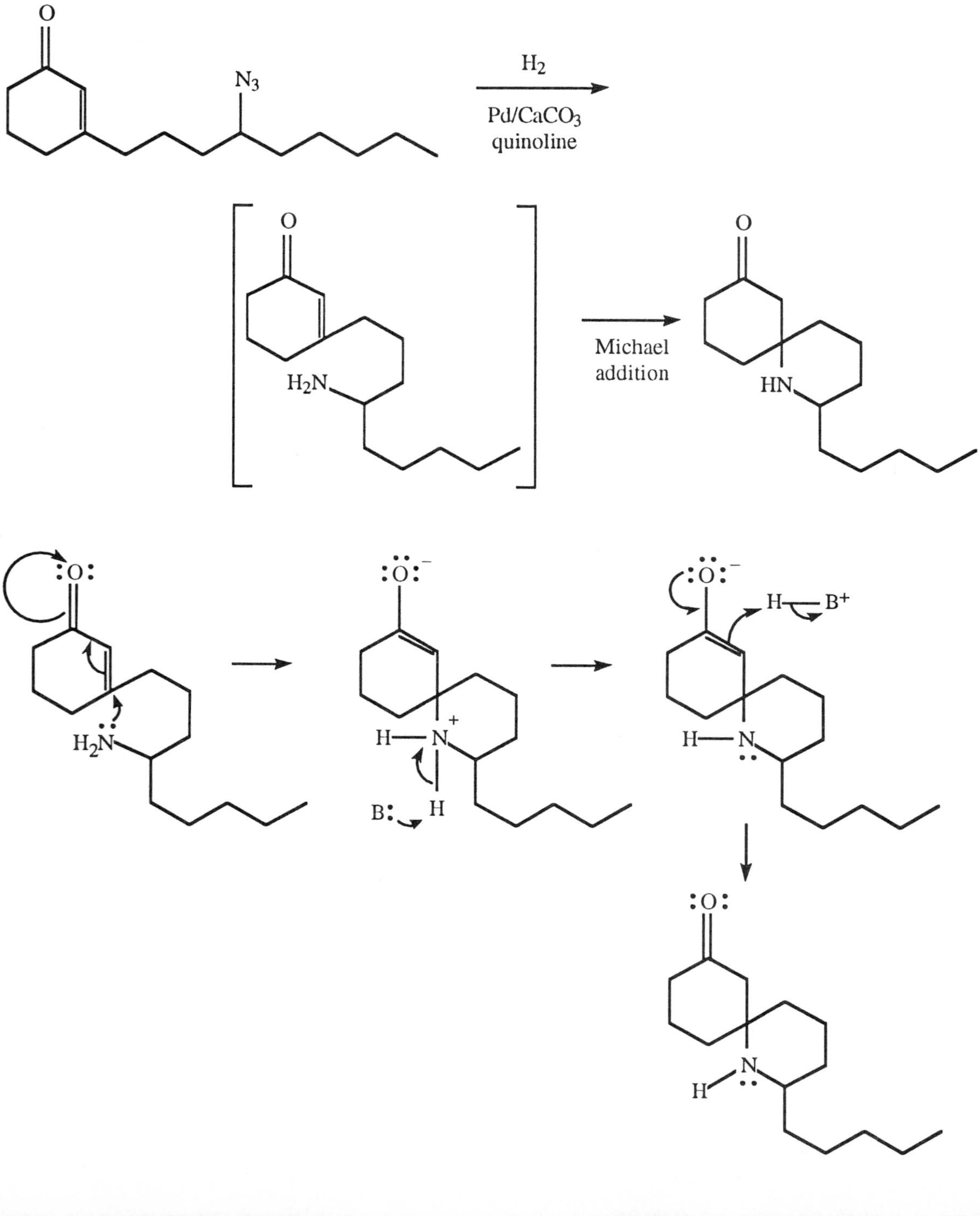

22.54   The leaving group must be N_2.

22.55   (a)

22.55 (cont)

(b)

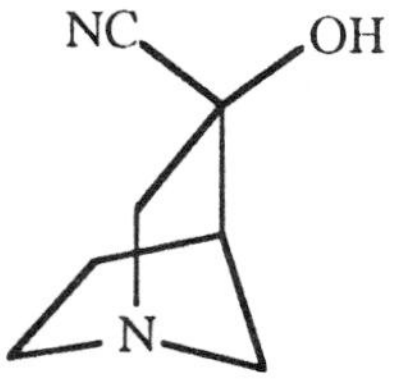

NC   OH

[structure of cyanohydrin with bicyclic amine N]

This cyanohydrin forms at higher temperatures; therefore, it must be the thermodynamic product. The other one forms faster at lower temperatures, so it is the kinetic product. The formation of a cyanohydrin must be reversible for the equilibrium to occur at higher temperatures.

22.56   The analysis of the proton magnetic resonance spectrum for Compound B is given below:

δ 3.2   (2H, singlet, —N$\underline{H}_2$)
δ 3.4   (3H, singlet, —OC$\underline{H}_3$)
δ 6.0 – 6.4   (4H, distorted para substitution pattern, Ar$\underline{H}$)

The distorted aromatic resonances and the relatively high chemical shift values tell us that the two groups para to each other on the aromatic ring are both electron-donating. Compound B is *p*-methoxyaniline. Since Compound B is made by reduction of Compound A using reducing agents such as tin in hydrochloric acid, Compound A must have a nitro group in place of the amino group. Compound A is *p*-nitroanisole (*p*-methoxynitrobenzene).

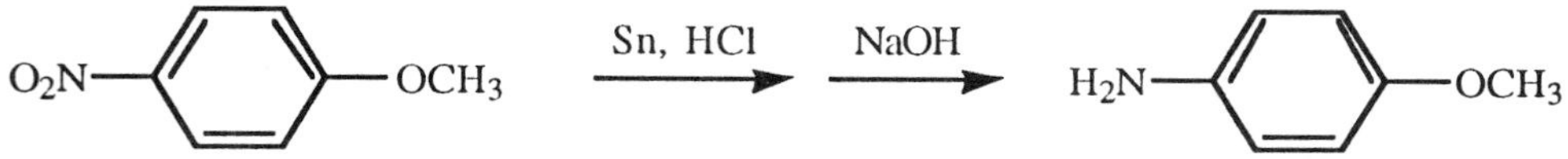

O_2N—⟨aromatic ring⟩—OCH_3 $\xrightarrow{\text{Sn, HCl}}$ $\xrightarrow{\text{NaOH}}$ H_2N—⟨aromatic ring⟩—OCH_3

Compound A   Compound B

22.57   The proton magnetic resonance spectrum of Compound C, $C_{10}H_{15}N$, shows a typical ethyl splitting pattern (a triplet at δ 1.1 and a quartet at δ 3.1). The chemical shift of the quartet tells us that the ethyl group is bonded to the nitrogen atom. There are also protons absorbing in the aromatic region (δ 6.0 – δ 6.9). The integration is 5:4:6 and tells us that there must be two ethyl groups bonded to the nitrogen. Compound C is *N,N*-diethylaniline.

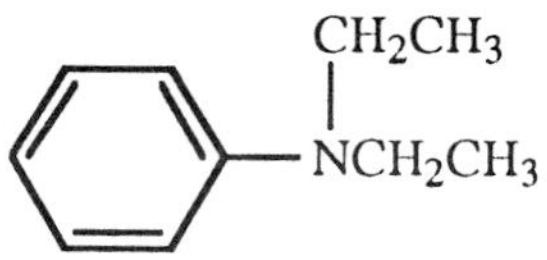

⟨aromatic ring⟩—NCH_2CH_3 with CH_2CH_3 branch

The proton magnetic resonance spectrum is summarized below:

δ 1.1   (6H, triplet, *J* 7 Hz, —CH$_2$C$\underline{H}_3$)
δ 3.1   (4H, quartet, *J* 7 Hz, —NC$\underline{H}_2$CH$_3$)
δ 6.0 – 6.9   (5H, Ar$\underline{H}$)

22.8 ppm   34.4 ppm
↓   ↓
$$CH_3 — \overset{\overset{\displaystyle O}{\|}}{C} — NHCH_2CH_3$$
↑   ↑
171.0 ppm   14.6 ppm

22.58

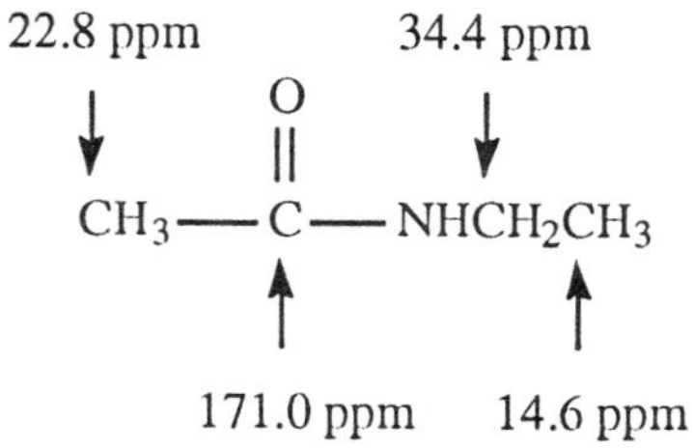

22.59   Compound D, C_3H_9N, has no units of unsaturation. The appearance of only two bands in the C-13 nuclear magnetic resonance spectrum tells us that Compound D is symmetrical. Compound D is isopropylamine (2-aminopropane).

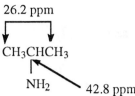

26.2 ppm

CH_3CHCH_3

NH_2 42.8 ppm

22.60   Compound E, $C_8H_{11}N$, has four units of unsaturation, which suggests the presence of an aromatic ring. This is confirmed by the presence of bands in the C-13 spectrum in the aromatic region. The solubility properties of Compound E tell us that it is an amine and the lack of reactivity with benzenesulfonyl chloride tells us that it is a tertiary amine. Six of the eight carbon atoms in Compound E are part of the benzene ring. Therefore, the remaining carbon atoms must be two methyl groups. Compound E is *N,N*-dimethylaniline.

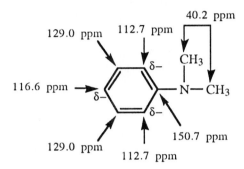

40.2 ppm

129.0 ppm   112.7 ppm

116.6 ppm

CH_3

$N-CH_3$

129.0 ppm   112.7 ppm   150.7 ppm

23

The Chemistry of Aromatic Compounds II. Synthetic Transformations

23.1

(a)

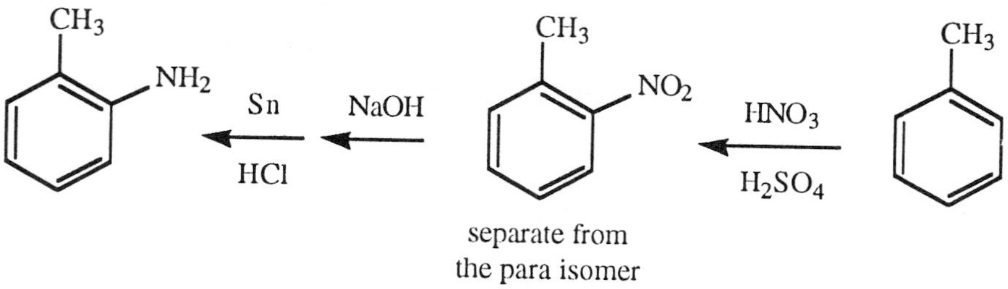

separate from
the para isomer

(b)

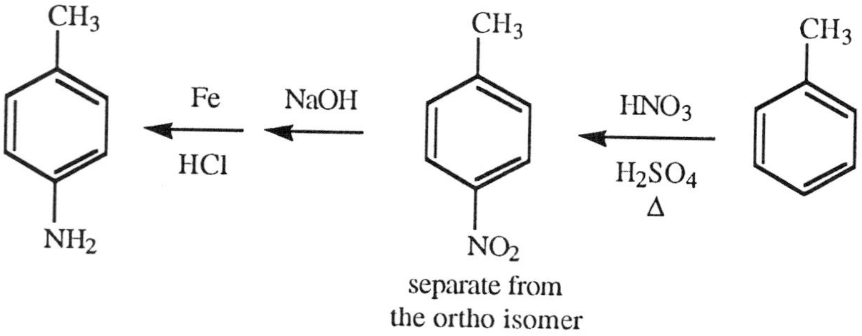

separate from
the ortho isomer

(c)

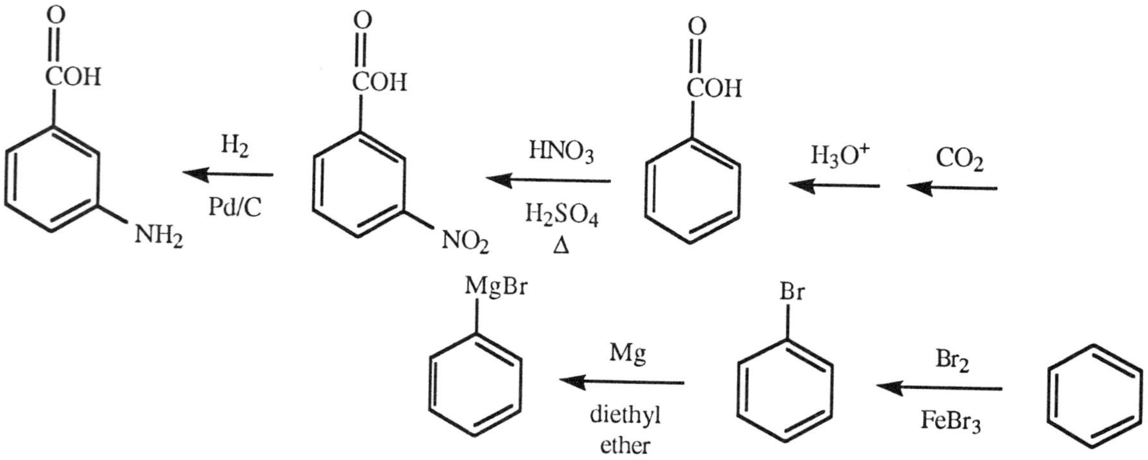

Concept Map 23.1   Diazonium ions in synthesis.

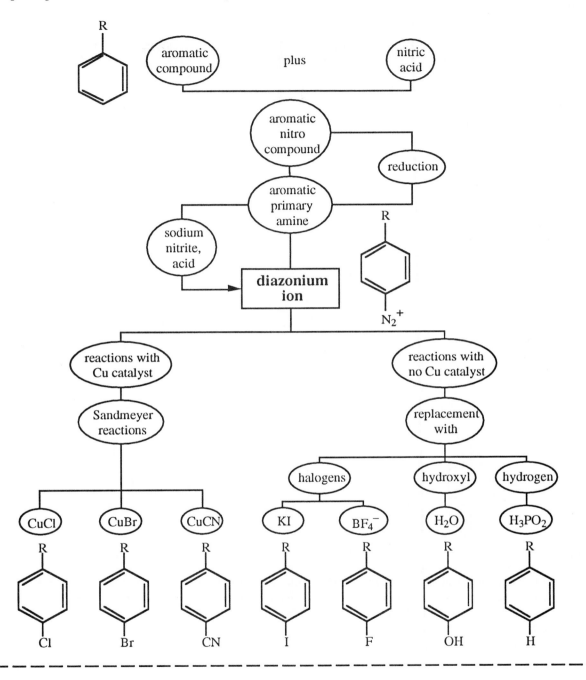

- -

23.2

23.2 (a) (cont)

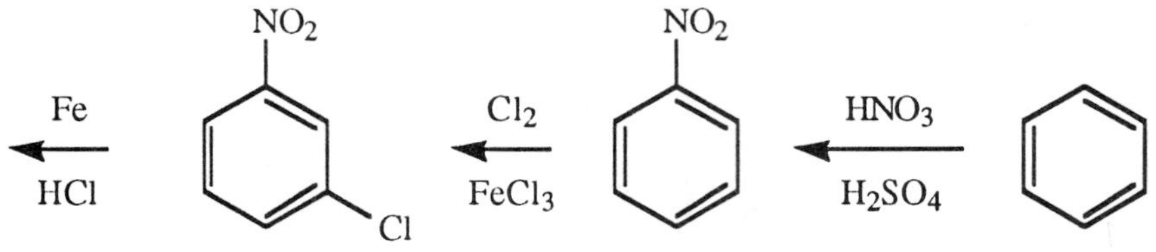

(b)

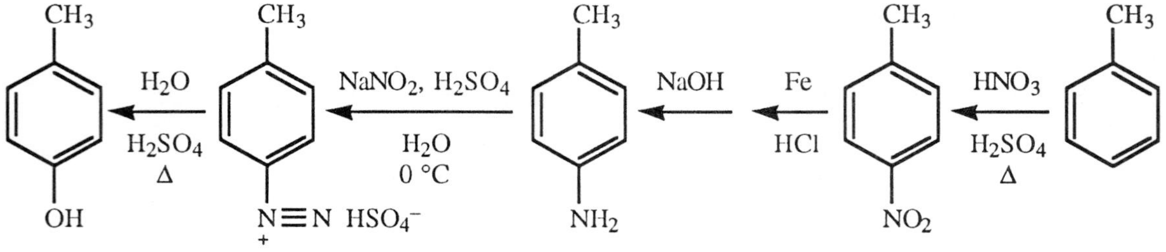

(c)

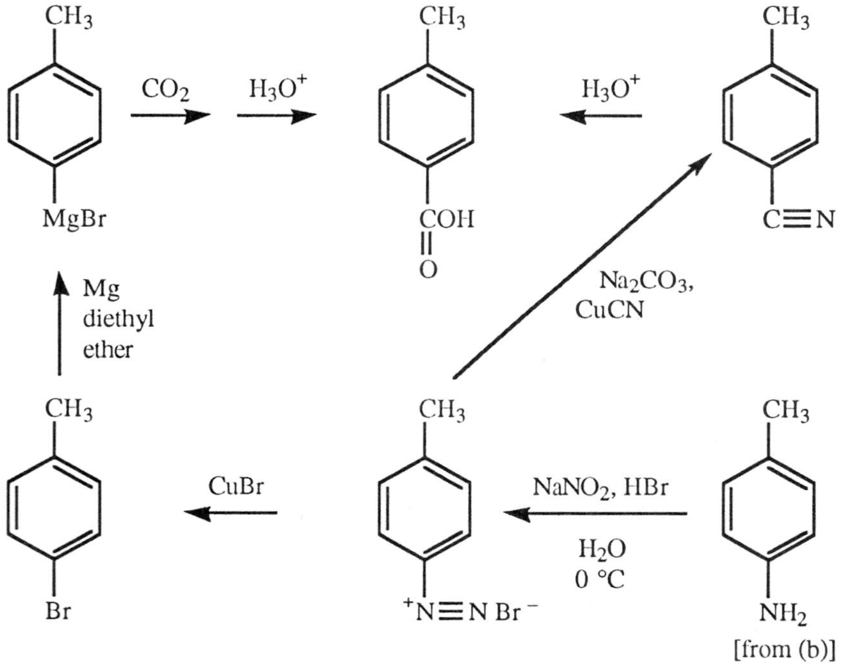

p-Bromotoluene can also be synthesized by treating toluene with bromine in the presence of ferric bromide, but it will be contaminated with o-bromotoluene. Separation of o-bromotoluene from p-bromotoluene is difficult because the two compounds differ by only 3° in boiling point. p-Nitrotoluene will also have the ortho isomer as a contaminant, but the two nitrotoluenes are easy to separate because the ortho isomer is a liquid at room temperature and the para isomer is a solid.

(d)

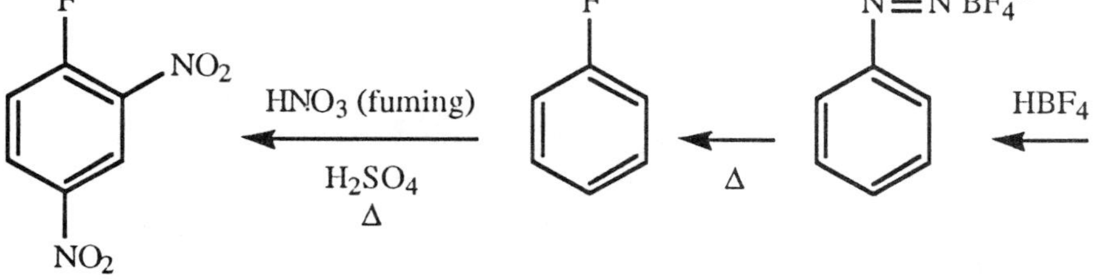

23.2 (d) (cont)

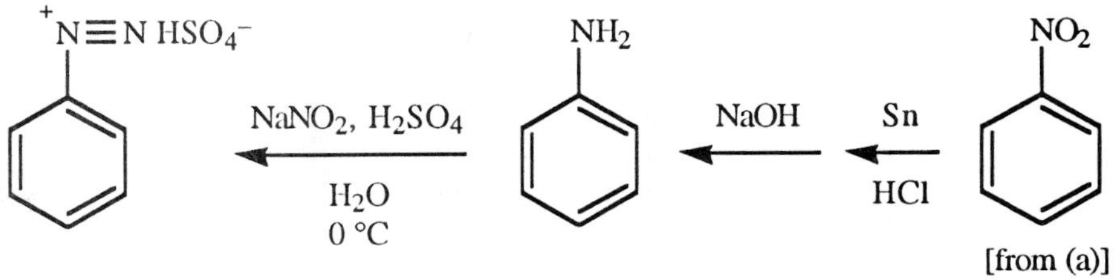

(e)

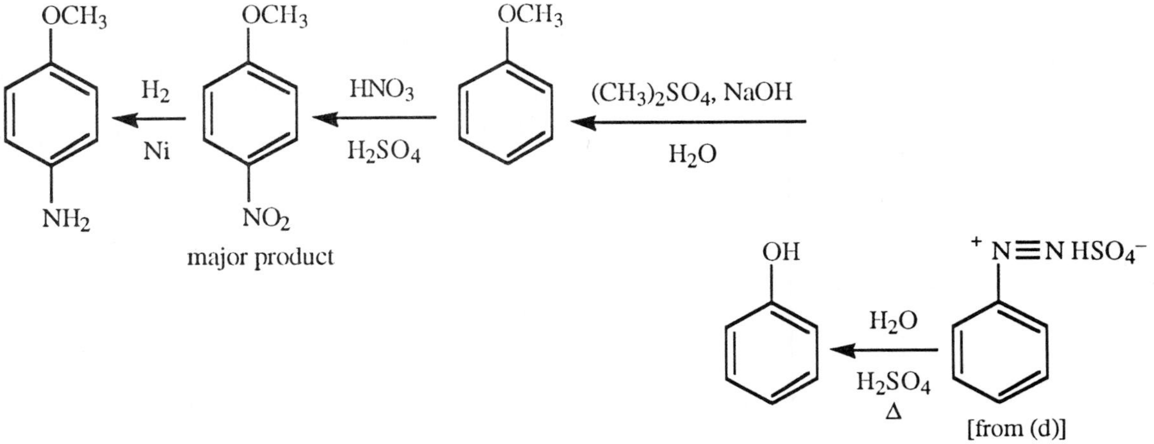

major product

(f)

major product

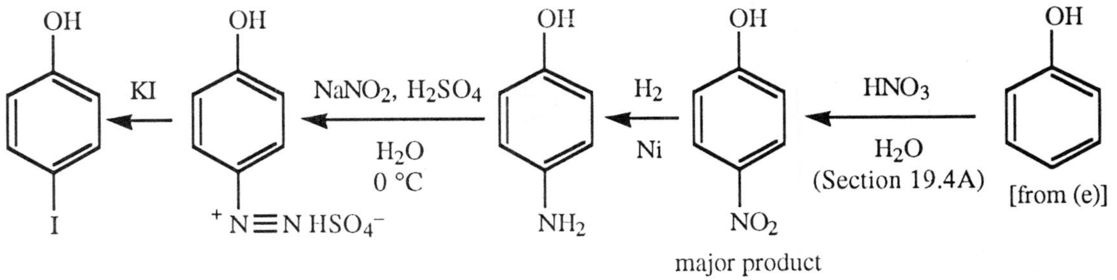

23.3

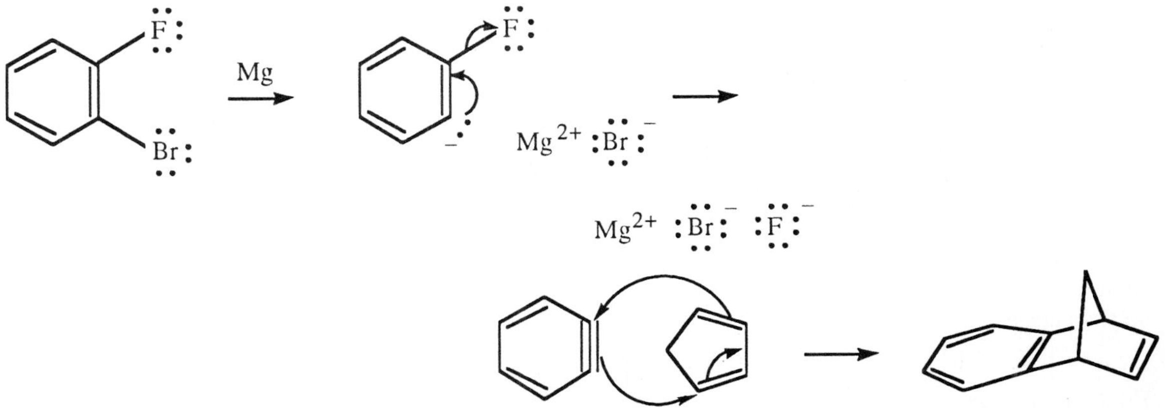

(See Section 13.6A for the formation of a Grignard reagent. The structure of the Grignard reagent is written so as to emphasize the carbanionic character of the carbon atom bonded to magnesium.)

Concept Map 23.2   Reactions of diazonium ions.

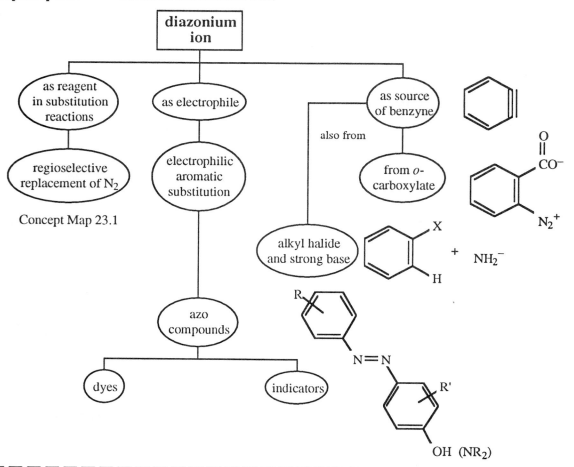

23.4

Protonation of this nitrogen atom gives a cation that has → no resonance stabilization.

Protonation of this nitrogen atom gives a cation that is stabilized by resonance; the positive charge is delocalized to the carbon atoms of an aromatic ring.

Protonation of this nitrogen atom gives a cation that is stabilized by resonance; the positive charge is delocalized to the carbon atoms of an aromatic ring and to the nitrogen atom of the amino group. Therefore this is the most stable cation of the three.

23.5

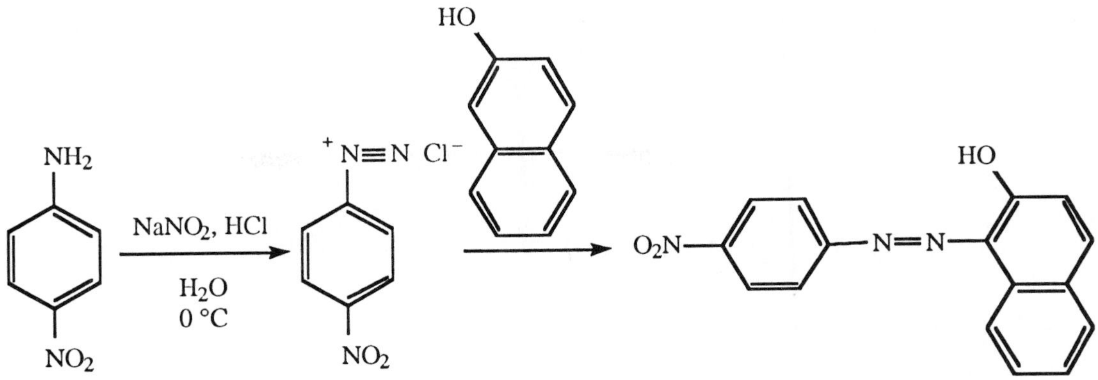

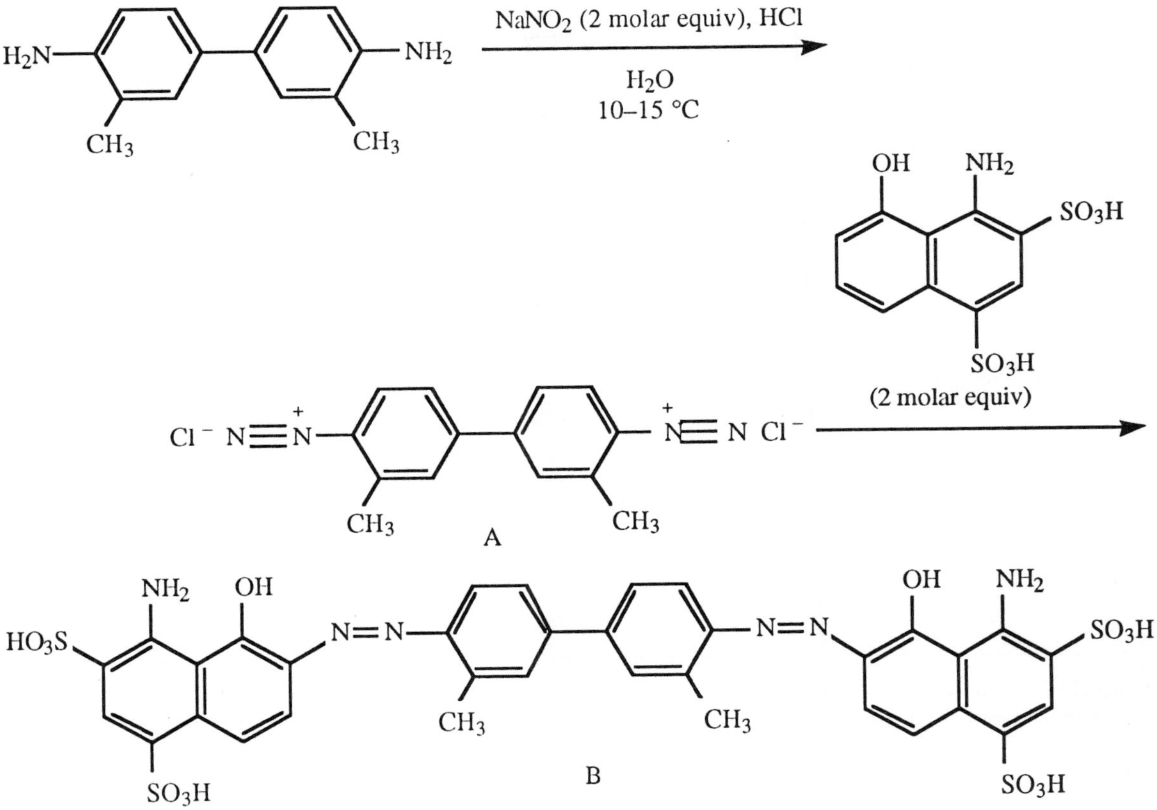

23.6

23.7

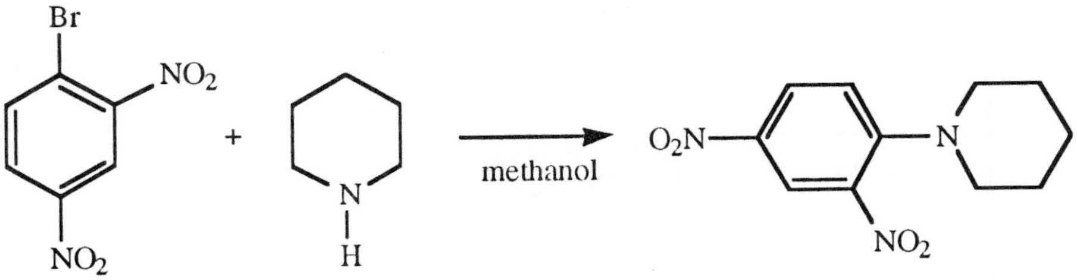

(a)

23.7 (cont)

(b)

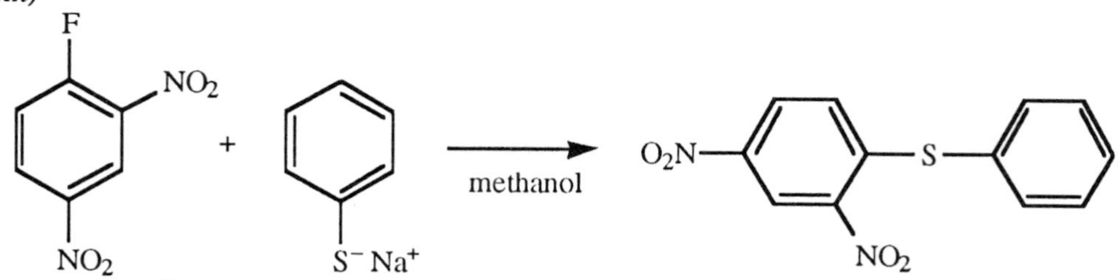

(c)

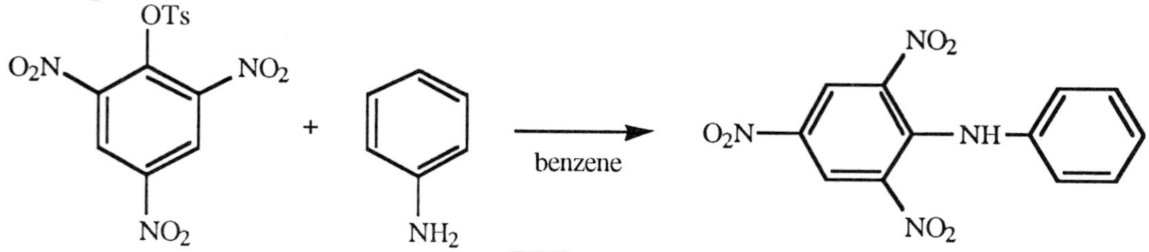

(d)

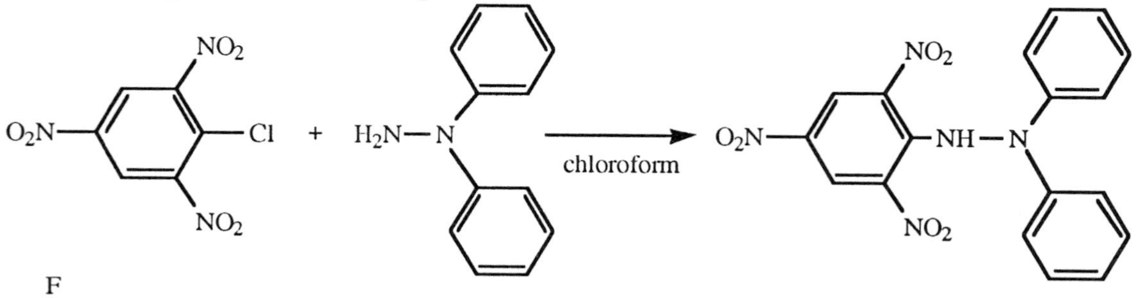

(e)

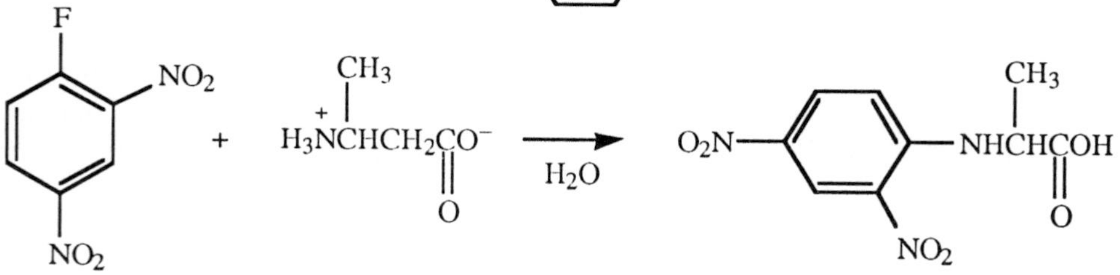

- -

Concept Map 23.3   (see p. 836)

- -

23.8

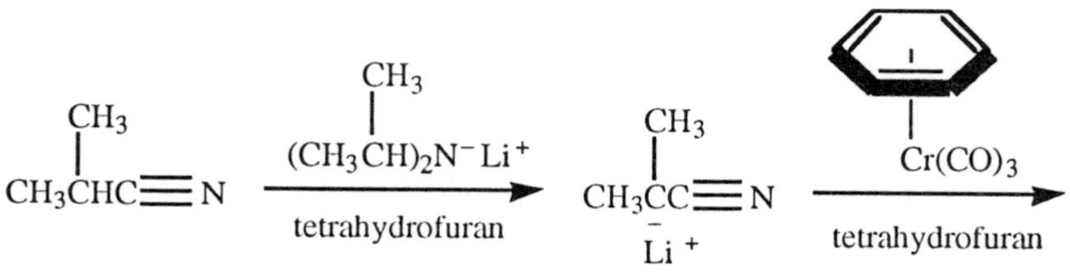

(a)

23.8 (a) (cont)

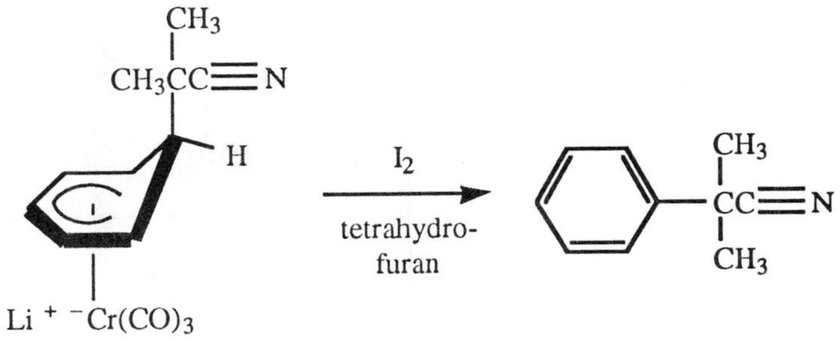

B C

(b)

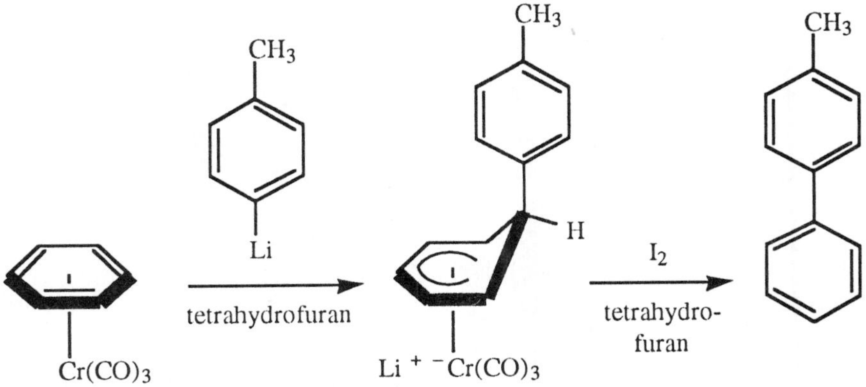

D E

(c)

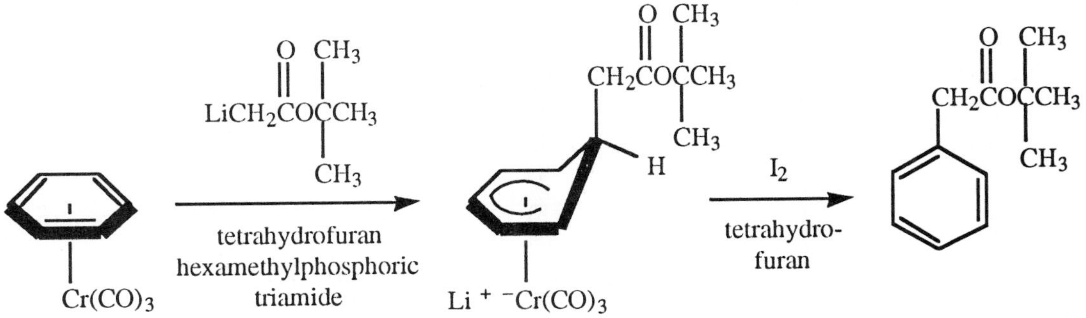

F G

(d)

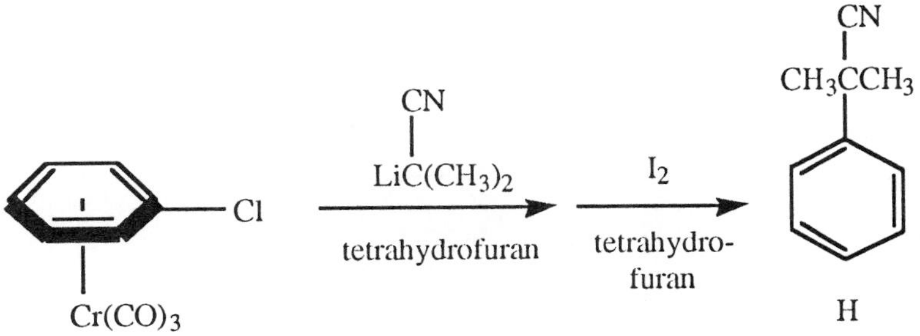

H

Concept Map 23.3   Nucleophilic aromatic substitution.

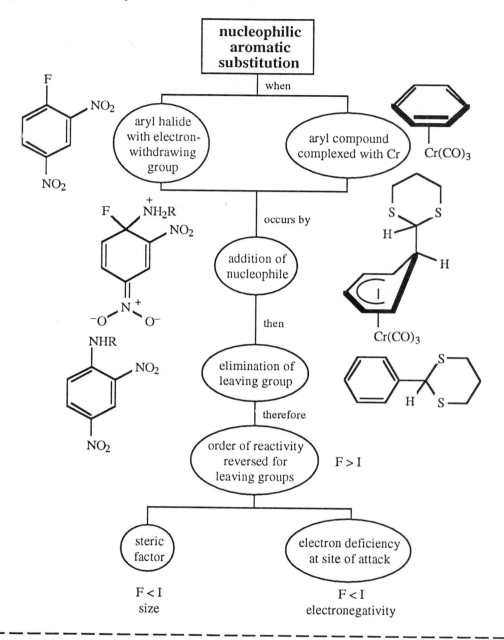

- -

23.9

23.9 (cont)

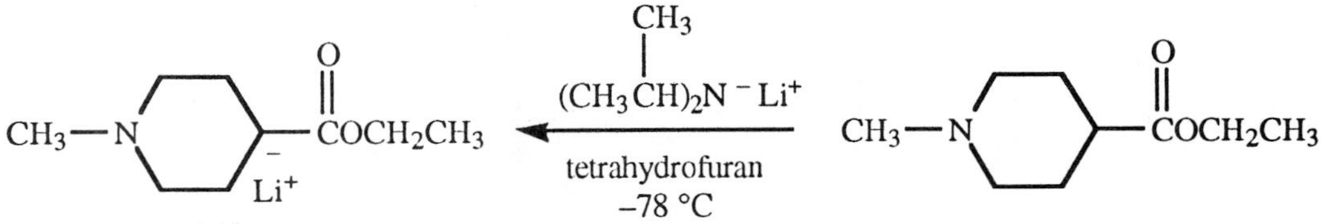

23.10

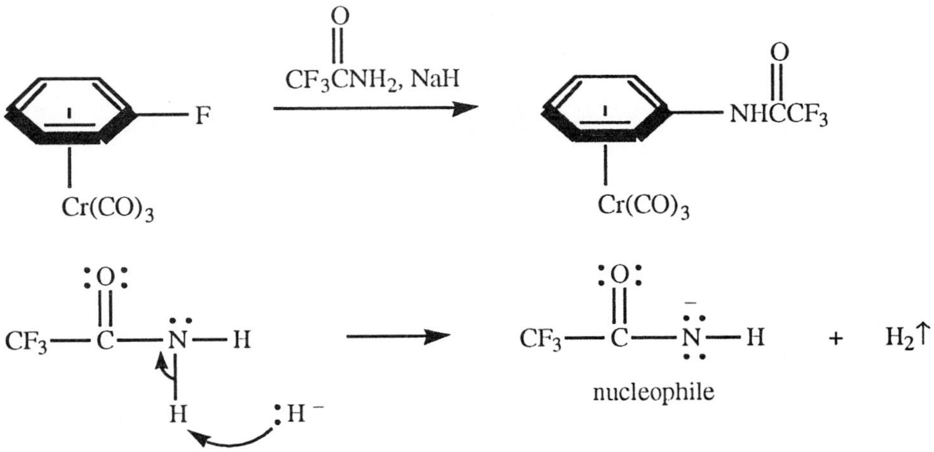

23.11

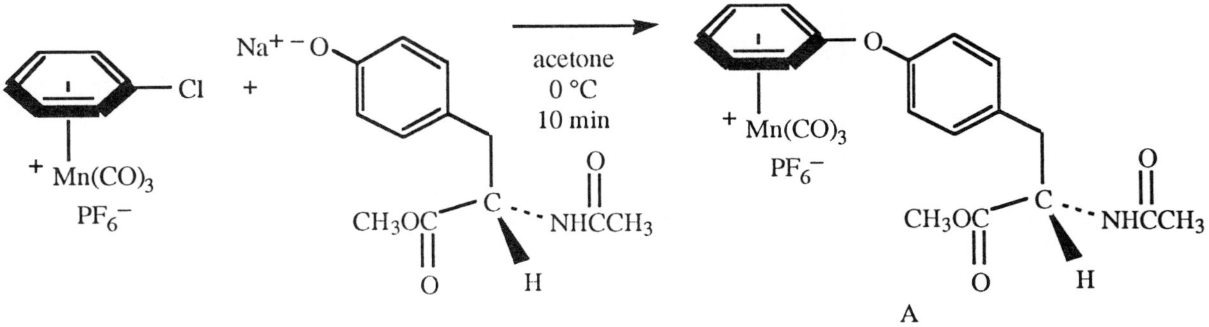

23.12

(a)

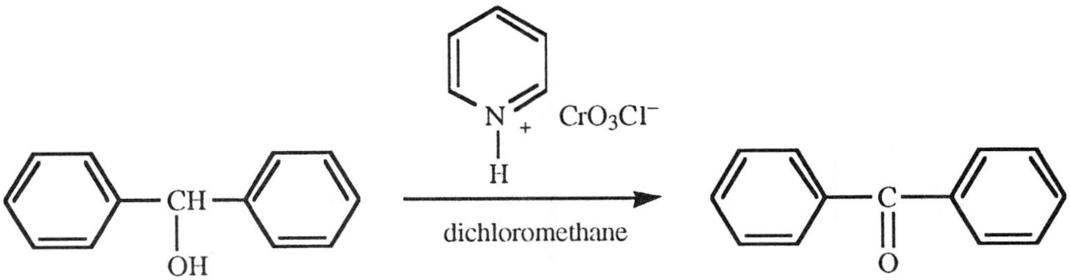

(b)

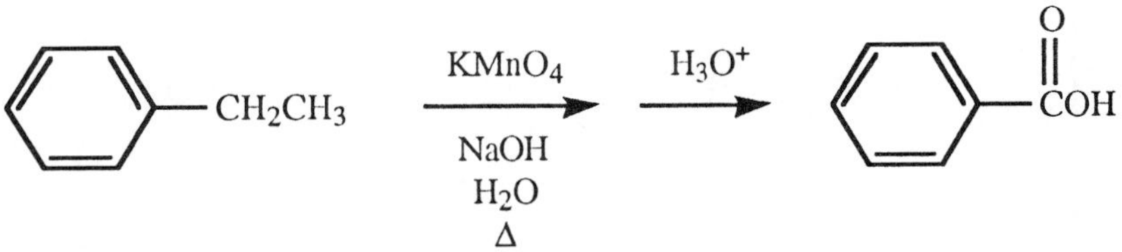

23.12 (cont)

(c)

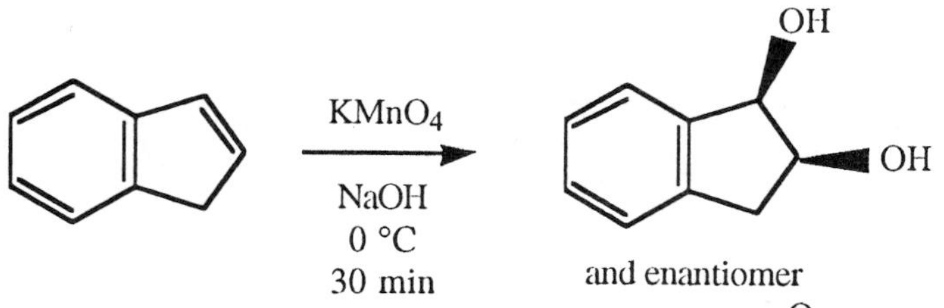

and enantiomer

(d)

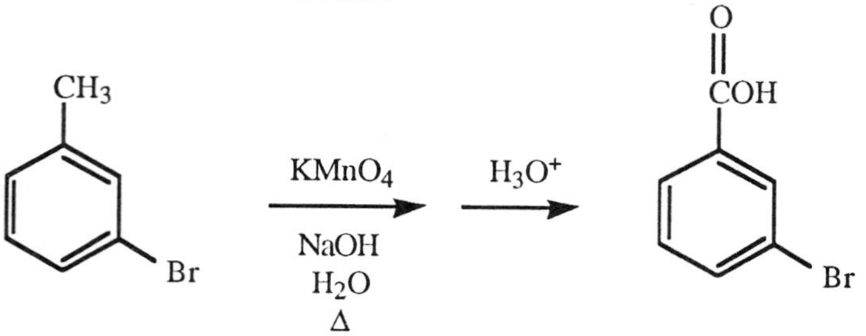

(e)

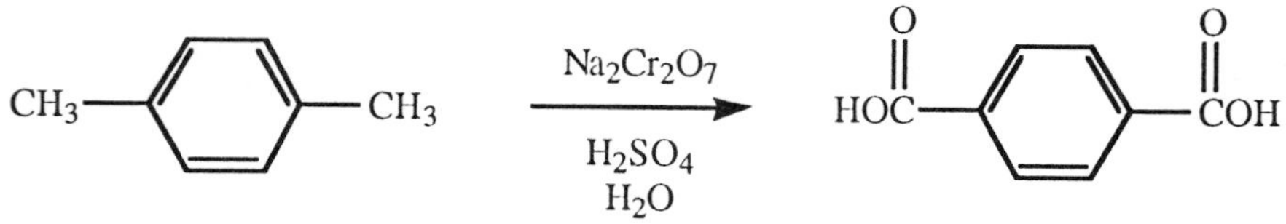

23.13 (a) Resonance contributors

1. *m*-nitrophenolate anion

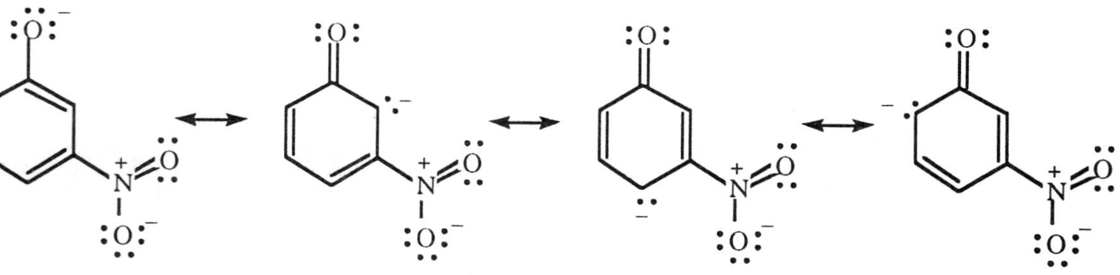

The negative charge is stabilized by the inductive effect of the electron-withdrawing nitro group, so *m*-nitrophenol is a stronger acid than phenol.

2. *p*-nitrophenolate anion

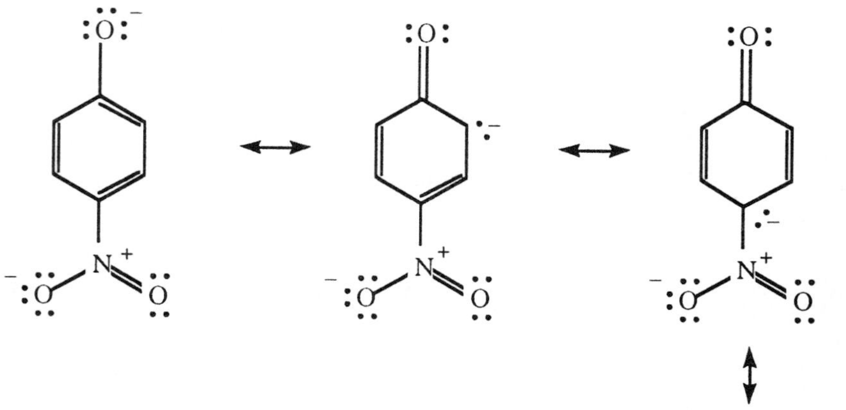

23.13 (a 2) (cont)

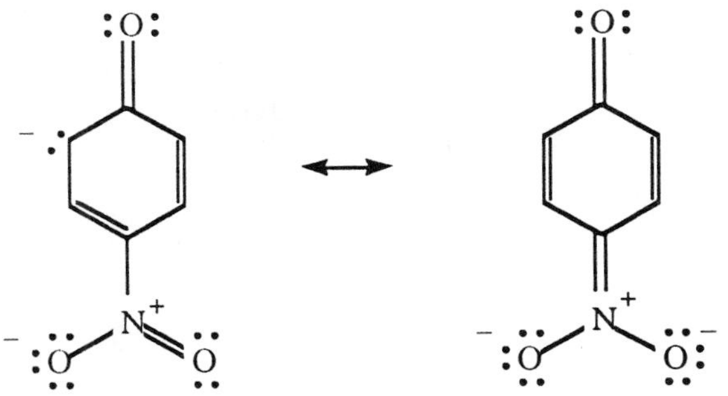

resonance contributor in which the negative charge is delocalized to the oxygen atoms of the nitro group

The negative charge of the *p*-nitrophenolate anion is stabilized not only by the inductive effect but by a resonance effect. Additional delocalization of charge, beyond what is available in the phenolate anion and in the *m*-nitrophenolate anion, is possible for the *p*-nitrophenolate anion. This extra stabilization of the conjugate base is reflected in the greater acidity of *p*-nitrophenol.

(b)

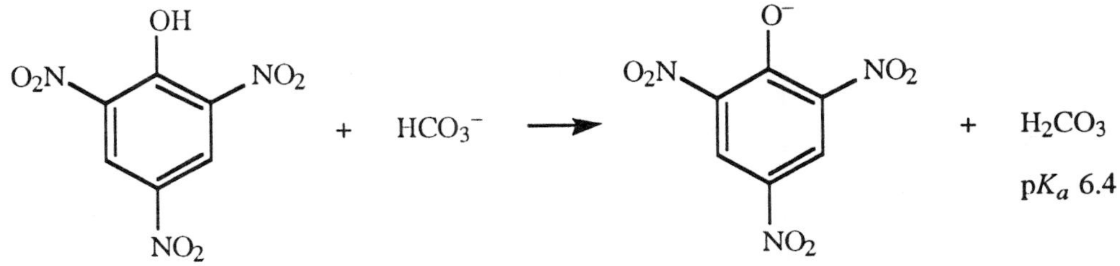

pK_a 0.25

Picric acid is a sufficiently strong acid to protonate the weak base bicarbonate anion. Although it is a phenol, the effect of the three nitro groups stabilizes the conjugate base of picric acid and thus increases the acidity of the compound so that it is more acidic than a simple carboxylic acid.

23.14 A carboxylic acid can usually be distinguished from a phenol by reaction with sodium bicarbonate, as long as the pK_a of the phenol is greater than that of carbonic acid. For example:

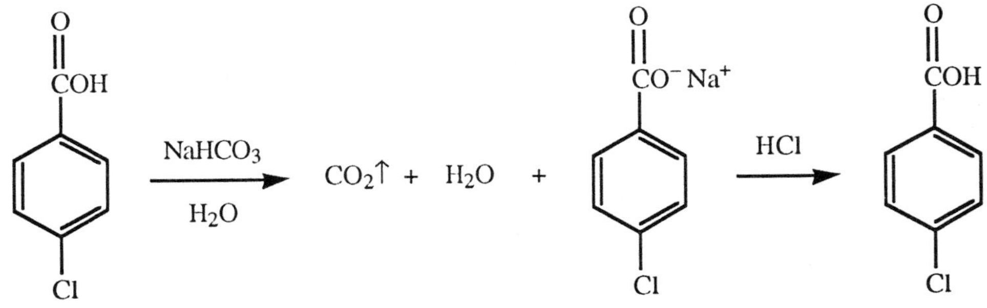

insoluble in water | soluble in water | insoluble in water

Aqueous sodium bicarbonate dissolves *p*-chlorobenzoic acid with the evolution of carbon dioxide gas. The carboxylic acid will be reprecipitated by strong acid. The phenol should not be soluble in sodium bicarbonate solution but should be soluble in a stronger base such as sodium hydroxide.

23.14 (cont)

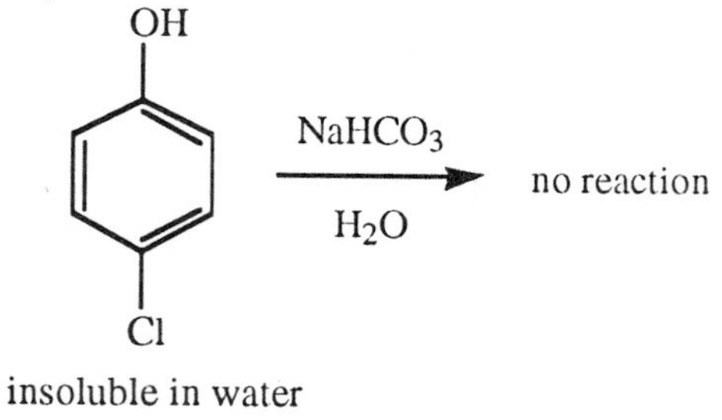

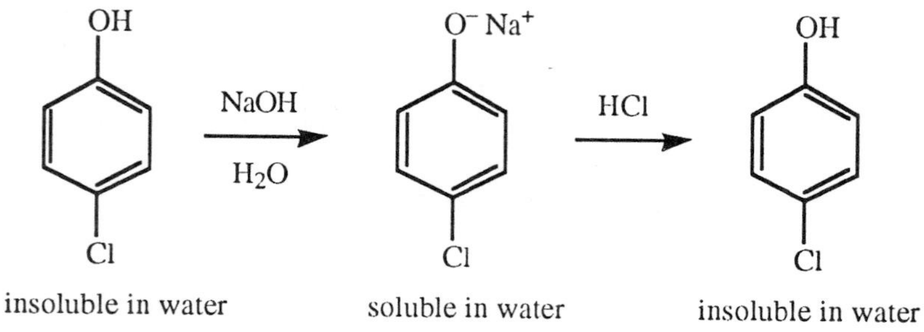

insoluble in water

insoluble in water   soluble in water   insoluble in water

23.15

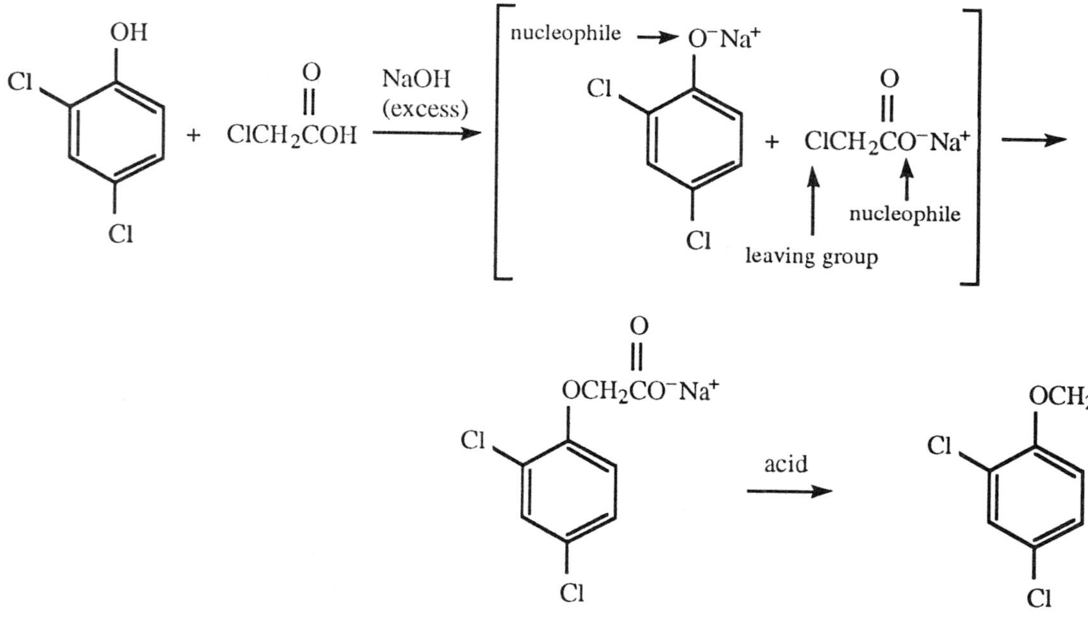

2,4-D

The phenolate anion is a stronger base and a better nucleophile than the chloroacetate anion. The oxygen atom of the phenolate anion will displace the chloride leaving group of the chloroacetate anion. Aromatic nucleophilic substitution reactions require strongly electron-withdrawing substituents on the benzene ring to stabilize the negative charge in the transition state. The chlorine atoms on the aromatic ring of the dichlorophenolate ion are not displaced in nucleophilic substitution reactions.

23.16   (1)

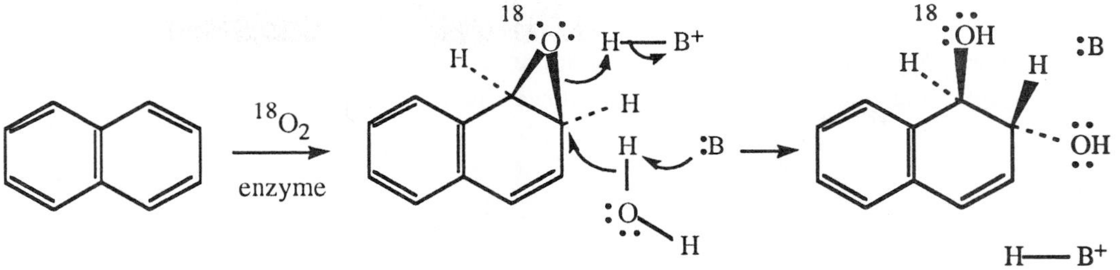

(2)

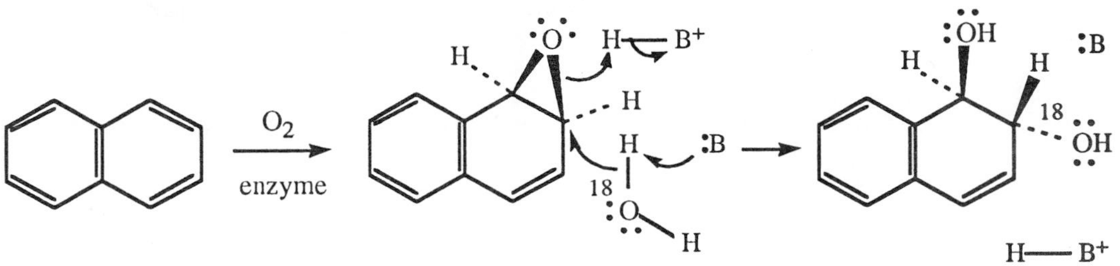

23.17

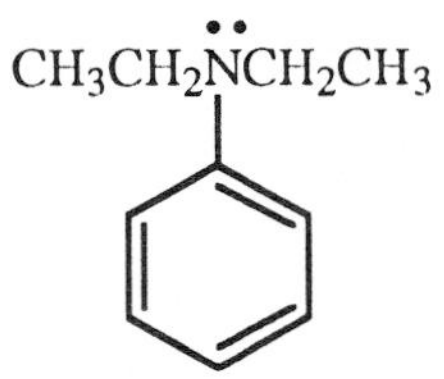

CH₃CH₂N̈CH₂CH₃

This compound has the nonbonding electrons on the nitrogen atom in conjugation with the aromatic ring. It is an arylamine and should have an ultraviolet spectrum resembling that of aniline.

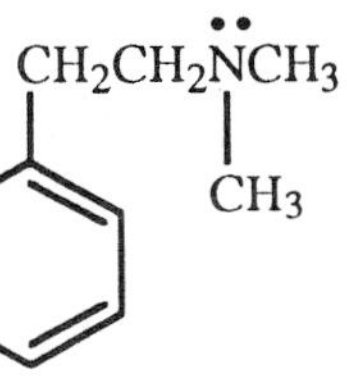

CH₂CH₂N̈CH₃

|

CH₃

In this compound the nitrogen atom is separated from the aromatic ring by sp^3-hybridized carbon atoms. The chromophore in this compound is an alkylbenzene, like toluene.

A comparison of the spectrum in Figure 23.7 with the spectra of aniline (Figure 23.6) and toluene (Figure 23.3) identifies Compound X as *N,N*-diethylaniline.

23.18   A shift of λ_{max} to a higher wavelength when potassium hydroxide is added to the solution points to a phenol. There are three possible phenols with the molecular formula C_7H_8O.

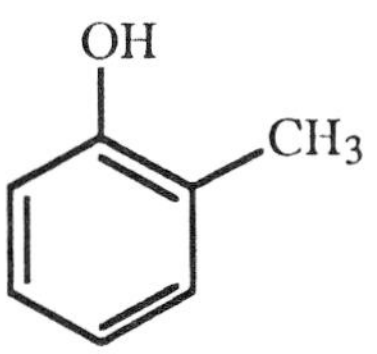

o-cresol

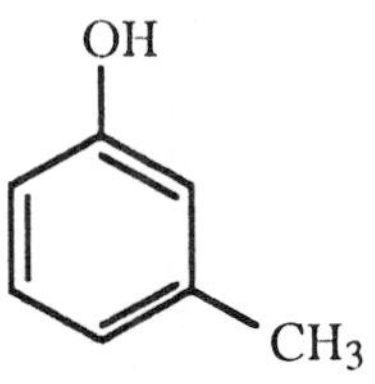

m-cresol

p-cresol

23.19 The absence of bands above 3000 cm⁻¹ in the infrared spectrum eliminates both a phenol and an alcohol. The absence of bands in the region between 2000 and 1600 cm⁻¹ eliminates a carbonyl compound such as an aldehyde or ketone. The ultraviolet spectrum is very similar to that of methoxybenzene (Figure 23.4), suggesting that Compound Z is a substituted aromatic ether. Possible structures are:

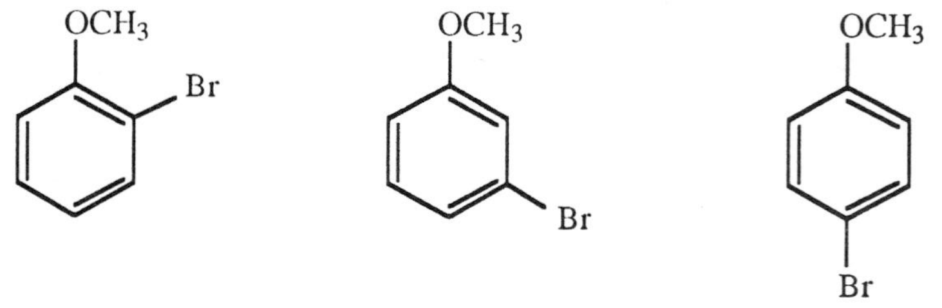

| *o*-bromomethoxy-
benzene | *m*-bromomethoxy-
benzene | *p*-bromomethoxy-
benzene |

23.20

(a)

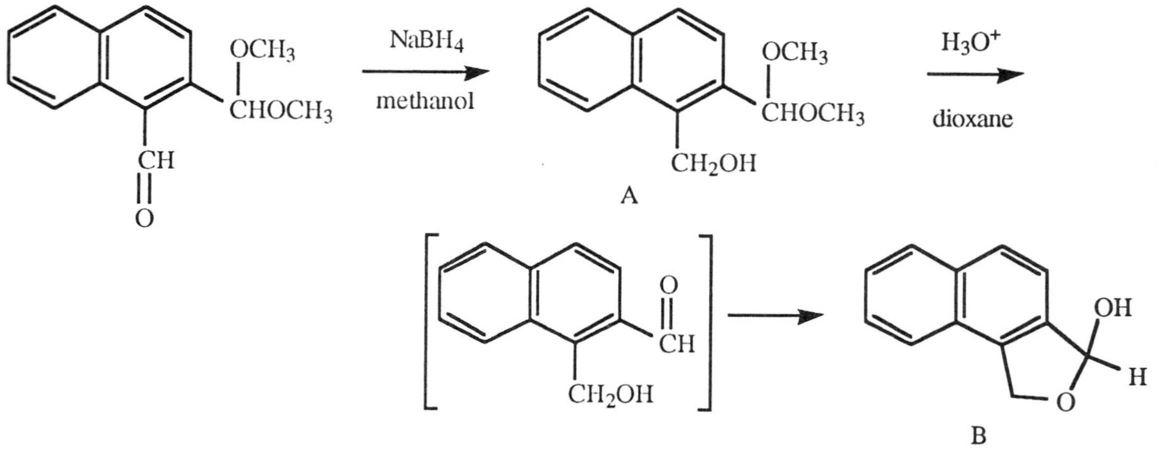

(b)

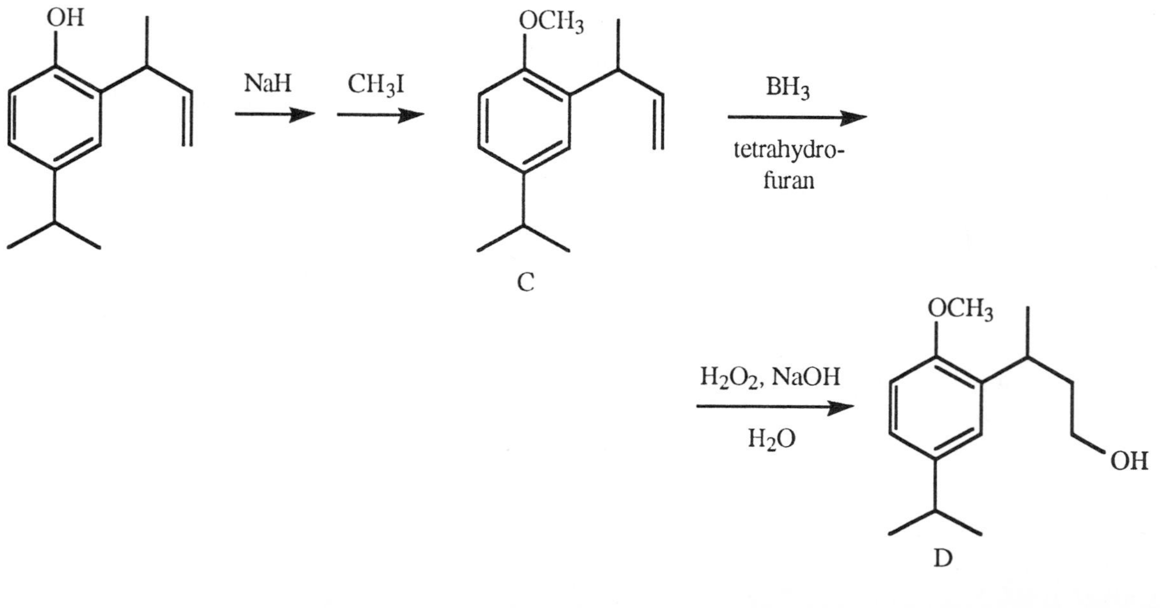

23.20 (cont)

(c)

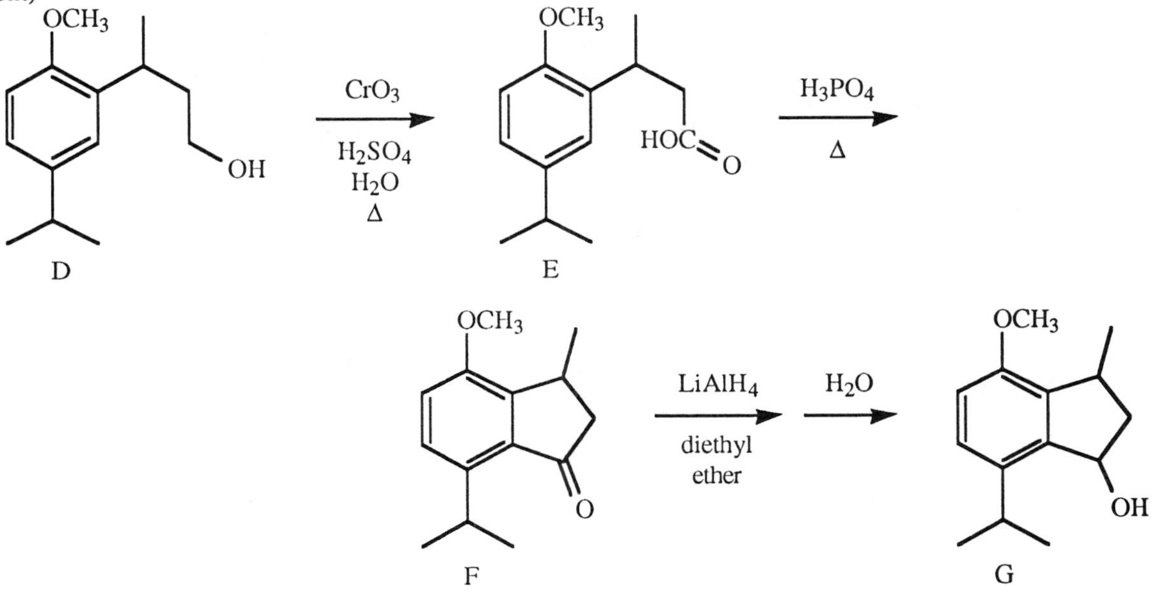

D → (CrO₃, H₂SO₄, H₂O, Δ) → E → (H₃PO₄, Δ)

F → (LiAlH₄, diethyl ether) → (H₂O) → G

(d)

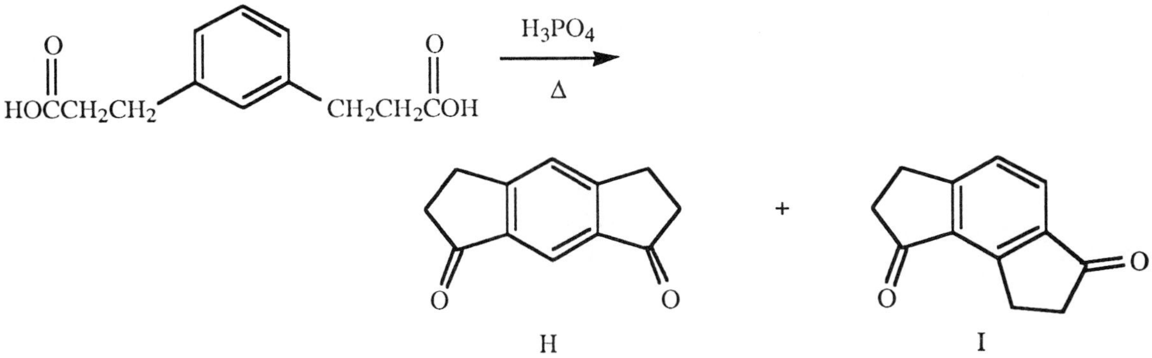

(H₃PO₄, Δ)

H + I

(e)

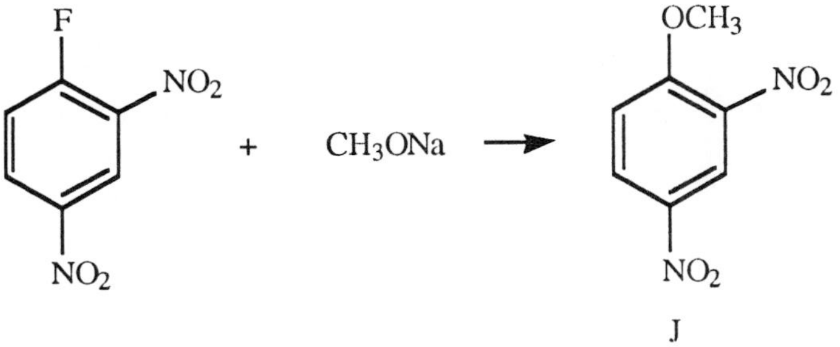

+ CH₃ONa → J

(f)

(KMnO₄, H₂O, Δ) → K

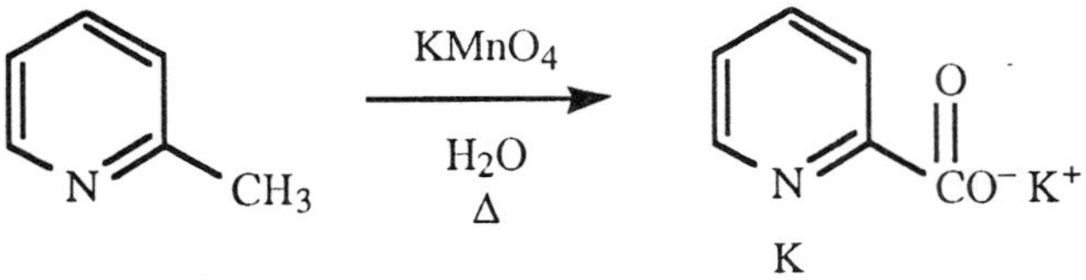

23.20 (cont)

(g)

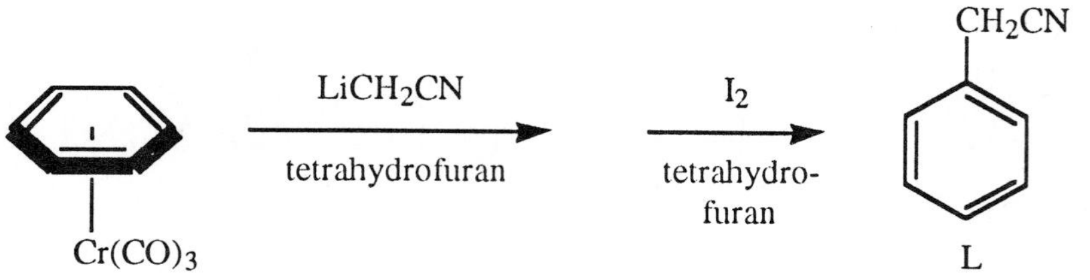

(h)

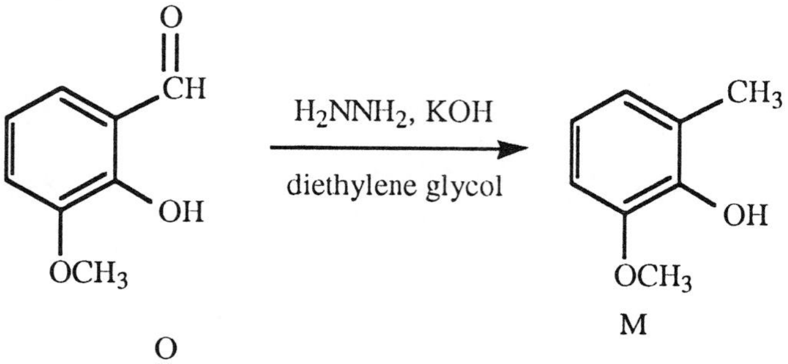

(i)

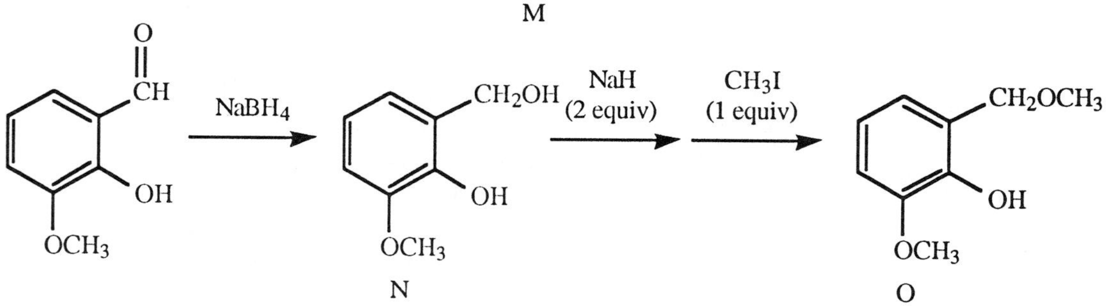

(an alkoxide ion is a better
nucleophile than a phenolate ion)

(j)

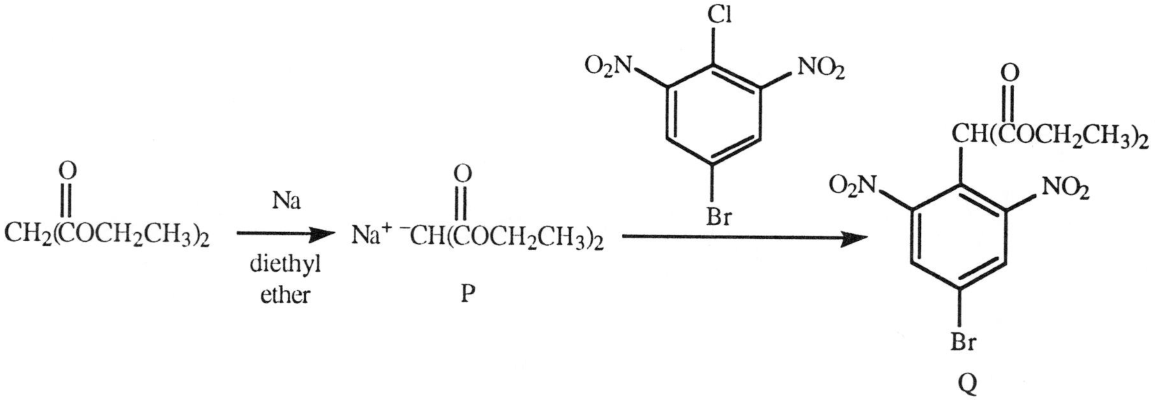

Chloride ion is a better leaving group than bromide ion for nucleophilic substitution reactions on aromatic rings (Section 23.3A). The location of the chlorine atom ortho to the nitro group is also essential to the reaction.

(k)

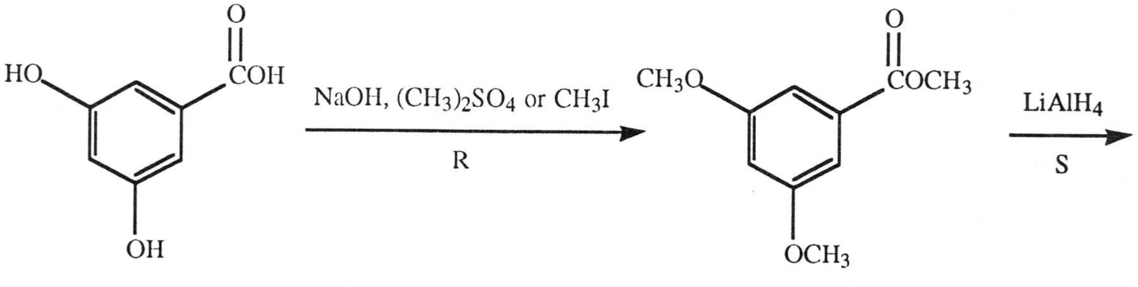

23.20 (k) (cont)

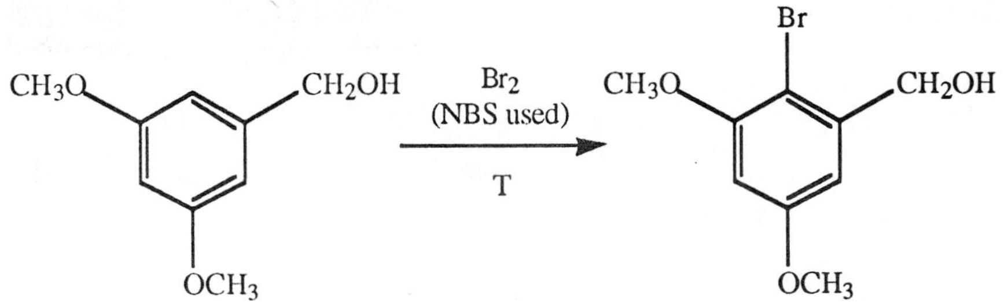

23.21

(a)

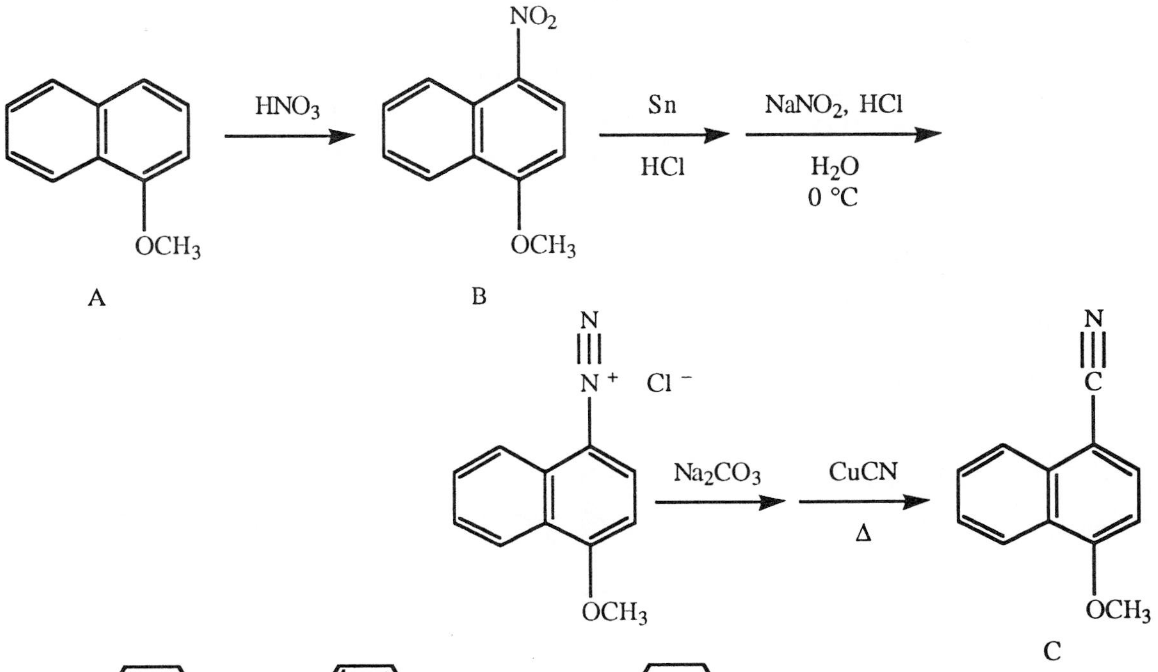

(b)

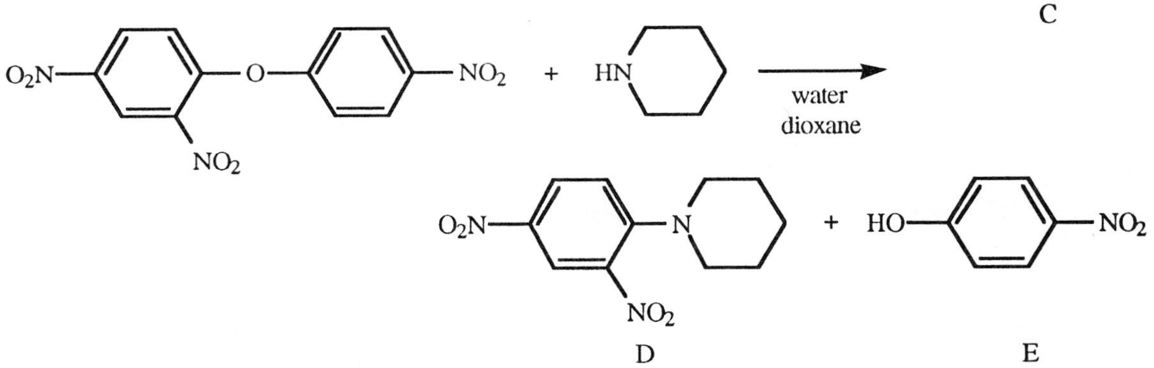

(c)

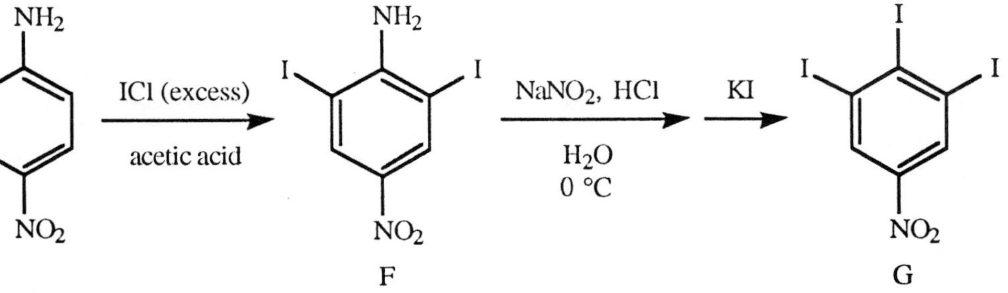

23.21 (cont)

(d)

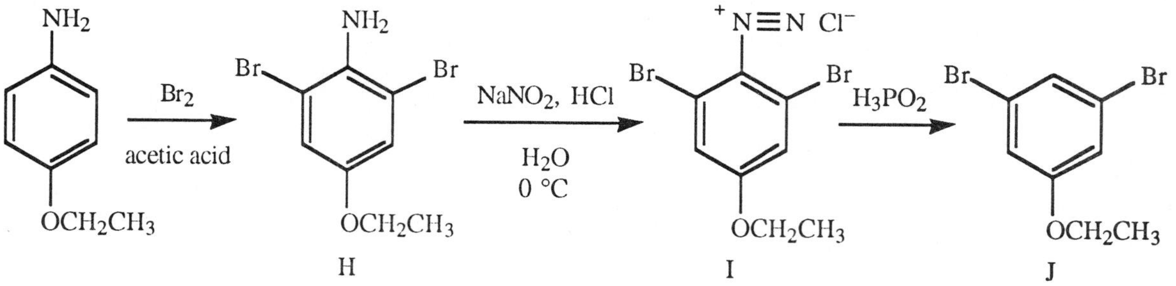

H

I

J

(e)

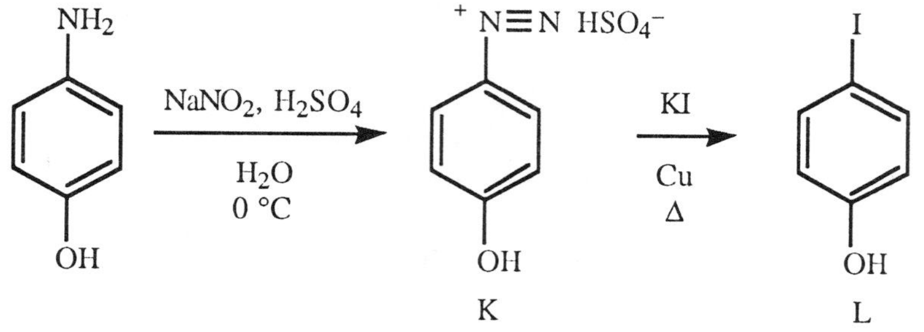

K

L

(f)

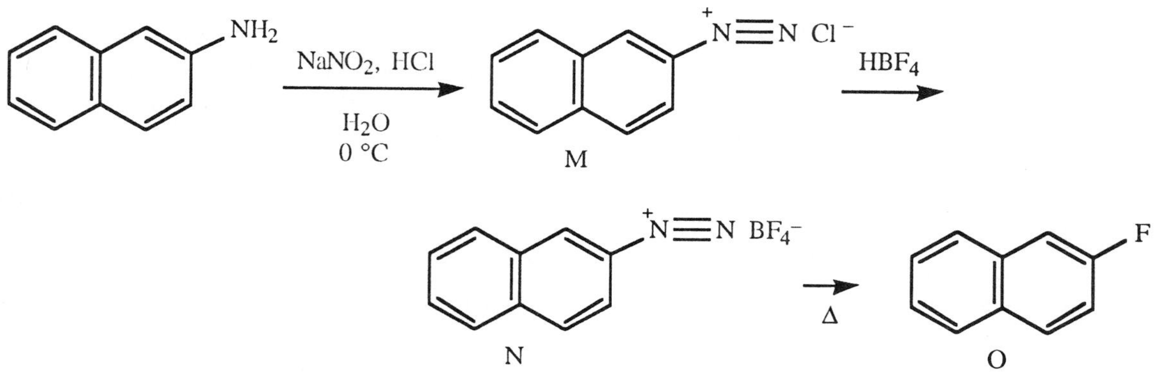

M

N

O

(g)

P

Q

R

S

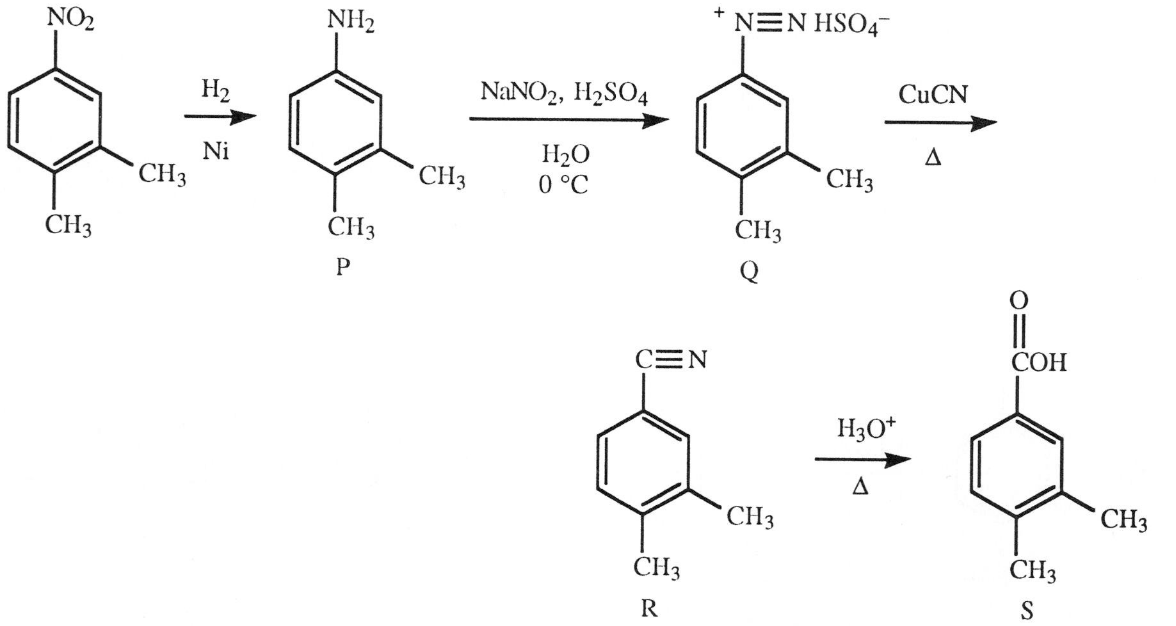

23.21 (cont)

(h)

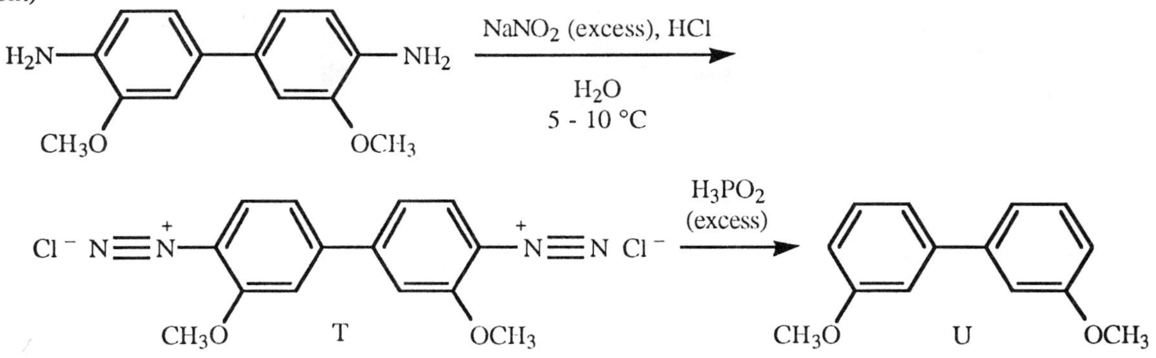

23.22

(a)

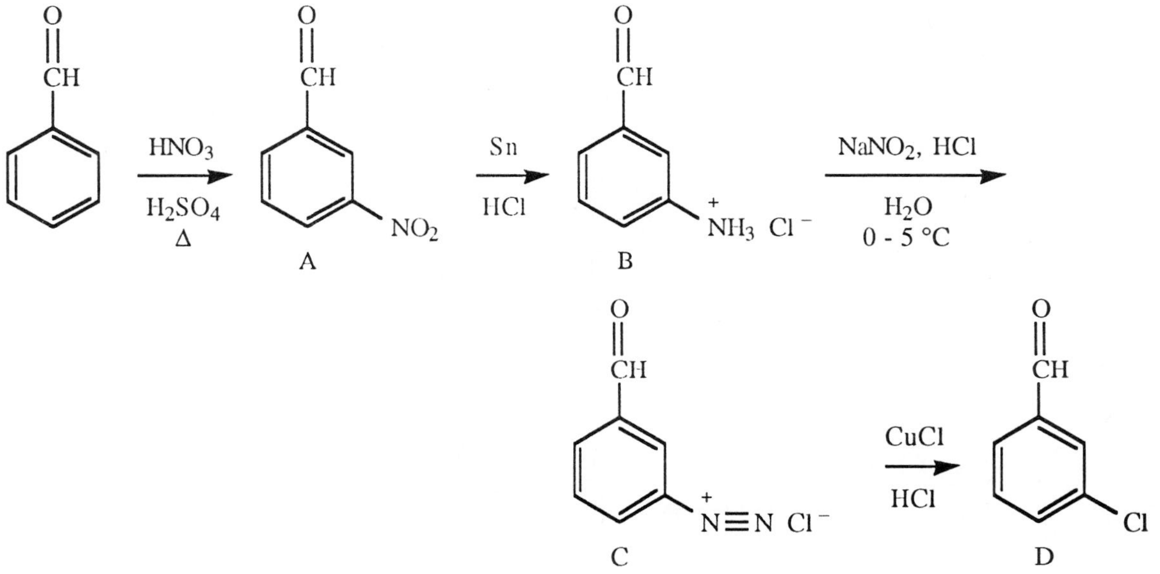

(b)

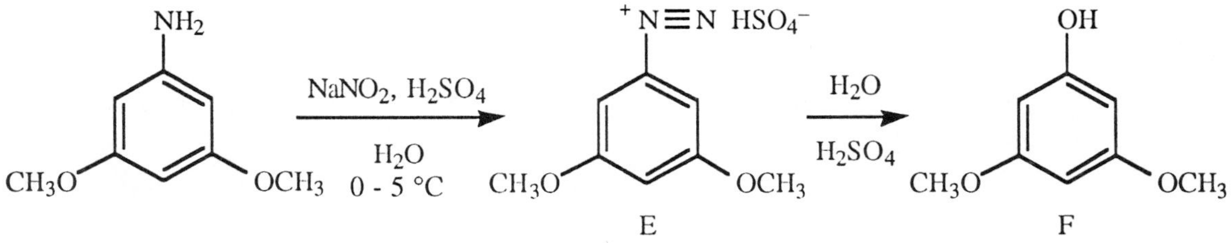

(c)

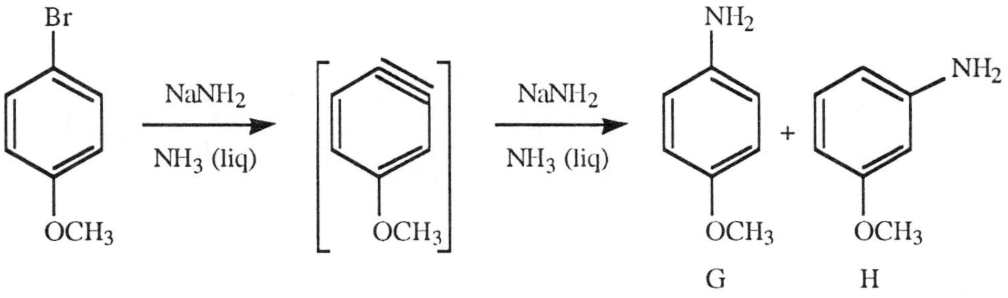

23.22 (cont)

(d)

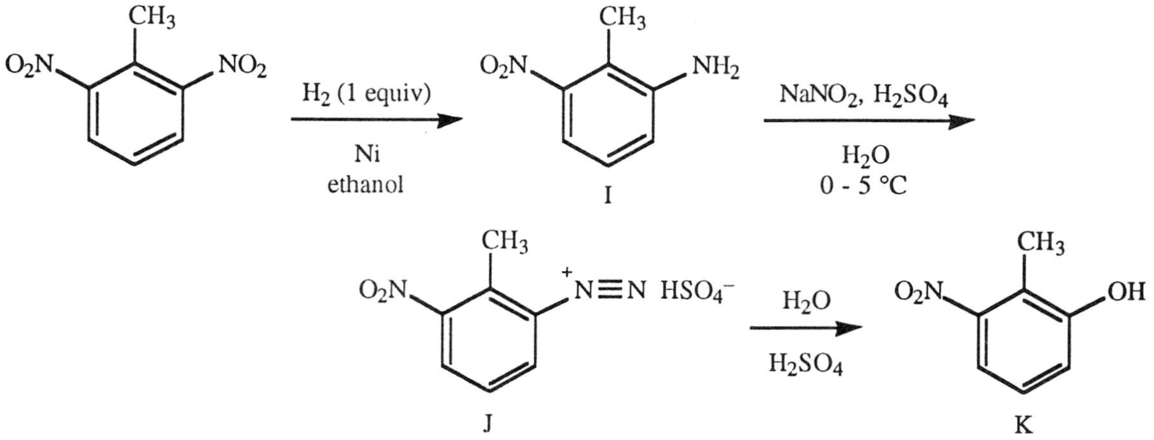

(e)

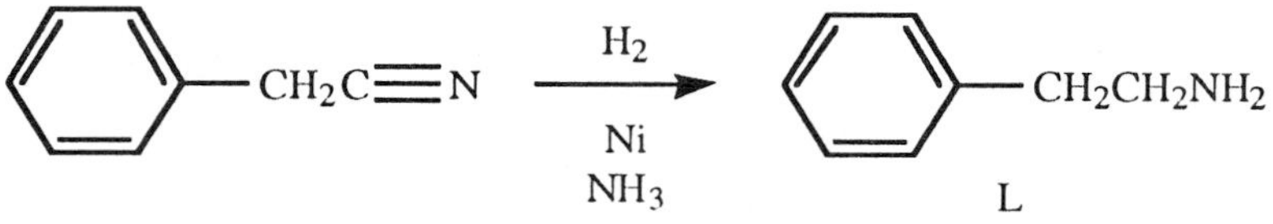

(f)

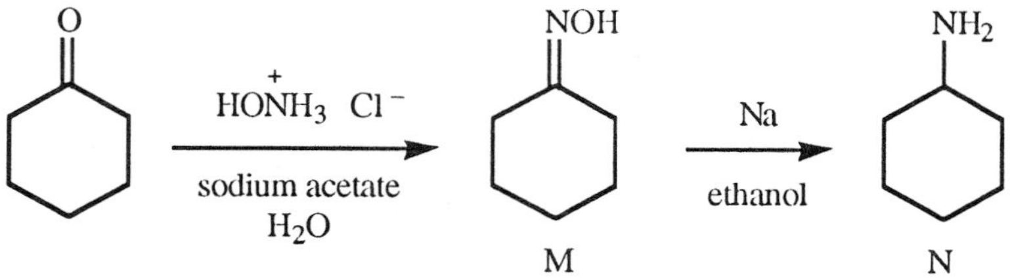

(g)

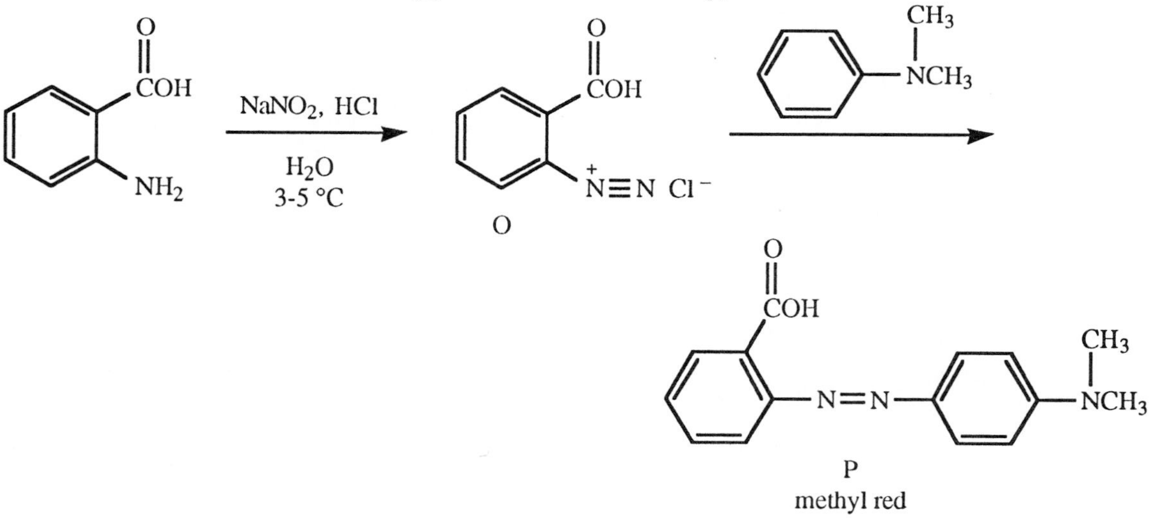

P
methyl red

(h)

Q

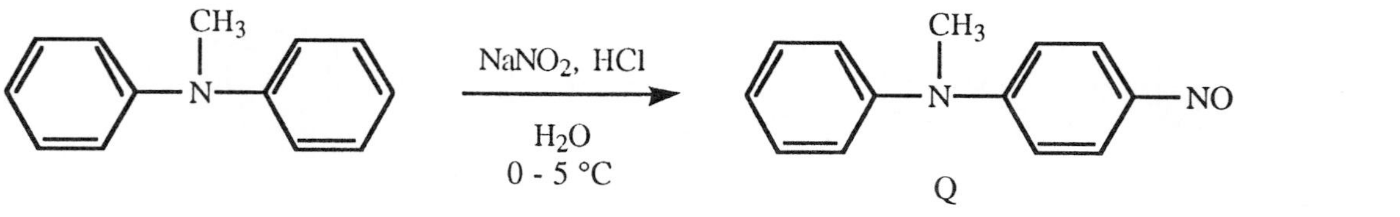

23.22 (cont)

(i)

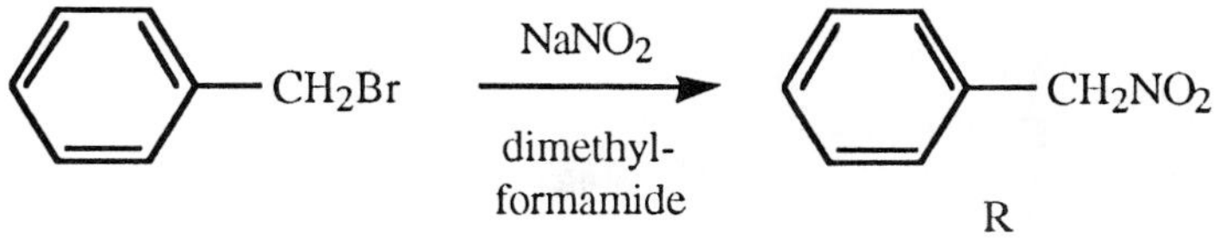

(j)

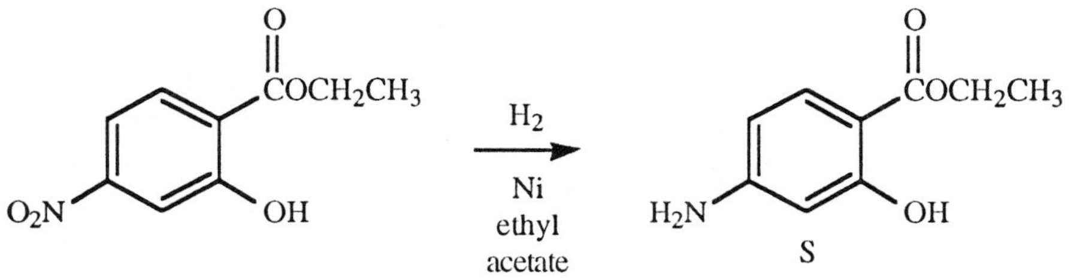

(k)

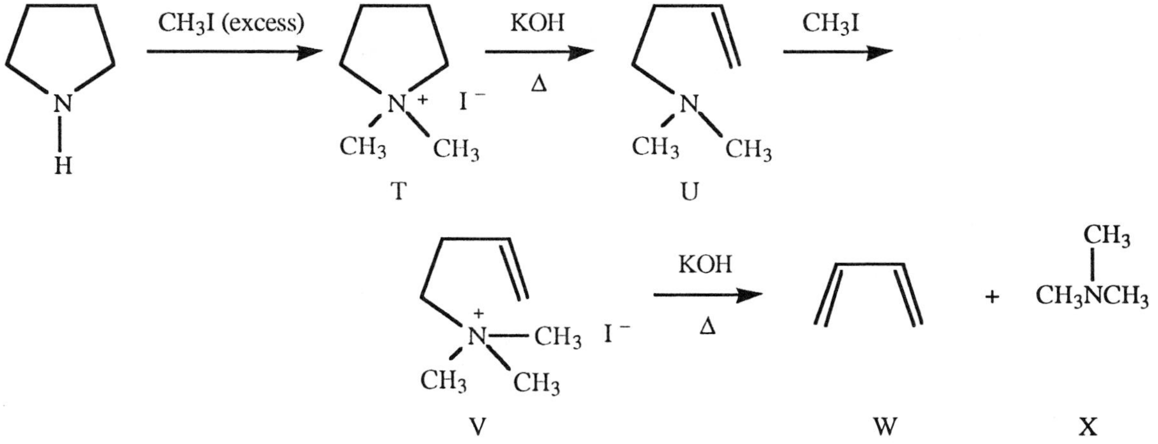

23.23

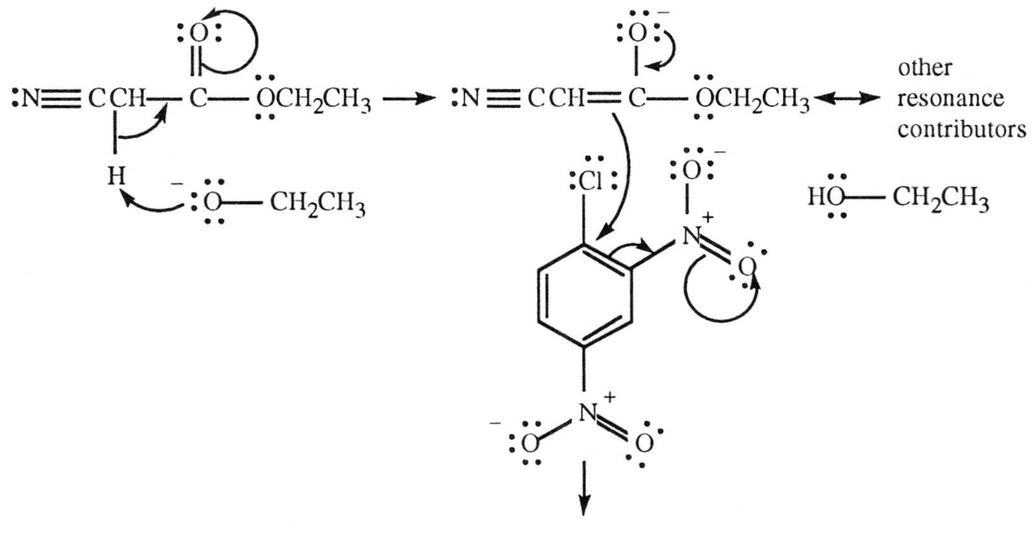

23.23 (cont)

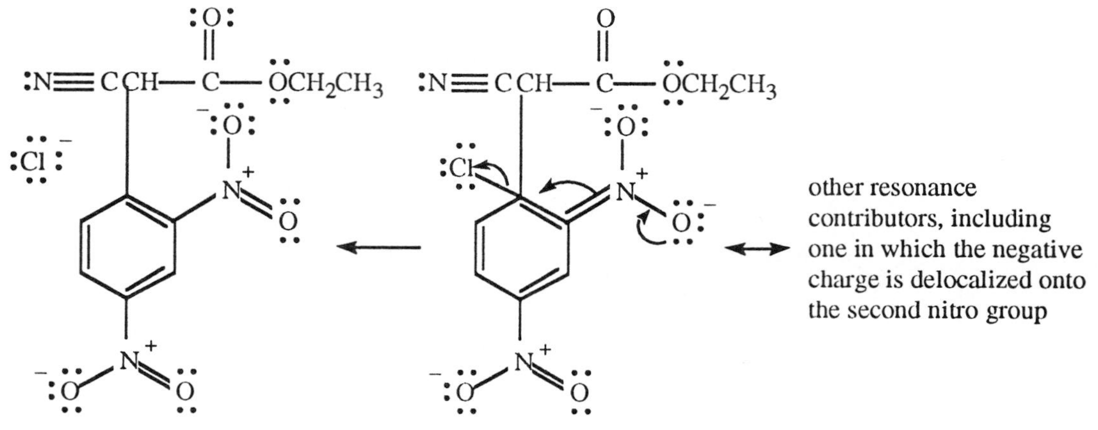
other resonance contributors, including one in which the negative charge is delocalized onto the second nitro group

23.24

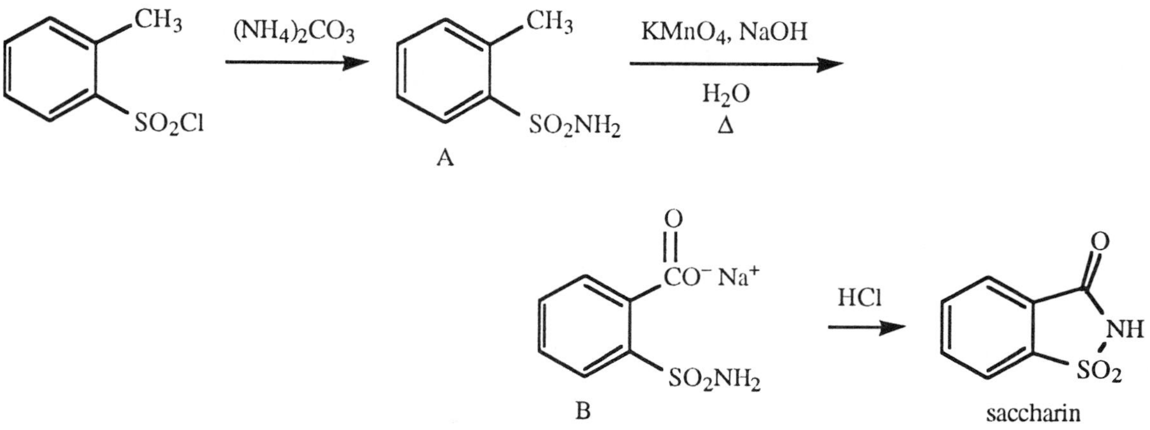
$(NH_4)_2CO_3$
A
$KMnO_4$, NaOH
H_2O
Δ
B
HCl
saccharin

23.25

(a)

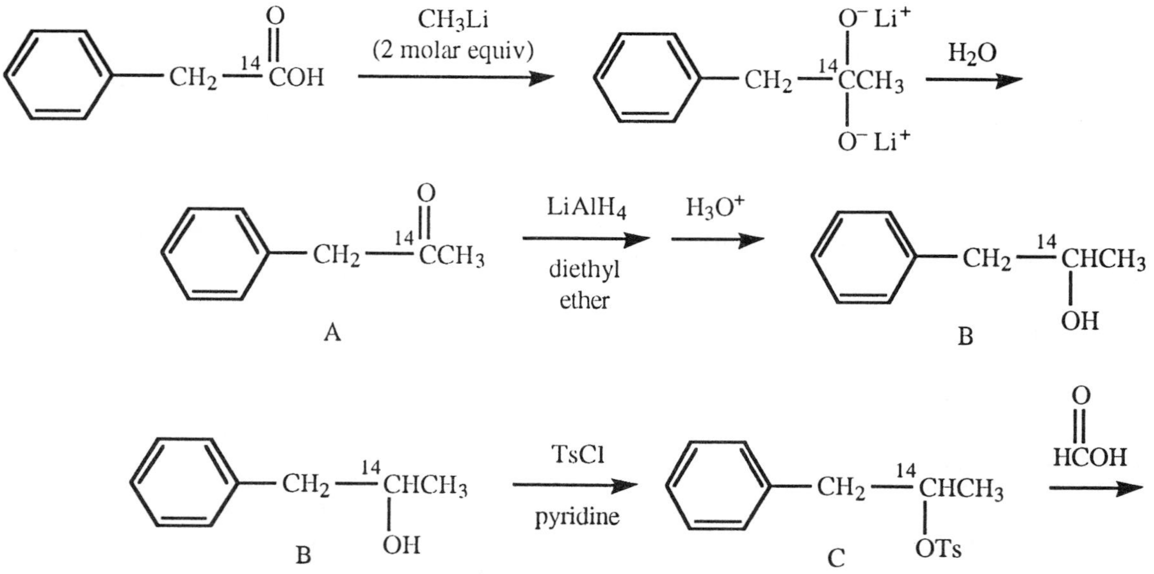
CH_3Li
(2 molar equiv)
H_2O
$LiAlH_4$
diethyl ether
H_3O^+
A
B
TsCl
pyridine
C

23.25 (a) (cont)

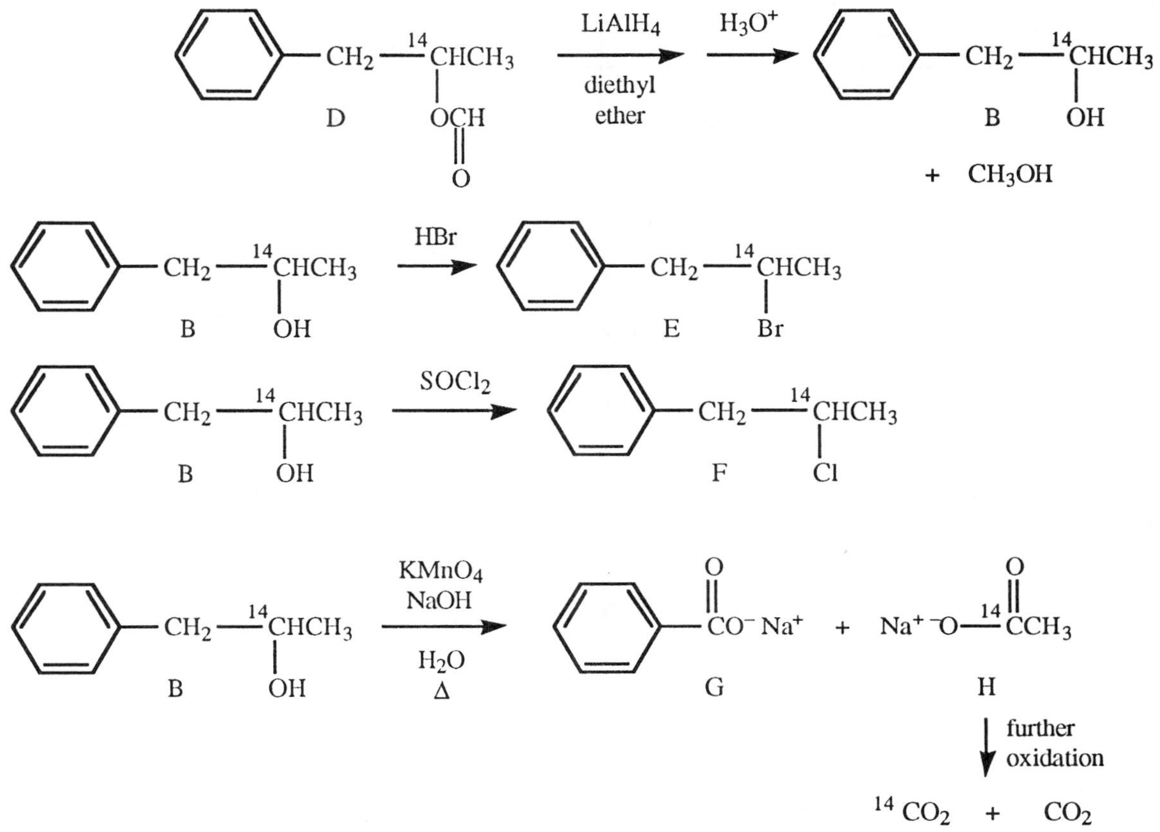

(b)

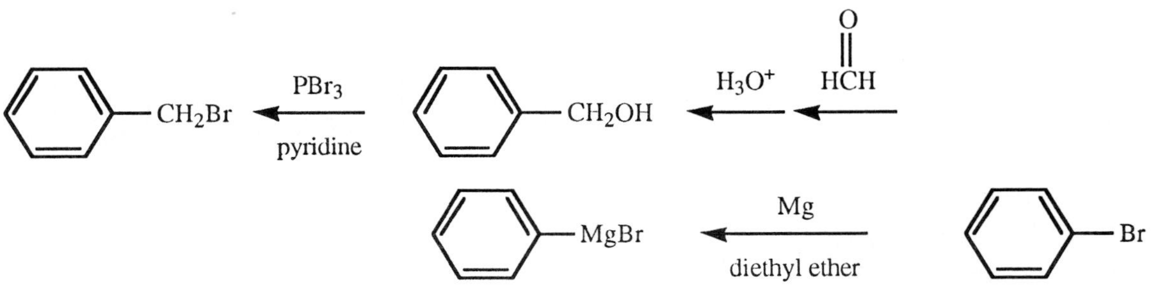

1.

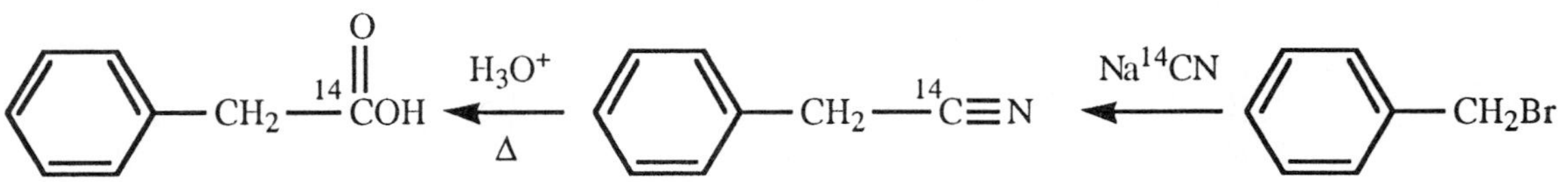

2.

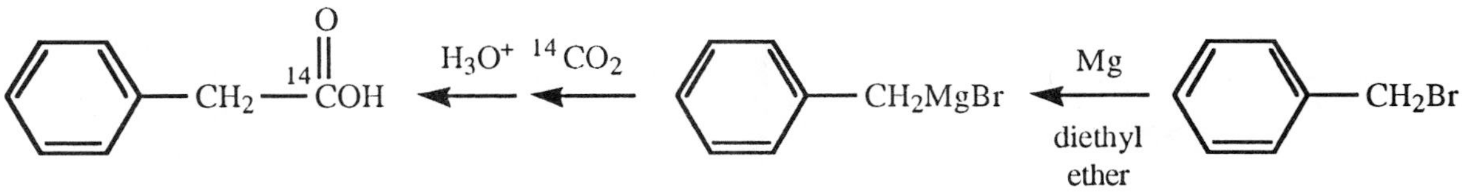

23.26

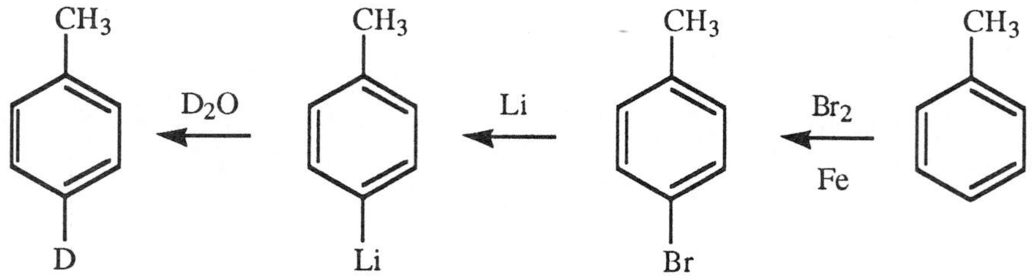

separate from
ortho isomer

23.27

most acidic   least acidic

(a)

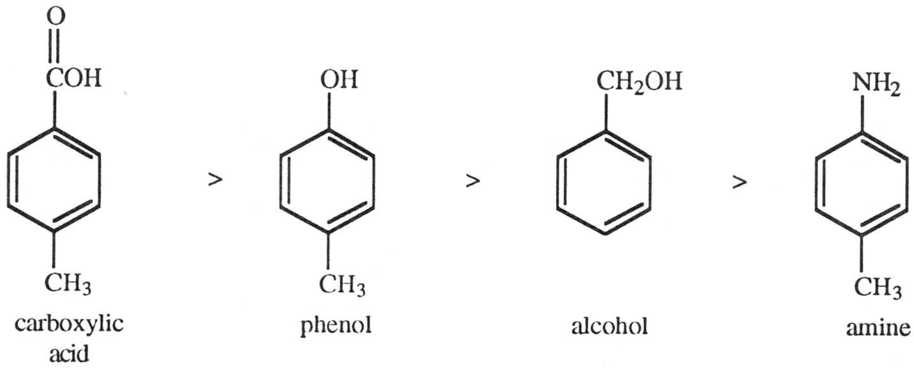

carboxylic
acid   phenol   alcohol   amine

(b)

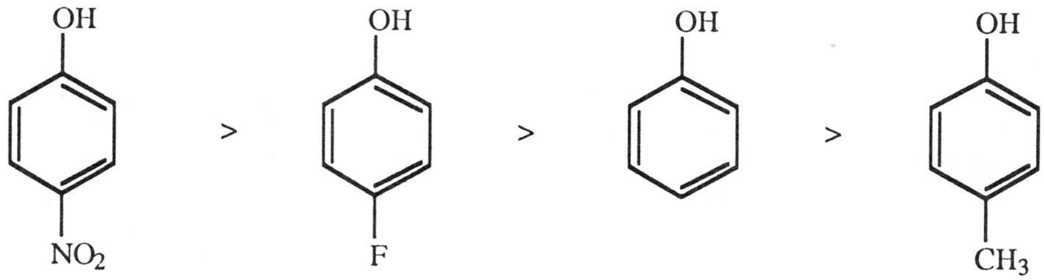

nitro group is
strongest electron-
withdrawing substituent

(c)

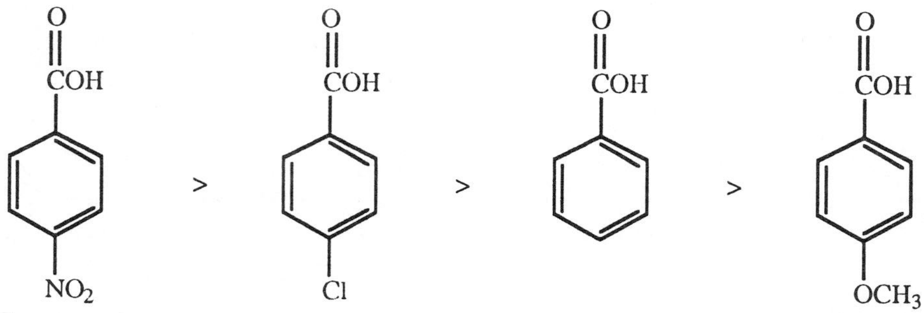

nitro group is
strongest electron-
withdrawing substituent

methoxy group
is electron-donating
in the para position

23.28

The reaction is a nucleophilic aromatic substitution reaction. The relative rates of reaction for the different nucleophiles reflect the relative nucleophilicities of the different species.

Methoxide is the best nucleophile in the series because it is by far the strongest base. The thiophenoxide anion, though a weaker base than the phenolate anion, is the better nucleophile because of the polarizability of the sulfur atom. Aniline is the weakest nucleophile. It is a weak base, meaning that the nonbonding electrons are not easily available for bond formation.

23.29

For Y: ——OCH$_3$ >   ——CH$_3$ >   ——H   >   ——Cl   >   ——NO$_2$

          <u>fastest</u>   <u>slowest</u>

The order of reactivity depends on the nucleophilicity of the nitrogen atom in the substituted aniline, which in turn is dependent on the availability of the nonbonding electrons. Electron-donating groups at the para position would increase the rate of the reaction, and electron-withdrawing groups would decrease the rate of the reaction.

The methoxyl group is strongly electron-donating by resonance.

The methyl group is weakly electron-donating by induction.

The chlorine is electron-withdrawing by induction.

23.29 (cont)

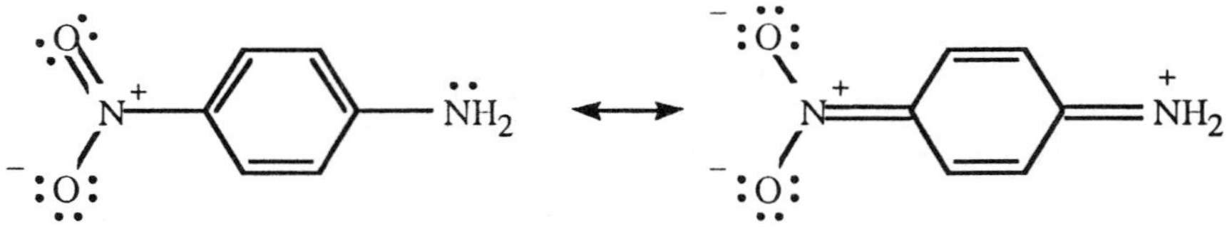

The nitro group is strongly electron-withdrawing by resonance.

23.30

(a)

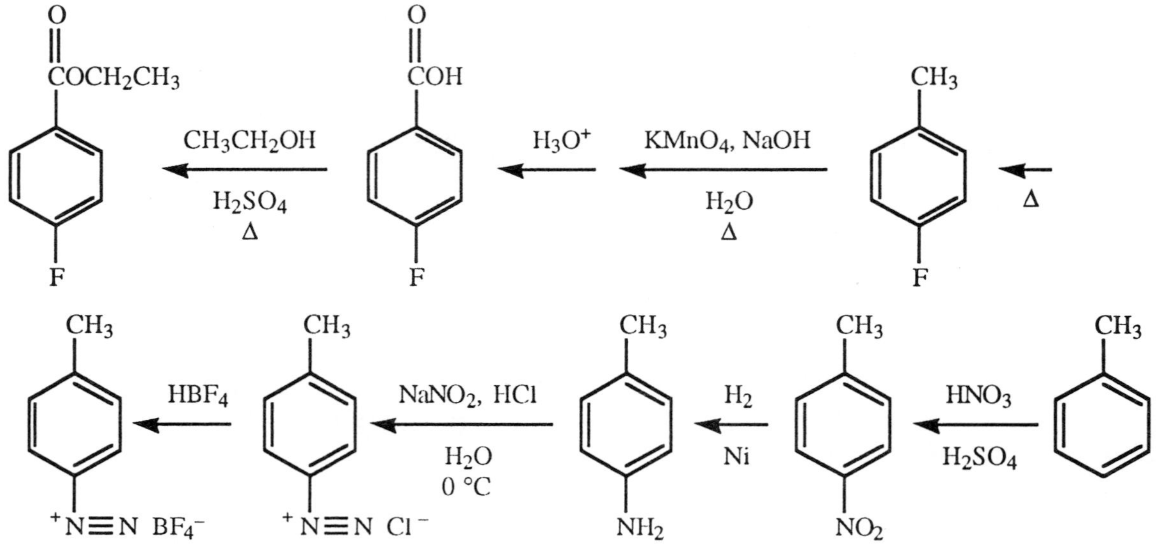

(b)

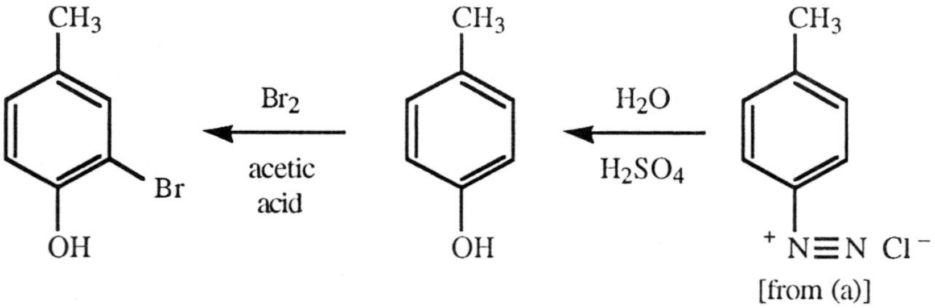

(c)

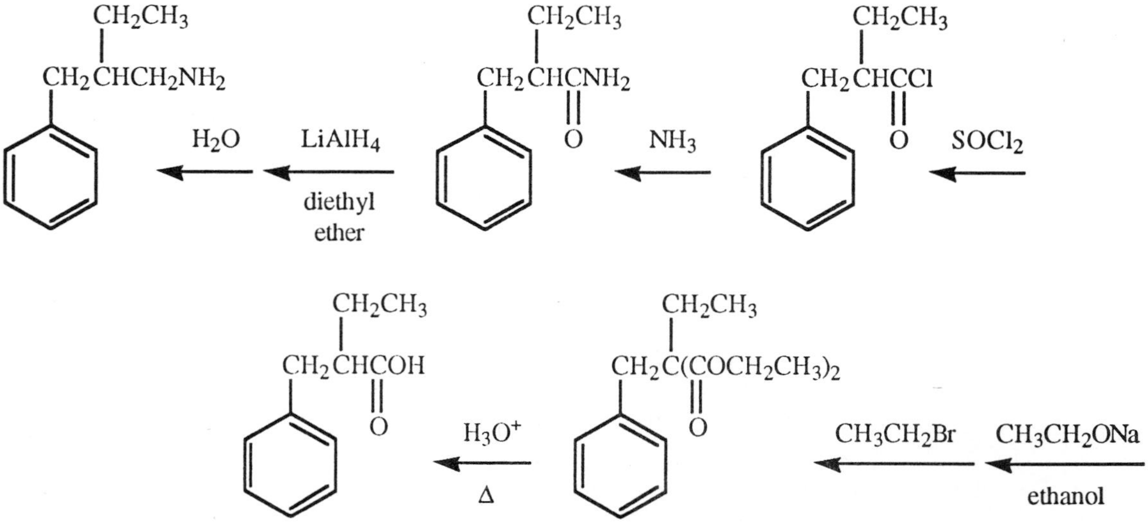

23.30 (c) (cont)

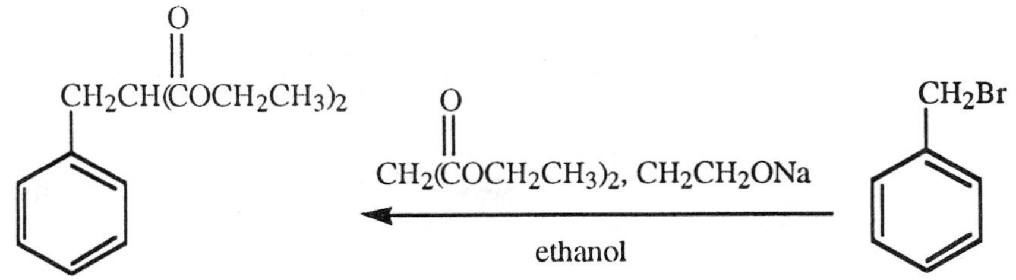

(d)

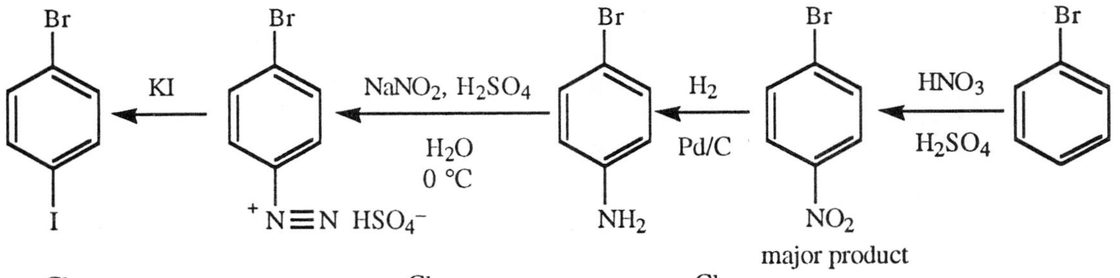

(e)

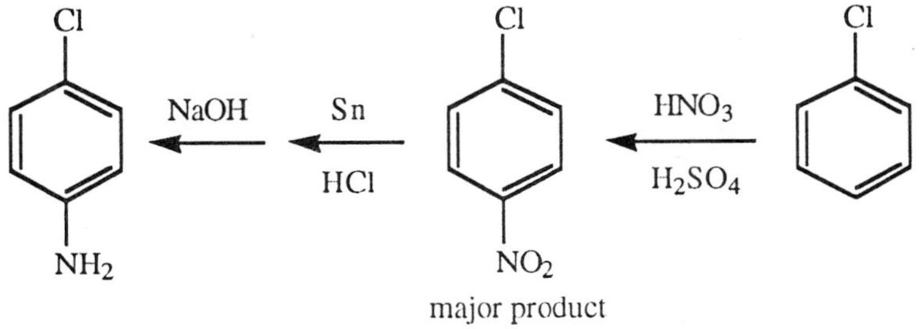

major product

(f)

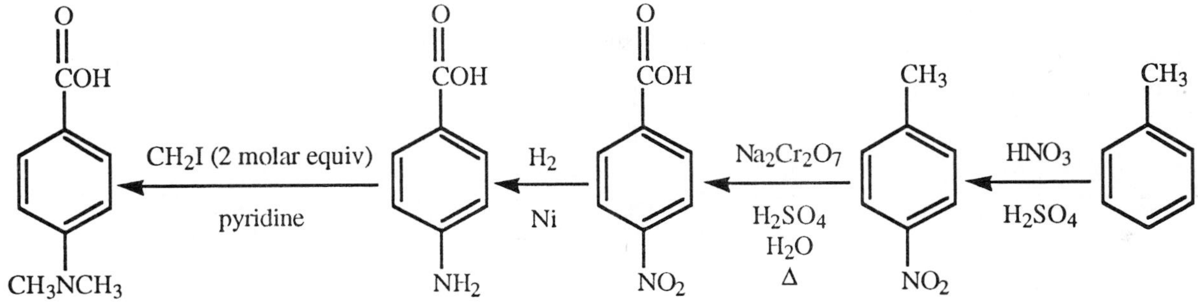

23.31

(not ⟍Cl; why?)

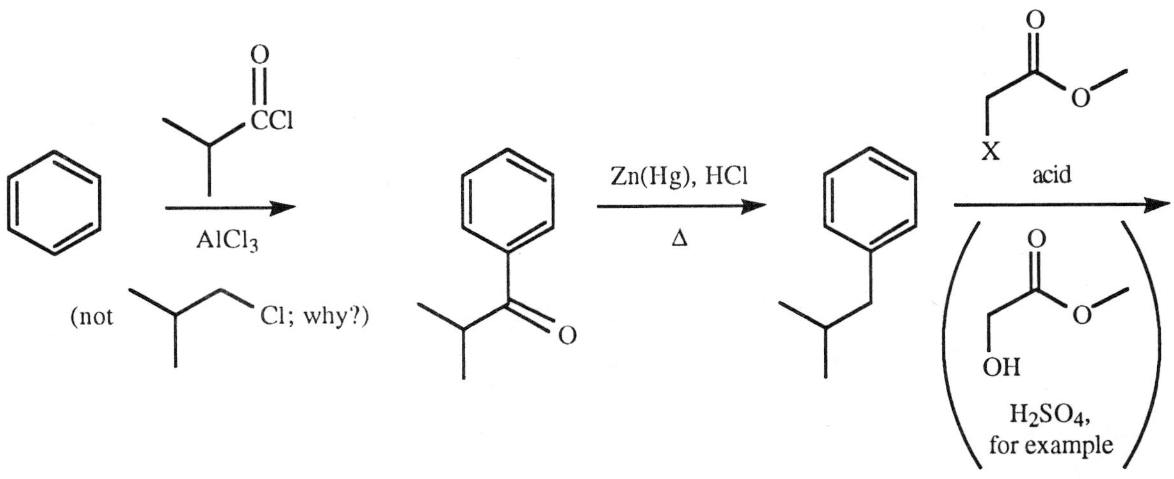

23.31 (cont)

23.32   This reaction goes by way of a tetrahedral anionic intermediate.

Delocalization of the negative charge onto the para nitro group requires that the *p* orbitals on nitrogen, oxygen and carbon be parallel to each other.

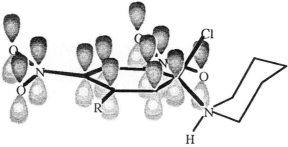

23.32 (cont)

As the R group gets larger [H > CH₃ > (CH₃)₃C], the nitro group can no longer remain in the plane of the ring, decreasing the p orbital overlap, and the stabilization of the intermediate anion. The large R groups ortho to the nitro group are said to cause **steric inhibition** of resonance, increasing the energy of activation, which in turn leads to a decrease in the rate of the reaction.

23.33   Iron has two electrons in the 4s orbital and six electrons in the 3d orbitals in its outermost shell. Iron(II) has lost the two electrons from the 4s orbital and has six electrons in the 3d orbitals. When it bonds to two cyclopentadienyl anions, each of which has six π electrons, iron has a total of eighteen electrons in its outermost shell, giving it the electronic configuration of the inert gas krypton, and leading to the stable compound ferrocene. Iron(III) has only five electrons in the 3d orbitals. Bonding with two cyclopentadienyl anions does not give iron(III) a filled shell, and, therefore, no stable compound forms.

23.34

(a)

23.34 (cont)

(b)

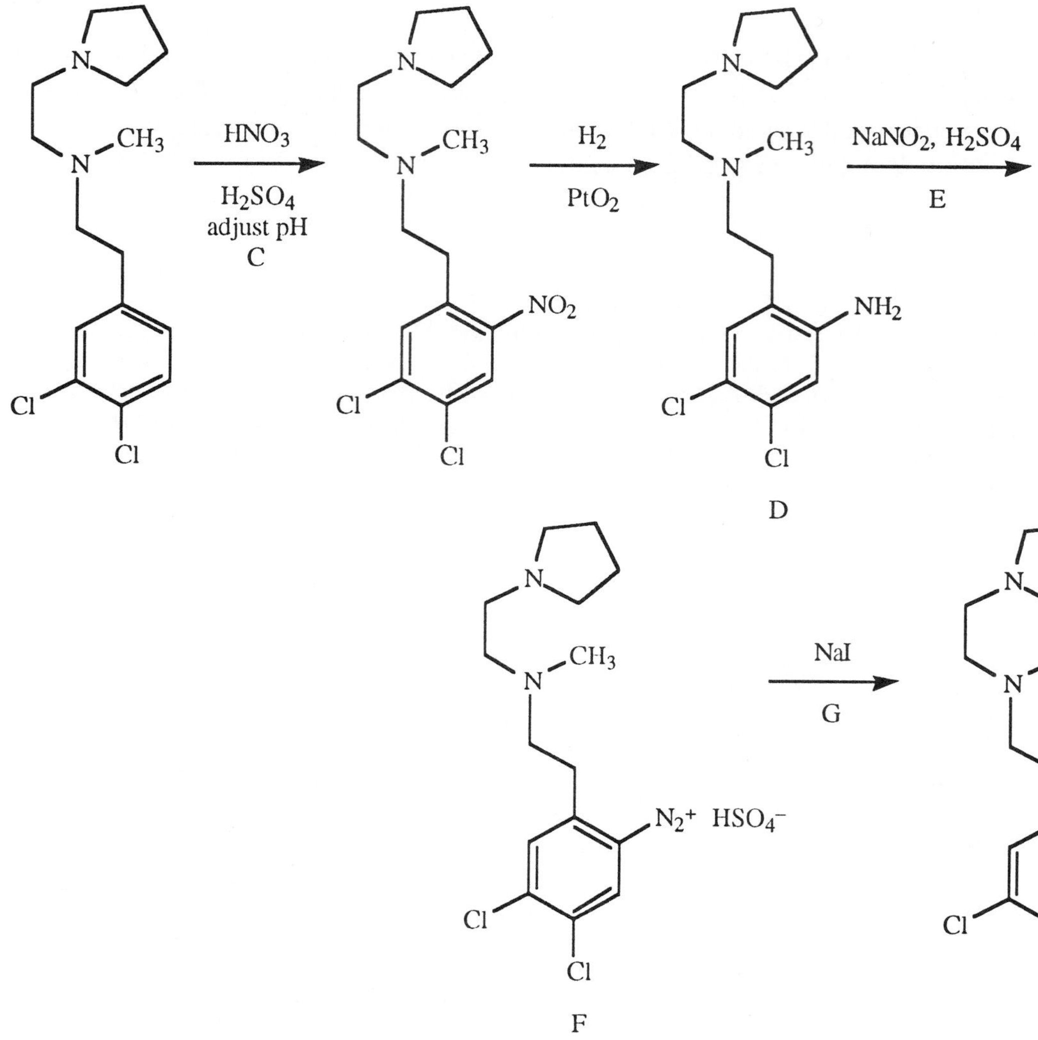

23.35

(a)

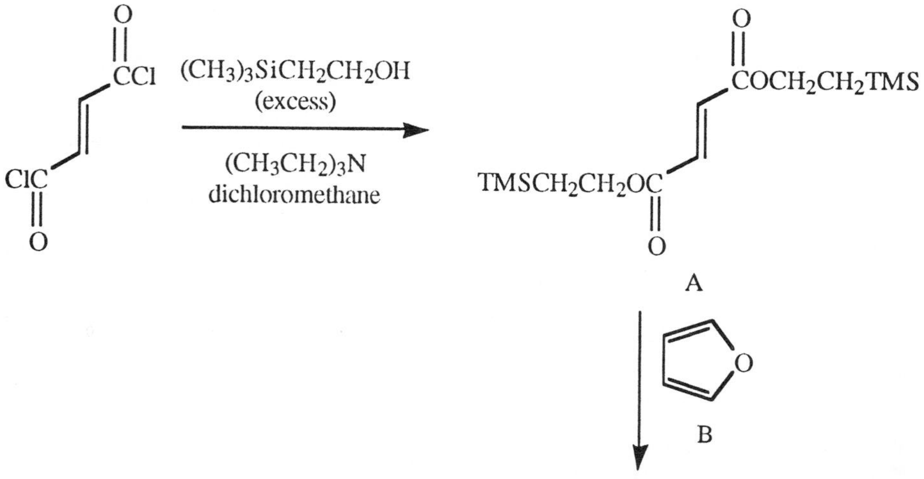

23.35 (cont)

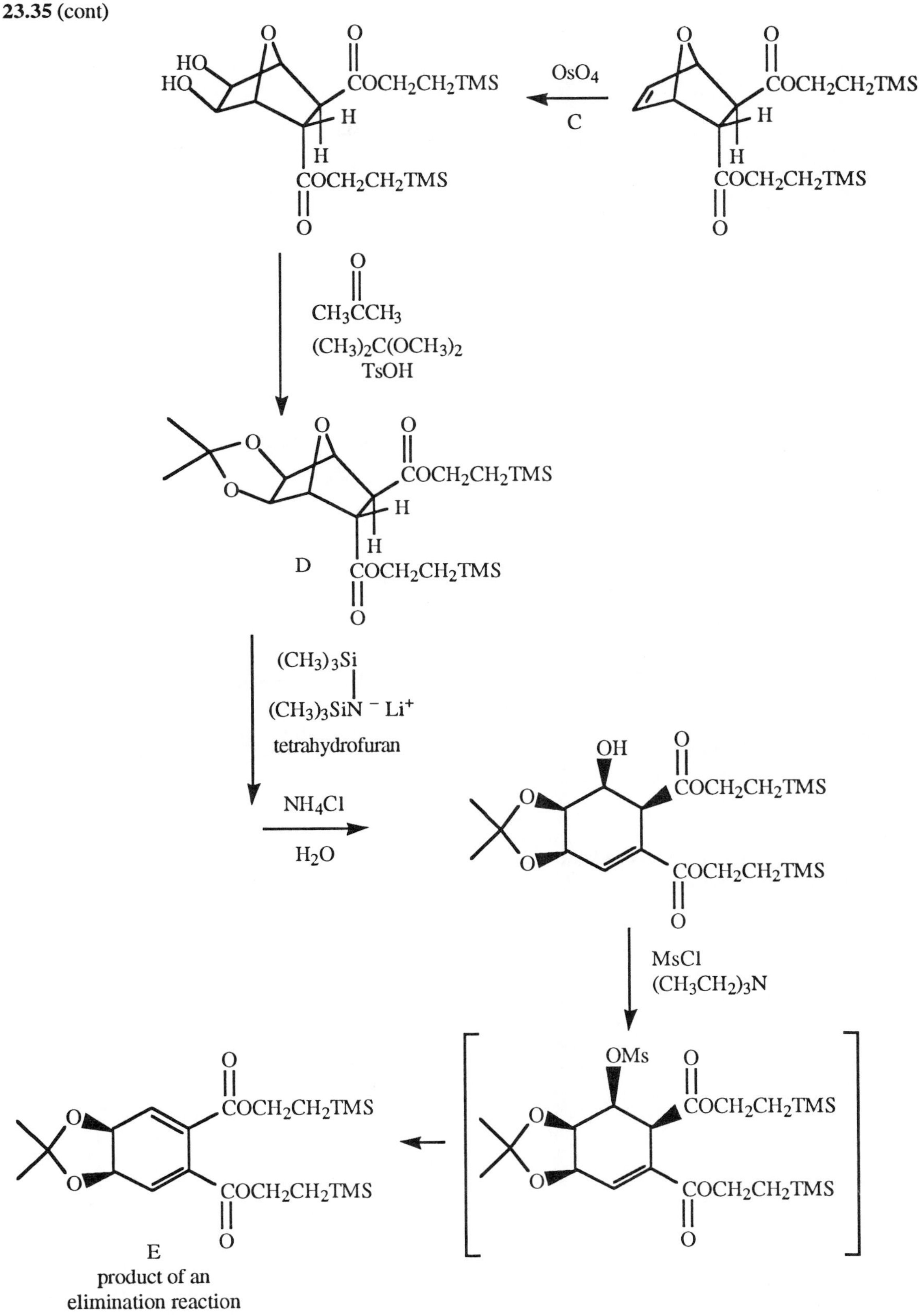

HO
HO
O
O
COCH2CH2TMS
H
H
COCH2CH2TMS
O
OsO4
C
CH3CCH3
(CH3)2C(OCH3)2
TsOH
D
(CH3)3Si
(CH3)3SiN − Li+
tetrahydrofuran
NH4Cl
H2O
OH
MsCl
(CH3CH2)3N
OMs
E
product of an
elimination reaction

23.35 (cont)

(b)

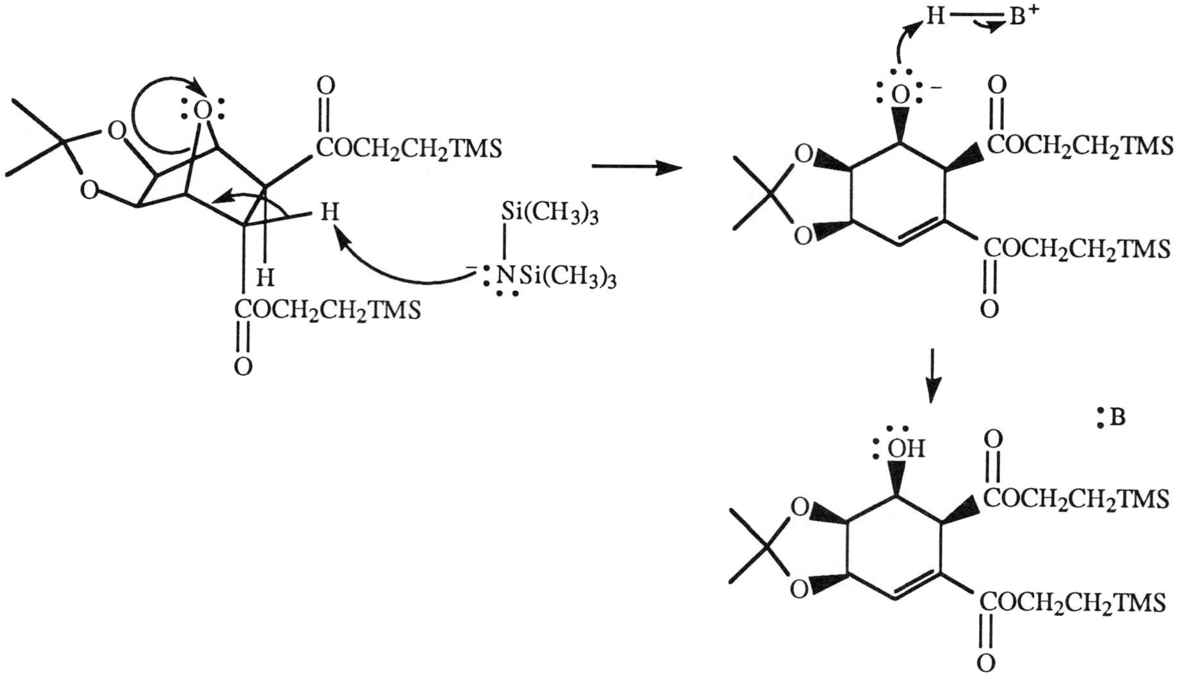

(c)

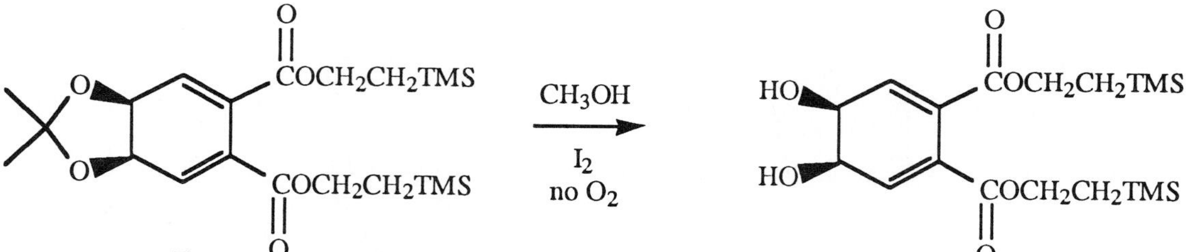

E $\xrightarrow[\substack{I_2 \\ \text{no } O_2}]{CH_3OH}$ F

The products at this stage are cyclohexadienes. They contain hydrogen atoms that are both allylic and adjacent to oxygen atoms, and, therefore, are very vulnerable to autooxidation reactions with oxygen (p. 902 in the text).

(d)

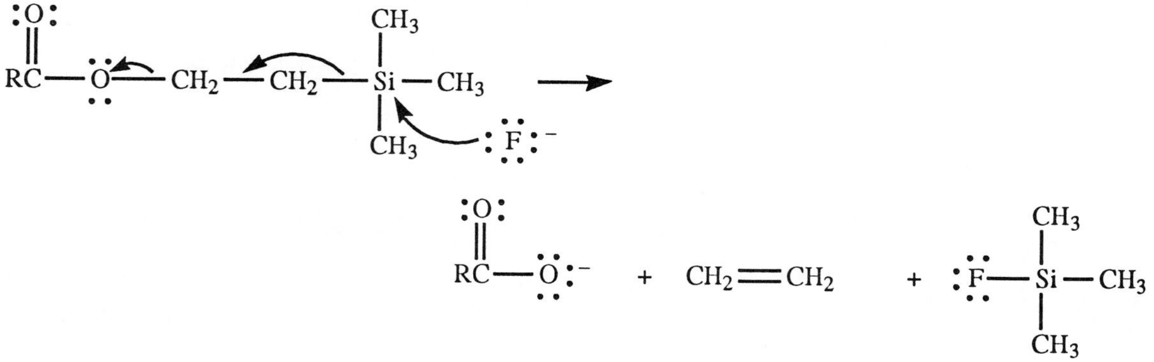

23.36

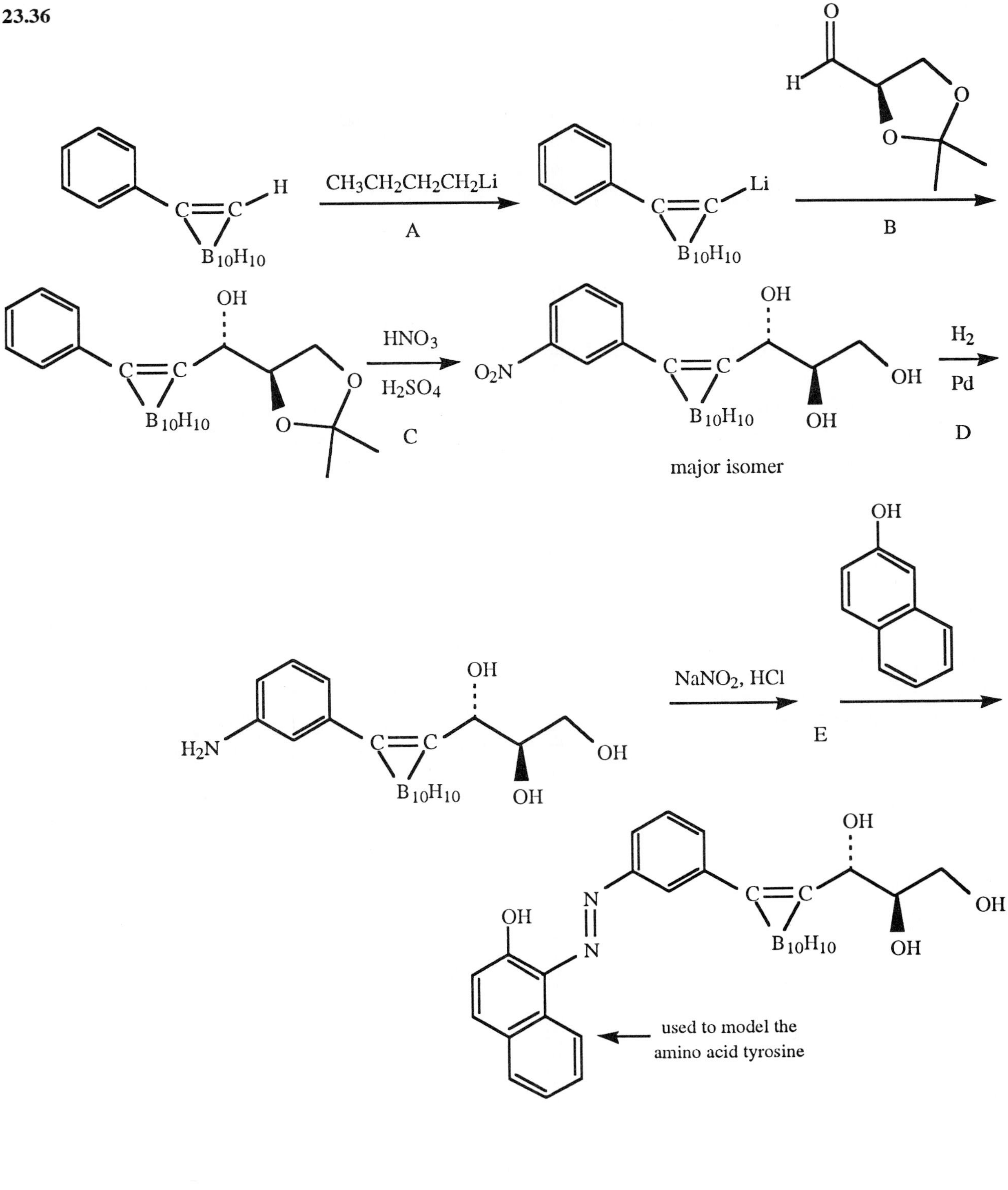

major isomer

used to model the
amino acid tyrosine

23.37   The infrared spectrum tells us that Compound A is an alcohol (3329 cm^{-1} is the O—H stretching frequency and 1046 cm^{-1} is the C—O stretching frequency) and Compound B is an ether (1245 and 1049 cm^{-1} are the C–O stretching frequencies). The proton magnetic resonance spectra are analyzed on the next page:

23.37 (cont)

Compound A:

δ 2.5	(2H, triplet, J 7 Hz, —CH$_2$C$\underline{H}_2$—)
δ 3.4	(2H, triplet, J 7 Hz, —OC$\underline{H}_2$CH$_2$—)
δ 3.7	(1H, singlet, —O$\underline{H}$)
δ 6.7	(5H, singlet, ArH)

Compound B:

δ 1.3	(3H, triplet, J 7 Hz, —CH$_2$C$\underline{H}_3$)
δ 3.7	(2H, quartet, J 7 Hz, —OC$\underline{H}_2$CH$_3$)
δ 6.3 – 7.0	(5H, multiplet, Ar$\underline{H}$)

The fragments for both Compound A and Compound B add up to $C_8H_{10}O$, which corresponds to a molecular weight of 122, which is where the molecular ion of each compound appears in the mass spectrum. We thus have the molecular formulas for Compounds A and B. Compound A is 2-phenylethanol. The base peak for Compound A, m/z 91, is the tropylium ion, which results from the following fragmentation:

Compound A
2-phenylethanol

$M \overset{+}{\underset{\cdot}{}}$
m/z 122

tropylium ion
m/z 91

From the infrared spectrum we know that Compound B is an ether. The quartet at δ 3.7, and the triplet at δ 1.3 in the proton magnetic resonance spectrum tells us that there is an isolated ethyl group in which the methylene group is attached to an oxygen atom. Compound B is ethyl phenyl ether (ethoxybenzene). The base peak in Compound B, m/z 94, is 28 mass units less than the molecular ion and corresponds to a rearrangement in which ethylene, C_2H_4, is lost.

23.37 (cont)

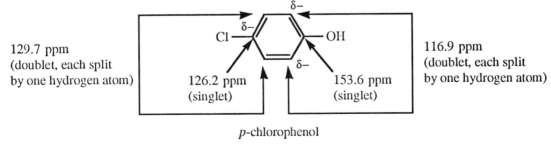

Compound B
ethyl phenyl ether
ethoxybenzene

M⁺
m/z 122

m/z 94

23.38   The increase in λ_{max} in the ultraviolet spectra in basic solution compared with the spectrum in acidic solution tells us that Compound C is a phenol. The presence of the hydroxyl group is confirmed by the broad band in the infrared spectrum at 3400 cm⁻¹.

The proton magnetic resonance data is summarized below:

δ 5.6   (1H, singlet, ArO<u>H</u>)
δ 6.1 – 6.9   (4H, para substitution pattern, Ar<u>H</u>)

The data so far tell us that Compound C is a para-substituted phenol. The C-13 nuclear magnetic resonance spectrum has only four bands instead of six in the aromatic region, suggesting that the benzene ring is symmetrically substituted, such as in para substitution.

The mass spectrum shows two molecular ions at m/z 128 and 130, with relative intensities of 3:1. The relative intensities suggest that Compound C contains chlorine. Putting together all the information gives us *p*-chlorophenol for the identity of Compound C.

The C-13 nuclear magnetic resonance spectrum is analyzed below:

129.7 ppm
(doublet, each split
by one hydrogen atom)

126.2 ppm
(singlet)

153.6 ppm
(singlet)

116.9 ppm
(doublet, each split
by one hydrogen atom)

p-chlorophenol

23.39   Integration of the bands in the proton magnetic resonance spectrum gives a ratio of 4:2:2:3. Disappearance of the broad band at δ 3.2 when Compound D is shaken with D_2O tells us that those hydrogens must be acidic protons attached to an oxygen or a nitrogen atom. The decrease in λ_{max} in the ultraviolet spectra in acid solution compared with basic solution tells us that Compound D is an aryl amine. The two bands in the infrared spectrum at 3480 and 3376 cm⁻¹ point to the presence of a primary amine, and the band at 1621 cm⁻¹ confirms the presence of the aromatic ring.

23.39 (cont)

The proton magnetic resonance data is summarized below:

δ 1.1	(3H, triplet, *J* 7 Hz, —CH$_2$CH$_3$)
δ 2.3	(2H, quartet, *J* 7 Hz, ArCH$_2$CH$_3$)
δ 3.2	(2H, broad singlet, —NH$_2$)
δ 6.0 – 6.7	(4H, ArH)

The data so far tell us that Compound D is an ethylaniline. The aromatic region in the proton magnetic resonance spectrum does not show the typical splitting pattern for para substitution. Therefore, Compound D must be either ortho- or meta-substituted. The C-13 nuclear magnetic resonance spectrum has eight bands, which also tells us that Compound D cannot be *p*-ethylaniline, which should have only six bands in the C-13 spectrum because of the symmetry of the molecule. The data, however, do not allow us to distinguish between *o*-ethylaniline and *m*-ethylaniline.

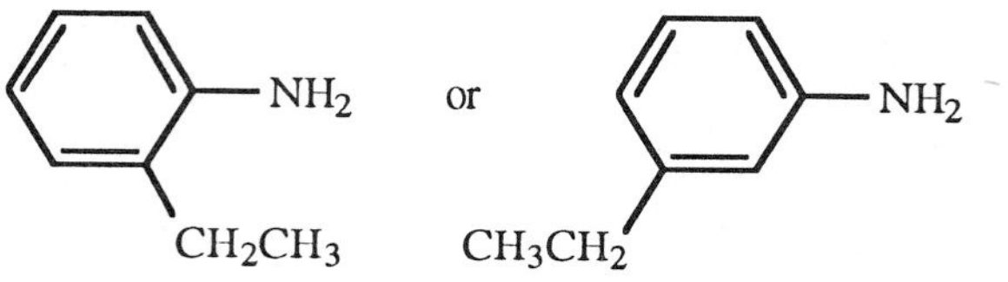

23.40 The absorption bands reported for the first compound are comparable to those for an alkylbenzene (compare with the spectrum of toluene, Figure 23.3). This suggests that the double bond in this compound is not conjugated with the aromatic ring. The first compound is 3-phenylpropene. In the spectrum of the second compound, the absorption bands are shifted to longer wavelengths and are also more intense, suggesting that this compound has a more extensive system of conjugation than the first one. The second compound is, therefore, 1-phenylpropene.

3-phenylpropene

no conjugation of the
double bond with the ring

1-phenylpropene

double bond in con-
jugation with the ring

23.41 We get the following information from the spectral data:

Mass spectrum: 756 – 720 = 36 amu, the gain in molecular weight of C$_{60}$. Therefore, 36 hydrogen atoms were added to C$_{60}$.

Proton NMR spectrum: The new compound contains hydrogen atoms. The chemical shifts indicate that there are two types of hydrogens. The broad multiplets indicate that they are coupled.

IR spectrum: The infrared spectrum confirms the presence of carbon-hydrogen bonds with bands at 2925 and 2855 cm^{-1}. Bands at 1620 and 675 cm^{-1} indicate the presence of double bonds.

UV spectrum: The ultraviolet spectrum of C$_{60}$ indicates that there is extensive conjugation of the double bonds. The new compound has lost that conjugation. The same conclusion can be drawn from the loss of the deep color of the C$_{60}$ solution during the reaction.

23.41 (cont)

All of the data suggest that 36 hydrogen atoms have added to C_{60}. They have done so in a way that breaks up the conjugation of the double bonds in C_{60}, leaving isolated double bonds in the compound, $C_{60}H_{36}$, that results. A look at the fragment of C_{60} shown on p. 839 in the text gives an idea of how this could happen.

23.42 (a) The highest wavelength of maximum absorption for Vitamin B_6 shifts from 292 nm to 315 nm when the solution is made basic. This change suggests that the vitamin contains a phenolic hydroxyl group. When the acidic proton of the hydroxyl group is replaced by a methyl group, the compound no longer ionizes at high pH and the absorption spectrum does not change with pH.

(b) The structure of Vitamin B_6 can be derived from its molecular formula, $C_6H_{11}NO_3$, and from the structures of the oxidation products of its methyl ether.

dicarboxylic acid
$C_9H_9NO_5$

precursor of
the lactone
$C_9H_{11}NO_4$

lactone

oxidation

Vitamin B_6
$C_8H_{11}NO_3$

methyl ether
of Vitamin B_6
$C_9H_{13}NO_3$

23.42 (b) (cont)

The side chains that are oxidized are the ones already bearing a hydroxyl group. Oxidation of an alcohol is easier than oxidation of an alkyl group. The reaction conditions, potassium permanganate at room temperature, are not vigorous enough to oxidize the alkyl group. The equations outlining the transformation of Vitamin B$_6$ are given on the previous page.

(c) Possible models for Vitamin B$_6$ should have a pyridine ring with a hydroxyl substituent. Alkyl substituents at different positions on the ring would also be helpful. Ultraviolet spectra of the model compounds should be taken under both acidic and basic conditions. Structures of some possible models are:

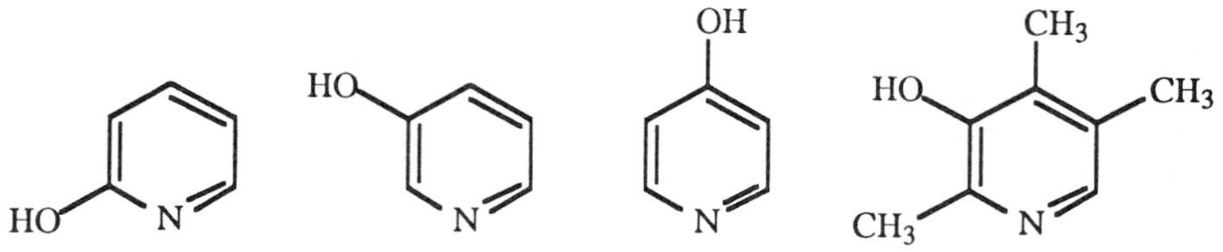

23.43

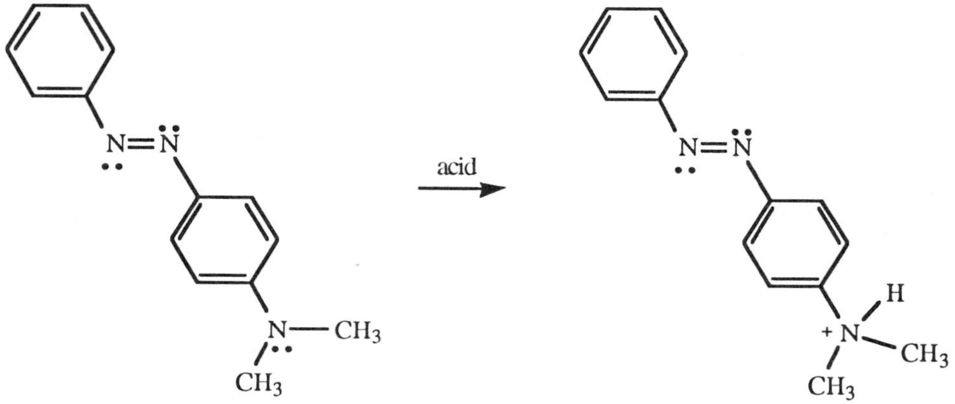

butter yellow
λ$_{max}$ 408 nm

acid

λ$_{max}$ 320 nm

Protonation of the nonbonding electrons on the amino group decreases the conjugation in the chromophore and, therefore, the wavelength of maximum absorption.

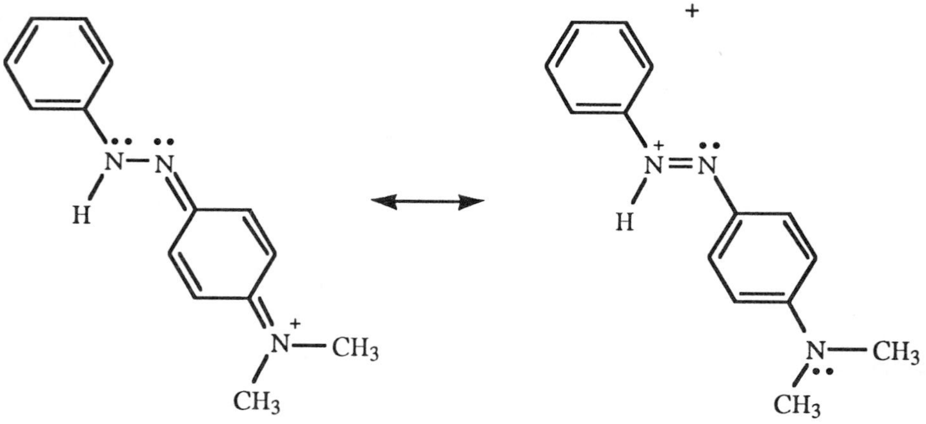

The chromophore in this species has quinoid character.

λ$_{max}$ 510 nm

24

The Chemistry of Heterocyclic Compounds

24.1

(a) 3-aminopyridine

(b) 2,5-dimethylthiophene

(c) ethyl 1-pyrrolecarboxylate

(d) 4-ethylisoquinoline

(e) 3-methyl-2-isoxazoline
(Note that the number 2 defines
the position of the double bond.)

(f) 1-ethylpyridine

(g) 4,4-dimethylazetidinone

(h) 3-indolecarboxylic acid

(i) 3-methylfuran

(j) 8-hydroxyquinoline

(k) 1,4-dimethylpyrazole

(l) 3-thiophenecarboxamide

(m) 2,3-diphenyloxazole

24.2

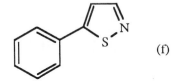

867

Concept Map 24.1   Classification of cyclic compounds.

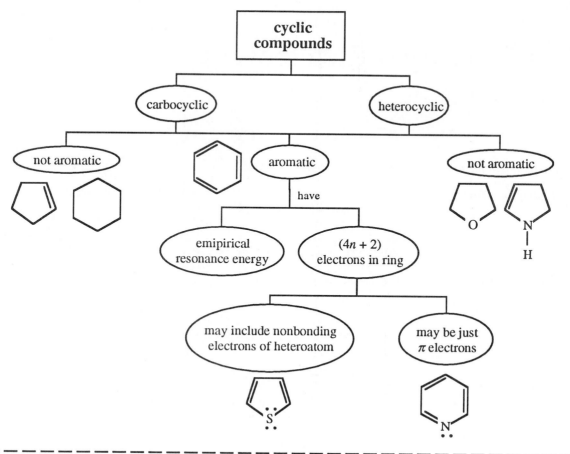

- -

24.3

24.4

no resonance stabilization;
localized charge

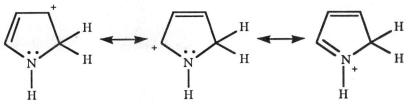

delocalization of charge by resonance

24.5 (a) The basicity of an amine is determined by the availability of the nonbonding electrons on the nitrogen atom. The pK_a values for their conjugate acids indicate that 2-methylpyridine is a slightly stronger base than pyridine, and 3-nitropyridine is a much weaker base than pyridine. The methyl group is weakly electron-donating; the nitro group is strongly electron-withdrawing.

pK_a of conjugate acid: 6.0   5.2
                          strongest base;
                          inductive effect of
                          the methyl group

pK_a of conjugate acid: 8.0

                                     weakest base;
                                     inductive and
                                     resonance effect of
                                     the nitro group

(b) The arguments used in part (a) of this problem can be applied to the compounds in this series also.

pK_a of conjugate acid: 2.0   1.3
                          strongest base;
                          inductive effect of
                          the methyl group

24.5 (b) (cont)

pK$_a$ of conjugate acid: −0.68

weakest base;
inductive and
resonance effect of
the carbonyl group

(c) The cyano group withdraws electrons from the ring and from the nitrogen atom of the substituted pyridines. Both inductive and resonance effects operate. The basicity of the pyridines appears to depend on the distance of the cyano group from the nitrogen atom in the ring.

pK$_a$ of conjugate acid: 1.9   1.5   −0.3

strongest base;   weakest base;
cyano group farthest   cyano group closest
from the nitrogen   to the nitrogen
atom in the ring   atom in the ring

24.6   (a) $CH_3CCH_2CH_2CCH_3$ $\xrightarrow[\Delta]{P_4S_{10}}$

(b) $HCCH_2CH_2CH$ $\xrightarrow{HCl}$

(c) $CH_3CCH_2CH_2CCH_3$ $\xrightarrow[\Delta]{P_4Se_{10}}$

selenium is in the same
family as oxygen and sulfur

(d) $CH_3CCH_2CH_2CCH_3$ $\xrightarrow[\Delta]{CH_3NH_2}$

24.7

24.8

(a)

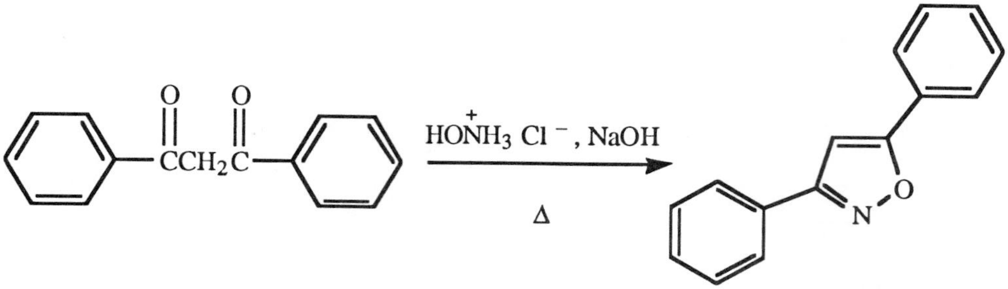

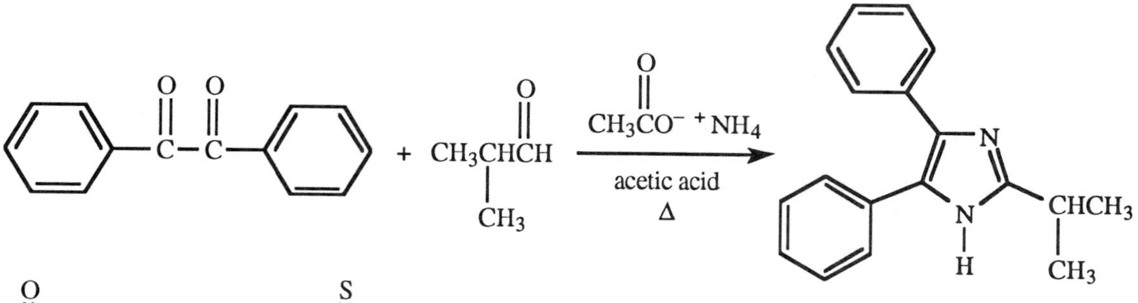

(b)

(c)

(d)

less hindered
carbonyl group

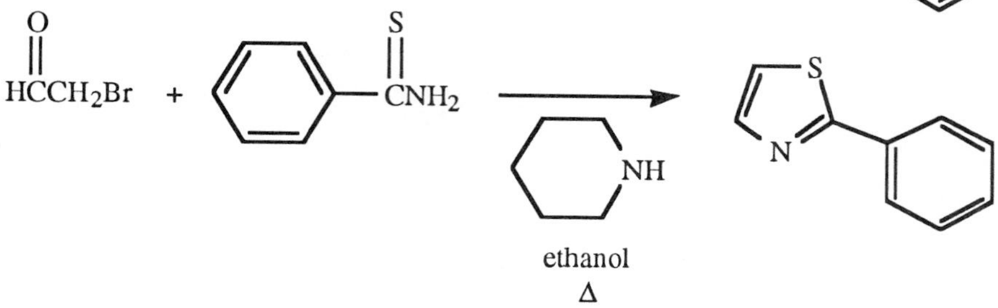

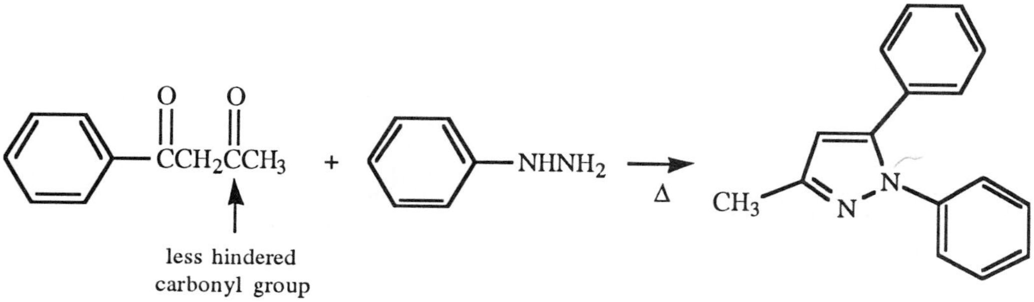

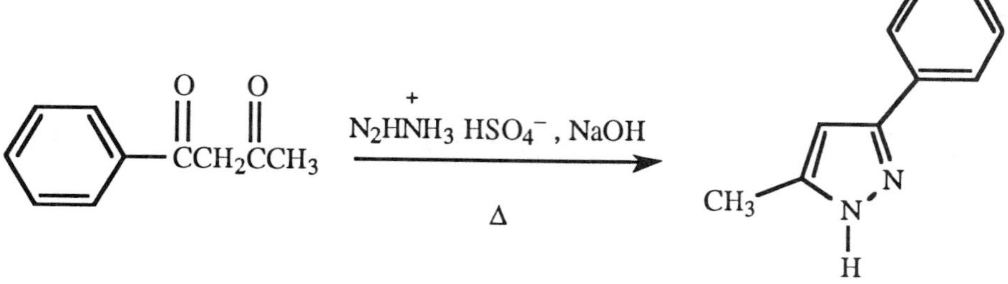

(e)

24.9

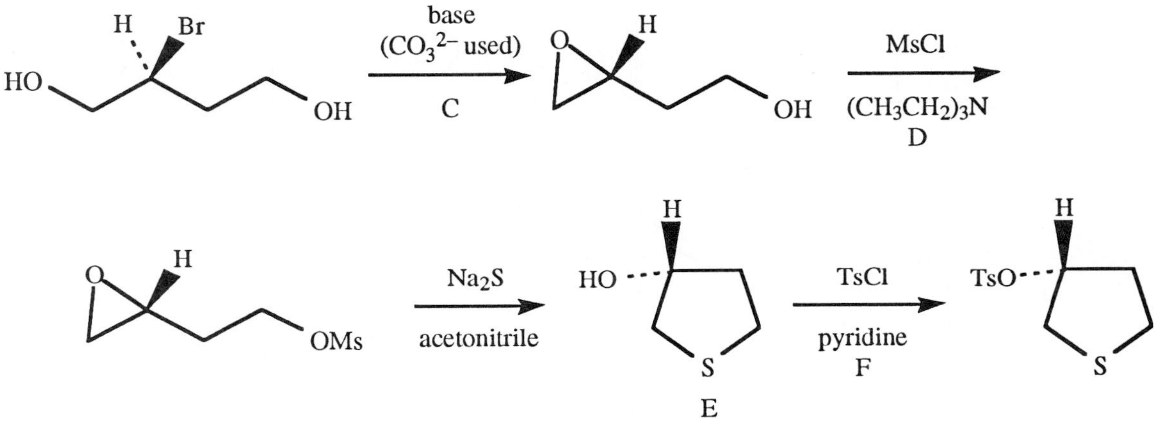

(a)

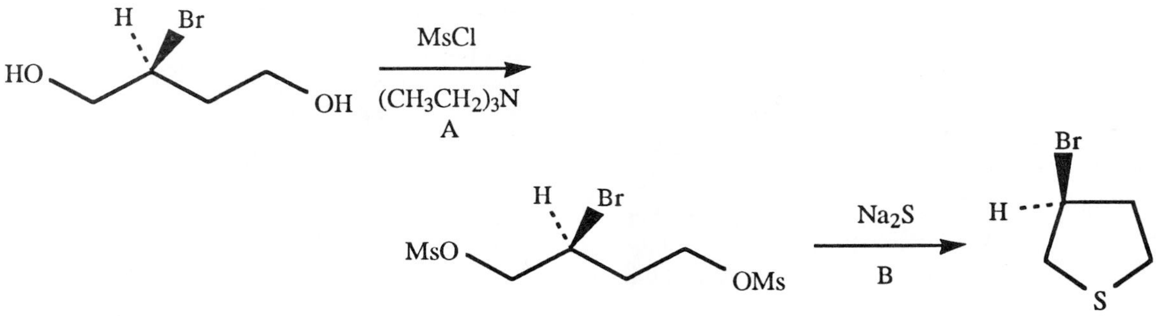

(b)

(c)

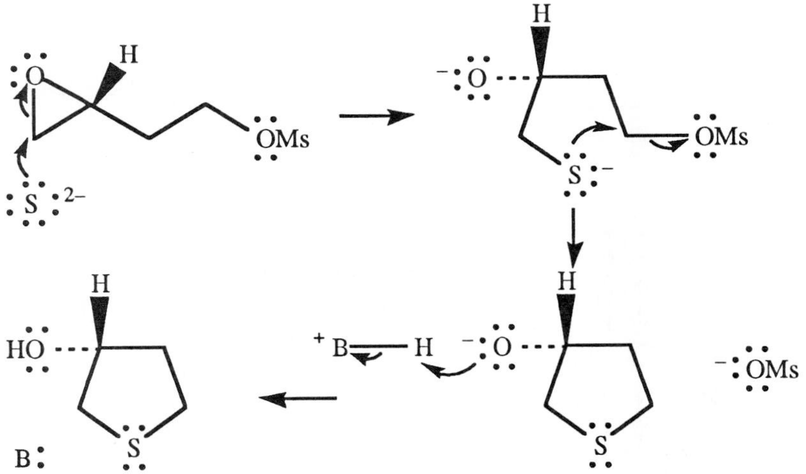

Concept Map 24.2   Synthesis of heterocycles from carbonyl compounds.

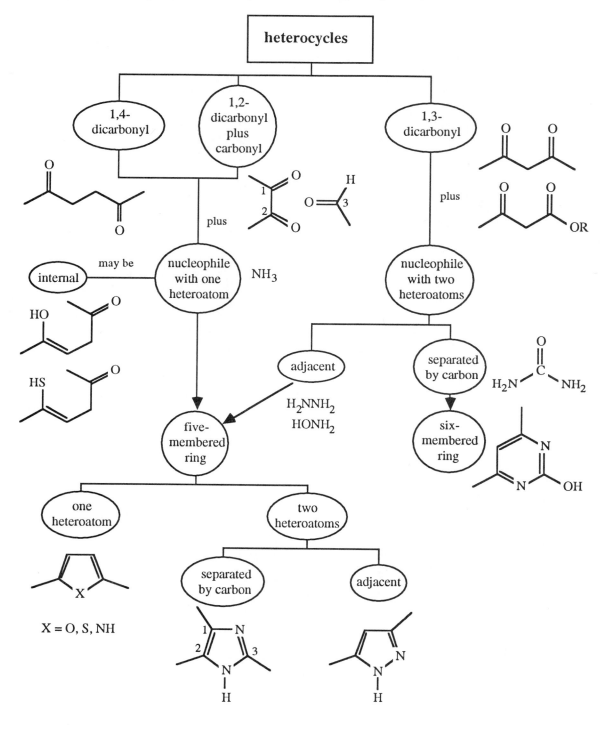

24.10

(a)

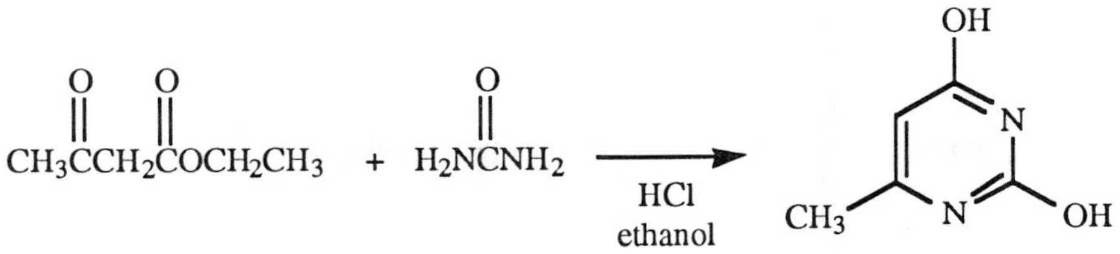

(b)

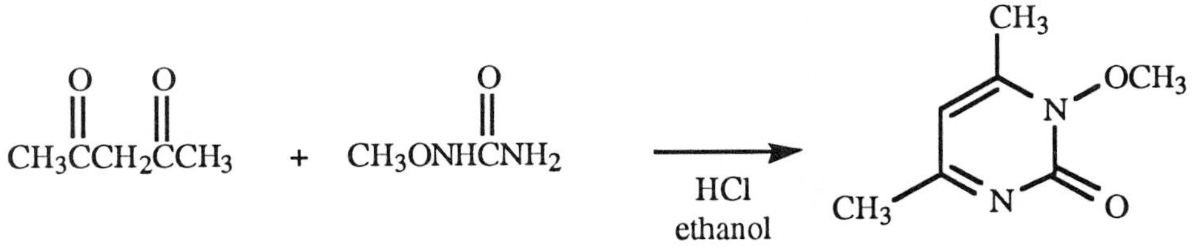

(c)

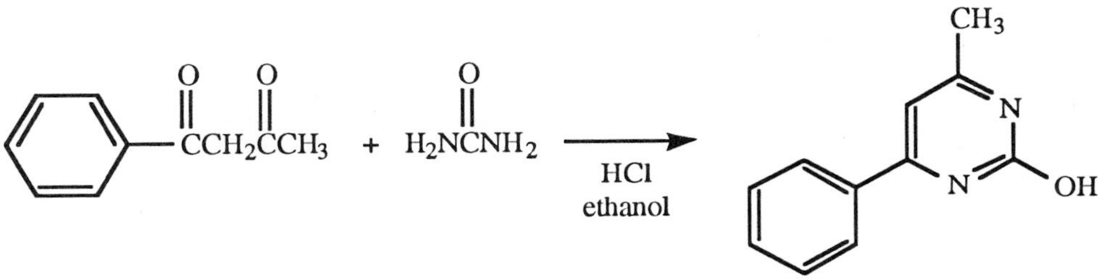

24.11

(a)

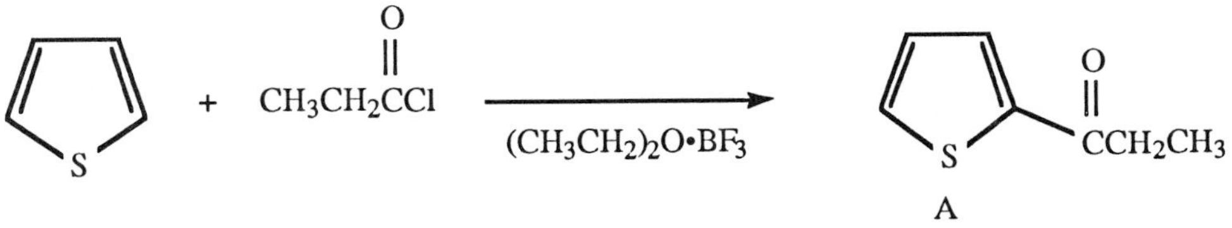

(b)

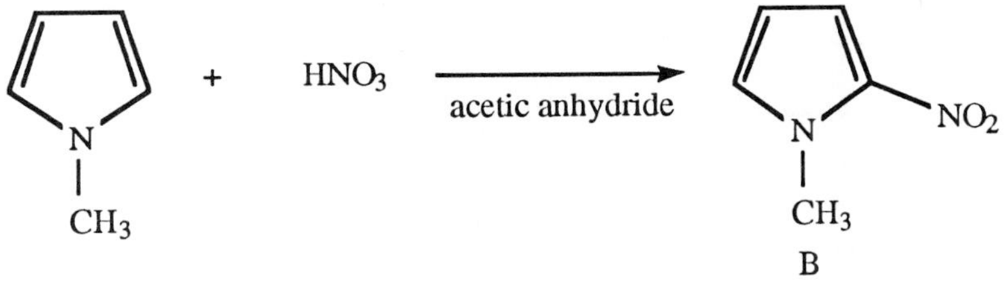

(c)

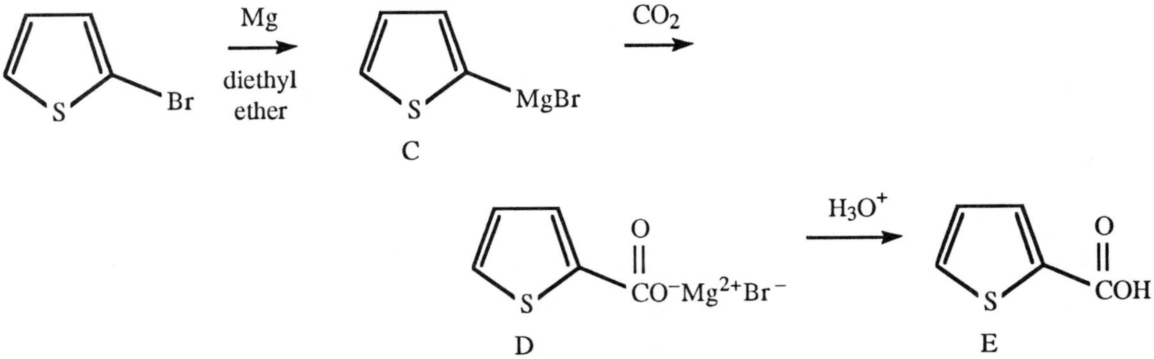

24.11 (cont)

(d)

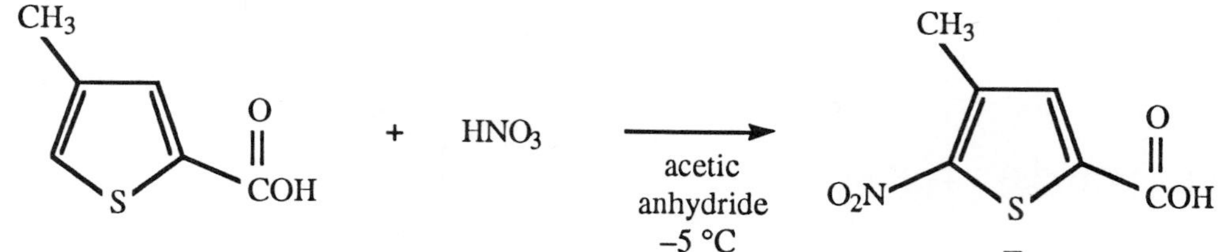

(e)

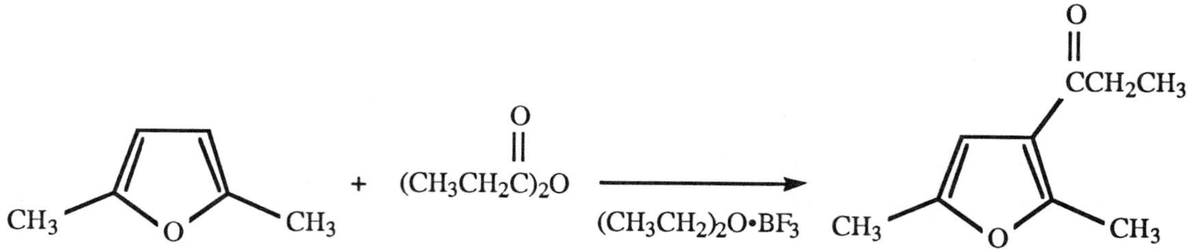

(f)

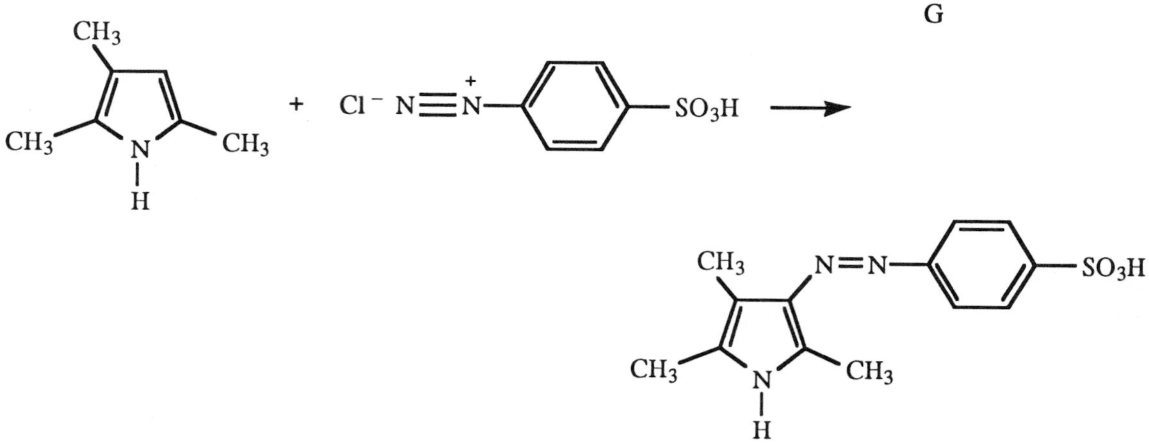

(g)

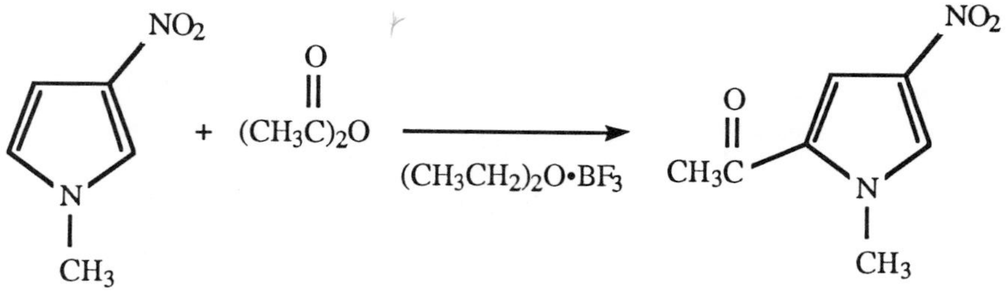

(h)

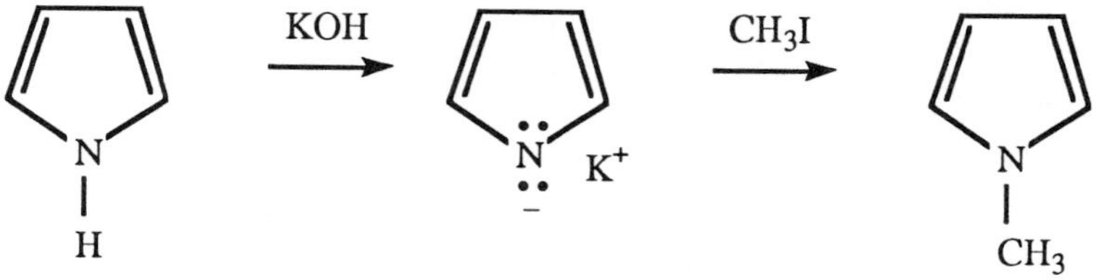

24.11 (cont)

(i)

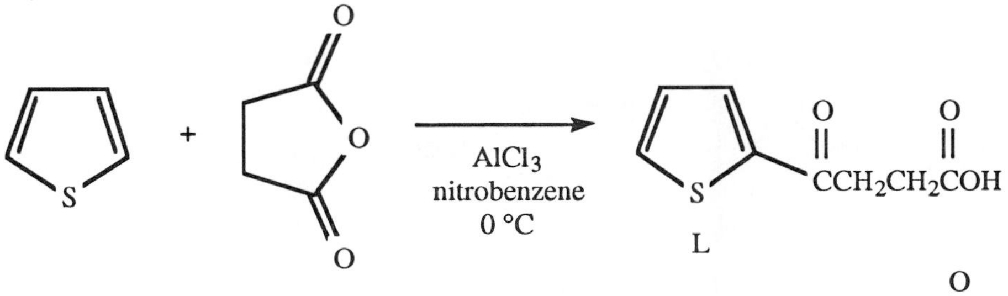

L

(j)

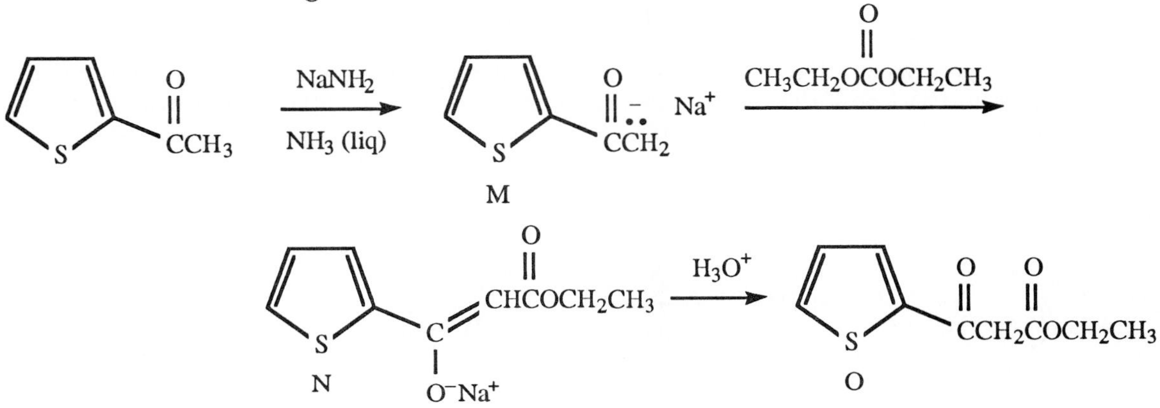

M

(k)

P

Q

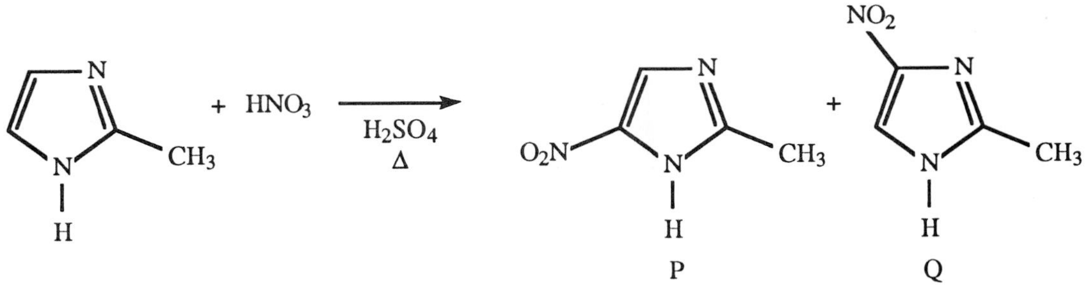

24.12 Three possible intermediates can form:

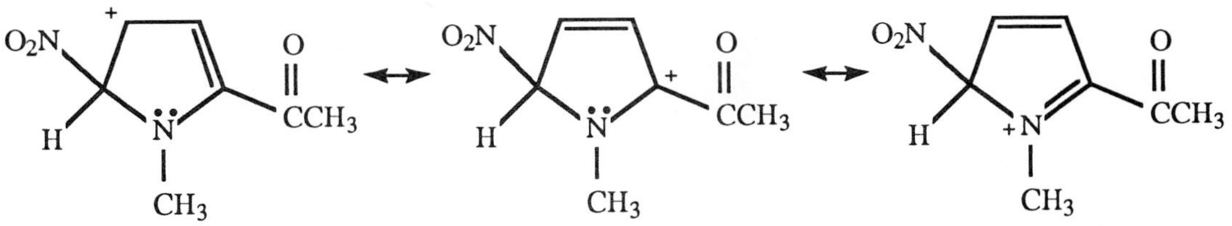

from attack at carbon 5

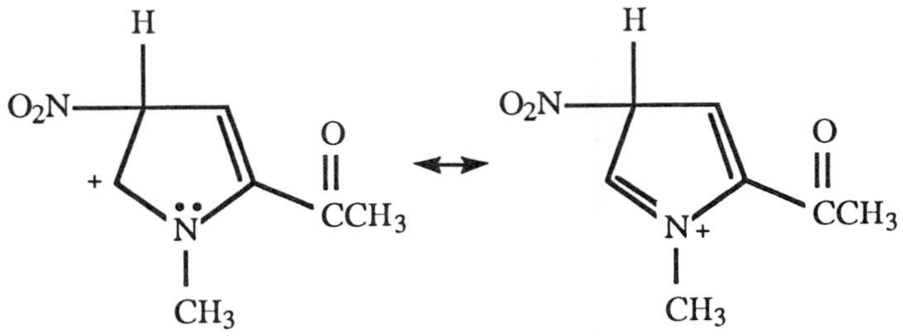

from attack at carbon 4

24.12 (cont)

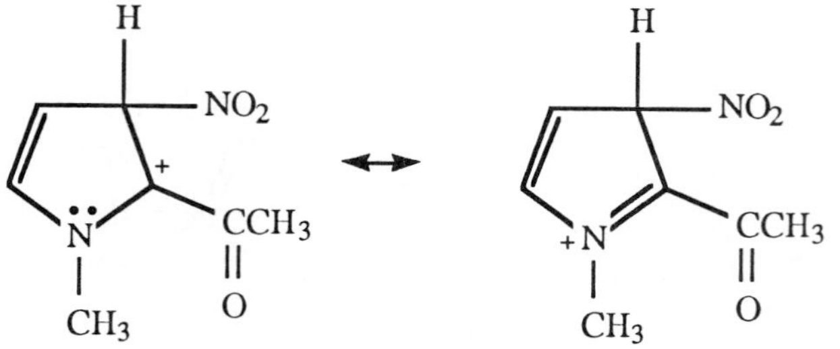

from attack at carbon 3

The intermediate of lowest energy is the one resulting from attack at the 4 position of the ring. In this intermediate the positive charge is delocalized to the nitrogen atom and no resonance contributor has a positive charge on the carbon atom bonded to the electron-withdrawing carbonyl group. Attack at carbon 5 is favored over attack at carbon 3 because there is greater delocalization of the positive charge in that intermediate.

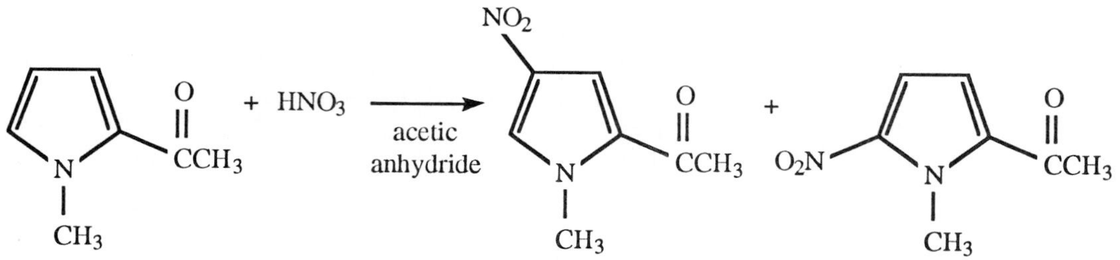

major product

24.13

(a)

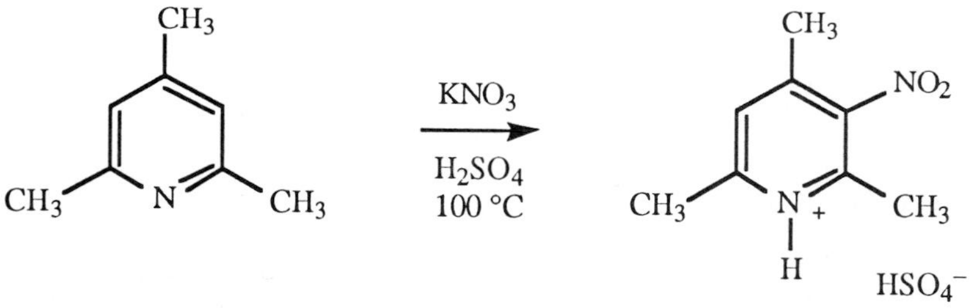

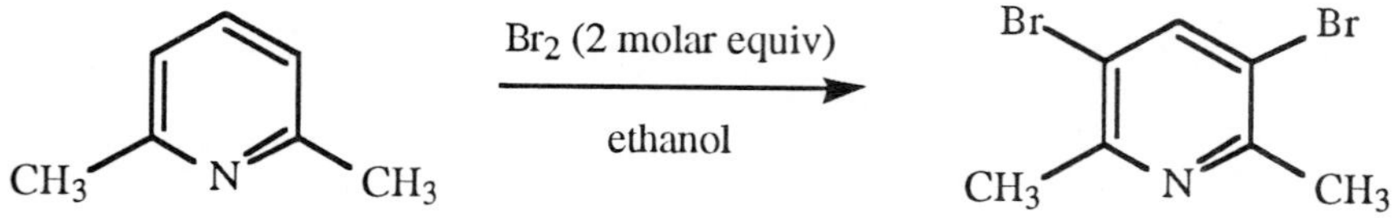

(b)

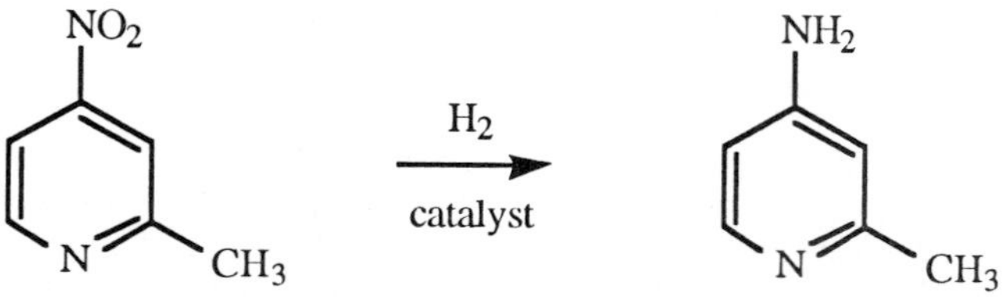

(c)

24.13 (cont)

(d)

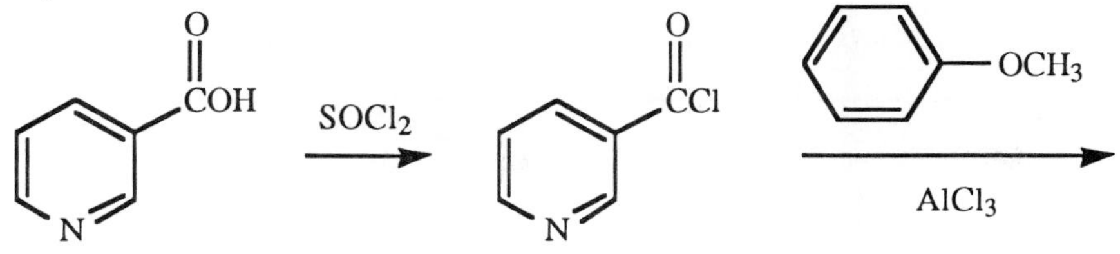

D

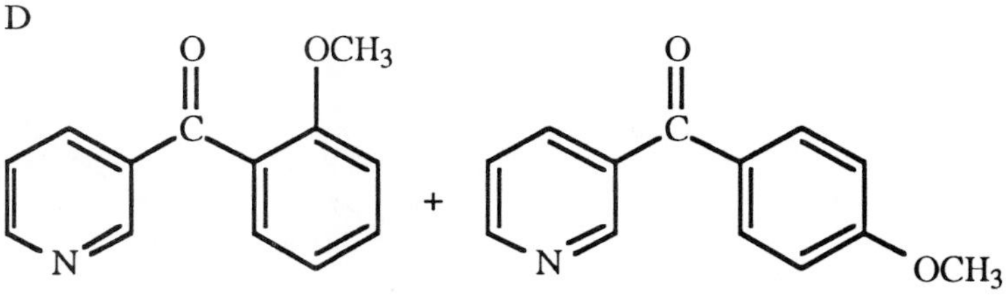

E

24.14

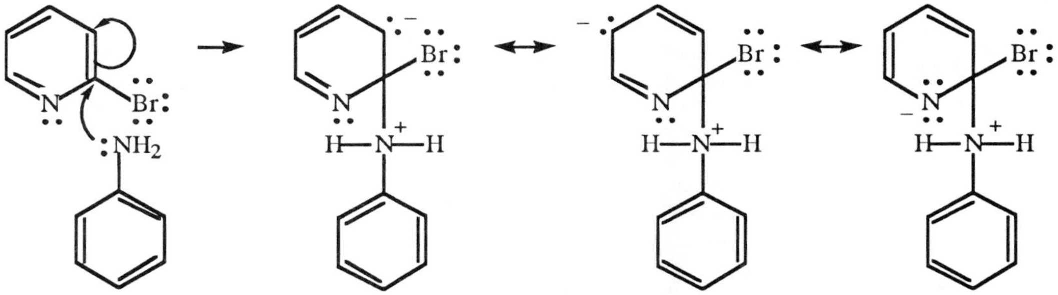

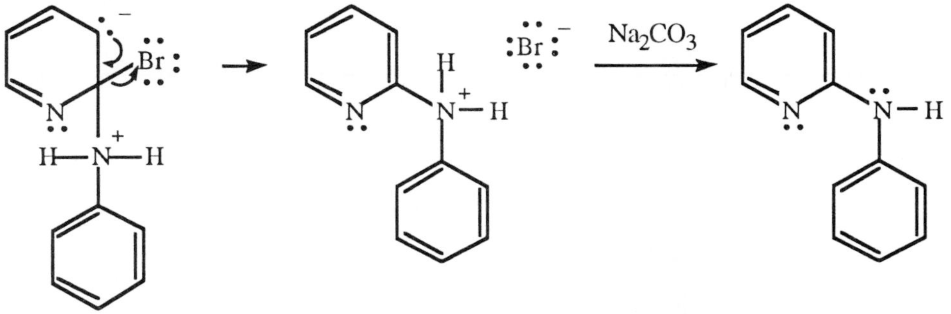

Concept Map 24.3   (see p. 880)

24.15

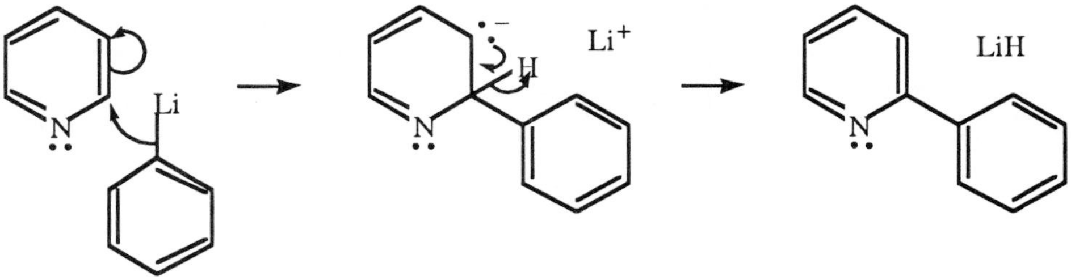

The reaction starts with attack of the organolithium reagent (a very strong nucleophile) on the carbon adjacent to the nitrogen atom to give a resonance-stabilized intermediate (see Problem 24.14). The leaving group is hydride ion. The driving force for the loss of hydride ion in this reaction and in the reaction of amide ion with pyridine is the recovery of aromaticity of the pyridine ring.

Concept Map 24.3   Substitution reactions of heterocycles.

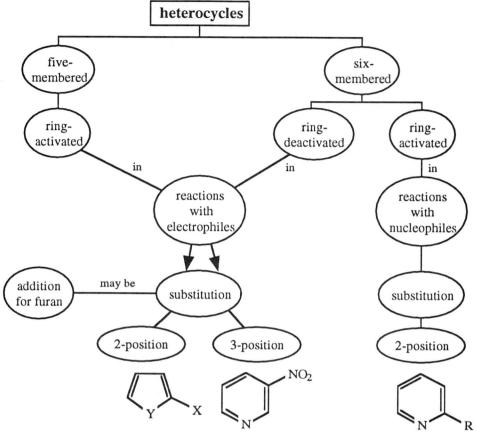

24.16

(a)

(b)

(c)

24.16 (cont)

(d)

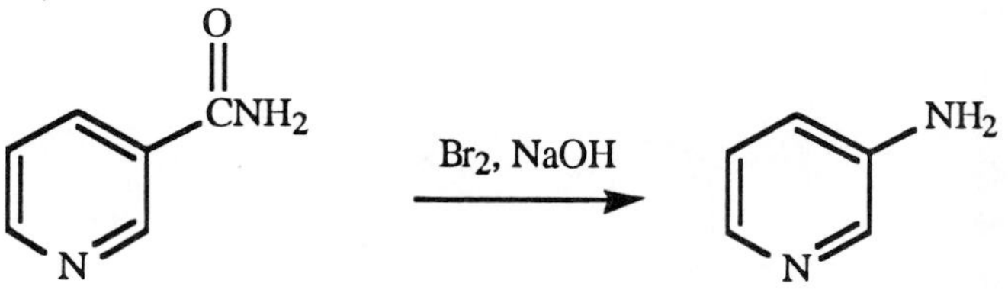

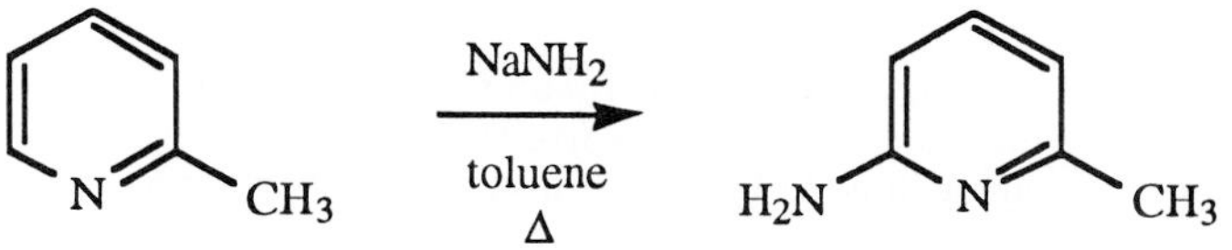

(e)

(f)

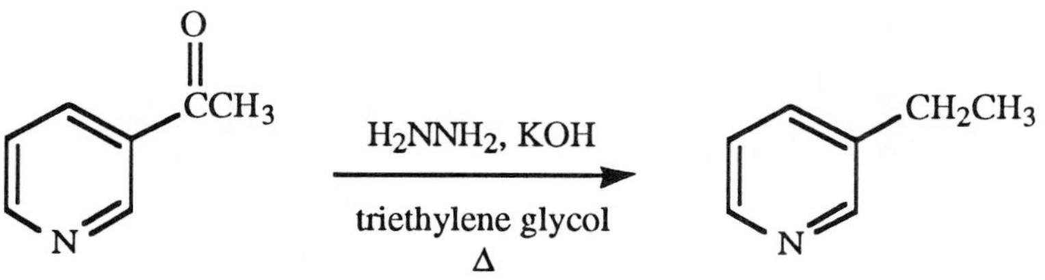

(g)

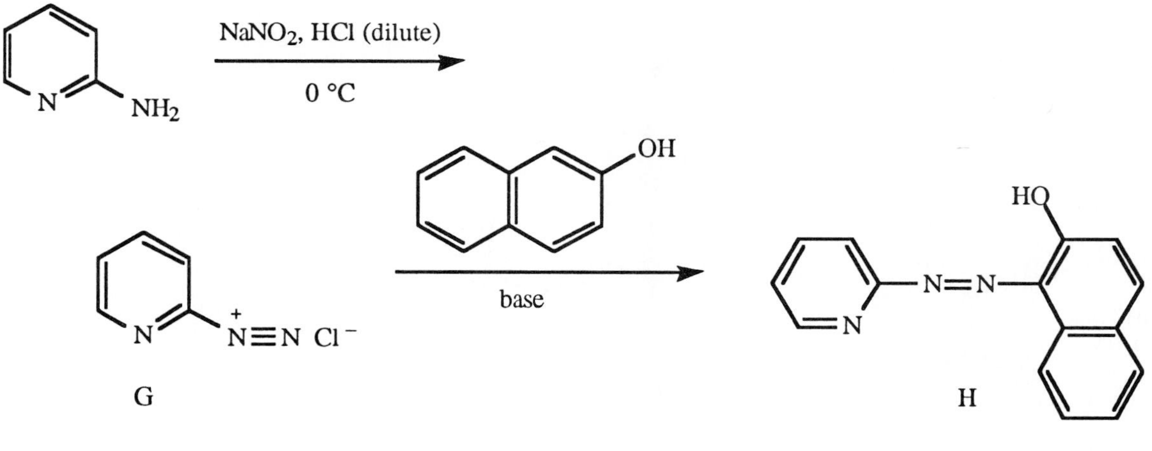

24.17

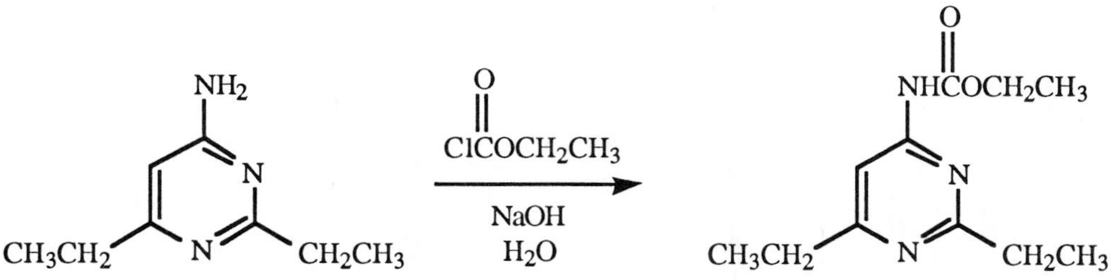

(a)

24.17 (cont)

(b)

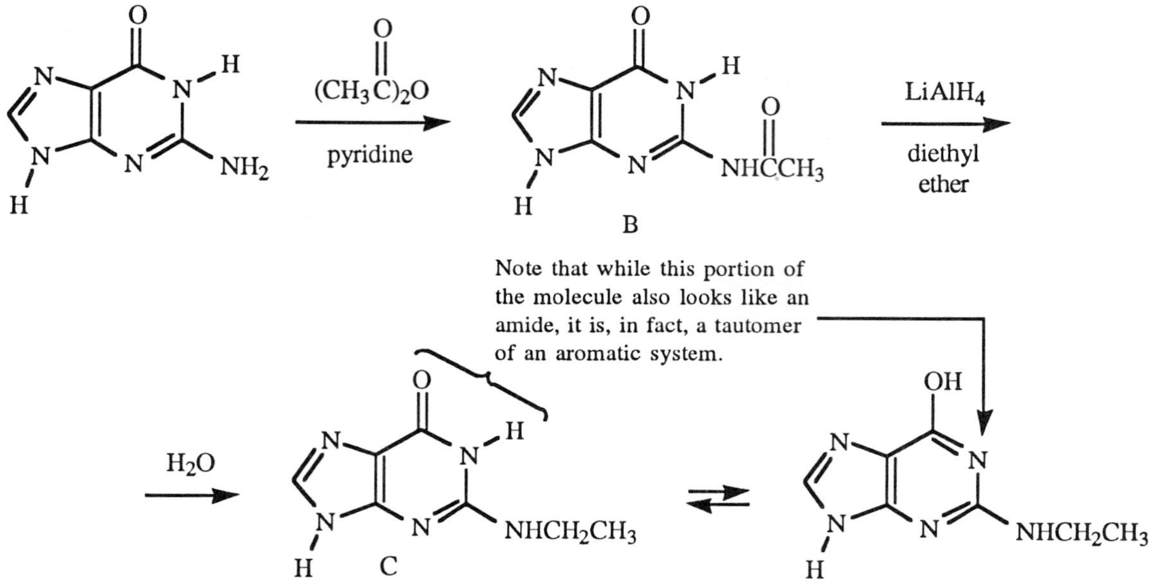

$(CH_3\overset{O}{\overset{||}{C}})_2O$
pyridine
B
LiAlH$_4$
diethyl ether
H$_2$O
C
Note that while this portion of the molecule also looks like an amide, it is, in fact, a tautomer of an aromatic system.
NHCH$_2$CH$_3$

(c)

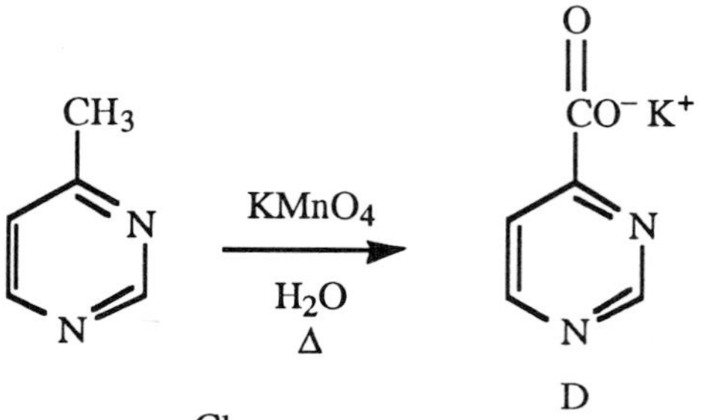

KMnO$_4$
H$_2$O
Δ
D

(d)

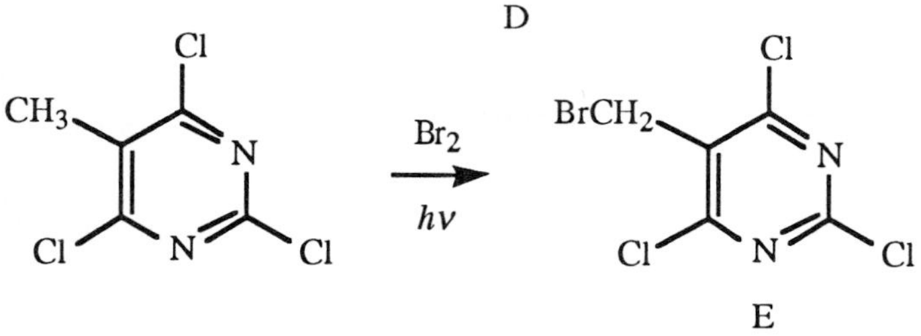

Br$_2$
hν
E

(e)

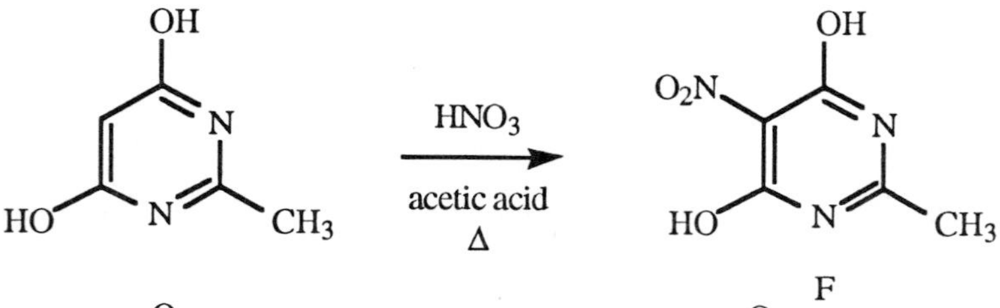

HNO$_3$
acetic acid
Δ
F

(f)

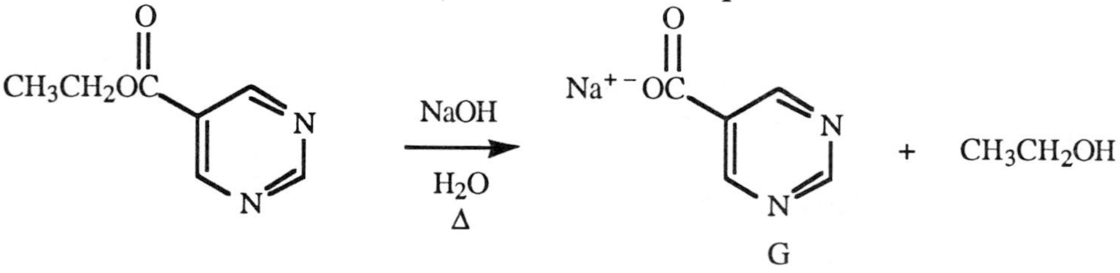

NaOH
H$_2$O
Δ
G
+ CH$_3$CH$_2$OH

24.17 (cont)

(g)

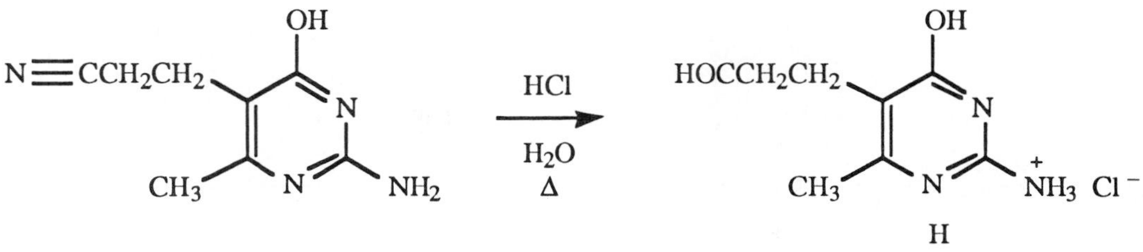

(h)

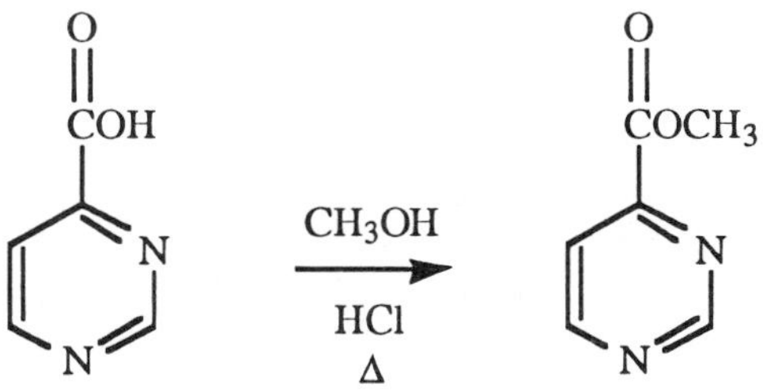

(i)

(j)

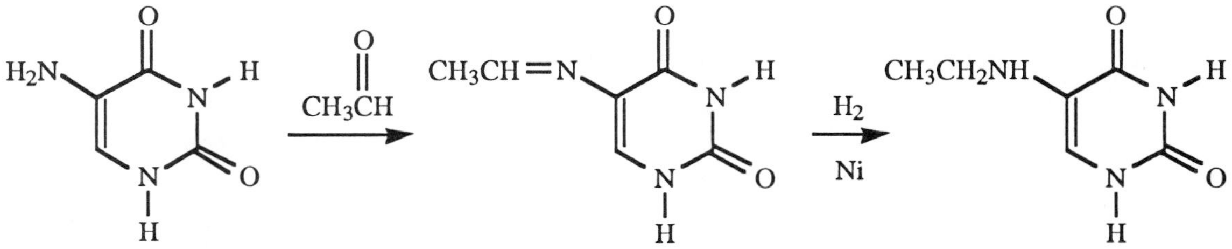

24.18   The adenine-thymine pair is held together by two hydrogen bonds, and the guanine-cytosine pair is held together by three hydrogen bonds. It takes more energy to pull apart a guanine-cytosine pair than an adenine-thymine pair. Therefore, the more guanine and cytosine a particular form of DNA contains, the higher the temperature required before the DNA melts.

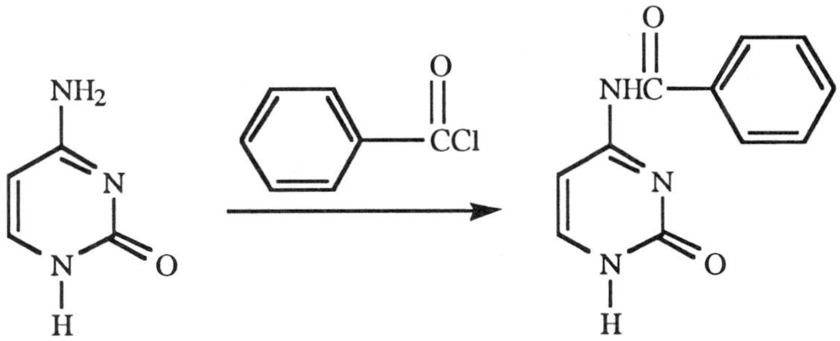

24.19   (a)

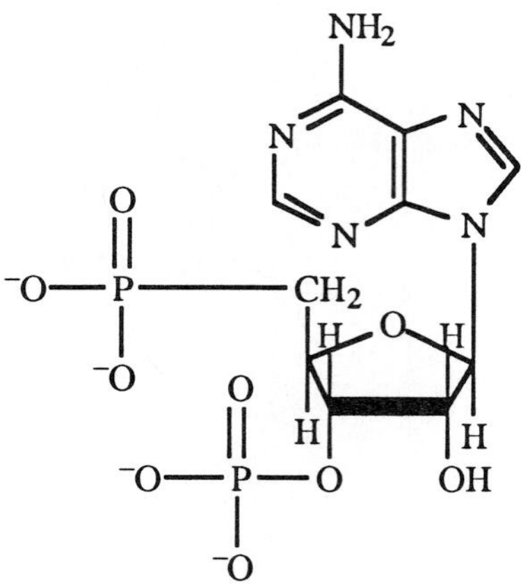

(b)

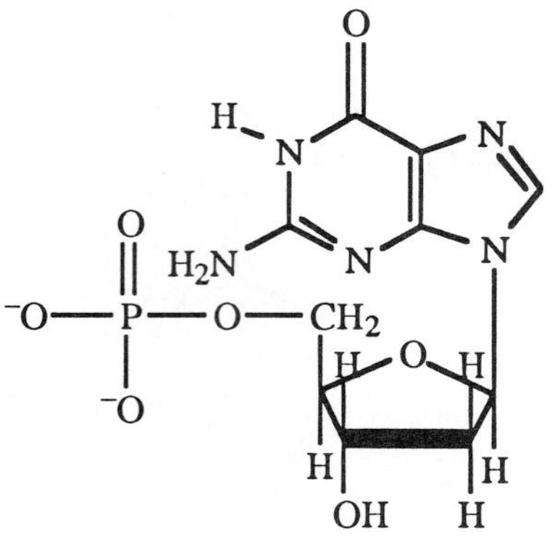

(c)

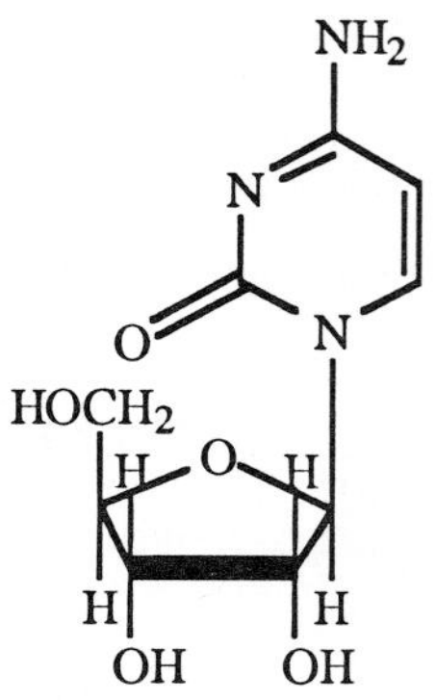

(d)

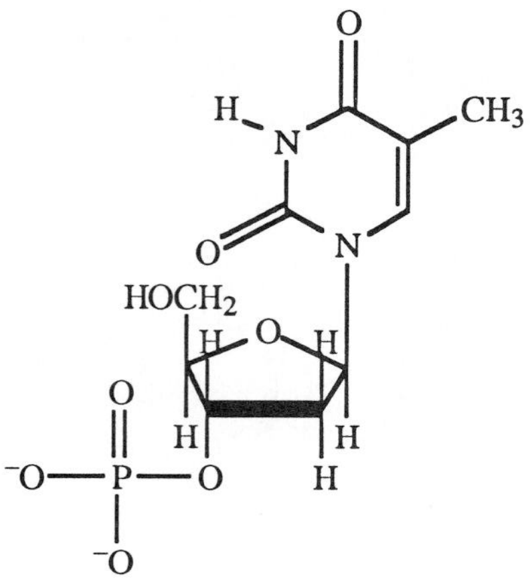

(e)

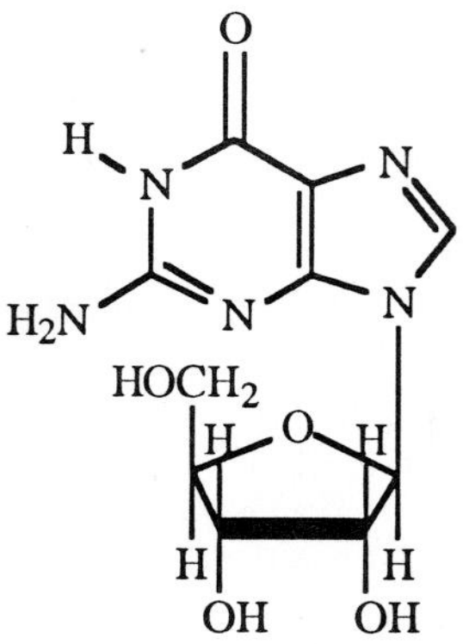

(f)

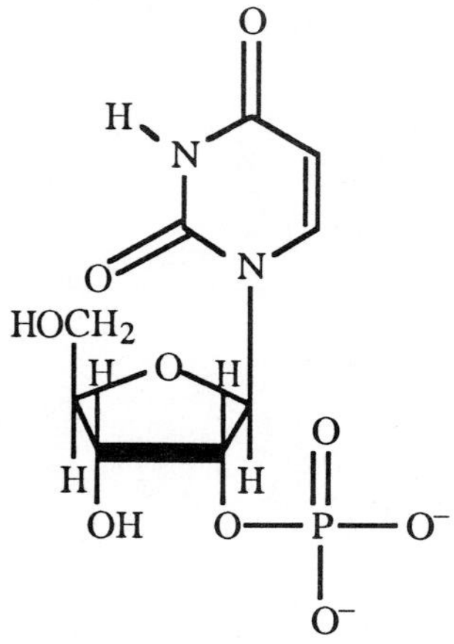

24.20

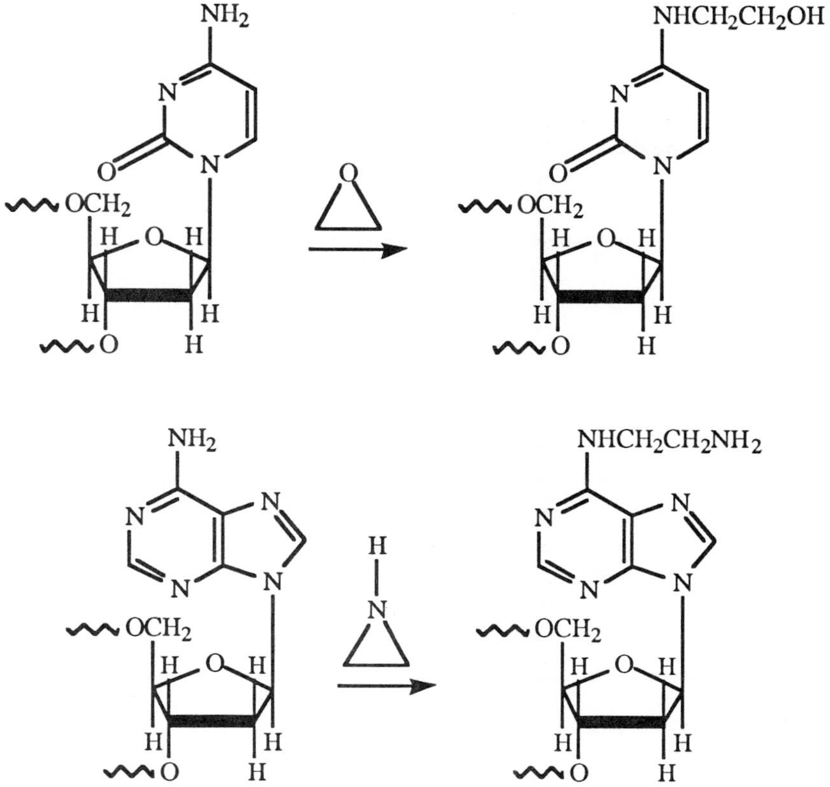

The heterocycles with three-membered rings are vulnerable to nucleophilic attack. The bases in DNA contain nucleophilic sites such as amino groups. Reaction of an amino group with an oxirane or an aziridine creates a new functionality at the sites in DNA that were previously involved in the hydrogen bonding so important to base pairing. Oxiranes and aziridines, therefore, disrupt the processes necessary for the transmission of genetic information and cause mutations.

24.21

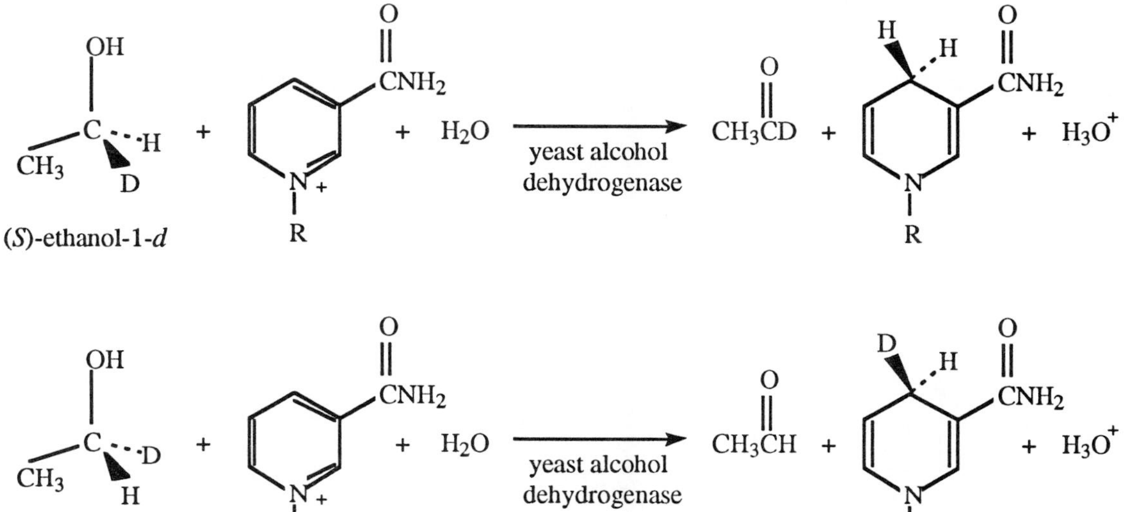

(S)-ethanol-1-d

(R)-ethanol-1-d

24.22

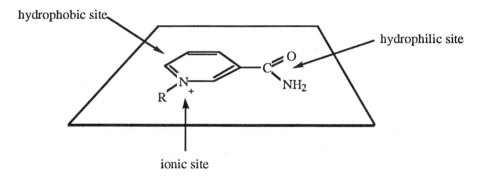

Nicotinamide fits the active site of the enzyme in this orientation.

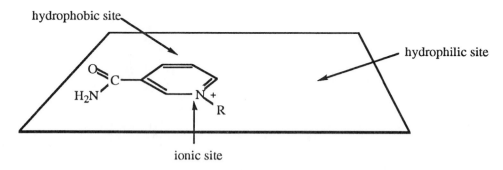

Nicotinamide does not fit the active site of the enzyme in this orientation.

The two orientations of the nicotinamide molecules as drawn are enantiomers of each other and are distinguished from each other by the enzyme.

24.23 (a)

$$CH_3 - \underset{\underset{CH_3}{|}}{\overset{\overset{CH_3}{|}}{C}} - H$$

no enantiotopic hydrogens

(b)

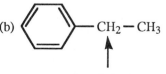

enantiotopic hydrogens

(c)

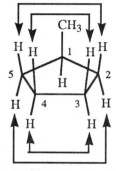

four sets of enantiotopic hydrogens;
the two sets of cis hydrogens on
carbons 2 and 5, and the two sets of cis
hydrogens on carbons 3 and 4

these two methyl groups are enantiotopic;
replacement of one gives rise to a stereocenter
on the adjacent carbon

(d)

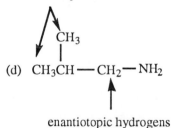

enantiotopic hydrogens

24.24

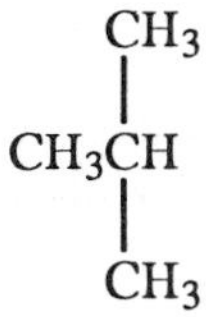

CH₃CH(CH₃)CH₃

no diasterotopic hydrogens

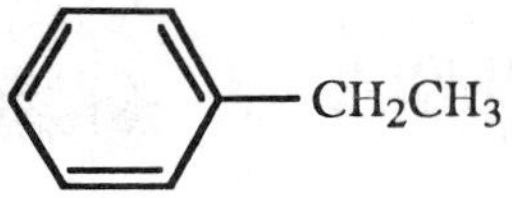

Ph—CH₂CH₃

no diasterotopic hydrogens

CH₃CH(CH₃)CH₂NH₂

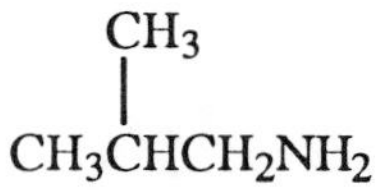

no diasterotopic hydrogens

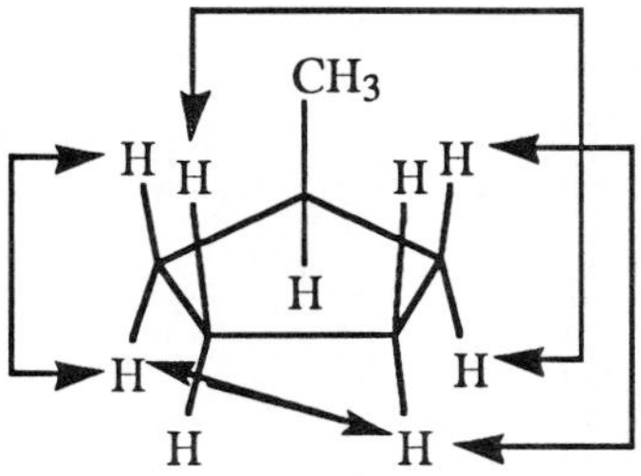

many sets of diastereotopic hydrogens,
four of which are shown

(a) 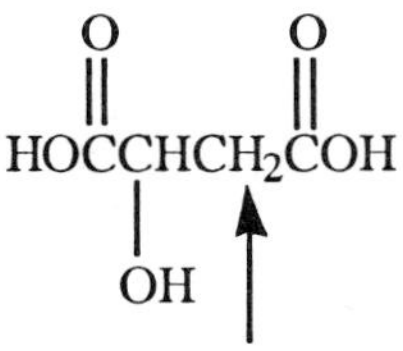 HOCCHCH₂COH
          |
          OH

diasterotopic
hydrogens

(b) 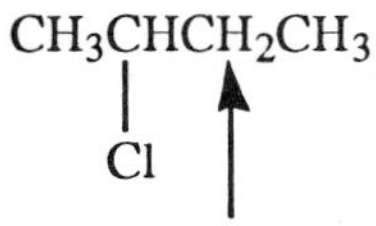 CH₃CHCH₂CH₃
          |
          Cl

diasterotopic
hydrogens

(c) 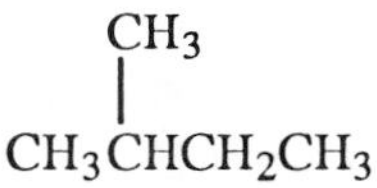 CH₃CHCH₂CH₃
          |
          CH₃

no diastereotopic
hydrogens

(d) 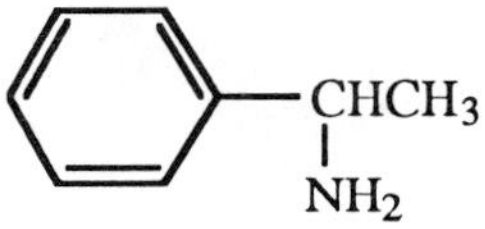 Ph—CHCH₃
          |
          NH₂

no diastereotopic
hydrogens

24.25

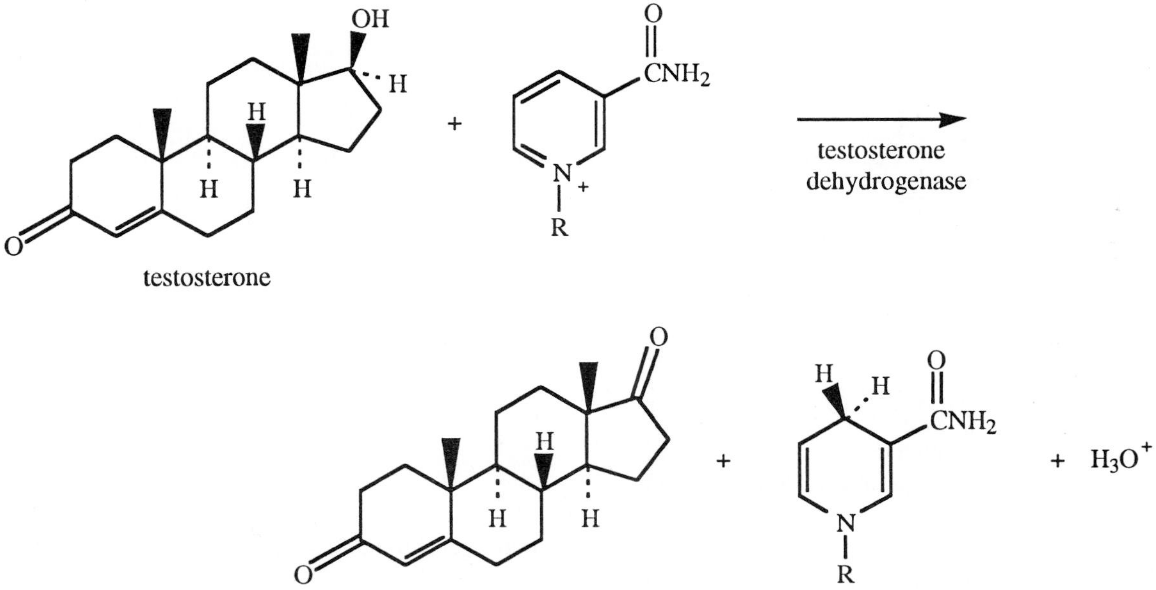

24.25 (cont)

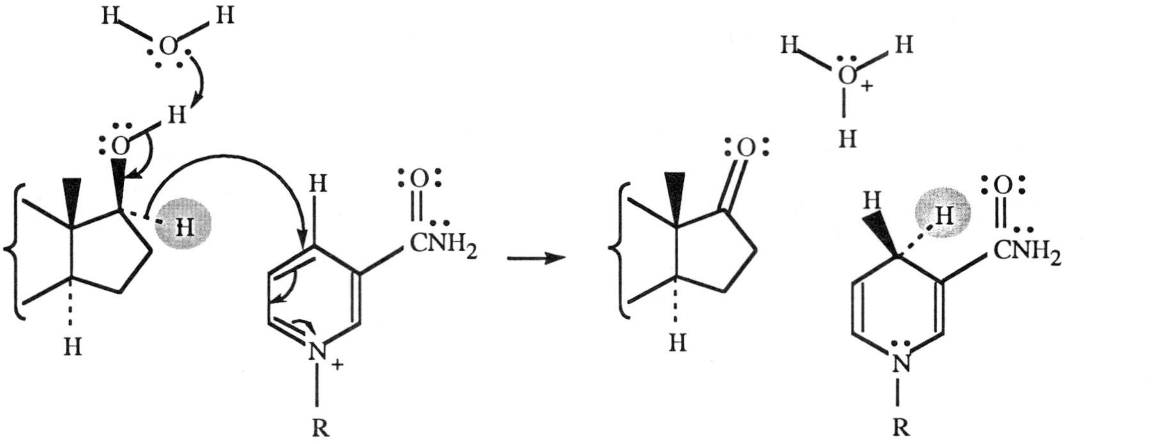

One hydrogen atom from testosterone goes to the solvent, the other to the back face of the nicotinamide group.

24.26

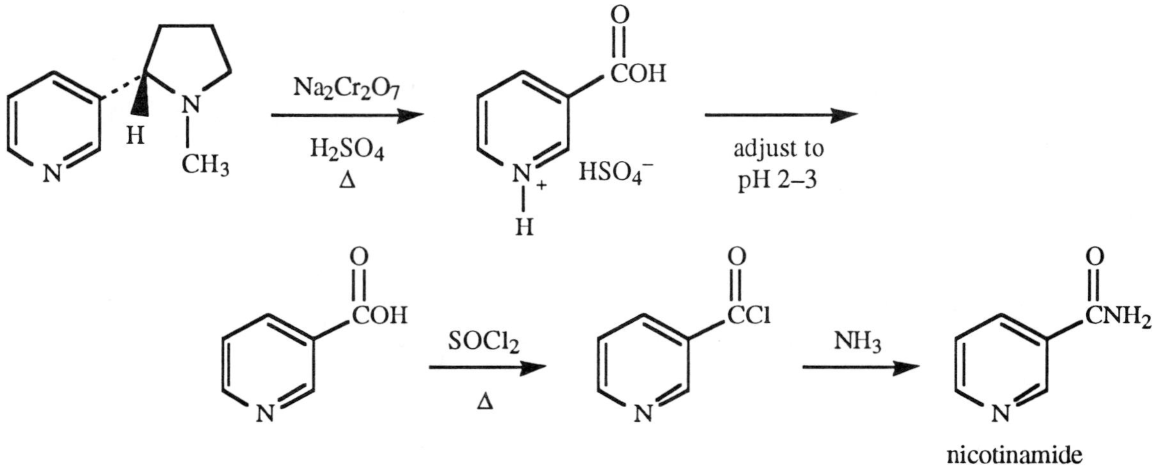

nicotinamide

24.27

intermediate from the addition of
the ylid from thymine to pyruvic acid

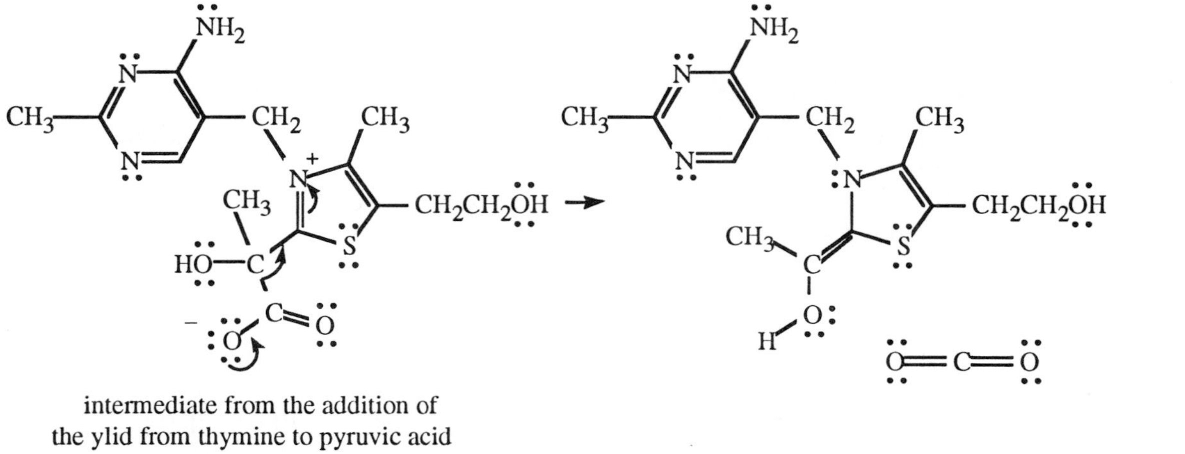

Decarboxylation occurs easily because the carbanion that is formed is stabilized by delocalization of charge to the positively charged nitrogen atom of the ylid. (See Problem 24.28 for the resonance contributors of the decarboxylation product.)

24.28

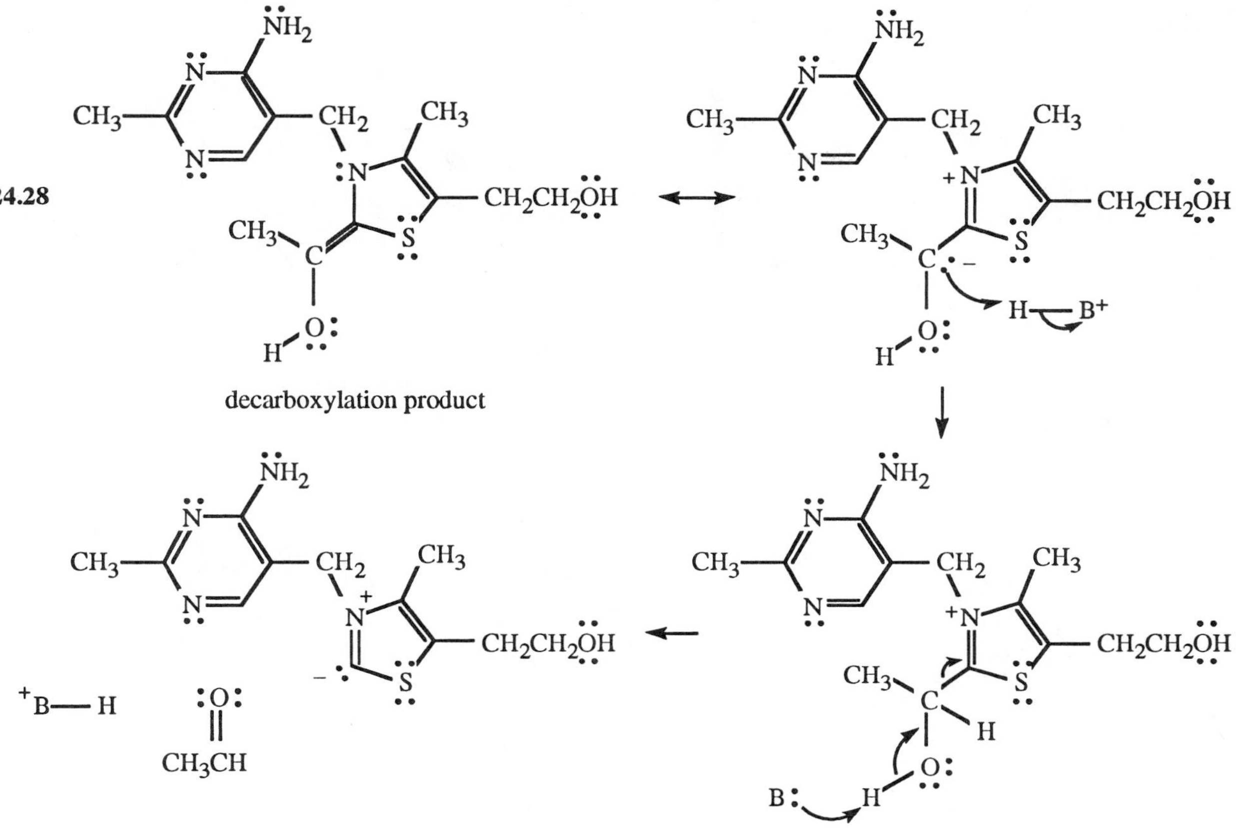
decarboxylation product

This resembles a retroaldol reaction.

24.29

(a)

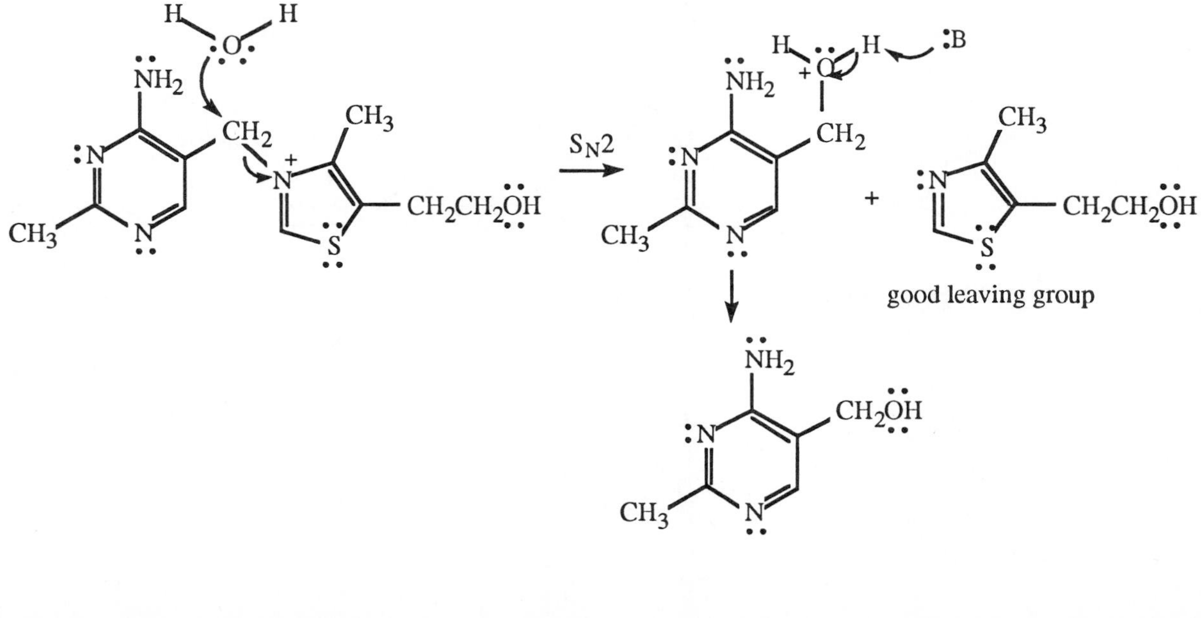
S_N2
good leaving group

24.29 (cont)

(b)

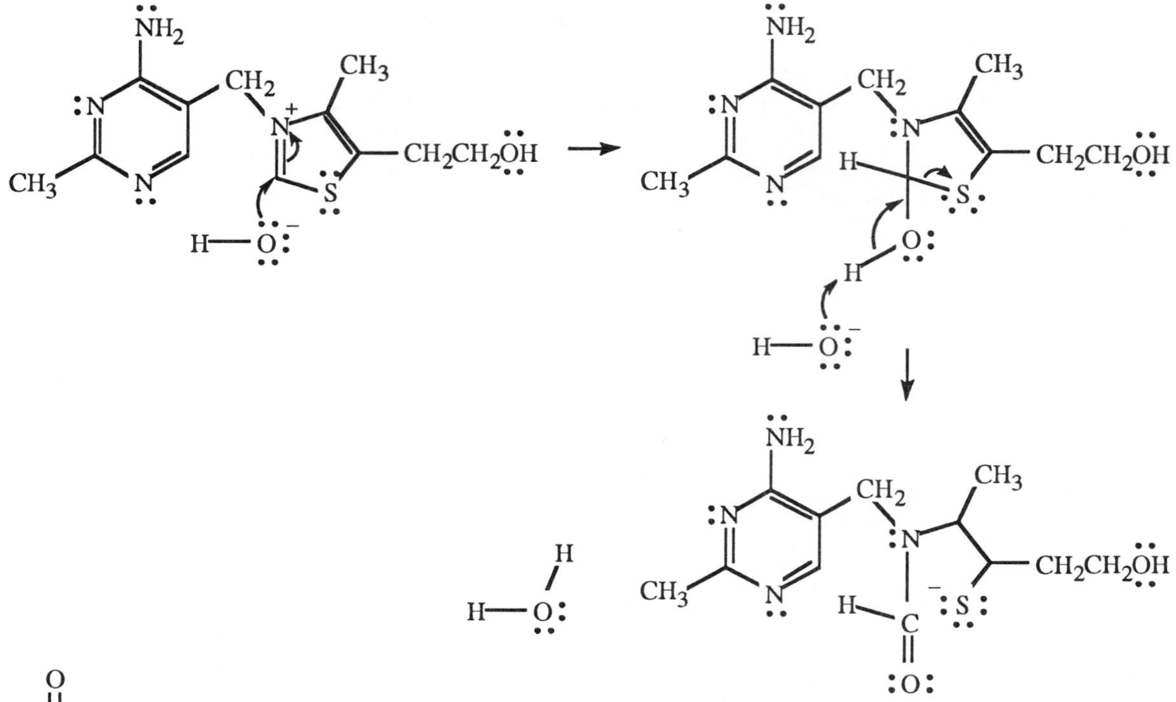

24.30

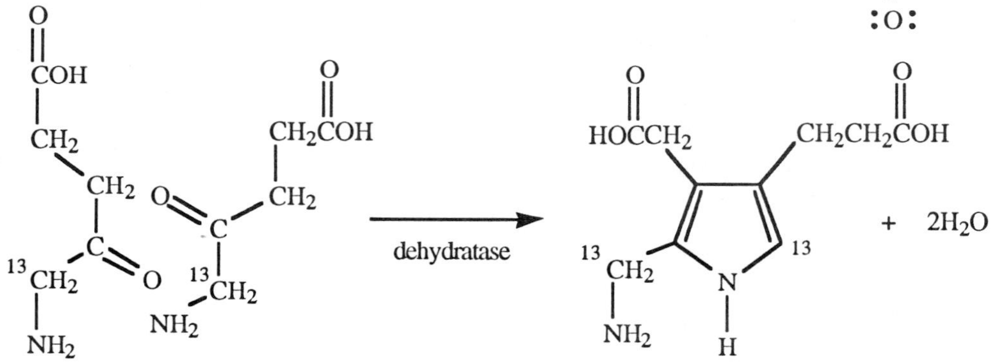

24.31

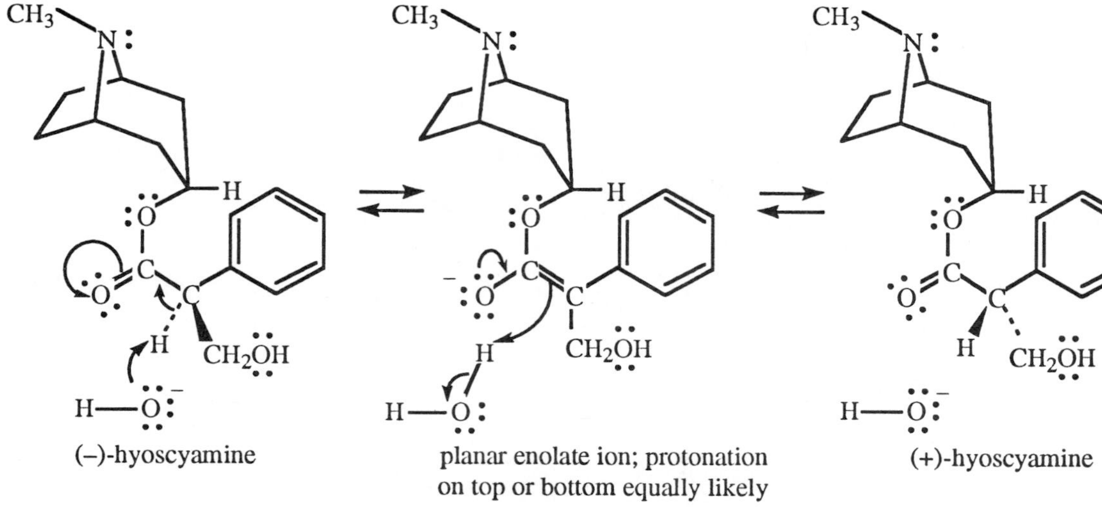

(−)-hyoscyamine   planar enolate ion; protonation   (+)-hyoscyamine
on top or bottom equally likely

Atropine is a racemic hyoscyamine, or the equilibrium mixture of the two enantiomers.

24.32

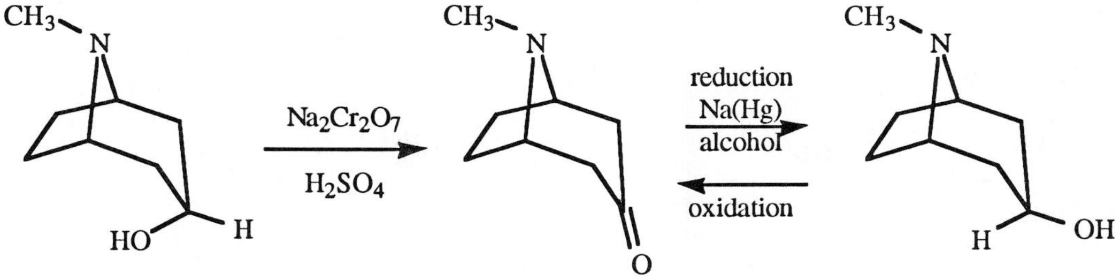

tropine   tropinone   ψ-tropine
                                                                    a stereoisomer of tropine

24.33

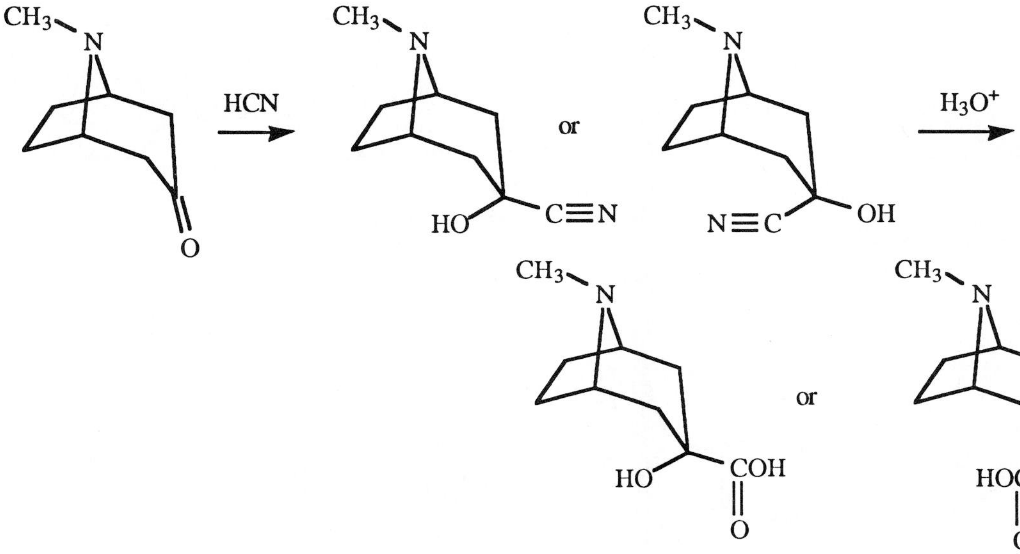

α-ecgonine

24.34

(1)

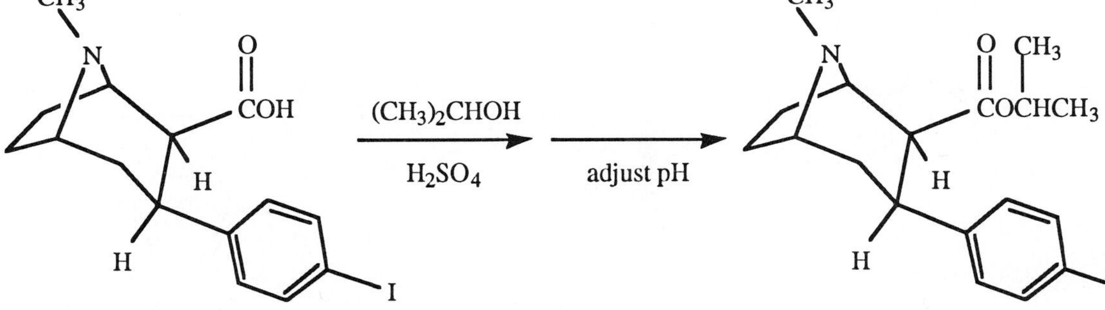

24.34 (cont)

(2)

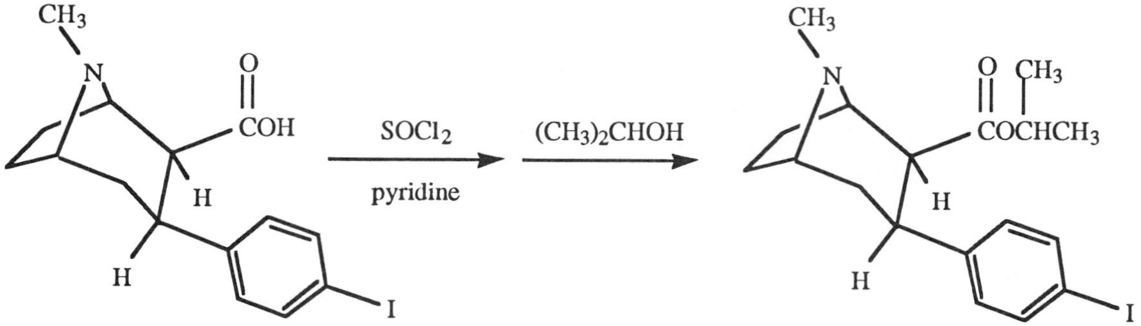

Method 2 is better since it does not depend on an equilibrium process for yield.

24.35

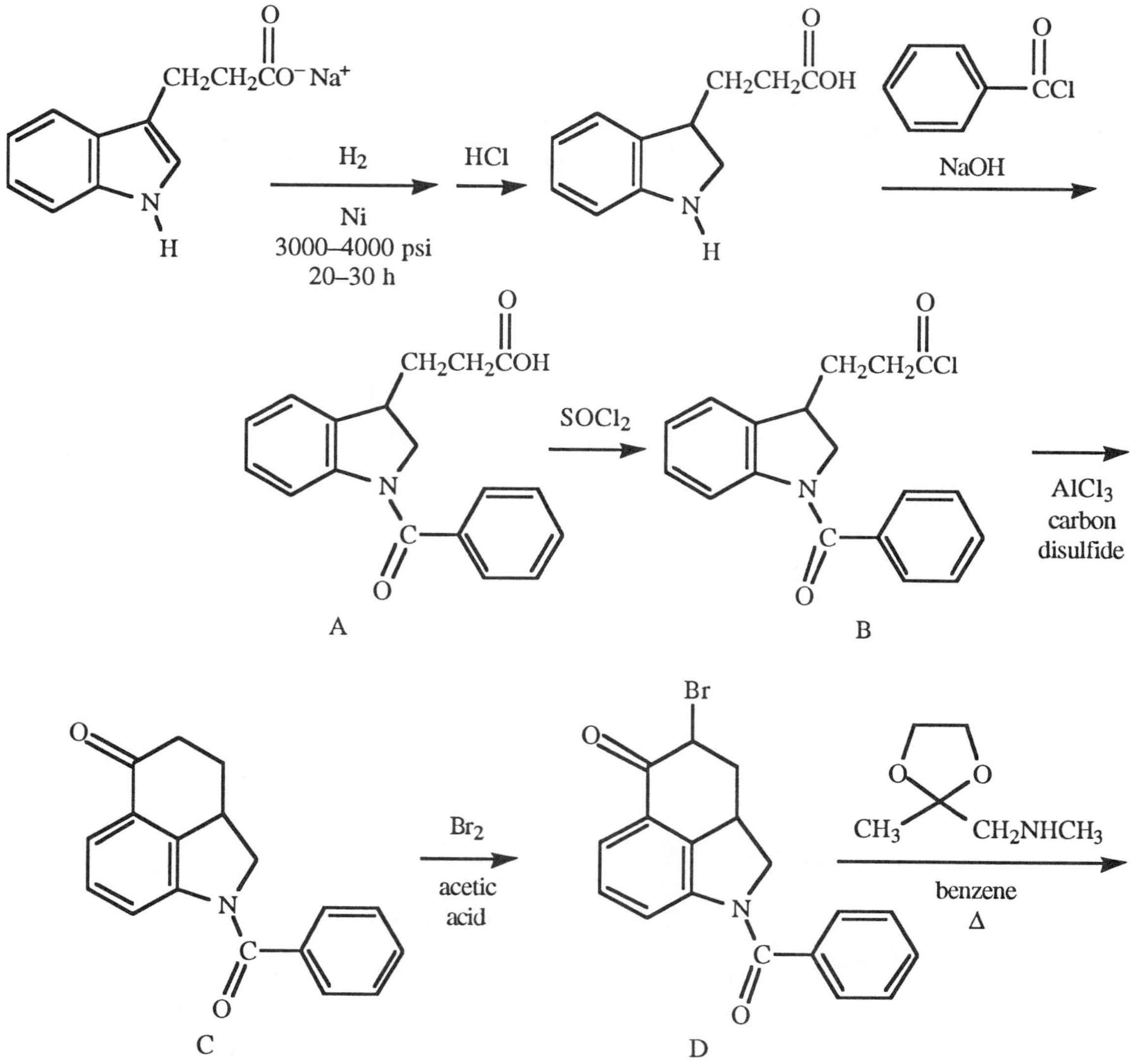

24.35 (cont)

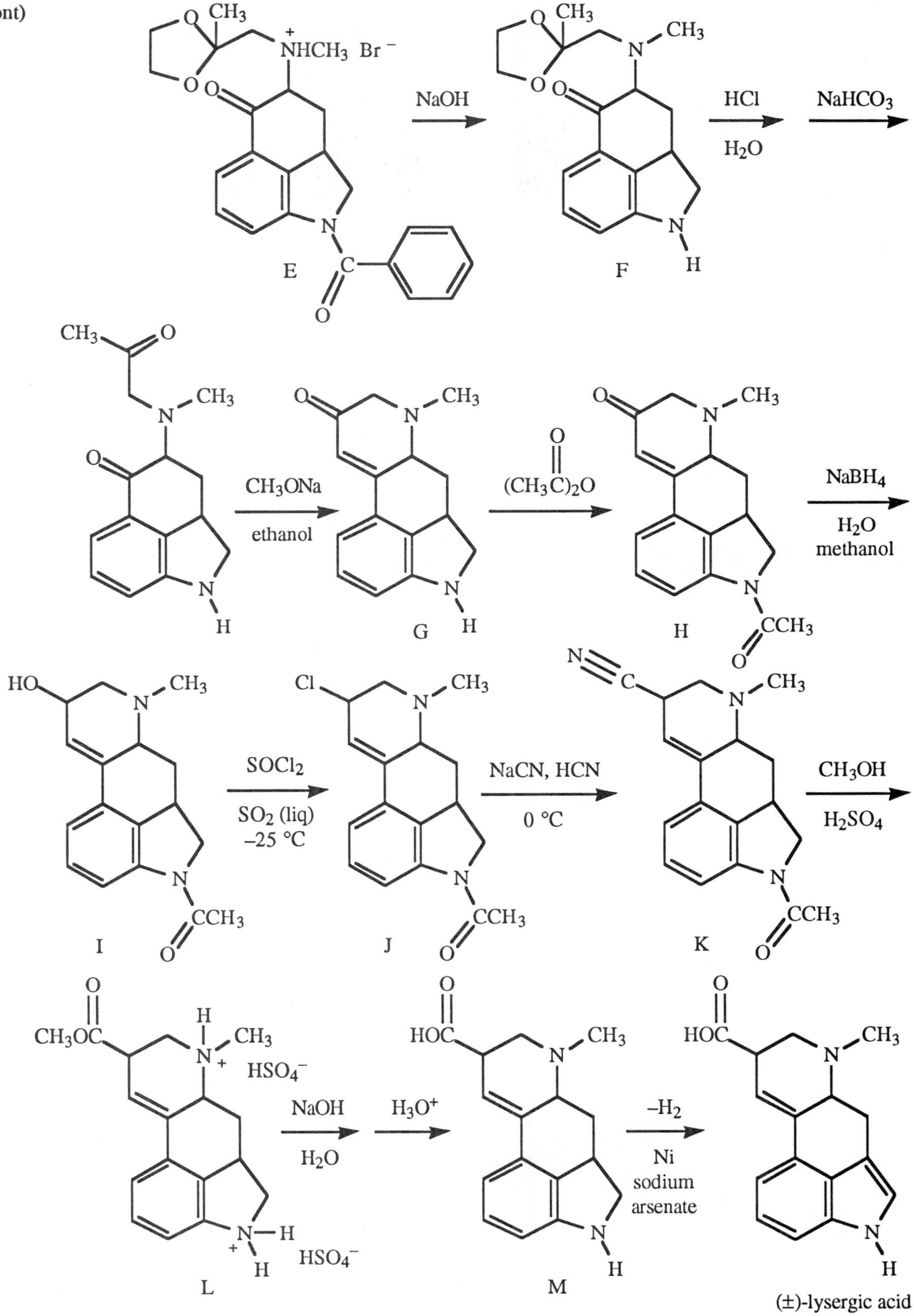

(±)-lysergic acid

24.36

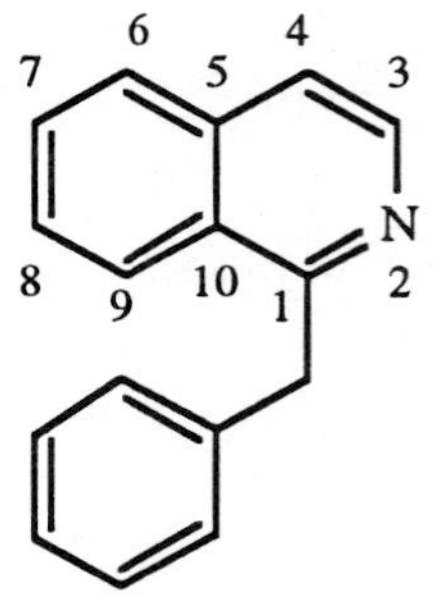

numbered around the isoquinoline
ring (*not* the numbering used in
naming the compound);
benzyl group on carbon 1

benzyl unit
on carbon 1

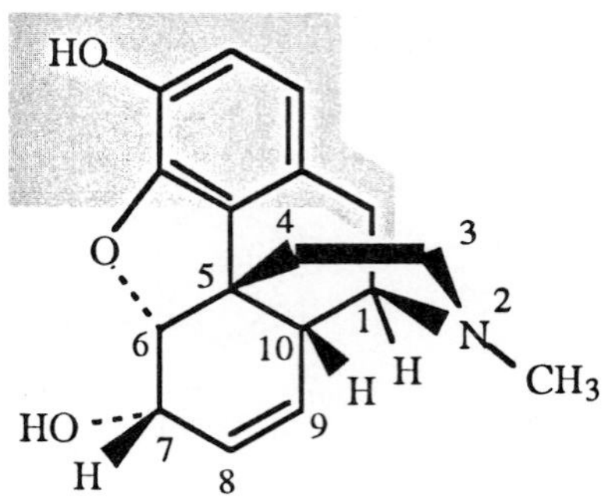

24.37

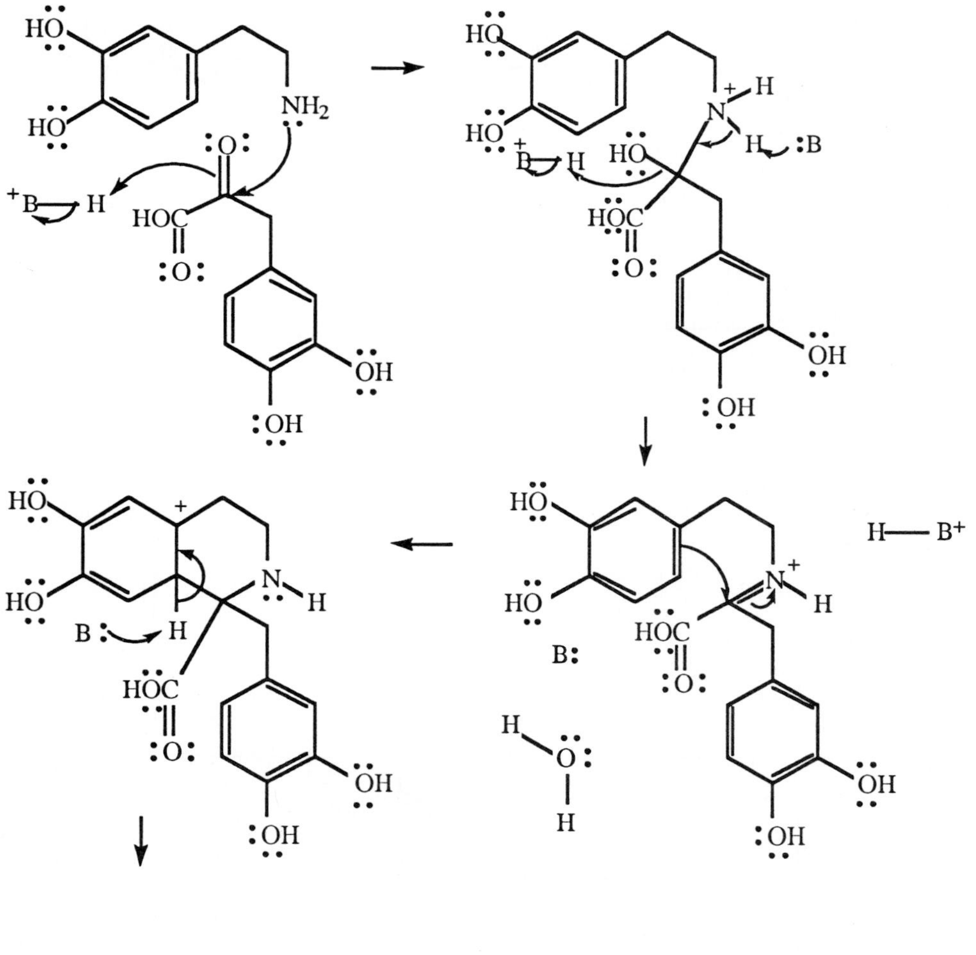

24.37 (cont)

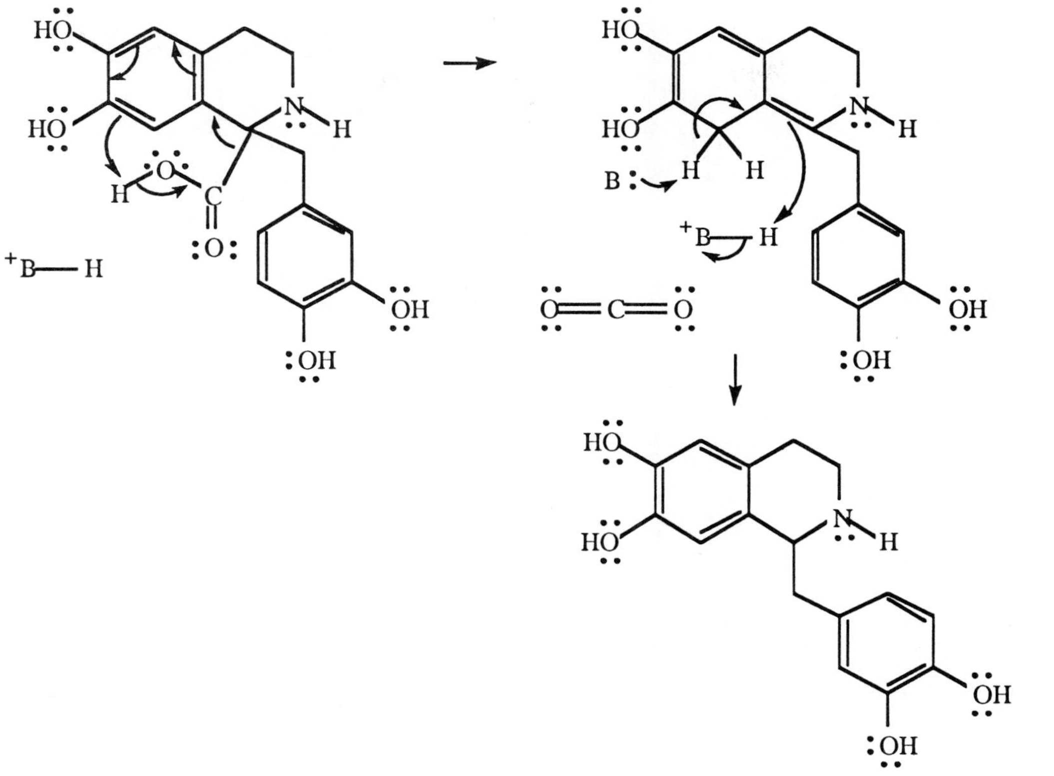

24.38

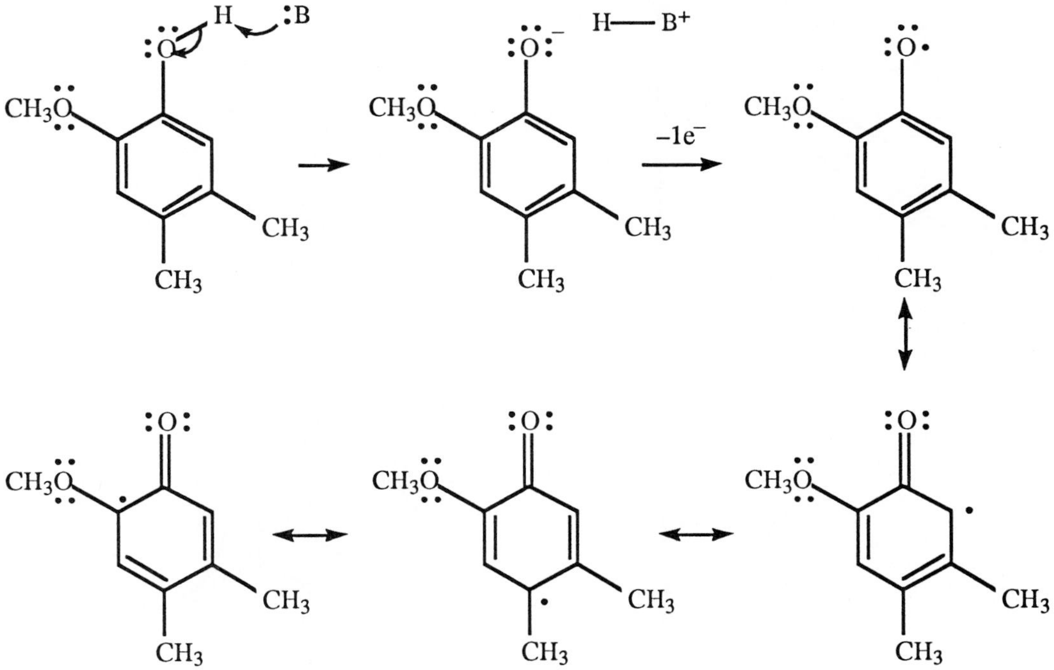

24.38 (cont)

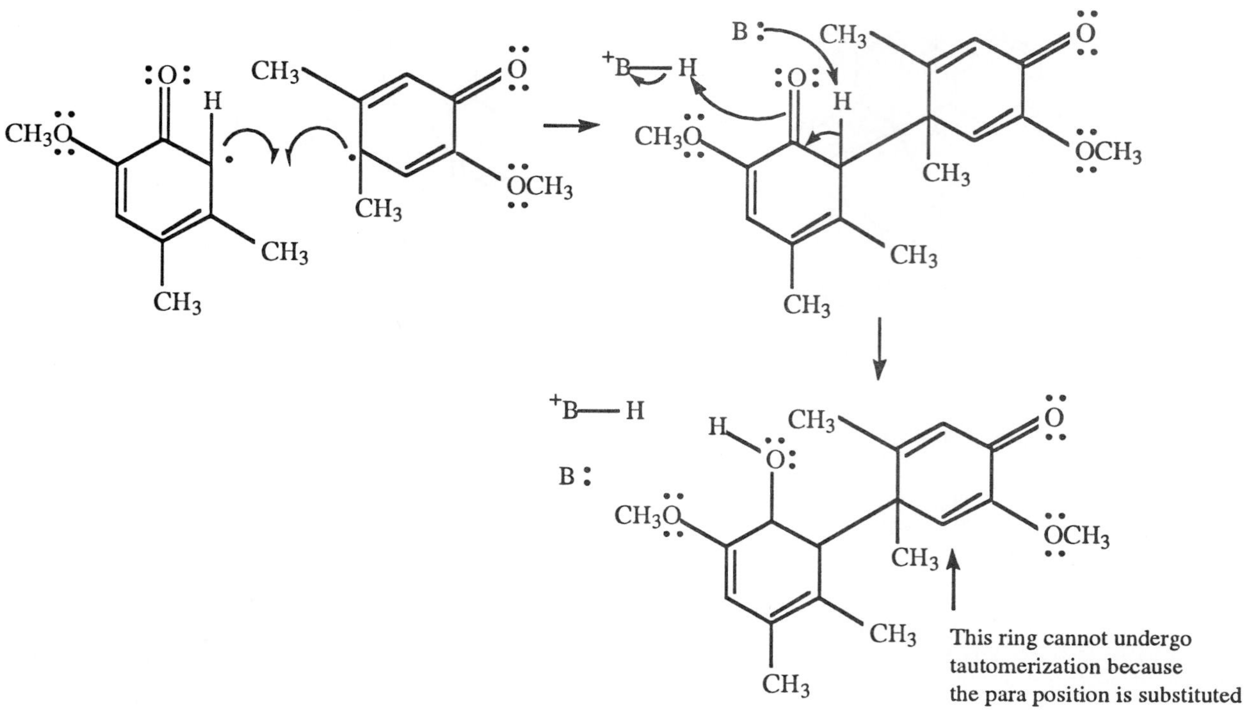

This ring cannot undergo
tautomerization because
the para position is substituted

24.39

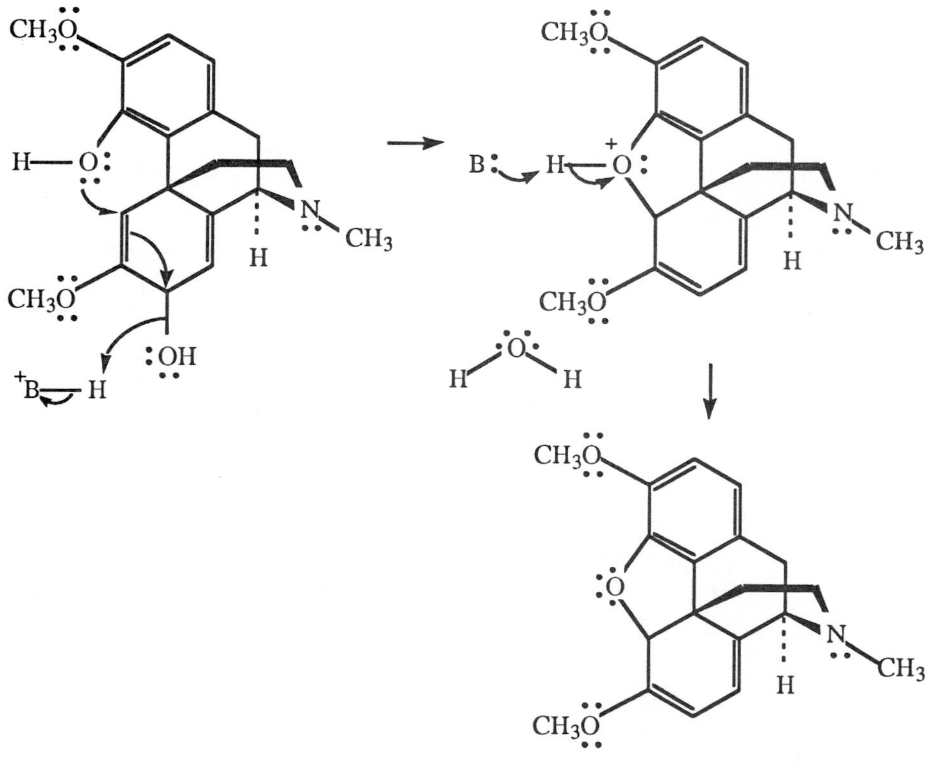

24.39 (cont)

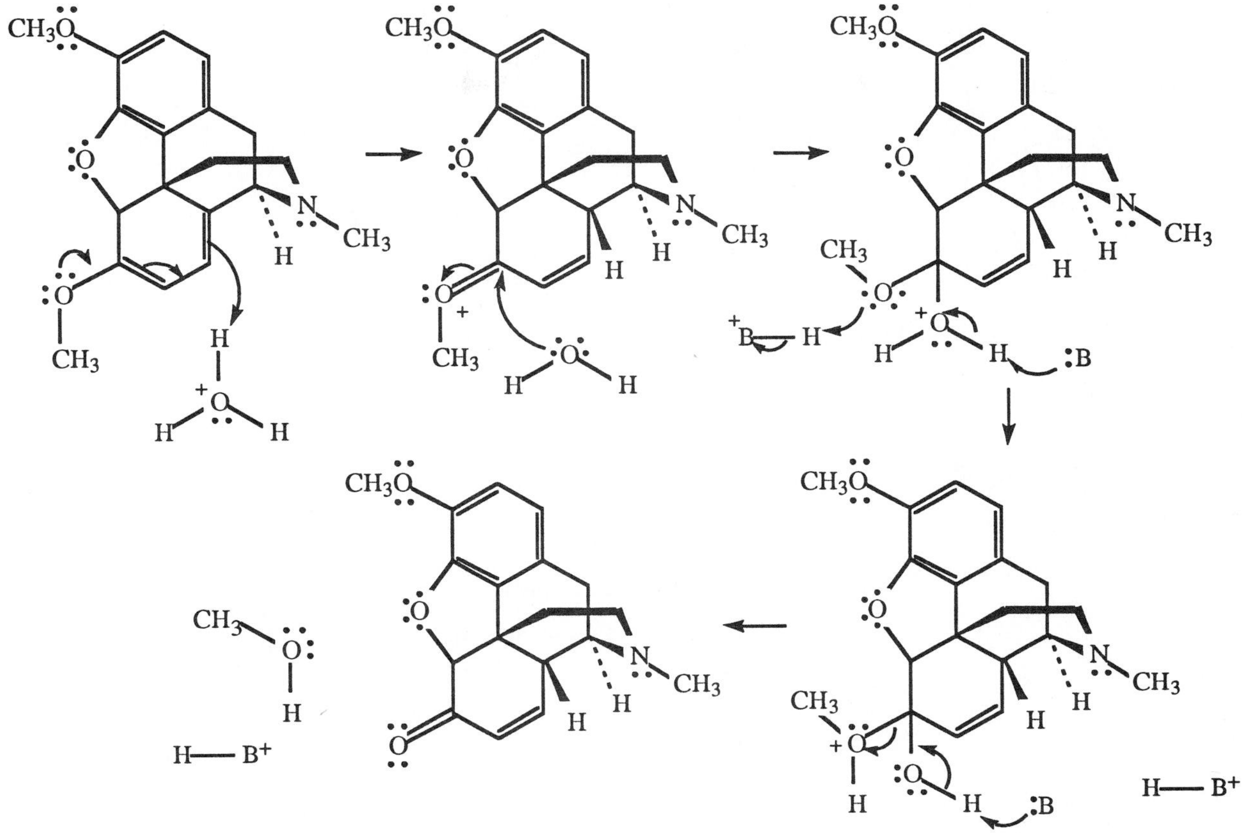

24.40

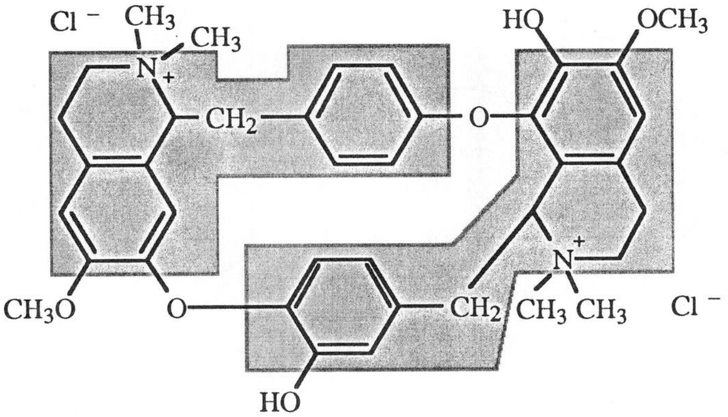

24.41

(a)

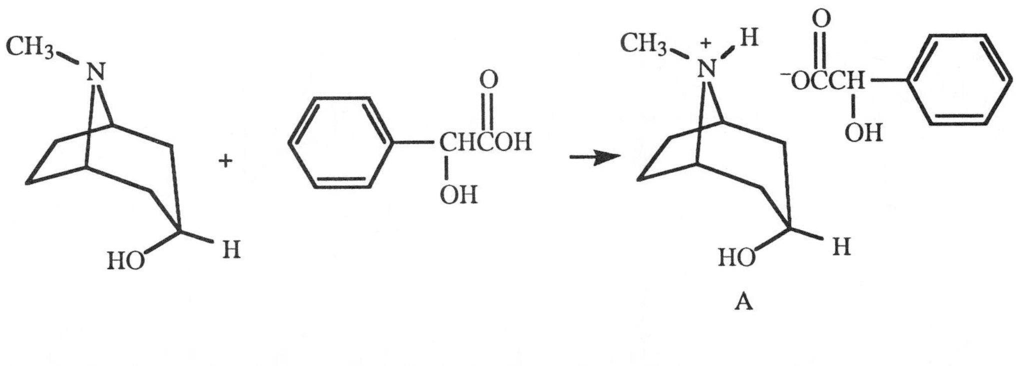

A

24.41 (cont)

(b)

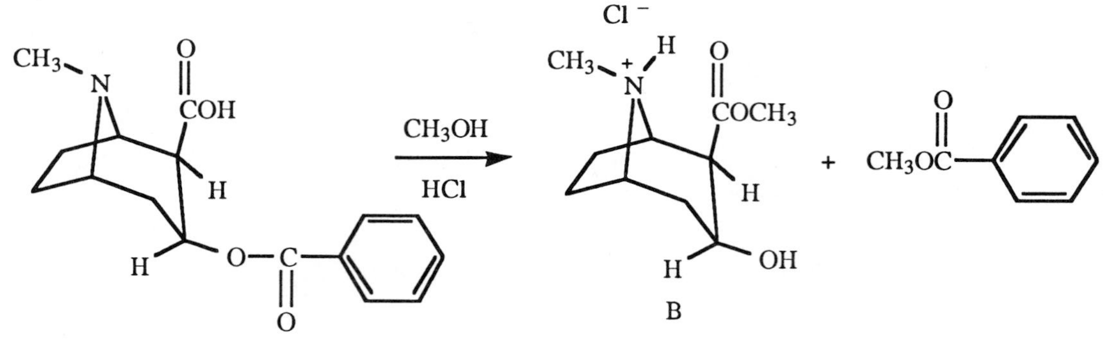

(c)

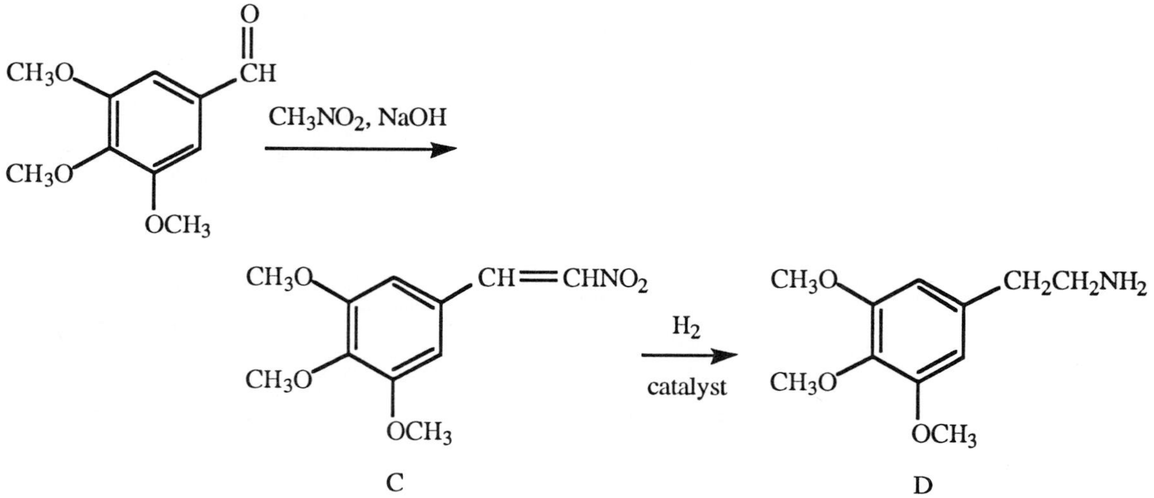

(d)

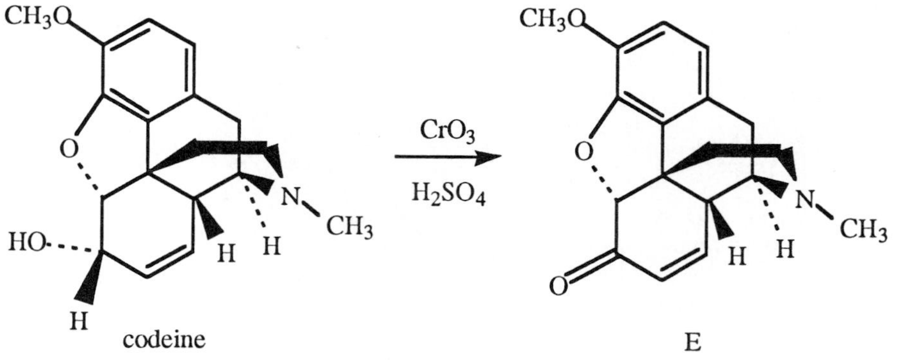

(e)

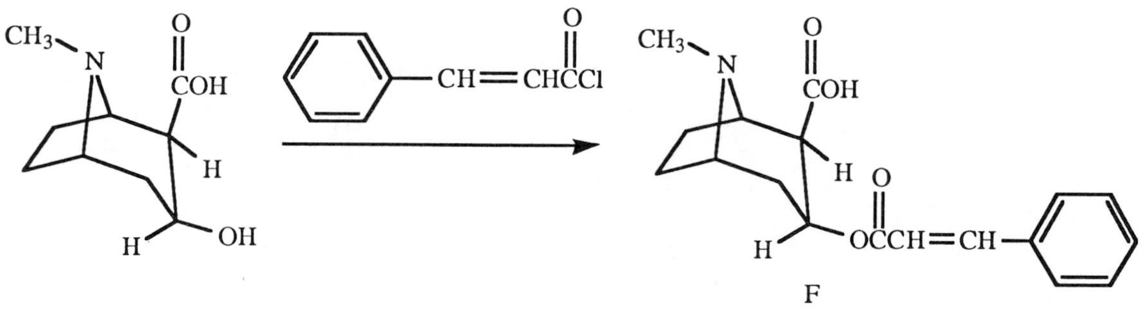

24.41 (cont)

(f)

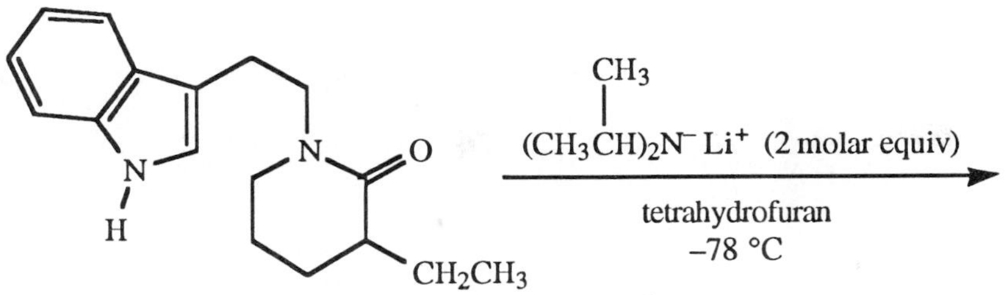

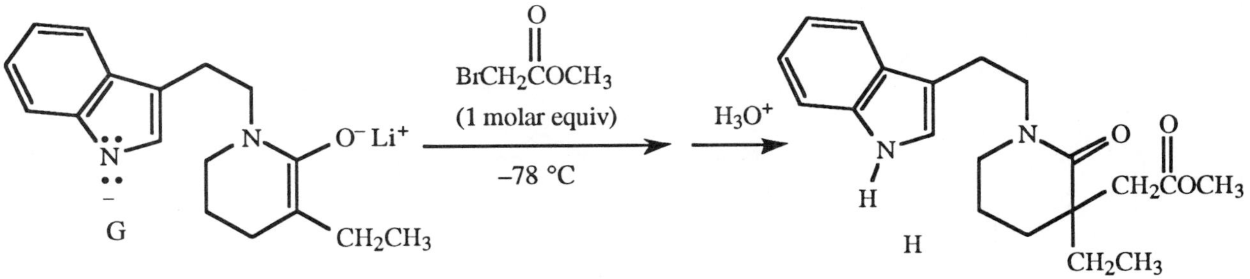

$$\xrightarrow[\substack{\text{tetrahydrofuran} \\ -78\ °C}]{\substack{CH_3 \\ | \\ (CH_3CH)_2N^-\ Li^+\ (2\ \text{molar equiv})}}$$

$$\xrightarrow[-78\ °C]{\substack{O \\ \parallel \\ BrCH_2COCH_3 \\ (1\ \text{molar equiv})}} \xrightarrow{H_3O^+}$$

G

H

24.42

(a)   3-isoquinolinecarboxylic acid   (b)   3-chloropyridine   (c)   2-aminopyrimidine

(d)   3-phenylisoxazolidine   (e)   4-nitropyrazole   (f)   6-chloro-3-phenylquinoline

(g)   methyl 3-furancarboxylate   (h)   1,4-dimethylimidazole   (i)   3-indolecarboxamide

(j)   2-phenylthiazole   (k)   2-thiophenecarboxylic acid

24.43

(a)

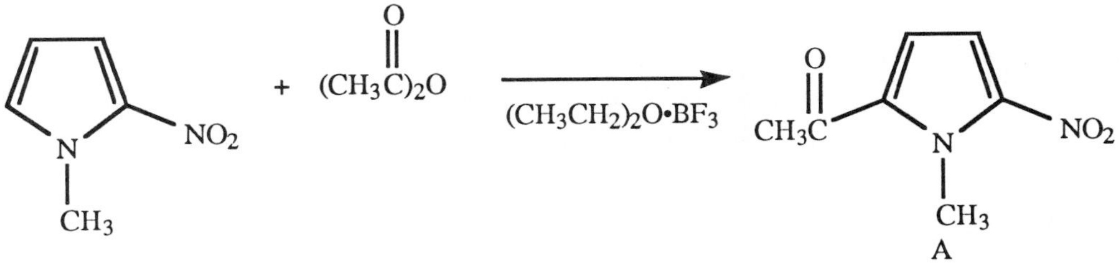

$$+ \ (CH_3C)_2O \ \xrightarrow{(CH_3CH_2)_2O\cdot BF_3}$$

A

(b)

$$+ \ H_2SO_4 \ \longrightarrow$$

B

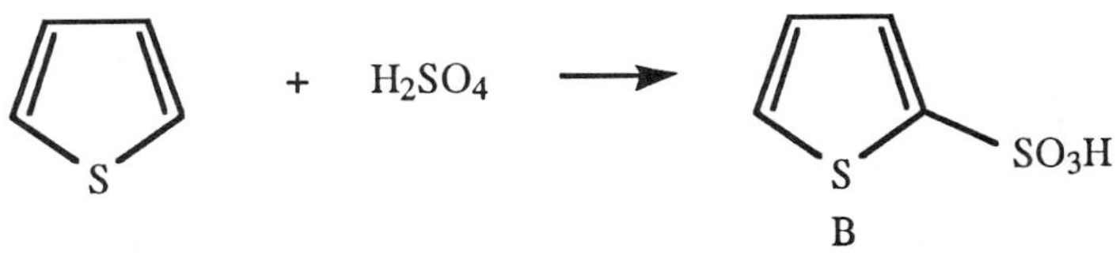

24.43 (cont)

(c)

$$\text{pyridine-COCH}_2\text{CH}_3 \xrightarrow[\text{diethyl ether}]{\text{LiAlH}_4} \xrightarrow{\text{H}_3\text{O}^+} \text{pyridine-CH}_2\text{OH (C)} + \text{CH}_3\text{CH}_2\text{OH}$$

(d)

$$\text{furan(CH}_3\text{O)(OCH}_3\text{)} + \text{H}_2 \xrightarrow{\text{Ni}} \text{tetrahydrofuran(CH}_3\text{O)(OCH}_3\text{) D}$$

(e)

$$\text{pyridine-CH}_2\text{CHCH}_3(\text{CH}_3) \xrightarrow[\substack{\text{H}_2\text{O} \\ \Delta}]{\text{KMnO}_4} \text{pyridine-CO}^-\text{K}^+ \text{(E)}$$

(f)

$$\text{thiophene} + (\text{CH}_3\text{CH}_2\text{CH}_2\overset{\text{O}}{\text{C}})_2\text{O} \xrightarrow[\text{acetic acid}]{\text{BF}_3} \text{thiophene-CCH}_2\text{CH}_2\text{CH}_3 \text{ (F)}$$

(g)

$$\text{O}_2\text{N-pyridine-NH}_2 \xrightarrow[\substack{\text{H}_2\text{O} \\ 0\ °\text{C}}]{\text{NaNO}_2,\ \text{HF}} \text{O}_2\text{N-pyridine-N}{\equiv}\text{N}^+\ \text{F}^- \text{ (G)}$$

(h)

$$\text{4-chloropyridine} + \text{aniline} \longrightarrow \text{pyridine-NH-phenyl (H)}$$

(i)

$$\text{pyrrole(N-H)} \xrightarrow{\text{KOH}} \text{pyrrole(N}^-\text{)K}^+ \text{ (I)} \xrightarrow{\overset{\text{O}}{\text{ClCOCH}_2\text{CH}_3}} \text{pyrrole(N-COCH}_2\text{CH}_3\text{) (J)}$$

24.43 (cont)

(j)

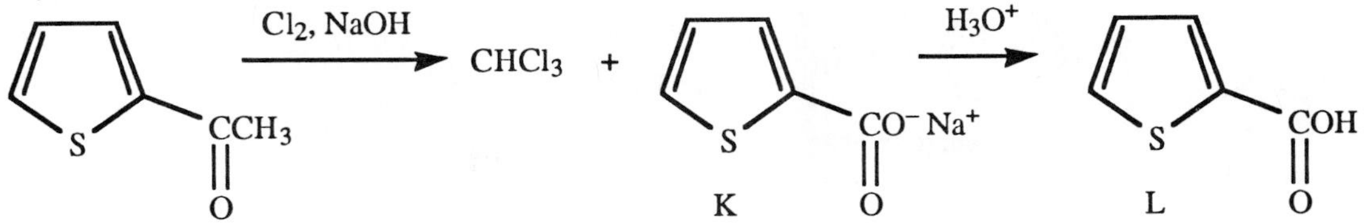

(k)

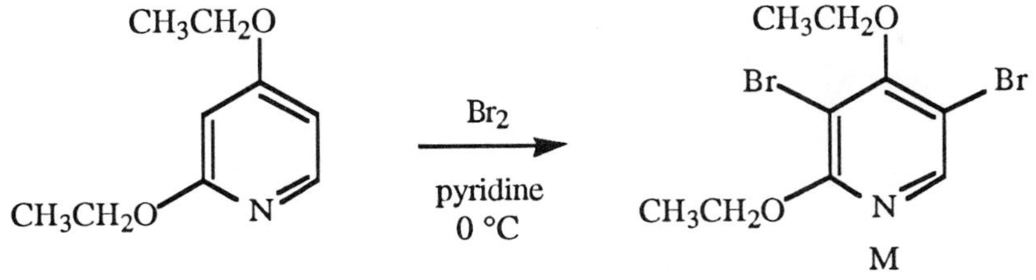

(l)

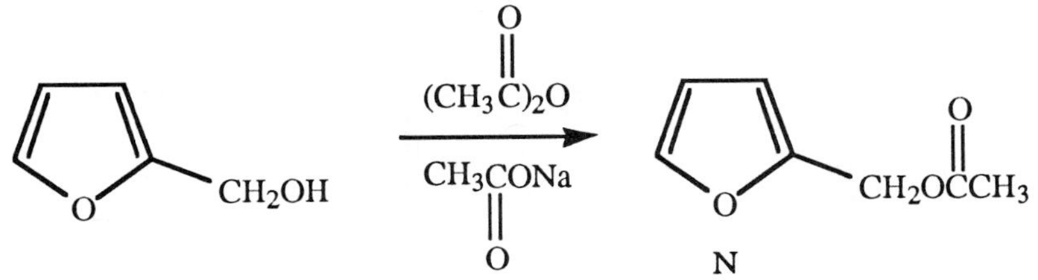

(m)

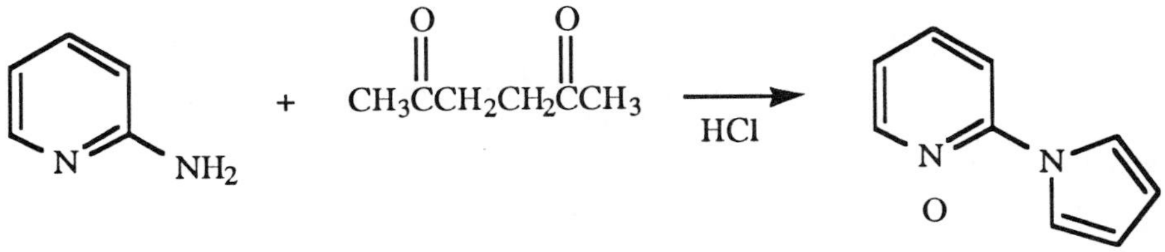

(n)

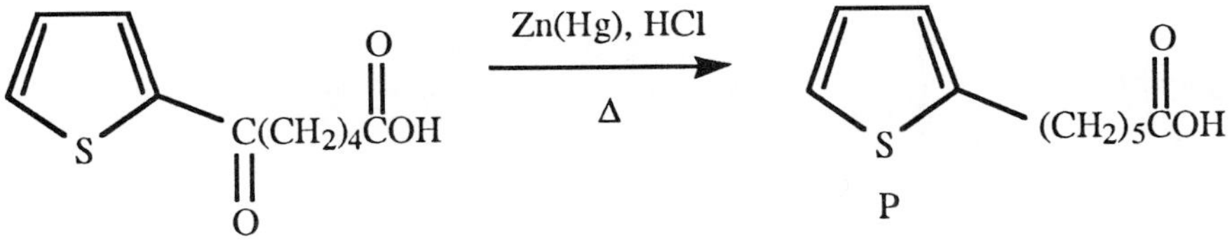

(o)

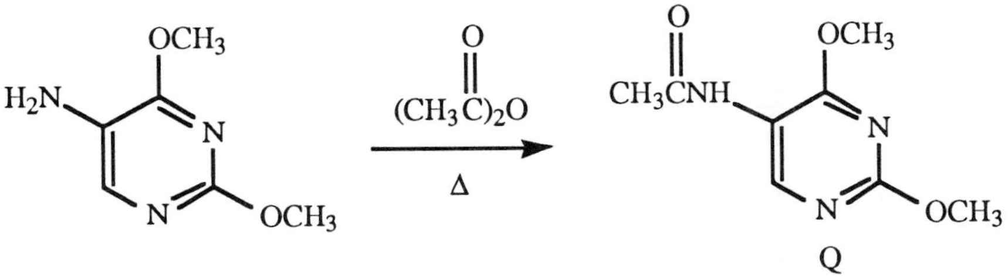

24.44

(a)

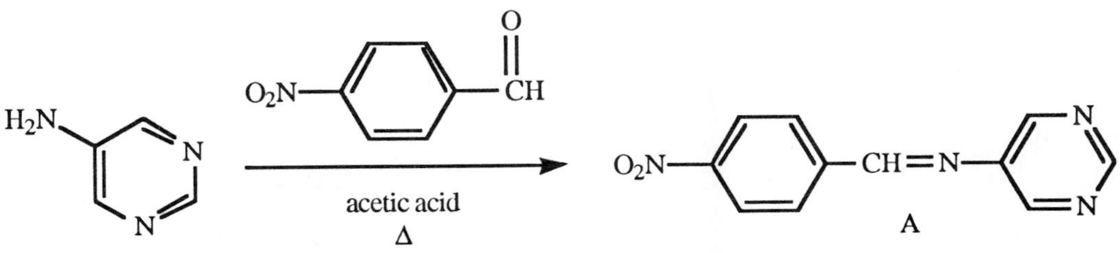

24.44 (cont)

(b)

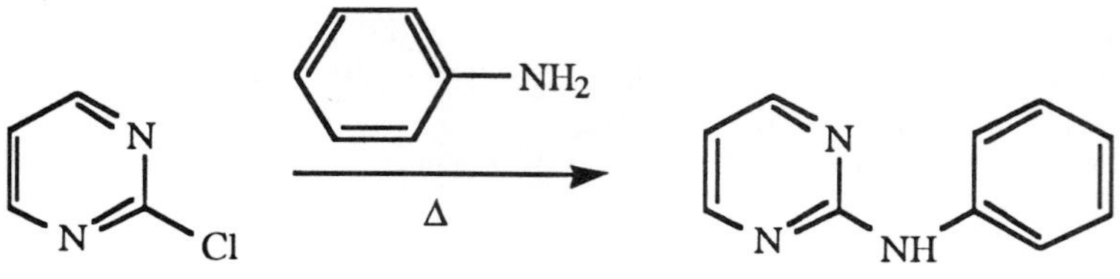

B

(c)

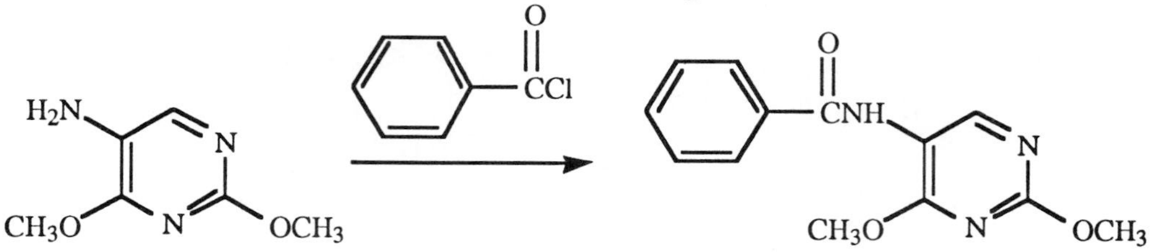

C

(d)

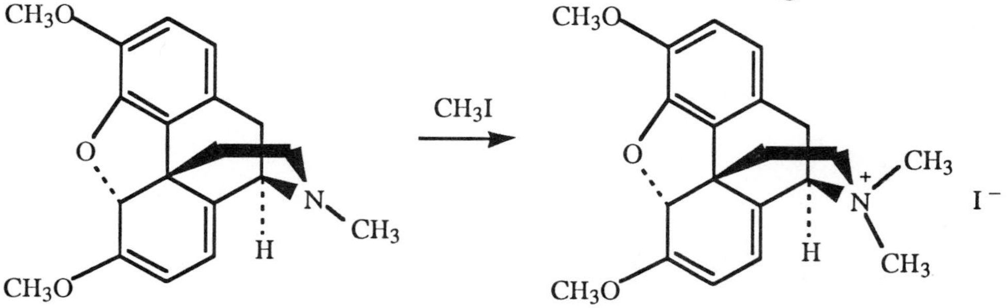

thebaine   D

(e)

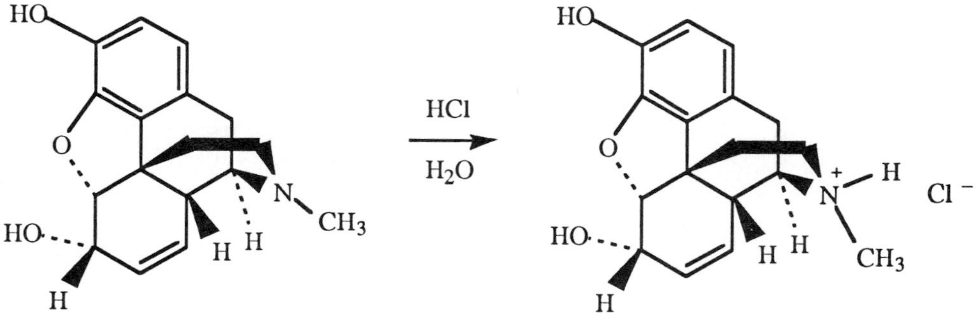

morphine   E

(f)

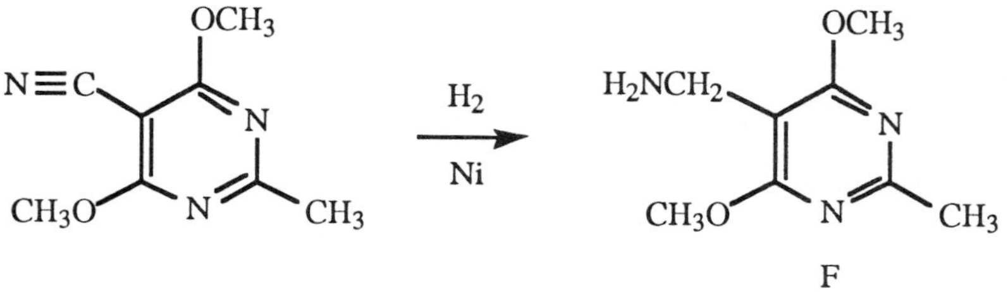

F

(g)

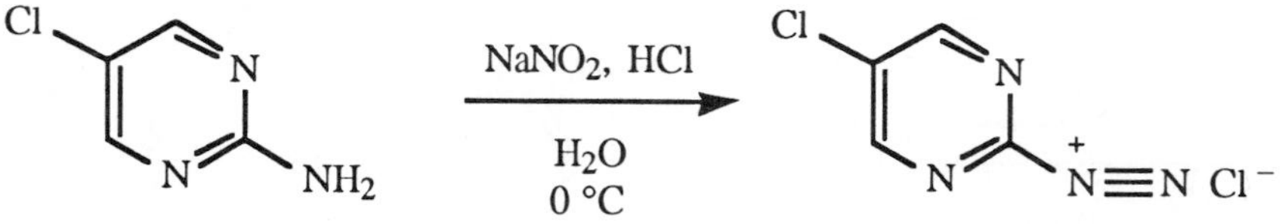

G

24.44 (cont)

(h)

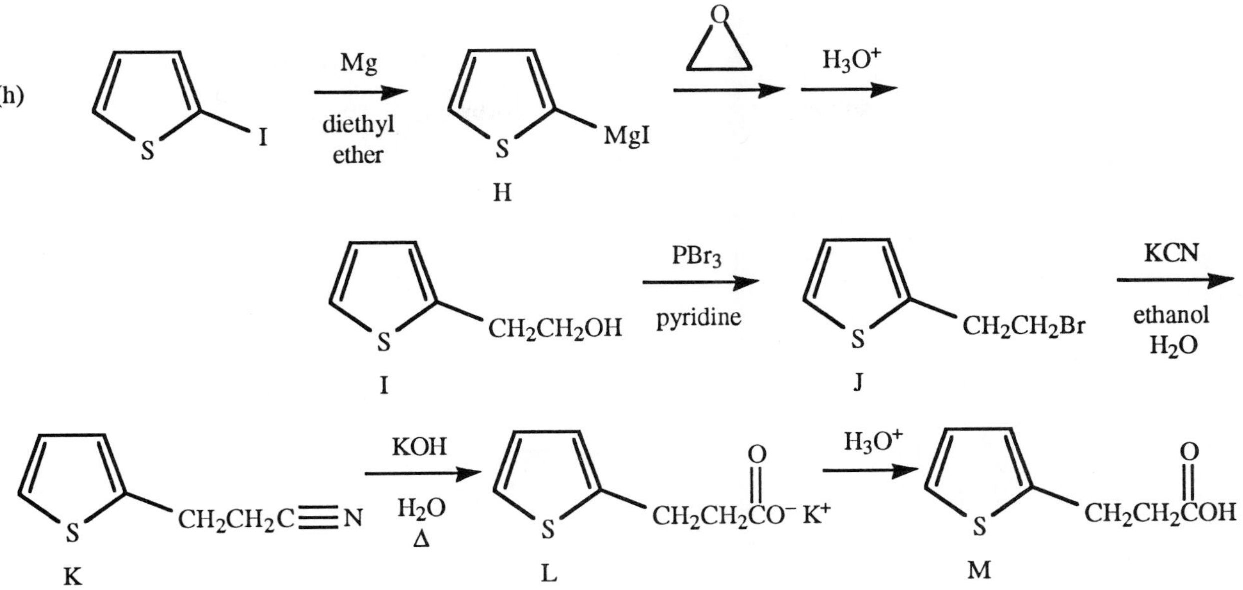

(i)

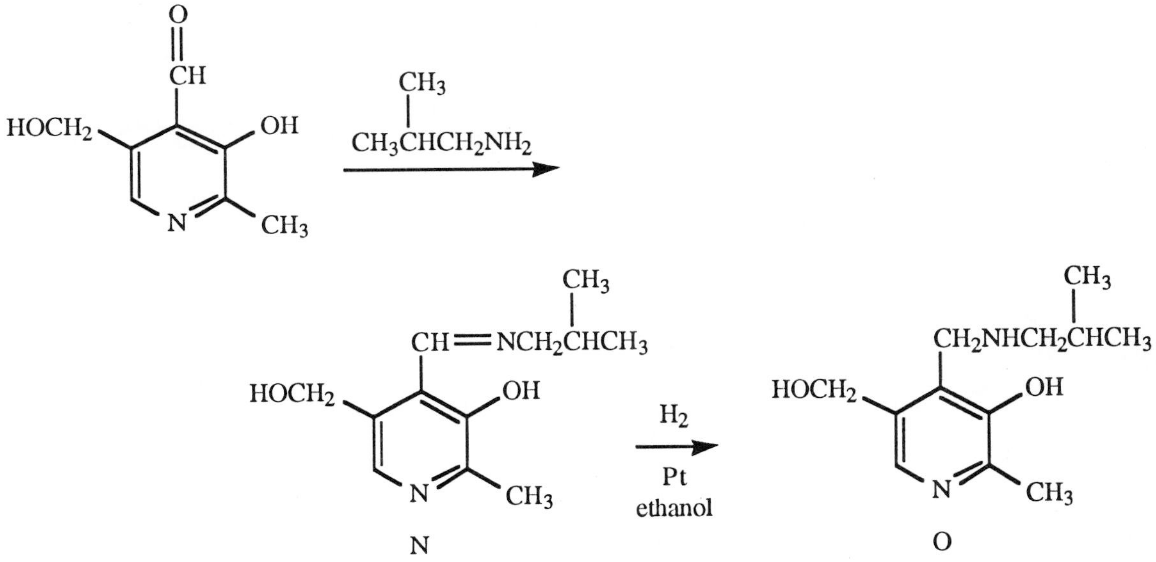

(j)

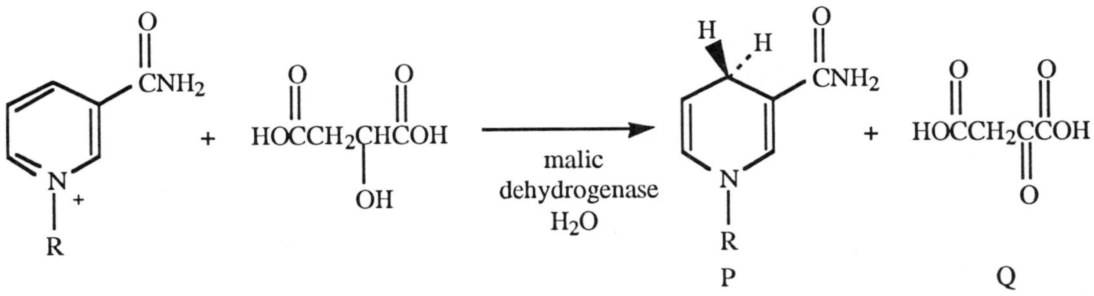

24.45

(a)

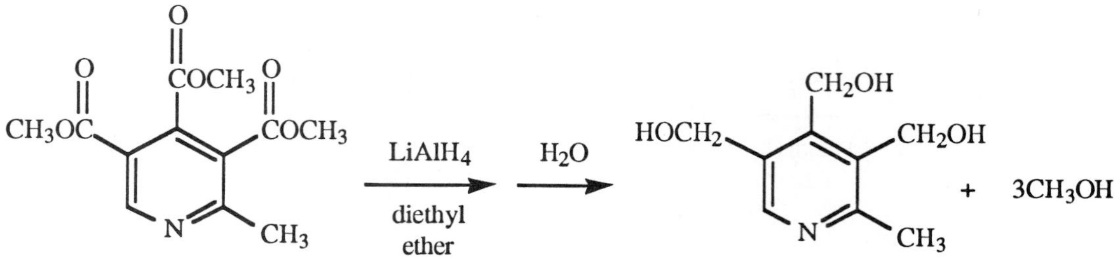

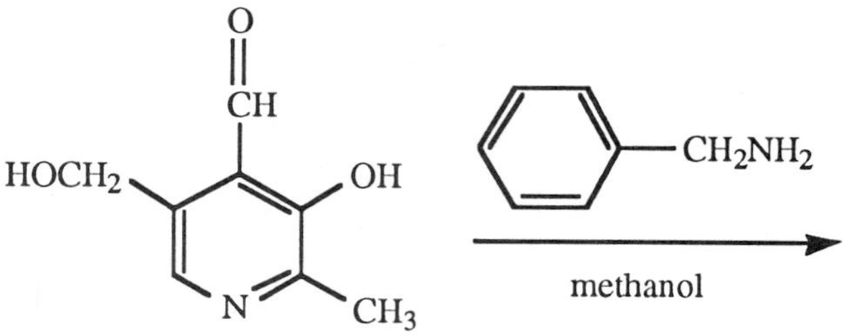

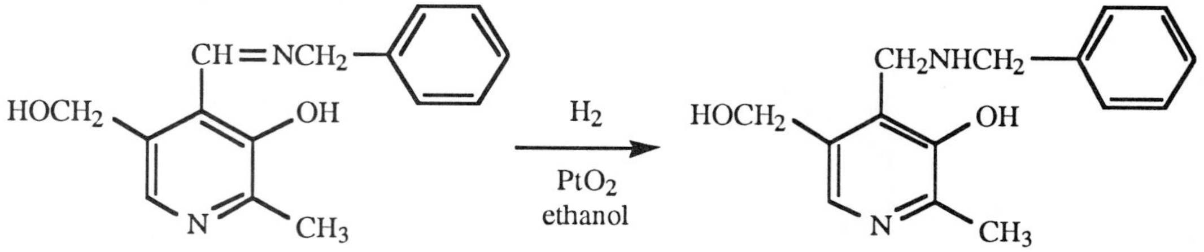

(b)

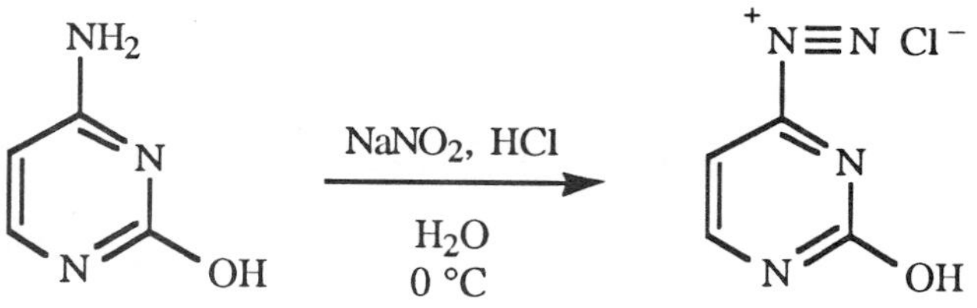

(c)

(d)

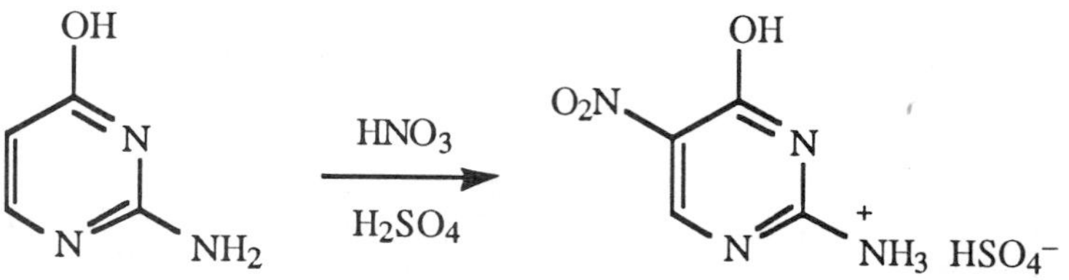

(e)

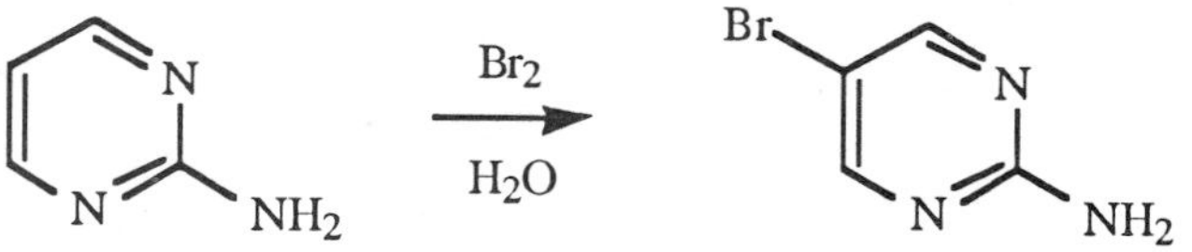

24.45 (cont)

(f)

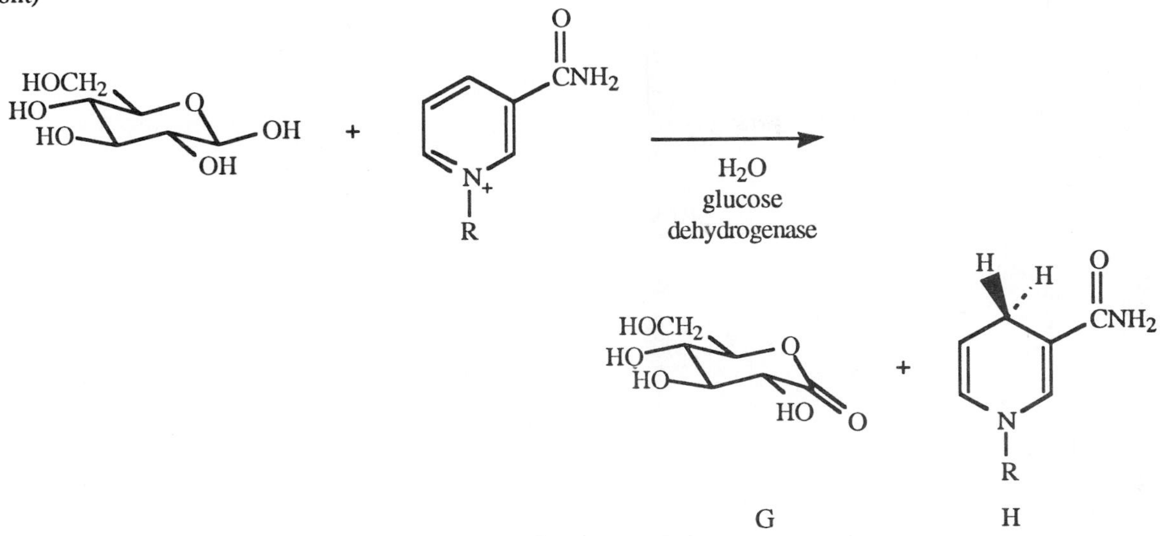

G 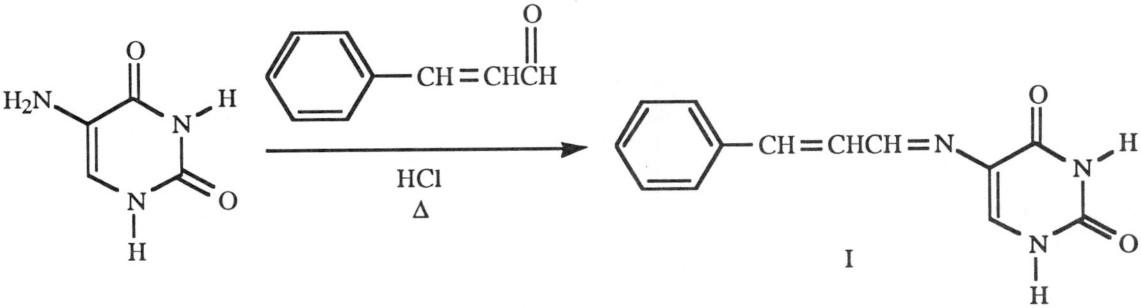   H

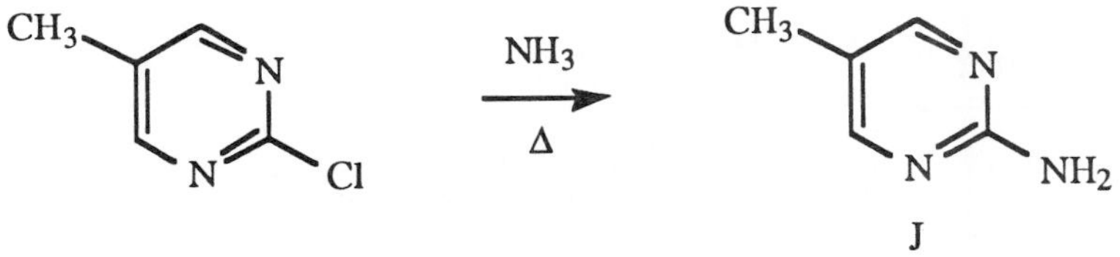

(g)

I

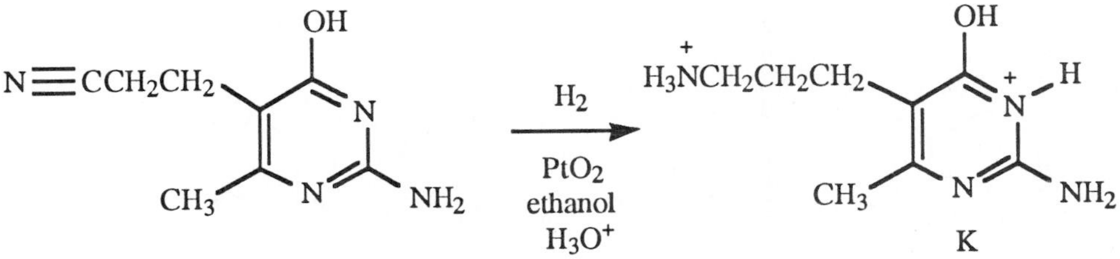

(h)

J

(i)

K

(j)

L

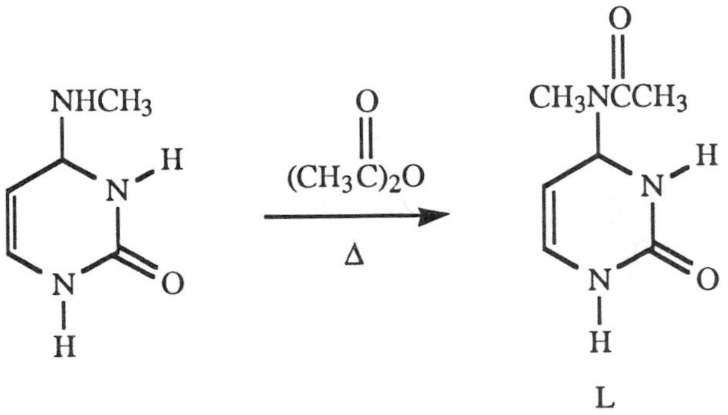

24.45 (cont)

(k)

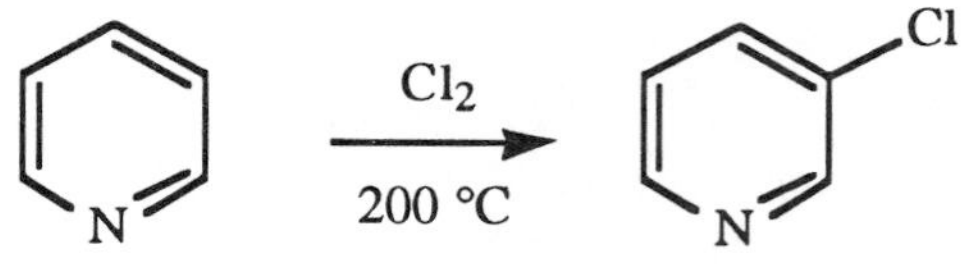

(l)

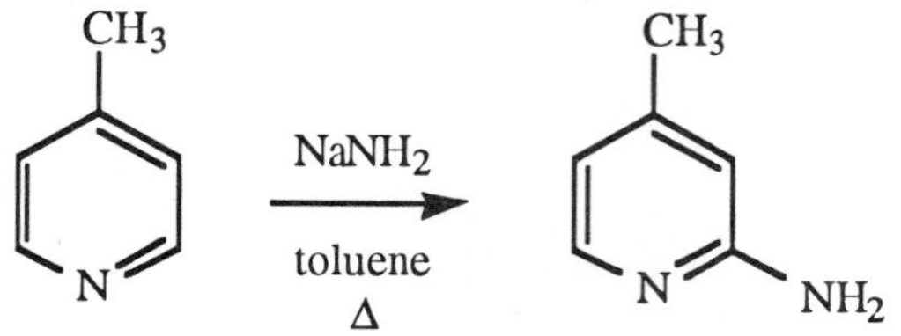

(m)

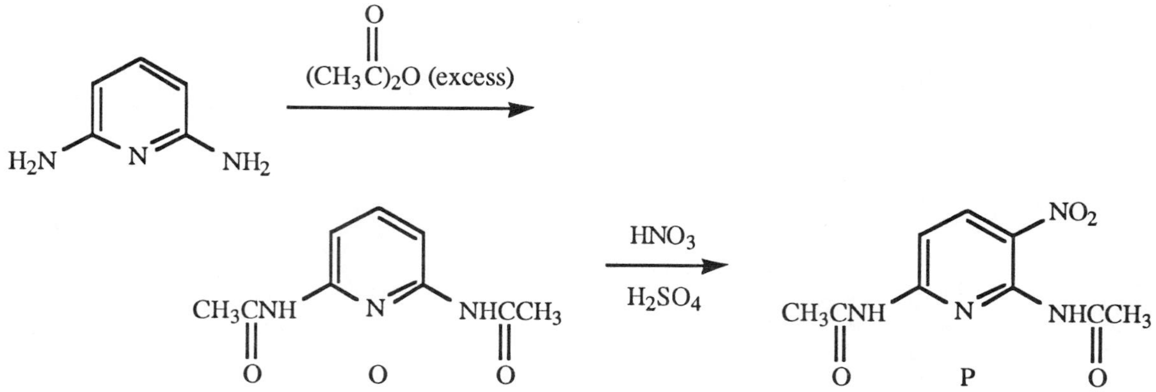

(n)

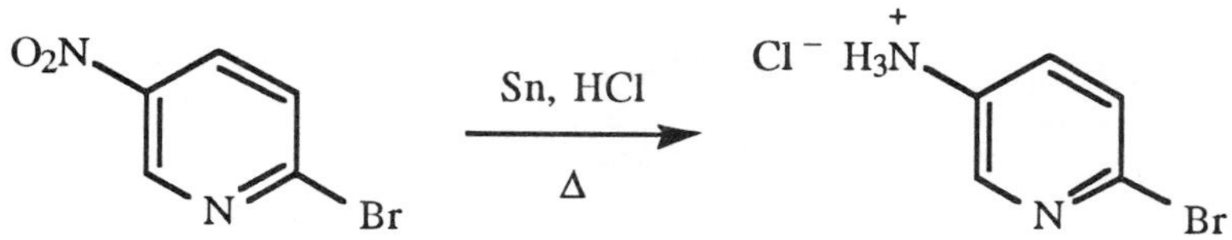

(o)

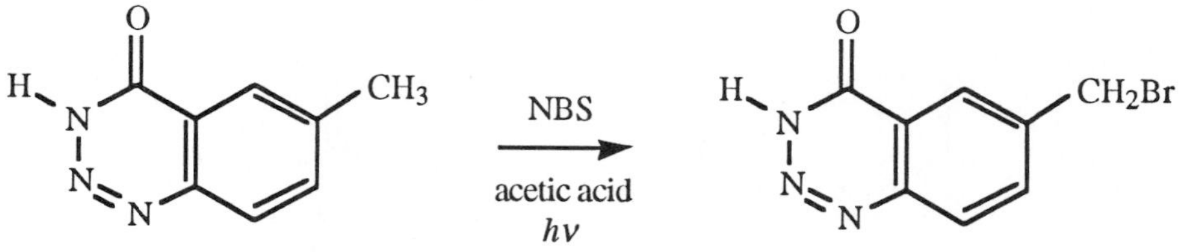

24.46

24.46 (cont)

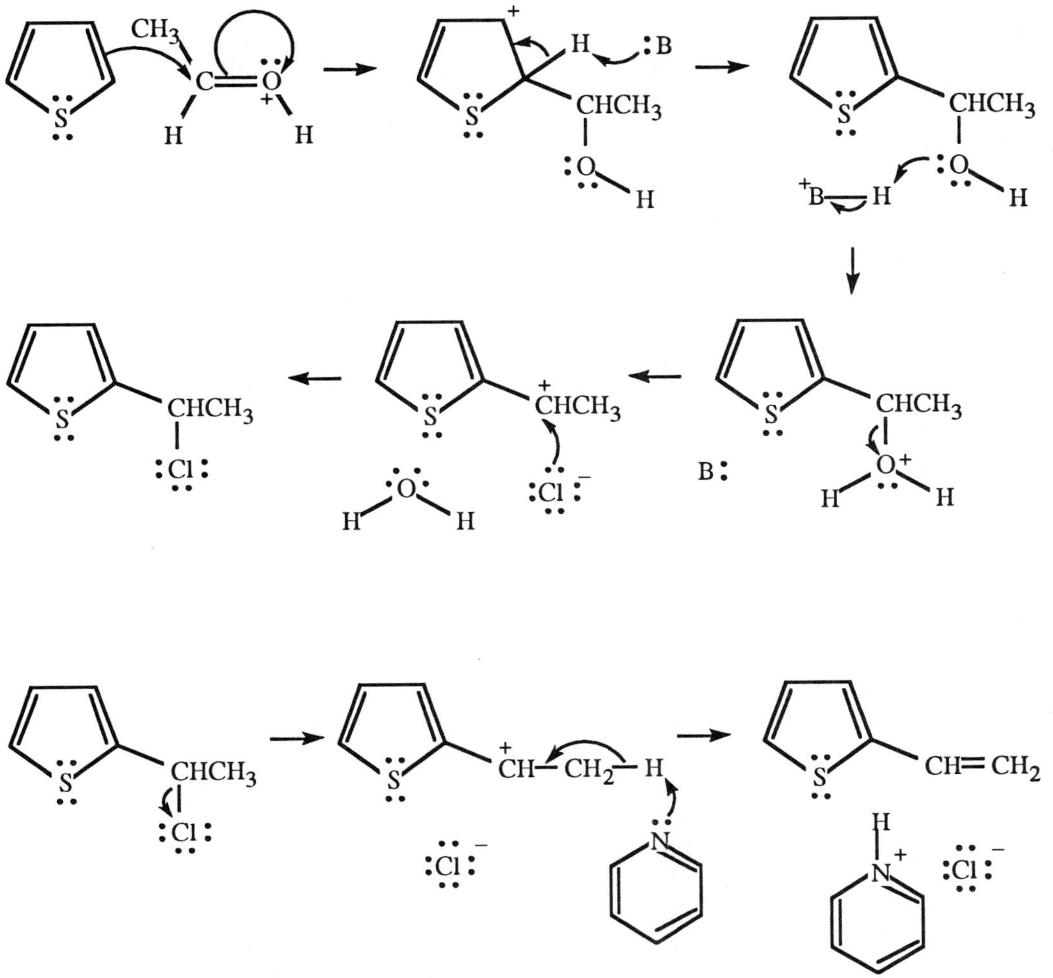

A carbocation adjacent to the thiophene ring is like a benzylic cation in stability so S_N1 and E1 reactions have been shown. S_N2 and E2 reactions are also a possibility.

24.47

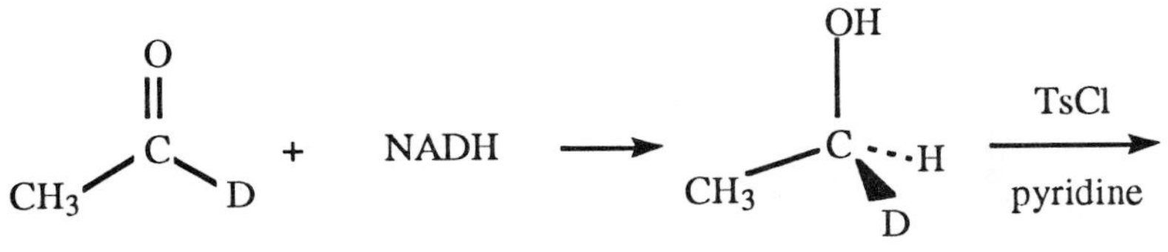

A
(*S*)-ethanol-1-*d*

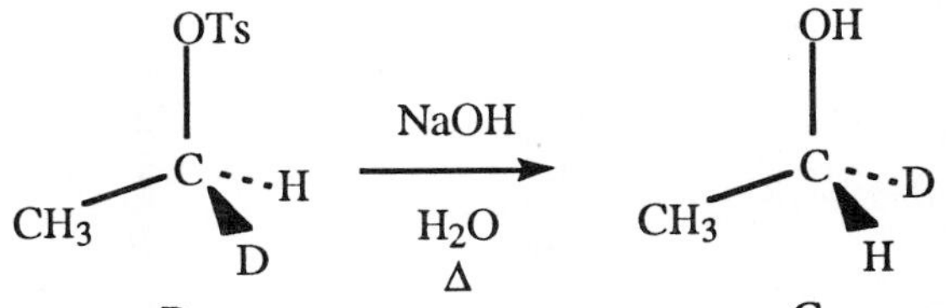

B

C
(*R*)-ethanol-1-*d*

24.47 (cont)

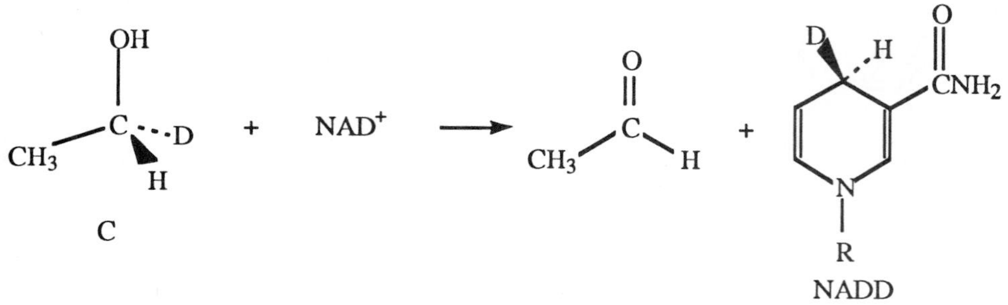

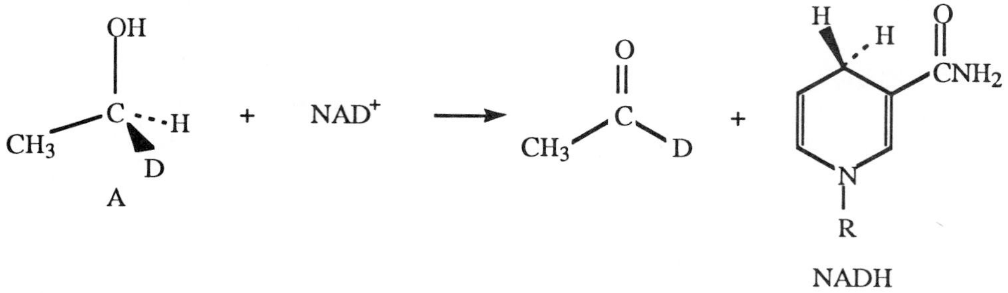

Alcohol A is converted to tosylate B with retention of configuration. The reaction of the tosylate B with hydroxide ion occurs by an S_N2 reaction with inversion of configuration. Thus an optically active alcohol is converted into its enantiomer. When the two alcohols, A and C, are oxidized with NAD$^+$, deuterium is removed from C and hydrogen from A.

These reactions are the reverse of those used to prepare the alcohols by enzymatic reductions. Reduction of acetaldehyde-1-d by NADH gave Alcohol A, while that of acetaldehyde by NADD gave Alcohol C (see p. 1091 in text).

The enzymatic oxidation-reduction reactions are stereoselective, so the proof that the enantiomeric alcohols lose deuterium and hydrogen selectively in the experiments outlined above also proves that the original reduction reactions give rise to enantiomeric ethanol-1-d species.

24.48

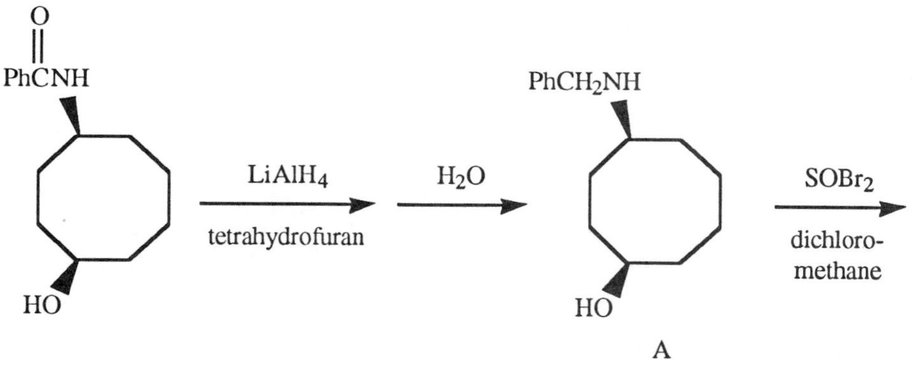

24.48 (cont)

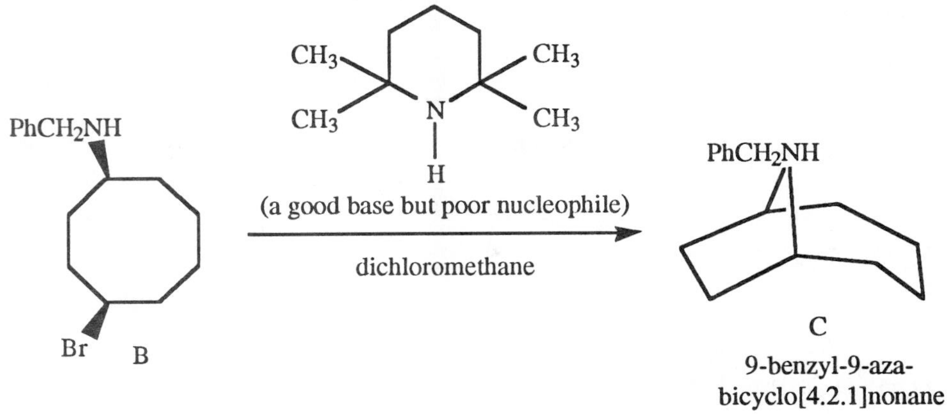

24.49

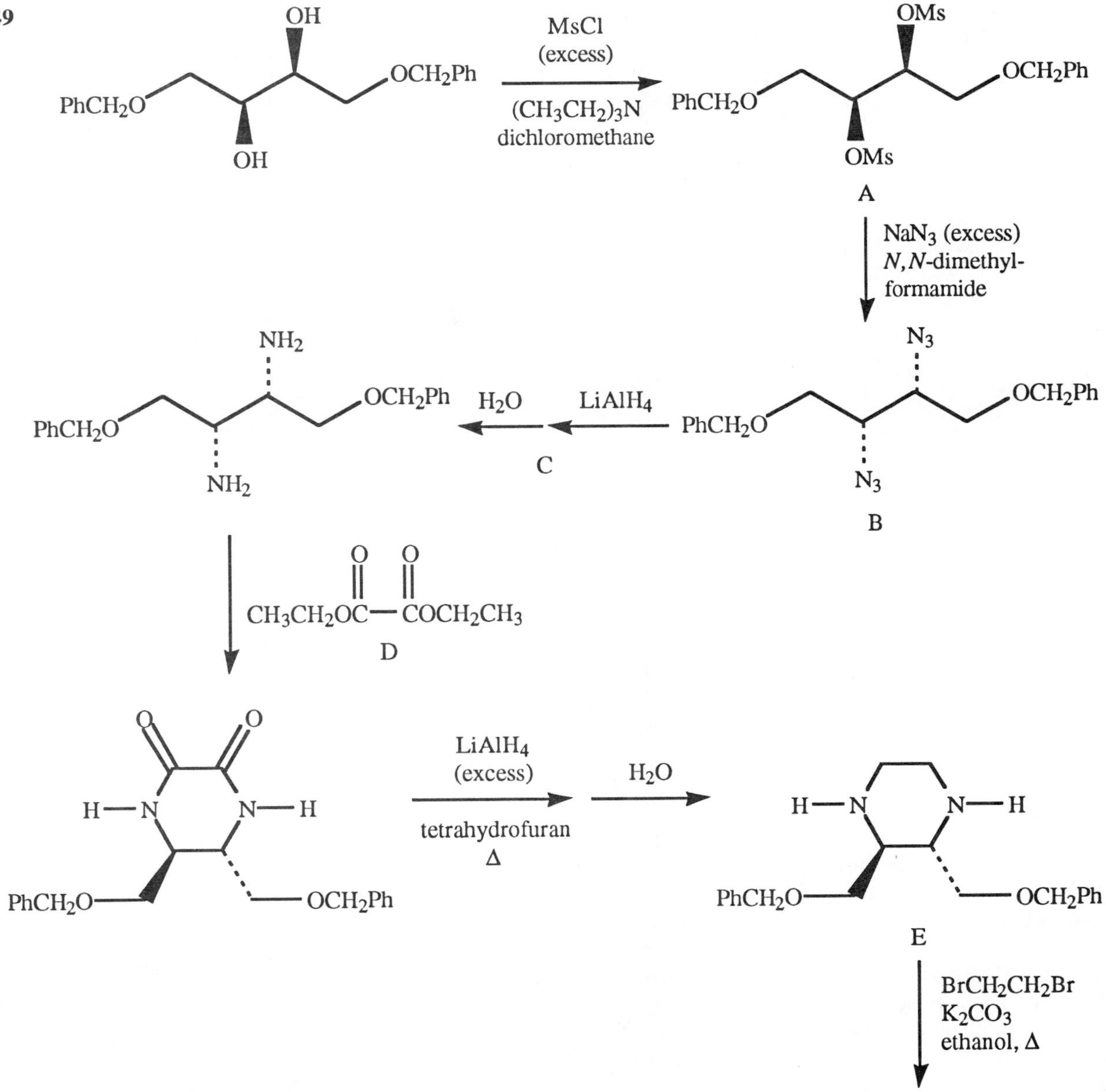

24.49 (cont)

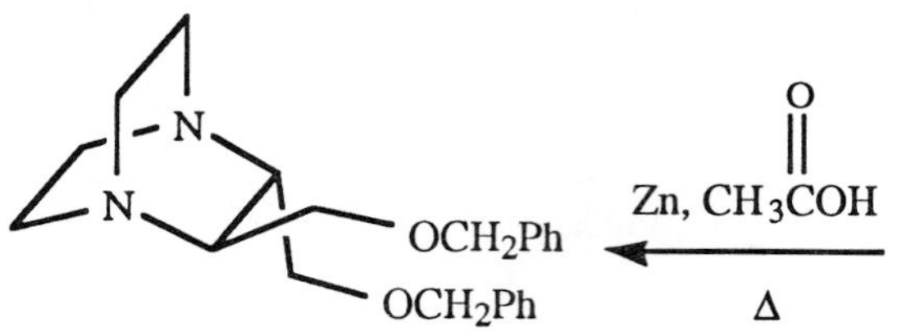

24.50

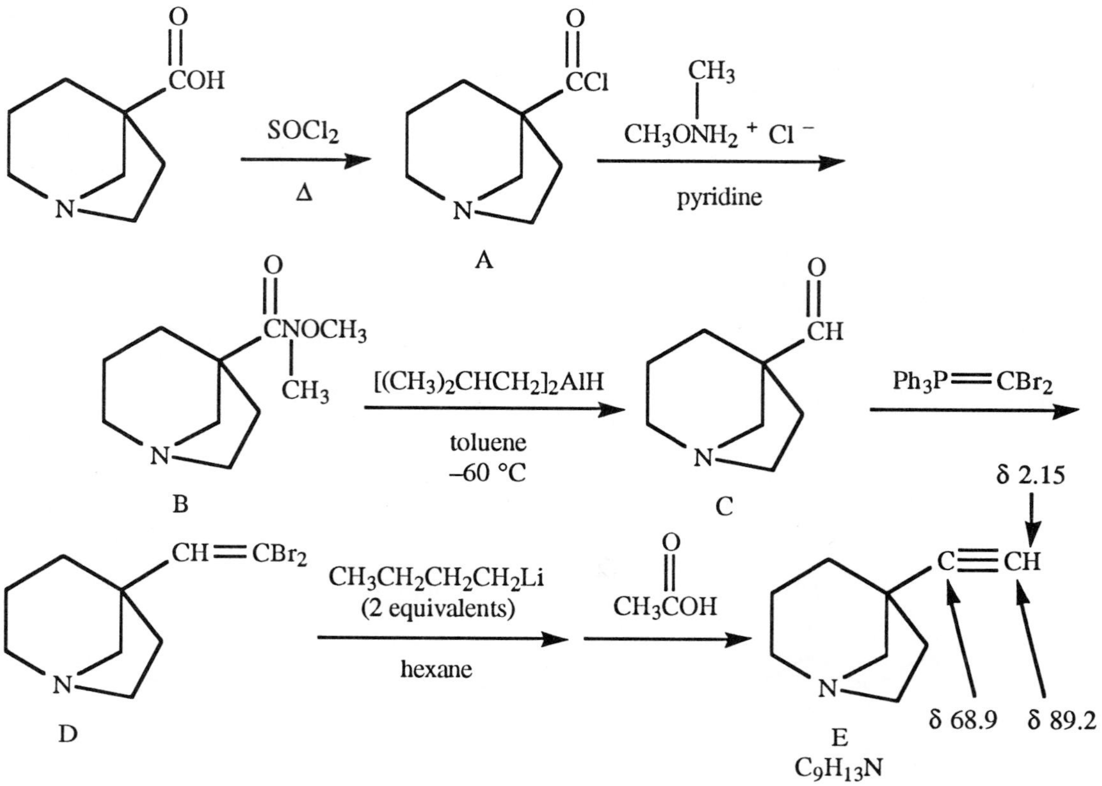

δ 2.15

δ 68.9   δ 89.2

E
$C_9H_{13}N$

24.51

| 4-methylimidazole | imidazole | benzimidazole | 4-nitroimidazole |
| least acidic | | | |

pK_a values   7.5   7.0   5.5   -0.1

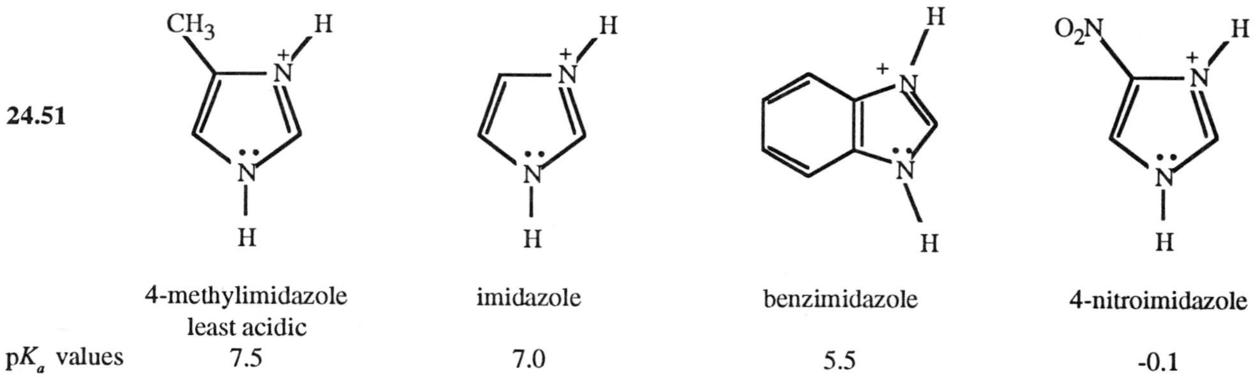

The acidity of each of these species is related to the basicity of the conjugate base, which, in turn, is related to the availability of nonbonding electrons on a nitrogen atom. In all of the imidazoles, basicity is derived from both of the nitrogen atoms in the ring because electrons from both are involved in stabilization of the conjugate acid by delocalization of charge. For example:

24.51 (cont)

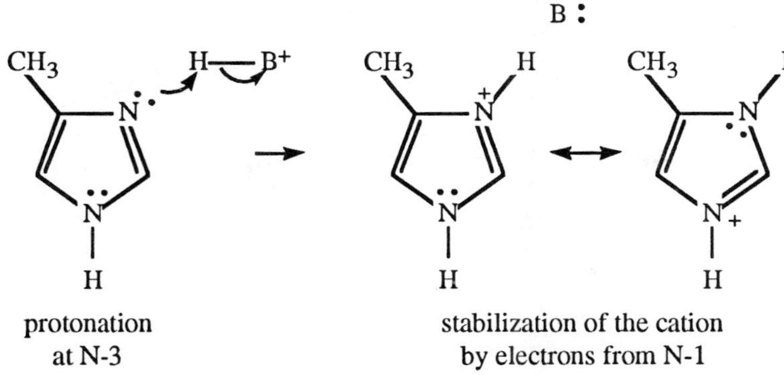
B:
CH3
H—B+
N
H
CH3
H
N
H
CH3
H
N
N+
H
protonation
at N-3
stabilization of the cation
by electrons from N-1

We expect 4-methylimidazole to be more basic than imidazole because the methyl group is electron-donating. In benzimidazole the nonbonding electrons are delocalized to the aromatic ring, making them less available at the nitrogen atoms.

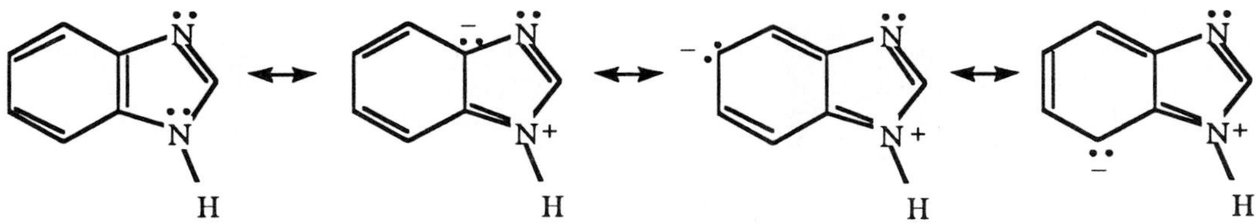
N
N
H
N
N+
H
N
N+
H
N
N+
H

In 4-nitroimidazole, the strong electron-withdrawing inductive effect of the nitro group combines with the resonance effect to greatly reduce electron density in the ring and at the nitrogen atoms.

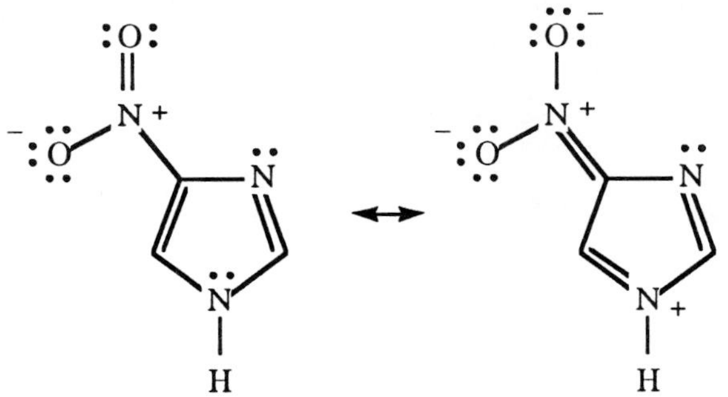
:O:
N+
O
N
N
H
O
N+
O
N
N+
H

24.52

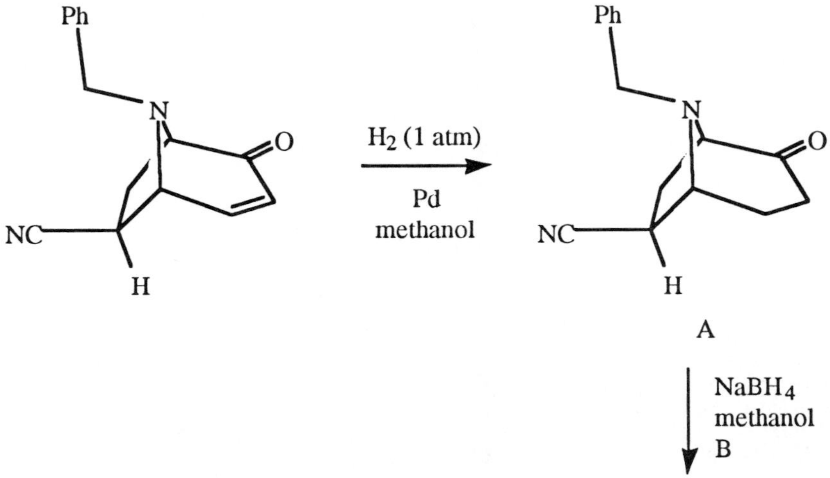
Ph
N
O
NC
H
H2 (1 atm)
Pd
methanol
Ph
N
O
NC
H
A
NaBH4
methanol
B

24.52 (cont)

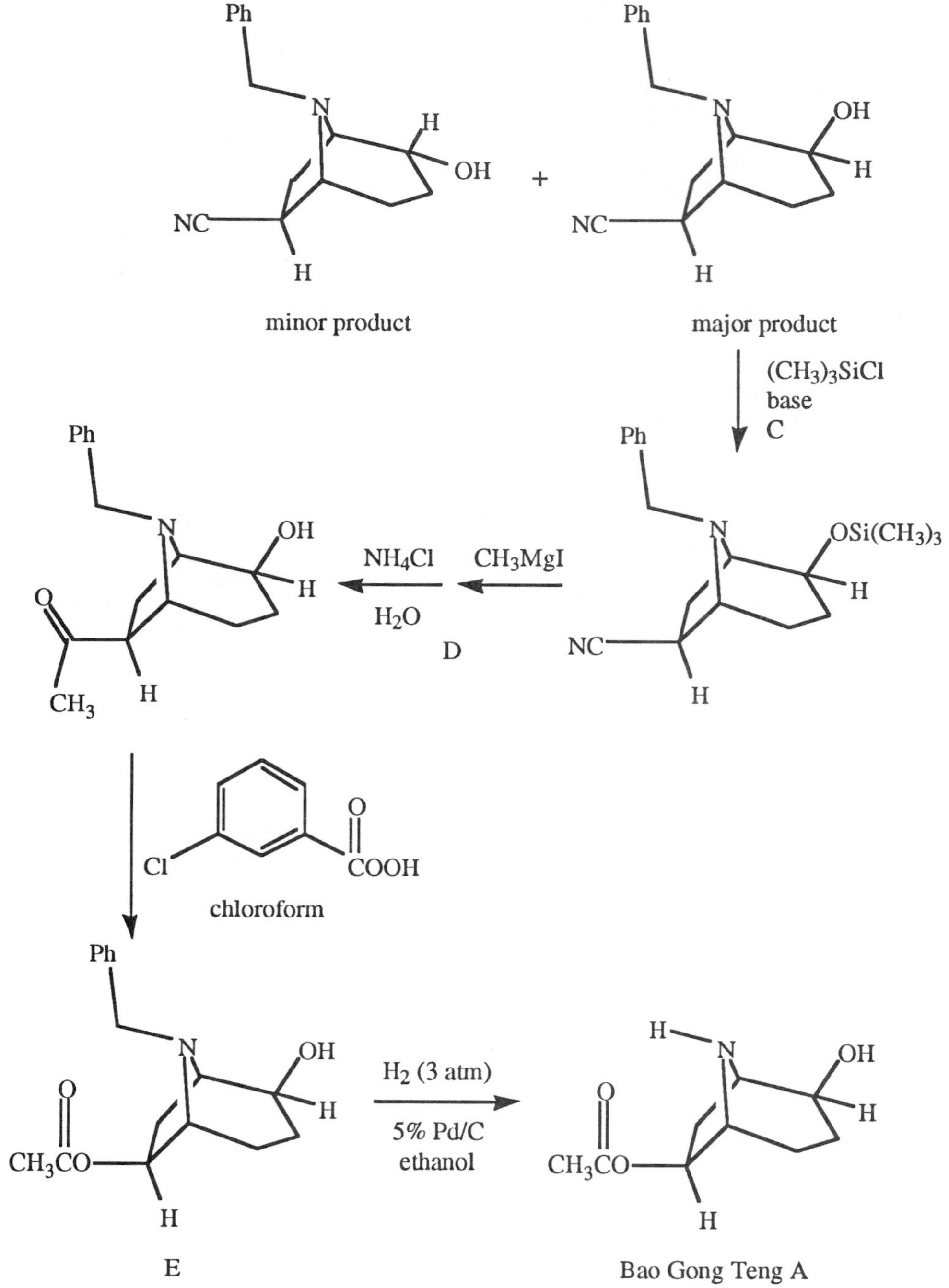

minor product   +   major product

$(CH_3)_3SiCl$
base
C

NH_4Cl CH_3MgI
H_2O
D

chloroform

H_2 (3 atm)
5% Pd/C
ethanol

E   Bao Gong Teng A

24.53

(a)

$AlCl_3$ $SOCl_2$
Δ

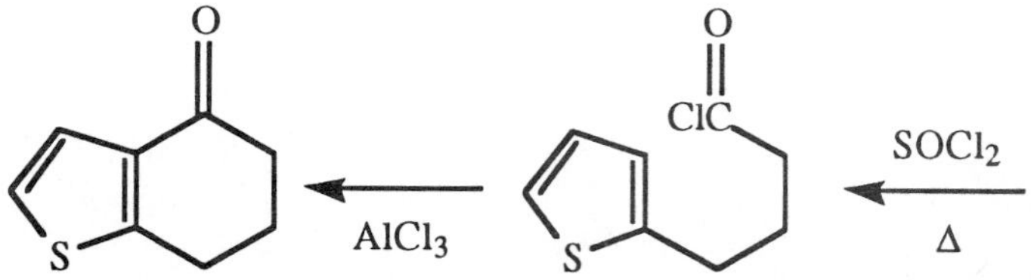

24.53 (a) (cont)

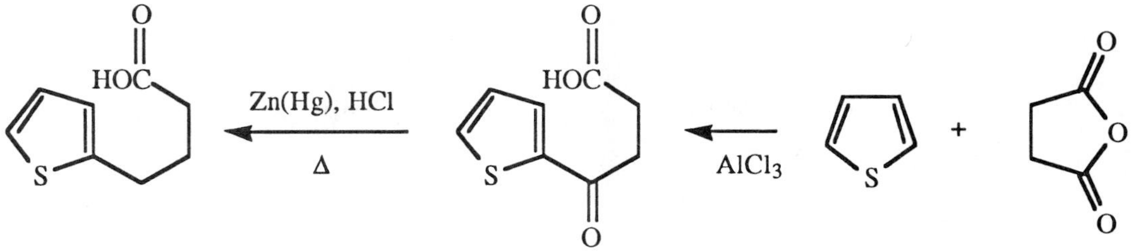

(b)

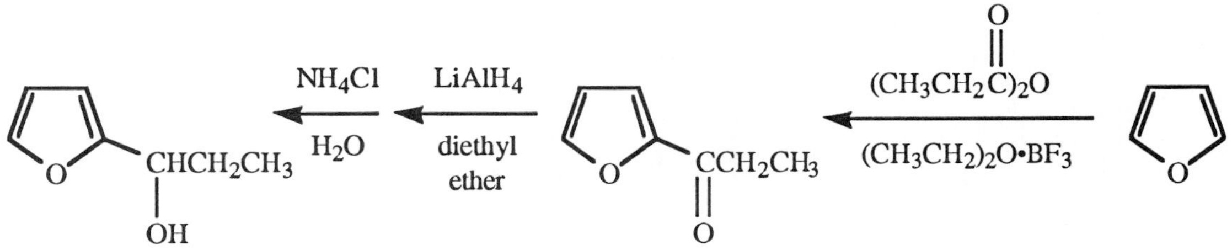

(c)

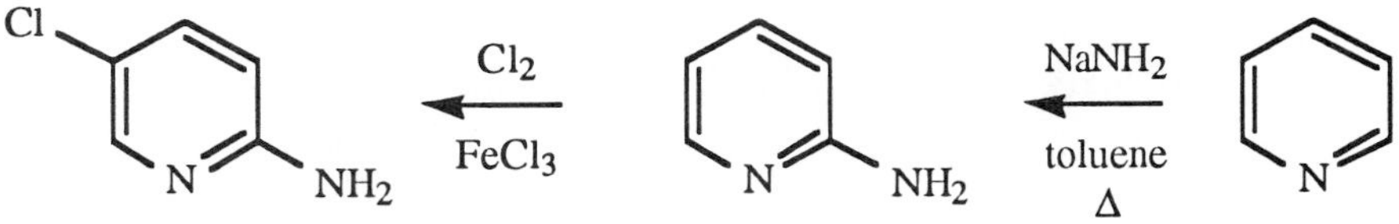

(d)

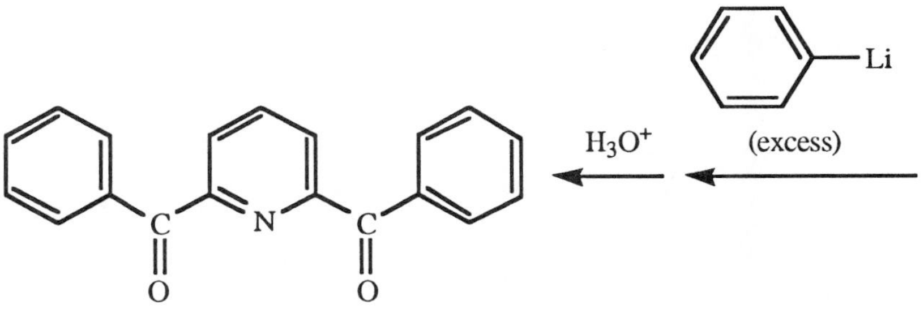

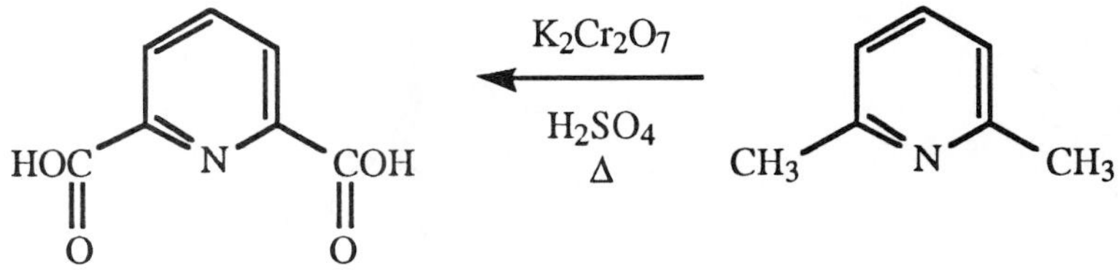

(e)

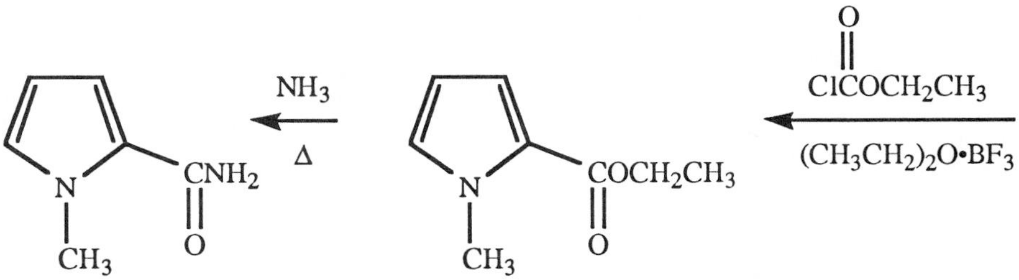

24.53 (e) (cont)

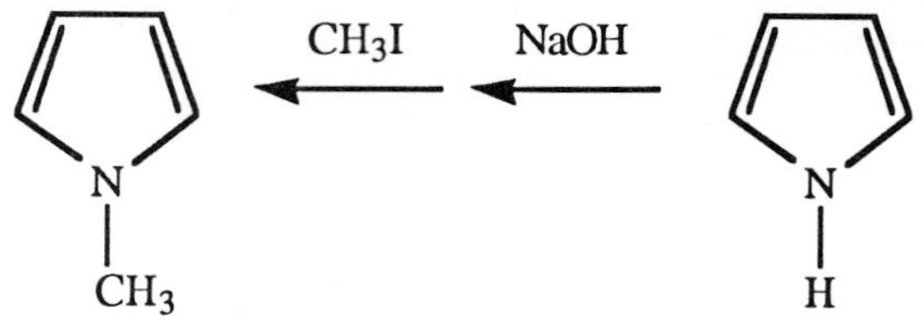

(f)

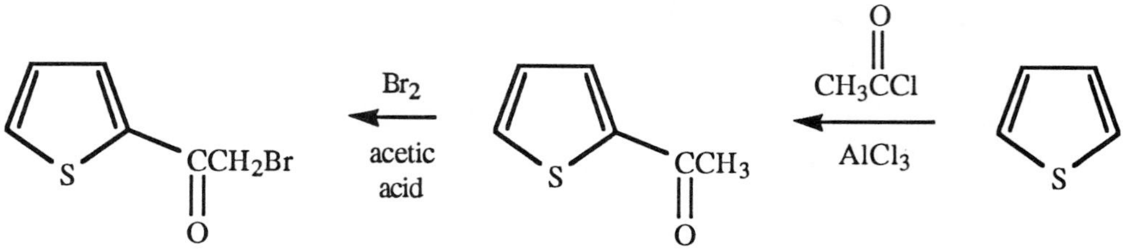

(g)

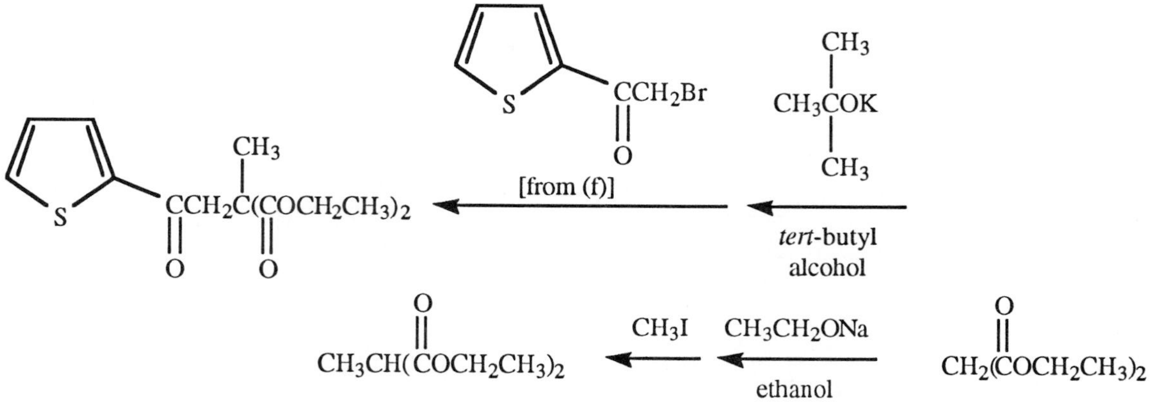

(h)

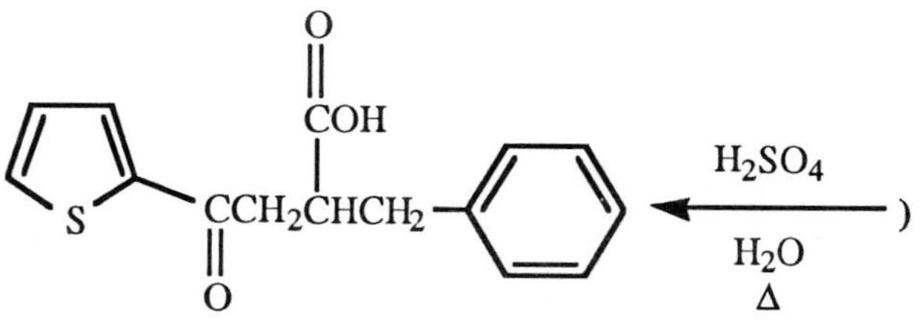

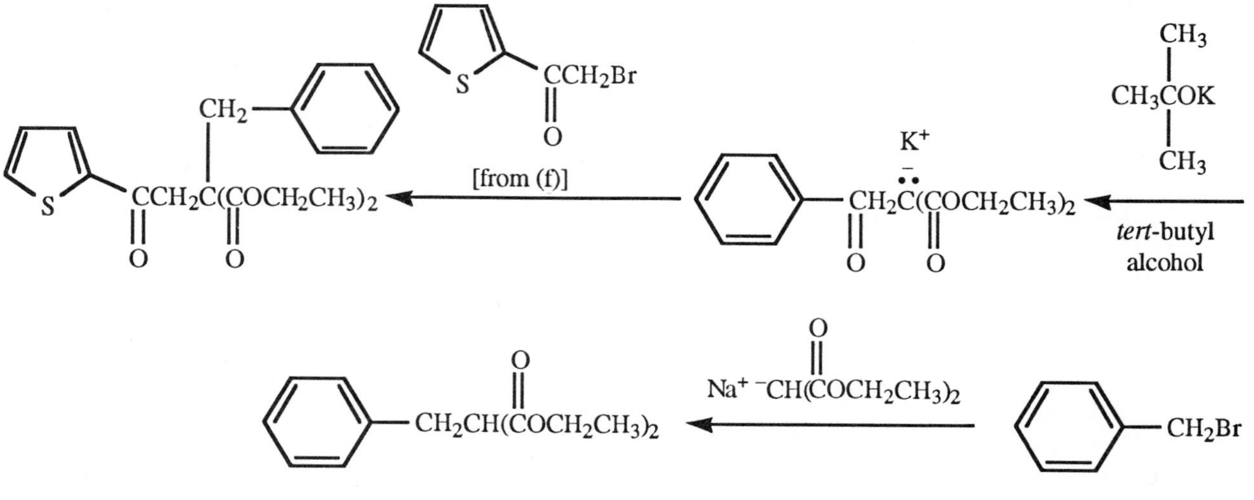

24.54

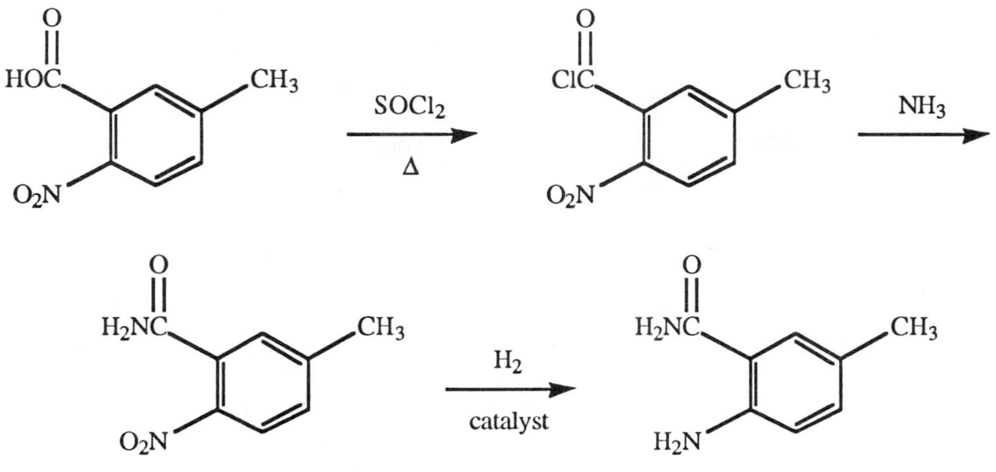

24.55

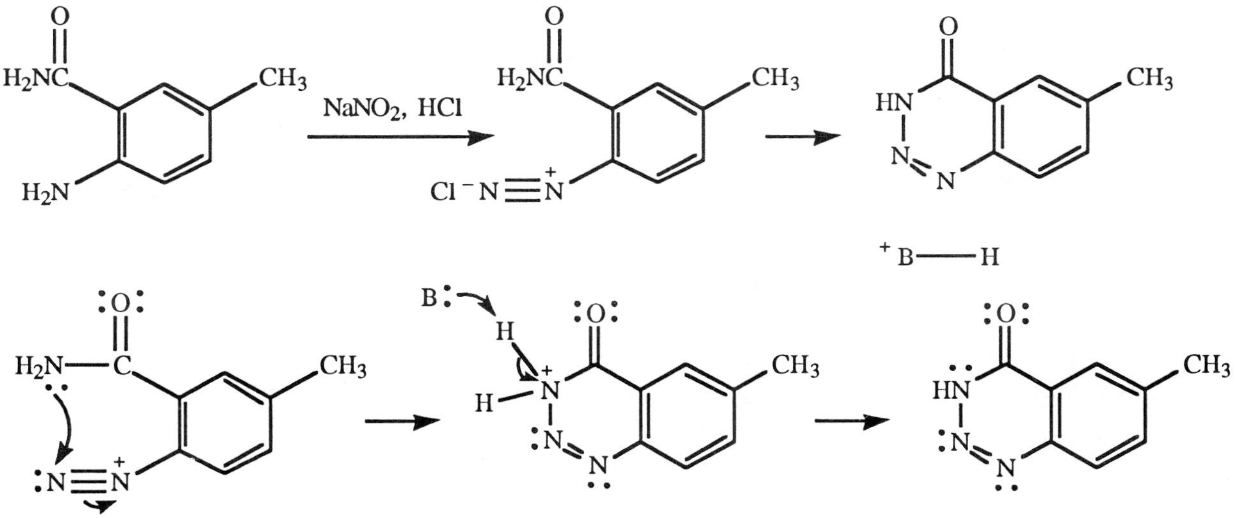

24.56

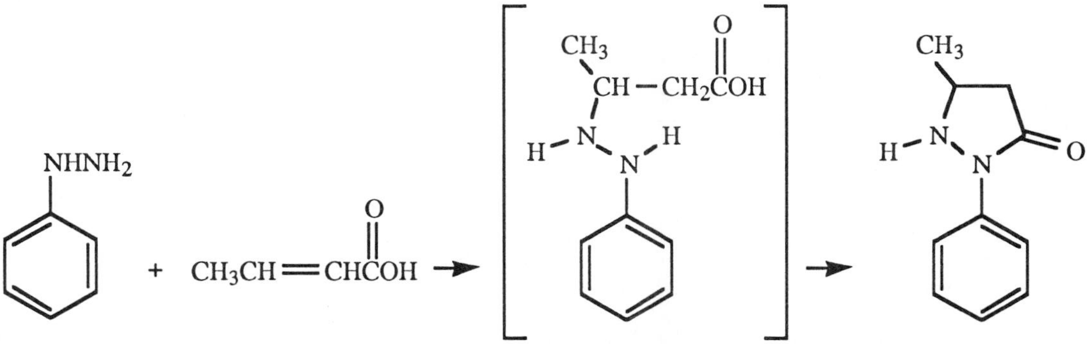

product from nucleophilic
addition to α,β-unsaturated
compound

$C_{10}H_{12}N_2O$

24.56 (cont)

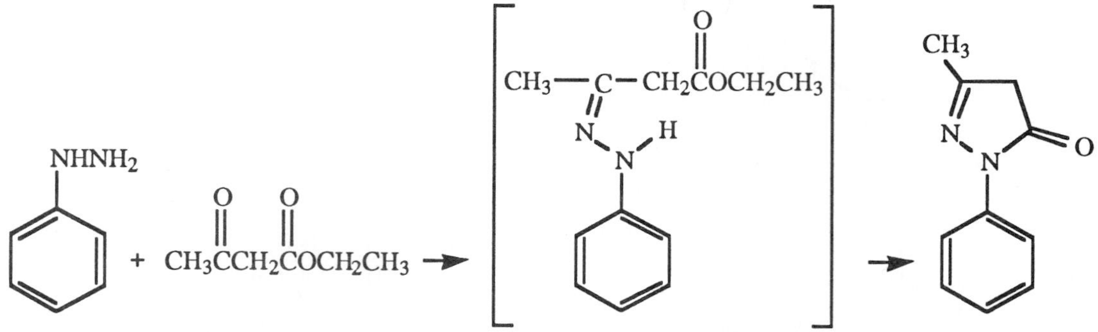

product from nucleophilic
addition to carbonyl of ketone
followed by loss of water to
give a phenylhydrazone

$C_{10}H_{10}N_2O$

24.57

(a)

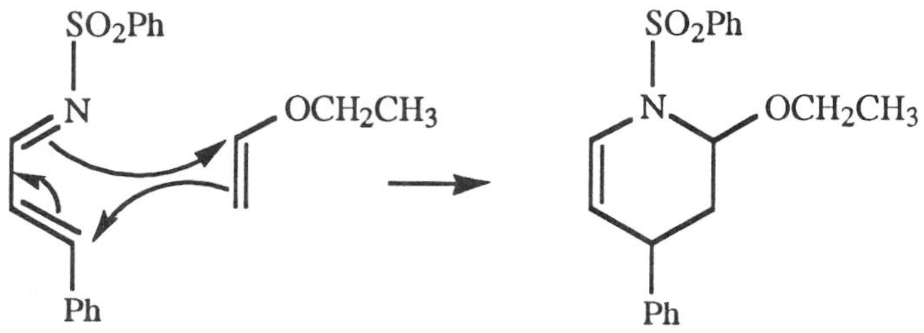

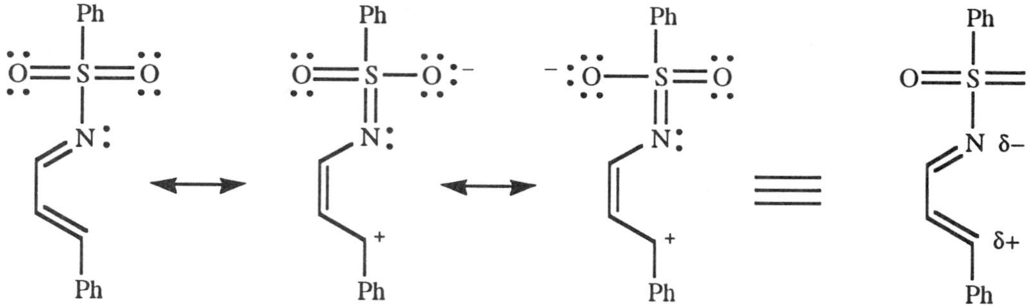

electron-withdrawing group
on the nitrogen polarizes the
diene by resonance

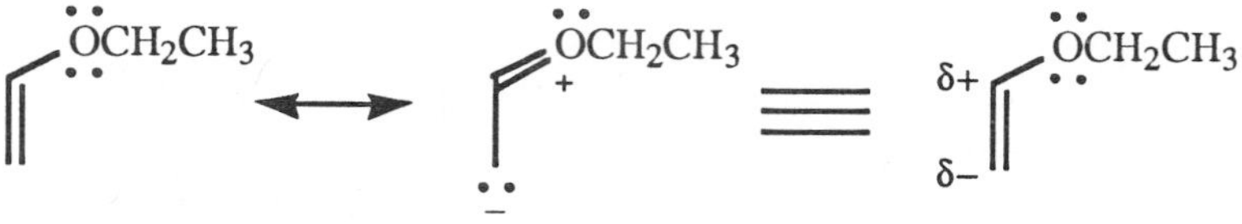

The dienophile, with an electron-rich substituent on it, is polarized by resonance. The diene and the dienophile react in a way that maximizes the interaction of the positive end of the dipole in one with the negative end of the dipole in the other. The regioselectivity of the reaction in which the more highly substituted adduct is formed can be explained in the same way.

24.57 (cont)

(b)   The addition of a second electron-withdrawing group to the diene increases the electron deficiency in the diene system, and, therefore, its reactivity toward an electron-rich dienophile.

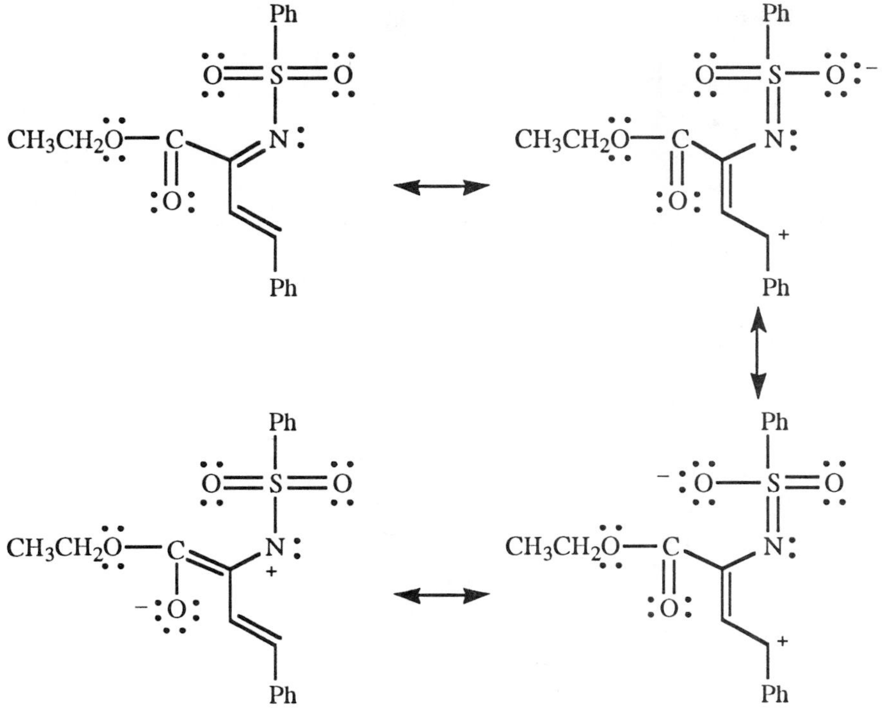

24.58   (1)   The bicyclic ring must be formed.

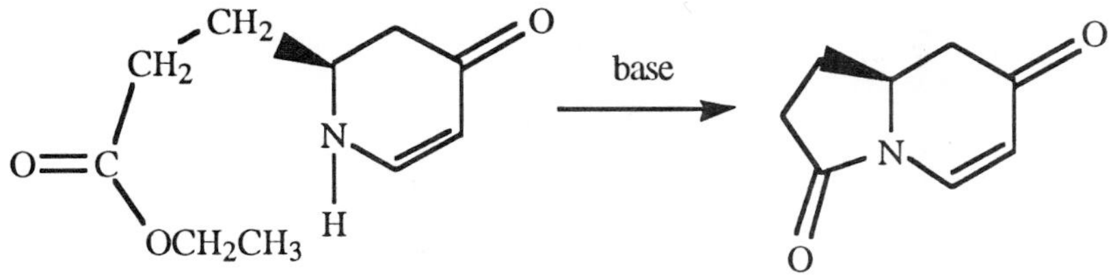

(2)   Carbonyl groups and the carbon-carbon double bond must be reduced.

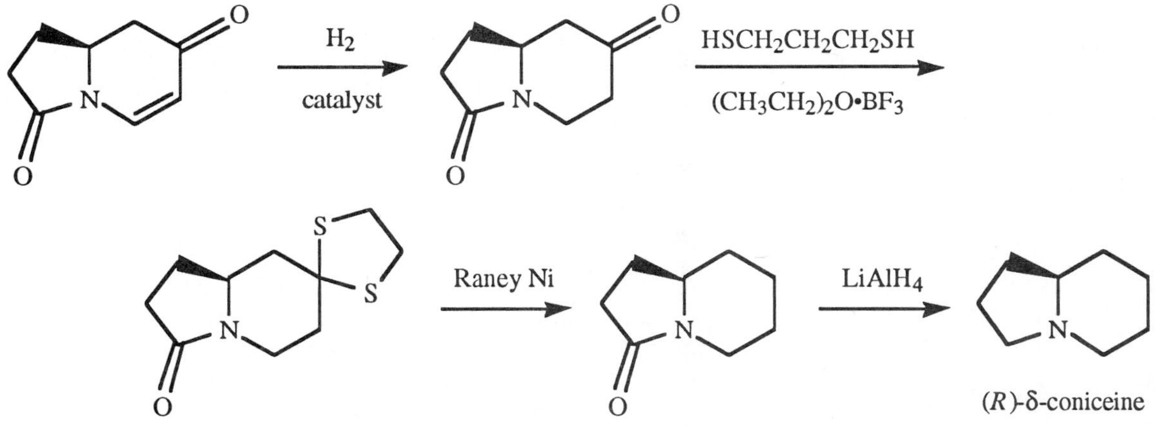

(*R*)-δ-coniceine

24.59

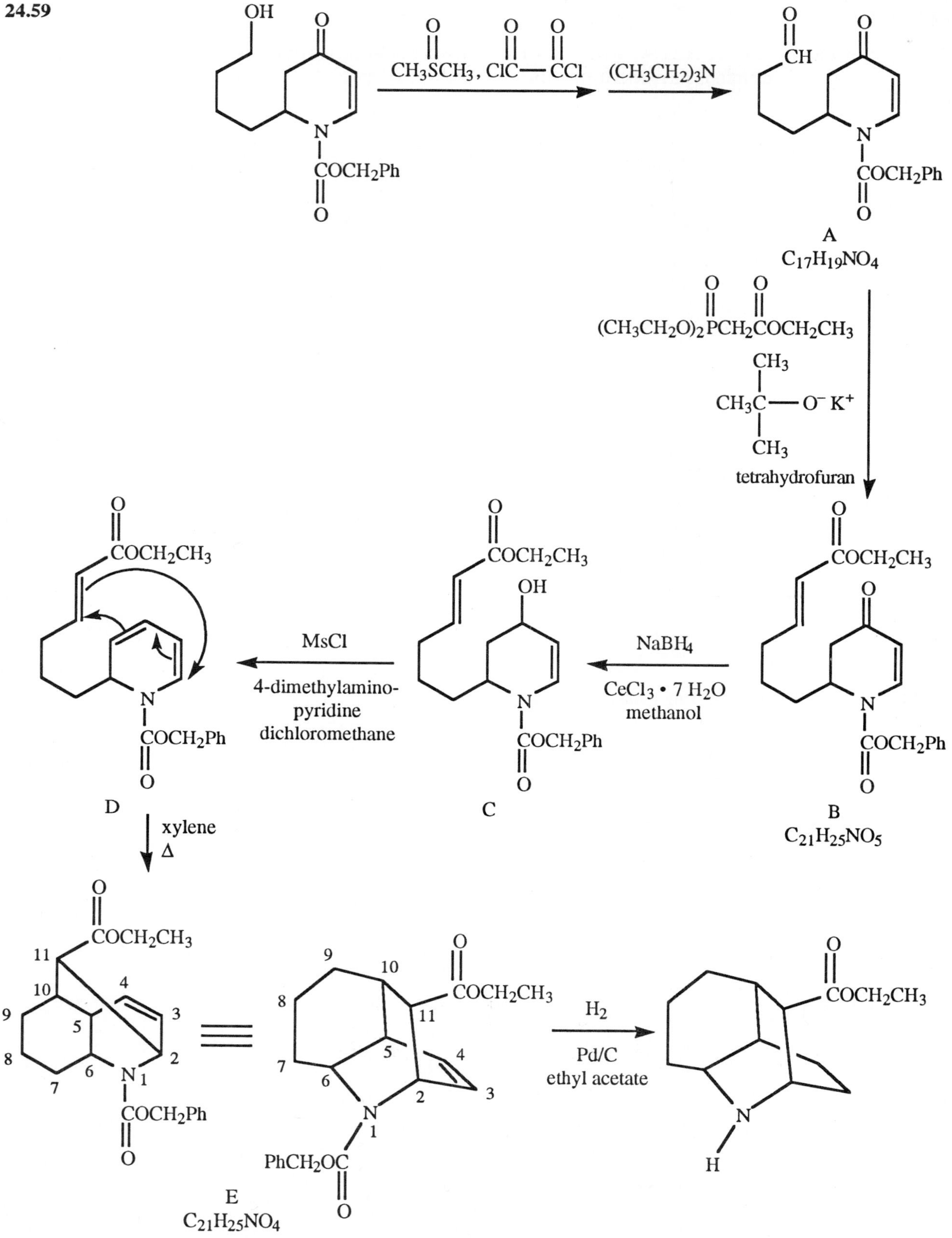

A
$C_{17}H_{19}NO_4$

B
$C_{21}H_{25}NO_5$

C

D

E
$C_{21}H_{25}NO_4$

25

Carbohydrates

25.1 A is a ketotriose
B is an aldotriose
C is an aldotetraose
D is a ketopentose

- -

Concept Map 25.1 Classification of carbohydrates.

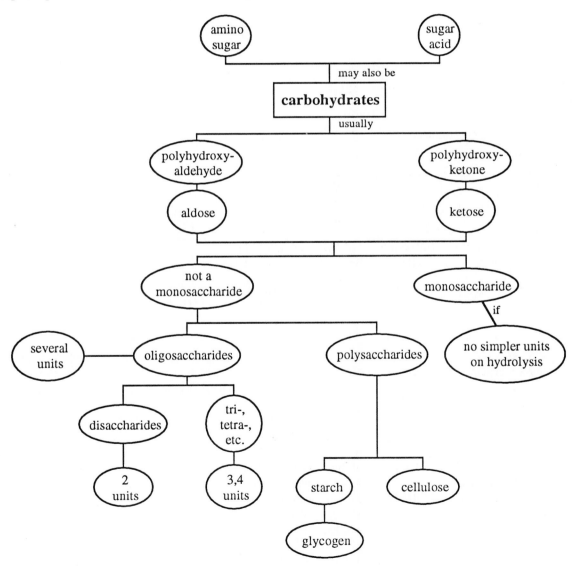

25.2   The configuration of (–)-lactic acid has been established as being *R* (Section 25.2A). Once we write the correct stereochemical structure for (*R*)-(–)-lactic acid, we can assign configuration to (–)-1-buten-3-ol and (–)-2-butanol also, because the reactions used for the interconversions do not break bonds to the stereocenter.

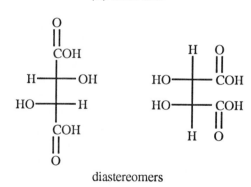

25.3

(a)

CH₃
H———Br
CH₂CH₃

Br
H———CH₃
CH₂CH₃

enantiomers

(*S*)-2-bromobutane   (*R*)-2-bromobutane

(b)

O
‖
COH
H———OH
CH₃

H   O
‖
HO———COH
CH₃

same

(*R*)-lactic acid

(c)

CH₃
H———Br
H———Br
CH₃

CH₃
H———Br
H———CH₃
Br

diastereomers

(2*S*, 3*R*)-2,3-
dibromobutane

(2*S*, 3*S*)-2,3-
dibromobutane

(d)

O
‖
COH
H———OH
HO———H
COH
‖
O

H   O
‖
HO———COH
HO———COH
‖
H   O

diastereomers

(2*R*, 3*R*)-2,3-
dihydroxy
butanedioic acid

(2*R*, 3*S*)-2,3-
dihydroxy
butanedioic acid

25.4   D-(+)-glyceraldehyde ≡ (*R*)-(+)-glyceraldehyde
D-(–)-lactic acid ≡ (*R*)-(–)-lactic acid
D-(–)-tartaric acid ≡ (2*S*, 3*S*)-(–)-2,3-dihydroxybutanedioic acid
D-(–)-ribose ≡ (2*R*, 3*R*, 4*R*)-(–)-2,3,4,5-tetrahydroxypentanal
D-(+)-glucose ≡ (2*R*, 3*S*, 4*R*, 5*R*)-(+)-2,3,4,5-pentahydroxyhexanal
D-(–)-fructose ≡ (3*S*, 4*R*, 5*R*)-(–)-1,3,4,5,6-pentahydroxy-2-hexanone
L-(–)-glyceraldehyde ≡ (*S*)-(–)-glyceraldehyde
L-(+)-lactic acid ≡ (*S*)-(+)-lactic acid
L-(+)-tartaric acid ≡ (2*R*, 3*R*)-(+)-2,3-dihydroxybutanedioic acid
L-(+)-alanine ≡ (*S*)-(+)-alanine
L-(+)-arabinose ≡ (2*R*, 3*S*, 4*S*)-(+)-2,3,4,5-tetrahydroxypentanal

25.5

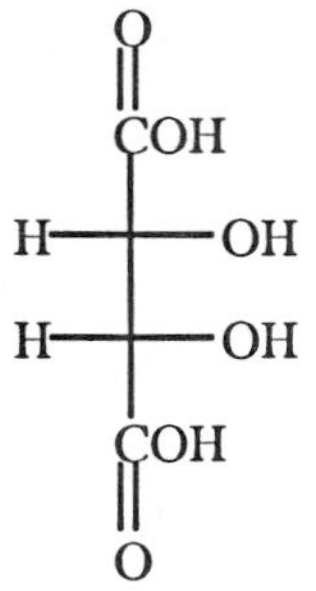

meso-tartaric acid

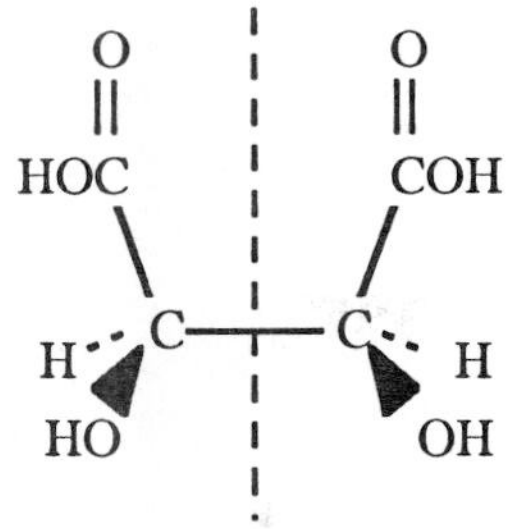

The plane of symmetry bisects
the central carbon-carbon bond.

25.6

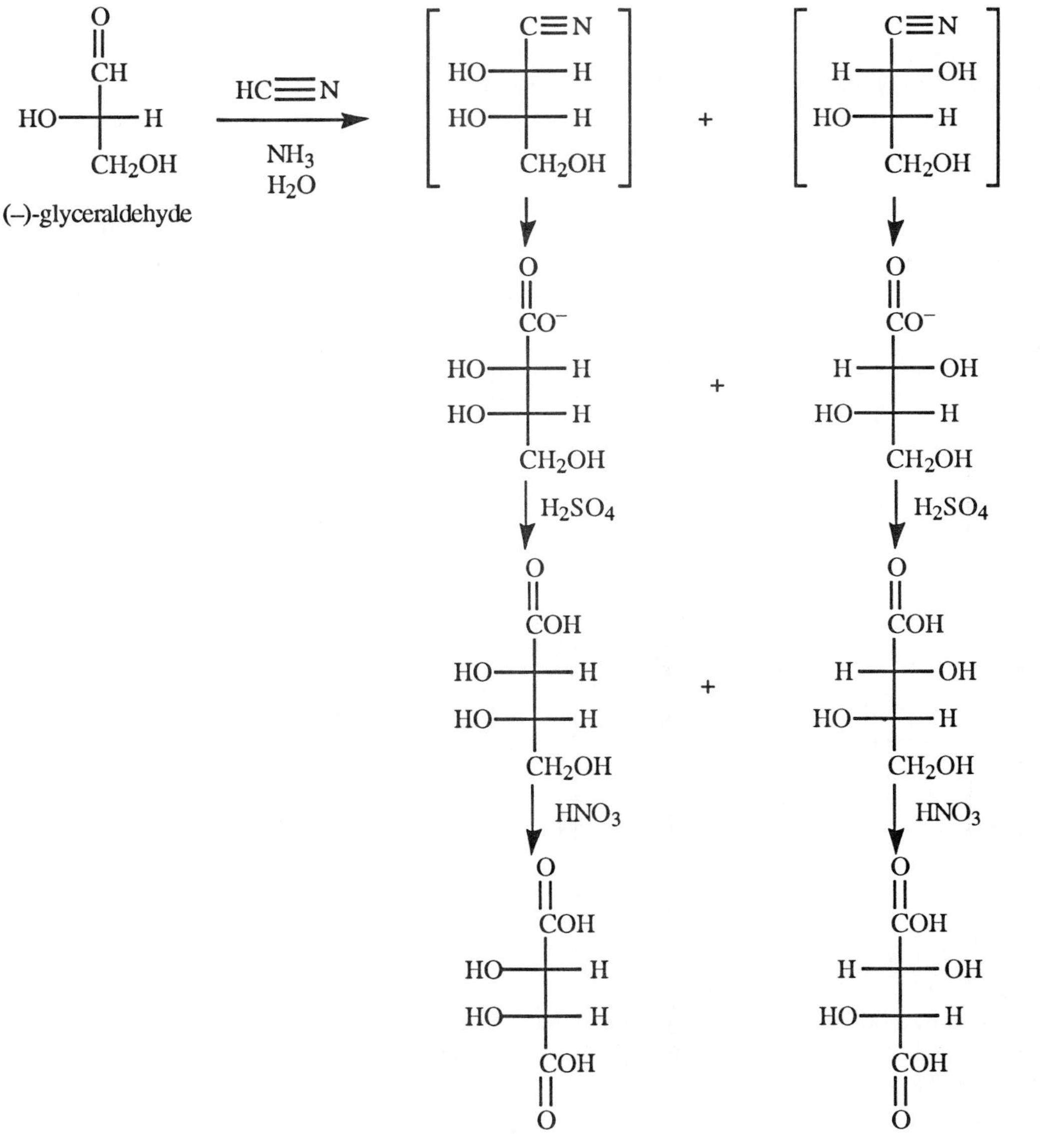

(–)-glyceraldehyde

(2*S*, 3*R*)-2,3- dihydroxy-
butanedioic acid

meso-tartaric acid
optically inactive

(2*R*, 3*R*)-2,3- dihydroxy–
butanedioic acid

L-(+)-tartaric acid
optically active

25.7

$$
\begin{array}{c}
\underset{\text{CH}}{\overset{\displaystyle\overset{\text{O}}{\|}}{}} \\
\text{HO}\!-\!\!-\!\text{H} \\
\text{H}\!-\!\!-\!\text{OH} \\
\text{HO}\!-\!\!-\!\text{H} \\
\text{HO}\!-\!\!-\!\text{H} \\
\text{CH}_2\text{OH}
\end{array}
$$

L-(–)-glucose

25.8   (a) D-*glycero*-D-*gluco*-heptose   (b) 4-deoxy-D-*glycero*-2-pentulose   (c) 3-deoxy-D-*ribo*-hexose

25.9   (a)

$$
\begin{array}{c}
\overset{\displaystyle\overset{\text{O}}{\|}}{\text{CH}} \\
\text{H}\!-\!\!-\!\text{OH} \\
\text{HO}\!-\!\!-\!\text{H} \\
\text{HO}\!-\!\!-\!\text{H} \\
\text{H}\!-\!\!-\!\text{OH} \\
\text{H}\!-\!\!-\!\text{OH} \\
\text{CH}_2\text{OH}
\end{array}
$$

(b)

$$
\begin{array}{c}
\overset{\displaystyle\overset{\text{O}}{\|}}{\text{CH}} \\
\text{HO}\!-\!\!-\!\text{H} \\
\text{H}\!-\!\!-\!\text{H} \\
\text{H}\!-\!\!-\!\text{OH} \\
\text{H}\!-\!\!-\!\text{OH} \\
\text{CH}_3
\end{array}
$$

(c)

$$
\begin{array}{c}
\overset{\displaystyle\overset{\text{O}}{\|}}{\text{CH}} \\
\text{H}\!-\!\!-\!\text{OH} \\
\text{HO}\!-\!\!-\!\text{H} \\
\text{H}\!-\!\!-\!\text{NHCCH}_3 \\
\text{H}\!-\!\!-\!\text{OH} \\
\text{CH}_3
\end{array}
$$

25.10

$$
\underset{\text{fructose}}{
\begin{array}{c}
\text{CH}_2\text{OH} \\
\text{C}\!=\!\text{O} \\
\text{CHOH} \\
\text{CHOH} \\
\text{CHOH} \\
\text{CH}_2\text{OH}
\end{array}}
\quad\xrightarrow[\text{H}_2\text{O}]{\text{HCN}}\quad
\underset{\text{no stereochemistry implied}}{
\left[
\begin{array}{c}
\text{CH}_2\text{OH} \\
\text{HO}\!-\!\text{C}\!-\!\text{C}\!\equiv\!\text{N} \\
\text{CHOH} \\
\text{CHOH} \\
\text{CHOH} \\
\text{CH}_2\text{OH}
\end{array}
\right]}
\quad\xrightarrow{\text{H}_2\text{O}}\quad
\begin{array}{c}
\text{CH}_2\text{OH} \\
\text{HO}\!-\!\text{C}\!-\!\text{COH} \\
\text{CHOH} \\
\text{CHOH} \\
\text{CHOH} \\
\text{CH}_2\text{OH}
\end{array}
\quad\xrightarrow[\text{P}]{\text{HI}}\quad
\underset{\substack{\text{2-methyl-}\\\text{hexanoic acid}}}{
\begin{array}{c}
\text{CH}_3 \\
\text{HO}\!-\!\text{C}\!-\!\text{COH} \\
\text{CH}_2 \\
\text{CH}_2 \\
\text{CH}_2 \\
\text{CH}_3
\end{array}}
$$

Because the addition of hydrogen cyanide to the carbonyl and the hydrolysis of the resulting nitrile gives a seven-carbon carboxylic acid with branching at the second carbon atom, fructose must be a ketose with the ketone function on carbon 2.

25.11   (a) The first step in the reduction is the protonation of the alcohol to give a good leaving group, followed by a nucleophilic substitution reaction (either S_N1 or S_N2, depending on the nature of the carbon atom) to form an alkyl halide.

25.11 (a) (cont)

$$R \!-\! OH \ + \ HI \ \longrightarrow \ R \!-\! I \ + \ H_2O$$

(b)

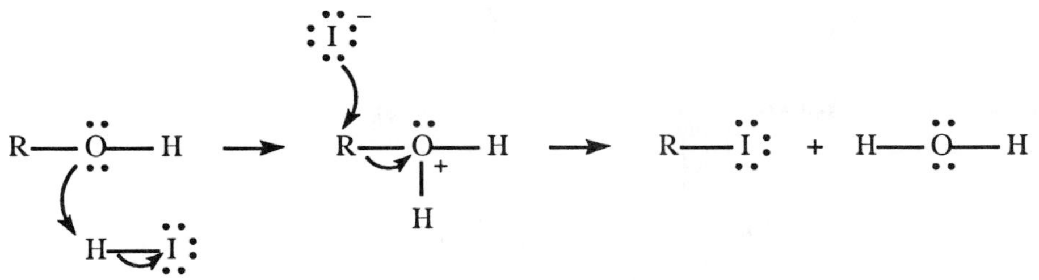

The rest of the reaction must be a free radical chain reaction.

initiation step:

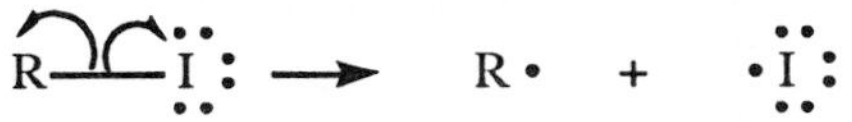

propagation steps:

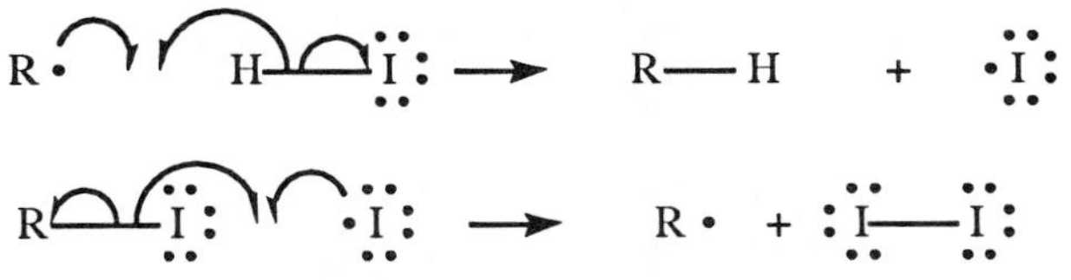

- -

Concept Map 25.2   (see p. 924)

25.12

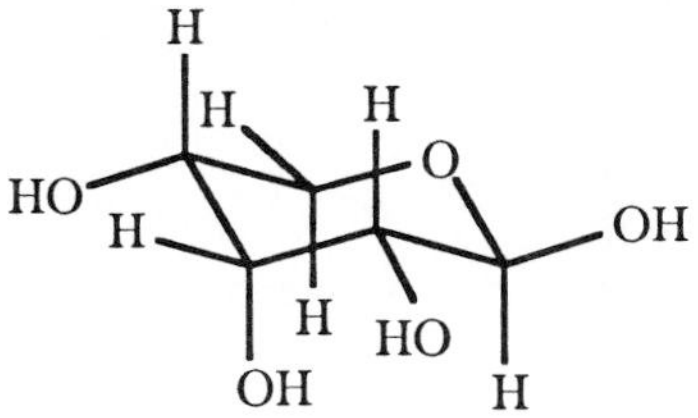

β-D-ribopyranose

25.13

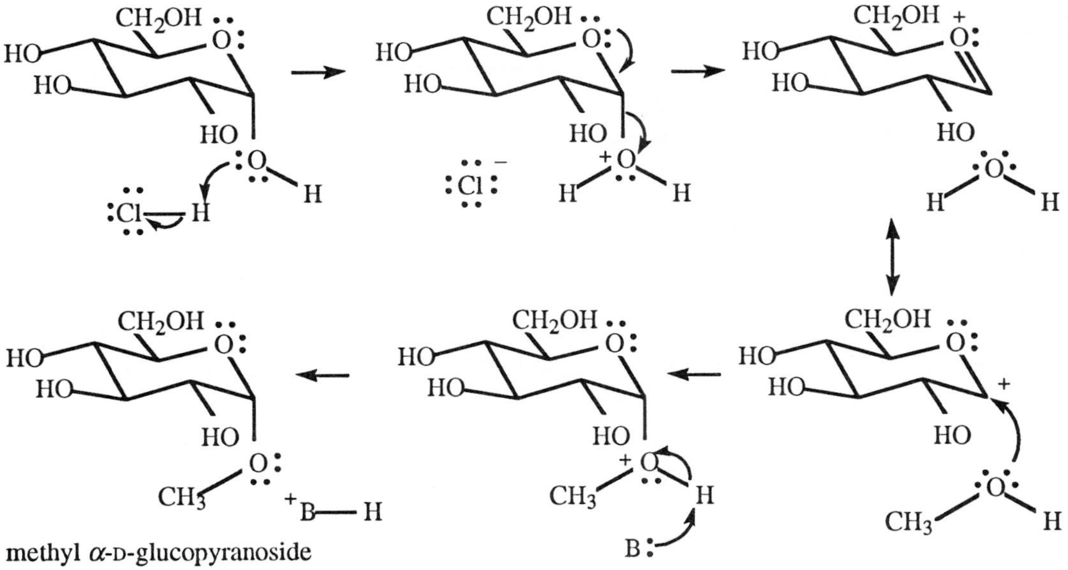

methyl α-D-glucopyranoside

Both α- and β-glucopyranose give the same carbocation intermediate. The intermediate reacts with methanol to give either the α-glucopyranoside or the β-glucopyranoside, depending on which side of the carbocation is attacked by the alcohol. The composition of the product mixture is determined by the relative stabilities of the two methyl glucopyranosides.

Concept Map 25.2   Structure of glucose.

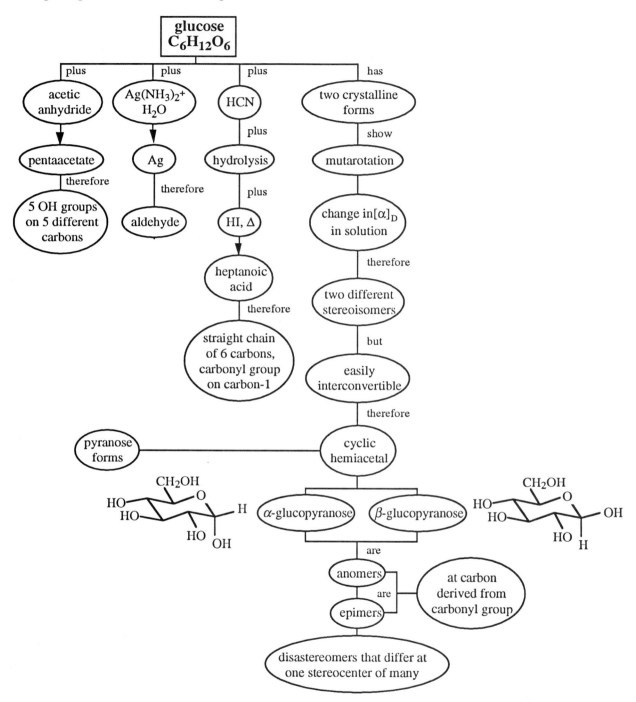

25.14

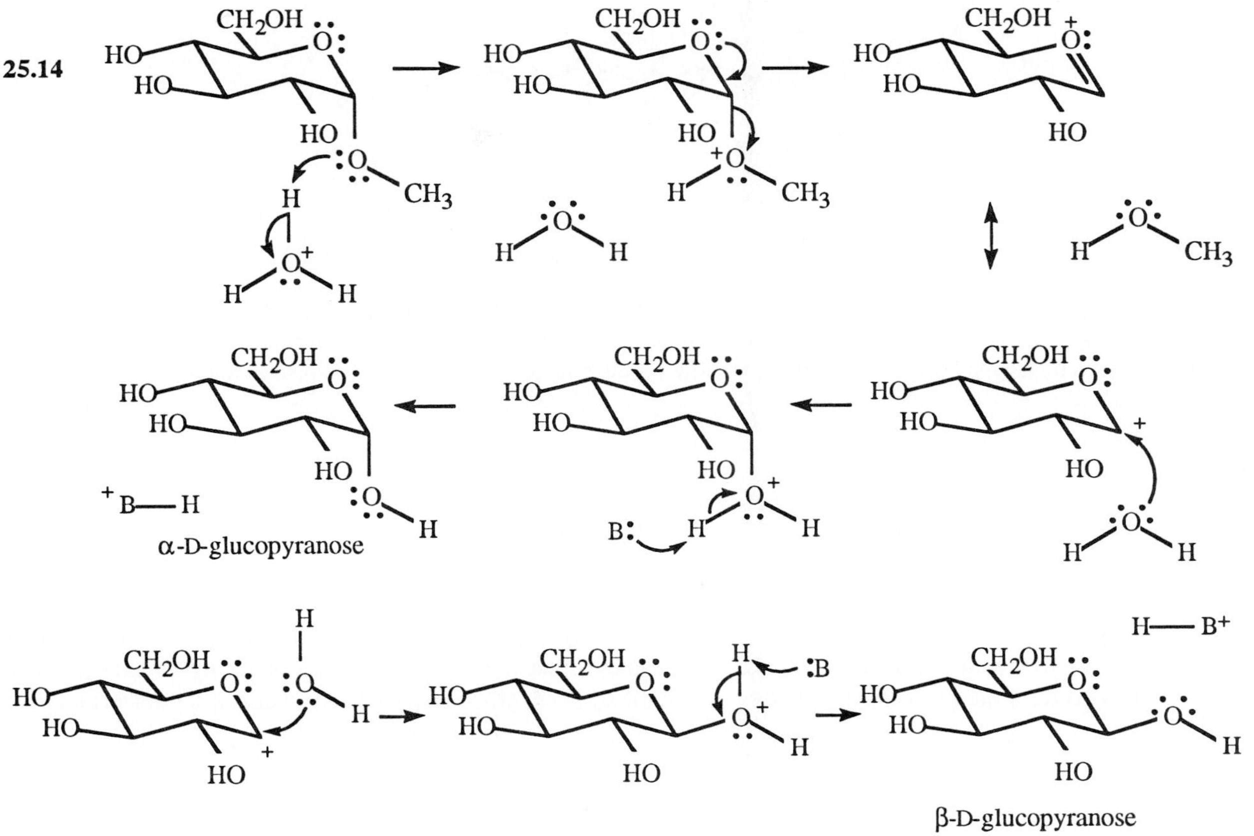

α-D-glucopyranose

β-D-glucopyranose

25.15

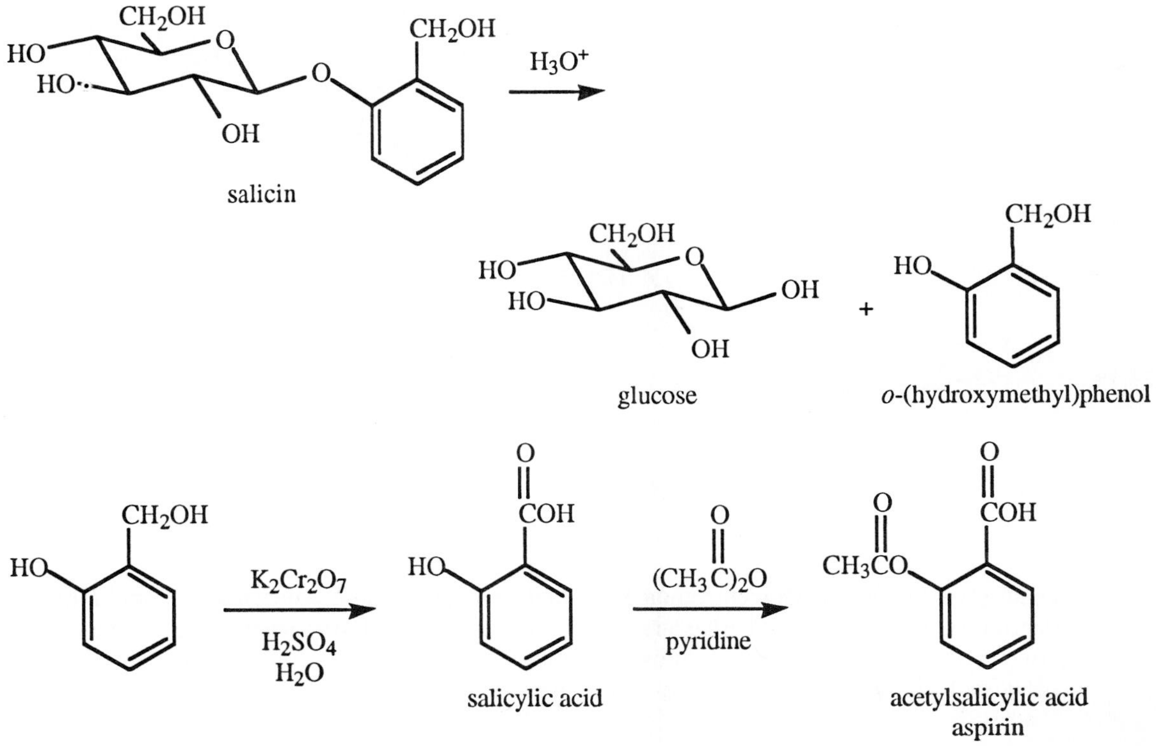

salicin

glucose   +   *o*-(hydroxymethyl)phenol

salicylic acid   acetylsalicylic acid
                                   aspirin

25.16

β-glucoside
cleaved by emulsin

arbutin

25.17 Intramolecular hydrogen bonding is important when deuteriochloroform is the solvent because hydrogen bonding with the solvent is not possible. The conformation in which the two hydroxy groups are axial allows them to hydrogen bond to each other, stabilizing the conformation with all axial substituents.

In D_2O, hydrogen bonding of the solvent to all of the groups is possible. Intramolecular hydrogen bonding is given up in favor of the more stable conformation in which all groups are equatorial and hydrogen bonded to the solvent. Note that the acidic hydrogens on the hydroxy groups have been exchanged for deuteriums.

25.18

Hydrogen bonding to the solvent is more important for the β-pyranose, which has an anomeric hydroxyl group, than for the β-pyranoside, which has an anomeric methoxyl group. Hydrogen bonding counteracts the anomeric effect, stabilizing the equatorial hydroxyl group relative to the axial one, more than it affects the equilibrium between the forms with the equatorial and axial methoxyl groups.

25.19 (a)

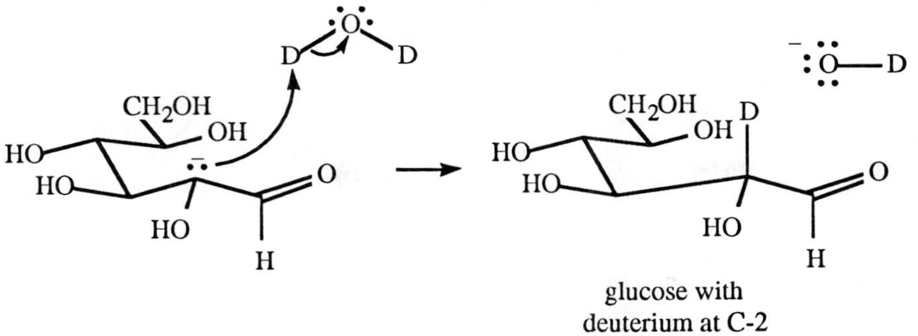

glucose with
deuterium at C-2

The carbanion at C-2 reacts with D_2O to give a deuterium atom bonded to carbon at that position in glucose or mannose.

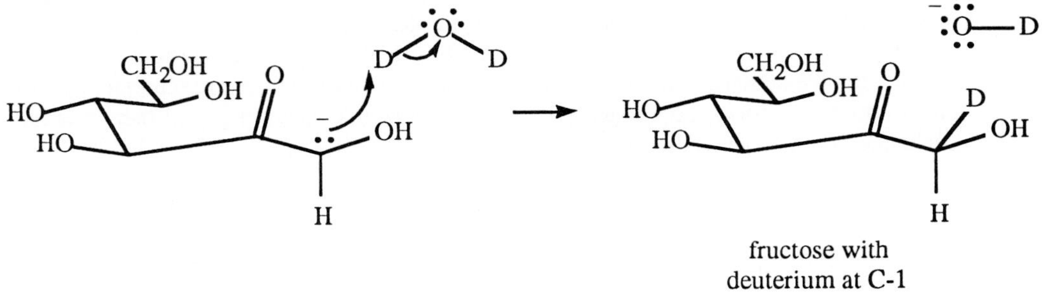

fructose with
deuterium at C-1

The carbanion at C-1 gives fructose with deuterium bonded to C-1.

(b) Fructose can pick up more than one deuterium atom because it can enolize again, losing a hydrogen atom from C-1. This enolate anion can be deuterated again at C-1.

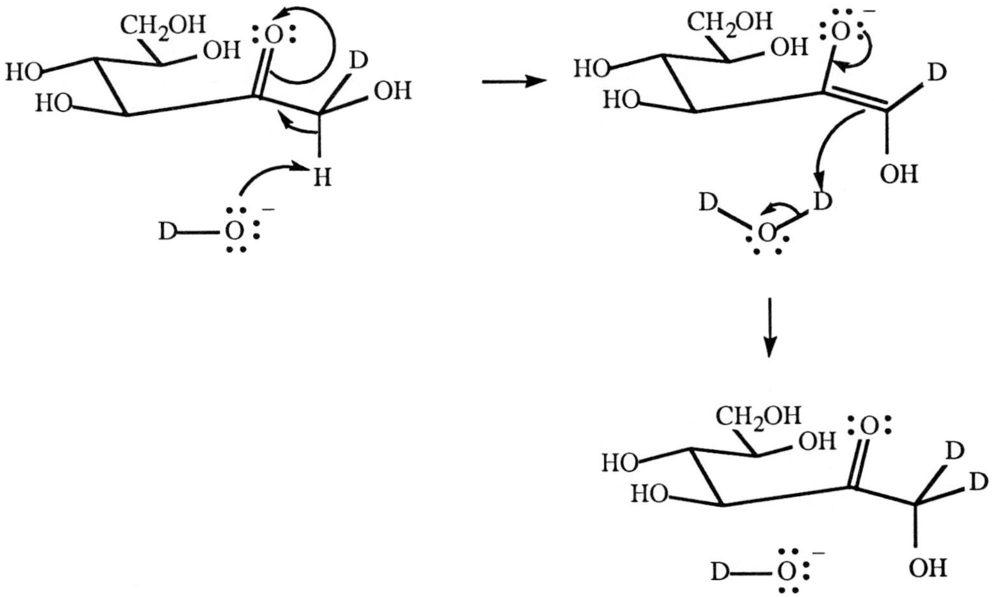

25.19 (b) (cont)

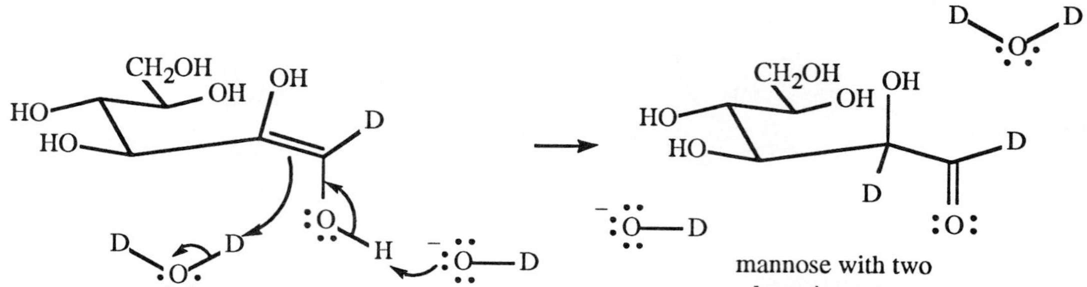

mannose with two
deuterium atoms
bonded to carbon

The presence, on average, of more than one deuterium atom in mannose suggests that some of the mannose is coming from fructose. (In the answer to this problem we have ignored the fact that all of the protons on the hydroxyl groups very rapidly exchange with D_2O and become —OD groups. They are converted back to H_2O, but the deuterium atoms bonded to carbon are much slower to exchange. The deuterium atoms on carbon can be located upon analysis so we have been concentrating on them and ignoring the rapid exchange taking place at the oxygen atoms.)

$$ROH \;+\; D_2O \;\rightleftharpoons\; ROD \;+\; HOD$$

$$ROD \;+\; H_2O \;\rightleftharpoons\; ROH \;+\; HOD$$

25.20

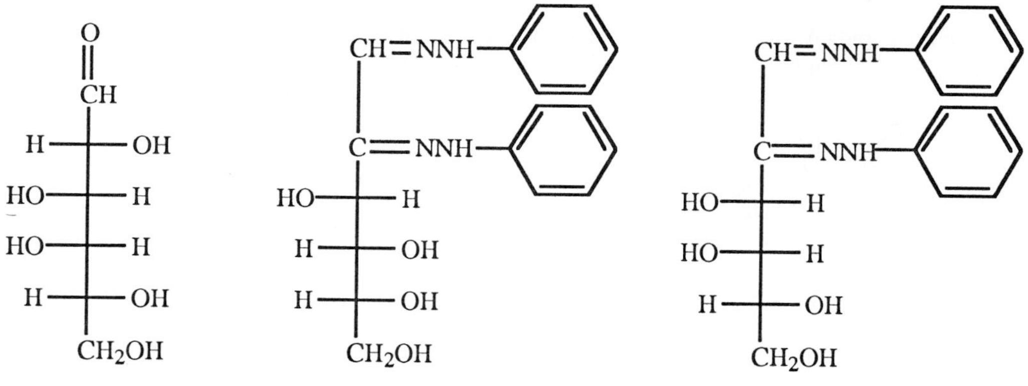

D-(+)-galactose

glucosazone

galactosazone

diastereomers
also epimers

25.21

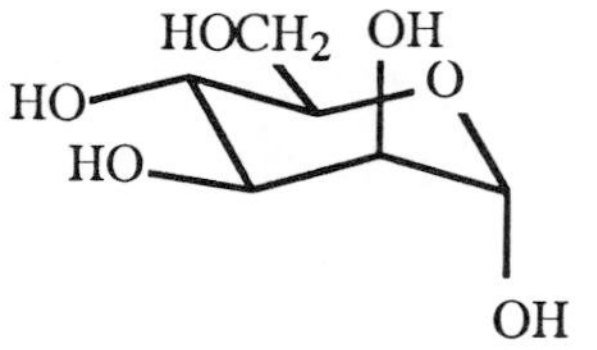

α-D-mannopyranose

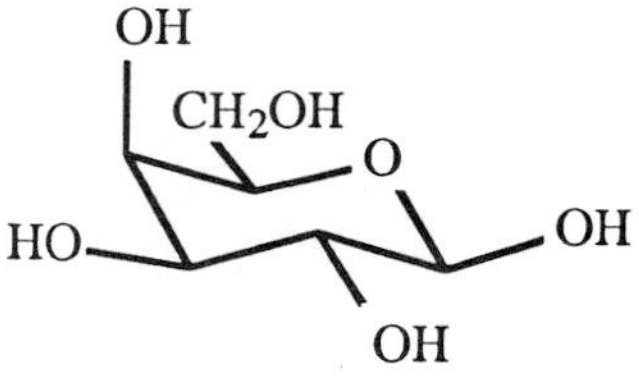

β-D-galactopyranose

25.22

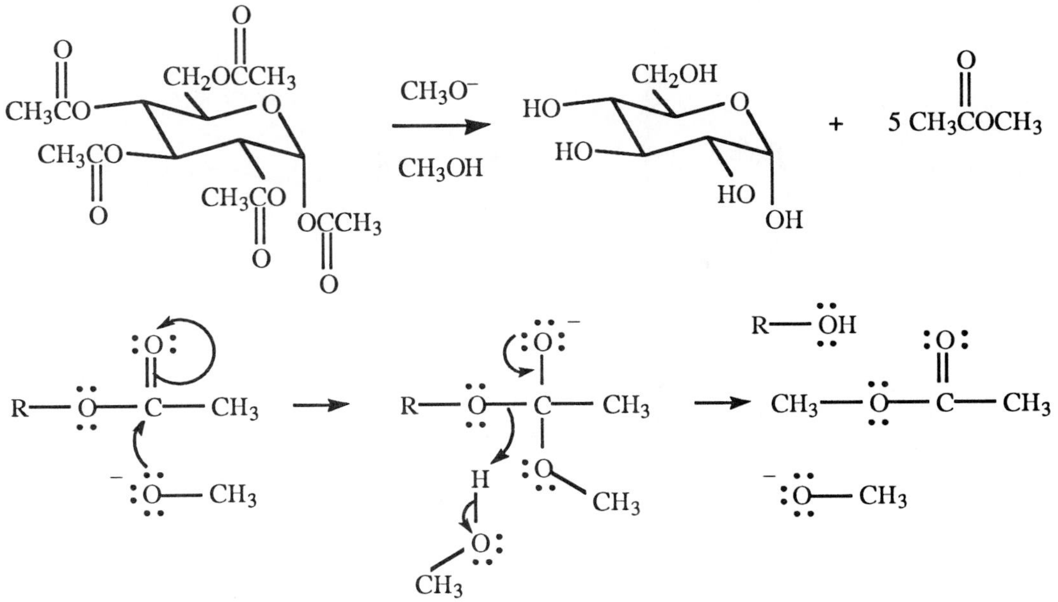

25.23

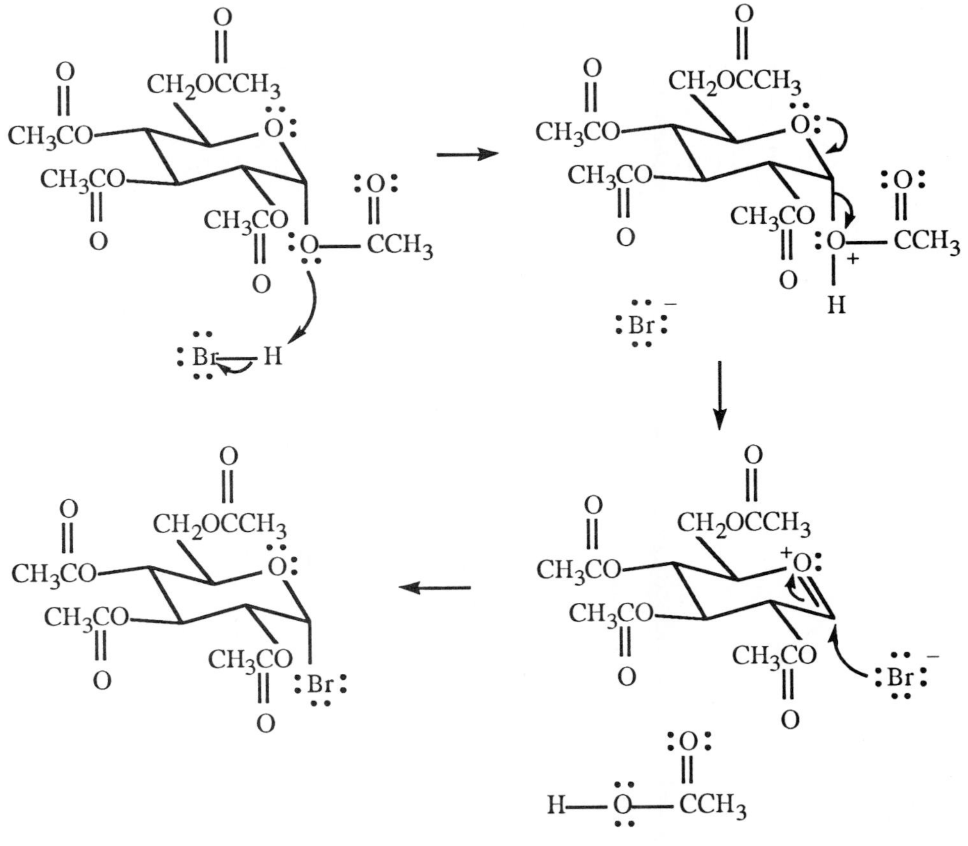

25.24   Acid hydrolysis would also cleave the glycosidic linkage between glucose and the phenol.

25.25

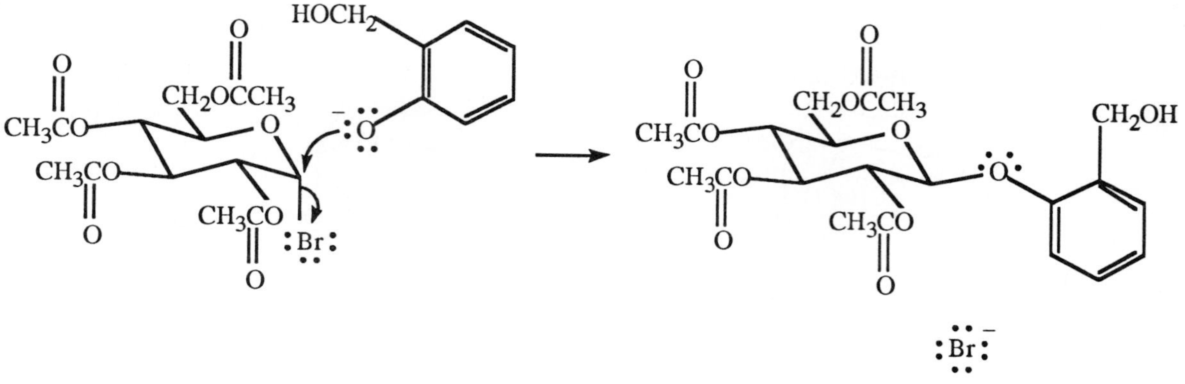

The reaction is an S_N2 reaction. It takes place with inversion of configuration.

25.26   Hydrolysis at C-1 gives a resonance-stabilized carbocation.

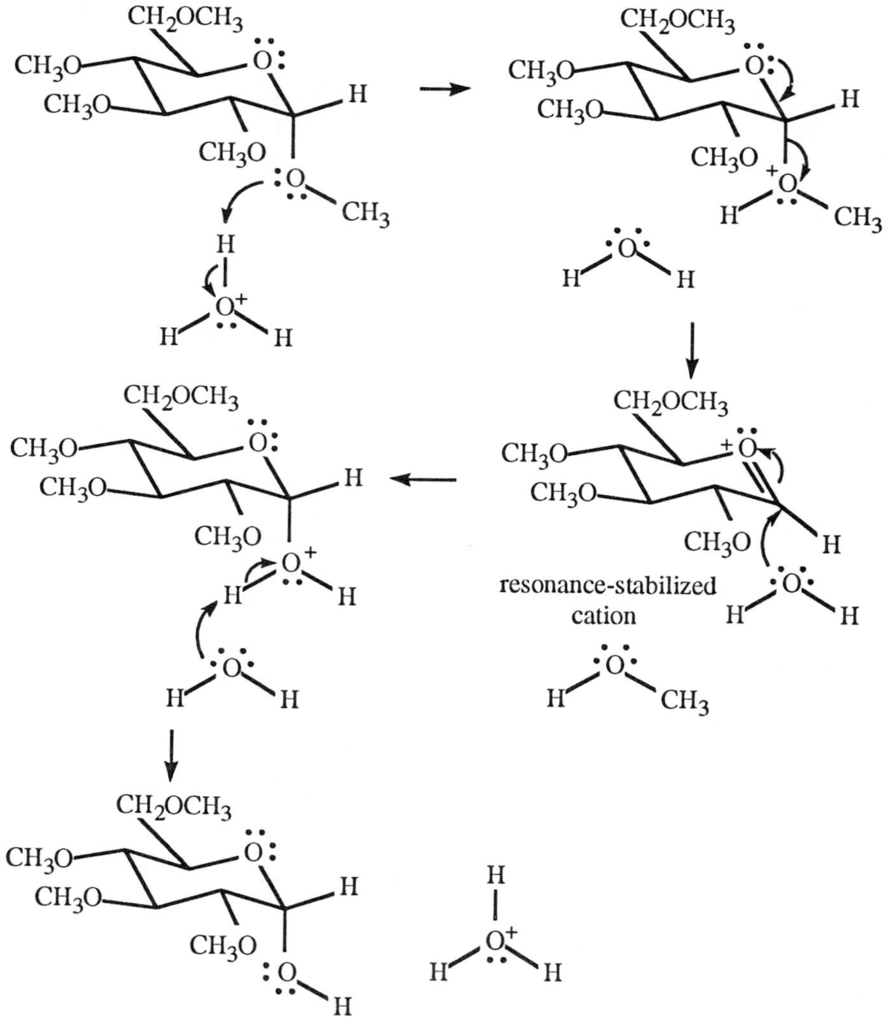

resonance-stabilized
cation

25.26 (cont)

Cleavage of the linkage at C-2 gives a secondary carbocation with no possibility of delocalization of charge.

secondary carbo-
cation at C-2 no
delocalization of
charge possible

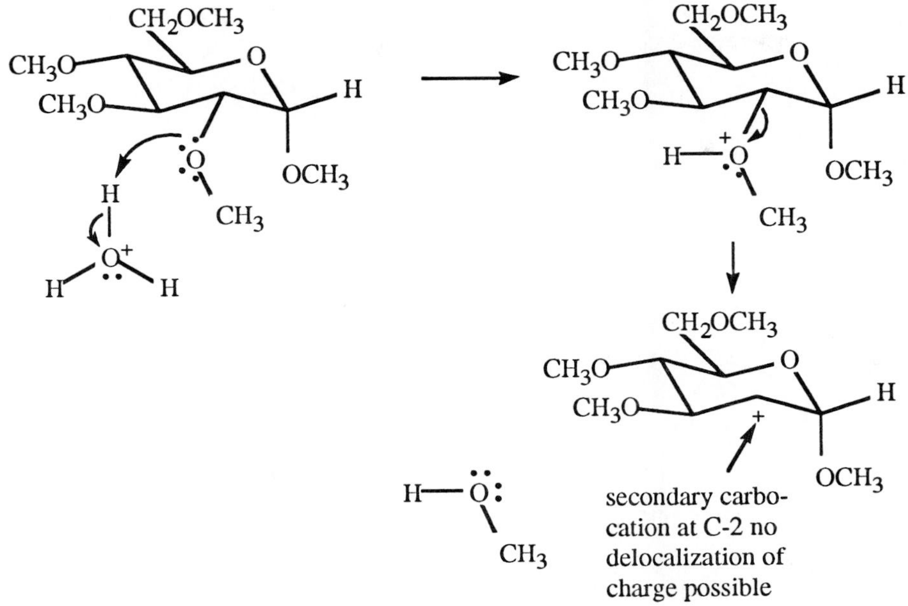

25.27

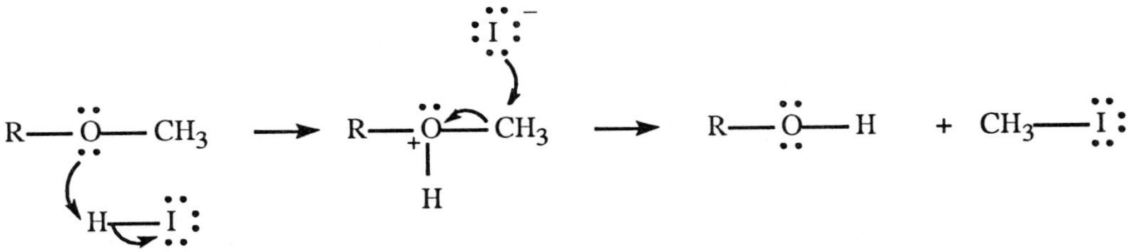

The concentrated acid protonates the ether, converting it to an oxonium ion with a good leaving group. Iodide ion displaces the leaving group. In dilute hydrochloric acid, water molecules compete with the ether for protons. The concentration of oxonium ions is much lower in dilute acid than in the concentrated acid. In addition, chloride ion is a weaker nucleophile than iodide ion, so the S_N2 reaction does not occur as readily. Therefore, it is possible to hydrolyze a glycoside linkage in dilute acid and leave untouched methyl ether functions at other hydroxyl groups in a sugar.

25.28

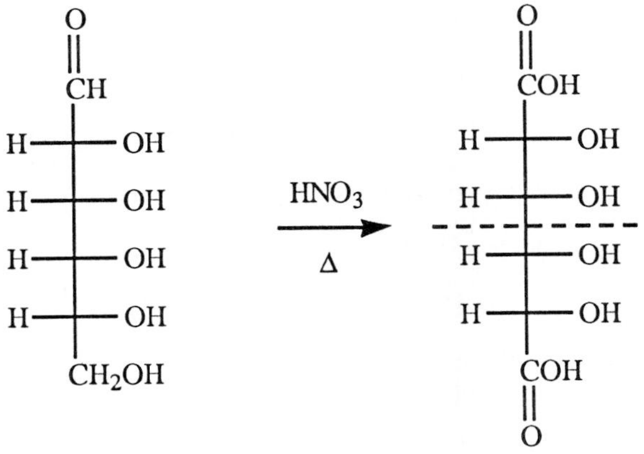

D-allose   allaric acid

25.29

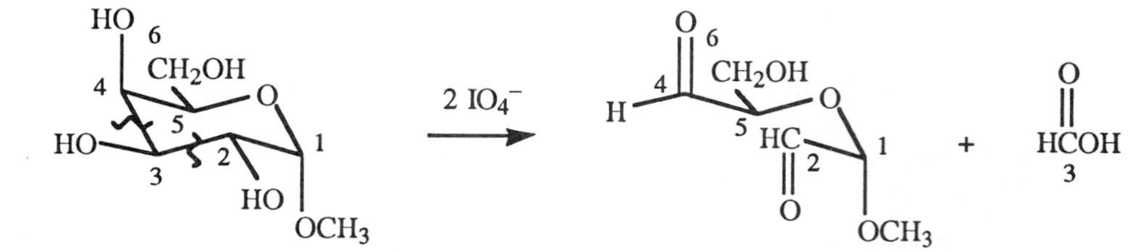

methyl α-D-galactopyranoside

25.30 They are diastereomers. They have the same configuration at the carbon derived from C-5 in glucose, and the opposite configuration at C-1.

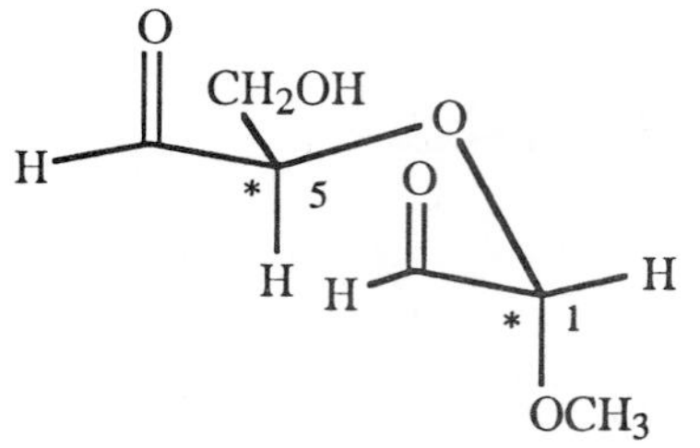

dialdehyde from methyl
α-D-glucopyranoside

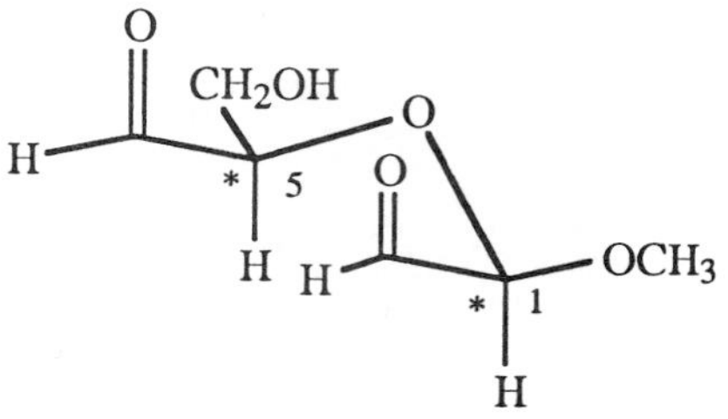

dialdehyde from methyl
β-D-glucopyranoside

Reduction with $NaBH_4$ reduces both aldehyde groups. This destroys the chirality at C-5, leaving only that at C-1. These compounds are enantiomers.

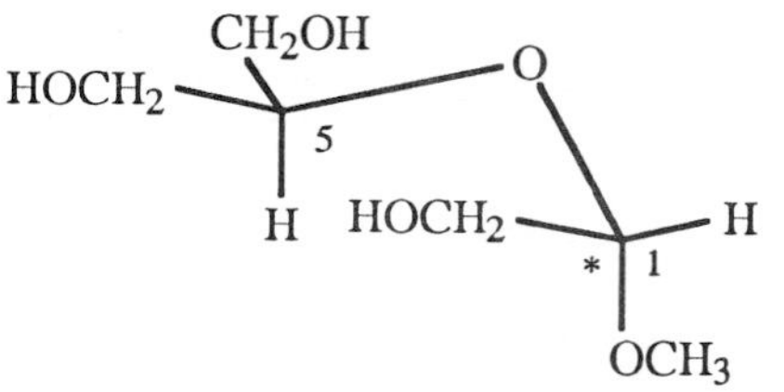

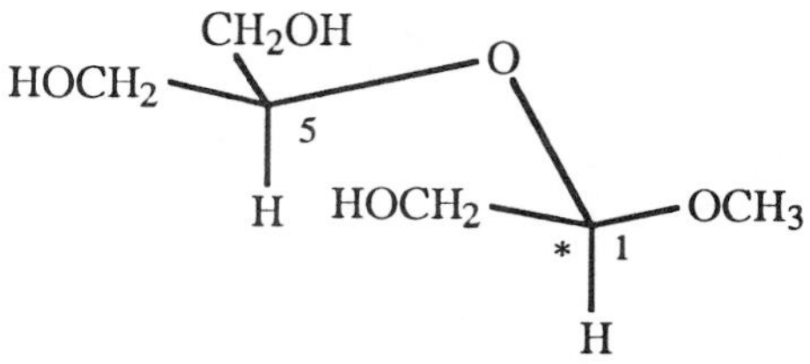

25.31

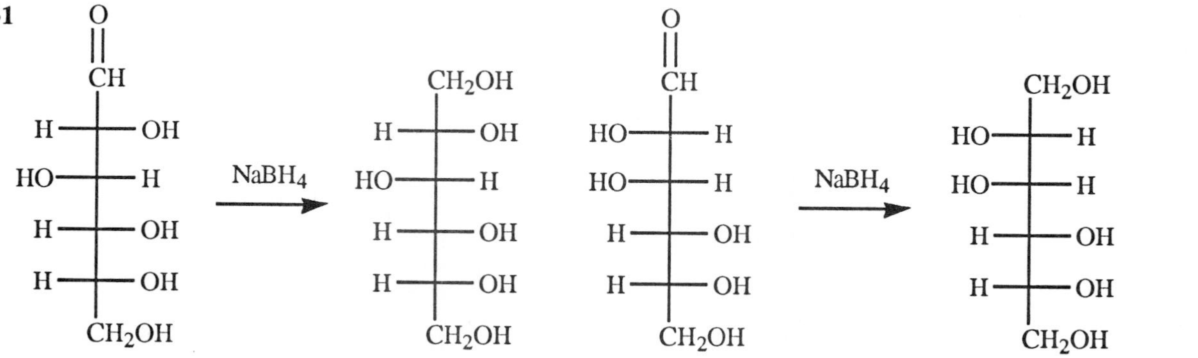

glucose

glucitol
single product; optically active

mannose

mannitol
single product; optically active

epimers

25.31 (cont)

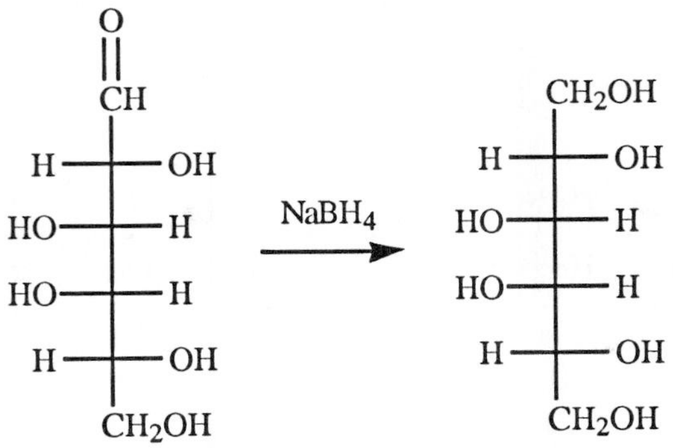

galactose

galactitol
single product;
a meso form; optically inactive

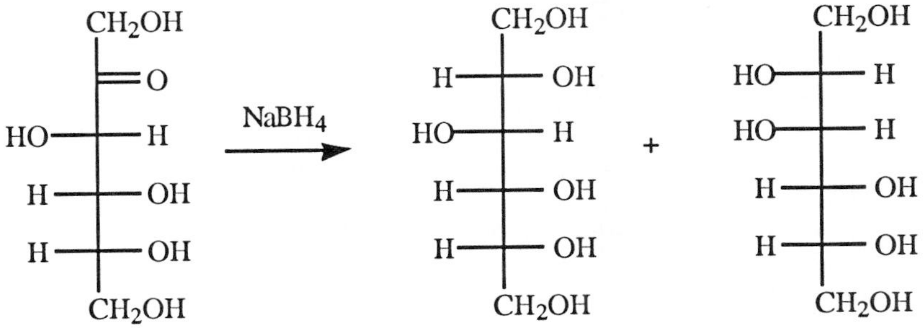

fructose

glucitol

mannitol

mixture of two products;
epimers; both optically active

Concept Map 25.3   (see p. 936)

25.32   Excess sodium borohydride would reduce the aldose to the alcohol.

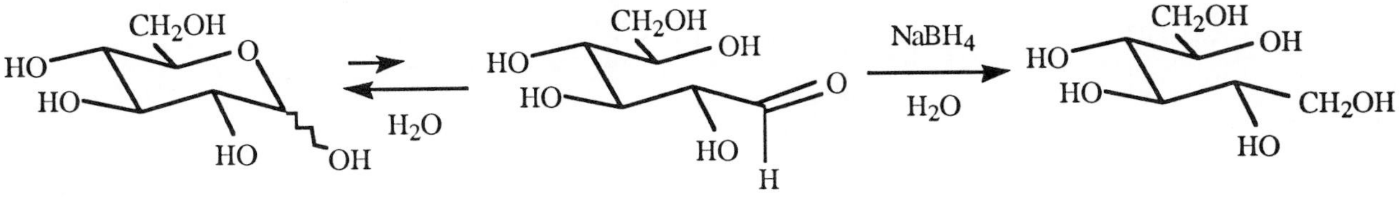

Concept Map 25.4   (see p. 939)

25.33

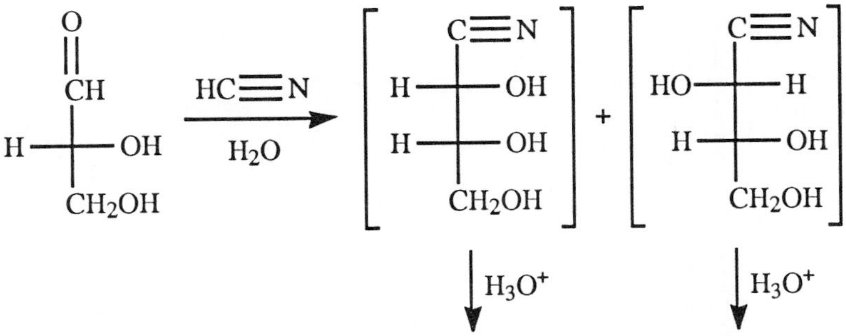

25.33 (cont)

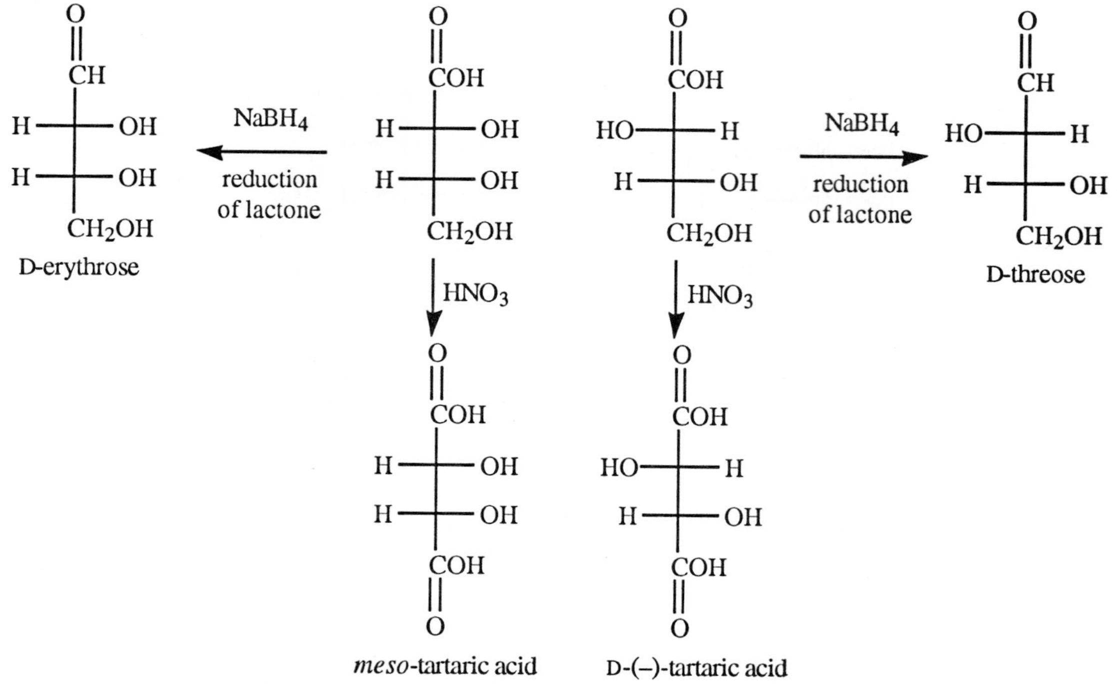

D-erythrose   NaBH₄ reduction of lactone

$$\text{NaBH}_4 \text{ reduction of lactone}$$

meso-tartaric acid   D-(−)-tartaric acid

D-threose   NaBH₄ reduction of lactone

25.34

D-ribose   NaBH₄ reduction of lactone

D-arabinose   NaBH₄ reduction of lactone

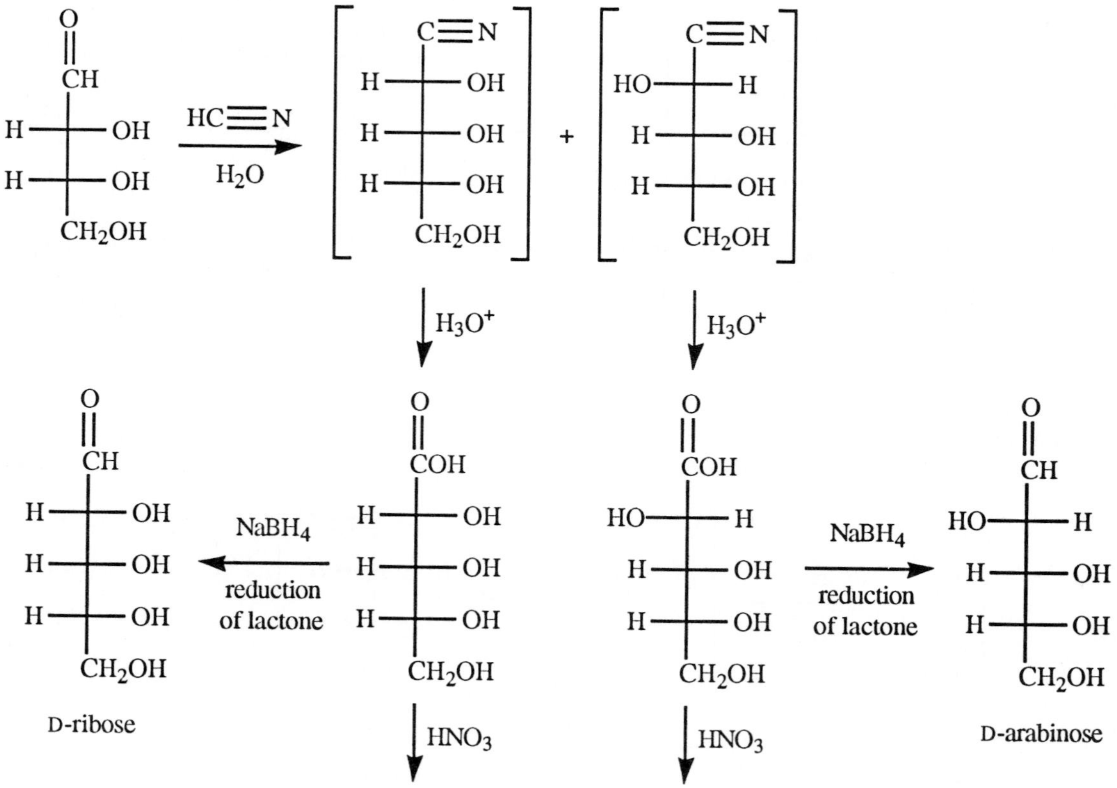

25.34 (cont)

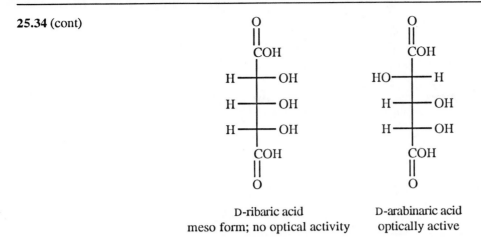

D-ribaric acid
meso form; no optical activity

D-arabinaric acid
optically active

D-Ribose must have all of the OH groups on the right in the Fischer projection formula, while D-arabinose must be the epimer of D-ribose at the new stereocenter created in this synthesis.

25.35

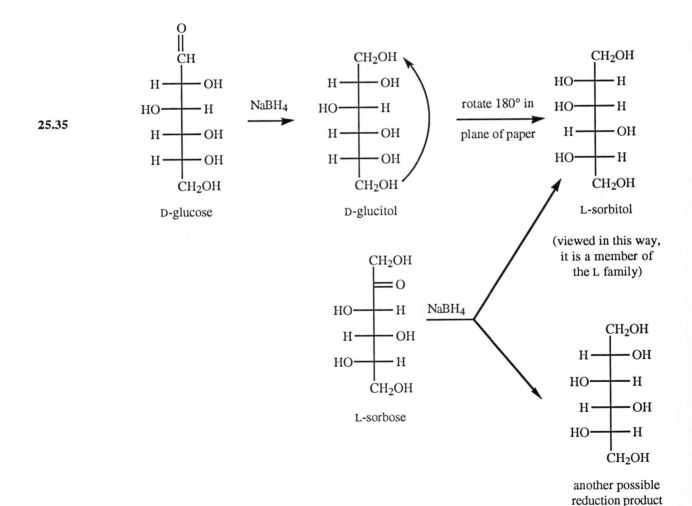

D-glucose

D-glucitol

L-sorbitol

(viewed in this way, it is a member of the L family)

L-sorbose

another possible reduction product of L-sorbose

Concept Map 25.3   Reactions of monosaccharides.

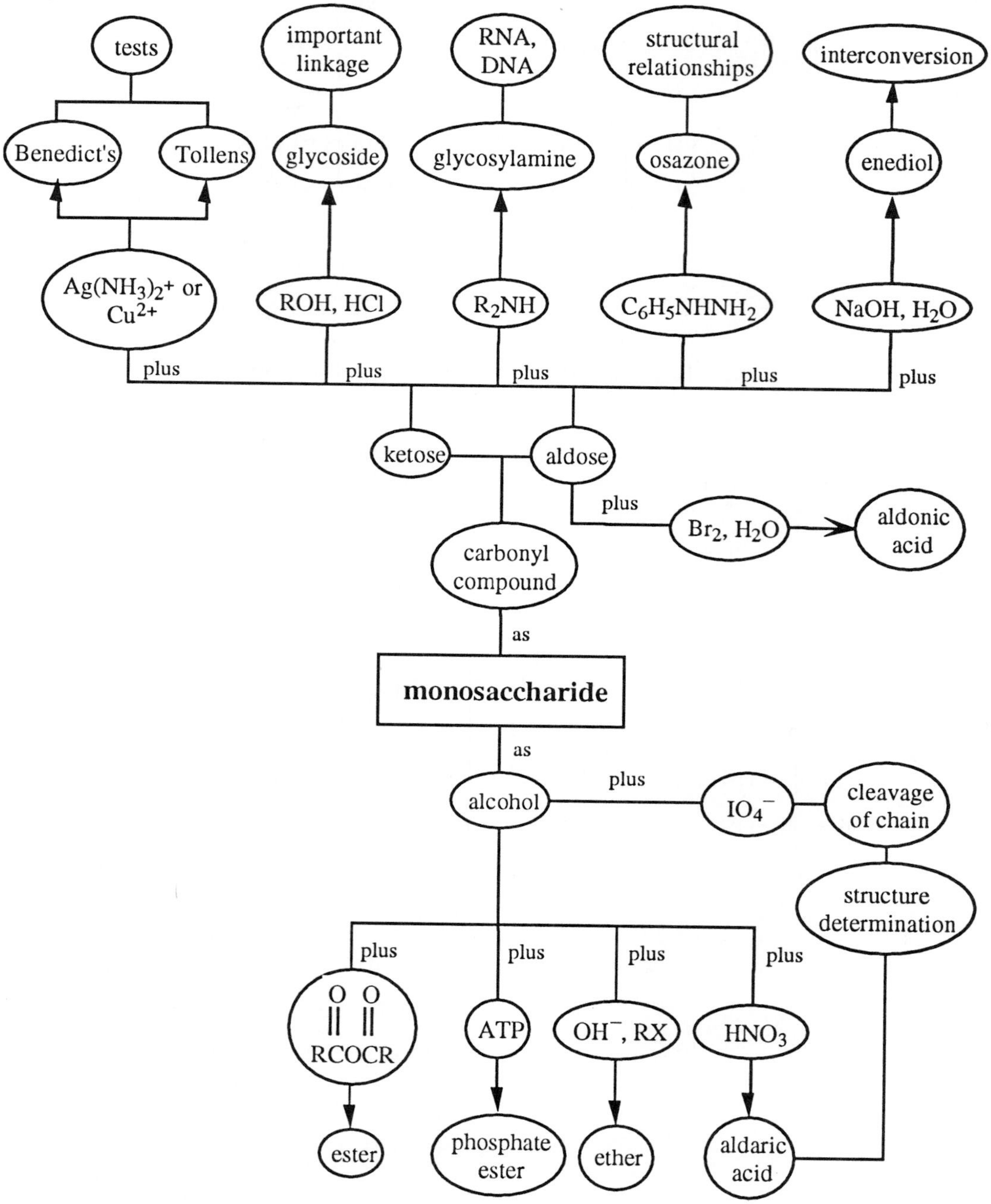

tests

Benedict's   Tollens

$Ag(NH_3)_2^+$ or Cu^{2+}

plus

important linkage

glycoside

ROH, HCl

plus

RNA, DNA

glycosylamine

R_2NH

plus

structural relationships

osazone

$C_6H_5NHNH_2$

plus

interconversion

enediol

$NaOH, H_2O$

plus

ketose   aldose

plus

Br_2, H_2O aldonic acid

carbonyl compound

as

monosaccharide

as

alcohol   plus IO_4^- cleavage of chain

structure determination

plus   plus   plus   plus

$\overset{O \quad O}{\overset{\| \quad \|}{RCOCR}}$ ATP OH^-, RX HNO_3

ester   phosphate ester   ether   aldaric acid

25.36

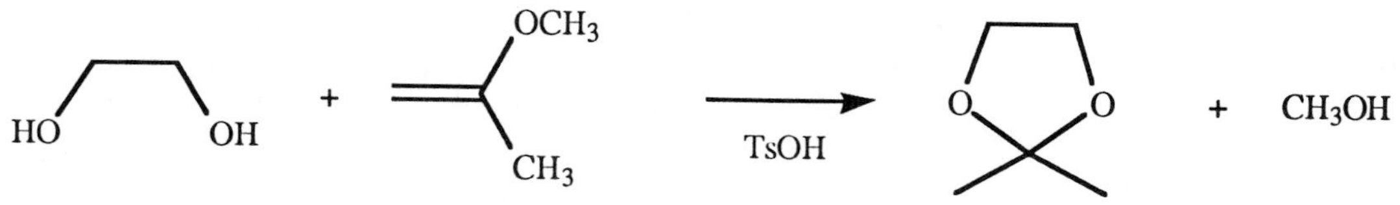

HO⁀OH + (structure with OCH₃ and CH₃) → (cyclic acetal structure with two O) + CH_3OH

TsOH

25.36 (cont)

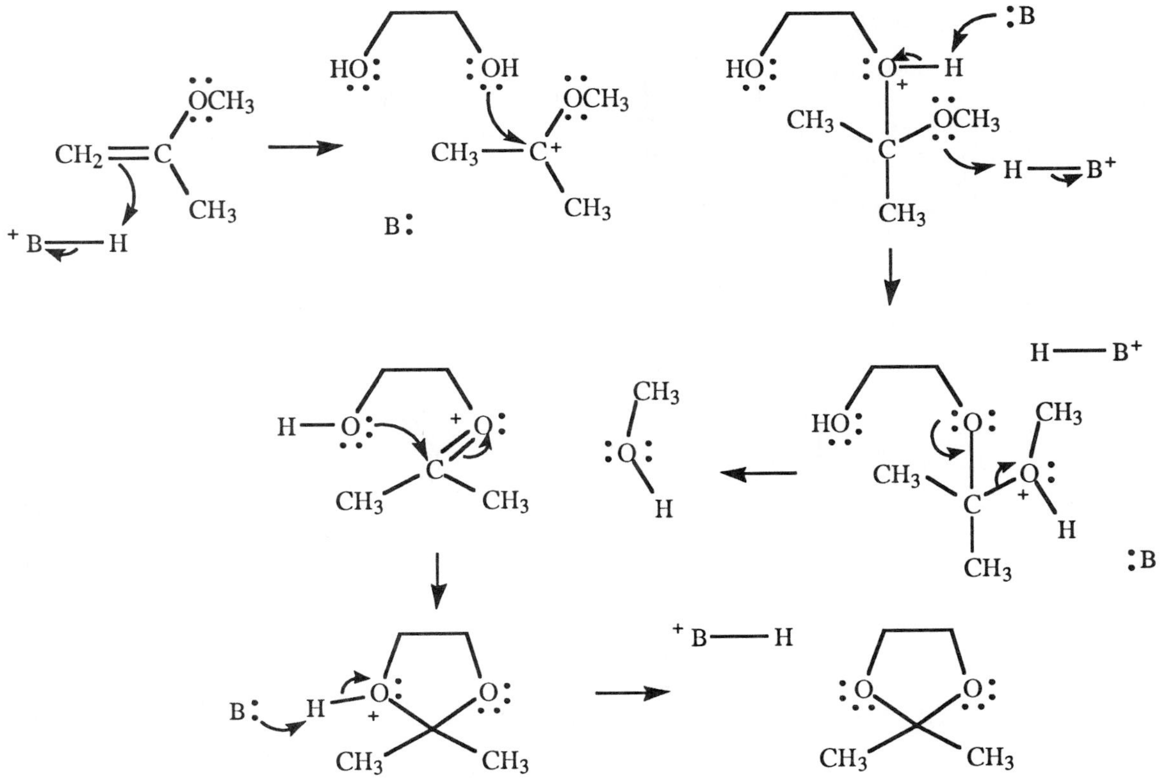

25.37

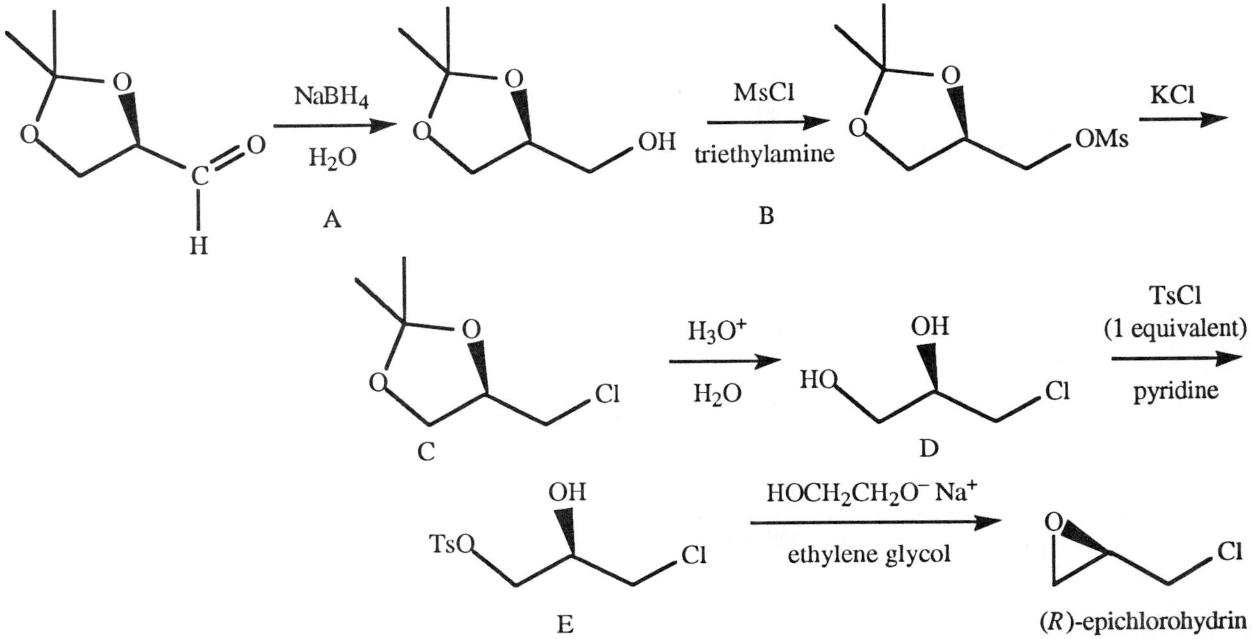

25.38

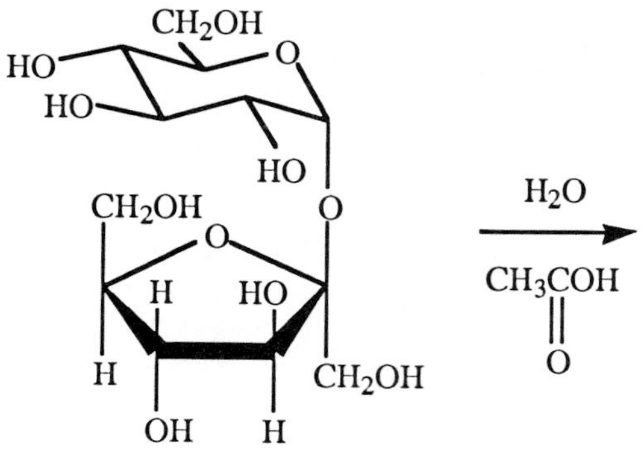

sucrose

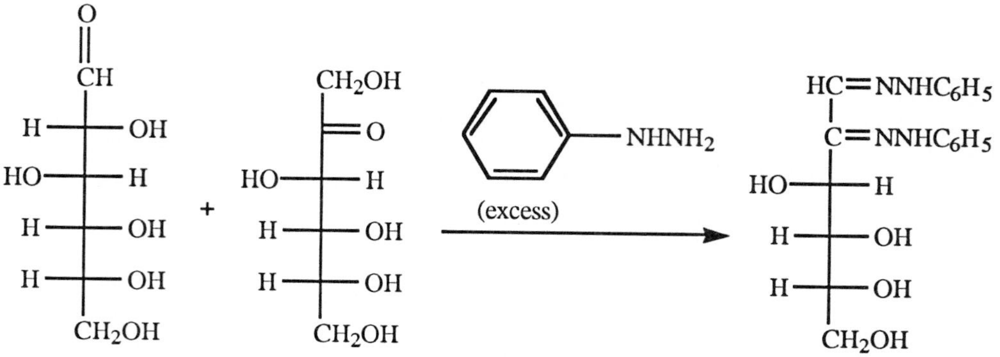

$$\xrightarrow[\underset{\underset{O}{\parallel}}{CH_3COH}]{H_2O}$$

glucose   fructose   (excess)   glucosazone

25.39

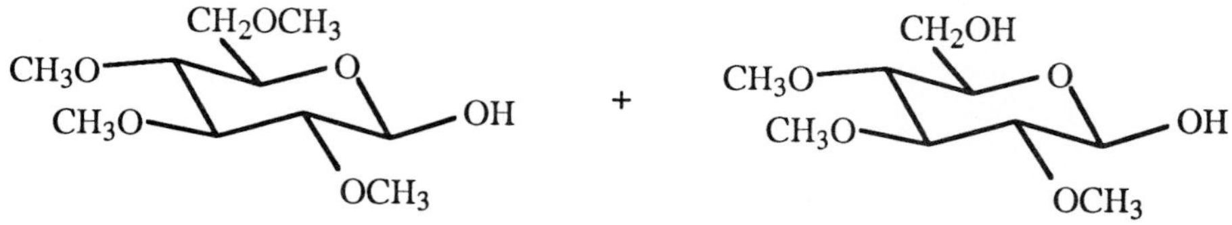

2,3,4,6-tetra-*O*-methyl-
D-glucopyranose
(therefore hydroxyl group
at C-1 in glycosidic linkage)

+

2,3,4-tri-*O*-methyl-
D-glucopyranose
(therefore hydroxyl group
at C-6 in glycosidic linkage;
anomeric carbon atom free because
gentiobiose is a reducing sugar)

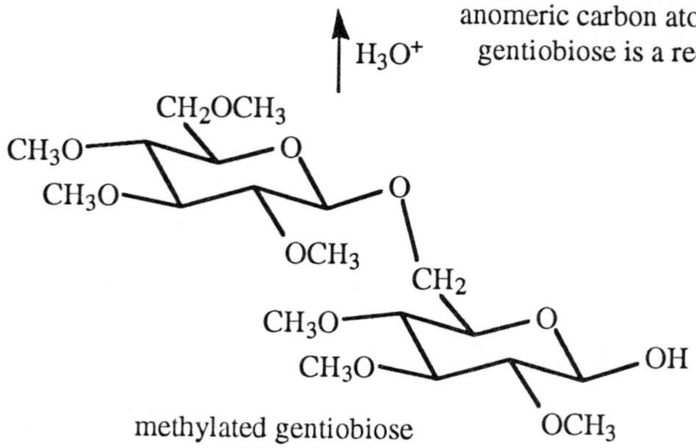

$\uparrow H_3O^+$

methylated gentiobiose

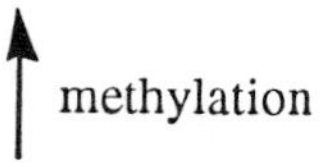

$\uparrow$ methylation

25.39 (cont)

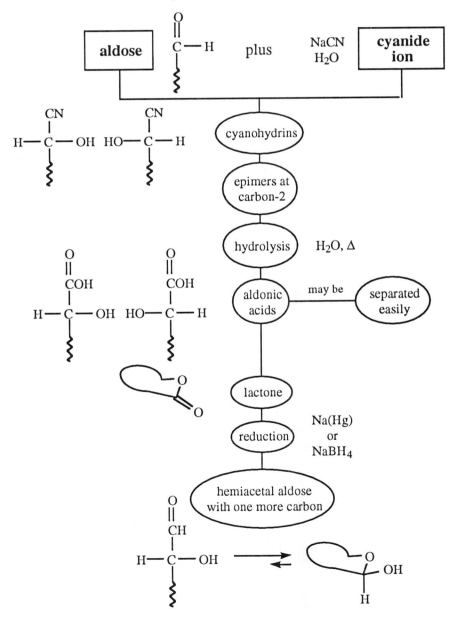

O-β-D-glucopyranosyl-(1 → 6)-D-glucopyranose

- -

Concept Map 25.4   The Kiliani-Fischer synthesis.

25.40

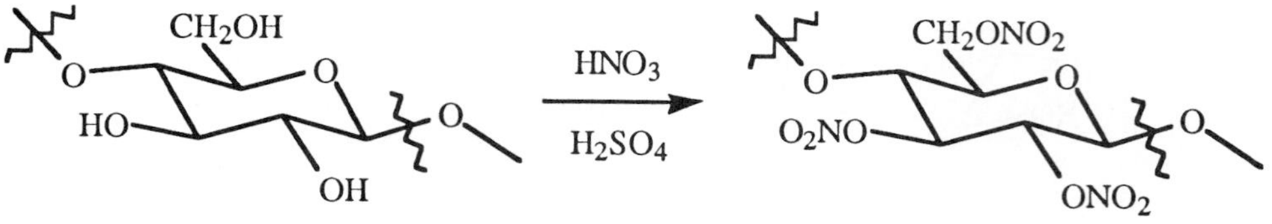

portion of cellulose   portion of cellulose trinitrate

25.41

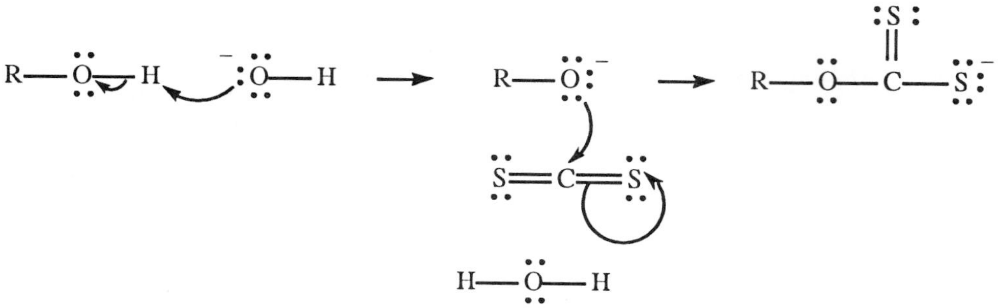

25.42

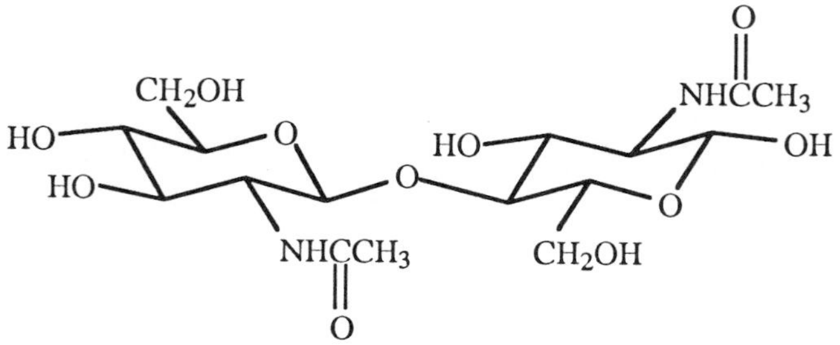

chitobiose

25.43

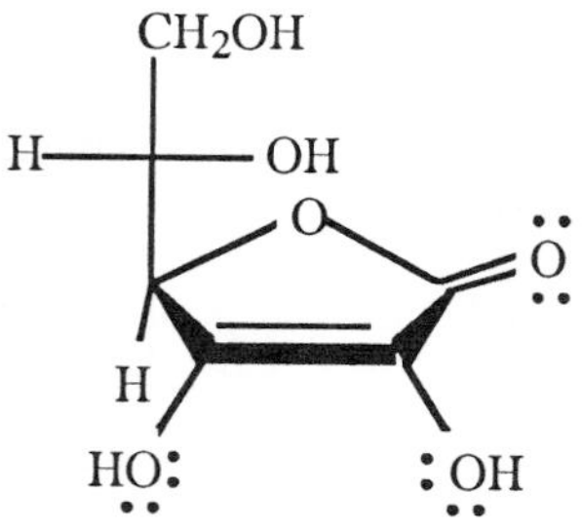

ascorbic acid

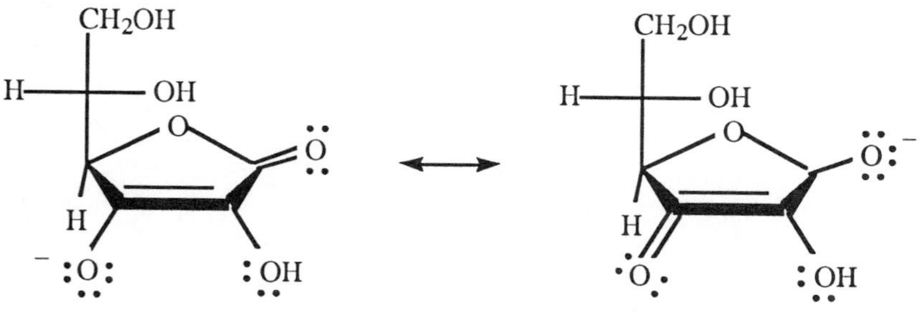

delocalization of charge to
oxygen atom of carbonyl group

25.43 (cont)

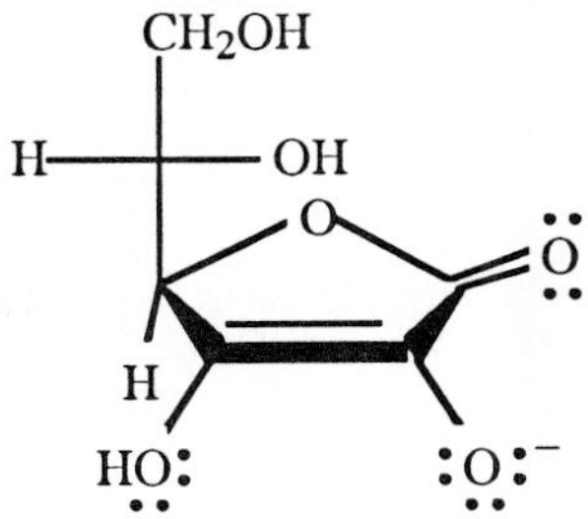

no possible delocalization of
charge to another oxygen atom

25.44

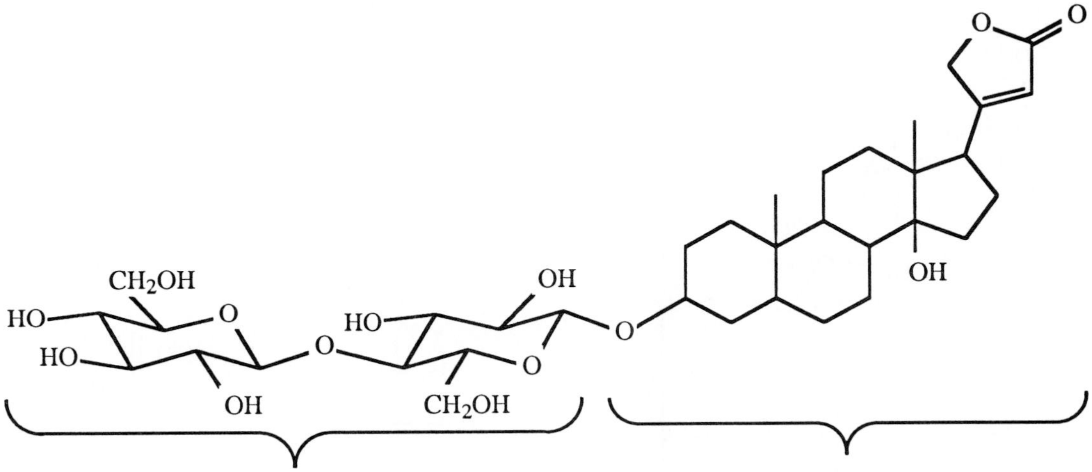

cellobiose   digitoxigenin

25.45

(a)

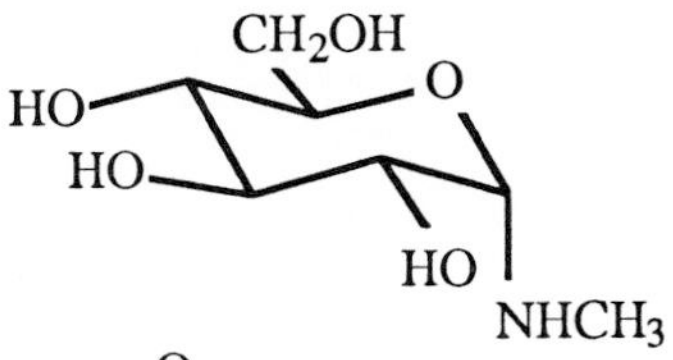

(b)

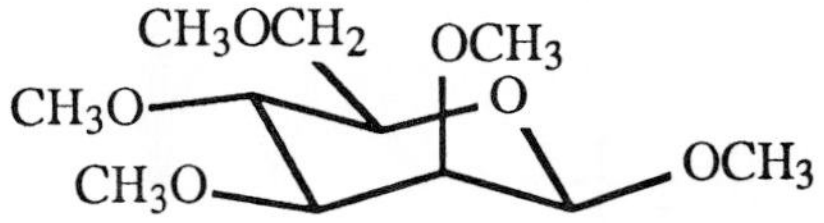

(c)

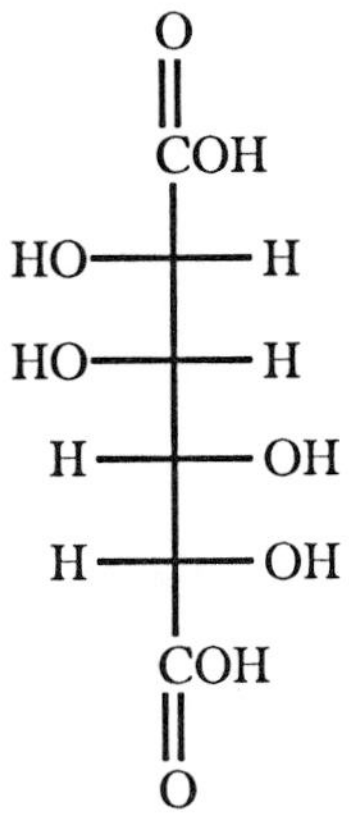

(d)

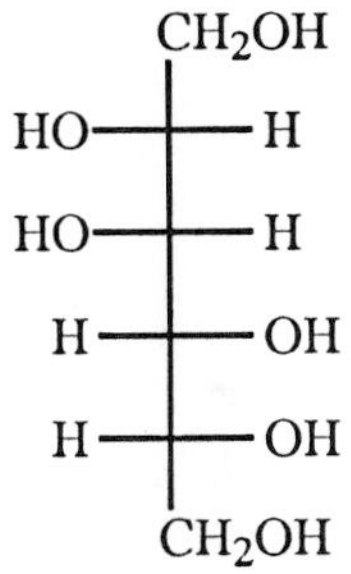

25.45 (cont)

(e)

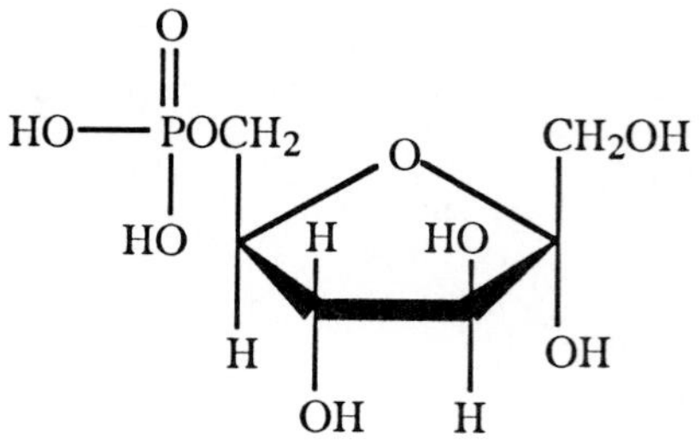

(f)

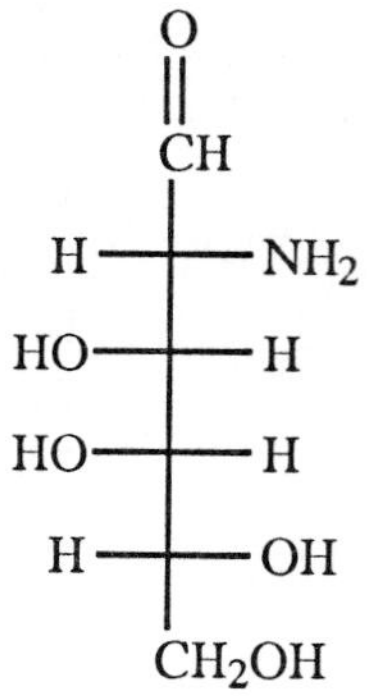

or

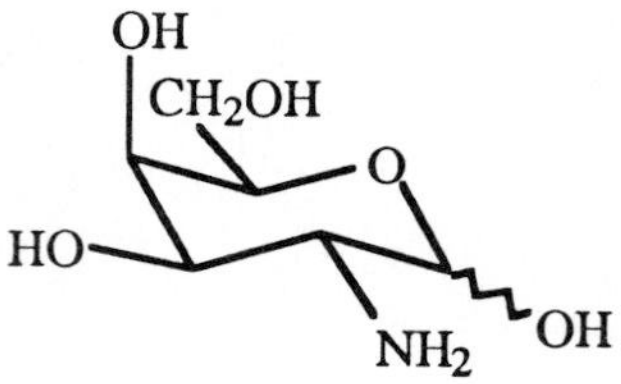

(g)

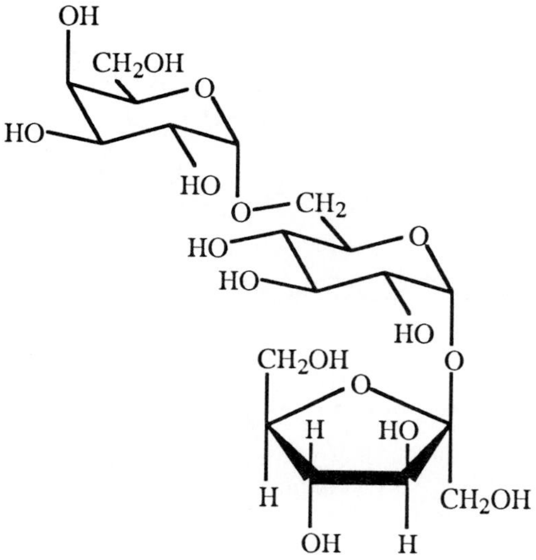

(h)

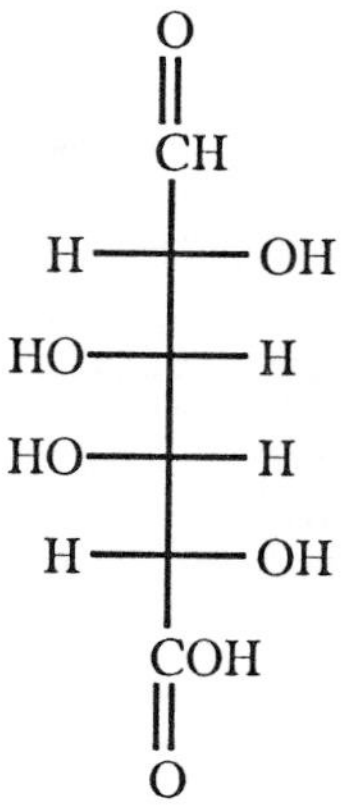

or

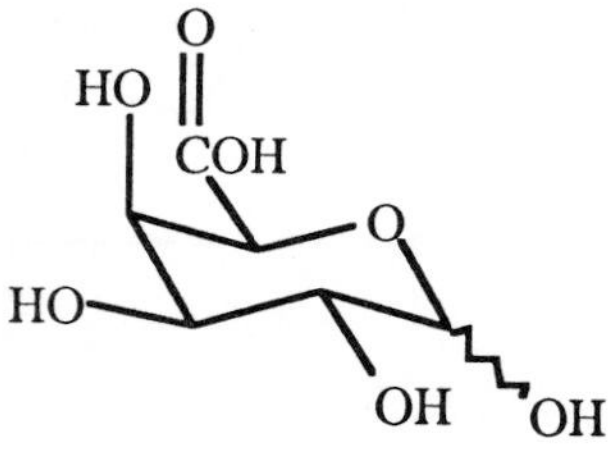

(i)

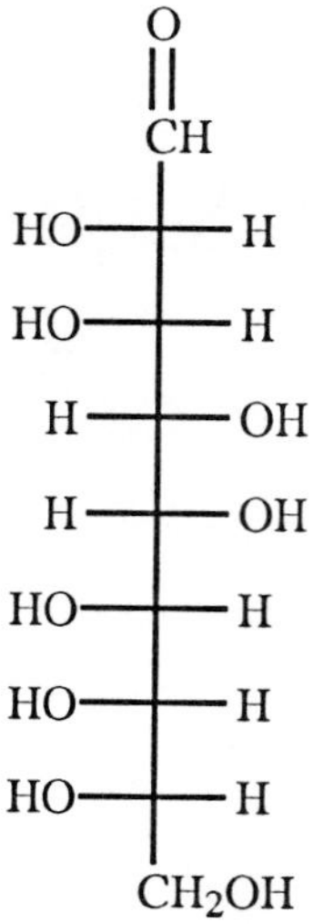

25.46 (a) *N*-acetyl-2-amino-2-deoxy-D-galactose
          2-acetamido-2-deoxy-D-galactose

       (b) α-D-glucopyranose-1-phosphate
            (shown in ionized form)

       (c) methyl β-D-2-deoxyribofuranoside

       (d) δ-galactonolactone

       (e) L-galactose

       (f) *N*-(β-D-ribofuranosyl)dimethylamine

       (g) α-D-galactopyranose pentaacetate

       (h) methyl 2,3,5-tri-*O*-methyl-β-D-
            ribofuranoside

       (i) L-*gluco*-2-heptulose

25.47

(a)

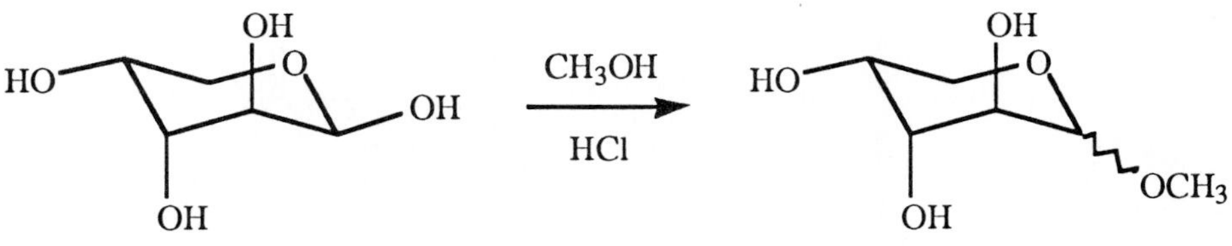

shown as β-D-arabinopyranose

mixture of methyl α- and
β-D-arabinopyranosides

(b)

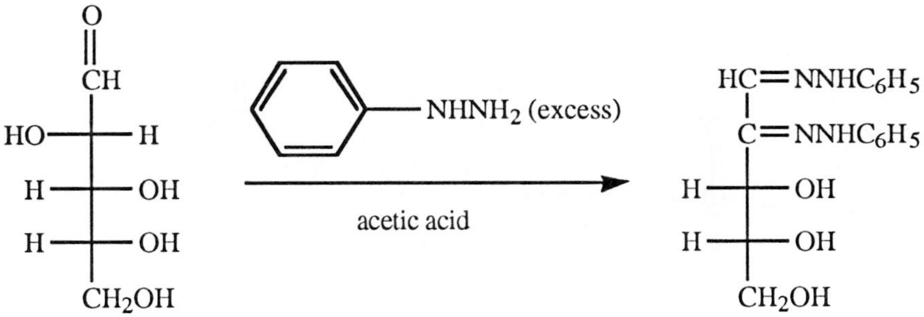

(c)

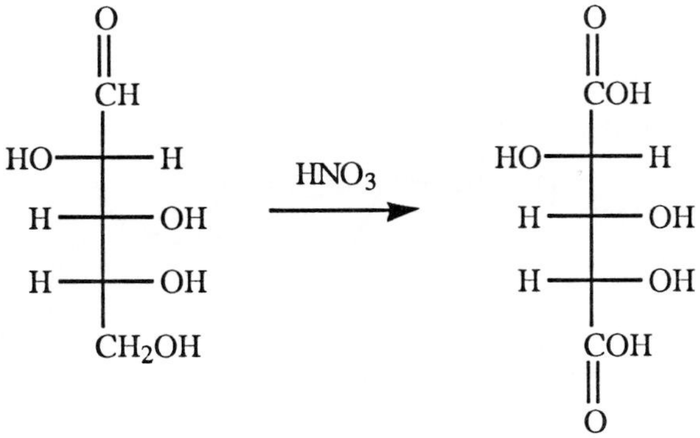

(d)

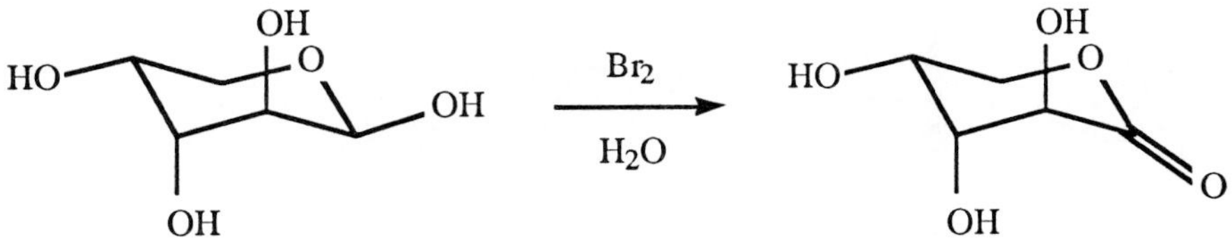

25.47 (cont)

(e)

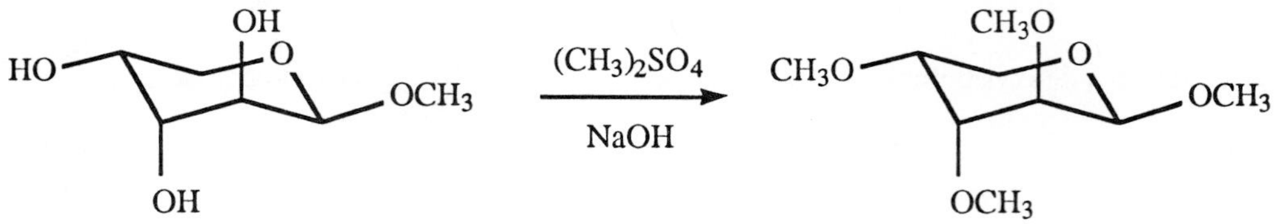

$$\xrightarrow[\text{NaOH}]{(CH_3)_2SO_4}$$

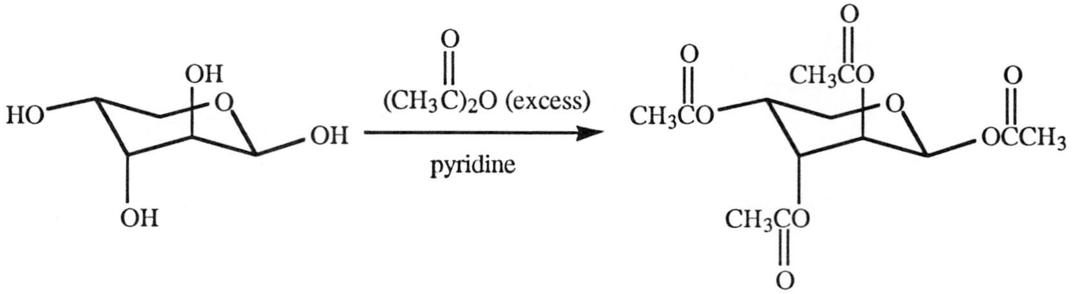

(f)

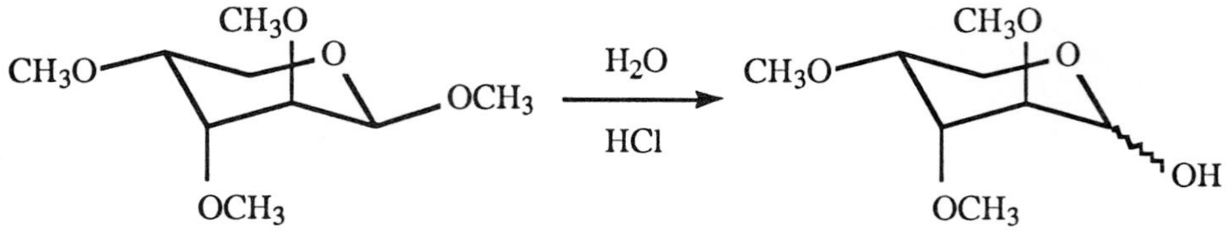

$$\xrightarrow[\text{HCl}]{H_2O}$$

(g)

$$\xrightarrow[\text{pyridine}]{(CH_3C)_2O \text{ (excess)}}$$

(h)

$$\xrightarrow[\text{cold}]{HBr}$$

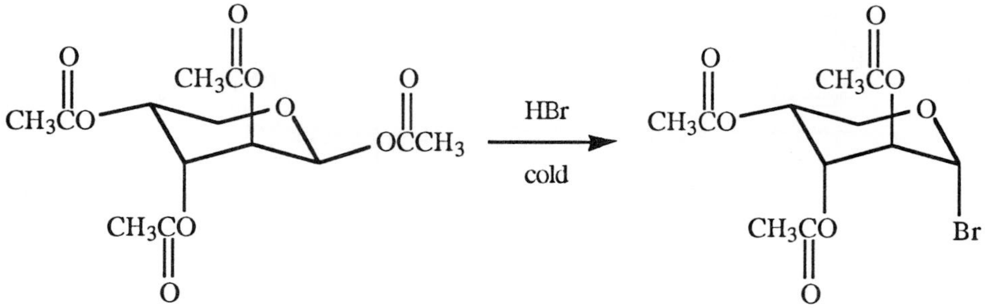

(i)

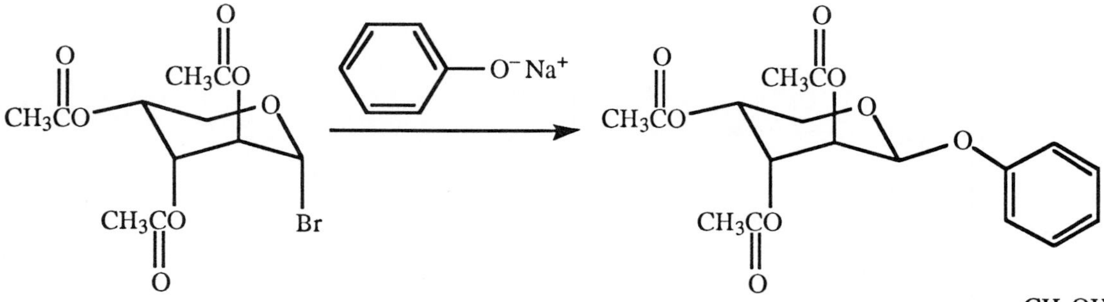

(j)

$$\xrightleftharpoons{NaOH, H_2O}$$

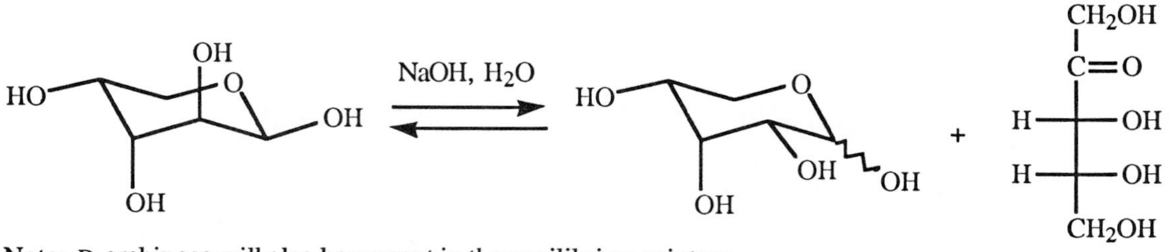

Note: D-arabinose will also be present in the equilibrium mixture.

25.47 (cont)

(k)

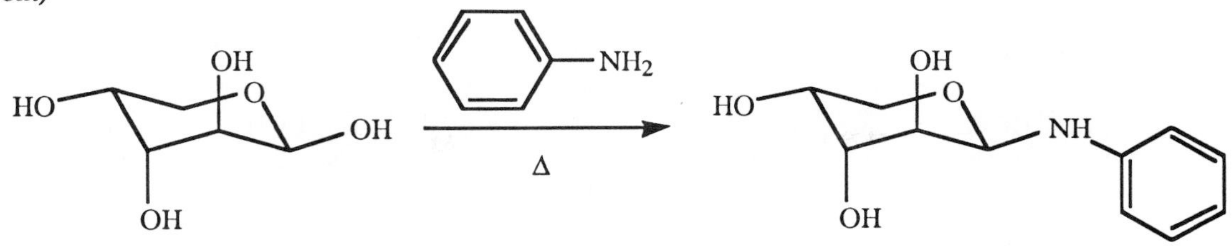

(l)

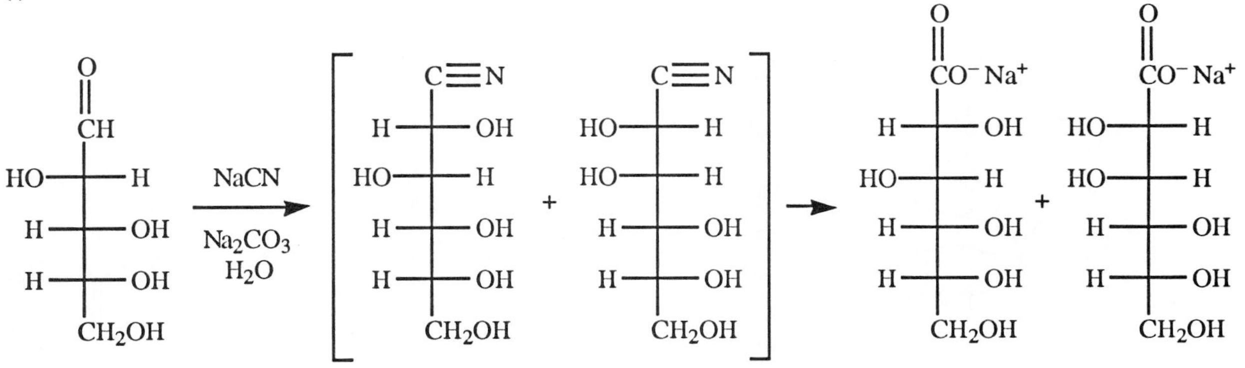

25.48

(a)

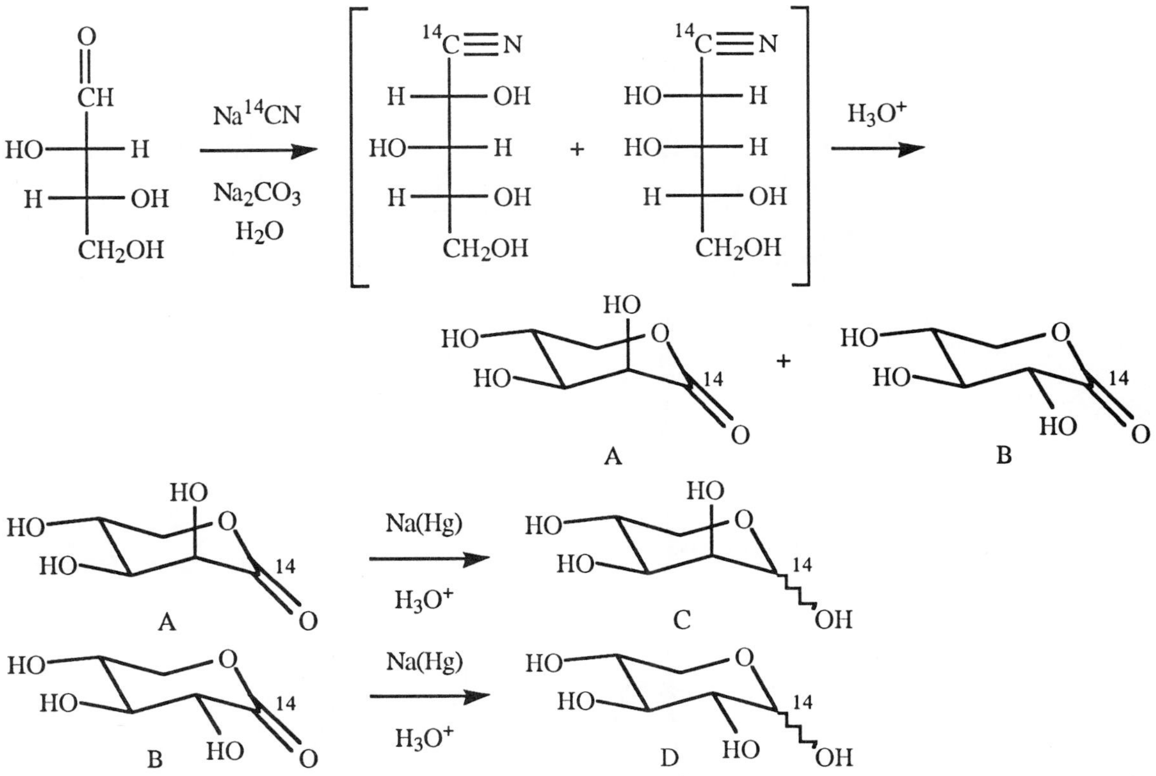

25.48 (cont)

(b)

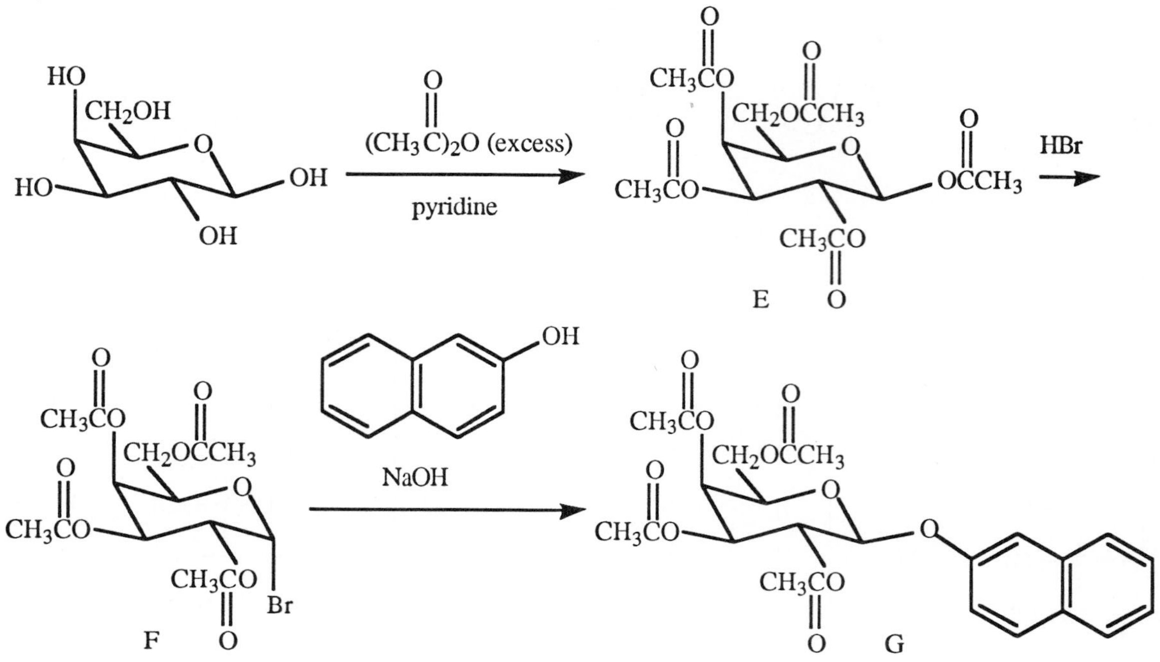

(c)

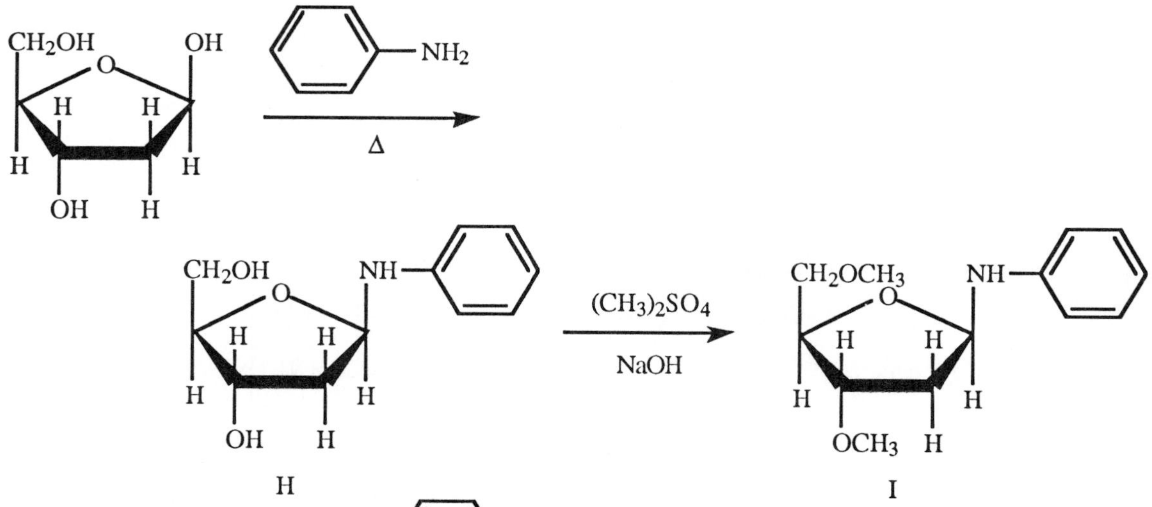

(d)

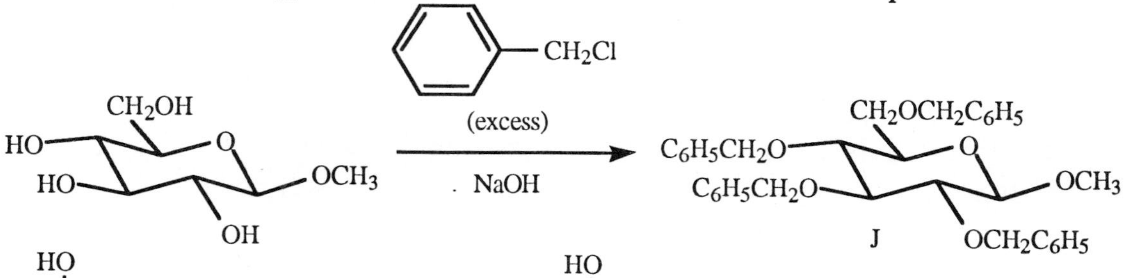

(e)

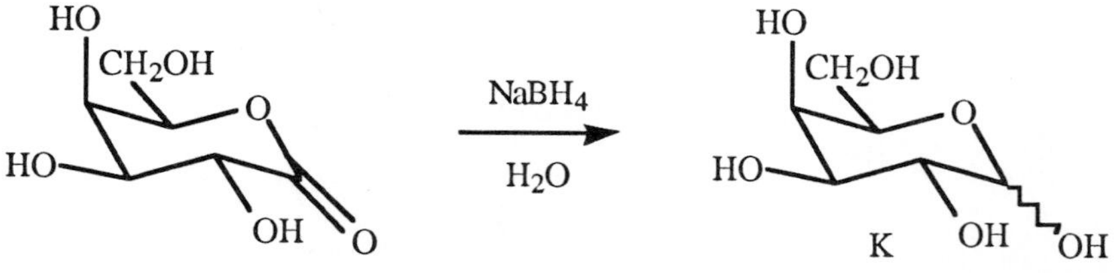

25.48 (cont)

(f)

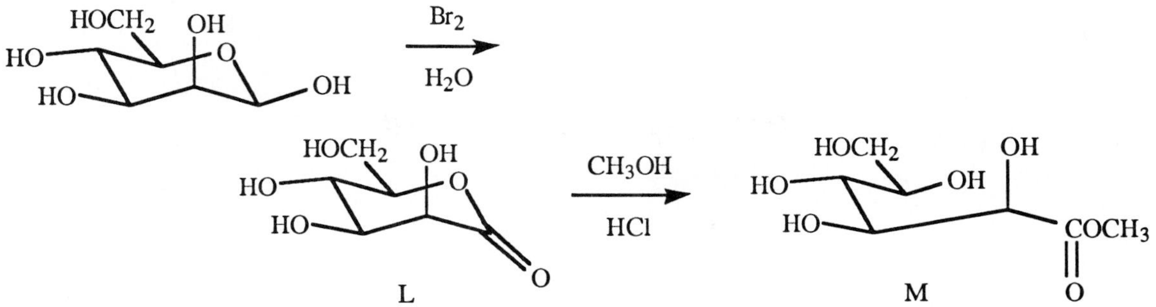

(g)

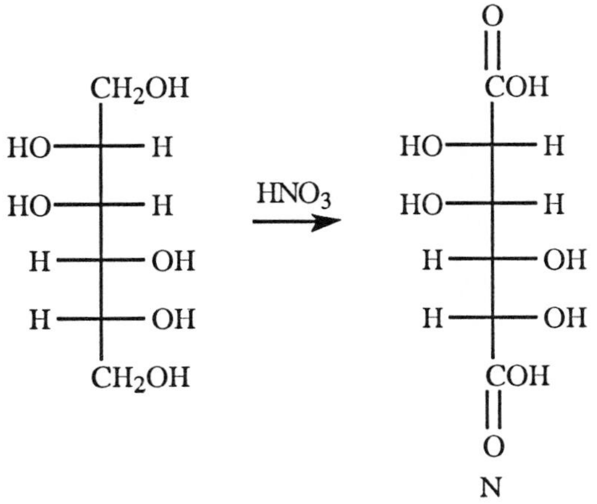

25.49

(a)

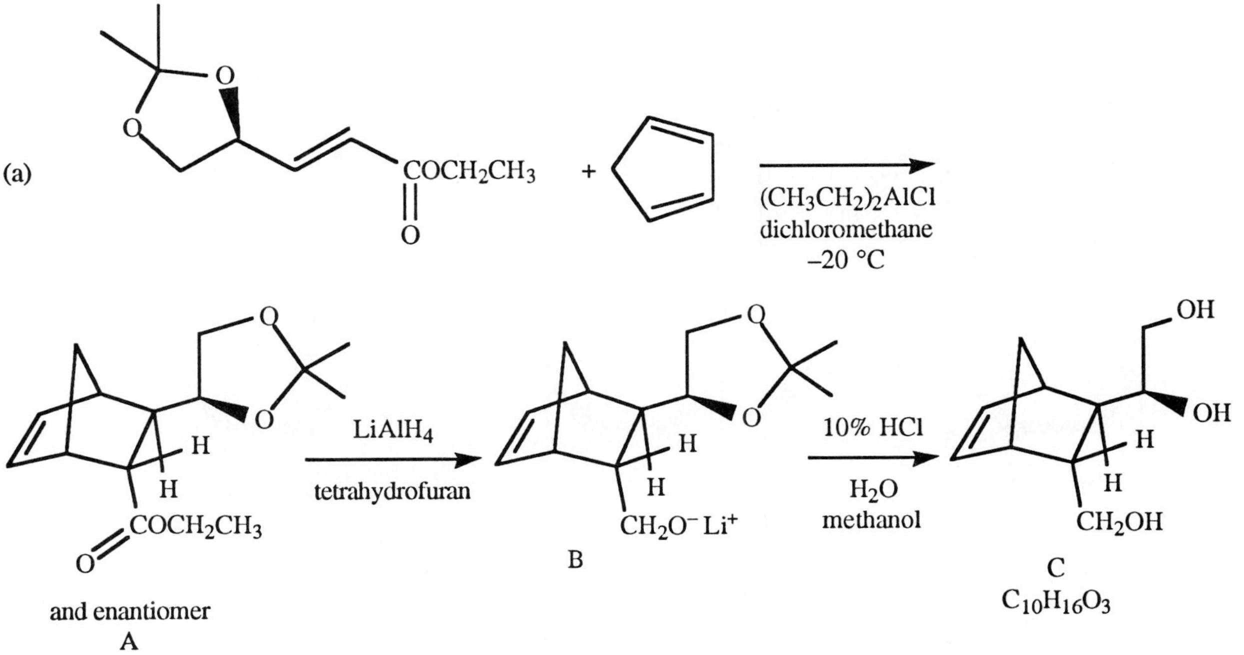

and enantiomer
A

25.49 (cont)

(b)

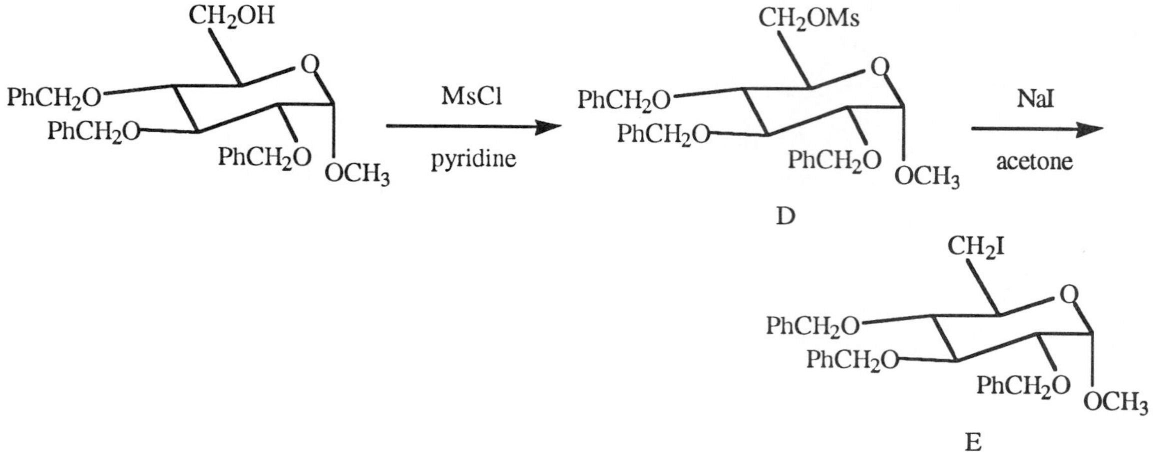

(c)

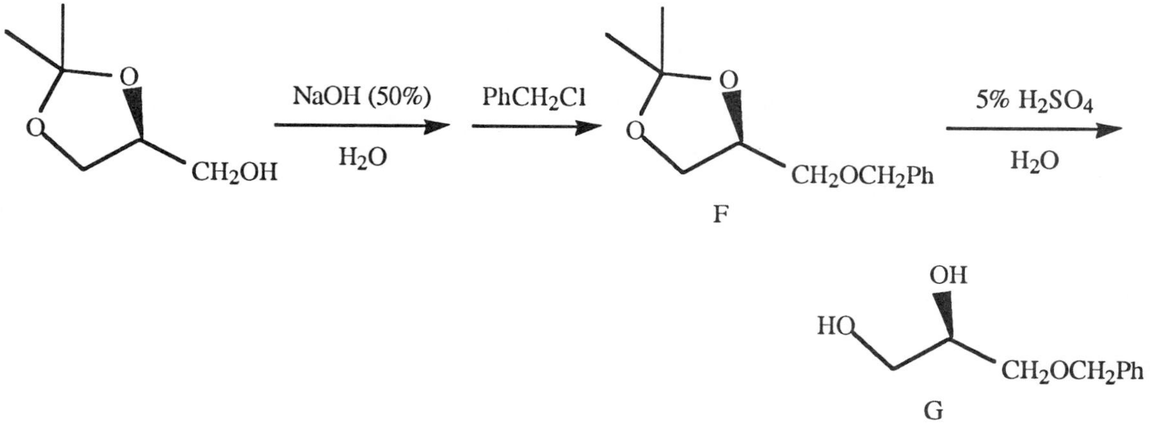

(d)

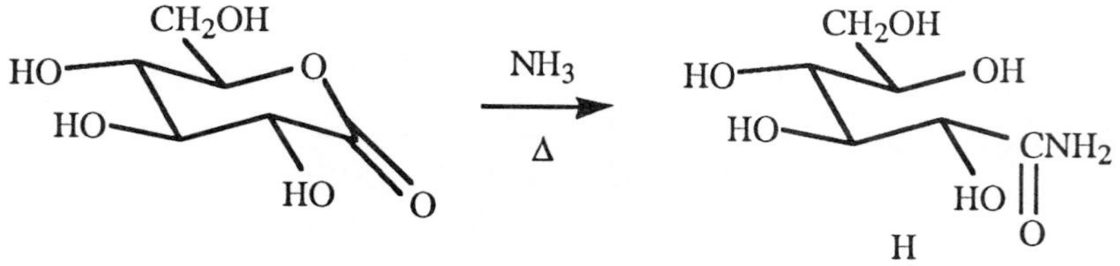

(e)

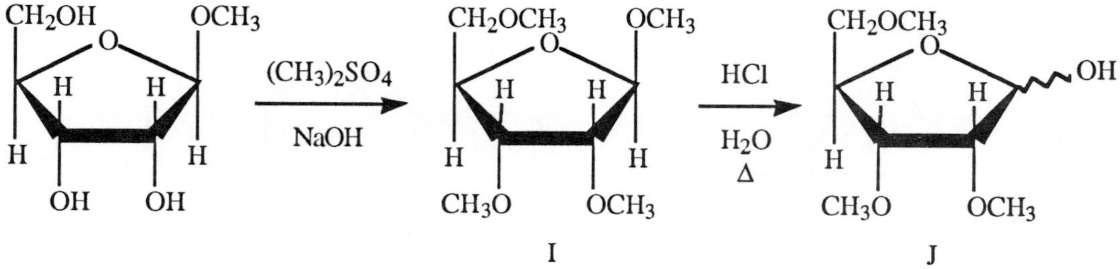

25.49 (cont)

(f)

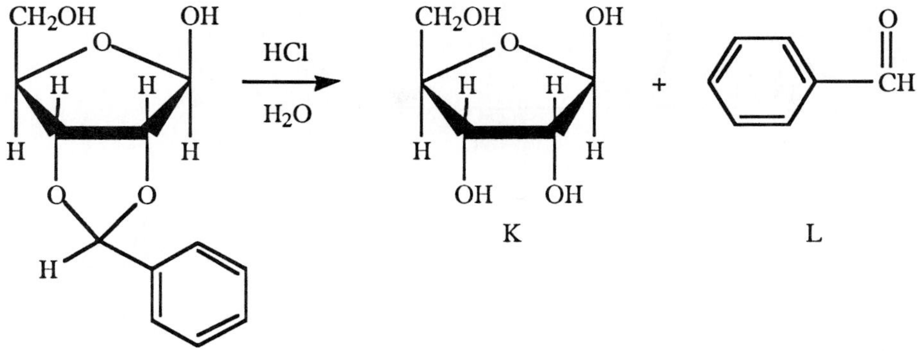

25.50

(a)

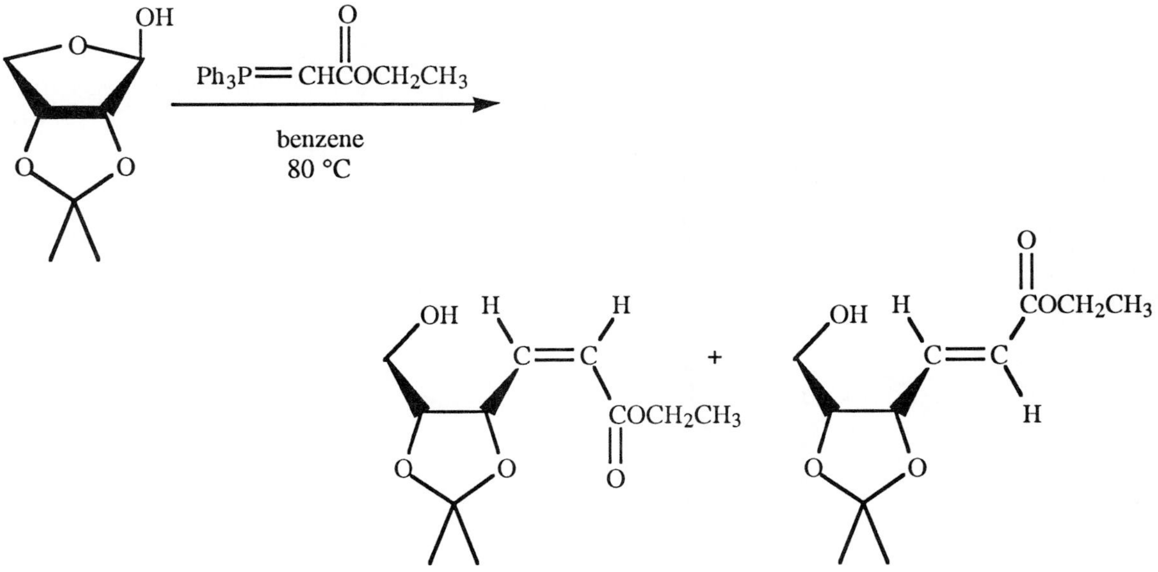

A and B are diastereomers

(b)

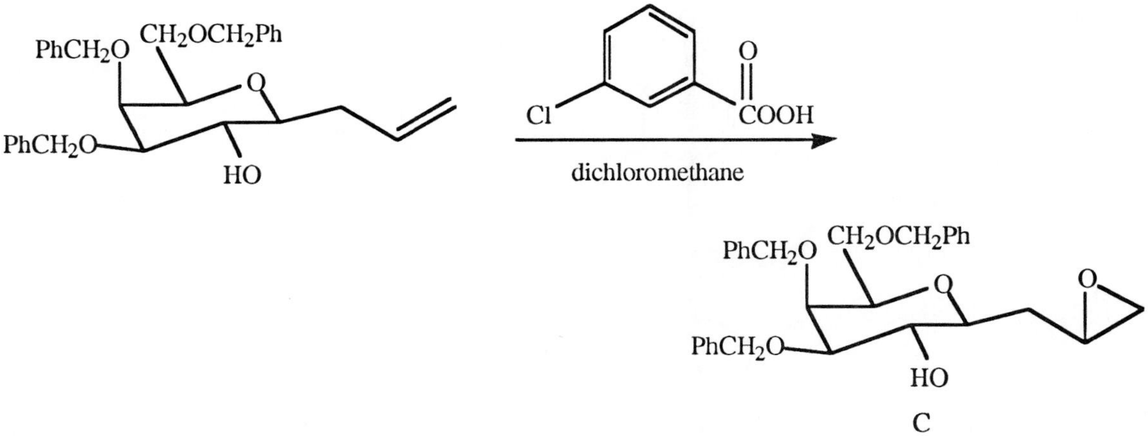

25.50 (cont)

(c)

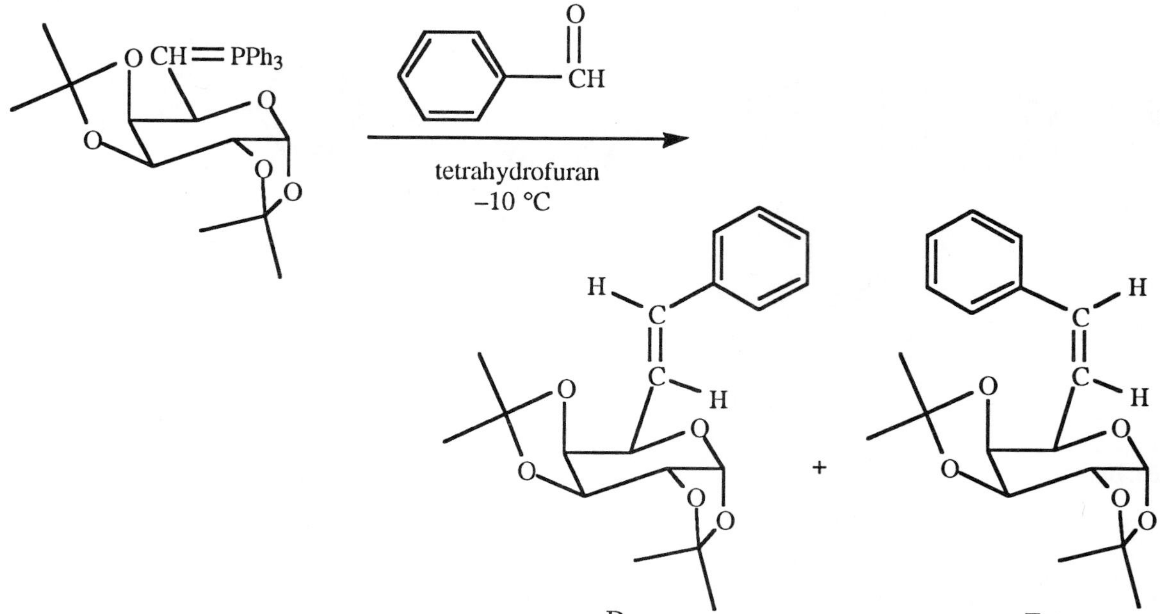

D   D and E are diastereomers   E

(d)

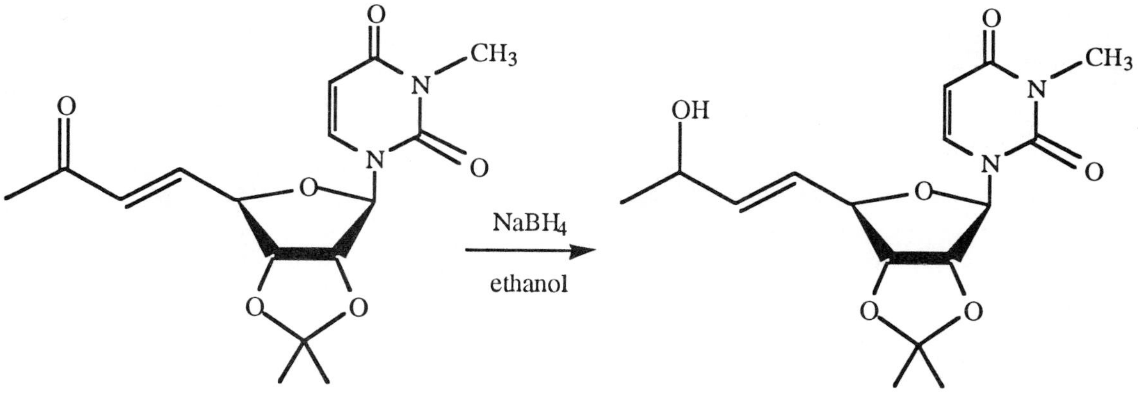

F

(e)

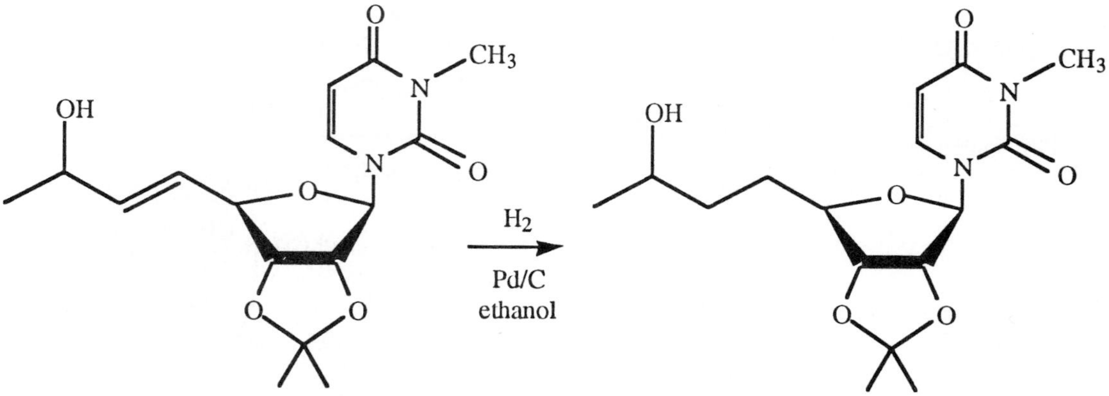

F   G

25.50 (e) (cont)

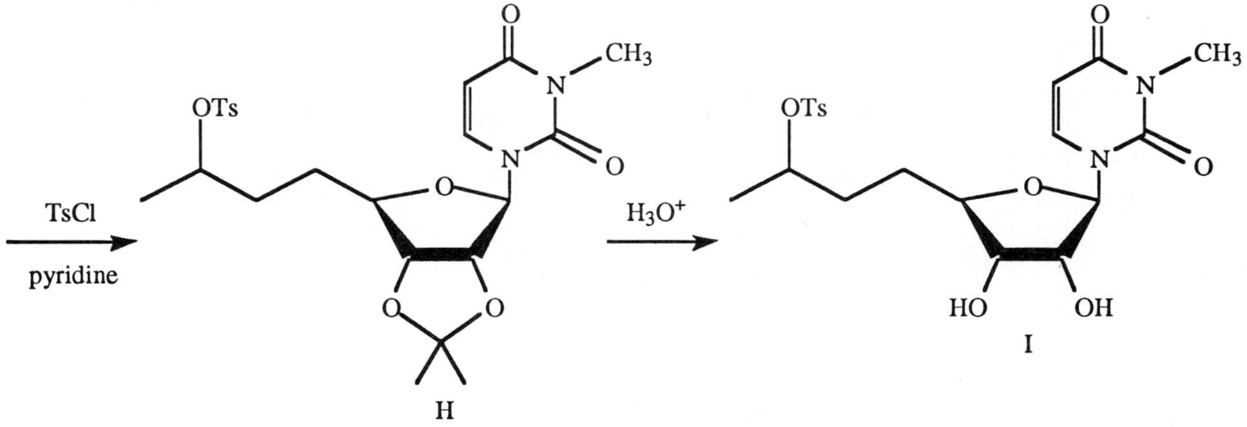

(f)

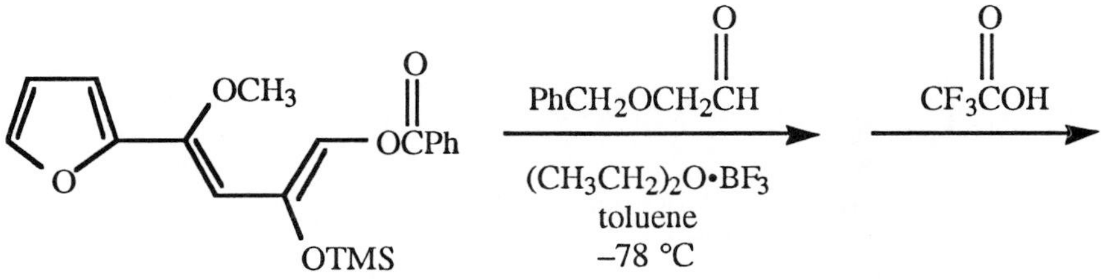

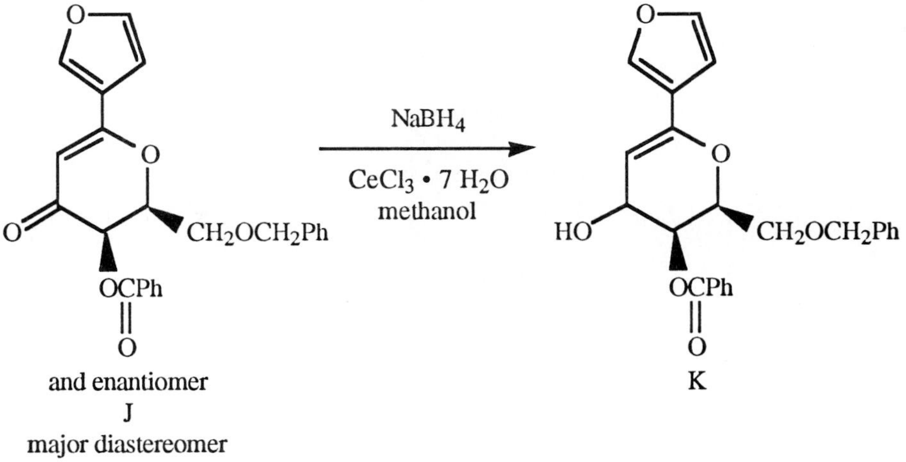

and enantiomer
J
major diastereomer

K

25.51

(+)-tartaric acid   A   B

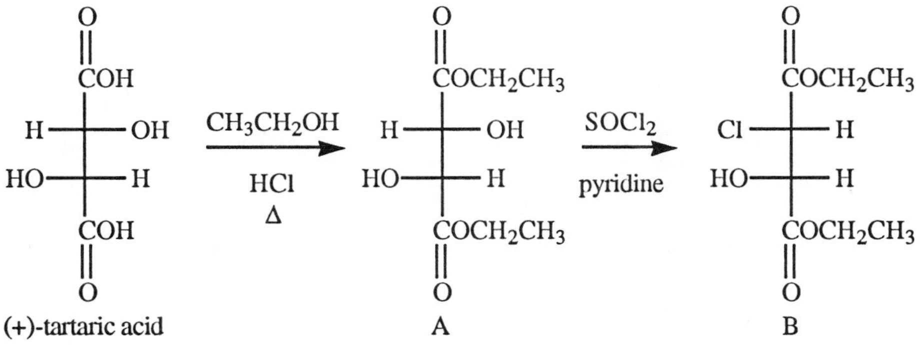

25.51 (cont)

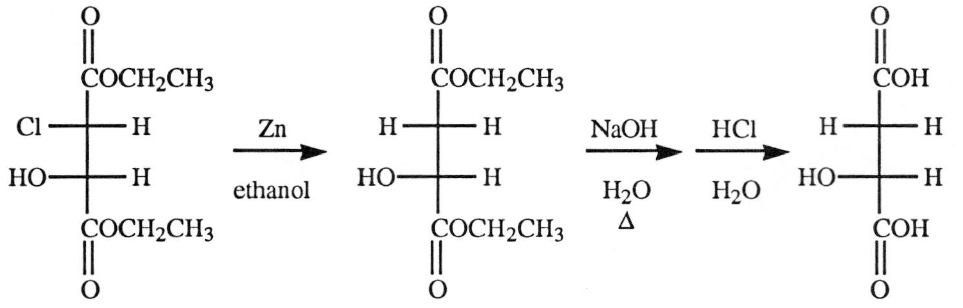

B C (+)-malic acid

Note: Conversion of either hydroxyl group in (+)-tartaric acid to the chloro compound and reduction of the carbon-chlorine bond gives (+)-malic acid.

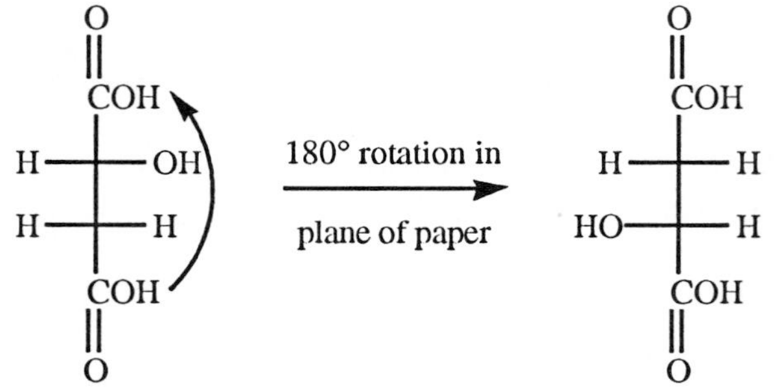

compound obtained if the lower hydroxyl group in (+)-tartaric acid is replaced by hydrogen

(+)-malic acid

25.52

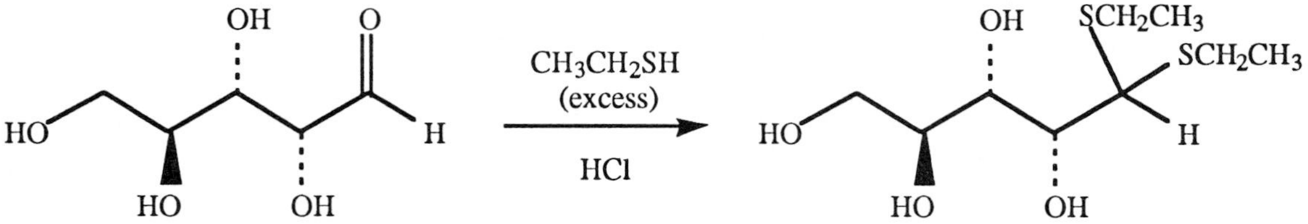

L-(+)-arabinose

A

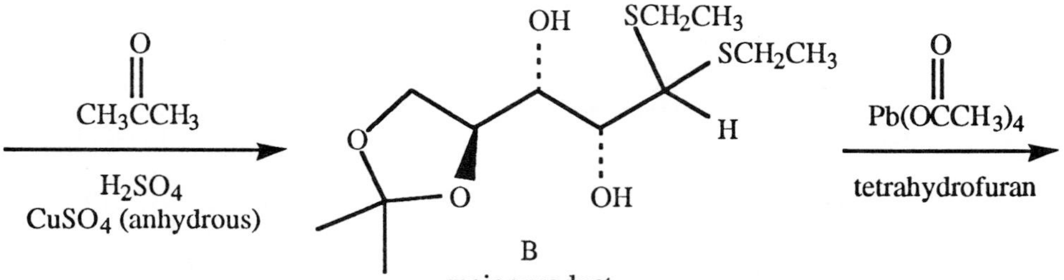

B
major product

25.52 (cont)

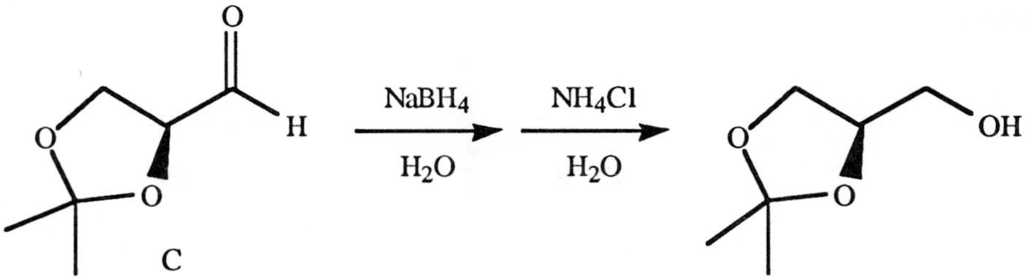

25.53

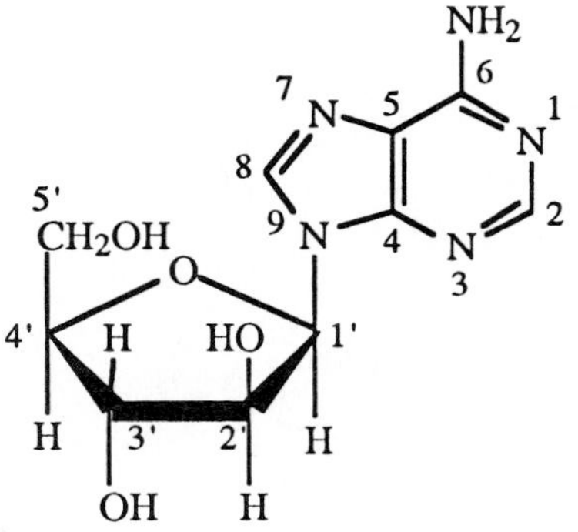

Vira-A

(Arabinose is epimeric with ribose at C-2, the 2' position in the sugar residue, so the structure of Vira-A is easily derived from that of adenosine.)

25.54

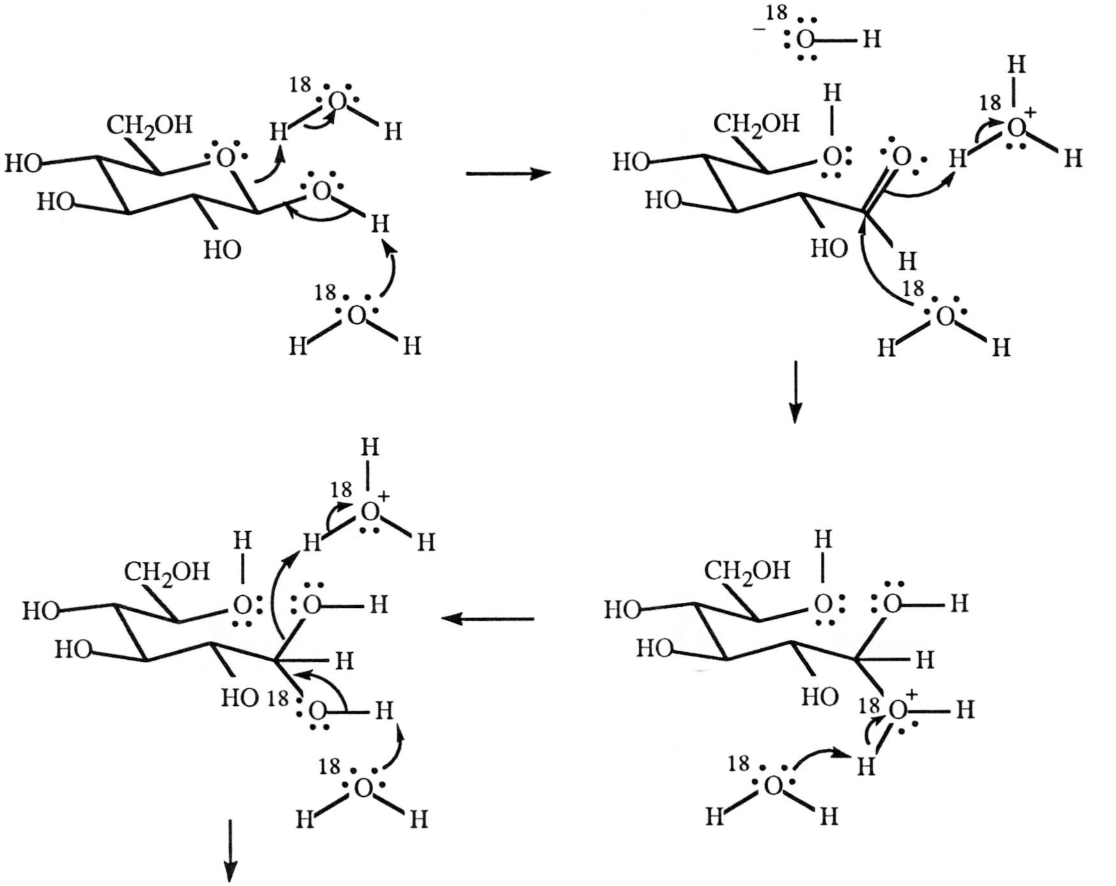

25.54 (cont)

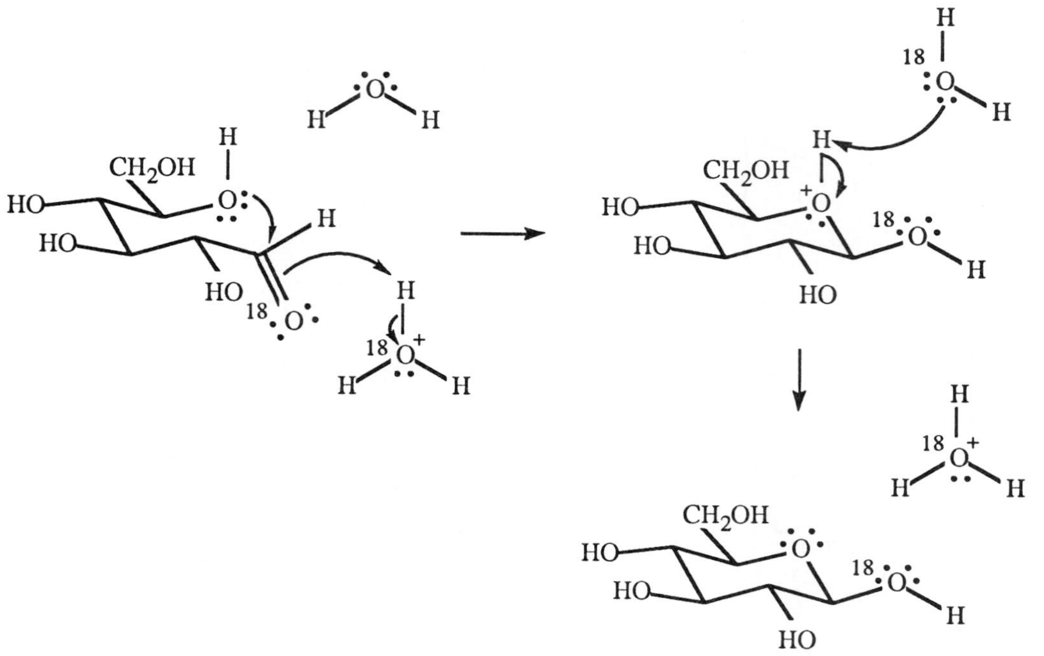

Exchange of oxygen atoms occurs at the anomeric carbon atom by way of the open-chain form of glucose.

25.55

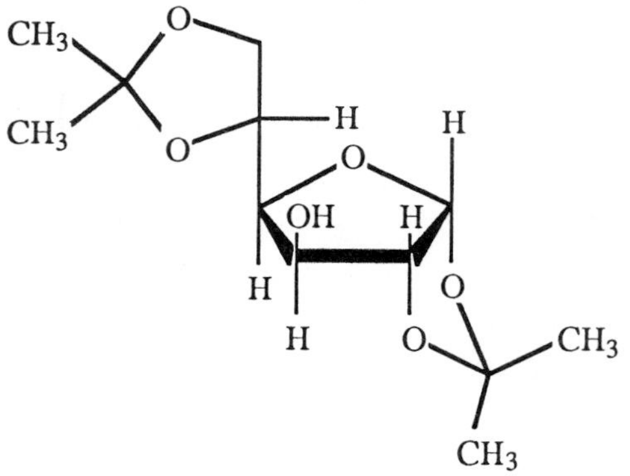

1,2,5,6-di-*O*-isopropylene-D-glucofuranose

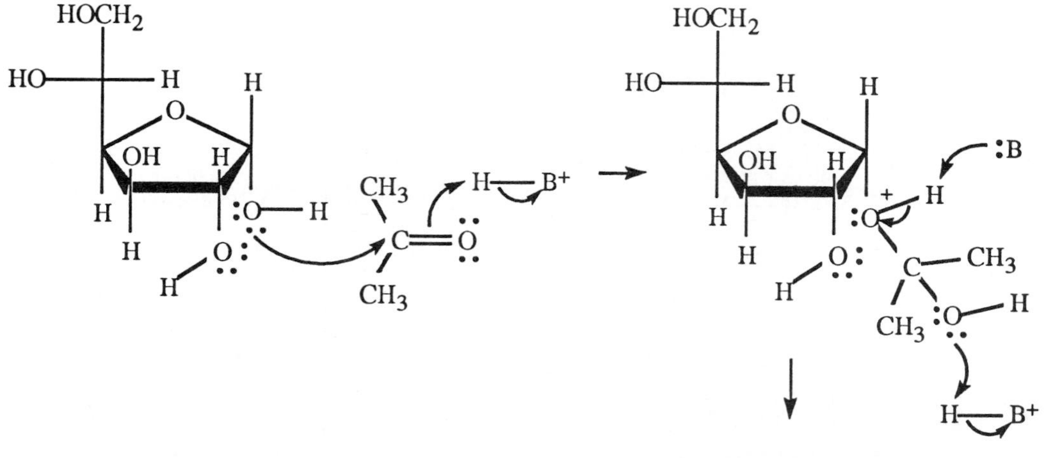

25.55 (cont)

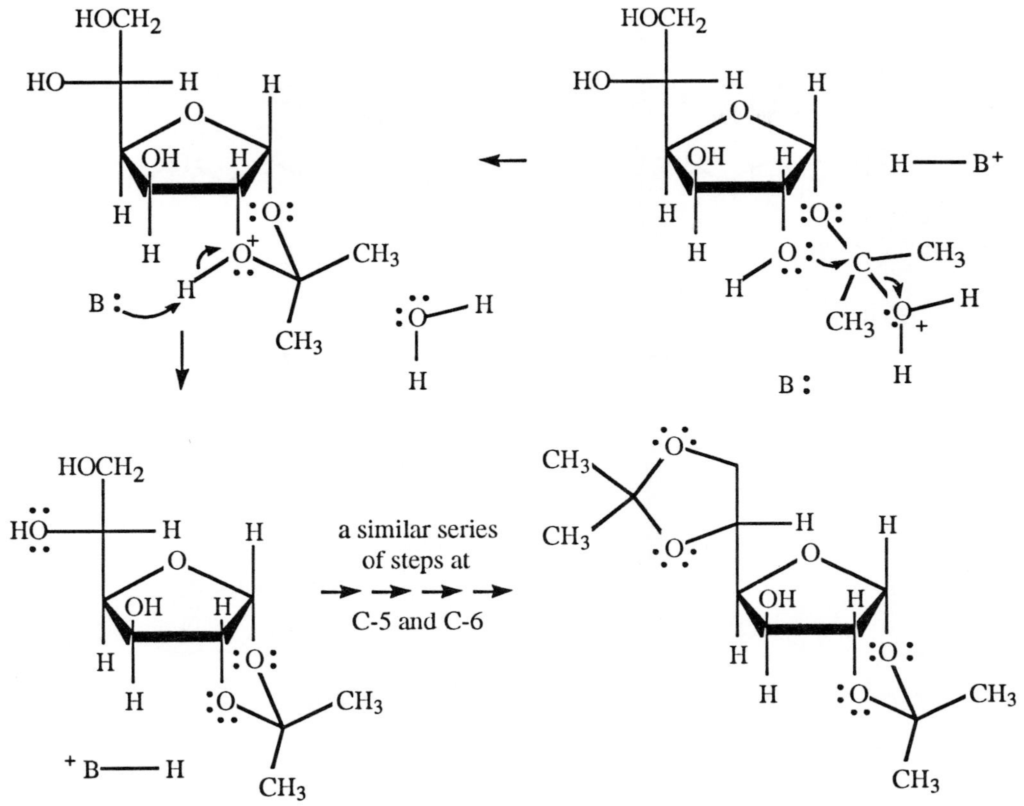
a similar series
of steps at
C-5 and C-6

25.56

(a)

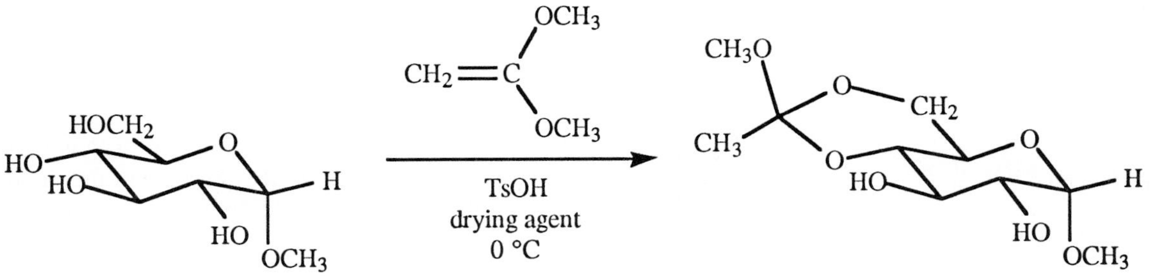
TsOH
drying agent
0 °C

(b)

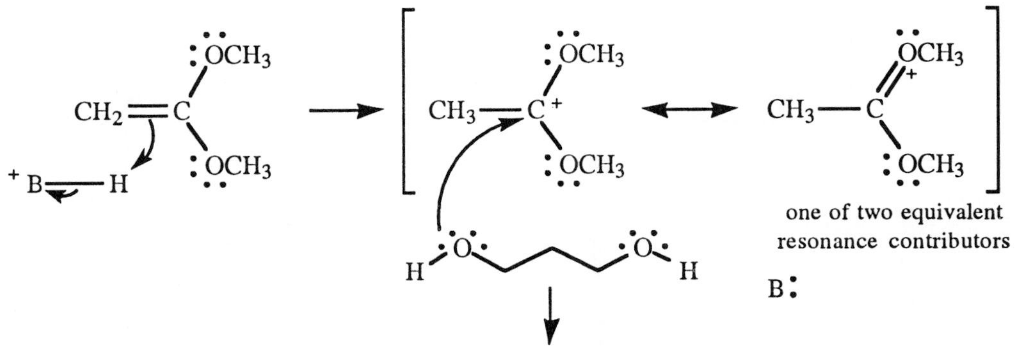
one of two equivalent
resonance contributors

25.56 (b) (cont)

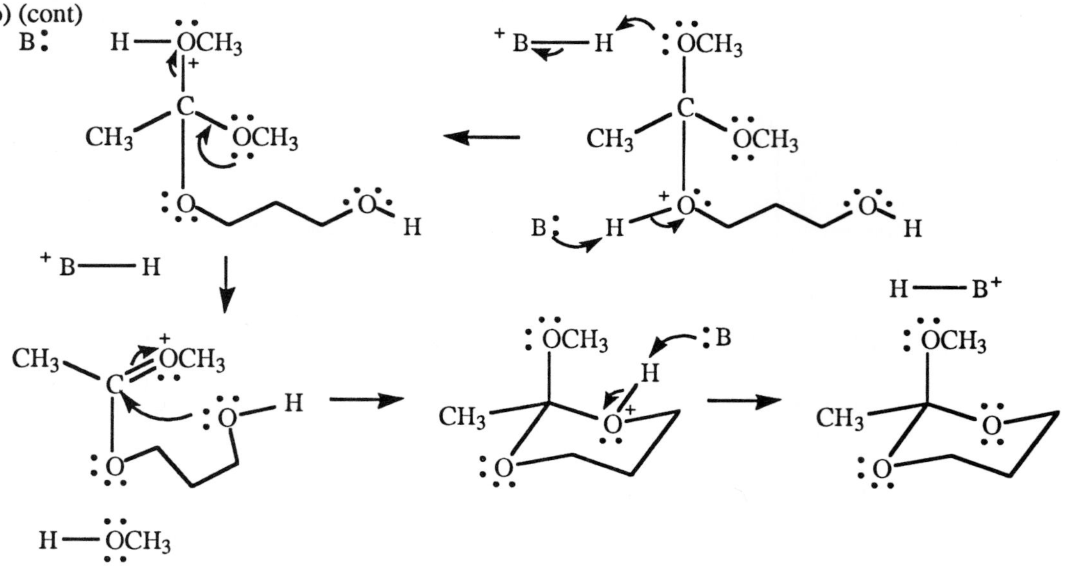

25.57

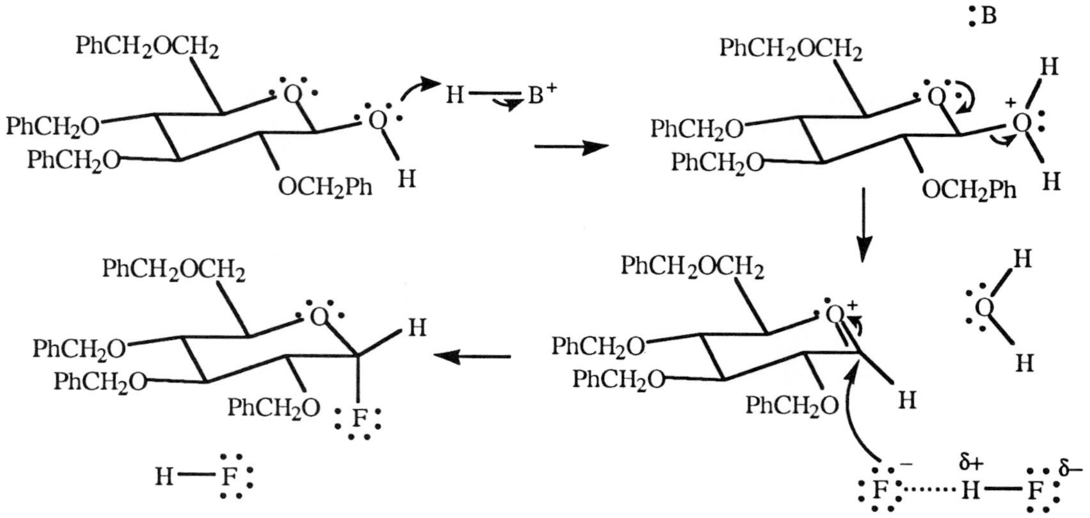

In solution, fluoride ion exists hydrogen bonded to hydrogen fluoride as HF_2^- because of the strength of the hydrogen bond to fluorine.

25.58

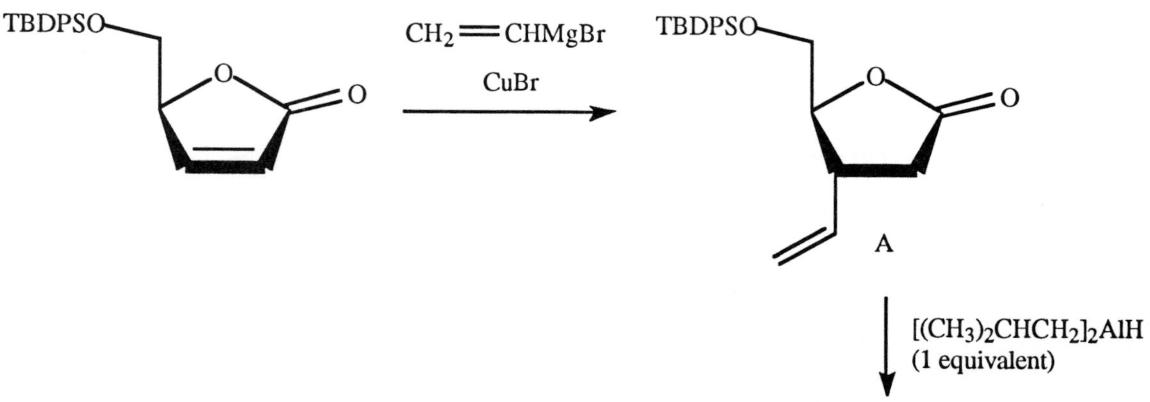

25.58 (cont)

25.59   The transformation from the starting material through A and B to the next structure shown needs to be analyzed before we can write structures for A and B.

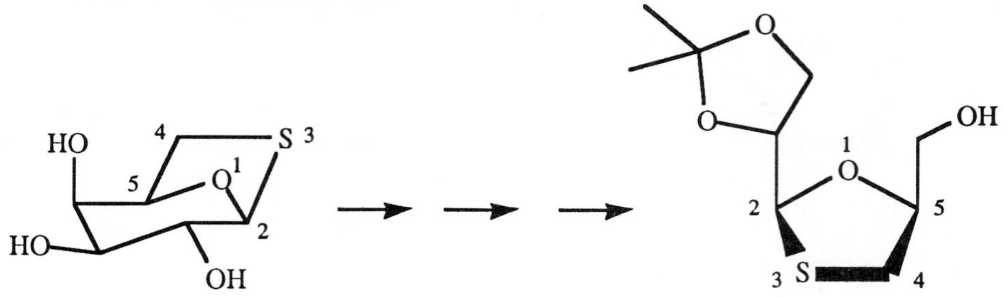

or

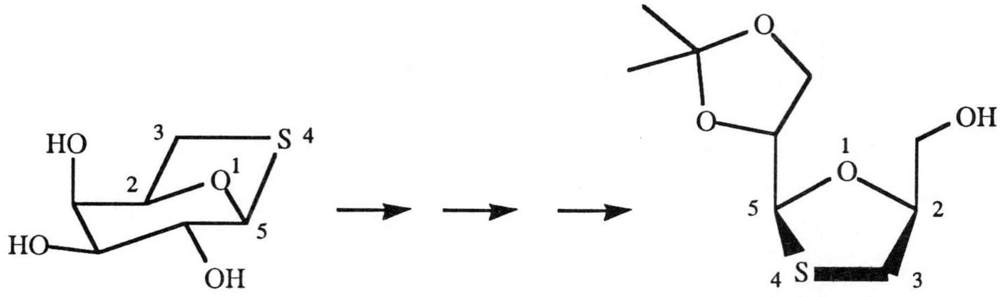

Either way we number the rings, cleavage of the carbohydrate ring appears to have taken place between the two hydroxyl groups that are cis to each other and can, therefore, easily form the cyclic periodate ester necessary for carbon-carbon bond cleavage.

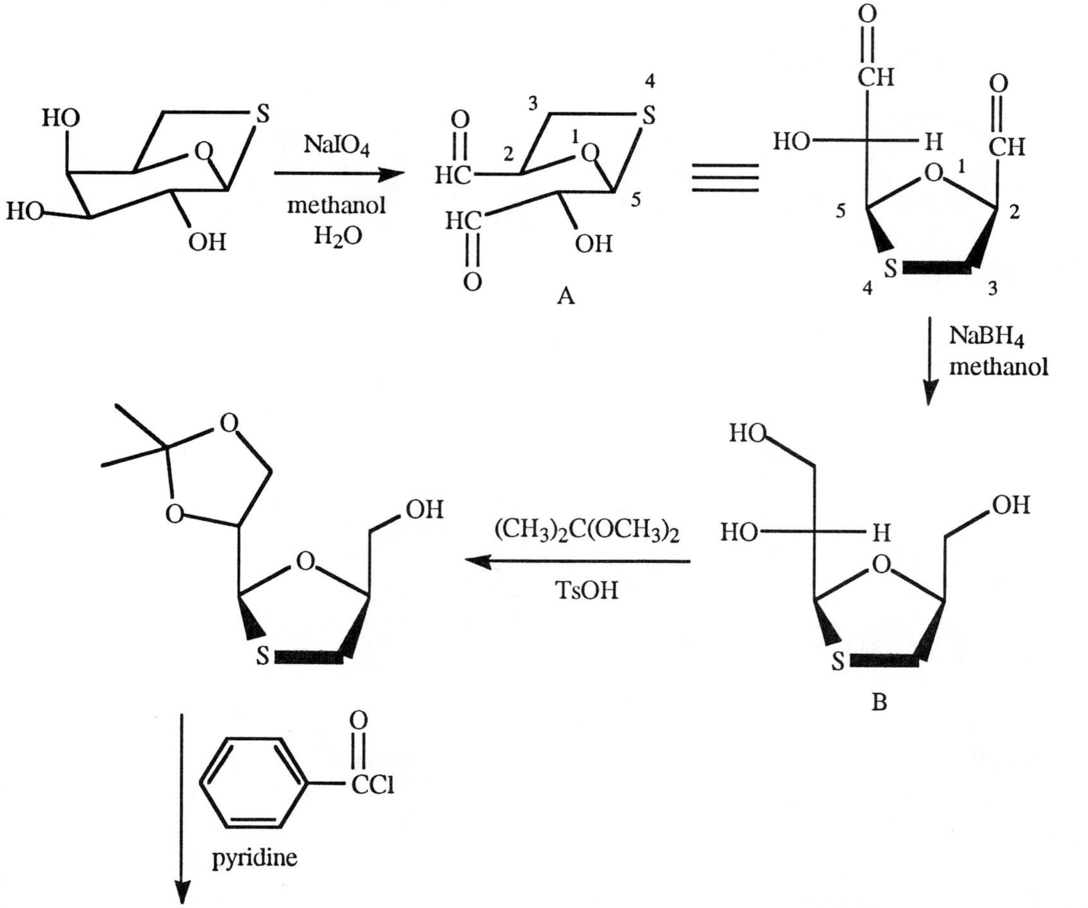

25.59 (cont)

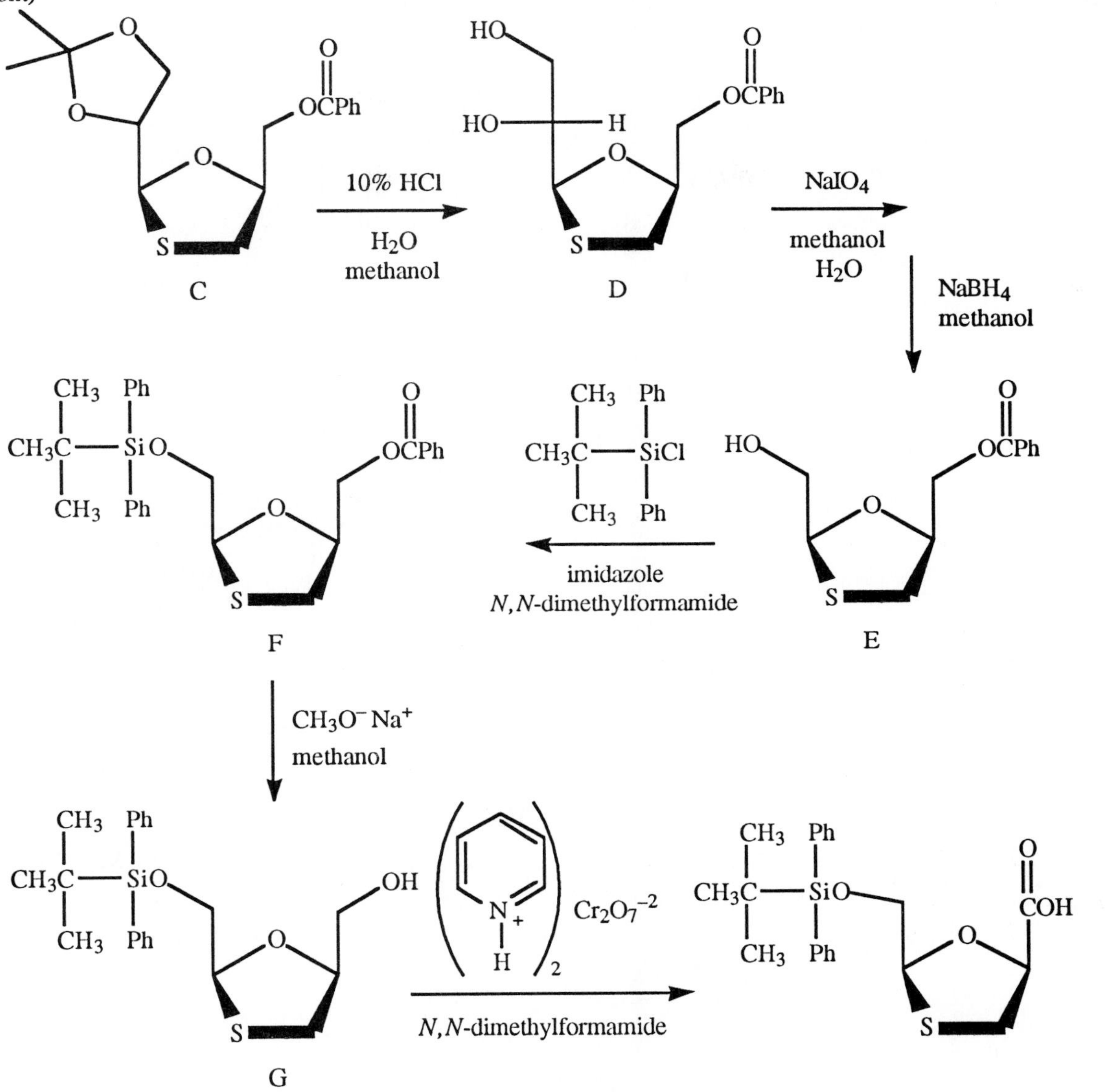

25.60   (a)

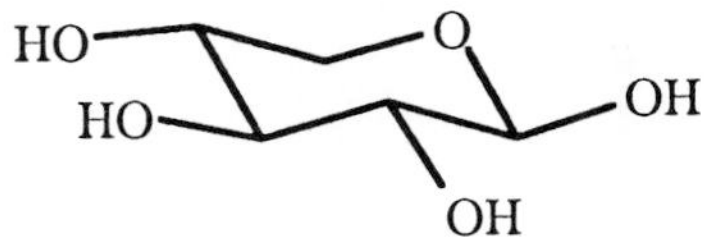

xylose
β-D-xylopyranose

(b) The glucose unit has the free hemiacetal function.

25.60 (cont)

(c)

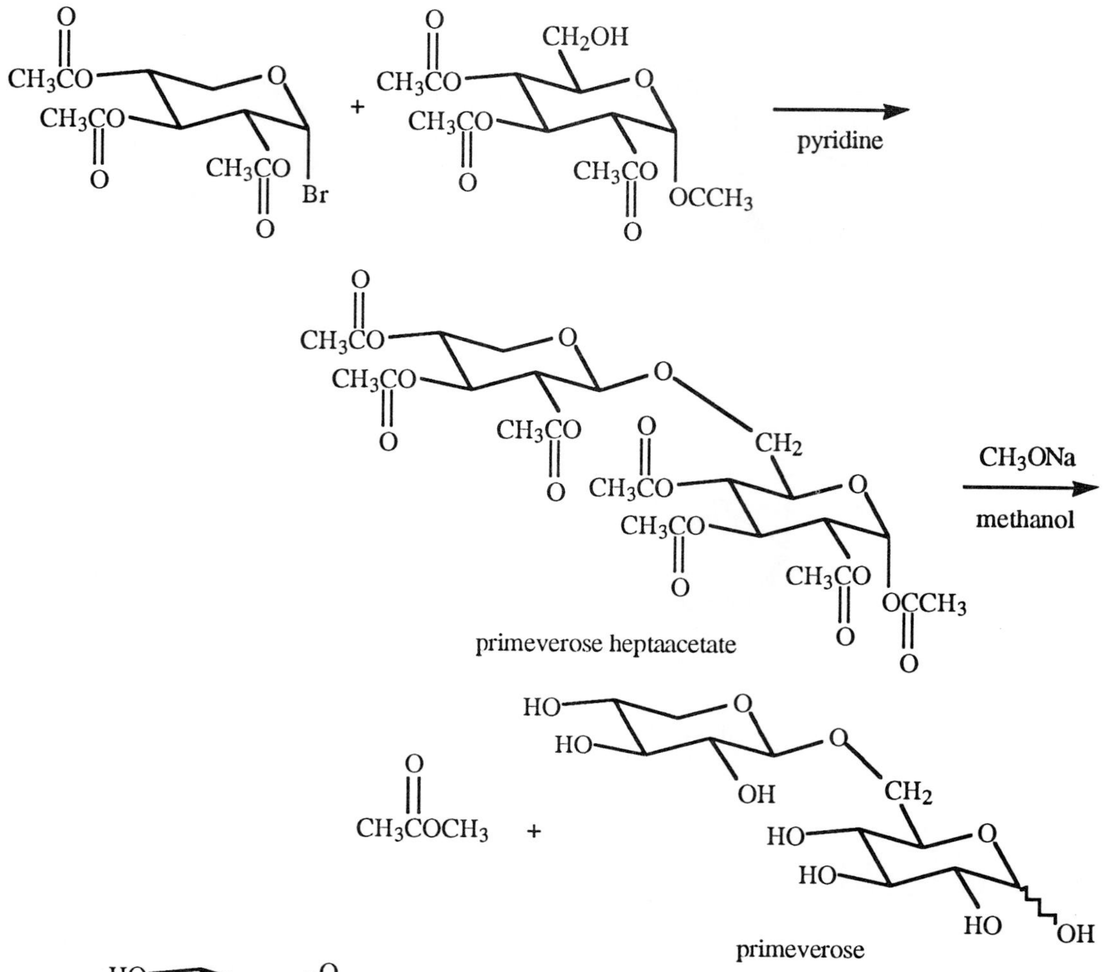
pyridine
CH3ONa
methanol
primeverose heptaacetate
primeverose

(d)

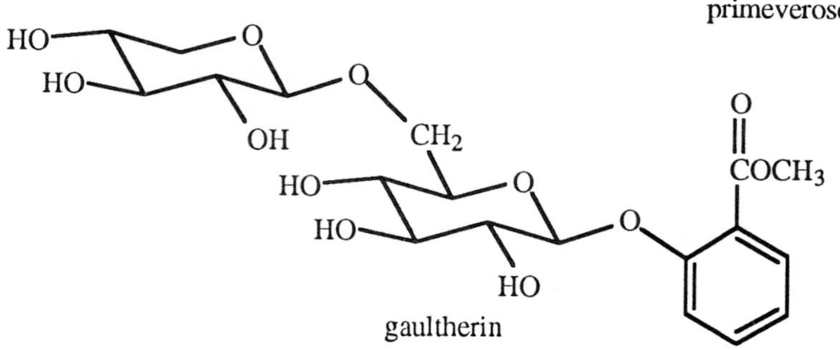

gaultherin

25.61

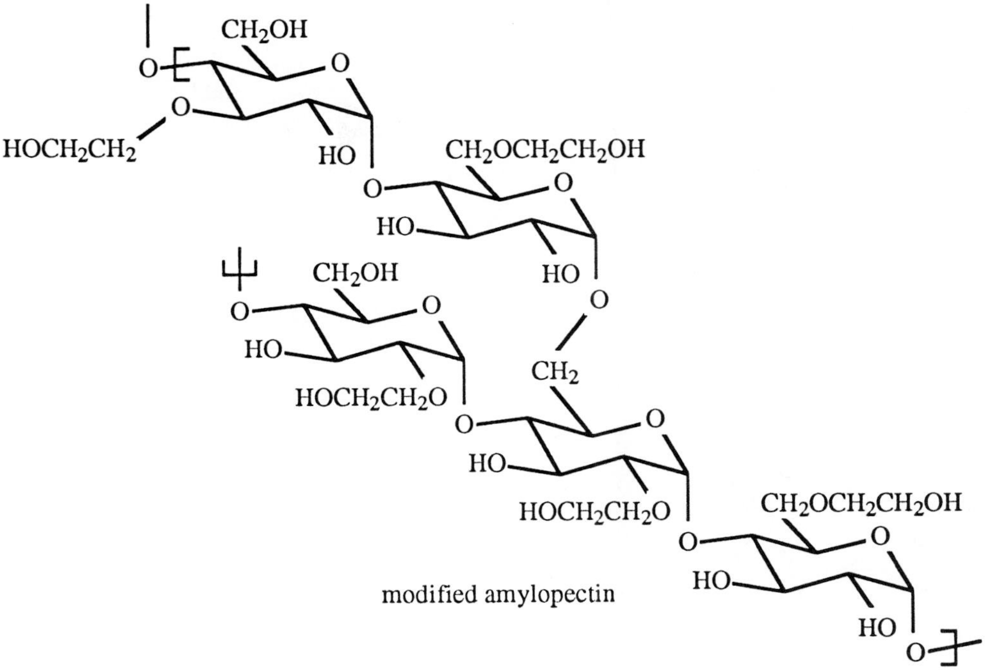

modified amylopectin

25.62

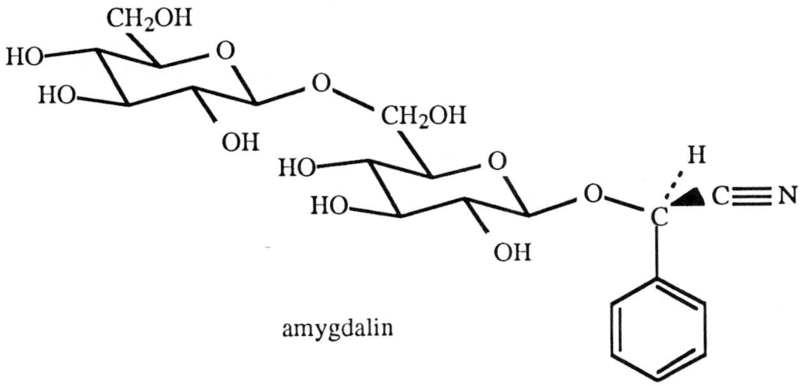

amygdalin

25.63

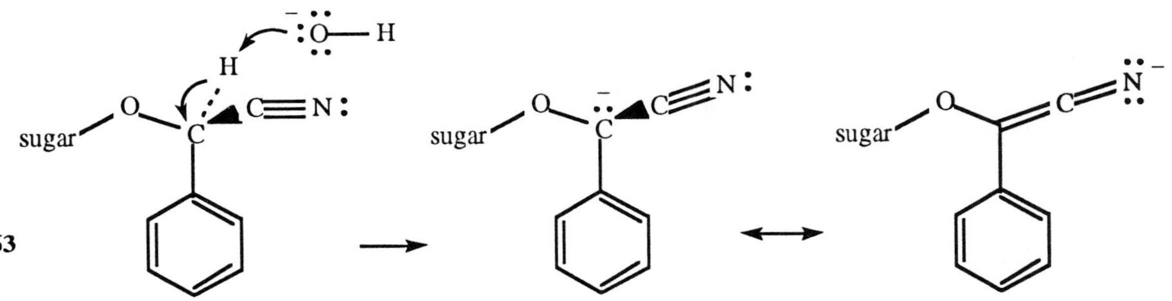

Loss of a proton gives a resonance-stabilized carbanion. The carbanion does not retain stereochemistry and can be protonated on either side.

25.63 (cont)

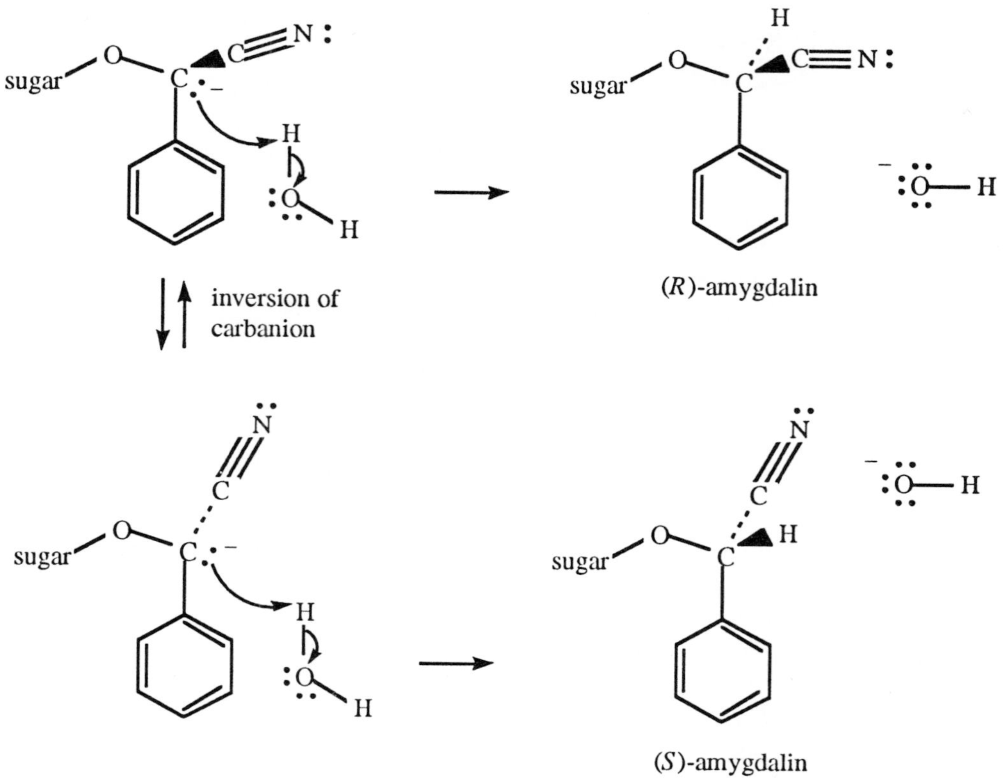

(R)-amygdalin

inversion of
carbanion

(S)-amygdalin

25.64   The reaction is an aldol condensation reaction between two glyceraldehyde molecules. The enediol intermediate that is necessary for the condensation reaction is formed faster from dihydroxyacetone (in which
deprotonation occurs at an O—H bond) than from glyceraldehyde (in which deprotonation must occur at a
C—H bond). The stereochemistry of the products is determined by the side of the carbonyl group that is attacked
by the enediol. The reaction is shown as taking place with the polar and bulky hydroxyl group of the enediol
pointing away from the oxygen atom of the carbonyl group undergoing nucleophilic attack.

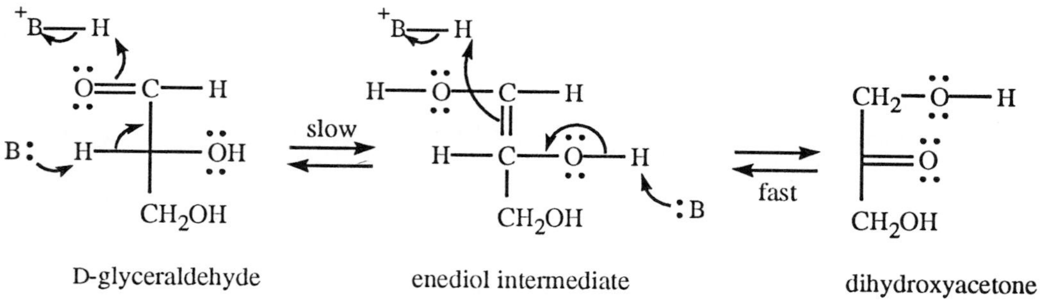

D-glyceraldehyde   enediol intermediate   dihydroxyacetone

25.64 (cont)

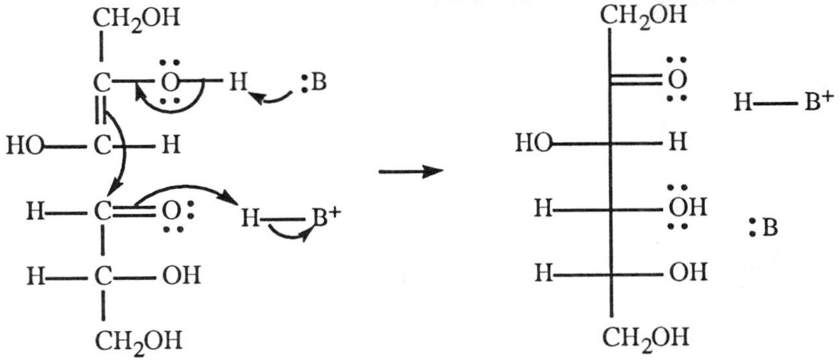

attack on one face of the
carbonyl group of D-
glyceraldehyde by the
enediol intermediate;
an aldol condensation

D-fructose

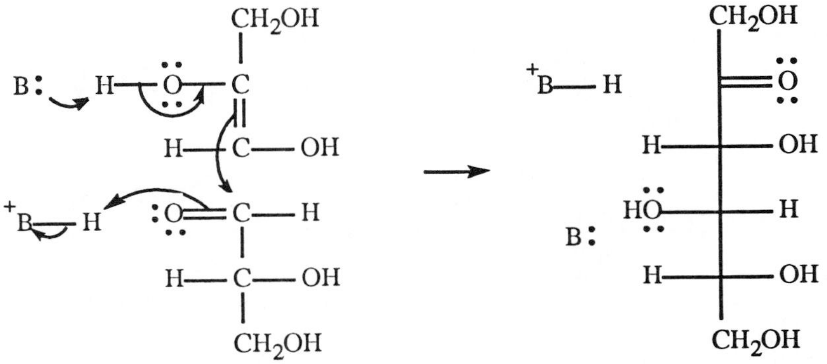

attack on the other face
of the carbonyl group of
D-glyceraldehyde by the
enediol intermediate

D-sorbose

25.65

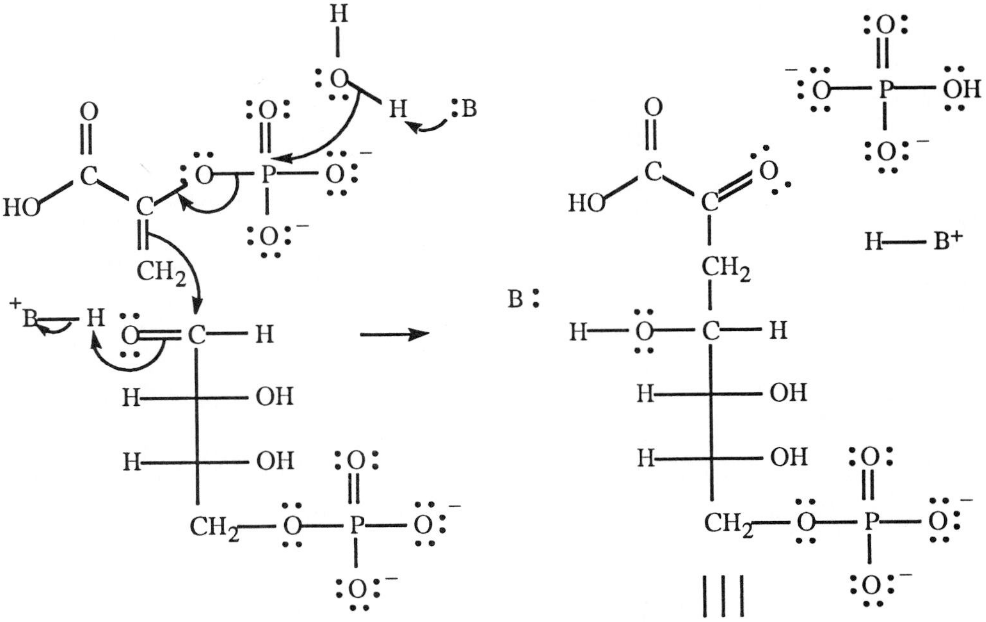

25.65 (cont)

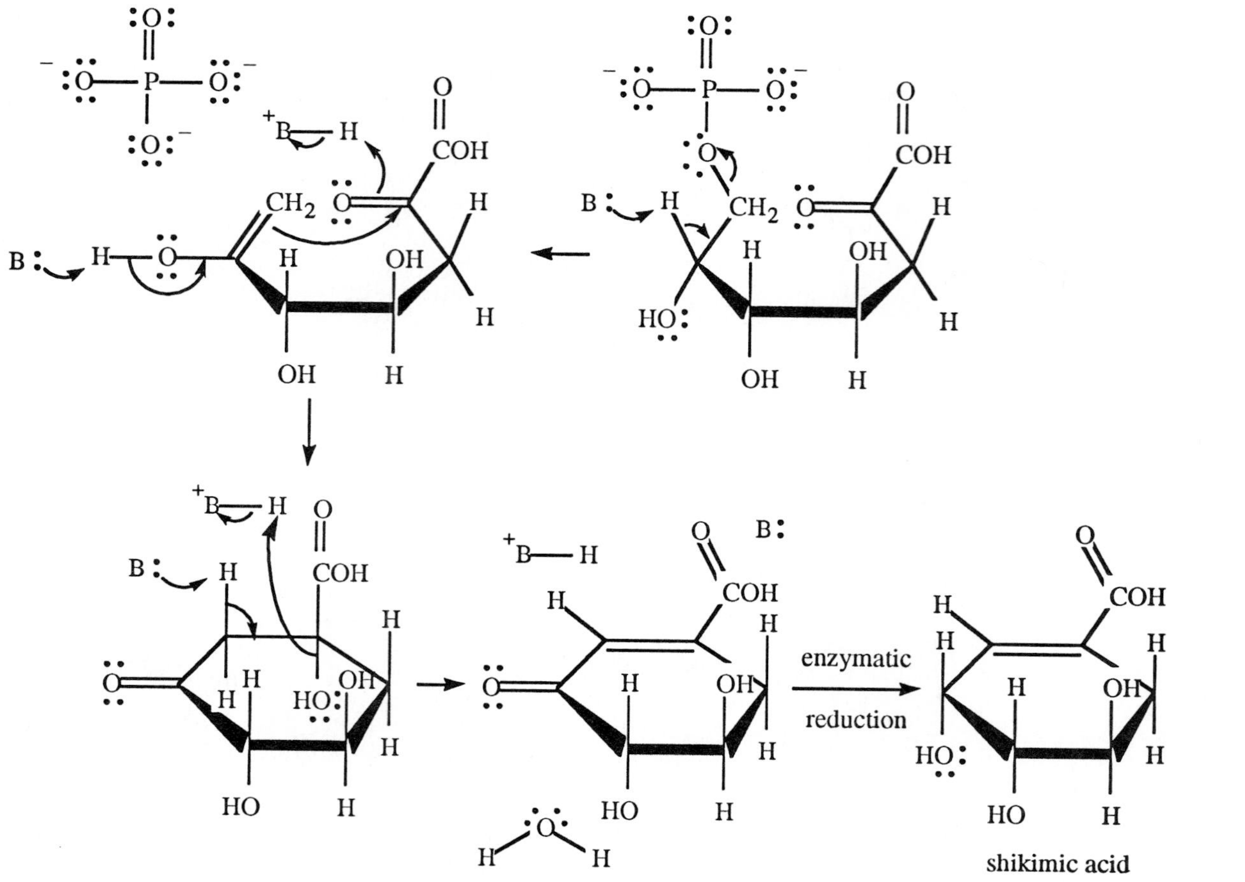

The transformations shown above occur in the presence of enzymes, and many bonds are shown being broken and made in one step. The stereoselectivity of an enzymatic reaction is demonstrated by the first transformation. In the laboratory, a mixture of diastereomers would be expected.

25.66

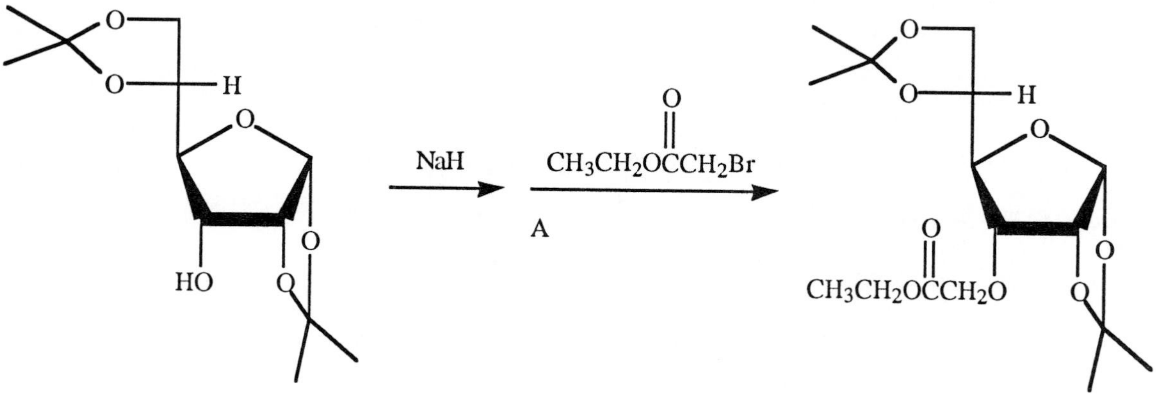

25.66 (cont)

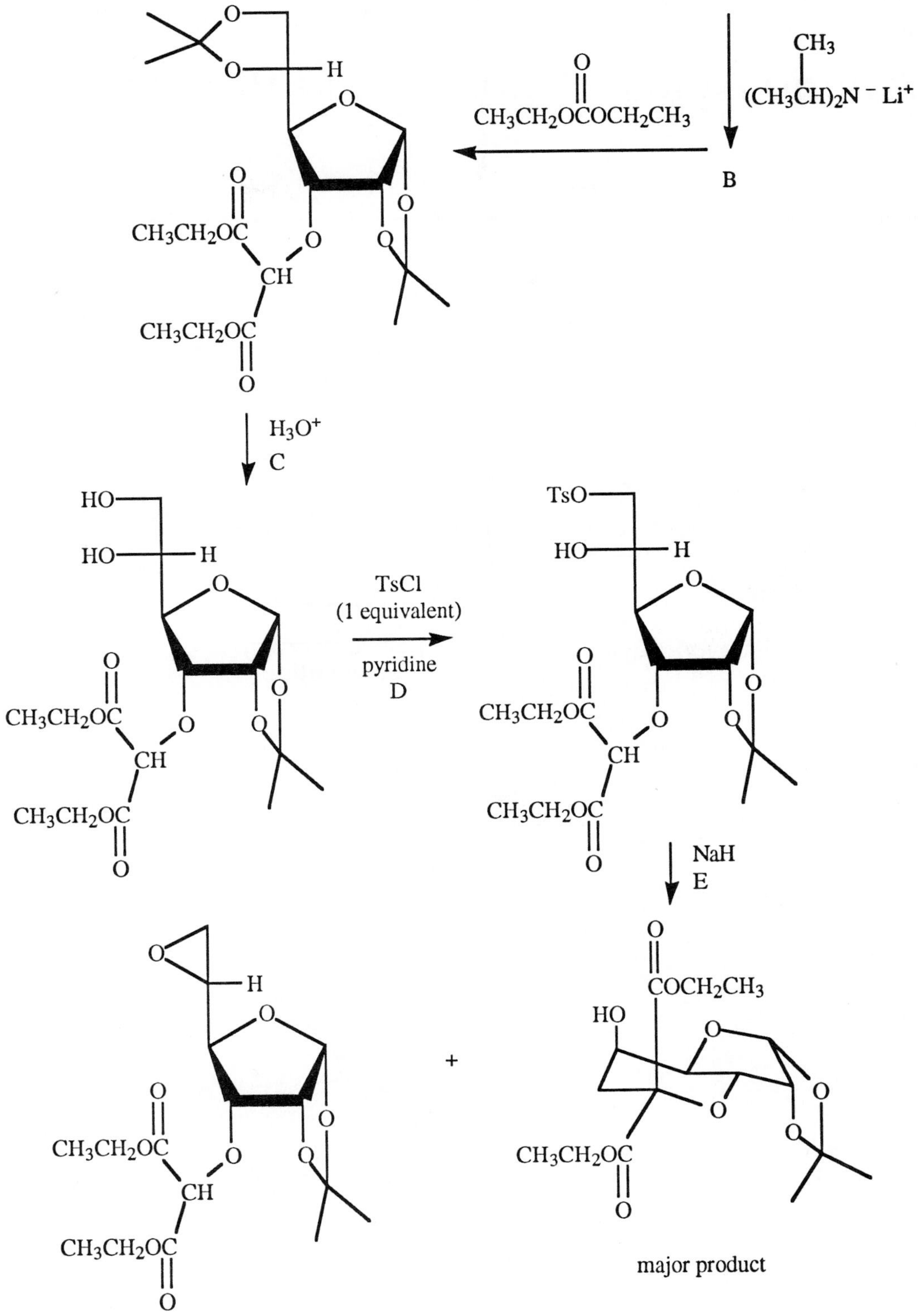

major product

26

Amino Acids, Peptides, and Proteins

Concept Map 26.1   Amino acids, polypeptides, proteins.

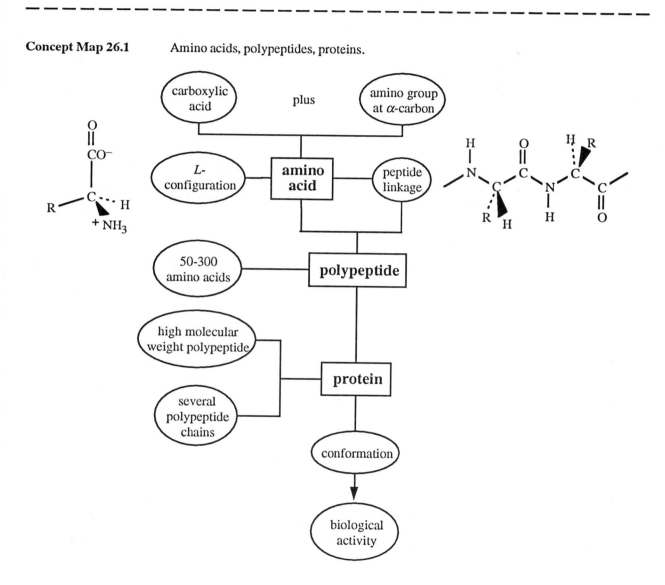

26.1

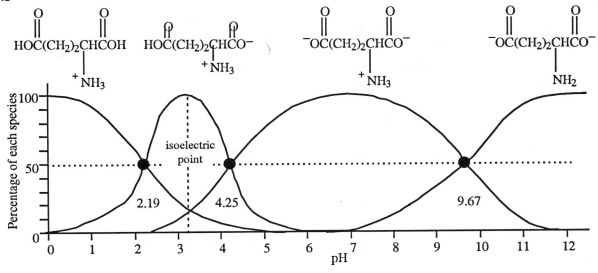

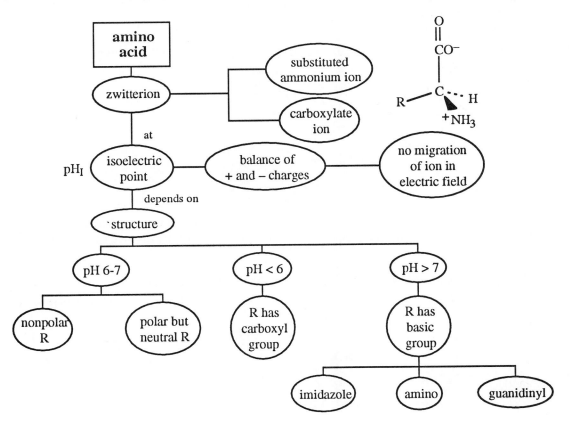

Concept Map 26.2   Acid-base properties of amino acids.

26.2   (a)

$$\underset{\overset{|}{+NH_3}}{HOCCH_2CHCO^-} \underset{H_3O^+}{\overset{OH^-}{\rightleftarrows}} \underset{\overset{|}{+NH_3}}{^-OCCH_2CHCO^-}$$

aspartic acid

$$\underset{\overset{|}{+NH_3}}{HOCCH_2CH_2CHCO^-} \underset{H_3O^+}{\overset{OH^-}{\rightleftarrows}} \underset{\overset{|}{+NH_3}}{^-OCCH_2CH_2CHCO^-}$$

glutamic acid

The carboxylic acid group on carbon 3 of aspartic acid is more acidic because it is closer to the electron-withdrawing effect of the ammonium ion on carbon 2 than is the acid group on carbon 4 of glutamic acid.

(b)   The carboxylic acid group on carbon 3 in aspartic acid is more acidic than the carboxylic acid group on carbon 4 in glutamic acid, therefore the carboxylic acid group in aspartic acid has a lower pK_a. The pH required to keep that group protonated and the molecule in the balanced zwitterionic state would also be lower than the pH required for glutamic acid.

26.3   All of them have the *S* configuration except for cysteine, which has the *R* configuration.

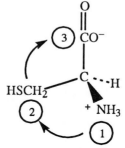

In all other amino acids the carboxylate group is of higher priority than the rest of the chain. The presence of sulfur in cysteine makes that group on the stereocenter of higher priority than the carboxylate group.

26.4   Isoleucine, threonine and hydroxyproline have more than one stereocenter.

Ile
configuration not
specified at carbon 3

Thr
(2*S*, 3*R*)-2-amino-3-
hydroxybutanoic acid

looks *S*, therefore *R*

Hypro
[2*S*, 4*R*]-4-
hydroxyproline

26.5   (a)

$$\overset{O}{\underset{\underset{\overset{|}{{}^{+}NH_3}}{|}}{\overset{||}{-OCCHCH_2CH_2CH_2CH_2\overset{\cdot\cdot}{N}H_2}}} + H-B^+ \;\rightleftharpoons\; \overset{O}{\underset{\underset{\overset{|}{{}^{+}NH_3}}{|}}{\overset{||}{-OCCHCH_2CH_2CH_2CH_2\overset{+}{N}H_3}}} + :B$$

lysine

When lysine is protonated at the ε-amino group, the resulting positive charge is localized at that nitrogen atom.

$$-OCCHCH_2CH_2CH_2NHCNH_2 + H-B^+ \;\rightleftharpoons\; -OCCHCH_2CH_2CH_2NHCNH_2 + B$$

arginine

When the guanidinyl group in arginine is protonated, the resulting cation is stabilized by delocalization of the charge to three nitrogen atoms. The guanidinyl group is, therefore, more basic, more easily protonated, than an amino group. The conjugate acid of guanidine, with a pK_a of 13.2, is a weaker acid than the conjugate acid of a primary amine, p$K_a \sim 10.5$.

(b)

tryptophan

indole ring
system

The nonbonding electrons on the nitrogen atom in the indole ring are part of an aromatic sextet, and are therefore not available for protonation. Protonation of that nitrogen atom would lead to loss of aromaticity, so the difference in energy between indole and its conjugate acid would be very large.

26.6

(a)

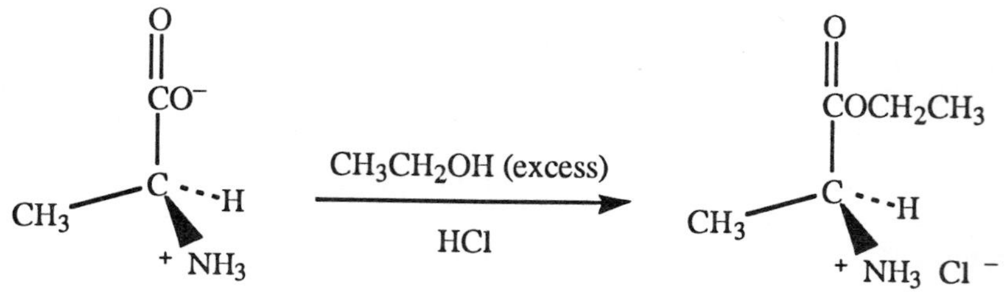

CH₃CH₂OH (excess)

HCl

(b)

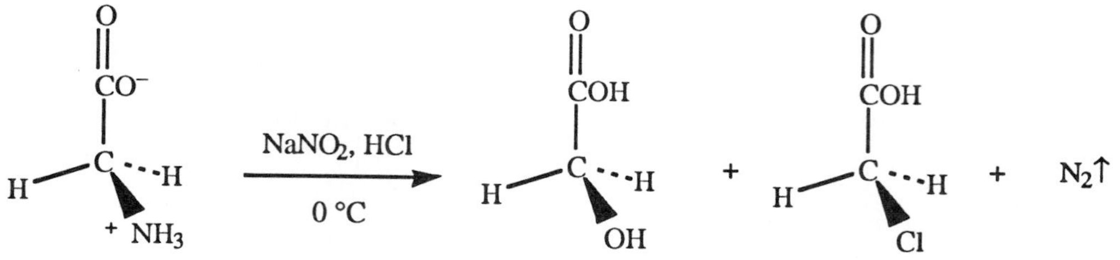

NaNO₂, HCl

0 °C

(c)

NaOH

SOCl₂

A

B

NH₃

C

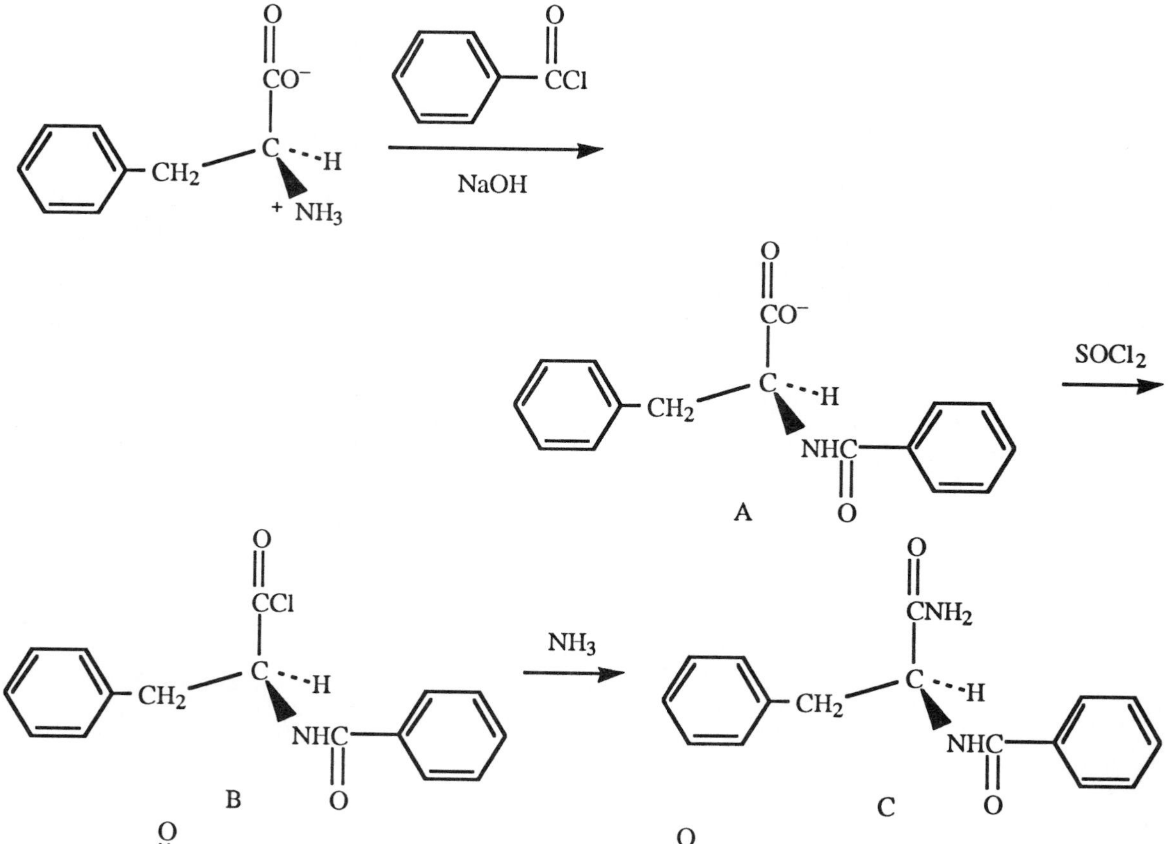

(d)

CH₃I (excess)

NaOH

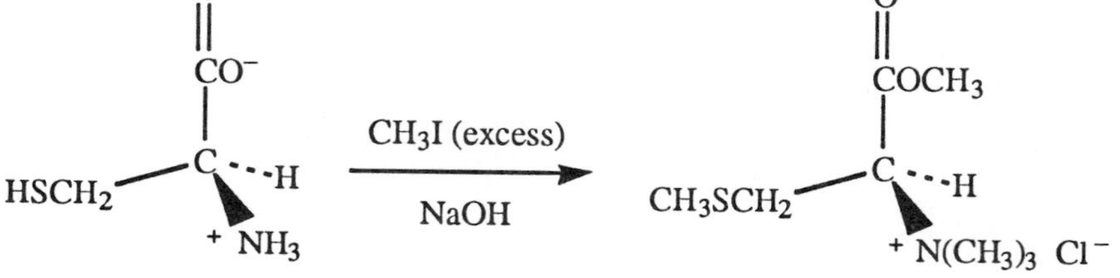

26.6

(e)

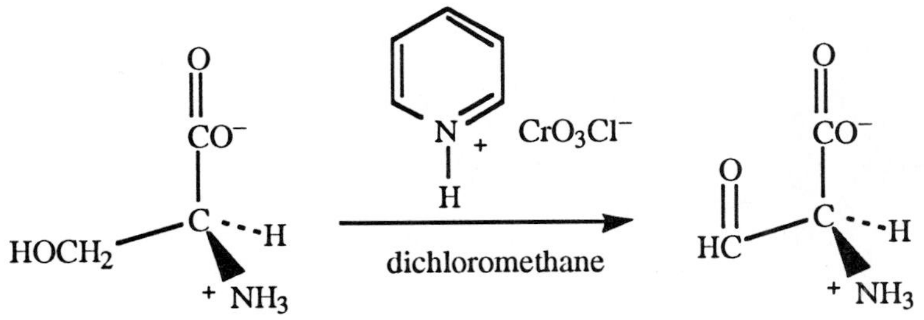

(f)

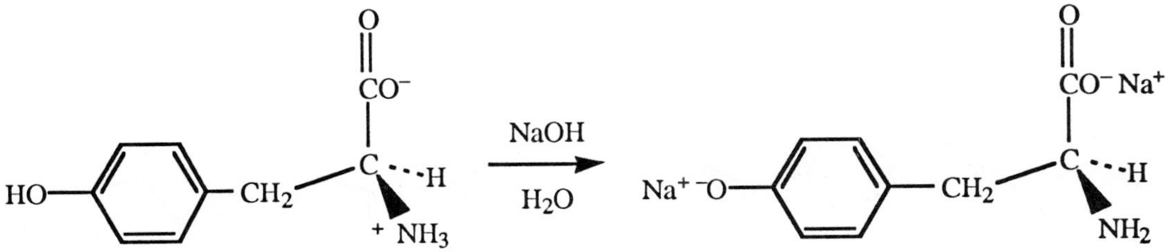

(g)

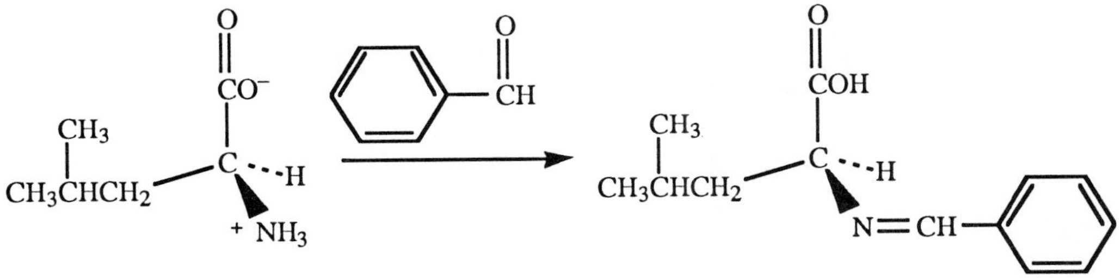

(h)

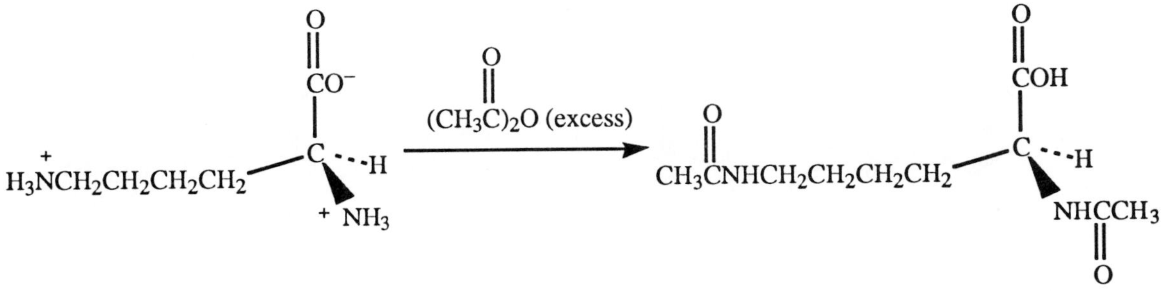

(i)

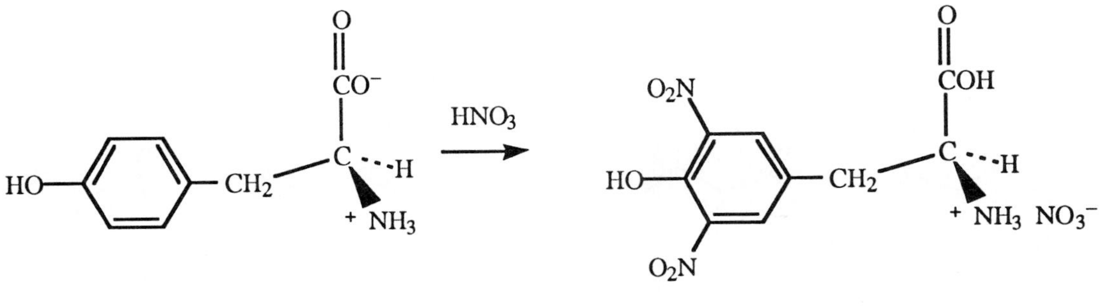

26.7

26.8   One possible peptide is:

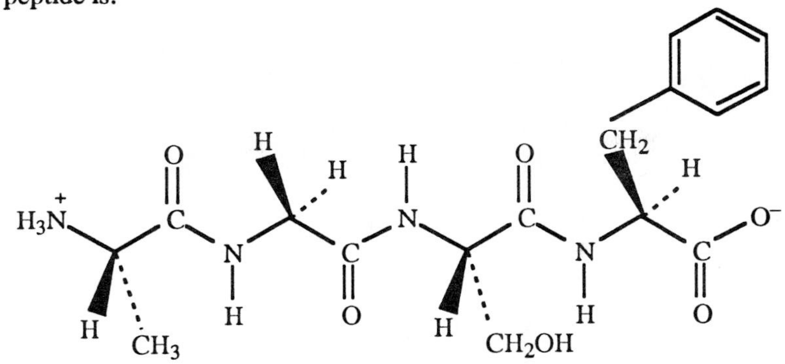

Ala-Gly-Ser-Phe

| *N*-terminal amino acid | *C*-terminal amino acid |

The other twenty-two, besides the one above and Ser-Ala-Phe-Gly, are:

Ala-Gly-Phe-Ser	Ser-Ala-Gly-Phe
Ala-Ser-Gly-Phe	Ser-Gly-Ala-Phe
Ala-Ser-Phe-Gly	Ser-Gly-Phe-Ala
Ala-Phe-Ser-Gly	Ser-Phe-Ala-Gly
Ala-Phe-Gly-Ser	Ser-Phe-Gly-Ala
Gly-Ala-Ser-Phe	Phe-Ala-Ser-Gly
Gly-Ala-Phe-Ser	Phe-Ala-Gly-Ser
Gly-Ser-Ala-Phe	Phe-Ser-Gly-Ala
Gly-Ser-Phe-Ala	Phe-Ser-Ala-Gly
Gly-Phe-Ala-Ser	Phe-Gly-Ser-Ala
Gly-Phe-Ser-Ala	Phe-Gly-Ala-Ser

| *N*-terminal amino acid | *C*-terminal amino acid | *N*-terminal amino acid | *C*-terminal amino acid |

26.9

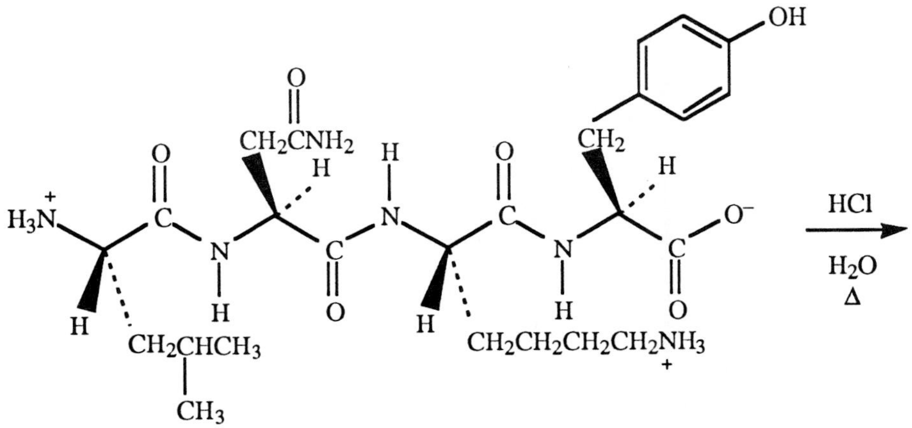

26.9 (cont)

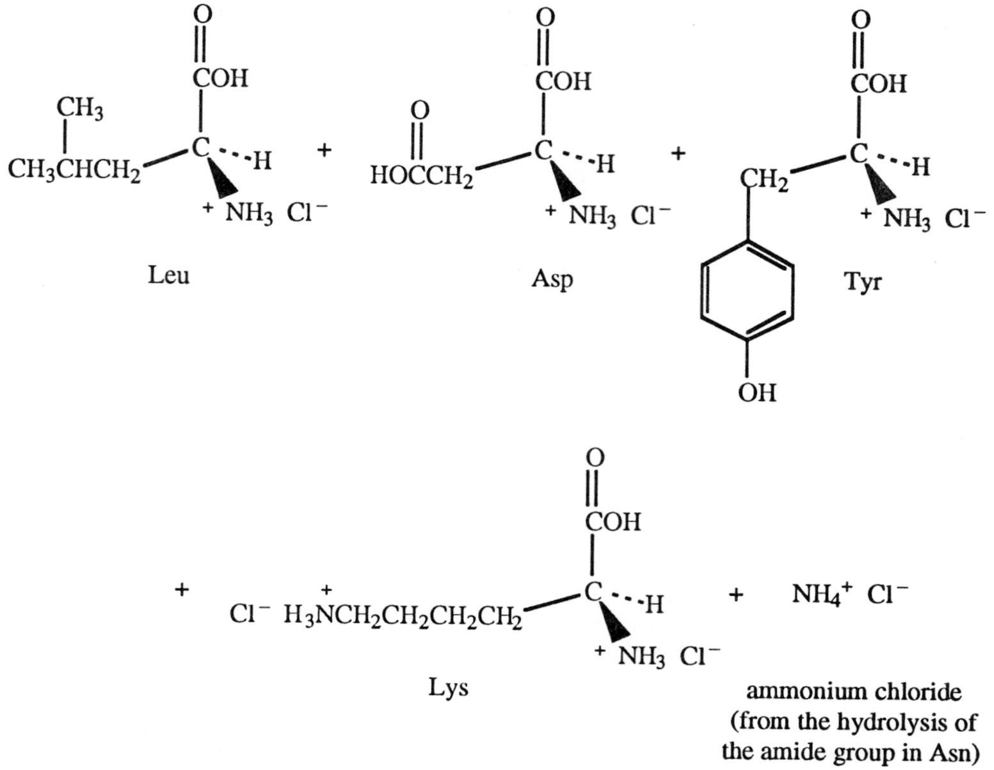
Leu
Asp
Tyr
Lys
ammonium chloride
(from the hydrolysis of
the amide group in Asn)

The hydrolysis was done in strong acid. Therefore, all amino acids are shown in their fully protonated forms.

26.10 (a) Gly-Phe-Thr-Lys will have $pH_I > 6$; it will move to the negative pole in an electric field at pH 6.

(b) Tyr-Ala-Val-Asn will have $pH_I \sim 6$; it will not move in an electric field at pH 6.

(c) Trp-Glu-Leu will have $pH_I < 6$; it will move to the positive pole in an electric field at pH 6.

(d) Pro-Hypro-Gly will have $pH_I \sim 6$; it will not move in an electric field at pH 6.

26.11

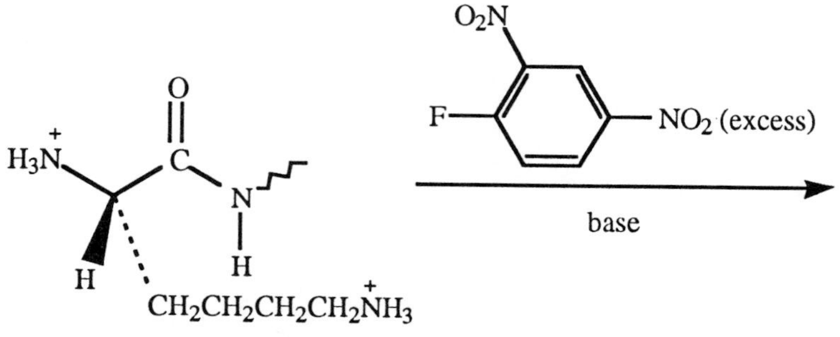
NO2 (excess)
base

N-terminal lysine

26.11 (cont)

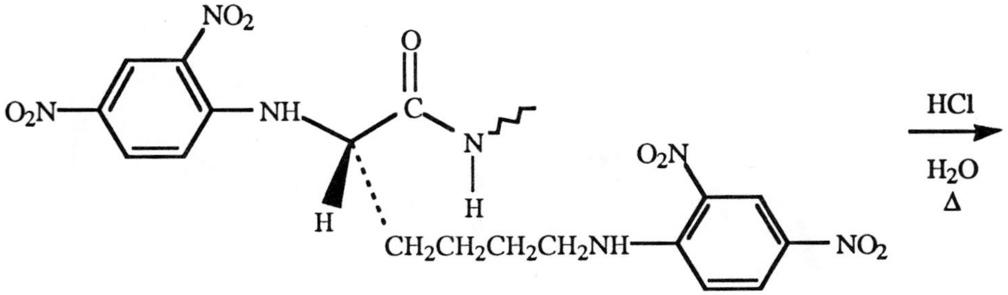

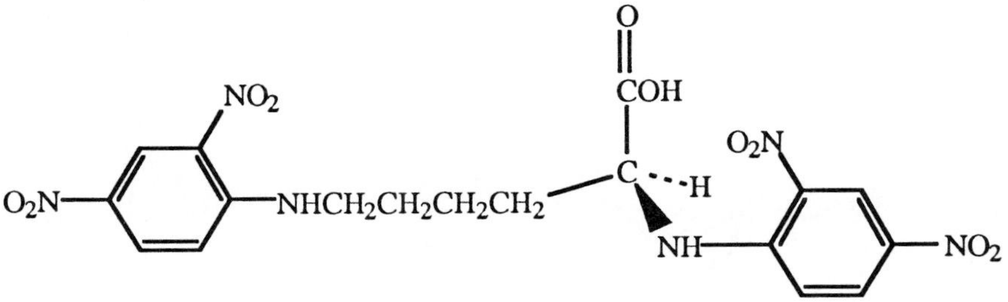

The *N*-terminal lysine
will be doubly labeled.

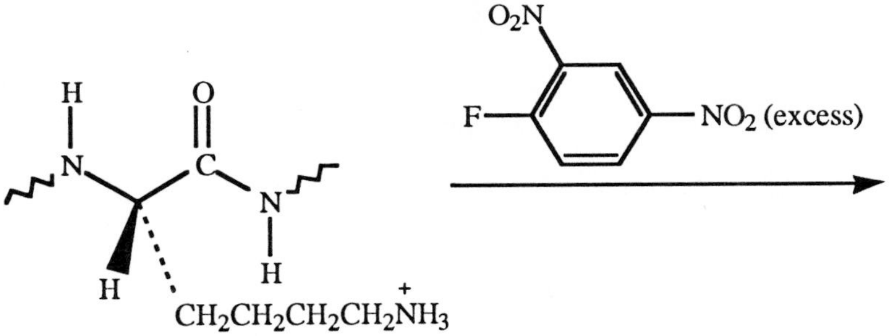

Lys in middle of chain

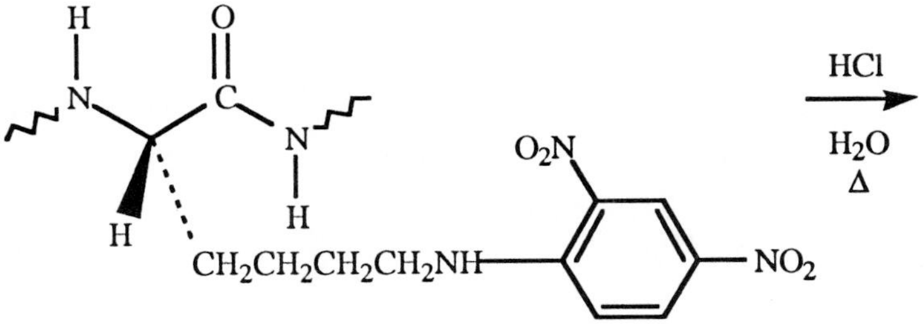

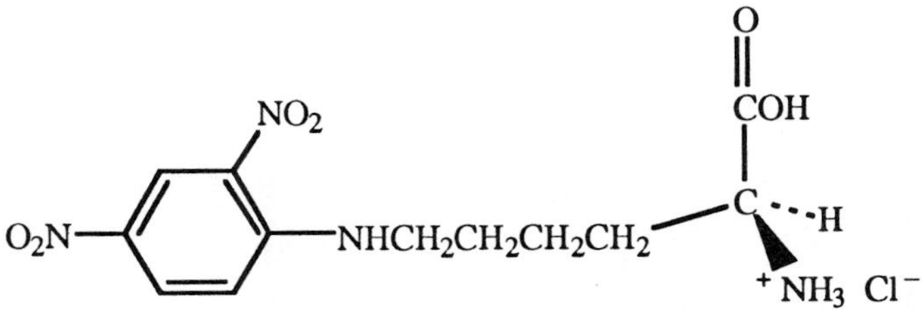

The lysine from the middle of the chain
will have a single dinitrophenyl label.

26.12

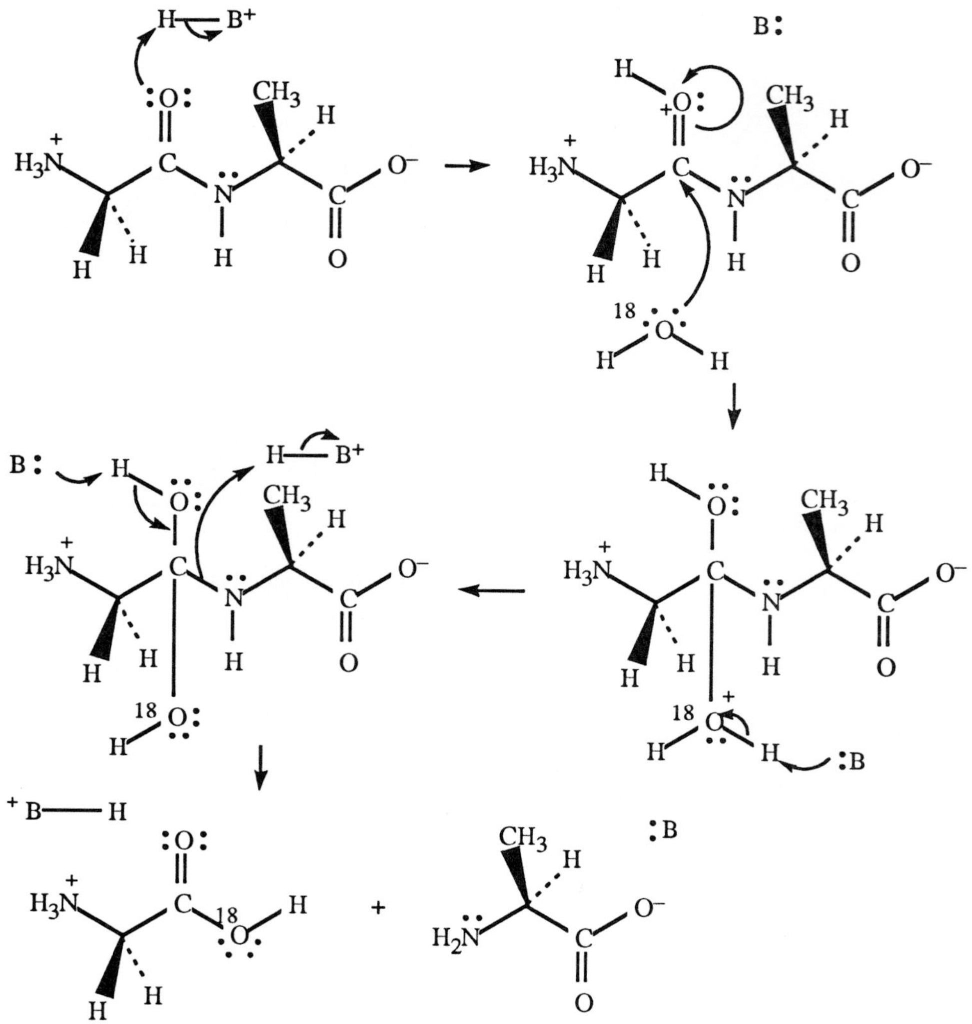

The carboxylic acid function of glycine, which was part of a peptide linkage, is labeled with ^{18}O.

No ^{18}O turns up in the carboxylic acid group of alanine, which was the *C*-terminal acid.

26.13 (a) The *N*-terminal amino acid is arginine.

(b) The fragments are shown lined up below:

        Arg–Gly–Pro
                Pro–Pro
                        Pro–Phe–Ile–Val
                        Pro–Phe–Ile

Therefore, the structure of the peptide is: Arg–Gly–Pro–Pro–Phe–Ile–Val.

Concept Map 26.3   Proof of structure of peptides and proteins.

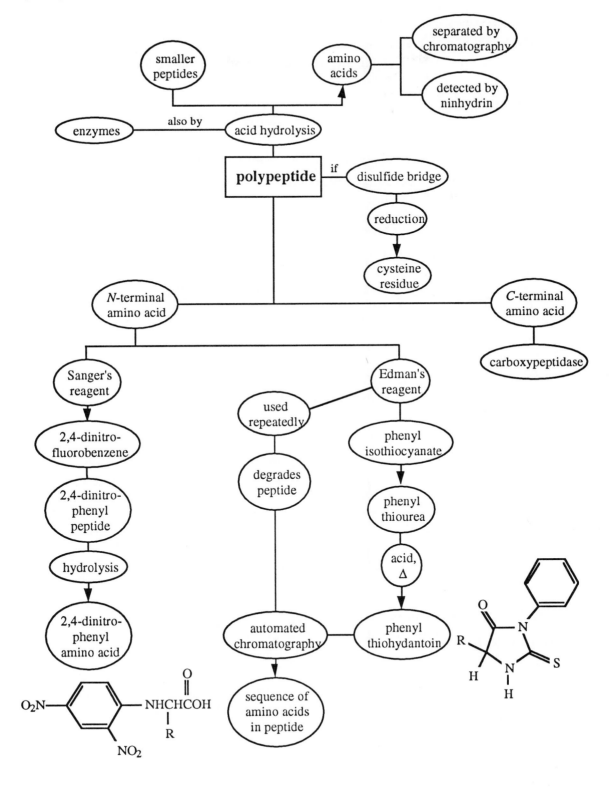

26.14   (a)   Cysteine is the *N*-terminal amino acid.
Glycinamide is the *C*-terminal amino acid.
Arginine is the second amino acid from the *C*-terminal end.

The peptide fragments are shown lined up below:

                    Cys–Pro–Arg–Gly (peptide of undetermined order*)
               Asp–Cys
          Glu–Asp–Cys
     Phe–Glu–Asp
     Phe–Glu
Cys–Try–Phe

*The information from the trypsin degradation gives the structure of the peptide of undetermined order.

Cys–Try–Phe–Glu–Asp–Cys---Pro–Arg–Gly is the order of amino acids from the degradation work.

   (b)   The full structure of the peptide must show a disulfide bridge and three amide groups to account for the three moles of ammonia lost in the hydrolysis. There are three free carboxylic acid groups in the structure shown in part (a): Glu, Asp, and the *C*-terminal Gly.

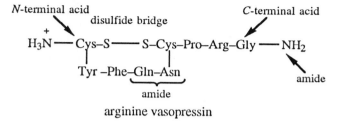

arginine vasopressin

- -

Concept Map 26.4   Protection of functional groups in peptide synthesis.

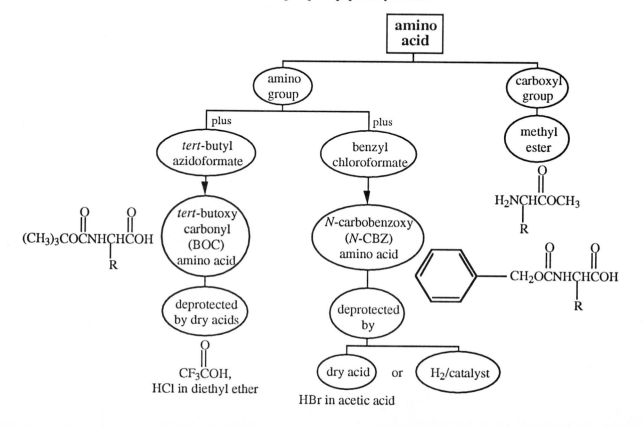

Concept Map 26.5   Activation of the carboxyl group in peptide synthesis.

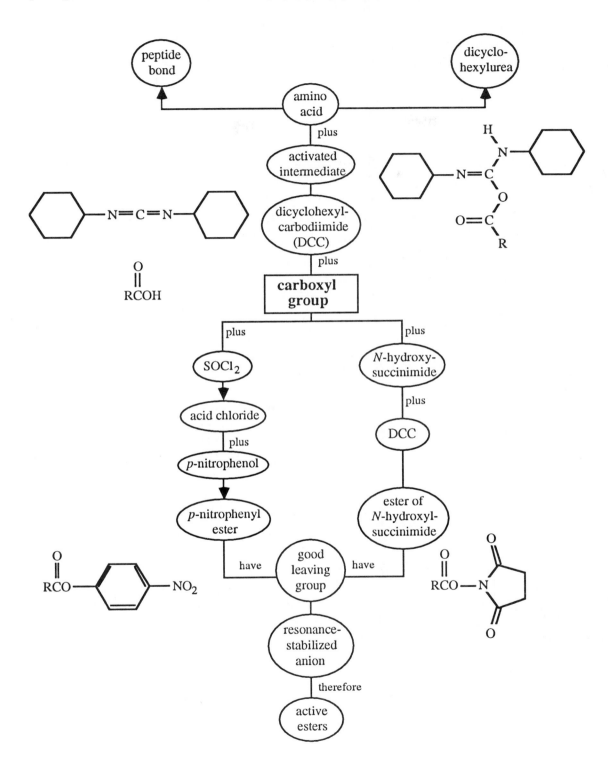

26.15

(a)

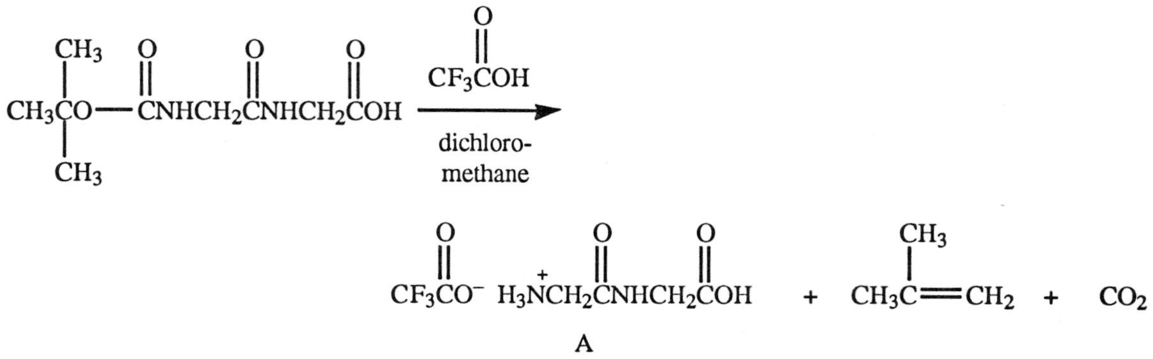

(b)

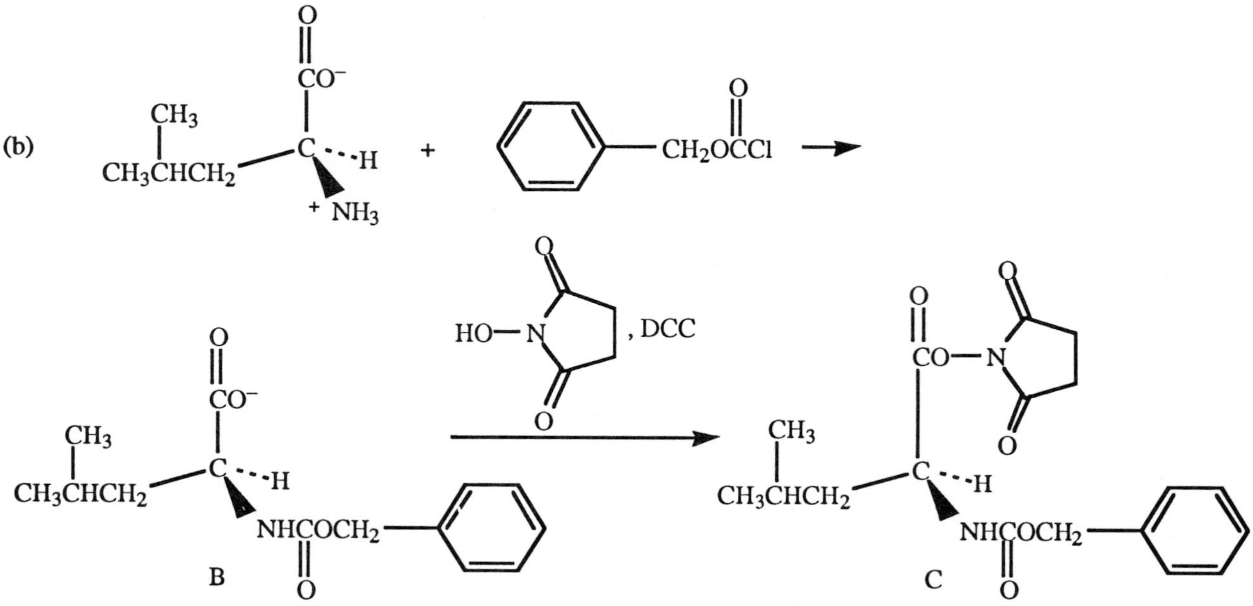

(c)

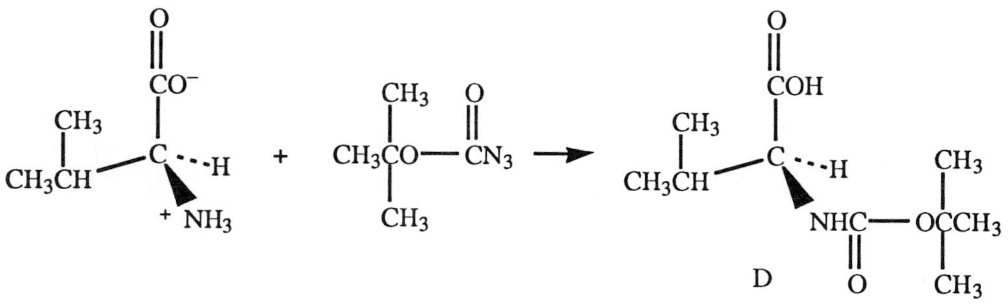

26.15 (cont)

(d)

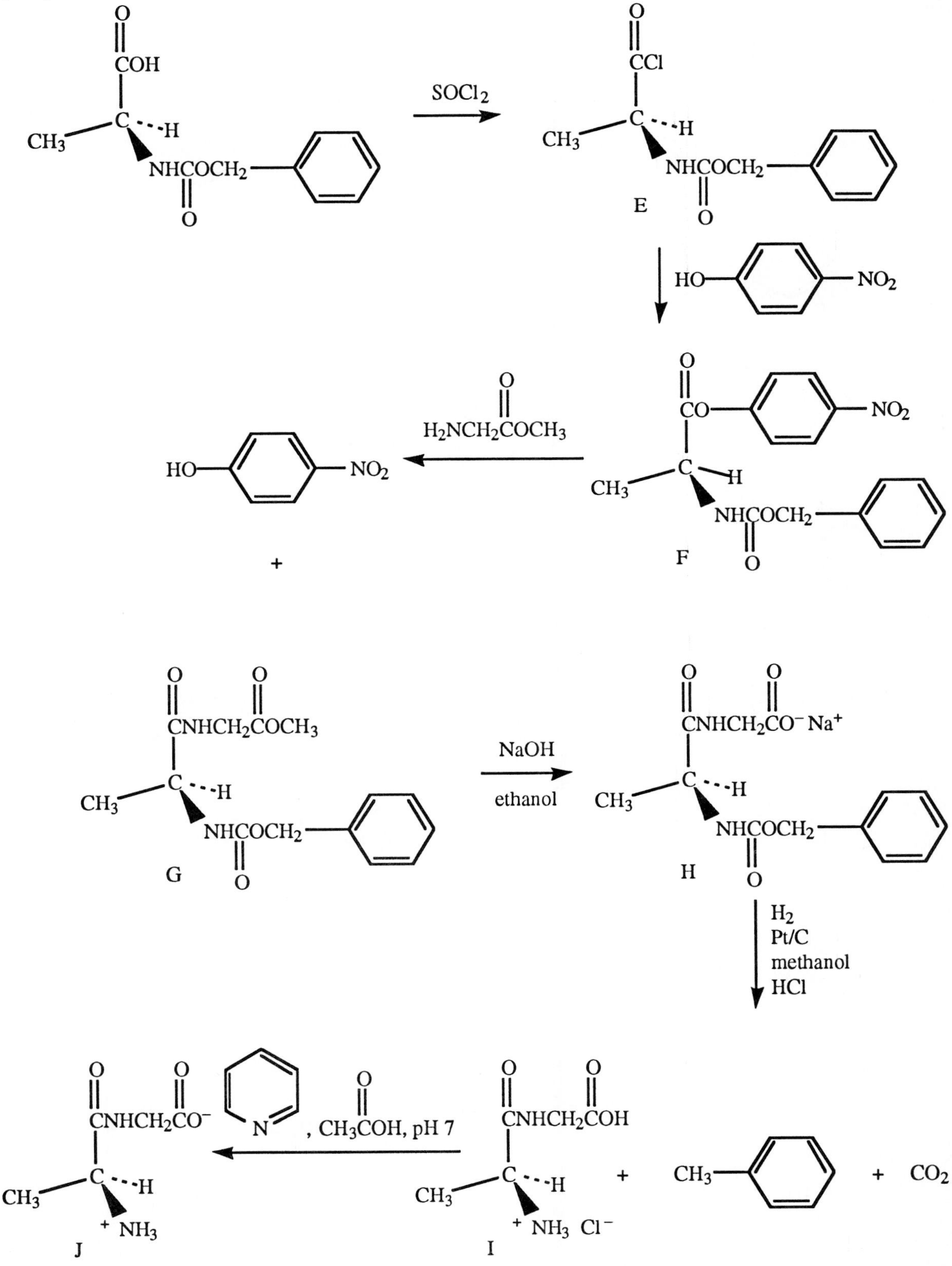

26.16

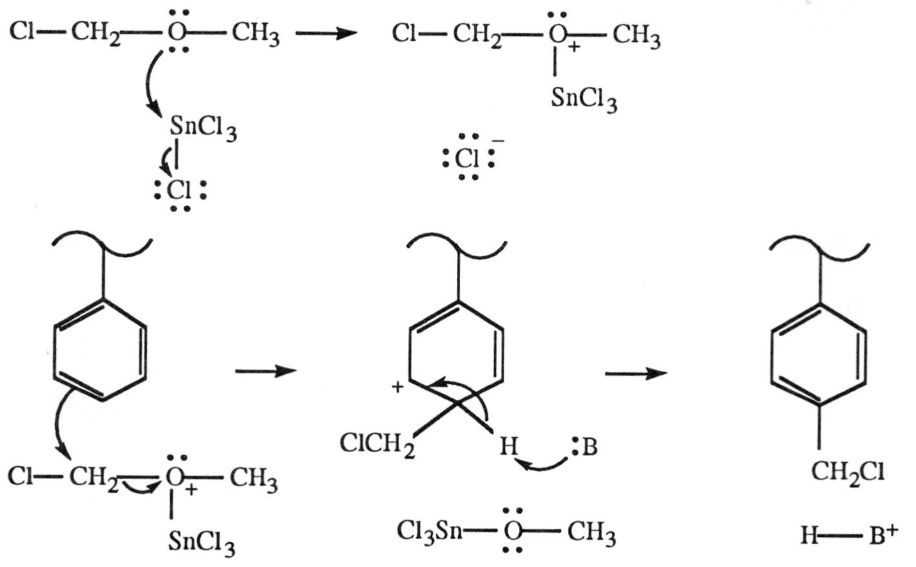

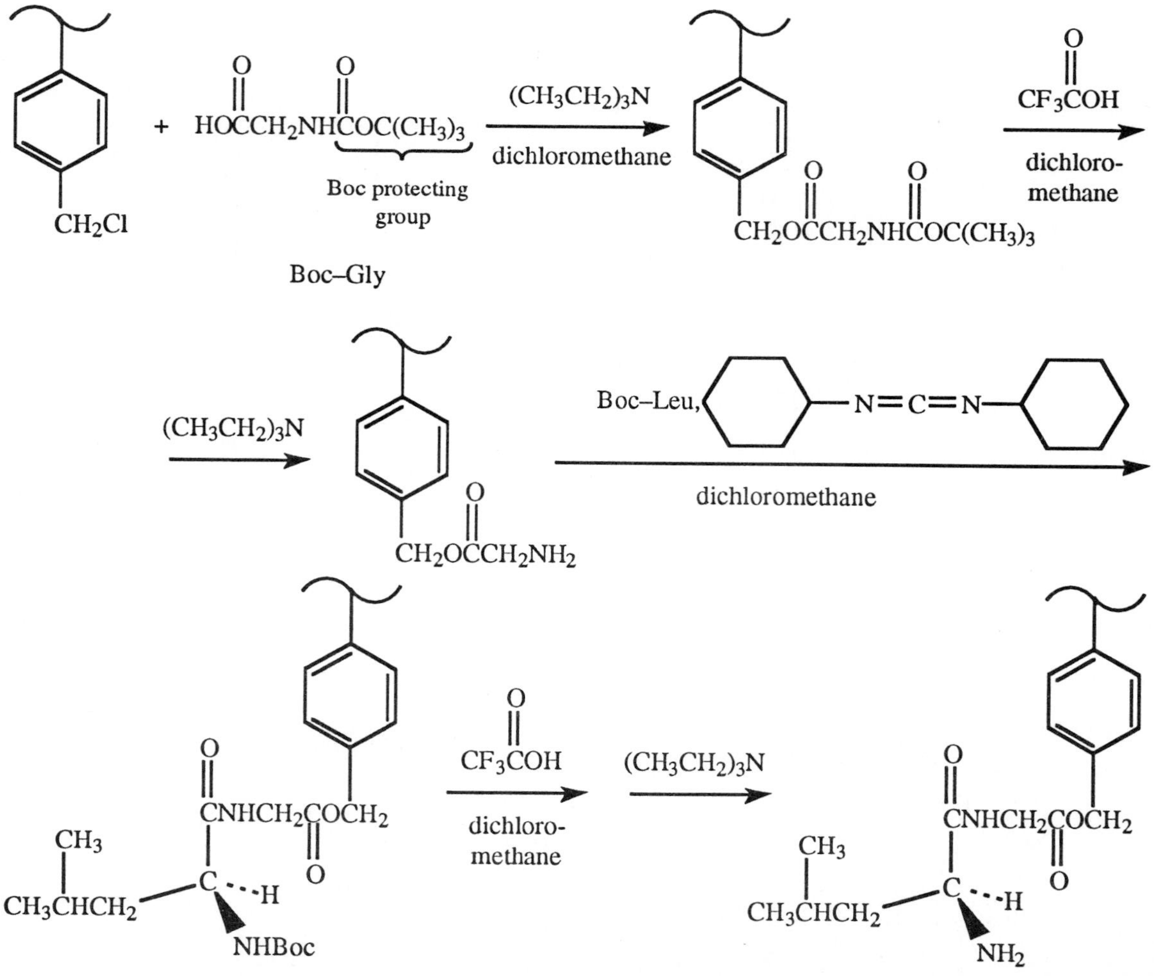

26.17

26.17 (cont)

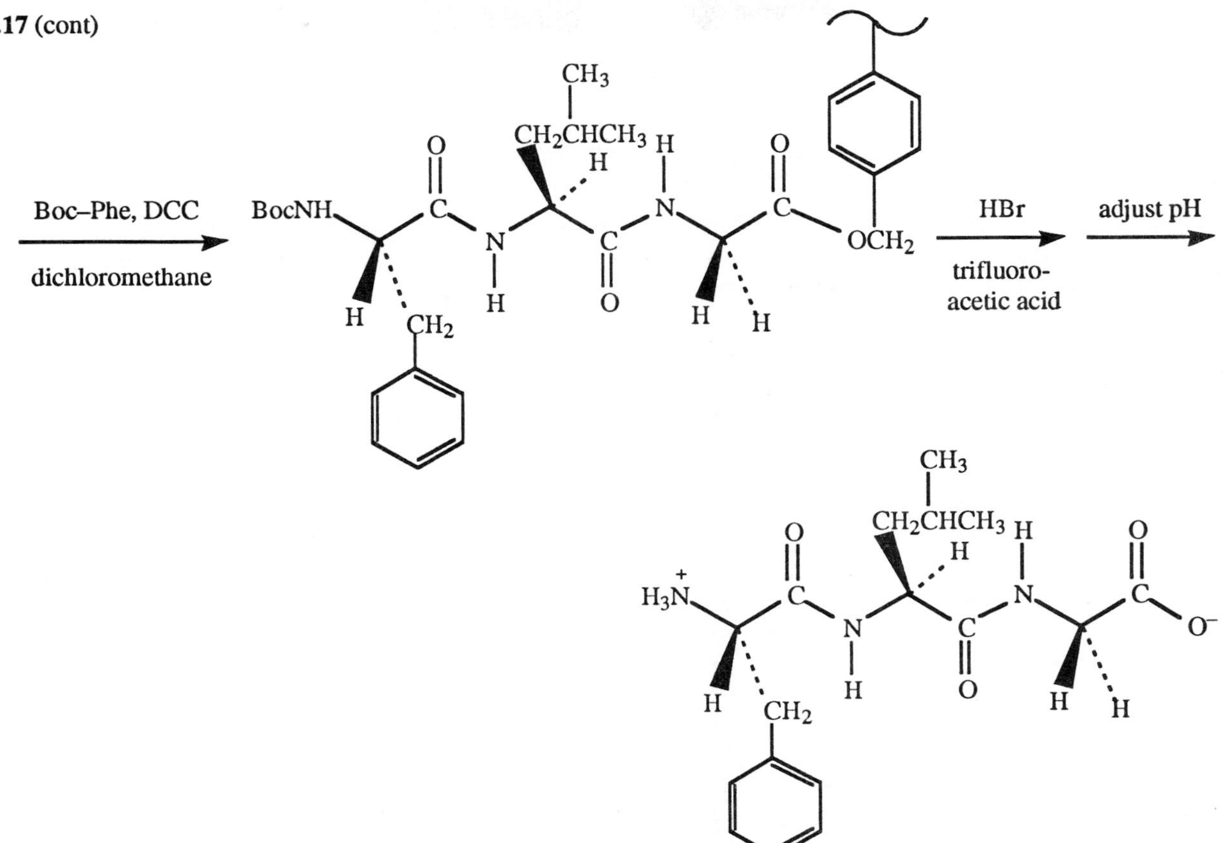

Phe–Leu–Gly

Concept Map 26.6   (see p. 984)

26.18

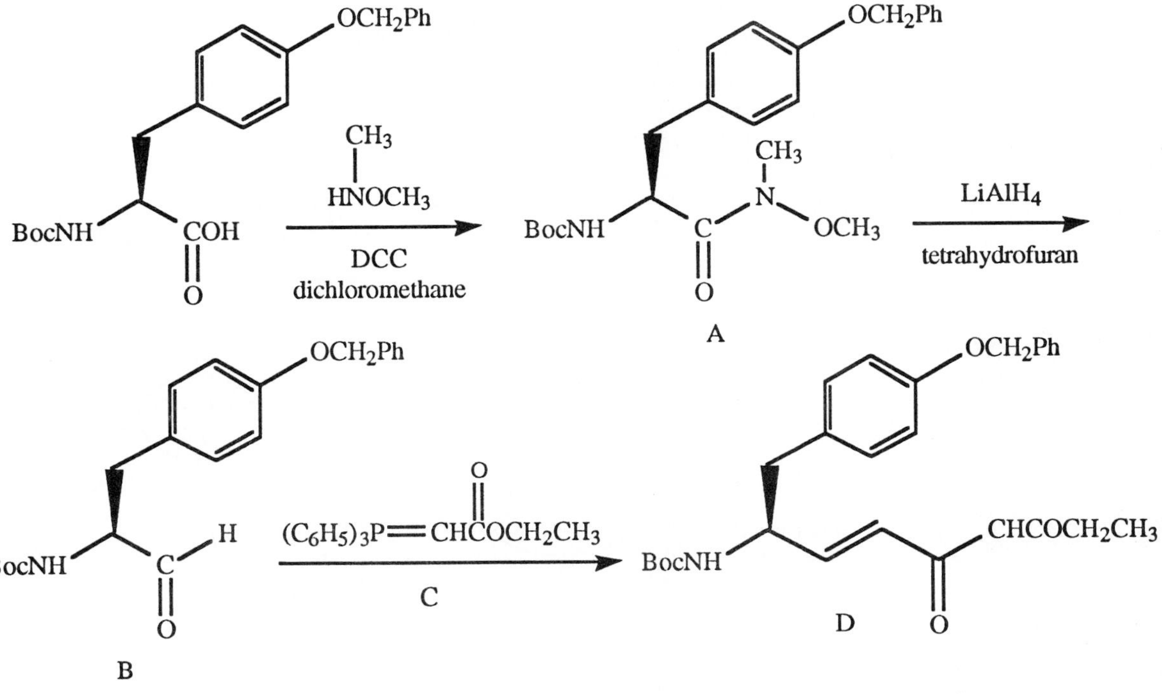

26.18 (cont)

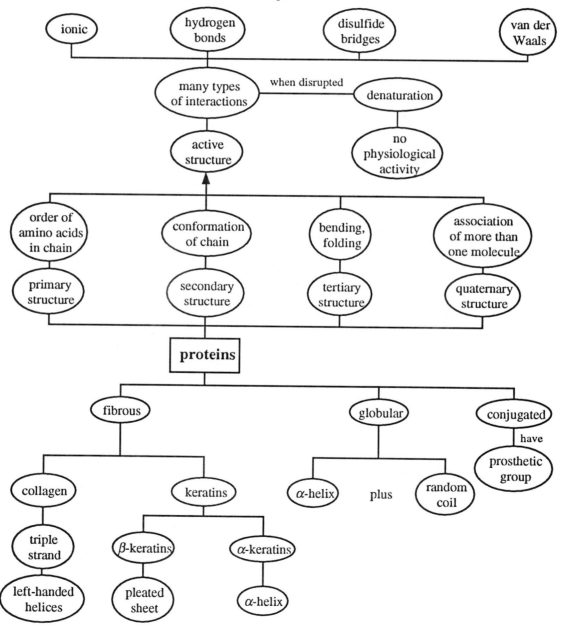

Concept Map 26.6   Conformation and structure in proteins.

26.19 (a)

$$H_3\overset{+}{N} - \underset{\underset{H}{\overset{|}{C}}}{\overset{O}{\underset{\parallel}{C}}} \cdots$$

(shown at pH 6)

(b)

$$CH_3 - \overset{\overset{\overset{O}{\parallel}}{COCH_2CH_3}}{\underset{NH_2}{\overset{|}{C}}} \cdots H$$

(c)

$$CH_3SCH_2CH_2 - \overset{\overset{\overset{O}{\parallel}}{COH}}{\underset{NHCOCH_2}{\overset{|}{C}}} \cdots H - \bigcirc$$

(d)

$$CH_3CH \underset{CH_3}{\overset{CH_3}{-}} \overset{\overset{\overset{O}{\parallel}}{COH}}{\underset{NHC-OCCH_3}{\overset{|}{C}}} \cdots H$$

(e)

$$CH_3CHCH_2 \underset{CH_3}{\overset{CH_3}{-}} \overset{\overset{\overset{O}{\parallel}}{COH}}{\underset{NH}{\overset{|}{C}}} \cdots H - \bigcirc - NO_2$$

(f)

$$HO - \bigcirc - CH_2 - \overset{\overset{\overset{O}{\parallel}}{CO}}{\underset{NH_2}{\overset{|}{C}}} \cdots H$$

26.20

(a) *N*-2,4-dinitrophenylphenylalanine   (b) glycylcysteinylhistidine   (c) *N*-benzoylglycine; hippuric acid

(d) *p*-nitrophenyl alaninate   (e) *N-tert*-butoxycarbonylleucine   (f) methyl cysteinate

26.21

(a)

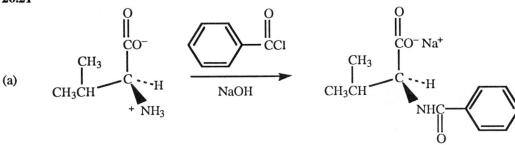

26.21 (cont)

(b)

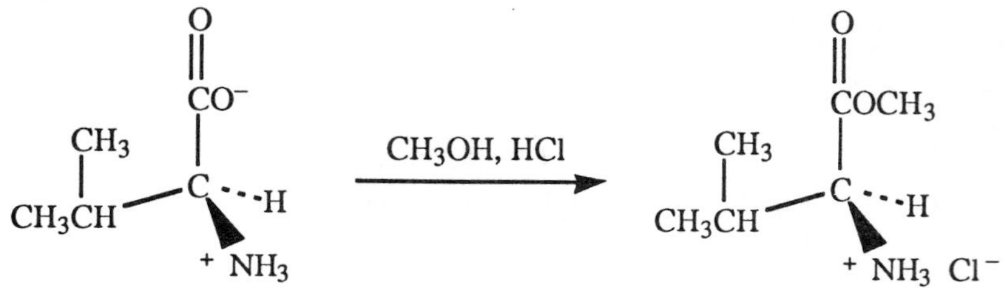

(c)

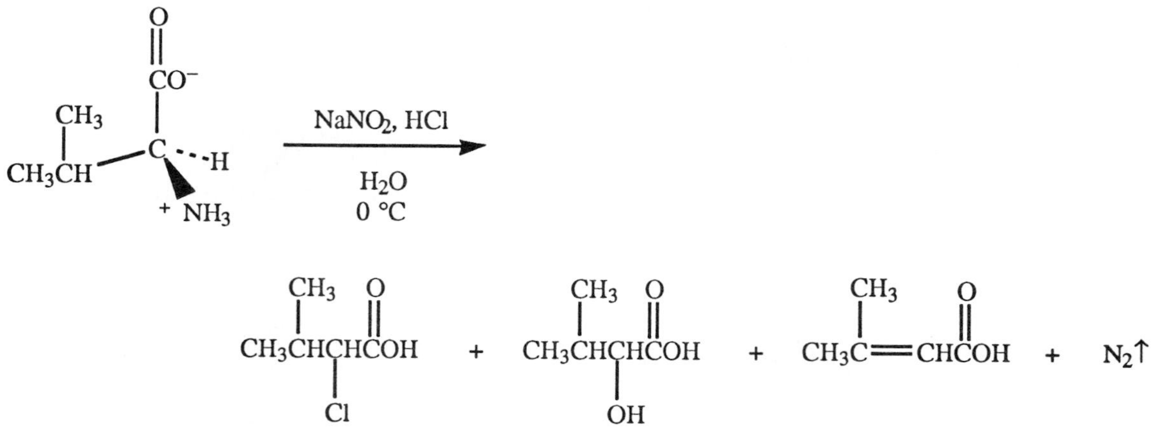

(d)

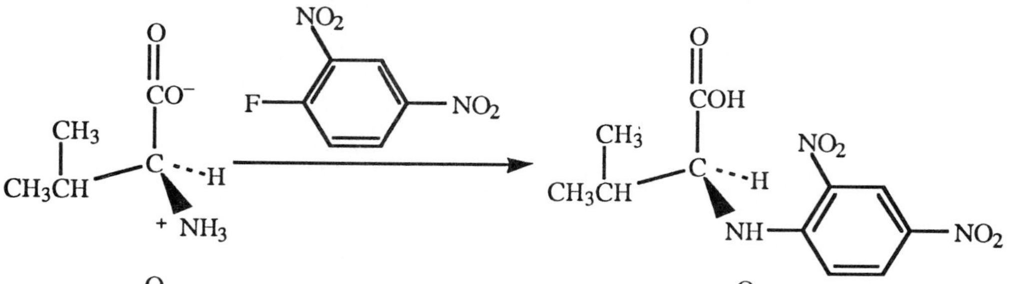

(e)

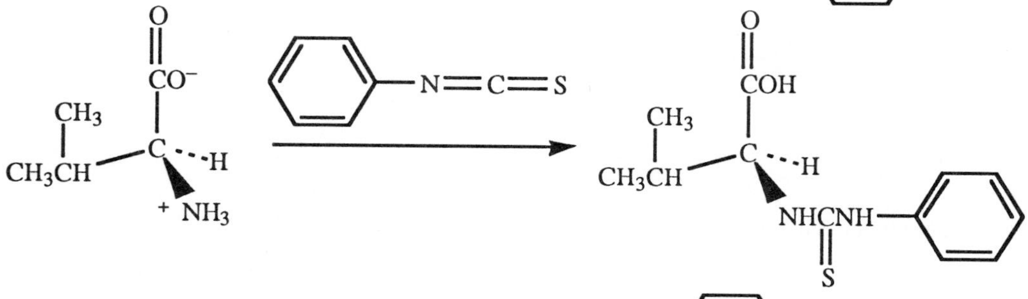

(f)

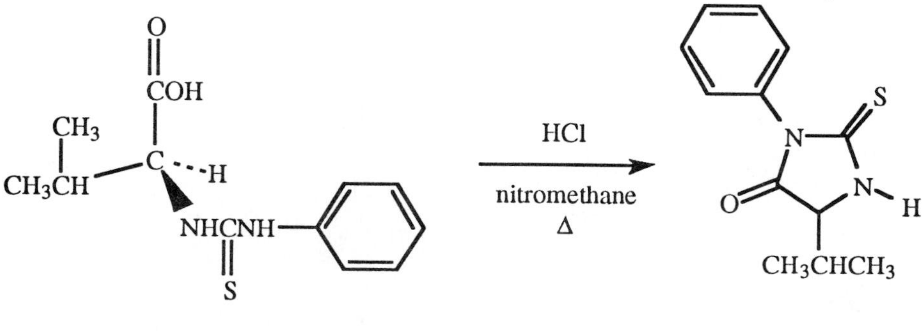

26.21 (cont)

(g)

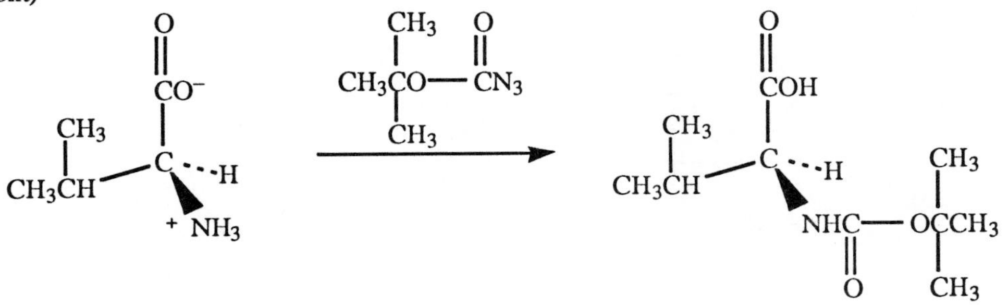

(h)

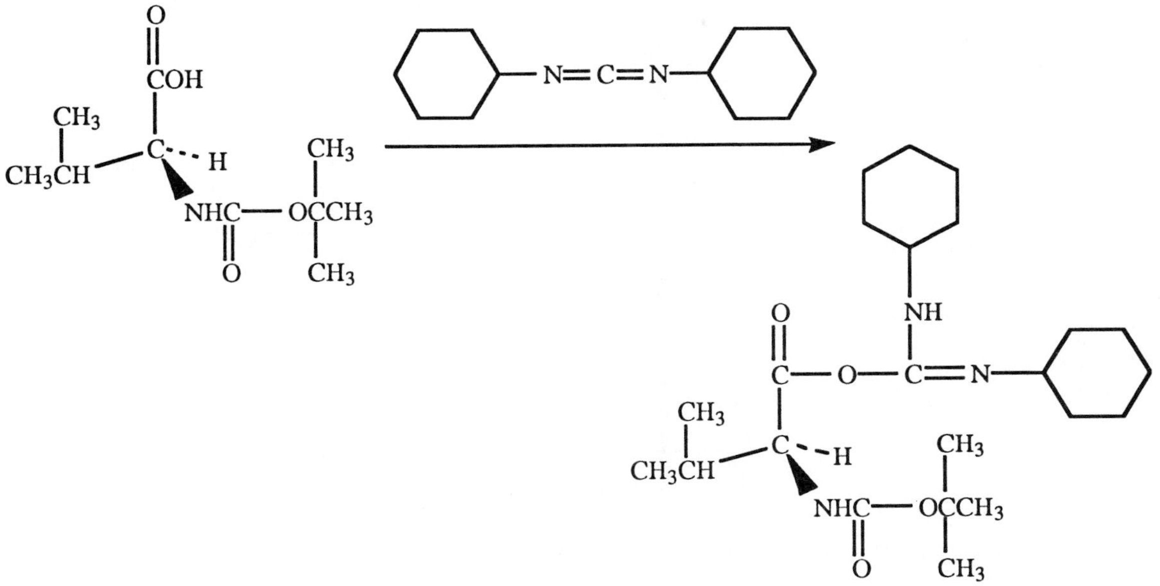

(i)

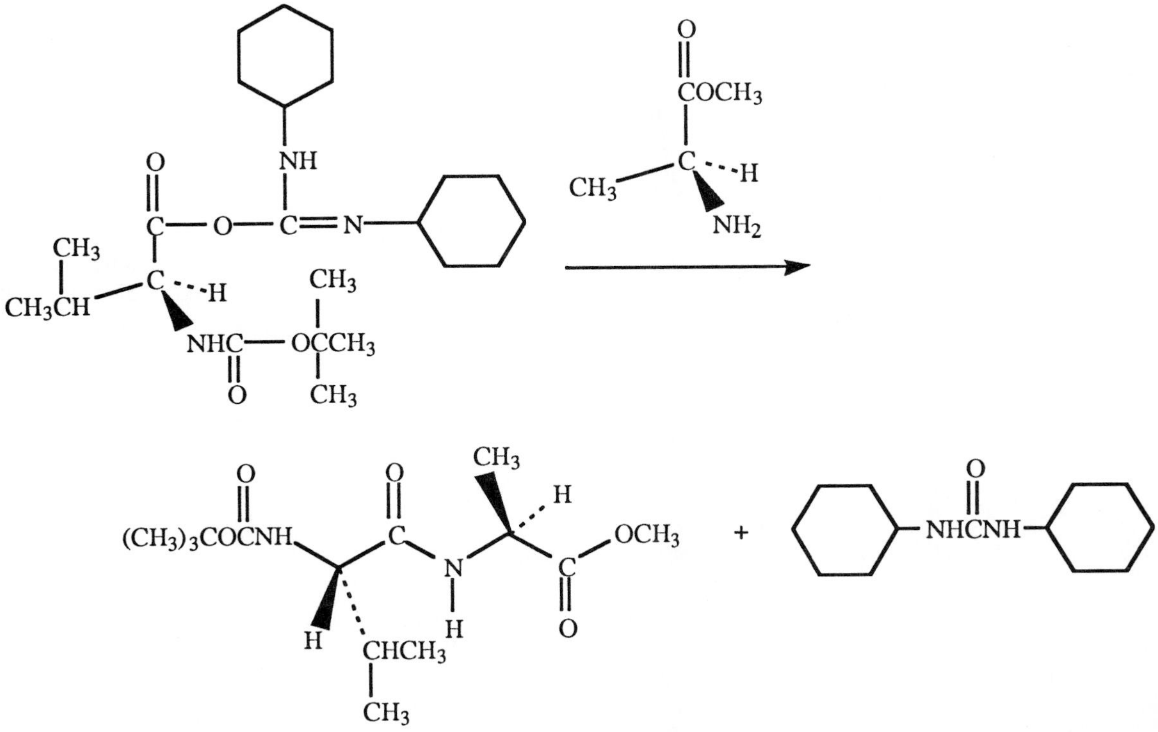

26.21 (cont)

(j)

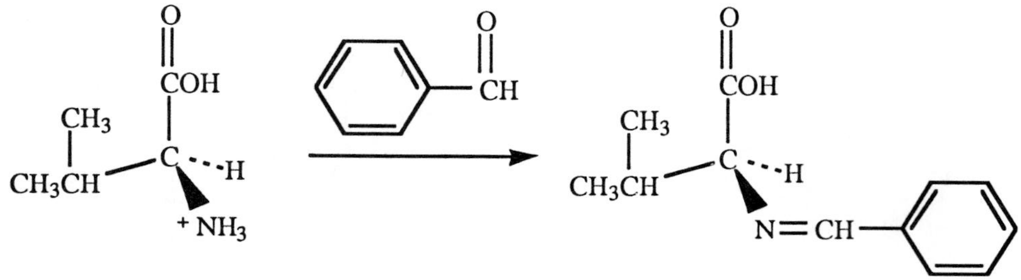

26.22

(a)

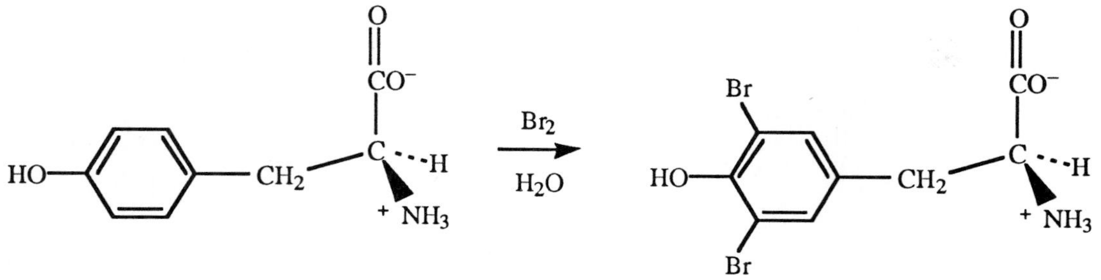

(b)

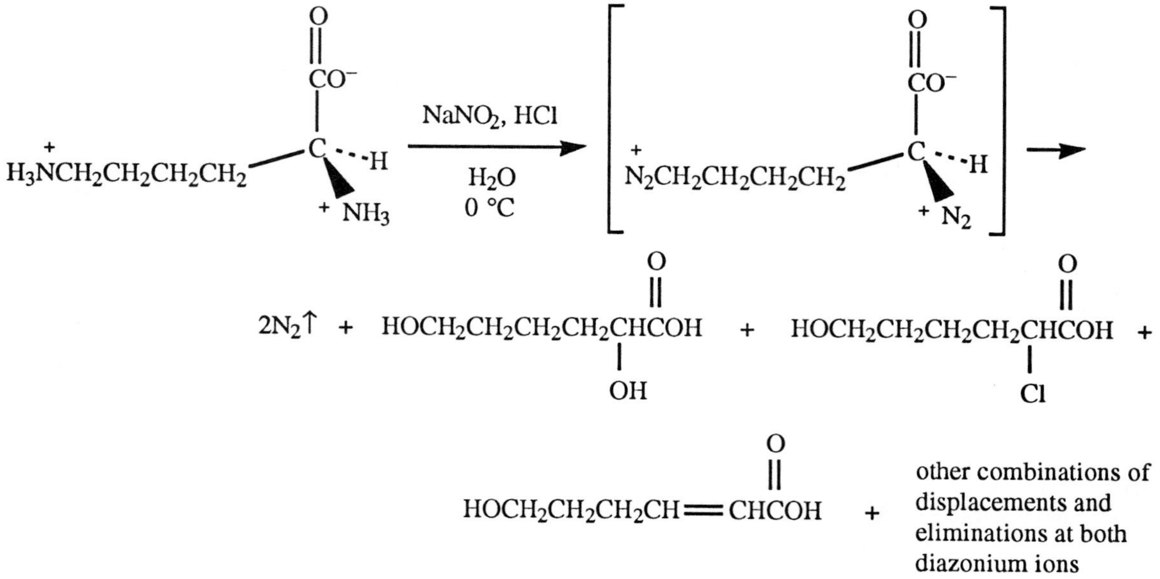

$2N_2\uparrow$ + $HOCH_2CH_2CH_2CH_2CHCOH$ + $HOCH_2CH_2CH_2CH_2CHCOH$ +
                        | |
                       OH   Cl

$HOCH_2CH_2CH_2CH=CHCOH$ +   other combinations of displacements and eliminations at both diazonium ions

(c)

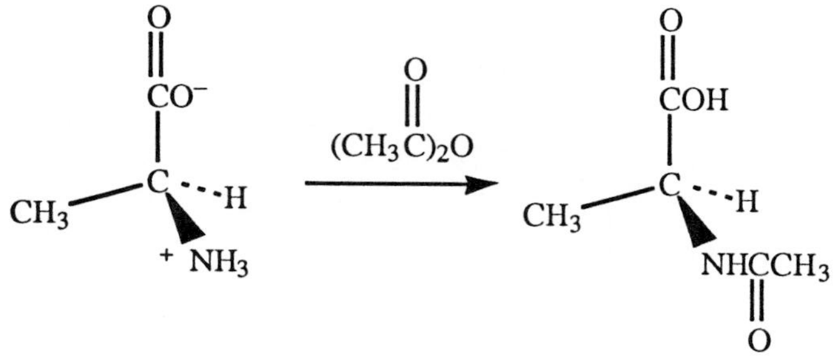

26.22 (cont)

(d)

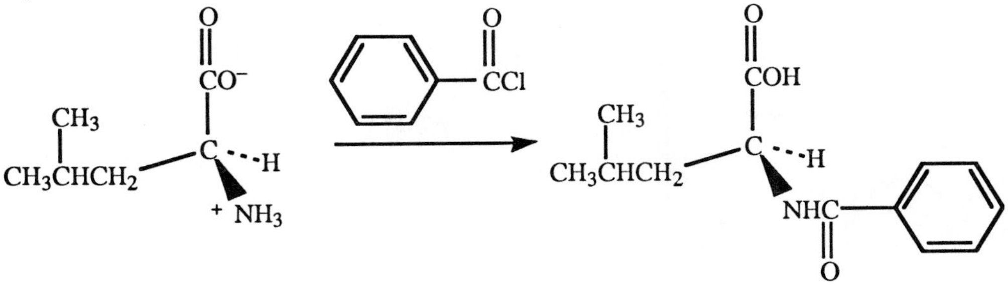

(e)

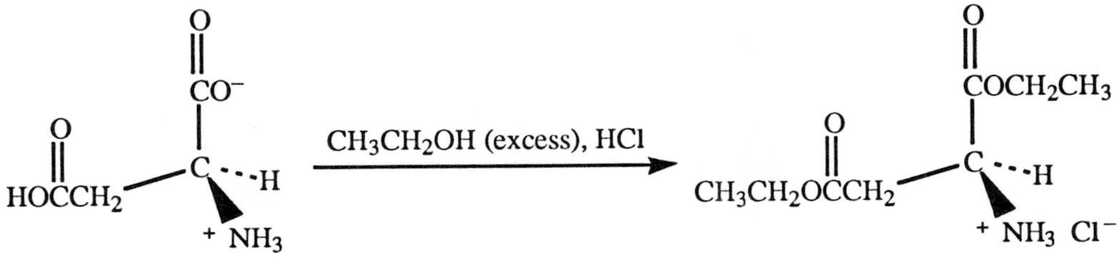

CH₃CH₂OH (excess), HCl

(f)

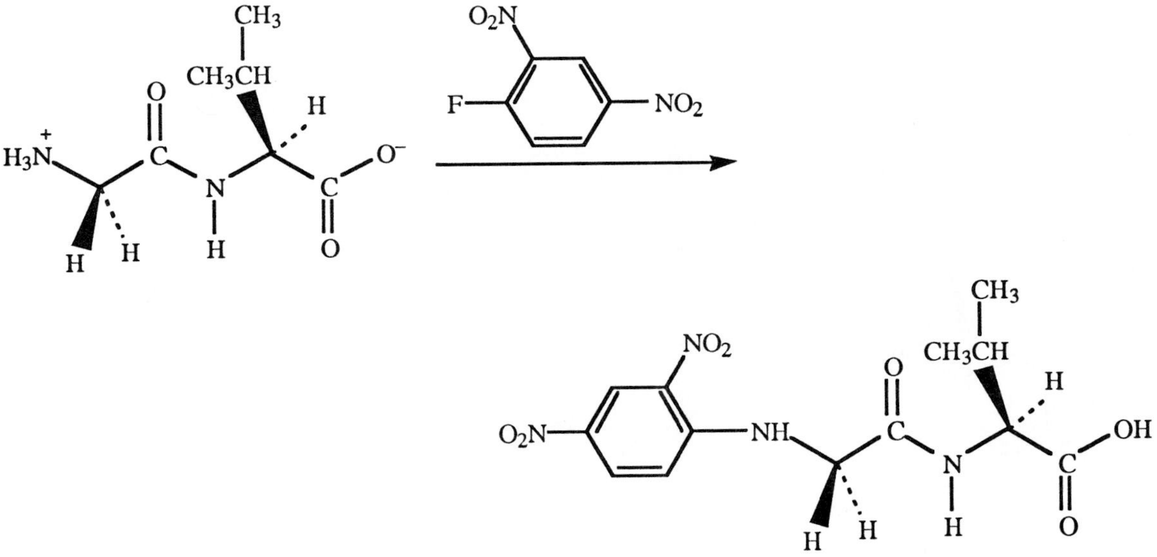

(g)   product of (f) $\xrightarrow[\Delta]{\text{H}_2\text{O, HCl}}$

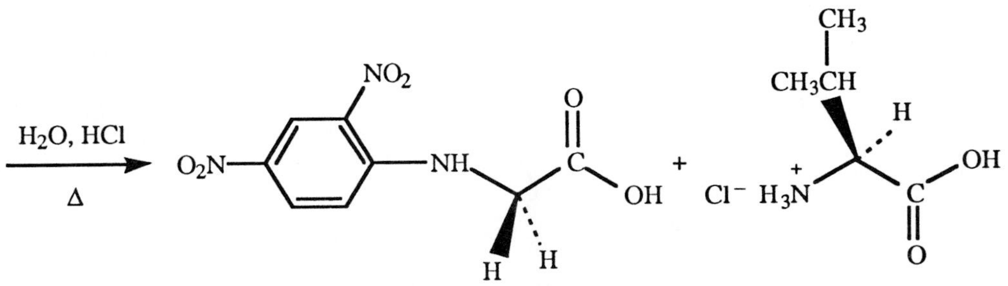

26.22 (cont)

(h)

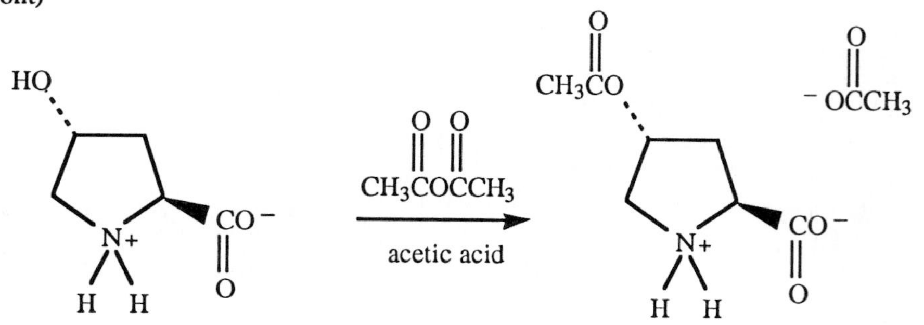

(i)

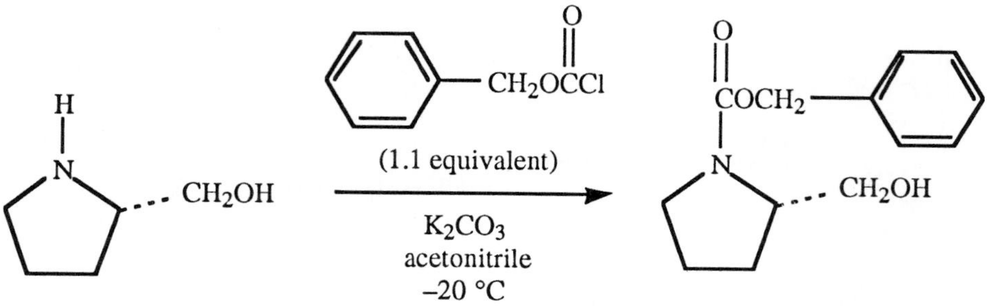

26.23

(a)

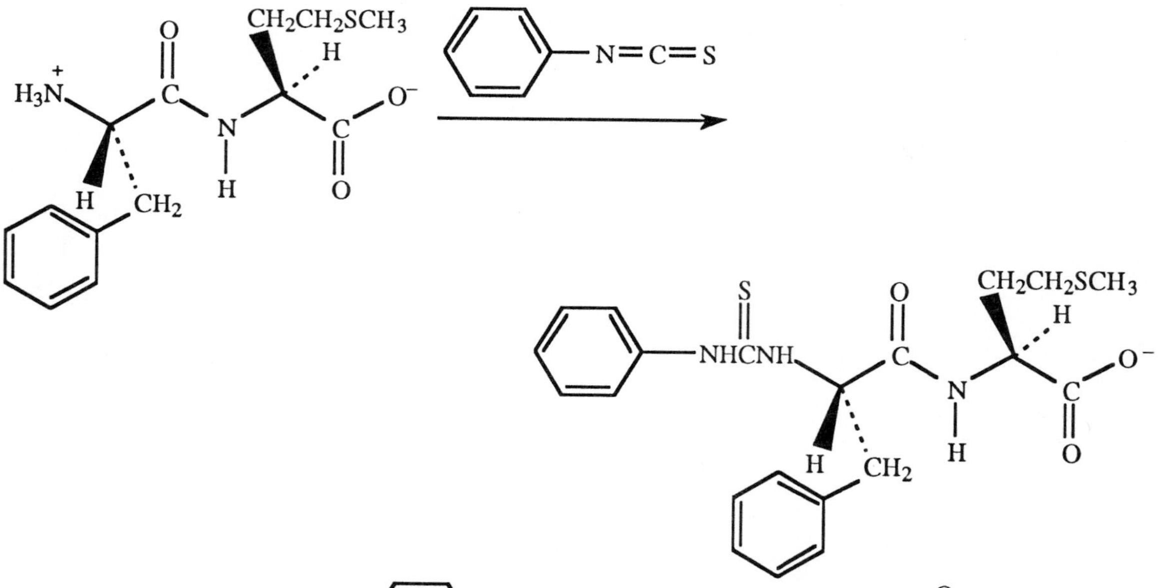

(b)   product of (a) $\xrightarrow[\substack{\text{nitromethane} \\ \Delta}]{\text{HCl}}$

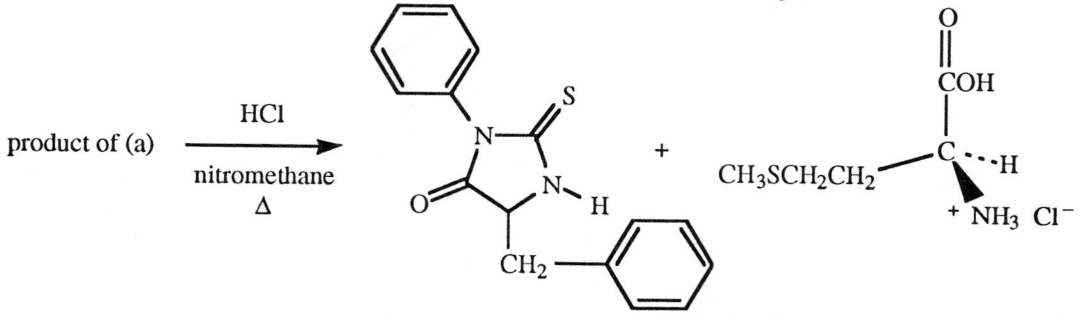

26.23 (cont)

(c)

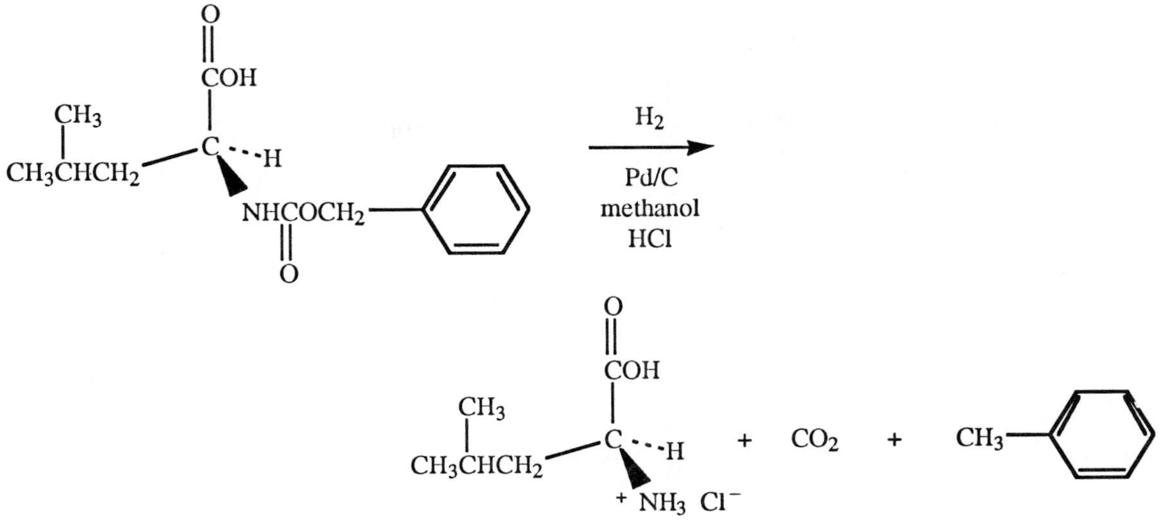

$$H_2 \quad \xrightarrow{\quad\quad} \quad Pd/C \quad methanol \quad HCl$$

$$+ \quad CO_2 \quad + \quad CH_3\text{—}\langle\text{phenyl}\rangle$$

(d)

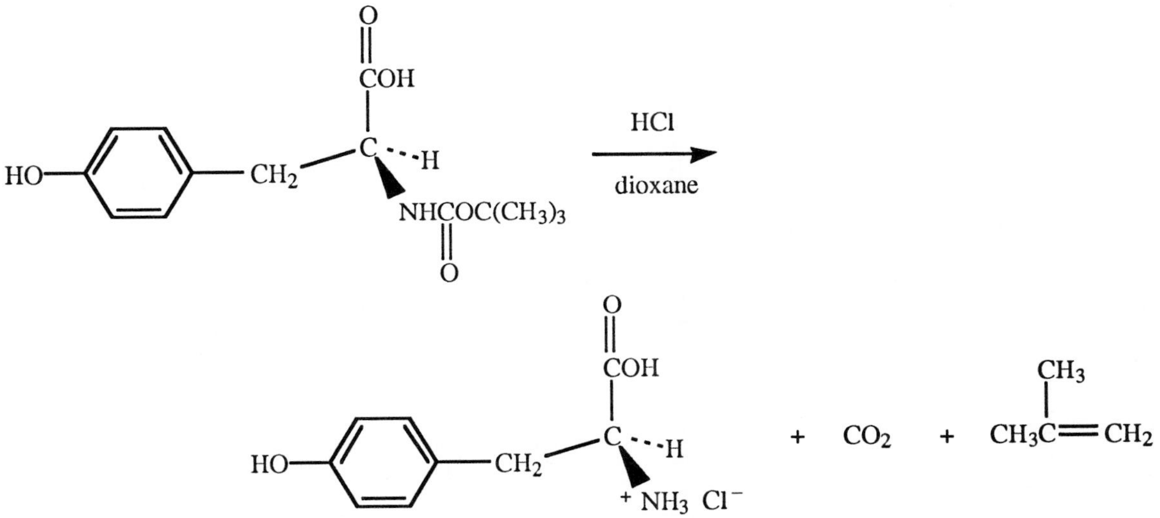

$$\xrightarrow[dioxane]{HCl}$$

$$+ \quad CO_2 \quad + \quad CH_3C(CH_3)\!=\!CH_2$$

(e)

$$HO\text{—}N\langle succinimide\rangle, \; DCC$$

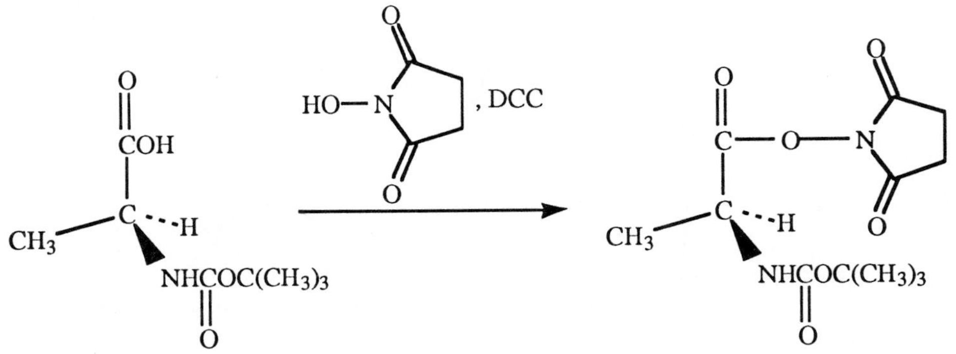

26.23 (cont)

(f)

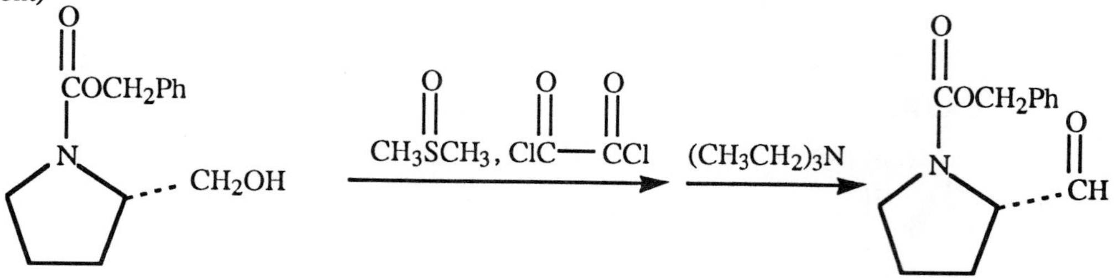

(g)

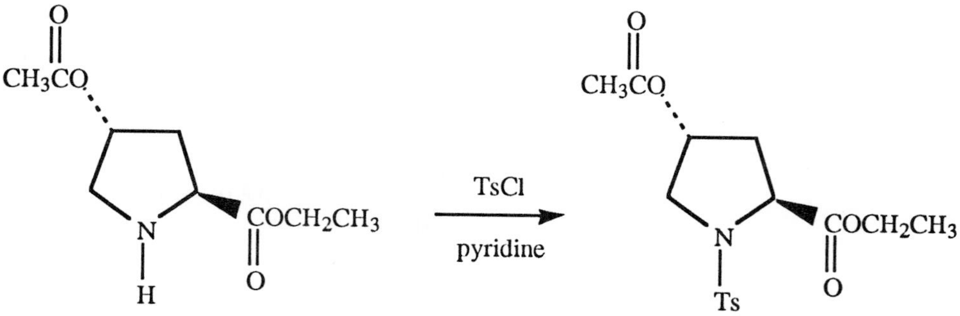

(h)

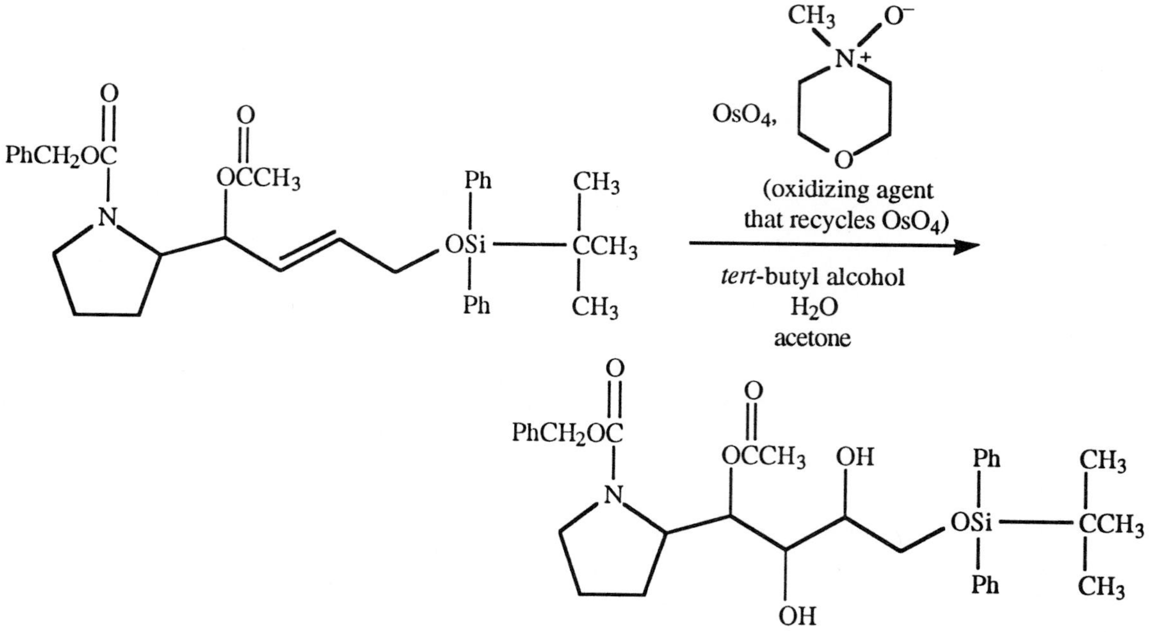

(i)

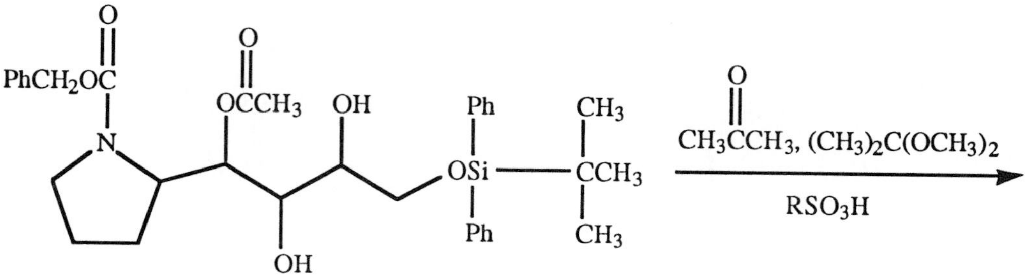

23.23 (cont)

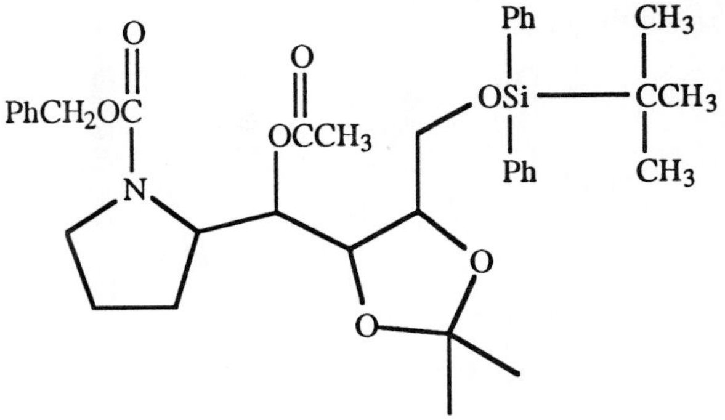

26.24

(a)

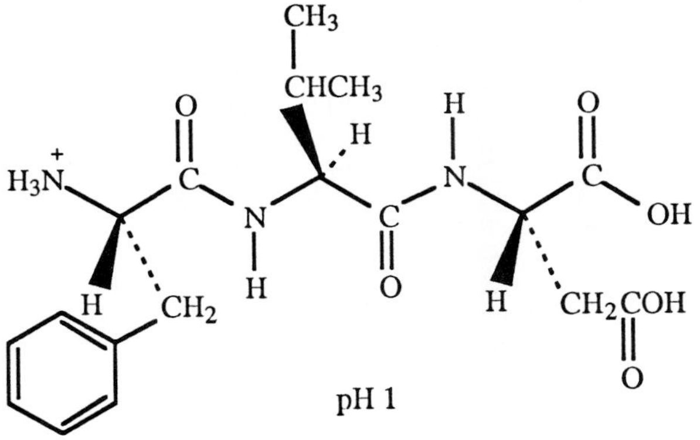

pH 1

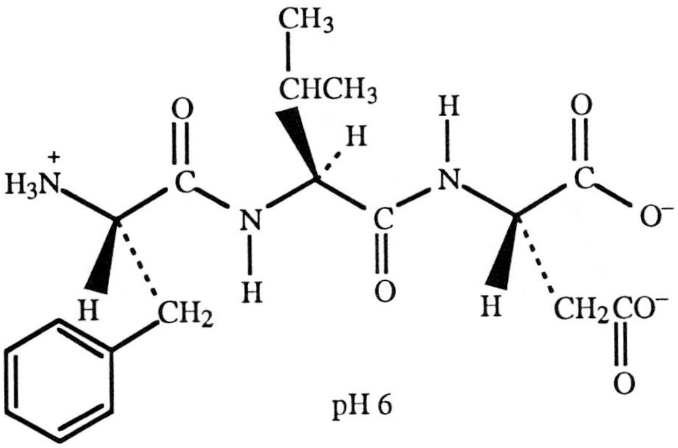

pH 6

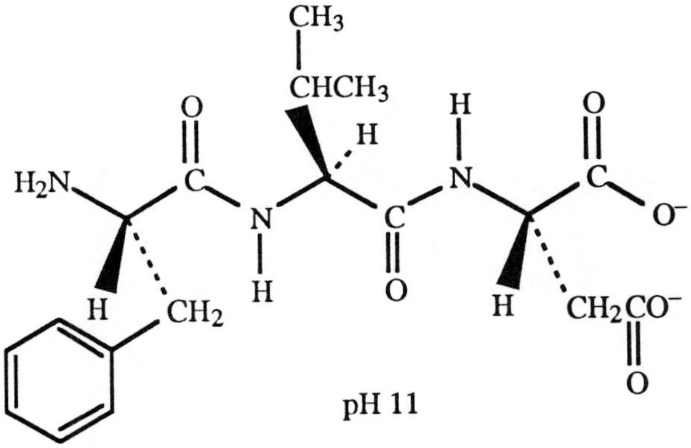

pH 11

(b)

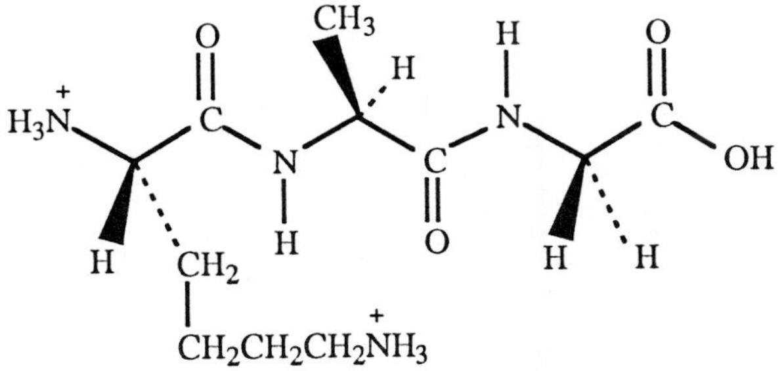

pH 1

26.24 (b) (cont)

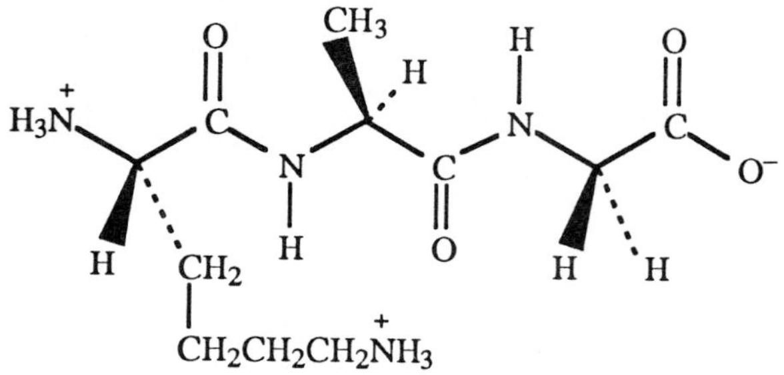

pH 6

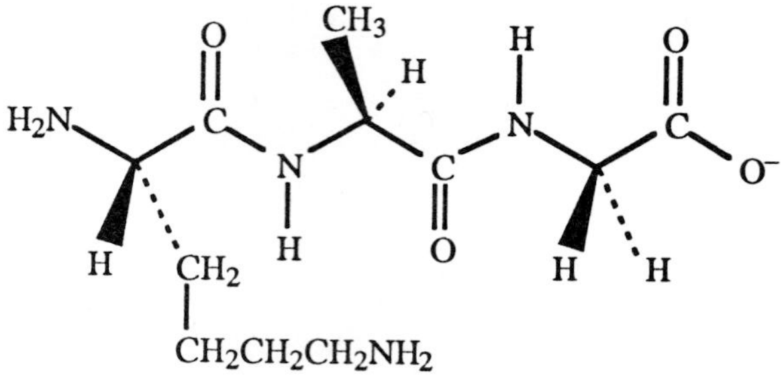

pH 11

(c)

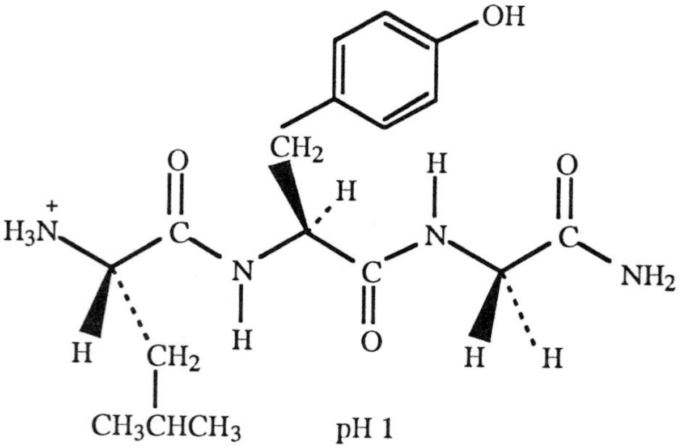

pH 1

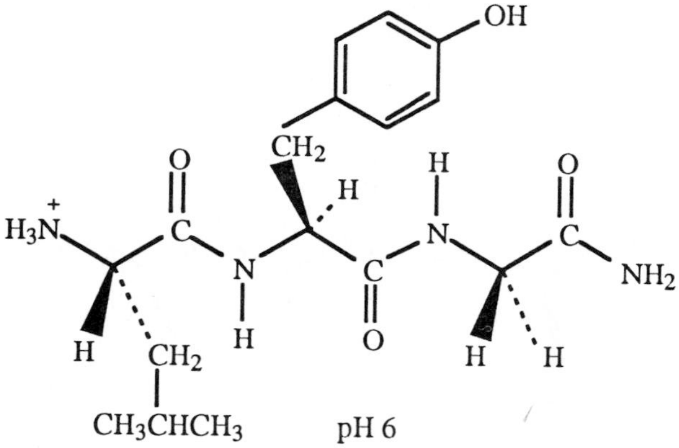

pH 6

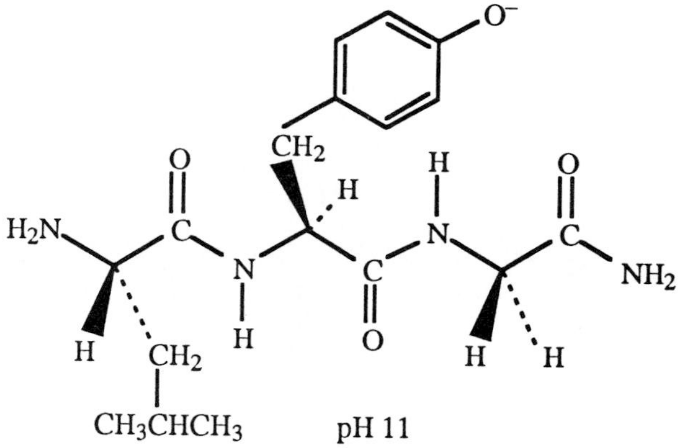

pH 11

(d)

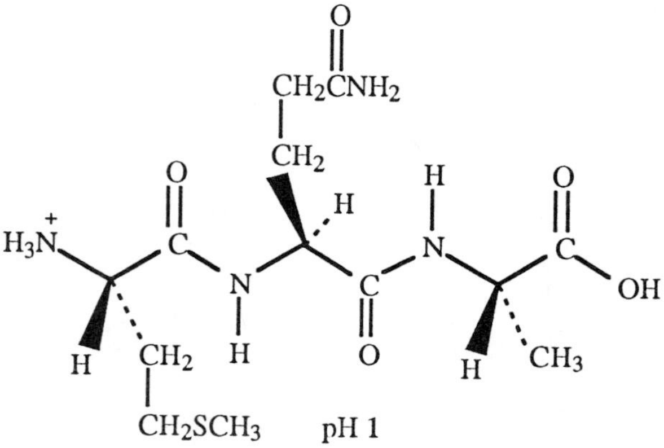

pH 1

26.24 (d) (cont)

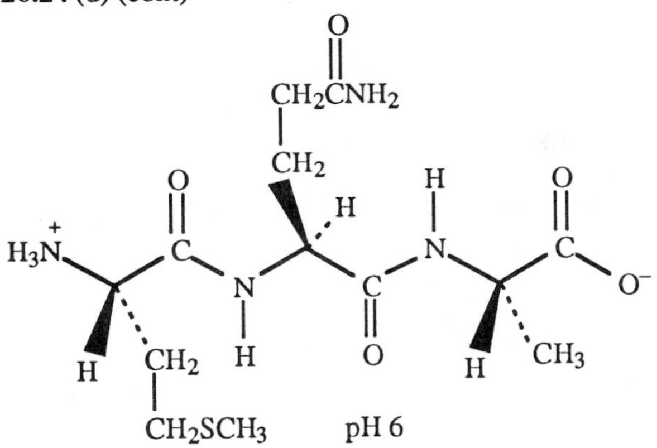

pH 6

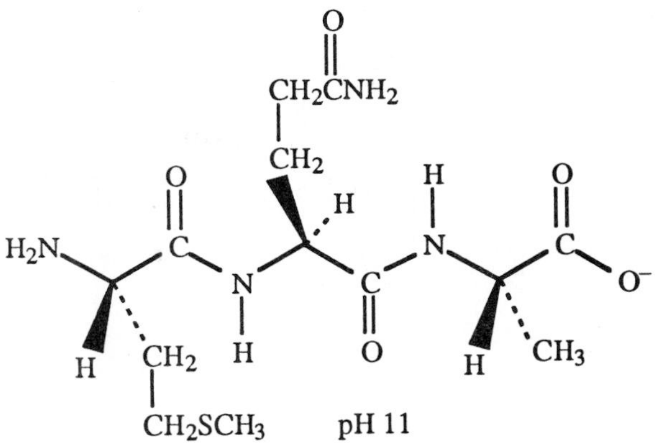

pH 11

26.25

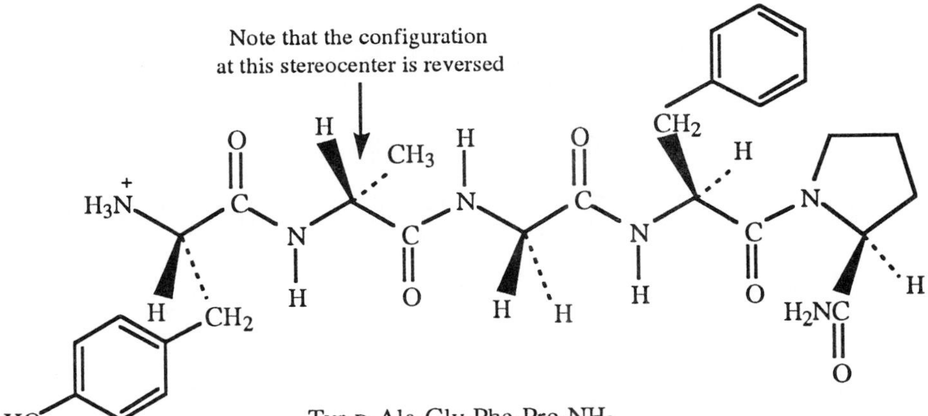

Note that the configuration
at this stereocenter is reversed

Tyr-D-Ala-Gly-Phe-Pro-NH₂

26.26

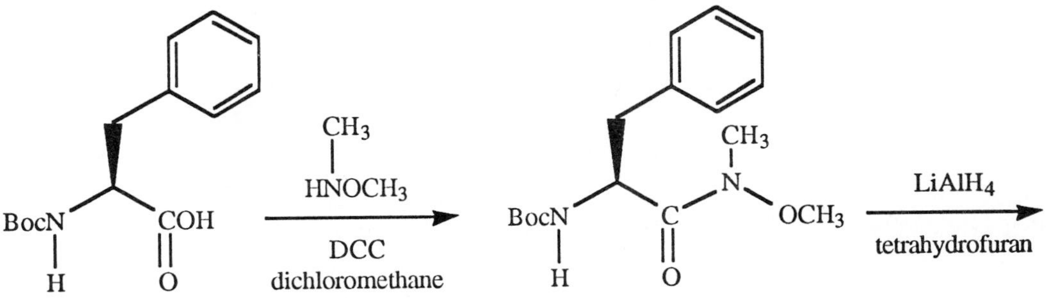

BocN   COH   →   BocN   →   LiAlH₄

HNOCH₃   DCC   dichloromethane   tetrahydrofuran

26.26 (cont)

26.27

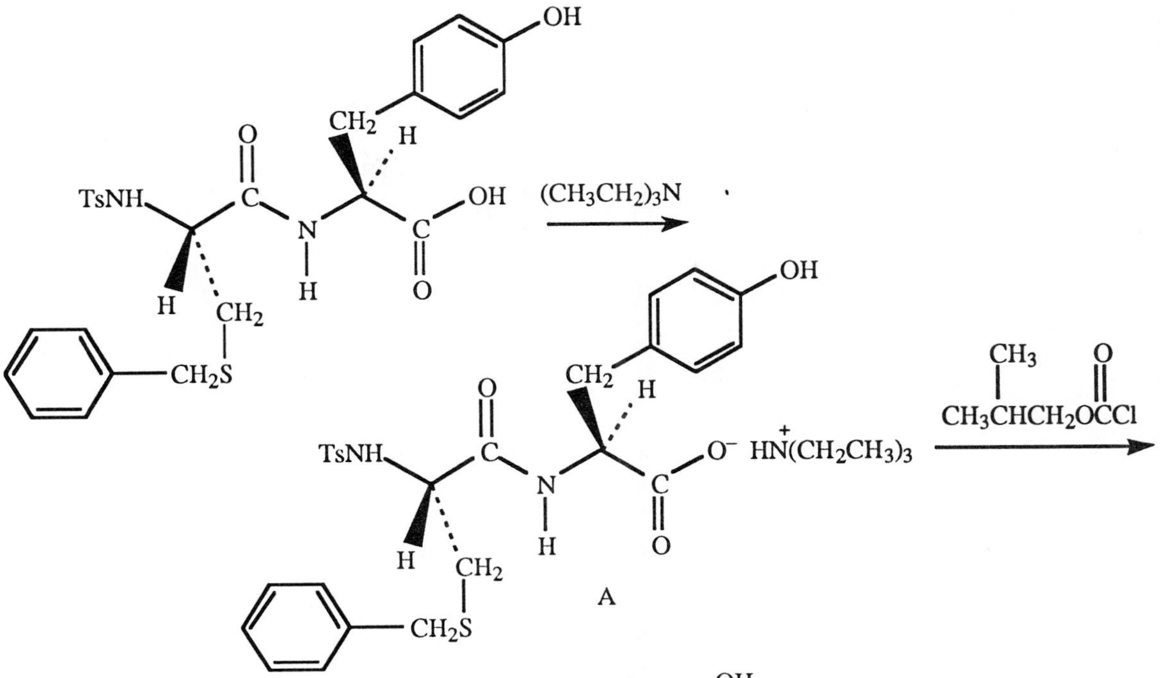

(CH₃CH₂)₃N

A

This functional group is like an anhydride in reactivity.

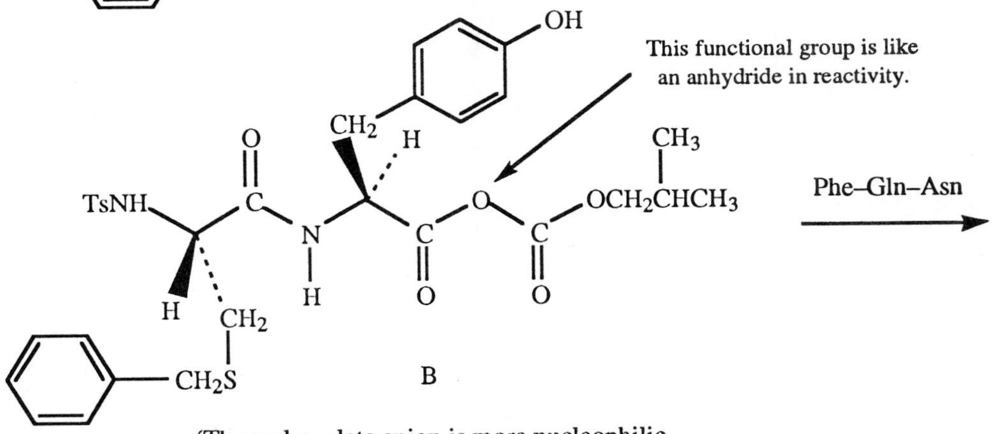

B

(The carboxylate anion is more nucleophilic than the hydroxyl group of the phenol.)

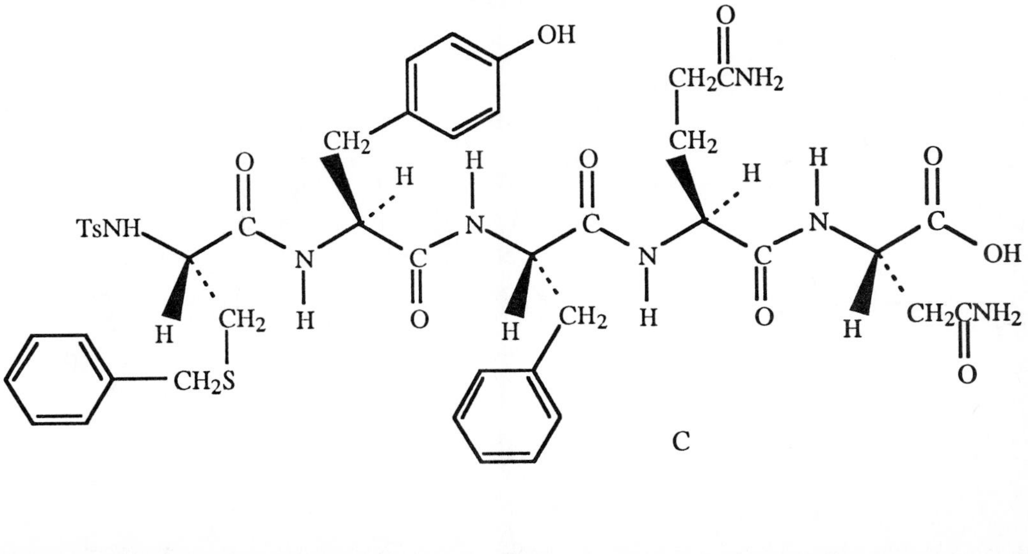

Phe–Gln–Asn

C

26.27 (cont)

26.27 (cont)

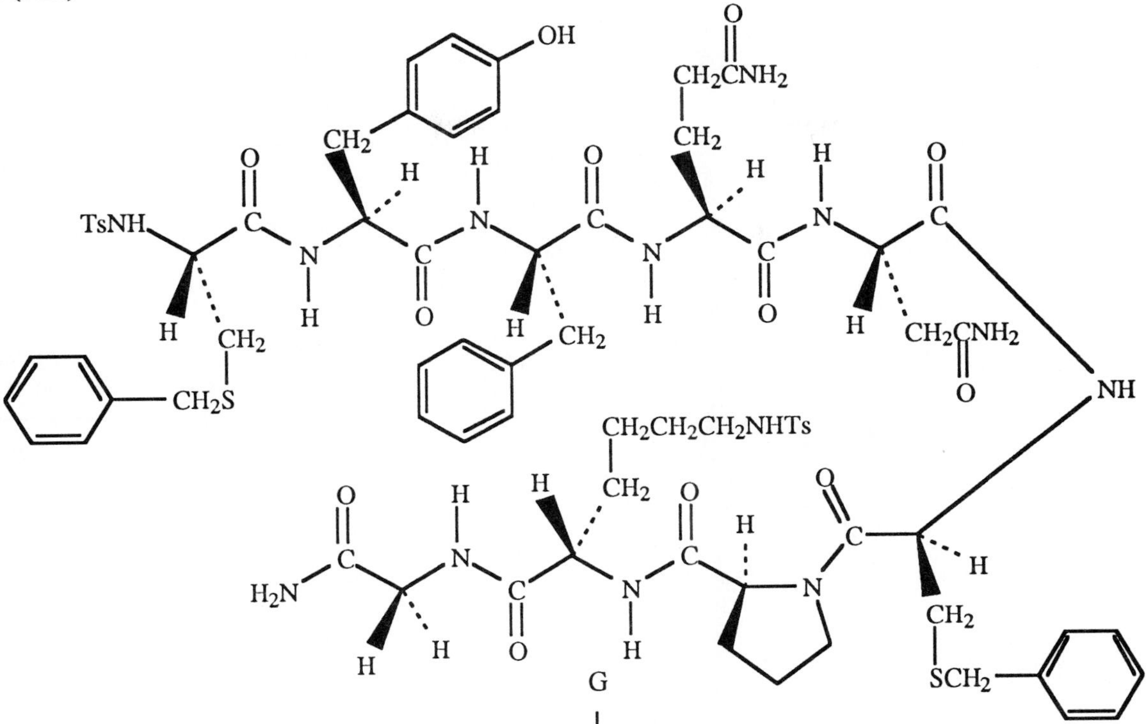

G

↓ Na, NH_3 (liq)

Cys–Tyr–Phe–Gln–Asn–Cys–Pro–Lys–Gly—NH_2

N-terminal end *C*-terminal end

H

↓ air oxidation

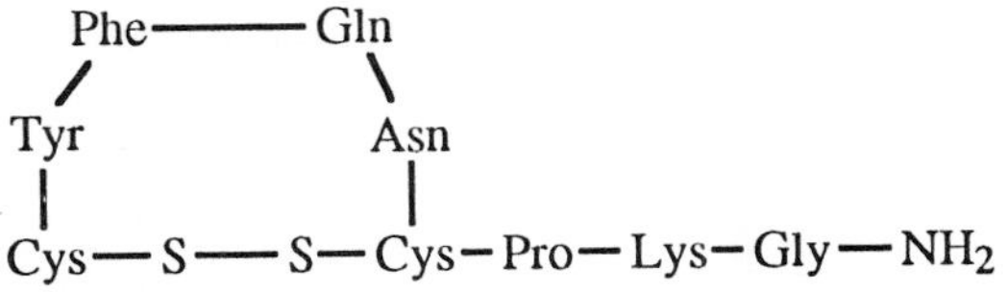

N-terminal end *C*-terminal end

I
lysine vasopressin

26.28

(a)

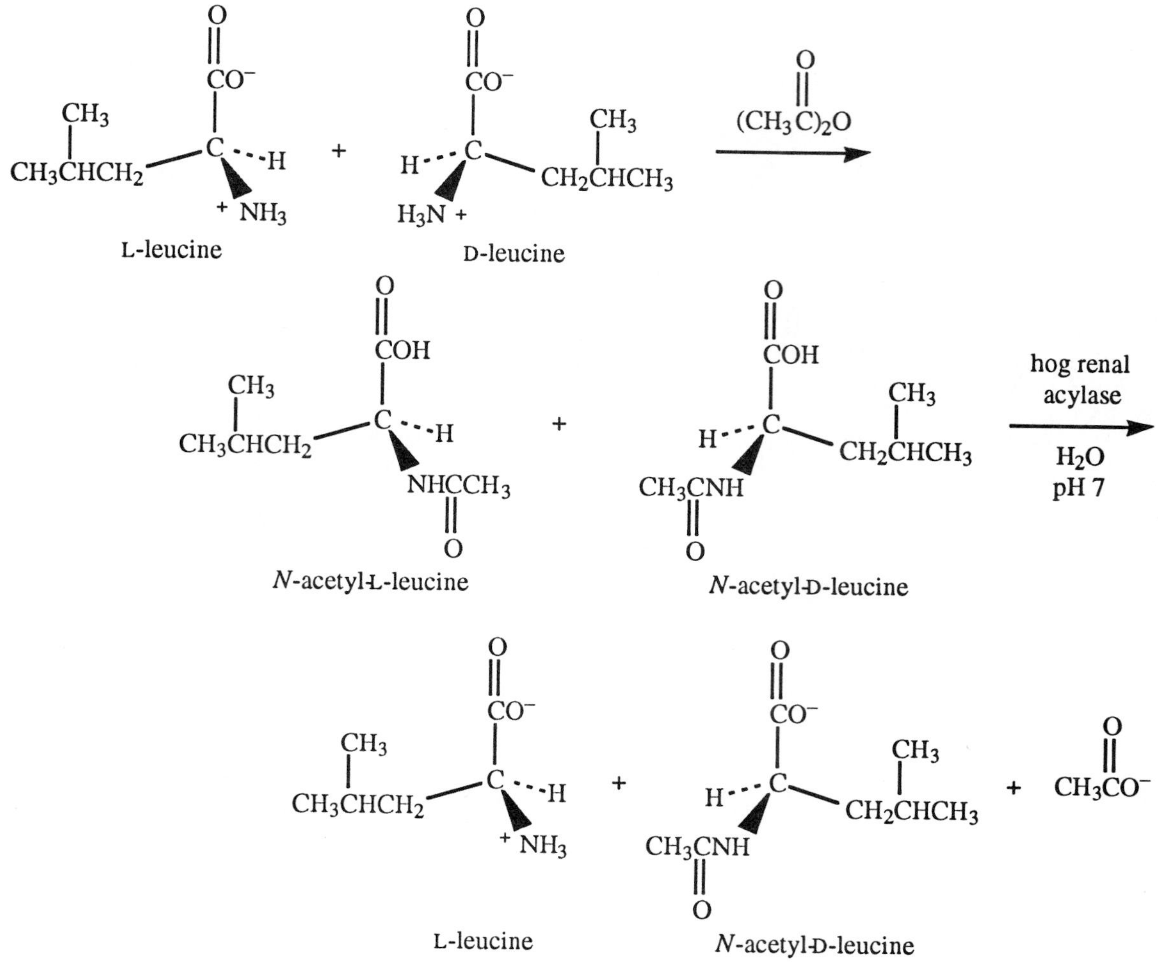

L-leucine + D-leucine → (CH₃CO)₂O

N-acetyl-L-leucine + N-acetyl-D-leucine → hog renal acylase, H₂O, pH 7

L-leucine + N-acetyl-D-leucine + CH₃CO⁻

(b) At pH 6, L-leucine will be its most insoluble as an internal salt or zwitterion. If the pH of the system is adjusted to 6, we can expect L-leucine to crystallize out.

The pH of the solution would have to be much lower for *N*-acetyl-D-leucine to precipitate, because the basicity of the amino group has been eliminated by conversion to an amide. The compound has the acidity of an alkyl carboxylic acid.

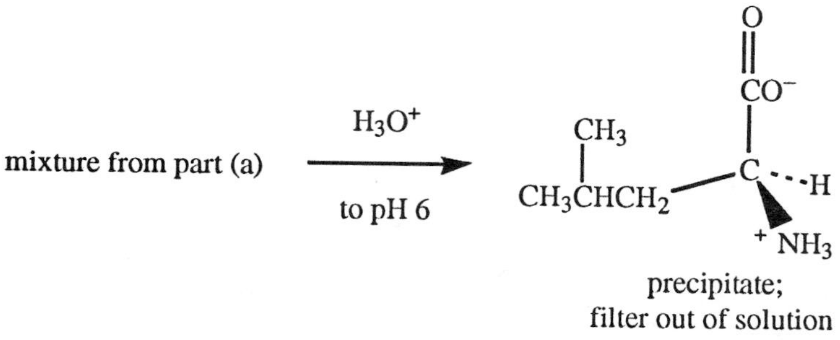

mixture from part (a) → H₃O⁺, to pH 6 →

precipitate;
filter out of solution

26.28 (b) (cont)

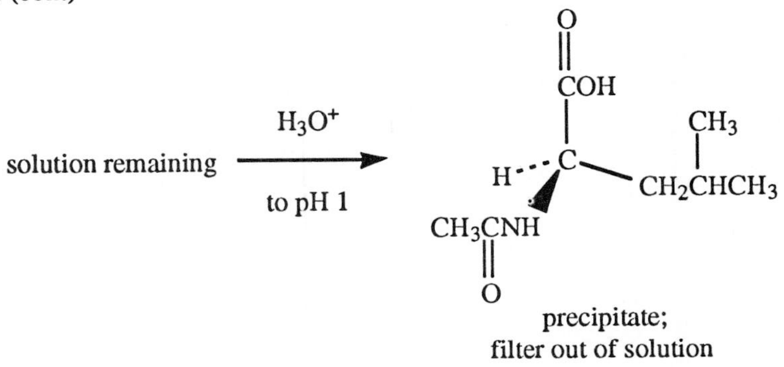

precipitate;
filter out of solution

(c)

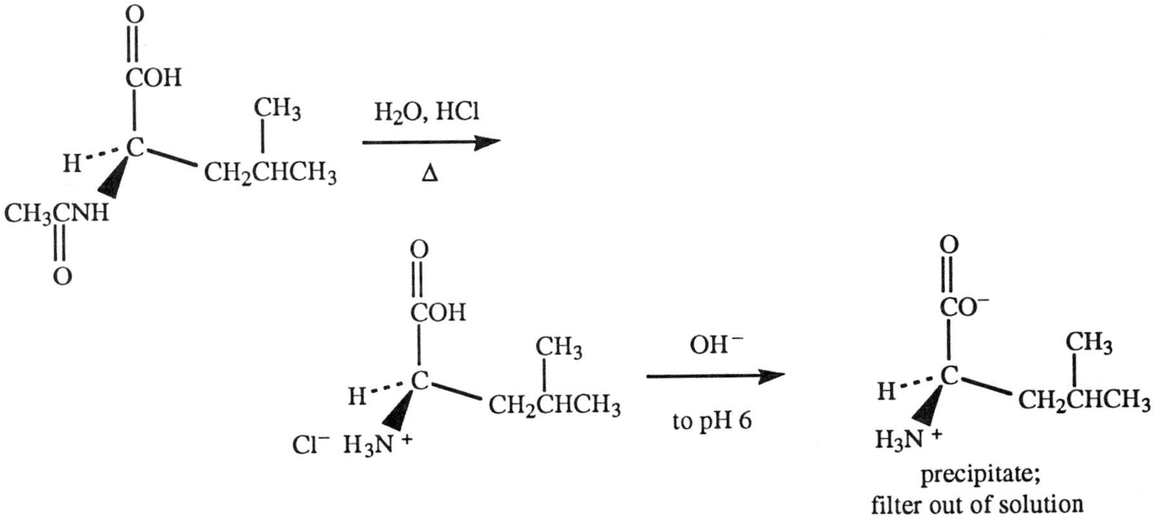

precipitate;
filter out of solution

26.29

(a)   (1)

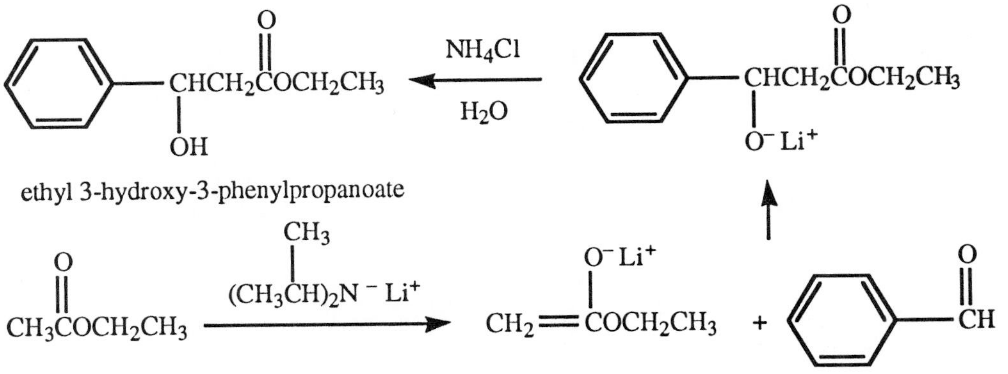

ethyl 3-hydroxy-3-phenylpropanoate

(2)

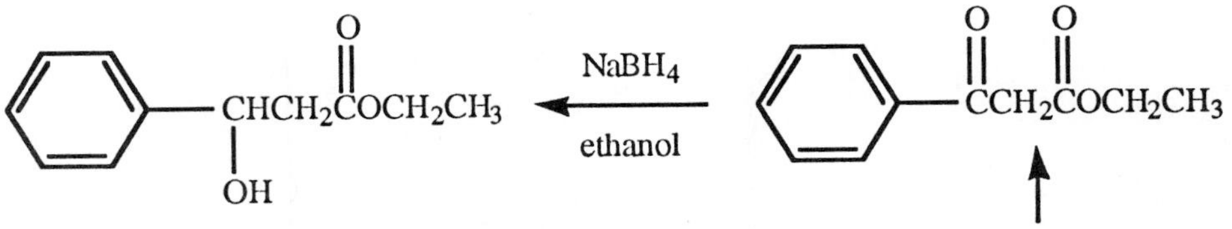

26.29 (a 2) (cont)

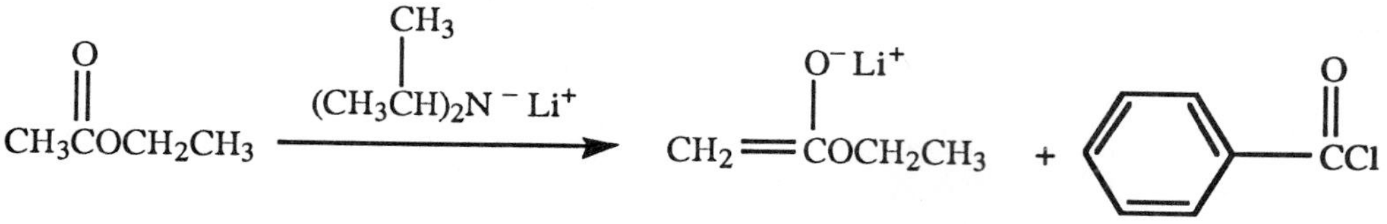

(b)

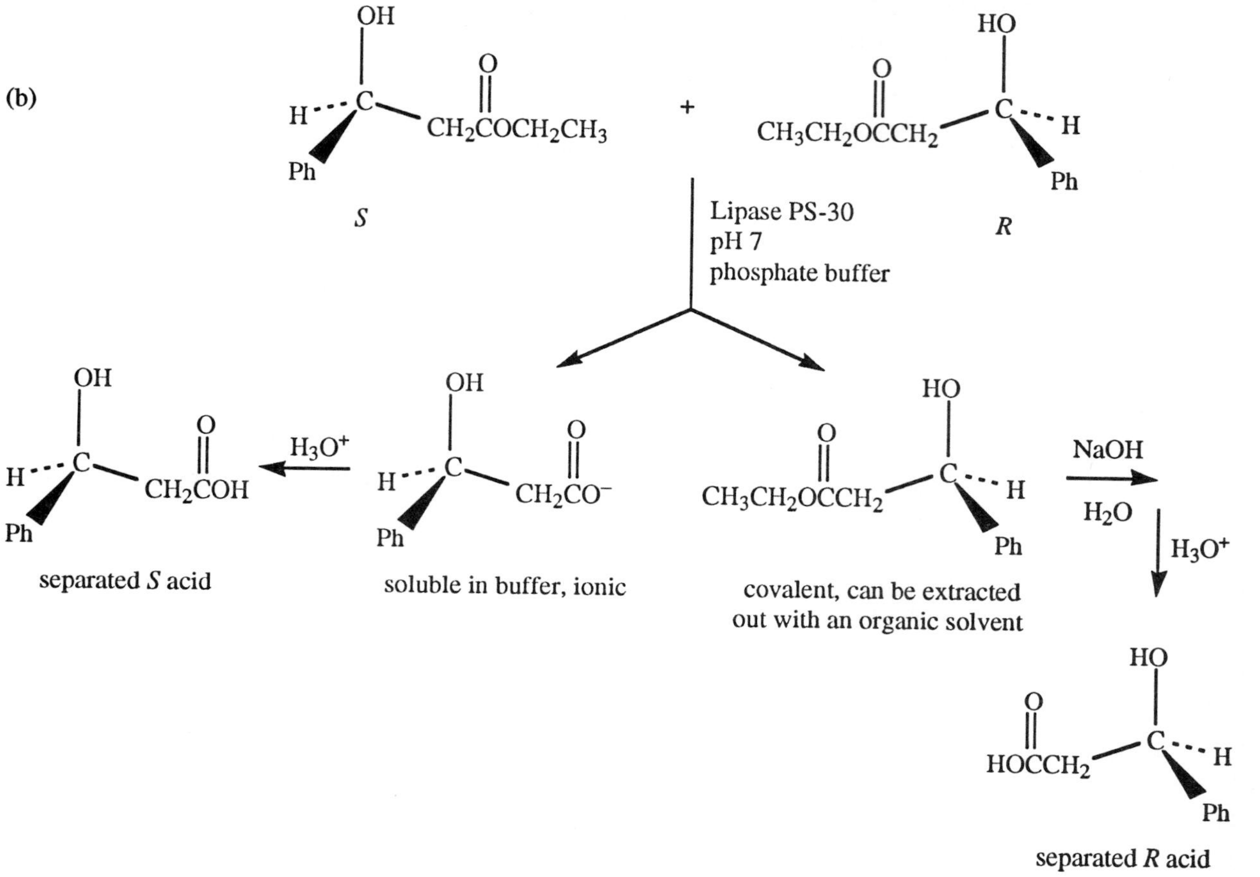

26.30

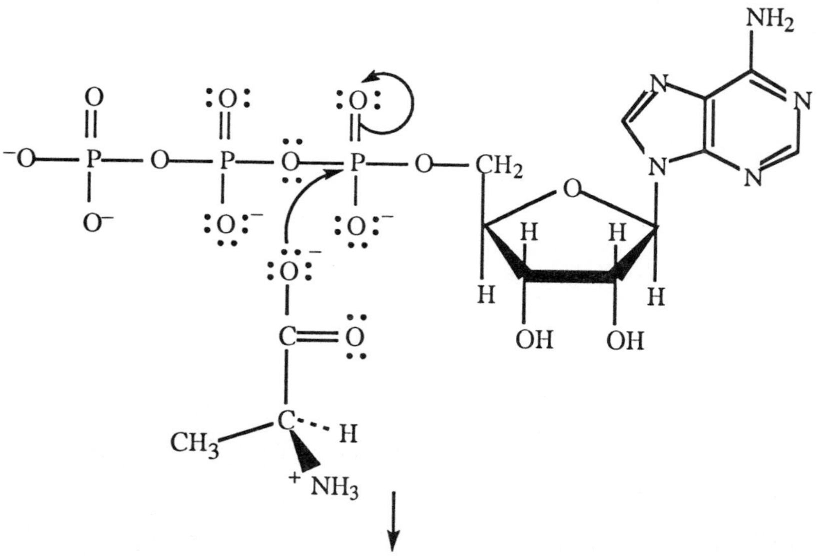

26.30 (cont)

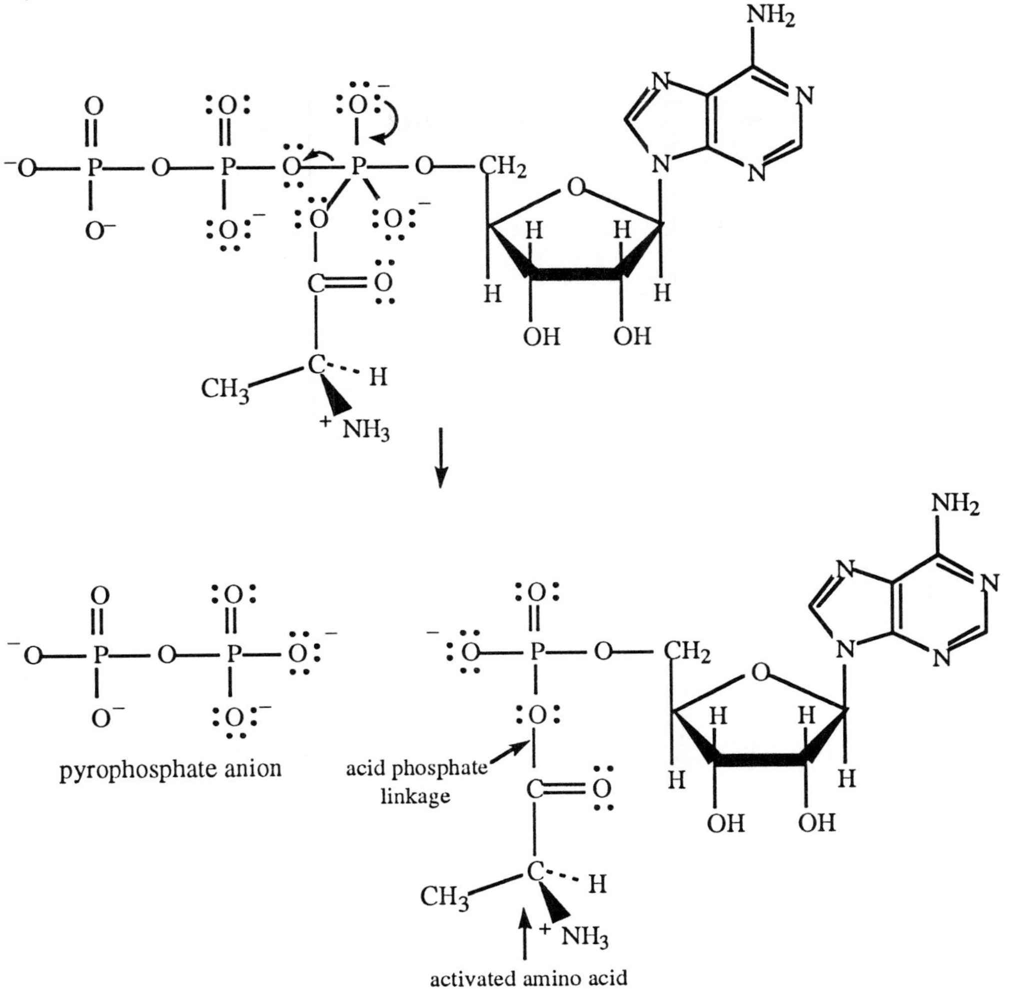

pyrophosphate anion

acid phosphate linkage

activated amino acid

26.31 xenopsin $\xrightarrow{\text{trypsin}}$ ⌐Glu—Gly—Lys—Arg + ⌐Glu—Gly—Lys—Arg—Pro—Trp

xenopsin $\xrightarrow{\text{chymotrypsin}}$ same peptides as above + Ile—Leu

N-terminal amino acid is ⌐Glu

C-terminal amino acid is Leu

Therefore, xenopsin is ⌐Glu—Gly—Lys—Arg—Pro—Trp—Ile—Leu

26.32 Combining the fragments from trypsin and chymotrypsin digestion gives the following three segments:

A peptide containing 17 amino acids from the *N*-terminal end of viscotoxin A_2:

Lys–Ser–Cys–Cys–Pro–Asn–Thr–Thr–Gly–Arg–Asn–Ile–Tyr–Asn–Thr–Cys–Arg

26.32 (cont)

A peptide containing 17 amino acids from the *C*-terminal end:

Ser–Gly–Cys–Lys–Ile–Ile–Ser–Ala–Ser–Thr–Cys–Pro–Ser–Tyr–Pro–Asp–Lys

A 12 amino acid peptide; no overlap with either of the *N*- or *C*-terminal fragments:

Phe–Gly–Gly–Gly–Ser–Arg–Glu–Val-Cys–Ala–Ser–Leu

The three peptides add up to 46 amino acids, the total number contained in viscotoxin A$_2$. The complete order of amino acids in viscotoxin A$_2$ is thus the order that exists in the sequence of the three peptide fragments.

Lys–Ser–Cys–Cys–Pro–Asn–Thr–Thr–Gly–Arg–Asn–Ile–Tyr–Asn–Thr

Ser–Leu–Ser–Ala–Cys–Val–Glu–Arg–Ser–Gly–Gly–Gly–Phe–Arg–Cys

Gly–Cys–Lys–Ile–Ile–Ser–Ala–Ser–Thr–Cys–Pro–Ser–Tyr–Pro–Asp–Lys

26.33 (a)

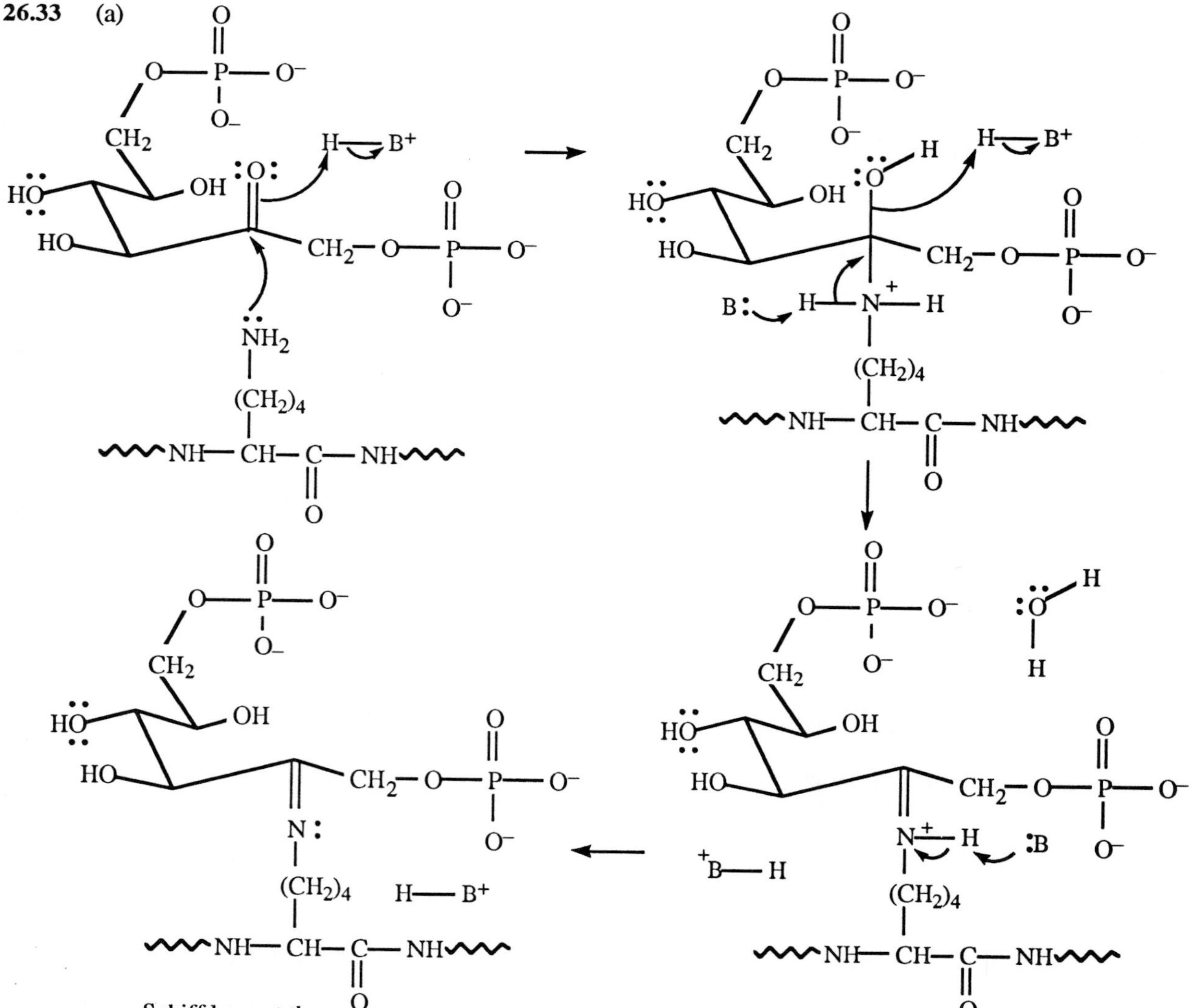

Schiff base at the active site of the enzyme

26.33 (cont)

(b)

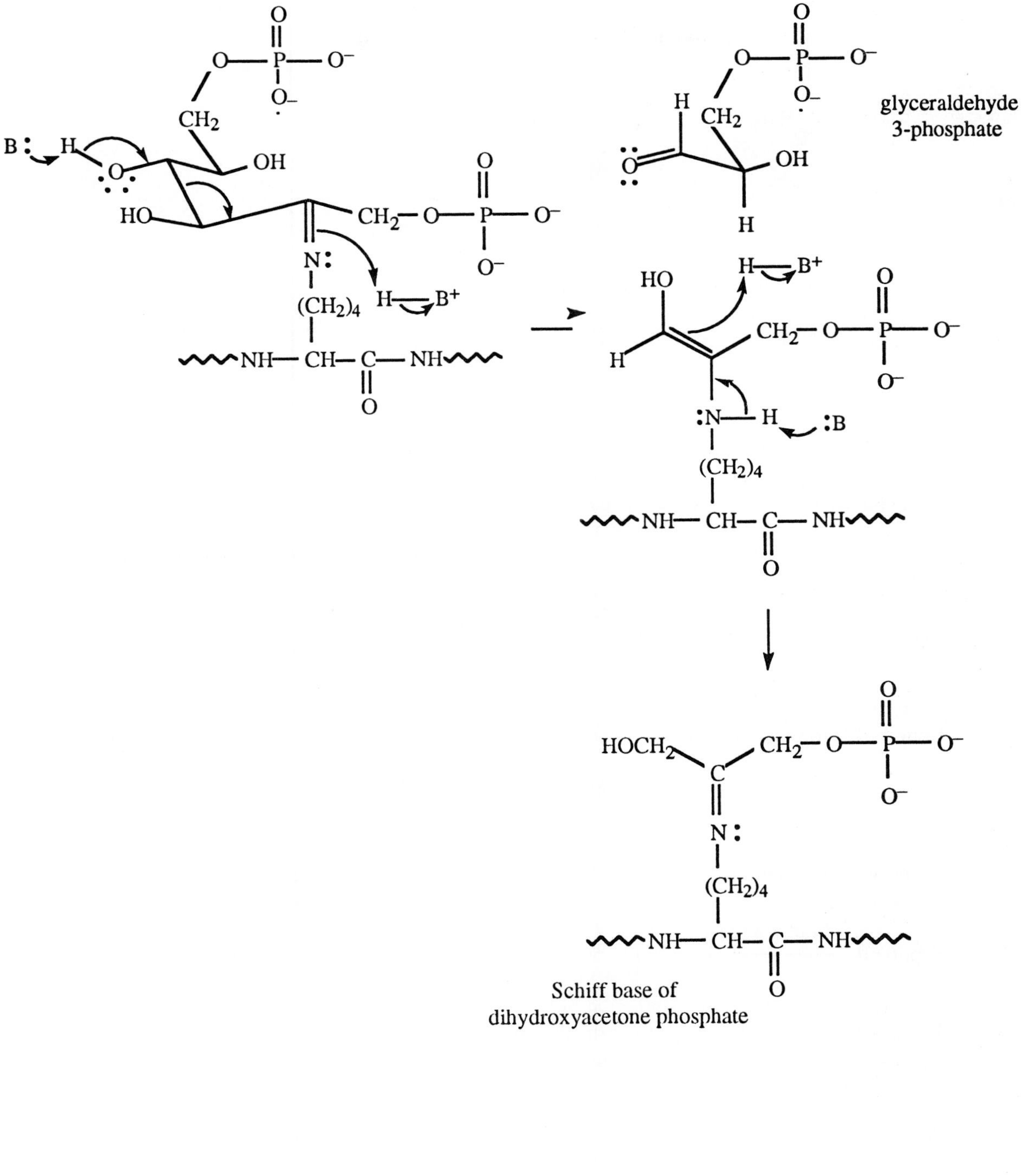

glyceraldehyde
3-phosphate

Schiff base of
dihydroxyacetone phosphate

26.33 (cont)

(c)

$$HOCH_2 - C - CH_2 - O - P - O^-$$

with N, (CH$_2$)$_4$, H groups and N–H, C=O

$$\xrightarrow[\text{H}_2\text{O}]{\text{NaBH}_4} \xrightarrow{\text{hydrolysis}}$$

$$CH_2OH$$
$$CHNHCH_2CH_2CH_2CH_2$$
$$CH_2OH$$

$$C - H \quad CO^- \quad +NH_3$$

lysine derivative of
dihydroxyacetone

26.34   Peptide P contains 11 amino acids:

Arg, Gln (2), Gly, Leu, Lys, Met, Phe (2), Pro (2)

Peptide A contains Arg, Gln (2), Lys, Phe, Pro (2). Therefore, Peptide B must contain Gly, Leu, Met, Phe.

$$\text{Peptide P (and Peptide A)} \xrightarrow{\text{Edman's reagent}} \begin{array}{l}\text{phenylthiohydantoins of}\\ \text{Arg, Pro, Lys, Pro, in that order}\end{array}$$

Therefore, the *N*-terminal amino acid of Peptide P (and Peptide A) is Arg. Peptide A contains the amino acids from the *N*-terminal end of P, and Peptide B must contain the amino acids from the *C*-terminal end of P.

$$\text{Peptide A} \xrightarrow{\text{carboxypeptidase}} \text{first Phe and then Gln}$$

Therefore, the *C*-terminal amino acid of A is Phe. Peptide A must have the structure Arg–Pro–Lys–Pro–Gln–Gln–Phe.

$$\text{Peptide B} \xrightarrow{\text{Edman's reagent}} \begin{array}{l}\text{phenylthiohydantoins of}\\ \text{Phe, Gly, Leu, in that order}\end{array}$$

Therefore, Peptide B must have the structure Phe–Gly–Leu–Met.

$$\text{Peptide P} \xrightarrow{\text{carboxypeptidase}} \text{no amino acid}$$

Therefore, the *C*-terminal amino acid of P does not have a free carboxylic acid group.

$$\text{Peptide P} \xrightarrow[\substack{\text{(hydrolysis of}\\ \text{amide bonds)}}]{\text{HCl (0.03 M)}} \text{Peptide C}$$

26.34 (cont)

Peptide C $\xrightarrow[\text{carboxypeptidase}]{}$ Gly, Met, Leu, Phe (order not determined)

Therefore, Peptide C has a *C*-terminal amino acid with a free carboxylic acid group. The same amino acids as those present in Peptides B and P are released to the solution, so Peptide C has the same sequence of amino acids as Peptide P. Both have Met as the *C*-terminal amino acid. In Peptide P, the *C*-terminal acid exists as the amide, Met—NH_2.

These conclusions can be summarized as answers to the subsections of the problem.

(a)   The *N*-terminal amino acid of Peptide P is Arg.

(b)   The *C*-terminal amino acid of Peptide P is Met.

(c)   The sequence of amino acids in Peptide A is Arg–Pro–Lys–Pro–Gln–Gln–Phe.

(d)   The sequence of amino acids in Peptide B is Phe–Gly–Leu–Met.

(e)   The complete structure of Peptide B is Phe–Gly–Leu–Met—NH_2.

(f)   The sequence of amino acids in Peptide C is Arg–Pro–Lys–Pro–Gln–Gln–Phe–Phe–Gly–Leu–Met.

(g)   The complete structure of Peptide P is Arg–Pro–Lys–Pro–Gln–Gln–Phe–Phe–Gly–Leu–Met—NH_2.

26.35   The solid state synthesis of Peptide P:

26.35 (cont)

(b)

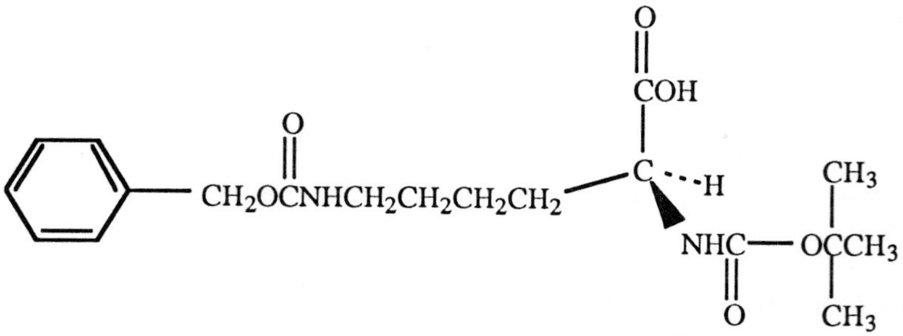

(c)

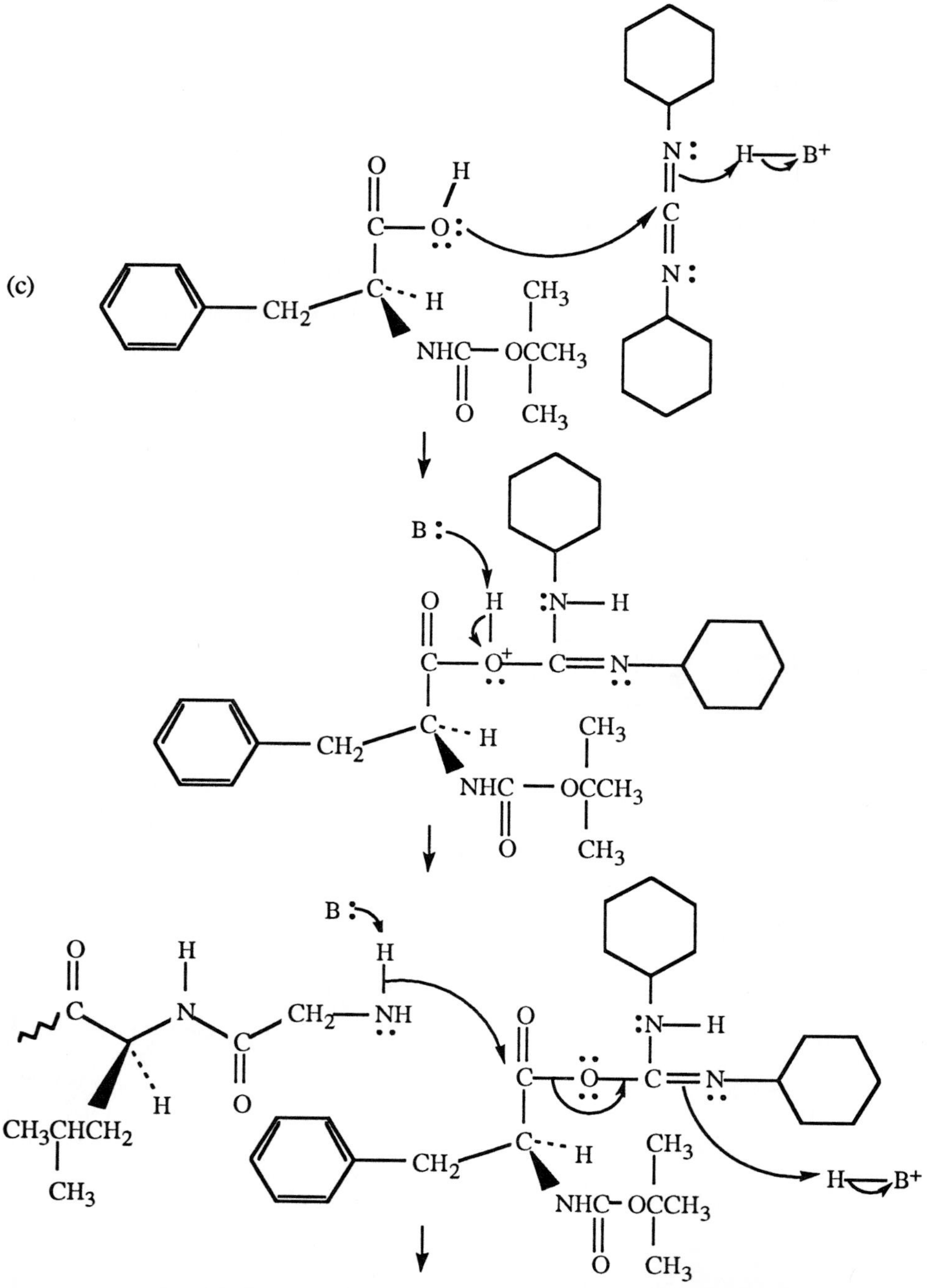

26.35 (c) (cont)

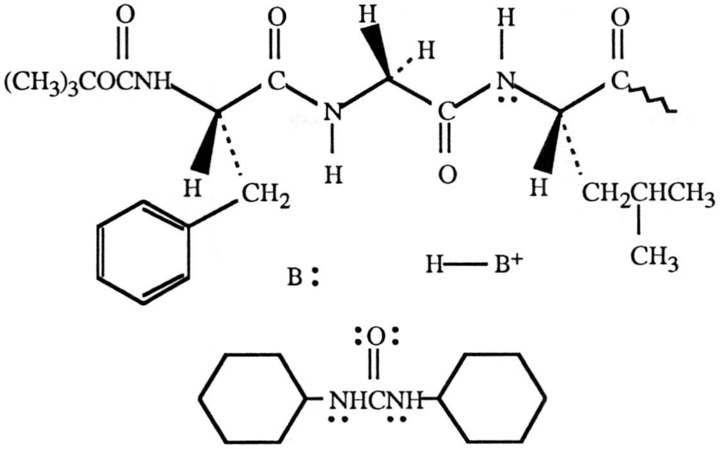

(d)

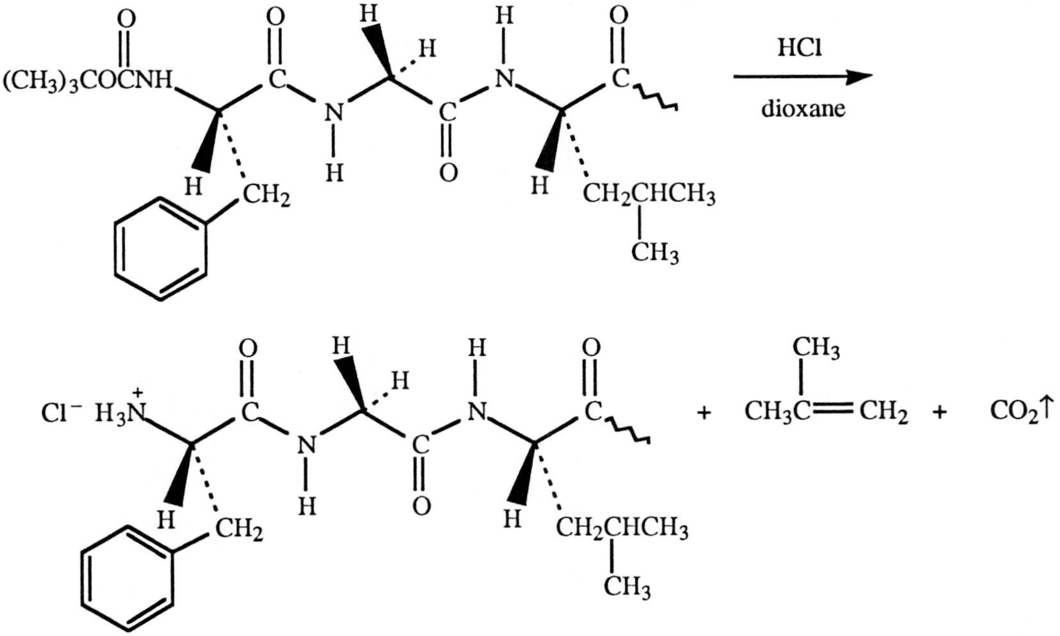

(e)

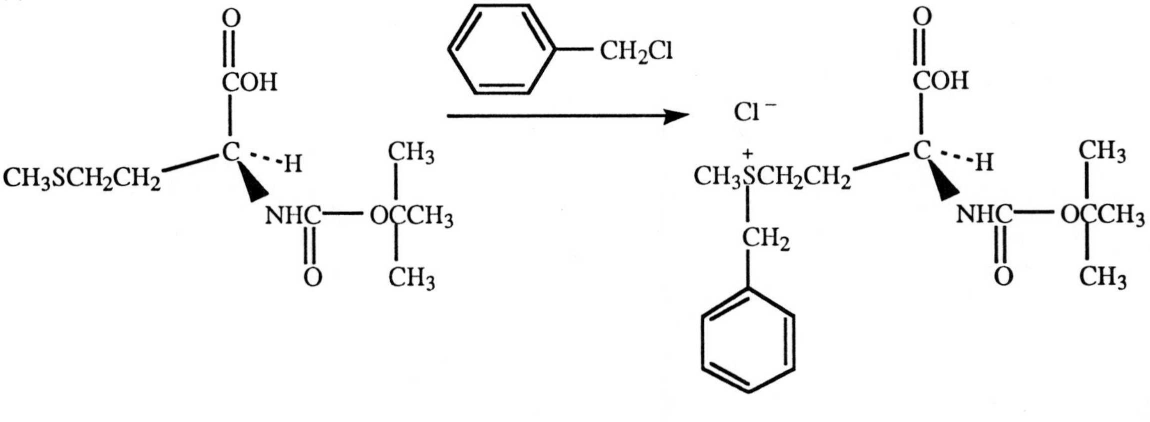

26.36

26.36 (cont)

26.37   (a)

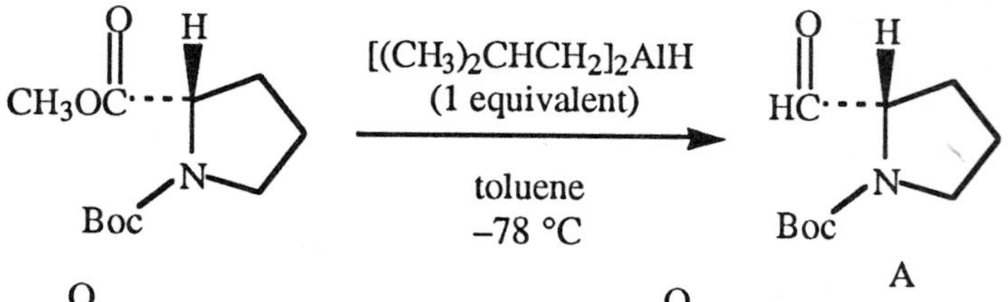

(b)

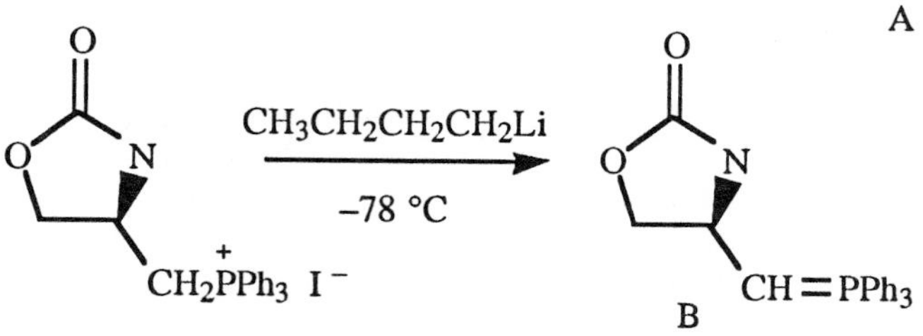

(c)

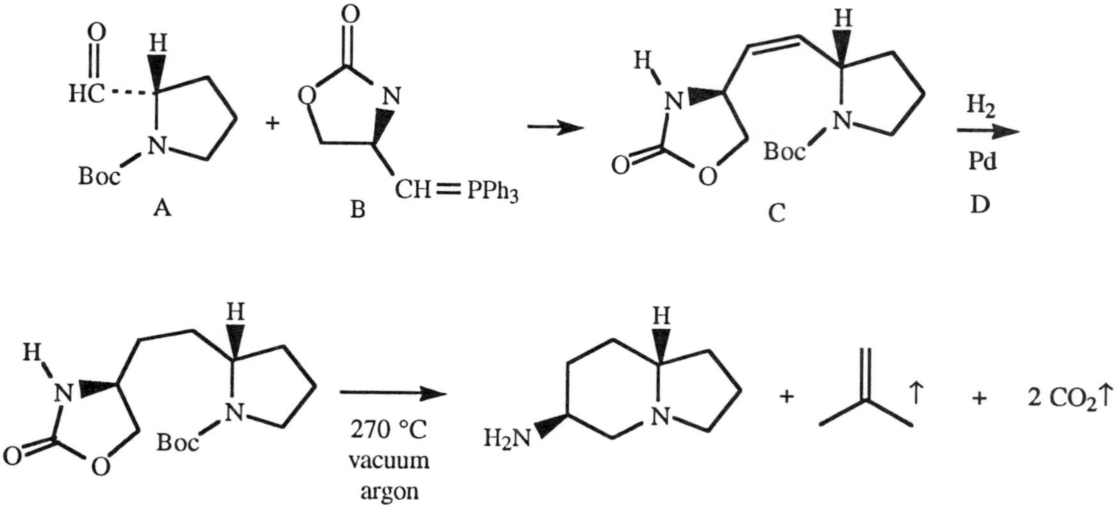

26.38

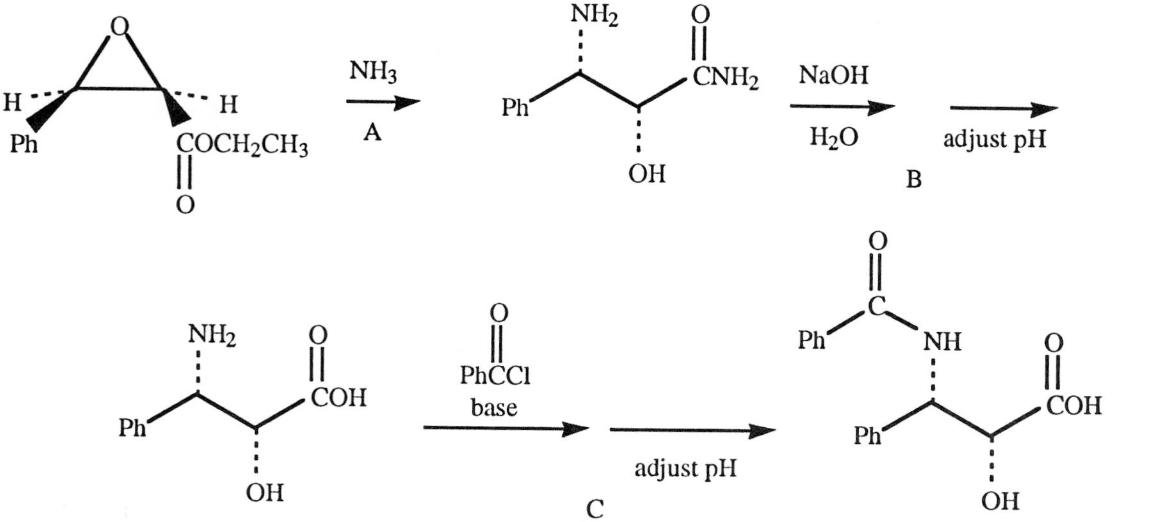

N-benzoyl-3-phenylisoserine

26.39

26.39 (cont)

26.39 (cont)

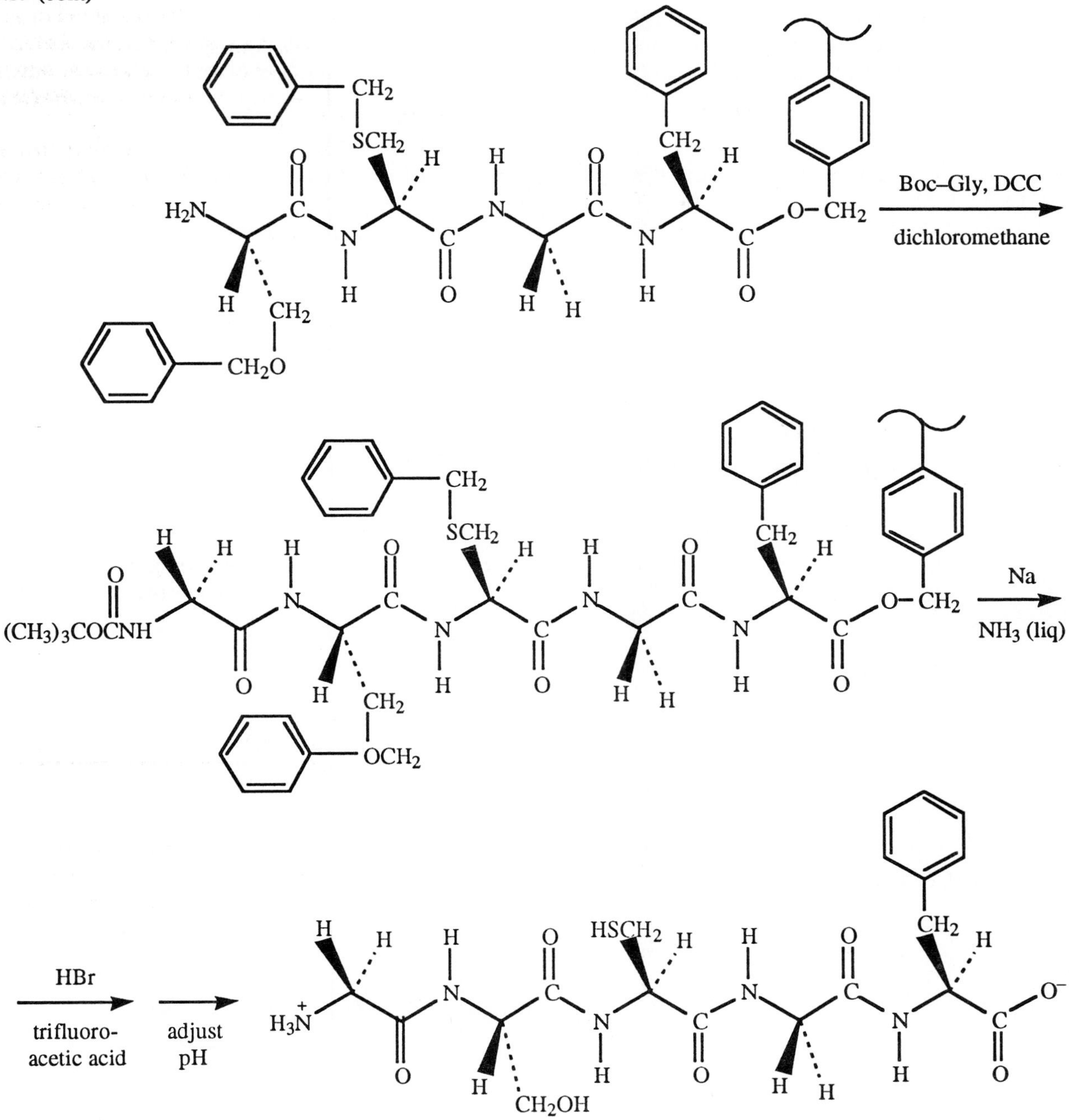

$\xrightarrow[\text{dichloromethane}]{\text{Boc–Gly, DCC}}$

$\xrightarrow[\text{NH}_3 \text{ (liq)}]{\text{Na}}$

$\xrightarrow[\substack{\text{trifluoro-} \\ \text{acetic acid}}]{\text{HBr}}$ $\xrightarrow[\text{pH}]{\text{adjust}}$

Gly–Ser–Cys–Gly–Phe

26.40   (a)   β-Carboxyaspartic acid is a malonic acid derivative. Malonic acid is stable when it is in basic solution, but it decarboxylates when it is heated with acid. β-Carboxyaspartic acid, therefore, survives basic hydrolysis of a peptide better than it survives acid-catalyzed hydrolysis. The decarboxylation product of β-carboxyaspartic acid is aspartic acid, which is then seen as the product of acid-catalyzed hydrolysis.

(b)

26.40 (b) (cont)

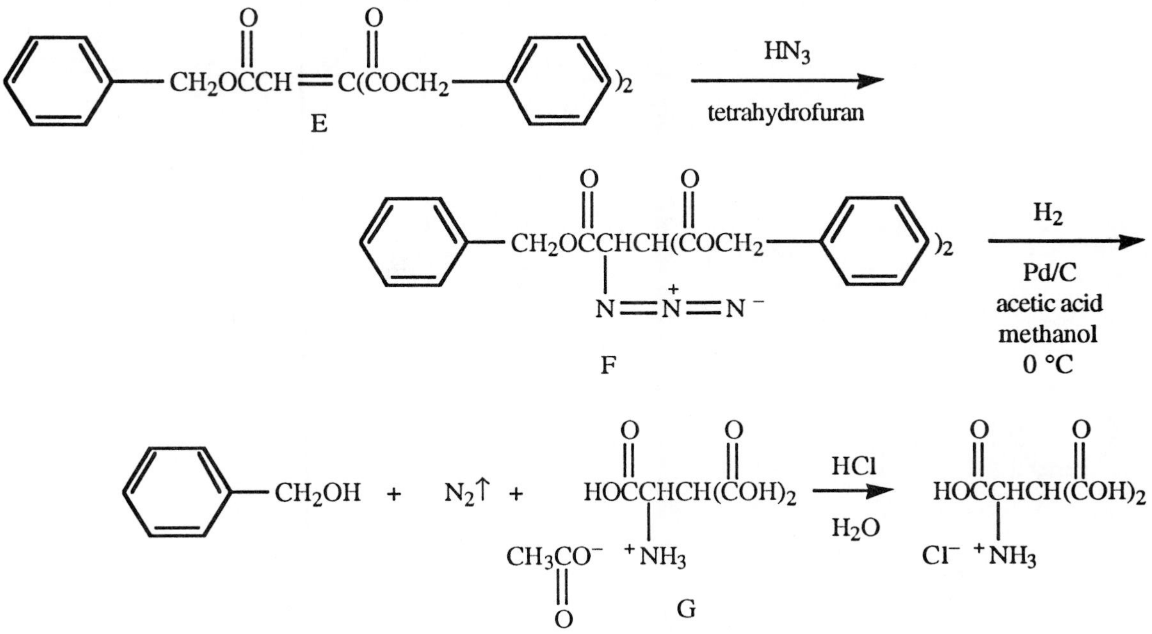

E

HN$_3$
tetrahydrofuran

F

H$_2$
Pd/C
acetic acid
methanol
0 °C

G

HCl
H$_2$O

(c)   Free radical substitution of the benzylic hydrogen atoms of the benzyl ethers can occur with Br$_2$ in the presence of light.

(d)   The ester groups on the carbon-carbon double bond are electron-withdrawing. The acid adds to give the more stable carbocation, the one with the positive charge on the carbon atom with only one ester group.

26.41

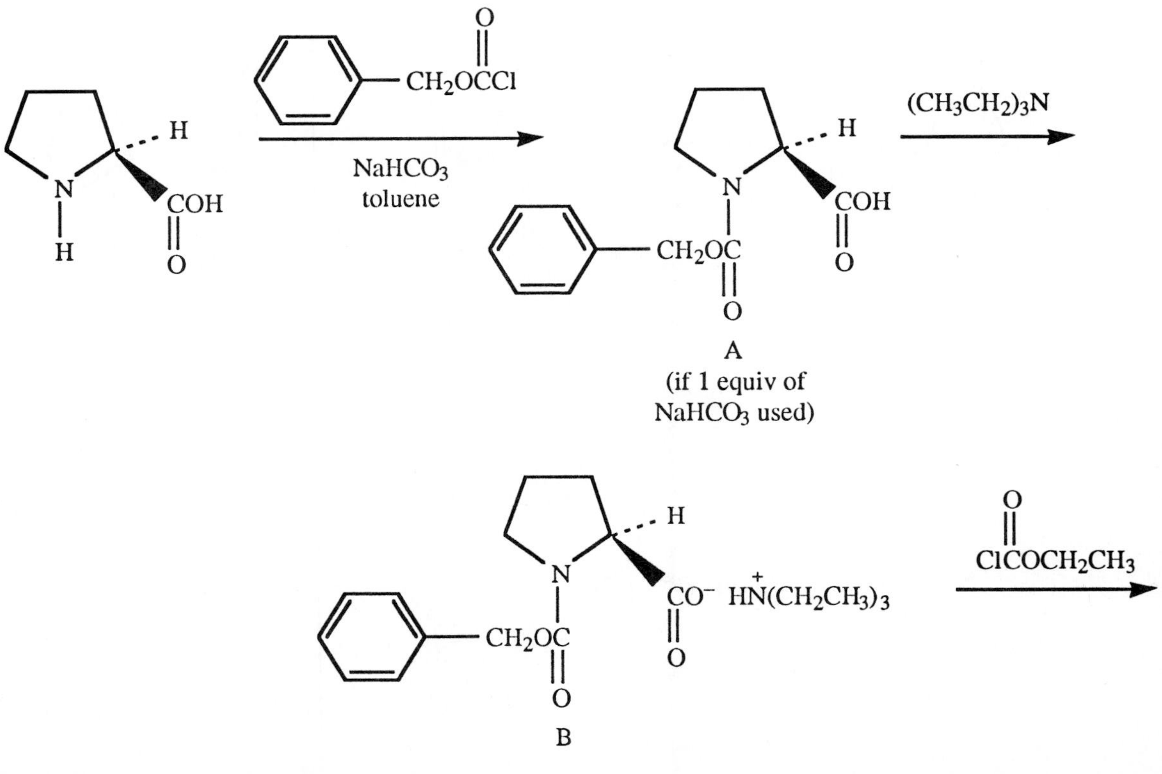

(CH$_3$CH$_2$)$_3$N

A
(if 1 equiv of
NaHCO$_3$ used)

B

O
‖
ClCOCH$_2$CH$_3$

26.41 (cont)

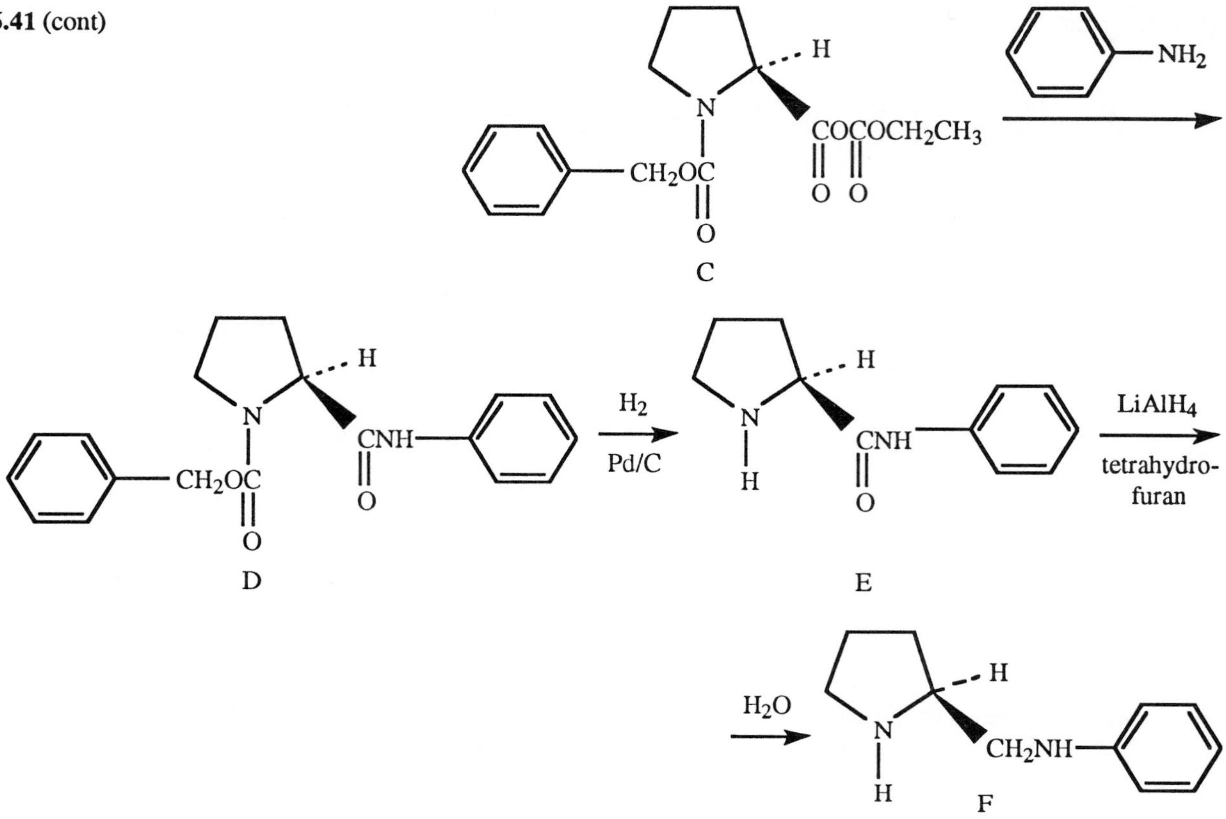

26.42

(a)

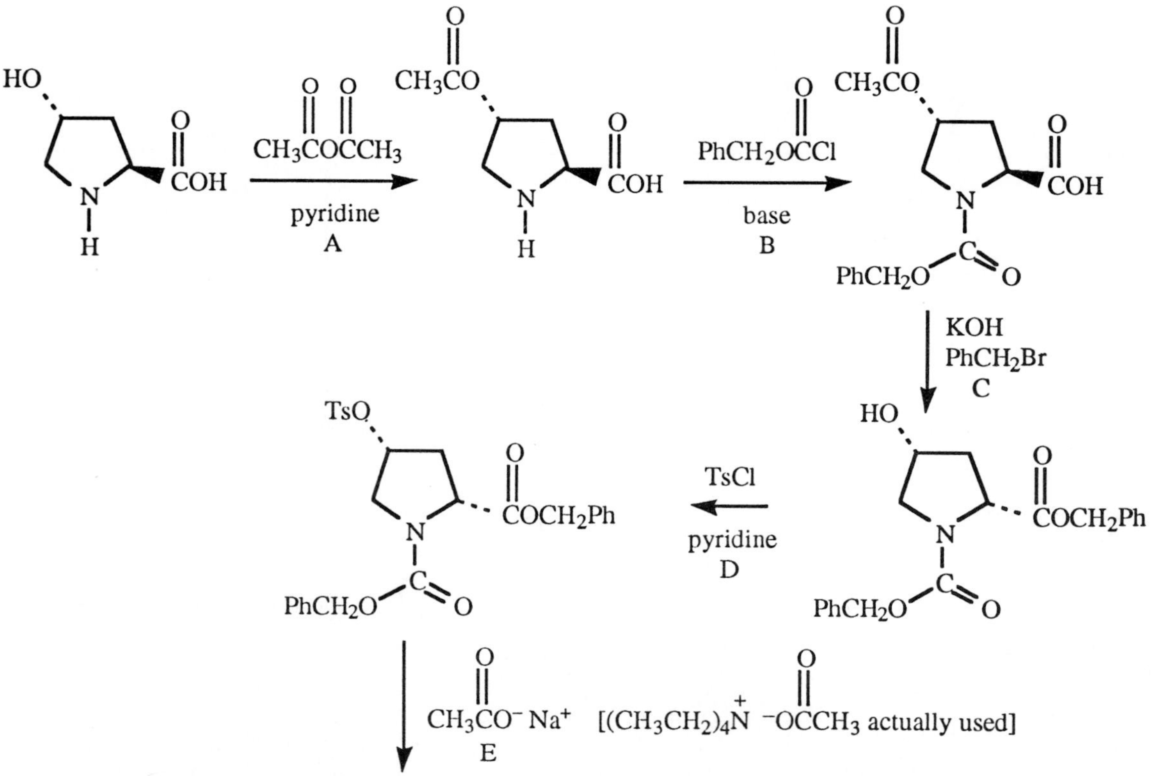

26.42 (a) (cont)

(b)   Enolization at the α-carbon atom of the amino acid in the strongly basic reaction mixture results in loss of stereochemistry. Reprotonation gives the other isomer.

26.42 (b) (cont)

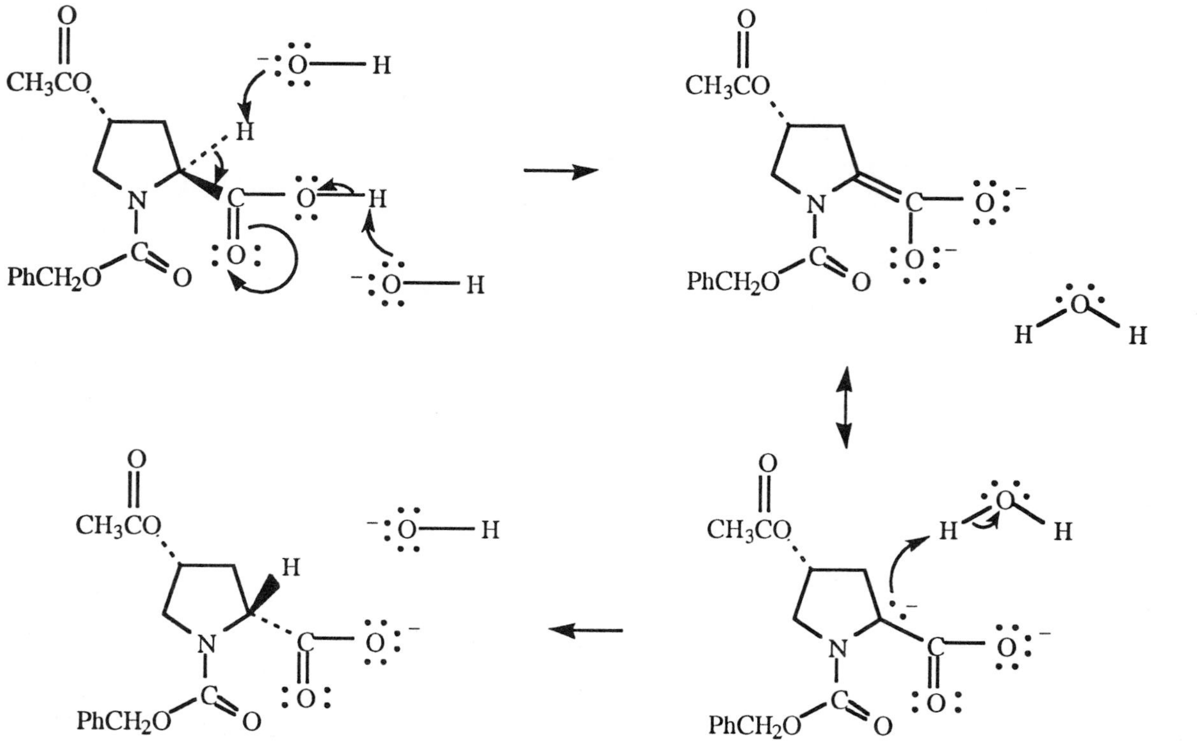

26.43

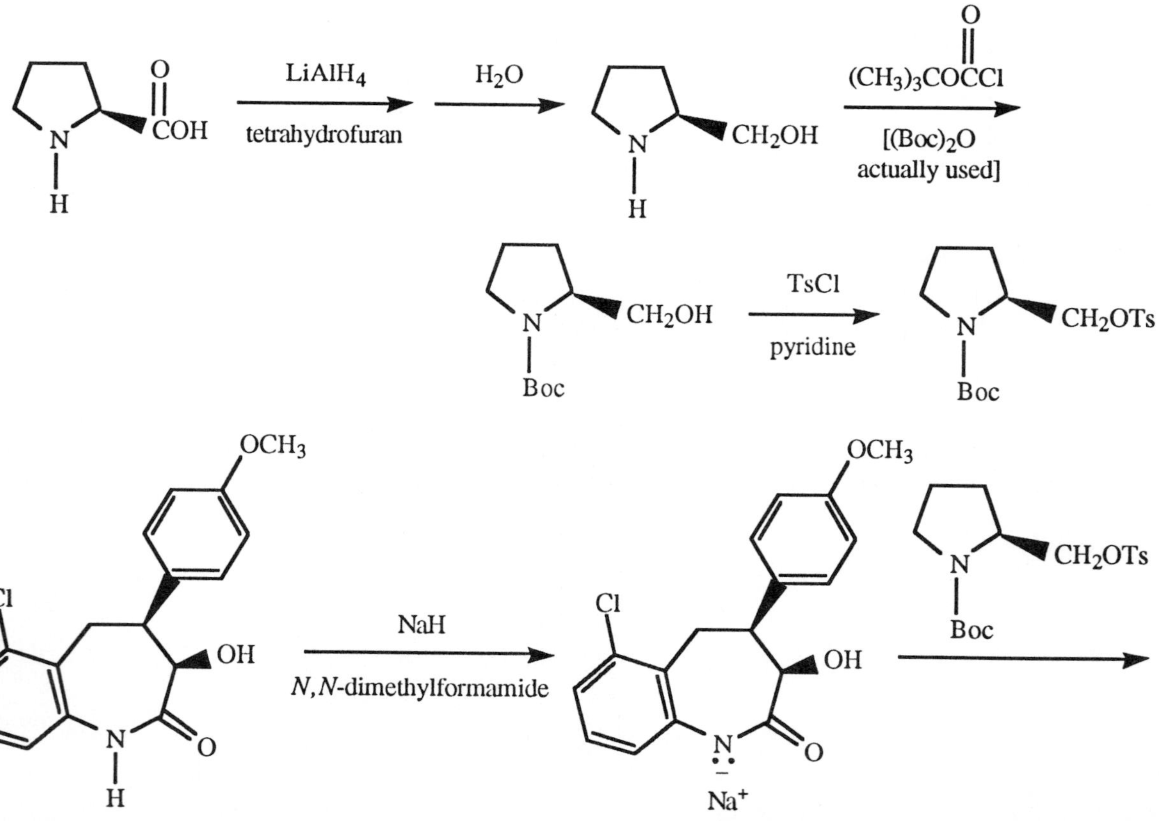

26.43 (cont)

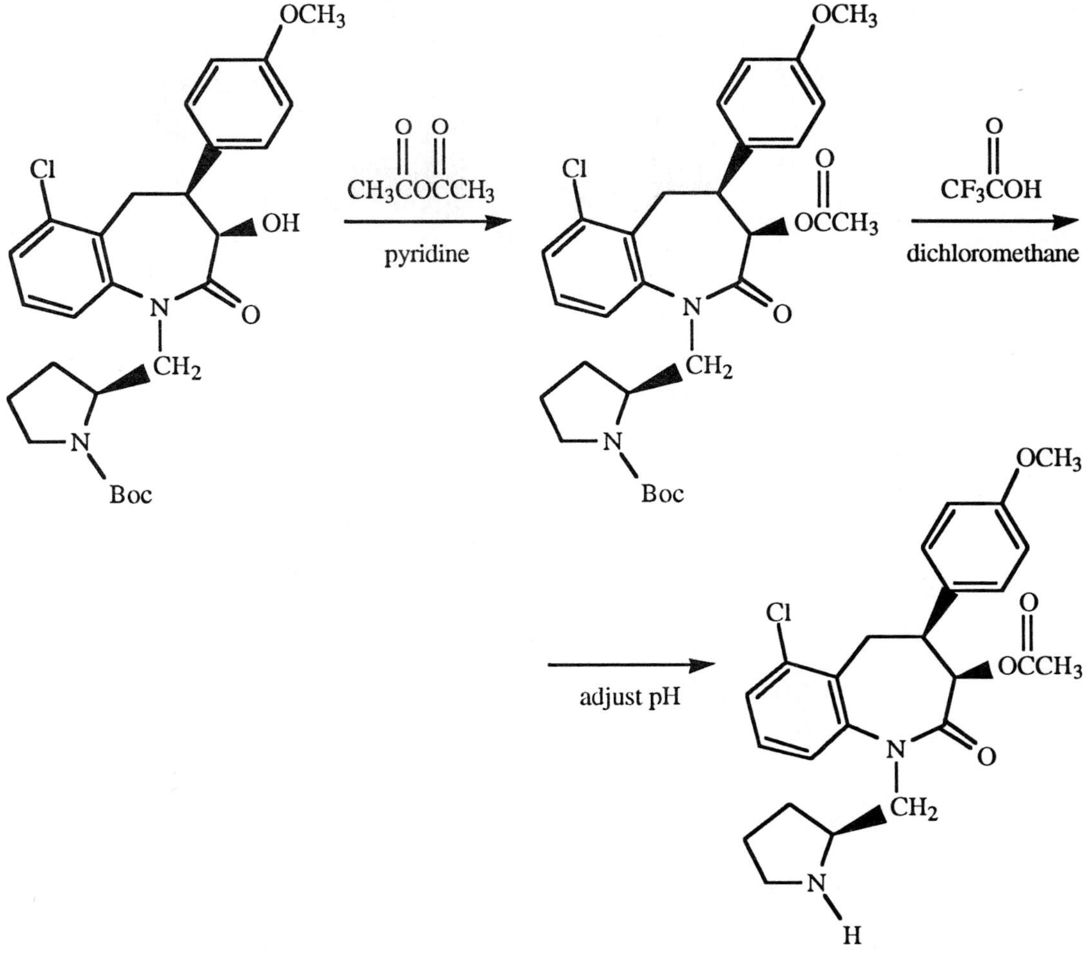

26.44

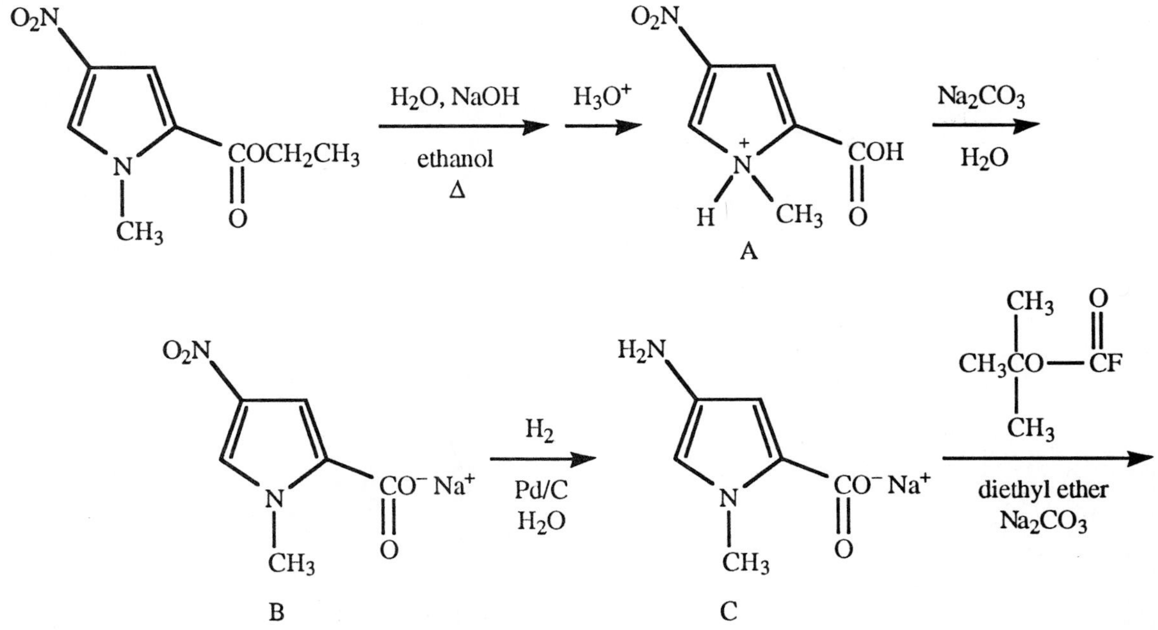

26.44 (cont)

Structure D, with reagent H₃O⁺ to pH 3 – 4, gives structure E, with reagent Cs₂CO₃, ethanol, H₂O.

Structure F, with reagent C₆H₅CH₂Br in dimethylformamide, gives structure G.

Structure G, with reagent CF₃COOH in dichloromethane, gives structure H, which with E, a carbodiimide, dimethylformamide, gives structure I.

Structure I, with reagent CF₃COOH in dichloromethane.

26.44 (cont)

26.44 (cont)

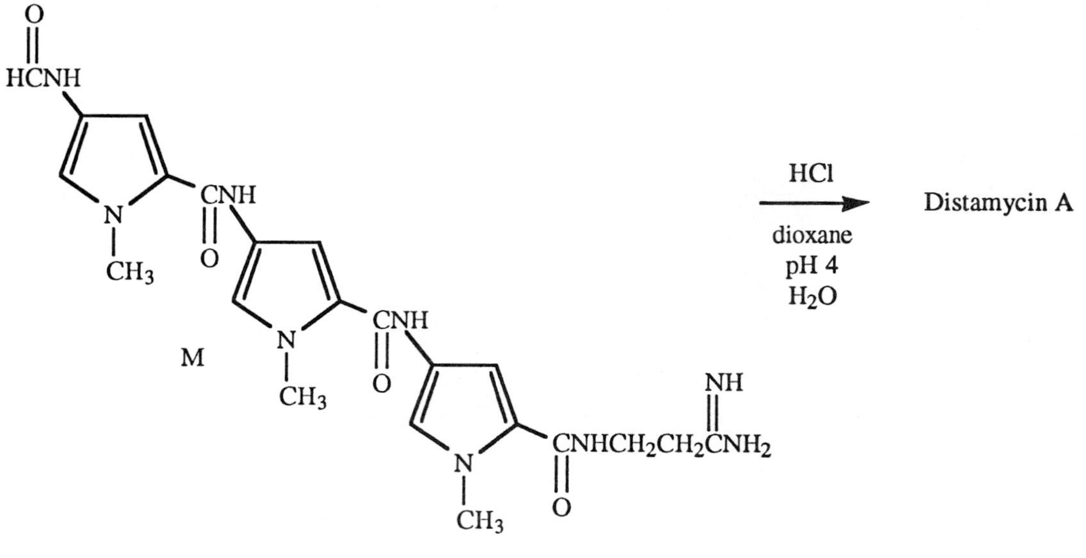

$$\text{M} \xrightarrow[\substack{\text{dioxane} \\ \text{pH 4} \\ \text{H}_2\text{O}}]{\text{HCl}} \text{Distamycin A}$$

27

Macromolecular Chemistry

Concept Map 27.1   Polymers.

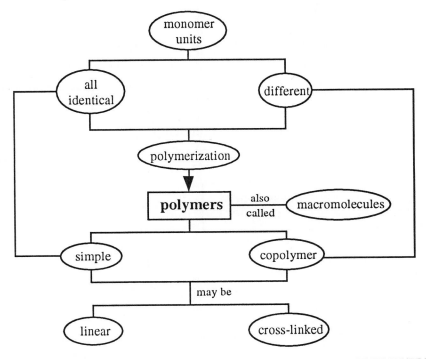

Concept Map 27.2   (see p. 1026)

27.1   Formation of the initiator:

$$CH_3(CH_2)_{10}C \overset{:O:}{\overset{\|}{}} \!\!\!-\!\! \ddot{O} \!\!-\!\! \ddot{O} \!\!-\!\! C(CH_2)_{10}CH_3 \quad \xrightarrow{\Delta} \quad 2\,CH_3(CH_2)_{10}C \overset{:O:}{\overset{\|}{}} \!\!\!-\!\! \ddot{O} \cdot$$

Initiation step:

$$CH_3(CH_2)_{10}C \overset{:O:}{\overset{\|}{}} \!\!\!-\!\! \ddot{O} \cdot \quad CH_2 \!\!=\!\! CH\ddot{Cl}: \quad \longrightarrow \quad CH_3(CH_2)_{10}C \overset{:O:}{\overset{\|}{}} \!\!\!-\!\! \ddot{O} \!\!-\!\! CH_2CH\ddot{Cl}:$$

1025

27.1 (cont)

Chain-propagation step:

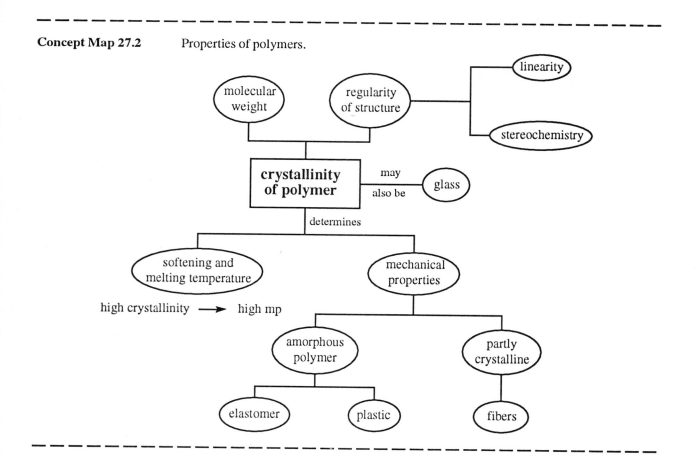

- -

Concept Map 27.2   Properties of polymers.

- -

27.2   (a) $HSCH_2COH$ (with O double-bonded to C)

(b) C_6H_5—$CH{=}CH_2$

(c) $CH_2{=}CHCH_3$

(d) $O{=}C{=}N$—C_6H_4—$N{=}C{=}O$ + $HOCH_2CH_2OH$

an isocyanate

(e) $CH_2{=}C(CH_3)COCH_3$ (with O double-bonded to C)

27.3   (a) $CH_2\!=\!\overset{\overset{\displaystyle CH_3}{|}}{C}$ ─⟨phenyl⟩   and $N\!\equiv\!CCH\!=\!CHC\!\equiv\!N$; a copolymer

(b) $CH_2\!=\!CCl_2$; a simple polymer

(c)   ⟨phenyl⟩─$CH\!=\!CH$─⟨phenyl⟩   and ⟨maleic anhydride⟩ ; a copolymer

(d) $CH_2\!=\!\overset{\overset{\displaystyle CH_3}{|}}{C}$ ─⟨phenyl⟩   and $CH_2\!=\!\overset{\overset{\displaystyle CH_3}{|}}{C}C\!\equiv\!N$; a copolymer

27.4 $\left[\!CH_2\!-\!CH\!-\!CH_2\!-\!CH\!-\!CH_2\!-\!CH\!\right]_n$
(with phenyl, phenyl (para), phenyl substituents)

$\xrightarrow[\Delta]{\substack{SO_3 \\ H_2SO_4}}$

$\left[\!CH_2\!-\!CH\!-\!CH_2\!-\!CH\!-\!CH_2\!-\!CH\!\right]_n$
(with phenyl-SO_3H, phenyl, phenyl substituents)

$\left[\!CH_2\!-\!CH\!-\!CH_2\!-\!CH\!-\!CH_2\!-\!CH\!\right]_m$
(with phenyl, phenyl substituents)

$\left[\!CH_2\!-\!CH\!-\!CH_2\!-\!CH\!-\!CH_2\!-\!CH\!\right]_m$
(with phenyl, phenyl-SO_3H substituents)

27.5

(a)   ⟨thiepanone (7-membered ring with S and C=O)⟩ $\xrightarrow{(CH_3)_3COK}$ $K^+ \;^-S(CH_2)_5\overset{\overset{\displaystyle O}{\|}}{C}S(CH_2)_5\overset{\overset{\displaystyle O}{\|}}{C}O\overset{\overset{\displaystyle CH_3}{|}}{\underset{\underset{\displaystyle CH_3}{|}}{C}}CH_3$ $\longrightarrow$ polymer

(b)   ⟨phenyl⟩─$CH\!=\!CH_2$ $\xrightarrow{\qquad\qquad}$

$\left[\text{naphthalene}\right]^{\overset{\displaystyle \cdot}{-}} Na^+$

tetrahydrofuran
$-80\ °C$

27.5 (b) (cont)

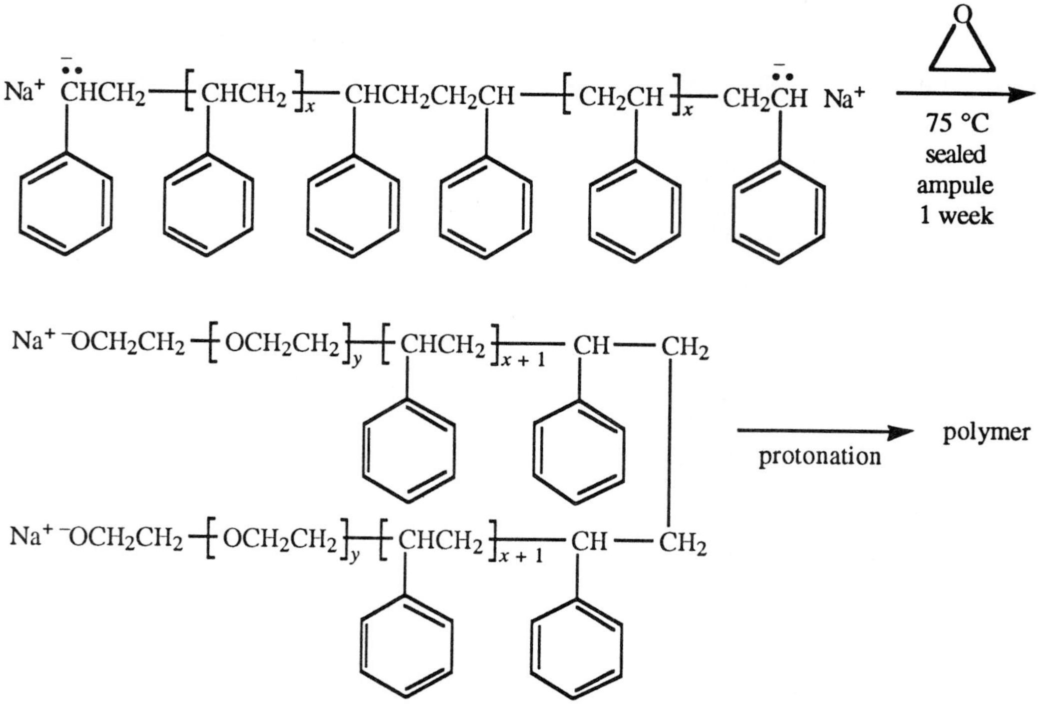

(c)

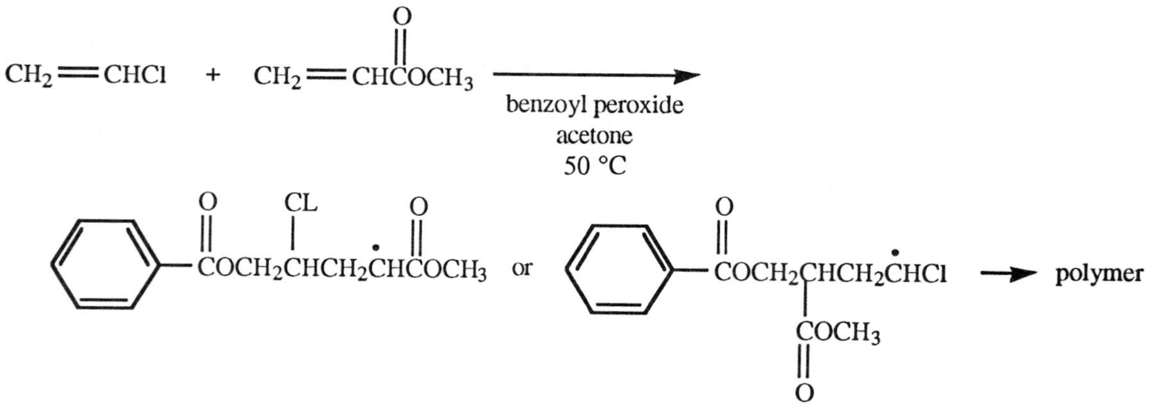

(d)

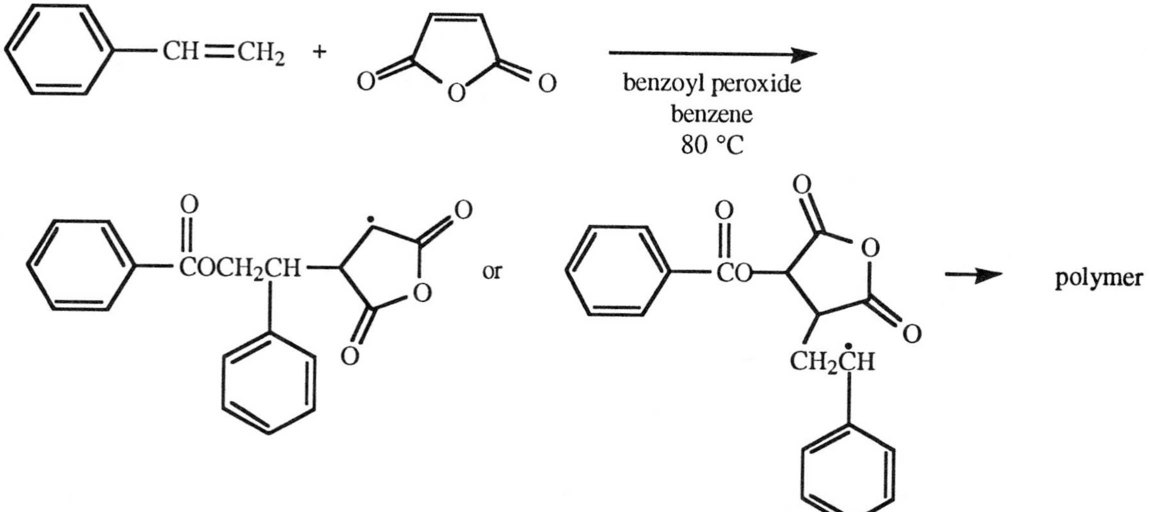

27.5 (cont)

(e)

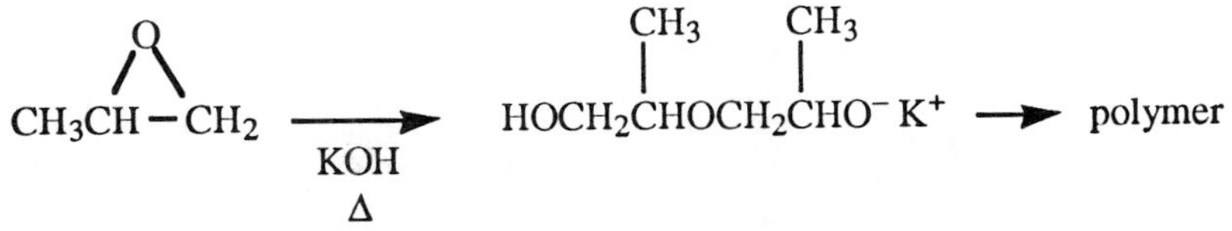

This is an S_N2 substitution reaction. The nucleophile reacts at the less hindered carbon atom of the oxirane.

27.6

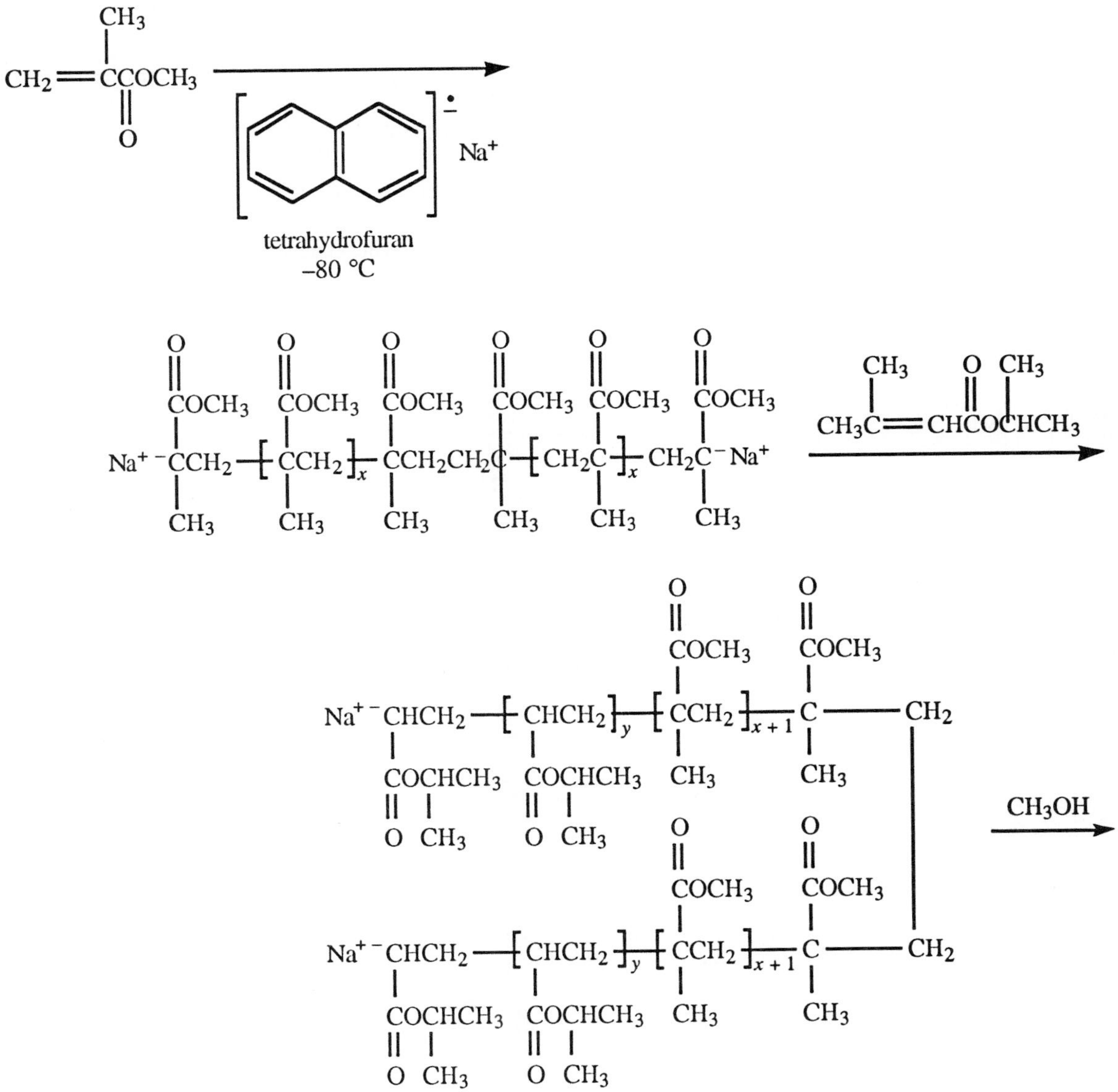

27.6 (cont)

$$CH_2CH_2 {-}{[}CHCH_2{]}_y{-}{[}CCH_2{]}_{x+1}{-}C{-}CH_2$$

(chain with pendant groups: $COCHCH_3$ / $O\ CH_3$; $COCHCH_3$ / $O\ CH_3$; CH_3; $COCH_3$ top and CH_3 bottom, $COCH_3$; two repeating chains shown cross-linked at CH_2)

A block copolymer can also be made by starting with isopropyl acrylate and then adding methyl methacrylate.

27.7 $AlCl_3 + CH_3C{-}Cl: \longrightarrow CH_3C{\equiv}\overset{+}{O}: + {^-}AlCl_4$

initiation

(tetrahydrofuran oxygen attacks $CH_3C{\equiv}\overset{+}{O}:$ giving oxocarbenium intermediate)

$$\overset{+}{O}{-}CCH_3 \longrightarrow \overset{+}{O}{-}(CH_2)_4{-}O{-}CCH_3$$

propagation

$$\overset{+}{O}{-}(CH_2)_4{-}O{-}(CH_2)_4{-}O{-}CCH_3$$

polymer $\longleftarrow$

27.8

(a)   (thiirane) $\xrightarrow{BF_3}$ $\overset{+}{S}{-}\overset{-}{B}F_3$ $\xrightarrow{\text{(thiirane)}}$ $\overset{+}{S}{-}CH_2CH_2\overset{-}{S}BF_3 \longrightarrow$ polymer

27.8 (cont)

(b)

$$\underset{\text{0.1 M HCl}}{\overset{\text{H}}{\underset{\Delta,\,2\text{ h}}{\longrightarrow}}}$$

(Nu represents the conjugate base of the acid initiator.)

(c) acid → $HO(CH_2)_4\overset{O}{\overset{\|}{C}}-Nu$ → $HO(CH_2)_4\overset{O}{\overset{\|}{C}}O(CH_2)_4\overset{O}{\overset{\|}{C}}-Nu$ → polymer

(Nu represents the conjugate base of the acid initiator.)

(d) base → $^-O(CH_2)_4\overset{O}{\overset{\|}{C}}-B$ → $^-O(CH_2)_4\overset{O}{\overset{\|}{C}}O(CH_2)_4\overset{O}{\overset{\|}{C}}-B$ → polymer

(B represents the basic initiator.)

(e)
solid state
(trace of acid
usually present)
→ $HOCH_2CH_2O\overset{O}{\overset{\|}{C}}-\overset{O}{\overset{\|}{C}}OH$ →

$HOCH_2CH_2O\overset{O}{\overset{\|}{C}}-\overset{O}{\overset{\|}{C}}OCH_2CH_2O\overset{O}{\overset{\|}{C}}-\overset{O}{\overset{\|}{C}}OH$ → → polymer

(f) $\underset{\underset{\text{H}_2\text{O (trace)}}{BF_3}}{\overset{\overset{\text{CH}_3}{|}}{CH_3C}=CH_2}$ → $\overset{\overset{\text{CH}_3}{|}}{\underset{+}{CH_3CCH_3}}$ $\overset{\overset{\text{CH}_3}{|}}{CH_3C=CH_2}$ → $\overset{\overset{\text{CH}_3\ \ \text{CH}_3}{|\ \ \ \ \ |}}{\underset{\underset{CH_3}{|}}{\underset{+}{CH_3CCH_2CCH_3}}}$ → polymer

27.9 $HO-\!\!\!\bigcirc\!\!\!-CH_2CH_2-\!\!\!\bigcirc\!\!\!-OH$ + $BrCH_2(CH_2)_6CH_2Br$ $\overset{NaOH}{\longrightarrow}$

1,2-bis(4-hydroxyphenyl)ethane   1,8-dibromooctane

$\left[\!\!-O-\!\!\!\bigcirc\!\!\!-CH_2CH_2-\!\!\!\bigcirc\!\!\!-OCH_2(CH_2)_6CH_2-\!\!\right]_n$

27.10

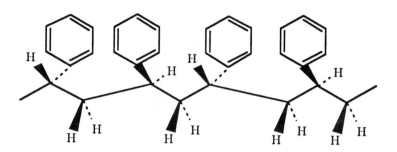

isotactic polystyrene

syndiotactic polystyrene

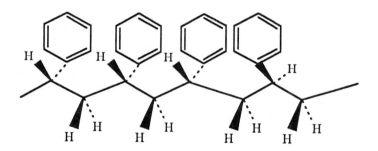

atactic polystyrene

fragment of the chain, showing only one of
several possible orientations of the phenyl
groups with respect to each other

27.11   In order for a polymer to be highly crystalline, it has to be highly ordered so it can pack easily into a crystal
structure. Isotactic and syndiotactic polymers, with their very regular primary structures, have strong
interactions between chains and form crystals. Atactic polymers, with their much less regular structures, do not
have as good interactions between chains and are more amorphous.

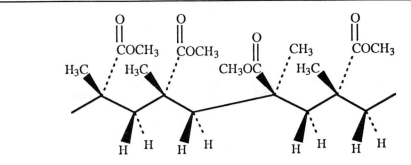

27.12

atactic poly(methyl methacrylate)

fragment of chain showing one of several possible
arrangements of the ester groups on the chain

Concept Map 27.3   Stereochemistry of polymers.

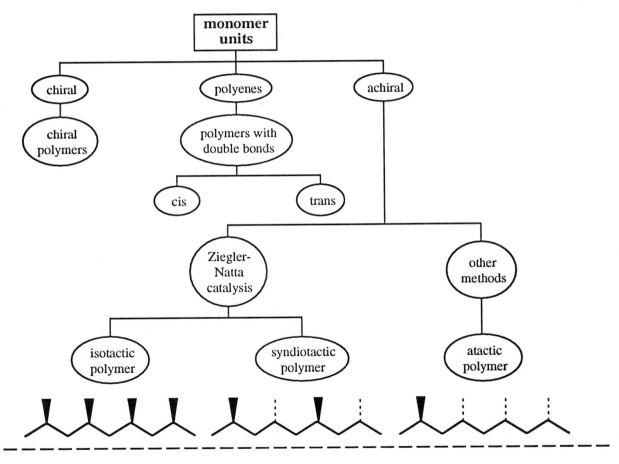

27.13

(a)

$$CH_3CH_2CHCH{=}CH_2 \xrightarrow[\substack{TiCl_4 \\ (CH_3CH_2)_2AlCl}]{} \xrightarrow{CH_3CH_2OH}$$

with CH₃ substituent on the CH

27.13 (a) (cont)

(b) $CH_2\!\!=\!\!CHOCH_3$ $\xrightarrow[\substack{VCl_3 \\ CH_3 \\ | \\ (CH_3CHCH_2)_3Al \\ heptane}]{}$ $\xrightarrow{CH_3CH_2OH}$

(c) $CH_2\!\!=\!\!CH_2$ + $\xrightarrow[\substack{VCl_3 \\ CH_3 \\ | \\ (CH_3CHCH_2)_2AlCl \\ heptane \\ -30\ °C}]{}$ $\xrightarrow{CH_3CH_2OH}$

(d) $CH_2\!\!=\!\!\overset{\overset{\displaystyle CH_3}{|}}{\underset{\underset{\displaystyle O}{||}}{C}}COCH_3$ $\xrightarrow[\substack{VCl_3 \\ CH_3 \\ | \\ (CH_3CHCH_2)_3Al \\ heptane}]{}$ $\xrightarrow{CH_3CH_2OH}$ $\left[\!CH_2\overset{\overset{\displaystyle CH_3}{|}}{\underset{\underset{\underset{\displaystyle O}{||}}{COCH_3}}{C}}\!\right]_n$

27.13 (cont)

(e)

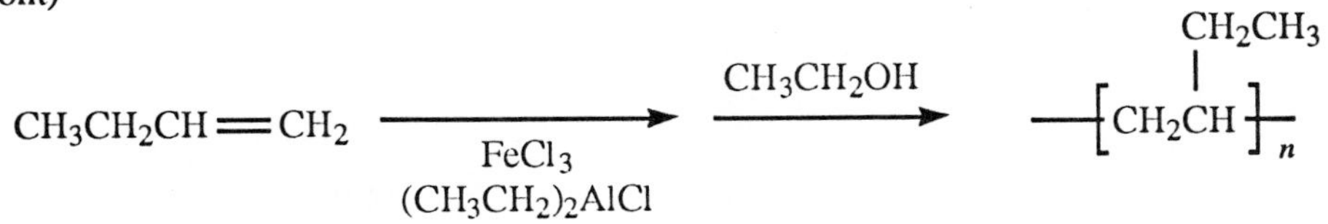

27.14

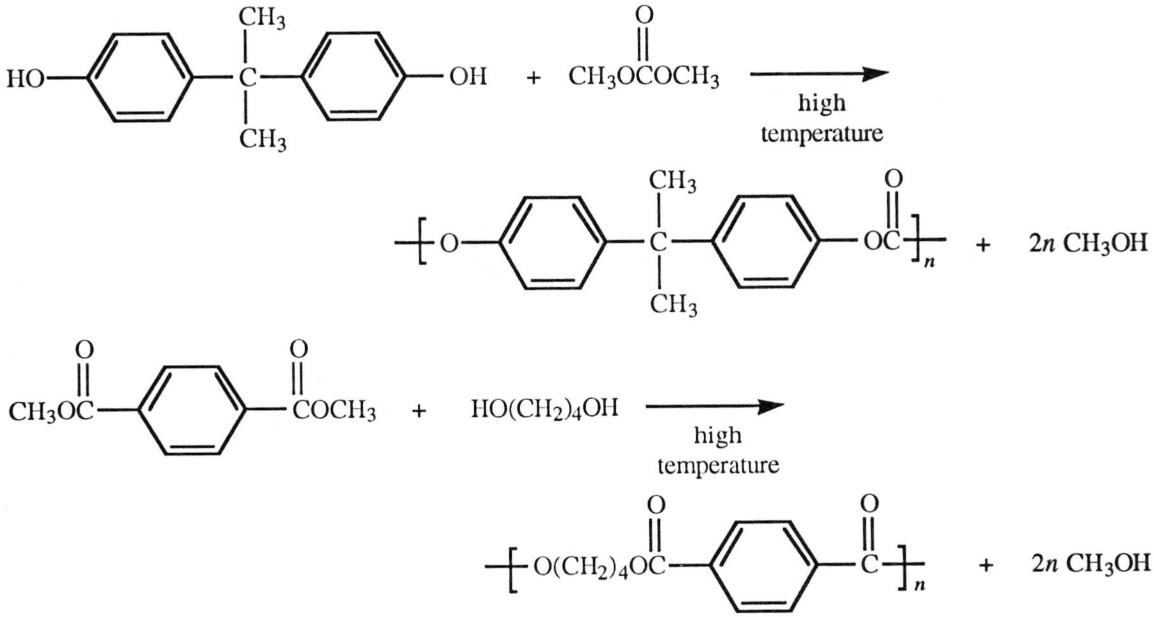

27.15 $H_2N(CH_2)_{10}\overset{O}{\overset{\|}{C}}OH \xrightarrow{\Delta} \left[NH(CH_2)_{10}\overset{O}{\overset{\|}{C}}\right]_n + 2n\ H_2O$

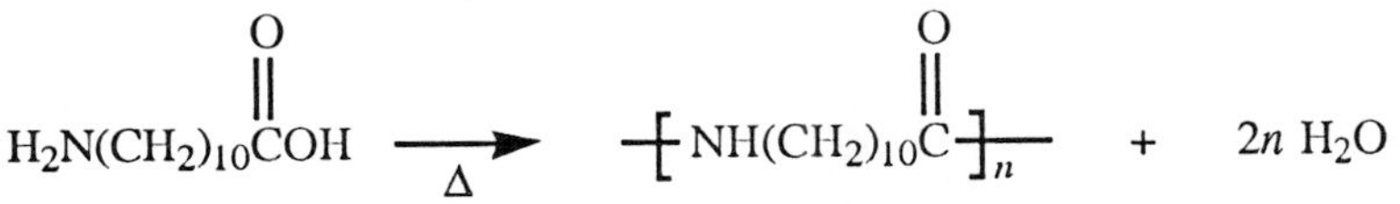

The commonly available lactams contain from four to seven atoms in the ring. The lactam of this monomer would contain twelve atoms in the ring and would not be readily available.

27.16

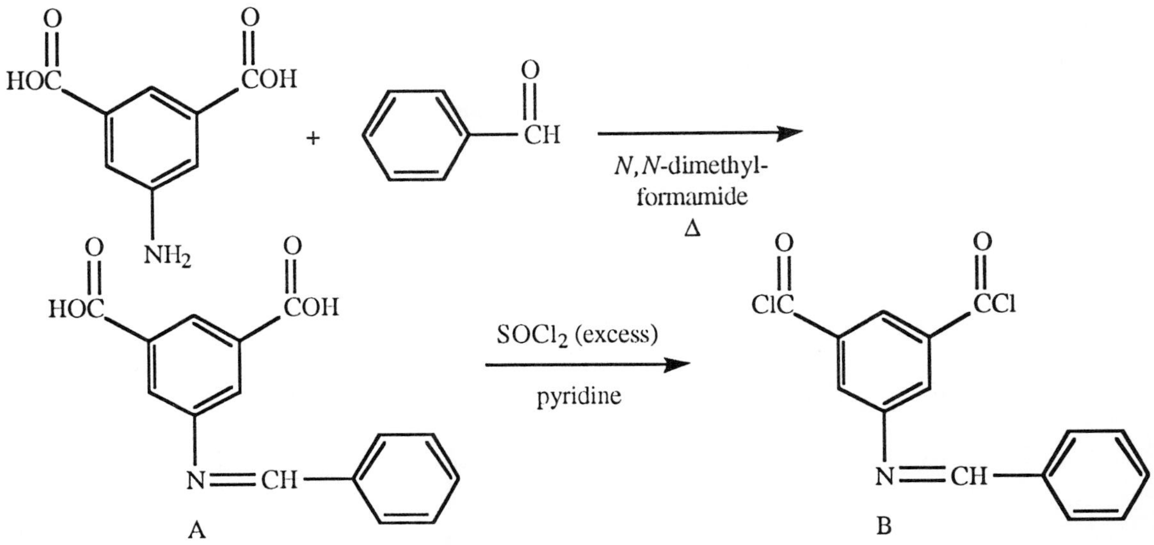

27.16 (cont)

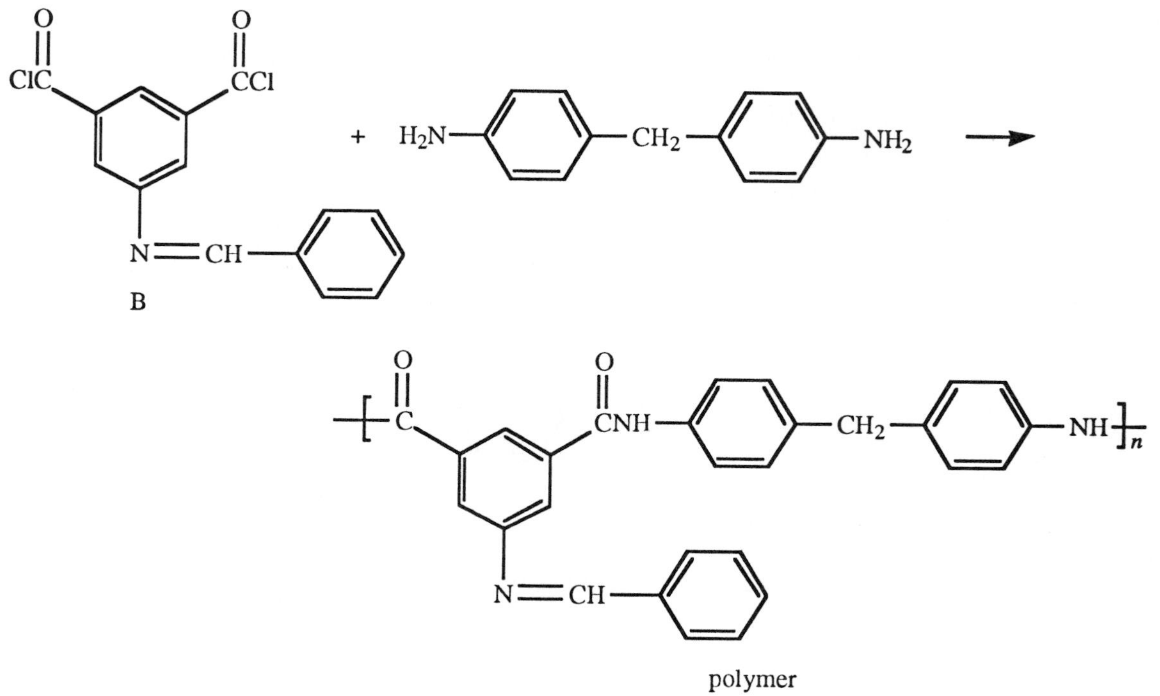

polymer

27.17

(a)

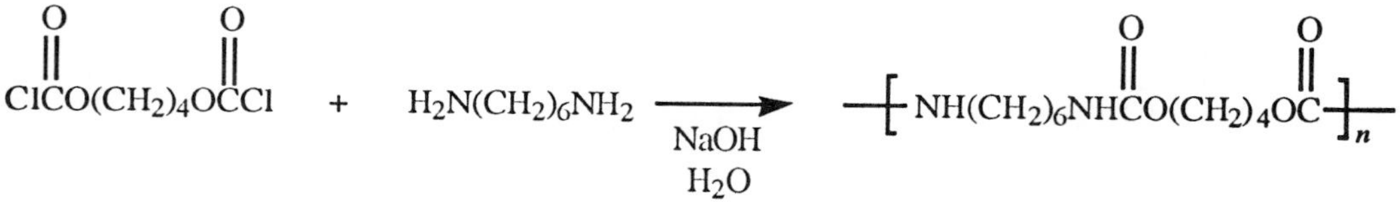

(b)

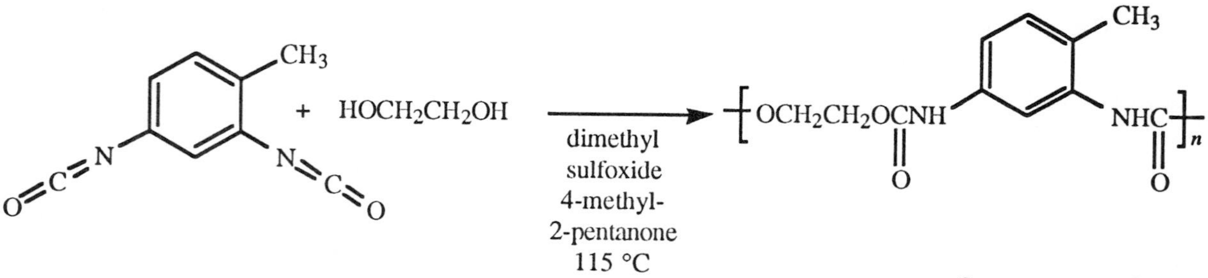

(c)

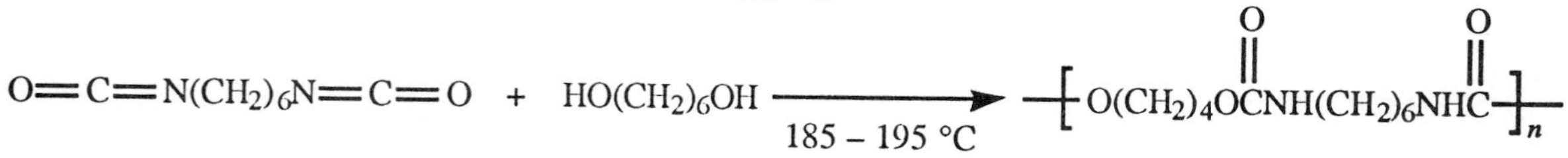

27.18

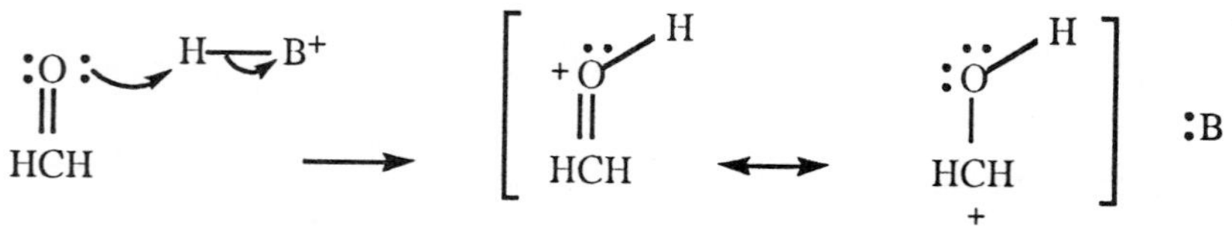

27.18 (cont)

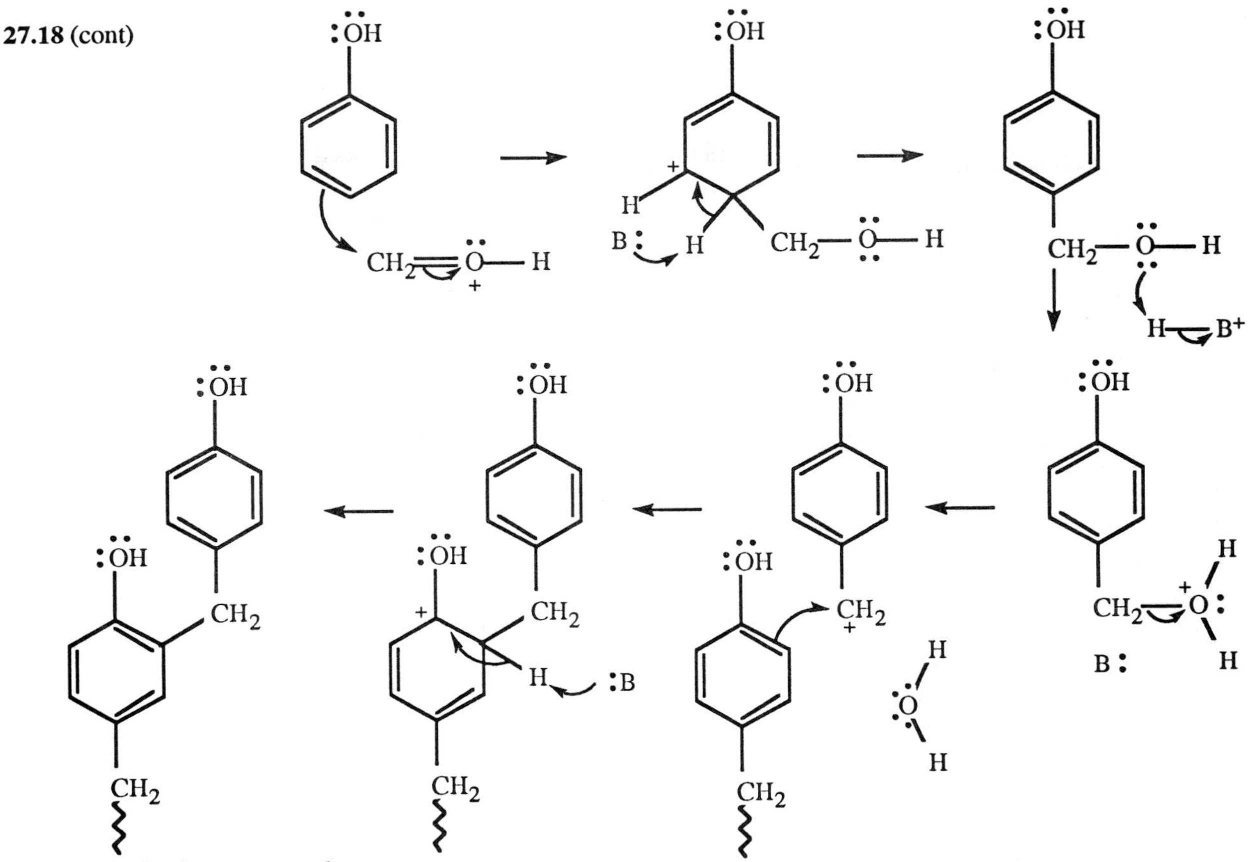

beginning of polymer network

27.19

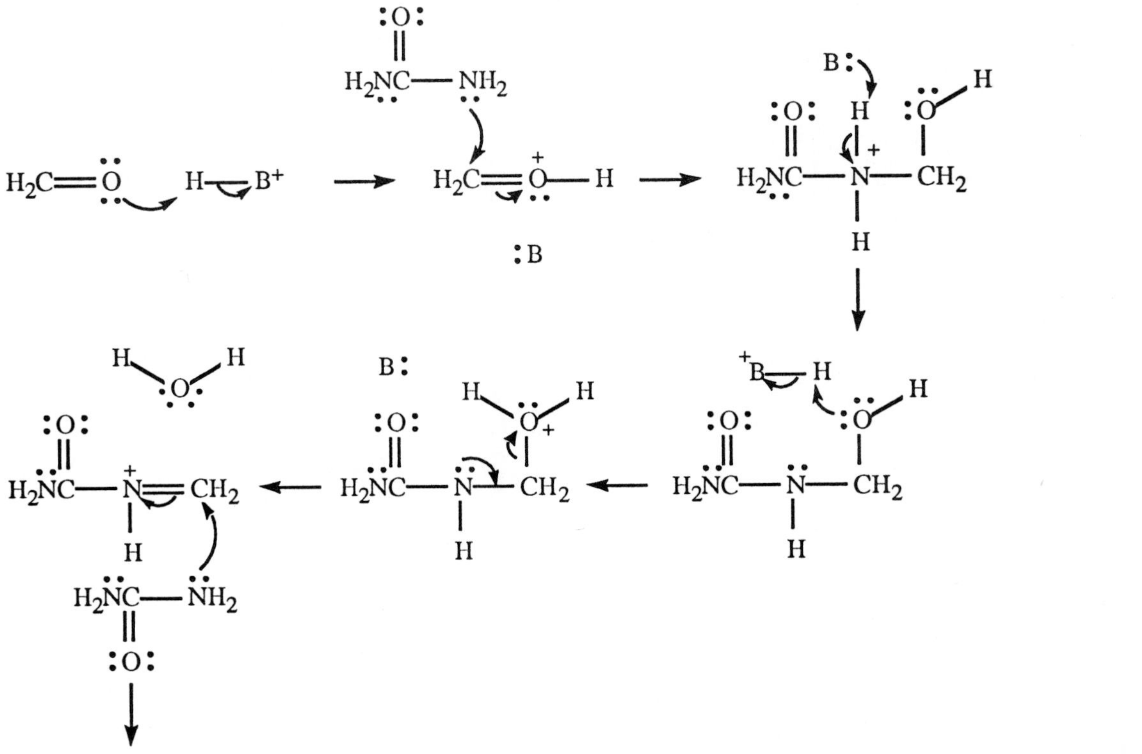

27.19 (cont)

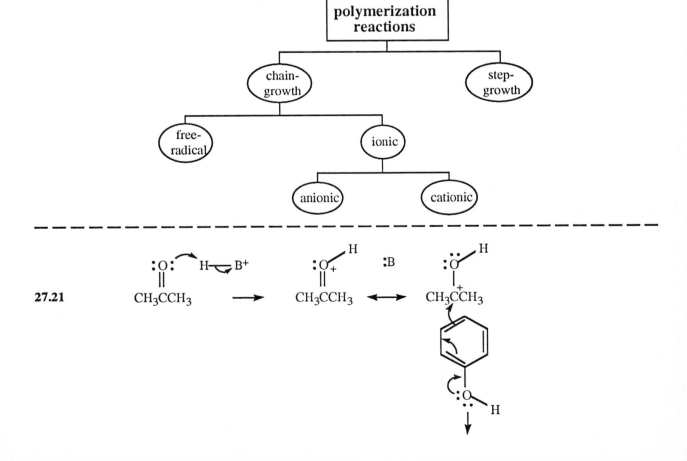

27.20

Concept Map 27.4   Types of polymerization reactions.

27.21

27.21 (cont)

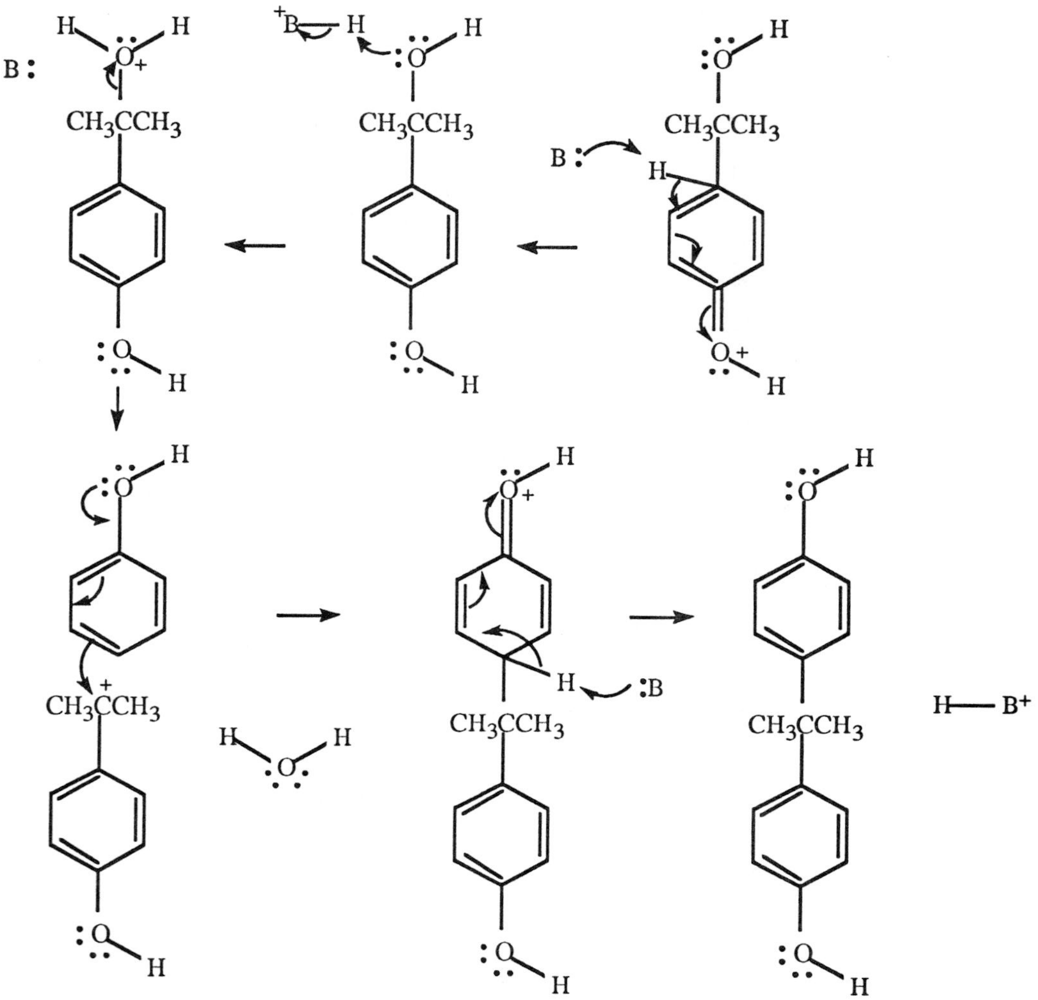

27.22 $CH_3CH = CH_2$ $\xrightarrow[\Delta]{Cl_2\ (g)}$ $ClCH_2CH = CH_2$ $\xrightarrow[H_2O]{Cl_2}$ $ClCH_2\overset{\displaystyle Cl}{\underset{\displaystyle |}{C}}HCH_2OH$

free radical
chlorination
at the allylic
position

electrophilic addition
of chlorine to the
double bond with water
as the electrophile

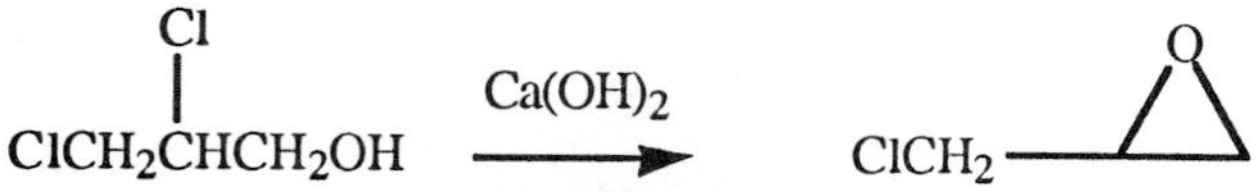

$ClCH_2\overset{\displaystyle Cl}{\underset{\displaystyle |}{C}}HCH_2OH$ $\xrightarrow{Ca(OH)_2}$ $ClCH_2 - \overset{\triangle}{}$ (epoxide)

intramolecular
S_N2 attack by
alkoxide ion

27.23

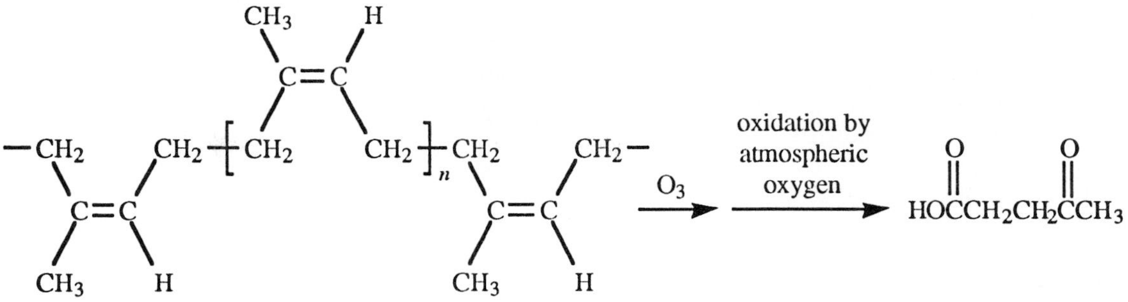

$$\text{polymer chain} \xrightarrow{\text{O}_3} \xrightarrow[\text{oxygen}]{\substack{\text{oxidation by} \\ \text{atmospheric}}} \text{HOCCH}_2\text{CH}_2\text{CCH}_3$$

27.24 $\text{CH}_2\!=\!\text{CHCH}\!=\!\text{CH}_2 \xrightarrow[\substack{\text{TiBr}_4 \\ \text{(CH}_3\text{CH}_2)_3\text{Al}}]{}$

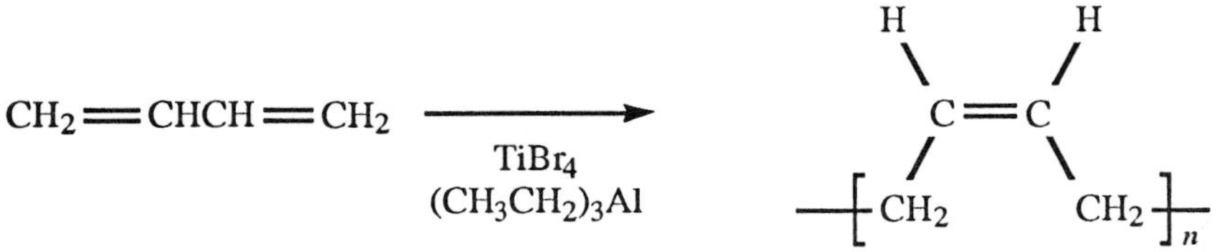

27.25 $\text{CH}_3\text{OCH}\!=\!\text{CHCH}\!=\!\text{CH}_2 \xrightarrow[\substack{\text{free radical} \\ \text{initiator}}]{}$

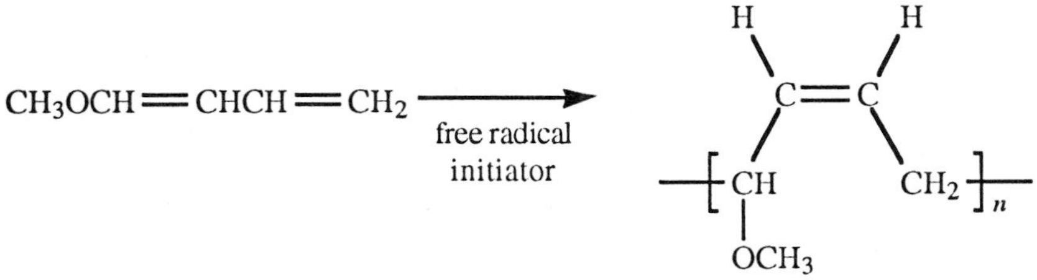

amorphous polymer;
cis double bonds

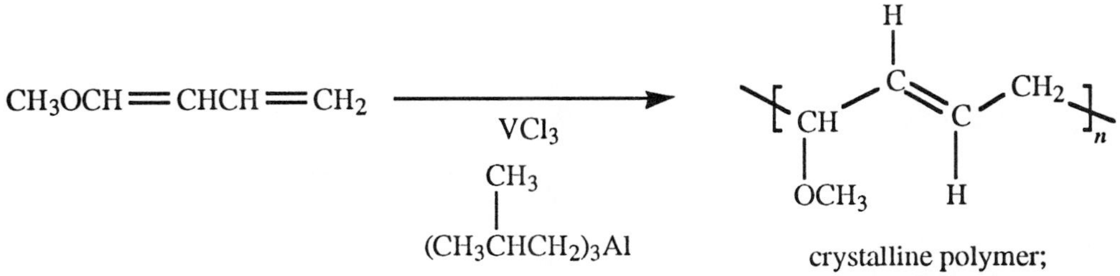

$\text{CH}_3\text{OCH}\!=\!\text{CHCH}\!=\!\text{CH}_2 \xrightarrow[\substack{\text{VCl}_3 \\ \text{(CH}_3\text{CHCH}_2)_3\text{Al}}]{}$

crystalline polymer;
trans double bonds

27.26 $\text{CH}\!=\!\text{CH}_2$ + $\text{CH}\!=\!\text{CH}_2$ + $\text{CH}_2\!=\!\text{CHCH}\!=\!\text{CH}_2 \xrightarrow[\text{temperature}]{\text{high}}$

or

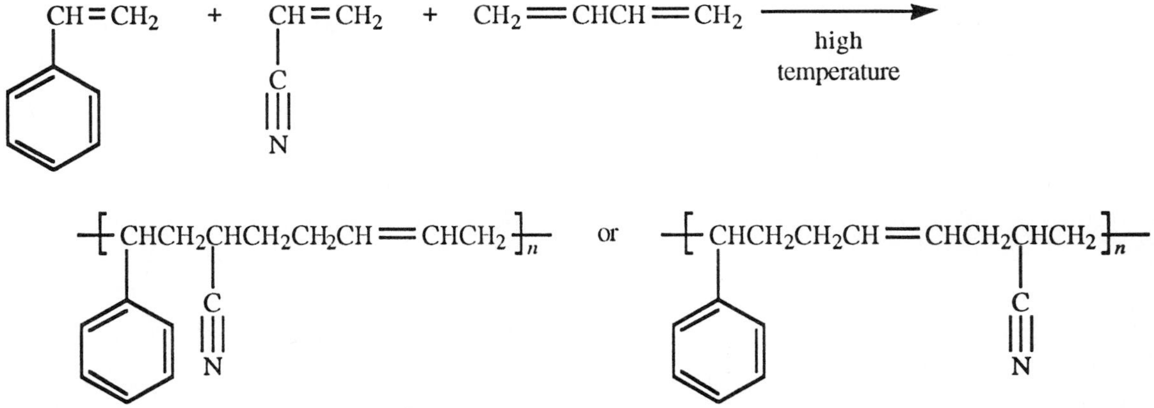

27.26 (cont)

There are many different copolymers that can be written for these three monomers, depending on the order in which they are arranged and on whether 1,2- or 1,4-addition to 1,3-butadiene takes place. Two of the possibilities are shown.

27.27

(a)

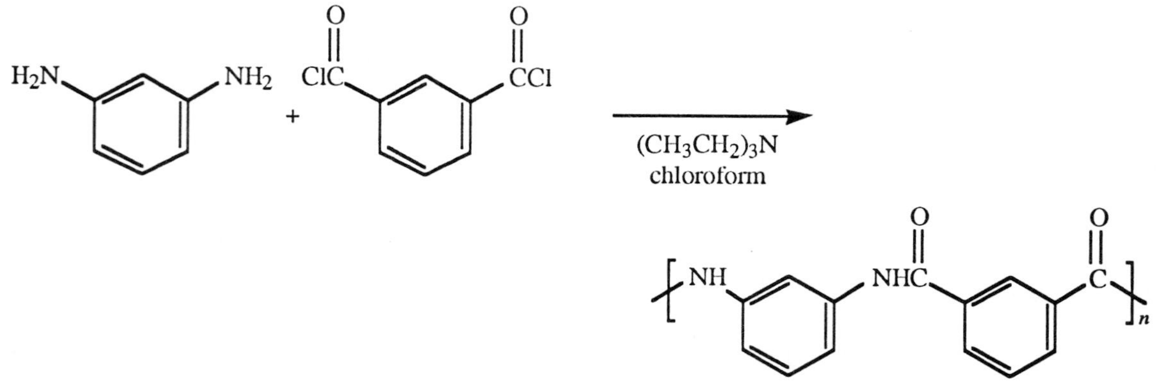

(b) $CH_2{=}CH_2$ + $CH_2{=}CCOCH_3$ $\xrightarrow{\text{peroxide}}$

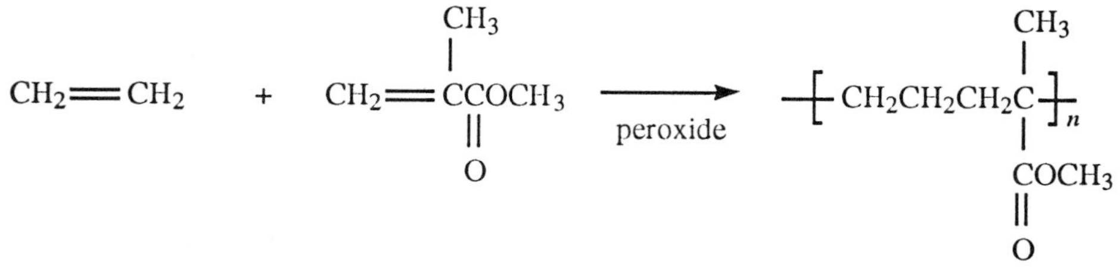

(c)

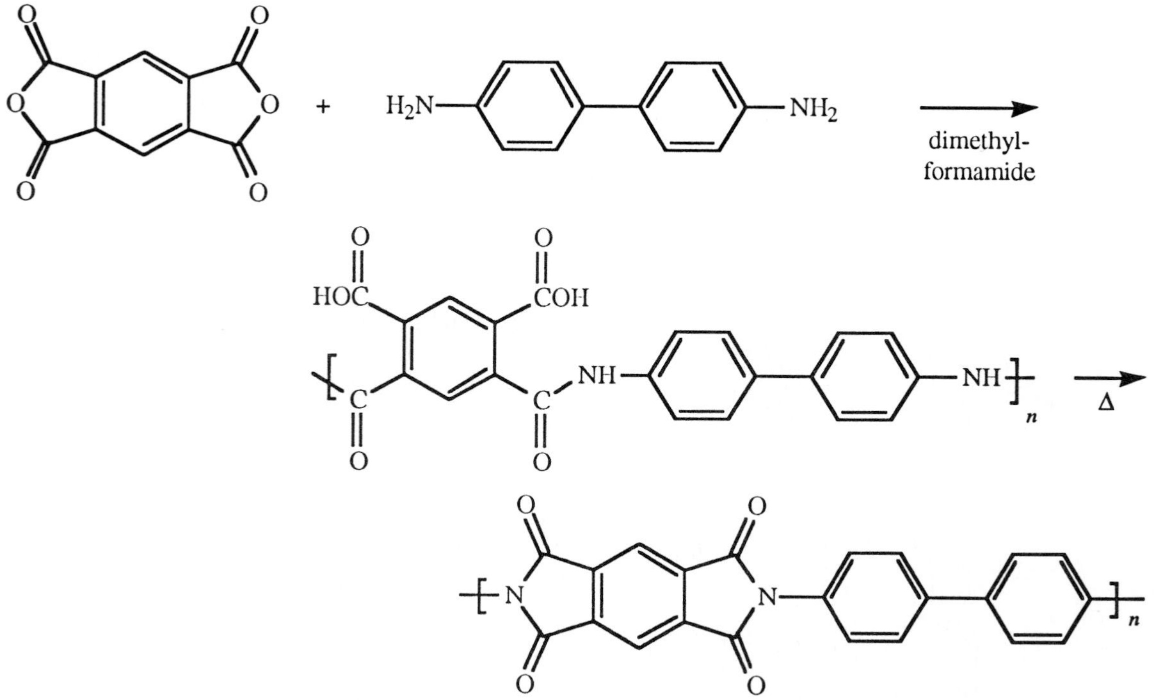

27.27 (cont)

(d)

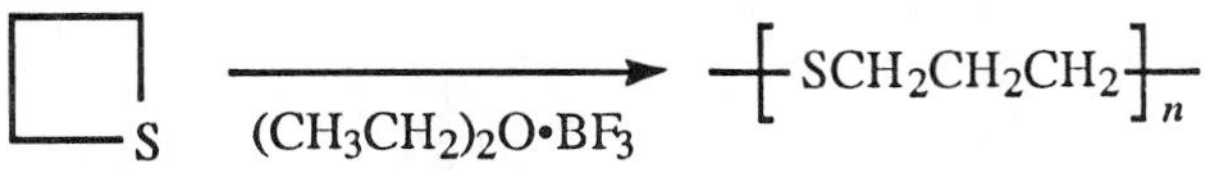

(e)

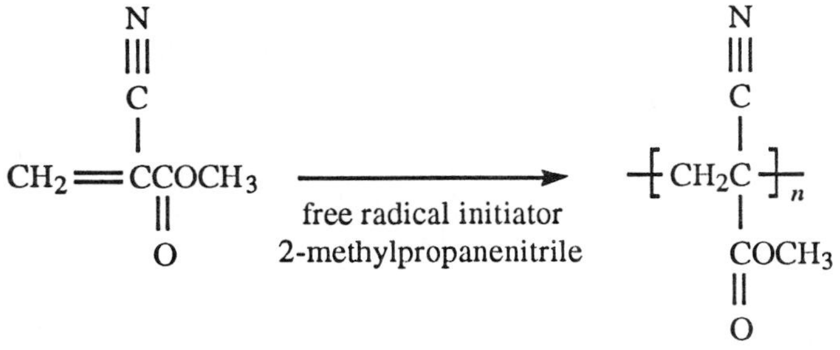

(f)

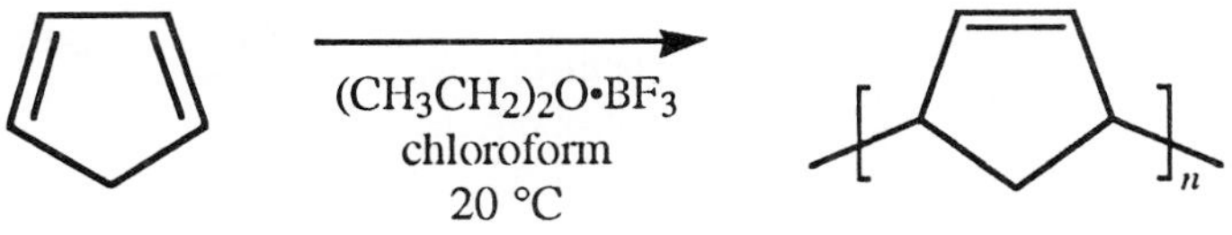

(g)

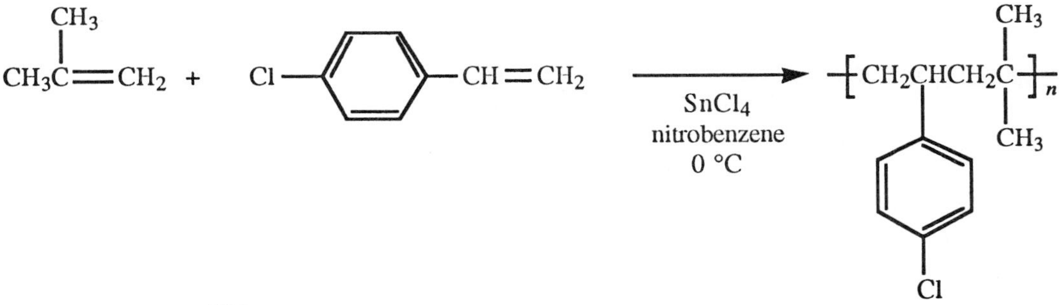

(h)

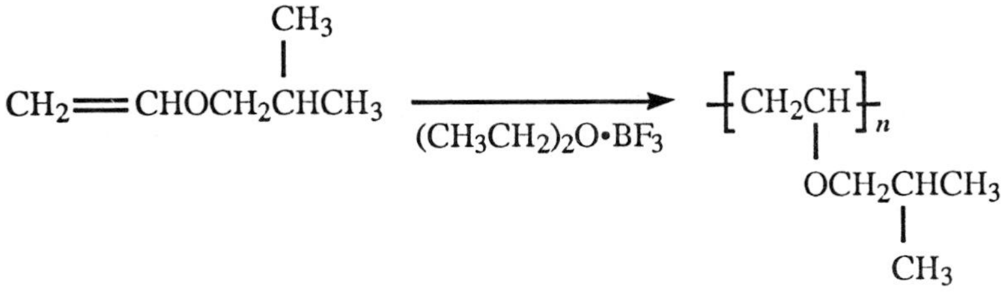

(i)

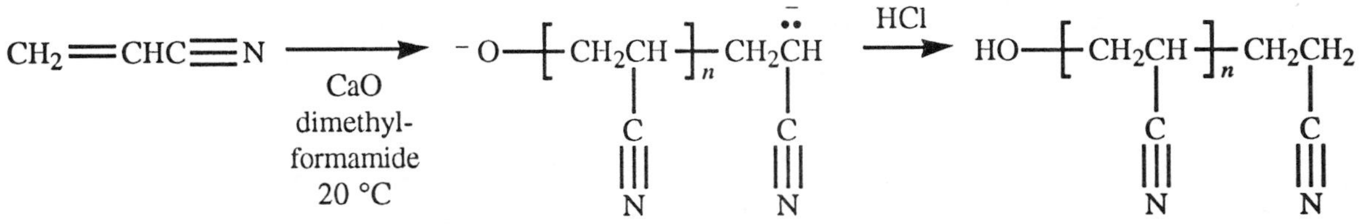

(j)

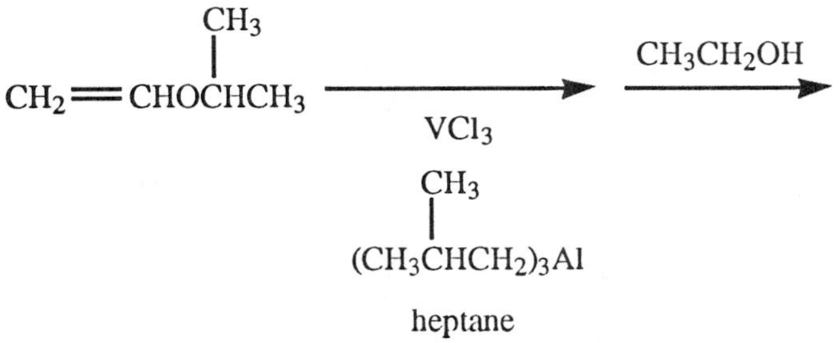

27.27 (cont)

$$\underset{\substack{|\\H}}{\overset{\substack{CH_3\\|\\OCHCH_3\\|}}{}} \cdots$$

(structure: stereochemical zig-zag polymer chain with OCHCH₃ (OCH(CH₃)) groups bearing CH₃, wedge/dash H substituents)

27.28

(a) $CH_3O{-}C_6H_4{-}CH{=}CH_2$ + $C_6H_5{-}CH{=}CH_2$ $\xrightarrow[\substack{\text{carbon}\\ \text{tetrachloride}\\ \text{nitrobenzene}\\ 0\,^\circ C}]{SnCl_4}$ $\left[CHCH_2{-}CHCH_2 \right]_n$

(p-methoxyphenyl group with OCH₃, and phenyl group as substituents on the polymer chain)

(b) $CH_2{=}CHCl$ $\xrightarrow{\text{benzoyl peroxide}}$ $\left[CH_2CH \right]_n$ with Cl substituent

(c) $CH_2{=}\underset{\substack{|\\O}}{\overset{\substack{C{\equiv}N\\|}}{C}}COCH_3$ $\xrightarrow[\substack{\text{methanol}\\ H_2O\\ 20\,^\circ C}]{}$ $\left[CH_2C \right]_n$ with $C{\equiv}N$ and $COCH_3$ (with =O) substituents

(d) $HO{-}C_6H_4{-}\underset{\substack{|\\CH_3}}{\overset{\substack{CH_3\\|}}{C}}{-}C_6H_4{-}OH$ + $\underset{\substack{||\\O}}{Cl\overset{O}{C}Cl}$ $\xrightarrow[\substack{H_2O}]{NaOH}$

$\left[O{-}C_6H_4{-}\underset{\substack{|\\CH_3}}{\overset{\substack{CH_3\\|}}{C}}{-}C_6H_4{-}OC \right]_n$

(e) $CH_2{=}CH_2$ + (cyclopentene) $\xrightarrow[\substack{VCl_4\\ (\text{hexyl})_3Al\\ \text{heptane}}]{CH_3CH_2OH}$ $\left[CH_2CH_2 \text{ (with fused cyclopentane ring)} \right]_n$

27.28 (cont)

(f)

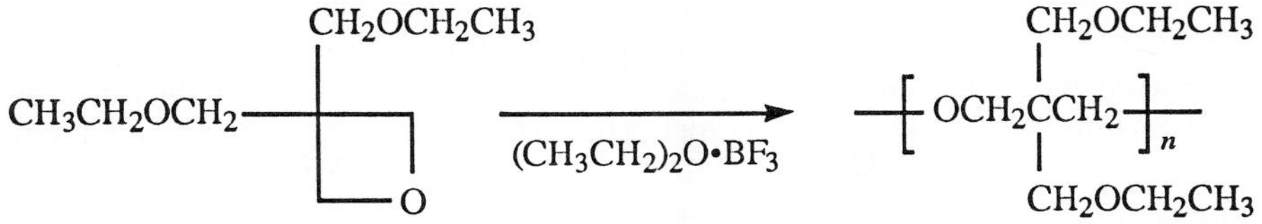

(g)

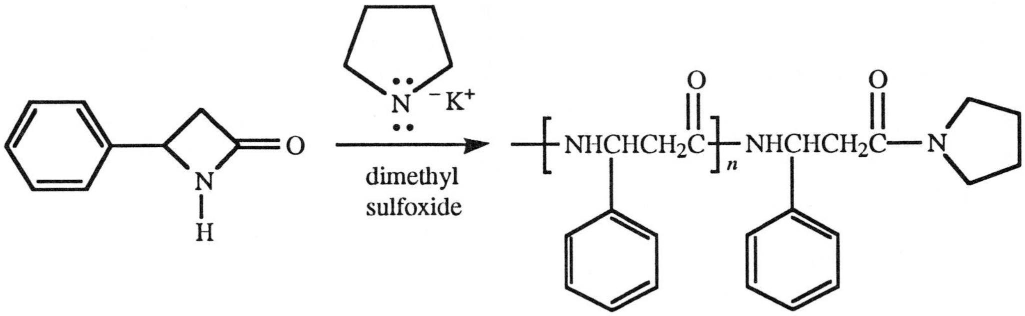

27.29   Cyano and ester groups are electron-withdrawing and stabilize a carbanionic intermediate by their inductive effects and also by resonance.

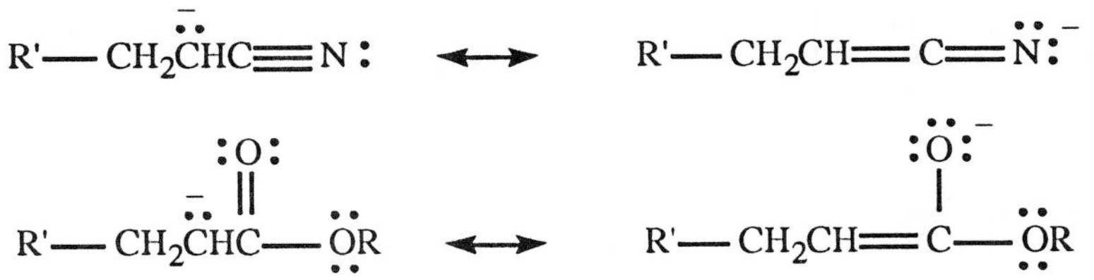

where R' is the initiator.

Radicals are also stabilized by groups such as the cyano group or carbonyl group.

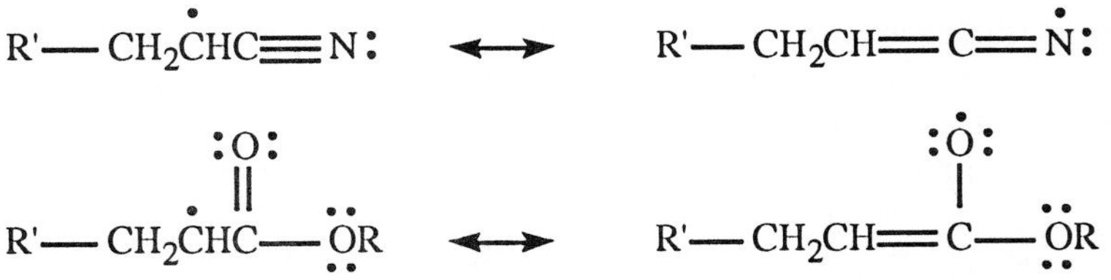

Reactions of these monomers that go through anionic or radical intermediates have low energies of activation and proceed easily.

Cationic intermediates are destabilized by the electron-withdrawing cyano and ester groups. Any resonance contributors that are written for such species delocalize the positive charge from carbon to the more electronegative nitrogen or oxygen atom.

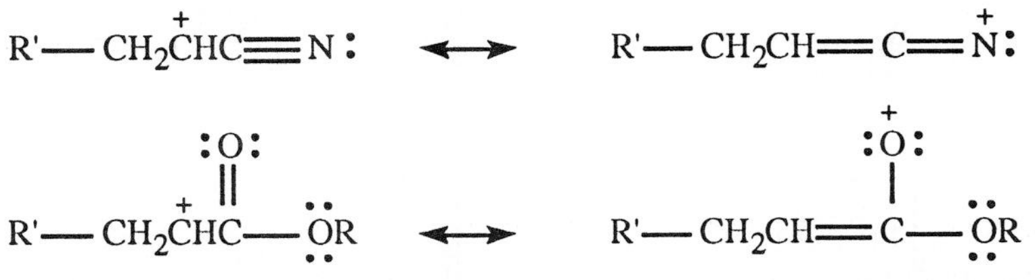

Such intermediates correspond to high-energy transition states and to high energies of activation for the reactions.

27.30　The rate of a reaction depends on the activation energy of the rate-determining step. The activation energy is the difference in energy between the reagents and the transition state. In rationalizing most relative rates, it is necessary only to look at the relative stabilization of the transition states. In general, the more stabilized the transition state is, the lower the activation energy and the faster the reaction. However, if the energy of a reagent is lowered even more by stabilization, the activation energy will be larger, and the rate correspondingly smaller.

In free radical polymerization the rate-determining step is the addition of an unstable radical to the alkene monomer. Substituents play an important role in stabilizing this radical, in which one species bears full radical character, and a less important role in stabilizing the transition state, in which radical character is distributed between two species. The consequences of this are illustrated graphically below.

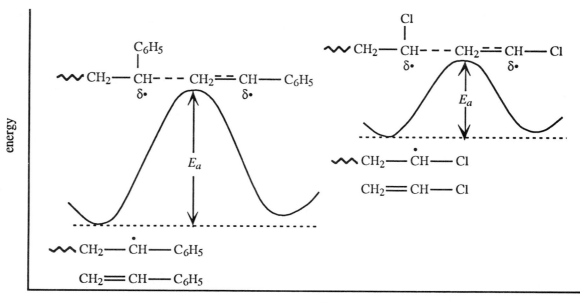

reaction coordinate

The phenyl group stabilizes the reagent radical more strongly than does the chlorine atom. Even though the transition state for the reaction of the vinyl chloride may be higher in energy in absolute terms than the transition state for the reaction of styrene, the energy of activation for the reaction of vinyl chloride is smaller. The relative rates of the polymerization reactions thus reflect the relative stabilization of radicals by the substituents on the double bond.

27.31

(a)

(b)

27.31 (b) (cont)

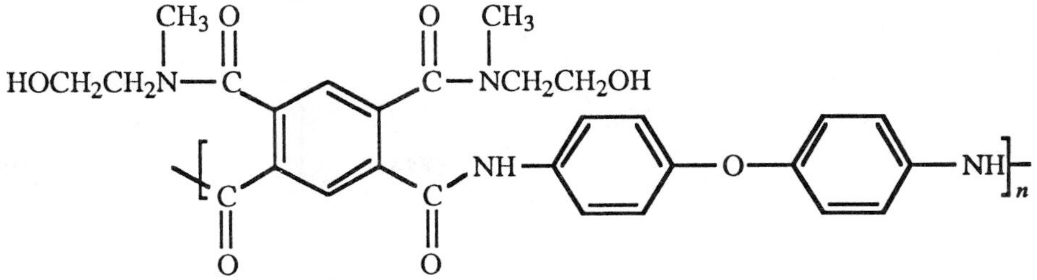

27.32   The first method would give a copolymer in which the amide bonds would be formed more or less randomly by the two different amines reacting with the acid chloride. There would be no pattern to the polymer. If we designate the siloxanediamine as H_2N—(Si)—NH_2 and the aryldiamine as H_2N—(Ar)—NH_2 , the first method would give polymers such as

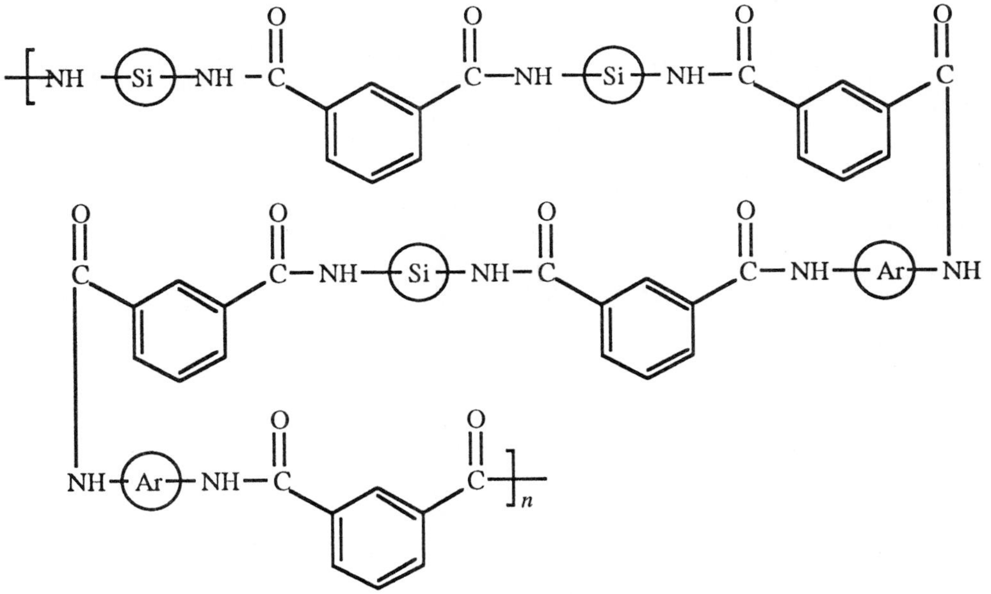

a random copolymer

The second method would give a block copolymer. It would be formed from the polymerization of oligomer A and oligomer B, each of which is a simple polymer. Oligomer A is an amine because a deficiency of acid chloride molecules is used in making it. Oligomer B is an acid chloride.

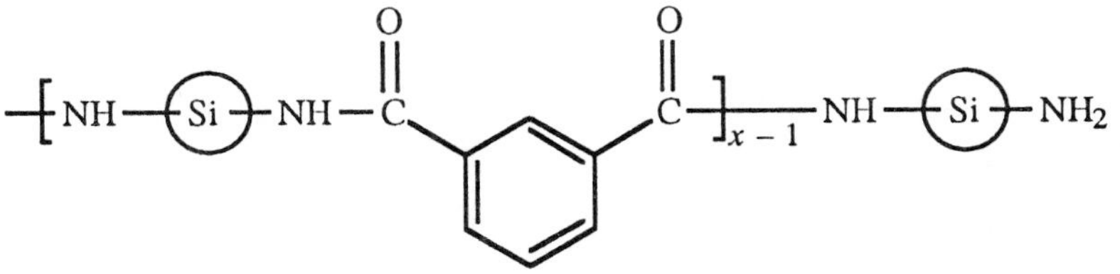

oligomer A

+

27.32 (cont)

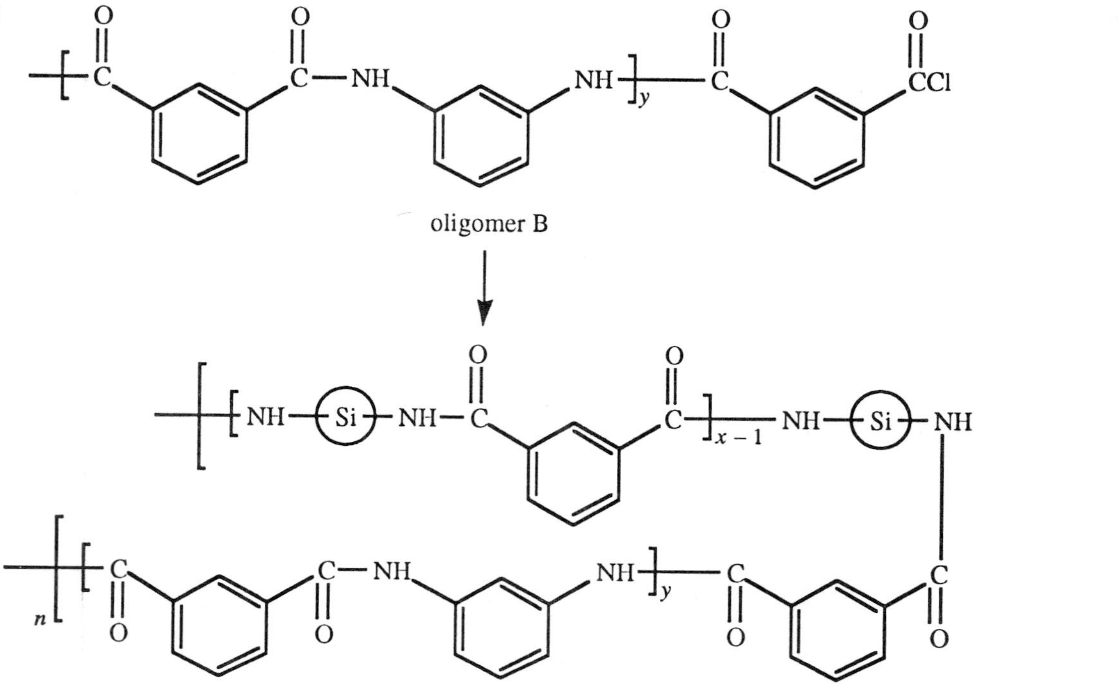
oligomer B

In the block copolymer, stretches of the polymer chain have structural regularity. The amine is either the siloxanediamine or the aryldiamine for many repeating units before a change occurs.

27.33

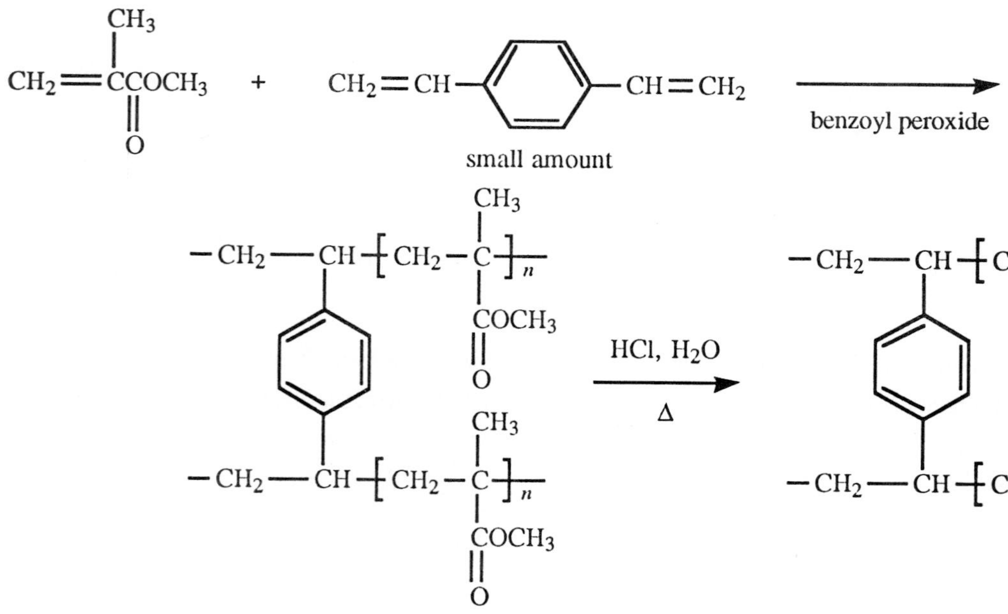
benzoyl peroxide
small amount
HCl, H₂O
Δ

27.34

1.

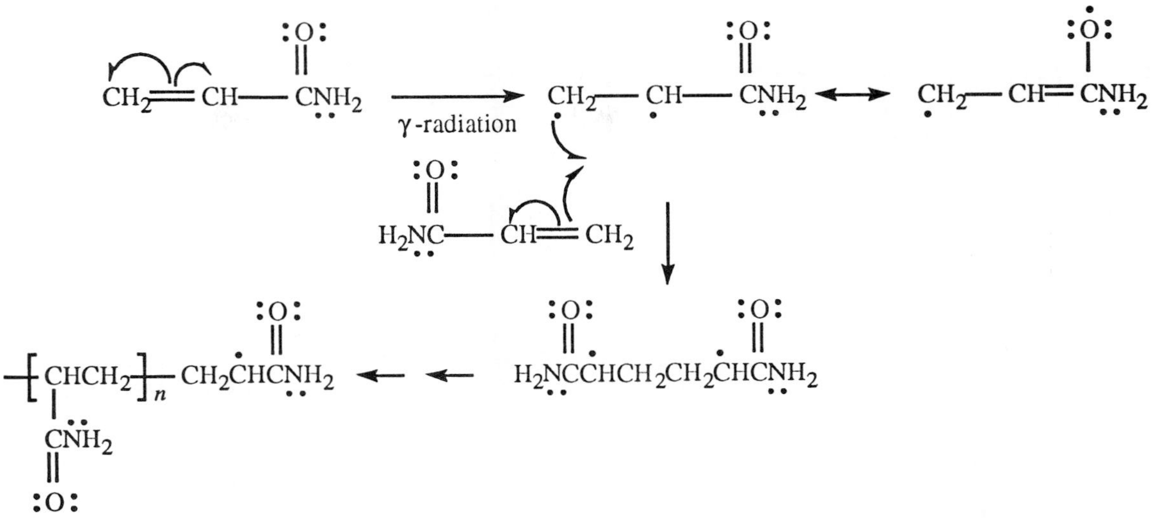

2.

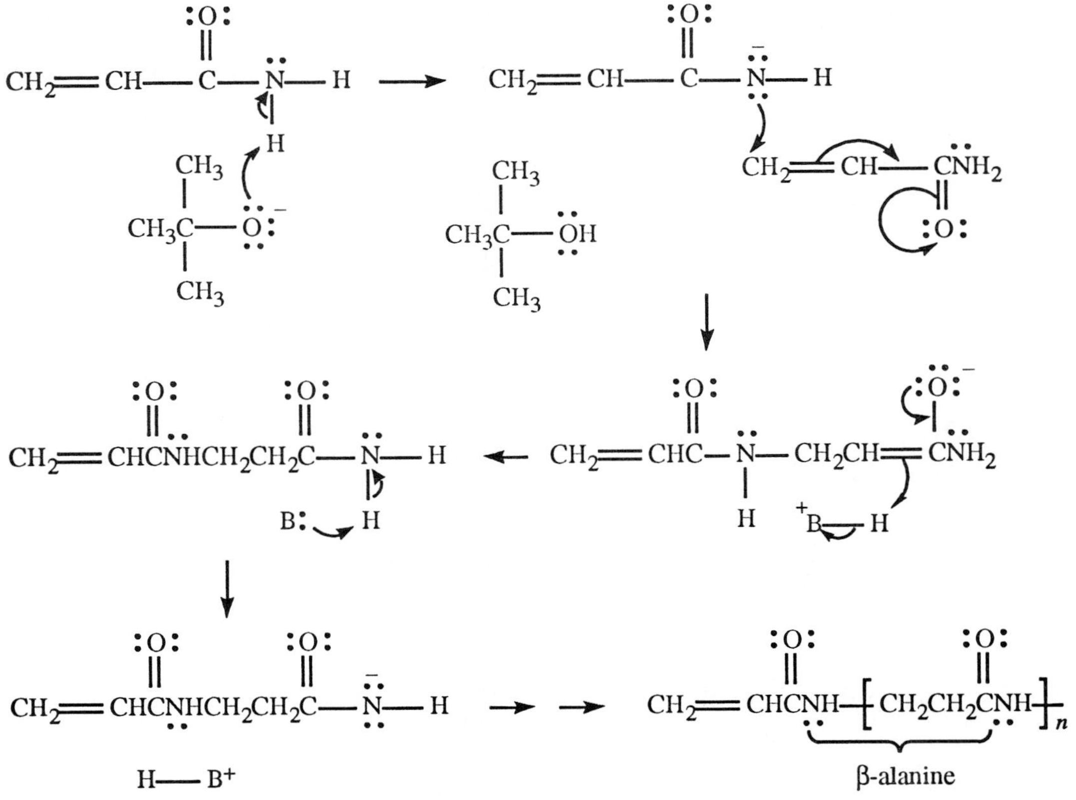

β-alanine

27.35

1.

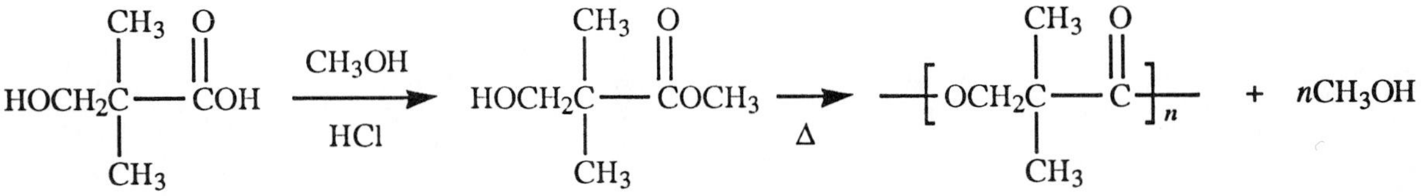

27.35 (cont)

2.

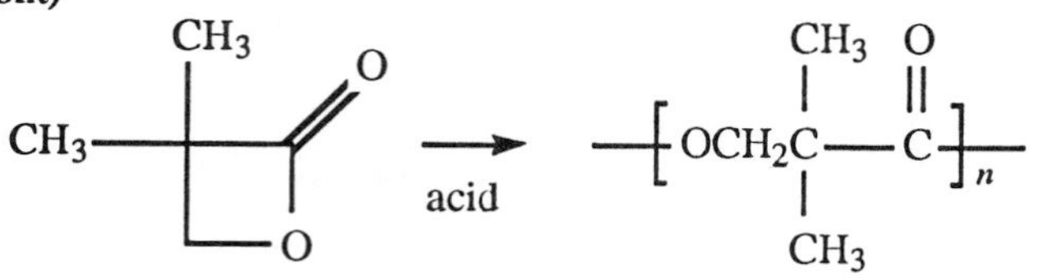

27.36   1.

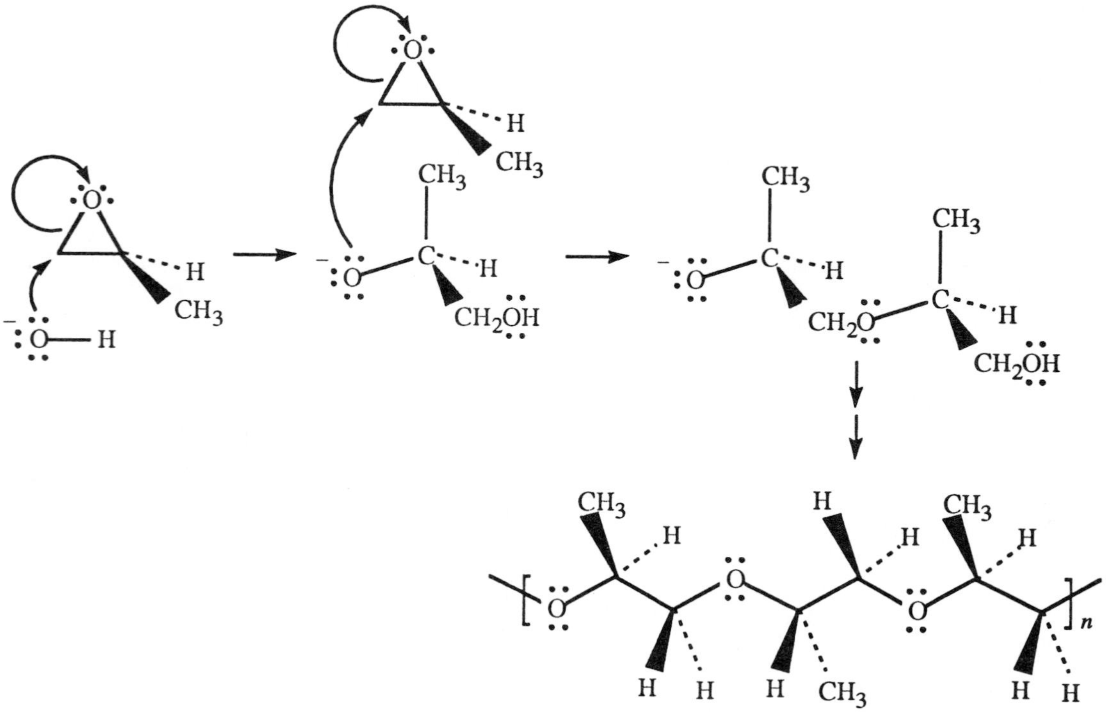

2.

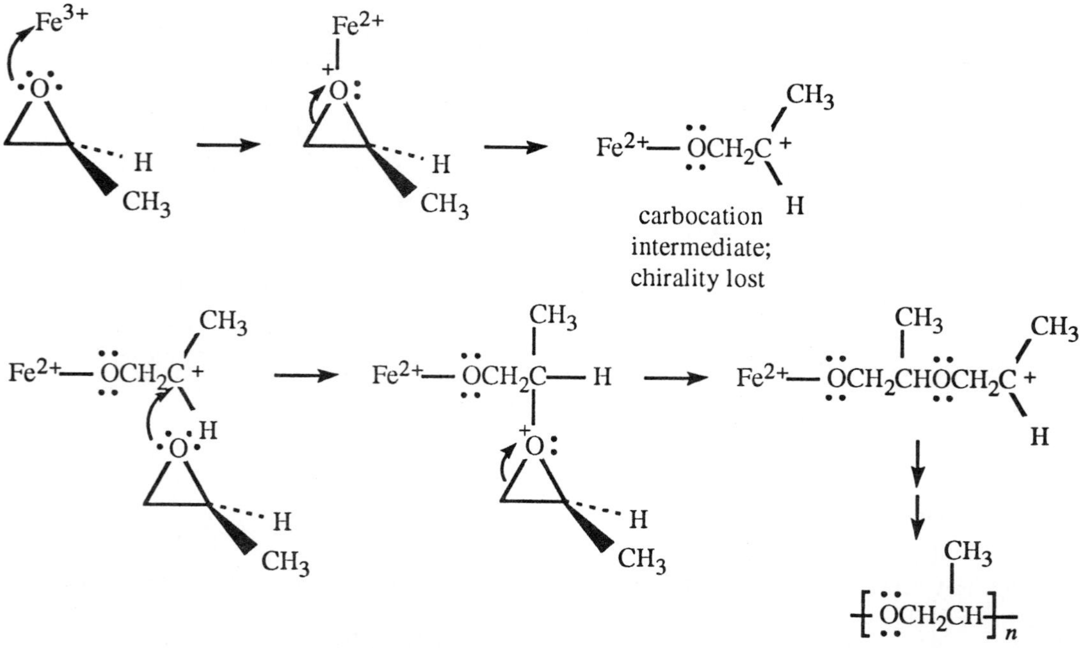

carbocation
intermediate;
chirality lost

27.37

(a)

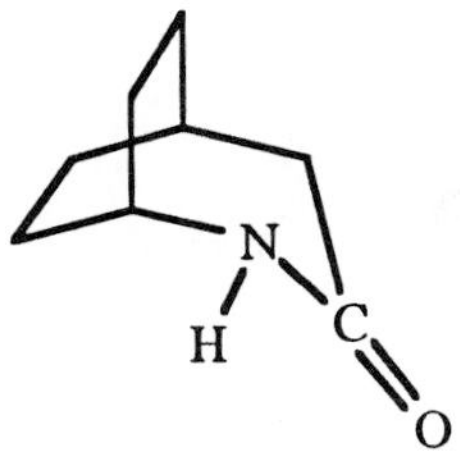

(b)

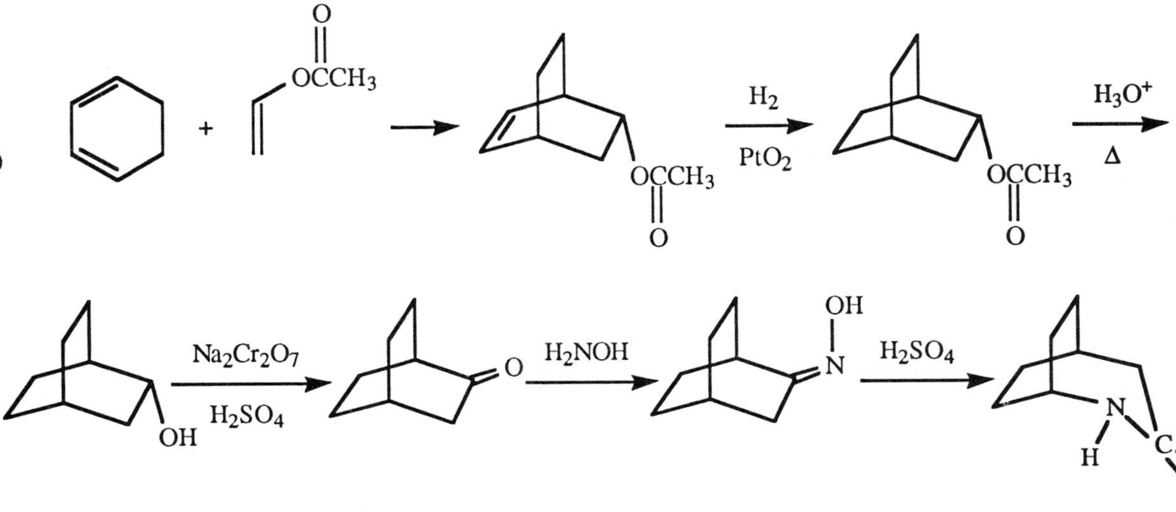

27.38

1.

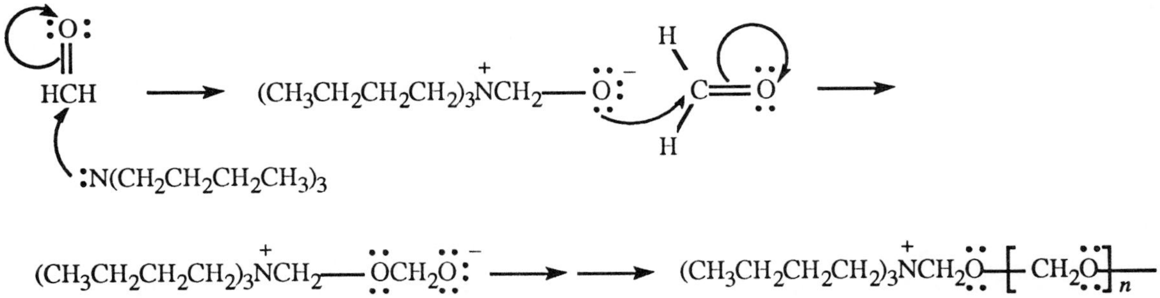

$(CH_3CH_2CH_2CH_2)_3\overset{+}{N}CH_2\!\!-\!\!\overset{..}{\underset{..}{O}}CH_2\overset{..}{\underset{..}{O}}\!\!:^- \longrightarrow \longrightarrow (CH_3CH_2CH_2CH_2)_3\overset{+}{N}CH_2\overset{..}{\underset{..}{O}}\!\!\left[\!\!-CH_2\overset{..}{\underset{..}{O}}\!\!\right]_n$

2.

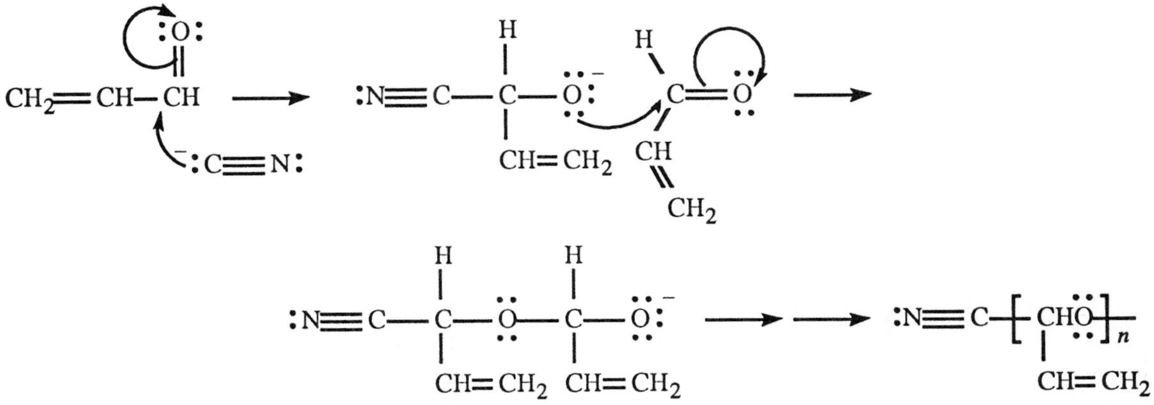

27.39

28

Concerted Reactions

Concept Map 28.1   Concerted reactions.

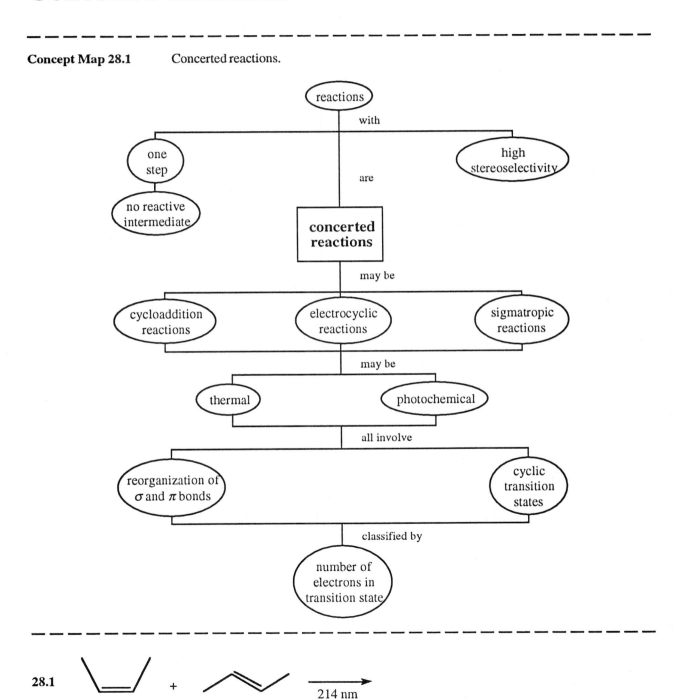

28.1   + $\xrightarrow{\text{214 nm}}$

28.1 (cont)

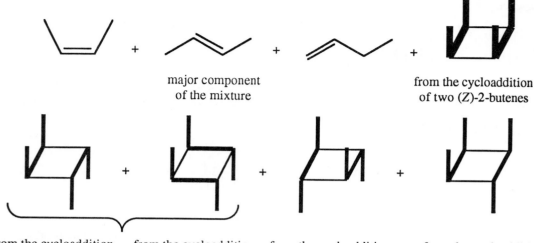

from the cycloaddition
of one (*E*)-2-butene
with one (*Z*)-2-butene;
not seen as a product in
the early stages of the
photochemical reaction
of either (*E*)-2-butene
or (*Z*)-2-butene

28.2

28.3

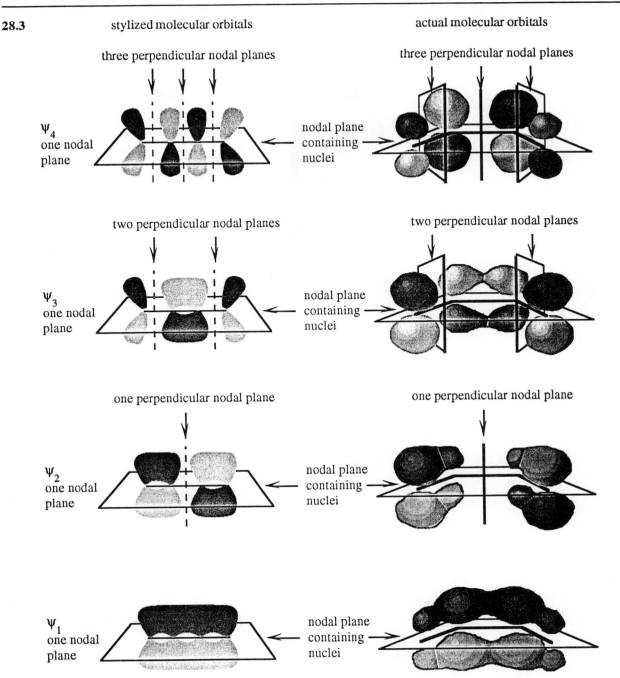

stylized molecular orbitals   actual molecular orbitals

three perpendicular nodal planes   three perpendicular nodal planes

Ψ_4
one nodal
plane

nodal plane
containing
nuclei

two perpendicular nodal planes   two perpendicular nodal planes

Ψ_3
one nodal
plane

nodal plane
containing
nuclei

one perpendicular nodal plane   one perpendicular nodal plane

Ψ_2
one nodal
plane

nodal plane
containing
nuclei

Ψ_1
one nodal
plane

nodal plane
containing
nuclei

28.4   The symmetry of a molecular orbital can be determined by looking at the symmetry of the atomic *p*-orbitals from which it is constructed since the molecular orbital retains that symmetry.

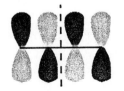

Ψ_4 antisymmetric

28.4 (cont)

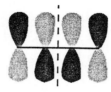

Ψ_3 symmetric

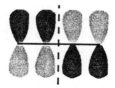

Ψ_2 antisymmetric

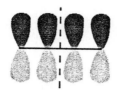

Ψ_1 symmetric

28.5

Ψ_4 ——————

Ψ_3 —————— LUMO

Ψ_2 —— ↑↓ —— HOMO

Ψ_1 —— ↑↓ ——

ground state of
1,3-butadiene

$\xrightarrow{h\nu}$

Ψ_4 —————— LUMO

Ψ_3 —— ↓ —— HOMO

Ψ_2 —— ↑ ——

Ψ_1 —— ↑↓ ——

excited state of
1,3-butadiene

28.6

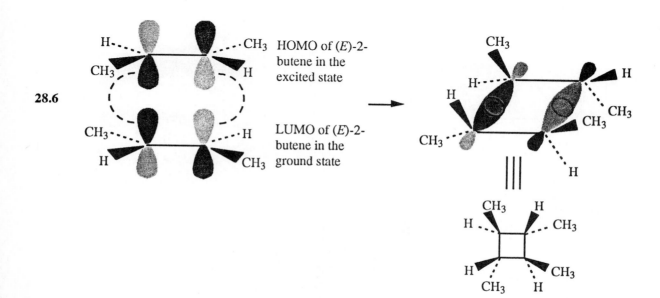

HOMO of (E)-2-
butene in the
excited state

LUMO of (E)-2-
butene in the
ground state

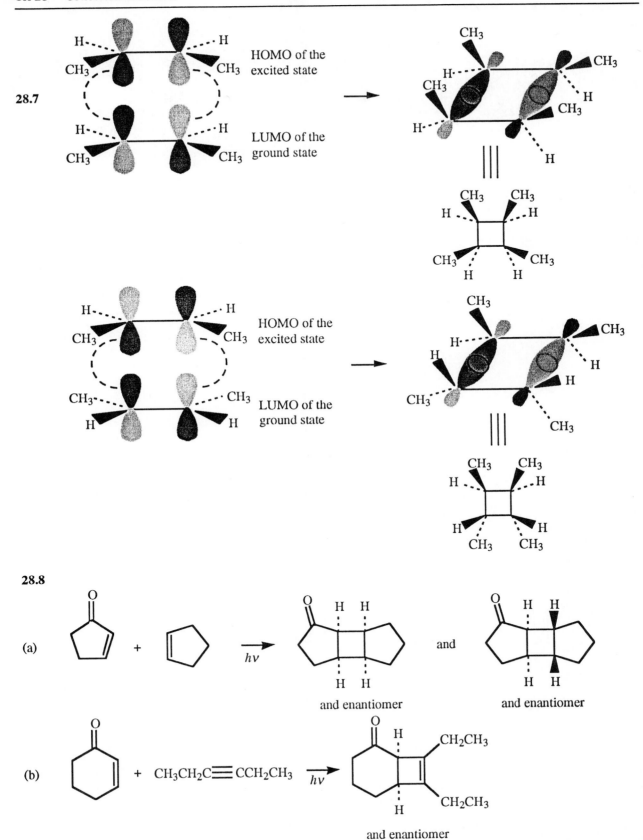

HOMO of the excited state

LUMO of the ground state

HOMO of the excited state

LUMO of the ground state

28.8

(a) and enantiomer   and   and enantiomer

(b) and enantiomer

28.8 (cont)

(c)

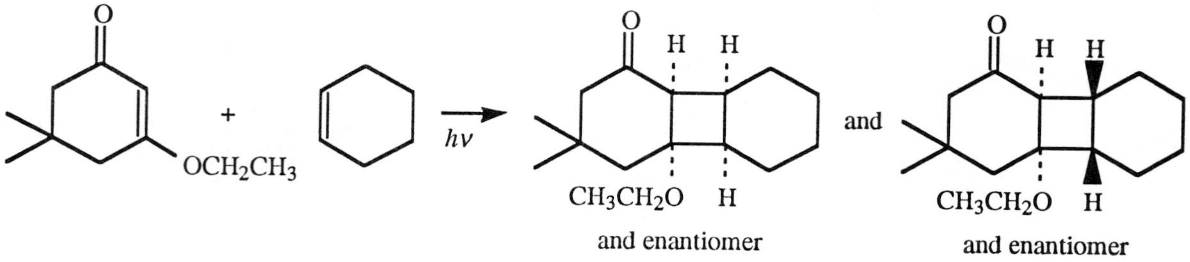

and

and enantiomer

and enantiomer

28.9

(a)

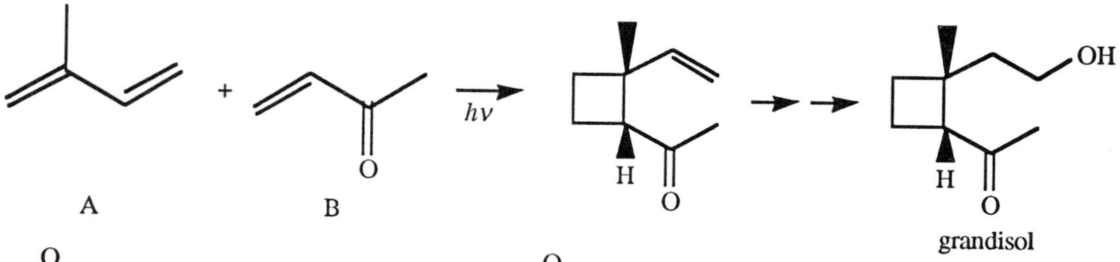

A

B

grandisol

(b)

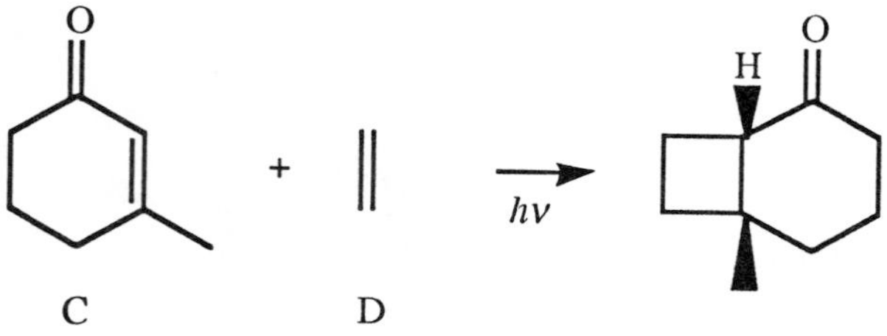

C

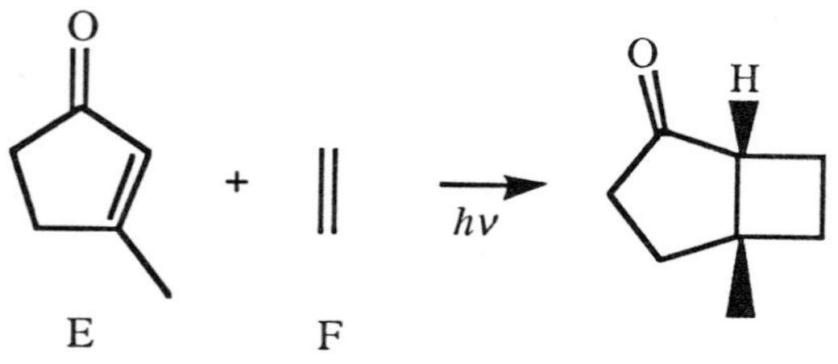

D

(c)

E

F

(d)

G

H

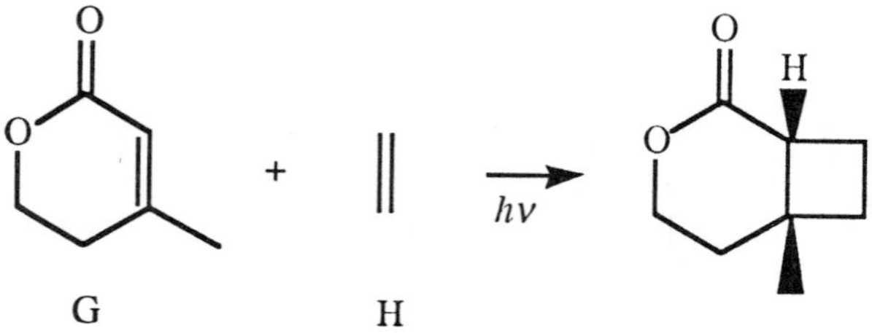

28.10

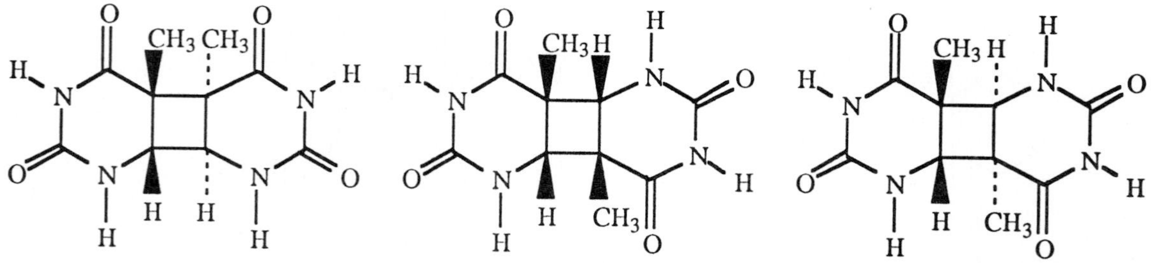

28.11

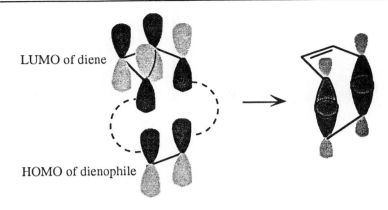

In this picture the lobes of the *p*-orbitals that are left over at carbons 2 and 3 of the diene will be bonding.

28.12

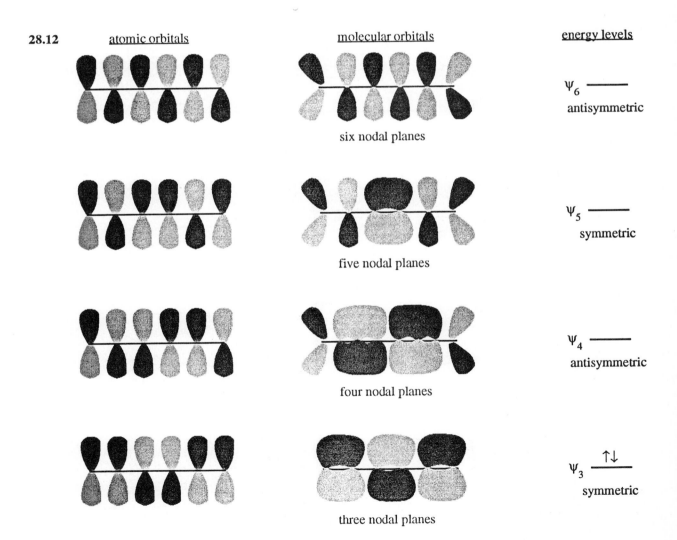

atomic orbitals	molecular orbitals	energy levels

six nodal planes — ψ_6 —— antisymmetric

five nodal planes — ψ_5 —— symmetric

four nodal planes — ψ_4 —— antisymmetric

three nodal planes — ψ_3 $\uparrow\downarrow$ symmetric

28.12 (cont)   <u>atomic orbitals</u>   <u>molecular orbitals</u>   <u>energy levels</u>

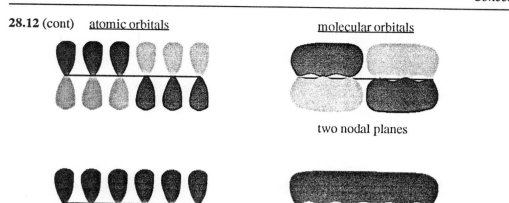

two nodal planes

Ψ_2 ⎯⎯ ↑↓
antisymmetric

one nodal plane

Ψ_1 ⎯⎯ ↑↓
symmetric

The lower three molecular orbitals are bonding; the upper three are antibonding. The HOMO is Ψ_3.

- -

Concept Map 28.2   Cycloaddition reactions.

```
         ┌──────────────┐                    ┌─────────────────┐
         │  alkene or   │       plus         │ other reactants │
         │   alkyne     │                    │  with π bonds   │
         └──────────────┘                    └─────────────────┘
                  │                                   │
                  └──────────────┬────────────────────┘
                                 ▼
                    ┌────────────────────────┐
                    │ cycloaddition reactions │
                    └────────────────────────┘
                            pictured as
                          ( interactions )
                    ┌──────────┴──────────────┐
            ┌──────────────┐          ┌─────────────────┐
            │   HOMO of    │          │    LUMO of      │
            │ one reactant │          │ other reactant  │
            └──────────────┘          └─────────────────┘
                    └───────────may be────────┘
            ( thermal )                  ( photochemical )
              when                           when
     ┌────────────────────┐        ┌────────────────────┐
     │ [4n + 2] π electrons│       │ [2 + 2] π electrons │
     │ in transition state │       │ in transition state │
     └────────────────────┘        └────────────────────┘
        ┌────────┴────────┐                  │
  ( six-membered )  ( five-membered )   ( four-membered )
  (     ring     )  (    ring      )    (     ring      )
```

(Section 28.4)

28.13

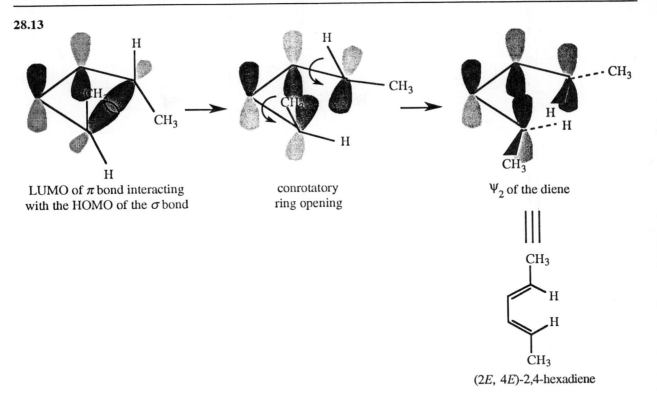

LUMO of π bond interacting
with the HOMO of the σ bond

conrotatory
ring opening

Ψ_2 of the diene

(2*E*, 4*E*)-2,4-hexadiene

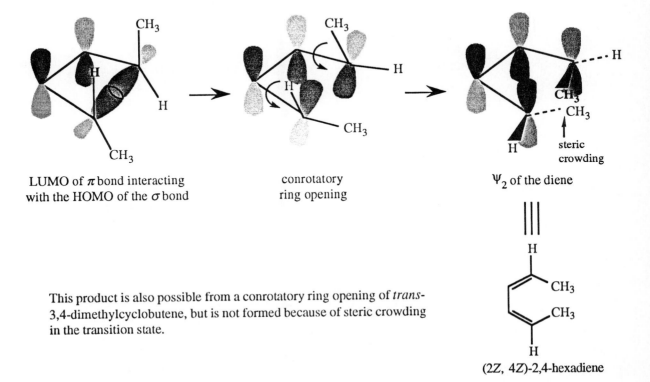

LUMO of π bond interacting
with the HOMO of the σ bond

conrotatory
ring opening

Ψ_2 of the diene

steric
crowding

(2*Z*, 4*Z*)-2,4-hexadiene

This product is also possible from a conrotatory ring opening of *trans*-
3,4-dimethylcyclobutene, but is not formed because of steric crowding
in the transition state.

28.14   The observed equilibrium involves two conrotatory electrocyclic reactions of a butadiene-cyclobutene system, one a ring closure and the other a ring opening.

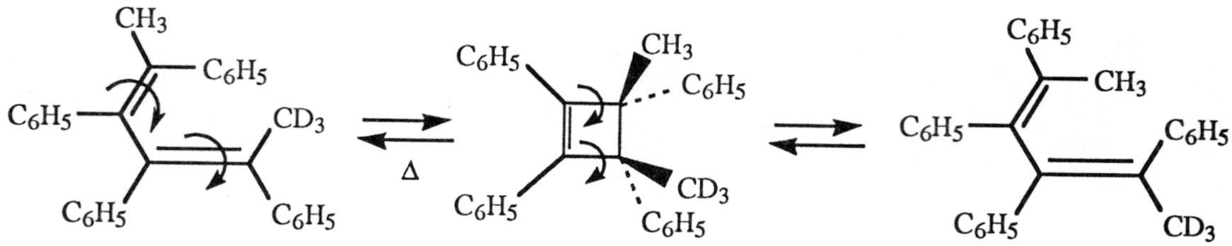

Other possible stereoisomers:

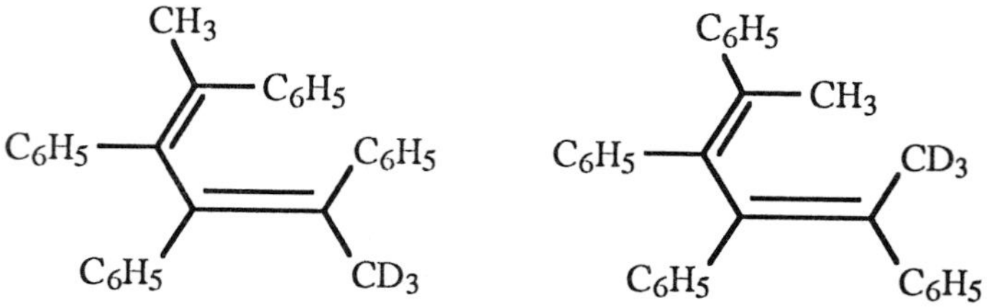

28.15

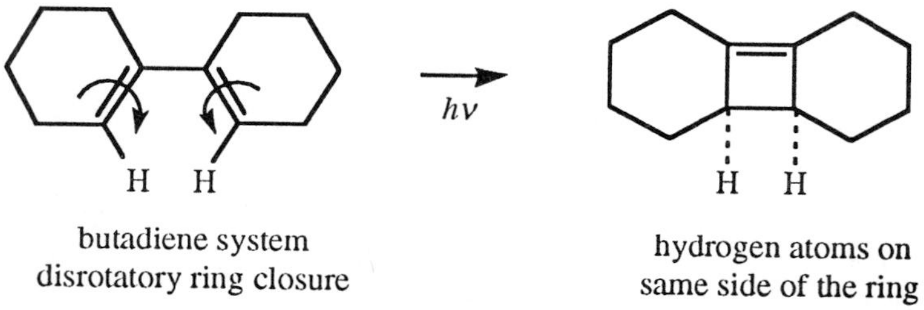

butadiene system
disrotatory ring closure

hydrogen atoms on
same side of the ring

28.16

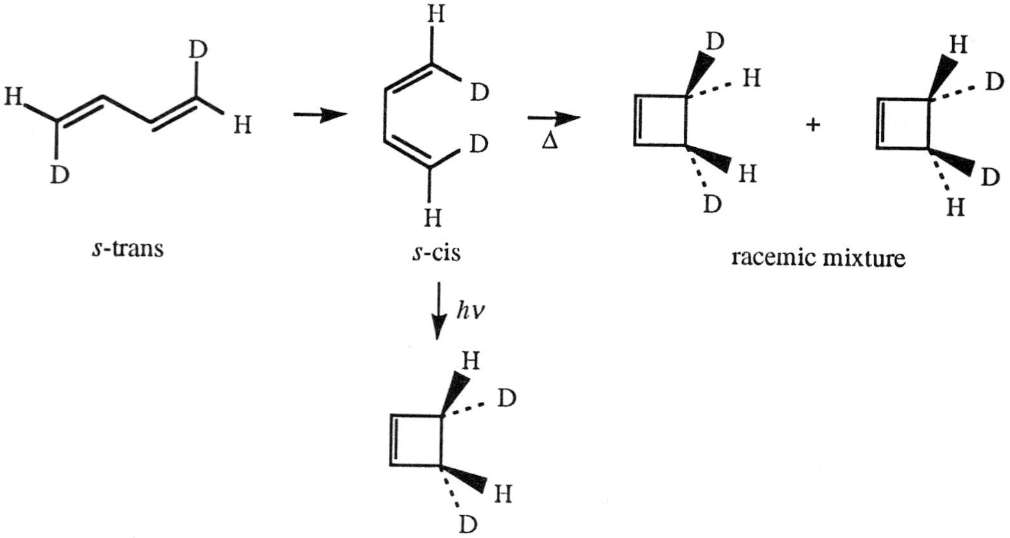

s-trans   *s*-cis   racemic mixture

28.16 (cont)

Thermal process:

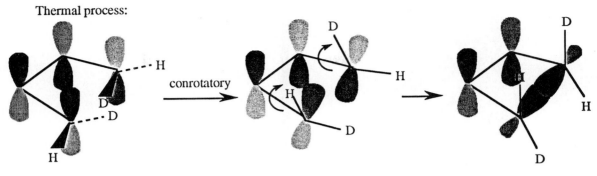

HOMO of diene
in ground state

Photochemical process:

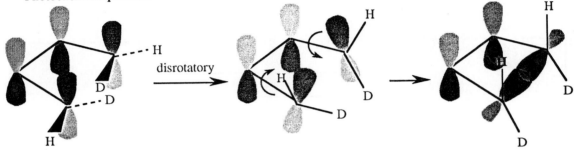

HOMO of diene
in excited state

 28.17

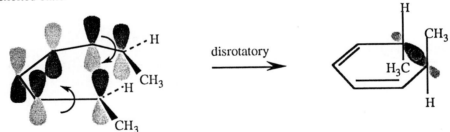

HOMO of the triene
in the ground state

(See Problem 28.12 for the molecular orbitals of the triene.)

 28.18

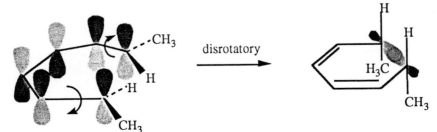

HOMO of the triene in the
ground state; thermal ring closure

same molecule as that
shown in Figure 28.11

28.18 (cont)

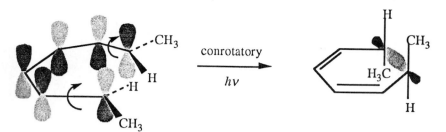

HOMO of the triene in the excited
state; photochemical ring closure

enantiomer of the molecule
shown in Figure 28.12

Concept Map 28.3   Electrocyclic reactions.

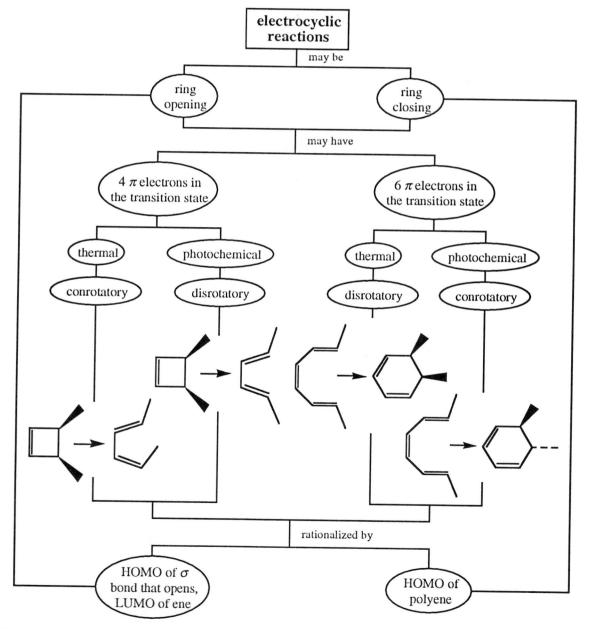

28.19   HOMO of triene in the ground state:

precalciferol → (outward disrotatory) → isopyrocalciferol

precalciferol → (inward disrotatory) → pyrocalciferol

28.20   allo-ocimene → ($h\nu$ conrotatory) → racemic mixture

Concept Map 28.4   Woodward-Hoffmann rules.

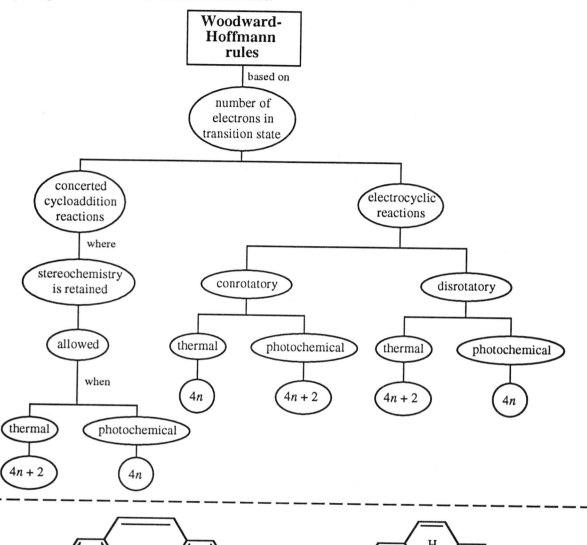

- -

28.21

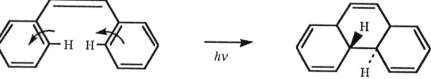

$4n + 2$ electrocyclic reaction; $n = 1$; conrotatory in the excited state; the hydrogen atoms at the ring junctions in dihydrophenanthrene will be trans to each other.

28.22

and enantiomer

$4n$ electrocyclic reaction, $n = 1$; conrotatory in ground state; the methyl groups on the new five-membered ring will be trans to each other.

28.23

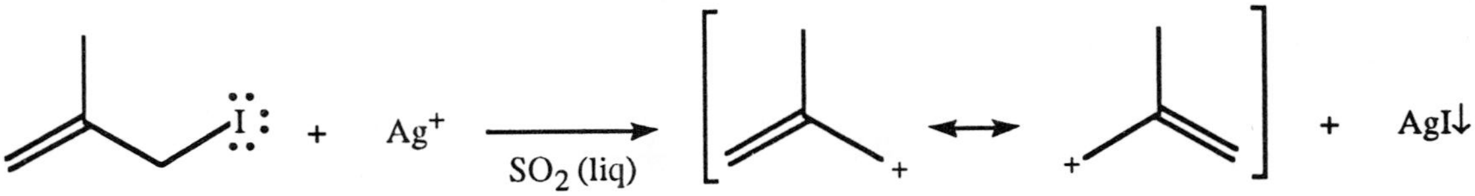
I
+
Ag^+
SO_2 (liq)
+
+
+
AgI↓

the dienophile is an allylic cation with two π electrons

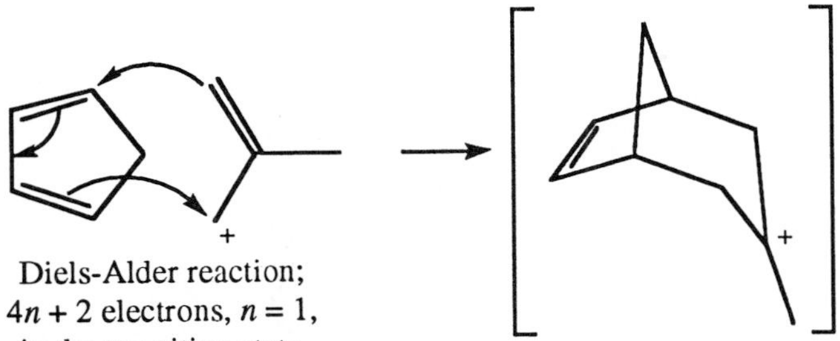
+
+

Diels-Alder reaction;
$4n + 2$ electrons, $n = 1$,
in the transition state

Concept Map 28.2 (see p. 1060)

28.24

(a)

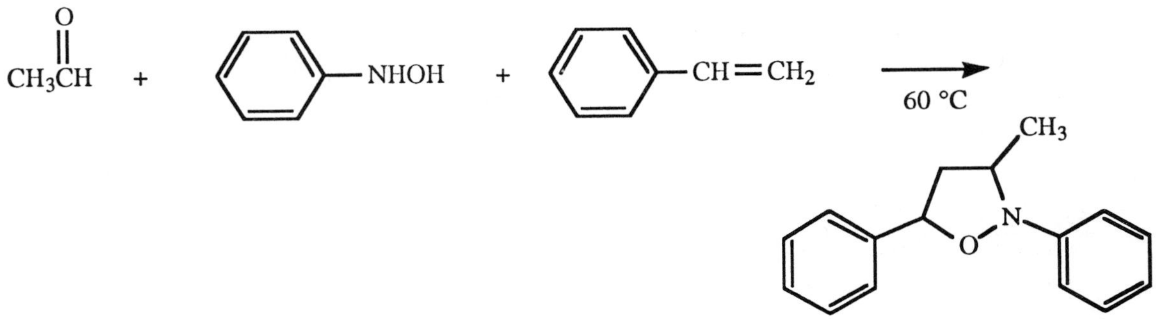
O
CH_3CH
+
NHOH
+
CH=CH_2
60 °C
CH_3
N
O

(b)

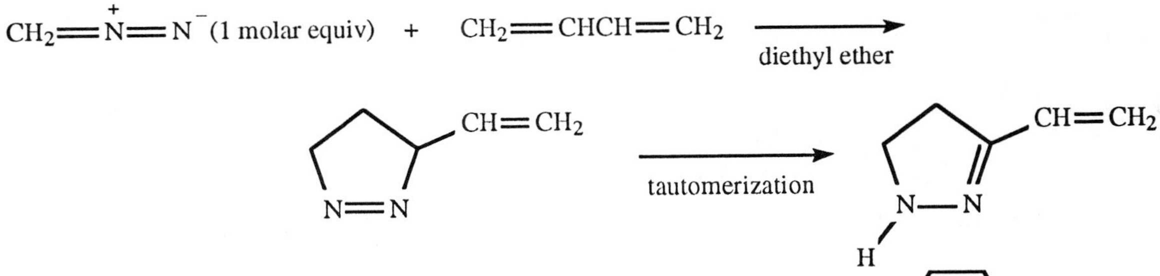
CH_2=$\overset{+}{N}$=$\overset{-}{N}$ (1 molar equiv)
+
CH_2=CHCH=CH_2
diethyl ether
CH=CH_2
N=N
tautomerization
CH=CH_2
N—N
H

(c)

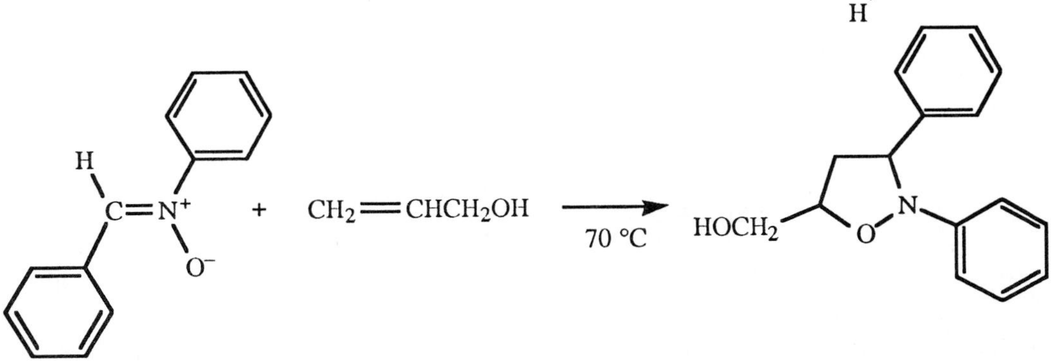
H
C=N^+
O^-
+
CH_2=$CHCH_2OH$
70 °C
$HOCH_2$
N
O

28.24 (cont)

(d) 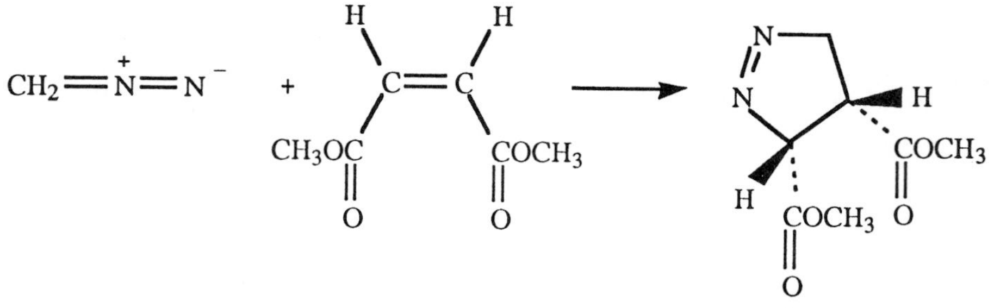 $CH_2{=}\overset{+}{N}{=}N^-$ +

and enantiomer

(e)

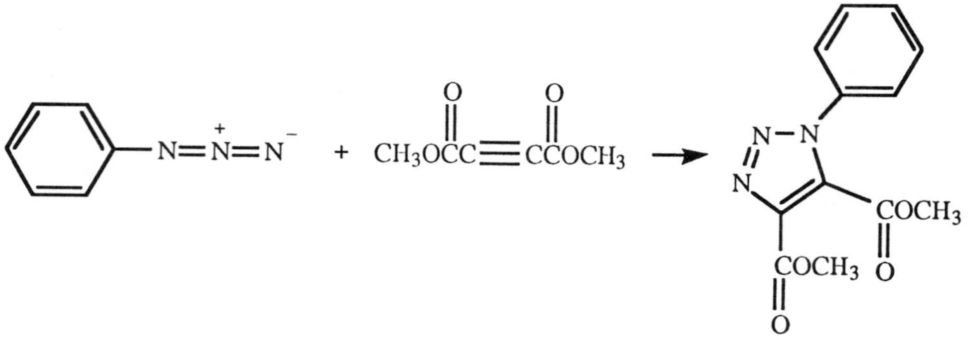

(f) 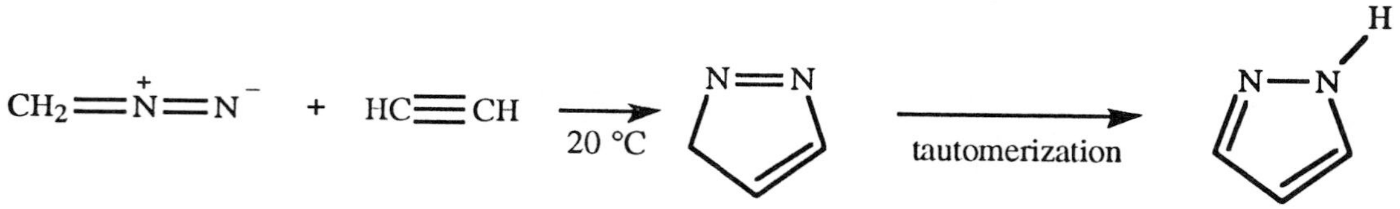 $CH_2{=}\overset{+}{N}{=}N^-$ + $HC{\equiv}CH$ $\xrightarrow{20\ ^\circ C}$ $\xrightarrow{\text{tautomerization}}$

more stable pyrazole

(g)

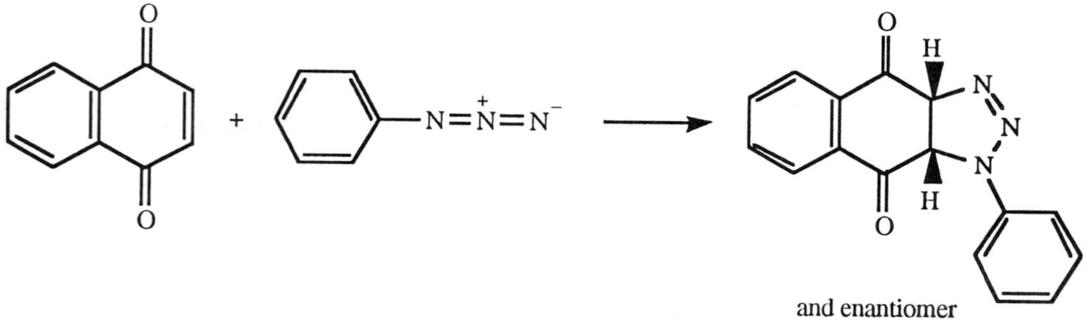

and enantiomer

(h)

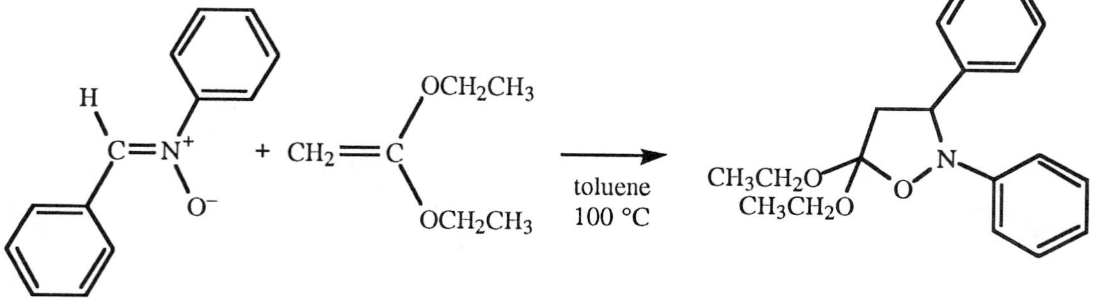

28.24 (cont)

(i)

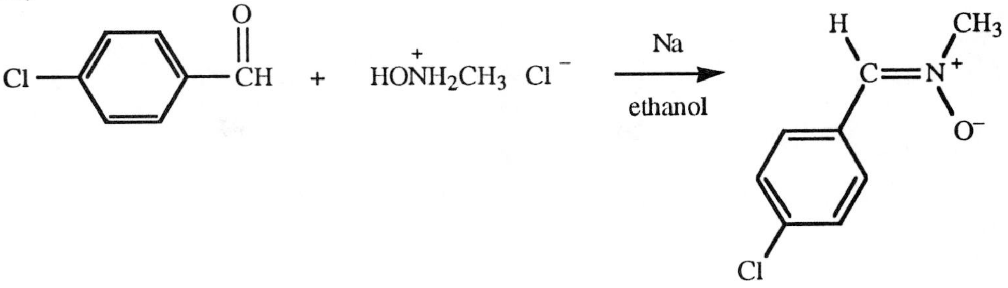

28.25

(a)

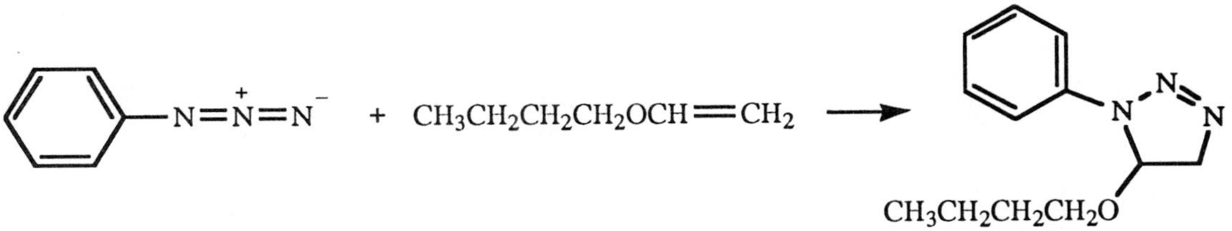

(b)

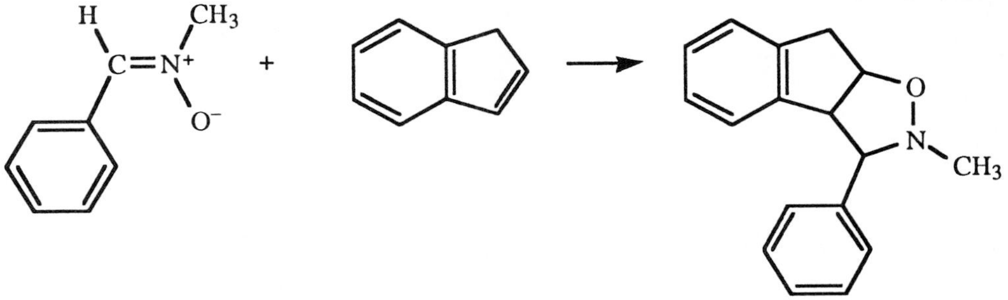

(c)

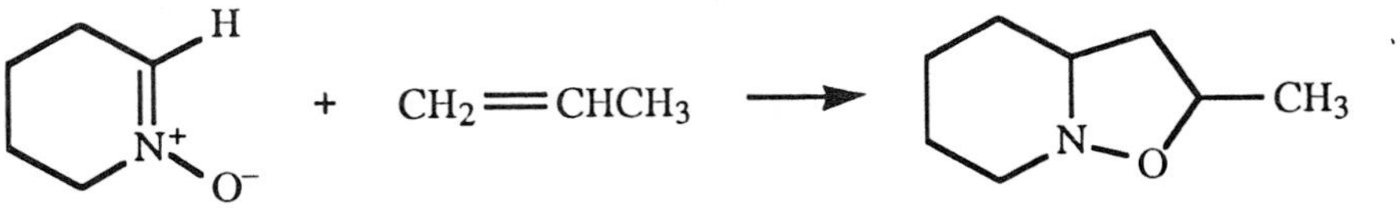

(d)

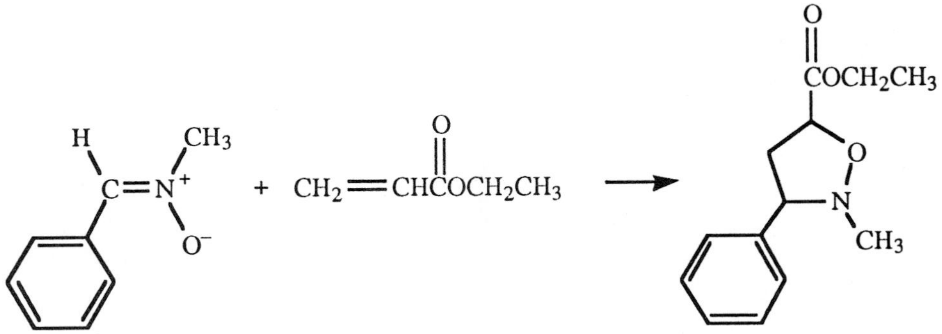

(e)

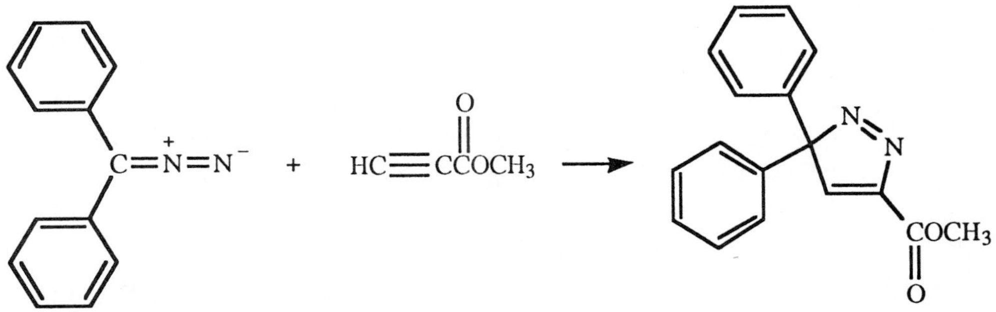

28.25 (cont)

(f)

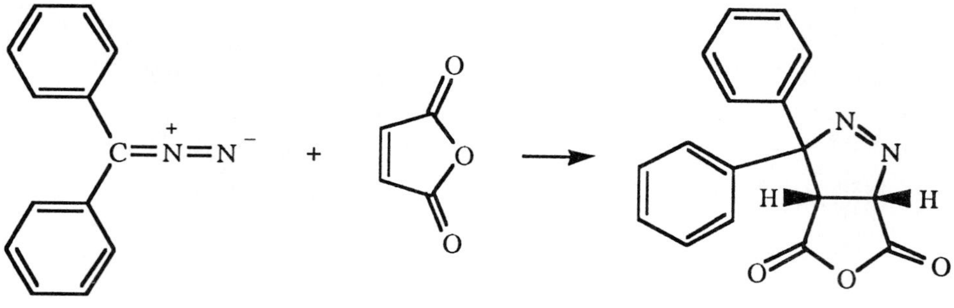

and enantiomer

(g)

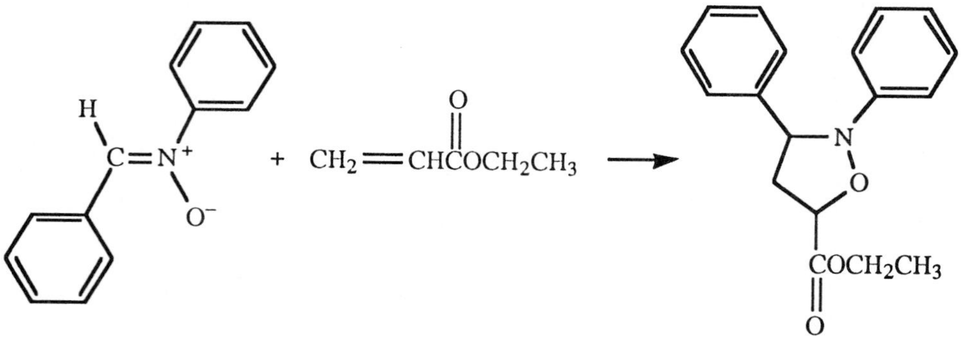

(h)

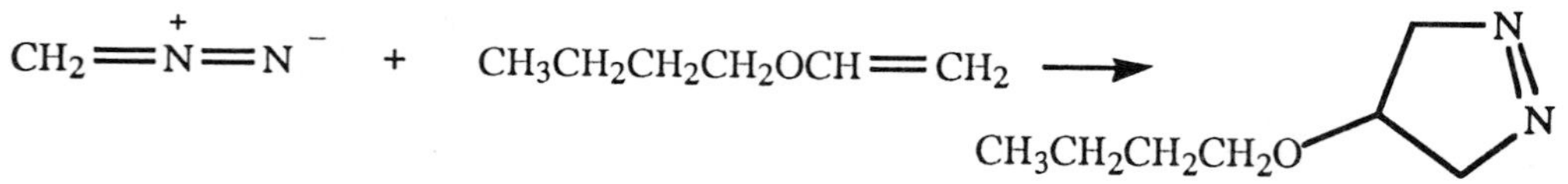

28.26

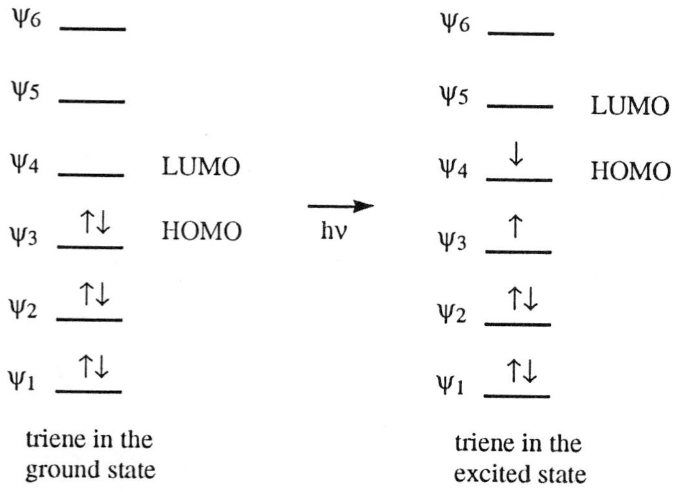

ψ_6 _____

ψ_5 _____

ψ_4 _____ LUMO

ψ_3 ↑↓ HOMO

ψ_2 ↑↓

ψ_1 ↑↓

triene in the
ground state

→ hν

ψ_6 _____

ψ_5 _____ LUMO

ψ_4 ↓ HOMO

ψ_3 ↑

ψ_2 ↑↓

ψ_1 ↑↓

triene in the
excited state

(See Problem 28.12 for pictures of the orbitals.)

28.26 (cont)

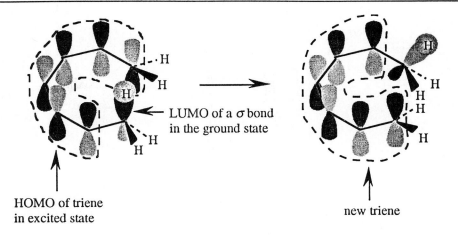

HOMO of triene
in excited state

← LUMO of a σ bond
in the ground state

new triene

28.27

(a)

CH₃CH₂OC — (structure with COCH₂CH₃) →[200 °C] product structure

(b)

phenyl structure →[180 °C] product structure

(c)

CH₃CH₂OC — (structure with N≡C) →[160 °C] product structure

(d)

(bicyclic structure with N≡C, C≡N) →[175 °C] product structure

28.28

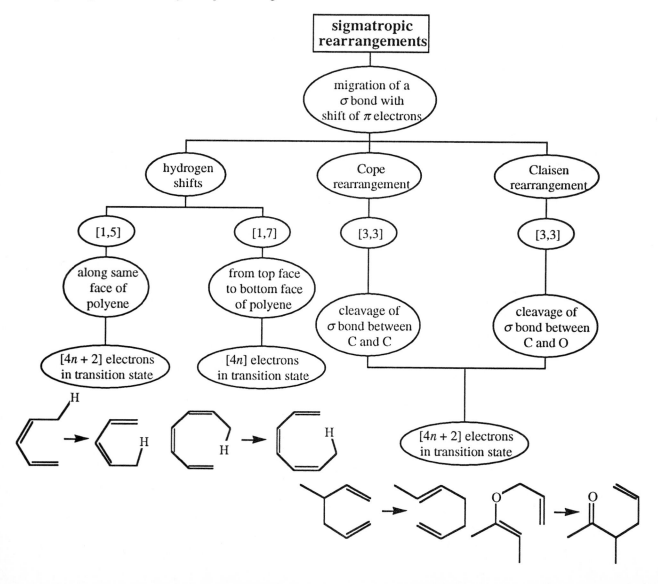

two isoprene units

desired product;
an unstable
intermediate

[3,3]-sigmatropic
rearrangement

The Cope rearrangement occurs easily because of relief of ring strain in going from a cyclobutane to a cyclooctadiene system.

Concept Map 28.5   Sigmatropic rearrangements.

28.29

(a) $CH_2\!=\!CHOCH_2CH\!=\!CH_2$ ≡   →
                                                                    255 °C

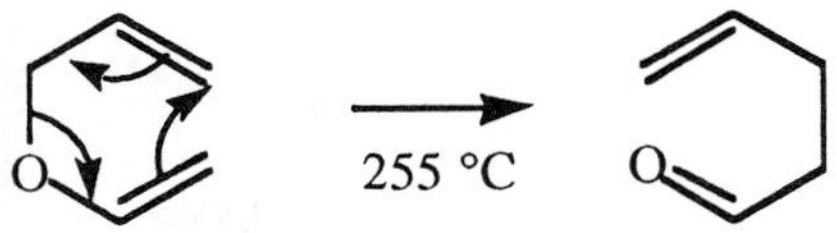

(b)   →
                                175 °C

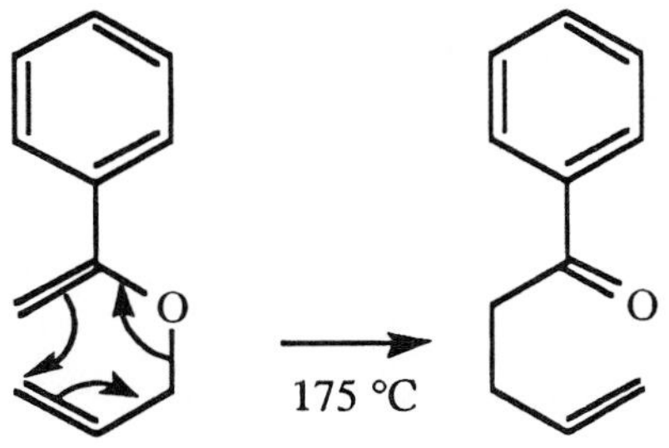

(c)

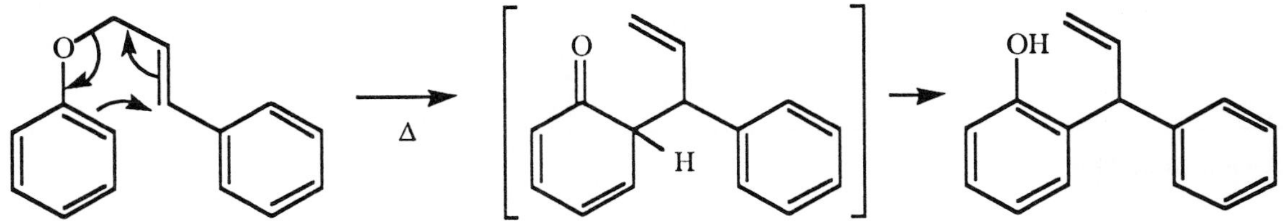

(d)
        255 °C

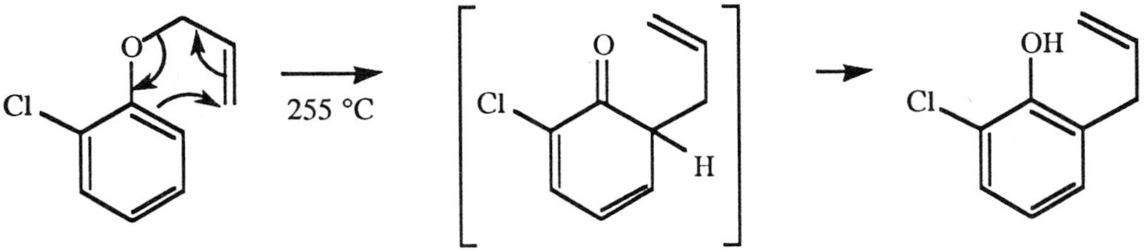

(e)
        N,N-dimethyl-
        aniline
        180 °C

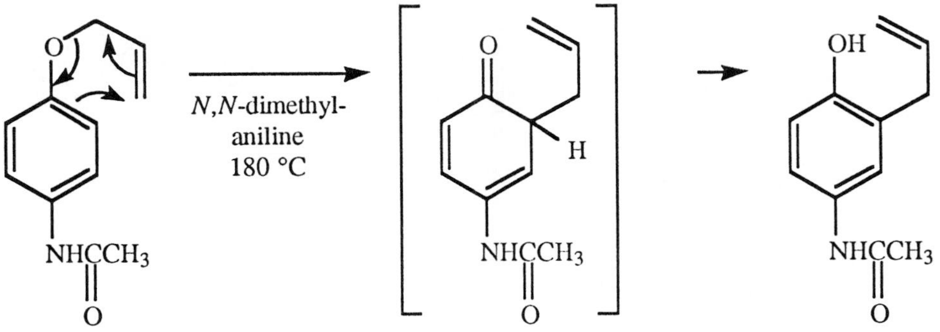

(f)
        N,N-dimethyl-
        aniline
        225 °C

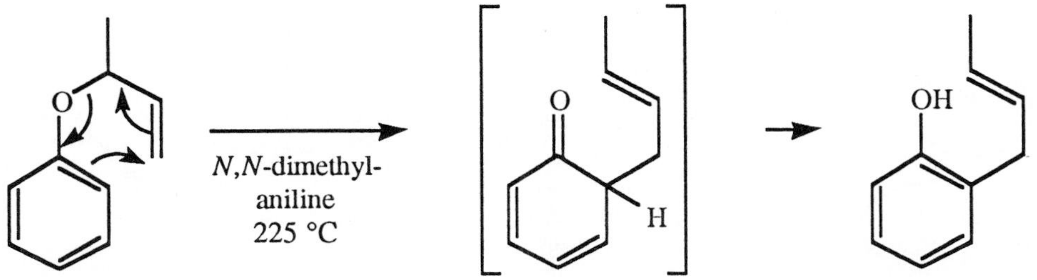

28.29 (cont)

(g)

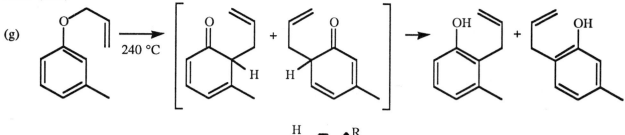

28.30

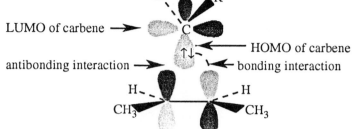

Concept Map 28.6   Carbenes.

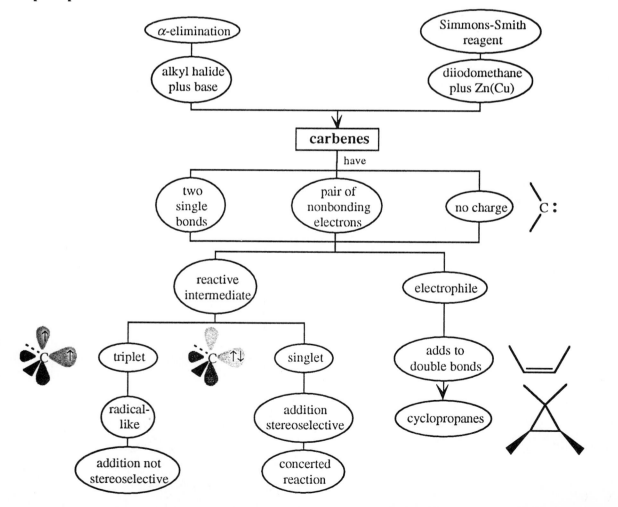

28.31

(a)

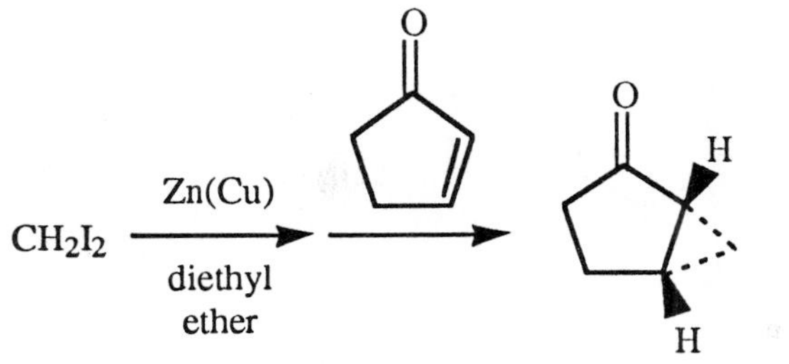
CH2I2
Zn(Cu)
diethyl
ether
O
O
H
H

and enantiomer

(b)

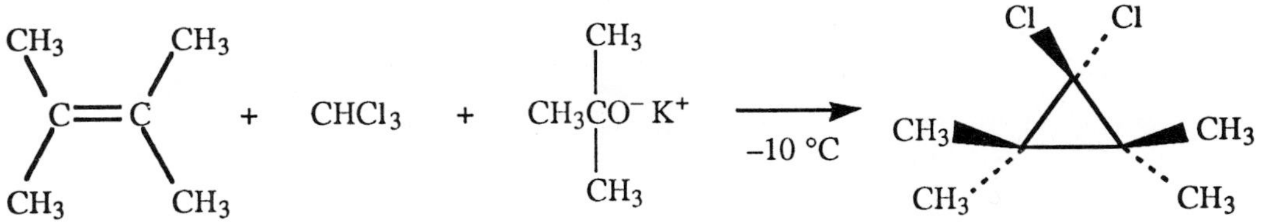
CH3
CH3
C=C
CH3
CH3
+
CHCl3
+
CH3
CH3CO⁻ K⁺
CH3
−10 °C
Cl
Cl
CH3
CH3
CH3
CH3

(c)

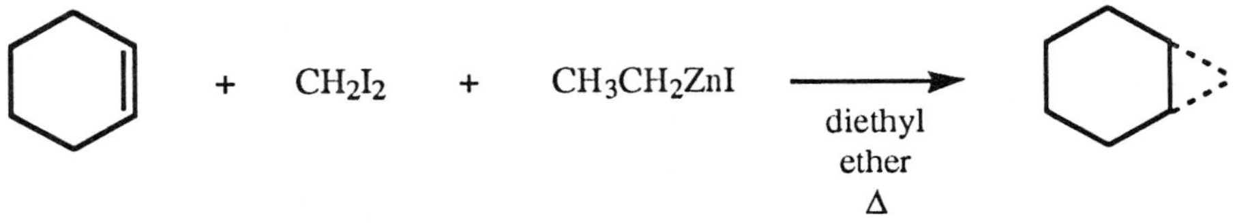
+
CH2I2
+
CH3CH2ZnI
diethyl
ether
Δ

(d)

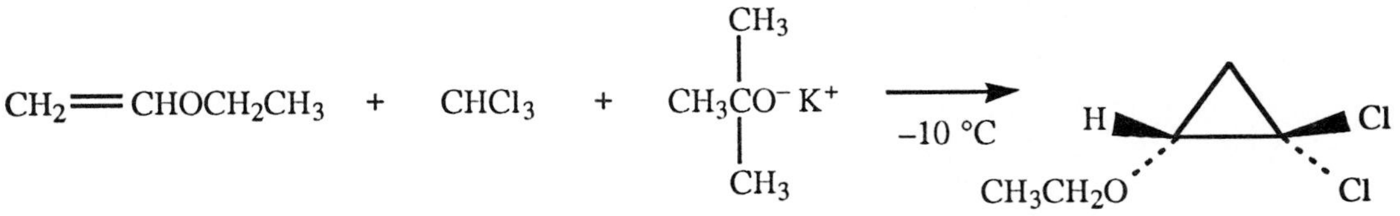
CH2=CHOCH2CH3
+
CHCl3
+
CH3
CH3CO⁻ K⁺
CH3
−10 °C
H
Cl
CH3CH2O
Cl

and enantiomer

(e)

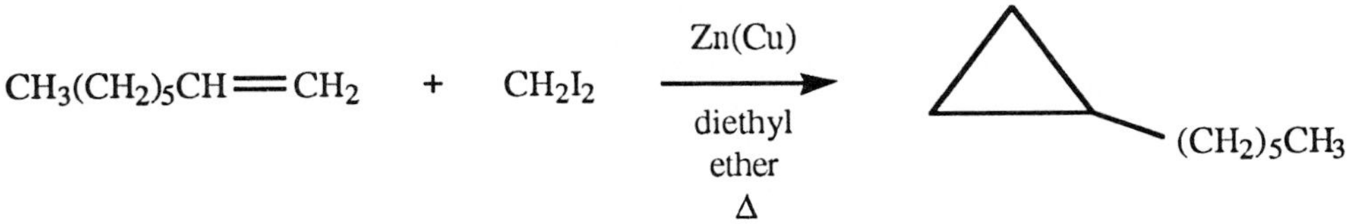
CH3(CH2)5CH=CH2
+
CH2I2
Zn(Cu)
diethyl
ether
Δ
(CH2)5CH3

(f)

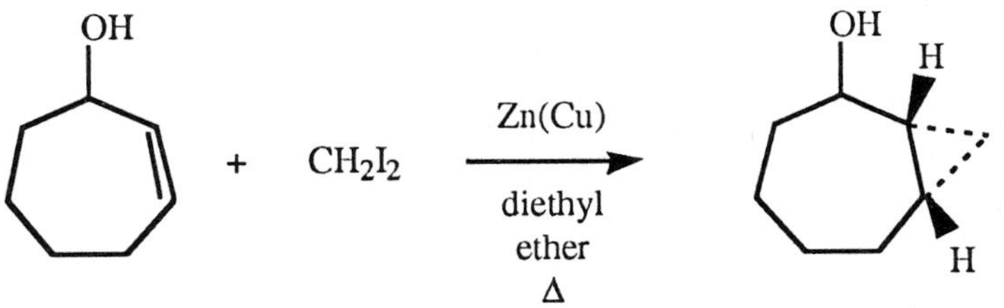
OH
+
CH2I2
Zn(Cu)
diethyl
ether
Δ
OH
H
H

and enantiomer

(g)

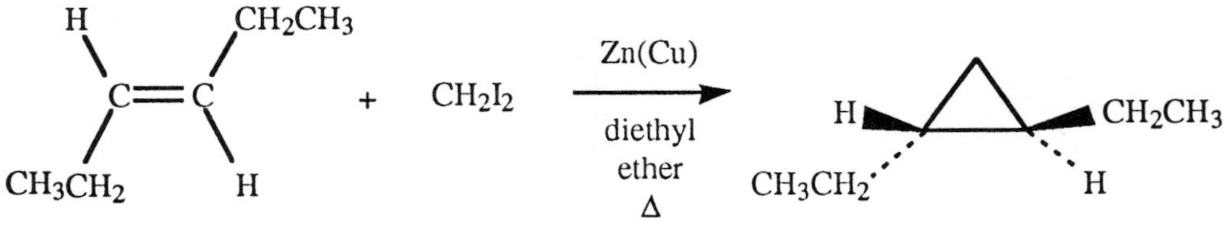
H
CH2CH3
C=C
CH3CH2
H
+
CH2I2
Zn(Cu)
diethyl
ether
Δ
H
CH2CH3
CH3CH2
H

and enantiomer

28.31 (cont)

(h)

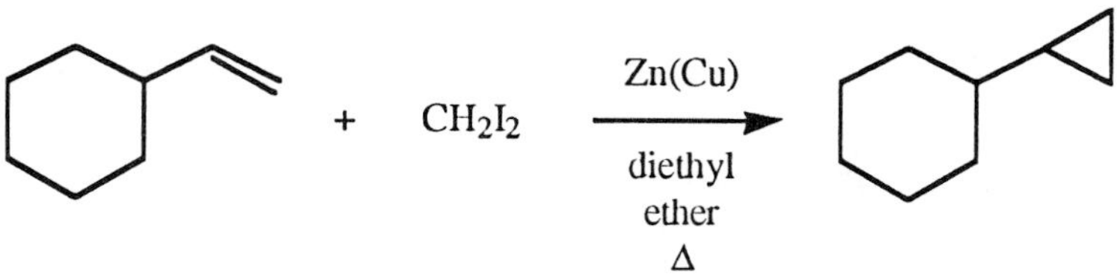

(i)

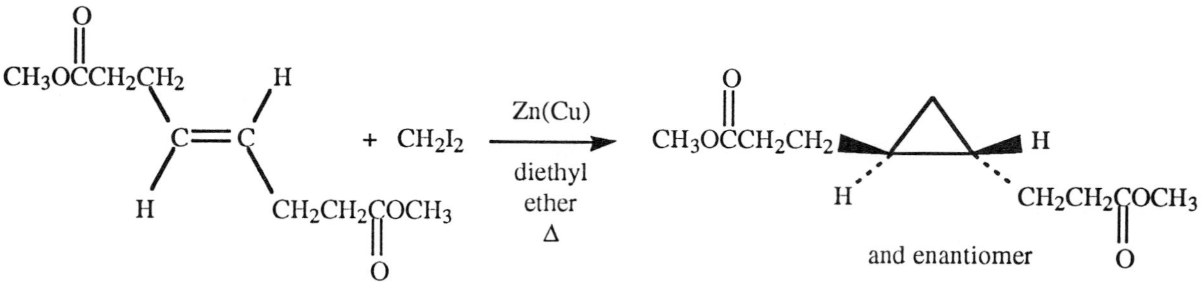

(j)

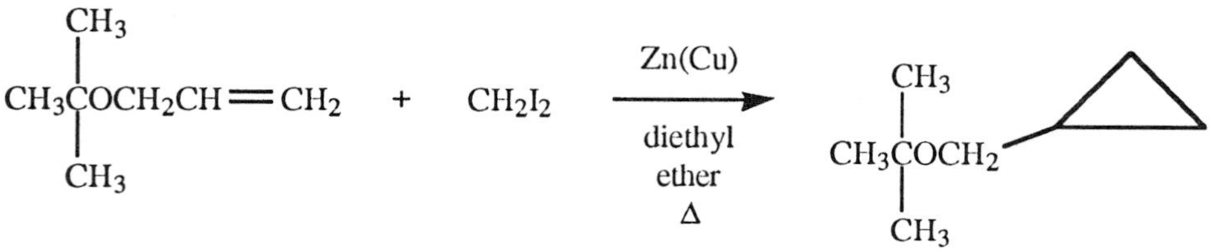

(k)

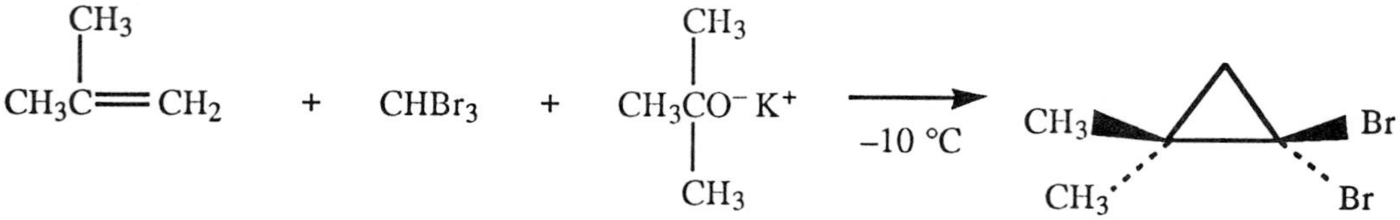

28.32

(a)

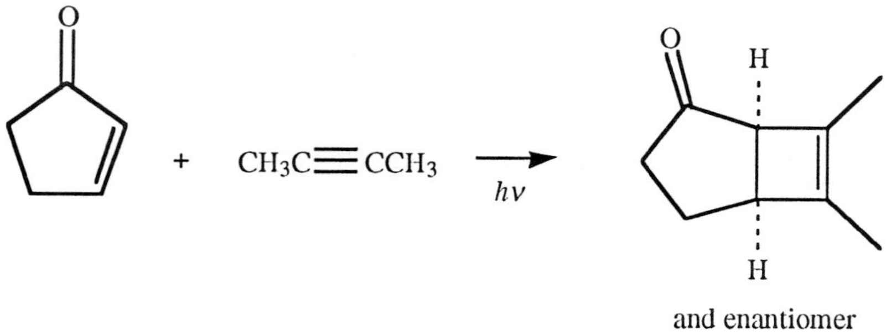

and enantiomer

(b)

and enantiomer

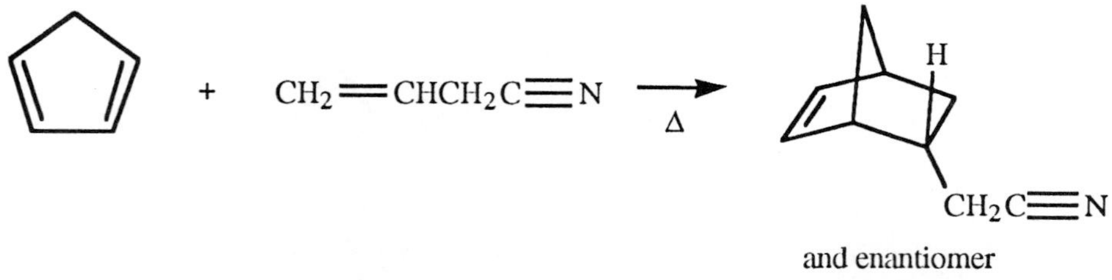

28.32 (cont)

(c)

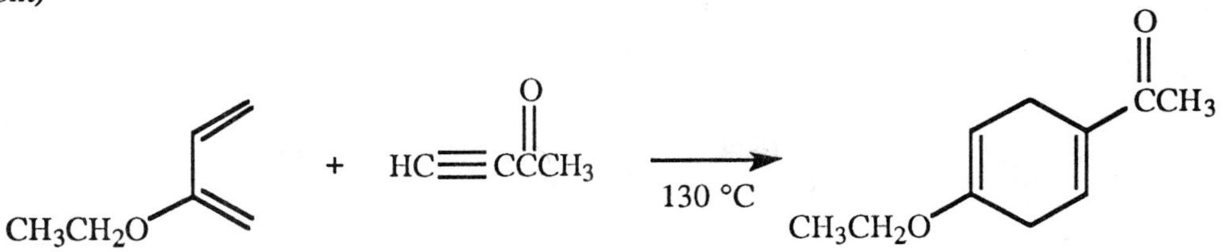

(d)

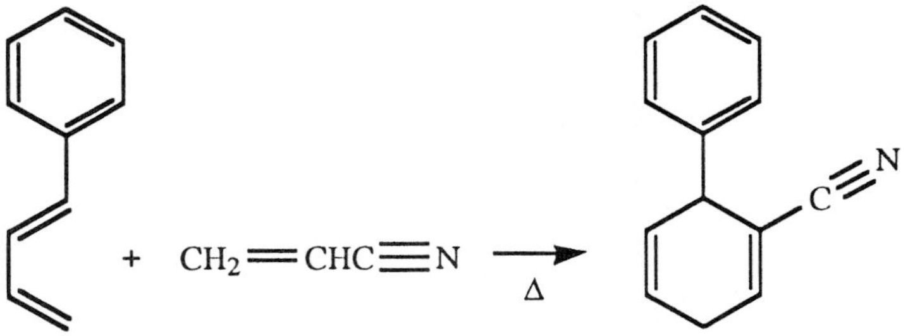

(e)

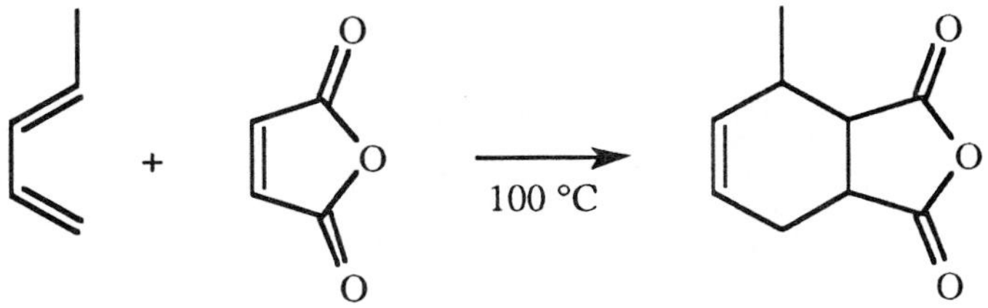

(f)

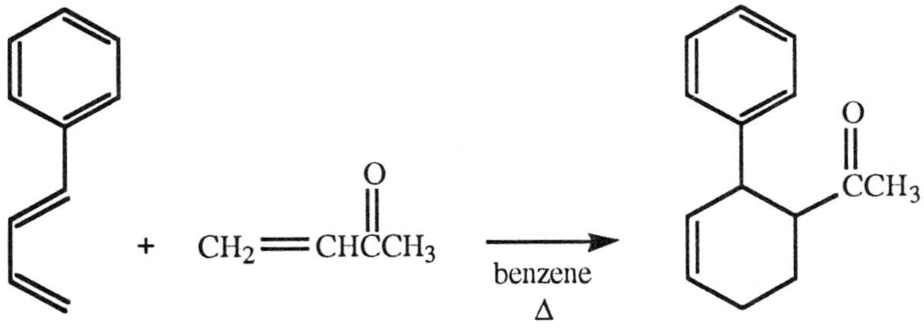

(g)

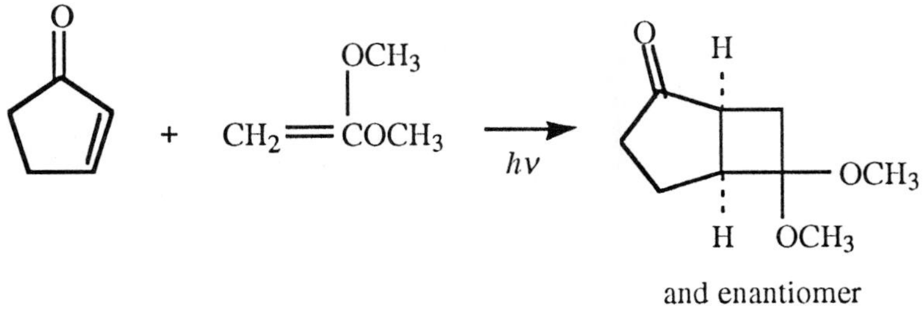

and enantiomer

(h)

and enantiomer

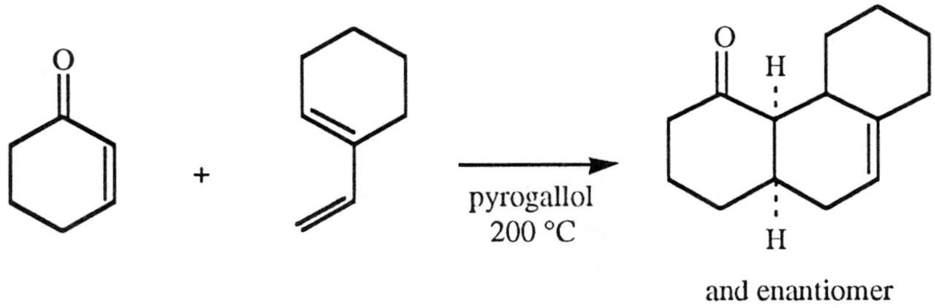

28.32 (cont)

(i)

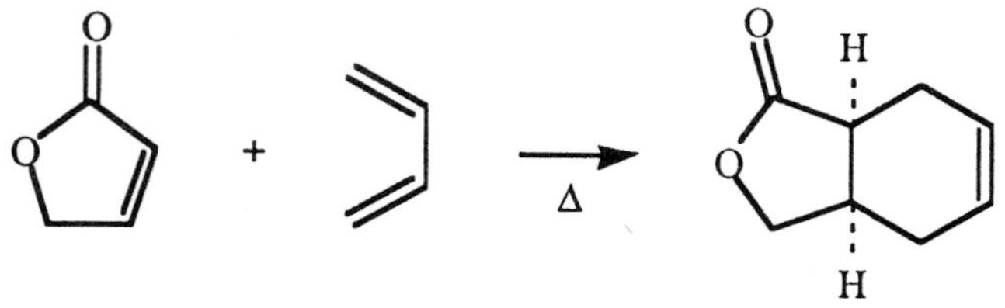

and enantiomer

(j)

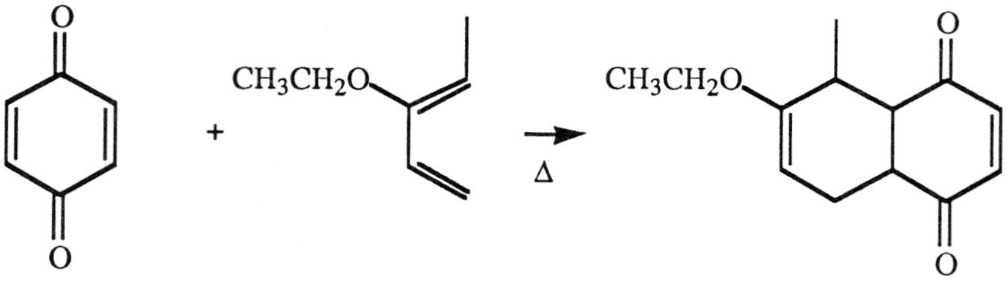

(k)

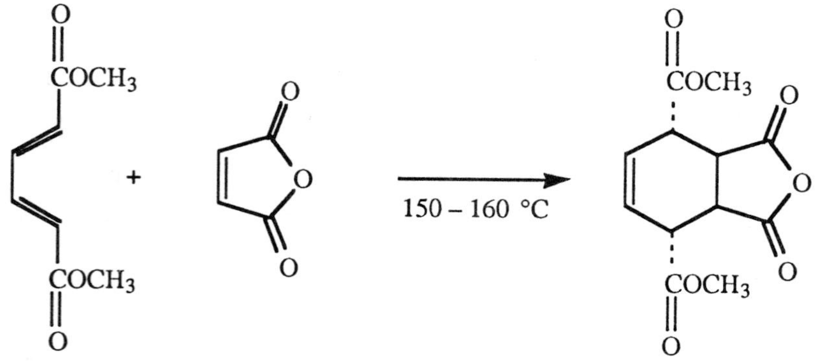

150 – 160 °C

(l)

$h\nu$

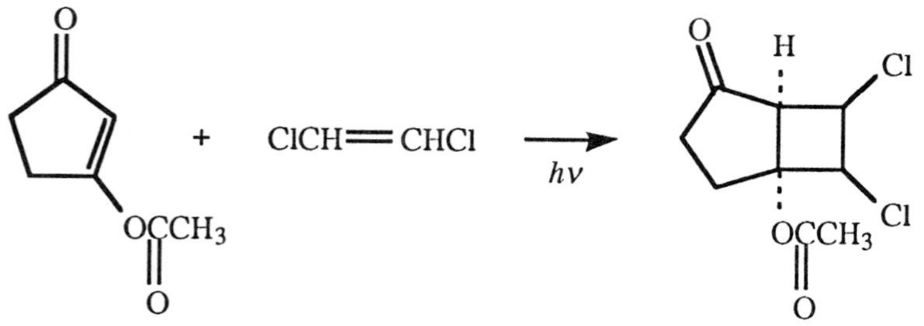

and enantiomer

28.33

(a)

Δ

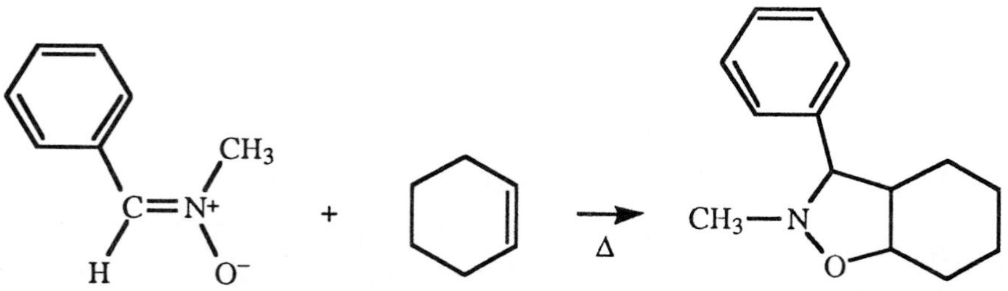

28.33 (cont)

(b)

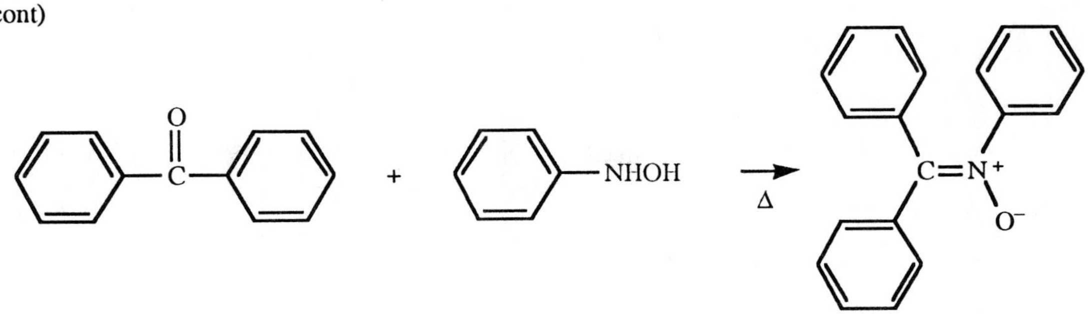

(c)

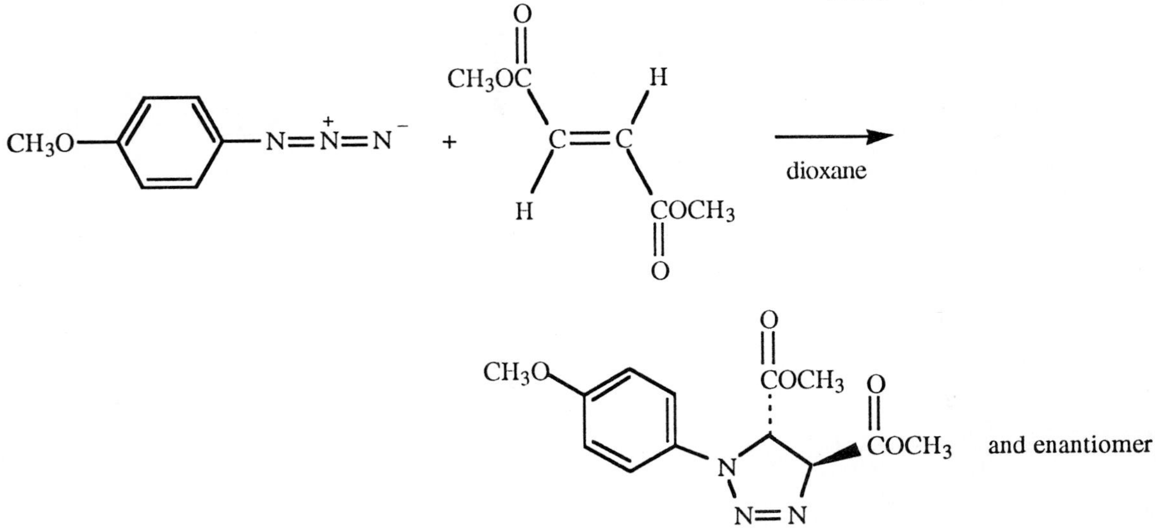

and enantiomer

(d)

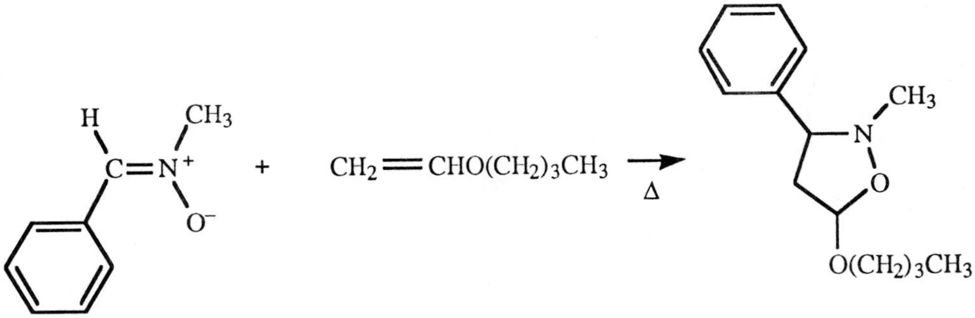

(e)

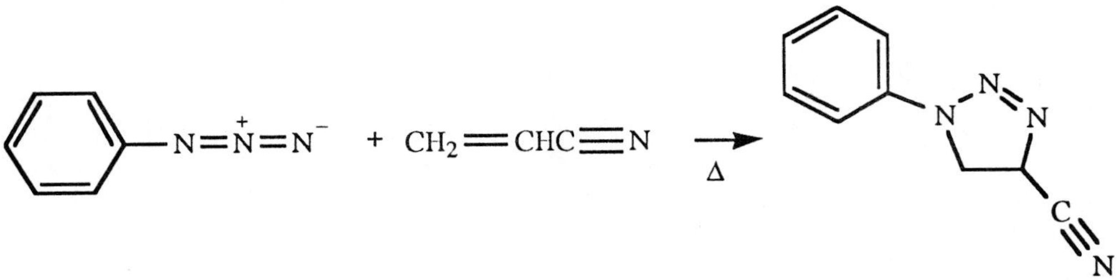

Note that the alkene in this problem is an electrophilic alkene. Accordingly, the orientation of the cycloaddition reaction is different from that seen in Problem 28.25 (a), in which phenyl azide adds to an alkene with an electron-donating group.

28.33 (cont)

(f)

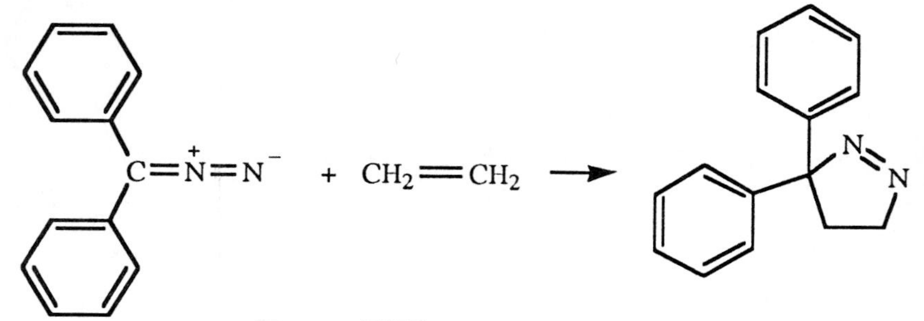

(g)

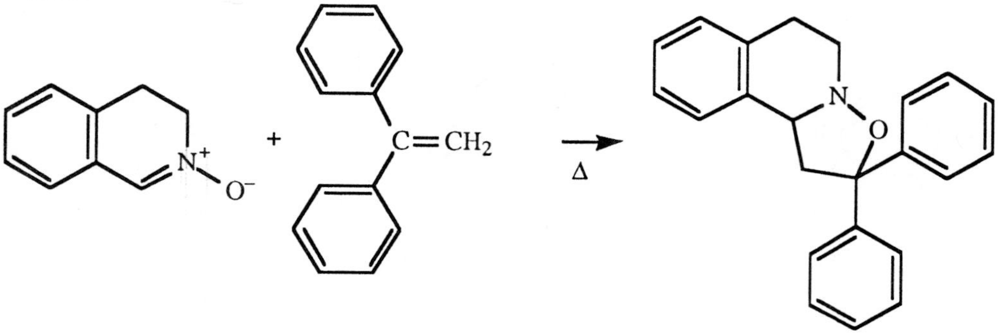

(h)

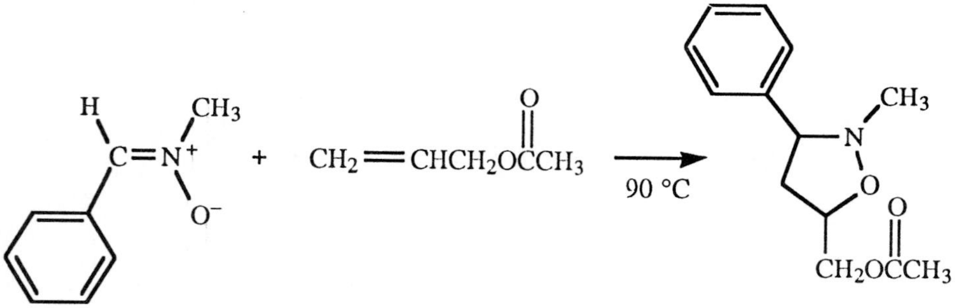

(i)

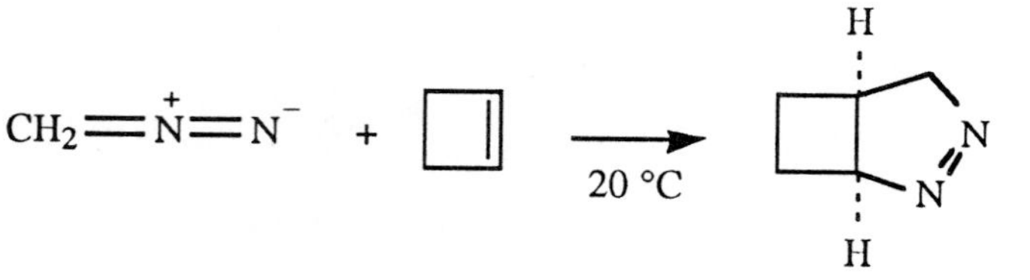

and enantiomer

28.34

(a)

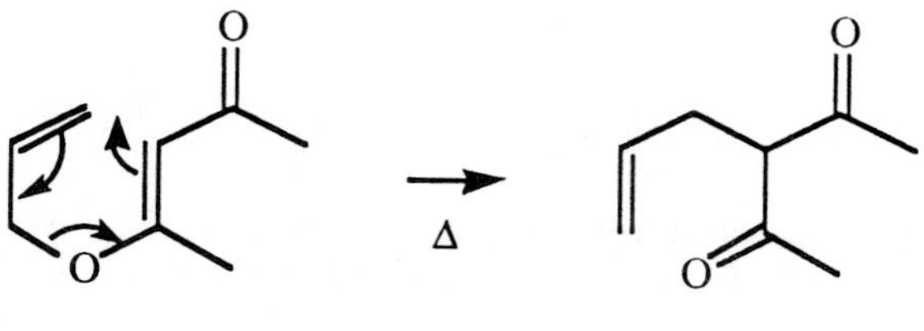

28.34 (cont)

(b)

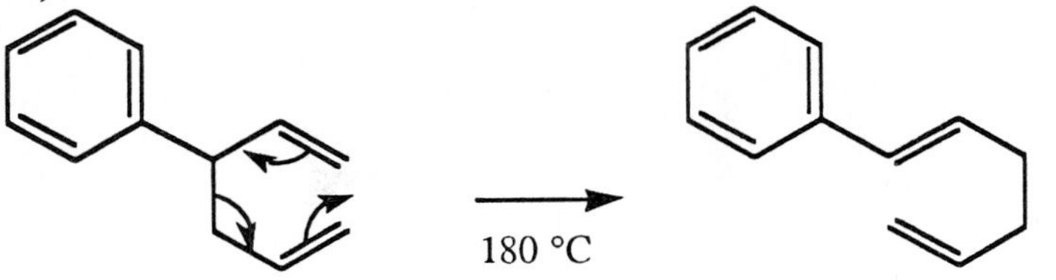

180 °C

(c)

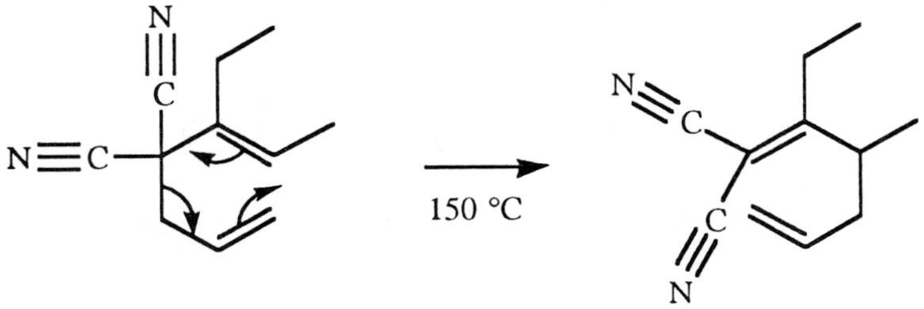

150 °C

(d)

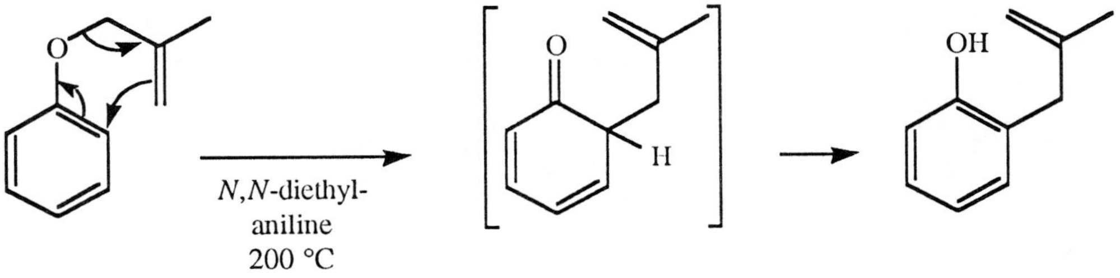

N,N-diethyl-
aniline
200 °C

(e)

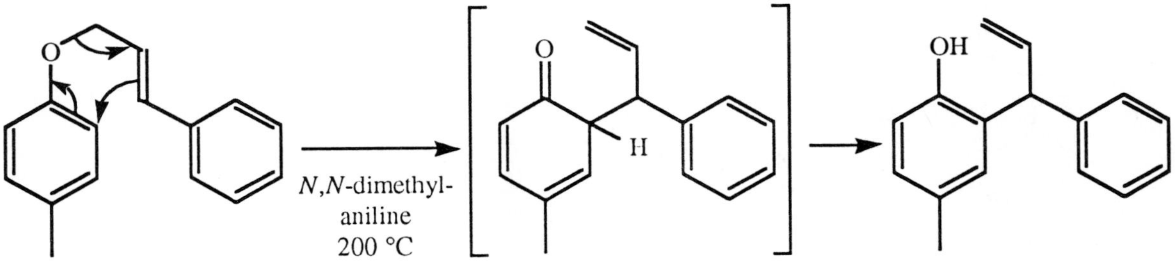

N,N-dimethyl-
aniline
200 °C

(f)

180 °C

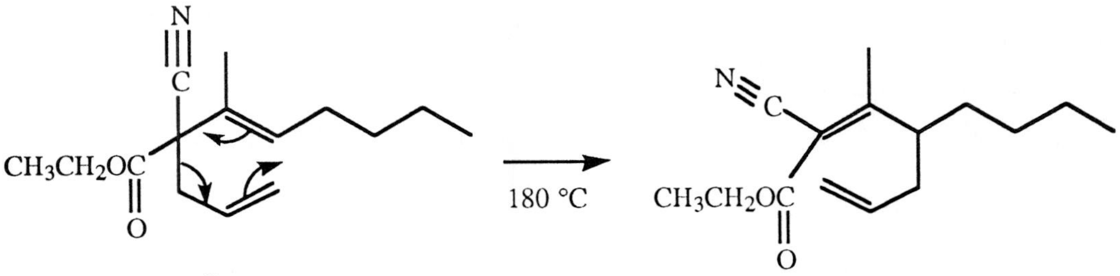

(g)

Δ

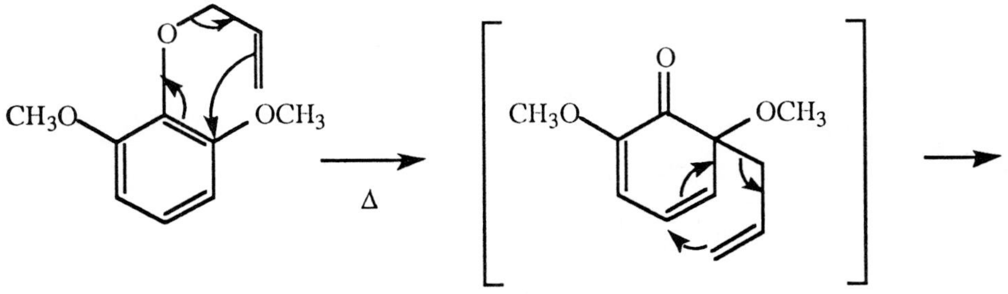

28.34 (g) (cont)

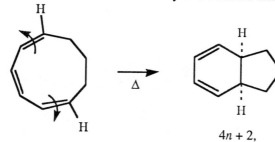

28.35

(a)

hv
diethyl
ether

$4n + 2$, conrotatory in the excited state

(b)

hv
diethyl
ether

$4n + 2$,
conrotatory in the excited state

(c)

Δ

$4n + 2$,
disrotatory in the ground state

(d)

Δ

$4n$,
conrotatory in the ground state

28.35 (cont)

(e)

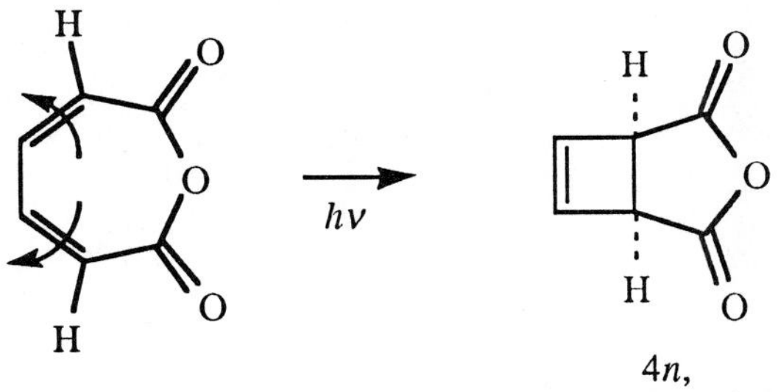

4n,
disrotatory in the ground state

(f)

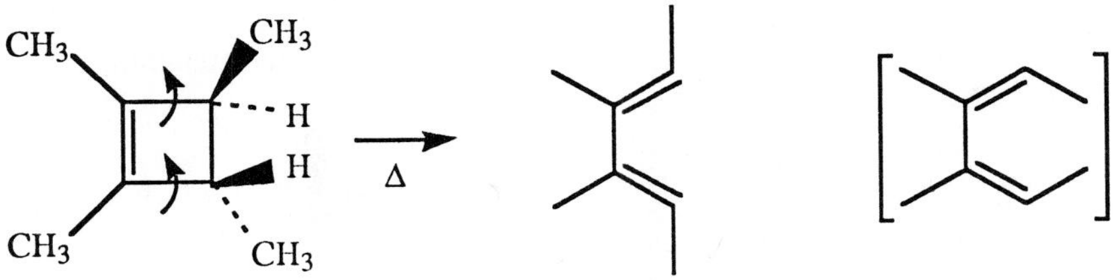

4n,
conrotatory in the ground state

too hindered
to form

28.36

(a)

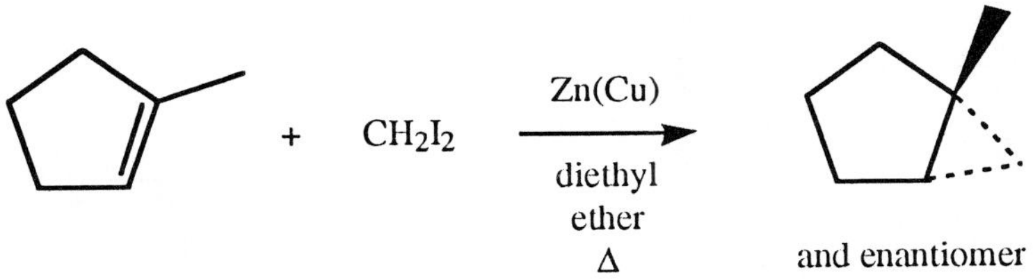

and enantiomer

(b)

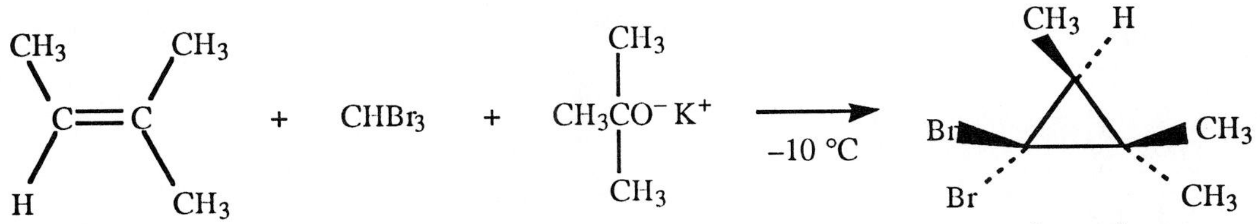

and enantiomer

(c) $CH_2{=}CHCCH_3$ + CH_2I_2

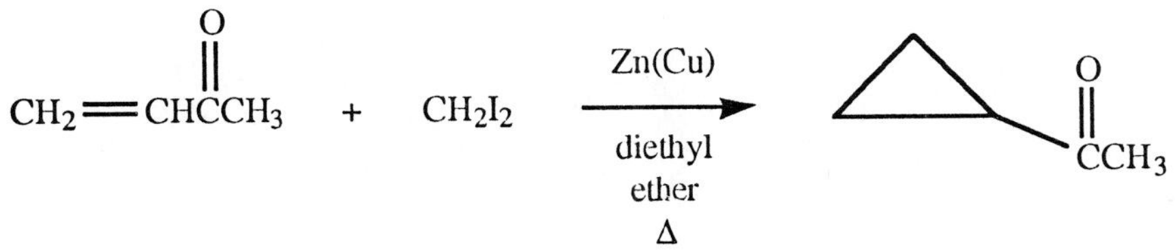

(d)

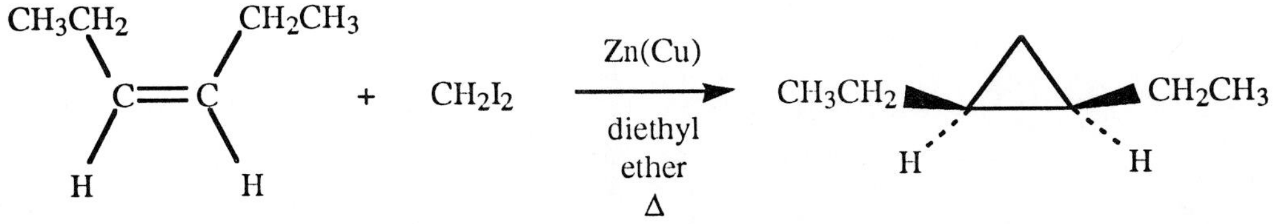

28.36 (cont)

(e)

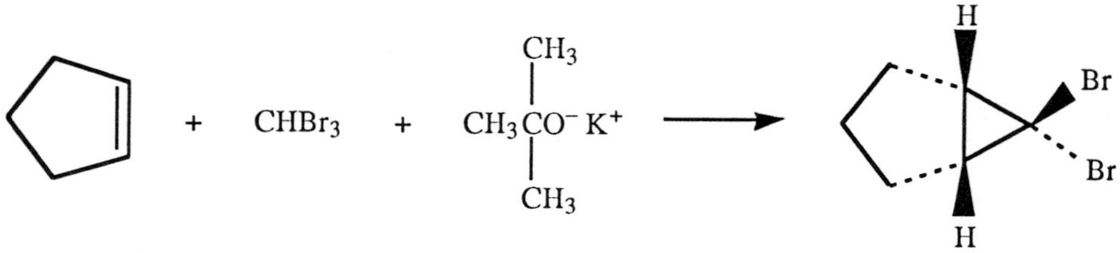

(f)

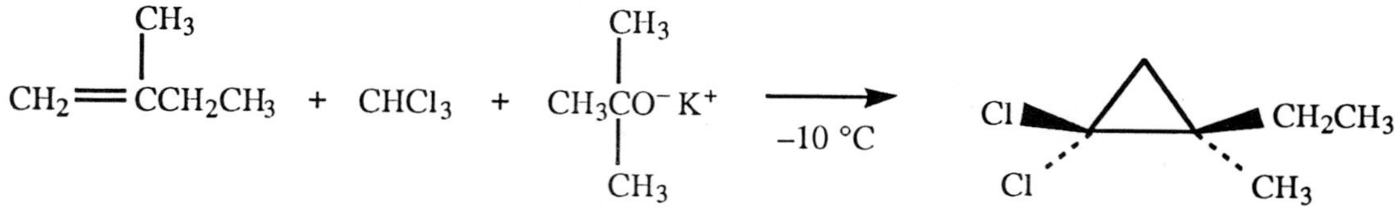

(g)

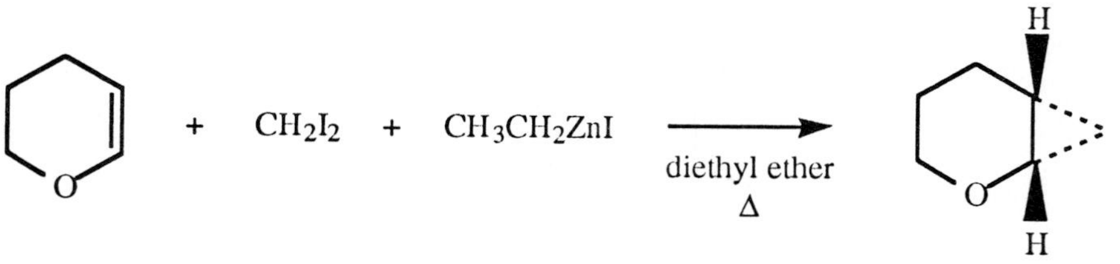

28.37

(a)

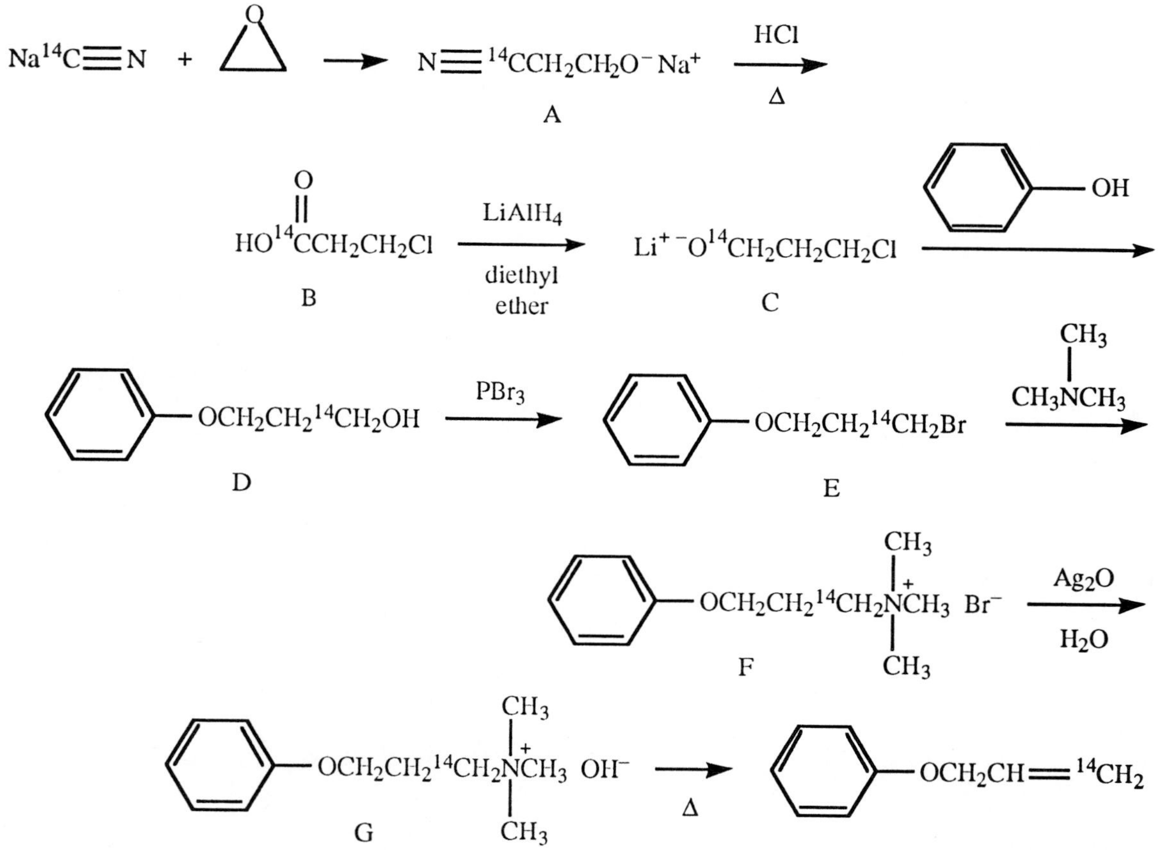

28.37 (cont)

(b)   Dehydration of an alcohol would lead to rearrangements, giving the more highly substituted alkene.

(c)

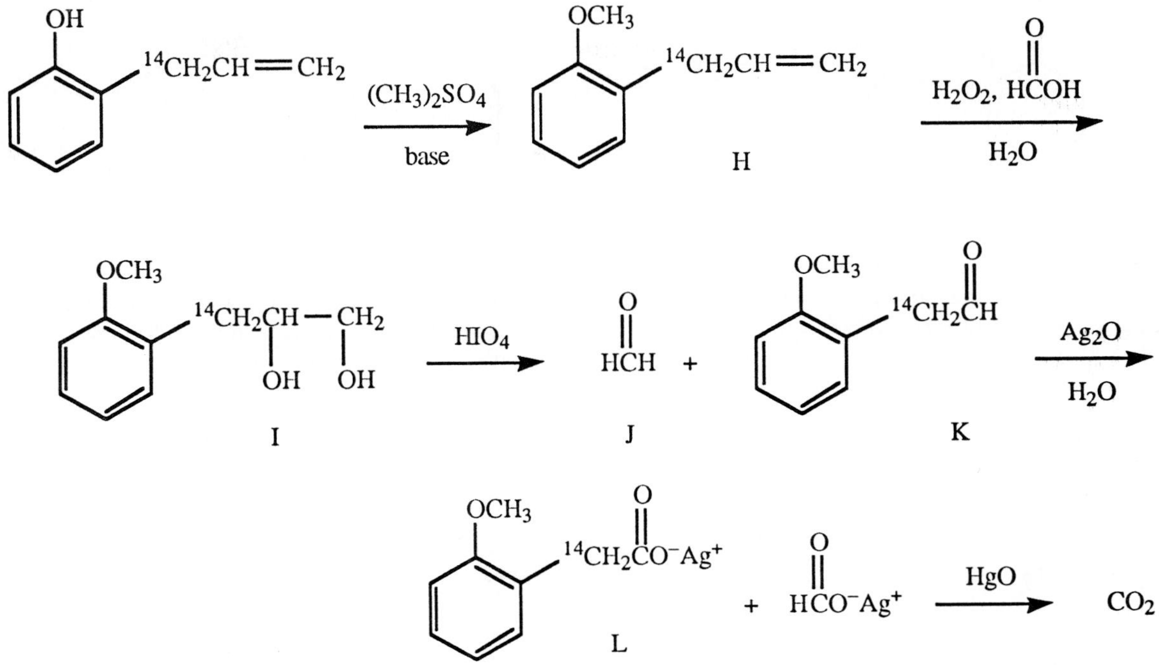

(d)   No. The ^{14}C label will be in the other product.

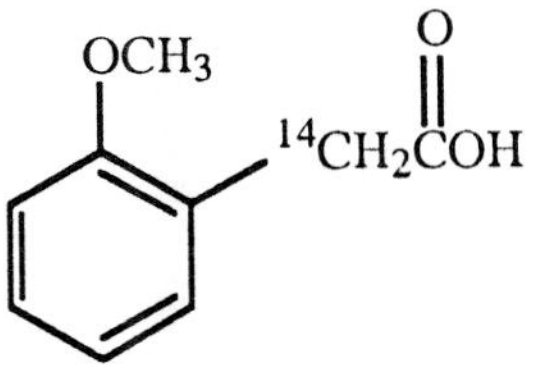

(e)   Yes.

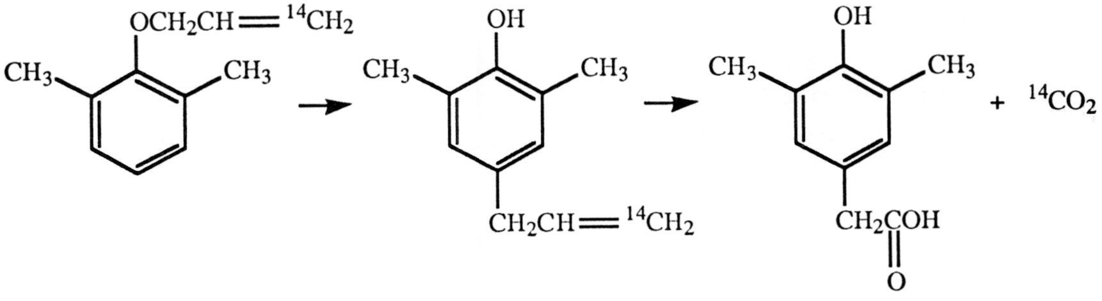

28.38

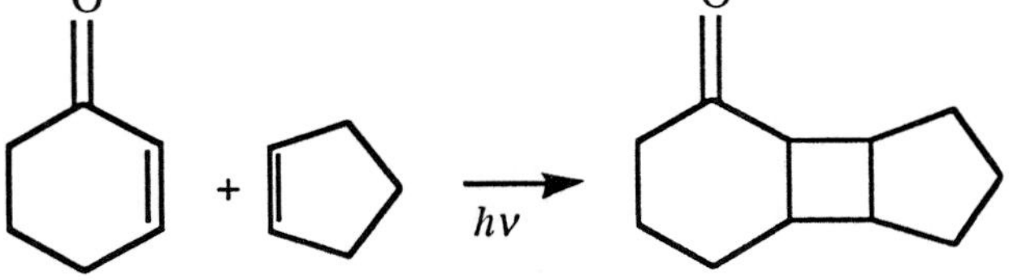

the only possible isomerism
is stereoisomerism

28.38 (cont)

The probable structures are shown below. Each structure also has an enantiomer.

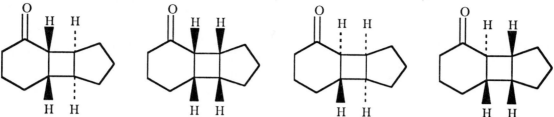

The ones in which the hydrogens are trans on the cyclopentane ring are not considered because of the mechanism of the reaction. It is the excited state of the cyclohexenone that reacts with the ground state of cyclopentene. The double bond in cyclohexenone may become twisted in the excited state, giving rise to a trans fusion of the six- and four-membered rings. The trans-fused compounds will isomerize in base through enolization and reprotonation to give the more stable cis-fused ring system. For example,

$$CH_3O^- Na^+ \xrightarrow{\text{methanol}} \quad \xrightarrow{CH_3OH} $$

This is the compound that is
formed in the largest amount in
the initial photochemical reaction.

28.39

4*n* + 2 system
n = 1
A

$\xrightarrow[\substack{\text{methanol} \\ \text{conrotatory} \\ \text{ring opening}}]{h\nu}$

B

cyclononane $\xleftarrow[\text{catalyst}]{H_2}$ B $\xrightarrow[\substack{\text{room temp-} \\ \text{erature;} \\ \text{disrotatory} \\ \text{ring closure}}]{}$ C

B is (1Z, 3Z, 5E)-1,3,5-cyclononatriene.

28.39 (cont)

A λ_{max} of 290 nm (ε 2050) in the ultraviolet spectrum points to the presence of a conjugated π system. The presence of bands in the infrared spectrum of B at 1645, 975, 960, and 670 cm^{-1} tells us that B has carbon-carbon double bonds. The band at 1645 cm^{-1} is the C=C stretching frequency. The bands at 975, 960 and 670 cm^{-1} are the bending frequencies for the C—H bonds on the double bonds.

The bands in the proton magnetic resonance spectrum of B correspond to the chemical shifts of vinyl hydrogen atoms (δ 4.7 – 6.2, 6H), and allylic and alkyl hydrogen atoms (δ 0.6 – 2.6, 6H).

28.40

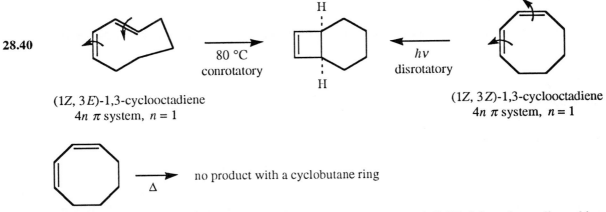

(1Z, 3E)-1,3-cyclooctadiene
4n π system, $n = 1$

(1Z, 3Z)-1,3-cyclooctadiene
4n π system, $n = 1$

no product with a cyclobutane ring

A 4n π system undergoes a thermal conrotatory ring closure, but for the (1Z, 3Z)-1,3-cyclooctadiene, this would lead to a trans ring junction between the cyclohexane ring and the cyclobutane ring. Such a trans fusion of the four-membered ring is highly strained.

>300 °C

This cannot be a concerted reaction because the concerted thermal ring opening of a 4n π system would be conrotatory and would give the (1Z, 3E)-1,3-cyclooctadiene (the reverse of the first reaction shown above).

28.41   Working backward from the structure of the isolated product, we get a cyclic diene for the structure of A.

A

To get to A from the starting material, we must open the cyclobutene ring in a conrotatory electrocyclic reaction.

28.41 (cont)

The product of the reaction, the cyclic diene, must come from an intramolecular cyclization reaction involving the carbonyl group of the aldehyde.

28.42   The formation of *trans*-5,6-dichloro-5,6-difluorobicyclo[2.2.1]hept-2-ene is a thermally allowed $[4n + 2]$ cycloaddition (a Diels-Alder) reaction, which occurs with high stereoselectivity.

The concerted thermal cycloaddition reaction for a system with $4n$ electrons is forbidden. The formation of 6,7-dichloro-6,7-difluorobicyclo[3.2.0]hept-2-ene goes by a radical mechanism, and gives mixtures of stereoisomers.

28.43

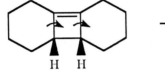

Δ
$4n$
conrotatory

Diels-Alder reaction

Δ
$4n + 2$
thermally allowed

28.44

Δ
conrotatory

$4n$ π system

A concerted reaction would produce a trans double bond in a six-membered ring. A six-membered ring containing a trans double bond is highly strained and unstable. Therefore, this cyclobutene does not open by a concerted reaction and would require high temperatures for reaction.

Δ
conrotatory

$4n$ π system

A concerted reaction would produce cis double bonds in both rings. This cyclobutene will open easily by a concerted mechanism.

28.45

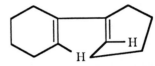

tautomerization

an enol, which is unstable;
no stereochemistry implied

28.46

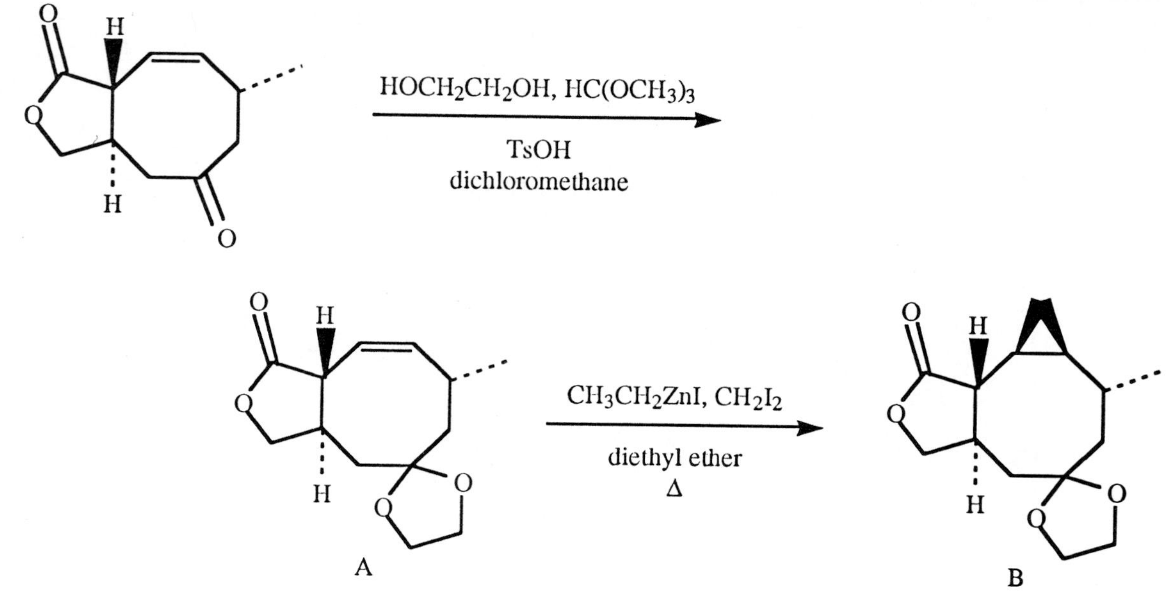

$$\xrightarrow[\substack{\text{TsOH} \\ \text{dichloromethane}}]{\text{HOCH}_2\text{CH}_2\text{OH, HC(OCH}_3)_3}$$

A

$$\xrightarrow[\substack{\text{diethyl ether} \\ \Delta}]{\text{CH}_3\text{CH}_2\text{ZnI, CH}_2\text{I}_2}$$

B

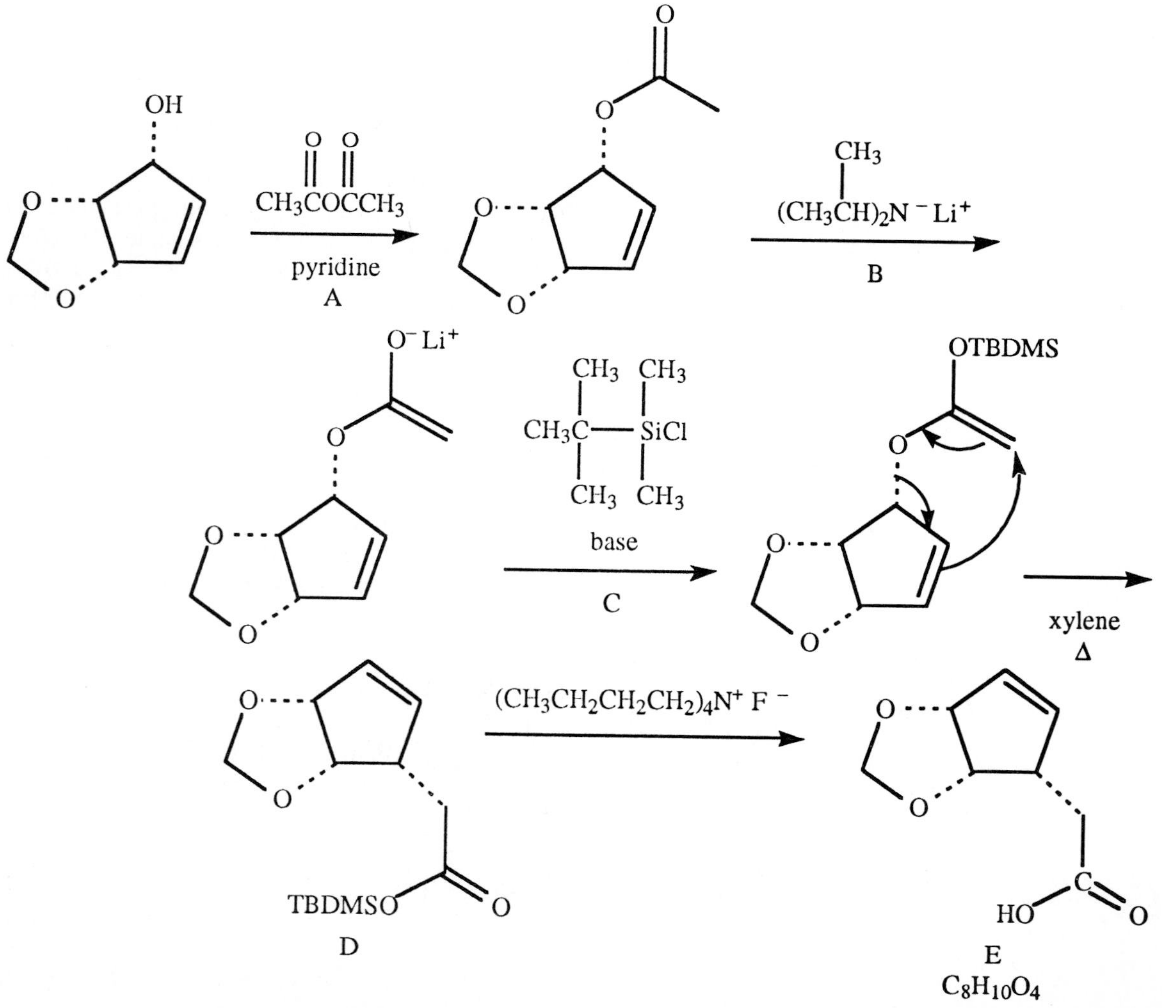

28.47

(a)

$$\xrightarrow[\substack{\text{pyridine} \\ A}]{\text{CH}_3\text{COCCH}_3}$$

$$\xrightarrow[B]{(CH_3CH)_2N^-Li^+ \;\; CH_3}$$

$$\xrightarrow[C]{\substack{\text{CH}_3\text{C}-\text{SiCl} \\ \text{base}}}$$

$$\xrightarrow[\substack{\text{xylene} \\ \Delta}]{}$$

D

$$\xrightarrow{(CH_3CH_2CH_2CH_2)_4N^+ F^-}$$

E

$C_8H_{10}O_4$

28.47 (cont)

(b)

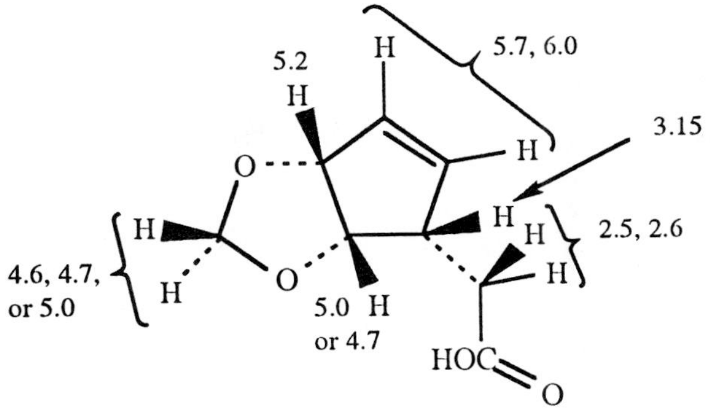

proton magnetic resonance assignments

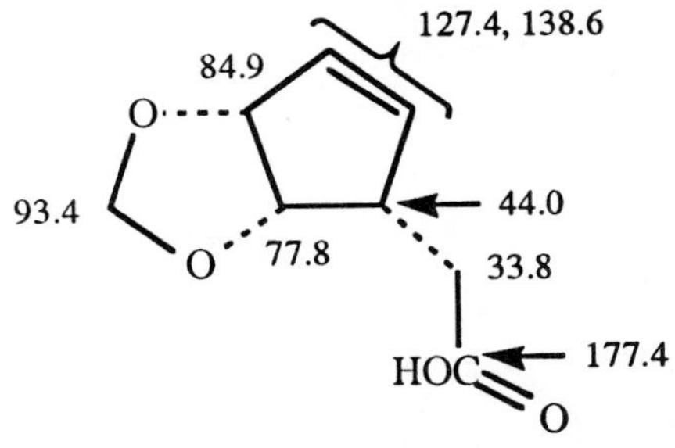

C-13 magnetic resonance assignments

28.48

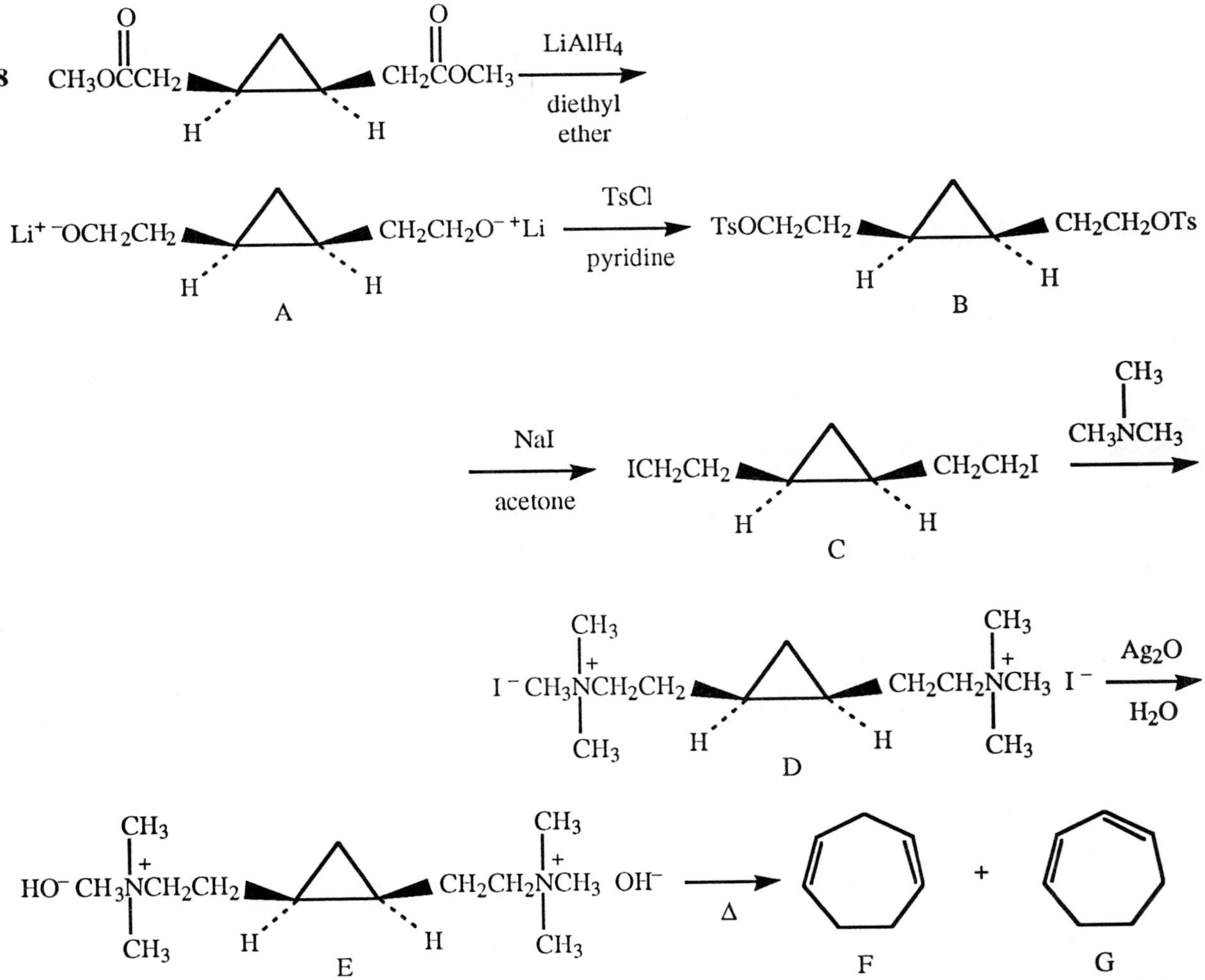

F
The ultraviolet spectrum shows the absence of a conjugated system.

G
The ultraviolet spectrum shows the presence of a conjugated system.

28.48 (cont)

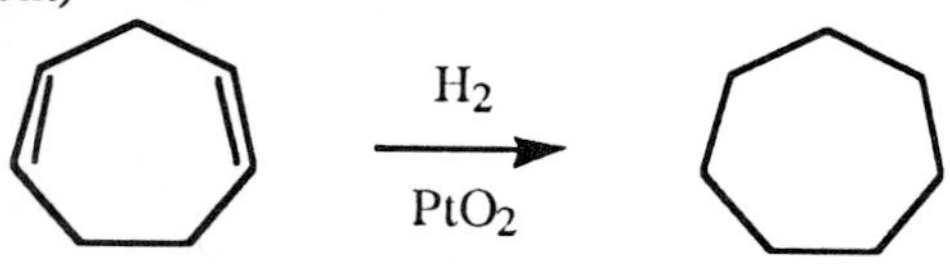

F

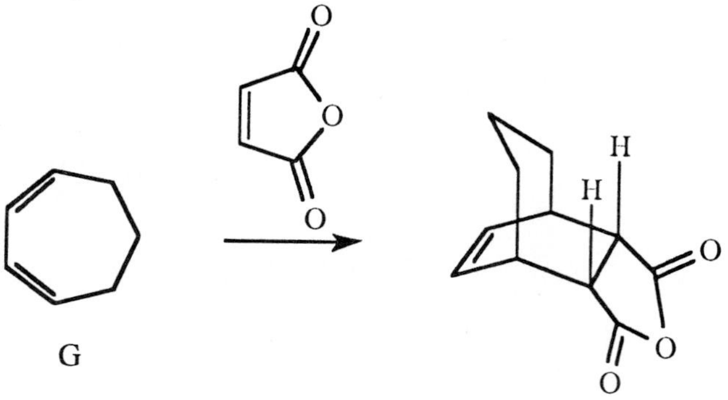

G

F is formed by a [3,3] sigmatropic rearrangement of the unstable intermediate formed when compound E is heated.

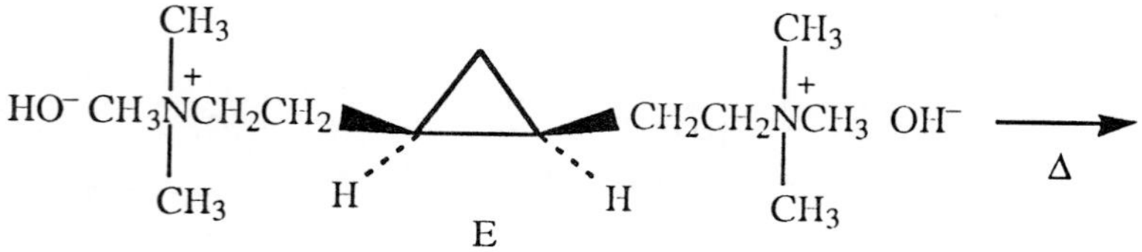

E

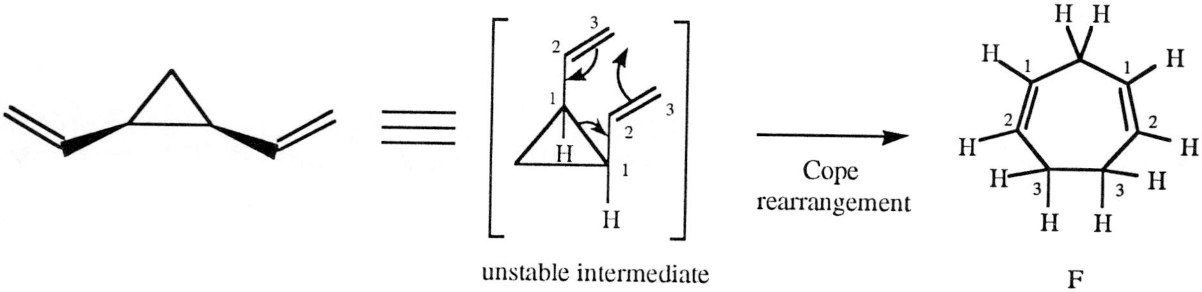

unstable intermediate

Cope rearrangement

F

No bands in the infrared spectrum for terminal alkenes; bands in the proton magnetic resonance spectrum for 4 vinyl hydrogen atoms, 4 methylene hydrogen atoms of one kind, and 2 methylene hydrogen atoms of another kind.

A λ_{max} of 248 nm in the ultraviolet spectrum of C points to the presence of a conjugated π system. G is formed from F in the presence of strong base.

28.48 (cont)

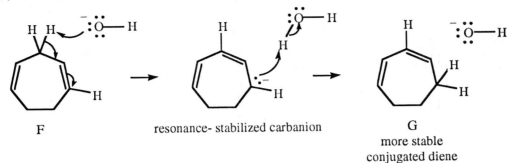

F   resonance- stabilized carbanion   G
                                                                          more stable
                                                                          conjugated diene

28.49

CH₃
₁H   NCH₃
     1   2
            3   [1,5] sigmatropic
                            shift   ──────────▶
     5   4

5-(*N*,*N*-dimethylamino)-
1,3-cycloheptadiene

CH₃NCH₃
      1   2
             3
¹H   5   4
     H

1-(*N*,*N*-dimethylamino)-
1,3-cycloheptadiene

(The numbers on the structure are to help
identify the sigmatropic rearrangement and
are not for nomenclature.)

28.50

◻ ─ COCH₃   +   ⬠   ────▶   H H
             ‖   hν   ◻⬠
             O   dichloro-   CH₃OC   H
                          methane   ‖
                                                O

                                              A
                                  product of a [2 + 2]
                                  cycloaddition reaction

────▶   180 °C
         3 h

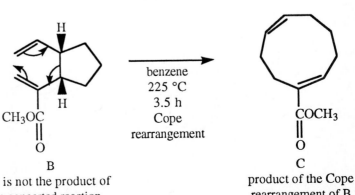

            H
        ⟍⟋⬠
      CH₃OC   H
           ‖
           O

           B
B is not the product of
a concerted reaction

benzene
225 °C
3.5 h
Cope
rearrangement

────▶

COCH₃
‖
O

C
product of the Cope
rearrangement of B

28.50 (cont)

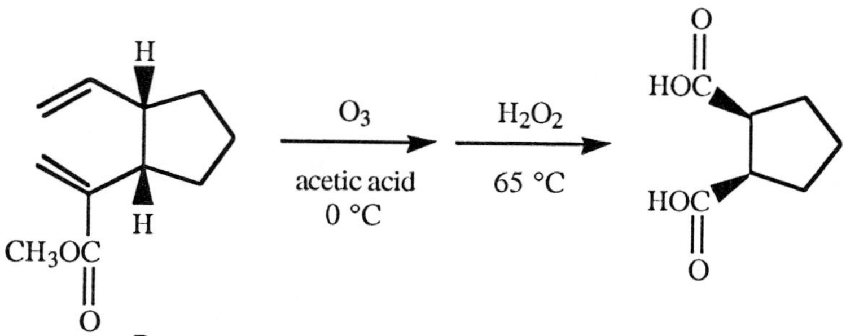
H
H
CH3OC
O
O3
acetic acid
0 °C
H2O2
65 °C
HOC
O
HOC
O

B

The assignment of the structure of B is supported by the results of the ozonolysis reaction (on the previous page) and by the proton magnetic resonance spectrum as shown below.

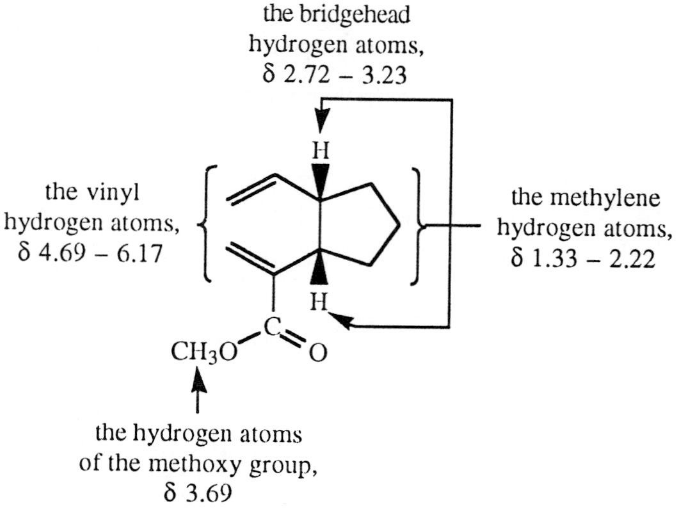
the bridgehead
hydrogen atoms,
δ 2.72 – 3.23
H
the vinyl
hydrogen atoms,
δ 4.69 – 6.17
the methylene
hydrogen atoms,
δ 1.33 – 2.22
H
CH3O
C
O
the hydrogen atoms
of the methoxy group,
δ 3.69

28.51

(a)

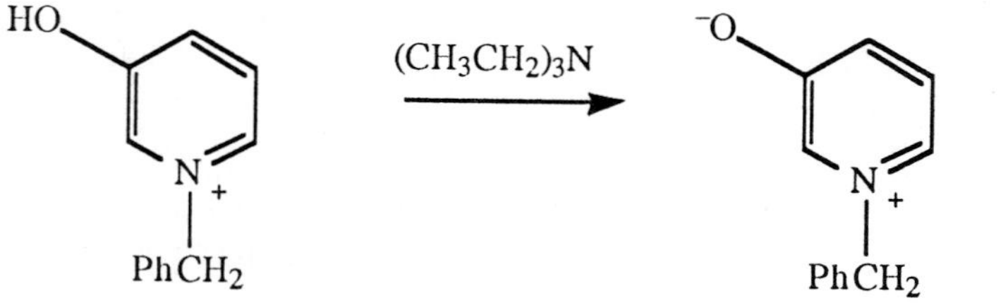
HO
N
+
PhCH2
(CH3CH2)3N
-O
N
+
PhCH2

A
1,3-dipolar species

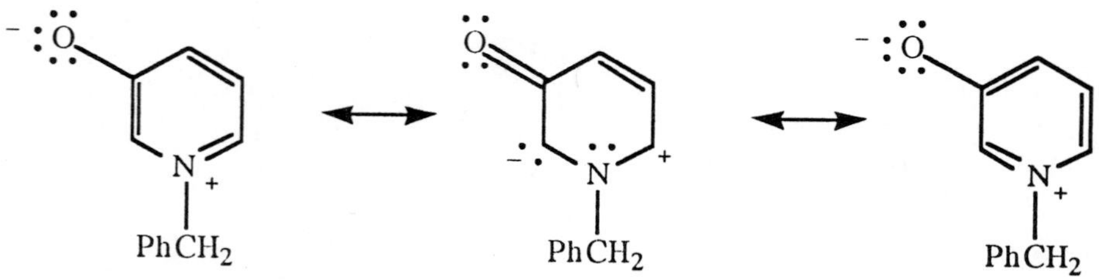
O
N
+
PhCH2
O
N
+
PhCH2
O
N
+
PhCH2

1,3-dipolar character

28.51 (cont)

(b)

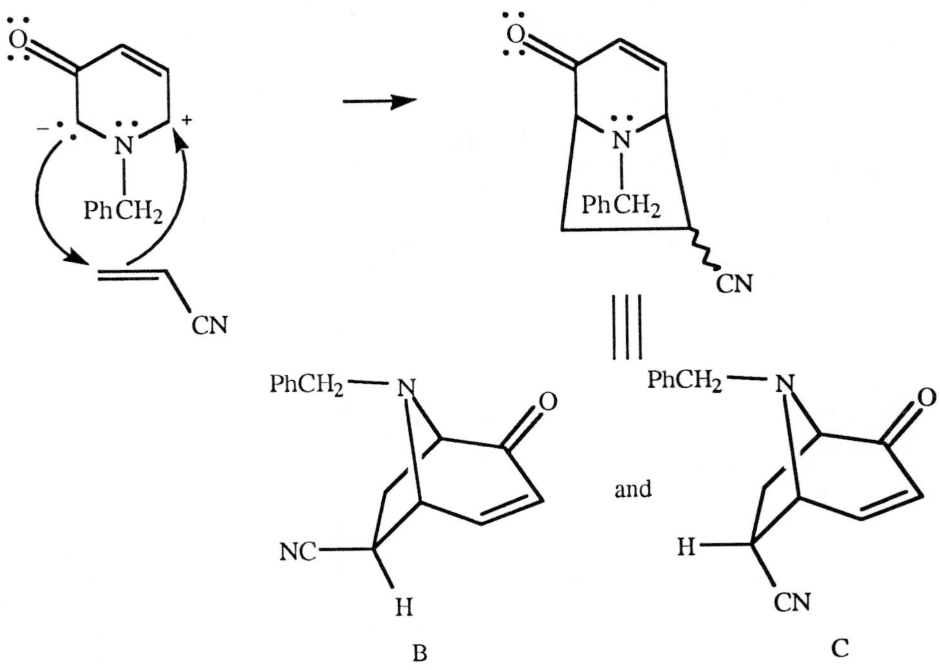

B is the structure that is related to Bao Gong Teng A.

28.52

(a)

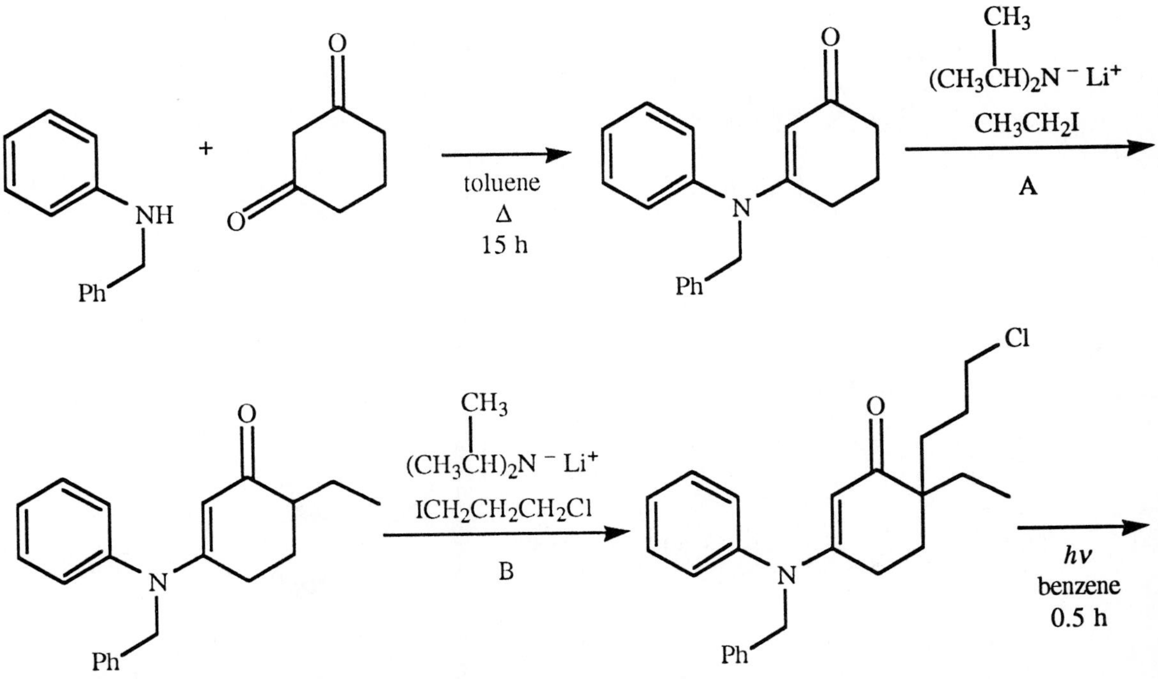

28.52 (a) (cont)

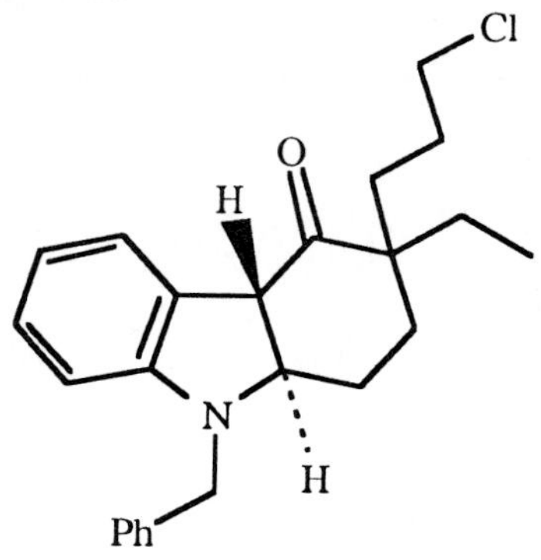

(b)   There are six electrons available for the reaction. The two nonbonding electrons on nitrogen participate. Conrotatory ring closure is expected.

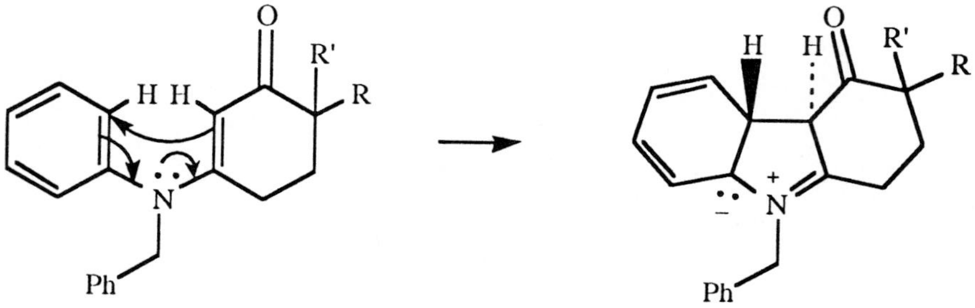

A pair of 1,2-hydrogen shifts completes the reaction, restoring aromaticity in one of the rings.

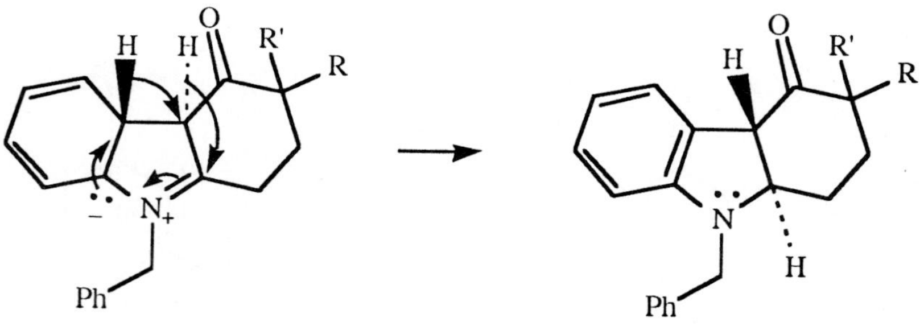

28.53

(a)   PhC≡CPh   +

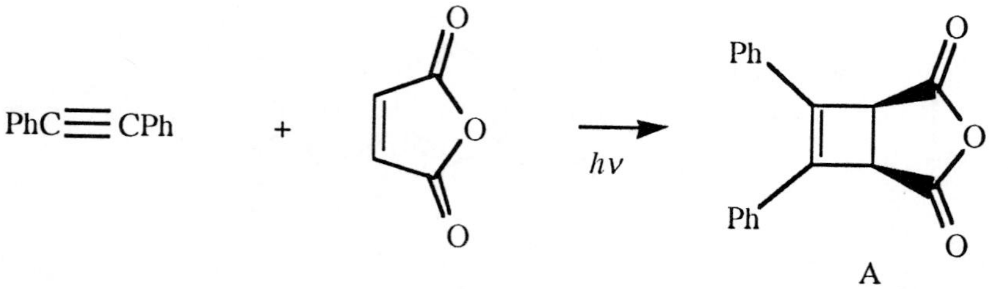

28.53 (cont)

(b)

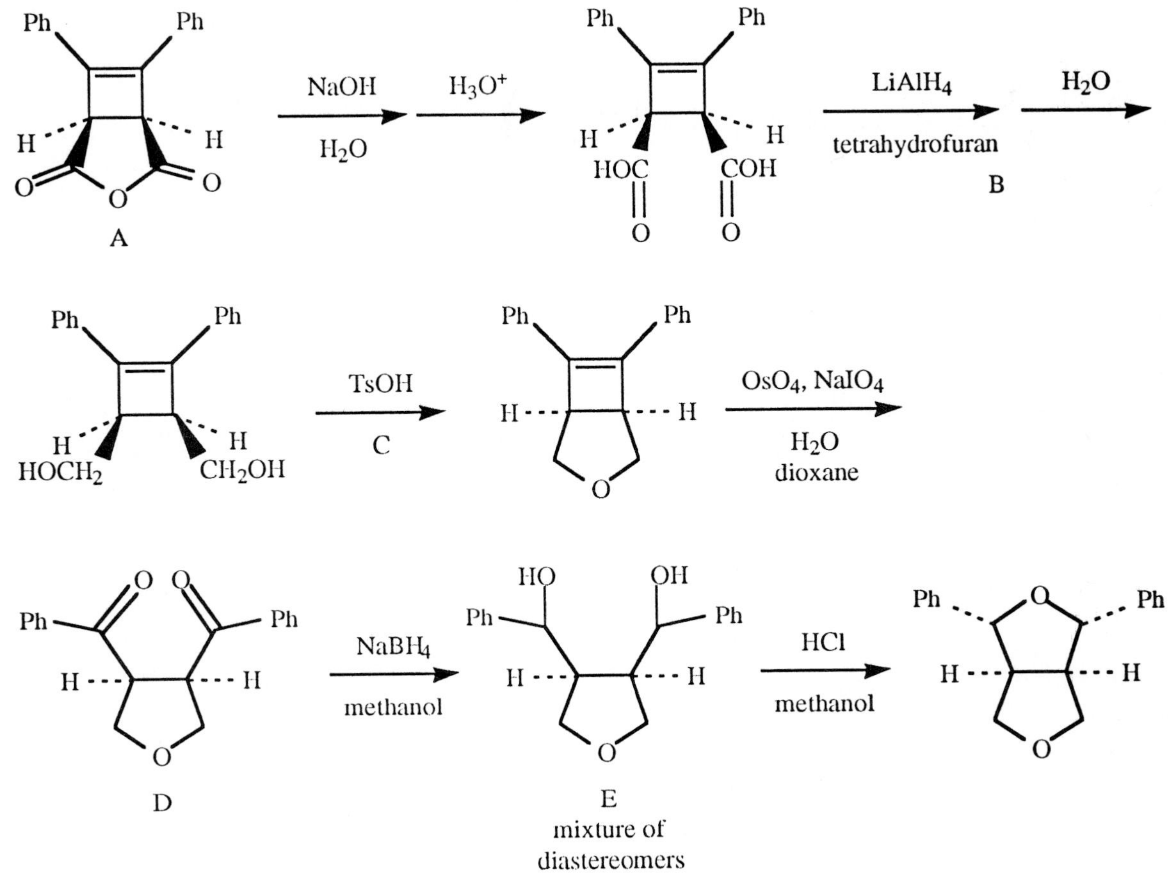

28.54

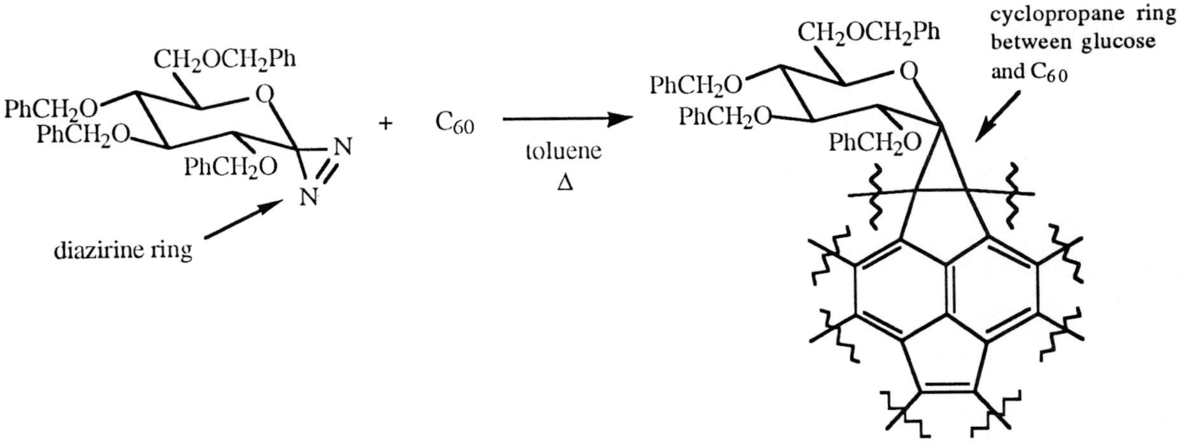

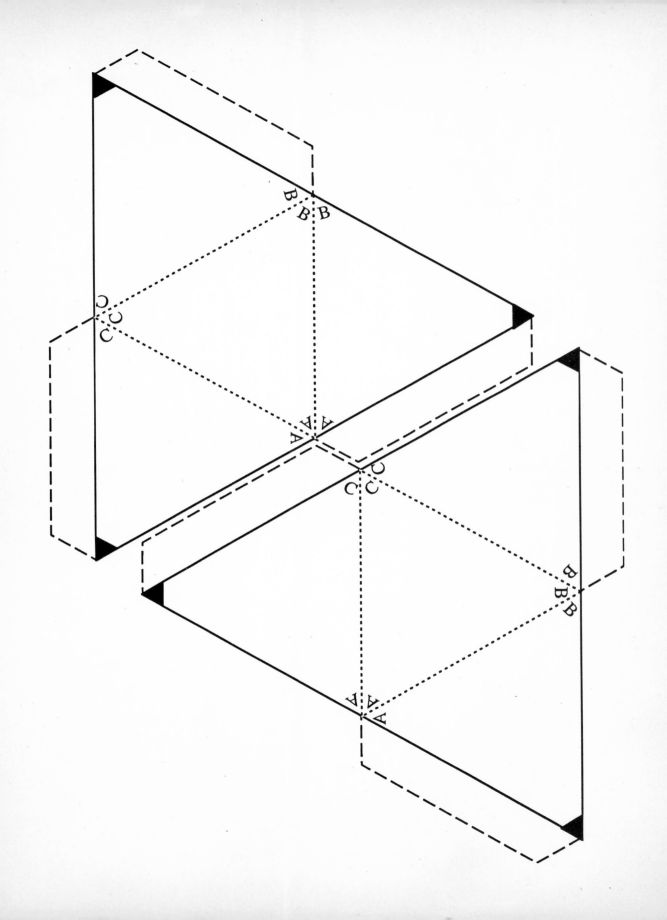